UNDERSTANDING NORMAL AND CLINICAL

Nutrition

Eighth Edition

SHARON RADY ROLFES | KATHRYN PINNA | ELLIE WHITNEY

WADSWORTH
CENGAGE Learning

Australia • Brazil • Japan • Korea • Mexico • Singapore • Spain • United Kingdom • United States

BP45

Understanding Normal and Clinical Nutrition, **Eighth Edition**
Sharon Rady Rolfes, Kathryn Pinna, Ellie Whitney

Publisher: Yolanda Cossio

Development Editor: Anna Lustig

Assistant Editor: Elesha Feldman

Editorial Assistant: Sarah Farrant

Technology Project Manager: Melinda Newfarmer

Marketing Communications Manager: Belinda Krohmer

Project Manager, Editorial Production: Trudy Brown

Creative Director: Rob Hugel

Art Director: John Walker

Print Buyer: Karen Hunt

Permissions Editor: Margaret Chamberlain-Gaston

Production Service: The Book Company

Text Designer: Dianne Beasley

Photo Researcher: Roman Barnes

Copy Editor: Mary Berry

Cover Designer: Dare Porter

Cover Image: Lisa Romerein, © 2008 Jupiterimages Corporation

Compositor: Lachina Publishing Services

For product information and technology assistance, contact us at
Cengage Learning Customer & Sales Support, 1-800-354-9706

For permission to use material from this text or product, submit all requests online at **cengage.com/permissions**
Further permissions questions can be emailed to **permissionrequest@cengage.com**

Library of Congress Control Number: 2008922067

ISBN-13: 978-0-495-55646-6

ISBN-10: 0-495-55646-7

Wadsworth
10 Davis Drive
Belmont, CA 94002-3098
USA

Cengage Learning is a leading provider of customized learning solutions with office locations around the globe, including Singapore, the United Kingdom, Australia, Mexico, Brazil, and Japan. Locate your local office at: **international.cengage.com/region**

Cengage Learning products are represented in Canada by Nelson Education, Ltd.

For your course and learning solutions, visit **academic.cengage.com**

Purchase any of our products at your local college store or at our preferred online store **www.ichapters.com**

Printed in Canada
1 2 3 4 5 6 7 12 11 10 09 08

5/11/09

To Ellie Whitney, my mentor, partner, and friend, with much appreciation for believing in me, sharing your wisdom, and giving me the opportunity to pursue a career more challenging and rewarding than any I could have imagined.

Sharon

To David Stone, for years of love, friendship, and assistance with numerous academic and musical pursuits.

Kathryn

To the memory of Gary Woodruff, the editor who first encouraged me to write.

Ellie

About the Authors

Sharon Rady Rolfes received her M.S. in nutrition and food science from Florida State University. She is a founding member of Nutrition and Health Associates, an information resource center that maintains a research database on over 1000 nutrition-related topics. Her other publications include the college textbooks *Understanding Nutrition* and *Nutrition for Health and Health Care* and a multimedia CD-ROM called *Nutrition Interactive*. In addition to writing, she occasionally teaches at Florida State University and serves as a consultant for various educational projects. Her volunteer work includes coordinating meals for the hungry and homeless and serving on the steering committee of Working Well, a community initiative designed to help local businesses improve the health and well-being of their employees. She maintains her registration as a dietitian and membership in the American Dietetic Association.

Kathryn Pinna received her M.S. and Ph.D. degrees in nutrition from the University of California at Berkeley. She has taught nutrition, food science, and biology courses in the San Francisco Bay Area for over 20 years. She has also worked as an outpatient dietitian, Internet consultant, and freelance writer. Her other publications include the textbooks *Nutrition for Health and Health Care* and *Nutrition and Diet Therapy*. She is a registered dietitian and a member of the American Society for Nutrition and the American Dietetic Association.

Ellie Whitney grew up in New York City and received her B.A. and Ph.D. degrees in English and Biology at Radcliffe/Harvard University and Washington University, respectively. She has lived in Tallahassee since 1970, has taught at both Florida State University and Florida A&M University, has written newspaper columns on environmental matters for the *Tallahassee Democrat*, and has authored almost a dozen college textbooks on nutrition, health, and related topics, many of which have been revised multiple times over the years. In addition to teaching and writing, she has spent the past three-plus decades exploring outdoor Florida and studying its ecology. Her latest book is *Priceless Florida: The Natural Ecosystems* (Pineapple Press, 2004).

Brief Contents

Contents

CHAPTER 5

The Lipids: Triglycerides, Phospholipids, and Sterols   138

CHAPTER 6

Protein: Amino Acids   180

CHAPTER 7

Metabolism: Transformations and Interactions   212

HOW TO BOXES

CASE STUDIES

Preface

Each year brings new discoveries in nutrition science. Staying current in this remarkable field remains a challenge for educators and health professionals alike. In this eighth edition of *Understanding Normal and Clinical Nutrition,* we present updated, comprehensive coverage of the fundamentals of nutrition and nutrition therapy for an introductory nutrition course. The early chapters focus on "normal" nutrition—recommendations about nutrition that are essential for maintaining health and preventing disease. The later chapters provide lessons in "clinical" nutrition—the pathophysiology and nutrition therapy for a wide range of medical conditions. As with previous editions, each chapter has been substantially revised and updated. New research topics, such as functional foods, probiotics, cytokines, and nutritional genomics, are introduced or more fully explored. The chapters include practical information and valuable resources to help readers apply nutrition knowledge and skills to their daily lives and the clinical setting.

Our goal in writing this book has always been to share our excitement about the field of nutrition in a manner that motivates students to study and learn. Moreover, we seek to provide accurate, current information that is meaningful to the student or health professional. Individuals who study nutrition often find nutritional science to be at once both fascinating and overwhelming; there are so many "details" to learn—new terms, new chemical structures, and new biological concepts. Taken one step at a time, however, the science of nutrition may seem less daunting and the "facts" more memorable. We hope that this book serves you well.

The Chapters Chapter 1 begins by exploring why we eat the foods we do and continues with a brief overview of the nutrients, the science of nutrition, recommended nutrient intakes, assessment, and important relationships between diet and health. Chapter 2 describes the diet-planning principles and food guides used to create diets that support good health and includes instructions on how to read a food label. In Chapter 3, readers follow the journey of digestion and absorption as the body transforms foods into nutrients. Chapters 4 through 6 describe carbohydrates, fats, and proteins—their chemistry, roles in the body, and places in the diet. Chapter 7 shows how the body derives energy from these three nutrients. Chapters 8 and 9 continue the story with a look at energy balance, the factors associated with overweight and underweight, and the benefits and dangers of weight loss and weight gain. Chapters 10 through 13 describe the vitamins, the minerals, and water—their roles in the body, deficiency and toxicity symptoms, and food sources. Chapters 14 through 16 complete the "normal" chapters by presenting the special nutrient needs of people who are at different phases of the life cycle—pregnancy and lactation, infancy, childhood, adolescence, and adulthood and the later years.

The remaining "clinical" chapters of the book focus on the nutrition care of individuals with health problems. Chapter 17 explains how illnesses and their treatments influence nutrient needs and describes the process of nutrition assessment. Chapter 18 discusses how nutrition care is implemented and introduces the different types of therapeutic diets used in patient care. Chapter 19 explores the potential interactions between nutrients and medications and examines the benefits and risks associated with herbal remedies. Chapters 20 and 21 describe special ways of feeding people who cannot eat conventional foods. Chapter 22 explains the inflammatory process and shows how metabolic and respiratory stress influence nutrient needs. Chapters 23 through 29 explore the pathology, medical treatment, and nutrition care associated with specific diseases, including gastrointestinal disorders, liver disease, diabetes mellitus, cardiovascular diseases, renal diseases, cancer, and HIV infection.

The Highlights Every chapter is followed by a highlight that provides readers with an in-depth look at a current, and often controversial, topic that relates to its companion chapter. New highlights in this edition feature foodborne illnesses and the role of probiotics in intestinal health.

Special Features The art and layout in this edition have been carefully designed to be inviting while enhancing student learning. In addition, special features help readers identify key concepts and apply nutrition knowledge. For example, when a new term is introduced, it is printed in bold type and a **definition** is provided. These definitions often include pronunciations and derivations to facilitate understanding. A glossary at the end of the book includes all defined terms.

> **definition** (DEF-eh-NISH-en): the meaning of a word.
> - **de** = from
> - **finis** = boundary

Nutrition in Your Life/Nutrition in the Clinical Setting

Chapters 1 through 16 begin with Nutrition in Your Life sections that introduce the essece of the chapter with a friendly and familiar scenario. Similiarly, Chapters 17 through 29 begin with Nutrition in the Clinical Setting sections, which introduce real-life concerns associated with diseases or their treatments.

Nutrition Portfolio/Clinical Portfolio

At the end of Chapters 1 through 16, a Nutrition Portfolio section revisits the messages introduced in the chapter and prompts readers to consider whether their personal choices meet the dietary goals discussed. Chapters 17 through 29 end with a Clinical Portfolio section, which enables readers to practice their clinical skills by addressing hypothetical clinical situations.

IN SUMMARY

Each major section within a chapter concludes with a summary paragraph that reviews the key concepts. Similarly, summary tables organize information in an easy-to-read format.

Also featured in this edition are the *Dietary Guidelines for Americans 2005* recommendations, which are introduced in Chapter 2 and presented throughout the text whenever their subjects are discussed. Look for the following design.

Dietary Guidelines for Americans 2005

These guidelines provide science-based advice to promote health and to reduce the risk of chronic disease through diet and physical activity.

HOW TO

Many of the chapters include "How to" sections that guide readers through problem-solving tasks. For example, the "How to" in Chapter 1 takes students through the steps of calculating energy intake from the grams of carbohydrate, fat, and protein in a food; another "How to" in Chapter 18 shows how to estimate the energy requirements of a hospital patient.

CASE STUDY

The clinical chapters include case studies that present problems and pose questions that allow readers to apply chapter material to hypothetical situations. Readers who successfully master these exercises will be better prepared to face "real-life" challenges that arise in the clinical setting.

NUTRITION ASSESSMENT CHECKLIST

The clinical chapters close with a Nutrition Assessment Checklist that helps readers evaluate how various disorders impair nutrition status. These sections highlight the medical, dietary, anthropometric, biochemical, and physical findings most relevant to patients with specific diseases.

Most of the clinical chapters also include a section on Diet-Drug Interactions that describes the nutrition-related concerns associated with the medications commonly used to treat the disorders described in the chapter.

NUTRITION CALCULATIONS

CENGAGENOW

Several of the early chapters close with a "Nutrition Calculation" section. These sections often reinforce the "How to" lessons and provide practice in doing nutrition-related calculations. The problems enable readers to practice their skills and then check their answers (found at the end of the chapter). Readers who successfully master these exercises will be well prepared for "real-life" nutrition-related problems.

NUTRITION ON THE NET

Each chapter and many highlights conclude with Nutrition on the Net—a list of websites for further study of topics covered in the accompanying text. These lists do not imply an endorsement of the organizations or their programs. We have tried to provide reputable sources, but cannot be responsible for the content of these sites. (Read Highlight 1 to learn how to find reliable information on the Internet.)

STUDY QUESTIONS

CENGAGENOW

Each chapter ends with study questions in essay and multiple-choice format. Study questions offer readers the opportunity to review the major concepts presented in the chapters in preparation for exams. The page numbers after each essay question refer readers to discussions that answer the question; multiple-choice answers appear at the end of the chapter.

The Appendixes The appendixes are valuable references for a number of purposes. Appendix A summarizes background information on the hormonal and nervous systems, complementing Appendixes B and C on basic chemistry, the chemical structures of nutrients, and major metabolic pathways. Appendix D describes measures of protein quality. Appendix E provides supplemental coverage of nutrition assessment. Appendix F presents the estimated energy requirements for men and women at various levels of physical activity. Appendix G presents the 2007 U.S. Exchange System. Appendix H is an 8000-item food composition table compiled from the latest nutrient database assembled by Axxya Systems. Appendix I presents recommendations from the World Health Organization (WHO) and information for Canadians—the 2005 *Beyond the Basics* meal-planning system and 2007 guidelines for healthy eating and physical activities. Appendix J presents the Healthy People 2010 nutrition-related objectives. Appendix K provides examples of commercial enteral formulas commonly used in tube feedings or to supplement oral diets.

The Inside Covers The inside covers put commonly used information at your fingertips. The front covers (pp. A, B, and C) present the current nutrient recommendations; the inside back cover (p. Y on the left) features the Daily Values used on food labels and a glossary of nutrient measures; and the inside back cover (p. Z on the right) shows suggested weight ranges for various heights (based on the Body Mass Index). The pages just prior to the back cover (pp. W–X) assist readers with calculations and conversions.

Closing Comments We have taken great care to provide accurate information and have included many references at the end of each chapter and highlight. To keep the number of references manageable, however, many statements that appeared in previous editions with references now appear without them. All statements reflect current nutrition knowledge, and the authors will supply references to back editions upon request. In addition to supporting text statements, the end-of-chapter references provide readers with resources for finding a good overview or more details on the subject. Nutrition is a fascinating subject, and we hope our enthusiasm for it comes through on every page.

Sharon Rady Rolfes
Kathryn Pinna
Ellie Whitney
May 2008

Acknowledgments

To produce a book requires the coordinated effort of a team of people—and, no doubt, each team member has another team of support people as well. We salute, with a big round of applause, everyone who has worked so diligently to ensure the quality of this book.

We thank our partners and friends, Linda DeBruyne and Fran Webb, for their valuable consultations and contributions; working together over the past 20+ years has been a most wonderful experience. We especially appreciate Linda's research assistance on several chapters. Special thanks go to our colleagues Gail Hammond for her Canadian perspective, Sylvia Crews for her revision of the Aids to Calculation section at the end of the book, and David Stone for his careful critique of several newly written sections in the clinical chapters. A thousand thank-yous to Beth Magana, Marni Jay Rolfes, and Alex Rodriguez for their careful attention to manuscript preparation and a multitude of other daily tasks.

We also thank the many people who have prepared the ancillaries that accompany this text: Harry Sitren and Ileana Trautwein for writing and enhancing the test bank; Gail Hammond, Melissa Langone, Barbara Quinn, Tania Rivera, Sharon Stewart, Lori Turner, and Daryle Wane for contributing to the instructor's manual; Connie Goff for preparing PowerPoint lecture presentations; and Celine Heskey for creating materials for Cengage Now. Thanks also to the folks at Axxya for their assistance in creating the food composition appendix and developing the computerized diet analysis program that accompanies this book.

Our special thanks to our editorial team for their hard work and enthusiasm— Peter Adams for his leadership and support; Anna Lustig for her efficient analysis of reviews and patience during manuscript preparation; Trudy Brown for her efforts in managing production; Mary Berry for her outstanding copyediting abilities, interest in accuracy, and eye for detail; Gary Kliewer of The Book Company for his diligent attention to the innumerable details involved in production; Roman Barnes for the extra care he took to locate meaningful photos; Pat Lewis for proofreading the final text pages; Elesha Feldman for her competent coordination of ancillaries and her work on the food composition appendix; and Erin Taylor for composing a thorough and useful index. We'd also like to thank Diane Beasley for creatively designing these pages, Cathy Leonard for coordinating artwork and page production, and Karyn Morrison and Margaret Chamberlain-Gaston for their assistance in obtaining permissions. To the many, many others involved in production and sales, we tip our hats in appreciation.

We are especially grateful to our associates, friends, and families for their continued encouragement and support. We also thank our many reviewers for their comments and contributions to this edition and all previous editions.

Reviewers of *Understanding Normal and Clinical Nutrition*

Melody Anacker
Montana State University

Janet Anderson
Utah State University

Judi Brooks
Eastern Michigan University

Richard S. Crow
University of Minnesota

Robert Davidson
Brigham Young University

Marguerite Dunne
Marist College

Brenda Eissenstat
Pennsylvania State University

Cindy Fitch
West Virginia University

Mary Flynn
Brown University

Gloria Gonzalez
Pensacola Junior College

Kathleen Gould
Townson University

Kathryn Henry
Hood College

Le Greta Hudson
University of Missouri–Columbia

Dale Larson
Johnson Community College

Katy Lenker
University of Central Oklahoma

Lorraine Lewis
Viterbo University

Kimberly Lower
Collin County Community College

Mary Maciolek
Middlesex County College

Kim McMahon
Utah State University

Steven Nizielski
Grand Valley State University

Anna Page
Johnson County Community College

Sarah Panarello
Yakima Valley Community College

Roman Pawlak
East Carolina University

Sue Roberts
Walla Walla Community College

Linda Shepherd
College of Saint Benedict, Saint John's University

Sandra Shortt
Cedarville University

Denise Signorelli
Community College of Southern Nevada

Mollie Smith
California State University, Fresno

Luann Soliah
Baylor University

Tammy Stephenson
University of Kentucky

Sherry Stewart
University of Texas at Dallas

Trinh Tran
City College of San Francisco

Eric Vlahov
University of Tampa

Janelle Walter
Baylor University

Stacie Wing-Gaia
University of Utah

Nutrition in Your Life

Believe it or not, you have probably eaten at least 20,000 meals in your life. Without any conscious effort on your part, your body uses the nutrients from those foods to make all its components, fuel all its activities, and defend itself against diseases. How successfully your body handles these tasks depends, in part, on your food choices. Nutritious food choices support healthy bodies.

The CengageNOW logo indicates an opportunity for online self-study, linking you to interactive tutorials and videos based on your level of understanding.

academic.cengage.com/login

How To: Practice Problems
Nutrition Portfolio Journal
Nutrition Calculations: Practice Problems

An Overview of Nutrition

Welcome to the world of **nutrition.** Although you may not always have been aware of it, nutrition has played a significant role in your life. And it will continue to affect you in major ways, depending on the **foods** you select.

Every day, several times a day, you make food choices that influence your body's health for better or worse. Each day's choices may benefit or harm your health only a little, but when these choices are repeated over years and decades, the rewards or consequences become major. That being the case, paying close attention to good eating habits now can bring you health benefits later. Conversely, carelessness about food choices can contribute to many chronic diseases ◆ prevalent in later life, including heart disease and cancer. Of course, some people will become ill or die young no matter what choices they make, and others will live long lives despite making poor choices. For the majority of us, however, the food choices we make each and every day will benefit or impair our health in proportion to the wisdom of those choices.

Although most people realize that their food habits affect their health, they often choose foods for other reasons. After all, foods bring to the table a variety of pleasures, traditions, and associations as well as nourishment. The challenge, then, is to combine favorite foods and fun times with a nutritionally balanced **diet.**

Food Choices

People decide what to eat, when to eat, and even whether to eat in highly personal ways, often based on behavioral or social motives rather than on an awareness of nutrition's importance to health. Many different food choices can support good health, and an understanding of nutrition helps you make sensible selections more often.

Personal Preference As you might expect, the number one reason people choose foods is taste—they like certain flavors. Two widely shared preferences are for the sweetness of sugar and the savoriness of salt. Liking high-fat foods also appears to be a universally common preference. Other preferences might be for the hot peppers

◆ In general, a **chronic** disease progresses slowly or with little change and lasts a long time. By comparison, an **acute** disease develops quickly, produces sharp symptoms, and runs a short course.
• **chronos** = time
• **acute** = sharp

nutrition: the science of foods and the nutrients and other substances they contain, and of their actions within the body (including ingestion, digestion, absorption, transport, metabolism, and excretion). A broader definition includes the social, economic, cultural, and psychological implications of food and eating.

foods: products derived from plants or animals that can be taken into the body to yield energy and nutrients for the maintenance of life and the growth and repair of tissues.

diet: the foods and beverages a person eats and drinks.

An enjoyable way to learn about other cultures is to taste their ethnic foods.

common in Mexican cooking or the curry spices of Indian cuisine. Some research suggests that genetics may influence people's food preferences.[1]

Habit People sometimes select foods out of habit. They eat cereal every morning, for example, simply because they have always eaten cereal for breakfast. Eating a familiar food and not having to make any decisions can be comforting.

Ethnic Heritage or Tradition

Among the strongest influences on food choices are ethnic heritage and tradition. People eat the foods they grew up eating. Every country, and in fact every region of a country, has its own typical foods and ways of combining them into meals. The "American diet" includes many ethnic foods from various countries, all adding variety to the diet. This is most evident when eating out: 60 percent of U.S. restaurants (excluding fast-food places) have an ethnic emphasis, most commonly Chinese, Italian, or Mexican.

Social Interactions Most people enjoy companionship while eating. It's fun to go out with friends for pizza or ice cream. Meals are social events, and sharing food is part of hospitality. Social customs invite people to accept food or drink offered by a host or shared by a group.

Availability, Convenience, and Economy People eat foods that are accessible, quick and easy to prepare, and within their financial means. Today's consumers value convenience and are willing to spend more than half of their food budget on meals that require little, if any, further preparation.[2] They frequently eat out, bring home ready-to-eat meals, or have food delivered. Even when they venture into the kitchen, they want to prepare a meal in 15 to 20 minutes, using less than a half dozen ingredients—and those "ingredients" are often semiprepared foods, such as canned soups. This emphasis on convenience limits food choices to the selections offered on menus and products designed for quick preparation. Whether decisions based on convenience meet a person's nutrition needs depends on the choices made. Eating a banana or a candy bar may be equally convenient, but the fruit offers more vitamins and minerals and less sugar and fat.

Positive and Negative Associations People tend to like particular foods associated with happy occasions—such as hot dogs at ball games or cake and ice cream at birthday parties. By the same token, people can develop aversions and dislike foods that they ate when they felt sick or that were forced on them.[3] By using foods as rewards or punishments, parents may inadvertently teach their children to like and dislike certain foods.

Emotional Comfort Some people cannot eat when they are emotionally upset. Others may eat in response to a variety of emotional stimuli—for example, to relieve boredom or depression or to calm anxiety.[4] A depressed person may choose to eat rather than to call a friend. A person who has returned home from an exciting evening out may unwind with a late-night snack. These people may find emotional comfort, in part, because foods can influence the brain's chemistry and the mind's response. Carbohydrates and alcohol, for example, tend to calm, whereas proteins and caffeine are more likely to activate. Eating in response to emotions can easily lead to overeating and obesity, but it may be appropriate at times. For example, sharing food at times of bereavement serves both the giver's need to provide comfort and the receiver's need to be cared for and to interact with others, as well as to take nourishment.

Values Food choices may reflect people's religious beliefs, political views, or environmental concerns. For example, many Christians forgo meat during Lent (the period prior to Easter), Jewish law includes an extensive set of dietary rules that govern the use of foods derived from animals, and Muslims fast between sunrise and sunset during Ramadan (the ninth month of the Islamic calendar). A concerned consumer may boycott fruit picked by migrant workers who have been exploited. People may buy vegetables from local farmers to save the fuel and environmental costs of foods shipped in from far away. They may also select foods packaged in containers that can be reused or recycled. Some consumers accept or reject foods that have been irradiated or genetically modified, depending on their approval of these processes.

Body Weight and Image Sometimes people select certain foods and supplements that they believe will improve their physical appearance and avoid those they believe might be detrimental. Such decisions can be beneficial when based on sound nutrition and fitness knowledge, but decisions based on fads or carried to extremes undermine good health, as pointed out in later discussions of eating disorders (Highlight 8).

Nutrition and Health Benefits Finally, of course, many consumers make food choices that will benefit health. Food manufacturers and restaurant chefs have responded to scientific findings linking health with nutrition by offering an abundant selection of health-promoting foods and beverages. Foods that provide health benefits beyond their nutrient contributions are called **functional foods.**[5] Whole foods—as natural and familiar as oatmeal or tomatoes—are the simplest functional foods. In other cases, foods have been modified to provide health benefits, perhaps by lowering the fat contents. In still other cases, manufacturers have fortified foods by adding nutrients or **phytochemicals** that provide health benefits (see Highlight 13). ◆ Examples of these functional foods include orange juice fortified with calcium to help build strong bones and margarine made with a plant sterol that lowers blood cholesterol.

Consumers typically welcome new foods into their diets, provided that these foods are reasonably priced, clearly labeled, easy to find in the grocery store, and convenient to prepare. These foods must also taste good—as good as the traditional choices. Of course, a person need not eat any of these "special" foods to enjoy a healthy diet; many "regular" foods provide numerous health benefits as well. In fact, "regular" foods such as whole grains; vegetables and legumes; fruits; meats, fish, and poultry; and milk products are among the healthiest choices a person can make.

To enhance your health, keep nutrition in mind when selecting foods.

◆ Functional foods may include whole foods, modified foods, or fortified foods.

IN SUMMARY

A person selects foods for a variety of reasons. Whatever those reasons may be, food choices influence health. Individual food selections neither make nor break a diet's healthfulness, but the balance of foods selected over time can make an important difference to health.[6] For this reason, people are wise to think "nutrition" when making their food choices.

The Nutrients

Biologically speaking, people eat to receive nourishment. Do you ever think of yourself as a biological being made of carefully arranged atoms, molecules, cells, tissues, and organs? Are you aware of the activity going on within your body even as you sit still? The atoms, molecules, and cells of your body continually move and change, even though the structures of your tissues and organs and your external appearance remain relatively constant. Your skin, which has covered you since your birth, is replaced entirely by new cells every seven years. The fat beneath your skin is not the

functional foods: foods that contain physiologically active compounds that provide health benefits beyond their nutrient contributions; sometimes called *designer foods* or *nutraceuticals.*

phytochemicals (FIE-toe-KEM-ih-cals): nonnutrient compounds found in plant-derived foods that have biological activity in the body.

• **phyto** = plant

Foods bring pleasure—and nutrients.

◆ As Chapter 5 explains, most lipids are fats.

same fat that was there a year ago. Your oldest red blood cell is only 120 days old, and the entire lining of your digestive tract is renewed every 3 to 5 days. To maintain your "self," you must continually replenish, from foods, the **energy** and the **nutrients** you deplete as your body maintains itself.

Nutrients in Foods and in the Body

Amazingly, our bodies can derive all the energy, structural materials, and regulating agents we need from the foods we eat. This section introduces the nutrients that foods deliver and shows how they participate in the dynamic processes that keep people alive and well.

Composition of Foods Chemical analysis of a food such as a tomato shows that it is composed primarily of water (95 percent). Most of the solid materials are carbohydrates, lipids, ◆ and proteins. If you could remove these materials, you would find a tiny residue of vitamins, minerals, and other compounds. Water, carbohydrates, lipids, proteins, vitamins, and some of the minerals found in foods are nutrients—substances the body uses for the growth, maintenance, and repair of its tissues.

This book focuses mostly on the nutrients, but foods contain other compounds as well—fibers, phytochemicals, pigments, additives, alcohols, and others. Some are beneficial, some are neutral, and a few are harmful. Later sections of the book touch on these compounds and their significance.

Composition of the Body A complete chemical analysis of your body would show that it is made of materials similar to those found in foods (see Figure 1-1). A healthy 150-pound body contains about 90 pounds of water and about 20 to 45 pounds of fat. The remaining pounds are mostly protein, carbohydrate, and the major minerals of the bones. Vitamins, other minerals, and incidental extras constitute a fraction of a pound.

FIGURE 1-1	Body Composition of Healthy-Weight Men and Women

The human body is made of compounds similar to those found in foods—mostly water (60 percent) and some fat (13 to 21 percent for young men, 23 to 31 percent for young women), with carbohydrate, protein, vitamins, minerals, and other minor constituents making up the remainder. (Chapter 8 describes the health hazards of too little or too much body fat.)

Key:

☐ % Carbohydrates, proteins, vitamins, minerals in the body

☐ % Fat in the body

☐ % Water in the body

energy: the capacity to do work. The energy in food is chemical energy. The body can convert this chemical energy to mechanical, electrical, or heat energy.

nutrients: chemical substances obtained from food and used in the body to provide energy, structural materials, and regulating agents to support growth, maintenance, and repair of the body's tissues. Nutrients may also reduce the risks of some diseases.

Chemical Composition of Nutrients The simplest of the nutrients are the minerals. Each mineral is a chemical element; its atoms are all alike. As a result, its identity never changes. For example, iron may have different electrical charges, but the individual iron atoms remain the same when they are in a food, when a person eats the food, when the iron becomes part of a red blood cell, when the cell is broken down, and when the iron is lost from the body by excretion. The next simplest nutrient is water, a compound made of two elements—hydrogen and oxygen. Minerals and water are **inorganic** nutrients—which means they do not contain carbon.

The other four classes of nutrients (carbohydrates, lipids, proteins, and vitamins) are more complex. In addition to hydrogen and oxygen, they all contain carbon, an element found in all living things. They are therefore called **organic** ◆ compounds (meaning, literally, "alive"). Protein and some vitamins also contain nitrogen and may contain other elements as well (see Table 1-1).

Essential Nutrients The body can make some nutrients, but it cannot make all of them. Also, it makes some in insufficient quantities to meet its needs and, therefore, must obtain these nutrients from foods. The nutrients that foods must supply are **essential nutrients.** When used to refer to nutrients, the word *essential* means more than just "necessary"; it means "needed from outside the body"—normally, from foods.

The Energy-Yielding Nutrients: Carbohydrate, Fat, and Protein

In the body, three organic nutrients can be used to provide energy: carbohydrate, fat, and protein. ◆ In contrast to these **energy-yielding nutrients**, vitamins, minerals, and water do not yield energy in the human body.

Energy Measured in kCalories The energy released from carbohydrates, fats, and proteins can be measured in **calories**—tiny units of energy so small that a single apple provides tens of thousands of them. To ease calculations, energy is expressed in 1000-calorie metric units known as kilocalories (shortened to kcalories, but commonly called "calories"). When you read in popular books or magazines that an apple provides "100 calories," it actually means 100 kcalories. This book uses the term kcalorie and its abbreviation kcal throughout, as do other scientific books and journals. ◆ The "How to" on p. 8 provides a few tips on "thinking metric."

◆ In agriculture, *organic* farming refers to growing crops and raising livestock according to standards set by the U.S. Department of Agriculture (USDA).

◆ Carbohydrate, fat, and protein are sometimes called **macronutrients** because the body requires them in relatively large amounts (many grams daily). In contrast, vitamins and minerals are **micronutrients**, required only in small amounts (milligrams or micrograms daily).

◆ The international unit for measuring food energy is the **joule**, a measure of *work* energy. To convert kcalories to kilojoules, multiply by 4.2; to convert kilojoules to kcalories, multiply by 0.24.

inorganic: not containing carbon or pertaining to living things.
- **in** = not

organic: in chemistry, a substance or molecule containing carbon-carbon bonds or carbon-hydrogen bonds. This definition excludes coal, diamonds, and a few carbon-containing compounds that contain only a single carbon and no hydrogen, such as carbon dioxide (CO_2), calcium carbonate ($CaCO_3$), magnesium carbonate ($MgCO_3$), and sodium cyanide (NaCN).

essential nutrients: nutrients a person must obtain from food because the body cannot make them for itself in sufficient quantity to meet physiological needs; also called **indispensable nutrients.** About 40 nutrients are currently known to be essential for human beings.

energy-yielding nutrients: the nutrients that break down to yield energy the body can use:
- Carbohydrate
- Fat
- Protein

calories: units by which energy is measured. Food energy is measured in **kilocalories** (1000 calories equal 1 kilocalorie), abbreviated **kcalories** or **kcal.** One kcalorie is the amount of heat necessary to raise the temperature of 1 kilogram (kg) of water 1°C. The scientific use of the term *kcalorie* is the same as the popular use of the term *calorie.*

| TABLE 1-1 | Elements in the Six Classes of Nutrients | | | | |

Notice that organic nutrients contain carbon.

	Carbon	Hydrogen	Oxygen	Nitrogen	Minerals
Inorganic nutrients					
Minerals					✓
Water		✓	✓		
Organic nutrients					
Carbohydrates	✓	✓	✓		
Lipids (fats)	✓	✓	✓		
Proteins[a]	✓	✓	✓	✓	
Vitamins[b]	✓	✓	✓		

[a] Some proteins also contain the mineral sulfur.
[b] Some vitamins contain nitrogen; some contain minerals.

HOW TO Think Metric

Like other scientists, nutrition scientists use metric units of measure. They measure food energy in kilocalories, people's height in centimeters, people's weight in kilograms, and the weights of foods and nutrients in grams, milligrams, or micrograms. For ease in using these measures, it helps to remember that the prefixes on the grams imply 1000. For example, a *kilo*gram is 1000 grams, a *milli*gram is 1/1000 of a gram, and a *micro*gram is 1/1000 of a milligram.

Most food labels and many recipe books provide "dual measures," listing both household measures, such as cups, quarts, and teaspoons, and metric measures, such as milliliters, liters, and grams. This practice gives people an opportunity to gradually learn to "think metric."

A person might begin to "think metric" by simply observing the measure—by noticing the amount of soda in a 2-liter bottle, for example. Through such experiences, a person can become familiar with a measure without having to do any conversions.

To facilitate communication, many members of the international scientific community have adopted a common system of measurement—the International System of Units (SI). In addition to using metric measures, the SI establishes common units of measurement. For example, the SI unit for measuring food energy is the joule (not the kcalorie). A joule is the amount of energy expended when 1 kilogram is moved 1 meter by a force of 1 newton. The joule is thus a measure of *work* energy, whereas the kcalorie is a measure of *heat* energy. While many scientists and journals report their findings in kilojoules (kJ), many others, particularly those in the United States, use kcalories (kcal). To convert energy measures from kcalories to kilojoules, multiply by 4.2. For example, a 50-kcalorie cookie provides 210 kilojoules:

$$50 \text{ kcal} \times 4.2 = 210 \text{ kJ}$$

Exact conversion factors for these and other units of measure are in the Aids to Calculation section on the last two pages of the book.

Volume: Liters (L)

1 L = 1000 milliliters (mL)
0.95 L = 1 quart
1 mL = 0.03 fluid ounces
240 mL = 1 cup

A liter of liquid is approximately one U.S. quart. (Four liters are only about 5 percent more than a gallon.)

One cup is about 240 milliliters; a half-cup of liquid is about 120 milliliters.

Weight: Grams (g)

1 g = 1000 milligrams (mg)
1 g = 0.04 ounce (oz)
1 oz = 28.35 g (or 30 g)
100 g = 3½ oz
1 kilogram (kg) = 1000 g
1 kg = 2.2 pounds (lb)
454 g = 1 lb

A kilogram is slightly more than 2 lb; conversely, a pound is about ½ kg.

A half-cup of vegetables weighs about 100 grams; one pea weighs about ½ gram.

A 5-pound bag of potatoes weighs about 2 kilograms, and a 176-pound person weighs 80 kilograms.

CENGAGENOW™
To practice thinking metrically, log on to **academic.cengage.com/login**, go to Chapter 1, then go to How To.

◆ Foods with a high energy density help with weight gain, whereas those with a low energy density help with weight loss.

energy density: a measure of the energy a food provides relative to the amount of food (kcalories per gram).

Energy from Foods The amount of energy a food provides depends on how much carbohydrate, fat, and protein it contains. When completely broken down in the body, a gram of carbohydrate yields about 4 kcalories of energy; a gram of protein also yields 4 kcalories; and a gram of fat yields 9 kcalories (see Table 1-2). Fat, therefore, has a greater **energy density** than either carbohydrate or protein. Figure 1-2 compares the energy density of two breakfast options, and later chapters describe how considering a food's energy density can help with weight management. ◆ The "How to" on p. 9 explains how to calculate the energy available from foods.

One other substance contributes energy—alcohol. Alcohol is not considered a nutrient because it interferes with the growth, maintenance, and repair of the body, but it does yield energy (7 kcalories per gram) when metabolized in the body. (Highlight 7 presents alcohol metabolism; Chapter 27 mentions the potential harmful role of alcohol in hypertension and the possible beneficial role in heart disease.)

FIGURE 1-2 Energy Density of Two Breakfast Options Compared

Gram for gram, ounce for ounce, and bite for bite, foods with a high energy density deliver more kcalories than foods with a low energy density. Both of these breakfast options provide 500 kcalories, but the cereal with milk, fruit salad, scrambled egg, turkey sausage, and toast with jam offers three times as much food as the doughnuts (based on weight); it has a lower energy density than the doughnuts. Selecting a variety of foods also helps to ensure nutrient adequacy.

LOWER ENERGY DENSITY

This 450-gram breakfast delivers 500 kcalories, for an energy density of 1.1 (500 kcal ÷ 450 g = 1.1 kcal/g).

HIGHER ENERGY DENSITY

This 144-gram breakfast delivers 500 kcalories, for an energy density of 3.5 (500 kcal ÷ 144 g = 3.5 kcal/g).

© Matthew Farruggio (both)

Most foods contain all three energy-yielding nutrients, as well as water, vitamins, minerals, and other substances. For example, meat contains water, fat, vitamins, and minerals as well as protein. Bread contains water, a trace of fat, a little protein, and some vitamins and minerals in addition to its carbohydrate. Only a few foods are exceptions to this rule, the common ones being sugar (pure carbohydrate) and oil (essentially pure fat).

Energy in the Body   The body uses the energy-yielding nutrients to fuel all its activities. When the body uses carbohydrate, fat, or protein for energy, the bonds between

HOW TO Calculate the Energy Available from Foods

To calculate the energy available from a food, multiply the number of grams of carbohydrate, protein, and fat by 4, 4, and 9, respectively. Then add the results together. For example, 1 slice of bread with 1 tablespoon of peanut butter on it contains 16 grams carbohydrate, 7 grams protein, and 9 grams fat:

16 g carbohydrate × 4 kcal/g = 64 kcal
7 g protein × 4 kcal/g = 28 kcal
9 g fat × 9 kcal/g = 81 kcal
Total = 173 kcal

From this information, you can calculate the percentage of kcalories each of the energy nutrients contributes to the total. To determine the percentage of kcalories from fat, for example, divide the 81 fat kcalories by the total 173 kcalories:

81 fat kcal ÷ 173 total kcal = 0.468 (rounded to 0.47)

Then multiply by 100 to get the percentage:

0.47 × 100 = 47%

Dietary recommendations that urge people to limit fat intake to 20 to 35 percent of kcalories refer to the day's total energy intake, not to individual foods. Still, if the proportion of fat in each food choice throughout a day exceeds 35 percent of kcalories, then the day's total surely will, too. Knowing that this snack provides 47 percent of its kcalories from fat alerts a person to the need to make lower-fat selections at other times that day.

CENGAGENOW™
To practice calculating the energy available from foods, log on to **academic.cengage.com/login**, go to Chapter 1, then go to How To.

TABLE 1-2 kCalorie Values of Energy Nutrients[a]

Nutrients	Energy (kcal/g)
Carbohydrate	4
Fat	9
Protein	4

NOTE: Alcohol contributes 7 kcalories per gram that can be used for energy, but it is not considered a nutrient because it interferes with the body's growth, maintenance, and repair.

[a]For those using kilojoules: 1 g carbohydrate = 17 kJ; 1 g protein = 17 kJ; 1 g fat = 37 kJ; and 1 g alcohol = 29 kJ.

◆ The processes by which nutrients are broken down to yield energy or used to make body structures are known as **metabolism** (defined and described further in Chapter 7).

the nutrient's atoms break. As the bonds break, they release energy. ◆ Some of this energy is released as heat, but some is used to send electrical impulses through the brain and nerves, to synthesize body compounds, and to move muscles. Thus the energy from food supports every activity from quiet thought to vigorous sports.

If the body does not use these nutrients to fuel its current activities, it rearranges them into storage compounds (such as body fat), to be used between meals and overnight when fresh energy supplies run low. If more energy is consumed than expended, the result is an increase in energy stores and weight gain. Similarly, if less energy is consumed than expended, the result is a decrease in energy stores and weight loss.

When consumed in excess of energy needs, alcohol, too, can be converted to body fat and stored. When alcohol contributes a substantial portion of the energy in a person's diet, the harm it does far exceeds the problems of excess body fat. (Highlight 7 describes the effects of alcohol on health and nutrition.)

Other Roles of Energy-Yielding Nutrients In addition to providing energy, carbohydrates, fats, and proteins provide the raw materials for building the body's tissues and regulating its many activities. In fact, protein's role as a fuel source is relatively minor compared with both the other two nutrients and its other roles. Proteins are found in structures such as the muscles and skin and help to regulate activities such as digestion and energy metabolism.

The Vitamins

The **vitamins** are also organic, but they do not provide energy. Instead, they facilitate the release of energy from carbohydrate, fat, and protein and participate in numerous other activities throughout the body.

Each of the 13 different vitamins has its own special roles to play.* One vitamin enables the eyes to see in dim light, another helps protect the lungs from air pollution, and still another helps make the sex hormones—among other things. When you cut yourself, one vitamin helps stop the bleeding and another helps repair the skin. Vitamins busily help replace old red blood cells and the lining of the digestive tract. Almost every action in the body requires the assistance of vitamins.

Vitamins can function only if they are intact, but because they are complex organic molecules, they are vulnerable to destruction by heat, light, and chemical agents. This is why the body handles them carefully, and why nutrition-wise cooks do, too. The strategies of cooking vegetables at moderate temperatures for short times and using small amounts of water help to preserve the vitamins.

The Minerals

In the body, some **minerals** are put together in orderly arrays in such structures as bones and teeth. Minerals are also found in the fluids of the body, which influences fluid properties. Whatever their roles, minerals do not yield energy.

Only 16 minerals are known to be essential in human nutrition.† Others are being studied to determine whether they play significant roles in the human body. Still other minerals are environmental contaminants that displace the nutrient minerals from their workplaces in the body, disrupting body functions. The problems caused by contaminant minerals are described in Chapter 13.

Because minerals are inorganic, they are indestructible and need not be handled with the special care that vitamins require. Minerals can, however, be bound by substances that interfere with the body's ability to absorb them. They can also be lost during food-refining processes or during cooking when they leach into water that is discarded.

vitamins: organic, essential nutrients required in small amounts by the body for health.

minerals: inorganic elements. Some minerals are essential nutrients required in small amounts by the body for health.

---

* The water-soluble vitamins are vitamin C and the eight B vitamins: thiamin, riboflavin, niacin, vitamins B_6 and B_{12}, folate, biotin, and pantothenic acid. The fat-soluble vitamins are vitamins A, D, E, and K. The water-soluble vitamins are the subject of Chapter 10 and the fat-soluble vitamins, of Chapter 11.
† The major minerals are calcium, phosphorus, potassium, sodium, chloride, magnesium, and sulfate. The trace minerals are iron, iodine, zinc, chromium, selenium, fluoride, molybdenum, copper, and manganese. Chapters 12 and 13 are devoted to the major and trace minerals, respectively.

Water

Water, indispensable and abundant, provides the environment in which nearly all the body's activities are conducted. It participates in many metabolic reactions and supplies the medium for transporting vital materials to cells and carrying waste products away from them. Water is discussed fully in Chapter 12, but it is mentioned in every chapter. If you watch for it, you cannot help but be impressed by water's participation in all life processes.

© Corbis

Water itself is an essential nutrient and naturally carries many minerals.

> **IN SUMMARY**
>
> Foods provide nutrients—substances that support the growth, maintenance, and repair of the body's tissues. The six classes of nutrients include:
>
> · Carbohydrates
> · Lipids (fats)
> · Proteins
> · Vitamins
> · Minerals
> · Water
>
> Foods rich in the energy-yielding nutrients (carbohydrates, fats, and proteins) provide the major materials for building the body's tissues and yield energy for the body's use or storage. Energy is measured in kcalories. Vitamins, minerals, and water facilitate a variety of activities in the body.

Without exaggeration, nutrients provide the physical and metabolic basis for nearly all that we are and all that we do. The next section introduces the science of nutrition with emphasis on the research methods scientists have used in uncovering the wonders of nutrition.

The Science of Nutrition

The science of nutrition is the study of the nutrients and other substances in foods and the body's handling of them. Its foundation depends on several other sciences, including biology, biochemistry, and physiology. As sciences go, nutrition is young, but as you can see from the size of this book, much has happened in nutrition's short life. And it is currently entering a tremendous growth spurt as scientists apply knowledge gained from sequencing the human **genome.** The integration of nutrition, genomics, and molecular biology has opened a whole new world of study called **nutritional genomics**—the science of how nutrients affect the activities of genes and how genes affect the interactions between diet and disease.[7] Highlight 6 describes how nutritional genomics is shaping the science of nutrition, and examples of nutrient–gene interactions appear throughout later sections of the book.

Conducting Research

Consumers may depend on personal experience or reports from friends ◆ to gather information on nutrition, but researchers use the scientific method to guide their work (see Figure 1-3 on p. 12). As the figure shows, research always begins with a problem or a question. For example, "What foods or nutrients might protect against the common cold?" In search of an answer, scientists make an educated guess **(hypothesis),** such as "foods rich in vitamin C reduce the number of common colds." Then they systematically conduct research studies to collect data that will test the hypothesis (see the glossary on p. 14 for definitions of research terms). Some examples of various types of research designs are presented in Figure 1-4 (p. 13). Each type of study has strengths and weaknesses (see Table 1-3 on p. 14). Consequently, some provide stronger evidence than others.

◆ A personal account of an experience or event is an **anecdote** and is not accepted as reliable scientific information.
  • **anekdotos** = unpublished

genome (GEE-nome): the full complement of genetic material (DNA) in the chromosomes of a cell. In human beings, the genome consists of 46 chromosomes. The study of genomes is called **genomics.**

nutritional genomics: the science of how nutrients affect the activities of genes **(nutrigenomics)** and how genes affect the interactions between diet and disease **(nutrigenetics)**.

FIGURE 1-3 The Scientific Method

Research scientists follow the scientific method. Note that most research generates new questions, not final answers. Thus the sequence begins anew, and research continues in a somewhat cyclical way.

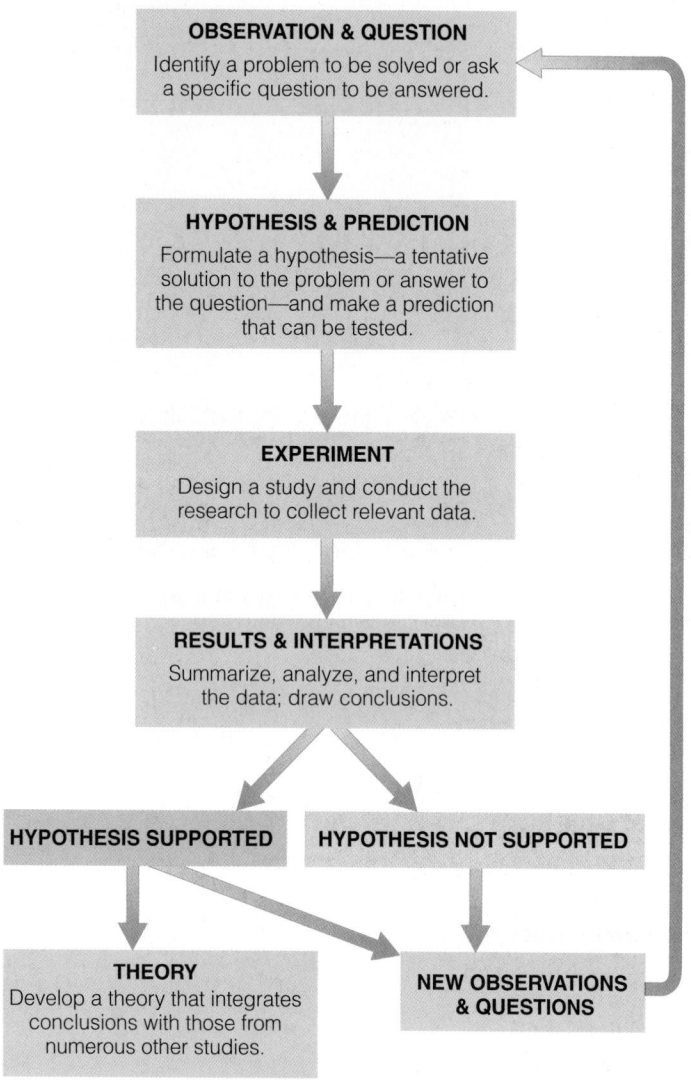

OBSERVATION & QUESTION
Identify a problem to be solved or ask a specific question to be answered.

HYPOTHESIS & PREDICTION
Formulate a hypothesis—a tentative solution to the problem or answer to the question—and make a prediction that can be tested.

EXPERIMENT
Design a study and conduct the research to collect relevant data.

RESULTS & INTERPRETATIONS
Summarize, analyze, and interpret the data; draw conclusions.

HYPOTHESIS SUPPORTED

HYPOTHESIS NOT SUPPORTED

THEORY
Develop a theory that integrates conclusions with those from numerous other studies.

NEW OBSERVATIONS & QUESTIONS

In attempting to discover whether a nutrient relieves symptoms or cures a disease, researchers deliberately manipulate one variable (for example, the amount of vitamin C in the diet) and measure any observed changes (perhaps the number of colds). As much as possible, all other conditions are held constant. The following paragraphs illustrate how this is accomplished.

Controls In studies examining the effectiveness of vitamin C, researchers typically divide the **subjects** into two groups. One group (the **experimental group**) receives a vitamin C supplement, and the other (the **control group**) does not. Researchers observe both groups to determine whether one group has fewer or shorter colds than the other. The following discussion describes some of the pitfalls inherent in an experiment of this kind and ways to avoid them.

In sorting subjects into two groups, researchers must ensure that each person has an equal chance of being assigned to either the experimental group or the control group. This is accomplished by **randomization;** that is, the subjects are chosen randomly from the same population by flipping a coin or some other method involving chance. Randomization helps to ensure that results reflect the treatment and not factors that might influence the grouping of subjects.

Importantly, the two groups of people must be similar and must have the same track record with respect to colds to rule out the possibility that observed differences in the rate, severity, or duration of colds might have occurred anyway. If, for example, the control group would normally catch twice as many colds as the experimental group, then the findings prove nothing.

In experiments involving a nutrient, the diets of both groups must also be similar, especially with respect to the nutrient being studied. If those in the experimental group were receiving less vitamin C from their usual diet, then any effects of the supplement may not be apparent.

Sample Size To ensure that chance variation between the two groups does not influence the results, the groups must be large. For example, if one member of a group of five people catches a bad cold by chance, he will pull the whole group's average toward bad colds; but if one member of a group of 500 catches a bad cold, she will not unduly affect the group average. Statistical methods are used to determine whether differences between groups of various sizes support a hypothesis.

Placebos If people who take vitamin C for colds *believe* it will cure them, their chances of recovery may improve. Taking anything believed to be beneficial may hasten recovery. This phenomenon, the result of expectations, is known as the **placebo effect.** In experiments designed to determine vitamin C's effect on colds, this mind-body effect must be rigorously controlled. Severity of symptoms is often a subjective measure, and people who believe they are receiving treatment may report less severe symptoms.

One way experimenters control for the placebo effect is to give pills to all participants. Those in the experimental group, for example, receive pills containing vitamin C, and those in the control group receive a **placebo**—pills of similar appearance and taste containing an inactive ingredient. This way, the expectations of both groups will be equal. It is not necessary to convince all subjects that they are receiving vitamin C, but the extent of belief or unbelief must be the same in both groups. A study conducted under these conditions is called a **blind exper-**

FIGURE 1-4 Examples of Research Designs

EPIDEMIOLOGICAL STUDIES

CROSS-SECTIONAL

Researchers observe how much and what kinds of foods a group of people eat and how healthy those people are. Their findings identify factors that might influence the incidence of a disease in various populations.

Example. The people of the Mediterranean region drink lots of wine, eat plenty of fat from olive oil, and have a lower incidence of heart disease than northern Europeans and North Americans.

CASE-CONTROL

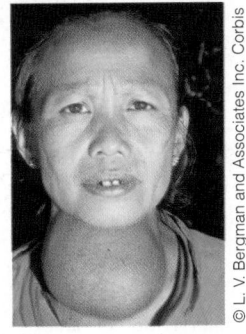

Researchers compare people who do and do not have a given condition such as a disease, closely matching them in age, gender, and other key variables so that differences in other factors will stand out. These differences may account for the condition in the group that has it.

Example. People with goiter lack iodine in their diets.

COHORT

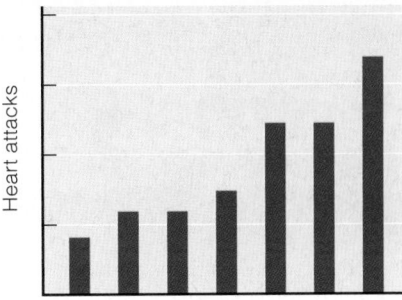

Researchers analyze data collected from a selected group of people (a cohort) at intervals over a certain period of time.

Example. Data collected periodically over the past several decades from over 5000 people randomly selected from the town of Framingham, Massachusetts, in 1948 have revealed that the risk of heart attack increases as blood cholesterol increases.

EXPERIMENTAL STUDIES

LABORATORY-BASED ANIMAL STUDIES

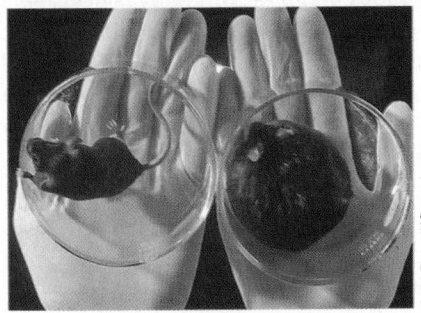

Researchers feed animals special diets that provide or omit specific nutrients and then observe any changes in health. Such studies test possible disease causes and treatments in a laboratory where all conditions can be controlled.

Example. Mice fed a high-fat diet eat less food than mice given a lower-fat diet, so they receive the same number of kcalories—but the mice eating the fat-rich diet become severely obese.

LABORATORY-BASED IN VITRO STUDIES

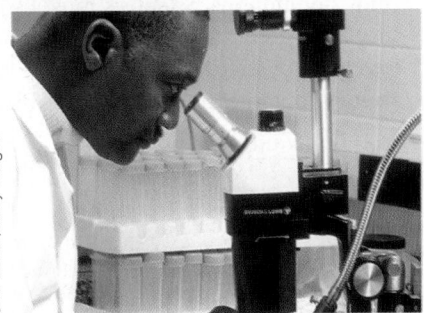

Researchers examine the effects of a specific variable on a tissue, cell, or molecule isolated from a living organism.

Example. Laboratory studies find that fish oils inhibit the growth and activity of the bacteria implicated in ulcer formation.

HUMAN INTERVENTION (OR CLINICAL) TRIALS

Researchers ask people to adopt a new behavior (for example, eat a citrus fruit, take a vitamin C supplement, or exercise daily). These trials help determine the effectiveness of such interventions on the development or prevention of disease.

Example. Heart disease risk factors improve when men receive fresh-squeezed orange juice daily for two months compared with those on a diet low in vitamin C—even when both groups follow a diet high in saturated fat.

iment—that is, the subjects do not know (are blind to) whether they are members of the experimental group (receiving treatment) or the control group (receiving the placebo).

Double Blind When both the subjects and the researchers do not know which subjects are in which group, the study is called a **double-blind experiment.** Being fallible human beings and having an emotional and sometimes financial investment

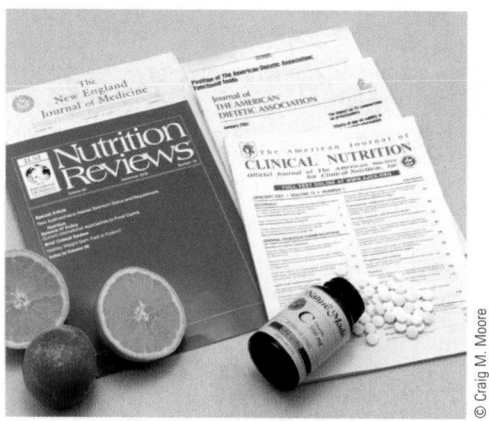

Knowledge about the nutrients and their effects on health comes from scientific study.

TABLE 1-3	Strengths and Weaknesses of Research Designs	
Type of Research	**Strengths**	**Weaknesses**
Epidemiological studies determine the incidence and distribution of diseases in a population. Epidemiological studies include cross-sectional, case-control, and cohort (see Figure 1-4).	• Can narrow down the list of possible causes • Can raise questions to pursue through other types of studies	• Cannot control variables that may influence the development or the prevention of a disease • Cannot prove cause and effect
Laboratory-based studies explore the effects of a specific variable on a tissue, cell, or molecule. Laboratory-based studies are often conducted in test tubes (in vitro) or on animals.	• Can control conditions • Can determine effects of a variable	• Cannot apply results from test tubes or animals to human beings
Human intervention or **clinical trials** involve human beings who follow a specified regimen.	• Can control conditions (for the most part) • Can apply findings to some groups of human beings	• Cannot generalize findings to all human beings • Cannot use certain treatments for clinical or ethical reasons

in a successful outcome, researchers might record and interpret results with a bias in the expected direction. To prevent such bias, the pills would be coded by a third party, who does not reveal to the experimenters which subjects were in which group until all results have been recorded.

Analyzing Research Findings

Research findings must be analyzed and interpreted with an awareness of each study's limitations. Scientists must be cautious about drawing any conclusions until they have accumulated a body of evidence from multiple studies that have used various types of research designs. As evidence accumulates, scientists begin to develop a **theory** that integrates the various findings and explains the complex relationships.

GLOSSARY OF RESEARCH TERMS

blind experiment: an experiment in which the subjects do not know whether they are members of the experimental group or the control group.

control group: a group of individuals similar in all possible respects to the experimental group except for the treatment. Ideally, the control group receives a placebo while the experimental group receives a real treatment.

correlation (CORE-ee-LAY-shun): the simultaneous increase, decrease, or change in two variables. If A increases as B increases, or if A decreases as B decreases, the correlation is **positive.** (This does not mean that A causes B or vice versa.) If A increases as B decreases, or if A decreases as B increases, the correlation is **negative.** (This does not mean that A prevents B or vice versa.) Some third factor may account for both A and B.

double-blind experiment: an experiment in which neither the subjects nor the researchers know which subjects are members of the experimental group and which are serving as control subjects, until after the experiment is over.

experimental group: a group of individuals similar in all possible respects to the control group except for the treatment. The experimental group receives the real treatment.

hypothesis (hi-POTH-eh-sis): an unproven statement that tentatively explains the relationships between two or more variables.

peer review: a process in which a panel of scientists rigorously evaluates a research study to assure that the scientific method was followed.

placebo (pla-SEE-bo): an inert, harmless medication given to provide comfort and hope; a sham treatment used in controlled research studies.

placebo effect: a change that occurs in reponse to expectations in the effectiveness of a treatment that actually has no pharmaceutical effects.

randomization (RAN-dom-ih-ZAY-shun): a process of choosing the members of the experimental and control groups without bias.

replication (REP-lih-KAY-shun): repeating an experiment and getting the same results. The skeptical scientist, on hearing of a new, exciting finding, will ask, "Has it been replicated yet?" If it hasn't, the scientist will withhold judgment regarding the finding's validity.

subjects: the people or animals participating in a research project.

theory: a tentative explanation that integrates many and diverse findings to further the understanding of a defined topic.

validity (va-LID-ih-tee): having the quality of being founded on fact or evidence.

variables: factors that change. A variable may depend on another variable (for example, a child's height depends on his age), or it may be independent (for example, a child's height does not depend on the color of her eyes). Sometimes both variables correlate with a third variable (a child's height and eye color both depend on genetics).

Correlations and Causes   Researchers often examine the relationships between two or more **variables**—for example, daily vitamin C intake and the number of colds or the duration and severity of cold symptoms. Importantly, researchers must be able to observe, measure, or verify the variables selected. Findings sometimes suggest no **correlation** between variables (regardless of the amount of vitamin C consumed, the number of colds remains the same). Other times, studies find either a **positive correlation** (the more vitamin C, the more colds) or a **negative correlation** (the more vitamin C, the fewer colds). Correlational evidence proves only that variables are associated, not that one is the cause of the other. People often jump to conclusions when they notice correlations, but their conclusions are often wrong. To actually prove that A causes B, scientists have to find evidence of the *mechanism*—that is, an explanation of how A might cause B.

Cautious Conclusions   When researchers record and analyze the results of their experiments, they must exercise caution in their interpretation of the findings. For example, in an epidemiological study, scientists may use a specific segment of the population—say, men 18 to 30 years old. When the scientists draw conclusions, they are careful not to generalize the findings to all people. Similarly, scientists performing research studies using animals are cautious in applying their findings to human beings. Conclusions from any one research study are always tentative and take into account findings from studies conducted by other scientists as well. As evidence accumulates, scientists gain confidence about making recommendations that affect people's health and lives. Still, their statements are worded cautiously, such as "A diet high in fruits and vegetables *may* protect against *some* cancers."

Quite often, as scientists approach an answer to one research question, they raise several more questions, so future research projects are never lacking. Further scientific investigation then seeks to answer questions such as "What substance or substances within fruits and vegetables provide protection?" If those substances turn out to be the vitamins found so abundantly in fresh produce, then, "How much is needed to offer protection?" "How do these vitamins protect against cancer?" "Is it their action as antioxidant nutrients?" "If not, might it be another action or even another substance that accounts for the protection fruits and vegetables provide against cancer?" (Highlight 11 explores the answers to these questions and reviews recent research on antioxidant nutrients and disease.)

Publishing Research

The findings from a research study are submitted to a board of reviewers composed of other scientists who rigorously evaluate the study to assure that the scientific method was followed—a process known as **peer review.** The reviewers critique the study's hypothesis, methodology, statistical significance, and conclusions. If the reviewers consider the conclusions to be well supported by the evidence—that is, if the research has **validity**—they endorse the work for publication in a scientific journal where others can read it. This raises an important point regarding information found on the Internet: much gets published without the rigorous scrutiny of peer review. Consequently, readers must assume greater responsibility for examining the data and conclusions presented—often without the benefit of journal citations.

Even when a new finding is published or released to the media, it is still only preliminary and not very meaningful by itself. Other scientists will need to confirm or disprove the findings through **replication.** To be accepted into the body of nutrition knowledge, a finding must stand up to rigorous, repeated testing in experiments performed by several different researchers. What we "know" in nutrition results from years of replicating study findings. Communicating the latest finding in its proper context without distorting or oversimplifying the message is a challenge for scientists and journalists alike.

With each report from scientists, the field of nutrition changes a little—each finding contributes another piece to the whole body of knowledge. People who know how science works understand that single findings, like single frames in a movie, are just small parts of a larger story. Over years, the picture of what is "true" in nutrition gradually changes, and dietary recommendations change to reflect the current understanding of scientific research. Highlight 5 provides a detailed look at how dietary fat recommendations have evolved over the past several decades as researchers have uncovered the relationships between the various kinds of fat and their roles in supporting or harming health.

> **IN SUMMARY**
>
> Scientists learn about nutrition by conducting experiments that follow the protocol of scientific research. Researchers take care to establish similar control and experimental groups, large sample sizes, placebos, and blind treatments. Their findings must be reviewed and replicated by other scientists before being accepted as valid.

The characteristics of well-designed research have enabled scientists to study the actions of nutrients in the body. Such research has laid the foundation for quantifying how much of each nutrient the body needs.

Dietary Reference Intakes

Using the results of thousands of research studies, nutrition experts have produced a set of standards that define the amounts of energy, nutrients, other dietary components, and physical activity that best support health. These recommendations are called **Dietary Reference Intakes (DRI),** and they reflect the collaborative efforts of researchers in both the United States and Canada.*[8] The inside front covers of this book provide a handy reference for DRI values.

Establishing Nutrient Recommendations

The DRI Committee consists of highly qualified scientists who base their estimates of nutrient needs on careful examination and interpretation of scientific evidence. These recommendations apply to healthy people and may not be appropriate for people with diseases that increase or decrease nutrient needs. The next several paragraphs discuss specific aspects of how the committee goes about establishing the values that make up the DRI:

- Estimated Average Requirements (EAR)
- Recommended Dietary Allowances (RDA)
- Adequate Intakes (AI)
- Tolerable Upper Intake Levels (UL)

Estimated Average Requirements (EAR) The committee reviews hundreds of research studies to determine the **requirement** for a nutrient—how much is needed in the diet. The committee selects a different criterion for each nutrient based on its various roles in performing activities in the body and in reducing disease risks.

An examination of all the available data reveals that each person's body is unique and has its own set of requirements. Men differ from women, and needs change as people grow from infancy through old age. For this reason, the committee clusters its recommendations for people into groups based on age and gender. Even so, the exact requirements for people of the same age and gender are likely to be different. For example, person A might need 40 units of a particular nutrient each day; person B might need 35; and person C, 57. Looking at enough people might reveal that their individual requirements fall into a symmetrical distribution,

Don't let the DRI "alphabet soup" of nutrient intake standards confuse you. Their names make sense when you learn their purposes.

© PhotoDisc/Getty Images

Dietary Reference Intakes (DRI): a set of nutrient intake values for healthy people in the United States and Canada. These values are used for planning and assessing diets and include:
- Estimated Average Requirements (EAR)
- Recommended Dietary Allowances (RDA)
- Adequate Intakes (AI)
- Tolerable Upper Intake Levels (UL)

requirement: the lowest continuing intake of a nutrient that will maintain a specified criterion of adequacy.

---

* The DRI reports are produced by the Food and Nutrition Board, Institute of Medicine of the National Academies, with active involvement of scientists from Canada.

with most near the midpoint and only a few at the extremes (see the left side of Figure 1-5). Using this information, the committee determines an **Estimated Average Requirement (EAR)** for each nutrient—the average amount that appears sufficient for half of the population. In Figure 1-5, the Estimated Average Requirement is shown as 45 units.

Recommended Dietary Allowances (RDA) Once a nutrient *requirement* is established, the committee must decide what intake to *recommend* for everybody—the **Recommended Dietary Allowance (RDA).** As you can see by the distribution in Figure 1-5, the Estimated Average Requirement (shown in the figure as 45 units) is probably closest to everyone's need. However, if people consumed exactly the average requirement of a given nutrient each day, half of the population would develop deficiencies of that nutrient—in Figure 1-5, for example, person C would be among them. Recommendations are therefore set high enough above the Estimated Average Requirement to meet the needs of most healthy people.

Small amounts above the daily requirement do no harm, whereas amounts below the requirement may lead to health problems. When people's nutrient intakes are consistently **deficient** (less than the requirement), their nutrient stores decline, and over time this decline leads to poor health and deficiency symptoms. Therefore, to ensure that the nutrient RDA meet the needs of as many people as possible, the RDA are set near the top end of the range of the population's estimated requirements.

In this example, a reasonable RDA might be 63 units a day (see the right side of Figure 1-5). Such a point can be calculated mathematically so that it covers about 98 percent of a population. Almost everybody—including person C whose needs were higher than the average—would be covered if they met this dietary goal. Relatively few people's requirements would exceed this recommendation, and even then, they wouldn't exceed by much.

Adequate Intakes (AI) For some nutrients, there is insufficient scientific evidence to determine an Estimated Average Requirement (which is needed to set an RDA). In these cases, the committee establishes an **Adequate Intake (AI)** instead of an RDA. An AI reflects the average amount of a nutrient that a group of healthy people consumes. Like the RDA, the AI may be used as nutrient goals for individuals.

FIGURE 1-5 Estimated Average Requirements (EAR) and Recommended Dietary Allowances (RDA) Compared

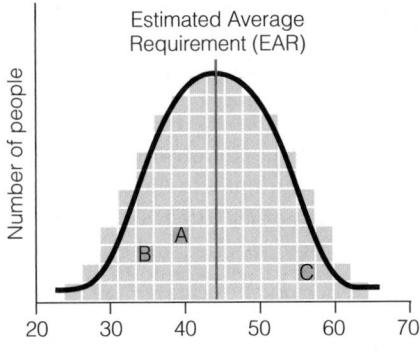
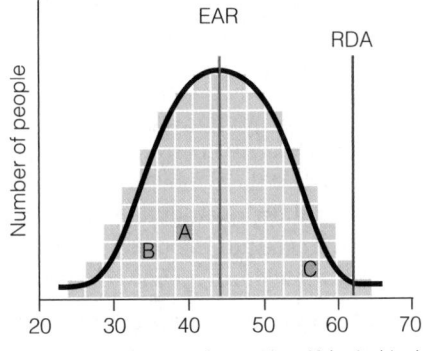

Each square in the graph above represents a person with unique nutritional requirements. (The text discusses three of these people—A, B, and C.) Some people require only a small amount of nutrient X and some require a lot. Most people, however, fall somewhere in the middle. This amount that covers half of the population is called the Estimated Average Requirement (EAR) and is represented here by the red line.

The Recommended Dietary Allowance (RDA) for a nutrient (shown here in purple) is set well above the EAR, covering about 98% of the population.

Estimated Average Requirement (EAR): the average daily amount of a nutrient that will maintain a specific biochemical or physiological function in half the healthy people of a given age and gender group.

Recommended Dietary Allowance (RDA): the average daily amount of a nutrient considered adequate to meet the known nutrient needs of practically all healthy people; a goal for dietary intake by individuals.

deficient: the amount of a nutrient below which almost all healthy people can be expected, over time, to experience deficiency symptoms.

Adequate Intake (AI): the average daily amount of a nutrient that appears sufficient to maintain a specified criterion; a value used as a guide for nutrient intake when an RDA cannot be determined.

FIGURE 1-6 Inaccurate versus Accurate View of Nutrient Intakes

The RDA or AI for a given nutrient represents a point that lies within a range of appropriate and reasonable intakes between toxicity and deficiency. Both of these recommendations are high enough to provide reserves in times of short-term dietary inadequacies, but not so high as to approach toxicity. Nutrient intakes above or below this range may be equally harmful.

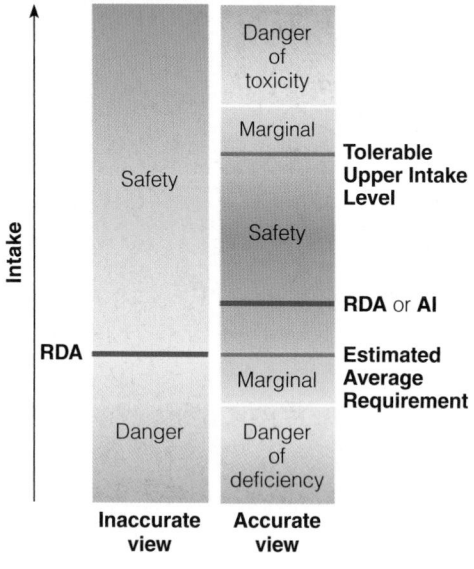

◆ Reference adults:
- Men: 19–30 yr, 5 ft 10 in., and 154 lb
- Women: 19–30 yr, 5 ft 4 in., and 126 lb

Tolerable Upper Intake Level (UL): the maximum daily amount of a nutrient that appears safe for most healthy people and beyond which there is an increased risk of adverse health effects.

Estimated Energy Requirement (EER): the average dietary energy intake that maintains energy balance and good health in a person of a given age, gender, weight, height, and level of physical activity.

Acceptable Macronutrient Distribution Ranges (AMDR): ranges of intakes for the energy nutrients that provide adequate energy and nutrients and reduce the risk of chronic diseases.

Although both the RDA and the AI serve as nutrient intake goals for individuals, their differences are noteworthy. An RDA for a given nutrient is based on enough scientific evidence to expect that the needs of almost all healthy people will be met. An AI, on the other hand, must rely more heavily on scientific judgments because sufficient evidence is lacking. The percentage of people covered by an AI is unknown; an AI is expected to exceed average requirements, but it may cover more or fewer people than an RDA would cover (if an RDA could be determined). For these reasons, AI values are more tentative than RDA. The table on the inside front cover identifies which nutrients have an RDA and which have an AI. Later chapters present the RDA and AI values for the vitamins and minerals.

Tolerable Upper Intake Levels (UL) As mentioned earlier, the recommended intakes for nutrients are generous, and they do not necessarily cover every individual for every nutrient. Nevertheless, it is probably best not to exceed these recommendations by very much or very often. Individual tolerances for high doses of nutrients vary, and somewhere above the recommended intake is a point beyond which a nutrient is likely to become toxic. This point is known as the **Tolerable Upper Intake Level (UL).** It is naive—and inaccurate—to think of recommendations as minimum amounts. A more accurate view is to see a person's nutrient needs as falling within a range, with marginal and danger zones both below and above it (see Figure 1-6).

Paying attention to upper levels is particularly useful in guarding against the overconsumption of nutrients, which may occur when people use large-dose supplements and fortified foods regularly. Later chapters discuss the dangers associated with excessively high intakes of vitamins and minerals, and the inside front cover (page C) presents tables that include the upper-level values for selected nutrients.

Establishing Energy Recommendations

In contrast to the RDA and AI values for nutrients, the recommendation for energy is not generous. Excess energy cannot be readily excreted and is eventually stored as body fat. These reserves may be beneficial when food is scarce, but they can also lead to obesity and its associated health consequences.

Estimated Energy Requirement (EER) The energy recommendation—called the **Estimated Energy Requirement (EER)**—represents the average dietary energy intake (kcalories per day) that will maintain energy balance in a person who has a healthy body weight and level of physical activity. ◆ Balance is key to the energy recommendation. Enough energy is needed to sustain a healthy and active life, but too much energy can lead to weight gain and obesity. Because *any* amount in excess of energy needs will result in weight gain, no upper level for energy has been determined.

Acceptable Macronutrient Distribution Ranges (AMDR) People don't eat energy directly; they derive energy from foods containing carbohydrate, fat, and protein. Each of these three energy-yielding nutrients contributes to the total energy intake, and those contributions vary in relation to each other. The DRI Committee has determined that the composition of a diet that provides adequate energy and nutrients and reduces the risk of chronic diseases is:

- 45–65 percent kcalories from carbohydrate
- 20–35 percent kcalories from fat
- 10–35 percent kcalories from protein

These values are known as **Acceptable Macronutrient Distribution Ranges (AMDR).**

Using Nutrient Recommendations

Although the intent of nutrient recommendations seems simple, they are the subject of much misunderstanding and controversy. Perhaps the following facts will help put them in perspective:

1. Estimates of adequate energy and nutrient intakes apply to *healthy* people. They need to be adjusted for malnourished people or those with medical problems who may require supplemented or restricted intakes.

2. *Recommendations* are not minimum requirements, nor are they necessarily optimal intakes for all individuals. Recommendations can only target "most" of the people and cannot account for individual variations in nutrient needs—yet. Given the recent explosion of knowledge about genetics, the day may be fast approaching when nutrition scientists will be able to determine an individual's optimal nutrient needs.[9] Until then, registered dietitians ◆ and other qualified health professionals can help determine if recommendations should be adjusted to meet individual needs.

3. Most nutrient goals are intended to be met through diets composed of a variety of *foods* whenever possible. Because foods contain mixtures of nutrients and nonnutrients, they deliver more than just those nutrients covered by the recommendations. Excess intakes of vitamins and minerals are unlikely when they come from foods rather than supplements.

4. Recommendations apply to *average* daily intakes. Trying to meet the recommendations for every nutrient every day is difficult and unnecessary. The length of time over which a person's intake can deviate from the average without risk of deficiency or overdose varies for each nutrient, depending on how the body uses and stores the nutrient. For most nutrients (such as thiamin and vitamin C), deprivation would lead to rapid development of deficiency symptoms (within days or weeks); for others (such as vitamin A and vitamin B_{12}), deficiencies would develop more slowly (over months or years).

5. Each of the DRI categories serves a unique purpose. For example, the Estimated Average Requirements are most appropriately used to develop and evaluate nutrition programs for *groups* such as schoolchildren or military personnel. The RDA (or AI if an RDA is not available) can be used to set goals for *individuals*. Tolerable Upper Intake Levels serve as a reminder to keep nutrient intakes below amounts that increase the risk of toxicity—not a common problem when nutrients derive from foods, but a real possibility for some nutrients if supplements are used regularly.

With these understandings, professionals can use the DRI for a variety of purposes.

◆ A **registered dietitian** is a college-educated food and nutrition specialist who is qualified to evaluate people's nutritional health and needs. See Highlight 1 for more on what constitutes a nutrition expert.

Comparing Nutrient Recommendations

At least 40 different nations and international organizations have published nutrient standards similar to those used in the United States and Canada. Slight differences may be apparent, reflecting differences both in the interpretation of the data from which the standards were derived and in the food habits and physical activities of the populations they serve.

Many countries use the recommendations developed by two international groups: FAO (Food and Agriculture Organization) and WHO (World Health Organization). ◆ The FAO/WHO recommendations are considered sufficient to maintain health in nearly all healthy people worldwide.

◆ Nutrient recommendations from FAO/WHO are provided in Appendix I.

IN SUMMARY

The Dietary Reference Intakes (DRI) are a set of nutrient intake values that can be used to plan and evaluate diets for healthy people. The Estimated Average Requirement (EAR) defines the amount of a nutrient that supports a specific function in the body for half of the population. The Recommended Dietary Allowance (RDA) is based on the Estimated Average Requirement and establishes a goal for dietary intake that will meet the needs of almost all healthy people. An Adequate Intake (AI) serves a similar purpose when an RDA cannot be determined. The Estimated Energy Requirement (EER) defines the average amount of energy intake needed to maintain energy balance, and the Acceptable Macronutrient Distribution Ranges (AMDR) define the proportions contributed by carbohydrate, fat, and protein to a healthy diet. The Tolerable Upper Intake Level (UL) establishes the highest amount that appears safe for regular consumption.

Nutrition Assessment

What happens when a person doesn't get enough or gets too much of a nutrient or energy? If the deficiency or excess is significant over time, the person exhibits signs of **malnutrition.** With a deficiency of energy, the person may display the symptoms of **undernutrition** by becoming extremely thin, losing muscle tissue, and becoming prone to infection and disease. With a deficiency of a nutrient, the person may experience skin rashes, depression, hair loss, bleeding gums, muscle spasms, night blindness, or other symptoms. With an excess of energy, the person may become obese and vulnerable to diseases associated with **overnutrition** such as heart disease and diabetes. With a sudden nutrient overdose, the person may experience hot flashes, yellowing skin, a rapid heart rate, low blood pressure, or other symptoms. Similarly, over time, regular intakes in excess of needs may also have adverse effects.

Malnutrition symptoms—such as diarrhea, skin rashes, and fatigue—are easy to miss because they resemble the symptoms of other diseases. But a person who has learned how to use assessment techniques to detect malnutrition can identify when these conditions are caused by poor nutrition and can recommend steps to correct it. This discussion presents the basics of nutrition assessment; many more details are offered in Chapter 17 and in Appendix E.

© Tom & Dee Ann Ann McCarthy/CORBIS

A peek inside the mouth provides clues to a person's nutrition status. An inflamed tongue may indicate a B vitamin deficiency, and mottled teeth may reveal fluoride toxicity, for example.

malnutrition: any condition caused by excess or deficient food energy or nutrient intake or by an imbalance of nutrients.

• **mal** = bad

undernutrition: deficient energy or nutrients.

overnutrition: excess energy or nutrients.

nutrition assessment: a comprehensive analysis of a person's nutrition status that uses health, socioeconomic, drug, and diet histories; anthropometric measurements; physical examinations; and laboratory tests.

Nutrition Assessment of Individuals

To prepare a **nutrition assessment,** a registered dietitian or other trained health care professional uses:

- Historical information
- Anthropometric data
- Physical examinations
- Laboratory tests

Each of these methods involves collecting data in various ways and interpreting each finding in relation to the others to create a total picture.

Historical Information   One step in evaluating nutrition status is to obtain information about a person's history with respect to health status, socioeconomic status, drug use, and diet. The health history reflects a person's medical record and may reveal a disease that interferes with the person's ability to eat or the body's use of nutrients. The person's family history of major diseases is also noteworthy, especially for conditions such as heart disease that have a genetic tendency to run in families. Economic circumstances may show a financial inability to buy foods or inadequate kitchen facilities in which to prepare them. Social factors such as marital status, ethnic background, and educational level also influence food choices and nutrition status. A drug history, including all prescribed and over-the-counter medications as well as illegal substances, may highlight possible interactions that lead to nutrient deficiencies (as described in Chapter 19). A diet history that examines a person's intake of foods, beverages, and supplements may reveal either a surplus or inadequacy of nutrients or energy.

To take a diet history, the assessor collects data about the foods a person eats. The data may be collected by recording the foods the person has eaten over a period of 24 hours, three days, or a week or more or by asking what foods the person typically eats and how much of each. The days in the record must be fairly typical of the person's diet, and portion sizes must be recorded accurately. To determine the amounts of nutrients consumed, the assessor usually enters the foods and their portion sizes into a computer using a diet analysis program. This step can also be done manually by looking up each food in a table of food composition such as Appendix H in this book. The assessor then compares the calculated nutrient intakes with the DRI to determine the probability of adequacy (see Figure 1-7).[10] Alternatively, the diet history might be compared against standards such as the USDA Food Guide or *Dietary Guidelines* (described in Chapter 2).

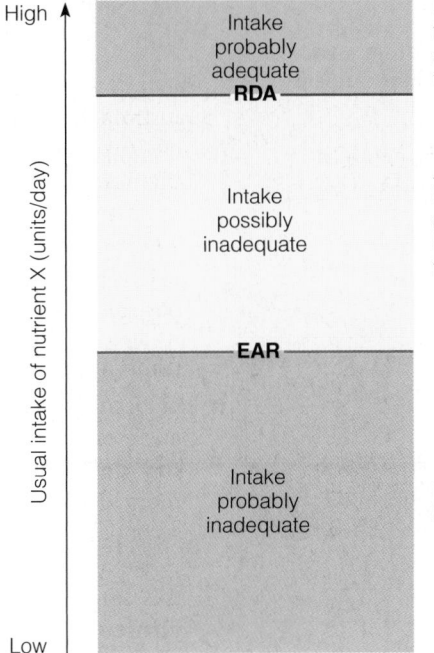

FIGURE 1-7 Using the DRI to Assess the Dietary Intake of a Healthy Individual

If a person's usual intake falls above the RDA, the intake is probably adequate because the RDA covers the needs of almost all people.

A usual intake that falls between the RDA and the EAR is more difficult to assess; the intake may be adequate, but the chances are greater or equal that it is inadequate.

If the usual intake falls below the EAR, it is probably inadequate.

An estimate of energy and nutrient intakes from a diet history, when combined with other sources of information, can help confirm or rule out the *possibility* of suspected nutrition problems. A sufficient intake of a nutrient does not guarantee adequacy, and an insufficient intake does not always indicate a deficiency. Such findings, however, warn of possible problems.

Anthropometric Data A second technique that may help to reveal nutrition problems is taking **anthropometric** measures such as height and weight. The assessor compares a person's measurements with standards specific for gender and age or with previous measures on the same individual. (Chapter 8 presents information on body weight and its standards.)

Measurements taken periodically and compared with previous measurements reveal patterns and indicate trends in a person's overall nutrition status, but they provide little information about specific nutrients. Instead, measurements out of line with expectations may reveal such problems as growth failure in children, wasting or swelling of body tissues in adults, and obesity—conditions that may reflect energy or nutrient deficiencies or excesses.

Physical Examinations A third nutrition assessment technique is a physical examination looking for clues to poor nutrition status. Every part of the body that can be inspected may offer such clues: the hair, eyes, skin, posture, tongue, fingernails, and others. The examination requires skill because many physical signs reflect more than one nutrient deficiency or toxicity—or even nonnutrition conditions. Like the other assessment techniques, a physical examination alone does not yield firm conclusions. Instead, physical examinations reveal possible imbalances that must be confirmed by other assessment techniques, or they confirm results from other assessment measures.

Laboratory Tests A fourth way to detect a developing deficiency, imbalance, or toxicity is to take samples of blood or urine, analyze them in the laboratory, and compare the results with normal values for a similar population. ◆ A goal of nutrition

◆ Assessment may one day depend on measures of how a nutrient influences genetic activity within the cells, instead of quantities in the blood or other tissues.

anthropometric (AN-throw-poe-MET-rick): relating to measurement of the physical characteristics of the body, such as height and weight.
• **anthropos** = human
• **metric** = measuring

FIGURE 1-8 Stages in the Development of a Nutrient Deficiency

Internal changes precede outward signs of deficiencies. However, outward signs of sickness need not appear before a person takes corrective measures. Laboratory tests can help determine nutrient status in the early stages.

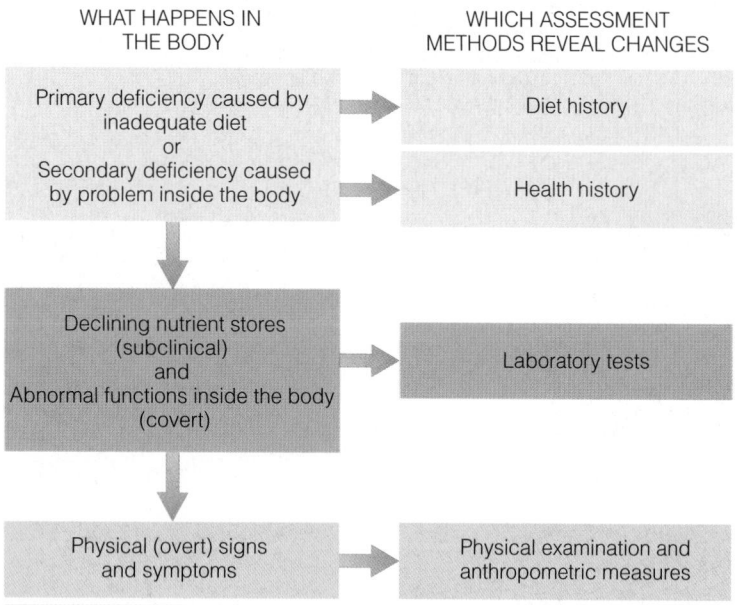

assessment is to uncover early signs of malnutrition before symptoms appear, and laboratory tests are most useful for this purpose. In addition, they can confirm suspicions raised by other assessment methods.

Iron, for Example The mineral iron can be used to illustrate the stages in the development of a nutrient deficiency and the assessment techniques useful in detecting them. The **overt,** or outward, signs of an iron deficiency appear at the end of a long sequence of events. Figure 1-8 describes what happens in the body as a nutrient deficiency progresses and shows which assessment methods can reveal those changes.

First, the body has too little iron—either because iron is lacking in the person's diet (a **primary deficiency**) or because the person's body doesn't absorb enough, excretes too much, or uses iron inefficiently (a **secondary deficiency**). A diet history provides clues to primary deficiencies; a health history provides clues to secondary deficiencies.

Next, the body begins to use up its stores of iron. At this stage, the deficiency might be described as **subclinical.** It exists as a **covert** condition, and although it might be detected by laboratory tests, no outward signs are apparent.

Finally, the body's iron stores are exhausted. Now, it cannot make enough iron-containing red blood cells to replace those that are aging and dying. Iron is needed in red blood cells to carry oxygen to all the body's tissues. When iron is lacking, fewer red blood cells are made, the new ones are pale and small, and every part of the body feels the effects of oxygen shortage. Now the overt symptoms of deficiency appear—weakness, fatigue, pallor, and headaches, reflecting the iron-deficient state of the blood. A physical examination will reveal these symptoms.

Nutrition Assessment of Populations

To assess a population's nutrition status, researchers conduct surveys using techniques similar to those used on individuals. The data collected are then used by various agencies for numerous purposes, including the development of national health goals.

◆ The new integrated survey is called *What We Eat in America.*

National Nutrition Surveys The National Nutrition Monitoring program coordinates the many nutrition-related surveys and research activities of various federal agencies. The integration of two major national surveys ◆ provides comprehensive data efficiently.[11] One survey collects data on the kinds and amounts of foods people eat.* Then researchers calculate the energy and nutrients in the foods and compare the amounts consumed with a standard. The other survey examines the people themselves, using anthropometric measurements, physical examinations, and laboratory tests.[†12] The data provide valuable information on several nutrition-related conditions, such as growth retardation, heart disease, and nutrient deficiencies. National nutrition surveys often oversample high-risk groups (low-income families, pregnant women, adolescents, the elderly, African Americans, and Mexican Americans) to glean an accurate estimate of their health and nutrition status.

The resulting wealth of information from the national nutrition surveys is used for a variety of purposes. For example, Congress uses this information to establish

overt (oh-VERT): out in the open and easy to observe.
• **ouvrir** = to open

primary deficiency: a nutrient deficiency caused by inadequate dietary intake of a nutrient.

secondary deficiency: a nutrient deficiency caused by something other than an inadequate intake such as a disease condition or drug interaction that reduces absorption, accelerates use, hastens excretion, or destroys the nutrient.

subclinical deficiency: a deficiency in the early stages, before the outward signs have appeared.

covert (KOH-vert): hidden, as if under covers.
• **couvrir** = to cover

---

* This survey was formerly called the Continuing Survey of Food Intakes by Individuals (CSFII), conducted by the U.S. Department of Agriculture (USDA).
† This survey is known as the National Health and Nutrition Examination Survey (NHANES), conducted by the U.S. Department of Health and Human Services (DHHS).

public policy on nutrition education, food assistance programs, and the regulation of the food supply. Scientists use the information to establish research priorities. The food industry uses these data to guide decisions in public relations and product development.[13] The Dietary Reference Intakes and other major reports that examine the relationships between diet and health depend on information collected from these nutrition surveys. These data also provide the basis for developing and monitoring national health goals.

National Health Goals **Healthy People** is a program that identifies the nation's health priorities and guides policies that promote health and prevent disease. At the start of each decade, the program sets goals for improving the nation's health during the following ten years. The goals of Healthy People 2010 focus on "improving the quality of life and eliminating disparity in health among racial and ethnic groups."[14] Nutrition is one of many focus areas, each with numerous objectives. Table 1-4 lists the nutrition and overweight objectives for 2010, and Appendix J includes a table of nutrition-related objectives from other focus areas.

At mid-decade, the nation's progress toward meeting its nutrition and overweight Healthy People 2010 goals was somewhat bleak. Trends in overweight and obesity worsened. Objectives to eat more fruits, vegetables, and whole grains and to increase physical activity showed little or no improvement. Clearly, "what we eat in America" must change if we hope to meet the Healthy People 2010 goals.

National Trends What do we eat in America and how has it changed over the past 30 years?[15] The short answer to both questions is "a lot." We eat more meals away from home, particularly at fast-food restaurants. We eat larger portions. We drink more sweetened beverages and eat more energy-dense, nutrient-poor foods such as candy and chips. We snack frequently. As a result of these dietary habits, our energy intake has risen and, consequently, so has the incidence of overweight and obesity. Overweight and obesity, in turn, profoundly influence our health—as the next section explains.

Surveys provide valuable information about the kinds of foods people eat.

TABLE 1-4	Healthy People 2010 Nutrition and Overweight Objectives

- Increase the proportion of adults who are at a *healthy weight.*
- Reduce the proportion of adults who are *obese.*
- Reduce the proportion of children and adolescents who are *overweight* or *obese.*
- Reduce *growth retardation* among low-income children under age 5 years.
- Increase the proportion of persons aged 2 years and older who consume at least two daily servings of *fruit.*
- Increase the proportion of persons aged 2 years and older who consume at least three daily servings of *vegetables,* with at least one-third being dark green or orange vegetables.
- Increase the proportion of persons aged 2 years and older who consume at least six daily servings of *grain products,* with at least three being whole grains.
- Increase the proportion of persons aged 2 years and older who consume less than 10 percent of kcalories from *saturated fat.*
- Increase the proportion of persons aged 2 years and older who consume no more than 30 percent of kcalories from *total fat.*

- Increase the proportion of persons aged 2 years and older who consume 2400 mg or less of *sodium.*
- Increase the proportion of persons aged 2 years and older who meet dietary recommendations for *calcium.*
- Reduce *iron deficiency* among young children, females of childbearing age, and pregnant females.
- Reduce *anemia* among low-income pregnant females in their third trimester.
- Increase the proportion of children and adolescents aged 6 to 19 years whose intake of *meals and snacks at school* contributes to good overall dietary quality.
- Increase the proportion of worksites that offer *nutrition or weight management classes or counseling.*
- Increase the proportion of physician office visits made by patients with a diagnosis of cardiovascular disease, diabetes, or hyperlipidemia that include *counseling or education related to diet and nutrition.*
- Increase *food security* among U.S. households and in so doing reduce hunger.

NOTE: "Nutrition and Overweight" is one of 28 focus areas, each with numerous objectives. Several of the other focus areas have nutrition-related objectives, and these are presented in Appendix J.
SOURCE: Healthy People 2010, **www.healthypeople.gov**

Healthy People: a national public health initiative under the jurisdiction of the U.S. Department of Health and Human Services (DHHS) that identifies the most significant preventable threats to health and focuses efforts toward eliminating them.

IN SUMMARY

People become malnourished when they get too little or too much energy or nutrients. Deficiencies, excesses, and imbalances of nutrients lead to malnutrition diseases. To detect malnutrition in individuals, health care professionals use four nutrition assessment methods. Reviewing dietary data and health information may suggest a nutrition problem in its earliest stages. Laboratory tests may detect it before it becomes overt, whereas anthropometrics and physical examinations pick up on the problem only after it causes symptoms. National surveys use similar assessment methods to measure people's food consumption and to evaluate the nutrition status of populations.

Diet and Health

Diet has always played a vital role in supporting health. Early nutrition research focused on identifying the nutrients in foods that would prevent such common diseases as rickets and scurvy, the vitamin D– and vitamin C–deficiency diseases. With this knowledge, developed countries have successfully defended against nutrient deficiency diseases. World hunger and nutrient deficiency diseases still pose a major health threat in developing countries, however, but not because of a lack of nutrition knowledge. More recently, nutrition research has focused on **chronic diseases** associated with energy and nutrient excesses. Once thought to be "rich countries' problems," chronic diseases have now become epidemic in developing countries as well—contributing to three out of five deaths worldwide.[16]

Chronic Diseases

Table 1-5 lists the ten leading causes of death in the United States. These "causes" are stated as if a single condition such as heart disease caused death, but most chronic diseases arise from multiple factors over many years. A person who died of heart disease may have been overweight, had high blood pressure, been a cigarette smoker, and spent years eating a diet high in saturated fat and getting too little exercise.

Of course, not all people who die of heart disease fit this description, nor do all people with these characteristics die of heart disease. People who are overweight might die from the complications of diabetes instead, or those who smoke might die of cancer. They might even die from something totally unrelated to any of these factors, such as an automobile accident. Still, statistical studies have shown that certain conditions and behaviors are linked to certain diseases.

Notice that Table 1-5 highlights five of the top six causes of death as having a link with diet or alcohol. During the past 30 years, as knowledge about these diet and disease relationships grew, the death rates for four of these—heart disease, cancers, strokes, and accidents—decreased.[17] Death rates for diabetes—a chronic disease closely associated with obesity—increased.

Risk Factors for Chronic Diseases

Factors that increase or reduce the *risk* of developing chronic diseases can be identified by analyzing statistical data. A strong association between a **risk factor** and a disease means that when the factor is present, the *likelihood* of developing the disease increases. It does not mean that all people with the risk factor will develop the disease. Similarly, a lack of risk factors does not guarantee freedom from a given disease. On the average, though, the more risk factors in a person's life, the greater that person's chances of developing the disease. Conversely, the fewer risk factors in a person's life, the better the chances for good health.

TABLE 1-5 Leading Causes of Death in the United States

	Percentage of Total Deaths
1. Heart disease	28.0
2. Cancers	22.7
3. Strokes	6.4
4. Chronic lung diseases	5.2
5. Accidents	4.5
6. Diabetes mellitus	3.0
7. Pneumonia and influenza	2.7
8. Alzheimer's disease	2.6
9. Kidney diseases	1.7
10. Blood infections	1.4

NOTE: The diseases highlighted in green have relationships with diet; yellow indicates a relationship with alcohol.
SOURCE: National Center for Health Statistics: **www.cdc.gov/nchs**

chronic diseases: diseases characterized by a slow progression and long duration. Examples include heart disease, cancer, and diabetes.

risk factor: a condition or behavior associated with an elevated frequency of a disease but not proved to be causal. Leading risk factors for chronic diseases include obesity, cigarette smoking, high blood pressure, high blood cholesterol, physical inactivity, and a diet high in saturated fats and low in vegetables, fruits, and whole grains.

Physical activity can be both fun and beneficial.

TABLE 1-6 Factors Contributing to Deaths in the United States

Factors	Percentage of Deaths
Tobacco	18
Poor diet/inactivity	15
Alcohol	4
Microbial agents	3
Toxic agents	2
Motor vehicles	2
Firearms	1
Sexual behavior	1
Illicit drugs	1

SOURCE: A. H. Mokdad and coauthors, Actual causes of death in the United States, 2000, *Journal of the American Medical Association* 291 (2004): 1238–1245, with corrections from *Journal of the American Medical Association* 293 (2005): 298.

Risk Factors Persist Risk factors tend to persist over time. Without intervention, a young adult with high blood pressure will most likely continue to have high blood pressure as an older adult, for example. Thus, to minimize the damage, early intervention is most effective.

Risk Factors Cluster Risk factors tend to cluster. For example, a person who is obese may be physically inactive, have high blood pressure, and have high blood cholesterol—all risk factors associated with heart disease. Intervention that focuses on one risk factor often benefits the others as well. For example, physical activity can help reduce weight. The physical activity and weight loss will, in turn, help to lower blood pressure and blood cholesterol.

Risk Factors in Perspective The most prominent factor contributing to death in the United States is tobacco use, ◆ followed closely by diet and activity patterns, and then alcohol use (see Table 1-6).[18] Risk factors such as smoking, poor dietary habits, physical inactivity, and alcohol consumption are personal behaviors that can be changed. Decisions to not smoke, to eat a well-balanced diet, to engage in regular physical activity, and to drink alcohol in moderation (if at all) improve the likelihood that a person will enjoy good health. Other risk factors, such as genetics, gender, and age, also play important roles in the development of chronic diseases, but they cannot be changed. Health recommendations acknowledge the influence of such factors on the development of disease, but they must focus on the factors that are changeable. For the two out of three Americans who do not smoke or drink alcohol excessively, the one choice that can influence long-term health prospects more than any other is diet.

◆ Cigarette smoking is responsible for almost one of every five deaths each year.

IN SUMMARY

Within the range set by genetics, a person's choice of diet influences long-term health. Diet has no influence on some diseases but is linked closely to others. Personal life choices, such as engaging in physical activity and using tobacco or alcohol, also affect health for the better or worse.

The next several chapters provide many more details about nutrients and how they support health. Whenever appropriate, the discussion shows how diet influences each of today's major diseases. Dietary recommendations appear again and again, as each nutrient's relationships with health is explored. Most people who follow the recommendations will benefit and can enjoy good health into their later years.

Nutrition Portfolio

Each chapter in this book ends with simple Nutrition Portfolio activities that invite you to review key messages and consider whether your personal choices are meeting the dietary goals introduced in the text. By keeping a journal of these Nutrition Portfolio assignments, you can examine how your knowledge and behaviors change as you progress in your study of nutrition.

Your food choices play a key role in keeping you healthy and reducing your risk of chronic diseases.

- Identify the factors that most influence your food choices for meals and snacks.

- List the chronic disease risk factors and conditions (listed in the definition of risk factors on p. 24) that you or members of your family have.

- Describe lifestyle changes you can make to improve your chances of enjoying good health.

NUTRITION ON THE NET

For further study of topics covered in this chapter, log on to **academic.cengage .com/nutrition/rolfes/UNCN8e**. Go to Chapter 1, then to Nutrition on the Net.

- Search for "nutrition" at the U.S. Government health and nutrition information sites: **www.healthfinder.gov** or **www.nutrition.gov**

- Learn more about basic science research from the National Science Foundation and Research!America: **www.nsf.gov** and **researchamerica.org**

- Review the Dietary Reference Intakes: **www.nap.edu**

- Review nutrition recommendations from the Food and Agriculture Organization and the World Health Organization: **www.fao.org** and **www.who.org**

- View Healthy People 2010: **www.healthypeople.gov**

- Visit the Food and Nutrition section of the Healthy Living area in Health Canada: **www.hc-sc.gc.ca**

- Learn about the national nutrition survey: **www.cdc.gov/nchs/nhanes.htm**

- Get information from the Food Surveys Research Group: **www.barc.usda.gov/bhnrc/foodsurvey**

- Visit the food and nutrition center of the Mayo Clinic: **www.mayohealth.org**

- Create a chart of your family health history at the U.S. Surgeon General's site: **familyhistory.hhs.gov**

NUTRITION CALCULATIONS

CENGAGENOW™ For additional practice, log on to **academic.cengage.com/login.** Go to Chapter 1, then to Nutrition Calculations.

Several chapters end with problems to give you practice in doing simple nutrition-related calculations. Although the situations are hypothetical, the numbers are real, and calculating the answers (check them on p. 29) provides a valuable nutrition lesson. Once you have mastered these examples, you will be prepared to examine your own food choices. Be sure to show your calculations for each problem.

1. Calculate the energy provided by a food's energy-nutrient contents. A cup of fried rice contains 5 grams protein, 30 grams carbohydrate, and 11 grams fat.
   a. How many kcalories does the rice provide from these energy nutrients?

   _____ = ___ kcal protein
   _____ = ___ kcal carbohydrate
   _____ = ___ kcal fat

   Total = ___ kcal

   b. What percentage of the energy in the fried rice comes from each of the energy-yielding nutrients?

   _____ = ___ % kcal from protein
   _____ = ___ % kcal from carbohydrate
   _____ = ___ % kcal from fat

   Total = ___ %

Note: The total should add up to 100%; 99% or 101% due to rounding is also acceptable.

   c. Calculate how many of the 146 kcalories provided by a 12-ounce can of beer come from alcohol, if the beer contains 1 gram protein and 13 grams carbohydrate. (Note: The remaining kcalories derive from alcohol.)

   1 g protein = ___ kcal protein
   13 g carbohydrate = ___ kcal carbohydrate
   = ___ kcal alcohol

   How many grams of alcohol does this represent?
   ___ g alcohol

2. Even a little nutrition knowledge can help you identify some bogus claims. Consider an advertisement for a new "super supplement" that claims the product provides 15 grams protein and 10 kcalories per dose. Is this possible? ___ Why or why not? _____ = ___ kcal

STUDY QUESTIONS

CENGAGENOW™

To assess your understanding of chapter topics, take the Student Practice Test and explore the modules recommended in your Personalized Study Plan. Log on to **academic.cengage.com/login.**

These questions will help you review this chapter. You will find the answers in the discussions on the pages provided.

1. Give several reasons (and examples) why people make the food choices that they do. (pp. 3–5)

2. What is a nutrient? Name the six classes of nutrients found in foods. What is an essential nutrient? (pp. 6–7)

3. Which nutrients are inorganic, and which are organic? Discuss the significance of that distinction. (pp. 7, 10)

4. Which nutrients yield energy, and how much energy do they yield per gram? How is energy measured? (pp. 7–10)

5. Describe how alcohol resembles nutrients. Why is alcohol not considered a nutrient? (pp. 8, 10)

6. What is the science of nutrition? Describe the types of research studies and methods used in acquiring nutrition information. (pp. 11–16)

7. Explain how variables might be correlational but not causal. (p. 15)

8. What are the DRI? Who develops the DRI? To whom do they apply? How are they used? In your description, identify the categories of DRI and indicate how they are related. (pp. 16–19)

9. What judgment factors are involved in setting the energy and nutrient recommendations? (pp. 17–18)

10. What happens when people get either too little or too much energy or nutrients? Define malnutrition, undernutrition, and overnutrition. Describe the four methods used to detect energy and nutrient deficiencies and excesses. (pp. 20–22)

11. What methods are used in nutrition surveys? What kinds of information can these surveys provide? (pp. 22–23)

12. Describe risk factors and their relationships to disease. (pp. 24–25)

These multiple choice questions will help you prepare for an exam. Answers can be found on p. 29.

1. When people eat the foods typical of their families or geographic region, their choices are influenced by:
   a. habit.
   b. nutrition.
   c. personal preference.
   d. ethnic heritage or tradition.

2. Both the human body and many foods are composed mostly of:
   a. fat.
   b. water.
   c. minerals.
   d. proteins.

3. The inorganic nutrients are:
   a. proteins and fats.
   b. vitamins and minerals.
   c. minerals and water.
   d. vitamins and proteins.

4. The energy-yielding nutrients are:
   a. fats, minerals, and water.
   b. minerals, proteins, and vitamins.
   c. carbohydrates, fats, and vitamins.
   d. carbohydrates, fats, and proteins.

5. Studies of populations that reveal correlations between dietary habits and disease incidence are:
   a. clinical trials.
   b. laboratory studies.
   c. case-control studies.
   d. epidemiological studies.

6. An experiment in which neither the researchers nor the subjects know who is receiving the treatment is known as:
   a. double blind.
   b. double control.
   c. blind variable.
   d. placebo control.

7. An RDA represents the:
   a. highest amount of a nutrient that appears safe for most healthy people.
   b. lowest amount of a nutrient that will maintain a specified criterion of adequacy.

   c. average amount of a nutrient considered adequate to meet the known nutrient needs of practically all healthy people.
   d. average amount of a nutrient that will maintain a specific biochemical or physiological function in half the people.

8. Historical information, physical examinations, laboratory tests, and anthropometric measures are:
   a. techniques used in diet planning.
   b. steps used in the scientific method.
   c. approaches used in disease prevention.
   d. methods used in a nutrition assessment.

9. A deficiency caused by an inadequate dietary intake is a(n):
   a. overt deficiency.
   b. covert deficiency.
   c. primary deficiency.
   d. secondary deficiency.

10. Behaviors such as smoking, dietary habits, physical activity, and alcohol consumption that influence the development of disease are known as:
    a. risk factors.
    b. chronic causes.
    c. preventive agents.
    d. disease descriptors.

REFERENCES

1. J. A. Mennella, M. Y. Pepino, and D. R. Reed, Genetic and environmental determinants of bitter perception and sweet preferences, *Pediatrics* 115 (2005): e216.
2. J. E. Tillotson, Our ready-prepared, ready-to-eat nation, *Nutrition Today* 37 (2002): 36–38.
3. D. Benton, Role of parents in the determination of the food preferences of children and the development of obesity, *International Journal of Obesity Related Metabolic Disorders* 28 (2004): 858–869.
4. L. Canetti, E. Bachar, and E. M. Berry, Food and emotion, *Behavioural Processes* 60 (2002): 157–164.
5. Position of the American Dietetic Association: Functional foods, *Journal of the American Dietetic Association* 104 (2004): 814–826.
6. Position of the American Dietetic Association: Total diet approach to communicating food and nutrition information, *Journal of the American Dietetic Association* 102 (2002): 100–108.
7. L. Afman and M. Müller, Nutrigenomics: From molecular nutrition to prevention of disease, *Journal of the American Dietetic Association* 106 (2006): 569–576; J. Ordovas and V. Mooser, Nutrigenomics and nutrigenetics, *Current Opinion in Lipidology* 15 (2005): 101–108; D. Shattuck, Nutritional genomics, *Journal of the American Dietetic Association* 103 (2003): 16, 18; P. Trayhurn, Nutritional genomics—"Nutrigenomics," *British Journal of Nutrition* 89 (2003): 1–2.
8. Committee on Dietary Reference Intakes, *Dietary Reference Intakes for Water, Potassium, Sodium, Chloride, and Sulfate* (Washington, D.C.: National Academies Press, 2005); Committee on Dietary Reference Intakes, *Dietary Reference Intakes for Energy, Carbohydrate, Fiber, Fat, Fatty Acids, Cholesterol, Protein, and Amino Acids* (Washington, D.C.: National Academies Press, 2005); Committee on Dietary Reference Intakes, *Dietary Reference Intakes for Vitamin A, Vitamin K, Arsenic, Boron, Chromium, Copper, Iodine, Iron, Manganese, Molybdenum, Nickel, Silicon, Vanadium, and Zinc* (Washington, D.C.: National Academy Press, 2001); Committee on Dietary Reference Intakes, *Dietary Reference Intakes for Vitamin C, Vitamin E, Selenium, and Carotenoids* (Washington, D.C.: National Academy Press, 2000); Committee on Dietary Reference Intakes, *Dietary Reference Intakes for Thiamin, Riboflavin, Niacin, Vitamin B6, Folate, Vitamin B12, Pantothenic Acid, Biotin, and Choline* (Washington, D.C.: National Academy Press, 1998); Committee on Dietary Reference Intakes, *Dietary Reference Intakes for Calcium, Phosphorus, Magnesium, Vitamin D, and Fluoride* (Washington, D.C.: National Academy Press, 1997).
9. Afman and Müller, 2006.
10. S. P. Murphy, S. I. Barr, and M. I. Poos, Using the new Dietary Reference Intakes to assess diets: A map to the maze, *Nutrition Reviews* 60 (2002): 267–275.
11. J. Dwyer and coauthors, Integration of the Continuing Survey of Food Intakes by Individuals and the National Health and Nutrition Examination Survey, *Journal of the American Dietetic Association* 101 (2001): 1142–1143.
12. J. Dwyer and coauthors, Collection of food and dietary supplement intake data: What we eat in America—NHANES, *Journal of Nutrition* 133 (2003): 590S–600S.
13. S. J. Crockett and coauthors, Nutrition monitoring application in the food industry, *Nutrition Today* 37 (2002): 130–135.
14. U.S. Department of Health and Human Services, *Healthy People 2010: Understanding and Improving Health*, January 2000.
15. R. R. Briefel and C. L. Johnson, Secular trends in dietary intake in the United States, *Annual Review of Nutrition* 24 (2004): 401–431.
16. B. M. Popkin, Global nutrition dynamics: The world is shifting rapidly toward a diet linked with noncommunicable diseases, *American Journal of Clinical Nutrition* 84 (2006): 289–298; D. Yach and coauthors, The global burden of chronic diseases: Overcoming impediments to prevention and control, *Journal of the American Medical Association* 291 (2004): 2616–2622.
17. A. Jemal and coauthors, Trends in the leading causes of death in the United States, 1970–2002, *Journal of the American Medical Association* 294 (2005): 1255–1259.
18. A. H. Mokdad and coauthors, Actual causes of death in the United States, 2000, *Journal of the American Medical Association* 291 (2004): 1238–1245.

ANSWERS

Nutrition Calculations

1. a. 5 g protein × 4 kcal/g = 20 kcal protein
30 g carbohydrate × 4 kcal/g = 120 kcal carbohydrate
11 g fat × 9 kcal/g = 99 kcal fat
Total = 239 kcal

b. 20 kcal ÷ 239 kcal × 100 = 8.4% kcal from protein
120 kcal ÷ 239 kcal × 100 = 50.2% kcal from carbohydrate
99 kcal ÷ 239 kcal × 100 = 41.4% kcal from fat
Total = 100%.

c. 1 g protein = 4 kcal protein
13 g carbohydrate = 52 kcal carbohydrate
146 total kcal − 56 kcal (protein + carbohydrate)
= 90 kcal alcohol
90 kcal alcohol ÷ 7 g/kcal = 12.9 g alcohol

2. No. 15 g protein × 4 kcal/g = 60 kcal

Study Questions (multiple choice)

1. d   2. b   3. c   4. d   5. d   6. a   7. c   8. d
9. c   10. a

Nutrition Information and Misinformation—On the Net and in the News

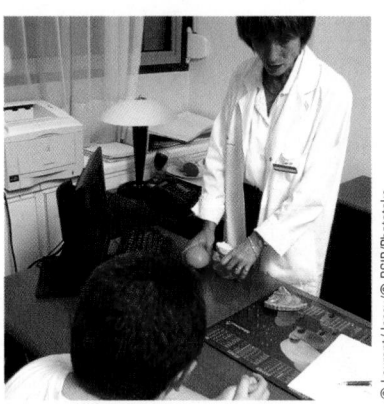

© Laurent/Jessy/© BSIP/Phototake

How can people distinguish valid nutrition information from misinformation? One excellent approach is to notice *who* is providing the information. The "who" behind the information is not always evident, though, especially in the world of electronic media. Keep in mind that *people* develop CD-ROMs and create websites on the Internet, just as people write books and report the news. In all cases, consumers need to determine whether the person is qualified to provide nutrition information.

This highlight begins by examining the unique potential as well as the problems of relying on the Internet and the media for nutrition information. It continues with a discussion of how to identify reliable nutrition information that applies to all resources, including the Internet and the news. (The glossary on p. 32 defines related terms.)

Nutrition on the Net

Got a question? The **Internet** has an answer. The Internet offers endless opportunities to obtain high-quality information, but it also delivers an abundance of incomplete, misleading, or inaccurate information.[1] Simply put: anyone can publish anything.

With hundreds of millions of **websites** on the **World Wide Web**, searching for nutrition information can be an overwhelming experience—much like walking into an enormous bookstore with millions of books, magazines, newspapers, and videos. And like a bookstore, the Internet offers no guarantees of the accuracy of the information found there—much of which is pure fiction.

When using the Internet, keep in mind that the quality of health-related information available covers a broad range.[2] You must evaluate websites for their accuracy, just like every other source. The accompanying "How to" provides tips for determining whether a website is reliable.

One of the most trustworthy sites used by scientists and others is the National Library of Medicine's PubMed, which provides free access to over 10 million abstracts (short descriptions) of research papers published in scientific journals around the world. Many abstracts provide links to websites where full articles are available. Figure H1-1 introduces this valuable resource.

Did you receive the e-mail warning about Costa Rican bananas causing the disease "necrotizing fasciitis"? If so, you've been scammed by Internet misinformation. When nutrition information arrives in unsolicited e-mails, be suspicious if:

- The person sending it to you didn't write it and you cannot determine who did or if that person is a nutrition expert
- The phrase "Forward this to everyone you know" appears
- The phrase "This is not a hoax" appears; chances are that it is
- The news is sensational and you've never heard about it from legitimate sources
- The language is emphatic and the text is sprinkled with capitalized words and exclamation marks
- No references are given or, if present, are of questionable validity when examined
- The message has been debunked on websites such as **www.quackwatch.org** or **www.urbanlegends.com**

Nutrition in the News

Consumers get much of their nutrition information from television news and magazine reports, which have heightened awareness of how diet influences the development of diseases. Consumers benefit from news coverage of nutrition when they learn to make lifestyle changes that will improve their health. Sometimes, however, when magazine articles or television programs report nutrition trends, they mislead consumers and create confusion. They often tell a lopsided story based on a few testimonials instead of presenting the results of research studies or a balance of expert opinions.

Tight deadlines and limited understanding sometimes make it difficult to provide a thorough report. Hungry for the latest news, the media often report scientific findings prematurely—without benefit of careful interpretation, replication, and peer review.[3] Usually, the reports present findings from a single, recently released study, making the news current and controversial. Consequently, the public receives diet and health news quickly, but not always in perspective. Reporters may twist inconclusive findings into "meaningful discoveries" when pressured to write catchy headlines and sensational stories.

As a result, "surprising new findings" seem to contradict one another, and consumers feel frustrated and betrayed. Occasionally, the reports are downright false, but more often the apparent contradictions are simply the normal result of science at work. A single study contributes to the big picture, but when viewed alone, it can easily distort the image. To be meaningful, the conclusions of any study must be presented cautiously within the context of other research findings.

Identifying Nutrition Experts

Regardless of whether the medium is electronic, print, or video, consumers need to ask whether the person behind the information is qualified to speak on nutrition. If the creator of an Internet website recommends eating three pineapples a day to lose weight, a trainer at the gym praises a high-protein diet, or a health-store clerk suggests an herbal supplement, should you believe these people? Can you distinguish between accurate news reports and infomercials on television? Have you noticed that many televised nutrition messages are presented by celebrities, fitness experts, psychologists, food editors, and chefs—that is, almost anyone except a **dietitian?** When you are confused or need sound dietary advice, whom should you ask?

Physicians and Other Health Care Professionals

Many people turn to physicians or other health care professionals for dietary advice, expecting them to know about all health-related matters. But are they the best sources of accurate and current information on nutrition? Only about 30 percent of all medical schools in the United States require students to take a separate nutrition course; less than half require the minimum 25 hours of nutrition instruction recommended by the National Academy of Sciences.[4] By comparison, most students reading this text are taking a nutrition class that provides an average of 45 hours of instruction.

The **American Dietetic Association (ADA)** asserts that standardized nutrition education should be included in the curricula for all health care professionals: physicians, nurses, physician's assistants, dental hygienists, physical and occupational therapists, social workers, and all others who provide services directly to clients. When these professionals understand the relevance of nutrition in the treatment and prevention of disease and have command of reliable nutrition information, then all the people they serve will also be better informed.

HOW TO Determine Whether a Website Is Reliable

To determine whether a website offers reliable nutrition information, ask the following questions:

- **Who?** Who is responsible for the site? Is it staffed by qualified professionals? Look for the authors' names and credentials. Have experts reviewed the content for accuracy?
- **When?** When was the site last updated? Because nutrition is an ever-changing science, sites need to be dated and updated frequently.
- **Where?** Where is the information coming from? The three letters following the dot in a Web address identify the site's affiliation. Addresses ending in "gov" (government), "edu" (educational institute), and "org" (organization) generally provide reliable information; "com" (commercial) sites represent businesses and, depending on their qualifications and integrity, may or may not offer dependable information.
- **Why?** Why is the site giving you this information? Is the site providing a public service or selling a product? Many commercial sites provide accurate information, but some do not. When money is the prime motivation, be aware that the information may be biased.

If you are satisfied with the answers to all of the questions above, then ask this final question:

- **What?** What is the message, and is it in line with other reliable sources? Information that contradicts common knowledge should be questioned. Many reliable sites provide links to other sites to facilitate your quest for knowledge, but this provision alone does not guarantee a reputable intention. Be aware that any site can link to any other site without permission.

FIGURE H1-1 PUBMED (www.pubmed.gov): Internet Resource for Scientific Nutrition References

The U.S. National Library of Medicine's PubMed website offers tutorials to help teach beginners to use the search system effectively. Often, simply visiting the site, typing a query in the "Search for" box, and clicking "Go" will yield satisfactory results.

For example, to find research concerning calcium and bone health, typing "calcium bone" nets over 30,000 results. Try setting limits on dates, types of articles, languages, and other criteria to obtain a more manageable number of abstracts to peruse.

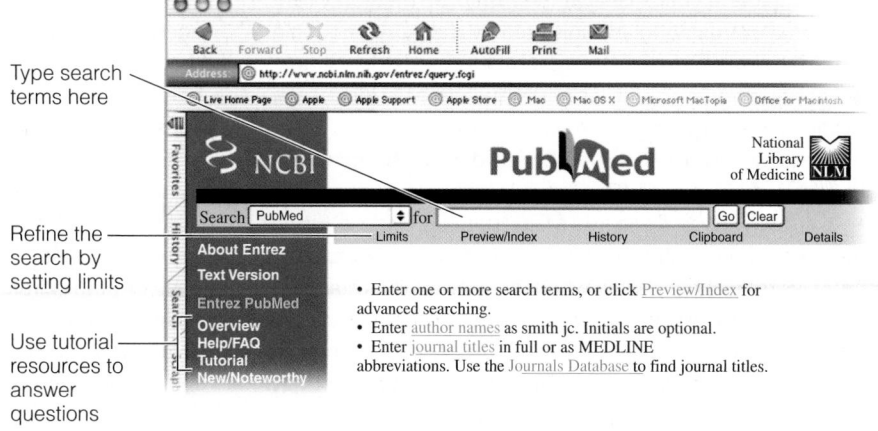

Type search terms here

Refine the search by setting limits

Use tutorial resources to answer questions

accredited: approved; in the case of medical centers or universities, certified by an agency recognized by the U.S. Department of Education.

American Dietetic Association (ADA): the professional organization of dietitians in the United States. The Canadian equivalent is Dietitians of Canada, which operates similarly.

certified nutritionists or **certified nutritional consultants** or **certified nutrition therapists:** a person who has been granted a document declaring his or her authority as a nutrition professional; see also *nutritionist.*

correspondence schools: schools that offer courses and degrees by mail. Some correspondence schools are accredited; others are not.

dietetic technician: a person who has completed a minimum of an associate's degree from an accredited university or college and an approved dietetic

technician program that includes a supervised practice experience. See also *dietetic technician, registered (DTR).*

dietetic technician, registered (DTR): a dietetic technician who has passed a national examination and maintains registration through continuing professional education.

dietitian: a person trained in nutrition, food science, and diet planning. See also *registered dietitian.*

DTR: see *dietetic technician, registered.*

fraudulent: the promotion, for financial gain, of devices, treatments, services, plans, or products (including diets and supplements) that alter or claim to alter a human condition without proof of safety or effectiveness. (The word *quackery* comes from the term *quacksalver,* meaning a person who quacks loudly about a miracle product— a lotion or a salve.)

Internet (the net): a worldwide network of millions of computers linked together to share information.

license to practice: permission under state or federal law, granted on meeting specified criteria, to use a certain title (such as dietitian) and offer certain services. **Licensed dietitians** may use the initials **LD** after their names.

misinformation: false or misleading information.

nutritionist: a person who specializes in the study of nutrition. Note that this definition does not specify qualifications and may apply not only to registered dietitians but also to self-described experts whose training is questionable. Most states have licensing laws that define the scope of practice for those calling themselves nutritionists.

public health dietitians: dietitians who specialize in providing nutrition services

through organized community efforts.

RD: see *registered dietitian.*

registered dietitian (RD): a person who has completed a minimum of a bachelor's degree from an accredited university or college, has completed approved course work and a supervised practice program, has passed a national examination, and maintains registration through continuing professional education.

registration: listing; with respect to health professionals, listing with a professional organization that requires specific course work, experience, and passing of an examination.

websites: Internet resources composed of text and graphic files, each with a unique URL (Uniform Resource Locator) that names the site (for example, www.usda.gov).

World Wide Web (the web, commonly abbreviated **www):** a graphical subset of the Internet.

Most health care professionals appreciate the connections between health and nutrition. Those who have specialized in clinical nutrition are especially well qualified to speak on the subject. Few, however, have the time or experience to develop diet plans and provide detailed diet instructions for clients. Often they wisely refer clients to a qualified nutrition expert—a **registered dietitian (RD).**

Registered Dietitians (RD)

A registered dietitian (RD) has the educational background necessary to deliver reliable nutrition advice and care.[5] To become an RD, a person must earn an undergraduate degree requiring about 60 semester hours in nutrition, food science, and other related subjects; complete a year's clinical internship or the equivalent; pass a national examination administered by the ADA; and maintain up-to-date knowledge and **registration** by participating in required continuing education activities such as attending seminars, taking courses, or writing professional papers.

Some states allow anyone to use the title dietitian or **nutritionist,** but others allow only an RD or people with specified qualifications to call themselves dietitians. Many states provide a further guarantee: a state registration, certification, or **license to practice.** In this way, states identify people who have met minimal standards of education and experience. Still, these state standards may fall short of those defining an RD. Similarly, some alternative educational programs qualify their graduates as **certified nutritionists, certified**

nutritional consultants, or **certified nutrition therapists**— terms that sound authoritative but lack the credentials of an RD.[6]

Dietitians perform a multitude of duties in many settings in most communities. They work in the food industry, pharmaceutical companies, home health agencies, long-term care institutions, private practice, public health departments, research centers, education settings, fitness centers, and hospitals. Depending on their work settings, dietitians can assume a number of different job responsibilities and positions. In hospitals, administrative dietitians manage the foodservice system; clinical dietitians provide client care; and nutrition support team dietitians coordinate nutrition care with other health care professionals. In the food industry, dietitians conduct research, develop products, and market services.

Public health dietitians who work in government-funded agencies play a key role in delivering nutrition services to people in the community. Among their many roles, public health dietitians help plan, coordinate, and evaluate food assistance programs; act as consultants to other agencies; manage finances; and much more.

Other Dietary Employees

In some facilities, a **dietetic technician** assists registered dietitians in both administrative and clinical responsibilities. A dietetic technician has been educated and trained to work under the guidance of a registered dietitian; upon passing a national examination, the title changes to **dietetic technician, registered (DTR).**

In addition to the dietetic technician, other dietary employees may include clerks, aides, cooks, porters, and other assistants. These dietary employees do not have extensive formal training in nutrition, and their ability to provide accurate information may be limited.

Identifying Fake Credentials

In contrast to registered dietitians, thousands of people obtain fake nutrition degrees and claim to be nutrition consultants or doctors of "nutrimedicine." These and other such titles may sound meaningful, but most of these people lack the established credentials and training of an ADA-sanctioned dietitian. If you look closely, you can see signs of their fake expertise.

Consider educational background, for example. The minimum standards of education for a dietitian specify a bachelor of science (BS) degree in food science and human nutrition or related fields from an **accredited** college or university.* Such a degree generally requires four to five years of study. In contrast, a fake nutrition expert may display a degree from a six-month correspondence course. Such a degree simply falls short. In some cases, businesses posing as legitimate **correspondence schools** offer even less—they sell certificates to anyone who pays the fees. To obtain these "degrees," a candidate need not attend any classes, read any books, or pass any examinations.

To safeguard educational quality, an accrediting agency recognized by the U.S. Department of Education (DOE) certifies that certain schools meet criteria established to ensure that an institution provides complete and accurate schooling. Unfortunately, fake nutrition degrees are available from schools "accredited" by more than 30 phony accrediting agencies. Acquiring false credentials is especially easy today, with **fraudulent** businesses operating via the Internet.

Knowing the qualifications of someone who provides nutrition information can help you determine whether that person's advice might be harmful or helpful. Don't be afraid to ask for credentials. The accompanying "How to" lists credible sources of nutrition information.

Red Flags of Nutrition Quackery

Figure H1-2 (p. 34) features eight red flags consumers can use to identify nutrition **misinformation.** Sales of unproven and dangerous products have always been a concern, but the Inter-

---

HOW TO Find Credible Sources of Nutrition Information

Government agencies, volunteer associations, consumer groups, and professional organizations provide consumers with reliable health and nutrition information. Credible sources of nutrition information include:

- Nutrition and food science departments at a university or community college
- Local agencies such as the health department or County Cooperative Extension Service
- Government health agencies such as:
  - Department of Agriculture (USDA)   www.usda.gov
  - Department of Health and Human Services (DHHS)   www.os.dhhs.gov
  - Food and Drug Administration (FDA)   www.fda.gov
  - Health Canada   www.hc-sc.gc.ca/nutrition
- Volunteer health agencies such as:
  - American Cancer Society   www.cancer.org
  - American Diabetes Association   www.diabetes.org
  - American Heart Association   www.americanheart.org
- Reputable consumer groups such as:
  - American Council on Science and Health   www.acsh.org
  - Federal Citizen Information Center   www.pueblo.gsa.gov
  - International Food Information Council   ific.org
- Professional health organizations such as:
  - American Dietetic Assocation   www.eatright.org
  - American Medical Association   www.ama-assn.org
  - Dietitians of Canada   www.dietitians.ca
- Journals such as:
  - *American Journal of Clinical Nutrition*   www.ajcn.org
  - *New England Journal of Medicine*   www.nejm.org
  - *Nutrition Reviews*   www.ilsi.org

---

net now provides merchants with an easy and inexpensive way to reach millions of customers around the world. Because of the difficulty in regulating the Internet, fraudulent and illegal sales of medical products have hit a bonanza. As is the case with the air, no one owns the Internet, and similarly, no one has control over the pollution. Countries have different laws regarding sales of drugs, dietary supplements, and other health products, but applying these laws to the Internet marketplace is almost impossible. Even if illegal activities could be defined and identified, finding the person responsible for a particular website is not always possible. Websites can open and close in a blink of a cursor. Now, more than ever, consumers must heed the caution "Buyer beware."

In summary, when you hear nutrition news, consider its source. Ask yourself these two questions: Is the person providing the information qualified to speak on nutrition? Is the information based on valid scientific research? If not, find a better source. After all, your health depends on it.

---

* To ensure the quality and continued improvement of nutrition and dietetics education programs, an ADA agency known as the Commission on Accreditation for Dietetics Education (CADE) establishes and enforces eligibility requirements and accreditation standards for programs preparing students for careers as registered dietitians or dietetic technicians. Programs meeting those standards are accredited by CADE.

FIGURE H1-2 Red Flags of Nutrition Quackery

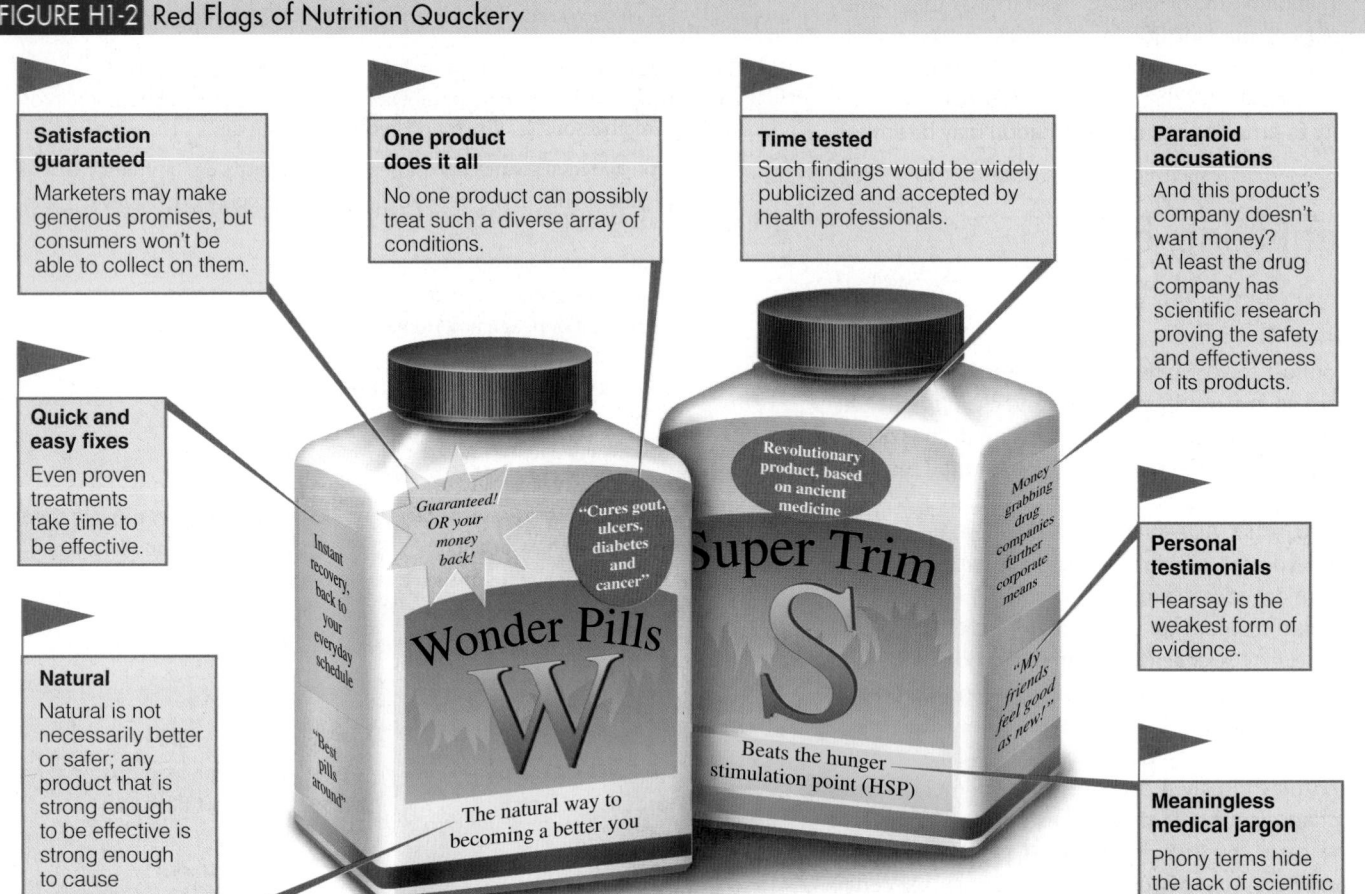

Satisfaction guaranteed
Marketers may make generous promises, but consumers won't be able to collect on them.

One product does it all
No one product can possibly treat such a diverse array of conditions.

Time tested
Such findings would be widely publicized and accepted by health professionals.

Paranoid accusations
And this product's company doesn't want money? At least the drug company has scientific research proving the safety and effectiveness of its products.

Quick and easy fixes
Even proven treatments take time to be effective.

Natural
Natural is not necessarily better or safer; any product that is strong enough to be effective is strong enough to cause side effects.

Personal testimonials
Hearsay is the weakest form of evidence.

Meaningless medical jargon
Phony terms hide the lack of scientific proof.

NUTRITION ON THE NET

For further study of topics covered in this chapter, log on to **academic.cengage.com/nutrition/rolfes/UNCN8e**. Go to Chapter 1, then to Nutrition on the Net.

- Visit the National Council Against Health Fraud: **www.ncahf.org**
- Find a registered dietitian in your area from the American Dietetic Association: **www.eatright.org**
- Find a nutrition professional in Canada from the Dietitians of Canada: **www.dietitians.ca**
- Find out whether a correspondence school is accredited from the Distance Education and Training Council's Accrediting Commission: **www.detc.org**
- Find useful and reliable health information from the Health on the Net Foundation: **www.hon.ch**
- Find out whether a school is properly accredited for a dietetics degree from the American Dietetic Association: **www.eatright.org/cade**
- Obtain a listing of accredited institutions, professionally accredited programs, and candidates for accreditation from the American Council on Education: **www.acenet.edu**
- Learn more about quackery from Stephen Barrett's Quackwatch: **www.quackwatch.org**
- Check out health-related hoaxes and urban legends: **www.cdc.gov/hoax_rumors.htm** and **www.urbanlegends.com/**
- Find reliable research articles: **www.pubmed.gov**

REFERENCES

1. Position of the American Dietetic Association: Food and nutrition misinformation, *Journal of the American Dietetic Association* 106 (2006): 601–607.
2. G. Eysenbach and coauthors, Empirical studies assessing the quality of health information for consumers on the World Wide Web: A systematic review, *Journal of the American Medical Association* 287 (2002): 2691–2700.
3. L. M. Schwartz, S. Woloshin, and L. Baczek, Media coverage of scientific meetings: Too much, too soon? *Journal of the American Medical Association* 287 (2002): 2859–2863.
4. K. M. Adams and coauthors, Status of nutrition education in medical schools, *American Journal of Clinical Nutrition* 83 (2006): 941S–944S.
5. Position of the American Dietetic Association: The roles of registered dieticians and dietetic technicians, registered in health promotion and disease prevention, *Journal of the American Dietetic Association* 106 (2006): 1875–1884.
6. Nutritionist imposters and how to spot them, *Nutrition and the M.D.*, September 2004, pp. 4–6.

Nutrition in Your Life

You make food choices—deciding what to eat and how much to eat—more than 1000 times every year. We eat so frequently that it's easy to choose a meal without giving any thought to its nutrient contributions or health consequences. Even when we want to make healthy choices, we may not know which foods to select or how much to consume. With a few tools and tips, you can learn to plan a healthy diet.

Planning a Healthy Diet

Chapter 1 explained that the body's many activities are supported by the nutrients delivered by the foods people eat. Food choices made over years influence the body's health, and consistently poor choices increase the risks of developing chronic diseases. This chapter shows how a person can select from the tens of thousands of available foods to create a diet that supports health. Fortunately, most foods provide several nutrients, so one trick for wise diet planning is to select a combination of foods that deliver a full array of nutrients. This chapter begins by introducing the diet-planning principles and dietary guidelines that assist people in selecting foods that will deliver nutrients without excess energy (kcalories).

Principles and Guidelines

How well you nourish yourself does not depend on the selection of any one food. Instead, it depends on the selection of many different foods at numerous meals over days, months, and years. Diet-planning principles and dietary guidelines are key concepts to keep in mind whenever you are selecting foods—whether shopping at the grocery store, choosing from a restaurant menu, or preparing a home-cooked meal.

Diet-Planning Principles

Diet planners have developed several ways to select foods. Whatever plan or combination of plans they use, though, they keep in mind the six basic diet-planning principles ◆ listed in the margin.

Adequacy   **Adequacy** means that the diet provides sufficient energy and enough of all the nutrients to meet the needs of healthy people. Take the essential nutrient iron, for example. Because the body loses some iron each day, people have to replace it by eating foods that contain iron. A person whose diet fails to provide enough iron-rich foods may develop the symptoms of iron-deficiency anemia: the person may feel weak, tired, and listless; have frequent headaches; and find that even the smallest amount of muscular work brings disabling fatigue. To prevent these deficiency symptoms, a person must include foods that supply adequate iron. The same is true for all the other essential nutrients introduced in Chapter 1.

◆ Diet-planning principles:
- **A**dequacy
- **B**alance
- k**C**alorie (energy) control
- Nutrient **D**ensity
- **M**oderation
- **V**ariety

adequacy (dietary): providing all the essential nutrients, fiber, and energy in amounts sufficient to maintain health.

© Polara Sutdios Inc.

To ensure an adequate and balanced diet, eat a variety of foods daily, choosing different foods from each group.

◆ Balance in the diet helps to ensure adequacy.

◆ Nutrient density promotes adequacy and kcalorie control.

Balance The art of balancing the diet involves consuming enough—but not too much—of each type of food. The essential minerals calcium and iron, taken together, illustrate the importance of dietary **balance.** Meats, fish, and poultry are rich in iron but poor in calcium. Conversely, milk and milk products are rich in calcium but poor in iron. Use some meat or meat alternates for iron; use some milk and milk products for calcium; and save some space for other foods, too, because a diet consisting of milk and meat alone would not be adequate. ◆ For the other nutrients, people need whole grains, vegetables, and fruits.

kCalorie (Energy) Control Designing an adequate diet without overeating requires careful planning. Once again, balance plays a key role. The amount of energy coming into the body from foods should balance with the amount of energy being used by the body to sustain its metabolic and physical activities. Upsetting this balance leads to gains or losses in body weight. The discussion of energy balance and weight control in Chapters 8 and 9 examines this issue in more detail, but the key to **kcalorie control** is to select foods of high **nutrient density.**

Nutrient Density To eat well without overeating, select foods that deliver the most nutrients for the least food energy. Consider foods containing calcium, for example. You can get about 300 milligrams of calcium from either $1^1/_2$ ounces of cheddar cheese or 1 cup of fat-free milk, but the cheese delivers about twice as much food energy (kcalories) as the milk. The fat-free milk, then, is twice as calcium dense as the cheddar cheese; it offers the same amount of calcium for half the kcalories. Both foods are excellent choices for adequacy's sake alone, but to achieve adequacy while controlling kcalories, ◆ the fat-free milk is the better choice. (Alternatively, a person could select a low-fat cheddar cheese.) The many bar graphs that appear in Chapters 10 through 13 highlight the most nutrient-dense choices, and the accompanying "How to" describes how to compare foods based on nutrient density.

balance (dietary): providing foods in proportion to each other and in proportion to the body's needs.

kcalorie (energy) control: management of food energy intake.

nutrient density: a measure of the nutrients a food provides relative to the energy it provides. The more nutrients and the fewer kcalories, the higher the nutrient density.

HOW TO Compare Foods Based on Nutrient Density

One way to evaluate foods is simply to notice their nutrient contribution *per serving:* 1 cup of milk provides about 300 milligrams of calcium, and $1/_2$ cup of fresh, cooked turnip greens provides about 100 milligrams. Thus a serving of milk offers three times as much calcium as a serving of turnip greens. To get 300 milligrams of calcium, a person could choose either 1 cup of milk or $1^1/_2$ cups of turnip greens.

Another valuable way to evaluate foods is to consider their nutrient density—their nutrient contribution *per kcalorie.* Fat-free milk delivers about 85 kcalories with its 300 milligrams of calcium. To calculate the nutrient density, divide milligrams by kcalories:

$$\frac{300 \text{ mg calcium}}{85 \text{ kcal}} = 3.5 \text{ mg per kcal}$$

Do the same for the fresh turnip greens, which provide 15 kcalories with the 100 milligrams of calcium:

$$\frac{100 \text{ mg calcium}}{15 \text{ kcal}} = 6.7 \text{ mg per kcal}$$

The more milligrams per kcalorie, the greater the nutrient density. Turnip greens are more calcium dense than milk. They provide more calcium *per kcalorie* than milk, but milk offers more calcium *per serving.* Both approaches offer valuable information, especially when combined with a realistic appraisal. What matters most is which are you more likely to consume—$1^1/_2$ cups of turnip greens or 1 cup of milk? You can get 300 milligrams of calcium from either, but the greens will save you about 40 kcalories (the savings would be even greater if you usually use whole milk).

Keep in mind, too, that calcium is only one of the many nutrients that foods provide. Similar calculations for protein, for example, would show that fat-free milk provides more protein both *per kcalorie* and *per serving* than turnip greens—that is, milk is more protein dense. Combining variety with nutrient density helps to ensure the adequacy of all nutrients.

Just like a person who has to pay for rent, food, clothes, and tuition on a limited budget, we have to obtain iron, calcium, and all the other essential nutrients on a limited energy allowance. Success depends on getting many nutrients for each kcalorie "dollar." For example, a can of cola and a handful of grapes may both provide about the same number of kcalories, but the grapes deliver many more nutrients. A person who makes nutrient-dense choices, such as fruit instead of cola, can meet daily nutrient needs on a lower energy budget. Such choices support good health.

Foods that are notably low in nutrient density—such as potato chips, candy, and colas—are sometimes called **empty-kcalorie foods.** The kcalories these foods provide are called "empty" because they deliver energy (from sugar, fat, or both) with little, or no, protein, vitamins, or minerals.

Moderation Foods rich in fat and sugar provide enjoyment and energy but relatively few nutrients. In addition, they promote weight gain when eaten in excess. A person practicing **moderation** ◆ eats such foods only on occasion and regularly selects foods low in solid fats and added sugars, a practice that automatically improves nutrient density. Returning to the example of cheddar cheese versus fat-free milk, the fat-free milk not only offers the same amount of calcium for less energy, but it also contains far less fat than the cheese.

◆ Moderation contributes to adequacy, balance, and kcalorie control.

Variety A diet may have all of the virtues just described and still lack **variety**, if a person eats the same foods day after day. People should select foods from each of the food groups daily and vary their choices within each food group from day to day for several reasons. First, different foods within the same group contain different arrays of nutrients. Among the fruits, for example, strawberries are especially rich in vitamin C while apricots are rich in vitamin A. Variety improves nutrient adequacy.[1] Second, no food is guaranteed entirely free of substances that, in excess, could be harmful. The strawberries might contain trace amounts of one contaminant, the apricots another. By alternating fruit choices, a person will ingest very little of either contaminant. Third, as the adage goes, variety is the spice of life. A person who eats beans frequently can enjoy pinto beans in Mexican burritos today, garbanzo beans in Greek salad tomorrow, and baked beans with barbecued chicken on the weekend. Eating nutritious meals need never be boring.

Dietary Guidelines for Americans

What should a person eat to stay healthy? The answers can be found in the *Dietary Guidelines for Americans 2005*. These guidelines provide science-based advice to promote health and to reduce risk of chronic diseases through diet and physical activity.[2] Table 2-1 presents the nine *Dietary Guidelines* topics with their key recommendations. These key recommendations, along with additional recommendations for specific population groups, also appear throughout the text as their subjects are discussed. The first three topics focus on choosing nutrient-dense foods within energy needs, maintaining a healthy body weight, and engaging in regular physical activity. The fourth topic, "Food Groups to Encourage," focuses on the selection of a variety of fruits and vegetables, whole grains, and milk. The next four topics advise people to choose sensibly in their use of fats, carbohydrates, salt, and alcoholic beverages (for those who partake). Finally, consumers are reminded to keep foods safe. Together, the *Dietary Guidelines* point the way toward better health. Table 2-2 presents Canada's *Guidelines for Healthy Eating*.

Some people might wonder why *dietary* guidelines include recommendations for physical activity. The simple answer is that most people who maintain a healthy body weight do more than eat right. They also exercise—the equivalent of 60 minutes or more of moderately intense physical activity daily. As you will see repeatedly throughout this text, food and physical activity choices are integral partners in supporting good health.

empty-kcalorie foods: a popular term used to denote foods that contribute energy but lack protein, vitamins, and minerals.

moderation (dietary): providing enough but not too much of a substance.

variety (dietary): eating a wide selection of foods within and among the major food groups.

TABLE 2-1 Key Recommendations of the *Dietary Guidelines for Americans 2005*

Adequate Nutrients within Energy Needs

- Consume a variety of nutrient-dense foods and beverages within and among the basic food groups; limit intakes of saturated and *trans* fats, cholesterol, added sugars, salt, and alcohol.
- Meet recommended intakes within energy needs by adopting a balanced eating pattern, such as the USDA Food Guide (see pp. 41–47).

Weight Management

- To maintain body weight in a healthy range, balance kcalories from foods and beverages with kcalories expended (see Chapters 8 and 9).
- To prevent gradual weight gain over time, make small decreases in food and beverage kcalories and increase physical activity.

Physical Activity

- Engage in regular physical activity and reduce sedentary activities to promote health, psychological well-being, and a healthy body weight.
- Achieve physical fitness by including cardiovascular conditioning, stretching exercises for flexibility, and resistance exercises or calisthenics for muscle strength and endurance.

Food Groups to Encourage

- Consume a sufficient amount of fruits, vegetables, milk and milk products, and whole grains while staying within energy needs.
- Select a variety of fruits and vegetables each day, including selections from all five vegetable subgroups (dark green, orange, legumes, starchy vegetables, and other vegetables) several times a week. Make at least half of the grain selections whole grains. Select fat-free or low-fat milk products.

Fats

- Consume less than 10 percent of kcalories from saturated fats and less than 300 milligrams of cholesterol per day, and keep *trans* fats consumption as low as possible (see Chapter 5).
- Keep total fat intake between 20 and 35 percent of kcalories; choose from mostly polyunsaturated and monounsaturated fat sources such as fish, nuts, and vegetable oils.
- Select and prepare foods that are lean, low fat, or fat-free and low in saturated and/or *trans* fats.

Carbohydrates

- Choose fiber-rich fruits, vegetables, and whole grains often.
- Choose and prepare foods and beverages with little added sugars (see Chapter 4).
- Reduce the incidence of dental caries by practicing good oral hygiene and consuming sugar- and starch-containing foods and beverages less frequently.

Sodium and Potassium

- Choose and prepare foods with little salt (less than 2300 milligrams sodium or approximately 1 teaspoon salt daily). At the same time, consume potassium-rich foods, such as fruits and vegetables (see Chapter 12).

Alcoholic Beverages

- Those who choose to drink alcoholic beverages should do so sensibly and in moderation (up to one drink per day for women and up to two drinks per day for men).
- Some individuals should not consume alcoholic beverages (see Highlight 7).

Food Safety

- To avoid microbial foodborne illness, keep foods safe: clean hands, food contact surfaces, and fruits and vegetables; separate raw, cooked, and ready-to-eat foods; cook foods to a safe internal temperature; chill perishable food promptly; and defrost food properly.
- Avoid unpasteurized milk and products made from it; raw or undercooked eggs, meat, poultry, fish, and shellfish; unpasteurized juices; raw sprouts.

NOTE: These guidelines are intended for adults and healthy children ages 2 and older.
SOURCE: The *Dietary Guidelines for Americans 2005*, available at **www.healthierus.gov/dietaryguidelines**.

TABLE 2-2 Canada's *Guidelines for Healthy Eating*

- Enjoy a variety of foods.
- Emphasize cereals, breads, other grain products, vegetables, and fruits.
- Choose lower-fat dairy products, leaner meats, and foods prepared with little or no fat.
- Achieve and maintain a healthy body weight by enjoying regular physical activity and healthy eating.
- Limit salt, alcohol, and caffeine.

SOURCE: These guidelines derive from *Action Towards Healthy Eating—Canada's Guidelines for Healthy Eating and Recommended Strategies for Implementation.*

IN SUMMARY

A well-planned diet delivers adequate nutrients, a balanced array of nutrients, and an appropriate amount of energy. It is based on nutrient-dense foods, moderate in substances that can be detrimental to health, and varied in its selections. The 2005 *Dietary Guidelines* apply these principles, offering practical advice on how to eat for good health.

Diet-Planning Guides

To plan a diet that achieves all of the dietary ideals just outlined, a person needs tools as well as knowledge. Among the most widely used tools for diet planning are **food group plans** that build a diet from clusters of foods that are similar in nutrient content. Thus each group represents a set of nutrients that differs somewhat from the nutrients supplied by the other groups. Selecting foods from each of the groups eases the task of creating an adequate and balanced diet.

USDA Food Guide

The 2005 *Dietary Guidelines* encourage consumers to adopt a balanced eating plan, such as the USDA's Food Guide (see Figure 2-1 on pp. 42–43). The USDA Food Guide assigns foods to five major groups ◆ and recommends daily amounts of foods from each group to meet nutrient needs. In addition to presenting the food groups, the figure lists the most notable nutrients of each group, the serving equivalents, and the foods within each group sorted by nutrient density. Chapter 15 provides a food guide for young children, and Appendix I presents Canada's food group plan, the *Food Guide to Healthy Eating*.

◆ Five food groups:
- Fruits
- Vegetables
- Grains
- Meat and legumes
- Milk

Dietary Guidelines for Americans 2005

Meet recommended intakes within energy needs by adopting a balanced eating pattern, such as the USDA Food Guide or the DASH eating plan. (The DASH eating plan is presented in Chapter 12.)

◆ Chapter 8 explains how to determine energy needs. For an approximation, turn to the DRI Estimated Energy Requirement (EER) on the inside front cover.

Recommended Amounts All food groups offer valuable nutrients, and people should make selections from each group daily. Table 2-3 specifies the amounts of foods from each group needed daily to create a healthful diet for several energy (kcalorie) levels. ◆ Estimated daily kcalorie needs for sedentary and active men and

food group plans: diet-planning tools that sort foods into groups based on nutrient content and then specify that people should eat certain amounts of foods from each group.

TABLE 2-3	Recommended Daily Amounts from Each Food Group							
	1600 kcal	1800 kcal	2000 kcal	2200 kcal	2400 kcal	2600 kcal	2800 kcal	3000 kcal
Fruits	1½ c	1½ c	2 c	2 c	2 c	2 c	2½ c	2½ c
Vegetables	2 c	2½ c	2½ c	3 c	3 c	3½ c	3½ c	4 c
Grains	5 oz	6 oz	6 oz	7 oz	8 oz	9 oz	10 oz	10 oz
Meat and legumes	5 oz	5 oz	5½ oz	6 oz	6½ oz	6½ oz	7 oz	7 oz
Milk	3 c	3 c	3 c	3 c	3 c	3 c	3 c	3 c
Oils	5 tsp	5 tsp	6 tsp	6 tsp	7 tsp	8 tsp	8 tsp	10 tsp
Discretionary kcalorie allowance	132 kcal	195 kcal	267 kcal	290 kcal	362 kcal	410 kcal	426 kcal	512 kcal

FIGURE 2-1 USDA Food Guide, 2005

Key:

● Foods generally high in nutrient density (choose most often)

▲ Foods lower in nutrient density (limit selections)

FRUITS

© Polara Studios, Inc.

Consume a variety of fruits and no more than one-third of the recommended intake as fruit juice.

These foods contribute folate, vitamin A, vitamin C, potassium, and fiber.

> ½ c fruit is equivalent to ½ c fresh, frozen, or canned fruit; 1 small fruit; ¼ c dried fruit; ½ c fruit juice.

● Apples, apricots, avocados, bananas, blueberries, cantaloupe, cherries, grapefruit, grapes, guava, kiwi, mango, oranges, papaya, peaches, pears, pineapples, plums, raspberries, strawberries, watermelon; dried fruit (dates, figs, raisins); unsweetened juices.

▲ Canned or frozen fruit in syrup; juices, punches, ades, and fruit drinks with added sugars; fried plantains.

VEGETABLES

© Polara Studios, Inc.

Choose a variety of vegetables from all five subgroups several times a week.

These foods contribute folate, vitamin A, vitamin C, vitamin K, vitamin E, magnesium, potassium, and fiber.

> ½ c vegetables is equivalent to ½ c cut-up raw or cooked vegetables; ½ c cooked legumes; ½ c vegetable juice; 1 c raw, leafy greens.

● Dark green vegetables: Broccoli and leafy greens such as arugula, beet greens, bok choy, collard greens, kale, mustard greens, romaine lettuce, spinach, and turnip greens.

● Orange and deep yellow vegetables: Carrots, carrot juice, pumpkin, sweet potatoes, and winter squash (acorn, butternut).

● Legumes: Black beans, black-eyed peas, garbanzo beans (chickpeas), kidney beans, lentils, navy beans, pinto beans, soybeans and soy products such as tofu, and split peas.

● Starchy vegetables: Cassava, corn, green peas, hominy, lima beans, and potatoes.

● Other vegetables: Artichokes, asparagus, bamboo shoots, bean sprouts, beets, brussels sprouts, cabbages, cactus, cauliflower, celery, cucumbers, eggplant, green beans, iceberg lettuce, mushrooms, okra, onions, peppers, seaweed, snow peas, tomatoes, vegetable juices, zucchini.

▲ Baked beans, candied sweet potatoes, coleslaw, French fries, potato salad, refried beans, scalloped potatoes, tempura vegetables.

GRAINS

© Polara Studios, Inc.

Make at least half of the grain selections whole grains.

These foods contribute folate, niacin, riboflavin, thiamin, iron, magnesium, selenium, and fiber.

> 1 oz grains is equivalent to 1 slice bread; ½ c cooked rice, pasta, or cereal; 1 oz dry pasta or rice; 1 c ready-to-eat cereal; 3 c popped popcorn.

● Whole grains (amaranth, barley, brown rice, buckwheat, bulgur, millet, oats, quinoa, rye, wheat) and whole-grain, low-fat breads, cereals, crackers, and pastas; popcorn.

● Enriched bagels, breads, cereals, pastas (couscous, macaroni, spaghetti), pretzels, rice, rolls, tortillas.

▲ Biscuits, cakes, cookies, cornbread, crackers, croissants, doughnuts, French toast, fried rice, granola, muffins, pancakes, pastries, pies, presweetened cereals, taco shells, waffles.

FIGURE 2-1 USDA Food Guide, 2005, continued

MEAT, POULTRY, FISH, LEGUMES, EGGS, AND NUTS

© Polara Studios, Inc.

Make lean or low-fat choices. Prepare them with little, or no, added fat.

Meat, poultry, fish, and eggs contribute protein, niacin, thiamin, vitamin B_6, vitamin B_{12}, iron, magnesium, potassium, and zinc; legumes and nuts are notable for their protein, folate, thiamin, vitamin E, iron, magnesium, potassium, zinc, and fiber.

> **1 oz meat is equivalent to 1 oz cooked lean meat, poultry, or fish; 1 egg; $\frac{1}{4}$ c cooked legumes or tofu; 1 tbs peanut butter; $\frac{1}{2}$ oz nuts or seeds.**

● Poultry (no skin), fish, shellfish, legumes, eggs, lean meat (fat-trimmed beef, game, ham, lamb, pork); low-fat tofu, tempeh, peanut butter, nuts (almonds, filberts, peanuts, pistachios, walnuts) or seeds (flaxseeds, pumpkin seeds, sunflower seeds).

△ Bacon; baked beans; fried meat, fish, poultry, eggs, or tofu; refried beans; ground beef; hot dogs; luncheon meats; marbled steaks; poultry with skin; sausages; spare ribs.

MILK, YOGURT, AND CHEESE

© Polara Studios, Inc.

Make fat-free or low-fat choices. Choose lactose-free products or other calcium-rich foods if you don't consume milk.

These foods contribute protein, riboflavin, vitamin B_{12}, calcium, magnesium, potassium, and, when fortified, vitamin A and vitamin D.

> **1 c milk is equivalent to 1 c fat-free milk or yogurt; $1\frac{1}{2}$ oz fat-free natural cheese; 2 oz fat-free processed cheese.**

● Fat-free milk and fat-free milk products such as buttermilk, cheeses, cottage cheese, yogurt; fat-free fortified soy milk.

△ 1% low-fat milk, 2% reduced-fat milk, and whole milk; low-fat, reduced-fat, and whole-milk products such as cheeses, cottage cheese, and yogurt; milk products with added sugars such as chocolate milk, custard, ice cream, ice milk, milk shakes, pudding, sherbet; fortified soy milk.

OILS

Matthew Farruggio

Select the recommended amounts of oils from among these sources.

These foods contribute vitamin E and essential fatty acids (see Chapter 5), along with abundant kcalories.

> **1 tsp oil is equivalent to 1 tbs low-fat mayonnaise; 2 tbs light salad dressing; 1 tsp vegetable oil; 1 tsp soft margarine.**

● Liquid vegetable oils such as canola, corn, flaxseed, nut, olive, peanut, safflower, sesame, soybean, and sunflower oils; mayonnaise, oil-based salad dressing, soft *trans*-free margarine.

● Unsaturated oils that occur naturally in foods such as avocados, fatty fish, nuts, olives, seeds (flaxseeds, sesame seeds), and shellfish.

SOLID FATS AND ADDED SUGARS

Matthew Farruggio

Limit intakes of food and beverages with solid fats and added sugars.

Solid fats deliver saturated fat and *trans* fat, and intake should be kept low. Solid fats and added sugars contribute abundant kcalories but few nutrients, and intakes should not exceed the discretionary kcalorie allowance—kcalories to meet energy needs after all nutrient needs have been met with nutrient-dense foods. Alcohol also contributes abundant kcalories but few nutrients, and its kcalories are counted among discretionary kcalories. See Table 2-3 for some discretionary kcalorie allowances.

△ Solid fats that occur in foods naturally such as milk fat and meat fat (see △ in previous lists).

△ Solid fats that are often added to foods such as butter, cream cheese, hard margarine, lard, sour cream, and shortening.

△ Added sugars such as brown sugar, candy, honey, jelly, molasses, soft drinks, sugar, and syrup.

△ Alcoholic beverages include beer, wine, and liquor.

TABLE 2-4	Estimated Daily kCalorie Needs for Adults	
	Sedentary[a]	Active[b]
Women		
19–30 yr	2000	2400
31–50 yr	1800	2200
51+ yr	1600	2100
Men		
19–30 yr	2400	3000
31–50 yr	2200	2900
51+ yr	2000	2600

[a]Sedentary describes a lifestyle that includes only the activities typical of day-to-day life.
[b]Active describes a lifestyle that includes physical activity equivalent to walking more than 3 miles per day at a rate of 3 to 4 miles per hour, in addition to the activities typical of day-to-day life. kCalorie values for active people reflect the midpoint of the range appropriate for age and gender, but within each group, older adults may need fewer kcalories and younger adults may need more.
NOTE: In addition to gender, age, and activity level, energy needs vary with height and weight (see Chapter 8 and Appendix F).

◆ Reminder: *Phytochemicals* are the nonnutrient compounds found in plant-derived foods that have biological activity in the body.

◆ The USDA nutrients of concern are fiber, vitamin A, vitamin C, vitamin E, and the minerals calcium, magnesium, and potassium.

legumes (lay-GYOOMS, LEG-yooms): plants of the bean and pea family, with seeds that are rich in protein compared with other plant-derived foods.

women are shown in Table 2-4. A sedentary young women needing 2000 kcalories a day, for example, would select 2 cups of fruit; 2¹/₂ cups of vegetables (dispersed among the vegetable subgroups); 6 ounces of grain foods (with at least half coming from whole grains); 5¹/₂ ounces of meat, poultry, or fish, or the equivalent of **legumes,** eggs, seeds, or nuts; and 3 cups of milk or yogurt, or the equivalent amount of cheese or fortified soy products. Additionally, a small amount of unsaturated oil, such as vegetable oil, or the oils of nuts, olives, or fatty fish, is required to supply needed nutrients.

All vegetables provide an array of vitamins, fiber, and the mineral potassium, but some vegetables are especially good sources of certain nutrients and beneficial phytochemicals. ◆ For this reason, the USDA Food Guide sorts the vegetable group into five subgroups. The dark green vegetables deliver the B vitamin folate; the orange vegetables provide vitamin A; legumes supply iron and protein; the starchy vegetables contribute carbohydrate energy; and the other vegetables fill in the gaps and add more of these same nutrients.

In a 2000-kcalorie diet, then, the recommended 2¹/₂ cups of daily vegetables should be varied among the subgroups over a week's time, as shown in Table 2-5. In other words, consuming 2¹/₂ cups of potatoes or even nutrient-rich spinach every day for seven days does *not* meet the recommended vegetable intakes. Potatoes and spinach make excellent choices when consumed in balance with vegetables from other subgroups. Intakes of vegetables are appropriately averaged over a week's time—it is not necessary to include every subgroup every day.

Notable Nutrients As Figure 2-1 notes, each food group contributes key nutrients. This feature provides flexibility in diet planning because a person can select any food from a food group and receive similar nutrients. For example, a person can choose milk, cheese, or yogurt and receive the same key nutrients. Importantly, foods provide not only these key nutrients, but small amounts of other nutrients and phytochemicals as well.

Because legumes contribute the same key nutrients—notably, protein, iron, and zinc—as meats, poultry, and fish, they are included in the same food group. For this reason, legumes are useful as meat alternatives, and they are also excellent sources of fiber and the B vitamin folate. To encourage frequent consumption, the USDA Food Guide also includes legumes as a subgroup of the vegetable group. Thus legumes count in either the vegetable group or the meat and legume group. In general, people who regularly eat meat, poultry, and fish count legumes as a vegetable, and vegetarians and others who seldom eat meat, poultry, or fish count legumes in the meat and legumes group.

The USDA Food Guide encourages greater consumption from certain food groups to provide the nutrients most often lacking ◆ in the diets of Americans. In general, most people need to eat:

- *More* dark green vegetables, orange vegetables, legumes, fruits, whole grains, and low-fat milk and milk products

| TABLE 2-5 | Recommended Weekly Amounts from the Vegetable Subgroups |

Table 2-3 specifies the recommended amounts of total vegetables per *day.* This table shows those amounts dispersed among five vegetable subgroups per *week.*

Vegetable Subgroups	1600 kcal	1800 kcal	2000 kcal	2200 kcal	2400 kcal	2600 kcal	2800 kcal	3000 kcal
Dark green	2 c	3 c	3 c	3 c	3 c	3 c	3 c	3 c
Orange and deep yellow	1½ c	2 c	2 c	2 c	2 c	2½ c	2½ c	2½ c
Legumes	2½ c	3 c	3 c	3 c	3 c	3½ c	3½ c	3½ c
Starchy	2½ c	3 c	3 c	6 c	6 c	7 c	7 c	9 c
Other	5½ c	6½ c	6½ c	7 c	7 c	8½ c	8½ c	10 c

- *Less* refined grains, total fats (especially saturated fat, *trans* fat, and cholesterol), added sugars, and total kcalories

Nutrient Density The USDA Food Guide provides a foundation for a healthy diet by emphasizing nutrient-dense options within each food group. By consistently selecting nutrient-dense foods, a person can obtain all the nutrients needed and still keep kcalories under control. In contrast, eating foods that are low in nutrient density makes it difficult to get enough nutrients without exceeding energy needs and gaining weight. For this reason, consumers should select low-fat foods from each group and foods without added fats or sugars—for example, fat-free milk instead of whole milk, baked chicken without the skin instead of hot dogs, green beans instead of French fries, orange juice instead of fruit punch, and whole-wheat bread instead of biscuits. Notice that the key in Figure 2-1 indicates which foods *within each group* are high or low in nutrient density. Oil is a notable exception: even though oil is pure fat and therefore rich in kcalories, a small amount of oil from sources such as nuts, fish, or vegetable oils is necessary every day to provide nutrients lacking from other foods. Consequently these high-fat foods are listed among the nutrient-dense foods (see Highlight 5 to learn why).

Dietary Guidelines for Americans 2005

Consume a variety of nutrient-dense foods and beverages within and among the basic food groups while choosing foods that limit the intake of saturated and *trans* fats, cholesterol, added sugars, salt, and alcohol.

Discretionary kCalorie Allowance At each kcalorie level, people who consistently choose nutrient-dense foods may be able to meet their nutrient needs without consuming their full allowance of kcalories. The difference between the kcalories needed to supply nutrients and those needed for energy—known as the **discretionary kcalorie allowance**—is illustrated in Figure 2-2. Table 2-3 (p. 41) includes the discretionary kcalorie allowance for several kcalorie levels. A person with discretionary kcalories available might choose to:

- Eat additional nutrient-dense foods, such as an extra serving of skinless chicken or a second ear of corn.
- Select a few foods with fats or added sugars, such as reduced-fat milk or sweetened cereal.
- Add a little fat or sugar to foods, such as butter or jelly on toast.
- Consume some alcohol. (Highlight 7 explains why this may not be a good choice for some individuals.)

Alternatively, a person wanting to lose weight might choose to:

- *Not* use the kcalories available from the discretionary kcalorie allowance.

Added fats and sugars are always counted as discretionary kcalories. The kcalories from the fat in higher-fat milks and meats are also counted among discretionary kcalories. It helps to think of fat-free milk as "milk" and whole milk or reduced-fat milk as "milk with added fat." Similarly, "meats" should be the leanest; other cuts are "meats with added fat." Puddings and other desserts made from whole milk provide discretionary kcalories from both the sugar added to sweeten them and the naturally occurring fat in the whole milk they contain. Even fruits, vegetables, and grains can carry discretionary kcalories into the diet in the form of peaches canned in syrup, scalloped potatoes, or high-fat crackers.

Discretionary kcalories must be counted separately from the kcalories of the nutrient-dense foods of which they may be a part. A fried chicken leg, for example, provides discretionary kcalories from two sources: the naturally occurring fat of the chicken skin and the added fat absorbed during frying. The kcalories of the skinless chicken underneath are not discretionary kcalories—they are necessary to provide the nutrients of chicken.

FIGURE 2-2 Discretionary kCalorie Allowance for a 2000-kCalorie Diet Plan

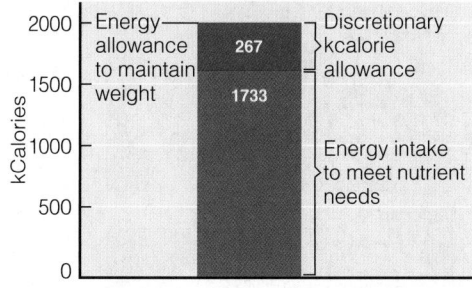

discretionary kcalorie allowance: the kcalories remaining in a person's energy allowance after consuming enough nutrient-dense foods to meet all nutrient needs for a day.

Serving Equivalents Recommended serving amounts for fruits, vegetables, and milk are measured in cups and those for grains and meats, in ounces. Figure 2-1 provides equivalent measures among the foods in each group specifying, for example, that 1 ounce of grains is equivalent to 1 slice of bread or $\frac{1}{2}$ cup of cooked rice.

A person using the USDA Food Guide can become more familiar with measured portions by determining the answers to questions such as these: ◆ What portion of a cup is a small handful of raisins? Is a "helping" of mashed potatoes more or less than a half-cup? How many ounces of cereal do you typically pour into the bowl? How many ounces is the steak at your favorite restaurant? How many cups of milk does your glass hold? Figure 2-1 (pp. 42–43) includes the serving sizes and equivalent amounts for foods within each group.

Mixtures of Foods Some foods—such as casseroles, soups, and sandwiches—fall into two or more food groups. With a little practice, users can learn to see these mixtures of foods as items from various food groups. For example, from the USDA Food Guide point of view, a taco represents four different food groups: the taco shell from the grains group; the onions, lettuce, and tomatoes from the "other vegetables" group; the ground beef from the meat group; and the cheese from the milk group.

Vegetarian Food Guide Vegetarian diets rely mainly on plant foods: grains, vegetables, legumes, fruits, seeds, and nuts. Some vegetarian diets include eggs, milk products, or both. People who do not eat meats or milk products can still use the USDA Food Guide to create an adequate diet.[3] ◆ The food groups are similar, and the amounts for each serving remain the same. Highlight 2 defines vegetarian terms and provides details on planning healthy vegetarian diets.

Ethnic Food Choices People can use the USDA Food Guide and still enjoy a diverse array of culinary styles by sorting ethnic foods into their appropriate food groups. For example, a person eating Mexican foods would find tortillas in the grains group, jicama in the vegetable group, and guava in the fruit group. Table 2-6 features ethnic food choices.

◆ For quick and easy estimates, visualize each portion as being about the size of a common object:
- 1 c fruit or vegetables = a baseball
- $\frac{1}{4}$ c dried fruit = a golf ball
- 3 oz meat = a deck of cards
- 2 tbs peanut butter = a marshmallow
- $1\frac{1}{2}$ oz cheese = 6 stacked dice
- $\frac{1}{2}$ c ice cream = a racquetball
- 4 small cookies = 4 poker chips

TABLE 2-6	Ethnic Food Choices				
	Grains	**Vegetables**	**Fruits**	**Meats and legumes**	**Milk**
Asian © Becky Luigart-Stayner/Corbis	Rice, noodles, millet	Amaranth, baby corn, bamboo shoots, chayote, bok choy, mung bean sprouts, sugar peas, straw mushrooms, water chestnuts, kelp	Carambola, guava, kumquat, lychee, persimmon, melons, mandarin orange	Soybeans and soy products such as soy milk and tofu, squid, duck eggs, pork, poultry, fish and other seafood, peanuts, cashews	Usually excluded
Mediterranean © Photo Disc/Getty Images	Pita pocket bread, pastas, rice, couscous, polenta, bulgur, focaccia, Italian bread	Eggplant, tomatoes, peppers, cucumbers, grape leaves	Olives, grapes, figs	Fish and other seafood, gyros, lamb, chicken, beef, pork, sausage, lentils, fava beans	Ricotta, provolone, parmesan, feta, mozzarella, and goat cheeses; yogurt
Mexican © Photo Disc/Getty Images	Tortillas (corn or flour), taco shells, rice	Chayote, corn, jicama, tomato salsa, cactus, cassava, tomatoes, yams, chilies	Guava, mango, papaya, avocado, plantain, bananas, oranges	Refried beans, fish, chicken, chorizo, beef, eggs	Cheese, custard

MyPyramid—Steps to a Healthier You The USDA created an educational tool called MyPyramid to illustrate the concepts of the *Dietary Guidelines* and the USDA Food Guide. Figure 2-3 presents a graphic image of MyPyramid, which was designed to encourage consumers to make healthy food and physical activity choices every day.

The abundant materials that support MyPyramid help consumers choose the kinds and amounts of foods to eat each day (**MyPyramid.gov**). In addition to creating a personal plan, consumers can find tips to help them improve their diet and lifestyle by "taking small steps each day."

◆ **MyPyramid.gov** offers information on vegetarian diets in its Tips & Resources section.

Exchange Lists

Food group plans are particularly well suited to help a person achieve dietary adequacy, balance, and variety. **Exchange lists** provide additional help in achieving kcalorie control and moderation. Originally developed for people with diabetes, exchange systems have proved useful for general diet planning as well.

Unlike the USDA Food Guide, which sorts foods primarily by their vitamin and mineral contents, the exchange system sorts foods according to their energy-nutrient contents. Consequently, foods do not always appear on the exchange list where you might first expect to find them. For example, cheeses are grouped with meats because, like meats, cheeses contribute energy from protein and fat but provide negligible carbohydrate. (In the USDA Food Guide presented earlier, cheeses are grouped with milk because they are milk products with similar calcium contents.)

exchange lists: diet-planning tools that organize foods by their proportions of carbohydrate, fat, and protein. Foods on any single list can be used interchangeably.

| FIGURE 2-3 | MyPyramid: Steps to a Healthier You |

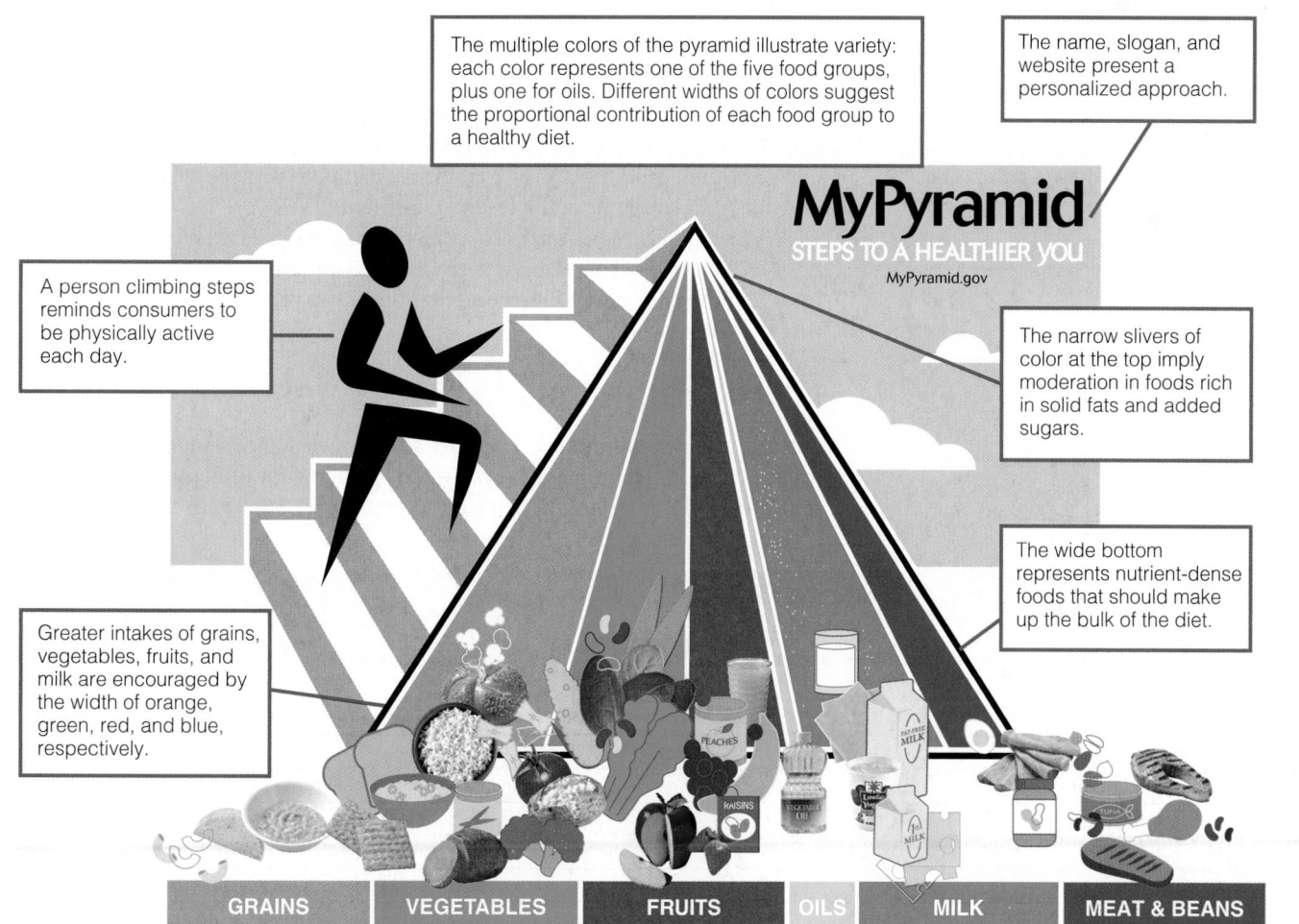

The multiple colors of the pyramid illustrate variety: each color represents one of the five food groups, plus one for oils. Different widths of colors suggest the proportional contribution of each food group to a healthy diet.

The name, slogan, and website present a personalized approach.

A person climbing steps reminds consumers to be physically active each day.

The narrow slivers of color at the top imply moderation in foods rich in solid fats and added sugars.

The wide bottom represents nutrient-dense foods that should make up the bulk of the diet.

Greater intakes of grains, vegetables, fruits, and milk are encouraged by the width of orange, green, red, and blue, respectively.

MyPyramid
STEPS TO A HEALTHIER YOU
MyPyramid.gov

GRAINS   VEGETABLES   FRUITS   OILS   MILK   MEAT & BEANS

SOURCE: USDA, 2005

Most bagels today weigh in at 4 ounces or more—meaning that a person eating one of these large bagels for breakfast is actually getting four or more grain servings, not one.

For similar reasons, starchy vegetables such as corn, green peas, and potatoes are listed with grains on the starch list in the exchange system, rather than with the vegetables. Likewise, olives are not classed as a "fruit" as a botanist would claim; they are classified as a "fat" because their fat content makes them more similar to oil than to berries. Bacon and nuts are also on the fat list to remind users of their high fat content. These groupings highlight the characteristics of foods that are significant to energy intake. To learn more about this useful diet-planning tool, study Appendix G, which gives details of the exchange system used in the United States, and Appendix I, which provides details of *Beyond the Basics,* a similar diet-planning system used in Canada.

Putting the Plan into Action

Familiarizing yourself with each of the food groups is the first step in diet planning. Table 2-7 shows how to use the USDA Food Guide to plan a 2000-kcalorie diet. The amounts listed from each of the food groups (see the second column of the table) were taken from Table 2-3 (p. 41). The next step is to assign the food groups to meals (and snacks), as in the remaining columns of Table 2-7.

Now, a person can begin to fill in the plan with real foods to create a menu. For example, the breakfast calls for 1 ounce grain, ¹/₂ cup fruit, and 1 cup milk. A person might select a bowl of cereal with banana slices and milk:

1 cup cereal = 1 ounce grain

1 small banana = ¹/₂ cup fruit

1 cup fat-free milk = 1 cup milk

Or ¹/₂ bagel and a bowl of cantaloupe pieces topped with yogurt:

¹/₂ small bagel = 1 ounce grain

¹/₂ cup melon pieces = ¹/₂ cup fruit

1 cup fat-free plain yogurt = 1 cup milk

Then the person can continue to create a diet plan by creating menus for lunch, dinner, and snacks. The final plan might look like the one in Figure 2-4. With the addition of a small amount of oils, this sample diet plan provides about 1850 kcalories and adequate amounts of the essential nutrients.

As you can see, we all make countless food-related decisions daily—whether we have a plan or not. Following a plan, such as the USDA Food Guide, that incorporates health recommendations and diet-planning principles helps a person make wise decisions.

From Guidelines to Groceries

Dietary recommendations emphasize nutrient-rich foods such as whole grains, fruits, vegetables, lean meats, fish, poultry, and low-fat milk products. You can design such a diet for yourself, but how do you begin? Start with the foods you enjoy

---

TABLE 2-7 Diet Planning Using the USDA Food Guide

This diet plan is one of many possibilities. It follows the amounts of foods suggested for a 2000-kcalorie diet as shown in Table 2-3 on p. 41 (with an extra ½ cup of vegetables).

Food Group	Amounts	Breakfast	Lunch	Snack	Dinner	Snack
Fruits	2 c	½ c		½ c	1 c	
Vegetables	2½ c		1 c		1½ c	
Grains	6 oz	1 oz	2 oz	½ oz	2 oz	½ oz
Meat and legumes	5½ oz		2 oz		3½ oz	
Milk	3 c	1 c		1 c		1 c
Oils	5½ tsp		1½ tsp		4 tsp	
Discretionary kcalorie allowance	267 kcal					

FIGURE 2-4 A Sample Diet Plan and Menu

This sample menu provides about 1850 kcalories and meets dietary recommendations to provide 45 to 65 percent of its kcalories from carbohydrate, 20 to 35 percent from fat, and 10 to 35 percent from protein. Some discretionary kcalories were spent on the fat in the low-fat cheese and in the sugar added to the graham crackers; about 150 discretionary kcalories remain available in this 2000-kcalorie diet plan.

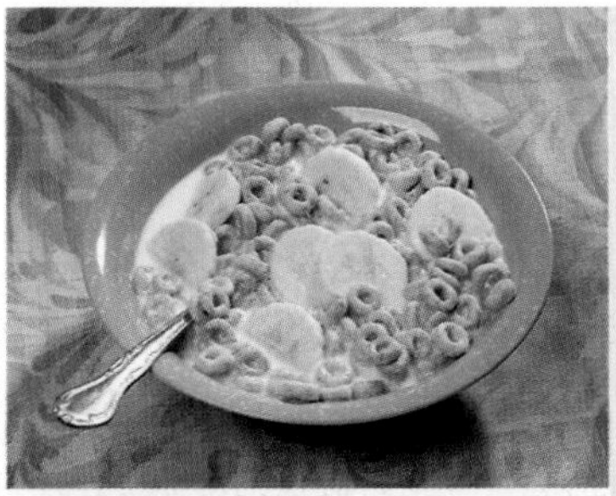

Amounts	❊ SAMPLE MENU ❊	Energy (kcal)
Breakfast		
1 oz whole grains	1 c whole-grain cereal	108
1 c milk	1 c fat-free milk	83
¹/₂ c fruit	1 small banana (sliced)	105
Lunch		
2 oz whole grains, 2 oz meats	1 turkey sandwich on roll	272
1¹/₂ tsp oils	1¹/₂ tbs low-fat mayonnaise	75
1 c vegetables	1 c vegetable juice	53
Snack		
¹/₂ oz whole grains	4 whole-wheat, reduced-fat crackers	86
1 c milk	1¹/₂ oz low-fat cheddar cheese	74
¹/₂ c fruit	1 small apple	72
Dinner		
¹/₂ c vegetables	1 c salad	8
1 oz meats	¹/₄ c garbanzo beans	71
2 tsp oils	2 tbs oil-based salad dressing and olives	81
¹/₂ c vegetables, 2¹/₂ oz meats, 2 oz enriched grains	Spaghetti with meat sauce	425
¹/₂ c vegetables	¹/₂ c green beans	22
2 tsp oils	2 tsp soft margarine	67
1 c fruit	1 c strawberries	49
Snack		
¹/₂ oz enriched grains	3 graham crackers	90
1 c milk	1 c fat-free milk	83

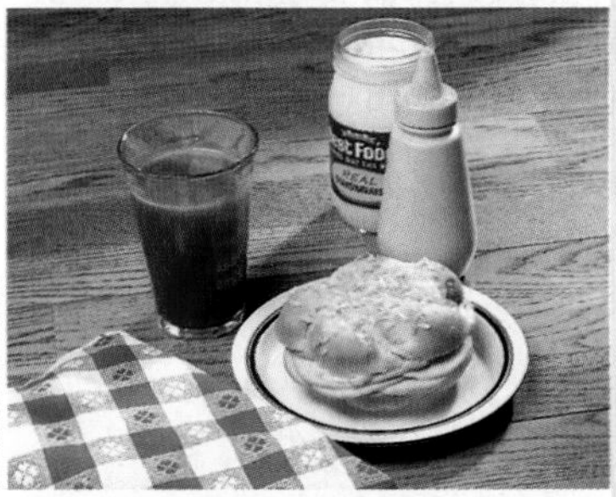

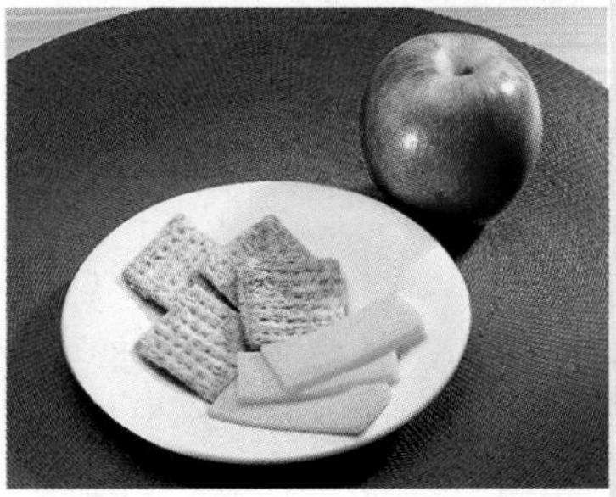

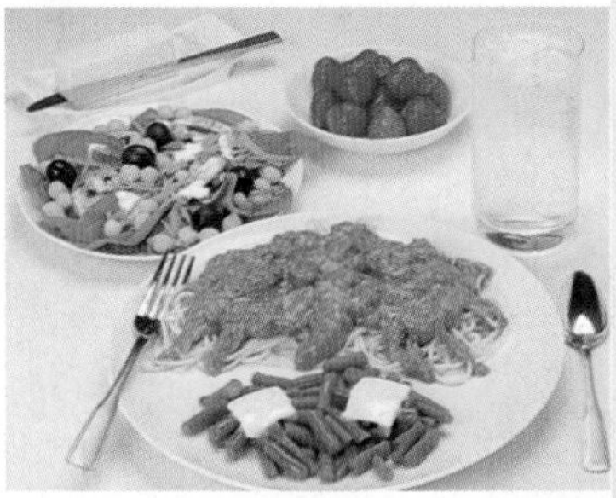

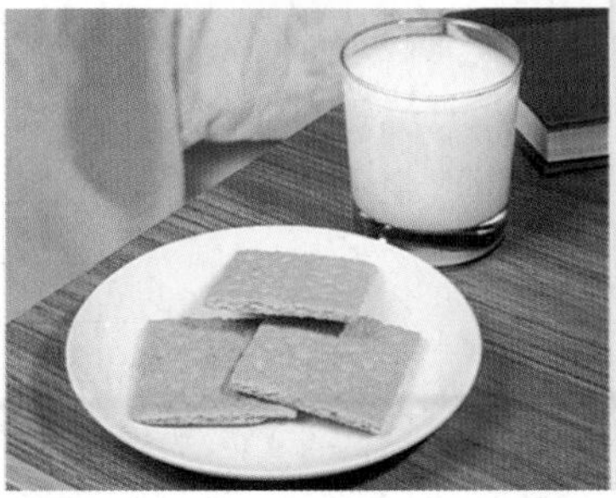

© Polara Studios, Inc.
© Polara Studios, Inc.
© Polara Studios, Inc.
© Quest
© Quest

eating. Then try to make improvements, little by little. When shopping, think of the food groups, and choose nutrient-dense foods within each group.

Be aware that many of the 50,000 food options available today are **processed foods** that have lost valuable nutrients and gained sugar, fat, and salt as they were transformed from farm-fresh foods to those found in the bags, boxes, and cans that line grocery-store shelves. Their value in the diet depends on the starting food and how it was prepared or processed. Sometimes these foods have been **fortified** to improve their nutrient contents.

Grains When shopping for grain products, you will find them described as *refined, enriched,* or *whole grain.* These terms refer to the milling process and the making of grain products, and they have different nutrition implications (see Figure 2-5). **Refined** foods may have lost many nutrients during processing; **enriched** products may have had some nutrients added back; and **whole-grain** products may be rich in fiber and all the nutrients found in the original grain. As such, whole-grain products support good health and should account for at least half of the grains daily.

When it became a common practice to refine the wheat flour used for bread by milling it and throwing away the bran and the germ, consumers suffered a tragic loss of many nutrients.[4] As a consequence, in the early 1940s Congress passed legislation requiring that all grain products that cross state lines be enriched with iron,

processed foods: foods that have been treated to change their physical, chemical, microbiological, or sensory properties.

fortified: the addition to a food of nutrients that were either not originally present or present in insignificant amounts. Fortification can be used to correct or prevent a widespread nutrient deficiency or to balance the total nutrient profile of a food.

refined: the process by which the coarse parts of a food are removed. When wheat is refined into flour, the bran, germ, and husk are removed, leaving only the endosperm.

enriched: the addition to a food of nutrients that were lost during processing so that the food will meet a specified standard.

whole grain: a grain milled in its entirety (all but the husk), not refined.

FIGURE 2-5 | A Wheat Plant

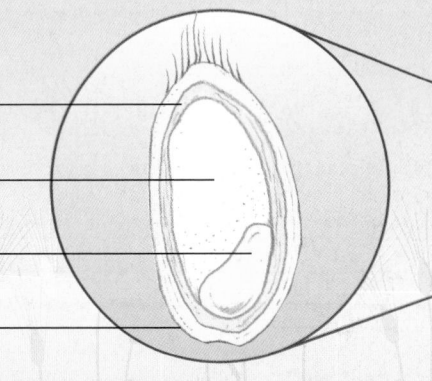

The protective coating of **bran** around the kernel of grain is rich in nutrients and fiber.

The **endosperm** contains starch and proteins.

The **germ** is the seed that grows into a wheat plant, so it is especially rich in vitamins and minerals to support new life.

The outer **husk** (or **chaff**) is the inedible part of a grain.

Whole-grain products contain much of the germ and bran, as well as the endosperm; that is why they are so nutritious.

Common types of flour:

- *Refined flour*—finely ground endosperm that is usually enriched with nutrients and bleached for whiteness; sometimes called *white flour.*
- *Wheat flour*—any flour made from the endosperm of the wheat kernel.
- *Whole-wheat flour*—any flour made from the entire wheat kernel.

The difference between *white flour* and *white wheat* is noteworthy. Typically, *white flour* refers to refined flour (as defined above). Most flour—whether refined, white, or whole wheat—is made from red wheat. Whole-grain products made from red wheat are typically brown and full flavored.

To capture the health benefits of whole grains for consumers who prefer white bread, manufacturers have been experimenting with an albino variety of wheat called *white wheat.* Whole-grain products made from white wheat provide the nutrients and fiber of a whole grain with a light color and natural sweetness. Read labels carefully—white bread is a whole-grain product only if it is made from whole white wheat.

Refined grain products contain only the endosperm. Even with nutrients added back, they are not as nutritious as whole-grain products, as the next figure shows.

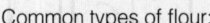

© Thomas Harm/Tom Peterson/Quest Photographic Inc.

 Dietary Guidelines for Americans 2005

Consume 3 or more ounce-equivalents of whole-grain products per day, with the rest of the recommended grains coming from enriched or whole-grain products. In general, at least half the grains should come from whole grains.

thiamin, riboflavin, and niacin. In 1996, this legislation was amended to include folate, a vitamin considered essential in the prevention of some birth defects. Most grain products that have been refined, such as rice, wheat pastas like macaroni and spaghetti, and cereals (both cooked and ready-to-eat types), have subsequently been enriched, ◆ and their labels say so.

Enrichment doesn't make a slice of bread rich in these added nutrients, but people who eat several slices a day obtain significantly more of these nutrients than they would from unenriched bread. Even though the enrichment of flour helps to prevent deficiencies of these nutrients, it fails to compensate for losses of many other nutrients and fiber. As Figure 2-6 shows, whole-grain items still outshine the enriched ones. Only *whole-grain* flour contains all of the nutritive portions of the grain. Whole-grain products, such as brown rice or oatmeal, provide more nutrients and fiber and contain less salt and sugar than flavored, processed rice or sweetened cereals.

Speaking of cereals, ready-to-eat breakfast cereals are the most highly fortified foods on the market. Like an enriched food, a *fortified* food has had nutrients added during processing, but in a fortified food, the added nutrients may not have been present in the original product. (The terms *fortified* and *enriched* may be used interchangeably.[5]) Some breakfast cereals made from refined flour and fortified with high doses of vitamins and minerals are actually more like supplements disguised

◆ Grain enrichment nutrients:
- Iron
- Thiamin
- Riboflavin
- Niacin
- Folate

FIGURE 2-6 Nutrients in Bread

Whole-grain bread is more nutritious than other breads, even enriched bread. For iron, thiamin, riboflavin, niacin, and folate, enriched bread provides about the same quantities as whole-grain bread and significantly more than unenriched bread. For fiber and the other nutrients (those shown here as well as those not shown), enriched bread provides less than whole-grain bread.

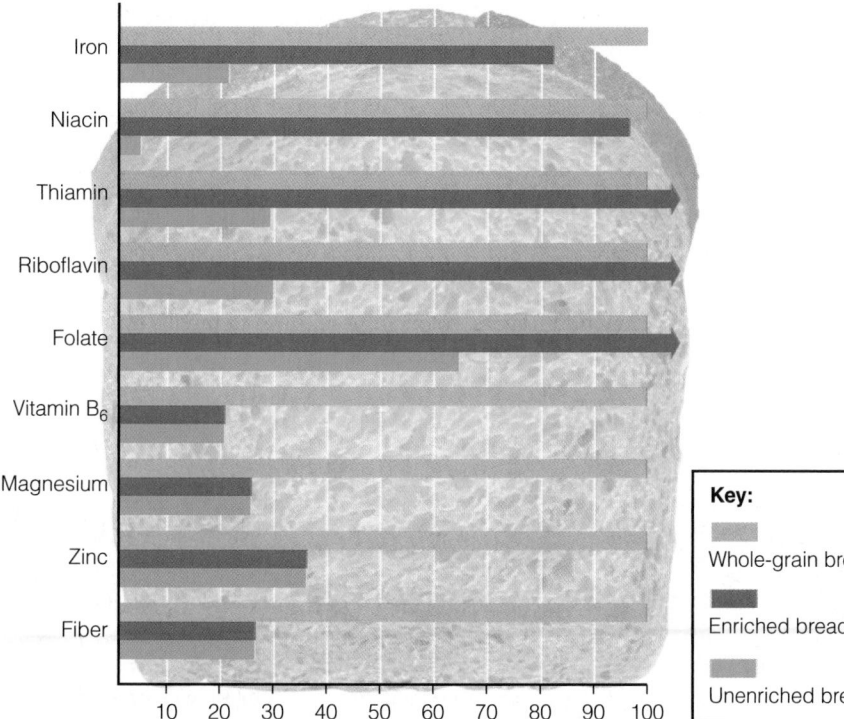

Percentage of nutrients as compared with whole-grain bread

Key:
Whole-grain bread
Enriched bread
Unenriched bread

When shopping for bread, look for the descriptive words *whole grain* or *whole wheat* and check the fiber contents on the Nutrition Facts panel of the label—the more fiber, the more likely the bread is a whole-grain product.

FIGURE 2-7 Eat 5 to 9 a Day for Better Health

The "5 to 9 a Day" campaign (**www.5aday.gov**) encourages consumers to eat a variety of fruits and vegetables. Because "everyone benefits from eating more," the campaign's slogan and messages are being revised to say *Fruits and Veggies—More Matters*.

as cereals than they are like whole grains. They may be nutritious—with respect to the nutrients added—but they still may fail to convey the full spectrum of nutrients that a whole-grain food or a mixture of such foods might provide. Still, fortified foods help people meet their vitamin and mineral needs.[6]

Vegetables Posters in the produce section of grocery stores encourage consumers to "eat 5 a day." Such efforts are part of a national educational campaign to increase fruit and vegetable consumption to 5 to 9 servings every day (see Figure 2-7). To help consumers remember to eat a variety of fruits and vegetables, the campaign provides practical tips, such as selecting from each of five colors.

Choose fresh vegetables often, especially dark green leafy and yellow-orange vegetables like spinach, broccoli, and sweet potatoes. Cooked or raw, vegetables are good sources of vitamins, minerals, and fiber. Frozen and canned vegetables without added salt are acceptable alternatives to fresh. To control fat, energy, and sodium intakes, limit butter and salt on vegetables.

Choose often from the variety of legumes available. ◆ They are an economical, low-fat, nutrient- and fiber-rich food choice.

◆ Legumes include a variety of beans and peas:

- Adzuki beans
- Black beans
- Black-eyed peas
- Fava beans
- Garbanzo beans
- Great northern beans
- Kidney beans
- Lentils
- Lima beans
- Navy beans
- Peanuts
- Pinto beans
- Soybeans
- Split peas

 Dietary Guidelines for Americans 2005

Choose a variety of fruits and vegetables each day. In particular, select from all five vegetable subgroups (dark green, orange, legumes, starchy vegetables, and other vegetables) several times a week.

Fruit Choose fresh fruits often, especially citrus fruits and yellow-orange fruits like cantaloupes and peaches. Frozen, dried, and canned fruits without added sugar are acceptable alternatives to fresh. Fruits supply valuable vitamins, minerals, fibers, and phytochemicals. They add flavors, colors, and textures to meals, and their natural sweetness makes them enjoyable as snacks or desserts.

Combining legumes with foods from other food groups creates delicious meals.

Add rice to red beans for a hearty meal.

Enjoy a Greek salad topped with garbanzo beans for a little ethnic diversity.

A bit of meat and lots of spices turn kidney beans into chili con carne.

Fruit juices are healthy beverages but contain little dietary fiber compared with whole fruits. Whole fruits satisfy the appetite better than juices, thereby helping people to limit food energy intakes. For people who need extra food energy, though, juices are a good choice. Be aware that sweetened fruit "drinks" or "ades" contain mostly water, sugar, and a little juice for flavor. Some may have been fortified with vitamin C or calcium but lack any other significant nutritional value.

 **Dietary** Guidelines for Americans 2005

Consume a sufficient amount of fruits and vegetables while staying within energy needs.

Meat, Fish, and Poultry Meat, fish, and poultry provide essential minerals, such as iron and zinc, and abundant B vitamins as well as protein. To buy and prepare these foods without excess energy, fat, and sodium takes a little knowledge and planning. When shopping in the meat department, choose fish, poultry, and lean cuts of beef and pork named "round" or "loin" (as in top round or pork tenderloin). As a guide, "prime" and "choice" cuts generally have more fat than "select" cuts. Restaurants usually serve prime cuts. Ground beef, even "lean" ground beef, derives most of its food energy from fat. Have the butcher trim and grind a lean round steak instead. Alternatively, **textured vegetable protein** can be used instead of ground beef in a casserole, spaghetti sauce, or chili, saving fat kcalories.

Weigh meat after it is cooked and the bones and fat are removed. In general, 4 ounces of raw meat is equal to about 3 ounces of cooked meat. Some examples of 3-ounce portions of meat include 1 medium pork chop, $^1/_2$ chicken breast, or 1 steak or hamburger about the size of a deck of cards. To keep fat intake moderate, bake, roast, broil, grill, or braise meats (but do not fry them in fat); remove the skin from poultry after cooking; trim visible fat before cooking; and drain fat after cooking. Chapter 5 offers many additional strategies for moderating fat intake.

Milk Shoppers find a variety of fortified foods in the dairy case. Examples are milk, to which vitamins A and D have been added, and soy milk, ◆ to which calcium, vitamin D, and vitamin B_{12} have been added. In addition, shoppers may find **imitation foods** (such as cheese products), **food substitutes** (such as egg substitutes), and functional foods ◆ (such as margarine with added plant sterols). As food technology advances, many such foods offer alternatives to traditional choices that may help people who want to reduce their fat and cholesterol intakes. Chapter 5 gives other examples.

When shopping, choose fat-free ◆ or low-fat milk, yogurt, and cheeses. Such selections help consumers meet their vitamin and mineral needs within their energy and fat allowances.[7] Milk products are important sources of calcium, but can provide too much sodium and fat if not selected with care.

◆ Be aware that not all soy milks have been fortified. Read labels carefully.

◆ Reminder: *Functional foods* contain physiologically active compounds that provide health benefits beyond basic nutrition.

◆ Milk descriptions:
  • **Fat-free** milk may also be called **nonfat**, **skim**, **zero-fat**, or **no-fat**.
  • **Low-fat** milk refers to 1% milk.
  • **Reduced-fat** milk refers to 2% milk; it may also be called **less-fat**.

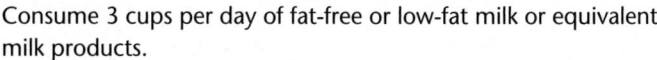 **Dietary** Guidelines for Americans 2005

Consume 3 cups per day of fat-free or low-fat milk or equivalent milk products.

IN SUMMARY

Food group plans such as the USDA Food Guide help consumers select the types and amounts of foods to provide adequacy, balance, and variety in the diet. They make it easier to plan a diet that includes a balance of grains, vegetables, fruits, meats, and milk products. In making any food choice, remember to view the food in the context of your total diet. The combination of many different foods provides the abundance of nutrients that is so essential to a healthy diet.

textured vegetable protein: processed soybean protein used in vegetarian products such as soy burgers.

imitation foods: foods that substitute for and resemble another food, but are nutritionally inferior to it with respect to vitamin, mineral, or protein content. If the substitute is not inferior to the food it resembles and if its name provides an accurate description of the product, it need not be labeled "imitation."

food substitutes: foods that are designed to replace other foods.

Food Labels

Many consumers read food labels to help them make healthy choices.[8] Food labels appear on virtually all processed foods, and posters or brochures provide similar nutrition information for fresh meats, fruits, and vegetables (see Figure 2-8). A few foods need not carry nutrition labels: those contributing few nutrients, such as plain coffee, tea, and spices; those produced by small businesses; and those prepared and sold in the same establishment. Producers of some of these items, however, voluntarily use labels. Even markets selling nonpackaged items voluntarily present nutrient information, either in brochures or on signs posted at the point of purchase. Restaurants need not supply complete nutrition information for menu items unless claims such as "low fat" or "heart healthy" have been made. When ordering such items, keep in mind that restaurants tend to serve extra-large portions—two to three times standard serving sizes. A "low-fat" ice cream, for example, may have only 3 grams of fat per ½ cup, but you may be served 2 cups for a total of 12 grams of fat and all their accompanying kcalories.

FIGURE 2-8 Example of a Food Label

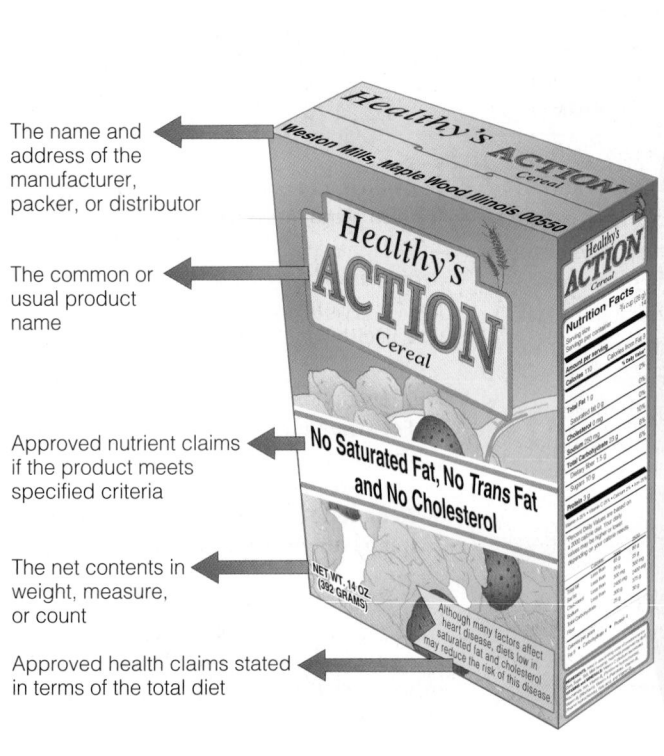

The name and address of the manufacturer, packer, or distributor

The common or usual product name

Approved nutrient claims if the product meets specified criteria

The net contents in weight, measure, or count

Approved health claims stated in terms of the total diet

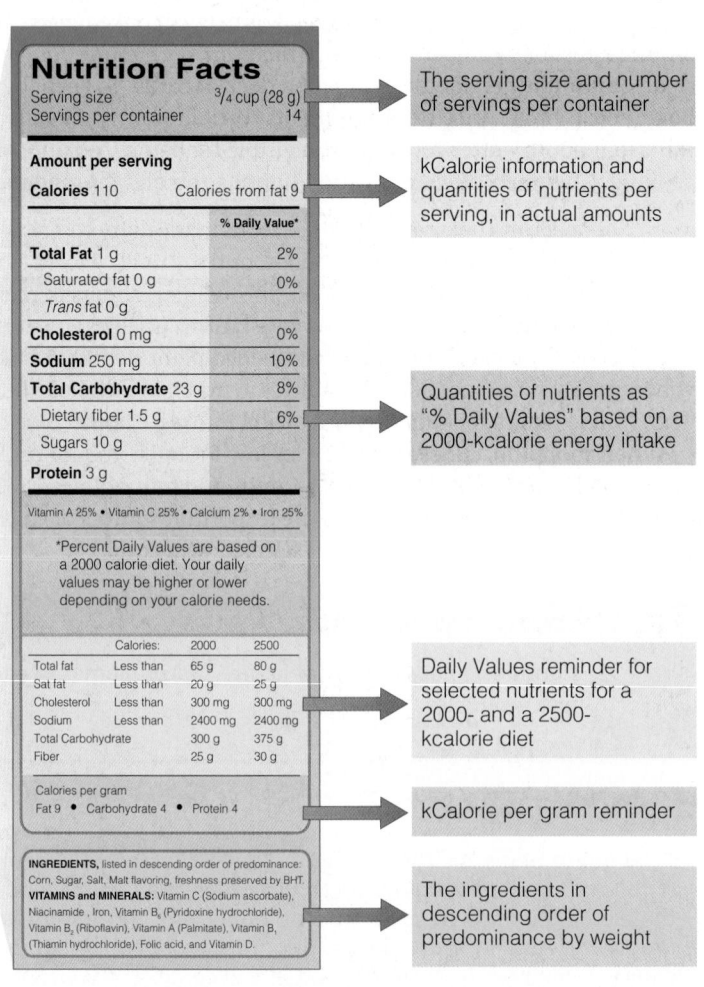

Nutrition Facts

Serving size   ³/₄ cup (28 g)
Servings per container   14

Amount per serving

Calories 110   Calories from fat 9

	% Daily Value*
Total Fat 1 g	2%
Saturated fat 0 g	0%
Trans fat 0 g	
Cholesterol 0 mg	0%
Sodium 250 mg	10%
Total Carbohydrate 23 g	8%
Dietary fiber 1.5 g	6%
Sugars 10 g	
Protein 3 g	

Vitamin A 25% • Vitamin C 25% • Calcium 2% • Iron 25%

*Percent Daily Values are based on a 2000 calorie diet. Your daily values may be higher or lower depending on your calorie needs.

	Calories:	2000	2500
Total fat	Less than	65 g	80 g
Sat fat	Less than	20 g	25 g
Cholesterol	Less than	300 mg	300 mg
Sodium	Less than	2400 mg	2400 mg
Total Carbohydrate		300 g	375 g
Fiber		25 g	30 g

Calories per gram
Fat 9 • Carbohydrate 4 • Protein 4

INGREDIENTS, listed in descending order of predominance: Corn, Sugar, Salt, Malt flavoring, freshness preserved by BHT. **VITAMINS and MINERALS:** Vitamin C (Sodium ascorbate), Niacinamide , Iron, Vitamin B₆ (Pyridoxine hydrochloride), Vitamin B₂ (Riboflavin), Vitamin A (Palmitate), Vitamin B₁ (Thiamin hydrochloride), Folic acid, and Vitamin D.

The serving size and number of servings per container

kCalorie information and quantities of nutrients per serving, in actual amounts

Quantities of nutrients as "% Daily Values" based on a 2000-kcalorie energy intake

Daily Values reminder for selected nutrients for a 2000- and a 2500-kcalorie diet

kCalorie per gram reminder

The ingredients in descending order of predominance by weight

The Ingredient List

All packaged foods must list all ingredients on the label in descending order of predominance by weight. Knowing that the first ingredient predominates by weight, consumers can glean much information. Compare these products, for example:

- A beverage powder that contains "sugar, citric acid, natural flavors . . ." versus a juice that contains "water, tomato concentrate, concentrated juices of carrots, celery . . ."
- A cereal that contains "puffed milled corn, sugar, corn syrup, molasses, salt . . ." versus one that contains "100 percent rolled oats"
- A canned fruit that contains "sugar, apples, water" versus one that contains simply "apples, water"

In each of these comparisons, consumers can see that the second product is the more nutrient dense.

Serving Sizes

Because labels present nutrient information *per serving,* they must identify the size of the serving. The Food and Drug Administration (FDA) has established specific serving sizes for various foods and requires that all labels for a given product use the same serving size. For example, the serving size for all ice creams is $1/2$ cup and for all beverages, 8 fluid ounces. This facilitates comparison shopping. Consumers can see at a glance which brand has more or fewer kcalories or grams of fat, for example. Standard serving sizes are expressed in both common household measures, such as cups, and metric measures, such as milliliters, to accommodate users of both types of measures (see Table 2-8).

When examining the nutrition facts on a food label, consumers need to compare the serving size on the label with how much they actually eat and adjust their calculations accordingly. For example, if the serving size is four cookies and you only eat two, then you need to cut the nutrient and kcalorie values in half; similarly, if you eat eight cookies, then you need to double the values. Notice, too, that small bags or individually wrapped items, such as chips or candy bars, may contain more than a single serving. The number of servings per container is listed just below the serving size.

Be aware that serving sizes on food labels are not always the same as those of the USDA Food Guide.[9] For example, a serving of rice on a food label is 1 cup, whereas in the USDA Food Guide it is $1/2$ cup. Unfortunately, this discrepancy, coupled with each person's own perception (oftentimes misperception) of standard serving sizes, sometimes creates confusion for consumers trying to follow recommendations.

Nutrition Facts

In addition to the serving size and the servings per container, the FDA requires that the "Nutrition Facts" panel on food labels present nutrient information in two ways—in quantities (such as grams) and as percentages of standards called the **Daily Values.** The Nutrition Facts panel must provide the nutrient amount, percent Daily Value, or both for the following:

- Total food energy (kcalories)
- Food energy from fat (kcalories)
- Total fat (grams and percent Daily Value)
- Saturated fat (grams and percent Daily Value)
- *Trans* fat (grams)
- Cholesterol (milligrams and percent Daily Value)
- Sodium (milligrams and percent Daily Value)

TABLE 2-8	Household and Metric Measures

- 1 teaspoon (tsp) = 5 milliliters (mL)
- 1 tablespoon (tbs) = 15 mL
- 1 cup (c) = 240 mL
- 1 fluid ounce (fl oz) = 30 mL
- 1 ounce (oz) = 28 grams (g)

NOTE: The Aids to Calculation section at the back of the book provides additional weights and measures.

Daily Values (DV): reference values developed by the FDA specifically for use on food labels.

Consumers read food labels to learn about the nutrient contents of a food or to compare similar foods.

- Total carbohydrate, which includes starch, sugar, and fiber (grams and percent Daily Value)
- Dietary fiber (grams and percent Daily Value)
- Sugars, which includes both those naturally present in and those added to the food (grams)
- Protein (grams)

The labels must also present nutrient content information as a percentage of the Daily Values for the following vitamins and minerals:

- Vitamin A
- Vitamin C
- Iron
- Calcium

The Daily Values

The FDA developed the Daily Values for use on food labels because comparing nutrient amounts against a standard helps make the numbers more meaningful to consumers. Table 2-9 presents the Daily Value standards for nutrients that are required to provide this information. Food labels list the amount of a nutrient in a product as a percentage of its Daily Value. A person reading a food label might wonder, for example, whether 1 milligram of iron or calcium is a little or a lot. As Table 2-9 shows, the Daily Value for iron is 18 milligrams, so 1 milligram of iron is enough to notice—it is more than 5 percent, and that is what the food label will say. But because the Daily Value for calcium on food labels is 1000 milligrams, 1 milligram of calcium is insignificant, and the food label will read "0%."

The Daily Values reflect dietary recommendations for nutrients and dietary components that have important relationships with health. The "% Daily Value" column on a label provides a ballpark estimate of how individual foods contribute to the total diet. It compares key nutrients in a serving of food with the goals of a person consuming 2000 kcalories per day. A 2000-kcalorie diet is considered about right for sedentary younger women, active older women, and sedentary older men.

| TABLE 2-9 | Daily Values for Food Labels |

Food labels must present the "% Daily Value" for these nutrients.

Food Component	Daily Value	Calculation Factors
Fat	65 g	30% of kcalories
Saturated fat	20 g	10% of kcalories
Cholesterol	300 mg	—
Carbohydrate (total)	300 g	60% of kcalories
Fiber	25 g	11.5 g per 1000 kcalories
Protein	50 g	10% of kcalories
Sodium	2400 mg	—
Potassium	3500 mg	—
Vitamin C	60 mg	—
Vitamin A	1500 µg	—
Calcium	1000 mg	—
Iron	18 mg	—

NOTE: Daily Values were established for adults and children over 4 years old. The values for energy-yielding nutrients are based on 2000 kcalories a day. For fiber, the Daily Value was rounded up from 23.

Young children and sedentary older women may need fewer kcalories. Most labels list, at the bottom, Daily Values for both a 2000-kcalorie and a 2500-kcalorie diet, but the "% Daily Value" column on all labels applies only to a 2000-kcalorie diet. A 2500-kcalorie diet is considered about right for many men, teenage boys, and active younger women. People who are exceptionally active may have still higher energy needs. Labels may also provide a reminder of the kcalories in a gram of carbohydrate, fat, and protein just below the Daily Value information (review Figure 2-8).

People who consume 2000 kcalories a day can simply add up all of the "% Daily Values" for a particular nutrient to see if their diet for the day fits recommendations. People who require more or less than 2000 kcalories daily must do some calculations to see how foods compare with their personal nutrition goals. They can use the calculation column in Table 2-9 or the suggestions presented in the accompanying "How to" feature.

Daily Values help consumers see easily whether a food contributes "a little" or "a lot" of a nutrient. ◆ For example, the "% Daily Value" column on a label of macaroni and cheese may say 20 percent for fat. This tells the consumer that each serving of this food contains about 20 percent of the day's allotted 65 grams of fat. A person consuming 2000 kcalories a day could simply keep track of the percentages of Daily Values from foods eaten in a day and try not to exceed 100 percent. Be aware that for some nutrients (such as fat and sodium) you will want to select foods with a low "% Daily Value" and for others (such as calcium and fiber) you will want a high "% Daily Value." To determine whether a particular food is a wise choice, a consumer needs to consider its place in the diet among all the other foods eaten during the day.

Daily Values also make it easy to compare foods. For example, a consumer might discover that frozen macaroni and cheese has a Daily Value for fat of 20 percent, whereas macaroni and cheese prepared from a boxed mix has a Daily Value of 15 percent. By comparing labels, consumers who are concerned about their fat intakes can make informed decisions.

The Daily Values used on labels are based in part on values from the 1968 Recommended Dietary Allowances. Since 1997, Dietary Reference Intakes that reflect scientific research on diet and health have been released. Efforts to update the Daily Values based on these current recommendations and to make labels more effective and easier to understand are underway.[10]

◆ % Daily Values:
- ≥ 20% = high or excellent source
- 10-19% = good source
- ≤ 5% = low

HOW TO Calculate Personal Daily Values

The Daily Values on food labels are designed for a 2000-kcalorie intake, but you can calculate a personal set of Daily Values based on your energy allowance. Consider a 1500-kcalorie intake, for example. To calculate a daily goal for fat, multiply energy intake by 30 percent:

1500 kcal × 0.30 kcal from fat
= 450 kcal from fat

The "kcalories from fat" are listed on food labels, so you can add all the "kcalories from fat" values for a day, using 450 as an upper limit. A person who prefers to count grams of fat can divide this 450 kcalories from fat by 9 kcalories per gram to determine the goal in grams:

450 kcal from fat ÷ 9 kcal/g
= 50 g fat

Alternatively, a person can calculate that 1500 kcalories is 75 percent of the 2000-kcalorie intake used for Daily Values:

1500 kcal ÷ 2000 kcal = 0.75
0.75 × 100 = 75%

Then, instead of trying to achieve 100 percent of the Daily Value, a person consuming 1500 kcalories will aim for 75 percent. Similarly, a person consuming 2800 kcalories would aim for 140 percent:

2800 kcal ÷ 2000 kcal = 1.40 or 140%

Table 2-9 includes a calculation column that can help you estimate your personal daily value for several nutrients.

CENGAGENOW
To calculate your personal daily values, log on to **academic.cengage.com/login**, then go to Chapter 2, then go to How To.

Nutrient Claims

Have you noticed phrases such as "good source of fiber" on a box of cereal or "rich in calcium" on a package of cheese? These and other **nutrient claims** may be used on labels as long as they meet FDA definitions, which include the conditions under which each term can be used. For example, in addition to having less than 2 milligrams of cholesterol, a "cholesterol-free" product may not contain more than 2 grams of saturated fat and *trans* fat combined per serving. The accompanying glossary defines nutrient terms on food labels, including criteria for foods described as "low," "reduced," and "free."

Some descriptions *imply* that a food contains, or does not contain, a nutrient. Implied claims are prohibited unless they meet specified criteria. For example, a claim that a product "contains no oil" *implies* that the food contains no fat. If the product is truly fat-free, then it may make the no-oil claim, but if it contains another source of fat, such as butter, it may not.

nutrient claims: statements that characterize the quantity of a nutrient in a food.

GLOSSARY OF TERMS ON FOOD LABELS

GENERAL TERMS

free: "nutritionally trivial" and unlikely to have a physiological consequence; synonyms include "without," "no," and "zero." A food that does not contain a nutrient naturally may make such a claim, but only as it applies to all similar foods (for example, "applesauce, a fat-free food").

good source of: the product provides between 10 and 19% of the Daily Value for a given nutrient per serving.

healthy: a food that is low in fat, saturated fat, cholesterol, and sodium and that contains at least 10% of the Daily Values for vitamin A, vitamin C, iron, calcium, protein, or fiber.

high: 20% or more of the Daily Value for a given nutrient per serving; synonyms include "rich in" or "excellent source."

less: at least 25% less of a given nutrient or kcalories than the comparison food (see individual nutrients); synonyms include "fewer" and "reduced."

light or **lite:** one-third fewer kcalories than the comparison food; 50% or less of the fat or sodium than the comparison food; any use of the term other than as defined must specify what it is referring to (for example, "light in color" or "light in texture").

low: an amount that would allow frequent consumption of a food without exceeding the Daily Value for the nutrient. A food that is naturally low in a nutrient may make such a claim, but only as it applies to all similar foods (for example, "fresh cauliflower, a low-sodium food"); synonyms include "little," "few," and "low source of."

more: at least 10% more of the Daily Value for a given nutrient than the comparison food; synonyms include "added" and "extra."

organic: on food labels, that at least 95% of the product's ingredients have been grown and processsed according to USDA regulations defining the use of fertilizers, herbicides, insecticides, fungicides, preservatives, and other chemical ingredients.

ENERGY

kcalorie-free: fewer than 5 kcal per serving.

low kcalorie: 40 kcal or less per serving.

reduced kcalorie: at least 25% fewer kcalories per serving than the comparison food.

FAT AND CHOLESTEROL[a]

percent fat-free: may be used only if the product meets the definition of *low fat* or *fat-free* and must reflect the amount of fat in 100 g (for example, a food that contains 2.5 g of fat per 50 g can claim to be "95 percent fat free").

fat-free: less than 0.5 g of fat per serving (and no added fat or oil); synonyms include "zero-fat," "no-fat," and "nonfat."

low fat: 3 g or less fat per serving.

less fat: 25% or less fat than the comparison food.

saturated fat-free: less than 0.5 g of saturated fat and 0.5 g of *trans* fat per serving.

low saturated fat: 1 g or less saturated fat and less than 0.5 g of *trans* fat per serving.

less saturated fat: 25% or less saturated fat and *trans* fat combined than the comparison food.

trans **fat-free:** less than 0.5 g of *trans* fat and less than 0.5 g of saturated fat per serving.

cholesterol-free: less than 2 mg cholesterol per serving and 2 g or less saturated fat and *trans* fat combined per serving.

low cholesterol: 20 mg or less cholesterol per serving and 2 g or less saturated fat and *trans* fat combined per serving.

less cholesterol: 25% or less cholesterol than the comparison food (reflecting a reduction of at least 20 mg per serving), and 2 g or less saturated fat and *trans* fat combined per serving.

extra lean: less than 5 g of fat, 2 g of saturated fat and *trans* fat combined, and 95 mg of cholesterol per serving and per 100 g of meat, poultry, and seafood.

lean: less than 10 g of fat, 4.5 g of saturated fat and *trans* fat combined, and 95 mg of cholesterol per serving and per 100 g of meat, poultry, and seafood.

CARBOHYDRATES: FIBER AND SUGAR

high fiber: 5 g or more fiber per serving. A high-fiber claim made on a food that contains more than 3 g fat per serving and per 100 g of food must also declare total fat.

sugar-free: less than 0.5 g of sugar per serving.

SODIUM

sodium-free and **salt-free:** less than 5 mg of sodium per serving.

low sodium: 140 mg or less per serving.

very low sodium: 35 mg or less per serving.

---

[a]Foods containing more than 13 grams total fat per serving or per 50 grams of food must indicate those contents immediately after a cholesterol claim. As you can see, all cholesterol claims are prohibited when the food contains more than 2 grams saturated fat and *trans* fat combined per serving.

Health Claims

Until 2003, the FDA held manufacturers to the highest standards of scientific evidence before approving **health claims** on food labels. Consumers reading "Diets low in sodium may reduce the risk of high blood pressure," for example, knew that the FDA had examined enough scientific evidence to establish a clear link between diet and health. Such reliable health claims make up the FDA's "A" list (see Table 2-10). The FDA refers to these health claims as "unqualified"—not that they lack the necessary qualifications, but that they can stand alone without further explanation or qualification.

These reliable health claims still appear on some food labels, but finding them may be difficult now that the FDA has created three additional categories of claims based on scientific evidence that is less conclusive (see Table 2-11). These categories were added after a court ruled: "Holding only the highest scientific standard for claims interferes with commercial free speech." Food manufacturers had argued that they should be allowed to inform consumers about possible benefits based on less than clear and convincing evidence. The FDA must allow manufacturers to provide information about nutrients and foods that show preliminary promise in preventing disease. These health claims are "qualified"—not that they meet the necessary qualifications, but that they require a qualifying explanation. For example, "Very limited and preliminary research suggests that eating one-half to one cup of tomatoes and/or tomato sauce a week may reduce the risk of prostate cancer. FDA concludes that there is little scientific evidence supporting the claim." Consumer groups argue that such information is confusing. Even with required disclaimers for health claims graded "B," "C," or "D," distinguishing "A" claims from others is difficult, as the next section shows. (Health claims on supplement labels are presented in Highlight 10.)

Structure-Function Claims

Unlike health claims, which require food manufacturers to collect scientific evidence and petition the FDA, **structure-function claims** can be made without any FDA approval. Product labels can claim to "slow aging," "improve memory," and "build strong bones" without any proof. The only criterion for a structure-function claim is that it must not mention a disease or symptom. Unfortunately, structure-function claims can be deceptively similar to health claims. Consider these statements:

- "May reduce the risk of heart disease."
- "Promotes a healthy heart."

Most consumers do not distinguish between these two types of claims.[11] In the statements above, for example, the first is a health claim that requires FDA approval and the second is an unproven, but legal, structure-function claim. Table 2-12 lists examples of structure-function claims.

TABLE 2-10 Food Label Health Claims—The "A" List

- Calcium and reduced risk of osteoporosis
- Sodium and reduced risk of hypertension
- Dietary saturated fat and cholesterol and reduced risk of coronary heart disease
- Dietary fat and reduced risk of cancer
- Fiber-containing grain products, fruits, and vegetables and reduced risk of cancer
- Fruits, vegetables, and grain products that contain fiber, particularly soluble fiber, and reduced risk of coronary heart disease
- Fruits and vegetables and reduced risk of cancer
- Folate and reduced risk of neural tube defects
- Sugar alcohols and reduced risk of tooth decay
- Soluble fiber from whole oats and from psyllium seed husk and reduced risk of heart disease
- Soy protein and reduced risk of heart disease
- Whole grains and reduced risk of heart disease and certain cancers
- Plant sterol and plant stanol esters and heart disease
- Potassium and reduced risk of hypertension and stroke

health claims: statements that characterize the relationship between a nutrient or other substance in a food and a disease or health-related condition.

structure-function claims: statements that characterize the relationship between a nutrient or other substance in a food and its role in the body.

TABLE 2-11 The FDA's Health Claims Report Card		
Grade	Level of Confidence in Health Claim	Required Label Disclaimers
A	High: Significant scientific agreement	These health claims do not require disclaimers; see Table 2-10 for examples.
B	Moderate: Evidence is supportive but not conclusive	"[Health claim.] Although there is scientific evidence supporting this claim, the evidence is not conclusive."
C	Low: Evidence is limited and not conclusive	"Some scientific evidence suggests [health claim]. However, FDA has determined that this evidence is limited and not conclusive."
D	Very low: Little scientific evidence supporting this claim	"Very limited and preliminary scientific research suggests [health claim]. FDA concludes that there is little scientific evidence supporting this claim."

TABLE 2-12 Examples of Structure-Function Claims

- Builds strong bones
- Promotes relaxation
- Improves memory
- Boosts the immune system
- Supports heart health
- Defends your health
- Slows aging
- Guards against colds
- Lifts your spirits

NOTE: Structure-function claims cannot make statements about diseases. See Table 2-10 on p. 59 for examples of health claims.

Consumer Education

Because labels are valuable only if people know how to use them, the FDA has designed several programs to educate consumers. Consumers who understand how to read labels are best able to apply the information to achieve and maintain healthful dietary practices.

Table 2-13 shows how the messages from the 2005 *Dietary Guidelines,* the USDA Food Guide, and food labels coordinate with each other. To promote healthy eating and physical activity, the "Healthier US Initiative" coordinates the efforts of national educational programs developed by government agencies.[12] The mission of this initiative is to deliver simple messages that will motivate consumers to make small changes in their eating and physical activity habits to yield big rewards.

TABLE 2-13 From Guidelines to Groceries

Dietary Guidelines	USDA Food Guide/MyPyramid	Food Labels
Adequate nutrients within energy needs	Select the recommended amounts from each food group at the energy level appropriate for your energy needs.	Look for foods that describe their vitamin, mineral, or fiber contents as a *good source* or *high.*
Weight management	Select nutrient-dense foods and beverages within and among the food groups. Limit high-fat foods and foods and beverages with added fats and sugars. Use appropriate portion sizes.	Look for foods that describe their kcalorie contents as *free, low, reduced, light,* or *less.*
Physical activity	Be phyisically active for at least 30 minutes most days of the week. Children and teenagers should be physically active for 60 minutes every day, or most days.	
Food groups to encourage	Select a variety of fruits each day. Include vegetables from all five subgroups (dark green, orange, legumes, starchy vegetables, and other vegetables) several times a week. Make at least half of the grain selections whole grains. Select fat-free or low-fat milk products.	Look for foods that describe their fiber contents as *good source* or *high.* Look for foods that provide at least 10% of the Daily Value for fiber, vitamin A, vitamin C, iron, and calcium from a variety of sources.
Fats	Choose foods within each group that are lean, low fat, or fat-free. Choose foods within each group that have little added fat.	Look for foods that describe their fat, saturated fat, *trans* fat, and cholesterol contents as *free, less, low, light, reduced, lean,* or *extra lean.* Look for foods that provide no more than 5% of the Daily Value for fat, saturated fat, and cholesterol.
Carbohydrates	Choose fiber-rich fruits, vegetables, and whole grains often. Choose foods and beverages within each group that have little added sugars.	Look for foods that describe their sugar contents as *free* or *reduced.* A food may be high in sugar if its ingredients list begins with or contains several of the following: *sugar, sucrose, fructose, maltose, lactose, honey, syrup, corn syrup, high-fructose corn syrup, molasses, evaporated cane juice,* or *fruit juice concentrate.*
Sodium and potassium	Choose foods within each group that are low in salt or sodium. Choose potassium-rich foods such as fruits and vegetables.	Look for foods that describe their salt and sodium contents as *free, low,* or *reduced.* Look for foods that provide no more than 5% of the Daily Value for sodium. Look for foods that provide at least 10% of the Daily Value for potassium.
Alcoholic beverages	Use sensibly and in moderation (no more than one drink a day for women and two drinks a day for men).	*Light* beverages contain fewer kcalories and less alcohol than regular versions.
Food safety		Follow the *safe handling instructions* on packages of meat and other safety instructions, such as *keep refrigerated,* on packages of perishable foods.

IN SUMMARY

Food labels provide consumers with information they need to select foods that will help them meet their nutrition and health goals. When labels contain relevant information presented in a standardized, easy-to-read format, consumers are well prepared to plan and create healthful diets.

This chapter provides the links to go from dietary guidelines to buying groceries and offers helpful tips for selecting nutritious foods. For information on foodborne illnesses, turn to Highlight 18.

Nutrition Portfolio

The secret to making healthy food choices is learning to incorporate the 2005 *Dietary Guidelines* and the USDA Food Guide into your decision-making process.

▧ Compare the foods you typically eat daily with the USDA Food Guide recommendations for your energy needs (see Table 2-3 on p. 41 and Table 2-4 on p. 44), making note of which food groups are usually over- or underrepresented.

▧ Describe your choices within each food group from day to day and include realistic suggestions for enhancing the variety in your diet.

▧ Write yourself a letter describing the dietary changes you can make to improve your chances of enjoying good health.

NUTRITION ON THE NET

For further study of topics covered in this chapter, log on to **academic.cengage.com/nutrition/rolfes/UNCN8e**. Go to Chapter 2, then to Nutrition on the Net.

- Search for "diet" and "food labels" at the U.S. Government health information site: **www.healthfinder.gov**
- Learn more about the *Dietary Guidelines for Americans:* **www.healthierus.gov/dietaryguidelines**
- Find Canadian information on nutrition guidelines and food labels at: **www.hc-sc.gc.ca**
- Learn more about the USDA Food Guide and MyPyramid: **mypyramid.gov**
- Visit the USDA Food Guide section (including its ethnic/cultural pyramids) of the U.S. Department of Agriculture: **www.nal.usda.gov/fnic**

- Visit the Traditional Diet Pyramids for various ethnic groups at Oldways Preservation and Exchange Trust: **www.oldwayspt.org**
- Search for "exchange lists" at the American Diabetes Association: **www.diabetes.org**
- Learn more about food labeling from the Food and Drug Administration: **www.cfsan.fda.gov**
- Search for "food labels" at the International Food Information Council: **www.ific.org**
- Assess your diet at the CNPP Interactive Healthy Eating Index: **www.usda.gov/cnpp**
- Get healthy eating tips from the "5 a day" programs: **www.5aday.gov** or **www.5aday.org**

NUTRITION CALCULATIONS

CENGAGENOW™ For additional practice log on to **academic.cengage.com/login**. Go to Chapter 2, then to Nutrition Calculations.

These problems will give you practice in doing simple nutrition-related calculations. Although the situations are hypothetical, the numbers are real, and calculating the answers (check them on p. 63) provides a valuable nutrition lesson. Be sure to show your calculations for each problem.

1. *Read a food label.* Look at the cereal label in Figure 2-8 and answer the following questions:
   a. What is the size of a serving of cereal?
   b. How many kcalories are in a serving?
   c. How much fat is in a serving?
   d. How many kcalories does this represent?
   e. What percentage of the kcalories in this product comes from fat?
   f. What does this tell you?
   g. What is the % Daily Value for fat?
   h. What does this tell you?
   i. Does this cereal meet the criteria for a low-fat product (refer to the glossary on p. 58)?
   j. How much fiber is in a serving?
   k. Read the Daily Value chart on the lower section of the label. What is the Daily Value for fiber?
   l. What percentage of the Daily Value for fiber does a serving of the cereal contribute? Show the calculation the label-makers used to come up with the % Daily Value for fiber.
   m. What is the predominant ingredient in the cereal?
   n. Have any nutrients been added to this cereal (is it fortified)?

2. *Calculate a personal Daily Value.* The Daily Values on food labels are for people with a 2000-kcalorie intake.
   a. Suppose a person has a 1600-kcalorie energy allowance. Use the calculation factors listed in Table 2-9 to calculate a set of personal "Daily Values" based on 1600 kcalories. Show your calculations.
   b. Revise the % Daily Value chart of the cereal label in Figure 2-8 based on your "Daily Values" for a 1600-kcalorie diet.

STUDY QUESTIONS

CENGAGENOW™

To assess your understanding of chapter topics, take the Student Practice Test and explore the modules recommended in your Personalized Study Plan. Log on to **academic.cengage.com/login**.

These questions will help you review this chapter. You will find the answers in the discussions on the pages provided.

1. Name the diet-planning principles and briefly describe how each principle helps in diet planning. (pp. 37–39)

2. What recommendations appear in the *Dietary Guidelines for Americans*? (pp. 39–40)

3. Name the five food groups in the USDA Food Guide and identify several foods typical of each group. Explain how such plans group foods and what diet-planning principles the plans best accommodate. How are food group plans used, and what are some of their strengths and weaknesses? (pp. 41–47)

4. Review the *Dietary Guidelines*. What types of grocery selections would you make to achieve those recommendations? (pp. 40, 48–53)

5. What information can you expect to find on a food label? How can this information help you choose between two similar products? (pp. 54–57)

6. What are the Daily Values? How can they help you meet health recommendations? (pp. 55–57)

7. Describe the differences between nutrient claims, health claims, and structure-function claims. (pp. 58–59)

These multiple choice questions will help you prepare for an exam. Answers can be found on p. 63.

1. The diet-planning principle that provides all the essential nutrients in sufficient amounts to support health is:
   a. balance.
   b. variety.
   c. adequacy.
   d. moderation.

2. A person who chooses a chicken leg that provides 0.5 milligram of iron and 95 kcalories instead of two table-spoons of peanut butter that also provide 0.5 milligram of iron but 188 kcalories is using the principle of nutrient:
   a. control.
   b. density.
   c. adequacy.
   d. moderation.

3. Which of the following is consistent with the *Dietary Guidelines for Americans*?
   a. Choose a diet restricted in fat and cholesterol.
   b. Balance the food you eat with physical activity.
   c. Choose a diet with plenty of milk products and meats.
   d. Eat an abundance of foods to ensure nutrient adequacy.

4. According to the USDA Food Guide, added fats and sugars are counted as:
   a. meats and grains.
   b. nutrient-dense foods.
   c. discretionary kcalories.
   d. oils and carbohydrates.

5. Foods within a given food group of the USDA Food Guide are similar in their contents of:
   a. energy.
   b. proteins and fibers.
   c. vitamins and minerals.
   d. carbohydrates and fats.

6. In the exchange system, each portion of food on any given list provides about the same amount of:
   a. energy.
   b. satiety.
   c. vitamins.
   d. minerals.

7. Enriched grain products are fortified with:
   a. fiber, folate, iron, niacin, and zinc.
   b. thiamin, iron, calcium, zinc, and sodium.
   c. iron, thiamin, riboflavin, niacin, and folate.
   d. folate, magnesium, vitamin B_6, zinc, and fiber.

8. Food labels list ingredients in:
    a. alphabetical order.
    b. ascending order of predominance by weight.
    c. descending order of predominance by weight.
    d. manufacturer's order of preference.

9. "Milk builds strong bones" is an example of a:
    a. health claim.
    b. nutrition fact.
    c. nutrient content claim.
    d. structure-function claim.

10. Daily Values on food labels are based on a:
    a. 1500-kcalorie diet.
    b. 2000-kcalorie diet.
    c. 2500-kcalorie diet.
    d. 3000-kcalorie diet.

REFERENCES

1. S. P. Murphy and coauthors, Simple measures of dietary variety are associated with improved dietary quality, *Journal of the American Dietetic Association* 106 (2006): 425–429.
2. U.S. Department of Agriculture and U.S. Department of Health and Human Services, *Dietary Guidelines for Americans, 2005*, available at www.healthierus.gov/dietaryguidelines.
3. Position of the American Dietetic Association and Dietitians of Canada: Vegetarian diets, *Journal of the American Dietetic Association* 103 (2003): 748–765.
4. J. R. Backstrand, The history and future of food fortification in the United States: A public health perspective, *Nutrition Reviews* 60 (2002): 15–26.
5. As cited in 21 Code of Federal Regulations—Food and Drugs, Section 104.20, 45 *Federal Register* 6323, January 25, 1980, as amended in 58 *Federal Register* 2228, January 6, 1993.
6. Position of the American Dietetic Association: Food fortification and nutritional supplements, *Journal of the American Dietetic Association* 105 (2005): 1300–1311.
7. R. Ranganathan and coauthors, The nutritional impact of dairy product consumption on dietary intakes of adults (1995–1996): The Bogalusa Heart Study, *Journal of the American Dietetic Association* 105 (2005): 1391–1400; L. G. Weinberg, L. A. Berner, and J. E. Groves, Nutrient contributions of dairy foods in the United States, Continuing Survey of Food Intakes by Individuals, 1994–1996, 1998, *Journal of the American Dietetic Association* 104 (2004): 895–902.
8. L. LeGault and coauthors, 2000–2001 Food Label and Package Survey: An update on prevalence of nutrition labeling and claims on processed, packaged foods, *Journal of the American Dietetic Association* 104 (2004): 952–958.
9. D. Herring and coauthors, Serving sizes in the Food Guide Pyramid and on the nutrition facts label: What's different and why? *Family Economics and Nutrition Review* 14 (2002): 71–73.
10. Dietary Reference Intakes (DRIs) for food labeling, *American Journal of Clinical Nutrition* 83 (2006): suppl; T. Philipson, Government perspective: Food labeling, *American Journal of Clinical Nutrition* 82 (2005): 262S–264S; The National Academy of Sciences, Dietary Reference Intakes: Guiding principles for nutrition labeling and fortification (2004), http://www.nap.edu/openbook/0309091438/html/R1.html.
11. P. Williams, Consumer understanding and use of health claims for foods, *Nutrition Reviews* 63 (2005): 256–264.
12. K. A. Donato, National health education programs to promote healthy eating and physical activity, *Nutrition Reviews* 64 (2006): S65–S70.

ANSWERS

Nutrition Calculations

1. a. ¾ cup (28 g)
   b. 110 kcalories
   c. 1 g fat
   d. 9 kcalories
   e. 9 kcal ÷ 110 kcal = 0.08
      0.08 × 100 = 8%
   f. This cereal derives 8 percent of its kcalories from fat
   g. 2%
   h. A serving of this cereal provides 2 percent of the 65 grams of fat recommended for a 2000-kcalorie diet
   i. Yes
   j. 1.5 g fiber
   k. 25 g
   l. 1.5 g ÷ 25 g = 0.06
      0.06 × 100 = 6%
   m. Corn
   n. Yes

2. a. Daily Values for 1600-kcalorie diet:
      Fat: 1600 kcal × 0.30 = 480 kcal from fat
         480 kcal ÷ 9 kcal/g = 53 g fat

   Saturated fat: 1600 kcal × 0.10 = 160 kcal from saturated fat
      160 kcal ÷ 9 kcal/g = 18 g saturated fat

   Cholesterol: 300 mg

   Carbohydrate: 1600 kcal × 0.60 = 960 kcal from carbohydrate
      960 kcal ÷ 4 kcal/g = 240 g carbohydrate

   Fiber: 1600 kcal ÷ 1000 kcal = 1.6
      1.6 × 11.5 g = 18.4 g fiber

   Protein: 1600 kcal × 0.10 = 160 kcal from protein
      160 kcal ÷ 4 kcal/g = 40 g protein

   Sodium: 2400 mg

   Potassium: 3500 mg

   b.

Total fat	2%	(1 g ÷ 53 g)
Saturated fat	0%	(0 g ÷ 18 g)
Cholesterol	0%	(no calculation needed)
Sodium	10%	(no calculation needed)
Total carbohydrate	10%	(23 g ÷ 240 g)
Dietary fiber	8%	(1.5 g ÷ 18.4 g)

Study Questions (multiple choice)

1. c   2. b   3. b   4. c   5. c   6. a   7. c   8. c
9. d   10. b

Vegetarian Diets

© Polora Studios, Inc.

The waiter presents this evening's specials: a fresh spinach salad topped with mandarin oranges, raisins, and sunflower seeds, served with a bowl of pasta smothered in a mushroom and tomato sauce and topped with grated parmesan cheese. Then this one: a salad made of chopped parsley, scallions, celery, and tomatoes mixed with bulgur wheat and dressed with olive oil and lemon juice, served with a spinach and feta cheese pie. Do these meals sound good to you? Or is something missing . . . a pork chop or ribeye, perhaps?

Would vegetarian fare be acceptable to you some of the time? Most of the time? Ever? Perhaps it is helpful to recognize that dietary choices fall along a continuum—from one end, where people eat no meat or foods of animal origin, to the other end, where they eat generous quantities daily. Meat's place in the diet has been the subject of much research and controversy, as this highlight will reveal. One of the missions of this highlight, in fact, is to identify the *range* of meat intakes most compatible with health. The health benefits of a primarily vegetarian diet seem to have encouraged many people to eat more vegetarian meals. The popular press refers to these "part-time vegetarians" who eat small amounts of meat from time to time as "flexitarians."

People who choose to exclude meat and other animal-derived foods from their diets today do so for many of the same reasons the Greek philosopher Pythagoras cited in the sixth century B.C.: physical health, ecological responsibility, and philosophical concerns. They might also cite world hunger issues, economic reasons, ethical concerns, or religious beliefs as motivating factors. Whatever their reasons—and even if they don't have a particular reason—people who exclude meat will be better prepared to plan well-balanced meals if they understand the nutrition and health implications of vegetarian diets.

Vegetarians generally are categorized, not by their motivations, but by the foods they choose to exclude (see the glossary below). Some people exclude red meat only; some also exclude chicken or fish; others also exclude eggs; and still others exclude milk and milk products as well. In fact, finding agreement on the definition of the term *vegetarian* is a challenge.[1]

As you will see, though, the foods a person *excludes* are not nearly as important as the foods a person *includes* in the diet. Vegetarian diets that include a variety of whole grains, vegetables, legumes, nuts, and fruits offer abundant complex carbohydrates and fibers, an assortment of vitamins and minerals, a mixture of phytochemicals, and little fat—characteristics that reflect current dietary recommendations aimed at promoting health and reducing obesity. Each of these foods—whole grains, vegetables, legumes, nuts, and fruits—independently reduces the risk for several chronic diseases.[2] This highlight examines the health benefits and potential problems of vegetarian diets and shows how to plan a well-balanced vegetarian diet.

GLOSSARY

lactovegetarians: people who include milk and milk products, but exclude meat, poultry, fish, seafood, and eggs from their diets.
- **lacto** = milk

lacto-ovo-vegetarians: people who include milk, milk products, and eggs, but exclude meat, poultry, fish, and seafood from their diets.
- **ovo** = egg

macrobiotic diets: extremely restrictive diets limited to a few grains and vegetables; based on metaphysical beliefs and not on nutrition. A macrobiotic diet might consist of brown rice, miso soup, and sea vegetables, for example.

meat replacements: products formulated to look and taste like meat, fish, or poultry; usually made of textured vegetable protein.

omnivores: people who have no formal restriction on the eating of any foods.
- **omni** = all
- **vores** = to eat

tempeh (TEM-pay): a fermented soybean food, rich in protein and fiber.

textured vegetable protein: processed soybean protein used in vegetarian products such as soy burgers; see also *meat replacements*.

tofu (TOE-foo): a curd made from soybeans, rich in protein and often fortified with calcium; used in many Asian and vegetarian dishes in place of meat.

vegans (VEE-gans): people who exclude all animal-derived foods (including meat, poultry, fish, eggs, and dairy products) from their diets; also called **pure vegetarians, strict vegetarians,** or **total vegetarians.**

vegetarians: a general term used to describe people who exclude meat, poultry, fish, or other animal-derived foods from their diets.

Health Benefits of Vegetarian Diets

Research on the health implications of vegetarian diets would be relatively easy if vegetarians differed from other people only in not eating meat. Many vegetarians, however, have also adopted lifestyles that may differ from many **omnivores:** they typically use no tobacco or illicit drugs, use little (if any) alcohol, and are physically active. Researchers must account for these lifestyle differences before they can determine which aspects of health correlate just with diet. Even then, *correlations* merely reveal what health factors *go with* the vegetarian diet, not what health effects may be *caused by* the diet. Despite these limitations, research findings suggest that well-planned vegetarian diets offer sound nutrition and health benefits to adults.[3] Dietary patterns that include very little, if any, meat may even increase life expectancy.[4]

Weight Control

In general, vegetarians maintain a lower and healthier body weight than nonvegetarians.[5] Vegetarians' lower body weights correlate with their high intakes of fiber and low intakes of fat. Because obesity impairs health in a number of ways, this gives vegetarians a health advantage.

Blood Pressure

Vegetarians tend to have lower blood pressure and lower rates of hypertension than nonvegetarians. Appropriate body weight helps to maintain a healthy blood pressure, as does a diet low in total fat and saturated fat and high in fiber, fruits, vegetables, and soy protein.[6] Lifestyle factors also influence blood pressure: smoking and alcohol intake raise blood pressure, and physical activity lowers it.

Heart Disease

The incidence of heart disease and related deaths is much lower for vegetarians than for meat eaters. The dietary factor most directly related to heart disease is saturated animal fat, and in general, vegetarian diets are lower in total fat, saturated fat, and cholesterol than typical meat-based diets.[7] The fats common in plant-based diets—the monounsaturated fats of olives, seeds, and nuts and the polyunsaturated fats of vegetable oils—are associated with a decreased risk of heart disease.[8] Furthermore, vegetarian diets are generally higher in dietary fiber, antioxidant vitamins, and phytochemicals—all factors that help control blood lipids and protect against heart disease.[9]

Many vegetarians include soy products such as **tofu** in their diets. Soy products may help to protect against heart disease because they contain polyunsaturated fats, fiber, vitamins, and minerals, and little saturated fat.[10] Even when intakes of energy, protein, carbohydrate, total fat, saturated fat, unsaturated fat, alcohol, and fiber are the same, people eating meals based on tofu have lower blood cholesterol and triglyceride levels than those eating meat. Some research suggests that soy protein and phytochemicals may be responsible for some of these health benefits (as Highlight 13 explains in greater detail).[11]

Cancer

Vegetarians have a significantly lower rate of cancer than the general population. Their low cancer rates may be due to their high intakes of fruits and vegetables (as Highlight 11 explains). In fact, the ratio of vegetables to meat may be the most relevant dietary factor responsible for cancer prevention.[12]

Some scientific findings indicate that vegetarian diets are associated not only with lower cancer mortality in general, but also with lower incidence of cancer at specific sites as well, most notably, colon cancer.[13] People with colon cancer seem to eat more meat, more saturated fat, and fewer vegetables than do people without colon cancer. High-protein, high-fat, low-fiber diets create an environment in the colon that promotes the development of cancer in some people. A high-meat diet has been associated with stomach cancer as well.[14]

Other Diseases

In addition to obesity, hypertension, heart disease, and cancer, vegetarian diets may help prevent diabetes, osteoporosis, diverticular disease, gallstones, and rheumatoid arthritis.[15] These health benefits of a vegetarian diet depend on wise diet planning.

Vegetarian Diet Planning

The vegetarian has the same meal-planning task as any other person—using a variety of foods to deliver all the needed nutrients within an energy allowance that maintains a healthy body weight (as discussed in Chapter 2). Vegetarians who include milk products and eggs can meet recommendations for most nutrients about as easily as nonvegetarians. Such diets provide enough energy, protein, and other nutrients to support the health of adults and the growth of children and adolescents.

Vegetarians who exclude milk products and eggs can select legumes, nuts, and seeds and products made from them, such as peanut butter, **tempeh,** and tofu, from the meat group. Those who do not use milk can use soy "milk"—a product made from soybeans that provides similar nutrients if fortified with calcium, vitamin D, and vitamin B_{12}.

The MyPyramid resources include tips for planning vegetarian diets using the USDA Food Guide. In addition, several food guides have been developed specifically for vegetarian diets.[16] They all address the particular nutrition concerns of vegetarians, but differ slightly. Figure H2-1 presents one version. When selecting from the vegetable and fruit groups, vegetarians should emphasize particularly good sources of calcium and iron, respectively. Green leafy vegetables, for example, provide almost five times as much calcium per serving as other vegetables. Similarly, dried fruits deserve special notice in the fruit group because they deliver six

FIGURE H2-1 An Example of a Vegetarian Food Pyramid

Review Figure 2–1 and Table 2–3 to find recommended daily amounts from each food group, serving size equivalents, examples of common foods within each group, and the most notable nutrients for each group. Tips for planning a vegetarian diet can be found at **MyPyramid.gov.**

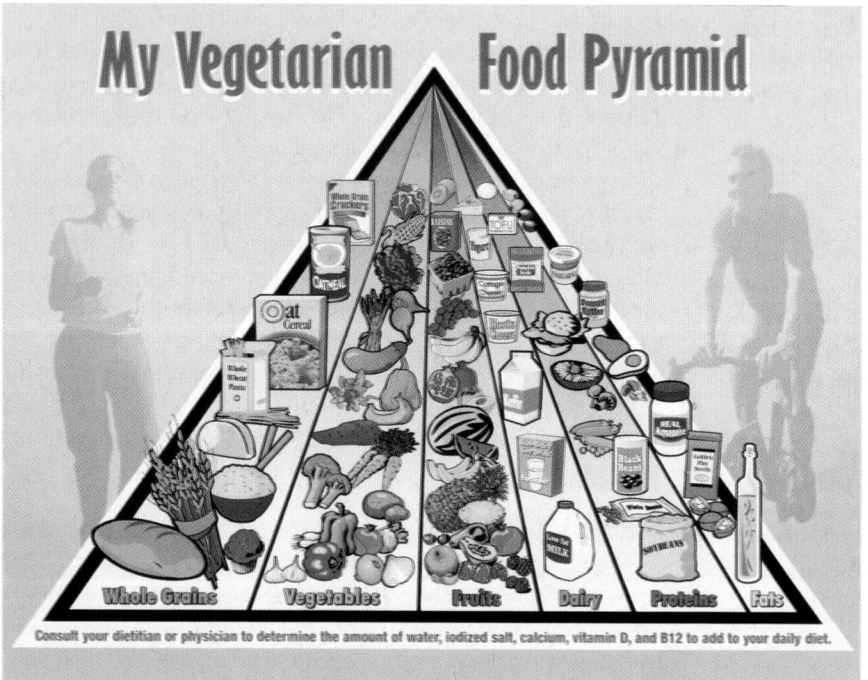

My Vegetarian Food Pyramid

Whole Grains Vegetables Fruits Dairy Proteins Fats

Consult your dietitian or physician to determine the amount of water, iodized salt, calcium, vitamin D, and B12 to add to your daily diet.

SOURCE: © GC Nutrition Council, 2006, adapted from USDA 2005 Dietary Guidelines and www.mypyramid.gov. Copies can be ordered from 301-680-6717.

times as much iron as other fruits. The milk group features fortified soy milks for those who do not use milk, cheese, or yogurt. The meat group is called "proteins" and includes legumes, soy products, nuts, and seeds. A group for oils encourages the use of vegetable oils, nuts, and seeds rich in unsaturated fats and omega-3 fatty acids. To ensure adequate intakes of vitamin B_{12}, vitamin D, and calcium, vegetarians need to select fortified foods or take supplements daily. The vegetarian food pyramid is flexible enough that a variety of people can use it: people who have adopted various vegetarian diets, those who want to make the transition to a vegetarian diet, and those who simply want to include more plant-based meals in their diet. Like MyPyramid, this vegetarian food pyramid also encourages physical activity.

Most vegetarians easily obtain large quantities of the nutrients that are abundant in plant foods: thiamin, folate, and vitamins B_6, C, A, and E. Vegetarian food guides help to ensure adequate intakes of the main nutrients vegetarian diets might otherwise lack: protein, iron, zinc, calcium, vitamin B_{12}, vitamin D, and omega-3 fatty acids.

Protein

The protein RDA for vegetarians is the same as for others, although some have suggested that it should be higher because of the lower digestibility of plant proteins.[17] **Lacto-ovo-vegetarians,** who use animal-derived foods such as milk and eggs, receive high-quality proteins and are likely to meet their protein needs. Even those who adopt only plant-based diets are likely to meet protein needs provided that their energy intakes are adequate and the protein sources varied.[18] The proteins of whole grains, legumes, seeds, nuts, and vegetables can provide adequate amounts of all the amino acids. An advantage of many vegetarian sources of protein is that they are generally lower in saturated fat than meats and are often higher in fiber and richer in some vitamins and minerals.

Vegetarians sometimes use **meat replacements** made of **textured vegetable protein** (soy protein). These foods are formulated to look and taste like meat, fish, or poultry. Many of these products are fortified to provide the vitamins and minerals found in animal sources of protein. A wise vegetarian learns to use a variety of whole, unrefined foods often and commercially prepared foods less frequently. Vegetarians may also use soy products such as tofu to bolster protein intake.

Iron

Getting enough iron can be a problem even for meat eaters, and those who eat no meat must pay special attention to their iron intake. The iron in plant foods such as legumes, dark green leafy vegetables, iron-fortified cereals, and whole-grain breads and cereals is poorly absorbed.[19] Because iron absorption from a vegetarian diet is low, the iron RDA for vegetarians is higher than for others (see Chapter 13 for more details).

Fortunately, the body seems to adapt to a vegetarian diet by absorbing iron more efficiently. Furthermore, iron absorption is enhanced by vitamin C, and vegetarians typically eat many vitamin C–rich fruits and vegetables. Consequently, vegetarians suffer no more iron deficiency than other people do.[20]

Zinc

Zinc is similar to iron in that meat is its richest food source, and zinc from plant sources is not well absorbed.[21] In addition, soy, which is commonly used as a meat alternative in vegetarian meals, interferes with zinc absorption. Nevertheless, most vegetarian adults are not zinc deficient. Perhaps the best advice to vegetarians regarding zinc is to eat a variety of nutrient-dense foods; include whole grains, nuts, and legumes such as black-eyed peas, pinto beans, and kidney beans; and maintain an adequate energy intake. For those who include seafood in their diets, oysters, crabmeat, and shrimp are rich in zinc.

Calcium

The calcium intakes of **lactovegetarians** are similar to those of the general population, but people who use no milk products risk deficiency. Careful planners select calcium-rich foods, such as calcium-fortified juices, soy milk, and breakfast cereals, in ample quantities regularly. This advice is especially important for children and adolescents. Soy formulas for infants are fortified with calcium and can be used in cooking, even for adults. Other good calcium sources include figs, some legumes, some green vegetables such as broccoli and turnip greens, some nuts such as almonds, certain seeds such as sesame seeds, and calcium-set tofu.* The choices should be varied because calcium absorption from some plant foods may be limited (as Chapter 12 explains).

Vitamin B$_{12}$

The requirement for vitamin B$_{12}$ is small, but this vitamin is found only in animal-derived foods. Consequently, vegetarians, in general, and **vegans** who eat no foods of animal original, in particular, may not get enough vitamin B$_{12}$ in their diets.[22] Fermented soy products such as tempeh may contain some vitamin B$_{12}$ from the bacteria, but unfortunately, much of the vitamin B$_{12}$ found in these products may be an inactive form. Seaweeds such as nori and chlorella supply some vitamin B$_{12}$, but not much, and excessive intakes of these foods can lead to iodine toxicity. To defend against vitamin B$_{12}$ deficiency, vegans must rely on vitamin B$_{12}$–fortified sources (such as soy milk or breakfast cereals) or supplements. Without vitamin B$_{12}$, the nerves suffer damage, leading to such health consequences as loss of vision.

Vitamin D

People who do not use vitamin D–fortified foods and do not receive enough exposure to sunlight to synthesize adequate vitamin D may need supplements to defend against bone loss. This is particularly important for infants, children, and older adults. In northern climates during winter months, young children on vegan diets can readily develop rickets, the vitamin D–deficiency disease.

Omega-3 Fatty Acids

Both Chapter 5 and Highlight 5 describe the health benefits of unsaturated fats, most notably the omega-3 fatty acids commonly found in fatty fish. To obtain sufficient amounts of omega-3 fatty acids, vegetarians need to consume flaxseed, walnuts, soybeans, and their oils.

Healthy Food Choices

In general, adults who eat vegetarian diets have lowered their risks of mortality and several chronic diseases, including obesity, high blood pressure, heart disease, and cancer. But there is nothing mysterious or magical about the vegetarian diet; vegetarianism is not a religion like Buddhism or Hinduism, but merely an eating plan that selects plant foods to deliver needed nutrients. The quality of the diet depends not on whether it includes meat, but on whether the other food choices are nutritionally sound. A diet that includes ample fruits, vegetables, whole grains, legumes, nuts, and seeds is higher in fiber, antioxidant vitamins, and phytochemicals, and lower in saturated fats than meat-based diets. Variety is key to nutritional adequacy in a vegetarian diet. Restrictive plans, such as **macrobiotic diets,** that limit selections to a few grains and vegetables cannot possibly deliver a full array of nutrients.

If not properly balanced, any diet—vegetarian or otherwise—can lack nutrients. Poorly planned vegetarian diets typically lack iron, zinc, calcium, vitamin B$_{12}$, and vitamin D; without planning, the meat eater's diet may lack vitamin A, vitamin C, folate, and fiber, among others. Quite simply, the negative health aspects of any diet, including vegetarian diets, reflect poor diet planning. Careful attention to energy intake and specific problem nutrients can ensure adequacy.

Keep in mind, too, that diet is only one factor influencing health. Whatever a diet consists of, its context is also important: no smoking, alcohol consumption in moderation (if at all), regular physical activity, adequate rest, and medical attention when needed all contribute to a healthy life. Establishing these healthy habits early in life seems to be the most important step one can take to reduce the risks of later diseases (as Highlight 15 explains).

---

* Calcium salts are often added during processing to coagulate the tofu.

NUTRITION ON THE NET

For further study of topics covered in this chapter, log on to **academic.cengage.com/nutrition/rolfes/UNCN8e**. Go to Chapter 2, then to Nutrition on the Net.

- Search for "vegetarian" at the Food and Drug Administration's site: **www.fda.gov**
- Visit the Vegetarian Resource Group: **www.vrg.org**
- Review another vegetarian diet pyramid developed by Oldways Preservation & Exchange Trust: **www.oldwayspt.org**

REFERENCES

1. S. I. Barr and G. E. Chapman, Perceptions and practices of self-defined current vegetarian, former vegetarian, and nonvegetarian women, *Journal of the American Dietetic Association* 102 (2002): 354–360.

2. J. Sabaté, The contribution of vegetarian diets to human health, *Forum of Nutrition* 56 (2003): 218–220.

3. Position of the American Dietetic Association and Dietitians of Canada: Vegetarian diets, *Journal of the American Dietetic Association* 103 (2003): 748–765; J. Sabaté, The contribution of vegetarian diets to health and disease: A paradigm shift? *American Journal of Clinical Nutrition* 78 (2003): 502S–507S.

4. P. N. Singh, J. Sabaté, and G. E. Fraser, Does low meat consumption increase life expectancy in humans? *American Journal of Clinical Nutrition* 78 (2003): 526S–532S.

5. P. K. Newby, K. L. Tucker, and A. Wolk, Risk of overweight and obesity among semivegetarian, lactovegetarian, and vegan women, *American Journal of Clinical Nutrition* 81 (2005): 1267–1274; N. Brathwaite and coauthors, Obesity, diabetes, hypertension, and vegetarian status among Seventh-Day Adventists in Barbados, *Ethnicity and Disease* 13 (2003): 34–39; E. H. Haddad and J. S. Tanzman, What do vegetarians in the United States eat? *American Journal of Clinical Nutrition* 78 (2003): 626S–632S.

6. S. E. Berkow and N. D. Barnard, Blood pressure regulation and vegetarian diets, *Nutrition Reviews* 63 (2005): 1–8; L. J. Appel, The effects of protein intake on blood pressure and cardiovascular disease, *Current Opinion in Lipidology* 14 (2003): 55–59.

7. J. E. Cade and coauthors, The UK Women's Cohort Study: Comparison of vegetarians, fish-eaters, and meat-eaters, *Public Health Nutrition* 7 (2004): 871–878; E. H. Haddad and J. S. Tanzman, What do vegetarians in the United States eat? *American Journal of Clinical Nutrition* 78 (2003): 626S–632S.

8. *Third Report of the National Cholesterol Education Program (NCEP) Expert Panel on Detection, Evaluation, and Treatment of High Blood Cholesterol in Adults (Adult Treatment Panel III)*, NIH publication no. 02-5215 (Bethesda, Md.: National Heart, Lung, and Blood Institute, 2002).

9. F. B. Hu, Plant-based foods and prevention of cardiovascular disease: An overview, *American Journal of Clinical Nutrition* 78 (2003): 544S–551S.

10. F. M. Sacks and coauthors, Soy protein, isoflavones, and cardiovascular health: An American Heart Association Science Advisory for professionals from the Nutrition Committee, *Circulation* 113 (2006): 1034–1044.

11. B. L. McVeigh and coauthors, Effect of soy protein varying in isoflavone content on serum lipids in healthy young men, *American Journal of Clinical Nutrition* 83 (2006): 244–251; D. Lukaczer and coauthors, Effect of a low glycemic index diet with soy protein and phytosterols on CVD risk factors in postmenopausal women, *Nutrition* 22 (2006): 104–113; M. S. Rosell and coauthors, Soy intake and blood cholesterol concentrations: A cross-sectional study of 1033 pre- and postmenopausal women in the Oxford arm of the European Prospective Investigation into Cancer and Nutrition, *American Journal of Clinical Nutrition* 80 (2004): 1391–1396; S. Tonstad, K. Smerud, and L. Hoie, A comparison of the effects of 2 doses of soy protein or casein on serum lipids, serum lipoproteins, and plasma total homocysteine in hypercholesterolemic subjects, *American Journal of Clinical Nutrition* 76 (2002): 78–84.

12. M. Kapiszewska, A vegetable to meat consumption ratio as a relevant factor determining cancer preventive diet: The Mediterranean versus other European countries, *Forum of Nutrition* 59 (2006): 130–153.

13. M. H. Lewin and coauthors, Red meat enhances the colonic formation of the DNA adduct O6-carboxymethyl guanine: Implications for colorectal cancer risk, *Cancer Research* 66 (2006): 1859–1865.

14. H. Chen and coauthors, Dietary patterns and adenocarcinoma of the esophagus and distal stomach, *American Journal of Clinical Nutrition* 75 (2002): 137–144.

15. C. Leitzmann, Vegetarian diets: What are the advantages? *Forum of Nutrition* 57 (2005): 147–156.

16. M. Virginia, V. Melina, and A. R. Mangels, A new food guide for North American vegetarians, *Journal of the American Dietetic Association* 103 (2003): 771–775; C. A. Venti and C. S. Johnston, Modified food guide pyramid for lactovegetarians and vegans, *Journal of Nutrition* 132 (2002): 1050–1054.

17. Venti and Johnston, 2002; V. Messina and A. R. Mangels, Considerations in planning vegan diets: Children, *Journal of the American Dietetic Association* 101 (2001): 661–669.

18. Position of the American Dietetic Association and Dietitians of Canada, 2003.

19. J. R. Hunt, Moving toward a plant-based diet: Are iron and zinc at risk? *Nutrition Reviews* 60 (2002): 127–134.

20. C. L. Larsson and G. K. Johansson, Dietary intake and nutritional status of young vegans and omnivores in Sweden, *American Journal of Clinical Nutrition* 76 (2002): 100–106.

21. Hunt, 2002.

22. W. Herrmann and coauthors, Vitamin B12 status, particularly holotranscobalamin II and methylmalonic acid concentrations, and hyperhomocysteinemia in vegetarians, *American Journal of Clinical Nutrition* 78 (2003): 131–136.

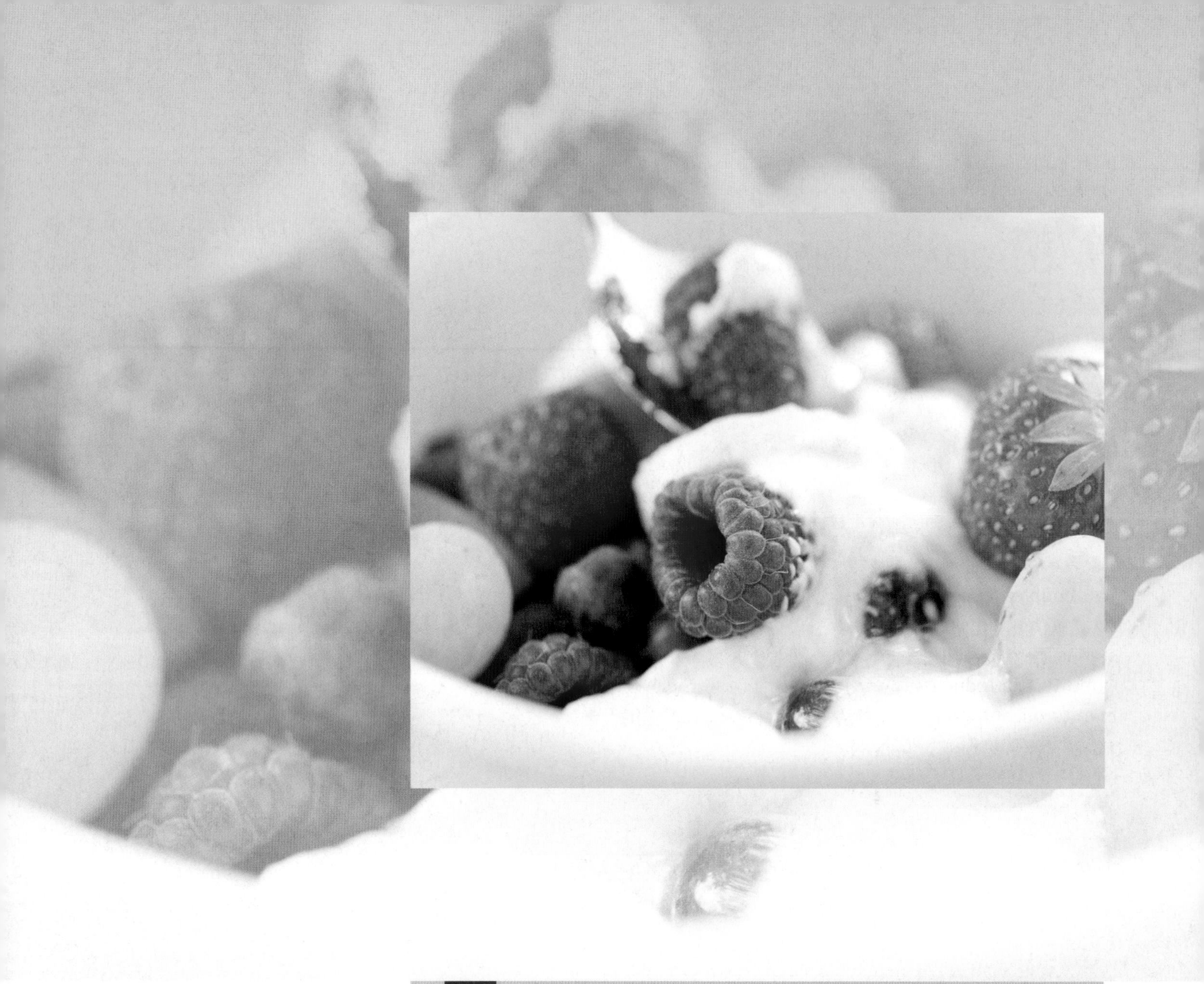

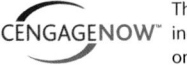

Foodcollection/Getty Images

CENGAGENOW™ The CengageNOW logo indicates an opportunity for online self-study, linking you to interactive tutorials and videos based on your level of understanding.

academic.cengage.com/login

Figure 3.8: Animated! The Digestive Fate of a Sandwich

Figure 3.11: Animated! The Vascular System

Nutrition Portfolio Journal

Nutrition in Your Life

Have you ever wondered what happens to the food you eat after you swallow it? Or how your body extracts nutrients from food? Have you ever marveled at how it all just seems to happen? Follow foods as they travel through the digestive system. Learn how a healthy digestive system transforms whatever food you give it—whether sirloin steak and potatoes or tofu and brussels sprouts—into the nutrients that will nourish the cells of your body.

Digestion, Absorption, and Transport

This chapter takes you on the journey that transforms the foods you eat into the nutrients featured in the later chapters. Then it follows the nutrients as they travel through the intestinal cells and into the body to do their work. This introduction presents a general overview of the processes common to all nutrients; later chapters discuss the specifics of digesting and absorbing individual nutrients.

Digestion

Digestion is the body's ingenious way of breaking down foods into nutrients in preparation for **absorption.** In the process, it overcomes many challenges without any conscious effort on your part. Consider these challenges:

1. Human beings breathe, eat, and drink through their mouths. Air taken in through the mouth must go to the lungs; food and liquid must go to the stomach. The throat must be arranged so that swallowing and breathing don't interfere with each other.

2. Below the lungs lies the diaphragm, a dome of muscle that separates the upper half of the major body cavity from the lower half. Food must pass through this wall to reach the stomach.

3. The materials within the digestive tract should be kept moving forward, slowly but steadily, at a pace that permits all reactions to reach completion.

4. To move through the system, food must be lubricated with fluids. Too much would form a liquid that would flow too rapidly; too little would form a paste too dry and compact to move at all. The amount of fluids must be regulated to keep the intestinal contents at the right consistency to move smoothly along.

5. When the digestive enzymes break food down, they need it in a finely divided form, suspended in enough liquid so that every particle is accessible. Once digestion is complete and the needed nutrients have been absorbed out of the tract and into the body, the system must excrete the remaining residue. Excreting all the water along with the solid residue, however, would be both wasteful and messy. Some water must be withdrawn to leave a paste just solid enough to be smooth and easy to pass.

6. The enzymes of the digestive tract are designed to digest carbohydrate, fat, and protein. The walls of the tract, composed of living cells, are also made of

digestion: the process by which food is broken down into absorbable units.
• **digestion** = take apart

absorption: the uptake of nutrients by the cells of the small intestine for transport into either the blood or the lymph.
• **absorb** = suck in

The process of digestion transforms all kinds of *foods* into *nutrients*.

◆ The process of chewing is called **mastication** (mass-tih-KAY-shun).

gastrointestinal (GI) tract: the digestive tract. The principal organs are the stomach and intestines.
• **gastro** = stomach
• **intestinalis** = intestine

carbohydrate, fat, and protein. These cells need protection against the action of the powerful digestive juices that they secrete.

7. Once waste matter has reached the end of the tract, it must be excreted, but it would be inconvenient and embarrassing if this function occurred continuously. Provision must be made for periodic, voluntary evacuation.

The following sections show how the body elegantly and efficiently handles these challenges.

Anatomy of the Digestive Tract

The **gastrointestinal (GI) tract** is a flexible muscular tube that extends from the mouth, through the esophagus, stomach, small intestine, large intestine, and rectum to the anus. Figure 3-1 traces the path followed by food from one end to the other. In a sense, the human body surrounds the GI tract. The inner space within the GI tract, called the **lumen**, is continuous from one end to the other. (GI anatomy terms appear in boldface type and are defined in the accompanying glossary.) Only when a nutrient or other substance finally penetrates the GI tract's wall does it enter the body proper; many materials pass through the GI tract without being digested or absorbed.

Mouth The process of digestion begins in the **mouth.** As you chew, ◆ your teeth crush large pieces of food into smaller ones, and fluids from foods, beverages, and salivary glands blend with these pieces to ease swallowing. Fluids also help dissolve the food so that you can taste it; only particles in solution can react with taste buds. When stimulated, the taste buds detect one, or a combination, of the four basic taste sensations: sweet, sour, bitter, and salty. Some scientists also include the flavor associated with monosodium glutamate, sometimes called *savory* or its Asian name, *umami* (oo-MOM-ee). In addition to these chemical triggers, aroma, texture, and temperature also affect a food's flavor. In fact, the sense of smell is thousands of times more sensitive than the sense of taste.

The tongue allows you not only to taste food, but also to move food around the mouth, facilitating chewing and swallowing. When you swallow a mouthful of

---

GLOSSARY OF GI ANATOMY TERMS

These terms are listed in order from start to end of the digestive system.

lumen (LOO-men): the space within a vessel, such as the intestine.

mouth: the oral cavity containing the tongue and teeth.

pharynx (FAIR-inks): the passageway leading from the nose and mouth to the larynx and esophagus, respectively.

epiglottis (epp-ih-GLOTT-iss): cartilage in the throat that guards the entrance to the trachea and prevents fluid or food from entering it when a person swallows.
• **epi** = upon (over)
• **glottis** = back of tongue

esophagus (ee-SOFF-ah-gus): the food pipe; the conduit from the mouth to the stomach.

sphincter (SFINK-ter): a circular muscle surrounding, and able to close, a body opening. Sphincters are found at specific points along

the GI tract and regulate the flow of food particles.
• **sphincter** = band (binder)

esophageal (ee-SOF-ah-GEE-al) **sphincter:** a sphincter muscle at the upper or lower end of the esophagus. The *lower esophageal sphincter* is also called the *cardiac sphincter.*

stomach: a muscular, elastic, saclike portion of the digestive tract that grinds and churns swallowed food, mixing it with acid and enzymes to form chyme.

pyloric (pie-LORE-ic) **sphincter:** the circular muscle that separates the stomach from the small intestine and regulates the flow of partially digested food into the small intestine; also called *pylorus* or *pyloric valve.*
• **pylorus** = gatekeeper

small intestine: a 10-foot length of small-diameter intestine that is the major site of digestion of

food and absorption of nutrients. Its segments are the duodenum, jejunum, and ileum.

gallbladder: the organ that stores and concentrates bile. When it receives the signal that fat is present in the duodenum, the gallbladder contracts and squirts bile through the bile duct into the duodenum.

pancreas: a gland that secretes digestive enzymes and juices into the duodenum. (The pancreas also secretes hormones into the blood that help to maintain glucose homeostasis.)

duodenum (doo-oh-DEEN-um, doo-ODD-num): the top portion of the small intestine (about "12 fingers' breadth" long in ancient terminology).
• **duodecim** = twelve

jejunum (je-JOON-um): the first two-fifths of the small intestine beyond the duodenum.

ileum (ILL-ee-um): the last segment of the small intestine.

ileocecal (ill-ee-oh-SEEK-ul) **valve:** the sphincter separating the small and large intestines.

large intestine or **colon** (COAL-un): the lower portion of intestine that completes the digestive process. Its segments are the ascending colon, the transverse colon, the descending colon, and the sigmoid colon.
• **sigmoid** = shaped like the letter S (sigma in Greek)

appendix: a narrow blind sac extending from the beginning of the colon that stores lymph cells.

rectum: the muscular terminal part of the intestine, extending from the sigmoid colon to the anus.

anus (AY-nus): the terminal outlet of the GI tract.

digestive system: all the organs and glands associated with the ingestion and digestion of food.

FIGURE 3–1 The Gastrointestinal Tract

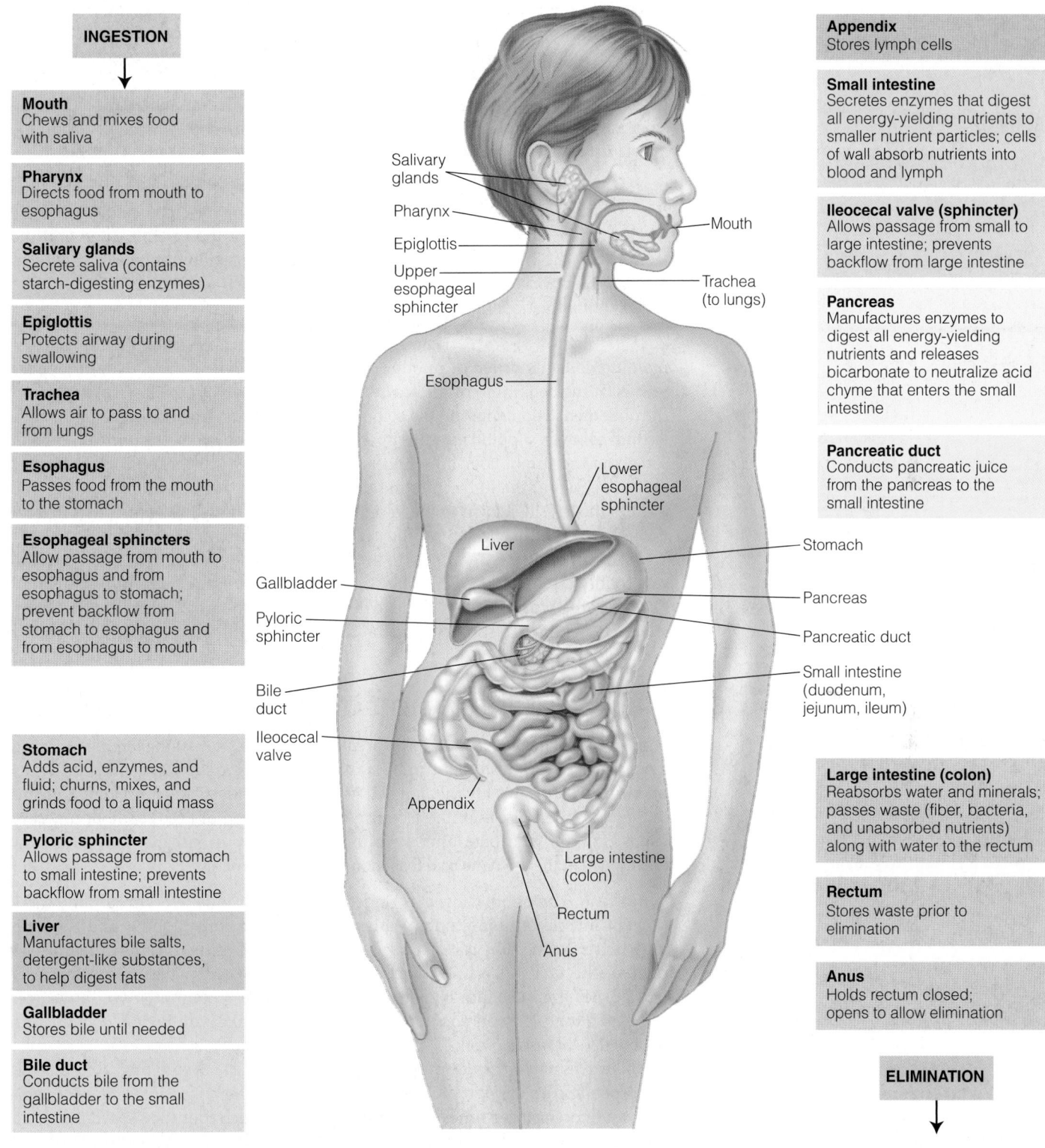

INGESTION

Mouth
Chews and mixes food with saliva

Pharynx
Directs food from mouth to esophagus

Salivary glands
Secrete saliva (contains starch-digesting enzymes)

Epiglottis
Protects airway during swallowing

Trachea
Allows air to pass to and from lungs

Esophagus
Passes food from the mouth to the stomach

Esophageal sphincters
Allow passage from mouth to esophagus and from esophagus to stomach; prevent backflow from stomach to esophagus and from esophagus to mouth

Stomach
Adds acid, enzymes, and fluid; churns, mixes, and grinds food to a liquid mass

Pyloric sphincter
Allows passage from stomach to small intestine; prevents backflow from small intestine

Liver
Manufactures bile salts, detergent-like substances, to help digest fats

Gallbladder
Stores bile until needed

Bile duct
Conducts bile from the gallbladder to the small intestine

Appendix
Stores lymph cells

Small intestine
Secretes enzymes that digest all energy-yielding nutrients to smaller nutrient particles; cells of wall absorb nutrients into blood and lymph

Ileocecal valve (sphincter)
Allows passage from small to large intestine; prevents backflow from large intestine

Pancreas
Manufactures enzymes to digest all energy-yielding nutrients and releases bicarbonate to neutralize acid chyme that enters the small intestine

Pancreatic duct
Conducts pancreatic juice from the pancreas to the small intestine

Large intestine (colon)
Reabsorbs water and minerals; passes waste (fiber, bacteria, and unabsorbed nutrients) along with water to the rectum

Rectum
Stores waste prior to elimination

Anus
Holds rectum closed; opens to allow elimination

ELIMINATION

Labels on figure:
Salivary glands
Pharynx
Epiglottis
Upper esophageal sphincter
Mouth
Trachea (to lungs)
Esophagus
Lower esophageal sphincter
Liver
Gallbladder
Pyloric sphincter
Bile duct
Ileocecal valve
Appendix
Stomach
Pancreas
Pancreatic duct
Small intestine (duodenum, jejunum, ileum)
Large intestine (colon)
Rectum
Anus

food, it passes through the **pharynx,** a short tube that is shared by both the **digestive system** and the respiratory system. To bypass the entrance to your lungs, the **epiglottis** closes off your air passages so that you don't choke when you swallow, thus resolving the first challenge. (Choking is discussed on pp. 92–93.) After a mouthful of food has been swallowed, it is called a **bolus.**

bolus (BOH-lus): a portion; with respect to food, the amount swallowed at one time.
• **bolos** = lump

Esophagus to the Stomach The **esophagus** has a **sphincter** muscle at each end. During a swallow, the upper **esophageal sphincter** opens. The bolus then slides down the esophagus, which passes through a hole in the diaphragm (challenge 2) to the **stomach.** The lower esophageal sphincter at the entrance to the stomach closes behind the bolus so that it proceeds forward and doesn't slip back into the esophagus (challenge 3). The stomach retains the bolus for a while in its upper portion. Little by little, the stomach transfers the food to its lower portion, adds juices to it, and grinds it to a semiliquid mass called **chyme.** Then, bit by bit, the stomach releases the chyme through the **pyloric sphincter,** which opens into the **small intestine** and then closes behind the chyme.

Small Intestine At the beginning of the small intestine, the chyme bypasses the opening from the common bile duct, which is dripping fluids (challenge 4) into the small intestine from two organs outside the GI tract—the **gallbladder** and the **pancreas.** The chyme travels on down the small intestine through its three segments—the **duodenum,** the **jejunum,** and the **ileum**—almost 10 feet of tubing coiled within the abdomen.*

Large Intestine (Colon) Having traveled the length of the small intestine, the remaining contents arrive at another sphincter (challenge 3 again): the **ileocecal valve,** at the beginning of the **large intestine (colon)** in the lower right side of the abdomen. Upon entering the colon, the contents pass another opening. Any intestinal contents slipping into this opening would end up in the **appendix,** a blind sac about the size of your little finger. The contents bypass this opening, however, and travel along the large intestine up the right side of the abdomen, across the front to the left side, down to the lower left side, and finally below the other folds of the intestines to the back of the body, above the **rectum.**

As the intestinal contents pass to the rectum, the colon withdraws water, leaving semisolid waste (challenge 5). The strong muscles of the rectum and anal canal hold back this waste until it is time to defecate. Then the rectal muscles relax (challenge 7), and the two sphincters of the **anus** open to allow passage of the waste.

The Muscular Action of Digestion

In the mouth, chewing, the addition of saliva, and the action of the tonguetransform food into a coarse mash that can be swallowed. After swallowing, you are generally unaware of all the activity that follows. As is the case with so much else that happens in the body, the muscles of the digestive tract meet internal needs without any conscious effort on your part. They keep things moving ◆ at just the right pace, slow enough to get the job done and fast enough to make progress.

Peristalsis The entire GI tract is ringed with circular muscles. Surrounding these rings of muscle are longitudinal muscles. When the rings tighten and the long muscles relax, the tube is constricted. When the rings relax and the long muscles tighten, the tube bulges. This action—called **peristalsis**—occurs continuously and pushes the intestinal contents along (challenge 3 again). (If you have ever watched a lump of food pass along the body of a snake, you have a good picture of how these muscles work.)

The waves of contraction ripple along the GI tract at varying rates and intensities depending on the part of the GI tract and on whether food is present. For example, waves occur three times per minute in the stomach, but they speed up to ten times per minute when chyme reaches the small intestine. When you have just eaten a meal, the waves are slow and continuous; when the GI tract is empty, the intestine is quiet except for periodic bursts of powerful rhythmic waves. Peristalsis,

◆ The ability of the GI tract muscles to move is called their **motility** (moh-TIL-ih-tee).

---

chyme (KIME): the semiliquid mass of partly digested food expelled by the stomach into the duodenum.
- **chymos** = juice

peristalsis (per-ih-STALL-sis): wavelike muscular contractions of the GI tract that push its contents along.
- **peri** = around
- **stellein** = wrap

---

* The small intestine is almost 2½ times shorter in living adults than it is at death, when muscles are relaxed and elongated.

along with sphincter muscles located at key places, keeps things moving along.

Stomach Action The stomach has the thickest walls and strongest muscles of all the GI tract organs. In addition to the circular and longitudinal muscles, it has a third layer of diagonal muscles that also alternately contract and relax (see Figure 3-2). These three sets of muscles work to force the chyme downward, but the pyloric sphincter usually remains tightly closed, preventing the chyme from passing into the duodenum of the small intestine. As a result, the chyme is churned and forced down, hits the pyloric sphincter, and remains in the stomach. Meanwhile, the stomach wall releases gastric juices. When the chyme is completely liquefied, the pyloric sphincter opens briefly, about three times a minute, to allow small portions of chyme to pass through. At this point, the chyme no longer resembles food in the least.

Segmentation The circular muscles of the intestines rhythmically contract and squeeze their contents (see Figure 3-3). These contractions,

FIGURE 3–2 Stomach Muscles

The stomach has three layers of muscles.

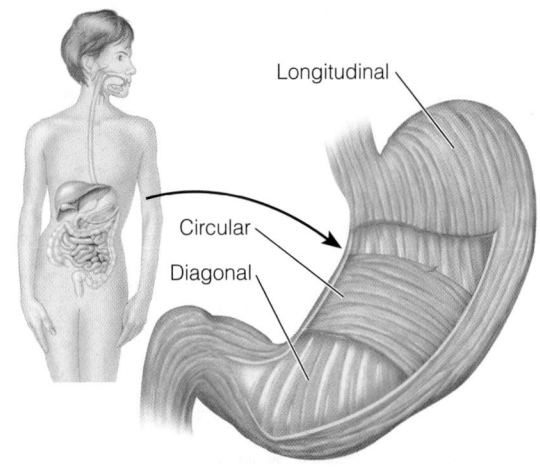

Longitudinal

Circular

Diagonal

FIGURE 3–3 Peristalsis and Segmentation

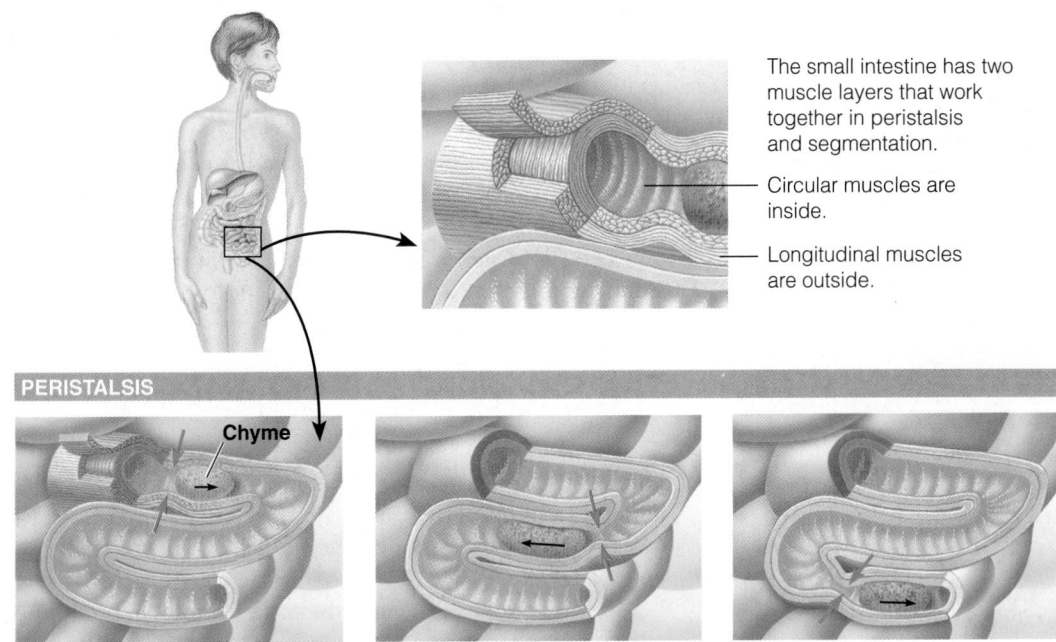

The small intestine has two muscle layers that work together in peristalsis and segmentation.

Circular muscles are inside.

Longitudinal muscles are outside.

PERISTALSIS

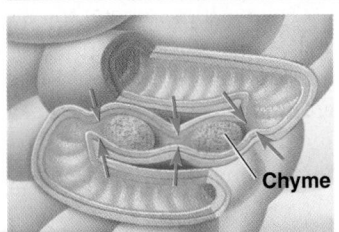

Chyme

The inner circular muscles contract, tightening the tube and pushing the food forward in the intestine.

When the circular muscles relax, the outer longitudinal muscles contract, and the intestinal tube is loose.

As the circular and longitudinal muscles tighten and relax, the chyme moves ahead of the constriction.

SEGMENTATION

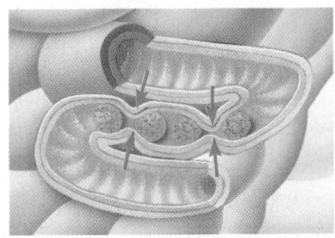

Chyme

Circular muscles contract, creating segments within the intestine.

As each set of circular muscles relaxes and contracts, the chyme is broken up and mixed with digestive juices.

These alternating contractions, occurring 12 to 16 times per minute, continue to mix the chyme and bring the nutrients into contact with the intestinal lining for absorption.

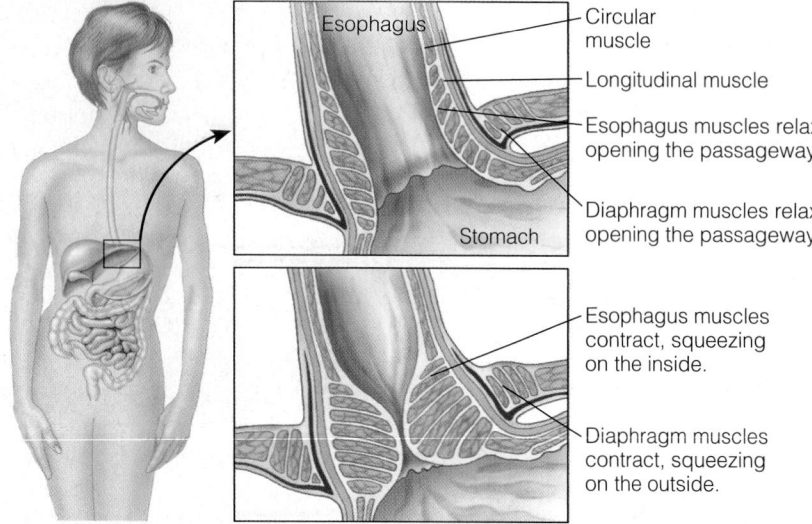

FIGURE 3–4 An Example of a Sphincter Muscle

When the circular muscles of a sphincter contract, the passage closes; when they relax, the passage opens.

Esophagus

Circular muscle

Longitudinal muscle

Esophagus muscles relax, opening the passageway.

Diaphragm muscles relax, opening the passageway.

Stomach

Esophagus muscles contract, squeezing on the inside.

Diaphragm muscles contract, squeezing on the outside.

FIGURE 3–5 The Salivary Glands

The salivary glands secrete saliva into the mouth and begin the digestive process. Given the short time food is in the mouth, salivary enzymes contribute little to digestion.

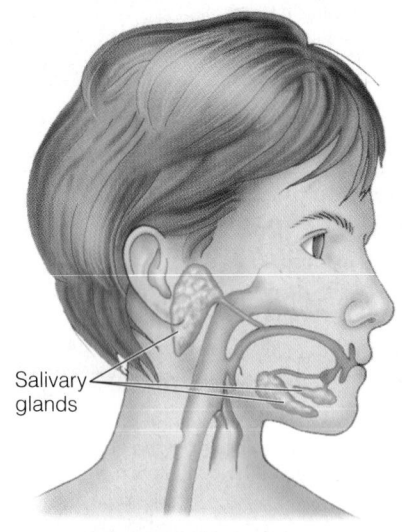

Salivary glands

segmentation (SEG-men-TAY-shun): a periodic squeezing or partitioning of the intestine at intervals along its length by its circular muscles.

reflux: a backward flow.
- **re** = back
- **flux** = flow

catalyst (CAT-uh-list): a compound that facilitates chemical reactions without itself being changed in the process.

called **segmentation**, mix the chyme and promote close contact with the digestive juices and the absorbing cells of the intestinal walls before letting the contents move slowly along. Figure 3-3 illustrates peristalsis and segmentation.

Sphincter Contractions Sphincter muscles periodically open and close, allowing the contents of the GI tract to move along at a controlled pace (challenge 3 again). At the top of the esophagus, the upper esophageal sphincter opens in response to swallowing. At the bottom of the esophagus, the lower esophageal sphincter (sometimes called the cardiac sphincter because of its proximity to the heart) prevents **reflux** of the stomach contents. At the bottom of the stomach, the pyloric sphincter, which stays closed most of the time, holds the chyme in the stomach long enough for it to be thoroughly mixed with gastric juice and liquefied. The pyloric sphincter also prevents the intestinal contents from backing up into the stomach. At the end of the small intestine, the ileocecal valve performs a similar function, allowing the contents of the small intestine to empty into the large intestine. Finally, the tightness of the rectal muscle is a kind of safety device; together with the two sphincters of the anus, it prevents elimination until you choose to perform it voluntarily (challenge 7). Figure 3-4 illustrates how sphincter muscles contract and relax to close and open passageways.

The Secretions of Digestion

The breakdown of food into nutrients requires secretions from five different organs: the salivary glands, the stomach, the pancreas, the liver (via the gallbladder), and the small intestine. These secretions enter the GI tract at various points along the way, bringing an abundance of water (challenge 3 again) and a variety of enzymes.

Enzymes are formally introduced in Chapter 6, but for now a simple definition will suffice. An enzyme is a protein that facilitates a chemical reaction—making a molecule, breaking a molecule apart, changing the arrangement of a molecule, or exchanging parts of molecules. As a **catalyst,** the enzyme itself remains unchanged. The enzymes involved in digestion facilitate a chemical reaction known as **hydrolysis**—the addition of water *(hydro)* to break *(lysis)* a molecule into smaller pieces. The glossary (p. 77) identifies some of the common **digestive enzymes** and related terms; later chapters introduce specific enzymes. When learning about enzymes, it helps to know that the word ending *-ase* denotes an enzyme. Enzymes are often identified by the organ they come from and the compounds they work on. *Gastric lipase*, for example, is a stomach enzyme that acts on lipids, whereas *pancreatic lipase* comes from the pancreas (and also works on lipids).

Saliva The **salivary glands,** shown in Figure 3-5, squirt just enough **saliva** to moisten each mouthful of food so that it can pass easily down the esophagus (challenge 4). (Digestive **glands** and their secretions are defined in the glossary on

p. 78.) The saliva contains water, salts, mucus, and enzymes that initiate the digestion of carbohydrates. Saliva also protects the teeth and the linings of the mouth, esophagus, and stomach from attack by substances that might harm them.

Gastric Juice In the stomach, **gastric glands** secrete **gastric juice,** a mixture of water, enzymes, and **hydrochloric acid,** which acts primarily in protein digestion. The acid is so strong that it causes the sensation of heartburn if it happens to reflux into the esophagus. Highlight 3, following this chapter, discusses heartburn, ulcers, and other common digestive problems.

The strong acidity of the stomach prevents bacterial growth and kills most bacteria that enter the body with food. It would destroy the cells of the stomach as well, but for their natural defenses. To protect themselves from gastric juice, the cells of the stomach wall secrete **mucus,** a thick, slippery, white substance that coats the cells, protecting them from the acid, enzymes, and disease-causing bacteria that might otherwise harm them (challenge 6).

Figure 3-6 shows how the strength of acids is measured—in **pH** ◆ units. Note that the acidity of gastric juice registers below "2" on the pH scale—stronger than vinegar. The stomach enzymes work most efficiently in the stomach's strong acid, but the salivary enzymes, which are swallowed with food, do not work in acid this strong. Consequently, the salivary digestion of carbohydrate gradually ceases when the stomach acid penetrates each newly swallowed bolus of food. When they enter the stomach, salivary enzymes become just other proteins to be digested.

Pancreatic Juice and Intestinal Enzymes By the time food leaves the stomach, digestion of all three energy nutrients (carbohydrates, fats, and proteins) has begun, and the action gains momentum in the small intestine. There the pancreas contributes digestive juices by way of ducts leading into the duodenum. The **pancreatic juice** contains enzymes that act on all three energy nutrients, and the cells of the intestinal wall also possess digestive enzymes on their surfaces.

In addition to enzymes, the pancreatic juice contains sodium **bicarbonate,** which is basic or alkaline—the opposite of the stomach's acid (review Figure 3-6). The pancreatic juice thus neutralizes the acidic chyme arriving in the small intestine from the stomach. From this point on, the chyme remains at a neutral or slightly alkaline pH. The enzymes of both the intestine and the pancreas work best in this environment.

Bile **Bile** also flows into the duodenum. The **liver** continuously produces bile, which is then concentrated and stored in the gallbladder. The gallbladder squirts

---

FIGURE 3–6 The pH Scale

A substance's acidity or alkalinity is measured in pH units. The pH is the negative logarithm of the hydrogen ion concentration. Each increment represents a tenfold increase in concentration of hydrogen particles. This means, for example, that a pH of 2 is 1000 times stronger than a pH of 5.

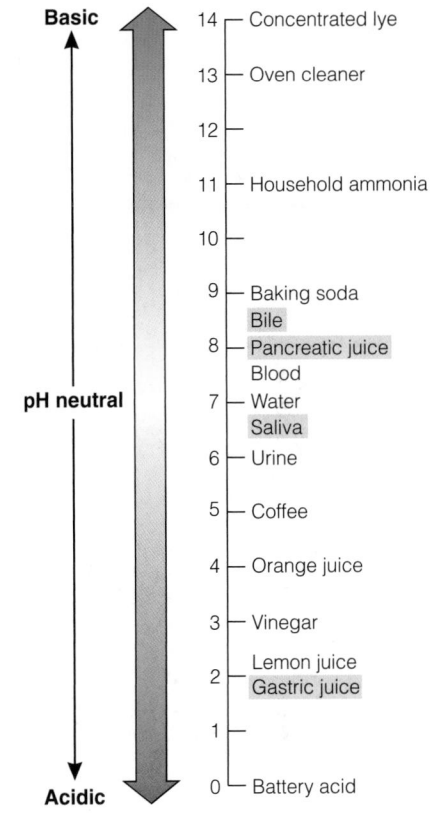

pH of common substances:

Basic	14	Concentrated lye
	13	Oven cleaner
	12	
	11	Household ammonia
	10	
	9	Baking soda
		Bile
	8	Pancreatic juice
		Blood
pH neutral	7	Water
		Saliva
	6	Urine
	5	Coffee
	4	Orange juice
	3	Vinegar
		Lemon juice
	2	Gastric juice
	1	
Acidic	0	Battery acid

◆ The lower the pH, the higher the H^+ ion concentration and the stronger the acid. A pH above 7 is alkaline, or base (a solution in which OH^- ions predominate).

GLOSSARY OF DIGESTIVE ENZYMES

digestive enzymes: proteins found in digestive juices that act on food substances, causing them to break down into simpler compounds.

-ase (ACE): a word ending denoting an enzyme. The word beginning often identifies the compounds the enzyme works on. Examples include:
- **carbohydrase** (KAR-boe-HIGH-drase), an enzyme that hydrolyzes carbohydrates.
- **lipase** (LYE-pase), an enzyme that hydrolyzes lipids (fats).

- **protease** (PRO-tee-ase), an enzyme that hydrolyzes proteins.

hydrolysis (high-DROL-ih-sis): a chemical reaction in which a major reactant is split into two products, with the addition of a hydrogen atom (H) to one and a hydroxyl group (OH) to the other (from water, H_2O). (The noun is **hydrolysis;** the verb is **hydrolyze.**)
- **hydro** = water
- **lysis** = breaking

pH: the unit of measure expressing a substance's acidity or alkalinity.

GLOSSARY OF DIGESTIVE GLANDS AND THEIR SECRETIONS

These terms are listed in order from start to end of the digestive tract.

glands: cells or groups of cells that secrete materials for special uses in the body. Glands may be **exocrine** (EKS-oh-crin) **glands,** secreting their materials "out" (into the digestive tract or onto the surface of the skin), or **endocrine** (EN-doe-crin) **glands,** secreting their materials "in" (into the blood).

• **exo** = outside
• **endo** = inside
• **krine** = to separate

salivary glands: exocrine glands that secrete saliva into the mouth.

saliva: the secretion of the salivary glands. Its principal enzyme begins carbohydrate digestion.

gastric glands: exocrine glands in the stomach wall that secrete gastric juice into the stomach.
• **gastro** = stomach

gastric juice: the digestive secretion of the gastric glands of the stomach.

hydrochloric acid: an acid composed of hydrogen and chloride atoms (HCl) that is normally produced by the gastric glands.

mucus (MYOO-kus): a slippery substance secreted by cells of the GI lining (and other body linings) that protects the cells from exposure to digestive juices (and other destructive agents). The lining of the GI tract with its coat of mucus is a **mucous membrane.** (The noun

is **mucus;** the adjective is **mucous.**)

liver: the organ that manufactures bile. (The liver's many other functions are described in Chapter 7.)

bile: an emulsifier that prepares fats and oils for digestion; an exocrine secretion made by the liver, stored in the gallbladder, and released into the small intestine when needed.

emulsifier (ee-MUL-sih-fire): a substance with both water-soluble and fat-soluble portions that promotes the mixing of oils and fats in a watery solution.

pancreatic (pank-ree-AT-ic) **juice:** the exocrine secretion of the pancreas, containing enzymes for the digestion of carbohydrate,

fat, and protein as well as bicarbonate, a neutralizing agent. The juice flows from the pancreas into the small intestine through the pancreatic duct. (The pancreas also has an endocrine function, the secretion of insulin and other hormones.)

bicarbonate: an alkaline compound with the formula HCO_3 that is secreted from the pancreas as part of the pancreatic juice. (Bicarbonate is also produced in all cell fluids from the dissociation of cabonic acid to help maintain the body's acid-base balance.)

---

the bile into the duodenum of the small intestine when fat arrives there. Bile is not an enzyme; it is an **emulsifier** that brings fats into suspension in water so that enzymes can break them down into their component parts. Thanks to all these secretions, the three energy-yielding nutrients are digested in the small intestine (the summary on p. 80 provides a table of digestive secretions and their actions).

The Final Stage

At this point, the three energy-yielding nutrients—carbohydrate, fat, and protein—have been disassembled and are ready to be absorbed. Most of the other nutrients—vitamins, minerals, and water—need no such disassembly; some vitamins and minerals are altered slightly during digestion, but most are absorbed as they are. Undigested residues, such as some fibers, are not absorbed. Instead, they continue through the digestive tract, providing a semisolid mass that helps exercise the muscles and keep them strong enough to perform peristalsis efficiently. Fiber also retains water, accounting for the pasty consistency of **stools,** and thereby carries some bile acids, some minerals, and some additives and contaminants with it out of the body.

By the time the contents of the GI tract reach the end of the small intestine, little remains but water, a few dissolved salts and body secretions, and undigested materials such as fiber. These enter the large intestine (colon).

In the colon, intestinal bacteria ferment some fibers, producing water, gas, and small fragments of fat that provide energy for the cells of the colon. The colon itself retrieves all materials that the body can recycle—water and dissolved salts (see Figure 3-7). The waste that is finally excreted has little or nothing of value left in it. The body has extracted all that it can use from the food. Figure 3-8 summarizes digestion by following a sandwich through the GI tract and into the body.

stools: waste matter discharged from the colon; also called **feces** (FEE-seez).

FIGURE 3–7 The Colon

The colon begins with the ascending colon rising upward toward the liver. It becomes the transverse colon as it turns and crosses the body toward the spleen. The descending colon turns downward and becomes the sigmoid colon, which extends to the rectum. Along the way, the colon mixes the intestinal contents, absorbs water and salts, and forms stools.

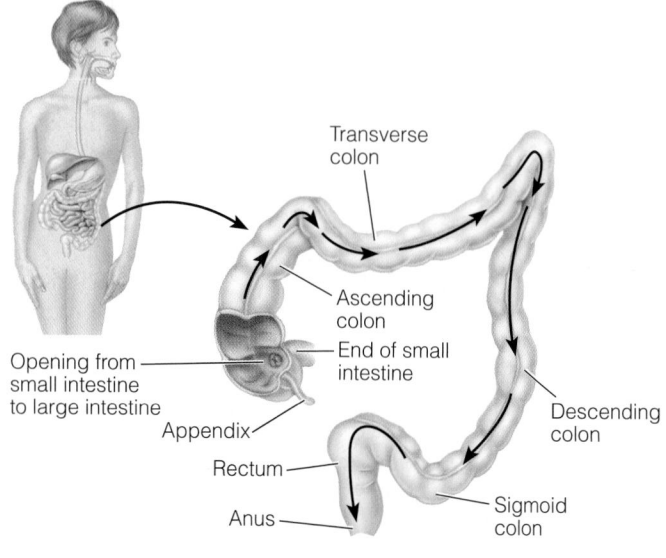

Transverse colon

Ascending colon

End of small intestine

Opening from small intestine to large intestine

Appendix

Rectum

Anus

Descending colon

Sigmoid colon

FIGURE 3-8   *Animated!* The Digestive Fate of a Sandwich

To review the digestive processes, follow a peanut butter and banana sandwich on whole-wheat, seasame seed bread through the GI tract. As the graph on the right illustrates, digestion of the energy nutrients begins in different parts of the GI tract, but all are ready for absorption by the time they reach the end of the small intestine.

CENGAGENOW™
To test your understanding
of these concepts, log on to
academic.cengage.com/login

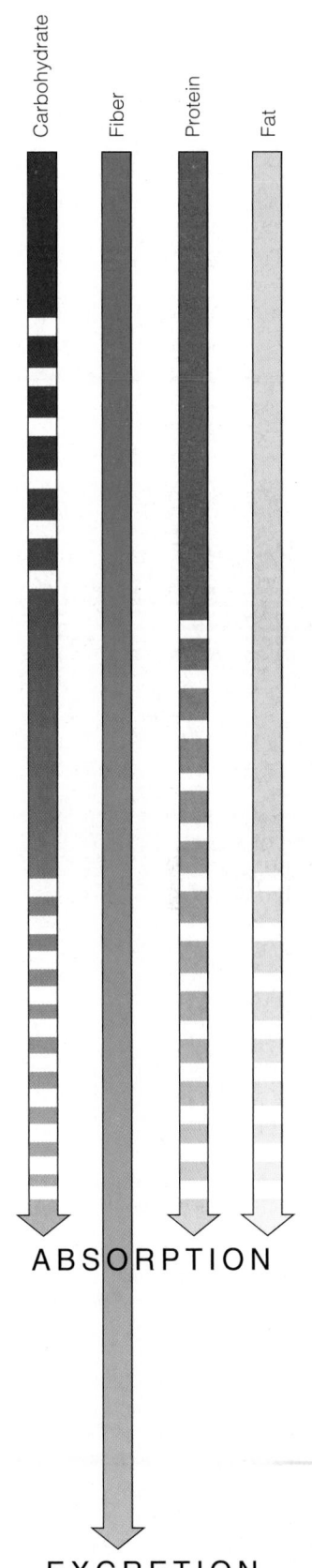

Carbohydrate

Fiber

Protein

Fat

ABSORPTION

EXCRETION

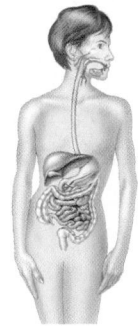

MOUTH: CHEWING AND SWALLOWING, WITH LITTLE DIGESTION

Carbohydrate digestion begins as the salivary enzyme starts to break down the starch from bread and peanut butter.
Fiber covering on the sesame seeds is crushed by the teeth, which exposes the nutrients inside the seeds to the upcoming digestive enzymes.

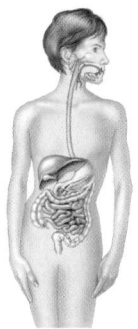

STOMACH: COLLECTING AND CHURNING, WITH SOME DIGESTION

Carbohydrate digestion continues until the mashed sandwich has been mixed with the gastric juices; the stomach acid of the gastric juices inactivates the salivary enzyme, and carbohydrate digestion ceases.
Proteins from the bread, seeds, and peanut butter begin to uncoil when they mix with the gastric acid, making them available to the gastric protease enzymes that begin to digest proteins.
Fat from the peanut butter forms a separate layer on top of the watery mixture.

SMALL INTESTINE: DIGESTING AND ABSORBING

Sugars from the banana require so little digestion that they begin to traverse the intestinal cells immediately on contact.
Starch digestion picks up when the pancreas sends pancreatic enzymes to the small intestine via the pancreatic duct. Enzymes on the surfaces of the small intestinal cells complete the process of breaking down starch into small fragments that can be absorbed through the intestinal cell walls and into the hepatic portal vein.
Fat from the peanut butter and seeds is emulsified with the watery digestive fluids by bile. Now the pancreatic and intestinal lipases can begin to break down the fat to smaller fragments that can be absorbed through the cells of the small intestinal wall and into the lymph.
Protein digestion depends on the pancreatic and intestinal proteases. Small fragments of protein are liberated and absorbed through the cells of the small intestinal wall and into the hepatic portal vein.
Vitamins and minerals are absorbed.

Note: Sugars and starches are members of the carbohydrate family.

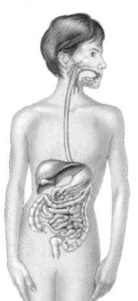

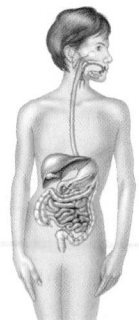

LARGE INTESTINE: REABSORBING AND ELIMINATING

Fluids and some minerals are absorbed.
Some fibers from the seeds, whole-wheat bread, peanut butter, and banana are partly digested by the bacteria living there, and some of these products are absorbed.
Most fibers pass through the large intestine and are excreted as feces; some fat, cholesterol, and minerals bind to fiber and are also excreted.

IN SUMMARY

As Figure 3-1 shows, food enters the mouth and travels down the esophagus and through the upper and lower esophageal sphincters to the stomach, then through the pyloric sphincter to the small intestine, on through the ileocecal valve to the large intestine, past the appendix to the rectum, ending at the anus. The wavelike contractions of peristalsis and the periodic squeezing of segmentation keep things moving at a reasonable pace. Along the way, secretions from the salivary glands, stomach, pancreas, liver (via the gallbladder), and small intestine deliver fluids and digestive enzymes.

Summary of Digestive Secretions and Their Major Actions

Organ or Gland	Target Organ	Secretion	Action
Salivary glands	Mouth	Saliva	Fluid eases swallowing; salivary enzyme breaks down **carbohydrate**.[a]
Gastric glands	Stomach	Gastric juice	Fluid mixes with bolus; hydrochloric acid uncoils **proteins**; enzymes break down proteins; mucus protects stomach cells.[a]
Pancreas	Small intestine	Pancreatic juice	Bicarbonate neutralizes acidic gastric juices; pancreatic enzymes break down **carbohydrates**, **fats**, and **proteins**.
Liver	Gallbladder	Bile	Bile stored until needed.
Gallbladder	Small intestine	Bile	Bile emulsifies **fat** so enzymes can attack.
Intestinal glands	Small intestine	Intestinal juice	Intestinal enzymes break down **carbohydrate**, **fat**, and **protein** fragments; mucus protects the intestinal wall.

[a] Saliva and gastric juices also contain lipases, but most fat breakdown occurs in the small intestines.

Food must first be digested and absorbed before the body can use it.

Foodcollection/Getty Images

villi (VILL-ee, VILL-eye): fingerlike projections from the folds of the small intestine; singular **villus.**

microvilli (MY-cro-VILL-ee, MY-cro-VILL-eye): tiny, hairlike projections on each cell of every villus that can trap nutrient particles and transport them into the cells; singular **microvillus.**

crypts (KRIPTS): tubular glands that lie between the intestinal villi and secrete intestinal juices into the small intestine.

goblet cells: cells of the GI tract (and lungs) that secrete mucus.

Absorption

Within three or four hours after you have eaten a dinner of beans and rice (or spinach lasagna, or steak and potatoes) with vegetable, salad, beverage, and dessert, your body must find a way to absorb the molecules derived from carbohydrate, protein, and fat digestion—and the vitamin and mineral molecules as well. Most absorption takes place in the small intestine, one of the most elegantly designed organ systems in the body. Within its 10-foot length, which provides a surface area equivalent to a tennis court, the small intestine engulfs and absorbs the nutrient molecules. To remove the molecules rapidly and provide room for more to be absorbed, a rush of circulating blood continuously washes the underside of this surface, carrying the absorbed nutrients away to the liver and other parts of the body. Figure 3-9 describes how nutrients are absorbed by simple diffusion, facilitated diffusion, or active transport. Later chapters provide details on specific nutrients. Before following nutrients through the body, we must look more closely at the anatomy of the absorptive system.

Anatomy of the Absorptive System

The inner surface of the small intestine looks smooth and slippery, but when viewed through a microscope, it turns out to be wrinkled into hundreds of folds. Each fold is contoured into thousands of fingerlike projections, as numerous as the hairs on velvet fabric. These small intestinal projections are the **villi.** A single villus, magnified still more, turns out to be composed of hundreds of cells, each covered with its own microscopic hairs, the **microvilli** (see Figure 3-10 on p. 82). In the crevices between the villi lie the **crypts**—tubular glands that secrete the intestinal juices into the small intestine. Nearby **goblet cells** secrete mucus.

FIGURE 3-9 | Absorption of Nutrients

Absorption of nutrients into intestinal cells typically occurs by simple diffusion, facilitated diffusion, or active transport.

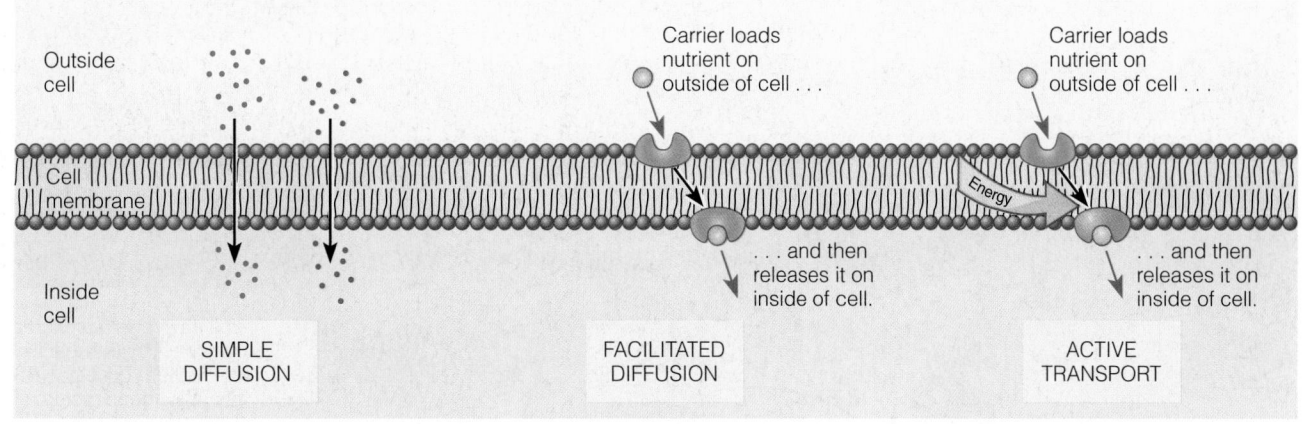

Some nutrients (such as water and small lipids) are absorbed by simple diffusion. They cross into intestinal cells freely.

Some nutrients (such as the water-soluble vitamins) are absorbed by facilitated diffusion. They need a specific carrier to transport them from one side of the cell membrane to the other. (Alternatively, facilitated diffusion may occur when the carrier changes the cell membrane in such a way that the nutrients can pass through.)

Some nutrients (such as glucose and amino acids) must be absorbed actively. These nutrients move against a concentration gradient, which requires energy.

The villi are in constant motion. Each villus is lined by a thin sheet of muscle, so it can wave, squirm, and wriggle like the tentacles of a sea anemone. Any nutrient molecule small enough to be absorbed is trapped among the microvilli that coat the cells and then drawn into the cells. Some partially digested nutrients are caught in the microvilli, digested further by enzymes there, and then absorbed into the cells.

A Closer Look at the Intestinal Cells

The cells of the villi are among the most amazing in the body, for they recognize and select the nutrients the body needs and regulate their absorption. As already described, each cell of a villus is coated with thousands of microvilli, which project from the cell's membrane (review Figure 3-10). In these microvilli, and in the membrane, lie hundreds of different kinds of enzymes and "pumps," which recognize and act on different nutrients. Descriptions of specific enzymes and "pumps" for each nutrient are presented in the following chapters where appropriate; the point here is that the cells are equipped to handle all kinds and combinations of foods and nutrients.

Specialization in the GI Tract A further refinement of the system is that the cells of successive portions of the intestinal tract are specialized to absorb different nutrients. The nutrients that are ready for absorption early are absorbed near the top of the tract; those that take longer to be digested are absorbed farther down. Registered dietitians and medical professionals who treat digestive disorders learn the specialized absorptive functions of different parts of the GI tract so that if one part becomes dysfunctional, the diet can be adjusted accordingly.

The Myth of "Food Combining" The idea that people should not eat certain food combinations (for example, fruit and meat) at the same meal, because the digestive system cannot handle more than one task at a time, is a myth. The art of "food combining" (which actually emphasizes "food separating") is based on this idea, and it represents faulty logic and a gross underestimation of the body's capabilities. In fact, the contrary is often true; foods eaten together can enhance each

FIGURE 3–10 The Small Intestinal Villi

Absorption of nutrients into intestinal cells typically occurs by simple diffusion or active transport.

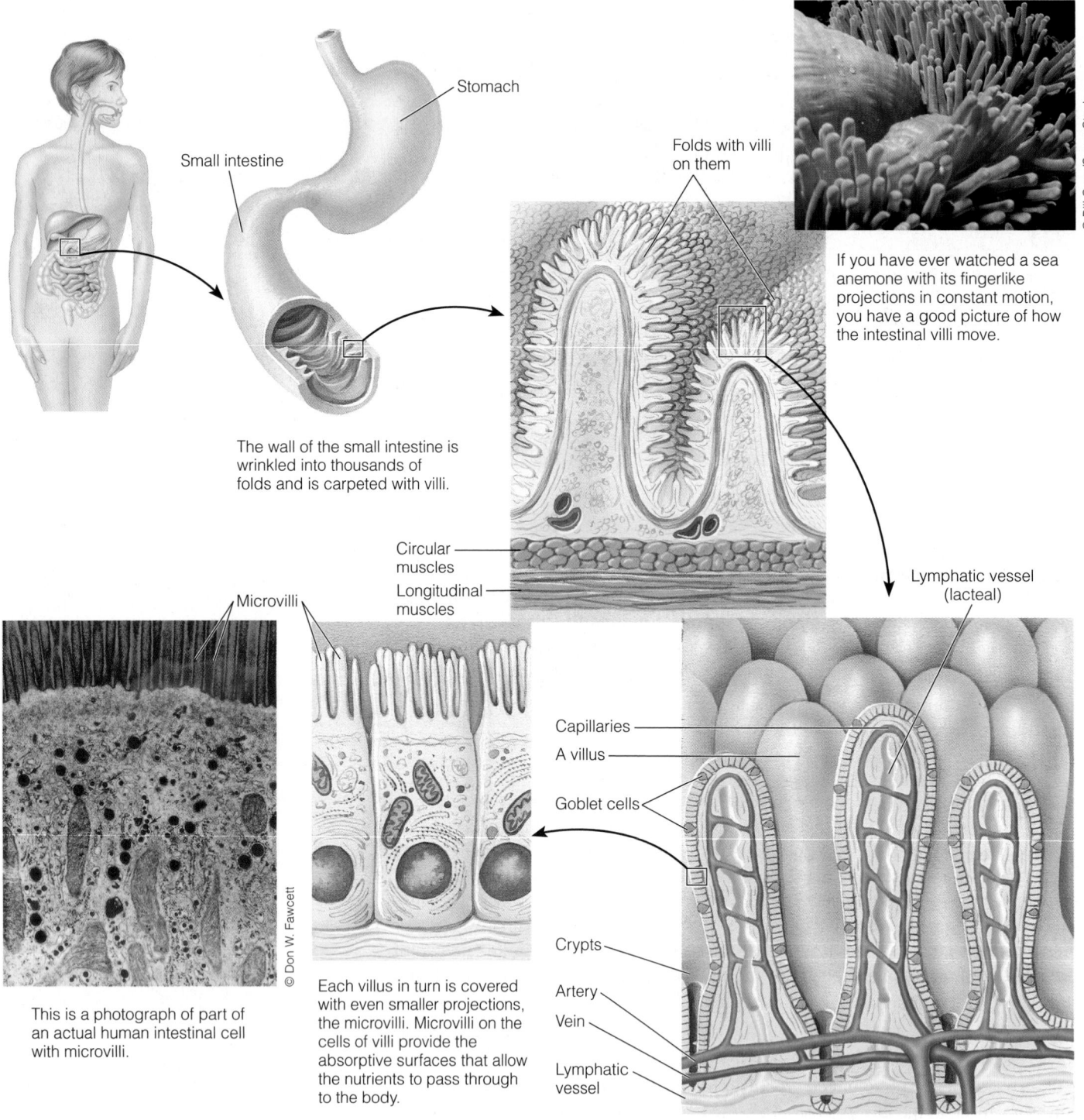

Stomach

Small intestine

Folds with villi on them

© Bill Crew/Super Stock

If you have ever watched a sea anemone with its fingerlike projections in constant motion, you have a good picture of how the intestinal villi move.

The wall of the small intestine is wrinkled into thousands of folds and is carpeted with villi.

Circular muscles

Longitudinal muscles

Lymphatic vessel (lacteal)

Microvilli

© Don W. Fawcett

This is a photograph of part of an actual human intestinal cell with microvilli.

Each villus in turn is covered with even smaller projections, the microvilli. Microvilli on the cells of villi provide the absorptive surfaces that allow the nutrients to pass through to the body.

Capillaries

A villus

Goblet cells

Crypts

Artery

Vein

Lymphatic vessel

other's use by the body. For example, vitamin C in a pineapple or other citrus fruit can enhance the absorption of iron from a meal of chicken and rice or other iron-containing foods. Many other instances of mutually beneficial interactions are presented in later chapters.

Preparing Nutrients for Transport When a nutrient molecule has crossed the cell of a villus, it enters either the bloodstream or the lymphatic system. Both transport systems supply vessels to each villus, as shown in Figure 3-10. The water-soluble nutrients and the smaller products of fat digestion are released directly into the bloodstream and guided directly to the liver where their fate and destination will be determined.

The larger fats and the fat-soluble vitamins are insoluble in water, however, and blood is mostly water. The intestinal cells assemble many of the products of fat digestion into larger molecules. These larger molecules cluster together with special proteins, forming chylomicrons. ◆ Because these chylomicrons cannot pass into the capillaries, they are released into the lymphatic system instead; the chylomicrons move through the lymph and later enter the bloodstream at a point near the heart, thus bypassing the liver at first. Details follow.

◆ Chylomicrons (kye-lo-MY-cronz) are described in Chapter 5.

IN SUMMARY

The many folds and villi of the small intestine dramatically increase its surface area, facilitating nutrient absorption. Nutrients pass through the cells of the villi and enter either the blood (if they are water soluble or small fat fragments) or the lymph (if they are fat soluble).

The Circulatory Systems

Once a nutrient has entered the bloodstream, it may be transported to any of the cells in the body, from the tips of the toes to the roots of the hair. The circulatory systems deliver nutrients wherever they are needed.

The Vascular System

The vascular, or blood circulatory, system is a closed system of vessels through which blood flows continuously, with the heart serving as the pump (see Figure 3-11, p. 84). As the blood circulates through this system, it picks up and delivers materials as needed.

All the body tissues derive oxygen and nutrients from the blood and deposit carbon dioxide and other wastes back into the blood. The lungs exchange carbon dioxide (which leaves the blood to be exhaled) and oxygen (which enters the blood to be delivered to all cells). The digestive system supplies the nutrients to be picked up. In the kidneys, wastes other than carbon dioxide are filtered out of the blood to be excreted in the urine.

Blood leaving the right side of the heart circulates through the lungs and then back to the left side of the heart. The left side of the heart then pumps the blood out of the **aorta** through **arteries** to all systems of the body. The blood circulates in the **capillaries**, where it exchanges material with the cells and then collects into **veins**, which return it again to the right side of the heart. In short, blood travels this simple route:

- Heart to arteries to capillaries to veins to heart

The routing of the blood leaving the digestive system has a special feature. The blood is carried to the digestive system (as to all organs) by way of an artery, which (as in all organs) branches into capillaries to reach every cell. Blood leaving the digestive system, however, goes by way of a vein. The **hepatic portal vein** directs blood not back to the heart, but to another organ—the liver. This vein *again* branches into *capillaries* so that every cell of the liver has access to the blood. Blood leaving the liver then *again* collects into a vein, called the **hepatic vein,** which returns blood to the heart.

The route is:

- Heart to arteries to capillaries (in intestines) to hepatic portal vein to capillaries (in liver) to hepatic vein to heart

aorta (ay-OR-tuh): the large, primary artery that conducts blood from the heart to the body's smaller arteries.

arteries: vessels that carry blood from the heart to the tissues.

capillaries (CAP-ill-aries): small vessels that branch from an artery. Capillaries connect arteries to veins. Exchange of oxygen, nutrients, and waste materials takes place across capillary walls.

veins (VANES): vessels that carry blood to the heart.

hepatic portal vein: the vein that collects blood from the GI tract and conducts it to capillaries in the liver.
- **portal** = gateway

hepatic vein: the vein that collects blood from the liver capillaries and returns it to the heart.
- **hepatic** = liver

FIGURE 3–11 *Animated!* The Vascular System

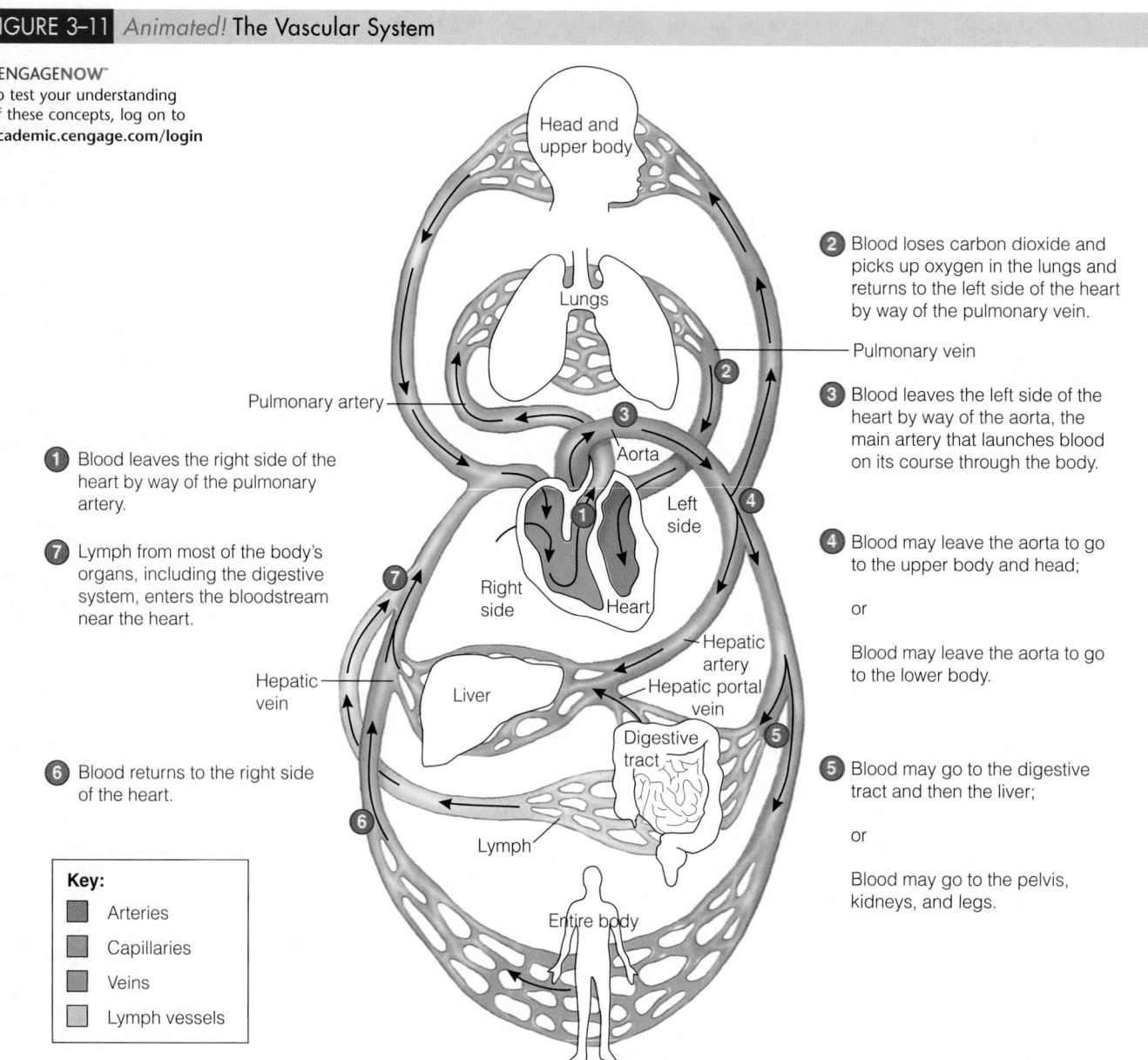

Head and upper body

Lungs

Pulmonary artery

Aorta

Left side

Right side

Heart

Hepatic vein

Liver

Hepatic artery

Hepatic portal vein

Digestive tract

Lymph

Entire body

2 Blood loses carbon dioxide and picks up oxygen in the lungs and returns to the left side of the heart by way of the pulmonary vein.

Pulmonary vein

3 Blood leaves the left side of the heart by way of the aorta, the main artery that launches blood on its course through the body.

4 Blood may leave the aorta to go to the upper body and head;

or

Blood may leave the aorta to go to the lower body.

5 Blood may go to the digestive tract and then the liver;

or

Blood may go to the pelvis, kidneys, and legs.

1 Blood leaves the right side of the heart by way of the pulmonary artery.

7 Lymph from most of the body's organs, including the digestive system, enters the bloodstream near the heart.

6 Blood returns to the right side of the heart.

Key:

■ Arteries

■ Capillaries

■ Veins

□ Lymph vessels

Figure 3-12 shows the liver's key position in nutrient transport. An anatomist studying this system knows there must be a reason for this special arrangement. The liver's placement ensures that it will be first to receive the nutrients absorbed from the GI tract. In fact, the liver has many jobs to do in preparing the absorbed nutrients for use by the body. It is the body's major metabolic organ.

You might guess that, in addition, the liver serves as a gatekeeper to defend against substances that might harm the heart or brain. This is why, when people ingest poisons that succeed in passing the first barrier (the intestinal cells), the liver quite often suffers the damage—from viruses such as hepatitis, from drugs such as barbiturates or alcohol, from toxins such as pesticide residues, and from contaminants such as mercury. Perhaps, in fact, you have been undervaluing your liver, not knowing what heroic tasks it quietly performs for you.

The Lymphatic System

The **lymphatic system** provides a one-way route for fluid from the tissue spaces to enter the blood. Unlike the vascular system, the lymphatic system has

lymphatic (lim-FAT-ic) **system:** a loosely organized system of vessels and ducts that convey fluids toward the heart. The GI part of the lymphatic system carries the products of fat digestion into the bloodstream.

FIGURE 3–12 The Liver

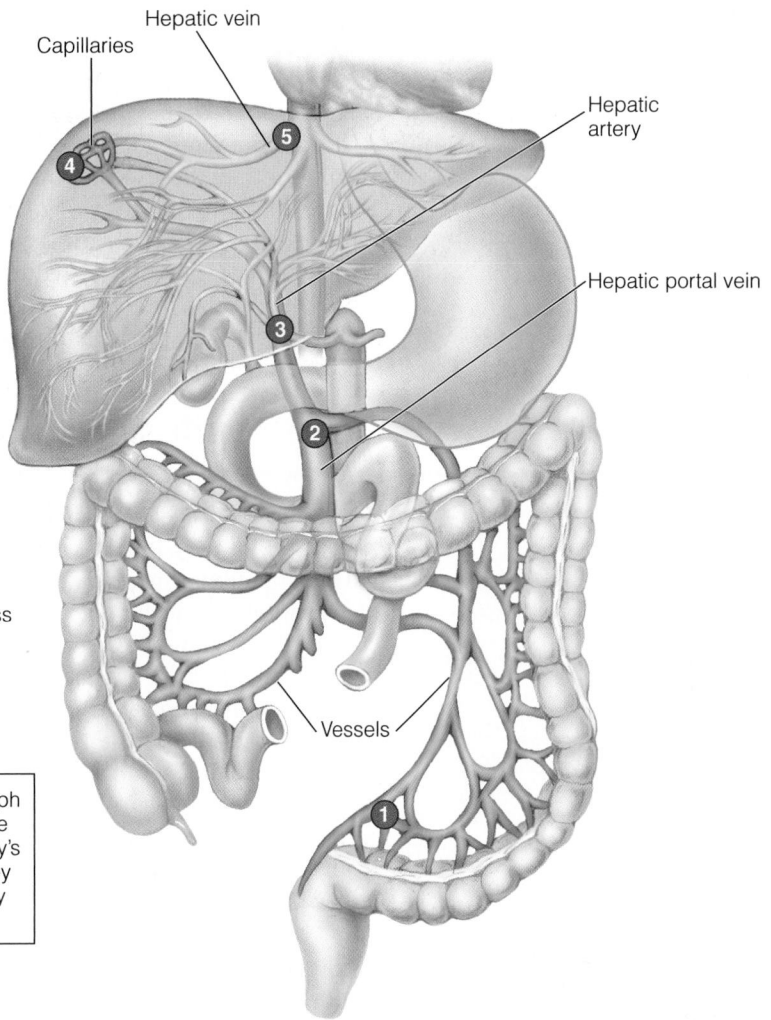

① Vessels gather up nutrients and reabsorbed water and salts from all over the digestive tract.

> Not shown here:
> Parallel to these vessels (veins) are other vessels (arteries) that carry oxygen-rich blood from the heart to the intestines.

② The vessels merge into the hepatic portal vein, which conducts all absorbed materials to the liver.

③ The hepatic artery brings a supply of freshly oxygenated blood (not loaded with nutrients) from the lungs to supply oxygen to the liver's own cells.

④ Capillaries branch all over the liver, making nutrients and oxygen available to all its cells and giving the cells access to blood from the digestive system.

⑤ The hepatic vein gathers up blood in the liver and returns it to the heart.

> In contrast, nutrients absorbed into lymph do not go to the liver first. They go to the heart, which pumps them to all the body's cells. The cells remove the nutrients they need, and the liver then has to deal only with the remnants.

Labels on figure: Capillaries, Hepatic vein, Hepatic artery, Hepatic portal vein, Vessels

no pump; instead, **lymph** circulates *between* the cells of the body and collects into tiny vessels. The fluid moves from one portion of the body to another as muscles contract and create pressure here and there. Ultimately, much of the lymph collects in the **thoracic duct** behind the heart. The thoracic duct opens into the **subclavian vein,** where the lymph enters the bloodstream. Thus nutrients from the GI tract that enter lymphatic vessels ◆ (large fats and fat-soluble vitamins) ultimately enter the bloodstream, circulating through arteries, capillaries, and veins like the other nutrients, with a notable exception—they bypass the liver at first.

Once inside the vascular system, the nutrients can travel freely to any destination and can be taken into cells and used as needed. What becomes of them is described in later chapters.

◆ The lymphatic vessels of the intestine that take up nutrients and pass them to the lymph circulation are called **lacteals** (LACK-tee-als).

lymph (LIMF): a clear yellowish fluid that is similar to blood except that it contains no red blood cells or platelets. Lymph from the GI tract transports fat and fat-soluble vitamins to the bloodstream via lymphatic vessels.

thoracic (thor-ASS-ic) **duct:** the main lymphatic vessel that collects lymph and drains into the left subclavian vein.

subclavian (sub-KLAY-vee-an) **vein:** the vein that provides passageway from the lymphatic system to the vascular system.

IN SUMMARY

Nutrients leaving the digestive system via the blood are routed directly to the liver before being transported to the body's cells. Those leaving via the lymphatic system eventually enter the vascular system but bypass the liver at first.

The Health and Regulation of the GI Tract

◆ Factors influencing GI function:
- Physical immaturity
- Aging
- Illness
- Nutrition

This section describes the bacterial conditions and hormonal regulation of a healthy GI tract, but many factors ◆ can influence normal GI function. For example, peristalsis and sphincter action are poorly coordinated in newborns, so infants tend to "spit up" during the first several months of life. Older adults often experience constipation, in part because the intestinal wall loses strength and elasticity with age, which slows GI motility. Diseases can also interfere with digestion and absorption and often lead to malnutrition. Lack of nourishment, in general, and lack of certain dietary constituents such as fiber, in particular, alter the structure and function of GI cells. Quite simply, GI tract health depends on adequate nutrition.

Gastrointestinal Bacteria

◆ Bacteria in the intestines are sometimes referred to as **flora** or **microflora.**

An estimated 10 trillion bacteria ◆ representing some 400 or more different species and subspecies live in a healthy GI tract. The prevalence of different bacteria in various parts of the GI tract depends on such factors as pH, peristalsis, diet, and other microorganisms. Relatively few microorganisms can live in the low pH of the stomach with its relatively rapid peristalsis, whereas the neutral pH and slow peristalsis of the lower small intestine and the large intestine permit the growth of a diverse and abundant bacterial population.[1]

Most of these bacteria normally do the body no harm and may actually do some good. Provided that the normal intestinal flora are thriving, infectious bacteria have a hard time establishing themselves to launch an attack on the system.

Diet is one of several factors that influence the body's bacterial population and environment. Consider **yogurt,** for example.[2] Yogurt contains *Lactobacillus* and other living bacteria. These microorganisms are considered **probiotics** because they change the conditions and native bacterial colonies in the GI tract in ways that seem to benefit health.[3] The potential GI health benefits of probiotics include helping to alleviate diarrhea, constipation, inflammatory bowel disease, ulcers, allergies, and lactose intolerance; enhance immune function; and protect against colon cancer.[4] Some probiotics may have adverse effects under certain circumstances.[5] Research studies continue to explore how diet influences GI bacteria and which foods—with their probiotics—affect GI health.

◆ Food components (such as fibers) that are not digested in the small intestine, but are used instead as food by bacteria to encourage their growth are called **prebiotics.**

GI bacteria also digest fibers and complex proteins.[6] ◆ In doing so, the bacteria produce nutrients such as short fragments of fat that the cells of the colon use for energy. Bacteria in the GI tract also produce several vitamins, ◆ including a significant amount of vitamin K, although the amount is insufficient to meet the body's total need for that vitamin.

◆ Vitamins produced by bacteria include:
- Biotin
- Folate
- Vitamin B_6
- Vitamin B_{12}
- Vitamin K

yogurt: milk product that results from the fermentation of lactic acid in milk by *Lactobacillus bulgaricus* and *Streptococcus thermophilus.*

probiotics: living microorganisms found in foods that, when consumed in sufficient quantities, are beneficial to health.
- **pro** = for
- **bios** = life

homeostasis (HOME-ee-oh-STAY-sis): the maintenance of constant internal conditions (such as blood chemistry, temperature, and blood pressure) by the body's control systems. A homeostatic system is constantly reacting to external forces to maintain limits set by the body's needs.
- **homeo** = the same
- **stasis** = staying

Gastrointestinal Hormones and Nerve Pathways

The ability of the digestive tract to handle its ever-changing contents routinely illustrates an important physiological principle that governs the way all living things function—the principle of **homeostasis.** Simply stated, survival depends on body conditions staying about the same; if they deviate too far from the norm, the body must "do something" to bring them back to normal. The body's regulation of digestion is one example of homeostatic regulation. The body also regulates its temperature, its blood pressure, and all other aspects of its blood chemistry in similar ways.

Two intricate and sensitive systems coordinate all the digestive and absorptive processes: the hormonal (or endocrine) system and the nervous system. Even before the first bite of food is taken, the mere thought, sight, or smell of food can trigger a response from these systems. Then, as food travels through the GI tract, it either stimulates or inhibits digestive secretions by way of messages that are carried from one section of the GI tract to another by both **hormones** ◆ and nerve pathways. (Appendix A presents a brief summary of the body's hormonal system and nervous system.)

Notice that the kinds of regulation described next are all examples of *feedback* mechanisms. A certain condition demands a response. The response changes that condition, and the change then cuts off the response. Thus the system is self-correcting. Examples follow:

- *The stomach normally maintains a pH between 1.5 and 1.7. How does it stay that way?* Food entering the stomach stimulates cells in the stomach wall to release the hormone **gastrin.** Gastrin, in turn, stimulates the stomach glands to secrete the components of hydrochloric acid. When pH 1.5 is reached, the acid itself turns off the gastrin-producing cells. They stop releasing gastrin, and the glands stop producing hydrochloric acid. Thus the system adjusts itself.

  Nerve receptors in the stomach wall also respond to the presence of food and stimulate the gastric glands to secrete juices and the muscles to contract. As the stomach empties, the receptors are no longer stimulated, the flow of juices slows, and the stomach quiets down.

- *The pyloric sphincter opens to let out a little chyme, then closes again. How does it know when to open and close?* When the pyloric sphincter relaxes, acidic chyme slips through. The cells of the pyloric muscle on the intestinal side sense the acid, causing the pyloric sphincter to close tightly. Only after the chyme has been neutralized by pancreatic bicarbonate and the juices surrounding the pyloric sphincter have become alkaline can the muscle relax again. This process ensures that the chyme will be released slowly enough to be neutralized as it flows through the small intestine. This is important because the small intestine has less of a mucous coating than the stomach does and so is not as well protected from acid.

- *As the chyme enters the intestine, the pancreas adds bicarbonate to it so that the intestinal contents always remain at a slightly alkaline pH. How does the pancreas know how much to add?* The presence of chyme stimulates the cells of the duodenum wall to release the hormone **secretin** into the blood. When secretin reaches the pancreas, it stimulates the pancreas to release its bicarbonate-rich juices. Thus, whenever the duodenum signals that acidic chyme is present, the pancreas responds by sending bicarbonate to neutralize it. When the need has been met, the cells of the duodenum wall are no longer stimulated to release secretin, the hormone no longer flows through the blood, the pancreas no longer receives the message, and it stops sending pancreatic juice. Nerves also regulate pancreatic secretions.

- *Pancreatic secretions contain a mixture of enzymes to digest carbohydrate, fat, and protein. How does the pancreas know how much of each type of enzyme to provide?* This is one of the most interesting questions physiologists have asked. Clearly, the pancreas does know what its owner has been eating, and it secretes enzyme mixtures tailored to handle the food mixtures that have been arriving recently (over the last several days). Enzyme activity changes proportionately in response to the amounts of carbohydrate, fat, and protein in the diet. If a person has been eating mostly carbohydrates, the pancreas makes and secretes mostly carbohydrases; if the person's diet has been high in fat, the pancreas produces more lipases; and so forth. Presumably, hormones from the GI tract, secreted in response to meals, keep the pancreas informed as to its digestive tasks. The day or two lag between the time a person's diet changes dramatically and the time digestion of the new diet becomes efficient explains why dietary changes can "upset digestion" and should be made gradually.

◆ In general, any gastrointestinal hormone may be called an **enterogastrone** (EN-ter-oh-GAS-trone), but the term refers specifically to any hormone that slows motility and inhibits gastric secretions.

hormones: chemical messengers. Hormones are secreted by a variety of glands in response to altered conditions in the body. Each hormone travels to one or more specific target tissues or organs, where it elicits a specific response to maintain homeostasis.

gastrin: a hormone secreted by cells in the stomach wall. Target organ: the glands of the stomach. Response: secretion of gastric acid.

secretin (see-CREET-in): a hormone produced by cells in the duodenum wall. Target organ: the pancreas. Response: secretion of bicarbonate-rich pancreatic juice.

◆ The inactive precursor of an enzyme is called a **proenzyme** or **zymogen** (ZYE-mo-jen).
- **pro** = before
- **zym** = concerning enzymes
- **gen** = to produce

• *Why don't the digestive enzymes damage the pancreas?* The pancreas protects itself from harm by producing an inactive form of the enzymes. ◆ It releases these proteins into the small intestine where they are activated to become enzymes. In pancreatitis, the digestive enzymes become active within the infected pancreas, causing inflammation and damaging the delicate pancreatic tissues.

• *When fat is present in the intestine, the gallbladder contracts to squirt bile into the intestine to emulsify the fat. How does the gallbladder get the message that fat is present?* Fat in the intestine stimulates cells of the intestinal wall to release the hormone **cholecystokinin (CCK).** This hormone, traveling by way of the blood to the gallbladder, stimulates it to contract, releasing bile into the small intestine. Cholecystokinin also travels to the pancreas, stimulates it to secrete its juices, releasing bicarbonate and enzymes into the small intestine. Once the fat in the intestine is emulsified and enzymes have begun to work on it, the fat no longer provokes release of the hormone, and the message to contract is canceled. (By the way, fat emulsification can continue even after a diseased gallbladder has been surgically removed because the liver can deliver bile directly to the small intestine.)

• *Fat and protein take longer to digest than carbohydrate does. When fat or protein is present, intestinal motility slows to allow time for its digestion. How does the intestine know when to slow down?* Cholecystokinin is released in response to fat or protein in the small intestine. In addition to its role in fat emulsification and digestion, cholecystokinin slows GI tract motility. Slowing the digestive process helps to maintain a pace that allows all reactions to reach completion. Hormonal and nervous mechanisms like these account for much of the body's ability to adapt to changing conditions.

Table 3-1 summarizes the actions of these GI hormones.

Once a person has started to learn the answers to questions like these, it may be hard to stop. Some people devote their whole lives to the study of physiology. For now, however, these few examples illustrate how all the processes throughout the digestive system are precisely and automatically regulated without any conscious effort.

IN SUMMARY

A diverse and abundant bacteria population support GI health. The regulation of GI processes depends on the coordinated efforts of the hormonal system and the nervous system; together, digestion and absorption transform foods into nutrients.

cholecystokinin (COAL-ee-SIS-toe-KINE-in), or **CCK:** a hormone produced by cells of the intestinal wall. Target organ: the gallbladder. Response: release of bile and slowing of GI motility.

The System at Its Best

This chapter describes the anatomy of the digestive tract on several levels: the sequence of digestive organs, the cells and structures of the villi, and the selective ma-

TABLE 3-1	The Primary Actions of GI Hormones			
Hormone:	Responds to:	Secreted from:	Stimulates:	Response:
Gastrin	Food in the stomach	Stomach wall	Stomach glands	Hydrochloric acid secreted into the stomach
Secretin	Acidic chyme in the small intestine	Duodenal wall	Pancreas	Bicarbonate-rich juices secreted into the small intestine
Cholecystokinin	Fat or protein in the small intestine	Intestinal wall	Gallbladder	Bile secreted into the duodenum
			Pancreas	Bicarbonate- and enzyme-rich juices secreted into the small intestine

chinery of the cell membranes. The intricate architecture of the digestive system makes it sensitive and responsive to conditions in its environment. Several different kinds of GI tract cells confer specific immunity against intestinal diseases such as inflammatory bowel disease. In addition, secretions from the GI tract—saliva, mucus, gastric acid, and digestive enzymes—not only help with digestion, but also defend against foreign invaders. Together the GI's team of bacteria, cells, and secretions defend the body against numerous challenges.[7] Knowing the optimal conditions will help you to make choices that promote the best functioning of the system.

Nourishing foods and pleasant conversations support a healthy digestive system.

One indispensable condition is good health of the digestive tract itself. This health is affected by such lifestyle factors as sleep, physical activity, and state of mind. Adequate sleep allows for repair and maintenance of tissue and removal of wastes that might impair efficient functioning. Activity promotes healthy muscle tone. Mental state influences the activity of regulatory nerves and hormones; for healthy digestion, you should be relaxed and tranquil at mealtimes.

Another factor in GI health is the kind of meals you eat. Among the characteristics of meals that promote optimal absorption of nutrients are those mentioned in Chapter 2: balance, moderation, variety, and adequacy. Balance and moderation require having neither too much nor too little of anything. For example, too much fat can be harmful, but some fat is beneficial in slowing down intestinal motility and providing time for absorption of some of the nutrients that are slow to be absorbed.

Variety is important for many reasons, but one is that some food constituents interfere with nutrient absorption. For example, some compounds common in high-fiber foods such as whole-grain cereals, certain leafy green vegetables, and legumes bind with minerals. To some extent, then, the minerals in those foods may become unavailable for absorption. These high-fiber foods are still valuable, but they need to be balanced with a variety of other foods that can provide the minerals.

As for adequacy—in a sense, this entire book is about dietary adequacy. But here, at the end of this chapter, is a good place to underline the interdependence of the nutrients. It could almost be said that every nutrient depends on every other. All the nutrients work together, and all are present in the cells of a healthy digestive tract. To maintain health and promote the functions of the GI tract, you should make balance, moderation, variety, and adequacy features of every day's menus.

Nutrition Portfolio

CENGAGENOW™
academic.cengage.com/login

A healthy digestive system can adjust to almost any diet and can handle any combination of foods with ease.

▪ Describe the physical and emotional environment that typically surrounds your meals, including how it affects you and how it might be improved.

▪ Detail any GI discomforts you may experience regularly and include suggestions to alleviate or prevent their occurrence (see Highlight 3).

▪ List any changes you can make in your eating habits to promote overall GI health.

NUTRITION ON THE NET

For further study of topics covered in this chapter, log on to **academic.cengage .com/nutrition/rolfes/UNCN8e**. Go to Chapter 3, then to Nutrition on the Net.

- Visit the Center for Digestive Health and Nutrition: **www.gihealth.com**

- Visit the patient information section of the American College of Gastroenterology: **www.acg.gi.org**

STUDY QUESTIONS

CENGAGE**NOW**

To assess your understanding of chapter topics, take the Student Practice Test and explore the modules recommended in your Personalized Study Plan. Log on to **academic.cengage.com/login.**

These questions will help you review this chapter. You will find the answers in the discussions on the pages provided.

1. Describe the challenges associated with digesting food and the solutions offered by the human body. (pp. 71–80)

2. Describe the path food follows as it travels through the digestive system. Summarize the muscular actions that take place along the way. (pp. 72–76)

3. Name five organs that secrete digestive juices. How do the juices and enzymes facilitate digestion? (pp. 76–78)

4. Describe the problems associated with absorbing nutrients and the solutions offered by the small intestine. (pp. 80–83)

5. How is blood routed through the digestive system? Which nutrients enter the bloodstream directly? Which are first absorbed into the lymph? (pp. 83–85)

6. Describe how the body coordinates and regulates the processes of digestion and absorption. (pp. 86–88)

7. How does the composition of the diet influence the functioning of the GI tract? (p. 89)

8. What steps can you take to help your GI tract function at its best? (p. 89)

These multiple choice questions will help you prepare for an exam. Answers can be found on p. 91.

1. The semiliquid, partially digested food that travels through the intestinal tract is called:
   a. bile.
   b. lymph.
   c. chyme.
   d. secretin.

2. The muscular contractions that move food through the GI tract are called:
   a. hydrolysis.
   b. sphincters.
   c. peristalsis.
   d. bowel movements.

3. The main function of bile is to:
   a. emulsify fats.

   b. catalyze hydrolysis.
   c. slow protein digestion.
   d. neutralize stomach acidity.

4. The pancreas neutralizes stomach acid in the small intestine by secreting:
   a. bile.
   b. mucus.
   c. enzymes.
   d. bicarbonate.

5. Which nutrient passes through the GI tract mostly undigested and unabsorbed?
   a. fat
   b. fiber
   c. protein
   d. carbohydrate

6. Absorption occurs primarily in the:
   a. mouth.
   b. stomach.
   c. small intestine.
   d. large intestine.

7. All blood leaving the GI tract travels first to the:
   a. heart.
   b. liver.
   c. kidneys.
   d. pancreas.

8. Which nutrients leave the GI tract by way of the lymphatic system?
   a. water and minerals
   b. proteins and minerals
   c. all vitamins and minerals
   d. fats and fat-soluble vitamins

9. Digestion and absorption are coordinated by the:
   a. pancreas and kidneys.
   b. liver and gallbladder.
   c. hormonal system and the nervous system.
   d. vascular system and the lymphatic system.

10. Gastrin, secretin, and cholecystokinin are examples of:
    a. crypts.
    b. enzymes.
    c. hormones.
    d. goblet cells.

REFERENCES

1. P. B. Eckburg and coauthors, Diversity of the human intestinal microbial flora, *Science* 308 (2005): 1635–1638; W. L. Hao and Y. K. Lee, Microflora of the gastrointestinal tract: A review, *Methods in Molecular Biology* 268 (2004): 491–502.

2. O. Adolfsson, S. N. Meydani, and R. M. Russell, Yogurt and gut function, *American Journal of Clinical Nutrition* 80 (2004): 245–256.

3. C. C. Chen and W. A. Walker, Probiotics and prebiotics: Role in clinical disease states, *Advances in Pediatrics* 52 (2005): 77–113; M. E. Sanders, Probiotics: Considerations for human health, *Nutrition Reviews* 61 (2003): 91–99; M. H. Floch and J. Hong-Curtiss, Probiotics and functional foods in gastrointestinal disorders, *Current Gastroenterology Reports* 3 (2001): 343–350; Probiotics and prebiotics, *American Journal of Clinical Nutrition (supp.)* 73 (2001): entire issue.

4. S. Santosa, E. Farnworth, and P. J. H. Jones, Probiotics and their potential health claims, *Nutrition Reviews* 64 (2006): 265–274; S. J. Salminen, M. Gueimonde, and E. Isolauri, Probiotics that modify disease risk, *American Society for Nutritional Sciences* 135 (2005): 1294–1298; F. Guarner and coauthors, Should yoghurt cultures be considered probiotic? *British Journal of Nutrition* 93 (2005): 783–786; J. M. Saavedra and A. Tschernia, Human studies with probiotics and prebiotics: Clinical implications, *British Journal of Nutrition* 87 (2002): S241–S246; P. Marteau and M. C. Boutron-Ruault, Nutritional advantages of probiotics and prebiotics, *British Journal of Nutrition* 87 (2002): S153–S157; G. T. Macfarlane and J. H. Cummings, Probiotics, infection and immunity, *Current Opinion in Infectious Diseases* 15 (2002): 501–506; L. Kopp-Hoolihan, Prophylactic and therapeutic uses of probiotics: A review, *Journal of the American Dietetic Association* 101 (2001): 229–238; M. B. Roberfroid, Prebiotics and probiotics: Are they functional foods? *American Journal of Clinical Nutrition* 71 (2000): 1682S–1687S.

5. J. Ezendam and H. van Loveren, Probiotics: Immunomodulation and evaluation of safety and efficacy, *Nutrition Reviews* 64 (2006): 1–14.

6. J. M. Wong and coauthors, Colonic health: Fermentation and short chain fatty acids, *Journal of Clinical Gastroenterology* 40 (2006): 235–243; S. Bengmark, Colonic food: Pre- and probiotics, *American Journal of Gastroenterology* 95 (2000): S5–S7.

7. P. Bourlioux and coauthors, The intestine and its microflora are partners for the protection of the host: Report on the Danone Symposium "The Intelligent Intestine," held in Paris, June 14, 2002, *American Journal of Clinical Nutrition* 78 (2003): 675–683.

ANSWERS

Study Questions (multiple choice)

1. c   2. c   3. a   4. d   5. b   6. c   7. b   8. d
9. c   10. c

Common Digestive Problems

© Corbis

The facts of anatomy and physiology presented in Chapter 3 permit easy understanding of some common problems that occasionally arise in the digestive tract. Food may slip into the air passages instead of the esophagus, causing choking. Bowel movements may be loose and watery, as in diarrhea, or painful and hard, as in constipation. Some people complain about belching, while others are bothered by intestinal gas. Sometimes people develop medical problems such as an ulcer. This highlight describes some of the symptoms of these common digestive problems and suggests strategies for preventing them (the glossary on p. 94 defines the relevant terms).

Choking

A person chokes when a piece of food slips into the **trachea** and becomes lodged so securely that it cuts off breathing (see Figure H3-1). Without oxygen, the person may suffer brain damage or die. For this reason, it is imperative that everyone learns to recognize a person grabbing his or her own throat as the international signal for choking (shown in Figure H3-2) and act promptly.

The choking scenario might read like this. A person is dining in a restaurant with friends. A chunk of food, usually meat, becomes lodged in his trachea so firmly that he cannot make a sound. No sound can be made because the **larynx** is in the trachea and makes sounds only when air is pushed across it. Often he chooses to suffer alone rather than "make a scene in public." If he tries to communicate distress to his friends, he must depend on pantomime. The friends are bewildered by his antics and become terribly worried when he "faints" after a few minutes without air. They call for an ambulance, but by the time it arrives, he is dead from suffocation.

To help a person who is choking, first ask this critical question: "Can you make any sound at all?" If so, relax. You have time to decide what you can do to help. Whatever you do, *do not* hit him on the back—the particle may become lodged more firmly in his air passage. If the person cannot make a sound, shout for help and perform the **Heimlich maneuver** (described in Figure H3-2). You would do well to take a life-saving course and practice these techniques because you will have no time for hesitation if you are called upon to perform this death-defying act.

Almost any food can cause choking, although some are cited more often than others: chunks of meat, hot dogs, nuts, whole grapes, raw carrots, marshmallows, hard or sticky candies, gum, popcorn, and peanut butter. These foods are particularly difficult for young children to safely chew and swallow. In 2000, more than 17,500 children (under 15 years old) in the United States choked; most of them choked on food, and 160 of them choked to death.[1] Always remain alert to the dangers of choking whenever young children are eating. To prevent choking, cut food into small pieces, chew thoroughly before swallowing, don't talk or laugh with food in your mouth, and don't eat when breathing hard.

Vomiting

Another common digestive mishap is **vomiting.** Vomiting can be a symptom of many different diseases or may arise in situations that upset the body's equilibrium, such as air or sea travel. For whatever reason, the contents of the stomach are propelled up through the esophagus to the mouth and expelled.

| FIGURE H3-1 | Normal Swallowing and Choking |

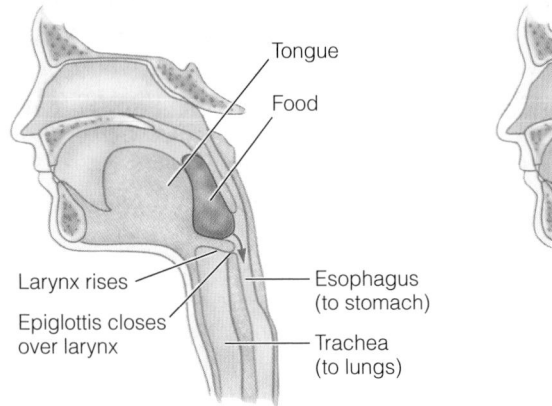
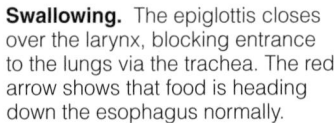
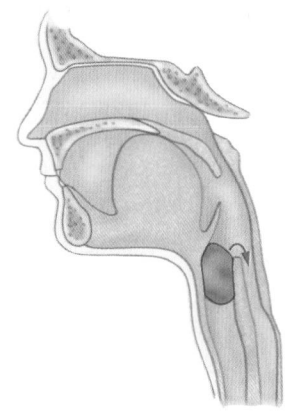

Swallowing. The epiglottis closes over the larynx, blocking entrance to the lungs via the trachea. The red arrow shows that food is heading down the esophagus normally.

Choking. A choking person cannot speak or gasp because food lodged in the trachea blocks the passage of air. The red arrow points to where the food should have gone to prevent choking.

FIGURE H3-2 First Aid for Choking

The first-aid strategy most likely to succeed is abdominal thrusts, sometimes called the Heimlich maneuver. Only if all else fails, open the person's mouth by grasping both his tongue and lower jaw and lifting. Then, and *only if* you can see the object, use your finger to sweep it out and begin rescue breathing.

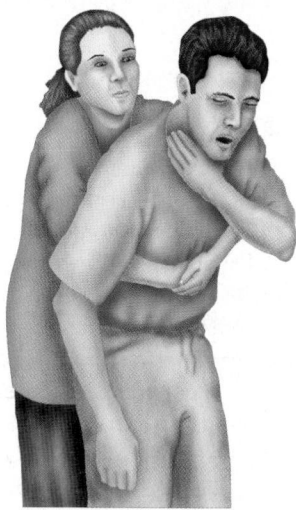

The universal signal for choking is when a person grabs his throat. It alerts others to the need for assistance. If this happens, stand behind the person, and wrap your arms around him. Place the thumb side of one fist snugly against his body, slightly above the navel and below the rib cage. Grasp your fist with your other hand and give him a sudden strong hug inward and upward. Repeat thrusts as necessary.

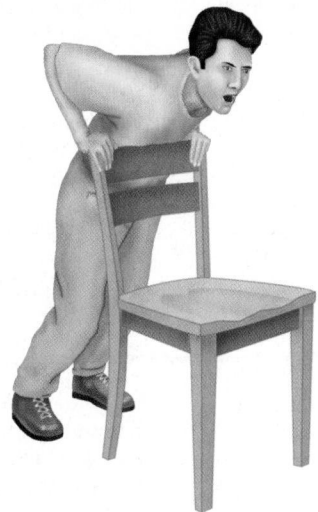

If you are choking and need to self-administer first aid, place the thumb side of one fist slightly above your navel and below your rib cage, grasp the fist with your other hand, and then press inward and upward with a quick motion. If this is unsuccessful, quickly press your upper abdomen over any firm surface such as the back of a chair, a countertop, or a railing.

If vomiting continues long enough or is severe enough, the muscular contractions will extend beyond the stomach and carry the contents of the duodenum, with its green bile, into the stomach and then up the esophagus. Although certainly unpleasant and wearying for the nauseated person, vomiting such as this is no cause for alarm. Vomiting is one of the body's adaptive mechanisms to rid itself of something irritating. The best advice is to rest and drink small amounts of liquids as tolerated until the nausea subsides.

A physician's care may be needed, however, when large quantities of fluid are lost from the GI tract, causing dehydration. With massive fluid loss from the GI tract, all of the body's other fluids redistribute themselves so that, eventually, fluid is taken from every cell of the body. Leaving the cells with the fluid are salts that are absolutely essential to the life of the cells, and they must be replaced. Replacement is difficult if the vomiting continues, and intravenous feedings of saline and glucose may be necessary while the physician diagnoses the cause of the vomiting and begins corrective therapy.

In an infant, vomiting is likely to become serious early in its course, and a physician should be contacted soon after onset. Infants have more fluid between their body cells than adults do, so more fluid can move readily into the digestive tract and be lost from the body. Consequently, the body water of infants becomes depleted and their body salt balance upset faster than in adults.

Self-induced vomiting, such as occurs in bulimia nervosa, also has serious consequences. In addition to fluid and salt imbalances, repeated vomiting can cause irritation and infection of the pharynx, esophagus, and salivary glands; erosion of the teeth and gums; and dental caries. The esophagus may rupture or tear, as may the stomach. Sometimes the eyes become red from pressure during vomiting. Bulimic behavior reflects underlying psychological problems that require intervention. (Bulimia nervosa is discussed fully in Highlight 8.)

Projectile vomiting is also serious. The contents of the stomach are expelled with such force that they leave the mouth in a wide arc like a bullet leaving a gun. This type of vomiting requires immediate medical attention.

Diarrhea

Diarrhea is characterized by frequent, loose, watery stools. Such stools indicate that the intestinal contents have moved too quickly through the intestines for fluid absorption to take place, or that water has been drawn from the cells lining the intestinal tract and added to the food residue. Like vomiting, diarrhea can lead to considerable fluid and salt losses, but the composition of the fluids is different. Stomach fluids lost in vomiting are highly acidic, whereas intestinal fluids lost in diarrhea are nearly neutral. When fluid losses require medical attention, correct replacement is crucial.

Diarrhea is a symptom of various medical conditions and treatments. It may occur abruptly in a healthy person as a result of infections (such as food poisoning) or as a side effect of medications. When used in large quantities, food ingredients such as the sugar alternative sorbitol and the fat alternative olestra may also cause diarrhea in some people. If a food is responsible, then that food must be omitted from the diet, at least temporarily. If medication is responsible, a different medicine, when possible, or a different form (injectable versus oral, for example) may alleviate the problem.

Diarrhea may also occur as a result of disorders of the GI tract, such as irritable bowel syndrome or colitis. **Irritable bowel syndrome** is one of the most common GI disorders and is characterized by a disturbance in the motility of the GI tract.[2] In most cases, GI contractions are stronger and last longer than normal, forcing intestinal contents through quickly and causing gas, bloating, and diarrhea. In some cases, however, GI contractions are weaker than normal, slowing the passage of intestinal contents and causing constipation. The exact cause of irritable bowel syndrome is not known, but researchers believe nerves and hormones are involved. The condition seems to worsen for some

GLOSSARY

acid controllers: medications used to prevent or relieve indigestion by suppressing production of acid in the stomach; also called **H2 blockers.** Common brands include Pepcid AC, Tagamet HB, Zantac 75, and Axid AR.

antacids: medications used to relieve indigestion by neutralizing acid in the stomach. Common brands include Alka-Seltzer, Maalox, Rolaids, and Tums.

belching: the expulsion of gas from the stomach through the mouth.

colitis (ko-LYE-tis): inflammation of the colon.

colonic irrigation: the popular, but potentially harmful practice of "washing" the large intestine with a powerful enema machine.

constipation: the condition of having infrequent or difficult bowel movements.

defecate (DEF-uh-cate): to move the bowels and eliminate waste.
• **defaecare** = to remove dregs

diarrhea: the frequent passage of watery bowel movements.

diverticula (dye-ver-TIC-you-la): sacs or pouches that develop in the weakened areas of the intestinal wall (like bulges in an inner tube where the tire wall is weak).
• **divertir** = to turn aside

diverticulitis (DYE-ver-tic-you-LYE-tis): infected or inflamed diverticula.
• **itis** = infection or inflammation

diverticulosis (DYE-ver-tic-you-LOH-sis): the condition of having diverticula. About one in every six people in Western countries develops diverticulosis in middle or later life.
• **osis** = condition

enemas: solutions inserted into the rectum and colon to stimulate a bowel movement and empty the lower large intestine.

gastroesophageal reflux: the backflow of stomach acid into the esophagus, causing damage to the cells of the esophagus and the sensation of heartburn. **Gastroesophageal reflux disease (GERD)** is characterized by symptoms of reflux occurring two or more times a week.

heartburn: a burning sensation in the chest area caused by backflow of stomach acid into the esophagus.

Heimlich (HIME-lick) **maneuver (abdominal thrust maneuver):** a technique for dislodging an object from the trachea of a choking person (see Figure H3-2); named for the physician who developed it.

hemorrhoids (HEM-oh-royds): painful swelling of the veins surrounding the rectum.

hiccups (HICK-ups): repeated cough-like sounds and jerks that are produced when an involuntary spasm of the diaphragm muscle sucks air down the windpipe; also spelled *hiccoughs.*

indigestion: incomplete or uncomfortable digestion, usually accompanied by pain, nausea, vomiting, heartburn, intestinal gas, or belching.
• **in** = not

irritable bowel syndrome: an intestinal disorder of unknown cause. Symptoms include abdominal discomfort and cramping, diarrhea, constipation, or alternating diarrhea and constipation.

larynx: the upper part of the air passageway that contains the vocal cords; also called the voice box (see Figure H3-1).

laxatives: substances that loosen the bowels and thereby prevent or treat constipation.

mineral oil: a purified liquid derived from petroleum and used to treat constipation.

peptic ulcer: a lesion in the mucous membrane of either the stomach (a gastric ulcer) or the duodenum (a duodenal ulcer).
• **peptic** = concerning digestion

trachea (TRAKE-ee-uh): the air passageway from the larynx to the lungs; also called the *windpipe.*

ulcer: a lesion of the skin or mucous membranes characterized by inflammation and damaged tissues. See also *peptic ulcer.*

vomiting: expulsion of the contents of the stomach up through the esophagus to the mouth.

---

people when they eat certain foods or during stressful events. These triggers seem to aggravate symptoms but not cause them. Dietary treatment hinges on identifying and avoiding individual foods that aggravate symptoms; small meals may also be beneficial. People with **colitis,** an inflammation of the large intestine, may also suffer from severe diarrhea. They often benefit from complete bowel rest and medication. If treatment fails, surgery to remove the colon and rectum may be necessary.

Treatment for diarrhea depends on cause and severity, but it always begins with rehydration.[3] Mild diarrhea may subside with simple rest and extra liquids (such as clear juices and soups) to replace fluid losses. However, call a physician if diarrhea is bloody or if it worsens or persists—especially in an infant, young child, elderly person, or person with a compromised immune system. Severe diarrhea can be life threatening.

Constipation

Like diarrhea, **constipation** describes a symptom, not a disease. Each person's GI tract has its own cycle of waste elimination, which depends on its owner's health, the type of food eaten, when it was eaten, and when the person takes time to **defecate.** What's normal for some people may not be normal for others. Some people have bowel movements three times a day; others

may have them three times a week. The symptoms of constipation include straining during bowel movements, hard stools, and infrequent bowel movements (fewer than three per week).[4] Ab-

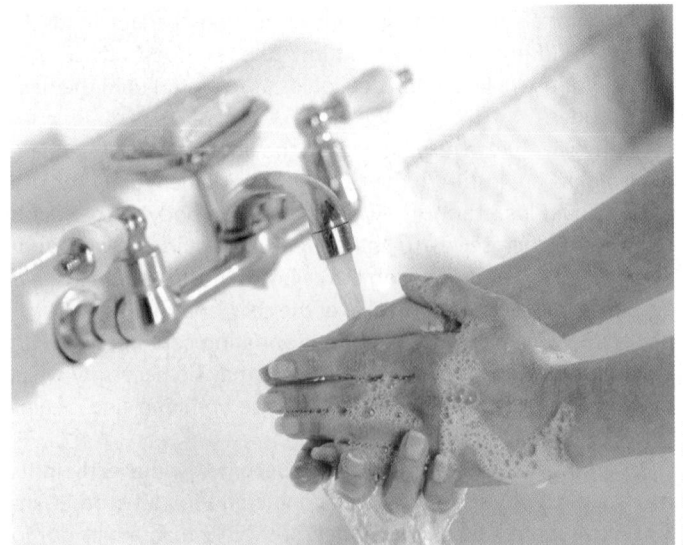

Personal hygiene (such as regular hand washing with soap and water) and safe food preparation (as described in Highlight 18) are easy and effective steps to take in preventing diarrheal diseases.

© Ariel Skelley/Corbis

dominal discomfort, headaches, backaches, and the passing of gas sometimes accompany constipation.

Often a person's lifestyle may cause constipation. Being too busy to respond to the defecation signal is a common complaint. If a person receives the signal to defecate and ignores it, the signal may not return for several hours. In the meantime, water continues to be withdrawn from the fecal matter, so when the person does defecate, the stools are dry and hard. In such a case, a person's daily regimen may need to be revised to allow time to have a bowel movement when the body sends its signal. One possibility is to go to bed earlier in order to rise earlier, allowing ample time for a leisurely breakfast and a movement.

Although constipation usually reflects lifestyle habits, in some cases it may be a side effect of medication or may reflect a medical problem such as tumors that are obstructing the passage of waste. If discomfort is associated with passing fecal matter, seek medical advice to rule out disease. Once this has been done, dietary or other measures for correction can be considered.

One dietary measure that may be appropriate is to increase dietary fiber to 20 to 25 grams per day over the course of a week or two. Fibers found in fruits, vegetables, and whole grains help to prevent constipation by increasing fecal mass. In the GI tract, fiber attracts water, creating soft, bulky stools that stimulate bowel contractions to push the contents along. These contractions strengthen the intestinal muscles. The improved muscle tone, together with the water content of the stools, eases elimination, reducing the pressure in the rectal veins and helping to prevent **hemorrhoids.** Chapter 4 provides more information on fiber's role in maintaining a healthy colon and reducing the risks of colon cancer and diverticulosis. **Diverticulosis** is a condition in which the intestinal walls develop bulges in weakened areas, most commonly in the colon (see Figure H3-3). These bulging pockets, known as **diverticula,** can worsen constipation, entrap feces, and become painfully infected and inflamed **(diverticulitis).** Treatment may require hospitalization, antibiotics, or surgery.

FIGURE H3-3 Diverticula in the Colon

Diverticula may develop anywhere along the GI tract, but they are most common in the colon.

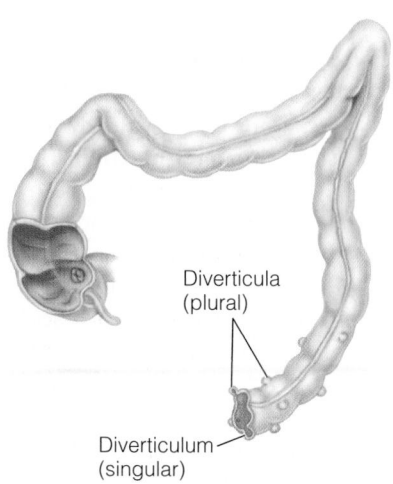

Diverticula
(plural)

Diverticulum
(singular)

Drinking plenty of water in conjunction with eating high-fiber foods also helps to prevent constipation. The increased bulk physically stimulates the upper GI tract, promoting peristalsis throughout. Similarly, physical activity improves the muscle tone and motility of the digestive tract. As little as 30 minutes of physical activity a day can help prevent or alleviate constipation.

Eating prunes—or "dried plums" as some have renamed them—can also be helpful. Prunes are high in fiber and also contain a laxative substance.* If a morning defecation is desired, a person can drink prune juice at bedtime; if the evening is preferred, the person can drink prune juice with breakfast.

These suggested changes in lifestyle or diet should correct chronic constipation without the use of **laxatives, enemas,** or **mineral oil,** although television commercials often try to persuade people otherwise. One of the fallacies often perpetrated by advertisements is that one person's successful use of a product is a good recommendation for others to use that product.

As a matter of fact, diet changes that relieve constipation for one person may increase the constipation of another. For instance, increasing fiber intake stimulates peristalsis and helps the person with a sluggish colon. Some people, though, have a spastic type of constipation, in which peristalsis promotes strong contractions that close off a segment of the colon and prevent passage; for these people, increasing fiber intake would be exactly the wrong thing to do.

A person who seems to need products such as laxatives frequently should seek a physician's advice. One potentially harmful but currently popular practice is **colonic irrigation**—the internal washing of the large intestine with a powerful enema machine. Such an extreme cleansing is not only unnecessary, but it can be hazardous, causing illness and death from equipment contamination, electrolyte depletion, and intestinal perforation. Less extreme practices can cause problems, too. Frequent use of laxatives and enemas can lead to dependency; upset the body's fluid, salt, and mineral balances; and, in the case of mineral oil, interfere with the absorption of fat-soluble vitamins. (Mineral oil dissolves the vitamins but is not itself absorbed. Instead, it leaves the body, carrying the vitamins with it.)

Belching and Gas

Many people complain of problems that they attribute to excessive gas. For some, **belching** is the complaint. Others blame intestinal gas for abdominal discomforts and embarrassment. Most people believe that the problems occur after they eat certain foods. This may be the case with intestinal gas, but belching results from swallowing air. The best advice for belching seems to be to eat slowly, chew thoroughly, and relax while eating.

Everyone swallows a little bit of air with each mouthful of food, but people who eat too fast may swallow too much air and then have to belch. Ill-fitting dentures, carbonated beverages, and chewing gum can also contribute to the swallowing of air with resultant belching. Occasionally, belching can be a sign of a more serious disorder, such as gallbladder disease or a peptic ulcer.

* This substance is dihydroxyphenyl isatin.

People troubled by gas need to determine which foods bother them and then eat those foods in moderation.

People who eat or drink too fast may also trigger **hiccups,** the repeated spasms that produce a cough-like sound and jerky movement. Normally, hiccups soon subside and are of no medical significance, but they can be bothersome. The most effective cure is to hold the breath for as long as possible, which helps to relieve the spasms of the diaphragm.

Although expelling gas can be a humiliating experience, it is quite normal. (People who experience painful bloating from mal-

absorption diseases, however, require medical treatment.) Healthy people expel several hundred milliliters of gas several times a day. Almost all (99 percent) of the gases expelled—nitrogen, oxygen, hydrogen, methane, and carbon dioxide—are odorless. The remaining "volatile" gases are the infamous ones.

Foods that produce gas usually must be determined individually. The most common offenders are foods rich in the carbohydrates—sugars, starches, and fibers. When partially digested carbohydrates reach the large intestine, bacteria digest them, giving off gas as a by-product. People can test foods suspected of forming gas by omitting them individually for a trial period to see if there is any improvement.

Heartburn and "Acid Indigestion"

Almost everyone has experienced **heartburn** at one time or another, usually soon after eating a meal. Medically known as **gastroesophageal reflux,** heartburn is the painful sensation a person feels behind the breastbone when the lower esophageal sphincter allows the stomach contents to reflux into the esophagus (see Figure H3-4). This may happen if a person eats or drinks too much (or both). Tight clothing and even changes of position (lying down, bending over) can cause it, too, as can some medications and smoking. Weight gain and overweight increase the frequency, severity, and duration of heartburn symptoms.[5] A defect of the sphincter muscle itself is a possible, but less common, cause.

If the heartburn is not caused by an anatomical defect, treatment is fairly simple. To avoid such misery in the future, the person needs to learn to eat less at a sitting, chew food more thoroughly, and eat it more slowly. Additional strategies are presented in Table H3-1 at the end of this highlight.

As far as "acid indigestion" is concerned, recall from Chapter 3 that the strong acidity of the stomach is a desirable condition—television commercials for **antacids** and **acid controllers** notwithstanding. People who overeat or eat too quickly are likely to suffer from **indigestion.** The muscular reaction of the stomach to unchewed lumps or to being overfilled may be so violent that it upsets normal peristalsis. When this happens, overeaters may taste the stomach acid and feel pain. Responding to advertisements, they may reach for antacids or acid controllers. Both of these drugs were originally designed to treat GI illnesses such as ulcers. As is true of most over-the-counter medicines, antacids and acid

---

FIGURE H3-4 Gastroesophageal Reflux

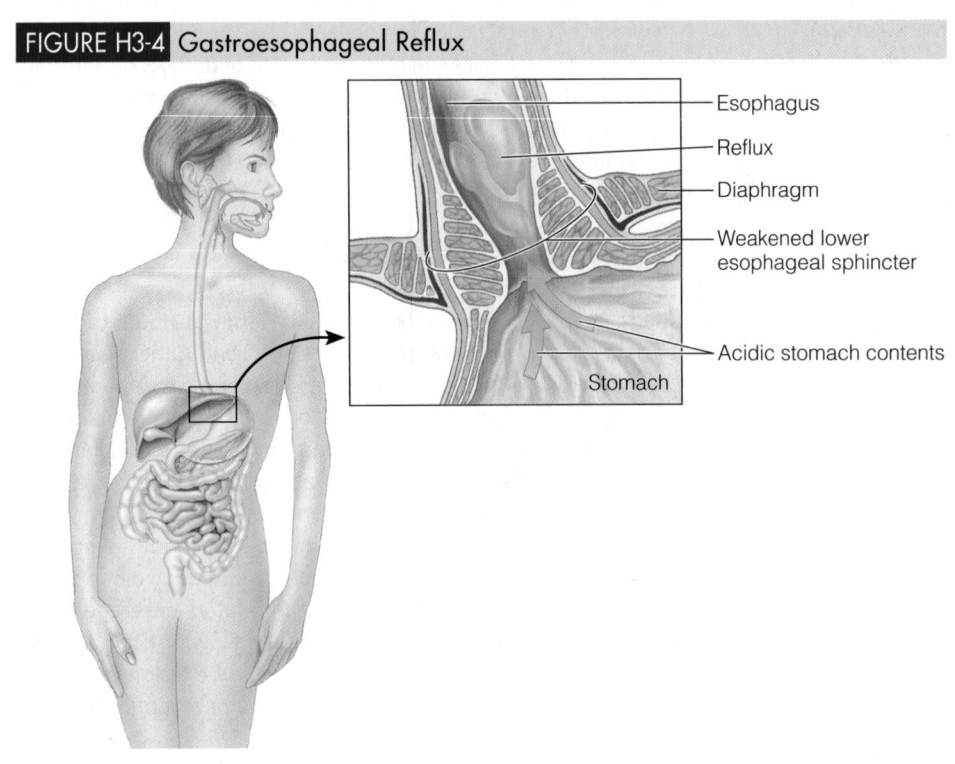

- Esophagus
- Reflux
- Diaphragm
- Weakened lower esophageal sphincter
- Acidic stomach contents

Stomach

controllers should be used only infrequently for occasional heartburn; they may mask or cause problems if used regularly. Acid-blocking drugs weaken the defensive mucous barrier of the GI tract, thereby increasing the risks of infections such as pneumonia, especially in vulnerable populations like the elderly.[6] Instead of self-medicating, people who suffer from frequent and regular bouts of heartburn and indigestion should try the strategies presented in the table below. If problems continue, they may need to see a physician, who can prescribe specific medication to control gastroesophageal reflux. Without treatment, the repeated splashes of acid can severely damage the cells of the esophagus, creating a condition known as Barrett's esophagus.[7] At that stage, the risk of cancer in the throat or esophagus increases dramatically. To repeat, if symptoms persist, see a doctor—don't self-medicate.

Ulcers

Ulcers are another common digestive problem. An **ulcer** is a lesion (a sore) and a **peptic ulcer** is a lesion in the lining of the stomach (gastric ulcers) or the duodenum of the small intestine (duodenal ulcers). The compromised lining is left unprotected and exposed to gastric juices, which can be painful. In some cases, ulcers can cause internal bleeding. If GI bleeding is excessive, iron deficiency may develop. Ulcers that perforate the GI lining can pose life-threatening complications.

Many people naively believe that an ulcer is caused by stress or spicy foods, but this is not the case. The stomach lining in a healthy person is well protected by its mucous coat. What, then, causes ulcers to form?

Three major causes of ulcers have been identified: bacterial infection with *Helicobacter pylori* (commonly abbreviated *H. pylori*);

the use of certain anti-inflammatory drugs such as aspirin, ibuprofen, and naproxen; and disorders that cause excessive gastric acid secretion. Most commonly, ulcers develop in response to *H. pylori* infection.[8] The cause of the ulcer dictates the type of medication used in treatment. For example, people with ulcers caused by infection receive antibiotics, whereas those with ulcers caused by medicines discontinue their use. In addition, all treatment plans aim to relieve pain, heal the ulcer, and prevent recurrence.

The regimen for ulcer treatment is to treat for infection, eliminate any food that routinely causes indigestion or pain, and avoid coffee and caffeine- and alcohol-containing beverages. Both regular and decaffeinated coffee stimulate acid secretion and so aggravate *existing* ulcers.

Ulcers and their treatments highlight the importance of not self-medicating when symptoms persist. People with *H. pylori* infection often take over-the-counter acid controllers to relieve the pain of their ulcers when, instead, they need physician-prescribed antibiotics. Suppressing gastric acidity not only fails to heal the ulcer, but it also actually worsens inflammation during an *H. pylori* infection. Furthermore, *H. pylori* infection has been linked with stomach cancer, making prompt diagnosis and appropriate treatment essential.[9]

Table H3-1 summarizes strategies to prevent or alleviate common GI problems. Many of these problems reflect hurried lifestyles. For this reason, many of their remedies require that people slow down and take the time to eat leisurely; chew food thoroughly to prevent choking, heartburn, and acid indigestion; rest until vomiting and diarrhea subside; and heed the urge to defecate. In addition, people must learn how to handle life's day-to-day problems and challenges without overreacting and becoming upset; learn how to relax, get enough sleep, and enjoy life. Remember, "what's eating you" may cause more GI distress than what you eat.

TABLE H3-1	Strategies to Prevent or Alleviate Common GI Problems		
GI Problem	**Strategies**	**GI Problem**	**Strategies**
Choking	• Take small bites of food. • Chew thoroughly before swallowing. • Don't talk or laugh with food in your mouth. • Don't eat when breathing hard.	Heartburn	• Eat small meals. • Drink liquids between meals. • Sit up while eating; elevate your head when lying down. • Wait 3 hours after eating before lying down. • Wait 2 hours after eating before exercising. • Refrain from wearing tight-fitting clothing. • Avoid foods, beverages, and medications that aggravate your heartburn. • Refrain from smoking cigarettes or using tobacco products. • Lose weight if overweight.
Diarrhea	• Rest. • Drink fluids to replace losses. • Call for medical help if diarrhea persists.		
Constipation	• Eat a high-fiber diet. • Drink plenty of fluids. • Exercise regularly. • Respond promptly to the urge to defecate.	Ulcer	• Take medicine as prescribed by your physician. • Avoid coffee and caffeine- and alcohol-containing beverages. • Avoid foods that aggravate your ulcer. • Minimize aspirin, ibuprofen, and naproxen use. • Refrain from smoking cigarettes.
Belching	• Eat slowly. • Chew thoroughly. • Relax while eating.		
Intestinal gas	• Eat bothersome foods in moderation.		

NUTRITION ON THE NET

For further study of topics covered in this chapter, log on to **academic.cengage.com/nutrition/rolfes/UNCN8e**. Go to Chapter 3, then to Nutrition on the Net.

- Search for "choking," "vomiting," "diarrhea," "constipation," "heartburn," "indigestion," and "ulcers" at the U.S. Government health information site: **www.healthfinder.gov**

- Visit the Center for Digestive Health and Nutrition: **www.gihealth.com**

- Visit the Digestive Diseases section of the National Institute of Diabetes, Digestive, and Kidney Diseases: **www.niddk.nih.gov/health/health.htm**

- Visit the patient information section of the American College of Gastroenterology: **www.acg.gi.org**

- Learn more about *H. pylori* from the Helicobacter Foundation: **www.helico.com**

REFERENCES

1. K. Gotsch, J. L. Annest, and P. Holmgreen, Nonfatal choking-related episodes among children-United States, 2001, *Morbidity and Mortality Weekly Report* 51 (2002): 945–948.
2. B. J. Horwitz and R. S. Fisher, The irritable bowel syndrome, *New England Journal of Medicine* 344 (2001): 1846–1850.
3. N. M. Thielman and R. L. Guerrant, Acute infectious diarrhea, *New England Journal of Medicine* 350 (2004): 38–47.
4. A. Lembo and M. Camilleri, Chronic constipation, *New England Journal of Medicine* 349 (2003): 1360–1368.

5. B. C. Jacobson and coauthors, Body-mass index and symptoms of gastroesophageal reflux in women, *New England Journal of Medicine* 354 (2006): 2340–2348.
6. R. J. F. Laheij and coauthors, Risk of community-acquired pneumonia and use of gastric acid-suppressive drugs, *Journal of the American Medical Association* 292 (2004): 1955–1960.
7. N. Shaheen and D. F. Ransohoff, Gastroesophageal reflux, Barrett's esophagus, and esophageal cancer: Scientific review, *Journal of the American Medical Association* 287 (2002): 1972–1981.

8. S. Suerbaum and P. Michetti, Helicobacter pylori infection, *New England Journal of Medicine* 347 (2002): 1175–1186.
9. N. Uemura and coauthors, Helicobacter pylori infection and the development of gastric cancer, *New England Journal of Medicine* 345 (2001): 784–789.

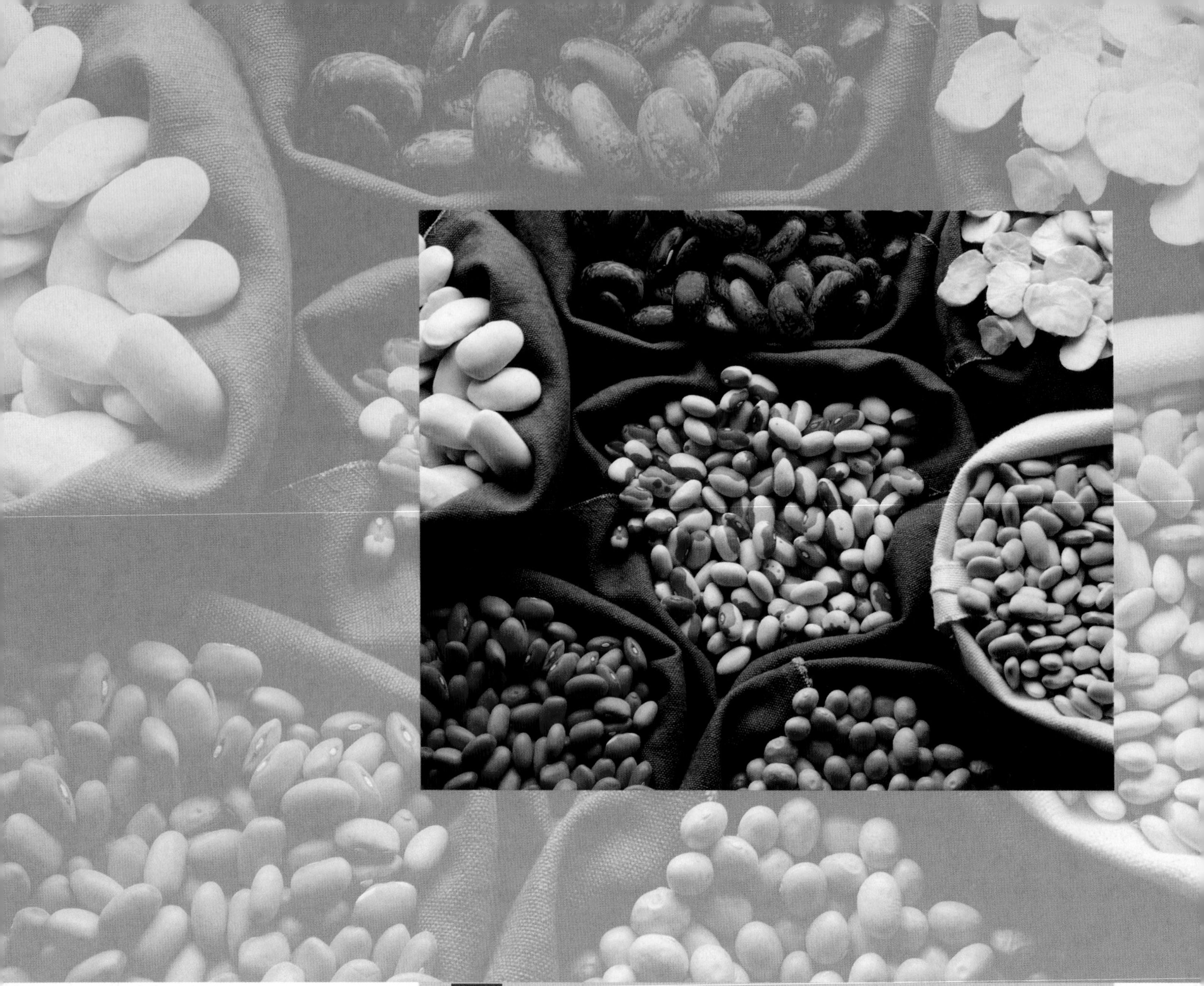

Kevin Summers/Getty Images

Nutrition in Your Life

Whether you are cramming for an exam or daydreaming about your next vacation, your brain needs carbohydrate to power its activities. Your muscles need carbohydrate to fuel their work, too, whether you are racing up the stairs to class or moving on the dance floor to your favorite music. Where can you get carbohydrate? And are some foods healthier choices than others? As you will learn from this chapter, whole grains, vegetables, legumes, and fruits naturally deliver ample carbohydrate and fiber with valuable vitamins and minerals and little or no fat. Milk products typically lack fiber, but they also provide carbohydrate along with an assortment of vitamins and minerals.

CENGAGENOW™ The CengageNOW logo indicates an opportunity for online self-study, linking you to interactive tutorials and videos based on your level of understanding.

academic.cengage.com/login

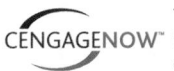

Figure 4.10: Animated! Carbohydrate Digestion in the GI Tract

Nutrition Portfolio Journal

Nutrition Calculations: Practice Problems

The Carbohydrates: Sugars, Starches, and Fibers

A student, quietly studying a textbook, is seldom aware that within his brain cells, billions of glucose molecules are splitting to provide the energy that permits him to learn. Yet glucose provides nearly all of the energy the human brain uses daily. Similarly, a marathon runner, bursting across the finish line in an explosion of sweat and triumph, seldom gives credit to the glycogen fuel her muscles have devoured to help her finish the race. Yet, together, these two **carbohydrates**—glucose and its storage form glycogen—provide about half of all the energy muscles and other body tissues use. The other half of the body's energy comes mostly from fat.

People don't eat glucose and glycogen directly. When they eat foods rich in carbohydrates, their bodies receive glucose for immediate energy and into glycogen for reserve energy. All plant foods—whole grains, vegetables, legumes, and fruits—provide ample carbohydrate. Milk also contains carbohydrates.

Many people mistakenly think of carbohydrates as "fattening" and avoid them when trying to lose weight. Such a strategy may be helpful if the carbohydrates are the simple sugars of soft drinks, candy, and cookies, but it is counterproductive if the carbohydrates are the complex carbohydrates of whole grains, vegetables, and legumes. As the next section explains, not all carbohydrates are created equal.

The Chemist's View of Carbohydrates

The dietary carbohydrate family includes the **simple carbohydrates** (the sugars) and the **complex carbohydrates** (the starches and fibers). The simple carbohydrates are those that chemists describe as:

- Monosaccharides—single sugars
- Disaccharides—sugars composed of pairs of monosaccharides

The complex carbohydrates are:

- Polysaccharides—large molecules composed of chains of monosaccharides

carbohydrates: compounds composed of carbon, oxygen, and hydrogen arranged as monosaccharides or multiples of monosaccharides. Most, but not all, carbohydrates have a ratio of one carbon molecule to one water molecule: $(CH_2O)_n$.
- **carbo** = carbon (C)
- **hydrate** = with water (H_2O)

simple carbohydrates (sugars): monosaccharides and disaccharides.

complex carbohydrates (starches and **fibers):** polysaccharides composed of straight or branched chains of monosaccharides.

FIGURE 4-1 Atoms and Their Bonds

The four main types of atoms found in nutrients are hydrogen (H), oxygen (O), nitrogen (N), and carbon (C).

$$H-\quad -O-\quad -N-\quad -\overset{|}{\underset{|}{C}}-$$

$$1\qquad 2\qquad 3\qquad 4$$

Each atom has a characteristic number of bonds it can form with other atoms.

$$H-\overset{\overset{H}{|}}{\underset{\underset{H}{|}}{C}}-\overset{\overset{H}{|}}{\underset{\underset{H}{|}}{C}}-O-H$$

Notice that in this simple molecule of ethyl alcohol, each H has one bond, O has two, and each C has four.

◆ Most of the monosaccharides important in nutrition are **hexoses**, simple sugars with six atoms of carbon and the formula C6H12O6.

• **hex** = six

FIGURE 4-2 Chemical Structure of Glucose

On paper, the structure of glucose has to be drawn flat, but in nature the five carbons and oxygen are roughly in a plane. The atoms attached to the ring carbons extend above and below the plane.

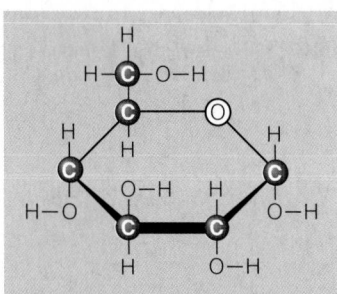

---

monosaccharides (mon-oh-SACK-uh-rides): carbohydrates of the general formula $C_nH_{2n}O_n$ that typically form a single ring. See Appendix C for the chemical structures of the monosaccharides.

• **mono** = one
• **saccharide** = sugar

glucose (GLOO-kose): a monosaccharide; sometimes known as blood sugar or **dextrose.**

• **ose** = carbohydrate
• ⬡ = glucose

To understand the structure of carbohydrates, look at the units of which they are made. The monosaccharides most important in nutrition ◆ each contain 6 carbon atoms, 12 hydrogens, and 6 oxygens (written in shorthand as $C_6H_{12}O_6$).

Each atom can form a certain number of chemical bonds with other atoms:

• Carbon atoms can form four bonds
• Nitrogen atoms, three
• Oxygen atoms, two
• Hydrogen atoms, only one

Chemists represent the bonds as lines between the chemical symbols (such as C, N, O, and H) that stand for the atoms (see Figure 4-1).

Atoms form molecules in ways that satisfy the bonding requirements of each atom. Figure 4-1 includes the structure of ethyl alcohol, the active ingredient of alcoholic beverages, as an example. The two carbons each have four bonds represented by lines; the oxygen has two; and each hydrogen has one bond connecting it to other atoms. Chemical structures bond according to these rules as dictated by nature.

IN SUMMARY

The carbohydrates are made of carbon (C), oxygen (O), and hydrogen (H). Each of these atoms can form a specified number of chemical bonds: carbon forms four, oxygen forms two, and hydrogen forms one.

The Simple Carbohydrates

The following list of the most important simple carbohydrates in nutrition symbolizes them as hexagons and pentagons of different colors.* Three are monosaccharides:

• Glucose
• Fructose
• Galactose

Three are disaccharides:

• Maltose (glucose + glucose)

• Sucrose (glucose + fructose)

• Lactose (glucose + galactose)

Monosaccharides

The three **monosaccharides** important in nutrition all have the same numbers and kinds of atoms, but in different arrangements. These chemical differences account for the differing sweetness of the monosaccharides. A pinch of purified glucose on the tongue gives only a mild sweet flavor, and galactose hardly tastes sweet at all. Fructose, however, is as intensely sweet as honey and, in fact, is the sugar primarily responsible for honey's sweetness.

Glucose Chemically, **glucose** is a larger and more complicated molecule than the ethyl alcohol shown in Figure 4-1, but it obeys the same rules of chemistry: each carbon atom has four bonds; each oxygen, two bonds; and each hydrogen, one bond. Figure 4-2 illustrates the chemical structure of a glucose molecule.

The diagram of a glucose molecule shows all the relationships between the atoms and proves simple on examination, but chemists have adopted even simpler ways to depict chemical structures. Figure 4-3 presents the chemical structure

* Fructose is shown as a pentagon, but like the other monosaccharides, it has six carbons (as you will see in Figure 4-4).

FIGURE 4-3 Simplified Diagrams of Glucose

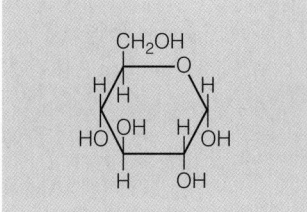

The lines representing some of the bonds and the carbons at the corners are not shown.

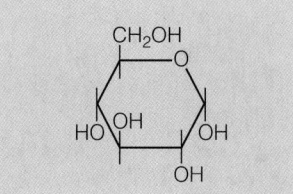

Now the single hydrogens are not shown, but lines still extend upward or downward from the ring to show where they belong.

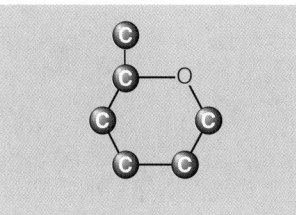

Another way to look at glucose is to notice that its six carbon atoms are all connected.

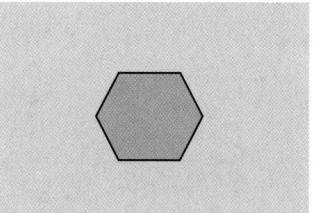

In this and other illustrations throughout this book, glucose is represented as a blue hexagon.

of glucose in a more simplified way by combining or omitting several symbols—yet it conveys the same information.

Commonly known as blood sugar, glucose serves as an essential energy source for all the body's activities. Its significance to nutrition is tremendous. Later sections explain that glucose is one of the two sugars in every disaccharide and the unit from which the polysaccharides are made almost exclusively. One of these polysaccharides, starch, is the chief food source of energy for all the world's people; another, glycogen, is an important storage form of energy in the body. Glucose reappears frequently throughout this chapter and all those that follow.

Fructose **Fructose** is the sweetest of the sugars. Curiously, fructose has exactly the same chemical *formula* as glucose—$C_6H_{12}O_6$—but its *structure* differs (see Figure 4-4). The arrangement of the atoms in fructose stimulates the taste buds on the tongue to produce the sweet sensation. Fructose occurs naturally in fruits and honey; other sources include products such as soft drinks, ready-to-eat cereals, and desserts that have been sweetened with high-fructose corn syrup (defined on p. 118).

Galactose The monosaccharide **galactose** occurs naturally as a single sugar in only a few foods. Galactose has the same numbers and kinds of atoms as glucose and fructose in yet another arrangement. Figure 4-5 shows galactose beside a molecule of glucose for comparison.

Disaccharides

The **disaccharides** are pairs of the three monosaccharides just described. Glucose occurs in all three; the second member of the pair is either fructose, galactose, or

fructose (FRUK-tose or FROOK-tose): a monosaccharide; sometimes known as fruit sugar or **levulose.** Fructose is found abundantly in fruits, honey, and saps.
- **fruct** = fruit
- ⬡ = fructose

galactose (ga-LAK-tose): a monosaccharide; part of the disaccharide lactose.
- ⬡ = galactose

disaccharides (dye-SACK-uh-rides): pairs of monosaccharides linked together. See Appendix C for the chemical structures of the disaccharides.
- **di** = two

FIGURE 4-4 Two Monosaccharides: Glucose and Fructose

Can you see the similarities? If you learned the rules in Figure 4-3, you will be able to "see" 6 carbons (numbered), 12 hydrogens (those shown plus one at the end of each single line), and 6 oxygens in both these compounds.

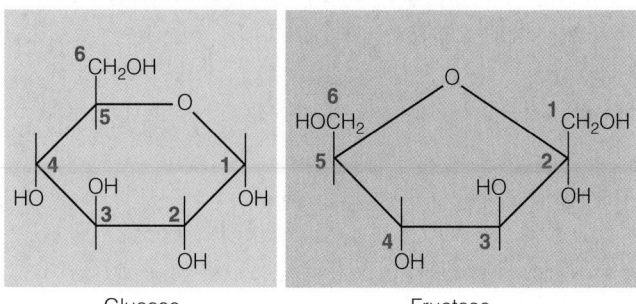

Glucose   Fructose

FIGURE 4-5 Two Monosaccharides: Glucose and Galactose

Notice the similarities and the difference (highlighted in red) between glucose and galactose. Both have 6 carbons, 12 hydrogens, and 6 oxygens, but the position of one OH group differs slightly.

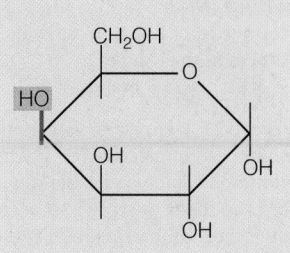

Glucose   Galactose

Fruits package their simple sugars with fibers, vitamins, and minerals, making them a sweet and healthy snack.

FIGURE 4-6 Condensation of Two Monosaccharides to Form a Disaccharide

Glucose + glucose ⟶ Maltose

An OH group from one glucose and an H atom from another glucose combine to create a molecule of H_2O.

The two glucose molecules bond together with a single O atom to form the disaccharide maltose.

another glucose. These carbohydrates—and all the other energy nutrients—are put together and taken apart by similar chemical reactions: condensation and hydrolysis.

Condensation To make a disaccharide, a chemical reaction known as **condensation** links two monosaccharides together (see Figure 4-6). A hydroxyl (OH) group from one monosaccharide and a hydrogen atom (H) from the other combine to create a molecule of water (H_2O). The two originally separate monosaccharides link together with a single oxygen (O).

Hydrolysis To break a disaccharide in two, a chemical reaction known as hydrolysis ◆ occurs (see Figure 4-7). A molecule of water splits to provide the H and OH needed to complete the resulting monosaccharides. Hydrolysis reactions commonly occur during digestion.

◆ Reminder: A *hydrolysis* reaction splits a molecule into two, with H added to one and OH to the other (from water); Chapter 3 explained that hydrolysis reactions break down molecules during digestion.

Maltose The disaccharide **maltose** consists of two glucose units. Maltose is produced whenever starch breaks down—as happens in human beings during carbohydrate digestion. It also occurs during the fermentation process that yields alcohol. Maltose is only a minor constituent of a few foods, most notably barley.

Sucrose Fructose and glucose together form **sucrose.** Because the fructose is accessible to the taste receptors, sucrose tastes sweet, accounting for some of the natural sweetness of fruits, vegetables, and grains. To make table sugar, sucrose is refined from the juices of sugarcane and sugar beets, then granulated. Depending on the extent to which it is refined, the product becomes the familiar brown, white, and powdered sugars available at grocery stores.

condensation: a chemical reaction in which two reactants combine to yield a larger product.

maltose (MAWL-tose): a disaccharide composed of two glucose units; sometimes known as malt sugar.
- ⬡⬡ = maltose

sucrose (SUE-krose): a disaccharide composed of glucose and fructose; commonly known as table sugar, beet sugar, or cane sugar. Sucrose also occurs in many fruits and some vegetables and grains.
- **sucro** = sugar
- ⬡⬡ = sucrose

FIGURE 4-7 Hydrolysis of a Disaccharide

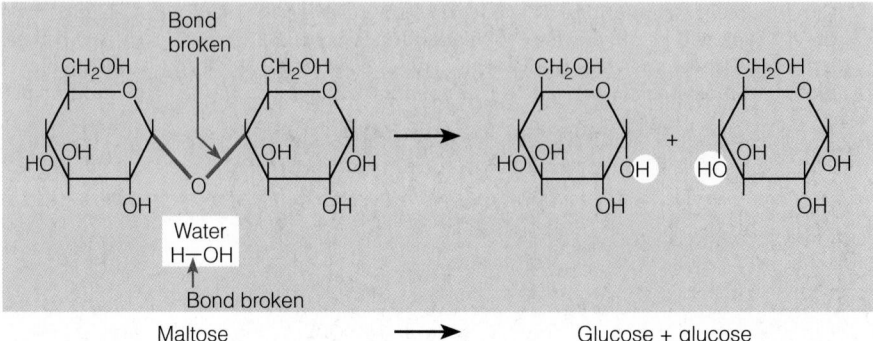

Bond broken

Water H—OH

Bond broken

Maltose ⟶ Glucose + glucose

The disaccharide maltose splits into two glucose molecules with H added to one and OH to the other (from the water molecule).

Lactose   The combination of galactose and glucose makes the disaccharide **lactose,** the principal carbohydrate of milk. Known as milk sugar, lactose contributes half of the energy (kcalories) provided by fat-free milk.

> **IN SUMMARY**
>
> Six simple carbohydrates, or sugars, are important in nutrition. The three monosaccharides (glucose, fructose, and galactose) all have the same chemical formula ($C_6H_{12}O_6$), but their structures differ. The three disaccharides (maltose, sucrose, and lactose) are pairs of monosaccharides, each containing a glucose paired with one of the three monosaccharides. The sugars derive primarily from plants, except for lactose and its component galactose, which come from milk and milk products. Two monosaccharides can be linked together by a condensation reaction to form a disaccharide and water. A disaccharide, in turn, can be broken into its two monosaccharides by a hydrolysis reaction using water.

The Complex Carbohydrates

The simple carbohydrates are the sugars just mentioned—the monosaccharides glucose, fructose, and galactose and the disaccharides maltose, sucrose, and lactose. In contrast, the complex carbohydrates contain many glucose units and, in some cases, a few other monosaccharides strung together as **polysaccharides.** Three types of polysaccharides are important in nutrition: glycogen, starches, and fibers.

Glycogen is a storage form of energy in the animal body; starches play that role in plants; and fibers provide structure in stems, trunks, roots, leaves, and skins of plants. Both glycogen and starch are built of glucose units; fibers are composed of a variety of monosaccharides and other carbohydrate derivatives.

Glycogen

Glycogen is found to only a limited extent in meats and not at all in plants.* For this reason, food is not a significant source of this carbohydrate. However, glycogen does perform an important role in the body. The human body stores glucose as glycogen—many glucose molecules linked together in highly branched chains (see the left side of Figure 4-8 on p. 106). This arrangement permits rapid hydrolysis. When the hormonal message "release energy" arrives at the glycogen storage sites in a liver or muscle cell, enzymes respond by attacking the many branches of glycogen simultaneously, making a surge of glucose available.†

Starches

The human body stores glucose as glycogen, but plant cells store glucose as **starches**—long, branched or unbranched chains of hundreds or thousands of glucose molecules linked together (see the middle and right side of Figure 4-8). These giant starch molecules are packed side by side in grains such as wheat or rice, in root crops and tubers such as yams and potatoes, and in legumes such as peas and beans. When you eat the plant, your body hydrolyzes the starch to glucose and uses the glucose for its own energy purposes.

All starchy foods come from plants. Grains are the richest food source of starch, providing much of the food energy for people all over the world—rice in Asia;

Major sources of starch include grains (such as rice, wheat, millet, rye, barley, and oats), legumes (such as kidney beans, black-eyed peas, pinto beans, navy beans, and garbanzo beans), tubers (such as potatoes), and root crops (such as yams and cassava).

> **lactose** (LAK-tose): a disaccharide composed of glucose and galactose; commonly known as milk sugar.
> • **lact** = milk
> • ◆◯◆ = lactose
>
> **polysaccharides:** compounds composed of many monosaccharides linked together. An intermediate string of three to ten monosaccharides is an **oligosaccharide.**
> • **poly** = many
> • **oligo** = few
>
> **glycogen** (GLY-ko-jen): an animal polysaccharide composed of glucose; manufactured and stored in the liver and muscles as a storage form of glucose. Glycogen is not a significant food source of carbohydrate and is not counted as one of the complex carbohydrates in foods.
> • **glyco** = glucose
> • **gen** = gives rise to
>
> **starches:** plant polysaccharides composed of glucose.

---

* Glycogen in animal muscles rapidly hydrolyzes after slaughter.
† Normally, only liver cells can produce glucose from glycogen to be sent *directly* to the blood; muscle cells can also produce glucose from glycogen, but must use it themselves. Muscle cells can restore the blood glucose level *indirectly*, however, as Chapter 7 explains.

FIGURE 4-8 Glycogen and Starch Molecules Compared (Small Segments)

Notice the more highly branched the structure, the greater the number of ends from which glucose can be released. (These units would have to be magnified millions of times to appear at the size shown in this figure. For details of the chemical structures, see Appendix C.)

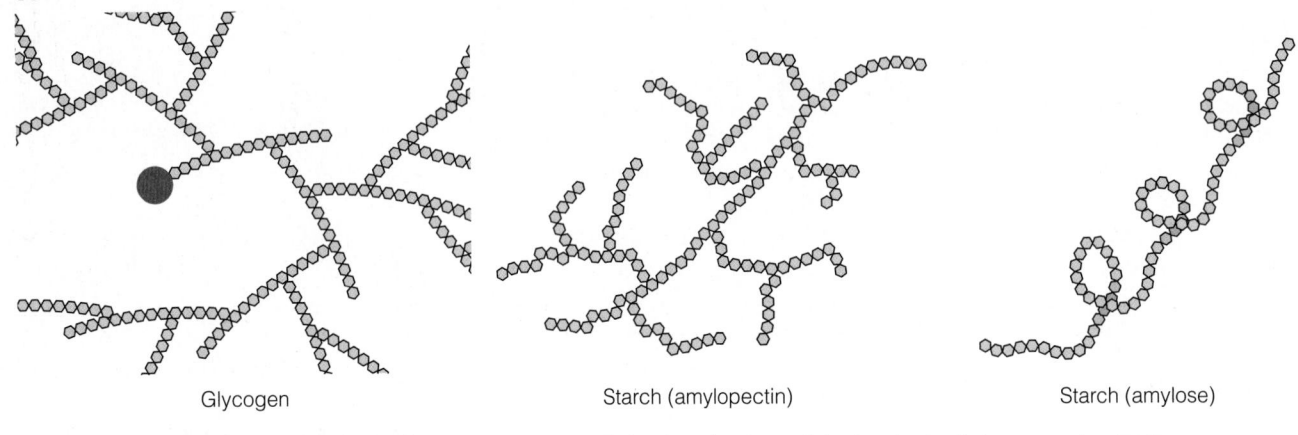

Glycogen

Starch (amylopectin)

Starch (amylose)

A glycogen molecule contains hundreds of glucose units in highly branched chains. Each new glycogen molecule needs a special protein for the attachment of the first glucose (shown here in red).

A starch molecule contains hundreds of glucose molecules in either occasionally branched chains (amylopectin) or unbranched chains (amylose).

wheat in Canada, the United States, and Europe; corn in much of Central and South America; and millet, rye, barley, and oats elsewhere. Legumes and tubers are also important sources of starch.

Fibers

Dietary fibers are the structural parts of plants and thus are found in all plant-derived foods—vegetables, fruits, whole grains, and legumes. Most dietary fibers are polysacharides. As mentioned earlier, starches are also polysacharides, but dietary fibers differ from starches in that the bonds between their monosaccharides cannot be broken down by digestive enzymes in the body. For this reason, dietary fibers are often described as *nonstarch polysaccharides*.* Figure 4-9 illustrates the difference in the bonds that link glucose molecules together in starch with those found in the fiber cellulose. Because dietary fibers pass through the body, they contribute no monosaccharides, and therefore little or no energy.

Even though most foods contain a variety of fibers, researchers often sort dietary fibers into two groups according to their solubility. Such distinctions help to explain their actions in the body.

Soluble Fibers Some dietary fibers dissolve in water **(soluble fibers)**, form gels **(viscous)**, and are easily digested by bacteria in the colon **(fermentable)**. Commonly found in oats, barley, legumes, and citrus fruits, soluble fibers are most often associated with protecting against heart disease and diabetes by lowering blood cholesterol and glucose levels, respectively.[1]

Insoluble Fibers Other fibers do not dissolve in water **(insoluble fibers)**, do not form gels (nonviscous), and are less readily fermented. Found mostly in whole grains (bran) and vegetables, insoluble fibers promote bowel movements and alleviate constipation.

Fiber Sources As mentioned, dietary fibers occur naturally in plants. When these fibers have been extracted from plants or manufactured and then added to foods or used in supplements they are called *functional fibers*—if they have beneficial health

dietary fibers: in plant foods, the *nonstarch polysaccharides* that are not digested by human digestive enzymes, although some are digested by GI tract bacteria. Dietary fibers include cellulose, hemicelluloses, pectins, gums, and mucilages and the nonpolysaccharides lignins, cutins, and tannins.

soluble fibers: indigestible food components that dissolve in water to form a gel. An example is pectin from fruit, which is used to thicken jellies.

viscous: a gel-like consistency.

fermentable: the extent to which bacteria in the GI tract can break down fibers to fragments that the body can use.†

insoluble fibers: indigestible food components that do not dissolve in water. Examples include the tough, fibrous structures found in the strings of celery and the skins of corn kernels.

* The nonstarch polysaccharide fibers include cellulose, hemicelluloses, pectins, gums, and mucilages. Fibers also include some *nonpolysaccharides* such as lignins, cutins, and tannins.
† Dietary fibers are fermented by bacteria in the colon to short-chain fatty acids, which are absorbed and metabolized by cells in the GI tract and liver (Chapter 5 describes fatty acids).

effects. Cellulose in cereals, for example, is a dietary fiber, but when consumed as a supplement to alleviate constipation, cellulose is considered a functional fiber. *Total fiber* refers to the sum of dietary fibers and functional fibers. These terms ♦ were created by the DRI Committee to accommodate products that may contain new fiber sources, but consumers may find them too confusing to be used on food labels.[2]

Resistant Starches   A few starches are classified as dietary fibers. Known as **resistant starches,** these starches escape digestion and absorption in the small intestine. Starch may resist digestion for several reasons, including the individual's efficiency in digesting starches and the food's physical properties. Resistant starch is common in whole legumes, raw potatoes, and unripe bananas.

Phytic Acid   Althought not classified as a dietary fiber, **phytic acid** is often found accompanying them in the same foods. Because of this close association, researchers have been unable to determine whether it is the dietary fiber, the phytic acid, or both, that binds with minerals, preventing their absorption. This binding presents a risk of mineral deficiencies, but the risk is minimal when total fiber intake is reasonable and mineral intake adequate. The nutrition consequences of such mineral losses are described further in Chapters 12 and 13.

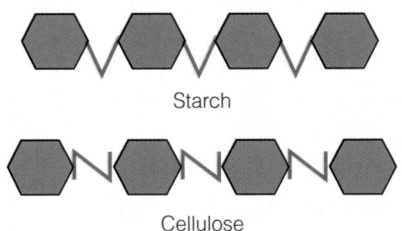

FIGURE 4-9   **Starch and Cellulose Molecules Compared (Small Segments)**

The bonds that link the glucose molecules together in cellulose are different from the bonds in starch (and glycogen). Human enzymes cannot digest cellulose. See Appendix C for chemical structures and descriptions of linkages.

Starch

Cellulose

♦ *Dietary fibers* occur naturally in intact plants. *Functional fibers* have been extracted from plants or manufactured and have beneficial effects in human beings. *Total fiber* is the sum of dietary fibers and functional fibers.

IN SUMMARY

The complex carbohydrates are the polysaccharides (chains of monosaccharides): glycogen, starches, and dietary fibers. Both glycogen and starch are storage forms of glucose—glycogen in the body, and starch in plants—and both yield energy for human use. The dietary fibers also contain glucose (and other monosaccharides), but their bonds cannot be broken by human digestive enzymes, so they yield little, if any, energy. The accompanying table summarizes the carbohydrate family of compounds.

The Carbohydrate Family	
Simple Carbohydrates (Sugars)	**Complex Carbohydrates**
• Monosaccharides:	• Polysaccharides:
Glucose	Glycogen[a]
Fructose	Starches
Galactose	Fibers
• Disaccharides:	
Maltose	
Sucrose	
Lactose	

[a]Glycogen is a complex carbohydrate (a polysaccharide) but not a *dietary* source of carbohydrate.

Digestion and Absorption of Carbohydrates

The ultimate goal of digestion and absorption of sugars and starches is to break them into small molecules—chiefly glucose—that the body can absorb and use. The large starch molecules require extensive breakdown; the disaccharides need only be broken once and the monosaccharides not at all. The initial splitting begins in the mouth; the final splitting and absorption occur in the small intestine; and conversion to a common energy currency (glucose) takes place in the liver. The details follow.

resistant starches: starches that escape digestion and absorption in the small intestine of healthy people.

phytic (FYE-tick) **acid:** a nonnutrient component of plant seeds; also called **phytate** (FYE-tate). Phytic acid occurs in the husks of grains, legumes, and seeds and is capable of binding minerals such as zinc, iron, calcium, magnesium, and copper in insoluble complexes in the intestine, which the body excretes unused.

◆ The short chains of glucose units that result from the breakdown of starch are known as **dextrins.** The word sometimes appears on food labels because dextrins can be used as thickening agents in processed foods.

◆ Reminder: A **bolus** is a portion of food swallowed at one time.

© Banana Stock/SuperStock

When a person eats carbohydrate-rich foods, the body receives a valuable commodity—glucose.

◆ Reminder: In general, the word ending –ase identifies an enzyme, and the beginning of the word identifies the molecule that the enzyme works on.

◆ Starches and sugars are called **available carbohydrates** because human digestive enzymes break them down for the body's use. In contrast, fibers are called **unavailable carbohydrates** because human digestive enzymes cannot break their bonds.

---

amylase (AM-ih-lace): an enzyme that hydrolyzes amylose (a form of starch). Amylase is a *carbohydrase,* an enzyme that breaks down carbohydrates.

satiety (sah-TIE-eh-tee): the feeling of fullness and satisfaction that occurs after a meal and inhibits eating until the next meal. Satiety determines how much time passes between meals.
- **sate** = to fill

maltase: an enzyme that hydrolyzes maltose

sucrase: an enzyme that hydrolyzes sucrose

lactase: an enzyme that hydrolyzes lactose

Carbohydrate Digestion

Figure 4-10 traces the digestion of carbohydrates through the GI tract. When a person eats foods containing starch, enzymes hydrolyze the long chains to shorter chains, ◆ the short chains to disaccharides, and, finally, the disaccharides to monosaccharides. This process begins in the mouth.

In the Mouth In the mouth, thoroughly chewing high-fiber foods slows eating and stimulates the flow of saliva. The salivary enzyme **amylase** starts to work, hydrolyzing starch to shorter polysaccharides and to the disaccharide maltose. In fact, you can taste the change if you hold a piece of starchy food like a cracker in your mouth for a few minutes without swallowing it—the cracker begins tasting sweeter as the enzyme acts on it. Because food is in the mouth for only a short time, very little carbohydrate digestion takes place there; it begins again in the small intestine.

In the Stomach The swallowed bolus ◆ mixes with the stomach's acid and protein-digesting enzymes, which inactivate salivary amylase. Thus the role of salivary amylase in starch digestion is relatively minor. To a small extent, the stomach's acid continues breaking down starch, but its juices contain no enzymes to digest carbohydrate. Fibers linger in the stomach and delay gastric emptying, thereby providing a feeling of fullness and **satiety.**

In the Small Intestine The small intestine performs most of the work of carbohydrate digestion. A major carbohydrate-digesting enzyme, pancreatic amylase, enters the intestine via the pancreatic duct and continues breaking down the polysaccharides to shorter glucose chains and maltose. The final step takes place on the outer membranes of the intestinal cells. There specific enzymes ◆ break down specific disaccharides:

- **Maltase** breaks maltose into two glucose molecules.
- **Sucrase** breaks sucrose into one glucose and one fructose molecule.
- **Lactase** breaks lactose into one glucose and one galactose molecule.

At this point, all polysaccharides and disaccharides have been broken down to monosaccharides—mostly glucose molecules, with some fructose and galactose molecules as well.

In the Large Intestine Within one to four hours after a meal, all the sugars and most of the starches have been digested. ◆ Only the fibers remain in the digestive tract. Fibers in the large intestine attract water, which softens the stools for passage without straining. Also, bacteria in the GI tract ferment some fibers. This process generates water, gas, and short-chain fatty acids (described in Chapter 5).* The colon uses these small fat molecules for energy. Metabolism of short-chain fatty acids also occurs in the cells of the liver. Fibers, therefore, can contribute some energy (1.5 to 2.5 kcalories per gram), depending on the extent to which they are broken down by bacteria and the fatty acids are absorbed.

Carbohydrate Absorption

Glucose is unique in that it can be absorbed to some extent through the lining of the mouth, but for the most part, nutrient absorption takes place in the small intestine. Glucose and galactose traverse the cells lining the small intestine by active transport; fructose is absorbed by facilitated diffusion, which slows its entry and produces a smaller rise in blood glucose. Likewise, unbranched chains of starch are digested slowly and produce a smaller rise in blood glucose than branched chains, which have many more places for enzymes to attack and release glucose rapidly.

As the blood from the intestines circulates through the liver, cells there take up fructose and galactose and convert them to other compounds, most often to glu-

---

* The short-chain fatty acids produced by GI bacteria are primarily acetic acid, propionic acid, and butyric acid.

FIGURE 4-10 *Animated!* Carbohydrate Digestion in the GI Tract

CENGAGENOW™
To test your understanding of these concepts, log on to
academic.cengage.com/login

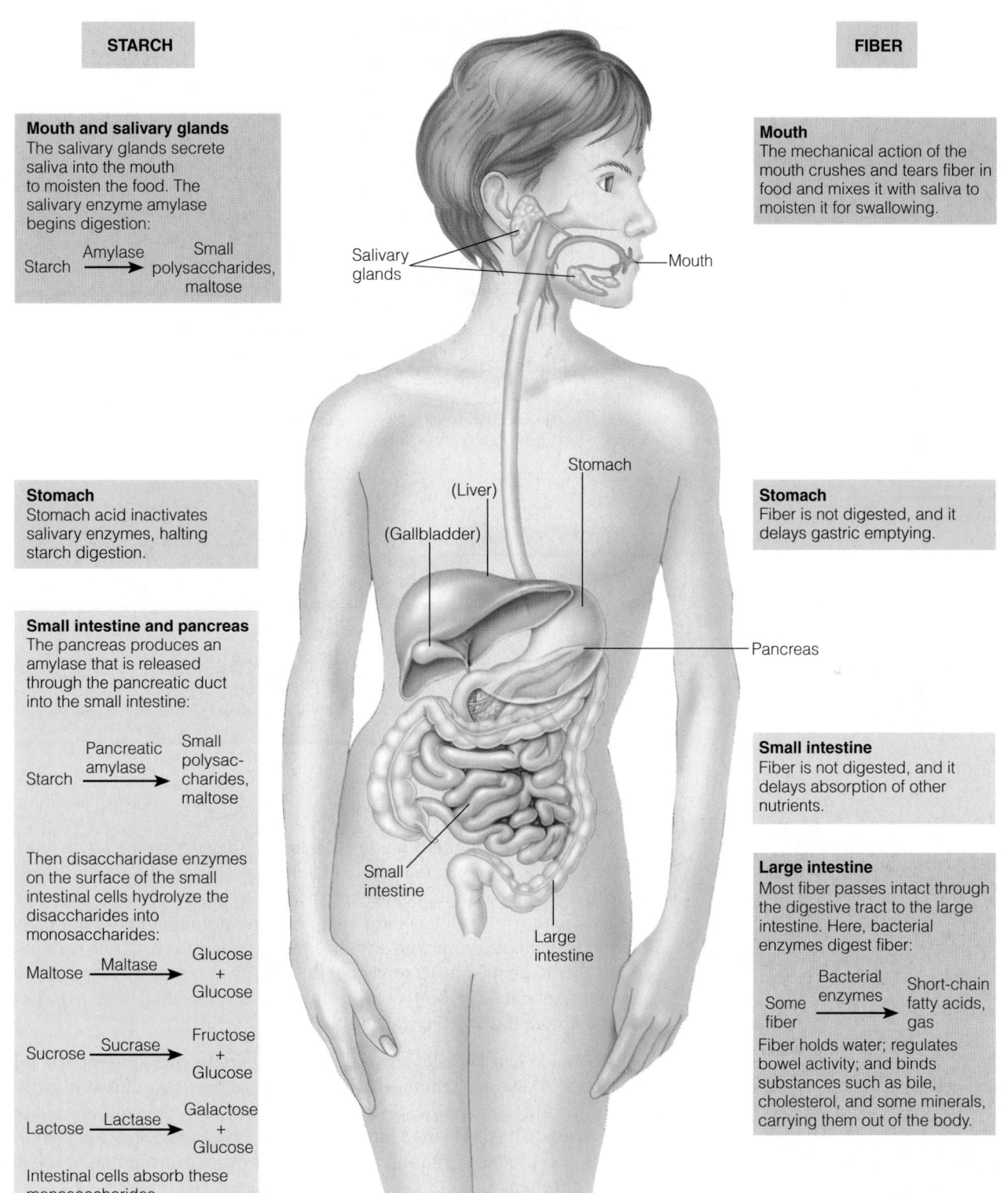

STARCH

Mouth and salivary glands
The salivary glands secrete saliva into the mouth to moisten the food. The salivary enzyme amylase begins digestion:

$$\text{Starch} \xrightarrow{\text{Amylase}} \text{Small polysaccharides, maltose}$$

Stomach
Stomach acid inactivates salivary enzymes, halting starch digestion.

Small intestine and pancreas
The pancreas produces an amylase that is released through the pancreatic duct into the small intestine:

$$\text{Starch} \xrightarrow[\text{amylase}]{\text{Pancreatic}} \text{Small polysaccharides, maltose}$$

Then disaccharidase enzymes on the surface of the small intestinal cells hydrolyze the disaccharides into monosaccharides:

$$\text{Maltose} \xrightarrow{\text{Maltase}} \text{Glucose} + \text{Glucose}$$

$$\text{Sucrose} \xrightarrow{\text{Sucrase}} \text{Fructose} + \text{Glucose}$$

$$\text{Lactose} \xrightarrow{\text{Lactase}} \text{Galactose} + \text{Glucose}$$

Intestinal cells absorb these monosaccharides.

FIBER

Mouth
The mechanical action of the mouth crushes and tears fiber in food and mixes it with saliva to moisten it for swallowing.

Stomach
Fiber is not digested, and it delays gastric emptying.

Small intestine
Fiber is not digested, and it delays absorption of other nutrients.

Large intestine
Most fiber passes intact through the digestive tract to the large intestine. Here, bacterial enzymes digest fiber:

$$\text{Some fiber} \xrightarrow[\text{enzymes}]{\text{Bacterial}} \text{Short-chain fatty acids, gas}$$

Fiber holds water; regulates bowel activity; and binds substances such as bile, cholesterol, and some minerals, carrying them out of the body.

Labels on figure: Salivary glands, Mouth, Stomach, (Liver), (Gallbladder), Pancreas, Small intestine, Large intestine

cose, as shown in Figure 4-11 (p. 110). Thus all disaccharides provide at least one glucose molecule directly, and they can provide another one indirectly—through the conversion of fructose and galactose to glucose.

FIGURE 4-11 Absorption of Monosaccharides

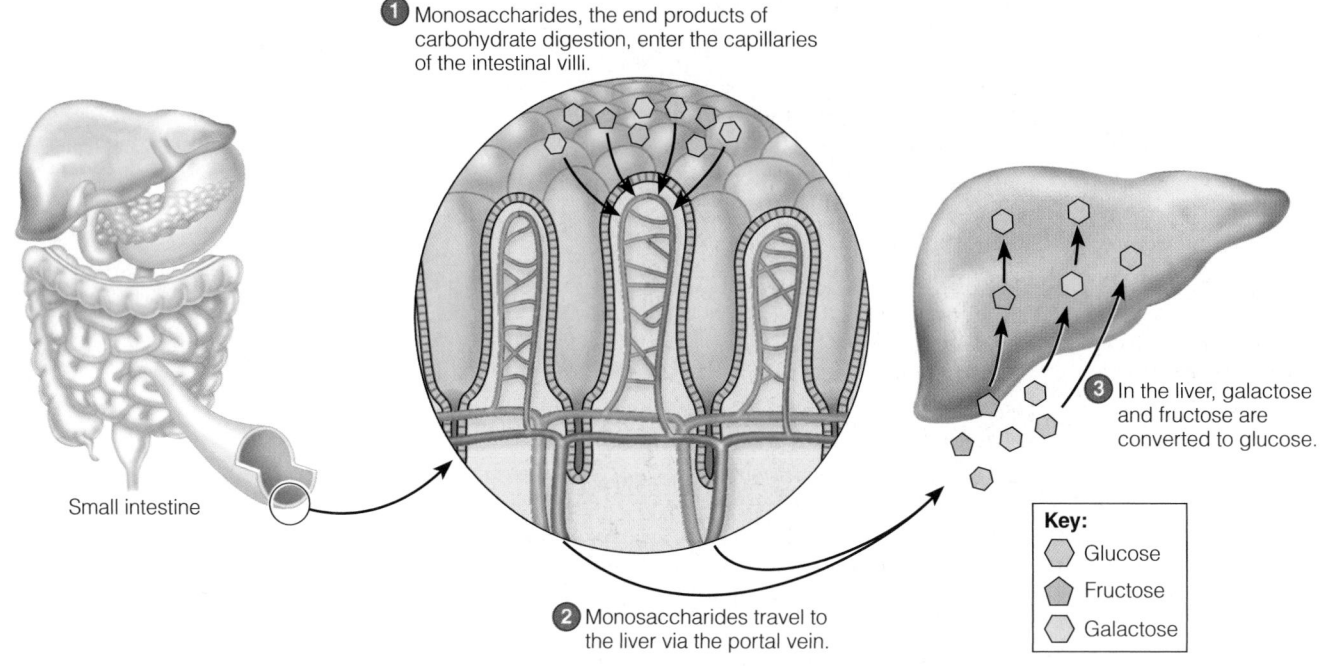

1 Monosaccharides, the end products of carbohydrate digestion, enter the capillaries of the intestinal villi.

Small intestine

2 Monosaccharides travel to the liver via the portal vein.

3 In the liver, galactose and fructose are converted to glucose.

Key:
- Glucose
- Fructose
- Galactose

IN SUMMARY

In the digestion and absorption of carbohydrates, the body breaks down starches into the disaccharide maltose. Maltose and the other disaccharides (lactose and sucrose) from foods are broken down into monosaccharides. Then monosaccharides are converted mostly to glucose to provide energy for the cells' work. The fibers help to regulate the passage of food through the GI system and slow the absorption of glucose, but they contribute little, if any, energy.

Lactose Intolerance

Normally, the intestinal cells produce enough of the enzyme lactase to ensure that the disaccharide lactose found in milk is both digested and absorbed efficiently. Lactase activity is highest immediately after birth, as befits an infant whose first and only food for a while will be breast milk or infant formula. In the great majority of the world's populations, lactase activity declines dramatically during childhood and adolescence to about 5 to 10 percent of the activity at birth. Only a relatively small percentage (about 30 percent) of the people in the world retain enough lactase to digest and absorb lactose efficiently throughout adult life.

Symptoms When more lactose is consumed than the available lactase can handle, lactose molecules remain in the intestine undigested, attracting water and causing bloating, abdominal discomfort, and diarrhea—the symptoms of **lactose intolerance.** The undigested lactose becomes food for intestinal bacteria, which multiply and produce irritating acid and gas, further contributing to the discomfort and diarrhea.

Causes As mentioned, lactase activity commonly declines with age. **Lactase deficiency** may also develop when the intestinal villi are damaged by disease, certain medicines, prolonged diarrhea, or malnutrition. Depending on the extent of the intestinal damage, lactose malabsorption may be temporary or permanent. In extremely rare cases, an infant is born with a lactase deficiency.

lactose intolerance: a condition that results from inability to digest the milk sugar lactose; characterized by bloating, gas, abdominal discomfort, and diarrhea. Lactose intolerance differs from milk allergy, which is caused by an immune reaction to the protein in milk.

lactase deficiency: a lack of the enzyme required to digest the disaccharide lactose into its component monosaccharides (glucose and galactose).

Prevalence The prevalence ◆ of lactose intolerance varies widely among ethnic groups, indicating that the trait is genetically determined. The prevalence of lactose intolerance is lowest among Scandinavians and other northern Europeans and highest among native North Americans and Southeast Asians.

Dietary Changes Managing lactose intolerance requires some dietary changes, although total elimination of milk products usually is not necessary. Excluding all milk products from the diet can lead to nutrient deficiencies because these foods are a major source of several nutrients, notably the mineral calcium, vitamin D, and the B vitamin riboflavin. Fortunately, many people with lactose intolerance can consume foods containing up to 6 grams of lactose ($^1/_2$ cup milk) without symptoms. The most successful strategies are to increase intake of milk products gradually, take them with other foods in meals, and spread their intake throughout the day. A change in the GI bacteria, not the reappearance of the missing enzyme, accounts for the ability to adapt to milk products. Importantly, most lactose-intolerant individuals need to *manage* their dairy consumption rather than *restrict* it.[3]

In many cases, lactose-intolerant people can tolerate fermented milk products such as yogurt and **kefir**.[4] The bacteria in these products digest lactose for their own use, thus reducing the lactose content. Even when the lactose content is equivalent to milk's, yogurt produces fewer symptoms. Hard cheeses, such as cheddar, and cottage cheese are often well tolerated because most of the lactose is removed with the whey during manufacturing. Lactose continues to diminish as cheese ages.

Many lactose-intolerant people use commercially prepared milk products that have been treated with an enzyme that breaks down the lactose. Alternatively, they take enzyme tablets with meals or add enzyme drops to their milk. The enzyme hydrolyzes much of the lactose in milk to glucose and galactose, which lactose-intolerant people can absorb without ill effects.

Because people's tolerance to lactose varies widely, lactose-restricted diets must be highly individualized. A completely lactose-free diet can be difficult because lactose appears not only in milk and milk products but also as an ingredient in many nondairy foods ◆ such as breads, cereals, breakfast drinks, salad dressings, and cake mixes. People on strict lactose-free diets need to read labels and avoid foods that include milk, milk solids, whey (milk liquid), and casein (milk protein, which may contain traces of lactose). They also need to check all medications with the pharmacist because 20 percent of prescription drugs and 5 percent of over-the-counter drugs contain lactose as a filler.

People who consume few or no milk products must take care to meet riboflavin, vitamin D, and calcium needs. Later chapters on the vitamins and minerals offer help with finding good nonmilk sources of these nutrients.

◆ Estimated prevalence of lactose intolerance:
>80% Southeast Asians
80% Native Americans
75% African Americans
70% Mediterranean peoples
60% Inuits
50% Hispanics
20% Caucasians
<10% Northern Europeans

◆ Lactose in selected foods:

Whole-wheat bread, 1 slice	0.5 g
Dinner roll, 1	0.5 g
Cheese, 1 oz	
Cheddar or American	0.5 g
Parmesan or cream	0.8 g
Doughnut (cake type), 1	1.2 g
Chocolate candy, 1 oz	2.3 g
Sherbet, 1 c	4.0 g
Cottage cheese (low-fat), 1 c	7.5 g
Ice cream, 1 c	9.0 g
Milk, 1 c	12.0 g
Yogurt (low-fat), 1 c	15.0 g

Note: Yogurt is often enriched with nonfat milk solids, which increase its lactose content to a level higher than milk's.

IN SUMMARY

Lactose intolerance is a common condition that occurs when there is insufficient lactase to digest the disaccharide lactose found in milk and milk products. Symptoms include GI distress. Because treatment requires limiting milk intake, other sources of riboflavin, vitamin D, and calcium must be included in the diet.

Glucose in the Body

The primary role of the available carbohydrates in human nutrition is to supply the body's cells with glucose for energy. Starch contributes most to the body's glucose supply, but as explained earlier, any of the monosaccharides can also provide glucose.

kefir (keh-FUR)**:** a fermented milk created by adding *Lactobacillus acidophilus* and other bacteria that break down lactose to glucose and galactose, producing a sweet, lactose-free product.

◆ The study of sugars is known as *glycobiology.*

◆ These combination molecules are known as *glycoproteins* and *glycolipids,* respectively.

Scientists have long known that providing energy is glucose's primary role in the body, but they have only recently uncovered additional roles that glucose and other sugars perform in the body.[5] ◆ Sugar molecules dangle from many of the body's protein and fat molecules, with dramatic consequences. Sugars attached to a protein change the protein's shape and function; when they bind to lipids in a cell's membranes, sugars alter the way cells recognize each other.[6] ◆ Cancer cells coated with sugar molecules, for example, are able to sneak by the cells of the immune system. Armed with this knowledge, scientists are now trying to use sugar molecules to create an anticancer vaccine. Further advances in knowledge are sure to reveal numerous ways these simple, yet remarkable, sugar molecules influence the health of the body.

A Preview of Carbohydrate Metabolism

Glucose plays the central role in carbohydrate metabolism. This brief discussion provides just enough information about carbohydrate metabolism to illustrate that the body needs and uses glucose as a chief energy nutrient. Chapter 7 provides a full description of energy metabolism.

Storing Glucose as Glycogen The liver stores about one-third of the body's total glycogen and releases glucose into the bloodstream as needed. After a meal, blood glucose rises, and liver cells link the excess glucose molecules by condensation reactions into long, branching chains of glycogen. When blood glucose falls, the liver cells break glycogen by hydrolysis reactions into single molecules of glucose and release them into the bloodstream. Thus glucose becomes available to supply energy to the brain and other tissues regardless of whether the person has eaten recently. Muscle cells can also store glucose as glycogen (the other two-thirds), but they hoard most of their supply, using it just for themselves during exercise. The brain maintains a small amount of glycogen, which is thought to provide an emergency energy reserve during times of severe glucose deprivation.[7]

The carbohydrates of grains, vegetables, fruits, and legumes supply most of the energy in a healthful diet.

Glycogen holds water and, therefore, is rather bulky. The body can store only enough glycogen to provide energy for relatively short periods of time—less than a day during rest and a few hours at most during exercise. For its long-term energy reserves, for use over days or weeks of food deprivation, the body uses its abundant, water-free fuel, fat, as Chapter 5 describes.

Using Glucose for Energy Glucose fuels the work of most of the body's cells. Inside a cell, enzymes break glucose in half. These halves can be put back together to make glucose, or they can be further broken down into even smaller fragments (never again to be reassembled to form glucose). The small fragments can yield energy when broken down completely to carbon dioxide and water (see Chapter 7).

As mentioned, the liver's glycogen stores last only for hours, not for days. To keep providing glucose to meet the body's energy needs, a person has to eat dietary carbohydrate frequently. Yet people who do not always attend faithfully to their bodies' carbohydrate needs still survive. How do they manage without glucose from dietary carbohydrate? Do they simply draw energy from the other two energy-yielding nutrients, fat and protein? They do draw energy from them, but not simply.

Making Glucose from Protein Glucose is the preferred energy source for brain cells, other nerve cells, and developing red blood cells. Body protein can be converted to glucose to some extent, but protein has jobs of its own that no other nutrient can do. Body fat cannot be converted to glucose to any significant extent. Thus, when a person does not replenish depleted glycogen stores by eating carbohydrate, body proteins are broken down to make glucose to fuel these special cells.

The conversion of protein to glucose is called **gluconeogenesis**—literally, the making of new glucose. Only adequate dietary carbohydrate can prevent this use of protein for energy, and this role of carbohydrate is known as its **protein-sparing action.**

gluconeogenesis (gloo-ko-nee-oh-JEN-ih-sis): the making of glucose from a noncarbohydrate source (described in more detail in Chapter 7).
- **gluco** = glucose
- **neo** = new
- **genesis** = making

protein-sparing action: the action of carbohydrate (and fat) in providing energy that allows protein to be used for other purposes.

Making Ketone Bodies from Fat Fragments An inadequate supply of carbohydrate can shift the body's energy metabolism in a precarious direction. With less carbohydrate providing glucose to meet the brain's energy needs, fat takes an alternative metabolic pathway; instead of entering the main energy pathway, fat fragments combine with each other, forming **ketone bodies.** Ketone bodies provide an alternate fuel source during starvation, but when their production exceeds their use, they accumulate in the blood, causing **ketosis,** a condition that disturbs the body's normal **acid-base balance,** as Chapter 7 describes. (Highlight 9 explores ketosis and the health consequences of low-carbohydrate diets further.)

To spare body protein and prevent ketosis, the body needs at least 50 to 100 grams of carbohydrate a day. Dietary recommendations urge people to select abundantly from carbohydrate-rich foods to provide for considerably more.

Using Glucose to Make Fat After meeting its energy needs and filling its glycogen stores to capacity, the body must find a way to handle any extra glucose. At first, energy metabolism shifts to use more glucose instead of fat. If that isn't enough to restore glucose balance, the liver breaks glucose into smaller molecules and puts them together into the more permanent energy-storage compound—fat. Thus when carbohydrate is abundant, fat is either conserved or created. The fat then travels to the fatty tissues of the body for storage. Unlike the liver cells, which can store only enough glycogen to meet less than a day's energy needs, fat cells can store seemingly unlimited quantities of fat.

The Constancy of Blood Glucose

Every body cell depends on glucose for its fuel to some extent, and the cells of the brain and the rest of the nervous system depend almost exclusively on glucose for their energy. The activities of these cells never cease, and they have limited ability to store glucose. Day and night, they continually draw on the supply of glucose in the fluid surrounding them. To maintain the supply, a steady stream of blood moves past these cells bringing more glucose from either the intestines (food) or the liver (via glycogen breakdown or gluconeogenesis).

Maintaining Glucose Homeostasis To function optimally, the body must maintain blood glucose within limits that permit the cells to nourish themselves. If blood glucose falls below normal, ◆ a person may become dizzy and weak; if it rises above normal, a person may become fatigued. Left untreated, fluctuations to the extremes—either high or low—can be fatal.

The Regulating Hormones Blood glucose homeostasis ◆ is regulated primarily by two hormones: *insulin,* which moves glucose from the blood into the cells, and *glucagon,* which brings glucose out of storage when necessary. Figure 4-12 (p. 114) depicts these hormonal regulators at work.

After a meal, as blood glucose rises, special cells of the pancreas respond by secreting **insulin** into the blood.* In general, the amount of insulin secreted corresponds with the rise in glucose. As the circulating insulin contacts the receptors on the body's other cells, the receptors respond by ushering glucose from the blood into the cells. Most of the cells take only the glucose they can use for energy right away, but the liver and muscle cells can assemble the small glucose units into long, branching chains of glycogen for storage. The liver cells can also convert glucose to fat for export to other cells. Thus elevated blood glucose returns to normal levels as excess glucose is stored as glycogen and fat.

When blood glucose falls (as occurs between meals), other special cells of the pancreas respond by secreting **glucagon** into the blood.† Glucagon raises blood glucose by signaling the liver to break down its glycogen stores and release glucose into the blood for use by all the other body cells.

---

* The *beta* (BAY-tuh) *cells,* one of several types of cells in the pancreas, secrete insulin in response to elevated blood glucose concentration.
† The *alpha cells* of the pancreas secrete glucagon in response to low blood glucose.

◆ Normal blood glucose (fasting): 70 to 100 mg/dL (published values vary slightly).

◆ Reminder: *Homeostasis* is the maintenance of constant internal conditions by the body's control systems.

ketone (KEE-tone) **bodies:** the product of the incomplete breakdown of fat when glucose is not available in the cells.

ketosis (kee-TOE-sis): an undesirably high concentration of ketone bodies in the blood and urine.

acid-base balance: the equilibrium in the body between acid and base concentrations (see Chapter 12).

insulin (IN-suh-lin): a hormone secreted by special cells in the pancreas in response to (among other things) increased blood glucose concentration. The primary role of insulin is to control the transport of glucose from the bloodstream into the muscle and fat cells.

glucagon (GLOO-ka-gon): a hormone that is secreted by special cells in the pancreas in response to low blood glucose concentration and elicits release of glucose from liver glycogen stores.

FIGURE 4-12 Maintaining Blood Glucose Homeostasis

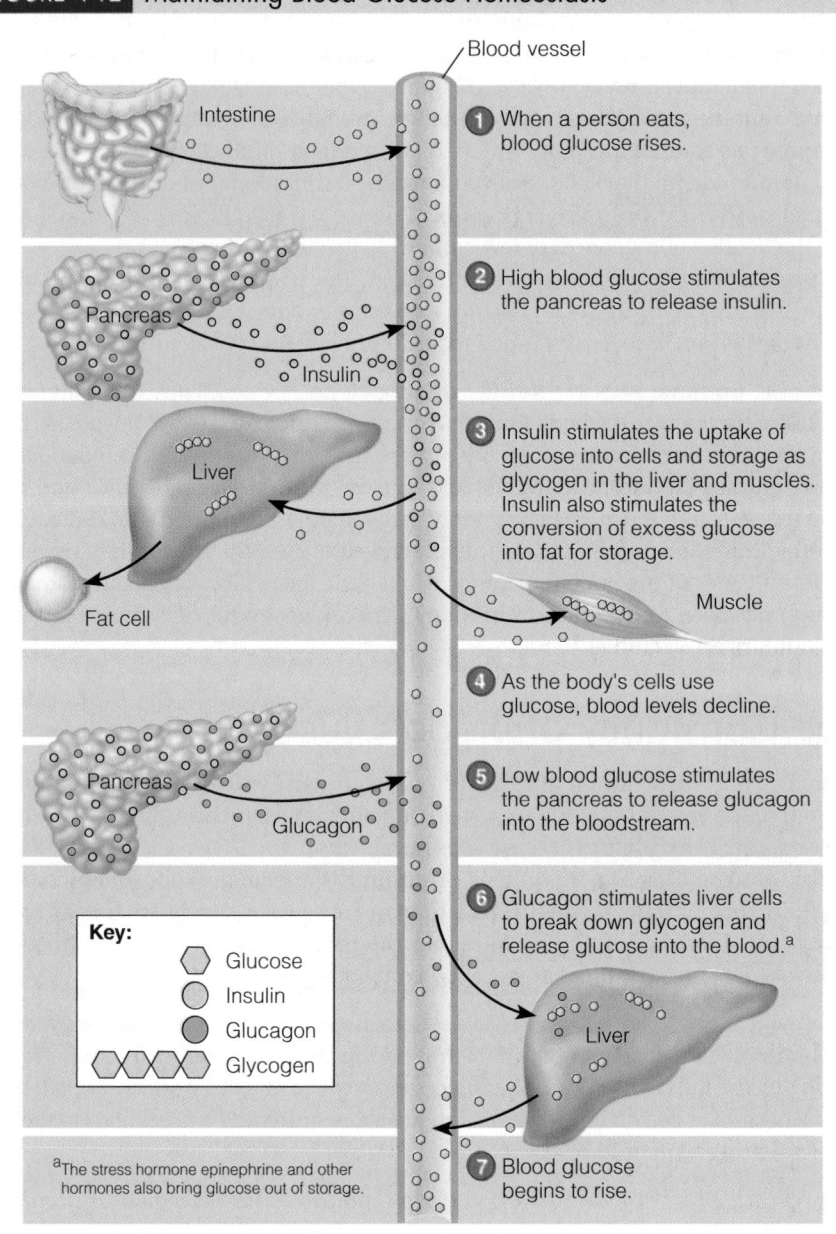

Blood vessel

Intestine

1 When a person eats, blood glucose rises.

Pancreas

Insulin

2 High blood glucose stimulates the pancreas to release insulin.

Liver

Fat cell

Muscle

3 Insulin stimulates the uptake of glucose into cells and storage as glycogen in the liver and muscles. Insulin also stimulates the conversion of excess glucose into fat for storage.

4 As the body's cells use glucose, blood levels decline.

Pancreas

Glucagon

5 Low blood glucose stimulates the pancreas to release glucagon into the bloodstream.

6 Glucagon stimulates liver cells to break down glycogen and release glucose into the blood.[a]

Liver

Key:
- Glucose
- Insulin
- Glucagon
- Glycogen

[a]The stress hormone epinephrine and other hormones also bring glucose out of storage.

7 Blood glucose begins to rise.

Another hormone that signals the liver cells to release glucose is the "fight-or-flight" hormone, **epinephrine.** When a person experiences stress, epinephrine acts quickly, ensuring that all the body cells have energy fuel in emergencies. Among its many roles in the body, epinephrine works to release glucose from liver glycogen to the blood.

Balancing within the Normal Range The maintenance of normal blood glucose ordinarily depends on two processes. When blood glucose falls below normal, food can replenish it, or in the absence of food, glucagon can signal the liver to break down glycogen stores. When blood glucose rises above normal, insulin can signal the cells to take in glucose for energy. Eating balanced meals at regular intervals helps the body maintain a happy medium between the extremes. Balanced meals that provide abundant complex carbohydrates, including fibers and a little fat, help to slow down the digestion and absorption of carbohydrate so that glucose enters the blood gradually, providing a steady, ongoing supply.

epinephrine (EP-ih-NEFF-rin): a hormone of the adrenal gland that modulates the stress response; formerly called **adrenaline.** When administered by injection, epinephrine counteracts anaphylactic shock by opening the airways and maintaining heartbeat and blood pressure.

Falling outside the Normal Range The influence of foods on blood glucose has given rise to the oversimplification that foods *govern* blood glucose concentrations. Foods do not; the body does. In some people, however, blood glucose regulation fails. When this happens, either of two conditions can result: diabetes or hypoglycemia. People with these conditions often plan their diets to help maintain their blood glucose within a normal range.

Diabetes In **diabetes,** blood glucose surges after a meal and remains above normal levels ◆ because insulin is either inadequate or ineffective. Thus *blood* glucose is central to diabetes, but *dietary* carbohydrates do not cause diabetes.

There are two main types of diabetes. In **type 1 diabetes,** the less common type, the pancreas fails to produce insulin. Although the exact cause is unclear, some research suggests that in genetically susceptible people, certain viruses activate the immune system to attack and destroy cells in the pancreas as if they were foreign cells. In **type 2 diabetes,** the more common type of diabetes, the cells fail to respond to insulin. ◆ This condition tends to occur as a consequence of obesity. As the incidence of obesity in the United States has risen in recent decades, the incidence of diabetes has followed. This trend is most notable among children and adolescents, as obesity among the nation's youth reaches epidemic proportions. Because obesity can precipitate type 2 diabetes, the best preventive measure is to maintain a healthy body weight. Concentrated sweets are not strictly excluded from the diabetic diet as they once were; they can be eaten in limited amounts with meals as part of a healthy diet. Chapter 14 describes the type of diabetes that develops in some women during pregnancy (gestational diabetes), and Chapter 26 gives full coverage to type 1 and type 2 diabetes and their associated problems.

Hypoglycemia In healthy people, blood glucose rises after eating and then gradually falls back into the normal range. The transition occurs without notice. Should blood glucose drop below normal, a person would experience the symptoms of **hypoglycemia:** weakness, rapid heartbeat, sweating, anxiety, hunger, and trembling. Most commonly, hypoglycemia is a consequence of poorly managed diabetes. Too much insulin, strenuous physical activity, inadequate food intake, or illness that causes blood glucose levels to plummet.

Hypoglycemia in healthy people is rare. Most people who experience hypoglycemia need only adjust their diets by replacing refined carbohydrates with fiber-rich carbohydrates and ensuring an adequate protein intake. In addition, smaller meals eaten more frequently may help. Hypoglycemia caused by certain medications, pancreatic tumors, overuse of insulin, alcohol abuse, uncontrolled diabetes, or other illnesses requires medical intervention.

The Glycemic Response The **glycemic response** refers to how quickly glucose is absorbed after a person eats, how high blood glucose rises, and how quickly it returns to normal. Slow absorption, a modest rise in blood glucose, and a smooth return to normal are desirable (a low glycemic response). Fast absorption, a surge in blood glucose, and an overreaction that plunges glucose below normal are less desirable (a high glycemic response). Different foods have different effects on blood glucose.

The rate of glucose absorption is particularly important to people with diabetes, who may benefit from limiting foods that produce too great a rise, or too sudden a fall, in blood glucose. To aid their choices, they may be able to use the **glycemic index,** a method of classifying foods according to their potential to raise blood glucose. ◆ Figure 4-13 (p. 116) ranks selected foods by their glycemic index.[8] Some studies have shown that selecting foods with a low glycemic index is a practical way to improve glucose control.[9]

Lowering the glycemic index of the *diet* may improve blood lipids and reduce the risk of heart disease as well.[10] A low glycemic diet may also help with weight management, although research findings are mixed.[11] Fibers and other slowly digested

◆ Blood glucose (fasting):
- Prediabetes: 100 to 125 mg/dL
- Diabetes: ≥ 126 mg/dL

◆ The condition of having blood glucose levels higher than normal, but below the diagnosis of diabetes, is sometimes called **prediabetes.**

◆ A related term, **glycemic load,** reflects both the glycemic index and the amount of carbohydrate.

diabetes (DYE-uh-BEET-eez): a chronic disorder of carbohydrate metabolism, usually resulting from insufficient or ineffective insulin.

type 1 diabetes: the less common type of diabetes in which the pancreas fails to produce insulin.

type 2 diabetes: the more common type of diabetes in which the cells fail to respond to insulin.

hypoglycemia (HIGH-po-gly-SEE-me-ah): an abnormally low blood glucose concentration.

glycemic (gly-SEEM-ic) **response:** the extent to which a food raises the blood glucose concentration and elicits an insulin response.

glycemic index: a method of classifying foods according to their potential for raising blood glucose.

FIGURE 4-13 Glycemic Index of Selected Foods

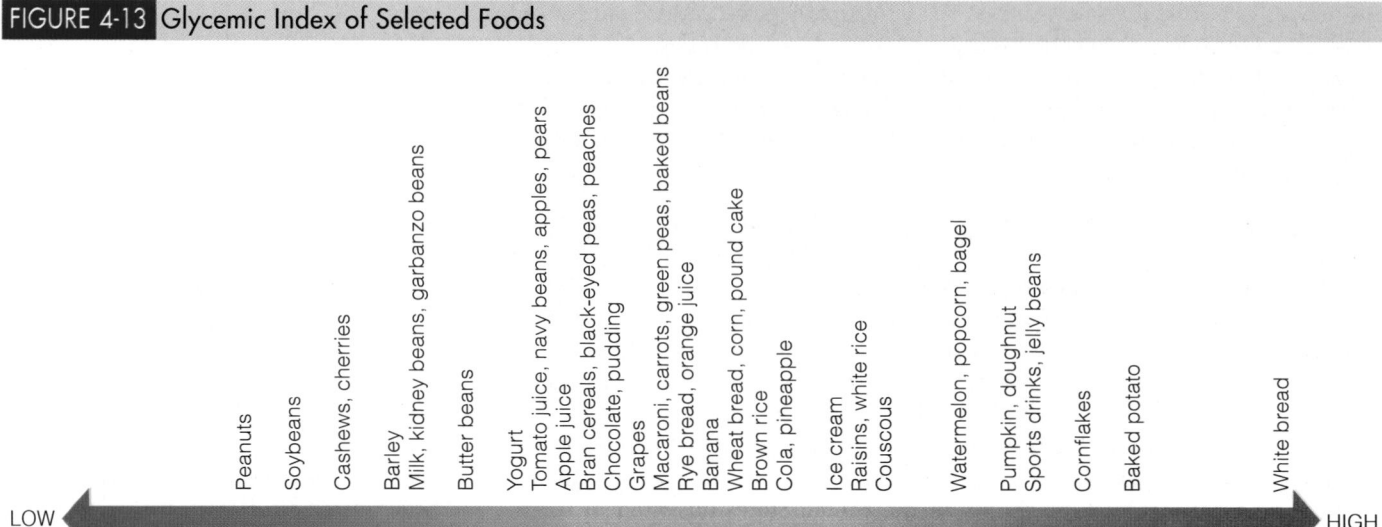

carbohydrates prolong the presence of foods in the digestive tract, thus providing greater satiety and diminishing the insulin response, which can help with weight control.[12] In contrast, the rapid absorption of glucose from a high glycemic diet seems to increase the risk of heart disease and promote overeating in some overweight people.[13]

Despite these possible benefits, the usefulness of the glycemic index is surrounded by controversy as researchers debate whether selecting foods based on the glycemic index is practical or offers any real health benefits.[14] Those opposing the use of the glycemic index argue that it is not sufficiently supported by scientific research.[15] The glycemic index has been determined for relatively few foods, and when the glycemic index has been established, it is based on an average of multiple tests with wide variations in their results. Values vary because of differences in the physical and chemical characteristics of foods, testing methods of laboratories, and digestive processes of individuals.

Furthermore, the practical utility of the glycemic index is limited because this information is neither provided on food labels nor intuitively apparent. Indeed, a food's glycemic index is not always what one might expect. Ice cream, for example, is a high-sugar food but produces less of a glycemic response than baked potatoes, a high-starch food. This effect is most likely because the fat in the ice cream slows GI motility and thus the rate of glucose absorption. Mashed potatoes produce more of a response than honey, probably because the fructose content of honey has little effect on blood glucose. In fact, sugars such as fructose generally have a moderate to low glycemic index.[16] Perhaps most relevant to real life, a food's glycemic effect differs depending on plant variety, food processing, cooking method, and whether it is eaten alone or with other foods.[17] Most people eat a variety of foods, cooked and raw, that provide different amounts of carbohydrate, fat, and protein—all of which influence the glycemic index of a meal.

Paying attention to the glycemic index may not be necessary because current guidelines already suggest many low glycemic index choices: whole grains, legumes, vegetables, fruits, and milk products. In addition, eating frequent, small meals spreads glucose absorption across the day and thus offers similar metabolic advantages to eating foods with a low glycemic response. People wanting to follow a low glycemic diet should be careful not to adopt a low-carbohydrate diet as well. The problems associated with a low-carbohydrate diet are addressed in Highlight 9.

IN SUMMARY

Dietary carbohydrates provide glucose that can be used by the cells for energy, stored by the liver and muscles as glycogen, or converted into fat if intakes exceed needs. All of the body's cells depend on glucose; those of the central nervous system are especially dependent on it. Without glucose, the body is forced to break down its protein tissues to make glucose and to alter energy metabolism to make ketone bodies from fats. Blood glucose regulation depends primarily on two pancreatic hormones: insulin to move glucose from the blood into the cells when levels are high and glucagon to free glucose from glycogen stores and release it into the blood when levels are low. The glycemic index measures how blood glucose responds to foods.

Health Effects and Recommended Intakes of Sugars

Ever since people first discovered honey and dates, they have enjoyed the sweetness of sugars. In the United States, the natural sugars of milk, fruits, vegetables, and grains account for about half of the sugar intake; the other half consists of sugars that have been refined and added to foods for a variety of purposes. ◆ The use of **added sugars** has risen steadily over the past several decades, both in the United States and around the world, with soft drinks and sugared fruit drinks accounting for most of the increase.[18] These added sugars assume various names on food labels: sucrose, invert sugar, corn sugar, corn syrups and solids, high-fructose corn syrup, and honey. A food is likely to be high in added sugars if its ingredient list starts with any of the sugars named in the glossary (p. 118) or if it includes several of them.

◆ As an additive, sugar:
  - Enhances flavor
  - Supplies texture and color to baked goods
  - Provides fuel for fermentation, causing bread to rise or producing alcohol
  - Acts as a bulking agent in ice cream and baked goods
  - Acts as a preservative in jams
  - Balances the acidity of tomato- and vinegar-based products

Health Effects of Sugars

In moderate amounts, sugars add pleasure to meals without harming health. In excess, however, they can be detrimental in two ways. One, sugars can contribute to nutrient deficiencies by supplying energy (kcalories) without providing nutrients. Two, sugars contribute to tooth decay.

Nutrient Deficiencies Empty-kcalorie foods that contain lots of added sugar such as cakes, candies, and sodas deliver glucose and energy with few, if any, other nutrients. By comparison, foods such as whole grains, vegetables, legumes, and fruits that contain some natural sugars and lots of starches and fibers deliver protein, vitamins, and minerals along with their glucose and energy.

A person spending 200 kcalories of a day's energy allowance on a 16-ounce soda gets little of value for those kcaloric "dollars." In contrast, a person using 200 kcalories on three slices of whole-wheat bread gets 9 grams of protein, 6 grams of fiber, plus several of the B vitamins with those kcalories. For the person who wants something sweet, a reasonable compromise might be two slices of bread with a teaspoon of jam on each. The amount of sugar a person can afford to eat depends on how many kcalories are available beyond those needed to deliver indispensable vitamins and minerals.

With careful food selections, a typical adult can obtain all the needed nutrients within an allowance of about 1500 kcalories. Some people have more generous energy allowances with which to "purchase" nutrients. For example, an active teenage boy may need as many as 3000 kcalories a day. If he eats mostly nutritious foods, then the "empty kcalories" of cola beverages

© Polara Studios Inc.

Over half of the added sugars in our diet come from soft drinks and table sugar, but baked goods, fruit drinks, ice cream, candy, and breakfast cereals also make substantial contributions.

added sugars: sugars and syrups used as an ingredient in the processing and preparation of foods such as breads, cakes, beverages, jellies, and ice cream as well as sugars eaten separately or added to foods at the table.

GLOSSARY OF ADDED SUGARS

brown sugar: refined white sugar crystals to which manufacturers have added molasses syrup with natural flavor and color; 91 to 96% pure sucrose.

confectioners' sugar: finely powdered sucrose, 99.9% pure.

corn sweeteners: corn syrup and sugars derived from corn.

corn syrup: a syrup made from cornstarch that has been treated with acid, high temperatures, and enzymes that produce glucose, maltose, and dextrins. See also *high-fructose corn syrup (HFCS)*.

dextrose: an older name for glucose.

granulated sugar: crystalline sucrose; 99.9% pure.

high-fructose corn syrup (HFCS): a syrup made from cornstarch that has been treated with an enzyme that converts some of the glucose to the sweeter fructose; made especially for use in processed foods and beverages, where it is the predominant sweetener. With a chemical structure similar to sucrose, HFCS has a fructose content of 42, 55, or 90%, with glucose making up the remainder.

honey: sugar (mostly sucrose) formed from nectar gathered by bees. An enzyme splits the sucrose into glucose and fructose. Composition and flavor vary, but honey always contains a mixture of sucrose, fructose, and glucose.

invert sugar: a mixture of glucose and fructose formed by the hydrolysis of sucrose in a chemical process; sold only in liquid form and sweeter than sucrose. Invert sugar is used as a food additive to help preserve freshness and prevent shrinkage.

levulose: an older name for fructose.

maple sugar: a sugar (mostly sucrose) purified from the concentrated sap of the sugar maple tree.

molasses: the thick brown syrup produced during sugar refining. Molasses retains residual sugar and other by-products and a few minerals; blackstrap molasses contains significant amounts of calcium and iron.

raw sugar: the first crop of crystals harvested during sugar processing. Raw sugar cannot be sold in the United States because it contains too much filth (dirt, insect fragments, and the like). Sugar sold as "raw sugar" domestically has actually gone through over half of the refining steps.

turbinado (ter-bih-NOD-oh) **sugar:** sugar produced using the same refining process as white sugar, but without the bleaching and anti-caking treatment. Traces of molasses give turbinado its sandy color.

white sugar: pure sucrose or "table sugar," produced by dissolving, concentrating, and recrystallizing raw sugar.

---

may be an acceptable addition to his diet. In contrast, an inactive older woman who is limited to fewer than 1500 kcalories a day can afford to eat only the most nutrient-dense foods.

Some people believe that because honey is a natural food, it is nutritious—or, at least, more nutritious than sugar.* A look at their chemical structures reveals the truth. Honey, like table sugar, contains glucose and fructose. The primary difference is that in table sugar the two monosaccharides are bonded together as a disaccharide, whereas in honey some of them are free. Whether a person eats monosaccharides individually, as in honey, or linked together, as in table sugar, they end up the same way in the body: as glucose and fructose.

Honey does contain a few vitamins and minerals, but not many, as Table 4-1 shows. Honey is denser than crystalline sugar, too, so it provides more energy per spoonful.

---

* Honey should never be fed to infants because of the risk of botulism. Chapters 16 and 19 provide more details.

TABLE 4-1 Sample Nutrients in Sugar and Other Foods

The indicated portion of any of these foods provides approximately 100 kcalories. Notice that for a similar number of kcalories and grams of carbohydrate, milk, legumes, fruits, grains, and vegetables offer more of the other nutrients than do the sugars.

	Size of 100 kcal Portion	Carbohydrate (g)	Protein (g)	Calcium (mg)	Iron (mg)	Vitamin A (µg)	Vitamin C (mg)
Foods							
Milk, 1% low-fat	1 c	12	8	300	0.1	144	2
Kidney beans	½ c	20	7	30	1.6	0	2
Apricots	6	24	2	30	1.1	554	22
Bread, whole-wheat	1½ slices	20	4	30	1.9	0	0
Broccoli, cooked	2 c	20	12	188	2.2	696	148
Sugars							
Sugar, white	2 tbs	24	0	trace	trace	0	0
Molasses, blackstrap	2½ tbs	28	0	343	12.6	0	0.1
Cola beverage	1 c	26	0	6	trace	0	0
Honey	1½ tbs	26	trace	2	0.2	0	trace

This is not to say that all sugar sources are alike, for some are more nutritious than others. Consider a fruit, say, an orange. The fruit may give you the same amounts of fructose and glucose and the same number of kcalories as a dose of sugar or honey, but the packaging is more valuable nutritionally. The fruit's sugars arrive in the body diluted in a large volume of water, packaged in fiber, and mixed with essential vitamins, minerals, and phytochemicals.

As these comparisons illustrate, the significant difference between sugar sources is not between "natural" honey and "purified" sugar but between concentrated sweets and the dilute, naturally occurring sugars that sweeten foods. You can suspect an exaggerated nutrition claim when someone asserts that one product is more nutritious than another because it contains honey.

Sugar can contribute to nutrient deficiencies only by displacing nutrients. For nutrition's sake, the appropriate attitude to take is not that sugar is "bad" and must be avoided, but that nutritious foods must come first. If nutritious foods crowd sugar out of the diet, that is fine—but not the other way around. As always, the goals to seek are balance, variety, and moderation.

Dental Caries Sugars from foods and from the breakdown of starches in the mouth can contribute to tooth decay. Bacteria in the mouth ferment the sugars and, in the process, produce an acid that erodes tooth enamel (see Figure 4-14), causing **dental caries,** or tooth decay. People can eat sugar without this happening, though, for much depends on how long foods stay in the mouth. Sticky foods stay on the teeth longer and continue to yield acid longer than foods that are readily cleared from the mouth. For that reason, sugar in a juice consumed quickly, for example, is less likely to cause dental caries than sugar in a pastry. By the same token, the sugar in sticky foods such as dried fruits can be more detrimental than its quantity alone would suggest.

Another concern is how often people eat sugar. Bacteria produce acid for 20 to 30 minutes after each exposure. If a person eats three pieces of candy at one time, the teeth will be exposed to approximately 30 minutes of acid destruction. But, if the person eats three pieces at half-hour intervals, the time of exposure increases to 90 minutes. Likewise, slowly sipping a sugary sports beverage may be more harmful than drinking quickly and clearing the mouth of sugar. Nonsugary foods can help remove sugar from tooth surfaces; hence, it is better to eat sugar with meals than between meals.[19] Foods such as milk and cheese may be particularly helpful in minimizing the effects of the acids and in restoring the lost enamel.[20]

Beverages such as soft drinks, orange juice, and sports drinks not only contain sugar but also have a low pH. These acidic drinks can erode tooth enamel and may explain why dental erosion is highly prevalent today.[21]

The development of caries depends on several factors: the bacteria that reside in **dental plaque,** the saliva that cleanses the mouth, the minerals that form the teeth, and the foods that remain after swallowing. For most people, good oral hygiene will prevent ◆ dental caries. In fact, regular brushing (twice a day, with a fluoride toothpaste) and flossing may be more effective in preventing dental caries than restricting sugary foods.

Dietary Guidelines for Americans 2005

Reduce the incidence of dental caries by practicing good oral hygiene and consuming sugar- and starch-containing foods and beverages less frequently.

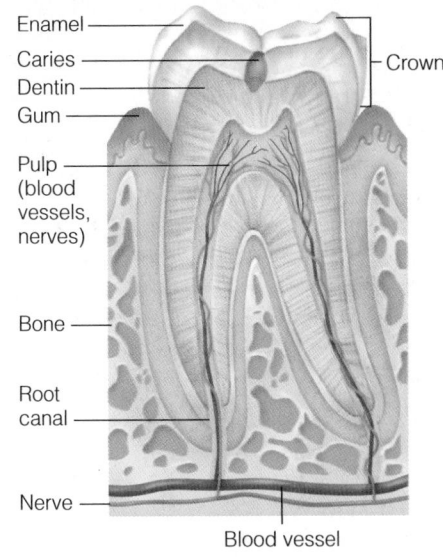

FIGURE 4-14 Dental Caries

Dental caries begins when acid dissolves the enamel that covers the tooth. If not repaired, the decay may penetrate the dentin and spread into the pulp of the tooth, causing inflammation, abscess, and possible loss of the tooth.

Enamel
Caries
Dentin
Gum
Crown
Pulp (blood vessels, nerves)
Bone
Root canal
Nerve
Blood vessel

◆ To prevent dental caries:
- Limit between-meal snacks containing sugars and starches.
- Brush and floss teeth regularly.
- If brushing and flossing are not possible, at least rinse with water.

dental caries: decay of teeth.
- **caries** = rottenness

dental plaque: a gummy mass of bacteria that grows on teeth and can lead to dental caries and gum disease.

Controversies Surrounding Sugars

Sugars have been blamed for a variety of other health problems.[22] The following paragraphs evaluate some of these controversies.

You receive about the same amount and kinds of sugars from an orange as from a tablespoon of honey, but the packaging makes a big nutrition difference.

Controversy: Does Sugar Cause Obesity? Over the past three decades, obesity rates have risen sharply in the United States. During the same period, consumption of added sugars has reached an all-time high—much of it because of the dramatic rise in high-fructose corn syrup used in beverages. Between 1977 and 2001, as people grew fatter, their intake of kcalories from fruit drinks and punches doubled and kcalories from soft drinks nearly tripled.[23] Although the use of this sweetener parallels unprecedented gains of body fatness, does it mean that the increasing sugar intakes are responsible for the increase in obesity?[24]

When eaten in excess of need, energy from added sugars contributes to body fat stores, just as excess energy from other sources does. Added sugars provide excess kcalories, raising the risk of weight gain and type 2 diabetes.[25] When total kcalorie intakes are controlled, however, *moderate* amounts of sugar do not cause obesity.[26]

People with diets *high* in added sugars often consume more kcalories each day than people with lower sugar intakes. Adolescents, for example, who drink as much as 26 ounces or more (about two cans) of sugar-sweetened soft drinks daily, consume 400 more kcalories a day than teens who don't. Overweight children and adolescents increase their risk of becoming obese by 60 percent with each additional syrup-sweetened drink they add to their daily diet. The liquid form of sugar in soft drinks makes it especially easy to overconsume kcalories.[27] Investigators are evaluating these and other possible links between fructose in the syrupy sweeteners of soft drinks and weight gain.[28] Research suggests that fructose from these added sugars favors the fat-making pathways.[29]

Limiting selections of foods and beverages high in added sugars can be an effective weight-loss strategy, especially for people whose excess kcalories come primarily from added sugars.[30] Replacing a can of cola with a glass of water every day, for example, can help a person lose a pound (or at least not gain a pound) in one month. That may not sound like much, but it adds up to more than 10 pounds a year, for very little effort.

Controversy: Does Sugar Cause Heart Disease? A diet high in added sugars can alter blood lipids to favor heart disease.[31] (Lipids include fats and cholesterol, as Chapter 5 explains.) This effect is most dramatic in people who respond to sucrose with abnormally high insulin secretions, which promote the making of excess fat.[32] For most people, though, moderate sugar intakes do *not* elevate blood lipids. To keep these findings in perspective, consider that heart disease correlates most closely with factors that have nothing to do with nutrition, such as smoking and genetics. Among dietary risk factors, several—such as saturated fats, *trans* fats, and cholesterol—have much stronger associations with heart disease than do sugar intakes.

Controversy: Does Sugar Cause Misbehavior in Children and Criminal Behavior in Adults? Sugar has been blamed for the misbehaviors of hyperactive children, delinquent adolescents, and lawbreaking adults. Such speculations have been based on personal stories and have not been confirmed by scientific research. No scientific evidence supports a relationship between sugar and hyperactivity or other misbehaviors. Chapter 15 provides accurate information on diet and children's behavior.

Controversy: Does Sugar Cause Cravings and Addictions? Foods in general, and carbohydrates and sugars more specifically, are not physically addictive in the ways that drugs are. Yet some people describe themselves as having "carbohydrate cravings" or being "sugar addicts." One frequently noted theory is that people seek carbohydrates as a way to increase their levels of the brain neurotransmitter **serotonin,** which elevates mood. Interestingly, when those with self-described carbohydrate cravings indulge, they tend to eat more of everything, but the percentage of energy from carbohydrates remains unchanged.[33] Alcohol also raises serotonin levels, and alcohol-dependent people who crave carbohydrates seem to handle sobriety better when given a high-carbohydrate diet.

One reasonable explanation for the carbohydrate cravings that some people experience involves the self-imposed labeling of a food as both "good" and "bad"—that is, one that is desirable but should be eaten with restraint. Chocolate is a familiar ex-

serotonin (SER-oh-TONE-in): a neurotransmitter important in sleep regulation, appetite control, intestinal motility, obsessive-compulsive behaviors, and mood disorders.

ample. Restricting intake heightens the desire further (a "craving"). Then "addiction" is used to explain why resisting the food is so difficult and, sometimes, even impossible. But the "addiction" is not pharmacological; a capsule of the psychoactive substances commonly found in chocolate, for example, does not satisfy the craving.

Recommended Intakes of Sugars

Because added sugars deliver kcalories but few or no nutrients, the 2005 *Dietary Guidelines* urge consumers to "choose and prepare foods and beverages with little added sugars." The USDA Food Guide counts these sugar kcalories (and those from solid fats and alcohol) as discretionary kcalories. Most people need to limit their use of added sugars. ◆ Estimates indicate that, on average, each person in the United States consumes about 105 pounds (almost 50 kilograms) of added sugar per year, or about 30 teaspoons (about 120 grams) of added sugar a day, an amount that exceeds these guidelines.[34]

◆ USDA Food Guide amounts of added sugars that can be included as discretionary kcalories when food choices are nutrient dense and fat ≤ 30% total kcal:
- 3 tsp for 1600 kcal diet
- 5 tsp for 1800 kcal diet
- 8 tsp for 2000 kcal diet
- 9 tsp for 2200 kcal diet
- 12 tsp for 2400 kcal diet

Dietary Guidelines for Americans 2005

Choose and prepare foods and beverages with little added sugars.

Estimating the *added* sugars in a diet is not always easy for consumers. Food labels list the total grams of sugar a food provides, but this total reflects both added sugars and those occurring naturally in foods. To help estimate sugar and energy intakes accurately, the list in the margin ◆ shows the amounts of concentrated sweets that are equivalent to 1 teaspoon of white sugar. These sugars all provide *about* 5 grams of carbohydrate and about 20 kcalories per teaspoon. Some are lower (16 kcalories for table sugar), and others are higher (22 kcalories for honey), but a 20-kcalorie average is an acceptable approximation. For a person who uses catsup liberally, it may help to remember that 1 tablespoon of catsup supplies about 1 teaspoon of sugar.

The DRI Committee did not set an upper level for sugar, but as mentioned, excessive intakes can interfere with sound nutrition and dental health. Few people can eat lots of sugary treats and still meet all of their nutrient needs without exceeding their kcalorie allowance. Specifically, the DRI suggests that added sugars should account for no more than 25 percent of the day's total energy intake.[35] When added sugars occupy this much of a diet, however, intakes from the five food groups fall below recommendations. For a person consuming 2000 kcalories a day, 25 percent represents 500 kcalories (that is, 125 grams, or 31 teaspoons) from concentrated sugars—and that's a lot of sugar. ◆ Perhaps an athlete in training whose energy needs are high can afford the added sugars from sports drinks without compromising nutrient intake, but most people do better by limiting their use of added sugars. The World Health Organization (WHO) and the Food and Agriculture Organization (FAO) suggest restricting consumption of added sugars to less than 10 percent of total energy.

◆ 1 tsp white sugar =
- 1 tsp brown sugar
- 1 tsp candy
- 1 tsp corn sweetener or corn syrup
- 1 tsp honey
- 1 tsp jam or jelly
- 1 tsp maple sugar or maple syrup
- 1 tsp molasses
- $1^1/_2$ oz carbonated soda
- 1 tbs catsup

◆ For perspective, each of these concentrated sugars provides about 500 kcal:
- 40 oz cola
- $^1/_2$ c honey
- 125 jelly beans
- 23 marshmallows
- 30 tsp sugar

How many kcalories from sugar does your favorite beverage or snack provide?

IN SUMMARY

Sugars pose no major health threat except for an increased risk of dental caries. Excessive intakes, however, may displace needed nutrients and fiber and may contribute to obesity when energy intake exceeds needs. A person deciding to limit daily sugar intake should recognize that not all sugars need to be restricted, just concentrated sweets, which are relatively empty of other nutrients and high in kcalories. Sugars that occur naturally in fruits, vegetables, and milk are acceptable.

Foods rich in starch and fiber offer many health benefits.

◆ Consuming 5 to 10 g of soluble fiber daily reduces blood cholesterol by 3 to 5%. For perspective, $1/2$ c dry oat bran provides 8 g of fiber, and 1 c cooked barley or $1/2$ c cooked legumes provides about 6 g of fiber.

Health Effects and Recommended Intakes of Starch and Fibers

Carbohydrates and fats are the two major sources of energy in the diet. When one is high, the other is usually low—and vice versa. A diet that provides abundant carbohydrate (45 to 65 percent of energy intake) and some fat (20 to 35 percent of energy intake) within a reasonable energy allowance best supports good health. To increase carbohydrate in the diet, focus on whole grains, vegetables, legumes, and fruits—foods noted for their starch, fibers, and naturally occurring sugars.

Health Effects of Starch and Fibers

In addition to starch, fibers, and natural sugars, whole grains, vegetables, legumes, and fruits supply valuable vitamins and minerals and little or no fat. The following paragraphs describe some of the health benefits of diets that include a variety of these foods daily.

Heart Disease High-carbohydrate diets, especially those rich in whole grains, may protect against heart disease and stroke, although sorting out the exact reasons why can be difficult.[36] Such diets are low in animal fat and cholesterol and high in fibers, vegetable proteins, and phytochemicals—all factors associated with a lower risk of heart disease. (The role of animal fat and cholesterol in heart disease is discussed in Chapter 5. The role of vegetable proteins in heart disease is presented in Chapter 6. The benefits of phytochemicals in disease prevention are featured in Highlight 13.)

Foods rich in soluble fibers (such as oat bran, barley, and legumes) lower blood cholesterol ◆ by binding with bile acids and thereby increasing their excretion. Consequently, the liver must use its cholesterol to make new bile acids. In addition, the bacterial by-products of fiber fermentation in the colon also inhibit cholesterol synthesis in the liver. The net result is lower blood cholesterol.[37]

Several researchers have speculated that fiber may also exert its effect by displacing fats in the diet. Whereas this is certainly helpful, even when dietary fat is low, high intakes of fibers exert a separate and significant cholesterol-lowering effect. In other words, a high-fiber diet helps to decrease the risk of heart disease independent of fat intake.[38]

Diabetes High-fiber foods—especially whole grains—play a key role in reducing the risk of type 2 diabetes.[39] When soluble fibers trap nutrients and delay their transit through the GI tract, glucose absorption is slowed, which helps to prevent the glucose surge and rebound that seem to be associated with diabetes onset.

GI Health Dietary fibers enhance the health of the large intestine. The healthier the intestinal walls, the better they can block absorption of unwanted constituents. Fibers such as cellulose (as in cereal brans, fruits, and vegetables) increase stool weight, easing passage, and reduce transit time. In this way, the fibers help to alleviate or prevent constipation.

Taken with ample fluids, fibers help to prevent several GI disorders. Large, soft stools ease elimination for the rectal muscles and reduce the pressure in the lower bowel, making it less likely that rectal veins will swell (hemorrhoids). Fiber prevents compaction of the intestinal contents, which could obstruct the appendix and permit bacteria to invade and infect it (appendicitis). In addition, fiber stimulates the GI tract muscles so that they retain their strength and resist bulging out into pouches known as diverticula (illustrated in Figure H3-3 on p. 95).[40]

Cancer Many, but not all, research studies suggest that increasing dietary fiber protects against colon cancer.[41] When the largest study of diet and cancer to date examined the diets of over a half million people in ten countries for four and a half years, the researchers found an inverse association between dietary fiber and colon cancer.[42] People who ate the most dietary fiber (35 grams per day) reduced their risk of colon cancer by 40 percent compared with those who ate the least fiber (15 grams per day). Importantly, the study focused on dietary fiber, not fiber supplements or additives, which lack valuable nutrients and phytochemicals that also help protect against cancer. Plant foods—vegetables, fruits, and whole-grain products—reduce the risks of colon and rectal cancers.[43]

Fibers may help prevent colon cancer by diluting, binding, and rapidly removing potential cancer-causing agents from the colon. In addition, soluble fibers stimulate bacterial fermentation of resistant starch and fiber in the colon, a process that produces short-chain fatty acids that lower the pH. These small fat molecules activate cancer-killing enzymes and inhibit inflammation in the colon.[44]

Weight Management High-fiber and whole-grain foods help a person to maintain a healthy body weight.[45] Foods rich in complex carbohydrates tend to be low in fat and added sugars and can therefore promote weight loss by delivering less energy ◆ per bite. In addition, as fibers absorb water from the digestive juices, they swell, creating feelings of fullness and delaying hunger.

Many weight-loss products on the market today contain bulk-inducing fibers such as methylcellulose, but buying pure fiber compounds like this is neither necessary nor advisable. Most experts agree that the health and weight management benefits attributed to fiber may come from other constituents of fiber-containing foods, and not from fiber alone.[46] For this reason, consumers should select whole grains, legumes, fruits, and vegetables instead of fiber supplements. High-fiber foods not only add bulk to the diet but are economical and nutritious as well. Table 4-2 summarizes fibers and their health benefits.

Harmful Effects of Excessive Fiber Intake Despite fibers' benefits to health, a diet high in fiber also has a few drawbacks. A person who has a small capacity and eats mostly high-fiber foods may not be able to take in enough food to meet energy or nutrient needs. The malnourished, the elderly, and young children adhering to all-plant (vegan) diets are especially vulnerable to this problem.

Launching suddenly into a high-fiber diet can cause temporary bouts of abdominal discomfort, gas, and diarrhea and, more seriously, can obstruct the GI tract. To

◆ Reminder:
 • Carbohydrate: 4 kcal/g
 • Fat: 9 kcal/g

TABLE 4-2 Dietary Fibers: Their Characteristics, Food Sources, and Health Effects in the Body

Fiber Characteristics	Major Food Sources	Actions in the Body	Health Benefits
Soluble, viscous, more fermentable • Gums and mucilages • Pectins • Psyllium[a] • Some hemicelluloses	Whole-grain products (barley, oats, oat bran, rye), fruits (apples, citrus), legumes, seeds and husks, vegetables; also extracted and used as food additives	• Lower blood cholesterol by binding bile • Slow glucose absorption • Slow transit of food through upper GI tract • Hold moisture in stools, softening them • Yield small fat molecules after fermentation that the colon can use for energy	• Lower risk of heart disease • Lower risk of diabetes
Insoluble, nonviscous, less fermentable • Cellulose • Lignins • Psyllium[a] • Resistant starch • Many hemicelluloses	Brown rice, fruits, legumes, seeds, vegetables (cabbage, carrots, brussels sprouts), wheat bran, whole grains; also extracted and used as food additives	• Increase fecal weight and speed fecal passage through colon • Provide bulk and feelings of fullness	• Alleviate constipation • Lower risks of diverticulosis, hemorrhoids, and appendicitis • May help with weight management

[a]Psyllium, a fiber laxative and cereal additive, has both soluble and insoluble properties.

prevent such complications, a person adopting a high-fiber diet can take the following precautions:

- Increase fiber intake gradually over several weeks to give the GI tract time to adapt.
- Drink plenty of liquids to soften the fiber as it moves through the GI tract.
- Select fiber-rich foods from a variety of sources—fruits, vegetables, legumes, and whole-grain breads and cereals.

Some fibers can limit the absorption of nutrients by speeding the transit of foods through the GI tract and by binding to minerals. When mineral intake is adequate, however, a reasonable intake of high-fiber foods does not seem to compromise mineral balance.

Clearly, fiber is like all the nutrients in that "more" is "better" only up to a point. Again, the key words are balance, moderation, and variety.

IN SUMMARY

Adequate intake of fiber:
- Fosters weight management
- Lowers blood cholesterol
- May help prevent colon cancer
- Helps prevent and control diabetes
- Helps prevent and alleviate hemorrhoids
- Helps prevent appendicitis
- Helps prevent diverticulosis

Excessive intake of fiber:
- Displaces energy- and nutrient-dense foods
- Causes intestinal discomfort and distention
- May interfere with mineral absorption

Recommended Intakes of Starch and Fibers

◆ The Aids to Calculations section at the end of this book explains how to solve such problems.

◆ RDA for carbohydrate:
- 130 g/day
- 45 to 65% of energy intake

◆ Daily Value:
- 300 g carbohydrate (based on 60% of 2000 kcal diet)

◆ To increase your fiber intake:
- Eat whole-grain cereals that contain ≥ 5 g fiber per serving for breakfast.
- Eat raw vegetables.
- Eat fruits (such as pears) and vegetables (such as potatoes) with their skins.
- Add legumes to soups, salads, and casseroles.
- Eat fresh and dried fruit for snacks.

◆ Daily Value:
- 25 g fiber (based on 11.5 g/1000 kcal)

Dietary recommendations suggest that carbohydrates provide about half (45 to 65 percent) of the energy requirement. A person consuming 2000 kcalories a day should therefore have 900 to 1300 kcalories of carbohydrate, or about 225 to 325 grams. ◆ This amount is more than adequate to meet the RDA ◆ for carbohydrate, which is set at 130 grams per day, based on the average minimum amount of glucose used by the brain.[47]

When it established the Daily Values that appear on food labels, the Food and Drug Administration (FDA) used a 60 percent of kcalories guideline in setting the Daily Value ◆ for carbohydrate at 300 grams per day. For most people, this means increasing total carbohydrate intake. To this end, the *Dietary Guidelines* encourage people to choose a variety of whole grains, vegetables, fruits, and legumes daily.

Dietary Guidelines for Americans 2005

Choose fiber-rich fruits, vegetables, and whole grains often.

Recommendations for fiber ◆ suggest the same foods just mentioned: whole grains, vegetables, fruits, and legumes, which also provide minerals and vitamins. The FDA set the Daily Value ◆ for fiber at 25 grams, rounding up from the recom-

TABLE 4-3 Fiber in Selected Foods

Grains

Whole-grain products provide about 1 to 2 grams (or more) of fiber per serving:

- 1 slice whole-wheat, pumpernickel, rye bread
- 1 oz ready-to-eat cereal (100% bran cereals contain 10 grams or more)
- ½ c cooked barley, bulgur, grits, oatmeal

Vegetable

Most vegetables contain about 2 to 3 grams of fiber per serving:

- 1 c raw bean sprouts
- ½ c cooked broccoli, brussels sprouts, cabbage, carrots, cauliflower, collards, corn, eggplant, green beans, green peas, kale, mushrooms, okra, parsnips, potatoes, pumpkin, spinach, sweet potatoes, swiss chard, winter squash
- ½ c chopped raw carrots, peppers

Fruit

Fresh, frozen, and dried fruits have about 2 grams of fiber per serving:

- 1 medium apple, banana, kiwi, nectarine, orange, pear
- ½ c applesauce, blackberries, blueberries, raspberries, strawberries
- Fruit juices contain very little fiber

Legumes

Many legumes provide about 6 to 8 grams of fiber per serving:

- ½ c cooked baked beans, black beans, black-eyed peas, kidney beans, navy beans, pinto beans

Some legumes provide about 5 grams of fiber per serving:

- ½ c cooked garbanzo beans, great northern beans, lentils, lima beans, split peas

NOTE: Appendix H provides fiber grams for over 2000 foods.

mended 11.5 grams per 1000-kcalories for a 2000-kcalorie intake. The DRI recommendation is slightly higher, at 14 grams per 1000-kcalorie intake. Similarly, the American Dietetic Association suggests 20 to 35 grams of dietary fiber daily, which is about two times higher than the average intake in the United States.[48] An effective way to add fiber while lowering fat is to substitute plant sources of proteins (legumes) for animal sources (meats). Table 4-3 presents a list of fiber sources.

As mentioned earlier, too much fiber is no better than too little. The World Health Organization recommends an upper limit of 40 grams of dietary fiber a day.

From Guidelines to Groceries

A diet following the USDA Food Guide, which includes several servings of fruits, vegetables, and grains daily, can easily supply the recommended amount of carbohydrates and fiber. In selecting high-fiber foods, keep in mind the principle of variety. The fibers in oats lower cholesterol, whereas those in bran help promote GI tract health. (Review Table 4-2 to see the diverse health effects of various fibers.)

Grains An ounce-equivalent of most foods in the grain group provides about 15 grams of carbohydrate, mostly as starch. Be aware that some foods in this group, especially snack crackers and baked goods such as biscuits, croissants, and muffins, contain added sugars, added fat, or both. When selecting from the grain group, be sure to include at least half as whole-grain products (see Figure 4-15, p. 126). The "3 are Key" message may help consumers to remember to choose a whole-grain cereal for breakfast, a whole-grain bread for lunch, and a whole-grain pasta or rice for dinner.

FIGURE 4-15 Bread Labels Compared

Food labels list the quantities of total carbohydrate, dietary fiber, and sugars. Total carbohydrate and dietary fiber are also stated as "% Daily Values." A close look at these two labels reveals that bread made from whole wheat-flour provides almost three times as much fiber as the one made mostly from refined wheat flour. When the words whole wheat or whole grain appear on the label, the bread inside contains all of the nutrients that bread can provide.

Whole Grain — WHOLE WHEAT

Nutrition Facts

Serving size 1 slice (30g)
Servings Per Container 15

Amount per serving

Calories 90	Calories from Fat 14

	% Daily Value*
Total Fat 1.5g	2%
Sodium 135mg	6%
Total Carbohydrate 15g	5%
Dietary fiber 2g	**8%**
Sugars 2g	
Protein 4g	

MADE FROM: UNBROMATED STONE GROUND 100% WHOLE WHEAT FLOUR, WATER, CRUSHED WHEAT, HIGH FRUCTOSE CORN SYRUP, PARTIALLY HYDROGENATED VEGETABLE SHORTENING (SOYBEAN AND COTTONSEED OILS), RAISIN JUICE CONCENTRATE, WHEAT GLUTEN, YEAST, WHOLE WHEAT FLAKES, UNSULPHURED MOLASSES, SALT, HONEY, VINEGAR, ENZYME MODIFIED SOY LECITHIN, CULTURED WHEY, UNBLEACHED WHEAT FLOUR AND SOY LECITHIN.

Natural — Wheat Bread

Nutrition Facts

Serving size 1 slice (30g)
Servings Per Container 15

Amount per serving

Calories 90	Calories from Fat 14

	% Daily Value*
Total Fat 1.5g	2%
Sodium 220mg	9%
Total Carbohydrate 15g	5%
Dietary fiber less than 1g	**2%**
Sugars 2g	
Protein 4g	

INGREDIENTS: UNBLEACHED ENRICHED WHEAT FLOUR [MALTED BARLEY FLOUR, NIACIN, REDUCED IRON, THIAMIN MONONITRATE (VITAMIN B1), RIBOFLAVIN (VITAMIN B2), FOLIC ACID], WATER, HIGH FRUCTOSE CORN SYRUP, MOLASSES, PARTIALLY HYDROGENATED SOYBEAN OIL, YEAST, CORN FLOUR, SALT, GROUND CARAWAY, WHEAT GLUTEN, CALCIUM PROPIONATE (PRESERVATIVE), MONOGLYCERIDES, SOY LECITHIN.

Vegetables The amount of carbohydrate a serving of vegetables provides depends primarily on its starch content. Starchy vegetables—a half-cup of cooked corn, peas, or potatoes—provide about 15 grams of carbohydrate per serving. A serving of most other *nonstarchy* vegetables—such as a half-cup of broccoli, green beans, or tomatoes—provides about 5 grams.

Fruits A typical fruit serving—a small banana, apple, or orange or a half-cup of most canned or fresh fruit—contains an average of about 15 grams of carbohydrate, mostly as sugars, including the fruit sugar fructose. Fruits vary greatly in their water and fiber contents and, therefore, in their sugar concentrations.

Milks and Milk Products A serving (a cup) of milk or yogurt provides about 12 grams of carbohydrate. Cottage cheese provides about 6 grams of carbohydrate per cup, but most other cheeses contain little, if any, carbohydrate.

Meats and Meat Alternates With two exceptions, foods in the meats and meat alternates group deliver almost no carbohydrate to the diet. The exceptions are nuts, which provide a little starch and fiber along with their abundant fat, and legumes, which provide an abundance of both starch and fiber. Just a half-cup serving of legumes provides about 20 grams of carbohydrate, a third from fiber.

Read Food Labels Food labels list the amount, in grams, of *total* carbohydrate—including starch, fibers, and sugars—per serving (review Figure 4-15). Fiber grams are also listed separately, as are the grams of sugars. (With this information, you can calculate starch grams ◆ by subtracting the grams of fibers and sugars from the total carbohydrate.) Sugars reflect both added sugars and those that occur naturally in foods. Total carbohydrate and dietary fiber are also expressed as "% Daily Values" for a person consuming 2000 kcalories; there is no Daily Value for sugars.

◆ To calculate starch grams using the first label in Figure 4-15:
15 g total − 4 g (dietary fiber + sugars) = 11 g starch

IN SUMMARY

Clearly, a diet rich in complex carbohydrates—starches and fibers—supports efforts to control body weight and prevent heart disease, cancer, diabetes, and GI disorders. For these reasons, recommendations urge people to eat plenty of whole grains, vegetables, legumes, and fruits—enough to provide 45 to 65 percent of the daily energy intake from carbohydrate.

In today's world, there is one other reason why plant foods rich in complex carbohydrates and natural sugars are a better choice than animal foods or foods high in concentrated sweets. In general, less energy and fewer resources are required to grow and process plant foods than to produce sugar or foods derived from animals.

Nutrition Portfolio

CENGAGENOW™
academic.cengage.com/login

Foods that derive from plants—whole grains, vegetables, legumes, and fruits—naturally provide ample carbohydrates and fiber with little or no fat. Refined foods often contain added sugars and fat.

▪ List the types and amounts of grain products you eat daily, making note of which are whole-grain or refined foods and how your choices could include more whole-grain options.

▪ List the types and amounts of fruits and vegetables you eat daily, making note of how many are dark-green, orange, or deep yellow, how many are starchy or legumes, and how your choices could include more of these options.

▪ Describe choices you can make in selecting and preparing foods and beverages to lower your intake of added sugars.

NUTRITION ON THE NET

For further study of topics covered in this chapter, log on to **academic.cengage.com/nutrition/rolfes/UNCN8e**. Go to Chapter 4, then to Nutrition on the Net.

• Search for "lactose intolerance" at the U.S. Government health information site: **www.healthfinder.gov**

• Search for "sugars" and "fiber" at the International Food Information Council site: **www.ific.org**

• Learn more about dental caries from the American Dental Association and the National Institute of Dental and

Craniofacial Research: **www.ada.org** and **www.nidcr.nih.gov**

• Learn more about diabetes from the American Diabetes Association, the Canadian Diabetes Association, and the National Institute of Diabetes and Digestive and Kidney Diseases: **www.diabetes.org**, **www.diabetes.ca**, and **www.niddk.nih.gov**

NUTRITION CALCULATIONS

CENGAGENOW™ For additional practice log on to **academic.cengage.com/login**. Go to Chapter 4, then to Nutrition Calculations.

These problems will give you practice in doing simple nutrition-related calculations. Although the situations are hypothetical, the numbers are real, and calculating the answers (check them on p. 131) provides a valuable lesson. Be sure to show your calculations for each problem.

Health recommendations suggest that 45 to 65 percent of the daily energy intake come from carbohydrates. Stating recommendations in terms of percentage of energy intake is meaningful only if energy intake is known. The following exercises illustrate this concept.

1. Calculate the carbohydrate intake (in grams) for a student who has a high carbohydrate intake (70 percent of energy intake) and a moderate energy intake (2000 kcalories a day).

   How does this carbohydrate intake compare to the Daily Value of 300 grams? To the 45 to 65 percent recommendation?

2. Now consider a professor who eats half as much carbohydrate as the student (in grams) and has the same energy intake. What percentage does carbohydrate contribute to the daily intake?

How does carbohydrate intake compare to the Daily Value of 300 grams? To the 45 to 65 percent recommendation?

3. Now consider an athlete who eats twice as much carbohydrate (in grams) as the student and has a much higher energy intake (6000 kcalories a day). What percentage does carbohydrate contribute to this person's daily intake?

   How does carbohydrate intake compare to the Daily Value of 300 grams? To the 45 to 65 percent recommendation?

4. One more example. In an attempt to lose weight, a person adopts a diet that provides 150 grams of carbohydrate per day and limits energy intake to 1000 kcalories. What percentage does carbohydrate contribute to this person's daily intake?

   How does this carbohydrate intake compare to the Daily Value of 300 grams? To the 45 to 65 percent recommendation?

These exercises should convince you of the importance of examining actual intake as well the percentage of energy intake.

STUDY QUESTIONS

CENGAGENOW™
To assess your understanding of chapter topics, take the Student Practice Test and explore the modules recommended in your Personalized Study Plan. Log on to **academic.cenage.com/login**.

These questions will help you review this chapter. You will find the answers in the discussions on the pages provided.

1. Which carbohydrates are described as simple and which are complex? (p. 101)

2. Describe the structure of a monosaccharide and name the three monosaccharides important in nutrition. Name the three disaccharides commonly found in foods and their component monosaccharides. In what foods are these sugars found? (pp. 102–105)

3. What happens in a condensation reaction? In a hydrolysis reaction? (p. 104)

4. Describe the structure of polysaccharides and name the ones important in nutrition. How are starch and glycogen similar, and how do they differ? How do the fibers differ from the other polysaccharides? (pp. 105–107)

5. Describe carbohydrate digestion and absorption. What role does fiber play in the process? (pp. 107–110)

6. What are the possible fates of glucose in the body? What is the protein-sparing action of carbohydrate? (pp. 111–113)

7. How does the body maintain its blood glucose concentration? What happens when the blood glucose concentration rises too high or falls too low? (pp. 113–117)

8. What are the health effects of sugars? What are the dietary recommendations regarding concentrated sugar intakes? (pp. 117–121)

9. What are the health effects of starches and fibers? What are the dietary recommendations regarding these complex carbohydrates? (pp. 122–125)

10. What foods provide starches and fibers? (pp. 125–126)

These multiple choice questions will help you prepare for an exam. Answers can be found on p. 131.

1. Carbohydrates are found in virtually all foods except:
   a. milks.
   b. meats.
   c. breads.
   d. fruits.

2. Disaccharides include:
   a. starch, glycogen, and fiber.
   b. amylose, pectin, and dextrose.
   c. sucrose, maltose, and lactose.
   d. glucose, galactose, and fructose.

3. The making of a disaccharide from two monosaccharides is an example of:
   a. digestion.
   b. hydrolysis.
   c. condensation.
   d. gluconeogenesis.

4. The storage form of glucose in the body is:
   a. insulin.
   b. maltose.
   c. glucagon.
   d. glycogen.

5. The significant difference between starch and cellulose is that:
   a. starch is a polysaccharide, but cellulose is not.
   b. animals can store glucose as starch, but not as cellulose.
   c. hormones can make glucose from cellulose, but not from starch.
   d. digestive enzymes can break the bonds in starch, but not in cellulose.

6. The ultimate goal of carbohydrate digestion and absorption is to yield:
   a. fibers.
   b. glucose.
   c. enzymes.
   d. amylase.

7. The enzyme that breaks a disaccharide into glucose and galactose is:
   a. amylase.
   b. maltase.
   c. sucrase.
   d. lactase.

8. With insufficient glucose in metabolism, fat fragments combine to form:
   a. dextrins.
   b. mucilages.
   c. phytic acids.
   d. ketone bodies.

9. What does the pancreas secrete when blood glucose rises? When blood glucose falls?
   a. insulin; glucagon
   b. glucagon; insulin
   c. insulin; glycogen
   d. glycogen; epinephrine

10. What percentage of the daily energy intake should come from carbohydrates?
    a. 15 to 20
    b. 25 to 30
    c. 45 to 50
    d. 45 to 65

REFERENCES

1. N. R. Sahyoan and coauthors, Whole-grain intake is inversely associated with metabolic syndrome and mortality in older adults, *American Journal of Clinical Nutrition* 83 (2006): 124–131; B. M. Davy and C. L. Melby, The effect of fiber-rich carbohydrates on features of Syndrome X, *Journal of the American Dietetic Association* 103 (2003): 86–96.

2. J. R. Jones, D. M. Lineback, and M. J. Levine, Dietary Reference Intakes: Implications for fiber labeling and consumption: A summary of the International Life Sciences Institute North America Fiber Workshop, June 1-2, 2004, Washington, DC, *Nutrition Reviews* 64 (2006): 31–38.

3. D. Savaiano, Lactose intolerance: A self-fulfilling prophecy leading to osteoporosis? *Nutrition Reviews* 61 (2003): 221–223.

4. S. R. Hertzler and S. M. Clancy, Kefir improves lactose digestion and tolerance in adults with lactose maldigestion, *Journal of the American Dietetic Association* 103 (2003): 582–587.

5. R. S. Haltiwanger and J. B. Lowe, Role of glycosylation in development, *Annual Review of Biochemistry* 73 (2004): 491–537; T. Maeder, Sweet medicines, *Scientific American* 287 (2002): 40–47; J. Travis, The true sweet science—Researchers develop a taster for the study of sugars, *Science News* 161 (2002): 232–233.

6. R. L. Schnaar, Glycolipid-mediated cell-cell recognition in inflammation and nerve regeneration, *Archives of Biochemistry and Biophysics* 426 (2004): 163–172.

7. R. Gruetter, Glycogen: The forgotten cerebral energy store, *Journal of Neuroscience Research* 74 (2003): 179–183.

8. K. Foster-Powell, S. H. A. Holt, and J. C. Brand-Miller, International table of glycemic index and glycemic load values: 2002, *American Journal of Clinical Nutrition* 76 (2002): 5–56.

9. A. M. Opperman and coauthors, Meta-analysis of the health effects of using the glycaemic index in meal-planning, *British Journal of Nutrition* 92 (2004): 367–381.

10. C. B. Ebbeling and coauthors, Effects of an ad libitum low-glycemic load diet on cardiovascular disease risk factors in obese young adults, *American Journal of Clinical Nutrition* 81 (2005): 976–982; S. Dickinson and J. Brand-Miller, Glycemic index, postprandial glycemia and cardiovascular disease, *Current Opinion in Lipidology* 16 (2005): 69–75; A. M. Opperman and coauthors, Meta-analysis of the health effects of using the glycaemic index in meal-planning, *British Journal of Nutrition* 92 (2004): 367–381; M. A. Pereira and coauthors, Effects of a low-glycemic load diet on resting energy expenditure and heart disease risk factors during weight loss, *Journal of the American Medical Association* 292 (2004): 2482–2490; T. M. S. Wolever, Carbohydrate and the regulation of blood glucose and metabolism, *Nutrition Reviews* 61 (2003): S40–S48; D. J. A. Jenkins and coauthors, Glycemic index: Overview of implications in health and disease, *American Journal of Clinical Nutrition* 76 (2002): 266S–273S.

11. G. Livesey, Low-glycemic diets and health: Implications for obesity, *Proceedings of the Nutrition Society* 64 (2005): 105–113; B. Sloth and coauthors, No difference in body weight decrease between a low-glycemic-index and a high-glycemic-index diet but reduced LDL cholesterol after 10-wk ad libitum intake of the low-glycemic-index diet, *American Journal of Clinical Nutrition* 80 (2004): 337–347; C. Bouché and coauthors, Five-week, low-glycemic index diet decreases total fat mass and improves plasma lipid profile in moderately overweight nondiabetic men, *Diabetes Care* 25 (2002): 822–828.

12. S. D. Ball and coauthors, Prolongation of satiety after low versus moderately high glycemic index meals in obese adolescents, *Pediatrics* 111 (2003): 488–494; S. B. Roberts, Glycemic index and satiety, *Nutrition in Clinical Care* 6 (2003): 20–26.

13. S. Liu and coauthors, Relation between a diet with a high glycemic load and plasma concentrations of high-sensitivity C-reactive protein in middle-aged women, *American Journal of Clinical Nutrition* 75 (2002): 492–498.

14. D. S. Ludwig, The glycemic index—Physiological mechanisms relating to obesity, diabetes, and cardiovascular disease, *Journal of the American Medical Association* 287 (2002): 2414–2423.

15. F. X. Pi-Sunyer, Glycemic index and disease, *American Journal of Clinical Nutrition* 76 (2002): 290S–298S.

16. D. R. Lineback and J. M. Jones, Sugars and health workshop: Summary and conclusions, *American Journal of Clinical Nutrition* 78 (2003): 893S–897S.

17. G. Fernandes, A. Velangi, and T. M. S. Wolever, Glycemic index of potatoes commonly consumed in North America, *Journal of the American Dietetic Association* 105 (2005): 557–562; E. M. Y. Chan and coauthors, Postprandial glucose response to Chinese foods in patients with type 2 diabetes, *Journal of the American Dietetic Association* 104 (2004): 1854–1858.

18. B. M. Popkin and S. J. Nielsen, The sweetening of the world's diet, *Obesity Research* 11 (2003): 1325–1332.

19. R. Touger-Decker and C. van Loveren, Sugars and dental caries, *American Journal of Clinical Nutrition* 78 (2003): 881S–892S.

20. S. Kashket and D. P. DePaola, Cheese consumption and the development and progression of dental caries, *Nutrition Reviews* 60 (2002): 97–103; Department of Health and Human Services, *Oral Health in America: A Report of the Surgeon General* (Rockville, Md.: National Institutes of Health, 2000), pp. 250–251.

21. S. Wongkhantee and coauthors, Effect of acidic food and drinks on surface hardness of enamel, dentine, and tooth-coloured filling materials, *Journal of Dentistry* 34 (2006): 214–220; W. K. Seow and K. M. Thong, Erosive effects of common beverages on extracted premolar teeth, *Australian Dental Journal* 50 (2005): 173–178.

22. J. M. Jones and K. Elam, Sugars and health: Is there an issue? *Journal of the American Dietetic Association* 103 (2003): 1058–1060.

23. G. A. Bray, S. J. Nielsen, and B. M. Popkin, Consumption of high-fructose corn syrup in beverages may play a role in the epidemic of obesity, *American Journal of Clinical Nutrition* 79 (2004): 537–543; S. J. Nielsen and B. M. Popkin, Changes in beverage intake between 1977 and 2001, *American Journal of Preventive Medicine* 27 (2004): 205–210.

24. A. M. Coulston and R. K. Johnson, Sugar and sugars: Myths and realities, *Journal of the American Dietetic Association* 102 (2002): 351–353.

25. M. B. Schulze and coauthors, Sugar-sweetened beverages, weight gain, and incidence of type 2 diabetes in young and middle-aged women, *Journal of the American Medical Association* 292 (2004): 927–934; L. S. Gross and coauthors, Increased consumption of refined carbohydrates and the epidemic of type 2 diabetes in the United States: An ecologic assessment, *American Journal of Clinical Nutrition* 79 (2004): 774–779.

26. Joint WHO/FAO Expert Consultation, 2003, pp. 57–58; S. H. F. Vermunt and coauthors, Effects of sugar intake on body weight: A review, *Obesity Reviews* 4 (2003): 91–99.

27. M. K. Hellerstein, Carbohydrate-induced hypertriglyceridemia: Modifying factors and implications for cardiovascular risk, *Current Opinion in Lipidology* 13 (2002): 33–40.

28. Bray, 2004; S. S. Elliott and coauthors, Fructose, weight gain, and the insulin resistance syndrome, *American Journal of Clinical Nutrition* 76 (2002): 911–922.

29. P. J. Havel, Dietary fructose: Implications for dysregulation of energy homeostasis and lipid/carbohydrate metabolism, *Nutrition Reviews* 63 (2005): 133–157; J. Wylie-Rosett, C. J. Segal-Isaacson, and A. Segal-Isaacson, Carbohydrates and increases in obesity: Does the type of carbohydrate make a difference? *Obesity Research* 12 (2004): 124S–129S; Elliott and coauthors, 2002.

30. J. James and coauthors, Preventing childhood obesity by reducing consumption of carbonated drinks: Cluster randomised controlled trial, *British Medical Journal* 10 (2004): 1136–1141.

31. S. K. Fried and S. P. Rao, Sugars, hyper-triglyceridemia, and cardiovascular disease, *American Journal of Clinical Nutrition* 78 (2003): 873S–880S; B. V. Howard and J. Wylie-Rosett, AHA Scientific Statement: Sugar and cardiovascular disease, *Circulation* 106 (2002): 523.

32. J. M. Schwarz and coauthors, Hepatic de novo lipogenesis in normoinsulinemic and hyperinsulinemic subjects consuming high-fat, low-carbohydrate and low-fat, high-carbohydrate isoenergetic diets, *American Journal of Clinical Nutrition* 77 (2003): 43–50.

33. S. Yanovski, Sugar and fat: Cravings and aversions, *Journal of Nutrition* 133 (2003): 835S–837S.

34. J. Putnam and S. Haley, Estimating consumption of caloric sweeteners, *Economic Research Service, Farm Service Agency, and Foreign Agricultural Service, USDA*, 2004.

35. Committee on Dietary Reference Intakes, *Dietary Reference Intakes: Energy, Carbohydrate, Fiber, Fat, Fatty Acids, Cholesterol, Protein, and Amino Acids* (Washington, D.C.: National Academies Press, 2005).

36. M. K. Jenson and coauthors, Whole grains, bran, and germ in relation to homocysteine and markers of glycemic control, lipids, and inflammation, *American Journal of Clinical Nutrition* 83 (2006): 275–283; M. K. Jensen and coauthors, Intakes of whole grains, bran, and germ and the risk of coronary heart disease in men, *American Journal of Clinical Nutrition* 80 (2004): 1492–1499; F. B. Hu and W. C. Willett, Optimal diets for prevention of coronary heart disease, *Journal of the American Medical Association* 288 (2002): 2569–2578; N. M. McKeown and coauthors, Whole-grain intake is favorably associated with metabolic risk factors for type 2 diabetes and cardiovascular disease in the Framingham Offspring Study, *American Journal of Clinical Nutrition* 76 (2002): 390–398; S. Liu, Intake of refined carbohydrates and whole grain foods in relation to risk of type 2 diabetes mellitus and coronary heart disease, *Journal of the American College of Nutrition* 21 (2002): 298–306.

37. K. M. Behall, D. J. Scholfield, and J. Hallfrisch, Diets containing barley significantly reduce lipids in mildly hypercholesterolemic men and women, *American Journal of Clinical Nutrition* 80 (2004): 1185–1193; B. M. Davy and coauthors, High-fiber oat cereal compared with wheat cereal consumption favorably alters LDL-cholesterol subclass and particle numbers in middle-aged and older men, *American Journal of Clinical Nutrition* 76 (2002): 351–358; D. J. A. Jenkins and coauthors, Soluble fiber intake at a dose approved by the US Food and Drug Administration for a claim of health benefits: Serum lipid risk factors for cardiovascular disease assessed in a randomized controlled crossover trial, *American Journal of Clinical Nutrition* 75 (2002): 834–839.

38. U. A. Ajani, E. S. Ford, and A. H. Mokdad, Dietary fiber and C-reactive protein: Findings from National Health and Nutrition Examination Survey Data, *Journal of Nutrition* 134 (2004): 1181–1185.

39. M. K. Jenson and coauthors, Whole grains, bran and germ in relation to homocysteine and markers of glycemic control, lipids, and inflammation, *American Journal of Clinical Nutrition* 83 (2006): 275–283; T. T. Fung and coauthors, Whole-grain intake and the risk of type 2 diabetes: A prospective study in men, *American Journal of Clinical Nutrition* 76 (2002): 535–540.

40. W. Aldoori and M. Ryan-Harshman, Preventing diverticular disease: Review of recent evidence on high-fibre diets, *Canadian Family Physician* 48 (2002): 1632–1637.

41. Y. Park and coauthors, Dietary fiber intake and risk of colorectal cancer, *Journal of the American Medical Association* 294 (2005): 2849–2857; T. Asano and R. S. McLeod, Dietary fibre for the prevention of colorectal adenomas and carcinomas, *Cochrane Database of Systematic Reviews* 2 (2002): CD003430.

42. S. A. Bingham and coauthors, Dietary fibre in food and protection against colorectal cancer in the European Prospective Investigation into Cancer and Nutrition (EPIC): An observational study, *Lancet* 361 (2003): 1496–1501.

43. M. L. Slattery and coauthors, Plant foods, fiber, and rectal cancer, *American Journal of Clinical Nutrition* 79 (2004): 274–281.

44. L. McMillan and coauthors, Opposing effects of butyrate and bile acids on apoptosis of human colon adenoma cells: Differential activation of PKC and MAP kinases, *British Journal of Cancer* 88 (2003): 748–753; M. E. Rodriguez-Cabezas and coauthors, Dietary fiber down-regulates colonic tumor necrosis factor alpha and nitric oxide production in trinitrobenzene-sulfonic acid-induced colitic rats, *Journal of Nutrition* 11 (2002): 3263–3271.

45. S. Liu and coauthors, Relation between changes in intakes of dietary fiber and grain products and changes in weight and development of obesity among middle-aged women, *American Journal of Clinical Nutrition* 78 (2003): 920–927.

46. P. Koh-Banerjee, Changes in whole-grain, bran, and cereal fiber consumption in relation to 8-y weight gain among men, *American Journal of Clinical Nutrition* 80 (2004): 1237–1245; Committee on Dietary Reference Intakes, 2002/2005, pp. 342–344.

47. Committee on Dietary Reference Intakes, 2005.

48. Position of the American Dietetic Association: Health implications of dietary fiber, *Journal of the American Dietetic Association* 102 (2002): 993–999.

ANSWERS

Nutrition Calculations

1. 0.7×2000 total kcal/day = 1400 kcal from carbohydrate/day

    1400 kcal from carbohydrate ÷ 4 kcal/g = 350 g carbohydrate

This carbohydrate intake is higher than the Daily Value and higher than the 45 to 65 percent recommendation.

2. 350 g carbohydrate ÷ 2 = 175 g carbohydrate/day

    175 g carbohydrate $\times$ 4 kcal/g = 700 kcal from carbohydrate

    700 kcal from carbohydrate ÷ 2000 total kcal/day = 0.35

    0.35×100 = 35% kcal from carbohydrate

This carbohydrate intake is lower than the Daily Value and lower than the 45 to 65 percent recommendation.

3. 350 g carbohydrate $\times$ 2 = 700 g carbohydrate/day

    700 g carbohydrate $\times$ 4 kcal/g = 2800 kcal from carbohydrate

    2800 kcal from carbohydrate ÷ 6000 total kcal/day = 0.47

    0.47×100 = 47% kcal from carbohydrate

This carbohydrate intake is higher than the Daily Value and meets the 45 to 65 percent recommendation.

4. 150 g carbohydrate $\times$ 4 kcal/g = 600 kcal from carbohydrate

    600 kcal from carbohydrate ÷ 1000 total kcal/day = 0.60

    0.60×100 = 60% kcal from carbohydrate

This carbohydrate intake is lower than the Daily Value and meets the 45 to 65 percent recommendation.

Study Questions (multiple choice)

1. b   2. c   3. c   4. d   5. d
6. b   7. d   8. d   9. a   10. d

Alternatives to Sugar

Funnette Division, Hoechst Celenese Corp.

Almost everyone finds pleasure in sweet foods—after all, the taste preference for sweets is inborn. To a child, the sweeter the food, the better. In adults, this preference is somewhat diminished, but most adults still enjoy an occasional sweet food or beverage. Because they want to control weight gain, blood glucose, and dental caries, many consumers turn to alternative sweeteners to help them limit kcalories and minimize sugar intake. In doing so, they encounter two sets of alternative sweeteners: **artificial sweeteners** and **sugar replacers.**

Artificial Sweeteners

The Food and Drug Administration (FDA) has approved the use of several artificial sweeteners—saccharin, aspartame, acesulfame potassium (acesulfame-K), sucralose, and neotame. Two others are awaiting FDA approval—alitame and cyclamate. Another—tagatose—did not need approval because it is generally recognized as a safe ingredient. These artificial sweeteners are sometimes called **nonnutritive sweeteners** because they provide virtually no energy. Table H4-1 and the accompanying glossary provide general

details about each of these sweeteners.

Saccharin, acesulfame-K, and sucralose are not metabolized in the body; in contrast, the body digests aspartame as a protein. In fact, aspartame yields energy (4 kcalories per gram, as does protein), but because so little is used, its energy contribution is negligible.

Some consumers have challenged the safety of using artificial sweeteners. Considering that all substances are toxic at some dose, it is little surprise that large doses of artificial sweeteners (or their components or metabolic by-products) have toxic effects. The question to ask is whether their ingestion is safe for human beings in quantities people normally use (and potentially abuse).

Saccharin

Saccharin, used for over 100 years in the United States, is currently used by some 50 million people—primarily in soft drinks, secondarily as a tabletop sweetener. Saccharin is rapidly excreted in the urine and does not accumulate in the body.

Questions about saccharin's safety surfaced in 1977, when experiments suggested that large doses of saccharin (equivalent to

GLOSSARY

Acceptable Daily Intake (ADI): the estimated amount of a sweetener that individuals can safely consume each day over the course of a lifetime without adverse effect.

acesulfame (AY-sul-fame) potassium: an artificial sweetener composed of an organic salt that has been approved for use in both the United States and Canada; also known as **acesulfame-K** because K is the chemical symbol for potassium.

alitame (AL-ih-tame): an artificial sweetener composed of two amino acids (alanine and aspartic acid); FDA approval pending.

artificial sweeteners: sugar substitutes that provide negligible, if any, energy; sometimes called **nonnutritive sweeteners.**

aspartame (ah-SPAR-tame or ASS-par-tame): an artificial sweetener composed of two amino acids (phenylalanine and aspartic acid); approved for use in both the United States and Canada.

cyclamate (SIGH-kla-mate): an artificial sweetener that is being considered for approval in the United States and is available in Canada as a tabletop sweetener, but not as an additive.

neotame (NEE-oh-tame): an artificial sweetener composed of two amino acids (phenylalanine and aspartic acid); approved for use in the United States.

nonnutritive sweeteners: sweeteners that yield no energy (or insignificant energy in the case of aspartame).

nutritive sweeteners: sweeteners that yield energy, including both sugars and sugar replacers.

saccharin (SAK-ah-ren): an artificial sweetener that has been approved for use in the United States. In Canada, approval for use in foods and beverages is pending; currently available only in pharmacies and only as a tabletop sweetener, not as an additive.

stevia (STEE-vee-ah): a South American shrub whose leaves are used as a sweetener; sold in the United States as a dietary supplement that provides sweetness without kcalories.

sucralose (SUE-kra-lose): an artificial sweetener approved for use in the United States and Canada.

sugar replacers: sugarlike compounds that can be derived from fruits or commercially produced from dextrose; also called **sugar alcohols** or **polyols.** Sugar alcohols are absorbed more slowly than other sugars and metabolized differently in the human body; they are not readily utilized by ordinary mouth bacteria. Examples are **maltitol, mannitol, sorbitol, xylitol, isomalt,** and **lactitol.**

tagatose (TAG-ah-tose): a monosaccharide structurally similar to fructose that is incompletely absorbed and thus provides only 1.5 kcalories per gram; approved for use as a "generally recognized as safe" ingredient.

TABLE H4-1 Sweeteners

Sweeteners	Relative Sweetness[a]	Energy (kcal/g)	Acceptable Daily Intake	Average Amount to Replace 1 tsp Sugar	Approved Uses
Approved Sweeteners (Trade Name)					
Saccharin (Sweet 'n Low)	450	0	5 mg/kg body weight	12 mg	Tabletop sweeteners, wide range of foods, beverages, cosmetics, and pharmaceutical products
Aspartame (Nutrasweet, Equal, NutraTaste)	200	4[b]	50 mg/kg body weight[c]	18 mg	General purpose sweetener in all foods and beverages Warning to people with PKU: Contains phenylalanine
Acesulfame potassium or Acesulfame-K (Sunette, Sweet One, Sweet 'n Safe)	200	0	15 mg/kg body weight[d]	25 mg	Tabletop sweeteners, puddings, gelatins, chewing gum, candies, baked goods, desserts, beverages
Sucralose (Splenda)	600	0	5 mg/kg body weight	6 mg	General purpose sweetener for all foods
Neotame	8000	0	18 mg/day	0.5 µg	Baked goods, nonalcoholic beverages, chewing gum, candies, frostings, frozen desserts, gelatins, puddings, jams and jellies, syrups
Tagatose (Nutralose)	0.8	1.5	7.5 g/day	1 tsp	Baked goods, beverages, cereals, chewing gum, confections, dairy products, dietary supplements, health bars, tabletop sweetener
Sweeteners with Approval Pending					**Proposed Uses**
Alitame	2000	4[e]	—		Beverages, baked goods, tabletop sweeteners, frozen desserts
Cyclamate	30	0	—		Tabletop sweeteners, baked goods

[a] Relative sweetness is determined by comparing the approximate sweetness of a sugar substitute with the sweetness of pure sucrose, which has been defined as 1.0. Chemical structure, temperature, acidity, and other flavors of the foods in which the substance occurs all influence relative sweetness.

[b] Aspartame provides 4 kcalories per gram, as does protein, but because so little is used, its energy contribution is negligible. In powdered form, it is sometimes mixed with lactose, however, so a 1-gram packet may provide 4 kcalories.

[c] Recommendations from the World Health Organization and in Europe and Canada limit aspartame intake to 40 milligrams per kilogram of body weight per day.

[d] Recommendations from the World Health Organization limit acesulfame-K intake to 9 milligrams per kilogram of body weight per day.

[e] Alitame provides 4 kcalories per gram, as does protein, but because so little is used, its energy contribution is negligible.

---

hundreds of cans of diet soda daily for a lifetime) increased the risk of bladder cancer in rats. The FDA proposed banning saccharin as a result. Public outcry in favor of saccharin was so loud, however, that Congress imposed a moratorium on the ban while additional safety studies were conducted. Products containing saccharin were required to carry a warning label until 2001, when studies concluded that saccharin did not cause cancer in humans.

Does saccharin cause cancer? The largest population study to date, involving 9000 men and women, showed that overall saccharin use did not increase the risk of cancer. Among certain small groups of the population, however, such as those who both smoked heavily and used saccharin, the risk of bladder cancer was slightly greater. Other studies involving more than 5000 people with bladder cancer showed no association between bladder cancer and saccharin use. In 2000, saccharin was removed from the list of suspected cancer-causing substances. Warning labels are no longer required.

Common sense dictates that consuming large amounts of any substance is probably not wise, but at current, moderate intake levels, saccharin appears to be safe for most people. It has been approved for use in more than 100 countries.

Aspartame

Aspartame is a simple chemical compound made of components common to many foods: two amino acids (phenylalanine and aspartic acid) and a methyl group (CH_3). Figure H4-1 (p. 134) shows its chemical structure. The flavors of the components give no clue to the combined effect; one of them tastes bitter, and the other is tasteless, but the combination creates a product that is 200 times sweeter than sucrose.

In the digestive tract, enzymes split aspartame into its three component parts. The body absorbs the two amino acids and uses them just as if they had come from food protein, which is made entirely of amino acids, including these two.

Because this sweetener contributes phenylalanine, products containing aspartame must bear a warning label for people with the inherited disease phenylketonuria (PKU). People with PKU are unable to dispose of any excess phenylalanine. The accumulation of phenylalanine and its by-products is toxic to the developing nervous system, causing irreversible brain damage. For this reason, all newborns in the United States are screened for PKU. The treatment for PKU is a special diet that must strike a balance,

FIGURE H4-1 Structure of Aspartame

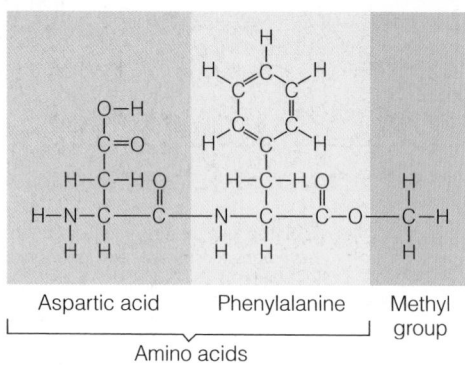

Aspartic acid | Phenylalanine | Methyl group

Amino acids

FIGURE H4-2 Metabolism of Aspartame

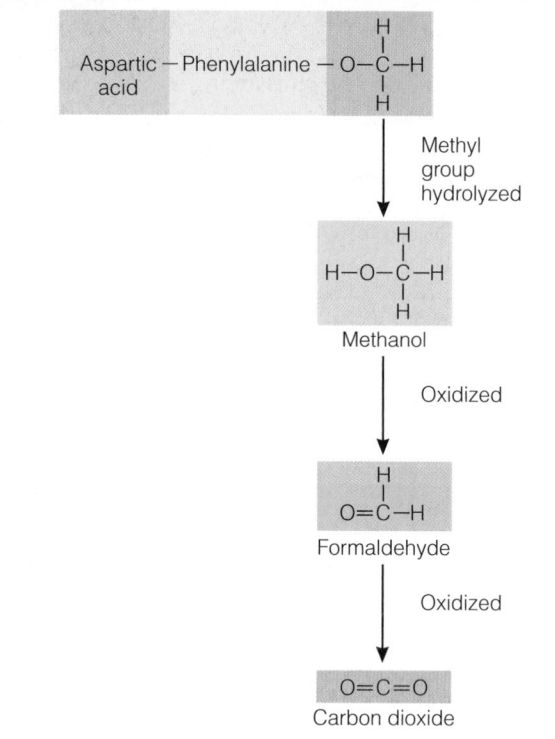

Aspartic acid — Phenylalanine — O—C—H

Methyl group hydrolyzed

Methanol

Oxidized

Formaldehyde

Oxidized

$O=C=O$
Carbon dioxide

providing enough phenylalanine to support normal growth and health but not enough to cause harm. The little extra phenylalanine from aspartame poses only a small risk, even in heavy users, but children with PKU need to get all their required phenylalanine from foods instead of from an artificial sweetener. The PKU diet excludes such protein- and nutrient-rich foods as milk, meat, fish, poultry, cheese, eggs, nuts, legumes, and many bread products. Consequently, these children have difficulty obtaining the many essential nutrients—such as calcium, iron, and the B vitamins—found along with phenylalanine in these foods. Children with PKU cannot afford to squander their limited phenylalanine allowance on the phenylalanine of aspartame, which contributes none of the associated vitamins or minerals essential for good health and normal growth.

During metabolism, the methyl group momentarily becomes methyl alcohol (methanol)—a potentially toxic compound (see Figure H4-2). This breakdown also occurs when aspartame-sweetened beverages are stored at warm temperatures over time. The amount of methanol produced may be safe to consume, but a person may not want to, considering that the beverage has lost its sweetness. In the body, enzymes convert methanol to formaldehyde, another toxic compound. Finally, formaldehyde is broken down to carbon dioxide. Before aspartame could be approved, the quantities of these products generated during metabolism had to be determined, and they were found to fall below the threshold at which they would cause harm. In fact, ounce for ounce, tomato juice yields six times as much methanol as a diet soda.

A recent Italian study found that aspartame caused cancer in female rats and fueled the controversies surrounding aspartame's safety.[1] Statements from the FDA and others, however, indicate that such a conclusion is not supported by the data.[2] The only valid scientific concern is that for people with epilepsy, excessive intake of aspartame may decrease their threshold for seizures; this does not appear to be a problem when intakes are within recommended amounts.[3]

Acesulfame-K

Because **acesulfame potassium (acesulfame-K)** passes through the body unchanged, it does not provide any energy nor does it increase the intake of potassium. Acesulfame-K is ap-

proved for use in the United States, Canada, and more than 60 other countries.

Sucralose

Sucralose is unique among the artificial sweeteners in that it is made from sugar that has had three of its hydroxyl (OH) groups replaced by chlorine atoms. The result is an exceptionally stable molecule that is much sweeter than sugar. Because the body does not recognize sucralose as a carbohydrate, it passes through the GI tract undigested and unabsorbed.

Neotame

Like aspartame, **neotame** also contains the amino acids phenylalanine and aspartic acid and a methyl group. Unlike aspartame, however, neotame has an additional side group attached. This simple difference makes all the difference to people with PKU because it blocks the digestive enzymes that normally separate phenylalanine and aspartic acid. Consequently, the amino acids are not absorbed and neotame need not carry a warning for people with PKU.

Tagatose

The FDA granted the fructose relative **tagatose** the status of "generally recognized as safe," making it available as a low-kcalorie sweetener for a variety of foods and beverages. This monosaccharide is naturally found in only a few foods, but it can be derived from lactose. Unlike fructose or lactose, however, 80 percent of tagatose remains unabsorbed until it reaches the large intestine. There, bacteria ferment tagatose, releasing gases and short chain fatty acids that are absorbed. As a result, tagatose provides 1.5 kcalories per gram. At high doses, tagatose causes flatulence, rumbling, and loose stools; otherwise, no adverse side effects have been noted. In fact, tagatose is a prebiotic that may benefit GI health. Unlike other sugars, tagatose does not promote dental caries and may carry a dental caries health claim.

Alitame and Cyclamate

FDA approval for **alitame** and **cyclamate** is still pending. To date, no safety issues have been raised for alitame, and it has been approved for use in other countries. In contrast, cyclamate has been battling safety issues for 50 years. Approved by the FDA in 1949, cyclamate was banned in 1969 principally on the basis of one study indicating that it caused bladder cancer in rats.

The National Research Council has reviewed dozens of studies on cyclamate and concluded that neither cyclamate nor its metabolites cause cancer. The council did, however, recommend further research to determine if heavy or long-term use poses risks. Although cyclamate does not *initiate* cancer, it may *promote* cancer development once it is started. The FDA currently has no policy on substances that enhance the cancer-causing activities of other substances, but it is unlikely to approve cyclamate soon, if at all. Agencies in more than 50 other countries, including Canada, have approved cyclamate.

Acceptable Daily Intake

The amount of artificial sweetener considered safe for daily use is called the **Acceptable Daily Intake (ADI).** The ADI represents the level of consumption that, if maintained every day throughout a person's life, would still be considered safe by a wide margin. It usually reflects an amount 100 times less than the level at which no observed effects occur in animal research studies.

The ADI for aspartame, for example, is 50 milligrams per kilogram of body weight. That is, the FDA approved aspartame based on the assumption that no one would consume more than 50 milligrams per kilogram of body weight in a day. This maximum daily intake is indeed high: for a 150-pound adult, it adds up to 97 packets of Equal or 20 cans of soft drinks sweetened only with aspartame. The company that produces aspartame estimates that if all the sugar and saccharin in the U.S. diet were replaced with aspartame, 1 percent of the population would be consuming the FDA maximum. Most people who use aspartame consume less than 5 milligrams per kilogram of body weight per day. But a young child who drinks four glasses of aspartame-sweetened beverages on a hot day and has five servings of other products with aspartame that day (such as pudding, chewing gum, cereal, gelatin, and frozen desserts) consumes the FDA maximum level. Although this intake presents no proven hazard, it seems wise to offer children other foods so as not to exceed the limit. Table H4-2 lists the average amounts of aspartame in some common foods.

For persons choosing to use artificial sweeteners, the American Dietetic Association wisely advises that they be used in moderation and only as part of a well-balanced nutritious diet.[4] The di-etary principles of moderation and variety help to reduce the possible risks associated with any food.

Artificial Sweeteners and Weight Control

The rate of obesity in the United States has been rising for decades. Foods and beverages sweetened with artificial sweeteners were among the first products developed to help people control their weight. Ironically, a few studies have reported that intense sweeteners, such as aspartame, may stimulate appetite, which could lead to weight gain. Contradicting these reports, most studies find no change in feelings of hunger and no change in food intakes or body weight. Adding to the confusion, some studies report lower energy intakes and greater weight losses when people eat or drink artificially sweetened products.[5]

When studying the effects of artificial sweeteners on food intake and body weight, researchers ask different questions and take different approaches. It matters, for example, whether the people used in a study are of a healthy weight and whether they are following a weight-loss diet. Motivations for using sweeteners differ, too, and this influences a person's actions. For example, one person might drink an artificially sweetened beverage now so as to be able to eat a high-kcalorie food later. This person's energy intake might stay the same or increase. A person trying to control food energy intake might drink an artificially sweetened beverage now and choose a low-kcalorie food later. This plan would help reduce the person's total energy intake.

In designing experiments on artificial sweeteners, researchers have to distinguish between the effects of sweetness and the effects of a particular substance. If a person is hungry shortly after eating an artificially sweetened snack, is that because the sweet taste (of all sweeteners, including sugars) stimulates appetite? Or is it because the artificial sweetener itself stimulates appetite? Research must also distinguish between the effects of food energy and the effects of the substance. If a person is hungry shortly after eating an artificially sweetened snack, is that because less food energy was available to satisfy hunger? Or is it because the artificial sweetener itself triggers hunger? Furthermore, if appetite is stimulated and a person feels hungry, does that actually lead to increased food intake?

Whether a person compensates for the energy reduction of artificial sweeteners either partially or fully depends on several factors. Using artificial sweeteners will not automatically lower energy intake; to control energy intake successfully, a person needs to make informed diet and activity decisions throughout the day (as Chapter 9 explains).

TABLE H4-2 Average Aspartame Contents of Selected Foods	
Food	Aspartame (mg)
12 oz diet soft drink	170
8 oz powdered drink	100
8 oz sugar-free fruit yogurt	124
4 oz gelatin dessert	80
1 packet sweetener	35

Stevia—An Herbal Alternative

The FDA has backed its approval or denial of artificial sweeteners with decades of extensive research. Such research is lacking for the herb **stevia,** a shrub whose leaves have long been used by the people of South America to sweeten their beverages. In the United States, stevia is sold in health-food stores as a dietary supplement. The FDA has reviewed the limited research on the use of stevia as an alternative to artificial sweeteners and found concerns regarding its effect on reproduction, cancer development, and energy metabolism. Used sparingly, stevia may do little harm, but the FDA could not approve its extensive and widespread use in the U.S. market. The European Union and the United Nations have reached similar conclusions. In Canada, provisional guidelines have been adopted for the use of stevia as a medicinal ingredient and as a sweetening agent. That stevia can be sold as a dietary supplement but not used as a food additive in the United States, highlights key differences in FDA regulations. Food additives must prove their safety and effectiveness before receiving FDA approval, whereas dietary supplements are not required to submit to any testing or receive any approval. (See Highlight 10 for information on dietary supplements and Chapter 19 for more on herbs.)

Sugar Replacers

Some "sugar-free" or reduced-kcalorie products contain sugar replacers.* The term *sugar replacers* describes the sugar alcohols—familiar examples include erythritol, mannitol, sorbitol, xylitol, maltitol, isomalt, and lactitol—that provide bulk and sweetness in cookies, hard candies, sugarless gums, jams, and jellies. These products claim to be "sugar-free" on their labels, but in this case, "sugar-free" does not mean free of kcalories. Sugar replacers do provide kcalories, but fewer than their carbohydrate cousins, the sugars. Because sugar replacers yield energy, they are sometimes referred to as **nutritive sweeteners.** Table H4-3 includes their energy values, but a simple estimate can help consumers: divide grams by 2. Sugar alcohols occur naturally in fruits and vegetables; manufacturers also use sugar alcohols as a low-energy bulk ingredient in many processed foods.

---

* To minimize confusion, the American Diabetes Association prefers the term sugar replacers instead of "sugar alcohols" (which connotes alcohol), "bulk sweeteners" (which connotes fiber), or "sugar substitutes" (which connotes aspartame and saccharin).

TABLE H4-3 Sugar Replacers

Sugar Alcohols	Relative Sweetness[a]	Energy (kcal/g)	Approved Uses
Erythritol	0.7	0.4	Beverages, frozen dairy desserts, baked goods, chewing gum, candies
Isomalt	0.5	2.0	Candies, chewing gum, ice cream, jams and jellies, frostings, beverages, baked goods
Lactitol	0.4	2.0	Candies, chewing gum, frozen dairy desserts, jams and jellies, frostings, baked goods
Maltitol	0.9	2.1	Particularly good for candy coating
Mannitol	0.7	1.6	Bulking agent, chewing gum
Sorbitol	0.5	2.6	Special dietary foods, candies, gums
Xylitol	1.0	2.4	Chewing gum, candies, pharmaceutical and oral health products

[a] Relative sweetness is determined by comparing the approximate sweetness of a sugar replacer with the sweetness of pure sucrose, which has been defined as 1.0. Chemical structure, temperature, acidity, and other flavors of the foods in which the substance occurs all influence relative sweetness.

Sugar alcohols evoke a low glycemic response. The body absorbs sugar alcohols slowly; consequently, they are slower to enter the bloodstream than other sugars. Side effects such as gas, abdominal discomfort, and diarrhea, however, make them less attractive than the artificial sweeteners. For this reason, regulations require food labels to state "Excess consumption may have a laxative effect" if reasonable consumption of that food could result in the daily ingestion of 50 grams of a sugar alcohol.

The real benefit of using sugar replacers is that they do not contribute to dental caries. Bacteria in the mouth cannot metabolize sugar alcohols as rapidly as sugar. They are therefore valuable in chewing gums, breath mints, and other products that people keep in their mouths for a while. Figure H4-3 presents labeling information for products using sugar alternatives.

The sugar replacers, like the artificial sweeteners, can occupy a place in the diet, and provided they are used in moderation, they will do no harm. In fact, they can help, both by providing an alternative to sugar for people with diabetes and by inhibiting caries-causing bacteria. People may find it appropriate to use all three sweeteners at times: artificial sweeteners, sugar replacers, and sugar itself.

FIGURE H4-3 | Sugar Alternatives on Food Labels

Products containing sugar replacers may claim to "not promote tooth decay" if they meet FDA criteria for dental plaque activity.

Products containing aspartame must carry a warning for people with phenylketonuria.

This ingredient list includes both sugar alcohols and artificial sweetenters.

INGREDIENTS: SORBITOL, MALTITOL, GUM BASE, MANNITOL, ARTIFICIAL AND NATURAL FLAVORING, ACACIA, SOFTENERS, TITANIUM DIOXIDE (COLOR), ASPARTAME, ACESULFAME POTASSIUM AND CANDELILLA WAX. PHENYLKETONURICS: CONTAINS PHENYLALANINE.

35% FEWER CALORIES THAN SUGARED GUM.

Nutrition Facts

Serving Size 2 pieces (3g)
Servings 6
Calories 5

Amount per serving	% DV*
Total Fat 0g	0%
Sodium 0mg	0%
Total Carb. 2g	1%
Sugars 0g	
Sugar Alcohol 2g	
Protein 0g	

*Percent Daily Values (DV) are based on a 2,000 calorie diet. Not a significant source of other nutrients.

Products containing less than 0.5 g of sugar per serving can claim to be "sugarless" or "sugar-free."

Products that claim to be "reduced kcalories" must provide at least 25% fewer kcalories per serving than the comparison item.

© Craig Moore

NUTRITION ON THE NET

For further study of topics covered in this chapter, log on to **academic.cengage.com/nutrition/rolfes/UNCN8e**. Go to Chapter 4, then to Nutrition on the Net.

- Search for "artificial sweeteners" at the U.S. Government health information site: **www.healthfinder.gov**

- Search for "sweeteners" at the International Food Information Council site: **www.ific.org**

REFERENCES

1. M. Soffritti and coauthors, Aspartame induces lymphomas and leukaemias in rats, *European Journal of Oncology* 10 (2005): 107–116.
2. U.S. Food and Drug Administration, FDA statement on European aspartame study, posted May 8, 2006, www.fda.gov; M. R. Weihrauch and V. Diehl, Artificial sweeteners—Do they bear a carcinogenic risk? *Annals of Oncology* 15 (2004): 1460–1465.
3. S. M. Jankovic, Controversies with aspartame, *Medicinski Pregled* 56 (2003): 27–29.
4. Position of the American Dietetic Association: Use of nutritive and nonnutritive sweeteners, *Journal of the American Dietetic Association* 104 (2004): 255–275.
5. S. H. F. Vermunt and coauthors, Effects of sugar intake on body weight: A review, *Obesity Reviews* 4 (2003): 91–99.

 The CengageNOW logo indicates an opportunity for online self-study, linking you to interactive tutorials and videos based on your level of understanding.

academic.cengage.com/login

Animated! Figure 5.17: Absorption of Fat

How To: Practice Problems

Nutrition Portfolio Journal

Nutrition Calculations: Practice Problems

Nutrition in Your Life

Most likely, you know what you don't like about body fat, but do you appreciate how it insulates you against the cold or powers your hike around a lake? And what about food fat? You're right to credit fat for providing the delicious flavors and aromas of buttered popcorn and fried chicken—and to criticize it for contributing to the weight gain and heart disease so common today. The challenge is to strike a healthy balance of enjoying some fat, but not too much. Learning which kinds of fats are most harmful will help you make wise decisions.

The Lipids: Triglycerides, Phospholipids, and Sterols

Most people are surprised to learn that fat has some virtues. Only when people consume either too much or too little fat, or too much of some kinds of fat, does poor health develop. It is true, though, that in our society of abundance, people are likely to consume too much fat.

Fat refers to the class of nutrients known as **lipids.** The lipid family includes triglycerides (**fats** and **oils**), phospholipids, and sterols. The triglycerides ◆ predominate, both in foods and in the body.

The Chemist's View of Fatty Acids and Triglycerides

Like carbohydrates, fatty acids and triglycerides are composed of carbon (C), hydrogen (H), and oxygen (O). Because these lipids have many more carbons and hydrogens in proportion to their oxygens, however, they can supply more energy per gram than carbohydrates can (Chapter 7 provides details).

The many names and relationships in the lipid family can seem overwhelming—like meeting a friend's extended family for the first time. To ease the introductions, this chapter first presents each of the lipids from a chemist's point of view using both words and diagrams. Then the chapter follows the lipids through digestion and absorption and into the body to examine their roles in health and disease. For people who think more easily in words than in chemical symbols, this *preview* of the upcoming chemistry may be helpful:

1. Every triglyceride contains one molecule of glycerol and three fatty acids (basically, chains of carbon atoms).

2. Fatty acids may be 4 to 24 (even numbers of) carbons long, the 18-carbon ones being the most common in foods and especially noteworthy in nutrition.

3. Fatty acids may be saturated or unsaturated. Unsaturated fatty acids may have one or more points of unsaturation. (That is, they may be monounsaturated or polyunsaturated.)

4. Of special importance in nutrition are the polyunsaturated fatty acids whose *first* point of unsaturation is next to the third carbon (known as omega-3 fatty acids) or next to the sixth carbon (omega-6).

5. The 18-carbon fatty acids that fit this description are linolenic acid (omega-3) and linoleic acid (omega-6). Each is the primary member of a family of longer-chain

◆ Of the lipids in foods, 95% are fats and oils (triglycerides); of the lipids stored in the body, 99% are triglycerides.

lipids: a family of compounds that includes triglycerides, phospholipids, and sterols. Lipids are characterized by their insolubility in water. (Lipids also include the fat-soluble vitamins, described in Chapter 11.)

fats: lipids that are solid at room temperature (77°F or 25°C).

oils: lipids that are liquid at room temperature (77°F or 25°C).

FIGURE 5-1 Acetic Acid

Acetic acid is a two-carbon organic acid.

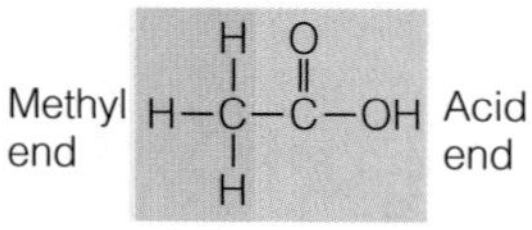

fatty acids that help to regulate blood pressure, blood clotting, and other body functions important to health.

The paragraphs, definitions, and diagrams that follow present this information again in much more detail.

Fatty Acids

A **fatty acid** is an organic acid—a chain of carbon atoms with hydrogens attached—that has an acid group (COOH) at one end and a methyl group (CH_3) at the other end. The organic acid shown in Figure 5-1 is acetic acid, the compound that gives vinegar its sour taste. Acetic acid is the shortest such acid, with a "chain" only two carbon atoms long.

The Length of the Carbon Chain   Most naturally occurring fatty acids contain even numbers of carbons in their chains—up to 24 carbons in length. This discussion begins with the 18-carbon fatty acids, which are abundant in our food supply. Stearic acid is the simplest of the 18-carbon fatty acids; the bonds between its carbons are all alike:

Stearic acid, an 18-carbon saturated fatty acid

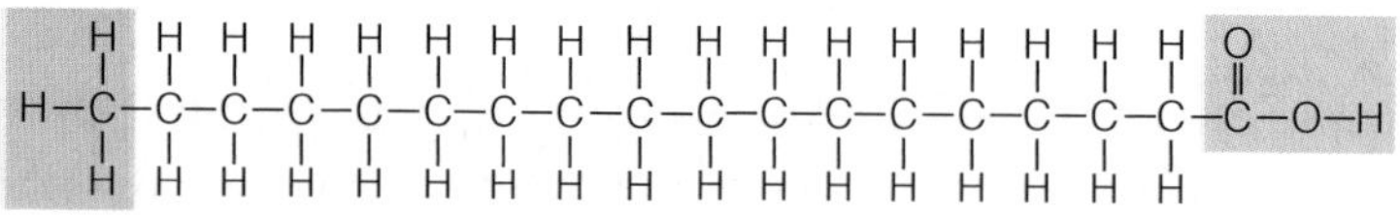

As you can see, stearic acid is 18 carbons long, and each atom meets the rules of chemical bonding described in Figure 4-1 on p. 102. The following structure also depicts stearic acid, but in a simpler way, with each "corner" on the zigzag line representing a carbon atom with two attached hydrogens:

Stearic acid (simplified structure)

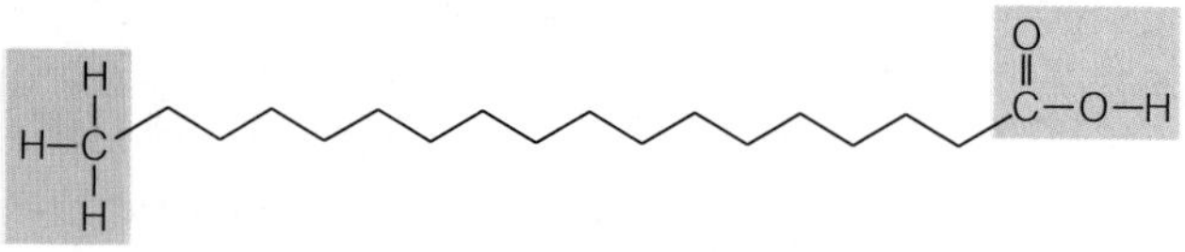

As mentioned, the carbon chains of fatty acids vary in length. The long-chain (12 to 24 carbons) fatty acids of meats, fish, and vegetable oils are most common in the diet. Smaller amounts of medium-chain (6 to 10 carbons) and short-chain (fewer than 6 carbons) fatty acids also occur, primarily in dairy products. (Tables C-1 and C-2 in Appendix C provide the names, chain lengths, and sources of fatty acids commonly found in foods.)

The Degree of Unsaturation   Stearic acid is a **saturated fatty acid** (terms that describe the saturation of fatty acids are defined in the accompanying glossary). A saturated fatty acid is fully loaded with hydrogen atoms and contains only single bonds between its carbon atoms. If two hydrogens were missing from the middle of the carbon chain, the remaining structure might be:

An impossible chemical structure

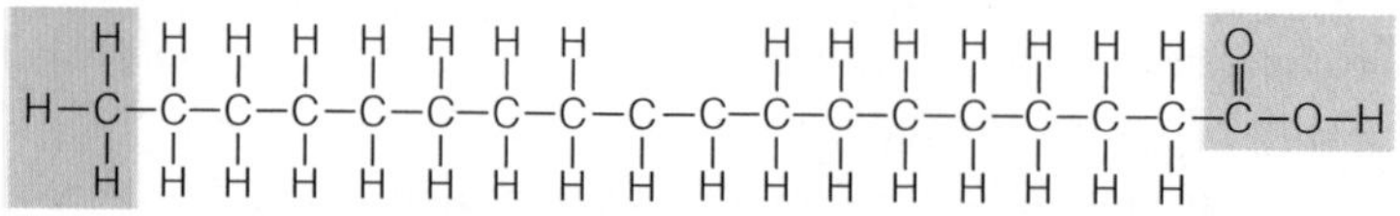

Such a compound cannot exist, however, because two of the carbons have only three bonds each, and nature requires that every carbon have four bonds. The two carbons therefore form a double bond:

Oleic acid, an 18-carbon monounsaturated fatty acid

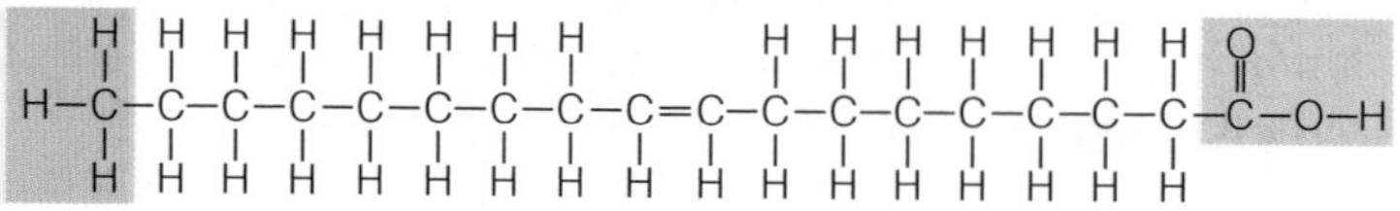

GLOSSARY OF FATTY ACID TERMS

fatty acid: an organic compound composed of a carbon chain with hydrogens attached and an acid group (COOH) at one end and a methyl group (CH_3) at the other end.

monounsaturated fatty acid (MUFA): a fatty acid that lacks two hydrogen atoms and has one double bond between carbons—for example, oleic acid. A **monounsaturated fat** is composed of triglycerides in which most of the fatty acids are monounsaturated.

• **mono** = one

point of unsaturation: the double bond of a fatty acid, where hydrogen atoms can easily be added to the structure.

polyunsaturated fatty acid (PUFA): a fatty acid that lacks four or more hydrogen atoms and has two or more double bonds between carbons—for

example, linoleic acid (two double bonds) and linolenic acid (three double bonds). A **polyunsaturated fat** is composed of triglycerides in which most of the fatty acids are polyunsaturated.

• **poly** = many

saturated fatty acid: a fatty acid carrying the maximum possible number of hydrogen atoms—for example, stearic acid. A

saturated fat is composed of triglycerides in which most of the fatty acids are saturated.

unsaturated fatty acid: a fatty acid that lacks hydrogen atoms and has at least one double bond between carbons (includes monounsaturated and polyunsaturated fatty acids). An **unsaturated fat** is composed of triglycerides in which most of the fatty acids are unsaturated.

The same structure drawn more simply looks like this: ◆

The double bond is a **point of unsaturation.** Hence, a fatty acid like this—with two hydrogens missing and a double bond—is an **unsaturated fatty acid.** This one is the 18-carbon **monounsaturated fatty acid** oleic acid, which is abundant in olive oil and canola oil.

A **polyunsaturated fatty acid** has two or more carbon-to-carbon double bonds. **Linoleic acid,** the 18-carbon fatty acid common in vegetable oils, lacks four hydrogens and has two double bonds:

Drawn more simply, linoleic acid looks like this (though the actual shape would kink at the double bonds):

A fourth 18-carbon fatty acid is **linolenic acid,** which has three double bonds. Table 5-1 presents the 18-carbon fatty acids. ◆

The Location of Double Bonds Fatty acids differ not only in the length of their chains and their degree of saturation, but also in the locations of their double bonds. Chemists identify polyunsaturated fatty acids by the position of the double bond nearest the methyl (CH_3) end of the carbon chain, which is described by an **omega** number. A polyunsaturated fatty acid with its first double bond three carbons away

Oleic acid (simplified structure)

◆ Remember that each "corner" on the zigzag line represents a carbon atom with two attached hydrogens. In addition, although drawn straight here, the actual shape kinks at the double bonds (as shown in the left side of Figure 5-8).

Linoleic acid, an 18-carbon polyunsaturated fatty acid

Linoleic acid (simplified structure)

◆ Chemists use a shorthand notation to describe fatty acids. The first number indicates the number of carbon atoms; the second, the number of the double bonds. For example, the notation for stearic acid is 18:0.

TABLE 5-1	18-Carbon Fatty Acids			
Name	Number of Carbon Atoms	Number of Double Bonds	Saturation	Common Food Sources
Stearic acid	18	0	Saturated	Most animal fats
Oleic acid	18	1	Monounsaturated	Olive, canola oils
Linoleic acid	18	2	Polyunsaturated	Sunflower, safflower, corn, and soybean oils
Linolenic acid	18	3	Polyunsaturated	Soybean and canola oils, flaxseed, walnuts

linoleic (lin-oh-LAY-ick) **acid:** an essential fatty acid with 18 carbons and two double bonds.

linolenic (lin-oh-LEN-ick) **acid:** an essential fatty acid with 18 carbons and three double bonds.

omega: the last letter of the Greek alphabet (ω), used by chemists to refer to the position of the first double bond from the methyl (CH_3) end of a fatty acid.

FIGURE 5-2 Omega-3 and Omega-6 Fatty Acids Compared

The omega number indicates the position of the first double bond in a fatty acid, counting from the methyl (CH_3) end. Thus an omega-3 fatty acid's first double bond occurs three carbons from the methyl end, and an omega-6 fatty acid's first double bond occurs six carbons from the methyl end. The members of an omega family may have different lengths and different numbers of double bonds, but the first double bond occurs at the same point in all of them. These structures are drawn linearly here to ease counting carbons and locating double bonds, but their shapes actually bend at the double bonds, as shown in Figure 5-8 (p. 145).

Linolenic acid, an omega-3 fatty acid

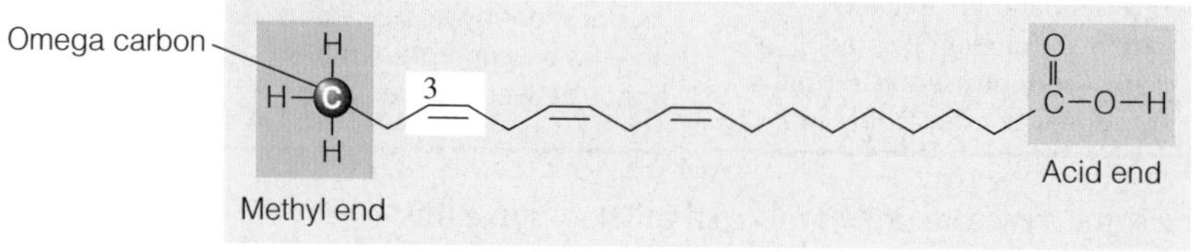

Linoleic acid, an omega-6 fatty acid

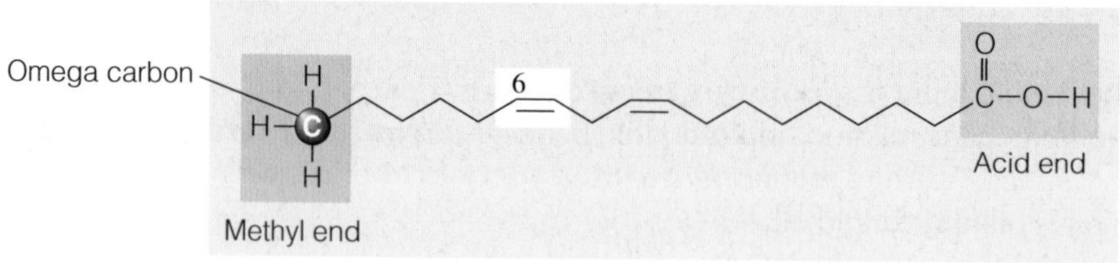

FIGURE 5-3 Glycerol

When glycerol is free, an OH group is attached to each carbon. When glycerol is part of a triglyceride, each carbon is attached to a fatty acid by a carbon-oxygen bond.

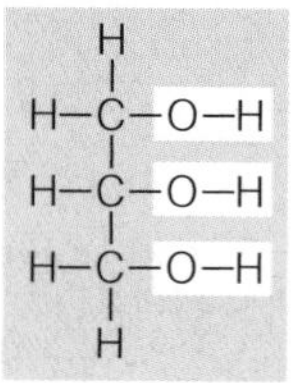

◆ The food industry often refers to these saturated vegetable oils as the "tropical oils."

omega-3 fatty acid: a polyunsaturated fatty acid in which the first double bond is three carbons away from the methyl (CH_3) end of the carbon chain.

omega-6 fatty acid: a polyunsaturated fatty acid in which the first double bond is six carbons from the methyl (CH_3) end of the carbon chain.

triglycerides (try-GLISS-er-rides): the chief form of fat in the diet and the major storage form of fat in the body; composed of a molecule of glycerol with three fatty acids attached; also called **triacylglycerols** (try-ay-seel-GLISS-er-ols).*
- **tri** = three
- **glyceride** = of glycerol
- **acyl** = a carbon chain

glycerol (GLISS-er-ol): an alcohol composed of a three-carbon chain, which can serve as the backbone for a triglyceride.
- **ol** = alcohol

from the methyl end is an **omega-3 fatty acid.** Similarly, an **omega-6 fatty acid** is a polyunsaturated fatty acid with its first double bond six carbons away from the methyl end. Figure 5-2 compares two 18-carbon fatty acids—linolenic acid (an omega-3 fatty acid) and linoleic acid (an omega-6 fatty acid).

Triglycerides

Few fatty acids occur free in foods or in the body. Most often, they are incorporated into **triglycerides**—lipids composed of three fatty acids attached to a **glycerol.** (Figure 5-3 presents a glycerol molecule.) To make a triglyceride, a series of condensation reactions combine a hydrogen atom (H) from the glycerol and a hydroxyl (OH) group from a fatty acid, forming a molecule of water (H_2O) and leaving a bond between the other two molecules (see Figure 5-4). Most triglycerides contain a mixture of more than one type of fatty acid (see Figure 5-5).

Degree of Unsaturation Revisited

The chemistry of a fatty acid—whether it is short or long, saturated or unsaturated, with its first double bond here or there—influences the characteristics of foods and the health of the body. A section later in this chapter explains how these features affect health; this section describes how the degree of unsaturation influences the fats and oils in foods.

Firmness The degree of unsaturation influences the firmness of fats at room temperature. Generally speaking, the polyunsaturated vegetable oils are liquid at room temperature, and the more saturated animal fats are solid. Not all vegetable oils are polyunsaturated, however. Cocoa butter, palm oil, palm kernel oil, and coconut oil ◆ are saturated even though they are of vegetable origin; they are firmer than most vegetable oils because of their saturation, but softer than most animal fats because of their shorter carbon chains (8 to 14 carbons long). Generally, the shorter the car-

---

* Research scientists commonly use the term *triacylglycerols;* this book continues to use the more familiar term *triglycerides,* as do many other health and nutrition books and journals.

FIGURE 5-4 Condensation of Glycerol and Fatty Acids to Form a Triglyceride

To make a triglyceride, three fatty acids attach to glycerol in condensation reactions.

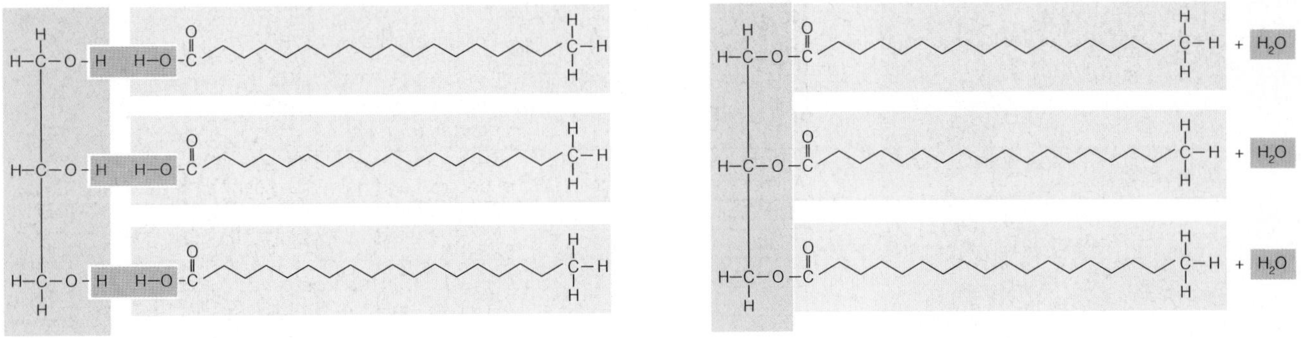

Glycerol + 3 fatty acids ⟶ Triglyceride + 3 water molecules

An H atom from glycerol and an OH group from a fatty acid combine to create water, leaving the O on the glycerol and the C at the acid end of each fatty acid to form a bond.

Three fatty acids attached to a glycerol form a triglyceride and yield water. In this example, all three fatty acids are stearic acid, but most often triglycerides contain mixtures of fatty acids (as shown in Figure 5-5).

bon chain, the softer the fat is at room temperature. Fatty acid compositions of selected fats and oils are shown in Figure 5-6 (p. 144), and Appendix H provides the fat and fatty acid contents of many other foods.

Stability Saturation also influences stability. All fats become spoiled when exposed to oxygen. Polyunsaturated fats spoil most readily because their double bonds are unstable; monounsaturated fats are slightly less susceptible. Saturated fats are most resistant to **oxidation** and thus least likely to become rancid. The oxidation of fats produces a variety of compounds that smell and taste rancid; other types of spoilage can occur due to microbial growth.

Manufacturers can protect fat-containing products against rancidity in three ways—none of them perfect. First, products may be sealed in air-tight, nonmetallic containers, protected from light, and refrigerated—an expensive and inconvenient storage system. Second, manufacturers may add **antioxidants** to compete for the oxygen and thus protect the oil (examples are the additives BHA and BHT and vitamin E).* Third, manufacturers may saturate some or all of the points of unsaturation by adding hydrogen molecules—a process known as hydrogenation.

Hydrogenation **Hydrogenation** offers two advantages. First, it protects against oxidation (thereby prolonging shelf life) by making polyunsaturated fats more saturated (see Figure 5-7, p. 144). Second, it alters the texture of foods by making liquid vegetable oils more solid (as in margarine and shortening). Hydrogenated fats make margarine spreadable, pie crusts flaky, and puddings creamy.

Trans-Fatty Acids Figure 5-7 illustrates the total hydrogenation of a polyunsaturated fatty acid to a saturated fatty acid, which rarely occurs during food processing. Most often, a fat is partially hydrogenated, and some of the double bonds that remain after processing change from *cis* to *trans*. In nature, most double bonds are *cis*—meaning that the hydrogens next to the double bonds are on the same side of the carbon chain. Only a few fatty acids (notably a small percentage of those found in milk and meat products) are ***trans*-fatty acids**—meaning that the hydrogens next to the double bonds are on opposite sides of the carbon chain (see Figure 5-8, p. 145).† These arrangements result in different configurations for the fatty acids, and this difference affects function: in the body, *trans*-fatty acids that derive from hydrogenation behave more like saturated fats than like unsaturated fats. The relationship between *trans*-fatty acids and heart disease has been the subject of much

FIGURE 5-5 A Mixed Triglyceride

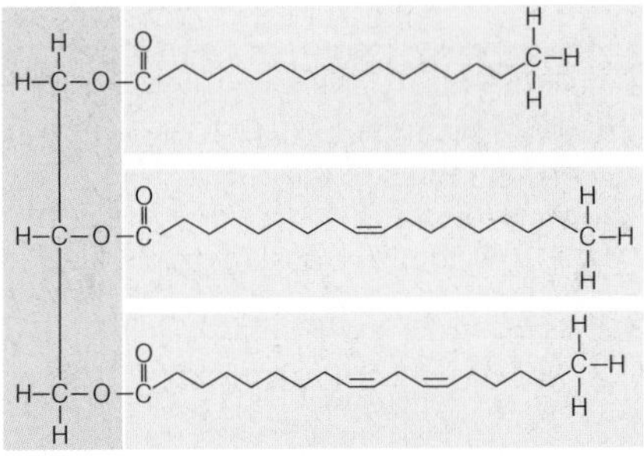

This mixed triglyceride includes a saturated fatty acid, a monounsaturated fatty acid, and a polyunsaturated fatty acid, respectively.

oxidation (OKS-ee-day-shun): the process of a substance combining with oxygen; oxidation reactions involve the loss of electrons.

antioxidants: as a food additive, preservatives that delay or prevent rancidity of fats in foods and other damage to food caused by oxygen.

hydrogenation (HIGH-dro-jen-AY-shun or high-DROJ-eh-NAY-shun): a chemical process by which hydrogens are added to monounsaturated or polyunsaturated fatty acids to reduce the number of double bonds, making the fats more saturated (solid) and more resistant to oxidation (protecting against rancidity). Hydrogenation produces *trans*-fatty acids.

***trans*-fatty acids:** fatty acids with hydrogens on opposite sides of the double bond.

* BHA is butylated hydroxyanisole; BHT is butylated hydroxytoluene.
† For example, most dairy products contain less than 0.5 grams *trans* fat per serving.

At room temperature, saturated fats (such as those commonly found in butter and other animal fats) are solid, whereas unsaturated fats (such as those found in vegetable oils) are usually liquid.

© Polara Studios Inc.

FIGURE 5-6 Comparison of Dietary Fats

Most fats are a mixture of saturated, monounsaturated, and polyunsaturated fatty acids.

Key:

■ Saturated		■ Polyunsaturated, omega-6	
■ Monounsaturated		■ Polyunsaturated, omega-3	

Animal fats and the tropical oils of coconut and palm are mostly **saturated** fatty acids.

Coconut oil	
Butter	
Beef tallow	
Palm oil	
Lard	

Some vegetable oils, such as olive and canola, are rich in **monounsaturated** fatty acids.

Olive oil	
Canola oil	
Peanut oil	

Many vegetable oils are rich in **polyunsaturated** fatty acids.

Safflower oil	
Flaxseed oil	
Walnut oil	
Sunflower oil	
Corn oil	

recent research, as a later section describes. In contrast, naturally occurring fatty acids, such as **conjugated linoleic acid**, that have a *trans* configuration may have health benefits.[1]

IN SUMMARY

The predominant lipids both in foods and in the body are triglycerides: glycerol backbones with three fatty acids attached. Fatty acids vary in the length of their carbon chains, their degrees of unsaturation, and the location of their double bond(s). Those that are fully loaded with hydrogens are saturated; those that are missing hydrogens and therefore have double bonds are unsaturated (monounsaturated or polyunsaturated). The vast majority of triglycerides contain more than one type of fatty acid. Fatty acid saturation affects fats' physical characteristics and storage properties. Hydrogenation, which makes polyunsaturated fats more saturated, gives rise to *trans*-fatty acids, altered fatty acids that may have health effects similar to those of saturated fatty acids.

FIGURE 5-7 Hydrogenation

Double bonds carry a slightly negative charge and readily accept positively charged hydrogen atoms, creating a saturated fatty acid. Most often, fat is partially hydrogenated, creating a *trans*-fatty acid (shown in Figure 5-8).

Polyunsaturated fatty acid → Hydrogenated (saturated) fatty acid

conjugated linoleic acid: a collective term for several fatty acids that have the same chemical formula as linoleic acid (18 carbons, two double bonds) but with different configurations.

FIGURE 5-8 *Cis-* and *Trans*-Fatty Acids Compared

This example shows the *cis* configuration for an 18-carbon monounsaturated fatty acid (oleic acid) and its corresponding *trans* configuration (elaidic acid).

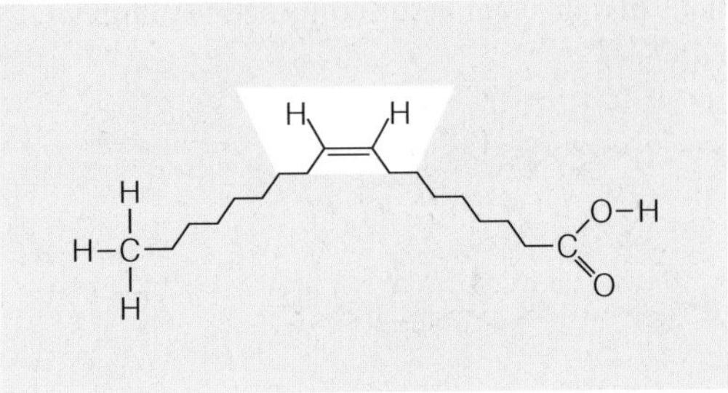

cis-fatty acid

A *cis*-fatty acid has its hydrogens on the same side of the double bond; *cis* molecules fold back into a U-like formation. Most naturally occuring unsaturated fatty acids in foods are *cis*.

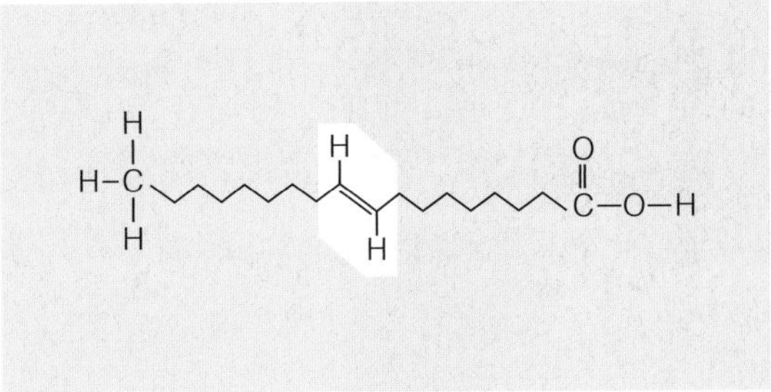

trans-fatty acid

A *trans*-fatty acid has its hydrogens on the opposite sides of the double bond; *trans* molecules are more linear. The *trans* form typically occurs in partially hydrogenated foods when hydrogen atoms shift around some double bonds and change the configuration from *cis* to *trans*.

The Chemist's View of Phospholipids and Sterols

The preceding pages have been devoted to one of the three classes of lipids, the triglycerides, and their component parts, the fatty acids. The other two classes of lipids, the phospholipids and sterols, make up only 5 percent of the lipids in the diet.

Phospholipids

The best-known **phospholipid** is **lecithin.** A diagram of a lecithin molecule is shown in Figure 5-9 (p. 146). Notice that lecithin has a backbone of glycerol with two of its three attachment sites occupied by fatty acids like those in triglycerides. The third site is occupied by a phosphate group and a molecule of **choline.** The fatty acids make phospholipids soluble in fat; the phosphate group allows them to dissolve in water. Such versatility enables the food industry to use phospholipids as emulsifiers ◆ to mix fats with water in such products as mayonnaise and candy bars.

Phospholipids in Foods In addition to the phospholipids used by the food industry as emulsifiers, phospholipids are also found naturally in foods. The richest food sources of lecithin are eggs, liver, soybeans, wheat germ, and peanuts.

Roles of Phospholipids The lecithins and other phospholipids are important constituents of cell membranes (see Figure 5-10, p. 146). Because phospholipids are soluble in both water and fat, they can help lipids move back and forth across the cell membranes into the watery fluids on both sides. Thus they enable fat-soluble substances, including vitamins and hormones, to pass easily in and out of cells. The phospholipids also act as emulsifiers in the body, helping to keep fats suspended in the blood and body fluids.

Lecithin periodically receives attention in the popular press. Its advocates claim that it is a major constituent of cell membranes (true), that cell membranes are essential to the integrity of cells (true), and that consumers must therefore take lecithin supplements (false). The liver makes from scratch all the lecithin a person needs. As for lecithin taken as a supplement, the digestive enzyme lecithinase ◆ in the intestine hydrolyzes most of it before it passes into the body, so little lecithin reaches the tissues intact. In other words, lecithin is *not an essential nutrient;* it is just another

◆ Reminder: *Emulsifiers* are substances with both water-soluble and fat-soluble portions that promote the mixing of oils and fats in watery solutions.

◆ Reminder: The word ending *-ase* denotes an enzyme. Hence, lecithinase is an enzyme that works on lecithin.

phospholipid (FOS-foe-LIP-id): a compound similar to a triglyceride but having a phosphate group (a phosphorus-containing salt) and choline (or another nitrogen-containing compound) in place of one of the fatty acids.

lecithin (LESS-uh-thin): one of the phospholipids. Both nature and the food industry use lecithin as an emulsifier to combine water-soluble and fat-soluble ingredients that do not ordinarily mix, such as water and oil.

choline (KOH-leen): a nitrogen-containing compound found in foods and made in the body from the amino acid methionine. Choline is part of the phospholipid lecithin and the neurotransmitter acetylcholine.

Matthew Farruggio

Without help from emulsifiers, fats and water don't mix.

FIGURE 5-10 Phospholipids of a Cell Membrane

A cell membrane is made of phospholipids assembled into an orderly formation called a bilayer. The fatty acid "tails" orient themselves away from the watery fluid inside and outside of the cell. The glycerol and phosphate "heads" are attracted to the watery fluid.

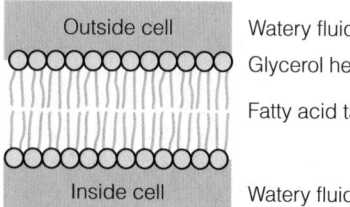

Outside cell — Watery fluid
Glycerol heads
Fatty acid tails
Inside cell — Watery fluid

◆ The chemical structure is the same, but cholesterol that is made in the body is called **endogenous** (en-DOGDE-eh-nus), whereas cholesterol from outside the body (from foods) is called **exogenous** (eks-ODGE-eh-nus).
- **endo** = within
- **gen** = arising
- **exo** = outside (the body)

sterols (STARE-ols or STEER-ols): compounds containing a four ring carbon structure with any of a variety of side chains attached.

cholesterol (koh-LESS-ter-ol): one of the sterols containing a four ring carbon structure with a carbon side chain.

FIGURE 5-9 Lecithin

Lecithin is one of the phospholipids. Notice that a molecule of lecithin is similar to a triglyceride but contains only two fatty acids. The third position is occupied by a phosphate group and a molecule of choline. Other phospholipids have different fatty acids at the upper two positions and different groups attached to phosphate.

From 2 fatty acids

The plus charge on the N is balanced by a negative ion—usually chloride.

From choline

From glycerol   From phosphate

lipid. Like other lipids, lecithin contributes 9 kcalories per gram—an unexpected "bonus" many people taking lecithin supplements fail to realize. Furthermore, large doses of lecithin may cause GI distress, sweating, and loss of appetite. Perhaps these symptoms can be considered beneficial—if they serve to warn people to stop self-dosing with lecithin.

IN SUMMARY

Phospholipids, including lecithin, have a unique chemical structure that allows them to be soluble in both water and fat. In the body, phospholipids are part of cell membranes; the food industry uses phospholipids as emulsifiers to mix fats with water.

Sterols

In addition to triglycerides and phospholipids, the lipids include the **sterols,** compounds with a multiple-ring structure.* The most famous sterol is **cholesterol**; Figure 5-11 (p. 147) shows its chemical structure.

Sterols in Foods Foods derived from both plants and animals contain sterols, but only those from animals contain significant amounts of cholesterol—meats, eggs, fish, poultry, and dairy products. Some people, confused about the distinction between dietary ◆ and blood cholesterol, have asked which foods contain the "good" cholesterol. "Good" cholesterol is not a type of cholesterol found in foods, but it refers to the way the body transports cholesterol in the blood, as explained later (p. 152).

Sterols other than cholesterol are naturally found in all plants. Being structurally similar to cholesterol, these plant sterols interfere with cholesterol absorption, thus lowering blood cholesterol levels.[2] Food manufacturers have fortified foods such as margarine with plant sterols, creating a functional food that helps to reduce blood cholesterol.

---

* The four-ring core structure identifies a steroid; sterols are alcohol derivatives with a steroid ring structure.

Roles of Sterols Many vitally important body compounds are sterols. Among them are bile acids, the sex hormones (such as testosterone), the adrenal hormones (such as cortisol), and vitamin D, as well as cholesterol itself. Cholesterol in the body can serve as the starting material for the synthesis of these compounds ◆ or as a structural component of cell membranes; more than 90 percent of all the body's cholesterol resides in the cells. Despite popular impressions to the contrary, cholesterol is not a villain lurking in some evil foods—it is a compound the body makes and uses. Right now, as you read, your liver is manufacturing cholesterol from fragments of carbohydrate, protein, and fat. In fact, the liver makes about 800 to 1500 milligrams of cholesterol per day, ◆ thus contributing much more to the body's total than does the diet.

Cholesterol's harmful effects in the body occur when it forms deposits in the artery walls. These deposits lead to **atherosclerosis,** a disease that causes heart attacks and strokes. (Chapter 27 provides many more details.)

IN SUMMARY

Sterols have a multiple-ring structure that differs from the structure of other lipids. In the body, sterols include cholesterol, bile, vitamin D, and some hormones. Animal-derived foods contain cholesterol. To summarize, the members of the lipid family include:

- **Triglycerides** (fats and oils), which are made of:
  - Glycerol (1 per triglyceride) and
  - Fatty acids (3 per triglyceride); depending on the number of double bonds, fatty acids may be:
    - *Saturated* (no double bonds)
    - *Monounsaturated* (one double bond)
    - *Polyunsaturated* (more than one double bond); depending on the location of the double bonds, polyunsaturated fatty acids may be:
      - *Omega-3* (first double bond 3 carbons away from methyl end)
      - *Omega-6* (first double bond 6 carbons away from methyl end)
- **Phospholipids** (such as lecithin)
- **Sterols** (such as cholesterol)

Digestion, Absorption, and Transport of Lipids

Each day, the GI tract receives, on average from the food we eat, 50 to 100 grams of triglycerides, 4 to 8 grams of phospholipids, and 200 to 350 milligrams of cholesterol. The body faces a challenge in digesting and absorbing these lipids: getting at them. Fats are **hydrophobic**—that is, they tend to separate from the watery fluids of the GI tract—whereas the enzymes for digesting fats are **hydrophilic.** The challenge is keeping the fats mixed in the watery fluids of the GI tract.

Lipid Digestion

The goal of fat digestion is to dismantle triglycerides into small molecules that the body can absorb and use—namely, **monoglycerides,** fatty acids, and glycerol. Figure 5-12 (p. 148) traces the digestion of triglycerides through the GI tract, and the following paragraphs provide the details.

In the Mouth Fat digestion starts off slowly in the mouth, with some hard fats beginning to melt when they reach body temperature. A salivary gland at the base of the tongue releases an enzyme (lingual lipase) ◆ that plays a minor role in fat

---

FIGURE 5-11 Cholesterol

The fat-soluble vitamin D is synthesized from cholesterol; notice the many structural similarities. The only difference is that cholesterol has a closed ring (highlighted in red), whereas vitamin D's is open, accounting for its vitamin activity. Notice, too, how different cholesterol is from the triglycerides and phospholipids.

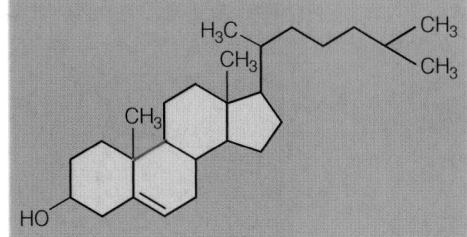

Cholesterol

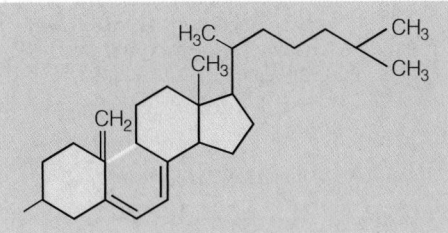

Vitamin D₃

◆ Compounds made from cholestrol:
- Bile acids
- Steroid hormones (testosterone, androgens, estrogens, progesterones, cortisol, cortisone, and aldosterone)
- Vitamin D

◆ For perspective, the Daily Value for cholesterol is 300 mg/day.

◆ Reminder: An enzyme that hydrolyzes lipids is called a *lipase; lingual* refers to the tongue.

atherosclerosis (ATH-er-oh-scler-OH-sis): a type of artery disease characterized by placques (accumulations of lipid-containing material) on the inner walls of the arteries (see Chapter 27).

hydrophobic (high-dro-FOE-bick): a term referring to water-fearing, or non-watersoluble, substances; also known as **lipophilic** (fat loving).
- **hydro** = water
- **phobia** = fear
- **lipo** = lipid
- **phile** = love

hydrophilic (high-dro-FIL-ick): a term referring to water-loving, or water-soluble, substances.

monoglycerides: molecules of glycerol with one fatty acid attached. A molecule of glycerol with two fatty acids attached is a **diglyceride.**
- **mono** = one
- **di** = two

FIGURE 5-12 Fat Digestion in the GI Tract

FAT

Mouth and salivary glands
Some hard fats begin to melt as they reach body temperature. The sublingual salivary gland in the base of the tongue secretes lingual lipase.

Stomach
The acid-stable lingual lipase initiates lipid digestion by hydrolyzing one bond of triglycerides to produce diglycerides and fatty acids. The degree of hydrolysis by lingual lipase is slight for most fats but may be appreciable for milk fats. The stomach's churning action mixes fat with water and acid. A gastric lipase accesses and hydrolyzes (only a very small amount of) fat.

Small intestine
Bile flows in from the gallbladder (via the common bile duct):

$$\text{Fat} \xrightarrow{\text{Bile}} \text{Emulsified fat}$$

Pancreatic lipase flows in from the pancreas (via the pancreatic duct):

$$\text{Emulsified fat} \xrightarrow[\text{lipase}]{\text{Pancreatic (and intestinal)}} \text{Monoglycerides, glycerol, fatty acids (absorbed)}$$
(triglycerides)

Large intestine
Some fat and cholesterol, trapped in fiber, exit in feces.

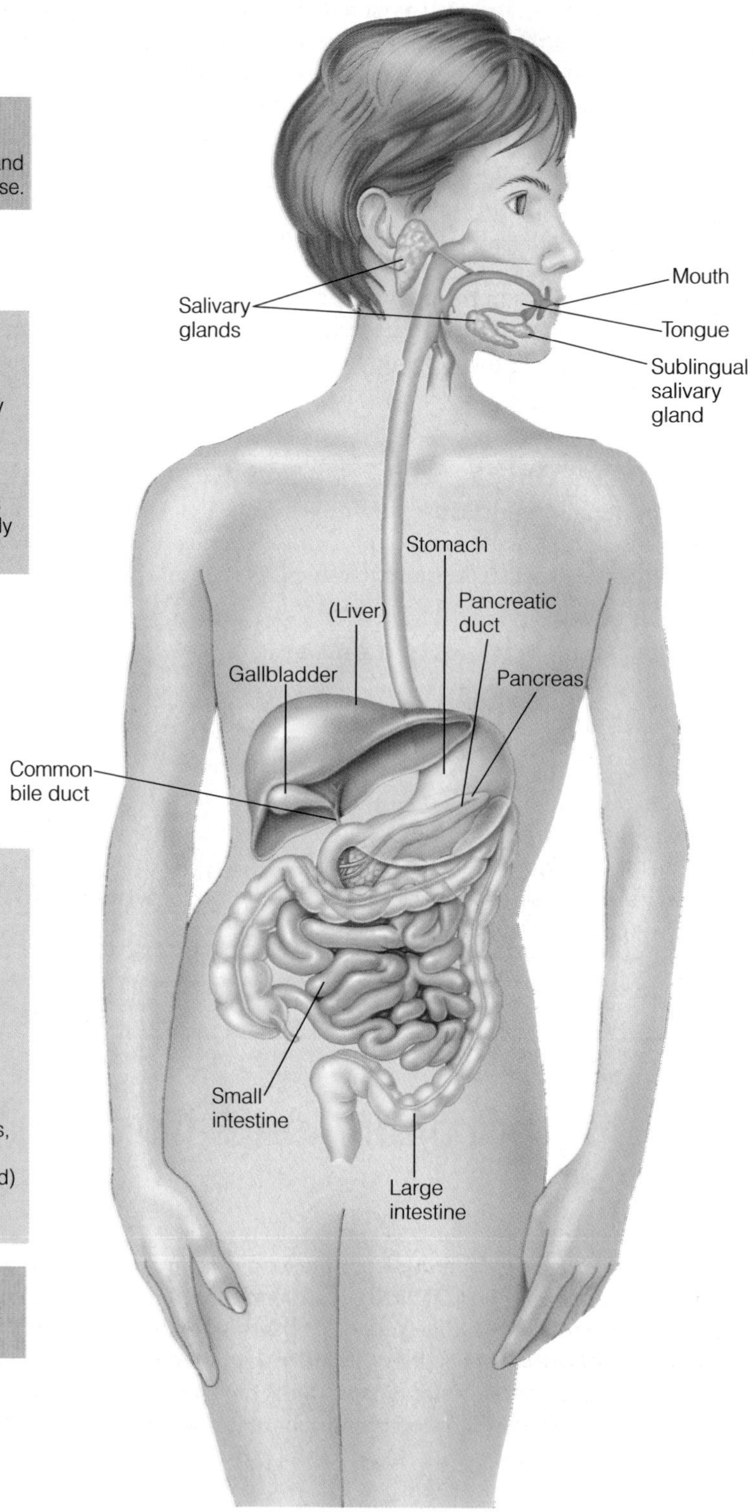

Salivary glands
Mouth
Tongue
Sublingual salivary gland
Stomach
(Liver)
Pancreatic duct
Gallbladder
Pancreas
Common bile duct
Small intestine
Large intestine

digestion in adults and an active role in infants. In infants, this enzyme efficiently digests the short- and medium-chain fatty acids found in milk.

In the Stomach In a quiet stomach, fat would float as a layer above the other components of swallowed food. But the strong muscle contractions of the stomach propel the stomach contents toward the pyloric sphincter. Some chyme passes through the pyloric sphincter periodically, but the remaining partially digested food is propelled back into the body of the stomach. This churning grinds the solid pieces to finer particles, mixes the chyme, and disperses the fat into smaller droplets. These actions help to expose the fat for attack by the gastric lipase enzyme—an enzyme that performs best in the acidic environment of the stomach. Still, little fat digestion takes place in the stomach; most of the action occurs in the small intestine.

In the Small Intestine When fat enters the small intestine, it triggers the release of the hormone cholecystokinin (CCK), which signals the gallbladder to release its stores of bile. (Remember that the liver makes bile, and the gallbladder stores it until it is needed.) Among bile's many ingredients ◆ are bile acids, which are made in the liver from cholesterol and have a similar structure. In addition, they often pair up with an amino acid (a building block of protein). The amino acid end is attracted to water, and the sterol end is attracted to fat (see Figure 5-13, p. 150). This structure improves bile's ability to act as an emulsifier, drawing fat molecules into the surrounding watery fluids. There, the fats are fully digested as they encounter lipase enzymes from the pancreas and small intestine. The process of emulsification is diagrammed in Figure 5-14 (p. 150).

Most of the hydrolysis of triglycerides occurs in the small intestine. The major fat-digesting enzymes are pancreatic lipases; some intestinal lipases are also active. These enzymes remove one, then the other, of each triglyceride's outer fatty acids, leaving a monoglyceride. Occasionally, enzymes remove all three fatty acids, leaving a free molecule of glycerol. Hydrolysis of a triglyceride is shown in Figure 5-15 (p. 151).

Phospholipids are digested similarly—that is, their fatty acids are removed by hydrolysis. The two fatty acids and the remaining phospholipid fragment are then absorbed. Most sterols can be absorbed as is; if any fatty acids are attached, they are first hydrolyzed off.

Bile's Routes After bile enters the small intestine and emulsifies fat, it has two possible destinations, illustrated in Figure 5-16 (p. 151). Most of the bile is reabsorbed from the intestine and recycled. The other possibility is that some of the bile can be trapped by dietary fibers in the large intestine and carried out of the body with the feces. Because cholesterol is needed to make bile, the excretion of bile effectively reduces blood cholesterol. As Chapter 4 explains, the dietary fibers most effective at lowering blood cholesterol this way are the soluble fibers commonly found in fruits, whole grains, and legumes.

Lipid Absorption

Figure 5-17 (p. 152) illustrates the absorption of lipids. Small molecules of digested triglycerides (glycerol and short- and medium-chain fatty acids) can diffuse easily into the intestinal cells; they are absorbed directly into the bloodstream. Larger molecules (the monoglycerides and long-chain fatty acids) merge into spherical complexes, known as **micelles.** Micelles are emulsified fat droplets formed by molecules of bile surrounding monoglycerides and fatty acids. This configuration permits solubility in the watery digestive fluids and transportation to the intestinal cells. Upon arrival, the lipid contents of the micelles diffuse into the intestinal cells. Once inside, the monoglycerides and long-chain fatty acids are reassembled into new triglycerides.

Within the intestinal cells, the newly made triglycerides and other lipids (cholesterol and phospholipids) are packed with protein into transport vehicles known as **chylomicrons.** The intestinal cells then release the chylomicrons into the lymphatic system. The chylomicrons glide through the lymph until they reach a point of entry into the bloodstream at the thoracic duct near the heart. (Recall from Chapter 3 that nutrients from the GI tract that enter the lymph system bypass the liver at first.) The blood carries these lipids to the rest of the body for immediate use

◆ In addition to bile acids and bile salts, bile contains cholesterol, phospholipids (especially lecithin), antibodies, water, electrolytes, and bilirubin and biliverdin (pigments resulting from the breakdown of heme).

micelles (MY-cells): tiny spherical complexes of emulsified fat that arise during digestion; most contain bile salts and the products of lipid digestion, including fatty acids, monoglycerides, and cholesterol.

chylomicrons (kye-lo-MY-cronz): the class of lipoproteins that transport lipids from the intestinal cells to the rest of the body.

FIGURE 5-13 A Bile Acid

This is one of several bile acids the liver makes from cholesterol. It is then bound to an amino acid to improve its ability to form micelles, spherical complexes of emulsified fat. Most bile acids occur as bile salts, usually in association with sodium, but sometimes with potassium or calcium.

Bile acid made from cholesterol (hydrophobic)	Bound to an amino acid from protein (hydrophilic)

---

or storage. A look at these lipids in the body reveals the kinds of fat the diet has been delivering.[3] The fat stores and muscle cells of people who eat a diet rich in unsaturated fats, for example, contain more unsaturated fats than those of people who select a diet high in saturated fats.

IN SUMMARY

The body makes special arrangements to digest and absorb lipids. It provides the emulsifier bile to make them accessible to the fat-digesting lipases that dismantle triglycerides, mostly to monoglycerides and fatty acids, for absorption by the intestinal cells. The intestinal cells assemble freshly absorbed lipids into chylomicrons, lipid packages with protein escorts, for transport so that cells all over the body may select needed lipids from them.

Lipid Transport

lipoproteins (LIP-oh-PRO-teenz): clusters of lipids associated with proteins that serve as transport vehicles for lipids in the lymph and blood.

The chylomicrons are only one of several clusters of lipids and proteins that are used as transport vehicles for fats. As a group, these vehicles are known as **lipoproteins,**

FIGURE 5-14 Emulsification of Fat by Bile

Like bile, detergents are emulsifiers and work the same way, which is why they are effective in removing grease spots from clothes. Molecule by molecule, the grease is dissolved out of the spot and suspended in the water, where it can be rinsed away.

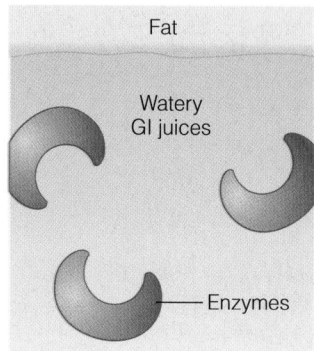

In the stomach, the fat and watery GI juices tend to separate. The enzymes in the GI juices can't get at the fat.

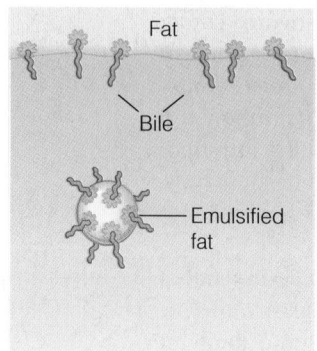

When fat enters the small intestine, the gallbladder secretes bile. Bile has an affinity for both fat and water, so it can bring the fat into the water.

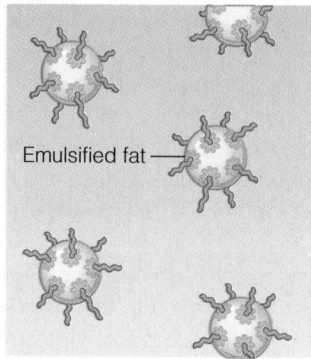

Bile's emulsifying action converts large fat globules into small droplets that repel each other.

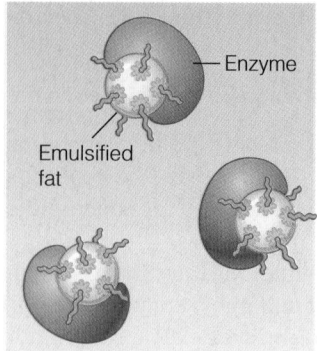

After emulsification, more fat is exposed to the enzymes, making fat digestion more efficient.

FIGURE 5-15 Digestion (Hydrolysis) of a Triglyceride

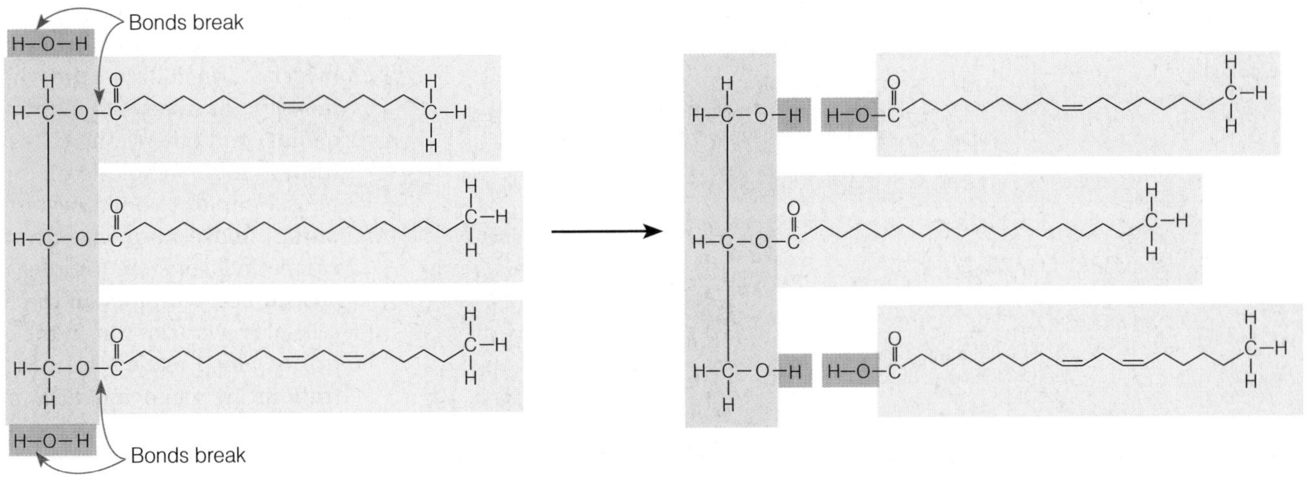

Bonds break

Bonds break

Triglyceride

The triglyceride and two molecules of water are split. The H and OH from water complete the structures of two fatty acids and leave a monoglyceride.

Monoglyceride + 2 fatty acids

These products may pass into the intestinal cells, but sometimes the monoglyceride is split with another molecule of water to give a third fatty acid and glycerol. Fatty acids, monoglycerides, and glycerol are absorbed into intestinal cells.

and they solve the body's problem of transporting fat through the watery bloodstream. The body makes four main types of lipoproteins, distinguished by their size and density.* Each type contains different kinds and amounts of lipids and proteins. ◆ Figure 5-18 (p. 153) shows the relative compositions and sizes of the lipoproteins.

Chylomicrons The chylomicrons are the largest and least dense of the lipoproteins. They transport *diet*-derived lipids (mostly triglycerides) from the intestine (via the lymph system) to the rest of the body. Cells all over the body remove triglycerides from the chylomicrons as they pass by, so the chylomicrons get smaller and smaller. Within 14 hours after absorption, most of the triglycerides have been depleted, and only a few remnants of protein, cholesterol, and phospholipid remain. Special protein receptors on the membranes of the liver cells recognize and remove these chylomicron remnants from the blood. After collecting the remnants, the liver cells first dismantle them and then either use or recycle the pieces.

VLDL (Very-Low-Density Lipoproteins) Meanwhile, in the liver—the most active site of lipid synthesis—cells are synthesizing other lipids. The liver cells use fatty acids arriving in the blood to make cholesterol, other fatty acids, and other compounds. At the same time, the liver cells may be making lipids from carbohydrates, proteins, or alcohol. Ultimately, the lipids made in the liver and those collected from chylomicron remnants are packaged with proteins as **VLDL (very-low-density lipoprotein)** and shipped to other parts of the body.

As the VLDL travel through the body, cells remove triglycerides, causing the VLDL to shrink. As a VLDL loses triglycerides, the proportion of lipids shifts, and the lipoprotein density increases. The remaining cholesterol-rich lipoprotein eventually becomes an **LDL (low-density lipoprotein).**[†] This transformation explains why LDL contain few triglycerides but are loaded with cholesterol.

---

* Chemists can identify the various lipoproteins by their density. They place a blood sample below a thick fluid in a test tube and spin the tube in a centrifuge. The most buoyant particles (highest in lipids) rise to the top and have the lowest density; the densest particles (highest in proteins) remain at the bottom and have the highest density. Others distribute themselves in between.
[†] Before becoming LDL, the VLDL are first transformed into intermediate-density lipoproteins (IDL), sometimes called VLDL remnants. Some IDL may be picked up by the liver and rapidly broken down; those IDL that remain in circulation continue to deliver triglycerides to the cells and eventually become LDL. Researchers debate whether IDL are simply transitional particles or a separate class of lipoproteins; normally, IDL do not accumulate in the blood. Measures of blood lipids include IDL with LDL.

FIGURE 5-16 Enterohepatic Circulation

Most of the bile released into the small intestine is reabsorbed and sent back to the liver to be reused. This cycle is called the **enterohepatic circulation** of bile. Some bile is excreted.

- **enteron** = intestine
- **hepat** = liver

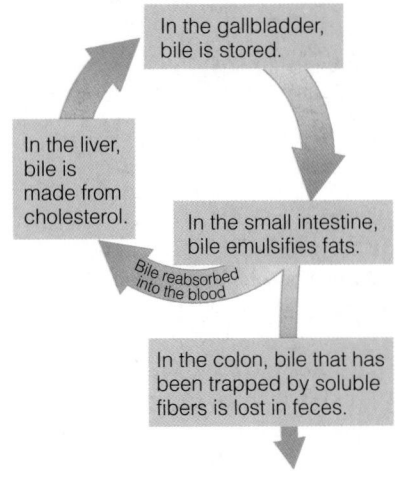

In the gallbladder, bile is stored.

In the liver, bile is made from cholesterol.

In the small intestine, bile emulsifies fats.

Bile reabsorbed into the blood

In the colon, bile that has been trapped by soluble fibers is lost in feces.

◆ The more lipids, the lower the density; the more proteins, the higher the density.

VLDL (very-low-density lipoprotein): the type of lipoprotein made primarily by liver cells to transport lipids to various tissues in the body; composed primarily of triglycerides.

LDL (low-density lipoprotein): the type of lipoprotein derived from very-low-density lipoproteins (VLDL) as VLDL triglycerides are removed and broken down; composed primarily of cholesterol.

FIGURE 5-17 *Animated!* Absorption of Fat

The end products of fat digestion are mostly monoglycerides, some fatty acids, and very little glycerol. Their absorption differs depending on their size. (In reality, molecules of fatty acid are too small to see without a powerful microscope, whereas villi are visible to the naked eye.)

CENGAGENOW™
To test your understanding of these concepts, log on to
academic.cengage.com/login

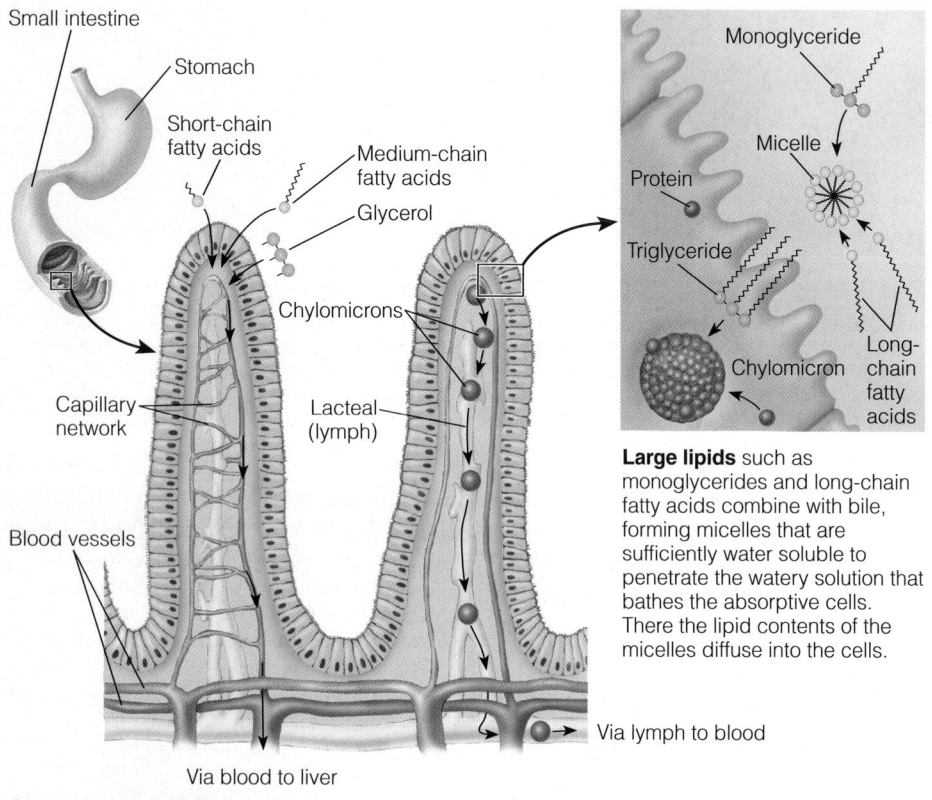

Large lipids such as monoglycerides and long-chain fatty acids combine with bile, forming micelles that are sufficiently water soluble to penetrate the watery solution that bathes the absorptive cells. There the lipid contents of the micelles diffuse into the cells.

Glycerol and small lipids such as short- and medium-chain fatty acids can move directly into the bloodstream.

LDL (Low-Density Lipoproteins)

The LDL circulate throughout the body, making their contents available to the cells of all tissues—muscles (including the heart muscle), fat stores, the mammary glands, and others. The cells take triglycerides, cholesterol, and phospholipids to build new membranes, make hormones or other compounds, or store for later use. Special LDL receptors on the liver cells play a crucial role in the control of blood cholesterol concentrations by removing LDL from circulation.

HDL (High-Density Lipoproteins)

Fat cells may release glycerol, fatty acids, cholesterol, and phospholipids to the blood. The liver makes **HDL (high-density lipoprotein)** to carry cholesterol from the cells back to the liver for recycling or disposal.

Health Implications

The distinction between LDL and HDL has implications for the health of the heart and blood vessels. The blood cholesterol linked to heart disease is LDL cholesterol. HDL also carry cholesterol, but elevated HDL represent cholesterol returning ◆ from the rest of the body to the liver for breakdown and excretion. High LDL cholesterol is associated with a high risk of heart attack, whereas high HDL cholesterol seems to have a protective effect. This is why some people refer to LDL as "bad," and HDL as "good," cholesterol. ◆ Keep in mind that the cholesterol itself is the same, and that the differences between LDL and HDL reflect the *proportions* and *types* of lipids and proteins within them—not the type of cholesterol. The margin ◆ lists factors that influence LDL and HDL, and Chapter 27 provides many more details.

Not too surprisingly, numerous genes influence how the body handles the uptake, synthesis, transport, and degradation of the lipoproteins. Much current research is focused on how nutrient-gene interactions may direct the progression of heart disease.

◆ The transport of cholesterol from the tissues to the liver is sometimes called the *scavenger pathway.*

◆ To help you remember, think of elevated **H**DL as **H**ealthy and elevated **L**DL as **L**ess healthy.

◆ Factors that lower LDL or raise HDL:
• Weight control
• Monounsaturated or polyunsaturated, instead of saturated, fat in the diet
• Soluble, viscous fibers (see Chapter 4)
• Phytochemicals (see Highlight 13)
• Moderate alcohol consumption
• Physical activity

HDL (high-density lipoprotein): the type of lipoprotein that transports cholesterol back to the liver from the cells; composed primarily of protein.

IN SUMMARY

The liver assembles lipids and proteins into lipoproteins for transport around the body. All four types of lipoproteins carry all classes of lipids (triglycerides, phospholipids, and cholesterol), but the chylomicrons are the largest and the highest in triglycerides; VLDL are smaller and are about half triglycerides; LDL are smaller still and are high in cholesterol; and HDL are the smallest and are rich in protein.

FIGURE 5-18 Sizes and Compositions of the Lipoproteins

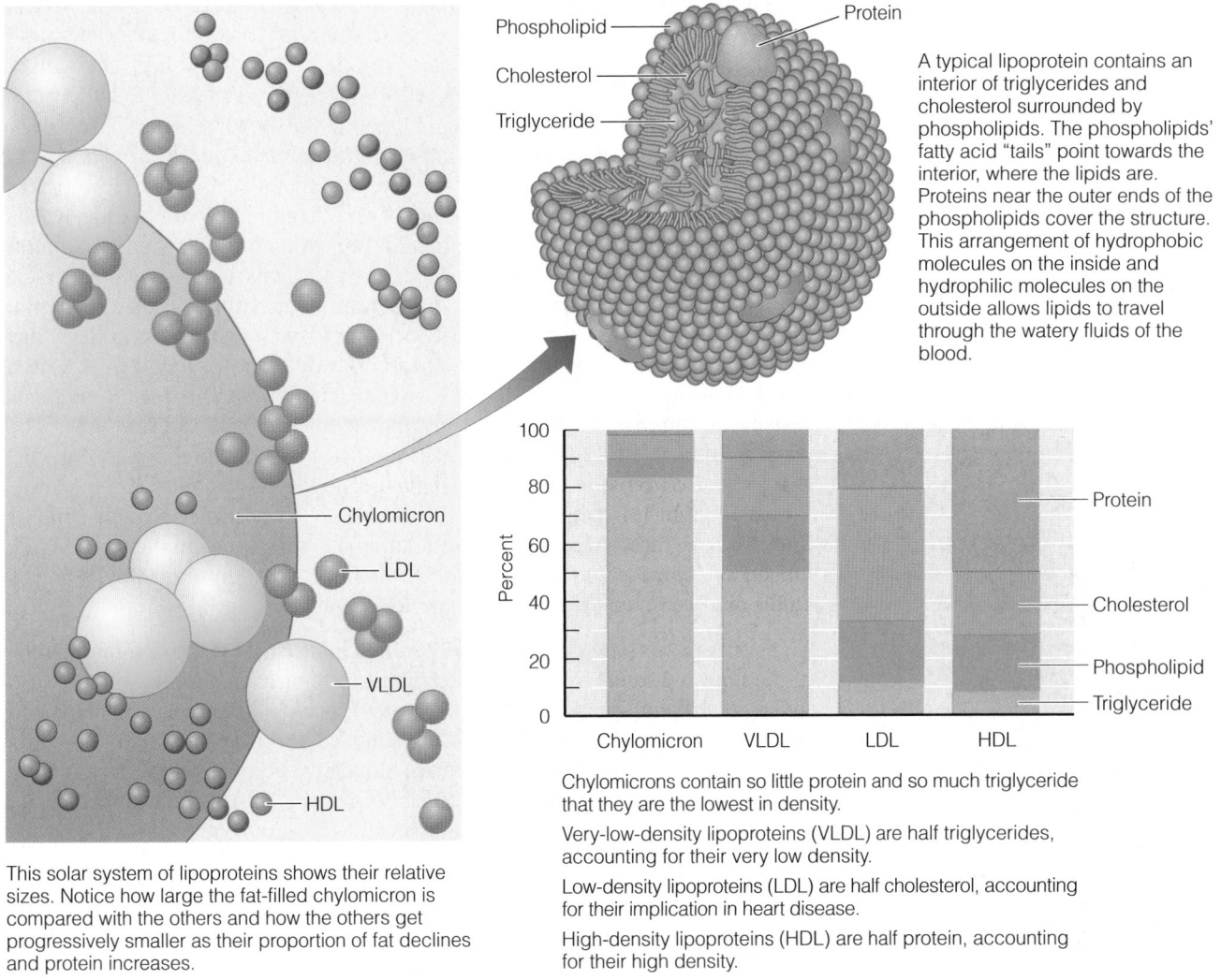

Phospholipid

Cholesterol

Triglyceride

Protein

A typical lipoprotein contains an interior of triglycerides and cholesterol surrounded by phospholipids. The phospholipids' fatty acid "tails" point towards the interior, where the lipids are. Proteins near the outer ends of the phospholipids cover the structure. This arrangement of hydrophobic molecules on the inside and hydrophilic molecules on the outside allows lipids to travel through the watery fluids of the blood.

Chylomicron

LDL

VLDL

HDL

This solar system of lipoproteins shows their relative sizes. Notice how large the fat-filled chylomicron is compared with the others and how the others get progressively smaller as their proportion of fat declines and protein increases.

Percent

Protein

Cholesterol

Phospholipid

Triglyceride

Chylomicron   VLDL   LDL   HDL

Chylomicrons contain so little protein and so much triglyceride that they are the lowest in density.

Very-low-density lipoproteins (VLDL) are half triglycerides, accounting for their very low density.

Low-density lipoproteins (LDL) are half cholesterol, accounting for their implication in heart disease.

High-density lipoproteins (HDL) are half protein, accounting for their high density.

Lipids in the Body

The blood carries lipids to various sites around the body. Once lipids arrive at their destinations, they can get to work providing energy, insulating against temperature extremes, protecting against shock, and maintaining cell membranes. This section provides an overview of the roles of triglycerides and fatty acids and then of the metabolic pathways they can follow within the body's cells.

Roles of Triglycerides

First and foremost, the triglycerides—either from food or from the body's fat stores—provide the body with energy. When a person dances all night, her dinner's triglycerides provide some of the fuel that keeps her moving. When a person loses his appetite, his stored triglycerides fuel much of his body's work until he can eat again.

Efficient energy metabolism depends on the energy nutrients—carbohydrate, fat, and protein—supporting each other. Glucose fragments combine with fat fragments during energy metabolism, and fat and carbohydrate help spare protein, providing energy so that protein can be used for other important tasks.

FIGURE 5-19 The Pathway from One Omega-6 Fatty Acid to Another

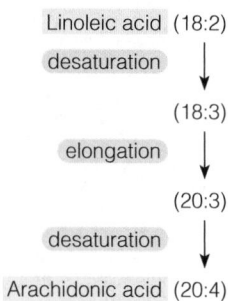

The first number indicates the number of carbons and the second, the number of double bonds. Similar reactions occur when the body makes the omega-3 fatty acids EPA and DHA from linolenic acid.

◆ A nonessential nutrient (such as arachidonic acid) that must be supplied by the diet in special circumstances (as in a linoleic acid deficiency) is considered *conditionally essential*.

essential fatty acids: fatty acids needed by the body but not made by it in amounts sufficient to meet physiological needs.

arachidonic (a-RACK-ih-DON-ic) **acid:** an omega-6 polyunsaturated fatty acid with 20 carbons and four double bonds; present in small amounts in meat and other animal products and synthesized in the body from linoleic acid.

eicosapentaenoic (EYE-cossa-PENTA-ee-NO-ick) **acid (EPA):** an omega-3 polyunsaturated fatty acid with 20 carbons and five double bonds; present in fish and synthesized in limited amounts in the body from linolenic acid.

docosahexaenoic (DOE-cossa-HEXA-ee-NO-ick) **acid (DHA):** an omega-3 polyunsaturated fatty acid with 22 carbons and six double bonds; present in fish and synthesized in limited amounts in the body from linolenic acid.

eicosanoids (eye-COSS-uh-noyds): derivatives of 20-carbon fatty acids; biologically active compounds that help to regulate blood pressure, blood clotting, and other body functions. They include *prostaglandins* (PROS-tah-GLAND-ins), *thromboxanes* (throm-BOX-ains), and *leukotrienes* (LOO-ko-TRY-eens).

Fat also insulates the body. Fat is a poor conductor of heat, so the layer of fat beneath the skin helps keep the body warm. Fat pads also serve as natural shock absorbers, providing a cushion for the bones and vital organs.

Essential Fatty Acids

The human body needs fatty acids, and it can make all but two of them—linoleic acid (the 18-carbon omega-6 fatty acid) and linolenic acid (the 18-carbon omega-3 fatty acid). These two fatty acids must be supplied by the diet and are therefore **essential fatty acids.** A simple definition of an essential nutrient has already been given: a nutrient that the body cannot make, or cannot make in sufficient quantities to meet its physiological needs. The cells do not possess the enzymes to make any of the omega-6 or omega-3 fatty acids from scratch, nor can they convert an omega-6 fatty acid to an omega-3 fatty acid or vice versa. Cells *can*, however, start with the 18-carbon member of an omega family and make the longer fatty acids of that family by forming double bonds (desaturation) and lengthening the chain two carbons at a time (elongation), as shown in Figure 5-19. This is a slow process because the omega-3 and omega-6 families compete for the same enzymes. Too much of a fatty acid from one family can create a deficiency of the other family's longer fatty acids, which is critical only when the diet fails to deliver adequate supplies. Therefore, the most effective way to maintain body supplies of all the omega-6 and omega-3 fatty acids is to obtain them directly from foods—most notably, from vegetable oils, seeds, nuts, fish, and other marine foods.

Linoleic Acid and the Omega-6 Family Linoleic acid is the primary member of the omega-6 family. When the body receives linoleic acid from the diet, it can make other members of the omega-6 family—such as the 20-carbon polyunsaturated fatty acid, **arachidonic acid.** If a linoleic acid deficiency should develop, arachidonic acid, and all other fatty acids that derive from linoleic acid, would also become essential and have to be obtained from the diet. ◆ Normally, vegetable oils and meats supply enough omega-6 fatty acids to meet the body's needs.

Linolenic Acid and the Omega-3 Family Linolenic acid is the primary member of the omega-3 family.* Like linoleic acid, linolenic acid cannot be made in the body and must be supplied by foods. Given this 18-carbon fatty acid, the body can make small amounts of the 20- and 22-carbon members of the omega-3 series, **eicosapentaenoic acid (EPA)** and **docosahexaenoic acid (DHA).** These omega-3 fatty acids are essential for normal growth and development, especially in the eyes and brain.[4] They may also play an important role in the prevention and treatment of heart disease.

Eicosanoids The body uses arachidonic acid and EPA to make substances known as **eicosanoids.** Eicosanoids are a diverse group of compounds that are sometimes described as "hormonelike," but they differ from hormones in important ways. For one, hormones are secreted in one location and travel to affect cells all over the body, whereas eicosanoids appear to affect only the cells in which they are made or nearby cells in the same localized environment. For another, hormones elicit the same response from all their target cells, whereas eicosanoids often have different effects on different cells.

The actions of various eicosanoids sometimes oppose each other. For example, one causes muscles to relax and blood vessels to dilate, whereas another causes muscles to contract and blood vessels to constrict. Certain eicosanoids participate in the immune response to injury and infection, producing fever, inflammation, and pain. One of the ways aspirin relieves these symptoms is by slowing the synthesis of these eicosanoids.

---

* This omega-3 linolenic acid is known as alpha-linolenic acid and is the fatty acid referred to in this chapter. Another fatty acid, also with 18 carbons and three double bonds, belongs to the omega-6 family and is known as gamma-linolenic acid.

Eicosanoids that derive from EPA differ from those that derive from arachidonic acid, with those from EPA providing greater health benefits.[5] The EPA eicosanoids help lower blood pressure, prevent blood clot formation, protect against irregular heartbeats, and reduce inflammation. Because the omega-6 and omega-3 fatty acids compete for the same enzymes to make arachidonic acid and EPA and to make the eicosanoids, the body needs these long-chain polyunsaturated fatty acids from the diet to make eicosanoids in sufficient quantities.[6]

Fatty Acid Deficiencies Most diets in the United States and Canada meet the minimum essential fatty acid requirement adequately. Historically, deficiencies have developed only in infants and young children who have been fed fat-free milk and low-fat diets or in hospital clients who have been mistakenly fed formulas that provided no polyunsaturated fatty acids for long periods of time. Classic deficiency symptoms include growth retardation, reproductive failure, skin lesions, kidney and liver disorders, and subtle neurological and visual problems.

Interestingly, a deficiency of omega-3 fatty acids (EPA and DHA) may be associated with depression.[7] Some neurochemical pathways in the brain become more active and others become less active.[8] It is unclear, however, which comes first—whether inadequate intake alters brain activity or depression alters fatty acid metabolism. To find the answers, researchers must untangle a multitude of confounding factors.

Double thanks: The body's fat stores provide energy for a walk, and the heel's fat pads cushion against the hard pavement.

IN SUMMARY

In the body, triglycerides:
- Provide an energy reserve when stored in the body's fat tissue
- Insulate against temperature extremes
- Protect against shock
- Help the body use carbohydrate and protein efficiently

Linoleic acid (18 carbons, omega-6) and linolenic acid (18 carbons, omega-3) are essential nutrients. They serve as structural parts of cell membranes and as precursors to the longer fatty acids that can make eicosanoids—powerful compounds that participate in blood pressure regulation, blood clot formation, and the immune response to injury and infection, among other functions. Because essential fatty acids are common in the diet and stored in the body, deficiencies are unlikely.

A Preview of Lipid Metabolism

The blood delivers triglycerides to the cells for their use. This is a preview of how the cells store and release energy from fat; Chapter 7 provides details.

Storing Fat as Fat The triglycerides, familiar as the fat in foods and as body fat, serve the body primarily as a source of fuel. Fat provides more than twice the energy of carbohydrate and protein, ◆ making it an extremely efficient storage form of energy. Unlike the liver's glycogen stores, the body's fat stores have virtually unlimited capacity, thanks to the special cells of the **adipose tissue.** Unlike most body cells, which can store only limited amounts of fat, the fat cells of the adipose tissue readily take up and store fat. An adipose cell is depicted in Figure 5-20.

To convert food fats to body fat, the body simply breaks them down, absorbs the parts, and puts them (and others) together again in storage. It requires very little energy to do this. An enzyme—**lipoprotein lipase (LPL)**—hydrolyzes triglycerides from lipoproteins, producing glycerol, fatty acids, and monoglycerides that enter the adipose cells. Inside the cells, other enzymes reassemble the pieces into triglycerides again for storage. Earlier, Figure 5-4 (p. 143) showed how the body can make a triglyceride from glycerol and fatty acids. Triglycerides fill the adipose cells, storing a lot of energy in a relatively small space. Adipose cells store fat

FIGURE 5-20 An Adipose Cell

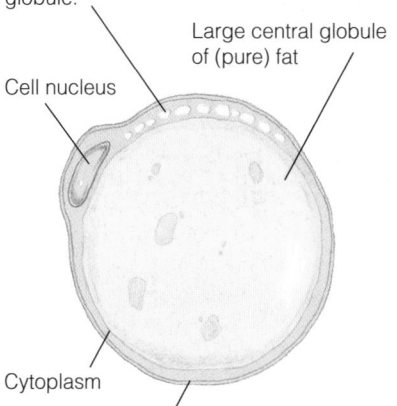

Newly imported triglycerides first form small droplets at the periphery of the cell, then merge with the large, central globule.

Large central globule of (pure) fat

Cell nucleus

Cytoplasm

As the central globule enlarges, the fat cell membrane expands to accommodate its swollen contents.

◆ Reminder: Gram for gram, fat provides more than twice as much energy (9 kcal) as carbohydrate or protein (4 kcal).

adipose (ADD-ih-poce) **tissue:** the body's fat tissue; consists of masses of triglyceride-storing cells.

lipoprotein lipase (LPL): an enzyme that hydrolyzes triglycerides passing by in the bloodstream and directs their parts into the cells, where they can be metabolized for energy or reassembled for storage.

Fat supplies most of the energy during a long-distance run.

◆ 1 lb body fat = 3500 kcal

◆ Desirable blood lipid profile:
- Total cholesterol: <200 mg/dL
- LDL cholesterol: <100 mg/dL
- HDL cholesterol: ≥60 mg/dL
- Triglycerides: <150 mg/dL

hormone-sensitive lipase: an enzyme inside adipose cells that responds to the body's need for fuel by hydrolyzing triglycerides so that their parts (glycerol and fatty acids) escape into the general circulation and thus become available to other cells for fuel. The signals to which this enzyme responds include epinephrine and glucagon, which oppose insulin (see Chapter 4).

blood lipid profile: results of blood tests that reveal a person's total cholesterol, triglycerides, and various lipoproteins.

after meals when a heavy traffic of chylomicrons and VLDL loaded with triglycerides passes by; they release it later whenever the other cells need replenishing.

Using Fat for Energy Fat supplies 60 percent of the body's ongoing energy needs during rest. During prolonged light to moderately intense exercise or extended periods of food deprivation, fat stores may make a slightly greater contribution to energy needs.

When cells demand energy, an enzyme (**hormone-sensitive lipase**) inside the adipose cells responds by dismantling stored triglycerides and releasing the glycerol and fatty acids directly into the blood. Energy-hungry cells anywhere in the body can then capture these compounds and take them through a series of chemical reactions to yield energy, carbon dioxide, and water.

A person who fasts (drinking only water) will rapidly metabolize body fat. A pound of body fat provides 3500 kcalories, ◆ so you might think a fasting person who expends 2000 kcalories a day could lose more than half a pound of body fat each day.* Actually, the person has to obtain some energy from lean tissue because the brain, nerves, and red blood cells need glucose. Also, the complete breakdown of fat requires carbohydrate or protein. Even on a total fast, a person cannot lose more than half a pound of pure fat per day. Still, in conditions of enforced starvation—say, during a siege or a famine—a fatter person can survive longer than a thinner person thanks to this energy reserve.

Although fat provides energy during a fast, it can provide very little glucose to give energy to the brain and nerves. Only the small glycerol molecule can be converted to glucose; fatty acids cannot be. (Figure 7-12 on p. 224 illustrates how only 3 of the 50 or so carbon atoms in a molecule of fat can yield glucose.) After prolonged glucose deprivation, brain and nerve cells develop the ability to derive about two-thirds of their minimum energy needs from the ketone bodies that the body makes from fat fragments. Ketone bodies cannot sustain life by themselves, however. As Chapter 7 explains, fasting for too long will cause death, even if the person still has ample body fat.

IN SUMMARY

The body can easily store unlimited amounts of fat if given excesses, and this body fat is used for energy when needed. (Remember that the liver can also convert excess carbohydrate and protein into fat.) Fat breakdown requires simultaneous carbohydrate breakdown for maximum efficiency; without carbohydrate, fats break down to ketone bodies.

Health Effects and Recommended Intakes of Lipids

Of all the nutrients, fat is most often linked with heart disease, some types of cancer, and obesity. Fortunately, the same recommendation can help with all of these health problems: choose a diet that is low in saturated fats, *trans* fats, and cholesterol and moderate in total fat.

Health Effects of Lipids

Hearing a physician say, "Your blood lipid profile looks fine," is reassuring. The **blood lipid profile** ◆ reveals the concentrations of various lipids in the blood,

---

* The reader who knows that 1 pound = 454 grams and that 1 gram of fat = 9 kcalories may wonder why a pound of body fat does not equal 4086 (9 × 454) kcalories. The reason is that body fat contains some cell water and other materials; it is not quite pure fat.

notably triglycerides and cholesterol, and their lipoprotein carriers (VLDL, LDL, and HDL). This information alerts people to possible disease risks and perhaps to a need for changing their exercise and eating habits. Both the amounts and types of fat in the diet influence people's risk for disease.[9]

Heart Disease Most people realize that elevated blood cholesterol is a major risk factor for **cardiovascular disease.** Cholesterol accumulates in the arteries, restricting blood flow and raising blood pressure. The consequences are deadly; in fact, heart disease is the nation's number one killer of adults. Blood cholesterol level is often used to predict the likelihood of a person's suffering a heart attack or stroke; the higher the cholesterol, the earlier and more likely the tragedy. Much of the effort to prevent heart disease focuses on lowering blood cholesterol.

Commercials advertise products that are low in cholesterol, and magazine articles tell readers how to cut the cholesterol from their favorite recipes. What most people don't realize, though, is that *food* cholesterol does not raise *blood* cholesterol as dramatically as *saturated fat* does.

Risks from Saturated Fats As mentioned earlier, LDL cholesterol raises the risk of heart disease. Saturated fats are most often implicated in raising LDL cholesterol. In general, the more saturated fat in the diet, the more LDL cholesterol in the body. Not all saturated fats have the same cholesterol-raising effect, however. Most notable among the saturated fatty acids that raise blood cholesterol are lauric, myristic, and palmitic acids (12, 14, and 16 carbons, respectively). In contrast, stearic acid (18 carbons) does not seem to raise blood cholesterol. However, making such distinctions may be impractical in diet planning because these saturated fatty acids typically appear together in the same foods.

Fats from animal sources are the main sources of saturated fats ◆ in most people's diets (see Figure 5-21). Some vegetable fats (coconut and palm) and hydrogenated fats provide smaller amounts of saturated fats. Selecting poultry or fish and fat-free milk products helps to lower saturated fat intake and heart disease risk. Using nonhydrogenated margarine and unsaturated cooking oil is another simple change that can dramatically lower saturated fat intake.

Risks from *Trans* Fats Research also suggests an association between dietary *trans*-fatty acids and heart disease.[10] In the body, *trans*-fatty acids alter blood cholesterol the same way some saturated fats do: they raise LDL cholesterol and, at high intakes, lower HDL cholesterol.[11] *Trans*-fatty acids also appear to increase inflammation and insulin resistance.[12] Limiting the intake of *trans*-fatty acids can improve blood cholesterol and lower the risk of heart disease. The estimated average intake of *trans*-fatty acids in the United States is about 5 grams per day—mostly from products that have been hydrogenated.[13] ◆

Reports on *trans*-fatty acids have raised consumer doubts about whether margarine is, after all, a better choice than butter for heart health. The American Heart Association has stated that because butter is rich in both saturated fat and cholesterol whereas margarine is made from vegetable fat with no dietary cholesterol, margarine is still preferable to butter. Be aware that soft margarines (liquid or tub) ◆ are less hydrogenated and relatively lower in *trans*-fatty acids; consequently, they do not raise blood cholesterol as much as the saturated fats of butter or the *trans* fats of hard (stick) margarines do. Some manufacturers are now offering nonhydrogenated margarines that are "*trans* fat free." The last section of this chapter describes how to read food labels and compares butter and margarines. Whichever you decide to use, remember to use them sparingly.

Risks from Cholesterol Although its effect is not as strong as that of saturated fat or *trans* fat, dietary cholesterol also raises blood cholesterol and increases the risk of heart disease. To maximize the effect on blood cholesterol, limit dietary cholesterol as well.

Recall that cholesterol is found in all foods derived from animals. Consequently, eating less fat from meats, eggs, and milk products helps lower dietary cholesterol intake ◆ (as well as total and saturated fat intakes). Figure 5-22 (p. 158) shows the

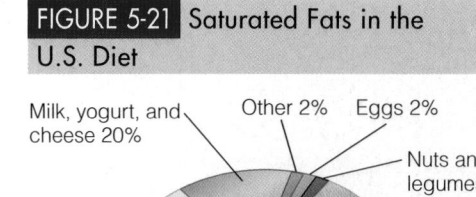

FIGURE 5-21 Saturated Fats in the U.S. Diet

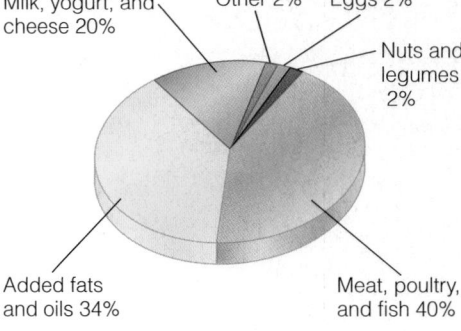

- Milk, yogurt, and cheese 20%
- Other 2%
- Eggs 2%
- Nuts and legumes 2%
- Added fats and oils 34%
- Meat, poultry, and fish 40%

Note that fruits, grains, and vegetables are insignificant sources, unless saturated fats are intentionally added to them during preparation.

◆ Major sources of saturated fats:
  - Whole milk, cream, butter, cheese
  - Fatty cuts of beef and pork
  - Coconut, palm, and palm kernel oils (and products containing them such as candies, pastries, pies, doughnuts, and cookies)

◆ Major sources of *trans* fats:
  - Deep-fried foods (vegetable shortening)
  - Cakes, cookies, doughnuts, pastry, crackers
  - Snack chips
  - Margarine
  - Imitation cheese
  - Meat and dairy products

◆ When selecting margarine, look for:
  - Soft (liquid or tub) instead of hard (stick)
  - ≤2 g saturated fat
  - Liquid vegetable oil (not hydrogenated or partially hydrogenated) as first ingredient
  - "*Trans* fat free"

◆ Major sources of cholesterol:
  - Eggs
  - Milk products
  - Meat, poultry, shellfish

cardiovascular disease (CVD): a general term for all diseases of the heart and blood vessels. Atherosclerosis is the main cause of CVD. When the arteries that carry blood to the heart muscle become blocked, the heart suffers damage known as **coronary heart disease (CHD).**
- **cardio** = heart
- **vascular** = blood vessels

cholesterol contents of selected foods. Many more foods, with their cholesterol contents, appear in Appendix H. For most people trying to lower blood cholesterol, however, limiting saturated fat is more effective than limiting cholesterol intake.

Most foods that are high in cholesterol are also high in saturated fat, but eggs are an exception. An egg contains only 1 gram of saturated fat but just over 200 milligrams of cholesterol—roughly two-thirds of the recommended daily limit. For people with a healthy lipid profile, eating one egg a day is not detrimental. People with high blood cholesterol, however, may benefit from limiting daily cholesterol intake to less that 200 milligrams.[14] When eggs are included in the diet, other sources of cholesterol may need to be limited on that day. Eggs are a valuable part of the diet because they are inexpensive, useful in cooking, and a source of high-quality protein and other nutrients. Low saturated fat, high omega-3 fat eggs are now available, and food manufacturers have produced several fat-free, cholesterol-free egg substitutes.

Benefits from Monounsaturated Fats and Polyunsaturated Fats Replacing both saturated and *trans* fats with monounsaturated ◆ and polyunsaturated ◆ fats may be the most effective dietary strategy in preventing heart disease. The lower rate of heart disease among people in the Mediterranean region of the world is often attributed to their liberal use of olive oil, a rich source of monounsaturated fatty acids. Olive oil also delivers valuable phytochemicals that help to protect against heart disease.[15] Replacing saturated fats with the polyunsaturated fatty acids of other vegetable oils also lowers blood cholesterol.[16] Highlight 5 examines various types of fats and their roles in supporting or harming heart health.

Benefits from Omega-3 Fats Research on the different types of fats has spotlighted the beneficial effects of the omega-3 ◆ polyunsaturated fatty acids in reducing the risks of heart disease and stroke.[17] Regular consumption of omega-3 fatty acids helps to prevent blood clots, protect against irregular heartbeats, and lower blood pressure, especially in people with hypertension or atherosclerosis.[18]

◆ Sources of monounsaturated fats:
- Olive oil, canola oil, peanut oil
- Avocados

◆ Sources of polyunsaturated fats:
- Vegetable oils (safflower, sesame, soy, corn, sunflower)
- Nuts and seeds

◆ Major sources of omega-3 fats:
- Vegetable oils (canola, soybean, flaxseed)
- Walnuts, flaxseeds
- Fatty fish (mackerel, salmon, sardines)

FIGURE 5-22 Cholesterol in Selected Foods

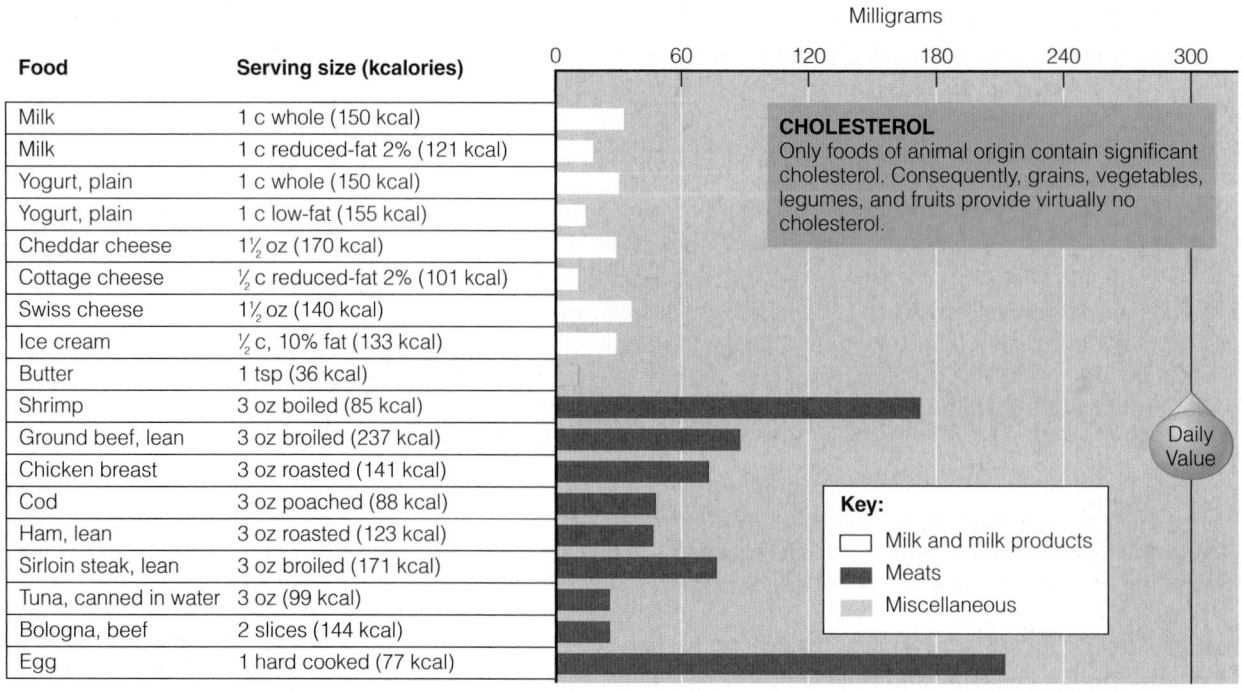

TABLE 5-2	Sources of Omega-3 and Omega-6 Fatty Acids
Omega-6	
Linoleic acid	Vegetable oils (corn, sunflower, safflower, soybean, cottonseed), poultry fat, nuts, seeds
Arachidonic acid	Meats, poultry, eggs (or can be made from linoleic acid)
Omega-3	
Linolenic acid	Oils (flaxseed, canola, walnut, wheat germ, soybean) Nuts and seeds (butternuts, flaxseeds, walnuts, soybean kernels) Vegetables (soybeans)
EPA and DHA	Human milk Pacific oysters and fish[a] (mackerel, salmon, bluefish, mullet, sablefish, menhaden, anchovy, herring, lake trout, sardines, tuna) (or can be made from linolenic acid)

[a]All fish contain some EPA and DHA; the amounts vary among species and within a species depending on such factors as diet, season, and environment. The fish listed here, except tuna, provide at least 1 gram of omega-3 fatty acids in 100 grams of fish (3.5 ounces). Tuna provides fewer omega-3 fatty acids, but because it is commonly consumed, its contribution can be significant.

Fatty fish are among the best sources of omega-3 fatty acids, and Highlight 5 features their role in supporting heart health. To maximize the benefits and minimize the risks, ◆ most healthy people should eat two servings of fish a week.[19]

Balance Omega-6 and Omega-3 Intakes Table 5-2 provides sources of omega-6 and omega-3 fatty acids. To obtain sufficient intakes and the right balance between omega-6 and omega-3 fatty acids, ◆ most people need to eat more fish and less meat.[20] The American Heart Association recommends two servings of fish a week, with an emphasis on fatty fish (salmon, herring, and mackerel, for example).[21] Eating fish instead of meat supports heart health, especially when combined with physical activity. Even one fish meal a month may be enough to make a difference.[22] When preparing fish, grill, bake, or broil, but do not fry. Fried fish from fast-food restaurants and frozen fried fish products are often low in omega-3 fatty acids and high in *trans-* and saturated fatty acids. Fish provides many minerals (except iron) and vitamins and is leaner than most other animal-protein sources. When used in a weight-loss program, eating fish improves blood lipids even more effectively than can be explained by losing weight or eating fish alone.

In addition to fish, other functional foods ◆ are being developed to help consumers improve their omega-3 fatty acid intake. For example, hens fed flaxseed produce eggs rich in omega-3 fatty acids. Including even one enriched egg in the diet daily can significantly increase a person's intake of omega-3 fatty acids. Another option may be to select wild game or pasture-fed cattle, which provide more omega-3 fatty acids and less saturated fat than grain-fed cattle.[23]

Omega-3 fatty acids are also available in capsules of fish oil supplements. Routine supplementation, however, is not recommended. High intakes of omega-3 polyunsaturated fatty acids may increase bleeding time, interfere with wound healing, raise LDL cholesterol, and suppress immune function.*[24] Such findings reinforce the concept that too much of a good thing can sometimes be harmful. People with heart disease, however, may benefit from doses greater than can be achieved through diet alone. They should always consult a physician first because including supplements as part of a treatment plan may be contraindicated for some patients.[25] Supplements may also provide relief for people with rheumatoid arthritis or asthma.[26]

Cancer The evidence for links between dietary fats and cancer ◆ is less convincing than for heart disease, but it does suggest possible associations between some types

◆ Fish relatively high in mercury:
  • Tilefish (also called golden snapper or golden bass), swordfish, king mackerel, shark
  Fish relatively low in mercury:
  • Cod, haddock, pollock, salmon, sole, tilapia
  • Most shellfish

◆ Recommended omega-6 to omega-3 ratio: 6 to 1

◆ Reminder: *Functional foods* contain physiologically active compounds that provide health benefits beyond basic nutrition (see Highlight 13 for a full discussion).

◆ Other risk factors for cancer include smoking, alcohol, and environmental contaminants. Chapter 29 provides more details about these risk factors and the development of cancer.

* Suppressed immune function is seen with daily intake of 0.9 to 9.4 grams EPA and 0.6 to 6.0 grams DHA for 3 to 24 weeks.

of fat and some types of cancers.[27] Dietary fat does not seem to *initiate* cancer development but, instead, may *promote* cancer once it has arisen.

The relationship between dietary fat and the risk of cancer differs for various types of cancers. In the case of breast cancer, evidence has been weak and inconclusive. Some studies indicate little or no association between dietary fat and breast cancer; others find that total *energy* intake and obesity contribute to the risk.[28] In the case of prostate cancer, some studies indicate a harmful association with total and saturated fat, although a specific type fatty acid has not yet been implicated.[29]

The relationship between dietary fat and the risk of cancer differs for various types of fats as well. The association between cancer and fat appears to be due primarily to saturated fats or dietary fat from meats (which is mostly saturated). Fat from milk or fish has not been implicated in cancer risk.[30] In fact, the omega-3 fatty acids of fish may protect against some cancers, although evidence does not support supplementation.[31] Thus dietary advice to reduce cancer risks parallels that given to reduce heart disease risks: reduce saturated fats and increase omega-3 fatty acids.

Obesity Fat contributes more than twice as many kcalories ◆ per gram as either carbohydrate or protein. Consequently, people who eat high-fat diets regularly may exceed their energy needs and gain weight, especially if they are inactive.[32] Because fat boosts energy intake, cutting fat from the diet can be an effective strategy in cutting kcalories. In some cases, though, choosing a fat-free food offers no kcalorie savings. Fat-free frozen desserts, for example, often have so much sugar added that the kcalorie count can be as high as in the regular-fat product. In this case, therefore, cutting fat and adding carbohydrate offers no kcalorie savings or weight-loss advantage. In fact, it may even raise energy intake and exacerbate weight problems. Later chapters revisit the role of dietary fat in the development of obesity.

◆ Fat is a more concentrated energy source than the other energy nutrients: 1 g carbohydrate or protein = 4 kcal, but 1 g fat = 9 kcal

IN SUMMARY

High blood LDL cholesterol poses a risk of heart disease, and high intakes of saturated and *trans* fats, specifically, contribute most to high LDL. Cholesterol in foods presents less of a risk. Omega-3 fatty acids appear to be protective.

Recommended Intakes of Fat

Some fat in the diet is essential for good health, but too much fat, especially saturated fat, increases the risks for chronic diseases. Defining the exact amount of fat, saturated fat, or cholesterol that benefits health or begins to harm health, however, is not possible. For this reason, no RDA or upper limit has been set. Instead, the DRI and 2005 *Dietary Guidelines* suggest a diet that is low in saturated fat, *trans* fat, and cholesterol and provides 20 to 35 percent of the daily energy intake from fat. ◆ The top end of this range is slightly higher than previous recommendations. This revision recognizes that diets with up to 35 percent of kcalories from fat can be compatible with good health if energy intake is reasonable and saturated fat intake is low. When total fat exceeds 35 percent, saturated fat increases to unhealthy levels.[33] For a 2000-kcalorie diet, 20 to 35 percent represents 400 to 700 kcalories from fat (roughly 45 to 75 grams). Part of this fat allowance should provide for the essential fatty acids—linoleic acid and linolenic acid. For this reason, an Adequate Intake (AI) has been established for these two fatty acids. Recommendations suggest that linoleic acid ◆ provide 5 to 10 percent of the daily energy intake and linolenic acid ◆ 0.6 to 1.2 percent.[34]

To help consumers meet the dietary fat goals, the Food and Drug Administration (FDA) established Daily Values ◆ on food labels using 30 percent of energy intake as the guideline for fat and 10 percent for saturated fat. The Daily Value for choles-

◆ DRI and 2005 *Dietary Guidelines* for fat:
- 20 to 35% of energy intake (from mostly polyunsaturated and monounsaturated fat sources such as fish, nuts, and vegetable oils)

◆ Linoleic acid (omega-6) AI:
Men:
- 19–50 yr: 17 g/day
- 51+ yr: 14 g/day
Women:
- 19–50 yr: 12 g/day
- 51+ yr: 11 g/day

◆ Linolenic acid (omega-3) AI:
- Men: 1.6 g/day
- Women: 1.1 g/day

◆ Daily Values:
- 65 g fat (based on 30% of 2000 kcal diet)
- 20 g saturated fat (based on 10% of 2000 kcal diet)
- 300 mg cholesterol

terol is 300 milligrams regardless of energy intake. There is no Daily Value for *trans* fat, but consumers should try to keep intakes as low as possible and within the 10 percent allotted for saturated fat. According to surveys, adults in the United States receive about 33 percent of their total energy from fat, with saturated fat contributing about 11 percent of the total. Cholesterol intakes in the United States average 190 milligrams a day for women and 290 for men. [35]

Dietary Guidelines for Americans 2005

Consume less than 10 percent of kcalories from saturated fatty acids and less than 300 mg/day of cholesterol, and keep *trans* fatty acid consumption as low as possible.

The fats of fish, nuts, and vegetable oils are not counted as discretionary kcalories because they provide valuable omega-3 fatty acids, essential fatty acids, and vitamin E. In contrast, solid fats ◆ deliver an abundance of saturated fatty acids; the USDA Food Guide counts them as discretionary kcalories. Discretionary kcalories may be used to add fats in cooking or at the table or to select higher fat items from the food groups. ◆

Although it is very difficult to do, some people actually manage to eat too little fat—to their detriment. Among them are people with eating disorders, described in Highlight 8, and athletes. Athletes following a diet too low in fat (less than 20 percent of total kcalories) fall short on energy, vitamins, minerals, and essential fatty acids as well as on performance.[36] As a practical guideline, it is wise to include the equivalent of at least a teaspoon of fat in every meal—a little peanut butter on toast or mayonnaise on tuna, for example. Dietary recommendations that limit fat were developed for healthy people over age two; Chapter 15 discusses the fat needs of infants and young children.

As the photos in Figure 5-23 show (p. 162), fat accounts for much of the energy in foods, and removing the fat from foods cuts energy and saturated fat intakes dramatically. To reduce dietary fat, eliminate fat as a seasoning and in cooking; remove the fat from high-fat foods; replace high-fat foods with low-fat alternatives; and emphasize whole grains, fruits, and vegetables. The remainder of this chapter identifies sources of fat in the diet, food group by food group.

From Guidelines to Groceries

Fats accompany protein in foods derived from animals, such as meat, fish, poultry, and eggs, and fats accompany carbohydrate in foods derived from plants, such as avocados and coconuts. Fats carry with them the four fat-soluble vitamins—A, D, E, and K—together with many of the compounds that give foods their flavor, texture, and palatability. Fat is responsible for the delicious aromas associated with sizzling bacon and hamburgers on the grill, onions being sautéed, or vegetables in a stir-fry. Of course, these wonderful characteristics lure people into eating too much from time to time. With careful selections, a diet following the USDA Food Guide can support good health and still meet fat recommendations (see the "How to" feature on p. 163).

Meats and Meat Alternates Many meats and meat alternates ◆ contain fat, saturated fat, and cholesterol but also provide high-quality protein and valuable vitamins and minerals. They can be included in a healthy diet if a person makes lean choices and prepares them using the suggestions outlined in the box on p. 163. Selecting "free-range" meats from grass-fed instead of grain-fed livestock offers the nutrient advantages of being lower in fat, and the fat has more polyunsaturated fatty acids, including the omega-3 type. Another strategy to lower blood cholesterol is to prepare meals using soy protein instead of animal protein.[37]

◆ Solid fats include meat and poultry fats (as in poultry skin, luncheon meats, sausage); milk fat (as in whole milk, cheese, butter); shortening (as in fried foods and baked goods); and hard margarines.

◆ The USDA Food Guide amounts of fats that can be included as discretionary kcalories when most food choices are nutrient dense and fat < 30% total kcal:
- 11 g for 1600 kcal diet
- 15 g for 1800 kcal diet
- 18 g for 2000 kcal diet
- 19 g for 2200 kcal diet
- 22 g for 2400 kcal diet

For perspective, 1 tsp oil = 5 g fat and provides about 45 kcal

◆ Very lean options:
- Chicken (white meat, no skin); cod, flounder, trout; tuna (canned in water); legumes

Lean options:
- Beef or pork "round" or "loin" cuts; chicken (dark meat, no skin); herring or salmon; tuna (canned in oil)

Medium-fat options:
- Ground beef, eggs, tofu

High-fat options:
- Sausage, bacon, luncheon meats, hot dogs, peanut butter, nuts

FIGURE 5-23 Cutting Fat Cuts kCalories—and Saturated Fat

Pork chop with fat (340 kcal, 19 g fat, 7 g saturated fat).

Potato with 1 tbs butter and 1 tbs sour cream (350 kcal, 14 g fat, 10 g saturated fat).

Whole milk, 1 c (150 kcal, 8 g fat, 5 g saturated fat).

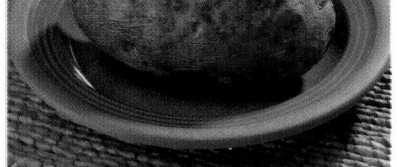

Pork chop with fat trimmed off (230 kcal, 9 g fat, 3 g saturated fat).

Plain potato (200 kcal, <1 g fat, 0 g saturated fat).

Fat-free milk, 1 c (90 kcal, <1 g fat, <1 g saturated fat).

© Polara Studios, Inc. (all)

Dietary Guidelines for Americans 2005

When selecting and preparing meat, poultry, and milk or milk products, make choices that are lean, low-fat, or fat-free.

◆ Fat-free and low-fat options:
  • Fat-free or 1% milk or yogurt (plain); fat-free and low-fat cheeses
  Reduced-fat options:
  • 2% milk, low-fat yogurt (plain)
  High-fat options:
  • Whole milk, regular cheeses

Milks and Milk Products Like meats, milks and milk products ◆ should also be selected with an awareness of their fat, saturated fat, and cholesterol contents. Fat-free and low-fat milk products provide as much or more protein, calcium, and other nutrients as their whole-milk versions—but with little or no saturated fat. Selecting fermented milk products, such as yogurt, may also help to lower blood cholesterol. These foods increase the population and activity of bacteria in the colon that ferment fibers. As Chapter 4 explained, this action lowers blood cholesterol as fibers bind with bile, thereby increasing excretion, and as bacteria produce short-chain fatty acids that inhibit cholesterol synthesis in the liver.[38]

Vegetables, Fruits, and Grains Choosing vegetables, fruits, whole grains, and legumes also helps lower the saturated fat, cholesterol, and total fat content of the diet. Most vegetables and fruits naturally contain little or no fat. Although avocados and olives are exceptions, most of their fat is unsaturated, which is not harmful to heart health. Most grains contain only small amounts of fat. Consumers need to read food labels, though, because some grain *products* such as fried taco shells, croissants, and biscuits are high in saturated fat, and pastries, crackers, and cookies may be high in *trans* fats. Similarly, many people add butter, margarine, or cheese sauce

HOW TO Make Heart-Healthy Choices—by Food Group

Breads and Cereals
- Select breads, cereals, and crackers that are low in saturated and *trans* fat (for example, bagels instead of croissants).
- Prepare pasta with a tomato sauce instead of a cheese or cream sauce.

Vegetables and Fruits
- Enjoy the natural flavor of steamed vegetables (without butter) for dinner and fruits for dessert.
- Eat at least two vegetables (in addition to a salad) with dinner.
- Snack on raw vegetables or fruits instead of high-fat items like potato chips.
- Buy frozen vegetables without sauce.

Milk and Milk Products
- Switch from whole milk to reduced-fat, from reduced-fat to low-fat, and from low-fat to fat-free (nonfat).
- Use fat-free and low-fat cheeses (such as part-skim ricotta and low-fat mozzarella) instead of regular cheeses.
- Use fat-free or low-fat yogurt or sour cream instead of regular sour cream.
- Use evaporated fat-free milk instead of cream.
- Enjoy fat-free frozen yogurt, sherbet, or ice milk instead of ice cream.

Meat and Legumes
- Fat adds up quickly, even with lean meat; limit intake to about 6 ounces (cooked weight) daily.
- Eat at least two servings of fish per week (particularly fish such as mackerel, lake trout, herring, sardines, and salmon).

- Choose fish, poultry, or lean cuts of pork or beef; look for unmarbled cuts named *round* or *loin* (eye of round, top round, bottom round, round tip, tenderloin, sirloin, center loin, and top loin).
- Choose processed meats such as lunch meats and hot dogs that are low in saturated fat and cholesterol.
- Trim the fat from pork and beef; remove the skin from poultry.
- Grill, roast, broil, bake, stir-fry, stew, or braise meats; don't fry. When possible, place food on a rack so that fat can drain.
- Use lean ground turkey or lean ground beef in recipes; brown ground meats without added fat, then drain off fat.
- Select tuna, sardines, and other canned meats packed in water; rinse oil-packed items with hot water to remove much of the fat.
- Fill kabob skewers with lots of vegetables and slivers of meat; create main dishes and casseroles by combining a little meat, fish, or poultry with a lot of pasta, rice, or vegetables.
- Use legumes often.
- Eat a meatless meal or two daily.
- Use egg substitutes in recipes instead of whole eggs or use two egg whites in place of each whole egg.

Fats and Oils
- Use butter or stick margarine sparingly; select soft margarines instead of hard margarines.
- Use fruit butters, reduced-kcalorie margarines, or butter replacers instead of butter.

- Use low-fat or fat-free mayonnaise and salad dressing instead of regular.
- Limit use of lard and meat fat.
- Limit use of products made with coconut oil, palm kernel oil, and palm oil (read labels on bakery goods, processed foods, popcorn oils, and nondairy creamers).
- Reduce use of hydrogenated shortenings and stick margarines and products that contain them (read labels on crackers, cookies, and other commercially prepared baked goods); use vegetable oils instead.

Miscellaneous
- Use a nonstick pan or coat the pan lightly with vegetable oil.
- Refrigerate soups and stews; when the fat solidifies, remove it.
- Use wine; lemon, orange, or tomato juice; herbs; spices; fruits; or broth instead of butter or margarine when cooking.
- Stir-fry in a small amount of oil; add moisture and flavor with broth, tomato juice, or wine.
- Use variety to enhance enjoyment of the meal: vary colors, textures, and temperatures—hot cooked versus cool raw foods—and use garnishes to complement food.
- Omit high-fat meat gravies and cheese sauces.

SOURCE: Adapted from *Third Report of the National Cholesterol Education Program (NCEP) Expert Panel on Detection, Evaluation, and Treatment of High Blood Cholesterol in Adults (Adult Treatment Panel III)*, NIH publication no. 02-5215 (Bethesda, Md.: National Heart, Lung, and Blood Institute, 2002), pp. V-25–V-27.

to grains and vegetables, which raises their saturated and *trans* fat contents. Because fruits are often eaten without added fat, a diet that includes several servings of fruit daily can help a person meet the dietary recommendations for fat.

A diet rich in vegetables, fruits, whole grains, and legumes also offers abundant vitamin C, folate, vitamin A, vitamin E, and dietary fiber—all important in supporting health. Consequently, such a diet protects against disease by reducing saturated fat, cholesterol, and total fat as well as by increasing nutrients. It also provides valuable phytochemicals that help defend against heart disease.

Invisible Fat *Visible* fat, such as butter and the fat trimmed from meat, is easy to see. *Invisible* fat is less apparent and can be present in foods in surprising amounts. Invisible fat "marbles" a steak or is hidden in foods like cheese. Any *fried* food contains abundant fat—potato chips, French fries, fried wontons, and fried fish. Many *baked* goods, too, are high in fat—pie crusts, pastries, crackers, biscuits, cornbread, doughnuts, sweet rolls, cookies, and cakes. Most chocolate bars deliver more kcalories from fat than from sugar. Even cream-of-mushroom soup prepared with water derives 66 percent of its energy from fat. Keep invisible fats in mind when making food selections.

Well-balanced, healthy meals provide some fat with an emphasis on monounsaturated and polyunsaturated fats.

Choose Wisely Consumers can find an abundant array of foods that are low in saturated fat, *trans* fat, cholesterol, and total fat. In many cases, they are familiar foods that are simply prepared with less fat. For example, fat can be removed by skimming milk or trimming meats. Manufacturers can dilute fat by adding water or whipping in air. They can use fat-free milk in creamy desserts and lean meats in frozen entrées. Sometimes manufacturers simply prepare the products differently. For example, fat-free potato chips may be baked instead of fried. Beyond lowering the fat content, manufacturers have developed margarines fortified with plant sterols that lower blood cholesterol.*[39] (Highlight 13 explores these and other functional foods designed to support health.) Such choices make heart-healthy eating easy.

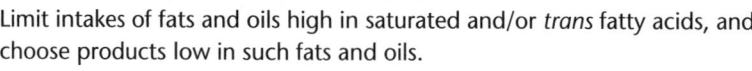

Dietary Guidelines for Americans 2005

Limit intakes of fats and oils high in saturated and/or *trans* fatty acids, and choose products low in such fats and oils.

To replace saturated fats with unsaturated fats, sauté foods in olive oil instead of butter, garnish salads with sunflower seeds instead of bacon, snack on mixed nuts instead of potato chips, use avocado instead of cheese on a sandwich, and eat salmon instead of steak. Table 5-3 shows how these simple substitutions can lower the saturated fat and raise the unsaturated fat in a meal. Highlight 5 provides more details about the benefits of healthy fats in the diet.

Fat Replacers Some foods are made with **fat replacers**—ingredients derived from carbohydrate, protein, or fat that can be used to replace some or all of the fat in foods. The body may digest and absorb some of these substances, so they may contribute some energy, although significantly less energy than fat's 9 kcalories per gram.

Fat replacers offering the sensory and cooking qualities of fats but none of the kcalories are called **artificial fats.** A familiar example of an artificial fat that has been approved for use in snack foods such as potato chips, crackers, and tortilla chips is **olestra.** Olestra's chemical structure is similar to that of a regular fat (a triglyceride) but with important differences. A triglyceride is composed of a glycerol molecule with three fatty acids attached, whereas olestra is made of a sucrose mol-

---

* Margarines that lower blood cholesterol contain plant sterols and are marketed under the brand names Benecol and Take Control.

fat replacers: ingredients that replace some or all of the functions of fat and may or may not provide energy.

artificial fats: zero-energy fat replacers that are chemically synthesized to mimic the sensory and cooking qualities of naturally occurring fats but are totally or partially resistant to digestion.

olestra: a synthetic fat made from sucrose and fatty acids that provides 0 kcalories per gram; also known as **sucrose polyester.**

TABLE 5-3   Choosing Unsaturated Fat instead of Saturated Fat

Portion sizes have been adjusted so that each of these foods provides approximately 100 kcalories. Notice that for a similar number of kcalories and grams of fat, the first choices offer less saturated fat and more unsaturated fat.

Foods (100 kcal portions)	Saturated Fat (g)	Unsaturated Fat (g)	Total Fat (g)
Olive oil (1 tbs) vs. butter (1 tbs)	2 vs. 7	9 vs. 4	11 vs. 11
Sunflower seeds (2 tbs) vs. bacon (2 slices)	1 vs. 3	7 vs. 6	8 vs. 9
Mixed nuts (2 tbs) vs. potato chips (10 chips)	1 vs. 2	8 vs. 5	9 vs. 7
Avocado (6 slices) vs. cheese (1 slice)	2 vs. 4	8 vs. 4	10 vs. 8
Salmon (2 oz) vs. steak (1½ oz)	1 vs. 2	3 vs. 3	4 vs. 5
Totals	**7 vs. 18**	**35 vs. 22**	**42 vs. 40**

ecule with six to eight fatty acids attached. Enzymes in the digestive tract cannot break the bonds of olestra, so unlike sucrose or fatty acids, olestra passes through the system unabsorbed.

The FDA's evaluation of olestra's safety addressed two questions. First, is olestra toxic? Research on both animals and human beings supports the safety of olestra as a partial replacement for dietary fats and oils, with no reports of cancer or birth defects. Second, does olestra affect either nutrient absorption or the health of the digestive tract? When olestra passes through the digestive tract unabsorbed, it binds with some of the fat-soluble vitamins A, D, E, and K and carries them out of the body, robbing the person of these valuable nutrients. To compensate for these losses, the FDA requires the manufacturer to fortify olestra with vitamins A, D, E, and K. Saturating olestra with these vitamins does not make the product a good source of vitamins, but it does block olestra's ability to bind with the vitamins from other foods. An asterisk in the ingredients list informs consumers that these added vitamins are "dietarily insignificant."

Some consumers experience digestive distress with olestra consumption, such as cramps, gas, bloating, and diarrhea. The FDA initially required a label warning stating that "olestra may cause abdominal cramping and loose stools" and that it "inhibits the absorption of some vitamins and other nutrients" but has since concluded that such a statement is no longer warranted.

Consumers need to keep in mind that low-fat and fat-free foods still deliver kcalories. Alternatives to fat can help to lower energy intake and support weight loss only when they actually *replace* fat and energy in the diet.[40]

Read Food Labels Labels list total fat, saturated fat, *trans* fat, and cholesterol contents of foods in addition to fat kcalories per serving (see Figure 5-24, p. 166). Because each package provides information for a single serving and because serving sizes are standardized, consumers can easily compare similar products.

Total fat, saturated fat, and cholesterol are also expressed as "% Daily Values" for a person consuming 2000 kcalories. People who consume more or less than 2000 kcalories daily can calculate their personal Daily Value for fat as described in the "How to" below. *Trans* fats do not have a Daily Value.

Beware of fast-food meals delivering too much fat, especially saturated fat. This double bacon cheeseburger, fries, and milkshake provide more than 1600 kcalories, with almost 90 grams of fat and over 30 grams of saturated fat—far exceeding dietary fat guidelines for the entire day.

HOW TO Calculate a Personal Daily Value for Fat

The % Daily Value for fat on food labels is based on 65 grams. To know how your intake compares with this recommendation, you can either count grams until you reach 65, or add the "% Daily Values" until you reach 100 percent—if your energy intake is 2000 kcalories a day. If your energy intake is more or less, you can calculate your personal daily fat allowance in grams. Suppose your energy intake is 1800 kcalories per day and your goal is 30 percent kcalories from fat. Multiply your total energy intake by 30 percent, then divide by 9:

1800 total kcal × 0.30 from fat = 540 fat kcal

540 fat kcal ÷ 9 kcal/g = 60 g fat

(In familiar measures, 60 grams of fat is about the same as ⅔ stick of butter or ¼ cup of oil.)

The accompanying table shows the numbers of grams of fat allowed per day for various energy intakes. With one of these numbers in mind, you can quickly evaluate the number of fat grams in foods you are considering eating.

CENGAGENOW™
To practice calculating a personal daily value for fat, log on to **academic.cengage.com/login**, go to Chapter 5, then go to How To.

Energy (kcal/day)	20% kCalories from Fat	35% kCalories from Fat	Fat (g/day)
1200	240	420	27–47
1400	280	490	31–54
1600	320	560	36–62
1800	360	630	40–70
2000	400	700	44–78
2200	440	770	49–86
2400	480	840	53–93
2600	520	910	58–101
2800	560	980	62–109
3000	600	1050	67–117

FIGURE 5-24 Butter and Margarine Labels Compared

Food labels list the kcalories from fat; the quantities and Daily Values for fat, saturated fat, and cholesterol; and the quantities for *trans* fat. Information on polyunsaturated and monounsaturated fats is optional. In this example, stick margarine has 2.5 g *trans* fat and tub margarine has 2 g *trans* fat. Products that contain 0.5 g or less of *trans* fat and 0.5 g or less of saturated fat may claim "no *trans* fat." Similarly, products that contain 2 mg or less of cholesterol and 2 g or less of saturated fat may claim to be "cholesterol-free."

If the list of ingredients includes hydrogenated oils, you know the food contains *trans* fat. Chapter 2 explained that foods list their ingredients in descending order of predominance by weight. As you can see from this example, the closer "partially hydrogenated oils" is to the beginning of the ingredients list, the more *trans* fats the product contains. Notice that most of the fat in butter is saturated, whereas most of the fat in margarine is unsaturated; partially hydrogenated margarines tend to have more *trans* fat than hydrogenated liquid margarines.

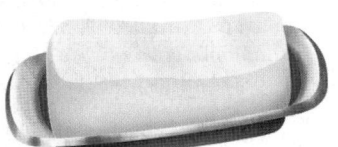

Butter

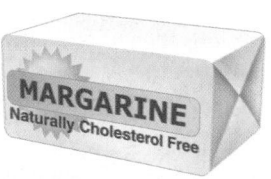

Margarine (stick)

Margarine (tub)

Margarine (liquid)

Nutrition Facts

Serving Size 1 Tbsp (14g)
Servings per container about 32

Amount per serving

Calories 100 Calories from Fat 100	
	%Daily Value*
Total Fat 11g	17%
Saturated Fat 7g	37%
Trans Fat 0g	
Cholesterol 30mg	10%
Sodium 95mg	4%
Total Carbohydrate 0g	0%
Protein 0g	

Vitamin A 8%

Not a significant source of dietary fiber, sugars, vitamin C, calcium, and iron.

*Percent Daily Values are based on a 2,000 calorie diet.

INGREDIENTS: Cream, salt.

Nutrition Facts

Serving Size 1 Tbsp (14g)
Servings per container about 32

Amount per serving

Calories 100 Calories from Fat 100	
	%Daily Value*
Total Fat 11g	17%
Saturated Fat 2g	11%
Trans Fat 2.5g	
Polyunsaturated Fat 3.5g	
Monounsaturated Fat 2.5g	
Cholesterol 0mg	0%
Sodium 105mg	4%
Total Carbohydrate 0g	0%
Protein 0g	

Vitamin A 10%

Not a significant source of dietary fiber, sugars, vitamin C, calcium, and iron.

*Percent Daily Values are based on a 2,000 calorie diet.

INGREDIENTS: Liquid soybean oil, partially hydrogenated soybean oil, water, buttermilk, salt, soy lecithin, sodium benzoate (as a preservative), vegetable mono and diglycerides, artificial flavor, vitamin A palmitate, colored with beta carotene (provitamin A).

Nutrition Facts

Serving size 1 Tbsp (14g)
Servings per container about 32

Amount per serving

Calories 100 Calories from Fat 100	
	%Daily Value*
Total Fat 11g	17%
Saturated Fat 2.5g	13%
Trans Fat 2g	
Polyunsaturated Fat 4g	
Monounsaturated Fat 2.5g	
Cholesterol 0mg	0%
Sodium 80mg	3%
Total Carbohydrate 0g	0%
Protein 0g	

Vitamin A 10%

Not a significant source of dietary fiber, sugars, vitamin C, calcium, and iron.

*Percent Daily Values are based on a 2,000 calorie diet.

INGREDIENTS: Liquid soybean oil, partially hydrogenated soybean oil, buttermilk, water, butter (cream, salt), salt, soy lecithin, vegetable mono and diglycerides, sodium benzoate added as a preservative, artificial flavor, vitamin A palmitate, colored with beta carotene.

Nutrition Facts

Serving size 1 Tbsp (14g)
Servings per container about 24

Amount per serving

Calories 70 Calories from Fat 70	
	%Daily Value*
Total Fat 8g	13%
Saturated Fat 1.5g	7%
Trans Fat 0g	
Polyunsaturated Fat 4.5g	
Monounsaturated Fat 2g	
Cholesterol 0mg	0%
Sodium 110mg	8%
Total Carbohydrate 0g	0%
Protein 0g	

Vitamin A 10%

Not a significant source of dietary fiber, sugars, vitamin C, calcium, and iron.

*Percent Daily Values are based on a 2,000 calorie diet.

INGREDIENTS: Liquid soybean oil, water, salt, hydrogenated cottonseed oil, vegetable monoglycerides and soy lecithin (emulsifiers), potassium sorbate and sodium benzoate (to preserve freshness), artificial flavor, phosphoric acid (acidulant), colored with beta carotene (source of vitamin A), vitamin A palmitate.

Be aware that the "% Daily Value" for fat is not the same as "% kcalories from fat." This important distinction is explained in the "How to" feature on p. 167. Because recommendations apply to average daily intakes rather than individual food items, food labels do not provide "% kcalories from fat." Still, you can get an idea of whether a particular food is high or low in fat.

CENGAGENOW™
To practice calculating % Daily Value and % kcalories from fat, log on to **academic.cengage.com/login**, go to Chapter 5, then go to How To.

HOW TO Understand "% Daily Value" and "% kCalories from Fat"

The "% Daily Value" that is used on food labels to describe the amount of fat in a food is not the same as the "% kcalories from fat" that is used in dietary recommendations to describe the amount of fat in the diet. They may appear similar, but their difference is worth understanding. Consider, for example, a piece of lemon meringue pie that provides 140 kcalories and 12 grams of fat. Because the Daily Value for fat is 65 grams for a 2000-kcalorie intake, 12 grams represent about 18 percent:

$$12 \text{ g} \div 65 \text{ g} = 0.18$$
$$0.18 \times 100 = 18\%$$

The pie's "% Daily Value" is 18 percent, or almost one-fifth, of the day's fat allowance.

Uninformed consumers may mistakenly believe that this food meets recommendations to limit fat to "20 to 35 percent kcalories," but it doesn't—for two reasons. First, the pie's 12 grams of fat contribute 108 of the 140 kcalories, for a total of 77 percent kcalories from fat:

$$12 \text{ g fat} \times 9 \text{ kcal/g} = 108 \text{ kcal}$$
$$108 \text{ kcal} \div 140 \text{ kcal} = 77\%$$

Second, the "percent kcalories from fat" guideline applies to a day's total intake, not to an individual food. Of course, if every selection throughout the day exceeds 35 percent kcalories from fat, you can be certain that the day's total intake will, too.

Whether a person's energy and fat allowance can afford a piece of lemon meringue pie depends on the other food and activity choices made that day.

IN SUMMARY

In foods, triglycerides:
- Deliver fat-soluble vitamins, energy, and essential fatty acids
- Contribute to the sensory appeal of foods and stimulate appetite

Although some fat in the diet is necessary, health authorities recommend a diet moderate in total fat and low in saturated fat, *trans* fat, and cholesterol. They also recommend replacing saturated fats with monounsaturated and polyunsaturated fats, particularly omega-3 fatty acids from foods such as fish, not from supplements. Many selection and preparation strategies can help bring these goals within reach, and food labels help to identify foods consistent with these guidelines.

If people were to make only one change in their diets, they would be wise to limit their intakes of saturated fat. Sometimes these choices can be difficult, though, because fats make foods taste delicious. To maintain good health, must a person give up all high-fat foods forever—never again to eat marbled steak, hollandaise sauce, or gooey chocolate cake? Not at all. These foods bring pleasure to a meal and can be enjoyed as part of a healthy diet when eaten occasionally in small quantities; but they should not be everyday foods. The key word for fat is *moderation,* not *deprivation.* Appreciate the energy and enjoyment that fat provides, but take care not to exceed your needs.

Nutrition Portfolio

To maintain good health, eat enough, but not too much, fat and select the right kinds.

- List the types and amounts of fats and oils you eat daily, making note of which ones are saturated, monounsaturated, or polyunsaturated and how your choices could include fewer saturated options.

- List the types and amounts of milk products, meats, fish, and poultry you eat daily, noting how your choices could include more low-fat options.

- Describe choices you can make in selecting and preparing foods to lower your intake of solid fats.

NUTRITION ON THE NET

For further study of topics covered in this chapter, log on to **academic.cengage .com/nutrition/rolfes/UNCN8e**. Go to Chapter 5, then to Nutrition on the Net.

- Search for "cholesterol" and "dietary fat" at the U.S. Government health information site: **www.healthfinder.gov**

- Search for "fat" at the International Food Information Council site: **www.ific.org**

- Find dietary strategies to prevent heart disease at the American Heart Association or National Heart, Lung, and Blood Institute: **www.americanheart.org** or **nhlbi.nih.gov**

NUTRITION CALCULATIONS

For additional practice log on to **academic.cengage.com/login.** Go to Chapter 5, then to Nutrition Calculations.

These problems will give you practice in doing simple nutrition-related calculations (see p. 171 for answers). Show your calculations for each problem.

1. Be aware of the fats in milks. Following are four categories of milk.

	Wt (g)	Fat (g)	Prot (g)	Carb (g)
Milk A (1 c)	244	8	8	12
Milk B (1 c)	244	5	8	12
Milk C (1 c)	244	3	8	12
Milk D (1 c)	244	0	8	12

   a. Based on *weight*, what percentage of each milk is fat (round off to a whole number)?

   b. How much energy from fat will a person receive from drinking 1 cup of each milk?

   c. How much total energy will the person receive from 1 cup of each milk?

   d. What percentage of the energy in each milk comes from fat?

   e. In the grocery store, how is each milk labeled?

2. Judge foods' fat contents by their labels.

   a. A food label says that one serving of the food contains 6.5 grams fat. What would the % Daily Value for fat be? What does the Daily Value you just calculated mean?

   b. How many kcalories from fat does a serving contain? (Round off to the nearest whole number.)

   c. If a *serving* of the food contains 200 kcalories, what percentage of the energy is from fat?

This example should show you how easy it is to evaluate foods' fat contents by reading labels and to see the difference between the % Daily Value and the percentage of kcalories from fat.

3. Now consider a piece of carrot cake. Remember that the Daily Value suggests 65 grams of fat as acceptable within a 2000-kcalorie diet. A serving of carrot cake provides 30 grams of fat. What percentage of the Daily Value is that? What does this mean?

STUDY QUESTIONS

CENGAGENOW

To assess your understanding of chapter topics, take the Student Practice Test and explore the modules recommended in your Personalized Study Plan. Log on to **academic.cengage.com/login.**

These questions will help you review this chapter. You will find the answers in the discussions on the pages provided.

1. Name three classes of lipids found in the body and in foods. What are some of their functions in the body? What features do fats bring to foods? (pp. 139, 145, 147, 153–155, 161)

2. What features distinguish fatty acids from each other? (pp. 139–142)

3. What does the term *omega* mean with respect to fatty acids? Describe the roles of the omega fatty acids in disease prevention. (pp. 141–142, 158–159)

4. What are the differences between saturated, unsaturated, monounsaturated, and polyunsaturated fatty acids? Describe the structure of a triglyceride. (pp. 140–143)

5. What does hydrogenation do to fats? What are *trans*-fatty acids, and how do they influence heart disease? (pp. 143–144, 157)

6. How do phospholipids differ from triglycerides in structure? How does cholesterol differ? How do these differences in structure affect function? (pp. 145–147)

7. What roles do phospholipids perform in the body? What roles does cholesterol play in the body? (pp. 145–147)

8. Trace the steps in fat digestion, absorption, and transport. Describe the routes cholesterol takes in the body. (pp. 147–153)

9. What do lipoproteins do? What are the differences among the chylomicrons, VLDL, LDL, and HDL? (pp. 150–153)

10. Which of the fatty acids are essential? Name their chief dietary sources. (pp. 154–155)

11. How does excessive fat intake influence health? What factors influence LDL, HDL, and total blood cholesterol? (pp. 156–160)

12. What are the dietary recommendations regarding fat and cholesterol intake? List ways to reduce intake. (pp. 160–165)

13. What is the Daily Value for fat (for a 2000-kcalorie diet)? What does this number represent? (pp. 165–167)

These multiple choice questions will help you prepare for an exam. Answers can be found on p. 171.

1. Saturated fatty acids:
    a. are always 18 carbons long.
    b. have at least one double bond.
    c. are fully loaded with hydrogens.
    d. are always liquid at room temperature.

2. A triglyceride consists of:
    a. three glycerols attached to a lipid.
    b. three fatty acids attached to a glucose.
    c. three fatty acids attached to a glycerol.
    d. three phospholipids attached to a cholesterol.

3. The difference between *cis*- and *trans*-fatty acids is:
    a. the number of double bonds.
    b. the length of their carbon chains.
    c. the location of the first double bond.
    d. the configuration around the double bond.

4. Which of the following is *not* true? Lecithin is:
    a. an emulsifier.
    b. a phospholipid.
    c. an essential nutrient.
    d. a constituent of cell membranes.

5. Chylomicrons are produced in the:
    a. liver.
    b. pancreas.
    c. gallbladder.
    d. small intestine.

6. Transport vehicles for lipids are called:
    a. micelles.
    b. lipoproteins.
    c. blood vessels.
    d. monoglycerides.

7. The lipoprotein most associated with a high risk of heart disease is:
    a. CHD.
    b. HDL.
    c. LDL.
    d. LPL.

8. Which of the following is *not* true? Fats:
    a. contain glucose.
    b. provide energy.
    c. protect against organ shock.
    d. carry vitamins A, D, E, and K.

9. The essential fatty acids include:
    a. stearic acid and oleic acid.
    b. oleic acid and linoleic acid.
    c. palmitic acid and linolenic acid.
    d. linoleic acid and linolenic acid.

10. A person consuming 2200 kcalories a day who wants to meet health recommendations should limit daily fat intake to:
    a. 20 to 35 grams.
    b. 50 to 85 grams.
    c. 75 to 100 grams.
    d. 90 to 130 grams.

REFERENCES

1. M. A. Zulet and coauthors, Inflammation and conjugated linoleic acid: Mechanisms of action and implications for human health, *Journal of Physiology and Biochemistry* 61 (2005): 483–494; M. A. Belury, Dietary conjugated linoleic acid in health: Physiological effects and mechanisms of action, *Annual Review of Nutrition* 22 (2002): 505–531.

2. K. A. Varady and coauthors, Plant sterols and endurance training combine to favorably alter plasma lipid profiles in previously sedentary hypercholesterolemic adults after 8 wk, *American Journal of Clinical Nutrition* 80 (2004): 1159–1166; M. Richelle and coauthors, Both free and esterified plant sterols reduce cholesterol absorption and the bioavailability of β-carotene and α-tocopherol in normocholesterolemic humans, *American Journal of Clinical Nutrition* 80 (2004): 171–177.

3. A. Andersson and coauthors, Fatty acid composition of skeletal muscle reflects dietary fat composition in humans, *American Journal of Clinical Nutrition* 76 (2002): 1222–1229; A. Baylin and coauthors, Adipose tissue biomarkers of fatty acid intake, *American Journal of Clinical Nutrition* 76 (2002): 750–757.

4 R. Uauy and A. D. Dangour, Nutrition in brain development and aging: Role of essential fatty acids, *Nutrition Reviews* 64 (2006): S24–S33; W. C. Heird and A. Lapillonne, The role of essential fatty acids in development, *Annual Review of Nutrition* 25 (2005): 549–571; J. M. Alessandri and coauthors, Polyunsaturated fatty acids in the central nervous system: Evolution of concepts and nutritional implications throughout life, *Reproduction, Nutrition, Development* 6 (2004): 509–538.

5. H. Tapiero and coauthors, Polyunsaturated fatty acids (PUFA) and eicosanoids in human health and pathologies, *Biomedicine and Pharmacotherapy* 56 (2002): 215–222.

6. M. T. Nakamura and T. Y. Nara, Structure, function, and dietary regulation of Δ6, Δ5, and Δ9 desaturases, *Annual Review of Nutrition* 24 (2004): 345–376.

7. J. R. Hibbeln, Seafood consumption, the DHA content of mothers' milk and prevalence rates of postpartum depression: A cross-national, ecological analysis, *Journal of Affective Disorders* 69 (2002): 15–29.

8. L. Zimmer and coauthors, The dopamine mesocorticolimbic pathway is affected by deficiency in n-3 polyunsaturated fatty acids, *American Journal of Clinical Nutrition* 75 (2002): 662–667.

9. P. J. Nestel and coauthors, Relation of diet to cardiovascular disease risk factors in subjects with cardiovascular disease in Australia and New Zealand: Analysis of the Long-Term Intervention with Pravastatin in Ischaemic Disease trial, *American Journal of Clinical Nutrition* 81 (2005): 1322–1329.

10. D. Mozaffarian and coauthors, Trans fatty acids and cardiovascular disease, *New England Journal of Medicine* 354 (2006): 1601–1613.

11. J. Dyerberg and coauthors, Effects of trans- and n-3 unsaturated fatty acids on cardiovascular risk markers in healthy males: An 8 weeks dietary intervention study, *European Journal of Clinical Nutrition* 58 (2004): 1062–1070; P. M. Clifton, J. B. Keogh, and M. Noakes, Trans fatty acids in adipose tissue and the food supply are associated with myocardial infarction, *Journal of Nutrition* 134 (2004): 874–879; N. M. deRoos, E. G. Schouten, and M. B. Katan, Trans fatty acids, HDL-cholesterol, and cardiovascular disease: Effects of dietary changes on vascular reactivity, *European Journal of Medical Research* 8 (2003): 355–357.

12. D. Mozaffarian and coauthors, Trans fatty acids and systemic inflammation in heart failure, *American Journal of Clinical Nutrition* 80 (2004): 1521–1525; D. J. Baer and coauthors, Dietary fatty acids affect plasma markers of inflammation in healthy men fed controlled diets: A randomized crossover study, *American Journal of Clinical Nutrition* 79 (2004): 969–973; D. Mozaffarian and coauthors, Dietary intake of *trans* fatty acids and systemic inflammation in women, *American Journal of Clinical Nutrition* 79 (2004): 606–612; G. A. Bray and coauthors, The influence of different fats and fatty acids on obesity, insulin resistance and inflammation, *Journal of Nutrition* 132 (2002): 2488–2491.

13. Federal Register 68, July 11, 2003, p. 41444.

14. Expert Panel on Detection, Evaluation, and Treatment of High Blood Cholesterol in Adults (Adult Treatment Panel III), *Third Report of the National Cholesterol Education Program (NCEP)*, NIH publication no. 02-5215 (Bethesda, Md.: National Heart, Lung, and Blood Institute, 2002), p. V-10.

15. A. H. Stark and Z. Madar, Olive oil as a functional food: Epidemiology and nutritional approaches, *Nutrition Reviews* 60 (2002): 170–176.

16. P. M. Kris-Etherton, K. D. Hecker, and A. E. Binkoski, Polyunsaturated fatty acids and cardiovascular health, *Nutrition Reviews* 62 (2004): 414–426.

17. J. L. Breslow, n-3 Fatty acids and cardiovascular disease, *American Journal of Clinical Nutrition* 83 (2006): 1477S–1482S; F. B. Hu and coauthors, Fish and omega-3 fatty acid intake and risk of coronary heart disease in women, *Journal of the American Medical Association* 287 (2002): 1815–1821; C. M. Albert and coauthors, Blood levels of long-chain n-3 fatty acids and the risk of sudden death, *New England Journal of Medicine* 346 (2002): 1113–1118.

18. Breslow, 2006; P. J. H. Jones and V. W. Y. Lau, Effect of n-3 polyunsaturated fatty acids on risk reduction of sudden death, *Nutrition Reviews* 60 (2002): 407–413.

19. M. C. Nesheim and A. L. Yaktine, eds., Seafood, *Seafood Choices: Balancing Benefits and Risks* (Washington, D. C.: National Academies Press, 2007), p. 12; C. W. Levenson and D. M. Axelrad, Too much of a good thing? Update on fish consumption and murcury exposure, *Nutrition Reviews* 64 (2006): 139–145; E. Guallar and coauthors, Mercury, fish oils, and the risk of myocardial infarction, *New England Journal of Medicine* 347 (2002): 1747–1754.

20. V. Wijendran and K. C. Hayes, Dietary n-6 and n-3 fatty acid balance and cardiovascular health, *Annual Review of Nutrition* 24 (2004): 597–615.

21. AHA Scientific statement: Diet and lifestyle recommendations revision 2006, *Circulation* 114 (2006): 82–96

22. K. He and coauthors, Fish consumption and risk of stroke in men, *Journal of the American Medical Association* 288 (2002): 3130–3136.

23. L. Cordain and coauthors, Fatty acid analysis of wild ruminant tissues: Evolutionary implications for reducing diet-related chronic disease, *European Journal of Clinical Nutrition* 56 (2002): 181–191.

24. S. Bechoua and coauthors, Influence of very low dietary intake of marine oil on some functional aspects of immune cells in healthy elderly people, *British Journal of Nutrition* 89 (2003): 523–532.

25. M. H. Raitt and coauthors, Fish oil supplementation and risk of ventricular tachycardia and ventricular fibrillation in patients with implantable defibrillators: A randomized control study, *Journal of the American Medical Association* 293 (2005): 2884–2891; P. M. Kris-Etherton and coauthors, AHA Scientific Statement: Fish consumption, fish oil, omega-3 fatty acids, and cardiovascular disease, *Circulation* 106 (2002): 2747–2757.

26. C. B. Stephensen, Fish oil and inflammatory disease: Is asthma the next target for n-3 fatty acid supplements? *Nutrition Reviews* 62 (2004): 486–489.

27. G. L. Khor, Dietary fat quality: A nutritional epidemiologist's view, *Asia Pacific Journal of Clinical Nutrition* 13 (2004): S22; R. Stoeckli and U. Keller, Nutritional fats and the risk of type 2 diabetes and cancer, *Physiology and Behavior* 83 (2004): 611–615.

28. M. D. Holmes and W. C. Willett, Does diet affect breast cancer risk? *Breast Cancer Research* 6 (2004): 170–178.

29. L. K. Dennis and coauthors, Problems with the assessment of dietary fat in prostate cancer studies, *American Journal of Epidemiology* 160 (2004): 436–444.

30. P. W. Parodi, Dairy product consumption and the risk of breast cancer, *Journal of the American College of Nutrition* 24 (2005): 556S–568S; J. Zhang and H. Kesteloot, Milk consumption in relation to incidence of prostate, breast, colon, and rectal cancers: Is there an independent effect? *Nutrition and Cancer* 53 (2005): 65–72.

31. C. H. MacLean and coauthors, Effects of omega-3 fatty acids on cancer risk-A systematic review, *Journal of the American Medical Association* 295 (2006): 403–415; W. E. Hardman, (n-3) Fatty acids and cancer therapy, *Journal of Nutrition* 134 (2004): 3427S–3430S; M. F. Leitzmann and coauthors, Dietary intake of n-3 and n-6 fatty acids and the risk of prostate cancer, *American Journal of Clinical Nutrition* 80 (2004): 204–216; S. C. Larsson and coauthors, Dietary long-chain n-3 fatty acids for the prevention of cancer: A review of potential mechanisms, *American Journal of Clinical Nutrition* 79 (2004): 935–945.

32. Committee on Dietary Reference Intakes, *Dietary Reference Intakes for Energy, Carbohydrate, Fiber, Fat, Fatty Acids, Cholesterol, Protein, and Amino Acids* (Washington, D.C.: National Academies Press, 2002/2005).

33. Committee on Dietary Reference Intakes, 2002/2005.

34. Committee on Dietary Reference Intakes, 2002/2005.
35. National Center for Health Statistics, *Chartbook on Trends in the Health of Americans, 2005,* www.cdc.gov/nchs, site visited on January 18, 2006; Committee on Dietary Reference Intakes, 2002/2005.
36. Position of the American Dietetic Association, Dietitians of Canada, and the American College of Sports Medicine: Nutrition and athletic performance, *Journal of the American Dietetic Association* 100 (2000): 1543–1556.
37. S. Tonstad, K. Smerud, and L. Høie, A comparison of the effects of 2 doses of soy

protein or casein on serum lipids, serum lipoproteins, and plasma total homocysteine in hypercholesterolemic subjects, *American Journal of Clinical Nutrition* 76 (2002): 78–84.
38. B. M. Davy and coauthors, High-fiber oat cereal compared with wheat cereal consumption favorably alters LDL-cholesterol subclass and particle numbers in middle-aged and older men, *American Journal of Clinical Nutrition* 76 (2002): 351–358; D. J. A. Jenkins and coauthors, Soluble fiber intake at a dose approved by the U.S. Food and Drug Administration for a claim of health benefits: Serum lipid risk factors for cardio-

vascular disease assessed in a randomized controlled crossover trial, *American Journal of Clinical Nutrition* 75 (2002): 834–839.
39. C. S. Patch, L. C. Tapsell, and P. G. Williams, Plant sterol/stanol prescription is an effective treatment strategy for managing hypercholesterolemia in outpatient clinical practice, *Journal of the American Dietetic Association* 105 (2005): 46–52.
40. Position of the American Dietetic Association: Fat replacers, *Journal of the American Dietetic Association* 105 (2005): 266–275.

ANSWERS

Nutrition Calculations

1. a. Milk A: 8 g fat ÷ 244 g total = 0.03; 0.03 × 100 = 3%
    Milk B: 5 g fat ÷ 244 g total = 0.02; 0.02 × 100 = 2%
    Milk C: 3 g fat ÷ 244 g total = 0.01; 0.01 × 100 = 1%
    Milk D: 0 g fat ÷ 244 g total = 0.00; 0.00 × 100 = 0%

   b. Milk A: 8 g fat × 9 kcal/g = 72 kcal from fat
    Milk B: 5 g fat × 9 kcal/g = 45 kcal from fat
    Milk C: 3 g fat × 9 kcal/g = 27 kcal from fat
    Milk D: 0 g fat × 9 kcal/g = 0 kcal from fat

   c. Milk A: (8 g fat × 9 kcal/g) + (8 g prot × 4 kcal/g) + (12 g carb × 4 kcal/g) = 152 kcal
    Milk B: (5 g fat × 9 kcal/g) + (8 g prot × 4 kcal/g) + (12 g carb × 4 kcal/g) = 125 kcal
    Milk C: (3 g fat × 9 kcal/g) + (8 g prot × 4 kcal/g) + (12 g carb × 4 kcal/g) = 107 kcal
    Milk D: (0 g fat × 9 kcal/g) + (8 g prot × 4 kcal/g) + (12 g carb × 4 kcal/g) = 80 kcal

   d. Milk A: 72 kcal from fat ÷ 152 total kcal = 0.47; 0.47 × 100 = 47%
    Milk B: 45 kcal from fat ÷ 125 total kcal = 0.36; 0.36 × 100 = 36%

   Milk C: 27 kcal from fat ÷ 107 total kcal = 0.25; 0.25 × 100 = 25%
   Milk D: 0 kcal from fat ÷ 80 total kcal = 0.00; 0.00 × 100 = 0%

   e. Milk A: whole
    Milk B: reduced-fat, 2%, or less-fat
    Milk C: low-fat or 1%
    Milk D: fat-free, nonfat, skim, zero-fat, or no-fat

2. a. 6.5 g ÷ 65 g = 0.1; 0.1 × 100 = 10%; a Daily Value of 10% means that one serving of this food contributes about 1/10 of the day's fat allotment

   b. 6.5 g × 9 kcal/g = 58.5, rounded to 59 kcal from fat

   c. (59 kcal from fat ÷ 200 kcal) × 100 = 30% kcalories from fat

3. (30 g fat ÷ 65 g fat) × 100 = 46% of the Daily Value for fat; this means that almost half of the day's fat allotment would be used in this one dessert

Study Questions (multiple choice)
1. c 2. c 3. d 4. c 5. d 6. b 7. c 8. a
9. d 10. b

High-Fat Foods—Friend or Foe?

© Philip Salverry/FoodPix/Jupiter Images

Eat less fat. Eat more fatty fish. Give up butter. Use margarine. Give up margarine. Use olive oil. Steer clear of saturated. Seek out omega-3. Stay away from *trans*. Stick with mono- and polyunsaturated. Keep fat intake moderate. Today's fat messages seem to be forever multiplying and changing. No wonder people feel confused about dietary fat. The confusion stems in part from the complexities of fat and in part from the nature of recommendations. As Chapter 5 explained, "dietary fat" refers to several kinds of fats. Some fats support health whereas others damage it, and foods typically provide a mixture of fats in varying proportions. Researchers have spent decades sorting through the relationships among the various kinds of fat and their roles in supporting or harming health. Translating these research findings into dietary recommendations is challenging. Too little information can mislead consumers, but too much detail can overwhelm them. As research findings accumulate, recommendations slowly evolve and become more refined. Fortunately, that's where we are with fat recommendations today—refining them from the general to the specific. Though they may seem to be "forever multiplying and changing," in fact, they are becoming more meaningful.

This highlight begins with a look at the dietary guidelines for fat intake. It continues by identifying which foods provide which fats and presenting the Mediterranean diet, an example of a food plan that embraces the heart-healthy fats. It closes with strategies to help consumers choose the right amounts of the right kinds of fats for a healthy diet.

Guidelines for Fat Intake

Dietary recommendations for fat have changed in recent years, shifting the emphasis from lowering total fat, in general, to limiting saturated and *trans* fat, specifically. For decades, health experts advised limiting intakes of total fat to 30 percent or less of energy intake. They recognized that saturated fats and *trans* fats are the fats that raise blood cholesterol but reasoned that by limiting total fat intake, saturated and *trans* fat intake would decline as well. People were simply advised to cut back on all fat and thereby they would cut back on saturated and *trans* fat. Such advice may have oversimplified the message and unnecessarily restricted total fat.

Low-fat diets have a place in treatment plans for people with elevated blood lipids or heart disease, but some researchers question the wisdom of such diets for healthy people as a means of controlling weight and preventing diseases. Several problems ac-

company low-fat diets. For one, many people find low-fat diets difficult to maintain over time. For another, low-fat diets are not necessarily low-kcalorie diets. If energy intake exceeds energy needs, weight gain follows, and obesity brings a host of health problems, including heart disease. For still another, diets extremely low in fat may exclude fatty fish, nuts, seeds, and vegetable oils—all valuable sources of many essential fatty acids, phytochemicals, vitamins, and minerals. Importantly, the fats from these sources protect against heart disease, as later sections of this highlight explain.

Instead of urging people to cut back on all fats, current recommendations suggest carefully replacing the "bad" saturated fats with the "good" unsaturated fats and enjoying them in moderation.[1] The goal is to create a diet moderate in kcalories that provides enough of the fats that support good health, but not too much of those that harm health. (Turn to pp. 156–160 for a review of the health consequences of each type of fat.)

With these findings and goals in mind, the DRI committee suggests a healthy range of 20 to 35 percent of energy intake from fat. This range appears to be compatible with low rates of heart disease, diabetes, obesity, and cancer.[2] Heart-healthy recommendations suggest that within this range, consumers should try to minimize their intakes of saturated fat, *trans* fat, and cholesterol and use monounsaturated and polyunsaturated fats instead.[3]

Asking consumers to limit their total fat intake was less than perfect advice, but it was straightforward—find the fat and cut back. Asking consumers to keep their intakes of saturated fats, *trans* fats, and cholesterol low and to use monounsaturated and polyunsaturated fats instead may be more on target with heart health, but it also makes diet planning more complicated. To make appropriate selections, consumers must first learn which foods contain which fats.

High-Fat Foods and Heart Health

Avocados, bacon, walnuts, potato chips, and mackerel are all high-fat foods, yet some of these foods have detrimental effects on heart health when consumed in excess, whereas others seem neutral or even beneficial. This section presents some of the accumulating evidence that helped to distinguish which high-fat foods belong in a healthy diet and which ones need to be kept to a minimum. As you will see, a little more fat in the diet may be compatible with heart health, but only if the great majority of it is the unsaturated kind.

Cook with Olive Oil

As it turns out, the traditional diets of Greece and other countries in the Mediterranean region offer an excellent example of eating patterns that use "good" fats liberally. Often, these diets are rich in olives and their oil. A classic study of the world's people, the Seven Countries Study, found that death rates from heart disease were strongly associated with diets high in saturated fats but only weakly linked with total fat.[4] In fact, the two countries with the highest fat intakes, Finland and the Greek island of Crete, had the highest (Finland) and lowest (Crete) rates of heart disease deaths. In both countries, the people consumed 40 percent or more of their kcalories from fat. Clearly, a high-fat diet was not the primary problem, so researchers refocused their attention on the *type* of fat. They began to notice the benefits of olive oil.

A diet that uses olive oil instead of other cooking fats, especially butter, stick margarine, and meat fats, may offer numerous health benefits.[5] Olive oil and other oils rich in mono- unsaturated fatty acids help to protect against heart disease by:

- Lowering total and LDL cholesterol and not lowering HDL cholesterol or raising triglycerides[6]
- Lowering LDL cholesterol susceptibility to oxidation[7]
- Lowering blood-clotting factors[8]
- Providing phytochemicals that act as antioxidants (see Highlight 11)[9]
- Lowering blood pressure[10]

When compared with other fats, olive oil seems to be a wise choice, but controlled clinical trials are too scarce to support population-wide recommendations to switch to a high-fat diet rich in olive oil. Importantly, olive oil is not a magic potion; drizzling it on foods does not make them healthier. Like other fats, olive oil delivers 9 kcalories per gram, which can contribute to weight gain in people who fail to balance their energy intake with their energy output. Its role in a healthy diet is to *replace* the saturated fats. Other vegetable oils, such as canola or safflower oil, are also generally low in saturated fats and high in unsaturated fats. For this reason, heart-healthy diets use these unsaturated vegetable oils as substitutes for the more saturated fats of butter, hydrogenated stick margarine, lard, or shortening. (Remember that the tropical oils—coconut, palm, and palm kernel—are too saturated to be included with the heart-healthy vegetable oils.)

Nibble on Nuts

Tree nuts and peanuts are traditionally excluded from low-fat diets, and for good reasons. Nuts provide up to 80 percent of their kcalories from fat, and a quarter cup (about an ounce) of mixed nuts provides over 200 kcalories. In a recent review of the literature, however, researchers found that people who ate a one-ounce serving of nuts on five or more days a week had a reduced risk of heart disease compared with people who consumed no nuts.[11] A smaller positive association was noted for

Olives and their oil may benefit heart health.

any amount greater than one serving of nuts a week. The nuts in this study were those commonly eaten in the United States: almonds, Brazil nuts, cashews, hazelnuts, macadamia nuts, pecans, pistachios, walnuts, and even peanuts. On average, these nuts contain mostly monounsaturated fat (59 percent), some polyunsaturated fat (27 percent), and little saturated fat (14 percent).

Research has shown a benefit from walnuts and almonds in particular. In study after study, walnuts, when substituted for other fats in the diet, produce favorable effects on blood lipids—even in people with elevated total and LDL cholesterol.[12] Results are similar for almonds. In one study, researchers gave men and women one of three kinds of snacks, all of equal kcalories: whole-wheat muffins, almonds (about 2¹/₂ ounces), or half muffins and half almonds.[13] At the end of a month, people receiving the full almond snack had the greatest drop in blood LDL cholesterol; those eating the half almond snack had a lesser, but still significant, drop in blood lipids; and those eating the muffin only snack had no change.

Studies on peanuts, macadamia nuts, pecans, and pistachios follow suit, indicating that including nuts may be a wise strategy against heart disease. Nuts may protect against heart disease because they provide:

- Monounsaturated and polyunsaturated fats in abundance, but few saturated fats
- Fiber, vegetable protein, and other valuable nutrients, including the antioxidant vitamin E (see Highlight 11)
- Phytochemicals that act as antioxidants (see Highlight 13)

Before advising consumers to include nuts in their diets, a caution is in order. As mentioned, most of the energy nuts provide comes from fats. Consequently, they deliver many kcalories per bite. In studies examining the effects of nuts on heart disease, researchers carefully adjust diets to make room for the nuts without

Matthew Farruggio

For heart health, snack on a few nuts instead of potato chips. Because nuts are energy dense (high in kcalories per ounce), it is especially important to keep portion size in mind when eating them.

increasing the total kcalories—that is, they use nuts *instead of, not in addition to,* other foods (such as meats, potato chips, oils, margarine, and butter). Consumers who do not make similar replacements could end up gaining weight if they simply add nuts on top of their regular diets. Weight gain, in turn, elevates blood lipids and raises the risks of heart disease.

Feast on Fish

Research into the health benefits of the long-chain omega-3 polyunsaturated fatty acids began with a simple observation: the native peoples of Alaska, northern Canada, and Greenland, who eat a diet rich in omega-3 fatty acids, notably EPA and DHA, have a remarkably low rate of heart disease even though their diets are relatively high in fat.[14] These omega-3 fatty acids help to protect against heart disease by:[15]

- Reducing blood triglycerides
- Preventing blood clots
- Protecting against irregular heartbeats
- Lowering blood pressure
- Defending against inflammation
- Serving as precursors to eicosanoids

For people with hypertension or atherosclerosis, these actions can be life saving.

Research studies have provided strong evidence that increasing omega-3 fatty acids in the diet supports heart health and lowers the rate of deaths from heart disease.[16] For this reason, the American Heart Association recommends including fish in a heart-healthy diet. People who eat some fish each week can lower their risks of heart attack and stroke. Table 5-2 on p. 159 lists fish that provide at least 1 gram of omega-3 fatty acids per serving.

Fish is the best source of EPA and DHA in the diet, but it is also a major source of mercury, an environmental contaminant. Most fish contain at least trace amounts of mercury, but tilefish (also known as

golden snapper or golden bass), swordfish, king mackerel, marlin, and shark have especially high levels. For this reason, the FDA advises pregnant and lactating women, women of childbearing age who may become pregnant, and young children to avoid:

- Tilefish (also called golden snapper or golden bass), swordfish, king mackeral, marlin, and shark

And to limit average weekly consumption of:

- A variety of fish and shellfish to 12 ounces (cooked or canned)
- White (albacore) tuna to 6 ounces (cooked or canned)

Commonly eaten seafood relatively low in mercury include shrimp, catfish, pollock, salmon, and canned light tuna.

In addition to the direct toxic effects of mercury, some (but not all) research suggests that mercury may diminish the health benefits of omega-3 fatty acids.[17] Such findings serve as a reminder that our health depends on the health of our planet. The protective effect of fish in the diet is available, provided that the fish and their surrounding waters are not heavily contaminated.

In an effort to limit exposure to pollutants, some consumers choose farm-raised fish. Compared with fish caught in the wild, farm-raised fish tend to be lower in mercury, but they are also lower in omega-3 fatty acids. When selecting fish, keep the diet strategies of variety and moderation in mind. Varying choices and eating moderate amounts helps to limit the intake of contaminants such as mercury.

© www.comstock.com

Fish is a good source of the omega-3 fatty acids.

High-Fat Foods and Heart Disease

The number one dietary determinant of LDL cholesterol is saturated fat. Figure H5-1 shows that each 1 percent increase in energy from saturated fatty acids in the diet may produce a 2 percent jump in heart disease risk by elevating blood LDL cholesterol. Conversely, reducing saturated fat intake by 1 percent can be expected to produce a 2 percent drop in heart disease risk by the same mechanism. Even a 2 percent drop in LDL represents a significant improvement for the health of the heart.[18] Like saturated fats, *trans* fats also raise heart disease risk by elevating LDL cholesterol. A heart-healthy diet limits foods rich in these two types of fat.

Limit Fatty Meats, Whole-Milk Products, and Tropical Oils

The major sources of saturated fats in the U.S. diet are fatty meats, whole milk products, tropical oils, and products made from any of these foods. To limit saturated fat intake, consumers must choose carefully among these high-fat foods. Over a third of the fat in most meats is saturated. Similarly, over half of the fat is saturated in whole milk and other high-fat dairy products, such as cheese, butter, cream, half-and-half, cream cheese, sour cream, and ice cream. The tropical oils of palm, palm kernel, and coconut, which are rarely used by consumers in the kitchen, are used heavily by food manufacturers, and are commonly found in many commercially prepared foods.

When choosing meats, milk products, and commercially prepared foods, look for those lowest in saturated fat. Labels provide a useful guide for comparing products in this regard, and Appendix H lists the saturated fat in several thousand foods.

Even with careful selections, a nutritionally adequate diet will provide some saturated fat. Zero saturated fat is not possible even when experts design menus with the mission to keep saturated fat as low as possible.[19] Because most saturated fats come from animal foods, vegetarian diets can, and usually do, deliver fewer saturated fats than mixed diets.

Limit Hydrogenated Foods

Chapter 5 explained that solid shortening and margarine are made from vegetable oil that has been hardened through hydrogenation. This process both saturates some of the unsaturated fatty acids and introduces *trans*-fatty acids. Many convenience foods contain *trans* fats, including:

- Fried foods such as French fries, chicken, and other commercially fried foods
- Commercial baked goods such as cookies, doughnuts, pastries, breads, and crackers
- Snack foods such as chips
- Imitation cheeses

To keep *trans* fat intake low, use these foods sparingly as an occasional taste treat.

Table H5-1 (p. 176) summarizes which foods provide which fats. Substituting unsaturated fats for saturated fats at each meal and snack can help protect against heart disease. Figure H5-2 (p. 176) compares two meals and shows how such substitutions can lower saturated fat and raise unsaturated fat—even when total fat and kcalories remain unchanged.

The Mediterranean Diet

The links between good health and traditional Mediterranean diets of the mid-1900s were introduced earlier with regard to olive

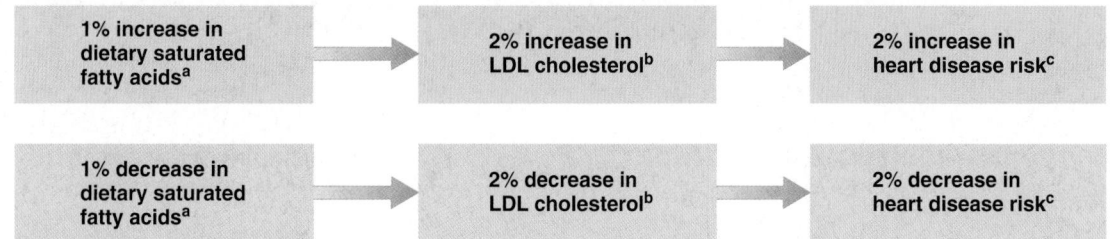

FIGURE H5-1 Potential Relationships among Dietary Saturated Fatty Acids, LDL Cholesterol, and Heart Disease Risk

[a]Percentage of change in total dietary energy from saturated fatty acids.
[b]Percentage of change in blood LDL cholesterol.
[c]Percentage of change in an individual's risk of heart disease; the percentage of change in risk may increase when blood lipid changes are sustained over time.

SOURCE: *Third Report of the National Cholesterol Education Program (NCEP) Expert Panel on Detection, Evaluation, and Treatment of High Blood Cholesterol in Adults (Adult Treatment Panel III)*, NIH publication no. 02-5215 (Bethesda, Md.: National Heart, Lung, and Blood Institute, 2002), p. V-8 and II-4.

TABLE H5-1   Major Sources of Various Fatty Acids

Healthful Fatty Acids

Monounsaturated	Omega-6 Polyunsaturated	Omega-3 Polyunsaturated
Avocado	Margarine (nonyhydrogenated)	Fatty fish (herring, mackerel, salmon, tuna)
Oils (canola, olive, peanut, sesame)	Oils (corn, cottonseed, safflower, soybean)	Flaxseed
Nuts (almonds, cashews, filberts, hazelnuts, macadamia nuts, peanuts, pecans, pistachios)	Nuts (pine nuts, walnuts)	Nuts (walnuts)
	Mayonnaise	
Olives	Salad dressing	
Peanut butter	Seeds (pumpkin, sunflower)	
Seeds (sesame)		

Harmful Fatty Acids

Saturated	*Trans*
Bacon	Fried foods (hydrogenated shortening)
Butter	Margarine (hydrogenated or partially hydrogenated)
Chocolate	
Coconut	Nondairy creamers
Cream cheese	Many fast foods
Cream, half-and-half	Shortening
Lard	Commercial baked goods (including doughnuts, cakes, cookies)
Meat	
Milk and milk products (whole)	Many snack foods (including microwave popcorn, chips, crackers)
Oils (coconut, palm, palm kernel)	
Shortening	
Sour cream	

NOTE: Keep in mind that foods contain a mixture of fatty acids.

FIGURE H5-2   Two Meals Compared: Replacing Saturated Fat with Unsaturated Fat

Examples of ways to replace saturated fats with unsaturated fats include sautéing vegetables in olive oil instead of butter, garnishing salads with avocado and sunflower seeds instead of bacon and blue cheese, and eating salmon instead of steak. Each of these meals provides roughly the same number of kcalories and grams of fat, but the one on the left has almost four times as much saturated fat and only half as many omega-3 fatty acids.

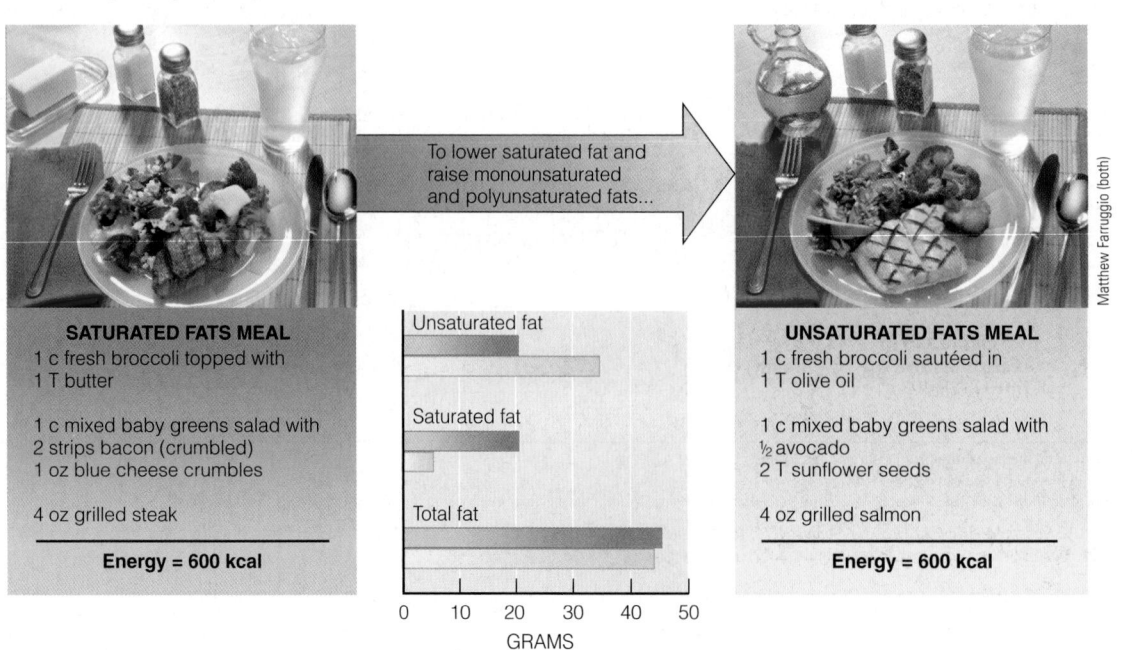

To lower saturated fat and raise monounsaturated and polyunsaturated fats...

SATURATED FATS MEAL
1 c fresh broccoli topped with 1 T butter

1 c mixed baby greens salad with 2 strips bacon (crumbled) 1 oz blue cheese crumbles

4 oz grilled steak

Energy = 600 kcal

Unsaturated fat

Saturated fat

Total fat

0   10   20   30   40   50
GRAMS

UNSATURATED FATS MEAL
1 c fresh broccoli sautéed in 1 T olive oil

1 c mixed baby greens salad with ½ avocado 2 T sunflower seeds

4 oz grilled salmon

Energy = 600 kcal

Matthew Farruggio (both)

oil. For people who eat these diets, the incidence of heart disease, some cancers, and other chronic diseases is low, and life expectancy is high.[20]

Although each of the many countries that border the Mediterranean Sea has its own culture, traditions, and dietary habits, their similarities are much greater than the use of olive oil alone. In fact, according to a recent study, no one factor alone can be credited with reducing disease risks—the association holds true only when the overall diet pattern is present.[21] Apparently, each of the foods contributes small benefits that harmonize to produce either a substantial cumulative or a synergistic effect.

The Mediterranean people focus their diets on crusty breads, whole grains, potatoes, and pastas; a variety of vegetables (including wild greens) and legumes; feta and mozzarella cheeses and yogurt; nuts; and fruits (especially grapes and figs). They eat some fish, other seafood, poultry, a few eggs, and little meat. Along with olives and olive oil, their principal sources of fat are nuts and fish; they rarely use butter or encounter hydrogenated fats. Consequently, traditional Mediterranean diets are:

- Low in saturated fat
- Very low in *trans* fat
- Rich in unsaturated fat
- Rich in complex carbohydrate and fiber
- Rich in nutrients and phytochemicals that support good health

People following the traditional Mediterranean diet can receive as much as 40 percent of a day's kcalories from fat, but their limited consumption of dairy products and meats provides less than 10 percent from saturated fats. In addition, because the animals in the Mediterranean region graze, the meat, dairy products, and eggs are richer in omega-3 fatty acids than those from animals fed grain. Other foods typical of the Mediterranean, such as wild plants and snails, provide omega-3 fatty acids as well. All in all, the traditional Mediterranean diet has gained a reputation for its health benefits as well as its delicious flavors, but beware of the typical Mediterranean-style cuisine available in U.S. restaurants. It has been adjusted to popular tastes, meaning that it is often much higher in saturated fats and meats—and much lower in the potentially beneficial constituents—than the traditional fare. Unfortunately, it appears that people in the Mediterranean region who are replacing some of their traditional dietary habits with those of the United States are losing the health benefits previously enjoyed.[22]

Conclusion

Are some fats "good," and others "bad" from the body's point of view? The saturated and *trans* fats indeed seem mostly bad for the health of the heart. Aside from providing energy, which unsaturated fats can do equally well, saturated and *trans* fats bring no indispensable benefits to the body. Furthermore, no harm can come from consuming diets low in them. Still, foods rich in these fats are often delicious, giving them a special place in the diet.

In contrast, the unsaturated fats are mostly good for the health of the heart when consumed in moderation. To date, their one proven fault seems to be that they, like all fats, provide abundant energy to the body and so may promote obesity if they drive kcalorie intakes higher than energy needs.[23] Obesity, in turn, often begets many body ills, as Chapter 8 makes clear.

When judging foods by their fatty acids, keep in mind that the fat in foods is a mixture of "good" and "bad," providing both saturated and unsaturated fatty acids. Even predominantly monounsaturated olive oil delivers some saturated fat. Consequently, even when a person chooses foods with mostly unsaturated fats, saturated fat can still add up if total fat is high. For this reason, fat must be kept below 35 percent of total kcalories if the diet is to be moderate in saturated fat. Even experts run into difficulty when attempting to create nutritious diets from a variety of foods that are low in saturated fats when kcalories from fat exceed 35 percent of the total.[24]

Does this mean that you must forever go without favorite cheeses, ice cream cones, or a grilled steak? The famous chef Julia Child made this point about moderation:

> An imaginary shelf labeled INDULGENCES is a good idea. It contains the best butter, jumbo-size eggs, heavy cream, marbled steaks, sausages and pâtés, hollandaise and butter sauces, French butter-cream fillings, gooey chocolate cakes, and all those lovely items that demand disciplined rationing. Thus, with these items high up and almost out of reach, we are ever conscious that they are not everyday foods. They are for special occasions, and when that occasion comes we can enjoy every mouthful.
> Julia Child, *The Way to Cook,* 1989

Additionally, food manufacturers have come to the assistance of consumers who wish to avoid the health threats from saturated and *trans* fats. Some margarine makers no longer offer products containing *trans* fats, and many snack manufacturers have reduced the saturated and *trans* fats in some products and now offer snack foods in 100-kcalorie packages. Other companies are following as consumers respond favorably.

Adopting some of the Mediterranean eating habits may serve those who enjoy a little more fat in the diet. Including vegetables, fruits, and legumes as part of a balanced daily diet is a good idea, as is *replacing* saturated fats such as butter, shortening, and meat fat with unsaturated fats like olive oil and the oils from nuts and fish. These foods provide vitamins, minerals, and phytochemicals—all valuable in protecting the body's health. The authors of this book do not stop there, however. They urge you to reduce fats from convenience foods and fast foods; choose small portions of meats, fish, and poultry; and include fresh foods from all the food groups each day. Take care to select portion sizes that will best meet your energy needs. Also, exercise daily.

REFERENCES

1. *Third Report of the National Cholesterol Education Program (NCEP) Expert Panel on Detection, Evaluation, and Treatment of High Blood Cholesterol in Adults (Adult Treatment Panel III)*, publication NIH no. 02-5215 (Bethesda, Md.: National Heart, Lung, and Blood Institute, 2002); Committee on Dietary Reference Intakes, *Dietary Reference Intakes for Energy, Carbohydrate, Fiber, Fat, Fatty Acids, Cholesterol, Protein, and Amino Acids* (Washington, D.C.: National Academies Press, 2002/2005).

2. Committee on Dietary Reference Intakes, 2002/2005, p. 769.

3. American Heart Association Scientific statement: Diet and lifestyle recommendations revision 2006, *Circulation* 114 (2006): 82–96; *Third Report of the National Cholesterol Education Program (NCEP) Expert Panel on Detection, Evaluation, and Treatment of High Blood Cholesterol in Adults (Adult Treatment Panel III)*, publication NIH no. 02-5215 (Bethesda, Md.: National Heart, Lung, and Blood Institute, 2002); Committee on Dietary Reference Intakes, *Dietary Reference Intakes for Energy, Carbohydrate, Fiber, Fat, Fatty Acids, Cholesterol, Protein, and Amino Acids* (Washington, D.C.: National Academies Press, 2002/2005).

4. A. Keys, *Seven Countries: A Multivariate Analysis of Death and Coronary Heart Disease* (Cambridge: Harvard University Press, 1980).

5. A. H. Stark and Z. Madar, Olive oil as a functional food: Epidemiology and nutritional approaches, *Nutrition Reviews* 60 (2002): 170–176.

6. M. I. Covas and coauthors, The effect of polyphenols in olive oil on heart disease risk factors, *Annals of Internal Medicine* 145 (2006): 333–341.

7. F. Visioli and coauthors, Virgin Olive Oil Study (VOLOS): Vasoprotective potential of extra virgin olive oil in mildly dislipidemic patients, *European Journal of Nutrition* 44 (2005): 121–127.

8. J. López-Miranda, Monounsaturated fat and cardiovascular risk, *Nutrition Reviews* 64 (2006): S2–S12.

9. F. Visioli and C. Galli, Biological properties of olive oil phytochemicals, *Critical Reviews in Food Science and Nutrition* 42 (2002): 209–221; M. N. Vissers and coauthors, Olive oil phenols are absorbed in humans, *Journal of Nutrition* 132 (2002): 409–417.

10. B. M. Rasmussen and coauthors, Effects of dietary saturated, monounsaturated, and n-3 fatty acids on blood pressure in healthy subjects, *American Journal of Clinical Nutrition* 83 (2006): 221–226; T. Psaltopoulou and coauthors, Olive oil, the Mediterranean diet, and arterial blood pressure: The Greek European Prospective Investigation into Cancer and Nutrition (EPIC) study, *American Journal of Clinical Nutrition* 80 (2004): 1012–1018.

11. J. H. Kelly and J. Sabate, Nuts and coronary heart disease: An epidemiological perspective, *British Journal of Nutrition* 96 (2006): S61–S67.

12. E. B. Feldman, The scientific evidence for a beneficial health relationship between walnuts and coronary heart disease, *Journal of Nutrition* 132 (2002): 1062S–1101S.

13. D. J. Jenkins and coauthors, Dose response of almonds on coronary heart disease risk factors: Blood lipids, oxidized low-density lipoproteins, lipoprotein (a), homocysteine, and pulmonary nitric oxide: A randomized, controlled, crossover trial, *Circulation* 106 (2002): 1327–1332.

14. E. Dewailly and coauthors, Cardiovascular disease risk factors and n-3 fatty acid status in the adult population of James Bay Cree, *American Journal of Clinical Nutrition* 76 (2002): 85–92.

15. J. L. Breslow, n-3 fatty acids and cardiovascular disease, *American Journal of Clinical Nutrition* 83 (2006): 1477S–1482S; P. J. H. Jones and V. W. Y. Lau, Effect of n-3 polyunsaturated fatty acids on risk reduction of sudden death, *Nutrition Reviews* 60 (2002): 407–413.

16. Breslow, 2006; F. B. Hu and coauthors, Fish and omega-3 fatty acid intake and risk of coronary heart disease in women, *Journal of the American Medical Association* 287 (2002): 1815–1821.

17. E. Guallar and coauthors, Mercury, fish oils, and the risk of myocardial infarction, *New England Journal of Medicine* 347 (2002): 1747–1754; K. Yoshizawa and coauthors, Mercury and the risk of coronary heart disease in man, *New England Journal of Medicine* 347 (2002): 1755–1760.

18. *Third Report of the National Cholesterol Education Program (NCEP) Expert Panel on Detection, Evaluation, and Treatment of High Blood Cholesterol in Adults (Adult Treatment Panel III)*, 2002, p.V-8.

19. Committee on Dietary Reference Intakes, 2002/2005, p. 835.

20. L. Serra-Majem, B. Roman, and R. Estruch, Scientific evidence of interventions using the Mediterranean diet: A systematic review, *Nutrition Reviews* 64 (2006): S27–S47; C. Pitsavos and coauthors, Adherence to the Mediterranean diet is associated with total antioxidant capacity in healthy adults: The ATTICA study, *American Journal of Clinical Nutrition* 82 (2005): 694–699; M. Meydani, A Mediterranean-style diet and metabolic syndrome, *Nutrition Reviews* 63 (2005): 312–314; D. B. Panagiotakos and coauthors, Can a Mediterranean diet moderate the development and clinical progression of coronary heart disease? A systematic review, *Medical Science Monitor* 10 (2004): RA193–RA198; K. T. B. Knoops and coauthors, Mediterranean diet, lifestyle factors, and 10-year mortality in elderly European men and women, *Journal of the American Medical Association* 292 (2004): 1433–1439; K. Esposito and coauthors, Effect of a Mediterranean-style diet on endothelial dysfunction and markers of vascular inflammation in the metabolic syndrome: A randomized study, *Journal of the American Medical Association* 292 (2004): 1440–1446.

21. A. Trichopoulou and coauthors, Adherence to a Mediterranean diet and survival in a Greek population, *New England Journal of Medicine* 348 (2003): 2599–2608.

22. F. Sofi and coauthors, Dietary habits, lifestyle, and cardiovascular risk factors in a clinically healthy Italian population: The "Florence" diet is not Mediterranean, *European Journal of Clinical Nutrition* 59 (2005): 584–591.

23. Committee on Dietary Reference Intakes, 2002/2005, pp. 796–797.

24. Committee on Dietary Reference Intakes, 2002/2005, pp. 799–802.

The CengageNOW logo indicates an opportunity for online self-study, linking you to interactive tutorials and videos based on your level of understanding.

academic.cengage.com/login

Nutrition in Your Life

Their versatility in the body is impressive. They help your muscles to contract, your blood to clot, and your eyes to see. They keep you alive and well by facilitating chemical reactions and defending against infections. Without them, your bones, skin, and hair would have no structure. No wonder they were named *proteins,* meaning "of prime importance." Does that mean proteins deserve top billing in your diet as well? Are the best sources of protein beef, beans, or broccoli? Learn which foods will supply you with enough, but not too much, high-quality protein.

Protein: Amino Acids

A few misconceptions surround the roles of protein in the body and the importance of protein in the diet. For example, people who associate meat with protein and protein with strength may eat steak to build muscles. Their thinking is only partly correct, however. Protein is a vital structural and working substance in all cells—not just muscle cells. To build strength, muscles cells need physical activity and all the nutrients—not just protein. Furthermore, protein is found in milk, eggs, legumes, and many grains and vegetables—not just meat. By overvaluing protein and overemphasizing meat in the diet, a person may mistakenly crowd out other, equally important nutrients and foods. As this chapter describes the various roles of protein in the body and food sources in the diet, keep in mind that protein is one of many nutrients needed to maintain good health.

The Chemist's View of Proteins

Chemically, **proteins** contain the same atoms as carbohydrates and lipids—carbon (C), hydrogen (H), and oxygen (O)—but proteins also contain nitrogen (N) atoms. These nitrogen atoms give the name *amino* (nitrogen containing) to the amino acids—the links in the chains of proteins.

Amino Acids

All **amino acids** have the same basic structure—a central carbon (C) atom with a hydrogen atom (H), an amino group (NH_2), and an acid group (COOH) attached to it. However, carbon atoms need to form four bonds, ◆ so a fourth attachment is necessary. This fourth site distinguishes each amino acid from the others. Attached to the carbon atom at the fourth bond is a distinct atom, or group of atoms, known as the *side group* or *side chain* (see Figure 6-1).

Unique Side Groups The side groups on amino acids vary from one amino acid to the next, making proteins more complex than either carbohydrates or lipids. A polysaccharide (starch, for example) may be several thousand units long, but each unit is a glucose molecule just like all the others. A protein, on the other hand, is

◆ Reminder:
  • H forms 1 bond
  • O forms 2 bonds
  • N forms 3 bonds
  • C forms 4 bonds

proteins: compounds composed of carbon, hydrogen, oxygen, and nitrogen atoms, arranged into amino acids linked in a chain. Some amino acids also contain sulfur atoms.

amino (a-MEEN-oh) **acids:** building blocks of proteins. Each contains an amino group, an acid group, a hydrogen atom, and a distinctive side group, all attached to a central carbon atom.
  • **amino** = containing nitrogen

FIGURE 6-1 | Amino Acid Structure

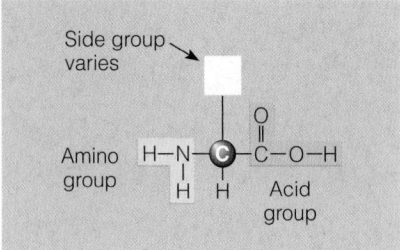

All amino acids have a carbon (known as the alpha-carbon), with an amino group (NH_2), an acid group (COOH), a hydrogen (H), and a side group attached. The side group is a unique chemical structure that differentiates one amino acid from another.

TABLE 6-1 | Amino Acids

Proteins are made up of about 20 common amino acids. The first column lists the essential amino acids for human beings (those the body cannot make—that must be provided in the diet). The second column lists the nonessential amino acids. In special cases, some nonessential amino acids may become conditionally essential (see the text). In a newborn, for example, only five amino acids are truly nonessential; the other nonessential amino acids are conditionally essential until the metabolic pathways are developed enough to make those amino acids in adequate amounts.

Essential Amino Acids		Nonessential Amino Acids	
Histidine	(HISS-tuh-deen)	Alanine	(AL-ah-neen)
Isoleucine	(eye-so-LOO-seen)	Arginine	(ARJ-ih-neen)
Leucine	(LOO-seen)	Asparagine	(ah-SPAR-ah-geen)
Lysine	(LYE-seen)	Aspartic acid	(ah-SPAR-tic acid)
Methionine	(meh-THIGH-oh-neen)	Cysteine	(SIS-teh-een)
Phenylalanine	(fen-il-AL-ah-neen)	Glutamic acid	(GLU-tam-ic acid)
Threonine	(THREE-oh-neen)	Glutamine	(GLU-tah-meen)
Tryptophan	(TRIP-toe-fan,	Glycine	(GLY-seen)
	TRIP-toe-fane)	Proline	(PRO-leen)
Valine	(VAY-leen)	Serine	(SEER-een)
		Tyrosine	(TIE-roe-seen)

made up of about 20 different amino acids, each with a different side group. Table 6-1 lists the amino acids most common in proteins.*

The simplest amino acid, glycine, has a hydrogen atom as its side group. A slightly more complex amino acid, alanine, has an extra carbon with three hydrogen atoms. Other amino acids have more complex side groups (see Figure 6-2 for examples). Thus, although all amino acids share a common structure, they differ in size, shape, electrical charge, and other characteristics because of differences in these side groups.

Nonessential Amino Acids More than half of the amino acids are *nonessential,* meaning that the body can synthesize them for itself. Proteins in foods usually deliver these amino acids, but it is not essential that they do so. The body can make all **nonessential amino acids,** given nitrogen to form the amino group and fragments from carbohydrate or fat to form the rest of the structure.

---

* Besides the 20 common amino acids, which can all be components of proteins, others do not occur in proteins, but can be found individually (for example, taurine and ornithine). Some amino acids occur in related forms (for example, proline can acquire an OH group to become hydroxyproline).

FIGURE 6-2 | Examples of Amino Acids

Note that all amino acids have a common chemical structure but that each has a different side group. Appendix C presents the chemical structures of the 20 amino acids most common in proteins.

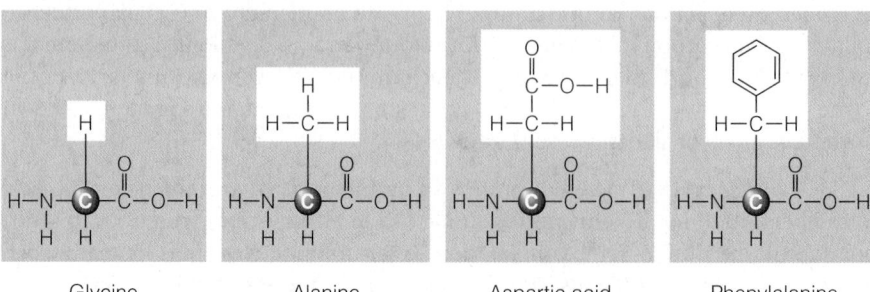

Glycine   Alanine   Aspartic acid   Phenylalanine

nonessential amino acids: amino acids that the body can synthesize (see Table 6-1).

Essential Amino Acids There are nine amino acids that the human body either cannot make at all or cannot make in sufficient quantity to meet its needs. These nine amino acids must be supplied by the diet; they are *essential.* ◆ The first column in Table 6-1 presents the **essential amino acids.**

Conditionally Essential Amino Acids Sometimes a nonessential amino acid becomes essential under special circumstances. For example, the body normally uses the essential amino acid phenylalanine to make tyrosine (a nonessential amino acid). But if the diet fails to supply enough phenylalanine, or if the body cannot make the conversion for some reason (as happens in the inherited disease phenylketonuria), then tyrosine becomes a **conditionally essential amino acid.**

◆ Some researchers refer to essential amino acids as **indispensable** and to nonessential amino acids as **dispensable.**

Proteins

Cells link amino acids end-to-end in a variety of sequences to form thousands of different proteins. A **peptide bond** unites each amino acid to the next.

Amino Acid Chains Condensation reactions connect amino acids, just as they combine monosaccharides to form disaccharides and fatty acids with glycerol to form triglycerides. Two amino acids bonded together form a **dipeptide** (see Figure 6-3). By another such reaction, a third amino acid can be added to the chain to form a **tripeptide.** As additional amino acids join the chain, a **polypeptide** is formed. Most proteins are a few dozen to several hundred amino acids long. Figure 6-4 (p. 184) provides an example—insulin.

Amino Acid Sequences If a person could walk along a carbohydrate molecule like starch, the first stepping stone would be a glucose. The next stepping stone would also be a glucose, and it would be followed by a glucose, and yet another glucose. But if a person were to walk along a polypeptide chain, each stepping stone would be one of 20 different amino acids. The first stepping stone might be the amino acid methionine. The second might be an alanine. The third might be a glycine, and the fourth a tryptophan, and so on. Walking along another polypeptide path, a person might step on a phenylalanine, then a valine, and a glutamine. In other words, amino acid sequences within proteins vary.

The amino acids can act somewhat like the letters in an alphabet. If you had only the letter G, all you could write would be a string of Gs: G–G–G–G–G–G–G. But with 20 different letters available, you can create poems, songs, and novels. Similarly, the 20 amino acids can be linked together in a variety of sequences—even more than are possible for letters in a word or words in a sentence. Thus the variety of possible sequences for polypeptide chains is tremendous.

essential amino acids: amino acids that the body cannot synthesize in amounts sufficient to meet physiological needs (see Table 6-1 on p. 182).

conditionally essential amino acid: an amino acid that is normally nonessential, but must be supplied by the diet in special circumstances when the need for it exceeds the body's ability to produce it.

peptide bond: a bond that connects the acid end of one amino acid with the amino end of another, forming a link in a protein chain.

dipeptide (dye-PEP-tide): two amino acids bonded together.
- **di** = two
- **peptide** = amino acid

tripeptide: three amino acids bonded together.
- **tri** = three

polypeptide: many (ten or more) amino acids bonded together.
- **poly** = many

FIGURE 6-3	Condensation of Two Amino Acids to Form a Dipeptide

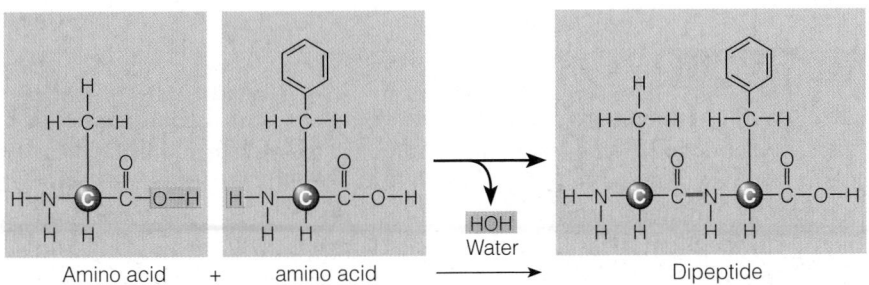

Amino acid + amino acid → Dipeptide

An OH group from the acid end of one amino acid and an H atom from the amino group of another join to form a molecule of water.

A peptide bond (highlighted in red) forms between the two amino acids, creating a dipeptide.

FIGURE 6-4 Amino Acid Sequence of Human Insulin

Human insulin is a relatively small protein that consists of 51 amino acids in two short polypeptide chains. (For amino acid abbreviations, see Appendix C.) Two bridges link the two chains. A third bridge spans a section within the short chain. Known as *disulfide bridges,* these links always involve the amino acid cysteine (Cys), whose side group contains sulfur (S). Cysteines connect to each other when bonds form between these side groups.

FIGURE 6-5 The Structure of Hemoglobin

Four highly folded polypeptide chains form the globular hemoglobin protein.

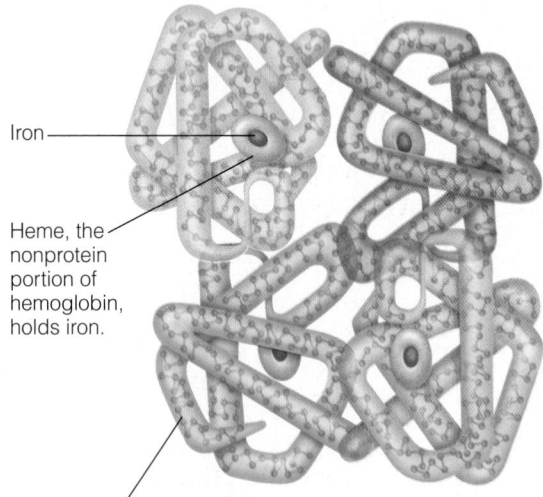

Iron

Heme, the nonprotein portion of hemoglobin, holds iron.

The amino acid sequence determines the shape of the polypeptide chain.

Protein Shapes Polypeptide chains twist into a variety of complex, tangled shapes, depending on their amino acid sequences. The unique side group of each amino acid gives it characteristics that attract it to, or repel it from, the surrounding fluids and other amino acids. Some amino acid side groups carry electrical charges that are attracted to water molecules; they are *hydrophilic.* Other side groups are neutral and are repelled by water; they are *hydrophobic.* As amino acids are strung together to make a polypeptide, the chain folds so that its charged hydrophilic side groups are on the outer surface near water; the neutral hydrophobic groups tuck themselves inside, away from water. The intricate, coiled shape the polypeptide finally assumes gives it maximum stability.

Protein Functions The extraordinary and unique shapes of proteins enable them to perform their various tasks in the body. Some form hollow balls that can carry and store materials within them, and some, such as those of tendons, are more than ten times as long as they are wide, forming strong, rod-like structures. Some polypeptides are functioning proteins just as they are; others need to associate with other polypeptides to form larger working complexes. Some proteins require minerals to activate them. One molecule of **hemoglobin**—the large, globular protein molecule that, by the billions, packs the red blood cells and carries oxygen—is made of four associated polypeptide chains, each holding the mineral iron (see Figure 6-5).

Protein Denaturation When proteins are subjected to heat, acid, or other conditions that disturb their stability, they undergo **denaturation**—that is, they uncoil and lose their shapes and, consequently, also lose their ability to function. Past a certain point, denaturation is irreversible. Familiar examples

hemoglobin (HE-moh-GLO-bin): the globular protein of the red blood cells that carries oxygen from the lungs to the cells throughout the body.
- **hemo** = blood
- **globin** = globular protein

denaturation (dee-NAY-chur-AY-shun): the change in a protein's shape and consequent loss of its function brought about by heat, agitation, acid, base, alcohol, heavy metals, or other agents.

IN SUMMARY

Chemically speaking, proteins are more complex than carbohydrates or lipids, being made of some 20 different amino acids, 9 of which the body cannot make; they are essential. Each amino acid contains an amino group, an acid group, a hydrogen atom, and a distinctive side group, all attached to a central carbon atom. Cells link amino acids together in a series of condensation reactions to create proteins. The distinctive sequence of amino acids in each protein determines its unique shape and function.

of denaturation include the hardening of an egg when it is cooked, the curdling of milk when acid is added, and the stiffening of egg whites when they are whipped.

Digestion and Absorption of Protein

Proteins in foods do not become body proteins directly. Instead, they supply the amino acids from which the body makes its own proteins. When a person eats foods containing protein, enzymes break the long polypeptide strands into shorter strands, the short strands into tripeptides and dipeptides, and, finally, the tripeptides and dipeptides into amino acids.

Protein Digestion

Figure 6-6 (p. 186) illustrates the digestion of protein through the GI tract. Proteins are crushed and moistened in the mouth, but the real action begins in the stomach.

In the Stomach The major event in the stomach is the partial breakdown (hydrolysis) of proteins. Hydrochloric acid uncoils (denatures) each protein's tangled strands so that digestive enzymes can attack the peptide bonds. The hydrochloric acid also converts the inactive form ◆ of the enzyme pepsinogen to its active form, **pepsin.** Pepsin cleaves proteins—large polypeptides—into smaller polypeptides and some amino acids.

◆ The inactive form of an enzyme is called a **proenzyme** or a **zymogen** (ZYE-moh-jen).

In the Small Intestine When polypeptides enter the small intestine, several pancreatic and intestinal **proteases** hydrolyze them further into short peptide chains, ◆ tripeptides, dipeptides, and amino acids. Then **peptidase** enzymes on the membrane surfaces of the intestinal cells split most of the dipeptides and tripeptides into single amino acids. Only a few peptides escape digestion and enter the blood intact. Figure 6-6 includes names of the digestive enzymes for protein and describes their actions.

◆ A string of four to nine amino acids is an **oligopeptide** (OL-ee-go-PEP-tide).
 • **oligo** = few

Protein Absorption

A number of specific carriers transport amino acids (and some dipeptides and tripeptides) into the intestinal cells. Once inside the intestinal cells, amino acids may be used for energy or to synthesize needed compounds. Amino acids that are not used by the intestinal cells are transported across the cell membrane into the surrounding fluid where they enter the capillaries on their way to the liver.

Consumers lacking nutrition knowledge may fail to realize that most proteins are broken down to amino acids before absorption. They may be mislead by advertisements urging them to "Eat enzyme A. It will help you digest your food." Or "Don't eat food B. It contains enzyme C, which will digest cells in your body." In reality, though, enzymes in foods are digested, just as all proteins are. Even the digestive enzymes—which function optimally at their specific pH—are denatured and digested when the pH of their environment changes. (For example, the enzyme pepsin, which works best in the low pH of the stomach becomes inactive and digested when it enters the higher pH of the small intestine.)

Another misconception is that eating predigested proteins (amino acid supplements) saves the body from having to digest proteins and keeps the digestive system from "overworking." Such a belief grossly underestimates the body's abilities. As a matter of fact, the digestive system handles whole proteins *better* than predigested ones because it dismantles and absorbs the amino acids at rates that are optimal for the body's use. (The last section of this chapter discusses amino acid supplements further.)

pepsin: a gastric enzyme that hydrolyzes protein. Pepsin is secreted in an inactive form, **pepsinogen,** which is activated by hydrochloric acid in the stomach.

proteases (PRO-tee-aces): enzymes that hydrolyze protein.

peptidase: a digestive enzyme that hydrolyzes peptide bonds. *Tripeptidases* cleave tripeptides; *dipeptidases* cleave dipeptides. *Endopeptidases* cleave peptide bonds within the chain to create smaller fragments, whereas *exopeptidases* cleave bonds at the ends to release free amino acids.
 • **tri** = three
 • **di** = two
 • **endo** = within
 • **exo** = outside

FIGURE 6-6 *Animated!* Protein Digestion in the GI Tract

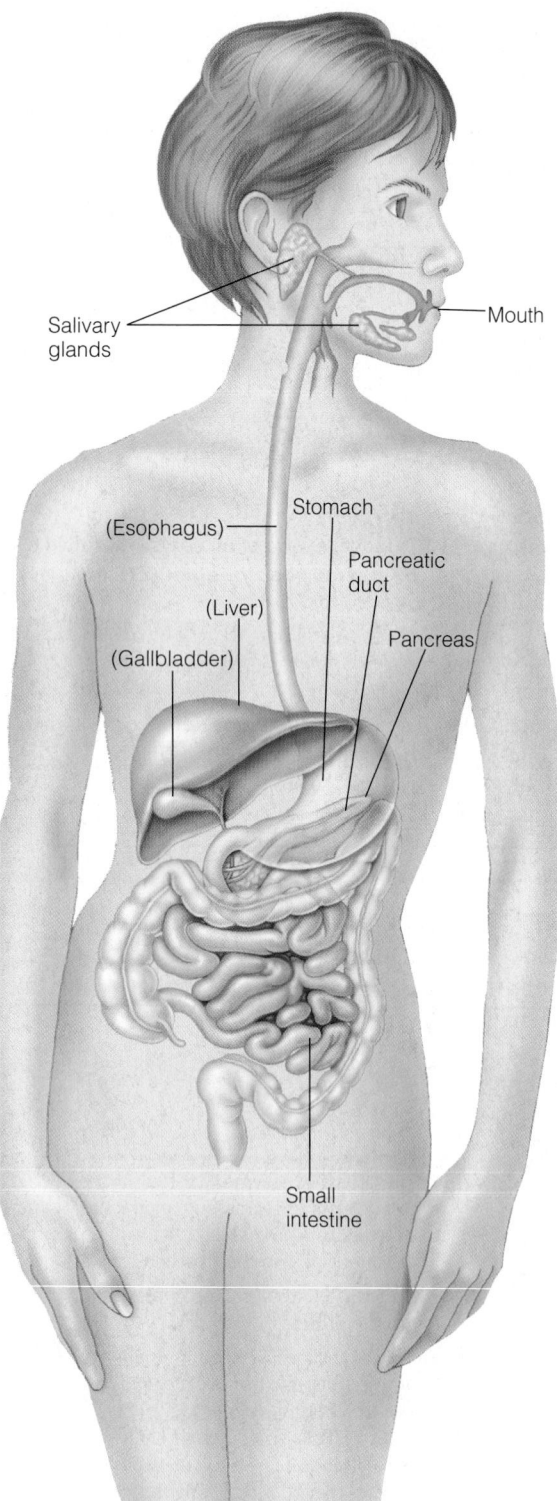

PROTEIN

Mouth and salivary glands
Chewing and crushing moisten protein-rich foods and mix them with saliva to be swallowed

Stomach
Hydrochloric acid (HCl) uncoils protein strands and activates stomach enzymes:

Protein $\xrightarrow{\text{pepsin, HCl}}$ smaller polypeptides

Small intestine and pancreas
Pancreatic and small intestinal enzymes split polypeptides further:

Poly-peptides $\xrightarrow{\text{pancreatic and intestinal proteases}}$ tripeptides, dipeptides, amino acids

Then enzymes on the surface of the small intestinal cells hydrolyze these peptides and the cells absorb them:

Peptides $\xrightarrow{\text{intestinal tripeptidases and dipeptidases}}$ amino acids (absorbed)

HYDROCHLORIC ACID AND THE DIGESTIVE ENZYMES

In the stomach:

Hydrochloric acid (HCl)
- Denatures protein structure
- Activates pepsinogen to pepsin

Pepsin
- Cleaves proteins to smaller polypeptides and some free amino acids
- Inhibits pepsinogen synthesis

In the small intestine:

Enteropeptidase[a]
- Converts pancreatic trypsinogen to trypsin

Trypsin
- Inhibits trypsinogen synthesis
- Cleaves peptide bonds next to the amino acids lysine and arginine
- Converts pancreatic procarboxypeptidases to carboxypeptidases
- Converts pancreatic chymotrypsinogen to chymotrypsin

Chymotrypsin
- Cleaves peptide bonds next to the amino acids phenylalanine, tyrosine, tryptophan, methionine, asparagine, and histidine

Carboxypeptidases
- Cleave amino acids from the acid (carboxyl) ends of polypeptides

Elastase and collagenase
- Cleave polypeptides into smaller polypeptides and tripeptides

Intestinal tripeptidases
- Cleave tripeptides to dipeptides and amino acids

Intestinal dipeptidases
- Cleave dipeptides to amino acids

Intestinal aminopeptidases
- Cleave amino acids from the amino ends of small polypeptides (oligopeptides)

[a]Enteropeptidase was formerly known as *enterokinase*.

IN SUMMARY

Digestion is facilitated mostly by the stomach's acid and enzymes, which first denature dietary proteins, then cleave them into smaller polypeptides and some amino acids. Pancreatic and intestinal enzymes split these polypeptides further, to oligo-, tri-, and dipeptides, and then split most of these to single amino acids. Then carriers in the membranes of intestinal cells transport the amino acids into the cells, where they are released into the bloodstream.

Proteins in the Body

The human body contains an estimated 30,000 different kinds of proteins. Of these, about 3000 have been studied, ◆ although with the recent surge in knowledge gained from sequencing the human genome, ◆ this number is growing rapidly. Only about 10 are described in this chapter—but these should be enough to illustrate the versatility, uniqueness, and importance of proteins. As you will see, each protein has a specific function, and that function is determined during protein synthesis.

◆ The study of the body's proteins is called **proteomics.**

◆ Reminder: The *human genome* is the full set of chromosomes, including all of the genes and associated DNA.

Protein Synthesis

Each human being is unique because of small differences in the body's proteins. These differences are determined by the amino acid sequences of proteins, which, in turn, are determined by genes. The following paragraphs describe in words the ways cells synthesize proteins; Figure 6-7 (p. 188) provides a pictorial description.

The instructions for making every protein in a person's body are transmitted by way of the genetic information received at conception. This body of knowledge, which is filed in the DNA (deoxyribonucleic acid) within the nucleus of every cell, never leaves the nucleus.

Delivering the Instructions Transforming the information in DNA into the appropriate sequence of amino acids needed to make a specific protein requires two major steps. In the first step, ◆ a stretch of DNA is used as a template to make a strand of RNA (ribonucleic acid) known as messenger RNA. Messenger RNA then carries the code across the nuclear membrane into the body of the cell. There it seeks out and attaches itself to one of the ribosomes (a protein-making machine, which is itself composed of RNA and protein), where the second step ◆ takes place. Situated on a ribosome, messenger RNA specifies the sequence in which the amino acids line up for the synthesis of a protein.

◆ This process of messenger RNA being made from a template of DNA is known as **transcription.**

◆ This process of messenger RNA directing the sequence of amino acids and synthesis of proteins is known as **translation.**

Lining Up the Amino Acids Other forms of RNA, called transfer RNA, collect amino acids from the cell fluid and bring them to the messenger. Each of the 20 amino acids has a specific transfer RNA. Thousands of transfer RNAs, each carrying its amino acid, cluster around the ribosomes, awaiting their turn to unload. When the messenger's list calls for a specific amino acid, the transfer RNA carrying that amino acid moves into position. Then the next loaded transfer RNA moves into place and then the next and the next. In this way, the amino acids line up in the sequence that is called for, and enzymes bind them together. Finally, the completed protein strand is released, and the transfer RNAs are freed to return for other loads of amino acids.

Sequencing Errors The sequence of amino acids in each protein determines its shape, which supports a specific function. If a genetic error alters the amino acid sequence of a protein, or if a mistake is made in copying the sequence, an altered protein will result, sometimes with dramatic consequences. The protein hemoglobin

FIGURE 6-7 *Animated!* Protein Synthesis

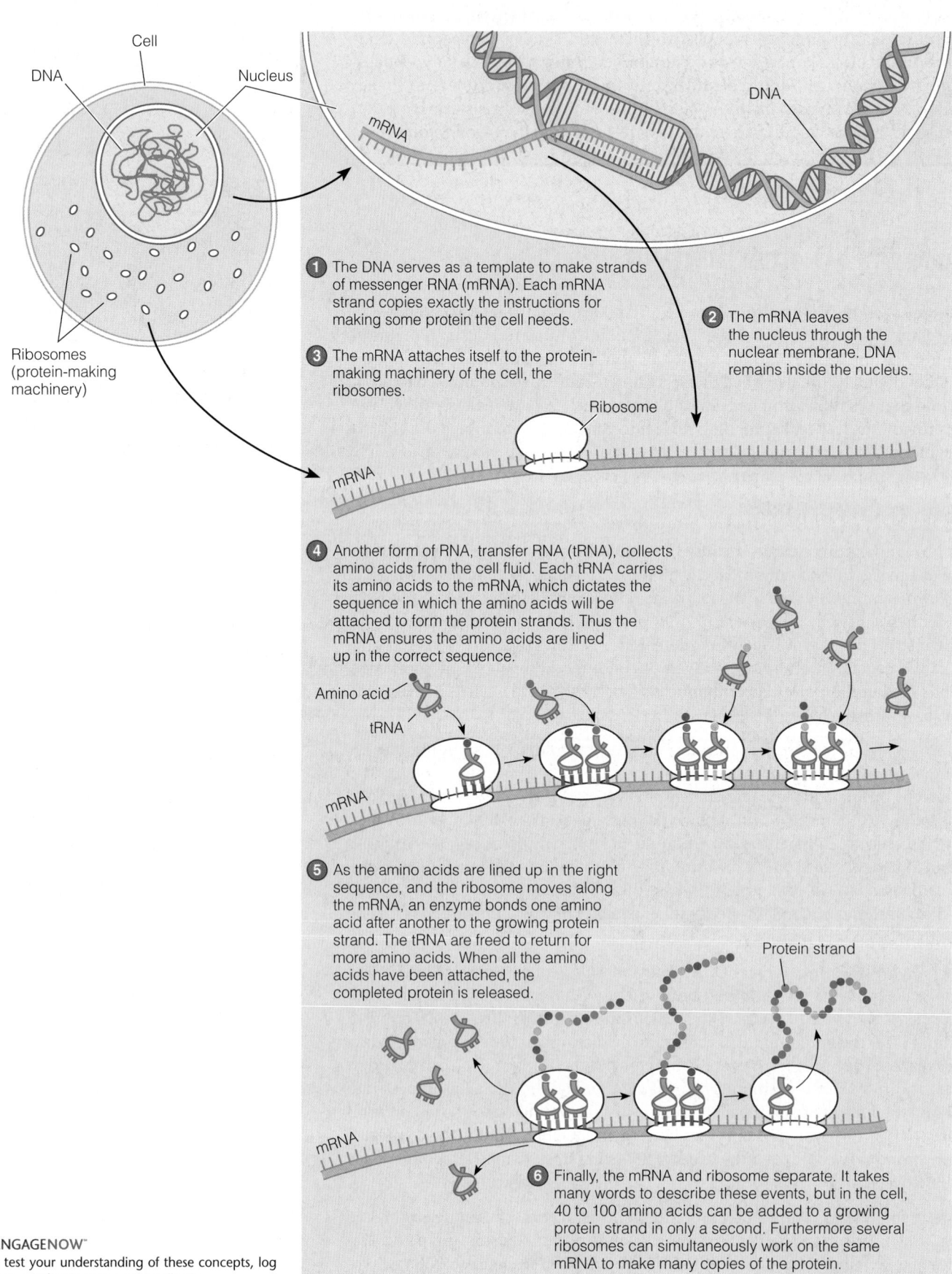

Cell

DNA

Nucleus

DNA

mRNA

Ribosomes (protein-making machinery)

1 The DNA serves as a template to make strands of messenger RNA (mRNA). Each mRNA strand copies exactly the instructions for making some protein the cell needs.

2 The mRNA leaves the nucleus through the nuclear membrane. DNA remains inside the nucleus.

3 The mRNA attaches itself to the protein-making machinery of the cell, the ribosomes.

Ribosome

mRNA

4 Another form of RNA, transfer RNA (tRNA), collects amino acids from the cell fluid. Each tRNA carries its amino acids to the mRNA, which dictates the sequence in which the amino acids will be attached to form the protein strands. Thus the mRNA ensures the amino acids are lined up in the correct sequence.

Amino acid

tRNA

mRNA

5 As the amino acids are lined up in the right sequence, and the ribosome moves along the mRNA, an enzyme bonds one amino acid after another to the growing protein strand. The tRNA are freed to return for more amino acids. When all the amino acids have been attached, the completed protein is released.

Protein strand

mRNA

6 Finally, the mRNA and ribosome separate. It takes many words to describe these events, but in the cell, 40 to 100 amino acids can be added to a growing protein strand in only a second. Furthermore several ribosomes can simultaneously work on the same mRNA to make many copies of the protein.

offers one example of such a genetic variation. In a person with **sickle-cell anemia,** ◆ two of hemoglobin's four polypeptide chains (described earlier on p. 184) have the normal sequence of amino acids, but the other two chains do not—they have the amino acid valine in a position that is normally occupied by glutamic acid (see Figure 6-8). This single alteration in the amino acid sequence changes the characteristics and shape of hemoglobin so much that it loses its ability to carry oxygen effectively. The red blood cells filled with this abnormal hemoglobin stiffen into elongated sickle, or crescent, shapes instead of maintaining their normal pliable disc shape—hence the name, sickle-cell anemia. Sickle-cell anemia raises energy needs, causes many medical problems, and can be fatal.[1] Caring for children with sickle-cell anemia includes diligent attention to their water needs; dehydration can trigger a crisis.

Nutrients and Gene Expression When a cell makes a protein as described earlier, scientists say that the gene for that protein has been "expressed." Cells can regulate **gene expression** to make the type of protein, in the amounts and at the rate, they need. Nearly all of the body's cells possess the genes for making all human proteins, but each type of cell makes only the proteins it needs. For example, cells of the pancreas express the gene for insulin; in other cells, that gene is idle. Similarly, the cells of the pancreas do not make the protein hemoglobin, which is needed only by the red blood cells.

Recent research has unveiled some of the fascinating ways nutrients regulate gene expression and protein synthesis (see Highlight 6). ◆ Because diet plays an ongoing role in our lives from conception to death, it has a major influence on gene expression and disease development.[2] The benefits of polyunsaturated fatty acids in defending against heart disease, for example, are partially explained by their role in influencing gene expression for lipid enzymes. Later chapters provide additional examples of relationships among nutrients, genes, and disease development.

| FIGURE 6-8 | Sickle Cell Compared with Normal Red Blood Cell |

Normally, red blood cells are disc-shaped, but in the inherited disorder sickle-cell anemia, red blood cells are sickle- or crescent-shaped. This alteration in shape occurs because valine replaces glutamic acid in the amino acid sequence of two of hemoglobin's polypeptide chains. As a result of this one alteration, the hemoglobin has a diminished capacity to carry oxygen.

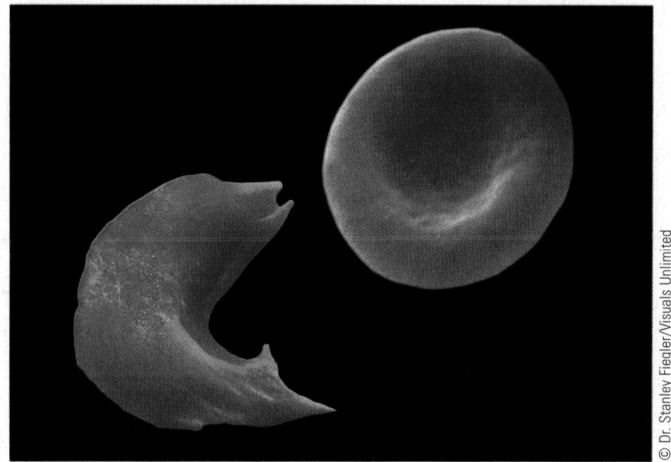

Sickle-shaped blood cell   Normal red blood cell

Amino acid sequence of normal hemoglobin:

Val—His—Leu—Thr—Pro—Glu—Glu

Amino acid sequence of sickle-cell hemoglobin:

Val—His—Leu—Thr—Pro—Val—Glu

© Dr. Stanley Flegler/Visuals Unlimited

◆ Anemia is not a disease, but a symptom of various diseases. In the case of sickle-cell anemia, a defect in the hemoglobin molecule changes the shape of the red blood cells. Later chapters describe the anemias of vitamin and mineral deficiencies. In all cases, the abnormal blood cells are unable to meet the body's oxygen demands.

◆ Nutrients can play key roles in activating or silencing genes. Switching genes on and off, without changing the genetic sequence itself, is known as **epigenetics.**
  • **epi** = among

sickle-cell anemia: a hereditary form of anemia characterized by abnormal sickle- or crescent-shaped red blood cells. Sickled cells interfere with oxygen transport and blood flow. Symptoms are precipitated by dehydration and insufficient oxygen (as may occur at high altitudes) and include hemolytic anemia (red blood cells burst), fever, and severe pain in the joints and abdomen.

gene expression: the process by which a cell converts the genetic code into RNA and protein.

IN SUMMARY

Cells synthesize proteins according to the genetic information provided by the DNA in the nucleus of each cell. This information dictates the order in which amino acids must be linked together to form a given protein. Sequencing errors occasionally occur, sometimes with significant consequences.

Roles of Proteins

Whenever the body is growing, repairing, or replacing tissue, proteins are involved. Sometimes their role is to facilitate or to regulate; other times it is to become part of a structure. Versatility is a key feature of proteins.

FIGURE 6-9 Enzyme Action

Each enzyme facilitates a specific chemical reaction. In this diagram, an enzyme enables two compounds to make a more complex structure, but the enzyme itself remains unchanged.

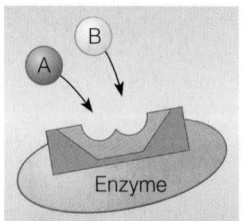
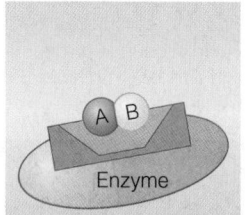
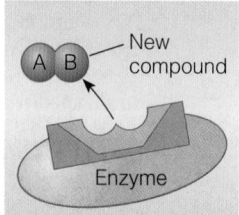

The separate compounds, A and B, are attracted to the enzyme's active site, making a reaction likely.

The enzyme forms a complex with A and B.

The enzyme is unchanged, but A and B have formed a new compound, AB.

◆ Breaking down reactions are **catabolic,** whereas building up reactions are **anabolic.** (Chapter 7 provides more details.)

◆ Recall from Chapter 5 that some hormones, such as estrogen and testosterone, derive from cholesterol.

matrix (MAY-tricks): the basic substance that gives form to a developing structure; in the body, the formative cells from which teeth and bones grow.

collagen (KOL-ah-jen): the protein from which connective tissues such as scars, tendons, ligaments, and the foundations of bones and teeth are made.

enzymes: proteins that facilitate chemical reactions without being changed in the process; protein catalysts.

fluid balance: maintenance of the proper types and amounts of fluid in each compartment of the body fluids (see also Chapter 12).

As Building Materials for Growth and Maintenance From the moment of conception, proteins form the building blocks of muscles, blood, and skin—in fact, of most body structures. For example, to build a bone or a tooth, cells first lay down a **matrix** of the protein **collagen** and then fill it with crystals of calcium, phosphorus, magnesium, fluoride, and other minerals.

Collagen also provides the material of ligaments and tendons and the strengthening glue between the cells of the artery walls that enables the arteries to withstand the pressure of the blood surging through them with each heartbeat. Also made of collagen are scars that knit the separated parts of torn tissues together.

Proteins are also needed for replacing dead or damaged cells. The life span of a skin cell is only about 30 days. As old skin cells are shed, new cells made largely of protein grow from underneath to replace them. Cells in the deeper skin layers synthesize new proteins to go into hair and fingernails. Muscle cells make new proteins to grow larger and stronger in response to exercise. Cells of the GI tract are replaced every few days. Both inside and outside, then, the body continuously deposits protein into the new cells that replace those that have been lost.

As Enzymes Some proteins act as **enzymes.** Digestive enzymes have appeared in every chapter since Chapter 3, but digestion is only one of the many processes facilitated by enzymes. Enzymes not only break down substances, but they also build substances (such as bone) ◆ and transform one substance into another (amino acids into glucose, for example). Figure 6-9 diagrams a synthesis reaction.

An analogy may help to clarify the role of enzymes. Enzymes are comparable to the clergy and judges who make and dissolve marriages. When a minister marries two people, they become a couple, with a new bond between them. They are joined together—but the minister remains unchanged. The minister represents enzymes that synthesize large compounds from smaller ones. One minister can perform thousands of marriage ceremonies, just as one enzyme can perform billions of synthetic reactions.

Similarly, a judge who lets married couples separate may decree many divorces before retiring. The judge represents enzymes that hydrolyze larger compounds to smaller ones; for example, the digestive enzymes. The point is that, like the minister and the judge, enzymes themselves are not altered by the reactions they facilitate. They are catalysts, permitting reactions to occur more quickly and efficiently than if substances depended on chance encounters alone.

As Hormones The body's many hormones are messenger molecules, and *some* hormones are proteins. ◆ Various endocrine glands in the body release hormones in response to changes that challenge the body. The blood carries the hormones from these glands to their target tissues, where they elicit the appropriate responses to restore and maintain normal conditions.

The hormone insulin provides a familiar example. When blood glucose rises, the pancreas releases its insulin. Insulin stimulates the transport proteins of the muscles and adipose tissue to pump glucose into the cells faster than it can leak out. (After acting on the message, the cells destroy the insulin.) Then, as blood glucose falls, the pancreas slows its release of insulin. Many other proteins act as hormones, regulating a variety of actions in the body (see Table 6-2 for examples).

As Regulators of Fluid Balance Proteins help to maintain the body's **fluid balance.** Figure 12-1 in Chapter 12 illustrates a cell and its associated fluids. As the figure explains, the body's fluids are contained inside the cells (intracellular) or outside the cells (extracellular). Extracellular fluids, in turn, can be found either in the spaces between the cells (interstitial) or within the blood vessels (intravascular). The fluid within the intravascular spaces is called plasma (essentially blood without its red blood cells). Fluids can flow freely between these compartments, but being large, proteins cannot. Proteins are trapped primarily within the cells and to a lesser extent in the plasma.

The exchange of materials between the blood and the cells takes place across the capillary walls, which allow the passage of fluids and a variety of materials—but usually not plasma proteins. Still some plasma proteins leak out of the capillaries into the interstitial fluid between the cells. These proteins cannot be reabsorbed back into the plasma; they normally reenter circulation via the lymph system. If plasma proteins enter the interstitial spaces faster than they can be cleared, fluid accumulates (because plasma proteins attract water) and causes swelling. Swelling due to an excess of interstitial fluid is known as **edema.** The protein-related causes of edema include:

- Excessive protein losses caused by kidney disease or large wounds (such as extensive burns)
- Inadequate protein synthesis caused by liver disease
- Inadequate dietary intake of protein

Whatever the cause of edema, the result is the same: a diminished capacity to deliver nutrients and oxygen to the cells and to remove wastes from them. As a consequence, cells fail to function adequately.

As Acid-Base Regulators Proteins also help to maintain the balance between **acids** and **bases** within the body fluids. Normal body processes continually produce acids and bases, which the blood carries to the kidneys and lungs for excretion. The challenge is to do this without upsetting the blood's acid-base balance.

In an acid solution, hydrogen ions (H^+) abound; the more hydrogen ions, the more concentrated the acid. Proteins, which have negative charges on their surfaces, attract hydrogen ions, which have positive charges. By accepting and releasing hydrogen ions, ◆ proteins maintain the acid-base balance of the blood and body fluids.

The blood's acid-base balance is tightly controlled. The extremes of **acidosis** and **alkalosis** lead to coma and death, largely because they denature working proteins. Disturbing a protein's shape renders it useless. To give just one example, denatured hemoglobin loses its capacity to carry oxygen.

As Transporters Some proteins move about in the body fluids, carrying nutrients and other molecules. The protein hemoglobin carries oxygen from the lungs to the cells. The lipoproteins transport lipids around the body. Special transport proteins carry vitamins and minerals.

The transport of the mineral iron provides an especially good illustration of these proteins' specificity and precision. When iron enters an intestinal cell after a meal has been digested and absorbed, it is captured by a protein. Before leaving the intestinal cell, iron is attached to another protein that carries it though the bloodstream to the cells. Once iron enters a cell, it is attached to a storage protein that will hold the iron until it is needed. When it is needed, iron is incorporated into proteins in the red blood cells and muscles that assist in oxygen transport and use. (Chapter 13 provides more details on how these protein carriers transport and store iron.)

Some transport proteins reside in cell membranes and act as "pumps," picking up compounds on one side of the membrane and releasing them on the other as needed. Each transport protein is specific for a certain compound or group of related compounds. Figure 6-10 (p. 192) illustrates how a membrane-bound transport protein helps to maintain the sodium and potassium concentrations in the fluids inside and outside cells. The balance of these two minerals is critical to nerve transmissions and muscle contractions; imbalances can cause irregular heartbeats, muscular weakness, kidney failure, and even death.

TABLE 6-2	Examples of Hormones and Their Actions
Hormones	**Actions**
Growth hormone	Promotes growth
Insulin and glucagon	Regulate blood glucose (see Chapter 4)
Thyroxin	Regulates the body's metabolic rate (see Chapter 8)
Calcitonin and parathyroid hormone	Regulate blood calcium (see Chapter 12)
Antidiuretic hormone	Regulates fluid and electrolyte balance (see Chapter 12)

NOTE: *Hormones* are chemical messengers that are secreted by endocrine glands in response to altered conditions in the body. Each travels to one or more specific target tissues or organs, where it elicits a specific response. For descriptions of many hormones important in nutrition, see Appendix A.

◆ Compounds that help keep a solution's acidity or alkalinity constant are called **buffers.**

edema (eh-DEEM-uh): the swelling of body tissue caused by excessive amounts of fluid in the interstitial spaces; seen in protein deficiency (among other conditions).

acids: compounds that release hydrogen ions in a solution.

bases: compounds that accept hydrogen ions in a solution.

acidosis (assi-DOE-sis): above-normal acidity in the blood and body fluids.

alkalosis (alka-LOE-sis): above-normal alkalinity (base) in the blood and body fluids.

FIGURE 6-10 *Animated!* An Example of a Transport Protein

This transport protein resides within a cell membrane and acts as a two-door passageway. Molecules enter on one side of the membrane and exit on the other, but the protein doesn't leave the membrane. This example shows how the transport protein moves sodium and potassium in opposite directions across the membrane to maintain a high concentration of potassium and a low concentration of sodium within the cell. This active transport system requires energy.

CENGAGENOW
To test your understanding of these concepts, log on to
academic.cengage.com/login

Key:
● Sodium
● Potassium

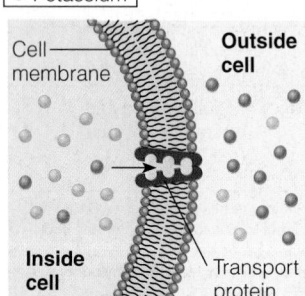
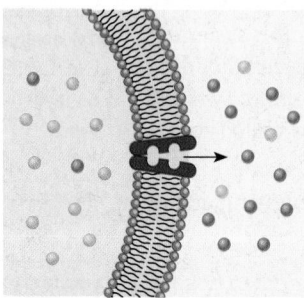
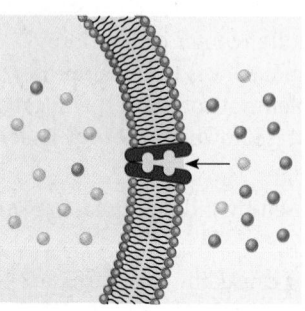
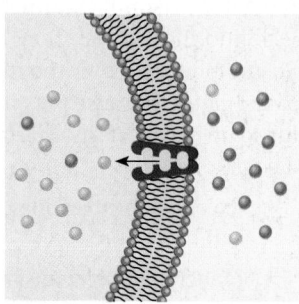

The transport protein picks up sodium from inside the cell.

The protein changes shape and releases sodium outside the cell.

The transport protein picks up potassium from outside the cell.

The protein changes shape and releases potassium inside the cell.

◆ Reminder: Protein provides 4 kcal/g. Return to p. 9 for a refresher on how to calculate the protein kcalories from foods.

◆ Reminder: The making of glucose from non-carbohydrate sources such as amino acids is *gluconeogenesis.*

antigens: substances that elicit the formation of antibodies or an inflammation reaction from the immune system. A bacterium, a virus, a toxin, and a protein in food that causes allergy are all examples of antigens.

antibodies: large proteins of the blood and body fluids, produced by the immune system in response to the invasion of the body by foreign molecules (usually proteins called *antigens*). Antibodies combine with and inactivate the foreign invaders, thus protecting the body.

immunity: the body's ability to defend itself against diseases (see also Highlight 17).

As Antibodies Proteins also defend the body against disease. A virus—whether it is one that causes flu, smallpox, measles, or the common cold—enters the cells and multiplies there. One virus may produce 100 replicas of itself within an hour or so. Each replica can then burst out and invade 100 different cells, soon yielding 10,000 virus particles, which invade 10,000 cells. Left free to do their worst, they will soon overwhelm the body with disease.

Fortunately, when the body detects these invading **antigens,** it manufactures **antibodies,** giant protein molecules designed specifically to combat them. The antibodies work so swiftly and efficiently that in a normal, healthy individual, most diseases never have a chance to get started. Without sufficient protein, though, the body cannot maintain its army of antibodies to resist infectious diseases.

Each antibody is designed to destroy a specific antigen. Once the body has manufactured antibodies against a particular antigen (such as the measles virus), it "remembers" how to make them. Consequently, the next time the body encounters that same antigen, it produces antibodies even more quickly. In other words, the body develops a molecular memory, known as **immunity.** (Chapter 15 describes food allergies—the immune system's response to food antigens.)

As a Source of Energy and Glucose Without energy, cells die; without glucose, the brain and nervous system falter. Even though proteins are needed to do the work that only they can perform, they will be sacrificed to provide energy ◆ and glucose ◆ during times of starvation or insufficient carbohydrate intake. The body will break down its tissue proteins to make amino acids available for energy or glucose production. In this way, protein can maintain blood glucose levels, but at the expense of losing lean body tissue. Chapter 7 provides many more details on energy metabolism.

Other Roles As mentioned earlier, proteins form integral parts of most body structures such as skin, muscles, and bones. They also participate in some of the body's most amazing activities such as blood clotting and vision. When a tissue is injured, a rapid chain of events leads to the production of fibrin, a stringy, insoluble mass of protein fibers that forms a solid clot from liquid blood. Later, more slowly, the protein collagen forms a scar to replace the clot and permanently heal the wound. The light-sensitive pigments in the cells of the eye's retina are molecules of the protein opsin. Opsin responds to light by changing its shape, thus initiating the nerve impulses that convey the sense of sight to the brain.

IN SUMMARY

The protein functions discussed here are summarized in the accompanying table. They are only a few of the many roles proteins play, but they convey some sense of the immense variety of proteins and their importance in the body.

Growth and maintenance	Proteins form integral parts of most body structures such as skin, tendons, membranes, muscles, organs, and bones. As such, they support the growth and repair of body tissues.
Enzymes	Proteins facilitate chemical reactions.
Hormones	Proteins regulate body processes. (Some, but not all, hormones are proteins.)
Fluid balance	Proteins help to maintain the volume and composition of body fluids.
Acid-base balance	Proteins help maintain the acid-base balance of body fluids by acting as buffers.
Transportation	Proteins transport substances, such as lipids, vitamins, minerals, and oxygen, around the body.
Antibodies	Proteins inactivate foreign invaders, thus protecting the body against diseases.
Energy and glucose	Proteins provide some fuel, and glucose if needed, for the body's energy needs.

© Ariel Skelley/Corbis

Growing children end each day with more bone, blood, muscle, and skin cells than they had at the beginning of the day.

A Preview of Protein Metabolism

This section previews protein metabolism; Chapter 7 provides a full description. Cells have several metabolic options, depending on their protein and energy needs.

Protein Turnover and the Amino Acid Pool Within each cell, proteins are continually being made and broken down, a process known as **protein turnover.** When proteins break down, they free amino acids. ◆ These amino acids mix with amino acids from dietary protein to form an **"amino acid pool"** within the cells and circulating blood. The rate of protein degradation and the amount of protein intake may vary, but the pattern of amino acids within the pool remains fairly constant. Regardless of their source, any of these amino acids can be used to make body proteins or other nitrogen-containing compounds, or they can be stripped of their nitrogen and used for energy (either immediately or stored as fat for later use).

Nitrogen Balance Protein turnover and **nitrogen balance** go hand in hand. In healthy adults, protein synthesis balances with degradation, and protein intake from food balances with nitrogen excretion in the urine, feces, and sweat. When nitrogen intake equals nitrogen output, the person is in nitrogen equilibrium, ◆ or zero nitrogen balance. Researchers use nitrogen balance studies to estimate protein requirements.[3]

If the body synthesizes more than it degrades and adds protein, nitrogen status becomes positive. Nitrogen status is positive in growing infants, children, adolescents, pregnant women, and people recovering from protein deficiency or illness; their nitrogen intake exceeds their nitrogen output. They are retaining protein in new tissues as they add blood, bone, skin, and muscle cells to their bodies.

If the body degrades more than it synthesizes and loses protein, nitrogen status becomes negative. Nitrogen status is negative in people who are starving or suffering other severe stresses such as burns, injuries, infections, and fever; their nitrogen

◆ Amino acids (or proteins) that derive from within the body are **endogenous** (en-DODGE-eh-nus). In contrast, those that derive from foods are **exogenous** (eks-ODGE-eh-nus).
- **endo** = within
- **gen** = arising
- **exo** = outside (the body)

◆ Nitrogen balance:
- Nitrogen equilibrium (zero nitrogen balance): N in = N out.
- Positive nitrogen: N in > N out.
- Negative nitrogen: N in < N out.

protein turnover: the degradation and synthesis of protein.

amino acid pool: the supply of amino acids derived from either food proteins or body proteins that collect in the cells and circulating blood and stand ready to be incorporated in proteins and other compounds or used for energy.

nitrogen balance: the amount of nitrogen consumed (N in) as compared with the amount of nitrogen excreted (N out) in a given period of time.*

---

* The genetic materials DNA and RNA contain nitrogen, but the quantity is insignificant compared with the amount in protein. Protein is 16 percent nitrogen. Said another way, the average protein weighs about 6.25 times as much as the nitrogen it contains, so scientists can estimate the amount of protein in a sample of food, body tissue, or other material by multiplying the weight of the nitrogen in it by 6.25.

output exceeds their nitrogen intake. During these times, the body loses nitrogen as it breaks down muscle and other body proteins for energy.

Using Amino Acids to Make Proteins or Nonessential Amino Acids As mentioned, cells can assemble amino acids into the proteins they need to do their work. If a particular nonessential amino acid is not readily available, cells can make it from another amino acid. If an essential amino acid is missing, the body may break down some of its own proteins to obtain it.

Using Amino Acids to Make Other Compounds Cells can also use amino acids to make other compounds. For example, the amino acid tyrosine is used to make the **neurotransmitters** norepinephrine and epinephrine, which relay nervous system messages throughout the body. Tyrosine can also be made into the pigment melanin, which is responsible for brown hair, eye, and skin color, or into the hormone thyroxin, which helps to regulate the metabolic rate. For another example, the amino acid tryptophan serves as a precursor for the vitamin niacin and for serotonin, a neurotransmitter important in sleep regulation, appetite control, and sensory perception.

Using Amino Acids for Energy and Glucose As mentioned earlier, when glucose or fatty acids are limited, cells are forced to use amino acids for energy and glucose. The body does not make a specialized storage form of protein as it does for carbohydrate and fat. Glucose is stored as glycogen in the liver and fat as triglycerides in adipose tissue, but protein in the body is available only from the working and structural components of the tissues. When the need arises, the body breaks down its tissue proteins and uses their amino acids for energy or glucose. Thus, over time, energy deprivation (starvation) always causes wasting of lean body tissue as well as fat loss. An adequate supply of carbohydrates and fats spares amino acids from being used for energy and allows them to perform their unique roles.

Deaminating Amino Acids When amino acids are broken down (as occurs when they are used for energy), they are first deaminated—stripped of their nitrogen-containing amino groups. **Deamination** produces ammonia, which the cells release into the bloodstream. The liver picks up the ammonia, converts it into urea (a less toxic compound), and returns the urea to the blood. The production of urea increases as dietary protein increases, until production hits its maximum rate at intakes approaching 250 grams per day. (Urea metabolism is described in Chapter 7.) The kidneys filter urea out of the blood; thus the amino nitrogen ends up in the urine. The remaining carbon fragments of the deaminated amino acids may enter a number of metabolic pathways—for example, they may be used for energy or for the production of glucose, ketones, cholesterol, or fat.*

Using Amino Acids to Make Fat Amino acids may be used to make fat when energy and protein intakes exceed needs and carbohydrate intake is adequate. The amino acids are deaminated, the nitrogen is excreted, and the remaining carbon fragments are converted to fat and stored for later use. In this way, protein-rich foods can contribute to weight gain.

IN SUMMARY

Proteins are constantly being synthesized and broken down as needed. The body's assimilation of amino acids into proteins and its release of amino acids via protein degradation and excretion can be tracked by measuring nitrogen balance, which should be positive during growth and steady in adulthood. An energy deficit or an inadequate protein intake may force the body to use amino acids as fuel, creating a negative nitrogen balance. Protein eaten in excess of need is degraded and stored as body fat.

neurotransmitters: chemicals that are released at the end of a nerve cell when a nerve impulse arrives there. They diffuse across the gap to the next cell and alter the membrane of that second cell to either inhibit or excite it.

deamination (dee-AM-ih-NAY-shun): removal of the amino (NH_2) group from a compound such as an amino acid.

* Chemists sometimes classify amino acids according to the destinations of their carbon fragments after deamination. If the fragment leads to the production of glucose, the amino acid is called *glucogenic;* if it leads to the formation of ketone bodies, fats, and sterols, the amino acid is called *ketogenic.* There is no sharp distinction between glucogenic and ketogenic amino acids, however. A few are both, most are considered glucogenic, only one (leucine) is clearly ketogenic.

Protein in Foods

In the United States and Canada, where nutritious foods are abundant, most people eat protein in such large quantities that they receive all the amino acids they need. In countries where food is scarce and the people eat only marginal amounts of protein-rich foods, however, the *quality* of the protein becomes crucial.

Protein Quality

The protein quality of the diet determines, in large part, how well children grow and how well adults maintain their health. Put simply, **high-quality proteins** provide enough of all the essential amino acids needed to support the body's work, and low-quality proteins don't. Two factors influence protein quality—the protein's digestibility and its amino acid composition.

Digestibility As explained earlier, proteins must be digested before they can provide amino acids. **Protein digestibility** depends on such factors as the protein's source and the other foods eaten with it. The digestibility of most animal proteins is high (90 to 99 percent); plant proteins are less digestible (70 to 90 percent for most, but over 90 percent for soy and legumes).

Amino Acid Composition To make proteins, a cell must have all the needed amino acids available simultaneously. The liver can produce any nonessential amino acid that may be in short supply so that the cells can continue linking amino acids into protein strands. If an essential amino acid is missing, though, a cell must dismantle its own proteins to obtain it. Therefore, to prevent protein breakdown, dietary protein must supply at least the nine essential amino acids plus enough nitrogen-containing amino groups and energy for the synthesis of the others. If the diet supplies too little of any essential amino acid, protein synthesis will be limited. The body makes whole proteins only; if one amino acid is missing, the others cannot form a "partial" protein. An essential amino acid supplied in less than the amount needed to support protein synthesis is called a **limiting amino acid.**

Reference Protein The quality of a food protein is determined by comparing its amino acid composition with the essential amino acid requirements of preschool-age children. Such a standard is called a **reference protein.** ◆ The rationale behind using the requirements of this age group is that if a protein will effectively support a young child's growth and development, then it will meet or exceed the requirements of older children and adults.

High-Quality Proteins As mentioned earlier, a high-quality protein contains all the essential amino acids in relatively the same amounts and proportions that human beings require; it may or may not contain all the nonessential amino acids. Proteins that are low in an essential amino acid cannot, by themselves, support protein synthesis. Generally, foods derived from animals (meat, fish, poultry, cheese, eggs, yogurt, and milk) provide high-quality proteins, although gelatin is an exception. (It lacks tryptophan and cannot support growth and health as a diet's sole protein.) Proteins from plants (vegetables, nuts, seeds, grains, and legumes) have more diverse amino acid patterns and tend to be limiting in one or more essential amino acids. Some plant proteins are notoriously low quality (for example, corn protein). A few others are high quality (for example, soy protein).

Researchers have developed several methods for evaluating the quality of food proteins and identifying high-quality proteins. Appendix D provides details.

Complementary Proteins In general, plant proteins are lower quality than animal proteins, and plants also offer less protein (per weight or measure of food). For this reason, many vegetarians improve the quality of proteins in their diets by combining plant-protein foods that have different but complementary amino acid patterns. This strategy yields **complementary proteins** that together contain all the

Black beans and rice, a favorite Hispanic combination, together provide a balanced array of amino acids.

◆ In the past, egg protein was commonly used as the reference protein. Table D-1 in Appendix D presents the amino acid profile of egg. As the reference protein, egg was assigned the value of 100; Table D-3 includes scores of other food proteins for comparison.

high-quality proteins: dietary proteins containing all the essential amino acids in relatively the same amounts that human beings require. They may also contain nonessential amino acids.

protein digestibility: a measure of the amount of amino acids absorbed from a given protein intake.

limiting amino acid: the essential amino acid found in the shortest supply relative to the amounts needed for protein synthesis in the body. Four amino acids are most likely to be limiting:

• Lysine
• Methionine
• Threonine
• Tryptophan

reference protein: a standard against which to measure the quality of other proteins.

complementary proteins: two or more dietary proteins whose amino acid assortments complement each other in such a way that the essential amino acids missing from one are supplied by the other.

FIGURE 6-11 Complementary Proteins

In general, legumes provide plenty of isoleucine (Ile) and lysine (Lys) but fall short in methionine (Met) and tryptophan (Trp). Grains have the opposite strengths and weaknesses, making them a perfect match for legumes.

	Ile	Lys	Met	Trp
Legumes	■	■		
Grains			■	■
Together	■	■	■	■

◆ Daily Value:
   • 50 g protein (based on 10% of 2000 kcal diet)

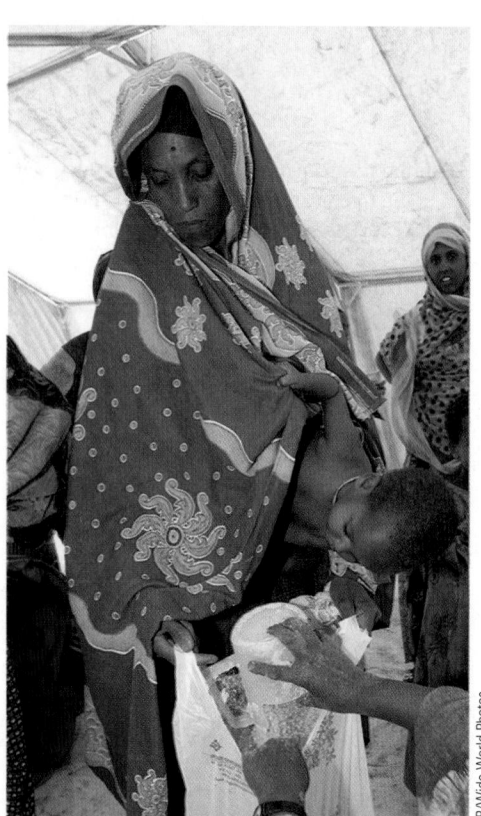

Donated food saves some people from starvation, but it is usually insufficient to meet nutrient needs or even to defend against hunger.

protein-energy malnutrition (PEM), also called **protein-kcalorie malnutrition (PCM)**: a deficiency of protein, energy, or both, including kwashiorkor, marasmus, and instances in which they overlap (see p. 198).

essential amino acids in quantities sufficient to support health. The protein quality of the combination is greater than for either food alone (see Figure 6-11).

Many people have long believed that combining plant proteins at every meal is critical to protein nutrition. For most healthy vegetarians, though, it is not necessary to balance amino acids at each meal if protein intake is varied and energy intake is sufficient.[4] Vegetarians can receive all the amino acids they need over the course of a day by eating a variety of whole grains, legumes, seeds, nuts, and vegetables. Protein deficiency will develop, however, when fruits and certain vegetables make up the core of the diet, severely limiting both the *quantity* and *quality* of protein. Highlight 2 describes how to plan a nutritious vegetarian diet.

IN SUMMARY

A diet that supplies all of the essential amino acids in adequate amounts ensures protein synthesis. The best guarantee of amino acid adequacy is to eat foods containing high-quality proteins or mixtures of foods containing complementary proteins that can each supply the amino acids missing in the other. In addition to its amino acid content, the quality of protein is measured by its digestibility and its ability to support growth. Such measures are of great importance in dealing with malnutrition worldwide, but in the United States and Canada, where protein deficiency is not common, protein quality scores of individual foods deserve little emphasis.

Protein Regulations for Food Labels

All food labels must state the quantity of protein in grams. The "% Daily Value" ◆ for protein is not mandatory on all labels but is required whenever a food makes a protein claim or is intended for consumption by children under four years old.* Whenever the Daily Value percentage is declared, researchers must determine the quality of the protein. Thus, when a % Daily Value is stated for protein, it reflects both quantity and quality.

Health Effects and Recommended Intakes of Protein

As you know by now, protein is indispensable to life. It should come as no surprise that protein deficiency can have devastating effects on people's health. But, like the other nutrients, protein in excess can also be harmful. This section examines the health effects and recommended intakes of protein.

Protein-Energy Malnutrition

When people are deprived of protein, energy, or both, the result is **protein-energy malnutrition (PEM)**. Although PEM touches many adult lives, it most often strikes early in childhood. It is one of the most prevalent and devastating forms of malnutrition in the world, afflicting one of every four children worldwide. Most of the 33,000 children who die each day are malnourished.[5]

Inadequate food intake leads to poor growth in children and to weight loss and wasting in adults. Children who are thin for their height may be suffering from

---

* For labeling purposes, the Daily Values for protein are as follows: for infants, 14 grams; for children under age four, 16 grams; for older children and adults, 50 grams; for pregnant women, 60 grams; and for lactating women, 65 grams.

acute PEM (recent severe food deprivation), whereas children who are short for their age have experienced **chronic PEM** (long-term food deprivation). Poor growth due to PEM is easy to overlook because a small child may look quite normal, but it is the most common sign of malnutrition.

PEM is most prevalent in Africa, Central America, South America, and East and Southeast Asia. In the United States, homeless people and those living in substandard housing in inner cities and rural areas have been diagnosed with PEM. In addition to those living in poverty, elderly people who live alone and adults who are addicted to drugs and alcohol are frequently victims of PEM. PEM can develop in young children when parents mistakenly provide "health-food beverages" ◆ that lack adequate energy or protein instead of milk, most commonly because of nutritional ignorance, perceived milk intolerance, or food faddism. Adult PEM is also seen in people hospitalized with infections such as AIDS or tuberculosis; these infections deplete body proteins, demand extra energy, induce nutrient losses, and alter metabolic pathways. Furthermore, poor nutrient intake during hospitalization worsens malnutrition and impairs recovery, whereas nutrition intervention often improves the body's response to other treatments and the chances of survival. PEM is also common in those suffering from the eating disorder anorexia nervosa (discussed in Highlight 8). Prevention emphasizes frequent, nutrient-dense, energy-dense meals and, equally important, resolution of the underlying causes of PEM—poverty, infections, and illness.

◆ Rice drinks are often sold as milk alternatives, but they fail to provide adequate protein, vitamins, and minerals.

Classifying PEM PEM occurs in two forms: marasmus and kwashiorkor, which differ in their clinical features (see Table 6-3). The following paragraphs present three clinical syndromes—marasmus, kwashiorkor, and the combination of the two.

Marasmus Appropriately named from the Greek word meaning "dying away," **marasmus** reflects a severe deprivation of food over a long time (chronic PEM). Put simply, the person is starving and suffering from an inadequate energy *and* protein intake (and inadequate essential fatty acids, vitamins, and minerals as well). Marasmus occurs most commonly in children from 6 to 18 months of age in all the overpopulated and impoverished areas of the world. Children in impoverished nations simply do not have enough to eat and subsist on diluted cereal drinks that supply scant energy and protein of low quality; such food can barely sustain life, much less support growth. Consequently, marasmic children look like little old people—just skin and bones.

acute PEM: protein-energy malnutrition caused by recent severe food restriction; characterized in children by thinness for height (wasting).

chronic PEM: protein-energy malnutrition caused by long-term food deprivation; characterized in children by short height for age (stunting).

marasmus (ma-RAZ-mus): a form of PEM that results from a severe deprivation, or impaired absorption, of energy, protein, vitamins, and minerals.

TABLE 6-3	Features of Marasmus and Kwashiorkor in Children

Separating PEM into two classifications oversimplifies the condition, but at the extremes, marasmus and kwashiorkor exhibit marked differences. Marasmus-kwashiorkor mix presents symptoms common to both marasmus and kwashiorkor. In all cases, children are likely to develop diarrhea, infections, and multiple nutrient deficiencies.

Marasmus	Kwashiorkor
Infancy (less than 2 yr)	Older infants and young children (1 to 3 yr)
Severe deprivation, or impaired absorption, of protein, energy, vitamins, and minerals	Inadequate protein intake or, more commonly, infections
Develops slowly; chronic PEM	Rapid onset; acute PEM
Severe weight loss	Some weight loss
Severe muscle wasting, with no body fat	Some muscle wasting, with retention of some body fat
Growth: <60% weight-for-age	Growth: 60 to 80% weight-for-age
No detectable edema	Edema
No fatty liver	Enlarged fatty liver
Anxiety, apathy	Apathy, misery, irritability, sadness
Good appetite possible	Loss of appetite
Hair is sparse, thin, and dry; easily pulled out	Hair is dry and brittle; easily pulled out; changes color; becomes straight
Skin is dry, thin, and easily wrinkles	Skin develops lesions

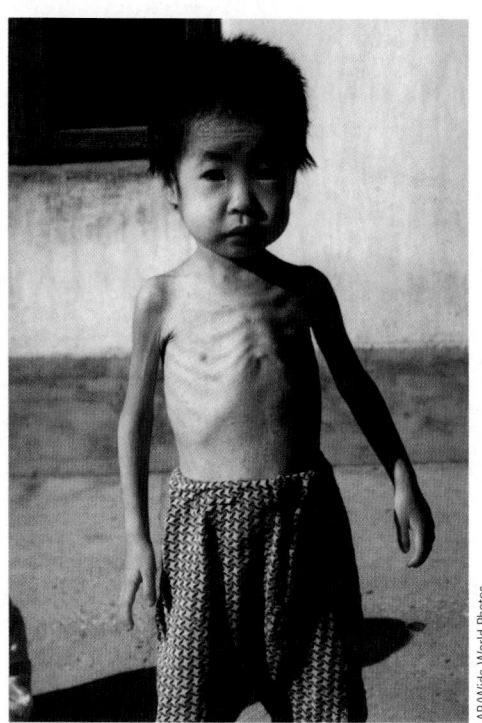

The extreme loss of muscle and fat characteristic of marasmus is apparent in this child's "matchstick" arms.

◆ For this reason, kwashiorkor is sometimes referred to as "wet" PEM and marasmus as "dry" PEM.

kwashiorkor (kwash-ee-OR-core, kwash-ee-or-CORE): a form of PEM that results either from inadequate protein intake or, more commonly, from infections.

dysentery (DISS-en-terry): an infection of the digestive tract that causes diarrhea.

Without adequate nutrition, muscles, including the heart, waste and weaken. Because the brain normally grows to almost its full adult size within the first two years of life, marasmus impairs brain development and learning ability. Reduced synthesis of key hormones slows metabolism and lowers body temperature. There is little or no fat under the skin to insulate against cold. Hospital workers find that children with marasmus need to be clothed, covered, and kept warm. Because these children often suffer delays in their mental and behavioral development, they also need loving care, a stimulating environment, and parental attention.

The starving child faces this threat to life by engaging in as little activity as possible—not even crying for food. The body musters all its forces to meet the crisis, so it cuts down on any expenditure of energy not needed for the functioning of the heart, lungs, and brain. Growth ceases; the child is no larger at age four than at age two. Enzymes are in short supply and the GI tract lining deteriorates. Consequently, the child can't digest and absorb what little food is eaten.

Kwashiorkor **Kwashiorkor** typically reflects a sudden and recent deprivation of food (acute PEM). Kwashiorkor is a Ghanaian word that refers to the birth position of a child and is used to describe the illness a child develops when the next child is born. When a mother who has been nursing her first child bears a second child, she weans the first child and puts the second one on the breast. The first child, suddenly switched from nutrient-dense, protein-rich breast milk to a starchy, protein-poor cereal, soon begins to sicken and die. Kwashiorkor typically sets in between 18 months and two years.

Kwashiorkor usually develops rapidly as a result of protein deficiency or, more commonly, is precipitated by an illness such as measles or other infection. Other factors, such as aflatoxins (a contaminant sometimes found in moldy grains), may also contribute to the development of, or symptoms that accompany, kwashiorkor.[6]

The loss of weight and body fat is usually not as severe in kwashiorkor as in marasmus, but some muscle wasting may occur. Proteins and hormones that previously maintained fluid balance diminish, and fluid leaks into the interstitial spaces. The child's limbs and abdomen become swollen with edema, a distinguishing feature of kwashiorkor. ◆ A fatty liver develops due to a lack of the protein carriers that transport fat out of the liver. The fatty liver lacks enzymes to clear metabolic toxins from the body, so their harmful effects are prolonged. Inflammation in response to these toxins and to infections further contributes to the edema that accompanies kwashiorkor. Without sufficient tyrosine to make melanin, the child's hair loses its color, and inadequate protein synthesis leaves the skin patchy and scaly, often with sores that fail to heal. The lack of proteins to carry or store iron leaves iron free. Unbound iron is common in children with kwashiorkor and may contribute to their illnesses and deaths by promoting bacterial growth and free-radical damage. (Free-radical damage is discussed fully in Highlight 11.)

Marasmus-Kwashiorkor Mix The combination of marasmus and kwashiorkor is characterized by the edema of kwashiorkor with the wasting of marasmus. Most often, the child suffers the effects of both malnutrition and infections. Some researchers believe that kwashiorkor and marasmus are two stages of the same disease. They point out that kwashiorkor and marasmus often exist side by side in the same community where children consume the same diet. They note that a child who has marasmus can later develop kwashiorkor. Some research indicates that marasmus represents the body's adaptation to starvation and that kwashiorkor develops when adaptation fails.

Infections In PEM, antibodies to fight off invading bacteria are degraded to provide amino acids for other uses, leaving the malnourished child vulnerable to infections. Blood proteins, including hemoglobin, are no longer synthesized, so the child becomes anemic and weak. **Dysentery,** an infection of the digestive tract, causes diarrhea, further depleting the body of nutrients and fluids. In the marasmic child, once infection sets in, kwashiorkor often follows, and the immune response weakens further.[7]

The combination of infections, fever, fluid imbalances, and anemia often leads to heart failure and occasionally sudden death. Infections combined with malnutrition are responsible for two-thirds of the deaths of young children in developing countries. Measles, which might make a healthy child sick for a week or two, kills a child with PEM within two or three days.

Rehabilitation   If caught in time, the life of a starving child may be saved with nutrition intervention. In severe cases, diarrhea will have incurred dramatic fluid and mineral losses that need to be replaced during the first 24 to 48 hours to help raise the blood pressure and strengthen the heartbeat. After that, protein and food energy may be given in *small* quantities, with intakes *gradually* increased as tolerated. Severely malnourished people, especially those with edema, recover better with an initial diet that is relatively low in protein (10 percent kcalories from protein).

Experts assure us that we possess the knowledge, technology, and resources to end hunger. Programs that tailor interventions to the local people and involve them in the process of identifying problems and devising solutions have the most success. To win the war on hunger, those who have the food, technology, and resources must make fighting hunger a priority (see Highlight 16 for more on hunger).

Health Effects of Protein

While many of the world's people struggle to obtain enough food energy and protein, in developed countries both are so abundant that problems of excess are seen. Overconsumption of protein offers no benefits and may pose health risks. High-protein diets have been implicated in several chronic diseases, including heart disease, cancer, osteoporosis, obesity, and kidney stones, but evidence is insufficient to establish an upper level.[8]

Researchers attempting to clarify the relationships between excess protein and chronic diseases face several obstacles. Population studies have difficulty determining whether diseases correlate with animal proteins or with their accompanying saturated fats, for example. Studies that rely on data from vegetarians must sort out the many lifestyle factors, in addition to a "no-meat diet," that might explain relationships between protein and health.

Heart Disease   A high-protein diet may contribute to the progression of heart disease. As Chapter 5 mentioned, foods rich in animal protein also tend to be rich in saturated fats. Consequently, it is not surprising to find a correlation between animal-protein intake (red meats and dairy products) and heart disease.[9] On the other hand, substituting vegetable protein for animal protein improves blood lipids and decreases heart disease mortality.[10]

Research suggests that elevated levels of the amino acid homocysteine may be an independent risk factor for heart disease, heart attacks, and sudden death in patients with heart disease.[11] Researchers do not yet fully understand the many factors—including a high protein diet—that can raise homocysteine in the blood or whether elevated levels are a cause or an effect of heart disease.[12] Until they can determine the exact role homocysteine plays in heart disease, researchers are following several leads in pursuit of the answers. Coffee's role in heart disease has been controversial, but research suggests it is among the most influential factors in raising homocysteine, which may explain some of the adverse health effects of heavy consumption.[13] Elevated homocysteine levels are among the many adverse health consequences of smoking cigarettes and drinking alcohol as well.[14] Homocysteine is also elevated with inadequate intakes of B vitamins and can usually be lowered with fortified foods or supplements of vitamin B_{12}, vitamin B_6, and folate.[15] Lowering homocysteine, however, may not help in preventing heart attacks.[16] Supplements of the B vitamins do not always benefit those with heart disease and in fact, may actually increase the risks.[17]

In contrast to homocysteine, the amino acid arginine may help protect against heart disease by lowering blood pressure and homocysteine levels.[18] Additional research is needed to confirm the benefits of arginine.[19] In the meantime, it is unwise

The edema characteristic of kwashiorkor is apparent in this child's swollen belly. Malnourished children commonly have an enlarged abdomen from parasites as well.

for consumers to use supplements of arginine, or any other amino acid for that matter (as pp. 202–203 explain). Physicians, however, may find it beneficial to add arginine supplements to their heart patients' treatment plan.[20]

Cancer As in heart disease, the effects of protein and fats on cancers cannot be easily separated. Population studies suggest a correlation between high intakes of animal proteins and some types of cancer (notably, cancer of the colon, breast, kidneys, pancreas, and prostate).

Adult Bone Loss (Osteoporosis) Chapter 12 presents calcium metabolism, and Highlight 12 elaborates on the main factors that influence osteoporosis. This section briefly describes the relationships between protein intake and bone loss. When protein intake is high, calcium excretion increases. Whether excess protein depletes the bones of their chief mineral may depend upon the ratio of calcium intake to protein intake. After all, bones need both protein and calcium. An ideal ratio has not been determined, but a young woman whose intake meets recommendations for both nutrients has a calcium-to-protein ratio of more than 20 to 1 (milligrams to grams), which probably provides adequate protection for the bones. For most women in the United States, however, average calcium intakes are lower and protein intakes are higher, yielding a 9-to-1 ratio, which may produce calcium losses significant enough to compromise bone health. In other words, the problem may reflect too little calcium, not too much protein.[21] In establishing recommendations, the DRI Committee considered protein's effect on calcium metabolism and bone health, but it did not find sufficient evidence to warrant an adjustment for calcium or an upper level for protein.[22]

Some (but not all) research suggests that animal protein may be more detrimental to calcium metabolism and bone health than vegetable protein.[23] A review of the topic, however, concludes that excess protein—whether from animal or vegetable sources—increases calcium excretion and, perhaps more importantly, that the other nutrients in the protein source may be equally, if not more, responsible for the effects on bone health.[24]

Inadequate intakes of protein may also compromise bone health.[25] Osteoporosis is particularly common in elderly women and in adolescents with anorexia nervosa—groups who typically receive less protein than they need. For these people, increasing protein intake may be just what they need to protect their bones.[26]

Weight Control Dietary protein may play a role in increasing body weight.[27] Protein-rich foods are often fat-rich foods that contribute to weight gain with its accompanying health risks. As Highlight 9 explains, weight-loss gimmicks that encourage a high-protein, low-carbohydrate diet may be temporarily effective, but only because they are low-kcalorie diets. Diets that provide adequate protein, moderate fat, and sufficient energy from carbohydrates can better support weight loss and good health. Including protein at each meal may help with weight loss by providing satiety.[28] Selecting too many protein-rich foods, such as meat and milk, may crowd out fruits, vegetables, and whole grains, making the diet inadequate in other nutrients.

Kidney Disease Excretion of the end products of protein metabolism depends, in part, on an adequate fluid intake and healthy kidneys. A high protein intake increases the work of the kidneys, but does not appear to diminish kidney function or cause kidney disease.[29] Restricting dietary protein, however, may help to slow the progression of kidney disease and limit the formation of kidney stones in people who have these conditions.

IN SUMMARY

Protein deficiencies arise from both energy-poor and protein-poor diets and lead to the devastating diseases of marasmus and kwashiorkor. Together, these diseases are known as PEM (protein-energy malnutrition), a major form of malnutrition causing death in children worldwide. Excesses of protein offer no advantage; in fact, overconsumption of protein-rich foods may incur health problems as well.

Recommended Intakes of Protein

As mentioned earlier, the body continuously breaks down and loses some protein and cannot store amino acids. To replace protein, the body needs dietary protein for two reasons. First, food protein is the only source of the *essential* amino acids, and second, it is the only practical source of *nitrogen* with which to build the nonessential amino acids and other nitrogen-containing compounds the body needs.

Given recommendations that people's fat intakes should contribute 20 to 35 percent of total food energy and carbohydrate intakes should contribute 45 to 65 percent, that leaves 10 to 35 percent for protein. In a 2000-kcalorie diet, that represents 200 to 700 kcalories from protein, or 50 to 175 grams. Average intakes in the United States and Canada fall within this range.

Protein RDA The protein RDA ◆ for adults is 0.8 grams per kilogram of healthy body weight per day. For infants and children, the RDA is slightly higher. The table on the inside front cover lists the RDA for males and females at various ages in two ways—grams per day based on reference body weights and grams per kilogram body weight per day.

The RDA generously covers the needs for replacing worn-out tissue, so it increases for larger people; it also covers the needs for building new tissue during growth, so it increases for infants, children, and pregnant women. The protein RDA is the same for athletes as for others, although some fitness authorities recommend a slightly higher intake.[30] The accompanying "How to" explains how to calculate your RDA for protein.

In setting the RDA, the DRI Committee assumes that people are healthy and do not have unusual metabolic needs for protein, that the protein eaten will be of mixed quality (from both high- and low-quality sources), and that the body will use the protein efficiently. In addition, the committee assumes that the protein is consumed along with sufficient carbohydrate and fat to provide adequate energy and that other nutrients in the diet are adequate.

Adequate Energy Note the qualification "adequate energy" in the preceding statement, and consider what happens if energy intake falls short of needs. An intake of 50 grams of protein provides 200 kcalories, which represents 10 percent of the total energy from protein, if the person receives 2000 kcalories a day. But if the person cuts energy intake drastically—to, say, 800 kcalories a day—then an intake of 200 kcalories from protein is suddenly 25 percent of the total; yet it's still the same amount of protein (number of grams). The protein intake is reasonable, but the energy intake is not. The low energy intake forces the body to use the protein to meet energy needs rather than to replace lost body protein. Similarly, if the person's energy intake is high—say, 4000 kcalories—the 50-gram protein intake represents only 5 percent of the total; yet it *still* is a reasonable protein intake. Again, the energy intake is unreasonable for most people, but in this case, it permits the protein to be used to meet the body's needs.

Be careful when judging protein (or carbohydrate or fat) intake as a percentage of energy. Always ascertain the number of grams as well, and compare it with the RDA or another standard stated in grams. A recommendation stated as a percentage of energy intake is useful only if the energy intake is within reason.

Protein in Abundance Most people in the United States and Canada receive more protein than they need. Even athletes in training typically don't need to increase their protein intakes because the additional foods they eat to meet their high energy needs deliver protein as well. That protein intake is high is not surprising considering the abundance of food eaten and the central role meats hold in the North American diet. A single ounce of meat (or $1/2$ cup legumes) delivers about 7 grams of protein, so 8 ounces of meat alone supplies more than the RDA for an average-size person. Besides meat, well-fed people eat many other nutritious foods, many of which also provide protein. A cup of milk provides 8 grams of protein.

HOW TO Calculate Recommended Protein Intakes

To figure your protein RDA:

- Look up the healthy weight for a person of your height (inside back cover). If your present weight falls within that range, use it for the following calculations. If your present weight falls outside the range, use the midpoint of the healthy weight range as your reference weight.
- Convert pounds to kilograms, if necessary (pounds divided by 2.2 equals kilograms).
- Multiply kilograms by 0.8 to get your RDA in grams per day. (Older teens 14 to 18 years old, multiply by 0.85.) Example:

Weight = 150 lb

150 lb ÷ 2.2 lb/kg = 68 kg (rounded off)

68 kg × 0.8 g/kg = 54 g protein (rounded off)

CENGAGENOW
To calculate recommended protein intakes, log on to **academic.cengage.com/login**, go to Chapter 6, then go to How To.

◆ RDA for protein:
- 0.8 g/kg/day
- 10 to 35% of energy intake

For many people, this 5-ounce steak provides almost all of the meat and much of the protein recommended for a day's intake.

© Polara Studios, Inc.

Vegetarians obtain their protein from whole grains, legumes, nuts, vegetables, and, in some cases, eggs and milk products.

Grains and vegetables provide small amounts of protein, but they can add up to significant quantities; fruits and fats provide no protein.

To illustrate how easy it is to overconsume protein, consider the amounts recommended by the USDA Food Guide for a 2000-kcalorie diet. Six ounces of grains provide about 18 grams of protein; 2^1/$_2$ cups of vegetables deliver about 10 grams; 3 cups of milk offer 24 grams; and 5^1/$_2$ ounces of meat supply 38 grams. This totals 90 grams of protein—higher than recommendations for most people and yet still lower than the average intake of people in the United States.

People in the United States and Canada get more protein than they need. If they have an adequate *food* intake, they have a more-than-adequate protein intake. The key diet-planning principle to emphasize for protein is moderation. Even though most people receive plenty of protein, some feel compelled to take supplements as well, as the next section describes.

IN SUMMARY

The optimal diet is adequate in energy from carbohydrate and fat and delivers 0.8 grams of protein per kilogram of healthy body weight each day. U.S. and Canadian diets are typically more than adequate in this respect.

Protein and Amino Acid Supplements

Websites, health-food stores, and popular magazine articles advertise a wide variety of protein supplements, and people take these supplements for many different reasons. Athletes take protein powders to build muscle. Dieters take them to spare their bodies' protein while losing weight. Women take them to strengthen their fingernails. People take individual amino acids, too—to cure herpes, to make themselves sleep better, to lose weight, and to relieve pain and depression.* Like many other magic solutions to health problems, protein and amino acid ◆ supplements don't work these miracles. Furthermore, they may be harmful.

Protein Powders Because the body builds muscle protein from amino acids, many athletes take protein powders with the false hope of stimulating muscle growth. Muscle work builds muscle; protein supplements do not, and athletes do not need them. Taking protein supplements does not improve athletic performance.[31] Protein powders can supply amino acids to the body, but nature's protein sources—lean meat, milk, eggs, and legumes—supply all these amino acids and more.

Whey protein appears to be particularly popular among athletes hoping to achieve greater muscle gains. A waste product of cheese manufacturing, whey protein is a common ingredient in many low-cost protein powders. When combined with strength training, whey supplements may increase protein synthesis slightly, but they do not seem to enhance athletic performance.[32] To build stronger muscles, athletes need to eat food with adequate energy and protein to support the weight-training work that does increase muscle mass. Those who still think they need more whey should pour a glass of milk; one cup provides 1.5 grams of whey.

Purified protein preparations contain none of the other nutrients needed to support the building of muscle, and the protein they supply is not needed by athletes who eat food. It is excess protein, and the body dismantles it and uses it for energy or stores it as body fat. The deamination of excess amino acids places an extra burden on the kidneys to excrete unused nitrogen.

Amino Acid Supplements Single amino acids do not occur naturally in foods and offer no benefit to the body; in fact, they may be harmful. The body was not designed to handle the high concentrations and unusual combinations of amino acids

◆ Use of amino acids as dietary supplements is *inappropriate*, especially for:
  • All women of childbearing age
  • Pregnant or lactating women
  • Infants, children, and adolescents
  • Elderly people
  • People with inborn errors of metabolism that affect their bodies' handling of amino acids
  • Smokers
  • People on low-protein diets
  • People with chronic or acute mental or physical illnesses who take amino acids without medical supervision

whey protein: a by-product of cheese production; falsely promoted as increasing muscle mass. Whey is the watery part of milk that separates from the curds.

* Canada only allows single amino acid supplements to be sold as drugs or used as food additives.

found in supplements. An excess of one amino acid can create such a demand for a carrier that it limits the absorption of another amino acid, presenting the possibility of a deficiency. Those amino acids winning the competition enter in excess, creating the possibility of toxicity. Toxicity of single amino acids in animal studies raises concerns about their use in human beings. Anyone considering taking amino acid supplements should check with a registered dietitian or physician first.

Most healthy athletes eating well-balanced diets do not need amino acid supplements. Advertisers point to research that identifies the **branched-chain amino acids** ◆ as the main ones used as fuel by exercising muscles. What the ads leave out is that compared to glucose and fatty acids, branched-chain amino acids provide very little fuel and that ordinary foods provide them in abundance anyway. Large doses of branched-chain amino acids can raise plasma ammonia concentrations, which can be toxic to the brain. Branched-chain amino acid supplements may be useful in conditions such as advanced liver failure, but otherwise, they are not routinely recommended.[33]

◆ The branched-chain amino acids are leucine, isoleucine, and valine.

In two cases, recommendations for single amino acid supplements have led to widespread public use—lysine to prevent or relieve the infections that cause herpes cold sores on the mouth or genital organs, and tryptophan to relieve pain, depression, and insomnia. In both cases, enthusiastic popular reports preceded careful scientific experiments and health recommendations. Research is insuffiencient to determine whether lysine suppresses herpes infections, but it appears safe (up to 3 grams per day) when taken in divided doses with meals.[34]

Tryptophan may be effective with respect to pain and sleep, but its use for these purposes is experimental. About 20 years ago, more than 1500 people who elected to take tryptophan supplements developed a rare blood disorder known as eosinophilia-myalgia syndrome (EMS). EMS is characterized by severe muscle and joint pain, extremely high fever, and, in over three dozen cases, death. Treatment for EMS usually involves physical therapy and low doses of corticosteroids to relieve symptoms temporarily. The Food and Drug Administration implicated impurities in the supplements, issued a recall of all products containing manufactured tryptophan, and warned that high-dose supplements of tryptophan might provoke EMS even in the absence of impurities.

IN SUMMARY

Normal, healthy people never need protein or amino acid supplements. It is safest to obtain lysine, tryptophan, and all other amino acids from protein-rich foods, eaten with abundant carbohydrate and some fat to facilitate their use in the body. With all that we know about science, it is hard to improve on nature.

Nutrition Portfolio

CENGAGENOW™
academic.cengage.com/login

Foods that derive from animals—meats, fish, poultry, eggs, and milk products—provide plenty of protein but are often accompanied by fat. Those that derive from plants— whole grains, vegetables, and legumes—may provide less protein but also less fat.

■ Calculate your daily protein needs and compare them with your protein intake. Consider whether you receive enough, but not too much, protein daily.

■ Describe your dietary sources of proteins and whether you use mostly plant-based or animal-based protein foods in your diet.

■ Debate the risks and benefits of taking protein or amino acid supplements.

branched-chain amino acids: the essential amino acids leucine, isoleucine, and valine, which are present in large amounts in skeletal muscle tissue; falsely promoted as fuel for exercising muscles.

NUTRITION ON THE NET

For further study of topics covered in this chapter, log on to **academic.cengage** **.com/nutrition/rolfes/UNCN8e**. Go to Chapter 6, then to Nutrition on the Net.

- Learn more about sickle-cell anemia from the National Heart, Lung, and Blood Institute or the Sickle Cell Disease Association of America: **www.nhlbi.nih.gov** or **www.sicklecelldisease.org**

- Learn more about protein-energy malnutrition and world hunger from the World Health Organization Nutrition

Programme or the National Institute of Child Health and Human Development: **www.who.int/nut** or **www.nichd.nih.gov**

- Highlight 16 offers many more websites on malnutrition and world hunger.

NUTRITION CALCULATIONS

CENGAGENOW˜ For additional practice, log on to **academic.cengage.com/login**. Go to Chapter 6, then to Nutrition Calculations.

These problems will give you practice in doing simple nutrition-related calculations using hypothetical situations (see p. 206 for answers). Once you have mastered these examples, you will be prepared to examine your own protein needs. Be sure to show your calculations for each problem.

1. Compute recommended protein intakes for people of different sizes. Refer to the "How to" on p. 201 and compute the protein recommendation for the following people. The intake for a woman who weighs 144 pounds is computed for you as an example.

    $$144 \text{ lb} \div 2.2 \text{ lb/kg} = 65 \text{ kg}$$

    $$0.8 \text{ g/kg} \times 65 \text{ kg} = 52 \text{ g protein per day}$$

    a. a woman who weighs 116 pounds
    b. a man (18 years) who weighs 180 pounds

2. The chapter warns that recommendations based on percentage of energy intake are not always appropriate. Consider a woman 26 years old who weighs 165 pounds. Her diet provides 1500 kcalories/day with 50 grams carbohydrate and 100 grams fat.
    a. What is this woman's protein intake? Show your calculations.
    b. Is her protein intake appropriate? Justify your answer.
    c. Are her carbohydrate and fat intakes appropriate? Justify your answer.

This exercise should help you develop a perspective on protein recommendations.

STUDY QUESTIONS

CENGAGENOW˜
To assess your understanding of chapter topics, take the Student Practice Test and explore the modules recommended in your Personalized Study Plan. Log on to **academic.cengage.com/login**.

These questions will help you review the chapter. You will find the answers in the discussions on the pages provided.

1. How does the chemical structure of proteins differ from the structures of carbohydrates and fats? (pp. 181–184)

2. Describe the structure of amino acids, and explain how their sequence in proteins affects the proteins' shapes. What are essential amino acids? (pp. 181–184)

3. Describe protein digestion and absorption. (pp. 185–186)

4. Describe protein synthesis. (pp. 187–189)

5. Describe some of the roles proteins play in the human body. (pp. 189–192)

6. What are enzymes? What roles do they play in chemical reactions? Describe the differences between enzymes and hormones. (p. 190)

7. How does the body use amino acids? What is deamination? Define nitrogen balance. What conditions are associated with zero, positive, and negative balance? (pp. 193–194)

8. What factors affect the quality of dietary protein? What is a high-quality protein? (pp. 195–196)

9. How can vegetarians meet their protein needs without eating meat? (pp. 195–196)

10. What are the health consequences of ingesting inadequate protein and energy? Describe marasmus and kwashiorkor. How can the two conditions be distinguished, and in what ways do they overlap? (pp. 196–199)

11. How might protein excess, or the type of protein eaten, influence health? (pp. 199–200)

12. What factors are considered in establishing recommended protein intakes? (pp. 201–202)

13. What are the benefits and risks of taking protein and amino acid supplements? (p. 202–203)

These multiple choice questions will help you prepare for an exam. Answers can be found on p. 206.

1. Which part of its chemical structure differentiates one amino acid from another?
   a. its side group
   b. its acid group
   c. its amino group
   d. its double bonds

2. Isoleucine, leucine, and lysine are:
   a. proteases.
   b. polypeptides.
   c. essential amino acids.
   d. complementary proteins.

3. In the stomach, hydrochloric acid:
   a. denatures proteins and activates pepsin.
   b. hydrolyzes proteins and denatures pepsin.
   c. emulsifies proteins and releases peptidase.
   d. condenses proteins and facilitates digestion.

4. Proteins that facilitate chemical reactions are:
   a. buffers.
   b. enzymes.
   c. hormones.
   d. antigens.

5. If an essential amino acid that is needed to make a protein is unavailable, the cells must:
   a. deaminate another amino acid.
   b. substitute a similar amino acid.
   c. break down proteins to obtain it.
   d. synthesize the amino acid from glucose and nitrogen.

6. Protein turnover describes the amount of protein:
   a. found in foods and the body.
   b. absorbed from the diet.
   c. synthesized and degraded.
   d. used to make glucose.

7. Which of the following foods provides the highest quality protein?
   a. egg
   b. corn
   c. gelatin
   d. whole grains

8. Marasmus develops from:
   a. too much fat clogging the liver.
   b. megadoses of amino acid supplements.
   c. inadequate protein and energy intake.
   d. excessive fluid intake causing edema.

9. The protein RDA for a healthy adult who weighs 180 pounds is:
   a. 50 milligrams/day.
   b. 65 grams/day.
   c. 180 grams/day.
   d. 2000 milligrams/day.

10. Which of these foods has the least protein per $\frac{1}{2}$ cup?
    a. rice
    b. broccoli
    c. pinto beans
    d. orange juice

REFERENCES

1. M. S. Buchowski and coauthors, Equation to estimate resting energy expenditure in adolescents with sickle cell anemia, *American Journal of Clinical Nutrition* 76 (2002): 1335–1344; Committee on Genetics, Health supervision for children with sickle cell disease, *Pediatrics* 109 (2002): 526–535.
2. J. M. Ordovas and D. Corella, Nutritional genomics, *Annual Review of Genomics and Human Genetics* 5 (2004): 71–118.
3. W. M. Rand, P. L. Pellett, and V. R. Young, Meta-analysis of nitrogen balance studies for estimating protein requirements in healthy adults, *American Journal of Clinical Nutrition* 77 (2003): 109–127.
4. Position of the American Dietetic Association and Dietitians of Canada: Vegetarian diets, *Journal of the American Dietetic Association* 103 (2003): 748–765.
5. Data from www.unicef.org, posted April 2005 and May 2006.
6. M. Krawinkel, Kwashiorkor is still not fully understood, *Bulletin of the World Health Organization* 81 (2003): 910–911.
7. M. Reid and coauthors, The acute-phase protein response to infection in edematous and nonedematous protein-energy malnutrition, *American Journal of Clinical Nutrition* 76 (2002): 1409–1415.
8. Committee on Dietary Reference Intakes, *Dietary Reference Intakes for Energy, Carbohydrate, Fiber, Fat, Fatty Acids, Cholesterol, Protein, and Amino Acids* (Washington, D.C.: National Academies Press, 2002/2005), p. 694.

9. L. E. Kelemen and coauthors, Associations of dietary protein with disease and mortality in a prospective study of postmenopausal women, *American Journal of Epidemiology* 161 (2005): 239–249.
10. B. L. McVeigh and coauthors, Effect of soy protein varying in isoflavone content on serum lipids in healthy young men, *American Journal of Clinical Nutrition* 83 (2006): 244–251; L. E. Kelemen and coauthors, Associations of dietary protein with disease and mortality in a prospective study of postmenopausal women, *American Journal of Epidemiology* 161 (2005): 239–249; S. Tonstad, K. Smerud, and L. Høie, A comparison of the effects of 2 doses of soy protein or casein on serum lipids, serum lipoproteins, and plasma total homocysteine in hypercholesterolemic subjects, *American Journal of Clinical Nutrition* 76 (2002): 78–84.
11. M. Haim and coauthors, Serum homocysteine and long-term risk of myocardial infarction and sudden death in patients with coronary heart disease, *Cardiology* 107 (2006): 52–56; M. B. Kazemi and coauthors, Homocysteine level and coronary artery disease, *Angiology* 57 (2006): 9–14; D. S. Wald, M. Law, and J. K. Morris, Homocysteine and cardiovascular disease: Evidence on causality from a meta-analysis, *British Medical Journal* 325 (2002): 1202–1217; The Homocysteine Studies Collaboration, Homocysteine and risk of ischemic heart

disease and stroke, *Journal of the American Medical Association* 288 (2002): 2015–2022.
12. J. Selhub, The many facets of hyperhomocysteinemia: Studies from the Framingham cohorts, *Journal of Nutrition* 136 (2006): 1726S–1730S; P. Verhoef and coauthors, A high-protein diet increases postprandial but not fasting plasma total homocysteine concentrations: A dietary controlled, crossover trial in healthy volunteers, *American Journal of Clinical Nutrition* 82 (2005): 553–558.
13. S. E. Chiuve and coauthors, Alcohol intake and methylenetetrahydrofolate reductase polymorphism modify the relation of folate intake to plasma homocysteine, *American Journal of Clinical Nutrition* 82 (2005): 155–162; P. Verhoef and coauthors, Contribution of caffeine to the homocysteine-raising effect of coffee: A randomized controlled trial in humans, *American Journal of Clinical Nutrition* 76 (2002): 1244–1248.
14. J. A. Troughton and coauthors, Homocysteine and coronary heart disease risk in the PRIME study, *Atherosclerosis* (2006); S. E. Chiuve and coauthors, Alcohol intake and methylenetetrahydrofolate reductase polymorphism modify the relation of folate intake to plasma homocysteine, *American Journal of Clinical Nutrition* 82 (2005): 155–162.
15. D. Genser and coauthors, Homocysteine, folate and vitamin B(12) in patients with coronary heart disease, *Annals of Nutrition &*

Metabolism 50 (2006): 413–419; Ø. Bleie and coauthors, Changes in basal and postmethionine load concentrations of total homocysteine and cystathionine after B vitamin intervention, *American Journal of Clinical Nutrition* 80 (2004): 641–648; E. Nurk and coauthors, Changes in lifestyle and plasma total homocysteine: The Hordaland Homocysteine Study, *American Journal of Clinical Nutrition* 79 (2004): 812–819; K. L. Tucker and coauthors, Breakfast cereal fortified with folic acid, vitamin B-6, and vitamin B-12 increases vitamin concentrations and reduces homocysteine concentrations: A randomized trial, *American Journal of Clinical Nutrition* 79 (2004): 805–811; J. F. Toole and coauthors, Lowering homocysteine in patients with ischemic stroke to prevent recurrent stroke, myocardial infarction, and death: The Vitamin Intervention for Stroke Prevention (VISP) randomized controlled trial, *Journal of the American Medical Association* 291 (2004): 565–575.

16. B-Vitamin Treatment Trialists' Collaboration, Homocysteine-lowering trials for prevention of cardiovascular events: A review of the design and power of the large randomized trials, *American Heart Journal* 151 (2006): 282–287.

17. E. Lonn and coauthors, Homocysteine lowering with folic acid and B vitamins in vascular disease, *New England Journal of Medicine* 354 (2006): 1567–1577; K. H. Bonaa and coauthors, Homocysteine lowering and cardiovascular events after acute myocardial infarction, *New England Journal of Medicine* 354 (2006): 1578–1588; G. Schnyder and coauthors, Effect of homocysteine-lowering therapy with folic acid, vitamin B12, and vitamin B6 on clinical outcome after percutaneous coronary intervention—The Swiss Heart Study: A randomized controlled trial, *Journal of the American Medical Association* 288 (2002): 973–979; B. J. Venn and coauthors, Dietary counseling to increase natural folate intake: A randomized, placebo-controlled trial in free-living subjects to assess effects on serum folate and plasma total homocysteine, *American Journal of Clinical Nutrition* 76 (2002): 758–765.

18. S. G. West and coauthors, Oral L-arginine improves hemodynamic responses to stress and reduces plasma homocysteine in hypercholesterolemic men, *Journal of Nutrition* 135 (2005): 212–217.

19. N. Gokce, L-arginine and hypertension, *Journal of Nutrition* 134 (2004): 2807S–2811S.

20. B. S. Kendler, Supplemental conditionally essential nutrients in cardiovascular disease therapy, *Journal of Cardiovascular Nursing* 21 (2006): 9–16.

21. B. Dawson-Hughes, Interaction of dietary calcium and protein in bone health in humans, *Journal of Nutrition* 133 (2003): 852S–854S.

22. Committee on Dietary Reference Intakes, 2002/2005, p. 841; Committee on Dietary Reference Intakes, *Dietary Reference Intakes for Calcium, Phosphorus, Magnesium, Vitamin D, and Fluoride* (Washington, D.C.: National Academy Press, 1997), pp. 75–76.

23. J. P. Bonjour, Dietary protein: An essential nutrient for bone health, *Journal of the American College of Nutrition* 24 (2005): 526S–536S; C. Weikert and coauthors, The relation between dietary protein, calcium and bone health in women: Results from the EPIC-Potsdam cohort, *Annals of Nutrition & Metabolism* 49 (2005): 312–318.

24. L. K. Massey, Dietary animal and plant protein and human bone health: A whole foods approach, *Journal of Nutrition* 133 (2003): 862S–865S.

25. F. Ginty, Dietary protein and bone health, *The Proceedings of the Nutrition Society* 62 (2003): 867–876; J. E. Kerstetter, K. O. O'Brien, and K. L. Insogna, Low protein intake: The impact on calcium and bone homeostasis in humans, *Journal of Nutrition* 133 (2003): 855S–861S.

26. A. Devine and coauthors, Protein consumption is an important predictor of lower limb bone mass in elderly women, *American Journal of Clinical Nutrition* 81 (2005): 1423–1428; J. Bell and S. J. Whiting, Elderly women need dietary protein to maintain bone mass, *Nutrition Reviews* 60 (2002): 337–341; M. T. Munoz and J. Argente, Anorexia nervosa in female adolescents: Endocrine and bone mineral density disturbances, *European Journal of Endocrinology* 147 (2002): 275–286.

27. A. Trichopoulou and coauthors, Lipid, protein and carbohydrate intake in relation to body mass index, *European Journal of Clinical Nutrition* 56 (2002): 37–43.

28. A. Astrup, The satiating power of protein—a key to obesity prevention? *American Journal of Clinical Nutrition* 82 (2005): 1–2; D. S. Weigle and coauthors, A high-protein diet induces sustained reductions in appetite, ad libitum caloric intake, and body weight despite compensatory changes in diurnal plasma leptin and ghrelin concentrations, *American Journal of Clinical Nutrition* 82 (2005): 41–48.

29. E. L. Knight and coauthors, The impact of protein intake on renal function decline in women with normal renal function or mild renal insufficiency, *Annals of Internal Medicine* 138 (2003): 460–467.

30. Position of the American Dietetic Association, Dietitians of Canada, and the American College of Sports Nutrition, Nutrition and athletic performance, *Journal of the American Dietetic Association* 100 (2000): 1543–1556.

31. L. L. Andersen and coauthors, The effect of resistance training combined with timed ingestion of protein on muscle fiber size and muscle strength, *Metabolism: Clinical and Experimental* 54 (2005): 151–156.

32. K. D. Tipton, Ingestion of casein and whey proteins result in muscle anabolism after resistance exercise, *Medicine and Science in Sports and Exercise* 36 (2004): 2073–2081.

33. R. Mascarenhas and S. Mobarhan, New support for branched-chain amino acid supplementation in advanced hepatic failure, *Nutrition Reviews* 62 (2004): 33–38.

34. M. M. Perfect and coauthors, Use of complementary and alternative medicine for the treatment of genital herpes, *Herpes* 12 (2005): 38–41.

ANSWERS

Nutrition Calculations

1. a. 116 lb ÷ 2.2 lb/kg = 53 kg

   0.8 g/kg × 53 kg = 42 g protein per day

   b. 180 lb ÷ 2.2 lb/kg = 82 kg

   He is 18 years old, so use 0.85 g/kg.

   0.85 g/kg × 82 kg = 70 g protein per day

2. a. 50 g carbohydrate × 4 kcal/g = 200 kcal from carbohydrate

   100 g fat × 9 kcal/g = 900 kcal from fat

   1500 kcal − (200 + 900 kcal) = 400 kcal from protein

   400 kcal ÷ 4 kcal/g = 100 g protein

   b. Using the RDA guideline of 0.8 g/kg, an appropriate protein intake for this woman would be 60 g protein/day (165 lb ÷ 2.2 lb/kg = 75 kg; 0.8 g/kg × 75 = 60 g/day). Her intake is higher than her RDA. Using the guideline that protein should contribute 10 to 35% of energy intake, her intake of 100 g protein on a 1500 kcal diet falls within the suggested range (400 kcal protein ÷ 1500 total kcal = 27%).

   c. Using the guideline that carbohydrate should contribute 45 to 65% and fat should contribute 20 to 35% of energy intake, her intake of 50 g carbohydrate is low (200 kcal carbohydrate ÷ 1500 total kcal = 13%), and her intake of 100 g fat is high (900 kcal fat ÷ 1500 total kcal = 60%).

Study Questions (multiple choice)

1. a   2. c   3. a   4. b   5. c   6. c   7. a
8. c   9. b   10. d

Nutritional Genomics

© Science VU/Visuals Unlimited

Imagine this scenario: A physician scrapes a sample of cells from inside your cheek and submits it to a **genomics** lab. The lab returns a report based on your genetic profile that reveals which diseases you are most likely to develop and makes recommendations for specific diet and lifestyle changes that can help you maintain good health. You may also be given a prescription for a dietary supplement that will best meet your personal nutrient requirements. Such a scenario may one day become reality as scientists uncover the genetic relationships between diet and disease. (Until then, however, consumers need to know that current genetic test kits commonly available on the Internet are unproven and quite likely fraudulent.)

How nutrients influence gene activity and how **genes** influence the activities of nutrients is the focus of a new field of study called **nutritional genomics** (see the accompanying glossary). Unlike sciences in the 20th century, nutritional genomics takes a comprehensive approach in analyzing information from several fields of study, providing an integrated understanding of the findings.[1] Consider how multiple disciplines contributed to our understanding of vitamin A over the past several decades, for example. Biochemistry revealed vitamin A's three chemical structures. Immunology identified the anti-infective properties of one

of these structures while physiology focused on another structure and it's role in vision. Epidemiology has reported improvements in the death rates and vision of malnourished children given vitamin A supplements, and biology has explored how such effects might be possible. The process was slow as researchers collected information on one gene, one action, and one nutrient at a time. Today's research in nutritional genomics involves all of the sciences, coordinating their multiple findings, and explaining their interactions among several genes, actions, and nutrients in relatively little time. As a result, nutrition knowledge is growing at an incredibly fast pace.

The recent surge in genomics research grew from the Human Genome Project, an international effort by industry and government scientists to identify and describe all of the genes in the **human genome**—that is, all the genetic information contained within a person's cells. Completed in 2003, this project developed many of the research technologies needed to study genes and genetic variation. Scientists are now working to identify the individual proteins made by the genes, the genes associated with diseases, and the dietary and lifestyle choices that most influence the expression of those genes. Such information will have major implications for society in general, and for health care in particular.[2]

GLOSSARY

chromosomes: structures within the nucleus of a cell made of DNA and associated proteins. Human beings have 46 chromosomes in 23 pairs. Each chromosome has many genes.

DNA (deoxyribonucleic acid): the double helix molecules of which genes are made.

epigenetics: the study of heritable changes in gene function that occur without a change in the DNA sequence.

gene expression: the process by which a cell converts the genetic code into RNA and protein.

genes: sections of chromosomes that contain the instructions

needed to make one or more proteins.

genetics: the study of genes and inheritance.

genomics: the study of all the genes in an organism and their interactions with environmental factors.

human genome (GEE-nome): the full complement of genetic material in the chromosomes of a person's cells.

microarray technology: research tools that analyze the expression of thousands of genes simultaneously and search for particular gene changes associated with a disease. DNA microarrays are also called *DNA chips.*

mutations: a permanent change in the DNA that can be inherited.

nucleotide bases: the nitrogen-containing building blocks of DNA and RNA—cytosine (C), thymine (T), uracil (U), guanine (G), and adenine (A). In DNA, the base pairs are A–T and C–G and in RNA, the base pairs are A–U and C–G.

nucleotides: the subunits of DNA and RNA molecules, composed of a phosphate group, a 5-carbon sugar (deoxyribose for DNA and ribose for RNA), and a nitrogen-containing base.

nutritional genomics: the science of how food (and its components) interacts with the

genome. The study of how nutrients affect the activities of genes is called *nutrigenomics.* The study of how genes affect the activities of nutrients is called *nutrigenetics.*

phenylketonuria (FEN-il-KEY-toe-NEW-ree-ah) or **PKU:** an inherited disorder characterized by failure to metabolize the amino acid phenylalanine to tyrosine.

RNA (ribonucleic acid): a compound similar to DNA, but RNA is a single strand with a ribose sugar instead of a deoxyribose sugar and uracil instead of thymine as one of its bases.

A Genomics Primer

Figure H6-1 shows the relationships among the materials that comprise the genome. As Chapter 6's discussion of protein synthesis pointed out, genetic information is encoded in DNA molecules within the nucleus of cells. The DNA molecules and associated proteins are packed within 46 **chromosomes.** The genes are segments of a DNA strand that can eventually be translated into one or more proteins. The sequence of **nucleotide bases** within each gene determines the amino acid sequence of a particular protein. Scientists currently estimate that there are between 20,000 and 25,000 genes in the human genome.

As Figure 6-7 (p. 188) explained, when cells make proteins, a **DNA** sequence is used to make messenger **RNA.** The **nucleotide** sequence in messenger RNA then determines the amino acid sequence to make a protein. This process—from genetic information to protein synthesis—is known as **gene expression.** Gene expression can be determined by measuring the amounts of messenger RNA in a tissue sample. **Microarray technology** (see photo on p. 207) allows researchers to detect messenger RNA and analyze the expression of thousands of genes simultaneously.

Simply having a certain gene does not determine that its associated trait will be expressed; the gene has to be activated. (Similarly, owning lamps does not ensure you will have light in your home unless you turn them on.) Nutrients are among many environmental factors that play key roles in either activating or silencing genes. Switching genes on and off does not change the DNA itself, but it can have dramatic consequences for a person's health.

The area of study that examines how environmental factors influence gene expression without changing the DNA is known as **epigenetics.** To turn genes on, enzymes attach proteins near the beginning of a gene. If enzymes attach a methyl group (CH_3) instead, the protein is blocked from binding to the gene and the gene remains switched off. Other factors influence gene expression as well, but methyl groups are currently the most well understood. They also are known to have dietary connections.

The accompanying photo of two mice illustrates epigenetics and how diet can influence genetic traits such as hair color and body weight. Both mice have a gene that tends to produce fat, yellow pups, but their mothers were given different diets. The mother of the mouse on the right was given a dietary supplement containing the B vitamins folate and vitamin B_{12}. These nutrients silenced the gene for "yellow and fat," resulting in brown pups with normal appetites. As Chapter 10 explains, one of the main roles of these B vitamins is to transfer methyl groups. In the case of the supplemented mice, methyl groups migrated onto DNA and shut off several genes, thus producing brown coats and protecting against the development of

FIGURE H6-1 | The Human Genome

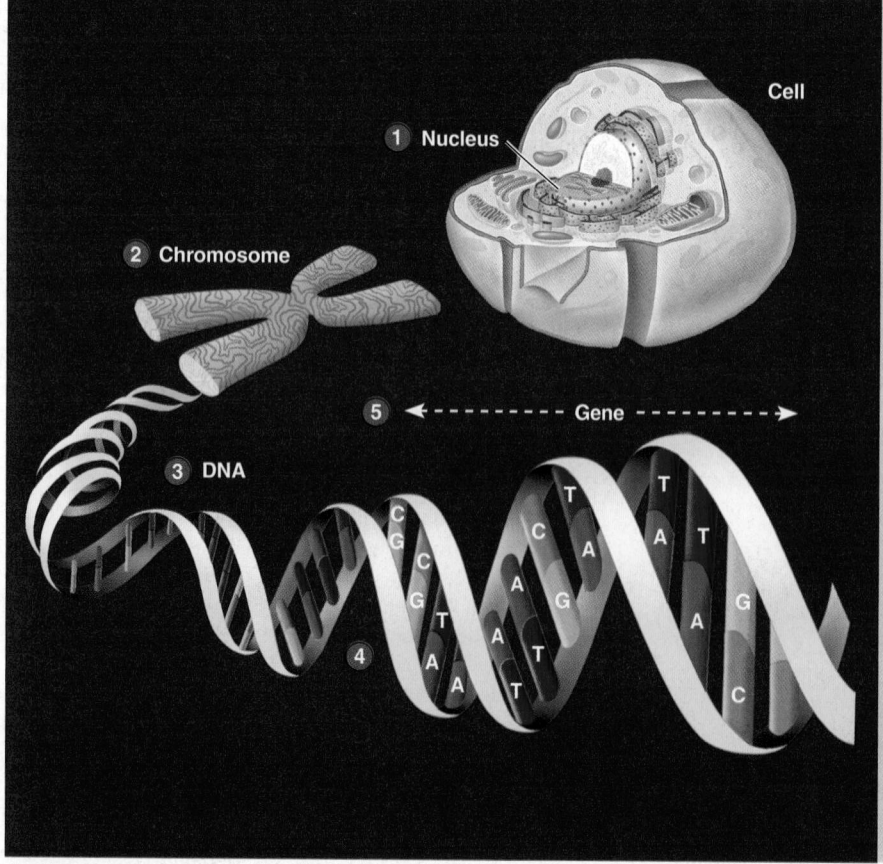

Cell

1 Nucleus

2 Chromosome

5 ◄ - - - - - - - - - Gene - - - - - - - - - ►

3 DNA

4

1 The human genome is a complete set of genetic material organized into 46 chromosomes, located within the nucleus of a cell.

2 A chromosome is made of DNA and associated proteins.

3 The double helical structure of a DNA molecule is made up of two long chains of nucleotides. Each nucleotide is composed of a phosphate group, a 5-carbon sugar, and a base.

4 The sequence of nucleotide bases (C, G, A, T) determines the amino acid sequence of proteins. These bases are connected by hydrogen bonding to form base pairs—adenine (A) with thymine (T) and guanine (G) with cytosine (C).

5 A gene is a segment of DNA that includes the information needed to synthesize one or more proteins.

Adapted from "A Primer: From DNA to Life," Human Genome Project, U.S. Department of Energy Office of Science; www.ornl.gov/sci/techresources/Human_Genome/

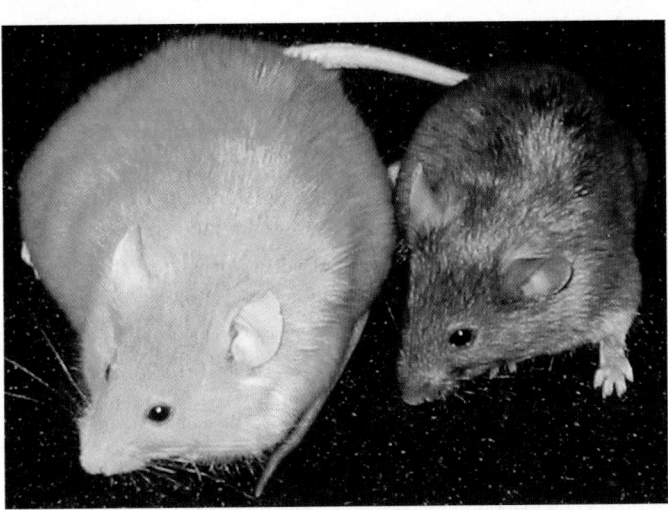

Both of these mice have the gene that tends to produce fat, yellow pups, but their mothers had different diets. The mother of the mouse on the right received a dietary supplement, which silenced the gene, resulting in brown pups with normal appetites.

obesity and some related diseases. Keep in mind that these changes occurred epigenetically. In other words, the DNA sequence within the genes of the mice remained the same.

Whether silencing or activating a gene is beneficial or harmful depends on what the gene does. Silencing a gene that stimulates cancer growth, for example, would be beneficial, but silencing a gene that suppresses cancer growth would be harmful. Similarly, activating a gene that defends against obesity would be beneficial, but activating a gene that promotes obesity would be harmful. Much research is under way to determine which nutrients activate or silence which genes.

Genetic Variation and Disease

Except for identical twins, no two persons are genetically identical. The variation in the genomes of any two persons, however, is only about 0.1 percent, a difference of only one nucleotide base in every 1000. Yet it is this incredibly small difference that makes each of us unique and explains why, given the same environmental influences, some of us develop certain diseases and others do not. Similarly, genetic variation explains why some of us respond to interventions such as diet and others do not. For example, following a diet low in saturated fats will significantly lower LDL cholesterol for most people, but the degree of change varies dramatically among individuals, with some people having only a small decrease or even a slight increase.[3] In other words, dietary factors may be more helpful or more harmful depending on a person's particular genetic variations.[4] (Such findings help to explain some of the conflicting results from research studies.) The goal of nutritional genomics is to custom design *specific* recommendations that fit the needs of *each* individual. Such personalized recommendations are expected to provide more effective disease prevention and treatment solutions.

Diseases characterized by a single-gene disorder are genetically predetermined, usually exert their effects early in life, and greatly affect those touched by them, but are relatively rare. The cause and effect of single-gene disorders is clear—those with the genetic defect get the disease and those without it don't. In contrast, the more common diseases, such as heart disease and cancer, are influenced by many genes and typically develop over several decades. These chronic diseases have multiple genetic components that *predispose* the prevention or development of a disease, depending on a variety of environmental factors (such as smoking, diet, and physical activity).[5] Both types are of interest to researchers in nutritional genomics.

Single-Gene Disorders

Some disorders are caused by **mutations** in single genes that are inherited at birth. The consequences of a missing or malfunctioning protein can seriously disrupt metabolism and may require significant dietary or medical intervention. A classic example of a diet-related, single-gene disorder is **phenylketonuria,** or **PKU.**

Approximately one in every 15,000 infants in the United States is born with PKU. PKU arises from mutations in the gene that codes for the enzyme that converts the essential amino acid phenylalanine to the amino acid tyrosine. Without this enzyme, phenylalanine and its metabolites accumulate and damage the nervous system, resulting in mental retardation, seizures, and behavior abnormalities. At the same time, the body cannot make tyrosine or compounds made from it (such as the neurotransmitter epinephrine). Consequently, tyrosine becomes an essential amino acid: because the body cannot make it, the diet must supply it.

Although the most debilitating effect is on brain development, other symptoms of PKU become evident if the condition is left untreated. Infants with PKU may have poor appetites and grow slowly. They may be irritable or have tremors or seizures. Their bodies and urine may have a musty odor. Their skin coloring may be unusually pale, and they may develop skin rashes.

The effect of nutrition intervention in PKU is remarkable. In fact, the only current treatment for PKU is a diet that restricts phenylalanine and supplies tyrosine to maintain blood levels of these amino acids within safe ranges. Because all foods containing protein provide phenylalanine, the diet must depend on a formula to supply a phenylalanine-free source of energy, protein, vitamins, and minerals. If the restricted diet is conscientiously followed, the symptoms can be prevented. Because phenylalanine is an essential amino acid, the diet cannot exclude it completely. Children with PKU need phenylalanine to grow, but they cannot handle excesses without detrimental effects. Therefore, their diets must provide enough phenylalanine to support normal growth and health but not enough to cause harm. The diet must also provide tyrosine. To ensure that blood concentrations of phenylalanine and tyrosine are close to normal, children and adults who have PKU must have blood tests periodically and adjust their diets as necessary.

Multigene Disorders

In multigene disorders, each of the genes can influence the progression of a disease, but no single gene causes the disease on its own. For this reason, genomics researchers must study the expression and interactions of *multiple* genes. Because multigene disorders are often sensitive to interactions with environmental influences, they are not as straightforward as single-gene disorders. Heart disease provides an example of a chronic disease with multiple gene and environmental influences. Consider that major risk factors for heart disease include elevated blood cholesterol levels, obesity, diabetes, and hypertension, yet the underlying genetic and environmental causes of any of these individual risk factors is not completely understood. Genomic research can reveal details about each of these risk factors. For example, tests could determine whether blood cholesterol levels are high due to increased cholesterol absorption or production or because of decreased cholesterol degradation.[6] This information could then guide physicians and dietitians to prescribe the most appropriate medical and dietary interventions from among many possible solutions.[7] Today's dietary recommendations advise a low-fat diet, which helps people with a small type of LDL but not those with the large type. In fact, a low-fat diet is actually more harmful for people with the large type. Finding the best option for each person will be a challenge given the many possible interactions between genes and environmental factors and the millions of possible gene variations in the human genome that make each individual unique.[8]

The results of genomic research are helping to explain findings from previous nutrition research. Consider dietary fat and heart disease, for example. As Highlight 5 explained, epidemiological and clinical studies have found that a diet high in unsaturated fatty acids often helps to maintain a healthy blood lipid profile. Now genetic studies offer an underlying explanation of this relationship: diets rich in polyunsaturated fatty acids activate genes responsible for making enzymes that break down fats and silence genes responsible for making enzymes that make fats.[9] Both actions change fat metabolism in the direction of lowering blood lipids.

To learn more about how individuals respond to diet, researchers examine the genetic differences between people. The most common genetic differences involve a change in a single nucleotide base located in a particular region of a DNA strand—thymine replacing cytosine, for example. Such variations are called single nucleotide polymorphisms (SNPs), and they commonly occur throughout the genome. Many SNPs (commonly pronounced "snips") have no effect on cell activity. In fact, SNPs are significant only if they affect the amino acid sequence of a protein in a way that alters its function *and* if that function is critical to the body's well-being. Research on a gene that plays a key role in lipid metabolism reveals differences in a person's response to diet depending on whether the gene has a common SNP. People with the SNP have lower LDL when eating a diet rich in polyunsaturated fatty acids—and higher LDL with a low intake—than those without the SNP.[10] These findings clearly show how diet (in this case, polyunsaturated fat) interacts with a gene (in this case, a fat metabolism gene with a SNP) to influence the development of a disease (changing blood lipids implicated in heart disease). The quest now is to identify the genetic characteristics that predict various responses to dietary recommendations.[11]

Clinical Concerns

Because multigene, chronic diseases are common, an understanding of the human genome will have widespread ramifications for health care. This new understanding of the human genome is expected to change health care by:

- Providing knowledge of an individual's genetic predisposition to specific diseases.

- Allowing physicians to develop "designer" therapies—prescribing the most effective schedule of screening, behavior changes (including diet), and medical interventions based on each individual's genetic profile.

- Enabling manufacturers to create new medications for each genetic variation so that physicians can prescribe the best medicine in the exact dose and frequency to enhance effectiveness and minimize the risks of side effects.

- Providing a better understanding of the nongenetic factors that influence disease development.

Enthusiasm surrounding genomic research needs to be put into perspective, however, in terms of the present status of clinical medicine as well as people's willingness to make difficult lifestyle choices. Critics have questioned whether genetic markers for disease would be more useful than simple clinical measurements, which reflect both genetic *and* environmental influences. In other words, knowing that a person is genetically predisposed to have high blood cholesterol is not necessarily more useful than knowing the person's actual blood cholesterol level.[12] Furthermore, if a disease has many genetic risk factors, each gene that contributes to susceptibility may have little influence on its own, so the benefits of identifying an individual genetic marker might be small. The long-range possibility is that many genetic markers will eventually be identified, and the hope is that the combined information will be a useful and accurate predictor of disease.

Having the knowledge to prevent disease and actually taking action do not always coincide. Despite the abundance of current dietary recommendations, people seem unwilling to make behavior changes known to improve their health. For example, it has been estimated that heart disease and type 2 diabetes are 90 percent preventable when people adopt an appropriate diet, maintain a healthy body weight, and exercise regularly.[13] Yet these two diseases remain among the leading causes of death. Given the difficulty that people have with current recommendations, it may be unrealistic to expect that many of them will enthusiastically adopt an even more detailed list of lifestyle modifications. Then again, compliance may be better when it is supported by information based on a person's own genetic profile.

The debate over nature versus nurture—whether genes or the environment are more influential—has quieted. The focus has shifted. Scientists acknowledge the important roles of each and understand the real answers lie within the myriad interactions. Current research is sorting through how nutrients (and other dietary factors) and genes confer health benefits or risks. Answers from genomic research may not become apparent for years to come, but the opportunities and rewards may prove well worth the efforts.[14]

NUTRITION ON THE NET

For further study of topics covered in this chapter, log on to **academic.cengage .com/nutrition/rolfes/UNCN8e**. Go to Chapter 6, then to Nutrition on the Net.

- Get information about human genomic discoveries and how they can be used to improve health from the Genomics and Disease Prevention site of the Centers for Disease Control: **www.cdc.gov/genomics**

REFERENCES

1. G. T. Keusch, What do –*omics* mean for the science and policy of the nutritional sciences? *American Journal of Clinical Nutrition* 83 (2006): 520S–522S.

2. N. Fogg-Johnson and J. Kaput, Nutrigenomics: An emerging scientific discipline, *Food Technology* 57 (2003): 60–67; R. Weinshilboum, Inheritance and drug response, *New England Journal of Medicine* 348 (2003): 529–537; A. E. Guttmacher and F. S. Collins, Genomic medicine—A primer, *New England Journal of Medicine* 347 (2002): 1512–1520.

3. D. Corella and J. M. Ordovas, Single nucleotide polymorphisms that influence lipid metabolism: Interaction with dietary factors, *Annual Review of Nutrition* 25 (2005): 341–390.

4. E. Trujillo, C. Davis, and J. Milner, Nutrigenomics, proteomics, metabolomics, and the practice of dietetics, *Journal of the American Dietetic Association* 106 (2006): 403–413.

5. J. Kaput and coauthors, The case for strategic international alliances to harness nutritional genomics for public and personal health, *British Journal of Nutrition* 94 (2005): 623–632; J. Kaput and R. L. Rodriguez, Nutritional genomics: The next frontier in the postgenome era, *Physiological Genomics* 16 (2004): 166–177.

6. J. B. German, M. A. Roberts, and S. M. Watkins, Personal metabolomics as a next generation nutritional assessment, *Journal of Nutrition* 133 (2003): 4260–4266.

7. R. M. DeBusk and coauthors, Nutritional genomics in practice: Where do we begin? *Journal of the American Dietetic Association* 105 (2005): 589–597.

8. J. M. Ordovas, Nutrigenetics, plasma lipids, and cardiovascular risk, *Journal of the American Dietetic Association* 106 (2006): 1074–1081.

9. H. Sampath and J. M. Ntambi, Polyunsaturated fatty acid regulation of genes of lipid metabolism, *Annual Review of Nutrition* 25 (2005): 317–340.

10. E. S. Tai and coauthors, Polyunsaturated fatty acids interact with PPARA–L162V polymorphism to affect plasma triglyceride apolipoprotein C-III concentrations in the Framingham Heart Study, *Journal of Nutrition* 135 (2005): 397–403.

11. J. M. Ordovas, The quest for cardiovascular health in the genomic era: Nutrigenetics and plasma lipoproteins, *Proceedings of the Nutrition Society* 63 (2004): 145–152.

12. W. C. Willett, Balancing life-style and genomics research for disease prevention, *Science* 296 (2002): 695–698.

13. S. Yusut and coauthors, Effect of potentially modifiable risk factors associated with myocardial infarction in 52 countries (the INTERHEART Study): Case-control study, *Lancet* 364 (2004): 937–952; Willett, 2002.

14. A. E. Guttmacher and F. S. Collins, Realizing the promise of genomics in biomedical research, *Journal of the American Medical Association* 294 (2005): 1399–1402; P. J. Stover, Nutritional genomics, *Physiological Genomics* 16 (2004): 161–165.

Nutrition in Your Life

You eat breakfast and hustle off to class. After lunch, you study for tomorrow's exam. Dinner is followed by an evening of dancing. Do you ever think about how the food you eat powers the activities of your life? What happens when you don't eat—or when you eat too much? Learn how the cells of your body transform carbohydrates, fats, and proteins into energy—and what happens when you give your cells too much or too little of any of these nutrients. Discover the metabolic pathways that lead to body fat and those that support physical activity. It's really quite fascinating.

CENGAGENOW™ The CengageNOW logo indicates an opportunity for online self-study, linking you to interactive tutorials and videos based on your level of understanding.

academic.cengage.com/login

Figure 7.5: Animated! Glycolysis: Glucose-to-Pyruvate

Figure 7.10: Animated! Fatty Acid-to-Acetyl CoA

Figure 7.18: Animated! The TCA Cycle

Figure 7.19: Animated! Electron Transport Chain and ATP Synthesis

Nutrition Portfolio Journal

Metabolism: Transformations and Interactions

Energy makes it possible for people to breathe, ride bicycles, compose music, and do everything else they do. All the energy that sustains human life initially comes from the sun—the ultimate source of energy. As Chapter 1 explained, *energy* is the capacity to do work. Although every aspect of our lives depends on energy, the concept of energy can be difficult to grasp because it cannot be seen or touched, and it manifests in various forms, including heat, mechanical, electrical, and chemical energy. In the body, heat energy maintains a constant body temperature, and electrical energy sends nerve impulses. Energy is stored in foods and in the body as chemical energy.

During **photosynthesis,** plants make simple sugars from carbon dioxide and capture the sun's light energy in the chemical bonds of those sugars. Then human beings eat either the plants or animals that have eaten the plants. These foods provide energy, but how does the body obtain that energy from foods? This chapter answers that question by following the nutrients that provide the body with **fuel** through a series of reactions that release energy from their chemical bonds. As the bonds break, they release energy in a controlled version of the same process by which wood burns in a fire. Both wood and food have the potential to provide energy. When wood burns in the presence of oxygen, it generates heat and light (energy), steam (water), and some carbon dioxide and ash (waste). Similarly, during **metabolism,** the body releases energy, water, and carbon dioxide.

By studying metabolism, you will understand how the body uses foods to meet its needs and why some foods meet those needs better than others. Readers who are interested in weight control will discover which foods contribute most to body fat and which to select when trying to gain or lose weight safely. Physically active readers will discover which foods best support endurance activities and which to select when trying to build lean body mass.

photosynthesis: the process by which green plants use the sun's energy to make carbohydrates from carbon dioxide and water.
• **photo** = light
• **synthesis** = put together (making)

fuel: compounds that cells can use for energy. The major fuels include glucose, fatty acids, and amino acids; other fuels include ketone bodies, lactate, glycerol, and alcohol.

metabolism: the sum total of all the chemical reactions that go on in living cells. Energy metabolism includes all the reactions by which the body obtains and expends the energy from food.
• **metaballein** = change

FIGURE 7-22   Feasting and Fasting

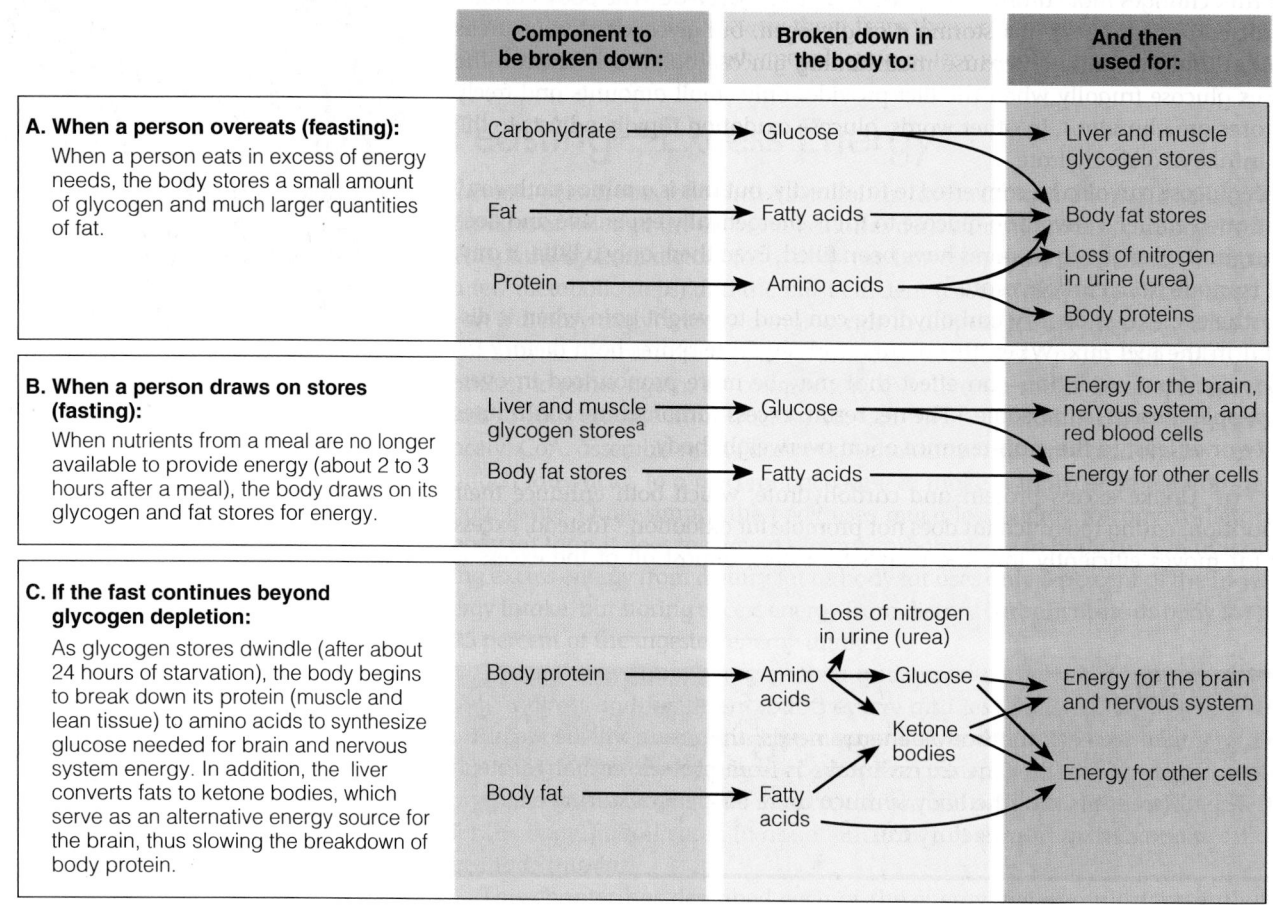

	Component to be broken down:	Broken down in the body to:	And then used for:
A. When a person overeats (feasting): When a person eats in excess of energy needs, the body stores a small amount of glycogen and much larger quantities of fat.	Carbohydrate → Fat → Protein →	Glucose Fatty acids Amino acids	Liver and muscle glycogen stores / Body fat stores / Loss of nitrogen in urine (urea) / Body proteins
B. When a person draws on stores (fasting): When nutrients from a meal are no longer available to provide energy (about 2 to 3 hours after a meal), the body draws on its glycogen and fat stores for energy.	Liver and muscle glycogen stores[a] → Body fat stores →	Glucose Fatty acids	Energy for the brain, nervous system, and red blood cells / Energy for other cells
C. If the fast continues beyond glycogen depletion: As glycogen stores dwindle (after about 24 hours of starvation), the body begins to break down its protein (muscle and lean tissue) to amino acids to synthesize glucose needed for brain and nervous system energy. In addition, the liver converts fats to ketone bodies, which serve as an alternative energy source for the brain, thus slowing the breakdown of body protein.	Body protein → Body fat →	Amino acids ↔ Glucose / Loss of nitrogen in urine (urea) / Ketone bodies / Fatty acids	Energy for the brain and nervous system / Energy for other cells

[a]The muscles' stored glycogen provides glucose only for the muscle in which the glycogen is stored.

cells, then breaking down to yield acetyl CoA, and finally delivering energy to power the cells' work. Several hours later, however, most of the glucose is used up—liver glycogen is exhausted and blood glucose begins to fall. Low blood glucose serves as a signal that promotes further fat breakdown and release of amino acids from muscles.

Glucose Needed for the Brain At this point, most of the cells are depending on fatty acids to continue providing their fuel. But red blood cells and the cells of the nervous system need glucose. Glucose is their primary energy fuel, and even when other energy fuels are available, glucose must be present to permit the energy-metabolizing machinery of the nervous system to work. Normally, the brain and nerve cells—which weigh only about three pounds—consume about half of the total *glucose* used each day (about 500 kcalories' worth). About one-fourth of the *energy* the adult body uses when it is at rest is spent by the brain; in children, it can be up to one-half.

Protein Meets Glucose Needs The red blood cells' and brain's special requirements for glucose pose a problem for the fasting body. The body can use its stores of fat, which may be quite generous, to furnish most of its cells with energy, but the red blood cells are completely dependent on glucose, ◆ and the brain and nerves prefer energy in the form of glucose. Amino acids that yield pyruvate can be used to make glucose, and to obtain the amino acids, body proteins must be broken down. For this reason, body protein tissues such as muscle and liver always break down to some extent during fasting. The amino acids that can't be used to make glucose are used as an energy source for other body cells.

The breakdown of body protein is an expensive way to obtain glucose. In the first few days of a fast, body protein provides about 90 percent of the needed glu-

◆ Red blood cells contain no mitochondria. Review Figure 7-1 (p. 214) to fully appreciate why red blood cells must depend on glucose for energy.

cose; glycerol, about 10 percent. If body protein losses were to continue at this rate, death would ensue within three weeks, regardless of the quantity of fat a person had stored. Fortunately, fat breakdown also increases with fasting—in fact, fat breakdown almost doubles, providing energy for other body cells and glycerol for glucose production.

The Shift to Ketosis As the fast continues, the body finds a way to use its fat to fuel the brain. It adapts by combining acetyl CoA fragments derived from fatty acids to produce an alternate energy source, ketone bodies (Figure 7-23). Normally produced and used only in small quantities, ketone bodies ◆ can provide fuel for some brain cells. Ketone body production rises until, after about ten days of fasting, it is meeting much of the nervous system's energy needs. Still, many areas of the brain rely exclusively on glucose, and to produce it, the body continues to sacrifice protein—albeit at a slower rate than in the early days of fasting.

When ketone bodies contain an acid group (COOH), they are called keto acids. Small amounts of keto acids are a normal part of the blood chemistry, but when their concentration rises, the pH of the blood drops. This is ketosis, a sign that the body's chemistry is going awry. Elevated blood ketones (ketonemia) are excreted in the urine (ketonuria). A fruity odor on the breath (known as acetone breath) develops, reflecting the presence of the ketone acetone.

Suppression of Appetite Ketosis also induces a loss of appetite. As starvation continues, this loss of appetite becomes an advantage to a person without access to food, because the search for food would be a waste of energy. When the person finds food and eats again, the body shifts out of ketosis, the hunger center gets the message that food is again available, and the appetite returns. Highlight 9 includes a discussion of the risks of ketosis-producing diets in its review of popular weight-loss diets.

Slowing of Metabolism In an effort to conserve body tissues for as long as possible, the hormones of fasting slow metabolism. As the body shifts to the use of ketone bodies, it simultaneously reduces its energy output and conserves both its fat and its lean tissue. Still the lean (protein-containing) organ tissues shrink in mass and perform less metabolic work, reducing energy expenditures. As the muscles waste, they can do less work and so demand less energy, reducing expenditures further. Although fasting may promote dramatic *weight* loss, a low-kcalorie diet better supports *fat* loss while retaining lean tissue.

◆ Reminder: *Ketone bodies* are compounds produced during the incomplete breakdown of fat when glucose is not available.

FIGURE 7-23 Ketone Body Formation

1 The first step in the formation of ketone bodies is the condensation of two molecules of acetyl CoA and the removal of the CoA to form a compound that is converted to the first ketone body.

Acetyl CoA   +   Acetyl CoA   + H_2O

2 CoA

A ketone, acetoacetate

2 This ketone body may lose a molecule of carbon dioxide to become another ketone.

CO_2

3 Or, the acetoacetate may add two hydrogens, becoming another ketone body (beta-hydroxybutyrate). See Appendix C for more details.

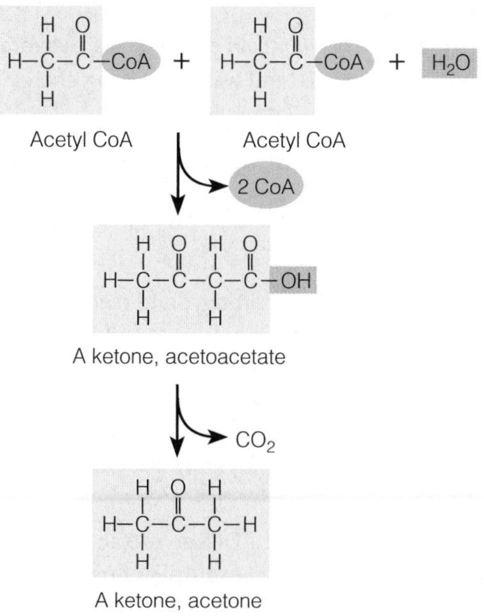

A ketone, acetone

Symptoms of Starvation The adaptations just described—slowing of energy output and reduction in fat loss—occur in the starving child, the hungry homeless adult, the fasting religious person, the adolescent with anorexia nervosa, and the malnourished hospital patient. Such adaptations help to prolong their lives and explain the physical symptoms of starvation: wasting; slowed heart rate, respiration, and metabolism; lowered body temperature; impaired vision; organ failure; and reduced resistance to disease.[10] Psychological effects of food deprivation include depression, anxiety, and food-related dreams.

The body's adaptations to fasting are sufficient to maintain life for a long time—up to two months. Mental alertness need not be diminished, and even some physical energy may remain unimpaired for a surprisingly long time. These remarkable adaptations, however, should not prevent anyone from recognizing the very real hazards that fasting presents.

IN SUMMARY

When fasting, the body makes a number of adaptations: increasing the breakdown of fat to provide energy for most of the cells, using glycerol and amino acids to make glucose for the red blood cells and central nervous system, producing ketones to fuel the brain, suppressing the appetite, and slowing metabolism. All of these measures conserve energy and minimize losses.

This chapter has probed the intricate details of metabolism at the level of the cells, exploring the transformations of nutrients to energy and to storage compounds. Several chapters and highlights build on this information. The highlight that follows this chapter shows how alcohol disrupts normal metabolism. Chapter 8 describes how a person's intake and expenditure of energy are reflected in body weight and body composition. Chapter 9 examines the consequences of unbalanced energy budgets—overweight and underweight. Chapter 10 shows the vital roles the B vitamins play as coenzymes assisting all the metabolic pathways described here.

CENGAGENOW™
academic.cengage.com/login

 Nutrition Portfolio

All day, every day, your cells dismantle carbohydrates, fats, and proteins, with the help of vitamins, minerals, and water, releasing energy to meet your body's immediate needs or storing it as fat for later use.

- Describe what types of foods best support aerobic and anaerobic activities.

- Consider whether you eat more protein, carbohydrate, or fat than your body needs.

- Explain how a low-carbohydrate diet forces your body into ketosis.

STUDY QUESTIONS

CENGAGENOW™
To assess your understanding of chapter topics, take the Student Practice Test and explore the modules recommended in your Personalized Study Plan. Log on to **academic.cengage.com/login**.

These questions will help you review the chapter. You will find the answers in the discussions on the pages provided.

1. Define metabolism, anabolism, and catabolism; give an example of each. (pp. 213–216)

2. Name one of the body's high-energy molecules, and describe how it is used. (pp. 216–217)

3. What are coenzymes, and what service do they provide in metabolism? (p. 216)

4. Name the four basic units, derived from foods, that are used by the body in metabolic transformations. How many carbons are in the "backbones" of each? (pp. 217–218)

5. Define aerobic and anaerobic metabolism. How does insufficient oxygen influence metabolism? (pp. 220–221)

6. How does the body dispose of excess nitrogen? (pp. 225–227)

7. Summarize the main steps in the metabolism of glucose, glycerol, fatty acids, and amino acids. (pp. 226–228)

8. Describe how a surplus of the three energy nutrients contributes to body fat stores. (pp. 219–226)

9. What adaptations does the body make during a fast? What are ketone bodies? Define ketosis. (pp. 233–236)

10. Distinguish between a loss of *fat* and a loss of *weight,* and describe how each might happen. (pp. 235–236)

These multiple choice questions will help you prepare for an exam. Answers can be found below.

1. Hydrolysis is an example of a(n):
   a. coupled reaction.
   b. anabolic reaction.
   c. catabolic reaction.
   d. synthesis reaction.

2. During metabolism, released energy is captured and transferred by:
   a. enzymes.
   b. pyruvate.
   c. acetyl CoA.
   d. adenosine triphosphate.

3. Glycolysis:
   a. requires oxygen.
   b. generates abundant energy.
   c. converts glucose to pyruvate.
   d. produces ammonia as a by-product.

4. The pathway from pyruvate to acetyl CoA:
   a. produces lactate.
   b. is known as gluconeogenesis.
   c. is metabolically irreversible.
   d. requires more energy than it produces.

5. For complete oxidation, acetyl CoA enters:
   a. glycolysis.
   b. the TCA cycle.
   c. the Cori cycle.
   d. the electron transport chain.

6. Deamination of an amino acid produces:
   a. vitamin B_6 and energy.
   b. pyruvate and acetyl CoA.
   c. ammonia and a keto acid.
   d. carbon dioxide and water.

7. Before entering the TCA cycle, each of the energy-yielding nutrients is broken down to:
   a. ammonia.
   b. pyruvate.
   c. electrons.
   d. acetyl CoA.

8. The body stores energy for future use in:
   a. proteins.
   b. acetyl CoA.
   c. triglycerides.
   d. ketone bodies.

9. During a fast, when glycogen stores have been depleted, the body begins to synthesize glucose from:
   a. acetyl CoA.
   b. amino acids.
   c. fatty acids.
   d. ketone bodies.

10. During a fast, the body produces ketone bodies by:
    a. hydrolyzing glycogen.
    b. condensing acetyl CoA.
    c. transaminating keto acids.
    d. converting ammonia to urea.

REFERENCES

1. R. H. Garrett and C. M. Grisham, *Biochemistry* (Belmont, Calif.: Thomson Brooks/ Cole, 2005), p. 73.
2. R. A. Robergs, F. Ghiasvand, and D. Parker, Biochemistry of exercise-induced metabolic acidosis, *American Journal of Physiology— Regulatory, Integrative and Comparative Physiology* 287 (2004): R502–R516.
3. T. H. Pederson and coauthors, Intracellular acidosis enhances the excitability of working muscle, *Science* 305 (2004): 1144–1147.
4. S. S. Gropper, J. L. Smith, and J. L. Groff, *Advanced Nutrition and Human Metabolism* (Belmont, Calif.: Wadsworth/Thomson Learning, 2005), p. 198.

5. Garrett and Grisham, 2005, p. 669.
6. M. K. Hellerstein, No common energy currency: De novo lipogenesis as the road less traveled, *American Journal of Clinical Nutrition* 74 (2001): 707–708.
7. R. M. Devitt and coauthors, De novo lipogenesis during controlled overfeeding with sucrose or glucose in lean and obese women, *American Journal of Clinical Nutrition* 74 (2001): 707–708.
8. I. Marques-Lopes and coauthors, Postprandial de novo lipogenesis and metabolic changes induced by a high-carbohydrate, low-fat meal in lean and overweight men,

American Journal of Clinical Nutrition 73 (2001): 253–261.
9. E. J. Parks, Macronutrient Metabolism Group Symposium on "Dietary fat: How low should we go?" Changes in fat synthesis influenced by dietary macronutrient content, *Proceedings of the Nutrition Society* 61 (2002): 281–286.
10. C. A. Jolly, Dietary restriction and immune function, *Journal of Nutrition* 134 (2004): 1853–1856.

ANSWERS

Study Questions (multiple choice)

1. c   2. d   3. c   4. c   5. b   6. c   7. d   8. c
9. b   10. b

Alcohol and Nutrition

With the understanding of metabolism gained from Chapter 7, you are in a position to understand how the body handles alcohol, how alcohol interferes with metabolism, and how alcohol impairs health and nutrition. Before examining alcohol's damaging effects, it may be appropriate to mention that drinking alcohol in *moderation* may have some health benefits, including reduced risks of heart attacks, strokes, dementia, diabetes, and osteoporosis.[1] Moderate alcohol consumption may lower mortality from all causes, but only in adults age 35 and older.[2] No health benefits are evident before middle age.[3] Importantly, any benefits of alcohol must be weighed against the many harmful effects described in this highlight, as well as the possibility of alcohol abuse.

Alcohol in Beverages

To the chemist, **alcohol** refers to a class of organic compounds containing hydroxyl (OH) groups (the accompanying glossary defines alcohol and related terms). The glycerol to which fatty acids are attached in triglycerides is an example of an alcohol to a chemist. To most people, though, *alcohol* refers to the intoxicating ingredient in **beer, wine,** and **distilled liquor (hard liquor).** The chemist's name for this particular alcohol is *ethyl alcohol,* or **ethanol.** Glycerol has 3 carbons with 3 hydroxyl groups attached; ethanol has only 2 carbons and 1 hydroxyl group (see Figure H7-1). The remainder of this highlight talks about the particular alcohol, ethanol, but refers to it simply as *alcohol.*

Alcohols affect living things profoundly, partly because they act as lipid solvents. Their ability to dissolve lipids out of cell membranes allows alcohols to penetrate rapidly into cells, destroying cell structures and thereby killing the cells. For this reason, most alcohols are toxic in relatively small amounts; by the same token, because they kill microbial cells, they are useful as disinfectants.

Ethanol is less toxic than the other alcohols. Sufficiently diluted and taken in small enough doses, its action in the brain produces an effect that people seek—not with zero risk, but with a low enough risk (if the doses are low enough) to be tolerable. Used in this way, alcohol is a **drug**—that is, a substance that modifies body functions. Like all drugs, alcohol both offers benefits and poses hazards. The 2005 *Dietary Guidelines* advise "those who choose to drink alcoholic beverages to do so sensibly and in moderation."

Dietary Guidelines for Americans 2005

- Those who choose to drink alcoholic beverages should do so sensibly and in moderation: up to one drink per day for women and two drinks per day for men.

- Alcoholic beverages should not be consumed by some individuals, including those who cannot restrict their alcohol intake, women of childbearing age who may become pregnant, pregnant and lactating women, children and adolescents, individuals taking medications that can interact with alcohol, and those with specific medical conditions.

- Alcoholic beverages should be avoided by individuals engaging in activities that require attention, skill, or coordination, such as driving or operating machinery.

The term **moderation** is important when describing alcohol use. How many drinks constitute moderate use, and how much is "a drink"? First, a **drink** is any alcoholic beverage that delivers ¹/₂ ounce of *pure ethanol:*

- 5 ounces of wine
- 10 ounces of wine cooler
- 12 ounces of beer
- 1¹/₂ ounces of distilled liquor (80 proof whiskey, scotch, rum, or vodka)

Beer, wine, and liquor deliver different amounts of alcohol. The amount of alcohol in distilled liquor is stated as **proof:** 100 proof liquor is 50 percent alcohol, 80 proof is 40 percent alcohol, and so forth. Wine and beer have less alcohol than distilled liquor, although some fortified wines and beers have more alcohol than the regular varieties (see photo caption on p. 239).

FIGURE H7-1 Two Alcohols: Glycerol and Ethanol

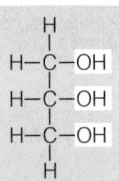

Glycerol is the alcohol used to make triglycerides.

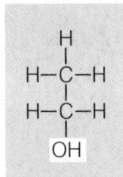

Ethanol is the alcohol in beer, wine, and distilled liquor.

GLOSSARY

acetaldehyde (ass-et-AL-duh-hide): an intermediate in alcohol metabolism.

alcohol: a class of organic compounds containing hydroxyl (OH) groups.

alcohol abuse: a pattern of drinking that includes failure to fulfill work, school, or home responsibilities; drinking in situations that are physically dangerous (as in driving while intoxicated); recurring alcohol-related legal problems (as in aggravated assault charges); or continued drinking despite ongoing social problems that are caused by or worsened by alcohol.

alcohol dehydrogenase (dee-high-DROJ-eh-nayz): an enzyme active in the stomach and the liver that converts ethanol to acetaldehyde.

alcoholism: a pattern of drinking that includes a strong craving for alcohol, a loss of control and an inability to stop drinking once begun, withdrawal symptoms (nausea, sweating, shakiness, and anxiety) after heavy drinking, and the need for increasing amounts of alcohol to feel "high."

antidiuretic hormone (ADH): a hormone produced by the pituitary gland in response to dehydration (or a high sodium concentration in the blood). It stimulates the kidneys to reabsorb more water and therefore prevents water loss in urine (also called *vasopressin*). (This ADH should not be confused with the enzyme alcohol dehydrogenase, which is also sometimes abbreviated ADH.)

beer: an alcoholic beverage brewed by fermenting malt and hops.

cirrhosis (seer-OH-sis): advanced liver disease in which liver cells turn orange, die, and harden, permanently losing their function; often associated with alcoholism.
• **cirrhos** = an orange

distilled liquor or **hard liquor:** an alcoholic beverage made by fermenting and distilling grains; sometimes called *distilled spirits*.

drink: a dose of any alcoholic beverage that delivers ½ oz of pure ethanol:
• 5 oz of wine
• 10 oz of wine cooler
• 12 oz of beer

• 1½ oz of hard liquor (80 proof whiskey, scotch, rum, or vodka)

drug: a substance that can modify one or more of the body's functions.

ethanol: a particular type of alcohol found in beer, wine, and distilled liquor; also called *ethyl alcohol* (see Figure H7-1). Ethanol is the most widely used—and abused—drug in our society. It is also the only legal, nonprescription drug that produces euphoria.

fatty liver: an early stage of liver deterioration seen in several diseases, including kwashiorkor and alcoholic liver disease. Fatty liver is characterized by an accumulation of fat in the liver cells.

fibrosis (fye-BROH-sis): an intermediate stage of liver deterioration seen in several diseases, including viral hepatitis and alcoholic liver disease. In fibrosis, the liver cells lose their function and assume the characteristics of connective tissue cells (fibers).

MEOS or **microsomal** (my-krow-SO-mal) **ethanol-oxidizing system:** a system of enzymes in the liver that oxidize not only

alcohol but also several classes of drugs.

moderation: in relation to alcohol consumption, not more than two drinks a day for the average-size man and not more than one drink a day for the average-size woman.

NAD (nicotinamide adenine dinucleotide): the main coenzyme form of the vitamin niacin. Its reduced form is NADH.

narcotic (nar-KOT-ic): a drug that dulls the senses, induces sleep, and becomes addictive with prolonged use.

proof: a way of stating the percentage of alcohol in distilled liquor. Liquor that is 100 proof is 50% alcohol; 90 proof is 45%, and so forth.

Wernicke-Korsakoff (VER-nee-key KORE-sah-kof) **syndrome:** a neurological disorder typically associated with chronic alcoholism and caused by a deficiency of the B vitamin thiamin; also called *alcohol-related dementia*.

wine: an alcoholic beverage made by fermenting grape juice.

12 oz beer — 10 oz wine cooler — 1½ oz liquor (80 proof whiskey, gin, brandy, rum, vodka) — 5 oz wine

Each of these servings equals one drink.

© Polara Studios, Inc.

Matthew Farruggio

Wines contain 7 to 24 percent alcohol by volume; those containing 14 percent or more must state their alcohol content on the label, whereas those with less than 14 percent may simply state "table wine" or "light wine." Beers typically contain less than 5 percent alcohol by volume and malt liquors, 5 to 8 percent; regulations vary, with some states requiring beer labels to show the alcohol content and others prohibiting such statements.

Second, because people have different tolerances for alcohol, it is impossible to name an exact daily amount of alcohol that is appropriate for everyone. Authorities have attempted to identify amounts that are acceptable for most healthy people. An accepted definition of moderation is up to two drinks per day for men and up to one drink per day for women. (Pregnant women are advised to abstain from alcohol, as Highlight 14 explains.) Notice that this advice is stated as a maximum, not as an average; seven drinks one night a week would not be considered moderate, even though one a day would be. Doubtless some people could consume slightly more; others could not handle nearly so much without risk. The amount a person can drink safely is highly individual, depending on genetics, health, gender, body composition, age, and family history.

Alcohol in the Body

From the moment an alcoholic beverage enters the body, alcohol is treated as if it has special privileges. Unlike foods, which require time for digestion, alcohol needs no digestion and is quickly absorbed across the walls of an empty stomach, reaching the brain within a few minutes. Consequently, a person can immediately feel euphoric when drinking, especially on an empty stomach.

When the stomach is full of food, alcohol has less chance of touching the walls and diffusing through, so its influence on the brain is slightly delayed. This information leads to a practical tip: eat snacks when drinking alcoholic beverages. Carbohydrate snacks slow alcohol absorption and high-fat snacks slow peristalsis, keeping the alcohol in the stomach longer. Salty snacks make a person thirsty; to quench thirst, drink water instead of more alcohol.

The stomach begins to break down alcohol with its **alcohol dehydrogenase** enzyme. Women produce less of this stomach enzyme than men; consequently, more alcohol reaches the intestine for absorption into the bloodstream. As a result, women absorb more alcohol than men of the same size who drink the same amount of alcohol. Consequently, they are more likely to become more intoxicated on less alcohol than men. Such differences between men and women help explain why women have a lower alcohol tolerance and a lower recommendation for moderate intake.

In the small intestine, alcohol is rapidly absorbed. From this point on, alcohol receives priority treatment: it gets absorbed and metabolized before most nutrients. Alcohol's priority status helps to ensure a speedy disposal and reflects two facts: alcohol cannot be stored in the body, and it is potentially toxic.

Alcohol Arrives in the Liver

The capillaries of the digestive tract merge into veins that carry the alcohol-laden blood to the liver. These veins branch and rebranch into capillaries that touch every liver cell. Liver cells are the only other cells in the body that can make enough of the alcohol dehydrogenase enzyme to oxidize alcohol at an appreciable rate. The routing of blood through the liver cells gives them the chance to dispose of some alcohol before it moves on.

Alcohol affects every organ of the body, but the most dramatic evidence of its disruptive behavior appears in the liver. If liver cells could talk, they would describe alcohol as demanding, egocentric, and disruptive of the liver's efficient way of running its business. For example, liver cells normally prefer fatty acids as their fuel, and they like to package excess fatty acids into triglycerides and ship them out to other tissues. When alcohol is present, however, the liver cells are forced to metabolize alcohol and let the fatty acids accumulate, sometimes in huge stockpiles. Alcohol metabolism can also permanently change liver cell structure, impairing the liver's ability to metabolize fats. As a result, heavy drinkers develop fatty livers.

The liver is the primary site of alcohol metabolism.[4] It can process about $1/2$ ounce of *ethanol* per hour (the amount in a typical drink), depending on the person's body size, previous drinking experience, food intake, and general health. This maximum rate of alcohol breakdown is set by the amount of alcohol dehydrogenase available. If more alcohol arrives at the liver than the enzymes can handle, the extra alcohol travels to all parts of the body, circulating again and again until liver enzymes are finally available to process it. Another practical tip derives from this information: drink slowly enough to allow the liver to keep up—no more than one drink per hour.

The amount of alcohol dehydrogenase enzyme present in the liver varies with individuals, depending on the genes they have inherited and on how recently they have eaten. Fasting for as little as a day forces the body to degrade its proteins, including the alcohol-processing enzymes, and this can slow the rate of alcohol metabolism by half. Drinking after not eating all day thus causes the drinker to feel the effects more promptly for two reasons: rapid absorption and slowed breakdown. By maintaining higher blood alcohol concentrations for longer times, alcohol can anesthetize the brain more completely (as described later in this highlight).

The alcohol dehydrogenase enzyme breaks down alcohol by removing hydrogens in two steps. (Figure H7-2 provides a simplified diagram of alcohol metabolism; Appendix C provides the chemical details.) In the first step, alcohol dehydrogenase oxidizes alcohol to **acetaldehyde.** High concentrations of acetaldehyde in the brain and other tissues are responsible for many of the damaging effects of **alcohol abuse.**

FIGURE H7-2 | Alcohol Metabolism

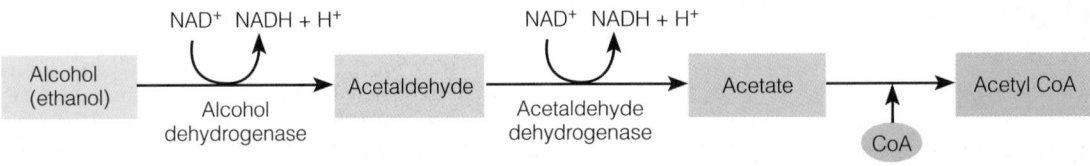

The conversion of alcohol to acetyl CoA requires the B vitamin niacin in its role as the coenzyme NAD. When the enzymes oxidize alcohol, they remove H atoms and attach them to NAD. Thus NAD is used up and NADH accumulates. (Note: More accurately, NAD+ is converted to NADH + H+.)

FIGURE H7-3 Alternate Route for Acetyl CoA: To Fat

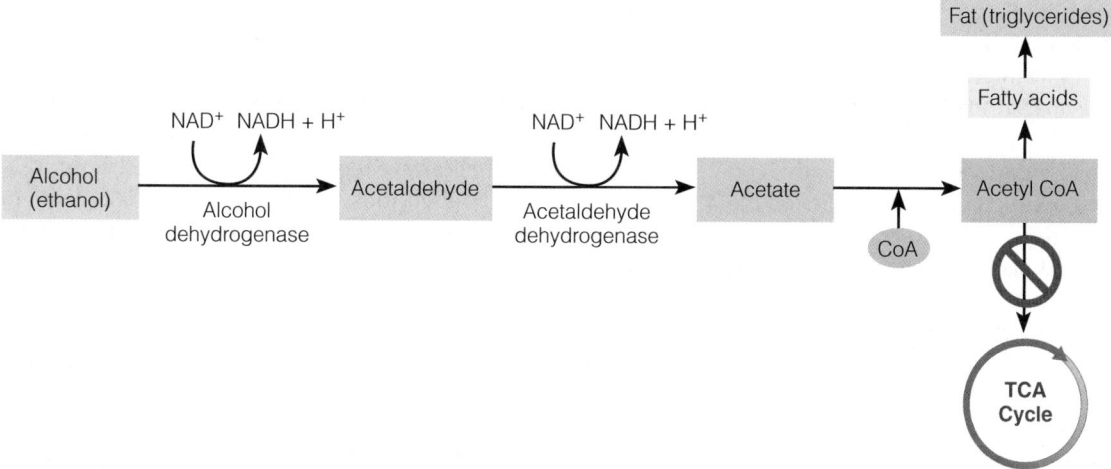

Acetyl CoA molecules are blocked from getting into the TCA cycle by the high level of NADH. Instead of being used for energy, the acetyl CoA molecules become building blocks for fatty acids.

In the second step, a related enzyme, acetaldehyde dehydrogenase, converts acetaldehyde to acetate, which is then converted to acetyl CoA—the "crossroads" compound introduced in Chapter 7 that can enter the TCA cycle to generate energy. These reactions produce hydrogen ions (H^+). The B vitamin niacin, in its role as the coenzyme **NAD (nicotinamide adenine dinucleotide),** helpfully picks up these hydrogen ions (becoming NADH). Thus, whenever the body breaks down alcohol, NAD diminishes and NADH accumulates. (Chapter 10 presents information on NAD and the other coenzyme roles of the B vitamins.)

Alcohol Disrupts the Liver

During alcohol metabolism, the multitude of other metabolic processes for which NAD is required, including glycolysis, the TCA cycle, and the electron transport chain, falter. Its presence is sorely missed in these energy pathways because it is the chief carrier of the hydrogens that travel with their electrons along the electron transport chain. Without adequate NAD, these energy pathways cannot function. Traffic either backs up, or an alternate route is taken. Such changes in the normal flow of energy pathways have striking physical consequences.

For one, the accumulation of hydrogen ions during alcohol metabolism shifts the body's acid-base balance toward acid. For another, the accumulation of NADH slows the TCA cycle, so pyruvate and acetyl CoA build up. Excess acetyl CoA then takes the route to fatty acid synthesis (as Figure H7-3 illustrates), and fat clogs the liver.

As you might expect, a liver overburdened with fat cannot function properly. Liver cells become less efficient at performing a number of tasks. Much of this inefficiency impairs a person's nutritional health in ways that cannot be corrected by diet alone. For example, the liver has difficulty activating vitamin D, as well as producing and releasing bile. To overcome such problems, a person needs to stop drinking alcohol.

The synthesis of fatty acids accelerates with exposure to alcohol. Fat accumulation can be seen in the liver after a single night of heavy drinking. **Fatty liver,** the first stage of liver deterioration seen in heavy drinkers, interferes with the distribution of nutrients and oxygen to the liver cells. Fatty liver is reversible with abstinence from alcohol. If fatty liver lasts long enough, however, the liver cells will die and form fibrous scar tissue. This second stage of liver deterioration is called **fibrosis.** Some liver cells can regenerate with good nutrition and abstinence from alcohol, but in the most advanced stage, **cirrhosis,** damage is the least reversible.

The fatty liver has difficulty generating glucose from protein. Without gluconeogenesis, blood glucose can plummet, leading to irreversible damage to the central nervous system.

The lack of glucose together with the overabundance of acetyl CoA sets the stage for ketosis. The body uses the acetyl CoA to make ketone bodies; their acidity pushes the acid-base balance further toward acid and suppresses nervous system activity.

Excess NADH also promotes the making of lactate from pyruvate. The conversion of pyruvate to lactate uses the hydrogens from NADH and restores some NAD, but a lactate buildup has serious consequences of its own—it adds still further to the body's acid burden and interferes with the excretion of another acid, uric acid, causing inflammation of the joints.

Alcohol alters both amino acid and protein metabolism. Synthesis of proteins important in the immune system slows down, weakening the body's defenses against infection. Protein deficiency can develop, both from a diminished synthesis of protein and from a poor diet. Normally, the cells would at least use the amino acids from the protein foods a person eats, but the drinker's liver deaminates the amino acids and uses the carbon fragments primarily to make fat or ketones. Eating well does not protect the drinker from protein depletion; a person has to stop drinking alcohol.

The liver's priority treatment of alcohol affects its handling of drugs as well as nutrients. In addition to the dehydrogenase enzyme already described, the liver possesses an enzyme system that metabolizes *both* alcohol and several other types of drugs. Called the **MEOS (microsomal ethanol-oxidizing system),** this system handles about one-fifth of the total alcohol a person consumes. At high blood concentrations or with repeated exposures, alcohol stimulates the synthesis of enzymes in the MEOS. The result is a more efficient metabolism of alcohol and tolerance to its effects.

As a person's blood alcohol rises, alcohol competes with—and wins out over—other drugs whose metabolism also relies on the MEOS. If a person drinks and uses another drug at the same time, the MEOS will dispose of alcohol first and metabolize the drug more slowly. While the drug waits to be handled later, the dose may build up so that its effects are greatly amplified—sometimes to the point of being fatal.

In contrast, once a heavy drinker stops drinking and alcohol is no longer competing with other drugs, the enhanced MEOS metabolizes drugs much faster than before. As a result, determining the correct dosages of medications can be challenging.

This discussion has emphasized the major way that the blood is cleared of alcohol—metabolism by the liver—but there is another way. About 10 percent of the alcohol leaves the body through the breath and in the urine. This is the basis for the breath and urine tests for drunkenness. The amounts of alcohol in the breath and in the urine are in proportion to the amount still in the bloodstream and brain. In nearly all states, legal drunkenness is set at 0.10 percent or less, reflecting the relationship between alcohol use and traffic and other accidents.

Alcohol Arrives in the Brain

Alcohol is a **narcotic.** People used it for centuries as an anesthetic because it can deaden pain. But alcohol was a poor anesthetic because one could never be sure how much a person would need and how much would be a fatal dose. Consequently, new, more predictable anesthetics have replaced alcohol. Nonetheless, alcohol continues to be used today as a kind of social anesthetic to help people relax or to relieve anxiety. People think that alcohol is a stimulant because it seems to relieve inhibitions. Actually, though, it accomplishes this by sedating *inhibitory* nerves, which are more numerous than excitatory nerves. Ultimately, alcohol acts as a depressant and affects all the nerve cells. Figure H7-4 describes alcohol's effects on the brain.

It is lucky that the brain centers respond to a rising blood alcohol concentration in the order described in Figure H7-4 because a person usually passes out before managing to drink a lethal dose. It is possible, though, to drink so fast that the effects of alcohol continue to accelerate after the person has passed out. Occasionally, a person dies from drinking enough to stop the heart before passing out. Table H7-1 shows the blood alcohol levels that correspond to progressively greater intoxication, and Table H7-2 shows the brain responses that occur at these blood levels.

Like liver cells, brain cells die with excessive exposure to alcohol. Liver cells may be replaced, but not all brain cells can regenerate. Thus some heavy drinkers suffer permanent brain damage.

FIGURE H7-4 Alcohol's Effects on the Brain

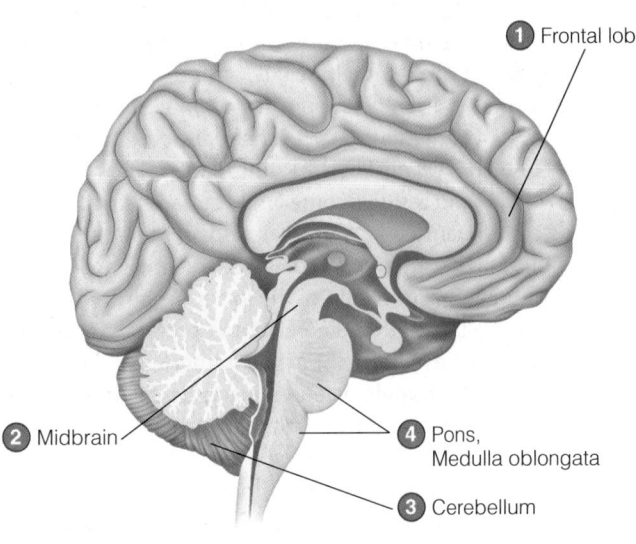

① Frontal lobe

② Midbrain

④ Pons, Medulla oblongata

③ Cerebellum

① Judgment and reasoning centers are most sensitive to alcohol. When alcohol flows to the brain, it first sedates the frontal lobe, the center of all conscious activity. As the alcohol molecules diffuse into the cells of these lobes, they interfere with reasoning and judgment.

② Speech and vision centers in the midbrain are affected next. If the drinker drinks faster than the rate at which the liver can oxidize the alcohol, blood alcohol concentrations rise: the speech and vision centers of the brain become sedated.

③ Voluntary muscular control is then affected. At still higher concentrations, the cells in the cerebellum responsible for coordination of voluntary muscles are affected, including those used in speech, eye-hand coordination, and limb movements. At this point people under the influence stagger or weave when they try to walk, or they may slur their speech.

④ Respiration and heart action are the last to be affected. Finally, the conscious brain is completely subdued, and the person passes out. Now the person can drink no more; this is fortunate because higher doses would anesthetize the deepest brain centers that control breathing and heartbeat, causing death.

TABLE H7-1 Alcohol Doses and Approximate Blood Level Percentages for Men and Women

Drinks[a]	Body Weight in Pounds—Men								
	100	120	140	160	180	200	220	240	ONLY SAFE DRIVING LIMIT
	00	00	00	00	00	00	00	00	
1	.04	.03	.03	.02	.02	.02	.02	.02	IMPAIRMENT BEGINS
2	.08	.06	.05	.05	.04	.04	.03	.03	
3	.11	.09	.08	.07	.06	.06	.05	.05	DRIVING SKILLS SIGNIFICANTLY AFFECTED
4	.15	.12	.11	.09	.08	.08	.07	.06	
5	.19	.16	.13	.12	.11	.09	.09	.08	
6	.23	.19	.16	.14	.13	.11	.10	.09	
7	.26	.22	.19	.16	.15	.13	.12	.11	
8	.30	.25	.21	.19	.17	.15	.14	.13	LEGALLY INTOXICATED
9	.34	.28	.24	.21	.19	.17	.15	.14	
10	.38	.31	.27	.23	.21	.19	.17	.16	

Drinks[a]	Body Weight in Pounds—Women									
	90	100	120	140	160	180	200	220	240	ONLY SAFE DRIVING LIMIT
	00	00	00	00	00	00	00	00	00	
1	.05	.05	.04	.03	.03	.03	.02	.02	.02	IMPAIRMENT BEGINS
2	.10	.09	.08	.07	.06	.05	.05	.04	.04	
3	.15	.14	.11	.10	.09	.08	.07	.06	.06	DRIVING SKILLS SIGNIFICANTLY AFFECTED
4	.20	.18	.15	.13	.11	.10	.09	.08	.08	
5	.25	.23	.19	.16	.14	.13	.11	.10	.09	
6	.30	.27	.23	.19	.17	.15	.14	.12	.11	
7	.35	.32	.27	.23	.20	.18	.16	.14	.13	
8	.40	.36	.30	.26	.23	.20	.18	.17	.15	LEGALLY INTOXICATED
9	.45	.41	.34	.29	.26	.23	.20	.19	.17	
10	.51	.45	.38	.32	.28	.25	.23	.21	.19	

NOTE: In some states, driving under the influence is proved when an adult's blood contains 0.08 percent alcohol, and in others, 0.10. Many states have adopted a "zero-tolerance" policy for drivers under age 21, using 0.02 percent as the limit.

[a] Taken within an hour or so; each drink equivalent to ½ ounce pure ethanol.

SOURCE: National Clearinghouse for Alcohol and Drug Information

TABLE H7-2 Alcohol Blood Levels and Brain Responses

Blood Alcohol Concentration	Effect on Brain
0.05	Impaired judgment, relaxed inhibitions, altered mood, increased heart rate
0.10	Impaired coordination, delayed reaction time, exaggerated emotions, impaired peripheral vision, impaired ability to operate a vehicle
0.15	Slurred speech, blurred vision, staggered walk, seriously impaired coordination and judgment
0.20	Double vision, inability to walk
0.30	Uninhibited behavior, stupor, confusion, inability to comprehend
0.40 to 0.60	Unconsciousness, shock, coma, death (cardiac or respiratory failure)

NOTE: Blood alcohol concentration depends on a number of factors, including alcohol in the beverage, the rate of consumption, the person's gender, and body weight. For example, a 100-pound female can become legally drunk (≥0.10 concentration) by drinking three beers in an hour, whereas a 220-pound male consuming that amount at the same rate would have a 0.05 blood alcohol concentration.

Whether alcohol impairs cognition in moderate drinkers is unclear.[5]

People who drink alcoholic beverages may notice that they urinate more, but they may be unaware of the vicious cycle that results. Alcohol depresses production of **antidiuretic hormone (ADH)**, a hormone produced by the pituitary gland that retains water—consequently, with less ADH, more water is lost. Loss of body water leads to thirst, and thirst leads to more drinking. Water will relieve dehydration, but the thirsty drinker may drink alcohol instead, which only worsens the problem. Such information provides another practical tip: drink water when thirsty and before each alcoholic drink. Drink an extra glass or two before going to bed. This strategy will help lessen the effects of a hangover.

Water loss is accompanied by the loss of important minerals. As Chapters 12 and 13 explain, these minerals are vital to the body's fluid balance and to many chemical reactions in the cells, including muscle action. Detoxification treatment includes restoration of mineral balance as quickly as possible.

Alcohol and Malnutrition

For many moderate drinkers, alcohol does not suppress food intake and may actually stimulate appetite. Moderate drinkers usually consume alcohol as *added* energy—on top of their normal food intake. In addition, alcohol in moderate doses is efficiently metabolized. Consequently, alcohol can contribute to body fat and weight gain—either by inhibiting oxidation or by being converted to fat.[6] Metabolically, alcohol is almost as efficient as fat in promoting obesity; each ounce of alcohol represents about a half-ounce of fat. Alcohol's contribution to body fat is most evident in the central obesity that commonly accompanies alcohol consumption, popularly—and appropriately—known as the "beer belly."[7] Alcohol in heavy doses, though, is not efficiently metabolized, generating more heat than fat. Heavy drinkers usually consume alcohol as *substituted* energy—instead of their normal food intake. They tend to eat poorly and suffer malnutrition.

Alcohol is rich in energy (7 kcalories per gram), but as with pure sugar or fat, the kcalories are empty of nutrients. The more alcohol people drink, the less likely that they will eat enough food to obtain adequate nutrients. The more kcalories spent on alcohol, the fewer kcalories available to spend on nutritious foods. Table H7-3 (p. 244) shows the kcalorie amounts of typical alcoholic beverages.

Chronic alcohol abuse not only displaces nutrients from the diet, but it also interferes with the body's metabolism of nutrients. Most dramatic is alcohol's effect on the B vitamin folate. The liver loses its ability to retain folate, and the kidneys increase their excretion of it. Alcohol abuse creates a folate deficiency that devastates digestive

TABLE H7-3 kCalories in Alcoholic Beverages and Mixers

Beverage	Amount (oz)	Energy (kcal)
Beer		
Regular	12	150
Light	12	78–131
Nonalcoholic	12	32–82
Distilled liquor (gin, rum, vodka, whiskey)		
80 proof	1½	100
86 proof	1½	105
90 proof	1½	110
Liqueurs		
Coffee liqueur, 53 proof	1½	175
Coffee and cream liqueur, 34 proof	1½	155
Crème de menthe, 72 proof	1½	185
Mixers		
Club soda	12	0
Cola	12	150
Cranberry juice cocktail	8	145
Diet drinks	12	2
Ginger ale or tonic	12	125
Grapefruit juice	8	95
Orange juice	8	110
Tomato or vegetable juice	8	45
Wine		
Dessert	3½	110–135
Nonalcoholic	8	14
Red or rosé	3½	75
White	3½	70
Wine cooler	12	170

system function. The intestine normally releases and retrieves folate continuously, but it becomes damaged by folate deficiency and alcohol toxicity, so it fails to retrieve its own folate and misses any that may trickle in from food as well. Alcohol also interferes with the action of folate in converting the amino acid homocysteine to methionine. The result is an excess of homocysteine, which has been linked to heart disease, and an inadequate supply of methionine, which slows the production of new cells, especially the rapidly dividing cells of the intestine and the blood. The combination of poor folate status and alcohol consumption has also been implicated in promoting colorectal cancer.

The inadequate food intake and impaired nutrient absorption that accompany chronic alcohol abuse frequently lead to a deficiency of another B vitamin—thiamin. In fact, the cluster of thiamin-deficiency symptoms commonly seen in chronic **alcoholism** has its own name—the **Wernicke-Korsakoff syndrome.** This syndrome is characterized by paralysis of the eye muscles, poor muscle coordination, impaired memory, and damaged nerves; it and other alcohol-related memory problems may respond to thiamin supplements.

Acetaldehyde, an intermediate in alcohol metabolism (review Figure H7-2, p. 240), interferes with nutrient use, too. For example, acetaldehyde dislodges vitamin B_6 from its protective binding protein so that it is destroyed, causing a vitamin B_6 deficiency and, thereby, lowered production of red blood cells.

Malnutrition occurs not only because of lack of intake and altered metabolism but because of direct toxic effects as well. Alcohol causes stomach cells to oversecrete both gastric acid and histamine, an immune system agent that produces inflammation. Beer in particular stimulates gastric acid secretion, irritating the linings of the stomach and esophagus and making them vulnerable to ulcer formation.

Overall, nutrient deficiencies are virtually inevitable in alcohol abuse, not only because alcohol displaces food but also because alcohol directly interferes with the body's use of nutrients, making them ineffective even if they are present. Intestinal cells fail to absorb B vitamins, notably, thiamin, folate, and vitamin B_{12}. Liver cells lose efficiency in activating vitamin D. Cells in the retina of the eye, which normally process the alcohol form of vitamin A (retinol) to its aldehyde form needed in vision (retinal), find themselves processing ethanol to acetaldehyde instead. Likewise, the liver cannot convert the aldehyde form of vitamin A to its acid form (retinoic acid), which is needed to support the growth of its (and all) cells.

Regardless of dietary intake, excessive drinking over a lifetime creates deficits of all the nutrients mentioned in this discussion and more. No diet can compensate for the damage caused by heavy alcohol consumption.

Alcohol's Short-Term Effects

The effects of abusing alcohol may be apparent immediately, or they may not become evident for years to come. Among the immediate consequences, all of the following involve alcohol use:[8]

- One-quarter of all emergency-room admissions
- One-third of all suicides
- One-half of all homicides
- One-half of all domestic violence incidents
- One-half of all traffic fatalities
- One-half of all fire victim fatalities

These statistics are sobering. The consequences of heavy drinking touch all races and all segments of society—men and women, young and old, rich and poor. One group particularly hard hit by heavy drinking is college students—not because they are prone to alcoholism, but because they live in an environment and are in a developmental stage of life in which heavy drinking is considered acceptable.[9]

Heavy drinking or binge drinking (defined as at least four drinks in a row for women and five drinks in a row for men) is widespread on college campuses and poses serious health and social consequences to drinkers and nondrinkers alike.*[10] In fact, binge drinking can kill: the respiratory center of the brain becomes anesthetized, and breathing stops. Acute alcohol intoxication can cause coronary artery spasms, leading to heart attacks.

Binge drinking is especially common among college students who live in a fraternity or sorority house, attend parties frequently, engage in other risky behaviors, and have a history of binge drinking in high school. Compared with nondrinkers or moderate drinkers, people who frequently binge drink (at least three times within two weeks) are more likely to engage in unpro-

---

* This definition of binge drinking, without specification of time elapsed, is consistent with standard practice in alcohol research.

tected sex, have multiple sex partners, damage property, and assault others.[11] On average, *every day* alcohol is involved in the:[12]

- Death of 5 college students
- Sexual assault of 266 college students
- Injury of 1641 college students
- Assault of 1907 college students

Binge drinkers skew the statistics on college students' alcohol use. The median number of drinks consumed by college students is 1.5 per week, but for binge drinkers, it is 14.5. Nationally, only 20 percent of all students are frequent binge drinkers; yet they account for two-thirds of all the alcohol students report consuming and most of the alcohol-related problems.

Binge drinking is not limited to college campuses, of course, but it is most common among 18- to 24-year-olds.[13] That age group and environment seem most accepting of such behavior despite its problems. Social acceptance may make it difficult for binge drinkers to recognize themselves as problem drinkers. For this reason, interventions must focus both on educating individuals and on changing the campus social environment.[14] The damage alcohol causes only becomes worse if the pattern is not broken. Alcohol abuse sets in much more quickly in young people than in adults. Those who start drinking at an early age more often suffer from alcoholism than people who start later on. Table H7-4 lists the key signs of alcoholism.

Alcohol's Long-Term Effects

The most devastating long-term effect of alcohol is the damage done to a child whose mother abused alcohol during pregnancy. The effects of alcohol on the unborn and the message that pregnant women should not drink alcohol are presented in Highlight 14.

For nonpregnant adults, a drink or two sets in motion many destructive processes in the body, but the next day's abstinence reverses them. As long as the doses are moderate, the time between them is ample, and nutrition is adequate, recovery is probably complete.

If the doses of alcohol are heavy and the time between them short, complete recovery cannot take place. Repeated onslaughts of alcohol gradually take a toll on all parts of the body (see Table H7-5, p. 246). Compared with nondrinkers and moderate drinkers, heavy drinkers have significantly greater risks of dying from all causes.[15] Excessive alcohol consumption is the third leading preventable cause of death in the United States.[16]

Personal Strategies

One obvious option available to people attending social gatherings is to enjoy the conversation, eat the food, and drink nonalcoholic beverages. Several nonalcoholic beverages are available that mimic the look and taste of their alcoholic counterparts. For those who enjoy champagne or beer, sparkling ciders and beers without alcohol are available. Instead of drinking a cocktail, a person can sip tomato juice with a slice of lime and a stalk of celery or just a plain cola beverage. Any of these drinks can ease conversation.

The person who chooses to drink alcohol should sip each drink slowly with food. The alcohol should arrive at the liver cells slowly enough that the enzymes can handle the load. It is best to space drinks, too, allowing about an hour or so to metabolize each drink.

If you want to help sober up a friend who has had too much to drink, don't bother walking arm in arm around the block. Walking muscles have to work harder, but muscle cells can't metabolize alcohol; only liver cells can. Remember that each person has a limited amount of the alcohol dehydrogenase enzyme that clears the blood at a steady rate. Time alone will do the job.

Nor will it help to give your friend a cup of coffee. Caffeine is a stimulant, but it won't speed up alcohol metabolism. The police say ruefully, "If you give a drunk a cup of coffee, you'll just have a wide-awake drunk on your hands." Table H7-6 (p. 246) presents other alcohol myths.

People who have passed out from drinking need 24 hours to sober up completely. Let them sleep, but watch over them. Encourage them to lie on their sides, instead of their backs. That way, if they vomit, they won't choke.

Don't drive too soon after drinking. The lack of glucose for the brain's function and the length of time needed to clear the blood of alcohol make alcohol's adverse effects linger long after its blood concentration has fallen. Driving coordination is still impaired the morning *after* a night of drinking, even if the drinking was moderate. Responsible aircraft pilots know that they must allow 24 hours for their bodies to clear alcohol completely, and they refuse to fly any sooner. The Federal Aviation Administration and major airlines enforce this rule.

TABLE H7-4 Signs of Alcoholism

- Tolerance—the person needs higher and higher intakes of alcohol to achieve intoxication
- Withdrawal—the person who stops drinking experiences anxiety, agitation, increased blood pressure, or seizures, or seeks alcohol to relieve these symptoms
- Impaired control—the person intends to have 1 or 2 drinks, but has 9 or 10 instead, or the person tries to control or quit drinking, but fails
- Disinterest—the person neglects important social, family, job, or school activities because of drinking
- Time—the person spends a great deal of time obtaining and drinking alcohol or recovering from excessive drinking
- Impaired ability—the person's intoxication or withdrawal symptoms interfere with work, school, or home
- Problems—the person continues drinking despite physical hazards or medical, legal, psychological, family, employment, or school problems

The presence of three or more of these conditions is required to make a diagnosis.

SOURCE: Adapted from *Diagnostic and Statistical Manual of Mental Disorders*, 4th ed. (Washington, D.C.: American Psychiatric Association, 1994).

TABLE H7-5 Health Effects of Heavy Alcohol Consumption

Health Problem	Effects of Alcohol
Arthritis	Increases the risk of inflamed joints
Cancer	Increases the risk of cancer of the liver, pancreas, rectum, and breast; increases the risk of cancer of the lungs, mouth, pharynx, larynx, and esophagus, where alcohol interacts synergistically with tobacco
Fetal alcohol syndrome	Causes physical and behavioral abnormalities in the fetus (see Highlight 14)
Heart disease	In heavy drinkers, raises blood pressure, blood lipids, and the risk of stroke and heart disease; when compared with those who abstain, heart disease risk is generally lower in light-to-moderate drinkers (see Chapter 27)
Hyperglycemia	Raises blood glucose
Hypoglycemia	Lowers blood glucose, especially in people with diabetes
Infertility	Increases the risks of menstrual disorders and spontaneous abortions (in women); suppresses luteinizing hormone (in women) and testosterone (in men)
Kidney disease	Enlarges the kidneys, alters hormone functions, and increases the risk of kidney failure
Liver disease	Causes fatty liver, alcoholic hepatitis, and cirrhosis
Malnutrition	Increases the risk of protein-energy malnutrition; low intakes of protein, calcium, iron, vitamin A, vitamin C, thiamin, vitamin B_6, and riboflavin; and impaired absorption of calcium, phosphorus, vitamin D, and zinc
Nervous disorders	Causes neuropathy and dementia; impairs balance and memory
Obesity	Increases energy intake, but is not a primary cause of obesity
Psychological disturbances	Causes depression, anxiety, and insomnia

NOTE: This list is by no means all-inclusive. Alcohol has direct toxic effects on all body systems.

TABLE H7-6 Myths and Truths Concerning Alcohol

Myth: Hard liquors such as rum, vodka, and tequila are more harmful than wine and beer.
Truth: The damage caused by alcohol depends largely on the *amount* consumed. Compared with hard liquor, beer and wine have relatively low percentages of alcohol, but they are often consumed in larger quantities.

Myth: Consuming alcohol with raw seafood diminishes the likelihood of getting hepatitis.
Truth: People have eaten contaminated oysters while drinking alcoholic beverages and not gotten as sick as those who were not drinking. But do not be misled: hepatitis is too serious an illness for anyone to depend on alcohol for protection.

Myth: Alcohol stimulates the appetite.
Truth: For some people, alcohol may stimulate appetite, but it seems to have the opposite effect in heavy drinkers. Heavy drinkers tend to eat poorly and suffer malnutrition.

Myth: Drinking alcohol is healthy.
Truth: Moderate alcohol consumption is associated with a lower risk for heart disease (see Chapter 27 for more details). Higher intakes, however, raise the risks for high blood pressure, stroke, heart disease, some cancers, accidents, violence, suicide, birth defects, and deaths in general. Furthermore, excessive alcohol consumption damages the liver, pancreas, brain, and heart. No authority recommends that nondrinkers begin drinking alcoholic beverages to obtain health benefits.

Myth: Wine increases the body's absorption of minerals.
Truth: Wine may increase the body's absorption of potassium, calcium, phosphorus, magnesium, and zinc, but the alcohol in wine also promotes the body's excretion of these minerals, so no benefit is gained.

Myth: Alcohol is legal and, therefore, not a drug.
Truth: Alcohol is legal for adults 21 years old and older, but it is also a drug—a substance that alters one or more of the body's functions.

Myth: A shot of alcohol warms you up.
Truth: Alcohol diverts blood flow to the skin making you *feel* warmer, but it actually cools the body.

Myth: Wine and beer are mild; they do not lead to alcoholism.
Truth: Alcoholism is not related to the kind of beverage, but rather to the quantity and frequency of consumption.

Myth: Mixing different types of drinks gives you a hangover.
Truth: Too much alcohol in any form produces a hangover.

Myth: Alcohol is a stimulant.
Truth: People think alcohol is a stimulant because it seems to relieve inhibitions, but it does so by depressing the activity of the brain. Alcohol is medically defined as a depressant drug.

Myth: Beer is a great source of carbohydrate, vitamins, minerals, and fluids.
Truth: Beer does provide some carbohydrate, but most of its kcalories come from alcohol. The few vitamins and minerals in beer cannot compete with rich food sources. And the diuretic effect of alcohol causes the body to lose more fluid in urine than is provided by the beer.

Look again at the drawing of the brain in Figure H7-4, and note that when someone drinks, judgment fails first. Judgment might tell a person to limit alcohol consumption to two drinks at a party, but if the first drink takes judgment away, many more drinks may follow. The failure to stop drinking as planned, on repeated occasions, is a danger sign warning that the person should not drink at all. The accompanying Nutrition on the Net provides websites for organizations that offer information about alcohol and alcohol abuse.

Ethanol interferes with a multitude of chemical and hormonal reactions in the body—many more than have been enumerated here. With heavy alcohol consumption, the potential for harm is great. The best way to escape the harmful effects of alcohol is, of course, to refuse alcohol altogether. If you do drink alcoholic beverages, do so with care, and in moderation.

NUTRITION ON THE NET

For further study of topics covered in this chapter, log on to **academic.cengage .com/nutrition/rolfes/UNCN8e**. Go to Chapter 7, then to Nutrition on the Net.

- Search for "alcohol" at the U.S. Government health site: **www.healthfinder.gov**

- Gather information on alcohol and drug abuse from the National Clearinghouse for Alcohol and Drug Information (NCADI): **ncadi.samhsa.gov**

- Learn more about alcoholism and drug dependence from the National Council on Alcoholism and Drug Dependence (NCADD): **www.ncadd.org**

- Visit the National Institute on Alcohol Abuse and Alcoholism: **www.collegedrinkingprevention.gov**

- Find help for a family alcohol problem from Alateen and Al-Anon Family support groups: **www.al-anon.alateen.org**

- Find help for an alcohol or drug problem from Alcoholics Anonymous (AA) or Narcotics Anonymous: **www.aa.org** or **www.wsoinc.com**

- Search for "party" to find tips for hosting a safe party from Mothers Against Drunk Driving (MADD): **www.madd.org**

REFERENCES

1. D. J. Meyerhoff and coauthors, Health risks of chronic moderate and heavy alcohol consumption: How much is too much? *Alcoholism, Clinical and Experimental Research* 29 (2005): 1334–1340; J. B. Standridge, R. G. Zylstra, and S. M. Adams, Alcohol consumption: An overview of benefits and risks, *Southern Medical Journal* 97 (2004): 664–672.
2. V. Arndt and coauthors, Age, alcohol consumption, and all-cause mortality, *Annals of Epidemiology* 14 (2004): 750–753.
3. J. Connor and coauthors, The burden of death, disease, and disability due to alcohol in New Zealand, *New Zealand Medical Journal* 118 (2005): U1412.
4. L. E. Nagy, Molecular aspects of alcohol metabolism: Transcription factors involved in early ethanol-induced liver injury, *Annual Review of Nutrition* 24 (2004): 55–78.
5. D. Krahn and coauthors, Alcohol use and cognition at mid-life: The importance of adjusting for baseline cognitive ability and educational attainment, *Alcoholism: Clinical and Experimental Research* 27 (2003): 1162–1166.
6. R. A. Breslow and B. A. Smothers, Drinking patterns and body mass index in never smokers: National Health Interview Survey, 1997–2001, *American Journal of Epidemiology* 161 (2005): 368–376; M. R. Yeomans, Effects of alcohol on food and energy intake in human subjects: Evidence for passive and active over-consumption of energy, *British Journal of Nutrition* 92 (2004): S31–S34; S. G.

Wannamethee and A. G. Shaper, Alcohol, body weight, and weight gain in middle-aged men, *American Journal of Clinical Nutrition* 77 (2003): 1312–1317; E. Jequier, Pathways to obesity, *International Journal of Obesity and Related Metabolic Disorders* 26 (2002): S12–S17.
7. S. G. Wannamethee, A. G. Shaper, and P. H. Whincup, Alcohol and adiposity: Effects of quantity and type of drink and time relation with meals, *International Journal of Obesity and Related Metabolic Disorders* 29 (2005): 1436–1444; J. M. Dorn and coauthors, Alcohol drinking patterns differentially affect central adiposity as measured by abdominal height in women and men, *Journal of Nutrition* 133 (2003): 2655–2662.
8. Position paper on drug policy: Physician Leadership on National Drug Policy (PLNDP), Brown University Center for Alcohol and Addiction Studies, 2000.
9. A. M. Brower, Are college students alcoholics? *Journal of American College Health* 50 (2002): 253–255.
10. R. D. Brewer and M. H. Swahn, Binge drinking and violence, *Journal of the American Medical Association* 294 (2005): 616–618; H. Wechsler and coauthors, Trends in college binge drinking during a period of increased prevention efforts—Findings from Harvard School of Public Health College Alcohol Study Surveys: 1993–2001, *Journal of American College Health* 50 (2002): 203–217.
11. Wechsler and coauthors, 2002.

12. R. W. Hingson and coauthors, Magnitude of alcohol-related mortality and morbidity among U.S. college students ages 18–24: Changes from 1998 to 2001, *Annual Review of Public Health* 26 (2005): 259–279.
13. National Center for Health Statistics, *Chartbook on Trends in the Health of Americans,* Alcohol consumption by adults 18 years of age and over, according to selected characteristics: United States, selected years 1997–2003, (2005): 264–266.
14. A. Ziemelis, R. B. Bucknam, and A. M. Elfessi, Prevention efforts underlying decreases in binge drinking at institutions of higher learning, *Journal of American College Health* 50 (2002): 238–252.
15. A. Y. Strandberg and coauthors, Alcohol consumption, 29-y total mortality, and quality of life in men in old age, *American Journal of Clinical Nutrition* 80 (2004): 1366–1371; I. R. White, D. R. Altmann, and K. Nanchahal, Alcohol consumption and mortality: Modeling risks for men and women at different ages, *British Medical Journal* 325 (2002): 191–197.
16. Centers for Disease Control, Alcohol-attributable deaths and years of potential life lost—United States, 2001, *Morbidity and Mortality Weekly Report* 53 (2004): 866–870.

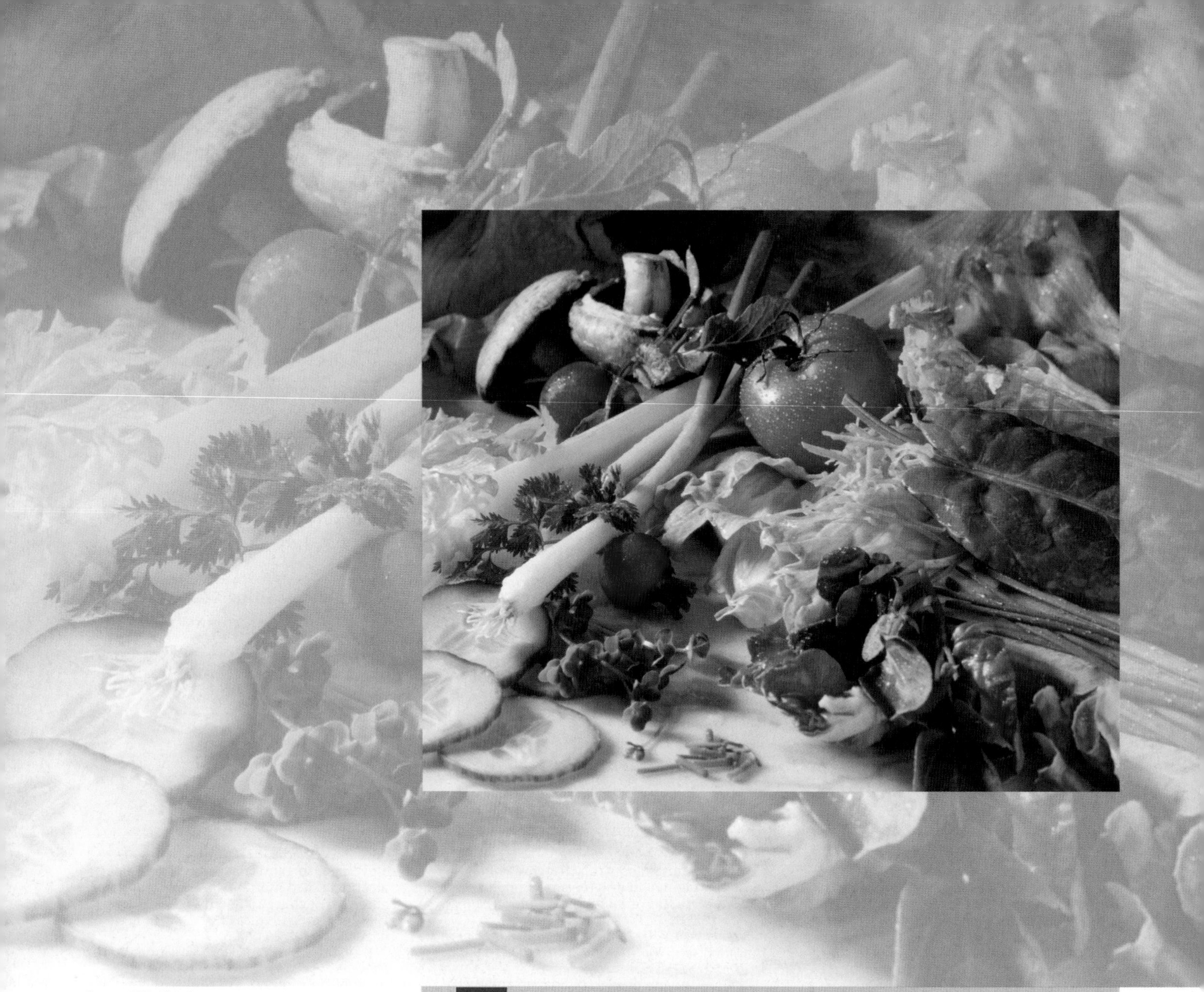

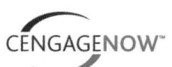

Rosemary Weller/Getty Images

The CengageNOW logo indicates an opportunity for online self-study, linking you to interactive tutorials and videos based on your level of understanding.

academic.cengage.com/login

How To: Practice Problems
Nutrition Portfolio Journal
Nutrition Calculations: Practice Problems

Nutrition in Your Life

It's a simple mathematical equation: energy in + energy out = energy balance. The reality, of course, is much more complex. One day you may devour a dozen doughnuts at midnight and sleep through your morning workout—tipping the scales toward weight gain. Another day you may snack on veggies and train for this weekend's 10K race—shifting the balance toward weight loss. Your body weight—especially as it relates to your body fat—and your level of fitness have consequences for your health. So, how are you doing? Are you ready to see how your "energy in" and "energy out" balance and whether your body weight and fat measures are consistent with good health?

Energy Balance and Body Composition

The body's remarkable machinery can cope with many extremes of diet. As Chapter 7 explained, both excess carbohydrate (glucose) and excess protein (amino acids) can contribute to body fat. To some extent, amino acids can be used to make glucose. To a very limited extent, even fat (the glycerol portion) can be used to make glucose. But a grossly unbalanced diet imposes hardships on the body. If energy intake is too low or if too little carbohydrate or protein is supplied, the body must degrade its own lean tissue to meet its glucose and protein needs. If energy intake is too high, the body stores fat.

Both excessive and deficient body fat result from an energy imbalance. The simple picture is as follows. People who have consumed more food energy than they have expended bank the surplus as body fat. To reduce body fat, they need to expend more energy than they take in from food. In contrast, people who have consumed too little food energy to support their bodies' activities have relied on their bodies' fat stores and possibly some of their lean tissues as well. To gain weight, these people need to take in more food energy than they expend. As you will see, though, the details of the body's weight regulation are quite complex.[1] This chapter describes energy balance and body composition and examines the health problems associated with having too much or too little body fat. The next chapter presents strategies toward resolving these problems.

Energy Balance

People expend energy continuously and eat periodically to refuel. Ideally, their energy intakes cover their energy expenditures without too much excess. Excess energy is stored as fat, and stored fat is used for energy between meals. The amount of body fat a person deposits in, or withdraws from, "storage" on any given day depends on the energy balance for that day—the amount consumed (energy in) versus the amount expended (energy out). When a person is maintaining weight, energy in equals energy out. When the balance shifts, weight changes. For each 3500 kcalories eaten in excess, a pound of body fat is stored; similarly, a pound of fat is lost for

When energy in balances with energy out, a person's body weight is stable.

◆ 1 lb body fat = 3500 kcal
Body fat, or adipose tissue, is composed of a mixture of mostly fat, some protein, and water. A pound of body fat (454 g) is approximately 87% fat, or (454 × 0.87) 395 g, and 395 g × 9 kcal/g = 3555 kcal.

each 3500 kcalories expended beyond those consumed. ◆ The fat stores of even a healthy-weight adult represent an ample reserve of energy—50,000 to 200,000 kcalories.

Quick changes in body weight are not simple changes in fat stores. Weight gained or lost rapidly includes some fat, large amounts of fluid, and some lean tissues such as muscle proteins and bone minerals. (Because water constitutes about 60 percent of an adult's body weight, retention or loss of water can greatly influence body weight.) Even over the long term, the composition of weight gained or lost is normally about 75 percent fat and 25 percent lean. During starvation, losses of fat and lean are about equal. (Recall from Chapter 7 that without adequate carbohydrate, protein-rich lean tissues break down to provide glucose.) Invariably, though, *fat* gains and losses are gradual. The next two sections examine the two sides of the energy-balance equation: energy in and energy out.

FIGURE 8-1 | Bomb Calorimeter

When food is burned, energy is released in the form of heat. Heat energy is measured in kcalories.

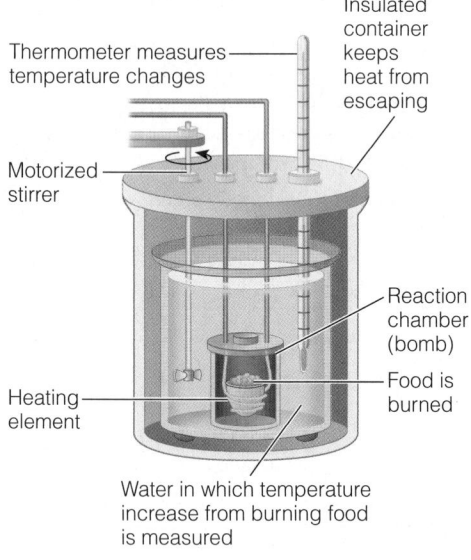

Thermometer measures temperature changes

Insulated container keeps heat from escaping

Motorized stirrer

Reaction chamber (bomb)

Food is burned

Heating element

Water in which temperature increase from burning food is measured

IN SUMMARY

When the energy consumed equals the energy expended, a person is in energy balance and body weight is stable. If more energy is taken in than is expended, a person gains weight. If more energy is expended than is taken in, a person loses weight.

Energy In: The kCalories Foods Provide

Foods and beverages provide the "energy in" part of the energy-balance equation. How much energy a person receives depends on the composition of the foods and beverages and on the amount the person eats and drinks.

Food Composition

To find out how many kcalories a food provides, a scientist can burn the food in a **bomb calorimeter** (see Figure 8-1). When the food burns, energy is released in the form of heat. The amount of heat given off provides a *direct* measure of the food's energy value (remember that kcalories are units of heat energy). In addition to releasing heat, these reactions generate carbon dioxide and water—just as the body's cells do when they metabolize the energy-yielding nutrients. When the food burns and the chemical bonds break, the carbons (C) and hydrogens (H) combine with oxygens (O) to form carbon dioxide (CO_2) and water (H_2O). The amount of oxygen consumed gives an *indirect* measure ◆ of the amount of energy released.

A bomb calorimeter measures the available energy in foods but overstates the amount of energy that the human body ◆ derives from foods. The body is less efficient than a calorimeter and cannot metabolize all of the energy-yielding nutrients in a food completely. Researchers can correct for this discrepancy mathematically to create useful tables of the energy values of foods (such as Appendix H). These

◆ Food energy values can be determined by:
  • **Direct calorimetry**, which measures the amount of heat released
  • **Indirect calorimetry**, which measures the amount of oxygen consumed

◆ The number of kcalories that the body derives from a food, in contrast to the number of kcalories determined by calorimetry, is the **physiological fuel value.**

bomb calorimeter (KAL-oh-RIM-eh-ter): an instrument that measures the heat energy released when foods are burned, thus providing an estimate of the potential energy of the foods.
  • **calor** = heat
  • **metron** = measure

values provide reasonable estimates, but they do not reflect the *precise* amount of energy a person will derive from the foods consumed.

The energy values of foods can also be computed from the amounts of carbohydrate, fat, and protein (and alcohol, if present) in the foods.* For example, a food ◆ containing 12 grams of carbohydrate, 5 grams of fat, and 8 grams of protein will provide 48 carbohydrate kcalories, 45 fat kcalories, and 32 protein kcalories, for a total of 125 kcalories. (To review how to calculate the energy available from foods, turn to p. 9.)

Food Intake

To achieve energy balance, the body must meet its needs without taking in too much or too little energy. Somehow the body decides how much and how often to eat—when to start eating and when to stop. As you will see, many signals initiate or delay eating. **Appetite** refers to the sensations of hunger, satiation, and satiety that prompt a person to eat or not eat.[2]

Hunger People eat for a variety of reasons, most obviously (although not necessarily most commonly) because they are hungry. Most people recognize **hunger** as an irritating feeling that prompts thoughts of food and motivates them to start eating. In the body, hunger is the physiological response to a need for food triggered by chemical messengers originating and acting in the brain, primarily in the **hypothalamus**.[3] Hunger can be influenced by the presence or absence of nutrients in the bloodstream, the size and composition of the preceding meal, customary eating patterns, climate (heat reduces food intake; cold increases it), exercise, hormones, and physical and mental illnesses. Hunger determines what to eat, when to eat, and how much to eat.

The stomach is ideally designed to handle periodic batches of food, and people typically eat meals at roughly four-hour intervals. Four hours after a meal, most, if not all, of the food has left the stomach. Most people do not feel like eating again until the stomach is either empty or almost so. Even then, a person may not feel hungry for quite a while.

Satiation During the course of a meal, as food enters the GI tract and hunger diminishes, **satiation** develops. As receptors in the stomach stretch and hormones such as cholecystokinin increase, the person begins to feel full.[4] The response: satiation occurs and the person stops eating.

Satiety After a meal, the feeling of **satiety** continues to suppress hunger and allows a person to not eat again for a while. Whereas *satiation* tells us to "stop eating," *satiety* reminds us to "not start eating again." Figure 8-2 (p. 252) summarizes the relationships among hunger, satiation, and satiety. Of course, people can override these signals, especially when presented with stressful situations or favorite foods.

Overriding Hunger and Satiety Not surprisingly, eating can be triggered by signals other than hunger, even when the body does not need food. Some people experience food cravings when they are bored or anxious. In fact, they may eat in response to any kind of stress, ◆ negative or positive. ("What do I do when I'm grieving? Eat. What do I do when I'm celebrating? Eat!") Many people respond to external cues such as the time of day ("It's time to eat") or the availability, sight, and taste of food ("I'd love a piece of chocolate even though I'm stuffed"). Environmental influences such as large portion sizes, favorite foods, or an abundance or variety of foods stimulate eating and increase energy intake.[5] These cognitive influences ◆ can easily lead to weight gain.

Eating can also be suppressed by signals other than satiety, even when a person is hungry. People with the eating disorder anorexia nervosa, for example, use

◆ Reminder:
  • 1 g carbohydrate = 4 kcal
  • 1 g fat = 9 kcal
  • 1 g protein = 4 kcal
  • 1 g alcohol = 7 kcal
As Chapter 1 mentioned, many scientists measure food energy in kilojoules instead. Conversion factors for these and other measures are in the Aids to Calculation section on the last two pages of the book.

◆ Eating in response to arousal is called **stress eating.**

◆ Cognitive influences include perceptions, memories, intellect, and social interactions.

appetite: the integrated response to the sight, smell, thought, or taste of food that initiates or delays eating.

hunger: the painful sensation caused by a lack of food that initiates food-seeking behavior.

hypothalamus (high-po-THAL-ah-mus): a brain center that controls activities such as maintenance of water balance, regulation of body temperature, and control of appetite.

satiation (say-she-AY-shun): the feeling of satisfaction and fullness that occurs during a meal and halts eating. Satiation determines how much food is consumed during a meal.

satiety: the feeling of fullness and satisfaction that occurs after a meal and inhibits eating until the next meal. Satiety determines how much time passes between meals.

---

* Some of the food energy values in the table of food composition in Appendix H were derived by bomb calorimetry, and many were calculated from their energy-yielding nutrient contents.

FIGURE 8-2 Hunger, Satiation, and Satiety

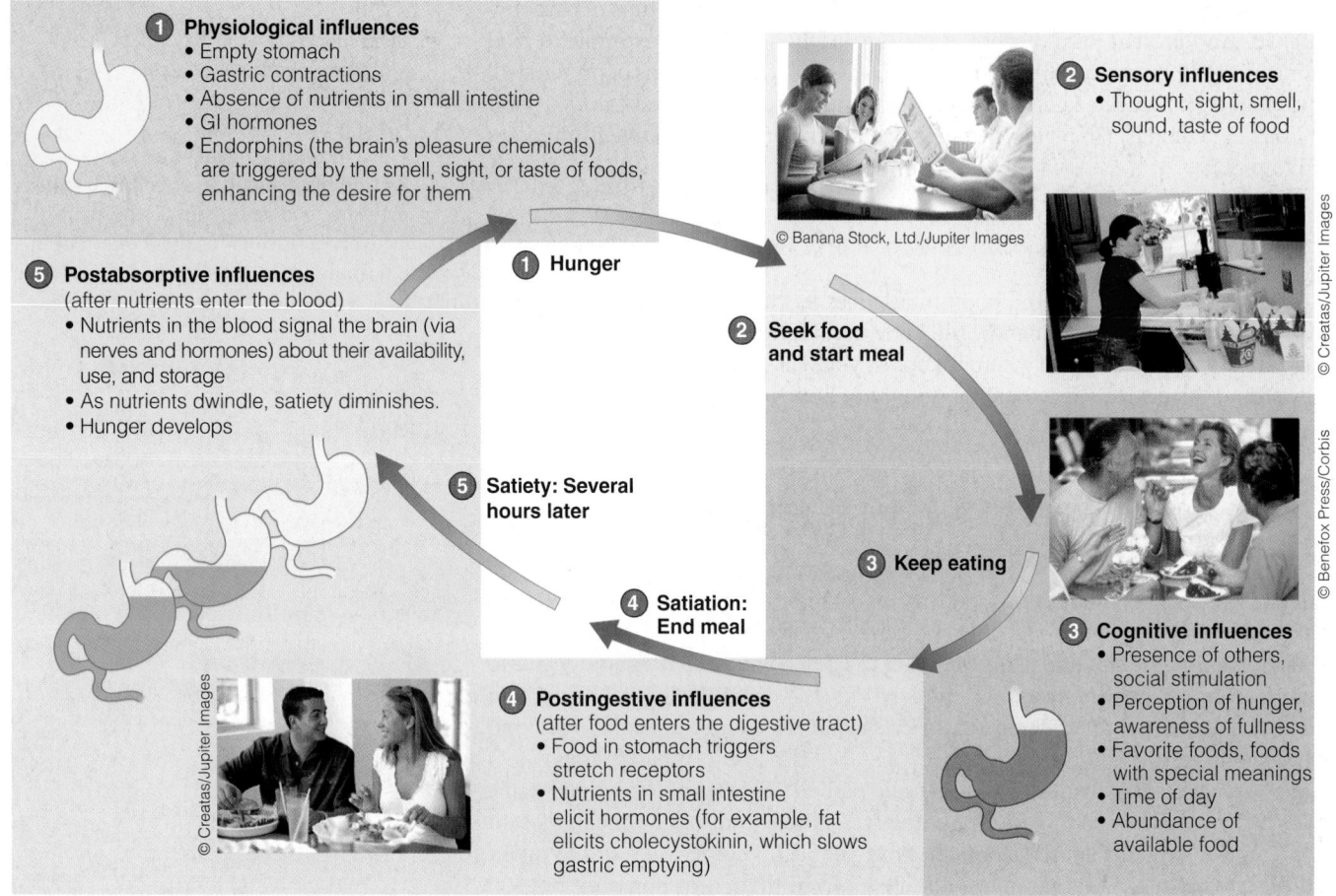

1 Physiological influences
- Empty stomach
- Gastric contractions
- Absence of nutrients in small intestine
- GI hormones
- Endorphins (the brain's pleasure chemicals) are triggered by the smell, sight, or taste of foods, enhancing the desire for them

2 Sensory influences
- Thought, sight, smell, sound, taste of food

5 Postabsorptive influences
(after nutrients enter the blood)
- Nutrients in the blood signal the brain (via nerves and hormones) about their availability, use, and storage
- As nutrients dwindle, satiety diminishes.
- Hunger develops

1 Hunger

2 Seek food and start meal

5 Satiety: Several hours later

3 Keep eating

4 Satiation: End meal

3 Cognitive influences
- Presence of others, social stimulation
- Perception of hunger, awareness of fullness
- Favorite foods, foods with special meanings
- Time of day
- Abundance of available food

4 Postingestive influences
(after food enters the digestive tract)
- Food in stomach triggers stretch receptors
- Nutrients in small intestine elicit hormones (for example, fat elicits cholecystokinin, which slows gastric emptying)

© Banana Stock, Ltd./Jupiter Images
© Creatas/Jupiter Images
© Benefox Press/Corbis
© Creatas/Jupiter Images

tremendous discipline to ignore the pangs of hunger. Some people simply cannot eat during times of stress, negative or positive. ("I'm too sad to eat." "I'm too excited to eat!") Why some people overeat in response to stress and others cannot eat at all remains a bit of a mystery, although researchers are beginning to understand the connections between stress hormones, brain activity, and "comfort foods."[6] Factors that appear to be involved include how the person perceives the stress and whether usual eating behaviors are restrained. (Highlight 8 features anorexia nervosa and other eating disorders.)

Sustaining Satiation and Satiety The extent to which foods produce satiation and sustain satiety depends in part on the nutrient composition of a meal.[7] Of the three energy-yielding nutrients, protein is considered the most **satiating.** Foods low in energy density are also more satiating.[8] High-fiber foods effectively provide satiation by filling the stomach and delaying the absorption of nutrients. For this reason, eating a large salad as a first course helps a person eat less during the meal.[9] In contrast, fat has a weak effect on satiation; consequently, eating high-fat foods may lead to passive overconsumption. High-fat foods are flavorful, which stimulates the appetite and entices people to eat more. High-fat foods are also energy dense; consequently, they deliver more kcalories per bite. (Chapter 1 introduced the concept of energy density, and Chapter 9 describes how considering a food's energy density can help with weight management.) Although fat provides little satiation during a meal, it produces strong satiety signals once it enters the intestine. Fat in the intestine triggers the release of cholecystokinin—a hormone that signals satiety and inhibits food intake.[10]

Eating high-fat foods while trying to limit energy intake requires small portion sizes, which can leave a person feeling unsatisfied. Portion size correlates directly with a food's satiety. Instead of eating small portions of high-fat foods and feeling

satiating: having the power to suppress hunger and inhibit eating.

deprived, a person can feel satisfied by eating large portions of high-protein and high-fiber foods. Figure 8-3 illustrates how fat influences portion size.

Message Central—The Hypothalamus As you can see, eating is a complex behavior controlled by a variety of psychological, social, metabolic, and physiological factors. The hypothalamus appears to be the control center, integrating messages about energy intake, expenditure, and storage from other parts of the brain and from the mouth, GI tract, and liver. Some of these messages influence satiation, which helps control the size of a meal; others influence satiety, which helps determine the frequency of meals.

Dozens of chemicals in the brain participate in appetite control and energy balance. By understanding the action of these brain chemicals, researchers may one day be able to control appetite. The greatest challenge now is to sort out the many actions of these brain chemicals. For example, one of these chemicals, **neuropeptide Y,** causes carbohydrate cravings, initiates eating, decreases energy expenditure, and increases fat storage—all factors favoring a positive energy balance and weight gain.

Regardless of hunger, people typically overeat when offered the abundance and variety of an "all you can eat" buffet.

IN SUMMARY

A mixture of signals governs a person's eating behaviors. Hunger and appetite initiate eating, whereas satiation and satiety stop and delay eating, respectively. Each responds to messages from the nervous and hormonal systems. Superimposed on these signals are complex factors involving emotions, habits, and other aspects of human behavior.

Energy Out: The kCalories the Body Expends

Chapter 7 explained that heat is released whenever the body breaks down carbohydrate, fat, or protein for energy and again when that energy is used to do work. The generation of heat, known as **thermogenesis,** can be measured to determine the amount of energy expended. ◆ The total energy a body expends reflects three main categories of thermogenesis:

• Energy expended for basal metabolism

• Energy expended for physical activity

◆ Energy expenditure, like food energy, can be determined by:

• **Direct calorimetry,** which measures the amount of heat released

• **Indirect calorimetry,** which measures the amount of oxygen consumed and carbon dioxide expelled

neuropeptide Y: a chemical produced in the brain that stimulates appetite, diminishes energy expenditure, and increases fat storage.

thermogenesis: the generation of heat; used in physiology and nutrition studies as an index of how much energy the body is expending.

FIGURE 8-3 How Fat Influences Portion Sizes

837 kcal
71 g fat

55 kcal
3 g fat

100 kcal
9 g fat

100 kcal
5 g fat

For the same size portion, peanuts deliver more than 15 times the kcalories and 20 times the fat of popcorn.

For the same number of kcalories, a person can have a few high-fat peanuts or almost 2 cups of high-fiber popcorn. (This comparison used oil-based popcorn; using air-popped popcorn would double the amount of popcorn in this example.)

At 6 feet 4 inches tall and 250 pounds (1.93 meters and 113 kilograms), this runner would be considered overweight by most standards. Yet he is clearly not overfat.

◆ In metric terms, a person 1.78 meters tall who weighs 68 kilograms may carry only about 14 of those kilograms as fat.

Body Weight, Body Composition, and Health

A person 5 feet 10 inches tall who weighs 150 pounds ◆ may carry only about 30 of those pounds as fat. The rest is mostly water and lean tissues—muscles, organs such as the heart and liver, and the bones of the skeleton. Direct measures of **body composition** are impossible in living human beings; instead, researchers assess body composition indirectly based on the following assumption:

$$\text{body weight} = \text{fat} + \text{lean tissue (including water)}$$

Weight gains and losses tell us nothing about how the body's composition may have changed, yet weight is the measure most people use to judge their "fatness." For many people, overweight means overfat, but this is not always the case. Athletes with dense bones and well-developed muscles may be overweight by some standards but have little body fat. Conversely, inactive people may seem to have acceptable weights, when, in fact, they may have too much body fat.

Defining Healthy Body Weight

How much should a person weigh? How can a person know if her weight is appropriate for her height? How can a person know if his weight is jeopardizing his health? Such questions seem so simple, yet the answers can be complex—and quite different depending on whom you ask.

The Criterion of Fashion In asking what is ideal, people often mistakenly turn to fashion for the answer. No doubt our society sets unrealistic ideals for body weight, especially for women. Miss America, our nation's icon of beauty, has never been overweight, and until recently, she has grown progressively thinner over the years (see Figure 8-5). Magazines, movies, and television all convey the message that to be thin is to be beautiful and happy. As a result, the media have a great influence on the weight concerns and dieting patterns of people of all ages, but most tragically on young, impressionable children and adolescents.[13] One-half of preteen girls and one-third of preteen boys are dissatisfied with their body weight and shape.[14]

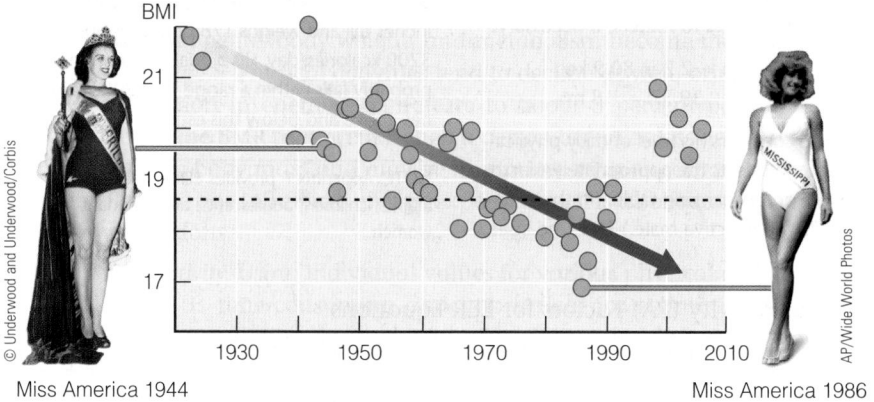

FIGURE 8-5 The Declining Weight of Miss America

Miss America 1944

Miss America 1986

As explained on p. 259, the body mass index (BMI) describes relative weight for height. Over the years, the BMI of Miss America has declined steadily. Since the mid-1960s, most have fallen below 18.5, the cutoff point indicating underweight with its associated health problems.

SOURCE: S. Rubenstein and B. Caballero, Is Miss America an undernourished role model? *Journal of the American Medical Association* 283 (2000): 1569.

body composition: the proportions of muscle, bone, fat, and other tissue that make up a person's total body weight.

TABLE 8-3	Tips for Accepting a Healthy Body Weight

- Value yourself and others for human attributes other than body weight. Realize that prejudging people by weight is as harmful as prejudging them by race, religion, or gender.
- Use positive, nonjudgmental descriptions of your body.
- Accept positive comments from others.
- Focus on your whole self including your intelligence, social grace, and professional and scholastic achievements.
- Accept that no magic diet exists.
- Stop dieting to lose weight. Adopt a lifestyle of healthy eating and physical activity permanently.

- Follow the USDA Food Guide. Never restrict food intake below the minimum levels that meet nutrient needs.
- Become physically active, not because it will help you get thin but because it will make you feel good and enhance your health.
- Seek support from loved ones. Tell them of your plan for a healthy life in the body you have been given.
- Seek professional counseling, *not* from a weight-loss counselor, but from someone who can help you make gains in self-esteem without weight as a factor.

Importantly, perceived body image has little to do with actual body weight or size. People of all shapes, sizes, and ages—including extremely thin fashion models with anorexia nervosa and fitness instructors with ideal body composition—have learned to be unhappy with their "overweight" bodies. Such dissatisfaction can lead to damaging behaviors, such as starvation diets, diet pill abuse, and health care avoidance.[15] The first step toward making healthy changes may be self-acceptance. Keep in mind that fashion is fickle; the body shapes valued by our society change with time. Furthermore, body shapes valued by our society differ from those of other societies. The standards defining "ideal" are subjective and frequently have little in common with health. Table 8-3 offers some tips for adopting health as an ideal, rather than society's misconceived image of beauty.

The Criterion of Health Even if our society were to accept fat as beautiful, obesity would still be a major risk factor for several life-threatening diseases. For this reason, the most important criterion for determining how much a person should weigh and how much body fat a person needs is not appearance but good health and longevity. Ideally, a person has enough fat to meet basic needs but not so much as to incur health risks. This range of healthy body weights has been identified using a common measure of weight and height—the body mass index.

Body Mass Index The **body mass index (BMI)** describes relative weight for height: ◆

$$BMI = \frac{weight\ (kg)}{height\ (m)^2} \quad or \quad \frac{weight\ (lb) \times 703}{height\ (in)^2}$$

Weight classifications based on BMI are presented in Figure 8-6 (p. 260). Notice that healthy weight falls between a BMI of 18.5 and 24.9, with **underweight** below 18.5, **overweight** above 25, and **obese** above 30. Well over half of adults in the United States have a BMI greater than 25, as Figure 8-7 (p. 260) shows.[16]

A BMI of 25 for adults represents a healthy target for overweight people to achieve or for others not to exceed. Obesity-related diseases and increased mortality become evident beyond a BMI of 25. The lower end of the healthy range may be a reasonable target for severely underweight people. BMI values slightly below the healthy range may be compatible with good health if food intake is adequate, but signs of illness, reduced work capacity, and poor reproductive function become apparent when BMI is below 17. The inside back cover presents weights and visual images associated with various BMI values. The "How to" on p. 261 describes how to determine an appropriate body weight based on BMI.

Keep in mind that BMI reflects height and weight measures and not body composition. Consequently, muscular athletes may be classified as over*weight* by BMI standards and not be over*fat*.[17] At the peak of his bodybuilding career, Arnold Schwarzenegger won the Mr. Olympia competition with a BMI of 31; the runner on p. 258 also has a BMI greater than 30. Yet neither would be considered obese. Striking differences in body composition are also apparent among people of various ethnic and racial groups, making standard BMI guidelines inappropriate for some

A healthy body contains enough lean tissue to support health and the right amount of fat to meet body needs.

◆ To convert pounds to kilograms:
lb × 2.2 lb/kg = kg
To convert inches to meters:
in × 39.37 in/m = m

body mass index (BMI): an index of a person's weight in relation to height; determined by dividing the weight (in kilograms) by the square of the height (in meters).

underweight: body weight below some standard of acceptable weight that is usually defined in relation to height (such as BMI); BMI below 18.5.

overweight: body weight above some standard of acceptable weight that is usually defined in relation to height (such as BMI); BMI 25 to 29.9.

obese: overweight with adverse health effects; BMI 30 or higher.

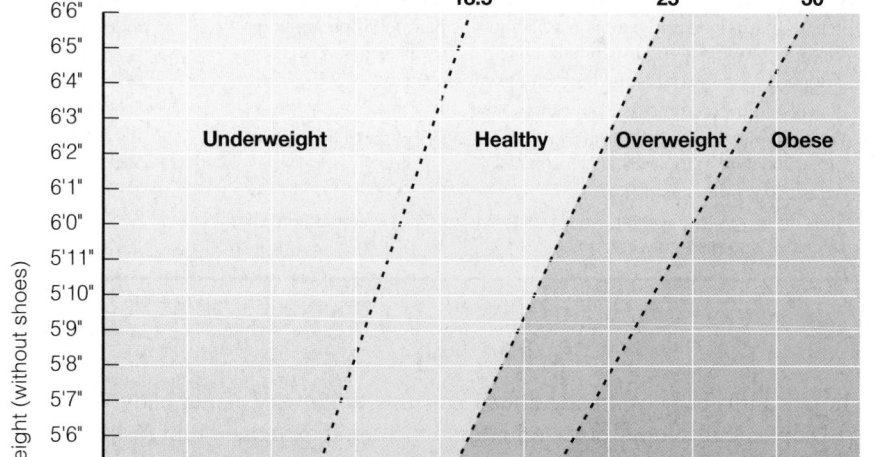

FIGURE 8-6   BMI Values Used to Assess Weight

NOTE: Chapter 15 presents BMI values for children and adolescents age 2 to 20.
SOURCE: U.S. Department of Agriculture and U.S. Department of Health and Human Services, *Nutrition and Your Health: Dietary Guidelines for Americans* (Washington, D.C.: 2000), p. 7.

FIGURE 8-7   Distribution of Body Weights in U.S. Adults

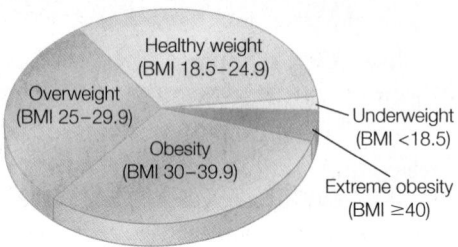

populations.[18] For example, blacks tend to have a greater bone density and protein content than whites; consequently, using BMI as the standard may overestimate the prevalence of obesity among blacks.

IN SUMMARY

Current standards for body weight are based on a person's weight in relation to height, called the body mass index (BMI), and reflect disease risks. To its disadvantage, BMI does not reflect body fat, and it may misclassify very muscular people as overweight.

Body Fat and Its Distribution

Although weight measures are inexpensive, easy to take, and highly accurate, they fail to reveal two valuable pieces of information in assessing disease risk: how much of the weight is fat and where the fat is located. The ideal amount of body fat depends partly on the person. A normal-weight man may have from 13 to 21 percent body fat; a woman, because of her greater quantity of essential fat, 23 to 31 percent. In general, health problems typically develop when body fat exceeds 22 percent in young men, 25 percent in men over age 40, 32 percent in young women, and 35 percent in women over age 40. Body fat may contribute as much as 70 percent in excessively obese adults. Figure 8-8 compares the body composition of healthy weight men and women.

CENGAGENOW™
To determine BMI, log on to **academic.cengage** **.com/login**, go to Chapter 8, then go to How To.

HOW TO Determine Body Weight Based on BMI

A person whose BMI reflects an unacceptable health risk can choose a desired BMI and then calculate an appropriate body weight. For example, a woman who is 5 feet 5 inches (1.65 meters) tall and weighs 180 pounds (82 kilograms) has a BMI of 30:

$$BMI = \frac{82 \text{ kg}}{1.65 \text{ m}^2} = 30$$

or

$$BMI = \frac{180 \text{ lb} \times 703}{65 \text{ in}^2} = 30$$

A reasonable target for most overweight people is a BMI 2 units below their current one. To determine a desired goal weight based on a BMI of 28, for example, the woman could divide the desired BMI by the factor appropriate for her height from the table below:

$$\text{desired BMI} \div \text{factor} = \text{goal weight}$$

$$28 \div 0.166 = 169 \text{ lb}$$

To reach a BMI of 28, this woman would need to lose 11 pounds. Such a calculation can help a person to determine realistic weight goals using health risk as a guide. Alternatively, a person could search the table on the inside back cover for the weight that corresponds to his or her height and the desired BMI.

Height	Factor	Height	Factor	Height	Factor
4'7" (1.40 m)	0.232	5'3" (1.60 m)	0.177	5'11" (1.80 m)	0.139
4'8" (1.42 m)	0.224	5'4" (1.63 m)	0.172	6'0" (1.83 m)	0.136
4'9" (1.45 m)	0.216	5'5" (1.65 m)	0.166	6'1" (1.85 m)	0.132
4'10" (1.47 m)	0.209	5'6" (1.68 m)	0.161	6'2" (1.88 m)	0.128
4'11" (1.50 m)	0.202	5'7" (1.70 m)	0.157	6'3" (1.90 m)	0.125
5'0" (1.52 m)	0.195	5'8" (1.73 m)	0.152	6'4" (1.93 m)	0.122
5'1" (1.55 m)	0.189	5'9" (1.75 m)	0.148	6'5" (1.96 m)	0.119
5'2" (1.57 m)	0.183	5'10" (1.78 m)	0.143	6'6" (1.98 m)	0.116

SOURCE: R. P. Abernathy, Body mass index: Determination and use, *Journal of the American Dietetic Association* 91 (1991): 843.

Some People Need Less Body Fat For many athletes, a lower percentage of body fat may be ideal—just enough fat to provide fuel, insulate and protect the body, assist in nerve impulse transmissions, and support normal hormone activity,

FIGURE 8-8 Male and Female Body Compositions Compared

The differences between male and female body compositions become apparent during adolescence. Lean body mass (primarily muscle) increases more in males than in females. Fat assumes a larger percentage of female body composition as essential body fat is deposited in the mammary glands and pelvic region in preparation for childbearing. Both men and women have essential fat associated with the bone marrow, the central nervous system, and the internal organs.

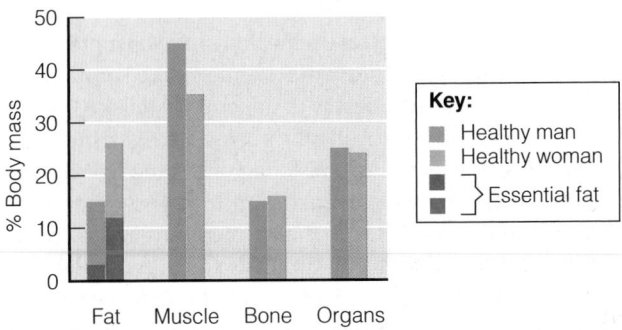

SOURCE: R. E. C. Wildman and D. M. Medeiros, *Advanced Human Nutrition* (Boca Raton, Fla.: CRC Press, 2000), pp. 321–323. Used with permission.

FIGURE 8-9 | Abdominal Fat

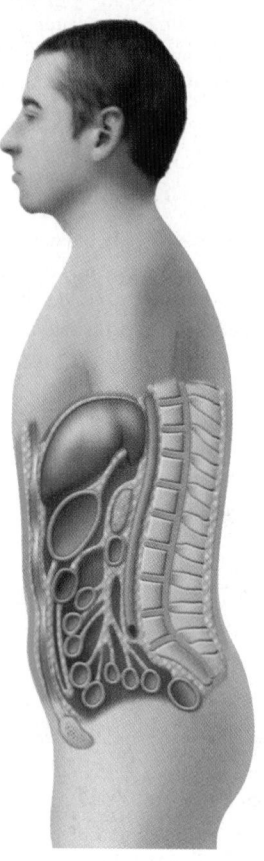
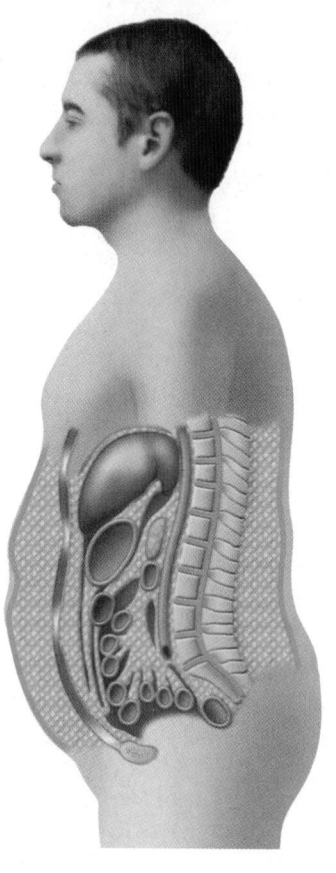

In healthy weight people, some fat is stored around the organs of the abdomen.

In overweight people, excess abdominal fat increases the risks of diseases.

but not so much as to burden the body with excess bulk. For some athletes, then, ideal body fat might be 5 to 10 percent for men and 15 to 20 percent for women. (Review the photo on p. 258 to appreciate what 8 percent body fat looks like.)

Some People Need More Body Fat For an Alaska fisherman, a higher percentage of body fat is probably beneficial because fat provides an insulating blanket to prevent excessive loss of body heat in cold climates. A woman starting a pregnancy needs sufficient body fat to support conception and fetal growth. Below a certain threshold for body fat, hormone synthesis falters, and individuals may become infertile, develop depression, experience abnormal hunger regulation, or become unable to keep warm. These thresholds differ for each function and for each individual; much remains to be learned about them.

Fat Distribution The distribution of fat on the body may be more critical than the total amount of fat alone. **Intra-abdominal fat** that is stored around the organs of the abdomen is referred to as **central obesity** or upper-body fat (see Figure 8-9). Independently of BMI or total body fat, central obesity is associated with increased risks of heart disease, stroke, diabetes, hypertension, gallstones, and some types of cancer.[19]

Abdominal fat is most common in men and to a lesser extent in women past menopause. Even when total body fat is similar, men have more abdominal fat than women. Regardless of gender, the risks of cardiovascular disease, diabetes, and mortality are increased for those with excessive abdominal fat. Interestingly, smokers tend to have more abdominal fat than nonsmokers even though they have lower BMI.[20]

Fat around the hips and thighs, sometimes referred to as lower-body fat, is most common in women during their reproductive years and seems relatively harmless. In fact, overweight people who do not have abdominal fat are less susceptible to

intra-abdominal fat: fat stored within the abdominal cavity in association with the internal abdominal organs, as opposed to the fat stored directly under the skin (subcutaneous fat).

central obesity: excess fat around the trunk of the body; also called **abdominal fat** or **upper-body fat.**

health problems than overweight people with abdominal fat. Figure 8-10 compares the body shapes of people with upper-body fat and lower-body fat.

Waist Circumference A person's **waist circumference** is the most practical indicator of fat distribution and central obesity. [21] In general, women with a waist circumference of greater than 35 inches (88 centimeters) and men with a waist circumference of greater than 40 inches (102 centimeters) have a high risk of central obesity-related health problems, such as diabetes and cardiovascular disease.[22] As waist circumference increases, disease risks increase.[23] Appendix E includes instructions for measuring waist circumference and assessing abdominal fat.

Some researchers use the waist-to-hip ratio when studying disease risks. The ratio requires another step or two (measuring the hips and comparing that measure to the waist measure), but it does not provide any additional information. Therefore, waist circumference alone is the preferred method for assessing abdominal fat in a clinical setting.*

Other Measures of Body Composition Health care professionals commonly use BMI and waist circumference measures because they are relatively easy and inexpensive. Together, these two measures prove most valuable in assessing a person's health risks and monitoring changes over time.[24] Researchers needing more precise measures of body composition may choose any of several other techniques to estimate body fat and its distribution (see Figure 8-11 on p. 264). Mastering these techniques requires proper instruction and practice to ensure reliability. In addition to the methods shown in Figure 8-11, researchers sometimes estimate body composition using these methods: total body water, radioactive potassium count, near-infrared spectrophotometry, ultrasound, computed tomography, and magnetic resonance imaging. Each method has advantages and disadvantages with respect to cost, technical difficulty, and precision of estimating body fat (see Appendix E for a comparison). Appendix E provides additional details and includes many of the tables and charts routinely used in assessment procedures.

IN SUMMARY

The ideal amount of body fat varies from person to person, but researchers have found that body fat in excess of 22 percent for young men and 32 percent for young women (the levels rise slightly with age) poses health risks. Central obesity, in which excess abdominal fat is distributed around the trunk of the body, presents greater health risks than excess fat distributed on the lower body.

Health Risks Associated with Body Weight and Body Fat

Body weight and fat distribution correlate with disease risks and life expectancy.[25] They indicate a greater *likelihood* of developing a chronic disease and shortening life expectancy. Not all overweight and underweight people will get sick and die before their time nor will all normal-weight people live long healthy lives. *Correlations* are not *causes*. For the most part, people with a BMI between 18.5 and 24.9 have relatively few health risks; risks increase as BMI falls below or rises above this range, indicating that both too little and too much body fat impair health.[26] Epidemiological data show a J- or U-shaped relationship between body weights and mortality (see Figure 8-12, p. 264).[27] People who are extremely underweight or extremely obese carry higher risks of early deaths than those whose weights fall within the acceptable range; these mortality risks decline with age.[28]

---

* The National Heart, Lung, and Blood Institute recommends using the waist circumference instead of the waist-to-hip ratio to assess obesity health risks.

FIGURE 8-10 "Apple" and "Pear" Body Shapes Compared

Popular articles sometimes call bodies with upper-body fat "apples" and those with lower-body fat, "pears." Researchers sometimes refer to upper-body fat as "android" (manlike) obesity and to lower-body fat as "gynoid" (womanlike) obesity.

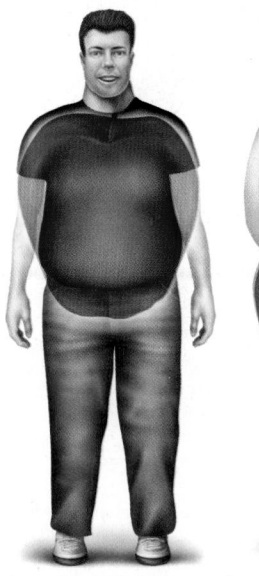
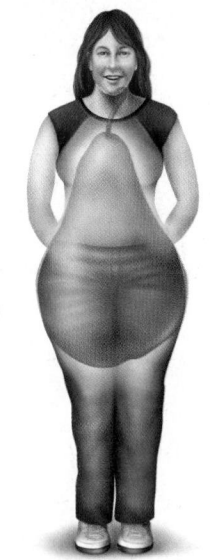
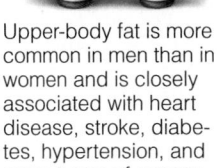

Upper-body fat is more common in men than in women and is closely associated with heart disease, stroke, diabetes, hypertension, and some types of cancer.

Lower body fat is more common in women than in men and is not usually associated with chronic diseases.

waist circumference: an anthropometric measurement used to assess a person's abdominal fat.

FIGURE 8-11 Common Methods Used to Assess Body Fat

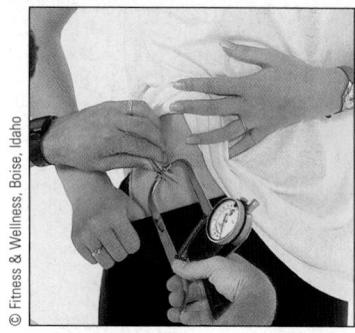

Skinfold measures estimate body fat by using a caliper to gauge the thickness of a fold of skin on the back of the arm (over the triceps), below the shoulder blade (subscapular), and in other places (including lower-body sites) and then comparing these measurements with standards.

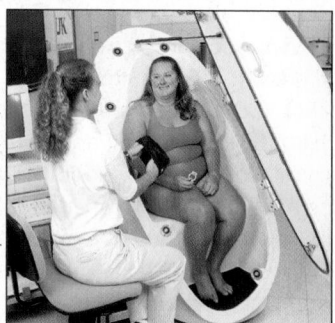

Air displacement plethysmography estimates body composition by having a person sit inside a chamber while computerized sensors determine the amount of air displaced by the person's body.

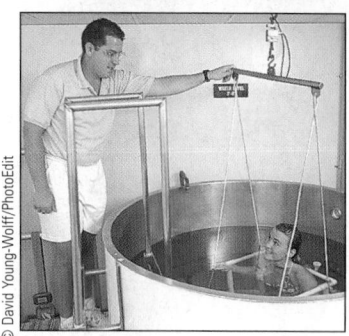

Hydrodensitometry measures body density by weighing the person first on land and then again while submerged in water. The difference between the person's actual weight and underwater weight provides a measure of the body's volume. A mathematical equation using the two measurements (volume and actual weight) determines body density, from which the percentage of body fat can be estimated.

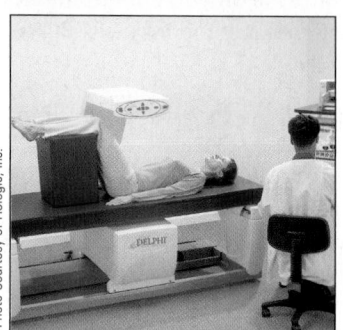

Dual energy X-ray absorptiometry (DEXA) uses two low-dose X-rays that differentiate among fat-free soft tissue (lean body mass), fat tissue, and bone tissue, providing a precise measurement of total fat and its distribution in all but extremely obese subjects.

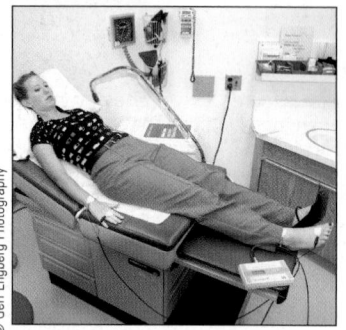

Bioelectrical impedance measures body fat by using a low-intensity electrical current. Because electrolyte-containing fluids, which readily conduct an electrical current, are found primarily in lean body tissues, the leaner the person, the less resistance to the current. The measurement of electrical resistance is then used in a mathematical equation to estimate the percentage of body fat.

FIGURE 8-12 BMI and Mortality

This J-shaped curve describes the relationship between body mass index (BMI) and mortality and shows that both underweight and overweight present risks of a premature death.

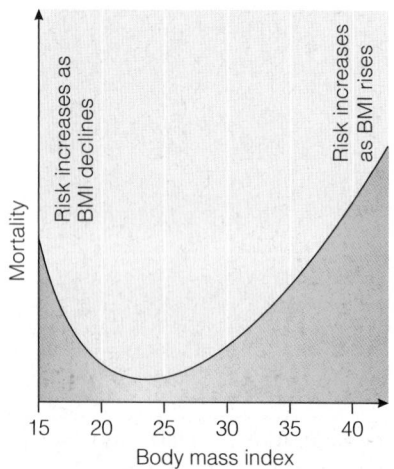

Independently of BMI, factors such as smoking habits raise health risks, and physical fitness lowers them.[29] A man with a BMI of 22 who smokes two packs of cigarettes a day is jeopardizing his health, whereas a woman with a BMI of 32 who walks briskly for an hour a day is improving her health.

Health Risks of Underweight Some underweight people enjoy an active, healthy life, but others are underweight because of malnutrition, smoking habits, substance abuse, or illnesses. Weight and fat measures alone would not reveal these underlying causes, but a complete assessment that includes a diet and medical history, physical examination, and biochemical analysis would.

An underweight person, especially an older adult, may be unable to preserve lean tissue during the fight against a wasting disease such as cancer or a digestive disorder, especially when the disease is accompanied by malnutrition. Without adequate nutrient and energy reserves, an underweight person will have a particularly tough battle against such medical stresses. Underweight women develop menstrual irregularities and become infertile. Exactly how infertility develops is unclear, but contributing factors include body weight as well as restricted energy and fat intake and depleted body fat stores. Those who do conceive may give birth to unhealthy infants. An underweight woman can improve her chances of having a healthy infant by gaining weight prior to conception, during pregnancy, or both. Underweight and significant weight loss are also associated with osteoporosis and bone fractures. For all these reasons, underweight people may benefit from enough of a weight gain to provide an energy reserve and protective amounts of all the nutrients that can be stored.

Health Risks of Overweight As for excessive body fat, the health risks are so many that it has been designated a disease—obesity. Among the health risks associated with obesity are diabetes, hypertension, cardiovascular disease, sleep apnea (abnormal ceasing of breathing during sleep), osteoarthritis, some cancers, gallbladder disease, kidney stones, respiratory problems (including Pickwickian syndrome, a breathing blockage linked with sudden death), and complications in pregnancy and surgery. Each year, these obesity-related illnesses cost our nation billions of dollars—in fact, as much as the medical costs of smoking.[30]

The cost in terms of lives is also great: an estimated 300,000 people die each year from obesity-related diseases. In fact, obesity is second only to tobacco in causing preventable illnesses and premature deaths. Mortality increases as excess weight increases; people with a BMI greater than 35 are more than twice as likely to die prematurely as others.[31] The risks associated with a high BMI appear to be greater for whites than for blacks; in fact, the health risks associated with obesity do not become apparent in black women until a BMI of 37.[32]

Equally important, both central obesity and weight gains of more than 20 pounds (9 kilograms) between early and middle adulthood correlate with increased disease risks.[33] Fluctuations in body weight, as typically occur with "yo-yo" dieting, may also increase the risks of chronic diseases and premature death. In contrast, sustained weight loss improves physical well-being, reduces disease risks, and increases life expectancy.

Cardiovascular Disease The relationship between obesity and cardiovascular disease risk is strong, with links to both elevated blood cholesterol and hypertension. Central obesity may raise the risk of heart attack and stroke as much as the three leading risk factors (high LDL cholesterol, hypertension, and smoking) do. ◆ In addition to body fat and its distribution, weight gain also increases the risk of cardiovascular disease. Weight loss, on the other hand, can effectively lower both blood cholesterol and blood pressure in obese people. Of course, lean and normal-weight people may also have high blood cholesterol and blood pressure, and these factors are just as dangerous in lean people as in obese people.

Diabetes Most adults with type 2 diabetes are overweight or obese.[34] Diabetes (type 2) is three times more likely to develop in an obese person than in a nonobese person. Furthermore, the person with type 2 diabetes often has central obesity. Central-body fat cells appear to be larger and more insulin-resistant than lower-body fat cells.[35] The association between **insulin resistance** and obesity is strong. Both are major risk factors for the development of type 2 diabetes.

Diabetes appears to be influenced by weight gains as well as by body weight. A weight gain of more than 10 pounds (4.5 kilograms) after the age of 18 doubles the risk of developing diabetes, even in women of average weight. In contrast, weight loss is effective in improving glucose tolerance and insulin resistance.[36]

Inflammation and the Metabolic Syndrome Chronic **inflammation** accompanies obesity, and inflammation contributes to chronic diseases.[37] As a person grows fatter, lipids first fill the adipose tissue and then migrate into other tissues such as the muscles and liver.[38] This accumulation of fat, especially in the abdominal region, changes the body's metabolism, resulting in insulin resistance, low HDL, high triglycerides, and high blood pressure.[39] This cluster of symptoms—collectively known as the metabolic syndrome—increases the risks for diabetes, hypertension, and atherosclerosis.◆ Fat accumulation, especially in the abdominal region, also activates genes that code for proteins ◆ involved in inflammation.[40] Furthermore, although relatively few immune cells are commonly found in adipose tissue, weight gain significantly increases their number and their role in inflammation.[41] Elevated blood lipids—whether due to obesity or to a high-fat diet—also promote inflammation.[42] Together, these factors help to explain why chronic inflammation accompanies obesity and how obesity contributes to the metabolic syndrome and the progression of

◆ Cardiovascular disease risk factors associated with obesity:
- High LDL cholesterol
- Low HDL cholesterol
- High blood pressure (hypertension)
- Diabetes

Chapter 27 provides many more details.

◆ Metabolic syndrome is a cluster of at least three of the following risk factors:
- High blood pressure
- High blood glucose
- High blood triglycerides
- Low HDL cholesterol
- High waist circumference

◆ Proteins released from adipose tissue signal changes in the body's fat and energy status and are called **adipokines.** Over 50 adipokines have been identified, some of which play a role in inflammation.

insulin resistance: the condition in which a normal amount of insulin produces a subnormal effect in muscle, adipose, and liver cells, resulting in an elevated fasting glucose; a metabolic consequence of obesity that precedes type 2 diabetes.

inflammation: an immunological response to cellular injury characterized by an increase in white blood cells.

Being active—even if overweight—is healthier than being sedentary. With a BMI of 36, aerobics instructor Jennifer Portnick is considered obese, but her daily workout routine helps to keep her in good health.

chronic diseases.[43] Even in healthy youngsters, body fat correlates positively with chronic inflammation.[44] As might be expected, weight loss reduces the number of immune cells in adipose tissue and changes gene expression to reduce inflammation.[45]

Cancer The risk of some cancers increases with both body weight and weight gain, but researchers do not fully understand the relationships. One possible explanation may be that obese people have elevated levels of hormones that could influence cancer development.[46] For example, adipose tissue is the major site of estrogen synthesis in women, obese women have elevated levels of estrogen, and estrogen has been implicated in the development of cancers of the female reproductive system—cancers that account for half of all cancers in women.

Fit and Fat versus Sedentary and Slim Importantly, BMI and weight gains and losses do not tell the whole story. Cardiorespiratory fitness also plays a major role in health and longevity, independently of BMI.[47] Normal-weight people who are fit have a lower risk of mortality than normal-weight people who are unfit. Furthermore, overweight but fit people have lower risks than normal-weight, unfit ones.[48] Clearly, a healthy body weight is good, but it may not be good enough. Fitness, in and of itself, offers many health benefits. The next chapter explores weight management and the benefits of achieving and maintaining a healthy weight.

IN SUMMARY

The weight appropriate for an individual depends largely on factors specific to that individual, including body fat distribution, family health history, and current health status. At the extremes, both overweight and underweight carry clear risks to health.

CENGAGENOW™
academic.cengage.com/login

Nutrition Portfolio

When combined with fitness, a healthy body weight will help you to defend against chronic diseases.

- Describe how your daily food intake and physical activity balance with each other.

- Calculate your estimated energy requirements.

- Describe any health risks that may be of concern for a person of your BMI and waist circumference.

NUTRITION ON THE NET

For further study of topics covered in this chapter, log on to **academic.cengage.com/nutrition/rolfes/UNCN8e**. Go to Chapter 8, then to Nutrition on the Net.

- Obtain food composition data from the USDA Nutrient Data Laboratory: **www.ars.usda.gov/ba/bhnrc/ndl**

- Learn about the 10,000 Steps Program at Shape Up America: **www.shapeup.org**

- Visit the special web pages and interactive applications for Healthy Weight: **www.nhlbi.nih.gov/subsites/index.htm**

NUTRITION CALCULATIONS

CENGAGENOW For additional practice log on to **academic.cengage.com/login**. Go to Chapter 8, then to Nutrition Calculations.

These problems give you practice in estimating energy needs. Once you have mastered these examples, you will be prepared to examine your own energy intakes and energy expenditures. Be sure to show your calculations for each problem and check p. 269 for answers.

1. Compare the energy a person might spend on various physical activities. Refer to Table 8-2 on p. 255, and compute how much energy a person who weighs 142 pounds (64.4 kilograms) would spend doing each of the following. You may want to compare various activities based on your weight.

   30 min vigorous aerobic dance:

   0.062 kcal/lb/min × 142 lb = 8.8 kcal/min
   (or 0.136 kcal/kg/min x 64.5 kg = 8.8 kcal/min)

   8.8 kcal/min × 30 min = 264 kcal

   a. 2 hours golf, carrying clubs
   b. 20 minutes running at 9 mph
   c. 45 minutes swimming at 20 yd/min
   d. 1 hour walking at 3.5 mph

2. Consider the effect of age on BMR. An infant who weighs 20 pounds (9.1 kilograms) has a BMR of 500 kcalories/day; an adult who weighs 170 pounds (77.3 kilograms) has a BMR of about 1500. Based on body weight, who has the faster BMR?

3. Compute daily energy needs for a woman, age 20, who is 5 feet 6 inches tall (1.68 meters), weighs 130 pounds (59 kilograms), and is lightly active.

4. Discover what weight is needed to achieve a desired BMI. Refer to the table on p. 261 and consider a person who is 5 feet 4 inches (1.63 meters) tall. Suppose this person wants to have a BMI of 21. What should this person weigh? Does this agree with the table on the inside back cover?

STUDY QUESTIONS

CENGAGENOW
To assess your understanding of chapter topics, take the Student Practice Test and explore the modules recommended in your Personalized Study Plan. Log on to **academic.cengage.com/login**.

These questions will help you review the chapter. You will find the answers in the discussions on the pages provided.

1. What are the consequences of an unbalanced energy budget? (pp. 249–250)

2. Define hunger, appetite, satiation, and satiety and describe how each influences food intake. (pp. 251–253)

3. Describe each component of energy expenditure. What factors influence each? How can energy expenditure be estimated? (pp. 253–257)

4. Distinguish between body weight and body composition. What assessment techniques are used to measure each? (pp. 258–264)

5. What problems are involved in defining "ideal" body weight? (pp. 258–259)

6. What is central obesity, and what is its relationship to disease? (pp. 262–265)

7. What risks are associated with excess body weight and excess body fat? (pp. 265–266)

These multiple choice questions will help you prepare for an exam. Answers can be found on p. 269.

1. A person who consistently consumes 1700 kcalories a day and spends 2200 kcalories a day for a month would be expected to:
   a. lose ½ to 1 pound.
   b. gain ½ to 1 pound.
   c. lose 4 to 5 pounds.
   d. gain 4 to 5 pounds.

2. A bomb calorimeter measures:
   a. physiological fuel.
   b. energy available from foods.
   c. kcalories a person derives from foods.
   d. heat a person releases in basal metabolism.

3. The psychological desire to eat that accompanies the sight, smell, or thought of food is known as:
   a. hunger.
   b. satiety.
   c. appetite.
   d. palatability.

4. A person watching television after dinner reaches for a snack during a commercial in response to:
   a. external cues.
   b. hunger signals.
   c. stress arousal.
   d. satiety factors.

5. The largest component of energy expenditure is:
   a. basal metabolism.
   b. physical activity.
   c. indirect calorimetry.
   d. thermic effect of food.

6. A major factor influencing BMR is:
   a. hunger.
   b. food intake.
   c. body composition.
   d. physical activity.

7. The thermic effect of an 800-kcalorie meal is about:
   a. 8 kcalories
   b. 80 kcalories
   c. 160 kcalories
   d. 200 kcalories

8. For health's sake, a person with a BMI of 21 might want to:
   a. lose weight
   b. maintain weight
   c. gain weight

9. Which of the following reflects height and weight?
   a. body mass index
   b. central obesity
   c. waist circumference
   d. body composition

10. Which of the following increases disease risks?
    a. BMI 19–21
    b. BMI 22–25
    c. lower-body fat
    d. central obesity

REFERENCES

1. G. A. Bray and C. M. Champagne, Beyond energy balance: There is more to obesity than kilocalories, *Journal of the American Dietetic Association* 105 (2005): S17–S23.
2. R. D. Mattes and coauthors, Appetite: Measurement and manipulation misgivings, *Journal of the American Dietetic Association* 105 (2005): S87–S97.
3. A. Del Parigi and coauthors, Sex differences in the human brain's response to hunger and satiation, *American Journal of Clinical Nutrition* 75 (2002): 1017–1022.
4. C. de Graaf and coauthors, Biomarkers of satiation and satiety, *American Journal of Clinical Nutrition* 79 (2004): 946–961; S. C. Woods, Gastrointestinal satiety signals: An overview of gastrointestinal signals that influence food intake, *American Journal of Physiology: Gastrointestinal and Liver Physiology* 286 (2004): G7–G13; T. H. Moran and K. P. Kinzig, Gastrointestinal satiety signals: Cholecystokinin, *American Journal of Physiology: Gastrointestinal and Liver Physiology* 286 (2004): G183–G188.
5. E. Kennedy, Dietary diversity, diet quality, and body weight regulation, *Nutrition Reviews* 62 (2004): S78–S81; B. Wansink, Environmental factors that increase the food intake and consumption volume of unknowing consumers, *Annual Review of Nutrition* 24 (2004): 455–479; B. J. Rolls, E. L. Morris, and L. S. Roe, Portion size of food affects energy intake in normal-weight and overweight men and women, *American Journal of Clinical Nutrition* 76 (2002): 1207–1213.
6. M. F. Dallman and coauthors, Chronic stress and obesity: A new view of "comfort food," *The Proceedings of the National Academy of Sciences* 100 (2003): 11696–11701.
7. D. E. Gerstein and coauthors, Clarifying concepts about macronutrients' effects on satiation and satiety, *Journal of the American Dietetic Association* 104 (2004): 1151–1153.
8. A. Drewnowski and coauthors, Dietary energy density and body weight: Is there a relationship? *Nutrition Reviews* 62 (2004): 403–413.
9. B. J. Rolls, L. S. Roe, and J. S. Meengs, Salad and satiety: Energy density and portion size of a first-course salad affect energy intake at lunch, *Journal of the American Dietetic Association* 104 (2004): 1570–1576.
10. B. Burton-Freeman, P. A. Davis, and B. O. Schneeman, Interaction of fat availability and sex on postprandial satiety and cholecystokinin after mixed-food meals, *American Journal of Clinical Nutrition* 80 (2004): 1207–1214.
11. G. P. Granata and L. J. Brandon, The thermic effect of food and obesity: Discrepant results and methodological variations, *Nutrition Reviews* 60 (2002): 223–233; L. Jonge and G. A. Bray, The thermic effect of food is reduced in obesity, *Nutrition Reviews* 60 (2002): 295–297.
12. N. Meunier and coauthors, Basal metabolic rate and thyroid hormones of late-middle-aged and older human subjects: The ZENITH study, *European Journal of Clinical Nutrition* 59 (2005): S53–S57; B. A. Parker and I. M. Chapman, Food intake and ageing—The role of the gut, *Mechanisms of Ageing and Development* 125 (2004): 859–866; I. M. Chapman, Endocrinology of anorexia of ageing, *Clinical Endocrinology and Metabolism* 18 (2004): 437–452.
13. J. Wardle, J. Waller, and E. Fox, Age of onset and body dissatisfaction in obesity, *Addictive Behaviors* 27 (2002): 561–573.
14. H. Truby and S. J. Paxton, Development of the Children's Body Image Scale, *British Journal of Clinical Psychology* 41 (2002): 185–203; H. A. Hausenblas and coauthors, Body image in middle school children, *Eating and Weight Disorders* 7 (2002): 244–248.
15. C. A. Drury and M. Louis, Exploring the association between body weight, stigma of obesity, and health care avoidance, *Journal of the American Academy of Nurse Practitioners* 14 (2002): 554–561.
16. K. M. Flegal and coauthors, Prevalence and trends in obesity among US adults, *Journal of the American Medical Association* 288 (2002): 1723–1727.
17. K. A. Witt and E. A. Bush, College athletes with an elevated body mass index often have a high upper arm muscle area, but not elevated triceps and subscapular skinfolds, *Journal of the American Dietetic Association* 105 (2005): 599–602.
18. R. P. Wildman and coauthors, Appropriate body mass index and waist circumference cutoffs for categorization of overweight and central adiposity among Chinese adults, *American Journal of Clinical Nutrition* 80 (2004): 1129–1136.
19. G. R. Dagenais and coauthors, Prognostic impact of body weight and abdominal obesity in women and men with cardiovascular disease, *American Heart Journal* 149 (2005): 54–60; Y. Wang and coauthors, Comparison of abdominal adiposity and overall obesity in predicting risk of type 2 diabetes among men, *American Journal of Clinical Nutrition* 81 (2005): 555–563; C. J. Tsai and coauthors, Prospective study of abdominal adiposity and gallstone disease in US men, *American Journal of Clinical Nutrition* 80 (2004): 38–44; T. B. Nguyen-Duy and coauthors, Visceral fat and liver fat are independent predictors of metabolic risk factors in men, *American Journal of Physiology: Endocrinology and Metabolism* (2003); G. Davì and coauthors, Platelet activation in obese women—Role of inflammation and oxidant stress, *Journal of the American Medical Association* 288 (2002): 2008–2014; J. M. Oppert and coauthors, Anthropometric estimates of muscle and fat mass in relation to cardiac and cancer mortality in men: The Paris Prospective Study, *American Journal of Clinical Nutrition* 75 (2002): 1107–1113.
20. D. Canoy and coauthors, Cigarette smoking and fat distribution in 21,828 British men and women: A population-based study, *Obesity Research* 13 (2005): 1466–1475.
21. Y. Wang and coauthors, Comparison of abdominal adiposity and overall obesity in predicting risk of type 2 diabetes among men, *American Journal of Clinical Nutrition* 81 (2005): 555–563; I. Janssen and coauthors, Body mass index and waist circumference independently contribute to the prediction of nonabdominal, abdominal subcutaneous, and visceral fat, *American Journal of Clinical Nutrition* 75 (2002): 683–688.
22. I. Lofgren and coauthors, Waist circumference is a better predictor than body mass index of coronary heart disease risk in overweight premenopausal women, *Journal of Nutrition* 134 (2004): 1071–1076; S. K. Zhu and coauthors, Waist circumference and obesity-associated risk factors among whites in the third National Health and Nutrition Examination Survey: Clinical action thresholds, *American Journal of Clinical Nutrition* 76 (2002): 743–749.
23. I. Janssen, P. T. Katzmarzyk, and R. Ross, Waist circumference and not body mass index explains obesity-related health risk, *American Journal of Clinical Nutrition* 79 (2004): 379–384.
24. G. A. Bray, Don't throw the baby out with the bath water, *American Journal of Clinical Nutrition* 79 (2004): 347–349.
25. A. H. Mokdad and coauthors, Prevalence of obesity, diabetes, and obesity-related health risk factors, 2001, *Journal of the American Medical Association* 289 (2003): 76–79; K. R. Fontaine and coauthors, Years of life lost due to obesity, *Journal of the American Medical Association* 289 (2003): 187–193.
26. D. M. Freedman and coauthors, Body mass index and all-cause mortality in a nationwide US cohort, *International Journal of Obesity* 30 (2006): 822–829; A. Thorogood and coauthors, Relation between body mass index and mortality in an unusually slim cohort, *Journal of Epidemiology and Community Health* 57 (2003) 130–133.
27. R. G. Rogers, R. A. Hummer, and P. M. Krueger, The effect of obesity on overall, circulatory disease- and diabetes-specific mortality, *Journal of Biosocial Science* 35 (2003): 107–129; D. B. Allison and coauthors, Differential associations of body mass index and adiposity with all-cause mortality among men in the first and second National Health and Nutrition Examination Surveys (NHANES I and NHANES II) follow-up

studies, *International Journal of Obesity and Related Metabolic Disorders* 26 (2002): 410–416; H. E. Meyer and coauthors, Body mass index and mortality: The influence of physical activity and smoking, *Medicine and Science in Sports and Exercise* 34 (2002): 1065–1070.

28. G. M. Price and coauthors, Weight, shape, and mortality risk in older persons: Elevated waist-hip ratio, not high body mass index, is associated with a greater risk of death, *American Journal of Clinical Nutrition* 84 (2006): 449–460; K. M. Flegal and coauthors, Excess deaths associated with underweight, overweight, and obesity, *Journal of the American Medical Association* 293 (2005): 1861–1867.

29. A. Peeters and coauthors, Obesity in adulthood and its consequences for life expectancy: A life-table analysis, *Annals of Internal Medicine* 138 (2003): 24–32.

30. E. A. Finkelstein, I. C. Fiebelkorn, and G. Wang, National medical expenditures attributable to overweight and obesity: How much, and who's paying? 2003, available at www.healthaffairs.org/WebExclusives/Finkelstein_Web_Excl_051403.htm.

31. K. M. Flegal and coauthors, Excess deaths associated with underweight, overweight, and obesity, *Journal of the American Medical Association* 293 (2005): 1861–1867.

32. J. E. Manson and S. S. Bassuk, Obesity in the United States: A fresh look at its high toll, *Journal of the American Medical Association* 289 (2003): 229–230; J. Stevens and coauthors, The effect of decision rules on the choice of a body mass index cutoff for obesity: Examples from African American and white women, *American Journal of Clinical Nutrition* 75 (2002): 986–992.

33. A. Schienkiewitz and coauthors, Body mass index history and risk of type 2 diabetes: Results from the European Prospective Investigation into Cancer and Nutrition (EPIC)—Potsdam Study, *American Journal of Clinical Nutrition* 84 (2006): 427–433.

34. Prevalence of overweight and obesity among adults with diagnosed diabetes—United States, 1988–1994 and 1999–2002, *Morbidity and Mortality Weekly Report* 53 (2004): 1066–1068.

35. E. H. Livingston, Lower body subcutaneous fat accumulation and diabetes mellitus risk, *Surgery for Obesity and Related Diseases* 2 (2006): 362–368.

36. G. M. Reaven, The insulin resistance syndrome: Definition and dietary approaches to treatment, *Annual Review of Nutrition* 25 (2005): 391–406; S. Klein and coauthors, Weight management through lifestyle modification for the prevention and management of type 2 diabetes: Rationale and strategies. A statement of the American Diabetes Association, the North American Association for the Study of Obesity, and the American Society for Clinical Nutrition, *American Journal of Clinical Nutrition* 80 (2004): 257–263.

37. R. DeCaterina and coauthors, Nutritional mechanisms that influence cardiovascular disease, *American Journal of Clinical Nutrition* 83 (2006): 421S–426S.

38. E. N. Hansen, A. Torquati, and N. N. Abumrad, Results of bariatric surgery, *Annual Review of Nutrition* 26 (2006): 481–511.

39. J. P. Despres, Is visceral obesity the cause of the metabolic syndrome, *Annals of Medicine* 38 (2006): 52–63.

40. P. Trayhurn, C. Bing, and I. S. Wood, Adipose tissue and adipokines—Energy regulation from the human perspective, *Journal of Nutrition* 136 (2006): 1935S–1939S; B. E. Wisse, The inflammatory syndrome: The role of adipose tissue cytokines in metabolic disorders linked to obesity, *Journal of the American Society of Nephrology* 15 (2004): 2792–2800.

41. A. H. Berg and P. E. Scherer, Adipose tissue, inflammation, and cardiovascular disease, *Circulation Research* 96 (2005): 939–968.

42. G. Boden, Fatty acid-induced inflammation and insulin resistance in skeletal muscle and liver, *Current Diabetes Reports* 6 (2006): 177–181.

43. D. C. W. Lau and coauthors, Adipokines: Molecular links between obesity and atherosclerosis, *American Journal of Physiology—Heart and Circulatory Physiology* 288 (2005): H2031–H2041.

44. A. Sbarbati and coauthors, Obesity and inflammation: Evidence for an elementary lesion, *Pediatrics* 117 (2006): 220–223; J. Warnberg and coauthors, Inflammatory proteins are related to total and abdominal adiposity in a healthy adolescent population: The AVENA Study, *American Journal of Clinical Nutrition* 84 (2006): 505–512.

45. J. P. Bastard and coauthors, Recent advances in the relationship between obesity, inflammation, and insulin resistance, *European Cytokine Network* 17 (2006): 4–12.

46. G. A. Bray, The underlying basis for obesity: Relationship to cancer, *Journal of Nutrition* 132 (2002): 3451S–3455S.

47. T. R. Wessel and coauthors, Relationship of physical fitness vs body mass index with coronary artery disease and cardiovascular events in women, *Journal of the American Medical Association* 292 (2004): 1179–1187; T. S. Church and coauthors, Exercise capacity and body composition as predictors of mortality among men with diabetes, *Diabetes Care* 27 (2004): 83–88; S. W. Farrell and coauthors, The relation of body mass index, cardiorespiratory fitness, and all-cause mortality in women, *Obesity Research* 10 (2002): 417–423; C. D. Lee and S. N. Blair, Cardiorespiratory fitness and smoking-related and total cancer mortality in men, *Medicine and Science in Sports and Exercise* 34 (2002): 735–739; C. D. Lee and S. N. Blair, Cardiorespiratory fitness and stroke mortality in men, *Medicine and Science in Sports and Exercise* 34 (2002): 592–595.

48. F. B. Hu and coauthors, Adiposity as compared with physical activity in predicting mortality among women, *New England Journal of Medicine* 351 (2004): 2694–2703.

ANSWERS

Nutrition Calculations

1. a. 0.045 kcal/lb/min × 142 lb = 6.4 kcal/min

    6.4 kcal/min × 120 min = 768 kcal

   b. 0.103 kcal/lb/min × 142 lb = 14.6 kcal/min

    14.6 kcal/min × 20 min = 292 kcal

   c. 0.032 kcal/lb/min × 142 lb = 4.5 kcal/min

    4.5 kcal/min × 45 min = 203 kcal

   d. 0.035 kcal/lb/min × 142 lb = 5 kcal/min

    5 kcal/min × 60 min = 300 kcal

2. The infant has the faster BMR (500 kcal/day ÷ 20 lb = 25 kcal/lb/day and 1500 kcal/day ÷ 170 lb = 8.8 kcal/lb/day). Because the infant has a BMR of 25 kcal/lb, whereas the adult has a BMR of 8.8 kcal/lb, the infant's BMR is almost 3 times faster than the adult's based on body weight.

3. EER = [354 − (6.91 × 20)] + 1.12 × [(9.36 × 59) + (726 × 1.68)]

   EER = (354 − 138.2) + 1.12 (552.24 + 1219.68)

   EER = (354 − 138.2) + 1.12 × 1771.9

   EER = 215.8 + 1984.6 = 2200 kcal/day

4. 21 ÷ 0.172 = 122 lb., yes

Study Questions (multiple choice)

1. c 2. b 3. c 4. a 5. a 6. c 7. b 8. b
9. a 10. d

Eating Disorders

For some people, low body weight becomes an obsessive goal, and they begin to view normal healthy body weight as being too fat. Their efforts to lose weight progress to a dangerously unhealthy point. An estimated 5 million people in the United States, primarily girls and young women, suffer from the **eating disorders** anorexia nervosa and bulimia nervosa (the accompanying glossary defines these and related terms).[1] Many more suffer from binge-eating disorders or other unspecified conditions that, even though they do not meet the strict criteria for anorexia nervosa or bulimia nervosa, imperil a person's well-being.

Why do so many people in our society suffer from eating disorders? Most experts agree that the causes include multiple factors: sociocultural, psychological, and perhaps neurochemical. Excessive pressure to be thin is at least partly to blame. Young people who attempt extreme weight loss may have learned to identify discomforts such as anger, jealousy, or disappointment with "feeling fat." They may also be depressed or suffer social anxiety. As weight loss becomes more of a focus, psychological problems worsen, and the likelihood of developing eating disorders intensifies. Athletes are among those most likely to develop eating disorders.

The Female Athlete Triad

At age 14, Suzanne was a top contender for a spot on the state gymnastics team. Each day her coach reminded team members that they must weigh no more than their assigned weights to qualify for competition. The coach chastised gymnasts who gained weight, and Suzanne was terrified of being singled out. Convinced that the less she weighed the better she would perform, Suzanne weighed herself several times a day to confirm that she had not exceeded her 80-pound limit. Driven to excel in her sport, Suzanne kept her weight down by eating very little and training very hard. Unlike many of her friends, Suzanne never began to menstruate. A few months before her fifteenth birthday, Suzanne's coach dropped her back to the second-level team. Suzanne blamed her poor performance on a slow-healing stress fracture. Mentally stressed and physically exhausted, she quit gymnastics and began overeating between periods of self-starvation. Suzanne had developed the dangerous combination of problems that characterize the **female athlete triad**—disordered eating, amenorrhea, and osteoporosis (see Figure H8-1).[2]

Disordered Eating

Part of the reason many athletes engage in **disordered eating** behaviors may be that they and their coaches have embraced unsuitable weight standards. An athlete's body must be heavier for a given height than a nonathlete's body because the athlete's body is dense, containing more healthy bone and muscle and less fat. When athletes rely only on the scales, they may mistakenly believe they are too fat because weight standards, such as the BMI, do not provide adequate information about body composition.

GLOSSARY

amenorrhea (ay-MEN-oh-REE-ah): the absence of or cessation of menstruation. **Primary amenorrhea** is menarche delayed beyond 16 years of age. **Secondary amenorrhea** is the absence of three to six consecutive menstrual cycles.

anorexia (an-oh-RECK-see-ah) **nervosa**: an eating disorder characterized by a refusal to maintain a minimally normal body weight and a distortion in perception of body shape and weight.

• **an** = without

• **orex** = mouth
• **nervos** = of nervous origin

binge-eating disorder: an eating disorder with criteria similar to those of bulimia nervosa, excluding purging or other compensatory behaviors.

bulimia (byoo-LEEM-ee-ah) **nervosa**: an eating disorder characterized by repeated episodes of binge eating usually followed by self-induced vomiting, misuse of laxatives or diuretics, fasting, or excessive exercise.

• **buli** = ox

cathartic (ka-THAR-tik): a strong laxative.

disordered eating: eating behaviors that are neither normal nor healthy, including restrained eating, fasting, binge eating, and purging.

eating disorders: disturbances in eating behavior that jeopardize a person's physical or psychological health.

emetic (em-ETT-ic): an agent that causes vomiting.

female athlete triad: a potentially fatal combination of three medical problems—

disordered eating, amenorrhea, and osteoporosis.

muscle dysmorphia (dis-MORE-fee-ah): a psychiatric disorder characterized by a preoccupation with building body mass.

stress fractures: bone damage or breaks caused by stress on bone surfaces during exercise.

unspecified eating disorders: eating disorders that do not meet the defined criteria for specific eating disorders.

FIGURE H8-1 The Female Athlete Triad

Eating Disorder
- Restrictive dieting (inadequate energy and nutrient intake)
- Overexercising
- Weight loss
- Lack of body fat

Osteoporosis
- Loss of calcium from bones

Amenorrhea
- Diminished hormones

Many young athletes severely restrict energy intakes to improve performance, enhance the aesthetic appeal of their performance, or meet the weight guidelines of their specific sports. They fail to realize that the loss of lean tissue that accompanies energy restriction actually impairs their physical performance. The increasing incidence of abnormal eating habits among athletes is cause for concern. Male athletes, especially wrestlers and gymnasts, are affected by these disorders as well, but females are most vulnerable. Risk factors for eating disorders among athletes include:

- Young age (adolescence)
- Pressure to excel at a chosen sport
- Focus on achieving or maintaining an "ideal" body weight or body fat percentage
- Participation in sports or competitions that emphasize a lean appearance or judge performance on aesthetic appeal such as gymnastics, wrestling, figure skating, or dance[3]
- Weight-loss dieting at an early age
- Unsupervised dieting

A few years ago, this Olympic gold medalist was weak and malnourished from anorexia nervosa. However, she recovered and set a world record in the cycling road race.

Amenorrhea

The prevalence of **amenorrhea** among premenopausal women in the United States is about 2 to 5 percent overall, but among female athletes, it may be as high as 66 percent. Contrary to previous notions, amenorrhea is *not* a normal adaptation to strenuous physical training: it is a symptom of something going wrong.[4] Amenorrhea is characterized by low blood estrogen, infertility, and often bone mineral losses. Excessive training, depleted body fat, low body weight, and inadequate nutrition all contribute to amenorrhea. However amenorrhea develops, it threatens the integrity of the bones. Bone losses remain significant even after recovery. (Women with bulimia frequently have menstrual irregularities, but because they rarely cease menstruating, they may be spared this loss of bone integrity.[5])

Osteoporosis

For most people, weight-bearing physical activity, dietary calcium, and (for women) the hormone estrogen protect against the bone loss of osteoporosis. For young women with disordered eating and amenorrhea, strenuous activity can impair bone health. Vigorous training combined with inadequate food intake disrupts metabolic and hormonal balances.[6] These disturbances compromise bone health, greatly increasing the risks of **stress fractures** today and of osteoporosis in later life. Stress fractures, a serious form of bone injury, commonly occur among dancers and other athletes with amenorrhea, low calcium intakes, and disordered eating. Many underweight young athletes have bones like those of postmenopausal women, and they may never recover their lost bone even after diagnosis and treatment—which makes prevention critical. Young athletes should be encouraged to consume 1300 milligrams of calcium each day, to eat nutrient-dense foods, and to obtain enough energy to support both weight gain and the energy expended in physical activity.

Other Dangerous Practices of Athletes

Only females face the threats of the female athlete triad, of course, but many male athletes face pressure to achieve a certain body weight and may develop eating disorders. Each week throughout the season, David drastically restricts his food and fluid intake before a wrestling match in an effort to "make weight." Wrestlers and their coaches believe that competing in a lower weight class will give them a competitive advantage over smaller opponents. To that end, David practices in rubber suits, sits in saunas, and takes diuretics to lose 4 to 6 pounds. He hopes to replenish the lost fluids, glycogen, and lean tissue during the hours between his weigh-in and competition, but the body needs days to correct this metabolic mayhem. Reestablishing fluid and electrolyte balances may take a day or two, replenishing glycogen stores may take two to three days, and replacing lean tissue may take even longer.

Ironically, the combination of food deprivation and dehydration impairs physical performance by reducing muscle strength, decreasing anaerobic power, and reducing endurance capacity. For optimal performance, wrestlers need to first achieve their competitive weight during the off-season and then eat well-balanced meals and drink plenty of fluids during the competitive season.

Some athletes go to extreme measures to bulk up and *gain* weight. People afflicted with **muscle dysmorphia** eat high-protein diets, take dietary supplements, weight train for hours at a time, and often abuse steroids in an attempt to bulk up. Their bodies are large and muscular, yet they see themselves as puny 90-pound weaklings. They are preoccupied with the idea that their bodies are too small or inadequately muscular. Like others with distorted body images, people with muscle dysmorphia weigh themselves frequently and center their lives on diet and exercise. Paying attention to diet and pumping iron for fitness is admirable, but obsessing over it can cause serious social, occupational, and physical problems.

Preventing Eating Disorders in Athletes

To prevent eating disorders in athletes and dancers, the performers, their coaches, and their parents must learn about inappropriate body weight ideals, improper weight-loss techniques, eating disorder development, proper nutrition, and safe weight-control methods. Young people naturally search for identity and will often follow the advice of a person in authority without question. Therefore, coaches and dance instructors should never encourage unhealthy weight loss to qualify for competition or to conform to distorted artistic ideals. Athletes who truly need to lose weight should try to do so during the off-season and under the supervision of a health care professional. Frequent weighings can push young people who are striving to lose weight into a cycle of starving to confront the scale, then bingeing uncontrollably afterward. The erosion of self-esteem that accompanies these events can interfere with normal psychological development and set the stage for serious problems later on.

Table H8-1 includes suggestions to help athletes and dancers protect themselves against developing eating disorders. The remaining sections describe eating disorders that anyone, athlete or nonathlete, may experience.

Anorexia Nervosa

Julie, 18 years old, is a superachiever in school. She watches her diet with great care, and she exercises daily, maintaining a rigorous schedule of self-discipline. She is thin, but she is determined to lose more weight. She is 5 feet 6 inches tall and weighs 85 pounds (roughly 1.68 meters and 39 kilograms). She has **anorexia nervosa.**

TABLE H8-1	Tips for Combating Eating Disorders

General Guidelines

- Never restrict food amounts to below those suggested for adequacy by the USDA Food Guide (see Table 2-3 on p. 41).

- Eat frequently. Include healthy snacks between meals. The person who eats frequently never gets so hungry as to allow hunger to dictate food choices.

- If not at a healthy weight, establish a reasonable weight goal based on a healthy body composition.

- Allow a reasonable time to achieve the goal. A reasonable loss of excess fat can be achieved at the rate of about 10 percent of body weight in six months.

- Establish a weight-maintenance support group with people who share interests.

Specific Guidelines for Athletes and Dancers

- Replace weight-based goals with performance-based goals.

- Restrict weight-loss activities to the off-season.

- Remember that eating disorders impair physical performance. Seek confidential help in obtaining treatment if needed.

- Focus on proper nutrition as an important facet of your training, as important as proper technique.

Characteristics of Anorexia Nervosa

Julie is unaware that she is undernourished, and she sees no need to obtain treatment. She developed amenorrhea several months ago and has become moody and chronically depressed. She insists that she is too fat, although her eyes are sunk in deep hollows in her face. Julie denies that she is ever tired, although she is close to physical exhaustion and no longer sleeps easily. Her family is concerned, and though reluctant to push her, they have finally insisted that she see a psychiatrist. Julie's psychiatrist has diagnosed anorexia nervosa (see Table H8-2) and prescribed group therapy as a start. If she does not begin to gain weight soon, she may need to be hospitalized.

As mentioned in the introduction, most anorexia nervosa victims are females; males account for only about 1 in 20 reported cases. Central to the diagnosis of anorexia nervosa is a distorted body image that overestimates personal body fatness. When Julie looks at herself in the mirror, she sees a "fat" 85-pound body. The more Julie overestimates her body size, the more resistant she is to treatment, and the more unwilling to examine her faulty values and misconceptions. Malnutrition is known to affect brain functioning and judgment in this way, causing lethargy, confusion, and delirium.

Anorexia nervosa cannot be self-diagnosed. Many people in our society are engaged in the pursuit of thinness, and denial runs high among people with anorexia nervosa. Some women have all the attitudes and behaviors associated with the condition, but without the dramatic weight loss.

Self-Starvation How can a person as thin as Julie continue to starve herself? Julie uses tremendous discipline against her hunger to strictly limit her portions of low-kcalorie foods. She will deny her hunger, and having adapted to so little food, she feels full af-

TABLE H8-2	Criteria for Diagnosis of Anorexia Nervosa

A person with anorexia nervosa demonstrates the following:

A. Refusal to maintain body weight at or above a minimal normal weight for age and height (e.g., weight loss leading to maintenance of body weight less than 85 percent of that expected; or failure to make expected weight gain during period of growth, leading to body weight less than 85 percent of that expected).

B. Intense fear of gaining weight or becoming fat, even though underweight.

C. Disturbance in the way in which one's body weight or shape is experienced, undue influence of body weight or shape on self-evaluation, or denial of the seriousness of the current low body weight.

D. In females past puberty, amenorrhea, i.e., the absence of at least three consecutive menstrual cycles. (A woman is considered to have amenorrhea if her periods occur only following hormone, e.g., estrogen, administration.)

Two types:

• *Restricting type:* During the episode of anorexia nervosa, the person does not regularly engage in binge eating or purging behavior (i.e., self-induced vomiting or the misuse of laxatives, diuretics, or enemas).

• *Binge eating/purging type:* During the episode of anorexia nervosa, the person regularly engages in binge eating or purging behavior (i.e., self-induced vomiting or the misuse of laxatives, diuretics, or enemas).

SOURCE: Reprinted with permission from American Psychiatric Association, *Diagnostic and Statistical Manual of Mental Disorders,* 4th ed. Text Revision. (Washington, D.C.: American Psychiatric Association, 2000).

ter eating only a half-dozen carrot sticks. She knows the kcalorie contents of dozens of foods and the kcalorie costs of as many exercises. If she feels that she has gained an ounce of weight, she runs or jumps rope until she is sure she has exercised it off. If she fears that the food she has eaten outweighs the exercise, she may take laxatives to hasten the passage of food from her system. She drinks water incessantly to fill her stomach, risking dangerous mineral imbalances. She is desperately hungry. In fact, she is starving, but she doesn't eat because her need for self-control dominates.

Many people, on learning of this disorder, say they wish they had "a touch" of it to get thin. They mistakenly think that people with anorexia nervosa feel no hunger. They also fail to recognize the pain of the associated psychological and physical trauma.

Physical Consequences The starvation of anorexia nervosa damages the body just as the starvation of war and poverty does. In fact, after a few months, most people with anorexia nervosa have protein-energy malnutrition (PEM) that is similar to marasmus (described in Chapter 6).[7] Their bodies have been depleted of both body fat and protein.[8] Victims are dying to be thin—quite literally. In young people, growth ceases and normal development falters. They lose so much lean tissue that basal metabolic rate slows. In addition, the heart pumps inefficiently and irregularly, the heart muscle becomes weak and thin, the chambers diminish in size, and the blood pressure falls.[9] Minerals that help to regulate heartbeat become unbalanced. Many deaths occur due to multiple organ system failure when the heart, kidneys, and liver cease to function.

Starvation brings other physical consequences as well, such as loss of brain tissue, impaired immune response, anemia, and a loss of digestive functions that worsens malnutrition. Peristalsis becomes sluggish, the stomach empties slowly, and the lining of the intestinal tract atrophies. The deteriorated GI tract fails to provide sufficient digestive enzymes and absorptive surfaces for handling any food that is eaten. The pancreas slows its production of digestive enzymes. The person may suffer from diarrhea, further worsening malnutrition.

Other effects of starvation include altered blood lipids, high blood vitamin A and vitamin E, low blood proteins, dry thin skin, abnormal nerve functioning, reduced bone density, low body temperature, low blood pressure, and the development of fine body hair (the body's attempt to keep warm). The electrical activity of the brain becomes abnormal, and insomnia is common. Both women and men lose their sex drives.

Women with anorexia nervosa develop amenorrhea. (It is one of the diagnostic criteria.) In young girls, the onset of menstruation is delayed. Menstrual periods typically resume with recovery, although some women never restart even after they have gained weight. Should an underweight woman with anorexia nervosa become pregnant, she is likely to give birth to an underweight baby—and low-birthweight babies face many health problems (as Chapter 14 explains). Mothers with anorexia nervosa may underfeed their children who then fail to grow and may also suffer the other consequences of starvation.

Treatment of Anorexia Nervosa

Treatment of anorexia nervosa requires a multidisciplinary approach.[10] Teams of physicians, nurses, psychiatrists, family therapists, and dietitians work together to resolve two sets of issues and behaviors: those relating to food and weight and those involving relationships with oneself and others. The first dietary objective is to stop weight loss while establishing regular eating patterns. Appropriate diet is crucial to recovery and must be tailored to individual client's needs. Because body weight is low and fear of weight gain is high, initial food intake may be small—perhaps only 1200 kcalories per day.[11] As eating becomes more comfortable, clients should gradually increase energy intake. Initially, clients may be unwilling to eat for themselves. Those who do eat will have a good chance of recovering without additional interventions. Even after recovery, however, energy intakes and eating behaviors may not fully return to normal.[12] Furthermore, weight gains may be slow because energy needs may be slightly elevated due to anxiety, abdominal pain, and cigarette smoking.[13]

Because anorexia nervosa is like starvation physically, health care professionals classify clients based on indicators of PEM.* Low-risk clients need nutrition counseling. Intermediate-risk clients may need supplements such as high-kcalorie, high-protein formulas in addition to regular meals. High-risk clients may require hospitalization and may need to be fed by tube at first to prevent death. This step may cause psychological trauma. Although drugs are commonly prescribed, they play a limited role in treatment.

---

* Indicators of protein-energy malnutrition: a low percentage of body fat, low serum albumin, low serum transferrin, and impaired immune reactions.

Denial runs high among those with anorexia nervosa. Few seek treatment on their own. About half of the women who are treated can maintain their body weight at 85 percent or more of a healthy weight, and at that weight, many of them begin menstruating again.[14] The other half have poor to fair treatment outcomes, relapse into abnormal eating behaviors, or die. Anorexia nervosa has one of the highest mortality rates among psychiatric disorders.[15] An estimated 1000 women die each year of anorexia nervosa—most commonly from cardiac complications due to malnutrition or by suicide.[16]

Before drawing conclusions about someone who is extremely thin or who eats very little, remember that diagnosis requires professional assessment. Several national organizations offer information for people who are seeking help with anorexia nervosa, either for themselves or for others.*

Bulimia Nervosa

Kelly is a charming, intelligent, 30-year-old flight attendant of normal weight who thinks constantly about food. She alternates between starving herself and secretly bingeing, and when she has eaten too much, she makes herself vomit. Most readers recognize these symptoms as those of **bulimia nervosa.**

Characteristics of Bulimia Nervosa

Bulimia nervosa is distinct from anorexia nervosa and is more prevalent, although the true incidence is difficult to establish because bulimia nervosa is not as physically apparent. More men suffer from bulimia nervosa than from anorexia nervosa, but bulimia nervosa is still more common in women than in men. The secretive nature of bulimic behaviors makes recognition of the problem difficult, but once it is recognized, diagnosis is based on the criteria listed in Table H8-3.

Like the typical person with bulimia nervosa, Kelly is single, female, and white. She is well educated and close to her ideal body weight, although her weight fluctuates over a range of 10 pounds or so every few weeks. She prefers to weigh less than the weight that her body maintains naturally.

Kelly seldom lets her eating disorder interfere with work or other activities, although a third of all bulimics do. From early childhood, she has been a high achiever and emotionally dependent on her parents. As a young teen, Kelly frequently followed severely restricted diets but could never maintain the weight loss. Kelly feels anxious at social events and cannot easily establish close personal relationships. She is usually depressed, is often impulsive, and has low self-esteem. When crisis hits, Kelly responds by replaying events, worrying excessively, and blaming herself but never asking for help—behaviors that interfere with effective coping.

Binge Eating Like the person with anorexia nervosa, the person with bulimia nervosa spends much time thinking about body weight and food. The preoccupation with food manifests itself in

---

* Internet sites are listed at the end of this highlight.

| TABLE H8-3 | Criteria for Diagnosis of Bulimia Nervosa |

A person with bulimia nervosa demonstrates the following:

A. Recurrent episodes of binge eating. An episode of binge eating is characterized by both of the following:

　1. Eating, in a discrete period of time (e.g., within any two-hour period), an amount of food that is definitely larger than most people would eat during a similar period of time and under similar circumstances.

　2. A sense of lack of control over eating during the episode (e.g., a feeling that one cannot stop eating or control what or how much one is eating).

B. Recurrent inappropriate compensatory behavior to prevent weight gain, such as self-induced vomiting; misuse of laxatives, diuretics, enemas, or other medications; fasting; or excessive exercise.

C. Binge eating and inappropriate compensatory behaviors both occur, on average, at least twice a week for three months.

D. Self-evaluation unduly influenced by body shape and weight.

E. The disturbance does not occur exclusively during episodes of anorexia nervosa.

Two types:

- *Purging type:* The person regularly engages in self-induced vomiting or the misuse of laxatives, diuretics, or enemas.

- *Nonpurging type:* The person uses other inappropriate compensatory behaviors, such as fasting or excessive exercise, but does not regularly engage in self-induced vomiting or the misuse of laxatives, diuretics, or enemas.

SOURCE: Reprinted with permission from American Psychiatric Association, *Diagnostic and Statistical Manual of Mental Disorders,* 4th ed. Text Revision. (Washington, D.C.: American Psychiatric Association, 2000).

---

secret binge-eating episodes, which usually progress through several emotional stages: anticipation and planning, anxiety, urgency to begin, rapid and uncontrollable consumption of food, relief and relaxation, disappointment, and finally shame or disgust.

A bulimic binge is characterized by a sense of lacking control over eating. During a binge, the person consumes food for its emotional comfort and cannot stop eating or control what or how much is eaten. A typical binge occurs periodically, in secret, usually at night, and lasts an hour or more. Because a binge frequently follows a period of rigid dieting, eating is accelerated by intense hunger. Energy restriction followed by bingeing can set in motion a pattern of weight cycling, which may make weight loss and maintenance more difficult over time.

During a binge, Kelly consumes thousands of kcalories of easy-to-eat, low-fiber, high-fat, and, especially, high-carbohydrate foods. Typically, she chooses cookies, cakes, and ice cream—and she eats the entire bag of cookies, the whole cake, and every last spoonful in a carton of ice cream. After the binge, Kelly pays the price with swollen hands and feet, bloating, fatigue, headache, nausea, and pain.

Purging To purge the food from her body, Kelly may use a **cathartic**—a strong laxative that can injure the lower intestinal tract. Or she may induce vomiting, with or without the use of an **emetic**—a drug intended as first aid for poisoning. These purging behaviors are often accompanied by feelings of shame or guilt. Hence a vicious cycle develops: negative self-perceptions

Bulimic binges are often followed by self-induced vomiting and feelings of shame or disgust.

followed by dieting, bingeing, and purging, which in turn lead to negative self-perceptions (see Figure H8-2).

On first glance, purging seems to offer a quick and easy solution to the problems of unwanted kcalories and body weight. Many people perceive such behavior as neutral or even positive, when, in fact, binge eating and purging have serious physical consequences. Signs of subclinical malnutrition are evident in a compromised immune system. Fluid and mineral imbalances caused by vomiting or diarrhea can lead to abnormal heart rhythms and injury to the kidneys. Urinary tract infections can lead to kidney failure. Vomiting causes irritation and infection of the pharynx, esophagus, and salivary glands; erosion of the teeth; and dental caries. The esophagus may rupture or tear, as may the

FIGURE H8-2 The Vicious Cycle of Restrictive Dieting and Binge Eating

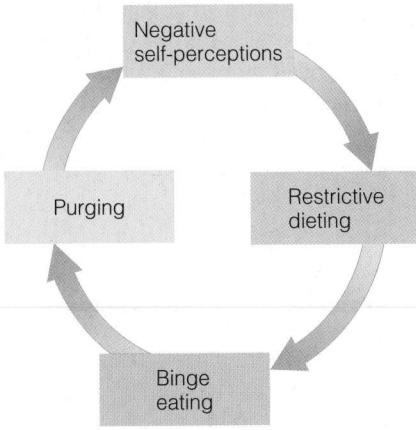

stomach. Sometimes the eyes become red from pressure during vomiting. The hands may be calloused or cut by the teeth while inducing vomiting. Overuse of emetics depletes potassium concentrations and can lead to death by heart failure.

Unlike Julie, Kelly is aware that her behavior is abnormal, and she is deeply ashamed of it. She wants to recover, and this makes recovery more likely for her than for Julie, who clings to denial. Feeling inadequate ("I can't even control my eating"), Kelly tends to be passive and to look to others for confirmation of her sense of worth. When she experiences rejection, either in reality or in her imagination, her bulimia nervosa becomes worse. If Kelly's depression deepens, she may seek solace in drug or alcohol abuse or in other addictive behaviors. Clinical depression is common in people with bulimia nervosa, and the rates of substance abuse are high.[17]

Treatment of Bulimia Nervosa

Kelly needs to establish regular eating patterns. She may also benefit from a regular exercise program.[18] Weight maintenance, rather than cyclic weight gains and losses, is the treatment goal. Major steps toward recovery include discontinuing purging and restrictive dieting habits and learning to eat three meals a day plus snacks.[19] Initially, energy intake should provide enough food to satisfy hunger and maintain body weight. Table H8-4 offers diet strategies to correct the eating problems of bulimia nervosa. About half of the women diagnosed with bulimia nervosa recover

TABLE H8-4 Diet Strategies for Combating Bulimia Nervosa

Planning Principles

- Plan meals and snacks; record plans in a food diary prior to eating.
- Plan meals and snacks that require eating at the table and using utensils.
- Refrain from finger foods.
- Refrain from "dieting" or skipping meals.

Nutrition Principles

- Eat a well-balanced diet and regularly timed meals consisting of a variety of foods.
- Include raw vegetables, salad, or raw fruit at meals to prolong eating times.
- Choose whole-grain, high-fiber breads, pasta, rice, and cereals to increase bulk.
- Consume adequate fluid, particularly water.

Other Tips

- Choose foods that provide protein and fat for satiety and bulky, fiber-rich carbohydrates for immediate feelings of fullness.
- Try including soups and other water-rich foods for satiety.
- Choose portions that meet the definition of "a serving" according to the Daily Food Guide (pp. 42–43).
- For convenience (and to reduce temptation) select foods that naturally divide into portions. Select one potato, rather than rice or pasta that can be overloaded onto the plate; purchase yogurt and cottage cheese in individual containers; look for small packages of precut steak or chicken; choose frozen dinners with measured portions.
- Include 30 minutes of physical activity every day—exercise may be an important tool in defeating bulimia.

completely after five to ten years, with or without treatment, but treatment probably speeds the recovery process.

A mental health professional should be on the treatment team to help clients with their depression and addictive behaviors. Some physicians prescribe the antidepressant drug fluoxetine in the treatment of bulimia nervosa.* Another drug that may be useful in the management of bulimia nervosa is naloxone, an opiate antagonist that suppresses the consumption of sweet and high-fat foods in binge-eaters.

Anorexia nervosa and bulimia nervosa are distinct eating disorders, yet they sometimes overlap in important ways. Anorexia victims may purge, and victims of both disorders may be overly concerned with body weight and have a tendency to drastically undereat. Many perceive foods as "forbidden" and "give in" to an eating binge. The two disorders can also appear in the same person, or one can lead to the other. Treatment is challenging and relapses are not unusual. Other people have **unspecified eating disorders** that fall short of the criteria for anorexia nervosa or bulimia nervosa but share some of their features. One such condition is binge-eating disorder.

Binge-Eating Disorder

Charlie is a 40-year-old schoolteacher who has been overweight all his life. His friends and family are forever encouraging him to lose weight, and he has come to believe that if he only had more willpower, dieting would work. He periodically gives dieting his best shot—restricting energy intake for a day or two only to succumb to uncontrollable cravings, especially for high-fat foods. Like Charlie, up to half of the obese people who try to lose weight periodically binge; unlike people with bulimia nervosa, however, they typically do not purge. Such an eating disorder does not meet the criteria for either anorexia nervosa or bulimia nervosa— yet such compulsive overeating is a problem and occurs in people of normal weight as well as those who are severely overweight. Table H8-5 lists criteria for unspecified eating disorders, including binge eating. Obesity alone is not an eating disorder.

Clinicians note differences between people with bulimia nervosa and those with binge-eating disorder.[20] People with **binge-eating disorder** consume less during a binge, rarely purge, and exert less restraint during times of dieting. Similarities also exist, including feeling out of control, disgusted, depressed, embarrassed, guilty, or distressed because of their self-perceived gluttony.[21]

---

* Fluoxetine is marketed under the trade name Prozac.

There are also differences between obese binge-eaters and obese people who do not binge. Those with the binge-eating disorder report higher rates of self-loathing, disgust about body size, depression, and anxiety. Their eating habits differ as well. Obese binge-eaters tend to consume more kcalories and more dessert and snack-type foods during regular meals and binges than obese people who do not binge.

Binge eating is a behavioral disorder that can be resolved with treatment. Resolving such behavior may not bring weight loss, but it may make participation in weight-control programs easier. It also improves physical health, mental health, and the chances of success in breaking the cycle of rapid weight losses and gains.

Eating Disorders in Society

Proof that society plays a role in eating disorders is found in their demographic distribution—they are known only in developed nations, and they become more prevalent as wealth increases and food becomes plentiful. Some people point to the vomitoriums of ancient times and claim that bulimia nervosa is not new, but the two are actually distinct. Ancient people were eating for pleasure, without guilt, and in the company of others; they vomited so that they could rejoin the feast. Bulimia nervosa is a disorder of isolation and is often accompanied by low self-esteem.

Chapter 8 described how our society sets unrealistic ideals for body weight, especially in women, and devalues those who do not conform to them. Anorexia nervosa and bulimia nervosa are not a form of rebellion against these unreasonable expectations, but rather an exaggerated acceptance of them. In fact, body dissatisfaction is a primary factor in the development of eating disorders.[22] Not everyone who is dissatisfied will develop an eating disorder, but everyone with an eating disorder is dissatisfied.

Characteristics of disordered eating such as restrained eating, fasting, binge eating, purging, fear of fatness, and distortion of body image are extraordinarily common among young girls. Most are "on diets," and many are poorly nourished. Some eat too little food to support normal growth; thus they miss out on their adolescent growth spurts and may never catch up. Many eat so little that hunger propels them into binge-purge cycles.

Perhaps a person's best defense against these disorders is to learn to appreciate his or her own uniqueness. When people discover and honor their body's real physical needs, they become unwilling to sacrifice health for conformity. To respect and value oneself may be lifesaving.

TABLE H8-5   Unspecified Eating Disorders, Including Binge-Eating Disorder

Criteria for Diagnosis of Unspecified Eating Disorders, in General

Many people have eating disorders but do not meet all the criteria to be classified as having anorexia nervosa or bulimia nervosa. Some examples include those who:

A. Meet all of the criteria for anorexia nervosa, except irregular menses.

B. Meet all of the criteria for anorexia nervosa, except that their current weights fall within the normal ranges.

C. Meet all of the criteria for bulimia nervosa, except that binges occur less frequently than stated in the criteria.

D. Are of normal body weight and who compensate inappropriately for eating small amounts of food (example: self-induced vomiting after eating two cookies).

E. Repeatedly chew food but spit it out without swallowing.

F. Have recurrent episodes of binge eating but do not compensate as do those with bulimia nervosa.

Criteria for Diagnosis of Binge-Eating Disorder, Specifically

A person with a binge-eating disorder demonstrates the following:

A. Recurrent episodes of binge eating. An episode of binge eating is characterized by both of the following:

   1. Eating, in a discrete period of time (e.g., within any two-hour period) an amount of food that is definitely larger than most people would eat in a similar period of time under similar circumstances.

   2. A sense of lack of control over eating during the episode (e.g., a feeling that one cannot stop eating or control what or how much one is eating).

B. Binge-eating episodes are associated with at least three of the following:

   1. Eating much more rapidly than normal.

   2. Eating until feeling uncomfortably full.

   3. Eating large amounts of food when not feeling physically hungry.

   4. Eating alone because of being embarrassed by how much one is eating.

   5. Feeling disgusted with oneself, depressed, or very guilty after overeating.

C. The binge eating causes marked distress.

D. The binge eating occurs, on average, at least twice a week for six months.

E. The binge eating is not associated with the regular use of inappropriate compensatory behaviors (e.g., purging, fasting, excessive exercise) and does not occur exclusively during the course of anorexia nervosa or bulimia nervosa.

SOURCE: Reprinted with permission from American Psychiatric Association, *Diagnostic and Statistical Manual of Mental Disorders*, 4th ed. Text Revision. (Washington, D.C.: American Psychiatric Association, 2000).

NUTRITION ON THE NET

For further study of topics covered in this chapter, log on to **academic.cengage.com/nutrition/rolfes/UNCN8e**. Go to Chapter 8, then to Nutrition on the Net.

- Search for "anorexia," "bulimia," and "eating disorders" at the U.S. Government health information site: **www.healthfinder.gov**

- Learn more about anorexia nervosa and related eating disorders from Anorexia Nervosa and Related Eating Disorders or the Academy of Eating Disorders: **www.anred.com** or **www.aedweb.org**

- Get facts about eating disorders from the National Institute of Mental Health: **www.nimh.nih.gov/publicat/eatingdisorders.cfm**

REFERENCES

1. Position of the American Dietetic Association: Nutrition intervention in the treatment of anorexia nervosa, bulimia nervosa, and eating disorders not otherwise specified (EDNOS), *Journal of the American Dietetic Association* 101 (2001): 810–819.
2. K. Kazis and E. Iglesias, The female athlete triad, *Adolescent Medicine* 14 (2003): 87–95; S. Sabatini, The female athlete triad, *American Journal of the Medical Sciences* 322 (2001): 193–195; Committee on Sports Medicine and Fitness, Medical concerns in the female athlete, *Pediatrics* 106 (2000): 610–613.
3. M. F. Reinking and L. E. Alexander, Prevalence of disordered-eating behaviors in undergraduate female collegiate athletes and nonathletes, *Journal of Athletic Training* 40 (2005): 47–51; M. K. Torstveit and J. Sundgot-Borgen, The female athlete triad: Are elite athletes at increased risk? *Medicine & Science in Sports and Exercise* 37 (2005): 184–193.
4. N. H. Golden, A review of the female athlete triad (amenorrhea, osteoporosis and disordered eating), *International Journal of Adolescent Medicine and Health* 14 (2002): 9–17.
5. S. J. Crow and coauthors, Long-term menstrual and reproductive function in patients with bulimia nervosa, *American Journal of Psychiatry* 159 (2002): 1048–1050.
6. C. L. Zanker and C. B. Cooke, Energy balance, bone turnover, and skeletal health in physically active individuals, *Medicine & Science in Sports and Exercise* 36 (2004): 1372–1381.
7. M. P. Fuhrman, P. Charney, and C. M. Mueller, Hepatic proteins and nutrition assessment, *Journal of the American Dietetic Association* 104 (2004): 1258–1264.
8. K. P. Kerruish and coauthors, Body composition in adolescents with anorexia nervosa, *American Journal of Clinical Nutrition* 75 (2002): 31–37.
9. C. Romano and coauthors, Reduced hemodynamic load and cardiac hypotrophy in patients with anorexia nervosa, *American Journal of Clinical Nutrition* 77 (2003): 308–312.
10. Committee on Adolescence, Identifying and treating eating disorders, *Pediatrics* 111 (2003): 204–211.
11. J. Yager and A. E. Andersen, Anorexia nervosa, *New England Journal of Medicine* 353 (2005): 1481–1488.
12. R. Sysko and coauthors, Eating behavior among women with anorexia nervosa, *American Journal of Clinical Nutrition* 82 (2005): 296–301; B. R. Carruth and J. D. Skinner, Dietary and physical activity patterns of young females with histories of eating disorders, *Topics in Clinical Nutrition* 16 (2000): 13–23.
13. V. van Wymelbeke and coauthors, Factors associated with the increase in resting energy expenditure during refeeding in malnourished anorexia nervosa patients, *American Journal of Clinical Nutrition* 80 (2004): 1469–1477.
14. H. C. Steinhausen, The outcome of anorexia nervosa in the 20th century, *American Journal of Psychiatry* 159 (2002): 1284–1293; B. Lowe and coauthors, Long-term outcome of anorexia nervosa in a prospective 21-year follow-up study, *Psychological Medicine* 31 (2001): 881–890.
15. P. K. Keel and coauthors, Predictors of mortality in eating disorders, *Archives of General Psychiatry* 60 (2003): 179–183.
16. M. B. Tamburrino and R. A. McGinnis, Anorexia nervosa: A review, *Panminerva Medica* 44 (2002): 301–311.
17. C. M. Bulik and coauthors, Alcohol use disorder comorbidity in eating disorders: A multicenter study, *Journal of Clinical Psychiatry* 65 (2004): 1000–1006.
18. J. Sundgot-Borgen and coauthors, The effect of exercise, cognitive therapy, and nutritional counseling in treating bulimia nervosa, *Medicine and Science in Sports and Exercise* 34 (2002): 190–195.
19. Position of the American Dietetic Association, 2001.
20. A. E. Dingemans, M. J. Bruna, and E. F. van Furth, Binge eating disorder: A review, *International Journal of Obesity and Related Metabolic Disorders* 26 (2002): 299–307.
21. D. M. Ackard and coauthors, Overeating among adolescents: Prevalence and associations with weight-related characteristics and psychological health, *Pediatrics* 111 (2003): 67–74.
22. J. Polivy and C. P. Herman, Causes of eating disorders, *Annual Review of Psychology* 53 (2002): 187–213.

Figure 9-1: Animated! Increasing Prevalence of Obesity among U.S. Adults

Figure 9-8: Animated! Influence of Physical Activity on Discretionary kCalorie Allowance

How To: Practice Problems

Nutrition Portfolio Journal

Nutrition Calculations: Practice Problems

Nutrition in Your Life

Are you pleased with your body weight? If so, you are a rare individual. Most people in our society think they should weigh more or less (mostly less) than they do. Usually, their primary concern is appearance, but they often understand that physical health is also somehow related to body weight. One does not necessarily cause the other—that is, an ideal body weight does not ensure good health. Instead, both depend on diet and physical activity. A well-balanced diet and active lifestyle support good health—and help maintain body weight within a reasonable range.

Weight Management: Overweight, Obesity, and Underweight

The previous chapter described how body weight is stable when energy in equals energy out. Weight gains occur when energy intake exceeds energy expended, and conversely, weight losses occur when energy expended exceeds energy intake. At the extremes, both overweight and underweight present health risks. **Weight management** is a key component of good health.

This chapter emphasizes overweight, partly because it has been more intensively studied and partly because it is a major health problem in the United States and a growing concern worldwide. Information on underweight is presented wherever appropriate. The highlight that follows this chapter examines fad diets.

Overweight and Obesity

Despite our preoccupation with body image and weight loss, the prevalence of overweight and obesity in the United States continues to rise dramatically.[1] In the past two decades, obesity increased in every state, in both genders, and across all ages, races, and educational levels (see Figure 9-1, p. 282). An estimated 66 percent of the adults in the United States are now considered overweight or obese, as defined by a BMI of 25 or greater.[2] ◆ The prevalence of overweight is especially high among women, the poor, blacks, and Hispanics.

The prevalence of overweight among children in the United States has also risen at an alarming rate. An estimated 33 percent of children and adolescents ages 2 to 19 years are either overweight or "at risk for overweight."[3] Chapter and Highlight 15 present information on overweight during childhood and adolescence.

◆ BMI:
  • Underweight: <18.5
  • Healthy weight: 18.5–24.9
  • Overweight: 25.0–29.9
  • Obese: ≥30

weight management: maintaining body weight in a healthy range by preventing gradual weight gain over time and losing weight if overweight.

FIGURE 9-1 *Animated!* Increasing Prevalence of Obesity (BMI ≥ 30) among U.S. Adults

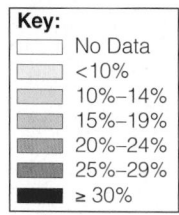

Key:
- No Data
- <10%
- 10%–14%
- 15%–19%
- 20%–24%
- 25%–29%
- ≥ 30%

CENGAGENOW™
To test your understanding of these concepts, log on to **academic.cengage.com/login.**

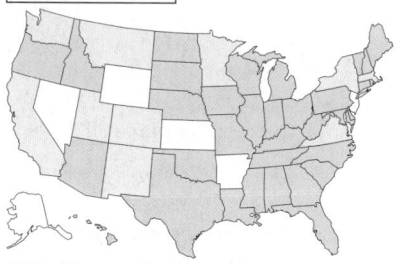

1990: No state had prevalence rates greater than or equal to 15 percent.

1995: Over half the states had prevalence rates greater than or equal to 15 percent, but no state had prevalence rates greater than or equal to 20 percent.

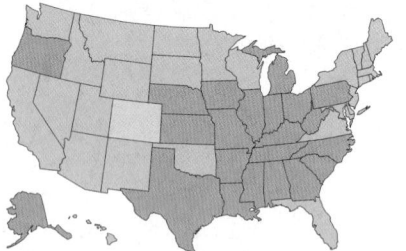

2000: Only one state had prevalence rates less than 15 percent, almost half of the states had prevalence rates greater than or equal to 20 percent, and no state had prevalence rates greater than or equal to 25 percent.

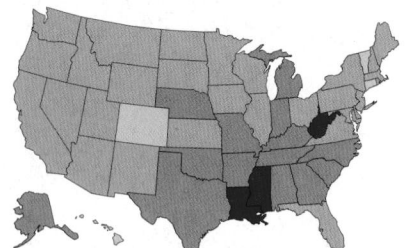

2005: Only four states had prevalence rates less than 20 percent, about one-third of the states had prevalence rates greater than or equal to 25 percent, with three states having prevalence rates greater than or equal to 30 percent.

SOURCE: www.cdc.nccdphp/dnpa/obesity/trend/maps/index.htm

Obesity is so widespread and its prevalence is rising so rapidly that many refer to it as an **epidemic.**[4] According to the World Health Organization, this epidemic of obesity has spread worldwide, affecting over 300 million adults. Contrary to popular opinion, obesity is not limited to industrialized nations; over 115 million people in developing countries suffer from obesity-related problems. Before examining the suspected causes of obesity and the various strategies used to treat it, it is helpful to understand the development and metabolism of body fat.

Fat Cell Development

When more energy is consumed than is expended, much of the excess energy is stored in the fat cells of adipose tissue. The amount of fat in a person's body reflects both the *number* and the *size* of the fat cells. The number of fat cells increases most rapidly during the growing years of late childhood and early puberty. After growth ceases, fat cell number may continue to increase whenever energy balance is positive. Obese people have more fat cells than healthy-weight people; their fat cells are also larger.

When energy intake exceeds expenditure, the fat cells accumulate triglycerides and expand in size (review Figure 5-20, p. 155). When the cells enlarge, they stimulate cell proliferation so that their numbers increase again.[5] Thus obesity develops ◆ when a person's fat cells increase in number, in size, or quite often both. Figure 9-2 illustrates fat cell development.

When energy out exceeds energy in, the size of fat cells dwindles, but not their number. People with extra fat cells tend to regain lost weight rapidly; with weight gain, their many fat cells readily fill. In contrast, people with an average number of enlarged fat cells may be more successful in maintaining weight losses; when their cells shrink, both cell size and number are normal. Prevention of obesity is most critical, then, during the growing years when fat cells increase in number.

As mentioned, excess fat is typically stored in adipose tissue. This stored fat may be well tolerated, but fat accumulation in organs such as the heart or liver clearly plays a key role in the development of diseases such as heart failure or fatty liver.[6] ◆

Fat Cell Metabolism

The enzyme lipoprotein lipase (LPL) ◆ promotes fat storage in both adipose and muscle cells. Obese people generally have much more LPL activity in their fat cells than lean people do (their muscle cell LPL activity is similar, though). This high LPL activity makes fat storage especially efficient. Consequently, even modest excesses in energy intake have a more dramatic impact on obese people than on lean people.

The activity of LPL is partially regulated by gender-specific hormones—estrogen in women and testosterone in men. In women, fat cells in the breasts, hips, and thighs produce abundant LPL, putting fat away in those body sites; in men, fat cells in the abdomen produce abundant LPL. This enzyme activity explains why men tend to develop central obesity around the abdomen (apple-shaped) whereas women more readily develop lower-body fat around the hips and thighs (pear-shaped).

Gender differences are also apparent in the activity of the enzymes controlling the release and breakdown of fat in various parts of the body. The release of lower-body fat is less active in women than in men, whereas the release of upper-body fat is similar. Furthermore, the rate of fat breakdown is lower in women than in men. Consequently, women may have a more difficult time losing fat in general, and from the hips and thighs in particular.

Enzyme activity may also explain why some people who lose weight regain it so easily. After weight loss, LPL activity increases, and it does so most dramatically in people who were fattest prior to weight loss. Apparently, weight loss serves as a signal to the gene that produces the LPL enzyme, saying "Make more of the enzyme that stores fat." People easily regain weight after having lost it because they are bat-

FIGURE 9-2 | Fat Cell Development

Fat cells are capable of increasing their size by 20-fold and their number by several thousandfold.

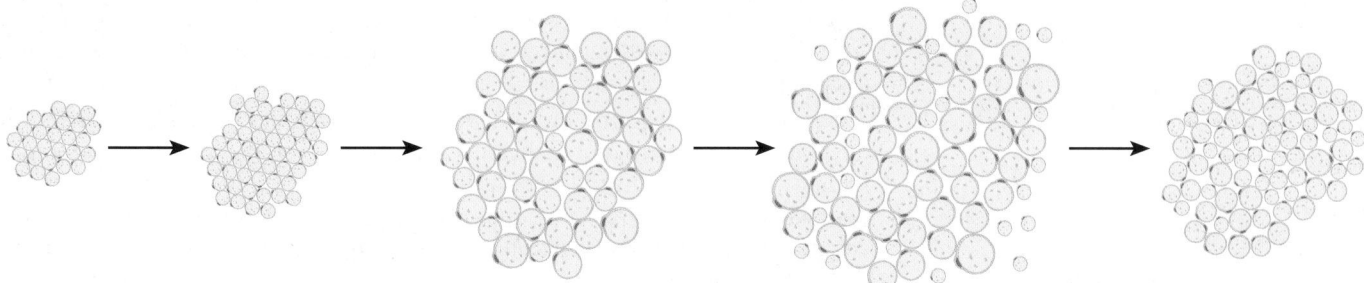

During growth, fat cells increase in number.

When energy intake exceeds expenditure, fat cells increase in size.

When fat cells have enlarged and energy intake continues to exceed energy expenditure, fat cells increase in number again.

With fat loss, the size of the fat cells shrinks but not the number.

tling against enzymes that want to store fat. The activities of these and other proteins provide an explanation for the observation that some inner mechanism seems to set a person's weight or body composition at a fixed point; the body will adjust to restore that **set point** if the person tries to change it.

Set-Point Theory

Many internal physiological variables, such as blood glucose, blood pH, and body temperature, remain fairly stable under a variety of conditions. The hypothalamus and other regulatory centers constantly monitor and delicately adjust conditions to maintain homeostasis. The stability of such complex systems may depend on set-point regulators that maintain variables within specified limits.

Researchers have confirmed that after weight gains or losses, the body adjusts its metabolism to restore the original weight. Energy expenditure increases after weight gain and decreases after weight loss. These changes in energy expenditure differ from those that would be expected based on body composition alone, and they help to explain why it is so difficult for an underweight person to maintain weight gains and an overweight person to maintain weight losses.

◆ Obesity due to an increase in the *number* of fat cells is **hyperplastic obesity.** Obesity due to an increase in the *size* of fat cells is **hypertrophic obesity.**

◆ The adverse effects of fat in nonadipose tissues are known as **lipotoxicity.**

◆ Reminder: *Lipoprotein lipase (LPL)* is an enzyme that hydrolyzes triglycerides passing by in the bloodstream and directs their parts into the cells, where they can be metabolized or reassembled for storage.

IN SUMMARY

Fat cells develop by increasing in number and size. Prevention of excess weight gain depends on maintaining a reasonable number of fat cells. With weight gains or losses, the body adjusts in an attempt to return to its previous status.

Causes of Overweight and Obesity

Why do people accumulate excess body fat? The obvious answer is that they take in more food energy than they expend. But that answer falls short of explaining why they do this. Is it genetic? Environmental? Cultural? Behavioral? Socioeconomic? Psychological? Metabolic? All of these? Most likely, obesity has many interrelated causes. Why an imbalance between energy intake and energy expenditure occurs remains a bit of a mystery; the next sections summarize possible explanations.

epidemic (ep-ih-DEM-ick): the appearance of a disease (usually infectious) or condition that attacks many people at the same time in the same region.
• **epi** = upon
• **demos** = people

set point: the point at which controls are set (for example, on a thermostat). The set-point theory that relates to body weight proposes that the body tends to maintain a certain weight by means of its own internal controls.

Genetics

Genetics plays a true causative role in relatively few cases of obesity, for example, in Prader-Willi syndrome—a genetic disorder characterized by excessive appetite, massive obesity, short stature, and often mental retardation. Most cases of obesity, however, do not stem from a genetic mutation, yet genetic influences do seem to be involved.

Researchers have found that adopted children tend to be similar in weight to their biological parents, not to their adoptive parents. Studies of twins yield similar findings: identical twins are twice as likely to weigh the same as fraternal twins—even when reared apart. These findings suggest an important role for genetics in determining a person's *susceptibility* to obesity. In other words, even if genes do not *cause* obesity, genetic factors interact with the food intake and activity patterns that lead to it and the metabolic pathways that maintain it.[7]

Clearly, something genetic makes a person more or less likely to gain or lose weight when overeating or undereating.[8] Some people gain more weight than others on comparable energy intakes. Given an extra 1000 kcalories a day for 100 days, some pairs of identical twins gain less than 10 pounds while others gain up to 30 pounds. Within each pair, the amounts of weight gained, percentages of body fat, and locations of fat deposits are similar. Similarly, some people lose more weight than others following comparable exercise routines.

Researchers have been examining several genes in search of answers to obesity questions. As the section on protein synthesis in Chapter 6 described, each cell expresses only the genes for the proteins it needs, and each protein performs a unique function. The following paragraphs describe some recent research involving proteins that might help explain appetite control, energy regulation, and obesity development.[9]

Leptin Researchers have identified an obesity gene, called *ob,* which is expressed primarily in the adipose tissue and codes for the protein **leptin.** Leptin acts as a hormone, primarily in the hypothalamus. Research suggests that leptin from adipose tissue signals sufficient energy stores and promotes a negative energy balance by suppressing appetite and increasing energy expenditure. Changes in energy expenditure primarily reflect changes in basal metabolism but may also include changes in physical activity patterns. Leptin is also released from stomach cells in response to the presence of food, suggesting a role for both short-term and long-term satiety regulation.[10]

Mice with a defective *ob* gene do not produce leptin and can weigh up to three times as much as normal mice and have five times as much body fat (see Figure 9-3). When injected with a synthetic form of leptin, the mice rapidly lose body fat. (Because leptin is a protein, it would be destroyed during digestion if given orally; consequently, it must be given by injection.) The fat cells not only lose fat, but they self-destruct (reducing cell number), which may explain why weight gains are delayed when the mice are fed again.

Although extremely rare, a genetic deficiency of leptin has been identified in human beings as well. An error in the gene that codes for leptin has been discovered in a few extremely obese children with barely detectable blood levels of leptin. Without leptin, the children have little appetite control; they are constantly hungry and eat considerably more than their siblings or peers. Given daily injections of leptin, these children lost a substantial amount of weight, confirming leptin's role in regulating appetite and body weight.[11]

Not too surprisingly, leptin injections are effective in suppressing appetite and supporting weight loss only when overeating and obesity are the result of a leptin deficiency. Very few obese people have a leptin deficiency, however. In fact, obese people generally have high leptin levels, and weight gain increases leptin concentrations. Researchers speculate that in obesity, leptin rises in an effort to overcome an insensitivity or resistance to leptin.

---

leptin: a protein produced by fat cells under direction of the *ob* gene that decreases appetite and increases energy expenditure; sometimes called the ***ob* protein.**

• **leptos** = thin

FIGURE 9-3 Mice with and without Leptin Compared

Both of these mice have a defective *ob* gene. Consequently, they do not produce leptin. They both became obese, but the one on the right received daily injections of leptin, which suppressed food intake and increased energy expenditure, resulting in weight loss.

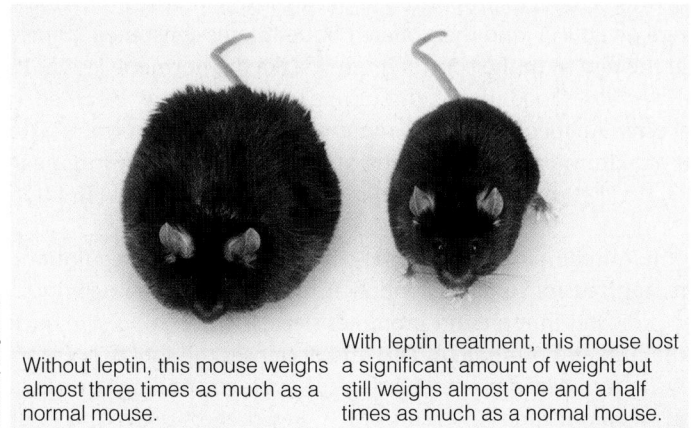

© Courtesy Amgen, Inc.

Without leptin, this mouse weighs almost three times as much as a normal mouse.

With leptin treatment, this mouse lost a significant amount of weight but still weighs almost one and a half times as much as a normal mouse.

Some researchers have reexamined the evidence on leptin from another point of view—one of undernutrition. Instead of focusing on leptin's role as a satiety signal that might help prevent obesity by regulating food intake, they view leptin as a starvation hormone that signals energy deficits.[12] When energy intake is low, leptin levels decline, and metabolism slows in an effort to reduce energy demands. Clearly, leptin plays a major role in energy regulation, but additional research is needed to clarify its actions when intake is either excessive or deficient.

In addition to its involvement in energy regulation, leptin plays several other roles in the body.[13] For example, leptin may inform the female reproductive system about body fat reserves; stimulate growth of new blood vessels, especially in the cornea of the eye; enhance the maturation of bone marrow cells; promote formation of red blood cells; and help support a normal immune response.[14] Elevated leptin levels may be partially responsible for the early maturation that commonly occurs in obese children.[15]

Ghrelin Leptin interacts with another protein that also acts as a hormone primarily in the hypothalamus.[16] Known as **ghrelin,** this protein is secreted primarily by the stomach cells and promotes a positive energy balance by stimulating appetite and promoting efficient energy storage.[17] The role ghrelin plays in regulating food intake and body weight is currently the subject of much intense research.[18]

Ghrelin triggers the desire to eat. Blood levels of ghrelin typically rise before and fall after a meal in proportion to the kcalories ingested—reflecting the hunger and satiety that precede and follow eating.[19] In general, fasting blood levels correlate inversely with body weight: lean people have high ghrelin levels and obese people have low levels.[20] Interestingly, although ghrelin levels are high in underweight people, they are exceptionally high in anorexia nervosa and return to normal with nutrition intervention—indicating that both body weight and nutrition status influence ghrelin levels.[21] Also noteworthy, ghrelin levels in Prader-Willi syndrome are markedly high and remain elevated even after a meal, which helps to explain the excessive appetite commonly seen in this disorder.[22] Similarly, ghrelin levels do not seem to decline as much after a meal in obese people or in people with binge-eating disorders as they do for lean people.[23]

Ghrelin fights to maintain a stable body weight.[24] In fact, some researchers speculate that its role is to maximize fat stores during times of famine.[25] On average, ghrelin levels are high whenever the body is in negative energy balance, as occurs during low-kcalorie diets, for example. This response may help explain why weight

ghrelin (GRELL-in): a protein produced by the stomach cells that enhances appetite and decreases energy expenditure.
- **ghre** = growth

loss is so difficult to maintain. Weight loss is more successful following gastric by-pass surgery, in part because ghrelin levels are abnormally low. (Why this is so remains unknown.)[26] Ghrelin levels decline again whenever the body is in positive energy balance, as occurs with weight gains.[27]

Ghrelin levels also decline in response to high levels of PYY, a peptide that the GI cells secrete after a meal in proportion to the kcalories ingested.[28] In one study, people who were given PYY and then offered buffet meals consumed 30 percent fewer kcalories in the day than the control group.[29] Like the hormone leptin, PYY signals satiety and decreases food intake, but unlike leptin, PYY may be an effective treatment for obesity. An ideal diet would maintain the satiating hormones (leptin, PYY, and cholecystokinin) and minimize the appetite stimulating hormone (ghrelin).[30] Fortunately, the diet that seems to do that best is one that is low in fat and rich in fiber.

Like leptin, ghrelin plays roles in the body beyond energy regulation. In fact, it was first recognized for its participation in growth hormone activity.[31] Some research also indicates that ghrelin promotes sleep.[32] Interestingly, a lack of sleep increases the hunger hormone ghrelin and decreases the satiety hormone leptin—which may help to explain epidemiological evidence finding an association between short sleep duration and high BMI.[33] Researchers are trying to understand the relationships among genes, sleep disorders, eating habits, and other related factors that may influence body weight and weight gain.[34]

Uncoupling Proteins Other genes code for proteins involved in energy metabolism. These proteins may influence the storing or expending of energy with different efficiencies or in different types of fat. The body has two types of fat: white and **brown adipose tissue.**[35] White adipose tissue stores fat for other cells to use for energy; brown adipose tissue releases stored energy as heat. Recall from Chapter 7 that when fat is oxidized, some of the energy is released in heat and some is captured in ATP. In brown adipose tissue, oxidation may be uncoupled ◆ from ATP formation, producing heat only.[36] By radiating energy away as heat, the body expends, rather than stores, energy. Brown fat and heat production is particularly important in newborns and in animals exposed to cold weather, especially those that hibernate.[37] They have plenty of brown adipose tissue. In contrast, most human adults have little brown fat—less than 1 percent of all fat cells and interspersed among the white fat cells.[38] The role of brown fat in body weight regulation, though probably minimal, is not yet understood.[39]

◆ Reminder: In *coupled reactions,* the energy released from the breakdown of one compound is used to create a bond in the formation of another compound. In *uncoupled reactions,* the energy is released as heat.

Uncoupling proteins are active not only in brown fat, but also in white fat and many other tissues. Their actions seem to influence the basal metabolic rate (BMR) and oppose the development of obesity. Animals with abundant amounts of these uncoupling proteins resist weight gain, whereas those with minimal amounts gain weight easily. Similarly, people with a genetic variant of an uncoupling protein have lower metabolic rates and are more overweight than others.[40] Whether the body dissipates the energy from an ice cream sundae as heat or stores it in body fat has major consequences for a person's body weight.

Environment

Although genetic studies indicate that body weight may be at least partially heritable, they do not fully explain obesity. In contrast to the studies mentioned earlier that found similar weights between identical twins, some identical twins have dramatically different body weights. With obesity rates rising over the past three decades and the **gene pool** remaining relatively unchanged, environment must also play a role in obesity. The *environment* includes all of the circumstances that we encounter daily that push us toward fatness or thinness. Keep in mind that genetic and environmental factors are not mutually exclusive; genes can influence eating behaviors, for example, and numerous eating behaviors influence body weight. A simple behavior, such as regularly skipping breakfast, for example, can contribute to obesity.[41]

brown adipose tissue: masses of specialized fat cells packed with pigmented mitochondria that produce heat instead of ATP.

gene pool: all the genetic information of a population at a given time.

Overeating One explanation for obesity is that overweight people overeat, although diet histories may not always reflect high intakes. Diet histories are not always accurate records of actual intakes; both normal-weight and obese people commonly misreport their dietary intakes.[42] Most importantly, current dietary intakes may not reflect the eating habits that led to obesity. Obese people who had a positive energy-balance for years and accumulated excess body fat may not currently have a positive energy balance. This reality highlights an important point: the energy-balance equation must consider time. Both present *and* past eating and activity patterns influence current body weight.

We live in an environment that exposes us to an abundance of high-kcalorie, high-fat foods that are readily available, relatively inexpensive, heavily advertised, ◆ and reasonably tasty.[43] Food is available everywhere, all the time—thanks largely to fast food. Our highways are lined with fast-food restaurants, and convenience stores and service stations offer fast food as well. Fast food is available in our schools, malls, and airports. It's convenient and it's available morning, noon, and night—and all times in between.

Most alarming are the extraordinarily large serving sizes and ready-to-go meals that offer supersize ◆ combinations. People buy the large sizes and combinations, perceiving them to be a good value, but then they eat more than they need—a bad deal. Large package or portion sizes can increase consumption—even when the food is not particularly appealing. Moviegoers given stale popcorn ate more when eating from a huge container than from a large container (both sizes were greater than anyone could finish).[44] Simply put, large portion sizes deliver more kcalories.[45] And portion sizes of virtually all foods and beverages have increased markedly in the past several decades, most notably at fast-food restaurants.[46] Not only have portion sizes increased over time, but they are now two to eight times larger than standard serving sizes.[47] The trend toward large portion sizes parallels the increasing prevalence of overweight and obesity in the United States, beginning in the 1970s, increasing sharply in the 1980s, and continuing today.[48]

Restaurant food, especially fast food, is a major player in the development of obesity.[49] Fast food is often high in fat.[50] Fat's 9 kcalories per gram quickly add up, amplifying people's energy intakes and enlarging their body fat stores. The combination of large portions and energy-dense foods is a double whammy.[51] Reducing portion sizes is somewhat helpful, but the real kcalorie savings come from lowering the energy density.[52] After all, large portions of foods with low energy density such as lean meats, fruits, and vegetables can help with weight loss. Unfortunately, these foods may not be as inexpensive, flavorful, and convenient as energy-dense foods.[53] Restaurants can help their customers eat healthfully by reducing portion sizes and offering more fruits, vegetables, legumes, and whole grains.[54]

Physical Inactivity Our environment fosters physical inactivity as well.[55] Life requires little exertion—escalators carry us up stairs, automobiles take us across town, buttons roll down windows, and remote controls change television channels from a distance. Modern technology has replaced physical activity at home, at work, and in transportation. Inactivity contributes to weight gain and poor health.[56] In turn, watching television, playing video games, and using the computer may contribute most to physical inactivity. The more time people spend in these sedentary activities, the more likely they are to be overweight.[57]

These sedentary activities contribute to weight gain in several ways. First, they require little energy beyond the resting metabolic rate. Second, they replace time spent in more vigorous activities. Third, watching television influences food purchases and correlates with between-meal snacking on the high-kcalorie, high-fat foods most heavily advertised.

People may be obese, therefore, not because they eat too much, but because they move too little—both in purposeful exercise and in the routines of daily life. One study reports that the differences in the time obese and lean people spent lying, sitting, standing, and moving accounts for about 350 kcalories a day.[58] Some obese people are so extraordinarily inactive that even when they eat less than lean people,

◆ The food industry spends $30 billion a year on advertising. The message? "Eat more."

◆ "Want fries with that?" A supersize portion delivers over 600 kcalories.

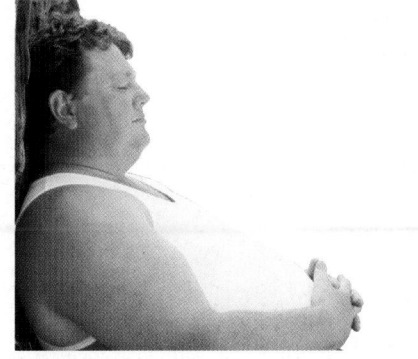

Lack of physical activity fosters obesity.

they still have an energy surplus. Reducing their food intake further would jeopardize health and incur nutrient deficiencies. Physical activity is a necessary component of nutritional health. People must be physically active if they are to eat enough food to deliver all the nutrients they need without unhealthy weight gain. In fact, to prevent weight gain, the DRI ◆ suggests an accumulation of 60 minutes of moderately intense physical activities every day in addition to the less intense activities of daily living.

◆ DRI for physical activity: 60 min/day (moderate intensity)

> ### IN SUMMARY
>
> Obesity has many causes and different combinations of causes in different people. Some causes, such as overeating and physical inactivity, may be within a person's control, and some, such as genetics, may be beyond it.

Problems of Overweight and Obesity

An estimated 35 to 45 percent of all U.S. women (and 20 to 30 percent of U.S. men) are trying to lose weight at any given time, spending up to $40 billion each year to do so.[59] Some of these people do not even need to lose weight. Others may benefit from weight loss, but they are not successful. Relatively few people succeed in losing weight, and even fewer succeed permanently. Whether an overweight person needs to lose weight is a question of health.

Health Risks

Chapter 8 described some of the health problems that commonly accompany obesity. In evaluating the risks to health from obesity, health care professionals use three indicators:[60]

◆ BMI 25.0–29.9 = overweight
BMI ≥30 = obese

◆ Men: >40 in (>102 cm)
Women: >35 in (>88 cm)

- Body mass index ◆ (BMI, as described in Chapter 8)
- Waist circumference ◆ (also described in Chapter 8)
- Disease risk profile, taking into account family history, life-threatening diseases, and common risk factors for chronic diseases[61]

The higher the BMI, the greater the waist circumference and the more risk factors—the greater the urgency to treat obesity.

People can best decide whether weight loss might be beneficial by considering their health status and motivation. People who are overweight by BMI standards, but otherwise in good health, might not benefit from losing weight; they might focus on preventing further weight gains instead. In contrast, those who are obese and suffering from a life-threatening disease such as diabetes might improve their health substantially by adopting a diet and exercise plan that supports weight loss. Motivation is a key component; to lose weight, a person needs to be ready and willing to make lifestyle changes for a lifetime.

◆ For reference, a woman with a BMI of 26 might be:
- 5 ft 3 in, 146 lb (1.60 m, 66.2 kg)
- 5 ft 5 in, 156 lb (1.65 m, 70.8 kg)
- 5 ft 7 in, 166 lb (1.70 m, 75.3 kg)

◆ Obese people and overweight people with two or more of these risk factors require aggressive treatment:
- Hypertension
- Cigarette smoking
- High LDL
- Low HDL
- Impaired glucose tolerance
- Family history of heart disease
- Men ≥45 yr; women ≥55 yr

Overweight in Good Health Often a person's motivations for weight loss have nothing to do with health. A healthy young woman with a BMI of 26 ◆ might want to lose a few pounds for spring break, but doing so might not improve her health. In fact, if she opts for a starvation diet or diet pills, she would be healthier *not* trying to lose weight.

◆ For reference, a man with a BMI of 28 might be:
- 5 ft 8 in, 184 lb (1.73 m, 83.5 kg)
- 5 ft 10 in, 195 lb (1.78m, 88.5 kg)
- 6 ft, 206 lb (1.83 m, 93.4 kg)

Obese or Overweight with Risk Factors Weight loss is recommended for people who are obese and those who are overweight (or who have a high waist circumference) with two or more risk factors for chronic diseases. ◆ A 50-year-old man with a BMI of 28 ◆ who has high blood pressure and a family history of heart disease can improve his health by adopting a diet low in saturated fat and a regular exercise plan.

Obese or Overweight with Life-Threatening Condition Weight loss is also recommended for a person who is either overweight or obese and suffering from a life-threatening condition such as heart disease, diabetes, or sleep apnea. ◆ The health benefits of weight loss are clear. For example, a 30-year-old man with a BMI of 40 ◆ might be able to prevent or control the diabetes that runs in his family by losing 75 pounds. Although the effort required to do so may be great, it may be no greater than the effort and consequences of living with diabetes.

Perceptions and Prejudices

Many people assume that every obese person can achieve slenderness and should pursue that goal. First consider that most obese people do not—for whatever reason—successfully lose weight and maintain their losses. Then consider the prejudice involved in that assumption. People come with varying weight tendencies, just as they come with varying potentials for height and degrees of health, yet we do not expect tall people to shrink or healthy people to get sick in an effort to become "normal."

Social Consequences Large segments of our society place such enormous value on thinness that obese people face prejudice and discrimination on the job, at school, and in social situations: they are judged on their appearance more than on their character.[62] Socially, obese people are stereotyped as lazy and lacking in self-control. Such a critical view of overweight is not prevalent in many other cultures, including segments of our own society. Instead, overweight is simply accepted or even embraced as a sign of robust health and beauty. Many overweight people today are tired of the obsession with weight control and simply want to be accepted as they are. To free society of its obsession with body weight and prejudice against obesity, people must first learn to judge others for who they are and not for what they weigh.

Psychological Problems Psychologically, obese people may suffer embarrassment when others treat them with hostility and contempt, and some have even come to view their own bodies as grotesque and loathsome. Parents and friends may scold them for lacking the discipline to resolve their weight problems. Health care professionals, including dietitians, are among the chief offenders. Criticism from others hurts self-esteem. Feelings of rejection, shame, or depression are common among obese people.

Most weight-loss programs assume that the problem can be solved simply by applying willpower and hard work. If determination were the only factor involved, though, the success rate would be far greater than it is. Overweight people may readily assume blame for failure to lose weight and maintain the losses when, in fact, it is the programs that have failed. Ineffective treatment and its associated sense of failure add to a person's psychological burden. Figure 9-4 illustrates how the devastating psychological effects of obesity and dieting perpetuate themselves.

Dangerous Interventions

People attach so many dreams of happiness to weight loss that they willingly risk huge sums of money for the slightest chance of success. As a result, weight-loss schemes flourish. Of the tens of thousands of claims, treatments, and theories for losing weight, few are effective—and many are downright dangerous. The negative effects must be carefully considered before embarking on any weight-loss program. Some interventions ◆ entail greater dangers than the risk of being overweight. Physical problems may arise from fad diets, "yo-yo" dieting, and drug use, and psychological problems may emerge from repeated "failures."

◆ Obese people and overweight people with any of these diseases require aggressive treatment:
- Heart disease
- Diabetes (type 2)
- Sleep apnea (a disturbance of breathing during sleep, including temporarily stopping)

◆ For reference, a man with a BMI of 40 might be:
- 5 ft 8 in, 265 lb (1.73 m, 120.2 kg)
- 5 ft 10 in, 280 lb (1.78 m, 127 kg)
- 6 ft, 295 lb (1.83 m, 133.8 kg)

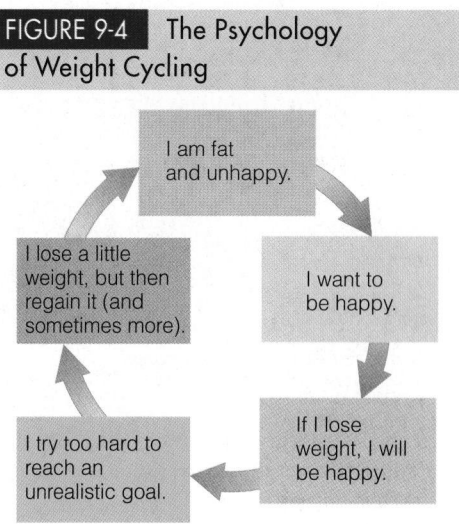

FIGURE 9-4 The Psychology of Weight Cycling

I am fat and unhappy.

I want to be happy.

If I lose weight, I will be happy.

I try too hard to reach an unrealistic goal.

I lose a little weight, but then regain it (and sometimes more).

◆ Scrutinize fad diets, magic potions, and wonder gizmos with a healthy dose of skepticism.

TABLE 9-1	Weight-Loss Consumer Bill of Rights (An Example)

1. *WARNING:* Rapid weight loss may cause serious health problems. Rapid weight loss is weight loss of more than 1½ to 2 pounds per week or weight loss of more than 1 percent of body weight per week after the second week of participation in a weight-loss program.

2. Consult your personal physician before starting any weight-loss program.

3. Only permanent lifestyle changes, such as making healthful food choices and increasing physical activity, promote long-term weight loss and successful maintenance.

4. Qualifications of this provider are available upon request.

5. *YOU HAVE A RIGHT TO:*

- Ask questions about the potential health risks of this program and its nutritional content, psychological support, and educational components.

- Receive an itemized statement of the actual or estimated price of the weight-loss program, including extra products, services, supplements, examinations, and laboratory tests.

- Know the actual or estimated duration of the program.

- Know the name, address, and qualifications of the dietitian or nutritionist who has reviewed and approved the weight-loss program.

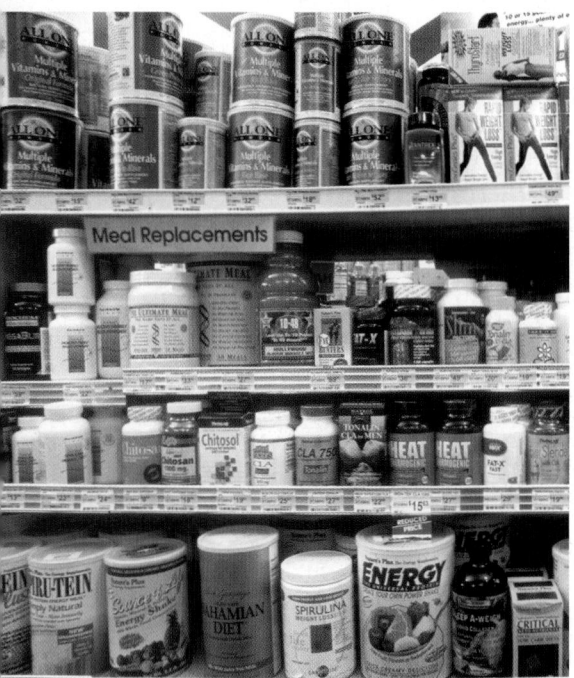

So many promises, so little success.

◆ Ephedrine is an amphetamine-like substance extracted from the Chinese ephedra herb *ma huang.*

fad diets: popular eating plans that promise quick weight loss. Most fad diets severely limit certain foods or overemphasize others (for example, never eat potatoes or pasta or eat cabbage soup daily).

serotonin (ser-oh-TONE-in): a neurotransmitter important in sleep regulation, appetite control, and sensory perception, among other roles. Serotonin is synthesized in the body from the amino acid tryptophan with the help of vitamin B$_6$.

Some of the nation's most popular diet books and weight-loss programs have misled consumers with unsubstantiated claims and deceptive testimonials. Furthermore, they fail to provide an assessment of the short- and long-term results of their treatment plans, even though such evaluations are possible and would permit consumers to make informed decisions. Of course, some weight-loss programs are better than others in terms of cost, approach, and customer satisfaction, but few are particularly successful in helping people keep lost weight off. Clients can expect reputable programs to abide by a consumer bill of rights that explains the risks associated with weight-loss programs and provides honest predictions of success (see Table 9-1).

Fad Diets **Fad diets** often sound good, but they typically fall short of delivering on their promises. They espouse exaggerated or false theories of weight loss and advise consumers to follow inadequate diets. Some fad diets are hazardous to health as Highlight 9 explains. Adverse reactions can be as minor as headaches, nausea, and dizziness or as serious as death. Table H9-4 (on p. 320) offers guidelines for identifying unsound weight-loss schemes and fad diets.

Weight-Loss Products Millions of people in the United States use nonprescription weight-loss products. Most of them are women, especially young overweight women, but almost 10 percent are of normal weight.

In their search for weight-loss magic, some consumers turn to "natural" herbal products and dietary supplements, even though few have proved to be effective. St. John's wort, for example, contains substances that inhibit the uptake of **serotonin** and thus suppress appetite. In addition to the many cautions that accompany the use of all herbal remedies, consumers should be aware that St. John's wort is often prepared in combination with the herbal stimulant ephedrine. ◆ Ephedrine-containing supplements promote modest short-term weight loss (about 2 pounds a month), but the associated risks are high.[63] These supplements have been implicated in several cases of heart attacks and seizures and have been linked to about 100 deaths. For this reason, the FDA has banned the sale of dietary supplements containing ephedra, but they are readily available on the Internet.* Table 9-2 presents the claims and the dangers behind ephedrine and several other common dietary supplements used for weight loss.[64]

Herbal laxatives containing senna, aloe, rhubarb root, cascara, castor oil, and buckthorn (or various combinations) are commonly sold as "dieter's tea." Such concoctions commonly cause nausea, vomiting, diarrhea, cramping, and fainting and may have contributed to the deaths of four women who had drastically reduced their food intakes. Consumers mistakenly believe that laxatives will diminish nutrient absorption and reduce kcalorie intake, but remember that absorption

---
* Ma huang (ephedrine) is illegal in Canada.

TABLE 9-2	Selected Herbal and Other Dietary Supplements Marketed for Weight Loss		
Product	**Manufacturers' Claims**	**Research Findings**	**Adverse Effects**
Bitter orange[a] (*Citrus aurantium,* a natural flavoring that contains synephrine, a compound structurally similiar to epinephrine)	Stimulates weight loss; provides an alternative to ephedra	Little evidence available	May increase blood pressure; may interact with drugs
Chitosan[b] (pronounced KITE-oh-san; derived from chitin, the substance that forms the hard shells of lobsters, crabs, and other crustaceans)	Binds to dietary fat, preventing digestion and absorption	Ineffective	Impaired absorption of fat-soluble vitamins
Chromium (trace mineral)	Eliminates body fat	Ineffective; weight gain reported when not accompanied by exercise	Headaches, sleep disturbances, and mood swings; hexavalent form is toxic and carcinogenic
Conjugated linoleic acid (CLA; a group of fatty acids related to linoleic acid, but with different *cis-* and *trans-*configurations)	Reduces body fat and suppresses appetite	Some evidence in animal studies, but ineffective in human studies	None known
Ephedrine[c] (amphetamine-like substance derived from the Chinese ephedra herb ma huang)	Speeds body's metabolism	Short-term weight loss and dangerous side effects	Insomnia, tremors, heart attacks, strokes, and death; FDA has banned the sale of these products
Hydroxycitric acid[d] (active ingredient derived from the rind of the tropical fruit *garcinia cambogia*)	Inhibits the enzyme that converts citric acid to fat; suppresses appetite	Ineffective	Toxicity symptoms reported in animal studies; headaches, respiratory and gastrointestinal distress in humans
Pyruvate[e] (3-carbon compound produced during glycolysis)	Speeds body's metabolism	Modest weight loss with high doses	GI distress
Yohimbine (derived from the bark of a West African tree)	Promotes weight loss	Ineffective	Nervousness, insomnia, anxiety, dizziness, tremors, headaches, nausea, vomiting, hypertension

NOTE: The FDA has not approved the use of any of these products; most products are used in conjunction with a 1000- to 1800-kcalorie diet.
[a]Marketed under the trade names Xenadrine EFX, Metabolife Ultra, NOW Diet Support.
[b]Marketed under the trade names Chitorich, Exofat, Fat Breaker, Fat Blocker, Fat Magnet, Fat Trapper, and Fatsorb.
[c]Marketed under the trade names Diet Fuel, Metabolife, and Nature's Nutrition Formula One.
[d]Marketed under the trade names Ultra Burn, Citralean, CitriMax, Citrin, Slim Life, Brindleslim, Medislim, and Beer Belly Busters.
[e]Marketed under the trade names Exercise in a Bottle, Pyruvate Punch, Pyruvate-c, and Provate.

occurs primarily in the small intestine and these laxatives act on the large intestine. Chapter 19 explores the possible benefits and potential dangers of herbal products. As it explains, current laws do not require manufacturers of dietary supplements to test the safety or effectiveness of any product. Consumers cannot assume that an herb or supplement of any kind is safe or effective just because it is available on the market. Supplements may contain contaminants and may not contain the amounts of active ingredients listed on the labels.[65] Anyone using dietary supplements for weight loss should first consult with a physician.

Other Gimmicks Other gimmicks don't help with weight loss either. Hot baths do not speed up metabolism so that pounds can be lost in hours. Steam and sauna baths do not melt the fat off the body, although they may dehydrate people so that they lose water weight. Brushes, sponges, wraps, creams, and massages intended to move, burn, or break up **"cellulite"** do nothing of the kind because there is no such thing as cellulite.

IN SUMMARY

The question of whether a person should lose weight depends on many factors: among them are the extent of overweight, age, health, and genetic makeup. Not all obesity will cause disease or shorten life expectancy. Just as there are unhealthy, normal-weight people, there are healthy, obese people. Some people may risk more in the process of losing weight than in remaining overweight. Fad diets and weight-loss supplements can be physically and psychologically damaging.

cellulite (SELL-you-light or SELL-you-leet): supposedly, a lumpy form of fat; actually, a fraud. Fatty areas of the body may appear lumpy when the strands of connective tissue that attach the skin to underlying muscles pull tight where the fat is thick. The fat itself is the same as fat anywhere else in the body. If the fat in these areas is lost, the lumpy appearance disappears.

Aggressive Treatments for Obesity

The appropriate strategies for weight reduction depend on the degree of obesity and the risk of disease. An overweight person in good health may need only to improve eating habits and increase physical activity, but someone with **clinically severe obesity** may need more aggressive treatment ◆ options—drugs or surgery.[66] Drugs appear to be modestly effective and safe, at least in the short term; surgery appears to be dramatically effective but can have severe complications, at least for some people.[67]

◆ The field of medicine that specializes in treating obesity is called **bariatrics**.
• **bar** = weight

Drugs

Based on new understandings of obesity's genetic basis and its classification as a chronic disease, much research effort has focused on drug treatments for obesity. Experts reason that if obesity is a chronic disease, it should be treated as such—and the treatment of most chronic diseases includes drugs. The challenge, then, is to develop an effective drug that can be used over time without adverse side effects or the potential for abuse.

Several drugs for weight loss have been tried over the years. When used as part of a long-term, comprehensive weight-loss program, drugs can help obese people to lose weight. Because weight regain commonly occurs with the discontinuation of drug therapy, treatment must be long term. Yet the long-term use of drugs poses risks. We don't yet know whether a person would be harmed more from maintaining a 100-pound excess or from taking a drug for a decade to keep the 100 pounds off. Physicians must prescribe drugs appropriately, inform consumers of the potential risks, and monitor side effects carefully. Two prescription drugs are currently on the market: sibutramine and orlistat. One reduces food intake; the other reduces nutrient absorption.[68]

Sibutramine Sibutramine suppresses appetite.* The drug is most effective when used in combination with a reduced-kcalorie diet and increased physical activity. Side effects include dry mouth, headache, constipation, rapid heart rate, and high blood pressure. The FDA warns those with high blood pressure not to use sibutramine and advises others to monitor their blood pressure.

Orlistat Orlistat takes a different approach to weight control.† It inhibits pancreatic lipase activity in the GI tract, thus blocking dietary fat digestion and absorption by about 30 percent. The drug is taken with meals and is most effective when accompanied by a reduced-kcalorie, low-fat diet. Side effects include gas, frequent bowel movements, and reduced absorption of fat-soluble vitamins. The FDA recently approved the over-the-counter sale of a low-dose version of orlistat.‡

Other Drugs Some physicians prescribe drugs that have not been approved for weight loss, a practice known as "off-label" use. These drugs have been approved for other conditions (such as seizures) and incidentally cause modest weight loss.[69] Physicians using off-label drugs must be well-informed of the drugs' use and effects and monitor their patients' responses closely.

◆ Surgery may be an option for people with all of the following conditions:
• Have tried diet and exercise programs without success
• Remain obese (BMI > 35)
• Have weight-related health problems

Surgery

Surgery ◆ as an approach to weight loss is justified in some specific cases of clinically severe obesity. Over 100,000 such surgeries are performed annually.[70] As Figure 9-5 shows, surgical procedures effectively limit food intake by reducing the capacity of the stomach. In addition, they suppress hunger by reducing production of the hormone ghrelin.[71] The results are dramatic: most people achieve a lasting weight loss of more than 50 percent of their excess body weight.[72] Importantly, most of them experience dramatic improvements in their diabetes, blood lipids, and blood pressure.[73]

clinically severe obesity: a BMI of 40 or greater or a BMI of 35 or greater with additional medical problems. A less preferred term used to describe the same condition is *morbid obesity*.

sibutramine (sigh-BYOO-tra-mean): a drug used in the treatment of obesity that slows the reabsorption of serotonin in the brain, thus suppressing appetite and creating a feeling of fullness.

orlistat (OR-leh-stat): a drug used in the treatment of obesity that inhibits the absorption of fat in the GI tract, thus limiting kcaloric intake.

* Sibutramine is marketed under the trade name Meridia.
† Orlistat is marketed under the trade name Xenical.
‡ The low-dose, over-the-counter version of orlistat is marketed under the trade name Alli (AL-eye).

FIGURE 9-5 Gastric Surgery Used in the Treatment of Severe Obesity

Both of these surgical procedures limit the amount of food that can be comfortably eaten.

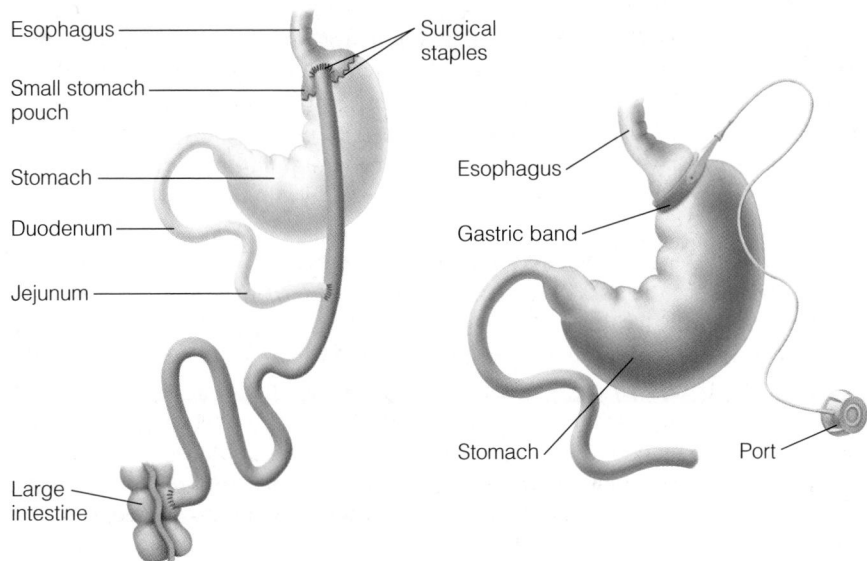

In gastric bypass, the surgeon constructs a small stomach pouch and creates an outlet directly to the small intestine, bypassing most of the stomach, the entire duodenum, and some of the jejunum. (Dark areas highlight the flow of food through the GI tract; pale areas indicate bypassed sections.)

In gastric banding, the surgeon uses a gastric band to reduce the opening from the esophagus to the stomach. The size of the opening can be adjusted by inflating or deflating the band by way of a port placed in the abdomen just beneath the skin.

Whether surgery is a reasonable option for obese teens is the subject of much debate among pediatricians and bariatric surgeons.[74] In addition to the criteria listed in the margin (p. 292) for adults considering surgery, teens must have a BMI greater than 40, and they must have attained skeletal maturity.[75] Considerations of the adolescent's physical growth, emotional development, family support, and ability to comply with dietary instructions weigh heavily in the decision.

The long-term safety and effectiveness of gastric surgery depend, in large part, on compliance with dietary instructions. Common immediate postsurgical complications include infections, nausea, vomiting, and dehydration. In the long term, deficiencies of iron, vitamin B_{12}, folate, calcium, and vitamin D are common.[76] Weight regain and psychological problems may also occur. Lifelong medical supervision is necessary for those who choose the surgical route, but in suitable candidates, the health benefits of weight loss may prove worth the risks.[77]

Another surgical procedure is used, not to treat obesity, but to remove the evidence. Plastic surgeons can extract some fat deposits by suction lipectomy, or "liposuction." This cosmetic procedure has little effect on body weight, but can alter body shape slightly in specific areas. Liposuction is a popular procedure in part because of its perceived safety, but, in fact, serious complications can occasionally result in death. Furthermore, removing adipose tissue by way of liposuction does not provide the health benefits that typically accompany weight loss.[78]

IN SUMMARY

Obese people with high risks of medical problems may need aggressive treatment, including drugs or surgery. Others may benefit most from improving eating and exercise habits.

Weight-Loss Strategies

Successful weight-loss strategies embrace small changes, moderate losses, and reasonable goals.[79] People who lose 10 to 20 pounds in a year by consistently choosing nutrient-dense foods and engaging in regular physical activity are much more likely to maintain the loss and reap health benefits than if they were to lose more weight in less time by adopting a radical fad diet. In keeping with this philosophy, the 2005 *Dietary Guidelines* advise those who need to lose weight to "aim for a slow, steady weight loss by decreasing kcalorie intake while maintaining an adequate nutrient intake and increasing physical activity." Even modest weight loss brings health benefits.

Modest weight loss, even when a person is still overweight, can improve control of diabetes and reduce the risks of heart disease by lowering blood pressure and blood cholesterol, especially for those with central obesity. Improvements in physical capabilities and bodily pain become evident with even a 5-pound weight loss. For these reasons, parameters such as blood pressure, blood cholesterol, or even vitality are more useful than body weight in marking success. People less concerned with disease risks may prefer to set goals for personal fitness, such as being able to play with children or climb stairs without becoming short of breath. Importantly, they can enjoy living a healthy life instead of focusing on the elusive goal of losing weight.

Whether the goal is health or fitness, expectations need to be reasonable. Unreachable targets ensure frustration and failure. When goals are achieved or exceeded, people enjoy rewards instead of finding disappointment.

Research findings highlight the great disparity between lofty expectations and reasonable success.[80] Before beginning a weight-loss program, obese women identified the weights they would describe as "dream," "happy," "acceptable," and "disappointing" (see Figure 9-6). All of these weights were below their starting weight. Their goal weights far exceeded the 5 to 10 percent recommended by experts, or even the 15 percent reported by the most successful weight-loss studies. Even their "disappointing" weights exceeded recommended goals. Close to a year later, and after an average loss of 35 pounds, almost half of the women did not achieve even their "disappointing" weights. They did, however, experience more physical, social, and psychological benefits than they had predicted for that weight. Still, in a cul-

FIGURE 9-6 Reasonable Weight Goals and Expectations Compared

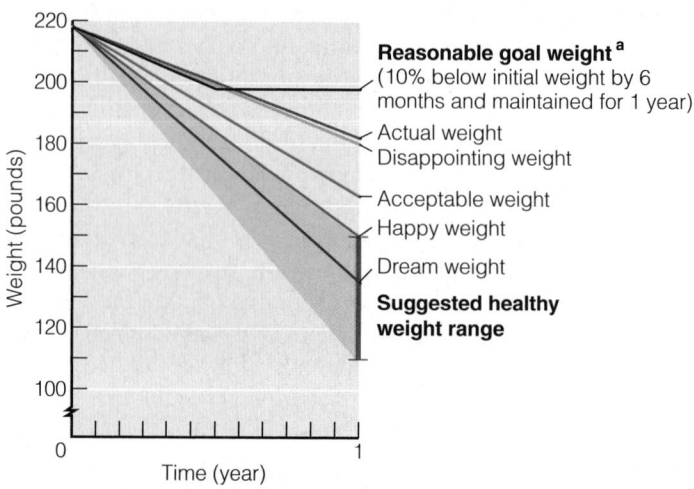

[a]Reasonable goal weights reflect pounds lost over time. Given more time, reasonable goals may eventually fall within the suggested healthy-weight range.

SOURCE: Adapted from G. D. Foster and coauthors, What is a reasonable weight loss? Patients' expectations and evaluations of obesity treatment outcomes, *Journal of Consulting and Clinical Psychology* 65 (1997): 79–85.

ture that overvalues thinness, these women were not satisfied with a 16 percent reduction in weight—not because their efforts were unsuccessful, but because their expectations were unrealistic.

Depending on initial body weight, a reasonable rate of weight loss for overweight people is $^1/_2$ to 2 pounds a week, ◆ or 10 percent of body weight over six months.[81] For a person weighing 250 pounds, a 10 percent loss is 25 pounds, or about 1 pound a week for six months. Such gradual weight losses are more likely to be maintained than rapid losses. Keep in mind that pursuing good health is a lifelong journey. Most adults are keenly aware of their body weights and shapes and realize that what they eat and what they do can make a difference to some extent. Those who are most successful at weight management seem to have fully incorporated healthful eating and physical activity into their daily lives.[82] Such advice—to reduce energy intake and increase physical activity—would hardly surprise anyone, yet relatively few people trying to control their weight follow these recommendations.

Eating Plans

Contrary to the claims of fad diets, no single food plan is magical, and no specific food must be included or avoided in a weight-management program. In designing a plan, people need only consider foods that they like or can learn to like, that are available, and that are within their means.

Be Realistic about Energy Intake The main characteristic of a weight-loss diet is that it provides less energy than the person needs to maintain present body weight. If food energy is restricted too severely, dieters may not receive sufficient nutrients and may lose lean tissue. Rapid weight loss usually means excessive loss of lean tissue, a lower BMR, and a rapid weight gain to follow. In addition, restrictive eating may set in motion the unhealthy behaviors of eating disorders as previously described in Highlight 8.

Table 9-3 outlines the recommendations of a weight-loss diet. Energy intake should provide nutritional adequacy without excess—that is, somewhere between

◆ Safe rate for weight loss:
- $^1/_2$ to 2 lb/week (0.2 to 0.9 kg)
- 10% body weight/6 mo
- For a person weighing 110 kg, a 10% loss is 11 kg, or about 0.5 kg a week for six months

TABLE 9-3	Recommendations for a Weight-Loss Diet
Nutrient	**Recommended Intake**
kCalories	
For people with BMI ≥ 35	Approximately 500 to 1000 kcalories per day reduction from usual intake
For people with BMI between 27 and 35	Approximately 300 to 500 kcalories per day reduction from usual intake
Total fat	30% or less of total kcalories
Saturated fatty acids[a]	8 to 10% of total kcalories
Monounsaturated fatty acids	Up to 15% of total kcalories
Polyunsaturated fatty acids	Up to 10% of total kcalories
Cholesterol[a]	300 mg or less per day
Protein[b]	Approximately 15% of total kcalories
Carbohydrate[c]	55% or more of total kcalories
Sodium chloride	No more than 2400 mg of sodium or approximately 6 g of sodium chloride (salt) per day
Calcium	1000 to 1500 mg per day
Fiber[c]	20 to 30 g per day

[a]People with high blood cholesterol should aim for less than 7 percent kcalories from saturated fat and 200 milligrams of cholesterol per day.
[b]Protein should be derived from plant sources and lean sources of animal protein.
[c]Carbohydrates and fiber should be derived from vegetables, fruits, and whole grains.
SOURCE: National Institutes of Health Obesity Education Initiative, *The Practical Guide: Identification, Evaluation, and Treatment of Overweight and Obesity in Adults* (Washington, D.C.: U.S. Department of Health and Human Services, 2000), p. 27.

deprivation and complete freedom to eat whatever, whenever. A reasonable suggestion is that an adult needs to increase activity and reduce food intake enough to create a deficit of 500 kcalories per day. Such a deficit produces a weight loss of about 1 pound per week—a rate that supports the loss of fat efficiently while retaining lean tissue. In general, weight-loss diets provide 1000 to 1200 kcalories per day for women and 1200 to 1600 kcalories a day for men.[83] Table 9-4 specifies the amounts of foods from each food group for these kcalorie levels.

Emphasize Nutritional Adequacy Nutritional adequacy is difficult to achieve on fewer than 1200 kcalories a day, and most healthy adults need never consume any less. A plan that provides an adequate intake supports a healthier and more successful weight loss than a restrictive plan that creates feelings of starvation and deprivation, which can lead to an irresistible urge to binge.

Table 9-4 includes the recommended amounts for diets providing 1000 to 1600 kcalories. Such an intake would allow most people to lose weight and still meet their nutrient needs with careful, nutrient-dense food selections. (Women might need iron supplements.) Keep in mind, too, that well-balanced diets that emphasize fruits, vegetables, whole grains, lean meats or meat alternates, and low-fat milk products offer many health rewards even when they don't result in weight loss. A supplement providing vitamins and minerals at or below 100 percent of the Daily Values can help people following low-kcalorie diets to achieve nutrient adequacy.[84]

Eat Small Portions As mentioned earlier, portion sizes at markets, at restaurants, and even at home have increased dramatically over the years.[85] We have come to expect large portions, and we have learned to clean our plates. Many of us pay more attention to these external cues defining how much to eat than to our internal cues of hunger and satiety.[86] For health's sake, we may need to learn to eat less food at each meal—one piece of chicken for dinner instead of two, a teaspoon of butter on vegetables instead of a tablespoon, and one cookie for dessert instead of six. The goal is to eat enough food for adequate energy, abundant vitamins and minerals, and some pleasure, but not more. This amount should leave a person feeling satisfied—not stuffed.

Keep in mind that even fat-free and low-fat foods can deliver a lot of kcalories when a person eats large quantities. A low-fat cookie or two can be a sweet treat even on a weight-loss diet, but larger portions defeat the savings.

Lower Energy Density Most people take their cues about how much to eat based on portion sizes, and the larger the portion size, the more they eat—even when the food is not particularly tasty.[87] To lower energy intake, a person can either reduce the portion size or reduce the energy density.[88] Selecting low-energy-dense foods seems to be more a successful strategy than restricting portion sizes.[89] Figure 9-7 illustrates how water, fiber, and fat influence energy density, and the accompanying "How to" feature compares foods based on their energy density. Foods containing water, those rich in fiber, and those low in fat help to lower energy density, providing more satiety for

TABLE 9-4	Daily Amounts from Each Food Group for 1000- to 1600-kCalorie Diets			
Food Group	1000 kcalories	1200 kcalories	1400 kcalories	1600 kcalories
Fruit	1 c	1 c	1½ c	1½ c
Vegetables	1 c	1½ c	1½ c	2 c
Grains	3 oz	4 oz	5 oz	5 oz
Meat and Legumes	2 oz	3 oz	4 oz	5 oz
Milk	3 c	3 c	3 c	3 c
Oils	3 tsp	3 tsp	3 tsp	4 tsp

NOTE: The USDA Food Guide patterns for 1000-, 1200-, and 1400-kcalories were designed for children and provided 2 cups milk. They were modified here to include an additional cup of milk, as 3 cups per day is recommended for all adults. The discretionary kcalorie allowance for these patterns is about 100 kcalories.

FIGURE 9-7 | Energy Density

Decreasing the energy density (kcal/g) of foods allows a person to eat satisfying portions while still reducing energy intake. To lower energy density, select foods high in water or fiber and low in fat.

Selecting grapes with their high water content instead of raisins increases the volume and cuts the energy intake in half.

Even at the same weight and similar serving sizes, the fiber-rich broccoli delivers twice the fiber of the potatoes for about one-fourth the energy.

By selecting the water-packed tuna (on the right) instead of the oil-packed tuna (on the left), a person can enjoy the same amount for fewer kcalories.

Matthew Farruggio (all)

fewer kcalories.[90] Because a low-energy-density diet is a low-fat, high-fiber diet rich in many vitamins and minerals, it supports good health in addition to weight loss.[91]

Remember Water Water helps with weight management in several ways. For one, foods with high water content (such as broth-based soups) increase fullness, reduce hunger, and consequently reduce energy intake. For another, drinking water fills the stomach between meals and satisfies thirst without adding kcalories. The average U.S. diet delivers an estimated 75 to 150 kcalories a day from sweetened beverages.[92] Simply replacing nutrient-poor, energy-dense beverages with water could save a person up to 15 pounds a year. Water also helps the GI tract adapt to a high-fiber diet.

Focus on Fiber Healthy meals and snacks center on high-fiber foods. Fresh fruits, vegetables, legumes, and whole grains offer abundant vitamins, minerals, and fiber but little fat. Consequently, high-carbohydrate diets rich in fiber tend to be relatively low in energy and high in nutrients.[93]

High-fiber foods also require effort to eat—an added bonus. Eating fiber-rich fruits and vegetables reduces energy density, lowers kcalorie intake, and promotes

HOW TO | Compare Foods Based on Energy Density

Chapter 2 described how to evaluate foods based on their nutrient density—their nutrient contribution per kcalorie. Another way to evaluate foods is to consider their energy density—their energy contribution per gram. This example compares carrot sticks with French fries. The conclusion is no surprise, but understanding the mathematics may offer valuable insight into the concept of energy density. A carrot weighing 72 grams delivers 31 kcalories. To calculate the energy density, divide kcalories by grams:

$$\frac{31 \text{ kcal}}{72 \text{ g}} = 0.43 \text{ kcal/g}$$

Do the same for French fries weighing 50 grams and contributing 167 kcalories:

$$\frac{167 \text{ kcal}}{50 \text{ g}} = 3.34 \text{ kcal/g}$$

The more kcalories per gram, the greater the energy density. French fries are more energy dense than carrots. They provide more energy per gram—and per bite. Considering a food's energy density is especially useful in planning diets for weight management. Foods with a high energy density help with weight gain, whereas foods with a low energy density help with weight loss.

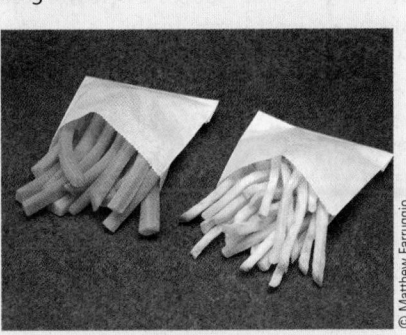

© Matthew Farruggio

CENGAGENOW™
To practice comparing foods based on energy density, log on **academic.cengage.com/login**, go to Chapter 9, then go to How To.

If you want to lose weight, steer clear of the empty kcalories in fancy coffee drinks. A 16-ounce café mocha delivers 400 kcalories—half of them from fat.

satiety.[94] The satiety signal indicating fullness is sent after a 20-minute lag, so a person who slows down and savors each bite eats less before the signal reaches the brain. Of course, much depends on whether the person pays attention to internal satiety signals and stops eating or, instead, responds to external cognitive influences and continues.

Choose Fats Sensibly Ideally, a weight-loss diet is both high in fiber and low in fat. Lowering the fat content of a food lowers its energy density—for example, selecting fat-free milk instead of whole milk. That way, a person can consume the usual amount (say, a cup of milk) at a lower energy intake (85 instead of 150 kcalories).

Fat has a weak satiating effect, and satiation plays a key role in determining food intake during a meal. Consequently, a person eating a high-fat meal raises energy intake by adding more food and more fat kcalories. For these reasons, measure fat with extra caution. Less fat in the diet means less fat in the body (review p. 163 for strategies to lower fat in the diet). Be careful not to take this advice to extremes, however; too little fat in the diet or in the body carries health risks as well, as Chapter 5 explained.

Whether a low-fat diet is the best option for weight loss is the subject of some controversy and much debate. An important point to notice in any discussion on weight-loss diets is total energy intake. *Low fat* simply means the energy derived from fat is relatively low compared with the total energy intake; it does not mean total energy intake is low. And reducing energy intake to less than expended is essential for weight loss. One way to lower energy intake is to lower fat intake. In these cases, adopting a low-fat diet can help with weight loss.[95]

Another currently popular way to lower energy intake is to lower carbohydrate intake. The highlight that follows this chapter discusses these diets fully, but findings from a recent study are worth mentioning here as well.[96] In this study, people were randomly assigned to one of two diets—either a low-carbohydrate diet or a low-fat diet. They were given descriptions of the diets and then fed themselves, as would be typical of many dieters. Both groups lost weight, but those on the low-carbohydrate diet lost more weight during the first six months; their diets produced a greater energy deficit. Interestingly, the differences in weight loss between the two groups disappeared by the end of one year. Between six months and one year, weight remained fairly stable in the low-fat group, but regains were evident in the low-carbohydrate group, suggesting that adhering to a low-carbohydrate diet for an extended length of time may be difficult. These findings highlight an important point: weight loss requires a commitment to long-term changes in food choices. They also confirm another critical point: weight loss depends on a low energy intake—not the proportion of energy nutrients.[97]

Watch for Other Empty kCalories A person trying to achieve or maintain a healthy weight needs to pay attention not only to fat, but to sugar and alcohol, too.[98] Using sugar or alcohol for pleasure on occasion is compatible with health as long as most daily choices are of nutrient-dense foods. Not only does alcohol add kcalories, but accompanying mixers can also add both kcalories and fat, especially in creamy drinks such as piña coladas (review Table H7-3 on p. 244). Furthermore, drinking alcohol reduces a person's inhibitions, which can sabotage weight-control efforts—at least temporarily.

IN SUMMARY

A person who adopts a lifelong "eating plan for good health" rather than a "diet for weight loss" will be more likely to keep the lost weight off. Table 9-5 provides several tips for successful weight management.

TABLE 9-5	Weight-Management Strategies

In General

- Focus on healthy eating and activity habits, not on weight losses or gains.
- Adopt reasonable expectations about health and fitness goals and about how long it will take to achieve them.
- Make nutritional adequacy a high priority.
- Learn, practice, and follow a healthful eating plan for the rest of your life.
- Participate in some form of physical activity regularly.
- Adopt permanent lifestyle changes to achieve and maintain a healthy weight.

For Weight Loss

- Energy out should exceed energy in by about 500 kcalories/day. Increase your physical activity enough to spend more energy than you consume from foods.
- Emphasize foods with a low energy density and a high nutrient density.
- Eat small portions. Share a restaurant meal with a friend or take home half for lunch tomorrow.
- Eat slowly.
- Limit high-fat foods. Make legumes, whole grains, vegetables, and fruits central to your diet plan.
- Limit low-fat treats to the serving size on the label.
- Limit concentrated sweets and alcoholic beverages.
- Drink a glass of water before you begin to eat and another while you eat. Drink plenty of water throughout the day (8 glasses or more a day).
- Keep a record of diet and exercise habits; it reveals problem areas, the first step toward improving behaviors.
- Learn alternative ways to deal with emotions and stresses.
- Attend support groups regularly or develop supportive relationships with others.

For Weight Gain

- Energy in should exceed energy out by at least 500 kcalories/day. Increase your food intake enough to store more energy than you spend in exercise. Exercise and eat to build muscles.
- Expect weight gain to take time (1 pound per month would be reasonable).
- Emphasize energy-dense foods.
- Eat at least three meals a day.
- Eat large portions of foods and expect to feel full.
- Eat snacks between meals.
- Drink plenty of juice and milk.

Physical Activity

The best approach to weight management includes physical activity.[99] Yet among people trying to lose weight, only half are physically active and only half of the active group meet minimal recommendations.[100] To prevent weight gains and support weight losses, current recommendations advise 60 minutes of moderately intense physical activity a day in addition to activities of daily life.[101] People who combine diet and exercise typically lose more fat, retain more muscle, and regain less weight than those who only follow a weight-loss diet. Even when people who include physical activity in their weight-management program do not lose more weight, they seem to follow their diet plans more closely and maintain their losses better than those who do not exercise. Consequently, they benefit from taking in a little less energy as well as from expending a little more energy in physical activity. Importantly, those who exercise reduce abdominal obesity and improve their blood pressure, insulin resistance, and cardiorespiratory fitness, regardless of weight loss.[102] Although there are many health benefits of physical activity, the focus here is on its role in weight management.

© Mike Chew/CORBIS

The key to good health is to combine sensible eating with regular exercise.

Dietary Guidelines for Americans 2005

To help manage body weight and prevent gradual, unhealthy body weight gain in adulthood, engage in approximately 60 minutes of moderate- to vigorous-intensity activity on most days of the week while not exceeding kcaloric intake requirements.

Activity and Energy Expenditure Table 8-2 (p. 255) shows how much energy each of several activities uses. The number of kcalories spent in an activity depends on body weight, intensity, and duration. For example, a person who weighs 150 pounds and walks 3¹/₂ miles in 60 minutes expends about 315 kcalories. That same person running 3 miles in 30 minutes uses a similar amount. By comparison, a 200-pound person running 3 miles in 30 minutes expends an additional 100 kcalories or so. The goal is to expend as much energy as your time allows. The greater the energy deficit created by exercise, the greater the fat loss. And be careful not to compensate for the energy spent in exercise by eating more food. Otherwise, energy balance won't shift and fat loss will be less significant.

Activity and Discretionary kCalorie Allowance Chapter 2 introduced the discretionary kcalorie allowance as the difference between the kcalories needed to supply nutrients and those needed to maintain energy balance. Because exercise expends energy, the energy allowance to maintain balance increases with increased physical activity—yet the energy needed to deliver needed nutrients remains about the same. In this way, physical activity increases the discretionary kcalorie allowance (see Figure 9-8). Having a larger discretionary kcalorie allowance puts a little more wiggle room in a weight-loss diet for such options as second helpings, sweet treats, or alcoholic beverages on occasion. Of course, selecting nutrient-dense foods and *not* using discretionary kcalories will maximize weight loss.

Activity and Metabolism Activity also contributes to energy expenditure in an indirect way—by speeding up metabolism. It does this both immediately and over the long term. On any given day, metabolism remains slightly elevated for several hours after intense and prolonged exercise. ◆ Over the long term, a person who engages in daily vigorous activity gradually develops more lean tissue. Metabolic rate rises accordingly, and this supports continued weight loss or maintenance.

Activity and Body Composition Physically active people have less body fat than sedentary people do—even if they have the same BMI. Physical activity, even

◆ This postexercise effect raises the energy expenditure of exercise by about 15 percent.

FIGURE 9-8   *Animated!* Influence of Physical Activity on Discretionary kCalorie Allowance

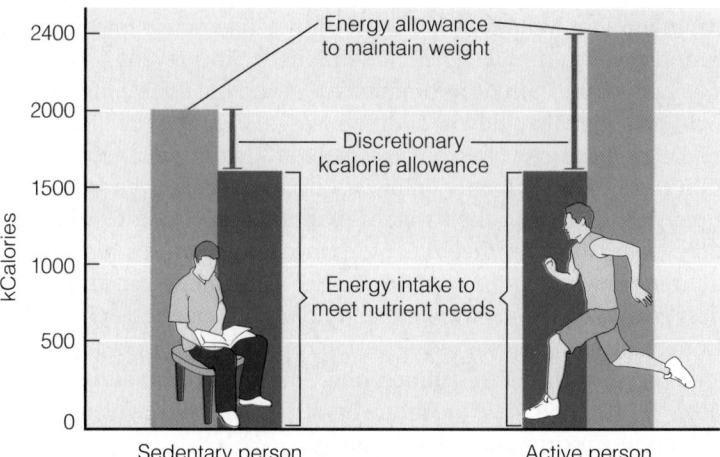

without weight loss, changes body composition: body fat decreases and lean body mass increases. Furthermore, exercise specifically decreases abdominal fat.[103]

Activity and Appetite Control Many people think that exercising will make them eat more, but this is not entirely true. Active people do have healthy appetites, but *immediately* after an intense workout, most people do not feel like eating. They may be thirsty and want to shower, but they are not hungry. The body has released fuels from storage to support the exercise, so glucose and fatty acids are abundant in the blood. At the same time, the body has suppressed its digestive functions. Hard physical work and eating are not compatible. A person must calm down, put energy fuels back in storage, and relax before eating. At that time, a physically active person may eat more than a sedentary person, but not so much as so fully compensate for the kcalories expended in exercise.[104]

Exercise may help curb the inappropriate appetite that accompanies boredom, anxiety, or depression. Weight-management programs encourage people who feel the urge to eat when not hungry to go out and exercise instead. The activity passes time, relieves anxiety, and prevents inappropriate eating.

Activity and Psychological Benefits Activity also helps reduce stress. Because stress itself cues inappropriate eating for many people, activity can help here, too. In addition, the fit person looks and feels healthy and, as a result, gains self-esteem. High self-esteem motivates a person to persist in seeking good health and fitness, which keeps the beneficial ◆ cycle going.

Choosing Activities Clearly, physical activity is a plus in a weight-management program. What kind of physical activity is best? People should choose activities that they enjoy and are willing to do regularly. What schedule of physical activity is best? ◆ It doesn't matter; whether a person chooses several short bouts of exercise or one continuous workout, the fitness and weight-loss benefits are the same—and any activity is better than being sedentary.

Health care professionals frequently advise people to engage in activities of low-to-moderate intensity for a long duration, such as an hour-long, fast-paced walk. The reasoning behind such advice is that people exercising at low-to-moderate intensity are more likely to stick with their activity for longer times and are less likely to injure themselves. A person who stays with an activity routine long enough to enjoy the rewards will be less inclined to give it up and will, over the long term, reap many health benefits. Activity of low-to-moderate intensity ◆ that expends at least 2000 kcalories per week is especially helpful for weight management. Higher levels produce even greater losses.[105]

In addition to exercise, a person can incorporate hundreds of energy-expending activities into daily routines: take the stairs instead of the elevator, walk to the neighbor's apartment instead of making a phone call, and rake the leaves instead of using a blower. Remember that sitting uses more kcalories than lying down, standing uses more kcalories than sitting, and moving uses more kcalories than standing. A 175-pound person who replaces a 30-minute television program with a 2-mile walk a day can expend enough energy to lose (or at least not gain) 18 pounds in a year. Meeting an activity goal of 10,000 steps a day helps to support a healthy BMI.[106] By wearing a pedometer, a person can easily track a day's activities without measuring miles or watching the clock. The point is to be active. Walk. Run. Swim. Dance. Cycle. Climb. Skip. Do whatever you enjoy doing—and do it often.

Spot Reducing People sometimes ask about "spot reducing." Unfortunately, muscles do not "own" the fat that surrounds them. Fat cells all over the body release fat in response to the demand of physical activity for use by whatever muscles are active. No exercise can remove the fat from any particular area.

Exercise can help with trouble spots in another way, though. The "trouble spot" for most men is the abdomen, their primary site of fat storage. During aerobic exercise, abdominal fat readily releases its stores, providing fuel to the physically active body. With regular exercise and weight loss, men will deplete these abdominal

◆ Benefits of physical activity in a weight-management program:
- Short-term increase in energy expenditure (from exercise and from a slight rise in metabolism)
- Long-term increase in BMR (from an increase in lean tissue)
- Improved body composition
- Appetite control
- Stress reduction and control of stress eating
- Physical, and therefore psychological, well-being
- Improved self-esteem

◆ For an active life, limit sedentary activities, engage in strength and flexibility activities, enjoy leisure activities often, engage in vigorous activities regularly, and be as active as possible every day.

◆ Estimated energy expended when walking at a moderate pace = 1 kcal/mi/kg body wt.

fat stores before those in the lower body. Women may also deplete abdominal fat with exercise, but their "trouble spots" are more likely to be their hips and thighs.

In addition to aerobic activity, strength training can help to improve the tone of muscles in a trouble area, and stretching to gain flexibility can help with associated posture problems. A combination of aerobic, strength, and flexibility workouts best improves fitness and physical appearance.

> ### IN SUMMARY
>
> Physical activity should be an integral part of a weight-control program. Physical activity can increase energy expenditure, improve body composition, help control appetite, reduce stress and stress eating, and enhance physical and psychological well-being.

Environmental Influences

Chapter 8 described how hormones regulate hunger, satiety, and satiation, but people don't always pay close attention to such internal signals. Instead, their eating behaviors are often dictated by environmental factors. Environmental factors include those surrounding the eating experience as well as those pertaining to the food itself.[107] Changing any of these factors can influence how much a person eats.[108]

Atmosphere The environment surrounding a meal or snack influences its duration. When the lighting, décor, aromas, and sounds of an environment are pleasant and comfortable, people tend to spend more time eating and thus eat more. A person needn't eat under neon lights with offensive music to eat less, of course. Instead, after completing a meal, remove food from the table and enjoy the ambience—without the presence of visual cues to stimulate additional eating.

Accessibility Among the strongest influences on how much we eat is the accessibility, ease, and convenience of obtaining food. In general, the less effort needed to obtain food, the more likely food will be eaten. Are you more likely to eat if half a leftover pizza is in your refrigerator or if you have to drive to the grocery store, buy a frozen pizza, and bake it for 45 minutes? Having food nearby and visible encourages eating. In one study, secretaries ate more chocolates when the candy was on their desks than when they had to walk six feet.[109] Interestingly, the secretaries underestimated the amount of chocolates they had eaten when the candy was on their desk and overestimated when it was a short distance away. The message is clear for people wanting to eat less candy (or any other tempting item)—keep it out of sight and in an inconvenient place (or don't even buy it).

Socializing People tend to eat more when socializing with others. Pleasant conversations extend the duration of a meal, allowing a person more time to eat more, and research confirms that the longer the meal, the greater the consumption.[110] In addition, by taking a visual cue from companions, a person might eat more when others at the table, clean their plates, or go to the buffet line for seconds. One way to eat less is to pace yourself with the person who seems to be eating the least and slowest. Social interactions also distract a person from paying attention to how much has been eaten. In some cases, socializing with friends during a meal may provide comfort and lower a person's motivation to limit consumption. In other cases, socializing with unfamiliar people during a meal—during a job interview or blind date, for example—may create stress and reduce food consumption. To eat less while socializing, pay attention to portion sizes.

Distractions Distractions influence food intake by initiating eating, interfering with internal controls to stop eating, and extending the duration of eating. Some people start eating dinner when a favorite television program comes on, regardless of hunger. Other people continue eating breakfast until they finish reading the newspaper. Such mindless eating can easily become overeating.

In addition to influencing the start and stop of a meal, distractions interfere with a person's ability to monitor and regulate how much is consumed.[111] Do you eat more popcorn when you are engrossed in a movie or if you are paying attention to how much popcorn you are eating? If distractions are a part of the eating experience, extra care is needed to control portion sizes.

Presence   The mere sight (or smell, or even thought) of a food can prompt a person to start eating—regardless of hunger. The chocolates in the clear candy dishes on the secretaries' desks were eaten much faster than those in opaque containers.[112]

Variety   When offered a variety of foods, or a variety of flavors of the same food, people tend to eat more. Interestingly, they tend to eat more even when variety is only *perceived.* Given six flavors of jelly beans, people will eat more when offered an assorted mixture than when presented with the exact same flavors and quantities sorted in a sectioned container.[113]

Variety is pleasing and distracting—two factors that slow the eating experience and delay satiation.[114] To limit intake, then, focus on a limited number of foods per meal. Be careful not to misunderstand and abandon variety in diet planning. Eating a variety of foods from each of the food groups is still a healthy plan—just not all at one meal.

Package and Portion Sizes   As noted earlier, the sizes of packages in grocery stores and portion sizes at restaurants and at home have increased dramatically in recent decades, contributing to the increase in obesity in the United States.[115] Put simply, we tend to clean our plates and finish the package. The larger the bag of potato chips, the greater the intake.[116] To keep from overeating, repackage snacks into smaller containers and eat them from a plate, not directly from the package.

Serving Containers   We often use plates, utensils, and glasses as visual cues to guide our decisions on how much to eat and drink.[117] If you plan to eat a bowl of ice cream, it matters whether the bowl you select holds 8 ounces or 24 ounces. Even the size of the serving container matters. Students took more—and ate more—snacks when serving from two large bowls instead of from four medium bowls.[118]

Large dinner plates and wide glasses create illusions and misperceptions about quantities consumed. A scoop of mashed potatoes on a small plate looks larger than the same-size scoop on a large plate, leading a person to underestimate the amount of food eaten.[119] To control portion sizes, use small bowls and plates, small serving spoons, and tall, narrow glasses.[120]

Eating from the package while distracted by television is a weight-gaining combination.

Donna Day/Getty Images

Behavior and Attitude

Behavior and attitude play important roles in supporting efforts to achieve and maintain appropriate body weight and composition. **Behavior modification** focuses on how to change behaviors to increase energy expenditure and decrease energy intake.[121] A person must commit to taking action.

Adopting a positive, matter-of-fact attitude helps to ensure success. Healthy eating and activity choices are an essential part of healthy living and should simply be incorporated into the day—much like brushing one's teeth or wearing a safety belt.

Become Aware of Behaviors   To solve a problem, a person must first identify all the behaviors that created the problem. Keeping a record will help to identify eating and exercise behaviors that may need changing (see Figure 9-9, p. 304). It will also establish a baseline against which to measure future progress.

Change Behaviors   Strategies ◆ focus on learning desired eating and exercise behaviors and eliminating unwanted behaviors. With so many possible behavior changes, a person can choose where to begin. Start simply and don't try to master

◆ Examples of behavioral strategies to support weight change:
- Do not grocery shop when hungry.
- Eat slowly (pause during meals, chew thoroughly, put down utensils between bites).
- Exercise when watching television.

behavior modification: the changing of behavior by the manipulation of antecedents (cues or environmental factors that trigger behavior), the behavior itself, and consequences (the penalties or rewards attached to behavior).

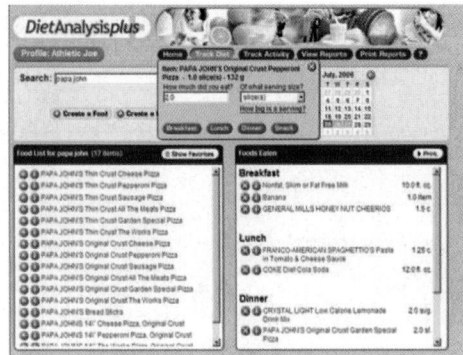

Diet analysis programs help people identify high-kcalorie foods and monitor their eating habits.

FIGURE 9-9 | Food Record

The entries in a food record should include the times and places of meals and snacks, the types and amounts of foods eaten, and a description of the individual's feelings when eating. The diary should also record physical activities: the kind, the intensity level, the duration, and the person's feelings about them.

Time	Place	Activity or food eaten	People present	Mood
10:30– 10:40	School vending machine	6 peanut butter crackers and 12 oz. cola	by myself	Starved
12:15– 12:30	Restaurant	Sub sandwich and 12 oz. cola	friends	relaxed & friendly
3:00– 3:45	Gym	Weight training	work out partner	tired
4:00– 4:10	Snack bar	Small frozen yogurt	by myself	OK

them all at once. Attempting too many changes at one time can be overwhelming. Pick one trouble area that is manageable and start there. Practice a desired behavior until it becomes routine. Then select another trouble area to work on, and so on. Another bit of advice along the same lines: don't try to tackle major changes during a particularly stressful time of life.

Personal Attitude For many people, overeating and being overweight have become an integral part of their identity. Those who fully understand their personal relationships with food are best prepared to make healthful changes in eating and exercise behaviors.

Sometimes habitual behaviors that are hazardous to health, such as smoking or drinking alcohol, contribute positively by helping people adapt to stressful situations. Similarly, many people overeat to cope with the stresses of life. To break out of that pattern, they must first identify the particular stressors that trigger the urge to overeat. Then, when faced with these situations, they must learn and practice problem-solving skills that will help them to respond appropriately.[122]

All this is not to imply that psychological therapy holds the magic answer to a weight problem. Still, efforts to improve one's general well-being may result in healthy eating and activity habits even when weight loss is not the primary goal. When the problems that trigger the urge to overeat are resolved in alternative ways, people may find they eat less. They may begin to respond appropriately to internal cues of hunger rather than inappropriately to external cues of stress. Sound emotional health supports a person's ability to take care of physical health in all ways—including nutrition, weight management, and fitness.

Support Groups Group support can prove helpful when making life changes. Some people find it useful to join a group such as Take Off Pounds Sensibly (TOPS), Weight Watchers (WW), Overeaters Anonymous (OA), or others. Some dieters prefer to form their own self-help groups or find support online. The Internet offers numerous opportunities for weight-loss education and counseling that may be effective alternatives to face-to-face programs.[123] As always, consumers need to choose wisely and avoid rip-offs.

IN SUMMARY

A surefire remedy for obesity has yet to be found, although many people find a combination of the approaches just described to be most effective. Diet and exercise shift energy balance so that more energy is being expended than is taken in. Physical activity increases energy expenditure, builds lean tissue, and improves health. Energy intake should be reduced by 500 to 1000 kcalories per day, depending on starting body weight and usual food intake. Behavior modification retrains habits to support a healthy eating and exercise plan. This treatment package requires time, individualization, and sometimes the assistance of a registered dietitian.

Weight Maintenance

People who are successful often experience much of their weight loss within half a year and then reach a plateau. This slowdown can be disappointing, but it should be recognized as an opportunity for the body to adjust to its new weight. Reaching a plateau provides a little relief from the distraction of weight-loss dieting. An appropriate goal at this point is to continue the eating and activity behaviors that will maintain weight. Attempting to lose additional weight at this point would require major effort and would almost certainly meet with failure.

The prevalence of **successful weight-loss maintenance** is difficult to determine, in part because researchers have used different criteria. Some look at success after one year and others after five years; some quantify success as 10 or more pounds lost and others as 5 or 10 percent of initial body weight lost. Furthermore, most research studies examine the success of one episode of weight loss in a structured program, but this scenario does not necessarily reflect the experiences of the general population. In reality, most people have lost weight several times in their lifetimes and did so on their own, not in a formal program. Almost 50 percent of people who intentionally lost weight have successfully maintained the loss for at least a year.[124]

Those who are successful in maintaining their weight loss have established vigorous exercise regimens and careful eating patterns, taking in less energy and a lower percentage of kcalories from fat than the national average.[125] Because these people are more efficient at storing fat, they do not have the same flexibility in their food and activity habits as their friends who have never been overweight. With weight loss, metabolism shifts downward so that formerly overweight people require less energy than might be expected given their current body weight and body composition. Consequently, to keep weight off, they must either eat less or exercise more than people the same size who have never been obese.

Physical activity plays a key role in maintaining weight.[126] Those who exercise vigorously are far more successful than those who are inactive. On average, weight maintenance requires a person to expend about 2000 kcalories in physical activity per week.[127] To accomplish this, a person might exercise either moderately (such as brisk walking) for 60 minutes a day or vigorously (such as fast bicycling) for 35 minutes a day, for example. Being active during both work hours and leisure time also helps a person to maintain weight loss.[128]

© Rick Gomez/Masterfile

Maintaining a healthy body weight requires maintaining the vigorous physical activities and careful eating habits that supported weight loss.

Dietary Guidelines for Americans 2005

To sustain weight loss in adulthood, participate in at least 60 to 90 minutes of moderately intense physical activity daily while not exceeding kcaloric intake requirements. Some people may need to consult with a healthcare provider before participating in this level of activity.

successful weight-loss maintenance: achieving a weight loss of at least 10 percent of initial body weight and maintaining the loss for at least one year.

In addition to limiting energy intake and exercising regularly, one other strategy may help with weight maintenance: frequent self-monitoring. People who weigh themselves periodically and monitor their eating and exercise habits regularly can detect weight gains in the early stages and promptly initiate changes to prevent relapse.[129]

Losing weight and maintaining the loss may not be as easy as gaining the weight in the first place, but it is possible. Those who have been successful find that it gets easier with time—the changes in diet and activity patterns become permanent.[130]

Prevention

Given the information presented up to this point in the chapter, the adage "An ounce of prevention is worth a pound of cure" seems particularly apropos. Preventing weight gain would benefit almost everybody.[131] Obesity is a major risk factor for numerous diseases, and losing weight is challenging and often temporary. Strategies for preventing weight gain ◆ are very similar to those for losing weight, with one exception: they begin early. Over the years, they become an integral part of a person's life. It is much easier for a person to resist doughnuts for breakfast if he rarely eats them. Similarly, a person will have little trouble walking each morning if she has always been active.

◆ To prevent weight gain:
  • Eat regular meals and limit snacking.
  • Drink water instead of high-kcalorie beverages.
  • Select sensible portion sizes and limit daily energy intake to no more than energy expended.
  • Become physically active and limit sedentary activities.

 **Dietary Guidelines for Americans 2005**

To prevent gradual weight gain over time, make small decreases in food and beverage kcalories and increase physical activity.

Public Health Programs

Has anyone in the United States *not* heard the message that obesity raises the risks of chronic diseases and that overweight people should aim for a healthy weight by eating sensibly and becoming physically active? Not likely. Yet implementing such advice is difficult in an environment of abundant food and physical inactivity. To successfully treat obesity, we may have to change the environment in which we live through public health law.[132] Table 9-6 provides examples of public health strategies that have been suggested to improve our nation's nutrition environment. Some of these strate-

TABLE 9-6	Suggested Public Health Strategies	
Strategies	**Examples of Suggested Nutritional Strategies**	**Examples of Successful Nonnutritional Strategies**
Impose safety standards to reduce the potential for harm.	• Regulate the energy or fat density of foods. • Regulate the size of packages of high-fat foods.	• Mandate safety glass in automobiles. • Regulate the lead content of paint.
Control commercial advertising to limit the influence of harmful products.	• Improve nutrition labeling and product packaging. • Restrict the promotion of high-fat foods (especially when directed at children).	• Restrict cigarette advertising (especially when directed at children). • Add health warnings to alcoholic beverages.
Control the conditions under which products are sold to limit exposure to hazardous substances.	• Remove high-fat, low–nutrient density foods from school vending machines. • Restrict the number of vendors licensed to sell high-fat foods.	• Mandate minimum-age laws for the use of tobacco, alcohol, and automobiles. • Restrict the number of vendors licensed to sell alcohol.
Control prices to reduce consumption.	• Tax soft drinks and other foods high in kcalories, fat, or sugar.	• Tax alcohol and tobacco.

SOURCES: Adapted from L. O. Gostin, Law as a tool to facilitate healthier lifestyles and prevent obesity, *Journal of the American Medical Association* 297 (2007): 87–90; M. Nestle and M. F. Jacobson, Halting the obesity epidemic: A public health policy approach, *Public Health Reports* 115 (2000): 12–24.

gies may seem radical, but dramatic measures may be needed if we are to curb the obesity epidemic that is sweeping across the nation.[133] Dozens of bills and resolutions are pending in Congress.[134] Whether changes in public policy—such as a tax on snack foods—will influence diet habits or simply generate revenues remains to be seen.[135]

IN SUMMARY

Preventing weight gains and maintaining weight losses require vigilant attention to diet and physical activity. Taking care of oneself is a lifelong responsibility.

Underweight

Underweight ◆ is a far less prevalent problem than overweight, affecting no more than 5 percent of U.S. adults (review Figure 8-7 on p. 260). Whether the underweight person needs to gain weight is a question of health and, like weight loss, a highly individual matter. People who are healthy at their present weight may stay there; there are no compelling reasons to try to gain weight. Those who are thin because of malnourishment or illness, however, might benefit from a diet that supports weight gain. Medical advice can help make the distinction.

Thin people may find gaining weight difficult. Those who wish to gain weight for appearance's sake or to improve their athletic performance need to be aware that healthful weight gains can be achieved only by physical conditioning combined with high energy intakes. On a high-kcalorie diet alone, a person may gain weight, but it will be mostly fat. Even if the gain improves appearance, it can be detrimental to health and might impair athletic performance. Therefore, in weight gain, as in weight loss, physical activity and energy intake are essential components of a sound plan.

◆ Reminder: *Underweight* is a body weight so low as to have adverse health effects; it is generally defined as BMI <18.5.

Problems of Underweight

The causes of underweight may be as diverse as those of overweight—genetic tendencies, hunger, appetite, and satiety irregularities; psychological traits; and metabolic factors. Habits learned early in childhood, especially food aversions, may perpetuate themselves.

The demand for energy to support physical activity and growth often contributes to underweight. An active, growing boy may need more than 4000 kcalories a day to maintain his weight and may be too busy to take time to eat adequately. Underweight people find it hard to gain weight due, in part, to their expenditure of energy in adaptive thermogenesis. So much energy may be expended adapting to a higher food intake that at first as many as 750 to 800 extra kcalories a day may be needed to gain a pound a week. Like those who want to lose weight, people who want to gain must learn new habits and learn to like new foods. They are also similarly vulnerable to potentially harmful schemes and would be wise to review the consumer bill of rights on p. 290, using "weight gain" instead of "weight loss" where appropriate.

As described in Highlight 8, the underweight condition anorexia nervosa sometimes develops in people who employ self-denial to control their weight. They go to such extremes that they become severely undernourished, achieving final body weights of 70 pounds or even less. The distinguishing feature of a person with anorexia nervosa, as opposed to other underweight people, is that the starvation is intentional. (See Highlight 8 for a review of anorexia nervosa and other eating disorders.)

Weight-Gain Strategies

Weight-gain strategies center on eating energy-dense foods that provide many kcalories in a small volume and exercising to build muscle. By using the USDA Food

Guide recommendations for the higher kcalorie levels (see Table 2-3 on p. 41), a person can gain weight while meeting nutrient needs.

Energy-Dense Foods Energy-dense foods (the very ones eliminated from a successful weight-loss diet) hold the key to weight gain. Pick the highest-kcalorie items from each food group—that is, milk shakes instead of fat-free milk, salmon instead of snapper, avocados instead of cucumbers, a cup of grape juice instead of a small apple, and whole-wheat muffins instead of whole-wheat bread. Because fat provides more than twice as many kcalories per teaspoon as sugar does, fat adds kcalories without adding much bulk.

Although eating high-kcalorie, high-fat foods is not healthy for most people, it may be essential for an underweight individual who needs to gain weight. An underweight person who is physically active and eating a nutritionally adequate diet can afford a few extra kcalories from fat. For health's sake, it is wise to select foods with monounsaturated and polyunsaturated fats instead of those with saturated or *trans* fats: for example, sautéing vegetables in olive oil instead of butter or hydrogenated margarine.

Regular Meals Daily People who are underweight need to make meals a priority and take the time to plan, prepare, and eat each meal. They should eat at least three healthy meals every day and learn to eat more food within the first 20 minutes of a meal. Another suggestion is to eat meaty appetizers or the main course first and leave the soup or salad until later.

Large Portions Underweight people need to learn to eat more food at each meal. For example, they can add extra slices of ham and cheese on the sandwich for lunch, drink milk from a larger glass, and eat cereal from a larger bowl.

The person should expect to feel full. Most underweight individuals are accustomed to small quantities of food. When they begin eating significantly more, they feel uncomfortable. This is normal and passes over time.

Extra Snacks Since a substantially higher energy intake is needed each day, in addition to eating more food at each meal, it is necessary to eat more frequently. Between-meal snacks do not interfere with later meals; they can readily lead to weight gains.[136] For example, a student might make three sandwiches in the morning and eat them between classes in addition to the day's three regular meals. Snacking on dried fruit, nuts, and seeds is also an easy way to add kcalories.

Juice and Milk Beverages provide an easy way to increase energy intake. Consider that 6 cups of cranberry juice add almost 1000 kcalories to the day's intake. kCalories can be added to milk by mixing in powdered milk or packets of instant breakfast.

For people who are underweight due to illness, concentrated liquid formulas are often recommended because a weak person can swallow them easily. A physician or registered dietitian can recommend high-protein, high-kcalorie formulas to help an underweight person maintain or gain weight. Used in addition to regular meals, these supplements can help considerably.

Exercising to Build Muscles To gain weight, use strength training primarily, and increase energy intake to support that exercise. Eating extra food will then support a gain of both muscle and fat. An additional 500 to 1000 kcalories a day above normal energy needs is enough to support the exercise as well as the building of muscle.[137]

IN SUMMARY

Both the incidence of underweight and the health problems associated with it are less prevalent than overweight and its associated problems. To gain weight, a person must train physically and increase energy intake by selecting energy-dense foods, eating regular meals, taking larger portions, and consuming extra snacks and beverages. Table 9-5 (p. 299) includes a summary of weight-gain strategies.

Nutrition Portfolio

CENGAGENOW™
academic.cengage.com/login

To enjoy good health and maintain a reasonable body weight, combine sensible eating habits and regular physical activity.

- Calculate your BMI and consider whether you need to lose or gain weight for the sake of good health.

- Reflect on your weight over the past year or so and explain any weight gains or loses.

- Describe the potential risks and possible benefits of fad diets and over-the-counter weight-loss drugs or herbal supplements.

NUTRITION ON THE NET

For further study of topics covered in this chapter, log on to **academic.cengage .com/nutrition/rolfes/UNCN8e**. Go to Chapter 9 then to Nutrition on the Net.

- Search for "obesity" and "weight control" at the U.S. Government health information site: **www.healthfinder.gov**

- Review the Clinical Guidelines on the Identification, Evaluation, and Treatment of Overweight and Obesity in Adults: **www.nhlbi.nih.gov/guidelines/obesity/ob_home.htm**

- Learn about the drugs used for weight loss from the Center for Drug Evaluation and Research: **www.fda.gov/cder**

- Learn about weight control and the WIN program from the Weight-control Information Network: **www.win.niddk.nih.gov**

- Visit weight-loss support groups, such as Take Off Pounds Sensibly (TOPS), Overeaters Anonymous (OA), and Weight Watchers: **www.tops.org**, **www.oa.org**, and **www .weightwatchers.com**

- See what the obesity professionals think at the North American Association for the Study of Obesity and the

- American Society for Bariatric Surgery: **www.naaso.org** and **www.asbs.org**

- Consider the nondietary approaches of HUGS International: **www.hugs.com**

- Learn about the 10,000 Step Program from Shape Up America!: **www.shapeup.org/10000steps.html**

- Find helpful information on achieving and maintaining a healthy weight from the Calorie Control Council: **www.caloriecontrol.org**

- Learn how to end size discrimination and improve the quality of life for fat people from the National Association to Advance Fat Acceptance: **www.naafa.org**

- Find good advice on starting a weight-loss program from the Partnership for Healthy Weight Management: **www.consumer.gov/weightloss**

- Consider ways to live a healthy life at any weight: **www.bodypositive.com**

NUTRITION CALCULATIONS

CENGAGENOW™ For additional practice log on to **academic.cengage.com/login**. Go to Chapter 9, then to Nutrition Calculations.

These problems give you practice in doing simple energy-balance calculations (see p. 314 for answers). Once you have mastered these examples, you will be prepared to examine your own food choices. Be sure to show your calculations for each problem.

1. Critique a commercial weight-loss plan. Consumers spend billions of dollars a year on weight-loss programs such as Slim-Fast, Sweet Success, Weight Watchers, Nutri/System, Jenny Craig, Optifast, Medifast, and Formula One. One such plan calls for a milk shake in the morning, at noon, and as an afternoon snack and "a sensible, balanced, low-fat dinner" in the evening. One shake mixed in 8 ounces of vitamin A– and D–fortified fat-free milk offers 190 kcalories; 32 grams of carbohydrate, 13 grams of protein, and 1 gram of fat; at least one-third of the Daily Value for all vitamins and minerals; plus 2 grams of fiber.

   a. Calculate the kcalories and grams of carbohydrate, protein, and fat that three shakes provide.

b. How do these values compare with the criteria listed in item 2 in Table H9-4 on p. 320?

c. Plan "a sensible, balanced, low-fat dinner" that will help make this weight-loss plan adequate and balanced. Now, how do the day's totals compare with the criteria in item 2 in Table H9-4 on p. 320?

d. Critique this plan using the other criteria described in Table H9-4 on p. 320 as a guide.

2. Evaluate a weight-gain attempt. People attempting to gain weight sometimes have a hard time because they choose low-kcalorie, high-bulk foods that make it hard to consume enough energy. Consider the following lunch: a chef's salad consisting of 2 cups iceberg lettuce, 1 whole tomato, 1 ounce swiss cheese, 1 ounce roasted ham (extra lean), 1 hard-boiled egg, ½ cup grated carrots, and ¼ cup Thousand Island salad dressing. If you weighed these foods, you'd find that they totaled 442 grams. This is a pretty filling meal.

a. The meal provides 459 kcalories. What is the energy density of this meal, expressed in kcalories per gram?

b. To gain weight, this person is advised to eat an additional 500 kcalories at this meal. Using foods with this same energy density, how much more chef's salad will this person have to eat?

c. Suppose a person simply can't do this. Try to reduce the bulk of this meal by replacing some of the lettuce with more energy-dense foods. Delete 1 cup lettuce from the salad and add another ounce roast ham and 1 ounce cheddar cheese. Show how these changes influence the weight and kcalories of this meal. (Use Appendix H.)

Item No./Food	Weight (g)	Energy (kcal)
Original totals:	442	459
Minus:		
#5083 Lettuce, 1 c	−_____	−_____
Plus:		
#12212 Roast ham, 1 oz	+_____	+_____
#1007 Cheddar cheese, 1 oz	+_____	+_____
Totals:	_____	_____

d. How many kcalories did the changes add?

e. How much more *weight* of food did these changes add?

This exercise should reveal why people attempting to gain weight are advised to add high-fat items, within reason, to their daily meals.

STUDY QUESTIONS

CENGAGENOW™

To assess your understanding of chapter topics, take the Student Practice Test and explore the modules recommended in your Personalized Study Plan. Log on to **academic.cengage.com/login.**

These questions will help you review the chapter. You will find the answers in the discussions on the pages provided.

1. Describe how body fat develops, and suggest some reasons why it is difficult for an obese person to maintain weight loss. (pp. 282–283)

2. What factors contribute to obesity? (pp. 283–288)

3. List several aggressive ways to treat obesity, and explain why such methods are not recommended for every overweight person. (pp. 292–293)

4. Discuss reasonable dietary strategies for achieving and maintaining a healthy body weight. (pp. 294–299)

5. What are the benefits of increased physical activity in a weight-loss program? (pp. 299–302)

6. Describe the behavioral strategies for changing an individual's dietary habits. What role does personal attitude play? (pp. 303–305)

7. Describe strategies for successful weight gain. (pp. 307–308)

These multiple choice questions will help you prepare for an exam. Answers can be found on p. 314.

1. With weight loss, fat cells:
   a. decrease in size only.
   b. decrease in number only.
   c. decrease in both number and size.
   d. decrease in number, but increase in size.

2. Obesity is caused by:
   a. overeating.
   b. inactivity.
   c. defective genes.
   d. multiple factors.

3. The protein produced by the fat cells under the direction of the *ob* gene is called:
   a. leptin.
   b. serotonin.
   c. sibutramine.
   d. phentermine.

4. The biggest problem associated with the use of drugs in the treatment of obesity is:
   a. cost.
   b. chronic dosage.
   c. ineffectiveness.
   d. adverse side effects.

5. A realistic goal for weight loss is to reduce body weight:
   a. down to the weight a person was at age 25.
   b. down to the ideal weight in the weight-for-height tables.
   c. by 10 percent over six months.
   d. by 15 percent over three months.

6. A nutritionally sound weight-loss diet might restrict daily energy intake to create a:
   a. 1000-kcalorie-per-month deficit.
   b. 500-kcalorie-per-month deficit.
   c. 500-kcalorie-per-day deficit.
   d. 3500-kcalorie-per-day deficit.

7. Successful weight loss depends on:
   a. avoiding fats and limiting water.
   b. taking supplements and drinking water.
   c. increasing proteins and restricting carbohydrates.
   d. reducing energy intake and increasing physical activity.

8. Physical activity does *not* help a person to:
   a. lose weight.
   b. retain muscle.
   c. maintain weight loss.
   d. lose fat in trouble spots.

9. Which strategy would *not* help an overweight person to lose weight?
   a. Exercise.
   b. Eat slowly.
   c. Limit high-fat foods.
   d. Eat energy-dense foods regularly.

10. Which strategy would *not* help an underweight person to gain weight?
    a. Exercise.
    b. Drink plenty of water.
    c. Eat snacks between meals.
    d. Eat large portions of foods.

REFERENCES

1. State-specific prevalence of obesity among adults—United States, 2005, *Morbidity and Mortality Weekly Report* 55 (2006): 985–988; C. L. Ogden and coauthors, Prevalence of overweight and obesity in the United States, 1999–2004, *Journal of the American Medical Association* 295 (2006): 1549–1555.
2. Ogden and coauthors, 2006; National Center for Health Statistics, *Chartbook on Trends in the Health of Americans, 2005*, www.cdc.gov/nchs, site visited on January 18, 2006.
3. Ogden and coauthors, 2006.
4. T. E. Kottke, L. A. Wu, and R. S. Hoffman, Economic and psychological implications of the obesity epidemic, *Mayo Clinic Proceedings* 78 (2003): 92–94; M. Kohn and M. Booth, The worldwide epidemic of obesity in adolescents, *Adolescent Medicine* 14 (2003): 1–9; G. du Toit and M. T. van der Merwe, The epidemic of childhood obesity, *South African Medical Journal* 93 (2003): 49–50; C. J. Schrodt, The obesity epidemic and physician responsibility, *Journal of the Kentucky Medical Association* 101 (2003): 27–28.
5. E. D. Rosen, The molecular control of adipogenesis with special reference to lymphatic pathology, *Annals of the New York Academy of Sciences* 979 (2002): 143–158.
6. J. E. Schaffer, Lipotoxicity: When tissues overeat, *Current Opinion in Lipidology* 14 (2003): 281–287.
7. R. J. F. Loos and T. Rankinen, Gene-diet interactions on body weight changes, *Journal of the American Dietetic Association* 105 (2005): S29–S34; S. Tholin and coauthors, Genetic and environmental influences on eating behavior: The Swedish Young Male Twins Study, *American Journal of Clinical Nutrition* 81 (2005): 564–569.
8. T. Rankinen and C. Bouchard, Genetics of food intake and eating behavior phenotypes in humans, *Annual Review of Nutrition* 26 (2006): 413–434; H. N. Lyon and J. N. Hirschhorn, Genetics of common forms of obesity: A brief overview, *American Journal of Clinical Nutrition* 82 (2005): 215S–217S.
9. H. K. Tiwari and coauthors, Is GAD2 on chromosome 10p12 a potential candidate gene for morbid obesity? *Nutrition Reviews* 63 (2005): 315–319; L. Bouchard and coauthors, Neuromedin: A strong candidate gene linking eating behaviors and susceptibility to obesity, *American Journal of Clinical Nutrition* 80 (2004): 1478–1486; E. Suviolahti and coauthors, The SLC6A14 gene shows evidence of association with obesity, *Journal of Clinical Investigation* 112 (2003): 1762–1772; D. E. Cummings and M. W. Schwartz, Genetics and pathophysiology of human obesity, *Annual Reviews of Medicine* 54 (2003): 453–471.

10. C. Pico and coauthors, Gastric leptin: A putative role in the short-term regulation of food intake, *British Journal of Nutrition* 90 (2003): 735–741.
11. S. O'Rahilly and coauthors, Minireview: Human obesity—Lessons from monogenic disorders, *Endocrinology* 144 (2003): 3757–3764.
12. A. M. Prentice and coauthors, Leptin and undernutrition, *Nutrition Reviews* 60 (2002): S56–S67.
13. J. Harvey and M. L. Ashford, Leptin in the CNS: Much more than a satiety signal, *Neuropharmacology* 44 (2003): 845–854.
14. P. Fietta, Focus on leptin, a pleiotropic hormone, *Minerva Medica* 96 (2005): 65–75; S. Takeda, F. Elefteriou, and G. Karsenty, Common endocrine control of body weight, reproduction, and bone mass, *Annual Review of Nutrition* 23 (2003): 403–411.
15. S. Shalitin and M. Phillip, Role of obesity and leptin in the pubertal process and pubertal growth—A review, *International Journal of Obesity Related Metabolic Disorders* 27 (2003): 869–874.
16. V. Popovic and L. H. Duntas, Brain somatic cross-talk: Ghrelin, leptin, and ultimate challengers of obesity, *Nutritional Neuroscience* 8 (2005): 1–5; J. Williams and S. Mobarhan, A critical interaction: Leptin and ghrelin, *Nutrition Reviews* 61 (2003): 391–393.
17. M. Kojima and K. Kangawa, Ghrelin, an orexigenic signaling molecule from the gastrointestinal tract, *Current Opinion in Pharmacology* 2 (2002): 665–668.
18. D. E. Cummings, K. E. Foster-Schubert, and J. Overduin, Gherlin and energy balance: Focus on current controversies, *Current Drug Targets* 6 (2005): 153–169; J. Eisenstein and A. Greenberg, Ghrelin: Update 2003, *Nutrition Reviews* 61 (2003): 101–104; O. Ukkola and S. Poykko, Ghrelin, growth and obesity, *Annals of Medicine* 34 (2002): 102–108.
19. H. S. Callahan and coauthors, Postprandial suppression of plasma ghrelin level is proportional to ingested caloric load but does not predict intermeal interval in humans, *Journal of Clinical Endocrinology and Metabolism* 89 (2003): 1319–1324; G. Schaller and coauthors, Plasma ghrelin concentrations are not regulated by glucose or insulin: A double-blind, placebo-controlled crossover clamp study, *Diabetes* 52 (2003): 16–20; G. Iniguez and coauthors, Fasting and postglucose ghrelin levels in SGA infants: Relationships with size and weight gain at one year of age, *Journal of Clinical Endocrinology and Metabolism* 87 (2002): 5830–5833.
20. M. Tanaka and coauthors, Habitual binge/purge behavior influences circulating ghrelin levels in eating disorders, *Journal of*

Psychiatric Research 37 (2003): 17–22; J. H. Lindeman and coauthors, Ghrelin and the hyposomatotropism of obesity, *Obesity Research* 10 (2002): 1161–1166.
21. M. Tanaka and coauthors, Effect of nutritional rehabilitation on circulating ghrelin and growth hormone levels in patients with anorexia nervosa, *Regulatory Peptides* 122 (2004): 163–168; V. Tolle and coauthors, Balance in ghrelin and leptin plasma levels in anorexia nervosa patients and constitutionally thin women, *Journal of Clinical Endocrinology and Metabolism* 88 (2003): 109–116; M. F. Saad and coauthors, Insulin regulates plasma ghrelin concentration, *Journal of Clinical Endocrinology and Metabolism* 87 (2002): 3997–4000.
22. A. M. Haqq and coauthors, Serum ghrelin levels are inversely correlated with body mass index, age, and insulin concentrations in normal children and are markedly increased in Prader-Willi syndrome, *Journal of Clinical Endocrinology and Metabolism* 88 (2003): 174–178; A. DelParigi and coauthors, High circulating ghrelin: A potential cause for hyperphagia and obesity in Prader-Willi syndrome, *Journal of Clinical Endocrinology and Metabolism* 87 (2002): 5461–5464.
23. A. Geliebter, M. E. Gluck, and S. A. Hashim, Plasma ghrelin concentrations are lower in binge-eating disorder, *Journal of Nutrition* 135 (2005): 1326–1330; P. J. English and coauthors, Food fails to suppress ghrelin levels in obese humans, *Journal of Clinical Endocrinology and Metabolism* 87 (2002): 2984.
24. D. E. Cummings and coauthors, Plasma ghrelin levels after diet-induced weight loss or gastric bypass surgery, *New England Journal of Medicine* 346 (2002): 1623–1630.
25. Cummings, Foster-Schubert, and Overduin, 2005.
26. Cummings, Foster-Schubert, and Overduin, 2005.
27. Iniguez and coauthors, 2002.
28. J. Korner and R. L. Leibel, To eat or not to eat—How the gut talks to the brain, *New England Journal of Medicine* 349 (2003): 926–930.
29. R. L. Batterham and coauthors, Inhibition of food intake in obese subjects by peptide YY3-36, *New England Journal of Medicine* 349 (2003): 941–948.
30. J. Orr and B. Davy, Dietary influences on peripheral hormones regulating energy intake: Potential applications for weight management, *Journal of the American Dietetic Association* 105 (2005): 1115–1124.
31. F. Broglio and coauthors, Ghrelin: Endocrine and non-endocrine actions, *Journal of Pediatric Endocrinology and Metabolism* 15 (2002): 1219–1227.

32. J. C. Weikel and coauthors, Ghrelin promotes slow-wave sleep in humans, American Journal of Physiology. *Endocrinology and Metabolism* 284 (2003): E407–E415.

33. N. D. Kohatsu and coauthors, Sleep duration and body mass index in rural population, *Archives of Internal Medicine* 166 (2006): 1701–1705; R. D. Verona and coauthors, Overweight and obese patients in a primary care population report less sleep than patients with a normal body mass index, *Archives of Internal Medicine* 165 (2005): 25–34; K. Spiegel and coauthors, Sleep curtailment in healthy young men is associated with decreased leptin levels, elevated ghrelin levels, and increased hunger and appetite, *Annals of Internal Medicine* 141 (2004): 846–850; G. Hasler and coauthors, The association between short sleep duration and obesity in young adults: A 13-year prospective study, *Sleep* 27 (2004): 661–666.

34. P. Hamet and J. Tremblay, Genetics of sleep-wake cycles and its disorders, *Metabolism* 55 (2006): S7–S12.

35. A. S. Avram, M. M. Avram, and W. D. James, Subcutaneous fat in normal and diseased states: 2. Anatomy and physiology of white and brown adipose tissue, *Journal of the American Academy of Dermatology* 53 (2005): 671–673.

36. P. Trayhurn, The biology of obesity, *Proceedings of the Nutrition Society* 64 (2005): 31–38; J. S. Kim-Han and L. L. Dugan, Mitochondrial uncoupling proteins in the central nervous system, *Antioxidants and Redox Signaling* 7 (2005): 1173–1181; R. J. F. Roos and T. Rankinen, Gene-diet interactions on body weight changes, *Journal of the American Dietetic Association* 105 (2005): S29–S34.

37. P. Laurberg, S. Andersen, and J. Karmisholt, Cold adaptation and thyroid hormone metabolism, *Hormone and Metabolic Research* 37 (2005): 545–549.

38. Avram, Avram, and James, 2005.

39. D. Ricquier, Respiration uncoupling and metabolism in the control of energy expenditure, *Proceedings of the Nutrition Society* 64 (2005): 47–52; W. D. van Marken Lichtenbelt and H. A. Daanen, Cold-induced metabolism, *Current Opinion in Clinical Nutrition and Metabolic Care* 6 (2003): 469–475.

40. S. Y. S. Kimm and coauthors, Racial differences in the relation between uncoupling protein genes and resting energy expenditure, *American Journal of Clinical Nutrition* 75 (2002): 714–719.

41. Y. Ma and coauthors, Association between eating patterns and obesity in a free-living US adult population, *American Journal of Epidemiology* 158 (2003): 85–92.

42. J. Maurer and coauthors, The psychological and behavioral characteristics related to energy misreporting, *Nutrition Reviews* 64 (2006): 53–66.

43. J. C. Peters, The challenge of managing body weight in the modern world, *Asia Pacific Journal of Clinical Nutrition* 11 (2002): S714–S717.

44. B. Wansink and J. Kim, Bad popcorn in big buckets: Portion size can influence intake as much as taste, *Journal of Nutrition Education and Behavior* 37 (2005): 242–245.

45. B. J. Rolls, E. L. Morris, and L. S. Roe, Portion size of food affects energy intake in normal-weight and overweight men and women, *American Journal of Clinical Nutrition* 76 (2002): 1207–1213.

46. S. J. Nielsen and B. M. Popkin, Patterns and trends in food portion sizes, 1977–1998, *Journal of the American Medical Association* 289 (2003): 450–453; H. Smiciklas-Wright and coauthors, Foods commonly eaten in the United States, 1989–1991 and 1994–1996: Are portion sizes changing? *Journal of the American Dietetic Association* 103 (2003): 41–47.

47. L. R. Young and M. Nestle, Expanding portion sizes in the US marketplace: Implications for nutrition counseling, *Journal of the American Dietetic Association* 103 (2003): 231–234.

48. L. R. Young and M. Nestle, The contribution of expanding portion sizes to the US obesity epidemic, *American Journal of Public Health* 92 (2002): 246–249.

49. J. E. Tillotson, America's obesity: Conflicting public policies, industrial economic development, and unintended human consequences, *Annual Review of Nutrition* 24 (2004): 617–643; Y. Ma and coauthors, Association between eating patterns and obesity in a free-living US adult population, *American Journal of Epidemiology* 158 (2003): 85–92.

50. S. A. Bowman and B. T. Vinyard, Fast food consumption of U.S. adults: Impact on energy and nutrient intakes and overweight status, *Journal of the American College of Nutrition* 23 (2004): 163–168; S. Paeratakul and coauthors, Fast-food consumption among US adults and children: Dietary and nutrient intake profile, *Journal of the American Dietetic Association* 103 (2003): 1332–1338.

51. B. J. Rolls, L. S. Roe, and J. S. Meengs, Reductions in portion size and energy density of foods are additive and lead to sustained decreases in energy intake, *American Journal of Clinical Nutrition* 83 (2006): 11–17; T. V. E. Kral, L. S. Roe, and B. J. Rolls, Combined effects of energy density and portion size on energy intake in women, *American Journal of Clinical Nutrition* 79 (2004): 962–968.

52. B. J. Rolls, The supersizing of America: Portion size and the obesity epidemic, *Nutrition Today* 38 (2003): 42–53.

53. A. Drewnowski and N. Darmon, The economics of obesity: Dietary energy density and energy cost, *American Journal of Clinical Nutrition* 82 (2005): 265S–273S.

54. C. H. Powers and M. A. Hess, A message to the restaurant industry: It's time to "step up to the plate," *Journal of the American Dietetic Association* 103 (2003): 1136–1138.

55. K. M. Booth, M. M. Pinkston, and W. S. C. Poston, Obesity and the built environment, *Journal of the American Dietetic Association* 105 (2005): S110–S117.

56. M. Lahti-Koski and coauthors, Associations of body mass index and obesity with physical activity, food choices, alcohol intake, and smoking in the 1982–1997 FINRISK Studies, *American Journal of Clinical Nutrition* 75 (2002): 809–817.

57. F. B. Hu and coauthors, Television watching and other sedentary behaviors in relation to risk of obesity and type 2 diabetes mellitus in women, *Journal of the American Medical Association* 289 (2003): 1785–1791.

58. J. A. Levine and coauthors, Interindividual variation in posture allocation: Possible role in human obesity, *Science* 307 (2005): 584–586.

59. *U.S. News and World Report*, June 16, 2003, p. 36; www.niddk.nih.gov/healthnutrit/pubs/statobes.htm.

60. R. F. Kushner and D. J. Blatner, Risk assessment of the overweight and obese patient, *Journal of the American Dietetic Association* 105 (2005): S53–S62; National Institutes of Health Obesity Education Initiative, *The Practical Guide: Identification, Evaluation, and Treatment of Overweight and Obesity in Adults*, NIH publication no. 00–4084 (Washington, D.C.: U.S. Department of Health and Human Services, 2000).

61. National Institutes of Health Obesity Education Initiative, 2000.

62. N. S. Wellman and B. Friedberg, Causes and consequences of adult obesity: Health, social and economic impacts in the United States, *Asia Pacific Journal of Clinical Nutrition* 11 (2002): S705–S709.

63. P. G. Shekelle and coauthors, Efficacy and safety of ephedra and ephedrine for weight loss and athletic performance: A meta-analysis, *Journal of the American Medical Association* 289 (2003): 1537–1545.

64. J. T. Dwyer, D. B. Allison, and P. M. Coates, Dietary supplements in weight reduction, *Journal of the American Dietetic Association* 105 (2005): S80–S86; R. B. Saper, D. M. Eisenberg, and R. S. Phillips, Common dietary supplements for weight loss, *American Family Physician* 70 (2004): 1731–1738; United States General Accounting Office, *Dietary Supplements for Weight Loss*, July 31, 2002.

65. S. P. Dolan and coauthors, Analysis of dietary supplements for arsenic, cadmium, mercury, and lead using inductively coupled plasma mass spectrometry, *Journal of Agricultural and Food Chemistry* 51 (2003): 1307–1312; A. H. Feifer, N. E. Fleshner, and L. Klotz, Analytical accuracy and reliability of commonly used nutritional supplements in prostate disease, *Journal of Urology* 168 (2002): 150–154.

66. S. Z. Yanovski and J. A. Yanovski, Obesity, *New England Journal of Medicine* 346 (2002): 591–602.

67. K. M. McTigue and coauthors, Screening and interventions for obesity in adults: Summary of the evidence for the U.S. Preventative Services Task Force, *Annals of Internal Medicine* 139 (2003): 933–949.

68. S. Schurgin and R. D. Siegel, Pharmacotherapy of obesity: An update, *Nutrition in Clinical Care* 6 (2003): 27–37.

69. S. B. Moyers, Medications as adjunct therapy for weight loss: Approved and off-label agents in use, *Journal of the American Dietetic Association* 105 (2005): 948–959.

70. R. Steinbrook, Surgery for severe obesity, *New England Journal of Medicine* 350 (2004): 1075–1079.

71. Cummings, Foster-Schubert, and Overduin, 2005.

72. G. L. Blackburn, Solutions in weight control: Lessons from gastric surgery, *American Journal of Clinical Nutrition* 82 (2005): 248S–252S; H. Buchwald and coauthors, Bariatric surgery: A systematic review and meta-analysis, *Journal of the American Medical Association* 292 (2004): 1724–1737.

73. E. N. Hansen, A. Torquati, and N. N. Abumrad, Results of bariatric surgery, *Annual Review of Nutrition* 26 (2006): 481–511; L. Sjöström and coauthors, Lifestyle, diabetes, and cardiovascular risk factors 10 years after bariatric surgery, *New England Journal of Medicine* 351 (2004): 2683–2693; Buchwald and coauthors, 2004.

74. S. E. Barlow, Bariatric surgery in adolescents: For treatment failures or health care system failures? *Pediatrics* 114 (2004): 252–253; H. Buchwald, Surgery for severely obese adolescents: Further insight from the American Society for Bariatric Surgery, *Pediatrics* 114 (2004): 253–254; B. M. Rodgers, Bariatric surgery for adolescents: A view from the American Pediatric Surgical Association, *Pediatrics* 144 (2004): 255–256.

75. T. H. Inge and coauthors, Bariatric surgery for severely overweight adolescents: Concerns and recommendations, *Pediatrics* 114 (2004): 217–223.

76. M. Shah, V. Simha, and A. Garg, Long-term impact of bariatric surgery on body weight, co-morbidities, and nutritional status: A review, *Journal of Clinical Endocrinology and Metabolism*, September 5, 2006.

77. R. E. Brolin, Bariatric surgery and long-term control of morbid obesity, *Journal of the American Medical Association* 288 (2002): 2793–2796.

78. S. Klein and coauthors, Absence of an effect of liposuction on insulin action and risk factors for coronary heart disease, *New England Journal of Medicine* 350 (2004): 2549–2557.

79. C. A. Nonas and G. D. Foster, Setting achievable goals for weight loss, *Journal of the American Dietetic Association* 105 (2005): S118–S123.

80. G. D. Foster and coauthors, Obese patients' perceptions of treatment outcomes and the factors that influence them, *Archives of Internal Medicine* 161 (2001): 2133–2139.

81. National Institutes of Health Obesity Education Initiative, 2000, p. 2.

82. Position of the American Dietetic Association: Weight management, *Journal of the American Dietetic Association* 102 (2002): 1145–1155.

83. National Institutes of Health Obesity Education Initiative, 2000, pp. 26–27.

84. J. T. Dwyer, D. B. Allison, and P. M. Coates, Dietary supplements in weight reduction, *Journal of the American Dietetic Association* 105 (2005): S80–S86.

85. Nielsen and Popkin, 2003; Smiciklas-Wright and coauthors, 2003; Young and Nestle, 2002.

86. B. J. Rolls and coauthors, Increasing the portion size of a sandwich increases energy intake, *Journal of the American Dietetic Association* 104 (2004): 367–372.

87. B. Wansink and J. Kim, Bad popcorn in big buckets: Portion size can influence intake as much as taste, *Journal of Nutrition Education and Behavior* 37 (2005): 242–245.

88. M. P. Mattson, Energy intake, meal frequency, and health: A neurobiological perspective, *Annual Review of Nutrition* 25 (2005): 237–260; T. V. Kral, L. S. Roe, and B. J. Rolls, Does nutrition information about the energy density of meals affect food intake in normal-weight women? *Appetite* 39 (2002): 137–145.

89. J. A. Ello-Martin, J. H. Ledikwe, and B. J. Rolls, The influence of food portion size and energy density on energy intake: Implications for weight management, *American Journal of Clinical Nutrition* 82 (2005): 236S–241S.

90. B. J. Rolls, A. Drewnowski, and J. H. Ledikwe, Changing the energy density of the diet as a strategy for weight management, *Journal of the American Dietetic Association* 105 (2005): S98–S103.

91. J. H. Ledikwe and coauthors, Low-energy-density diets are associated with high diet quality in adults in the United States, *Journal of the American Dietetic Association* 106 (2006): 1172–1180.

92. B. M. Popkin and coauthors, A new proposed guidance system for beverage consumption in the United States, *American Journal of Clinical Nutrition* 83 (2006): 529–542.

93. S. A. Bowman and J. T. Spence, A comparison of low-carbohydrate vs. high-carbohydrate diets: Energy restriction, nutrient quality and correlation to body mass index, *Journal of the American College of Nutrition* 21 (2002): 268–274.

94. B. J. Rolls, J. A. Ello-Martin, and B. C. Tohill, What can intervention studies tell us about the relationship between fruit and vegetable consumption and weight management? *Nutrition Reviews* 62 (2004): 1–17.

95. A. Astrup and coauthors, Low-fat diets and energy balance: How does the evidence stand in 2002? *Proceedings of the Nutrition Society* 61 (2002): 299–309; S. D. Poppitt and coauthors, Long-term effects of ad libitum low-fat, high-carbohydrate diets on weight and serum lipids in overweight subjects with metabolic syndrome, *American Journal of Clinical Nutrition* 75 (2002): 11–20.

96. G. D. Foster and coauthors, A randomized trial of a low-carbohydrate diet for obesity, *New England Journal of Medicine* 348 (2003): 2082–2090.

97. D. K. Layman and coauthors, A reduced ratio of dietary carbohydrate to protein improves body composition and blood lipid profiles during weight loss in adult women, *Journal of Nutrition* 133 (2003): 411–417; D. M. Bravata and coauthors, Efficacy and safety of low-carbohydrate diets, *Journal of the American Medical Association* 289 (2003): 1837–1850; S. Pirozzo and coauthors, Advice on low-fat diets for obesity, *Cochrane Database of Systematic Review* (2002), available at www.update-software.com/abstracts/ab003640.htm.

98. Astrup and coauthors, 2002.

99. J. M. Jakicic and A. D. Otto, Treatment and prevention of obesity: What is the role of exercise? *Nutrition Reviews* 64 (2006): S57–S61.

100. J. Kruger and coauthors, Physical activity profiles of U.S. adults trying to lose weight: NHIS 1998, *Medicine and Science in Sports and Exercise* 37 (2005): 364–368.

101. Committee on Dietary Reference Intakes, *Dietary Reference Intakes for Energy, Carbohydrate, Fiber, Fat, Fatty Acids, Cholesterol, Protein, and Amino Acids,* (Washington, D.C.: National Academies Press, 2002/2005).

102. L. L. Frank and coauthors, Effects of exercise on metabolic risk variables in overweight postmenopausal women: A randomized clinical trial, *Obesity Research* 13 (2005): 615–625; J. F. Carroll and C. K. Kyser, Exercise training in obesity lowers blood pressure independent of weight change, *Medicine and Science in Sports and Exercise* 34 (2002): 596–601; B. Gutin and coauthors, Effects of exercise intensity on cardiovascular fitness, total body composition, and visceral adiposity of obese adolescents, *American Journal of Clinical Nutrition* 75 (2002): 818–826.

103. C. A. Holcomb, D. L. Heim, and T. M. Loughin, Physical activity minimizes the association of body fatness with abdominal obesity in white, premenopausal women: Results from the Third National Health and Nutrition Examination Survey, *Journal of the American Dietetic Association* 104 (2004): 1859–1862; Gutin and coauthors, 2002.

104. M. Pomerleau and coauthors, Effects of exercise intensity on food intake and appetite in women, *American Journal of Clinical Nutrition* 80 (2004): 1230–1236.

105. R. W. Jeffery and coauthors, Physical activity and weight loss: Does prescribing higher physical activity goals improve outcome? *American Journal of Clinical Nutrition* 78 (2003): 684–689; C. A. Slentz and coauthors, Effects of the amount of exercise on body weight, body composition, and measures of central obesity: STRRIDE—A randomized controlled study, *Archives of Internal Medicine* 164 (2004): 31–39.

106. D. L. Thompson, J. Rakow, and S. M. Perdue, Relationship between accumulated walking and body composition in middle-aged women, *Medicine and Science in Sports and Exercise* 36 (2004): 911–914; H. R. Wyatt and coauthors, A Colorado statewide survey of walking and its relation to excessive weight, *Medicine and Science in Sports and Exercise* 37 (2005): 724–730.

107. B. Wansink, Environmental factors that increase the food intake and consumption volume of unknowing consumers, *Annual Review of Nutrition* 24 (2004): 455–479.

108. N. Stroebele and J. M. DeCastro, Effect of ambience on food intake and food choice, *Nutrition* 20 (2004): 821–838.

109. B. Wansink, J. E. Painter, and Y. K. Lee, The office candy dish: Proximity's influence on estimated and actual consumption, *International Journal of Obesity* 30 (2006): 871–875.

110. P. Pliner and coauthors, Meal duration mediates the effect of "social facilitation" on eating in humans, *Appetite* 46 (2006): 189–198.

111. J. M. Poothullil, Recognition of oral sensory satisfaction and regulation of the volume of intake in humans, *Nutritional Neuroscience* 8 (2005): 245–250.

112. Wansink, Painter, and Lee, 2006.

113. B. E. Kahn and B. Wansink, The influence of assortment structure on perceived variety and consumption quantities, *Journal of Consumer Research* 30 (2004): 519–533.

114. M. M. Hetherington and coauthors, Understanding variety: Tasting different foods delays satiation, *Physiology and Behavior* 87 (2006): 263–271.

115. S. J. Nielsen and B. M. Popkin, Patterns and trends in food portion sizes, 1977–1998, *Journal of the American Medical Association* 289 (2003): 450–453; H. Smiciklas-Wright and coauthors, Foods commonly eaten in the United States, 1989–1991 and 1994–1996: Are portion sizes changing? *Journal of the American Dietetic Association* 103 (2003): 41–47; L. R. Young and M. Nestle, Expanding portion sizes in the US marketplace: Implications for nutrition counseling, *Journal of the American Dietetic Association* 103 (2003): 231–234;

116. B. J. Rolls and coauthors, Increasing the portion size of a packaged snack increases energy intake in men and women, *Appetite* 42 (2004): 63–69.

117. B. Wansink, J. E. Painter, and J. North, Bottomless bowls: Why visual cues of portion size may influence intake, *Obesity Research* 13 (2005): 93–100.

118. B. Wansink and M. M. Cheney, Super bowls: Serving bowl size and food consumption, *Journal of the American Medical Association* 293 (2005): 1727–1728.

119. B. Wansink, K. van Ittersum, and J. E. Painter, Ice cream illusions bowls, spoons, and self-served portion sizes, *American Journal of Preventive Medicine* 31 (2006): 240–243.

120. B. Wansink and K. van Ittersum, Shape of glass and amount of alcohol poured: Comparative study of effect of practice and concentration, *British Medical Journal* 331 (2005): 1512–1514.

121. L. A. Berkel and coauthors, Behavioral interventions for obesity, *Journal of the American Dietetic Association* 105 (2005): S35–S43; G. D. Foster, A. P. Makris, and B. A. Bailer, Behavioral treatment of obesity, *American Journal of Clinical Nutrition* 82 (2005): 230S–235S.

122. S. M. Byrne, Psychological aspects of weight maintenance and relapse in obesity, *Journal of Psychosomatic Research* 53 (2002): 1029–1036.

123. D. F. Tate, E. H. Jackvony, and R. R. Wing, Effects of Internet behavioral counseling on weight loss in adults at risk for type 2 diabetes: A randomized trial, *Journal of the American Medical Association* 289 (2003): 1833–1836.

124. G. L. Blackburn, and B. A. Waltman, Expanding the limits of treatment—New strategic initiatives, *Journal of the American Dietetic Association* 105 (2005): S131–S135.

125. M. S. Leser, S. Z. Yanovski, and J. A. Yanovski, A low-fat intake and greater activity level are associated with lower weight regain 3 years after completing a very-low-calorie diet, *Journal of the American Dietetic Association* 102 (2002): 1252–1256.

126. R. L. Weinsier and coauthors, Free-living activity energy expenditure in women successful and unsuccessful at maintaining a normal body weight, *American Journal of Clinical Nutrition* 75 (2002): 499–504.

127. American College of Sports Medicine, 2001.
128. M. A. van Baak and coauthors, Leisure-time activity is an important determinant of long-term weight maintenance after weight loss in the Sibutramine Trial on Obesity Reduction and Maintenance (STORM trial), *American Journal of Clinical Nutrition* 78 (2003): 209–214.
129. R. R. Wing and coauthors, A self-regulation program for maintenance of weight loss, *New England Journal of Medicine* 355 (2006): 1563–1571.
130. R. R. Wing and S. Phelan, Long-term weight loss maintenance, *American Journal of Clinical Nutrition* 82 (2005): 222S–225S.
131. J. O. Hill, H. Thompson, and H. Wyatt, Weight maintenance: What's missing? *Journal of the American Dietetic Association* 105 (2005): S63–S66.

132. L. O. Gostin, Law as a tool to facilitate healthier lifestyles and prevent obesity, *Journal of the American Medical Association* 297 (2007): 87–90; M. M. Mello, D. M. Studdert, and T. A. Brennan, Obesity—The new frontier of public health law, *New England Journal of Medicine* 354 (2006): 2601–2610; R. E. Killingsworth, Health promoting community design: A new paradigm to promote healthy and active communities, *American Journal of Health Promotion* 17 (2003): 169–170.
133. S. L. Mercer and coauthors, Possible lessons from the tobacco experience for obesity control, *American Journal of Clinical Nutrition* 77 (2003): 1073S–1082S.
134. R. Smith, Passing an effective obesity bill, *Journal of the American Dietetic Association* 106 (2006): 1349–1350.

135. F. Kuchler, A. Tegene, and J. M. Harris, Taxing snack foods: What to expect for diet and tax revenues, *Current Issues in Economics of Food Markets*, Agriculture Information Bulletin No. 747–08, August 2004.
136. C. Marmonier and coauthors, Snacks consumed in a nonhungry state have poor satiating efficiency: Influence of snack composition on substrate utilization and hunger, *American Journal of Clinical Nutrition* 76 (2002): 518–528.
137. Position Paper: Nutrition and athletic performance—Position of the American Dietetic Association, Dietitians of Canada, and the American College of Sports Medicine, *Journal of the American Dietetic Association* 100 (2000): 1543–1556.

ANSWERS

Nutrition Calculations

1. a. Three milk shakes provide: 3×190 kcal $= 570$ kcal; 3×32 g carbohydrate $= 96$ g carbohydrate; 3×13 g protein $= 39$ g protein; and 3×1 g fat $= 3$ g fat.

   b. To meet this criteria, the plan needs *at least* an additional 430 kcalories (1000 kcal − 570 kcal = 430 kcal); an additional 7 to 17 grams of protein, depending on the person's RDA based on gender and age (56 g − 39 g = 17 g and 46 g − 39 g = 7 g); an additional 4 grams of carbohydrate (100 g − 96 g = 4 g); and some additional fat.

   c. Of course, there are many possible dinners that you could plan. One might be:

   Salad made with 1 c lettuce, 1 c chopped tomatoes and onions, ¼ c garbanzo beans, and 2 tbs low-fat dressing

   4 oz grilled chicken

   1 medium baked potato

   1 c summer squash and zucchini

   1 c melon cubes

   This meal brings the day's totals to 1215 kcalories, 90 g of protein, 192 g of carbohydrate, and 13 g of fat, which meets the goals for kcalories, protein, and carbohydrate. Because the milk shake has been fortified, all vitamin and mineral needs are covered as well. The only possible dietary shortcoming is that the day's percent kcalories from fat is low (only 10%), but because energy and nutrient recommendations have been met and the goal is weight loss, this may be acceptable.

   d. This weight-loss plan uses a liquid formula rather than foods, making clients dependent on a special device (the formula) rather than teaching them how to make good choices from the conventional food supply. It provides no information about dropout rates, the long-term success of clients, or weight maintenance after the program ends.

2. a. 459 kcal ÷ 442 g = 1.04 kcal/g

   b. More than another whole salad (1.04 kcal/g × 500 kcal = 520 g)

   c.

Item No./Food	Weight (g)	Energy (kcal)
Original totals:	442	459
Minus:		
#5083 Lettuce, 1 c	−55	−6
Plus:		
#12212 Roast ham, 1 oz	+28	+41
#1007 Cheddar cheese, 1 oz	+28	+113
Totals:	443 g	607 kcal

   d. 607 kcal − 459 kcal = 148 kcal added

   e. 443 g − 442 g = 1 g added

Study Questions (multiple choice)

1. a 2. d 3. a 4. d 5. c 6. c 7. d 8. d 9. d 10. b

The Latest and Greatest Weight-Loss Diet—Again

© Geri Engberg

To paraphrase William Shakespeare, "a fad diet by any other name would still be a fad diet." And the names are legion: the Atkins Diet, the Calories Don't Count diet, the Protein Power diet, the Carbohydrate Addict's diet, the Lo-Carbo diet, the South Beach diet, the Zone diet.* Year after year, "new and improved" diets appear on bookstore shelves and circulate among friends. People of all sizes eagerly try the best diet on the market ever, hoping that this one will really work. Sometimes these diets seem to work for a while, but more often than not, their success is short-lived. Then another diet takes the spotlight. Here's how Dr. K. Brownell, an obesity researcher at Yale University, describes this phenomenon: "When I get calls about the latest diet fad, I imagine a trick birthday cake candle that keeps lighting up and we have to keep blowing it out."

Realizing that fad diets do not offer a safe and effective plan for weight loss, health professionals speak out, but they never get the candle blown out permanently. New fad diets can keep making outrageous claims because no one requires their advocates to prove what they say. Fad diet gurus do not have to conduct credible research on the benefits or dangers of their diets. They can simply make recommendations and then later, if questioned, search for bits and pieces of research that support the conclusions they have already reached. That's backwards. Diet and health recommendations should *follow* years of sound research that has been reviewed by panels of scientists *before* being offered to the public.

Because anyone can publish anything—in books or on the Internet—peddlers of fad diets can make unsubstantiated statements that fall far short of the truth but sound impressive to the uninformed. They often offer distorted bits of legitimate research. They may start with one or more actual facts but then leap from one erroneous conclusion to the next. Anyone who wants to believe these claims has to wonder how the thousands of scientists working on obesity research over the past century could possibly have missed such obvious connections. Table H9-1 (p. 316) presents some of the claims and truths of fad diets.

Fad diets come in almost as many shapes and sizes as the people who search them out. Some restrict fats or carbohydrates, some limit portion sizes, some focus on food combinations, and some claim that a person's genetic type or blood type determines the foods best suited to manage weight and prevent disease. Table H9-2 (p. 317) compares some of today's more popular diets. Regardless of their names, many popular diets espouse a carbohydrate-restricted or carbohydrate-modified diet. Some diets claim that all or some types of carbohydrates are bad. Some go so far as to equate carbohydrates with toxic poisons or addictive drugs. "Bad" carbohydrates—such as sugar, white flour, and potatoes—are considered evil because they are absorbed easily and raise blood glucose. The pancreas then responds by secreting insulin—and insulin is touted as the real villain responsible for our nation's epidemic of obesity. Whether restricting overall carbohydrate intake or replacing certain "bad" carbohydrates with "good" carbohydrates, these diets tend to overemphasize protein. This highlight examines some of the science and the science fiction behind a few carbohydrate-restricted or carbohydrate-modified, high-protein fad diets.

The Diet's Appeal

Perhaps the greatest appeal of fad diets such as the Atkins Diet is that it turns nutrient recommendations upside down. Foods such as meats and milk products that need to be selected carefully to limit saturated fat can be eaten with abandon on this diet. Grains, legumes, vegetables, and fruits that consumers are told to eat in abundance can now be ignored. For some people, this is a dream come true: steaks without the potatoes, ribs without the coleslaw, and meatballs without the pasta. Who can resist the promise of weight loss while eating freely from a list of favorite foods?

To lure dieters in, proponents of fad diets often blame current recommendations for our obesity troubles. They claim that the incidence of obesity is rising because we are eating less fat. Such a claim may impress the naive, but it sends skeptical people running for the facts. True, the incidence of obesity has risen dramatically over the past two decades.[1] True, our intake of fat has dropped from 35 to 33 percent of daily energy intake.[2] Such facts might seem to imply that lowering fat intake leads to obesity, but this is an erroneous conclusion. The *percentage* declined only because average energy intakes increased by almost 200 kcalories a

* The following sources offer comparisons and evaluations of various fad diets for your review: Battle of the diet books II, *Nutrition Action Healthletter,* July/August 2006, pp. 10–11; B. Liebman, Weighing the diet books, *Nutrition Action Healthletter,* January/February 2004, pp. 1–8; S. T. St. Jeor and coauthors, Dietary protein and weight reduction: A statement for healthcare professionals from the nutrition committee of the Council on Nutrition, Physical Activity, and Metabolism of the American Heart Association, *Circulation* 104 (2001): 1869–1874.

TABLE H9-1 The Claims and Truths of Fad Diets

The Claim:	You can lose weight "easily."
The Truth:	Most fad diet plans have complicated rules that require you to calculate protein requirements, count carbohydrate grams, combine certain foods, time meal intervals, purchase special products, plan daily menus, and measure serving sizes.
The Claim:	You can lose weight by eating a specific ratio of carbohydrates, protein, and fat.
The Truth:	Weight loss depends on spending more energy than you take in, not on the proportion of energy nutrients.
The Claim:	This "revolutionary diet" can "reset your genetic code."
The Truth:	You inherited your genes and cannot alter your genetic code.
The Claim:	High-protein diets are popular, selling more than 20 million books, because they work.
The Truth:	Weight-loss books are popular because people grasp for quick fixes and simple solutions to their weight problems. If book sales were an indication of weight-loss success, we would be a lean nation—but they're not, and neither are we.
The Claim:	People gain weight on low-fat diets.
The Truth:	People can gain weight on low-fat diets if they overindulge in carbohydrates and proteins while cutting fat; low-fat diets are not necessarily low-kcalorie diets. But people can also lose weight on low-fat diets if they cut kcalories as well as fat.
The Claim:	High-protein diets energize the brain.
The Truth:	The brain depends on glucose for its energy; the primary dietary source of glucose is carbohydrate, not protein.
The Claim:	Thousands of people have been successful with this plan.
The Truth:	Authors of fad diets have not published their research findings in scientific journals. Success stories are anecdotal and failures are not reported.
The Claim:	Carbohydrates raise blood glucose levels, triggering insulin production and fat storage.
The Truth:	Insulin promotes fat storage when energy intake exceeds energy needs. Furthermore, insulin is only one hormone involved in the complex processes of maintaining the body's energy balance and health.
The Claim:	Eat protein and lose weight.
The Truth:	For every complicated problem, there is a simple—and wrong—solution.

Low-carbohydrate meals overemphasize meat, fish, poultry, eggs, and cheeses, and shun breads, pastas, fruits, and vegetables.

day (from 1878 kcalories a day to 2056). Actual fat intake *increased* by 3 grams a day (from 73 grams to 76). Furthermore, fewer than half of us engage in regular physical activity.[3] Obesity experts blame our high energy intakes and low energy outputs for the increase in obesity. Weight loss, after all, depends on a negative energy balance. To their credit, some of these diet plans recommend exercise—and regular physical activity is an integral component of successful weight loss.[4]

Dieters are also lured into fad diets by sophisticated—yet often erroneous—explanations of the metabolic consequences of eating certain foods. Terms such as *eicosanoids* and *de novo lipogen-*

esis are scattered about, often intimidating readers into believing that the authors must be right given their brilliance in understanding the body. Several of the latest fad diets hold *insulin* responsible for the obesity problem and the *glycemic index* as the weight loss solution. Yet, among nutrition researchers, controversy continues to surround the questions of whether insulin promotes weight gain or a low glycemic diet fosters weight loss.[5]

What does insulin do? Among its roles, insulin facilitates the transport of glucose into the cells, the storage of fatty acids as fat, and the synthesis of cholesterol. It is an anabolic hormone that builds and stores. True—but there's more to the story. Insulin is only one of many factors involved in the body's metabolism of nutrients and regulation of body weight. Furthermore, as Chapter 4's discussion of the glycemic index pointed out, blood glucose and insulin do not always respond to foods as might be expected. The glycemic effect of a food depends on how the food is ripened, processed, and cooked; the time of day the food is eaten; the other foods eaten with it; and the presence or absence of certain diseases such as type 2 diabetes in the person eating the food.[6] Thus the glycemic effect of a particular food varies—fad diet books mislead people by claiming that each food has a set glycemic effect. Many carbohydrates—fruits, vegetables, legumes, and whole grains—are rich in fibers that slow glucose absorption and moderate insulin response. Furthermore, there is no clear evidence that elevated blood insulin concentrations promote weight gain in healthy people or that foods with a low glycemic effect promote weight loss.[7] A review of the evidence thus far concludes that the ideal long-term study has not yet been conducted.[8]

Most importantly, insulin is critical to maintaining health, as any person with type 1 diabetes can attest. Insulin causes problems only when a person develops insulin resistance—that is, when the body's cells do not respond to the large quantities of insulin that the pancreas continues to pump out in an effort to get a response. Insulin resistance is a major health problem—but it is not caused by carbohydrate, or by protein, or by fat. It results from being overweight. When a person loses weight, insulin response improves.

TABLE H9-2 Popular Diets Compared

Diet	Major Premise Promoted	Strong Point(s)	Weak Point(s)
High-Carbohydrate, Low-Fat			
Ornish Diet	• By strictly limiting fat (both animal and vegetable), you eat fewer kcalories without eating less food.	• High-fiber, low-fat foods in this plan can lower blood cholesterol and blood pressure.	• So little fat that essential fatty acids may be lacking. • Limits fish, nuts, and olive oil which may protect against heart disease.
Pritikin Program	• By eating low-fat, mainly plant-based foods, you can eat more food and still feel satisfied.	• No food group is completely eliminated in this high-fiber, low-fat diet program. • Some use of foods rich in omega-3 fatty acids are encouraged.	• For some people, very low-fat diets may be unsatisfying and therefore difficult to adhere to.
Low-Carbohydrate, High-Protein			
Atkins Diet	• People are overweight or obese because they have metabolic imbalances caused by eating too many carbohydrates; by restricting carbohydrates, these imbalances can be corrected. • You can lose weight without lowering kcalorie intake.	• Quick, short-term weight loss is achieved.	• Restricts carbohydrates to a level that induces ketosis. • Ketosis can cause nausea, light-headedness, and fatigue. • Ketosis can worsen existing medical problems such as kidney disease. • A diet high in fat such as Atkins can increase the risk of heart disease and some cancers.
Low-Carbohydrate			
Zone Diet	• Eating the correct proportions of carbohydrates, fat, and protein leads to hormonal balance, weight loss, disease prevention, and increased vitality.	• Promotes weight loss because it is a low-kcalorie diet.	• The diet is rigid, restrictive, and complicated, making it difficult for most people to follow accurately. • The overblown health claims of the diet's proponents are based on misinterpreted science and remain unsubstantiated.
Carbohydrate-Modified			
South Beach Diet	• Eating "good carbohdrates" such as vegetables, whole-wheat pastas, and brown rice will maintain satiety and resist cravings for "bad carbohydrates" such as white rice and potatoes.	• Encourages consumption of vegetables, lean meats, and fish, and the use of unsaturated oils when cooking. • Restricts fatty meats and cheeses as well as sweets.	• Starchy carbohydrates and all fruits are completely excluded during the first two weeks.
The Ultimate Weight Solution Diet	• Foods that require great effort to prepare and eat are nutrient-dense; eating these kinds of foods (raw vegetables, vegetable soups, whole grains, beans, meats, poultry, and fish) will lead to weight loss. • Foods that take little effort to prepare and eat provide excess kcalories relative to nutrients; eating these kinds of foods (fast foods, puddings, high-kcalorie convenience foods, processed foods) leads to uncontrolled eating and weight gain.	• Encourages consumption of lean meats and fish; whole grains; vegetables; fruit; and low-fat milk, yogurt, and cheese. • Restricts fatty meats and cheeses as well as sweets. • Encourages exercise.	• Confusing as to exactly what to eat or how much.
Metabolic Type			
Eat Right 4 Your Type	• Your blood type determines which foods you should eat or not eat.	None	• Food groups or individual foods are excluded, depending on blood type. • No scientific data on the relationship between blood type and food choices.

Another distortion of the facts is the claim that high-protein foods expend more energy. As Chapter 8 mentioned, the thermic effect of food for protein is higher than for carbohydrate or fat, but the increase is still insignificant—perhaps the equivalent of two pounds per year, at most.

If low-carbohydrate, high-protein diets were as successful as some people claim, then consumers who tried them would lose weight, and their obesity problems would be solved. But this is not the case. Similarly, if high-protein diets were as worthless as others claim, then consumers would eventually stop pursuing them. Clearly, this is not happening either. These diets have enough going for them that they work for some people at least for a short time, but they fail to produce long-lasting results for most people. Studies report that people following high-protein, low-carbohydrate diets do lose weight.[9] In fact, they lose more than people following conventional high-carbohydrate, low-fat diets—but only for the first six months. Their later gains make up the difference, so total weight loss is no different after one year.[10] The following sections examine some of the apparent achievements and shortcomings of high-protein diets.[11]

The Diet's Achievements

With over half of our nation's adults overweight and many more concerned about their weight, the market for a weight-loss book, product, or program is huge (no pun intended). Americans spend an estimated $33 billion a year on weight-loss books and products. Even a plan that offers only minimal weight-loss success easily attracts a following. Carbohydrate-modified and high-protein, low-carbohydrate diet plans offer a little success to some people for a short time. Here's why.

Don't Count kCalories

Who wants to count kcalories? Even experienced dieters find counting kcalories burdensome, not to mention timeworn. They want a new, easy way to lose weight, and high-protein diet plans seem to offer this boon. But, though these diets often claim to disregard kcalories, their design typically ensures a low energy intake. Most of the sample menu plans provided by these diets, especially in the early stages, are designed to deliver an average of 1200 kcalories a day.

Even when counting kcalories is truly not necessary, the total for these diets tends to be low simply because food intake is so limited. Without its refried beans, tortilla wrapping, and chopped vegetables, a burrito is reduced to a pile of ground beef. Without the baked potato, there's no need for butter and sour cream. Weight loss occurs because of the low energy intake—not the proportion of energy nutrients.[12] Success, then, depends on the restricted intake, not on protein's magical powers or carbohydrate's evil forces. This is an important point. Any diet can produce weight loss, at least temporarily, if intake is restricted. The real value of a diet is determined by its ability to maintain weight loss and support good health over the long term. The goal is not simply weight loss, but health gains—and whether carbohydrate-modified or high-protein, low-carbohydrate diets can support optimal health over time remains unknown.

Satisfy Hunger

Protein may promote weight loss by providing satiety.[13] As Chapter 8 mentioned, of the three energy-yielding nutrients, protein is the most satiating. High-protein meals suppress hunger and delay the start of the next meal. Furthermore, people tend to eat less after a high-protein meal than after a low-protein one. In one study, when protein intake increased from 15 percent of total energy to 30 percent but carbohydrate was held constant at 50 percent of total energy, people decreased their energy intakes and lost body weight and body fat.[14] This research suggests that less emphasis should be placed on carbohydrate restriction.

In real-life situations, there is a strong association between a person's protein intake and BMI—the higher the intake, the higher the BMI.[15] This association remains apparent even after adjusting for energy intake and physical activity. All meals—whether designed for weight loss or not—should include enough protein to satisfy hunger, but not so much as to contribute to weight gain.

Follow a Plan

Most people need specific instructions and examples to make dietary changes. Popular diets offer dieters a plan. The user doesn't have to decide what foods to eat, how to prepare them, or how much to eat. Unfortunately, these instructions only serve short-term weight-loss needs. They do not provide for long-term changes in lifestyle that will support weight maintenance or health goals.

The success of any weight-loss diet depends on the person adopting the plan and sticking with it. People who prefer the high-protein, low-carbohydrate diet over the high-carbohydrate, low-fat diet may have more success at sticking with it. Again, weight loss occurs because of the duration of a low-kcalorie plan—not the proportion of energy nutrients.[16]

Limit Choices

Diets that omit hundreds of foods and several food groups limit a person's options and lack variety. Chapter 2 praised variety as a valuable way to ensure an adequate intake of nutrients, but variety also entices people to eat more food and gain more weight. Without variety, some people lose interest in eating, which further reduces energy intake. Even if the allowed foods are favorites, eating the same foods week after week can become monotonous.

The Diet's Shortcomings

Most of the foods that fad diets promote are healthy foods—lean meats, fat-free or low-fat milk and yogurt, vegetables, whole grains, beans, and fruit. The *Dietary Guidelines for Americans 2005* encourage consumers to eat the same foods. The *Dietary Guidelines* also advise consumers to eat less saturated fat, however, and some carbohydrate-restricted or carbohydrate-modified diets can be high in saturated fat. Like some of the carbohydrate-modified diets, the *Dietary Guidelines* also encourage people to eat a diet high in fiber-rich carbohydrate foods. Fad diet claims that people lose weight because they switch from eating "bad" carbohydrates to eating "good" ones, however, are misleading; in truth, people lose weight on these diets because they are eating fewer kcalories not because they are eating different kinds of kcalories. Still, people who have followed carbohydrate-restricted or carbohydrate-modified diet plans for several months have lost weight. Can these diets be harmful?

Too Much Fat

Some fad diets focus so intently on promoting protein and curbing carbohydrate that they fail to account for the fat that accompanies many high-protein foods. A breakfast of bacon and eggs, lunch of ham and cheese, and dinner of barbecued short ribs would provide 100 grams of protein—and 121 grams of fat! Yet this day's meals, even with a snack of peanuts, provide only 1600 kcalories. Without careful selection, protein-rich diets can be extraordinarily high in saturated fat and cholesterol—dietary factors that raise LDL cholesterol and the risks for heart disease.

Overall, studies report that people following high-protein, low-carbohydrate diets have little or no change in blood pressure or blood lipids—risk factors for heart disease.[17] Some researchers speculate that the weight loss that occurs on these diets offsets the adverse effects of a diet high in saturated fat and low in fruits and vegetables.[18] Others point out that different sources of protein have different effects on risk factors for heart disease.[19] For example, the effects of white meat from chicken or fish differ from those of red meat. Diets containing large amounts of red meat appear to increase the risk of heart disease. In contrast, replacing animal sources of protein with plant sources of protein may benefit health.

Too Much Protein

Moderation has been a recurring theme throughout this text, with recommendations to get enough, but not too much of anything, and cautions that too much can be as harmful as too little. Too much protein can contribute to weight gain just as too much carbohydrate or fat can. As mentioned earlier, protein intake is positively associated with BMI.[20] The DRI Committee did not establish an upper level for protein, but it does recognize that high-protein diets have been implicated in chronic diseases such as osteoporosis, kidney stones and kidney disease, some cancers, heart disease, and obesity.[21] Health recommendations typically advise a protein intake of 50 to 100 grams per day and within the range of 10 to 35 percent of energy intake.[22] This range allows for flexibility without risk of harm. By comparison, popular high-protein diets suggest a protein intake of 70 to 160 grams per day, representing 25 to 65 percent of energy intake.[23]

Guidelines from the DRI committee include higher protein intakes (10 to 35 percent of total energy) than recommended previously, but long-term studies of high-protein intakes are needed to ascertain the health consequences of such diets. One such study is currently under way: The DiOGenes (Diet, Obesity, and Genes) project is examining the interactions among a high dietary protein intake, the glycemic effect of foods, and genetic and behavioral factors in preventing weight gain and regain.[24] The study focuses on about 700 overweight or obese adults and their children in eight different countries across Europe and may involve the United States as well.

Too Little Everything Else

The quality of the diet suffers when carbohydrates are restricted.[25] Without fruits, vegetables, and whole grains, high-protein diets lack not only carbohydrate, but fiber, vitamins, minerals, and phytochemicals as well—all dietary factors protective against disease.[26] To help shore up some of these inadequacies, fad diets often recommend a dietary supplement. Conveniently, many of the companies selling fad diets also peddle these supplements. But as Highlights 10 and 11 explain, foods offer many more health benefits than any supplement can provide. Quite simply, if the diet is inadequate, it needs to be improved, not supplemented.

The Body's Perspective

When a person consumes a low-carbohydrate diet, a metabolism similar to that of fasting prevails. (See Chapter 7 for a review of fasting.) With little dietary carbohydrate coming in, the body uses its glycogen stores to provide glucose for the cells of the brain, nerves, and blood. Once the body depletes its glycogen reserves, it begins making glucose from the amino acids of protein (gluconeogenesis). A low-carbohydrate diet may provide abundant protein from food, but the body still uses some protein from body tissues.

Dieters can know glycogen depletion has occurred and gluconeogenesis has begun by monitoring their urine. Whenever glycogen or protein is broken down, water is released and urine production increases. Low-carbohydrate diets also induce ketosis, and ketones can be detected in the urine. Ketones form whenever glucose is lacking and fat breakdown is incomplete.

Many fad diets regard ketosis as the key to losing weight, but studies comparing weight-loss diets find no relation between ketosis and weight loss.[27] People in ketosis may experience a loss of

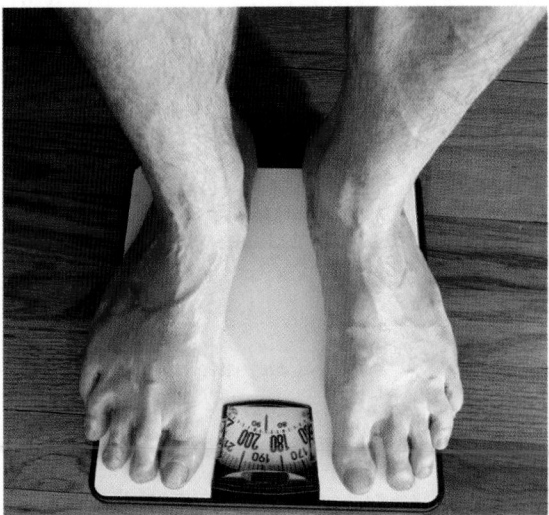

The wise consumer distinguishes between loss of fat and loss of weight.

TABLE H9-3 Adverse Side Effects of Low-Carbohydrate, Ketogenic Diets

- Nausea
- Fatigue (especially if physically active)
- Constipation
- Low blood pressure
- Elevated uric acid (which may exacerbate kidney disease and cause inflammation of the joints in those predisposed to gout)
- Stale, foul taste in the mouth (bad breath)
- In pregnant women, fetal harm and stillbirth

appetite and a dramatic weight loss within the first few days. They should know that much of this weight loss reflects the loss of glycogen and protein together with large quantities of body fluids and important minerals.[28] They need to appreciate the difference between loss of *fat* and loss of *weight*. Fat losses on ketogenic diets are no greater than on other diets providing the same number of kcalories. Once the dieter returns to well-balanced meals that provide adequate energy, carbohydrate, fat, protein, vitamins, and minerals, the body avidly retains these needed nutrients. The weight will return, quite often to a level higher than the starting point. Table H9-3 lists other consequences of a ketogenic diet.

Table H9-4 offers guidelines for identifying fad diets and other weight-loss scams; it includes the hallmarks of a reasonable weight-loss program as well. Diets that overemphasize protein and fall short on carbohydrate may not harm healthy people if used for only a little while, but they cannot support optimal health for long. Chapter 9 includes reasonable approaches to weight management and concludes that the ideal diet is one you can live with for the rest of your life. Keep that criterion in mind when you evaluate the next "latest and greatest weight-loss diet" that comes along.

TABLE H9-4 Guidelines for Identifying Fad Diets and Other Weight-Loss Scams

Fad Diets and Weight-Loss Scams	Healthy Diet Guidelines
1. They promise dramatic, rapid weight loss.	1. Weight loss should be gradual and not exceed 2 pounds per week.
2. They promote diets that are nutritionally unbalanced or extremely low in kcalories.	2. Diets should provide: • A reasonable number of kcalories (not fewer than 1000 kcalories per day for women and 1200 kcalories per day for men) • Enough, but not too much, protein (between the RDA and twice the RDA) • Enough, but not too much, fat (between 20 and 35% of daily energy intake from fat) • Enough carbohydrates to spare protein and prevent ketosis (at least 100 grams per day) and 20 to 30 grams of fiber from food sources • A balanced assortment of vitamins and minerals from a variety of foods from each of the food groups • At least 1 liter (about 1 quart) of water daily or 1 milliliter per kcalorie daily—whichever is more.
3. They use liquid formulas rather than foods.	3. Foods should accommodate a person's ethnic background, taste preferences, and financial means.
4. They attempt to make clients dependent upon special foods or devices.	4. Programs should teach clients how to make good choices from the conventional food supply.
5. They fail to encourage permanent, realistic lifestyle changes.	5. Programs should teach physical activity plans that involve spending at least 300 kcalories a day and behavior-modification strategies that help to correct poor eating habits.
6. They misrepresent salespeople as "counselors" supposedly qualified to give guidance in nutrition and/or general health.	6. Even if adequately trained, such "counselors" would still be objectionable because of the obvious conflict of interest that exists when providers profit directly from products they recommend and sell.
7. They collect large sums of money at the start or require that clients sign contracts for expensive, long-term programs.	7. Programs should be reasonably priced and run on a pay-as-you-go basis.
8. They fail to inform clients of the risks associated with weight loss in general or the specific program being promoted.	8. They should provide information about dropout rates, the long-term success of their clients, and possible diet side effects.
9. They promote unproven or spurious weight-loss aids such as human chorionic gonadotropin hormone (HCG), starch blockers, diuretics, sauna belts, body wraps, passive exercise, ear stapling, acupuncture, electric muscle-stimulating (EMS) devices, spirulina, amino acid supplements (e.g. arginine, ornithine), glucomannan, methylcellulose (a "bulking agent"), "unique" ingredients, and so forth.	9. They should focus on nutrient-rich foods and regular exercise.
10. They fail to provide for weight maintenance after the program ends.	10. They should provide a plan for weight maintenance after successful weight loss.

SOURCES: Adapted from American College of Sports Medicine, *ACSM's Guidelines for Exercise Testing and Prescription* (Baltimore: Williams & Wilkins, 1995), pp. 218–219; J. T. Dwyer, Treatment of obesity: Conventional programs and fad diets, in *Obesity,* ed. P. Björntorp and B.N. Brodoff (Philadelphia: J.B. Lippincott, 1992), p. 668; *National Council Against Health Fraud Newsletter,* March/April 1987, National Council Against Health Fraud, Inc.

REFERENCES

1. C. L. Ogden and coauthors, Prevalence of overweight and obesity in the United States, 1999–2004, *Journal of the American Medical Association* 295 (2006): 1549–1555.
2. P. Chanmugam and coauthors, Did fat intake in the United States really decline between 1989–1991 and 1994–1996? *Journal of the American Dietetic Association* 103 (2003): 867–872.
3. P. M. Barnes and C. A. Schoenborn, *Physical Activity among Adults: United States, 2000,* 2003, available at www.cdc.gov/nchs/about/major/nhis/released200306.htm#7.
4. J. M. Jakicic and A. D. Otto, Physical activity considerations for the treatment and prevention of obesity, *American Journal of Clinical Nutrition* 82 (2005): 226S–229S.
5. R. Clemens and P. Pressman, Clinical value of glycemic index unclear, *Food Technology* 58 (2004): 18; M. A. Pereira and coauthors, Effects of a low-glycemic load diet on resting energy expenditure and heart disease risk factors during weight loss, *Journal of the American Medical Association* 292 (2004): 2482–2490; A. Raben, Should obese patients be counselled to follow a low-glycaemic index diet? No, *Obesity Reviews* 3 (2002): 245–256; D. B. Pawlak, C. B. Ebbeling, and D. S. Ludwig, Should obese patients be counselled to follow a low-glycaemic index diet? Yes, *Obesity Reviews* 3 (2002): 235–243.
6. F. X. Pi-Sunyer, Glycemic index and disease, *American Journal of Clinical Nutrition* 76 (2002): 290S–298S.
7. Pi-Sunyer, 2002.
8. A. G. Pittas and S. B. Roberts, Dietary composition and weight loss: Can we individualize dietary prescriptions according to insulin sensitivity or secretion status? *Nutrition Reviews* 64 (2006): 435–448; Raben, 2002.
9. A. Astrup, T. M. Larsen, and A. Harper, Atkins and other low-carbohydrate diets: Hoax or an effective tool for weight loss? *The Lancet* 364 (2004): 897–899; E. C. Westman and coauthors, Effect of 6-month adherence to a very low carbohydrate diet program, *American Journal of Medicine* 113 (2002): 30–36.
10. L. Stern and coauthors, The effects of low-carbohydrate versus conventional weight loss diets in severely obese adults: One-year follow-up of a randomized trial, *Annals of Internal Medicine* 140 (2004): 778–785; G. D. Foster and coauthors, A randomized trial of a low-carbohydrate diet for obesity, *New England Journal of Medicine* 348 (2003): 2082–2090.
11. J. Eisenstein and coauthors, High-protein weight-loss diets: Are they safe and do they work? A review of the experimental and epidemiologic data, *Nutrition Reviews* 60 (2002): 189–200.
12. D. K. Layman and coauthors, A reduced ratio of dietary carbohydrate to protein improves body composition and blood lipid profiles during weight loss in adult women, *Journal of Nutrition* 133 (2003): 411–417; D. M. Bravata and coauthors, Efficacy and safety of low-carbohydrate diets, *Journal of the American Medical Association* 289 (2003): 1837–1850.
13. D. A. Schoeller and A. C. Buchholz, Energetics of obesity and weight control: Does diet composition matter? *Journal of the American Dietetic Association* 105 (2005): S24–S28; S. M. Nickols-Richardson and coauthors, Perceived hunger is lower and weight loss is greater in overweight premenopausal women consuming a low-carbohydrate/high-protein vs high-carbohydrate/low-fat diet, *Journal of the American Dietetic Association* 105 (2005): 1433–1437.
14. D. S. Weigle and coauthors, A high-protein diet induces sustained reductions in appetite, ad libitum caloric intake, and body weight despite compensatory changes in diurnal plasma leptin and ghrelin concentrations, *American Journal of Clinical Nutrition* 82 (2005): 41–48.
15. A. Trichopoulou and coauthors, Lipid, protein and carbohydrate intake in relation to body mass index, *European Journal of Clinical Nutrition* 56 (2002): 37–43.
16. Bravata and coauthors, 2003.
17. Bravata and coauthors, 2003.
18. Foster and coauthors, 2003.
19. F. B. Hu, Protein, body weight, and cardiovascular health, *American Journal of Clinical Nutrition* 82 (2005): 242S–247S.
20. A. Trichopoulou and coauthors, Lipid, protein and carbohydrate intake in relation to body mass index, *European Journal of Clinical Nutrition* 56 (2002): 37–43.
21. Committee on Dietary Reference Intakes, *Dietary Reference Intakes for Energy, Carbohydrate, Fiber, Fat, Fatty Acids, Cholesterol, Protein, and Amino Acids* (Washington, D.C.: National Academies Press, 2002/2005).
22. Committee on Dietary Reference Intakes, 2002/2005; S. T. St. Jeor and coauthors, Dietary protein and weight reduction: A statement for healthcare professionals from the nutrition committee of the Council on Nutrition, Physical Activity, and Metabolism of the American Heart Association, *Circulation* 104 (2001): 1869–1874.
23. St. Jeor and coauthors, 2001.
24. W. H. M. Saris and A. Harper, DiOGenes: A multidisciplinary offensive focused on the obesity epidemic, *Obesity Reviews* 6 (2005): 175–176.
25. L. S. Greene-Finestone and coauthors, Adolescents' low-carbohydrate-density diets are related to poorer dietary intakes, *Journal of the American Dietetic Association* 105 (2005): 1783–1788; E. T. Kennedy and coauthors, Popular diets: Correlation to health, nutrition, and obesity, *Journal of the American Dietetic Association* 101 (2001): 411–420.
26. W. Cunningham and D. Hyson, The skinny on high-protein, low-carbohydrate diets, *Preventive Cardiology* 9 (2006): 166–171.
27. M. D. Coleman and S. M. Nickols-Richardson, Urinary ketones reflect serum ketone concentration but do not relate weight loss in overweight premenopausal women following a low-carbohydrate/high-protein diet, *Journal of the American Dietetic Association* 105 (2005): 608–611; Foster and coauthors, 2003.
28. St. Jeor and coauthors, 2001.

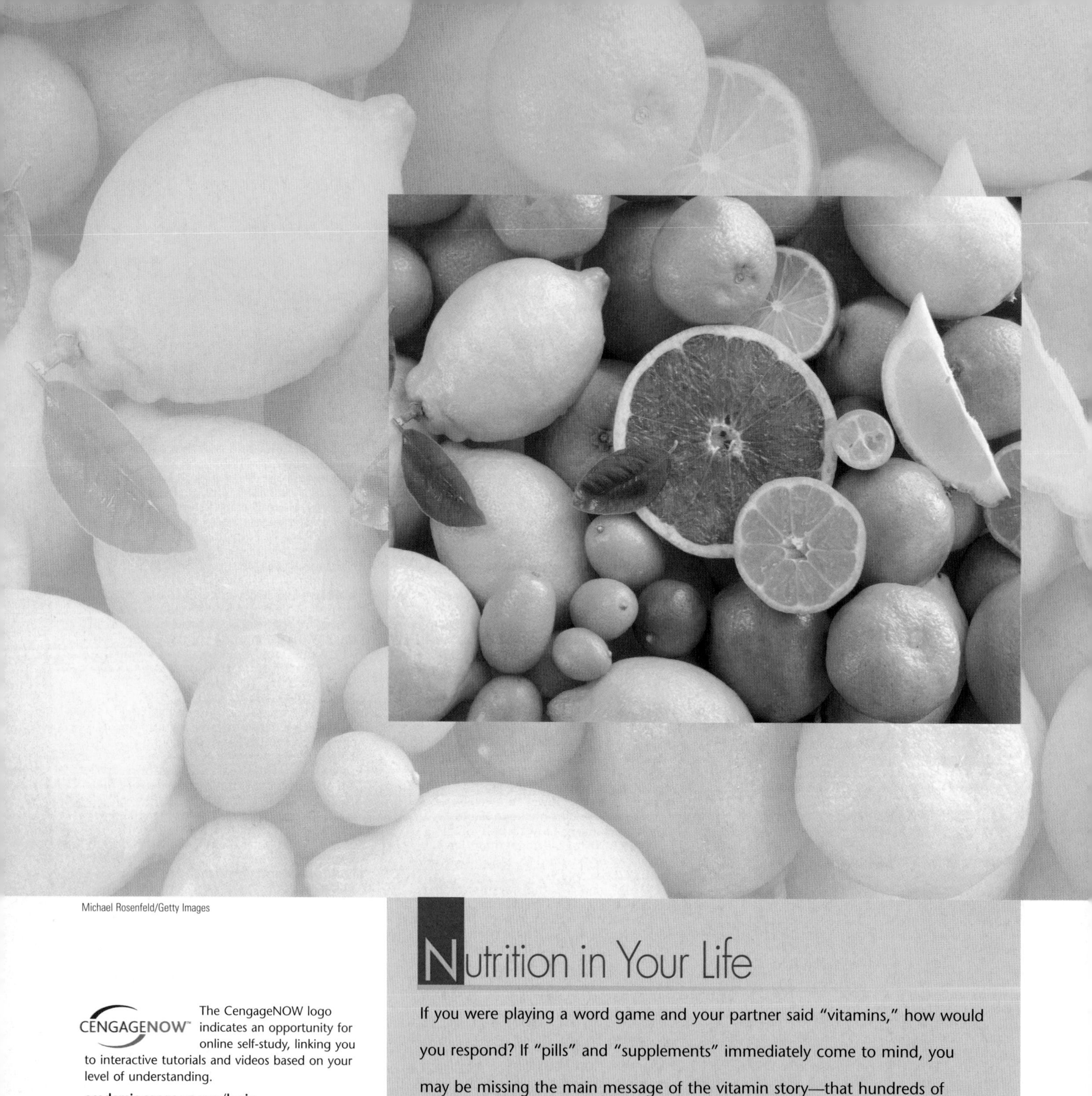

The CengageNOW logo indicates an opportunity for online self-study, linking you to interactive tutorials and videos based on your level of understanding.

academic.cengage.com/login

Figure 10-1: Animated! Coenzyme Action

Figure 10-13: Animated! Metabolic Pathways Involving B Vitamins

How To: Practice Problems

Nutrition Portfolio Journal

Nutrition Calculations: Practice Problems

Nutrition in Your Life

If you were playing a word game and your partner said "vitamins," how would you respond? If "pills" and "supplements" immediately come to mind, you may be missing the main message of the vitamin story—that hundreds of foods deliver over a dozen vitamins that participate in thousands of activities throughout your body. Quite simply, foods supply vitamins to support all that you are and all that you do—and supplements of any one of them, or even a combination of them, can't compete with foods in keeping you healthy.

The Water Soluble Vitamins: B Vitamins and Vitamin C

Earlier chapters focused on the energy-yielding nutrients, which play leading roles in the body. The vitamins and minerals are their supporting cast. This chapter begins with an overview of the vitamins and then examines each of the water-soluble vitamins and a nonvitamin relative named choline; the next chapter features the fat-soluble vitamins. Chapters 12 and 13 present the minerals.

The Vitamins—An Overview

Researchers first recognized that foods contain substances that are "vital to life" in the early 1900s. Since then, the world of vitamins has opened up dramatically. The vitamins ◆ are powerful substances, as their *absence* attests. Vitamin A deficiency can cause blindness; a lack of the B vitamin niacin can cause dementia; and a lack of vitamin D can retard bone growth. The consequences of deficiencies are so dire, and the effects of restoring the needed vitamins so dramatic, that people spend billions of dollars every year in the belief that vitamin pills will cure a host of ailments (see Highlight 10). Vitamins certainly support sound nutritional health, but they do not cure all ills. Furthermore, vitamin supplements do not offer the many benefits that come from vitamin-rich foods.

The *presence* of the vitamins also attests to their power. The B vitamin folate helps to prevent birth defects. Vitamin C seems to protect against certain types of cancer. Similarly, vitamin E seems to help protect against some facets of cardiovascular disease. As you will see, the vitamins' roles in supporting optimal health extend far beyond preventing deficiency diseases. In fact, some of the credit given to low-fat diets in preventing disease actually belongs to the vitamins found in vegetables, fruits, and whole grains (see Highlight 11 for more on vitamins in disease prevention).

The vitamins differ from carbohydrates, fats, and proteins in the following ways:

- *Structure.* Vitamins are individual units; they are not linked together (as are molecules of glucose or amino acids). Appendix C presents the chemical structure for each of the vitamins.

- *Function.* Vitamins do not yield usable energy when broken down; they assist the enzymes that release energy from carbohydrates, fats, and proteins.

- *Food contents.* The amounts of vitamins people ingest daily from foods and the amounts they require are measured in *micrograms* (µg) or *milligrams* (mg), rather than grams (g).◆

◆ Reminder: The *vitamins* are organic, essential nutrients required in tiny amounts to perform specific functions that promote growth, reproduction, or the maintenance of health and life.
- **vita** = life
- **amine** = containing nitrogen (the first vitamins discovered contained nitrogen)

◆ 1 g = 1000 mg
1 mg = 1000 µg
For perspective, a dollar bill weighs about 1 g.

Polara Studios, Inc.

To minimize vitamin losses, wrap cut fruits and vegetables or store them in airtight containers.

The vitamins are similar to the energy-yielding nutrients, though, in that they are vital to life, organic, and available from foods.

Bioavailability The amount of vitamins available from foods depends not only on the quantity provided by a food but also on the amount absorbed and used by the body—referred to as the vitamins' **bioavailability.** The quantity of vitamins in a food can be determined relatively easily. Researchers analyze foods to determine their vitamin contents and publish the results in tables of food composition such as Appendix H. Determining the bioavailability of a vitamin is a more complex task because it depends on many factors, including:

- Efficiency of digestion and time of transit through the GI tract
- Previous nutrient intake and nutrition status
- Other foods consumed at the same time (Chapters 10–13 describe factors that inhibit or enhance the absorption of individual vitamins and minerals.)
- Method of food preparation (raw, cooked, or processed)
- Source of the nutrient (synthetic, fortified, or naturally occurring)

Experts consider these factors when estimating recommended intakes.

Precursors Some of the vitamins are available from foods in inactive forms known as **precursors,** or provitamins. Once inside the body, the precursor is converted to an active form of the vitamin. Thus, in measuring a person's vitamin intake, it is important to count both the amount of the active vitamin and the potential amount available from its precursors. The discussions and summary tables throughout this chapter and the next indicate which vitamins have precursors.

Organic Nature Being organic, vitamins can be destroyed and left unable to perform their duties. Therefore, they must be handled with care during storage and in cooking. Prolonged heating may destroy much of the thiamin in food. Because riboflavin can be destroyed by the ultraviolet rays of the sun or by fluorescent light, foods stored in transparent glass containers are most likely to lose riboflavin. Oxygen destroys vitamin C, so losses occur when foods are cut, processed, and stored; these losses may be enough to reduce its action in the body.[1] Table 10-1 summarizes ways to minimize nutrient losses in the kitchen.

Solubility As you may recall, carbohydrates and proteins are hydrophilic and lipids are hydrophobic. The vitamins divide along the same lines—the hydrophilic, water-soluble ones ◆ are the eight B vitamins and vitamin C; the hydrophobic, fat-soluble ones are vitamins A, D, E, and K. As each vitamin was discovered, it was given a name and sometimes a letter and number as well. Many of the water-soluble vitamins have multiple names, which has led to some confusion. The margin lists the standard names, and summary tables throughout this chapter provide the common alternative names.

Solubility is apparent in the food sources of the different vitamins, and it affects their absorption, transport, storage, and excretion by the body. The water-soluble vitamins are found in the watery compartments of foods; the fat-soluble vitamins

◆ **Water-soluble vitamins:**
- B vitamins:
 Thiamin
 Riboflavin
 Niacin
 Biotin
 Pantothenic acid
 Vitamin B_6
 Folate
 Vitamin B_{12}
- Vitamin C

Fat-soluble vitamins:
- Vitamin A
- Vitamin D
- Vitamin E
- Vitamin K

bioavailability: the rate at and the extent to which a nutrient is absorbed and used.

precursors: substances that precede others; with regard to vitamins, compounds that can be converted into active vitamins; also known as **provitamins.**

TABLE 10-1	Minimizing Nutrient Losses

- To slow the degradation of vitamins, refrigerate (most) fruits and vegetables.
- To minimize the oxidation of vitamins, store fruits and vegetables that have been cut in airtight wrappers, and store juices that have been opened in closed containers (and refrigerate them).
- To prevent losses during washing, rinse fruits and vegetables before cutting.
- To minimize losses during cooking, use a microwave oven or steam vegetables in a small amount of water. Add vegetables after water has come to a boil. Use the cooking water in mixed dishes such as casseroles and soups. Avoid high temperatures and long cooking times.

usually occur together in the fats and oils of foods. On being absorbed, the water-soluble vitamins move directly into the blood. Like fats, however, the fat-soluble vitamins must first enter the lymph, then the blood. Once in the blood, many of the water-soluble vitamins travel freely, whereas many of the fat-soluble vitamins require protein carriers for transport. Upon reaching the cells, water-soluble vitamins freely circulate in the water-filled compartments of the body, but fat-soluble vitamins are held in fatty tissues and the liver until needed. The kidneys, monitoring the blood that flows through them, detect and remove small excesses of water-soluble vitamins (large excesses, however, may overwhelm the system, creating adverse effects). Fat-soluble vitamins tend to remain in fat-storage sites in the body rather than being excreted, and so are more likely to reach toxic levels when consumed in excess.

Because the body stores fat-soluble vitamins, they can be eaten in large amounts once in a while and still meet the body's needs over time. Water-soluble vitamins are retained for varying periods in the body. Although a single day's omission from the diet does not bring on a deficiency, the water-soluble vitamins must still be eaten more regularly than the fat-soluble vitamins.

Toxicity Knowledge about some of the amazing roles of vitamins has prompted many people to assume that "more is better" and take vitamin supplements. But just as an inadequate intake can cause harm, so can an excessive intake. Even some of the water-soluble vitamins have adverse effects when taken in large doses.

That a vitamin can be both essential and harmful may seem surprising, but the same is true of most nutrients. The effects of every substance depend on its dose, and this is one reason consumers should not self-prescribe supplements for their ailments. See the "How to" below for a perspective on doses.

The Committee on Dietary Reference Intakes (DRI) addresses the possibility of adverse effects from high doses of nutrients by establishing Tolerable Upper Intake Levels. An Upper Level defines the highest amount of a nutrient that is likely not to cause harm for most healthy people when consumed daily. The risk of harm increases as intakes rise above the Upper Level. Of the nutrients discussed in this chapter, niacin, vitamin B_6, folate, choline, and vitamin C have Upper Levels, and these values are presented in their respective summary tables. Data are lacking to establish Upper Levels for the remaining B vitamins, but this does not mean that

---

HOW TO Understand Dose Levels and Effects

A substance may have a beneficial or harmful effect, but a critical thinker would not conclude that the substance itself was beneficial or harmful without first asking what dose was used. The accompanying figure shows three possible relationships between dose levels and effects. The third diagram represents the situation with nutrients—more is better up to a point, but beyond that point, still more can be harmful.

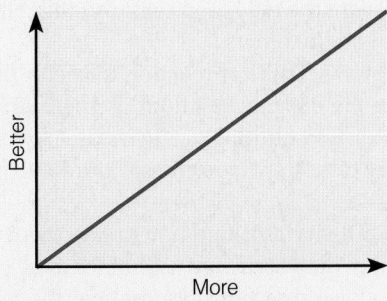

As you progress in the direction of more, the effect gets better and better, with no end in sight (real life is seldom, if ever, like this).

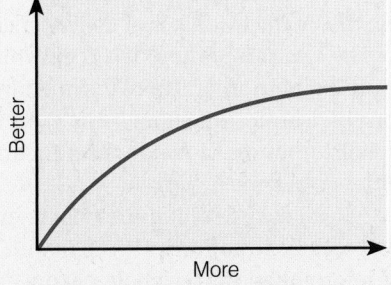

As you progress in the direction of more, the effect reaches a maximum and then a plateau, becoming no better with higher doses.

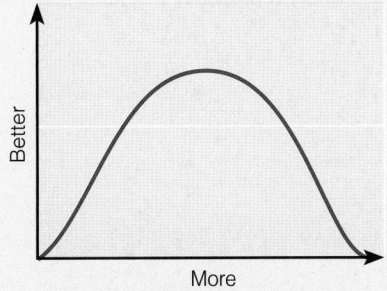

As you progress in the direction of more, the effect reaches an optimum at some intermediate dose and then declines, showing that more is better up to a point and then harmful. That too much can be as harmful as too little represents the situation with most nutrients.

excessively high intakes would be without risk. (The inside front cover pages present Upper Levels for the vitamins and minerals.)

IN SUMMARY

The vitamins are essential nutrients needed in tiny amounts in the diet both to prevent deficiency diseases and to support optimal health. The water-soluble vitamins are the B vitamins and vitamin C; the fat-soluble vitamins are vitamins A, D, E, and K. The accompanying table summarizes the differences between the water-soluble and fat-soluble vitamins.

	Water-Soluble Vitamins: B Vitamins and Vitamin C	Fat-Soluble Vitamins: Vitamins A, D, E, and K
Absorption	Directly into the blood	First into the lymph, then the blood
Transport	Travel freely	Many require protein carriers
Storage	Circulate freely in water-filled parts of the body	Stored in the cells associated with fat
Excretion	Kidneys detect and remove excess in urine	Less readily excreted; tend to remain in fat-storage sites
Toxicity	Possible to reach toxic levels when consumed from supplements	Likely to reach toxic levels when consumed from supplements
Requirements	Needed in frequent doses (perhaps 1 to 3 days)	Needed in periodic doses (perhaps weeks or even months)

NOTE: Exceptions occur, but these differences between the water-soluble and fat-soluble vitamins are valid generalizations.

The discussion of B vitamins that follows begins with a brief description of each of them, then offers a look at the ways they work together. Thus, a preview of the individual vitamins is followed by a survey of how they work together, in concert.

The B Vitamins—As Individuals

Despite supplement advertisements that claim otherwise, the vitamins do not provide the body with fuel for energy. It is true, though, that without B vitamins the body would lack energy. The energy-yielding nutrients—carbohydrate, fat, and protein—are used for fuel; the B vitamins help the body to use that fuel. Several of the B vitamins—thiamin, riboflavin, niacin, pantothenic acid, and biotin—form part of the coenzymes ◆ that assist certain enzymes in the release of energy from carbohydrate, fat, and protein. Other B vitamins play other indispensable roles in metabolism. Vitamin B_6 assists enzymes that metabolize amino acids; folate and vitamin B_{12} help cells to multiply. Among these cells are the red blood cells and the cells lining the GI tract—cells that deliver energy to all the others.

The vitamin portion of a coenzyme allows a chemical reaction to occur; the remaining portion of the coenzyme binds to the enzyme. Without its coenzyme, an enzyme cannot function. Thus symptoms of B vitamin deficiencies directly reflect the disturbances of metabolism incurred by a lack of coenzymes. Figure 10-1 illustrates coenzyme action.

The following sections describe individual B vitamins and note many coenzymes and metabolic pathways. Keep in mind that a later discussion assembles these pieces of information into a whole picture. The following sections also present the recommendations, deficiency and toxicity symptoms, and food sources for each vitamin. The recommendations for the B vitamins and vitamin C reflect the 1998 and 2000 DRI, respectively.[2] For thiamin, riboflavin, niacin, vitamin B_6, folate, vitamin B_{12} and vitamin C, sufficient data were available to establish an RDA; for biotin, pantothenic acid, and choline, an Adequate Intake (AI) was set; only niacin, vitamin B_6, folate, choline, and vitamin C have Tolerable Upper Intake Levels. These values appear in the summary tables and figures that follow and on the pages of the inside front cover.

◆ Reminder: A *coenzyme* is a small organic molecule that associates closely with certain enzymes; many B vitamins form an integral part of coenzymes.

FIGURE 10-1 *Animated!* Coenzyme Action

Some vitamins form part of the coenzymes that enable enzymes either to synthesize compounds (as illustrated by the lower enzymes in this figure) or to dismantle compounds (as illustrated by the upper enzymes).

CENGAGENOW™
To test your understanding of these concepts, log on to **academic.cengage.com/login.**

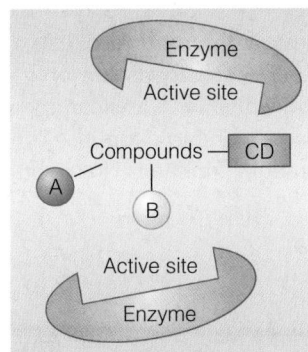
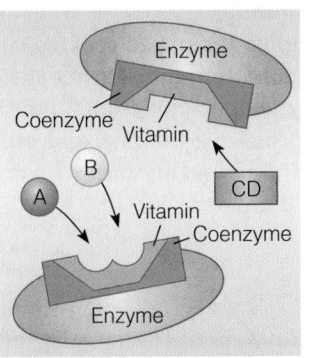
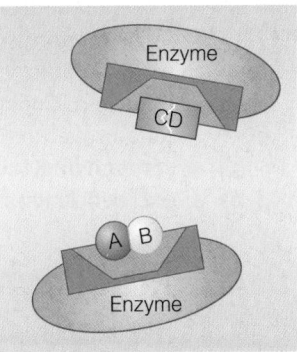
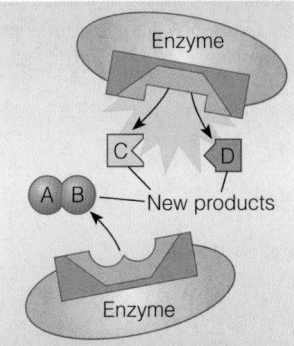

Without coenzymes, compounds A, B, and CD don't respond to their enzymes.

With the coenzymes in place, compounds are attracted to their sites on the enzymes . . .

. . . and the reactions proceed instantaneously. The coenzymes often donate or accept electrons, atoms, or groups of atoms.

The reactions are completed with either the formation of a new product, AB, or the breaking apart of a compound into two new products, C and D, and the release of energy.

Thiamin

Thiamin is the vitamin part of the coenzyme TPP (thiamin pyrophosphate), which assists in energy metabolism. The TPP coenzyme participates in the conversion of pyruvate to acetyl CoA (described in Chapter 7). The reaction removes one carbon from the 3-carbon pyruvate to make the 2-carbon acetyl CoA and carbon dioxide (CO_2). Later, TPP participates in a similar step in the TCA cycle where it helps convert a 5-carbon compound to a 4-carbon compound. Besides playing these pivotal roles in the energy metabolism of all cells, thiamin occupies a special site on the membranes of nerve cells. Consequently, processes in nerves and in their responding tissues, the muscles, depend heavily on thiamin.

Thiamin Recommendations Dietary recommendations are based primarily on thiamin's role in enzyme activity. Generally, thiamin needs will be met if a person eats enough food to meet energy needs—if that energy comes from nutritious foods. The average thiamin intake in the United States and Canada meets or exceeds recommendations.

Thiamin Deficiency and Toxicity People who fail to eat enough food to meet energy needs risk nutrient deficiencies, including thiamin deficiency. Inadequate thiamin intakes have been reported among the nation's malnourished and homeless people. Similarly, people who derive most of their energy from empty-kcalorie items risk thiamin deficiency. Alcohol ◆ is a good example. It contributes energy but provides few, if any, nutrients and often displaces food. In addition, alcohol impairs thiamin absorption and enhances thiamin excretion in the urine, doubling the risk of deficiency. An estimated four out of five alcoholics are thiamin deficient.

Prolonged thiamin deficiency can result in the disease **beriberi,** which was first observed in Indonesia when the custom of polishing rice became widespread.[3] Rice provided 80 percent of the energy intake of the people of that area, and the germ and bran of the rice grain was their principal source of thiamin. When the germ and bran were removed in the preparation of white rice, beriberi spread like wildfire. The symptoms of beriberi include damage to the nervous system as well as to the heart and other muscles. Figure 10-2 presents one of the symptoms of beriberi. No adverse effects have been associated with excesses of thiamin; no Upper Level has been determined.

◆ Severe thiamin deficiency in alcohol abusers is called the **Wernicke-Korsakoff** (VER-nee-key KORE-sah-kof) **syndrome.** Symptoms include disorientation, loss of short-term memory, jerky eye movements, and staggering gait.

thiamin (THIGH-ah-min): a B vitamin. The coenzyme form is **TPP (thiamin pyrophosphate).**

beriberi: the thiamin-deficiency disease.
- **beri** = weakness
- **beriberi** = "I can't, I can't"

FIGURE 10-2 Thiamin-Deficiency Symptom—The Edema of Beriberi

Beriberi may be characterized as "wet" (referring to edema) or "dry" (with muscle wasting, but no edema). Physical examination confirms that this person has wet beriberi. Notice how the impression of the physician's thumb remains on the leg.

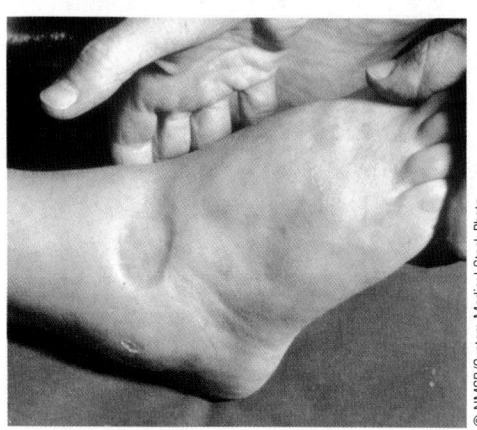

© NMSB/Custom Medical Stock Photo

Thiamin Food Sources Before examining Figure 10-3, you may want to read the accompanying "How to," which describes the many features found in this and similar figures in this chapter and the next three chapters. When you look at Figure 10-3, notice that thiamin occurs in small quantities in many nutritious foods. The long red bar near the bottom of the graph shows that meats in the pork family are exceptionally rich in thiamin. Yellow bars confirm that enriched grains are a reliable source of thiamin.

As mentioned earlier, prolonged cooking can destroy thiamin. Also, like other water-soluble vitamins, thiamin leaches into water when foods are boiled or blanched. Cooking methods that require little or no water such as steaming and microwave heating conserve thiamin and other water-soluble vitamins. The accompanying table (p. 329) summarizes thiamin's main functions, food sources, and deficiency symptoms.

Riboflavin

Like thiamin, **riboflavin** serves as a coenzyme in many reactions, most notably in the release of energy from nutrients in all body cells. The coenzyme forms of riboflavin are FMN (flavin mononucleotide) and FAD (flavin adenine dinucleotide); both can accept and then donate two hydrogens (see Figure 10-4, p. 330). During energy metabolism, FAD picks up two hydrogens (with their electrons) from the TCA cycle and delivers them to the electron transport chain (described in Chapter 7).

FIGURE 10-3 Thiamin in Selected Foods

See the "How to" section on the next page for more information on using this figure.

Food	Serving size (kcalories)
Bread, whole wheat	1 oz slice (70 kcal)
Cornflakes, fortified	1 oz (110 kcal)
Spaghetti pasta	½ c cooked (99 kcal)
Tortilla, flour	1 10"-round (234 kcal)
Broccoli	½ c cooked (22 kcal)
Carrots	½ c shredded raw (24 kcal)
Potato	1 medium baked w/skin (133 kcal)
Tomato juice	¾ c (31 kcal)
Banana	1 medium raw (109 kcal)
Orange	1 medium raw (62 kcal)
Strawberries	½ c fresh (22 kcal)
Watermelon	1 slice (92 kcal)
Milk	1 c reduced-fat 2% (121 kcal)
Yogurt, plain	1 c low-fat (155 kcal)
Cheddar cheese	1½ oz (171 kcal)
Cottage cheese	½ c low-fat 2% (101 kcal)
Pinto beans	½ c cooked (117 kcal)
Peanut butter	2 tbs (188 kcal)
Sunflower seeds	1 oz dry (165 kcal)
Tofu (soybean curd)	½ c (76 kcal)
Ground beef, lean	3 oz broiled (244 kcal)
Chicken breast	3 oz roasted (140 kcal)
Tuna, canned in water	3 oz (99 kcal)
Egg	1 hard cooked (78 kcal)
Excellent, and sometimes unusual, sources:	
Pork chop, lean	3 oz broiled (169 kcal)
Soy milk	1 c (81 kcal)
Squash, acorn	½ c baked (69 kcal)

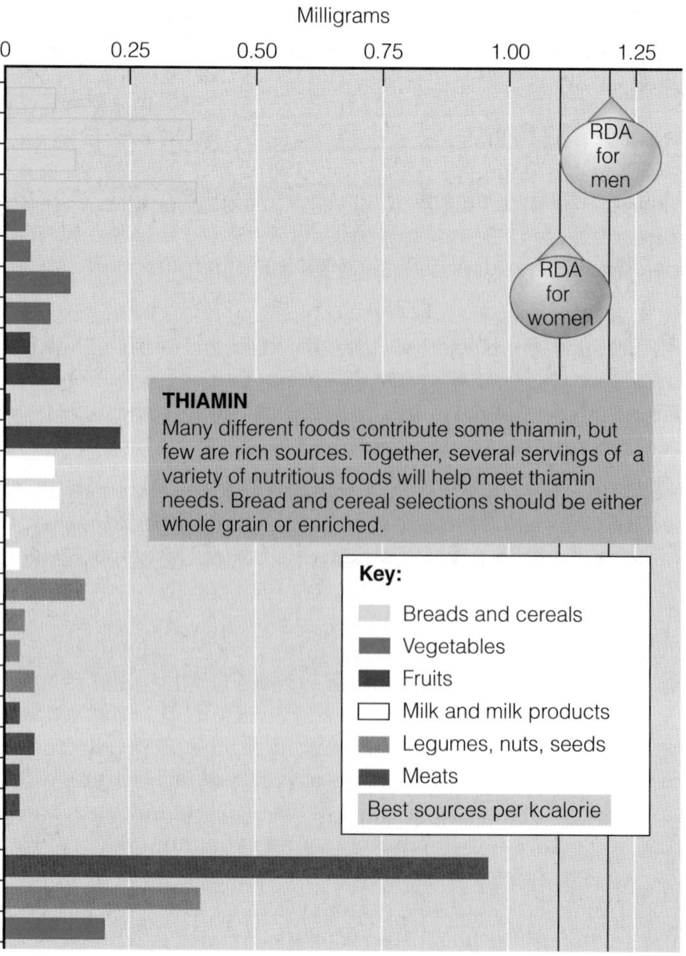

Milligrams: 0, 0.25, 0.50, 0.75, 1.00, 1.25

RDA for men

RDA for women

THIAMIN
Many different foods contribute some thiamin, but few are rich sources. Together, several servings of a variety of nutritious foods will help meet thiamin needs. Bread and cereal selections should be either whole grain or enriched.

Key:
- Breads and cereals
- Vegetables
- Fruits
- Milk and milk products
- Legumes, nuts, seeds
- Meats
- Best sources per kcalorie

HOW TO | Evaluate Foods for Their Nutrient Contributions

Figure 10-3 is the first of a series of figures in this and the next three chapters that present the vitamins and minerals in foods. Each figure presents the same 24 foods, which were selected to ensure a variety of choices representative of each of the food groups as suggested by the USDA Food Guide. For example, a bread, a cereal, and a pasta were chosen from the grain group. The suggestion to include a variety of vegetables was also considered: dark green, leafy vegetables (broccoli); deep orange and yellow vegetables (carrots); starchy vegetables (potatoes); legumes (pinto beans); and other vegetables (tomato juice). The selection of fruits followed suggestions to use whole fruits (bananas); citrus fruits (oranges); melons (watermelon); and berries (strawberries). Items were selected from the milk and meat groups in a similar way. In addition to the 24 foods that appear in all of the figures, three different foods

were selected for each of the nutrients to add variety and often reflect excellent, and sometimes unusual, sources.

Notice that the figures list the food, the serving size, and the food energy (kcalories) on the left. The amount of the nutrient per serving is presented in the graph on the right along with the RDA (or AI) for adults, so you can see how many servings would be needed to meet recommendations.

The colored bars show at a glance which food groups best provide a nutrient: yellow for breads and cereals; green for vegetables; purple for fruits; white for milk and milk products; brown for legumes; and red for meat, fish, and poultry. Because the USDA Food Guide mentions legumes with both the meat group and the vegetable group and because legumes are especially rich in many vitamins and minerals, they have been given their own color to high-

light their nutrient contributions.

Notice how the bar graphs shift in the various figures. Careful study of all of the figures taken together will confirm that variety is the key to nutrient adequacy.

Another way to evaluate foods for their nutrient contributions is to consider their nutrient density (their thiamin *per 100 kcalories*, for example). Quite often, vegetables rank higher on a nutrient-per-kcalorie list than they do on a nutrient-per-serving list (see p. 38 to review how to evaluate foods based on nutrient density). The left column in the figure highlights about five foods that offer the best deal for your energy "dollar" (the kcalorie). Notice how many of them are vegetables.

Realistically, people cannot eat for single nutrients. Fortunately, most foods deliver more than one nutrient, allowing people to combine foods into nourishing meals.

IN SUMMARY | Thiamin

Other Names

Vitamin B₁

RDA

Men: 1.2 mg/day

Women: 1.1 mg/day

Chief Functions in the Body

Part of coenzyme TPP (thiamin pyrophosphate) used in energy metabolism

Significant Sources

Whole-grain, fortified, or enriched grain products; moderate amounts in all nutritious food; pork

Easily destroyed by heat

Deficiency Disease

Beriberi (wet, with edema; dry, with muscle wasting)

Deficiency Symptomsᵃ

Enlarged heart, cardiac failure; muscular weakness; apathy, poor short-term memory, confusion, irritability; anorexia, weight loss

Toxicity Symptoms

None reported

ᵃSevere thiamin deficiency is often related to heavy alcohol consumption with limited food consumption (Wernicke-Korsakoff syndrome).

© Polara Studios Inc.

Pork is the richest source of thiamin, but enriched or whole-grain products typically make the greatest contribution to a day's intake because of the quantities eaten. Legumes such as split peas are also valuable sources of thiamin.

Riboflavin Recommendations Like thiamin's RDA, riboflavin's RDA is based primarily on its role in enzyme activity. Most people in the United States and Canada meet or exceed riboflavin recommendations.

Riboflavin Deficiency and Toxicity Riboflavin deficiency ◆ most often accompanies other nutrient deficiencies. Lack of the vitamin causes inflammation of the membranes of the mouth, skin, eyes, and GI tract. Excesses of riboflavin appear to cause no harm; no Upper Level has been established.

Riboflavin Food Sources The greatest contributions of riboflavin come from milk and milk products (see Figure 10-5, p. 331). Whole-grain or enriched bread and cereal products are also valuable sources because of the quantities typically consumed.

◆ Riboflavin deficiency is called **ariboflavinosis** (ay-RYE-boh-FLAY-vin-oh-sis).
- **a** = not
- **osis** = condition

riboflavin (RYE-boh-flay-vin): a B vitamin. The coenzyme forms are **FMN (flavin mononucleotide)** and **FAD (flavin adenine dinucleotide)**.

FIGURE 10-4 Riboflavin Coenzyme, Accepting and Donating Hydrogens

This figure shows the chemical structure of the riboflavin portion of the coenzyme only; the remainder of the coenzyme structure is represented by dotted lines (see Appendix C for the complete chemical structures of FAD and FMN). The reactive sites that accept and donate hydrogens are highlighted in white.

FAD

FADH$_2$

During the TCA cycle, compounds release hydrogens, and the riboflavin coenzyme FAD picks up two of them. As it accepts two hydrogens, FAD becomes FADH$_2$.

FADH$_2$ carries the hydrogens to the electron transport chain. At the end of the electron transport chain, the hydrogens are accepted by oxygen, creating water, and FADH$_2$ becomes FAD again. For every FADH$_2$ that passes through the electron transport chain, 2 ATP are generated.

◆ Turn to p. 38 for a review of how to evaluate foods based on nutrient density (per kcalorie).

When riboflavin sources are ranked by nutrient density (per kcalorie), ◆ many dark green, leafy vegetables (such as broccoli, turnip greens, asparagus, and spinach) appear high on the list. Vegans and others who don't use milk must rely on ample servings of dark greens and enriched grains for riboflavin. Nutritional yeast is another good source.

Ultraviolet light and irradiation destroy riboflavin. For these reasons, milk is sold in cardboard or opaque plastic containers, and precautions are taken when vitamin D is added to milk by irradiation.* In contrast, riboflavin is stable to heat, so cooking does not destroy it. The following summary table lists riboflavin's chief functions, food sources, and deficiency symptoms.

---

* Vitamin D can be added to milk by feeding cows irradiated yeast or by irradiating the milk itself.

All of these foods are rich in riboflavin, but milk and milk products provide much of the riboflavin in the diets of most people.

IN SUMMARY Riboflavin

Other Names

Vitamin B$_2$

RDA

Men: 1.3 mg/day

Women: 1.1 mg/day

Chief Functions in the Body

Part of coenzymes FMN (flavin mononucleotide) and FAD (flavin adenine dinucleotide) used in energy metabolism

Significant Sources

Milk products (yogurt, cheese); whole-grain, fortified, or enriched grain products; liver

Easily destroyed by ultraviolet light and irradiation

Deficiency Disease

Ariboflavinosis (ay-RYE-boh-FLAY-vin-oh-sis)

Deficiency Symptoms

Sore throat; cracks and redness at corners of mouth;[a] painful, smooth, purplish red tongue;[b] inflammation characterized by skin lesions covered with greasy scales

Toxicity Symptoms

None reported

[a]Cracks at the corners of the mouth are called *angular stomatitis* or *cheilosis* (kye-LOH-sis or kee-LOH-sis).
[b]Smoothness of the tongue is caused by loss of its surface structures and is termed *glossitis* (gloss-EYE-tis).

© Polara Studios Inc.

FIGURE 10-5   Riboflavin in Selected Foods

See the "How to" section on p. 329 for more information on using this figure.

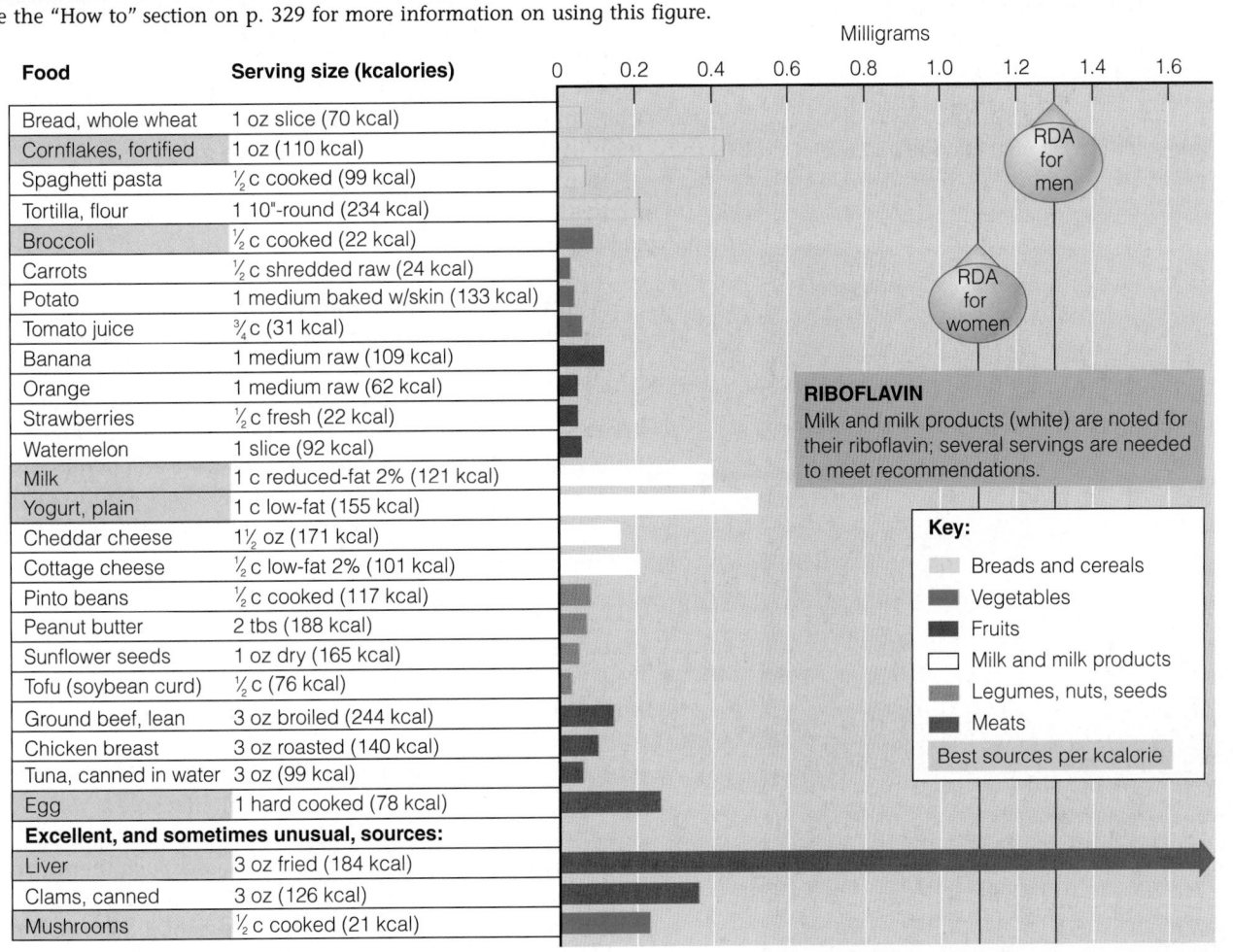

Food	Serving size (kcalories)
Bread, whole wheat	1 oz slice (70 kcal)
Cornflakes, fortified	1 oz (110 kcal)
Spaghetti pasta	½ c cooked (99 kcal)
Tortilla, flour	1 10"-round (234 kcal)
Broccoli	½ c cooked (22 kcal)
Carrots	½ c shredded raw (24 kcal)
Potato	1 medium baked w/skin (133 kcal)
Tomato juice	¾ c (31 kcal)
Banana	1 medium raw (109 kcal)
Orange	1 medium raw (62 kcal)
Strawberries	½ c fresh (22 kcal)
Watermelon	1 slice (92 kcal)
Milk	1 c reduced-fat 2% (121 kcal)
Yogurt, plain	1 c low-fat (155 kcal)
Cheddar cheese	1½ oz (171 kcal)
Cottage cheese	½ c low-fat 2% (101 kcal)
Pinto beans	½ c cooked (117 kcal)
Peanut butter	2 tbs (188 kcal)
Sunflower seeds	1 oz dry (165 kcal)
Tofu (soybean curd)	½ c (76 kcal)
Ground beef, lean	3 oz broiled (244 kcal)
Chicken breast	3 oz roasted (140 kcal)
Tuna, canned in water	3 oz (99 kcal)
Egg	1 hard cooked (78 kcal)
Excellent, and sometimes unusual, sources:	
Liver	3 oz fried (184 kcal)
Clams, canned	3 oz (126 kcal)
Mushrooms	½ c cooked (21 kcal)

RIBOFLAVIN
Milk and milk products (white) are noted for their riboflavin; several servings are needed to meet recommendations.

Key:
- Breads and cereals
- Vegetables
- Fruits
- Milk and milk products
- Legumes, nuts, seeds
- Meats
- Best sources per kcalorie

Niacin

The name **niacin** describes two chemical structures: nicotinic acid and nicotinamide (also known as niacinamide). The body can easily convert nicotinic acid to nicotinamide, which is the major form of niacin in the blood.

The two coenzyme forms of niacin, NAD (nicotinamide adenine dinucleotide) and NADP (the phosphate form), participate in numerous metabolic reactions. They are central in energy-transfer reactions, especially the metabolism of glucose, fat, and alcohol. NAD is similar to the riboflavin coenzymes in that it carries hydrogens (and their electrons) during metabolic reactions, including the pathway from the TCA cycle to the electron transport chain.

Niacin Recommendations   Niacin is unique among the B vitamins in that the body can make it from the amino acid tryptophan. To make 1 milligram of niacin requires approximately 60 milligrams of dietary tryptophan. For this reason, recommended intakes are stated in **niacin equivalents (NE).** ◆ A food containing 1 milligram of niacin and 60 milligrams of tryptophan provides the equivalent of 2 milligrams of niacin, or 2 niacin equivalents. The RDA for niacin allows for this conversion and is stated in niacin equivalents; average niacin intakes in the United States and Canada exceed recommendations.

Niacin Deficiency   The niacin-deficiency disease, **pellagra**, produces the symptoms of diarrhea, dermatitis, dementia, and eventually death (often called "the four Ds").

◆ 1 NE = 1 mg niacin or 60 mg tryptophan

niacin (NIGH-a-sin): a B vitamin. The coenzyme forms are **NAD (nicotinamide adenine dinucleotide)** and **NADP** (the phosphate form of NAD). Niacin can be eaten preformed or made in the body from its precursor, tryptophan, one of the amino acids.

niacin equivalents (NE): the amount of niacin present in food, including the niacin that can theoretically be made from its precursor, tryptophan, present in the food.

pellagra (pell-AY-gra): the niacin-deficiency disease.
- **pellis** = skin
- **agra** = rough

FIGURE 10-6 Niacin-Deficiency Symptom—The Dermatitis of Pellagra

In the dermatitis of pellagra, the skin darkens and flakes away as if it were sunburned. The protein-deficiency disease kwashiorkor also produces a "flaky paint" dermatitis, but the two are easily distinguished. The dermatitis of pellagra is bilateral and symmetrical and occurs only on those parts of the body exposed to the sun.

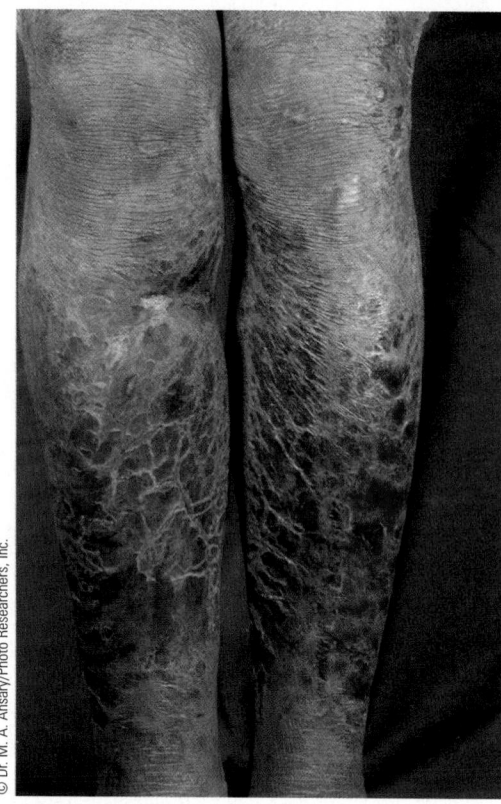

© Dr. M. A. Ansary/Photo Researchers, Inc.

◆ When a normal dose of a nutrient (levels commonly found in foods) provides a normal blood concentration, the nutrient is having a *physiological* effect. When a large dose (levels commonly available only from supplements) overwhelms some body system and acts like a drug, the nutrient is having a *pharmacological* effect.
• **physio** = natural
• **pharma** = drug

niacin flush: a temporary burning, tingling, and itching sensation that occurs when a person takes a large dose of nicotinic acid; often accompanied by a headache and reddened face, arms, and chest.

In the early 1900s, pellagra caused widespread misery and some 87,000 deaths in the U.S. South, where many people subsisted on a low-protein diet centered on corn. This diet supplied neither enough niacin nor enough tryptophan. At least 70 percent of the niacin in corn is bound to complex carbohydrates and small peptides, making it unavailable for absorption. Furthermore, corn is high in the amino acid leucine, which interferes with the tryptophan-to-niacin conversion, thus further contributing to the development of pellagra. Figure 10-6 illustrates the dermatitis of pellagra.

Pellagra was originally believed to be caused by an infection. Medical researchers spent many years and much effort searching for infectious microbes until they realized that the problem was not what was *present* in the food but what was *absent* from it. That a disease such as pellagra could be caused by diet—and not by germs—was a groundbreaking discovery. It contradicted commonly held medical opinions that diseases were caused only by infectious agents. By carefully following the scientific method (as described in Chapter 1), researchers advanced the science of nutrition dramatically.*

Niacin Toxicity Naturally occurring niacin from foods ◆ causes no harm, but large doses from supplements or drugs produce a variety of adverse effects, most notably **"niacin flush."** Niacin flush occurs when nicotinic acid is taken in doses only three to four times the RDA. It dilates the capillaries and causes a tingling sensation that can be painful. The nicotinamide form does not produce this effect—nor does it lower blood cholesterol.

Large doses of nicotinic acid have been used to help lower blood cholesterol and prevent heart disease. Such therapy must be closely monitored. People with the following conditions may be particularly susceptible to the toxic effects of niacin: liver disease, diabetes, peptic ulcers, gout, irregular heartbeats, inflammatory bowel disease, migraine headaches, and alcoholism.

Niacin Food Sources Tables of food composition typically list preformed niacin only, but as mentioned, niacin can also be made in the body from the amino acid tryptophan. Dietary tryptophan could meet about half the daily niacin need for most people, but the average diet easily supplies enough preformed niacin. The "How to" on p. 333 shows how to estimate the total amount of niacin available from both tryptophan and preformed niacin in the diet.

Figure 10-7 (p. 334) presents niacin in selected foods. Meat, poultry, legumes, and enriched and whole grains contribute about half the niacin people consume. Mushrooms, potatoes, and tomatoes are among the richest vegetable sources, and they can provide abundant niacin when eaten in generous amounts.

Niacin is less vulnerable to losses during food preparation and storage than other water-soluble vitamins. Being fairly heat-resistant, niacin can withstand reasonable cooking times, but like other water-soluble vitamins, it will leach into cooking water. The summary table includes food sources as well as niacin's various names, functions, and deficiency and toxicity symptoms.

---

* Dr. Joseph Goldberger, a physician for the U.S. government, headed the investigations that determined that pellagra was a dietary disorder, not an infectious disease. He died several years before Conrad Elevjhem discovered that a deficiency of niacin caused pellagra.

HOW TO Estimate Niacin Equivalents

To estimate niacin equivalents:

- Calculate total protein consumed (grams).
- Assuming that the RDA amount of protein will be used first to make body protein, subtract the RDA to obtain "leftover" protein available to make niacin (grams). (Actually, the RDA provides a generous protein allowance, so "leftover" protein may be even greater than this.)
- About 1 gram of every 100 grams of high-quality protein is tryptophan, so divide by 100 to obtain the tryptophan in this leftover protein (grams).
- Multiply by 1000 to express this amount of tryptophan in milligrams.
- Divide by 60 to get niacin equivalents (milligrams).
- Finally, add the amount of preformed niacin obtained in the diet (milligrams).

For example, suppose that a 19-year-old woman who weighs 130 pounds consumes 75 grams of protein in a day. To calculate her protein RDA, first convert pounds to kilo-grams if necessary, and then multiply by 0.8 g/kg:

$$130 \text{ lb} \div 2.2 \text{ lb/kg} = 59 \text{ kg}$$
$$59 \text{ kg} \times 0.8 \text{ g/kg} = 47 \text{ g}$$

Then determine her leftover protein by subtracting her RDA from her intake:

$$75 \text{ g protein intake} - 47 \text{ g protein RDA} = 28 \text{ g protein leftover}$$

Next calculate the amount of tryptophan in this leftover protein:

$$28 \text{ g protein} \div 100 = 0.28 \text{ g tryptophan}$$
$$0.28 \text{ g tryptophan} \times 1000 = 280 \text{ mg tryptophan}$$

Then convert milligrams of tryptophan to niacin equivalents:

$$280 \text{ mg tryptophan} \div 60 = 4.7 \text{ mg NE}$$

To determine the total amount of niacin available from the diet, add the amount available from tryptophan (4.7 mg NE) to the amount of preformed niacin obtained from the diet.

CENGAGENOW
To practice estimating niacin requirements, log on to **academic.cengage.com/login**, go to Chapter 10, then go to How To.

IN SUMMARY Niacin

Other Names

Nicotinic acid, nicotinamide, niacinamide, vitamin B_3; precursor is dietary tryptophan (an amino acid)

RDA

Men: 16 mg NE/day

Women: 14 mg NE/day

Upper Level

Adults: 35 mg/day

Chief Functions in the Body

Part of coenzymes NAD (nicotinamide adenine dinucleotide) and NADP (its phosphate form) used in energy metabolism

Significant Sources

Milk, eggs, meat, poultry, fish; whole-grain, fortified, and enriched grain products; nuts and all protein-containing foods

Deficiency Disease

Pellagra

Deficiency Symptoms

Diarrhea, abdominal pain, vomiting; inflamed, swollen, smooth, bright red tongue;[a] depression, apathy, fatigue, loss of memory, headache; bilateral symmetrical rash on areas exposed to sunlight

Toxicity Symptoms

Painful flush, hives, and rash ("niacin flush"); nausea and vomiting; liver damage, impaired glucose tolerance

[a]Smoothness of the tongue is caused by loss of its surface structures and is termed *glossitis* (gloss-EYE-tis).

Protein-rich foods such as meat, fish, poultry, and peanut butter contribute much of the niacin in people's diets. Enriched breads and cereals and a few vegetables are also rich in niacin.

© Polara Studios, Inc.

Biotin

Biotin plays an important role in metabolism as a coenzyme that carries activated carbon dioxide. This role is critical in the TCA cycle: biotin delivers a carbon

biotin (BY-oh-tin): a B vitamin that functions as a coenzyme in metabolism.

FIGURE 10-7 Niacin in Selected Foods

See the "How to" section on p. 329 for more information on using this figure.

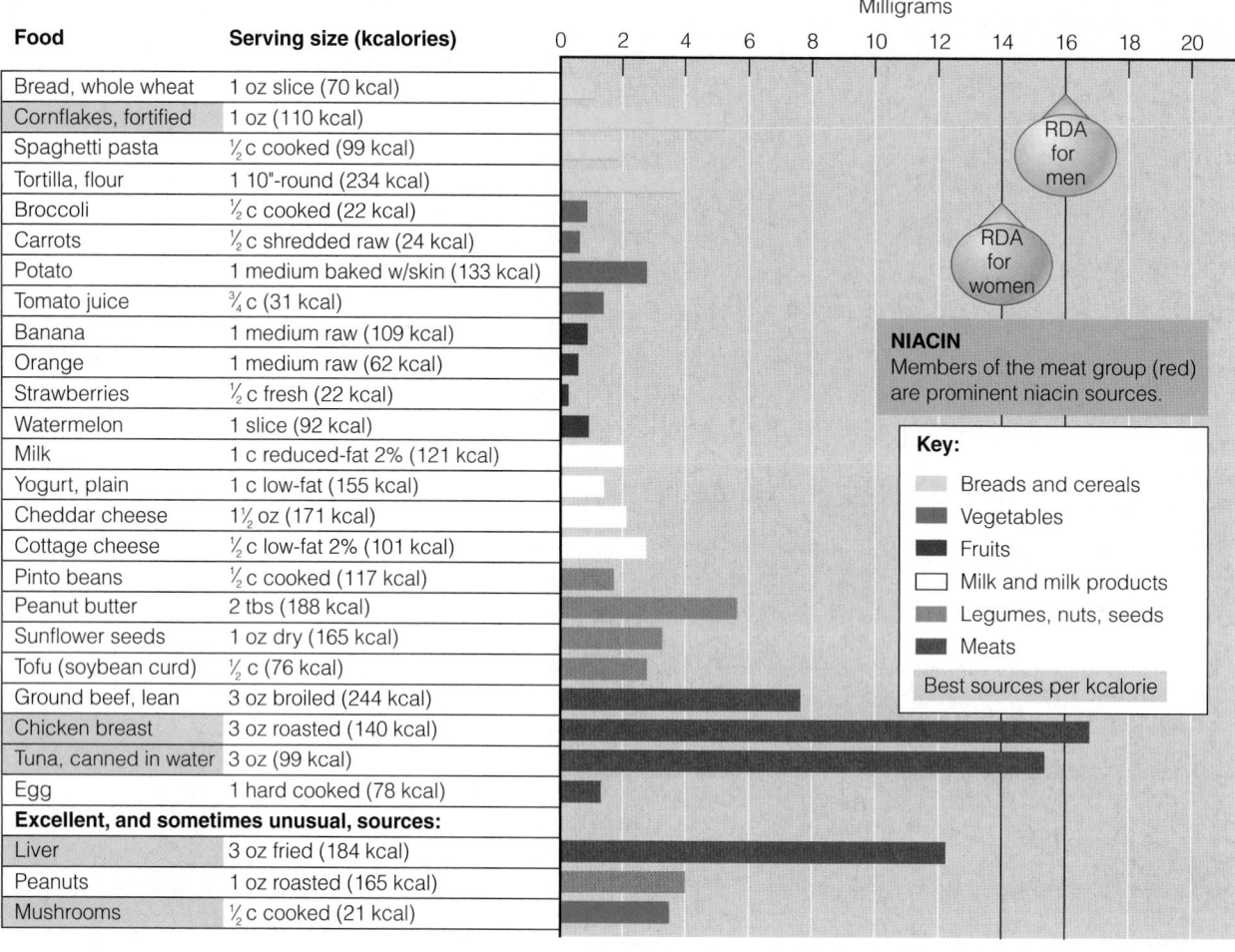

Milligrams

Food	Serving size (kcalories)
Bread, whole wheat	1 oz slice (70 kcal)
Cornflakes, fortified	1 oz (110 kcal)
Spaghetti pasta	½ c cooked (99 kcal)
Tortilla, flour	1 10"-round (234 kcal)
Broccoli	½ c cooked (22 kcal)
Carrots	½ c shredded raw (24 kcal)
Potato	1 medium baked w/skin (133 kcal)
Tomato juice	¾ c (31 kcal)
Banana	1 medium raw (109 kcal)
Orange	1 medium raw (62 kcal)
Strawberries	½ c fresh (22 kcal)
Watermelon	1 slice (92 kcal)
Milk	1 c reduced-fat 2% (121 kcal)
Yogurt, plain	1 c low-fat (155 kcal)
Cheddar cheese	1½ oz (171 kcal)
Cottage cheese	½ c low-fat 2% (101 kcal)
Pinto beans	½ c cooked (117 kcal)
Peanut butter	2 tbs (188 kcal)
Sunflower seeds	1 oz dry (165 kcal)
Tofu (soybean curd)	½ c (76 kcal)
Ground beef, lean	3 oz broiled (244 kcal)
Chicken breast	3 oz roasted (140 kcal)
Tuna, canned in water	3 oz (99 kcal)
Egg	1 hard cooked (78 kcal)
Excellent, and sometimes unusual, sources:	
Liver	3 oz fried (184 kcal)
Peanuts	1 oz roasted (165 kcal)
Mushrooms	½ c cooked (21 kcal)

RDA for men
RDA for women

NIACIN
Members of the meat group (red) are prominent niacin sources.

Key:
- Breads and cereals
- Vegetables
- Fruits
- Milk and milk products
- Legumes, nuts, seeds
- Meats

Best sources per kcalorie

◆ Reminder: *Gluconeogenesis* is the synthesis of glucose from noncarbohydrate sources such as amino acids or glycerol.

◆ The protein **avidin** (AV-eh-din) in egg whites binds biotin.
• **avid** = greedy

to 3-carbon pyruvate, thus replenishing oxaloacetate, the 4-carbon compound needed to combine with acetyl CoA to keep the TCA cycle turning. The biotin coenzyme also participates in gluconeogenesis, ◆ fatty acid synthesis, and the breakdown of certain fatty acids and amino acids. Recent research has uncovered roles for biotin in gene expression.[4]

Biotin Recommendations Biotin is needed in very small amounts. Instead of an RDA, an Adequate Intake (AI) has been determined.

Biotin Deficiency and Toxicity Biotin deficiencies rarely occur. Researchers can induce a biotin deficiency in animals or human beings by feeding them raw egg whites, which contain a protein ◆ that binds biotin and thus prevents its absorption. Biotin-deficiency symptoms include skin rash, hair loss, and neurological impairment. More than two dozen egg whites must be consumed daily for several months to produce these effects, however, and the eggs have to be raw; cooking denatures the binding protein. No adverse effects from high biotin intakes have been reported, but some research indicates that biotin supplementation damages DNA.[5] Biotin does not have an Upper Level.

Biotin Food Sources Biotin is widespread in foods (including egg yolks), so eating a variety of foods protects against deficiencies. Some biotin is also synthesized by GI tract bacteria, but this amount may not contribute much to the biotin absorbed. A review of biotin facts is provided in the summary table.

IN SUMMARY	Biotin
Adequate Intake (AI)	**Deficiency Symptoms**
Adults: 30 µg/day	Depression, lethargy, hallucinations, numb or tingling sensation in the arms and legs; red, scaly rash around the eyes, nose, and mouth; hair loss
Chief Functions in the Body	
Part of a coenzyme used in energy metabolism, fat synthesis, amino acid metabolism, and glycogen synthesis	**Toxicity Symptoms**
	None reported
Significant Sources	
Widespread in foods; liver, egg yolks, soybeans, fish, whole grains; also produced by GI bacteria	

Pantothenic Acid

Pantothenic acid is part of the chemical structure of coenzyme A—the same CoA that forms acetyl CoA, the "crossroads" compound in several metabolic pathways, including the TCA cycle. (Appendix C presents the chemical structures of these two molecules and shows that coenzyme A is made up in part of pantothenic acid.) As such, it is involved in more than 100 different steps in the synthesis of lipids, neurotransmitters, steroid hormones, and hemoglobin.

Pantothenic Acid Recommendations An Adequate Intake (AI) for pantothenic acid has been set. It reflects the amount needed to replace daily losses.

Pantothenic Acid Deficiency and Toxicity Pantothenic acid deficiency is rare. Its symptoms involve a general failure of all the body's systems and include fatigue, GI distress, and neurological disturbances. The "burning feet" syndrome that affected prisoners of war in Asia during World War II is thought to have been caused by pantothenic acid deficiency. No toxic effects have been reported, and no Upper Level has been established.

Pantothenic Acid Food Sources Pantothenic acid is widespread in foods, and typical diets seem to provide adequate intakes. Beef, poultry, whole grains, potatoes, tomatoes, and broccoli are particularly good sources. Losses of pantothenic acid during food production can be substantial because it is readily destroyed by the freezing, canning, and refining processes. The following summary table presents pantothenic acid facts.

IN SUMMARY	Pantothenic Acid
Adequate Intake (AI)	**Deficiency Symptoms**
Adults: 5 mg/day	Vomiting, nausea, stomach cramps; insomnia, fatigue, depression, irritability, restlessness, apathy; hypoglycemia, increased sensitivity to insulin; numbness, muscle cramps, inability to walk
Chief Functions in the Body	
Part of coenzyme A, used in energy metabolism	
Significant Sources	**Toxicity Symptoms**
Widespread in foods; chicken, beef, potatoes, oats, tomatoes, liver, egg yolk, broccoli, whole grains	None reported
Easily destroyed by food processing	

pantothenic (PAN-toe-THEN-ick) **acid:** a B vitamin. The principal active form is part of coenzyme A, called "CoA" throughout Chapter 7.
• **pantos** = everywhere

Vitamin B$_6$

Vitamin B$_6$ occurs in three forms—pyridoxal, pyridoxine, and pyridoxamine. All three can be converted to the coenzyme PLP (pyridoxal phosphate), which is active in amino acid metabolism. Because PLP can transfer amino groups (NH$_2$) from an amino acid to a keto acid, the body can make nonessential amino acids (review Figure 7-15, p. 226). The ability to add and remove amino groups makes PLP valuable in protein and urea metabolism as well. The conversions of the amino acid tryptophan to niacin or to the neurotransmitter serotonin ◆ also depend on PLP as does the synthesis of heme (the nonprotein portion of hemoglobin), nucleic acids (such as DNA and RNA), and lecithin.

A surge of research in the last decade has revealed that vitamin B$_6$ influences cognitive performance, immune function, and steroid hormone activity. Unlike other water-soluble vitamins, vitamin B$_6$ is stored extensively in muscle tissue.

Vitamin B$_6$ Recommendations Because the vitamin B$_6$ coenzymes play many roles in amino acid metabolism, previous RDA were expressed in terms of protein intakes; the current RDA for vitamin B$_6$, however, is not. Research does not support claims that large doses of vitamin B$_6$ enhance muscle strength or physical endurance. Vitamin supplements cannot compete with a nutritious diet and physical training.

Vitamin B$_6$ Deficiency Without adequate vitamin B$_6$, synthesis of key neurotransmitters diminishes, and abnormal compounds produced during tryptophan metabolism accumulate in the brain. Early symptoms of vitamin B$_6$ deficiency include depression and confusion; advanced symptoms include abnormal brain wave patterns and convulsions.

Alcohol contributes to the destruction and loss of vitamin B$_6$ from the body. As Highlight 7 described, when the body breaks down alcohol, it produces acetaldehyde. If allowed to accumulate, acetaldehyde dislodges the PLP coenzyme from its enzymes; once loose, PLP breaks down and is excreted. Low concentrations of PLP increase the risk of heart disease.[6]

Another drug that acts as a vitamin B$_6$ **antagonist** is INH, a medication that inhibits the growth of the tuberculosis bacterium.* This drug has saved countless lives, but as a vitamin B$_6$ antagonist, INH binds and inactivates the vitamin, inducing a deficiency. Whenever INH is used to treat tuberculosis, vitamin B$_6$ supplements must be given to protect against deficiency.

Vitamin B$_6$ Toxicity The first major report of vitamin B$_6$ toxicity appeared in the early 1980s. Until that time, everyone (including researchers and dietitians) believed that, like the other water-soluble vitamins, vitamin B$_6$ could not reach toxic concentrations in the body. The report described neurological damage in people who had been taking more than 2 *grams* of vitamin B$_6$ daily (20 times the current Upper Level of 100 *milligrams* per day) for two months or more.

Some people have taken vitamin B$_6$ supplements in an attempt to cure **carpal tunnel syndrome** and sleep disorders even though such treatment seems to be ineffective or at least inconclusive.[7] Self-prescribing is ill-advised because large doses of vitamin B$_6$ taken for months or years may cause irreversible nerve degeneration.

Vitamin B$_6$ Food Sources As you can see from the colors in Figure 10-8 (p. 337), meats, fish, and poultry (red bars), potatoes and a few other vegetables (green bars), and fruits (purple bars) offer vitamin B$_6$. As is true of most of the other vitamins, fruits and vegetables would rank considerably higher if foods were judged by nutrient density (vitamin B$_6$ per kcalorie). Several servings of vitamin B$_6$–rich foods are needed to meet recommended intakes.

Foods lose vitamin B$_6$ when heated. Information is limited, but vitamin B$_6$ bioavailability from plant-derived foods seems to be lower than from animal-

---

◆ Reminder: *Serotonin* is a neurotransmitter important in appetite control, sleep regulation, and sensory perception, among other roles; it is synthesized in the body from the amino acid tryptophan with the help of vitamin B$_6$.

vitamin B$_6$: a family of compounds— pyridoxal, pyridoxine, and pyridoxamine. The primary active coenzyme form is **PLP (pyridoxal phosphate).**

antagonist: a competing factor that counteracts the action of another factor. When a drug displaces a vitamin from its site of action, the drug renders the vitamin ineffective and thus acts as a vitamin antagonist.

carpal tunnel syndrome: a pinched nerve at the wrist, causing pain or numbness in the hand. It is often caused by repetitive motion of the wrist.

---

* INH stands for isonicotinic acid hydrazide.

FIGURE 10-8 Vitamin B₆ in Selected Foods

See the "How to" section on p. 329 for more information on using this figure.

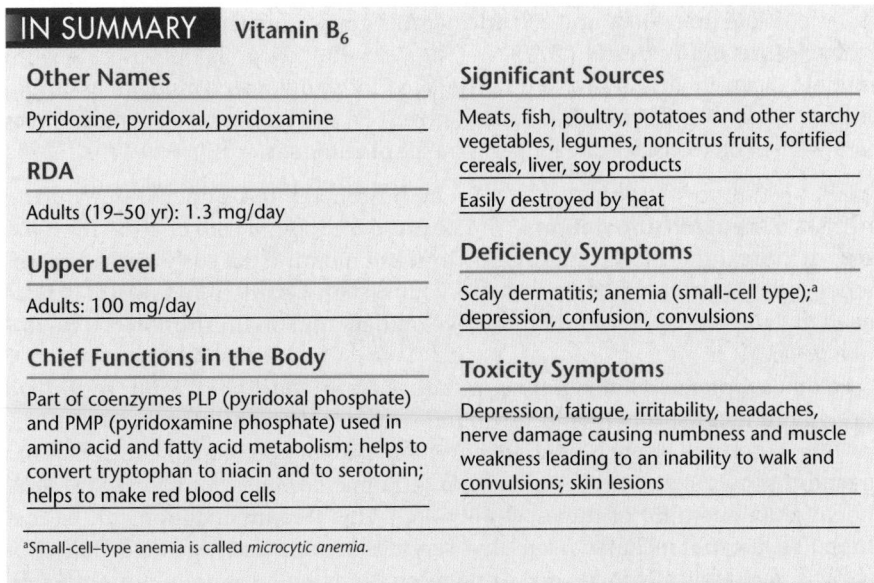

Food	Serving size (kcalories)
Bread, whole wheat	1 oz slice (70 kcal)
Cornflakes, fortified	1 oz (110 kcal)
Spaghetti pasta	½ c cooked (99 kcal)
Tortilla, flour	1 10"-round (234 kcal)
Broccoli	½ c cooked (22 kcal)
Carrots	½ c shredded raw (24 kcal)
Potato	1 medium baked w/skin (133 kcal)
Tomato juice	¾ c (31 kcal)
Banana	1 medium raw (109 kcal)
Orange	1 medium raw (62 kcal)
Strawberries	½ c fresh (22 kcal)
Watermelon	1 slice (92 kcal)
Milk	1 c reduced-fat 2% (121 kcal)
Yogurt, plain	1 c low-fat (155 kcal)
Cheddar cheese	1½ oz (171 kcal)
Cottage cheese	½ c low-fat 2% (101 kcal)
Pinto beans	½ c cooked (117 kcal)
Peanut butter	2 tbs (188 kcal)
Sunflower seeds	1 oz dry (165 kcal)
Tofu (soybean curd)	½ c (76 kcal)
Ground beef, lean	3 oz broiled (244 kcal)
Chicken breast	3 oz roasted (140 kcal)
Tuna, canned in water	3 oz (99 kcal)
Egg	1 hard cooked (78 kcal)
Excellent, and sometimes unusual, sources:	
Prune juice	¾ c (137 kcal)
Bluefish	3 oz baked (135 kcal)
Squash, acorn	½ c baked (69 kcal)

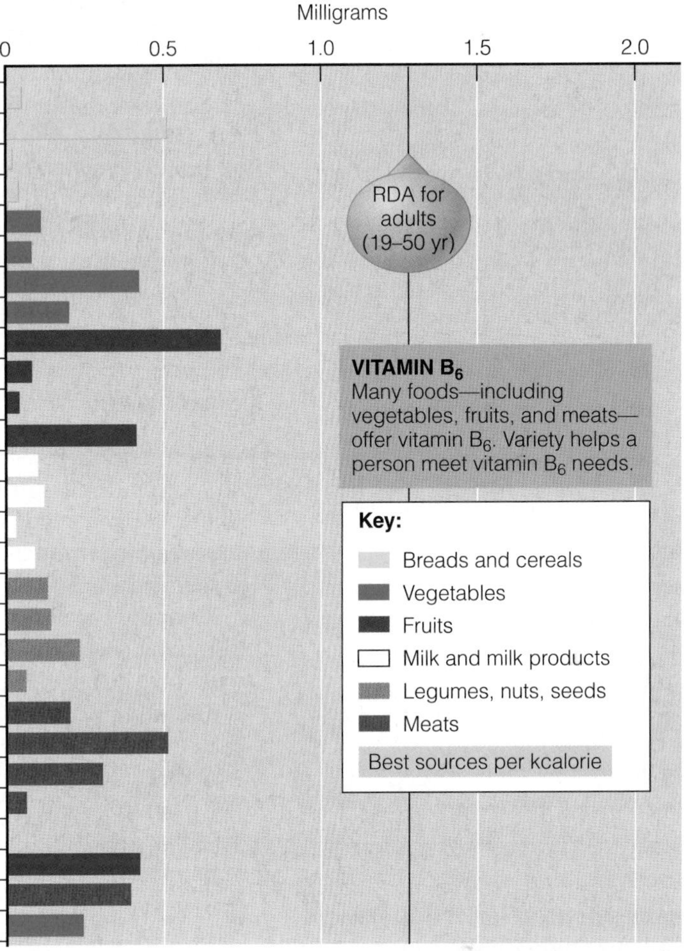

RDA for adults (19–50 yr)

VITAMIN B₆
Many foods—including vegetables, fruits, and meats—offer vitamin B₆. Variety helps a person meet vitamin B₆ needs.

Key:
- ▢ Breads and cereals
- ▮ Vegetables
- ■ Fruits
- ▢ Milk and milk products
- ▮ Legumes, nuts, seeds
- ▮ Meats
- Best sources per kcalorie

derived foods. Fiber does not appear to interfere with absorption of vitamin B₆. The summary table lists food sources of vitamin B₆ as well as its chief functions in the body and the common symptoms of deficiency and toxicity.

IN SUMMARY Vitamin B₆

Other Names

Pyridoxine, pyridoxal, pyridoxamine

RDA

Adults (19–50 yr): 1.3 mg/day

Upper Level

Adults: 100 mg/day

Chief Functions in the Body

Part of coenzymes PLP (pyridoxal phosphate) and PMP (pyridoxamine phosphate) used in amino acid and fatty acid metabolism; helps to convert tryptophan to niacin and to serotonin; helps to make red blood cells

Significant Sources

Meats, fish, poultry, potatoes and other starchy vegetables, legumes, noncitrus fruits, fortified cereals, liver, soy products

Easily destroyed by heat

Deficiency Symptoms

Scaly dermatitis; anemia (small-cell type);[a] depression, confusion, convulsions

Toxicity Symptoms

Depression, fatigue, irritability, headaches, nerve damage causing numbness and muscle weakness leading to an inability to walk and convulsions; skin lesions

[a]Small-cell–type anemia is called *microcytic anemia.*

Most protein-rich foods such as meat, fish, and poultry provide ample vitamin B₆; some vegetables and fruits are good sources, too.

© Polara Studios Inc.

Folate

Folate, also known as folacin or folic acid, has a chemical name that would fit a flying dinosaur: pteroylglutamic acid (PGA for short). Its primary coenzyme form, THF (tetrahydrofolate), serves as part of an enzyme complex that transfers one-carbon compounds that arise during metabolism. This action helps convert vitamin B_{12} to one of its coenzyme forms and helps synthesize the DNA required for all rapidly growing cells.

Foods deliver folate mostly in the "bound" form—that is, combined with a string of amino acids (glutamate), known as polyglutamate. (See Appendix C for the chemical structure.) The small intestine prefers to absorb the "free" folate form—folate with only one glutamate attached (the monoglutamate form).[8] Enzymes on the intestinal cell surfaces hydrolyze the polyglutamate to monoglutamate and several glutamates. Then the monoglutamate is attached to a methyl group (CH_3). Special transport systems deliver the monoglutamate with its methyl group to the liver and other body cells.

For the folate coenzyme to function, the methyl group must be removed by an enzyme that requires the help of vitamin B_{12}. Without that help, folate becomes trapped inside cells in its methyl form, unavailable to support DNA synthesis and cell growth. Figure 10-9 summarizes the process of folate's absorption and activation.

To dispose of excess folate, the liver secretes most of it into bile and ships it to the gallbladder. Thus folate returns to the intestine in an enterohepatic circulation route like that of bile itself (review Figure 5-16, p. 151).

This complicated system for handling folate is vulnerable to GI tract injuries. Because folate is actively secreted back into the GI tract with bile, it has to be reabsorbed repeatedly. If the GI tract cells are damaged, then folate is rapidly lost from the body. Such is the case in alcohol abuse; folate deficiency rapidly develops and, ironically, further damages the GI tract. The folate coenzymes, remember, are active in cell multiplication—and the cells lining the GI tract are among the most rapidly renewed cells in the body. When unable to make new cells, the GI tract deteriorates and not only loses folate, but also fails to absorb other nutrients.

Folate Recommendations The bioavailability of folate ranges from 50 percent for foods to 100 percent for supplements taken on an empty stomach. These differences in bioavailability were considered when establishing the folate RDA. Naturally occurring folate from foods is given full credit. Synthetic folate from fortified foods and supplements is given extra credit because, on average, it is 1.7 times more available than naturally occurring food folate. Thus a person consuming 100 micrograms of folate from foods and 100 micrograms from a supplement receives 270 **dietary folate equivalents (DFE).** ◆ (The "How to" on p. 339 describes how to estimate dietary folate equivalents.) The need for folate rises considerably during pregnancy and whenever cells are multiplying, so the recommendations for pregnant women are considerably higher than for other adults.

Folate and Neural Tube Defects Folate has proven to be critical in reducing the risks of **neural tube defects.**[9] ◆ The brain and spinal cord develop from the **neural tube,** and defects in its orderly formation during the early weeks of pregnancy may result in various central nervous system disorders and death. (Chapter 14 includes photos of neural tube development and an illustration of a neural tube defect.)

Folate supplements taken one month before conception and continued throughout the first trimester of pregnancy can help prevent neural tube defects. For this reason, all women of childbearing age ◆ who are capable of becoming pregnant should consume 0.4 milligram (400 micrograms) of folate daily, ◆ although only one-third of them actually do.[10] This recommendation can be met through a diet that includes at least five servings of fruits and vegetables daily, but many women typically fail to do so and receive only half this amount from foods.

◆ To calculate DFE:

DFE = μg food folate + (1.7 × μg synthetic folate)

Using the example in the text:

$$\begin{array}{r} 100 \text{ μg food} \\ + \ 170 \text{ μg supplement } (1.7 \times 100 \text{ μg}) \\ \hline 270 \text{ μg DFE} \end{array}$$

◆ The two main types of neural tube defects are **spina bifida** (literally, "split spine") and **anencephaly** ("no brain").

◆ Women of childbearing age (15 to 45 yr) should:
- Eat folate-rich foods
- Eat folate-fortified foods
- Take a multivitamin daily (most provide 400 μg folate)

◆ Reminder: A milligram (mg) is one-thousandth of a gram. A microgram (μg) is one-thousandth of a milligram (or one-millionth of a gram).
- 0.4 mg = 400 μg

---

folate (FOLE-ate): a B vitamin; also known as folic acid, folacin, or pteroylglutamic (tare-o-EEL-glue-TAM-ick) acid (PGA). The coenzyme forms are **DHF (dihydrofolate)** and **THF (tetrahydrofolate).**

dietary folate equivalents (DFE): the amount of folate available to the body from naturally occurring sources, fortified foods, and supplements, accounting for differences in the bioavailability from each source.

neural tube defects: malformations of the brain, spinal cord, or both during embryonic development that often result in lifelong disability or death.

neural tube: the embryonic tissue that forms the brain and spinal cord.

FIGURE 10-9 Folate's Absorption and Activation

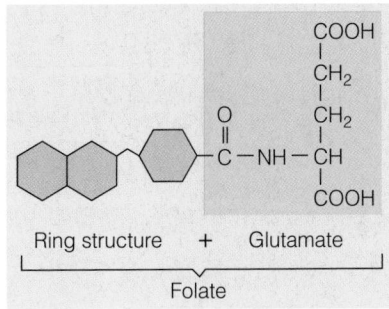

Ring structure + Glutamate

Folate

In foods, folate naturally occurs as polyglutamate. (Folate occurs as monoglutamate in fortified foods and supplements.)

Spinach

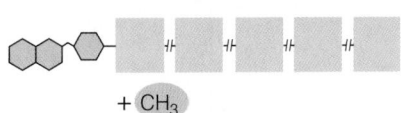

In the intestine, digestion breaks glutamates off . . . and adds a methyl group. Folate is absorbed and delivered to cells.

Intestine

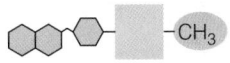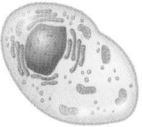

In the cells, folate is trapped in its inactive form.

Cell

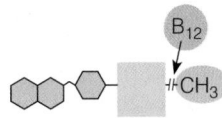

To activate folate, vitamin B_{12} removes and keeps the methyl group, which activates vitamin B_{12}.

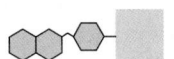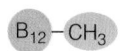

Both the folate coenzyme and the vitamin B_{12} coenzyme are now active and available for DNA synthesis.

DNA

HOW TO Estimate Dietary Folate Equivalents

Folate is expressed in terms of DFE (dietary folate equivalents) because synthetic folate from supplements and fortified foods is absorbed at almost twice (1.7 times) the rate of naturally occurring folate from other foods. Use the following equation to calculate:

DFE = μg food folate + (1.7 × μg synthetic folate)

Consider, for example, a pregnant woman who takes a supplement and eats a bowl of fortified cornflakes, 2 slices of fortified bread, and a cup of fortified pasta. From the supplement and fortified foods, she obtains synthetic folate:

Supplement	100 μg folate
Fortified cornflakes	100 μg folate
Fortified bread	40 μg folate
Fortified pasta	60 μg folate
	300 μg folate

To calculate the DFE, multiply the amount of synthetic folate by 1.7:

300 μg × 1.7 = 510 μg DFE

Now add the naturally occurring folate from the other foods in her diet—in this example, another 90 μg of folate.

510 μg DFE + 90 μg = 600 μg DFE

Notice that if we had not converted synthetic folate from supplements and fortified foods to DFE, then this woman's intake would appear to fall short of the 600 μg recommendation for pregnancy (300 μg + 90 μg = 390 μg). But as our example shows, her intake does meet the recommendation. At this time, supplement and fortified food labels list folate in μg only, not μg DFE, making such calculations necessary.

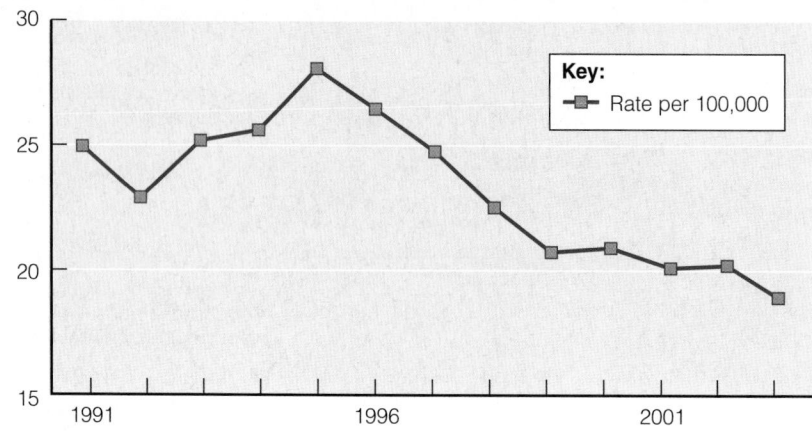

FIGURE 10-10 Decreasing Spina Bifida Rates since Folate Fortification

Neural tube defects have declined since folate fortification began in 1996.

SOURCE: National Vital Statistics System, National Center for Health Statistics, Centers for Disease Control.

Furthermore, because of the enhanced bioavailability of synthetic folate, supplementation or fortification improves folate status significantly. Women who have given birth to infants with neural tube defects previously should consume 4 milligrams of folate daily before conception and throughout the first trimester of pregnancy.

Because half of the pregnancies each year are unplanned and because neural tube defects occur early in development before most women realize they are pregnant, the Food and Drug Administration (FDA) has mandated that grain products be fortified to deliver folate to the U.S. population.* Labels on fortified products may claim that "adequate intake of folate has been shown to reduce the risk of neural tube defects." Fortification has improved folate status in women of childbearing age and lowered the number of neural tube defects that occur each year, as Figure 10-10 shows.[11] Whether additional fortification will help save even more infants is a topic of current debate.[12]

Folate fortification raises safety concerns as well, especially because folate intakes from fortified foods are more than twice as high as originally predicted.[13] Because high intakes of folate complicate the diagnosis of a vitamin B_{12} deficiency, folate consumption should not exceed 1 milligram daily without close medical supervision.[14]

Some research suggests a relationship between abnormal folate metabolism and non-neural tube birth defects such as Down syndrome.[15] Folate's exact role, however, remains unclear, and supplementation does not appear to decrease the prevalence of Down syndrome.[16] Some women whose infants develop these defects are *not* deficient in folate, and others with severe folate deficiencies do *not* give birth to infants with birth defects.[17] Researchers continue to look for other factors that must also be involved.

Folate and Heart Disease The FDA's decision to fortify grain products with folate was strengthened by research indicating an important role for folate in defending against heart disease. As Chapter 6 mentioned, research indicates that high levels of the amino acid homocysteine and low levels of folate increase the risk of fatal heart disease.[18] One of folate's key roles in the body is to break down homocys-

---

* Bread products, flour, corn grits, cornmeal, farina, rice, macaroni, and noodles must be fortified with 140 micrograms of folate per 100 grams of grain. For perspective, 100 grams is roughly 3 slices of bread; 1 cup of flour; 1/2 cup of corn grits, cornmeal, farina, or rice; or 3/4 cup of macaroni or noodles.

teine. Without folate, homocysteine accumulates, which seems to enhance blood clot formation and arterial wall deterioration. Fortified foods and folate supplements raise blood folate and reduce blood homocysteine levels to an extent that may help to prevent heart disease.[19] Supplements do not seem to reduce the risk of death from cardiovascular causes.[20]

Folate and Cancer Folate may also play a role in preventing cancer.[21] Notably, folate may be most effective in protecting those most likely to develop cancers: men who smoke (against pancreatic cancer) and women who drink alcohol (against breast cancer).[22]

Folate Deficiency Folate deficiency impairs cell division and protein synthesis—processes critical to growing tissues. In a folate deficiency, the replacement of red blood cells and GI tract cells falters. Not surprisingly, then, two of the first symptoms of a folate deficiency are **anemia** and GI tract deterioration.

The anemia of folate deficiency is characterized by large, ◆ immature red blood cells. Without folate, DNA damage destroys many of the red blood cells as they attempt to divide and mature.[23] The result is fewer, but larger, red blood cells that cannot carry oxygen or travel through the capillaries as efficiently as normal red blood cells.

Folate deficiencies may develop from inadequate intake and have been reported in infants who were fed goat's milk, which is notoriously low in folate. Folate deficiency may also result from impaired absorption or an unusual metabolic need for the vitamin. Metabolic needs increase in situations where cell multiplication must speed up, such as pregnancies involving twins and triplets; cancer; skin-destroying diseases such as chicken pox and measles; and burns, blood loss, GI tract damage, and the like.

Of all the vitamins, folate appears to be most vulnerable to interactions with drugs, which can lead to a secondary deficiency. Some medications, notably anticancer drugs, have a chemical structure similar to folate's structure and can displace the vitamin from enzymes and interfere with normal metabolism. Like all cells, cancer cells need the real vitamin to multiply—without it, they die. Unfortunately, these drugs affect both cancerous cells and healthy cells, and they create a folate deficiency for all cells. (Chapter 19 discusses nutrient-drug interactions and includes a figure illustrating the similarities between the vitamin folate and the anticancer drug methotrexate.)

Aspirin and antacids also interfere with the body's handling of folate. Healthy adults who use these drugs to relieve an occasional headache or upset stomach need not be concerned, but people who rely heavily on aspirin or antacids should be aware of the nutrition consequences. Oral contraceptives may also impair folate status, as may smoking.[24]

Folate Toxicity Naturally occurring folate from foods alone appears to cause no harm. Excess folate from fortified foods or supplements, however, can reach levels that are high enough to obscure a vitamin B_{12} deficiency and delay diagnosis of neurological damage. For this reason, an Upper Level has been established for folate from fortified foods or supplements (see the inside front cover).

Folate Food Sources Figure 10-11 (p. 342) shows that folate is especially abundant in legumes, fruits, and vegetables. The vitamin's name suggests the word *foliage,* and indeed, leafy green vegetables are outstanding sources. With fortification, grain products also contribute folate. The small red and white bars in Figure 10-11 indicate that meats, milk, and milk products are poor folate sources. Heat and oxidation during cooking and storage can destroy as much as half of the folate in foods. The table on the next page provides a summary of folate information.

Leafy dark green vegetables (such as spinach and broccoli), legumes (such as black beans, kidney beans, and black-eyed peas), liver, and some fruits (notably citrus fruits and juices) are naturally rich in folate.

◆ Large-cell anemia is known as **macrocytic** or **megaloblastic anemia**.
- **macro** = large
- **cyte** = cell
- **mega** = large

anemia (ah-NEE-me-ah): literally, "too little blood." Anemia is any condition in which too few red blood cells are present, or the red blood cells are immature (and therefore large) or too small or contain too little hemoglobin to carry the normal amount of oxygen to the tissues. It is not a disease itself but can be a symptom of many different disease conditions, including many nutrient deficiencies, bleeding, excessive red blood cell destruction, and defective red blood cell formation.
- **an** = without
- **emia** = blood

TABLE H10-1 Vitamin and Mineral Intakes for Adults

Nutrient	Tolerable Upper Intake Levels[a]	Daily Values	Typical Multivitamin-Mineral Supplement	Average Single-Nutrient Supplement
Vitamins				
Vitamin A	3000 µg (10,000 IU)	5000 IU	5000 IU	8000 to 10,000 IU
Vitamin D	50 µg (2000 IU)	400 IU	400 IU	400 IU
Vitamin E	1000 mg (1500 to 2200 IU)[b]	30 IU	30 IU	100 to 1000 IU
Vitamin K	—[c]	80 µg	40 µg	—[e]
Thiamin	—[c]	1.5 mg	1.5 mg	50 mg
Riboflavin	—[c]	1.7 mg	1.7 mg	25 mg
Niacin (as niacinamide)	35 mg[b]	20 mg	20 mg	100 to 500 mg
Vitamin B6	100 mg	2 mg	2 mg	100 to 200 mg
Folate	1000 µg[b]	400 µg	400 µg	400 µg
Vitamin B12	—[c]	6 µg	6 µg	100 to 1000 µg
Pantothenic acid	—[c]	10 mg	10 mg	100 to 500 mg
Biotin	—[c]	300 µg	30 µg	300 to 600 µg
Vitamin C	2000 mg	60 mg	10 mg	500 to 2000 mg
Choline	3500 mg	—	10 mg	250 mg
Minerals				
Calcium	2500 mg	1000 mg	160 mg	250 to 600 mg
Phosphorus	4000 mg	1000 mg	110 mg	—[e]
Magnesium	350 mg[d]	400 mg	100 mg	250 mg
Iron	45 mg	18 mg	18 mg	18 to 30 mg
Zinc	40 mg	15 mg	15 mg	10 to 100 mg
Iodine	1100 µg	150 µg	150 µg	—[e]
Selenium	400 µg	70 µg	10 µg	50 to 200 µg
Fluoride	10 mg	—	—	—[e]
Copper	10 mg	2 mg	0.5 mg	—[e]
Manganese	11 mg	2 mg	5 mg	—[e]
Chromium	—[c]	120 µg	25 µg	200 to 400 µg
Molybdenum	2000 µg	75 µg	25 µg	—[e]

[a] Unless otherwise noted, Upper Levels represent total intakes from food, water, and supplements.
[b] Upper Levels represent intakes from supplements, fortified foods, or both.
[c] These nutrients have been evaluated by the DRI Committee for Tolerable Upper Intake Levels, but none were established because of insufficient data. No adverse effects have been reported with intakes of these nutrients at levels typical of supplements, but caution is still advised, given the potential for harm that accompanies excessive intakes.
[d] Upper Levels represent intakes from supplements only.
[e] Available as a single supplement by prescription.

of accidental ingestion fatalities among children. Even mild overdoses cause GI distress, nausea, and black diarrhea that reflects gastric bleeding. Severe overdoses result in bloody diarrhea, shock, liver damage, coma, and death.

Life-Threatening Misinformation

Another problem arises when people who are ill come to believe that high doses of vitamins or minerals can be therapeutic. Not only can high doses be toxic, but the person may take them instead of seeking medical help. Furthermore, there are no guarantees that the supplements will be effective. Marketing materials for supplements often make health statements that are required to be "truthful and not misleading," but they often fall far short of both. Chapter 19 revisits this topic and includes a discussion of herbal preparations.

Unknown Needs

Another argument against the use of supplements is that no one knows exactly how to formulate the "ideal" supplement. What nutrients should be included? Which, if any, of the phytochemicals should be included? How much of each? On whose needs should the choices be based? Surveys have repeatedly shown little relationship between the supplements people take and the nutrients they actually need.

False Sense of Security

Another argument against supplement use is that it may lull people into a false sense of security. A person might eat irresponsibly, thinking, "My supplement will cover my needs." Or, experiencing a warning symptom of a disease, a person might postpone seeking a diagnosis, thinking, "I probably just need a supplement to make this go away." Such self-diagnosis is potentially dangerous.

Other Invalid Reasons

Other invalid reasons people might use for taking supplements include:

- The belief that the food supply or soil contains inadequate nutrients
- The belief that supplements can provide energy
- The belief that supplements can enhance athletic performance or build lean body tissues without physical work or faster than work alone
- The belief that supplements will help a person cope with stress
- The belief that supplements can prevent, treat, or cure conditions ranging from the common cold to cancer

Ironically, people with health problems are more likely to take supplements than other people, yet today's health problems are more likely to be due to overnutrition and poor lifestyle choices than to nutrient deficiencies. The truth—that most people would benefit from improving their eating and exercise habits—is harder to swallow than a supplement pill.

Bioavailability and Antagonistic Actions

In general, the body absorbs nutrients best from foods in which the nutrients are diluted and dispersed among other substances that may facilitate their absorption. Taken in pure, concentrated form, nutrients are likely to interfere with one another's absorption or with the absorption of nutrients in foods eaten at the same time. Documentation of these effects is particularly extensive for minerals: zinc hinders copper and calcium absorption, iron hinders zinc absorption, calcium hinders magnesium and iron absorption, and magnesium hinders the absorption of calcium and iron. Similarly, binding agents in supplements limit mineral absorption.

Although minerals provide the most familiar and best-documented examples, interference among vitamins is now being seen as supplement use increases. The vitamin A precursor beta-carotene, long thought to be nontoxic, interferes with vitamin E metabolism when taken over the long term as a dietary supplement. Vitamin E, on the other hand, antagonizes vitamin K activity and so should not be used by people being treated for blood-clotting disorders. Consumers who want the benefits of optimal absorption of nutrients should eat ordinary foods, selected for nutrient density and variety.

Whenever the diet is inadequate, the person should first attempt to improve it so as to obtain the needed nutrients from foods. If that is truly impossible, then the person needs a multivitamin-mineral supplement that supplies between 50 and 150 percent of the Daily Value for each of the nutrients. These amounts reflect the ranges commonly found in foods and therefore are compatible with the body's normal handling of nutrients (its physiologic tolerance). The next section provides some pointers to assist in the selection of an appropriate supplement.

Selection of Supplements

Whenever a physician or registered dietitian recommends a supplement, follow the directions carefully. When selecting a supplement yourself, look for a single, balanced vitamin-mineral supplement. Supplements with a USP verification logo have been tested by the U.S. Pharmacopeia (USP) to assure that the supplement:

- Contains the declared ingredients and amounts listed on the label
- Does not contain harmful levels of contaminants
- Will disintegrate and release ingredients in the body
- Was made under safe and sanitary conditions

If you decide to take a vitamin-mineral supplement, ignore the eye-catching art and meaningless claims. Pay attention to the form the supplements are in, the list of ingredients, and the price. Here's where the truth lies, and from it you can make a rational decision based on facts. You have two basic questions to answer.

Form

The first question: What form do you want—chewable, liquid, or pills? If you'd rather drink your supplements than chew them, fine. (If you choose a chewable form, though, be aware that chewable vitamin C can dissolve tooth enamel.) If you choose pills, look for statements about the disintegration time. The USP suggests that supplements should completely disintegrate within 30 to 45 minutes.* Obviously, supplements that don't dissolve have little chance of entering the bloodstream, so look for a brand that claims to meet USP disintegration standards.

Contents

The second question: What vitamins and minerals do *you* need? Generally, an appropriate supplement provides vitamins and minerals in amounts that do not exceed recommended intakes. Avoid supplements that, in a daily dose, provide more than the Tolerable Upper Intake Level for *any* nutrient. Avoid preparations with more than 10 milligrams of iron per dose, except as prescribed by a physician. Iron is hard to get rid of once it's in the body, and an excess of iron can cause problems, just as a deficiency can (see Chapter 13).

Misleading Claims

Be aware that "organic" or "natural" supplements are no more effective than others and often cost more. The word *synthetic*

---

* The USP establishes standards for quality, strength, and purity of supplements.

may sound like "fake," but to synthesize just means to put together.

Avoid products that make **"high potency"** claims. More is not better (review the "How to" on p. 325). Remember that foods are also providing these nutrients. Nutrients can build up and cause unexpected problems. For example, a man who takes vitamins and begins to lose his hair may think his hair loss means he needs *more* vitamins, when in fact it may be the early sign of a vitamin A overdose. (Of course, it may be completely unrelated to nutrition as well.)

Be wise to fake vitamins and preparations that contain items not needed in human nutrition, such as carnitine and inositol. Such ingredients reveal a marketing strategy aimed at your pocket, not at your health. The manufacturer wants you to believe that its pills contain the latest "new" nutrient that other brands omit, but in reality, these substances are not known to be needed by human beings.

Realize that the claim that supplements "relieve stress" is another marketing ploy. If you give even passing thought to what people mean by "stress," you'll realize manufacturers could never design a supplement to meet everyone's needs. Is it stressful to take an exam? Well, yes. Is it stressful to survive a major car wreck with third-degree burns and multiple bone fractures? Definitely, yes. The body's responses to these stresses are different. The body does use vitamins and minerals in mounting a stress response, but a body fed a well-balanced diet can meet the needs of most minor stresses. For the major ones, medical intervention is needed. In any case, taking a vitamin supplement won't make life any less stressful.

Other marketing tricks to sidestep are "green" pills that contain dehydrated, crushed parsley, alfalfa, and other fruit and vegetable extracts. The nutrients and phytochemicals advertised can be obtained from a serving of vegetables more easily and for less money. Such pills may also provide enzymes, but enzymes are inactivated in the stomach during protein digestion.

Be aware that some geriatric "tonics" are low in vitamins and minerals and may be high in alcohol. The liquids designed for infants offer a more complete option.

Recognize the latest nutrition buzzwords. Manufacturers were marketing "antioxidant" supplements before the print had time to dry on the first scientific reports of antioxidant vitamins' action in preventing cancer and cardiovascular disease. Remember, too, that high doses can alter a nutrient's action in the body. An antioxidant in physiological quantities may be beneficial, but in pharmacological quantities, it may act as a prooxidant and produce harmful by-products. Highlight 11 explores antioxidants and supplement use in more detail.

Finally, be aware that advertising on the Internet is cheap and not closely regulated. Promotional e-mails can be sent to millions of people in an instant. Internet messages can easily cite references and provide links to other sites, implying an endorsement when in fact none has been given.[10] Be cautious when examining unsolicited information and search for a balanced perspective.

Cost

When shopping for supplements, remember that local or store brands may be just as good as nationally advertised brands. If they are less expensive, it may be because the price does not have to cover the cost of national advertising.

Regulation of Supplements

The Dietary Supplement Health and Education Act of 1994 was intended to enable consumers to make informed choices about nutrient supplements. The act subjects supplements to the same general labeling requirements that apply to foods. Specifically:

- Nutrition labeling for dietary supplements is required.
- Labels may make nutrient claims (as "high" or "low") according to specific criteria (for example, "an excellent source of vitamin C").
- Labels may claim that the lack of a nutrient can cause a deficiency disease, but if they do, they must also include the prevalence of that deficiency disease in the United States.
- Labels may make health claims that are supported by significant scientific agreement and are not brand specific (for example, "folate protects against neural tube defects").
- Labels may claim to diagnose, treat, cure, or relieve common complaints such as menstrual cramps or memory loss, but may *not* make claims about specific diseases (except as noted above).
- Labels may make structure-function claims about the role a nutrient plays in the body, how the nutrient performs its function, and how consuming the nutrient is associated with general well-being. These claims must be accompanied by an **FDA** disclaimer statement: "This statement has not been evaluated by the Food and Drug Administration. This product is not intended to diagnose, treat, cure or prevent any disease." Figure H10-1 provides an example of a supplement label that complies with the requirements.

The multibillion-dollar-a-year supplement industry spends much money and effort influencing these regulations. The net effect of the Dietary Supplement Health and Education Act was a deregulation of the supplement industry. Unlike food additives or

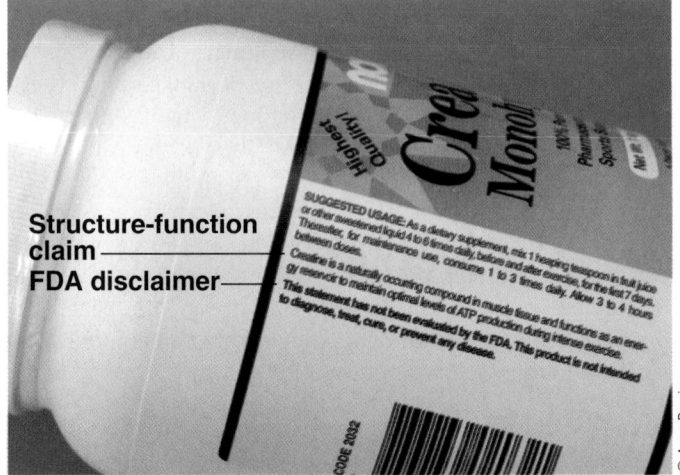

Structure-function claim

FDA disclaimer

Structure-function claims do not need FDA authorization, but they must be accompanied by a disclaimer.

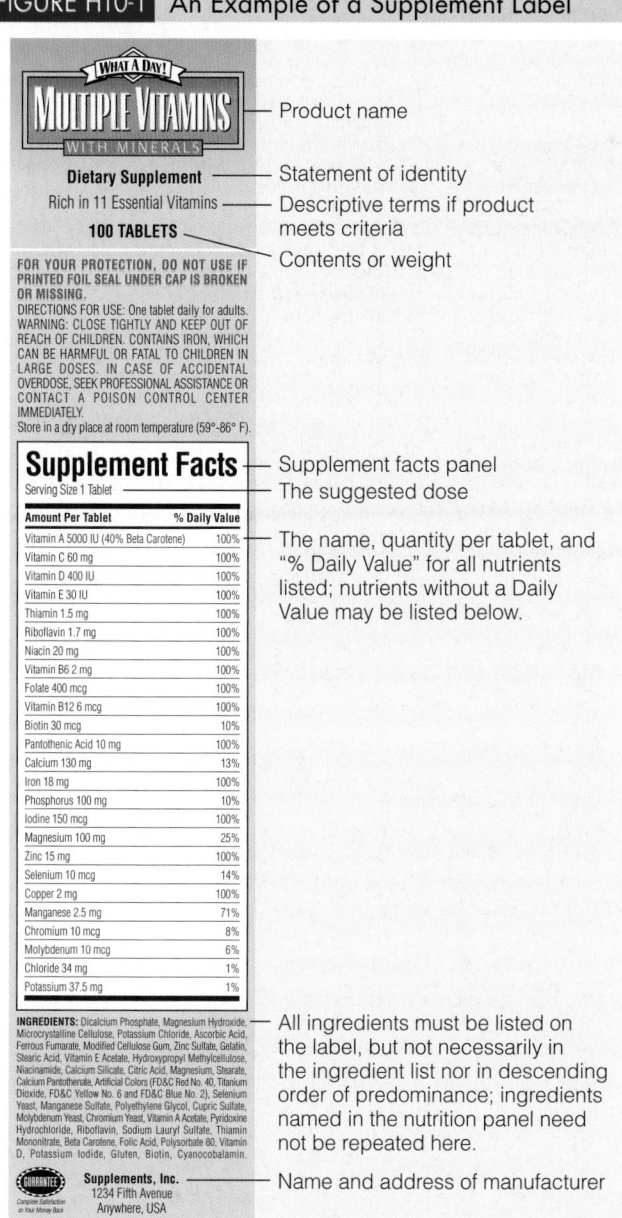

FIGURE H10-1 An Example of a Supplement Label

— Product name

— Statement of identity

— Descriptive terms if product meets criteria

— Contents or weight

— Supplement facts panel

— The suggested dose

— The name, quantity per tablet, and "% Daily Value" for all nutrients listed; nutrients without a Daily Value may be listed below.

— All ingredients must be listed on the label, but not necessarily in the ingredient list nor in descending order of predominance; ingredients named in the nutrition panel need not be repeated here.

— Name and address of manufacturer

drugs, supplements do not need to be proved safe and effective, nor do they need the FDA's approval before being marketed. Furthermore, there are no standards for potency or dosage and no requirements for providing warnings of potential side effects. Should a problem arise, the burden falls to the FDA to prove that the supplement poses a "significant or unreasonable risk of illness or injury."[11] Only then would it be removed from the market. When asked, most Americans express support for greater regulation of dietary supplements. Health professionals agree.[12]

If all the nutrients we need can come from food, why not just eat food? Foods have so much more to offer than supplements do. Nutrients in foods come in an infinite variety of combinations with a multitude of different carriers and absorption enhancers. They come with water, fiber, and an array of beneficial phytochemicals. Foods stimulate the GI tract to keep it healthy. They provide energy, and as long as you need energy each day, why not have nutritious foods deliver it? Foods offer pleasure, satiety, and opportunities for socializing while eating. In no way can nutrient supplements hold a candle to foods as a means of meeting human health needs. For further proof, read Highlight 11.

NUTRITION ON THE NET

For further study of topics covered in this chapter, log on to **academic.cengage.com/nutrition/rolfes/UNCN8e**. Go to Chapter 10, then to Nutrition on the Net.

- Gather information from the Office of Dietary Supplements or Health Canada: **dietary-supplements.info.nih.gov** or **www.hc-sc.gc.ca**

- Report adverse reactions associated with dietary supplements to the FDA's MedWatch program: **www.fda.gov/medwatch**

- Search for "supplements" at the American Dietetic Association: **www.eatright.org**

- Learn more about supplements from the FDA Center for Food Safety and Applied Nutrition: **www.cfsan.fda.gov/~dms/supplmnt.html**

- Obtain consumer information on dietary supplements from the U.S. Pharmacopeia: **www.usp.org**

- Review the Federal Trade Commission policies for dietary supplement advertising: **www.ftc.gov/bcp/conline/pubs/buspubs/dietsupp.htm**

REFERENCES

1. National Institutes of Health, Multivitamin/mineral supplements and chronic disease prevention, *Annals of Internal Medicine* 145 (2006): 364–371; A. E. Millen, K. W. Dodd, and A. F. Subar, Use of vitamin, mineral, nonvitamin, and nonmineral supplements in the United States: The 1987, 1992, and 2000 National Health Interview Survey results, *Journal of the American Dietetic Association* 104 (2004): 942–950.

2. Position of the American Dietetic Association: Fortification and nutritional supplements, *Journal of the American Dietetic Association* 105 (2005): 1300–1311.

3. Practice Paper of the American Dietetic Association: Dietary supplements, *Journal of the American Dietetic Association* 105 (2005): 460–470; J. R. Hunt, Tailoring advice on dietary supplements: An opportunity for dietetics professionals, *Journal of the American Dietetic Association* 102 (2002): 1754–1755; C. Thomson and coauthors, Guidelines regarding the recommendation and sale of dietary supplements, *Journal of the American Dietetic Association* 102 (2002): 1158–1164.

4. D. E. Wildish, An evidence-based approach for dietitian prescription of multiple vitamins with minerals, *Journal of the American Dietetic Association* 104 (2004): 779–786.

5. K. M. Fairfield and R. H. Fletcher, Vitamins for chronic disease prevention in adults: Scientific review, *Journal of the American Medical Association* 287 (2002): 3116–3126.

6. R. H. Fletcher and K. M. Fairfield, Vitamins for chronic disease prevention in adults: Clinical applications, *Journal of the American Medical Association* 287 (2002): 3127–3129.

7. B. N. Ames, The metabolic tune-up: Metabolic harmony and disease prevention, *Journal of Nutrition* 133 (2003): 1544S–1548S.

8. P. M. Kris-Etherton and coauthors, Antioxidant vitamin supplements and cardiovascular disease, *Circulation* 110 (2004): 637–641; C. D. Morris and S. Carson, Routine vitamin supplementation to prevent cardiovascular disease: A summary of the evidence for the U.S. Preventive Services Task Force, *Annals of Internal Medicine* 139 (2003): 56–70; B. Hasanain and A. D. Mooradian, Antioxidant vitamins and their influence in diabetes mellitus, *Current Diabetes Reports* 2 (2002): 448–456.

9. National Institutes of Health, 2006.

10. J. M. Drazen, Inappropriate advertising of dietary supplements, *New England Journal of Medicine* 348 (2003): 777–778.

11. Institute of Medicine and National Research Council, *Dietary supplements: A framework for evaluating safety,* (Washington, D.C.: National Academy Press, 2004); C. L. Taylor, Regulatory frameworks for functional foods and dietary supplements, *Nutrition Reviews* 62 (2004): 55–59.

12. P. B. Fontanarosa, D. Rennie, and C. D. DeAngelis, The need for regulation of dietary supplements—Lessons from ephedra, *Journal of the American Medical Association* 289 (2003): 1568–1570.

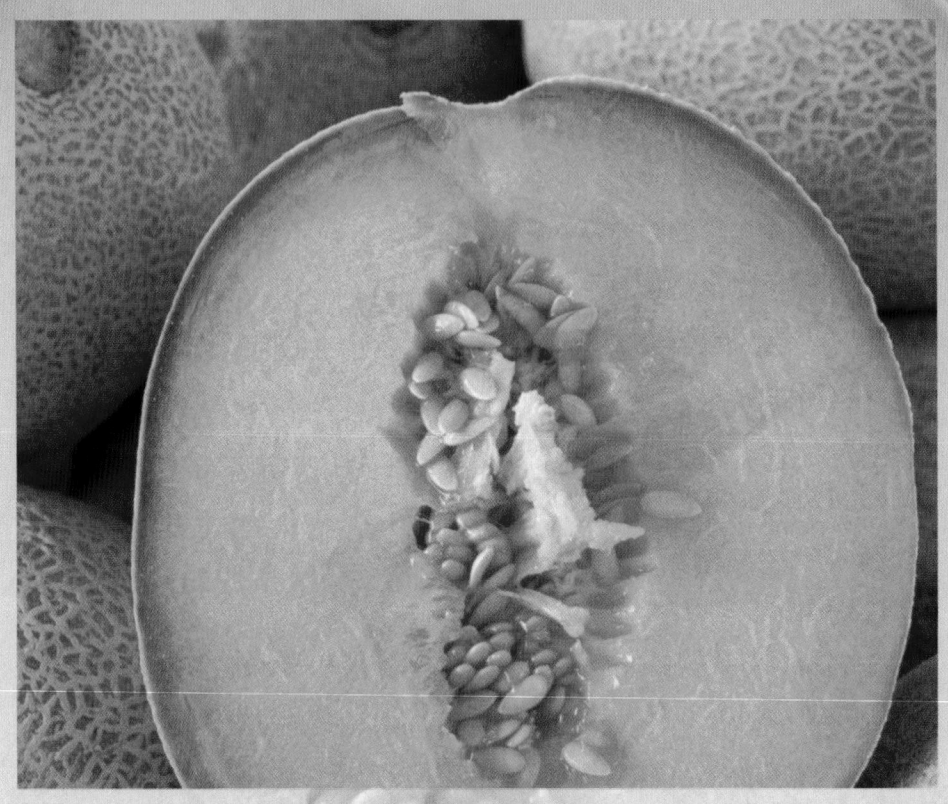

The CengageNOW logo indicates an opportunity for online self-study, linking you to interactive tutorials and videos based on your level of understanding.

academic.cengage.com/login

Figure 11-3: Animated! Vitamin A's Role in Vision
Figure 11-9: Animated! Vitamin D Synthesis and Activation

Nutrition Portfolio Journal

Nutrition Calculations: Practice Problems

Nutrition in Your Life

Realizing that vitamin A from vegetables participates in vision, a mom encourages her children to "eat your carrots" because "they're good for your eyes." A dad takes his children outside to "enjoy the fresh air and sunshine" because they need the vitamin D that is made with the help of the sun. A physician recommends that a patient use vitamin E to slow the progression of heart disease. Another physician gives a newborn a dose of vitamin K to protect against life-threatening blood loss. These common daily occurrences highlight some of the heroic work of the fat-soluble vitamins.

The Fat-Soluble Vitamins: A, D, E, and K

The fat-soluble vitamins A, D, E, and K differ from the water-soluble vitamins in several significant ways (review the table on p. 326). Being insoluble in the watery GI juices, the fat-soluble vitamins require bile for their absorption. Upon absorption, fat-soluble vitamins travel through the lymphatic system within chylomicrons before entering the bloodstream, where many of them require protein carriers for transport. The fat-soluble vitamins participate in numerous activities throughout the body, but excesses are stored primarily in the liver and adipose tissue. The body maintains blood concentrations by retrieving these vitamins from storage as needed; thus people can eat less than their daily need for days, weeks, or even months or years without ill effects. They need only ensure that, over time, *average* daily intakes approximate recommendations. By the same token, because fat-soluble vitamins are not readily excreted, the risk of toxicity is greater than it is for the water-soluble vitamins.

Vitamin A and Beta-Carotene

Vitamin A was the first fat-soluble vitamin to be recognized. Almost a century later, vitamin A and its precursor, **beta-carotene,** ◆ continue to intrigue researchers with their diverse roles and profound effects on health.

Three different forms of vitamin A are active in the body: retinol, retinal, and retinoic acid. Collectively, these compounds are known as **retinoids.** Foods derived from animals provide compounds (retinyl esters) that are readily digested and absorbed as retinol in the intestine.[1] Foods derived from plants provide **carotenoids,** ◆ some of which have **vitamin A activity.*** The most studied of the carotenoids is beta-carotene, which can be split to form retinol in the intestine and liver. Beta-carotene's absorption and conversion are significantly less efficient than those of the retinoids.[2] Figure 11-1 (p. 370) illustrates the structural similarities and differences of these vitamin A compounds and the cleavage of beta-carotene.

The cells can convert retinol and retinal to the other active forms of vitamin A as needed. The conversion of retinol to retinal is reversible, but the further conversion of

◆ A compound that can be converted into an active vitamin is called a *precursor.*

◆ Carotenoids are among the best-known phytochemicals.

vitamin A: all naturally occurring compounds with the biological activity of retinol (RET-ih-nol), the alcohol form of vitamin A.

beta-carotene (BAY-tah KARE-oh-teen): one of the carotenoids; an orange pigment and vitamin A precursor found in plants.

retinoids (RET-ih-noyds): chemically related compounds with biological activity similar to that of retinol; metabolites of retinol.

carotenoids (kah-ROT-eh-noyds): pigments commonly found in plants and animals, some of which have vitamin A activity. The carotenoid with the greatest vitamin A activity is beta-carotene.

vitamin A activity: a term referring to both the active forms of vitamin A and the precursor forms in foods without distinguishing between them.

---

* Carotenoids with vitamin A activity include alpha-carotene, beta-carotene, and beta-cryptoxanthin; carotenoids with no vitamin A activity include lycopene, lutein, and zeaxanthin.

FIGURE 11-1 Forms of Vitamin A

In this diagram, corners represent carbon atoms, as in all previous diagrams in this book. A further simplification here is that methyl groups (CH_3) are understood to be at the ends of the lines extending from corners. (See Appendix C for complete structures.)

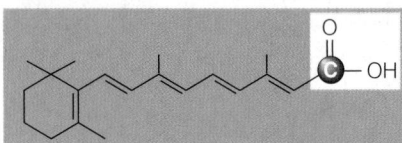

Retinol, the alcohol form

Retinal, the aldehyde form

Retinoic acid, the acid form

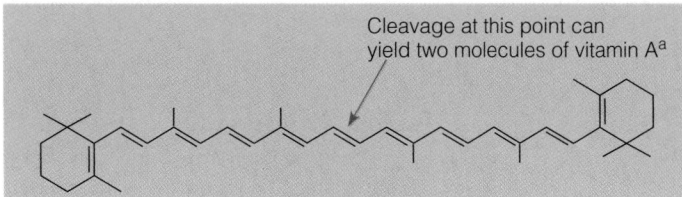

Cleavage at this point can yield two molecules of vitamin A[a]

Beta-carotene, a precursor

*Sometimes cleavage occurs at other points as well, so that one molecule of beta-carotene may yield only one molecule of vitamin A. Furthermore, not all beta-carotene is converted to vitamin A, and absorption of beta-carotene is not as efficient as that of vitamin A. For these reasons, 12 µg of beta-carotene are equivalent to 1 µg of vitamin A. Conversion of other carotenoids to vitamin A is even less efficient.

retinal to retinoic acid is irreversible (see Figure 11-2). This irreversibility is significant because each form of vitamin A performs a function that the others cannot.

Several proteins participate in the digestion and absorption of vitamin A.[3] After absorption via the lymph system, vitamin A eventually arrives at the liver, where it is stored. There, a special transport protein, **retinol-binding protein (RBP),** picks up vitamin A from the liver and carries it in the blood. Cells that use vitamin A have special protein receptors for it, as if the vitamin were fragile and had to be passed carefully from hand to hand without being dropped. Each form of vitamin A has its own receptor protein (retinol has several) within the cells.

Roles in the Body

Vitamin A is a versatile vitamin, known to influence over 500 genes.[4] Its major roles include:

- Promoting vision
- Participating in protein synthesis and cell differentiation (and thereby maintaining the health of epithelial tissues and skin)
- Supporting reproduction and growth

As mentioned, each form of vitamin A performs specific tasks. Retinol supports reproduction and is the major transport and storage form of the vitamin. Retinal is active in vision and is also an intermediate in the conversion of retinol to retinoic acid (review Figure 11-2). Retinoic acid acts like a hormone, regulating cell differentiation, growth, and embryonic development.[5] Animals raised on retinoic acid

FIGURE 11-2 Conversion of Vitamin A Compounds

Notice that the conversion from retinol to retinal is reversible, whereas the pathway from retinal to retinoic acid is not.

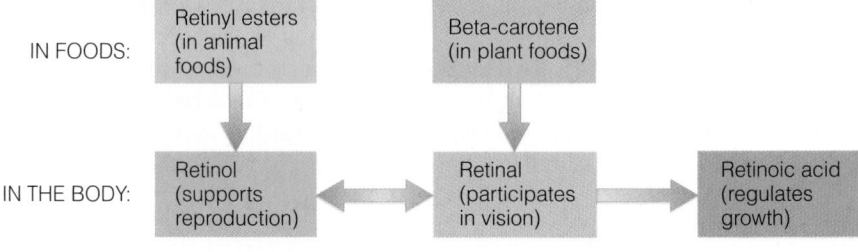

retinol-binding protein (RBP): the specific protein responsible for transporting retinol.

as their sole source of vitamin A can grow normally, but they become blind because retinoic acid cannot be converted to retinal (review Figure 11-2).

Vitamin A in Vision Vitamin A plays two indispensable roles in the eye: it helps maintain a crystal-clear outer window, the **cornea,** and it participates in the conversion of light energy into nerve impulses at the **retina** (see Figure 11-3 for details). The cells of the retina contain **pigment** molecules called **rhodopsin;** each rhodopsin molecule is composed of a protein called **opsin** bonded to a molecule of retinal. ◆ When light passes through the cornea of the eye and strikes the cells of the retina, rhodopsin responds by changing shape and becoming bleached. As it does, the retinal shifts from a *cis* to a *trans* configuration, just as fatty acids do during hydrogenation (see pp. 143–145). The *trans*-retinal cannot remain bonded to opsin. When retinal is released, opsin changes shape, thereby disturbing the membrane of the cell and generating an electrical impulse that travels along the cell's length. At the other end of the cell, the impulse is transmitted to a nerve cell, which conveys the message to the brain. Much of the retinal is then converted back to its active *cis* form and combined with the opsin protein to regenerate the pigment rhodopsin. Some retinal, however, may be oxidized to retinoic acid, a biochemical dead end for the visual process. Visual activity leads to repeated small losses of retinal, necessitating its constant replenishment either directly from foods or indirectly from retinol stores.

Vitamin A in Protein Synthesis and Cell Differentiation Despite its important role in vision, only one-thousandth of the body's vitamin A is in the retina. Much more is in the cells lining the body's surfaces. There, the vitamin participates in protein synthesis and **cell differentiation,** a process by which each type of cell develops to perform a specific function. Its role in cell differentiation helps explain how vitamin A may prevent cancer.[6]

All body surfaces, both inside and out, are covered by layers of cells known as **epithelial cells.** The **epithelial tissue** on the outside of the body is, of course, the skin—and vitamin A helps to protect against skin damage from sunlight.[7] The epithelial tissues that line the inside of the body are the **mucous membranes:** the linings of the mouth, stomach, and intestines; the linings of the lungs and the passages leading to them; the linings of the urinary bladder and urethra; the linings of the uterus and vagina; and the linings of the eyelids and sinus passageways. Within the body, the mucous membranes of the GI tract alone line an area larger than a quarter of a football field, and vitamin A helps to maintain their integrity (see Figure 11-4, p. 372).

Vitamin A promotes differentiation of epithelial cells and goblet cells, one-celled glands that synthesize and secrete mucus. Mucus coats and protects the epithelial cells from invasive microorganisms and other harmful substances, such as gastric juices.

◆ More than 100 million cells reside in the retina, and each contains about 30 million molecules of vitamin A-containing visual pigments.

cornea (KOR-nee-uh): the transparent membrane covering the outside of the eye.

retina (RET-in-uh): the layer of light-sensitive nerve cells lining the back of the inside of the eye; consists of rods and cones.

pigment: a molecule capable of absorbing certain wavelengths of light so that it reflects only those that we perceive as a certain color.

rhodopsin (ro-DOP-sin): a light-sensitive pigment of the retina; contains the retinal form of vitamin A and the protein opsin.
- **rhod** = red (pigment)
- **opsin** = visual protein

opsin (OP-sin): the protein portion of the visual pigment molecule.

cell differentiation (DIF-er-EN-she-AY-shun): the process by which immature cells develop specific functions different from those of the original that are characteristic of their mature cell type.

epithelial (ep-i-THEE-lee-ul) **cells**: cells on the surface of the skin and mucous membranes.

epithelial tissue: the layer of the body that serves as a selective barrier between the body's interior and the environment. (Examples are the cornea of the eyes, the skin, the respiratory lining of the lungs, and the lining of the digestive tract.)

mucous (MYOO-kus) **membranes:** the membranes, composed of mucus-secreting cells, that line the surfaces of body tissues.

FIGURE 11-3 *Animated!* **Vitamin A's Role in Vision**

CENGAGENOW
To test your understanding of these concepts, log on to **academic.cengage.com/login.**

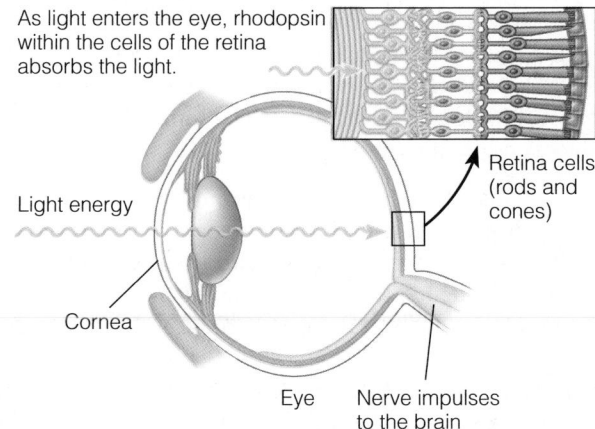

As light enters the eye, rhodopsin within the cells of the retina absorbs the light.

Light energy

Retina cells (rods and cones)

Cornea

Eye   Nerve impulses to the brain

The cells of the retina contain rhodopsin, a molecule composed of opsin (a protein) and *cis*-retinal (vitamin A).

cis-Retinal   *trans*-Retinal

As rhodopsin absorbs light, retinal changes from *cis* to *trans*, which triggers a nerve impulse that carries visual information to the brain.

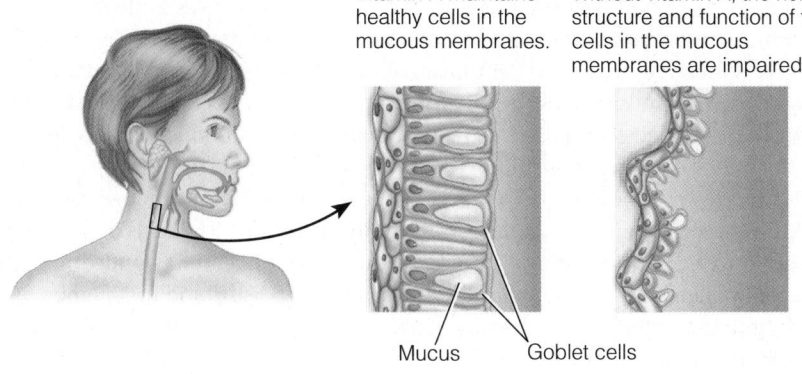

FIGURE 11-4 Mucous Membrane Integrity

Vitamin A maintains healthy cells in the mucous membranes.

Without vitamin A, the normal structure and function of the cells in the mucous membranes are impaired.

Mucus   Goblet cells

◆ The cells that destroy bone during growth are **osteoclasts;** those that build bone are **osteoblasts.**
- **osteo** = bone
- **clast** = break
- **blast** = build

◆ The sacs of degradative enzymes are **lysosomes** (LYE-so-zomes).

◆ Key antioxidant nutrients:
- Vitamin C, vitamin E, beta-carotene
- Selenium

Vitamin A in Reproduction and Growth As mentioned, vitamin A also supports reproduction and growth. In men, retinol participates in sperm development, and in women, vitamin A supports normal fetal development during pregnancy. Children lacking vitamin A fail to grow. When given vitamin A supplements, these children gain weight and grow taller.

The growth of bones illustrates that growth is a complex phenomenon of **remodeling.** To convert a small bone into a large bone, the bone-remodeling cells must "undo" some parts of the bone as they go, ◆ and vitamin A participates in the dismantling. The cells that break down bone contain sacs of degradative enzymes. ◆ With the help of vitamin A, these enzymes eat away at selected sites in the bone, removing the parts that are not needed.

Beta-Carotene as an Antioxidant In the body, beta-carotene serves primarily as a vitamin A precursor.[8] Not all dietary beta-carotene is converted to active vitamin A, however. Some beta-carotene may act as an antioxidant ◆ capable of protecting the body against disease. (See Highlight 11 for details.)

Vitamin A Deficiency

Vitamin A status depends mostly on the adequacy of vitamin A stores, 90 percent of which are in the liver. Vitamin A status also depends on a person's protein status because retinol-binding proteins serve as the vitamin's transport carriers inside the body.

If a person were to stop eating vitamin A–containing foods, deficiency symptoms would not begin to appear until after stores were depleted—one to two years for a healthy adult but much sooner for a growing child. Then the consequences would be profound and severe. Vitamin A deficiency is uncommon in the United States, but it is one of the developing world's major nutrition problems. More than 100 million children worldwide have some degree of vitamin A deficiency and thus are vulnerable to infectious diseases and blindness.

Infectious Diseases In developing countries around the world, measles is a devastating infectious disease, killing as many as two million children each year. The severity of the illness often correlates with the degree of vitamin A deficiency; deaths are usually due to related infections such as pneumonia and severe diarrhea. Providing large doses of vitamin A reduces the risk of dying from these infections.

The World Health Organization (WHO) and UNICEF (the United Nations International Children's Emergency Fund) have made the control of vitamin A deficiency a major goal in their quest to improve child health and survival throughout

---

remodeling: the dismantling and reformation of a structure, in this case, bone.

FIGURE 11-5 Vitamin A–Deficiency Symptom—Night Blindness

These photographs illustrate the eyes' slow recovery in response to a flash of bright light at night. In animal research studies, the response rate is measured with electrodes.

In dim light, you can make out the details in this room. You are using your rods for vision.

A flash of bright light momentarily blinds you as the pigment in the rods is bleached.

You quickly recover and can see the details again in a few seconds.

With inadequate vitamin A, you do not recover but remain blinded for many seconds.

the developing world. They recommend routine vitamin A supplementation for all children with measles in areas where vitamin A deficiency is a problem or where the measles death rate is high. In the United States, the American Academy of Pediatrics recommends vitamin A supplementation for certain groups of measles-infected infants and children. Vitamin A supplementation also protects against the complications of other life-threatening infections, including malaria, lung diseases, and HIV (human immunodeficiency virus, the virus that causes AIDS).[9]

Night Blindness **Night blindness** is one of the first detectable signs of vitamin A deficiency and permits early diagnosis. In night blindness, the retina does not receive enough retinal to regenerate the visual pigments bleached by light. The person loses the ability to recover promptly from the temporary blinding that follows a flash of bright light at night or to see after the lights go out. In many parts of the world, after the sun goes down, vitamin A–deficient people become night-blind: children cannot find their shoes or toys, and women cannot fetch water or wash dishes. They often cling to others or sit still, afraid that they may trip and fall or lose their way if they try to walk alone. In many developing countries, night blindness due to vitamin A deficiency is so common that the people have special words to describe it. In Indonesia, the term is *buta ayam,* which means "chicken eyes" or "chicken blindness." (Chickens do not have the cells of the retina that respond to dim light and therefore cannot see at night.) Figure 11-5 shows the eyes' slow recovery in response to a flash of bright light in night blindness.

Blindness (Xerophthalmia) Beyond night blindness is total blindness—failure to see at all. Night blindness is caused by a lack of vitamin A at the back of the eye, the retina; total blindness is caused by a lack at the front of the eye, the cornea. Severe vitamin A deficiency is the major cause of childhood blindness in the world, causing more than half a million preschool children to lose their sight each year. Blindness due to vitamin A deficiency, known as **xerophthalmia**, develops in stages. At first, the cornea becomes dry and hard, a condition known as **xerosis.** Then, corneal xerosis can quickly progress to **keratomalacia,** the softening of the cornea that leads to irreversible blindness.

Keratinization Elsewhere in the body, vitamin A deficiency affects other surfaces. On the body's outer surface, the epithelial cells change shape and begin to secrete the protein **keratin**—the hard, inflexible protein of hair and nails. As Figure 11-6 shows, the skin becomes dry, rough, and scaly as lumps of keratin accumulate (**keratinization**). Without vitamin A, the goblet cells in the GI tract diminish in number and activity, limiting the secretion of mucus. With less mucus, normal digestion and absorption of nutrients falter, and this, in turn, worsens malnutrition by limiting the absorption of whatever nutrients the diet may deliver. Similar changes in the cells of

FIGURE 11-6 Vitamin A–Deficiency Symptom—The Rough Skin of Keratinization

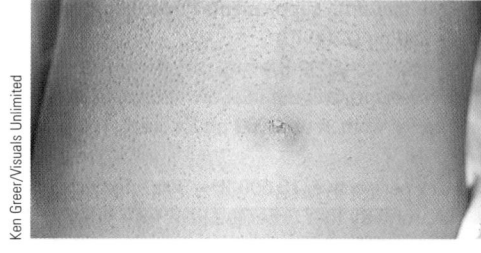

In vitamin A deficiency, the epithelial cells secrete the protein keratin in a process known as *keratinization.* (Keratinization doesn't occur in the GI tract, but mucus-producing cells dwindle and mucus production declines.) The extreme of this condition is *hyperkeratinization* or *hyperkeratosis.* When keratin accumulates around hair follicles, the condition is known as *follicular hyperkeratosis.*

night blindness: slow recovery of vision after flashes of bright light at night or an inability to see in dim light; an early symptom of vitamin A deficiency.

xerophthalmia (zer-off-THAL-mee-uh): progressive blindness caused by severe vitamin A deficiency.

• **xero** = dry

• **ophthalm** = eye

xerosis (zee-ROW-sis): abnormal drying of the skin and mucous membranes; a sign of vitamin A deficiency.

keratomalacia (KARE-ah-toe-ma-LAY-shuh): softening of the cornea that leads to irreversible blindness; seen in severe vitamin A deficiency.

keratin (KARE-uh-tin): a water-insoluble protein; the normal protein of hair and nails.

keratinization: accumulation of keratin in a tissue; a sign of vitamin A deficiency.

The carotenoids in foods bring colors to meals; the retinoids in our eyes allow us to see them.

© Polara Studios Inc.

compound. The beta-carotene in dark green, leafy vegetables is abundant but masked by large amounts of the green pigment **chlorophyll.**

Bright color is not always a sign of vitamin A activity, however. Beets and corn, for example, derive their colors from the red and yellow **xanthophylls,** which have no vitamin A activity. As for white plant foods such as potatoes, cauliflower, pasta, and rice, they also offer little or no vitamin A.

Vitamin A–Poor Fast Foods Fast foods often lack vitamin A. Anyone who dines frequently on hamburgers, French fries, and colas is wise to emphasize colorful vegetables and fruits at other meals.

Vitamin A–Rich Liver People sometimes wonder if eating liver too frequently can cause vitamin A toxicity. Liver is a rich source because vitamin A is stored in the livers of animals, just as in humans.* Arctic explorers who have eaten large quantities of polar bear liver have become ill with symptoms suggesting vitamin A toxicity, as have young children who regularly ate a chicken liver spread that provided three times their daily recommended intake. Liver offers many nutrients, and eating it periodically may improve a person's nutrition status. But caution is warranted not to eat too much too often, especially for pregnant women. With one ounce of beef liver providing more than three times the RDA for vitamin A, intakes can rise quickly.

IN SUMMARY

Vitamin A is found in the body in three forms: retinol, retinal, and retinoic acid. Together, they are essential to vision, healthy epithelial tissues, and growth. Vitamin A deficiency is a major health problem worldwide, leading to infections, blindness, and keratinization. Toxicity can also cause problems and is most often associated with supplement abuse. Animal-derived foods such as liver and whole or fortified milk provide retinoids, whereas brightly colored plant-derived foods such as spinach, carrots, and pumpkins provide beta-carotene and other carotenoids. In addition to serving as a precursor for vitamin A, beta-carotene may act as an antioxidant in the body. The accompanying table summarizes vitamin A's functions in the body, deficiency symptoms, toxicity symptoms, and food sources.

Vitamin A

Other Names

Retinol, retinal, retinoic acid; precursors are carotenoids such as beta-carotene

Beta-carotene: spinach and other dark leafy greens; broccoli, deep orange fruits (apricots, cantaloupe) and vegetables (squash, carrots, sweet potatoes, pumpkin)

RDA

Men: 900 µg RAE/day

Women: 700 µg RAE/day

Deficiency Disease

Hypovitaminosis A

Deficiency Symptoms

Night blindness, corneal drying (xerosis), triangular gray spots on eye (Bitot's spots), softening of the cornea (keratomalacia), and corneal degeneration and blindness (xerophthalmia); impaired immunity (infectious diseases); plugging of hair follicles with keratin, forming white lumps (hyperkeratosis)

Upper Level

Adults: 3000 µg/day

Chief Functions in the Body

Vision; maintenance of cornea, epithelial cells, mucous membranes, skin; bone and tooth growth; reproduction; immunity

Toxicity Disease

Hypervitaminosis A[a]

Significant Sources

Retinol: fortified milk, cheese, cream, butter, fortified margarine, eggs, liver

(continued)

[a]A related condition, *hypercarotenemia*, is caused by the accumulation of too much of the vitamin A precursor beta-carotene in the blood, which turns the skin noticeably yellow. Hypercarotenemia is not, strictly speaking, a toxicity symptom.

chlorophyll (KLO-row-fil): the green pigment of plants, which absorbs light and transfers the energy to other molecules, thereby initiating photosynthesis.

xanthophylls (ZAN-tho-fills): pigments found in plants; responsible for the color changes seen in autumn leaves.

* The liver is not the only organ that stores vitamin A. The kidneys, adrenals, and other organs do, too, but the liver stores the most and is the most commonly eaten organ meat.

Vitamin A (continued)	
Chronic Toxicity Symptoms	**Acute Toxicity Symptoms**
Increased activity of osteoclasts[b] causing reduced bone density; liver abnormalities; birth defects	Blurred vision, nausea, vomiting, vertigo; increase of pressure inside skull, mimicking brain tumor; headaches; muscle incoordination

[b] *Osteoclasts* are the cells that destroy bone during its growth. Those that build bone are *osteoblasts*.

Vitamin D

Vitamin D (calciferol) ◆ is different from all the other nutrients in that the body can synthesize it, with the help of sunlight, from a precursor that the body makes from cholesterol. Therefore, vitamin D is not an essential nutrient; given enough time in the sun, people need no vitamin D from foods.

Figure 11-9 diagrams the pathway for making and activating vitamin D. Ultraviolet rays from the sun hit the precursor in the skin and convert it to previtamin D_3. This compound works its way into the body and slowly, over the next 36 hours, is converted to its active form with the help of the body's heat. The biological activity of the active vitamin is 500- to 1000-fold greater than that of its precursor.

Regardless of whether the body manufactures vitamin D_3 or obtains it directly from foods, two hydroxylation reactions must occur before the vitamin becomes fully active.[15] First, the liver adds an OH group, and then the kidneys add another OH group to produce the active vitamin. A review of Figure 11-9 reveals how diseases affecting either the liver or the kidneys can interfere with the activation of vitamin D and produce symptoms of deficiency.

Roles in the Body

Though called a vitamin, vitamin D is actually a hormone—a compound manufactured by one part of the body that causes another part to respond. Like vitamin A, vitamin D has a binding protein that carries it to the target organs—most notably, the intestines, the kidneys, and the bones. All respond to vitamin D by making the minerals needed for bone growth and maintenance available.

Vitamin D in Bone Growth Vitamin D is a member of a large and cooperative bone-making and maintenance team ◆ composed of nutrients and other compounds, including vitamins A, C, and K; hormones (parathyroid hormone and calcitonin); the protein collagen; and the minerals calcium, phosphorus, magnesium, and fluoride. Vitamin D's special role in bone growth is to maintain blood concentrations of calcium and phosphorus. The bones grow denser and stronger as they absorb and deposit these minerals.

Vitamin D raises blood concentrations of these minerals in three ways. It enhances their absorption from the GI tract, their reabsorption by the kidneys, and their mobilization from the bones into the blood.[16] The vitamin may work alone, as it does in the GI tract, or in combination with parathyroid hormone, as it does in the bones and kidneys. Vitamin D is the director, but the star of the show is calcium. Details of calcium balance appear in Chapter 12.

Vitamin D in Other Roles Scientists have discovered many other vitamin D target tissues, including cells of the immune system, brain and nervous system, pancreas, skin, muscles and cartilage, and reproductive organs. Because vitamin D has numerous functions, it may be valuable in treating a number of disorders. Recent evidence suggests that vitamin D may protect against tuberculosis, gum inflammation, multiple sclerosis, and some cancers.[17]

◆ Vitamin D comes in many forms, the two most important being a plant version called **vitamin D_2** or **ergocalciferol** (ER-go-kal-SIF-er-ol) and an animal version called **vitamin D_3** or **cholecalciferol** (KO-lee-kal-SIF-er-ol).

◆ Key bone nutrients:
 • Vitamin D, vitamin K, vitamin A
 • Calcium, phosphorus, magnesium, fluoride

FIGURE 11-9 *Animated!* **Vitamin D Synthesis and Activation**

The precursor of vitamin D is made in the liver from cholesterol (see Figure 5-11 on p. 147 and Appendix C). The activation of vitamin D is a closely regulated process. The final product, active vitamin D, is also known as 1,25-dihydroxycholecalciferol (or calcitriol).

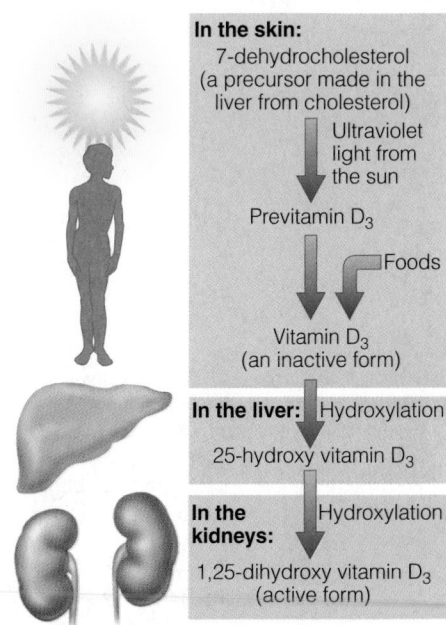

In the skin:
7-dehydrocholesterol (a precursor made in the liver from cholesterol)
↓ Ultraviolet light from the sun
Previtamin D_3
↓ Foods
Vitamin D_3 (an inactive form)
In the liver: Hydroxylation
25-hydroxy vitamin D_3
In the kidneys: Hydroxylation
1,25-dihydroxy vitamin D_3 (active form)

Vitamin D Deficiency

Factors that contribute to vitamin D deficiency include dark skin, breastfeeding without supplementation, lack of sunlight, and not using fortified milk. In vitamin D deficiency, production of the protein that binds calcium in the intestinal cells slows. Thus, even when calcium in the diet is adequate, it passes through the GI tract unabsorbed, leaving the bones undersupplied. Consequently, a vitamin D deficiency creates a calcium deficiency and increases the risks of several chronic diseases, most notably osteoporosis.[18] Vitamin D–deficient adolescents may not reach their peak bone mass.[19]

Rickets Worldwide, the vitamin D–deficiency disease **rickets** still afflicts many children.[20] In the United States, rickets is not common, but when it occurs, young, breastfed, black children are the ones most likely to be affected.[21] In rickets, the bones fail to calcify normally, causing growth retardation and skeletal abnormalities. The bones become so weak that they bend when they have to support the body's weight (see Figure 11-10). A child with rickets who is old enough to walk characteristically develops bowed legs, often the most obvious sign of the disease. Another sign is the beaded ribs ◆ that result from the poorly formed attachments of the bones to the cartilage.

◆ Because the poorly formed rib attachments resemble rosary beads, this symptom is commonly known as **rachitic** (ra-KIT-ik) **rosary** ("the rosary of rickets").

Osteomalacia In adults, the poor mineralization of bone results in the painful bone disease **osteomalacia**.[22] The bones become increasingly soft, flexible, brittle, and deformed.

Osteoporosis Any failure to synthesize adequate vitamin D or obtain enough from foods sets the stage for a loss of calcium from the bones, which can result in fractures. Highlight 12 describes the many factors that lead to osteoporosis, a condition of reduced bone density.

The Elderly Vitamin D deficiency is especially likely in older adults for several reasons. For one, the skin, liver, and kidneys lose their capacity to make and activate vi-

rickets: the vitamin D–deficiency disease in children characterized by inadequate mineralization of bone (manifested in bowed legs or knock-knees, outward-bowed chest, and knobs on ribs). A rare type of rickets, not caused by vitamin D deficiency, is known as *vitamin D–refractory rickets.*

osteomalacia (OS-tee-oh-ma-LAY-shuh): a bone disease characterized by softening of the bones. Symptoms include bending of the spine and bowing of the legs. The disease occurs most often in adult women.
• **osteo** = bone
• **malacia** = softening

FIGURE 11-10 Vitamin D–Deficiency Symptoms—Bowed Legs and Beaded Ribs of Rickets

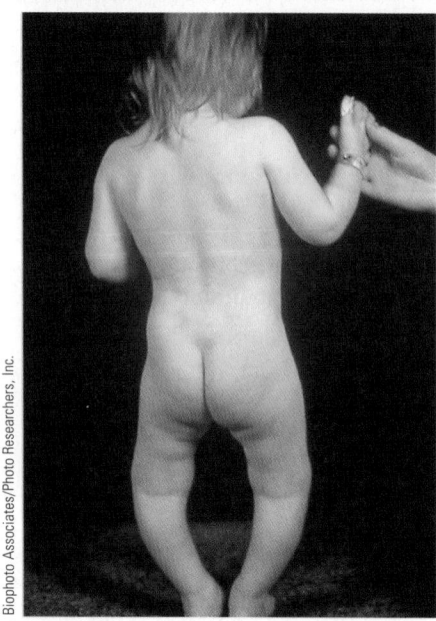

Biophoto Associates/Photo Researchers, Inc.

Bowed legs. In rickets, the poorly formed long bones of the legs bend outward as weight-bearing activities such as walking begin.

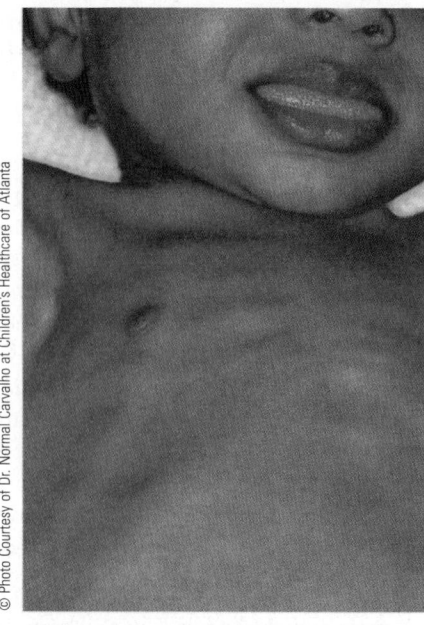

© Photo Courtesy of Dr. Normal Carvalho at Children's Healthcare of Atlanta

Beaded ribs. In rickets, a series of "beads" develop where the cartilages and bones attach.

tamin D with advancing age. For another, older adults typically drink little or no milk—the main dietary source of vitamin D. And finally, older adults typically spend much of the day indoors, and when they do venture outside, many of them cautiously wear protective clothing or apply sunscreen to all sun-exposed areas of their skin. Dark-skinned people living in northern regions are particularly vulnerable.[23] All of these factors increase the likelihood of vitamin D deficiency and its consequences: bone losses and fractures. Vitamin D supplementation helps to reduce the risks of falls and fractures in elderly persons.[24]

Vitamin D Toxicity

Vitamin D clearly illustrates how nutrients in optimal amounts support health, but both inadequacies and excesses cause trouble. Vitamin D is the most likely of the vitamins to have toxic effects when consumed in excessive amounts. The amounts of vitamin D made by the skin and found in foods are well within the safe limits set by the Upper Level, but supplements containing the vitamin in concentrated form should be kept out of the reach of children and used cautiously, if at all, by adults.

Excess vitamin D raises the concentration of blood calcium. ◆ Excess blood calcium tends to precipitate in the soft tissue, forming stones, especially in the kidneys where calcium is concentrated in the effort to excrete it. Calcification may also harden the blood vessels and is especially dangerous in the major arteries of the heart and lungs, where it can cause death.

◆ High blood calcium is known as **hypercalcemia** and may develop from a variety of disorders, including vitamin D toxicity. It does *not* develop from a high calcium intake.

Vitamin D Recommendations and Sources

Only a few foods contain vitamin D naturally. Fortunately, the body can make vitamin D with the help of a little sunshine. In setting dietary recommendations, however, the DRI Committee assumed that no vitamin D was available from skin synthesis. Current recommendations may be insufficient, however, given recent research showing numerous health benefits and safety of higher intakes.[25]

Vitamin D in Foods Most adults, especially in sunny regions, need not make special efforts to obtain vitamin D from food. People who are not outdoors much or who live in northern or predominantly cloudy or smoggy areas are advised to drink at least 2 cups of vitamin D–fortified milk a day. The fortification of milk with vitamin D is the best guarantee that people will meet their needs and underscores the importance of milk in a well-balanced diet.* Despite vitamin D fortification, the average intake in the United States falls short of recommendations.[26]

Without adequate sunshine, fortification, or supplementation, a vegan diet cannot meet vitamin D needs. Vegetarians who do not include milk in their diets may use vitamin D–fortified soy milk and cereals. Importantly, feeding infants and young children nonfortified "health beverages" instead of milk or infant formula can create severe nutrient deficiencies, including rickets.

Vitamin D from the Sun Most of the world's population relies on natural exposure to sunlight to maintain adequate vitamin D nutrition. The sun imposes no risk of vitamin D toxicity; prolonged exposure to sunlight degrades the vitamin D precursor in the skin, preventing its conversion to the active vitamin. Even lifeguards on southern beaches are safe from vitamin D toxicity from the sun.

Prolonged exposure to sunlight does, however, prematurely wrinkle the skin and present the risk of skin cancer. Sunscreens help reduce these risks, but unfortunately, sunscreens with sun protection factors (SPF) of 8 and higher also prevent vitamin D synthesis. A strategy to avoid this dilemma is to apply sunscreen after enough time has elapsed to provide sufficient vitamin D synthesis. For

A cold glass of milk refreshes as it replenishes vitamin D and other bone-building nutrients.

---

* Vitamin D fortification of milk in the United States is 10 micrograms cholecalciferol (400 IU) per quart; in Canada, it is 9 to 12 micrograms (350 to 470 IU) per liter, with a current proposal to raise it slightly.

The sunshine vitamin—vitamin D.

© Fotographia/CORBIS

FIGURE 11-11 Vitamin D Synthesis and Latitude

Above 40° north latitude (and below 40° south latitude in the southern hemisphere), vitamin D synthesis essentially ceases for the four months of winter. Synthesis increases as spring approaches, peaks in summer, and declines again in the fall. People living in regions of extreme northern (or extreme southern) latitudes may miss as much as six months of vitamin D production.

most people, exposing hands, face, and arms on a clear summer day for 5 to 10 minutes two or three times a week should be sufficient to maintain vitamin D nutrition.[27]

The pigments of dark skin provide some protection from the sun's damage, but they also reduce vitamin D synthesis. Dark-skinned people require longer sunlight exposure than light-skinned people: heavily pigmented skin achieves the same amount of vitamin D synthesis in three hours as fair skin in 30 minutes. Latitude, season, and time of day ◆ also have dramatic effects on vitamin D synthesis (see Figure 11-11). Heavy clouds, smoke, or smog block the ultraviolet (UV) rays of the sun that promote vitamin D synthesis. Differences in skin pigmentation, latitude, and smog may account for the finding that African American people, especially those in northern, smoggy cities, are most likely to be vitamin D deficient and develop rickets.[28] To ensure an adequate vitamin D status, supplements may be needed.[29] The body's vitamin D stores from summer synthesis alone are insufficient to meet winter needs.[30]

◆ Factors that may limit sun exposure and, therefore, vitamin D synthesis:
- Geographic location
- Season of the year
- Time of day
- Air pollution
- Clothing
- Tall buildings
- Indoor living
- Sunscreens

 **Dietary Guidelines for Americans 2005**

People with dark skin and those with insufficient exposure to sunlight should consume extra vitamin D from vitamin D–fortified foods and/or supplements.

Depending on the radiation used, the UV rays from tanning lamps and tanning beds may also stimulate vitamin D synthesis and increase bone density.[31] The potential hazards of skin damage, however, may outweigh any possible benefits.* The Food and Drug Administration (FDA) warns that if the lamps are not properly filtered, people using tanning booths risk burns, damage to the eyes and blood vessels, and skin cancer.

* The best wavelengths for vitamin D synthesis are UV-B rays between 290 and 310 nanometers. Some tanning parlors advertise "UV-A rays only, for a tan without the burn," but UV-A rays can damage the skin.

IN SUMMARY

Vitamin D can be synthesized in the body with the help of sunlight or obtained from fortified milk. It sends signals to three primary target sites: the GI tract to absorb more calcium and phosphorus, the bones to release more, and the kidneys to retain more. These actions maintain blood calcium concentrations and support bone formation. A deficiency causes rickets in childhood and osteomalacia in later life. The table below summarizes vitamin D facts.

Vitamin D

Other Names

Calciferol (kal-SIF-er-ol), 1,25-dihydroxy vitamin D (calcitriol); the animal version is vitamin D_3 or cholecalciferol; the plant version is vitamin D_2 or ergocalciferol; precursor is the body's own cholesterol

Adequate Intake (AI)

Adults: 5 µg/day (19–50 yr)

    10 µg/day (51–70 yr)

    15 µg/day (>70 yr)

Upper Level

Adults: 50 µg/day

Chief Functions in the Body

Mineralization of bones (raises blood calcium and phosphorus by increasing absorption from digestive tract, withdrawing calcium from bones, stimulating retention by kidneys)

Significant Sources

Synthesized in the body with the help of sunlight; fortified milk, margarine, butter, juices, cereals, and chocolate mixes; veal, beef, egg yolks, liver, fatty fish (herring, salmon, sardines) and their oils

Deficiency Symptoms

Rickets in Children

Inadequate calcification, resulting in misshapen bones (bowing of legs); enlargement of ends of long bones (knees, wrists); deformities of ribs (bowed, with beads or knobs);[a] delayed closing of fontanel, resulting in rapid enlargement of head (see figure below); lax muscles resulting in protrusion of abdomen; muscle spasms

Osteomalacia or Osteoporosis in Adults

Loss of calcium, resulting in soft, flexible, brittle, and deformed bones; progressive weakness; pain in pelvis, lower back, and legs

Toxicity Disease

Hypervitaminosis D

Toxicity Symptoms

Elevated blood calcium; calcification of soft tissues (blood vessels, kidneys, heart, lungs, tisues around joints)

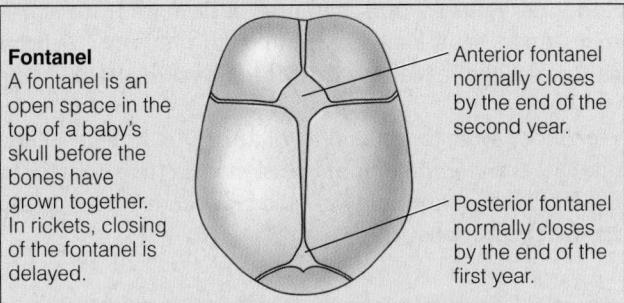

Fontanel
A fontanel is an open space in the top of a baby's skull before the bones have grown together. In rickets, closing of the fontanel is delayed.

Anterior fontanel normally closes by the end of the second year.

Posterior fontanel normally closes by the end of the first year.

[a]Bowing of the ribs causes the symptoms known as *pigeon breast*. The beads that form on the ribs resemble rosary beads; thus this symptom is known as *rachitic* (ra-KIT-ik) *rosary* ("the rosary of rickets").

Vitamin E

Researchers discovered a component of vegetable oils necessary for reproduction in rats and named this antisterility factor **tocopherol,** which means "to bring forth offspring." When chemists isolated four different tocopherol compounds, they designated them by the first four letters of the Greek alphabet: alpha, beta, gamma, and delta. The tocopherols consist of a complex ring structure and a long saturated side chain. (Appendix C provides the chemical structures.) The positions of methyl

tocopherol (tuh-KOFF-er-ol): a general term for several chemically related compounds, one of which has vitamin E activity. (See Appendix C for chemical structures.)

groups (CH_3) on the side chain and their chemical rotations distinguish one tocopherol from another. **Alpha-tocopherol** is the only one with vitamin E activity in the human body.[32] The other tocopherols are not readily converted to alpha-tocopherol in the body, nor do they perform the same roles. Whether these other tocopherols might be beneficial in other ways is the subject of current research.[33]

Vitamin E as an Antioxidant

Vitamin E is a fat-soluble antioxidant ◆ and one of the body's primary defenders against the adverse effects of free radicals. Its main action is to stop the chain reaction of free radicals producing more free radicals (see Highlight 11). In doing so, vitamin E protects the vulnerable components of the cells and their membranes from destruction. Most notably, vitamin E prevents the oxidation of the polyunsaturated fatty acids, but it protects other lipids and related compounds (for example, vitamin A) as well.

Accumulating evidence suggests that vitamin E may reduce the risk of heart disease by protecting low-density lipoproteins (LDL) against oxidation and reducing inflammation.[34] The oxidation of LDL and inflammation have been implicated as key factors in the development of heart disease. Highlight 11 provides many more details on how vitamin E and other antioxidants protect against chronic diseases, such as heart disease and cancer.

Vitamin E Deficiency

In human beings, a primary deficiency of vitamin E (from poor dietary intake) is rare; deficiency is usually associated with diseases of fat malabsorption such as cystic fibrosis. Without vitamin E, the red blood cells break open and spill their contents, probably due to oxidation of the polyunsaturated fatty acids in their membranes. This classic sign of vitamin E deficiency, known as **erythrocyte hemolysis,** is seen in premature infants, born before the transfer of vitamin E from the mother to the infant that takes place in the last weeks of pregnancy. Vitamin E treatment corrects **hemolytic anemia.**

Prolonged vitamin E deficiency also causes neuromuscular dysfunction involving the spinal cord and retina of the eye. Common symptoms include loss of muscle coordination and reflexes and impaired vision and speech. Vitamin E treatment corrects these neurological symptoms of vitamin E deficiency, but it does *not* prevent or cure the hereditary **muscular dystrophy** that afflicts young children.

Two other conditions seem to respond to vitamin E treatment, although results are inconsistent. One is a nonmalignant breast disease **(fibrocystic breast disease),** and the other is an abnormality of blood flow that causes cramping in the legs **(intermittent claudication).**

Vitamin E Toxicity

Vitamin E supplement use has risen in recent years as its protective actions against chronic diseases have been recognized. Still, toxicity is rare, and it appears safe across a broad range of intakes.[35] The Upper Level for vitamin E (1000 milligrams) is more than 65 times greater than the recommended intake for adults (15 milligrams). Extremely high doses of vitamin E may interfere with the blood-clotting action of vitamin K and enhance the effects of drugs used to oppose blood clotting, causing hemorrhage.

Vitamin E Recommendations

The current RDA for vitamin E is based on the alpha-tocopherol form only. As mentioned earlier, the other tocopherols cannot be converted to alpha-tocopherol, nor

---

◆ Key antioxidant nutrients:
  • Vitamin C, vitamin E, beta-carotene
  • Selenium

---

alpha-tocopherol: the active vitamin E compound.

erythrocyte (eh-RITH-ro-cite) **hemolysis** (he-MOLL-uh-sis): the breaking open of red blood cells (erythrocytes); a symptom of vitamin E–deficiency disease in human beings.
  • **erythro** = red
  • **cyte** = cell
  • **hemo** = blood
  • **lysis** = breaking

hemolytic (HE-moh-LIT-ick) **anemia:** the condition of having too few red blood cells as a result of erythrocyte hemolysis.

muscular dystrophy (DIS-tro-fee): a hereditary disease in which the muscles gradually weaken. Its most debilitating effects arise in the lungs.

fibrocystic (FYE-bro-SIS-tik) **breast disease:** a harmless condition in which the breasts develop lumps, sometimes associated with caffeine consumption. In some, it responds to abstinence from caffeine; in others, it can be treated with vitamin E.
  • **fibro** = fibrous tissue
  • **cyst** = closed sac

intermittent claudication (klaw-dih-KAY-shun): severe calf pain caused by inadequate blood supply. It occurs when walking and subsides during rest.
  • **intermittent** = at intervals
  • **claudicare** = to limp

can they perform the same metabolic roles in the body. A person who consumes large quantities of polyunsaturated fatty acids needs more vitamin E. Fortunately, vitamin E and polyunsaturated fatty acids tend to occur together in the same foods. Current research suggests that most adults in the United States fall short of recommended intakes for vitamin E and that smokers may have a higher requirement.[36]

Fat-soluble vitamin E is found predominantly in vegetable oils, seeds, and nuts.

Vitamin E in Foods

Vitamin E is widespread in foods. Much of the vitamin E in the diet comes from vegetable oils and products made from them, such as margarine and salad dressings. Wheat germ oil is especially rich in vitamin E.

Because vitamin E is readily destroyed by heat processing (such as deep-fat frying) and oxidation, fresh or lightly processed foods are preferable sources. Most processed and convenience foods do not contribute enough vitamin E to ensure an adequate intake.

Prior to 2000, values of the vitamin E in foods reflected all of the tocopherols and were expressed in "milligrams of tocopherol equivalents." ◆ These measures overestimated the amount of alpha-tocopherol. To estimate the alpha-tocopherol content of foods stated in tocopherol equivalents, multiply by 0.8.[37]

◆ Appendix H accurately presents vitamin E data in milligrams of alpha-tocopherol.

IN SUMMARY

Vitamin E acts as an antioxidant, defending lipids and other components of the cells against oxidative damage. Deficiencies are rare, but they do occur in premature infants, the primary symptom being erythrocyte hemolysis. Vitamin E is found predominantly in vegetable oils and appears to be one of the least toxic of the fat-soluble vitamins. The summary table reviews vitamin E's functions, deficiency symptoms, toxicity symptoms, and food sources.

Vitamin E

Other Names	Significant Sources
Alpha-tocopherol	Polyunsaturated plant oils (margarine, salad dressings, shortenings), leafy green vegetables, wheat germ, whole grains, liver, egg yolks, nuts, seeds, fatty meats
RDA	
Adults: 15 mg/day	Easily destroyed by heat and oxygen
Upper Level	**Deficiency Symptoms**
Adults: 1000 mg/day	Red blood cell breakage,[a] nerve damage
Chief Functions in the Body	**Toxicity Symptoms**
Antioxidant (stabilization of cell membranes, regulation of oxidation reactions, protection of polyunsaturated fatty acids [PUFA] and vitamin A)	Augments the effects of anticlotting medication

[a]The breaking of red blood cells is called *erythrocyte hemolysis.*

Vitamin K

Like vitamin D, vitamin K can be obtained from a nonfood source. Bacteria in the GI tract synthesize vitamin K that the body can absorb. Vitamin K ◆ acts primarily in blood clotting, where its presence can make the difference between life and death. Blood has a remarkable ability to remain liquid, but it can turn solid within seconds

◆ K stands for the Danish word *koagulation* ("coagulation" or "clotting").

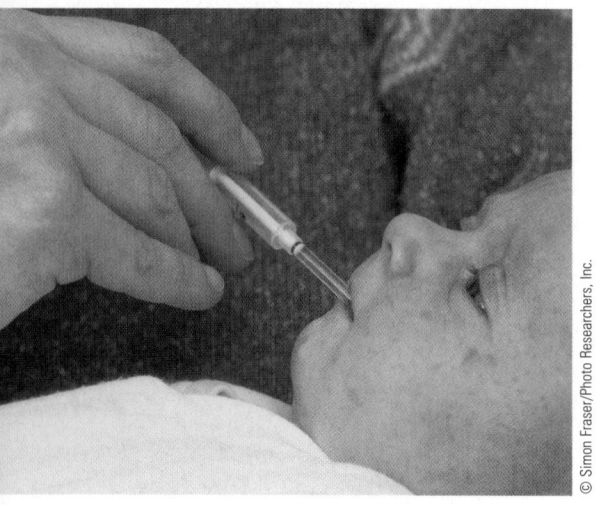

© Simon Fraser/Photo Researchers, Inc.

Soon after birth, newborn infants receive a dose of vitamin K to prevent hemorrhagic disease.

◆ Hemophilia is caused by a genetic defect and has no relation to vitamin K.

◆ Reminder: A *primary deficiency* develops in response to an inadequate dietary intake whereas a *secondary deficiency* occurs for other reasons.

when the integrity of that system is disturbed. (If blood did not clot, a single pinprick could drain the entire body of all its blood.)

Roles in the Body

More than a dozen different proteins and the mineral calcium are involved in making a blood clot. Vitamin K is essential for the activation of several of these proteins, among them prothrombin, made by the liver as a precursor of the protein thrombin (see Figure 11-12). When any of the blood-clotting factors is lacking, **hemorrhagic disease** results. If an artery or vein is cut or broken, bleeding goes unchecked. (Of course, this is not to say that hemorrhaging is always caused by vitamin K deficiency. Another cause is the hereditary disorder **hemophilia,** ◆ which is not curable with vitamin K.)

Vitamin K also participates in the synthesis of bone proteins. Without vitamin K, the bones produce an abnormal protein that cannot bind to the minerals that normally form bones, resulting in low bone density.[38] An adequate intake of vitamin K helps to make the bone protein correctly, decreases bone turnover, and protects against hip fractures.[39]

Vitamin K is historically known for its role in blood clotting, and more recently for its participation in bone building, but researchers continue to discover proteins needing vitamin K's assistance.[40] These proteins have been identified in the plaques of atherosclerosis, the kidneys, and the nervous system.

Vitamin K Deficiency

A primary deficiency ◆ of vitamin K is rare, but a secondary deficiency may occur in two circumstances. First, whenever fat absorption falters, as occurs when bile production fails, vitamin K absorption diminishes. Second, some drugs disrupt vitamin K's synthesis and action in the body: antibiotics kill the vitamin K–producing bacteria in the intestine, and anticoagulant drugs interfere with vitamin K metabolism and activity. When vitamin K deficiency does occur, it can be fatal.

Newborn infants present a unique case of vitamin K nutrition because they are born with a **sterile** intestinal tract, and the vitamin K–producing bacteria take weeks to establish themselves. At the same time, plasma prothrombin concentrations are low. (This reduces the likelihood of fatal blood clotting during the stress of

hemorrhagic (hem-oh-RAJ-ik) **disease:** a disease characterized by excessive bleeding.

hemophilia (HE-moh-FEEL-ee-ah): a hereditary disease in which the blood is unable to clot because it lacks the ability to synthesize certain clotting factors.

sterile: free of microorganisms, such as bacteria.

FIGURE 11-12 Blood-Clotting Process

When blood is exposed to air, foreign substances, or secretions from injured tissues, platelets (small, cell-like structures in the blood) release a phospholipid known as thromboplastin. Thromboplastin catalyzes the conversion of the inactive protein prothrombin to the active enzyme thrombin. Thrombin then catalyzes the conversion of the precursor protein fibrinogen to the active protein fibrin that forms the clot.

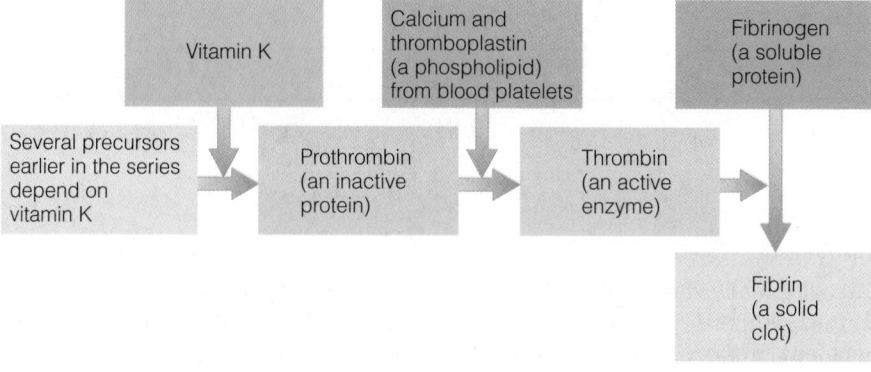

birth.) To prevent hemorrhagic disease in the newborn, a single dose of vitamin K
◆ (usually as the naturally occurring form, phylloquinone) is given at birth either
orally or by intramuscular injection. Concerns that vitamin K given at birth raises
the risks of childhood cancer are unproved and unlikely.

Vitamin K Toxicity

Toxicity is not common, and no adverse effects have been reported with high intakes
of vitamin K. Therefore, an Upper Level has not been established. High doses of vitamin K can reduce the effectiveness of anticoagulant drugs used to prevent blood
clotting.[41] People taking these drugs should eat vitamin K–rich foods in moderation
and keep their intakes consistent from day to day.

Vitamin K Recommendations and Sources

As mentioned earlier, vitamin K is made in the GI tract by the billions of bacteria
that normally reside there. Once synthesized, vitamin K is absorbed and stored in
the liver. This source provides only about half of a person's needs. Vitamin K–rich
foods such as green vegetables and vegetable oils can easily supply the rest.

◆ The natural form of vitamin K is **phylloquinone** (FILL-oh-KWIN-own); the synthetic form is **menadione** (men-uh-DYE-
own). See Appendix C for the chemistry of
these structures.

Notable food sources of vitamin K include
green vegetables such as collards, spinach, bib
lettuce, brussels sprouts, and cabbage and vegetable oils such as soybean oil and canola oil.

© Matthew Farruggio

IN SUMMARY

Vitamin K helps with blood clotting, and its deficiency causes hemorrhagic disease (uncontrolled bleeding). Bacteria in the GI tract can make the vitamin;
people typically receive about half of their requirements from bacterial synthesis and half from foods such as green vegetables and vegetable oils. Because
people depend on bacterial synthesis for vitamin K, deficiency is most likely in
newborn infants and in people taking antibiotics. The accompanying table
provides a summary of vitamin K facts.

Vitamin K

Other Names	Significant Sources
Phylloquinone, menaquinone, menadione, naphthoquinone	Bacterial synthesis in the digestive tract;[a] liver; leafy green vegetables, cabbage-type vegetables; milk

Adequate Intakes (AI)	Deficiency Symptoms
Men: 120 µg/day	Hemorrhaging
Women: 90 µg/day	

Chief Functions in the Body	Toxicity Symptoms
Synthesis of blood-clotting proteins and bone proteins	None known

[a]Vitamin K needs cannot be met from bacterial synthesis alone; however, it is a potentially important source in the small intestine, where absorption efficiency ranges from 40 to 70 percent.

The Fat-Soluble Vitamins— In Summary

The four fat-soluble vitamins play many specific roles in the growth and maintenance of the body. Their presence affects the health and function of the eyes, skin,
GI tract, lungs, bones, teeth, nervous system, and blood; their deficiencies become
apparent in these same areas. Toxicities of the fat-soluble vitamins are possible,
especially when people use supplements, because the body stores excesses.

As with the water-soluble vitamins, the function of one fat-soluble vitamin often depends on the presence of another. Recall that vitamin E protects vitamin A from oxidation. In vitamin E deficiency, vitamin A absorption and storage are impaired. Three of the four fat-soluble vitamins—A, D, and K—play important roles in bone growth and remodeling. As mentioned, vitamin K helps synthesize a specific bone protein, and vitamin D regulates that synthesis. Vitamin A, in turn, may control which bone-building genes respond to vitamin D.

Fat-soluble vitamins also interact with minerals. Vitamin D and calcium cooperate in bone formation, and zinc is required for the synthesis of vitamin A's transport protein, retinol-binding protein. Zinc also assists the enzyme that regenerates retinal from retinol in the eye.

The roles of the fat-soluble vitamins differ from those of the water-soluble vitamins, and they appear in different foods—yet they are just as essential to life. The need for them underlines the importance of eating a wide variety of nourishing foods daily. The following table condenses the information on fat-soluble vitamins into a short summary.

IN SUMMARY   The Fat-Soluble Vitamins

Vitamin and Chief Functions	Deficiency Symptoms	Toxicity Symptoms	Significant Sources
Vitamin A Vision; maintenance of cornea, epithelial cells, mucous membranes, skin; bone and tooth growth; reproduction; immunity	Infectious diseases, night blindness, blindness (xerophthalmia), keratinization	Reduced bone mineral density, liver abnormalities, birth defects	Retinol: milk and milk products Beta-carotene: dark green leafy and deep yellow/orange vegetables
Vitamin D Mineralization of bones (raises blood calcium and phosphorus by increasing absorption from digestive tract, withdrawing calcium from bones, stimulating retention by kidneys)	Rickets, osteomalacia	Calcium imbalance (calcification of soft tissues and formation of stones)	Synthesized in the body with the help of sunshine; fortified milk
Vitamin E Antioxidant (stabilization of cell membranes, regulation of oxidation reactions, protection of polyunsaturated fatty acids [PUFA] and vitamin A)	Erythrocyte hemolysis, nerve damage	Hemorrhagic effects	Vegetable oils
Vitamin K Synthesis of blood-clotting proteins and bone proteins	Hemorrhage	None known	Synthesized in the body by GI bacteria; green leafy vegetables

CENGAGENOW™
academic.cengage.com/login

Nutrition Portfolio

For the fat-soluble vitamins, select colorful fruits and vegetables, fortified milk or soy products, and vegetable oils; use supplements with caution, if at all.

■ Examine your weekly choices of vegetables and evaluate whether you meet the recommendations for dark green or orange and deep yellow vegetables.

■ Consider whether you drink enough vitamin D–fortified milk or go outside in the sunshine regularly.

■ Describe the vegetable oils you use when you cook and their vitamin contributions.

NUTRITION ON THE NET

For further study of topics covered in this chapter, log on to **academic.cengage .com/nutrition/rolfes/UNCN8e**. Go to Chapter 11, then to Nutrition on the Net.

- Search for "vitamins" at the American Dietetic Association: **www.eatright.org**
- Review the Dietary Reference Intakes for vitamins A, D, E, and K and the carotenoids by searching for "DRI": **www.nap.edu**

- Visit the World Health Organization to learn about "vitamin deficiencies" around the world: **www.who.int**
- Search for "vitamins" at the U.S. Government health information site: **www.healthfinder.gov**
- Learn how fruits and vegetables support a healthy diet rich in vitamins from the 5 A Day for Better Health program: **www.5aday.com** or **www.5aday.gov**

NUTRITION CALCULATIONS

CENGAGENOW™ For additional practice log on to **academic.cengage.com/login**. Go to Chapter 11, then to Nutrition Calculations.

These exercises will help you learn the best food sources for the vitamins and prepare you to examine your own food choices. See p. 389 for answers.

1. Review the units in which vitamins are measured (a spot check). For each of these vitamins, note the unit of measure:

   Vitamin A   Vitamin D

   Vitamin E   Vitamin K

2. Analyze the vitamin contents of foods. Review the figures, photos, and food sources sections in Chapters 10 and 11 and list the food group(s) that contributed the

most of each vitamin. Which food groups offer the most thiamin? The most riboflavin? The most niacin? The most vitamin B_6? The most folate? The most vitamin B_{12}? The most vitamin C? The most vitamin A? The most vitamin D? The most vitamin E?

List the groups that provide "the most" and compare them with the USDA Food Guide in Chapter 2.

This exercise should convince you that each of the food groups provides some, but not all, of the vitamins needed daily. For a full array, a person needs to eat a variety of foods from each of the food groups regularly.

STUDY QUESTIONS

CENGAGENOW™

To assess your understanding of chapter topics, take the Student Practice Test and explore the modules recommended in your Personalized Study Plan. Log on to **academic.cengage.com/login**.

These questions will help you review the chapter. You will find the answers in the discussions on the pages provided.

1. List the fat-soluble vitamins. What characteristics do they have in common? How do they differ from the water-soluble vitamins? (p. 369)

2. Summarize the roles of vitamin A and the symptoms of its deficiency. (pp. 370–374)

3. What are vitamin precursors? Name the precursors of vitamin A, and tell in what classes of foods they are located. Give examples of foods with high vitamin A activity. (pp. 369, 374–376)

4. How is vitamin D unique among the vitamins? What is its chief function? What are the richest sources of this vitamin? (pp. 377, 379–380)

5. Describe vitamin E's role as an antioxidant. What are the chief symptoms of vitamin E deficiency? (p. 382)

6. What is vitamin K's primary role in the body? What conditions may lead to vitamin K deficiency? (pp. 384–385)

These multiple choice questions will help you prepare for an exam. Answers can be found on p. 389.

1. Fat-soluble vitamins:
   a. are easily excreted.
   b. seldom reach toxic levels.
   c. require bile for absorption.
   d. are not stored in the body's tissues.

2. The form of vitamin A active in vision is:
   a. retinal.
   b. retinol.
   c. rhodopsin.
   d. retinoic acid.

3. Vitamin A–deficiency symptoms include:
   a. rickets and osteomalacia.
   b. hemorrhaging and jaundice.
   c. night blindness and keratomalacia.
   d. fibrocystic breast disease and erythrocyte hemolysis.

4. Good sources of vitamin A include:
   a. oatmeal, pinto beans, and ham.
   b. apricots, turnip greens, and liver.
   c. whole-wheat bread, green peas, and tuna.
   d. corn, grapefruit juice, and sunflower seeds.

5. To keep minerals available in the blood, vitamin D targets:
   a. the skin, the muscles, and the bones.
   b. the kidneys, the liver, and the bones.
   c. the intestines, the kidneys, and the bones.
   d. the intestines, the pancreas, and the liver.

6. Vitamin D can be synthesized from a precursor that the body makes from:
   a. bilirubin.
   b. tocopherol.
   c. cholesterol.
   d. beta-carotene.

7. Vitamin E's most notable role is to:
   a. protect lipids against oxidation.
   b. activate blood-clotting proteins.
   c. support protein and DNA synthesis.
   d. enhance calcium deposits in the bones.

8. The classic sign of vitamin E deficiency is:
   a. rickets.
   b. xeropthalmia.
   c. muscular dystrophy.
   d. erythrocyte hemolysis.

9. Without vitamin K:
   a. muscles atrophy.
   b. bones become soft.
   c. skin rashes develop.
   d. blood fails to clot.

10. A significant amount of vitamin K comes from:
    a. vegetable oils.
    b. sunlight exposure.
    c. bacterial synthesis.
    d. fortified grain products.

REFERENCES

1. E. H. Harrison, Mechanisms of digestion and absorption of dietary vitamin A, *Annual Review of Nutrition* 25 (2005): 87–103.
2. S. J. Hickenbottom and coauthors, Variability in conversion of β-carotene to vitamin A in men as measured by using a double-tracer study design, *American Journal of Clinical Nutrition* 75 (2002): 900–907; K. J. Yeum and R. M. Russell, Carotenoid bioavailability and bioconversion, *Annual Review of Nutrition* 22 (2002): 483–504.
3. E. H. Harrison, Mechanisms of digestion and absorption of dietary vitamin A, *Annual Review of Nutrition* 25 (2005): 87–103.
4. J. Bastien and C. Rochette-Egly, Nuclear retinoid receptors and the transcription of retinoid-target genes, *Gene* 328 (2004): 1–16; J. E. Balmer and R. Blomhoff, Gene expression regulation by retinoic acid, *Journal of Lipid Research* 43 (2002): 1773–1808.
5. M. Clagett-Dame and H. F. DeLuca, The role of vitamin A in mammalian reproduction and embryonic development, *Annual Review of Nutrition* 22 (2002): 347–381.
6. T. Oren, J. A. Sher, and T. Evans, Hematopoiesis and retinoids: Development and disease, *Leukemia Lymphoma* 44 (2003): 1881–1891; A. C. Ross, Advances in retinoid research: Mechanisms of cancer chemoprevention symposium introduction, *Journal of Nutrition* 133 (2003): 271S–272S.
7. H. Sies and W. Stahl, Nutritional protection against skin damage from sunlight, *Annual Review of Nutrition* 24 (2004): 173–200.
8. Committee on Dietary Reference Intakes, *Dietary Reference Intakes for Vitamin C, Vitamin E, Selenium, and Carotenoids* (Washington, D.C.: National Academy Press, 2000).
9. E. Villamor and coauthors, Vitamin A supplements ameliorate the adverse effect of HIV-1, malaria, and diarrheal infections on child growth, *Pediatrics* 109 (2002): e6.
10. K. L. Penniston and S. A. Tanumihardjo, The acute and chonic toxic effects of vitamin A, *American Journal of Clinical Nutrition* 83 (2006): 191–201.
11. A. Mazzone and A. dal Canton, Images in clinical medicine—Hypercarotenemia, *New England Journal of Medicine* 346 (2002): 821.
12. P. S. Genaro and L. A. Martini, Vitamin A supplementation and risk of skeletal fracture, *Nutrition Reviews* 62 (2004): 65–72; P. Lips, Hypervitaminosis A and fractures, *New England Journal of Medicine* 348 (2003): 347–349; K. Michaëlsson and coauthors, Serum retinol levels and the risk of fracture, *New England Journal of Medicine* 348 (2003): 287–294; D. Feskanich and coauthors, Vitamin A intake and hip fractures among postmenopausal women, *Journal of the American Medical Association* 287 (2002): 47–54.
13. H. A. Jackson and A. H. Sheehan, Effect of vitamin A on fracture risk, *The Annals of Pharmacotherapy* 39 (2005): 2086–2090.
14. M. J. Brown and coauthors, Carotenoid bioavailability is higher from salads ingested with full-fat than with fat-reduced salad dressings as measured with electrochemical detection, *American Journal of Clinical Nutrition* 80 (2004): 396–403.
15. P. Lips, Vitamin D physiology, *Progress in Biophysics and Molecular Biology* 92 (2006): 4–8.
16. H. F. DeLuca, Overview of general physiologic features and functions of vitamin D, *American Journal of Clinical Nutrition* 80 (2004): 1689S–1696S.
17. P. T. Liu and coauthors, Toll-like receptor triggering of a vitamin D-mediated human antimicrobial response, *Science* 311 (2006): 1770–1773; T. Dietrich and coauthors, Association between serum concentrations of 25-hydroxyvitamin D and gingival inflammation, *American Journal of Clinical Nutrition* 82 (2005): 575–580; J. Welsh, Vitamin D and breast cancer: Insights from animal models, *American Journal of Clinical Nutrition* 80 (2004): 1721S–1724S; I. A. van der Mei and coauthors, Past exposure to sun, skin phenotype, and risk of multiple sclerosis: Case-control study, *British Medical Journal* 327 (2003): 316–321.
18. M. F. Holick, Vitamin D: Importance in the prevention of cancers, type 1 diabetes, heart disease, and osteoporosis, *American Journal of Clinical Nutrition* 79 (2004): 362–371.
19. M. K. M. Lehtonen-Veromaa and coauthors, Vitamin D and attainment of peak bone mass among peripubertal Finnish girls: A 3-y prospective study, *American Journal of Clinical Nutrition* 76 (2002): 1446–1453.
20. S. A. Abrams, Nutritional rickets: An old disease returns, *Nutrition Reviews* 60 (2002): 111–115.

21. P. Weisberg and coauthors, Nutritional rickets among children in the United States: Review of cases reported between 1986 and 2003, *American Journal of Clinical Nutrition* 80 (2004): 1697S–1705S.

22. M. F. Holick, High prevalence of vitamin D inadequacy and implications for health, *Mayo Clinic Proceedings* 81 (2006): 353–373.

23. M. S. Calvo and S. J. Whiting, Prevalence of vitamin D insufficiency in Canada and the United States: Importance to health status and efficacy of current food fortification and dietary supplement use, *Nutrition Reviews* 61 (2003): 107–113.

24. H. A. Bischoff-Ferrari and coauthors, Fracture prevention with vitamin D supplementation: A meta-analysis of randomized controlled trials, *Journal of the American Medical Association* 293 (2005): 2257–2264; H. A. Bischoff-Ferrari and coauthors, Effect of vitamin D on falls: A meta-analysis, *Journal of the American Medical Association* 291 (2004): 1999–2006.

25. H. A. Bischoff-Ferrari and coauthors, Estimation of optimal serum concentrations of 25-hydroxyvitamin D for multiple health outcomes, *American Journal of Clinical Nutrition* 84 (2006): 18–28.

26. M. S. Calvo, S. L. Whiting, and C. N. Barton, Vitamin D fortification in the United States and Canada: Current status and data needs, *American Journal of Clinical Nutrition* 80 (2004): 1710S–1716S; C. Moore and coauthors, Vitamin D intake in the United States, *Journal of the American Dietetic Association* 104 (2004): 980–983.

27. M. F. Holick, Sunlight and vitamin D for bone health and prevention of autoimmune diseases, cancers, and cardiovascular disease, *American Journal of Clinical Nutrition* 80 (2004): 1678S–1688S.

28. T. A. Sentongo and coauthors, Vitamin D status in children, adolescents, and young adults with Crohn disease, *American Journal*

of Clinical Nutrition 76 (2002): 1077–1081; S. Nesby-O'Dell and coauthors, Hypovitaminosis D prevalence and determinants among African American and white women of reproductive age: Third National Health and Nutrition Examination Survey, 1988–1994, *American Journal of Clinical Nutrition* 76 (2002): 187–192.

29. L. Steingrimsdottir and coauthors, Relationship between serum parathyroid hormone levels, vitamin D sufficiency, and calcium intake, *Journal of the American Medical Association* 294 (2005): 2336–2341.

30. R. P. Heaney and coauthors, Human serum 25-hydroxycholecalciferol response to extended oral dosing with cholecalciferol, *American Journal of Clinical Nutrition* 77 (2003): 204–210.

31. V. Tangpricha and coauthors, Tanning is associated with optimal vitamin D status (serum 25-hydroxyvitamin D concentration) and higher bone mineral density, *American Journal of Clinical Nutrition* 80 (2004): 1645–1649.

32. Committee on Dietary Reference Intakes, 2000.

33. S. Devaraj and I. Jialal, Failure of vitamin E in clinical trials: Is gamma-tocopherol the answer? *Nutrition Reviews* 63 (2005): 290–293; M. C. Morris and coauthors, Relation of the tocopherol forms to incident Alzheimer disease and to cognitive change, *American Journal of Clinical Nutrition* 81 (2005): 508–514; A. M. Papas, Beyond α-tocopherol: The role of the other tocopherols and tocotrienols, in M. S. Meskin and coeditors, *Phytochemicals in Nutrition and Health* (Boca Raton, Fla.: CRC Press, 2002), pp. 61–77; Q. Jiang and coauthors, γ-Tocopherol, the major form of vitamin E in the US diet, deserves more attention, *American Journal of Clinical Nutrition* 74 (2001): 714–722.

34. U. Singh, S. Devaraj, and I. Jialal, Vitamin E, oxidative stress, and inflammation, *Annual Review of Nutrition* 25 (2005): 151–174.

35. J. N. Hathcock and coauthors, Vitamins E and C are safe across a broad range of intakes, *American Journal of Clinical Nutrition* 81 (2005): 736–745.

36. R. S. Bruno and coauthors, α-Tocopherol disappearance is faster in cigarette smokers and is inversely related to their ascorbic acid status, *American Journal of Clinical Nutrition* 81 (2005): 95–103; J. Maras and coauthors, Intake of α-Tocopherol is limited among US adults, *Journal of the American Dietetic Association* 104 (2004): 567–575.

37. Committee on Dietary Reference Intakes, 2000.

38. S. L. Booth and coauthors, Vitamin K intake and bone mineral density in women and men, *American Journal of Clinical Nutrition* 77 (2003): 512–516.

39. K. D. Cashman, Vitamin K status may be an important determinant of childhood bone health, *Nutrition Reviews* 63 (2005): 284–293; H. J. Kalkwarf and coauthors, Vitamin K, bone turnover, and bone mass in girls, *American Journal of Clinical Nutrition* 80 (2004): 1075–1080; N. C. Binkley and coauthors, A high phylloquinone intake is required to achieve maximal osteocalcin γ-carboxylation, *American Journal of Clinical Nutrition* 76 (2002): 1055–1060.

40. K. L. Berkner, The vitamin K-dependent carboxylase, *Annual Review of Nutrition* 25 (2005): 127–149.

41. M. A. Johnson, Influence of vitamin K on anticoagulant therapy depends on vitamin K status and the source and chemical forms of vitamin K, *Nutrition Reviews* 63 (2005): 91–100.

ANSWERS

Nutrition Calculations

1. Vitamin A: µg RAE Vitamin D: µg
   Vitamin E: mg Vitamin K: µg

2. Thiamin: Legumes and grains
   Riboflavin: Milks, grains, and meats
   Niacin: Meats and grains
   Vitamin B$_6$: Meats
   Folate: Legumes and vegetables
   Vitamin B$_{12}$: Meats and milks
   Vitamin C: Vegetables and fruits

Vitamin A: Vegetables, fruits, and milks

Vitamin D: Milks

Vitamin E: Legumes and oils

Taken together, "the most" groups form the USDA Food Guide—grains, vegetables, legumes, fruits, milks, meats, and oils.

Study Questions (multiple choice)

1. c 2. a 3. c 4. b 5. c 6. c 7. a 8. d
9. d 10. c

Antioxidant Nutrients in Disease Prevention

© Nick Clements/Taxi/Getty Images

Count on supplement manufacturers to exploit the day's hot topics in nutrition. The moment bits of research news surface, new supplements appear—and terms like "antioxidants" and "lycopene" become household words. Friendly faces in TV commercials try to persuade us that these supplements hold the magic in the fight against aging and disease. New supplements hit the market and cash registers ring. Vitamin C, for years the leading single nutrient supplement, gains new popularity, and sales of lutein, beta-carotene, and vitamin E supplements soar as well.

In the meantime, scientists and medical experts around the world continue their work to clarify and confirm the roles of antioxidants in preventing chronic diseases. This highlight summarizes some of the accumulating evidence. It also revisits the advantages of foods over supplements. But first it is important to introduce the troublemakers—the **free radicals.** (The accompanying glossary defines free radicals and related terms.)

Free Radicals and Disease

Chapter 7 described how the body's cells use oxygen in metabolic reactions. In the process, oxygen sometimes reacts with body compounds and produces highly unstable molecules known as free radicals. In addition to normal body processes, environmental factors such as ultraviolet radiation, air pollution, and tobacco smoke generate free radicals.

A free radical is a molecule with one or more unpaired electrons.* An electron without a partner is unstable and highly reactive. To regain its stability, the free radical quickly finds a stable but vulnerable compound from which to steal an electron.

With the loss of an electron, the formerly stable molecule becomes a free radical itself and steals an electron from another nearby molecule. Thus, an electron-snatching chain reaction is under way with free radicals producing more free radicals. Antioxidants neutralize free radicals by donating one of their own electrons, thus ending the chain reaction. When they lose electrons, antioxidants do not become free radicals because they are stable in either form. (Review Figure 10-15 on p. 351 to see how ascorbic acid can give up two hydrogens with their electrons and become dehydroascorbic acid.)

Once formed, free radicals attack. Occasionally, these free-radical attacks are helpful. For example, cells of the immune system use free radicals as ammunition in an "oxidative burst" that demolishes disease-causing viruses and bacteria. Most often, however, free-radical attacks cause widespread damage. They commonly damage the polyunsaturated fatty acids in lipoproteins and in cell membranes, disrupting the transport of substances into and out of cells. Free radicals also alter DNA, RNA, and proteins, creating excesses and deficiencies of specific proteins, impairing cell functions, and eliciting an inflammatory response. All of these actions contribute to cell damage, disease progression, and aging (see Figure H11-1).

---

* Many free radicals exist, but oxygen-derived free radicals are most common in the human body. Examples of oxygen-derived free radicals include superoxide radical ($O_2\cdot^-$), hydroxyl radical ($OH\cdot$), and nitric oxide ($NO\cdot$). (The dots in the symbols represent the unpaired electrons.) Technically, hydrogen peroxide (H_2O_2) and singlet oxygen are not free radicals because they contain paired electrons, but the unstable conformation of their electrons makes radical-producing reactions likely. Scientists sometimes use the term *reactive oxygen species (ROS)* to describe all of these compounds.

GLOSSARY

free radicals: unstable and highly reactive atoms or molecules that have one or more unpaired electrons in the outer orbital. (See Appendix B for a review of basic chemistry concepts.)

oxidants (OKS-ih-dants): compounds (such as oxygen itself) that oxidize other

compounds. Compounds that prevent oxidation are called *antioxidants*, whereas those that promote it are called *prooxidants*.
• **anti** = against
• **pro** = for

prooxidants: substances that significantly induce oxidative stress.

Reminders: **Dietary antioxidants** are substances typically found in foods that significantly decrease the adverse effects of free radicals on normal functions in the body. **Nonnutrients** are compounds in foods that do not fit into the six classes of nutrients.

Phytochemicals are nonnutrient compounds found in plant-derived foods that have biological activity in the body.
Oxidative stress is a condition in which the production of oxidants and free radicals exceeds the body's ability to handle them and prevent damage.

FIGURE H11-1   Free Radical Damage

Free radicals are highly reactive. They might attack the polyunsaturated fatty acids in a cell membrane, which generates lipid radicals that damage cells and accelerate disease progression. Free radicals might also attack and damage DNA, RNA, and proteins, which interferes with the body's ability to maintain normal cell function, causing disease and premature aging.

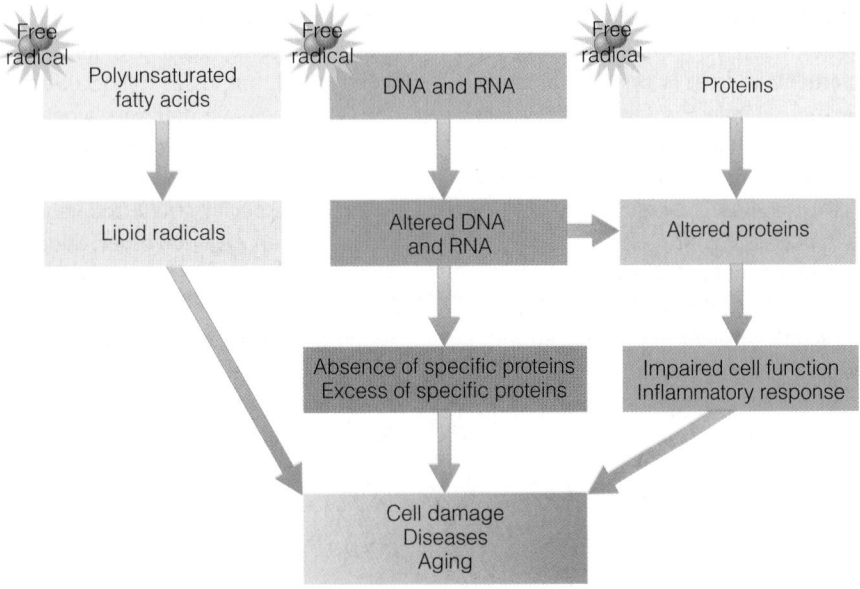

The body's natural defenses and repair systems try to control the destruction caused by free radicals, but these systems are not 100 percent effective. In fact, they become less effective with age, and the unrepaired damage accumulates. To some extent, **dietary antioxidants** defend the body against **oxidative stress,** but if antioxidants are unavailable or if free-radical production becomes excessive, health problems may develop.[1] Oxygen-derived free radicals may cause diseases, not only by indiscriminately destroying the valuable components of cells, but also by serving as signals for specific activities within the cells. Scientists have identified oxidative stress as a causative factor and antioxidants as a protective factor in cognitive performance and the aging process as well as in the development of diseases such as cancer, arthritis, cataracts, diabetes, and heart disease.[2]

Defending against Free Radicals

The body maintains a couple lines of defense against free-radical damage. A system of enzymes disarms the most harmful **oxidants.*** The action of these enzymes depends on the minerals selenium, copper, manganese, and zinc. If the diet fails to provide adequate supplies of these minerals, this line of defense weakens. The body also uses the antioxidant vitamins—vitamin E, beta-

carotene, and vitamin C. Vitamin E defends the body's lipids (cell membranes and lipoproteins, for example) by efficiently stopping the free-radical chain reaction. Beta-carotene also acts as an antioxidant in lipid membranes. Vitamin C protects other tissues, such as the skin and fluid of the blood, against free-radical attacks.[3] Vitamin C seems especially adept at neutralizing free radicals from polluted air and cigarette smoke; it may also restore oxidized vitamin E to its active state.

Dietary antioxidants may also include **nonnutrients**—some of the **phytochemicals** (featured in Highlight 13). Together, nutrients and phytochemicals with antioxidant activity minimize damage in the following ways:

- Limiting free-radical formation
- Destroying free radicals or their precursors
- Stimulating antioxidant enzyme activity
- Repairing oxidative damage
- Stimulating repair enzyme activity

These actions play key roles in defending the body against cancer and heart disease.

Defending against Cancer

Cancers arise when cellular DNA is damaged—sometimes by free-radical attacks. Antioxidants may reduce cancer risks by protecting DNA from this damage. Many researchers have reported low rates of cancer in people whose diets include abundant vegetables and fruits, rich in antioxidants.[4] Preliminary reports suggest an inverse relationship between DNA damage and vegetable

* These enzymes include glutathione peroxidase, thioredoxin reductase, superoxide dismutase, and catalase.

intake and a positive relationship with beef and pork intake. Laboratory studies with animals and with cells in tissue culture also seem to support such findings.

Foods rich in vitamin C seem to protect against certain types of cancers, especially those of the mouth, larynx, esophagus, and stomach. Such a correlation may reflect the benefits of a diet rich in fruits and vegetables and low in fat; it does not necessarily support taking vitamin C supplements to treat or prevent cancer.

Researchers hypothesize that vitamin E might inhibit cancer formation by attacking free radicals that damage DNA. Evidence that vitamin E helps guard against cancer, however, is contradictory and inconclusive.[5]

Several studies report a cancer-preventing benefit of vegetables and fruits rich in beta-carotene and the other carotenoids as well. Carotenoids seem to protect against oxidative damage to DNA.[6] High concentrations of beta-carotene are associated with a lower mortality from all causes and lower rates of cancer.[7]

Defending against Heart Disease

High blood cholesterol carried in LDL is a major risk factor for cardiovascular disease, but how do LDL exert their damage? One scenario is that free radicals within the arterial walls oxidize LDL, changing their structure and function. The oxidized LDL then accelerate the formation of artery-clogging plaques.[8] These free radicals also oxidize the polyunsaturated fatty acids of the cell membranes, sparking additional changes in the arterial walls, which impede the flow of blood. Susceptibility to such oxidative damage within the arterial walls is heightened by a diet high in saturated fat or cigarette smoke. In contrast, diets that include plenty of fruits and vegetables, especially when combined with little saturated fat, strengthen antioxidant defenses against LDL oxidation. Antioxidant nutrients taken as supplements also seem to slow the early progression of atherosclerosis.[9]

Antioxidants, especially vitamin E, may protect against cardiovascular disease.[10] Epidemiological studies suggest that people who eat foods rich in vitamin E have relatively few atherosclerotic plaques and low rates of death from heart disease.[11] Similarly, large doses of vitamin E supplements may slow the progression of heart disease. Among its many protective roles, vitamin E defends against LDL oxidation, inflammation, arterial injuries, and blood clotting.[12] Less clear is whether vitamin E supplements benefit people who already have heart disease or multiple risk factors for it. Antioxidant supplements may not be beneficial and, in fact, may even be harmful for these people.[13]

Vitamin C supplements may reduce the risk of heart disease.[14] Some studies suggest that vitamin C protects against LDL oxidation, raises HDL, lowers total cholesterol, and improves blood pressure. Vitamin C may also minimize inflammation and the free-radical action within the arterial wall.[15]

Foods, Supplements, or Both?

In the process of scavenging and quenching free radicals, antioxidants themselves become oxidized. To some extent, they can be regenerated, but losses still occur and free radicals attack continuously. To maintain defenses, a person must replenish dietary antioxidants regularly. But should antioxidants be replenished from foods or from supplements?

Foods—especially fruits and vegetables—offer not only antioxidants, but an array of other valuable vitamins and minerals as well. Importantly, deficiencies of these nutrients can damage DNA as readily as free radicals can. Eating fruits and vegetables in abundance protects against both deficiencies and diseases. A major review of the evidence gathered from metabolic studies, epidemiologic studies, and dietary intervention trials identified three dietary strategies most effective in preventing heart disease:[16]

- Use unsaturated fats (that have not been hydrogenated) instead of saturated or *trans* fats (see Highlight 5).
- Select foods rich in omega-3 fatty acids (see Chapter 5).
- Consume a diet high in fruits, vegetables, nuts, and whole grains and low in refined grain products.

Such a diet combined with exercise, weight control, and not smoking serves as the best prescription for health. Notably, taking supplements is not among these disease-prevention recommendations.

Some research suggests a protective effect from as little as a daily glass of orange juice or carrot juice (rich sources of vitamin C and beta-carotene, respectively). Other intervention studies, however, have used levels of nutrients that far exceed current recommendations and can be achieved only by taking supplements. In making their recommendations for the antioxidant nutrients, members of the DRI Committee considered whether these studies support substantially higher intakes to help protect against chronic diseases. They did raise the recommendations for vitamins C and E, but they do not support taking vitamin pills over eating a healthy diet.

While awaiting additional research, should people anticipate the "go-ahead" and start taking antioxidant supplements now? Most scientists agree that the evidence is insufficient for such a recommendation.[17] Though fruits and vegetables containing many antioxidant nutrients and phytochemicals have been associated with a diminished risk of many cancers, supplements have not always proved beneficial. In fact, sometimes the benefits are more apparent when the vitamins come from foods rather than from supplements. In other words, the antioxidant actions of fruits and vegetables are greater than their nutrients alone can explain.[18] Without data to confirm the benefits of supplements, we cannot accept the potential risks.[19] And the risks are real.

Consider the findings from meta-analysis studies of the relationships between daily supplements of vitamin E, beta-carotene, or both and total mortality. Researchers concluded that supplements either had *no benefit* or *increased* mortality and should be avoided.[20]

Even if research clearly proves that a particular nutrient is the ultimate protective ingredient in foods, supplements would not be the answer because their contents are limited. Vitamin E supplements, for example, usually contain alpha-tocopherol, but foods provide an assortment of tocopherols among other nutrients, many of which provide valuable protection against free-radical damage. In addition to a full array of nutrients, foods provide phytochemicals that also fight against many diseases.[21] Supplements shortchange users. Furthermore, supplements should only be used as an adjunct to other measures such as smoking cessation, weight control, physical activity, and medication as needed.[22]

Clearly, much more research is needed to define optimal and dangerous levels of intake. This much we know: antioxidants behave differently under various conditions. At physiological levels typical of a healthy diet, they act as antioxidants, but at pharmacological doses typical of supplements, they may act as **prooxidants,** stimulating the production of free radicals and altering metabolism in a way that may promote disease. A high intake of vitamin C from supplements, for example, may *increase* the risk of heart disease in women with diabetes.[23] High doses (more than 400 IU per day) of vitamin E supplements may increase mortality.[24] Until the optimum intake of antioxidant nutrients can be determined, the risks of supplement use remain unclear. The best way to add antioxidants to the diet is to eat generous servings of fruits and vegetables daily.

It should be clear by now that we cannot know the identity and action of every chemical in every food. Even if we did, why create a supplement to replicate a food? Why not eat foods and

Many cancer-fighting products are available now at your local produce counter.

enjoy the pleasure, nourishment, and health benefits they provide? The beneficial constituents in foods are widespread among plants. Among the fruits, pomegranates, berries, and citrus rank high in antioxidants; top antioxidant vegetables include kale, spinach, and brussels sprouts; millet and oats contain the most antioxidants among the grains; pinto beans and soybeans are the outstanding legumes; and walnuts outshine the other nuts.[25] But don't try to single out one particular food for its magic nutrient, antioxidant, or phytochemical. Instead, eat a wide variety of fruits, vegetables, grains, legumes, and nuts every day—and get *all* the magic compounds these foods have to offer.

REFERENCES

1. A. J. McEligot, S. Yang, and F. L. Meyskens, Redox regulation by intrinsic species and extrinsic nutrients in normal and cancer cells, *Annual Review of Nutrition* 25 (2005): 261–295; S. F. Clark, The biochemistry of antioxidants revisited, *Nutrition in Clinical Practice* 17 (2002): 5–17.
2. J. L. Evans and coauthors, Are oxidative stress-activated signaling pathways mediators of insulin resistance and beta-cell dysfunction? *Diabetes* 52 (2003): 1–8; F. Grodstein, J. Chen, and W. C. Willett, High-dose antioxidant supplements and cognitive function in community-dwelling elderly women, *American Journal of Clinical Nutrition* 77 (2003): 975–984; M. J. Engelhart and coauthors, Dietary intake of antioxidants and risk of Alzheimer disease, *Journal of the American Medical Association* 287 (2002): 3223–3229.
3. M. V. Catani and coauthors, Biological role of vitamin C in keratinocytes, *Nutrition Reviews* 63 (2005): 81–90.
4. D. P. Hayes, The protective role of fruits and vegetables against radiation-induced cancer, *Nutrition Reviews* 63 (2005): 303–311; A. Martin and coauthors, Roles of vitamins E and C on neurodegenerative diseases and

cognitive performance, *Nutrition Reviews* 60 (2002): 308–326; H. Chen and coauthors, Dietary patterns and adenocarcinoma of the esophagus and distal stomach, *American Journal of Clinical Nutrition* 75 (2002): 137–144.
5. D. Q. Pham and R. Plakogiannis, Vitamin E supplementation in cardiovascular disease and cancer prevention: Part 1, *Annals of Pharmacotherapy* 39 (2005): 1870–1878.
6. X. Zhao and coauthors, Modification of lymphocyte DNA damage by carotenoid supplementation in postmenopausal women, *American Journal of Clinical Nutrition* 83 (2006): 163–169.
7. B. Buijsse and coauthors, Plasma carotene and α-tocopherol in relation to 10-y all-cause and cause-specific mortality in European elderly: The Survey in Europe on Nutrition and the Elderly, a Concerted Action (SENECA), *American Journal of Clinical Nutrition* 82 (2005): 879–886.
8. G. A. A. Ferns and D. J. Lamb, What does the lipoprotein oxidation phenomenon mean? *Biochemical Society Transactions* 32 (2004): 160–163; W. Jessup, L. Kritharides, and R. Stocker, Lipid oxidation in atherogenesis: An overview, *Biochemical Society Transactions* 32 (2004): 134–138.

9. L. Liu and M. Meydani, Combined vitamin C and E supplementation retards early progression of arteriosclerosis in heart transplant patients, *Nutrition Reviews* 60 (2002): 368–371; H. Y. Huang and coauthors, Effects of vitamin C and vitamin E on in vivo lipid peroxidation: Results of a randomized controlled trial, *American Journal of Clinical Nutrition* 76 (2002): 549–555.
10. E. K. Kabagambe and coauthors, Some dietary and adipose tissue carotenoids are associated with the risk of nonfatal acute myocardial infarction in Costa Rica, *Journal of Nutrition* 135 (2005): 1763–1769; A. Iannuzzi and coauthors, Dietary and circulating antioxidant vitamins in relation to carotid plaques in middle-aged women, *American Journal of Clinical Nutrition* 76 (2002): 582–587.
11. A. Iannuzzi and coauthors, Dietary and circulating antioxidant vitamins in relation to carotid plaques in middle-aged women, *American Journal of Clinical Nutrition* 76 (2002): 582–587.
12. U. Singh, S. Devaraj, and I. Jialal, Vitamin E, oxidative stress, and inflammation, *Annual Review of Nutrition* 25 (2005): 151–174; S. Devaraj, A. Harris, and I. Jialal, Modulation of monocyte-macrophage function with

α-tocopherol: Implications for atherosclerosis, *Nutrition Reviews* 60 (2002): 8–14; L. J. van Tits and coauthors, α-Tocopherol supplementation decreases production of superoxide and cytokines by leukocytes ex vivo in both normolipidemic and hypertriglyceridemic individuals, *American Journal of Clinical Nutrition* 71 (2000): 458–464; M. Meydani, Vitamin E and prevention of heart disease in high-risk patients, *Nutrition Reviews* 58 (2000): 278–281.

13. The HOPE and HOPE-TOO Investigators, Effects of long-term vitamin E supplementation on cardiovascular events and cancer: A randomized controlled trial, *Journal of the American Medical Association* 293 (2005): 1338–1347; D. D. Waters and coauthors, Effects of hormone replacement therapy and antioxidant vitamin supplements on coronary atherosclerosis in postmenopausal women: A randomized controlled trial, *Journal of the American Medical Association* 288 (2002): 2432–2440.

14. P. Knekt and coauthors, Antioxidant vitamins and coronary heart disease risk: A pooled analysis of 9 cohorts, *American Journal of Clinical Nutrition* 80 (2004): 1508–1520.

15. S. G. Wannamethee and coauthors, Associations of vitamin C status, fruit and vegetable intakes, and markers of inflammation and hemostasis, *American Journal of Clinical Nutrition* 83 (2006): 567–574.

16. F. B. Hu and W. C. Willett, Optimal diets for prevention of coronary heart disease, *Journal of the American Medical Association* 288 (2002): 2569–2578.

17. H. Y. Huang and coauthors, The efficacy and safety of multivitamin and mineral supplement use to prevent cancer and chronic disease in adults: A systematic review for a National Institutes of Health state-of-the-science conference, *Annals of Internal Medicine* 145 (2006): 372–385; P. M. Kris-Etherton and coauthors, Antioxidant vitamin supplements and cardiovascular disease, *Circulation* 110 (2004): 637–641.

18. L. O. Dragsted and coauthors, The 6-a-day study: Effects of fruit and vegetables on markers of oxidative stress and antioxidative defense in healthy nonsmokers, *American Journal of Clinical Nutrition* 79 (2004): 1060–1072.

19. S. Hercberg, The history of β-carotene and cancers: From observational to intervention studies. What lessons can be drawn for future research on polyphenols? *American Journal of Clinical Nutrition* 81 (2005): 218S–222S.

20. E. R. Miller and coauthors, Meta-analysis: High-dosage vitamin E supplementation may increase all-cause mortality, *Annals of Internal Medicine* 142 (2005): 37–46; I. Lee and coauthors, Vitamin E in the primary prevention of cardiovascular disease and cancer—The Women's Health Study: A randomized controlled trial, *Journal of the American Medical Association* 294 (2005): 56–65; D. P. Vivekananthan and coauthors, Use of antioxidant vitamins for the prevention of cardiovascular disease: Meta-analysis of randomised trials, *Lancet* 361 (2003): 2017–2023.

21. P. M. Kris-Etherton and coauthors, Bioactive compounds in nutrition and health-research methodologies for establishing biological function: The antioxidant and anti-inflammatory effects of flavonoids on atherosclerosis, *Annual Review of Nutrition* 24 (2004): 511–538.

22. J. E. Manson, S. S. Bassuk, and M. J. Stampfer, Does vitamin E supplementation prevent cardiovascular events? *Journal of Womens Health* 12 (2003): 123–136.

23. D. H. Lee and coauthors, Does supplemental vitamin C increase cardiovascular disease risk in women with diabetes? *American Journal of Clinical Nutrition* 80 (2004): 1194–1200.

24. E. R. Miller and coauthors, Meta-analysis: High-dosage vitamin E supplementation may increase all-cause mortality, *Annals of Internal Medicine* 142 (2005): 37–46.

25. B. L. Halvorsen and coauthors, A systematic screening of total antioxidants in dietary plants, *Journal of Nutrition* 132 (2002): 461–471.

The CengageNOW logo indicates an opportunity for online self-study, linking you to interactive tutorials and videos based on your level of understanding.

academic.cengage.com/login

Nutrition in Your Life

What's your beverage of choice? If you said water, then congratulate yourself for recognizing its importance in maintaining your body's fluid balance. If you answered milk, then pat yourself on the back for taking good care of your bones. Faced with a lack of water, you would realize within days how vital it is to your very survival. The consequences of a lack of milk (or other calcium-rich foods) are also dramatic, but may not become apparent for decades. Water, calcium, and all the other major minerals support fluid balance and bone health. Before getting too comfortable reading this chapter, you might want to get yourself a glass of water or milk. Your body will thank you.

Water and the Major Minerals

Water is an essential nutrient, more important to life than any of the others. The body needs more water each day than any other nutrient. Furthermore, you can survive only a few days without water, whereas a deficiency of the other nutrients may take weeks, months, or even years to develop.

This chapter begins with a look at water and the body's fluids. The body maintains an appropriate balance and distribution of fluids with the help of another class of nutrients—the minerals. In addition to introducing the minerals that help regulate body fluids, this chapter describes many of the other important functions minerals perform in the body.

Water and the Body Fluids

Water constitutes about 60 percent of an adult's body weight and a higher percentage of a child's (see Figure 1–1, p. 6). Because water makes up about three-fourths of the weight of lean tissue and less than one-fourth of the weight of fat, a person's body composition influences how much of the body's weight is water. The proportion of water is generally smaller in females, obese people, and the elderly because of their smaller proportion of lean tissue.

In the body, water is the fluid in which all life processes occur. The water in the body fluids:

- Carries nutrients and waste products throughout the body

- Maintains the structure of large molecules such as proteins and glycogen

- Participates in metabolic reactions

- Serves as the solvent for minerals, vitamins, amino acids, glucose, and many other small molecules so that they can participate in metabolic activities

- Acts as a lubricant and cushion around joints and inside the eyes, the spinal cord, and, in pregnancy, the amniotic sac surrounding the fetus in the womb

- Aids in the regulation of normal body temperature (Evaporation of sweat from the skin removes excess heat from the body.)

- Maintains blood volume

© Michael Pole/CORBIS

Water is the most indispensable nutrient.

◆ Water balance: intake = output

◆ Fluids in the body:
  • Intracellular (inside cells)
  • Extracellular (outside cells)
    • Interstitial (between cells)
    • Intravascular (inside blood vessels)

◆ Reminder: The *hypothalamus* is a brain center that controls activities such as maintenance of water balance, regulation of body temperature, and control of appetite.

To support these and other vital functions, the body actively maintains an appropriate **water balance.** ◆

Water Balance and Recommended Intakes

Every cell contains fluid of the exact composition that is best for that cell **(intracellular fluid)** and is bathed externally in another such fluid **(interstitial fluid).** Interstitial fluid is the largest component of **extracellular fluid.** ◆ Figure 12-1 illustrates a cell and its associated fluids. These fluids continually lose and replace their components, yet the composition in each compartment remains remarkably constant under normal conditions. Because imbalances can be devastating, the body quickly responds by adjusting both water intake and excretion as needed. Consequently, the entire system of cells and fluids remains in a delicate, but controlled, state of homeostasis.

Water Intake **Thirst** and satiety influence water intake, apparently in response to changes sensed by the mouth, hypothalamus, ◆ and nerves. When water intake is inadequate, the blood becomes concentrated (having lost water but not the dissolved substances within it), the mouth becomes dry, and the hypothalamus initiates drinking behavior. When water intake is excessive, the stomach expands and stretch receptors send signals to stop drinking. Similar signals are sent from receptors in the heart as blood volume increases.

Thirst drives a person to seek water, but it lags behind the body's need. When too much water is lost from the body and not replaced, **dehydration** develops. A first sign of dehydration is thirst, the signal that the body has already lost some of its fluid. If a person is unable to obtain fluid or, as in many elderly people, fails to perceive the thirst message, the symptoms of dehydration may progress rapidly from thirst to weakness, exhaustion, and delirium—and end in death if not corrected (see Table 12-1). Dehydration may easily develop with either water deprivation or excessive water losses.

Water intoxication, on the other hand, is rare but can occur with excessive water ingestion and kidney disorders that reduce urine production. The symptoms may include confusion, convulsions, and even death in extreme cases. Excessive water ingestion (10 to 20 liters) within a few hours contributes to the dangerous condition known as hyponatremia, sometimes seen in endurance athletes. For this reason, guidelines suggest limiting fluid intake during times of heavy sweating to 1 to 1.5 liters per hour.[1]

Water Sources The obvious dietary sources of water are water itself and other beverages, but nearly all foods also contain water. Most fruits and vegetables contain up to 90 percent water, and many meats and cheeses contain at least 50 percent. (See Table 12-2 for selected foods and Appendix H for many more.) Also, water is generated during metabolism. Recall from Chapter 7 that when the energy-yielding

water balance: the balance between water intake and output (losses).

intracellular fluid: fluid within the cells, usually high in potassium and phosphate. Intracellular fluid accounts for approximately two-thirds of the body's water.
• **intra** = within

interstitial (IN-ter-STISH-al) **fluid:** fluid between the cells (intercellular), usually high in sodium and chloride. Interstitial fluid is a large component of extracellular fluid.
• **inter** = in the midst, between

extracellular fluid: fluid outside the cells. Extracellular fluid includes two main components—the interstitial fluid and plasma. Extracellular fluid accounts for approximately one-third of the body's water.
• **extra** = outside

thirst: a conscious desire to drink.

dehydration: the condition in which body water output exceeds water input. Symptoms include thirst, dry skin and mucous membranes, rapid heartbeat, low blood pressure, and weakness.

water intoxication: the rare condition in which body water contents are too high in all body fluid compartments.

TABLE 12-1	Signs of Dehydration	
Body Weight Lost (%)	**Symptoms**
1–2	Thirst, fatigue, weakness, vague discomfort, loss of appetite
3–4	Impaired physical performance, dry mouth, reduction in urine, flushed skin, impatience, apathy
5–6	Difficulty concentrating, headache, irritability, sleepiness, impaired temperature regulation, increased respiratory rate
7–10	Dizziness, spastic muscles, loss of balance, delirium, exhaustion, collapse

NOTE: The onset and severity of symptoms at various percentages of body weight lost depend on the activity, fitness level, degree of acclimation, temperature, and humidity. If not corrected, dehydration can lead to death.

TABLE 12-2 Percentage of Water in Selected Foods

100%	Water
90–99%	Fat-free milk, strawberries, watermelon, lettuce, cabbage, celery, spinach, broccoli
80–89%	Fruit juice, yogurt, apples, grapes, oranges, carrots
70–79%	Shrimp, bananas, corn, potatoes, avocados, cottage cheese, ricotta cheese
60–69%	Pasta, legumes, salmon, ice cream, chicken breast
50–59%	Ground beef, hot dogs, feta cheese
40–49%	Pizza
30–39%	Cheddar cheese, bagels, bread
20–29%	Pepperoni sausage, cake, biscuits
10–19%	Butter, margarine, raisins
1–9%	Crackers, cereals, pretzels, taco shells, peanut butter, nuts
0%	Oils, sugars

FIGURE 12-1 One Cell and Its Associated Fluids

Fluids are found within the cells (intracellular) or outside the cells (extracellular). Extracellular fluids include plasma (the fluid portion of blood in the intravascular spaces of blood vessels) and interstitial fluids (the tissue fluid that fills the intercellular spaces between the cells).

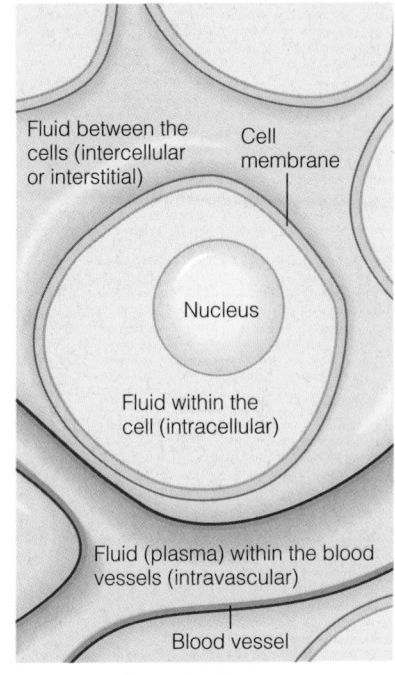

nutrients break down, their carbons and hydrogens combine with oxygen to yield carbon dioxide (CO_2) and water (H_2O). As Table 12-3 shows, the water derived daily from these three sources averages about $2^1/_2$ liters (roughly $2^1/_2$ quarts or $10^1/_2$ cups).

Water Losses The body must excrete a minimum of about 500 milliliters (about 2 cups) of water each day ◆ as urine—enough to carry away the waste products generated by a day's metabolic activities. Above this amount, excretion adjusts to balance intake. If a person drinks more water, the kidneys excrete more urine, and the urine becomes more dilute. In addition to urine, water is lost from the lungs as vapor and from the skin as sweat; some is also lost in feces.* The amount of fluid lost from each source varies, depending on the environment (such as heat or humidity) and physical conditions (such as exercise or fever). On average, daily losses total about $2^1/_2$ liters. Table 12-3 shows how water excretion balances intake; maintaining this balance requires healthy kidneys and an adequate intake of fluids.

Water Recommendations Because water needs vary depending on diet, activity, environmental temperature, and humidity, a general water requirement is difficult to establish. Recommendations ◆ are sometimes expressed in proportion to the amount of energy expended under average environmental conditions.[2] The recommended water intake for a person who expends 2000 kcalories a day, for example, is 2 to 3 liters of water (about 8 to 12 cups). This recommendation is in line with the Adequate Intake (AI) for *total* water set by the DRI Committee. ◆ Total water includes not only drinking water, but water in other beverages and in foods as well.

◆ The amount of water the body has to excrete each day to dispose of its wastes is the **obligatory** (ah-BLIG-ah-TORE-ee) **water excretion**—about 500 mL (about 2 c, or a pint).

◆ Water recommendation:
• 1.0 to 1.5 mL/kcal expended (adults)[†]
• 1.5 mL/kcal expended (infants and athletes)
Conversion factors:
• 1 mL = 0.03 fluid ounce
• 125 mL ≈ $^1/_2$ c
Easy estimation: $^1/_2$ c per 100 kcal expended

◆ AI for *total* water:
• Men: 3.7 L/day
• Women: 2.7 L/day
Conversion factors:
• 1 L ≈ 1 qt ≈ 32 oz ≈ 4 c

TABLE 12-3 Water Balance

Water Sources	Amount (mL)	Water Losses	Amount (mL)
Liquids	550 to 1500	Kidneys (urine)	500 to 1400
Foods	700 to 1000	Skin (sweat)	450 to 900
Metabolic water	200 to 300	Lungs (breath)	350
		GI tract (feces)	150
Total	1450 to 2800	Total	1450 to 2800

NOTE: For perspective, 100 mL is a little less than $^1/_2$ cup and 1000 mL is a little more than 1 quart (1 mL = 0.03 oz).

---

* Water lost from the lungs and skin accounts for almost one-half of the daily losses even when a person is not visibly perspiring; these losses are commonly referred to as *insensible water losses.*
† For those using kilojoules: 4.2 to 6.3 mL/kJ expended.

Because a wide range of water intakes will prevent dehydration and its harmful consequences, the AI is based on average intakes. People who are physically active or who live in hot environments may need more.[3]

Which beverages are best? Any beverage can readily meet the body's fluid needs, but those with few or no kcalories do so without contributing to weight gain. Given that obesity is a major health problem and that beverages currently represent over 20 percent of the total energy intake in the United States, most people would do well to select water as their preferred beverage. Other choices include tea, coffee, nonfat and low-fat milk and soymilk, artificially sweetened beverages, fruit and vegetable juices, sports drinks, and lastly, sweetened nutrient-poor beverages.[4]

Some research indicates that people who drink caffeinated beverages lose a little more fluid than when drinking water because caffeine acts as a diuretic. The DRI Committee considered such findings in their recommendations for water intake and concluded: "Caffeinated beverages contribute to the daily total water intake similar to that contributed by non-caffeinated beverages."[5] In other words, it doesn't seem to matter whether people rely on caffeine-containing beverages or other beverages to meet their fluid needs.

As Highlight 7 explained, alcohol acts as a diuretic, and it has many adverse effects on health and nutrition status. Alcohol should not be used to meet fluid needs.

Health Effects of Water In addition to meeting the body's fluid needs, drinking plenty of water may protect against urinary stones and constipation.[6] Even mild dehydration seems to interfere with daily tasks involving concentration, alertness, and short-term memory.[7]

The kind of water a person drinks may also make a difference to health. Water is usually either hard or soft. **Hard water** has high concentrations of calcium and magnesium; sodium or potassium is the principal mineral of **soft water.** (See the accompanying glossary for these and other common terms used to describe water.) In practical terms, soft water makes more bubbles with less soap; hard water leaves a ring on the tub, a crust of rocklike crystals in the teakettle, and a gray residue in the laundry.

Soft water may seem more desirable around the house, and some homeowners purchase water softeners that replace magnesium and calcium with sodium. In the

GLOSSARY OF WATER TERMS

artesian water: water drawn from a well that taps a confined aquifer in which the water is under pressure.

bottled water: drinking water sold in bottles.

carbonated water: water that contains carbon dioxide gas, either naturally occurring or added, that causes bubbles to form in it; also called *bubbling* or *sparkling water.* Seltzer, soda, and tonic waters are legally soft drinks and are not regulated as water.

distilled water: water that has been vaporized and recondensed, leaving it free of dissolved minerals.

filtered water: water treated by filtration, usually through *activated carbon filters* that reduce the lead in tap water, or by *reverse osmosis* units that force pressurized water across a membrane removing lead, arsenic, and some microorganisms from tap water.

hard water: water with a high calcium and magnesium content.

mineral water: water from a spring or well that typically contains 250 to 500 parts per million (ppm) of minerals. Minerals give water a distinctive flavor. Many mineral waters are high in sodium.

natural water: water obtained from a spring or well that is certified to be safe and sanitary. The mineral content may not be changed, but the water may be treated in other ways such as with ozone or by filtration.

public water: water from a municipal or county water system that has been treated and disinfected.

purified water: water that has been treated by distillation or other physical or chemical processes that remove dissolved solids. Because purified water contains no minerals or contaminants, it is useful for medical and research purposes.

soft water: water with a high sodium or potassium content.

spring water: water originating from an underground spring or well. It may be bubbly (carbonated), or "flat" or "still," meaning not carbonated. Brand names such as "Spring Pure" do not necessarily mean that the water comes from a spring.

well water: water drawn from ground water by tapping into an aquifer.

body, however, soft water with sodium may aggravate hypertension and heart disease. In contrast, the minerals in hard water may benefit these conditions.

Soft water also more easily dissolves certain contaminant minerals, such as cadmium and lead, from old plumbing pipes. As Chapter 13 explains, these contaminant minerals harm the body by displacing the nutrient minerals from their normal sites of action. People who live in old buildings should run the cold water tap a minute to flush out harmful minerals whenever the water faucet has been off for more than six hours. Many people select **bottled water,** believing it to be safer than tap water and therefore worth its substantial cost.

IN SUMMARY

Water makes up about 60 percent of the adult body's weight. It assists with the transport of nutrients and waste products throughout the body, participates in chemical reactions, acts as a solvent, serves as a shock absorber, and regulates body temperature. To maintain water balance, intake from liquids, foods, and metabolism must equal losses from the kidneys, skin, lungs, and GI tract. The amount and type of water a person drinks may have positive or negative health effects.

Blood Volume and Blood Pressure

Fluids maintain the blood volume, which in turn influences blood pressure. The kidneys are central to the regulation of blood volume and blood pressure.[8] All day, every day, the kidneys reabsorb needed substances and water and excrete wastes with some water in the urine (see Figure 12-2 on p. 402). The kidneys meticulously adjust the volume and the concentration of the urine to accommodate changes in the body, including variations in the day's food and beverage intakes. Instructions on whether to retain or release substances or water come from ADH, renin, angiotensin, and aldosterone.

ADH and Water Retention Whenever blood volume or blood pressure falls too low, or whenever the extracellular fluid becomes too concentrated, the hypothalamus signals the pituitary gland to release antidiuretic hormone (ADH). ◆ ADH is a water-conserving hormone ◆ that stimulates the kidneys to reabsorb water. Consequently, the more water you need, the less your kidneys excrete. These events also trigger thirst. Drinking water and retaining fluids raise the blood volume and dilute the concentrated fluids, thus helping to restore homeostasis.

Renin and Sodium Retention Cells in the kidneys respond to low blood pressure by releasing an enzyme called **renin.** Through a complex series of events, renin causes the kidneys to reabsorb sodium. Sodium reabsorption, in turn, is always accompanied by water retention, which helps to restore blood volume and blood pressure.

Angiotensin and Blood Vessel Constriction In addition to its role in sodium retention, renin converts the blood protein angiotensinogen to its active form—**angiotensin.** Angiotensin is a powerful **vasoconstrictor** that narrows the diameters of blood vessels, thereby raising the blood pressure.

Aldosterone and Sodium Retention In addition to acting as a vasoconstrictor, angiotensin stimulates the release of the hormone **aldosterone** from the **adrenal glands.** Aldosterone signals the kidneys to retain more sodium, and therefore water, because when sodium moves, fluids follow. Again, the effect is that when more water is needed, less is excreted.

All of these actions are presented in Figure 12-3 (p. 403) and help to explain why high-sodium diets aggravate conditions such as hypertension or edema. Too much

◆ Reminder: *Antidiuretic hormone (ADH)* is a hormone produced by the pituitary gland in response to dehydration (or a high sodium concentration in the blood). It stimulates the kidneys to reabsorb more water and therefore to excrete less.

◆ Recall from Highlight 7 that alcohol depresses ADH activity, thus promoting fluid losses and dehydration. In addition to its antidiuretic effect, ADH elevates blood pressure and so is also called **vasopressin** (VAS-oh-PRES-in).
  • **vaso** = vessel
  • **press** = pressure

---

renin (REN-in): an enzyme from the kidneys that activates angiotensin.

angiotensin (AN-gee-oh-TEN-sin): a hormone involved in blood pressure regulation. Its precursor protein is called *angiotensinogen;* it is activated by *renin,* an enzyme from the kidneys.

vasoconstrictor (VAS-oh-kon-STRIK-tor): a substance that constricts or narrows the blood vessels.

aldosterone (al-DOS-ter-own): a hormone secreted by the adrenal glands that regulates blood pressure by increasing the reabsorption of sodium by the kidneys. Aldosterone also regulates chloride and potassium concentrations.

adrenal glands: glands adjacent to, and just above, each kidney.

FIGURE 12-2 *Animated!* A Nephron, One of the Kidney's Many Functioning Units

CENGAGENOW™
To test your understanding of these concepts, log on to **academic.cengage.com/login.**

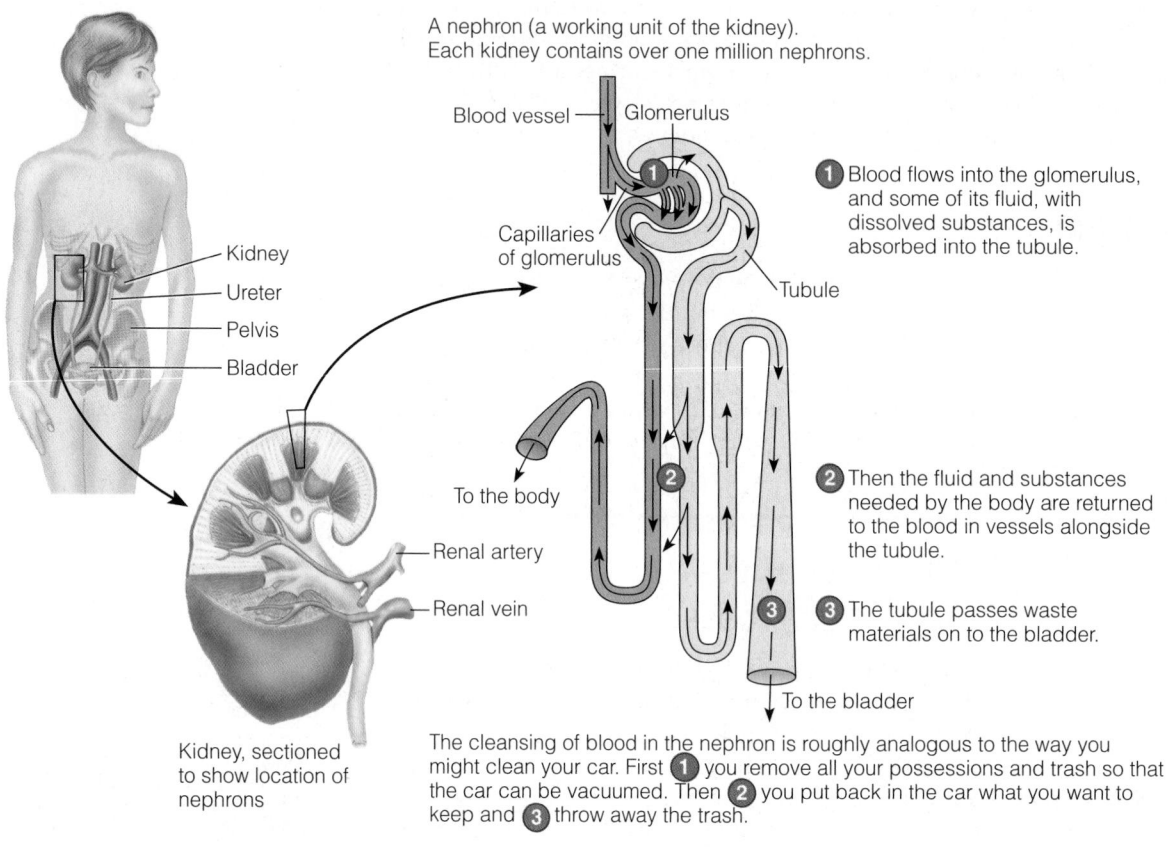

A nephron (a working unit of the kidney).
Each kidney contains over one million nephrons.

Blood vessel — Glomerulus

Capillaries of glomerulus

Tubule

1 Blood flows into the glomerulus, and some of its fluid, with dissolved substances, is absorbed into the tubule.

Kidney
Ureter
Pelvis
Bladder

To the body

Renal artery

Renal vein

2 Then the fluid and substances needed by the body are returned to the blood in vessels alongside the tubule.

3 The tubule passes waste materials on to the bladder.

To the bladder

Kidney, sectioned to show location of nephrons

The cleansing of blood in the nephron is roughly analogous to the way you might clean your car. First **1** you remove all your possessions and trash so that the car can be vacuumed. Then **2** you put back in the car what you want to keep and **3** throw away the trash.

sodium causes water retention and an accompanying rise in blood pressure or swelling in the interstitial spaces. Chapter 27 discusses hypertension in detail.

IN SUMMARY

In response to low blood volume, low blood pressure, or highly concentrated body fluids, these actions combine to effectively restore homeostasis:

• ADH retains water.
• Renin retains sodium.
• Angiotensin constricts blood vessels.
• Aldosterone retains sodium.

These actions can maintain water balance only if a person drinks enough water.

◆ The major minerals:
  • Sodium
  • Chloride
  • Potassium
  • Calcium
  • Phosphorus
  • Magnesium
  • Sulfur

Fluid and Electrolyte Balance

Maintaining a balance of about two-thirds of the body fluids inside the cells and one-third outside is vital to the life of the cells. If too much water were to enter the cells, they might rupture; if too much water were to leave, they would collapse. To control the movement of water, the cells direct the movement of the major minerals. ◆

FIGURE 12-3 *Animated!* How the Body Regulates Blood Volume

CENGAGENOW˙
To test your understanding of these concepts, log on to **academic.cengage.com/login.**

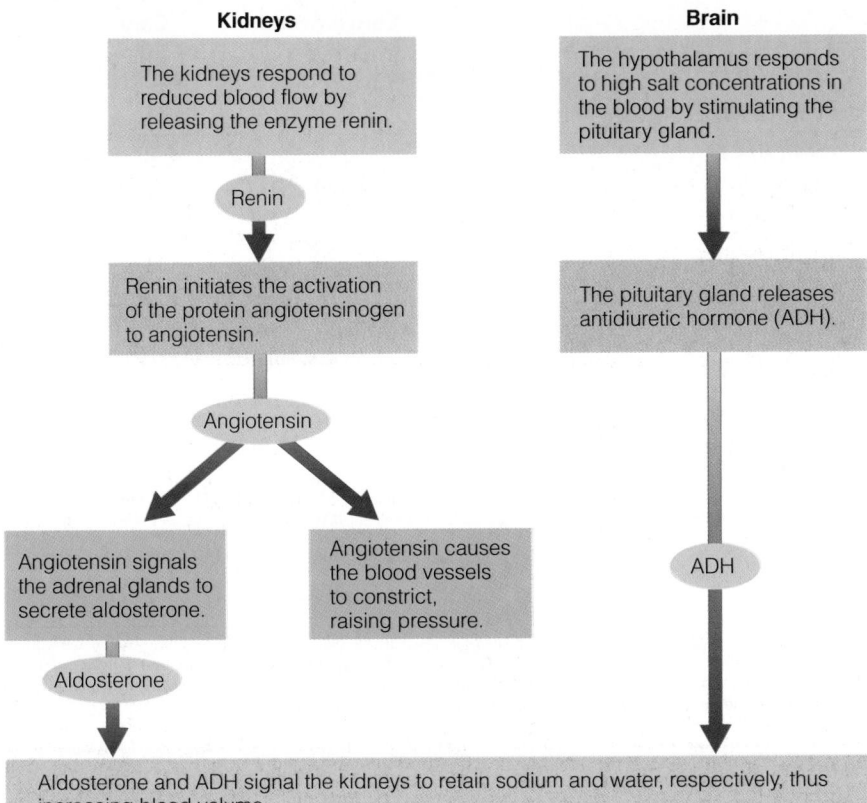

Kidneys

The kidneys respond to reduced blood flow by releasing the enzyme renin.

Renin

Renin initiates the activation of the protein angiotensinogen to angiotensin.

Angiotensin

Angiotensin signals the adrenal glands to secrete aldosterone.

Angiotensin causes the blood vessels to constrict, raising pressure.

Aldosterone

Brain

The hypothalamus responds to high salt concentrations in the blood by stimulating the pituitary gland.

The pituitary gland releases antidiuretic hormone (ADH).

ADH

Aldosterone and ADH signal the kidneys to retain sodium and water, respectively, thus increasing blood volume.

◆ To remember the difference between cations and anions, think of the "t" in cations as a "plus" (+) sign and the "n" in anions as "negative."

◆ A neutral molecule, such as water, that has opposite charges spatially separated within the molecule is **polar.** See Appendix B for more details.

salt: a compound composed of a positive ion other than H⁺ and a negative ion other than OH⁻. An example is sodium chloride (Na⁺ Cl⁻).
- **Na** = sodium
- **Cl** = chloride

dissociates (dis-SO-see-aites): physically separates.

ions (EYE-uns): atoms or molecules that have gained or lost electrons and therefore have electrical charges. Examples include the positively charged sodium ion (Na⁺) and the negatively charged chloride ion (Cl⁻). For a closer look at ions, see Appendix B.

cations (CAT-eye-uns): positively charged ions.

anions (AN-eye-uns): negatively charged ions.

electrolytes: salts that dissolve in water and dissociate into charged particles called ions.

electrolyte solutions: solutions that can conduct electricity.

milliequivalents (mEq): the concentration of electrolytes in a volume of solution. Milliequivalents are a useful measure when considering ions because the number of charges reveals characteristics about the solution that are not evident when the concentration is expressed in terms of weight.

Dissociation of Salt in Water When a mineral **salt** such as sodium chloride (NaCl) dissolves in water, it separates **(dissociates)** into **ions**—positively and negatively charged particles (Na⁺ and Cl⁻). The positive ions are **cations;** the negative ones are **anions.** ◆ Unlike pure water, which conducts electricity poorly, ions dissolved in water carry electrical current. For this reason, salts that dissociate into ions are called **electrolytes,** and fluids that contain them are **electrolyte solutions.**

In all electrolyte solutions, anion and cation concentrations are balanced (the number of negative and positive charges are equal). If a fluid contains 1000 negative charges, it must contain 1000 positive charges, too. If an anion enters the fluid, a cation must accompany it or another anion must leave so that electrical neutrality will be maintained. Thus, whenever sodium (Na⁺) ions leave a cell, potassium (K⁺) ions enter, for example. In fact, it's a good bet that whenever Na⁺ and K⁺ ions are moving, they are going in opposite directions.

Table 12-4 (p. 404) shows that, indeed, the positive and negative charges inside and outside cells are perfectly balanced even though the numbers of each kind of ion differ over a wide range. Inside the cells, the positive charges total 202 and the negative charges balance these perfectly. Outside the cells, the amounts and proportions of the ions differ from those inside, but again the positive and negative charges balance. (Scientists count these charges in **milliequivalents, mEq.**)

Electrolytes Attract Water Electrolytes attract water. Each water molecule has a net charge of zero, ◆ but the oxygen side of the molecule has a slight negative charge, and the hydrogens have a slight positive charge. Figure 12-4 (p. 404) shows the result in an electrolyte solution: both positive and negative ions attract clusters of water

TABLE 12-4 Important Body Electrolytes		
Electrolytes	Intracellular (inside cells) Concentration (mEq/L)	Extracellular (outside cells) Concentration (mEq/L)
Cations (positively charged ions)		
Sodium (Na$^+$)	10	142
Potassium (K$^+$)	150	5
Calcium (Ca^{++})	2	5
Magnesium (Mg^{++})	40	3
	202	155
Anions (negatively charged ions)		
Chloride (Cl$^-$)	2	103
Bicarbonate (HCO$_3$$^-$)	10	27
Phosphate (HPO$_4$$^=$)	103	2
Sulfate (SO$_4$$^=$)	20	1
Organic acids (lactate, pyruvate)	10	6
Proteins	57	16
	202	155

NOTE: The numbers of positive and negative charges in a given fluid are the same. For example, in extracellular fluid, the cations and anions both equal 155 milliequivalents per liter (mEq/L). Of the cations, sodium ions make up 142 mEq/L; and potassium, calcium, and magnesium ions make up the remainder. Of the anions, chloride ions number 103 mEq/L; bicarbonate ions number 27; and the rest are provided by phosphate ions, sulfate ions, organic acids, and protein.

molecules around them. This attraction dissolves salts in water and enables the body to move fluids into appropriate compartments.

Water Follows Electrolytes As Figure 12-5 shows, some electrolytes reside primarily outside the cells (notably, sodium and chloride), whereas others reside predominantly inside the cells (notably, potassium, magnesium, phosphate, ◆ and sulfate). Cell membranes are *selectively permeable,* meaning that they allow the pas-

◆ The word ending *-ate* denotes a salt of the mineral. Thus, phosphate is the salt form of the mineral phosphorus, and sulfate is the salt form of sulfur.

FIGURE 12-4 Water Dissolves Salts and Follows Electrolytes

The structural arrangement of the two hydrogen atoms and one oxygen atom enables water to dissolve salts. Water's role as a solvent is one of its most valuable characteristics.

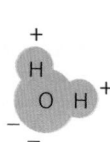

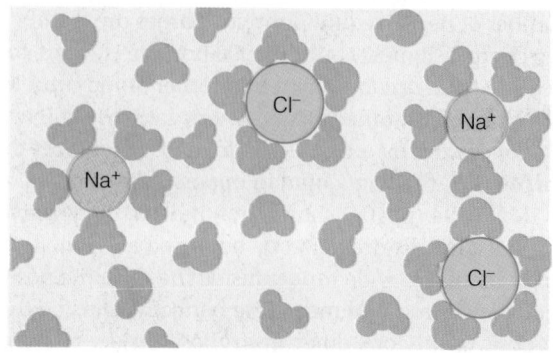

The negatively charged electrons that bond the hydrogens to the oxygen spend most of their time near the oxygen atom. As a result, the oxygen is slightly negative, and the hydrogens are slightly positive (see Appendix B).

In an electrolyte solution, water molecules are attracted to both anions and cations. Notice that the negative oxygen atoms of the water molecules are drawn to the sodium cation (Na$^+$), whereas the positive hydrogen atoms of the water molecules are drawn to the chloride ions (Cl$^-$).

FIGURE 12-5 A Cell and Its Electrolytes

All of these electrolytes are found both inside and outside the cells, but each can be found mostly on one side or the other of the cell membrane.

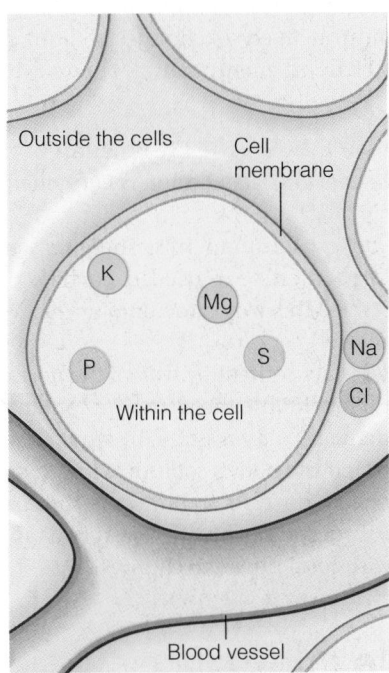

Outside the cells

Cell membrane

Within the cell

Blood vessel

Chemical symbols:
- K = potassium
- P = phosphorus
- Mg = magnesium
- S = sulfate
- Na = sodium
- Cl = chloride

Key:

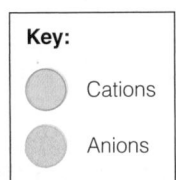 Cations

Anions

When immersed in water, raisins become plump because water moves toward the higher concentration of sugar inside the raisins.

When sprinkled with salt, vegetables "sweat" because water moves toward the higher concentration of salt outside the eggplant.

sage of some molecules, but not others. Whenever electrolytes move across the membrane, water follows.

The movement of water across a membrane toward the more concentrated **solutes** is called **osmosis.** The amount of pressure needed to prevent the movement of water across a membrane is called the **osmotic pressure.** Figure 12-6 presents osmosis, and the photos of salted eggplant and rehydrated raisins provide familiar examples.

Proteins Regulate Flow of Fluids and Ions Chapter 6 described how proteins attract water and help to regulate fluid movement. In addition, transport proteins in

FIGURE 12-6 Osmosis

Water flows in the direction of the more highly concentrated solution.

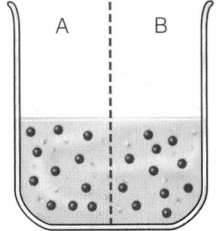 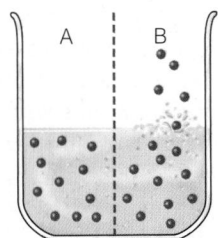 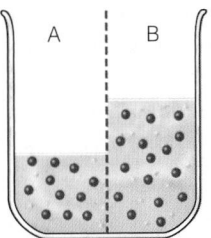

1 With equal numbers of solute particles on both sides of the semipermeable membrane, the concentrations are equal, and the tendency of water to move in either direction is about the same.

2 Now additional solute is added to side B. Solute cannot flow across the divider (in the case of a cell, its membrane).

3 Water can flow both ways across the divider, but has a greater tendency to move from side A to side B, where there is a greater concentration of solute. The volume of water becomes greater on side B, and the concentrations on side A and B become equal.

solutes (SOLL-yutes): the substances that are dissolved in a solution. The number of molecules in a given volume of fluid is the **solute concentration.**

osmosis: the movement of water across a membrane *toward* the side where the solutes are more concentrated.

osmotic pressure: the amount of pressure needed to prevent the movement of water across a membrane.

Physically active people must remember to replace their body fluids.

the cell membranes regulate the passage of positive ions and other substances from one side of the membrane to the other. Negative ions follow positive ions, and water flows toward the more concentrated solution.

A protein that regulates the flow of fluids and ions in and out of cells is the sodium-potassium pump. The pump actively exchanges sodium for potassium across the cell membrane, using ATP as an energy source. Figure 6-10 on p. 192 illustrates this action.

Regulation of Fluid and Electrolyte Balance The amounts of various minerals in the body must remain nearly constant. Regulation occurs chiefly at two sites: the GI tract and the kidneys.

The digestive juices of the GI tract contain minerals. These minerals and those from foods are reabsorbed in the large intestine as needed. Each day, 8 liters of fluids and associated minerals are recycled this way, providing ample opportunity for the regulation of electrolyte balance.

The kidneys' control of the body's *water* content by way of the hormone ADH has already been described (see p. 401). To regulate the *electrolyte* contents, the kidneys depend on the adrenal glands, which send out messages by way of the hormone aldosterone (also explained on p. 401). If the body's sodium is low, aldosterone stimulates sodium reabsorption from the kidneys. As sodium is reabsorbed, potassium (another positive ion) is excreted in accordance with the rule that total positive charges must remain in balance with total negative charges.

Fluid and Electrolyte Imbalance

Normally, the body defends itself successfully against fluid and electrolyte imbalances. Certain situations and some medications, however, may overwhelm the body's ability to compensate. Severe, prolonged vomiting and diarrhea as well as heavy sweating, burns, and traumatic wounds may incur such great fluid and electrolyte losses as to precipitate a medical emergency.

Different Solutes Lost by Different Routes Different solutes are lost depending on why fluid is lost. If fluid is lost by vomiting or diarrhea, sodium is lost indiscriminately. If the adrenal glands oversecrete aldosterone, as may occur when they develop a tumor, the kidneys may excrete too much potassium. Also, the person with uncontrolled diabetes may lose glucose, a solute not normally excreted, and large amounts of fluid with it. Each situation results in dehydration, but drinking water alone cannot restore electrolyte balance. Medical intervention is required.

Replacing Lost Fluids and Electrolytes In many cases, people can replace the fluids and minerals lost in sweat or in a temporary bout of diarrhea by drinking plain cool water and eating regular foods. Some cases, however, demand rapid replacement of fluids and electrolytes—for example, when diarrhea threatens the life of a malnourished child. Caregivers around the world have learned to use simple formulas ◆ to treat mild-to-moderate cases of diarrhea. These lifesaving formulas do not require hospitalization and can be prepared from ingredients available locally. Caregivers need only learn to measure ingredients carefully and use sanitary water. Once rehydrated, a person can begin eating foods.

Acid-Base Balance

The body uses its ions not only to help maintain fluid and electrolyte balance, but also to regulate the acidity (pH) ◆ of its fluids. The pH scale introduced in Chapter 3 is repeated here, in Figure 12-7, with the normal and abnormal pH ranges of the blood added. As you can see, the body must maintain the pH within a narrow range to avoid life-threatening consequences. Slight deviations in either direction can denature proteins, causing metabolic mayhem. Enzymes

◆ Health care workers use **oral rehydration therapy (ORT)**—a simple solution of sugar, salt, and water, taken by mouth—to treat dehydration caused by diarrhea. A simple ORT recipe (cool before giving):
• $^1/_2$ L boiling water
• A small handful of sugar (4 tsp)
• 3 pinches of salt ($^1/_2$ tsp)

◆ Reminder: *pH* is the unit of measure expressing a substance's acidity or alkalinity.

FIGURE 12-7 The pH Scale

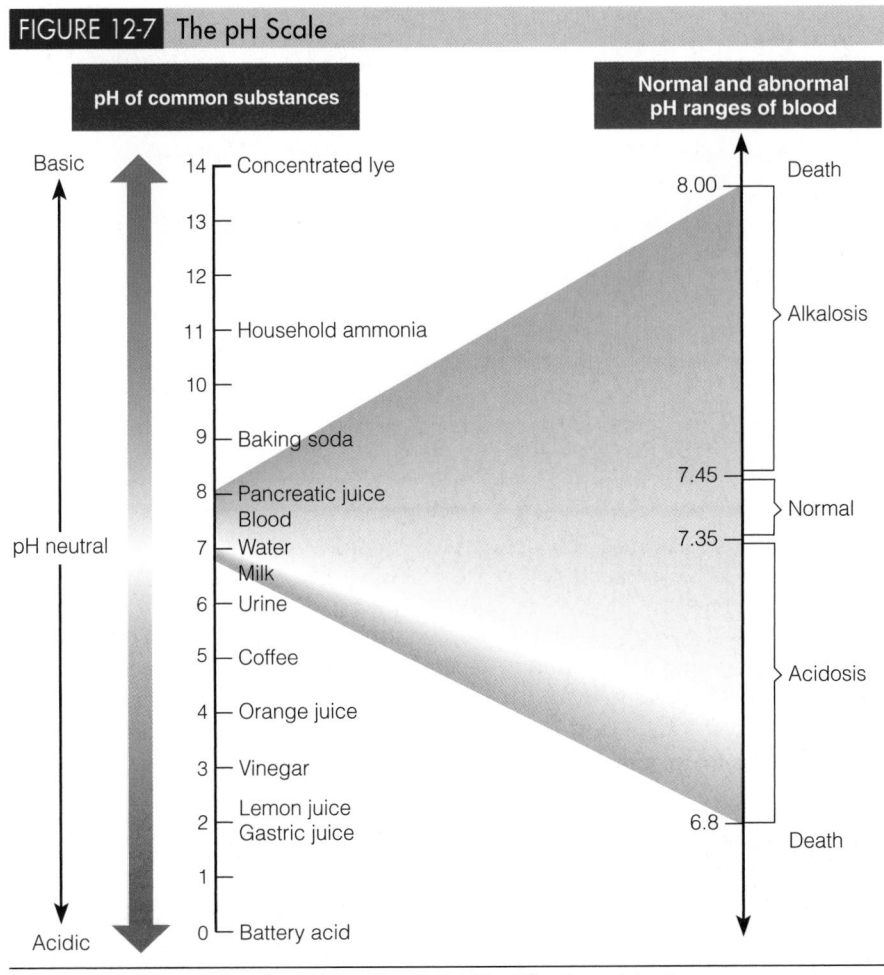

NOTE: Each step is ten times as concentrated in base ($^1/_{10}$ as much acid, or H$^+$) as the one below it.

couldn't catalyze reactions and hemoglobin couldn't carry oxygen—to name just two examples.

The acidity of the body's fluids is determined by the concentration of hydrogen ions (H$^+$). ◆ A high concentration of hydrogen ions is very acidic. Normal energy metabolism generates hydrogen ions, as well as many other acids, that must be neutralized. Three systems defend the body against fluctuations in pH—buffers in the blood, respiration in the lungs, and excretion in the kidneys.

Regulation by the Buffers Bicarbonate ◆ (a base) and **carbonic acid** (an acid) in the body fluids (as well as some proteins) protect the body against changes in acidity by acting as buffers—substances that can neutralize acids or bases. Figure 12-8 (p. 408) presents the chemical reactions of this buffer system, which is primarily under the control of the lungs and kidneys.

Carbon dioxide, which is formed all the time during energy metabolism, dissolves in water to form carbonic acid in the blood. Carbonic acid, in turn, dissociates to form hydrogen ions and bicarbonate ions. The appropriate balance between carbonic acid and bicarbonate is essential to maintaining optimal blood pH.

Regulation in the Lungs The lungs control the concentration of carbonic acid by raising or slowing the respiration rate, depending on whether the pH needs to be increased or decreased. If too much carbonic acid builds up, the respiration rate speeds up; this hyperventilation increases the amount of carbon dioxide exhaled, thereby lowering the carbonic acid concentration and restoring homeostasis. Conversely, if bicarbonate builds up, the respiration rate slows; carbon dioxide is retained and forms more carbonic acid. Again, homeostasis is restored.

◆ The lower the pH, the higher the H$^+$ ion concentration and the stronger the acid. A pH above 7 is alkaline, or base (a solution in which OH$^-$ ions predominate).

◆ Reminder: *Bicarbonate* is an alkaline compound with the formula HCO$_3$. It is produced in all cell fluids from the dissociation of carbonic acid to help maintain the body's acid-base balance. (Bicarbonate is also secreted from the pancreas during digestion as part of the pancreatic juice.)

carbonic acid: a compound with the formula H$_2$CO$_3$ that results from the combination of carbon dioxide (CO$_2$) and water (H$_2$O); of particular importance in maintaining the body's acid-base balance.

FIGURE 12-8 Bicarbonate-Carbonic Acid Buffer System

The reversible reactions of the bicarbonate-carbonic acid buffer system help to regulate the body's pH. Recall from Chapter 7 that carbon dioxide and water are formed during energy metabolism.

Carbon dioxide (CO_2) is a volatile gas that quickly dissolves in water (H_2O), forming carbonic acid (H_2CO_3):

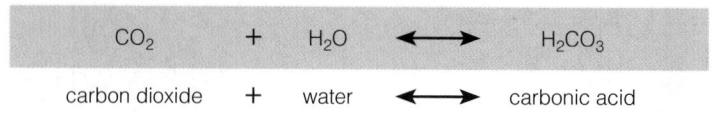

carbon dioxide + water ⟷ carbonic acid

Carbonic acid readily dissociates to a hydrogen ion (H^+) and a bicarbonate ion (HCO_3^-):

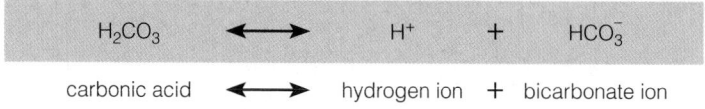

carbonic acid ⟷ hydrogen ion + bicarbonate ion

Regulation in the Kidneys The kidneys control the concentration of bicarbonate by either reabsorbing or excreting it, depending on whether the pH needs to be increased or decreased, respectively. Their work is complex, but the net effect is easy to sum up. The *body's* total acid burden remains nearly constant; the acidity of the *urine* fluctuates to accommodate that balance.

IN SUMMARY

Electrolytes (charged minerals) in the fluids help distribute the fluids inside and outside the cells, thus ensuring the appropriate water balance and acid-base balance to support all life processes. Excessive losses of fluids and electrolytes upset these balances, and the kidneys play a key role in restoring homeostasis.

The Minerals—An Overview

Figure 12-9 (p. 409) shows the amounts of the **major minerals** found in the body and, for comparison, some of the trace minerals. The distinction between the major and trace minerals does not mean that one group is more important than the other—all minerals are vital. The major minerals are so named because they are present, and needed, in larger amounts in the body. They are shown at the top of the figure and are discussed in this chapter. The trace minerals (shown at the bottom) are discussed in Chapter 13. A few generalizations pertain to all of the minerals and distinguish them from the vitamins. Especially notable is their chemical nature.

Inorganic Elements Unlike the organic vitamins, which are easily destroyed, minerals are inorganic elements ◆ that always retain their chemical identity. Once minerals enter the body proper, they remain there until excreted; they cannot be changed into anything else. Iron, for example, may temporarily combine with other charged elements in salts, but it is always iron. Neither can minerals be destroyed by heat, air, acid, or mixing. Consequently, little care is needed to preserve minerals during food preparation. In fact, the ash that remains when a food is burned contains all the minerals that were in the food originally. Minerals can be lost from food only when they leach into cooking water that is then poured down the drain.

◆ Reminder: An *inorganic* substance does not contain carbon.

major minerals: essential mineral nutrients found in the human body in amounts larger than 5 g; sometimes called **macrominerals.**

FIGURE 12-9 Minerals in a 60-kilogram (132-pound) Human Body

Not only are the major minerals present in the body in larger amounts than the trace minerals, but they are also needed by the body in larger amounts. Recommended intakes for the major minerals are stated in *hundreds of milligrams* or *grams,* whereas those for the trace minerals are listed in *tens of milligrams* or even *micrograms.*

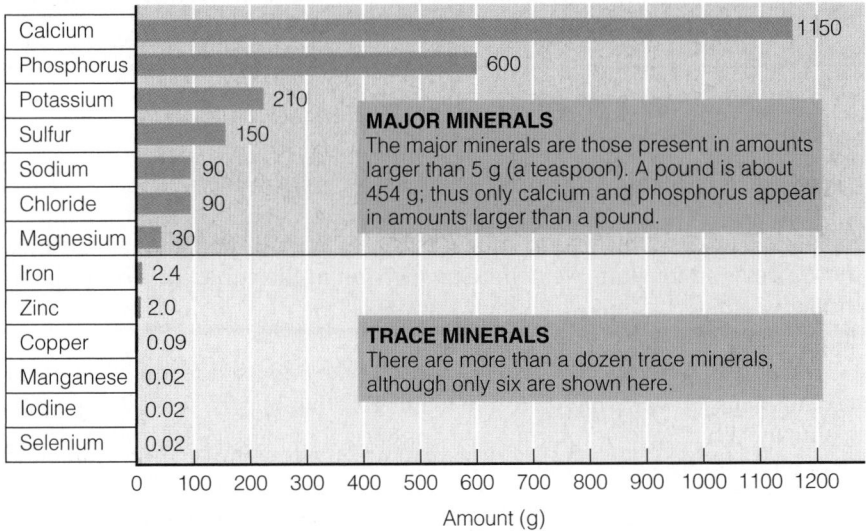

MAJOR MINERALS
The major minerals are those present in amounts larger than 5 g (a teaspoon). A pound is about 454 g; thus only calcium and phosphorus appear in amounts larger than a pound.

TRACE MINERALS
There are more than a dozen trace minerals, although only six are shown here.

Mineral	Amount (g)
Calcium	1150
Phosphorus	600
Potassium	210
Sulfur	150
Sodium	90
Chloride	90
Magnesium	30
Iron	2.4
Zinc	2.0
Copper	0.09
Manganese	0.02
Iodine	0.02
Selenium	0.02

The Body's Handling of Minerals The minerals also differ from the vitamins in the amounts the body can absorb and in the extent to which they must be specially handled. Some minerals, such as potassium, are easily absorbed into the blood, transported freely, and readily excreted by the kidneys, much like the water-soluble vitamins. Other minerals, such as calcium, are more like fat-soluble vitamins in that they must have carriers to be absorbed and transported. And, like some of the fat-soluble vitamins, minerals taken in excess can be toxic.

Variable Bioavailability The bioavailability ◆ of minerals varies. Some foods contain **binders** that combine chemically with minerals, preventing their absorption and carrying them out of the body with other wastes. Examples of binders include phytates, which are found primarily in legumes and grains, and oxalates, which are present in rhubarb and spinach, among other foods. These foods contain more minerals than the body actually receives for use.

Nutrient Interactions Chapter 10 described how the presence or absence of one vitamin can affect another's absorption, metabolism, and excretion. The same is true of the minerals. The interactions between sodium and calcium, for example, cause both to be excreted when sodium intakes are high. Phosphorus binds with magnesium in the GI tract, so magnesium absorption is limited when phosphorus intakes are high. These are just two examples of the interactions involving minerals featured in this chapter. Discussions in both this chapter and the next point out additional problems that arise from such interactions. Notice how often they reflect an excess of one mineral creating an inadequacy of another and how supplements—not foods—are most often to blame.

Varied Roles Although all the major minerals help to maintain the body's fluid balance as described earlier, sodium, chloride, and potassium are most noted for that role.◆ For this reason, these three minerals are discussed first here. Later sections describe the minerals most noted for their roles in bone growth and health—calcium, phosphorus, and magnesium.

◆ Reminder: *Bioavailability* refers to the rate at and the extent to which a nutrient is absorbed and used.

◆ Key fluid balance nutrients:
• Sodium, potassium, chloride

binders: chemical compounds in foods that combine with nutrients (especially minerals) to form complexes the body cannot absorb. Examples include **phytates** (FYE-tates) and **oxalates** (OCK-sa-lates).

IN SUMMARY

The major minerals are found in larger quantities in the body, whereas the trace minerals occur in smaller amounts. Minerals are inorganic elements that retain their chemical identities. They usually receive special handling and regulation in the body, and they may bind with other substances or interact with other minerals, thus limiting their absorption.

Sodium

People have held salt (sodium chloride) in high regard throughout recorded history. We describe someone we admire as "the salt of the earth" and someone we consider worthless as "not worth their salt." Even the word *salary* comes from the Latin word for salt.

Cultures vary in their use of salt, but most people find its taste innately appealing. Salt brings its own tangy taste and enhances other flavors, most likely by suppressing the bitter flavors. You can taste this effect for yourself: tonic water with its bitter quinine tastes sweeter with a little salt added.

Sodium Roles in the Body **Sodium** is the principal cation of the extracellular fluid and the primary regulator of its volume. Sodium also helps maintain acid-base balance and is essential to nerve impulse transmission and muscle contraction.*

Sodium is readily absorbed by the intestinal tract and travels freely in the blood until it reaches the kidneys, which filter all the sodium out of the blood. Then, with great precision, the kidneys return to the bloodstream the exact amount of sodium the body needs. Normally, the amount excreted is approximately equal to the amount ingested on a given day. When blood sodium rises, as when a person eats salted foods, thirst signals the person to drink until the appropriate sodium-to-water ratio is restored. Then the kidneys excrete both the excess water and the excess sodium together.

Sodium Recommendations Diets rarely lack sodium, and even when intakes are low, the body adapts by reducing sodium losses in urine and sweat, thus making deficiencies unlikely. Sodium recommendations ◆ are set low enough to protect against high blood pressure, but high enough to allow an adequate intake of other nutrients with a typical diet. Because high sodium intakes correlate with high blood pressure, the Upper Level for adults is set at 2300 milligrams per day, slightly lower than the Daily Value used on food labels (2400 milligrams). The average sodium intake for adults in the United States exceeds the Upper Level—and most adults will develop hypertension at some point in their lives.

Sodium and Hypertension For years, a high *sodium* intake was considered the primary factor responsible for high blood pressure. Then research pointed to *salt* (sodium chloride) as the dietary culprit. Salt has a greater effect on blood pressure than either sodium or chloride alone or in combination with other ions.

For some individuals, blood pressure increases in response to excesses in salt intake. People most likely to have a **salt sensitivity** include those whose parents had high blood pressure, those with chronic kidney disease or diabetes, African Americans, and people over 50 years of age.[†] Overweight people also appear to be particularly sensitive to the effect of salt on blood pressure. For them, a high salt intake correlates strongly with heart disease, and salt restriction helps to lower their blood pressure.

In fact, a salt-restricted diet lowers blood pressure in people without hypertension as well. Because reducing salt intake causes no harm and diminishes the risk

◆ AI for sodium:
- 1500 mg/day (19–50 yr)
- 1300 mg/day (51–70 yr)
- 1200 mg/day (>70 yr)

sodium: the principal cation in the extracellular fluids of the body; critical to the maintenance of fluid balance, nerve impulse transmissions, and muscle contractions.

salt sensitivity: a characteristic of individuals who respond to a high salt intake with an increase in blood pressure or to a low salt intake with a decrease in blood pressure.

---

* One of the ways the kidneys regulate acid-base balance is by excreting hydrogen ions (H^+) in exchange for sodium ions (Na^+).
† Compared with others, salt-sensitive individuals have elevated concentrations of renin in their blood.

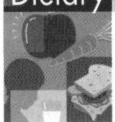

Dietary Guidelines for Americans 2005

Consume less than 2300 mg (approximately 1 tsp of salt) of sodium per day.

◆ Salt (sodium chloride) is about 40% sodium.
1 g salt contributes 400 mg sodium
5 g salt = 1 tsp
1 tsp salt contributes 2000 mg sodium

of hypertension and heart disease, the 2005 *Dietary Guidelines* advise limiting daily *salt* intake to about 1 teaspoon ◆ (the equivalent of 2.3 grams or 2300 milligrams of *sodium*). Higher intakes seem to be well tolerated in most healthy people, however. The accompanying "How to" offers strategies for cutting salt (and therefore sodium) intake.

One diet plan, known as the DASH (Dietary Approaches to Stop Hypertension) diet, also lowers blood pressure. The DASH approach emphasizes fruits, vegetables, and low-fat milk products; includes whole grains, nuts, poultry, and fish; and calls for reduced intakes of red meat, butter, and other high-fat foods. The DASH diet in combination with a reduced sodium intake is even more effective in lowering blood pressure than either strategy alone. Chapter 27 offers a complete discussion of hypertension and the dietary recommendations for its prevention and treatment.

Sodium and Bone Loss (Osteoporosis) A high salt intake is also associated with increased calcium excretion, but its influence on bone loss is less clear.[9] In addition, potassium may prevent the increase in calcium excretion caused by a high-salt diet.[10] For these reasons, dietary advice to prevent bone loss parallel those suggested for hypertension—a DASH diet that is low in sodium and abundant in potassium-rich fruits and vegetables and calcium-rich low-fat milk products.[11]

Sodium in Foods In general, processed foods have the most sodium, whereas unprocessed foods such as fresh fruits, vegetables, milk, and meats have the least. In fact, as much as 75 percent of the sodium in people's diets comes from salt added to foods by manufacturers; about 15 percent comes from salt added during cooking and at the table; and only 10 percent comes from the natural content in foods.

HOW TO Cut Salt (and Sodium) Intake

Most people eat more salt (and therefore sodium) than they need. Some people can lower their blood pressure by avoiding highly salted foods and removing the salt-shaker from the table. Foods eaten without salt may seem less tasty at first, but with repetition, people can learn to enjoy the natural flavors of many unsalted foods. Strategies to cut salt intake include:

- Select fresh, unprocessed foods.
- Cook with little or no added salt.
- Prepare foods with sodium-free spices such as basil, bay leaves, curry, garlic, ginger, mint, oregano, pepper, rosemary, and thyme; lemon juice; vinegar; or wine.
- Add little or no salt at the table; taste foods before adding salt.
- Read labels with an eye open for sodium. (See the glossary on p. 58 for terms used to describe the sodium contents of foods on labels.)

- Select low-salt or salt-free products when available.

Use these foods sparingly:

- Foods prepared in brine, such as pickles, olives, and sauerkraut
- Salty or smoked meats, such as bologna, corned or chipped beef, bacon, frank-furters, ham, lunch meats, salt pork, sausage, and smoked tongue
- Salty or smoked fish, such as anchovies, caviar, salted and dried cod, herring, sardines, and smoked salmon
- Snack items such as potato chips, pretzels, salted popcorn, salted nuts, and crackers
- Condiments such as bouillon cubes; seasoned salts; MSG; soy, teriyaki, Worcestershire, and barbeque sauces; prepared horseradish, catsup, and mustard
- Cheeses, especially processed types
- Canned and instant soups

Fresh herbs add flavor to a recipe without adding salt.

Because processed foods may contain sodium without chloride, as in additives such as sodium bicarbonate or sodium saccharin, they do not always taste salty. Most people are surprised to learn that 1 ounce of cornflakes contains more sodium than 1 ounce of salted peanuts—and that 1/2 cup of instant chocolate pudding contains still more. (The peanuts taste saltier because the salt is all on the surface, where the tongue's sensors immediately pick it up.)

Figure 12-10 shows that processed foods not only contain more sodium than their less processed counterparts but also have less potassium. Low potassium may be as significant as high sodium when it comes to blood pressure regulation, so processed foods have two strikes against them.

Dietary Guidelines for Americans 2005

Choose and prepare foods with little salt. At the same time, consume potassium-rich foods, such as fruits and vegetables.

Sodium Deficiency If blood sodium drops, as may occur with vomiting, diarrhea, or heavy sweating, both sodium and water must be replenished. Under normal conditions of sweating due to physical activity, salt losses can easily be replaced later in the day with ordinary foods. Salt tablets are not recommended because too much salt, especially if taken with too little water, can induce dehydration. During intense activities, such as ultra-endurance events, athletes can lose so much sodium and drink so much water that they develop hyponatremia—the dangerous condition of having too little sodium in the blood.

FIGURE 12-10 | What Processing Does to the Sodium and Potassium Contents of Foods

People who eat foods high in salt often happen to be eating fewer potassium-containing foods at the same time. Notice how potassium is lost and sodium is gained as foods become more processed, causing the potassium-to-sodium ratio to fall dramatically. Even when potassium isn't lost, the addition of sodium still lowers the potassium-to-sodium ratio. Limiting sodium intake may help in two ways, then—by lowering blood pressure in salt-sensitive individuals and by indirectly raising potassium intakes in all individuals.

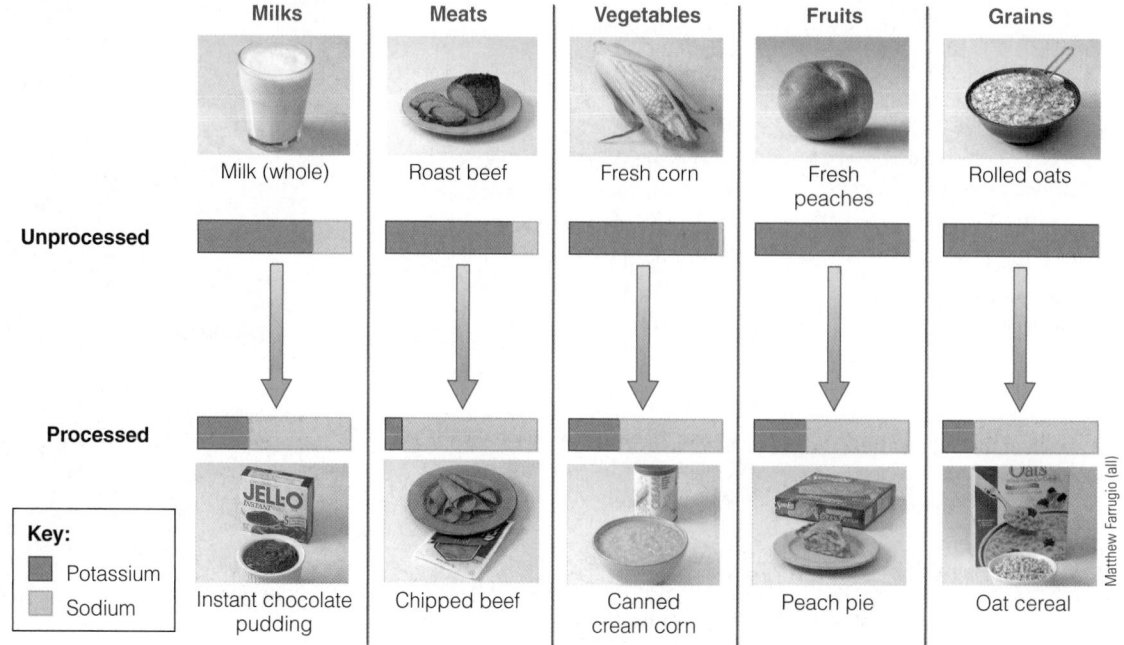

Matthew Farrugio (all)

Sodium Toxicity and Excessive Intakes The immediate symptoms of acute sodium toxicity are edema and hypertension, but such toxicity poses no problem as long as water needs are met. Prolonged excessive sodium intake ◆ may contribute to hypertension in some people, as explained earlier.

◆ UL for sodium: 2300 mg/day

IN SUMMARY

Sodium is the main cation outside cells and one of the primary electrolytes responsible for maintaining fluid balance. Dietary deficiency is rare, and excesses may aggravate hypertension in some people. For this reason, health professionals advise a diet moderate in salt and sodium. The accompanying table summarizes information about sodium.

Sodium

Adequate Intake (AI)	Deficiency Symptoms
Adults: 1500 mg/day (19–50 yr) 1300 mg/day (51–70 yr) 1200 mg/day (>70 yr)	Muscle cramps, mental apathy, loss of appetite
	Toxicity Symptoms
Upper Level	Edema, acute hypertension
Adults: 2300 mg/day	**Significant Sources**
Chief Functions in the Body	Table salt, soy sauce; moderate amounts in meats, milks, breads, and vegetables; large amounts in processed foods
Maintains normal fluid and electrolyte balance; assists in nerve impulse transmission and muscle contraction	

Chloride

The element *chlorine* (Cl$_2$) is a poisonous gas. When chlorine reacts with sodium or hydrogen, however, it forms the negative chloride ion (Cl$^-$). *Chloride*, an essential nutrient, is required in the diet.

Chloride Roles in the Body **Chloride** is the major anion of the extracellular fluids (outside the cells), where it occurs mostly in association with sodium. Chloride moves passively across membranes through channels and so also associates with potassium inside cells. Like sodium and potassium, chloride maintains fluid and electrolyte balance.

In the stomach, the chloride ion is part of hydrochloric acid, which maintains the strong acidity of the gastric juice. One of the most serious consequences of vomiting is the loss of this acid ◆ from the stomach, which upsets the acid-base balance.* Such imbalances are commonly seen in bulimia nervosa, as described in Highlight 8.

◆ Reminder: The loss of acid can lead to *alkalosis,* an above-normal alkalinity in the blood and body fluids.

Chloride Recommendations and Intakes Chloride is abundant in foods (especially processed foods) as part of sodium chloride and other salts. Because the proportion of chloride in salt is greater than sodium, ◆ chloride recommendations are slightly higher than, but still equivalent to, those of sodium. In other words, $^3/_4$ teaspoon of salt will deliver some sodium, more chloride, and still meet the AI for both.

◆ Salt (sodium chloride) is about 60% chloride.
1 g salt contributes 600 mg chloride
5 g salt = 1 tsp
1 tsp salt contributes 3000 mg chloride

Chloride Deficiency and Toxicity Diets rarely lack chloride. Chloride losses may occur in conditions such as heavy sweating, chronic diarrhea, and vomiting. The only known cause of high blood chloride concentrations is dehydration due to

chloride (KLO-ride): the major anion in the extracellular fluids of the body. Chloride is the ionic form of chlorine, Cl$^-$. See Appendix B for a description of the chlorine-to-chloride conversion.

---

* Hydrochloric acid secretion into the stomach involves the addition of bicarbonate ions (base) to the plasma. These bicarbonate ions (HCO$_3^-$) are neutralized by hydrogen ions (H$^+$) from the gastric secretions that are reabsorbed into the plasma. When hydrochloric acid is lost during vomiting, these hydrogen ions are no longer available for reabsorption, and so, in effect, the concentrations of bicarbonate ions in the plasma are increased. In this way, excessive vomiting of acidic gastric juices leads to *metabolic alkalosis.*

water deficiency. In both cases, consuming ordinary foods and beverages can restore chloride balance.

IN SUMMARY

Chloride is the major anion outside cells, and it associates closely with sodium. In addition to its role in fluid balance, chloride is part of the stomach's hydrochloric acid. The accompanying table summarizes information on chloride.

Chloride

Adequate Intake (AI)	Deficiency Symptoms
Adults: 2300 mg/day (19–50 yr) 2000 mg/day (51—70 yr) 1800 mg/day (>70 yr)	Do not occur under normal circumstances

	Toxicity Symptoms
Upper Level	Vomiting
Adults: 3600 mg/day	

	Significant Sources
Chief Functions in the Body	Table salt, soy sauce; moderate amounts in meats, milks, eggs; large amounts in processed foods
Maintains normal fluid and electrolyte balance; part of hydrochloric acid found in the stomach, necessary for proper digestion	

Potassium

Like sodium, **potassium** is a positively charged ion. In contrast to sodium, potassium is the body's principal intracellular cation, *inside* the body cells.

Potassium Roles in the Body Potassium plays a major role in maintaining fluid and electrolyte balance and cell integrity. During nerve impulse transmission and muscle contraction, potassium and sodium briefly trade places across the cell membrane. The cell then quickly pumps them back into place. Controlling potassium distribution is a high priority for the body because it affects many aspects of homeostasis, including a steady heartbeat.

Potassium Recommendations and Intakes Potassium is abundant in all living cells, both plant and animal. Because cells remain intact unless foods are processed, the richest sources of potassium are *fresh* foods—as Figure 12-11 (p. 415) shows. In contrast, most processed foods such as canned vegetables, ready-to-eat cereals, and luncheon meats contain less potassium—and more sodium (recall Figure 12-10, p. 412). To meet the AI for potassium, most people need to increase their intake of fruits and vegetables to five to nine servings daily.

Potassium and Hypertension Diets low in potassium seem to play an important role in the development of high blood pressure. Low potassium intakes raise blood pressure, whereas high potassium intakes, especially when combined with low sodium intakes, appear to both prevent and correct hypertension.[12] ◆ Potassium-rich fruits and vegetables also appear to reduce the risk of stroke—more so than can be explained by the reduction in blood pressure alone.

Potassium Deficiency Potassium deficiency is characterized by an increase in blood pressure, salt sensitivity, kidney stones, and bone turnover. As deficiency progresses, symptoms include irregular heartbeats, muscle weakness, and glucose intolerance.

Potassium Toxicity Potassium toxicity does not result from overeating foods high in potassium; therefore an Upper Level was not set. It can result from overconsumption of potassium salts or supplements (including some "energy fitness shakes") and from certain diseases or treatments. Given more potassium than the body needs, the

◆ Reminder: The DASH diet, used to lower blood pressure, emphasizes potassium-rich foods such as fruits and vegetables.

potassium: the principal cation within the body's cells; critical to the maintenance of fluid balance, nerve impulse transmissions, and muscle contractions.

FIGURE 12-11 Potassium in Selected Foods

See the "How to" on p. 329 for more information on using this figure.

Food	Serving size (kcalories)	Milligrams
Bread, whole wheat	1 oz slice (70 kcal)	
Cornflakes, fortified	1 oz (110 kcal)	
Spaghetti pasta	½ c cooked (99 kcal)	
Tortilla, flour	1 10"-round (234 kcal)	
Broccoli	½ c cooked (22 kcal)	
Carrots	½ c shredded raw (24 kcal)	
Potato	1 medium baked w/skin (133 kcal)	
Tomato juice	¾ c (31 kcal)	
Banana	1 medium raw (109 kcal)	
Orange	1 medium raw (62 kcal)	
Strawberries	½ c fresh (22 kcal)	
Watermelon	1 slice (92 kcal)	
Milk	1 c reduced-fat 2% (121 kcal)	
Yogurt, plain	1 c low-fat (155 kcal)	
Cheddar cheese	1½ oz (171 kcal)	
Cottage cheese	½ c low-fat 2% (101 kcal)	
Pinto beans	½ c cooked (117 kcal)	
Peanut butter	2 tbs (188 kcal)	
Sunflower seeds	1 oz dry (165 kcal)	
Tofu (soybean curd)	½ c (76 kcal)	
Ground beef, lean	3 oz broiled (244 kcal)	
Chicken breast	3 oz roasted (140 kcal)	
Tuna, canned in water	3 oz (99 kcal)	
Egg	1 hard cooked (78 kcal)	
Excellent, and sometimes unusual, sources:		
Squash, acorn	½ c baked (69 kcal)	
Soybeans	½ c cooked (149 kcal)	
Artichoke	1 (60 kcal)	

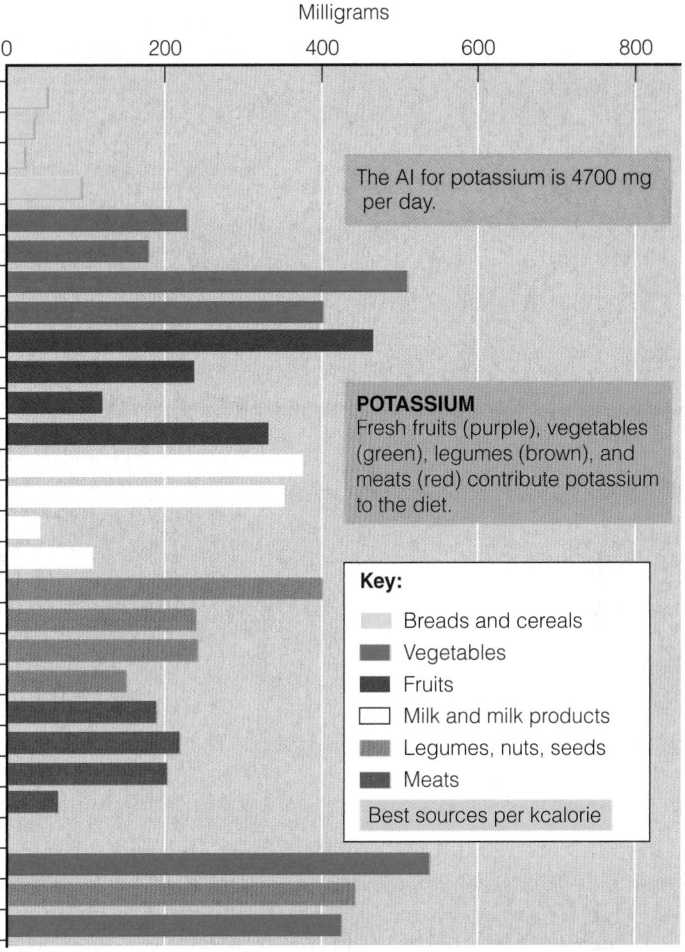

The AI for potassium is 4700 mg per day.

POTASSIUM
Fresh fruits (purple), vegetables (green), legumes (brown), and meats (red) contribute potassium to the diet.

Key:
- Breads and cereals
- Vegetables
- Fruits
- Milk and milk products
- Legumes, nuts, seeds
- Meats

Best sources per kcalorie

kidneys accelerate their excretion. If the GI tract is bypassed, however, and potassium is injected directly into a vein, it can stop the heart.

IN SUMMARY

Potassium, like sodium and chloride, is an electrolyte that plays an important role in maintaining fluid balance. Potassium is the primary cation inside cells; fresh foods, notably fruits and vegetables, are its best sources. The table below summarizes facts about potassium.

Potassium

Adequate Intake (AI)

Adults: 4700 mg/day

Chief Functions in the Body

Maintains normal fluid and electrolyte balance; facilitates many reactions; supports cell integrity; assists in nerve impulse transmission and muscle contractions

Deficiency Symptoms[a]

Irregular heatbeat, muscular weakness, glucose intolerance

Toxicity Symptoms

Muscular weakness; vomiting; if given into a vein, can stop the heart

Significant Sources

All whole foods: meats, milks, fruits, vegetables, grains, legumes

[a]Deficiency accompanies dehydration.

© Polara Studios Inc.

Fresh foods, especially fruits and vegetables, provide potassium in abundance.

Calcium

Calcium is the most abundant mineral in the body. It receives much emphasis in this chapter and in the highlight that follows because an adequate intake helps grow a healthy skeleton in early life and minimize bone loss in later life.

Calcium Roles in the Body

Ninety-nine percent of the body's calcium is in the bones (and teeth), where it plays two roles. First, it is an integral part of bone structure, providing a rigid frame that holds the body upright and serves as attachment points for muscles, making motion possible. Second, it serves as a calcium bank, offering a readily available source of the mineral to the body fluids should a drop in blood calcium occur.

Calcium in Bones As bones begin to form, calcium salts form crystals, called **hydroxyapatite**, on a matrix of the protein collagen. During **mineralization,** as the crystals become denser, they give strength and rigidity to the maturing bones. As a result, the long leg bones of children can support their weight by the time they have learned to walk.

Many people have the idea that once a bone is built, it is inert like a rock. Actually, the bones are gaining and losing minerals continuously in an ongoing process of remodeling. Growing children gain more bone than they lose, and healthy adults maintain a reasonable balance. When withdrawals substantially exceed deposits, problems such as osteoporosis develop (as described in Highlight 12).

The formation of teeth follows a pattern similar to that of bones. The turnover of minerals in teeth is not as rapid as in bone, however; fluoride hardens and stabilizes the crystals of teeth, opposing the withdrawal of minerals from them.

Calcium in Body Fluids Although only 1 percent of the body's calcium circulates in the extracellular and intracellular fluids, its presence there is vital to life. Many of its actions help to maintain normal blood pressure.

Cells throughout the body can detect calcium in the extracellular fluids and respond accordingly. For example, when the extracellular fluid contains too little calcium, the parathyroid glands release parathyroid hormone and the kidneys reabsorb calcium—all in an effort to raise calcium levels. Extracellular calcium also participates in blood clotting.

The calcium in intracellular fluids binds to proteins within the cells and activates them. ◆ These proteins participate in the regulation of muscle contractions, the transmission of nerve impulses, the secretion of hormones, and the activation of some enzyme reactions.

Calcium and Disease Prevention Calcium may protect against hypertension. For this reason, restricting sodium to treat hypertension is narrow advice, especially considering the success of the DASH diet in lowering blood pressure. The DASH diet is not particularly low in sodium, but it is rich in calcium, as well as in magnesium and potassium. As mentioned earlier, the DASH diet, together with a reduced sodium intake, is more effective in lowering blood pressure than either strategy alone. Some research also suggests protective relationships between dietary calcium and blood cholesterol, diabetes, and colon cancer.[13] Highlight 12 explores calcium's role in preventing osteoporosis.

Calcium and Obesity Calcium may also play a role in maintaining a healthy body weight.[14] Analyses of national survey data as well as small clinical studies show an inverse relationship between calcium intake and body fatness: the higher the calcium intake, the lower the body fatness.[15] In particular, calcium from dairy foods, but *not* from supplements, seems to influence body weight.[16] An adequate dietary calcium intake may help prevent excessive fat accumulation by stimulating hormonal action that targets the breakdown of stored fat.[17] Not all research suggests that calcium or dairy foods are associated with body weight.[18] Large, well-designed clinical studies are needed to clarify the effects of dietary calcium intake on body weight.

◆ An example of a protein that calcium binds with and activates is **calmodulin** (cal-MOD-you-lin). One of calmodulin's roles is to activate the enzymes involved in breaking down glycogen, which releases energy for muscle contractions.

calcium: the most abundant mineral in the body; found primarily in the body's bones and teeth.

hydroxyapatite (high-drox-ee-APP-ah-tite): crystals made of calcium and phosphorus.

mineralization: the process in which calcium, phosphorus, and other minerals crystallize on the collagen matrix of a growing bone, hardening the bone.

FIGURE 12-12 *Animated!* Calcium Balance

Blood calcium is regulated in part by vitamin D and two hormones—calcitonin and parathyroid hormone. Bone serves as a reservoir when blood calcium is high and as a source of calcium when blood calcium is low. Osteoclasts break down bone and release calcium into the blood; osteoblasts build new bone using calcium from the blood.

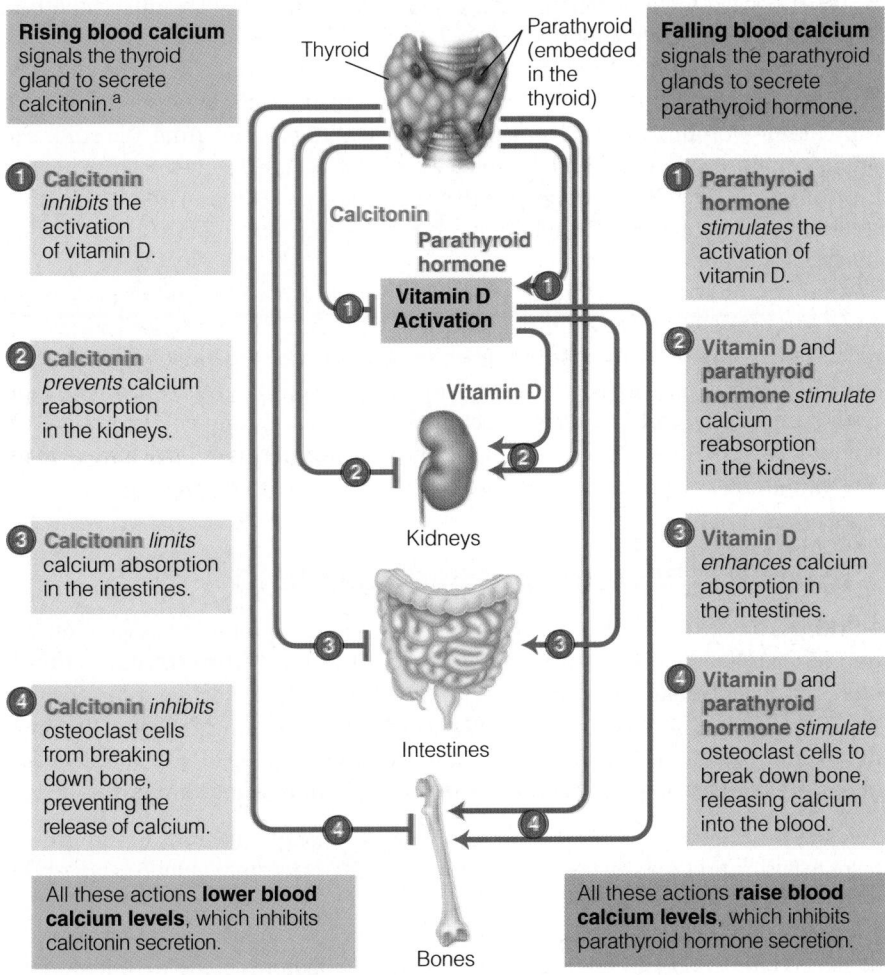

Rising blood calcium signals the thyroid gland to secrete calcitonin.[a]

1 **Calcitonin** *inhibits* the activation of vitamin D.

2 **Calcitonin** *prevents* calcium reabsorption in the kidneys.

3 **Calcitonin** *limits* calcium absorption in the intestines.

4 **Calcitonin** *inhibits* osteoclast cells from breaking down bone, preventing the release of calcium.

All these actions **lower blood calcium levels**, which inhibits calcitonin secretion.

Falling blood calcium signals the parathyroid glands to secrete parathyroid hormone.

1 **Parathyroid hormone** *stimulates* the activation of vitamin D.

2 **Vitamin D** and **parathyroid hormone** *stimulate* calcium reabsorption in the kidneys.

3 **Vitamin D** *enhances* calcium absorption in the intestines.

4 **Vitamin D** and **parathyroid hormone** *stimulate* osteoclast cells to break down bone, releasing calcium into the blood.

All these actions **raise blood calcium levels**, which inhibits parathyroid hormone secretion.

Thyroid — Parathyroid (embedded in the thyroid)

Calcitonin · Parathyroid hormone · **Vitamin D Activation** · Vitamin D · Kidneys · Intestines · Bones

[a]Calcitonin plays a major role in defending infants and young children against the dangers of rising blood calcium that can occur when regular feedings of milk deliver large quantities of calcium to a small body. In contrast, calcitonin plays a relatively minor role in adults because their absorption of calcium is less efficient and their bodies are larger, making elevated blood calcium unlikely.

Calcium Balance Calcium homeostasis involves a system of hormones and vitamin D. Whenever blood calcium falls too low or rises too high, three organ systems respond: the intestines, bones, and kidneys. Figure 12-12 illustrates how vitamin D and two hormones—**parathyroid hormone** and **calcitonin**—return blood calcium to normal.

The calcium in bone provides a nearly inexhaustible bank of calcium for the blood. The blood borrows and returns calcium as needed so that even with a dietary deficiency, *blood* calcium remains normal—even as *bone* calcium diminishes (see Figure 12-13, p. 418). Blood calcium changes only in response to abnormal regulatory control, not to diet. A person can have an inadequate calcium intake for years and suffer no noticeable symptoms. Only later in life does it become apparent that bone integrity has been compromised.

Blood calcium above normal results in **calcium rigor:** the muscles contract and cannot relax. Similarly, blood calcium below normal causes **calcium tetany**—also characterized by uncontrolled muscle contraction. These conditions do *not* reflect a *dietary* excess or lack of calcium; they are caused by a lack of vitamin D or by abnormal secretion of the regulatory hormones. A chronic *dietary* deficiency of calcium, or a chronic deficiency due to poor absorption over the years,

parathyroid hormone: a hormone from the parathyroid glands that regulates blood calcium by raising it when levels fall too low; also known as **parathormone** (PAIR-ah-THOR-moan).

calcitonin (KAL-seh-TOE-nin): a hormone secreted by the thyroid gland that regulates blood calcium by lowering it when levels rise too high.

calcium rigor: hardness or stiffness of the muscles caused by high blood calcium concentrations.

calcium tetany (TET-ah-nee): intermittent spasm of the extremities due to nervous and muscular excitability caused by low blood calcium concentrations.

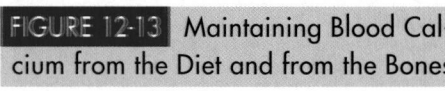

FIGURE 12-13 Maintaining Blood Calcium from the Diet and from the Bones

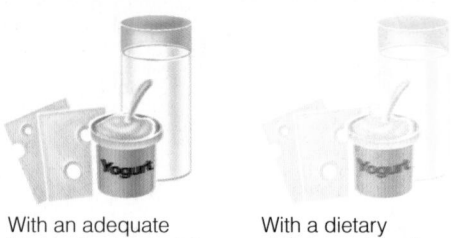

With an adequate intake of calcium-rich food, blood calcium remains normal . . .

With a dietary deficiency, blood calcium still remains normal . . .

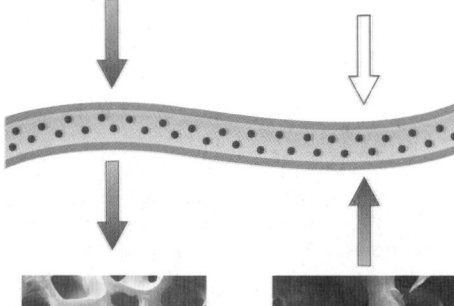

. . . and bones deposit calcium. The result is strong, dense bones.

. . . because bones give up calcium to the blood. The result is weak, osteoporotic bones.

© David Dempster from J Bone Miner Res, 1986 (both)

♦ Factors that *enhance* calcium absorption:
- Stomach acid
- Vitamin D
- Lactose (in infants only)

Factors that *inhibit* calcium absorption:
- Lack of stomach acid
- Vitamin D deficiency
- High phosphorus intake
- Phytates (in seeds, nuts, grains)
- Oxalates (in beet greens, rhubarb, spinach, sweet potatoes)

♦ Suggested daily amounts:
- Young children (2 to 8 yr): 2 c
- Older children, teenagers, and all adults: 3 c

calcium-binding protein: a protein in the intestinal cells, made with the help of vitamin D, that facilitates calcium absorption.

peak bone mass: the highest attainable bone density for an individual, developed during the first three decades of life.

depletes the savings account in the bones. Again: the *bones,* not the blood, are robbed by a calcium deficiency.

Calcium Absorption Many factors affect calcium absorption, but on average, adults absorb about 25 percent of the calcium they ingest. The stomach's acidity helps to keep calcium soluble, and vitamin D helps to make the **calcium-binding protein** needed for absorption. (This explains why calcium-rich milk is the best food for vitamin D fortification.)

Whenever calcium is needed, the body increases its production of the calcium-binding protein to improve calcium absorption. The result is obvious in the case of a pregnant woman, who absorbs 50 percent of the calcium from the milk she drinks. Similarly, growing children and teens absorb 50 to 60 percent of the calcium they consume. Then, when bone growth slows or stops, absorption falls to the adult level of about 25 percent. In addition, absorption becomes more efficient during times of inadequate intakes.

Many of the conditions that enhance calcium absorption inhibit its absorption when they are absent. For example, sufficient vitamin D supports absorption, and a deficiency impairs it. In addition, fiber, in general, and the binders phytate and oxalate, in particular, interfere with calcium absorption, but their effects are relatively minor in typical U.S. diets. Vegetables with oxalates and whole grains with phytates are nutritious foods, of course, but they are not useful calcium sources. The margin note ♦ presents factors that influence calcium balance.

Calcium Recommendations and Sources

Calcium is unlike most other nutrients in that hormones maintain its *blood* concentration regardless of dietary intake. As Figure 12-13 shows, when calcium intake is high, the *bones* benefit; when intake is low, the *bones* suffer. Calcium recommendations are therefore based on the amount needed to retain the most calcium in bones. By retaining the most calcium possible, the bones can develop to their fullest potential in size and density—their **peak bone mass**—within genetic limits.

Calcium Recommendations Because obtaining enough calcium during growth helps to ensure that the skeleton will be strong and dense, recommendations have been set high at 1300 milligrams daily for adolescents up to the age of 18 years. Between the ages of 19 and 50, recommendations are lowered to 1000 milligrams a day; for older adults, recommendations are raised again to 1200 milligrams a day to minimize the bone loss that tends to occur later in life. Some authorities advocate as much as 1500 milligrams a day for women over 50. Many people in the United States and Canada, particularly women, have calcium intakes far below current recommendations. High intakes of calcium from supplements may have adverse effects such as kidney stone formation.[19] For this reason, an Upper Level has been established (see inside front cover).

High intakes of both dietary protein and sodium increase calcium losses, but whether these losses impair bone development remains unclear. In the case of protein, high intakes of either animal or plant proteins may be problematic, but the effects are minimized by the beneficial effects of other nutrients in the food and diet—for example, by the potassium in legumes and the calcium in milk.[20] In establishing an Adequate Intake (AI) for calcium, the DRI Committee considered these nutrient interactions and did not adjust dietary recommendations based on this information.

Calcium in Milk Products Figure 12-14 shows that calcium is found most abundantly in a single class of foods—milk. ♦ The person who doesn't like to drink milk may prefer to eat cheese or yogurt. Alternatively, milk and milk products can be concealed in foods. Powdered fat-free milk can be added to casseroles, soups, and other mixed dishes during preparation; 5 heaping tablespoons offer the equivalent of 1 cup of milk. This simple step is an excellent way for older women not only to obtain extra calcium, but more protein, vitamins, and minerals as well.

It is especially difficult for children who don't drink milk to meet their calcium needs.[21] Children who don't drink milk have lower calcium intakes and poorer bone

FIGURE 12-14 Calcium in Selected Foods

See the "How to" on p. 329 for more information on using this figure.

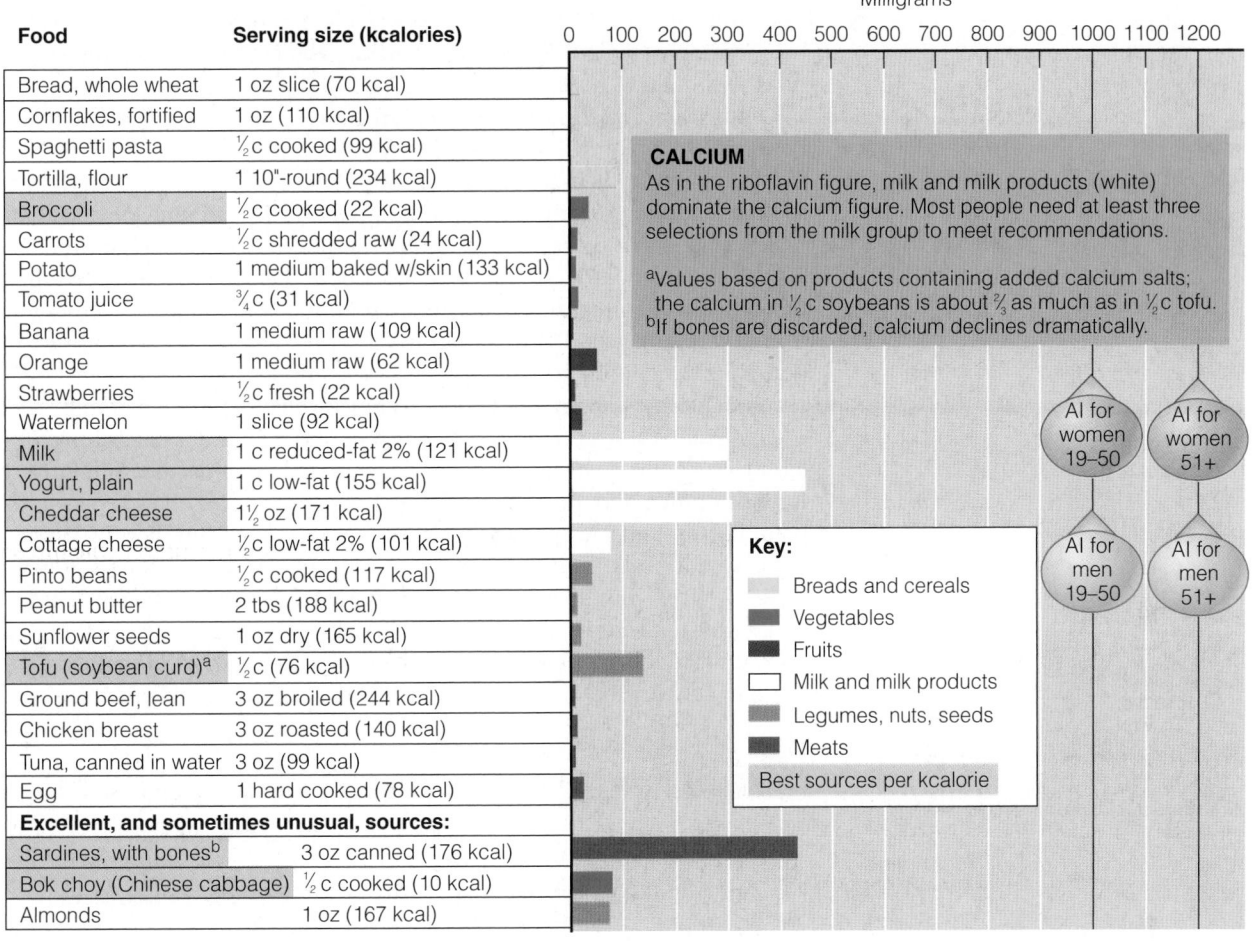

health than those who drink milk regularly.[22] The consequences of drinking too little milk during childhood and adolescence persist into adulthood. Women who seldom drank milk as children or teenagers have lower bone density and greater risk of fractures than those who drank milk regularly.[23] It is possible for people who do not drink milk to obtain adequate calcium, but only if they carefully select other calcium-rich foods.

Calcium in Other Foods Many people, for a variety of reasons, cannot or do not drink milk. Some cultures do not use milk in their cuisines; some vegetarians exclude milk as well as meat; and some people are allergic to milk protein or are lactose intolerant. ◆ Others simply do not enjoy the taste of milk. These people need to find non-milk sources of calcium to help meet their calcium needs. Some brands of tofu, corn tortillas, some nuts (such as almonds), and some seeds (such as sesame seeds) can supply calcium for the person who doesn't use milk products. A slice of most breads contains only about 5 to 10 percent of the calcium found in milk, but it can be a major source for people who eat many slices because the calcium is well absorbed.

Among the vegetables, mustard and turnip greens, bok choy, kale, parsley, watercress, and broccoli are good sources of available calcium. So are some seaweeds such as the nori popular in Japanese cooking. Some dark green, leafy vegetables—notably spinach and Swiss chard—appear to be calcium-rich but actually provide little, if any, calcium to the body because of the binders they contain. It would take 8 cups of spinach—containing six times as much calcium as 1 cup of milk—to deliver the equivalent in *absorbable* calcium.

◆ People with lactose intolerance may be able to consume small quantities of milk, as Chapter 4 explains.

FIGURE 12-15 Bioavailability of Calcium from Selected Foods

≥50% absorbed	Cauliflower, watercress, brussels sprouts, rutabaga, kale, mustard greens, bok choy, broccoli, turnip greens
≈30% absorbed	Milk, calcium-fortified soy milk, calcium-set tofu, cheese, yogurt, calcium-fortified foods and beverages
≈20% absorbed	Almonds, sesame seeds, pinto beans, sweet potatoes
≤5% absorbed	Spinach, rhubarb, Swiss chard

CENGAGENOW™
To practice estimating calcium intake, log on to **academic.cengage.com/login**, go to Chapter 12, then go to How To.

Milk and milk products are notorious for their calcium, but calcium-set tofu, bok choy, kale, calcium-fortified orange juice, and broccoli are also rich in calcium.

Matthew Farruggio

With the exception of foods such as spinach that contain calcium binders, however, the calcium content of foods is usually more important than bioavailability. Consequently, recognizing that people eat a variety of foods containing calcium, the DRI Committee did not consider calcium bioavailability when setting recommendations. Figure 12-15 ranks selected foods according to their calcium bioavailability.

Oysters are also a rich source of calcium, as are small fish eaten with their bones, such as canned sardines. Many Asians prepare a stock from bones that helps account for their adequate calcium intake without the use of milk. They soak the cracked bones from chicken, turkey, pork, or fish in vinegar and then slowly boil the bones until they become soft. The bones release calcium into the acidic broth, and most of the vinegar boils off. Cooks then use the stock, which contains more than 100 milligrams of calcium per tablespoon, in place of water to prepare soups, vegetables, and rice. Similarly, cooks in the Navajo tribe use an ash prepared from the branches and needles of the juniper tree in their recipes. One teaspoon of juniper ash provides about as much calcium as a cup of milk.

Some mineral waters provide as much as 500 milligrams of calcium per liter, offering a convenient way to meet both calcium and water needs.[24] Similarly, calcium-fortified orange juice and other fruit and vegetable juices allow a person to obtain both calcium and vitamins easily. Other examples of calcium-fortified foods include high-calcium milk (milk with extra calcium added) and calcium-fortified cereals. Fortified juices and foods help consumers increase calcium intakes, but depending on the calcium sources, the bioavailability may be significantly less than quantities listed on food labels.[25] The "How to" below describes a shortcut method for estimating your calcium intake. Highlight 12 discusses calcium supplements.

HOW TO Estimate Your Calcium Intake

Most dietitians have developed useful shortcuts to help them estimate nutrient intakes and "see" inadequacies in the diet. They can tell at a glance whether a day's meals fall short of calcium recommendations, for example.

To estimate calcium intakes, keep two bits of information in mind:

- A cup of milk provides about 300 milligrams of calcium.
- Adults need between 1000 and 1200 milligrams of calcium per day, which represents 3 to 4 cups of milk—or the equivalent:

$$1000 \text{ mg} \div 300 \text{ mg/c} = 3\tfrac{1}{3} \text{ c}$$
$$1200 \text{ mg} \div 300 \text{ mg/c} = 4 \text{ c}$$

If a person drinks 3 to 4 cups of milk a day, it's easy to see that calcium needs are being met. If not, it takes some detective work to identify the other sources and estimate total calcium intake.

To estimate a person's daily calcium intake, use this shortcut, which compares the calcium in calcium-rich foods to the calcium content of milk. The calcium in a cup of milk is assigned 1 point, and the goal is to attain 3 to 4 points per day. Foods are given points as follows:

- 1 c milk, yogurt, or fortified soy milk or 1½ oz cheese = 1 point

- 4 oz canned fish with bones (sardines) = 1 point
- 1 c ice cream, cottage cheese, or calcium-rich vegetable (see the text) = ½ point

Then, because other foods also contribute small amounts of calcium, together they are given a point.

- Well-balanced diet containing a variety of foods = 1 point

Now consider a day's meals with calcium in mind. Cereal with 1 cup of milk for breakfast (1 point for milk), a ham and cheese sub sandwich for lunch (1 point for cheese), and a cup of broccoli and lasagna for dinner (½ point for calcium-rich vegetable and 1 point for cheese in lasagna)—plus 1 point for all other foods eaten that day—adds up to 4½ points. This shortcut estimate indicates that calcium recommendations have been met, and a diet analysis of these few foods reveals a calcium intake of over 1000 milligrams. By knowing the best sources of each nutrient, you can learn to scan the day's meals and quickly see if you are meeting your daily goals.

A generalization that has been gaining strength throughout this book is supported by the information given here about calcium. A balanced diet that supplies a variety of foods is the best plan to ensure adequacy for all essential nutrients. All food groups should be included, and none should be overemphasized. In our culture, calcium intake is usually inadequate wherever milk is lacking in the diet—whether through ignorance, poverty, simple dislike, fad dieting, lactose intolerance, or allergy. By contrast, iron is usually lacking whenever milk is overemphasized, as Chapter 13 explains.

Calcium Deficiency

A low calcium intake during the growing years limits the bones' ability to reach their optimal mass and density. Most people achieve a peak bone mass by their late 20s, and dense bones best protect against age-related bone loss and fractures (see Figure 12-16). All adults lose bone as they grow older, beginning between the ages of 30 and 40. When bone losses reach the point of causing fractures under common, everyday stresses, the condition is known as **osteoporosis.** Osteoporosis affects more than 44 million people in the United States, mostly older women.

Unlike many diseases that make themselves known through symptoms such as pain, shortness of breath, skin lesions, tiredness, and the like, osteoporosis is silent. The body sends no signals saying bones are losing their calcium and, as a result, their integrity. Blood samples offer no clues because blood calcium remains normal regardless of bone content, and measures of bone density are not routinely taken. Highlight 12 suggests strategies to protect against bone loss, of which eating calcium-rich foods is only one.

FIGURE 12-16 Phases of Bone Development throughout Life

The active growth phase occurs from birth to approximately age 20. The next phase of peak bone mass development occurs between the ages of 12 and 30. The final phase, when bone resorption exceeds formation, begins between the ages of 30 and 40 and continues through the remainder of life.

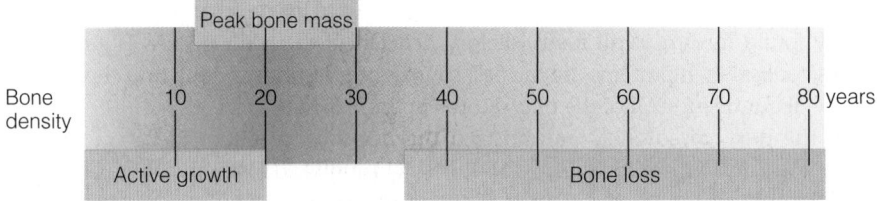

(continued)

IN SUMMARY

Most of the body's calcium is in the bones where it provides a rigid structure and a reservoir of calcium for the blood. Blood calcium participates in muscle contraction, blood clotting, and nerve impulses, and it is closely regulated by a system of hormones and vitamin D. Calcium is found predominantly in milk and milk products, but some other foods including certain vegetables and tofu also provide calcium. Even when calcium intake is inadequate, blood calcium remains normal, but at the expense of bone loss, which can lead to osteoporosis. Calcium's roles, deficiency symptoms, and food sources are summarized on the next page.

---

osteoporosis (OS-tee-oh-pore-OH-sis): a disease in which the bones become porous and fragile due to a loss of minerals; also called **adult bone loss.**
- **osteo** = bone
- **porosis** = porous

Calcium (continued)

Adequate Intake (AI)

Adults: 1000 mg/day (19–50 yr)
        1200 mg/day (>51 yr)

Upper Level

Adults: 2500 mg/day

Chief Functions in the Body

Mineralization of bones and teeth; also involved in muscle contraction and relaxation, nerve functioning, blood clotting, blood pressure

Deficiency Symptoms

Stunted growth in children; bone loss (osteoporosis) in adults

Toxicity Symptoms

Constipation; increased risk of urinary stone formation and kidney dysfunction; interference with absorption of other minerals

Significant Sources

Milk and milk products, small fish (with bones), calcium-set tofu, greens (bok choy, broccoli, chard, kale), legumes

Phosphorus

Phosphorus is the second most abundant mineral in the body. About 85 percent of it is found combined with calcium in the hydroxyapatite crystals of bones and teeth.

Phosphorus Roles in the Body Phosphorus salts (phosphates) are found not only in bones and teeth, but in all body cells as part of a major buffer system (phosphoric acid and its salts). Phosphorus is also part of DNA and RNA and is therefore necessary for all growth.

Phosphorus assists in energy metabolism. Many enzymes and the B vitamins become active only when a phosphate group is attached. ATP itself, the energy currency of the cells, uses three phosphate groups to do its work.

Lipids containing phosphorus as part of their structures (phospholipids) help to transport other lipids in the blood. Phospholipids are also the major structural components of cell membranes, where they control the transport of nutrients into and out of the cells. Some proteins, such as the casein in milk, contain phosphorus as part of their structures (phosphoproteins).

Phosphorus Recommendations and Intakes Because phosphorus is commonly found in almost all foods, dietary deficiencies are unlikely. As Figure 12-17 shows, foods rich in proteins are the best sources of phosphorus. Milk and cheese contribute about one-fourth of the phosphorus in the U.S. diet.

In the past, researchers emphasized the importance of an ideal calcium-to-phosphorus ratio from the diet to support calcium metabolism, but there is little or no evidence to support this concept. The quantities of calcium and phosphorus in the diet are far more important than their ratio to each other. A high phosphorus intake has been blamed for bone loss when, in fact, a low calcium intake—not a phosphorus toxicity or an improper ratio—is responsible. Research shows that the displacement of milk in the diet by cola drinks, not the phosphoric acid content of the beverages, has adverse effects on bone. No adverse effects of high dietary phosphorus intakes have been reported; still, an Upper Level has been established (see inside front cover).

IN SUMMARY

Phosphorus accompanies calcium both in the crystals of bone and in many foods such as milk. Phosphorus is also important in energy metabolism, as part of phospholipids, and as part of the genetic materials DNA and RNA. The summary table on the next page lists functions of, and other information about, phosphorus.

(continued)

phosphorus: a major mineral found mostly in the body's bones and teeth.

FIGURE 12-17 Phosphorus in Selected Foods

See the "How to" on p. 329 for more information on using this figure.

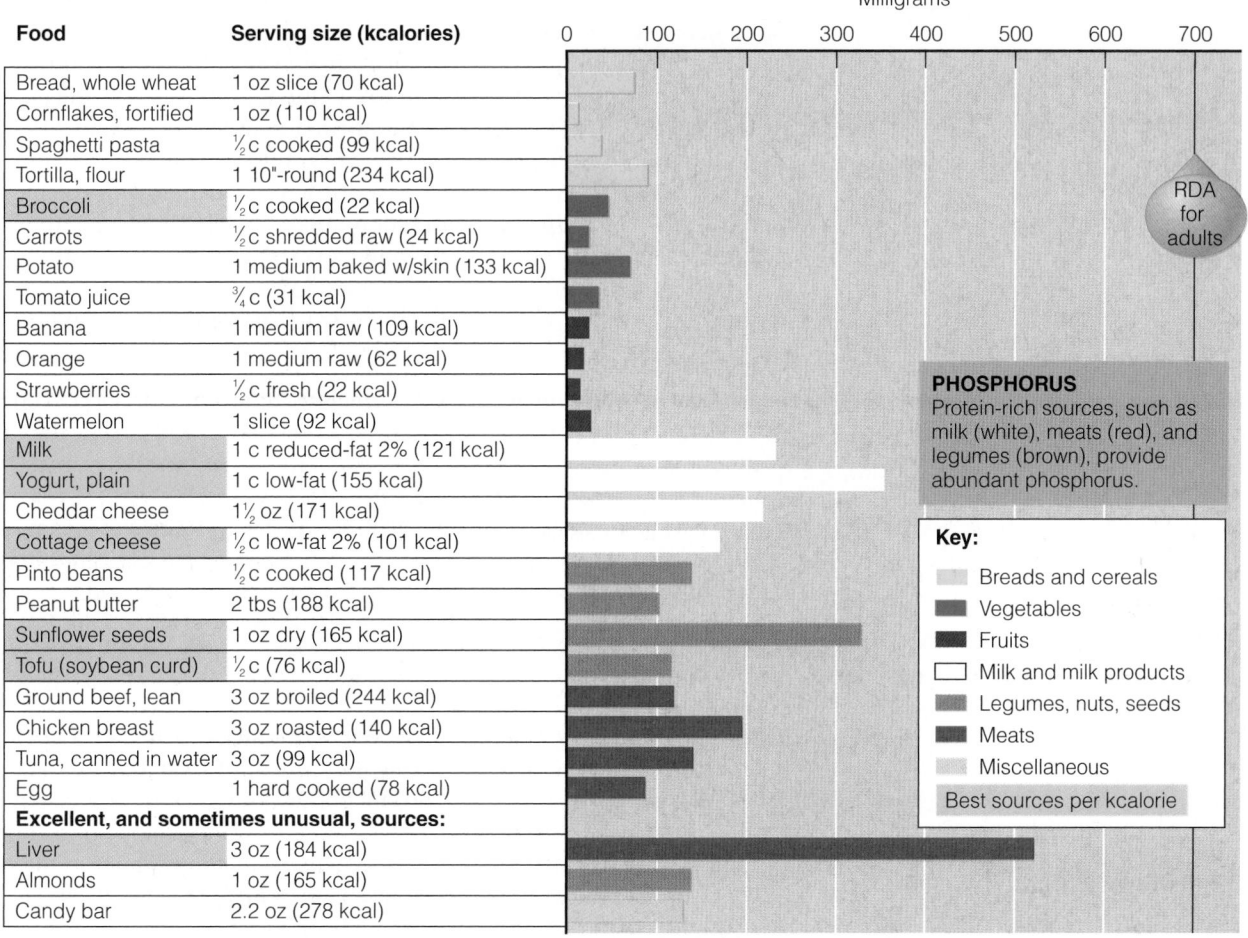

Food	Serving size (kcalories)
Bread, whole wheat	1 oz slice (70 kcal)
Cornflakes, fortified	1 oz (110 kcal)
Spaghetti pasta	½ c cooked (99 kcal)
Tortilla, flour	1 10"-round (234 kcal)
Broccoli	½ c cooked (22 kcal)
Carrots	½ c shredded raw (24 kcal)
Potato	1 medium baked w/skin (133 kcal)
Tomato juice	¾ c (31 kcal)
Banana	1 medium raw (109 kcal)
Orange	1 medium raw (62 kcal)
Strawberries	½ c fresh (22 kcal)
Watermelon	1 slice (92 kcal)
Milk	1 c reduced-fat 2% (121 kcal)
Yogurt, plain	1 c low-fat (155 kcal)
Cheddar cheese	1½ oz (171 kcal)
Cottage cheese	½ c low-fat 2% (101 kcal)
Pinto beans	½ c cooked (117 kcal)
Peanut butter	2 tbs (188 kcal)
Sunflower seeds	1 oz dry (165 kcal)
Tofu (soybean curd)	½ c (76 kcal)
Ground beef, lean	3 oz broiled (244 kcal)
Chicken breast	3 oz roasted (140 kcal)
Tuna, canned in water	3 oz (99 kcal)
Egg	1 hard cooked (78 kcal)
Excellent, and sometimes unusual, sources:	
Liver	3 oz (184 kcal)
Almonds	1 oz (165 kcal)
Candy bar	2.2 oz (278 kcal)

RDA for adults

PHOSPHORUS
Protein-rich sources, such as milk (white), meats (red), and legumes (brown), provide abundant phosphorus.

Key:
- Breads and cereals
- Vegetables
- Fruits
- Milk and milk products
- Legumes, nuts, seeds
- Meats
- Miscellaneous
- Best sources per kcalorie

Phosphorus

RDA

Adults: 700 mg/day

Upper Level

Adults (19–70 yr): 4000 mg/day

Chief Functions in the Body

Mineralization of bones and teeth; part of every cell; important in genetic material, part of phospholipids, used in energy transfer and in buffer systems that maintain acid-base balance

Deficiency Symptoms

Muscular weakness, bone pain[a]

Toxicity Symptoms

Calcification of nonskeletal tissues, particularly the kidneys

Significant Sources

All animal tissues (meat, fish, poultry, eggs, milk)

[a]Dietary deficiency rarely occurs, but some drugs can bind with phosphorus making it unavailable and resulting in bone loss that is characterized by weakness and pain.

Magnesium

Magnesium barely qualifies as a major mineral: only about 1 ounce of magnesium is present in the body of a 130-pound person. Over half of the body's magnesium is in the bones. Much of the rest is in the muscles and soft tissues, with only 1

magnesium: a cation within the body's cells, active in many enzyme systems.

◆ Reminder: A *catalyst* is a compound that facilitates chemical reactions without itself being changed in the process.

percent in the extracellular fluid. As with calcium, bone magnesium may serve as a reservoir to ensure normal blood concentrations.

Magnesium Roles in the Body In addition to maintaining bone health, magnesium acts in all the cells of the soft tissues, where it forms part of the protein-making machinery and is necessary for energy metabolism. It participates in hundreds of enzyme systems. A major role of magnesium is as a catalyst ◆ in the reaction that adds the last phosphate to the high-energy compound ATP, making it essential to the body's use of glucose; the synthesis of protein, fat, and nucleic acids; and the cells' membrane transport systems. Together with calcium, magnesium is involved in muscle contraction and blood clotting: calcium promotes the processes, whereas magnesium inhibits them. This dynamic interaction between the two minerals helps regulate blood pressure and lung function. Like many other nutrients, magnesium supports the normal functioning of the immune system.

Magnesium Intakes Average dietary magnesium estimates for U.S. adults fall below recommendations. Dietary intake data, however, do not include the contribution made by water. In areas with hard water, the water contributes both calcium and magnesium to daily intakes. Mineral waters noted earlier for their calcium content may also be magnesium-rich and can be important sources of this mineral for those who drink them.[26] Bioavailability of magnesium from mineral water is about 50 percent, but it improves when the water is consumed with a meal.[27]

The brown bars in Figure 12-18 indicate that legumes, seeds, and nuts make significant magnesium contributions. Magnesium is part of the chlorophyll molecule, so leafy green vegetables are also good sources.

Magnesium Deficiency Even with average magnesium intakes below recommendations, deficiency symptoms rarely appear except with diseases. Magnesium deficiency may develop in cases of alcohol abuse, protein malnutrition, kidney disorders, and prolonged vomiting or diarrhea. People using diuretics may also show symptoms. A severe magnesium deficiency causes a tetany similar to the calcium tetany described earlier. Magnesium deficiencies also impair central nervous system activity and may be responsible for the hallucinations experienced during alcohol withdrawal.

Magnesium and Hypertension Magnesium is critical to heart function and seems to protect against hypertension and heart disease.[28] Interestingly, people living in areas of the country with hard water, which contains high concentrations of calcium and magnesium, tend to have low rates of heart disease. With magnesium deficiency, the walls of the arteries and capillaries tend to constrict—a possible explanation for the hypertensive effect.

Magnesium Toxicity Magnesium toxicity is rare, but it can be fatal. The Upper Level for magnesium applies only to nonfood sources such as supplements or magnesium salts.

IN SUMMARY

Like calcium and phosphorus, magnesium supports bone mineralization. Magnesium is also involved in numerous enzyme systems and in heart function. It is found abundantly in legumes and leafy green vegetables and, in some areas, in water. The table below offers a summary.

Magnesium

RDA	Upper Level
Men (19–30 yr): 400 mg/day	Adults: 350 mg nonfood magnesium/day
Women (19–30 yr): 310 mg/day	

(continued)

FIGURE 12-18 Magnesium in Selected Foods

See the "How to" on p. 329 for more information on using this figure.

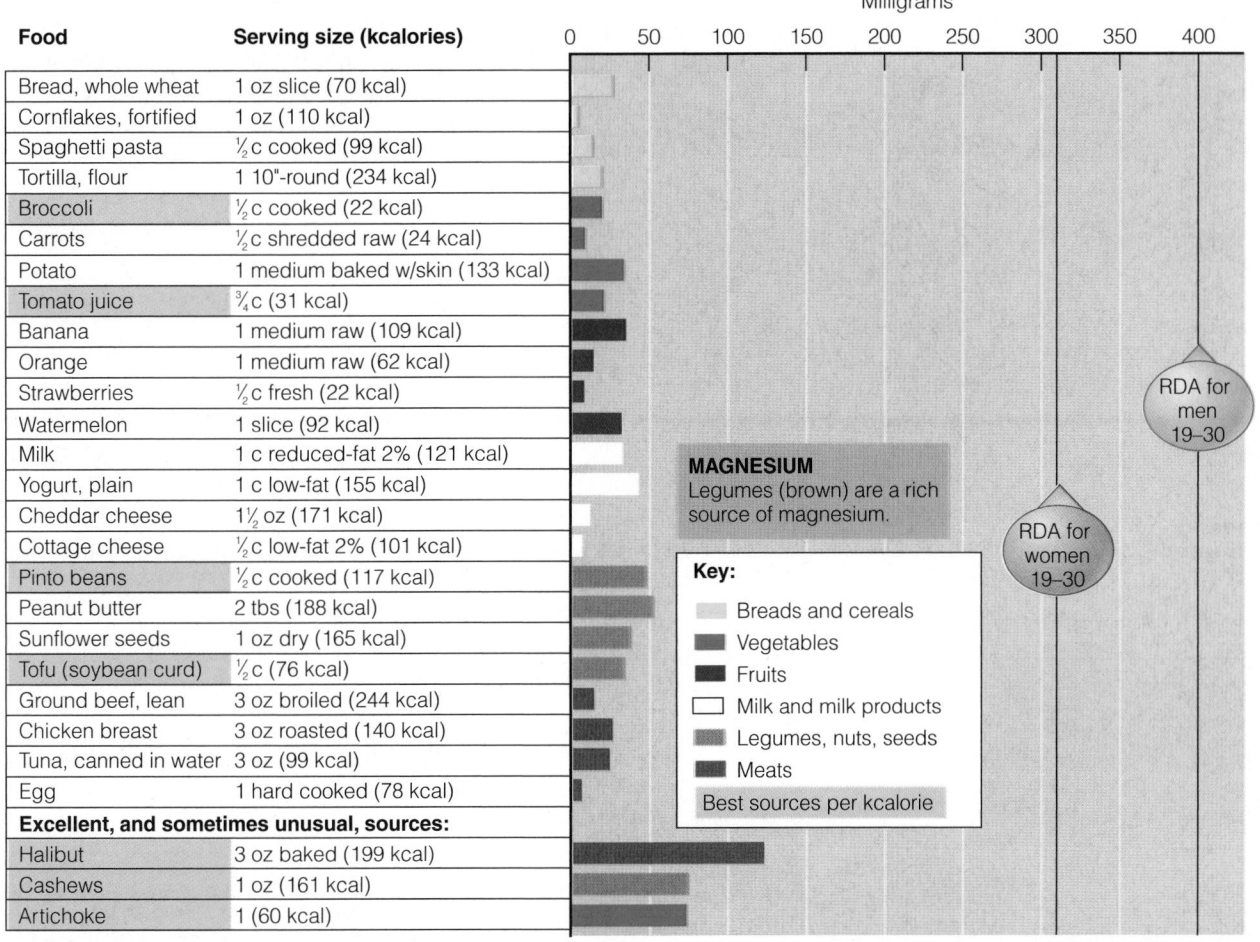

Food	Serving size (kcalories)
Bread, whole wheat	1 oz slice (70 kcal)
Cornflakes, fortified	1 oz (110 kcal)
Spaghetti pasta	½ c cooked (99 kcal)
Tortilla, flour	1 10"-round (234 kcal)
Broccoli	½ c cooked (22 kcal)
Carrots	½ c shredded raw (24 kcal)
Potato	1 medium baked w/skin (133 kcal)
Tomato juice	¾ c (31 kcal)
Banana	1 medium raw (109 kcal)
Orange	1 medium raw (62 kcal)
Strawberries	½ c fresh (22 kcal)
Watermelon	1 slice (92 kcal)
Milk	1 c reduced-fat 2% (121 kcal)
Yogurt, plain	1 c low-fat (155 kcal)
Cheddar cheese	1½ oz (171 kcal)
Cottage cheese	½ c low-fat 2% (101 kcal)
Pinto beans	½ c cooked (117 kcal)
Peanut butter	2 tbs (188 kcal)
Sunflower seeds	1 oz dry (165 kcal)
Tofu (soybean curd)	½ c (76 kcal)
Ground beef, lean	3 oz broiled (244 kcal)
Chicken breast	3 oz roasted (140 kcal)
Tuna, canned in water	3 oz (99 kcal)
Egg	1 hard cooked (78 kcal)
Excellent, and sometimes unusual, sources:	
Halibut	3 oz baked (199 kcal)
Cashews	1 oz (161 kcal)
Artichoke	1 (60 kcal)

MAGNESIUM
Legumes (brown) are a rich source of magnesium.

Key:
- Breads and cereals
- Vegetables
- Fruits
- Milk and milk products
- Legumes, nuts, seeds
- Meats
- Best sources per kcalorie

RDA for men 19–30

RDA for women 19–30

Magnesium (continued)

Chief Functions in the Body

Bone mineralization, building of protein, enzyme action, normal muscle contraction, nerve impulse transmission, maintenance of teeth, and functioning of immune system

Deficiency Symptoms

Weakness; confusion; if extreme, convulsions, bizarre muscle movements (especially of eye and face muscles), hallucinations, and difficulty in swallowing; in children, growth failure[a]

Toxicity Symptoms

From nonfood sources only; diarrhea, alkalosis, dehydration

Significant Sources

Nuts, legumes, whole grains, dark green vegetables, seafood, chocolate, cocoa

[a]A still more severe deficiency causes tetany, an extreme, prolonged contraction of the muscles similar to that caused by low blood calcium.

Sulfate

Sulfate is the oxidized form of the mineral **sulfur,** as it exists in food and water. The body's need for sulfate is easily met by a variety of foods and beverages. In addition, the body receives sulfate from the amino acids methionine and cysteine found in dietary proteins. These sulfur-containing amino acids help determine the contour of protein molecules. The sulfur-containing side chains in cysteine molecules can link to each other, forming disulfide bridges, which stabilize the protein structure. (See the drawing of insulin with its disulfide bridges on p. 184.) Skin, hair, and nails contain some of the body's more rigid proteins, which have a high sulfur content.

Because the body's sulfate needs are easily met with normal protein intakes, there is no recommended intake for sulfate. Deficiencies do not occur when diets contain protein. Only when people lack protein to the point of severe deficiency will they lack the sulfur-containing amino acids.

sulfate: the oxidized form of sulfur.

sulfur: a mineral present in the body as part of some proteins.

IN SUMMARY

Like the other nutrients, minerals' actions are coordinated to get the body's work done. The major minerals, especially sodium, chloride, and potassium, influence the body's fluid balance; whenever an anion moves, a cation moves—always maintaining homeostasis. Sodium, chloride, potassium, calcium, and magnesium are key members of the team of nutrients that direct nerve impulse transmission and muscle contraction. They are also the primary nutrients involved in regulating blood pressure. Phosphorus and magnesium participate in many reactions involving glucose, fatty acids, amino acids, and the vitamins. Calcium, phosphorus, and magnesium combine to form the structure of the bones and teeth. Each major mineral also plays other specific roles in the body. (See the summary table below.)

The Major Minerals

Mineral and Chief Functions	Deficiency Symptoms	Toxicity Symptoms	Significant Sources
Sodium Maintains normal fluid and electrolyte balance; assists in nerve impulse transmission and muscle contraction	Muscle cramps, mental apathy, loss of appetite	Edema, acute hypertension	Table salt, soy sauce; moderate amounts in meats, milks, breads, and vegetables; large amounts in processed foods
Chloride Maintains normal fluid and electrolyte balance; part of hydrochloric acid found in the stomach, necessary for proper digestion	Do not occur under normal circumstances	Vomiting	Table salt, soy sauce; moderate amounts in meats, milks, eggs; large amounts in processed foods
Potassium Maintains normal fluid and electrolyte balance; facilitates many reactions; supports cell integrity; assists in nerve impulse transmission and muscle contractions	Irregular heartbeat, muscular weakness, glucose intolerance	Muscular weakness; vomiting; if given into a vein, can stop the heart	All whole foods; meats, milks, fruits, vegetables, grains, legumes
Calcium Mineralization of bones and teeth; also involved in muscle contraction and relaxation, nerve functioning, blood clotting, and blood pressure	Stunted growth in children; bone loss (osteoporosis) in adults	Constipation; increased risk of urinary stone formation and kidney dysfunction; interference with absorption of other minerals	Milk and milk products, small fish (with bones), tofu, greens (bok choy, broccoli, chard), legumes
Phosphorus Mineralization of bones and teeth; part of every cell; important in genetic material, part of phospholipids, used in energy transfer and in buffer systems that maintain acid-base balance	Muscular weakness, bone pain[a]	Calcification of nonskeletal tissues, particularly the kidneys	All animal tissues (meat, fish, poultry, eggs, milk)
Magnesium Bone mineralization, building of protein, enzyme action, normal muscle contraction, nerve impulse transmission, maintenance of teeth, and functioning of immune system	Weakness; confusion; if extreme, convulsions, bizarre muscle movements (especially of eye and face muscles), hallucinations, and difficulty in swallowing; in children, growth failure[b]	From nonfood sources only; diarrhea, alkalosis, dehydration	Nuts, legumes, whole grains, dark green vegetables, seafood, chocolate, cocoa
Sulfate As part of proteins, stabilizes their shape by forming disulfide bridges; part of the vitamins biotin and thiamin and the hormone insulin	None known; protein deficiency would occur first	Toxicity would occur only if sulfur-containing amino acids were eaten in excess; this (in animals) suppresses growth	All protein-containing foods (meats, fish, poultry, eggs, milk, legumes, nuts)

[a]Dietary deficiency rarely occurs, but some drugs can bind with phosphorus making it unavailable and resulting in bone loss that is characterized by weakness and pain.
[b]A still more severe deficiency causes tetany, an extreme, prolonged contraction of the muscles similar to that caused by low blood calcium.

With all of the tasks these minerals perform, they are of great importance to life. Consuming enough of each of them every day is easy, given a variety of foods from each of the food groups. Whole-grain breads supply magnesium; fruits, vegetables, and legumes provide magnesium and potassium, too; milks offer calcium and phosphorus; meats offer phosphorus and sulfate as well; all foods provide sodium and chloride, with excesses being more problematic than inadequacies. The message is quite simple and has been repeated throughout this text: for an adequate intake of all the nutrients, including the major minerals, choose different foods from each of the five food groups. And drink plenty of water.

Nutrition Portfolio

CENGAGENOW
academic.cengage.com/login

Many people may miss the mark when it comes to drinking enough water to keep their bodies well hydrated or obtaining enough calcium to promote strong bones; in contrast, sodium intakes often exceed those recommended for health.

- Describe your strategy for ensuring that you drink plenty of water—about 8 glasses—every day.

- Explain the importance of selecting and preparing foods with less salt.

- Determine whether you drink at least 3 glasses of milk—or get the equivalent in calcium—every day.

NUTRITION ON THE NET

For further study of topics covered in this chapter, log on to **academic.cengage .com/nutrition/rolfes/UNCN8e**. Go to Chapter 12, then to Nutrition on the Net.

- Search for "minerals" at the American Dietetic Association site: **www.eatright.org**

- Learn about sodium in foods and on food labels from the Food and Drug Administration: **www.fda.gov/fdac/ foodlabel/sodium.html**

- Find tips and recipes for including more milk in the diet: **www.whymilk.com**

- Learn about the benefits of calcium from the National Dairy Council: **www.nationaldairycouncil.org**

NUTRITION CALCULATIONS

CENGAGENOW For additional practice log on to **academic.cengage.com/login**. Go to Chapter 12, then to Nutrition Calculations.

These problems give you an appreciation for the minerals in foods. Be sure to show your calculations (see p. 430 for answers).

1. For each of these minerals, note the unit of measure:
   Calcium   Magnesium   Phosphorus
   Potassium   Sodium

2. Learn to appreciate calcium-dense foods. The foods in the accompanying table are ranked in order of their calcium contents per serving.
   a. Which foods offer the most calcium per kcalorie? To calculate calcium density, divide calcium (mg) by energy (kcal). Record your answer in the table (round your answers); the first one is done for you.

b. The top five items ranked in order of calcium contents per serving are sardines > milk > cheese > salmon > broccoli. What are the top five items in order of calcium content per kcalorie?

Food	Calcium (mg)	Energy (kcal)	Calcium Density (mg/kcal)
Sardines, 3 oz canned	325	176	1.85
Milk, fat-free, 1 c	301	85	
Cheddar cheese, 1 oz	204	114	
Salmon, 3 oz canned	182	118	
Broccoli, cooked from fresh, chopped, ½ c	36	22	
Sweet potato, baked in skin, 1 ea	32	140	
Cantaloupe melon, ½	29	93	
Whole-wheat bread, 1 slice	21	64	
Apple, 1 medium	15	125	
Sirloin steak, lean, 3 oz	9	171	

This information should convince you that milk, milk products, fish eaten with their bones, and dark green vegetables are the best choices for calcium.

3. a. Consider how the rate of absorption influences the amount of calcium available for the body's use. Use Figure 12-15 on p. 420 to determine how much calcium the body actually receives from the foods listed in the accompanying table by multiplying the milligrams of calcium in the food by the percentage absorbed. The first one is done for you.

b. To appreciate how the absorption rate influences the amount of calcium available to the body, compare broccoli with almonds. Which provides more calcium in foods and to the body?

c. To appreciate how the calcium content of foods influences the amount of calcium available to the body, compare cauliflower with milk. How much cauliflower would a person have to eat to receive an equivalent amount of calcium as from 1 cup of milk? How does your answer change when you account for differences in their absorption rates?

Food	Calcium in the Food (mg)	Absorption Rate (%)	Calcium in the Body (mg)
Cauliflower, ½ c cooked, fresh	10	≥50	≥5
Broccoli, ½ c cooked, fresh	36		
Milk, 1 c 1% low-fat	300		
Almonds, 1 oz	75		
Spinach, 1 c raw	55		

STUDY QUESTIONS

CENGAGENOW™
To assess your understanding of chapter topics, take the Student Practice Test and explore the modules recommended in your Personalized Study Plan. Log on to **academic.cengage.com/login**.

These questions will help you review the chapter. You will find the answers in the discussions on the pages provided.

1. List the roles of water in the body. (p. 397)

2. List the sources of water intake and routes of water excretion. (pp. 398–399)

3. What is ADH? Where does it exert its action? What is aldosterone? How does it work? (p. 401)

4. How does the body use electrolytes to regulate fluid balance? (pp. 402–406)

5. What do the terms *major* and *trace* mean when describing the minerals in the body? (pp. 408–409)

6. Describe some characteristics of minerals that distinguish them from vitamins. (pp. 408–409)

7. What is the major function of sodium in the body? Describe how the kidneys regulate blood sodium. Is a dietary deficiency of sodium likely? Why or why not? (pp. 410–413)

8. List calcium's roles in the body. How does the body keep blood calcium constant regardless of intake? (pp. 416–418)

9. Name significant food sources of calcium. What are the consequences of inadequate intakes? (pp. 408–421)

10. List the roles of phosphorus in the body. Discuss the relationships between calcium and phosphorus. Is a dietary deficiency of phosphorus likely? Why or why not? (pp. 422–423)

11. State the major functions of chloride, potassium, magnesium, and sulfur in the body. Are deficiencies of these nutrients likely to occur in your own diet? Why or why not? (pp. 413–415, 423–425)

These multiple choice questions will help you prepare for an exam. Answers can be found on p. 430.

1. The body generates water during the:
   a. buffering of acids.
   b. dismantling of bone.
   c. metabolism of minerals.
   d. breakdown of energy nutrients.

2. Regulation of fluid and electrolyte balance and acid-base balance depends primarily on the:
   a. kidneys.
   b. intestines.
   c. sweat glands.
   d. specialized tear ducts.

3. The distinction between the major and trace minerals reflects the:
   a. ability of their ions to form salts.
   b. amounts of their contents in the body.
   c. importance of their functions in the body.
   d. capacity to retain their identity after absorption.

4. The principal cation in extracellular fluids is:
   a. sodium.
   b. chloride.
   c. potassium.
   d. phosphorus.

5. The role of chloride in the stomach is to help:
   a. support nerve impulses.
   b. convey hormonal messages.
   c. maintain a strong acidity.
   d. assist in muscular contractions.

6. Which would provide the most potassium?
   a. bologna
   b. potatoes
   c. pickles
   d. whole-wheat bread

7. Calcium homeostasis depends on:
   a. vitamin K, aldosterone, and renin.
   b. vitamin K, parathyroid hormone, and renin.
   c. vitamin D, aldosterone, and calcitonin.
   d. vitamin D, calcitonin, and parathyroid hormone.

8. Calcium absorption is hindered by:
   a. lactose.
   b. oxalates.
   c. vitamin D.
   d. stomach acid.

9. Phosphorus assists in many activities in the body, but *not*:
   a. energy metabolism.
   b. the clotting of blood.
   c. the transport of lipids.
   d. bone and teeth formation.

10. Most of the body's magnesium can be found in the:
    a. bones.
    b. nerves.
    c. muscles.
    d. extracellular fluids.

REFERENCES

1. J. W. Gardner, Death by water intoxication, *Military Medicine* 167 (2002): 432–434.
2. F. Manz and A. Wentz, Hydration status in the United States and Germany, *Nutrition Reviews* 63 (2005): S55–S62.
3. M. N. Sawka, S. N. Cheuvront, and R. Carter III, Human water needs, *Nutrition Reviews* 63 (2005): S30–S39.
4. B. M. Popkin and coauthors, A new proposed guidance system for beverage consumption in the United States, *American Journal of Clinical Nutrition* 83 (2006): 529–542.
5. Committee on Dietary Reference Intakes, *Dietary Reference Intakes for Water, Potassium, Sodium, Chloride, and Sulfate* (Washington, D.C.: National Academies Press, 2004), pp. 120–121.
6. F. Manz and A. Wentz, The importance of good hydration for the prevention of chronic diseases, *Nutrition Reviews* 63 (2005): S2–S5.
7. P. Ritz and G. Berrut, The importance of good hydration for day-to-day health, *Nutrition Reviews* 63 (2005): S6–S13.
8. K. M. O'Shaughnessy and F. E. Karet, Salt handling and hypertension, *Annual Review of Nutrition* 26 (2006): 343–365.
9. M. Harrington and K. D. Cashman, High salt intake appears to increase bone resorption in postmenopausal women but high potassium intake ameliorates this adverse effect, *Nutrition Reviews* 61 (2003): 179–183.
10. D. E. Sellmeyer, M. Schloetter, and A. Sebastin, Potassium citrate prevents increased urine calcium excretion and bone resorption induced by a high sodium chloride diet, *Journal of Clinical Endocrinology and Metabolism* 87 (2002): 2008–2012.
11. P. Lin and coauthors, The DASH diet and sodium reduction improve markers of bone turnover and calcium metabolism in adults, *Journal of Nutrition* 133 (2003): 3130–3136.
12. C. A. Nowson and coauthors, Blood pressure response to dietary modifications in free-living individuals, *Journal of Nutrition* 134 (2004): 2322–2329.
13. S. C. Larsson and coauthors, Calcium and dairy food intakes are inversely associated with colorectal cancer risk in the Cohort of Swedish Men, *American Journal of Clinical Nutrition* 83 (2006): 667–673; A. Flood and coauthors, Calcium from diet and supplements is associated with reduced risk of colorectal cancer in a prospective cohort of women, *Cancer Epidemiology, Biomarkers, and Prevention* 14 (2005): 126–132; U. Peters and coauthors, Calcium intake and colorectal adenoma in a US colorectal cancer early detection program, *American Journal of Clinical Nutrition* 80 (2004): 1358–1365; E. Cho and coauthors, Dairy foods, calcium, and colorectal cancer: A pooled analysis of 10 cohort studies, *Journal of the National Cancer Institute* 96 (2004): 1015–1022; M. Jacqmain and coauthors, Calcium intake, body composition, and lipoprotein-lipid concentrations, *American Journal of Clinical Nutrition* 77 (2003): 1448–1452.
14. S. J. Parikh and J. A. Yanovski, Calcium and adiposity, *American Journal of Clinical Nutrition* 77 (2003): 281–287; D. Teegarden, Calcium intake and reduction in weight or fat mass, *Journal of Nutrition* 133 (2003): 249S–251S; R. P. Heaney, K. M. Davies, and M. J. Barger-Lux, Calcium and weight: Clinical studies, *Journal of the American College of Nutrition* 21 (2002): 152–155.
15. R. J. Loos and coauthors, Calcium intake is associated with adiposity in black and white men and white women of the HERITAGE Family Study, *Journal of Nutrition* 134 (2004): 1772–1778; Jacqmain and coauthors, 2003; Teegarden, 2003; Heaney, Davies, and Barger-Lux, 2002.
16. J. K. Lorenzen and coauthors, Calcium supplementation for 1 y does not reduce body weight or fat mass in young girls, *American Journal of Clinical Nutrition* 83 (2006): 18–23; M. B. Zemel and coauthors, Dietary calcium and dairy products accelerate weight and fat loss during energy restriction in obese adults, *American Journal of Clinical Nutrition* 75 (2002): 342S.
17. M. B. Zemel, Mechanisms of dairy modulation of adiposity, *Journal of Nutrition* 133 (2003): 252S–256S.
18. S. N. Rajpathak and coauthors, Calcium and dairy intakes in relation to long-term weight gain in US men, *American Journal of Clinical Nutrition* 83 (2006): 559–566; C. W. Gunther and coauthors, Dairy products do not lead to alterations in body weight or fat mass in young women in a 1-y intervention, *American Journal of Clinical Nutrition* 81 (2005): 751–756; S. I. Barr, Increased dairy product or calcium intake: Is body weight or composition affected in humans? *Journal of Nutrition* 133 (2003): 245S–248S.
19. R. D. Jackson and coauthors, Calcium plus vitamin D supplementation and the risk of fractures, *New England Journal of Medicine* 354 (2006): 669–683.

20. L. K. Massey, Dietary animal and plant protein and human bone health: A whole foods approach, *Journal of Nutrition* 133 (2003): 862S–865S.
21. X. Gao and coauthors, Meeting adequate intake for dietary calcium without dairy foods in adolescents aged 9 to 18 years (National Health and Nutrition Examination Survey 2001–2002), *Journal of the American Dietetic Association* 106 (2006): 1759–1765.
22. R. E. Black and coauthors, Children who avoid drinking cow milk have low dietary calcium intakes and poor bone health, *American Journal of Clinical Nutrition* 76 (2002): 675–680.
23. F. R. Greer, N. F. Krebs, and the Committee on Nutrition, Optimizing bone health and calcium intakes of infants, children, and adolescents, *Pediatrics* 117 (2006) 578–585; H. J. Kalkwarf, J. C. Khoury, and B. P. Lanphear, Milk intake during childhood and adolescence, adult bone density, and osteoporotic fractures in US women, *American Journal of Clinical Nutrition* 77 (2003): 257–265.
24. R. P. Heaney, Absorbability and utility of calcium in mineral waters, *American Journal of Clinical Nutrition* 84 (2006): 371–374.
25. R. P. Heaney and coauthors, Calcium fortification systems differ in bioavailability, *Journal of the American Dietetic Association* 105 (2005): 807–809.
26. Galan and coauthors, 2002.
27. M. Sabatier and coauthors, Meal effect on magnesium bioavailability from mineral water in healthy women, *American Journal of Clinical Nutrition* 75 (2002): 65–71.
28. S. H. Jee and coauthors, The effect of magnesium supplementation on blood pressure: A meta-analysis of randomized clinical trials, *American Journal of Hypertension* 15 (2002): 691–696.

ANSWERS

Nutrition Calculations

1. Calcium: mg   Magnesium: mg   Phosphorus: mg

   Potassium: mg   Sodium: mg

2. a.

Food	Calcium Density (mg/kcal)
Sardines, 3 oz canned	325 mg ÷ 176 kcal = 1.85 mg/kcal
Milk, fat-free, 1 c	301 mg ÷ 85 kcal = 3.54 mg/kcal
Cheddar cheese, 1 oz	204 mg ÷ 114 kcal = 1.79 mg/kcal
Salmon, 3 oz canned	182 mg ÷ 118 kcal = 1.54 mg/kcal
Broccoli, cooked from fresh, chopped, ½ c	36 mg ÷ 22 kcal = 1.64 mg/kcal
Sweet potato, baked in skin, 1 ea	32 mg ÷ 140 kcal = 0.23 mg/kcal
Cantaloupe melon, ½	29 mg ÷ 93 kcal = 0.31 mg/kcal
Whole-wheat bread, 1 slice	21 mg ÷ 64 kcal = 0.33 mg/kcal
Apple, 1 medium	15 mg ÷ 125 kcal = 0.12 mg/kcal
Sirloin steak, lean, 3 oz	9 mg ÷ 171 kcal = 0.05 mg/kcal

    b. Ranked by calcium density (calcium per kcalorie):
       milk > sardines > cheese > broccoli > salmon

3. a.

Food	Calcium in Food (mg) × Absorption rate (%) = Calcium in the Body (mg)
Cauliflower, ½ c cooked, fresh	10 mg × 0.50 = 5 mg (or more)
Broccoli, ½ c cooked, fresh	36 mg × 0.50 = 18 mg (or more)
Milk, 1 c 1% low-fat	300 mg × 0.30 = 90 mg
Almonds, 1 oz	75 mg × 0.20 = 15 mg
Spinach, 1 c raw	55 mg × 0.05 = 3 mg (or less)

    b. The almonds offer more than twice as much calcium per serving, but an equivalent amount after absorption.

    c. To equal the 300 milligrams provided by milk, a person would need to eat 15 cups of cauliflower (300 mg/c milk 4 10 mg/½ c cauliflower 5 30 ½ c or 15 c). After considering the better absorption rate of cauliflower, a person would need to eat 9 cups of cauliflower (5 mg/½ c or 10 mg/c; 90 mg 4 10 mg/c 5 9 c) to match the 90 milligrams available to the body from milk after absorption. The better absorption rate reduced the quantity of cauliflower significantly, but that's still a lot of cauliflower.

Study Questions (multiple choice)

1. d   2. a   3. b   4. a   5. c   6. b   7. d   8. b

9. b   10. a

Osteoporosis and Calcium

© Photo Disc/Getty Images

Osteoporosis becomes apparent during the later years, but it develops much earlier—and without warning. Few people are aware that their bones are being robbed of their strength. The problem often first becomes evident when someone's hip suddenly gives way. People say, "She fell and broke her hip," but in fact the hip may have been so fragile that it broke *before* she fell. Even bumping into a table may be enough to shatter a porous bone into fragments so numerous and scattered that they cannot be reassembled. Removing them and replacing them with an artificial joint requires major surgery. An estimated 300,000 people in the United States are hospitalized each year because of hip fractures related to osteoporosis. About a fourth die of complications within a year. A fourth of those who survive will never walk or live independently again. Their quality of life slips downward.

This highlight examines osteoporosis, one of the most prevalent diseases of aging, affecting more than 44 million people in the United States—most of them women over 50.[1] It reviews the many factors that contribute to the 1.5 million breaks in the bones of the hips, vertebrae, wrists, arms, and ankles each year. And it presents strategies to reduce the risks, paying special attention to the role of dietary calcium.

Bone Development and Disintegration

Bone has two compartments: the outer, hard shell of **cortical bone** and the inner, lacy matrix of **trabecular bone.** (The glossary defines these and other bone-related terms.) Both can lose minerals, but in different ways and at different rates. The photograph on p. 432 shows a human leg bone sliced lengthwise, exposing the lacy, calcium-containing crystals of trabecular bone. These crystals give up calcium to the blood when the diet runs short, and they take up calcium again when the supply is plentiful (review Figure 12-13 on p. 418). For people who have eaten calcium-rich foods throughout the bone-forming years of their youth, these deposits make bones dense and provide a rich reservoir of calcium.

Surrounding and protecting the trabecular bone is a dense, ivorylike exterior shell—the cortical bone. Cortical bone composes the shafts of the long bones, and a thin cortical shell caps the end of the bone, too. Both compartments confer strength on bone: cortical bone provides the sturdy outer wall, and trabecular bone provides support along the lines of stress.

The two types of bone play different roles in calcium balance and osteoporosis. Supplied with blood vessels and metabolically

GLOSSARY

bone meal or **powdered bone:** crushed or ground bone preparations intended to supply calcium to the diet. Calcium from bone is not well absorbed and is often contaminated with toxic minerals such as arsenic, mercury, lead, and cadmium.

bone density: a measure of bone strength. When minerals fill the bone matrix (making it dense), they give it strength.

cortical bone: the very dense bone tissue that forms the outer shell surrounding trabecular bone

and comprises the shaft of a long bone.

dolomite: a compound of minerals (calcium magnesium carbonate) found in limestone and marble. Dolomite is powdered and is sold as a calcium-magnesium supplement. However, it may be contaminated with toxic minerals, is not well absorbed, and interacts adversely with absorption of other esssential minerals.

oyster shell: a product made from the powdered shells of oysters

that is sold as a calcium supplement, but it is not well absorbed by the digestive system.

trabecular (tra-BECK-you-lar) **bone:** the lacy inner structure of calcium crystals that supports the bone's structure and provides a calcium storage bank.

type I osteoporosis: osteoporosis characterized by rapid bone losses, primarily of trabecular bone.

type II osteoporosis: osteoporosis characterized by gradual losses of

both trabecular and cortical bone.

Reminder: **Osteoporosis** is a disease characterized by porous and fragile bones.

Antacids are medications used to relieve indigestion by neutralizing acid in the stomach. Calcium-containing preparations (such as Tums) contain available calcium. Antacids with aluminum or magnesium hydroxides (such as Rolaids) can accelerate calcium losses.

active, trabecular bone is sensitive to hormones that govern day-to-day deposits and withdrawals of calcium. It readily gives up minerals whenever blood calcium needs replenishing. Losses of trabecular bone start becoming significant for men and women in their 30s, although losses can occur whenever calcium withdrawals exceed deposits.

Cortical bone also gives up calcium, but slowly and at a steady pace. Cortical bone losses typically begin at about age 40 and continue slowly but surely thereafter.

Losses of trabecular and cortical bone reflect two types of osteoporosis, which cause two types of bone breaks. **Type I osteoporosis** involves losses of trabecular bone (see Figure H12-1). These losses sometimes exceed three times the expected rate, and bone breaks may occur suddenly. Trabecular bone becomes so fragile that even the body's own weight can overburden the spine—vertebrae may suddenly disintegrate and crush down, painfully pinching major nerves. Wrists may break as bone ends weaken, and teeth may loosen or fall out as the trabecular bone of the jaw recedes. Women are most often the victims of this type of osteoporosis, outnumbering men six to one.

In **type II osteoporosis,** the calcium of both cortical and trabecular bone is drawn out of storage, but slowly over the years. As old age approaches, the vertebrae may compress into wedge shapes, forming what is often called a "dowager's hump," the posture many older people assume as they "grow shorter." Figure H12-2 (p. 433) shows the effect of compressed spinal bone on a woman's height and posture. Because both the cortical shell and the trabecular interior weaken, breaks most often occur in the hip, as mentioned in the introductory paragraph. A woman is twice as likely as a man to suffer type II osteoporosis.

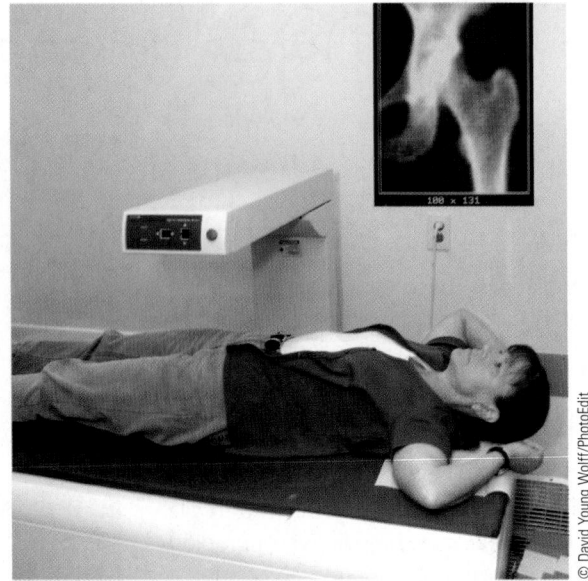

Using a DEXA (dual-energy X-ray absorpiometry) scan to measure bone mineral density identifies osteoporosis, determines risks for fractures, and tracks responses to treatment.

Table H12-1 summarizes the differences between the two types of osteoporosis. Physicians can diagnose osteoporosis and assess the risk of bone fractures by measuring **bone density** using dual-energy X-ray absorptiometry (DEXA scan) or ultrasound. They also consider risk factors that predict bone fractures, including age, personal and family history of fracture, BMI, and physical inactivity.[2] Table H12-2 summarizes the major risk factors and protective factors for osteo-

FIGURE H12-1 Healthy and Osteoporotic Trabecular Bones

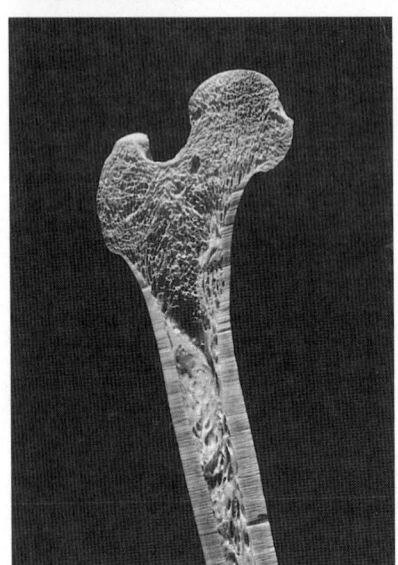

Trabecular bone is the lacy network of calcium-containing crystals that fills the interior. Cortical bone is the dense, ivorylike bone that forms the exterior shell.

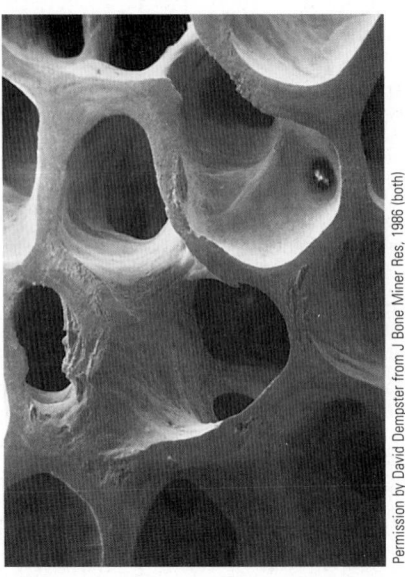

Electron micrograph of healthy trabecular bone.

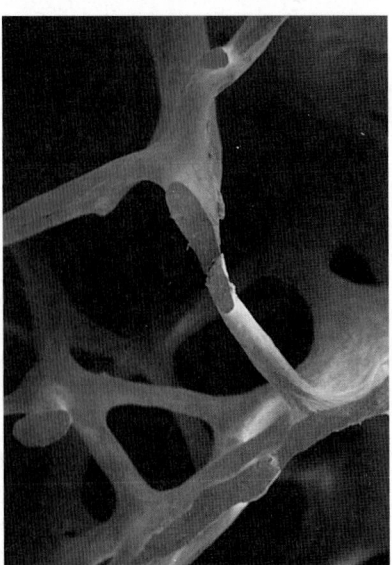

Electron micrograph of trabecular bone affected by osteoporosis.

FIGURE H12-2 Loss of Height in a Woman Caused by Osteoporosis

The woman on the left is about 50 years old. On the right, she is 80 years old. Her legs have not grown shorter. Instead, her back has lost length due to collapse of her spinal bones (vertebrae). Collapsed vertebrae cannot protect the spinal nerves from pressure that causes excruciating pain.

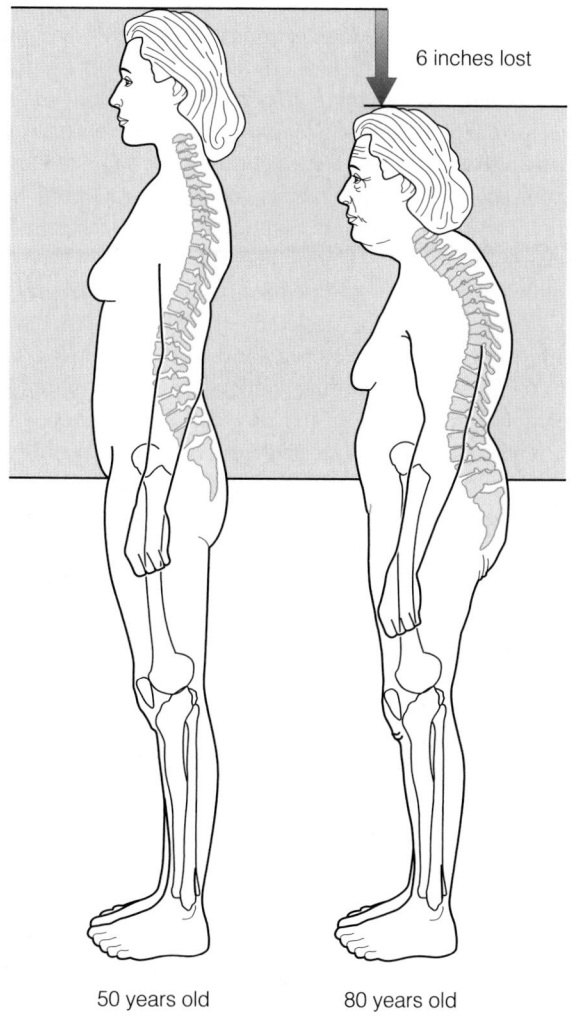

6 inches lost

50 years old   80 years old

porosis. The more risk factors that apply to a person, the greater the chances of bone loss. Notice that several risk factors that are influential in the development of osteoporosis—such as age, gender, and genetics—cannot be changed. Other risk factors—such as diet, physical activity, body weight, smoking, and alcohol use—are personal behaviors that can be changed. By eating a calcium-rich, well-balanced diet, being physically active, abstaining from smoking, and drinking alcohol in moderation (if at all), people can defend themselves against osteoporosis. These decisions are particularly important for those with other risk factors that cannot be changed.

Whether a person develops osteoporosis seems to depend on the interactions of several factors, including nutrition. The strongest predictor of bone density is age: osteoporosis is responsible for 90 percent of the hip fractures in women and 80 percent in men over the age of 65.

Age and Bone Calcium

Two major stages of life are critical in the development of osteoporosis. The first is the bone-acquiring stage of childhood and adolescence. The second is the bone-losing decades of late adulthood (especially in women after menopause). The bones gain strength and density all through the growing years and into young adulthood. As people age, the cells that build bone gradually become less active, but those that dismantle bone continue working. The result is that bone loss exceeds bone formation. Some bone loss is inevitable, but losses can be curtailed by maximizing bone mass.

TABLE H12-1 Types of Osteoporosis Compared

	Type I	Type II
Other name	Postmenopausal osteoporosis	Senile osteoporosis
Age of onset	50 to 70 years old	70 years and older
Bone loss	Trabecular bone	Both trabecular and cortical bone
Fracture sites	Wrist and spine	Hip
Gender incidence	6 women to 1 man	2 women to 1 man
Primary causes	Rapid loss of estrogen in women following menopause; loss of testosterone in men with advancing age	Reduced calcium absorption, increased bone mineral loss, increased propensity to fall

TABLE H12-2 Risk Factors and Protective Factors for Osteoporosis

Risk Factors	Protective Factors
• Older age	• Younger age
• Low BMI	• High BMI
• Caucasian, Asian, or Hispanic heritage	• African American heritage
• Cigarette smoking	• No smoking
• Alcohol consumption in excess	• Alcohol consumption in moderation
• Sedentary lifestyle	• Regular weight-bearing exercise
• Use of glucocorticoids or anticonvulsants	• Use of diuretics
• Female gender	• Male gender
• Maternal history of osteoporosis fracture or personal history of fracture	• Bone density assessment and treatment (if necessary)
• Estrogen deficiency in women (amenorrhea or menopause, especially early or surgically induced); testosterone deficiency in men	• Use of estrogen therapy
• Lifetime diet inadequate in calcium and vitamin D	• Lifetime diet rich in calcium and vitamin D

Maximizing Bone Mass

To maximize bone mass, the diet must deliver an adequate supply of calcium during the first three decades of life. Children and teens who get enough calcium and vitamin D have denser bones than those with inadequate intakes.[3] With little or no calcium from the diet, the body must depend on bone to supply calcium to the blood—bone mass diminishes, and bones lose their density and strength. When people reach the bone-losing years of middle age, those who formed dense bones during their youth have the advantage. They simply have more bone starting out and can lose more before suffering ill effects. Figure H12-3 demonstrates this effect.

Minimizing Bone Loss

Not only does dietary calcium build strong bones in youth, but it remains important in protecting against losses in the later years. Unfortunately, calcium intakes of older adults are typically low, and calcium absorption declines after menopause.[4] The kidneys do not activate vitamin D as well as they did earlier (recall that active vitamin D enhances calcium absorption). Also, sunlight is needed to form vitamin D, and many older people spend little or no time outdoors in the sunshine. For these reasons, and because intakes of vitamin D are typically low anyway, blood vitamin D declines.

Some of the hormones that regulate bone and calcium metabolism also change with age and accelerate bone mineral withdrawal.* Together, these age-related factors contribute to bone loss: inefficient bone remodeling, reduced calcium intakes, impaired calcium absorption, poor vitamin D status, and hormonal changes that favor bone mineral withdrawal.

FIGURE H12-3 Bone Losses over Time Compared

Peak bone mass is achieved by age 30. Women gradually lose bone mass until menopause, when losses accelerate dramatically and then gradually taper off.

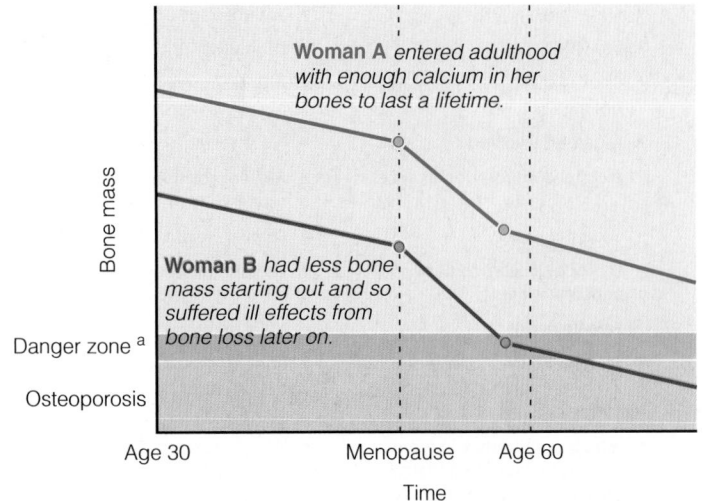

Woman A *entered adulthood with enough calcium in her bones to last a lifetime.*

Woman B *had less bone mass starting out and so suffered ill effects from bone loss later on.*

Bone mass

Danger zone [a]

Osteoporosis

Age 30   Menopause   Age 60

Time

[a]People with a moderate degree of bone mass reduction are said to have *osteopenia* and are at increased risk of fractures.
SOURCE: Data from Committee on Dietary Reference Intakes, *Dietary Reference Intakes for Calcium, Phosphorus, Magnesium, Vitamin D, and Fluoride* (Washington, D.C.: National Academy Press, 1997), pp. 71–145.

Gender and Hormones

After age, gender is the next strongest predictor of osteoporosis. Men have greater bone density than women at maturity, and women have greater losses than men in later life. Consequently, men develop bone problems about 10 years later than women, and women account for four out of five cases of osteoporosis.[5] Menopause imperils women's bones. Bone dwindles rapidly when the hormone estrogen diminishes and menstruation ceases. Women may lose up to 20 percent of their bone mass during the six to eight years following menopause. Eventually, losses taper off so that women again lose bone at the same rate as men their age. Losses of bone minerals continue throughout the remainder of a woman's lifetime, but not at the free-fall pace of the menopause years (review Figure H12-3).

Rapid bone losses also occur when *young* women's ovaries fail to produce enough estrogen, causing menstruation to cease. In some cases, diseased ovaries are to blame and must be removed; in others, the ovaries fail to produce sufficient estrogen because the women suffer from anorexia nervosa and have unreasonably restricted their body weight (see Highlight 8). The amenorrhea and low body weights explain much of the bone loss seen in these young women, even years after diagnosis and treatment. Estrogen therapy can help nonmenstruating women prevent further bone loss and reduce the incidence of fractures.[6] Because estrogen therapy may increase the risks for breast cancer, women must carefully weigh any potential benefits against the possible dangers.[7] The two main classes of drugs used to prevent or treat osteoporosis are antiresorptive agents that block bone resorption by inhibiting osteoclast activity (examples include raloxifene, alendronate, risedronate, and calcitonin) and anabolic agents that stimulate bone formation by acting on osteoblasts (an example is parathyroid hormone).[†8] A combination of these drugs or of hormone replacement and a drug may be most beneficial.[9]

Some women who choose not to use estrogen therapy turn to soy as an alternative treatment. Interestingly, the phytochemicals commonly found in soybeans mimic the actions of estrogen in the body. When natural estrogen is lacking, as after menopause, these phytochemicals may step in to stimulate estrogen-sensitive tissues. By way of this action, soy and its phytochemicals may help to prevent the rapid bone losses of the menopause years.[10] Research is far from conclusive, but some evidence suggests that soy may indeed offer some protection.[11]

If estrogen deficiency is a major cause of osteoporosis in women, what is the cause of bone loss in men? The male sex hormone testosterone appears to play a role. Men with low levels of testosterone, as occurs after re-

---

* Among the hormones suggested as influential are parathyroid hormone, calcitonin, and estrogen.
† Raloxifene (rah-LOX-ih-feen) is a selective estrogen-receptor modulator (SERM), marketed as Evista; alendronate (a-LEN-droe-nate) is a bisphosphonate, marketed as Fosamax; risedronate (rih-SEH-droe-nate) is a bisphosphonate, marketed as Actonel; and calcitonin is a hormone, marketed as Calcimar and Miacalcin.

moval of diseased testes or when testes lose function with aging, suffer more fractures. Treatment for men with osteoporosis includes testosterone replacement therapy. Thus both male and female sex hormones participate in the development and treatment of osteoporosis.

Genetics and Ethnicity

Osteoporosis may, in part, be hereditary, and family history of osteoporosis or fracture is a risk factor. The exact role of genetics is unclear, but it most likely influences both the peak bone mass achieved during growth and the bone loss incurred during the later years. The extent to which a given genetic potential is realized, however, depends on many outside factors. Diet and physical activity, for example, can maximize peak bone density during growth, whereas alcohol and tobacco abuse can accelerate bone losses later in life.

Risks of osteoporosis appear to run along racial lines and reflect genetic differences in bone development. African Americans, for example, seem to use and conserve calcium more efficiently than Caucasians.[12] Consequently, even though their calcium intakes are typically lower, black people have denser bones than white people do. Greater bone density expresses itself in less bone loss, fewer fractures, and a lower rate of osteoporosis among blacks.[13] Fractures, for example, are about twice as likely in white women age 65 or older as in black women.

Other ethnic groups have a high risk of osteoporosis. Asians from China and Japan, Mexican Americans, Hispanic people from Central and South America, and Inuit people from St. Lawrence Island typically have lower bone density than Caucasians. One might expect that these groups would suffer more bone fractures, but this is not always the case. Again, genetic differences may explain why. Asians, for example, generally have small, compact hips, which makes them less susceptible to fractures.

Findings from around the world demonstrate that although a person's genes may lay the groundwork for bone health, environmental factors influence the genes' ultimate expression. Diet in general, and calcium in particular, are among those environmental factors. Others include physical activity, body weight, smoking, and alcohol. Importantly, all of these factors are within a person's control.

Physical Activity and Body Weight

Physical activity may be the single most important factor supporting bone growth during adolescence.[14] Muscle strength and bone strength go together. When muscles work, they pull on the bones, stimulating them to develop more trabeculae and grow denser. The hormones that promote new muscle growth also favor the building of bone. As a result, active bones are denser and stronger than sedentary bones.[15]

To keep bones healthy, a person should engage in weight training or weight-bearing endurance activities (such as tennis

and jogging or vigorous walking) regularly.[16] Regular physical activity combined with an adequate calcium intake helps to maximize bone density in adolescence.[17] Adults can also maximize and maintain bone density with a regular program of weight training. Even past menopause, when most women are losing bone, weight training improves bone density.[18]

Strength training helps to build strong bones.

Heavier body weights and weight gains place a similar stress on the bones and promote their density. In fact, weight losses reduce bone density and increase the risk of fractures—in part because energy restriction diminishes calcium absorption and compromises calcium balance.[19] As mentioned in Highlight 8, the combination of underweight, severely restricted energy intake, extreme daily exercise, and amenorrhea reliably predicts bone loss.

Smoking and Alcohol

Add bone damage to the list of ill consequences associated with smoking. The bones of smokers are less dense than those of nonsmokers—even after controlling for differences in age, body weight, and physical activity habits.[20] Fortunately, the damaging effects can be reversed with smoking cessation. Blood indicators of beneficial bone activity are apparent six weeks after a person stops smoking.[21] In time, bone density is similar for former smokers and nonsmokers.

People who abuse alcohol often suffer from osteoporosis and experience more bone breaks than others. Several factors appear to be involved. Alcohol enhances fluid excretion, leading to excessive calcium losses in the urine; upsets the hormonal balance required for healthy bones; slows bone formation, leading to lower bone density; stimulates bone breakdown; and increases the risk of falling.

Dietary Calcium

Bone strength later in life depends most on how well the bones were built during childhood and adolescence. Adequate calcium nutrition during the growing years is essential to achieving optimal peak bone mass. Simply put, growing children who do not get enough calcium do not have strong bones.[22] Neither do adults who did not get enough calcium during their childhood and adolescence.[23] To that end, the DRI Committee recommends 1300 milligrams of calcium per day for everyone 9 through 18 years of age. Unfortunately, few girls meet the recommendations for calcium during these bone-forming years. (Boys generally obtain intakes close to those recommended because they eat more food.) Consequently, most girls start their adult years with less-than-optimal bone density. As adults, women rarely meet their recommended intakes of 1000 to 1200 milligrams from food. Some authorities suggest 1500 milligrams of calcium for post-menopausal women who are not receiving estrogen, but they warn that intakes exceeding 2500 milligrams a day could cause health problems.

Other Nutrients

Much research has focused on calcium, but other nutrients support bone health, too.[24] Adequate protein protects bones and reduces the likelihood of hip fractures.[25] As mentioned earlier, vitamin D is needed to maintain calcium metabolism and optimal bone health.[26] Supplementation with vitamin D reduces bone loss and the risk of fractures.[27] Vitamin K decreases bone turnover and protects against hip fractures.[28] The minerals magnesium and potassium also help to maintain bone mineral density. Vitamin A is needed in the bone-remodeling process, but too much vitamin A may be associated with osteoporosis.[29] Omega-3 fatty acids may help preserve bone integrity.[30] Additional research points to the bone benefits not of a specific nutrient, but of a diet rich in fruits and vegetables.[31] In contrast, diets containing too much salt are associated with bone losses.[32] Clearly, a well-balanced diet that depends on all the food groups to supply a full array of nutrients is central to bone health.

A Perspective on Supplements

Bone health depends, in part, on calcium. People who do not consume milk products or other calcium-rich foods in amounts that provide even half the recommended calcium should consider consulting a registered dietitian who can assess the diet and suggest food choices to correct any inadequacies. For those who are unable to consume enough calcium-rich foods, taking calcium supplements may help to enhance bone density and protect against bone loss.[33]

Selecting a calcium supplement requires a little investigative work to sort through the many options. Before examining calcium supplements, recognize that multivitamin-mineral pills contain little or no calcium. The label may list a few milligrams of calcium, but remember that the recommended intake is a gram or more for adults.

Calcium supplements are typically sold as compounds of calcium carbonate (common in **antacids** and fortified chocolate candies), citrate, gluconate, lactate, malate, or phosphate. These supplements often include magnesium, vitamin D, or both. In addition, some calcium supplements are made from **bone meal,** **oyster shell,** or **dolomite** (limestone). Many calcium supplements, especially those derived from these natural products, contain lead—which impairs health in numerous ways, as Chapter 13 points out.[34] Fortunately, calcium interferes with the absorption and action of lead in the body.

The first question to ask is how much calcium the supplement provides. Most calcium supplements provide between 250 and 1000 milligrams of calcium. To be safe, total calcium intake from both foods and supplements should not exceed 2500 milligrams a day. Read the label to find out how much a dose supplies. Unless the label states otherwise, supplements of calcium carbonate are 40 percent calcium; those of calcium citrate are 21 percent; lactate, 13 percent; and gluconate, 9 percent. Select a low-dose supplement, and take it several times a day rather than taking a large-dose supplement all at once. Taking supplements in doses of 500 milligrams or less improves absorption. Small doses also help ease the GI distress (constipation, intestinal bloating, and excessive gas) that sometimes accompanies calcium supplement use.

The next question to ask is how well the body absorbs and uses the calcium from various supplements. Most healthy people absorb calcium equally well (and as well as from milk) from any of these supplements: calcium carbonate, citrate, or phosphate. More important than supplement solubility is tablet disintegration. When manufacturers compress large quantities of calcium into small pills, the stomach acid has difficulty penetrating the pill. To test a supplement's ability to dissolve, drop it into a 6-ounce cup of vinegar, and stir occasionally. A high-quality formulation will dissolve within half an hour.

Finally, people who choose supplements must take them regularly. Furthermore, consideration should be given to the best time to take the supplements. To circumvent adverse nutrient interactions, take calcium supplements between, not with, meals. (Importantly, do not take calcium supplements with iron supplements or iron-rich meals; calcium inhibits iron absorption.) To enhance calcium absorption, take supplements with meals. If such contradictory advice drives you crazy, reconsider the benefits of food sources of calcium. Most experts agree that foods are the best source of most nutrients.

Some Closing Thoughts

Unfortunately, many of the strongest risk factors for osteoporosis are beyond people's control: age, gender, and genetics. But several strategies are still effective for prevention.[35] First, ensure an optimal peak bone mass during childhood and adolescence by eating a balanced diet rich in calcium and engaging in regular physical activity. Then, maintain that bone mass by continuing those healthy diet and activity habits, abstaining from cigarette smoking, and using alcohol moderately, if at all. Finally, minimize bone loss by maintaining an adequate nutrition and exercise regimen, and, for women, consult a physician about calcium supplements or other drug therapies that may be effective both in preventing bone loss and in restoring lost bone. The reward is the best possible chance of preserving bone health throughout life.

NUTRITION ON THE NET

For further study of topics covered in this chapter, log on to **academic.cengage .com/nutrition/rolfes/UNCN8e**. Go to Chapter 12, then to Nutrition on the Net.

• Search for "falls and fractures" at the National Institute on Aging: **www.nih.gov/nia**

• Visit the National Institutes of Health Osteoporosis and Related Bone Diseases' National Resource Center: **www.osteo.org**

• Obtain additional information from the National Osteoporosis Foundation: **www.nof.org**

REFERENCES

1. U.S. Department of Health and Human Services, *Bone Health and Osteoporosis: A report of the Surgeon General,* (Rockville, Md.: U.S. Department of Health and Human Services, Office of the Surgeon General, 2004).

2. L. G. Raisz, Screening for osteoporosis, *New England Journal of Medicine* 353 (2005): 164–171.

3. F. R. Greer, N. F. Krebs, and the Committee on Nutrition, Optimizing bone health and calcium intakes of infants, children, and adolescents, *Pediatrics* 117 (2006) 578–585.

4. B. E. C. Nordin and coauthors, Effect of age on calcium absorption in postmenopausal women, *American Journal of Clinical Nutrition* 80 (2004): 998–1002.

5. J. M. Campion and M. J. Maricic, Osteoporosis in men, *American Family Physician* 67 (2003): 1521–1526.

6. H. J. Kloosterboer and A. G. Ederveen, Pros and cons of existing treatment modalities in osteoporosis: A comparison between tibolone, SERMs and estrogen (+/- progestogen) treatments, *Journal of Steroid Biochemistry and Molecular Biology* 83 (2002): 157–165; R. A. Sayegh and P. G. Stubblefield, Bone metabolism and the perimenopause overview, risk factors, screening, and osteoporosis preventive measures, *Obstetrics and Gynecology Clinics of North America* 29 (2002): 495–510.

7. R. T. Chlebowski and coauthors, Influence of estrogen plus progestin on breast cancer and mammography in healthy postmenopausal women: The Women's Health Initiative Randomized Trial, *Journal of the American Medical Association* 289 (2003): 3243–3253; C. G. Solomon and R. G. Dluhy, Rethinking postmenopausal hormone therapy, *New England Journal of Medicine* 348 (2003): 579–580; Writing Group for the Women's Health Initiative Investigators, Risks and benefits of estrogen plus progestin in healthy postmenopausal women: Principal results from the Women's Health Initiative Randomized Controlled Trial, *Journal of the American Medical Association* 288 (2002): 321–333; O. Ylikorkala and M. Metsaheikkila, Hormone replacement therapy in women with a history of breast cancer, *Gynecological Endocrinology* 16 (2002): 469–478.

8. C. J. Rosen, Postmenopausal osteoporosis, *New England Journal of Medicine* 353 (2005): 595–603; J. F. Whitfield, How to grow bone to treat osteoporosis and mend fractures, *Current Rheumatology Reports* 5 (2003): 45–56.

9. R. P. Heaney and R. R. Recker, Combination and sequential therapy of osteoporosis, *New England Journal of Medicine* 353 (2005): 624–625; S. L. Greenspan, N. M. Resnick, and R. A. Parker, Combination therapy with hormone replacement and alendronate for

prevention of bone loss in elderly women: A randomized controlled trial, *Journal of the American Medical Association* 289 (2003): 2525–2533.

10. C. Atkinson and coauthors, The effects of phytoestrogen isoflavones on bone density in women: A double-blind, randomized, placebo-controlled trial, *American Journal of Clinical Nutrition* 79 (2004): 326–333; R. Brynin, Soy and its isoflavones: A review of their effects on bone density, *Alternative Medicine Review* 7 (2002): 317–327.

11. B. H. Arjmandi and coauthors, Soy protein has a greater effect on bone in postmenopausal women not on hormone replacement therapy, as evidenced by reducing bone resorption and urinary calcium excretion, *Journal of Clinical Endocrinology and Metabolism* 88 (2003): 1048–1054; T. Uesugi, Y. Fukui, and Y. Yamori, Beneficial effects of soybean isoflavone supplementation on bone metabolism and serum lipids in postmenopausal Japanese women: A four-week study, *Journal of the American College of Nutrition* 21 (2002): 97–102.

12. K. Wigertz and coauthors, Racial differences in calcium retention in response to dietary salt in adolescent girls, *American Journal of Clinical Nutrition* 81 (2005): 845–850.

13. J. A. Cauley and coauthors, Longitudinal study of changes in hip bone mineral density in Caucasian and African-American women, *Journal of the American Geriatrics Society* 53 (2005): 183–189; J. A. Cauley and coauthors, Bone mineral density and the risk of incident nonspinal fractures in black and white women, *Journal of the American Medical Association* 293 (2005): 2102–2108.

14. A. J. Lanou, S. E. Berkow, and N. D. Barnard, Calcium, dairy products, and bone health in children and young adults: A reevaluation of the evidence, *Pediatrics* 115 (2005): 736–743.

15. F. R. Greer, Bone health: It's more than calcium intake, *Pediatrics* 115 (2005): 792–794.

16. American College of Sports Medicine Position Stand, Physical activity and bone health, *Medicine and Science in Sports and Exercise* 36 (2004): 1985–1996.

17. J. M. Welch and C. M. Weaver, Calcium and exercise affect the growing skeleton, *Nutrition Reviews* 63 (2005): 361–373; T. Lloyd and coauthors, Lifestyle factors and the development of bone mass and bone strength in young women, *Journal of Pediatrics* 144 (2004): 776–782; M. C. Wang and coauthors, Diet in midpuberty and sedentary activity in prepuberty predict peak bone mass, *American Journal of Clinical Nutrition* 77 (2003): 495–503; S. J. Stear and coauthors, Effect of a calcium and exercise intervention on the bone mineral status of 16-18-y-old adolescent girls, *American Journal of Clinical Nutrition* 77 (2003): 985–992.

18. E. C. Cussler and coauthors, Weight lifted in strength training predicts bone change in postmenopausal women, *Medicine and Science in Sports and Exercise* 35 (2003): 10–17.

19. M. Cifuentes and coauthors, Weight loss and calcium intake influence calcium absorption in overweight postmenopausal women, *American Journal of Clinical Nutrition* 80 (2004): 123–130; T. L. Radak, Caloric restriction and calcium's effect on bone metabolism and body composition in overweight and obese premenopausal women, *Nutrition Reviews* 62 (2004): 468–481.

20. P. Gerdhem and K. J. Obrant, Effects of cigarette-smoking on bone mass as assessed by dual-energy X-ray absorptiometry and ultrasound, *Osteoporosis International* 13 (2002): 932–936.

21. C. Oncken and coauthors, Effects of smoking cessation or reduction on hormone profiles and bone turnover in postmenopausal women, *Nicotine and Tobacco Research* 4 (2002): 451–458.

22. R. E. Black and coauthors, Children who avoid drinking cow milk have low dietary calcium intakes and poor bone health, *American Journal of Clinical Nutrition* 76 (2002): 675–680.

23. H. J. Kalkwarf, J. C. Khoury, and B. P. Lanphear, Milk intake during childhood and adolescence, adult bone density, and osteoporotic fractures in US women, *American Journal of Clinical Nutrition* 77 (2003): 257–265.

24. J. W. Nieves, Osteoporosis: The role of micronutrients, *American Journal of Clinical Nutrition* 81 (2005): 1232S–1239S.

25. J. Bell, Elderly women need dietary protein to maintain bone mass, *Nutrition Reviews* 60 (2002): 337–341; B. Dawson-Hughes and S. S. Harris, Calcium intake influences the association of protein intake with rates of bone loss in elderly men and women, *American Journal of Clinical Nutrition* 75 (2002): 773–779; J. H. E. Promislow and coauthors, Protein consumption and bone mineral density in the elderly: The Rancho Bernardo Study, *American Journal of Epidemiology* 155 (2002): 636–644.

26. L. Steingrimsdottir and coauthors, Relationship between serum parathyroid hormone levels, vitamin D sufficiency, and calcium intake, *Journal of the American Medical Association* 294 (2005): 2336–2341.

27. H. A. Bischoff-Ferrari and coauthors, Fracture prevention with vitamin D supplementation: A meta-analysis of randomized controlled trials, *Journal of the American Medical Association* 293 (2005): 2257–2264; D. Feskanich, W. C. Willett, and G. A. Colditz, Calcium, vitamin D, milk consumption, and hip fractures: A prospective study among postmenopausal women,

American Journal of Clinical Nutrition 77 (2003): 504–511.

28. H. J. Kalkwarf and coauthors, Vitamin K, bone turnover, and bone mass in girls, *American Journal of Clinical Nutrition* 80 (2004): 1075–1080; N. C. Binkley and coauthors, A high phylloquinone intake is required to achieve maximal osteocalcin γ-carboxylation, *American Journal of Clinical Nutrition* 76 (2002): 1055–1060.

29. K. Michaelsson and coauthors, Serum retinol levels and the risk of fractures, *New England Journal of Medicine* 348 (2003): 287–294; D. Feskanich and coauthors, Vitamin A intake and hip fractures among postmenopausal women, *Journal of the American Medical Association* 287 (2002): 47–54; S. Johnasson and coauthors, Subclinical hypervitaminosis A causes fragile bones in rats, *Bone* 31 (2002): 685–689.

30. L. A. Weiss, E. Barrett-Connor, and D. von Mühlen, Ratio of n -6 to n -3 fatty acids and bone mineral density in older adults: The Rancho Bernardo Study, *American Journal of Clinical Nutrition* 81 (2005): 934–938.

31. H. Vatanparast and coauthors, Positive effects of vegetable and fruit consumption and calcium intake on bone mineral accrual in boys during growth from childhood to adolescence: The University of Saskatchewan Pediatric Bone Mineral Accrual Study, *American Journal of Clinical Nutrition* 82 (2005): 700–706; C. P. McGartland and coauthors, Fruit and vegetable consumption and bone mineral density: The Northern Ireland Young Hearts Project, *American Journal of Clinical Nutrition* 80 (2004): 1019–1023; L. Doyle and K. D. Cashman, The DASH diet may have beneficial effects on bone health, *Nutrition Reviews* 62 (2004): 215–220; K. L. Tucker and coauthors, Bone mineral density and dietary patterns in older adults: The Framingham Osteoporosis Study, *American Journal of Clinical Nutrition* 76 (2002): 245–252.

32. M. Harrington and K. D. Cashman, High salt intake appears to increase bone resorption in postmenopausal women but high potassium intake ameliorates this adverse effect, *Nutrition Reviews* 61 (2003): 179–183; Tucker and coauthors, 2002.

33. V. Matkovic and coauthors, Calcium supplementation and bone mineral density in females from childhood to young adulthood: A randomized controlled trial, *American Journal of Clinical Nutrition* 81 (2005): 175–188; R. P. Dodiuk-Gad and coauthors, Sustained effect of short-term calcium supplementation on bone mass in adolescent girls with low calcium intake, *American Journal of Clinical Nutrition* 81 (2005): 168–174; L. D. McCabe and coauthors, Dairy intakes affect bone density in the elderly, *American Journal of Clinical Nutrition* 80 (2004): 1066–1074.

34. E. A. Ross, N. J. Szabo, and I. R. Tebbett, Lead content of calcium supplements, *Journal of the American Medical Association* 284 (2000): 1425–1429.

35. NIH Consensus Development Panel on Osteoporosis Prevention, Diagnosis, and Therapy, Osteoporosis prevention, diagnosis, and therapy, *Journal of the American Medical Association* 285 (2001): 785–795.

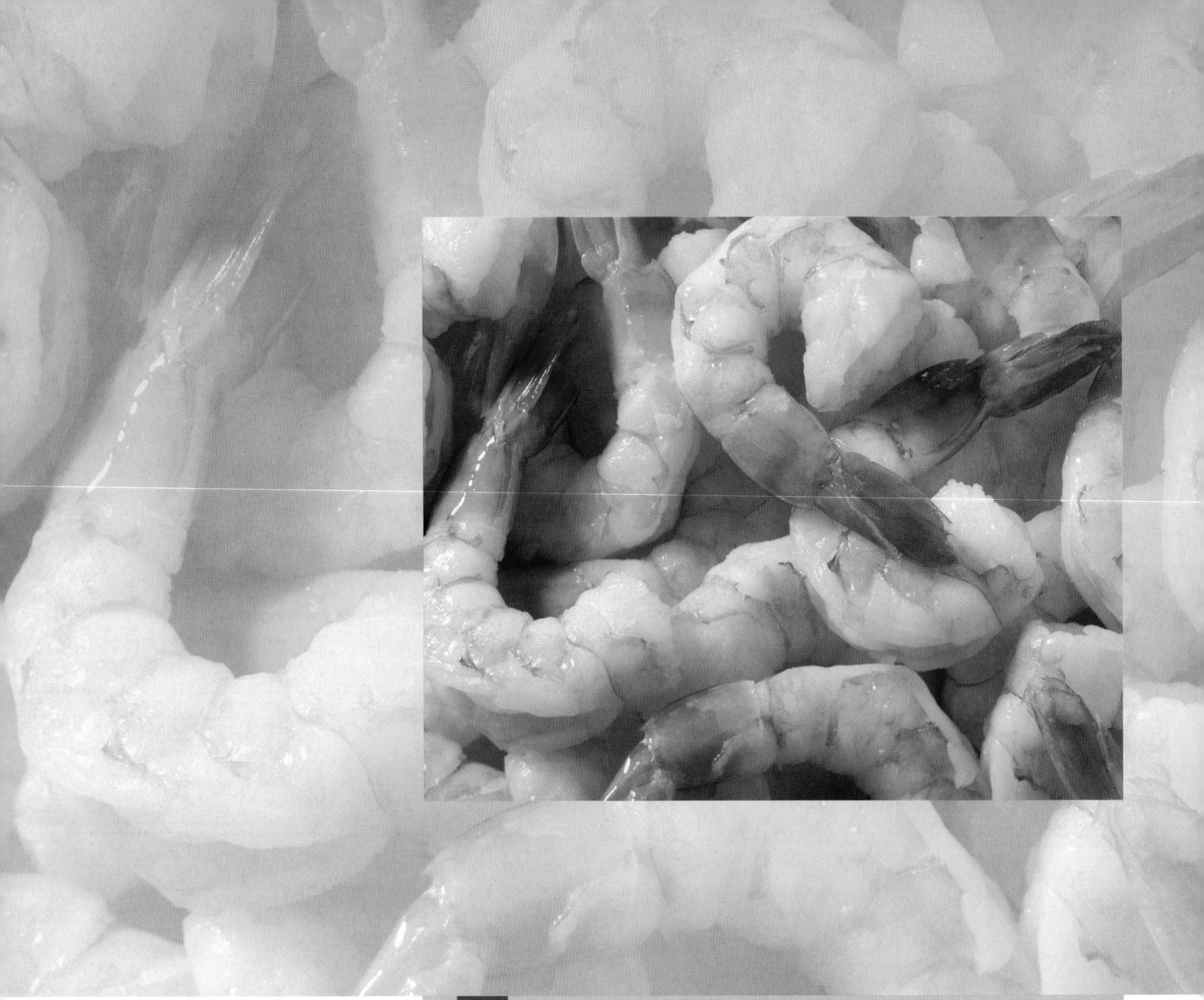

Food Image Source/Getty Images

CENGAGENOW™ The CengageNOW logo indicates an opportunity for online self-study, linking you to interactive tutorials and videos based on your level of understanding.

academic.cengage.com/login

Figure 13-3: Animated! Iron Recycled in the Body

Figure 13-6: Animated! Enteropancreatic Circulation of Zinc

Nutrition Portfolio Journal

Nutrition Calculations: Practice Problems

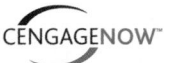

Nutrition in Your Life

Trace—barely a perceptible amount. But the trace minerals tackle big jobs. Your blood can't carry oxygen without iron, and insulin can't deliver glucose without chromium. Teeth become decayed without fluoride, and thyroid glands develop goiter without iodine. Together, the trace minerals—iron, zinc, iodine, selenium, copper, manganese, fluoride, chromium, and molybdenum—keep you healthy and strong. Where can you get these amazing minerals? A variety of foods, especially those from the meat and meat alternate group, sprinkled with a little iodized salt and complemented by a glass of fluoridated water will do the trick. It's remarkable what your body can do with only a few milligrams—or even micrograms—of the trace minerals.

The Trace Minerals

Figure 12-9 in the last chapter (p. 409) showed the tiny quantities of **trace minerals** in the human body. The trace minerals are so named because they are present, and needed, in relatively small amounts in the body. All together, they would produce only a bit of dust, hardly enough to fill a teaspoon. Yet they are no less important than the major minerals or any of the other nutrients. Each of the trace minerals performs a vital role. A deficiency of any of them may be fatal, and an excess of many is equally deadly. Remarkably, people's diets normally supply just enough of these minerals to maintain health.

The Trace Minerals—An Overview

The body requires the trace minerals in minuscule quantities. They participate in diverse tasks all over the body, each having special duties that only it can perform.

Food Sources The trace mineral contents of foods depend on soil and water composition and on how foods are processed. Furthermore, many factors in the diet and within the body affect the minerals' bioavailability. ◆ Still, outstanding food sources for each of the trace minerals, just like those for the other nutrients, include a wide variety of foods, especially unprocessed, whole foods.

Deficiencies Severe deficiencies of the better-known minerals are easy to recognize. Deficiencies of the others may be harder to diagnose, and for all minerals, mild deficiencies are easy to overlook. Because the minerals are active in all the body systems—the GI tract, cardiovascular system, blood, muscles, bones, and central nervous system—deficiencies can have wide-reaching effects and can affect people of all ages. The most common result of a deficiency in children is failure to grow and thrive.

Toxicities Some of the trace minerals are toxic at intakes not far above the estimated requirements. Thus it is important not to habitually exceed the Upper Level of recommended intakes. Many vitamin-mineral supplements contain trace minerals, making it easy for users to exceed their needs. Highlight 10 discusses supplement use and some of the regulations included in the Dietary Supplement Health

◆ Reminder: *Bioavailability* refers to the rate at and the extent to which a nutrient is absorbed and used.

trace minerals: essential mineral nutrients found in the human body in amounts smaller than 5 g; sometimes called **microminerals.**

and Education Act. As that discussion notes, the Food and Drug Administration (FDA) has no authority to limit the amounts of trace minerals in supplements; consumers have demanded the freedom to choose their own doses of nutrients.* Individuals who take supplements must therefore be aware of the possible dangers and select supplements that contain no more than 100 percent of the Daily Value. It would be easier and safer to meet nutrient needs by selecting a variety of foods than by combining an assortment of supplements (see Highlight 10).

Interactions Interactions among the trace minerals are common and often well coordinated to meet the body's needs. For example, several of the trace minerals support insulin's work, influencing its synthesis, storage, release, and action.

At other times, interactions lead to nutrient imbalances. An excess of one may cause a deficiency of another. (A slight manganese overload, for example, may aggravate an iron deficiency.) A deficiency of one may interfere with the work of another. (A selenium deficiency halts the activation of the iodine-containing thyroid hormones.) A deficiency of a trace mineral may even open the way for a contaminant mineral to cause a toxic reaction. (Iron deficiency, for example, makes the body vulnerable to lead poisoning.) These examples reinforce the need to balance intakes and to use supplements wisely, if at all. A good food source of one nutrient may be a poor food source of another, and factors that enhance the action of some trace minerals may interfere with others. (Meats are a good source of iron but a poor source of calcium; vitamin C enhances the absorption of iron but hinders that of copper.) Research on the trace minerals is active, suggesting that we have much more to learn about them.

IN SUMMARY

Although the body uses only tiny amounts of the trace minerals, they are vital to health. Because so little is required, the trace minerals can be toxic at levels not far above estimated requirements—a consideration for supplement users. Like the other nutrients, the trace minerals are best obtained by eating a variety of whole foods.

Iron

Iron is an essential nutrient, vital to many of the cells' activities, but it poses a problem for millions of people. Some people simply don't eat enough iron-containing foods to support their health optimally, whereas others absorb so much iron that it threatens their health. Iron exemplifies the principle that both too little and too much of a nutrient in the body can be harmful. In its wisdom, the body has several ways to achieve iron homeostasis, protecting against both deficiency and overload.[1]

Iron Roles in the Body

◆ Iron's two ionic states:
• Ferrous iron (reduced): Fe^{++}
• Ferric iron (oxidized): Fe^{+++}

◆ Reminder: A *cofactor* is a substance that works with an enzyme to facilitate a chemical reaction.

Iron has the knack of switching back and forth between two ionic states. ◆ In the reduced state, iron has lost two electrons and therefore has a net positive charge of two; it is known as *ferrous iron*. In the oxidized state, iron has lost a third electron, has a net positive charge of three, and is known as *ferric iron*. Ferrous iron can be oxidized to ferric iron, and ferric iron can be reduced to ferrous iron. Thus iron can serve as a cofactor ◆ to enzymes involved in oxidation-reduction reactions—reactions so widespread in metabolism that they occur in all cells. Enzymes involved in making amino acids, collagen, hormones, and neurotransmitters all require iron. (For details about ions, oxidation, and reduction, see Appendix B.)

---

* Canada regulates the amounts of trace minerals in supplements.

Iron forms a part of the electron carriers that participate in the electron transport chain (discussed in Chapter 7).* In this pathway, these carriers transfer hydrogens and electrons to oxygen, forming water, and in the process, make ATP for the cells' energy use.

Most of the body's iron is found in two proteins: hemoglobin ◆ in the red blood cells and **myoglobin** in the muscle cells. In both, iron helps accept, carry, and then release oxygen.

Iron Absorption and Metabolism

The body conserves iron. Because it is difficult to excrete iron once it is in the body, balance is maintained primarily through absorption. More iron is absorbed when stores are empty and less is absorbed when stores are full.[2]

Iron Absorption Special proteins help the body absorb iron from food (see Figure 13-1). One protein, called mucosal **ferritin,** receives iron from food and stores it in the mucosal cells ◆ of the small intestine. When the body needs iron, mucosal ferritin releases some iron to another protein, called mucosal **transferrin.** Mucosal transferrin transfers the iron to another protein, *blood transferrin,* which transports the iron to the rest of the body. If the body does not need iron, it is carried out when the intestinal cells are shed and excreted in the feces; intestinal cells are replaced about every three to five days. By holding iron temporarily, these cells control iron absorption by either delivering iron when the day's intake falls short or disposing of it when intakes exceed needs.

Heme and Nonheme Iron Iron absorption depends in part on its dietary source.[3] Iron occurs in two forms in foods: as **heme** iron, which is found only in foods derived from the flesh of animals, such as meats, poultry, and fish and as nonheme iron, which is found in both plant-derived and animal-derived foods (see Figure 13-2, p. 444). On average, heme iron represents about 10 percent of the iron a

---

* The iron-containing electron carriers of the electron transport chain are known as *cytochromes.* See Appendix C for details of this pathway.

◆ Reminder: *Hemoglobin* is the oxygen-carrying protein of the red blood cells that transports oxygen from the lungs to tissues throughout the body; hemoglobin accounts for 80% of the body's iron.

◆ A mucous membrane such as the one that lines the GI tract is sometimes called the **mucosa** (mu-KO-sa). The adjective of mucosa is **mucosal** (mu-KO-sal).

> **myoglobin:** the oxygen-holding protein of the muscle cells.
> • **myo** = muscle
>
> **ferritin** (FAIR-ih-tin): the iron storage protein.
>
> **transferrin** (trans-FAIR-in): the iron transport protein.
>
> **heme** (HEEM): the iron-holding part of the hemoglobin and myoglobin proteins. About 40% of the iron in meat, fish, and poultry is bound into heme; the other 60% is **nonheme** iron.

FIGURE 13-1 | Iron Absorption

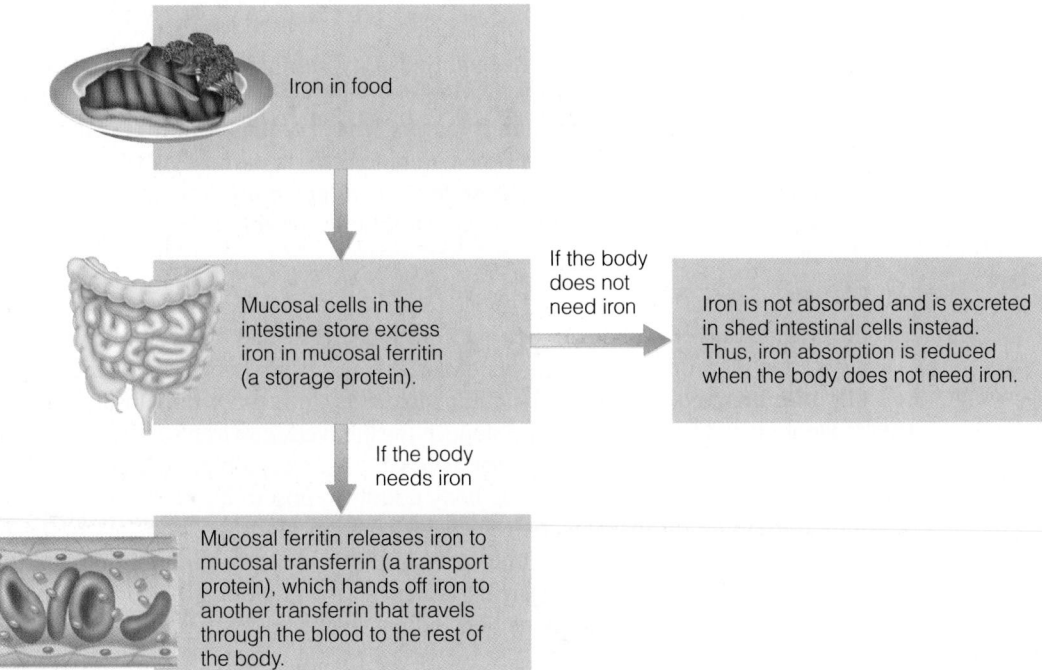

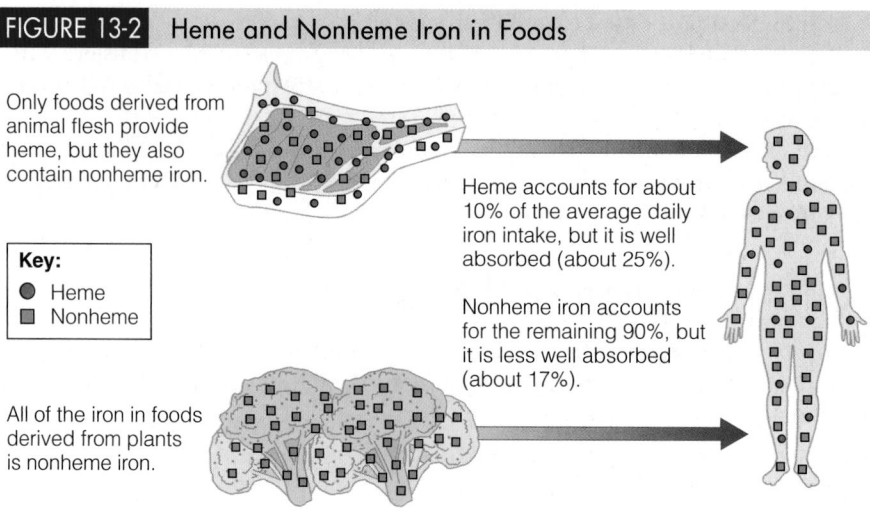

FIGURE 13-2 Heme and Nonheme Iron in Foods

Only foods derived from animal flesh provide heme, but they also contain nonheme iron.

Key:
● Heme
■ Nonheme

Heme accounts for about 10% of the average daily iron intake, but it is well absorbed (about 25%).

Nonheme iron accounts for the remaining 90%, but it is less well absorbed (about 17%).

All of the iron in foods derived from plants is nonheme iron.

This chili dinner provides several factors that may enhance iron absorption: heme and nonheme iron and MFP from meat, nonheme iron from legumes, and vitamin C from tomatoes.

◆ Factors that *enhance* nonheme iron absorption:
 • MFP factor
 • Vitamin C (ascorbic acid)

◆ Factors that *inhibit* nonheme iron absorption:
 • Phytates (legumes, grains, and rice)
 • Vegetable proteins (soybeans, legumes, nuts)
 • Calcium (milk)
 • Tannic acid (and other polyphenols in tea and coffee)

MFP factor: a peptide released during the digestion of **m**eat, **f**ish, and **p**oultry that enhances nonheme iron absorption.

person consumes in a day. Even though heme iron accounts for only a small proportion of the intake, it is so well absorbed that it contributes significant iron. About 25 percent of heme iron and 17 percent of nonheme iron is absorbed, depending on dietary factors and the body's iron stores.[4] In iron deficiency, absorption increases. In iron overload, absorption declines.[5] Researchers disagree as to whether heme iron absorption responds to iron stores as sensitively as nonheme iron absorption does.

Absorption-Enhancing Factors Meat, fish, and poultry contain not only the well-absorbed heme iron, but also a peptide (called the **MFP factor)** that promotes the absorption of nonheme iron ◆ from other foods eaten at the same meal.[6] Vitamin C also enhances nonheme iron absorption from foods eaten in the same meal by capturing the iron and keeping it in the reduced ferrous form, ready for absorption. Some acids and sugars also enhance nonheme iron absorption.

Absorption-Inhibiting Factors Some dietary factors bind with nonheme iron, inhibiting absorption. ◆ These factors include the phytates in legumes, whole grains, and rice; the vegetable proteins in soybeans, other legumes, and nuts; the calcium in milk; and the polyphenols (such as tannic acid) in tea, coffee, grain products, oregano, and red wine.

Dietary Factors Combined The many dietary enhancers, inhibitors, and their combined effects make it difficult to estimate iron absorption. Most of these factors exert a strong influence individually, but not when combined with the others in a meal. Furthermore, the impact of the combined effects diminishes when a diet is evaluated over several days. When multiple meals are analyzed together, three factors appear to be most relevant: MFP and vitamin C as enhancers and phytates as inhibitors.

Individual Variation Overall, about 18 percent of dietary iron is absorbed from mixed diets and only about 10 percent from vegetarian diets.[7] As you might expect, vegetarian diets do not have the benefit of easy-to-absorb heme iron or the help of MFP in enhancing absorption. In addition to dietary influences, iron absorption also depends on an individual's health, stage in the life cycle, and iron status. Absorption can be as low as 2 percent in a person with GI disease or as high as 35 percent in a rapidly growing, healthy child. The body adapts to absorb more iron when a person's iron stores fall short or when the need increases for any reason (such as pregnancy). The body makes more mucosal transferrin to absorb more iron from the intestines and more blood transferrin to carry more iron around the body. Similarly, when iron stores are sufficient, the body adapts to absorb less iron.

Iron Transport and Storage Blood transferrin delivers iron to the bone marrow and other tissues. The bone marrow uses large quantities to make new red blood cells, whereas other tissues use less. Surplus iron is stored in the protein ferritin, primarily in the liver, but also in the bone marrow and spleen. When dietary iron has been plentiful, ferritin is constantly and rapidly made and broken down, providing an ever-ready supply of iron. When iron concentrations become abnormally high, the liver converts some ferritin into another storage protein called **hemosiderin.** Hemosiderin releases iron more slowly than ferritin does. By storing excess iron, the body protects itself: free iron acts as a free radical, attacking cell lipids, DNA, and protein. (See Highlight 11 for more information on free radicals and the damage they can cause.)

Iron Recycling The average red blood cell lives about four months; then the spleen and liver cells remove it from the blood, take it apart, and prepare the degradation products for excretion or recycling. The iron is salvaged: the liver attaches it to blood transferrin, which transports it back to the bone marrow to be reused in making new red blood cells. Thus, although red blood cells live for only about four months, the iron recycles through each new generation of cells (see Figure 13-3). The body loses some iron daily via the GI tract and, if bleeding occurs, in blood. Only tiny amounts of iron are lost in urine, sweat, and shed skin.*

Iron Balance Maintaining iron balance depends on the careful regulation of iron absorption, transport, storage, recycling, and losses. The hormone **hepcidin** is central to the regulation of iron balance.[8] Produced by the liver, hepcidin helps to maintain blood iron within the normal range by inhibiting absorption from the intestines and transport out of storage as needed.

Iron Deficiency

Worldwide, **iron deficiency** is the most common nutrient deficiency, affecting more than 1.2 billion people.[9] In developing countries, almost half of preschool

---

* Adults lose about 1.0 milligram of iron per day. Women lose additional iron in menses. Menstrual losses vary considerably, but over a month, they average about 0.5 milligram per day.

hemosiderin (heem-oh-SID-er-in): an iron storage protein primarily made in times of iron overload.

hepcidin: a hormone produced by the liver that regulates iron balance.

iron deficiency: the state of having depleted iron stores.

FIGURE 13-3 *Animated!* **Iron Recycled in the Body**

Once iron enters the body, most of it is recycled. Some is lost with body tissues and must be replaced by eating iron-containing food.

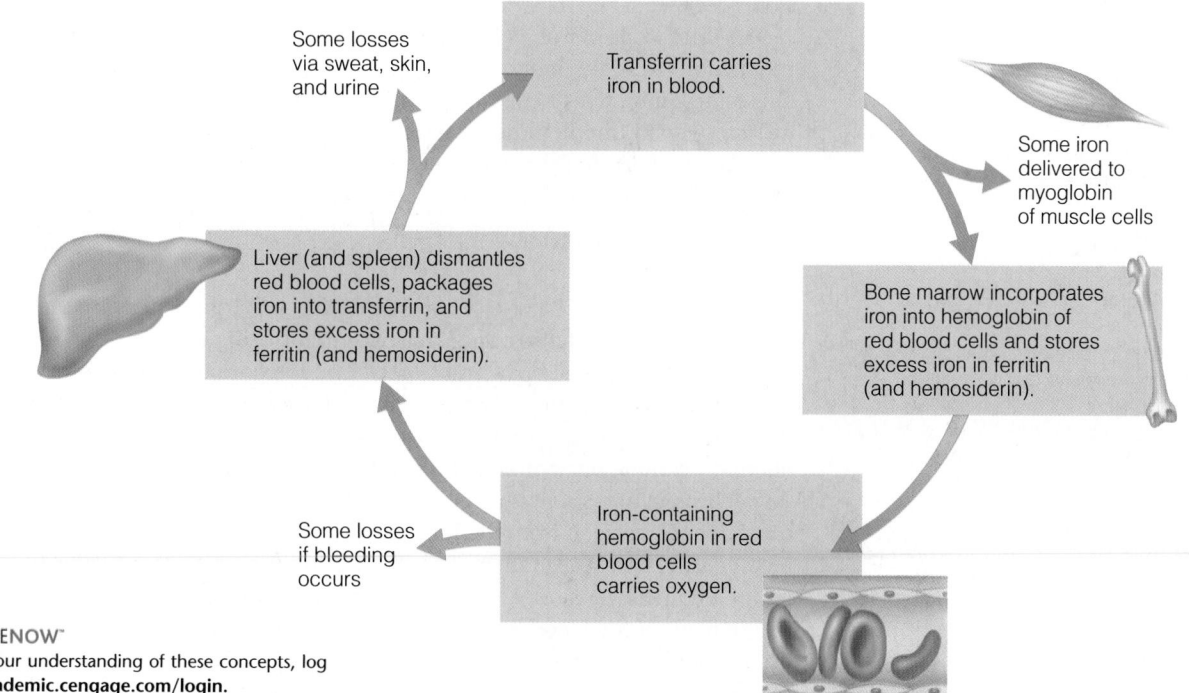

Some losses via sweat, skin, and urine

Transferrin carries iron in blood.

Some iron delivered to myoglobin of muscle cells

Liver (and spleen) dismantles red blood cells, packages iron into transferrin, and stores excess iron in ferritin (and hemosiderin).

Bone marrow incorporates iron into hemoglobin of red blood cells and stores excess iron in ferritin (and hemosiderin).

Some losses if bleeding occurs

Iron-containing hemoglobin in red blood cells carries oxygen.

CENGAGENOW™
To test your understanding of these concepts, log on to **academic.cengage.com/login.**

children and pregnant women suffer from **iron-deficiency anemia.**[10] In the United States, iron deficiency is less prevalent, but it still affects 10 percent of toddlers, adolescent girls, and women of childbearing age. Iron deficiency is also relatively common among overweight children and adolescents compared with those who are normal weight.[11] Preventing and correcting iron deficiency are high priorities.[12]

Vulnerable Stages of Life Some stages of life ◆ demand more iron but provide less, making deficiency likely. Women in their reproductive years are especially prone to iron deficiency because of repeated blood losses during menstruation. Pregnancy demands additional iron to support the added blood volume, growth of the fetus, and blood loss during childbirth. Infants and young children receive little iron from their high-milk diets, yet need extra iron to support their rapid growth. Iron deficiency among toddlers in the United States is common.[13] The rapid growth of adolescence, especially for males, and the menstrual losses of females also demand extra iron that a typical teen diet may not provide. An adequate iron intake is especially important during these stages of life.

◆ High risk for iron deficiency:
- Women in their reproductive years
- Pregnant women
- Infants and young children
- Teenagers

Blood Losses Bleeding ◆ from any site incurs iron losses. In some cases, such as an active ulcer, the bleeding may not be obvious, but even small chronic blood losses significantly deplete iron reserves. In developing countries, blood loss is often brought on by malaria and parasitic infections of the GI tract. People who donate blood regularly also incur losses and may benefit from iron supplements. As mentioned, menstrual losses can be considerable as they tap women's iron stores regularly.

◆ The iron content of blood is about 0.5 mg/100 mL blood. A person donating a pint of blood (approximately 500 mL) loses about 2.5 mg of iron.

Assessment of Iron Deficiency Iron deficiency develops in stages. ◆ This section provides a brief overview of how to detect these stages, and Appendix E provides more details. In the first stage of iron deficiency, iron stores diminish. Measures of serum ferritin (in the blood) reflect iron stores and are most valuable in assessing iron status at this earliest stage.

The second stage of iron deficiency is characterized by a decrease in transport iron: serum iron falls, and the iron-carrying protein transferrin *increases* (an adaptation that enhances iron absorption). Together, measurements of serum iron and transferrin can determine the severity of the deficiency—the more transferrin and the less iron in the blood, the more advanced the deficiency is. Transferrin saturation—the percentage of transferrin that is saturated with iron—decreases as iron stores decline.

◆ Stages of iron deficiency:
- Iron stores diminish
- Transport iron decreases
- Hemoglobin production declines

The third stage of iron deficiency occurs when the lack of iron limits hemoglobin production. Now the hemoglobin precursor, **erythrocyte protoporphyrin,** begins to accumulate as hemoglobin and **hematocrit** values decline.

Hemoglobin and hematocrit tests are easy, quick, and inexpensive, so they are the tests most commonly used in evaluating iron status. Their usefulness in detecting iron deficiency is limited, however, because they are late indicators. Furthermore, other nutrient deficiencies and medical conditions can influence their values.

◆ Iron-deficiency anemia is a **microcytic** (my-cro-SIT-ic) **hypochromic** (high-po-KROME-ic) **anemia.**
- **micro** = small
- **cytic** = cell
- **hypo** = too little
- **chrom** = color

Iron Deficiency and Anemia Iron deficiency and iron-deficiency anemia are not the same: people may be iron deficient without being anemic. The term *iron deficiency* refers to depleted body iron stores without regard to the degree of depletion or to the presence of anemia. The term *iron-deficiency anemia* refers to the severe depletion of iron stores that results in a low hemoglobin concentration. In iron-deficiency anemia, hemoglobin synthesis decreases, resulting in red blood cells that are pale (hypochronic) and small (microcytic), ◆ as shown in Figure 13-4.[14] These cells can't carry enough oxygen from the lungs to the tissues. Without adequate iron, energy metabolism in the cells falters. The result is fatigue, weakness, headaches, apathy, pallor, and poor resistance to cold temperatures. Because hemoglobin is the bright red pigment of the blood, the skin of a fair person who is anemic may become noticeably pale. In a dark-skinned person, the tongue and eye lining, normally pink, is very pale.

iron-deficiency anemia: severe depletion of iron stores that results in low hemoglobin and small, pale red blood cells. Anemias that impair hemoglobin synthesis are **microcytic** (small cell).
- **micro** = small
- **cytic** = cell

erythrocyte protoporphyrin (PRO-toe-PORE-fe-rin): a precursor to hemoglobin.

hematocrit (hee-MAT-oh-krit): measurement of the volume of the red blood cells packed by centrifuge in a given volume of blood.

FIGURE 13-4 | Normal and Anemic Blood Cells

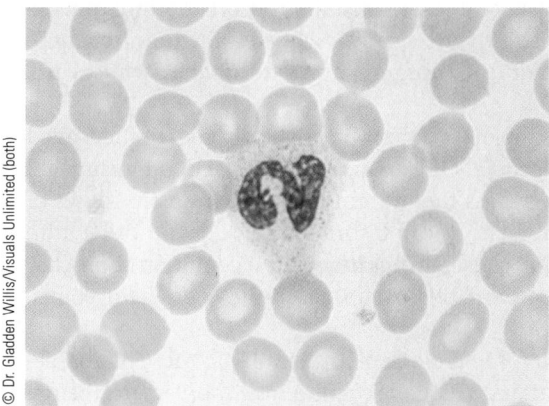

Both size and color are normal in these blood cells.

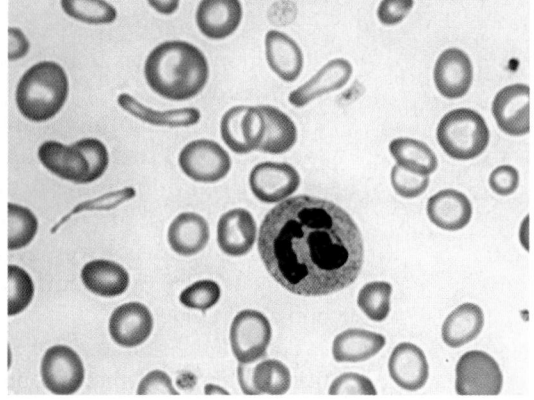

Blood cells in iron-deficiency anemia are small (microcytic) and pale (hypochromic) because they contain less hemoglobin.

The fatigue that accompanies iron-deficiency anemia differs from the tiredness a person experiences from a simple lack of sleep. People with anemia feel fatigue only when they exert themselves. Iron supplementation can relieve the fatigue and improve the body's response to physical activity.[15]

Iron Deficiency and Behavior Long before the red blood cells are affected and anemia is diagnosed, a developing iron deficiency affects behavior. Even at slightly lowered iron levels, energy metabolism is impaired and neurotransmitter synthesis is altered, reducing physical work capacity and mental productivity.[16] Without the physical energy and mental alertness to work, plan, think, play, sing, or learn, people simply do these things less. They have no obvious deficiency symptoms; they just appear unmotivated, apathetic, and less physically fit. Work productivity and voluntary activities decline.

Many of the symptoms associated with iron deficiency are easily mistaken for behavioral or motivational problems. A restless child who fails to pay attention in class might be thought contrary. An apathetic homemaker who has let housework pile up might be thought lazy. No responsible dietitian would ever claim that all behavioral problems are caused by nutrient deficiencies, but poor nutrition is always a possible contributor to problems like these. When investigating a behavioral problem, check the adequacy of the diet and seek a routine physical examination before undertaking more expensive, and possibly harmful, treatment options. (The effects of iron deficiency on children's behavior are discussed further in Chapter 15.)

Iron Deficiency and Pica A curious behavior seen in some iron-deficient people, especially in women and children of low-income groups, is **pica**—an appetite for ice, clay, paste, and other nonfood substances. These substances contain no iron and cannot remedy a deficiency; in fact, clay actually inhibits iron absorption, which may explain the iron deficiency that accompanies such behavior.

Iron Toxicity

In general, even a diet that includes fortified foods poses no special risk for iron toxicity.[17] The body normally absorbs less iron when its stores are full, but some individuals are poorly defended against excess iron. Once considered rare, **iron overload** has emerged as an important disorder of iron metabolism and regulation.

pica (PIE-ka): a craving for nonfood substances. Also known as **geophagia** (gee-oh-FAY-gee-uh) when referring to clay eating and **pagophagia** (pag-oh-FAY-gee-uh) when referring to ice craving.

iron overload: toxicity from excess iron.

Iron Overload The iron overload disorder known as **hemochromatosis** is usually caused by a genetic failure to prevent unneeded iron in the diet from being absorbed.[18] Recent research suggests that just as insulin supports normal glucose homeostasis and its absence or ineffectiveness causes diabetes, the hormone hepcidin supports iron homeostasis and its absence or ineffectiveness causes hemochromatosis.

Hereditary hemochromatosis is the most common genetic disorder in the United States, affecting some 1.5 million people. Other causes of iron overload include repeated blood transfusions (which bypass the intestinal defense), massive doses of supplementary iron (which overwhelm the intestinal defense), and other rare metabolic disorders. Excess iron may cause **hemosiderosis,** a condition characterized by deposits of the iron storage protein hemosiderin in the liver, heart, joints, and other tissues.

Some of the signs and symptoms of iron overload are similar to those of iron deficiency: apathy, lethargy, and fatigue. Therefore, taking iron supplements before assessing iron status is clearly unwise; hemoglobin tests alone would fail to make the distinction because excess iron accumulates in storage. Iron overload assessment tests measure transferrin saturation and serum ferritin.

Iron overload is characterized by tissue damage, especially in iron-storing organs such as the liver. Infections are likely because bacteria thrive on iron-rich blood. Symptoms are most severe in alcohol abusers because alcohol damages the intestine, further impairing its defenses against absorbing excess iron. Untreated hemochromatosis increases the risks of diabetes, liver cancer, heart disease, and arthritis.

Iron overload is more common in men than in women and is twice as prevalent among men as iron deficiency. The widespread fortification of foods with iron makes it difficult for people with hemochromatosis to follow a low-iron diet, and greater dangers lie in the indiscriminate use of iron and vitamin C supplements. Vitamin C not only enhances iron absorption, but also releases iron from ferritin, allowing free iron to wreak the damage typical of free radicals. Thus vitamin C acts as a *pro*oxidant when taken in high doses. (See Highlight 11 for a discussion of free radicals and their effects on disease development.)

Iron and Heart Disease Some research suggests a link between heart disease and iron, especially when accompanied by alcohol consumption.[19] As mentioned, free radicals can attack ferritin, causing it to release iron from storage. Free iron, in turn, acts as an oxidant that can generate more free radicals. Whether iron's role in oxidative stress contributes to the development of diseases is unclear.[20]

Iron and Cancer There may be an association between iron and some cancers.[21] Explanations for how iron might be involved in causing cancer focus on its free-radical activity, which can damage DNA (see Highlight 11). One of the benefits of a high-fiber diet may be that the accompanying phytates bind iron, making it less available for such reactions.

Iron Poisoning Large doses of iron supplements cause GI distress, including constipation, nausea, vomiting, and diarrhea. These effects may not be as serious as other consequences of iron toxicity, but they are consistent enough to establish an Upper Level of 45 milligrams per day for adults.

Ingestion of iron-containing supplements remains a leading cause of accidental poisoning in small children. Symptoms of toxicity include nausea, vomiting, diarrhea, a rapid heartbeat, a weak pulse, dizziness, shock, and confusion. As few as five iron tablets containing as little as 200 milligrams of iron have caused the deaths of dozens of young children. The exact cause of these deaths is uncertain, but excessive free-radical damage is thought to play a role in heart failure and respiratory distress. Autopsy reports reveal iron deposits and cell death in the stomach, small intestine, liver, and blood vessels (which can cause internal bleeding). Keep iron-containing tablets out of the reach of children. If you suspect iron poisoning, call the nearest poison control center or a physician immediately.

hemochromatosis (HE-moh-KRO-ma-toe-sis): a genetically determined failure to prevent absorption of unneeded dietary iron that is characterized by iron overload and tissue damage.

hemosiderosis (HE-moh-sid-er-OH-sis): a condition characterized by the deposition of hemosiderin in the liver and other tissues.

Iron Recommendations and Sources

To obtain enough iron, people must first select iron-rich foods and then take advantage of factors that maximize iron absorption. This discussion begins by identifying iron-rich foods and then reviews the factors affecting absorption.

Recommended Iron Intakes The usual diet in the United States provides about 6 to 7 milligrams of iron for every 1000 kcalories. The recommended daily intake for men is 8 milligrams, and because most men eat more than 2000 kcalories a day, they can meet their iron needs with little effort. Women in their reproductive years, however, need 18 milligrams a day. The accompanying "How to" explains how to calculate the recommended intake.

Vegetarians need 1.8 times as much iron ◆ to make up for the low bioavailability typical of their diets.[22] To maximize iron absorption, vegetarians should incorporate iron-rich foods into a diet that is low in inhibitors (foods such as leavened breads and fermented soy products such as miso and tempeh) and high in enhancers (foods rich in vitamin C and the organic acids found in fruits and vegetables). Good vegetarian sources of iron include soy foods (such as soybeans and tofu), legumes (such as lentils and kidney beans), nuts (such as cashews and almonds), seeds (such as pumpkin seeds and sunflower seeds), cereals (such as cream of wheat and oatmeal), dried fruit (such as apricots and raisins), vegetables (such as mushrooms and potatoes), and blackstrap molasses.

Because women have higher iron needs and lower energy needs, they sometimes have trouble obtaining enough iron. On average, women receive only 12 to 13 milligrams of iron per day, which is not enough iron for women until after menopause. To meet their iron needs from foods, premenopausal women need to select iron-rich foods at every meal.

Dietary Guidelines for Americans 2005
Women of childbearing age who may become pregnant should eat foods high in heme-iron and/or consume iron-rich plant foods or iron-fortified foods with an enhancer of iron absorption, such as vitamin C–rich foods.

Iron in Foods Figure 13-5 (p. 450) shows the amounts of iron in selected foods. Meats, fish, and poultry contribute the most iron per serving; other protein-rich foods such as legumes and eggs are also good sources. Although an indispensable part of the diet, foods in the milk group are notoriously poor in iron. Grain products vary, with whole-grain, enriched, and fortified breads and cereals contributing significantly to iron intakes. Finally, dark greens (such as broccoli) and dried fruits (such as raisins) contribute some iron.

◆ To calculate the RDA for vegetarians, multiply by 1.8:
- 8 mg × 1.8 = 14 mg/day (vegetarian men)
- 18 mg × 1.8 = 32 mg/day (vegetarian women, 19 to 50 yr)

When the label on a grain product says "enriched," it means iron and several B vitamins have been added.

HOW TO Estimate the Recommended Daily Intake for Iron

To calculate the recommended daily iron intake, the DRI Committee considers a number of factors. For example, for a woman of childbearing age (19 to 50):

- Losses from feces, urine, sweat, and shed skin: 1.0 milligram
- Losses through menstruation: 0.5 milligram (about 14 milligrams total averaged over 28 days)

These losses reflect an average daily need (total) of 1.5 milligrams of *absorbed* iron.

An estimated average requirement is determined based on the daily need and the assumption that an average of 18 percent of ingested iron is absorbed:

1.5 mg iron (needed)
÷ 0.18 (percent iron absorbed)
= 8 mg iron (estimated average requirement)

Then, a margin of safety is added to cover the needs of essentially all women of childbearing age, and the RDA is set at 18 milligrams.

FIGURE 13-5 Iron in Selected Foods

See the "How to" section on p. 329 for more information on using this figure.

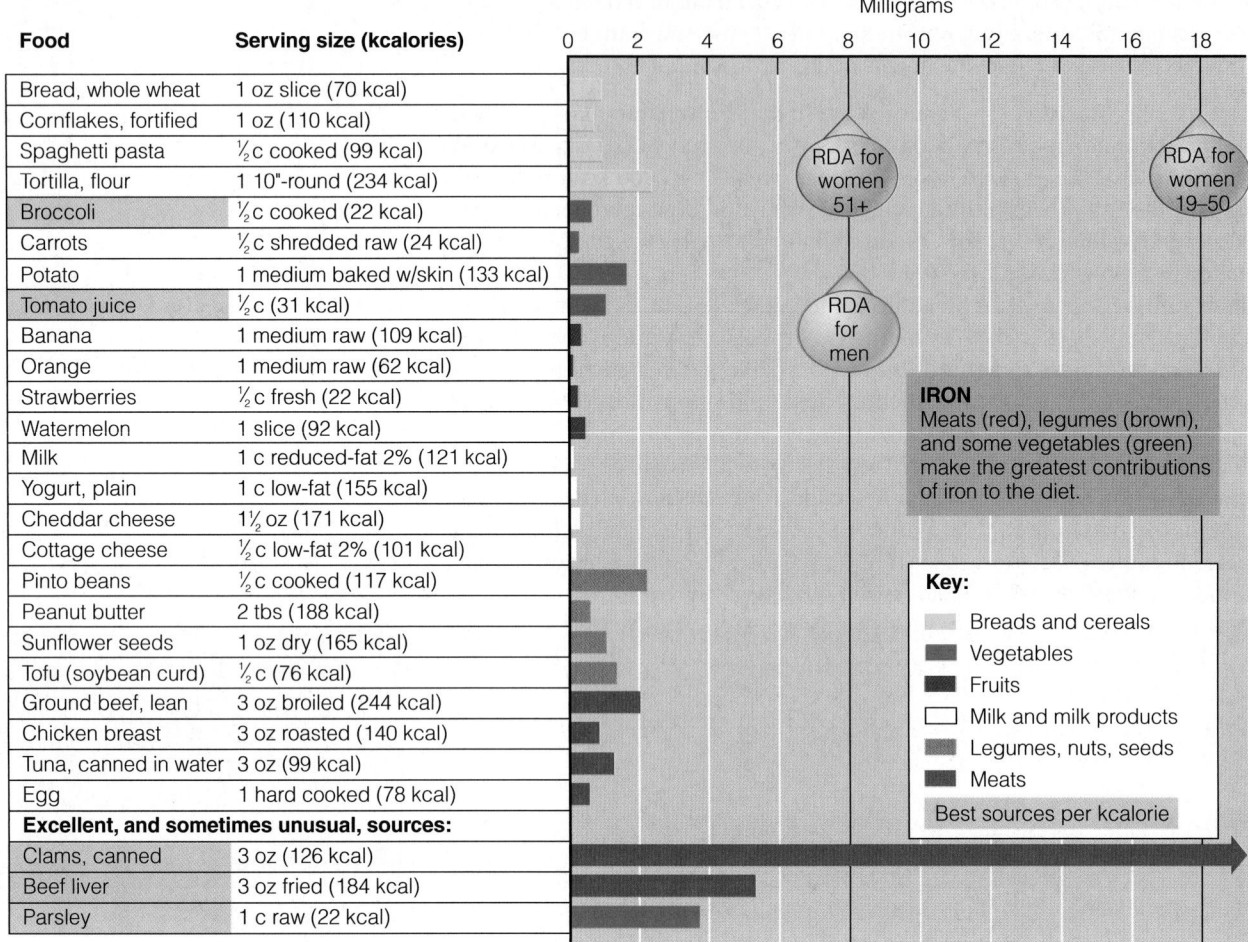

Food	Serving size (kcalories)
Bread, whole wheat	1 oz slice (70 kcal)
Cornflakes, fortified	1 oz (110 kcal)
Spaghetti pasta	½ c cooked (99 kcal)
Tortilla, flour	1 10"-round (234 kcal)
Broccoli	½ c cooked (22 kcal)
Carrots	½ c shredded raw (24 kcal)
Potato	1 medium baked w/skin (133 kcal)
Tomato juice	½ c (31 kcal)
Banana	1 medium raw (109 kcal)
Orange	1 medium raw (62 kcal)
Strawberries	½ c fresh (22 kcal)
Watermelon	1 slice (92 kcal)
Milk	1 c reduced-fat 2% (121 kcal)
Yogurt, plain	1 c low-fat (155 kcal)
Cheddar cheese	1½ oz (171 kcal)
Cottage cheese	½ c low-fat 2% (101 kcal)
Pinto beans	½ c cooked (117 kcal)
Peanut butter	2 tbs (188 kcal)
Sunflower seeds	1 oz dry (165 kcal)
Tofu (soybean curd)	½ c (76 kcal)
Ground beef, lean	3 oz broiled (244 kcal)
Chicken breast	3 oz roasted (140 kcal)
Tuna, canned in water	3 oz (99 kcal)
Egg	1 hard cooked (78 kcal)
Excellent, and sometimes unusual, sources:	
Clams, canned	3 oz (126 kcal)
Beef liver	3 oz fried (184 kcal)
Parsley	1 c raw (22 kcal)

IRON
Meats (red), legumes (brown), and some vegetables (green) make the greatest contributions of iron to the diet.

Key:
- Breads and cereals
- Vegetables
- Fruits
- Milk and milk products
- Legumes, nuts, seeds
- Meats

Best sources per kcalorie

Iron-Enriched Foods Iron is one of the enrichment nutrients for grain products. One serving of enriched bread or cereal provides only a little iron, but because people eat many servings of these foods, the contribution can be significant. Iron added to foods is not absorbed as well as naturally occurring iron, but when eaten with absorption-enhancing foods, enrichment iron can make a difference. In cases of iron overload, enrichment may exacerbate the problem.[23]

Maximizing Iron Absorption In general, the bioavailability of iron is high in meats, fish, and poultry, intermediate in grains and legumes, and low in most vegetables, especially those containing oxalates such as spinach. As mentioned earlier, the amount of iron ultimately absorbed from a meal depends on the combined effects of several enhancing and inhibiting factors. For maximum absorption of non-heme iron, eat meat for MFP and fruits or vegetables for vitamin C. The iron of baked beans, for example, will be enhanced by the MFP in a piece of ham served with them. The iron of bread will be enhanced by the vitamin C in a slice of tomato on a sandwich.

Iron Contamination and Supplementation

In addition to the iron from foods, **contamination iron** from nonfood sources of inorganic iron salts can contribute to the day's intakes. People can also get iron from supplements.

contamination iron: iron found in foods as the result of contamination by inorganic iron salts from iron cookware, iron-containing soils, and the like.

Contamination Iron Foods cooked in iron cookware take up iron salts. The more acidic the food and the longer it is cooked in iron cookware, the higher the iron content. The iron content of eggs can triple in the time it takes to scramble them in an iron pan. Admittedly, the absorption of this iron may be poor (perhaps only 1 to 2 percent), but every little bit helps a person who is trying to increase iron intake.

Iron Supplements People who are iron deficient may need supplements as well as an iron-rich, absorption-enhancing diet. Many physicians routinely recommend iron supplements to pregnant women, infants, and young children. Iron from supplements is less well absorbed than that from food, so the doses must be high. The absorption of iron taken as ferrous sulfate or as an iron **chelate** is better than that from other iron supplements. Absorption also improves when supplements are taken between meals, at bedtime on an empty stomach, and with liquids (other than milk, tea, or coffee, which inhibit absorption). Taking iron supplements in a single dose instead of several doses per day is equally effective and may improve a person's willingness to take it regularly.

There is no benefit to taking iron supplements with orange juice because vitamin C does not enhance absorption from supplements as it does from foods. (Vitamin C enhances iron absorption by converting insoluble ferric iron in foods to the more soluble ferrous iron, and supplemental iron is already in the ferrous form.) Constipation is a common side effect of iron supplementation; drinking plenty of water may help to relieve this problem.

An old-fashioned iron skillet adds iron to foods.

IN SUMMARY

Most of the body's iron is in hemoglobin and myoglobin where it carries oxygen for use in energy metabolism; some iron is also required for enzymes involved in a variety of reactions. Special proteins assist with iron absorption, transport, and storage—all helping to maintain an appropriate balance, because both too little and too much iron can be damaging. Iron deficiency is most common among infants and young children, teenagers, women of childbearing age, and pregnant women. Symptoms include fatigue and anemia. Iron overload is most common in men. Heme iron, which is found only in meat, fish, and poultry, is better absorbed than nonheme iron, which occurs in most foods. Nonheme iron absorption is improved by eating iron-containing foods with foods containing the MFP factor and vitamin C; absorption is limited by phytates and oxalates. The summary table presents a few iron facts.

Iron

RDA

| Men: 8 mg/day |
| Women: 18 mg/day (19–50 yr) |
| 8 mg/day (51+) |

Upper Level

| Adults: 45 mg/day |

Chief Functions in the Body

Part of the protein hemoglobin, which carries oxygen in the blood; part of the protein myoglobin in muscles, which makes oxygen available for muscle contraction; necessary for the utilization of energy as part of the cells' metabolic machinery

Significant Sources

Red meats, fish, poultry, shellfish, eggs, legumes, dried fruits

Deficiency Symptoms

Anemia: weakness, fatigue, headaches; impaired work performance and cognitive function; impaired immunity; pale skin, nailbeds, mucous membranes, and palm creases; concave nails; inability to regulate body temperature; pica

Toxicity Symptoms

GI distress
Iron overload: infections, fatigue, joint pain, skin pigmentation, organ damage

chelate (KEY-late): a substance that can grasp the positive ions of a mineral.
• **chele** = claw

Zinc

◆ Reminder: A *cofactor* is a substance that works with an enzyme to facilitate a chemical reaction.

Zinc is a versatile trace element required as a cofactor ◆ by more than 100 enzymes. Virtually all cells contain zinc, but the highest concentrations are found in muscle and bone.[24]

Zinc Roles in the Body

◆ Metalloenzymes that require zinc:
- Help make parts of the genetic materials DNA and RNA
- Manufacture heme for hemoglobin
- Participate in essential fatty acid metabolism
- Release vitamin A from liver stores
- Metabolize carbohydrates
- Synthesize proteins
- Metabolize alcohol in the liver
- Dispose of damaging free radicals

Zinc supports the work of numerous proteins in the body, such as the **metalloenzymes,** ◆ which are involved in a variety of metabolic processes, including the regulation of gene expression.* In addition, zinc stabilizes cell membranes, helping to strengthen their defense against free-radical attacks. Zinc also assists in immune function and in growth and development. Zinc participates in the synthesis, storage, and release of the hormone insulin in the pancreas, although it does not appear to play a direct role in insulin's action. Zinc interacts with platelets in blood clotting, affects thyroid hormone function, and influences behavior and learning performance. It is needed to produce the active form of vitamin A (retinal) in visual pigments and the retinol-binding protein that transports vitamin A. It is essential to normal taste perception, wound healing, the making of sperm, and fetal development. A zinc deficiency impairs all these and other functions, underlining the vast importance of zinc in supporting the body's proteins.

Zinc Absorption and Metabolism

The body's handling of zinc resembles that of iron in some ways and differs in others. A key difference is the circular passage of zinc from the intestine to the body and back again.

Zinc Absorption The rate of zinc absorption varies from about 15 to 40 percent, depending on a person's zinc status—if more is needed, more is absorbed. Also, dietary factors influence zinc absorption. For example, phytates bind zinc, thus limiting its bioavailability.[25]

Upon absorption into an intestinal cell, zinc has two options. It may become involved in the metabolic functions of the cell itself. Alternatively, it may be retained within the cell by **metallothionein,** a special binding protein similar to the iron storage protein, mucosal ferritin.

Metallothionein in the intestinal cells helps to regulate zinc absorption by holding it in reserve until the body needs zinc. Then metallothionein releases zinc into the blood where it can be transported around the body. Metallothionein in the liver performs a similar role, binding zinc until other body tissues signal a need for it.

Zinc Recycling Some zinc eventually reaches the pancreas, where it is incorporated into many of the digestive enzymes that the pancreas releases into the intestine at mealtimes. The intestine thus receives two doses of zinc with each meal—one from foods and the other from the zinc-rich pancreatic secretions. The recycling of zinc in the body from the pancreas to the intestine and back to the pancreas is referred to as the **enteropancreatic circulation** of zinc. As this zinc circulates through the intestine, it may be excreted in shed intestinal cells or absorbed into the body on any of its times around (see Figure 13-6). The body loses zinc primarily in feces. Smaller losses occur in urine, shed skin, hair, sweat, menstrual fluids, and semen.

Zinc Transport Zinc's main transport vehicle in the blood is the protein albumin. Some zinc also binds to transferrin—the same transferrin that carries iron in the

metalloenzymes (meh-TAL-oh-EN-zimes): enzymes that contain one or more minerals as part of their structures.

metallothionein (meh-TAL-oh-THIGH-oh-neen): a sulfur-rich protein that avidly binds with and transports metals such as zinc.
- **metallo** = containing a metal
- **thio** = containing sulfur
- **ein** = a protein

enteropancreatic (EN-ter-oh-PAN-kree-AT-ik) **circulation:** the circulatory route from the pancreas to the intestine and back to the pancreas.

---

* Among the metalloenzymes requiring zinc are carbonic anhydrase, deoxythymidine kinase, DNA and RNA polymerase, and alkaline phosphatase.

FIGURE 13-6 *Animated!* Enteropancreatic Circulation of Zinc

Some zinc from food is absorbed by the small intestine and sent to the pancreas to be incorporated into digestive enzymes that return to the small intestine. This cycle is called the enteropancreatic circulation of zinc.

CENGAGENOW™
To test your understanding
of these concepts, log on to
academic.cengage.com/login.

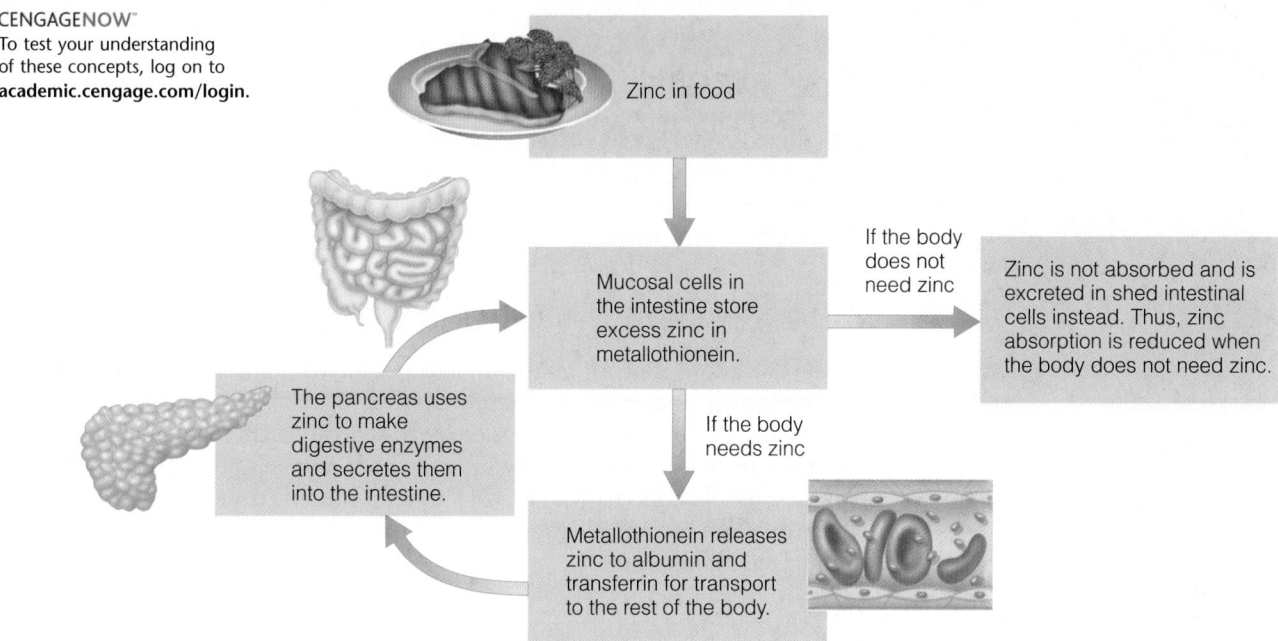

Zinc in food

Mucosal cells in the intestine store excess zinc in metallothionein.

If the body does not need zinc

Zinc is not absorbed and is excreted in shed intestinal cells instead. Thus, zinc absorption is reduced when the body does not need zinc.

The pancreas uses zinc to make digestive enzymes and secretes them into the intestine.

If the body needs zinc

Metallothionein releases zinc to albumin and transferrin for transport to the rest of the body.

blood. In healthy individuals, transferrin is usually less than 50 percent saturated with iron, but in iron overload, it is more saturated. Diets that deliver more than twice as much iron as zinc leave too few transferrin sites available for zinc. The result is poor zinc absorption. The converse is also true: large doses of zinc inhibit iron absorption.

Large doses of zinc create a similar problem with another essential mineral, copper. These nutrient interactions highlight one of the many reasons why people should use supplements conservatively, if at all: supplementation can easily create imbalances.

Zinc Deficiency

Severe zinc deficiencies are not widespread in developed countries, but they do occur in vulnerable groups—pregnant women, young children, the elderly, and the poor. Human zinc deficiency was first reported in the 1960s in children and adolescent boys in Egypt, Iran, and Turkey. Children have especially high zinc needs because they are growing rapidly and synthesizing many zinc-containing proteins, and the native diets among those populations were not meeting these needs. Middle Eastern diets are typically low in the richest zinc source, meats, and the staple foods are legumes, unleavened breads, and other whole-grain foods—all high in fiber and phytates, which inhibit zinc absorption.*

Figure 13-7 shows the severe growth retardation and mentions the immature sexual development characteristic of zinc deficiency. In addition, zinc deficiency hinders digestion and absorption, causing diarrhea, which worsens malnutrition not only for zinc, but for all nutrients. It also impairs the immune response, making infections likely—among them, GI tract infections, which worsen malnutrition, including zinc malnutrition (a classic downward spiral of events).[26] Chronic zinc deficiency damages the central nervous system and brain and may lead to poor motor development and cognitive performance. Because zinc deficiency directly impairs vitamin A metabolism, vitamin A–deficiency symptoms often appear. Zinc

FIGURE 13-7 Zinc-Deficiency Symptom—The Stunted Growth of Dwarfism

The growth retardation, known as dwarfism, is rightly ascribed to zinc deficiency because it is partially reversible when zinc is restored to the diet.

© H. Sanstead, University of Texas at Galveston

The Egyptian man on the right is an adult of average height. The Egyptian boy on the left is 17 years old but is only 4 feet tall, like a 7-year-old in the United States. His genitalia are like those of a 6-year-old.

---

* Unleavened bread contains no yeast, which normally breaks down phytates during fermentation.

Zinc is highest in protein-rich foods such as oysters, beef, poultry, legumes, and nuts.

deficiency also disturbs thyroid function and the metabolic rate. It alters taste, causes loss of appetite, and slows wound healing—in fact, its symptoms are so pervasive that generalized malnutrition and sickness are more likely to be the diagnosis than simple zinc deficiency.

Zinc Toxicity

High doses (over 50 milligrams) of zinc may cause vomiting, diarrhea, headaches, exhaustion, and other symptoms. An Upper Level for adults was set at 40 milligrams based on zinc's interference in copper metabolism—an effect that, in animals, leads to degeneration of the heart muscle.

Zinc Recommendations and Sources

Figure 13-8 shows zinc amounts in foods per serving. Zinc is highest in protein-rich foods such as shellfish (especially oysters), meats, poultry, milk, and cheese. Legumes and whole-grain products are good sources of zinc if eaten in large quantities; in typical U.S. diets, the phytate content of grains is not high enough to impair zinc absorption. Vegetables vary in zinc content depending on the soil in which they are grown. Average intakes in the United States are slightly higher than recommendations.

FIGURE 13-8 Zinc in Selected Foods

See the "How to" section on p. 329 for more information on using this figure.

Food	Serving size (kcalories)
Bread, whole wheat	1 oz slice (70 kcal)
Cornflakes, fortified	1 oz (110 kcal)
Spaghetti pasta	½ c cooked (99 kcal)
Tortilla, flour	1 10"-round (234 kcal)
Broccoli	½ c cooked (22 kcal)
Carrots	½ c shredded raw (24 kcal)
Potato	1 medium baked w/skin (133 kcal)
Tomato juice	¾ c (31 kcal)
Banana	1 medium raw (109 kcal)
Orange	1 medium raw (62 kcal)
Strawberries	½ c fresh (22 kcal)
Watermelon	1 slice (92 kcal)
Milk	1 c reduced-fat 2% (121 kcal)
Yogurt, plain	1 c low-fat (155 kcal)
Cheddar cheese	1½ oz (171 kcal)
Cottage cheese	½ c low-fat 2% (101 kcal)
Pinto beans	½ c cooked (117 kcal)
Peanut butter	2 tbs (188 kcal)
Sunflower seeds	1 oz dry (165 kcal)
Tofu (soybean curd)	½ c (76 kcal)
Ground beef, lean	3 oz broiled (244 kcal)
Chicken breast	3 oz roasted (140 kcal)
Tuna, canned in water	3 oz (99 kcal)
Egg	1 hard cooked (78 kcal)
Excellent, and sometimes unusual, sources:	
Oysters	3 oz cooked (139 kcal)
Sirloin steak, lean	3 oz broiled (172 kcal)
Crab	3 oz cooked (94 kcal)

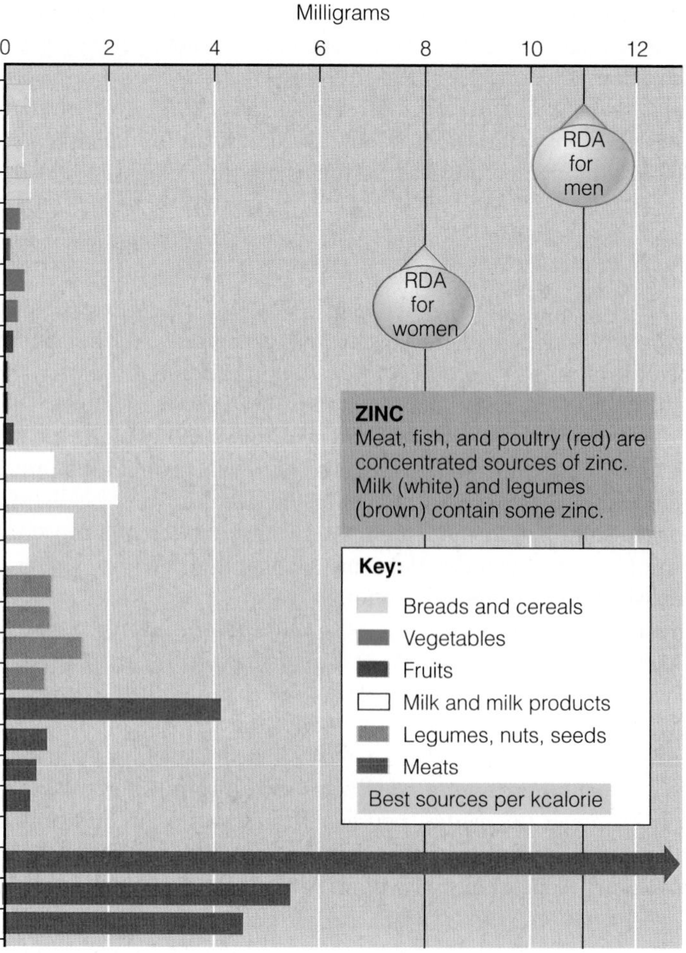

Milligrams scale: 0, 2, 4, 6, 8, 10, 12

RDA for men (~11 mg); RDA for women (~8 mg)

ZINC
Meat, fish, and poultry (red) are concentrated sources of zinc. Milk (white) and legumes (brown) contain some zinc.

Key:
- Breads and cereals
- Vegetables
- Fruits
- Milk and milk products
- Legumes, nuts, seeds
- Meats
- Best sources per kcalorie

Zinc Supplementation

In developed countries, most people obtain enough zinc from the diet without resorting to supplements. In developing countries, zinc supplements play a major role in the treatment of childhood infectious diseases. Zinc supplements effectively reduce the incidence of disease and death associated with diarrhea.[27]

The use of zinc lozenges to treat the common cold has been controversial and inconclusive, with some studies finding them effective and others not.[28] The different study results may reflect the effectiveness of various zinc compounds. Some studies using zinc gluconate report shorter duration of cold symptoms, whereas most studies using other combinations of zinc report no effect. Common side effects of zinc lozenges include nausea and bad taste reactions.

IN SUMMARY

Zinc-requiring enzymes participate in a multitude of reactions affecting growth, vitamin A activity, and pancreatic digestive enzyme synthesis, among others. Both dietary zinc and zinc-rich pancreatic secretions (via enteropancreatic circulation) are available for absorption. Absorption is monitored by a special binding protein (metallothionein) in the intestine. Protein-rich foods derived from animals are the best sources of bioavailable zinc. Fiber and phytates in cereals bind zinc, limiting absorption. Growth retardation and sexual immaturity are hallmark symptoms of zinc deficiency. These facts and others are included in the following table.

Zinc

RDA

Men: 11 mg/day

Women: 8 mg/day

Significant Sources

Protein-containing foods: red meats, shellfish, whole grains; some fortified cereals

Upper Level

Adults: 40 mg/day

Deficiency Symptoms[a]

Growth retardation, delayed sexual maturation, impaired immune function, hair loss, eye and skin lesions, loss of appetite

Chief Functions in the Body

Part of many enzymes; associated with the hormone insulin; involved in making genetic material and proteins, immune reactions, transport of vitamin A, taste perception, wound healing, the making of sperm, and the normal development of the fetus

Toxicity Symptoms

Loss of appetite, impaired immunity, low HDL, copper and iron deficiencies

[a]A rare inherited disease of zinc malabsorption, *acrodermatitis* (AK-roh-der-ma-TIE-tis) *enteropathica* (EN-ter-oh-PATH-ick-ah), causes additional and more severe symptoms.

Iodine

Traces of the iodine ion (called iodide) ♦ are indispensable to life. In the GI tract, iodine from foods becomes iodide. This chapter uses the term *iodine* when referring to the nutrient in foods and *iodide* when referring to it in the body. Iodide occurs in the body in minuscule amounts, but its principal role in the body and its requirement are well established.

Iodide Roles in the Body Iodide is an integral part of the thyroid hormones ♦ that regulate body temperature, metabolic rate, reproduction, growth, blood cell production, nerve and muscle function, and more. By controlling the rate at which the cells use oxygen, these hormones influence the amount of energy released during basal metabolism.

♦ The ion form of *iodine* is called *iodide*.

♦ The thyroid gland releases tetraiodothyronine (T_4), commonly known as **thyroxine** (thigh-ROCKS-in), to its target tissues. Upon reaching the cells, T_4 is deiodinated to tri-iodothyronine (T_3), which is the active form of the hormone.

FIGURE 13-9 Iodine-Deficiency Symptom—The Enlarged Thyroid of Goiter

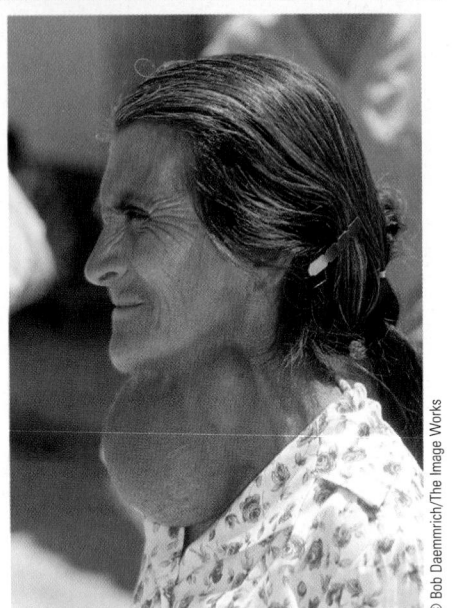

In iodine deficiency, the thyroid gland enlarges—a condition known as simple goiter.

◆ Thyroid-stimulating hormone is also called *thyrotropin*.

◆ Examples of goitrogen-containing foods:
  • Cabbage, spinach, radishes, rutabagas
  • Soybeans, peanuts
  • Peaches, strawberries

◆ The underactivity of the thyroid gland is known as *hypothyroidism* and may be caused by iodine deficiency or any number of other causes. Without treatment, an infant with *congenital hypothyroidism* will develop the physical and mental retardation of *cretinism*.

◆ Iodized salt contains about 60 µg iodine per gram salt.

◆ On average, ½ tsp iodized salt provides the RDA for iodine.

goiter (GOY-ter): an enlargement of the thyroid gland due to an iodine deficiency, malfunction of the gland, or overconsumption of a goitrogen. Goiter caused by iodine deficiency is **simple goiter.**

goitrogen (GOY-troh-jen): a substance that enlarges the thyroid gland and causes **toxic goiter.** Goitrogens occur naturally in such foods as cabbage, kale, brussels sprouts, cauliflower, broccoli, and kohlrabi.

cretinism (CREE-tin-ism): a congenital disease characterized by mental and physical retardation and commonly caused by maternal iodine deficiency during pregnancy.

Iodine Deficiency The hypothalamus regulates thyroid hormone production by controlling the release of the pituitary's thyroid-stimulating hormone (TSH). ◆ With iodine deficiency, thyroid hormone production declines, and the body responds by secreting more TSH in a futile attempt to accelerate iodide uptake by the thyroid gland. If a deficiency persists, the cells of the thyroid gland enlarge to trap as much iodide as possible. Sometimes the gland enlarges until it makes a visible lump in the neck, a simple **goiter** (shown in Figure 13-9).

Goiter afflicts about 200 million people the world over, many of them in South America, Asia, and Africa. In all but 4 percent of these cases, the cause is iodine deficiency. As for the 4 percent (8 million), most have goiter because they regularly eat excessive amounts of foods ◆ that contain an antithyroid substance (**goitrogen**) whose effect is not counteracted by dietary iodine. The goitrogens present in plants remind us that even natural components of foods can cause harm when eaten in excess.

Goiter may be the earliest and most obvious sign of iodine deficiency, but the most tragic and prevalent damage occurs in the brain. Children with even a mild iodine deficiency typically have goiters and perform poorly in school. With sustained treatment, however, mental performance in the classroom as well as thyroid function improves.[29]

A severe iodine deficiency during pregnancy causes the extreme and irreversible mental and physical retardation known as **cretinism.** ◆ Cretinism affects approximately 6 million people worldwide and can be averted by the early diagnosis and treatment of maternal iodine deficiency. A worldwide effort to provide iodized salt to people living in iodine-deficient areas has been dramatically successful. Because iron deficiency is common among people with iodine deficiency and because iron deficiency reduces the effectiveness of iodized salt, dual fortification with both iron and iodine may be most beneficial.[30]

Iodine Toxicity Excessive intakes of iodine can interfere with thyroid function and enlarge the gland, just as deficiency can.[31] During pregnancy, exposure to excessive iodine from foods, prenatal supplements, or medications is especially damaging to the developing infant. An infant exposed to toxic amounts of iodine during gestation may develop a goiter so severe as to block the airways and cause suffocation. The Upper Level is 1100 micrograms per day for an adult—several times higher than average intakes.

Iodine Recommendations and Sources The ocean is the world's major source of iodine. In coastal areas, seafood, water, and even iodine-containing sea mist are dependable iodine sources. Further inland, the amount of iodine in foods is variable and generally reflects the amount present in the soil in which plants are grown or on which animals graze. Landmasses that were once under the ocean have soils rich in iodine; those in flood-prone areas where water leaches iodine from the soil are poor in iodine. In the United States and Canada, the iodization of salt ◆ has eliminated the widespread misery caused by iodine deficiency during the 1930s, but iodized salt is not available in many parts of the world. Some countries add iodine to bread, fish paste, or drinking water instead.

Although average consumption of iodine in the United States exceeds recommendations, it falls below toxic levels. Some of the excess iodine in the U.S. diet stems from fast foods, which use iodized salt liberally. Some iodine comes from bakery products and from milk. The baking industry uses iodates (iodine salts) as dough conditioners, and most dairies feed cows iodine-containing medications and use iodine to disinfect milking equipment. Now that these sources have been identified, food industries have reduced their use of these compounds, but the sudden emergence of this problem points to a need for continued surveillance of the food supply. Processed foods in the United States use regular salt, not iodized salt.

The recommended intake of iodine for adults is a minuscule amount. The need for iodine is easily met by consuming seafood, vegetables grown in iodine-rich soil, and iodized salt. ◆ In the United States, labels indicate whether salt is iodized; in Canada, all table salt is iodized.

IN SUMMARY

Iodide, the ion of the mineral iodine, is an essential component of the thyroid hormone. An iodine deficiency can lead to simple goiter (enlargement of the thyroid gland) and can impair fetal development, causing cretinism. Iodization of salt has largely eliminated iodine deficiency in the United States and Canada. The table provides a summary of iodine.

Iodine

RDA	**Deficiency Disease**
Adults: 150 µg/day	Simple goiter, cretinism
Upper Level	**Deficiency Symptoms**
1100 µg/day	Underactive thyroid gland, goiter, mental and physical retardation in infants (cretinism)
Chief Functions in the Body	**Toxicity Symptoms**
A component of two thyroid hormones that help to regulate growth, development, and metabolic rate	Underactive thyroid gland, elevated TSH, goiter
Significant Sources	
Iodized salt, seafood, bread, dairy products, plants grown in iodine-rich soil and animals fed those plants	

Only "iodized salt" has had iodine added.

© Craig M. Moore

Selenium

The essential mineral **selenium** shares some of the chemical characteristics of the mineral sulfur. This similarity allows selenium to substitute for sulfur in the amino acids methionine, cysteine, and cystine.[32]

Selenium Roles in the Body Selenium is one of the body's antioxidant nutrients, ◆ working primarily as a part of proteins—most notably, the enzyme glutathione peroxidase.[33] Glutathione peroxidase and vitamin E work in tandem. Glutathione peroxidase prevents free-radical formation, thus blocking the chain reaction before it begins; if free radicals do form and a chain reaction starts, vitamin E stops it. (Highlight 11 describes free-radical formation, chain reactions, and antioxidant action in detail.) Another enzyme that converts the thyroid hormone to its active form also contains selenium.

◆ Key antioxidant nutrients:
  • Vitamin C, vitamin E, beta-carotene
  • Selenium

Selenium Deficiency Selenium deficiency is associated with a heart disease ◆ that is prevalent in regions of China where the soil and foods lack selenium. Although the primary cause of this heart disease is probably a virus, selenium deficiency appears to predispose people to it, and adequate selenium seems to prevent it.

◆ The heart disease associated with selenium deficiency is named **Keshan** (KESH-an or ka-SHAWN) **disease** for one of the provinces of China where it was studied. Keshan disease is characterized by heart enlargement and insufficiency; fibrous tissue replaces the muscle tissue that normally composes the middle layer of the walls of the heart.

Selenium and Cancer Some research suggests that selenium may protect against some types of cancers.[34] Given the potential for harm and the lack of conclusive evidence, however, recommendations to take selenium supplements would be premature—and perhaps ineffective as well. Selenium from foods appears to be more effective in inhibiting cancer growth than selenium from supplements. Such a finding reinforces a theme that has been repeated throughout this text—foods offer many more health benefits than supplements.

Selenium Recommendations and Sources Selenium is found in the soil, and therefore in the crops grown for consumption.[35] People living in regions with selenium-poor soil may still get enough selenium, partly because they eat vegetables

selenium (se-LEEN-ee-um): a trace element.

and grains transported from other regions and partly because they eat meats and other animal products, which are reliable sources of selenium. Average intakes in the United States and Canada are above the RDA, which is based on the amount needed to maximize glutathione peroxidase activity.

Selenium Toxicity Because high doses of selenium are toxic, an Upper Level has been set. Selenium toxicity causes loss and brittleness of hair and nails, garlic breath odor, and nervous system abnormalities.

IN SUMMARY

Selenium is an antioxidant nutrient that works closely with the glutathione peroxidase enzyme and vitamin E. Selenium is found in association with protein in foods. Deficiencies are associated with a predisposition to a type of heart abnormality known as Keshan disease. See the table below for a summary of selenium.

Selenium

RDA

Adults: 55 µg/day

Upper Level

Adults: 400 µg/day

Chief Functions in the Body

Defends against oxidation; regulates thyroid hormone

Significant Sources

Seafood, meat, whole grains, fruits, and vegetables (depending on soil content)

Deficiency Symptoms

Predisposition to heart disease characterized by cardiac tissue becoming fibrous (Keshan disease)

Toxicity Symptoms

Loss and brittleness of hair and nails; skin rash, fatigue, irritability, and nervous system disorders; garlic breath odor

Copper

The body contains about 100 milligrams of copper. It is found in a variety of cells and tissues.

Copper Roles in the Body Copper serves as a constituent of several enzymes. The copper-containing enzymes have diverse metabolic roles with one common characteristic: all involve reactions that consume oxygen or oxygen radicals. For example, copper-containing enzymes catalyze the oxidation of ferrous iron to ferric iron.*[36] Copper's role in iron metabolism makes it a key factor in hemoglobin synthesis. Two copper- and zinc-containing enzymes participate in the body's natural defense against free radicals.† Still another copper enzyme helps to manufacture collagen and heal wounds.‡ Copper, like iron, is needed in many of the metabolic reactions related to the release of energy.§

Copper Deficiency and Toxicity Typical U.S. diets provide adequate amounts of copper and deficiency is rare. In animals, copper deficiency raises blood cholesterol and damages blood vessels, raising questions about whether low dietary copper might contribute to cardiovascular disease in humans.

---

* The copper-containing enzyme *ceruloplasmin* participates in the oxidation of ferrous iron to ferric iron.
† Two copper-containing *superoxide dismutase* enzymes defend against free radicals.
‡ The copper-containing enzyme *lysyl oxidase* helps synthesize connective tissues.
§ The copper-containing enzyme *cytochrome C oxidase* participates in the electron transport chain.

Some genetic disorders create a copper toxicity, but excessive intakes from foods are unlikely. Excessive intakes from supplements may cause liver damage, and therefore an Upper Level has been set.

Two rare genetic disorders affect copper status in opposite directions. In Menkes disease, the intestinal cells absorb copper, but cannot release it into circulation, causing a life-threatening deficiency. In Wilson's disease, copper accumulates in the liver and brain, creating a life-threatening toxicity. Wilson's disease can be controlled by reducing copper intake, using chelating agents such as penicillamine, and taking zinc supplements, which interfere with copper absorption.

Copper Recommendations and Sources The richest food sources of copper are legumes, whole grains, nuts, shellfish, and seeds. Over half of the copper from foods is absorbed, and the major route of elimination appears to be bile. Water may also provide copper, depending on the type of plumbing pipe and the hardness of the water.

IN SUMMARY

Copper is a component of several enzymes, all of which are involved in some way with oxygen or oxidation. Some act as antioxidants; others are essential to iron metabolism. Legumes, whole grains, and shellfish are good sources of copper. See the table for a summary of copper facts.

Copper

RDA	Significant Sources
Adults: 900 µg/day	Seafood, nuts, whole grains, seeds, legumes

Upper Level	Deficiency Symptoms
Adults: 10,000 µg/day (10 mg/day)	Anemia, bone abnormalities

Chief Functions in the Body	Toxicity Symptoms
Necessary for the absorption and use of iron in the formation of hemoglobin; part of several enzymes	Liver damage

---

Manganese

The human body contains a tiny 20 milligrams of manganese. Most of it can be found in the bones and metabolically active organs such as the liver, kidneys, and pancreas.

Manganese Roles in the Body Manganese acts as a cofactor for many enzymes that facilitate the metabolism of carbohydrate, lipids, and amino acids. In addition, manganese-containing metalloenzymes assist in bone formation and the conversion of pyruvate to a TCA cycle compound.

Manganese Deficiency and Toxicity Manganese requirements are low, and many plant foods contain significant amounts of this trace mineral, so deficiencies are rare. As is true of other trace minerals, however, dietary factors such as phytates inhibit its absorption. In addition, high intakes of iron and calcium limit manganese absorption, so people who use supplements of those minerals regularly may impair their manganese status.

Toxicity is more likely to occur from an environment contaminated with manganese than from dietary intake.[37] Miners who inhale large quantities of manganese dust on the job over prolonged periods show symptoms of a brain disease,

c. Name three foods that are higher on the second list than they were on the first list.

d. What do these foods have in common?

Food	Iron (mg)	Energy (kcal)	Iron Density (mg/kcal)
Milk, fat-free, 1 c	0.10	85	
Cheddar cheese, 1 oz	0.19	114	
Broccoli, cooked from fresh, chopped, 1 c	1.31	44	
Sweet potato, baked in skin, 1 ea	0.51	117	
Cantaloupe melon, ½	0.56	93	
Carrots, from fresh, ½ c	0.48	35	
Whole-wheat bread, 1 slice	0.87	64	
Green peas, cooked from frozen, ½ c	1.26	62	
Apple, medium	0.38	125	
Sirloin steak, lean, 4 oz	3.81	228	
Pork chop, lean, broiled, 1 ea	0.66	166	

STUDY QUESTIONS

CENGAGENOW

To assess your understanding of chapter topics, take the Student Practice Test and explore the modules recommended in your Personalized Study Plan. Log on to **academic.cengage.com/login.**

These questions will help you review the chapter. You will find the answers in the discussions on the pages provided.

1. Distinguish between heme and nonheme iron. Discuss the factors that enhance iron absorption. (pp. 443–444)

2. Distinguish between iron deficiency and iron-deficiency anemia. What are the symptoms of iron-deficiency anemia? (pp. 445–447)

3. What causes iron overload? What are its symptoms? (p. 448)

4. Describe the similarities and differences in the absorption and regulation of iron and zinc. (pp. 443–445, 452–453)

5. Discuss possible reasons for a low intake of zinc. What factors affect the bioavailability of zinc? (p. 454)

6. Describe the principal functions of iodide, selenium, copper, manganese, fluoride, chromium, and molybdenum in the body. (pp. 455–462)

7. What public health measure has been used in preventing simple goiter? What measure has been recommended for protection against tooth decay? (pp. 456–457, 460–461)

8. Discuss the importance of balanced and varied diets in obtaining the essential minerals and avoiding toxicities. (pp. 463–465)

9. Describe some of the ways trace minerals interact with each other and with other nutrients. (p. 463)

These multiple choice questions will help you prepare for an exam. Answers can be found on p. 468.

1. Iron absorption is impaired by:
   a. heme.
   b. phytates.
   c. vitamin C.
   d. MFP factor.

2. Which of these people is *least* likely to develop an iron deficiency?
   a. 3-year-old boy
   b. 52-year-old man
   c. 17-year-old girl
   d. 24-year-old woman

3. Which of the following would *not* describe the blood cells of a severe iron deficiency?
   a. anemic
   b. microcytic
   c. pernicious
   d. hypochromic

4. Which provides the most absorbable iron?
   a. 1 apple
   b. 1 c milk
   c. 3 oz steak
   d. ½ c spinach

5. The intestinal protein that helps to regulate zinc absorption is:
   a. albumin.
   b. ferritin.
   c. hemosiderin.
   d. metallothionein.

6. A classic sign of zinc deficiency is:
   a. anemia.
   b. goiter.
   c. mottled teeth.
   d. growth retardation.

7. Cretinism is caused by a deficiency of:
   a. iron.
   b. zinc.
   c. iodine.
   d. selenium.

8. The mineral best known for its role as an antioxidant is:
   a. copper.
   b. selenium.
   c. manganese.
   d. molybdenum.

9. Fluorosis occurs when fluoride:
   a. is excessive.
   b. is inadequate.
   c. binds with phosphorus.
   d. interacts with calcium.

10. Which mineral enhances insulin activity?
    a. zinc
    b. iodine
    c. chromium
    d. manganese

REFERENCES

1. M. W. Hentze, M. U. Muckenthaler, and N. C. Andrews, Molecular control of mammalian iron metabolism, *Cell* 117 (2004): 285–297.
2. R. E. Fleming and B. R. Bacon, Orchestration of iron homeostasis, *New England Journal of Medicine* 352 (2005): 1741–1744; Chung and Wessling-Resnick, Lessons learned from genetic and nutritional iron deficiencies, *Nutrition Reviews* 62 (2004): 212–220.
3. E. G. Theil, Iron, ferritin, and nutrition, *Annual Review of Nutrition* 24 (2004): 327–343.
4. Committee on Dietary Reference Intakes, *Dietary Reference Intakes for Vitamin A, Vitamin K, Arsenic, Boron, Chromium, Copper, Iodine, Iron, Manganese, Molybdenum, Nickel, Silicon, Vanadium, and Zinc* (Washington, D.C.: National Academy Press, 2001), p. 315.
5. S. Miret, R. J. Simpson, and A. T. McKie, Physiology and molecular biology of dietary iron absorption, *Annual Review of Nutrition* 23 (2003): 283–301.
6. R. F. Hurrell and coauthors, Meat protein fractions enhance nonheme iron absorption in humans, *Journal of Nutrition* 136 (2006): 2808–2812.
7. Committee on Dietary Reference Intakes, 2001, p. 351.
8. E. Nemeth and T. Ganz, Regulation of iron metabolism by hepcidin, *Annual Review of Nutrition* 26 (2006): 323–342.
9. J. L. Beard and J. R. Connor, Iron status and neural functioning, *Annual Review of Nutrition* 23 (2003): 41–58.
10. World Health Organization, www.who.int/nut/ida.htm.
11. K. G. Nead and coauthors, Overweight children and adolescents: A risk group for iron deficiency, *Pediatrics* 114 (2004): 104–108.
12. Iron deficiency—United States, 1999–2000, *Morbidity and Mortality Weekly Report* 51 (2002): 897–899.
13. K. C. White, Anemia is a poor predictor of iron deficiency among toddlers in the United States: For heme the bell tolls, *Pediatrics* 115 (2005): 315–320.
14. M. J. Koury and P. Ponka, New insights into erythropoiesis: The roles of folate, vitamin B12, and iron, *Annual Review of Nutrition* 24 (2004): 105–131.
15. T. Brownlie and coauthors, Marginal iron deficiency without anemia impairs aerobic adaptation among previously untrained women, *American Journal of Clinical Nutrition* 75 (2002): 734–742.
16. J. Beard, Iron deficiency alters brain development and functioning, *Journal of Nutrition* 133 (2003): 1468S–1472S; E. M. Ross, Evaluation and treatment of iron deficiency in adults, *Nutrition in Clinical Care* 5 (2002): 220–224.

17. P. C. Adams and coauthors, Hemochromatosis and iron-overload screening in a racially diverse population, *New England Journal of Medicine* 352 (2005): 1769–1778; A. L. M. Heath and S. J. Fairweather-Tait, Health implications of iron overload: The role of diet and genotype, *Nutrition Reviews* 61 (2003): 45–62.
18. A. Pietrangelo, Hereditary hemochromatosis, *Annual Review of Nutrition* 26 (2006): 251–270.
19. D. Lee, A. R. Folsom, and D. R. Jacobs, Iron, zinc, and alcohol consumption and mortality from cardiovascular diseases: The Iowa Women's Health Study, *American Journal of Clinical Nutrition* 81 (2005): 787–791; U. Ramakrishnan, E. Kuklina, and A. D. Stein, Iron stores and cardiovascular disease risk factors in women of reproductive age in the United States, *American Journal of Clinical Nutrition* 76 (2002): 1256–1260.
20. M. B. Reddy and L. Clark, Iron, oxidative stress, and disease risk, *Nutrition Reviews* 62 (2004): 120–124; J. L. Derstine and coauthors, Iron status in association with cardiovascular disease risk in 3 controlled feeding studies, *American Journal of Clinical Nutrition* 77 (2003): 56–62.
21. A. G. Mainous and coauthors, Iron, lipids, and risk of cancer in the Framingham Offspring Cohort, *American Journal of Epidemiology* 160 (2005): 1115–1122.
22. Committee on Dietary Reference Intakes, 2001, p. 351.
23. J. R. Backstrand, The history and future of food fortification in the United States: A public health perspective, *Nutrition Reviews* 60 (2002): 15–26.
24. H. Tapiero and K. D. Tew, Trace elements in human physiology and pathology: Zinc and metallothioneins, *Biomedicine and Pharmacotherapy* 57 (2003): 399–411.
25. C. L. Adams and coauthors, Zinc absorption from a low-phytic acid maize, *American Journal of Clinical Nutrition* 76 (2002): 556–559.
26. C. F. Walker and R. E. Black, Zinc and the risk for infectious disease, *Annual Review of Nutrition* 24 (2004): 255–275.
27. J. M. M. Gardner and coauthors, Zinc supplementation and psychosocial stimulation: Effects on the development of undernourished Jamaican children, *American Journal of Clinical Nutrition* 82 (2005): 399–405; T. A. Strand and coauthors, Effectiveness and efficacy of zinc for the treatment of acute diarrhea in young children, *Pediatrics* 109 (2002): 898–903; N. Bhandari and coauthors, Substantial reduction in severe diarrheal morbidity by daily zinc supplementation in young North Indian children, *Pediatrics* 109 (2002): e86.

28. G. A. Eby and W. W. Halcomb, Ineffectiveness of zinc gluconate nasal spray and zinc orotate lozenges in common-cold treatment: A double-blind placebo-controlled clinical trial, *Alternative Therapies in Health and Medicine* 12 (2006): 34–48; B. Arroll, Non-antibiotic treatments for upper-respiratory tract infections (common cold), *Respiratory Medicine* 99 (2005): 1477–1484; B. H. McElroy and S. P. Miller, Effectiveness of zinc gluconate glycine lozenges (Cold-Eeze) against the common cold in school-aged subjects: A retrospective chart review, *American Journal of Therapeutics* 9 (2002): 472–475.
29. M. B. Zimmermann and coauthors, Rapid relapse of thyroid dysfunction and goiter in school-age children after discontinuation of salt iodization, *American Journal of Clinical Nutrition* 79 (2004): 642–645.
30. M. B. Zimmerman, The influence of iron status on iodine utilization and thyroid function, *Annual Review of Nutrition* 26 (2006): 367–389.
31. W. Teng and coauthors, Effect of iodine intake on thyroid diseases in China, *New England Journal of Medicine* 354 (2006): 2783–2793.
32. D. M. Driscoll and P. R. Copeland, Mechanism and regulation of selenoprotein synthesis, *Annual Review of Nutrition* 23 (2003): 17–40.
33. R. F. Burk and K. E. Hill, Selenoprotein P: An extracellular protein with unique physical characteristics and a role in selenium homeostasis, *Annual Review of Nutrition* 25 (2005): 215–235.
34. A. J. Duffield-Lillico, I. Shureiqi, and S. M. Lippman, Can selenium prevent colorectal cancer? A signpost from epidemiology, *Journal of the National Cancer Institute* 96 (2004): 1645–1647.
35. J. W. Finley, Selenium accumulation in plant foods, *Nutrition Reviews* 63 (2005): 196–202.
36. N. E. Hellman and J. D. Gitlin, Ceruloplasmin metabolism and function, *Annual Review of Nutrition* 22 (2002): 439–458.
37. J. W. Finley, Does environmental exposure to manganese pose a health risk to healthy adults? *Nutrition Reviews* 62 (2004): 148–153.
38. Populations receiving optimally fluoridated public drinking water—United States, 2000, *Morbidity and Mortality Weekly Report* 51 (2002): 144–147.
39. Position of the American Dietetic Association: The impact of fluoride on health, *Journal of the American Dietetic Association* 105 (2005): 1620–1628.
40. Surveillance for dental caries, dental sealants, tooth retention, edentulism, and enamel fluorosis—United States, 1988–1994 and 1999–2002, *Morbidity and Mortality Weekly Report* 54 (2005): 1–44.

41. Populations receiving optimally fluoridated public drinking water—United States, 2000, 2002.

42. M. D. Althuis and coauthors, Glucose and insulin responses to dietary chromium supplements: A meta-analysis, *American Journal of Clinical Nutrition* 76 (2002): 148–155.

43. T. A. Devirian and S. L. Volpe, The physiological effects of dietary boron, *Critical Reviews in Food and Science Nutrition* 43 (2003): 219–231.

44. Committee on Environmental Health, Lead exposure in children: Prevention, detection, and management, *Pediatrics* 116 (2005):

1036–1046; Blood lead levels—United States, 1999–2002, *Morbidity and Mortality Weekly Report* 54 (2005): 513–527.

45. D. C. Bellinger, Lead, *Pediatrics* 113 (2004): 1016–1022.

46. S. G. Selevan and coauthors, Blood lead concentration and delayed puberty in girls, *New England Journal of Medicine* 348 (2003): 1527–1536.

47. R. J. Billings, R. J. Berkowitz, and G. Watson, Teeth, *Pediatrics* 113 (2004): 1120–1127.

48. R. L. Canfield and coauthors, Intellectual impairment in children with blood lead concentrations below 10 μg per deciliter,

New England Journal of Medicine 348 (2003): 1517–1526.

49. A. R. Kemper and coauthors, Follow-up testing among children with elevated screening blood lead levels, *Journal of the American Medical Association* 293 (2005): 2232–2237.

50. K. Kalia and S. J. Flora, Strategies for safe and effective therapeutic measures for chronic arsenic and lead poisoning, *Journal of Occupational Health* 47 (2005): 1–21; S. P. Murphy and coauthors, Simple measures of dietary variety are associated with improved dietary quality, *Journal of the American Dietetic Association* 106 (2006): 425–429.

ANSWERS

Nutrition Calculations

1. Iron: mg   Selenium: μg   Fluoride: mg

   Zinc: mg   Copper: μg   Chromium: μg

   Iodine: μg   Manganese: mg   Molybdenum: μg

2. a. Ranked by iron per serving: sirloin steak > broccoli > green peas > bread > pork chop > cantaloupe > sweet potato > carrots > apple > cheese > milk

  b.

Food	Iron Density (mg/kcal)
Milk, fat-free, 1 c	0.10 mg ÷ 85 kcal = 0.0012 mg/kcal
Cheddar cheese, 1 oz	0.19 mg ÷ 114 kcal = 0.0017 mg/kcal
Broccoli, cooked from fresh, chopped, 1 c	1.31 mg ÷ 44 kcal = 0.0298 mg/kcal
Sweet potato, baked in skin, 1 ea	0.51 mg ÷ 117 kcal = 0.0044 mg/kcal
Cantaloupe melon, ½	0.56 mg ÷ 93 kcal = 0.0060 mg/kcal
Carrots, from fresh, ½ c	0.48 mg ÷ 35 kcal = 0.0137 mg/kcal
Whole-wheat bread, 1 slice	0.87 mg ÷ 64 kcal = 0.0136 mg/kcal
Green peas, cooked from frozen, ½ c	1.26 mg ÷ 62 kcal = 0.0203 mg/kcal
Apple, medium	0.38 mg ÷ 125 kcal = 0.0030 mg/kcal
Sirloin steak, lean, 4 oz	3.81 mg ÷ 228 kcal = 0.0167 mg/kcal
Pork chop, lean broiled, 1 ea	0.66 mg ÷ 166 kcal = 0.0040 mg/kcal

  Ranked by iron density (iron per kcalorie): broccoli > green peas > sirloin steak > carrots > bread > cantaloupe > sweet potato > pork chop > apple > cheese > milk

  c. Broccoli, green peas, and carrots are all higher on the per-kcalorie list.

  d. They are all vegetables.

Study Questions (multiple choice)

1. b   2. b   3. c   4. c   5. d   6. d   7. c   8. b   9. a   10. c

Phytochemicals and Functional Foods

© John E. Kelly/FoodPix/Jupiter Images

Chapter 13 completes the introductory discussions on the six classes of nutrients—carbohydrates, lipids, proteins, vitamins, minerals, and water. In addition to these nutrients, foods contain thousands of nonnutrient compounds, including the phytochemicals. Chapter 1 introduced the **phytochemicals** as compounds found in plant-derived foods (*phyto* means plant) that have biological activity in the body. Research on phytochemicals is unfolding daily, adding to our knowledge of their roles in human health, but there are still many questions and only tentative answers. Just a few of the tens of thousands of phytochemicals have been researched at all, and only a sampling are mentioned in this highlight—enough to illustrate their wide variety of food sources and roles in supporting health.

The concept that foods provide health benefits beyond those of the nutrients emerged from numerous epidemiological studies showing the protective effects of plant-based diets on cancer and heart disease. People have been using foods to maintain health and prevent disease for years, but now these foods have been given a name—they are called **functional foods.** (The accompanying glossary defines this and other terms.) As Chapter 1 explained, functional foods include all foods (whole, fortified, or modified foods) that have a potentially beneficial effect on health.[1] Much of this text touts the benefits of nature's functional foods—grains rich in dietary fibers, fish rich in omega-3 fatty acids, and fruits rich in phytochemicals, for example. This highlight begins with a look at some of these familiar functional foods, the phytochemicals they contain, and their roles in disease prevention. Then the discussion turns to examine the most controversial of functional foods—novel foods to which phytochemicals have been added to promote health. How these foods fit into a healthy diet is still unclear.[2]

The Phytochemicals

In foods, phytochemicals impart tastes, aromas, colors, and other characteristics. They give hot peppers their burning sensation, garlic its pungent flavor, and tomatoes their dark red color. In the body, phytochemicals can have profound physiological effects, acting as antioxidants, mimicking hormones, and suppressing the development of diseases.[3] Table H13-1 (p. 470) presents the names, possible effects, and food sources of some of the better-known phytochemicals.

Defending against Cancer

A variety of phytochemicals from a variety of foods appear to protect against DNA damage and defend the body against cancer. A few examples follow.

Soybeans and products made from them correlate with low rates of some cancers.[4] Soybeans—as well as other legumes, **flaxseeds,** whole grains, fruits, and vegetables—are a rich source of an array of phytochemicals, among them the **phytoestrogens.** Because the chemical structure of these phytochemicals is similar to the steroid hormone estrogen, they can weakly mimic or modulate the effects of estrogen in the body.[5] They also have antioxidant

GLOSSARY

flavonoids (FLAY-von-oyds): yellow pigments in foods; phytochemicals that may exert physiological effects on the body.

flaxseeds: the small brown seeds of the flax plant; valued as a source of linseed oil, fiber, and omega-3 fatty acids.

lignans: phytochemicals present in flaxseed, but not in flax oil, that are converted to phytosterols by intestinal bacteria and are under study as

possible anticancer agents.

lutein (LOO-teen): a plant pigment of yellow hue; a phytochemical believed to play roles in eye functioning and health.

lycopene (LYE-koh-peen): a pigment responsible for the red color of tomatoes and other red-hued vegetables; a phytochemical that may act as an antioxidant in the body.

phytoestrogens: plant-derived compounds that have structural and functional similarities to human estrogen. Phytoestrogens include the isoflavones genistein, daidzein, and glycitein.

phytosterols: plant-derived compounds that have structural similarities to cholesterol and lower blood cholesterol by competing with cholesterol for absorption. Phytosterols include sterol esters and stanol esters.

Reminders: **Phytochemicals** are nonnutrient compounds found in plant-derived foods that have biological activity in the body.

Functional foods are foods that contain physiologically active compounds that provide health benefits beyond basic nutrition.

TABLE H13-1   Phytochemicals—Their Food Sources and Actions

Name	Possible Effects	Food Sources
Alkylresorcinols[a]	May contribute to the protective effect of grains in reducing the risks of diabetes, heart disease, and some cancers.	Whole grain wheat and rye
Capsaicin	Modulates blood clotting, possibly reducing the risk of fatal clots in heart and artery disease.	Hot peppers
Carotenoids (include beta-carotene, lycopene, lutein, and hundreds of related compounds)[b]	Act as antioxidants, possibly reducing risks of cancer and other diseases.	Deeply pigmented fruits and vegetables (apricots, broccoli, cantaloupe, carrots, pumpkin, spinach, sweet potatoes, tomatoes)
Curcumin	May inhibit enzymes that activate carcinogens.	Tumeric, a yellow-colored spice
Flavonoids (include flavones, flavonols, isoflavones, catechins, and others)[a,c]	Act as antioxidants; scavenge carcinogens; bind to nitrates in the stomach, preventing conversion to nitrosamines; inhibit cell proliferation.	Berries, black tea, celery, citrus fruits, green tea, olives, onions, oregano, purple grapes, purple grape juice, soybeans and soy products, vegetables, whole wheat, wine
Indoles[d]	May trigger production of enzymes that block DNA damage from carcinogens; may inhibit estrogen action.	Broccoli and other cruciferous vegetables (brussels sprouts, cabbage, cauliflower), horseradish, mustard greens
Isothiocyanates (including sulforaphane)	Inhibit enzymes that activate carcinogens; trigger production of enzymes that detoxify carcinogens.	Broccoli and other cruciferous vegetables (brussels sprouts, cabbage, cauliflower), horseradish, mustard greens
Lignans[e]	Block estrogen activity in cells, possibly reducing the risk of cancer of the breast, colon, ovaries, and prostate.	Flaxseed and its oil, whole grains
Monoterpenes (include limonene)	May trigger enzyme production to detoxify carcinogens; inhibit cancer promotion and cell proliferation.	Citrus fruit peels and oils
Organosulfur compounds	May speed production of carcinogen-destroying enzymes; slow production of carcinogen-activating enzymes.	Chives, garlic, leeks, onions
Phenolic acids[a]	May trigger enzyme production to make carcinogens water soluble, facilitating excretion.	Coffee beans, fruits (apples, blueberries, cherries, grapes, oranges, pears, prunes), oats, potatoes, soybeans
Phytic acid	Binds to minerals, preventing free-radical formation, possibly reducing cancer risk.	Whole grains
Phytoestrogens (genistein and daidzein)	Estrogen inhibition may produce these actions: inhibit cell replication in GI tract; reduce risk of breast, colon, ovarian, prostate, and other estrogen-sensitive cancers; reduce cancer cell survival. Estrogen mimicking may reduce risk of osteoporosis.	Soybeans, soy flour, soy milk, tofu, textured vegetable protein, other legume products
Protease inhibitors	May suppress enzyme production in cancer cells, slowing tumor growth; inhibit hormone binding; inhibit malignant changes in cells.	Broccoli sprouts, potatoes, soybeans and other legumes, soy products
Resveratrol	Offsets artery-damaging effects of high-fat diets.	Red wine, peanuts
Saponins	May interfere with DNA replication, preventing cancer cells from multiplying; stimulate immune response.	Alfalfa sprouts, other sprouts, green vegetables, potatoes, tomatoes
Tannins[a]	May inhibit carcinogen activation and cancer promotion; act as antioxidants.	Black-eyed peas, grapes, lentils, red and white wine, tea

[a]A subset of the larger group *phenolic phytochemicals*.
[b]Other carotenoids include alpha-carotene, beta-cryptoxanthin, and zeaxanthin.
[c]Other flavonoids of interest include ellagic acid and ferulic acid; see also *phytoestrogens*.
[d]Indoles include dithiothiones, isothiocyantes, and others.
[e]Lignans act as phytosterols and phytoestrogens, but their food sources are limited.

activity that appears to slow the growth of breast and prostate cancers.[6] However, the use of phytoestrogen supplements is ill-advised as they may stimulate the growth of estrogen-dependent cancers (such as breast cancer).[7] Even the role of soy foods for breast cancer survivors is uncertain. Soy foods may be most effective when consumed in moderation throughout life. The American Cancer Society recommends: "Breast cancer survivors should consume only moderate amounts of soy foods as part of a healthy plant-based diet and should not intentionally ingest very high levels of soy products."[8]

Tomatoes seem to offer protection against cancers of the esophagus, lungs, prostate, and stomach. Among the phytochemicals responsible for this effect is **lycopene,** one of beta-carotene's many carotenoid relatives. Lycopene is the pigment that gives apricots, guava, papaya, pink grapefruits, and watermelon their red color—and it is especially abundant in tomatoes and cooked tomato products. Lycopene is a powerful antioxidant that seems to inhibit the growth of cancer cells.[9] Importantly, these benefits are seen when people eat *foods* containing lycopene.[10]

Soybeans and tomatoes are only two of the many fruits and vegetables credited with providing anticancer activity. Strong and convincing evidence shows that the risk of many cancers, and perhaps of cancer in general, decreases when diets include an abundance of fruits and vegetables.[11] To that end, current recommendations urge consumers to eat five to nine servings of fruits and vegetables a day.

Defending against Heart Disease

Diets based primarily on unprocessed foods appear to support heart health better than those founded on highly refined foods—perhaps because of the abundance of nutrients, fiber, or phytochemicals such as the **flavonoids**.[12] Flavonoids, a large group of phytochemicals known for their health-promoting qualities, are found in whole grains, legumes, soy, vegetables, fruits, herbs, spices, teas, chocolate, nuts, olive oil, and red wines.[13] Flavonoids are powerful antioxidants that may help to protect LDL cholesterol against oxidation and reduce blood platelet stickiness, making blood clots less likely.[14] An abundance of flavonoid-containing *foods* in the diet lowers the risks of chronic diseases.[15] Importantly, no claims can be made for flavonoids themselves as the protective factor, particularly when they are extracted from foods and sold as supplements.[16]

In addition to flavonoids, fruits and vegetables are rich in carotenoids. Studies suggest that a diet rich in carotenoids is also associated with a lower risk of heart disease.[17] Notable among the carotenoids that may defend against heart disease are **lutein** and lycopene.[18]

The **phytosterols** of soybeans and the **lignans** of flaxseed may also protect against heart disease.[19] These cholesterol-like molecules are naturally found in all plants and inhibit cholesterol absorption in the body. As a result, blood cholesterol levels decline.[20] These phytochemicals also seem to protect against heart disease by acting as antioxidants and lowering blood pressure.[21]

The Phytochemicals in Perspective

Because foods deliver thousands of phytochemicals in addition to dozens of nutrients, researchers must be careful in giving credit for particular health benefits to any one compound. Diets rich in whole grains, legumes, vegetables, fruits, and nuts seem to protect against heart disease and cancer, but identifying *the* specific foods or components of foods that are responsible is difficult.[22] Each food possesses a unique array of phytochemicals—citrus fruits provide monoterpenes; grapes, resveratrol; and flaxseed, lignans. (Review Table H13-1 for the possible effects and other food sources of these phytochemicals.) Broccoli may contain as many as 10,000 different phytochemicals—each with the potential to influ-

Nature offers a variety of functional foods that provide us with many health benefits.

ence some action in the body. Beverages such as wine, spices such as oregano, and oils such as olive oil (especially virgin olive oil) contain many phytochemicals that may explain, in part, why people who live in the Mediterranean region have reduced risks of heart disease and cancer.[23] Phytochemicals might also explain why the DASH diet is so effective in lowering blood pressure and blood lipids.[24] Even identifying all of the phytochemicals and their effects doesn't answer all the questions because the actions of phytochemicals may be complementary or overlapping—which reinforces the principle of variety in diet planning. For an appreciation of the array of phytochemicals offered by a variety of fruits and vegetables, see Figure H13-1 (p. 472).

Functional Foods

Because foods naturally contain thousands of phytochemicals that are biologically active in the body, virtually all of them have some special value in supporting health. In other words, even simple, whole foods, in reality, are functional foods. Cranberries may help protect against urinary tract infections; garlic may lower blood cholesterol; and tomatoes may protect against some cancers, just to name a few examples.[25] But that hasn't stopped food manufacturers from trying to create functional foods as well. The creation of more functional foods has become the fastest-growing trend and the greatest influence transforming the American food supply.[26]

Many processed foods become functional foods when they are fortified with nutrients or enhanced with phytochemicals or herbs (calcium-fortified orange juice, for example). Less frequently, an entirely new food is created, as in the case of a meat substitute made of mycoprotein—a protein derived from a fungus.*[27] This functional food not only provides dietary fiber, polyunsaturated fats, and high-quality protein, but it lowers LDL cholesterol, raises HDL cholesterol, improves glucose response, and prolongs satiety after a meal. Such a novel functional food raises the question—is it a food or a drug?

Foods as Pharmacy

Not too long ago, most of us could agree on what was a food and what was a drug. Today, functional foods blur the distinctions.[28] They have characteristics similar to both foods and drugs, but do not fit neatly into either category. Consider margarine, for example.

Eating nonhydrogenated margarine sparingly instead of butter generously may lower blood cholesterol slightly over several months and clearly falls into the food category. Taking the drug Lipitor, on the other hand, lowers blood cholesterol significantly within weeks and clearly falls into the drug category. But margarine enhanced with a phytosterol that lowers blood cholesterol is in a gray area between the two. The margarine looks and tastes like a food, but it acts like a drug.

The use of functional foods as drugs creates a whole new set of diet-planning challenges. Not only must foods provide an adequate intake of all the nutrients to support good health, but they must

---

* This mycoprotein product is marketed under the trade name Quorn (pronounced KWORN).

© Courtesy of Brassica Protection Products, © 2001 PhotoDisc, © Eye Wire, Inc., Courtesy of Flax Council of Canada, PhotoDisc/Getty Images, Matthew Farruggio

FIGURE H13-1 An Array of Phytochemicals in a Variety of Fruits and Vegetables

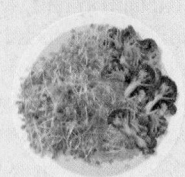

Broccoli and broccoli sprouts contain an abundance of the cancer-fighting phytochemical sulforaphane.

An apple a day—rich in flavonoids—may protect against lung cancer.

The phytoestrogens of soybeans seem to starve cancer cells and inhibit tumor growth; the phytosterols may lower blood cholesterol and protect

Garlic, with its abundant organosulfur compounds, may lower blood cholesterol and protect against stomach cancer.

The phytochemical resveratrol found in grapes (and nuts) protects against cancer by inhibiting cell growth and against heart disease by limiting clot formation and inflammation.

The ellagic acid of strawberries may inhibit certain types of cancer.

Tomatoes, with their abundant lycopene, may defend against cancer by protecting DNA from oxidative damage.

The monoterpenes of citrus fruits (and cherries) may inhibit cancer growth.

The flavonoids in black tea may protect against heart disease, whereas those in green tea may defend against cancer.

The flavonoids in cocoa and chocolate defend against oxidation and reduce the tendency of blood to clot.

Spinach and other colorful vegetables contain the carotenoids lutein and zeaxanthin, which help protect the eyes against macular degeneration.

Flaxseed, the richest source of lignans, may prevent the spread of cancer.

Blueberries, a rich source of flavonoids, improve memory in animals.

also deliver drug-like ingredients to protect against disease. Like drugs used to treat chronic diseases, functional foods may need to be eaten several times a day for several months or years to have a beneficial effect. Sporadic users may be disappointed in the results. Margarine enriched with 2 to 3 grams of phytosterols may reduce cholesterol by up to 15 percent, much more than regular margarine does, but not nearly as much as the more than 30 percent reduction seen with cholesterol-lowering drugs.[29] For this reason, functional foods may be more useful for prevention and mild cases of disease than for intervention and more severe cases.

Foods and drugs differ dramatically in cost as well. Functional foods such as fruits and vegetables incur no added costs, of course, but foods that have been manufactured with added phytochemicals can be expensive, costing up to six times as much as

their conventional counterparts. The price of functional foods typically falls between that of traditional foods and medicines.

Unanswered Questions

To achieve a desired health effect, which is the better choice: to eat a food designed to affect some body function or simply to adjust the diet? Does it make more sense to use a margarine enhanced with a phytosterol that lowers blood cholesterol or simply to limit the amount of butter eaten?* Is it smarter to eat eggs enriched with omega-3 fatty acids or to restrict egg consumption?

---

* Margarine products that lower blood cholesterol contain either sterol esters from vegetable oils, soybeans, and corn or stanol esters from wood pulp.

Might functional foods offer a sensible solution for improving our nation's health—if done correctly? Perhaps so, but the problem is that the food industry is moving too fast for either scientists or the Food and Drug Administration to keep up. Consumers were able to buy soup with St. John's wort that claimed to enhance mood and fruit juice with echinacea that was supposed to fight colds while scientists were still conducting their studies on these ingredients. Research to determine the safety and effectiveness of these substances is still in progress. Until this work is complete, consumers are on their own in finding the answers to the following questions:

- *Does it work?* Research is generally lacking and findings are often inconclusive.

- *How much does it contain?* Food labels are not required to list the quantities of added phytochemicals. Even if they were, consumers have no standard for comparison and cannot deduce whether the amounts listed are a little or a lot. Most importantly, until research is complete, food manufacturers do not know what amounts (if any) are most effective—or most toxic.

- *Is it safe?* Functional foods can act like drugs. They contain ingredients that can alter body functions and cause allergies, drug interactions, drowsiness, and other side effects. Yet, unlike drug labels, food labels do not provide instructions for the dosage, frequency, or duration of treatment.

- *Is it healthy?* Adding phytochemicals to a food does not magically make it a healthy choice. A candy bar may be fortified with phytochemicals, but it is still made mostly of sugar and fat.

Critics suggest that the designation "functional foods" may be nothing more than a marketing tool. After all, even the most experienced researchers cannot yet identify the perfect combination of nutrients and phytochemicals to support optimal health. Yet manufacturers are freely experimenting with various concoctions as if they possessed that knowledge. Is it okay for them to sprinkle phytochemicals on fried snack foods or caramel candies and label them "functional," thus implying health benefits?

Future Foods

Nature has elegantly designed foods to provide us with a complex array of dozens of nutrients and thousands of additional

Functional foods currently on the market promise to "enhance mood," "promote relaxation and good karma," "increase alertness," and "improve memory," among other claims.

compounds that may benefit health—most of which we have yet to identify or understand. Over the years, we have taken those foods, deconstructed them, and then reconstructed them in an effort to "improve" them. With new scientific understandings of how nutrients—and the myriad other compounds in foods—interact with genes, we may someday be able to design foods to meet the *exact* health needs of *each* individual.[30] Indeed, our knowledge of the human genome and of human nutrition may well merge to allow specific recommendations for individuals based on their predisposition to diet-related diseases.

If the present trend continues, someday physicians may be able to prescribe the perfect foods to enhance your health, and farmers will be able to grow them. Scientists have already developed gene technology to alter the composition of food crops. They can grow rice enriched with vitamin A and tomatoes containing a hepatitis vaccine, for example. It seems quite likely that foods can be created to meet every possible human need. But then, in a sense, that was largely true 100 years ago when we relied on the bounty of nature.

NUTRITION ON THE NET

For further study of topics covered in this chapter, log on to **academic.cengage.com/nutrition/rolfes/UNCN8e**. Go to Chapter 13, then to Nutrition on the Net.

- Search for "functional foods" at the International Food Information Council: **www.ific.org**

- Search for "functional foods" at the Center for Science in the Public Interest: **www.cspinet.org**

- Find out if warnings have been issued for any food ingredients at the FDA website: **www.fda.gov**

REFERENCES

1. Position of the American Dietetic Association: Functional foods, *Journal of the American Dietetic Association* 104 (2004): 814–826.

2. C. H. Halsted, Dietary supplements and functional foods: 2 sides of a coin? *American Journal of Clinical Nutrition* 77 (2003): 1001S–1007S.

3. C. Manach and coauthors, Polyphenols: Food sources and bioavailability, *American Journal of Clinical Nutrition* 79 (2004): 727–747; P. M. Kris-Etherton and coauthors, Bioactive compounds in foods: Their role in the prevention of cardiovascular disease and cancer, *American Journal of Medicine* 113 (2002): 71S–88S.

4. M. B. Schabath and coauthors, Dietary phytoestrogens and lung cancer risk, *Journal of the American Medical Association* 294 (2005): 1493–1504; W. H. Xu and coauthors, Soya food intake and risk of endometrial cancer among Chinese women in Shanghai: Population based case-control study, *British Medical Journal* 328 (2004): 1285–1288.

5. I. C. Munro and coauthors, Soy isoflavones: A safety review, *Nutrition Reviews* 61 (2003): 1–33.

6. T. A. Ryan-Borchers and coauthors, Soy isoflavones modulate immune function in healthy postmenopausal women, *American Journal of Clinical Nutrition* 83 (2006): 1118–1125; C. A. Lamartiniere and coauthors, Genistein chemoprevention: Timing and mechanisms of action in murine mammary and prostate, *Journal of Nutrition* 132 (2002): 552S–558S.

7. M. Messina, W. McCaskill-Stevens, J. W. Lampe, Addressing the soy and breast cancer relationship: Review, commentary, and workshop proceedings, *Journal of the National Cancer Institute* 98 (2006): 1275–1284.

8. G. Maskarinec, Soy foods for breast cancer survivors and women at high risk for breast cancer? *Journal of the American Dietetic Association* 105 (2005): 1524–1528.

9. A. Basu and V. Imrhan, Tomatoes versus lycopene in oxidative stress and carcinogenesis: Conclusions from clinical trials, *European Journal of Clinical Nutrition* (2006); D. Heber and Q. Y. Lu, Overview of mechanisms of action of lycopene, *Experimental Biology and Medicine* 227 (2002): 920–923; T. M. Vogt and coauthors, Serum lycopene, other serum carotenoids, and risk of prostate cancer in US blacks and whites, *American Journal of Epidemiology* 155 (2002): 1023–1032.

10. S. Ellinger, J. Ellinger, and P. Stehle, Tomatoes, tomato products and lycopene in the prevention and treatment of prostate cancer: Do we have the evidence from intervention studies? *Current Opinion in Clinical Nutrition and Metabolic Care* 9 (2006): 722–727; E. Giovannucci and coauthors, A prospective study of tomato products, lycopene, and prostate cancer risk, *Journal of the National Cancer Institute* 94 (2002): 391–398.

11. C. A. Gonzalez, Nutrition and cancer: The current epidemiological evidence, *British Journal of Nutrition* 96 (2006): S42–S45; H. Vainio and E. Weiderpass, Fruit and vegetables in cancer prevention, *Nutrition and Cancer* 54 (2006): 111–142.

12. J. A. Ross and C. M. Kasum, Dietary flavonoids: Bioavailability, metabolic effects, and safety, *Annual Review of Nutrition* 22 (2002): 19–34.

13. M. B. Engler and M. M. Engler, The emerging role of flavonoid-rich cocoa and chocolate in cardiovascular health and disease, *Nutrition Reviews* 64 (2006): 109–118; M. W. Ariefdjohan and D. A. Savaiano, Chocolate and cardiovascular health: Is it too good to be true? *Nutrition Reviews* 63 (2005): 427–430; F. M. Steinberg, M. M. Bearden, and C. L. Keen, Cocoa and chocolate flavonoids: Implications for cardiovascular health, *Journal of the American Dietetic Association* 103 (2003): 215–223; F. Visioli and C. Galli, Biological properties of olive oil phytochemicals, *Critical Reviews in Food Science and Nutrition* 42 (2002): 209–221; Y. J. Surh, Anti-tumor promoting potential of selected spice ingredients with antioxidative and anti-inflammatory activities: A short review, *Food and Chemical Toxicology* 40 (2002): 1091–1097; J. M. Geleijnse and coauthors, Inverse association of tea and flavonoid intakes with incident myocardial infarction: The Rotterdam Study, *American Journal of Clinical Nutrition* 75 (2002): 880–886.

14. M. Messina, C. Gardner, and S. Barnes, Gaining insight into the health effects of soy but a long way still to go: Commentary on the Fourth International Symposium on the Role of Soy in Preventing and Treating Chronic Disease, *Journal of Nutrition* 132 (2002): 547S–551S; P. Knekt and coauthors, Flavonoid intake and risk of chronic diseases, *American Journal of Clinical Nutrition* 76 (2002): 560–568.

15. Ross and Kasum, 2002.

16. S. K. Osganian and coauthors, Dietary carotenoids and risk of coronary artery disease in women, *American Journal of Clinical Nutrition* 77 (2003): 1390–1399; S. Liu and coauthors, Intake of vegetables rich in carotenoids and risk of coronary heart disease in men: The Physicians' Heart Study, *International Journal of Epidemiology* 30 (2001): 130–135.

17. T. H. Rissanen and coauthors, Serum lycopene concentrations and carotid atherosclerosis: The Kuopio Ischaemic Heart Disease Risk Factor Study, *American Journal of Clinical Nutrition* 77 (2003): 133–138; Heber and Lu, 2002.

18. L. T. Bloedon and P. O. Szapary, Flaxseed and cardiovascular risk, *Nutrition Reviews* 62 (2004): 18–27; X. Zhang and coauthors, Soy food consumption is associated with lower risk of coronary heart disease in Chinese women, *Journal of Nutrition* 133 (2003): 2874–2878; R. E. Ostlund, Jr., Phytosterols in human nutrition, *Annual Review of Nutrition* 22 (2002): 533–549.

19. V. W. Y. Lau, M. Journoud, and P. J. H. Jones, Plant sterols are efficacious in lowering plasma LDL and non-HDL cholesterol in hypercholesterolemic type 2 diabetic and nondiabetic persons, *American Journal of Clinical Nutrition* 81 (2005): 1351–1358; S. Zhan and S. C. Ho, Meta-analysis of the effects of soy protein containing isoflavones on the lipid profile, *American Journal of*

Clinical Nutrition 81 (2005): 397–408; E. A. Lucas and coauthors, Flaxseed improves lipid profile without altering biomarkers of bone metabolism in postmenopausal women, *Journal of Clinical Endocrinology and Metabolism* 87 (2002): 1527–1532; C. A. Vanstone and coauthors, Unesterified plant sterols and stanols lower LDL-cholesterol concentrations equivalently in hypercholesterolemic persons, *American Journal of Clinical Nutrition* 76 (2002): 1272–1278.

20. L. T. Bloedon and P. O. Szapary, Flaxseed and cardiovascular risk, *Nutrition Reviews* 62 (2004): 18–27; M. Rivas and coauthors, Soy milk lowers blood pressure in men and women with mild to moderate essential hypertension, *Journal of Nutrition* 132 (2002): 1900–1902.

21. M. I. Covas and coauthors, The effect of polyphenols in olive oil on heart disease risk factors: A randomized trial, *Annals of Internal Medicine* 145 (2006): 333–341; Y. Z. H-Y. Hashim and coauthors, Components of olive oil and chemoprevention of colorectal cancer, *Nutrition Reviews* 63 (2005): 374–386; F. Visioli, A. Poli, and C. Gall, Antioxidant and other biological activities of phenols from olives and olive oil, *Medicinal Research Reviews* 22 (2002): 65–75.

22. M. M. Most, Estimated phytochemical content of the Dietary Approaches to Stop Hypertension (DASH) Diet is higher than in the control study diet, *Journal of the American Dietetic Association* 104 (2004): 1725–1727.

23. A. B. Howell and B. Foxman, Cranberry juice and adhesion of antibiotic resistant uropathogens, *Journal of the American Medical Association* 287 (2002): 3082–3083; C. W. Hadley and coauthors, Tomatoes, lycopene, and prostate cancer: Progress and promise, *Experimental Biology and Medicine* 227 (2002): 869–880.

24. Position of the American Dietetic Association, 2004.

25. T. Peregrin, Mycoprotein: Is America ready for a meat substitute derived from a fungus? *Journal of the American Dietetic Association* 102 (2002): 628.

26. C. L. Taylor, Regulatory frameworks for functional foods and dietary supplements, *Nutrition Reviews* 62 (2004): 55–59.

27. C. S. Patch, L. C. Tapsell, and P. G. Williams, Plant sterol/stanol prescription is an effective treatment strategy for managing hypercholesterolemia in outpatient clinical practice, *Journal of the American Dietetic Association* 105 (2005): 46–52; D. A. J. M. Kerckhoffs and coauthors, Effects on the human serum lipoprotein profile of β-glucan, soy protein and isoflavones, plant sterols and stanols, garlic and tocotrienols, *Journal of Nutrition* 132 (2002): 2494–2505; L. A. Simons, Additive effect of plant sterol-ester margarine and cerivastatin in lowering low-density lipoprotein cholesterol in primary hypercholesterolemia, *American Journal of Cardiology* 90 (2002): 737–740.

28. J. A. Milner, Functional foods and health: A US perspective, *British Journal of Nutrition* 88 (2002): S151–158.

The CengageNOW logo indicates an opportunity for online self-study, linking you to interactive tutorials and videos based on your level of understanding.

academic.cengage.com/login

Nutrition Portfolio Journal

Nutrition in Your Life

Food choices have consequences. Sometimes they happen immediately, as when you get heartburn after eating a pepperoni and jalapeño pizza. Other times they sneak up on you, as when you gain weight after repeatedly overindulging in double hot fudge sundaes. Quite often, they are temporary and easily resolved, as when hunger pangs strike after you drink only a diet cola for lunch. During pregnancy, however, the consequences of a woman's food choices are dramatic. They affect not just her health, but also the growth and development of another human being—and not just for today, but for years to come. Making smart food choices is a huge responsibility, but fortunately, it's fairly simple.

Life Cycle Nutrition: Pregnancy and Lactation

All people—pregnant and lactating women, infants, children, adolescents, and adults—need the same nutrients, but the amounts they need vary depending on their stage of life. This chapter focuses on nutrition in preparation for, and support of, pregnancy and lactation. The next two chapters address the needs of infants, children, adolescents, and older adults.

Nutrition prior to Pregnancy

A section on nutrition prior to pregnancy must, by its nature, focus mainly on women. Both a man's and a woman's nutrition may affect **fertility** and possibly the genetic contributions they make to their children, but it is the woman's nutrition that has the most direct influence on the developing fetus. Her body provides the environment for the growth and development of a new human being. Prior to pregnancy, a woman has a unique opportunity to prepare herself physically, mentally, and emotionally for the many changes to come. In preparation for a healthy pregnancy, a woman can establish the following habits:[1]

- *Achieve and maintain a healthy body weight.* Both underweight and overweight are associated with infertility.[2] Overweight and obese men have low sperm counts and hormonal changes that reduce fertility.[3] Excess body fat in women disrupts menstrual regularity and ovarian hormone production.[4] Should a pregnancy occur, mothers, both underweight and overweight, and their newborns, face increased risks of complications.

- *Choose an adequate and balanced diet.* Malnutrition reduces fertility and impairs the early development of an infant should a woman become pregnant.

- *Be physically active.* A woman who wants to be physically active when she is pregnant needs to become physically active beforehand.

- *Receive regular medical care.* Regular health care visits can help ensure a healthy start to pregnancy.

- *Manage chronic conditions.* Diseases such as diabetes, HIV/AIDS, PKU, and sexually transmitted diseases can adversely affect a pregnancy and need close medical attention to help ensure a healthy outcome.

fertility: the capacity of a woman to produce a normal ovum periodically and of a man to produce normal sperm; the ability to reproduce.

Young adults can prepare for a healthy pregnancy by taking care of themselves today.

• *Avoid harmful influences.* Both maternal and paternal ingestion of harmful substances (such as cigarettes, alcohol, drugs, or environmental contaminants) can cause abnormalities, alter genes or their expression, and interfere with fertility.

Young adults who nourish and protect their bodies do so not only for their own sakes, but also for future generations.[5]

Dietary Guidelines for Americans 2005

• Women of childbearing age who may become pregnant should eat foods high in heme-iron and/or consume iron-rich plant foods or iron-fortified foods with an enhancer of iron absorption, such as vitamin C–rich foods.

• Women of childbearing age who may become pregnant should consume adequate synthetic folate daily from fortified foods or supplements in addition to naturally occurring folate from a variety of foods.

Growth and Development during Pregnancy

A whole new life begins at **conception.** Organ systems develop rapidly, and nutrition plays many supportive roles. This section describes placental development and fetal growth, paying close attention to times of intense developmental activity.

Placental Development

In the early days of pregnancy, a spongy structure known as the **placenta** develops in the **uterus.** Two associated structures also form (see Figure 14-1). One is the **amniotic sac,** a fluid-filled balloonlike structure that houses the developing fetus. The other is the **umbilical cord,** a ropelike structure containing fetal blood vessels that extends through the fetus's "belly button" (the umbilicus) to the placenta. These three structures play crucial roles during pregnancy and then are expelled from the uterus during childbirth.

The placenta develops as an interweaving of fetal and maternal blood vessels embedded in the uterine wall. The maternal blood transfers oxygen and nutrients to the fetus's blood and picks up fetal waste products. By exchanging oxygen, nutrients, and waste products, the placenta performs the respiratory, absorptive, and excretory functions that the fetus's lungs, digestive system, and kidneys will provide after birth.

The placenta is a versatile, metabolically active organ. Like all body tissues, the placenta uses energy and nutrients to support its work. Like a gland, it produces an array of hormones that maintain pregnancy and prepare the mother's breasts for lactation (making milk). A healthy placenta is essential for the developing fetus to attain its full potential.[6]

Fetal Growth and Development

Fetal development begins with the fertilization of an **ovum** by a **sperm.** Three stages follow: the zygote, the embryo, and the fetus (see Figure 14-2).

The Zygote The newly fertilized ovum, or **zygote,** begins as a single cell and divides to become many cells during the days after fertilization. Within two weeks, the zygote embeds itself in the uterine wall—a process known as **implantation.** Cell division continues as each set of cells divides into many other cells. As development proceeds, the zygote becomes an embryo.

conception: the union of the male sperm and the female ovum; fertilization.

placenta (plah-SEN-tuh): the organ that develops inside the uterus early in pregnancy, through which the fetus receives nutrients and oxygen and returns carbon dioxide and other waste products to be excreted.

uterus (YOU-ter-us): the muscular organ within which the infant develops before birth.

amniotic (am-nee-OTT-ic) **sac:** the "bag of waters" in the uterus, in which the fetus floats.

umbilical (um-BILL-ih-cul) **cord:** the ropelike structure through which the fetus's veins and arteries reach the placenta; the route of nourishment and oxygen to the fetus and the route of waste disposal from the fetus. The scar in the middle of the abdomen that marks the former attachment of the umbilical cord is the **umbilicus** (um-BILL-ih-cus), commonly known as the "belly button."

ovum (OH-vum): the female reproductive cell, capable of developing into a new organism upon fertilization; commonly referred to as an egg.

sperm: the male reproductive cell, capable of fertilizing an ovum.

zygote (ZY-goat): the product of the union of ovum and sperm; so-called for the first two weeks after fertilization.

implantation: the stage of development in which the zygote embeds itself in the wall of the uterus and begins to develop; occurs during the first two weeks after conception.

FIGURE 14-1 | The Placenta and Associated Structures

To understand how placental villi absorb nutrients without maternal and fetal blood interacting directly, think of how the intestinal villi work. The GI side of the intestinal villi is bathed in a nutrient-rich fluid (chyme). The intestinal villi absorb the nutrient molecules and release them into the body via capillaries. Similarly, the maternal side of the placental villi is bathed in nutrient-rich maternal blood. The placental villi absorb the nutrient molecules and release them to the fetus via fetal capillaries.

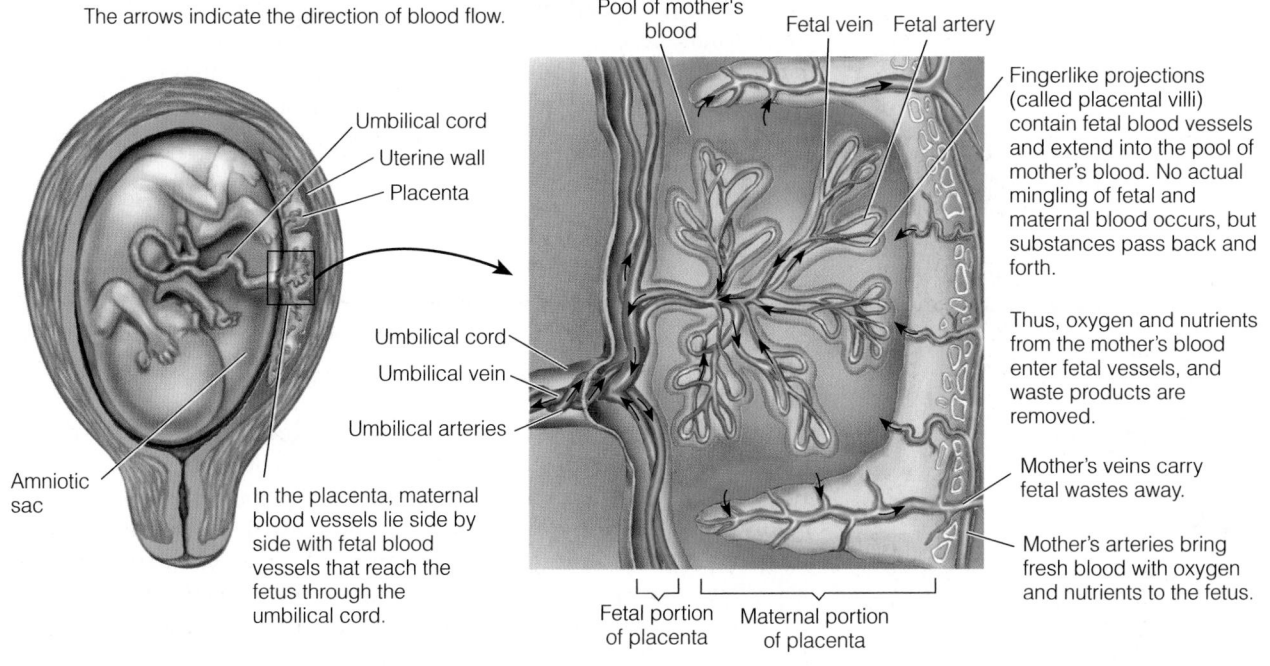

The arrows indicate the direction of blood flow.

Umbilical cord
Uterine wall
Placenta

Amniotic sac

In the placenta, maternal blood vessels lie side by side with fetal blood vessels that reach the fetus through the umbilical cord.

Umbilical cord
Umbilical vein
Umbilical arteries

Pool of mother's blood
Fetal vein   Fetal artery

Fingerlike projections (called placental villi) contain fetal blood vessels and extend into the pool of mother's blood. No actual mingling of fetal and maternal blood occurs, but substances pass back and forth.

Thus, oxygen and nutrients from the mother's blood enter fetal vessels, and waste products are removed.

Mother's veins carry fetal wastes away.

Mother's arteries bring fresh blood with oxygen and nutrients to the fetus.

Fetal portion of placenta   Maternal portion of placenta

FIGURE 14-2 | Stages of Embryonic and Fetal Development

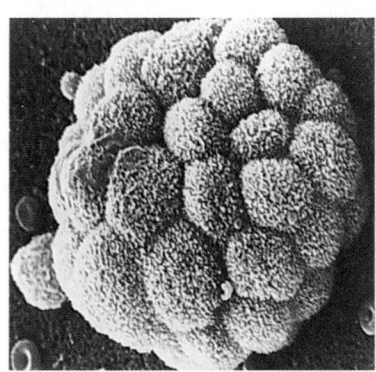

1 A newly fertilized ovum is about the size of a period at the end of this sentence. This **zygote** at less than one week after fertilization is not much bigger and is ready for implantation.

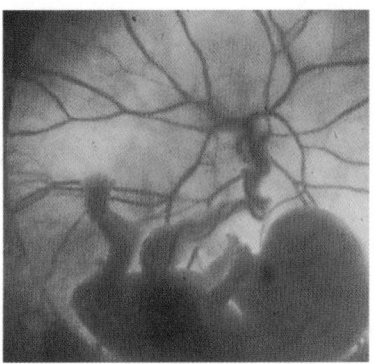

3 A **fetus** after 11 weeks of development is just over an inch long. Notice the umbilical cord and blood vessels connecting the fetus with the placenta.

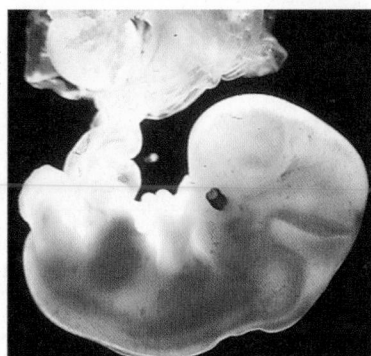

2 After implantation, the placenta develops and begins to provide nourishment to the developing embryo. An **embryo** 5 weeks after fertilization is about ¹/₂ inch long.

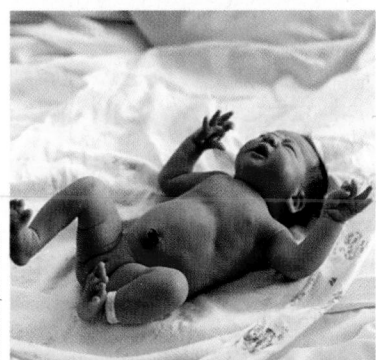

4 A **newborn infant** after nine months of development measures close to 20 inches in length. From 8 weeks to term, this infant grew 20 times longer and 50 times heavier.

FIGURE 14-3 The Concept of Critical Periods in Fetal Development

Critical periods occur early in fetal development. An adverse influence felt early in pregnancy can have a much more severe and prolonged impact than one felt later on.

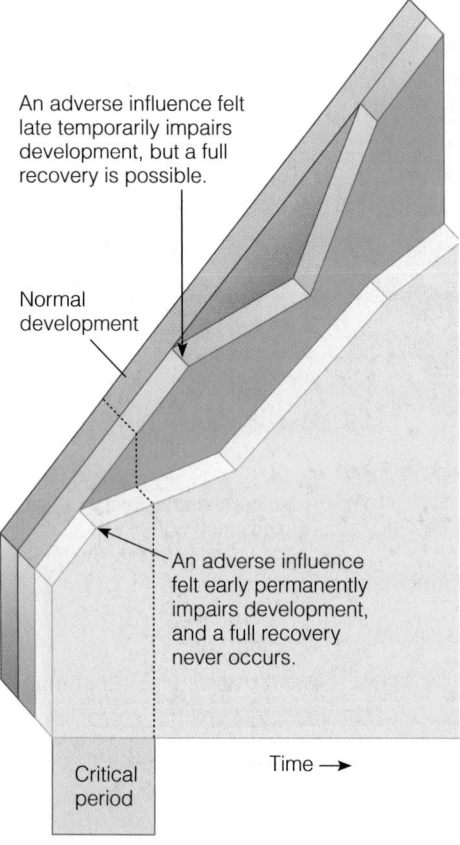

An adverse influence felt late temporarily impairs development, but a full recovery is possible.

Normal development

An adverse influence felt early permanently impairs development, and a full recovery never occurs.

Time →

Critical period

◆ Reminder: The *neural tube* is the structure that eventually becomes the brain and spinal cord.

embryo (EM-bree-oh): the developing infant from two to eight weeks after conception.

fetus (FEET-us): the developing infant from eight weeks after conception until term.

critical periods: finite periods during development in which certain events occur that will have irreversible effects on later developmental stages; usually a period of rapid cell division.

gestation (jes-TAY-shun): the period from conception to birth. For human beings, the average length of a healthy gestation is 40 weeks. Pregnancy is often divided into three-month periods, called **trimesters.**

The Embryo The **embryo** develops at an amazing rate. At first, the number of cells in the embryo doubles approximately every 24 hours; later the rate slows, and only one doubling occurs during the final 10 weeks of pregnancy. At 8 weeks, the 1¼-inch embryo has a complete central nervous system, a beating heart, a digestive system, well-defined fingers and toes, and the beginnings of facial features.

The Fetus The **fetus** continues to grow during the next 7 months. Each organ grows to maturity according to its own schedule, with greater intensity at some times than at others. As Figure 14-2 (p. 479) shows, fetal growth is phenomenal: weight increases from less than an ounce to about 7½ pounds (3500 grams). Most successful pregnancies last 38 to 42 weeks and produce a healthy infant weighing between 6½ and 9 pounds.

Critical Periods

Times of intense development and rapid cell division are called **critical periods**—critical in the sense that those cellular activities can occur only at those times. If cell division and number are limited during a critical period, full recovery is not possible (see Figure 14-3).

The development of each organ and tissue is most vulnerable to adverse influences (such as nutrient deficiencies or toxins) during its own critical period (see Figure 14-4). The critical period for neural tube ◆ development, for example, is from 17 to 30 days **gestation.** Consequently, neural tube development is most vulnerable to nutrient deficiencies, nutrient excesses, or toxins during this critical time—when most women do not even realize that they are pregnant. Any abnormal development of the neural tube or its failure to close completely can cause a major defect in the central nervous system. Figure 14-5 shows photos of neural tube development in the early weeks of gestation.

FIGURE 14-4 Critical Periods of Development

During embryonic development (from 2 to 8 weeks), many of the tissues are in their critical periods (purple area of the bars); events occur that will have irreversible effects on the development of those tissues. In the later stages of development (green area of the bars), the tissues continue to grow and change, but the events are less critical in that they are relatively minor or reversible.

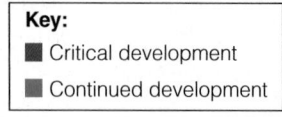

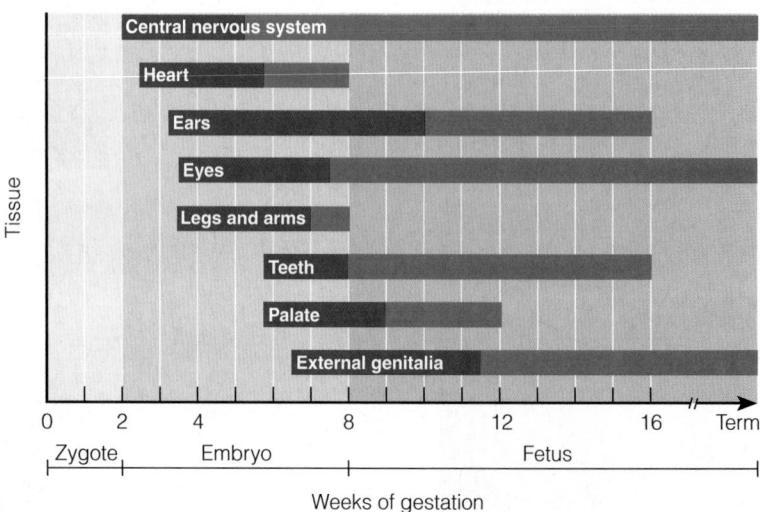

SOURCE: Adapted from *Before We Are Born: Essentials of Embryology and Birth Defects* by K. L. Moore. and T. V. N. Persaud: W. B. Saunders, 2003.

FIGURE 14-5 Neural Tube Development

The neural tube is the beginning structure of the brain and spinal cord. Any failure of the neural tube to close or to develop normally results in central nervous system disorders such as spina bifida and anencephaly. Successful development of the neural tube depends, in part, on the vitamin folate.

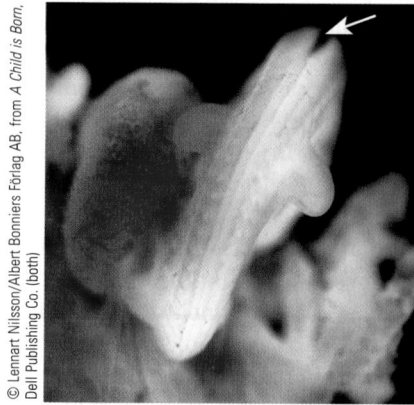
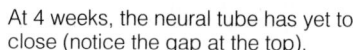
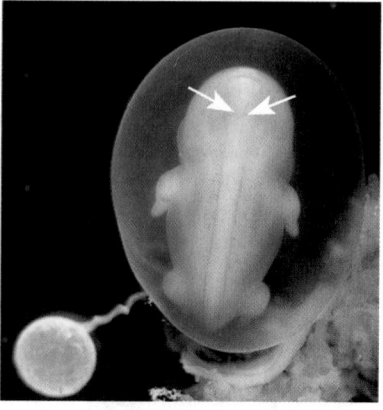

© Lennart Nilsson/Albert Bonniers Förlag AB, from *A Child is Born*, Dell Publishing Co. (both)

At 4 weeks, the neural tube has yet to close (notice the gap at the top).

At 6 weeks, the neural tube (outlined by the delicate red vertebral arteries) has successfully closed.

Neural Tube Defects In the United States, approximately 30 of every 100,000 newborns are born with a neural tube defect; ◆ some 1000 or so infants are affected each year.* Many other pregnancies with neural tube defects end in abortions or stillbirths.

The two most common types of neural tube defects are anencephaly and spina bifida. In **anencephaly,** the upper end of the neural tube fails to close. Consequently, the brain is either missing or fails to develop. Pregnancies affected by anencephaly often end in miscarriage; infants born with anencephaly die shortly after birth.

Spina bifida is characterized by incomplete closure of the spinal cord and its bony encasement (see Figure 14-6 on p. 482). The meninges membranes covering the spinal cord often protrude as a sac, which may rupture and lead to meningitis, a life-threatening infection. Spina bifida is accompanied by varying degrees of paralysis, depending on the extent of the spinal cord damage. Mild cases may not even be noticed, but severe cases lead to death. Common problems include clubfoot, dislocated hip, kidney disorders, curvature of the spine, muscle weakness, mental handicaps, and motor and sensory losses.

The cause of neural tube defects is unknown, but researchers are examining several gene-gene, gene-nutrient, and gene-environment interactions.[7] A pregnancy affected by a neural tube defect can occur in any woman, but these factors make it more likely:

- A previous pregnancy affected by a neural tube defect
- Maternal diabetes (type 1)
- Maternal use of antiseizure medications
- Maternal obesity
- Exposure to high temperatures early in pregnancy (prolonged fever or hot-tub use)
- Race/ethnicity (more common among whites and Hispanics than others)
- Low socioeconomic status

Folate supplementation reduces the risk.

◆ Reminder: A *neural tube defect* is a malformation of the brain, spinal cord, or both during embryonic development. The two main types of neural tube defects are **spina bifida** (literally, "split spine") and **anencephaly** ("no brain").

anencephaly (AN-en-SEF-a-lee): an uncommon and always fatal type of neural tube defect; characterized by the absence of a brain.
- **an** = not (without)
- **encephalus** = brain

spina (SPY-nah) **bifida** (BIFF-ih-dah): one of the most common types of neural tube defects; characterized by the incomplete closure of the spinal cord and its bony encasement.
- **spina** = spine
- **bifida** = split

* Worldwide, some 300,000 to 400,000 infants are born with neural tube defects each year.

Energy and Nutrient Needs during Pregnancy

From conception to birth, all parts of the infant—bones, muscles, organs, blood cells, skin, and other tissues—are made from nutrients in the foods the mother eats. For most women, nutrient needs during pregnancy and lactation ◆ are higher than at any other time (see Figure 14-10). To meet the high nutrient demands of pregnancy, a woman will need to make careful food choices, but her body will also help by maximizing absorption and minimizing losses.

◆ The Dietary Reference Intakes (DRI) table on the inside front cover provides separate listings for women during pregnancy and lactation, reflecting their heightened nutrient needs. Chapters 10–13 presented details on the vitamins and minerals.

Energy The enhanced work of pregnancy raises the basal metabolic rate dramatically and demands extra energy.[28] Energy needs of pregnant women are greater than those of nonpregnant women—an additional 340 kcalories per day during the second trimester and an extra 450 kcalories per day during the third. ◆ A woman can easily get these added kcalories with nutrient dense selections from the five food groups. See Table 2-3 (p. 41) for suggested dietary patterns for several kcalorie levels and Figure 14-11 (p. 490) for a sample menu for pregnant and lactating women.

◆ Energy requirement during pregnancy:
• 2nd trimester: + 340 kcal/day
• 3rd trimester: + 450 kcal/day

For a 2000-kcalorie daily intake, these added kcalories represent about 15 to 20 percent more food energy than before pregnancy. The increase in nutrient needs is often greater than this, so nutrient-dense foods should be chosen to supply the extra kcalories: foods such as whole-grain breads and cereals, legumes, dark green vegetables, citrus fruits, low-fat milk and milk products, and lean meats, fish, poultry, and eggs. Ample carbohydrate (ideally, 175 grams or more per day and certainly no less than 135 grams) is necessary to fuel the fetal brain. Sufficient carbohydrate ensures that the protein needed for growth will not be broken down and used to make glucose.

◆ Protein RDA during pregnancy:
• + 25 g/day

Protein The protein RDA ◆ for pregnancy is an additional 25 grams per day higher than for nonpregnant women. Pregnant women can easily meet their protein needs by selecting meats, milk products, and protein-containing plant foods such as legumes, whole grains, nuts, and seeds. Because use of high-protein supplements during pregnancy may be harmful to the infant's development, it is discouraged.

Essential Fatty Acids The high nutrient requirements of pregnancy leave little room in the diet for excess fat, but the essential long-chain polyunsaturated fatty acids are particularly important to the growth and development of the fetus. The brain is largely made of lipid material, and it depends heavily on the long-chain omega-3 and omega-6 fatty acids for its growth, function, and structure.[29] (See Table 5-2 on p. 159 for a list of good food sources of the omega fatty acids.)

Nutrients for Blood Production and Cell Growth New cells are laid down at a tremendous pace as the fetus grows and develops. At the same time, the mother's red blood cell mass expands. All nutrients are important in these processes, but for folate, vitamin B_{12}, iron, and zinc, the needs are especially great due to their key roles in the synthesis of DNA and new cells.

◆ Folate RDA during pregnancy:
• 600 µg/day

The requirement for folate increases dramatically during pregnancy. ◆ It is best to obtain sufficient folate from a combination of supplements, fortified foods, and a diet that includes fruits, juices, green vegetables, and whole grains.[30] The "How to" feature in Chapter 10 on p. 339 described how folate from each of these sources contributes to a day's intake.

◆ Vitamin B_{12} RDA during pregnancy:
• 2.6 µg/day

The pregnant woman also has a slightly greater need for the B vitamin that activates the folate enzyme—vitamin B_{12}. ◆ Generally, even modest amounts of meat, fish, eggs, or milk products together with body stores easily meet the need for vitamin B_{12}. Vegans who exclude all foods of animal origin, however, need daily supplements of vitamin B_{12} or vitamin B_{12}–fortified foods to prevent the neurological complications of a deficiency.

FIGURE 14-10 Comparison of Nutrient Recommendations for Nonpregnant, Pregnant, and Lactating Women

For actual values, turn to the table on the inside front cover.

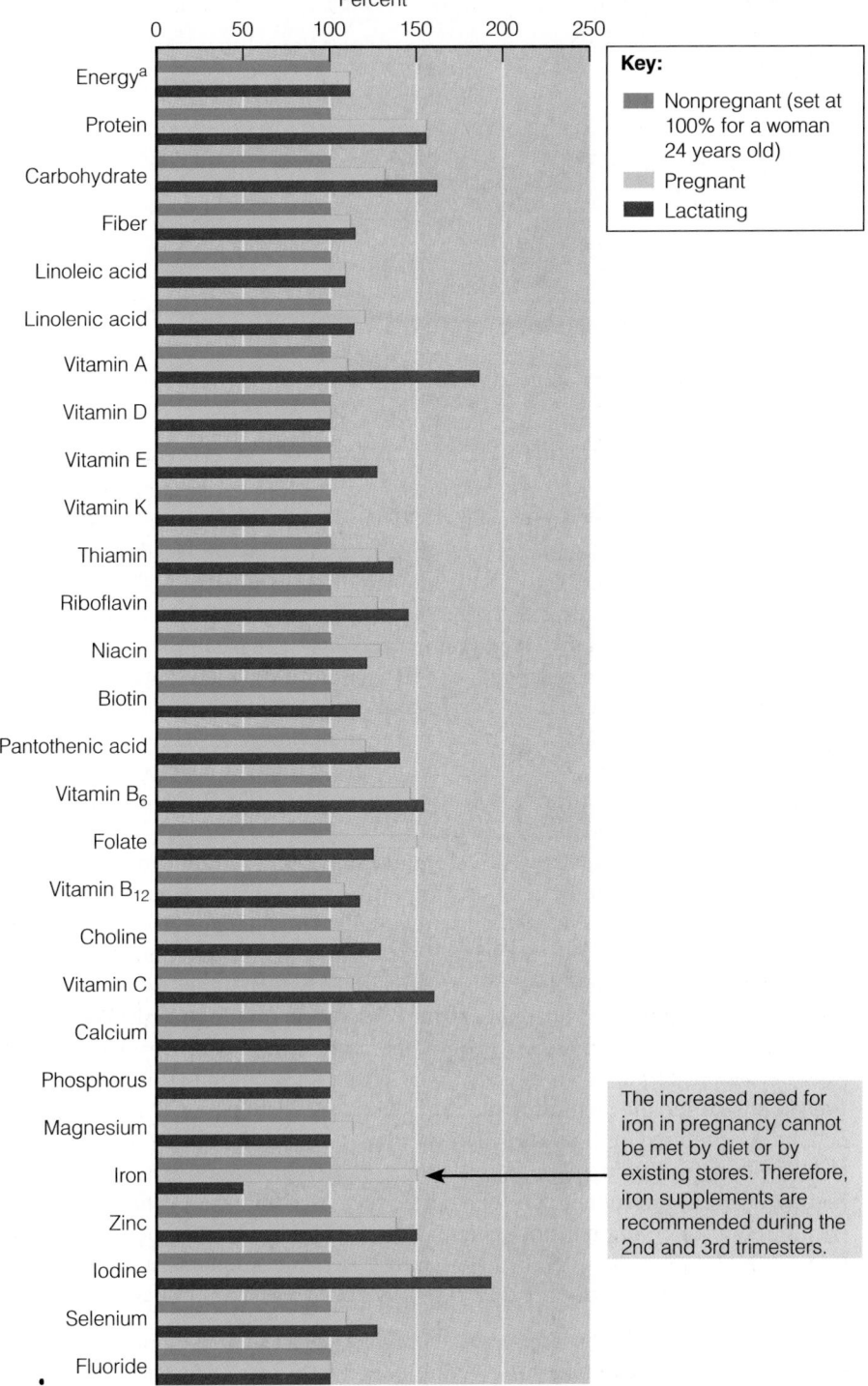

Key:
- Nonpregnant (set at 100% for a woman 24 years old)
- Pregnant
- Lactating

The increased need for iron in pregnancy cannot be met by diet or by existing stores. Therefore, iron supplements are recommended during the 2nd and 3rd trimesters.

[a]Energy allowance during pregnancy is for 2nd trimester; energy allowance during the 3rd trimester is slightly higher; no additional allowance is provided during the 1st trimester. Energy allowance during lactation is for the first 6 months; energy allowance during the second 6 months is slightly higher.

FIGURE 14-11 Daily Food Choices for Pregnant and Lactating Women

SAMPLE MENU

Breakfast
1 whole-wheat English muffin
2 tbs peanut butter
1 c low-fat vanilla yogurt
½ c fresh strawberries
1 c orange juice

Midmorning snack
½ c cranberry juice
1 oz pretzels

Lunch
Sandwich (tuna salad on whole-wheat bread)
½ carrot (sticks)
1 c low-fat milk

Dinner
Chicken cacciatore
   3 oz chicken
   ½ c stewed tomatoes
1 c rice
½ c summer squash
1½ c salad (spinach, mushrooms, carrots)
1 tbs salad dressing
1 slice Italian bread
2 tsp soft margarine
1 c low-fat milk

NOTE: This sample meal plan provides about 2500 kcalories (55% from carbohydrate, 20% from protein, and 25% from fat) and meets most of the vitamin and mineral needs of pregnant and lactating women.

◆ Iron RDA during pregnancy:
• 27 mg/day

◆ Zinc RDA during pregnancy:
• 12 mg/day (≤18 yr)
• 11 mg/day (19–50 yr)

◆ The AI for vitamin D does not increase during pregnancy.

Pregnant women need iron ◆ to support their enlarged blood volume and to provide for placental and fetal needs.[31] The developing fetus draws on maternal iron stores to create sufficient stores of its own to last through the first four to six months after birth. Even women with inadequate iron stores transfer significant amounts of iron to the fetus, suggesting that the iron needs of the fetus have priority over those of the mother.[32] In addition, blood losses are inevitable at birth, especially during a cesarean section, and can further drain the mother's supply.*

During pregnancy, the body makes several adaptations to help meet the exceptionally high need for iron. Menstruation, the major route of iron loss in women, ceases, and iron absorption improves thanks to an increase in blood transferrin, the body's iron-absorbing and iron-carrying protein. Without sufficient intake, though, iron stores would quickly dwindle.

Few women enter pregnancy with adequate iron stores, so a daily iron supplement is recommended during the second and third trimesters for all pregnant women. For this reason, most prenatal supplements provide 30 to 60 milligrams of iron a day. To enhance iron absorption, the supplement should be taken between meals or at bedtime and with liquids other than milk, coffee, or tea, which inhibit iron absorption. Drinking orange juice does not enhance iron absorption from supplements as it does from foods; vitamin C enhances iron absorption by converting iron from ferric to ferrous, but supplemental iron is already in the ferrous form. Vitamin C is helpful, however, in preventing the premature rupture of amniotic membranes.[33]

Zinc ◆ is required for DNA and RNA synthesis and thus for protein synthesis and cell development. Typical zinc intakes for pregnant women are lower than recommendations, but fortunately, zinc absorption increases when zinc intakes are low.[34] Routine supplementation is not advised.[35] Women taking iron supplements (more than 30 milligrams per day), however, may need zinc supplementation because large doses of iron can interfere with the body's absorption and use of zinc.

Nutrients for Bone Development Vitamin D and the bone-building minerals calcium, phosphorus, magnesium, and fluoride are in great demand during pregnancy. Insufficient intakes may produce abnormal fetal bones and teeth.

Vitamin D ◆ plays a vital role in calcium absorption and utilization. Consequently, severe maternal vitamin D deficiency interferes with normal calcium metabolism, resulting in rickets in the infant and osteomalacia in the mother.[36] Regular exposure to sunlight and consumption of vitamin D–fortified milk are usually sufficient to provide the recommended amount of vitamin D during preg-

---

* On average, almost twice as much blood is lost during a cesarean delivery as during the average vaginal delivery of a single fetus.

nancy, although some researchers question whether current recommendations are adequate.[37] Routine supplementation is not recommended because of the toxicity risk. Vegans who avoid milk, eggs, and fish may receive enough vitamin D from regular exposure to sunlight and from fortified soy milk.

Calcium absorption and retention increases dramatically in pregnancy, helping the mother to meet the calcium needs of pregnancy.[38] ◆ During the last trimester, as the fetal bones begin to calcify, over 300 milligrams a day are transferred to the fetus. Recommendations to ensure an adequate calcium intake during pregnancy help to conserve maternal bone while supplying fetal needs.[39]

Calcium intakes for pregnant women ◆ typically fall below recommendations. Because bones are still actively depositing minerals until about age 30, adequate calcium is especially important for young women. Pregnant women under age 25 who receive less than 600 milligrams of dietary calcium daily need to increase their intake of milk, cheese, yogurt, and other calcium-rich foods. Alternatively, and less preferably, they may need a daily supplement of 600 milligrams of calcium.

Other Nutrients The nutrients mentioned here are those most intensely involved in blood production, cell growth, and bone growth. Of course, other nutrients are also needed during pregnancy to support the growth and health of both fetus and mother. Even with adequate nutrition, repeated pregnancies, less than a year apart, deplete nutrient reserves. When this happens, fetal growth may be compromised, and maternal health may decline. The optimal interval between pregnancies is 18 to 23 months.

Nutrient Supplements Pregnant women who make wise food choices can meet most of their nutrient needs, with the possible exception of iron. Even so, physicians routinely recommend daily multivitamin-mineral supplements for pregnant women. Prenatal supplements typically contain greater amounts of folate, iron, and calcium than regular vitamin-mineral supplements. These supplements are particularly beneficial for women who do not eat adequately and for those in high-risk groups: women carrying multiple fetuses, cigarette smokers, and alcohol and drug abusers. The use of prenatal supplements may help reduce the risks of preterm delivery, low infant birthweights, and birth defects. Supplement use *prior* to conception also seems to reduce the risk of preterm births.[40] Figure 14-12 presents a label from a standard prenatal supplement.

◆ The AI for calcium does not increase during pregnancy.

◆ The USDA Food Guide suggests consuming 3 cups per day of fat-free or low-fat milk or the equivalent in milk products.

FIGURE 14-12 Example of a Prenatal Supplement

Supplement Facts
Serving Size 1 Tablet

Amount Per Tablet	% Daily Value for Pregnant/ Lactating Women
Vitamin A 4000 IU	50%
Vitamin C 100 mg	167%
Vitamin D 400 IU	100%
Vitamin E 11 IU	37%
Thiamin 1.84 mg	108%
Riboflavin 1.7 mg	85%
Niacin 18 mg	90%
Vitamin B6 2.6 mg	104%
Folate 800 mcg	100%
Vitamin B12 4 mcg	50%
Calcium 200 mg	15%
Iron 27 mg	150%
Zinc 25 mg	167%

INGREDIENTS: calcium carbonate, microcrystalline cellulose, dicalcium phosphate, ascorbic acid, ferrous fumarate, zinc oxide, acacia, sucrose ester, niacinamide, modified cellulose gum, di-alpha tocopheryl acetate, hydroxypropyl methylcellulose, hydroxypropyl cellulose, artificial colors (FD&C blue no. 1 lake, FD&C red no. 40 lake, FD&C yellow no. 6 lake, titanium dioxide), polyethylene glycol, starch, pyridoxine hydrochloride, vitamin A acetate, riboflavin, thiamin mononitrate, folic acid, beta carotene, cholecalciferol, maltodextrin, gluten, cyanocobalamin, sodium bisulfite.

Vegetarian Diets during Pregnancy and Lactation

In general, a vegetarian diet can support a healthy pregnancy and successful lactation if it provides adequate energy; includes milk and milk products; and contains a wide variety of legumes, cereals, fruits, and vegetables.[41] Many vegetarian women are well nourished, with nutrient intakes from diet alone exceeding the RDA for all vitamins and minerals except iron, which is low for most women. In contrast, vegan women who restrict themselves to an exclusively plant-based diet generally have low food energy intakes and are thin. For pregnant women, this can be a problem. Women with low prepregnancy weights and small weight gains during pregnancy jeopardize a healthy pregnancy.

Vegan diets may require supplementation with vitamin B_{12}, calcium, and vitamin D, or the addition of foods fortified with these nutrients. Infants of vegan parents may suffer spinal cord damage and develop severe psychomotor retardation due to a lack of vitamin B_{12} in the mother's diet during pregnancy. Breastfed infants of vegan mothers have been reported to develop vitamin B_{12} deficiency and severe movement disorders. Giving the infants vitamin B_{12} supplements corrects the blood and neurological symptoms of deficiency, as well as the structural abnormalities, but cognitive and language development delays may persist. A vegan mother needs a regular source of vitamin B_{12}–fortified foods or a supplement that provides 2.6 micrograms daily.

A pregnant woman who cannot meet her calcium needs through diet alone may need 600 milligrams of supplemental calcium daily, taken with meals. Pregnant women who do not receive sufficient dietary vitamin D or enough exposure to sunlight may need a supplement that provides 10 micrograms daily.

Common Nutrition-Related Concerns of Pregnancy

Nausea, constipation, heartburn, and food sensitivities are common nutrition-related concerns during pregnancy. A few simple strategies can help alleviate maternal discomforts (see Table 14-2).

Nausea Not all women have queasy stomachs in the early months of pregnancy, but many do. The nausea of "morning sickness" may actually occur anytime and ranges from mild queasiness to debilitating nausea and vomiting. Severe and continued vomiting may require hospitalization if it results in acidosis, dehydration, or excessive weight loss. The hormonal changes of early pregnancy seem to be responsible for a woman's sensitivities to the appearance, texture, or smell of foods. Traditional strategies for quelling nausea are listed in Table 14-2, but some women

TABLE 14-2 Strategies to Alleviate Maternal Discomforts

To Alleviate the Nausea of Pregnancy	To Prevent or Alleviate Constipation	To Prevent or Relieve Heartburn
• On waking, arise slowly. • Eat dry toast or crackers. • Chew gum or suck hard candies. • Eat small, frequent meals. • Avoid foods with offensive odors. • When nauseated, drink carbonated beverages instead of citrus juice, water, milk, coffee, or tea.	• Eat foods high in fiber (fruits, vegetables, and whole-grain cereals). • Exercise regularly. • Drink at least eight glasses of liquids a day. • Respond promptly to the urge to defecate. • Use laxatives only as prescribed by a physician; do not use mineral oil, because it interferes with absorption of fat-soluble vitamins.	• Relax and eat slowly. • Chew food thoroughly. • Eat small, frequent meals. • Drink liquids between meals. • Avoid spicy or greasy foods. • Sit up while eating; elevate the head while sleeping. • Wait an hour after eating before lying down. • Wait two hours after eating before exercising.

benefit most from resting when nauseous and simply eating the foods they want when they feel like eating. They may also find comfort in a cleaner, quieter, and more temperate environment.

Constipation and Hemorrhoids As the hormones of pregnancy alter muscle tone and the growing fetus crowds intestinal organs, an expectant mother may experience constipation. She may also develop hemorrhoids (swollen veins of the rectum). Hemorrhoids can be painful, and straining during bowel movements may cause bleeding. She can gain relief by following the strategies listed in Table 14-2.

Heartburn Heartburn is another common complaint during pregnancy. The hormones of pregnancy relax the digestive muscles, and the growing fetus puts increasing pressure on the mother's stomach. This combination allows stomach acid to back up into the lower esophagus, creating a burning sensation near the heart. Tips to help relieve heartburn are included in Table 14-2.

Food Cravings and Aversions Some women develop cravings for, or aversions to, particular foods and beverages during pregnancy. **Food cravings** and **food aversions** are fairly common, but they do not seem to reflect real physiological needs. In other words, a woman who craves pickles does not necessarily need salt. Similarly, cravings for ice cream are common in pregnancy but do not signify a calcium deficiency. Cravings and aversions that arise during pregnancy are most likely due to hormone-induced changes in sensitivity to taste and smell.

Nonfood Cravings Some pregnant women develop cravings for nonfood items ◆ such as freezer frost, laundry starch, clay, soil, or ice—a practice known as pica. Pica is a cultural phenomenon that reflects a society's folklore; it is especially common among African American women.[42] Pica is often associated with iron-deficiency anemia, but whether iron deficiency leads to pica or pica leads to iron deficiency is unclear. Eating clay or soil may interfere with iron absorption and displace iron-rich foods from the diet.

◆ Reminder: *Pica* is the general term for eating nonfood items. The specific craving for nonfood items that come from the earth, such as clay or dirt, is known as *geophagia*.

> **IN SUMMARY**
>
> Energy and nutrient needs are high during pregnancy. A balanced diet that includes an extra serving from each of the five food groups can usually meet these needs, with the possible exception of iron and folate (supplements are recommended). The nausea, constipation, and heartburn that sometimes accompany pregnancy can usually be alleviated with a few simple strategies. Food cravings do not typically reflect physiological needs.

◆ Nutrition advice in prenatal care:
 • Eat well-balanced meals.
 • Gain enough weight to support fetal growth.
 • Take prenatal supplements as prescribed.
 • Stop drinking alcohol.

High-Risk Pregnancies

Some pregnancies jeopardize the life and health of the mother and infant. Table 14-3 (p. 494) identifies several characteristics of a **high-risk pregnancy**. A woman with none of these risk factors is said to have a **low-risk pregnancy**. The more factors that apply, the higher the risk. All pregnant women, especially those in high-risk categories, need prenatal care, including dietary ◆ advice.

The Infant's Birthweight

A high-risk pregnancy is likely to produce an infant with **low birthweight**. Low-birthweight infants, defined as infants who weigh $5^1/2$ pounds or less, are classified according to their gestational age. Preterm infants are born before they are fully developed; they are often underweight and have trouble breathing because their lungs are immature. Preterm infants may be small, but if their size and

food cravings: strong desires to eat particular foods.

food aversions: strong desires to avoid particular foods.

high-risk pregnancy: a pregnancy characterized by indicators that make it likely the birth will be surrounded by problems such as premature delivery, difficult birth, retarded growth, birth defects, and early infant death.

low-risk pregnancy: a pregnancy characterized by indicators that make a normal outcome likely.

low birthweight (LBW): a birthweight of $5^1/2$ lb (2500 g) or less; indicates probable poor health in the newborn and poor nutrition status in the mother during pregnancy, before pregnancy, or both. Normal birthweight for a full-term baby is $6^1/2$ to $8^3/4$ lb (about 3000 to 4000 g).

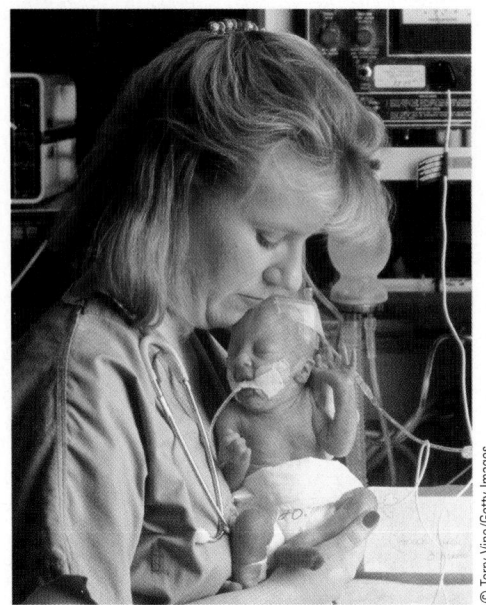

Low-birthweight babies need special care and nourishment.

TABLE 14-3	High-Risk Pregnancy Factors	
Factor	**Condition That Raises Risk**
Maternal weight	
• Prior to pregnancy	Prepregnancy BMI either <18.5 or >25
• During pregnancy	Insufficient or excessive pregnancy weight gain
Maternal nutrition	Nutrient deficiencies or toxicities; eating disorders
Socioeconomic status	Poverty, lack of family support, low level of education, limited food available
Lifestyle habits	Smoking, alcohol or other drug use
Age	Teens, especially 15 years or younger; women 35 years or older
Previous pregnancies	
• Number	Many previous pregnancies (3 or more to mothers under age 20; 4 or more to mothers age 20 or older)
• Interval	Short or long intervals between pregnancies (<18 months or >59 months)
• Outcomes	Previous history of problems
• Multiple births	Twins or triplets
• Birthweight	Low- or high-birthweight infants
Maternal health	
• High blood pressure	Development of pregnancy-related hypertension
• Diabetes	Development of gestational diabetes
• Chronic diseases	Diabetes; heart, respiratory, and kidney disease; certain genetic disorders; special diets and medications

◆ The weight of some preterm infants is **appropriate for gestational age (AGA)**; others are **small for gestational age (SGA)**, often reflecting malnutrition.

weight are appropriate for their age, ◆ they can catch up in growth given adequate nutrition support. In contrast, small-for-gestational-age infants have suffered growth failure in the uterus and do not catch up as well. For the most part, survival improves with increased gestational age and birthweight.

Low-birthweight infants are more likely to experience complications during delivery than normal-weight babies. They also have a statistically greater chance of having physical and mental birth defects, contracting diseases, and dying early in life. Of infants who die before their first birthdays, about two-thirds were low-birthweight newborns. Very-low-birthweight infants ($3\frac{1}{2}$ pounds or less) struggle not only for their immediate physical health and survival, but for their future cognitive development and abilities as well.

A strong relationship is evident between socioeconomic disadvantage and low birthweight. Low socioeconomic status impairs fetal development by causing stress and by limiting access to medical care and to nutritious foods. Low socioeconomic status often accompanies teen pregnancies, smoking, and alcohol and drug abuse—all predictors of low birthweight.

Malnutrition and Pregnancy

Good nutrition clearly supports a pregnancy. In contrast, malnutrition interferes with the ability to conceive, the likelihood of implantation, and the subsequent development of a fetus should conception and implantation occur.[43]

◆ Reminder: *Amenorrhea* is the temporary or permanent absence of menstrual periods. Amenorrhea is normal before puberty, after menopause, during pregnancy, and during lactation; otherwise it is abnormal.

Malnutrition and Fertility The nutrition habits and lifestyle choices people make can influence the course of a pregnancy they are not even planning at the time. Severe malnutrition and food deprivation can reduce fertility because women may develop amenorrhea, ◆ and men may be unable to produce viable sperm. Furthermore, both men and women lose sexual interest during times of starvation. Starvation arises predictably during famines, wars, and droughts, but it can also occur amidst peace and plenty. Many young women who diet excessively are starving and suffering from malnutrition (see Highlight 8).

Malnutrition and Early Pregnancy   If a malnourished woman does become pregnant, she faces the challenge of supporting both the growth of a baby and her own health with inadequate nutrient stores. Malnutrition prior to and around conception prevents the placenta from developing fully. A poorly developed placenta cannot deliver optimum nourishment to the fetus, and the infant will be born small and possibly with physical and cognitive abnormalities. If this small infant is a female, she may develop poorly and have an elevated risk of developing a chronic condition that could impair her ability to give birth to a healthy infant. Thus a woman's malnutrition can adversely affect not only her children but her *grandchildren*.

Malnutrition and Fetal Development   Without adequate nutrition during pregnancy, fetal growth and infant health are compromised. In general, consequences of malnutrition during pregnancy include fetal growth retardation, congenital malformations (birth defects), spontaneous abortion and stillbirth, preterm birth, and low infant birthweight. Preterm birth and low infant birthweight, in turn, predict the risk of stillbirth in a *subsequent* pregnancy.[44] Malnutrition, coupled with low birthweight, is a factor in more than half of all deaths of children under four years of age worldwide.

Food Assistance Programs

Women in high-risk pregnancies can find assistance from the WIC program—a high-quality, cost-effective health care and nutrition services program for women, infants, and children in the United States. Formally known as the Special Supplemental Nutrition Program for Women, Infants, and Children, WIC provides nutrition education and nutritious foods to infants, children up to age five, and pregnant and breastfeeding women who qualify financially and have a high risk of medical or nutritional problems. ◆ The program is both remedial and preventive: services include health care referrals, nutrition education, and food packages or vouchers for specific foods. These foods supply nutrients known to be lacking in the diets of the target population—most notably, protein, calcium, iron, vitamin A, and vitamin C. WIC-sponsored foods include tuna fish, carrots, eggs, milk, iron-fortified cereal, vitamin C–rich juice, cheese, legumes, peanut butter, and infant formula.

◆ WIC participants:
- $^1/_3$ of all pregnant women
- $^1/_2$ of all infants
- $^1/_4$ of all children ages 1–4 yr

More than 7 million people—most of them young children—receive WIC benefits each month. Prenatal WIC participation can effectively reduce infant mortality, low birthweight, and maternal and newborn medical costs. In 2003, Congress appropriated over $4.5 billion for WIC. For every dollar spent on WIC, an estimated three dollars in medical costs are saved in the first two months after birth.

Maternal Health

Medical disorders can threaten the life and health of both mother and fetus. If diagnosed and treated early, many diseases can be managed to ensure a healthy outcome—another strong argument for early prenatal care. Furthermore, the changes in pregnancy can reveal disease risks, making screening important and early intervention possible.[45]

Preexisting Diabetes   Whether diabetes presents risks depends on how well it is controlled before and during pregnancy. Without proper management of maternal diabetes, women face high infertility rates, and those who do conceive may experience episodes of severe hypoglycemia or hyperglycemia, spontaneous abortions, and pregnancy-related hypertension. Infants may be large, suffer physical and mental abnormalities, and experience other complications such as severe hypoglycemia or respiratory distress, both of which can be fatal. Ideally, a woman with diabetes will receive the prenatal care needed to achieve glucose control before conception and continued glucose control throughout pregnancy.

◆ Risk factors for gestational diabetes:
  - Age 25 or older
  - BMI >25 or excessive weight gain
  - Complications in previous pregnancies, including gestational diabetes or high-birthweight infant
  - Prediabetes or symptoms of diabetes
  - Family history of diabetes
  - Hispanic, black, Native American, South or East Asian, Pacific Islander, or indigenous Australian

◆ The hypertensive diseases of pregnancy are sometimes called **toxemia.**

◆ The normal edema of pregnancy responds to gravity; fluid pools in the ankles. The edema of preeclampsia is a generalized edema. The differences between these two types of edema help with the diagnosis of preeclampsia.

◆ Warning signs of preeclampsia:
  - Hypertension
  - Protein in the urine
  - Upper abdominal pain
  - Severe and constant headaches
  - Swelling, especially of the face
  - Dizziness
  - Blurred vision
  - Sudden weight gain (1 lb/day)
  - Fetal growth retardation

---

gestational diabetes: abnormal glucose tolerance during pregnancy.

transient hypertension of pregnancy: high blood pressure that develops in the second half of pregnancy and resolves after childbirth, usually without affecting the outcome of the pregnancy.

preeclampsia (PRE-ee-KLAMP-see-ah): a condition characterized by hypertension, fluid retention, and protein in the urine; formerly known as *pregnancy-induced hypertension.*[a]

eclampsia (eh-KLAMP-see-ah): a severe stage of preeclampsia characterized by convulsions.

---

[a] The Working Group on High Blood Pressure in Pregnancy, convened by the National High Blood Pressure Education Program of the National Heart, Lung, and Blood Institute, suggested abandoning the term *pregnancy-induced hypertension* because it failed to differentiate between the mild, transient hypertension of pregnancy and the life-threatening hypertension of preeclampsia.

Gestational Diabetes For every 14 women entering pregnancy without diabetes, one will develop a condition known as **gestational diabetes** during pregnancy. Gestational diabetes usually develops during the second half of pregnancy, with subsequent return to normal after childbirth. Some women with gestational diabetes, however, develop diabetes (usually type 2) after pregnancy, especially if they are overweight. For this reason, health care professionals strongly advise against excessive weight gain during pregnancy.

The most common consequences of gestational diabetes are complications during labor and delivery and a high infant birthweight.[46] Birth defects associated with gestational diabetes include heart damage, limb deformities, and neural tube defects. To ensure that the problems of gestational diabetes are dealt with promptly, physicians screen for the risk factors ◆ listed in the margin and test high-risk women for glucose intolerance immediately and average-risk women between 24 and 28 weeks gestation.[47] Dietary recommendations should meet the needs of pregnancy and maternal blood glucose goals.[48] To maintain normal blood glucose levels, carbohydrates should be restricted to 35 to 40 percent of energy intake. To limit excessive weight gain, obese women should limit energy intake to about 25 kcalories per kilogram body weight. Diet and moderate exercise may control gestational diabetes, but if blood glucose fails to normalize, insulin or other drugs may be required. Importantly, treatment reduces birth complications, infant deaths, and maybe even postpartum depression.[49]

Preexisting Hypertension Hypertension complicates pregnancy and affects its outcome in different ways, depending on when the hypertension first develops and on how severe it becomes. In addition to the threats hypertension always carries (such as heart attack and stroke), high blood pressure increases the risks of a low-birthweight infant or the separation of the placenta from the wall of the uterus before the birth, resulting in stillbirth. Ideally, before a woman with hypertension becomes pregnant, her blood pressure is under control.

Transient Hypertension of Pregnancy Some women develop hypertension during the second half of pregnancy.* Most often, the rise in blood pressure is mild and does not affect the pregnancy adversely. Blood pressure usually returns to normal during the first few weeks after childbirth. This **transient hypertension of pregnancy** differs from the life-threatening hypertensive diseases ◆ of pregnancy —preeclampsia and eclampsia.

Preeclampsia and Eclampsia Hypertension may signal the onset of **preeclampsia,** a condition characterized not only by high blood pressure but also by protein in the urine and fluid retention (edema). The edema ◆ of preeclampsia is a whole-body edema, distinct from the localized fluid retention women normally experience late in pregnancy.

The cause of preeclampsia remains unclear, but it usually occurs with first pregnancies ◆ and most often after 20 weeks gestation.[50] Symptoms typically regress within two days of delivery. Both men and women who were born of pregnancies complicated by preeclampsia are more likely to have a child born of a pregnancy complicated by preeclampsia, suggesting a genetic predisposition. Black women have a much greater risk of preeclampsia than white women.

Preeclampsia affects almost all of the mother's organs—the circulatory system, liver, kidneys, and brain. Blood flow through the vessels that supply oxygen and nutrients to the placenta diminishes. For this reason, preeclampsia often retards fetal growth. In some cases, the placenta separates from the uterus, resulting in preterm birth or stillbirth.

Preeclampsia can progress rapidly to **eclampsia**—a condition characterized by convulsive seizures and coma. Maternal death during pregnancy and childbirth is

---

* Blood pressure of 140/90 millimeters mercury or greater during the second half of pregnancy in a woman who has not previously exhibited hypertension indicates high blood pressure. So does a rise in systolic blood pressure of 30 millimeters or in diastolic blood pressure of 15 millimeters on at least two occasions more than six hours apart. By this rule, an apparently "normal" blood pressure of 120/85 is high for a woman whose normal value is 90/70.

extremely rare in developed countries, but when it does occur, eclampsia is a common cause. The rate of death for black women with eclampsia is more than four times the rate for white women.

Preeclampsia demands prompt medical attention. Treatment focuses on controlling blood pressure and preventing convulsions. If preeclampsia develops early and is severe, induced labor or cesarean section may be necessary, regardless of gestational age. The infant will be preterm, with all of the associated problems, including poor lung development and special care needs. Several dietary factors have been studied, but none have proved conclusive in preventing preeclampsia. Limited research suggests that exercise may protect against preeclampsia by stimulating placenta growth and vascularity and reducing oxidative stress.[51]

The Mother's Age

Maternal age also influences the course of a pregnancy. Compared with women of the physically ideal childbearing age of 20 to 25, both younger and older women face more complications of pregnancy.

Pregnancy in Adolescents Many adolescents become sexually active before age 19, and approximately 900,000 adolescent girls face pregnancies each year in the United States; slightly more than half of them give birth.[52] Nourishing a growing fetus adds to a teenage girl's nutrition burden, especially if her growth is still incomplete. Simply being young increases the risks of pregnancy complications independently of important socioeconomic factors.

Common complications among adolescent mothers include iron-deficiency anemia (which may reflect poor diet and inadequate prenatal care) and prolonged labor (which reflects the mother's physical immaturity). On a positive note, maternal death is lowest for mothers under age 20.

Pregnant teenagers have higher rates of stillbirths, preterm births, and low-birthweight infants than do adult women. Many of these infants suffer physical problems, require intensive care, and die within the first year. The care of infants born to teenagers costs our society an estimated $1 billion annually. Because teenagers have few financial resources, they cannot pay these costs. Furthermore, their low economic status contributes significantly to the complications surrounding their pregnancies. At a time when prenatal care is most important, it is less accessible. And the pattern of teenage pregnancies continues from generation to generation, with almost 40 percent of the daughters born to teenage mothers becoming teenage mothers themselves. Clearly, teenage pregnancy is a major public health problem.

To support the needs of both mother and fetus, young teenagers (13 to 16 years old) are encouraged to strive for the highest weight gains recommended for pregnancy. For a teen who enters pregnancy at a healthy body weight, a weight gain of approximately 35 pounds is recommended; this amount minimizes the risk of delivering a low-birthweight infant. Gaining less weight may limit fetal growth. Pregnant and lactating teenagers can use the USDA Food Guide presented in Table 2-3 and Figure 2-1 (pp. 41–43), making sure to select a high enough kcalorie level to support adequate weight gain.

Without the appropriate economic, social, and physical support, a young mother will not be able to care for herself during her pregnancy and for her child after the birth. To improve her chances for a successful pregnancy and a healthy infant, she must seek prenatal care. WIC helps pregnant teenagers obtain adequate food for themselves and their infants. (WIC is introduced on p. 495.)

Pregnancy in Older Women In the last several decades, many women have delayed childbearing while they pursue education and careers. As a result, the number of first births to women 35 and older has increased dramatically. Most of these women, even those over age 50, have healthy pregnancies.[53]

The few complications associated with later childbearing often reflect chronic conditions such as hypertension and diabetes, which can complicate an otherwise healthy pregnancy. These complications may result in a cesarean section, which is twice as common in women over 35 as among younger women. For all these reasons, maternal death rates are higher in women over 35 than in younger women.

The babies of older mothers face problems of their own including higher rates of preterm births and low birthweight.[54] Their rates of birth defects are also high. Because 1 out of 50 pregnancies in older women produces an infant with genetic abnormalities, obstetricians routinely screen women older than 35. For a 40-year-old mother, the risk of having a child with **Down syndrome,** for example, is about 1 in 100 compared with 1 in 300 for a 35-year-old and 1 in 10,000 for a 20-year-old. In addition, fetal death is twice as high for women 35 years and older than for younger women. Why this is so remains a bit of a mystery. One possibility is that the uterine blood vessels of older women may not fully adapt to the increased demands of pregnancy.

Practices Incompatible with Pregnancy

Besides malnutrition, a variety of lifestyle factors can have adverse effects on pregnancy, and some may be teratogenic. ◆ People who are planning to have children can make the choice to practice healthy behaviors.

Alcohol One out of ten pregnant women drinks alcohol at some time during her pregnancy; 1 out of 50 drinks frequently.[55] Alcohol consumption during pregnancy can cause irreversible mental and physical retardation of the fetus—fetal alcohol syndrome (FAS). Of the leading causes of mental retardation, FAS is the only one that is totally *preventable*. To that end, the surgeon general urges all pregnant women to refrain from drinking alcohol. Fetal alcohol syndrome is the topic of Highlight 14, which includes mention of how alcohol consumption by men may also affect fertility and fetal development.

Medicinal Drugs Drugs other than alcohol can also cause complications during pregnancy, problems in labor, and serious birth defects. For these reasons, pregnant women should not take any medicines without consulting their physicians, who must weigh the benefits against the risks.

Herbal Supplements Similarly, pregnant women should seek a physician's advice before using herbal supplements. Women sometimes seek herbal preparations during their pregnancies to quell nausea, induce labor, aid digestion, promote water loss, support restful sleep, and fight depression. As Chapter 19 explains, some herbs may be safe, but many others are definitely harmful.

Illicit Drugs The recommendation to avoid drugs during pregnancy also includes illicit drugs, of course. Unfortunately, use of illicit drugs, such as cocaine and marijuana, is common among some pregnant women.

Drugs of abuse, such as cocaine, easily cross the placenta and impair fetal growth and development. Furthermore, they are responsible for preterm births, low-birthweight infants, perinatal deaths, ◆ and sudden infant deaths. If these newborns survive, central nervous system damage is evident: their cries, sleep, and behaviors early in life are abnormal, and their cognitive development later in life is impaired.[56] They may be hypersensitive or underaroused; those who test positive for drugs suffer the greatest effects of toxicity and withdrawal.

Smoking and Chewing Tobacco Unfortunately, an estimated one out of nine pregnant women in the United States smokes, with higher rates for older teens.[57] Smoking cigarettes and chewing tobacco at any time exert harmful effects, and pregnancy dramatically magnifies the hazards of these practices. Smoking restricts the blood supply to the growing fetus and thus limits oxygen and nutrient delivery and waste removal. A mother who smokes is more likely to have a complicated birth and a low-birthweight infant. Indeed, of all preventable causes of low birthweight in the United States, smoking is at the top of the list. Although, most infants born to cigarette smokers are low birthweight, some are not, suggesting that the effect of smoking on birthweight also depends, in part, on genes involved in the metabolism of smoking toxins.[58]

◆ Reminder: The word *teratogenic* describes a factor that causes abnormal fetal development and birth defects.

◆ The word *perinatal* refers to the time between the 28th week of gestation and 1 month after birth.

Down syndrome: a genetic abnormality that causes mental retardation, short stature, and flattened facial features.

In addition to contributing to low birthweight, smoking interferes with lung growth and increases the risks of respiratory infections and childhood asthma.[59] It can also cause death in an otherwise healthy fetus or newborn. A positive relationship exists between **sudden infant death syndrome (SIDS)** and both cigarette smoking during pregnancy and postnatal exposure to passive smoke.[60] Smoking during pregnancy may even harm the intellectual and behavioral development of the child later in life. The margin ◆ lists other complications of smoking during pregnancy.

Infants of mothers who chew tobacco also have low birthweights and high rates of fetal deaths. Any woman who smokes cigarettes or chews tobacco and is considering pregnancy or who is already pregnant should try to quit.

Environmental Contaminants Proving that environmental contaminants cause reproductive damage is difficult, but evidence in wildlife is established and seems likely for human beings.[61] Infants and young children of pregnant women exposed to environmental contaminants such as lead show signs of delayed mental and psychomotor development. During pregnancy, lead readily moves across the placenta, inflicting severe damage on the developing fetal nervous system.[62] In addition, infants exposed to even low levels of lead during gestation weigh less at birth and consequently struggle to survive. For these reasons, it is particularly important that pregnant women receive foods and beverages grown and prepared in environments free of contamination. A diet high in calcium will help to defend against lead contamination, and breastfeeding may help to counterbalance developmental damage incurred from contamination during pregnancy.[63]

Mercury is among the contaminants of concern. As Chapter 5 mentioned, fatty fish are a good source of omega-3 fatty acids, but some fish contain large amounts of the pollutant mercury, which can harm the developing brain and nervous system.[64] Because the benefits of moderate fish consumption outweigh the risks, pregnant (and lactating) women should do the following:[65]

- Avoid shark, swordfish, king mackerel, and tilefish (also called golden snapper or golden bass).
- Limit average weekly consumption to 12 ounces (cooked or canned) of seafood *or* to 6 ounces (cooked or canned) of white (albacore) tuna.

Supplements of fish oil are not recommended because they may contain concentrated toxins and because their effects on pregnancy remain unknown.

Foodborne Illness As Highlight 18 explains, foodborne illnesses arise when people eat foods that contain infectious microbes or microbes that produce toxins. At best, the vomiting and diarrhea associated with these illnesses can leave a pregnant woman exhausted and dehydrated; at worse, foodborne illnesses can cause meningitis, pneumonia, or even fetal death. Pregnant women are about 20 times more likely than other healthy adults to get the foodborne illness **listeriosis.** The margin ◆ presents tips to prevent listeriosis, and Highlight 18 includes precautions to minimize the risks of other common foodborne illness.

Dietary Guidelines for Americans 2005

- Pregnant women should not eat or drink unpasteurized milk, milk products, or juices; raw or undercooked eggs, meat, or poultry; or raw sprouts.
- Pregnant women should only eat certain deli meats and frankfurters that have been reheated to steaming hot.

Vitamin-Mineral Megadoses The pregnant woman who is trying to eat well may mistakenly assume that more is better when it comes to vitamin-mineral supplements. This is simply not true; many vitamins and minerals are toxic when taken

◆ Complications associated with smoking during pregnancy:
- Fetal growth retardation
- Low birthweight
- Complications at birth (prolonged final stage of labor)
- Mislocation of the placenta
- Premature separation of the placenta
- Vaginal bleeding
- Spontaneous abortion
- Fetal death
- Sudden infant death syndrome (SIDS)
- Middle ear diseases
- Cardiac and respiratory diseases

◆ Listeriosis can be prevented in the following ways:
- Use only pasteurized juices and dairy products; avoid Mexican soft cheeses, feta cheese, brie, Camembert, and blue-veined cheeses such as Roquefort.
- Thoroughly cook meat, poultry, eggs, and seafood.
- Thoroughly reheat hot dogs, luncheon meats, and deli meats, including cured meats such as salami.
- Wash all fruits and vegetables.
- Avoid refrigerated pâté, meat spreads, smoked seafood such as salmon or trout, and any fish labeled "nova," "lox," or "kippered," unless prepared in a cooked dish.

sudden infant death syndrome (SIDS): the unexpected and unexplained death of an apparently well infant; the most common cause of death of infants between the second week and the end of the first year of life; also called *crib death*.

listeriosis: an infection caused by eating food contaminated with the bacterium *Listeria monocytogenes*, which can be killed by pasteurization and cooking but can survive at refrigerated temperatures; certain ready-to-eat foods, such as hot dogs and deli meats, may become contaminated after cooking or processing, but before packaging.

in excess. Excessive vitamin A is particularly infamous for its role in malformations of the cranial nervous system. Intakes before the seventh week appear to be the most damaging. (Review Figure 14-4 on p. 480 to see how many tissues are in their critical periods prior to the seventh week.) For this reason, vitamin A supplements are not given during pregnancy unless there is specific evidence of deficiency, which is rare. A pregnant woman can obtain all the vitamin A and most of the other vitamins and minerals she needs by making wise food choices. She should take supplements only on the advice of a registered dietitian or physician.

Caffeine Caffeine crosses the placenta, and the developing fetus has a limited ability to metabolize it. Research studies have not proved that caffeine (even in high doses) causes birth defects in human infants (as it does in animals), but some evidence suggests that heavy use increases the risk of fetal death.[66] (In these studies, heavy caffeine use is defined as the equivalent of eight or more cups of coffee a day.) All things considered, it is most sensible to limit caffeine consumption to the equivalent of a cup of coffee or two 12-ounce cola beverages a day. (The caffeine contents of selected beverages, foods, and drugs are listed at the beginning of Appendix H.)

Weight-Loss Dieting Weight-loss dieting, even for short periods, is hazardous during pregnancy. Low-carbohydrate diets or fasts that cause ketosis deprive the fetal brain of needed glucose and may impair cognitive development. Such diets are also likely to lack other nutrients vital to fetal growth. Regardless of prepregnancy weight, pregnant women should never intentionally lose weight.

Sugar Substitutes Artificial sweeteners have been extensively investigated and found to be acceptable during pregnancy if used within the FDA's guidelines (presented in Highlight 4).[67] Still, it is prudent for pregnant women to use sweeteners in moderation and within an otherwise nutritious and well-balanced diet. Women with phenylketonuria should not use aspartame, as Highlight 4 explained.

IN SUMMARY

High-risk pregnancies, especially for teenagers, threaten the life and health of both mother and infant. Proper nutrition and abstinence from smoking, alcohol, and other drugs improve the outcome. In addition, prenatal care includes monitoring pregnant women for gestational diabetes and preeclampsia.

In general, the following guidelines will allow most women to enjoy a healthy pregnancy:[68]

- Get prenatal care.
- Eat a balanced diet, safely prepared.
- Take prenatal supplements as prescribed.
- Gain a healthy amount of weight.
- Refrain from cigarettes, alcohol, and drugs (including herbs, unless prescribed by a physician).

Childbirth marks the end of pregnancy and the beginning of a new set of parental responsibilities—including feeding the newborn.

Nutrition during Lactation

◆ To learn about breastfeeding, a pregnant woman can read at least one of the many books available. At the end of this chapter, Nutrition on the Net provides a list of resources, including LaLeche League International.

Before the end of her pregnancy, a woman needs to consider whether to feed her infant breast milk, ◆ infant formula, or both. These options are the only recommended foods for an infant during the first four to six months of life. The rate of breastfeeding is close to the Healthy People 2010 goal of 75 percent at birth, but it falls far short of goals at six months and a year.[69] This section focuses on how the mother's nutrition supports the making of breast milk, and the next chapter describes how the infant benefits from drinking breast milk.

In many countries around the world, a woman breastfeeds her newborn without considering the alternatives or making a conscious decision. In other parts of the world, a woman feeds her newborn formula simply because she knows so little about breastfeeding. She may have misconceptions or feel uncomfortable about a process she has never seen or experienced. Breastfeeding offers many health benefits to both mother and infant, and every pregnant woman should seriously consider it (see Table 14-4).[70] Even so, there are sometimes valid reasons for not breastfeeding, and formula-fed infants grow and develop into healthy children.

Lactation: A Physiological Process

Lactation naturally follows pregnancy, as the mother's body continues to nourish the infant. The **mammary glands** secrete milk for this purpose. The mammary glands develop during puberty but remain fairly inactive until pregnancy. During pregnancy, hormones promote the growth and branching of a duct system in the breasts and the development of the milk-producing cells.

The hormones **prolactin** and **oxytocin** finely coordinate lactation. The infant's demand for milk stimulates the release of these hormones, which signal the mammary glands to supply milk. Prolactin is responsible for milk production. As long as the infant is nursing, prolactin concentrations remain high, and milk production continues.

The hormone oxytocin causes the mammary glands to eject milk into the ducts, a response known as the **let-down reflex.** The mother feels this reflex as a contraction of the breast, followed by the flow of milk and the release of pressure. By relaxing and eating well, the nursing mother promotes easy let-down of milk and greatly enhances her chances of successful lactation.

A women who decides to breastfeed offers her infant a full array of nutrients and protective factors to support optimal health and development.

TABLE 14-4 Benefits of Breastfeeding

For Infants:

- Provides the appropriate composition and balance of nutrients with high bioavailability
- Provides hormones that promote physiological development
- Improves cognitive development
- Protects against a variety of infections
- May protect against some chronic diseases, such as diabetes (both types), obesity, atherosclerosis, asthma, and hypertension, later in life
- Protects against food allergies

For Mothers:

- Contracts the uterus
- Delays the return of regular ovulation, thus lengthening birth intervals (is not, however, a dependable method of contraception)
- Conserves iron stores (by prolonging amenorrhea)
- May protect against breast and ovarian cancer and reduce the risk of diabetes (type 2)

Other:

- Cost savings from not needing medical treatment for childhood illnesses or time off work to care for them
- Cost savings from not needing to purchase formula (even after adjusting for added foods in the diet of a lactating mother)[a]
- Environmental savings to society from not needing to manufacture, package, and ship formula and dispose of the packaging
- Convenience of not having to shop for and prepare formula

[a]A nursing mother produces more than 35 gallons of milk during the first six months, saving roughly $450 in formula costs.

---

lactation: production and secretion of breast milk for the purpose of nourishing an infant.

mammary glands: glands of the female breast that secrete milk.

prolactin (pro-LAK-tin): a hormone secreted from the anterior pituitary gland that acts on the mammary glands to promote the production of milk. The release of prolactin is mediated by **prolactin-inhibiting hormone (PIH).**

- **pro** = promote
- **lacto** = milk

oxytocin (OCK-see-TOH-sin): a hormone that stimulates the mammary glands to eject milk during lactation and the uterus to contract during childbirth.

let-down reflex: the reflex that forces milk to the front of the breast when the infant begins to nurse.

Breastfeeding: A Learned Behavior

Lactation is an automatic physiological process that virtually all mothers are capable of doing. Breastfeeding, on the other hand, is a learned behavior that not all mothers decide to do. Of women who do breastfeed, those who receive early and repeated information and support breastfeed their infants longer than others. Health care professionals  play an important role in providing encouragement and accurate information on breastfeeding.[71] Women who have been successful breastfeeding can offer advice and dispel misperceptions about lifestyle issues. Table 14-5 lists ten steps maternity facilities and health care professionals can take to promote successful breastfeeding among new mothers.[72]

◆ Some hospitals employ *certified lactation consultants* who specialize in helping new mothers establish a healthy breastfeeding relationship with their newborn. These consultants are often registered nurses with specialized training in breast and infant anatomy and physiology.

The mother's partner also plays an important role in encouraging breastfeeding.[73] When partners support the decision, mothers are more likely to start and continue breastfeeding. Clearly, educating those closest to the mother could change attitudes and promote breastfeeding.

Most healthy women who want to breastfeed can do so with a little preparation. Physical obstacles to breastfeeding are rare, although most nursing mothers quit before the recommended six months because of perceived difficulties.[74] Obese mothers seem to have a particularly difficult time, perhaps because of reduced prolactin levels.[75] Successful breastfeeding requires adequate nutrition and rest. This, plus the support of all who care, will help to enhance the well-being of mother and infant.

Maternal Energy and Nutrient Needs during Lactation

Ideally, the mother who chooses to breastfeed her infant will continue to eat nutrient-dense foods throughout lactation. An adequate diet is needed to support the stamina, patience, and self-confidence that nursing an infant demands.

Energy Intake and Exercise A nursing mother produces about 25 ounces of milk per day, with considerable variation from woman to woman and in the same woman from time to time, depending primarily on the infant's demand for milk. To produce an adequate supply of milk, a woman needs extra energy—almost 500

A jog through the park provides an opportunity for physical activity and fresh air.

TABLE 14-5	Ten Steps to Successful Breastfeeding

To promote breastfeeding, every maternity facility should:

- Develop a written breastfeeding policy that is routinely communicated to all health care staff
- Train all health care staff in the skills necessary to implement the breastfeeding policy
- Inform all pregnant women about the benefits and management of breastfeeding
- Help mothers initiate breastfeeding within ½ hour of birth
- Show mothers how to breastfeed and how to maintain lactation, even if they need to be separated from their infants
- Give newborn infants no food or drink other than breast milk, unless medically indicated
- Practice rooming-in, allowing mothers and infants to remain together 24 hours a day
- Encourage breastfeeding on demand
- Give no artificial nipples or pacifiers to breastfeeding infants[a]
- Foster the establishment of breastfeeding support groups and refer mothers to them at discharge from the facility

[a]Compared with nonusers, infants who use pacifiers breastfeed less frequently and stop breastfeeding at a younger age. C. G. Victora and coauthors, Pacifier use and short breastfeeding duration: Cause, consequence, or coincidence? *Pediatrics* 99 (1997): 445–453.
SOURCE: United Nations Children's Fund and World Health Organization, *Protecting, Promoting and Supporting Breastfeeding: The Special Role of Maternity Services.*

kcalories a day above her regular need during the first six months of lactation. To meet this energy need, ◆ she can eat an extra 330 kcalories of food each day and let the fat reserves she accumulated during pregnancy provide the rest. Most women need at least 1800 kcalories a day to receive all the nutrients required for successful lactation. Severe energy restriction may hinder milk production.

After the birth of the infant, many women actively try to lose the extra weight and body fat they accumulated during pregnancy.[76] Opinions differ as to whether breastfeeding helps with postpartum weight loss. Lactating women may lose body *fat* more slowly than nonlactating women, but the rate of *weight* loss is about the same.[77] In general, most women lose one to two pounds a month during the first four to six months of lactation; some may lose more, and others may maintain or even gain weight. Neither the quality nor the quantity of breast milk is adversely affected by moderate weight loss, and infants grow normally.

◆ Energy requirement during lactation:
• 1st 6 mo: +330 kcal/day
• 2nd 6 mo: +400 kcal/day

Dietary Guidelines for Americans 2005

Moderate weight reduction is safe for breastfeeding women and does not compromise weight gain of the nursing infant.

Women often exercise to lose weight and improve fitness, and this is compatible with breastfeeding and infant growth. Because intense physical activity can raise the lactate concentration of breast milk and influence the milk's taste, some infants may prefer milk produced prior to exercise. In these cases, mothers can either breastfeed before exercise or express their milk before exercise for use afterward.

Dietary Guidelines for Americans 2005

Neither acute nor regular exercise adversely affects the mother's ability to successfully breastfeed.

Energy Nutrients Recommendations for protein and fatty acids intakes remain about the same during lactation as during pregnancy, but they increase for carbohydrates and fibers. Nursing mothers need additional carbohydrate to replace the glucose used to make the lactose in breast milk. The fiber recommendation is 1 gram higher simply because it is based on kcalorie intake, which increases during lactation.

Vitamins and Minerals A question often raised is whether a mother's milk may lack a nutrient if she fails to get enough in her diet. The answer differs from one nutrient to the next, but in general, nutritional inadequacies reduce the *quantity*, not the *quality*, of breast milk. Women can produce milk with adequate protein, carbohydrate, fat, and most minerals, even when their own supplies are limited. For these nutrients and for the vitamin folate as well, milk quality is maintained at the expense of maternal stores. This is most evident in the case of calcium: dietary calcium has no effect on the calcium concentration of breast milk, but maternal bones lose some density during lactation if calcium intakes are inadequate.[78] Bone density increases again when lactation ends; breastfeeding has no long-term harmful effects on bones.[79] The nutrients in breast milk that are most likely to decline in response to prolonged inadequate intakes are the vitamins—especially vitamins B_6, B_{12}, A, and D. Review Figure 14-10 (p. 489) to compare a lactating woman's nutrient needs with those of pregnant and nonpregnant women.

Water Despite misconceptions, a mother who drinks more fluid does not produce more breast milk. To protect herself from dehydration, however, a lactating woman

Nutritious foods support successful lactation.

◆ AI for *total* water (including drinking water, other beverages, and foods) during lactation: 3.8 L/day

needs to drink plenty of fluids. ◆ A sensible guideline is to drink a glass of milk, juice, or water at each meal and each time the infant nurses.

Nutrient Supplements Most lactating women can obtain all the nutrients they need from a well-balanced diet without taking vitamin-mineral supplements. Nevertheless, some may need iron supplements, not to enhance the iron in their breast milk, but to refill their depleted iron stores. The mother's iron stores dwindle during pregnancy as she supplies the developing fetus with enough iron to last through the first four to six months of the infant's life. In addition, childbirth may have incurred blood losses. Thus a woman may need iron supplements during lactation even though, until menstruation resumes, her iron requirement is about half that of other nonpregnant women her age.

Food Assistance Programs In general, women most likely to participate in the food assistance program WIC—those who are poor and have little education—are less likely to breastfeed. Furthermore, WIC provides infant formula at no cost. Because WIC recognizes the many benefits of breastfeeding, efforts are made to overcome this dilemma. In addition to nutrition education, breastfeeding mothers receive the following WIC incentives:

- Higher priority in certification into WIC
- Longer eligibility to participate in WIC
- More foods and larger quantities
- Breast pumps and other support materials

Together, these efforts help to provide nutrition support and encourage WIC mothers to breastfeed.

Particular Foods Foods with strong or spicy flavors (such as garlic) may alter the flavor of breast milk. A sudden change in the taste of the milk may annoy some infants. Familiar flavors may enhance enjoyment.

Infants who develop symptoms of food allergy may be more comfortable if the mother's diet excludes the most common offenders—cow's milk, eggs, fish, peanuts, and tree nuts. Generally, infants with a strong family history of food allergies benefit from breastfeeding.

A nursing mother can usually eat whatever nutritious foods she chooses. If she suspects a particular food is causing the infant discomfort, her physician may recommend a dietary challenge: eliminate the food from the diet to see if the infant's reactions subside; then return the food to the diet, and again monitor the infant's reactions. If a food must be eliminated for an extended time, appropriate substitutions must be made to ensure nutrient adequacy.

Maternal Health

If a woman has an ordinary cold, she can continue nursing without worry. If susceptible, the infant will catch it from her anyway. (Thanks to the immunological protection of breast milk, the baby may be less susceptible than a formula-fed baby would be.) With appropriate treatment, a woman who has an infectious disease such as tuberculosis or hepatitis can breastfeed; transmission is rare.[80] Women with HIV (human immunodeficiency virus) infections, however, should consider other options.

HIV Infection and AIDS Mothers with HIV infections can transmit the virus (which causes AIDS) to their infants through breast milk, especially during the early months of breastfeeding.[81] Where safe alternatives are available, HIV-positive women should *not* breastfeed their infants. In developing countries, where the feeding of inappropriate or contaminated formulas causes 1.5 million infant deaths each year, the decision is less obvious. To prevent the mother-to-child transmission of HIV, WHO and UNICEF urge mothers in developing countries *not* to breastfeed. However, they stress the importance of finding suitable feeding alternatives to prevent the malnutrition, disease, and death that commonly occur when women in these countries do not breastfeed.

Diabetes Women with diabetes (type 1) may need careful monitoring and counseling to ensure successful lactation. These women need to adjust their energy intakes and insulin doses to meet the heightened needs of lactation. Maintaining good glucose control helps to initiate lactation and support milk production.

Postpartum Amenorrhea Women who breastfeed experience prolonged **postpartum amenorrhea.** Absent menstrual periods, however, do not protect a woman from pregnancy. To prevent pregnancy, a couple must use some form of contraception. Breastfeeding women who use oral contraceptives should use progestin-only agents for at least the first six months.[82] Estrogen-containing oral contraceptives reduce the volume and the protein content of breast milk.

Breast Health Some women fear that breastfeeding will cause their breasts to sag. The breasts do swell and become heavy and large immediately after the birth, but even when they produce enough milk to nourish a thriving infant, they eventually shrink back to their prepregnant size. Given proper support, diet, and exercise, breasts often return to their former shape and size when lactation ends. Breasts change their shape as the body ages, but breastfeeding does not accelerate this process.

Whether the physical and hormonal events of pregnancy and lactation protect women from later breast cancer is an area of active research.[83] Some research suggests no association between breastfeeding and breast cancer, whereas other research suggests a protective effect. Protection against breast cancer is most apparent for premenopausal women who were young when they breastfed and who breastfed for a long time.

Practices Incompatible with Lactation

Some substances impair milk production or enter breast milk and interfere with infant development. This section discusses practices that a breastfeeding mother should avoid.

Alcohol Alcohol easily enters breast milk, and its concentration peaks within an hour of ingestion. Infants drink less breast milk when their mothers have consumed even small amounts of alcohol (equivalent to a can of beer). Three possible reasons, acting separately or together, may explain why. For one, the alcohol may have altered the flavor of the breast milk and thereby the infants' acceptance of it. For another, because infants metabolize alcohol inefficiently, even low doses may be potent enough to suppress their feeding and cause sleepiness. Third, the alcohol may have interfered with lactation by inhibiting the hormone oxytocin.

In the past, alcohol has been recommended to mothers to facilitate lactation despite a lack of scientific evidence that it does so. The research summarized here suggests that alcohol actually hinders breastfeeding. An occasional alcoholic beverage may be within safe limits, but breastfeeding should be avoided for at least two hours afterwards.

Medicinal Drugs Most medicines are compatible with breastfeeding, but some are contraindicated, either because they suppress lactation or because they are secreted into breast milk and can harm the infant.[84] As a precaution, a nursing mother should consult with her physician prior to taking any drug, including herbal supplements.

Illicit Drugs Illicit drugs, of course, are harmful to the physical and emotional health of both the mother and the nursing infant. Breast milk can deliver such high doses of illicit drugs as to cause irritability, tremors, hallucinations, and even death in infants. Women whose infants have overdosed on illicit drugs contained in breast milk have been convicted of murder.

postpartum amenorrhea: the normal temporary absence of menstrual periods immediately following childbirth.

Smoking Because cigarette smoking reduces milk volume, smokers may produce too little milk to meet their infants' energy needs. The milk they do produce contains nicotine, which alters its smell and flavor. Consequently, infants of breastfeeding mothers who smoke gain less weight than infants of those who do not smoke. Furthermore, infant exposure to passive smoke negates the protective effect breastfeeding offers against SIDS and increases the risks dramatically.

Environmental Contaminants Some environmental contaminants in the food supply, such as DDT, PCBs, and dioxin, can find their way into breast milk. Inuit mothers living in Arctic Québec who eat seal and beluga whale blubber have high concentrations of DDT and PCBs in their breast milk, but the impact on infant development is unclear. Preliminary studies indicate that the children of these Inuit mothers are developing normally. Researchers speculate that the abundant omega-3 fatty acids of the Inuit diet may protect against damage to the central nervous system. Breast milk tainted with dioxins interferes with tooth development during early infancy, producing soft, mottled teeth that are vulnerable to dental caries. To limit mercury intake, lactating women should heed the fish restrictions mentioned earlier for pregnant women (see p. 499).

Caffeine Caffeine enters breast milk and may make an infant irritable and wakeful. As during pregnancy, caffeine consumption should be moderate—the equivalent of one to two cups of coffee a day. Larger doses of caffeine may interfere with the bioavailability of iron from breast milk and impair the infant's iron status.

IN SUMMARY

The lactating woman needs extra fluid and enough energy and nutrients to produce about 25 ounces of milk a day. Breastfeeding is contraindicated for those with HIV/AIDS. Alcohol, other drugs, smoking, and contaminants may reduce milk production or enter breast milk and impair infant development.

This chapter has focused on the nutrition needs of the mother during pregnancy and lactation. The next chapter explores the dietary needs of infants, children, and adolescents.

CENGAGENOW™
academic.cengage.com/login

 Nutrition Portfolio

The choices a woman makes in preparation for, and in support of, pregnancy and lactation can influence both her health and her infant's development—today and for decades to come.

■ For women of childbearing age, determine whether you consume at least 400 micrograms of dietary folate equivalents daily.

■ For women who are pregnant, evaluate whether you are meeting your nutrition needs and gaining the amount of weight recommended.

■ For women who are about to give birth, carefully consider all the advantages of breastfeeding your infant and obtain the needed advice to support you.

NUTRITION ON THE NET

For further study of topics covered in this chapter, log on to **academic.cengage.com/nutrition/rolfes/UNCN8e**. Go to Chapter 14, then to Nutrition on the Net.

- Visit the pregnancy and child health center of the Mayo Clinic: **www.mayohealth.org**

- Learn more about having a healthy baby and about birth defects from the March of Dimes and the National Center on Birth Defects and Developmental Disabilities: **www.modimes.org** and **www.cdc.gov/ncbddd**

- Learn more about neural tube defects from the Spina Bifida Association of America: **www.sbaa.org**

- Search for "birth defects," "pregnancy," "adolescent pregnancy," "maternal and infant health," and "breastfeeding" at the U.S. Government health information site: **www.healthfinder.gov**

- Search for "pregnancy" at the American Dietetic Association site: **www.eatright.org**

- Learn more about the WIC program: **www.fns.usda.gov/fns**

- Visit the American College of Obstetricians and Gynecologists: **www.acog.org**

- Learn more about gestational diabetes from the American Diabetes Association: **www.diabetes.org**

- Learn more about breastfeeding from LaLeche League International: **www.lalecheleague.org**

- Obtain prenatal nutrition guidelines from Health Canada: **www.hc-sc.gc.ca**

STUDY QUESTIONS

CENGAGENOW™
To assess your understanding of chapter topics, take the Student Practice Test and explore the modules recommended in your Personalized Study Plan. Log on to **academic.cengage.com/login.**

These questions will help you review the chapter. You will find the answers in the discussions on the pages provided.

1. Describe the placenta and its function. (p. 478–479)

2. Describe the normal events of fetal development. How does malnutrition impair fetal development? (pp. 478–480, 495)

3. Define the term *critical period*. How do adverse influences during critical periods affect later health? (pp. 480–483)

4. Explain why women of childbearing age need folate in their diets. How much is recommended, and how can women ensure that these needs are met? (pp. 481–483)

5. What is the recommended pattern of weight gain during pregnancy for a woman at a healthy weight? For an underweight woman? For an overweight woman? (pp. 484–486)

6. What does a pregnant woman need to know about exercise? (pp. 486–487)

7. Which nutrients are needed in the greatest amounts during pregnancy? Why are they so important? Describe wise food choices for the pregnant woman. (pp. 487–491)

8. Define low-risk and high-risk pregnancies. What is the significance of infant birthweight in terms of the child's future health? (pp. 493–495)

9. Describe some of the special problems of the pregnant adolescent. Which nutrients are needed in increased amounts? (p. 497)

10. What practices should be avoided during pregnancy? Why? (pp. 498–500)

11. How do nutrient needs during lactation differ from nutrient needs during pregnancy? (pp. 489, 502–504)

These multiple choice questions will help you prepare for an exam. Answers can be found on p. 510.

1. The spongy structure that delivers nutrients to the fetus and returns waste products to the mother is called the:
   a. embryo.
   b. uterus.
   c. placenta.
   d. amniotic sac.

2. Which of these strategies is *not* a healthy option for an overweight woman?
   a. Limit weight gain during pregnancy.
   b. Postpone weight loss until after pregnancy.
   c. Follow a weight-loss diet during pregnancy.
   d. Try to achieve a healthy weight before becoming pregnant.

3. A reasonable weight gain during pregnancy for a normal-weight woman is about:
   a. 10 pounds.
   b. 20 pounds.
   c. 30 pounds.
   d. 40 pounds.

4. Energy needs during pregnancy increase by about:
   a. 100 kcalories/day.
   b. 300 kcalories/day.
   c. 500 kcalories/day.
   d. 700 kcalories/day.

5. To help prevent neural tube defects, grain products are now fortified with:
   a. iron.
   b. folate.
   c. protein.
   d. vitamin C.

6. Pregnant women should *not* take supplements of:
   a. iron.
   b. folate.
   c. vitamin A.
   d. vitamin C.

7. The combination of high blood pressure, protein in the urine, and edema signals:
   a. jaundice.
   b. preeclampsia.
   c. gestational diabetes.
   d. gestational hypertension.

8. To facilitate lactation, a mother needs:
   a. about 5000 kcalories a day.
   b. adequate nutrition and rest.
   c. vitamin and mineral supplements.
   d. a glass of wine or beer before each feeding.

9. A breastfeeding woman should drink plenty of water to:
   a. produce more milk.
   b. suppress lactation.
   c. prevent dehydration.
   d. dilute nutrient concentration.

10. A woman may need iron supplements during lactation:
    a. to enhance the iron in her breast milk.
    b. to provide iron for the infant's growth.
    c. to replace the iron in her body's stores.
    d. to support the increase in her blood volume.

REFERENCES

1. Recommendations to improve preconception health and health care—United States, *Morbidity and Mortality Weekly Report* 55 (2006): 1–23.
2. M. J. Davies, Evidence for effects of weight on reproduction in women, *Reproductive Biomedicine Online* 12 (2006): 552–561.
3. M. Sallmen and coauthors, Reduced fertility among overweight and obese men, *Epidemiology* 17 (2006): 520–523; H. I. Kort and coauthors, Impact of body mass index values on sperm quantity and quality, *Journal of Andrology* 27 (2006): 450–452.
4. R. Pasquali and A. Gambineri, Metabolic effects of obesity on reproduction, *Reproductive Biomedicine Online* 12 (2006): 542–551.
5. R. M. Sharpe and S. Franks, Environment, lifestyle and infertility—An inter-generational issue, *Nature Cell Biology* 4 (2002): s33–s40.
6. J. C. Cross and L. Mickelson, Nutritional influences on implantation and placental development, *Nutrition Reviews* 64 (2006): S12–S18.
7. R. Padmanabhan, Etiology, pathogenesis and prevention of neural tube defects, *Congenital Anomalies* 46 (2006): 55–67.
8. K. A. Bol, J. S. Collins, and R. S. Kirby, Survival of infants with neural tube defects in the presence of folic acid fortification, *Pediatrics* 117 (2006): 803–813; T. Tamura and M. F. Picciano, Folate and human reproduction, *American Journal of Clinical Nutrition* 83 (2006): 993–1016; L. B. Bailey and R. J. Berry, Folic acid supplementation and the occurrence of congenital heart defects, orofacial clefts, multiple births, and miscarriage, *American Journal of Clinical Nutrition* 81 (2005): 1213S–1217S.
9. Spina bifida and anencephaly before and after folic acid mandate—United States, 1995–1996 and 1999–2000, *Morbidity and Mortality Weekly Report* 53 (2004): 362–365; J. Erickson, Folic acid and prevention of spina bifida and anencephaly, *Morbidity and Mortality Weekly Report* 51 (2002): 1–3.
10. R. L. Brent and G. P. Oakley, The folate debate, *Pediatrics* 117 (2006): 1418–1419; J. I. Rader and B. O. Schneeman, Prevalence of neural tube defects, folate status, and folate fortification of enriched cereal-grain products in the United States, *Pediatrics* 117 (2006): 1394–1399.
11. M. Hanson and coauthors, Report on the 2nd World Congress on Fetal Origins of Adult Disease, Brighton, U.K., June 7–10, 2003, *Pediatric Research* 55 (2004): 894–897; G. Wu and coauthors, Maternal nutrition and fetal development, *Journal of Nutrition* 134 (2004): 2169–2172; C. N. Hales and S. E. Ozanne, For debate: Fetal and early postnatal growth restriction lead to diabetes, the metabolic syndrome and renal failure, *Diabetologia* 46 (2003): 1013–1019.
12. S. E. Moore and coauthors, Birth weight predicts response to vaccination in adults born in an urban slum in Lahore, Pakistan, *American Journal of Clinical Nutrition* 80 (2004): 453–459; B. E. Birgisdottir and coauthors, Size at birth and glucose intolerance in a relatively genetically homogeneous, high-birth weight population, *American Journal of Clinical Nutrition* 76 (2002): 399–403.
13. R. C. Painter and coauthors, Early onset of coronary artery disease after prenatal exposure to the Dutch famine, *American Journal of Clinical Nutrition* 84 (2006): 322–327; O. A. Kensara and coauthors, Fetal programming of body composition: Relation between birth weight and body composition measured with dual-energy X-ray absorptiometry and anthropometric methods in older Englishmen, *American Journal of Clinical Nutrition* 82 (2005): 980–987; P. Szitányi, J. Janda, and R. Poledne, Intrauterine undernutrition and programming as a new risk of cardiovascular disease in later life, *Physiological Research* 52 (2003): 389–395; A. Singhal and coauthors, Programming of lean body mass: A link between birth weight, obesity, and cardiovascular disease? *American Journal of Clinical Nutrition* 77 (2003): 726–730.
14. G. Wolf, Adult type 2 diabetes induced by intrauterine growth retardation, *Nutrition Reviews* 61 (2003): 176–179.
15. P. L. Hofman and coauthors, Premature birth and later insulin resistance, *New England Journal of Medicine* 351 (2004): 2179–2186; M. A. Sperling, Prematurity—A window of opportunity? *New England Journal of Medicine* 351 (2004): 2229–2231.
16. Hofman and coauthors, 2004; Sperling, 2004.
17. L. Adair and D. Dahly, Developmental determinants of blood pressure in adults, *Annual Review of Nutrition* 25 (2005): 407–434; K. M. Moritz, M. Dodic, and E. M. Wintour, Kidney development and the fetal programming of adult disease, *Bioessays* 25 (2003): 212–220; M. Symonds and coauthors, Maternal nutrient restriction during placental growth, programming of fetal adiposity and juvenile blood pressure control, *Archives of Physiology and Biochemistry* 111 (2003): 45–52.
18. Adair and Dahly, 2005; C. M. Law and coauthors, Fetal, infant, and childhood growth and adult blood pressure: A longitudinal study from birth to 22 years of age, *Circulation* 105 (2002): 1088–1092.
19. R. A. Waterland and R. L. Jirtle, Transposable elements: Targets for early nutritional effects on epigenetic gene regulation, *Molecular and Cellular Biology* 23 (2003): 5293–5300.
20. A. J. Drake and B. R. Walker, The intergenerational effects of fetal programming: Nongenomic mechanisms for the inheritance of low birth weight and cardiovascular risk, *Journal of Endocrinology* 180 (2004): 1–16.
21. D. B. Sarwer and coauthors, Pregnancy and obesity: A review and agenda for future research, *Journal of Women's Health* 15 (2006): 720–733; T. Henriksen, Nutrition and pregnancy outcome, *Nutrition Reviews* 64 (2006): S19–S23; J. C. King, Maternal obesity, metabolism, and pregnancy outcomes, *Annual Review of Nutrition* 26 (2006): 271–291.
22. T. K. Young and B. Woodmansee, Factors that are associated with cesarean delivery in a large private practice: The importance of pregnancy body mass index and weight gain, *American Journal of Obstetrics and Gynecology* 187 (2002): 312–318.
23. J. C. King, Maternal obesity, metabolism, and pregnancy outcomes, *Annual Review of Nutrition* 26 (2006): 271–291.
24. M. L. Watkins and coauthors, Maternal obesity and risk for birth defects, *Pediatrics* 111 (2003): 1152–1158.
25. M. E. Roselló-Soberón, L. Fuentes-Chaparro, and E. Casanueva, Twin pregnancies: Eating for three? Maternal nutrition update, *Nutrition Reviews* 63 (2005): 295–302.
26. D. A. Krummel, Postpartum weight control: A vicious cycle, *Journal of the American Dietetic Association* 107 (2007): 37–40; E. Villamor and S. Cnattingius, Interpregnancy weight change and risk of adverse pregnancy outcomes: A population-based study, *Lancet* 368 (2006): 1164–1170.

27. R. Artal and M. O'Toole, Guidelines of the American College of Obstetricians and Gynecologists for exercise during pregnancy and the postpartum period, *British Journal of Sports Medicine* 37 (2003): 6–12; Committee of Obstetric Practice, Exercise during pregnancy and the postpartum period, *Obstetrics and Gynecology* 99 (2002): 171–173.

28. M. Lof and coauthors, Changes in basal metabolic rate during pregnancy in relation to changes in body weight and composition, cardiac output, insulin-like growth factor I, and thyroid hormones and in relation to fetal growth, *American Journal of Clinical Nutrition* 81 (2005): 678–685; N. F. Butte and coauthors, Energy requirements during pregnancy based on total energy expenditure and energy deposition, *American Journal of Clinical Nutrition* 79 (2004): 1078–1087.

29. R. Uauy and A. D. Dangour, Nutrition in brain development and aging: Role of essential fatty acids, *Nutrition Reviews* 64 (2006): S24–S33.

30. Committee on Dietary Reference Intakes, *Dietary Reference Intakes for Thiamin, Riboflavin, Niacin, Vitamin B6, Folate, Vitamin B12, Pantothenic Acid, Biotin, and Choline* (Washington, D.C.: National Academy Press, 1998), pp. 196–305.

31. T. O. Scholl, Iron status during pregnancy: Setting the stage for mother and infant, *American Journal of Clinical Nutrition* 81 (2005): 1218S–1222S.

32. K. O. O'Brien and coauthors, Maternal iron status influences iron transfer to the fetus during the third trimester of pregnancy, *American Journal of Clinical Nutrition* 77 (2003): 924–930.

33. E. Casanueva and coauthors, Vitamin C supplementation to prevent premature rupture of the chorioamniotic membranes: A randomized trial, *American Journal of Clinical Nutrition* 81 (2005): 859–863.

34. C. M. Donangelo and coauthors, Zinc absorption and kinetics during pregnancy and lactation in Brazilian women, *American Journal of Clinical Nutrition* 82 (2005): 118–124.

35. D. Shah and H. P. S. Sachdev, Zinc deficiency in pregnancy and fetal outcome, *Nutrition Reviews* 64 (2006): 15–30.

36. N. Pawley and N. J. Bishop, Prenatal and infant predictors of bone health: The influence of vitamin D, *American Journal of Clinical Nutrition* 80 (2004): 1748S–1751S.

37. B. W. Hollis and C. L. Wagner, Assessment of dietary vitamin D requirements during pregnancy and lactation, *American Journal of Clinical Nutrition* 79 (2004): 717–726.

38. C. L. V. Zapata and coauthors, Calcium homeostasis during pregnancy and lactation in Brazilian women with low calcium intakes: A longitudinal study, *American Journal of Clinical Nutrition* 80 (2004): 417–422.

39. K. O. O'Brien and coauthors, Bone calcium turnover during pregnancy and lactation in women with low calcium diets is associated with calcium intake and circulating insulin-like growth factor 1 concentrations, *American Journal of Clinical Nutrition* 83 (2006): 317–323.

40. A. Vahratian and coauthors, Multivitamin use and the risk of preterm birth, *American Journal of Epidemiology* 160 (2004): 886–892.

41. Position of the American Dietetic Association and Dietitians of Canada: Vegetarian diets, *Journal of the American Dietetic Association* 103 (2003): 748–765.

42. R. W. Corbett, C. Ryan, and S. P. Weinrich, Pica in pregnancy: Does it affect pregnancy outcomes? *American Journal of Maternal Child Nursing* 28 (2003): 183–189.

43. L. H. Allen, Multiple micronutrients in pregnancy and lactation: An overview, *American Journal of Clinical Nutrition* 81 (2005): 1206S–1212S.

44. P. J. Surkan and coauthors, Previous preterm and small-for-gestational-age births and the subsequent risk of stillbirth, *New England Journal of Medicine* 350 (2004): 777–785.

45. R. J. Kaaja and I. A. Greer, Manifestations of chronic disease during pregnancy, *Journal of the American Medical Association* 294 (2005): 2751–2757.

46. W. van Wootten and R. E. Turner, Macrosomia in neonates of mothers with gestational diabetes is associated with body mass index and previous gestational diabetes, *Journal of the American Dietetic Association* 102 (2002): 241–243.

47. American Diabetes Association, Diagnosis and classification of diabetes mellitus, *Diabetes Care* 29 (2006): S43–S48; Report of the Expert Committee on the Diagnosis and Classification of Diabetes Mellitus, *Diabetes Care* 26 (2003): S5–S20.

48. Position statement from the American Diabetes Association: Gestational diabetes mellitus, *Diabetes Care* 26 (2003): S103–S105.

49. C. A. Crowther and coauthors, Effect of treatment of gestational diabetes mellitus on pregnancy outcomes, *New England Journal of Medicine* 352 (2005): 2477–2486; O. Langer and coauthors, Overweight and obese in gestational diabetes: The impact on pregnancy outcomes, *American Journal of Obstetrics and Gynecology* 192 (2005): 1768–1776.

50. C. G. Solomon and E. W. Seely, Preeclampsia—Searching for the cause, *New England Journal of Medicine* 350 (2004): 641–642.

51. C. B. Rudra and coauthors, Perceived exertion during prepregnancy physical activity and preeclampsia risk, *Medicine and Science in Sports and Exercise* 37 (2005): 1836–1841; T. L. Weissgerber, L. A. Wolfe, and G. A. L. Davies, The role of regular physical activity in preeclampsia prevention, *Medicine and Science in Sports and Exercise* 36 (2004): 2024–2031.

52. J. D. Klein and the Committee on Adolescence, Adolescent pregnancy: Current trends and issues, *Pediatrics* 116 (2005): 281–286.

53. R. J. Paulson and coauthors, Pregnancy in the sixth decade of life—Obstetric outcomes in women of advanced reproductive age, *Journal of the American Medical Association* 288 (2002): 2320–2323.

54. S. C. Tough and coauthors, Delayed childbearing and its impact on population rate changes in lower birth weight, multiple birth, and preterm delivery, *Pediatrics* 109 (2002): 399–403.

55. Alcohol consumption among women who are pregnant or who might become pregnant—United States, 2002, *Morbidity and Mortality Weekly Report* 53 (2004): 1178–1181.

56. L. T. Singer and coauthors, Cognitive and motor outcomes of cocaine-exposed infants, *Journal of the American Medical Association* 287 (2002): 1952–1960.

57. Smoking during pregnancy—United States, 1990–2002, *Morbidity and Mortality Weekly Report* 53 (2004): 911–915.

58. X. Wang and coauthors, Maternal cigarette smoking, metabolic gene polymorphism, and infant birth weight, *Journal of the American Medical Association* 287 (2002): 195–202.

59. J. R. DiFranza, C. A. Aligne, and M. Weitzman, Prenatal and postnatal environmental tobacco smoke exposure and children's health, *Pediatrics* 113 (2004): 1007–1015.

60. DiFranza, Aligne, and Weitzman, 2004.

61. R. L. Brent, S. Tanski, and M. Weitzman, A pediatric perspective on the unique vulnerability and resilience of the embryo and the child to environmental toxicants: The importance of rigorous research concerning age and agent, *Pediatrics* 113 (2004): 935–944; R. M. Sharpe and D. S. Irvine, How

strong is the evidence of a link between environmental chemicals and adverse effects on human reproductive health? *British Medical Journal* 328 (2004): 447–451.

62. A. Gomaa and coauthors, Maternal bone lead as an independent risk factor for fetal neurotoxicity: A prospective study, *Pediatrics* 110 (2002): 110–118.

63. N. Ribas-Fitó and coauthors, Breastfeeding, exposure to organochlorine compounds, and neurodevelopment in infants, *Pediatrics* 111 (2003): e580–e585.

64. S. E. Schober and coauthors, Blood mercury levels in US children and women of childbearing age, 1999–2000, *Journal of the American Medical Association* 289 (2003): 1667–1674.

65. D. Mozaffarian and E. B. Rimm, Fish intake, contaminants, and human health: Evaluating the risks and the benefits, *Journal of the American Medical Association* 296 (2006): 1885–1899; Institute of Medicine report brief, *Seafood Choices: Balancing Benefits and Risks,* October 2006.

66. B. H. Bech and coauthors, Coffee and fetal death: A cohort study with prospective data, *American Journal of Epidemiology* 162 (2005): 983–990.

67. Position of the American Dietetic Association: Use of nutritive and nonnutritive sweeteners, *Journal of the American Dietetic Association* 104 (2004): 255–275.

68. Position of the American Dietetic Association: Nutrition and lifestyle for a healthy pregnancy outcome, *Journal of the American Dietetic Association* 102 (2002): 1479–1490.

69. R. Li and coauthors, Breastfeeding rates in the United States by characteristics of the child, mother, or family: The 2002 National Immunization Survey, *Pediatrics* 115 (2005): e31; R. Li and coauthors, Prevalence of breastfeeding in the United States: The 2001 National Immunization Survey, *Pediatrics* 111 (2003): 1198–1201; A. S. Ryan, Z. Wenjun, and A. Acosta, Breastfeeding continues to increase into the new millennium, *Pediatrics* 110 (2002): 1103–1109.

70. American Academy of Pediatrics, Breastfeeding and the use of human milk, *Pediatrics* 115 (2005): 496–506; Position of the American Dietetic Association: Promoting and supporting breastfeeding, *Journal of the American Dietetic Association* 105 (2005): 810–818.

71. K. A. Bonuck and coauthors, Randomized, controlled trial of a prenatal and postnatal lactation consultant intervention on duration and intensity of breastfeeding up to 12 months, *Pediatrics* 116 (2005): 1413–1426; J. Labarere and coauthors, Efficacy of breastfeeding support provided by trained clinicians during an early, routine, preventive visit: A prospective, randomized, open trial of 226 Mother-infant pairs, *Pediatrics* 115 (2005): e139; E. M. Taveras and coauthors, Mothers' and clinicians' perspectives on breastfeeding counseling during routine preventive visits, *Pediatrics* 113 (2004): e405.

72. S. Merten, J. Dratva, and U. Ackermann-Liebrich, Do baby-friendly hospitals influence breastfeeding duration on a national level? *Pediatrics* 116 (2005): e702; A. Merewood and coauthors, Breastfeeding rates in US baby-friendly hospitals: Results of a national survey, *Pediatrics* 116 (2005): 628–634.

73. A. Pisacane and coauthors, A controlled trial of the father's role in breastfeeding promotion, *Pediatrics* 116 (2005): e494; C. L. Dennis, Breastfeeding initiation and duration: A 1990–2000 literature review, *Journal of Obstetric, Gynecologic and Neonatal Nursing* 31 (2002): 12–32.

74. Dennis, 2002.

75. C. A. Lovelady, Is maternal obesity a cause of poor lactation performance? *Nutrition Reviews* 63 (2005): 352–355.

76. D. A. Krummel and coauthors, Stages of change for weight management in postpartum women, *Journal of the American Dietetic Association* 104 (2004): 1102–1108.

77. K. S. Wosje and H. J. Kalkwarf, Lactation, weaning, and calcium supplementation: Effects on body composition in postpartum women, *American Journal of Clinical Nutrition* 80 (2004): 423–429.

78. K. O. O'Brien and coauthors, Bone calcium turnover during pregnancy and lactation in women with low calcium diets is associated with calcium intake and circulating insulin-like growth factor 1 concentrations, *American Journal of Clinical Nutrition* 83 (2006): 317–323.

79. F. F. Bezerra and coauthors, Bone mass is recovered from lactation to postweaning in adolescent mothers with low calcium intakes, *American Journal of Clinical Nutrition* 80 (2004): 1322–1326; L. M. Paton and coauthors, Pregnancy and lactation have no long-term deleterious effect on measures of bone mineral in healthy women: A twin study, *American Journal of Clinical Nutrition* 77 (2003): 707–714.

80. J. S. Wang, Q. R. Zhu, and X. H. Wang, Breastfeeding does not pose any additional risk of immunoprophylaxis failure on infants of HBV carrier mothers, *International Journal of Clinical Practice* 57 (2003): 100–102; J. B. Hill and coauthors, Risk of hepatitis B transmission in breast-fed infants of chronic hepatitis B carriers, *Obstetrics and Gynecology* 99 (2002): 1049–1052; M. L. Newell and L. Pembrey, Mother-to-child transmission of hepatitis C virus infection, *Drugs of Today* 38 (2002): 321–337.

81. J. S. Read and the Committee on Pediatric AIDS, human milk, breastfeeding, and transmission of human immunodeficiency virus type 1 in the United States, *Pediatrics* 112 (2003): 1196–1205.

82. R. Lesnewski and L. Prine, Initiating hormonal contraception, *American Family Physician* 74 (2006): 105–112.

83. S. Cnattingius and coauthors, Pregnancy characteristics and maternal risk of breast cancer, *Journal of the American Medical Association* 294 (2005): 2474–2480.

84. S. Ito and A. Lee, Drug excretion into breast milk—Overview, *Advanced Drug Delivery Reviews* 55 (2003): 617–627.

ANSWERS

1. c 2. c 3. c 4. b 5. b 6. c 7. b 8. b 9. c 10. c

Fetal Alcohol Syndrome

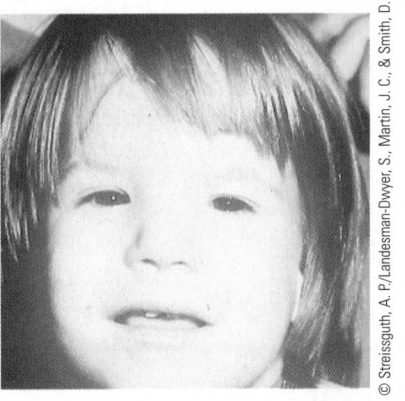

© Streissguth, A. P./Landesman-Dwyer, S., Martin, J. C., & Smith, D. W.

As Chapter 14 mentioned, drinking alcohol during pregnancy endangers the fetus. Alcohol crosses the placenta freely and deprives the developing fetus of both nutrients and oxygen. The damaging effects of alcohol on the developing fetus cover a range of abnormalities referred to as **fetal alcohol spectrum disorder** (see the glossary on p. 512).[1] Those at the most severe end of the spectrum are described as having **fetal alcohol syndrome (FAS)**, a cluster of physical, mental, and neurobehavioral symptoms that includes:

- Prenatal and postnatal growth retardation
- Impairment of the brain and central nervous system, with consequent mental retardation, poor motor skills and coordination, and hyperactivity
- Abnormalities of the face and skull (see Figure H14-1)
- Increased frequency of major birth defects: cleft palate, heart defects, and defects in ears, eyes, genitals, and urinary system

Tragically, the damage evident at birth persists: children with FAS never fully recover.[2]

Each year, as many as 6000 infants are born with FAS because their mothers drank too much alcohol during pregnancy.[3] In addition, some 4 million infants are born with **prenatal alcohol exposure.** The cluster of mental problems associated with prenatal alcohol exposure is known as **alcohol-related neurodevelopmental disorder (ARND),** and the physical malformations are referred to as **alcohol-related birth defects (ARBD).** Some children with ARBD and ARND have no outward signs; others may be short or have only minor facial abnormalities. They often go undiagnosed even when they develop learning difficulties in the early school years. Mood disorders and problem behaviors, such as aggression, are common.[4]

The surgeon general states that pregnant women should abstain from alcohol. Abstinence from alcohol is the best policy for pregnant women both because alcohol consumption during pregnancy has such severe consequences and because FAS can only be prevented—it cannot be treated. Further, because the most severe damage occurs around the time of conception—*before a woman may even realize that she is pregnant*—the warning to abstain includes women who may become pregnant.

Drinking during Pregnancy

As mentioned in Chapter 14, 1 out of 10 pregnant women drinks alcohol at some time during her pregnancy; 1 out of 50 uses alcohol frequently and admits to binge drinking.[5] When a woman drinks during pregnancy, she causes damage in two ways: directly, by intoxication, and indirectly, by malnutrition. Prior to the

FIGURE H14-1 Typical Facial Characteristics of FAS

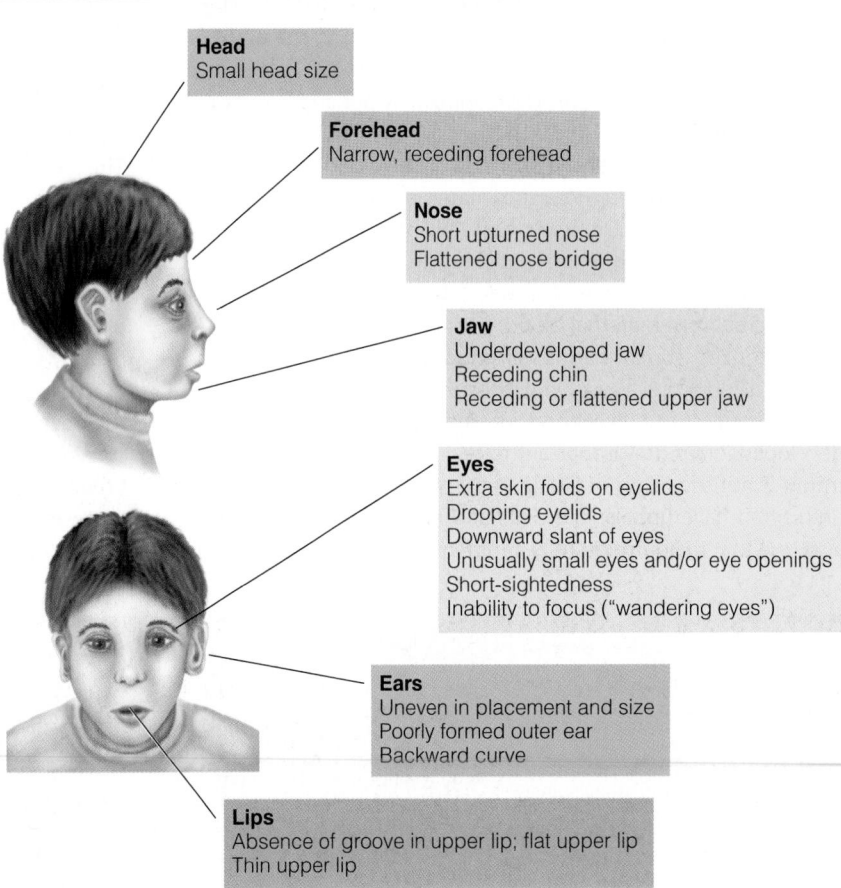

Head
Small head size

Forehead
Narrow, receding forehead

Nose
Short upturned nose
Flattened nose bridge

Jaw
Underdeveloped jaw
Receding chin
Receding or flattened upper jaw

Eyes
Extra skin folds on eyelids
Drooping eyelids
Downward slant of eyes
Unusually small eyes and/or eye openings
Short-sightedness
Inability to focus ("wandering eyes")

Ears
Uneven in placement and size
Poorly formed outer ear
Backward curve

Lips
Absence of groove in upper lip; flat upper lip
Thin upper lip

Characteristic facial features may diminish with time, but children with FAS typically continue to be short and underweight for their age.

complete formation of the placenta (approximately 12 weeks), alcohol diffuses directly into the tissues of the developing embryo, causing incredible damage. (Review Figure 14-4 on p. 480 and note that the critical periods for most tissues occur during embryonic development.) Alcohol interferes with the orderly development of tissues during their critical periods, reducing the number of cells and damaging those that are produced. The damage of alcohol toxicity during brain development is apparent in its reduced size and impaired function.[6]

When alcohol crosses the placenta, fetal blood alcohol rises until it reaches equilibrium with maternal blood alcohol. The mother may not even appear drunk, but the fetus may be poisoned. The fetus's body is small, its detoxification system is immature, and alcohol remains in fetal blood long after it has disappeared from maternal blood.

A pregnant woman harms her unborn child not only by consuming alcohol but also by not consuming food. This combination enhances the likelihood of malnutrition and a poorly developed infant. It is important to realize, however, that malnutrition is not the cause of FAS. It is true that mothers of FAS children often have unbalanced diets and nutrient deficiencies. It is also true that malnutrition may augment the clinical signs seen in

these children, but it is the *alcohol* that causes the damage. An adequate diet alone will not prevent FAS if alcohol abuse continues.

How Much Is Too Much?

A pregnant woman need not have an alcohol-abuse problem to give birth to a baby with FAS. She need only drink in excess of her liver's capacity to detoxify alcohol. Even one drink a day threatens neurological development and behaviors.[7] Four drinks a day dramatically increase the risk of having an infant with physical malformations.

In addition to total alcohol intake, drinking patterns play an important role. Most FAS studies report their findings in terms of average intake per day, but people usually drink more heavily on some days than on others. For example, a woman who drinks an *average* of 1 ounce of alcohol (2 drinks) a day may not drink at all during the week, but then have 10 drinks on Saturday night, exposing the fetus to extremely toxic quantities of alcohol. Whether various drinking patterns incur damage depends on the frequency of consumption, the quantity consumed, and the stage of fetal development at the time of each drinking episode.

An occasional drink may be innocuous, but researchers are unable to say how much alcohol is safe to consume during pregnancy. For this reason, health care professionals urge women to stop drinking alcohol as soon as they realize they are pregnant or better, as soon as they *plan* to become pregnant. Why take any risk? Only the woman who abstains is sure of protecting her infant from FAS.

When Is the Damage Done?

The first month or two of pregnancy is a critical period of fetal development. Because pregnancy usually cannot be confirmed before five to six weeks, a woman may not even realize she is pregnant during that critical time. Therefore, it is advisable for women who are trying to conceive, or who suspect they might be pregnant, to abstain or curtail their alcohol intakes to ensure a healthy start.

The type of abnormality observed in an FAS infant depends on the developmental events occurring at the times of alcohol exposure. During the first trimester, developing organs such as the brain, heart, and kidneys may be malformed. During the second trimester, the risk of spontaneous abortion increases. During the third trimester, body and brain growth may be retarded.

GLOSSARY

alcohol-related birth defects (ARBD): malformations in the skeletal and organ systems (heart, kidneys, eyes, ears) associated with prenatal alcohol exposure.

alcohol-related neurodevelopmental disorder (ARND): abnormalities in the central nervous system and cognitive development associated with prenatal alcohol exposure.

fetal alcohol spectrum disorder: a range of physical, behavioral, and cognitive abnormalities caused by prenatal alcohol exposure.

fetal alcohol syndrome (FAS): a cluster of physical, behavioral, and cognitive abnormalities associated with prenatal alcohol exposure, including facial malformations, growth retardation, and central nervous system disorders.

prenatal alcohol exposure: subjecting a fetus to a pattern of excessive alcohol intake characterized by substantial regular use or heavy episodic drinking.

NOTE: See Highlight 7 for other alcohol-related terms and information.

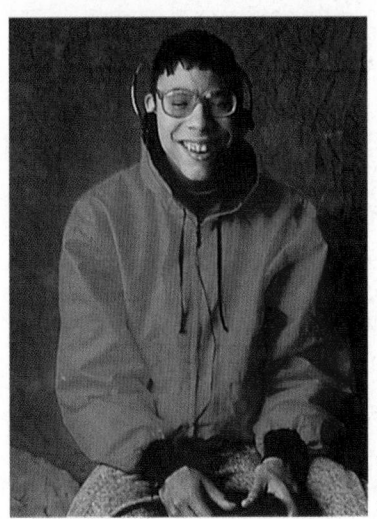

Children born with FAS must live with the long-term consequences of prenatal brain damage.

© 1995 George Steinmetz

Male alcohol ingestion may also affect fertility and fetal development.[8] Animal studies have found smaller litter sizes, lower birthweights, reduced survival rates, and impaired learning ability in the offspring of males consuming alcohol prior to conception. An association between paternal alcohol intake one month prior to conception and low infant birthweight is also apparent in human beings. (Paternal alcohol intake was defined as an average of 2 or more drinks daily or at least 5 drinks on one occasion.) This relationship was independent of either parent's smoking and of the mother's use of alcohol, caffeine, or other drugs.

In view of the damage caused by FAS, prevention efforts focus on educating women not to drink during pregnancy.[9] Everyone should know of the potential dangers. Women who drink alcohol and who are sexually active may benefit from counseling and ef-

All containers of beer, wine, and liquor warn women not to drink alcoholic beverages during pregnancy because of the risk of birth defects.

Matthew Farruggio

fective contraception to prevent pregnancy.[10] Almost half of all pregnancies are unintended, with many conceived during a binge-drinking episode.[11]

Public service announcements and alcohol beverage warning labels help to raise awareness. Everyone should hear the message loud and clear: Don't drink alcohol prior to conception or during pregnancy.

NUTRITION ON THE NET

For further study of topics covered in this chapter, log on to **academic.cengage.com/nutrition/rolfes/UNCN8e**. Go to Chapter 14, then to Nutrition on the Net.

- Visit the National Organization on Fetal Alcohol Syndrome: **www.nofas.org**
- Search for "fetal alcohol syndrome" at the U.S. Government health information site: **www.healthfinder.gov**

- Request information on fetal alcohol syndrome from the National Clearinghouse for Alcohol and Drug Information: **ncadi.samsha.gov**
- Request information on drinking during pregnancy from the National Institute on Alcohol Abuse and Alcoholism: **www.niaaa.nih.gov**
- Gather facts on fetal alcohol syndrome from the March of Dimes: **www.modimes.org**

REFERENCES

1. H. E. Hoyme and coauthors, A practical clinical approach to diagnosis of fetal alcohol spectrum disorders: Clarification of the 1996 Institute of Medicine Criteria, *Pediatrics* 115 (2005): 39–47.
2. N. L. Day and coauthors, Prenatal alcohol exposure predicts continued deficits in offspring size at 14 years of age, *Alcoholism: Clinical and Experimental Research* 26 (2002): 1584–1591; M. D. Cornelius and coauthors, Alcohol, tobacco and marijuana use among pregnant teenagers: 6-year follow-up of offspring growth effects, *Neurotoxicology and Teratology* 24 (2002): 703–710.
3. Guidelines for identifying and referring persons with fetal alcohol syndrome, *Morbidity and Mortality Weekly Report* 54 (2005): 1–10.

4. M. J. O'Connor and coauthors, Psychiatric illness in a clinical sample of children with prenatal alcohol exposure, *American Journal of Drug and Alcohol Abuse* 28 (2002): 743–754.
5. Alcohol consumption among women who are pregnant or who might become pregnant—United States, 2002, *Morbidity and Mortality Weekly Report* 53 (2004): 1178–1181.
6. J. W. Olney and coauthors, The enigma of fetal alcohol neurotoxicity, *Annals of Medicine* 34 (2002): 109–119.
7. S. W. Jacobson and coauthors, Validity of maternal report of prenatal alcohol, cocaine, and smoking in relation to neurobehavioral outcome, *Pediatrics* 109 (2002): 815–825.
8. H. Klonoff-Cohen, P. Lam-Kruglick, and C. Gonzalez, Effects of maternal and paternal

alcohol consumption on the success rates of in vitro fertilization and gamete intrafallopian transfer, *Fertility and Sterility* 79 (2003): 330–339.
9. J. R. Hankin, Fetal alcohol syndrome prevention research, *Alcohol Research and Health* 26 (2002): 58–65.
10. The Project CHOICES Intervention Research Group, Reducing the risk of alcohol-exposed pregnancies: A study of a motivational intervention in community settings, *Pediatrics* 111 (2003): 1131–1135.
11. T. S. Naimi and coauthors, Binge drinking in the preconception period and the risk of unintended pregnancy: Implications for women and their children, *Pediatrics* 111 (2003): 1136–1141.

Nutrition in Your Life

Much of this book has focused on you—your food choices and how they might affect your health. This chapter shifts the focus from you the recipient to you the caregiver. One day (if not already), children will depend on you to feed them well and teach them wisely. The responsibility of nourishing children can seem overwhelming at times, but the job is fairly simple. Offer children a variety of nutritious foods to support their growth, and teach them how to make healthy food and activity choices. Presenting foods in a relaxed and supportive environment nourishes both physical and emotional well-being.

 The CengageNOW logo indicates an opportunity for online self-study, linking you to interactive tutorials and videos based on your level of understanding.

academic.cengage.com/login

How To: Practice Problems
Nutrition Portfolio Journal

Life Cycle Nutrition: Infancy, Childhood, and Adolescence

The first year of life is a time of phenomenal growth and development. After the first year, a child continues to grow and change, but more slowly. Still, the cumulative effects over the next decade are remarkable. Then, as the child enters the teen years, the pace toward adulthood accelerates dramatically. This chapter examines the special nutrient needs of infants, children, and adolescents.

Nutrition during Infancy

Initially, the infant drinks only breast milk or formula but later begins to eat some foods, as appropriate. Common sense in the selection of infant foods along with a nurturing, relaxed environment support an infant's health and well-being.

Energy and Nutrient Needs

An infant grows fast during the first year, as Figure 15-1 shows. Growth directly reflects nutrient intake and is an important parameter in assessing the nutrition status of infants and children. Health care professionals measure the heights and weights of infants and children at intervals and compare the measurements with standard growth curves for gender and age and with previous measures of each child (see the "How to," p. 516).

Energy Intake and Activity A healthy infant's birthweight doubles by about five months of age and triples by one year, typically reaching 20 to 25 pounds. The infant's length changes more slowly than weight, increasing about 10 inches from birth to one year. By the end of the first year, infant growth slows considerably; during the second year, an infant typically gains less than 10 pounds and grows about 5 inches in height.

Not only do infants grow rapidly, but their energy requirement is remarkably high—about twice that of an adult, based on body weight. A newborn baby requires about 450 kcalories per day, whereas most adults require about 2000 kcalories per day. In terms of body weight, the difference is remarkable. Infants require about 100 kcalories per kilogram of body weight per day, whereas most adults need fewer than 40 (see Table 15-1, p. 516). If an infant's energy needs were applied to an adult, a 170-pound adult would require over 7000 kcalories a day. After six months, the infant's energy needs decline as the growth rate slows, but some of the energy saved by slower growth is spent in increased activity.

FIGURE 15-1 Weight Gain of Infants in Their First Five Years of Life

In the first year, an infant's birthweight may triple, but over the following several years, the rate of weight gain gradually diminishes.

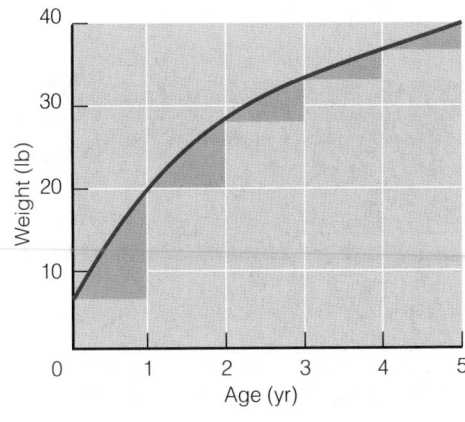

HOW TO Plot Measures on a Growth Chart

You can assess the growth of infants and children by plotting their measurements on a percentile graph. Percentile graphs divide the measures of a population into 100 equal divisions so that half of the population falls at or above the 50th percentile and half falls below. Using percentiles allows for comparisons among people of the same age and gender.

To plot measures on a growth chart, follow these steps:

- Select the appropriate chart based on age and gender. For this example, use the accompanying chart, which gives percentiles for weight for girls from birth to 36 months. (Appendix E provides other growth charts for both boys and girls of various ages.)
- Locate the infant's age along the horizontal axis at the bottom of the chart (in this example, 6 months).
- Locate the infant's weight in pounds or kilograms along the vertical axis of the chart (in this example, 17 pounds or 7.7 kilograms).
- Mark the chart where the age and weight lines intersect (shown here with a red dot), and follow the curved line to find the percentile.

This six-month-old infant is at the 75th percentile. Her pediatrician will weigh her again over the next few months and expect the growth curve to follow the same percentile throughout the first year. In general, dramatic changes or measures much above the 80th percentile or much below the 10th percentile may be cause for concern.

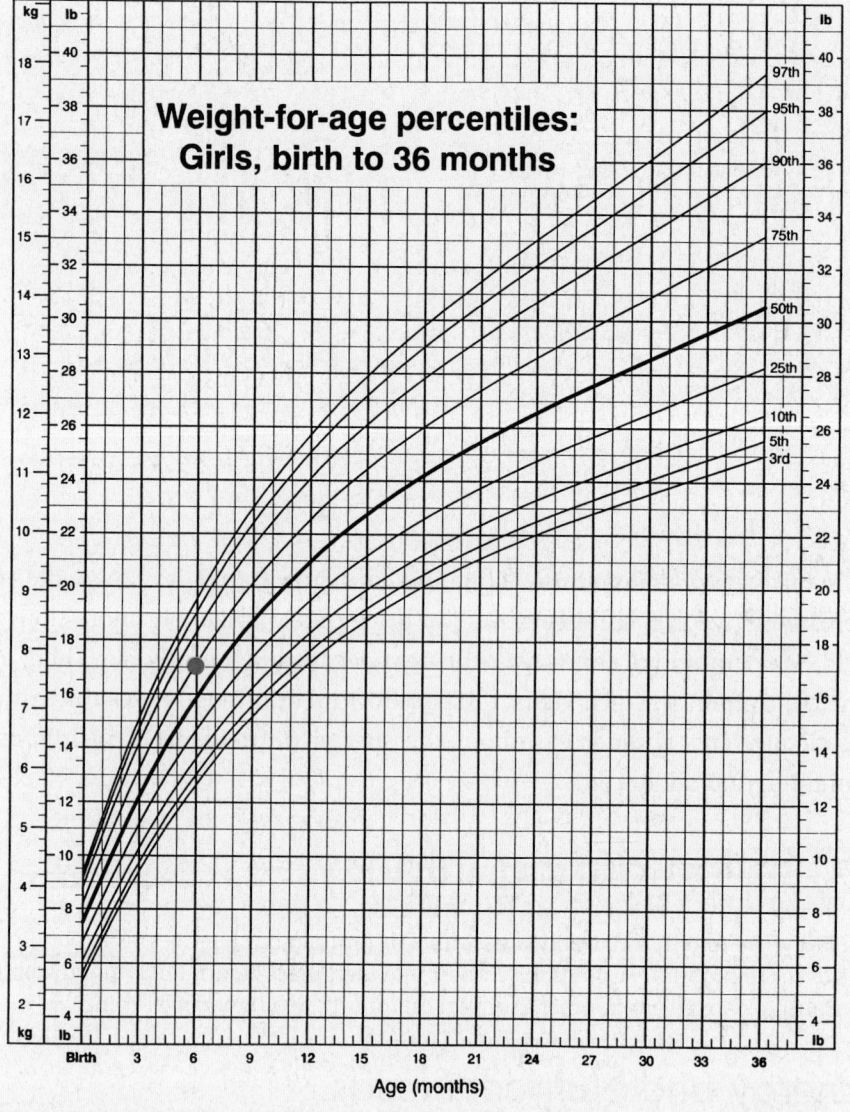

SOURCE: Developed by the National Center for Health Statistics in collaboration with the National Center for Chronic Disease Prevention and Health Promotion (2000).

CENGAGENOW™
To practice plotting measures on a growth chart, log on to **academic.cengage.com/login**, go to Chapter 15, then go to How To.

TABLE 15-1 Infant and Adult Heart Rate, Respiration Rate, and Energy Needs Compared

	Infants	Adults
Heart rate (beats/minute)	120 to 140	70 to 80
Respiration rate (breaths/minute)	20 to 40	15 to 20
Energy needs (kcal/body weight)	45/lb (100/kg)	<18/lb (<40/kg)

Energy Nutrients Recommendations for the energy nutrients—carbohydrate, fat, and protein—during the first six months of life are based on the average intakes of healthy, full-term infants fed breast milk.[1] During the second six months of life, recommendations reflect typical intakes from solid foods as well as breast milk.

As discussed in Chapter 4, carbohydrates provide energy to all the cells of the body, especially those in the brain, which depend primarily on glucose to fuel activities. Relative to the size of the body, an infant's brain is larger and uses relatively more glucose—about 60 percent of the day's total energy intake.[2]

Fat provides most of the energy in breast milk and standard infant formula. Its high energy density supports the rapid growth of early infancy.

No single nutrient is more essential to growth than protein. All of the body's cells and most of its fluids contain protein; it is the basic building material of the body's tissues. Chapter 6 detailed the problems inadequate protein can cause. Excess dietary protein can cause problems, too, especially in a small infant. Too much protein stresses the liver and kidneys, which have to metabolize and excrete the excess nitrogen. Signs of protein overload include acidosis, dehydration, diarrhea, elevated blood ammonia, elevated blood urea, and fever. Such problems are not common, but they have been observed in infants fed inappropriate foods, such as fat-free milk or concentrated formula.

Vitamins and Minerals As with the energy nutrients, the recommendations for the vitamins and minerals are based on the average amount of nutrients consumed by thriving infants breastfed by well-nourished mothers. An infant's needs for most of these nutrients, in proportion to body weight, are more than double those of an adult. Figure 15-2 illustrates this by comparing a five-month-old infant's needs per unit of body weight with those of an adult man. Some of the differences are extraordinary.

Water One of the most essential nutrients for infants, as for everyone, is water. The younger the infant, the greater the percentage of body weight is water. During early infancy, breast milk or infant formula normally provides enough water to replace fluid losses in a healthy infant. Even in hot, dry climates, neither breastfed nor bottle-fed infants need supplemental water.[3] Because much of the fluid in an infant's body is located *outside* the cells—between the cells and in the blood vessels—rapid fluid losses and the resulting dehydration can be life-threatening. Conditions that cause rapid fluid loss, such as diarrhea or vomiting, require treatment with an electrolyte solution designed for infants.

After six months, energy saved by slower growth is spent in increased activity.

FIGURE 15-2 Recommended Intakes of an Infant and an Adult Compared on the Basis of Body Weight

Because infants are small, they need smaller total amounts of the nutrients than adults do, but when comparisons are based on body weight, infants need more than twice as much of many nutrients. Infants use large amounts of energy and nutrients, in proportion to their body size, to keep all their metabolic processes going.

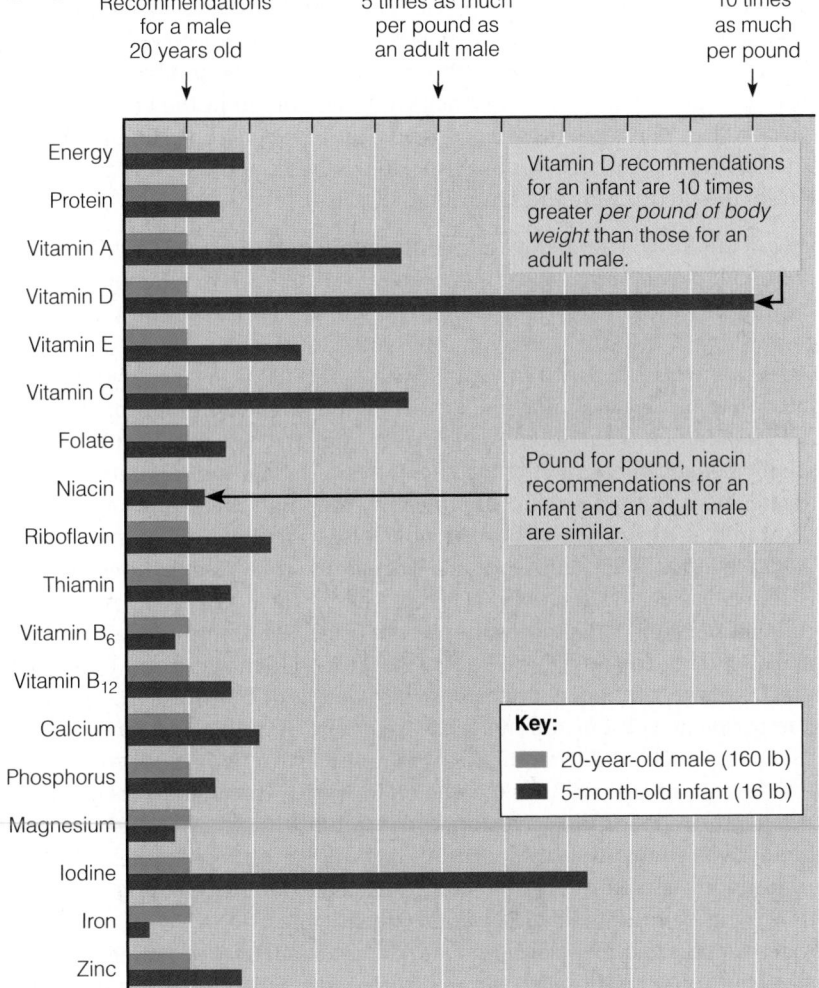

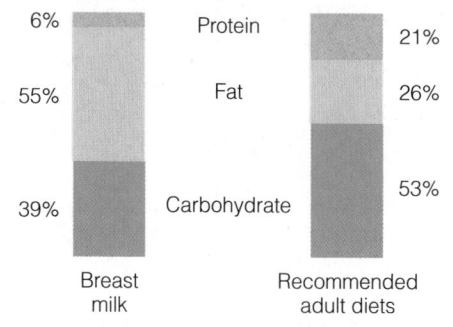

FIGURE 15-3 Percentages of Energy-Yielding Nutrients in Breast Milk and in Recommended Adult Diets

The proportions of energy-yielding nutrients in human breast milk differ from those recommended for adults.[a]

	Breast milk	Recommended adult diets
Protein	6%	21%
Fat	55%	26%
Carbohydrate	39%	53%

[a]The values listed for adults represent approximate midpoints of the acceptable ranges for protein (10 to 35 percent), fat (20 to 35 percent), and carbohydrate (45 to 65 percent).

◆ Chapter 14 discussed breastfeeding, breastfeeding support, reasons why some women choose not to breastfeed, and contraindications to breastfeeding.

alpha-lactalbumin (lact-AL-byoo-min): a major protein in human breast milk, as opposed to **casein** (CAY-seen), a major protein in cow's milk.

Breast Milk

In the United States and Canada, the two dietary practices that have the most significant effect on an infant's nutrition are the milk the infant receives and the age at which solid foods are introduced. A later section discusses the introduction of solid foods, but as to the milk, both the American Academy of Pediatrics (AAP) and the Canadian Paediatric Society strongly recommend breastfeeding for healthy full-term infants, except where specific contraindications exist. The American Dietetic Association (ADA) also advocates breastfeeding for the nutritional health it confers on the infant as well as for the many other benefits it provides both infant and mother (review Table 14-4, p. 501).[4]

Breast milk excels as a source of nutrients for infants. Its unique nutrient composition and protective factors promote optimal infant health and development throughout the first year of life. The AAP, the Canadian Paediatric Society, and the ADA recommend exclusive breastfeeding for 6 months, and breastfeeding with complementary foods for at least 12 months for infants.[5] Experts add, though, that iron-fortified formula, which imitates the nutrient composition of breast milk, is an acceptable alternative. After all, the primary goal is to provide the infant nourishment in a relaxed and loving environment. ◆

Frequency and Duration of Breastfeeding Breast milk is more easily and completely digested than formula, so breastfed infants usually need to eat more frequently than formula-fed infants do. During the first few weeks, approximately 8 to 12 feedings a day, on demand, as soon as the infant shows early signs of hunger such as increased alertness, activity, or suckling motions, promote optimal milk production and infant growth.[6] Crying is a late indicator of hunger. An infant who nurses every two to three hours and sleeps contentedly between feedings is adequately nourished. As the infant gets older, stomach capacity enlarges and the mother's milk production increases, allowing for longer intervals between feedings.

Even though the infant obtains about half the milk from the breast during the first two or three minutes of sucking, breastfeeding is encouraged for about 10 to 15 minutes on each breast. The infant's sucking, as well as the complete removal of milk from the breast, stimulates lactation.

Energy Nutrients The energy-nutrient composition of breast milk differs dramatically from that recommended for adult diets (see Figure 15-3). Yet for infants, breast milk is nature's most nearly perfect food, providing the clear lesson that people at different stages of life have different nutrient needs.

The carbohydrate in breast milk (and infant formula) is the disaccharide lactose. In addition to being easily digested, lactose enhances calcium absorption.

The amount of protein in breast milk is less than in cow's milk, but this quantity is actually beneficial because it places less stress on the infant's immature kidneys to excrete the major end product of protein metabolism, urea. Much of the protein in breast milk is **alpha-lactalbumin**, which is efficiently digested and absorbed.

As for the lipids, breast milk contains a generous proportion of the essential fatty acids linoleic acid and linolenic acid, as well as their longer-chain derivatives arachidonic acid and DHA (docosahexaenoic acid). Infant formula used to provide only linoleic acid and linolenic acid, but now arachidonic acid and DHA are also included.[7] Infants can make arachidonic acid and DHA from linoleic and linolenic acid, respectively, but some infants may need more than they can make.

Arachidonic acid and DHA are found abundantly in both the retina of the eye and the brain, and research has focused on the visual and mental development of breastfed infants and infants fed standard formula without DHA and arachidonic acid added.[8] Breastfed infants generally score higher on tests of mental development than formula-fed infants do, and researchers are investigating whether this difference can be attributed to DHA and arachidonic acid in breast milk.[9] In one study, researchers found no developmental or visual differences between infants fed standard formula and those fed formula with added DHA and arachidonic acid.[10] In two other studies, however, infants fed the formula fortified with DHA and arachidonic acid formula had sharper vision at one year of age than those who were fed standard formula.[11]

Vitamins With the exception of vitamin D, the vitamins in breast milk are ample to support infant growth. The vitamin D in breast milk is low, and vitamin D deficiency impairs bone mineralization. Vitamin D deficiency is most likely in infants who are not exposed to sunlight daily, have darkly pigmented skin, and receive breast milk without vitamin D supplementation.[12] Reports of infants in the United States developing the vitamin D–deficiency disease rickets and recommendations by the AAP to keep infants under six months of age out of direct sunlight have prompted updated vitamin D guidelines. The AAP now recommends a vitamin D supplement for all infants who are breastfed exclusively, and for any infants who do not receive at least 500 milliliters (15 ounces) per day of vitamin D–fortified formula.[13]

Minerals The calcium content of breast milk is ideal for infant bone growth, and the calcium is well absorbed. Breast milk contains relatively small amounts of iron, but the iron has a high bioavailability. Zinc also has a high bioavailability, thanks to the presence of a zinc-binding protein. Breast milk is low in sodium, another benefit for immature kidneys. Fluoride promotes the development of strong teeth, but breast milk is not a good source.

Supplements Pediatricians may routinely prescribe liquid supplements containing vitamin D, iron, and fluoride. Table 15-2 offers a schedule of supplements during infancy. In addition, the AAP recommends giving a single dose of vitamin K to infants at birth to protect them from bleeding to death. (See Chapter 11 for a description of vitamin K's role in blood clotting.)

Immunological Protection In addition to nutritional benefits, breast milk offers immunological protection. Not only is breast milk sterile, but it actively fights disease and protects infants from illnesses.[14] Such protection is most valuable during the first year, when the infant's immune system is not fully prepared to mount a response against infection.

During the first two or three days after delivery, the breasts produce **colostrum,** a premilk substance containing mostly serum with antibodies and white blood cells. Colostrum (like breast milk) helps protect the newborn from infections against which the mother has developed immunity. The maternal antibodies swallowed with the milk inactivate disease-causing bacteria within the digestive tract before they can start infections. This explains, in part, why breastfed infants have fewer intestinal infections than formula-fed infants.

In addition to antibodies, colostrum and breast milk provide other powerful agents ◆ that help to fight against bacterial infection. Among them are **bifidus factors,** which favor the growth of the "friendly" bacterium *Lactobacillus bifidus*

Women are encouraged to breastfeed whenever possible because breast milk offers infants many nutrient and health advantages.

◆ Protective factors in breast milk:
- Antibodies
- Bifidus factors
- Lactoferrin
- Lactadherin
- Growth factor
- Lipase enzyme

TABLE 15-2 Supplements for Full-Term Infants			
	Vitamin D[a]	Iron[b]	Fluoride[c]
Breastfed infants:			
Birth to six months of age	✔		
Six months to one year	✔	✔	✔
Formula-fed infants:			
Birth to six months of age			
Six months to one year		✔	✔

[a]Vitamin D supplements are recommended for all infants who are exclusively breastfed and for any infants who do not receive at least 500 milliliters (15 ounces) of vitamin D–fortified formula.
[b]Infants four to six months of age need additional iron, preferably in the form of iron-fortified cereal for both breastfed and formula-fed infants and iron-fortified infant formula for formula-fed infants.
[c]At six months of age, breastfed infants and formula-fed infants who receive ready-to-use formulas (these are prepared with water low in fluoride) or formula mixed with water that contains little or no fluoride (less than 0.3 ppm) need supplements.
SOURCE: Adapted from Committee on Nutrition, American Academy of Pediatrics, *Pediatric Nutrition Handbook,* 5th ed., ed. R. E. Kleinman (Elk Grove Village, Ill.: American Academy of Pediatrics, 2004).

colostrum (ko-LAHS-trum): a milklike secretion from the breast, present during the first day or so after delivery before milk appears; rich in protective factors.

bifidus (BIFF-id-us, by-FEED-us) **factors:** factors in colostrum and breast milk that favor the growth of the "friendly" bacterium *Lactobacillus* (lack-toh-ba-SILL-us) *bifidus* in the infant's intestinal tract, so that other, less desirable intestinal inhabitants will not flourish.

in the infant's digestive tract, so that other, harmful bacteria cannot become established. An iron-binding protein in breast milk, **lactoferrin,** keeps bacteria from getting the iron they need to grow, helps absorb iron into the infant's bloodstream, and kills some bacteria directly.[15] The protein **lactadherin** in breast milk binds to, and inhibits replication of, the virus that causes most infant diarrhea.[16] Breastfeeding also protects against other common illnesses of infancy such as middle ear infection and respiratory illness.[17] In addition, a growth factor that is present in breast milk stimulates the development and maintenance of the infant's digestive tract and its protective factors. Several breast milk enzymes such as lipase also help protect the infant against infection. Clearly, breast milk is a very special substance.

Allergy and Disease Protection In addition to protection against infection, breast milk may offer protection against the development of allergies. Compared with formula-fed infants, breastfed infants have a lower incidence of allergic reactions, such as asthma, recurrent wheezing, and skin rash.[18] This protection is especially noticeable among infants with a family history of allergies.[19] Similarly, breast milk may offer protection against the development of cardiovascular disease. Compared with formula-fed infants, breastfed infants have lower blood pressure and lower blood cholesterol as adults.[20]

Other Potential Benefits Breastfeeding may also help protect against excessive weight gain later. A review of more than 60 published studies investigating the relationship between infant feeding and obesity suggests that initial breastfeeding protects against obesity in later life.[21] A well-controlled survey of more than 15,000 adolescents and their mothers indicated that those who were mostly breastfed for the first six months of life were less likely to become overweight than those who were fed formula.[22] A study of much younger children (three to five years of age), however, found no clear evidence that breastfeeding influences body weight.[23] These researchers noted that other factors, especially the mother's weight, strongly predict overweight in children.

Many studies suggest a beneficial effect of breastfeeding on intelligence, but when subjected to strict standards of methodology (for example, large sample size and appropriate intelligence testing), the evidence is less convincing.[24] Nevertheless, the possibility that breastfeeding may positively affect later intelligence is intriguing. It may be that some specific component of breast milk, such as DHA, stimulates brain development or that certain factors associated with the feeding process itself promote intellect. Most likely, a combination of factors are involved. More large, well-controlled studies are needed to confirm the effects, if any, of breastfeeding on later intelligence.

Breast Milk Banks Similar to blood banks that collect blood from individuals to give to others in need, **breast milk banks** receive milk from lactating women who have an abundant supply to give to infants whose own mothers' milk is unavailable or insufficient. The women who donate breast milk are carefully screened to exclude those who smoke cigarettes, use illegal drugs, take medications (including high doses of dietary supplements), drink more than two alcoholic beverages a day, or have communicable diseases. The breast milk from several donors is pooled to ensure an even distribution of all components, pasteurized to destroy bacteria, checked for contamination, and frozen before being shipped overnight to hospitals, where it is dispensed by physician prescription. In the absence of mother's own breast milk, donor milk may be the life-saving solution for fragile infants, most notably those with very low birthweight or unusual medical conditions.[25]

Infant Formula

A woman who breastfeeds for a year can **wean** her infant to cow's milk, bypassing the need for infant formula. However, a woman who decides to feed her infant for-

---

lactoferrin (lack-toh-FERR-in): a protein in breast milk that binds iron and keeps it from supporting the growth of the infant's intestinal bacteria.

lactadherin (lack-tad-HAIR-in): a protein in breast milk that attacks diarrhea-causing viruses.

breast milk bank: a service that collects, screens, processes, and distributes donated human milk.

wean: to gradually replace breast milk with infant formula or other foods appropriate to an infant's diet.

mula from birth, to wean to formula after less than a year of breastfeeding, or to substitute formula for breastfeeding on occasion must select an appropriate infant formula and learn to prepare it.

Infant Formula Composition Formula manufacturers attempt to copy the nutrient composition of breast milk as closely as possible. Figure 15-4 illustrates the energy-nutrient balance of both. The AAP recommends that all formula-fed infants receive iron-fortified infant formulas. The increasing use of iron-fortified formulas during the past few decades is a major reason for the decline in iron-deficiency anemia among U.S. infants.

Risks of Formula Feeding Infant formulas contain no protective antibodies for infants, but in general, vaccinations, purified water, and clean environments in developed countries help protect infants from infections. Formulas can be prepared safely by following the rules of proper food handling and by using water that is free of contamination. Of particular concern is lead-contaminated water, a major source of lead poisoning in infants. Because the first water drawn from the tap each day is highest in lead, a person living in a house with old, lead-soldered plumbing should let the water run a few minutes before drinking or using it to prepare formula or food.

In developing countries and in poor areas of the United States, formula may be unavailable, prepared with contaminated water, or overdiluted in an attempt to save money. Contaminated formulas often cause infections, leading to diarrhea, dehydration, and malabsorption. Without sterilization and refrigeration, formula is an ideal breeding ground for bacteria. Whenever such risks are present, breast-feeding can be a life-saving option: breast milk is sterile, and its antibodies enhance an infant's resistance to infections.

Infant Formula Standards National and international standards have been set for the nutrient contents of infant formulas. In the United States, the standard developed by the AAP reflects "human milk taken from well-nourished mothers during the first or second month of lactation, when the infant's growth rate is high." The Food and Drug Administration (FDA) mandates the safety and nutritional quality of infant formulas. Formulas meeting these standards have similar nutrient compositions. Small differences among formulas are sometimes confusing, but they are usually unimportant.

Special Formulas Standard cow's milk-based formulas are inappropriate for some infants. Special formulas have been designed to meet the dietary needs of infants with specific conditions such as prematurity or inherited diseases. Infants allergic to milk protein can drink special **hypoallergenic formulas** or formulas based on soy protein.[26] Soy formulas also use cornstarch and sucrose instead of lactose and so are recommended for infants with lactose intolerance as well. They are also useful as an alternative to milk-based formulas for vegan families. Despite these limited uses, soy formulas account for one-fourth of the infant formulas sold today. While soy formulas support the normal growth and development of infants, for infants who don't need them, they offer no advantage over milk formulas.

Inappropriate Formulas Caregivers must use only products designed for infants; soy *beverages,* for example, are nutritionally incomplete and inappropriate for infants. Goat's milk is also inappropriate for infants in part because of its low folate content. An infant receiving goat's milk is likely to develop "goat's milk anemia," an anemia characteristic of folate deficiency.

Nursing Bottle Tooth Decay An infant cannot be allowed to sleep with a bottle because of the potential damage to developing teeth. Salivary flow, which normally cleanses the mouth, diminishes as the infant falls asleep. Prolonged sucking on a bottle of formula, milk, or juice bathes the upper teeth in a carbohydrate-rich fluid that nourishes decay-producing bacteria. (The tongue covers and protects most of the lower teeth, but they, too, may be affected.) The result is extensive and rapid tooth decay (see Figure 15-5, p. 522). To prevent **nursing bottle tooth decay,** no infant should be put to bed with a bottle of nourishing fluid.

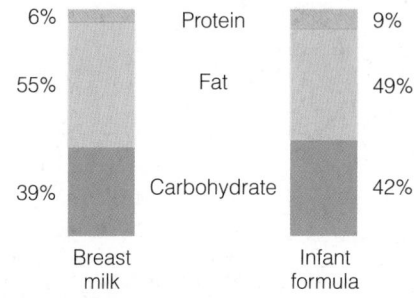

FIGURE 15-4 Percentages of Energy-Yielding Nutrients in Breast Milk and in Infant Formula

The average proportions of energy-yielding nutrients in human breast milk and formula differ slightly. In contrast, cow's milk provides too much protein (20%) and too little carbohydrate (30%).

6%	Protein	9%
55%	Fat	49%
39%	Carbohydrate	42%
Breast milk		Infant formula

The infant thrives on infant formula offered with affection.

© Jon Feingersh/Corbis

hypoallergenic formulas: clinically tested infant formulas that support infant growth and development but do not provoke reactions in 90% of infants or children with confirmed cow's milk allergy.

nursing bottle tooth decay: extensive tooth decay due to prolonged tooth contact with formula, milk, fruit juice, or other carbohydrate-rich liquid offered to an infant in a bottle.

FIGURE 15-5 Nursing Bottle Tooth Decay

This child was frequently put to bed sucking on a bottle filled with apple juice, so the teeth were bathed in carbohydrate for long periods of time—a perfect medium for bacterial growth. The upper teeth show signs of decay.

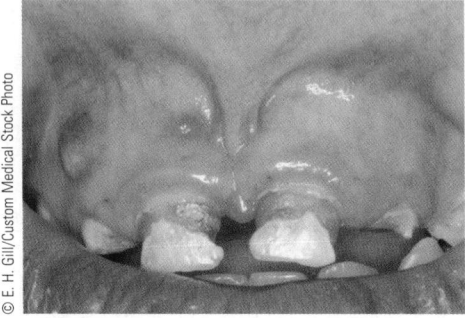

Special Needs of Preterm Infants

An estimated one out of eight pregnancies in the United States results in a preterm birth.[27] The terms *preterm* and *premature* imply incomplete fetal development, or immaturity, of many body systems. As might be expected, preterm birth is a leading cause of infant deaths. Preterm infants face physical independence from their mothers before some of their organs and body tissues are ready. The rate of weight gain in the fetus is greater during the last trimester of gestation than at any other time. Therefore, a preterm infant is most often a low-birthweight infant as well. A premature birth deprives the infant of the nutritional support of the placenta during a time of maximal growth.

The last trimester of gestation is also a time of building nutrient stores. Being born with limited nutrient stores intensifies the already precarious situation for the infant. The physical and metabolic immaturity of preterm infants further compromises their nutrition status. Nutrient absorption, especially of fat and calcium, from an immature GI tract is limited. Consequently, preterm, low-birthweight infants are candidates for nutrient imbalances. Deficiencies of the fat-soluble vitamins, calcium, iron, and zinc are common.

Preterm breast milk is well suited to meet a preterm infant's needs. During early lactation, preterm milk contains higher concentrations of protein and is lower in volume than term milk. The low milk volume is advantageous because preterm infants consume small quantities of milk per feeding, and the higher protein concentration allows for better growth. In many instances, supplements of nutrients specifically designed for preterm infants are added to the mother's expressed breast milk and fed to the infant from a bottle. When fortified with a preterm supplement, preterm breast milk supports growth at a rate that approximates the growth rate that would have occurred within the uterus.

Introducing Cow's Milk

The age at which whole cow's milk should be introduced to the infant's diet has long been a source of controversy. The AAP advises that whole cow's milk is not appropriate during the first year.[28] Children one to two years of age should not be given reduced-fat, low-fat, or fat-free milk routinely; they need the fat of whole milk. Between the ages of two and five years, a gradual transition from whole milk to the lower-fat milks can take place, but care should be taken to avoid excessive restriction of dietary fat.

Dietary Guidelines for Americans 2005

Children two to eight years should consume 2 cups per day of fat-free or low-fat milk or equivalent milk products.

In some infants, particularly those younger than six months of age, whole cow's milk may cause intestinal bleeding, which can lead to iron deficiency. Cow's milk is also a poor source of iron. Consequently, it both causes iron loss and fails to replace iron. Furthermore, the bioavailability of iron from infant cereal and other foods is reduced when cow's milk replaces breast milk or iron-fortified formula during the first year. Compared with breast milk or iron-fortified formula, cow's milk is higher in calcium and lower in vitamin C, characteristics that reduce iron absorption. Furthermore, the higher protein concentration of cow's milk can stress the infant's kidneys. In short, cow's milk is a poor choice during the first year of life; infants need breast milk or iron-fortified infant formula.

Introducing Solid Foods

The high nutrient needs of infancy are met first by breast milk or formula only and then by the limited addition of selected foods over time. Infants gradually develop the ability to chew, swallow, and digest the wide variety of foods available to adults. The caregiver's selection of appropriate foods at the appropriate stages of development is prerequisite to the infant's optimal growth and health.

When to Begin In addition to breast milk or formula, an infant can begin eating solid foods between four and six months.[29] The AAP supports exclusive breastfeeding for six months but recognizes that infants are often developmentally ready to accept complementary foods between four and six months of age.[30] The main purpose of introducing solid foods is to provide needed nutrients that are no longer supplied adequately by breast milk or formula alone. The foods chosen must be those ◆ that the infant is developmentally capable of handling both physically and metabolically. The exact timing depends on the individual infant's needs and developmental readiness (see Table 15-3), which vary from infant to infant because of differences in growth rates, activities, and environmental conditions. ◆ In short, the addition of foods to an infant's diet should be governed by three considerations: the infant's nutrient needs, the infant's physical readiness to handle different forms of foods, and the need to detect and control allergic reactions.

Food Allergies To prevent allergy and to facilitate its prompt identification should it occur, experts recommend introducing single-ingredient foods, one at a time, in small portions, and waiting four to five days before introducing the next new food.[31] For example, rice cereal is usually the first cereal introduced because it

◆ The German word **beikost** (BYE-cost) describes any nonmilk foods given to an infant.

◆ Digestive secretions gradually increase throughout the first year of life, making the digestion of solid foods more efficient.

TABLE 15-3 Infant Development and Recommended Foods

Because each stage of development builds on the previous stage, the foods from an earlier stage continue to be included in all later stages.

Age (mo)	Feeding Skill	Appropriate Foods Added to the Diet
0–4	Turns head toward any object that brushes cheek. Initially swallows using back of tongue; gradually begins to swallow using front of tongue as well. Strong reflex (extrusion) to push food out during first 2 to 3 months.	Feed breast milk or infant formula.
4–6	Extrusion reflex diminishes, and the ability to swallow nonliquid foods develops. Indicates desire for food by opening mouth and leaning forward. Indicates satiety or disinterest by turning away and leaning back. Sits erect with support at 6 months. Begins chewing action. Brings hand to mouth. Grasps objects with palm of hand.	Begin iron-fortified cereal mixed with breast milk, formula, or water. Begin pureed vegetables and fruits.
6–8	Able to self-feed finger foods. Develops pincer (finger to thumb) grasp. Begins to drink from cup.	Begin textured vegetables and fruits. Begin unsweetened, diluted fruit juices from cup.
8–10	Begins to hold own bottle. Reaches for and grabs food and spoon. Sits unsupported.	Begin breads and cereals from table. Begin yogurt. Begin pieces of soft, cooked vegetables and fruit from table. Gradually begin finely cut meats, fish, casseroles, cheese, eggs, and mashed legumes.
10–12	Begins to master spoon, but still spills some.	Add variety. Gradually increase portion sizes.[a]

[a] Portion sizes for infants and young children are smaller than those for an adult. For example, a grain serving might be ½ slice of bread instead of 1 slice, or ¼ cup rice instead of ½ cup.

SOURCE: Adapted in part from Committee on Nutrition, American Academy of Pediatrics, *Pediatric Nutrition Handbook*, 5th ed., ed. R. E. Kleinman (Elk Grove Village, Ill.: American Academy of Pediatrics, 2004), pp. 103–115.

Foods such as iron-fortified cereals and formulas, mashed legumes, and strained meats provide iron.

is the least allergenic. When it is clear that rice cereal is not causing an allergy, another grain, perhaps barley or oats, is introduced. Wheat cereal is offered last because it is the most common offender. If a cereal causes an allergic reaction such as a skin rash, digestive upset, or respiratory discomfort, it should be discontinued before introducing the next food. A later section in this chapter offers more information about food allergies.

Choice of Infant Foods Infant foods should be selected to provide variety, balance, and moderation. Commercial baby foods offer a wide variety of palatable, nutritious foods in a safe and convenient form. Homemade infant foods can be as nutritious as commercially prepared ones, as long as the cook minimizes nutrient losses during preparation. Ingredients for homemade foods should be fresh, whole foods without added salt, sugar, or seasonings. Pureed food can be frozen in ice cube trays, providing convenient-sized blocks of food that can be thawed, warmed, and fed to the infant. To guard against foodborne illnesses, hands and equipment must be kept clean.

Because recommendations to restrict fat do not apply to children under age two, labels on foods for children under two (such as infant meats and cereals) cannot carry information about fat. Fat information is omitted from infant food labels to prevent parents from restricting fat in infants' diets. Fearing that their infant will become overweight, parents may unintentionally malnourish the infant by limiting fat. In fact, infants and young children, because of their rapid growth, need more fat than older children and adults.

Foods to Provide Iron Rapid growth demands iron. At about four to six months, the infant begins to need more iron than body stores plus breast milk or iron-fortified formula can provide. In addition to breast milk or iron-fortified formula, infants can receive iron from iron-fortified cereals and, once they readily accept solid foods, from meat or meat alternates such as legumes. Iron-fortified cereals contribute a significant amount of iron to an infant's diet, but the iron's bioavailability is poor.[32] Caregivers can enhance iron absorption from iron-fortified cereals by serving vitamin C–rich foods with meals.

Foods to Provide Vitamin C The best sources of vitamin C are fruits and vegetables (see pp. 354–355 in Chapter 10). It has been suggested that infants who are introduced to fruits before vegetables may develop a preference for sweets and find the vegetables less palatable, but there is no evidence to support offering these foods in a particular order.[33]

Fruit juice is a good source of vitamin C, but drinking too much juice can lead to diarrhea in infants and young children.[34] AAP recommendations limit juice consumption for infants and young children (one to six years of age) to between 4 and 6 ounces per day.[35] Beyond these limits, fruit juices contribute excessive kcalories and displace other nutrient-rich foods. Fruit juices should be diluted and served in a cup, not a bottle, once the infant is six months of age or older.

Foods to Omit Concentrated sweets, including baby food "desserts," have no place in an infant's diet. They convey no nutrients to support growth, and the extra food energy can promote obesity. Products containing sugar alcohols such as sorbitol should also be limited, as they may cause diarrhea. Canned vegetables are also inappropriate for infants, as they often contain too much sodium. Honey and corn syrup should never be fed to infants because of the risk of **botulism.*** Infants and young children are vulnerable to foodborne illnesses, and the *Dietary Guidelines 2005* address this risk.

---

botulism (BOT-chew-lism): an often fatal foodborne illness caused by the ingestion of foods containing a toxin produced by bacteria that grow without oxygen.

---

* In infants, but not in older individuals, ingestion of *Clostridium botulinum* spores can cause illness when the spores germinate in the intestine and produce a toxin, which is absorbed. Symptoms include poor feeding, constipation, loss of tension in the arteries and muscles, weakness, and respiratory compromise. Infant botulism has been implicated in 5 percent of cases of sudden infant death syndrome (SIDS).

Dietary Guidelines for Americans 2005

Infants and young children should not eat or drink unpasteurized milk, milk products, or juices; raw or undercooked eggs, meat, poultry, fish, or shellfish; or raw sprouts.

Ideally, a one-year-old eats many of the same foods as the rest of the family.

Infants and even young children cannot safely chew and swallow any of the foods listed in the margin; ◆ they can easily choke on these foods, a risk not worth taking. Nonfood items may present even greater choking hazards to infants and young children.[36] Parents and caregivers must pay careful attention to eliminate choking hazards in children's environments.

Vegetarian Diets during Infancy The newborn infant is a lactovegetarian. As long as the infant has access to sufficient quantities of either infant formula or breast milk (plus a vitamin D supplement) from a mother who eats an adequate diet, the infant will thrive during the early months. "Health-food beverages," such as rice milk, are inappropriate choices because they lack the protein, vitamins, and minerals infants and toddlers need; in fact, their use can lead to severe nutritional deficiencies.

Infants beyond about six months of age present a greater challenge in terms of meeting nutrient needs by way of vegetarian and, especially, vegan diets. Continued breastfeeding or formula feeding is recommended, but supplementary feedings are necessary to ensure adequate energy and iron intakes. Infants and young children in vegetarian families should be given iron-fortified infant cereals well into the second year. Mashed or pureed legumes, tofu, and cooked eggs can be added to their diets in place of meat.

The risks of malnutrition in infants increase with weaning and reliance on table foods. Infants who receive a well-balanced vegetarian diet that includes milk products and a variety of other foods can easily meet their nutritional requirements for growth. This is not always true for vegan infants; the growth of vegan infants slows significantly around the time of transition from breast milk to solid foods. Protein-energy malnutrition and deficiencies of vitamin D, vitamin B_{12}, iron, and calcium have been reported in infants fed vegan diets. Vegan diets that are high in fiber, other complex carbohydrates, and water will fill infants' stomachs before meeting their energy needs. This problem can be partially alleviated by providing more energy-dense foods, such as nut butters, legumes, dried fruit spreads, and mashed avocado. Using soy formulas (or milk) fortified with calcium, vitamin B_{12}, and vitamin D and including vitamin C–containing foods at meals to enhance iron absorption will help prevent other nutrient deficiencies in vegan diets. Parents or caregivers who choose to feed their infants vegan diets should consult with their pediatrician and a registered dietitian frequently to ensure a nutritionally adequate diet that will support growth.

Foods at One Year At one year of age, whole cow's milk can become a primary source of most of the nutrients an infant needs; 2 to 3 cups a day meets those needs sufficiently. Ingesting more milk than this can displace iron-rich foods, which can lead to **milk anemia.** If powdered milk is used, it should contain fat.

Other foods—meats, iron-fortified cereals, enriched or whole-grain breads, fruits, and vegetables—should be supplied in variety and in amounts sufficient to round out total energy needs. Ideally, a one-year-old will sit at the table, eat many of the same foods everyone else eats, and drink liquids from a cup, not a bottle. Figure 15-6 shows a meal plan that meets a one-year-old's requirements.

◆ To prevent choking, do not give infants or young children:
- Raw carrots
- Cherries
- Gum
- Hard or gel-type candies
- Hot dog slices
- Marshmallows
- Nuts
- Peanut butter
- Popcorn
- Raw celery
- Whole beans
- Whole grapes

Keep these nonfood items out of their reach:
- Coins
- Small balls
- Balloons
- Pen tops

FIGURE 15-6 Sample Meal Plan for a One-Year-Old

SAMPLE MENU

Breakfast	½ c iron-fortified, unsweetened breakfast cereal
	¼ c whole milk (with cereal)
	½ c orange juice
Morning snack	½ c yogurt
	½ c fruit[a]
Lunch	½ sandwich: 1 slice bread with 2 tbs tuna salad or egg salad
	½ c vegetables[b] (steamed carrots)
	½ c whole milk
Afternoon snack	½ slice whole-wheat toast
	1 tbs apple butter
	½ c whole milk
Dinner	1 oz chopped meat or ¼ c well-cooked mashed legumes
	¼ c potato, rice, or pasta
	½ c vegetables[b] (chopped broccoli)
	½ c whole milk

[a]Include citrus fruits, melons, and berries.
[b]Include dark green, leafy and deep yellow vegetables.

Mealtimes with Toddlers

The nurturing of a young child involves more than nutrition. Those who care for young children are responsible not only for providing nutritious milk, foods, and water, but also a safe, loving, secure environment in which the children may grow

milk anemia: iron-deficiency anemia that develops when an excessive milk intake displaces iron-rich foods from the diet.

Toddlers need vitamin A– and vitamin D–fortified whole milk.

and develop. In light of toddlers' developmental and nutrient needs and their often contrary and willful behavior, a few feeding guidelines may be helpful:

- Discourage unacceptable behavior, such as standing at the table or throwing food, by removing the young child from the table to wait until later to eat. Be consistent and firm, not punitive. The child will soon learn to sit and eat.
- Let toddlers explore and enjoy food, even if this means eating with fingers for a while. Learning to use a spoon will come in time.
- Don't force food on children. Rejecting new foods is normal and acceptance is more likely as children become familiar with new foods through repeated opportunities to taste them.
- Provide nutritious foods, and let children choose which ones, and how much, they will eat. Gradually, they will acquire a taste for different foods.
- Limit sweets. Infants and young children have little room for empty-kcalorie foods in their daily energy allowance. Do not use sweets as a reward for eating meals.
- Don't turn the dining table into a battleground. Make mealtimes enjoyable. Teach healthy food choices and eating habits in a pleasant environment.

IN SUMMARY

The primary food for infants during the first 12 months is either breast milk or iron-fortified formula. In addition to nutrients, breast milk also offers immunological protection. At about four to six months, infants should gradually begin eating solid foods. By one year, they are drinking from a cup and eating many of the same foods as the rest of the family.

Nutrition during Childhood

Each year from age one to adolescence, a child typically grows taller by 2 to 3 inches and heavier by 5 to 6 pounds. Growth charts provide valuable clues to a child's health. Weight gains out of proportion to height gains may reflect overeating and inactivity, whereas measures significantly below the standard suggest malnutrition.

Increases in height and weight are only two of the many changes growing children experience (see Figure 15-7). At age one, children can stand alone and are beginning to toddle; by two, they can walk and are learning to run; and by three, they can jump and climb with confidence. Bones and muscles increase in mass and density to make these accomplishments possible. Thereafter, lengthening of the long bones and increases in musculature proceed unevenly and more slowly until adolescence.

Energy and Nutrient Needs

Children's appetites begin to diminish around one year, consistent with the slowing growth. Thereafter, children spontaneously vary their food intakes to coincide with their growth patterns; they demand more food during periods of rapid growth than during slow growth. Sometimes they seem insatiable, and other times they seem to live on air and water.

Children's energy intakes also vary widely from meal to meal. Even so, their total daily intakes remain remarkably constant.[37] If children eat less at one meal, they typically eat more at the next, and vice versa. Overweight children are exceptions: they do not always adjust their energy intakes appropriately and may eat in response to external cues, disregarding hunger and satiety signals.

Energy Intake and Activity Individual children's energy needs vary widely, depending on their growth and physical activity. A one-year-old child needs about 800 kcalo-

FIGURE 15-7 Body Shape of One-Year-Old and Two-Year-Old Compared

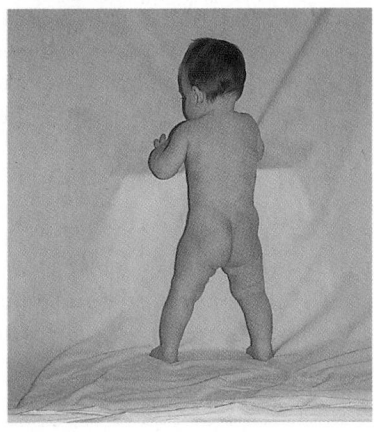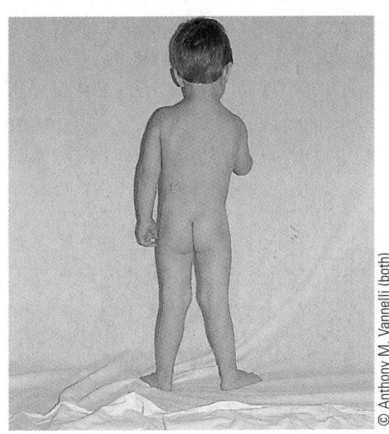

The body shape of a one-year-old (left) changes dramatically by age two (right). The two-year-old has lost much of the baby fat; the muscles (especially in the back, buttocks, and legs) have firmed and strengthened; and the leg bones have lengthened.

ries a day; an active six-year-old needs twice as many kcalories a day. By age ten, an active child needs about 2000 kcalories a day. Total energy needs increase slightly with age, but energy needs per kilogram of body weight actually decline gradually.

Physically active children of any age need more energy because they expend more, and inactive children can become obese even when they eat less food than the average. Unfortunately, our nation's children are becoming less and less active, with young girls showing a marked reduction in their physical activity. Schools would serve our children well by offering activities to promote physical fitness.[38] Children who learn to enjoy physical play and exercise, both at home and at school, are best prepared to maintain active lifestyles as adults.

Dietary Guidelines for Americans 2005

Children should engage in at least 60 minutes of physical activity on most, preferably all, days of the week.

Some children, notably those adhering to a vegan diet, may have difficulty meeting their energy needs. Grains, vegetables, and fruits provide plenty of fiber, adding bulk, but may provide too few kcalories to support growth. Soy products, other legumes, and nut or seed butters offer more concentrated sources of energy to support optimal growth and development.[39]

Carbohydrate and Fiber Carbohydrate recommendations are based on glucose use by the brain. After one year of age, brain glucose use remains fairly constant and is within the adult range. Carbohydrate recommendations for children from the age of one year on are therefore the same as for adults (see inside front cover).[40]

Fiber recommendations ◆ derive from adult intakes shown to reduce the risk of coronary heart disease and are based on energy intakes. Consequently, fiber recommendations for younger children with low energy intakes are less than those for older ones with high energy intakes.[41]

Dietary Guidelines for Americans 2005

Children and adolescents should consume whole-grain products often, and at least half of the grains should be whole grains.

◆ Fiber recommendations for children:

Age (yr)	AI (g/day)
1–3	19
4–8	25
9–13	
Boys	31
Girls	26
14–18	
Boys	38
Girls	26

Fat and Fatty Acids No RDA for total fat has been established, but the DRI Committee recommends a fat intake of 30 to 40 percent of energy for children 1 to 3 years of age and 25 to 35 percent for children 4 to 18 years of age.[42] As long as children's energy intakes are adequate, fat intakes below 30 percent of total energy do not impair growth.[43] Children who eat low-fat diets, however, tend to have low intakes of some vitamins and minerals. Recommended intakes of the essential fatty acids are based on average intakes (see inside front cover).

Dietary Guidelines for Americans 2005

Keep total fat intake between 30 to 35 percent of kcalories for children 2 to 3 years of age and between 25 and 35 percent of kcalories for children and adolescents 4 to 18 years of age, with most fats coming from sources of polyunsaturated and monounsaturated fatty acids, such as fish, nuts, and vegetable oils.

Protein Like energy needs, total protein needs increase slightly with age, but when the child's body weight is considered, the protein requirement actually declines slightly (see inside front cover). Protein recommendations must consider the requirements for maintaining nitrogen balance, the quality of protein consumed, and the added needs of growth.

Vitamins and Minerals The vitamin and mineral needs of children increase with age (see inside front cover). A balanced diet of nutritious foods can meet children's needs for these nutrients, with the notable exception of iron. Iron-deficiency anemia is a major problem worldwide, as well as being prevalent among U.S. and Canadian children, especially toddlers one to two years of age.[44] During the second year of life, toddlers progress from a diet of iron-rich infant foods such as breast milk, iron-fortified formula, and iron-fortified infant cereal to a diet of adult foods and iron-poor cow's milk. In addition, their appetites often fluctuate—some become finicky about the foods they eat, and others prefer milk and juice to solid foods.[45] All of these situations can interfere with children eating iron-rich foods at a critical time for brain growth and development.

To prevent iron deficiency, children's foods must deliver 7 to 10 milligrams of iron per day. To achieve this goal, snacks and meals should include iron-rich foods, and milk intake should be reasonable so that it will not displace lean meats, fish, poultry, eggs, legumes, and whole-grain or enriched products. (Chapter 13 described iron-rich foods and ways to maximize iron absorption.)

Supplements With the exception of specific recommendations for fluoride, iron, and vitamin D during infancy and childhood, the AAP and other professional groups agree that well-nourished children do not need vitamin and mineral supplements. Despite this, many children and adolescents take supplements.[46] Ironically, children with poor nutrient intakes typically do not receive supplements, and those who do take supplements typically receive extra nutrients they do not need.[47] Furthermore, researchers are still studying the safety of supplement use by children.[48] The Federal Trade Commission has warned parents about giving supplements advertised to prevent or cure childhood illnesses such as colds, ear infections, or asthma. Dietary supplements on the market today include many herbal products that have not been tested for safety and effectiveness in children.

Planning Children's Meals To provide all the needed nutrients, children's meals should include a variety of foods from each food group—in amounts suited to their

◆ www.MyPyramid.gov/kids

appetites and needs. Figure 15-8 presents MyPyramid ◆ designed for children 6 to 11 years of age and includes the recommended amounts of food for an 1800-kcalorie intake. Table 15-4 (p. 530) lists amounts of food for several kcalorie levels below 1800 kcalories, which are appropriate for most younger children and sedentary older children. Review Table 2-3 on page 41 for recommended daily amounts of foods from each group for higher kcalorie levels, which are appropriate for active older children.

FIGURE 15-8 Food Guide Pyramid for Young Children

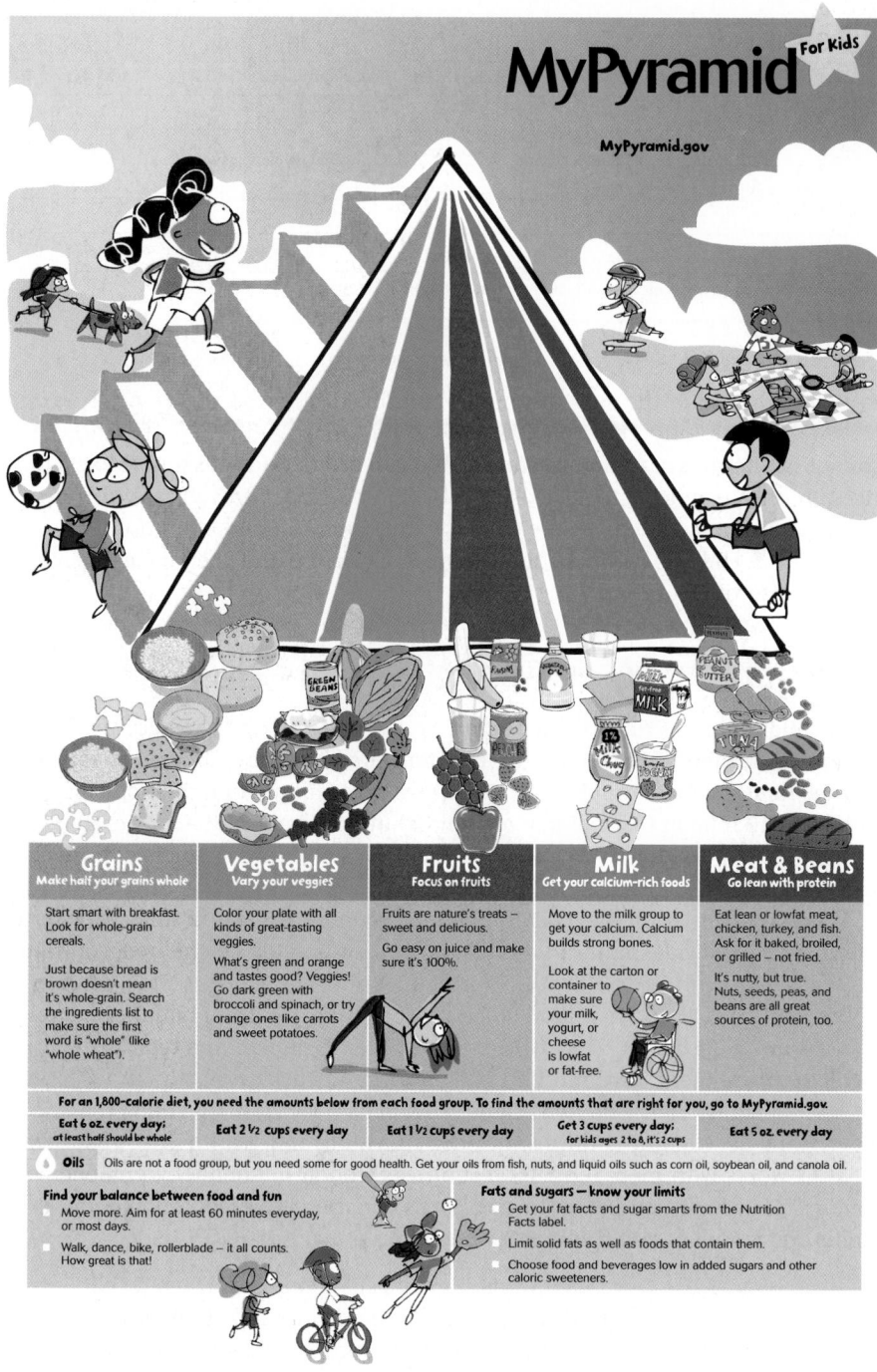

Estimated daily kcalorie needs for active and sedentary children of various ages are shown in Table 15-5 (p. 530).

Children whose diets follow the pattern presented in Figure 15-8 meet their nutrient needs fully, but few children eat according to these recommendations. Based on an analysis of the most recent national food intake data, the USDA found that most (81 percent) children between two and nine years of age have diets that need substantial improvement.[49] A comprehensive survey, called the Feeding Infants and Toddlers Study (FITS), assessed the food and nutrient intakes of more than 3000 infants and toddlers.[50] The survey found that fruit and vegetable intakes of infants and toddlers are limited, and in fact, about 25 percent of infants and toddlers older than 9 months did not eat a single serving of fruits or vegetables in a day.[51] By 15

TABLE 15-4 Recommended Daily Amounts from Each Food Group (1000 to 1600 kCalories)

Food Group	1000 kcal	1200 kcal	1400 kcal	1600 kcal
Fruits	1 c	1 c	1½ c	1½ c
Vegetables	1 c	1½ c	1½ c	2 c
Grains	3 oz	4 oz	5 oz	5 oz
Meat and legumes	2 oz	3 oz	4 oz	5 oz
Milk	2 c	2 c	2 c	3 c
Oils	3 tsp	3 tsp	3 tsp	4 tsp

NOTE: The discretionary kcalorie allowance for these patterns is about 100 kcalories.

TABLE 15-5 Estimated Daily kCalorie Needs for Children

Children	Sedentary[a]	Active[b]
2 to 3 yr	1000	1400
Females		
4 to 8 yr	1200	1800
9 to 13 yr	1600	2200
Males		
4 to 8 yr	1400	2000
9 to 13 yr	1800	2600

[a] *Sedentary* describes a lifestyle that includes only the activities typical of day-to-day life.
[b] *Active* describes a lifestyle that includes at least 60 minutes per day of moderate physical activity (equivalent to walking more than 3 miles per day at 3 to 4 miles per hour) in addition to the activities of day-to-day life.

to 18 months of age, the most commonly consumed vegetable was French fries and the most commonly consumed fruit was bananas—neither particularly rich sources of vitamins or minerals. Parents and caregivers of infants and toddlers thus need to offer a much greater variety of nutrient-dense vegetables and fruits at meals and snacks to help ensure adequate nutrition. Among other nutrition concerns for U.S. children are inadequate intakes of calcium and fiber and excessive intakes of saturated fat.[52]

Hunger and Malnutrition in Children

Most children in the United States and Canada have access to regular meals, but hunger and malnutrition do appear in certain circumstances. Children in very low-income families, for example, are more likely to be hungry and malnourished. An estimated 12 million U.S. children are hungry at least some of the time and are living in poverty.[53] Highlight 16 examines the causes and consequences of hunger in the United States.

When hunger is chronic, children become malnourished and suffer growth retardation. Worldwide, malnutrition takes a devastating toll on children, contributing to nearly half of the deaths of children under four years old. Vitamin A deficiency afflicts 3 to 10 million children worldwide, inducing blindness, stunted growth, and infections.[54] Zinc deficiency also retards growth and typically accompanies protein-energy malnutrition and vitamin A deficiency.

The United Nations Children's Fund, known as UNICEF, helps children living in poverty in developing countries get the nutrition and health care they need. UNICEF works with more than 160 countries through national governments, private-sector partners, and other international agencies to protect children and their rights and to reduce childhood death and illness.

Hunger and Behavior Even when hunger is temporary, as when a child misses one meal, behavior and academic performance are affected. Children who eat nutritious breakfasts improve their school performance and are tardy or absent significantly less often than their peers who do not. A nutritious breakfast is a central feature of a diet that meets the needs of children and supports their healthy growth and development.[55] Children who skip breakfast typically do not make up the deficits at later meals—they simply have lower intakes of energy, vitamins, and minerals than those who eat breakfast. Without breakfast, children perform poorly in tasks requiring concentration, their attention spans are shorter, and they even score lower on intelligence tests than their well-fed peers. Malnourished children are particularly vulnerable. Common sense dictates that it is unreasonable to expect anyone to learn and perform without fuel. For the child who hasn't had breakfast, the morning's lessons may be lost altogether. Even if a child has eaten breakfast, discomfort from hunger may become distracting by late morning. Teachers aware of the late-morning slump in their classrooms wisely request that midmorning snacks be provided; snacks improve classroom performance all the way to lunchtime.

HOW TO Protect against Lead Toxicity

Researchers simultaneously made three major discoveries about lead toxicity: lead poisoning has *subtle* effects, the effects are *permanent,* and they occur at *low levels of exposure.* The amount of lead recognized to cause harm is only 10 micrograms per 100 milliliters of blood. Some research shows that blood lead concentrations *below* this amount may adversely affect children's scores on intelligence tests.[a] Consequently, consumers should take ultraconservative measures to protect themselves, and especially their infants and young children, from lead poisoning. The American Academy of Pediatrics and the Centers for Disease Control recommend screening in communities with a substantial number of houses built before 1950 and in those with a substantial number of children with elevated lead levels. In addition to screening children most likely to be exposed, pediatricians should alert all parents to the possible dangers of lead exposure and explain prevention strategies.

Preventive strategies include:

- In contaminated environments, keep small children from putting dirty or old painted objects in their mouths, and make sure children wash their hands before eating. Similarly, keep small children from eating any nonfood items. Lead poisoning has been reported in young children who have eaten crayons or pool cue chalk.

- Wet-mop floors and damp-sponge walls regularly. Children's blood lead levels decline when the homes they live in are cleaned regularly.

- Be aware that other countries do not have the same regulations protecting consumers against lead. Children have been poisoned by eating crayons made in China and drinking fruit juice canned in Mexico.

- Do not use lead-contaminated water to make infant formula.

- Once you have opened canned food, store it in a lead-free container to prevent lead migration into the food.

- Do not store acidic foods or beverages (such as vinegar or orange juice) in ceramic dishware or alcoholic beverages in pewter or crystal decanters.

- Many manufacturers are now making lead-safe products. Old, handmade, or imported ceramic cups and bowls may contain lead and should not be used to heat coffee or tea or acidic foods such as tomato soup.

- U.S. wineries have stopped using lead in their foil seals, but older bottles may still be around, and other countries may still use lead. To be safe, wipe the foil-sealed rim of a wine bottle with a clean wet cloth before removing the cork.

- Feed children nutritious meals regularly.

- Before using your newspaper to wrap food, mulch garden plants, or add to your compost, confirm with the publisher that the paper uses no lead in its ink.

The Environmental Protection Agency (EPA) also publishes a booklet, *Lead and Your Drinking Water,* in which the following cautions appear:

- Have the water in your home tested by a competent laboratory.

- Use only cold water for drinking, cooking, and making formula (cold water absorbs less lead).

- When water has been standing in pipes for more than two hours, flush the cold-water pipes by running water through them for 30 seconds before using it for drinking, cooking, or mixing formulas.

- If lead contamination of your water supply seems probable, obtain additional information and advice from the EPA and your local public health agency.

By taking these steps, parents can protect themselves and their children from this preventable danger.[b]

[a]R. L. Canfield and coauthors, Intellectual impairment in children with blood lead concentrations below 10 μg per deciliter, *New England Journal of Medicine* 348 (2003): 1517–1526.

[b]Call the National Lead Information Center hotline at (800) 424-LEAD (424-5323) for general information.

Misbehaving Even a child who is not truly hyperactive can be difficult to manage at times. Michael may act unruly out of a desire for attention, Jessica may be cranky because of a lack of sleep, Christopher may react violently after watching too much television, and Sheila may be unable to sit still in class due to a lack of exercise. All of these children may benefit from more consistent care—regular hours of sleep, regular mealtimes, and regular outdoor activity.

Food Allergy and Intolerance

Food allergy is frequently blamed for physical and behavioral abnormalities in children, but just 6 percent of children are diagnosed with true food allergies.[64] Food allergies diminish with age, until in adulthood they affect only about 1 or 2 percent of the population.[65]

A true food allergy occurs when fractions of a food protein or other large molecule are absorbed into the blood and elicit an immunologic response. (Recall that proteins are normally dismantled in the digestive tract to amino acids that are absorbed without such a reaction.) The body's immune system reacts to these large food molecules as it does to other antigens—by producing antibodies, histamines, or other defensive agents.

Detecting Food Allergy Allergies may have one or two components. They always involve antibodies, but they may or may not involve symptoms. ◆ This means

◆ A person who produces antibodies *without* having any symptoms has an **asymptomatic allergy;** a person who produces antibodies *and* has symptoms has a **symptomatic allergy.**

food allergy: an adverse reaction to food that involves an immune response; also called **food-hypersensitivity reaction.**

These normally wholesome foods may cause life-threatening symptoms in people with allergies.

◆ Symptoms of impending anaphylactic shock:
- Tingling sensation in mouth
- Swelling of the tongue and throat
- Irritated, reddened eyes
- Difficulty breathing, asthma
- Hives, swelling, rashes
- Vomiting, abdominal cramps, diarrhea
- Drop in blood pressure
- Loss of consciousness
- Death

◆ Reminder: *Epinephrine* is a hormone of the adrenal gland that modulates the stress response; formerly called **adrenaline.** When administered by injection, epinephrine counteracts anaphylactic shock by opening the airways and maintaining heartbeat and blood pressure.

---

anaphylactic (ana-fill-LAC-tic) **shock:** a life-threatening, whole-body allergic reaction to an offending substance.

adverse reactions: unusual responses to food (including intolerances and allergies).

food intolerances: adverse reactions to foods that do not involve the immune system.

---

that allergies can be diagnosed only by testing for antibodies. Even symptoms exactly like those of an allergy may not be caused by an allergy. However, once a food allergy has been diagnosed, the required treatment is strict elimination of the offending food. Children with allergies, like all children, need all their nutrients, so it is important to include other foods that offer the same nutrients as the omitted foods.[66]

Allergic reactions to food may be immediate or delayed. In either case, the antigen interacts immediately with the immune system, but the timing of symptoms varies from minutes to 24 hours after consumption of the antigen. Identifying the food that causes an immediate allergic reaction is fairly easy because the symptoms appear shortly after the food is eaten. Identifying the food that causes a delayed reaction is more difficult because the symptoms may not appear until much later. By this time, many other foods may have been eaten, complicating the picture.

Anaphylactic Shock The life-threatening food allergy reaction of **anaphylactic shock** is most often caused by peanuts, tree nuts, milk, eggs, wheat, soybeans, fish, or shellfish. Among these foods, eggs, milk, soy, and peanuts most often cause problems in children. Children are more likely to outgrow allergies to eggs, milk, and soy than allergies to peanuts. Peanuts cause more life-threatening reactions than do all other food allergies combined. Research is currently under way to help people with peanut allergies tolerate small doses, thus saving lives and minimizing reactions.[67] One possible solution depends on finding a natural, hypoallergenic peanut among the 14,000 varieties of peanuts. Families of children with a life-threatening food allergy and school personnel who supervise them must guard them against any exposure to the allergen. The child must learn to identify which foods pose a problem and then learn and use refusal skills for all foods that may contain the allergen.

Parents of children with allergies can pack safe foods for lunches and snacks and ask school officials to strictly enforce a "no swapping" policy in the lunchroom. The child must be able to recognize the symptoms of impending anaphylactic shock, ◆ such as a tingling of the tongue, throat, or skin, or difficulty breathing. Any person with food allergies severe enough to cause anaphylactic shock should wear a medical alert bracelet or necklace. Finally, the responsible child and the school staff should be prepared with injections of epinephrine, ◆ which prevents anaphylaxis after exposure to the allergen. Many preventable deaths occur each year when people with food allergies accidentally ingest the allergen but have no epinephrine available.

Food Labeling As of 2006, food labels must list the presence of common allergens in plain language, using the names of the eight most common allergy-causing foods.[68] For example, a food containing "textured vegetable protein" must say "soy" on its label. Similarly, "casein" must be identified as "milk," and so forth. Food producers must also prevent cross-contamination during production and clearly label foods in which it is likely to occur.[69] For example, equipment used for making peanut butter must be scrupulously clean before being used to pulverize cashew nuts for cashew butter to protect unsuspecting cashew butter consumers from peanut allergens.

Technology may soon offer new solutions. New drugs are being developed that may interfere with the immune response that causes allergic reactions.[70] Also, through genetic engineering, scientists may one day create allergen-free peanuts, soybeans, and other foods to make them safer.

Food Intolerances Not all **adverse reactions** to foods are food allergies, although even physicians may describe them as such. Signs of adverse reactions to foods include stomachaches, headaches, rapid pulse rate, nausea, wheezing, hives, bronchial irritation, coughs, and other such discomforts. Among the causes may be reactions to chemicals in foods, such as the flavor enhancer monosodium glutamate (MSG), the natural laxative in prunes, or the mineral sulfur; digestive diseases, such as obstructions or injuries; enzyme deficiencies, such as lactose intolerance; and even psychological aversions. These reactions involve symptoms but no antibody production. Therefore, they are **food intolerances,** not allergies.

Pesticides on produce may also cause adverse reactions. Pesticides that were applied in the fields may linger on the foods. Health risks from pesticide exposure may be low for healthy adults, but children are vulnerable. Therefore, government agencies have set a **tolerance level** for each pesticide by first identifying foods that children commonly eat in large amounts and then considering the effects of pesticide exposure during each developmental stage.

Hunger, lead poisoning, hyperactivity, and allergic reactions can all adversely affect a child's nutrition status and health. Fortunately, each of these problems has solutions. They may not be easy solutions, but at least we have a reasonably good understanding of the problems and ways to correct them. Such is not the case with the most pervasive health problem for children in the United States—obesity.

Childhood Obesity

The number of overweight children has increased dramatically over the past three decades (see Figure 15-9). Like their parents, children in the United States are becoming fatter. An estimated 17 percent of U.S. children and adolescents 2 to 19 years of age are overweight.[71] Based on data from the BMI-for-age growth charts, children and adolescents are categorized as *at risk of overweight* above the 85th percentile and as *overweight* at the 95th percentile and above. Prevalence data reflect only children and adolescents in the overweight category. If those at risk of overweight were also included, the estimated 17 percent would likely double. Figure 15-10 (p. 536) presents the BMI for children and adolescents, indicating cutoff points for overweight and at risk of overweight.

The use of the term *overweight* instead of *obese* when referring to children with a BMI above the age- and gender-specific 95th percentiles is controversial. Some experts think it is best not to label children as obese, whereas others think it important to recognize the full extent of the problem. The Institute of Medicine's Committee on Prevention of Obesity in Children and Youth acknowledges the use of the term *overweight* to describe obese children but asserts that *obese* conveys the seriousness, urgency, medical nature, and need for immediate action more effectively than the term *overweight* does.[72]

The problem of obesity in children is especially troubling because overweight children have the potential of becoming obese adults with all the social, economic, and medical ramifications that often accompany obesity. They have additional problems, too, arising from differences in their growth, physical health, and psychological development. In trying to explain the rise in childhood obesity, researchers point to both genetic and environmental factors.

Genetic and Environmental Factors Parental obesity predicts an early increase in a young child's BMI, and it more than doubles the chances that a young child will become an obese adult. Children with neither parent obese have a less than 10 percent chance of becoming obese in adulthood, whereas overweight teens with at least one obese parent have a greater than 80 percent chance of being obese adults. Also, as children grow older, their body weight becomes an important factor in determining their obesity as adults.[73] The link between parental and child obesity reflects both genetic and environmental factors (as described in Chapter 9).

Diet and physical inactivity must also play a role in explaining why children are heavier today than they were 30 or so years ago. As the prevalence of childhood obesity throughout the United States has more than doubled for young children and adolescents, and tripled for children 6 to 11 years of age, the society our children live in has changed considerably.[74] In many families today, both parents work outside the home and work longer hours; more emphasis is placed on convenience foods and foods eaten away from home; meal choices at school are more diverse and often less nutritious; sedentary activities such as watching television and playing video or computer games occupy much of children's free time; and opportunities for

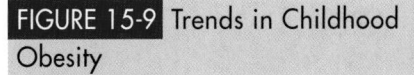

FIGURE 15-9 Trends in Childhood Obesity

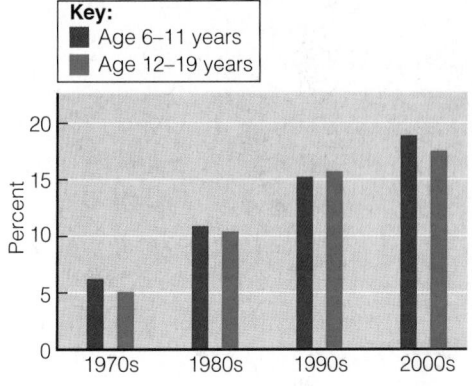

Key:
■ Age 6–11 years
■ Age 12–19 years

tolerance level: the maximum amount of residue permitted in a food when a pesticide is used according to the label directions.

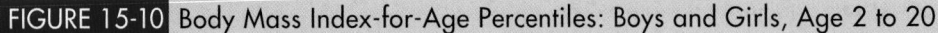

FIGURE 15-10 Body Mass Index-for-Age Percentiles: Boys and Girls, Age 2 to 20

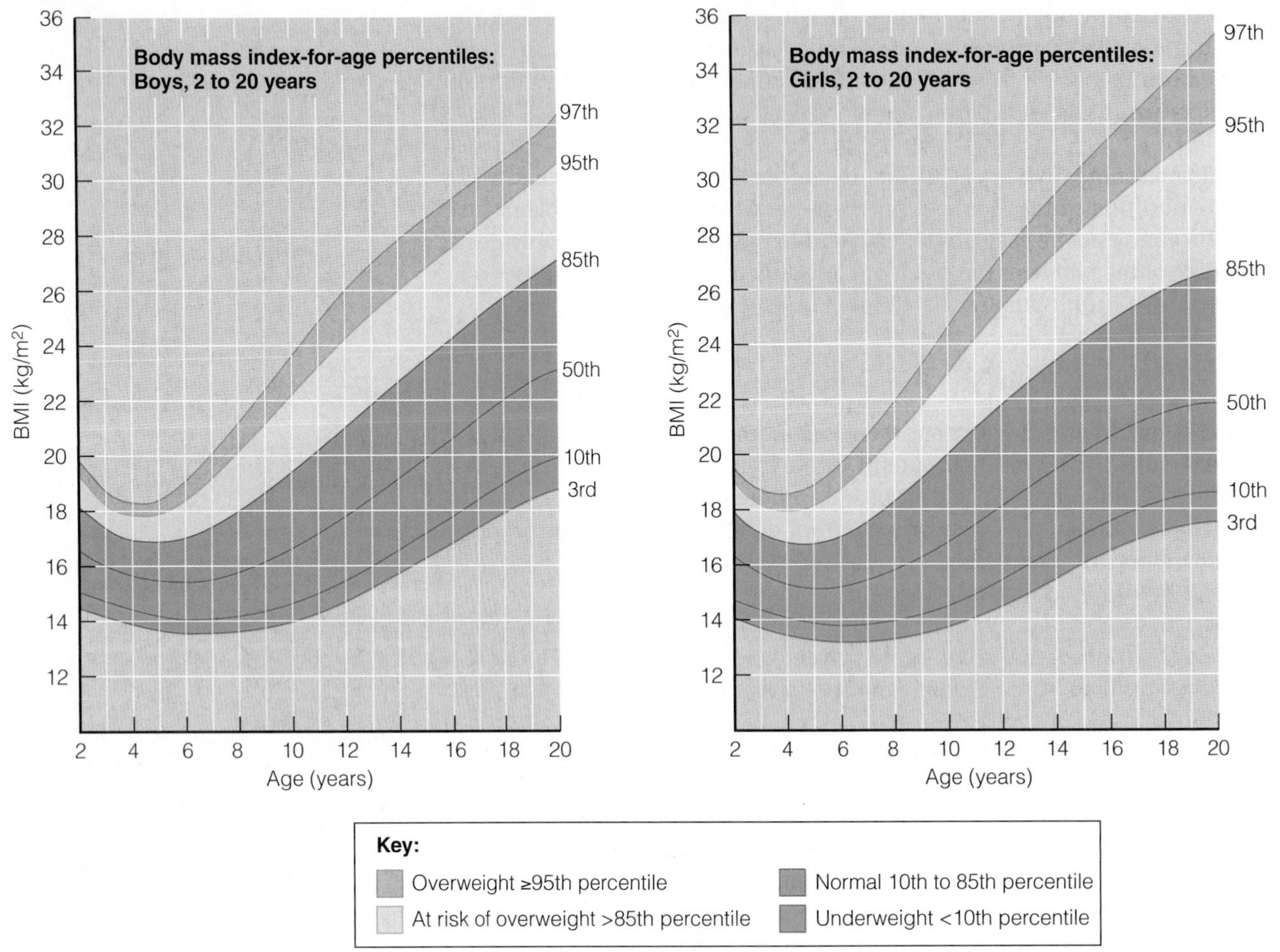

physical activity and outdoor play both during and after school have declined.[75] All of these factors—and many others—influence children's eating and activity patterns.

Children learn food behaviors from their families, and research confirms the significant roles parents play in teaching their children about healthy food choices, providing nutrient-dense foods, and serving as role models.[76] When parents eat fruits and vegetables frequently, their children do, too.[77] The more fruits and vegetables children eat, the more vitamins, minerals, and fibers, and the less saturated fat in their diets.

Research shows that one in four toddlers (19 to 24 months of age) exceeds estimated energy requirements as a result of eating such foods as candy, pizza, chicken nuggets, soda, sweet tea, and salty snacks like cheese puffs and chips.[78] Thus, when researchers ask, "Are today's children eating more kcalories than those of 30 years ago?" the answer is, "Yes." Some researchers report an increase of 100 to 200 kcalories a day for all age groups, enough to account for significant weight gains.[79]

Coincidentally or not, as the prevalence of obesity among both children and adults has surged over the past three decades, so has the consumption of added sugars and, especially, high-fructose corn syrup—the easily consumed, energy-dense liquid sugar added to soft drinks. Each 12-ounce can of soft drink provides the equivalent of about 10 teaspoons of sugar and 150 kcalories. More than half of children in school consume at least one soft drink each day at school; adolescent males consume the most—four or more cans daily.[80] According to one estimate, the risk of obesity increases by 60 percent with each sugared soft drink consumed daily.[81]

No doubt, the tremendous increase in soft drink consumption plays a role, but much of the obesity epidemic can be explained by lack of physical activity. Children have become more sedentary, and sedentary children are more often overweight.[82] Television watching ◆ may contribute most to physical inactivity. A child who spends more than an hour or two each day in front of a television, computer monitor, or other media can become overweight and develop unhealthy blood lipids even while eating fewer kcalories than a more active child.[83]

Children who have television sets in their bedrooms spend more time watching TV and are more likely to be overweight than children who do not have televisions in their rooms.[84] Children who watch a great deal of television are most likely to be overweight and least likely to eat family meals or fruits and vegetables.[85] They often snack on the nutrient-poor, energy-dense foods that are advertised.[86] The average child sees an estimated 30,000 TV commercials a year—many peddling foods high in sugar, saturated fat, and salt such as sugar-coated breakfast cereals, candy bars, chips, fast foods, and carbonated beverages. More than half of all food advertisements are aimed specifically at children and market their products as fun and exciting.[87] Not surprisingly, the more time children spend watching television, the more they request these advertised foods and beverages—and they get their requests about half of the time.[88] The most popular foods and beverages are marketed to children and adolescents on the Internet as well, using "advergaming" (advertised product as part of a game), cartoon characters or "spokes-characters," and designated children's areas.[89]

The physically inactive time spent watching television is second only to time spent sleeping. Children also spend more time playing video games. These activities use no more energy than resting, displace participation in more vigorous activities, and foster snacking on high-fat foods.[90] Simply reducing the amount of time spent watching television (and playing video games) can improve a child's BMI. The American Academy of Pediatrics (AAP) now recommends limiting television and video time to two hours per day as a strategy to help prevent childhood obesity.[91]

Growth Overweight children develop a characteristic set of physical traits. They typically begin puberty earlier and so grow taller than their peers at first, but then they stop growing at a shorter height. They develop greater bone and muscle mass in response to the demand of having to carry more weight—both fat and lean weight. Consequently, they appear "stocky" even when they lose their excess fat.

Physical Health Like overweight adults, overweight children display a blood lipid profile indicating that atherosclerosis is beginning to develop—high levels of total cholesterol, triglycerides, and LDL cholesterol. Overweight children also tend to have high blood pressure; in fact, obesity is a leading cause of pediatric hypertension.[92] Their risks for developing type 2 diabetes and respiratory diseases (such as asthma) are also exceptionally high.[93] These relationships between childhood obesity and chronic diseases are discussed fully in Highlight 15.

Psychological Development In addition to the physical consequences, childhood obesity brings a host of emotional and social problems.[94] Because people frequently judge others on appearance more than on character, overweight children are often victims of prejudice. Many suffer discrimination by adults and rejection by their peers. They may have poor self-images, a sense of failure, and a passive approach to life. Television shows, which are a major influence in children's lives, often portray the fat person as the bumbling misfit. Overweight children may come to accept this negative stereotype in themselves and in others, which can lead to additional emotional and social problems. Researchers investigating children's reactions to various body types find that both normal-weight and underweight children respond unfavorably to overweight bodies.

Prevention and Treatment of Obesity Medical science has worked wonders in preventing or curing many of even the most serious childhood diseases, but obesity remains a challenge.[95] Once excess fat has been stored, it is challenging to lose. In light of all this, parents are encouraged to make major efforts to prevent childhood

◆ TV fosters obesity because it:
- Requires no energy beyond basal metabolism
- Replaces vigorous activities
- Encourages snacking
- Promotes a sedentary lifestyle

Playing video games influences children's activity patterns similarly.

Television watching influences children's eating habits and activity patterns.

obesity or to begin treatment early—before adolescence.[96] Treatment must consider the many aspects of the problem and possible solutions. The most successful approach integrates diet, physical activity, psychological support, and behavioral changes.[97]

Diet The initial goal for overweight children is to reduce the rate of weight gain; that is, to maintain weight while the child grows taller. Continued growth will then accomplish the desired change in weight for height. Weight loss is usually not recommended because diet restriction can interfere with growth and development. Intervention for some older, overweight children with accompanying medical conditions may warrant weight loss, but this treatment requires an individualized approach based on the degree of overweight and severity of the medical conditions.[98] Whether the goal is to treat or prevent obesity, the following strategies may be helpful:

- Serve family meals that reflect kcalorie control both in the foods offered and in the ways foods are prepared.
- Involve children in shopping for food and preparing meals.
- Encourage children to eat only when they are hungry, to eat slowly, to pause and enjoy their table companions, and to stop eating when they are full.
- Teach them how to select nutrient-dense foods (low-fat and non-fat milk and milk products for children 3 years of age and older, fruits and vegetables, whole grains, legumes, fish, and lean meat) that will meet their nutrient needs within their energy allowances. Also, teach them to serve themselves appropriate portions at meals; the amount of food offered influences the amount of food eaten.[99]
- Limit foods high in saturated and *trans* fats (see Table H5-1 in Highlight 5) and high-sugar foods, including sugar-sweetened soft drinks.
- Never force children to clean their plates.
- Plan for snack times and provide a variety of nutritious snacks (see Table 15-8 later in this chapter).
- Discourage eating while watching TV.

Dietary Guidelines for Americans 2005

Help overweight children reduce the rate of body weight gain while allowing growth and development. Consult a health care provider before placing a child on a weight-reduction diet.

Physical Activity The many benefits of physical activity are well known but often are not enough to motivate overweight people, especially children. Yet regular vigorous activity can improve a child's weight, body composition, and physical fitness.[100] Ideally, parents will limit sedentary activities and encourage daily physical activity to promote strong skeletal, muscular, and cardiovascular development and instill in their children the desire to be physically active throughout life. Most importantly, parents need to set a good example. Physical activity is a natural and lifelong behavior of healthy living. It can be as simple as riding a bike, playing tag, jumping rope, or doing chores. It need not be an organized sport; it just needs to be some activity on a regular basis. The AAP supports the efforts of schools to include more physical activity in the curriculum and encourages parents to support their children's participation.[101]

Psychological Support Weight-loss programs that involve parents and other caregivers in treatment report greater success than those without parental involvement. Because obesity in parents and their children tends to be positively correlated, both benefit when parents participate in a weight-loss program. Parental attitudes about food greatly influence children's eating behavior, so it is important that the influence be positive. Otherwise, eating problems may become exacerbated.

Behavioral Changes In contrast to traditional weight-loss programs that focus on *what* to eat, behavioral programs focus on *how* to eat. These techniques involve changing learned habits that lead a child to eat excessively.

Obesity is prevalent in our society. Because treatment of obesity is frequently unsuccessful, it is most important to prevent its onset. Above all, be sensible in teaching children how to maintain appropriate body weight. Children can easily get the impression that their worth is tied to their body weight. Parents and the media are most influential in shaping self-concept, weight concerns, and dieting practices.[102] Some parents fail to realize that society's ideal of slimness can be perilously close to starvation and that a child encouraged to "diet" cannot obtain the energy and nutrients required for normal growth and development. Even healthy children without diagnosable eating disorders have been observed to limit their growth through "dieting." Weight gain in truly overweight children can be managed without compromising growth, but it should be overseen by a health care professional.

Eating is more fun for children when friends are there.

Mealtimes at Home

Traditionally, parents served as **gatekeepers,** determining what foods and activities were available in their children's lives. Then the children made their own selections. Gatekeepers who wanted to promote nutritious choices and healthful habits provided access to nutrient-dense, delicious foods and opportunities for active play at home.

In today's consumer-oriented society, children have greater influence over family decisions concerning food—the fast-food restaurant the family chooses when eating out, the type of food the family eats at home, and the specific brands the family purchases at the grocery store. Parental guidance in food choices is still necessary, but teaching children consumer skills to help them make informed choices is equally important.

Honoring Children's Preferences Researchers attempting to explain children's food preferences encounter contradictions. Children say they like colorful foods, yet they most often reject green and yellow vegetables in favor of brown peanut butter and white potatoes, apple wedges, and bread. They seem to like raw vegetables better than cooked ones, so it is wise to offer vegetables that are raw or slightly undercooked, served separately, and easy to eat. Foods should be warm, not hot, because a child's mouth is much more sensitive than an adult's. The flavor should be mild because a child has more taste buds, and smooth foods such as mashed potatoes or split-pea soup should contain no lumps (a child wonders, with some disgust, what the lumps might be). Children prefer foods that are familiar, so offer various foods regularly.

Make mealtimes fun for children. Young children like to eat at little tables and to be served small portions of food. They like sandwiches cut in different geometric shapes and common foods called silly names. They also like to eat with other children, and they tend to eat more when in the company of their friends. Children are also more likely to give up their prejudices against foods when they see their peers eating them.

Learning through Participation Allowing children to help plan and prepare the family's meals provides enjoyable learning experiences and encourages children to eat the foods they have prepared. Vegetables are pretty, especially when fresh, and provide opportunities for children to learn about color, seeds, growing vegetables, and shapes and textures—all of which are fascinating to young children. Measuring, stirring, washing, and arranging foods are skills that even a young child can practice with enjoyment and pride (see Table 15-7).

Avoiding Power Struggles Problems over food often arise during the second or third year, when children begin asserting their independence. Many of these problems stem from the conflict between children's developmental stages and capabilities and parents who, in attempting to do what they think is best for their children,

TABLE 15-7 Food Skills of Preschool Children[a]

Age 1 to 2 years, when large muscles develop:
- Uses short-shanked spoon
- Helps feed self
- Lifts and drinks from cup
- Helps scrub, tear, break, or dip foods

Age 3 years, when medium hand muscles develop:
- Spears food with fork
- Feeds self independently
- Helps wrap, pour, mix, shake, or spread foods
- Helps crack nuts with supervision

Age 4 years, when small finger muscles develop:
- Uses all utensils and napkin
- Helps roll, juice, mash, or peel foods
- Cracks egg shells

Age 5 years, when fine coordination of fingers and hands develops:
- Helps measure, grind, grate, and cut (soft foods with dull knife)
- Uses hand mixer with supervision

[a]These ages are approximate. Healthy, normal children develop at their own pace.

gatekeepers: with respect to nutrition, key people who control other people's access to foods and thereby exert profound impacts on their nutrition. Examples are the spouse who buys and cooks the food, the parent who feeds the children, and the caregiver in a day-care center.

Children enjoy eating the foods they help to prepare.

try to control every aspect of eating. Such conflicts can disrupt children's abilities to regulate their own food intakes or to determine their own likes and dislikes. For example, many people share the misconception that children must be persuaded or coerced to try new foods. In fact, the opposite is true. When children are forced to try new foods, even by way of rewards, they are less likely to try those foods again than are children who are left to decide for themselves. Similarly, when children are restricted from eating their favorite foods, they are more likely to want those foods.[103] Wise parents provide healthful foods and allow their child to determine *how much* and even *whether* to eat.

When introducing new foods, offer them one at a time and only in small amounts such as one bite at first. The more often a food is presented to a young child, the more likely the child will accept that food. Offer the new food at the beginning of the meal, when the child is hungry, and allow the child to make the decision to accept or reject it. Never make an issue of food acceptance.

Choking Prevention Parents must always be alert to the dangers of choking. A choking child is silent, so an adult should be present whenever a child is eating. Make sure the child sits when eating; choking is more likely when a child is running or falling. (See p. 525 for a list of foods and nonfood items most likely to cause choking.)

Playing First Children may be more relaxed and attentive at mealtime if outdoor play or other fun activities are scheduled before, rather than immediately after, mealtime. Otherwise children "hurry up and eat" so that they can go play.

Snacking Parents may find that when their children snack, they aren't hungry at mealtimes. Instead of teaching children *not* to snack, parents are wise to teach them *how* to snack. Provide snacks that are as nutritious as the foods served at mealtime. Snacks can even be mealtime foods served individually over time, instead of all at once on one plate. When providing snacks to children, think of the five food groups and offer such snacks as pieces of cheese, tangerine slices, and egg salad on whole-wheat crackers (see Table 15-8). Snacks that are easy to prepare should be readily available to children, especially if they arrive home from school before their parents.

To ensure that children have healthy appetites and plenty of room for nutritious foods when they are hungry, parents and teachers must limit access to candy, soft drinks, and other concentrated sweets. Limiting access includes limiting the amount of pocket money children have to buy such foods themselves.[104] If these foods are permitted in large quantities, the only possible outcomes are nutrient deficiencies, obesity, or both. The preference for sweets is innate; most children do not naturally select nutritious foods on the basis of taste. When children are allowed to create meals freely from a variety of foods, they typically select foods that provide a lot of sugar. When their parents are watching, or even when they only think their parents are watching, children improve their selections.

Sweets need not be banned altogether. Children who are exceptionally active can enjoy high-kcalorie foods such as ice cream or pudding from the milk group or pancakes from the bread group. Sedentary children need to become more active so they can also enjoy some of these foods without unhealthy weight gain.

Preventing Dental Caries Children frequently snack on sticky, sugary foods that stay on the teeth and provide an ideal environment for the growth of bacteria that cause dental caries. Teach children to brush and floss after meals, to brush or rinse after eating snacks, to avoid sticky foods, and to select crisp or fibrous foods frequently.

Serving as Role Models In an effort to practice these many tips, parents may overlook perhaps the single most important influence on their children's food habits—themselves.[105] Parents who don't eat carrots shouldn't be surprised when their children refuse to eat carrots. Likewise, parents who comment negatively on the smell of brussels sprouts may not be able to persuade children to try them. Children learn much through imitation. It is not surprising that children prefer the foods other family members enjoy and dislike foods that are never offered to them.[106] Parents, older

TABLE 15-8 Healthful Snack Ideas—Think Food Groups, Alone and in Combination

Selecting two or more foods from different food groups adds variety and nutrient balance to snacks. The combinations are endless, so be creative. Whenever possible, choose whole grains, low-fat or reduced-fat milk products, and lean meats.

Grains

Grain products are filling snacks, especially when combined with other foods:

- Cereal with fruit and milk
- Crackers and cheese
- Whole-grain toast with peanut butter
- Popcorn with grated cheese
- Oatmeal raisin cookies with milk

Vegetables

Cut-up, fresh, raw vegetables make great snacks alone or in combination with foods from other food groups:

- Celery with peanut butter
- Broccoli, cauliflower, and carrot sticks with a flavored cottage cheese dip

Fruits

Fruits are delicious snacks and can be eaten alone—fresh, dried, or juiced—or combined with other foods:

- Apples and cheese
- Bananas and peanut butter
- Peaches with yogurt
- Raisins mixed with sunflower seeds or nuts

Meats and Legumes

Meats and legumes add protein to snacks:

- Refried beans with nachos and cheese
- Tuna on crackers
- Luncheon meat on whole-grain bread

Milk and Milk Products

Milk can be used as a beverage with any snack, and many other milk products, such as yogurt and cheese, can be eaten alone or with other foods as listed above.

---

siblings, and other caregivers set an irresistible example by sitting with younger children, eating the same foods, and having pleasant conversations during mealtimes.

While serving and enjoying food, caregivers can promote both physical and emotional growth at every stage of a child's life. They can help their children develop both a positive self-concept and a positive attitude toward food. With good beginnings, children will grow without the conflicts and confusions about food that can lead to nutrition and health problems.

Nutrition at School

While parents are doing what they can to establish good eating habits in their children at home, others are preparing and serving foods to their children at day-care centers and schools. In addition, children begin to learn about food and nutrition in the classroom. Meeting the nutrition and education needs of children is critical to supporting their healthy growth and development.[107] ◆

Meals at School The U.S. government assists schools financially so that every student can receive nutritious meals at school. Both the School Breakfast Program and

◆ The American Dietetic Association has set nutrition standards for child-care programs. Among them, meal plans should include the following:

- Be nutritionally adequate and consistent with the *Dietary Guidelines for Americans*
- Involve parents in planning
- Follow recommended meal patterns that balance energy and nutrients with children's ages, appetites, activity levels, and special needs while respecting cultural and ethnic differences
- Minimize added fat, sugar, and sodium
- Emphasize fresh fruit, fresh and frozen vegetables, and whole grains
- Provide furniture and eating utensils that are age appropriate and developmentally suitable to encourage children to accept and enjoy mealtime

School lunches provide children with nourishment at little or no charge.

◆ The school breakfast must contain at a minimum:
- One serving of fluid milk
- One serving of fruit or vegetable or full-strength juice
- Two servings of bread or bread alternates; or two servings of meat or meat alternates; or one of each

the National School Lunch Program provide meals at a reasonable cost to children from families with the financial means to pay. Meals are available free or at reduced cost to children from low-income families. In addition, schools can obtain food commodities. Nationally, the U.S. Department of Agriculture (USDA) administers the programs; on the state level, state departments of education operate them.* The programs usually cost local school districts little, but the educational rewards are great. Several studies have reported that children who participate in school food programs perform better in the classroom.[108]

More than 28 million children receive lunches through the National School Lunch Program—half of them free or at a reduced price.[109] School lunches offer a variety of food choices and help children meet at least one-third of their recommended intakes for energy, protein, vitamin A, vitamin C, iron, and calcium. Table 15-9 shows school lunch patterns for children of different ages and specifies the numbers of servings of milk, protein-rich foods (meat, poultry, fish, cheese, eggs, legumes, or peanut butter), vegetables, fruits, and breads or other grain foods. In an effort to help reduce disease risk, all government-funded meals served at schools must follow the *Dietary Guidelines for Americans.*

Parents often rely on school lunches to meet a significant part of their children's nutrient needs on school days. Indeed, students who regularly eat school lunches have higher intakes of many nutrients and fiber than students who do not.[110]

The School Breakfast Program ◆ is available in more than 80 percent of the nation's schools that offer school lunch, and close to 9 million children participate in it.[111] Nevertheless, for many children who need it, the School Breakfast Program is either unavailable, or the children do not participate in it.[112] The majority of children who eat school breakfasts are from low-income families. As research results continue to emphasize the positive impact breakfast has on school performance and health, vigorous campaigns to expand school breakfast programs are under way.

---

* School lunches in Canada are administered locally and therefore vary from area to area.

Food Group	Preschool (Age)		Grade School through High School (Grade)		
	1 to 2	3 to 4	K to 3	4 to 6	7 to 12
Meat or meat alternate 1 serving:					
Lean meat, poultry, or fish	1 oz	1½ oz	1½ oz	2 oz	3 oz
Cheese	1 oz	1½ oz	1½ oz	2 oz	3 oz
Large egg(s)	½	¾	¾	1	1½
Cooked dry beans or peas	¼ c	⅜ c	⅜ c	½ c	¾ c
Peanut butter	2 tbs	3 tbs	3 tbs	4 tbs	6 tbs
Yogurt	½ c	¾ c	¾ c	1 c	1½ c
Peanuts, soynuts, tree nuts, or seeds[b]	½ oz	¾ oz	¾ oz	1 oz	1½ oz
Vegetable and/or fruit 2 or more servings, both to total	½ c	½ c	½ c	¾ c	¾ c
Bread or bread alternate[c] Servings	5/week	8/week	8/week	8/week	10/week
Milk 1 serving of fluid milk	¾ c	¾ c	1 c	1 c	1 c

TABLE 15-9 School Lunch Patterns for Different Ages[a]

[a]The quantities listed represent per-lunch minimums for each age and grade except those for the oldest group, which are recommendations. Schools unable to serve the recommended quantities for grades 7 to 12 must provide at least the amount shown for grades 4 to 6.
[b]These meat alternates may be used to meet no more than half of the meat or meat alternate requirement; therefore, they must be used in a meal with another meat or meat alternate.
[c]Schools must serve daily at least ½ serving of bread or bread alternate to the youngest age group and at least 1 serving to older children.
SOURCE: U.S. Department of Agriculture, National School Lunch Program Regulations, revised January 1, 1998.

Another federal program, the Child and Adult Care Food Program (CACFP), operates similarly and provides funds to organized child-care programs. All eligible children, centers, and family day-care homes may participate. Sponsors are reimbursed for most meal costs and may also receive USDA commodity foods.

Competing Influences at School Serving healthful lunches is only half the battle; students need to eat them, too. Short lunch periods and long waiting lines prevent some students from eating a school lunch and leave others with too little time to complete their meals.[113] Nutrition efforts at schools are also undermined when students can buy what the USDA labels "competitive foods"—meals from fast-food restaurants or a la carte foods such as pizza or snack foods and carbonated beverages from snack bars, school stores, and vending machines.[114] In one study, students who selected competitive foods in addition to, or instead of, school meals consumed more energy and fat and less calcium and vitamin A than those who selected only the school lunch.[115]

Increasingly, school-based nutrition issues are being addressed by legislation. Some states restrict the sale of competitive foods and have higher rates of participation in school meal programs than the national average. Federal legislation mandates that all school districts that participate in the USDA's National School Lunch Program develop and put in place a local wellness policy.[116] Nutrition professionals advocate further legislative measures that would prohibit sales of food and beverages from vending machines or school stores in middle and high schools until 30 minutes after the end of the last meal unless they are part of the school foodservice and meet *Dietary Guidelines* standards.[117] Reducing the prices of nutritious foods also greatly increases the likelihood that students will purchase them.[118]

IN SUMMARY

Children's appetites and nutrient needs reflect their stage of growth. Those who are chronically hungry and malnourished suffer growth retardation; when hunger is temporary and nutrient deficiencies are mild, the problems are usually more subtle—such as poor academic performance. Iron deficiency is widespread and has many physical and behavioral consequences. "Hyper" behavior is not caused by poor nutrition; misbehavior may be due to lack of sleep, too little physical activity, or too much television, among other things. Childhood obesity has become a major health problem. Adults at home and at school need to provide children with nutrient-dense foods and teach them how to make healthful diet and activity choices.

Nutrition during Adolescence

Teenagers make many more choices for themselves than they did as children. They are not fed, they eat; they are not sent out to play, they choose to go. At the same time, social pressures thrust choices at them, such as whether to drink alcoholic beverages and whether to develop their bodies to meet extreme ideals of slimness or athletic prowess. Their interest in nutrition—both valid information and misinformation—derives from personal, immediate experiences. They are concerned with how diet can improve their lives now—they engage in fad dieting in order to fit into a new bathing suit, avoid greasy foods in an effort to clear acne, or eat a pile of spaghetti to prepare for a big sporting event. In presenting information on the nutrition and health of adolescents, this section includes many topics of interest to teens.

Growth and Development

With the onset of **adolescence,** the steady growth of childhood speeds up abruptly and dramatically, and the growth patterns of female and male become distinct.

adolescence: the period from the beginning of puberty until maturity.

Nutritious snacks contribute valuable nutrients to an active teen's diet.

Hormones direct the intensity of the adolescent growth spurt, profoundly affecting every organ of the body, including the brain. After two to three years of intense growth and a few more at a slower pace, physically mature adults emerge.

In general, the adolescent growth spurt begins at age 10 or 11 for females and at 12 or 13 for males. It lasts about two and a half years. Before **puberty,** male and female body compositions differ only slightly, but during the adolescent spurt, differences between the genders become apparent in the skeletal system, lean body mass, and fat stores. In females, fat assumes a larger percentage of the total body weight, and in males, the lean body mass—principally muscle and bone—increases much more than in females (review Figure 8-8, p. 261). On average, males grow 8 inches taller, and females, 6 inches taller. Males gain approximately 45 pounds, and females, about 35 pounds.

Energy and Nutrient Needs

Energy and nutrient needs are greater during adolescence than at any other time of life, except pregnancy and lactation. In general, nutrient needs rise throughout childhood, peak in adolescence, and then level off or even diminish as the teen becomes an adult.

Energy Intake and Activity The energy needs of adolescents vary greatly, depending on their current rate of growth, gender, body composition, and physical activity.[119] Boys' energy needs may be especially high; they typically grow faster than girls and, as mentioned, develop a greater proportion of lean body mass. An exceptionally active boy of 15 may need 3500 kcalories or more a day just to maintain his weight. Girls start growing earlier than boys and attain shorter heights and lower weights, so their energy needs peak sooner and decline earlier than those of their male peers. A sedentary girl of 15 whose growth is nearly at a standstill may need fewer than 1800 kcalories a day if she is to avoid excessive weight gain. Thus adolescent girls need to pay special attention to being physically active and selecting foods of high nutrient density so as to meet their nutrient needs without exceeding their energy needs.

Dietary Guidelines for Americans 2005

Adolescents should engage in at least 60 minutes of physical activity on most, preferably all, days of the week.

The insidious problem of obesity becomes ever more apparent in adolescence and often continues into adulthood. The problem is most evident in females of African American descent and in Hispanic children of both genders. Without intervention, overweight adolescents face numerous physical and socioeconomic consequences for years to come. The consequences of obesity are so dramatic and our society's attitude toward obese people is so negative that even teens of normal or below-normal weight may perceive a need to lose weight. When taken to extremes, restrictive diets bring dramatic physical consequences of their own, as Highlight 8 explained.

Vitamins The RDA (or AI) for most vitamins increases during the adolescent years (see the table on the inside front cover). Several of the vitamin recommendations for adolescents are similar to those for adults, including the recommendation for vitamin D. During puberty, both the activation of vitamin D and the absorption of calcium are enhanced, thus supporting the intense skeletal growth of the adolescent years without additional vitamin D.

Iron The need for iron increases during adolescence for both females and males, but for different reasons. Iron needs increase for females as they start to menstruate

puberty: the period in life in which a person becomes physically capable of reproduction.

and for males as their lean body mass develops. Hence, the RDA increases at age 14 for both males and females. For females, the RDA remains high into late adulthood. For males, the RDA returns to preadolescent values in early adulthood.

In addition, iron needs increase when the adolescent growth spurt begins, whether that occurs before or after age 14. Therefore, boys in a growth spurt need an additional 2.9 milligrams of iron per day above the RDA for their age; girls need an additional 1.1 milligrams per day.[120]

Furthermore, iron recommendations for girls before age 14 do not reflect the iron losses of menstruation. The average age of menarche (first menstruation) in the United States is 12.5 years, however.[121] Therefore, for girls under the age of 14 who have started to menstruate, an additional 2.5 milligrams of iron per day is recommended.[122] Thus the RDA for iron depends not only on age and gender but also on whether the individual is in a growth spurt or has begun to menstruate, as listed in the margin. ◆

Iron intakes often fail to keep pace with increasing needs, especially for females, who typically consume less iron-rich foods such as meat and fewer total kcalories than males. Not surprisingly, iron deficiency is most prevalent among adolescent girls. Iron-deficient children and teens score lower on standardized tests than those who are not iron deficient.

Calcium Adolescence is a crucial time for bone development, and the requirement for calcium reaches its peak during these years.[123] Unfortunately, low calcium intakes among adolescents have reached crisis proportions: 90 percent of females and 70 percent of males ages 12 to 19 years have calcium intakes below recommendations.[124] Low calcium intakes during times of active growth, especially if paired with physical inactivity, can compromise the development of peak bone mass, which is considered the best protection against adolescent fractures and adult osteoporosis. Increasing milk products in the diet to meet calcium recommendations greatly increases bone density.[125] Once again, however, teenage girls are most vulnerable, for their milk—and therefore their calcium—intakes begin to decline at the time when their calcium needs are greatest.[126] Furthermore, women have much greater bone losses than men in later life. In addition to dietary calcium, bones grow stronger with physical activity. However, because few high schools require students to attend physical education classes, most adolescents must make a point to be physically active during leisure time.

◆ Iron RDA for males:
- 9–13 yr: 8 mg/day
- 9–13 yr in growth spurt: 10.9 mg/day
- 14–18 yr: 11 mg/day
- 14–18 yr in growth spurt: 13.9 mg/day

Iron RDA for females:
- 9–13 yr: 8 mg/day
- 9–13 yr in menarche: 10.5 mg/day
- 9–13 yr in menarche and growth spurt: 11.6 mg/day
- 14–18 yr: 15 mg/day
- 14–18 yr in growth spurt: 16.1 mg/day

Dietary Guidelines for Americans 2005

Children 9 years of age and older should consume 3 cups per day of fat-free or low-fat milk or equivalent milk products.

Food Choices and Health Habits

Teenagers like the freedom to come and go as they choose. They eat what they want if it is convenient and if they have the time.[127] With a multitude of afterschool, social, and job activities, they almost inevitably fall into irregular eating habits. At any given time on any given day, a teenager may be skipping a meal, eating a snack, preparing a meal, or consuming food prepared by a parent or restaurant. Adolescents who frequently eat meals with their families, however, eat more fruits, vegetables, grains, and calcium-rich foods, and drink fewer soft drinks, than those who seldom eat with their families.[128] Furthermore, the more often teenagers eat dinner with their families, the less likely they are to smoke, drink, or use drugs.[129] Many adolescents also begin to skip breakfast on a regular basis, missing out on important nutrients that are not made up at later meals during the day. Compared with those who skip breakfast, teenagers who do eat breakfast have higher intakes of vitamins A, C, and riboflavin, as well as calcium, iron, and zinc.[130] Teenagers who eat breakfast are therefore more likely to meet their nutrient intake recommendations.

Because their lunches rarely include fruits, vegetables, or milk, many teens fail to get all the vitamins and minerals they need each day.

◆ For perspective, caffeine-containing soft drinks typically deliver between 30 and 55 mg of caffeine per 12-ounce can. A pharmacologically active dose of caffeine is defined as 200 mg. Appendix H starts with a table listing the caffeine contents of selected foods, beverages, and drugs.

Ideally, in light of adolescents' busy schedules and desire for freedom, the adult continues to play the role of gatekeeper, controlling the type and availability of food in the teenager's environment. Teenagers should find plenty of nutritious, easy-to-grab foods in the refrigerator (meats for sandwiches; low-fat cheeses; fresh, raw vegetables and fruits; fruit juices; and milk) and more in the cabinets (whole-grain breads, peanut butter, nuts, popcorn, and cereal). In many households today, the adults work outside the home, and teenagers perform some of the gatekeepers' roles, such as shopping for groceries or choosing fast or prepared foods.

Snacks Snacks typically provide at least a fourth of the average teenager's daily food energy intake. Most often, favorite snacks are too high in saturated fat and sodium and too low in fiber to support the future health of the arteries.[131] Table 15-8, p. 541 shows how to combine foods from different food groups to create healthy snacks.

Beverages Most frequently, adolescents drink soft drinks instead of fruit juice or milk with lunch, supper, and snacks. About the only time they select fruit juices is at breakfast. When teens drink milk, they are more likely to consume it with a meal (especially breakfast) than as a snack. Soft drinks, when chosen as the primary beverage, may affect bone density because they displace milk from the diet.[132] Because of their greater food intakes, boys are more likely than girls to drink enough milk to meet their calcium needs.

Over the past three decades, teens (especially girls) have been drinking more soft drinks and less milk.[133] Adolescents who drink soft drinks regularly have a higher energy intake and a lower calcium intake than those who do not; they are also more likely to be overweight.[134]

Soft drinks containing caffeine present a different problem if caffeine ◆ intake becomes excessive. Caffeine seems to be relatively harmless when used in moderate doses (the equivalent of fewer than three 12-ounce cola beverages a day). In greater amounts, however, it can cause the symptoms associated with anxiety, such as sweating, tenseness, and inability to concentrate.

Eating Away from Home Adolescents eat about one-third of their meals away from home, and their nutritional welfare is enhanced or hindered by the choices they make. A lunch consisting of a hamburger, a chocolate shake, and French fries supplies substantial quantities of many nutrients at a kcalorie cost of about 800, an energy intake some adolescents can afford. When they eat this sort of lunch, teens can adjust their breakfast and dinner choices to include fruits and vegetables for vitamin A, vitamin C, folate, and fiber and lean meats and legumes for iron and zinc. (See Appendix H for the nutrient contents of fast foods.) Fortunately, many fast-food restaurants are offering more nutritious choices than the standard hamburger meal.

Peer Influence Many of the food and health choices adolescents make reflect the opinions and actions of their peers. When others perceive milk as "babyish," a teen may choose soft drinks instead; when others skip lunch and hang out in the parking lot, a teen may join in for the camaraderie, regardless of hunger. Adults need to remember that adolescents have the right to make their own decisions—even if they are contrary to the adults' views. Gatekeepers can set up the environment so that nutritious foods are available and can stand by with reliable nutrition information and advice, but the rest is up to the adolescents. Ultimately, they make the choices. (Highlight 8 examines the influence of social pressures on the development of eating disorders.)

Problems Adolescents Face

Physical maturity and growing independence present adolescents with new choices. The consequences of those choices will influence their nutritional health both today and throughout life. Some teenagers begin using drugs, alcohol, and tobacco; others wisely refrain. Information about the use of these substances is presented here because most people are first exposed to them during adolescence, but it actually applies to people of all ages.

Marijuana Almost half of the high school students in the United States report having at least tried marijuana.[135] Marijuana is unique among drugs in that it seems to enhance the enjoyment of eating, especially of sweets, a phenomenon commonly known as "the munchies." Prolonged use of marijuana, however, does not seem to bring about a weight gain.

Cocaine, Crack, and Methamphetamine Cocaine, crack, and methamphetamine stimulate the nervous system and elicit the stress response—constricted blood vessels, raised blood pressure, dilated pupils, and increased body temperature. These drugs also drive away feelings of fatigue. They occasionally cause immediate death—usually by heart attack, stroke, or seizure in an already damaged body system. During prolonged episodes of drug use, abusers suffer dehydration and electrolyte imbalances. Decreases in appetite, weight loss, and malnutrition are common. Notably, the craving for these drugs replaces hunger; rats given unlimited cocaine will choose it over food until they starve to death. Thus, unlike marijuana use, cocaine, crack, and methamphetamine use has major nutritional consequences.

Ecstasy The club drug ecstasy has become alarmingly popular in recent years. Ecstasy signals the nerve cells to dump all their stored serotonin ◆ at once and then prevents its reabsorption. The rush of serotonin flooding the gap between the nerve cells (the *synapse*) alters a person's mood, but it may also damage nerve cells and impair memory. Because serotonin helps to regulate body temperature, overheating is a common and potentially dangerous side effect of this drug. People who use ecstasy regularly tend to lose weight.

◆ Reminder: *Serotonin* is a neurotransmitter important in the regulation of appetite, sleep, and body temperature.

Drug Abuse, in General The nutrition problems associated with other drugs vary in degree, but drug abusers in general face multiple nutrition problems. ◆ During withdrawal from drugs, an important part of treatment is to identify and correct nutrient deficiencies.

Alcohol Abuse Sooner or later all teenagers face the decision of whether to drink alcohol. The law forbids the sale of alcohol to people under 21, but most adolescents who want it can get it. By the end of high school, 77 percent of students have tried alcohol, and about half have been drunk at least once.[136] Highlight 7 describes how alcohol affects nutrition status. To sum it up, alcohol provides energy but no nutrients, and it can displace nutritious foods from the diet. Alcohol alters nutrient absorption and metabolism, so imbalances develop. People who cannot keep their alcohol use moderate must abstain to maintain their health. Highlight 7 lists resources for people with alcohol-related problems.

◆ Nutrition problems of drug abusers:
- They buy drugs with money that could be spent on food.
- They lose interest in food during "highs."
- They use drugs that suppress appetite.
- Their lifestyle fails to promote good eating habits.
- If they use intravenous (IV) drugs, they may contract AIDS, hepatitis, or other infectious diseases, which increase their nutrient needs. Hepatitis also causes taste changes and loss of appetite.
- Medicines used to treat drug abuse may alter nutrition status.

Smoking Slightly less than 30 percent of U.S. high school students report smoking a cigarette in the previous month.[137] This is the lowest rate of smoking among high school students since 1991. Cigarette smoking is a pervasive health problem causing thousands of people to suffer from cancer and diseases of the cardiovascular, digestive, and respiratory systems. These effects are beyond the scope of nutrition, but smoking cigarettes does influence hunger, body weight, and nutrient status.

Smoking a cigarette eases feelings of hunger. When smokers receive a hunger signal, they can quiet it with cigarettes instead of food. Such behavior ignores body signals and postpones energy and nutrient intake. Indeed, smokers tend to weigh less than nonsmokers and to gain weight when they stop smoking. People contemplating giving up cigarettes should know that the average weight gain is about 10 pounds in the first year. Smokers wanting to quit should prepare for the possibility of weight gain and adjust their diet and activity habits so as to maintain weight during and after quitting. Smoking cessation programs need to include strategies for weight management.

Nutrient intakes of smokers and nonsmokers differ. Smokers tend to have lower intakes of dietary fiber, vitamin A, beta-carotene, folate, and vitamin C. The association between smoking and low intakes of fruits and vegetables rich in these nutrients may be noteworthy, considering their protective effect against lung cancer (see Highlight 11).

◆ The vitamin C requirement for people who regularly smoke cigarettes is an additional 35 mg/day.

Compared with nonsmokers, smokers require more vitamin C ◆ to maintain steady body pools. Oxidants in cigarette smoke accelerate vitamin C metabolism and deplete smokers' body stores of this antioxidant. This depletion is even evident to some degree in nonsmokers who are exposed to passive smoke.[138]

Beta-carotene enhances the immune response and protects against some cancer activity. Specifically, the risk of lung cancer is greatest for smokers who have the lowest intakes. Of course, such evidence should not be misinterpreted. It does not mean that as long as people eat their carrots, they can safely use tobacco. Nor does it mean that beta-carotene *supplements* are beneficial; smokers taking beta-carotene supplements actually had a higher incidence of lung cancer and risk of death than those taking a placebo. (See Highlight 11 for more details.) Smokers are ten times more likely to get lung cancer than nonsmokers. Both smokers and non-smokers, however, can reduce their cancer risks by eating fruits and vegetables rich in antioxidants. (See Highlight 11 for details on antioxidant nutrients and disease prevention.)

Smokeless Tobacco Like cigarettes, smokeless tobacco use is linked to many health problems, from minor mouth sores to tumors in the nasal cavities, cheeks, gums, and throat. The risk of mouth and throat cancers is even greater than for smoking to-bacco. Other drawbacks to tobacco chewing and snuff dipping include bad breath, stained teeth, and blunted senses of smell and taste. Tobacco chewing also damages the gums, tooth surfaces, and jawbones, making teeth loss later in life likely.

The nutrition and lifestyle choices people make as children and adolescents have long-term, as well as immediate, effects on their health. Highlight 15 describes how sound choices and good habits during childhood and adolescence can help prevent chronic diseases later in life.

Nutrition Portfolio

Encouraging children to eat nutritious foods today helps them learn how to make healthy food choices tomorrow.

▓ If there are children in your life, think about the food they eat and consider whether they receive enough food for healthy growth, but not so much as to lead to obesity.

▓ Describe the advantages of physical activity to children's health and well-being.

▓ Plan a day's menu for a child 4 to 8 years of age, making sure to include foods that provide enough calcium and iron.

NUTRITION ON THE NET

• Learn more about breast milk banks from the Human Milk Banking Association of North America: **www.hmbana.com**

• Search for "infants," "baby bottle tooth decay," "premature birth," "hyperactivity," "food allergies," and "adolescent health," at the U.S. Government health information site: **www.healthfinder.gov**

• Learn how to care for infants, children, and adolescents from the American Academy of Pediatrics and the Canadian Paediatric Society: **www.aap.org** and **www.cps.ca**

• Download the current growth charts and learn about their most recent revision: **www.cdc.gov/growthcharts**

• Get information on the Food Guide Pyramid for young children from the USDA: **www.MyPyramid.gov/kids**

• Get tips for feeding children from the American Dietetic Association: **www.eatright.org**

- Get tips for keeping children healthy from the Nemours Foundation: **www.kidshealth.org**

- Visit the National Center for Education in Maternal & Child Health and the National Institute of Child Health and Human Development: **www.ncemch.org** and **www.nichd.nih.gov**

- Learn about the Child Nutrition Programs: **www.fns.usda.gov/fns**

- Learn how UNICEF works to protect children: **www.unicef.org**

- Learn how to reduce lead exposure in your home from the U.S. Department of Housing and Urban Development Office of Lead Hazard Control: **www.hud.gov/lead**

- Learn more about food allergies from the American Academy of Allergy, Asthma, and Immunology; the Food Allergy Network; and the International Food Information Council: **www.aaaai.org, www.foodallergy.org,** and **www.ific.org**

- Learn more about hyperactivity from Children and Adults with Attention Deficit/Hyperactivity Disorders: **www.chadd.org**

- Visit the Milk Matters section of the National Institute of Child Health and Human Development (NICHD): **www.nichd.nih.gov**

- Learn more about caffeine from the International Food Information Council: **www.ific.org**

- To learn about healthy foods and to find recipes and ideas for physical activities, visit: **www.kidnetic.com**

- Get weight-loss tips for children and adolescents: **www.shapedown.com**

- Learn about nondietary approaches to weight loss from HUGS International: **www.hugs.com**

- Read the message for parents and teens on the risks of tobacco use from the American Academy of Pediatrics: **www.aap.org**

- Get help quitting smoking at QuitNet: **www.quitnet.com**

- Visit the Tobacco Information and Prevention Source (TIPS) of the Centers for Disease Control and Prevention: **www.cdc.gov/tobacco/sgr/sgr_2000**

STUDY QUESTIONS

CENGAGENOW™

To assess your understanding of chapter topics, take the Student Practice Test and explore the modules recommended in your Personalized Study Plan. Log on to **academic.cengage.com/login.**

These questions will help you review the chapter. You will find the answers in the discussions on the pages provided.

1. Describe some of the nutrient and immunological attributes of breast milk. (pp. 518–520)

2. What are the appropriate uses of formula feeding? What criteria would you use in selecting an infant formula? (pp. 520–521)

3. Why are solid foods not recommended for an infant during the first few months of life? When is an infant ready to start eating solid food? (pp. 523–525)

4. Identify foods that are inappropriate for infants and explain why they are inappropriate. (pp. 522, 524–525)

5. What nutrition problems are most common in children? What strategies can help prevent these problems? (pp. 530–532)

6. Describe the relationships between nutrition and behavior. How does television influence nutrition? (pp. 532–533, 537)

7. Describe a true food allergy. Which foods most often cause allergic reactions? How do food allergies influence nutrition status? (pp. 533–535)

8. Describe the problems associated with childhood obesity and the strategies for prevention and treatment. (pp. 535–539)

9. List strategies for introducing nutritious foods to children. (pp. 539–541)

10. What impact do school meal programs have on the nutrition status of children? (pp. 541–543)

11. Describe the changes in nutrient needs from childhood to adolescence. Why is an adolescent girl more likely to develop an iron deficiency than is a boy? (pp. 543–545)

12. How do adolescents' eating habits influence their nutrient intakes? (pp. 545–546)

13. How does the use of illicit drugs influence nutrition status? (p. 547)

14. How do the nutrient intakes of smokers differ from those of nonsmokers? What impacts can those differences exert on health? (pp. 547–548)

These multiple choice questions will help you prepare for an exam. Answers can be found on p. 553.

1. A reasonable weight for a healthy five-month-old infant who weighed 8 pounds at birth might be:
   a. 12 pounds.
   b. 16 pounds.
   c. 20 pounds.
   d. 24 pounds.

2. Dehydration can develop quickly in infants because:
   a. much of their body water is extracellular.
   b. they lose a lot of water through urination and tears.
   c. only a small percentage of their body weight is water.
   d. they drink lots of breast milk or formula, but little water.

3. An infant should begin eating solid foods between:
   a. 2 and 4 weeks.
   b. 1 and 3 months.
   c. 4 and 6 months.
   d. 8 and 10 months.

4. Among U.S. and Canadian children, the most prevalent nutrient deficiency is of:
   a. iron.
   b. folate.
   c. protein.
   d. vitamin D.

5. A true food allergy always:
   a. elicits an immune response.
   b. causes an immediate reaction.
   c. creates an aversion to the offending food.
   d. involves symptoms such as headaches or hives.

6. Which of the following strategies is *not* effective?
   a. Play first, eat later.
   b. Provide small portions.
   c. Encourage children to help prepare meals.
   d. Use dessert as a reward for eating vegetables.

7. To help teenagers consume a balanced diet, parents can:
   a. monitor the teens' food intake.
   b. give up—parents can't influence teenagers.
   c. keep the pantry and refrigerator well stocked.
   d. forbid snacking and insist on regular, well-balanced meals.

8. During adolescence, energy and nutrient needs:
   a. reach a peak.
   b. fall dramatically.
   c. rise, but do not peak until adulthood.
   d. fluctuate so much that generalizations can't be made.

9. The nutrients most likely to fall short in the adolescent diet are:
   a. sodium and fat.
   b. folate and zinc.
   c. iron and calcium.
   d. protein and vitamin A.

10. To balance the day's intake, an adolescent who eats a hamburger, fries, and cola at lunch might benefit most from a dinner of:
    a. fried chicken, rice, and banana.
    b. ribeye steak, baked potato, and salad.
    c. pork chop, mashed potatoes, and apple juice.
    d. spaghetti with meat sauce, broccoli, and milk.

REFERENCES

1. Committee on Dietary Reference Intakes, *Dietary Reference Intakes for Energy, Carbohydrate, Fiber, Fat, Fatty Acids, Cholesterol, Protein, and Amino Acids* (Washington, D.C.: National Academies Press, 2005).
2. Committee on Dietary Reference Intakes, 2005, pp. 280–281.
3. Committee on Nutrition, American Academy of Pediatrics, *Pediatric Nutrition Handbook*, 5th ed., ed. R. E. Kleinman (Elk Grove Village, Ill.: American Academy of Pediatrics, 2004), pp. 103–115.
4. Position of the American Dietetic Association: Promoting and supporting breastfeeding, *Journal of the American Dietetic Association* 105 (2005): 810–818.
5. American Academy of Pediatrics, Policy statement: Breastfeeding and the use of human milk, *Pediatrics* 115 (2005): 496–506; M. Boland, Exclusive breastfeeding should continue to six months, *Paediatrics and Child Health* 10 (2005): 148–149; Position of the American Dietetic Association, 2005.
6. American Academy of Pediatrics, 2005.
7. J. D. Carver, Advances in nutritional modifications of infant formulas, *American Journal of Clinical Nutrition* 77 (2003): 1550S–1554S.
8. W. C. Heird and A. Lapillonne, The role of essential fatty acids in development, *Annual Review of Nutrition* 25 (2005): 549–571; J. C. McCann and B. N. Ames, Is docosahexaenoic acid, an n-3 long-chain polyunsaturated fatty acid, required for development of normal brain function? An overview of evidence from cognitive and behavioral tests in humans and animals, *American Journal of Clinical Nutrition* 82 (2005): 281–295; N. Auestad and coauthors, Visual, cognitive, and language assessments at 39 months: A follow-up study of children fed formulas containing long-chain polyunsaturated fatty acids to 1 year of age, *Pediatrics* 112 (2003): e177–183.
9. C. L. Cheatham, J. Columbo, and S. E. Carlson, n-3 Fatty acids and cognitive and visual acuity development: Methodologic and conceptual considerations, *American Journal of Clinical Nutrition* 83 (2006): 1458S–1466S; W. W. Koo, Efficacy and safety of docosahexaenoic acid and arachidonic acid addition to infant formulas: Can one buy better vision and intelligence? *Journal of the American College of Nutrition* 22 (2003): 101–107; E. E. Birch and coauthors, A randomized controlled trial of long-chain polyunsaturated fatty acid supplementation of formula in term infants after weaning at 6 wk of age, *American Journal of Clinical Nutrition* 75 (2002): 570–580.
10. Auestad and coauthors, 2003.
11. E. E. Birch and coauthors, Visual maturation of term infants fed long-chain polyunsaturated fatty acid-supplemented or control formula for 12 mo, *American Journal of Clinical Nutrition* 81 (2005): 871–879; E. E. Birch and coauthors, A randomized controlled trial of long-chain polyunsaturated fatty acid supplementation of formula in term infants after weaning at 6 wk of age, *American Journal of Clinical Nutrition* 75 (2002): 570–580.
12. L. M. Gartner, F. R. Greer, and the Section on Breastfeeding and Committee on Nutrition, Prevention of rickets and vitamin D deficiency: New guidelines for vitamin D intake, *Pediatrics* 111 (2003): 908–910.
13. Gartner, Greer, and the Section on Breastfeeding and Committee on Nutrition, 2003.
14. American Academy of Pediatrics, 2005; Position of the American Dietetic Association, 2005.
15. B. Lönnerdal, Nutritional and physiologic significance of human milk proteins, *American Journal of Clinical Nutrition* 77 (2003): 1537S–1543S.
16. D. S. Newburg, G. M. Ruiz-Palacios, and A. L. Morrow, Human milk glycans protect infants against enteric pathogens, *Annual Review of Nutrition* 25 (2005): 37–58.
17. C. J. Chantry, C. R. Howard, and P. Auinger, Full breastfeeding duration and associated decrease in respiratory tract infection in US children, *Pediatrics* 117 (2006): 425–432; American Academy of Pediatrics, 2005; Position of the American Dietetic Association, 2005.
18. R. S. Zeiger and N. J. Friedman, The relationship of breastfeeding to the development of atopic disorders, *Nestle Nutrition Workshop Series: Pediatric Program* 57 (2006): 93–108.
19. M. Gdalevich, D. Mimouni, and M. Mimouni, Breastfeeding and the risk of bronchial asthma in childhood: A systematic review with meta-analysis of prospective studies, *Journal of Pediatrics* 139 (2001): 261–266.
20. A. Singhal, Early nutrition and long-term cardiovascular health, *Nutrition Reviews* 64 (2006): S44–S49; R. M. Martin, D. Gunnell, and G. D. Smith, Breastfeeding in infancy

and blood pressure in later life: Systematic review and meta-analysis, *American Journal of Epidemiology* 161 (2005): 15–26; C. G. Owen and coauthors, Infant feeding and blood cholesterol: A study in adolescents and systematic review, *Pediatrics* 110 (2002): 597–608.

21. C. G. Owen and coauthors, Effect of infant feeding on the risk of obesity across the life course: A quantitative review of published evidence, *Pediatrics* 115 (2005): 1367–1377.

22. M. W. Gillman and coauthors, Risk of overweight among adolescents who were breastfed as infants, *Journal of the American Medical Association* 285 (2001): 2461–2467.

23. M. L. Hediger and coauthors, Association between infant breastfeeding and overweight in young children, *Journal of the American Medical Association* 285 (2001): 2453–2460.

24. M. C. Daniels and L. S. Adair, Breastfeeding influences cognitive development in Filipino children, *Journal of Nutrition* 135 (2005): 2589–2595; E. L. Mortensen and coauthors, The association between duration of breastfeeding and adult intelligence, *Journal of the American Medical Association* 287 (2002): 2365–2371; A. Jain, J. Concato, and J. M. Leventhal, How good is the evidence linking breastfeeding and intelligence? *Pediatrics* 109 (2002): 1044–1053.

25. M. R. Tully, L. Lockhart-Borman, and K. Updegrove, Stories of success: The use of donor milk is increasing in North America, *Journal of Human Lactation* 20 (2004): 75–77.

26. L. Seppo and coauthors, A follow-up study of nutrient intake, nutritional status, and growth in infants with cow milk allergy fed either a soy formula or an extensively hydrolyzed whey formula, *American Journal of Clinical Nutrition* 82 (2005): 140–145; Committee on Nutrition, American Academy of Pediatrics, *Pediatric Nutrition Handbook*, 5th ed., ed. R. Kleinman (Elk Grove Village, Ill.: American Academy of Pediatrics, 2004), pp. 87–97.

27. D. Hoyert and coauthors, Annual summary of vital statistics: 2004, *Pediatrics* 117 (2006): 168–183.

28. Committee on Nutrition, American Academy of Pediatrics, 2004, p. 111.

29. American Academy of Pediatrics, Breastfeeding and the use of human milk, *Pediatrics* 115 (2005): 496–506.

30. Committee on Nutrition, American Academy of Pediatrics, 2004, pp. 105–108.

31. A. Fiocchi, A. Assa'ad, and S. Bahna, Food allergy and the introduction of solid foods to infants: A consensus document, *Annals of Allergy, Asthma and Immunology* 97 (2006): 10–21.

32. L. Hallberg and coauthors, The role of meat to improve the critical iron balance during weaning, *Pediatrics* 111 (2003): 864–870.

33. Committee on Nutrition, American Academy of Pediatrics, 2004, pp. 105–108.

34. Committee on Nutrition, American Academy of Pediatrics, 2004, pp. 103–115.

35. Committee on Nutrition, American Academy of Pediatrics, 2004, pp. 103–115.

36. Centers for Disease Control and Prevention, Nonfatal choking-related episodes among children—United States, 2001, *Morbidity and Mortality Weekly Report* 51 (2002): 945–948.

37. M. K. Fox and coauthors, Relationship between portion size and energy intake among infants and toddlers: Evidence of self-regulation, *Journal of the American Dietetic Association* 106 (2006): S77–S83.

38. American Academy of Pediatrics, Council on Sports Medicine and Fitness and Council on School Health, Active healthy living: Prevention of childhood obesity through increased physical activity, *Pediatrics* 117 (2006): 1834–1842.

39. V. Messina and A. R. Mangels, Considerations in planning vegan diets: Children, *Journal of the American Dietetic Association* 101 (2001): 661–669.

40. Committee on Dietary Reference Intakes, 2005, Chapter 6.

41. Committee on Dietary Reference Intakes, 2005, Chapter 7.

42. Committee on Dietary Reference Intakes, 2005, Chapter 11.

43. Committee on Dietary Reference Intakes, 2005, Chapter 8.

44. K. C. White, Anemia is a poor predictor of iron deficiency among toddlers in the United States: For heme the bell tolls, *Pediatrics* 115 (2005): 315–320; Centers for Disease Control and Prevention, Iron deficiency—United States, 1999–2000, *Morbidity and Mortality Weekly Report* 51 (2002): 897–899.

45. S. L. Johnson, Children's food acceptance patterns: The interface of ontogeny and nutrition needs, *Nutrition Reviews* 60 (2002): S91–S94.

46. R. E. Kleinman, Current approaches to standards of care for children: How does the pediatric community currently approach this issue? *Nutrition Today* 37 (2002): 177–178.

47. R. Briefel and coauthors, Feeding Infants and Toddlers Study: Do vitamin and mineral supplements contribute to nutrient adequacy or excess among US infants and toddlers? *Journal of the American Dietetic Association* 106 (2006): S52–S65.

48. D. J. Raiten, M. F. Picciano, and P. Coates, Dietary supplement use in children: Who, what, why, and where do we go from here: Executive summary, *Nutrition Today* 37 (2002): 167–169.

49. M. Lino, and coauthors, U.S. Department of Agriculture, Center for Nutrition Policy and Promotion, The quality of young children's diets, *Family Economics and Nutrition Review* 14 (2002): 52–59.

50. P. Ziegler and coauthors, Feeding infants and toddlers study (FITS): Development of the FITS Survey in comparison to other dietary survey methods, *Journal of the American Dietetic Association* 106 (2006): S12–S27.

51. J. Stang, Improving the eating patterns of infants and toddlers, *Journal of the American Dietetic Association* 106 (2006): S7–S9; M. K. Fox and coauthors, Feeding infants and toddlers study: What foods are infants and toddlers eating? *Journal of the American Dietetic Association* 104 (2004): S22–S30.

52. Position of the American Dietetic Association: Dietary guidance for healthy children ages 2 to 11 years, *Journal of the American Dietetic Association* 104 (2004): 660–677.

53. M. Nord, M. Andrews, and S. Carlson, Household food security in the United States, 2005, November 2006 available at www.ers.usda.gov/publications/err29.

54. Committee on Dietary Reference Intakes, *Dietary Reference Intakes for Vitamin A, Vitamin K, Arsenic, Boron, Chromium, Copper, Iodine, Iron, Manganese, Molybdenum, Nickel, Silicon, Vanadium, and Zinc* (Washington, D.C.: National Academy Press, 2001), pp. 82–161.

55. G. C. Rampersaud and coauthors, Breakfast habits, nutritional status, body weight, and academic performance in children and adolescents, *Journal of the American Dietetic Association* 105 (2005): 743–760; S. G. Affenito and coauthors, Breakfast consumption by African-American and white adolescent girls correlates positively with calcium and fiber intake and negatively with body mass index, *Journal of the American Dietetic Association* 105 (2005): 938–945; Position of the American Dietetic Association, 2004.

56. J. L. Beard and J. R. Connor, Iron status and neural functioning, *Annual Review of Nutrition* 23 (2003): 41–58; Committee on Dietary Reference Intakes, 2001, pp. 290–393.

57. B. Lozhoff and coauthors, Long-lasting neural and behavioral effects of iron deficiency in infancy, *Nutrition Reviews* 64 (2006): S34–S43; Beard and Connor, 2003.

58. Centers for Disease Control and Prevention, Blood lead levels—United States, 1999–2002, *Morbidity and Mortality Weekly Report* 54 (2005): 513–616.

59. Committee on Environmental Health, American Academy of Pediatrics, Policy statement: Lead exposure in children: Prevention, detection, and management, *Pediatrics* 116 (2005): 1036–1046.

60. Committee on Environmental Health, American Academy of Pediatrics, 2005; X. Liu and coauthors, Do children with falling blood lead levels have improved cognition? *Pediatrics* 110 (2002): 787–791.

61. Centers for Disease Control and Prevention, 2005.

62. E. Romano and coauthors, Development and prediction of hyperactive symptoms from 2 to 7 years in a population-based sample, *Pediatrics* 117 (2006): 2101–2109; L. T. Blanchard, M. J. Gurka, and J. A. Blackman, Emotional, developmental, and behavioral health of American children and their families: A report from the 2003 National Survey of Children's Health, *Pediatrics* 117 (2006): e1202–1212.

63. M. L. Wolraich and coauthors, Attention-deficit/hyperactivity disorder among adolescents: A review of the diagnosis, treatment, and clinical implications, *Pediatrics* 115 (2006): 1734–1746; S. Parmet, C. Lynm, and R. M. Glass, Attention-deficit/hyperactivity disorder, *Journal of the American Medical Association* 288 (2002): 1804; Subcommittee on Attention-Deficit/Hyperactivity Disorder, American Academy of Pediatrics, Clinical practice guideline: Treatment of the school-aged child with attention-deficit/hyperactivity disorder, *Pediatrics* 108 (2001): 1033–1044.

64. U. S. Department of Health and Human Services, National Institutes of Health, National Institute of Allergy and Infectious Diseases, *Food Allergy: An Overview*, NIH publication no. 04-5518 (July 2004), available at www.niaid.nih.gov; R. Formanek, Food allergies: When food becomes the enemy, *FDA Consumer*, July/August 2001, pp. 10–16.

65. Formanek, 2001.

66. L. Christie and coauthors, Food allergies in children affect nutrient intake and growth, *Journal of the American Dietetic Association* 102 (2002): 1648–1651.

67. H. Metzger, Two approaches to peanut allergy, *New England Journal of Medicine* 348 (2003): 1046–1048.

68. Food Allergen Labeling and Consumer Protection Act of 2004, available at http://thomas.loc.gov/cgi-bin/query/F?c108:6:./temp/~c108Dz8zuL:e48634.

69. Formanek, 2001.

70. B. Merz, Studying peanut anaphylaxis, *New England Journal of Medicine* 348 (2003): 975–976; Metzger, 2003; X. M. Li and coauthors, Persistent protective effect of heat-killed *Escherichia coli* producing "engineered," recombinant peanut proteins in a murine model of peanut allergy, *Journal of Allergy and Clinical Immunology* 112 (2003): 159–167.

71. C. L. Ogden and coauthors, Prevalence of overweight and obesity in the United States, 1999–2004, *Journal of the American Medical Association* 295 (2006): 1549–1555.

72. J. P. Koplan, C. T. Liverman, and V. I. Kraak, eds., *Preventing Childhood Obesity: Health in the Balance* (Washington, D.C.: National Academies Press, 2005), pp. 79–123.

73. A. Must, Does overweight in childhood have an impact on adult health? *Nutrition Reviews* 61 (2003): 139–142; S. S. Guo and coauthors, Predicting overweight and obe-

sity in adulthood from body mass index values in childhood and adolescence, *American Journal of Clinical Nutrition* 76 (2002): 653–658; A. D. Salbe and coauthors, Assessing risk factors for obesity between childhood and adolescence: I. Birth weight, childhood adiposity, parental obesity, insulin, and leptin, *Pediatrics* 110 (2002): 299–306.

74. Koplan Liverman, and Kraak, 2005.

75. American Academy of Pediatrics, Council on Sports Medicine and Fitness and Council on School Health, Active healthy living: Prevention of childhood obesity through increased physical activity, *Pediatrics* 117 (2006): 1834–1842; Koplan, Liverman, and Kraak, 2005.

76. D. Benton, Role of parents in the determination of the food preferences of children and the development of obesity, *International Journal of Obesity Related Metabolic Disorders* 28 (2004): 858–869.

77. J. O. Fisher and coauthors, Parental influences on young girls' fruit and vegetable, micronutrient, and fat intakes, *Journal of the American Dietetic Association* 102 (2002): 58–64.

78. S. A. Lederman and coauthors, Summary of the presentations at the Conference on Preventing Childhood Obesity, December 8, 2003, *Pediatrics* 114 (2004): 1146–1173.

79. S. Kranz, A. M. Siega-Riz, and A. H. Herring, Changes in diet quality of American preschoolers between 1977 and 1998, *American Journal of Public Health* 94 (2004): 1525–1530; S. J. Nielsen, A. M. Siega-Riz, and B. M. Popkin, Trends in energy intake in U.S. between 1977 and 1996: Similar shifts seen across age groups, *Obesity Research* 10 (2002): 370–378.

80. Committee on School Health, American Academy of Pediatrics, Soft drinks in schools, *Pediatrics* 113 (2004): 152–154.

81. D. S. Ludwig, K. E. Peterson, and L. S. Gortmaker, Relation between consumption of sugar-sweetened drinks and childhood obesity: A prospective, observational analysis, *Lancet* 357 (2001): 505–508.

82. American Academy of Pediatrics, Council on Sports Medicine and Fitness and Council on School Health, Active healthy living: Prevention of childhood obesity through increased physical activity, *Pediatrics* 117 (2006): 1834–1842.

83. M. H. Proctor and coauthors, Television viewing and change in body fat from preschool to early adolescence: The Framingham Children's Study, *International Journal of Obesity and Related Metabolic Disorders* 27 (2003): 827–833.

84. B. A. Dennison, T. A. Erb, and P. L. Jenkins, Television viewing and television in bedroom associated with overweight risk among low-income preschool children, *Pediatrics* 109 (2002): 1028–1035.

85. S. Gable, Y. Chang, and J. L. Krull, Television watching and frequency of family meals are predictive of overweight onset and persistance in a national sample of school-aged children, *Journal of the American Dietetic Association* 107 (2007): 53–61; K. A. Coon and coauthors, Relationships between use of television during meals and children's food consumption patterns, *Pediatrics* 107 (2001): e71.

86. J. L. Wiecha and coauthors, When children eat what they watch: Impact of television viewing on dietary intake in youth, *Archives of Pediatrics & Adolescent Medicine* 160 (2006): 436–442; S. C. Folta and coauthors, Food advertising targeted at school-age children: A content analysis, *Journal of Nutrition Education and Behavior* 38 (2006): 244–248.

87. S. M. Connor, Food-related advertising on preschool television: Building brand recognition in young viewers, *Pediatrics* 118 (2006): 1478–1485; Folta and coauthors, 2006.

88. L. J. Chamberlain, Y. Wang, and T. N. Robinson, Does children's screen time predict requests for advertised products? Cross-sectional and prospective analyses, *Archives of Pediatrics & Adolescent Medicine* 160 (2006): 363–368; Y. Aktas-Arnas, The effects of television food advertisement on children's food purchasing requests, *Pediatrics International* 48 (2006): 138–145; M. O'Dougherty, M. Story, and J. Stang, Observations of parent-child co-shoppers in supermarkets: Children's involvement in food selections, parental yielding, and refusal strategies, *Journal of Nutrition Education and Behavior* 38 (2006): 183–188.

89. K. Weber, M. Story, and L. Harnack, Internet food marketing strategies aimed at children and adolescents: A content analysis of food and beverage brand web sites, *Journal of the American Dietetic Association* 106 (2006): 1463–1466.

90. J. Utter and coauthors, Couch potatoes or french fries: Are sedentary behaviors associated with body mass index, physical activity, and dietary behaviors among adolescents? *Journal of the American Dietetic Association* 103 (2003): 1298–1305.

91. American Academy of Pediatrics, Committee on Nutrition, Prevention of pediatric overweight and obesity, *Pediatrics* 112 (2003): 424–430.

92. R. Jago and coauthors, Prevalence of abnormal lipid and blood pressure values among an ethnically diverse population of eighth-grade adolescents and screening implications, *Pediatrics* 117 (2006): 2065–2073.

93. M. L. Cruz and coauthors, Pediatric obesity and insulin resistance: Chronic disease risk and implications for treatment and prevention beyond body weight modification, *Annual Review of Nutrition* 25 (2005): 435–468; A. Must and S. E. Anderson, Effects of obesity on morbidity in children and adolescents, *Nutrition in Clinical Care* 6 (2003): 4–12.

94. J. B. Schwimmer, T. M. Burwinkle, and J. W. Varni, Health-related quality of life of severely obese children and adolescents, *Journal of the American Medical Association* 289 (2003): 1813–1819.

95. S. Caprio and M. Genel, Confronting the epidemic of childhood obesity, *Pediatrics* 115 (2005): 494–495.

96. Position of the American Dietetic Association: Individual-, family-, school-, and community-based interventions for pediatric overweight, *Journal of the American Dietetic Association* 106 (2006): 925–945; S. Kirk, B. J. Scott, and S. R. Daniels, Pediatric obesity epidemic: Treatment options, *Journal of the American Dietetic Association* 105 (2005): S44–S51.

97. Kirk, Scott, and Daniels, 2005.

98. Kirk, Scott, and Daniels, 2005.

99. K. L. McConahy and coauthors, Portion size of common foods predicts energy intake among preschool-aged children, *Journal of the American Dietetic Association* 104 (2004): 975–979; B. J. Rolls, D. Engell, and L. L. Birch, Serving portion size influences 5-year-old but not 3-year-old children's food intakes, *Journal of the American Dietetic Association* 100 (2000): 232–234.

100. American Academy of Pediatrics, Council on Sports Medicine and Fitness and Council on School Health, Active healthy living: Prevention of childhood obesity through increased physical activity, *Pediatrics* 117 (2006): 1834–1842.

101. American Academy of Pediatrics, 2006.

102. Position of the American Dietetic Association: Individual-, family-, school-, and community-based interventions for pediatric overweight, *Journal of the American Dietetic Association* 106 (2006): 925–945;

American Academy of Pediatrics, Committee on Nutrition, Prevention of pediatric overweight and obesity, *Pediatrics* 112 (2003): 424–430; D. Spruijt-Metz and coauthors, Relation between mothers' child-feeding practices and children's adiposity, *American Journal of Clinical Nutrition* 75 (2002): 581–586; A. E. Field and coauthors, Peer, parent, and media influences on the development of weight concerns and frequent dieting among preadolescent and adolescent girls and boys, *Pediatrics* 107 (2001): 54–60.

103. D. Benton, Role of parents in the determination of the food preferences of children and the development of obesity, *International Journal of Obesity and Related Metabolic Disorders* 28 (2004): 858–869.

104. B. P. Roberts, A. S. Blinkhorn, and J. T. Duxbury, The power of children over adults when obtaining sweet snacks, *International Journal of Paediatric Dentistry* 13 (2003): 76–84.

105. J. Wardle, S. Carnell, and L. Cooke, Parental control over feeding and children's fruit and vegetable intake: How are they related? *Journal of the American Dietetic Association* 105 (2005): 227–232; A. T. Galloway and coauthors, Parental pressure, dietary patterns, and weight status among girls who are "picky eaters," *Journal of the American Dietetic Association* 105 (2005): 541–548; L. J. Cooke and coauthors, Demographic, familial and trait predictors of fruit and vegetable consumption by pre-school children, *Public Health Nutrition* 2 (2004): 251–252.

106. J. D. Skinner and coauthors, Children's food preferences: A longitudinal analysis, *Journal of the American Dietetic Association* 102 (2002): 1638–1647.

107. Position of the American Dietetic Association: Benchmarks for nutrition programs in child care settings, *Journal of the American Dietetic Association* 105 (2005): 979–986; Position of the American Dietetic Association, Society of Nutrition Education, and American School Food Service Association-Nutrition services: An essential component of comprehensive school health programs, *Journal of the American Dietetic Association* 103 (2003): 505–514.

108. Position of the American Dietetic Association, Society of Nutrition Education, and American School Food Service Association, 2003.

109. Position of the American Dietetic Association: Local support for nutrition integrity in schools, *Journal of the American Dietetic Association* 106 (2006): 122–133.

110. Position of the American Dietetic Association: Dietary guidance for healthy children ages 2 to 11 years, *Journal of the American Dietetic Association* 104 (2004): 660–677; 2004; K. W. Cullen and I. Zakeri, Fruits, vegetables, milk, and sweetened beverages consumption and access to a la carte/snack bar meals at school, *American Journal of Public Health* 94 (2004): 463–467; P. M. Gleason and C. W. Suitor, Eating at school: How the National School Lunch Program affects children's diets, *American Journal of Agricultural Economics* 85 (2003): 1047–1051.

111. Position of the American Dietetic Association, 2006.

112. Position of the American Dietetic Association, 2006.

113. Position of the American Dietetic Association, 2006.

114. Position of the American Dietetic Association, 2006; C. Probart and coauthors, Competitive foods available in Pennsylvania public high schools, *Journal of the American Dietetic Association* 105 (2005): 1243–1249; Cullen and Zakeri, 2004; Committee on School Health, American Academy of Pediatrics, Soft drinks in schools, *Pediatrics* 113 (2004): 152–154.

115. S. B. Templeton and coauthors, Competitive foods increase the intake of energy and decrease the intake of certain nutrients by adolescents consuming school lunch, *Journal of the American Dietetic Association* 105 (2005): 215–220.

116. Position of the American Dietetic Association, 2006.

117. Position of the American Dietetic Association, Society for Nutrition Education, and American School Food Service Association, Nutrition Services: An essential component of comprehensive school health programs, *Journal of the American Dietetic Association* 103 (2003): 505–514.

118. S. A. French, Pricing effects on food choices, *Journal of Nutrition* 133 (2003): 841S–843S.

119. Committee on Dietary Reference Intakes, 2005, Chapter 5.

120. Committee on Dietary Reference Intakes, 2001, pp. 290–393.

121. W. C. Chumlea and coauthors, Age at menarche and racial comparisons in US girls, *Pediatrics* 111 (2003): 110–113.

122. Committee on Dietary Reference Intakes, 2001, pp. 290–393.

123. F. R. Greer, N. F. Krebs, and the Committee on Nutrition, American Academy of Pediatrics, Optimizing bone health and calcium intakes of infants, children, and adolescents, *Pediatrics* 117 (2006): 578–585.

124. Greer, Krebs, and the Committee on Nutrition, 2006.

125. H. J. Kalkwarf, J. C. Khoury, and B. P. Lanphear, Milk intake during childhood and adolescence, adult bone density, and osteoporotic fractures in US women, *American Journal of Clinical Nutrition* 77 (2003): 257–265.

126. S. A. Bowman, Beverage choices of young females: Changes and impact on nutrient intakes, *Journal of the American Dietetic Association* 102 (2002): 1234–1239.

127. M. Story, D. Neumark-Sztainer, and S. French, Individual and environmental influences on adolescent eating behaviors, *Journal of the American Dietetic Association* 102 (2002): S40–S51.

128. D. Neumark-Sztainer and coauthors, Family meal patterns: Associations with sociodemographic characteristics and improved dietary intake among adolescents, *Journal of the American Dietetic Association* 103 (2003): 317–322.

129. National Center on Addiction and Substance Abuse (CASA) at Columbia University, *The Importance of Family Dinners,* September, 2003.

130. Rampersaud and coauthors, 2005.

131. American Heart Association, S. S. Gidding and coauthors, Dietary recommendations for children and adolescents: A guide for practitioners, *Pediatrics* 117 (2006): 544–559.

132. Greer, Krebs, and the Committee on Nutrition, 2006; H. Vatanparast and coauthors, Positive effects of vegetable and fruit consumption and calcium intake on bone mineral accrual in boys during growth from childhood to adolescence: The University of Saskatchewan Pediatric Bone Mineral Accrual Study, *American Journal of Clinical Nutrition* 82 (2005): 700–706; G. Mrdjenovic and D. A. Levitsky, Nutritional and energetic consequences of sweetened drink consumption in 6- to 13-year-old children, *Journal of Pediatrics* 142 (2003): 604–610.

133. S. A. French, B. H. Lin, and J. F. Guthrie, National trends in soft drink consumption among children and adolescents age 6 to 17 years: Prevalence, amounts, and sources, 1997/1978 to 1994/1998, *Journal of the American Dietetic Association* 103 (2003): 1326–1331; Bowman, 2002.

134. J. James and coauthors, Preventing childhood obesity by reducing consumption of carbonated drinks: Cluster randomised controlled trial, *British Medical Journal* 328 (2004): 1237.

135. American Academy of Pediatrics, J. W. King and Committee on Substance Abuse, Tobacco, Alcohol, and Other Drugs: The role of the pediatrician in prevention, identification, and management of substance abuse, *Pediatrics* 115 (2005): 816–821.

136. American Academy of Pediatrics, King and Committee on Substance Abuse, 2005.

137. Centers for Disease Control and Prevention, Youth tobacco surveillance—United States, 2001–2002, *Morbidity and Mortality Weekly Report* 55 (2006): entire supplement.

138. A. M. Preston and coauthors, Influence of environmental tobacco smoke on vitamin C status in children, *American Journal of Clinical Nutrition* 77 (2003): 167–172.

ANSWERS

Study Questions (multiple choice)

1. b   2. a   3. c   4. a   5. a   6. d   7. c   8. a   9. c   10. d

Childhood Obesity and the Early Development of Chronic Diseases

When people think about the health problems of children and adolescents, they typically think of ear infections, colds, and acne—not heart disease, diabetes, or hypertension. Today, however, unprecedented numbers of U.S. children are being diagnosed with obesity and the serious "adult diseases," such as type 2 diabetes, that accompany overweight.[1] When type 2 diabetes develops before the age of 20, the incidence of diabetic kidney disease and death in middle age increases dramatically, largely because of the long duration of the disease.[2] For children born in the United States in the year 2000, the risk of developing type 2 diabetes sometime in their lives is estimated to be 30 percent for boys and 40 percent for girls.[3] U.S. children are not alone—rapidly rising rates of obesity threaten the health of an alarming number of children around the globe.[4] Without immediate intervention, millions of children are destined to develop type 2 diabetes and hypertension in childhood followed by **cardiovascular disease (CVD)** in early adulthood.

This highlight focuses on efforts to prevent childhood obesity and the development of heart disease and type 2 diabetes, but the benefits extend to other obesity-related diseases as well. The years of childhood (ages 2 to 18) are emphasized here, because the earlier in life health-promoting habits become established, the better they will stick. Chapters 26 and 27 fill in the rest of the story of nutrition's role in the development and treatment of diabetes and heart disease, respectively.

Invariably, questions arise as to what extent genetics is involved in disease development. For heart disease and type 2 diabetes, genetics does not appear to play a *determining* role; that is, a person is not simply destined at birth to develop these diseases. Instead, genetics appears to play a *permissive* role—the potential is inherited and will develop if given a push by poor health choices such as excessive weight gain, poor diet, sedentary lifestyle, and cigarette smoking.

Many experts agree that preventing or treating obesity in childhood will reduce the rate of chronic diseases in adulthood. Without intervention, most overweight children become overweight adolescents who become overweight adults, and being overweight exacerbates every chronic disease that adults face.[5]

Early Development of Type 2 Diabetes

In recent years, type 2 diabetes, a chronic disease closely linked with obesity, has been on the rise among children and adolescents as the prevalence of obesity in U.S. youth has increased.[6] Obesity is the most important risk factor for type 2 diabetes—most of the children diagnosed with it are obese.[7] Most are diagnosed during puberty, but as children become more obese and less active, the trend is shifting to younger children. Type 2 diabetes is most likely to occur in those who are obese and sedentary and have a family history of diabetes.

In type 2 diabetes, the cells become insulin-resistant—that is, the cells become less sensitive to insulin, reducing the amount of glucose entering the cells from the blood. The combination of obesity and insulin resistance produces a cluster of symptoms, including high blood cholesterol and high blood pressure, which, in turn, promotes the development of atherosclerosis and the early development of CVD.[8] Other common problems evident by early adulthood include kidney disease, blindness, and miscarriages. The complications of diabetes, especially when encountered at a young age, can shorten life expectancy.

Prevention and treatment of type 2 diabetes depend on weight management, which can be particularly difficult in a youngster's world of food advertising, video games, and pocket money for candy bars. The activity and dietary suggestions to help defend against heart disease later in this highlight apply to type 2 diabetes as well.

Early Development of Heart Disease

Most people consider heart disease to be an adult disease because its incidence rises with advancing age, and symptoms rarely appear before age 30. The disease process actually begins much earlier.

Atherosclerosis

Most cardiovascular disease involves **atherosclerosis** (see the glossary, p. 555 for this and related terms). Atherosclerosis develops when regions of an artery's walls become progressively thickened with **plaque**—an accumulation of fatty deposits, smooth muscle cells, and fibrous connective tissue. If it progresses, atherosclerosis may eventually block the flow of blood to the heart and cause a heart attack or cut off blood flow to the brain and cause a stroke. Infants are born with healthy, smooth, clear arteries, but within the first decade of life, **fatty streaks** may begin to appear

GLOSSARY

atherosclerosis (ATH-er-oh-scler-OH-sis): a type of artery disease characterized by plaques (accumulations of lipid-containing material) on the inner walls of the arteries (see Chapter 27).
- **athero** = porridge or soft

- **scleros** = hard
- **osis** = condition

cardiovascular disease (CVD): a general term for all diseases of the heart and blood vessels. Atherosclerosis is the main cause of CVD. When the arteries that carry blood to the heart muscle

become blocked, the heart suffers damage known as **coronary heart disease (CHD).**
- **cardio** = heart
- **vascular** = blood vessels

fatty streaks: accumulations of cholesterol and other lipids along the walls of the arteries.

plaque (PLACK): an accumulation of fatty deposits, smooth muscle cells, and fibrous connective tissue that develops in the artery walls in atherosclerosis. Plaque associated with atherosclerosis is known as **atheromatous** (ATH-er-OH-ma-tus) **plaque.**

(see Figure H15-1). During adolescence, these fatty streaks may begin to accumulate fibrous connective tissue. By early adulthood, the fibrous plaques may begin to calcify and become raised lesions, especially in boys and young men. As the lesions grow more numerous and enlarge, the heart disease rate begins to rise, most dramatically at about age 45 in men and 55 in women. From this point on, arterial damage and blockage progress rapidly, and heart attacks and strokes threaten life. In short, the consequences of atherosclerosis, which become apparent only in adulthood, have their beginnings in the first decades of life.[9]

Atherosclerosis is not inevitable; people can grow old with relatively clear arteries. Early lesions may either progress or regress, depending on several factors, many of which reflect lifestyle behaviors. Smoking, for example, is strongly associated with the prevalence of fatty streaks and raised lesions, even in young adults.

Blood Cholesterol

As blood cholesterol rises, atherosclerosis worsens. Cholesterol values at birth are similar in all populations; differences emerge in early childhood. Standard values for cholesterol in children and adolescents (ages 2 to 18 years) are listed in Table H15-1 (p. 556).

In general, blood cholesterol tends to rise as dietary saturated fat intakes increase. Blood cholesterol also correlates with childhood obesity, especially abdominal obesity.[10] LDL cholesterol rises with obesity, and HDL declines. These relationships are apparent throughout childhood, and their magnitude increases with age.

Children who are both overweight and have high blood cholesterol are likely to have parents who develop heart disease early.[11] For this reason, selective screening is recommended for children and adolescents whose parents (or grandparents) have heart disease; those whose parents have elevated blood cholesterol; and those whose family history is unavailable, especially if other risk factors are evident.[12] Because blood cholesterol in children is a good predictor of adult values, some experts recommend universal screening for all children, and particularly for those who are overweight, smoke, are sedentary, or consume diets high in saturated fat.

Early—but not advanced—atherosclerotic lesions are reversible, making screening and education a high priority. Both those with family histories of heart disease and those with multiple risk factors need intervention. Children with the highest risks of developing heart disease are sedentary and obese, with high blood pressure and high blood cholesterol.[13] In contrast, children with the lowest risks of heart disease are physically active and of normal weight, with low blood pressure and favorable lipid profiles. Routine pediatric care should identify these known risk factors and provide intervention when needed.

FIGURE H15-1 The Formation of Plaques in Atherosclerosis

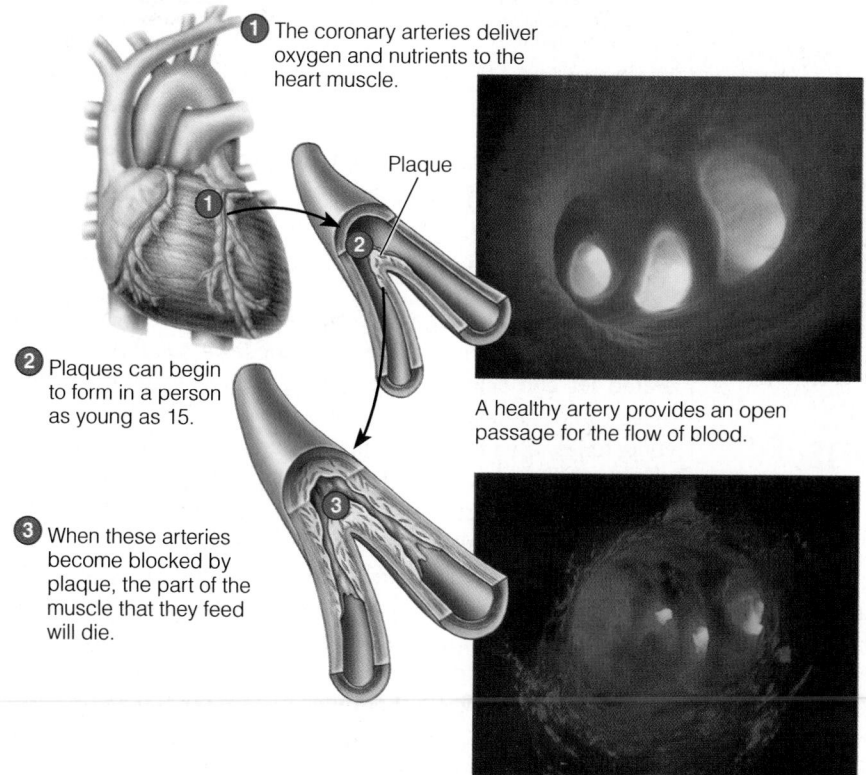

1 The coronary arteries deliver oxygen and nutrients to the heart muscle.

Plaque

2 Plaques can begin to form in a person as young as 15.

3 When these arteries become blocked by plaque, the part of the muscle that they feed will die.

© Courtesy of Zeneca Pharmaceutical Division, Cheshire, England (both)

A healthy artery provides an open passage for the flow of blood.

Plaques form along the artery's inner wall, reducing blood flow. Clots can form, aggravating the problem.

Blood Pressure

Pediatricians routinely monitor blood pressure in children and adolescents. High blood pressure may signal an underlying disease or the early onset of hypertension. Hypertension accelerates the development of atheroscerlosis.[14]

TABLE H15-1 Cholesterol Values for Children and Adolescents

Disease Risk	Total Cholesterol (mg/dL)	LDL Cholesterol (mg/dL)
Acceptable	<170	<100
Borderline	170–199	100–129
High	≥200	≥130

NOTE: Adult values appear in Chapter 27.

TABLE H15-2 American Heart Association Dietary Guidelines and Strategies for Children[a]

- Balance dietary kcalories with physical activity to maintain normal growth.
- Every day, engage in 60 minutes of moderate to vigorous play or physical activity.
- Eat vegetables and fruits daily. Use fresh, frozen, and canned vegetables and fruits and serve at every meal; limit those with added fats, salt, and sugar.
- Limit juice intake (4 to 6 ounces per day for children 1 to 6 years of age, 8 to 12 ounces for children 7 to 18 years of age).
- Use vegetable oils (canola, soybean, olive, safflower, or other unsaturated oils) and soft margarines low in saturated fat and *trans* fatty acids instead of butter or most other animal fats in the diet.
- Choose whole-grain breads and cereals rather than refined products; read labels and make sure that "whole grain" is the first ingredient.
- Reduce the intake of sugar-sweetened beverages and foods.
- Consume low-fat and nonfat milk and milk products daily.
- Include 2 servings of fish per week, especially fatty fish such as broiled or baked salmon.
- Choose legumes and tofu in place of meat for some meals.
- Choose only lean cuts of meat and reduced-fat meat products; remove the skin from poultry.
- Use less salt, including salt from processed foods. Breads, breakfast cereals, and soups may be high in salt and/or sugar so read food labels and choose high-fiber, low-salt, low-sugar alternatives.
- Limit the intake of high-kcalorie add-ons such as gravy, Alfredo sauce, cream sauce, cheese sauce, and hollandaise sauce.
- Serve age-appropriate portion sizes on appropriately sized plates and bowls.

[a] These guidelines are for children 3 years of age and older.
SOURCE: Adapted from American Heart Association, Samuel S. Gidding, and coauthors, Dietary recommendations for children and adolescents: A guide for practitioners, *Pediatrics* 117 (2006): 544–559.

Like atherosclerosis and high blood cholesterol, hypertension may develop in the first decades of life, especially among obese children, and worsen with time.[15] Children can control their hypertension by participating in regular aerobic activity and by losing weight or maintaining their weight as they grow taller. Evidence is needed to clarify whether restricting sodium in children's and adolescent's diets lowers blood pressure.

Physical Activity

Research has also confirmed an association between blood lipids and physical activity in children, similar to that seen in adults. Physically active children have a better lipid profile and lower blood pressure than physically inactive children, and these positive findings often persist into adulthood.

Just as blood cholesterol and obesity track over the years, so does a youngster's level of physical activity. Those who are inactive now are likely to still be inactive years later. Similarly, those who are physically active now tend to remain so. Compared with inactive teens, those who are physically active weigh less, smoke less, eat a diet lower in saturated fats, and have better blood lipid profiles. Both obesity and blood cholesterol correlate with the inactive pastime of watching television. The message is clear: physical activity offers numerous health benefits, and children who are active today are most likely to be active for years to come.

Dietary Recommendations for Children

Regardless of family history, experts agree that all children over age two should eat a variety of foods and maintain desirable weight (see Table H15-2). Children (4 to 18 years of age) should receive at least 25 percent and no more than 35 percent of total energy from fat, less than 10 percent from saturated fat, and less than 300 milligrams of cholesterol per day.[16] Recommendations limiting fat and cholesterol are not intended for infants or children under two years old. Infants and toddlers need a higher percentage of fat to support their rapid growth.

Moderation, Not Deprivation

Healthy children over age two can begin the transition to eating according to recommendations by eating fewer foods high in saturated fat and selecting more fruits and vegetables. Healthy meals can occasionally include moderate amounts of a child's favorite foods, even if they are high in saturated fat such as French fries and ice cream. A steady diet of offerings from some "children's menus" in restaurants such as chicken nuggets, hot dogs, and French fries, easily exceeds a prudent intake of saturated fat, *trans* fat, and kcalories, however, and invites both nutrient shortages and weight gains.[17] Fortunately, most restaurants chains are changing children's menus to include steamed vegetables, fruit cups, and broiled or grilled poultry—additions welcomed by busy parents who often dine out or purchase take-out foods.

Other fatty foods, such as nuts, vegetable oils, and some varieties of fish such as light canned tuna or salmon, are important for their essential fatty acids. Low-fat milk and milk products also deserve special attention in a child's diet for the needed calcium and other nutrients they supply.[18]

Parents and caregivers play a key role in helping children establish healthy eating habits. Balanced meals need to provide lean meat, poultry, fish, and legumes; fruits and vegetables; whole grains; and low-fat milk products. Such meals can provide enough energy and nutrients to support growth and maintain blood cholesterol within a healthy range.

Pediatricians warn parents to avoid extremes. Although intentions may be good, excessive food restriction may create nutrient deficiencies and impair growth. Furthermore, parental control over eating may instigate battles and foster attitudes about foods that can lead to inappropriate eating behaviors.

Diet First, Drugs Later

Experts agree that children with high blood cholesterol should first be treated with diet. If blood cholesterol remains high in children ten years and older after 6 to 12 months of dietary intervention, then drugs may be necessary to lower blood cholesterol. Drugs can effectively lower blood cholesterol without interfering with adolescent growth or development.[19]

Smoking

Even though the focus of this text is nutrition, another risk factor for heart disease that starts in childhood and carries over into adulthood must also be addressed—cigarette smoking. Each day 3000 children light up for the first time—typically in grade school. Among high school students, almost two out of three have tried smoking, and one in five smokes regularly.[20] Approximately 80 percent of all adult smokers began smoking before the age of 18.

Of those teenagers who continue smoking, half will eventually die of smoking-related causes. Efforts to teach children about the dangers of smoking need to be aggressive. Children are not likely to consider the long-term health consequences of tobacco use. They are more likely to be struck by the immediate health

Cigarette smoking is the number one preventable cause of deaths.

consequences, such as shortness of breath when playing sports, or social consequences, such as having bad breath. Whatever the context, the message to all children and teens should be clear: don't start smoking. If you've already started, quit.

In conclusion, *adult* heart disease is a major *pediatric* problem. Without intervention, some 60 million children are destined to suffer its consequences within the next 30 years. Optimal prevention efforts focus on children, especially on those who are overweight.[21]

Just as young children receive vaccinations against infectious diseases, they need screening for, and education about, chronic diseases. Many health education programs have been implemented in schools around the country. These programs are most effective when they include education in the classroom, heart-healthy meals in the lunchroom, fitness activities on the playground, and parental involvement at home.

NUTRITION ON THE NET

For further study of topics covered in this chapter, log on to **academic.cengage .com/nutrition/rolfes/UNCN8e**. Go to Chapter 15, then to Nutrition on the Net.

- Get weight-loss tips for children and adolescents: **www.shapedown.com**

- Learn about nondietary approaches to weight loss from HUGS International: **www.hugs.com**
- Visit the Nemours Foundation: **www.kidshealth.org**
- Find information on diabetes in children at the American Diabetes Association and Juvenile Diabetes Research Foundation: **www.diabetes.org** and **www.jdrf.org**

REFERENCES

1. C. L. Ogden and coauthors, Prevalence of overweight and obesity in the United States, 1999–2004, *Journal of the American Medical Association* 295 (2006): 1549–1555; J. P. Kaplan, C. T. Liverman, and V. I. Kraak, eds., *Preventing Childhood Obesity: Health in the Balance* (Washington, D.C.: National Academies Press, 2005), pp. 1–20; M. L. Cruz and coauthors, Pediatric obesity and insulin resistance: Chronic disease risk and implications for treatment and prevention beyond body weight modification, *Annual Review of Nutrition* 25 (2005): 435–468; T. Lobstein, L. Baur, and R. Uauy, Obesity in children and young people: A crisis in public health, *Obesity Reviews* 5 (2004): 4–85.

2. M. E. Pavkov and coauthors, Effect of youth-onset type 2 diabetes mellitus on incidence of end-stage renal disease and mortality in young and middle-aged Pima Indians, *Journal of the American Medical Association* 296 (2006): 421–426.

3. Kaplan, Liverman, and Kraak, 2005.

4. Cruz and coauthors, 2005; M. Kohn and M. Booth, The worldwide epidemic of obesity in adolescents, *Adolescent Medicine* 14 (2003): 1–9; L. S. Lieberman, Dietary, evolutionary, and modernizing influences on the prevalence of type 2 diabetes, *Annual Review of Nutrition* 23 (2003): 345–377; Committee on Nutrition, American Academy of Pediatrics, Prevention of pediatric overweight and obesity, *Pediatrics* 112 (2003): 424–430.

5. A. Must, Does overweight in childhood have an impact on adult health? *Nutrition Reviews* 61 (2003): 139–142; D. S. Freedman, Clustering of coronary heart disease risk factors among obese children, *Journal of Pediatric Endocrinology and Metabolism* 15 (2002): 1099–1108.

6. T. S. Hannon, G. Rao, and S. A. Arslanian, Childhood obesity and type 2 diabetes melli-

tus, *Pediatrics* 116 (2005): 473–480; Cruz and coauthors, 2005.

7. Hannon, Rao, and Arslanian, 2005; Cruz and coauthors, 2005.

8. G. S. Boyd and coauthors, Effect of obesity and high blood pressure on plasma lipid levels in children and adolescents, *Pediatrics* 116 (2005): 473–480; R. Kohen-Avramoglu, A. Theriault, and K. Adeli, Emergence of the metabolic syndrome in childhood: An epidemiological overview and mechanistic link to dyslipidemia, *Clinical Biochemistry* 36 (2003): 413–420.

9. S. Li and coauthors, Childhood cardiovascular risk factors and carotid vascular changes in adulthood: The Bogalusa Heart Study, *Journal of the American Medical Association* 290 (2003): 2271–2276; K. B. Keller and L. Lemberg, Obesity and the metabolic syndrome, *American Journal of Clinical Care* 12 (2003): 167–170.

10. O. Fiedland and coauthors, Obesity and lipid profiles in children and adolescents, *Journal of Pediatric Endocrinology and Metabolism* 15 (2002): 1011–1016; T. Dwyer and coauthors, Syndrome X in 8-y-old Australian children: Stronger associations with current body fatness than with infant size or growth, *International Journal of Obesity and Related Metabolic Disorders* 26 (2002): 1301–1309.

11. B. Glowinska, M. Urban, and A. Koput, Cardiovascular risk factors in children with obesity, hypertension and diabetes: Lipoprotein (a) levels and body mass index correlate with family history of cardiovascular disease, *European Journal of Pediatrics* 161 (2002): 511–518.

12. A. Wiegman and coauthors, Family history and cardiovascular risk in familial hypercholesterolemia: Data in more than 1000 children, *Circulation* 107 (2003): 1473–1478.

13. V. N. Muratova and coauthors, The relation of obesity to cardiovascular risk factors among children: The CARDIAC project, *West Virginia Medical Journal* 98 (2002): 263–267.

14. National High Blood Pressure Education Program Working Group on High Blood Pressure in Children and Adolescents, The Fourth Report on the Diagnosis, Evaluation, and Treatment of High Blood Pressure in Children and Adolescents, *Pediatrics* 114 (2004): 555S–576S.

15. Dwyer and coauthors, 2002.

16. Committee on Dietary Reference Intakes, *Dietary Reference Intakes for Energy, Carbohydrate, Fiber, Fat, Fatty Acids, Cholesterol, Protein, and Amino Acids* (Washington, D.C.: National Academies Press, 2005), pp. 769–879.

17. J. Hurley and B. Liebman, Kids' cuisine: "What would you like with your fries?" *Nutrition Action Healthletter* 31 (2004): 12–15.

18. F. R. Greer, N. F. Krebs, and the Committee on Nutrition, American Academy of Pediatrics, Optimizing bone health and calcium intakes of infants, children, and adolescents, *Pediatrics* 117 (2006): 578–585.

19. S. de Jongh and coauthors, Efficacy and safety of statin therapy in children with familial hypercholesterolemia: A randomized, double-blind, placebo-controlled trial with simvastatin, *Circulation* 106 (2002): 2231–2237.

20. Centers for Disease Control and Prevention, Youth tobacco surveillance—United States, 2001–2002, *Morbidity and Mortality Weekly Report* 55 (2006): entire supplement.

21. Committee on Nutrition, American Academy of Pediatrics, Prevention of pediatric overweight and obesity, *Obesity* 112 (2003): 424–430.

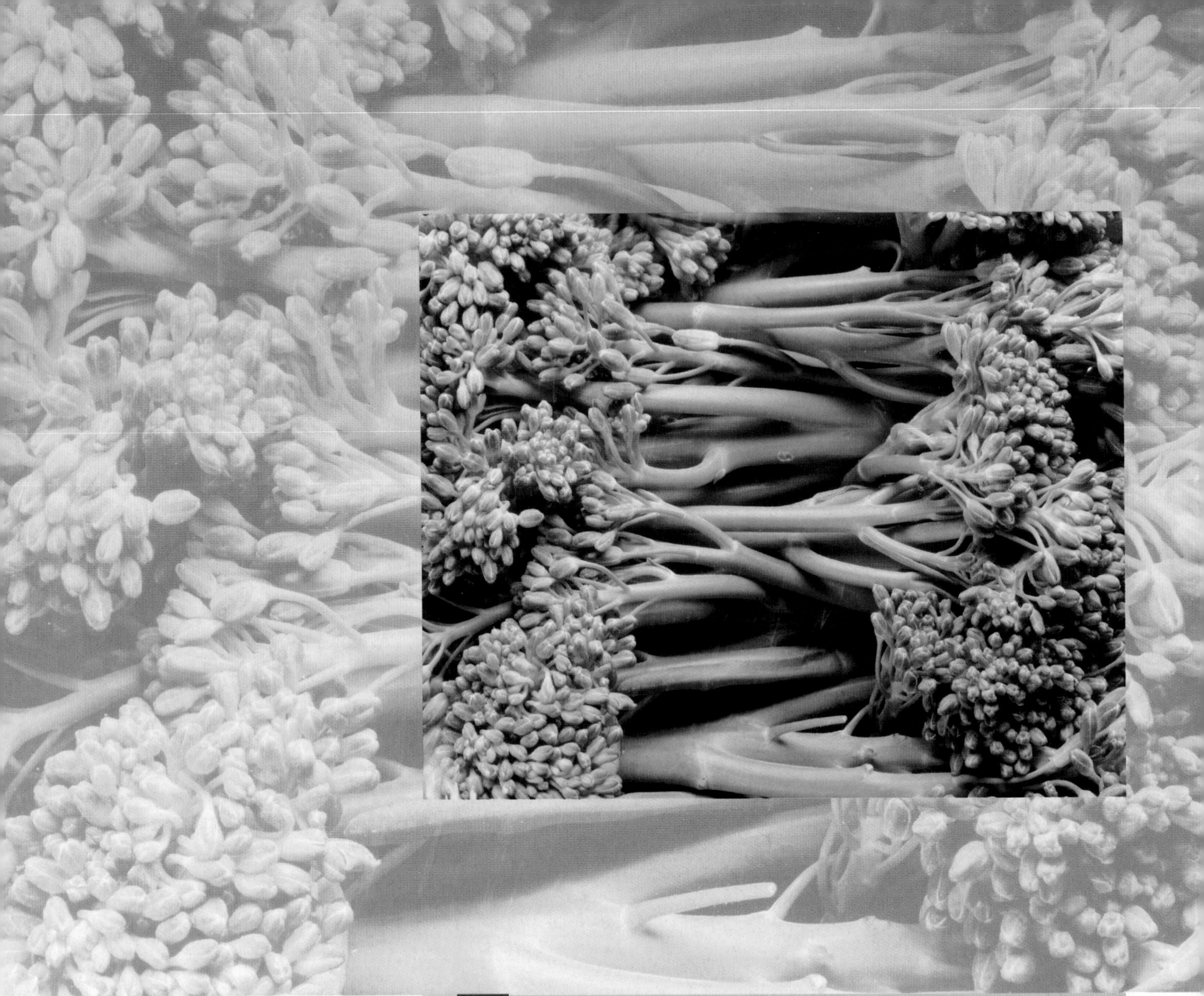

Still Images/Getty Images

CENGAGENOW™ The CengageNOW logo indicates an opportunity for online self-study, linking you to interactive tutorials and videos based on your level of understanding.

academic.cengage.com/login

How To: Practice Problems

Nutrition Portfolio Journal

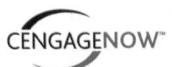

Nutrition in Your Life

Take a moment to envision yourself 20, 40, or even 60 years from now. Are you physically fit and healthy? Can you see yourself walking on the beach with friends or tossing a ball with children? Are you able to climb stairs and carry your own groceries? Importantly, are you enjoying life? If you're lucky, you will grow old with good health, but much of that depends on your actions today—and every day from now until then. Making nutritious foods and physical activities a priority in your life can help bring rewards of continued health and enjoyment in later life.

Life Cycle Nutrition: Adulthood and the Later Years

Wise food choices, made throughout adulthood, can support a person's ability to meet physical, emotional, and mental challenges and to enjoy freedom from disease. Two goals motivate adults to pay attention to their diets: promoting health and slowing aging. Much of this text has focused on nutrition to support health, and later chapters feature dietary treatments of diseases. This chapter focuses on aging and the nutrition needs of older adults.

The U.S. population is growing older. The majority is now middle-aged, and the ratio of old people to young is increasing, as Figure 16-1 (p. 562) shows. In 1900, only 1 out of 25 people was 65 or older. In 2000, 1 out of 8 had reached age 65. Projections for 2030 are 1 out of 5.

Our society uses the arbitrary age of 65 years to define the transition point between middle age and old age, but growing "old" happens day by day, with changes occurring gradually over time. Since 1950 the population of those over 65 has almost tripled. Remarkably, the fastest-growing age group has been people over 85 years; since 1950 their numbers have increased sevenfold. The number of people in the United States age 100 or older doubled in the last decade. Similar trends are occurring in populations worldwide.[1]

Life expectancy in the United States for white women is 81 years and for black women, 76 years; for white men, it is 75 years and for black men, 69 years—all record highs and much higher than the average life expectancy of 47 years in 1900.[2] Women who live to 80 can expect to survive an additional 9 years, on average; men, an additional 7 years. Advances in medical science—antibiotics and other treatments—are largely responsible for almost doubling the life expectancy in the 20th century. Improved nutrition and an abundant food supply have also contributed to lengthening life expectancy. The **life span** has not lengthened as dramatically; human **longevity** appears to have an upper limit. The potential human life span is currently 130 years. With recent advances in medical technology

life expectancy: the average number of years lived by people in a given society.

life span: the maximum number of years of life attainable by a member of a species.

longevity: long duration of life.

FIGURE 16-1 The Aging of the U.S. Population

In general, the percentage of older people in the population has increased over the decades whereas the percentage of younger people has decreased.

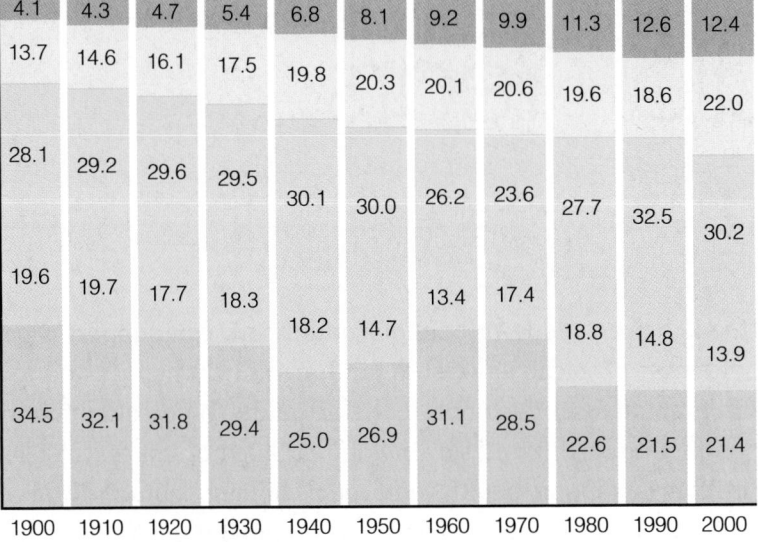

Key:
- ≥65 years
- 45–64 years
- 25–44 years
- 15–24 years
- >15 years

Year	≥65	45–64	25–44	15–24	>15
1900	4.1	13.7	28.1	19.6	34.5
1910	4.3	14.6	29.2	19.7	32.1
1920	4.7	16.1	29.6	17.7	31.8
1930	5.4	17.5	29.5	18.3	29.4
1940	6.8	19.8	30.1	18.2	25.0
1950	8.1	20.3	30.0	14.7	26.9
1960	9.2	20.1	26.2	13.4	31.1
1970	9.9	20.6	23.6	17.4	28.5
1980	11.3	19.6	27.7	18.8	22.6
1990	12.6	18.6	32.5	14.8	21.5
2000	12.4	22.0	30.2	13.9	21.4

SOURCE: U.S. Census Bureau, Decennial census of population, 1900 to 2000.

and genetic knowledge, however, researchers may one day be able to extend the life span even further by slowing, or perhaps preventing, aging and its accompanying diseases.[3]

Nutrition and Longevity

Research in the field of aging is active—and difficult. Researchers are challenged by the diversity of older adults. When older adults experience health problems, it is hard to know whether to attribute these problems to genetics, aging, or other environmental factors such as nutrition. The idea that nutrition can influence the aging process is particularly appealing because people can control and change their eating habits. The questions being asked include:

- To what extent is aging inevitable, and can it be slowed through changes in lifestyle and environment?
- What role does nutrition play in the aging process, and what role can it play in slowing aging?

With respect to the first question, it seems that aging is an inevitable, natural process, programmed into the genes at conception.[4] People can, however, slow the process within genetic limits by adopting healthy lifestyle habits such as eating nutritious food and engaging in physical activity. In fact, an estimated 70 to 80 percent of the average person's life expectancy may depend on individual health-related behaviors; genes determine the remaining 20 to 30 percent.[5]

With respect to the second question, good nutrition helps to maintain a healthy body and can therefore ease the aging process in many significant ways. Clearly, nutrition can improve the **quality of life** in the later years.

quality of life: a person's perceived physical and mental well-being.

Observation of Older Adults

The strategies adults use to meet the two goals mentioned at the start of this chapter—promoting health and slowing aging—are actually very much the same. What to eat, when to sleep, how physically active to be, and other lifestyle choices greatly influence both physical health and the aging process.

Healthy Habits A person's **physiological age** reflects his or her health status and may or may not reflect the person's **chronological age.** Quite simply, some people seem younger, and others older, than their years. Six lifestyle behaviors seem to have the greatest influence on people's health and therefore on their physiological age:

- Sleeping regularly and adequately
- Eating well-balanced meals, including breakfast, regularly
- Engaging in physical activity regularly
- Not smoking
- Not using alcohol, or using it in moderation
- Maintaining a healthy body weight

Over the years, the effects of these lifestyle choices accumulate—that is, people who follow most of these practices live longer and have fewer disabilities as they age. They are in better health, even when older in chronological age, than people who do not adopt these behaviors. Even though people cannot change their birth dates, they may be able to add years to, and enhance the quality of, their lives. Physical activity seems to be most influential in preventing or slowing the many changes that define a stereotypical "old" person. After all, many of the physical limitations that accompany aging occur because people become inactive, not because they become older.

Physical Activity The many remarkable benefits of regular physical activity are not limited to the young. Compared with those who are inactive, older adults who are active weigh less; have greater flexibility, more endurance, better balance, and better health; and live longer.[6] They reap additional benefits from various activities as well: aerobic activities improve cardiorespiratory endurance, blood pressure, and blood lipid concentrations; moderate endurance activities improve the quality of sleep; and strength training improves posture and mobility. In fact, regular physical activity is the most powerful predictor of a person's mobility in the later years. Physical activity also increases blood flow to the brain, thereby preserving mental ability, alleviating depression, supporting independence, and improving quality of life.[7]

Dietary Guidelines for Americans 2005

Older adults should participate in regular physical activity to reduce the functional declines associated with aging and to achieve the other benefits of physical activity identified for all adults.

Muscle mass and muscle strength tend to decline with aging, making older people vulnerable to falls and immobility. Falls are a major cause of fear, injury, disability, and even death among older adults.[8] Many lose their independence as a result of falls. Regular physical activity tones, firms, and strengthens muscles, helping to improve confidence, reduce the risk of falling, and lessen the risk of injury should a fall occur.

Even without a fall, older adults may become so weak that they can no longer perform life's daily tasks, such as climbing stairs, carrying packages, and opening jars. By improving muscle strength, which allows a person to perform these tasks, strength training helps to maintain independence. Even in frail, elderly people over

physiological age: a person's age as estimated from her or his body's health and probable life expectancy.

chronological age: a person's age in years from his or her date of birth.

Regular physical activity promotes a healthy, independent lifestyle.

85 years of age, strength training not only improves balance, muscle strength, and mobility, but it also increases energy expenditure and energy intake, thereby enhancing nutrient intakes. This finding highlights another reason to be physically active: a person spending energy can afford to eat more food and thus receives more nutrients. People who are committed to an ongoing fitness program can benefit from higher energy and nutrient intakes and still maintain their body weights.

Ideally, physical activity should be part of each day's schedule and should be intense enough to prevent muscle atrophy and to speed up the heartbeat and respiration rate. Although aging reduces both speed and endurance to some degree, older adults can still train and achieve exceptional performances. Healthy older adults who have not been active can ease into a suitable routine. They can start by walking short distances until they are walking at least 10 minutes continuously and then gradually increase their distance to a 30- to 45-minute workout at least 5 days a week. Table 16-1 provides exercise guidelines for seniors. People with medical conditions should check with a physician before beginning an exercise routine, as should sedentary men over 40 and sedentary women over 50 who want to participate in a vigorous program.

Manipulation of Diet

In their efforts to understand longevity, researchers have not only observed people, but they have also manipulated influencing factors, such as diet, in animals. This research has given rise to some interesting and suggestive findings.

Energy Restriction in Animals   Animals live longer and have fewer age-related diseases when their energy intakes are restricted. These life-prolonging benefits become evident when the diet provides enough food to prevent malnutrition and an energy intake of about 70 percent of normal. Exactly how energy restriction prolongs life remains largely unexplained, although gene activity appears to play a key role. The genetic activity of old mice differs from that of young mice, with some genes becoming more active with age and others less active. With an energy-restricted diet, many of the genetic activities of older mice revert to those of younger mice. These "slow-aging" genetic changes are apparent in as little as one month on an energy-restricted, but still nutritionally adequate, diet.

TABLE 16-1	Exercise Guidelines for Older Adults			
	Endurance	**Strength**	**Balance**	**Flexibility**
Examples				
Start easy	Be active 5 minutes on most or all days.	Using 0- to 2-pound weights, do 1 set of 8 repetitions twice a week.	Hold onto table or chair with one hand, then with one finger.	Hold stretch 10 seconds; do each stretch 3 times.
Progress gradually to goal	Be active 30 minutes (minimum) on most or all days.	Increase weight as able; do 2 sets of 8–15 repetitions twice a week.	Do not hold onto table or chair; then close eyes.	Hold stretch 30 seconds; do each stretch 5 times.
Cautions and comments	Stop if you are breathing so hard you can't talk or if you feel dizziness or chest pain.	Breathe out as you contract and in as you relax (do not hold breath); use smooth, steady movements.	Incorporate balance techniques with strength exercises as you progress.	Stretch after strength and endurance exercises for 20 minutes, 3 times a week; use slow, steady movements; bend joints slightly.

SOURCE: *Exercise: A Guide from the National Institute on Aging*, www.nia.nih.gov.

The consequences of energy restriction in animals include a delay in the onset, or prevention, of diseases such as atherosclerosis; prolonged growth and development; and improved blood glucose, insulin sensitivity, and blood lipids. In addition, energy metabolism slows and body temperature drops—indications of a reduced rate of oxygen consumption. As Highlight 11 explained, the use of oxygen during energy metabolism produces free radicals, which have been implicated in the aging process. Restricting energy intake in animals not only produces fewer free radicals, but also increases antioxidant activity and enhances DNA repair. Reducing oxidative stress may at least partially explain how restricting energy intake lengthens life expectancy.[9]

Interestingly, longevity appears to depend on restricting energy intake and not on the amount of body fat. Genetically obese rats live longer when given a restricted diet even though their body fat is similar to that of other rats allowed to eat freely.

Energy Restriction in Human Beings Research on a variety of animals ◆ confirms the relationship between energy restriction and longevity. Applying the results of animal studies to human beings is problematic, however, and conducting studies on human beings raises numerous questions—beginning with how to define energy restriction.[10] Does it mean eating less or just weighing less? Is it less than you want or less than the average? Does eating less have to result in weight loss? Does it matter whether weight loss results from more exercise or from less food? Or whether weight loss is intentional or unintentional? Answers await research.

Extreme starvation to extend life, like any extreme, is rarely, if ever, worth the price. Moderation, on the other hand, may be valuable. Many of the physiological responses to energy restriction seen in animals also occur in people whose intakes are *moderately* restricted. When people cut back on their usual energy intake by 10 to 20 percent, ◆ body weight, body fat, and blood pressure drop, and blood lipids and insulin response improve—favorable changes for preventing chronic diseases.[11] Some research suggests that fasting on alternative days may provide similar benefits.[12] The reduction in oxidative damage that occurs with energy restriction in animals also occurs in people whose diets include antioxidant nutrients and phytochemicals. Diets, such as the Mediterranean diet, which include an abundance of fruits, vegetables, olive oil, and red wine—with their array of antioxidants and phytochemicals—support good health and long life.[13] Clearly, nutritional adequacy is essential to living a long and healthy life.

◆ kCalorie-restricted research has been conducted on various species, including mice, rats, rhesus monkeys, cynomolgus monkeys, spiders, and fish.

◆ For perspective, a person with a usual energy intake of 2000 kcalories might cut back to 1600 to 1800 kcalories.

IN SUMMARY

Life expectancy in the United States increased dramatically in the 20th century. Factors that enhance longevity include limited or no alcohol use, regular balanced meals, weight control, adequate sleep, abstinence from smoking, and regular physical activity. Energy restriction in animals seems to lengthen their lives. Whether such dietary intervention in human beings is beneficial remains unknown. At the very least, nutrition—especially when combined with regular physical activity—can influence aging and longevity in human beings by supporting good health and preventing disease.

The Aging Process

As people get older, each person becomes less and less like anyone else. The older people are, the more time has elapsed for such factors as nutrition, genetics, physical activity, and everyday **stress** to influence physical and psychological aging.

Stress promotes the early onset of age-related diseases.[14] Both physical **stressors** (such as alcohol abuse, other drug abuse, smoking, pain, and illness) and

stress: any threat to a person's well-being; a demand placed on the body to adapt.

stressors: environmental elements, physical or psychological, that cause stress.

psychological stressors (such as exams, divorce, moving, and the death of a loved one) elicit the body's **stress response.** The body responds to such stressors with an elaborate series of physiological steps, as the nervous and hormonal systems bring about defensive readiness in every body part. These effects favor physical action—the classic fight-or-flight response. Prolonged or severe stress can drain the body of its reserves and leave it weakened, aged, and vulnerable to illness, especially if physical action is not taken. As people age, they lose their ability to adapt to both external and internal disturbances. When disease strikes, the reduced ability to adapt makes the aging individual more vulnerable to death than a younger person.

Because the stress response is mediated by hormones, it differs between men and women.[15] The fight-or-flight response may be more typical of men than of women. Women's reactions to stress more typically follow a pattern of "tend-and-befriend."[16] Women *tend* by nurturing and protecting themselves and their children. These actions promote safety and reduce stress. Women *befriend* by creating and maintaining a social group that can help in the process.

Highlight 11 described the oxidative stresses and cellular damage that occur when free radicals exceed the body's ability to defend itself. Increased free-radical activity and decreased antioxidant protection are common features of aging—and antioxidants seem to help slow the aging process.[17] Such findings seem to suggest that the fountain of youth may actually be a cornucopia of fruits and vegetables rich in antioxidants. (Return to Highlight 11 for more details on the antioxidant action of fruits and vegetables in defending against oxidative stress.)

Physiological Changes

As aging progresses, inevitable changes in each of the body's organs contribute to the body's declining function. These physiological changes influence nutrition status, just as growth and development do in the earlier stages of the life cycle.

Body Weight Two-thirds of older adults in the United States are now considered overweight or obese. Chapter 8 presented the many health problems that accompany obesity and the BMI guidelines for a healthy body weight (18.5 to 24.9). These guidelines apply to all adults, regardless of age, but they may be too restrictive for older adults. The importance of body weight in defending against chronic diseases differs for older adults. Being moderately *overweight* may not be harmful. For adults over 65, health risks do not become apparent until BMI reaches at least 27—and the relationship tends to diminish with age until it disappears by age 75. Older adults who are *obese*, however, face serious medical complications and can significantly improve their quality of life with weight loss.[18]

For some older adults, a low body weight may be more detrimental than a high one. Low body weight often reflects malnutrition and the trauma associated with a fall. Many older adults experience unintentional weight loss, in large part because of an inadequate food intake. Without adequate nutrient reserves, an underweight person may be unprepared to fight against diseases. For underweight people, even a slight weight loss (5 percent) increases the likelihood of disease and premature death, making every meal a life-saving event.

Body Composition In general, older people tend to lose bone and muscle and gain body fat. Many of these changes occur because some hormones that regulate appetite and metabolism become less active with age, whereas others become more active.*[19]

Loss of muscle, known as **sarcopenia,** can be significant in the later years, and its consequences can be quite dramatic (see Figure 16-2).[20] As muscles di-

---

stress response: the body's response to stress, mediated by both nerves and hormones.

sarcopenia (SAR-koh-PEE-nee-ah): loss of skeletal muscle mass, strength, and quality.
- **sarco** = flesh
- **penia** = loss or lack

---

* Causes of diminished appetite in older adults include increased cholecystokinin, leptin, and cytokines and decreased ghrelin and testosterone. Additional examples of hormones that change with age include growth hormone and androgens, which decline with advancing age, thus contributing to the decrease in lean body mass, and prolactin, which increases with age, helping to maintain body fat. Insulin sensitivity also diminishes as people grow older, most likely because of increases in body fat and decreases in physical activity.

FIGURE 16-2 Sarcopenia

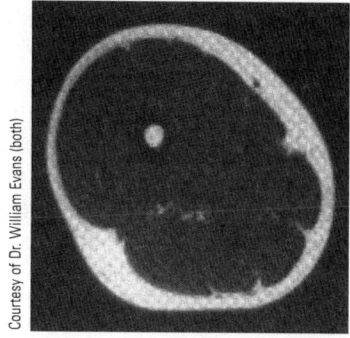 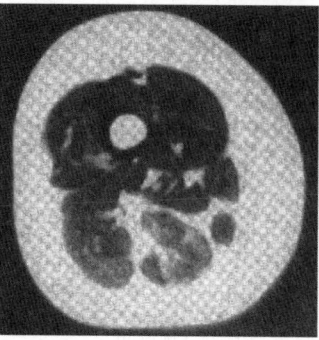

Courtesy of Dr. William Evans (both)

These cross sections of two women's thighs may appear to be about the same size from the outside, but the 20-year-old woman's thigh (left) is dense with muscle tissue. The 64-year-old woman's thigh (right) has lost muscle and gained fat, changes that may be largely preventable with strength-building physical activities.

minish and weaken, people lose the ability to move and maintain balance—making falls likely. The limitations that accompany the loss of muscle mass and strength play a key role in the diminishing health that often accompanies aging.[21] Optimal nutrition and regular physical activity can help maintain muscle mass and strength and minimize the changes in body composition associated with aging.[22]

Risk factors for sarcopenia include weight loss, little physical activity, and cigarette smoking.[23] Obesity and the inflammation that accompanies it may also contribute to sarcopenia.[24]

Immune System Changes in the immune system also bring declining function with age. In addition, the immune system is compromised by nutrient deficiencies. Thus the combination of age and malnutrition makes older people vulnerable to infectious diseases. Adding insult to injury, antibiotics often are not effective against infections in people with compromised immune systems. Consequently, infectious diseases are a major cause of death in older adults. Older adults may improve their immune system responses by exercising regularly.

GI Tract In the GI tract, the intestinal wall loses strength and elasticity with age, and GI hormone secretions change. All of these actions slow motility. Constipation is much more common in the elderly than in the young. Changes in GI hormone secretions also diminish appetite, leading to decreased energy intake and weight loss.[25]

Atrophic gastritis, a condition that affects almost one-third of those over 60, is characterized ◆ by an inflamed stomach, bacterial overgrowth, and a lack of hydrochloric acid and intrinsic factor. All of these can impair the digestion and absorption of nutrients, most notably, vitamin B_{12}, but also biotin, folate, calcium, iron, and zinc.

Difficulty in swallowing, medically known as **dysphagia,** occurs in all age groups, but especially in the elderly. Being unable to swallow a mouthful of food can be scary, painful, and dangerous. Even swallowing liquids can be a problem for some people. Consequently, the person may eat less food and drink fewer beverages, resulting in weight loss, malnutrition, and dehydration. Dietary intervention for dysphagia is highly individualized based on the person's abilities and tolerances. The diet typically provides moist, soft-textured, tender-cooked, or pureed foods and thickened liquids.

Tooth Loss Regular dental care over a lifetime protects against tooth loss and gum disease, which are common in old age. These conditions make chewing difficult or

◆ Consequences of atrophic gastritis:
• Inflamed stomach
• Increased bacterial growth
• Reduced hydrochloric acid
• Reduced intrinsic factor
• Increased risk of nutrient deficiencies, notably of vitamin B_{12}

dysphagia (dis-FAY-jah): difficulty in swallowing.

◆ The medical term for lack of teeth is **edentulous** (ee-DENT-you-lus).
  • **e** = without
  • **dens** = teeth

◆ Conditions requiring dental care:
  Dry mouth
  Eating difficulty
  No dental care within two years
  Tooth or mouth pain
  Altered food selections
  Lesions, sores, or lumps in mouth

painful. Dentures, even when they fit properly, are less effective than natural teeth, and inefficient chewing can cause choking. People with tooth loss, ◆ gum disease, and ill-fitting dentures tend to limit their food selections to soft foods. If foods such as corn on the cob, apples, and hard rolls are replaced by creamed corn, applesauce, and rice, then nutrition status may not be greatly affected. However, when food groups are eliminated and variety is limited, poor nutrition follows. People without teeth typically eat fewer fruits and vegetables and have less variety in their diets.[26] Consequently, they have low intakes of fiber and vitamins, which exacerbates their dental and overall health problems.[27] To determine whether a visit to the dentist is needed, an older adult can check the conditions listed in the margin. ◆

Sensory Losses and Other Physical Problems Sensory losses and other physical problems can also interfere with an older person's ability to obtain adequate nourishment. Failing eyesight, for example, can make driving to the grocery store impossible and shopping for food a frustrating experience. It may become so difficult to read food labels and count money that the person doesn't buy needed foods. Carrying bags of groceries may be an unmanageable task. Similarly, a person with limited mobility may find cooking and cleaning up too hard to do. Not too surprisingly, the prevalence of undernutrition is high among those who are homebound.

Sensory losses can also interfere with a person's ability or willingness to eat. Taste and smell sensitivities tend to diminish with age and may make eating less enjoyable. If a person eats less, then weight loss and nutrient deficiencies may follow. Loss of vision and hearing may contribute to social isolation, and eating alone may lead to poor intake.

Other Changes

In addition to the physiological changes that accompany aging, adults change in many other ways that influence their nutrition status.[28] Psychological, economic, and social factors play big roles in a person's ability and willingness to eat.

Psychological Changes Although not an inevitable component of aging, depression is common among older adults.[29] Depressed people, even those without disabilities, lose their ability to perform simple physical tasks. They frequently lose their appetite and the motivation to cook or even to eat. An overwhelming sense of grief and sadness at the death of a spouse, friend, or family member may leave a person, especially an elderly person, feeling powerless to overcome depression. When a person is suffering the heartache and loneliness of bereavement, cooking meals may not seem worthwhile. The support and companionship of family and friends, especially at mealtimes, can help overcome depression and enhance appetite.

Economic Changes Overall, older adults today have higher incomes than their cohorts of previous generations. Still, 10 percent of the people over age 65 live in poverty. Factors such as living arrangements and income make significant differences in the food choices, eating habits, and nutrition status of older adults, especially those over age 80. People of low socioeconomic means are likely to have inadequate food and nutrient intakes. Only about one-third of the needy elderly receive assistance from federal programs.

Social Changes Malnutrition among older adults is most common in hospitals and nursing homes.[30] In the community, malnutrition is most likely to occur among those living alone, especially men; those with the least education; those living in federally funded housing (an indicator of low income); and those who have recently experienced a change in lifestyle. Adults who live alone do not necessarily make poor food choices, but they often consume too little food. Loneliness is directly related to nutritional inadequacies, especially of energy intake.

© Bob Thomas/Stone/Getty Images

Shared meals can brighten the day and enhance the appetite.

IN SUMMARY

Many changes that accompany aging can impair nutrition status. Among physiological changes, hormone activity alters body composition, immune system changes raise the risk of infections, atrophic gastritis interferes with digestion and absorption, and tooth loss limits food choices. Psychological changes such as depression, economic changes such as loss of income, and social changes such as loneliness contribute to poor food intake.

Energy and Nutrient Needs of Older Adults

Growing old can be enjoyable for people who take care of their health and live each day fully.

Knowledge about the nutrient needs and nutrition status of older adults has grown considerably in recent years. The Dietary Reference Intakes (DRI) cluster people over 50 into two age categories—one group of 51 to 70 years and one of 71 and older. Increasingly, research is showing that the nutrition needs of people 50 to 70 years old differ from those of people over 70.

Setting standards for older people is difficult because individual differences become more pronounced as people grow older.[31] People start out with different genetic predispositions and ways of handling nutrients, and the effects of these differences become magnified with years of unique dietary habits. For example, one person may tend to omit fruits and vegetables from his diet, and by the time he is old, he may have a set of nutrition problems associated with a lack of fiber and antioxidants. Another person may have omitted milk and milk products all her life—her nutrition problems may be related to a lack of calcium. Also, as people age, they suffer different chronic diseases and take various medicines—both of which will affect nutrient needs. For all of these reasons, researchers have difficulty even defining "healthy aging," a prerequisite to developing recommendations to meet the "needs of practically all healthy persons." The following discussion gives special attention to the nutrients of greatest concern.

Water

Despite real fluid needs, many older people do not seem to feel thirsty or notice mouth dryness. Many nursing home employees say it is hard to persuade their elderly clients to drink enough water and fruit juices. Older adults may find it difficult and bothersome to get a drink or to get to a bathroom. Those who have lost bladder control may be afraid to drink too much water.

Dehydration is a risk for older adults.[32] Total body water decreases as people age, so even mild stresses such as fever or hot weather can precipitate rapid dehydration in older adults. Dehydrated older adults seem to be more susceptible to urinary tract infections, pneumonia, **pressure ulcers,** and confusion and disorientation. To prevent dehydration, older adults need to drink *at least* six glasses of water a day. ◆

◆ Beverage recommendation for adults 51+ yr:
- Men: 13 c/day
- Women: 9 c/day

◆ When using the tables in Appendix F to estimate energy requirements:
- Men: Subtract 10 kcal/day for each year of age above 19
- Women: Subtract 7 kcal/day for each year of age above 19

pressure ulcers: damage to the skin and underlying tissues as a result of compression and poor circulation; commonly seen in people who are bedridden or chairbound.

Energy and Energy Nutrients

On average, energy needs decline an estimated 5 percent per decade. One reason is that people usually reduce their physical activity as they age, although they need not do so. Another reason is that basal metabolic rate declines 1 to 2 percent per decade in part because lean body mass and thyroid hormones diminish.[33]

The lower energy expenditure of older adults means that they need to eat less food to maintain their weights. Accordingly, the estimated energy requirements ◆ for adults decrease steadily after age 19, as the "How to" on p. 570 explains.

HOW TO Estimate Energy Requirements for Older Adults

The "How to" on p. 257 described how to estimate the energy requirements for adults using an equation that accounts for age, physical activity, weight, and height. Alternatively, energy requirements for older adults can be "guesstimated" by using the values listed in the tables in Appendix F for adults 30 years of age and subtracting 7 kcalories for women and 10 kcalories for men per day for each year over 30.

For example, Table F-4 lists 2556 kcalories per day for a woman who is 5 feet 5 inches tall,

weighs 150 pounds, and has a low activity level. To estimate the energy requirements of a similar 50-year-old woman, subtract 7 kcalories per day for each year over 30:

$$50 - 30 = 20 \text{ yr}$$
$$20 \text{ yr} \times 7 \text{ kcal/day} = 140 \text{ kcal/day}$$
$$2556 \text{ kcal/day (at age 30)} - 140 \text{ kcal/day}$$
$$= 2416 \text{ kcal/day (at age 50)}$$

Similarly, using Table F-5 to estimate the energy requirements of a sedentary 65-year-old man who is 5 feet 11 inches tall and

weighs 250 pounds, subtract 10 kcalories per day for each year over 30:

$$65 - 30 = 35 \text{ yr}$$
$$35 \text{ yr} \times 10 \text{ kcal/day} = 350 \text{ kcal/day}$$
$$3088 \text{ kcal/day (at age 30)} - 350 \text{ kcal/day}$$
$$= 2738 \text{ kcal/day (at age 65)}$$

(Adults between the ages of 19 and 30 can also use the values listed in the tables in Appendix F by adding 7 kcalories for women and 10 kcalories for men per day for each year below 30.)

CENGAGENOW
To practice estimating energy requirements for older adults, log on to **academic.cengage.com/login**, go to Chapter 16, then go to How To.

On limited energy allowances, people must select mostly nutrient-dense foods. There is little leeway for added sugars, solid fats, or alcohol. The USDA Food Guide (pp. 41–47) offers a dietary framework for adults of all ages.

Protein Because energy needs decrease, protein must be obtained from low-kcalorie sources of high-quality protein, such as lean meats, poultry, fish, and eggs; fat-free and low-fat milk products; and legumes. Protein is especially important for the elderly to support a healthy immune system, prevent muscle wasting, and optimize bone mass.

Underweight or malnourished older adults need protein- and energy-dense snacks such as hard-boiled eggs, tuna fish and crackers, peanut butter on wheat toast, and hearty soups. Drinking liquid nutritional formulas between meals can also boost energy and nutrient intakes.[34] Importantly, the diet should provide enjoyment as well as nutrients.[35]

Carbohydrate and Fiber As always, abundant carbohydrate is needed to protect protein from being used as an energy source. Sources of complex carbohydrates such as legumes, vegetables, whole grains, and fruits are also rich in fiber and essential vitamins and minerals. Average fiber intakes among older adults are lower than current recommendations (14 grams per 1000 kcalories).[36] Eating high-fiber foods and drinking water can alleviate constipation—a condition common among older adults, especially nursing home residents. Physical inactivity and medications also contribute to the high incidence of constipation.

Fat As is true for people of all ages, fat intake needs to be moderate in the diets of most older adults—enough to enhance flavors and provide valuable nutrients, but not so much as to raise the risks of cancer, atherosclerosis, and other degenerative diseases. This recommendation should not be taken too far; limiting fat too severely may lead to nutrient deficiencies and weight loss—two problems that carry greater health risks in the elderly than overweight.

Vitamins and Minerals

Most people can achieve adequate vitamin and mineral intakes simply by including foods from all food groups in their diets, but older adults often omit fruits and vegetables. Similarly, few older adults consume the recommended amounts of milk or milk products.

◆ Reminder: *Atrophic gastritis* is a chronic inflammation of the stomach characterized by inadequate hydrochloric acid and intrinsic factor—two key players in vitamin B_{12} absorption.

Vitamin B_{12} An estimated 10 to 30 percent of adults over 50 have atrophic gastritis. ◆ As Chapter 10 explained, people with atrophic gastritis are particularly vulnerable to vitamin B_{12} deficiency. The bacterial overgrowth that accompanies this condition uses up the vitamin, and without hydrochloric acid and intrinsic factor, digestion and absorption of vitamin B_{12} are inefficient. Given the poor cognition, anemia, and devastating neurological effects associated with a vitamin B_{12} deficiency, an adequate intake is imperative.[37] The RDA for older adults is the same as for younger adults, but with the added suggestion to obtain most of a day's intake from vitamin B_{12}–fortified foods and supplements.[38] The bioavailability of vitamin B_{12} from these sources is better than from foods.

Dietary Guidelines for Americans 2005

People over age 50 should consume vitamin B_{12} from fortified foods or supplements.

Vitamin D Vitamin D deficiency is a problem among older adults. Only vitamin D–fortified milk provides significant vitamin D, and many older adults drink little or no milk. Further compromising the vitamin D status of many older people, especially those in nursing homes, is their limited exposure to sunlight. Finally, aging reduces the skin's capacity to make vitamin D and the kidneys' ability to convert it to its active form. Not only are older adults not getting enough vitamin D, but they may actually need more to improve both muscle and bone strength.[39] To prevent bone loss and to maintain vitamin D status, especially in those who engage in minimal outdoor activity, adults 51 to 70 years old need 10 micrograms daily, and those over 70 need 15 micrograms.[40]

Dietary Guidelines for Americans 2005

Older adults should consume extra vitamin D from vitamin D–fortified foods and/or supplements.

Calcium Both Chapter 12 and Highlight 12 emphasized the importance of abundant dietary calcium throughout life, especially for women after menopause, to protect against osteoporosis. The DRI Committee recommends 1200 milligrams of calcium daily, but the calcium intakes of older people in the United States are well below recommendations.[41] Some older adults avoid milk and milk products because they dislike these foods or associate them with stomach discomfort. Simple solutions include using calcium-fortified juices, adding powdered milk to recipes, and taking supplements. Chapter 12 offered many other strategies for including nonmilk sources of calcium for those who do not drink milk.

Iron The iron needs of men remain unchanged throughout adulthood. For women, iron needs decrease substantially when blood loss through menstruation ceases. Consequently, iron-deficiency anemia is less common in older adults than in younger people. In fact, elevated iron stores are more likely than deficiency in older people, especially those who take iron supplements, eat red meat regularly, and include vitamin C–rich fruits in their daily diet.[42]

Nevertheless, iron deficiency may develop in older adults, especially when their food energy intakes are low. Aside from diet, two other factors may lead to iron deficiency in older people: chronic blood loss from diseases and medicines and poor iron absorption due to reduced stomach acid secretion and antacid use. Iron deficiency impairs immunity and leaves older adults vulnerable to infectious diseases.[43] Anyone concerned with older people's nutrition should keep these possibilities in mind.

Taking time to nourish your body well is a gift you give yourself.

Nutrient Supplements

People judge for themselves how to manage their nutrition, and more than half of older adults turn to dietary supplements.[44] When recommended by a physician or registered dietitian, vitamin D and calcium supplements for osteoporosis or vitamin B_{12} for pernicious anemia may be beneficial. Many health care professionals recommend a daily multivitamin-mineral supplement that provides 100 percent or less of the Daily Value for the listed nutrients.[45] They reason that such a supplement is more likely to be beneficial than to cause harm.

People with small energy allowances would do well to become more active so they can afford to eat more food. Food is the best source of nutrients for everybody. Supplements are just that—supplements to foods, not substitutes for them. For anyone who is motivated to obtain the best possible health, it is never too late to learn to eat well, drink water, exercise regularly, and adopt other lifestyle habits such as quitting smoking and moderating alcohol use.

IN SUMMARY

The table below summarizes the nutrient concerns of aging. Although some nutrients need special attention in the diet, supplements are not routinely recommended. The ever-growing number of older people creates an urgent need to learn more about how their nutrient requirements differ from those of others and how such knowledge can enhance their health.

Nutrient	Effect of Aging	Comments
Water	Lack of thirst and decreased total body water make dehydration likely.	Mild dehydration is a common cause of confusion. Difficulty obtaining water or getting to the bathroom may compound the problem.
Energy	Need decreases as muscle mass decreases (sarcopenia).	Physical activity moderates the decline.
Fiber	Likelihood of constipation increases with low intakes and changes in the GI tract.	Inadequate water intakes and lack of physical activity, along with some medications, compound the problem.
Protein	Needs may stay the same or increase slightly.	Low-fat, high-fiber legumes and grains meet both protein and other nutrient needs.
Vitamin B_{12}	Atrophic gastritis is common.	Deficiency causes neurological damage; supplements may be needed.
Vitamin D	Increased likelihood of inadequate intake; skin synthesis declines.	Daily sunlight exposure in moderation or supplements may be beneficial.
Calcium	Intakes may be low; osteoporosis is common.	Stomach discomfort commonly limits milk intake; calcium substitutes or supplements may be needed.
Iron	In women, status improves after menopause; deficiencies are linked to chronic blood losses and low stomach acid output.	Adequate stomach acid is required for absorption; antacid or other medicine use may aggravate iron deficiency; vitamin C and meat increase absorption.

Nutrition-Related Concerns of Older Adults

Nutrition may play a greater role than has been realized in preventing many changes once thought to be inevitable consequences of growing older. The following discussions of vision, arthritis, and the aging brain show that nutrition may provide at least some protection against some of the conditions associated with aging.

Vision

One key aspect of healthy aging is maintaining good vision.[46] Age-related eye diseases that impair vision, such as cataract and macular degeneration, correlate with poor survival that cannot be explained by other risk factors.[47] Following a healthy diet as described by the *Dietary Guidelines for Americans* is one way to protect against these age-related vision problems.[48]

Cataracts **Cataracts** are age-related thickenings in the lenses of the eyes that impair vision. If not surgically removed, they ultimately lead to blindness. Cataracts

cataracts (KAT-ah-rakts): thickenings of the eye lenses that impair vision and can lead to blindness.

occur even in well-nourished individuals as a result of ultraviolet light exposure, oxidative stress, injury, viral infections, toxic substances, and genetic disorders. Many cataracts, however, are vaguely called senile cataracts—meaning "caused by aging." In the United States, more than half of all adults 65 and older have a cataract.

Oxidative stress appears to play a significant role in the development of cataracts, and the antioxidant nutrients may help minimize the damage. Studies have reported an inverse relationship between cataracts and dietary intakes of vitamin C, vitamin E, and carotenoids; taking supplements or eating fruits and vegetables rich in these antioxidant nutrients seems to slow the progression or reduce the risk of developing cataracts.[49]

One other diet-related factor may play a role in the development of cataracts—obesity.[50] Obesity appears to be associated with cataracts, but its role has not been identified. Risk factors that typically accompany overweight, such as inactivity, diabetes, or hypertension, do not explain the association.

Macular Degeneration The leading cause of visual loss among older people is age-related **macular degeneration,** a deterioration of the macular region of the retina.[51] As with cataracts, risk factors for age-related macular degeneration include oxidative stress from sunlight, and preventive factors may include supplements of antioxidant vitamins plus zinc and the carotenoids lutein and zeaxanthin.[52] Total dietary fat may also be a risk factor for macular degeneration, but the omega-3 fatty acids of fish may be protective.

Arthritis

More than 40 million people in the United States have some form of **arthritis.** As the population ages, it is expected that the prevalence will increase to 60 million by 2020.

Osteoarthritis The most common type of arthritis that disables older people is **osteoarthritis,** a painful deterioration of the cartilage in the joints. During movement, the ends of bones are normally protected from wear by cartilage and by small sacs of fluid that act as a lubricant. With age, the cartilage sometimes disintegrates, and the joints become malformed and painful to move.

One known connection between osteoarthritis ◆ and nutrition is overweight. Weight loss may relieve some of the pain for overweight persons with osteoarthritis, partly because the joints affected are often weight-bearing joints that are stressed and irritated by having to carry excess pounds. Interestingly, though, weight loss often relieves much of the pain of arthritis in the hands as well, even though they are not weight-bearing joints. Jogging and other weight-bearing exercises do not worsen arthritis. In fact, both aerobic activity and strength training offer improvements in physical performance and pain relief, especially when accompanied by even modest weight loss.[53]

Rheumatoid Arthritis Another type of arthritis known as **rheumatoid arthritis** has possible links to diet through the immune system.[54] In rheumatoid arthritis, the immune system mistakenly attacks the bone coverings as if they were made of foreign tissue. In some individuals, certain foods, notably vegetables and olive oil, may moderate the inflammatory response and provide some relief.[55]

The omega-3 fatty acids commonly found in fish oil reduce joint tenderness and improve mobility in some people with rheumatoid arthritis.[56] The same diet recommended for heart health—one low in saturated fat from meats and milk products and high in omega-3 fats from fish—helps prevent or reduce the inflammation in the joints that makes arthritis so painful.

Another possible link between nutrition and rheumatoid arthritis involves the oxidative damage to the membranes within joints that causes inflammation and swelling. The antioxidant vitamins C and E and the carotenoids defend against oxidation, and increased intakes of these nutrients may help prevent or relieve the pain of rheumatoid arthritis.[57]

◆ Risk factors for osteoarthritis:
- Age
- Smoking
- High BMI at age 40
- Lack of hormone therapy (in women)

macular (MACK-you-lar) **degeneration:** deterioration of the macular area of the eye that can lead to loss of central vision and eventual blindness. The **macula** is a small, oval, yellowish region in the center of the retina that provides the sharp, straight-ahead vision so critical to reading and driving.

arthritis: inflammation of a joint, usually accompanied by pain, swelling, and structural changes.

osteoarthritis: a painful, degenerative disease of the joints that occurs when the cartilage in a joint deteriorates; joint structure is damaged, with loss of function; also called **degenerative arthritis.**

rheumatoid (ROO-ma-toyd) **arthritis:** a disease of the immune system involving painful inflammation of the joints and related structures.

Gout Another form of arthritis, which most commonly affects men, is **gout**, a condition characterized by deposits of uric acid crystals in the joints. Uric acid derives from the breakdown of **purines**, primarily from those made by the body but also from those found in foods.[58] Foods such as meat and seafood that are rich in purines increase uric acid levels and the risk of gout, whereas milk products seem to lower uric acid levels and the risk of gout.[59]

Treatment Treatment for arthritis—dietary or otherwise—may help relieve discomfort and improve mobility, but it does not cure the condition. Traditional medical intervention for arthritis includes medication and surgery. Alternative therapies to treat arthritis abound, but none have proved safe and effective in scientific studies. Popular supplements—glucosamine, chondroitin, or a combination—may relieve pain and improve mobility as well as over-the-counter pain relievers, but mixed reports from studies emphasize the need for additional research.[60] The regular use of drugs and supplements can impose nutrition risks; many affect appetite and alter the body's use of nutrients, as Chapter 19 explains.

The Aging Brain

The brain, like all of the body's organs, responds to both genetic and environmental factors that can enhance or diminish its amazing capacities. One of the challenges researchers face when studying the human brain is to distinguish among normal age-related physiological changes, changes caused by diseases, and changes that result from cumulative, environmental factors such as diet.

The brain normally changes in some characteristic ways as it ages. For one thing, its blood supply decreases. For another, the number of **neurons,** the brain cells that specialize in transmitting information, diminishes as people age. When the number of nerve cells in one part of the cerebral cortex diminishes, hearing and speech are affected. Losses of neurons in other parts of the cortex can impair memory and cognitive function. When the number of neurons in the hindbrain diminishes, balance and posture are affected. Losses of neurons in other parts of the brain affect still other functions. Some of the cognitive loss and forgetfulness generally attributed to aging may be due in part to environmental, and therefore controllable, factors—including nutrient deficiencies.

Nutrient Deficiencies and Brain Function Nutrients influence the development and activities of the brain. The ability of neurons to synthesize specific neurotransmitters depends in part on the availability of precursor nutrients that are obtained from the diet.[61] The neurotransmitter serotonin, for example, derives from the amino acid tryptophan. To function properly, the enzymes involved in neurotransmitter synthesis require vitamins and minerals. Thus nutrient deficiencies may contribute to the loss of memory and cognition that some older adults experience. Such losses may be preventable or at least diminished or delayed through diet and exercise.[62] Table 16-2 summarizes some of the better-known connections between brain function and nutrients.

In some instances, the degree of cognitive loss is extensive. Such **senile dementia** may be attributable to a specific disorder such as a brain tumor or Alzheimer's disease. Table 16-3 lists common signs of dementia.

Alzheimer's Disease Much attention has focused on the *abnormal* deterioration of the brain called **Alzheimer's disease,** which affects 10 percent of U.S. adults by age 65 and 30 percent of those over 85. Diagnosis of Alzheimer's disease depends on its characteristic symptoms: the victim gradually loses memory and reasoning, the ability to communicate, physical capabilities, and eventually life itself.[63] Nerve cells in the brain die, and communication between the cells breaks down.

Researchers are closing in on the exact cause of Alzheimer's disease.* Clearly, genetic factors are involved.[64] Free radicals and oxidative stress also seem to be in-

---

TABLE 16-2 Summary of Nutrient-Brain Relationships

Brain Function	Depends on an Adequate Intake of:
Short-term memory	Vitamin B_{12}, vitamin C, vitamin E
Performance in problem-solving tests	Riboflavin, folate, vitamin B_{12}, vitamin C
Mental health	Thiamin, niacin, zinc, folate
Cognition	Folate, vitamin B_6, vitamin B_{12}, iron, vitamin E
Vision	Essential fatty acids, vitamin A
Neurotransmitter synthesis	Tyrosine, tryptophan, choline

gout (GOWT): a common form of arthritis characterized by deposits of uric acid crystals in the joints.

purines: compounds of nitrogen-containing bases such as adenine, guanine, and caffeine. Purines that originate from the body are *endogenous* and those that derive from foods are *exogenous*.

neurons: nerve cells; the structural and functional units of the nervous system. Neurons initiate and conduct nerve impulse transmissions.

senile dementia: the loss of brain function beyond the normal loss of physical adeptness and memory that occurs with aging.

Alzheimer's disease: a degenerative disease of the brain involving memory loss and major structural changes in neuron networks; also known as *senile dementia of the Alzheimer's type (SDAT)*, *primary degenerative dementia of senile onset*, or *chronic brain syndrome*.

---

*A report on the genetic and other aspects of Alzheimer's is available from Alzheimer's Disease Education and Referral Center, P.O. Box 8250, Silver Springs, MD 20907-8250.

volved.[65] Nerve cells in the brains of people with Alzheimer's disease show evidence of free-radical attack—damage to DNA, cell membranes, and proteins. They also show evidence of the minerals that trigger free-radical attacks—iron, copper, zinc, and aluminum. Some research suggests that the antioxidant nutrients can limit free-radical damage and delay or prevent Alzheimer's disease.[66]

In Alzheimer's disease, the brain develops **senile plaques** and **neurofibrillary tangles.** Senile plaques are clumps of a protein fragment called beta-amyloid, whereas neurofibrillary tangles are snarls of the fibers that extend from the nerve cells. Both seem to occur in response to oxidative stress.[67] Researchers question whether these characteristics are the cause or the result of Alzheimer's disease.[68] In fact, scientists are unsure whether these plaques and tangles are causing the damage, serving as markers, or even protecting by sequestering the proteins that begin the dementia process.[69]

Late in the course of the disease there is a decline in the activity of the enzyme that assists in the production of the neurotransmitter acetylcholine from choline and acetyl CoA. Acetylcholine is essential to memory, but supplements of choline (or of lecithin, which contains choline) have no effect on memory or on the progression of the disease. Drugs that inhibit the breakdown of acetylcholine, on the other hand, have proved beneficial.

Research suggests that cardiovascular disease risk factors such as high blood pressure, diabetes, and elevated levels of homocysteine may be related to the development of Alzheimer's disease.[70] Diets designed to support a healthy heart, including omega-3 fatty acids and light-to-moderate alcohol intake, may benefit a healthy brain as well.[71]

Treatment for Alzheimer's disease involves providing care to clients and support to their families. Drugs are used to improve or at least to slow the loss of short-term memory and cognition, but they do not cure the disease. Other drugs may be used to control depression, anxiety, and behavior problems.

Maintaining appropriate body weight may be the most important nutrition concern for the person with Alzheimer's disease. Depression and forgetfulness can lead to changes in eating behaviors and poor food intake. Furthermore, changes in the body's weight-regulation system may contribute to weight loss. Perhaps the best that a caregiver can do nutritionally for a person with Alzheimer's disease is to supervise food planning and mealtimes. Providing well-liked and well-balanced meals and snacks in a cheerful atmosphere encourages food consumption. To minimize confusion, offer a few ready-to-eat foods, in bite-size pieces, with seasonings and sauces. To avoid mealtime disruptions, control distractions such as music, television, children, and the telephone.

TABLE 16-3	Common Signs of Dementia

- Agitated behavior
- Becoming lost in familiar surroundings or circumstances
- Confusion
- Delusions
- Loss of interest in daily activities
- Loss of memory
- Loss of problem-solving skills
- Unclear thinking

Both foods and mental challenges nourish the brain.

IN SUMMARY

Senile dementia and other losses of brain function afflict millions of older adults, and others face loss of vision due to cataracts or macular degeneration or cope with the pain of arthritis. As the number of people over age 65 continues to grow, the need for solutions to these problems becomes urgent. Some problems may be inevitable, but others are preventable and good nutrition may play a key role.

Food Choices and Eating Habits of Older Adults

Older people are an incredibly diverse group, and for the most part, they are independent, socially sophisticated, mentally lucid, fully participating members of society who report themselves to be happy and healthy. In fact, the quality of life

senile plaques: clumps of the protein fragment beta-amyloid on the nerve cells, commonly found in the brains of people with Alzheimer's dementia.

neurofibrillary tangles: snarls of the threadlike strands that extend from the nerve cells, commonly found in the brains of people with Alzheimer's dementia.

among the elderly has improved, and their chronic disabilities have declined dramatically in recent years.[72] By practicing stress-management skills, maintaining physical fitness, participating in activities of interest, and cultivating spiritual health, as well as obtaining adequate nourishment, people can support a high quality of life into old age (see Table 16-4 for some strategies).

Older people spend more money per person on foods to eat at home than other age groups and less money on foods away from home. Manufacturers would be wise to cater to the preferences of older adults by providing good-tasting, nutritious foods in easy-to-open, single-serving packages with labels that are easy to read. Such services enable older adults to maintain their independence and to feel a sense of control and involvement in their own lives. Another way older adults can take care of themselves is by remaining or becoming physically active. As mentioned earlier, physical activity helps preserve one's ability to perform daily tasks and so promotes independence.

Familiarity, taste, and health beliefs are most influential on older people's food choices. Eating foods that are familiar, especially ethnic foods that recall family meals and pleasant times, can be comforting. People 65 and over are less likely to diet to lose weight than younger people are, but they are more likely to diet in pursuit of medical goals such as controlling blood glucose and cholesterol.

Food Assistance Programs

The Nutrition Screening Initiative is part of a national effort to identify and treat nutrition problems in older persons; it uses a screening checklist. To *determine* the risk of malnutrition in older clients, health care professionals can keep in mind the characteristics and questions listed in Table 16-5.

An integral component of the Older Americans Act (OAA) is the OAA Nutrition Program, formerly known as the Elderly Nutrition Program. Its services are designed to improve older people's nutrition status and enable them to avoid medical problems, continue living in communities of their own choice, and stay out of institutions. Its specific goals are to provide low-cost, nutritious meals; opportunities for social interaction; homemaker education and shopping assistance; counseling and referral to social services; and transportation. The program's mission has always been to provide "more than a meal."

TABLE 16-4 Strategies for Growing Old Healthfully

- Choose nutrient-dense foods.
- Be physically active. Walk, run, dance, swim, bike, or row for aerobic activity. Lift weights, do calisthenics, or pursue some other activity to tone, firm, and strengthen muscles. Practice balancing on one foot or doing simple movements with your eyes closed. Modify activities to suit changing abilities and tastes.
- Maintain appropriate body weight.
- Reduce stress (cultivate self-esteem, maintain a positive attitude, manage time wisely, know your limits, practice assertiveness, release tension, and take action).
- For women, discuss with a physician the risks and benefits of estrogen replacement therapy.
- For people who smoke, discuss with a physician strategies and programs to help you quit.
- Expect to enjoy sex, and learn new ways of enhancing it.
- Use alcohol only moderately, if at all; use drugs only as prescribed.
- Take care to prevent accidents.
- Expect good vision and hearing throughout life; obtain glasses and hearing aids if necessary.
- Take care of your teeth; obtain dentures if necessary.

- Be alert to confusion as a disease symptom, and seek diagnosis.
- Take medications as prescribed; see a physician before self-prescribing medicines or herbal remedies and a registered dietitian before self-prescribing supplements.
- Control depression through activities and friendships; seek professional help if necessary.
- Drink 6 to 8 glasses of water every day.
- Practice mental skills. Keep on solving math problems and crossword puzzles, playing cards or other games, reading, writing, imagining, and creating.
- Make financial plans early to ensure security.
- Accept change. Work at recovering from losses; make new friends.
- Cultivate spiritual health. Cherish personal values. Make life meaningful.
- Go outside for sunshine and fresh air as often as possible.
- Be socially active—play bridge, join an exercise or dance group, take a class, teach a class, eat with friends, volunteer time to help others.
- Stay interested in life—pursue a hobby, spend time with grandchildren, take a trip, read, grow a garden, or go to the movies.
- Enjoy life.

TABLE 16-5	Risk Factors for Malnutrition in Older Adults

	These questions help *determine* the risk of malnutrition in older adults:
Disease	• Do you have an illness or condition that changes the types or amounts of foods you eat?
Eating poorly	• Do you eat fewer than two meals a day? Do you eat fruits, vegetables, and milk products daily?
Tooth loss or mouth pain	• Is it difficult or painful to eat?
Economic hardship	• Do you have enough money to buy the food you need?
Reduced social contact	• Do you eat alone most of the time?
Multiple medications	• Do you take three or more different prescribed or over-the-counter medications daily?
Involuntary weight loss or gain	• Have you lost or gained 10 pounds or more in the last six months?
Needs assistance	• Are you physically able to shop, cook, and feed yourself?
Elderly person	• Are you older than 80?

Social interactions at a congregate meal site can be as nourishing as the foods served.

The OAA Nutrition Program provides for **congregate meals** at group settings such as community centers. Administrators try to select sites for congregate meals where as many eligible people as possible can participate. Volunteers may also deliver meals to those who are homebound either permanently or temporarily; these home-delivered meals are known as **Meals on Wheels.** Although the home-delivery program ensures nutrition, its recipients miss out on the social benefits of the congregate meals. Therefore, every effort is made to persuade older people to come to the shared meals, if they can. All persons aged 60 years and older and their spouses are eligible to receive meals from these programs, regardless of their income. Priority is given to those who are economically and socially needy. An estimated 3 million of our nation's older adults benefit from these meals.

These programs provide at least one meal a day that meets a third of the RDA for this age group, and they must operate five or more days a week. Many programs voluntarily offer additional services designed to appeal to older adults: provisions for special diets (to meet medical needs or religious preferences), food pantries, ethnic meals, and delivery of meals to the homeless. Adding breakfast to the service increases energy and nutrient intakes, which helps to relieve hunger and depression.[73]

Older adults can also take advantage of the Senior Farmers Market Nutrition Program, which provides low-income older adults with coupons that can be exchanged for fresh fruits, vegetables, and herbs at community-supported farmers' markets and roadside stands. This program increases fresh fruit and vegetable consumption, provides nutrition information, and even reaches the homebound elderly, a group of people who normally do not have access to farmers' markets.

Older adults can learn about the available programs in their communities by looking in the Yellow Pages of the telephone book under "Social Services" or "Senior Citizens' Organizations."* In addition, the local senior center and hospital can usually direct people to programs that provide nutrition and other health-related services.

Meals for Singles

Many older adults live alone, and singles of all ages face challenges in purchasing, storing, and preparing food. Large packages of meat and vegetables are often intended for families of four or more, and even a head of lettuce can spoil before one

---

* To find a local provider, call Eldercare Locator at (800) 677-1116.

congregate meals: nutrition programs that provide food for the elderly in conveniently located settings such as community centers.

Meals on Wheels: a nutrition program that delivers food for the elderly to their homes.

person can use it all. Many singles live in small dwellings and have little storage space for foods. A limited income presents additional obstacles. This section offers suggestions that can help to solve some of the problems singles face, beginning with a special note about the dangers of foodborne illness.

Foodborne Illness The risk of older adults getting a foodborne illness is greater than for other adults. The consequences of an upset stomach, diarrhea, fever, vomiting, abdominal cramps, and dehydration are oftentimes more severe, sometimes leading to paralysis, meningitis, or even death. For these reasons, older adults need to carefully follow the food safety suggestions presented in Highlight 18.

Dietary Guidelines for Americans 2005

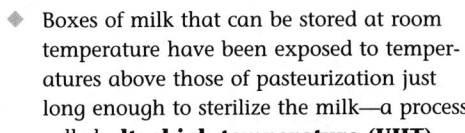

- Older adults should not eat or drink unpasteurized milk, milk products, or juices; raw or undercooked eggs, meat, poultry, fish, or shellfish; or raw sprouts.
- Older adults should only eat certain deli meats and frankfurters that have been reheated to steaming hot.

Spend Wisely People who have the means to shop and cook for themselves can cut their food bills simply by being wise shoppers. Large supermarkets are usually less expensive than convenience stores. A grocery list helps reduce impulse buying, and specials and coupons can save money when the items featured are those that the shopper needs and uses.

Buying the right amount so as not to waste any food is a challenge for people eating alone. They can buy fresh milk in the size best suited for personal needs. Pint-size and even cup-size boxes ◆ of milk are available and can be stored unopened on a shelf for as long as three months without refrigeration.

◆ Boxes of milk that can be stored at room temperature have been exposed to temperatures above those of pasteurization just long enough to sterilize the milk—a process called **ultrahigh temperature (UHT)**.

Many foods that offer a variety of nutrients for practically pennies have a long shelf life; staples such as rice, pastas, dry powdered milk, and dried legumes can be purchased in bulk and stored for months at room temperature. Other foods that are usually a good buy include whole pieces of cheese rather than sliced or shredded cheese, fresh produce in season, variety meats such as chicken livers, and cereals that require cooking instead of ready-to-serve cereals.

A person who has ample freezer space can buy large packages of meat, such as pork chops, ground beef, or chicken, when they are on sale. Then the meat can be immediately wrapped into individual servings for the freezer. All the individual servings can be put in a bag marked appropriately with the contents and the date.

Frozen vegetables are more economical in large bags than in small boxes. After the amount needed is taken out, the bag can be closed tightly with a twist tie or rubber band. If the package is returned quickly to the freezer each time, the vegetables will stay fresh for a long time.

Finally, breads and cereals usually must be purchased in larger quantities. Again the amount needed for a few days can be taken out and the rest stored in the freezer.

Grocers will break open a package of wrapped meat and rewrap the portion needed. Similarly, eggs can be purchased by the half-dozen. Eggs do keep for long periods, though, if stored properly in the refrigerator.

Fresh fruits and vegetables can be purchased individually. A person can buy fresh fruit at various stages of ripeness: a ripe one to eat right away, a semiripe one to eat soon after, and a green one to ripen on the windowsill. If vegetables are packaged in large quantities, the grocer can break open the package so that a smaller amount can be purchased. Small cans of fruits and vegetables, even though they are more expensive per unit, are a reasonable alternative, considering that it is expensive to buy a regular-size can and let the unused portion spoil.

Be Creative Creative chefs think of various ways to use foods when only large amounts are available. For example, a head of cauliflower can be divided into

Buy only what you will use.

© Lonnie Duka/Index Stock Imagery

thirds. Then one-third is cooked and eaten hot. Another third is put into a vinegar and oil marinade for use in a salad. And the last third can be used in a casserole or stew.

A variety of vegetables and meats can be enjoyed stir-fried; inexpensive vegetables such as cabbage, celery, and onion are delicious when crisp cooked in a little oil with herbs or lemon added. Interesting frozen vegetable mixtures are available in larger grocery stores. Cooked, leftover vegetables can be dropped in at the last minute. A bonus of a stir-fried meal is that there is only one pan to wash. Similarly, a microwave oven allows a chef to use fewer pots and pans. Meals and leftovers can also be frozen or refrigerated in microwavable containers to reheat as needed.

Many frozen dinners offer nutritious options. Adding a fresh salad, a whole-wheat roll, and a glass of milk can make a nutritionally balanced meal.

Also, single people shouldn't hesitate to invite someone to share meals with them whenever there is enough food. It's likely that the person will return the invitation, and both parties will get to enjoy companionship and a meal prepared by others.

Invite guests to share a meal.

IN SUMMARY

Older people can benefit from both the nutrients provided and the social interaction available at congregate meals. Other government programs deliver meals to those who are homebound. With creativity and careful shopping, those living alone can prepare nutritious, inexpensive meals. Physical activity, mental challenges, stress management, and social activities can also help people grow old comfortably.

Nutrition Portfolio

By eating a balanced diet, maintaining a healthy body weight, and engaging in a variety of physical, social, and mental activities, you can enjoy good health in later life.

- Visit older adults in your community and consider whether they have the financial means, physical ability, and social support they need to eat adequately.

- Note whether they have experienced an unintentional loss of weight recently.

- Discuss how they occupy their time physically, socially, and mentally.

NUTRITION ON THE NET

For further study of topics covered in this chapter, log on to **academic.cengage .com/nutrition/rolfes/UNCN8e**. Go to Chapter 16, then to Nutrition on the Net.

- Search for "aging," "arthritis," and "Alzheimer's" on the U.S. Government health information site: **www.healthfinder.gov**

- Visit the National Aging Information Center of the Administration on Aging: **www.aoa.gov**

- Visit the American Geriatrics Society: **www.americangeriatrics.org**

- Visit the National Institute on Aging: **www.nia.nih.gov**

- Visit the American Association of Retired Persons: **www.aarp.org**

- Get nutrition tips for growing older in good health from the American Dietetic Association: **www.eatright.org**

- Learn more about cataracts and macular degeneration from the National Eye Institute, the Macular Degeneration Partnership, and the American Society of Cataract and Refractive Surgery: **www.nei.nih.gov**, **www.macd.net**, and **www.ascrs.org**

- Learn more about arthritis from the Arthritis Society, the Arthritis Foundation, and the National Institute of Arthritis and Musculoskeletal and Skin Diseases: **www.arthritis.ca, www.arthritis.org,** and **www.niams.nih.gov**

- Learn more about Alzheimer's disease from the NIA Alzheimer's Disease Education and Referral Center and the

Alzheimer's Association: **www.alzheimers.org** and **www.alz.org**

- Find out about federal government programs designed to help senior citizens maintain good health: **www.seniors.gov**

STUDY QUESTIONS

CENGAGENOW™
To assess your understanding of chapter topics, take the Student Practice Test and explore the modules recommended in your Personalized Study Plan. Log onto **academic.cengage.com/login.**

These questions will help you review the chapter. You will find the answers in the discussions on the pages provided.

1. What roles does nutrition play in aging, and what roles can it play in retarding aging? (pp. 562–565)

2. What are some of the physiological changes that occur in the body's systems with aging? To what extent can aging be prevented? (pp. 565–568)

3. Why does the risk of dehydration increase as people age? (p. 569)

4. Why do energy needs usually decline with advancing age? (p. 569)

5. Which vitamins and minerals need special consideration for the elderly? Explain why. Identify some factors that complicate the task of setting nutrient standards for older adults. (pp. 570–571)

6. Discuss the relationships between nutrition and cataracts and between nutrition and arthritis. (pp. 572–574)

7. What characteristics contribute to malnutrition in older people? (pp. 568–569, 575–577)

These multiple choice questions will help you prepare for an exam. Answers can be found on p. 582.

1. Life expectancy in the United States is about:
    a. 48 to 60 years.
    b. 58 to 70 years.
    c. 68 to 80 years.
    d. 78 to 90 years.

2. The human life span is about:
    a. 85 years.
    b. 100 years.
    c. 115 years.
    d. 130 years.

3. A 72-year-old person whose physical health is similar to that of people 10 years younger has a(n):
    a. chronological age of 62.
    b. physiological age of 72.
    c. physiological age of 62.
    d. absolute age of minus 10.

4. Rats live longest when given diets that:
    a. eliminate all fat.
    b. provide lots of protein.
    c. allow them to eat freely.
    d. restrict their energy intakes.

5. Which characteristic is *not* commonly associated with atrophic gastritis?
    a. inflamed stomach
    b. vitamin B_{12} toxicity
    c. bacterial overgrowth
    d. lack of intrinsic factor

6. On average, adult energy needs:
    a. decline 5 percent per year.
    b. decline 5 percent per decade.
    c. remain stable throughout life.
    d. rise gradually throughout life.

7. Which nutrients seem to protect against cataract development?
    a. minerals
    b. lecithins
    c. antioxidants
    d. amino acids

8. The best dietary advice for a person with osteoarthritis might be to:
    a. avoid milk products.
    b. take fish oil supplements.
    c. take vitamin E supplements.
    d. lose weight, if overweight.

9. Congregate meal programs are preferable to Meals on Wheels because they provide:
    a. nutritious meals.
    b. referral services.
    c. social interactions.
    d. financial assistance.

10. The Elderly Nutrition Program is available to:
    a. all people 65 years and older.
    b. all people 60 years and older.
    c. homebound people only, 60 years and older.
    d. low-income people only, 60 years and older..

REFERENCES

1. Trends in aging—United States and worldwide, *Morbidity and Mortality Weekly Report* 52 (2003): 101–106.

2. D. L. Hoyert and coauthors, Annual summary of vital statistics: 2004, *Pediatrics* 117 (2006): 168–183.

3. Living well to 100: Nutrition, genetics, inflammation, supplement to *American Journal of Clinical Nutrition* 83 (2006): 401S–490S.

4. W. S. Browner and coauthors, The genetics of human longevity, *American Journal of Medicine* 117 (2004): 851–860.

5. T. Perls, Genetic and environmental influences on exceptional longevity and the AGE nomogram, *Annals of the New York Academy of Sciences* 959 (2002): 1–13.

6. M. E. Cress and coauthors, Physical activity programs and behavior counseling in older adult populations, *Medicine and Science in Sports and Exercise* 36 (2004): 1997–2003; E. W. Gregg and coauthors, Relationship of changes in physical activity and mortality among older women, *Journal of the American Medical Association* 289 (2003): 2379–2386.

7. Cress and coauthors, 2004; C. H. Hillman and coauthors, Physical activity and executive control: Implications for increased cognitive health during older adulthood, *Research Quarterly for Exercise and Sport* 75 (2004): 176–185; W. J. Strawbridge and coauthors, Physical activity reduces the risk of subsequent depression for older adults, *American Journal of Epidemiology* 156 (2002): 328–334; A. J. Schuit and coauthors, Physical activity and cognitive decline, the role of the apolipoprotein e4 allele, *Medicine and Science in Sports and Exercise* 33 (2001): 772–777.

8. Fatalities and injuries from falls among older adults—United States, 1993–2003 and 2001–2005, *Morbidity and Mortality Weekly Report* 55 (2006): 1221–1224.

9. G. Barja, Endogenous oxidative stress: Relationship to aging, longevity and caloric restriction, *Ageing Research and Reviews* 1 (2002): 397–411; B. J. Merry, Molecular mechanisms linking calorie restriction and longevity, *International Journal of Biochemistry and Cell Biology* 34 (2002): 1340–1354.

10. L. K. Heilbronn and E. Ravussin, Calorie restriction and aging: Review of the literature and implications for studies in humans, *American Journal of Clinical Nutrition* 78 (2003): 361–369.

11. G. Wolf, Calorie restriction increases life span: A molecular mechanism, *Nutrition Reviews* 64 (2006): 89–92; L. Fontana and coauthors, Long-term calorie restriction is highly effective in reducing the risk for atherosclerosis in humans, *Proceedings of the National Academy of Sciences* 101 (2004): 6659–6663.

12. L. K. Heilbronn and coauthors, Alternate-day fasting in nonobese subjects: Effects on body weight, body composition, and energy metabolism, *American Journal of Clinical Nutrition* 81 (2005): 69–73; M. P. Mattson, Energy intake, meal frequency, and health: A neurobiological perspective, *Annual Review of Nutrition* 25 (2005): 237–260; R. M. Anson and coauthors, Intermittent fasting dissociates beneficial effects of dietary restriction on glucose metabolism and neuronal resistance to injury from calorie intake, *Proceeding of the National Academy of Sciences* 100 (2003): 6216–6220.

13. K. T. B. Knoops and coauthors, Mediterranean diet, lifestyle factors, and 10-year mortality in elderly European men and women: The HALE Project, *Journal of the American Medical Association* 292 (2004): 1433–1439.

14. E. S. Epel and coauthors, Accelerated telomere shortening in response to life stress, *Proceedings of the National Academy of Sciences* 101 (2004): 17312–17315.

15. S. A. Motzer and V. Hertig, Stress, stress response, and health, *Nursing Clinics of North America* 39 (2004): 1–17.

16. S. E. Taylor and coauthors, Biobehavioral responses to stress in females: Tend-and-befriend, not fight-or-flight, *Psychological Review* 107 (2000): 411–429.

17. B. P. Yu and H. Y. Chung, Adaptive mechanisms to oxidative stress during aging, *Mechanisms of Ageing and Development* 127 (2006): 436–443; F. Sierra, Is (your cellular response to) stress killing you? *Journals of Gerontology. Series A, Biological Sciences and Medical Sciences* 61 (2006): 557–561; D. P. Jones, Extracellular redox state: Refining the definition of oxidative stress in aging, *Rejuvenation Research* 9 (2006): 169–181.

18. D. T. Villareal and coauthors, Obesity in older adults: Technical review and position statement of the American Society for Nutrition and NAASO, The Obesity Society, *American Journal of Clinical Nutrition* 82 (2005): 923–934.

19. I. M. Chapman, Endocrinology of anorexia of ageing, Best Practice and Research. *Clinical Endocrinology and Metabolism* 18 (2004): 437–452.

20. H. K. Kamel, Sarcopenia and aging, *Nutrition Reviews* 61 (2003): 157–167; C. W. Bales and C. S. Ritchie, Sarcopenia, weight loss, and nutritional frailty in the elderly, *Annual Review of Nutrition* 22 (2002): 309–323.

21. M. Cesari and coauthors, Frailty syndrome and skeletal muscle: Results from the Invecchiare in Chianti study, *American Journal of Clinical Nutrition* 83 (2006): 1142–1148.

22. K. S. Nair, Aging muscle, *American Journal of Clinical Nutrition* 81 (2005): 953–963; R. D. Hansen and B. J. Allen, Habitual physical activity, anabolic hormones, and potassium content of fat-free mass in postmenopausal women, *American Journal of Clinical Nutrition* 75 (2002): 314–320.

23. A. B. Newman and coauthors, Weight change and the conservation of lean mass in old age: The Health, Aging and Body Composition Study, *American Journal of Clinical Nutrition* 82 (2005): 872–878; P. Szulc and coauthors, Hormonal and lifestyle determinants of appendicular skeletal muscle mass in men: The MINOS study, *American Journal of Clinical Nutrition* 80 (2004): 496–503.

24. M. Cesari and coauthors, Sarcopenia, obesity, and inflammation-Results from the Trial of Angiotensin Converting Enzyme Inhibition and Novel Cardiovascular Risk Factors study, *American Journal of Clinical Nutrition* 82 (2005): 428–434.

25. B. A. Parker and I. M. Chapman, Food intake and ageing—The role of the gut, *Mechanisms of Ageing and Development* 125 (2004): 859–866.

26. N. R. Sahyoun, C. L. Lin, and E. Krall, Nutritional status of the older adult is associated with dentition status, *Journal of the American Dietetic Association* 103 (2003): 61–66.

27. R. L. Bailey and coauthors, Persistent oral health problems associated with comorbidity and impaired diet quality in older adults, *Journal of the American Dietetic Association* 104 (2004): 1273–1276.

28. Position paper of the American Dietetic Association: Nutrition across the spectrum of aging, *Journal of the American Dietetic Association* 105 (2005): 616–633.

29. D. G. Blazer, Depression in late life: Review and commentary, *Journals of Gerontology: Series A, Biological Sciences and Medical Sciences* 58 (2003): 249–265; J. Unutzer, M. L. Bruce, and NIMH Affective Disorders Workgroup, The elderly, *Mental Health Services Research* 4 (2002): 245–247.

30. N. Kagansky and coauthors, Poor nutritional habits are predictors of poor outcome in very old hospitalized patients, *American Journal of Clinical Nutrition* 82 (2005): 784–791; N. L. Crogan and A. Pasvogel, The influence of protein-calorie malnutrition on quality of life in nursing homes, *Journals of Gerontology: Series A, Biological Sciences and Medical Sciences* 58 (2003): 159–164; Y. Guigoz, S. Lauque, and B. J. Vellas, Identifying the elderly at risk for malnutrition. The Mini Nutritional Assessment, *Clinics in Geriatric Medicine* 18 (2002): 737–757.

31. R. Chernoff, Micronutrient requirements in older women, *American Journal of Clinical Nutrition* 81 (2005): 1240S–1245S.

32. M. Ferry, Strategies for ensuring good hydration in the elderly, *Nutrition Reviews* 63 (2005): S22–S29.

33. N. Meunier and coauthors, Basal metabolic rate and thyroid hormones of late-middle-aged and older human subjects: The ZENITH study, *European Journal of Clinical Nutrition* 59 (2005): S53–S57.

34. M. M. G. Wilson, R. Purushothaman, and J. E. Morley, Effect of liquid dietary supplements on energy intake in the elderly, *American Journal of Clinical Nutrition* 75 (2002): 944–947.

35. Position of the American Dietetic Association: Liberalization of the diet prescription improves quality of life for older adults in long-term care, *Journal of the American Dietetic Association* 105 (2005): 1955–1965.

36. Committee on Dietary Reference Intakes, *Dietary Reference Intakes for Energy, Carbohydrate, Fiber, Fat, Fatty Acids, Cholesterol, Protein, and Amino Acids* (Washington, D.C.: National Academies Press, 2002).

37. M. A. Johnson and coauthors, Hyperhomocysteinemia and vitamin B-12 deficiency in elderly using Title IIIc nutrition services, *American Journal of Clinical Nutrition* 77 (2003): 211–220.

38. Committee on Dietary Reference Intakes, *Dietary Reference Intakes for Thiamin, Riboflavin, Niacin, Vitamin B6, Folate, Vitamin B12, Pantothenic Acid, Biotin, and Choline,* (Washington, D.C.: National Academy Press, 2000), p. 338.

39. H. A. Bischoff-Ferrari and coauthors, Higher 25-hydroxyvitamin D concentrations are associated with better lower-extremity function in both active and inactive persons aged ≥60 y, *American Journal of Clinical Nutrition* 80 (2004): 752–758.

40. Committee on Dietary Reference Intakes, *Dietary Reference Intakes for Calcium, Phosphorus, Magnesium, Vitamin D, and Fluoride* (Washington, D.C.: National Academy Press, 1997).

41. Committee on Dietary Reference Intakes, 1997.

42. D. J. Fleming and coauthors, Dietary factors associated with the risk of high iron stores in the elderly Framingham Heart Study cohort, *American Journal of Clinical Nutrition* 76 (2002): 1375–1384.

43. N. Ahluwalia and coauthors, Immune function is impaired in iron-deficient, homebound, older women, *American Journal of Clinical Nutrition* 79 (2004): 516–521.

44. R. Costello and coauthors, Executive summary: Conference on dietary supplement use in the elderly—Proceedings of the conference held January 14–15, 2003, Natcher Auditorium, National Institutes of Health, Bethesda, Md., *Nutrition Reviews* 62 (2004): 160–175.

45. R. H. Fletcher and K. M. Fairfield, Vitamins for chronic disease prevention in adults, *Journal of the American Medical Association* 287 (2002): 3127–3129; W. C. Willett and M. J. Stampfer, What vitamins should I be taking,

doctor? *New England Journal of Medicine* 345 (2001): 1819–1824.

46. T. Ostbye and coauthors, Ten dimensions of health and their relationships with overall self-reported health and survival in a predominately religiously active elderly population: The Cache County memory study, *Journal of the American Geriatrics Society* 54 (2006): 199–209.

47. M. D. Knudtson, B. E. Klein, and R. Klein, Age-related disease, visual impairment, and survival: The Beaver Dam Eye Study, *Archives of Ophthalmology* 124 (2006): 243–249.

48. S. M. Moeller and coauthors, Overall adherence to the Dietary Guidelines for Americans is associated with reduced prevalence of early age-related nuclear lens opacities in women, *Journal of Nutrition* 134 (2004): 1812–1819.

49. W. G. Christen and coauthors, Fruit and vegetable intake and the risk of cataract in women, *American Journal of Clinical Nutrition* 81 (2005): 1417–1422; C. Chitchumroonchokchai and coauthors, Xanthophylls and α-tocopherol decrease UVB-induced lipid peroxidation and stress signaling in human lens epithelial cells, *Journal of Nutrition* 134 (2004): 3225–3232; A. Taylor and coauthors, Long-term intake of vitamins and carotenoids and odds of early age-related cortical and posterior subcapsular lens opacities, *American Journal of Clinical Nutrition* 75 (2002): 540–549; The REACT Group, The Roche European American Cataract Trial (REACT): A randomized clinical trial to investigate the efficacy of an oral antioxidant micronutrient mixture to slow progression of age-related cataract, *Ophthalmic Epidemiology* 9 (2002): 49–80.

50. J. M. Weintraub and coauthors, A prospective study of the relationship between body mass index and cataract extraction among US women and men, *International Journal of Obesity and Related Metabolic Disorders* 26 (2002): 1588–1595.

51. The Eye Diseases Prevalence Research Group, Age-related macular degeneration is the leading cause of blindness. . . , *Archives of Ophthalmology* 122 (2004): 564–572; J. L. Gottlieb, Age-related macular degeneration, *Journal of the American Medical Association* 288 (2002): 2233–2236.

52. P. R. Trumbo and K. C. Ellwood, Lutein and zeaxanthin intakes and risk of age-related macular degeneration and cataracts: An evaluation using the Food and Drug Administration's evidence-based review system for health claims, *American Journal of Clinical Nutrition* 84 (2006): 971–974; R. van Leeuwen and coauthors, Dietary intake of antioxidants and risk of age-related macular degeneration, *Journal of the American Medical Association* 294 (2005): 3101–3107; D. Hartmann and coauthors, Plasma kinetics of zeaxanthin and 3′-dehydro-lutein after multiple oral doses of synthetic zeaxanthin, *American Journal of Clinical Nutrition* 79 (2004): 410–417; N. I. Krinsky, J. T. Landrum, and R. A. Bone, Biologic mechanisms of the protective role of lutein and zeaxanthin in the eye, *Annual Review of Nutrition* 23 (2003): 171–201.

53. L. Devos-Comby, T. Cronan, and S. C. Roesch, Do exercise and self-management interventions benefit patients with osteoarthritis of the knee? A metaanalytic review, *Journal of Rheumatology* 33 (2006): 744–756; S. P. Messier and coauthors, Exercise and dietary weight loss in overweight and obese older adults with knee osteoarthritis: The Arthritis, Diet, and Activity Promotion Trial, *Arthritis Rheumatism* 50 (2004): 1501–1510.

54. D. J. Pattison, D. P. Symmons, and A. Young, Does diet have a role in the aetiology of rheumatoid arthritis? *Proceedings of the Nutrition Society* 63 (2004): 137–143.

55. L. Skoldstam, L. Hagfors, and G. Johansson, An experimental study of a Mediterranean diet intervention for patients with rheumatoid arthritis, *Annals of the Rheumatic Diseases* 62 (2003): 208–214.

56. O. Adam, Dietary fatty acids and immune reactions in synovial tissue, *European Journal of Medical Research* 8 (2003): 381–387; L. Cleland, M. James, and S. Proudman, The role of fish oils in the treatment of rheumatoid arthritis, *Drugs* 63 (2003): 845–853.

57. D. J. Pattison and coauthors, Dietary ß-cryptoxanthin and inflammatory polyarthritis: Results from a population-based prospective study, *American Journal of Clinical Nutrition* 82 (2005): 451–455; J. R. Cerhan and coauthors, Antioxidant micronutrients and risk of rheumatoid arthritis in a cohort of older women, *American Journal of Epidemiology* 157 (2003): 345–354.

58. N. Schlesinger, Dietary factors and hyperuricaemia, *Current Pharmaceutical Design* 11 (2005): 4133–4138.

59. H. K. Choi, S. Liu, and G. Curhan, Intake of purine-rich foods, protein, and dairy products and relationship to serum levels of uric acid: The Third National Health and Nutrition Examination Survey, *Arthritis and Rheumatism* 52 (2005): 283–289; H. K. Choi and coauthors, Purine-rich foods, dairy and protein intake, and the risk of gout in men, *New England Journal of Medicine* 350 (2004): 1093–1103.

60. D. O. Clegg and coauthors, Glucosamine, chondroitin sulfate, and the two in combination for painful knee osteoarthritis, *New England Journal of Medicine* 354 (2006): 795–808; F. Richy and coauthors, Structural and symptomatic efficacy of glucosamine and chondroitin in knee osteoarthritis: A comprehensive meta-analysis, *Archives of Internal Medicine* 163 (2003): 1514–1522.

61. R. J. Wurtman and coauthors, Effects of normal meals rich in carbohydrates or proteins on plasma tryptophan and tyrosine ratios, *American Journal of Clinical Nutrition* 77 (2003): 128–132.

62. R. D. Abbott and coauthors, Walking and dementia in physically capable elderly men, *Journal of the American Medical Association* 292 (2004): 1447–1453; J. Weuve and coauthors, Physical activity, including walking, and cognitive function in older women, *Journal of the American Medical Association* 292 (2004): 1454–1461.

63. J. L. Cummings and G. Cole, Alzheimer disease, *Journal of the American Medical Association* 287 (2002): 2335–2338.

64. T. D. Bird, Genetic factors in Alzheimer's disease, *New England Journal of Medicine* 352 (2005): 862–864.

65. P. I. Moreira and coauthors, Oxidative stress: The old enemy in Alzheimer's disease pathophysiology, *Current Alzheimer Research* 2 (2005): 403–408.

66. M. C. Morris and coauthors, Relation of the tocopherol forms to incident Alzheimer disease and to cognitive change, *American Journal of Clinical Nutrition* 81 (2005): 508–514; P. P. Zandi and coauthors, Reduced risk of Alzheimer disease in users of antioxidant vitamin supplements: The Cache County Study, *Archives of Neurology* 61 (2004): 82–88; M. J. Engelhart and coauthors, Dietary intake of antioxidants and risk of Alzheimer disease, *Journal of the American Medical Association* 287 (2002): 3223–3229; M. C. Morris, Dietary intake of antioxidant nutrients and the risk of incident Alzheimer disease in a biracial community study, *Journal of the American Medical Association* 287 (2002): 3230–3237.

67. R. J. Castellani and coauthors, Antioxidant protection and neurodegenerative disease: The role of amyloid-beta and tau, *American Journal of Alzheimer's Disease and Other Dementias* 21 (2006): 126–130; P. Zafrilla and coauthors, Oxidative stress in Alzheimer patients in different stages of the disease, *Current Medicinal Chemistry* 13 (2006): 1075–1083.

68. R. A. Armstrong, Plaques and tangles and the pathogenesis of Alzheimer's disease, *Folia Neuropathologica* 44 (2006): 1–11; G. L. Wenk, Neuropathologic changes in Alzheimer's disease: Potential targets for treatment, *Journal of Clinical Psychiatry* 67 (2006): 3–7.

69. A. Nunomura and coauthors, Neuropathology in Alzheimer's disease: Awaking from a hundred-year-old dream, *Science of Aging Knowledge Environment* (2006): pe10; R. E. Tanzi, Tangles and neurodegenerative disease—A surprising twist, *New England Journal of Medicine* 353 (2005): 1853–1855.

70. G. Ravaglia and coauthors, Homocysteine and folate as risk factors for dementia and Alzheimer disease, *American Journal of Clinical Nutrition* 82 (2005): 636–643; K. L. Tucker and coauthors, High homocysteine and low B vitamins predict cognitive decline in aging men: The Veterans Affairs Normative Aging Study, *American Journal of Clinical Nutrition* 82 (2005): 627–635; P. Quadri and coauthors, Homocysteine, folate, and vitamin B-12 in mild cognitive impairment, Alzheimer disease, and vascular dementia, *American Journal of Clinical Nutrition* 80 (2004): 114–122; S. Seshadri and coauthors, Plasma homocysteine as a risk factor for dementia and Alzheimer's disease, *New England Journal of Medicine* 346 (2002): 476–483.

71. R. Uauy and A. D. Dangour, Nutrition in brain development and aging: Role of essential fatty acids, *Nutrition Reviews* 64 (2006): S24–S33; T. den Heijer and coauthors, Alcohol intake in relation to brain magnetic resonance imaging findings in older persons without dementia, *American Journal of Clinical Nutrition* 80 (2004): 992–997; F. Calon and coauthors, Docosahexaenoic acid protects from dendritic pathology in an Alzheimer's disease mouse model, *Neuron* 43 (2004): 633–645.

72. V. A. Freedman, L. G. Martin, and R. F. Schoeni, Recent trends in disability and functioning among older adults in the United States: A systematic review, *Journal of the American Medical Association* 288 (2002): 3137–3146.

73. E. A. Gollub and D. O. Weddle, Improvements in nutritional intake and quality of life among frail homebound older adults receiving home-delivered breakfast and lunch, *Journal of the American Dietetic Association* 104 (2004): 1227–1235.

ANSWERS

Study Questions (multiple choice)

1. c   2. d   3. c   4. d   5. b   6. b   7. c   8. d   9. c   10. b

Hunger and Community Nutrition

One person in every eight worldwide experiences persistent hunger—not the healthy appetite triggered by anticipation of a hearty meal but the painful sensation caused by a lack of food. The physical feelings are the same, but the hunger described in this highlight takes on greater meaning because the lack of food is recurrent and involuntary. Hunger deprives a person of the physical and mental energy needed to enjoy a full life and often leads to severe malnutrition and death. Tens of thousands of people—one child every five seconds—die of starvation each day.[1]

Resolving the hunger problem may seem at first beyond the influence of the ordinary person. Can one person's choice to volunteer at a food recovery program make a difference? In truth, such choices do produce several benefits. For one, a person's action may influence many other people over time. For another, an action repeated becomes a habit, with compounded benefits. For still another, making choices with an awareness of the consequences gives a person a sense of personal control, hope, and effectiveness. The daily actions of many concerned people can help solve the problems of hunger in their own neighborhoods or on the other side of the world.

Hunger in the United States

Ideally, all people at all times would have access to enough food to support an active, healthy life; in other words, they would experience **food security.** Unfortunately, over 38 million people in the United States, including 14 million children, live in poverty and cannot afford to buy enough food to maintain good health.[2] Said another way, one out of ten households

experiences hunger or the threat of hunger. Given the agricultural bounty and enormous wealth in this country, do these numbers surprise you? The limited or uncertain availability of nutritionally adequate and safe foods is known as **food insecurity** and is a major social problem in our nation today. Inadequate diets lead to poor health in adults and impaired physical, psychological, and cognitive development in children.

The accompanying "How to" (p. 584) presents the questions used in national surveys to identify food insecurity in the United States, and Figure H16-1 shows the most recent findings. Responses to these questions provide crude, but necessary, data to estimate the degree of hunger in this country.[3]

Defining Hunger in the United States

At its most extreme, people experience hunger because they have absolutely no food. More often, they have too little food **(food insufficiency)** and try to stretch their limited resources by eating small meals or skipping meals—often for days at a time. Sometimes hungry people obtain enough food to satisfy their hunger, perhaps by seeking food assistance or finding food through socially unacceptable ways—begging from strangers, stealing from markets, or scavenging through garbage cans, for example. Sometimes obtaining food raises concerns for food safety—for example, when rot, slime, mold, or insects have damaged foods or when people eat others' leftovers or meat from roadkill.[4]

Hunger has many causes, but in developed countries, the primary cause is **food poverty.** People are hungry not because there is no food nearby to purchase but because they lack

---

GLOSSARY

emergency shelters: facilities that are used to provide temporary housing.

food bank: a facility that collects and distributes food donations to authorized organizations feeding the hungry.

food insecurity: limited or uncertain access to foods of sufficient quality or quantity to sustain a healthy and active life.

food insufficiency: an inadequate amount of food due to a lack of resources.

food pantries: programs that provide groceries to be prepared and eaten at home.

food poverty: hunger resulting from inadequate access to available food for various reasons, including inadequate resources, political obstacles, social disruptions, poor weather conditions, and lack of transportation.

food recovery: collecting wholesome food for distribution to low-income people who are hungry. Four common methods of food recovery are:

- *field gleaning:* collecting crops from fields that either have already been harvested or are not profitable to harvest.

- *nonperishable food collection:* collecting processed foods from wholesalers and markets.

- *perishable food rescue or salvage:* collecting perishable produce from wholesalers and markets.

- *prepared food rescue:* collecting prepared foods from commercial kitchens.

food security: certain access to enough food for all people at all times to sustain a healthy and active life.

soup kitchens: programs that provide prepared meals to be eaten on site.

HOW TO Identify Food Insecurity in a U.S. Household

To determine the extent of food insecurity in a household, surveys ask questions about behaviors and conditions known to characterize households having difficulty meeting basic food needs during the past 12 months. Most often, adults tend to protect their children from hunger. In the most severe cases, children also suffer from hunger and eat less.

1. Did you worry whether food would run out before you got money to buy more?
2. Did you find that the food you bought just didn't last and you didn't have money to buy more?
3. Were you unable to afford to eat balanced meals?
4. Did you or other adults in your household ever cut the size of your meals or skip meals because there wasn't enough food?
5. Did this happen in 3 or more months during the previous year?
6. Did you ever eat less than you felt you should because there wasn't enough money for food?
7. Were you ever hungry but didn't eat because you couldn't afford enough food?

8. Did you ever lose weight because you didn't have enough money to buy food?
9. Did you or other adults in your household ever not eat for a whole day because you were running out of money to buy food?
10. Did this happen in 3 or more months during the previous year?
11. Did you rely on only a few kinds of low-cost food to feed your children because you were running out of money to buy food?
12. Were you unable to feed your children a balanced meal because you couldn't afford it?
13. Were your children not eating enough because you just couldn't afford enough food?
14. Did you ever cut the size of your children's meals because there wasn't enough money for food?
15. Were your children ever hungry but you just couldn't afford enough food?
16. Did your children ever skip a meal because there wasn't enough money for food?
17. Did this happen in 3 or more months during the previous year?

18. Did your children ever not eat for a whole day because there wasn't enough money for food?

The more positive responses, the greater the food insecurity. Households with children answer all of the questions and are categorized as follows:

≤ 2 positive responses = food secure

3–7 positive responses = low food security

≥ 8 positive responses = very low food security

Households without children answer the first 10 questions and are categorized as follows:

≤ 2 positive responses = food secure

3–5 positive responses = low food security

≥ 6 positive responses = very low food security

Figure H16-1 (below) shows the results of the 2005 surveys.

SOURCE: United States Department of Agriculture, *Household Food Security in the United States, 2005,* available at www.ers.usda.gov/publications/err29.

FIGURE H16-1 Prevalence of Food Insecurity and Hunger in U.S. Households, 2005

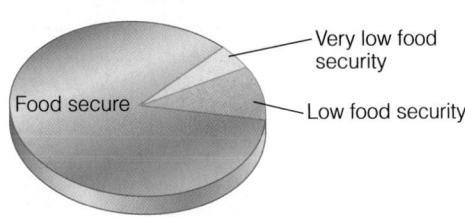

SOURCE: Economic Research Service, U.S. Department of Agriculture, www.ers.usda.gov/publications/, posted November 2006 and visited December 7, 2006.

money. In the United States, an estimated one out of eight people (including one out of six children) lives in poverty. Even those above the poverty line may not have food security. Physical and mental illnesses and disabilities, unemployment, low-paying jobs, unexpected or ongoing medical expenses, and high living expenses threaten their financial stability. When money is tight, people are forced to choose between food and life's other necessities—utilities, housing, and medical care. Food costs are more variable and flexible; people can choose to

purchase fewer groceries to lower the monthly food bill, but they usually can't decide to pay only a portion of the bills for electricity, rent, or medication. Further contributing to food poverty are other problems such as abuse of alcohol and other drugs; lack of awareness of available food assistance programs; and the reluctance of people, particularly the elderly, to accept what they perceive as "welfare" or "charity." Lack of resources remains the major cause of food poverty, and solving this problem would do a lot to relieve hunger.

In the United States, poverty and hunger reach across various segments of society, most often touching single parents living in households with their children, Hispanics and African Americans, and individuals living in the inner cities. People living in poverty are simply unable to buy sufficient amounts of nourishing foods, even if they are wise shoppers. For many of the children in these families, school lunch (and breakfast, where available) may be the only nourishment for the day. Otherwise they go hungry, waiting for an adult to find money for food. Not surprisingly, these children are more likely to have health problems than those who eat regularly.[5] They also tend to perform poorly in school and in social situations.[6]

Ironically, hunger and obesity exist side by side in the United States—sometimes within the same person. It may seem paradoxical that in the same individual hunger reflects an inadequate food intake while obesity implies an excessive intake, but research studies have confirmed the relationship.[7] The highest rates of obesity occur among those living in the greatest poverty—the same people who live with food insecurity.[8] Unfortunately, many healthful food choices cost more than energy-dense foods that foster weight gain but offer few, if any, nutrients.[9] Foods such as doughnuts, pizzas, and hamburgers provide the most energy and satiety for the least cost. Furthermore, people who are unsure about their next meal are likely to overeat when food or money is available. Interestingly, food-insecure people who do not participate in food assistance programs have a greater risk of obesity than those who do participate—illustrating that providing food actually helps to prevent obesity.[10] The accompanying "How to" offers shopping tips for those on a limited budget.

Relieving Hunger in the United States

The American Dietetic Association (ADA) calls for aggressive action to bring an end to domestic food insecurity and hunger and to achieve food and nutrition security for everybody living in the United States.[11] Many federal and local programs aim to prevent or relieve malnutrition and hunger in the United States.

Adequate nutrition and food security are essential in supporting good health and achieving the public health goals of the United States. To that end, an extensive network of federal assistance programs provides life-giving food to millions of U.S. citizens daily. One out of every six Americans receives food assistance of some kind, at a total cost of over $40 billion per year. Even so, the programs are not fully successful in preventing hunger, but they do seem to improve the nutrient intakes of those who participate. Programs described in earlier chapters include the WIC program for low-income pregnant women, breastfeeding mothers, and their young children (Chapter 14); the school lunch, breakfast, and child-care food programs for children (Chapter 15); and the food assistance programs for older adults such as congregate meals and Meals on Wheels (Chapter 16).

AP/Wide World Photos

The fight against hunger depends on the helping hands of caring volunteers.

The Food Stamp Program, administered by the U.S. Department of Agriculture (USDA), is the largest of the federal food assistance programs, both in amount of money spent and in number of people served. It provides assistance to almost 26 million people at a cost of over $25 billion per year; more than half of the recipients are children.[12] The USDA issues debit cards through state agencies to households—people who buy and prepare food together. The amount a household receives depends on its size, resources, and income. The average monthly benefit is about $86 per person.[13] Recipients may use the cards to purchase food and food-bearing plants and seeds,

---

HOW TO Plan Healthy, Thrifty Meals

Chapter 2 introduced the USDA MyPyramid Food Guide and principles for planning a healthy diet. Meeting that goal on a limited budget adds to the challenge. To save money and spend wisely, plan and shop for healthy meals with the following tips in mind:

Planning

- Make a grocery list before going to the store to avoid expensive "impulse" items.
- Do not shop when hungry.
- Use leftovers.
- Center meals on rice, noodles, and other grains.
- Use small quantities of meat, poultry, fish, or eggs.
- Use legumes instead of meat, poultry, fish, or eggs several times a week.
- Use cooked cereals such as oatmeal instead of ready-to-eat breakfast cereals.

- Cook large quantities when time and money allow.
- Check for sales and clip coupons for products you need; plan meals to take advantage of sale items.

Shopping

- Buy day-old bread and other products from the bakery outlet.
- Select whole foods instead of convenience foods (potatoes instead of instant mashed potatoes, for example).
- Try store brands.
- Buy fresh produce that is in season; buy canned or frozen items at other times.
- Buy only the amount of fresh foods that you will eat before it spoils. Buy large bags of frozen items or dry goods; when cooking, take out the amount needed and store the remainder.

- Buy fat-free dry milk; mix and refrigerate quantities needed for a day or two. Buy fresh milk by the gallon or half-gallon.
- Buy less expensive cuts of meat. Chuck and bottom round roast are usually inexpensive; cover during cooking and cook long enough to make meat tender. Buy whole chickens instead of pieces.
- Compare the unit price (cost per ounce, for example) of similar foods so that you can select the least expensive brand or size.
- Buy nonfood items such as toilet paper and laundry detergent at discount stores instead of grocery stores.

For daily menus and recipes for healthy, thrifty meals, visit the USDA Center for Nutrition Policy and Promotion: www.usda.gov/cnpp.

but not to buy tobacco, cleaning items, alcohol, or other nonfood items.

The Food Stamp Program improves nutrient intakes significantly, but hunger continues to plague the United States. Of the estimated 2 million homeless people in the United States who are eligible for food assistance, only 15 percent of single adults and 50 percent of families receive food stamps.

Efforts to resolve the problem of hunger in the United States do not depend solely on federal assistance programs. National **food recovery** programs have made a dramatic difference. The largest program, Second Harvest, coordinates the efforts of more than 40,000 **food pantries, emergency shelters,** and **soup kitchens** that feed over 25 million people a year.

Each year, an estimated one-fifth of our food supply is wasted in fields, commercial kitchens, grocery stores, and restaurants—that's enough food to feed 49 million people. Food recovery programs collect and distribute good food that would otherwise go to waste. Volunteers might pick corn left in an already harvested field, a grocer might deliver ripe bananas to a local **food bank,** and a caterer might take leftover chicken salad to a community shelter, for example. All of these efforts help to feed the hungry in the United States.

Food recovery programs depend on volunteers. Concerned citizens work through local agencies and churches to feed the hungry. Community-based food pantries provide groceries, and soup kitchens serve prepared meals. These meals often deliver adequate nourishment, but most homeless people receive fewer than one and a half meals a day, so many are still inadequately nourished. Combinations of various strategies help to build food security in a community.[14]

Solutions

Every segment of our society can join in the fight against hunger and poverty and help improve the political and environmental policies that surround them. The federal government, the states, local communities, big business and small companies, educators, and all individuals, including dietitians and foodservice managers, have many opportunities to resolve these problems.

Government policies that encourage sustainable food production protect the environment while using less energy and fewer resources. A healthy environment enhances our ability to produce enough food to meet the needs of all hungry people.

Businesses can take the initiative to help; some already have. Several large corporations are major supporters of antihunger programs. Many grocery stores and restaurants participate in food recovery programs by giving their leftover foods to community distribution centers.

Educators, including nutrition educators, can teach others about the underlying social and political causes of poverty, the root cause of hunger. At the college level, they can teach the relationships between hunger and population, hunger and envi-

Each person's choice to get involved and be heard can help lead to needed change.

ronmental degradation, hunger and the status of women, and hunger and global economics. They can advocate legislation to address these problems. They can teach the poor to develop and run nutrition programs in their own communities and to fight on their own behalf for antipoverty, antihunger legislation.

Dietitians and foodservice managers have a special role to play, and their efforts can make an impressive difference. Their professional organization, the ADA, urges members to conserve resources and minimize waste in both their professional and their personal lives.[15] In addition, the ADA urges its members to educate themselves and others on hunger, its consequences, and programs to fight it; to conduct research on the effectiveness and benefits of programs; and to serve as advocates on the local, state, and national levels to help end hunger in the United States.[16] Globally, the ADA supports programs that combat malnutrition, provide food security, promote self-sufficiency, respect local cultures, protect the environment, and sustain the economy.[17]

Individuals can assist the global community in solving its poverty and hunger problems by joining and working for hunger-relief organizations (see Table H16-1). They can also support organizations that lobby for the needed changes in U.S. economic policies toward developing countries.

"Be part of the solution, not part of the problem," an adage says. In other words, don't waste time or energy moaning and groaning about how bad things are: do something to improve them. This adage is as applicable to today's global environmental problems as it is to an unwashed dish in the kitchen sink. They are our problems: human beings created them, and human beings must solve them.

TABLE H16-1 Hunger-Relief Organizations

Action without Borders
79 Fifth Ave., 17th Floor
New York, NY 10118
(212) 843-3973
www.idealist.org

Bread for the World
50 F St. NW, Suite 500
Washington, DC 20001
(800) 82-BREAD or (800) 822-7323
(202) 639-9400; fax (202) 639-9401
www.bread.org

Center on Hunger and Poverty
Brandeis University
Mailstop 077
Waltham, MA 02454
(781) 736-8885
www.centeronhunger.org

Community Food Security Coalition
P.O. Box 909
Venice, CA 90294
(310) 822-5410
www.foodsecurity.org

Congressional Hunger Center
229½ Pennsylvania Ave.
Washington, DC 20003
(202) 547-7022
www.hungercenter.org

Food and Agriculture
  Organization (FAO) of the
  United Nations
2175 K St. NW, Suite 300
Washington, DC 20437
(202) 653-2400
www.fao.org

HungerWeb
Tufts Nutrition
nutrition.tufts.edu/academic/
hungerweb/overview

Oxfam America
26 West St.
Boston, MA 02111-1206
(800) 77-OXFAM or
(800) 776-9326
www.oxfamamerica.org

Pan American Health Organization
525 23rd St. NW
Washington, DC 20037
(202) 974-3000
www.paho.org

Second Harvest
35 E. Wacker Dr., #2000
Chicago, IL 60601
(800) 771-2303
www.secondharvest.org

Society of St. Andrew
3383 Sweet Hollow Rd.
Big Island, VA 24526
(800) 333-4597
www.endhunger.org

United Nations Children's Fund
  (UNICEF)
3 United Nations Plaza
New York, NY 10017-4414
(212) 326-7035
www.unicef.org

World Food Program
Via Vittorio Emanuele Orlando, 83
Rome, Italy 00148
www.wfp.org

World Health Organization (WHO)
525 23rd St. NW
Washington, DC 20037
(202) 974-3000
www.who.org

World Hunger Year (WHY)
505 Eighth Ave., 21st Floor
New York, NY 10018-6582
(800) GleanIt
www.worldhungeryear.org

NUTRITION ON THE NET

For further study of topics covered in this chapter, log on to **academic.cengage .com/nutrition/rolfes/UNCN8e**. Go to Chapter 16, then to Nutrition on the Net.

- Explore the problems of hunger, malnutrition, and food insecurity at the Feeding Minds, Fighting Hunger site: **www.feedingminds.org**

- Learn about constructive, community-based solutions to the problems of poverty and hunger within and between the public and private sectors from the National Hunger Clearinghouse: **www.worldhungeryear.org/nhc**

- Visit the USDA Food Stamp Program: **www.fns.usda.gov/fsp**

- Download recipes, sample menus, and numerous tips for planning, shopping for, and cooking healthy meals on a tight budget from the USDA cookbook entitled "Recipes and Tips for Healthy, Thrifty Meals": **www.usda.gov/cnpp**

- Review the Best Practices Manual for Food Recovery and Gleaning at the USDA Food and Nutrition Service site: **www.fns.usda.gov/fdd/gleaning/gleanintro.htm**

- Find information on feeding the hungry from the Emergency Food and Shelter Program: **www.efsp.unitedway.org**

- Donate free food at The Hunger Site: **www.thehungersite .com**

- See Table H16-1 (above) for additional websites.

REFERENCES

1. Food and Agriculture Organization of the United Nations, *State of Food Insecurity in the World 2005*.
2. United States Department of Agriculture, *Household Food Security in the United States, 2004*, ERS Research Briefs, available at www.ers.usda.gov/publications.
3. J. S. Hampl and R. Hill, Dietetic approaches to US hunger and food insecurity, *Journal of the American Dietetic Association* 102 (2002): 919–923.
4. K. M. Kempson and coauthors, Food management practices used by people with limited resources to maintain food sufficiency as reported by nutrition educators, *Journal of the American Dietetic Association* 102 (2002): 1795–1799.
5. J. T. Cook and coauthors, Food insecurity is associated with adverse health outcomes among human infants and toddlers, *Journal of Nutrition* 134 (2004): 1432–1438.
6. D. F. Jyoti, E. A. Frongillo, and S. J. Jones, Food insecurity affects school children's academic performance, weight gain, and social skills, *Journal of Nutrition* 135 (2005): 2831–2839.

7. L. M. Scheier, What is the hunger-obesity paradox? *Journal of the American Dietetic Association* 105 (2005): 883–886.
8. P. E. Wilde and J. N. Peterman, Individual weight change is associated with household food security status, *Journal of Nutrition* 136 (2006): 1395–1400; A. Drewnowski and S. E. Specter, Poverty and obesity: The role of energy density and energy costs, *American Journal of Clinical Nutrition* 79 (2004): 6–16; E. J. Adams, L. Grummer-Strawn, and G. Chavez, Food insecurity is associated with increased risk of obesity in California women, *Journal of Nutrition* 133 (2003): 1070–1074.
9. Drewnowski and Specter, 2004.
10. S. J. Jones and E. A. Frongillo, The modifying effects of Food Stamp Program participation on the relation between food insecurity and weight change in women, *Journal of Nutrition* 136 (2006): 1091–1094; S. J. Jones and coauthors, Lower risk of overweight in school-aged food insecure girls who participate in food assistance, *Archives of Pediatrics and Adolescent Medicine* 157 (2003): 780–784.

11. Position of the American Dietetic Association: Food insecurity and hunger in the United States, *Journal of the American Dietetic Association* 106 (2006): 446–458.
12. USDA Food and Nutrition Service, www.fns.usda.gov/fsp, site visited August 30, 2006.
13. USDA Food and Nutrition Service, www.fns.usda.gov/fsp, site visited August 30, 2006.
14. C. McCullum and coauthors, Evidence-based strategies to build community food security, *Journal of the American Dietetic Association* 105 (2005): 278–283.
15. Position of the American Dietetic Association: Dietetic professionals can implement practices to conserve natural resources and protect the environment, *Journal of the American Dietetic Association* 101 (2001): 1221–1227.
16. Position of the American Dietetic Association, 2006.
17. Position of the American Dietetic Association: Addressing world hunger, malnutrition, and food insecurity, *Journal of the American Dietetic Association* 103 (2003): 1046–1057.

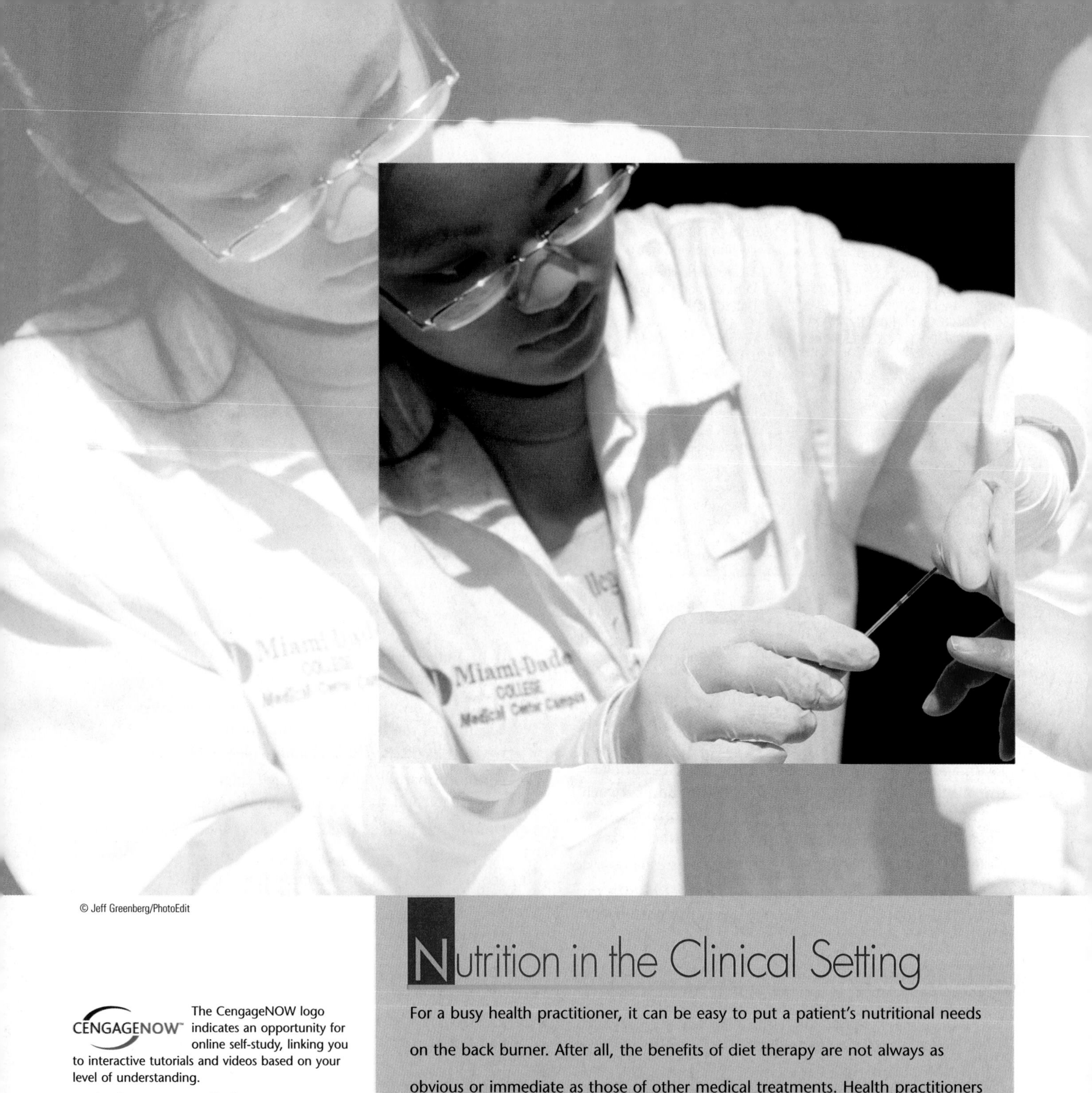

The CengageNOW logo indicates an opportunity for online self-study, linking you to interactive tutorials and videos based on your level of understanding.

academic.cengage.com/login

Nutrition in the Clinical Setting

For a busy health practitioner, it can be easy to put a patient's nutritional needs on the back burner. After all, the benefits of diet therapy are not always as obvious or immediate as those of other medical treatments. Health practitioners who want to provide the best care for their patients, however, soon learn that an appropriate diet can improve both short-term and long-term outcomes of many disease treatments. Moreover, patients are often concerned about the impact their diet has on their disease condition. The remaining chapters of this book will show how dietary treatments can contribute to medical care and improve the quality of life for people who have become ill.

Nutrition Care and Assessment

Previous chapters of this book introduced the nutrients and described how the appropriate dietary choices can support good health. Turning now to clinical nutrition, the remaining chapters describe how various medical disorders can affect nutrition status and nutrient needs. This chapter introduces the process that is used for providing nutrition care and the strategies used for assessing nutrition status.

Nutrition in Health Care

Malnutrition is frequently reported in patients hospitalized with acute illness, and acutely ill individuals without nutrition problems on admission often exhibit a subsequent decline in nutrition status. In the past few decades, estimates of malnutrition in hospital patients have ranged from 38 to 62 percent.[1] Poor nutrition status weakens immune function and compromises a person's healing ability, influencing both the course of disease and the body's response to treatment. Thus, preventing and correcting nutrition problems can improve the outcome of disease treatments and can also help to prevent complications.

Effects of Illness on Nutrition Status

An illness, its symptoms, and its treatments can lead to malnutrition by reducing food intake, interfering with digestion and absorption, or altering nutrient metabolism and excretion (see Figure 17-1 on p. 590). For example, the nausea associated with some illnesses and disease treatments can diminish appetite and reduce food intake; similarly, an inflamed mouth or esophagus can make the physical act of eating uncomfortable. Certain medications can cause anorexia or gastrointestinal discomfort or can interfere with nutrient function and metabolism. Prolonged bed rest often results in **pressure sores,** which increase metabolic stress and raise protein and energy needs. ◆

The dietary changes required during an acute illness are usually temporary and can be tailored to accommodate an individual's preferences and lifestyle.

◆ Chapter 22 discusses the nutrition needs of patients undergoing acute metabolic stress.

pressure sores: regions of damaged skin and tissue due to prolonged pressure on the affected area by an external object, such as a bed, wheelchair, or cast; vulnerable areas of the body include buttocks, hips, and heels. Also called *decubitus* (deh-KYU-bih-tus) *ulcers.*

FIGURE 17-1 Ways in Which Illness Can Affect Nutrition Status

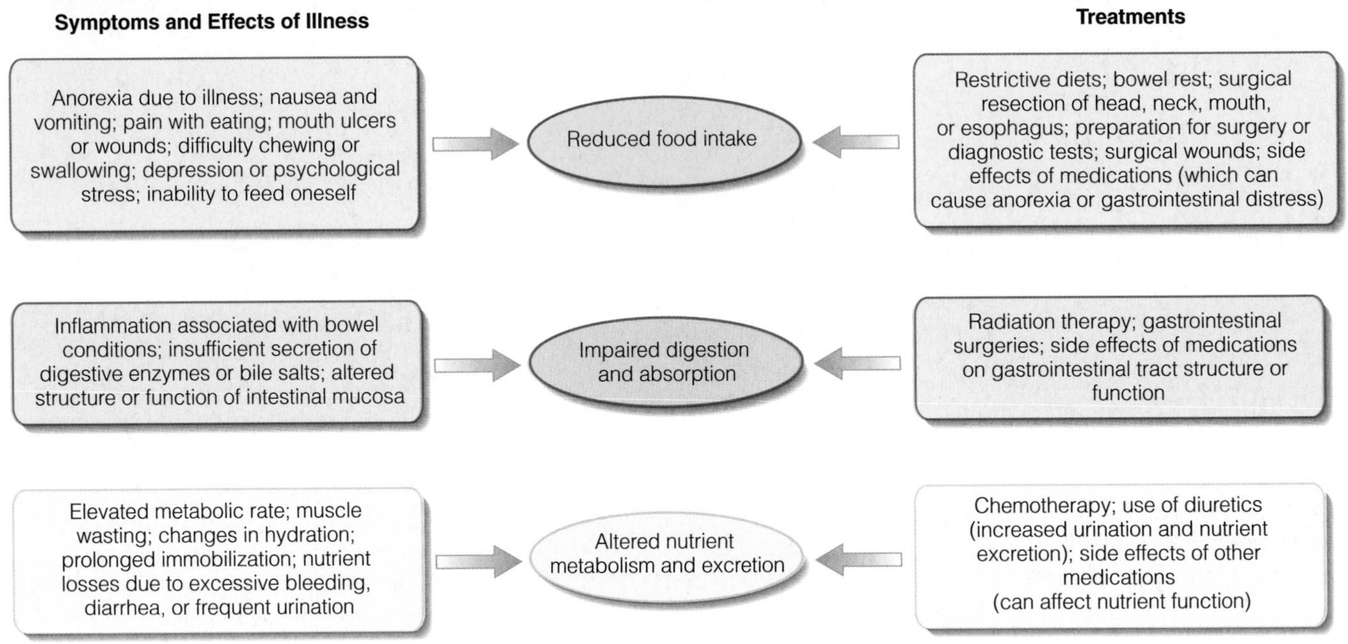

Symptoms and Effects of Illness

Treatments

Anorexia due to illness; nausea and vomiting; pain with eating; mouth ulcers or wounds; difficulty chewing or swallowing; depression or psychological stress; inability to feed oneself

→ Reduced food intake ←

Restrictive diets; bowel rest; surgical resection of head, neck, mouth, or esophagus; preparation for surgery or diagnostic tests; surgical wounds; side effects of medications (which can cause anorexia or gastrointestinal distress)

Inflammation associated with bowel conditions; insufficient secretion of digestive enzymes or bile salts; altered structure or function of intestinal mucosa

→ Impaired digestion and absorption ←

Radiation therapy; gastrointestinal surgeries; side effects of medications on gastrointestinal tract structure or function

Elevated metabolic rate; muscle wasting; changes in hydration; prolonged immobilization; nutrient losses due to excessive bleeding, diarrhea, or frequent urination

→ Altered nutrient metabolism and excretion ←

Chemotherapy; use of diuretics (increased urination and nutrient excretion); side effects of other medications (can affect nutrient function)

Conversely, chronic illnesses may necessitate long-term dietary modifications. For example, diabetes treatment requires lifelong changes in diet and lifestyle that some people may find difficult to maintain. The challenge for health professionals is to help their patients appreciate the potential benefits of treatment and accept dietary changes that can improve their health.

Responsibility for Nutrition Care

The members of a health care team work together to ensure that the nutritional needs of patients are met during illness. The roles of these team members in nutrition care may overlap, and job descriptions in different institutions may vary somewhat. In some cases, nutrition care is incorporated into the medical care plan developed by the entire health care team. Such plans, called **critical pathways,** outline coordinated plans of care for specific medical diagnoses, treatments, or procedures.

Physicians Physicians are responsible for meeting all of a patient's medical needs, including nutrition. They prescribe **diet orders** and other orders related to nutrition care, including referrals for medical nutrition therapy and dietary counseling. Physicians rely on nurses, registered dietitians, and other health professionals to alert them to nutrition problems, suggest strategies for handling these problems, and provide nutrition services.

Registered Dietitians Registered dietitians ◆ are food and nutrition experts who are uniquely qualified to provide **medical nutrition therapy.** They conduct nutrition and dietary assessments; diagnose nutrition problems; develop, implement, and evaluate **nutrition care plans** (described in a later section); plan and approve menus; and provide nutrition education. Registered dietitians may also work as managers of food and cafeteria services in health care institutions.

Registered Dietetic Technicians Registered dietetic technicians often work in partnership with registered dietitians and assist in the implementation and monitoring of nutrition services. Depending on their background and experience, they may screen patients for nutrition problems, provide patient education and counseling, de-

◆ Reminder: A *registered dietitian (RD)* has successfully completed the education and training specified by the the American Dietetic Association (or Dietitians of Canada), including an undergraduate degree in nutrition or dietetics, a supervised internship, and a national examination.

critical pathways: coordinated programs of treatment that merge the care plans of different health practitioners; also called *clinical pathways.*

diet orders: specific instructions regarding dietary management; also called *diet prescriptions.*

medical nutrition therapy: nutrition care provided by a registered dietitian; includes assessing nutrition status, diagnosing nutrition problems, and providing nutrition care.

nutrition care plans: strategies for meeting an individual's nutritional needs.

velop menus and recipes, ensure appropriate meal delivery, and monitor patients' food choices and intakes. Dietetic technicians sometimes supervise foodservice operations and may have roles in purchasing, inventory, quality control, sanitation, or safety.

Nurses Nurses interact closely with patients and thus are in an ideal position to identify people who would benefit from nutrition services. They often screen patients for nutrition problems and may participate in nutrition assessments. Nurses also provide direct nutrition care, such as encouraging patients to eat, finding practical solutions to food-related problems, recording a patient's food intake, and answering questions about special diets. As members of **nutrition support teams,** nurses are responsible for administering tube and intravenous feedings. In facilities that do not employ registered dietitians, nurses often assume responsibility for much of the nutrition care. Table 17-1 provides examples of **nursing diagnoses** that are likely to be associated with nutrition problems.

Other Health Care Professionals Other health care professionals may assist with nutrition care. Pharmacists, physical therapists, occupational therapists, speech therapists, social workers, nursing assistants, and home health care aides can be instrumental in alerting dietitians or nurses to nutrition problems or may share relevant information about a patient's health status or personal needs.

Nutrition Screening

To identify patients who are malnourished or at risk for malnutrition, a **nutrition screening** is conducted within 24 hours of a patient's admission to a hospital or other extended-care facility. It may also be included in outpatient services and community health programs. ◆ A nutrition screening involves collecting a limited amount of health-related information that can identify malnutrition (most often, protein-energy malnutrition, or PEM ◆). Although the screening should be sensitive enough to identify the patients who require nutrition care, it must be simple enough to be completed within 10 to 15 minutes. Usually a nurse, nursing assistant, registered dietitian, or dietetic technician performs and documents the screening. In some instances, a screening may be repeated (or followed up with a more comprehensive screening) during a patient's stay.

The information gathered during a nutrition screening varies according to the patient population, the type of care offered by the health care facility, and the patient's medical problem. Usually included are the admitting diagnosis, physical measurements, laboratory test results, and information about diet and health provided by the patient or caregiver (see Table 17-2 for examples). A number of screening tools that use different combinations of these variables have become popular in recent years; these tools include the *Mini Nutritional Assessment* and the *Subjective Global Assessment*, outlined in Tables 17-3 (p. 592) and 17-4 (p. 593). The Mini

TABLE 17-1 Nursing Diagnoses with Nutritional Implications

- Chronic confusion
- Chronic pain
- Constipation
- Diarrhea
- Disturbed body image
- Feeding self-care deficit
- Imbalanced nutrition: less than body requirements
- Imbalanced nutrition: more than body requirements
- Impaired dentition
- Impaired oral mucous membrane
- Impaired physical mobility
- Impaired swallowing
- Ineffective infant feeding pattern
- Nausea
- Readiness for enhanced nutrition
- Risk for aspiration
- Risk for deficient fluid volume
- Risk for unstable blood glucose

SOURCE: NANDA International, *Nursing Diagnoses: Definitions and Classification 2007-2008* (Philadelphia: NANDA International, 2007).

◆ Reminder: The Nutrition Screening Initiative, which addresses malnutrition risk in older adults, is described in Chapter 16 (pp. 576–577).

◆ Reminder: *Protein-energy malnutrition (PEM)* is a deficiency of protein and food energy and is characterized by weight loss and loss of muscle tissue.

TABLE 17-2 Information Included in a Nutrition Screening

- Age, medical diagnosis, severity of illness
- Height and weight, BMI, unintentional weight changes
- Tissue wasting, loss of subcutaneous fat
- Changes in appetite or food intake
- Problems that interfere with food intake (such as chewing or swallowing difficulty, or nausea and vomiting)
- Food allergies or intolerances, extensive dietary restrictions
- Laboratory test results that indicate poor health status
- History of diabetes, renal disease, or other chronic illness
- Presence of anemia or pressure sores
- Use of medications that can impair nutrition status
- Depression, social isolation, dementia

nutrition support teams: health care professionals responsible for the provision of nutrients by tube feeding or intravenous infusion.

nursing diagnoses: clinical judgments about actual or potential health problems that provide the basis for selecting appropriate nursing interventions.

nutrition screening: a brief assessment of health-related variables to identify patients who are malnourished or at risk for malnutrition.

TABLE 17-3 Mini Nutritional Assessment[a]

The Mini Nutritional Assessment is a two-part screening tool consisting of a brief screening step and, if warranted, a more detailed assessment. A referral for nutrition care is warranted if the combined scores from Step 1 and Step 2 indicate that the patient is malnourished or at risk for malnutrition.

Step 1: Screening

The patient or caregiver provides answers to questions A through D, and the screener determines answers to questions E and F. If the sum of points is 11 or less, the patient is at risk for malnutrition and Step 2 is conducted. A sum of 12 or more suggests that the patient is not at risk. Point values are shown below each question.

A. Has food intake declined in the past 3 months due to loss of appetite, digestive problems, or chewing or swallowing difficulties?

   0 = severe loss of appetite; 1 = moderate loss of appetite; 2 = no loss of appetite

B. Any weight loss during the past 3 months?

   0 = weight loss >3 kg; 1 = does not know; 2 = weight loss of 1 to 3 kg; 3 = no weight loss

C. Any mobility problems?

   0 = cannot leave bed or chair; 1 = cannot leave house; 2 = able to leave house

D. Any psychological stress or acute disease in the past 3 months?

   0 = yes; 2 = no

E. Any neuropsychological problems?

   0 = severe dementia or depression; 1 = mild dementia; 2 = no psychological problems

F. Healthy BMI?

   0 = BMI is <19; 1 = BMI is 19 to <21; 2 = BMI is 21 to <23; 3 = BMI is ≥23

Step 2: Assessment

The patient or caregiver provides answers to questions G through P, and the screener determines answers to questions Q and R. The total score includes the sum of points from both Step 1 and Step 2. A total score of less than 17 points suggests that the patient is malnourished; a score of 17 to 23.5 points indicates risk of malnutrition.

G. Lives independently? 0 = no; 1 = yes

H. Takes more than 3 prescription drugs per day? 0 = yes; 1 = no

I. Presence of pressure sores or skin ulcers? 0 = yes; 1 = no

J. Number of full meals consumed daily? 0 = 1 meal; 1 = 2 meals; 2 = 3 meals

K. Intake of protein-containing foods?

   • At least 1 serving of milk products per day? (yes/no)

   • 2 or more servings of legumes or eggs per week? (yes/no)

   • Meat, fish, or poultry every day? (yes/no)

   0 = 0 or 1 "yes" response; 0.5 = 2 "yes" responses; 1 = 3 "yes" responses

L. At least 2 servings of fruits and vegetables per day? 0 = no; 1 = yes

M. Amount of fluid consumed daily?

   0 = <3 cups; 0.5 = 3 to 5 cups; 1 = >5 cups

N. Needs assistance with feeding?

   0 = needs assistance; 1 = self-fed with some difficulty; 2 = no assistance needed

O. Patient's view of own nutritional status?

   0 = feels malnourished; 1 = uncertain; 2 = feels adequately nourished

P. Patient's view of own health status in comparison with others?

   0 = not as good; 0.5 doesn't know; 1 = as good as others; 2 = better than others

Q. Midarm circumference in centimeters? 0 = <21; 0.5 = 21 to <22; 1 = ≤22

R. Calf circumference in centimeters? 0 = <31; 1 = ≥31

[a]An interactive version is available at www.mna-elderly.com/clinical-practice.htm.
SOURCE: M. C. Murphy and coauthors, The use of the Mini-Nutritional Assessment (MNA) tool in elderly orthopaedic patients, *European Journal of Clinical Nutrition* 54 (2000): 555–562. Reprinted by permission

Nutritional Assessment was developed to detect malnutrition in adults over 65 years of age, whereas the Subjective Global Assessment has been found to be applicable to a variety of patient populations. Briefer screening methods may use just two or three variables; for example, a tool called the *Nutrition Risk Index* determines

TABLE 17-4 Subjective Global Assessment

The Subjective Global Assessment rates features of the medical history and physical examination. Each variable is given an A, B, or C rating: A for well nourished, B for potential or mild malnutrition, and C for severe malnutrition. Patients are classified according to the final numbers of A, B, and C rankings.

Medical History

- Body weight changes: percentage change in past 6 months; weight change in past 2 weeks
- Dietary changes: suboptimal, low-kcalorie, liquid diet, or starvation
- GI symptoms: nausea, diarrhea, vomiting, or anorexia for more than 2 weeks
- Functional ability: full capacity versus suboptimal, walking versus bedridden
- Degree of disease-related metabolic stress: low, medium, or high

Physical Examination

- Subcutaneous fat loss (triceps or chest)
- Muscle loss (quadriceps or deltoids)
- Ankle edema
- Sacral (lower spine) edema
- Ascites (abdominal edema)

Classification:

A: Well nourished: if no significant loss of weight, fat, or muscle tissue and no dietary difficulties, functional impairments, or GI symptoms; also applies to patients with recent weight gain and improved appetite, functioning, or medical prognosis

B: Moderate malnutrition: if 5 to 10 percent weight loss, mild loss of muscle or fat tissue, decreased food intake, and digestive or functional difficulties that impair food intake; the B classification usually applies to patients with an even mix of A, B, and C ratings

C: Severe malnutrition: if more than 10 percent weight loss, severe loss of muscle or fat tissue, edema, multiple GI symptoms, and functional impairments

SOURCES: R. S. Gibson, *Principles of Nutritional Assessment* (New York: Oxford University Press, 2005), pp. 809–826; A. S. Detsky and coauthors, What is subjective global assessment of nutritional status? *Journal of Parenteral and Enteral Nutrition* 11 (1987): 8–13.

malnutrition risk by evaluating weight loss and serum albumin levels.[2] Note that there is no single screening method that is universally accepted, and health care institutions often develop specific techniques that meet their particular needs.

A nutrition or health screening may lead to a referral for nutrition care. The following section describes the next stage of the process: the method used by dietitians to address nutritional concerns.

The Nutrition Care Process

Registered dietitians use a systematic approach to medical nutrition therapy called the **nutrition care process.** ◆ Figure 17-2 presents the four distinct, yet interrelated, steps of the nutrition care process:[3]

1. Nutrition assessment
2. Nutrition diagnosis
3. Nutrition intervention
4. Nutrition monitoring and evaluation

Although the nutrition care process is easiest to visualize as a series of steps, the steps are frequently revisited in order to reassess and revise diagnoses and intervention strategies. Note that each step of the nutrition care process must be documented in the medical record, providing a record for future reference and facilitating communication among members of the health care team. Chapter 18 provides additional information about documentation.

Nutrition Assessment A nutrition assessment involves the collection and analysis of health-related information for the purpose of identifying specific nutrition

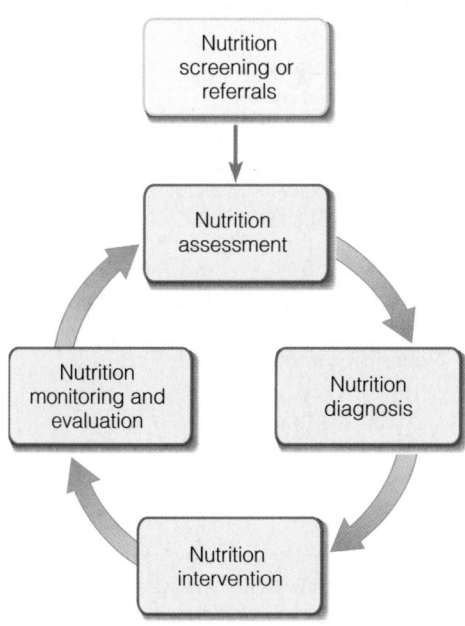

FIGURE 17-2 The Nutrition Care Process

Nutrition screening or referrals → Nutrition assessment → Nutrition diagnosis → Nutrition intervention → Nutrition monitoring and evaluation

◆ As a comparison, the *nursing process* consists of five steps:
1. Assessment
2. Nursing diagnosis
3. Outcome identification and planning
4. Implementation
5. Evaluation

nutrition care process: a problem-solving method that dietetics professionals use to evaluate and treat nutrition-related problems.

TABLE 17-5 Examples of Nutrition Diagnoses

Intake Diagnoses

- Excessive alcohol intake
- Inadequate energy intake
- Inadequate fluid intake
- Inappropriate infusion of parenteral nutrition
- Increased calcium needs
- Inconsistent carbohydrate intake

Clinical Diagnoses

- Altered blood potassium levels
- Altered GI function (constipation)
- Breastfeeding difficulty
- Food-medication interaction
- Involuntary weight gain
- Swallowing difficulty

Behavioral-Environmental Diagnoses

- Disordered eating pattern
- Impaired ability to prepare meals
- Limited access to food
- Physical inactivity
- Self-feeding difficulty
- Undesirable food choices

SOURCE: American Dietetic Association, *Nutrition Diagnosis and Intervention: Standardized Language for the Nutrition Care Process* (Chicago: American Dietetic Association, 2007).

◆ This format is called a *PES statement* because it includes the *Problem*, the *Etiology*, and the *Signs* and symptoms.

◆ The American Dietetic Association maintains an Evidence Analysis Library to keep members updated about recent developments in nutrition and dietetics research.

problems and their underlying causes. A well-conducted assessment allows the dietitian to devise a plan of action to prevent or correct nutrient imbalances or to evaluate whether a particular care plan is working. An assessment typically includes information from medical, personal, social, and food/nutrition histories; anthropometric and biochemical analyses; medical tests; and a physical examination. When appropriate, assessment data are compared with reliable standards to help with their interpretation. The second half of this chapter describes the components of nutrition assessment in detail.

Nutrition Diagnosis After completing a nutrition assessment, the dietitian identifies existing and potential nutrition problems, a step that requires a careful and objective analysis of the patterns and relationships among the assessment data. Each nutrition problem receives a separate diagnosis, which includes the specific problem, etiology or cause, and signs and symptoms that provide evidence of the problem.[4] ◆ For example, a potential nutrition diagnosis might be "Involuntary weight gain *(the problem)* related to chronic use of a medication (corticosteroids) that causes weight gain *(the etiology or cause)* as evidenced by an unintentional weight gain of 10 percent of body weight over the past six months *(the sign or symptom)."* Note that a nutrition diagnosis is likely to change over the course of illness, due to either a successful nutrition intervention or resolution of the medical problem.

Nutrition diagnoses fall into three main categories: *intake, clinical,* and *behavioral-environmental.* Intake-related diagnoses involve either the inadequate or excessive ingestion of nutrients, energy, fluid, alcohol, dietary supplements, and food ingredients. Clinical diagnoses involve medical or physical conditions that disrupt nutrition status, such as disruptions in physiological or mechanical functioning, altered nutrient metabolism, and body weight problems. Behavioral-environmental diagnoses include problems related to the patient's knowledge, attitudes, or beliefs; the physical environment; access to food; and food safety. Table 17-5 lists examples of nutrition diagnoses in each of these categories.

Nutrition Intervention After nutrition problems are identified, the appropriate nutrition care is planned and implemented. Nutrition interventions attempt to modify dietary and lifestyle practices or environmental conditions that interfere with nutrition status or health. When possible, the intervention targets the etiology or cause of the problem as identified in the nutrition diagnosis. Nutrition interventions often include both counseling and education components, which require close interaction with the patient. To be successful, the intervention must carefully consider the patient's dietary and lifestyle preferences. Note that nutrition interventions used by dietitians are *evidence based;* that is, they are based on scientific rationale and supported by the results of high-quality research.[5] ◆

The goals of nutrition interventions are stated in terms of measurable outcomes, such as the results of laboratory or anthropometric tests. For example, goals for an overweight person with diabetes might include target ranges for blood glucose levels and body weight. Other desirable outcomes include positive changes in dietary behaviors and lifestyle; for example, the diabetes patient may need to learn how to control carbohydrate intake or portion sizes and may benefit from regular exercise. These outcomes can be assessed during an interview with the patient.

Although many aspects of nutrition care fall within the scope of dietetics practice, others require the assistance of other health professionals. For example, a physician's help would be required if a medication interfered with food intake; the nursing or foodservice staff might be involved if the feeding environment or meal delivery required adjustment. Chapter 18 provides additional information about nutrition intervention.

Nutrition Monitoring and Evaluation The effectiveness of the nutrition care plan must be evaluated periodically: the original goals and outcome measures should be reviewed and compared with updated assessment data, and the patient's progress should be monitored carefully. Sometimes a change in a person's situation alters nutritional needs; for example, a change in the medical treatment or a new medication may alter a person's tolerance to certain foods. A nutrition care plan must be flexible enough to adapt to the new situation.

If progress is slow or a patient is unable or unwilling to make the suggested changes, the care plan should be redesigned and take into account the reasons why the earlier plan was not successful. The new plan may need to include motivational techniques or additional patient education. If the patient remains unwilling to modify behaviors despite the expected benefits, the health care provider can try again at a later time when the patient may be more receptive.

IN SUMMARY

Illnesses and their treatments can affect food intake and nutrient needs, leading to malnutrition. In turn, poor nutrition can reduce the effectiveness of medical treatments. The combined efforts of each member of the health care team ensure that patients receive optimal nutrition care. Nutrition screening identifies individuals who can benefit from nutrition assessment and follow-up nutrition care. The nutrition care process includes four interrelated steps: nutrition assessment, nutrition diagnosis, nutrition intervention, and nutrition monitoring and evaluation.

Nutrition Assessment

As described earlier in this chapter, a nutrition assessment provides the information needed for diagnosing nutrition problems and designing a nutrition care plan, and follow-up assessments help to determine whether the care plan has been effective. Ideally, the assessment should be sensitive enough to detect subtle nutrition problems and specific enough to identify problem nutrients. For most nutrient imbalances, a variety of tests are necessary to identify nutrition problems. The remainder of this chapter describes the types of information and measures that are most commonly obtained in a nutrition assessment.

Historical Information

Historical information provides clues about the patient's nutrition status and nutrient needs and uncovers personal preferences that need to be considered when developing a nutrition care plan. Table 17-6 summarizes the various types of historical information that contribute to a nutrition assessment.[6] This information can be obtained from the medical record or by interviewing the patient or caregiver. ◆

◆ The terms used in this section are based on the standardized language developed by the American Dietetic Association; other resources may use different terminology.

TABLE 17-6 Historical Information Used in Nutrition Assessment[a]

Medical/Health History	Medication/Supplement History	Personal/Social History	Food/Nutrition History
Current complaint(s)	Prescription drugs	Age	Food intake
Past medical conditions	Over-the-counter drugs	Occupation	Food allergies and intolerances
Surgical history	Dietary/herbal supplements	Educational level	Nutrition/health knowledge
Family medical history	Alcohol intake	Socioeconomic status	Food availability
Chronic disease risk	Cigarettes/tobacco use	Cultural/ethnic identity	Physical activity and exercise patterns
Mental/emotional health status	Illegal drug use	Religious beliefs	
		Home/family situation	
		Cognitive abilities	

[a]Historical information can be categorized in different ways; the categories shown here conform to the standardized language used in nutrition diagnoses.
SOURCE: American Dietetic Association, *Nutrition Diagnosis and Intervention: Standardized Language for the Nutrition Care Process* (Chicago: American Dietetic Association, 2007).

TABLE 17-7 Medical Problems Often Associated with Malnutrition

- Acquired immune deficiency syndrome (AIDS)
- Alcoholism
- Anorexia nervosa
- Bulimia nervosa
- Burns (extensive or severe)
- Cancer and cancer treatments
- Cardiovascular diseases
- Celiac disease
- Chewing/swallowing difficulties
- Chronic kidney disease
- Dementia
- Diabetes mellitus
- Feeding disabilities
- Infections
- Inflammatory bowel diseases
- Liver disease
- Malabsorption
- Mental illness
- Pressure sores
- Surgery (major)
- Vomiting (prolonged or severe)

Medical/Health History A substantial number of medical problems and their treatments may either interfere with food intake or require dietary changes; Table 17-7 lists examples. The medical history generally includes the family medical history as well; this information reveals a person's genetic susceptibilities for diseases that can potentially be prevented with dietary and lifestyle changes. (Note that the family medical history refers to blood relatives only.)

Medication/Supplement History A number of medications have detrimental effects on nutrition status, and various dietary components can alter drug absorption or metabolism. Chapter 19 provides examples of notable diet-drug interactions that may need consideration when planning nutrition care. Use of alcohol, tobacco, or illegal drugs may alter food intake and ultimately have disruptive effects on health and nutrition status.

Personal/Social History Personal and social factors influence food choices, as well as a person's ability to manage health and nutrition problems. For example, financial concerns may restrict access to health care and nutritious foods. Cultural background or religious beliefs can affect food preferences. Certain family members may be responsible for preparing or procuring food. An individual with cognitive disabilities may eat poorly or be unable to follow complex dietary instructions.

Food/Nutrition History A food/nutrition history (often called a *diet history*) is a detailed account of a person's dietary practices. It includes food intake data, lifestyle habits, and information about the various factors that may influence dietary choices, such as the person's knowledge or beliefs about nutrition and health. Although different methods are used for collecting food intake data, the procedure often includes an interview about recent food intake (for example, a *24-hour recall*) and a survey of usual food choices (such as a *food frequency questionnaire*). A good food/nutrition history can uncover current or potential nutrition problems, as well as patterns of behavior that contribute to health problems. The following section describes the most common methods of gathering food intake information.

Food Intake Data

Obtaining accurate food intake data is challenging, and results may vary depending on the individual's memory and honesty and the assessor's skill and training. In addition, each method has its own strengths and weaknesses, so best results are obtained by using a combination of methods. Table 17-8 summarizes the methods commonly used and each method's advantages and disadvantages.

Once food intake data are collected, the nutrient intake can be estimated using dietary analysis software or a table of food composition (such as that in Appendix H) and nutrient intake levels can be compared with RDA and AI values. Another option is to compare the food list with a diet-planning guide such as the USDA Food Guide (see Chapter 2). The food list also reveals a person's food preferences, which are helpful for developing an appropriate nutrition care plan, planning menus, or providing dietary counseling.

The 24-Hour Recall The **24-hour recall** is a guided interview in which an individual recounts all of the foods and beverages consumed in the past 24 hours or during the previous day. The interviewer includes questions about the times when meals or snacks were eaten, amounts consumed, and ways in which foods were prepared. Accuracy can be improved by prompting the respondent to recall food items that are often forgotten, such as snack foods, beverages, and condiments.

In a typical interview, the assessor may begin by asking: "What is the first thing you ate or drank yesterday morning?" After the first food items are described, the follow-up questions might be: "What time was that?" and "How much did you eat?" Questioning continues until the intake record for the day is complete. Food models or measuring cups and spoons can be used to help the individual visualize and describe the amounts consumed. After the day's intake is recounted, the interviewer asks whether the intake that day is fairly typical and, if not, how it varies

24-hour recall: a record of foods consumed in the previous 24 hours; sometimes modified to include foods consumed in a typical day.

TABLE 17-8	Methods for Obtaining Food Intake Data		
Method	**Description**	**Advantages**	**Disadvantages**
24-hour recall	Guided interview in which the foods and beverages consumed in a 24-hour period are described in detail.	• Results are not dependent on literacy or educational level of respondent. • Interview occurs after food is consumed, so it does not interfere with food choices. • It is a relatively easy and quick assessment method.	• Process is reliant on memory. • Food items that cause embarrassment (alcohol, desserts) may be omitted. • Underestimation and overestimation of food intakes are common. • Skill of interviewer affects outcome. • Data from a single day cannot represent the respondent's usual intake accurately. • Seasonal variations may not be addressed.
Food frequency questionnaire	Written survey of food consumption during a specific period of time, often a one-year period.	• Process examines long-term food intake, so day-to-day and seasonal variability should not affect results. • It is completed after food is consumed, so it does not interfere with food choices. • It is a low-cost method.	• Process is reliant on memory. • It is not good for monitoring short-term changes in food intake. • Serving sizes are often difficult for respondents to evaluate without assistance. • Calculated nutrient intakes may not be accurate. • Food lists include common foods only. • Food lists for the general population are of limited value in special populations.
Food record	Written account of food consumed during a specified period, usually several consecutive days. Accuracy is improved by including weights or measures of foods.	• Process does not rely on memory. • Recording foods as they are consumed improves likelihood of obtaining accurate food intake data. • It is useful for controlling intake because keeping records can increase awareness of food choices.	• Recording process itself influences food intake. • Process is time-consuming and burdensome for respondent; requires high degree of motivation. • Underreporting is common. • It requires literacy and the physical ability to write. • Seasonal changes in diet are not taken into account.
Direct observation	Observation of meal trays or shelf inventories before and after eating; possible only in residential facilities.	• Process does not rely on memory. • It does not interfere with person's food intake. • It can be used to evaluate acceptability of prescribed diet.	• Process is possible only in residential situations. • It is labor intensive.

from the person's usual intake. A recall interview may be conducted on several nonconsecutive days to obtain a better representation of a person's usual diet.

A recall interview often yields useful data for developing an acceptable nutrition care plan and identifying food items that may need to be restricted due to illness. It is a poor technique, however, for determining the adequacy of a diet, because it does not take into account fluctuations in food intake or seasonal variations. Moreover, food intakes are often underestimated because the process relies on an individual's memory and reporting accuracy. People often forget to mention alcohol, soft drinks, snack foods, and desserts unless specifically prompted to do so, and some individuals find it embarrassing to report consumption of foods such as chocolate, butter, and red meat.[7]

Food Frequency Questionnaire A **food frequency questionnaire** surveys the foods and beverages regularly consumed during a specific time period. Some questionnaires are qualitative only: food lists contain common foods, organized by food group, with check boxes to indicate frequency of consumption. Other types of questionnaires provide semiquantitative information by including portion sizes as well. Figure 17-3 shows a sample section of a semiquantitative questionnaire that surveys fruit intake over the previous year. Because the respondent is often asked to estimate food intakes over a one-year period, the results should not be affected by seasonal changes in diet. Conversely, a disadvantage of this method is its inability to determine recent changes in food intake. In addition, food frequency questionnaires typically list only common foods, so food intake data are not as accurate as those obtained by dietary recall methods.

Simple versions of food frequency questionnaires focus on food categories relevant to a person's medical condition. For example, a questionnaire designed to evaluate calcium intake may include only milk products, fortified foods, certain fruits and vegetables, and dietary supplements that contain calcium. A computer analysis can then quickly estimate the individual's calcium intake and compare it with recommendations.

Food Record A **food record** is a written account of foods and beverages consumed during a specified time period, usually several consecutive days. Foods are recorded as they are consumed in order to obtain the most complete and accurate record possible; thus the process does not rely on memory. A detailed food record includes the types and amounts of foods and beverages consumed, times of consumption, and methods of preparation. For weight-management purposes, it may also include information about a person's mood (happy, stressed), the occasion (party, family meal), activities engaged in while eating (watching TV, driving to work), and daily physical activity. For establishing blood glucose control, the record may include information about medications, physical activity, and the results of blood glucose monitoring.

The food record provides valuable information about food intake, as well as a person's response to and compliance with medical nutrition therapy. Unfortunately,

food frequency questionnaire: a survey of foods routinely consumed. Some questionnaires ask about the types of food eaten and yield only qualitative information; others include questions about portions consumed and yield semiquantitative data as well.

food record: a detailed log of food eaten during a specified time period, usually several days; also called a *food diary*. A food record may also include information regarding disease symptoms, physical activity, and medication use.

FIGURE 17-3 Sample Section of a Food Frequency Questionnaire

FRUIT	HOW OFTEN									HOW MUCH			
	Never or less than once per month	1 per mon.	2–3 per mon.	1 per week	2 per week	3–4 per week	5–6 per week	Every day	MEDIUM SERVING	YOUR SERVING SIZE		
										S	M	L
EXAMPLE: Bananas	○	○	○	●	○	○	○	○	1 medium	○ 1/2	● 1	○ 2
Bananas	○	○	○	○	○	○	○	○	1 medium	○ 1/2	○ 1	○ 2
Apples, applesauce	○	○	○	○	○	○	○	○	1 medium or 1/2 cup	○ 1/2	○ 1	○ 2
Oranges (not including juice)	○	○	○	○	○	○	○	○	1 medium	○ 1/2	○ 1	○ 2
Grapefruit (not including juice)	○	○	○	○	○	○	○	○	1/2 medium	○ 1/4	○ 1/2	○ 1
Cantaloupe	○	○	○	○	○	○	○	○	1/4 medium	○ 1/8	○ 1/4	○ 1/2
Peaches, apricots (fresh, in season)	○	○	○	○	○	○	○	○	1 medium	○ 1/2	○ 1	○ 2
Peaches, apricots (canned or dried)	○	○	○	○	○	○	○	○	1 medium or 1/2 cup	○ 1/2	○ 1	○ 2
Prunes, or prune juice	○	○	○	○	○	○	○	○	1/2 cup	○ 1/4	○ 1/2	○ 1
Watermelon (in season)	○	○	○	○	○	○	○	○	1 slice	○ 1/2	○ 1	○ 2
Strawberries, other berries (in season)	○	○	○	○	○	○	○	○	1/2 cup	○ 1/4	○ 1/2	○ 1
Any other fruit, including kiwi, fruit cocktail, grapes, raisins, mangoes	○	○	○	○	○	○	○	○	1/2 cup	○ 1/4	○ 1/2	○ 1

food records require a great deal of time to complete, and people need to be highly motivated to keep accurate records. Another drawback is that the recording process itself may influence food intake. Furthermore, a reliable estimate of a person's overall nutrient intake is difficult to obtain when data are collected for just a few days or even a week, because of day-to-day and seasonal variations in food intake.

Direct Observation In facilities that serve meals, food intakes can be directly observed and analyzed. This method can also reveal a person's food preferences, changes in appetite, and any problems with a prescribed diet. Health practitioners use direct observation to conduct patients' **kcalorie counts** to determine the food energy (and often, protein) consumed by patients during a single day or several consecutive days. To perform a kcalorie count, the clinician estimates food intake by recording the dietary items that a patient is given at meals and subtracting the amounts remaining after meals are completed; this procedure allows an estimate of the kcaloric content of foods and beverages actually consumed. Although a useful means of discerning patients' intakes, direct observation requires regular and careful documentation and can be labor intensive and costly.

Food models and measuring utensils can help an individual visualize portion sizes.

Anthropometric Data

Anthropometric data ◆ can reveal problems related to both overnutrition and protein-energy malnutrition. Height (or length) and weight are the most widely used anthropometric measurements and help to evaluate growth in children and nutrition status in adults. Other helpful data include body composition tests (described in Chapter 8 and Appendix E) and circumferences of the head, waist, and limbs.

◆ Reminder: *Anthropometric* refers to physical measurements of the body.

Height (or Length) Poor growth in children can signify malnutrition. In adults, height measurements alone do not reflect current nutrition status but can be used for estimating a person's energy needs or appropriate body weight. Length is measured in infants and children younger than 24 months of age, and height is usually measured in older children and adults. ◆ Length can also be measured in adults and children who are unable to stand unassisted due to physical or medical reasons. The "How to" on p. 600 describes some standard techniques for measuring length and height.

◆ *Length* is measured while a person is recumbent (lying down), whereas *height* is measured while a person is standing upright.

In adults who are unable to stand, height can be estimated from equations that include either the knee height or the full-arm span, both of which correlate well with height.[8] Knee height extends from the top of the knee to the heel and is measured with a knee-height caliper while the individual is in a sitting position or lying flat on the back with the knee bent at a 90-degree angle. The full-arm span is the distance from the tip of one middle finger to the other when the arms are extended horizontally; it is measured most accurately when the measuring device is affixed to the wall and the subject is standing with the back to the wall and arms fully extended. In children with disabilities that affect stature, alternative measures of linear growth include the full-arm span, lower-leg lengths (knee to heel, similar to the knee-height measure), and upper-arm lengths (shoulder to elbow), which are compared with reference percentiles.

Body Weight During clinical care, health care providers must monitor body weights carefully: weight changes may reflect changes in hydration status, and involuntary weight losses may signify PEM. Body weights are typically compared with healthy ranges on height-weight tables and growth charts or used to calculate the body mass index (BMI). ◆ The "How to" on p. 600 includes suggestions for improving the accuracy of weight measurements.

◆ Reminder: $BMI = \dfrac{weight\ (kg)}{height\ (m)^2}$
A healthy BMI typically falls between 18.5 and 25.

Head Circumference A head circumference measurement helps to assess brain growth and malnutrition in children up to three years of age, although this measure is not necessarily reduced in a malnourished child. Head circumference values can also track brain development in premature and small-for-gestational-age infants. To measure head circumference, the assessor encircles the largest circumference measure

kcalorie counts: the determination of food energy (and often, protein) consumed by patients for one or more days.

HOW TO Measure Length and Height

To improve the accuracy of length and height measurements, keep the following in mind:

- Always measure—never ask! Self-reported heights are less accurate than measured heights. If height is not measured, document that the height is self-reported.
- Measure the length of infants and young children by using a measuring board with a fixed headboard and a movable footboard. It generally takes two people to measure length. One person gently holds the infant's head against the headboard; the other straightens the infant's legs and moves the footboard to the bottom of the infant's feet.
- Measure height next to a wall on which a nonstretchable measuring tape or board has been fixed. Ask the person to stand erect without shoes and with heels together. The person's eyes and head should be facing forward, with heels, buttocks, and shoulder blades touching the wall. Place a ruler or other flat, stiff object on the top of the head at a right angle to the wall, and carefully note the height measurement.
- Immediately record length and height measurements to the nearest $1/8$ inch or 0.1 centimeter.
- For evaluating growth rate in young children, use the appropriate growth chart (Appendix E) when plotting results. If length is measured, use the growth chart for children between 0 and 36 months; if height is measured, use the chart for individuals between 2 and 20 years.
- Higher values are obtained from supine measurements than from vertical height measurements due to gravity.

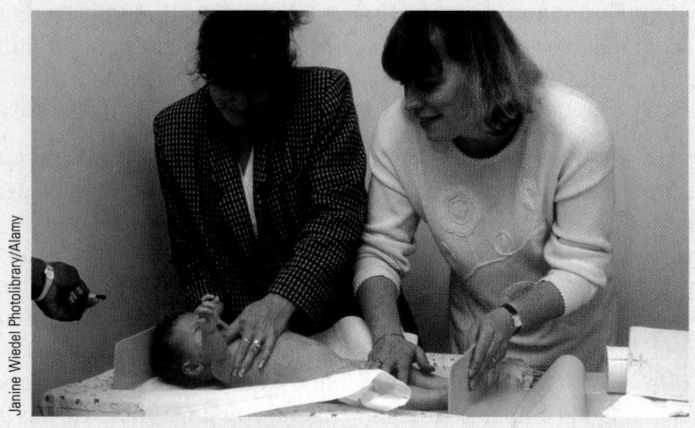

It takes two people to measure the length of an infant.

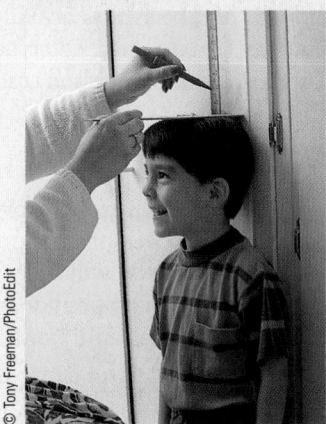

Standing erect allows for an accurate height measurement.

HOW TO Measure Weight

Tips for measuring weight include:

- Always measure—never ask! Self-reported weights are often inaccurate. If weight is not measured, document that the weight is self-reported.
- Valid weight measurements require scales that have been carefully maintained, calibrated, and checked for accuracy at regular intervals. Beam balance and electronic scales are the most accurate. Bathroom scales are inaccurate and inappropriate in the clinical setting.
- Measure an infant's weight with a scale that allows the infant to sit or lie down. The tray should be large enough to support an infant or young child up to 40 pounds, and the scale should weigh in $1/2$-ounce or 10-gram increments. For accurate results, weigh infants without clothes or diapers. Excessive movement by the infant can reduce accuracy.
- Children who can stand are weighed in the same way as adults, using beam balance or electronic scales with platforms large enough for standing comfortably. If repeated weight measurements are needed, each weighing should take place at the same time of day (preferably before breakfast), in the same amount of clothing, after the person has voided, and on the same scale. Record weights to the nearest $1/4$ pound or 0.1 kilogram.
- Special scales and hospital beds with built-in scales are available for weighing people who are bedridden.

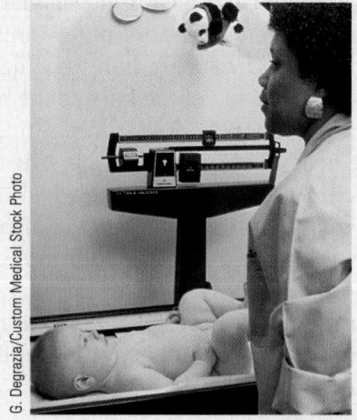

Infants are weighed on scales that allow them to sit or lie down.

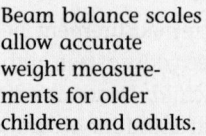
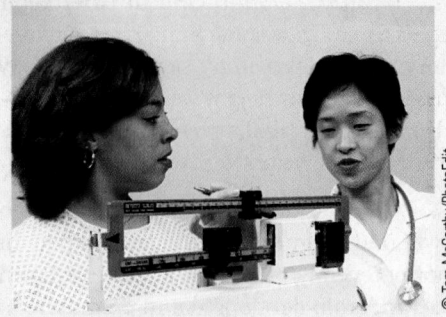

Beam balance scales allow accurate weight measurements for older children and adults.

of a child's head with a nonstretchable measuring tape: the tape is placed just above the eyebrows and ears and around the occipital prominence at the back of the head (see the photo). The measurement is read to the nearest $^1/_8$ inch or 0.1 centimeter.

Circumferences of Waist and Limbs The waist circumference correlates with abdominal fat and can help in assessing overnutrition (see Chapter 8, p. 263). Circumferences of the mid-upper arm, mid-thigh, and mid-calf regions can help in evaluating the effects of illness, aging, and PEM on skeletal muscle tissue. For improved accuracy, circumference measurements are often used together with skinfold measurements to correct for the subcutaneous fat in limbs.

Anthropometric Assessment in Infants and Children To evaluate growth patterns, periodic measurements of height (or length), weight, and head circumference are plotted on growth charts, such as those provided in Appendix E. ◆ The most commonly used growth charts compare height (or length) to age, weight to age, head circumference to age, weight to length, and BMI to age. Although individual growth patterns vary, a child's growth will generally stay at about the same percentile throughout childhood; a sharp drop in a previously steady growth pattern suggests malnutrition. Growth patterns that fall below the 5th percentile may also be cause for concern, although genetic influences must be considered when interpreting low values. Growth charts with BMI-for-age percentiles can be used to assess risk of underweight and overweight in children over two years of age: the 10th and 85th percentiles are used as cutoffs to identify children who may be malnourished or overweight, respectively.[9] Chapter 15 provides additional information about growth during infancy and childhood.

Anthropometric Assessment in Adults Weight and height are the primary anthropometric values monitored in adults. As mentioned previously, weight changes must be evaluated carefully during illness: although unintentional weight *loss* can indicate malnutrition, weight *gain* may result from fluid retention rather than overnutrition. Fluid retention often accompanies worsening disease in patients with heart failure, liver cirrhosis, and kidney failure, and it can mask the weight loss associated with PEM. In assessing the significance of weight loss, the rate should be considered as well as the amount: an involuntary weight loss of more than 10 percent within a six-month period suggests risk of PEM.[10] Note that certain types of medications can also contribute to weight changes.

To assess the degree of nutritional risk associated with illness, weight data are often expressed as "percent of ideal body weight" (%IBW) or "percent of usual body weight" (%UBW). Because a healthy body weight usually falls within a BMI range of 18.5 and 24.9, an "ideal weight" can be found using a BMI table or graph (see Figure 8-6, p. 260 or the inside back cover of this book).[11] The %IBW is not as useful as the %UBW for interpreting weight changes that occur in underweight, overweight, or obese individuals. In underweight patients, the %IBW can overestimate the degree of weight loss due to illness. In overweight or obese patients, the weight loss resulting from illness may be overlooked. ◆ General guidelines for estimating and evaluating %IBW and %UBW are provided in Table 17-9 and in the "How to" on p. 602.

Many of the illnesses discussed in later chapters are associated with losses in muscle tissue that resist nutrition intervention. In addition, losses in both muscle tissue and height are common with aging even though body weights may remain stable. Thus, health practitioners may utilize skinfold and limb circumference measurements to evaluate body composition changes that need to be addressed in the treatment plan.

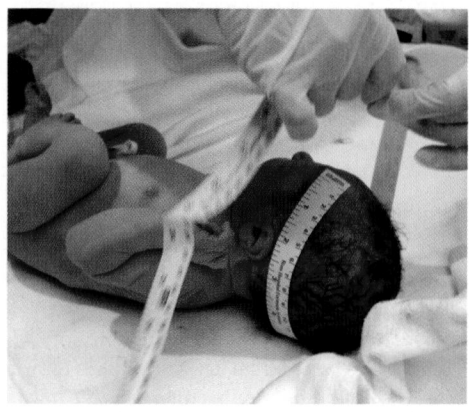

Head circumference measurements can help to assess brain growth.

◆ The Centers for Disease Control and Prevention provides complete sets of growth charts at its website: **cdc.gov/growthcharts/**.

◆ Reminder: Guidelines for assessing overweight and obesity were described in Chapter 8, p. 259.

Biochemical Data

Biochemical data provide information about protein-energy nutrition, vitamin and mineral status, fluid and electrolyte balance, and organ function. Most tests are based on analyses of blood or urine samples, which contain proteins, nutrients, and

TABLE 17-9	Use of Body Weight for Assessing Nutritional Risk	
%IBW	**%UBW**	**Nutritional Risk**
80–89	85–95	Risk of mild malnutrition
70–79	75–84	Risk of moderate malnutrition
<70	<75	Risk of severe malnutrition

◆ Blood test results are reported in terms of either *plasma* or *serum* levels. *Plasma* is the yellow fluid that remains after cells are removed; it still contains clotting factors. *Serum* is the fluid remaining after both cells and clotting factors are removed.

◆ Fluid retention can result in lab results that are deceptively low. Dehydration may cause lab results to be deceptively high.

◆ The term *half-life* defines the length of time that a substance remains in plasma. The albumin in plasma has a 3-week half-life, meaning that half of the amount circulating in plasma is degraded in a 3-week period.

metabolites that reflect nutrition and health status. Table 17-10 lists and describes common blood tests ◆ that have nutritional implications. Laboratory tests relevant to specific diseases will be discussed in the chapters that follow.

Interpreting laboratory values can be challenging because a number of factors influence the test results. For example, serum protein values can be affected by fluid imbalances, ◆ pregnancy, medications, and exercise. Similarly, serum levels of vitamins and minerals are often poor indicators of nutrient deficiency because the values are affected by multiple variables; therefore, a variety of tests are generally needed to diagnose a nutrition problem. Taken together with other assessment data, however, laboratory test results help to present a clearer picture than is possible to obtain otherwise.

Plasma Proteins Plasma protein levels can help in the assessment of protein-energy status, but the levels may fluctuate for other reasons as well.[12] For example, plasma proteins are synthesized in the liver, so plasma concentrations can reflect liver function. Metabolic stress alters plasma protein levels, because the liver increases its synthesis of some plasma proteins and reduces the synthesis of others. Values may also be influenced by pregnancy, kidney function, zinc status, and some medications. Because plasma proteins are affected by so many factors, their values must be considered along with other data to evaluate nutrition status. The following paragraphs describe several of the plasma proteins commonly measured during illness.

Albumin Albumin is the most abundant plasma protein, and its levels are routinely monitored during illness. Although many medical conditions influence albumin, it is slow to reflect changes in nutrition status because of its large body pool and slow rate of degradation. ◆ In people with chronic PEM, albumin levels remain normal for long periods of time despite depletion of body proteins; levels fall only after prolonged malnutrition. Likewise, when malnutrition is treated, albumin concentrations increase slowly, so albumin is not a sensitive indicator of effective treatment.

HOW TO Estimate and Evaluate %IBW and %UBW

To estimate %IBW, compare an individual's current weight with a reasonable (ideal) weight from a BMI table or other appropriate reference:

$$\%IBW = \frac{\text{current weight}}{\text{ideal weight}} \times 100$$

For example, suppose you wish to calculate the %IBW for a woman who is 5 feet 8 inches tall and weighs 116 pounds. The midpoint of the healthy BMI range is approximately 22, so using a BMI table (as shown on the inside back cover of this book), you estimate that a reasonable weight for this woman would be about 144 pounds:

$$\%IBW = \frac{116}{144} \times 100 = 80.6\%$$

The woman in this example weighs about 80.6 percent of her ideal body weight. A look at Table 17-9 indicates that at 80.6 percent of IBW, she may be mildly malnourished. Keep in mind that the calculation of "ideal body weight" is somewhat arbitrary, because the BMI table and various

other references provide a range of weights for individuals of a given height.

To estimate %UBW, compare a person's current weight with the weight that the person generally maintains:

$$\%UBW = \frac{\text{current weight}}{\text{usual weight}} \times 100$$

For example, if a man loses 32 pounds during illness and his usual weight is 180 pounds, his current weight would be 148 pounds. These values can be incorporated into the above equation:

$$\%UBW = \frac{148}{180} \times 100 = 82.2\%$$

The man in this example weighs 82.2 percent of his usual weight. A look at Table 17-9 shows that a person at 82 percent of UBW may be moderately malnourished.

TABLE 17-10 Routine Laboratory Tests with Nutritional Implications

This table presents a partial listing of some uses of commonly performed lab tests that have implications for nutritional problems.

Laboratory Test	Acceptable Range	Description
Hematology		
Red blood cell (RBC) count	Male: 4.3–5.7 million/μL Female: 3.8–5.1 million/μL	Number of RBC; aids anemia diagnosis.
Hemoglobin (Hb)	Male: 13.5–17.5 g/dL Female: 12.0–16.0 g/dL	Hemoglobin content of RBC; aids anemia diagnosis.
Hematocrit (Hct)	Male: 39–49% Female: 35–45%	Percentage RBC in total blood volume; aids anemia diagnosis.
Mean corpuscular volume (MCV)	80–100 fL	RBC size; helps to distinguish between microcytic and macrocytic anemias.
Mean corpuscular hemoglobin concentration (MCHC)	31–37% Hb/cell	Hb concentration within RBC; helps to distinguish iron-deficiency anemia.
White blood cell (WBC) count	4500–11,000 cells/μL	Number of WBC; general assessment of immunity.
Blood Chemistry		
Serum Proteins		
Total protein	6.4–8.3 g/dL	Protein levels are not specific to disease or highly sensitive; they can reflect body protein, illness or infections, changes in hydration or metabolism, pregnancy, or medications.
Albumin	3.4–4.8 g/dL	May reflect illness or PEM; slow to respond to improvement or worsening of disease.
Transferrin	200–400 mg/dL >60 yr: 180–380 mg/dL	May reflect illness, PEM, or iron deficiency; slightly more sensitive to changes than albumin.
Prealbumin (transthyretin)	10–40 mg/dL	May reflect illness or PEM; more responsive to health status changes than albumin or transferrin.
C-reactive protein	68–8200 ng/mL	Indicator of inflammation or disease.
Serum Enzymes		
Creatine kinase (CK)	Male: 38–174 U/L Female: 26–140 U/L	Different forms of CK are found in muscle, brain, and heart. High levels in blood may indicate heart attack, brain tissue damage, or skeletal muscle injury.
Lactate dehydrogenase (LDH)	208–378 U/L	LDH is found in many tissues. Specific types may be elevated after heart attack, lung damage, or liver disease.
Alkaline phosphatase	25–100 U/L	Found in many tissues; often measured to evaluate liver function.
Aspartate aminotransferase (AST, formerly SGOT)	10–30 U/L	Usually monitored to assess liver damage; elevated in most liver diseases. Levels are somewhat increased after muscle injury.
Alanine aminotransferase (ALT, formerly SGPT)	Male: 10–40 U/L Female: 7–35 U/L	Usually monitored to assess liver damage; elevated in most liver diseases. Levels are somewhat increased after muscle injury.
Serum Electrolytes		
Sodium	136–146 mEq/L	Helps to evaluate hydration status or neuromuscular, kidney, and adrenal functions.
Potassium	3.5–5.1 mEq/L	Helps to evaluate acid-base balance and kidney function; can detect potassium imbalances.
Chloride	98–106 mEq/L	Helps to evaluate hydration status and detect acid-base and electrolyte imbalances.
Other		
Glucose (fasting)[a]	74–106 mg/dL >60 yr: 80–115 mg/dL	Detects risk of glucose intolerance, diabetes mellitus, and hypoglycemia; helps to monitor diabetes treatment.
Glycosylated hemoglobin (HbA$_{1c}$)	5.0–7.5% of Hb	Used to monitor long-term blood glucose control (approximately 1 to 3 months prior).
Blood urea nitrogen (BUN)	6–20 mg/dL	Primarily used to monitor kidney function; value is altered by liver failure, dehydration, or shock.
Uric acid	Male: 3.5–7.2 mg/dL Female: 2.6–6.0 mg/dL	Used for detecting gout or changes in kidney function; levels affected by age and diet; varies among different ethnic groups.
Creatinine (serum or plasma)	Male: 0.7–1.3 mg/dL Female: 0.6–1.1 mg/dL	Used to monitor renal function.

[a]Fasting glucose levels that repeatedly exceed 100 mg/dL suggest prediabetes.

NOTE: μL = microliter; dL = deciliter; fL = femtoliter; ng = nanogram; U/L = units per liter; mEq = milliequivalents.

SOURCE: L. Goldman and D. Ausiello, eds., *Cecil Textbook of Medicine* (Philadelphia: Saunders, 2004).

◆ Transferrin's half-life in plasma is approximately 8 to 10 days.

◆ Half-lives of prealbumin and retinol-binding protein are 2 days and 12 hours, respectively.

Transferrin Transferrin is an iron-transport protein, and its concentrations can respond to both iron deficiency and PEM. Transferrin levels rise as iron status worsens and fall as iron status improves, so using transferrin values to evaluate protein-energy status is difficult if an iron deficiency is also present. Transferrin is degraded more rapidly than albumin, ◆ but its levels change relatively slowly in response to nutrition therapy.

Prealbumin and Retinol-Binding Protein Levels of prealbumin (also called transthyretin) and retinol-binding protein decrease rapidly during PEM and respond quickly to improved protein intakes. ◆ Thus these proteins are more sensitive than albumin to changes in protein status. Like other plasma proteins, their usefulness in nutrition assessment is limited because they are affected by a number of different factors, including metabolic stress, zinc deficiency, and various medical conditions. Prealbumin and retinol-binding protein are more expensive to measure than albumin, so they are not routinely included during nutrition assessment.

Medical Tests and Procedures

Nutrient deficiencies often impair physiological functions, so medical tests and procedures are sometimes helpful for evaluating the physiological changes that accompany malnutrition or disease.[13] For example, both PEM and zinc deficiency can depress immunity, which can be assessed by testing the skin's response to antigens that ordinarily cause redness and swelling when immune function is adequate. Alterations in metabolic rate associated with disease or malnutrition can be assessed using indirect calorimetry (see Chapter 8). Muscle weakness due to **wasting,** or loss of muscle tissue, can be assessed by hand-grip strength. The chapters that follow provide additional examples of medical tests or procedures that suggest declines in physiological functioning.

Physical Examinations

As with other assessment methods, interpreting physical signs of malnutrition requires skill and clinical judgment. Most physical signs are nonspecific; they can reflect any of several nutrient deficiencies, as well as conditions unrelated to nutrition. For example, cracked lips may be caused by several B vitamin deficiencies but may also be caused by sunburn, windburn, or dehydration. Dietary and laboratory data are usually needed as additional evidence to confirm suspected nutrient deficiencies.

Clinical Signs of Malnutrition Signs of malnutrition tend to appear most often in parts of the body where cell replacement occurs at a rapid rate, such as the hair, skin, and digestive tract (including the mouth and tongue). Table 17-11 lists some clinical signs of nutrient deficiencies. Many of the symptoms listed occur only in advanced stages of deficiency. The summary tables in Chapters 10 through 13 provide additional examples of the physical signs of nutrient imbalances.

Hydration State As mentioned previously, either fluid retention or dehydration may accompany some illnesses and may also result from the use of certain medications. Therefore, recognizing the patient's hydration state is necessary for the correct interpretation of blood tests and the body weight measurement.

Fluid retention (also called *edema*) can accompany malnutrition, infection, or injury. It can be caused by impaired blood circulation, and it frequently accompanies disorders of the heart and blood vessels, kidneys, liver, and lungs. Physical signs of fluid retention include weight gain, facial puffiness, swelling of limbs, abdominal distention, and tight-fitting shoes.

Dehydration can result from vomiting, diarrhea, sweating, fever, excessive urination, and skin injuries or burns (due to fluid loss through skin lesions). Dehydration risk is greatest in older adults, who have reduced thirst responses to water deprivation. Symptoms include thirst, dry skin or mouth, and reduced skin tension;

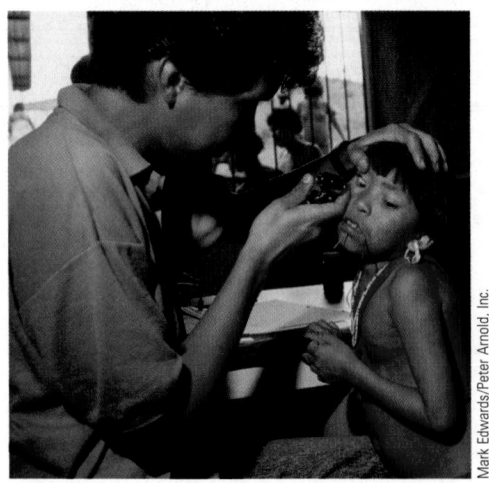

Mark Edwards/Peter Arnold, Inc.

Physical signs of malnutrition are often evident in parts of the body where the cells are replaced at a rapid rate.

wasting: the gradual atrophy (loss) of body tissues; associated with protein-energy malnutrition or chronic illness.

TABLE 17-11	Clinical Signs of Nutrient Deficiencies		
Body System	Acceptable	Signs of Malnutrition	Other Possible Causes
Hair	Shiny, firm in scalp	Dull, brittle, dry, loose; falls out (PEM); corkscrew hair (copper)	Excessive hair bleaching; hair loss from aging, chemotherapy, or radiation therapy
Eyes	Bright, clear pink membranes; adjust easily to light	Pale membranes (iron); spots, dryness, nightblindness (vitamin A); redness at corners of eyes (B vitamins)	Anemia, unrelated to nutrition; eye disorders; allergies
Lips	Smooth	Dry, cracked, or with sores in the corner of the lips (B vitamins)	Sunburn, windburn, excessive salivation from ill-fitting dentures or other disorders
Mouth and gums	Red tongue without swelling, normal sense of taste; teeth without caries; gums without bleeding, swelling, or pain	Smooth or magenta tongue (B vitamins), decreased taste sensations (zinc); swollen, bleeding gums (vitamin C)	Medications, periodontal disease (poor oral hygiene)
Skin	Smooth, firm, good color	Poor wound healing (PEM, vitamin C, zinc); dry, rough, lack of fat under skin (essential fatty acids, PEM, B vitamins); bruising, bleeding under skin (vitamins C and K)	Poor skin care, diabetes mellitus, aging, medications
Nails	Smooth, firm, pink	Ridged (PEM); spoon shaped, pale (iron)	
Other	—	Dementia, peripheral neuropathy (B vitamins); swollen glands at front of neck (PEM, iodine); bowed legs (vitamin D)	Disorders of aging (dementia), diabetes mellitus (peripheral neuropathy)

the urine color may be dark yellow or amber, and urine volume may be unusually low. The Case Study below can help you review the different components of a nutrition assessment.

CASE STUDY — Nutrition Screening and Assessment

Elise Walden is an 85-year-old retired businesswoman who has been a widow for 10 years. She uses a walker and has poorly fitting dentures. She was recently admitted to the hospital with pneumonia and also has congestive heart failure and diabetes. She routinely takes several medications to control blood glucose, hypertension, and heart function, and, in addition to these, the physician ordered antibiotics to treat the pneumonia. During an initial nutrition screening, Mrs. Walden stated that she had been eating very poorly over the past two weeks. She said that she usually weighs about 125 pounds—a fact that was documented in her medical chart from a previous visit. Although she felt she was losing weight, she didn't know how much weight she may have lost or when she started losing weight. Upon admission to the hospital, Mrs. Walden weighed 110 pounds and was 5 feet, 3 inches tall. Her serum albumin level was 3.0 grams per deciliter. A physical exam revealed edema, and several other laboratory tests confirmed that she was retaining fluid. As a result of the nutrition screening, Mrs. Walden was referred to a registered dietitian for a complete nutrition assessment.

1. From the brief description provided, which items in Mrs. Walden's medical, social, and diet histories might alert the dietitian that this patient is at risk of malnutrition?
2. Identify a healthy body weight for Mrs. Walden, and calculate her %IBW and %UBW. What do the results reveal? What effect does fluid retention have on Mrs. Walden's weight?
3. How can fluid retention alter Mrs. Walden's serum protein levels? What physical symptoms may have suggested that she was retaining excess fluid?
4. What tools can be used to estimate Mrs. Walden's usual food intake? What medical, physical, and social factors are likely to affect her dietary intake?
5. Describe other types of assessment information the dietitian may need before developing a nutrition care plan.

IN SUMMARY

Nutrition assessments typically include historical information, anthropometric and biochemical data, medical tests and procedures, and physical examinations. Health care providers assess food intake using 24-hour recall interviews, food frequency questionnaires, food records, and direct observation. Anthropometric measurements help to evaluate growth patterns, overnutrition and undernutrition, and body composition. Biochemical analyses help in the assessment of nutrient imbalances but are influenced by various other medical problems. Physical examinations can help the assessor detect signs of nutrient deficiency and fluid imbalances.

CENGAGENOW™
academic.cengage.com/login

Clinical Portfolio

■ Describe the potential nutritional implications of these findings from a patient's medical, personal, and social histories: age 78, lives alone, recently lost spouse, uses a walker, has no natural teeth or dentures, has a history of hypertension and diabetes, uses medications that cause frequent urination.

■ Calculate the %IBW and %UBW for a man who is 5 feet 11 inches tall with a current weight of 150 pounds and a usual body weight of 180 pounds. What additional information do you need to interpret the implications of his weight loss?

■ Nurses often shoulder much of the responsibility for collecting food intake data for kcalorie counts because they typically deliver food trays and snacks and later retrieve them. Why is it important to verify and record both what the patient receives (foods and amounts) and the foods that remain uneaten? When might patients be enlisted in the collection of food intake data, and when might such a course be unwise?

NUTRITION ON THE NET

For further study of topics covered in this chapter, log on to **academic.cengage.com/nutrition/rolfes/UNCN8e**. Go to Chapter 17, then to Nutrition on the Net.

• Learn about careers in nutrition and dietetics at the website of the American Dietetic Association: **www.eatright.org**

• Try the interactive version of the Mini Nutritional Assessment: **www.mna-elderly.com/clinical-practice.htm**

• Obtain food composition data from the USDA Nutrient Data Laboratory: **www.nal.usda.gov/fnic/foodcomp/search/**

• Analyze your diet by following the process described at this website: **www.mypyramid.gov**

STUDY QUESTIONS

CENGAGENOW™

To assess your understanding of chapter topics, take the Student Practice Test and explore the modules recommended in your Personalized Study Plan. Log on to **academic.cengage.com/login.**

These questions will help you review the chapter. You will find the answers in the discussions on the pages provided.

1. In what ways can illnesses affect nutrition status? Contrast the roles of the different health care professionals in providing nutrition care. (pp. 589–591)

2. Give examples of the methods used for screening patients for malnutrition. Explain how nutrition screening differs from a complete nutrition assessment. (pp. 591–593)

3. Discuss each of the steps of the nutrition care process. (pp. 593–595)

4. Give examples of the types of information included in medical, social, and diet histories. (pp. 595–596)

5. Describe the methods of gathering food intake data, and indicate the advantages and disadvantages of each process. (pp. 596–599)

6. What types of anthropometric measurements are included in nutrition assessments? Explain how these measurements help in the evaluation of nutrition status. (pp. 599–601)

7. How do biochemical analyses help to assess nutrition status? Give examples. What confounding factors may influence the results of blood tests? (pp. 601–605)

8. Give examples of physical signs that can suggest malnutrition. Describe the signs and symptoms that can result from fluid retention and dehydration. (pp. 604–605)

These multiple choice questions will help you prepare for an exam. Answers can be found on p. 608.

1. Mr. Hom experiences loss of appetite, difficulty swallowing, and mouth pain as a consequence of illness. Mr. Hom is at risk of malnutrition due to:
   a. altered metabolism.
   b. reduced food intake.
   c. altered excretion of nutrients.
   d. altered digestion and absorption.

2. The central role of nurses in health care makes them well positioned for:
   a. calculating patients' nutrient needs.
   b. providing medical nutrition therapy.
   c. conducting complete nutrition assessments.
   d. identifying patients at risk for malnutrition.

3. Of the following data collected during a nutrition screening, which item does not place the person at risk for malnutrition?
   a. having a health problem that is frequently associated with PEM
   b. the use of prescription medications that affect nutrient needs

   c. residing with a spouse in a middle-income neighborhood
   d. a significant reduction in food intake over the past five or more days

4. The nutrition care process is a systematic approach for:
   a. identifying the nutrient content of foods.
   b. ordering special diets.
   c. conducting nutrition screening.
   d. meeting the nutrition needs of patients.

5. To conduct complete nutrition assessments, dietitians rely on several sources of information, which include all of the following *except:*
   a. nutrition care plans.
   b. body measurements.
   c. medical, medication, and social histories.
   d. biochemical data.

6. Which dietary assessment method does a health practitioner use to conduct a kcalorie count?
   a. 24-hour recall interview
   b. food frequency questionnaire
   c. food record
   d. direct observation

7. The %UBW of a person who weighs 135 pounds and has a usual body weight of 150 pounds is:
   a. 111 percent.
   b. 90 percent.
   c. 86 percent.
   d. 74 percent.

8. A malnourished patient has just begun to eat after days without significant amounts of food. Which of the following blood tests would change most quickly as the patient's nutrition status improves?
   a. albumin
   b. transferrin
   c. serum electrolytes
   d. retinol-binding protein

9. Which sign of PEM would be unlikely to show up in a physical examination?
   a. low plasma protein levels
   b. dull, brittle hair
   c. poor wound healing
   d. wasted appearance

10. Fluid retention may cause all of the following effects *except:*
    a. lab results that are deceptively high.
    b. facial puffiness.
    c. lab results that are deceptively low.
    d. tight-fitting shoes.

REFERENCES

1. D. C. Heimburger, Adulthood, in M. E. Shils and coeditors, *Modern Nutrition in Health and Disease* (Baltimore: Lippincott Williams & Wilkins, 2006), pp. 830–842.
2. R. S. Gibson, *Principles of Nutritional Assessment* (New York: Oxford University Press, 2005), pp. 809–826.
3. K. Lacey and E. Pritchett, Nutrition care process and model: ADA adopts road map to quality care and outcomes management, *Journal of the American Dietetic Association* 103 (2003): 1061–1072.
4. American Dietetic Association, *Nutrition Diagnosis and Intervention: Standardized Language for the Nutrition Care Process* (Chicago: American Dietetic Association, 2007).
5. American Dietetic Association, 2007.
6. American Dietetic Association, 2007.
7. Gibson, 2005; L. C. Tapsell, V. Brenninger, and J. Barnard, Applying conversation analysis to foster accurate reporting in the diet history interview, *Journal of the American Dietetic Association* 100 (2000): 818–824.
8. Gibson, 2005.
9. K. M. Flegal, R. Wei, and C. Ogden, Weight-for-stature compared with body mass index-for-age growth charts for the United States from the Centers for Disease Control and Prevention, *American Journal of Clinical Nutrition* 75 (2002): 761–766.
10. S. B. Heymsfield and R. N. Baumgartner, Body composition and anthropometry, in M. E. Shils and coeditors, *Modern Nutrition in Health and Disease* (Baltimore: Lippincott Williams & Wilkins, 2006), pp. 751–770.
11. B. Shah, K. Sucher, and C. B. Hollenbeck, Comparison of ideal body weight equations and published height-weight tables with body mass index tables for healthy adults in the United States, *Nutrition in Clinical Practice* 21 (2006): 312–319.
12. H. P. Fuhrman, P. Charney, and C. M. Mueller, Hepatic proteins and nutrition assessment, *Journal of the American Dietetic Association* 104 (2004): 1258–1264.
13. Gibson, 2005.

ANSWERS

Study Questions (multiple choice)

1. b   2. d   3. c   4. d   5. a   6. d   7. b   8. d   9. a   10. a

Nutrition and Immunity

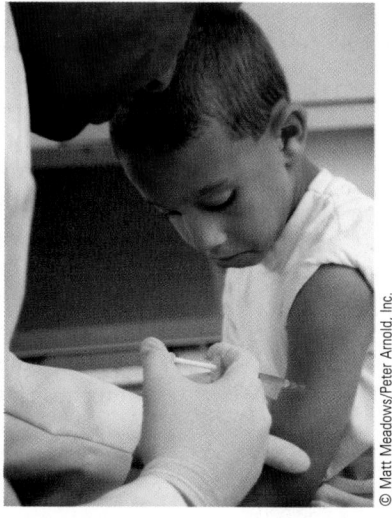

The **immune system** protects the body by fighting infectious agents and eliminating abnormal or "worn-out" cells. Its elaborate network of interacting cells and molecules works to block invading organisms from entering the body and destroys those that do gain entry. Substances that elicit an immune response are called antigens; common examples include foreign proteins produced by bacteria, viruses, parasites, or fungi. Because the immune system can usually distinguish between the body's cells and proteins and those of invading organisms, the body's own tissues are protected.

This highlight introduces the immune system and its relationships to malnutrition and illness; the glossary below defines relevant terms. Later chapters examine some of the relationships between specific illnesses and immune processes. Some diseases result from inadequate immune responses, as when infections spread, causing sepsis (Chapter 22), or when malignant cells develop into tumors (Chapter 29). Other conditions, such as inflammatory bowel diseases (Chapter 24) and atherosclerosis (Chapter 27), result from **inflammation** (Chapter 22). Most of the time, however, the immune system's carefully orchestrated actions are quietly working to preserve health.

GLOSSARY

acute-phase proteins: plasma proteins released from the liver at the onset of acute infection. An example is **C-reactive protein,** which is considered one of the main indicators of severe infection and has antimicrobial effects.

adaptive immunity: immunity that is specific for particular antigens; it adapts to antigens in an individual's environment and is characterized by "memory" for particular antigens. Also called **acquired immunity.**

allergen: any substance that triggers an inappropriate immune response.

allergy: an excessive and inappropriate immune reaction to a harmless substance.

autoimmune diseases: diseases characterized by an attack of immune defenses on the body's own cells.

B cell: a lymphocyte that produces antibodies.

cell-mediated immunity: immunity conferred by T cells and macrophages.

complement: a group of plasma proteins that assist the activities of antibodies.

cytokines (SIGH-toe-kines): signaling proteins produced by the body's cells; those produced by white blood cells regulate immune cell development and immune responses.

humoral immunity: immunity conferred by B cells, which produce and release antibodies into body fluids.

• **humor** = fluid

hypersensitivity: immune responses that are excessive or inappropriate. One type of hypersensitivity is *allergy.*

immune system: the body's defense system against foreign substances.

immunoglobulins (IM-you-no-GLOB-you-linz): large globular proteins produced by B cells that function as antibodies.

inflammation: a nonspecific response to injury or infection; a type of innate immune response.

innate immunity: immunity that is present at birth, unchanging throughout life, and nonspecific for particular antigens; also called **natural immunity.**

leukocytes: blood cells that function in immunity; also called **white blood cells.**

lymph (LIMF): the body fluid carried in lymphatic vessels; lymph is collected from the extracellular fluids of body tissues and ultimately transported to the bloodstream.

lymphatic vessels: vessels through which lymph travels.

lymphocytes (LIM-foe-sites): white blood cells that recognize specific antigens and therefore function in adaptive immunity; include *T cells* and *B cells.*

lymphoid tissues: tissues that have roles in immunity.

lysozyme (LYE-so-zyme): an enzyme with antibacterial properties; found in immune cells and body secretions such as tears, saliva, and sweat.

macrophages (MAK-roe-fay-jez): monocytes that have left circulation and settled in a tissue, where they serve as scavengers and activate the immune response.

monocytes (MON-oh-sites): cells released from the bone marrow that move into tissues and mature into macrophages.

natural killer cells: lymphocytes that confer nonspecific immunity by destroying a wide array of viruses and tumor cells.

neutrophils (NEW-tro-fills): the most common type of white blood cell. Neutrophils destroy antigens by phagocytosis.

phagocytes (FAG-oh-sites): white blood cells (primarily neutrophils and macrophages) that have the ability to engulf and destroy antigens.

• **phagein** = to eat

phagocytosis (FAG-oh-sigh-TOE-sis): the process by which phagocytes engulf and destroy antigens.

T cell: a lymphocyte that attacks antigens; functions in cell-mediated immunity.

FIGURE H17-1 The Lymphatic System

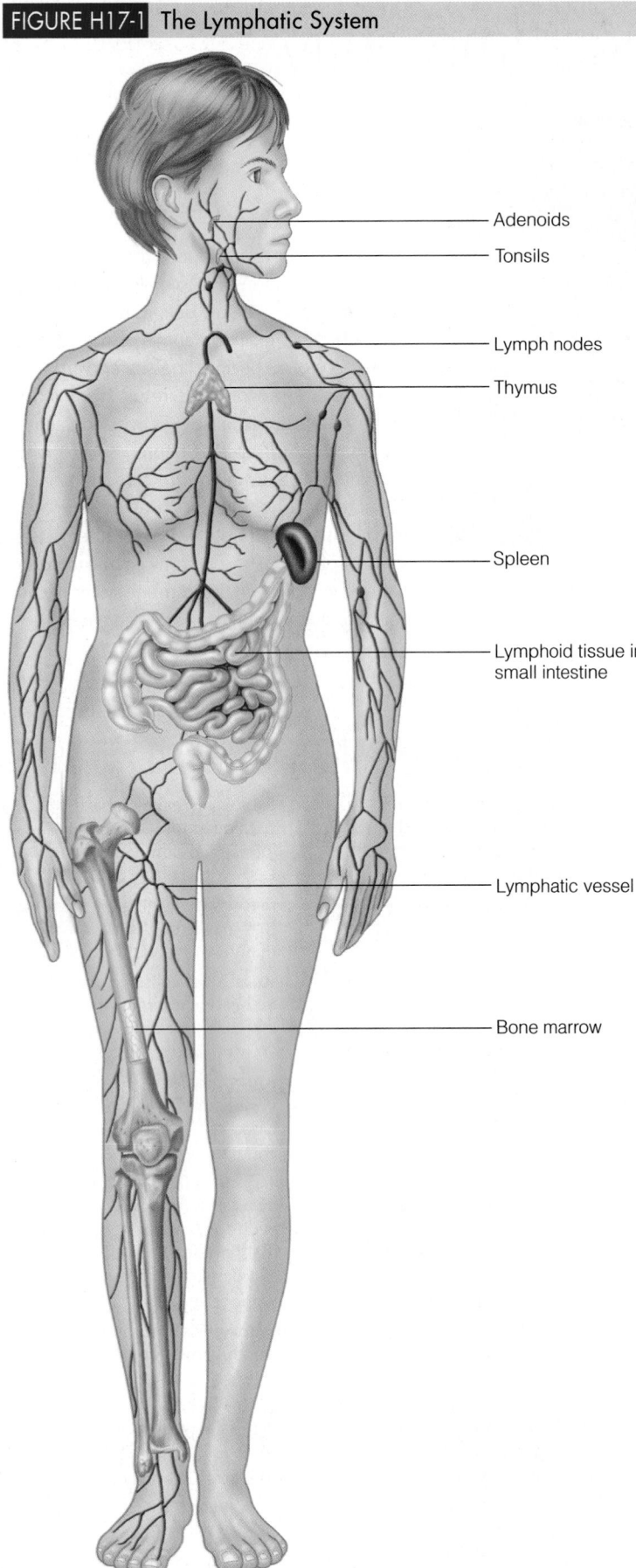

FIGURE H17-1 The Lymphatic System

- Adenoids
- Tonsils
- Lymph nodes
- Thymus
- Spleen
- Lymphoid tissue in small intestine
- Lymphatic vessel
- Bone marrow

Tissues of the Immune System

The immune system resides in no single organ, but depends on the physical and chemical interactions of a loosely organized network of cells and tissues scattered throughout the body.[1] The tissues and organs involved in immunity are collectively known as the lymphatic system (see Figure H17-1). **Lymphoid tissues** include the thymus gland and bone marrow (where **lymphocytes** are made), and the spleen, tonsils, adenoids, and lymph nodes (where foreign materials and debris are filtered out and discarded). Additional lymphoid tissue is dispersed in various locations throughout the body—especially within the mucosal linings of the gastrointestinal tract, the respiratory tract, and the genitourinary tract, where antigens are most likely to enter the body.

The cells active in immunity are the **leukocytes** (commonly known as **white blood cells**) and several types of accessory cells, as described in Table H17-1 and discussed in the following pages. These cells act by releasing chemicals such as enzymes, prostaglandins, and histamine, as well as proteins called **cytokines** that bind to receptors on target cells. White blood cells travel between the tissues and blood in **lymph,** a body fluid carried by the **lymphatic vessels.** Lymph is collected from the extracellular fluids that bathe tissues and is eventually transported to the bloodstream.

Examples of Innate Immunity

The immune protection present at birth is called **innate,** or **natural, immunity.** Innate immunity is nonspecific—it deters and destroys a wide range of pathogens. Nonspecific defenses include physical barriers to invading organisms, actions of defensive proteins, and activities of phagocytes and natural killer cells.

Physical Barriers to Infection

The body's first line of defense—the skin and mucous membranes—prevents the entry of infectious agents, which might otherwise gain easy access to tissues and blood. Skin not only provides an impenetrable physical barrier but also contains its own lymphoid tissue and a variety of immune cells interspersed in its outer layers. Mucous membranes lining the gastrointestinal, respiratory, and genitourinary tracts also act as barriers to infection: the mucous layers of these tissues trap microorganisms and prevent them from attaching to tissue surfaces.[2]

Microbes that arrive in the stomach face possible destruction from acidic gastric juices and enzymes. Those that survive enter the small intestine, where digestive secretions and

TABLE H17-1 Cells of the Immune System

White Blood Cells	Cell Type	Function
Lymphocytes	T cells	Activate macrophages Assist B cells Destroy virally infected cells
	B cells	Produce and secrete antibodies
	Natural killer cells	Destroy virally infected cells
Phagocytes	Monocytes/macrophages[a]	Present antigen fragments to T cells Engulf pathogens and cellular debris
	Neutrophils	Engulf pathogens and cellular debris
	Eosinophils	Release proteins that damage parasites Suppress inflammatory reactions
Accessory Cells	**Cell Type**	**Function**
Inflammatory mediators	Basophils	Release mediators that regulate inflammation
	Mast cells	Release mediators that regulate inflammation
	Platelets	Have primary role in blood clotting Release mediators that regulate inflammation

[a]Monocytes circulate in blood and become macrophages after they enter tissues.

specialized cells, antibodies, and lymphoid tissue protect against infection. The large intestine also contains defensive cells and antibodies, as well as stable bacterial populations that help to maintain mucosal tissue and create a hostile environment for invasive bacteria.[3]

Defensive Proteins

Proteins contribute to nonspecific immune defenses by serving as enzymes or signaling molecules. The liver releases **acute-phase proteins** in response to trauma, infection, or inflammation. Some acute-phase proteins, such as **C-reactive protein,** have antimicrobial activities that destroy some types of bacteria. C-reactive protein is considered a "marker" of acute inflammation and becomes elevated only when the body is fighting disease. Other acute-phase proteins include **complement,** a group of about 25 plasma proteins, so named because the proteins "complement" the activities of antibodies. When an antibody interacts with an antigen, a complex is formed that starts a series of reactions between the complement proteins. These actions may render microbes more susceptible to phagocytosis (described later), puncture a target cell's membrane, or help rid the body of antigen-antibody complexes. Another protein, **lysozyme,** attacks bacteria by breaking down carbohydrates on bacterial cell walls, causing the bacteria to burst.

Phagocytes

Upon entering the body, pathogens may encounter **phagocytes,** the scavenger cells of the immune system. Phagocytes engulf and digest bacteria, cellular debris (from damaged cells), and foreign particles in a process called **phagocytosis.** Phagocytes are attracted to their targets by the presence of common microbial products, complement fragments, or chemical signals produced by cells. They pull in their prey by extending pseudopods ("false feet") and then douse it with a mix of potent chemicals that include hydrolytic enzymes, lysozyme, and free radicals.

The two main types of phagocytes are neutrophils and macrophages. **Neutrophils** are the predominant leukocytes in the blood, making up about 50 to 65 percent of the total. They also have the shortest life spans, surviving only a day or two after they are released from bone marrow. Neutrophils migrate into tissues in response to injury or infection and accumulate in large numbers during the inflammatory process (discussed in Chapter 22). **Macrophages** are initially released from bone marrow as **monocytes;** after about a day in circulation, a monocyte migrates into one particular tissue, where it develops into a macrophage and survives for several months or longer. Each tissue has its own resident macrophages, and although their names may vary, they have similar functions in all tissues in which they reside. Examples of tissue macrophages include the Langerhans cells in the skin and the Kupffer cells in the liver.

Macrophages move and kill bacteria more slowly than neutrophils, but they are larger and can engulf larger targets, such as the body's dead and damaged cells. They also have the additional ability to display fragments of engulfed antigens on their cell surfaces for lymphocytes to recognize. This action triggers the immune responses of the lymphocytes, as described in a later section.

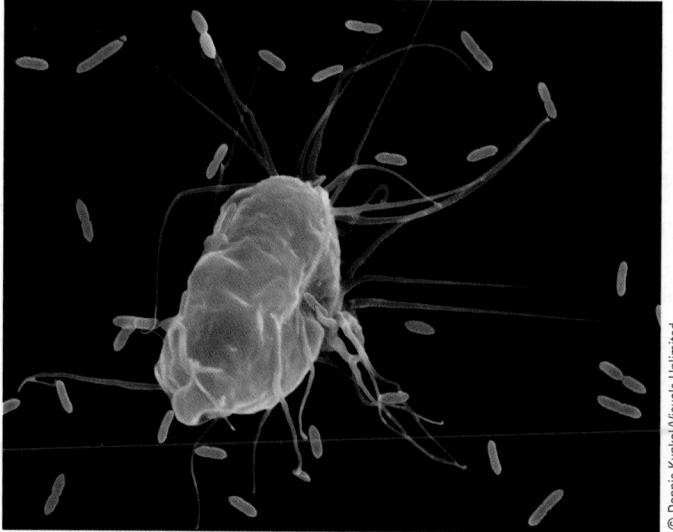

A macrophage extends pseudopods to pull in and engulf bacteria.

© Dennis Kunkel/Visuals Unlimited

Natural Killer Cells

Natural killer cells, which are members of the lymphocyte family, recognize and destroy virus-infected cells and tumor cells. These killer cells produce pore-forming proteins (called perforins) that puncture their target cells' membranes. The killer cells then transfer destructive enzymes into the damaged cells, further destroying their structures and encouraging self-destruction. Finally, phagocytes arrive at the scene to remove fragments left behind by the newly destroyed cells. The next section discusses the other members of the lymphocyte family, the B cells and T cells, which have critical roles in adaptive immunity.

Examples of Adaptive Immunity

In **adaptive,** or **acquired, immunity,** immune cells and proteins recognize *specific* pathogens as being foreign. Each **B cell** or **T cell** generates antibodies or receptors that can recognize only one type of antigen. Once activated, a lymphocyte produces other cells just like itself so that the newly formed army can attack the invading antigens and combat the infection. The lymphocytes are able to recognize a huge and diverse number of foreign molecules. Some of the lymphocytes serve as "memory cells," which survive for many years, enabling the immune system to respond rapidly if the same infection recurs.

B Cells

The B cells confer **humoral immunity,** so named because the cells' secretions, not the cells themselves, mount the defense within bodily fluids. B cells respond to antigens by producing antibodies that travel in the blood or tissue fluids to the site of infection. Antibodies, also known as **immunoglobulins,** are literally large globular proteins that provide immune protection. Each B cell expresses thousands of identical antibodies on its surface. Once an antigen binds, the B cell multiplies. Its daughter cells produce large numbers of the same antibody and secrete them into the surrounding fluids. The free antibodies then attach to the surfaces of antigens to neutralize them or make them an easy target for attack by phagocytes. The antibodies can also bind to viral proteins to prevent viruses from entering cells.

T Cells

T cells participate in **cell-mediated immunity,** so named because the cells themselves direct an immune response. A T cell has thousands of identical receptors on its cell surface (called T-cell receptors) that can recognize only one type of antigen. The antigens are displayed on the surfaces of antigen-presenting cells, specialized cells designed for this task (such as macrophages and B cells). After a *helper T cell* binds to an antigen fragment on an antigen-presenting cell, it recruits a *cytotoxic T cell* to the region to attack and destroy the local antigens. The actions of cytotoxic T cells are similar to those of natural killer cells: they perforate cell

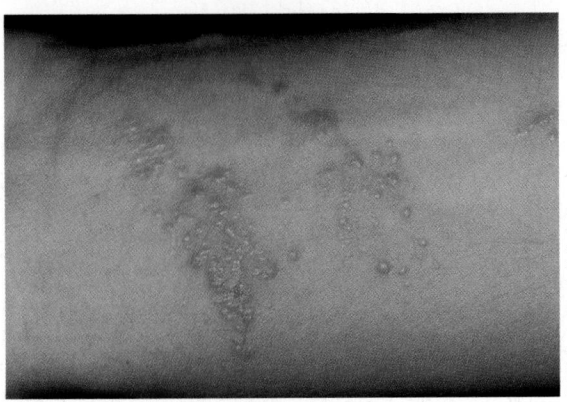

The rash that appears after contact with poison oak is an example of skin hypersensitivity.

© Bill Beatty/Visuals Unlimited

membranes and deliver powerful chemicals that eventually lead to a cell's destruction. Helper T cells can also activate B cells to produce antibodies and can activate macrophages to destroy the pathogens they have engulfed.

Undesirable Effects of Immunity

The body's immune function can sometimes create problems. Exaggerated or inappropriate immune reactions, referred to as **hypersensitivity,** can lead to discomfort or illness. **Allergy** is an example of an exaggerated response to an **allergen,** a harmless protein that may be eaten or inhaled. (Food allergy was introduced in Chapter 15 and is discussed further in Highlight 25.) As another example, the immune complexes formed from antigens and antibodies can cause damage to tissues if not readily cleared by phagocytes. **Autoimmune diseases,** including such familiar diseases as type 1 diabetes mellitus and pernicious anemia, develop when immune responses are mounted against the body's own cells. Although the effects of the immune system are lifesaving when directed at harmful pathogens, they can be life threatening when turned against the body.

Malnutrition and Immunity

Malnutrition affects all aspects of immunity, including both innate and adaptive immune defenses. For example, both PEM and vitamin A deficiency can result in damage to skin and mucous membranes, allowing microorganisms easier entry into the body. Deficiencies of protein and various micronutrients can affect the synthesis of hydrolytic enzymes, complement, antibodies, and other proteins important for immune function. Cell-mediated immunity is impaired in numerous ways by both PEM and zinc deficiencies.[4]

Because PEM is usually associated with multiple micronutrient deficiencies, it has been difficult for researchers to separate out the influences of individual nutrients. Nevertheless, zinc, iron, and vitamin A deficiencies are among the predominant micronutrient deficiencies worldwide, and a large body of research has demonstrated that each has a strong, independent influence on immunity. Notably, supplementation of various micronutrients (especially zinc and vitamin A) in malnourished populations has been found to reduce the incidence and severity of illness.[5]

Malnutrition and Infection

Because malnutrition impairs immune function in numerous ways, it is frequently associated with an increased risk of infection. In addition, an infection itself can worsen malnutrition (as described in the following section). The result is a downward spiral in immunity and overall health. Malnourished populations have higher-than-normal incidences of infectious diseases such as measles, malaria, acute respiratory infections, diarrheal diseases, and tuberculosis.[6] These diseases are major causes of morbidity and mortality in developing countries.

Infection and Nutrition Status

As mentioned, the effects of infection can be detrimental to nutrition status.[7] Anorexia often develops and is worse when an infection is severe, resulting in weight loss, negative nitrogen balance, and delayed growth and healing. Intestinal infections can cause nutrient malabsorption, atrophy of intestinal tissue, substantial blood loss, and diarrhea. Furthermore, infections generally stimulate metabolic processes, raising metabolic rate and nutrient needs as well. Chapter 22 delves further into the consequences of severe infection and discusses the nutrient needs of individuals who suffer from these conditions.

REFERENCES

1. P. C. Calder, Immunological parameters: What do they mean? *Journal of Nutrition* 137 (2007): 773S–780S.
2. P. Winkler and coauthors, Molecular and cellular basis of microflora-host interactions, *Journal of Nutrition* 137 (2007): 756S–772S.
3. Winkler and coauthors, 2007.
4. G. Fernandes, C. A. Jolly, and R. A. Lawrence, Nutrition and the immune system, in M. E. Shils and coeditors, *Modern Nutrition in Health and Disease* (Baltimore: Lippincott Williams & Wilkins, 2006), pp. 670–684; R. D. Semba, Nutrition and infection, in M. E. Shils and coeditors, *Modern Nutrition in Health and Disease* (Baltimore: Lippincott Williams & Wilkins, 2006), pp. 1401–1413.
5. N. S. Scrimshaw, Historical concepts of interactions, synergism and antagonism between nutrition and infection, *Journal of Nutrition* 133 (2003): 316S–321S; K. H. Brown, Diarrhea and malnutrition, *Journal of Nutrition* 133 (2003): 328S–332S.
6. Semba, 2006.
7. U. E. Schaible and S. H. E. Kaufmann, Malnutrition and infection: Complex mechanisms and global impacts, *PLoS Medicine* 4 (2007) e115, available at doi:10.1371/journal.pmed.0040115; G. T. Keusch, The history of nutrition: Malnutrition, infection and immunity, *Journal of Nutrition* 133 (2003): 336S–340S; Scrimshaw, 2003.

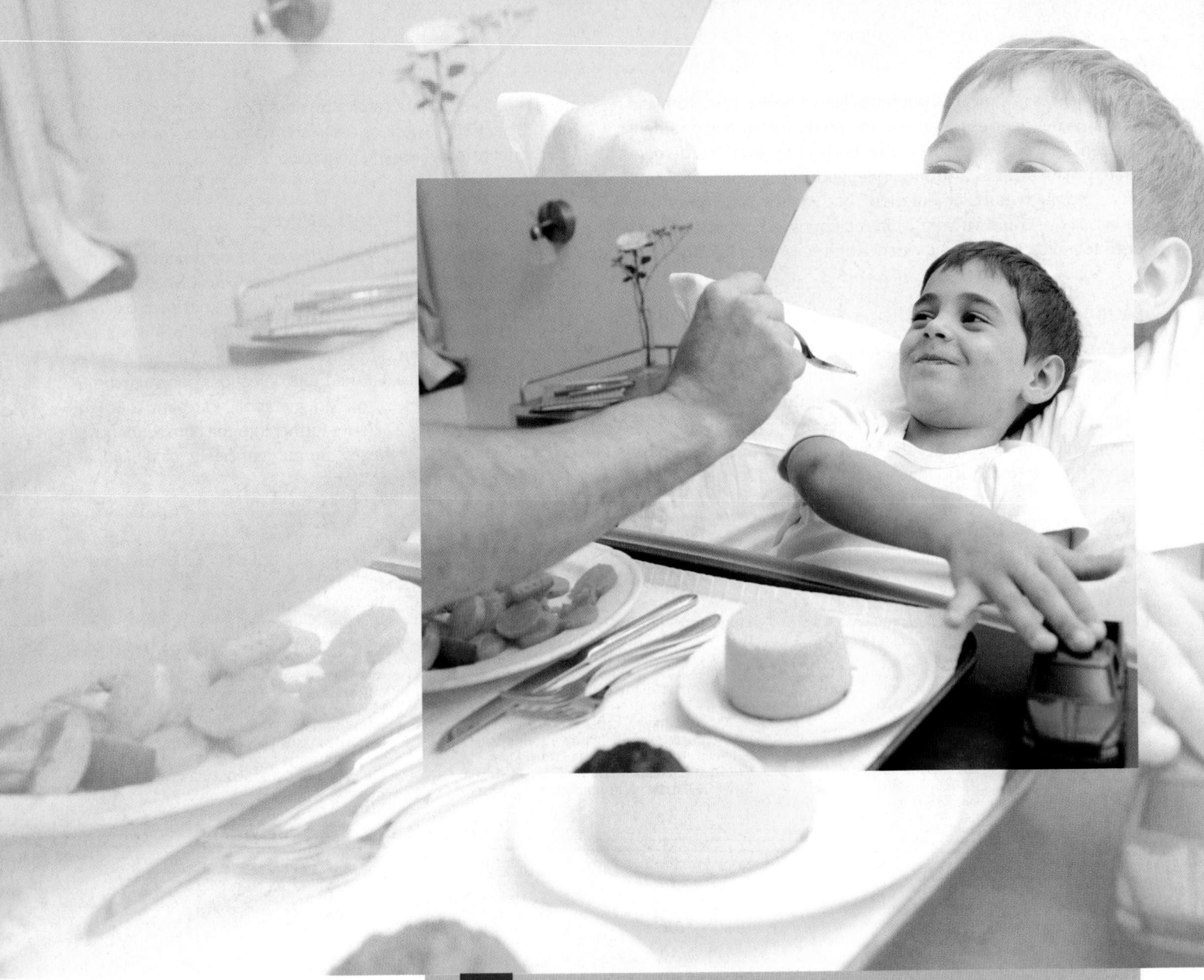

Nutrition in the Clinical Setting

When working with patients, remember to establish a caring environment. Use familiar language, maintain eye contact, and be a good listener. Showing your interest can go a long way toward winning a patient's trust. A patient may greet even an ideal dietary plan with resentment and bitterness, for it may restrict favorite foods and make it more difficult to forget about an illness. When their interactions with health practitioners are positive and encouraging, individuals are more likely to make the dietary changes that benefit health.

Nutrition Intervention

Chapter 17 discussed the interactions between illness and nutrition status and described the process of nutrition assessment. The results of the nutrition assessment allow dietitians to diagnose actual and potential nutrition problems. ◆ This chapter explains how dietitians and other health care professionals address nutrition problems and provide nutrition care. Ensuring that dietary needs are met is a key part of this process, so the chapter also describes methods for estimating energy requirements, common dietary modifications, and the foodservice provided in health care facilities.

◆ Nutrition diagnoses were discussed in Chapter 17, p. 594.

Implementing Nutrition Care

After formulating nutrition diagnoses, the dietitian determines the appropriate nutrition interventions. Table 18-1 on p. 616 shows how nutrition interventions are categorized.[1] Most nutrition interventions include a **nutrition prescription,** which provides specific dietary recommendations regarding food, nutrient, or energy intake or feeding method. Many interventions include nutrition education and counseling, which provide the knowledge, skills, and motivation that enable the patient to make necessary dietary and lifestyle changes. Some nutrition interventions require coordination with a number of other health professionals or facilities.

A nutrition intervention always includes two interrelated components: the planning process and the plan's implementation.[2] As Table 18-2 on p. 616 shows, the planning phase includes prioritizing the nutrition problems that were identified, determining their proper treatments, and setting goals. Implementing the plan involves communication with the patient, caregiver, and colleagues; carrying out the necessary treatments; and adjusting the plan when necessary.

Documenting Nutrition Care

Each step of the nutrition care process must be documented in the patient's medical record. The entries should be as succinct as possible so that they can be easily read

nutrition prescription: specific dietary recommendations related to food, nutrient, or energy intake or feeding method.

TABLE 18-1 Examples of Nutrition Interventions

Intervention	Examples
Food and/or nutrient delivery	Providing appropriate meals, snacks, and dietary supplements
	Providing specialized nutrition support (tube and intravenous feedings)
	Determining need for feeding assistance or adjustment in feeding environment
	Managing nutrition-related medication problems
Nutrition education	Providing basic nutrition-related instruction
	Providing in-depth training to increase dietary knowledge or skills
Nutrition counseling	Helping individual set priorities and establish goals
	Motivating individual to change behaviors
	Solving problems that interfere with the nutrition care plan
Coordination of nutrition care	Providing referrals or consulting other health professionals or agencies that can assist with treatment
	Organizing treatments that involve other health professionals or health care facilities
	Arranging transfer of nutrition care to another professional or location

SOURCE: American Dietetic Association, *Nutrition Diagnosis and Intervention: Standardized Language for the Nutrition Care Process* (Chicago: American Dietetic Association, 2007).

TABLE 18-2 Elements of Nutrition Interventions

Planning Nutrition Care

- Prioritizing nutrition diagnoses
- Consulting dietetics practice guidelines
- Reviewing the policies of the health care facility
- Determining specific dietary recommendations
- Conferring with the patient or caregivers
- Establishing goals and expected outcomes

Implementing the Nutrition Care Plan

- Documenting the nutrition care plan in the medical record
- Discussing the nutrition care plan with the patient or caregivers
- Individualizing treatment as warranted
- Continuing data collection and documentation
- Revising the nutrition care plan as warranted

SOURCE: American Dietetic Association, *Nutrition Diagnosis and Intervention: Standardized Language for the Nutrition Care Process* (Chicago: American Dietetic Association, 2007).

and quickly understood by the other members of the health care team. In addition, electronic (computerized) data systems, which have been widely adopted in the past decade, have standardized templates that require concise language. Before making entries in patients' medical records, health care professionals need to learn the particular charting methods preferred by their medical facility. The following sections describe some popular formats used for documenting nutrition care.[3]

ADIME Format The ADIME format closely reflects the steps of the nutrition care process. Each letter represents one of the steps: *Assessment, Diagnosis, Intervention, Monitoring,* and *Evaluation.* Using this format, the nutrition care plan would be recorded as follows:

- *Assessment.* The assessment section summarizes relevant assessment results, such as the medical problem, historical information, height, weight, BMI, laboratory test results, and relevant symptoms.
- *Diagnosis.* In the diagnosis section, the nutrition diagnoses are listed and prioritized.
- *Intervention.* The intervention section describes treatment goals and expected outcomes, specific interventions, and the patient's responses to nutrition care.
- *Monitoring and evaluation.* The monitoring and evaluation sections record the patient's progress, changes in the patient's condition, and adjustments in the care plan.

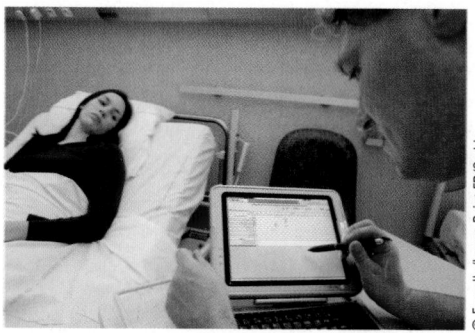

Many health care facilities maintain computerized medical records, which have standardized templates that require concise language.

SOAP Format The SOAP format is the oldest method used for documenting nutrition care and is still in popular use. The letters refer to Subjective, Objective, Assessment, and the Plan for care, and the nutrition care plan would be documented as follows:

- *Subjective.* The "Subjective" section includes assessment data obtained from the medical record and an interview with the patient or caregiver, including the chief medical problem and relevant symptoms.
- *Objective.* The "Objective" section includes objective assessment data, such as the results of biochemical analyses, anthropometric tests, medical procedures, and physical examinations.
- *Assessment.* A brief evaluation of the subjective and objective data and the nutrition diagnoses are presented in the "Assessment" section.
- *Plan.* Under "Plan" are recommendations that can help solve the problem, including the nutrition prescription, plans for nutrition education and counseling, and referrals to other professionals or agencies.

Figure 18-1 on p. 618 shows an example of a SOAP note, although there are many possible variations.

PES Statement The PES statement, introduced in Chapter 17 (see p. 594), is the general structure used for formatting nutrition diagnoses and can be used in any formatting style. The PES statement is so named because it includes the *Problem, Etiology* or cause of the problem, and the *Signs* and symptoms that provide evidence for the problem. The SOAP note in Figure 18-1 includes two PES statements.

Approaches to Nutrition Care

A nutrition care plan often involves significant dietary modifications. To ensure better compliance, the plan needs to be compatible with the desires and abilities of the person it is designed to help. The challenge is greater if dietary changes are required for extended periods.

Long-Term Dietary Intervention When long-term changes are necessary, a care plan must take into account a person's current food habits, lifestyle, and degree of motivation. Behavior change is a process that occurs in stages; therefore, more than one consultation is usually necessary. The following approaches may be helpful in implementing long-term dietary changes:[4]

- *Determine the individual's readiness for change.* Some people have little desire to change their dietary behaviors, and even individuals who are willing may not be fully prepared to take the necessary steps. The health practitioner needs to consider a patient's readiness to adopt new dietary behaviors before attempting to implement an ambitious care plan.
- *Emphasize what to eat, rather than what not to eat.* Emphasizing foods to include in the diet, rather than those to restrict, can make dietary changes

FIGURE 18-1 Example of a SOAP Note

SOAP NOTE

Patient Name: James Steiner   Date: Sept. 15, 2008

Age: 58   Gender: Male   Medical diagnosis: Hypercholesterolemia

Subjective:

Mr. Steiner recently learned of his hypercholesterolemia; wants to try dietary/lifestyle changes to reduce need for the medication. Reports frequent snacking and little time for exercise. Willing to attempt weight loss.

Objective:

Total cholesterol: 288 mg/dL   Height: 6'1"; Weight: 268 lb.
LDL-C: 214 mg/dL; HDL-C: 48 mg/dL   BMI: 35.4
Triglycerides: 132 mg/dL   Waist circumference: 45"

Assessment:

Abdominal obesity; analysis reveals intake of approximately 4200 kcal per day, about 1500 kcal above estimated needs; snack food choices are high in kcal and saturated fat.
Nutrition Diagnoses:
1. Obesity related to excess energy intake of 1500 kcal/day and physical inactivity as evidenced by BMI of 35.4
2. Undesirable food choices related to inadequate access to appropriate foods at work as evidenced by elevated body weight and LDL cholesterol

Plan:

Goal: 15 lb. weight loss over next 6 months.
Mr. Steiner to start 45-minute walking program, evenings.
Nutrition prescription: reduction of food intake to about 2400 kcal per day with about 30% kcal from fat, and 7% of kcal from saturated fat.
Initial education: appropriate food portions, low-kcal foods and snacks, food sources of saturated fat, pre-planning lunches at work.
Referral: Heart-healthy workshop on Sept. 22 (one week); Mr. Steiner to attend with wife.
Follow-up visit: Oct 15 (one month); Mr. Steiner to keep 3-day food record before visit; will identify appropriate food portions and between-meal snacks.

Form completed by: Genevieve Johnson, MPH, RD Position: Dietitian, Nutrition Services

more appealing. For example, encouraging additional fruits and vegetables is a more attractive message than telling the patient to restrict butter, cream sauces, and ice cream.

- *Suggest only one or two changes at a time.* People are more likely to adopt a dietary plan that does not deviate too much from their usual diet. If they succeed in adopting one or two changes, they are more likely to stick to the plan and be open to additional suggestions. Stricter plans may yield quicker results but are useful only for highly motivated people.

Nutrition Education Nutrition education allows patients to learn about the dietary factors that affect their particular medical condition. Ideally, this knowledge can motivate them to change their diet and lifestyle in order to improve their health status.

A nutrition education program should be tailored to a person's age, level of literacy, and cultural background. Learning style must also be considered: some people learn best by discussion supplemented with written materials, whereas others prefer visual examples, such as food models and measuring devices.[5] Information is provided in either one-on-one sessions or group discussions. The meeting should include an assessment of the person's understanding of the material and commitment to making changes. Follow-up sessions can reveal whether the person has successfully adopted a dietary plan. For example, a dietitian who counsels a woman who is lactose intolerant and hesitant to use milk products might proceed as follows:

- The dietitian provides sample menus of a nutritionally adequate diet that limits milk and milk products. Together, the dietitian and the woman design menus that consider her food preferences.

- The dietitian describes the types and amounts of milk products that are unlikely to cause symptoms and explains how to gradually incorporate these foods into the diet.

- Using diet analysis software, the dietitian demonstrates how altering intakes of calcium-containing foods changes a meal's calcium content.

- The dietitian explains how to use the Daily Values on food labels to estimate the calcium content of packaged foods.

- The dietitian provides information about the advantages and disadvantages of different calcium supplements.

- The dietitian assesses the woman's understanding by having her identify nonmilk products that are high in calcium.

Ideally, the dietitian would be able to monitor the woman's progress in a subsequent counseling session.

Follow-up Care For optimal success, dietitians should try to monitor the patient's progress and periodically evaluate the effectiveness of the nutrition care plan. Doing so usually involves comparing relevant outcome measures (such as the results of blood tests) with initial values and meeting with the patient to learn whether the plan has been satisfactory from the patient's point of view. Such follow-up efforts can reveal whether the care plan needs to be revised, as is often the case when a person's situation changes. For example, after a pregnant woman delivers her baby, she may need instructions on how to feed her infant or how to modify her diet to support lactation (if she is breastfeeding) so that she can return to a healthy body weight. If a follow-up meeting with a dietitian is not possible, a dietetic technician or other qualified health practitioner should provide additional guidance and education.

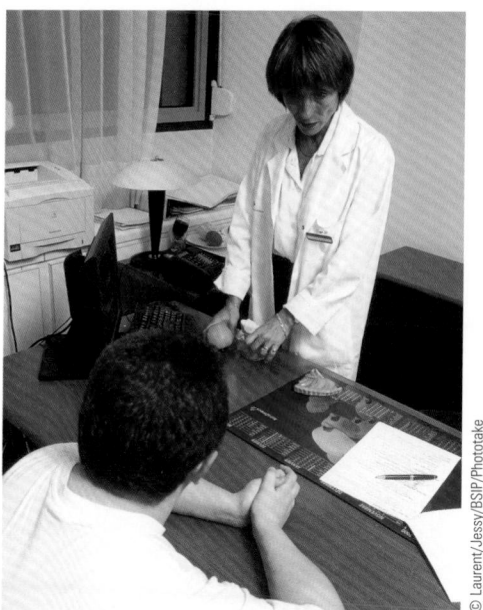

Dietary counseling requires sensitivity to cultural orientation, educational background, and motivation for change.

IN SUMMARY

Nutrition interventions are designed to correct the nutrition problems associated with illness. An intervention should take into account a person's food habits, lifestyle, cultural orientation, educational background, and degree of motivation. Each step of nutrition care should be clearly documented in the medical record; the ADIME and SOAP formats are popular styles of documentation. Nutrition education may be provided in an individual counseling session or group workshop. Nutrition care can be evaluated by reviewing relevant outcome measures of health status and determining the patient's understanding and acceptance of the intervention.

Determining Energy Requirements

To determine energy requirements for hospital patients, clinicians typically estimate the resting metabolic rate (RMR), and then adjust the RMR value with "stress factors" that account for medical problems and, in some cases, medical treatments. In ambulatory patients, a factor for activity level may also be necessary. The standard clinical procedure for determining RMR is indirect calorimetry, which measures oxygen consumption and carbon dioxide production (thereby determining kcalories burned) during a period of rest.[6] The procedure is labor intensive, so clinicians more often use

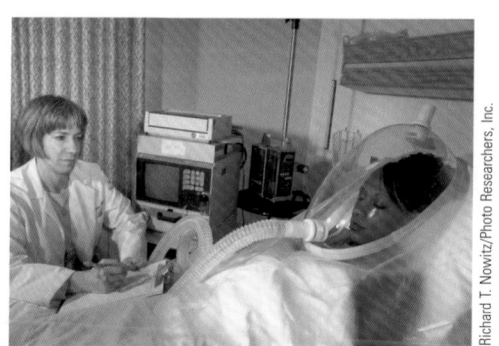

Indirect calorimetry is performed using equipment that analyzes the oxygen and carbon dioxide content of inhaled and exhaled air.

TABLE 18-3 Selected Equations for Estimating Resting Metabolic Rate (RMR)

Harris-Benedict[a]

Women:

RMR = 655.1 + [9.563 × weight (kg)] + [1.85 × height (cm)] − [4.676 × age (years)]

Men:

RMR = 66.5 + [13.75 × weight (kg)] + [5.003 × height (cm)] − [6.755 × age (years)]

Mifflin-St. Jeor

Women:

RMR = [9.99 × weight (kg)] + [6.25 × height (cm)] − [4.92 × age (years)] − 161

Men:

RMR = [9.99 × weight (kg)] + [6.25 × height (cm)] − [4.92 × age (years)] + 5

WHO/FAO/UNU[a]

Girls and women (age range, years):

10–18: RMR = [7.4 × weight (kg)] + [482 × height (m)] + 217

18–30: RMR = [13.3 × weight (kg)] + [334 × height (m)] + 35

30–60: RMR = [8.7 × weight (kg)] − [25 × height (m)] + 865

>60: RMR = [9.2 × weight (kg)] + [637 × height (m)] − 302

Men and boys (age range, years):

10–18: RMR = [16.6 × weight (kg)] + [77 × height (m)] + 572

18–30: RMR = [15.4 × weight (kg)] − [27 × height (m)] + 717

30–60: RMR = [11.3 × weight (kg)] + [16 × height (m)] + 901

>60: RMR = [8.8 × weight (kg)] + [1128 × height (m)] − 1071

[a]Although these equations are sometimes used for estimating basal metabolic rate (BMR), they were derived from data measured during resting conditions in most cases.

SOURCES: M. D. Mifflin and coauthors, A new predictive equation for resting energy expenditure in healthy individuals, *American Journal of Clinical Nutrition* 51 (1990): 241–247; World Health Organization, *Energy and Protein Requirements: Report of a Joint FAOAVHO/UNU Expert Consultation* (Geneva: World Health Organization, 1985); J. A. Harris and F. G. Benedict, A biometric study of human basal metabolism, *Proceedings of the National Academy of Sciences USA* 4 (1918): 370–373.

predictive equations that yield similar results. Table 18-3 lists several RMR equations in common use; the "How to" on p. 621 presents an example of this method.

In overweight and obese individuals who are not critically ill, the Mifflin-St. Jeor equation has been found to yield the most accurate results.[7] In other equations, adjusted body weights are sometimes used in place of actual body weights in an attempt to improve accuracy. For example, some research studies have suggested that the Harris-Benedict equation may be more appropriate for obese patients if the body weight used in the equation falls between an estimated ideal weight and the patient's actual weight. ◆ Other studies, however, have been unable to confirm the usefulness of body weight adjustments in predictive equations.[8]

Critical care patients may have energy needs that are considerably higher than normal due to fever, mechanical ventilation, restlessness, or the presence of open wounds. Patients who are critically ill are usually bedridden and inactive, however, so the energy needed for physical activity is minimal. Energy requirements for critical care patients are discussed further in Chapter 22.

◆ A method sometimes used for adjusting body weight:
Adjusted body weight = ideal weight + 0.25 (actual weight − ideal weight)

HOW TO Estimate the Energy Requirements of a Hospital Patient

To determine the energy requirements of a hospital patient, the health practitioner calculates the patient's resting metabolic rate (RMR) and then applies a "stress factor" to accommodate the additional energy needs imposed by illness. The stress factor 1.25 has been shown to be reasonably accurate for many hospitalized patients; other examples are listed in Table 22-2 on p. 713.

The following example uses the WHO/FAO/UNU equation (shown in Table 18-3) and the stress factor 1.25 to determine the energy needs of a 54-year-old female patient who is 5 feet 3 inches tall, weighs 115 pounds, and is confined to bed.

Step 1: The patient's weight and height are converted to the units used in the equation:

Weight in kilograms = 115 lb ÷ 2.2 lb/kg = 52.3 kg
Height in meters = 63 in. × 0.0254 m/in. = 1.6 m

Step 2: Using the WHO/FAO/UNU equation for estimating the RMR in women between 30 and 60 years old:

$$RMR = [8.7 \times weight (kg)] - [25 \times height (m)] + 865$$
$$= (8.7 \times 52.3) - (25 \times 1.6) + 865$$
$$= 455 - 40 + 865 = 1280 \text{ kcal}$$

Step 3: The RMR value is multiplied by the appropriate stress factor:

$$RMR \times stress factor = 1280 \times 1.25 = 1600 \text{ kcal}$$

Thus, an appropriate energy intake for this patient would be approximately 1600 kcal. Her weight should be monitored to determine if her actual needs are higher or lower.

For a patient who is not confined to bed, an additional activity factor can be applied to accommodate the extra energy needs. For example, if the patient in the example begins limited activity while in the hospital, an activity factor of 1.2 can be multiplied by the results obtained in Step 3:

$$1600 \times activity factor = 1600 \times 1.2 = 1920 \text{ kcal}$$

The activity factor for a hospitalized patient often falls between 1.1 and 1.4, and it is likely to change as the patient's condition improves.

IN SUMMARY

Energy requirements are typically estimated by multiplying a person's resting metabolic rate (RMR) by factors that account for the medical condition, medical treatments, and activity level. The RMR value can be obtained from indirect calorimetry or a predictive equation. Energy needs of critical care patients can be increased by fever, mechanical ventilation, restlessness, and open wounds.

Dietary Modifications

During illness, many patients can meet energy and nutrient needs by following a **standard diet.** Other patients may require a **modified diet,** which is altered by changing food consistency or nutrient content or by including or eliminating specific foods. If a patient's medical condition makes it difficult to meet nutrient needs orally, two options remain: *tube feedings* and *intravenous feedings.* This section introduces the use of modified diets and alternative feeding routes in clinical care. Later chapters describe other types of modified diets and additional dietary strategies for treating nutritional problems.

Modified Diets

Table 18-4 on p. 622 lists examples of modified diets that are often prescribed during illness.[9] Diets with altered texture and consistency are often prescribed for individuals with chewing and swallowing difficulties. Diets with modified nutrient or food content are frequently used to relieve disease symptoms or reduce the risk of developing complications. Some patients may have several medical problems and need a number of dietary changes. Keep in mind that modified diets should be adjusted to satisfy individual preferences and tolerances and may also need to be altered as a patient's condition changes.

standard diet: a diet that includes all foods and meets the nutrient needs of healthy people; also called a *regular diet.*

modified diet: a diet that is altered by changing food consistency or nutrient content or by including or eliminating specific foods; also called a *therapeutic diet.*

TABLE 18-4 Examples of Modified Diets

Type of Diet	Description of Diet	Appropriate Uses
Modified Texture and Consistency		
Mechanically altered diets	Contain foods that are modified in texture. Pureed diets include only pureed foods; mechanical soft diets may include solid foods that are mashed, minced, ground, or soft.	Pureed diets are used for people with swallowing difficulty, poor lip and tongue control, or oral hypersensitivity. Mechanical soft diets are appropriate for people with limited chewing ability or certain swallowing impairments.
Blenderized liquid diet	Contains fluids and foods that are blenderized to liquid form.	For people who cannot chew, swallow easily, or tolerate solid foods.
Clear liquid diet	Contains clear fluids or foods that are liquid at room temperature and leave minimal residue in the colon.	For preparation for bowel surgery or colonoscopy, for acute GI disturbances (such as after GI surgeries), or as a transition diet after intravenous feeding. For short-term use only.
Therapeutic Diets		
Fat-restricted diet	Restricts fat to low (<50 g/day) or very low (<25 g/day) levels in the diet.	For people who have certain malabsorptive disorders or symptoms of diarrhea, flatulence, or steatorrhea (fecal fat) resulting from dietary fat intolerance.
Fiber-restricted diet	Restricts fiber to low levels in the diet (<10 g/day).	For acute phases of intestinal disorders or to reduce fecal output before surgery. Not recommended for long-term use.
Sodium-restricted diet	Restricts sodium; degree of restriction depends on symptoms and disease severity.	To prevent fluid retention or induce fluid loss; used in hypertension, congestive heart failure, renal disease, and liver disease.
High-kcalorie, high-protein diet	Contains foods that are kcalorie and protein dense.	Used for increased kcalorie and protein requirements (in cancer, AIDS, burns, trauma, and other illnesses); also used to reverse malnutrition, improve nutrition status, or promote weight gain.

SOURCES: American Dietetic Association, *Nutrition Care Manual* (Chicago: American Dietetic Association, 2005); American Dietetic Association, *Manual of Clinical Dietetics* (Chicago: American Dietetic Association, 2000).

Mechanically Altered Diets Mechanically altered diets are helpful for individuals who have difficulty chewing or swallowing. Chewing difficulties usually result from dental problems. Impaired swallowing, or **dysphagia,** may result from neurological disorders, surgical procedures involving the head and neck, and various physiological or anatomical abnormalities that restrict the movement of food within the throat or esophagus. Dysphagia diets are highly individualized because swallowing problems can vary greatly. Furthermore, patients must be monitored regularly because swallowing ability can fluctuate over time. Chapter 23 provides details about the specific diets used for treating dysphagia.

Table 18-5 lists examples of foods often included in mechanically altered diets. Some diets may contain mostly pureed foods (*pureed diet*), whereas a less restrictive diet may include moist, soft-textured foods that easily form a bolus (*mechanical soft diet,* or simply, *soft diet*). Diets for people with chewing problems typically include foods that are ground or minced (*ground/minced diet*). Note that the foods used in these diets can overlap, and individual tolerances should ultimately determine whether foods are included or excluded.

Blenderized Liquid Diet Blenderized diets may be prescribed following oral or facial surgeries (for example, jaw wiring) or be recommended to individuals with chewing problems. Soft or tender foods that can be blenderized (often with added liquid) are available from all food groups, and they include cereals and breads; cooked vegetables; fresh or cooked fruits without skins and seeds; cooked, tender meats and fish; and potatoes, rice, and pasta. Foods that do not blend well should be excluded; exam-

dysphagia: difficulty swallowing.

TABLE 18-5	Foods Included in Mechanically Altered Diets

Depending on the feeding problem, a mechanically altered diet may include foods that are pureed, mashed, ground, minced, or soft textured. Foods vary according to tolerance.

Pureed Diets	Mechanical Soft Diets
Milk products: Milk, smooth yogurt, pudding	**Milk products:** Milk, yogurt with soft fruit, pudding, cottage cheese
Fruits: Pureed fruits and juices without pulp, skin, seeds, or chunks; well-mashed fresh bananas; applesauce	**Fruits:** Canned or cooked fruits without seeds or skin, fruit juices with small amounts of pulp, ripe bananas
Vegetables: Pureed cooked vegetables without seeds or skins, mashed potatoes, pureed potatoes with gravy	**Vegetables:** Soft, well-cooked vegetables that are not rubbery or fibrous; well-cooked, moist potatoes
Meats and meat substitutes: Pureed meats (with gravy), pureed casseroles (with broth), hummus or other pureed legume spread	**Meats and meat substitutes:** Ground, minced, or tender meat, poultry, or fish with gravy or sauce; tofu; well-cooked legumes; scrambled eggs
Breads and cereals: Smooth cooked cereals such as cream of wheat, slurried breads and pancakes,[a] pureed rice and pasta	**Breads and cereals:** Cooked cereals or moistened dry cereals with minimal texture, soft pancakes or breads, well-cooked noodles or dumplings in sauce or gravy

[a] Slurried foods are mixed with liquid until the consistency is appropriate; they may be gelled and shaped to improve their appearance.
SOURCE: American Dietetic Association, *Nutrition Care Manual* (Chicago: American Dietetic Association, 2005).

ples include nuts and seeds, dried fruits, sausage and frankfurters, hard cheeses, raw vegetables, corn, and celery.

Clear Liquid Diet Clear liquids, which require minimal digestion and are easily tolerated by the gastrointestinal (GI) tract, are often the foods recommended before some GI procedures (such as GI examinations, X-rays, or surgeries), after GI surgery, or after fasting or intravenous feeding. The **clear liquid diet** consists of clear fluids and foods that are liquid at body temperature and leave little undigested material (called *residue*) in the colon. Permitted foods include clear or pulp-free fruit juices, carbonated beverages, clear meat and vegetable broths (such as consommé and bouillon), fruit-flavored or unflavored gelatin, fruit ices made from clear juices, frozen juice bars, and plain hard candy. Although the clear liquid diet provides fluid and electrolytes, its nutrient and energy contents are extremely limited. If used for longer than a day or two, this diet should be supplemented with commercially prepared low-residue formulas that provide required nutrients. Figure 18-2 on p. 624 gives an example of a one-day clear liquid menu.

Sometimes a **full liquid diet,** a liquid diet that is not limited to clear liquids, is used as a transitional diet between liquids and solid foods. In addition to clear liquids, a full liquid diet may include milk, eggnog, cream soups, and thin cereal gruels. Because the diet contains milk products, it may be inappropriate for patients with significant lactose intolerance. Moreover, a gradual progression from clear liquids to solid foods is generally unnecessary, ◆ so the usefulness of this diet is in question.

Fat-Restricted Diet A fat-restricted diet is recommended for reducing the symptoms of fat malabsorption, which frequently accompanies diseases of the liver, gallbladder, pancreas, and intestines. Fat restriction may also alleviate the symptoms of heartburn. Although fat intake is occasionally limited to as little as 25 grams daily, it should not be restricted more than necessary, because fat is an important source of kcalories. Chapter 24 provides additional information about fat-restricted diets.

Most foods included in a fat-restricted diet provide less than 1 gram of fat per serving. The diet includes fat-free milk products, most breads and cooked grains, fat-free broths and soups, vegetables prepared without fats, most fruits, and fat-free candies and sweets (see Table 24-5 on p. 762). Restricted foods include low-fat and whole-milk products, baked products with added fat (like muffins), and most

◆ A change in diet as a patient's food tolerance improves is called *diet progression.*

clear liquid diet: a diet that consists of foods that are liquid at body temperature, require minimal digestion, and contribute limited residue (undigested material) in the colon.

full liquid diet: a liquid diet that includes clear liquids, milk, yogurt, ice cream, and liquid nutritional supplements (such as Ensure).

FIGURE 18-2 Menu—Clear Liquid Diet

✻ SAMPLE MENU ✻

Breakfast	Strained orange juice
	Flavored gelatin
	Ginger ale
	Coffee or tea, sugar
Lunch	Bouillon or consommé
	Flavored gelatin
	Frozen juice bars
	Apple or grape juice
	Coffee or tea, sugar
Supper	Bouillon or consommé
	Flavored gelatin
	Fruit ice
	Cranberry juice
	Coffee or tea, sugar
Snacks	Soft drinks
	Fruit ices
	Hard candy

◆ Specific information about the fiber content of foods can be found in Chapter 4 (pp. 122–125) and Appendix H.

◆ The average sodium intake in the United States is approximately 3400 mg per day. The sodium UL was set at 2300 mg to help prevent hypertension.

prepared desserts. Lean meat and meat substitutes are permitted but may be restricted to 4 to 6 ounces per day, depending on the degree of restriction. Some patients with malabsorptive conditions cannot tolerate large amounts of lactose or dietary fiber, so foods that include these substances may also need to be excluded from the diet.

Fiber-Restricted Diet Fiber restriction is recommended during acute phases of intestinal disorders, when the presence of fiber may exacerbate intestinal discomfort or cause diarrhea or blockages. Fiber-restricted diets are sometimes used before surgery to minimize fecal volume and after surgery during transition to a regular diet. Long-term fiber restriction is discouraged, however, because it is associated with constipation, diverticulosis, and other problems.

Fiber-restricted diets often eliminate whole-grain breads and cereals, nuts and seeds, raw and dried fruits, berries, dried beans and peas, chunky peanut butter, winter squash, and most raw vegetables. ◆ If required, even greater reductions in colonic residue can be achieved by following a **low-residue diet,** which excludes most fruits and vegetables, foods high in resistant starch (see p. 107), milk products that contain significant lactose, and foods that contain fructose or sugar alcohols (such as sorbitol). These foods contribute to colonic residue because some of their nutrients may be poorly digested (such as the lactose in milk) or poorly absorbed (such as sorbitol and fructose). Note that the terms "low-fiber diet" and "low-residue diet" are often used interchangeably.

Sodium-Restricted Diet Sodium restriction can help to prevent or correct fluid retention and is often recommended for treatment of hypertension, congestive heart failure, kidney disease, and liver disease. The degree of restriction depends on the illness, the severity of symptoms, and the specific drug treatment prescribed. In most cases, sodium is restricted to 2000 or 3000 milligrams daily, although more severe restrictions may be used in the hospital setting. Many patients find it difficult to comply with sodium restrictions, so while the sodium recommendation is an attempt to improve the patient's medical problem, it may still exceed the tolerable upper intake level (UL) for sodium of 2300 milligrams. ◆

A sodium-restricted diet limits the use of salt (both in cooking and at the table), eliminates most prepared foods and condiments, and limits consumption of milk and milk products (if excessive). Because so many processed foods are high in sodium, people following a sodium-restricted diet should check food labels and consume only low-sodium products. Sodium restriction is difficult to implement on a long-term basis because many people find low-sodium diets unpalatable and fail to adhere to them. Additional information about sodium restriction is provided in Chapters 27 and 28.

High-kCalorie, High-Protein Diet The high-kcalorie, high-protein diet is used to increase kcalorie and protein intakes in patients who have unusually high requirements or in those who are eating poorly. High-fat foods are added to increase energy intakes; consequently, the diet may exceed 35 percent kcalories from fat. Consuming small, frequent meals and commercial liquid supplements (such as Ensure or Boost) can also help a patient meet increased energy, protein, and nutrient needs.

Examples of foods included in high-kcalorie, high-protein diets are listed in Table 18-6. Some of these foods are high in saturated fat, which is limited in heart-healthy diets. These foods are used liberally in diets for malnourished patients to help correct their immediate nutrition problems—weight loss and muscle wasting. Chapter 29 offers additional suggestions for increasing the kcalorie and protein contents of meals.

Alternative Feeding Routes

In most cases, patients meet their nutrient needs by consuming regular foods. If their nutrient needs are high or their appetites poor, liquid supplements can be added to their diets to improve their intakes. Sometimes, however, a person's med-

low-residue diet: a diet low in fiber and other food constituents that contribute to colonic residue.

TABLE 18-6 Foods Included in High-kCalorie, High-Protein Diets

Milk products	Whole milk, half-and-half, cream Cheese Milk shakes, eggnog Ice cream, whipped cream
Fruits	Dried fruit Canned fruit in heavy syrup Avocado
Vegetables	Vegetables prepared with butter, margarine, sour cream, mayonnaise, or salad dressing Cream of vegetable soups

Meats and high-protein foods	All meats, fish, and poultry, including bacon, frankfurters,
	    and luncheon meats; eggs
	All meats, prepared fried or covered in cream sauces and
	    gravies
	Nuts and seeds, peanut and other nut butters, coconut

Breads and cereals	Granola and dry cereals prepared with whole milk or cream and dried fruit Hot cereals with whole milk or cream, or added fat Pasta, rice, and potatoes with added fat Pancakes, waffles, French toast

ical condition makes it difficult to meet nutrient needs orally. Two options remain: **tube feedings** and **intravenous feedings,** described more fully in Chapters 20 and 21.

- *Tube feedings.* Nutritionally complete formulas can be delivered through a tube placed directly into the stomach or intestine. Tube feedings are preferred to intravenous feedings if the GI tract is functioning. For example, a person in a coma is unable to eat but may be able to digest foods and absorb nutrients normally. In this situation, a tube feeding would be the appropriate option.
- *Intravenous feedings.* A person's medical condition sometimes prohibits the use of the GI tract to deliver nutrients. If the person is malnourished and the GI tract cannot be used for a significant period of time, intravenous feedings (also called *parenteral nutrition*) can meet nutritional needs.

Nothing by Mouth (NPO)

An order to not give a patient anything at all—food, beverages, or medications—is indicated by NPO, an abbreviation for *non per os,* meaning "nothing by mouth." For example, an order may read "NPO for 24 hours" or "NPO until after X-ray." The NPO order is commonly used during certain acute illnesses or diagnostic tests involving the GI tract.

IN SUMMARY

Diets prescribed during illness can be modified in consistency, nutrient content, or food content. Mechanically altered diets are used for people with swallowing and chewing difficulties. Clear liquid diets may be used briefly after acute gastrointestinal disturbances or intravenous feedings or before various diagnostic tests. Some medical conditions may require the restriction of specific nutrients, such as fat, fiber, or sodium. A high-kcalorie, high-protein diet may help to prevent or reverse malnutrition, improve nutrition status, or promote weight gain. In some cases, nutrients need to be delivered via tube feedings or intravenously.

tube feedings: liquid formulas delivered through a tube placed in the stomach or intestine.

intravenous feedings: the provision of nutrients through a vein, bypassing the intestine; also called **parenteral nutrition.**

Foodservice departments strive to prepare appetizing and nutritious meals and may accommodate dozens of special diets.

Foodservice

The work of a foodservice department can appear deceptively simple: appropriate foods are delivered to patients who need specific types of diets. Behind the scenes, however, a complex system is at work. A foodservice department faces a daily challenge in planning, producing, and delivering hundreds of nutritious meals and accommodating dozens of special diets and food preferences.

Although this discussion focuses on the foodservice in hospitals, much of the information applies to the foodservice in any health care facility, including nursing homes, assisted living centers, rehabilitation centers, and residential mental health care facilities. An important difference between hospitals and long-term health care facilities deserves mention, however. When patients in hospitals eat poorly, they can make up for nutrient deficits by eating well when they return home. Residents of a long-term care facility do not have that option. For this reason, foodservice departments in long-term care facilities must make even greater efforts to ensure that their patients receive and consume nutritious foods.

Menu Planning

When designing menus for modified diets, the dietary and foodservice personnel refer to a **diet manual,** which details the exact foods or preparation methods to include or exclude in a modified diet. The diet manual may also outline the rationale and indications for use of the diets and include sample menus. The manual may be compiled by the dietetics staff or adopted from another health care facility or a dietetics organization.

Food Selection

Most hospitals provide **selective menus** from which patients can select their meals. A patient who must follow a modified diet receives menus that include only the foods specified in the hospital's diet manual for that particular diet. By allowing a choice, however, this system ensures that patients will receive the foods they prefer and are most likely to eat. An added advantage is that patients can become familiar with the modified diets as they select foods from the appropriate menus. Examples of selective menus are shown in Figure 18-3.

In hospitals that provide selective menus, patients may need to make menu selections a day or two in advance so that the foodservice department can estimate the amounts and types of food they need to purchase and prepare. Each menu identifies the patient and room number, the meal (breakfast, lunch, or supper), type of diet, and the day the food will be served. Menus are usually color-coded by diet, which helps to ensure that foodservice employees put the right foods on food trays. Color-coding also helps the person delivering the tray confirm that the right diet was delivered.

Sometimes a menu is not marked correctly or is misplaced, in which case the patient may receive a meal selected by the foodservice department. Other potential problems that may arise when selective menus are used include the following:

- Patients may have difficulty seeing, reading, understanding, or physically marking menus.
- Patients may not understand that their selections will be for the next (or another) day.

diet manual: a resource that specifies the foods allowed and restricted in modified diets and provides sample menus.

selective menus: menus that provide choices in some or all menu categories.

FIGURE 18-3 Sample Lunch Menus

LOW-FAT/LOW CHOLESTEROL/CARDIAC SUNDAY

✵∽ Lunch ∽✵

LF = Low Fat   LSLF = Low Sodium, Low Fat

Meats

LSLF Baked chicken   LSLF Baked fish (cod)

Starchy Vegetables

LSLF Rice   LSLF Boiled potatoes

Vegetables

LSLF Baby carrots   LSLF Green beans

Soup/Salad/Juice	Dressings
LSLF Coleslaw	Diet French
Gelatin	Diet Thousand Island
Tomato soup	Diet Italian
Tossed salad	

Desserts

Pears   Fresh fruit

Breads

LF Dinner roll	Bran bread
White bread	LS Crackers
Wheat bread	

Beverages & Condiments

Coffee	Creamer
Decaf. coffee	Sugar
Hot tea	Sugar substitute
Decaf. hot tea	Herb seasoning
Iced tea	Lemon
Buttermilk	Margarine
Fat-free milk	Mustard
	Diet mayonnaise
	Catsup

Name _____   **Room** _____

LOW SODIUM	SUNDAY

✵∽ Lunch ∽✵

LF = Low Fat   LSLF = Low Sodium, Low Fat

Meats

LSLF Baked chicken   LSLF Baked fish (cod)

Starchy Vegetables

LSLF Rice   LSLF Boiled potatoes

Vegetables

LSLF Baby carrots   LSLF Green beans

Soup/Salad/Juice	Dressings
LSLF Coleslaw	Diet French
LS Chicken broth	Diet Thousand Island
Apple juice	Diet Italian
Tossed salad	

Desserts

Pears   Fresh fruit

Breads

Dinner roll	Bran bread
White bread	LS Crackers
Wheat bread	

Beverages & Condiments

Coffee	Sugar
Decaf. coffee	Sugar substitute
Hot tea	Creamer
Decaf. hot tea	Lemon
Iced tea	Herb seasoning
Whole milk	Margarine
2% milk	Diet mustard
Fat-free milk	Diet mayonnaise
No salt	Diet catsup

Name _____   **Room** _____

RENAL	SUNDAY

✵∽ Lunch ∽✵

LF = Low Fat   LSLF = Low Sodium, Low Fat

Meats

LSLF Baked chicken   LSLF Baked fish

Starchy Vegetables

LSLF Rice   LSLF Dialyzed potatoes

Vegetables

LSLF Baby carrots   LSLF Green beans

Soup/Salad/Juice	Dressings
Lemonade	Diet French
LSLF Coleslaw	Diet Thousand Island
Tossed salad	Diet Italian
(no tomato)	

Desserts

Pears   Apple pie

Breads

Dinner roll	Bran bread
White bread	LS Crackers
Wheat bread	

Beverages & Condiments

Coffee	Sugar
Decaf. coffee	Sugar substitute
Hot tea	Creamer
Decaf. hot tea	Lemon
Iced tea	Margarine
	Diet mustard
	Mayonnaise
No salt	

Name _____   **Room** _____

- Patients may be out of their rooms (for tests, procedures, or physical activity) or asleep when the menus arrive and may miss the menu pickup time.
- Patients may be too ill or too disinterested in food to make menu selections.

Problems with menu procedures can often be corrected by explaining the system or by taking the time to help patients mark menus.

Some hospitals do not offer selective menus. Instead, they may provide **nonselective menus** (menus with preselected food items) or menus that include some elements of both systems **(semiselective menus).** Nonselective menus have been gaining popularity in hospital foodservice because they simplify operations and may help to cut costs.

Food Preparation and Delivery

The responsibility for budgeting, purchasing, planning, preparing, and serving appropriate meals rests with either an administrative dietitian or a foodservice director. In some facilities, foodservice companies from outside the hospital are contracted to perform these duties.

The logistics of preparing foods tailored to each modified diet can be overwhelming. For this reason, foodservice departments use systems designed to limit costs and minimize errors. Meals may be produced in a central kitchen and delivered directly to patients' rooms, using serving equipment that keeps hot foods hot and cold foods cold. Another popular practice is to produce meals beforehand,

nonselective menus: menus that do not allow choices and list only preselected food items.

semiselective menus: menus that combine aspects of both selective and nonselective menus.

Foodborne Illnesses

© Digital Imagery/PhotoDisc/Getty Images

Preparing meals to meet the special dietary needs of patients during times of sickness requires careful attention to food safety. An estimated 76 million people experience **foodborne illness** each year in the United States.[1] For some 5000 people each year, the adverse effects of the illness can be so severe as to cause death. Most vulnerable are pregnant women; very young, very old, sick, or malnourished people; and individuals with a weakened immune system (as in people with AIDS). By taking the proper precautions, people can minimize their chances of contracting foodborne illnesses.

Foodborne Infections and Food Intoxications

Foodborne illness can be caused by either infection or intoxication. Table H18-1 summarizes the most common or severe foodborne illnesses, along with their food sources, general symptoms, and prevention methods.

Foodborne Infections

Foodborne infections are caused by eating foods contaminated by microbes that can cause disease. The most common foodborne **pathogen** is *Salmonella,* which enters the GI tract in contaminated foods such as undercooked poultry and unpasteurized milk. Symptoms generally include abdominal cramps, fever, vomiting, and diarrhea.

Food Intoxications

Food intoxications are caused by eating foods containing natural toxins or, more likely, microbes that produce toxins. The most common food toxin is produced by *Staphylococcus aureus;* it affects more than one million people each year. Less common, but more infamous, is *Clostridium botulinum,* an organism that produces a deadly toxin in anaerobic conditions such as improperly canned (especially home-canned) foods and homemade garlic or herb-flavored oils stored at room temperature. Because the toxin paralyzes muscles, a person with botulism—*C. botulinum* intoxication—has difficulty seeing, speaking, swallowing, and breathing.[2] Because death can occur within 24 hours of onset, botulism demands immediate medical attention. Even then, survivors may suffer the effects for months or years.

Food Safety in the Marketplace

Transmission of foodborne illness has changed along with our food supplies and lifestyles.[3] In the past, foodborne illness was caused by one person's error in a small setting, such as improperly refrigerated egg salad at a family picnic, and it affected only a few victims. Today, we eat more foods that have been prepared and packaged by others for mass consumption. Consequently, when a food manufacturer or restaurant chef makes an error, foodborne illness can become epidemic. An estimated 80 percent of reported foodborne illnesses are caused by errors in a commercial setting, such as the improper **pasteurization** of milk at a large dairy.

GLOSSARY

cross-contamination: the contamination of food by bacteria that occurs when the food comes into contact with surfaces previously touched by raw meat, poultry, or seafood.

foodborne illness: illness transmitted to human beings through food and water, caused by either an infectious agent (foodborne infection) or a poisonous substance (food intoxication); commonly known as **food poisoning.**

pasteurization: heat processing of food that inactivates some,

but not all, microorganisms in the food; not a sterilization process. Bacteria that cause spoilage are still present.

pathogen (PATH-oh-jen): a microorganism capable of producing disease.

sushi: vinegar-flavored rice and seafood, typically wrapped in seaweed and stuffed with colorful vegetables. Some sushi is stuffed with raw fish; other varieties contain cooked seafood.

TABLE H18-1 Foodborne Illnesses

Disease and Organism That Causes It	Most Frequent Food Sources	Onset and General Symptoms	Prevention Methods[a]
Foodborne Infections			
Campylobacteriosis (KAM-pee-loh-BAK-ter-ee-OH-sis) *Campylobacter* bacterium	Raw and undercooked poultry, unpasteurized milk, contaminated water	Onset: 2 to 5 days. Diarrhea, vomiting, abdominal cramps, fever; sometimes bloody stools; lasts 2 to 10 days.	Cook foods thoroughly; use pasteurized milk; use sanitary food-handling methods.
Cryptosporidiosis (KRIP-toe-spo-rid-ee-OH-sis) *Crytosporidium parvum* parasite	Commonly contaminated swimming or drinking water, even from treated sources; highly chlorine-resistant; contaminated raw produce and unpasteurized juices and ciders	Onset: 2 to 10 days. Diarrhea, stomach cramps, upset stomach, slight fever; symptoms may come and go for weeks or months.	Wash all raw vegetables and fruits before peeling; use pasteurized milk and juice; do not swallow drops of water while using pools, hot tubs, ponds, lakes, rivers, or streams for recreation.
Cyclosporiasis (sigh-clo-spore-EYE-uh-sis) *Cyclospora cayetanensis* parasite	Contaminated water, contaminated fresh produce	Onset: 1 to 14 days. Watery diarrhea, loss of appetite, weight loss, stomach cramps, nausea, vomiting, fatigue; symptoms may come and go for weeks or months.	Use treated, boiled, or bottled water; cook foods thoroughly; peel fruits.
E. coli infection *Escherichia coli*[b] bacterium	Undercooked ground beef, unpasteurized milk and juices, raw fruits and vegetables, contaminated water, person-to-person contact	Onset: 1 to 8 days. Severe bloody diarrhea, abdominal cramps, vomiting; lasts 5 to 10 days.	Cook ground beef thoroughly; use pasteurized milk; use sanitary food-handling methods; use treated, boiled, or bottled water.
Gastroenteritis[c] Norwalk virus	Person-to-person contact; raw foods, salads, sandwiches	Onset: 1 to 2 days. Vomiting; lasts 1 to 2 days.	Use sanitary food-handling methods.
Giardiasis (JYE-are-DYE-ah-sis) *Giardia intestinalis* parasite	Contaminated water; uncooked foods	Onset: 7 to 14 days. Diarrhea (but occasionally constipation), abdominal pain, gas.	Use sanitary food-handling methods; avoid raw fruits and vegetables where parasites are endemic; dispose of sewage properly.
Hepatitis (HEP-ah-TIE-tis) Hepatitis A virus	Undercooked or raw shellfish	Onset: 15 to 50 days (28 days average). Diarrhea, dark urine, fever, headache, nausea, abdominal pain, jaundice (yellowed skin and eyes from buildup of wastes); lasts 2 to 12 weeks.	Cook foods thoroughly.
Listeriosis (lis-TER-ee-OH-sis) *Listeria monocytogenes* bacterium	Unpasteurized milk; fresh soft cheeses; luncheon meats, hot dogs	Onset: 1 to 21 days. Fever, muscle aches; nausea, vomiting, blood poisoning, complications in pregnancy, and meningitis (stiff neck, severe headache, and fever).	Use sanitary food-handling methods; cook foods thoroughly; use pasteurized milk.
Perfringens (per-FRINGE-enz) **food poisoning** *Clostridium perfringens* bacterium	Meats and meat products stored at between 120°F and 130°F	Onset: 8 to 16 hr. Abdominal pain, diarrhea, nausea; lasts 1 to 2 days.	Use sanitary food-handling methods; cook foods thoroughly; refrigerate foods promptly and properly.
Salmonellosis (sal-moh-neh-LOH-sis) *Salmonella* bacteria (>2300 types)	Raw or undercooked eggs, meats, poultry, raw milk and other dairy products, shrimp, frog legs, yeast, coconut, pasta, and chocolate	Onset: 1 to 3 days. Fever, vomiting, abdominal cramps, diarrhea; lasts 4 to 7 days; can be fatal.	Use sanitary food-handling methods; use pasteurized milk; cook foods thoroughly; refrigerate foods promptly and properly.
Shigellosis (shi-gel-LOH-sis) *Shigella* bacteria (>30 types)	Person-to-person contact, raw foods, salads, sandwiches, contaminated water	Onset: 1 to 2 days. Bloody diarrhea, cramps, fever; lasts 4 to 7 days.	Use sanitary food-handling methods; cook foods thoroughly; use proper refrigeration.
Vibrio (VIB-ree-oh) **infection** *Vibrio vulnificus*[d] bacterium	Raw or undercooked seafood, contaminated water	Onset: 1 to 7 days. Diarrhea, abdominal cramps, nausea, vomiting; lasts 2 to 5 days; can be fatal.	Use sanitary food-handling methods; cook foods thoroughly.
Yersiniosis (yer-SIN-ee-OH-sis) *Yersinia enterocolitica* bacterium	Raw and undercooked pork, unpasteurized milk	Onset: 1 to 2 days. Diarrhea, vomiting, fever, abdominal pain; lasts 1 to 3 weeks.	Cook foods throughly; use pasteurized milk; use treated, boiled, or bottled water.
Food Intoxications			
Botulism (BOT-chew-lizm) Botulinum toxin [produced by *Clostridium botulinum* bacterium, which grows without oxygen, in low-acid foods, and at temperatures between 40°F and 120°F; the **botulinum** (BOT-chew-line-um) **toxin** responsible for botulism is called **botulin** (BOT-chew-lin)]	Anaerobic environment of low acidity (canned corn, peppers, green beans, soups, beets, asparagus, mushrooms, ripe olives, spinach, tuna, chicken, chicken liver, liver pâté, luncheon meats, ham, sausage, stuffed eggplant, lobster, and smoked and salted fish)	Onset: 4 to 36 hr. Nervous system symptoms, including double vision, inability to swallow, speech difficulty, and progressive paralysis of the respiratory system; often fatal; leaves prolonged symptoms in survivors.	Use proper canning methods for low-acid foods; refrigerate homemade garlic and herb oils; avoid commercially prepared foods with leaky seals or with bent, bulging, or broken cans.
Staphylococcal (STAF-il-oh-KOK-al) **food poisoning** Staphylococcal toxin (produced by *Staphylococcus aureus* bacterium)	Toxin produced in improperly refrigerated meats; egg, tuna, potato, and macaroni salads; cream-filled pastries	Onset: 1 to 6 hr. Diarrhea, nausea, vomiting, abdominal cramps, fever; lasts 1 to 2 days.	Use sanitary food-handling methods; cook food thoroughly; refrigerate foods promptly and properly; use proper home-canning methods.

NOTE: Travelers' diarrhea is most commonly caused by *E. coli, Campylobacter jejuni, Shigella,* and *Salmonella.*

[a]The "How to" on pp. 636–637 provides more details on the proper handling, cooking, and refrigeration of foods.

[b]The most serious strain is *E. coli* STEC O157.

[c]Gastroenteritis refers to an inflammation of the stomach and intestines but is the most common name used for illnesses caused by Norwalk viruses.

[d]Most cases of *Vibrio vulnificus* infection occur in persons with underlying illness, particularly those with liver disorders, diabetes, cancer, and AIDS, and those who require long-term steroid use. The fatality rate is 50 percent for this population.

To prevent food intoxication from homemade flavored oils, wash and dry the herbs before adding them to the oil and keep the oil refrigerated.

In the mid-1990s, when a fast-food restaurant served undercooked burgers tainted with an infectious strain of *Escherichia coli,* hundreds of patrons became ill, and at least 3 people died. In the early 2000s, a national food company had to recall more than 4 million pounds of poultry products after *Listeria* poisoning killed 7 people and made more than 50 others sick. In the 2006 *E. coli* outbreak due to contaminated fresh spinach, nearly 200 people became sick, and 2 elderly women and a 2-year-old boy died before consumers got the FDA message not to eat fresh spinach. These incidents and others have focused the national spotlight on two important safety issues: disease-causing organisms are commonly found in raw foods, and thorough cooking kills most of these foodborne pathogens. This heightened awareness sparked a much needed overhaul of national food safety programs.

Industry Controls

To make our food supply safe for consumers, the USDA, the FDA, and the food-processing industries have developed and implemented programs to control foodborne illness.* The Hazard Analysis Critical Control Points (HACCP) system requires food manufacturers to identify points of contamination and implement controls to prevent foodborne disease. For example, after tracing two large outbreaks of salmonellosis to imported cantaloupe, producers began using chlorinated water to wash the melons and to make ice for packing and shipping. Safety procedures such as this prevent hundreds of thousands of foodborne illnesses each year and are responsible for the decline in infections over the past decade.[4]

This example raises another issue regarding the safety of imported foods. FDA inspectors cannot keep pace with the increasing numbers of imported foods; they inspect fewer than 2

---

* In addition to the HACCP, these programs include the Emerging Infections Program (EIP), the Foodborne Diseases Active Surveillance Network (Food-Net), and the Food Safety Inspection Service (FSIS).

percent of the almost 3 million shipments of fruits, vegetables, and seafood coming into more than 300 ports in the United States each year. The FDA is working with other countries to adopt the safe food-handling practices used in the United States.

Consumer Awareness

Canned and packaged foods sold in grocery stores are easily controlled, but rare accidents do happen. Batch numbering makes it possible to recall contaminated foods through public announcements via newspapers, television, and radio. In the grocery store, consumers can buy items before the "sell by" date and inspect the safety seals and wrappers of packages. A broken seal, bulging can lid, or mangled package alerts consumers to the possibility of microbe or insect contamination, spoilage, or even vandalism.

State and local health regulations provide guidelines on the cleanliness of facilities and the safe preparation of foods for restaurants, cafeterias, and fast-food establishments. Even so, consumers can also take these actions to help prevent foodborne illnesses when dining out:

- Wash hands with hot, soapy water before meals.
- Expect clean tabletops, dinnerware, utensils, and food preparation areas.
- Expect cooked foods to be served piping hot and salads to be fresh and cold.
- Refrigerate doggy bags within two hours.

Improper handling of foods can occur anywhere along the line, from commercial manufacturers to large supermarkets to small restaurants to private homes. Maintaining a safe food supply requires everyone's efforts (see Figure H18-1).

Food Safety in the Kitchen

Whether microbes multiply and cause illness depends, in part, on a few key food-handling behaviors in the kitchen—whether the kitchen is one in your home, a hospital cafeteria, a gourmet restaurant, or a canning manufacturer.[5] Figure H18-2 summarizes the four simple things that can help most to prevent foodborne illness:

- *Keep a clean, safe kitchen.* Wash countertops, cutting boards, hands, sponges, and utensils in hot, soapy water before and after each step of food preparation.
- *Avoid cross-contamination.* Keep raw eggs, meat, poultry, and seafood separate from other foods. Wash all utensils and surfaces (such as cutting boards or platters) that have been in contact with these foods with hot, soapy water before using them again. Bacteria inevitably left on the surfaces from the raw meat can recontaminate the cooked meat or other foods—a problem known as **cross-contamination.** Washing raw eggs, meat, and poultry is not recommended because the extra handling increases the risk of cross-contamination.
- *Keep hot foods hot.* Cook foods long enough to reach internal temperatures that will kill microbes, and maintain ade-

FIGURE H18-1 Food Safety from Farms to Consumers

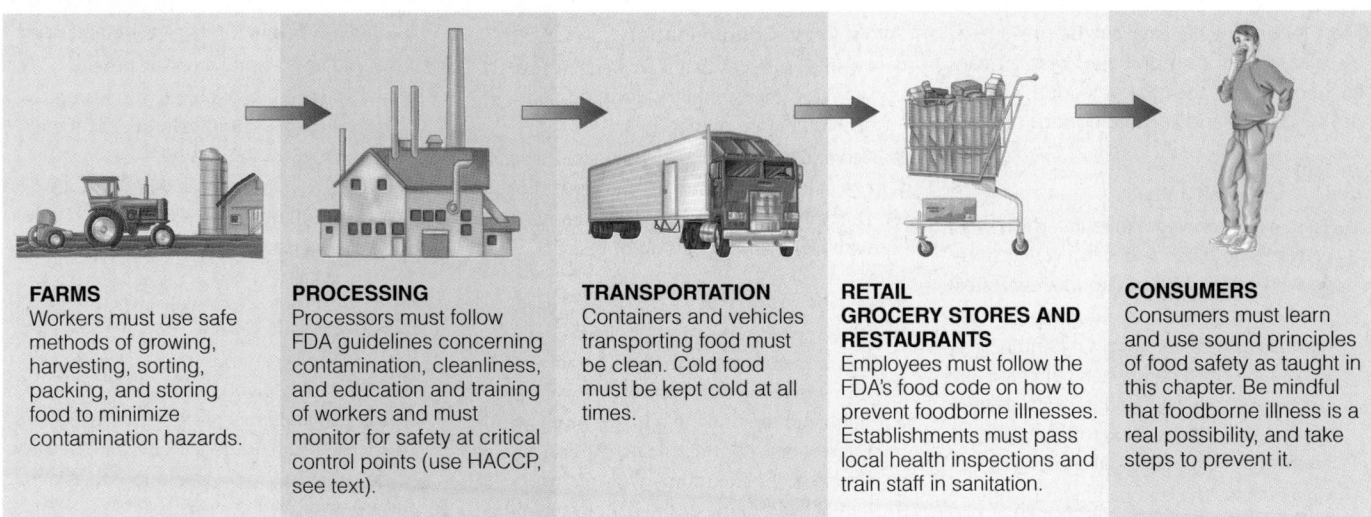

FARMS
Workers must use safe methods of growing, harvesting, sorting, packing, and storing food to minimize contamination hazards.

PROCESSING
Processors must follow FDA guidelines concerning contamination, cleanliness, and education and training of workers and must monitor for safety at critical control points (use HACCP, see text).

TRANSPORTATION
Containers and vehicles transporting food must be clean. Cold food must be kept cold at all times.

RETAIL GROCERY STORES AND RESTAURANTS
Employees must follow the FDA's food code on how to prevent foodborne illnesses. Establishments must pass local health inspections and train staff in sanitation.

CONSUMERS
Consumers must learn and use sound principles of food safety as taught in this chapter. Be mindful that foodborne illness is a real possibility, and take steps to prevent it.

FIGURE H18-2 Fight Bac!

Four ways to keep food safe. The Fight Bac! website is at **www.fightbac.org.**

Dietary Guidelines for Americans 2005

To avoid microbial foodborne illness:

- Clean hands, food contact surfaces, and fruits and vegetables.
- Separate raw, cooked, and ready-to-eat foods while shopping, preparing, or storing foods.
- Do *not* wash or rinse meat and poultry.
- Cook foods to a safe temperature to kill microorganisms.
- Chill (refrigerate) perishable food promptly, and defrost foods properly.

quate temperatures to prevent bacterial growth until the foods are served.

- *Keep cold foods cold.* Go directly home upon leaving the grocery store, and place foods in the refrigerator or freezer right away. After a meal, refrigerate any leftovers immediately.

Unfortunately, consumers commonly fail to follow these simple food-handling recommendations.[6] See the "How to" on pp. 636–637 for additional food safety tips.

Safe Handling of Meats and Poultry

Meats and poultry contain bacteria and provide a moist, nutrient-rich environment that favors microbial growth. Ground meat is especially susceptible because it receives more handling than other kinds of meat and has more surface exposed to bacterial

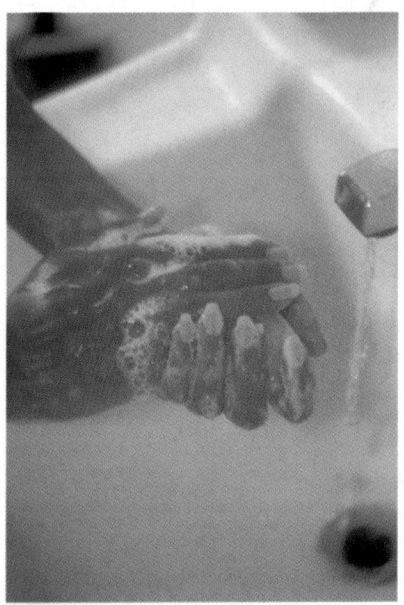

Wash your hands with warm water and soap for at least 20 seconds before preparing or eating food to reduce the chance of microbial contamination.

HOW TO | Prevent Foodborne Illness

Most foodborne illnesses can be prevented by following four simple rules: keep a clean kitchen, avoid cross-contamination, keep hot foods hot, and keep cold foods cold.

Keep a Clean Kitchen

- Wash fruits and vegetables in a clean sink with a scrub brush and warm water; store washed and unwashed produce separately.
- Use hot, soapy water to wash hands, utensils, dishes, nonporous cutting boards, and countertops before handling food and between tasks when working with different foods. Use a bleach solution on cutting boards (one capful per gallon of water).
- Cover cuts with clean bandages before food preparation; dirty bandages carry harmful microorganisms.
- Mix foods with utensils, not hands; keep hands and utensils away from mouth, nose, and hair.
- Anyone may be a carrier of bacteria and should avoid coughing or sneezing over food. A person with a skin infection or infectious disease should not prepare food.
- Wash or replace sponges and towels regularly.
- Clean up food spills and crumb-filled crevices.

Avoid Cross-Contamination

- Wash all surfaces that have been in contact with raw meats, poultry, eggs, fish, and shellfish before reusing.
- Serve cooked foods on a clean plate. Separate raw foods from those that have been cooked.
- Don't use marinade that was in contact with raw meat for basting or sauces.

Keep Hot Foods Hot

- When cooking meats or poultry, use a thermometer to test the internal temperature. Insert the thermometer between the thigh and the body of a turkey or into the thickest part of other meats, making sure the tip of the thermometer is not in contact with bone or the pan. Cook to the temperature indicated for that particular meat (see Figure H18-4 on p. 637); cook hamburgers to at least medium well-done. If you have safety questions, call the USDA Meat and Poultry Hotline: (800) 535-4555.
- Cook stuffing separately, or stuff poultry just prior to cooking.
- Do not cook large cuts of meat or turkey in a microwave oven; it leaves some parts undercooked while overcooking others.
- Cook eggs before eating them (soft-boiled for at least 3½ minutes; scrambled until set, not runny; fried for at least 3 minutes on one side and 1 minute on the other).
- Cook seafood thoroughly. If you have safety questions about seafood, call the FDA hotline: (800) FDA-4010.
- When serving foods, maintain temperatures at 140°F or higher.
- Heat leftovers thoroughly to at least 165°F.

Keep Cold Foods Cold

- When running errands, stop at the grocery store last. When you get home, refrigerate the perishable groceries (such as meats and dairy products) immediately. Do not leave perishables in the car any longer than it takes for ice cream to melt.
- Put packages of raw meat, fish, or poultry on a plate before refrigerating to prevent juices from dripping on food stored below.
- Buy only foods that are solidly frozen in store freezers.
- Keep cold foods at 40°F or less; keep frozen foods at 0°F or less (keep a thermometer in the refrigerator).
- Marinate meats in the refrigerator, not on the counter.

---

contamination. Consumers cannot detect the harmful bacteria in or on meat. Figure H18-3 presents label instructions for the safe handling of meat and poultry and two types of USDA seals. For safety's sake, cook meat thoroughly, using a thermometer to test the internal temperature (see Figure H18-4).

Mad Cow Disease

Reports on mad cow disease from dozens of countries, including Canada and the United States, have sparked consumer concerns.[7] Mad cow disease is a slowly progressive, fatal condition that affects the central nervous system of cattle.* A similar disease develops in people who have eaten contaminated beef from infected cows (milk products appear to be safe).† Approximately 150 cases have been reported worldwide, primarily in the United Kingdom. The USDA has taken numerous steps to prevent the transmission of mad cow

---

* Mad cow disease is technically known as bovine spongiform encephalopathy (BSE).
† The human form of BSE is called variant Creutzfeldt-Jakob disease (vCJD).

FIGURE H18-3 | Meat and Poultry Safety, Grading, and Inspection Seals

Inspection is mandatory; grading is voluntary. Neither guarantees that the product will not cause foodborne illnesses, but consumers can help to prevent foodborne illnesses by following the safe handling instructions.

The voluntary "Graded by USDA" seal indicates that the product has been graded for tenderness, juiciness, and flavor. Beef is graded Prime (abundant marbling of the meat muscle), Choice (less marbling), and Select (lean). Similarly, poultry is graded A, B, and C.

The mandatory "Inspected and Passed by the USDA" seal ensures that meat and poultry products are safe, wholesome, and correctly labeled. Inspection does not guarantee that the meat is free of potentially harmful bacteria.

Safe Handling Instructions

THIS PRODUCT WAS PREPARED FROM INSPECTED AND PASSED MEAT AND/OR POULTRY. SOME FOOD PRODUCTS MAY CONTAIN BACTERIA THAT CAN CAUSE ILLNESS IF THE PRODUCT IS MISHANDLED OR COOKED IMPROPERLY. FOR YOUR PROTECTION, FOLLOW THESE SAFE HANDLING INSTRUCTIONS.

KEEP REFRIGERATED OR FROZEN. THAW IN REFRIGERATOR OR MICROWAVE.

KEEP RAW MEAT AND POULTRY SEPARATE FROM OTHER FOODS. WASH WORKING SURFACES (INCLUDING CUTTING BOARDS), UTENSILS, AND HANDS AFTER TOUCHING RAW MEAT OR POULTRY.

COOK THOROUGHLY.

KEEP HOT FOODS HOT. REFRIGERATE LEFTOVERS IMMEDIATELY OR DISCARD.

The USDA requires that safe handling instructions appear on all packages of meat and poultry.

HOW TO Prevent Foodborne Illness, *continued*

- Refrigerate leftovers promptly; use shallow containers to cool foods faster; use leftovers within 3 to 4 days.
- Thaw meats or poultry in the refrigerator, not at room temperature. If you must hasten thawing, use cool water (changed every 30 minutes) or a microwave oven.
- Freeze meat, fish, or poultry immediately if not planning to use within a few days.

In General

- Do not reuse disposable containers; use nondisposable containers or recycle instead.
- Do not taste food that is suspect. "If in doubt, throw it out."
- Throw out foods with danger-signaling odors. Be aware, though, that most food-poisoning bacteria are odorless, colorless, and tasteless.
- Do not buy or use items that have broken seals or mangled packaging; such containers cannot protect against microbes, insects, spoilage, or even vandalism. Check safety seals, buttons, and expiration dates.
- Follow label instructions for storing and preparing packaged and frozen foods; throw out foods that have been thawed or refrozen.

- Discard foods that are discolored, moldy, or decayed or that have been contaminated by insects or rodents.

For Specific Food Items

- *Canned goods.* Carefully discard food from cans that leak or bulge so that other people and animals will not accidentally ingest it; before canning, seek professional advice from the USDA Extension Service (check your phone book under U.S. government listings, or ask directory assistance).
- *Milk and cheeses.* Use only pasteurized milk and milk products. Aged cheeses, such as cheddar and Swiss, do well for an hour or two without refrigeration, but they should be refrigerated or stored in an ice chest for longer periods.
- *Eggs.* Use clean eggs with intact shells. Do not eat eggs, even pasteurized eggs, raw; raw eggs are commonly found in Caesar salad dressing, eggnog, cookie dough, hollandaise sauce, and key lime pie. Cook eggs until whites are firmly set and yolks begin to thicken.
- *Honey.* Honey may contain dormant bacterial spores, which can awaken in the human body to produce botulism. In adults, this poses little hazard, but infants

under one year of age should never be fed honey. Honey can accumulate enough toxin to kill an infant; it has been implicated in several cases of sudden infant death. (Honey can also be contaminated with environmental pollutants picked up by the bees.)
- *Mayonnaise.* Commercial mayonnaise may actually help a food to resist spoilage because of the acid content. Still, keep it cold after opening.
- *Mixed salads.* Mixed salads of chopped ingredients spoil easily because they have extensive surface area for bacteria to invade, and they have been in contact with cutting boards, hands, and kitchen utensils that easily transmit bacteria to food (regardless of their mayonnaise content). Chill them well before, during, and after serving.
- *Picnic foods.* Choose foods that last without refrigeration, such as fresh fruits and vegetables, breads and crackers, and canned spreads and cheeses that can be opened and used immediately. Pack foods cold, layer ice between foods, and keep foods out of water.
- *Seafood.* Buy only fresh seafood that has been properly refrigerated or iced. Cooked seafood should be stored separately from raw seafood to avoid cross-contamination.

FIGURE H18-4 Recommended Safe Temperatures (Fahrenheit)

Bacteria multiply rapidly at temperatures between 40°F and 140°F. Cook foods to the temperatures shown on this thermometer and hold them at 140°F or higher.

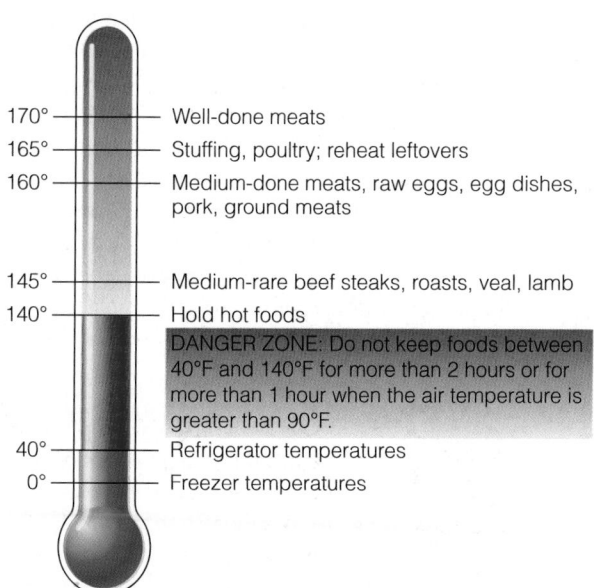

170° — Well-done meats
165° — Stuffing, poultry; reheat leftovers
160° — Medium-done meats, raw eggs, egg dishes, pork, ground meats
145° — Medium-rare beef steaks, roasts, veal, lamb
140° — Hold hot foods
DANGER ZONE: Do not keep foods between 40°F and 140°F for more than 2 hours or for more than 1 hour when the air temperature is greater than 90°F.
40° — Refrigerator temperatures
0° — Freezer temperatures

disease in cattle, and if these measures are followed, then risks from eating meat from U.S. cattle are low.[8] Because the infectious agents occur in the intestines, central nervous system, and other organs, but not in muscle meat, concerned consumers may want to select whole cuts of meat instead of ground beef or sausage. A few recent reports of hunters developing fatal neurological disorders have raised concerns about a similar disease in wild game. Hunters and consumers who regularly eat elk, deer, or antelope should check the advisories of their state department of agriculture.

Avian Influenza

Avian influenza (bird flu) is a very contagious and life-threatening viral infection that naturally occurs among birds, including chickens, ducks, and turkeys. The risk of bird flu in human beings is relatively low, and most cases have resulted from direct contact with infected birds or their contaminated environment. Because the virus can change easily,

Cook hamburgers to 160°F; color alone cannot determine doneness. Some burgers will turn brown before reaching 160°F, whereas others may retain some pink color, even when cooked to 175°F.

Eating raw seafood is a risky proposition.

scientists are concerned that it could infect people and spread rapidly from person to person, creating a pandemic. It is important to note that bird flu is not transmitted by eating poultry.

Safe Handling of Seafood

Most seafood available in the United States and Canada is safe, but eating it undercooked or raw can cause severe illnesses—hepatitis, worm or parasite infestation, viral intestinal disorders, and other diseases.* Rumor has it that freezing fish will make it safe to eat raw, but this is only partly true. Commercial freezing kills mature parasitic worms, but only cooking can kill all worm eggs and other microorganisms that can cause illness. For safety's sake, all seafood should be cooked until it is opaque. Even **sushi** can be safe to eat when chefs combine cooked seafood and other ingredients into these delicacies.

Eating raw oysters can be dangerous for anyone, but people with liver disease and weakened immune systems are most vulnerable. At least 10 species of bacteria found in raw oysters can cause serious illness and even death.† Raw oysters may also carry the hepatitis A virus, which can cause liver disease. Some hot sauces can kill many of these bacteria, but not the virus; alcohol may also protect some people against some oyster-borne illnesses, but not enough to guarantee protection (or to recommend drinking alcohol). Pasteurization of raw oysters—holding them at a specified temperature for a specified time—holds promise for killing bacteria without cooking the oysters or altering their texture or flavor.

As population density increases along the shores of seafood-harvesting waters, pollution inevitably invades the sea life there. Preventing seafood-borne illness is in large part a task of controlling water pollution. To help ensure a safe seafood market, the FDA requires processors to adopt food safety practices based on the HACCP system mentioned earlier.

Chemical pollution and microbial contamination lurk not only in the water but also in the boats and warehouses where seafood is cleaned, prepared, and refrigerated. Because seafood is one of the most perishable foods, time and temperature are critical to its freshness, flavor, and safety. To keep seafood as fresh as possible, people in the industry must "keep it cold, keep it clean, and keep it moving." Wise consumers eat it cooked.

Other Precautions and Procedures

Fresh food generally smells fresh. Not all types of food poisoning are detectable by odor, but some bacterial wastes produce "off" odors. If an abnormal odor exists, the food is spoiled. Throw it out or, if it was recently purchased, return it to the grocery store. Do not taste it. Table H18-2 lists safe refrigerator storage times for selected foods.

Local health departments and the USDA Extension Service can provide additional information about food safety. If precautions fail and a mild foodborne illness develops, drink clear liquids to replace fluids lost through vomiting and diarrhea. If serious foodborne illness is suspected, first call a physician. Then wrap the remainder of the suspected food and label the container so that the food cannot be mistakenly eaten, place it in the refrigerator, and hold it for possible inspection by health authorities.

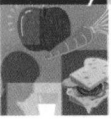

Dietary Guidelines for Americans 2005

- To avoid microbial foodborne illness, avoid raw (unpasteurized) milk or any products made from unpasteurized milk, raw or partially cooked eggs or foods containing raw eggs, raw or undercooked meat and poultry, unpasteurized juices, and raw sprouts.

---

* Diseases caused by toxins from the sea include ciguatera, scombroid poisoning, and paralytic and neurotoxic shellfish poisoning.
† Raw oysters can carry the bacterium *Vibrio vulnificus*; see Table H18-1 for details.

TABLE H18-2 Safe Refrigerator Storage Times (≤40°F)

1 to 2 Days

Raw ground meats, breakfast or other raw sausages, raw fish or poultry; gravies

3 to 5 Days

Raw steaks, roasts, or chops; cooked meats, poultry, vegetables, and mixed dishes; lunch meats (packages opened); mayonnaise salads (chicken, egg, pasta, tuna)

1 Week

Hard-cooked eggs, bacon or hot dogs (opened packages); smoked sausages or seafood

2 to 4 Weeks

Raw eggs (in shells); lunch meats, bacon, or hot dogs (packages unopened); dry sausages (pepperoni, hard salami); most aged and processed cheeses (Swiss, brick)

2 Months

Mayonnaise (opened jar); most dry cheese (Parmesan, Romano)

Millions of people suffer mild to life-threatening symptoms caused by foodborne illnesses (review Table H18-1). As the "How to" on pp. 636–637 describes, most of these illnesses can be prevented by storing and cooking foods at their proper temperatures and by preparing them in sanitary conditions.

NUTRITION ON THE NET

For further study of topics covered in this chapter, log on to **academic.cengage.com/nutrition/rolfes/UNCN8e**. Go to Chapter 18, then to Nutrition on the Net.

- Get food safety tips from the Government Food Safety Information site or from the Fight BAC! Campaign of the Partnership for Food Safety Education: **www.foodsafety.gov** or **www.fightbac.org**

- Learn more about foodborne illnesses from the National Center for Infectious Diseases at the Centers for Disease Control and Prevention. Search for foodborne illness at: **www.cdc.gov**

- Learn about the various types of food thermometers and how and when to use them from the USDA Thermy Campaign: **www.fsis.usda.gov/thermy**

REFERENCES

1. Centers for Disease Control and Prevention, *FoodNet Surveillance Report for 2004* (Atlanta, Ga.: Centers for Disease Control and Prevention, June 2006).
2. E. A. Coleman and M. E. Yergler, Botulism, *American Journal of Nursing* 102 (2002): 44–47.
3. Position of the American Dietetic Association: Food and water safety, *Journal of the American Dietetic Association* 103 (2003): 1203–1218.
4. Centers for Disease Control and Prevention, June 2006.
5. B. J. McCabe-Sellers and S. E. Beattie, Food safety: Emerging trends in foodborne illness surveillance and prevention, *Journal of the American Dietetic Association* 104 (2004): 1709–1717.
6. J. B. Anderson and coauthors, A camera's view of consumer food-handling behaviors, *Journal of the American Dietetic Association* 104 (2004): 186–191.
7. U.S. Food and Drug Administration, Consumer asked questions about BSE in products regulated by FDA's Center for Food Safety and Applied Nutrition (CFSAN), available at www.cfsan.fda.gov/~comm/
bsefaq.html, site updated September 14, 2005, and visited December 6, 2006; U.S. Department of Agriculture, Bovine spongiform encephalopathy (BSE) Q & A's, available at www.aphis.usda.gov/lpa/issues/bse/bse_q&a.html, site updated January 21, 2004, and visited December 6, 2006.
8. C. A. Donnelly, Bovine spongiform encephalopathy in the United States: An epidemiologist's view, *New England Journal of Medicine* 350 (2004): 539–542; T. Hampton, What now, mad cow? Experts put risk to US public in perspective, *Journal of the American Medical Association* 291 (2004): 543–549.

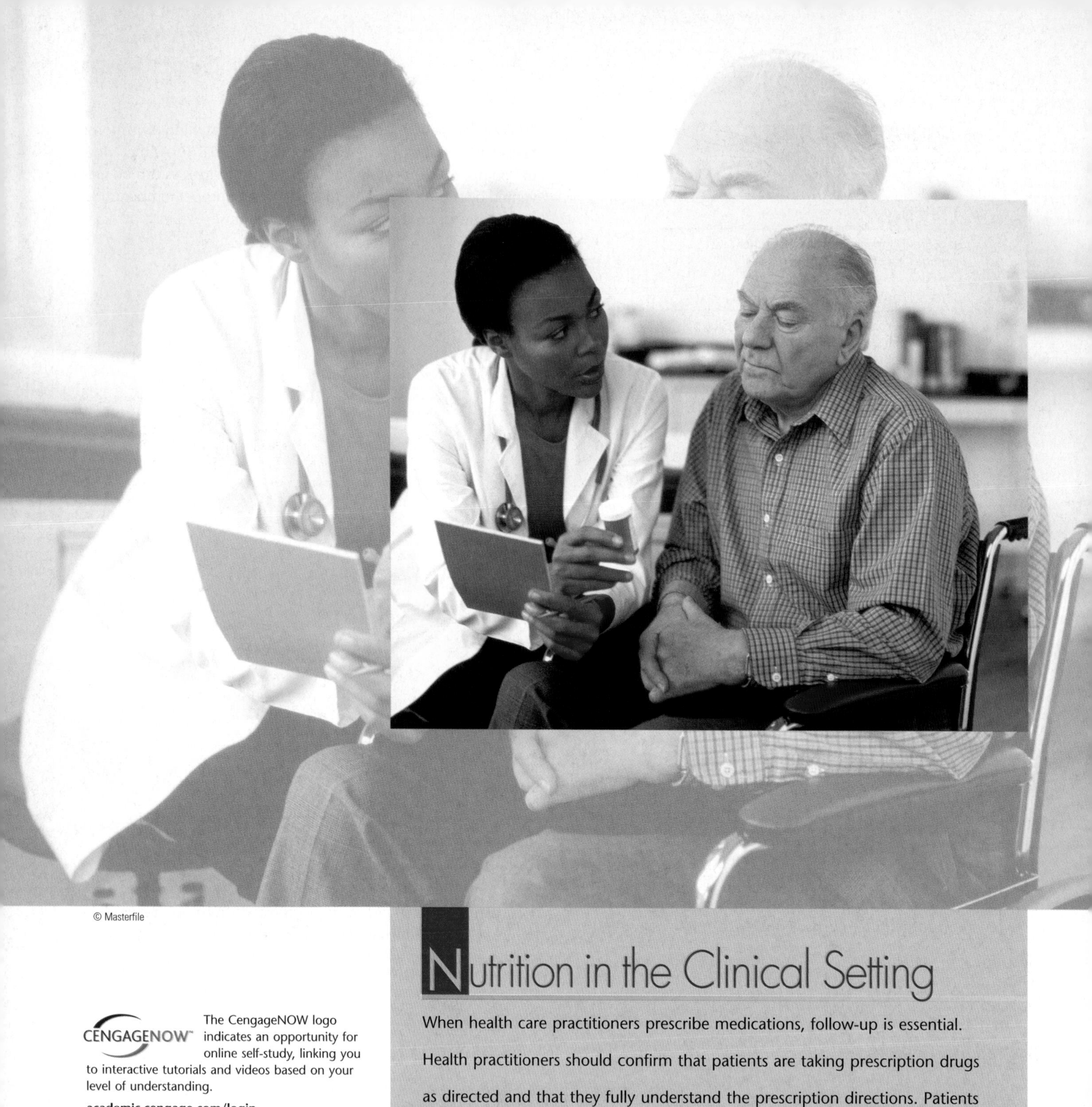

© Masterfile

Nutrition in the Clinical Setting

When health care practitioners prescribe medications, follow-up is essential. Health practitioners should confirm that patients are taking prescription drugs as directed and that they fully understand the prescription directions. Patients may feel uncomfortable admitting their uncertainty about the directions, intolerance to side effects, or inability to purchase the drugs they need. Others may feel hesitant to discuss their use of herbal products. Patients are more likely to confide in health care providers who take the time to discuss these difficulties.

Medications, Herbal Products, and Diet-Drug Interactions

People frequently rely on medications to prevent and treat health problems. They may also use herbal products, which have become a popular alternative therapy. Because any ingested chemical can affect metabolism and potentially disrupt body processes, both medications and herbal products may produce adverse effects. Serious side effects may occur when medications interact with each other (drug-drug interactions) or with nutrients and other dietary components (diet-drug interactions). This chapter discusses the uses of medications and herbal supplements and describes potential diet-drug interactions.

Medications in Disease Treatment

Medications must be proved to be safe and effective before they can be marketed in the United States. The Food and Drug Administration (FDA) is responsible for approving sales of new drugs and inspecting facilities where drugs are manufactured. It also oversees the advertising of prescription drugs and labeling of over-the-counter drugs.[1]

Prescription Drugs Prescription drugs are usually given to treat serious conditions and may cause severe side effects. For these reasons, they are sold by prescription only, which ensures that a physician has evaluated the patient's medical condition and determined that the benefits of using the medication outweigh the risks of incurring side effects.

Over-the-Counter Drugs Over-the-counter (OTC) drugs are those that can be used safely and effectively without medical supervision. People usually use them to treat less serious illnesses that are easily self-diagnosed. Examples include aspirin to treat headaches or pain, decongestants to relieve stuffy noses, and antacids to combat indigestion. Prescription drugs considered safe enough for self-medication are frequently switched to OTC status, sometimes in smaller doses than are available by prescription. Labels on OTC drugs are regulated and provide information about the drugs' appropriate uses, dosages, and potential adverse effects.

Although over-the-counter drugs are considered safe enough for self-medication, they can cause adverse effects when used inappropriately.

◆ Consumers can search for generic equivalents to brand-name drugs at **www.fda.gov/cder/ob/default.htm.**

◆ The FDA's MedWatch program encourages health professionals and consumers to report medication problems by mail, fax, telephone, or the Internet (**www.fda.gov/ medwatch**).

Health care practitioners should caution patients that adverse effects may occur if OTC drugs are used inappropriately. Under certain circumstances, the active ingredients in these drugs can worsen medical conditions, produce complications, and interact with other medications. Furthermore, people using products with several active ingredients may inadvertently take toxic amounts of a substance when using several drugs simultaneously. For example, a person with a cold may take one medication to treat a cough and another medication for a headache without realizing that both contain an analgesic (pain medication).

Generic Drugs Brand-name drugs are usually given patent protection for 20 years after the patent is submitted. After the patent protection expires, generic versions of the drugs can be sold. To gain FDA approval, a generic drug must have the same biological effects as the original drug: it must contain the same active ingredients; be identical in strength, dosage form, and route of administration; and meet the same requirements for purity and quality. Thus, consumers can be confident that generic drugs are as safe and effective as the brand-name products they replace. ◆ The advantage is a substantial savings—generic drugs typically cost 20 to 75 percent less than their brand-name counterparts.

Risks from Medications

The risk of an adverse reaction always accompanies the use of a medicine. Thus, a medication should be used only when the benefits of using it outweigh the potential risks. Before prescribing drug treatments, health professionals should discuss with patients the potential side effects of using the medications and should also advise them if alternative treatments are available. The risks associated with medications become greater when a drug is incorrectly prescribed or administered. This section discusses the types of risks associated with drugs and suggests some steps for managing risk.[2]

Side Effects By the time a drug reaches the marketplace, large-scale clinical trials have revealed the majority of side effects associated with its use. However, rare side effects are sometimes detected only after a drug has been more widely used. In some instances, these effects occur because drugs are used for longer periods or in different circumstances than originally anticipated. The FDA monitors adverse events after drugs are marketed. Manufacturers are required to submit periodic reports, and individuals using the drugs are encouraged to report unexpected effects directly to the FDA. ◆ In some cases, the FDA may decide to change labeling information or even withdraw drugs from the marketplace due to their unacceptable risks to health.

Drug-Drug Interactions When a person uses multiple drugs, one drug may alter the effects of another drug, and the risk of side effects increases. These problems are common in older adults, who are likely to use several medications daily over long periods. Primary care physicians often supervise medication use, but some individuals use drugs prescribed by a number of different physicians. Others may use OTC medications and dietary supplements in addition to prescription drugs without being aware of the risks associated with certain combinations.

Diet-Drug Interactions Substances in the diet may alter the effectiveness of drugs, and drugs may affect food intake, digestion, absorption, metabolism, or excretion of nutrients. The second half of this chapter describes these interactions and explains how they may affect nutrition status.

Medication Errors A medication error is any preventable action that causes inappropriate drug use or patient harm due to mistakes made by a health professional or patient. Many errors leading to patient harm involve the use of incorrect drugs or improper dosages.[3] The wrong drug is sometimes administered when two different drugs have names that look or sound alike or have similar packaging. In other cases, the physician's prescription is misread or misinterpreted; for example, a patient died after receiving 10 milliliters of morphine solution instead of 10 milligrams—a 20-fold overdose.

Several policy changes and programs are helping to reduce medication errors. For example, the bar codes currently used on medications and patient identification bracelets allow health practitioners to verify that the correct medication and dosage is administered: error messages alert personnel if the drug, dose, or timing of administration is inappropriate. In addition, a national education campaign is attempting to eliminate one of the most common but preventable sources of medication errors—the use of ambiguous medical abbreviations (see the examples in Table 19-1). Because terms such as these are easily misread or misinterpreted, they can no longer be used in clinical documentation related to patient care.

Patients at High Risk of Adverse Effects

Health care professionals should be aware that some patients are more vulnerable than others to adverse effects from drugs. This category includes the populations that rarely participate in clinical trials that determine product safety: pregnant and lactating women, children, and people with medical conditions that are not the main focus of the study. In these groups, side effects may be discovered only after a drug has been marketed. Children may react in different ways to drugs than adults do, and the appropriate dosage for their age and size is often unknown. Also, limited data are available on drug safety in older adults. Elderly people with chronic

TABLE 19-1 Terms Prohibited on Clinical Documentation			
Prohibited Terms	**Intended Meaning**	**Potential Problem**	**Correct Term for Documentation**
U	Unit	Can be misread as the number 0 or 4; may cause 10-fold overdose or higher.	Write out "unit."
IU	International unit	Can be misread as IV (intravenous) or 10.	Write out "international unit."
Trailing zero (1.0 mg) or lack of leading zero (.1 mg)	1 mg; 0.1 mg	Decimal point can be missed, leading to 10-fold error in dosages.	Never use zero by itself after a decimal. Always use zero before a decimal point.
HS, hs	*HS* means "half-strength"; *hs* means "bedtime" (abbreviation for "hours of sleep").	Can be mistaken for one another.	Write out "half-strength" or "bedtime."
µg	Microgram	Can be misread as mg (milligram).	Write "mcg."
A.S., A.D., A.U.	Abbreviations of the Latin for left ear, right ear, and both ears	Can be misread as O.S., O.D., and O.U., meaning left eye, right eye, and both eyes.	Write out full words.
T.I.W.	Three times a week	Can be mistaken for "three times a day" or "twice weekly."	Write out "3 times weekly."
Q.D. (q.d.), Q.O.D (q.o.d.)	*Q.D.* means "every day"; *Q.O.D.* means "every other day."	Can be mistaken for one another or misread as "q.i.d." (four times daily).	Write out "daily" or "every other day."
q1d	Daily	Can be misread as q.i.d. (four times daily).	Write out "daily."

diseases that require multiple medications are especially susceptible to adverse effects. They are also more likely to have impaired function of the liver or kidneys—the two organs critical to metabolizing and eliminating drugs from the body. The "How to" below provides suggestions that may help to reduce the risks of adverse effects from medications.

Elderly people using multiple medications are especially susceptible to adverse effects from drugs.

◆ Reminder: Highlight 10 discusses vitamin and mineral supplements.

IN SUMMARY

Both prescription and OTC drugs must be shown to be safe and effective before they are sold. The benefits of using a medication should be greater than the risks associated with its use. Potential risks include side effects, drug-drug and diet-drug interactions, and medication errors. The most common types of medication errors involve incorrect dosages or use of the wrong drug. Patients at highest risk of experiencing adverse effects from medications include pregnant and nursing women, children, and the elderly. Health professionals should discuss the risks and benefits of medications with patients and alert them to potential dangers and possible solutions.

Herbal Products

Use of herbal products has grown rapidly in the past decade. In a 2004 research study of 61,587 older adults (ages 50 to 76 years), researchers found that one-third of the participants used herbal and other specialty (nonplant) supplements.[4] ◆ Consumers use these products in the hope of improving their general health and preventing or treating specific diseases. The top-selling herbal supplements include ginseng, echinacea, *Ginkgo biloba*, garlic, and St. John's wort.[5] Table 19-2 lists these and other popular herbs along with their common uses and potential risks associated with their use.

Effectiveness and Safety of Herbal Products

Despite the popularity of herbal products in the United States, the benefits of their use are uncertain. The marketplace is currently inundated with herbal "remedies" of dubious effectiveness. There is no question that many medicinal herbs contain naturally occurring compounds that can exert physiological effects. Few herbal products, however, have been rigorously tested, many make unfounded claims, and some contain contaminants or produce toxic effects.[6]

HOW TO Reduce the Risks of Adverse Effects from Medications

To limit the risks of adverse effects from medications, health care providers can take the following steps:

- Advise the patient that drugs should not be taken unless absolutely necessary. Discuss dietary or lifestyle practices that may have benefits similar to those of drugs. For example, individuals may not need to take laxatives if they increase consumption of foods high in fiber and begin exercising regularly.
- Request a complete list of prescription medications, over-the-counter drugs, and dietary supplements that a patient is taking. Make sure that at least one physician is coordinating the patient's drug use.
- Verify that the patient understands how to take medications properly. Alert the patient to potential interactions between drugs and between drugs and dietary substances.

- Encourage the patient to keep track of adverse effects. Inform patients that new or unusual symptoms may be the result of a medication rather than their medical condition. In some cases, other medications that treat the same condition may have fewer side effects.
- Encourage the patient to purchase all medications at the same pharmacy so that the pharmacist can alert the physician and patient to potential problems.
- Include information about drug safety when providing health education.

TABLE 19-2 Popular Herbs, Their Common Uses, and Adverse Effects

Herb	Scientific Name	Common Uses	Adverse Effects
Black cohosh	Cimicifuga racemosa	Relief of menopausal symptoms	Rare; occasional stomach upset, headache, weight gain
Chaparral	Larrea tridentata	General tonic; treatment of infection, cancer, and arthritis	Hepatitis, liver failure
Comfrey	Symphytum officinale	Wound healing (topical use), treatment of lung and GI disorders	Liver damage
Echinacea	Echinacea augustifolia, E. pallida, E. purpurea	Prevention and treatment of upper respiratory infections	Rare; occasional allergic reactions
Feverfew	Tanacetum parthenium	Prevention of migraine headache	Mouth and tongue sores, swelling of lips, GI upset
Garlic	Allium sativum	Reduction of blood clotting, athero-sclerosis, blood pressure, and blood cholesterol	Halitosis (bad breath), body odor; occasional dyspepsia, flatulence, excessive bleeding, anorexia, allergic reactions
Ginger	Zingiber officinale	Prevention and treatment of nausea and motion sickness	Rare; occasional heartburn
Ginkgo	Ginkgo biloba	Treatment of dementia, memory defects, and circulatory impairment	Rare; occasional stomach upset, headache, skin hypersensitivity, excessive bleeding
Ginseng	Panax ginseng, P. quinquefolius	General tonic, reduction of blood glucose levels	Rare
Kava	Piper methysticum	Treatment of anxiety, stress, and insomnia	Dyspepsia, restlessness, drowsiness, tremor, headache, dermatitis (with heavy use), occasional hepatitis and liver failure
St. John's wort	Hypericum perforatum	Treatment of mild to moderate depression	Rare; occasional stomach upset, fatigue, dizziness, headache, dry mouth, dermatitis, skin photosensitivity
Saw palmetto	Serenoa repens	Reduction of symptoms associated with enlarged prostate	Rare
Valerian	Valeriana officinalis	Sedation, treatment of insomnia	Rare
Yohimbe	Pausinystalia yohimbe	Treatment of erectile dysfunction	Anxiety, headache, dizziness, nausea, rapid heartbeat, hypertension, increased urinary frequency; isolated reports of renal failure, blood disorders, and airway constriction

SOURCE: M. Rotblatt and I. Ziment, *Evidence-Based Herbal Medicine* (Philadelphia: Hanley & Belfus, 2002).

Efficacy Herbs have been used for centuries to treat medical conditions, and many have acquired reputations for being beneficial for people with specific diseases. Unfortunately, only a limited number of clinical studies support the traditional uses, and the results of studies that suggest little or no benefit are rarely publicized by the supplement industry. The National Center for Complementary and Alternative Medicine (a division of the National Institutes of Health) is currently funding large, controlled trials of several popular herbal treatments in an effort to obtain reliable efficacy and safety data.

Although labels on herbal products cannot make claims about preventing or treating specific diseases, suggestive statements are common. For example, a label may claim that an herb "promotes restful sleep" but cannot state that it cures insomnia. Stores often shelve herbal products by health condition; for example, posted signs may indicate the supplements suggested for "liver health" or "men's health." The reading materials positioned close to those shelves often suggest that the products can improve one's health.

Consistency of Herbal Ingredients Herbs contain numerous compounds, and it is often unclear which of these ingredients, if any, might produce the implied beneficial effects. Because the compounds in herbs vary among species and are affected by a plant's growing conditions, different samples of an herb can have different chemical compositions. The preparation method may also cause variations in the composition of an herbal product. Some manufacturers attempt to standardize the herbal

Despite the popularity of echinacea, its benefits for treating the common cold have not been supported by some well-designed clinical studies.

◆ Although the FDA ruling to ban ephedra was overturned by a Utah district court in 2005, the ruling was upheld by an appellate court in August 2006.

extracts they sell so that the compound believed to be beneficial is more likely to be obtained from each dose.

Even when the active ingredients in an herbal supplement have been shown to be safe and effective, the dosage suggested on the label might not provide the quantity of active ingredients found to be effective. For example, a consumer group (ConsumerLab.com) tested nine *Ginkgo biloba* products and found that seven of the products, when consumed at the recommended dosage, lacked adequate levels of one or more compounds believed to be helpful.[7] In some cases, the desired herbal product is not even present. In a university study of echinacea preparations, 10 percent of the 59 products tested contained no measurable echinacea, and only 52 percent of the samples contained the variety of echinacea listed on the label.[8]

Safety Issues Unlike drugs, herbal products do not need FDA approval before they are marketed. According to the Dietary Supplement Health and Education Act (DSHEA) of 1994, the companies that produce or distribute dietary (including herbal) supplements are responsible for determining their safety, but these companies are not required to provide any evidence. If a company receives reports of illness or injury related to the use of its products, it does not need to submit this information to the FDA. In addition, the FDA must show that a supplement is unsafe before it can take action to remove the product from the marketplace.

Consumers of herbal supplements often assume that because plants are "natural," herbal products must be harmless. Many herbal remedies have toxic effects, however. The most common adverse effects of herbs include diarrhea, nausea, and vomiting.[9] The popular herbs kava, chaparral, and comfrey have caused liver damage. The use of yohimbe (promoted for bodybuilding) has been linked to renal failure, seizures, and heart palpitations. In 2004, the FDA removed the herb ephedra (also known as *ma huang*) from the market, advising that its side effects (which include elevated blood pressure and rapid heartbeat) could cause heart attack or stroke. ◆ Note that the adverse effects of herbs are seldom listed on supplement labels.

Like drugs, herbs may either intensify or interfere with the effects of other herbs and drugs. Information about herb-drug interactions is limited, however, and much of what is known has been obtained from case studies rather than from controlled clinical trials. An herb may increase or decrease the effects of medications, or it may raise the risk of toxicity. For example, garlic, ginkgo, and ginseng may increase the risk of bleeding when used with anticoagulant drugs.[10] St. John's wort has been found to inhibit the actions of oral contraceptives, anticoagulants, and other drugs. Individuals may be more susceptible to the adverse effects of a drug if the herb they are using has a similar effect; for example, ginger and ginseng contain compounds that raise blood pressure and may increase the toxicity of drugs that have a similar side effect.[11] Table 19-3 provides examples of herb-drug interactions.

Contamination of herbal products is another safety concern. Some products have been found to contain lead and other toxic metals in excessive amounts.[12] Other contaminants frequently found in herbal products include molds, bacteria, and pesticides that have been banned for use on food crops.[13] Adulteration of imported products has been a serious concern: several studies found that some herbal products imported from China and India contained synthetic drugs that were not declared on the label.[14] There have also been reports of serious illnesses and fatalities occurring from the intentional or accidental substitutions of one plant species for another.[15]

Use of Herbal Products in Illness

When people self-medicate or ask the advice of store clerks instead of seeking effective medical treatment, the consequences are sometimes serious and irreversible. A

TABLE 19-3	Examples of Herb-Drug Interactions	
Herb	**Drug**	**Interaction**
American ginseng	Estrogens, corticosteroids	Enhances hormonal response
American ginseng	Breast cancer therapeutic agent	Synergistically inhibits cancer cell growth
American ginseng, karela	Blood glucose regulators	Affect blood glucose levels
Echinacea (possible immunostimulant)	Cyclosporine and corticosteroids (immunosuppressants)	May reduce drug effectiveness
Evening primrose oil, borage	Anticonvulsants	Lower seizure threshold
Feverfew	Aspirin, ibuprofen, and other nonsteroidal anti-inflammatory drugs	Negates the effect of the herb in treating migraine headaches
Feverfew, garlic, ginkgo, ginger, and Asian ginseng	Warfarin, coumarin (anticlotting drugs, "blood thinners")	Prolong bleeding time; increase likelihood of hemorrhage
Garlic	Protease inhibitor (HIV drug)	May reduce drug effectiveness
Kava, valerian	Anesthetics	May enhance drug action
Kelp (iodine source)	Synthroid or other thyroid hormone replacers	Interferes with drug action
Kyushin, licorice, plantain, uzara root, hawthorn, Asian ginseng	Digoxin (cardiac antiarrhythmic drug derived from the herb foxglove)	Interfere with drug action and monitoring
St. John's wort, saw palmetto, black tea	Iron	Tannins in herbs inhibit iron absorption
Valerian	Barbiturates	Causes excessive sedation

visit to the health-food store for an herbal remedy may be less stressful than a visit to the doctor, but it may delay getting an appropriate treatment and allow an illness to progress. Consumers should inform their health care providers about their use of herbal supplements so that a comprehensive care plan can be developed and potential problems can be averted.

Many people are unaware that herbal products can interact with medications. Because research on herbs is often lacking, assessing potential interactions is difficult for health care professionals and patients alike. Some pharmacology textbooks and handbooks now contain information about herbal products and potential herb-drug interactions, and various consumer websites and periodicals provide information about the safety of brand-name products. Health professionals should turn to these resources when a patient reports use of herbal products.

IN SUMMARY

Herbal products are not reliable treatments for medical conditions; there is little evidence demonstrating their effectiveness and safety, and the concentrations of active ingredients may vary greatly. Manufacturers and distributors of herbal supplements are responsible for determining product safety, and the FDA is unable to take action against unsafe supplements before they reach the market. Safety concerns include adverse effects, herb-drug interactions, and contamination. Consumers using herbs may delay getting an appropriate treatment for their condition.

Diet-Drug Interactions

When working with patients, medical personnel should be alert to possible interactions between drugs and dietary substances. These interactions can raise health care costs and result in serious, and sometimes fatal, complications. With hundreds of diet-drug interactions known and more to be identified in the future, health professionals must learn to take steps to prevent or lessen their adverse consequences. Diet-drug interactions generally fall into the following categories:

- Drugs can alter food intake by reducing the appetite or by causing complications that make food consumption difficult or unpleasant. Some drugs may increase appetite and cause weight gain.

- Drugs can alter the absorption, metabolism, and excretion of nutrients. Conversely, nutrients and other food components can alter the absorption, metabolism, and excretion of drugs.

- Some interactions between dietary components and drugs can be toxic.

Examples of these types of diet-drug interactions are shown in Table 19-4. The remaining sections of this chapter describe some specific diet-drug interactions and their clinical significance.[16]

Drug Effects on Food Intake

Some drugs can make food intake difficult or unpleasant: they may suppress the appetite, alter taste sensations, induce nausea or vomiting, cause mouth dryness, or lead to inflammation or lesions in the mouth or GI tract. Certain side effects, including abdominal discomfort, constipation, and diarrhea, may be worsened by food consumption. Medications that cause drowsiness, such as sedatives and some painkillers, can make a person too tired to eat.

Drug complications that reduce food intake are significant only when they continue for a long period. Although many drugs can cause nausea in some individuals, the nausea often subsides after the first few doses of the medication and thus has little effect on nutrition status. If side effects persist, other medications may be used to treat them; for example, antinauseants and antiemetics may help to reduce nausea and vomiting and thereby improve food intake.

Some medications stimulate food intake and encourage weight gain. Unintentional weight gain may result from the use of some antipsychotics, antidepressants, and corticosteroids (for example, prednisone). People using these drugs may be unable to feel satiated and may gain 40 to 60 pounds in just a few months. For some conditions, however, weight gain is desirable. Patients with diseases that cause wasting, such as cancer or AIDS, are sometimes prescribed appetite enhancers such as megestrol acetate, a progesterone analog, or dronabinol, which is derived from the active ingredient in marijuana.

Drug Effects on Nutrient Absorption

The medications that most often cause widespread nutrient malabsorption are those that damage the intestinal mucosa. Antineoplastic and antiretroviral drugs ◆ are especially detrimental, although nonsteroidal anti-inflammatory drugs (NSAIDs) and some antibiotics can have similar, though milder, effects. This section describes additional ways in which medications may alter nutrient absorption.

◆ *Antineoplastic drugs* combat tumor growth. *Antiretroviral drugs* treat HIV infection.

Drug-Nutrient Binding Some medications bind nutrients in the GI tract, preventing their absorption. For example, bile acid binders, which reduce cholesterol levels, may bind to fat-soluble vitamins. Some antibiotics, notably tetracycline and ciprofloxacin, bind to the calcium in foods and supplements, reducing the absorption

TABLE 19-4	Examples of Diet-Drug Interactions

Drugs May Alter Food Intake by:

Altering the appetite (amphetamines suppress appetite; corticosteroids increase appetite).

Interfering with taste or smell (amphetamines change taste perceptions).

Inducing nausea or vomiting (digitalis may do both).

Interfering with oral function (some antidepressants may cause dry mouth).

Causing sores or inflammation in the mouth (methotrexate may cause painful mouth ulcers).

Drugs May Alter Nutrient Absorption by:

Changing the acidity of the digestive tract (antacids may interfere with iron and folate absorption).

Damaging mucosal cells (cancer chemotherapy may damage mucosal cells).

Binding to nutrients (bile acid binders bind to fat-soluble vitamins).

Foods and Nutrients May Alter Drug Absorption by:

Stimulating secretion of gastric acid (the antifungal agent ketoconazole is absorbed better with meals due to increased acid secretion).

Altering rate of gastric emptying (intestinal absorption of drugs may be delayed when they are taken with food).

Binding to drugs (calcium binds to tetracycline, reducing drug and calcium absorption).

Competing for absorption sites in the intestine (dietary amino acids interfere with levodopa absorption).

Drugs and Nutrients May Interact and Alter Metabolism by:

Acting as structural analogs (as do warfarin and vitamin K).

Using similar enzyme systems (phenobarbital induces liver enzymes that increase metabolism of folate, vitamin D, and vitamin K).

Competing for transport on plasma proteins (fatty acids and drugs may compete for the same sites on the plasma protein albumin).

Drugs May Alter Nutrient Excretion by:

Altering reabsorption in the kidneys (some diuretics increase the excretion of sodium and potassium).

Causing diarrhea or vomiting (diarrhea and vomiting may cause electrolyte losses).

Food May Alter Medication Excretion by:

Inducing activities of liver enzymes that metabolize drugs to allow their excretion (components of charcoal-broiled meats increase metabolism of warfarin, theophylline, and acetaminophen).

Toxicity May Occur from Combining Foods and Drugs by:

Increasing side effects of the drug (caffeine in beverages can increase adverse effects of stimulants).

Increasing drug action to excessive levels (grapefruit components may block metabolism of drugs and enhance drugs' actions and side effects).

---

of both the calcium and the antibiotic. Other minerals, such as iron, magnesium, and zinc, can also bind to some antibiotics. Consumers are advised to use dairy products and all mineral supplements at least 2 hours apart from these medications.

Altered Stomach Acidity Medications that reduce stomach acidity can impair the absorption of vitamin B_{12}, folate, and iron. Examples include antacids, which neutralize stomach acid by acting as weak bases, and antiulcer drugs (such as proton pump inhibitors and H2 blockers), which interfere with acid secretion.

Direct Inhibition Several drugs impede nutrient absorption by interfering with their intestinal metabolism or transport into mucosal cells. For example, the antibiotics trimethoprim and pyrimethamine compete with folate for absorption into intestinal cells.

To help prevent diet-drug interactions, ask about *all* of the drugs and supplements a patient takes, including prescription and over-the-counter drugs, herbal products, and other dietary supplements.

◆ Reminder: *Phytates* are compounds found in many plant foods, including whole grains and legumes. Phytates can bind to minerals and other substances and reduce their absorption.

◆ Cortisol is a steroid hormone secreted by the adrenal cortex as part of the body's stress response.

Dietary Effects on Drug Absorption

Most drugs are absorbed in the upper small intestine. Major influences on drug absorption include the stomach-emptying rate, the level of acidity in the stomach, and direct interactions with dietary components. The drug's formulation may also influence its absorption. The instructions included with medications typically advise whether food should be included or avoided with use.

Stomach-Emptying Rate Drugs reach the small intestine more quickly when the stomach is empty. Therefore, taking a medication with meals may delay its absorption, although the total amount absorbed may not be lower. As an example, aspirin works faster when taken on an empty stomach, although taking it with food is often encouraged to reduce stomach irritation.

Slow stomach emptying can enhance drug absorption, because the drug's absorption sites in the small intestine do not become saturated. Slow drug absorption due to slow stomach emptying can be a problem, however, if high drug concentrations are needed for effectiveness, as when a hypnotic is taken to induce sleep.

Stomach Acidity Some drugs are better absorbed in an acidic environment, so conditions that are too alkaline may reduce their absorption. Hence, antacid medications may reduce the absorption of these drugs. Other drugs can be damaged by acid and are often available in coated forms that resist the stomach's acidity.

Interactions with Dietary Components Some dietary substances can bind to drugs and inhibit their absorption. For example, the phytates in foods can bind to digoxin, a drug prescribed for heart disease. ◆ High-fiber diets may decrease the absorption of some tricyclic antidepressants. As mentioned earlier, calcium can bind to some antibiotics, reducing absorption of both the calcium and the drug.

Drug Effects on Nutrient Metabolism

Drugs and nutrients share similar enzyme systems in the small intestine and liver. Consequently, some drugs may enhance or inhibit the activities of enzymes needed for nutrient metabolism. For example, the anticonvulsants phenobarbital and phenytoin increase levels of the liver enzymes that metabolize folate, vitamin D, and vitamin K; therefore, persons using these drugs may require supplements of these vitamins.

The drug methotrexate, used to treat cancer and inflammatory conditions, acts by interfering with folate metabolism and thus depriving rapidly dividing cancer cells of the folate they need to multiply. Methotrexate resembles folate in structure (see Figure 19-1) and competes with folate for the enzyme that converts folate to its active form. The adverse effects of using methotrexate therefore include symptoms of folate deficiency. These adverse effects can be reduced by using a preactivated form of folate (called leucovorin), which is often prescribed along with methotrexate to ensure that the body's rapidly dividing cells (such as cells of the digestive tract, skin cells, and red blood cells) receive adequate folate.

Isoniazid (INH), an antituberculosis drug, sometimes induces vitamin B_6 deficiency. Isoniazid is similar in structure to vitamin B_6 and can therefore interfere with the vitamin's conversion to its active form. Because the drug must be taken for at least six months to treat infection, vitamin B_6 supplements are often given to prevent deficiency.[17]

Corticosteroids, used as anti-inflammatory agents and immunosuppressants, have actions that mimic those of the hormone cortisol. ◆ Long-term corticosteroid use can have broad effects on nutritional health and may cause weight gain, muscle wasting, bone loss, and hyperglycemia, with eventual development of osteoporosis and diabetes.

FIGURE 19-1 Folate and Methotrexate

By competing for the enzyme that activates folate, methotrexate prevents cancer cells from obtaining the folate they need to multiply. In the process, normal cells are also deprived of the folate they need.

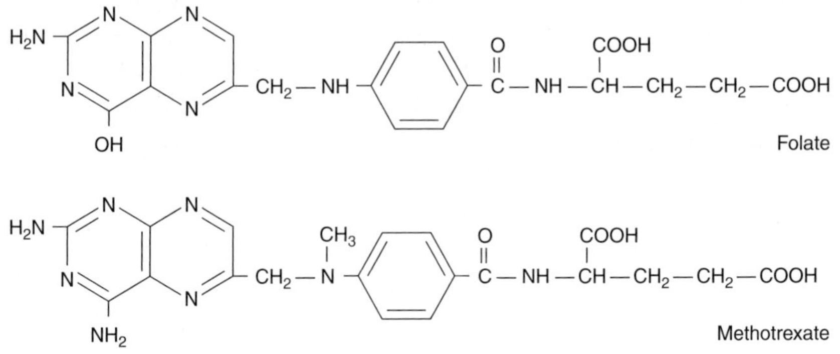

Dietary Effects on Drug Metabolism

Some food components alter the activities of enzymes that metabolize drugs or may counteract drug effects in other ways. Compounds in grapefruit juice (and whole grapefruit) have been found to inhibit or inactivate enzymes that metabolize a number of different drugs. As a result of the reduced enzyme action, blood concentrations of the drugs increase, leading to stronger physiological effects. The effect of the grapefruit juice lasts for a substantial period after the juice is consumed; for example, in experiments with a drug prescribed for heart disease, the juice's effect had an estimated half-life of 12 hours.[18] ◆ Table 19-5 on p. 652 provides examples of drugs that interact with grapefruit juice, as well as some common drugs that are unaffected.

A number of dietary substances can alter the activity of the anticoagulant drug warfarin. The most important interaction is with vitamin K, which is structurally similar to warfarin. Warfarin acts by blocking the enzyme that activates vitamin K, thereby preventing the synthesis of blood-clotting factors. ◆ The amount of warfarin prescribed is dependent, in part, on how much vitamin K is in the diet. If vitamin K consumption from foods or supplements increases substantially, it can weaken the effect of the drug. Individuals using warfarin are advised to consume similar amounts of vitamin K daily to keep warfarin activity stable. The dietary sources highest in vitamin K are green leafy vegetables.

Several popular herbs contain natural compounds that may enhance the activity of warfarin and therefore should be avoided during warfarin treatment. These herbs include St. John's wort, garlic, ginseng, dong quai, danshen, and others.[19]

Drug Effects on Nutrient Excretion

Drugs that enhance urinary excretion can interfere with nutrient reabsorption in the kidneys, ◆ resulting in greater urinary losses of the nutrients. For example, some diuretics can accelerate the excretion of calcium, potassium, magnesium, and thiamin, and dietary supplements may be necessary to avoid deficiency. Risk of nutrient depletion is highest if multiple drugs with the same effect are used, if kidney function is impaired, or if the medications are used for a long time. Note that some diuretics can cause certain minerals to be retained, rather than excreted.[20]

◆ Reminder: The term *half-life* can be used to define the time period of a chemical effect. If the grapefruit effect has a 12-hour half-life, this means that after 12 hours, its biological effect is half of the maximum effect measured.

◆ Reminder: Vitamin K is required for the synthesis of prothrombin and several other blood-clotting proteins.

◆ When the kidneys reabsorb a substance, they retain it in the blood. Substances that are not reabsorbed are excreted in urine.

TABLE 19-5 Examples of Grapefruit Juice–Drug Interactions

Drug Category	Drugs Affected by Grapefruit Juice	Drugs Unaffected by Grapefruit Juice
Cardiovascular drugs	Felodipine	Amlodipine
	Nicardipine	Diltiazem
	Nifedipine	Propafenone
	Verapamil	Quinidine
Cholesterol-lowering drugs	Atorvastatin	Pravastatin
	Lovastatin	
	Simvastatin	
Central nervous system drugs	Buspirone	Clomipramine
	Carbamazepine	Haloperidol
	Diazepam	
	Triazolam	
Anti-infective drugs	Saquinavir	Clarithromycin
		Itraconazole
Estrogens	Ethinylestradiol	17-β-estradiol
Anticoagulants	—	Acenocoumarol
		Warfarin
Immunosuppressants	Cyclosporine	Prednisone
	Tacrolimus	
Antiasthmatic drugs	—	Theophylline

SOURCE: D. G. Bailey, M. O. Arnold, and J. D. Spence, Inhibitors in the diet: Grapefruit juice–drug interactions, in R. H. Levy and coeditors, *Metabolic Drug Interactions* (Phildelphia: Lippincott Williams & Wilkins, 2000), pp. 661–669.

Dietary Effects on Drug Excretion

Inadequate excretion of medications can cause toxicity, whereas excessive losses may reduce the amount available for therapeutic effect. Some food components can alter drug reabsorption by the kidneys. For example, the amount of the medication lithium ◆ reabsorbed by the kidneys is similar to the amount of sodium reabsorbed. Thus, both sodium depletion and dehydration, which increase sodium reabsorption, can result in lithium retention. Similarly, a person with a high sodium intake will excrete more sodium in the urine, and therefore more lithium. Individuals using lithium are advised to maintain a consistent sodium intake from day to day in order to maintain a stable blood level of lithium.[21]

Urine acidity can affect drug excretion due to the effects of pH on a compound's ionic (chemical) form. The medication quinidine, used to treat arrhythmias, is excreted more readily in acidic urine. Foods or drugs that cause urine to become more alkaline may reduce quinidine excretion and raise blood levels of the medication.

Diet-Drug Interactions and Toxicity

Interactions between food components and drugs can cause toxicity or exacerbate a drug's side effects. The combination of tyramine, a food component, and monoamine oxidase (MAO) inhibitors, which include some antidepressants and a medication that treats Parkinson's disease, can be fatal. MAO inhibitors block an enzyme that normally inactivates tyramine, as well as the hormones epinephrine

◆ Lithium is used to prevent mood swings in patients with bipolar disorder.

and norepinephrine. When people who take MAO inhibitors consume excessive tyramine, the increased tyramine in the blood can induce a sudden release of accumulated norepinephrine. This surge in norepinephrine results in severe headaches, rapid heartbeat, and a dangerous rise in blood pressure. For this reason, people taking MAO inhibitors are advised to restrict their intakes of foods rich in tyramine.

Tyramine occurs naturally in foods and is also formed when bacteria degrade the protein in foods. Thus, the tyramine content of a food usually increases when a food ages or spoils. Individuals at risk of tyramine toxicity are advised to buy mainly fresh foods and consume them promptly.[22] Foods that often contain substantial amounts of tyramine are listed in Table 19-6.

Considering the number of medications available and the many ways in which drugs and dietary substances can interact, it should not be surprising that serious side effects are increasingly recognized. Health professionals should attempt to understand the mechanisms of diet-drug interactions, identify them when they occur, and prevent them whenever possible. The "How to" below offers some practical advice about preventing diet-drug interactions.

TABLE 19-6 Examples of Foods with a High Tyramine Content[a]

- Aged cheeses
- Aged meats
- Alcoholic beverages (beer, wine)
- Anchovies
- Caviar
- Fava beans
- Fermented foods (sauerkraut, sausages)
- Feta cheese
- Lima beans
- Mushrooms
- Pickled fish or meat
- Prepared soy foods (miso, tempeh, tofu)
- Smoked fish or meat
- Soy sauce
- Yeast extract (Marmite)

[a]The tyramine content of foods depends on storage conditions and processing; thus the amounts in similar products can vary substantially.

IN SUMMARY

Medications can alter food intake and affect the absorption, metabolism, and excretion of nutrients. Components of foods can similarly affect the absorption, metabolism, and excretion of medications. Some drugs may reduce appetite and cause damage to the GI tract. Binding between drugs and nutrients may inhibit their absorption. Drugs and nutrients may interfere with each other's metabolism because they use similar enzymes in the small intestine and liver. Diet-drug interactions may cause excessive losses of nutrients and alter the urinary excretion of medications.

HOW TO Prevent Diet-Drug Interactions

The Joint Commission, an accreditation agency for health care organizations, has recommended that all patients be educated about potential diet-drug interactions. Health professionals can help by informing patients of precautions related to medications and watching for signs of problems that may arise.

To prevent diet-drug interactions, first list the types and amounts of over-the-counter drugs, prescription medications, and dietary supplements that the patient uses on a regular basis. Look up each medication in a drug reference and make a note of:

- The appropriate method of administration (twice daily or at bedtime, for example).
- How the medication should be administered with respect to foods, beverages, and specific nutrients (for example, take on an empty stomach, take with food, do not take with milk, or do not drink alcoholic beverages while using the medication).
- How the medication should be used with respect to other medications.
- The side effects that may affect food intake (nausea and vomiting, constipation or diarrhea, or sedation, for example) or nutrient needs (interference with nutrient absorption or metabolism, for example).

A similar process can be used to review the dietary supplements that a person is taking. A reliable reference may list their appropriate uses, possible side effects, and potential interactions with food and medications.

Patients who take multiple medications may need help learning when to take each medication to avoid drug-drug or diet-drug interactions. The health practitioner can use information from a patient's diet history (see Chapter 17) to help the patient coordinate meals and drugs so as to avoid interactions.

Some medications have well-known effects on nutrition status. The health practitioner should remain alert for signs of problems, especially when:

- Nutritional problems are a frequent result of using the medication.
- A patient uses multiple medications.
- The patient is in a high-risk group, for example, a child, a pregnant or lactating woman, an older adult, or a person who is malnourished, abuses alcohol, or has impaired liver or kidney function.
- The patient needs to use the medication for a long period of time.

Check with the pharmacist for additional information about medications and potential interactions.

Clinical Portfolio

▨ A patient mentions that he regularly takes six to seven dietary and herbal supplements and that he has not told the physician that he uses them. His prescription medications include an antihypertensive agent (to reduce blood pressure) and warfarin. What approach might you take to learn the details about the patient's supplement use and his reasons for taking them? If you discover that some of the supplements may pose a risk for diet-drug or herb-drug interactions with his prescription medications, what steps should you take?

▨ An elderly woman in a residential home has been losing weight since her arrival there. She has been taking several medications to treat both a heart problem and a mild case of bronchitis. You notice that she eats only a few bites at mealtimes and seems disinterested in food. Describe several steps you can take to learn whether the medications are interfering with her food intake in some way.

NUTRITION ON THE NET

- Visit the home page of the U.S. Food and Drug Administration (FDA), the agency that regulates all drugs in the United States: **www.fda.gov**

- The FDA provides safety information about drugs and other medical products on the MedWatch website: **www.fda.gov/ medwatch**

- The Medline Plus website provides information about drugs, dietary supplements, and herbal products: **www.nlm.nih .gov/medlineplus/druginformation.html**

- Find information about dietary supplements at the Office of Dietary Supplements, a division of the National Institutes of Health: **dietary-supplements.info.nih.gov/**

STUDY QUESTIONS

These questions will help you review the chapter. You will find the answers in the discussions on the pages provided.

1. Describe similarities and differences among prescription medications, over-the-counter drugs, and generic versions of drugs. (pp. 641–642)

2. Identify factors that frequently cause medication errors. What recommendations have been proposed to reduce the incidences of these errors? (pp. 642–643)

3. Explain why some patient populations are at high risk for adverse effects from drugs. (pp. 643–644)

4. Explain why herbal products are not dependable treatments for medical conditions. Describe possible dangers associated with the use of these products. (pp. 644–647)

5. Discuss ways in which medications can affect food intake. (pp. 648–649)

6. Describe how medications can interfere with nutrient absorption and how dietary factors can affect drug absorption. (pp. 648–650)

7. Explain how drugs and nutrients may influence each other's metabolism, and provide examples. (pp. 650–651)

8. Discuss diet-drug interactions that can alter the excretion of nutrients or medications. Explain why tyramine intake must be monitored by people using monoamine oxidase (MAO) inhibitors. (pp. 651–653)

These multiple choice questions will help you prepare for an exam. Answers can be found on p. 656.

1. Over-the-counter drugs are:
   a. unlikely to cause adverse effects.
   b. unlikely to interact with dietary components.
   c. generally used for longer periods of time than prescription medications.
   d. used to treat illnesses that are normally self-diagnosed and self-treated.

2. Recommendations for reducing incidences of medication errors include:
   a. physician supervision whenever drugs are administered.
   b. advising patients to take only one medication at a time.
   c. requiring that prescriptions be typed instead of handwritten.
   d. avoiding the use of confusing terms on clinical documents.

3. Adverse drug effects are most likely when:
   a. multiple medications are used.
   b. generic drugs are substituted for brand-name drugs.
   c. patients begin using a new medication.
   d. medications are taken for just one or two days.

4. An important difference between medications and herbal products that reach the marketplace is that:
   a. medications that cause adverse effects cannot be sold.
   b. medications are subject to contamination with toxic metals, molds, and bacteria.
   c. herbal products are not required to prove safety and effectiveness.
   d. herbal products must provide standard amounts of active ingredients.

5. An important step that health practitioners can take to limit the risk of medication-related side effects is to:
   a. recommend use of over-the-counter drugs instead of prescription medications.
   b. encourage use of herbal supplements rather than prescription medications.
   c. advise patients to take medications separately from meals.
   d. ask patients to fully describe the types and amounts of medications and dietary supplements they are using.

6. Examples of medication-related symptoms that can significantly limit food intake include:
   a. ringing in the ears.
   b. persistent nausea and vomiting.
   c. insomnia.
   d. skin rash.

7. Factors that typically interfere with drug absorption include:
   a. binding between drugs and food components.
   b. use of antacid therapies.
   c. a rapid stomach-emptying rate.
   d. all of the above.

8. Compounds in grapefruit juice:
   a. bind to antibiotics, reducing absorption.
   b. cause excessive drug excretion.
   c. strengthen the effects of certain drugs.
   d. alter acidity in the stomach, impairing drug absorption.

9. Vitamin K consumption should be consistent in patients using:
   a. tetracycline.
   b. isoniazid.
   c. warfarin.
   d. lithium.

10. People who use MAO inhibiters must limit consumption of:
    a. whole milk and yogurt.
    b. aged cheeses.
    c. dark green leafy vegetables.
    d. grapefruit juice.

REFERENCES

1. U.S. Food and Drug Administration, *CDER 2005 Report to the Nation: Improving Public Health Through Human Drugs* (Rockville, Md.: U.S. Food and Drug Administration, 2005).
2. U.S. Food and Drug Administration, 2005.
3. Institute of Medicine, Committee on Identifying and Preventing Medication Errors, *Preventing Medication Errors* (Washington, D.C.: National Academy Press, 2007).
4. S. Gunther and coauthors, Demographic and health-related correlates of herbal and specialty supplement use, *Journal of the American Dietetic Association* 104 (2004): 27–34.
5. J. Kennedy, Herb and supplement use in the U.S. adult population, *Clinical Therapeutics* 27 (2005): 1847–1858.
6. J. J. Mucksavage and L.-N. Chan, Dietary supplement interactions with medication, in J. I. Boullata and V. T. Armenti, eds., *Handbook of Drug-Nutrient Interactions* (Totowa, N.J.: Humana Press, 2004), pp. 217–233; C. H. Halsted, Dietary supplements and functional foods: 2 sides of a coin? *American Journal of Clinical Nutrition* 77 (2003): 1001S–1007S; J. Barnes, L. A. Anderson, and J. D. Phillipson, *Herbal Medicines: A Guide for Healthcare Professionals* (Chicago: Pharmaceutical Press, 2002); M. Rotblatt and I. Ziment, eds., *Evidence-Based Herbal Medicine* (Philadelphia: Hanley & Belfus, 2002).
7. ConsumerLab.com, Product review: Memory enhancement supplements (ginkgo, huperzine A, phosphatidylserine, and acetyl-L-carnitine), available at www.consumerlab.com/results/ginkgobiloba.asp, posted November 30, 2005; site visited September 1, 2007.
8. C. M. Gilroy and coauthors, Echinacea and truth in labeling, *Archives of Internal Medicine* 163 (2003): 699–704.
9. Halsted, 2003.
10. Mucksavage and Chan, 2004; B. I. Gurley and D. W. Hagan, Herbal and dietary supplement interactions with drugs, in B. J. McCabe, E. H. Frankel, and J. J. Wolfe, eds., *Handbook of Food-Drug Interactions* (Boca Raton, Fla.: CRC Press, 2003), pp. 259–293.
11. Barnes, Anderson, and Phillipson, 2002.
12. ConsumerLab.com, 2005.
13. V. H. Tournas, E. Katsoudas, and E. J. Miracco, Moulds, yeasts, and aerobic plate counts in ginseng supplements, *International Journal of Food Microbiology* 108 (2006): 178–181; K. S. Leung and coauthors, Systematic evaluation of organochlorine pesticide residues in Chinese materia medica, *Phytotherapy Research* 19 (2005): 514–518.
14. R. J. Ko, A U.S. perspective on the adverse reactions from traditional Chinese medicines, *Journal of the Chinese Medical Association* 67 (2004): 109–116; E. Ernst, Toxic heavy metals and undeclared drugs in Asian herbal medicines, *Trends in Pharmacological Sciences* 23 (2002): 136–139.
15. Barnes, Anderson, and Phillipson, 2002.
16. L.-N. Chan, Drug-nutrient interactions, in M. E. Shils and coeditors, *Modern Nutrition in Health and Disease* (Baltimore: Lippincott

Williams & Wilkins, 2006), pp. 1539–1553.

17. I. F. Btaiche and M. D. Kraft, Nutrients that may optimize drug effects, in J. I. Boullata and V. T. Armenti, eds., *Handbook of Drug-Nutrient Interactions* (Totowa, N.J.: Humana Press, 2004), pp. 195–216.

18. D. G. Bailey, Grapefruit juice–drug interaction issues, in J. I. Boullata and V. T. Armenti, eds., *Handbook of Drug-Nutrient Interactions* (Totowa, N.J.: Humana Press, 2004), pp. 175–194.

19. M. L. Chavez, M. A. Jordan, and P. I. Chavez, Evidence-based drug–herbal interactions, *Life Sciences* 78 (2006): 2146–2157.

20. S. A. Shapses, Y. R. Schlussel, and M. Cifuentes, Drug-nutrient interactions that impact mineral status, in J. I. Boullata and V. T. Armenti, eds., *Handbook of Drug-Nutrient Interactions* (Totowa, N.J.: Humana Press, 2004), pp. 301–328; B. J. McCabe, E. H. Frankel, and J. J. Wolfe, Monitoring nutritional status in drug regimens, in B. J. McCabe, E. H. Frankel, and J. J. Wolfe, eds., *Handbook of Food-Drug Interactions* (Boca Raton, Fla.: CRC Press, 2003), pp. 73–108.

21. McCabe, Frankel, and Wolfe, Monitoring nutritional status in drug regimens, 2003.

22. B. J. McCabe, Dietary counseling to prevent food-drug interactions, in B. J. McCabe, E. H. Frankel, and J. J. Wolfe, eds., *Handbook of Food-Drug Interactions* (Boca Raton, Fla.: CRC Press, 2003), pp. 295–324.

ANSWERS

Study Questions (multiple choice)

1. d   2. d   3. a   4. c   5. d   6. b   7. d   8. c   9. c   10. b

Anemia in Illness

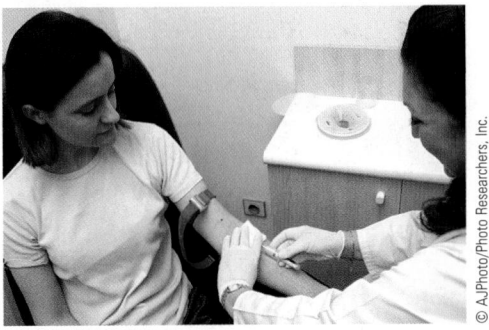

© AJPhoto/Photo Researchers, Inc.

Anemia—a reduction of red blood cells that lowers the oxygen-carrying capacity of the blood—is frequently the first sign of illness and may be the disorder that initially drives an individual to seek medical attention. Anemia is associated with a great number of diseases and is common among hospital patients: some 20 to 40 percent exhibit some degree of anemia.[1] Earlier chapters in this textbook described some of the relationships between anemia and nutrient deficiencies. This highlight explains how and why anemia develops during the course of illness. The glossary below defines relevant terms.

Overview of Anemia

Anemia develops when red blood cells (also called *erythrocytes*) are unable to be produced in sufficient numbers, are too quickly destroyed, or are lost due to bleeding. Because red blood cells contain the hemoglobin that supplies oxygen to tissues, their absence can result in fatigue and reduced stamina. The deficiency of oxygen in tissues is the main stimulus for the production of additional red blood cells.

Red Blood Cell Production

The production of red blood cells **(erythropoiesis)** takes place in the bone marrow, which is a soft tissue found in certain types of bone. The process begins when kidney cells sense the low oxygen content of blood and release the hormone **erythropoietin** (see Figure H19-1 on p. 658). Erythropoietin travels to the bone marrow, where it stimulates precursor cells (stem cells) to divide and differentiate into red blood cells. The cells that are released

from the bone marrow are immature red blood cells called **reticulocytes.** Reticulocytes develop into mature red blood cells over a 24- to 48-hour period while they circulate in the bloodstream.

Nutritional Anemias

The nutrient deficiencies that most frequently upset red blood cell production are those of iron, folate, and vitamin B_{12}. Iron is required for hemoglobin production, and deficiency results in **microcytic anemia,** characterized by small, hypochromic cells (see p. 446). Vitamin B_{12} and folate participate in DNA synthesis, and deficiency of either nutrient leads to **macrocytic anemia,** characterized by large immature cells (see pp. 343–344).

Other nutrient deficiencies may cause anemia, although not as frequently. Vitamin E helps to maintain cell membrane integrity, and its deficiency is associated with **hemolytic anemia** (red blood cell breakdown). Vitamin B_6 plays a role in hemoglobin production, and a deficiency can sometimes cause microcytic anemia. Vitamin C supports blood vessel integrity; fragile and bleeding capillaries may result from deficiency. Protein-energy malnutrition leads to anemia because red blood cell development depends on protein synthesis. Although nutrient deficiencies may result from dietary inadequacy, they can also arise during the course of illness due to the effects of disease on intestinal absorption, nutrient metabolism, and nutrient losses.

Identifying Causes of Anemia

Identifying the cause of anemia is sometimes quite challenging. In some cases, anemia may be a well-known consequence of disease, as when renal failure impairs the synthesis of the hormone

GLOSSARY

anemia of chronic disease: anemia that develops in persons with chronic illness; may resemble iron-deficiency anemia even though iron stores are often adequate.

aplastic anemia: anemia characterized by the inability of bone marrow to produce adequate numbers of blood cells. Causes include genetic

defects, viruses, radiation treatment, and drug toxicity.

erythropoiesis (eh-RIH-throh-poy-EE-sis): production of red blood cells within the bone marrow.

erythropoietin (eh-RIH-throh-POY-eh-tin): a hormone produced by kidney cells that stimulates red blood cell production.

hemolytic (hee-moe-LIH-tic) **anemia:** anemia characterized by the breakdown of red blood cells.

macrocytic anemia: anemia characterized by large red blood cells, as occurs in folate and vitamin B_{12} deficiency; also called **megaloblastic anemia.**

microcytic anemia: anemia characterized by small, hypochromic (pale) red blood

cells, as occurs in iron deficiency.

peripheral blood smear: a blood sample spread on a glass slide and stained for analysis under a microscope. *Peripheral* refers to the use of circulating blood rather than tissue blood.

reticulocytes: immature red blood cells released into blood by the bone marrow.

FIGURE H19-1 Erythropoiesis

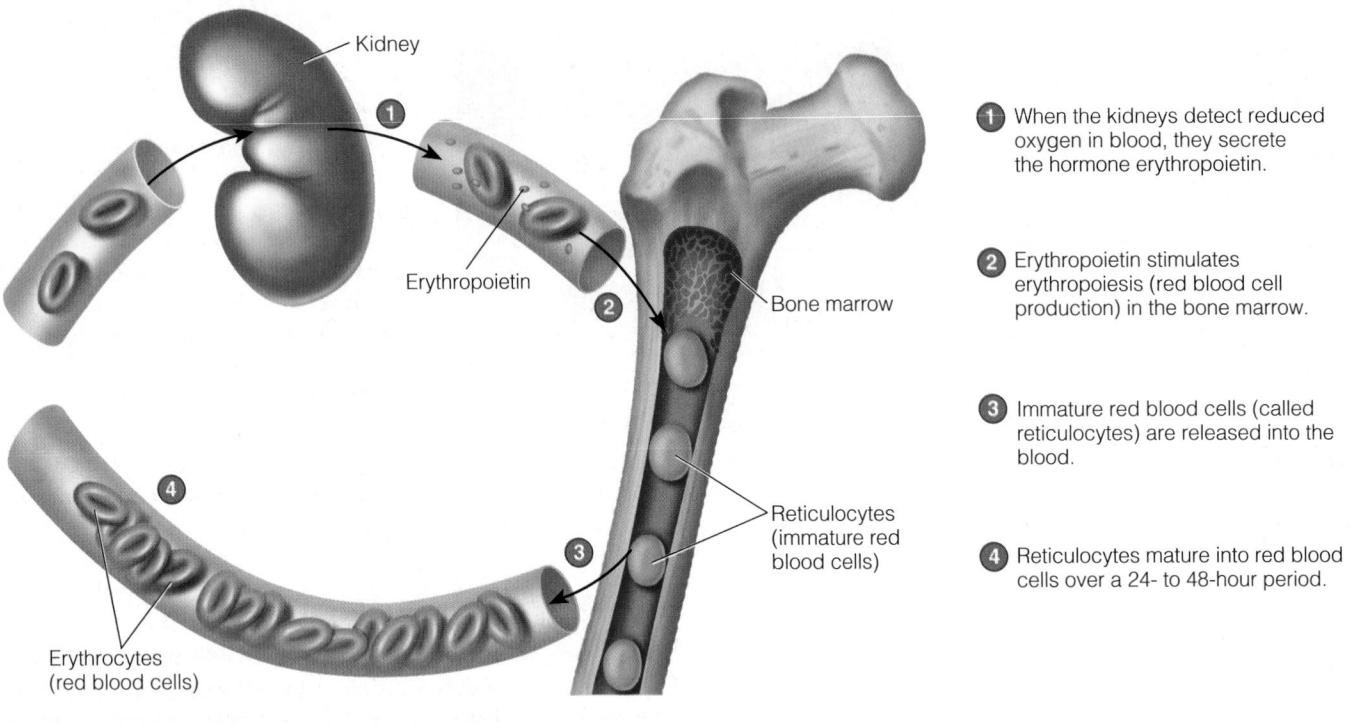

Kidney

Erythropoietin

Bone marrow

Reticulocytes
(immature red
blood cells)

Erythrocytes
(red blood cells)

1. When the kidneys detect reduced oxygen in blood, they secrete the hormone erythropoietin.

2. Erythropoietin stimulates erythropoiesis (red blood cell production) in the bone marrow.

3. Immature red blood cells (called reticulocytes) are released into the blood.

4. Reticulocytes mature into red blood cells over a 24- to 48-hour period.

SOURCE: Reprinted with permission from L. Sherwood, *Human Physiology,* 5th ed., (Brooks/Cole, 2004), Figure 11-4, p. 395.

erythropoietin. When anemia develops rapidly, blood loss is often the cause, whereas a more gradual onset suggests malnutrition, chronic illness, or slow, chronic bleeding. The results of laboratory tests provide valuable clues, although conditions such as dehydration and inflammation can influence the values. Laboratory results are especially difficult to analyze if several disturbances are present simultaneously. A **peripheral blood smear** (see the photo) is often used to study abnormalities in red blood cell shape and may also reveal an underlying cause.

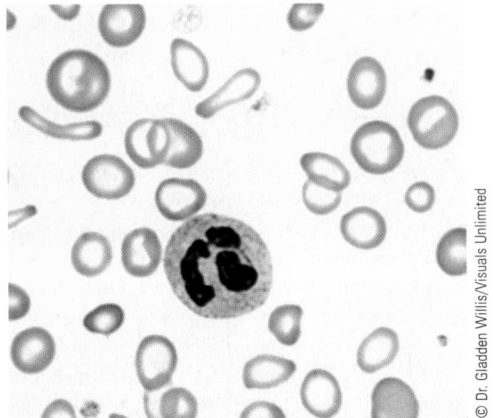

A peripheral blood smear provides information about the number and shape of blood cells.

Nutritional Anemias in Illness

There are various ways in which illnesses can lead to iron, folate, or vitamin B_{12} deficiencies, the main causes of the nutritional anemias. Blood loss, common to many illnesses, is a primary cause of iron deficiency. Some illnesses may result in a reduction in food intake, as discussed in Chapter 17. The liver's stores of iron and vitamin B_{12} are often adequate to prevent deficiencies during transient illnesses, but reserves of folate are limited; thus a folate deficiency can develop within a few months if dietary intakes are low. If several nutrient deficiencies occur simultaneously, it may be difficult to identify the cause of anemia using standard blood tests (see Appendix E) because both macrocytic and microcytic anemia may be present.

Blood Loss

As mentioned, blood loss can eventually lead to iron deficiency. Unfortunately, slow, chronic bleeding may be difficult to identify before anemia develops.[2] Gastrointestinal conditions often cause bleeding; examples include peptic ulcers, inflammatory bowel conditions, and gastrointestinal varices (enlarged veins) that develop in advanced liver disease. Excessive bleeding can accompany coagulation disorders, which are usually due to liver disease, genetic defects, or vitamin K deficiency. Frequent blood draws or surgical procedures also contribute to blood loss.

Nutrient Malabsorption

Chapter 24 explains how disorders that damage the small intestine can lead to nutrient malabsorption. Diseases like Crohn's disease and celiac disease can destroy intestinal mucosa and reduce the absorption of all nutrients. Iron is primarily absorbed in the duodenum and upper jejunum, and its absorption is impaired by conditions that reduce hydrochloric acid secretion or result in surgical resection (removal) of the upper intestine. Resection of the stomach or ileum can hasten the onset of vitamin B_{12} deficiency because both organs have roles in vitamin B_{12} absorption: you may recall from Chapter 10 that the stomach produces a protein (called *intrinsic factor*) needed for vitamin B_{12} absorption and that the ileum is the site of vitamin B_{12} absorption.

Anemia of Chronic Disease

Chronic disease itself can cause anemia, and anemia is sometimes the initial sign that chronic disease is present.[3] In fact, the **anemia of chronic disease** is the most common type of anemia affecting hospitalized patients and patients with chronic illnesses.[4] This type of anemia usually occurs in individuals who have chronic infections, inflammatory conditions, autoimmune disorders, or cancer. Although often a mild form of anemia, it can progress and become severe enough to require blood transfusions.[5]

The anemia of chronic disease is characterized by alterations both in the distribution of iron among tissues and in the rates of red blood cell production and destruction. As a result of the inflammatory response, macrophages in the liver, spleen, and bone marrow sequester iron, making it unavailable for erythropoiesis and hence slowing the rate of production of new red blood cells. In addition, red blood cells are destroyed more rapidly than usual, and the reduced production of red blood cells cannot keep pace. Finally, iron absorption is impaired, possibly because intestinal cells inhibit iron's release into blood. Eventu-

ally, outright iron deficiency may result from inadequate iron absorption.[6]

Blood tests help to distinguish between the anemia of chronic disease and iron-deficiency anemia (see Table H19-1). The combination of low serum iron and low total iron-binding capacity suggests the anemia of chronic disease rather than iron deficiency. In addition, serum ferritin levels are normal or elevated, whereas they are typically low in iron deficiency. Diagnosis is more complicated if both types of anemia are present.[7]

Medications and Anemia

Anemia is among the adverse effects that may result from medication use. Medications can alter nutrient metabolism, can impair blood coagulation and erythropoiesis, and sometimes lead to increased red blood cell destruction. Because the life span of red blood cells is about 120 days, the long-term use of medications is more likely to result in anemia than is short-term use.

Drug-Nutrient Interactions

As Chapter 19 described, there are numerous ways in which medications can alter nutrient metabolism; the most common are listed in Table 19-4 on p. 649. A number of medications are known to influence the absorption or metabolism of folate and lead to macrocytic anemia. Sulfasalazine (used for ulcerative colitis) and some anticonvulsant drugs inhibit folate absorption, and methotrexate (an immunosuppressive), triamterene (a diuretic), and pyrimethamine (an antimalarial) interfere with folate metabolism.[8] If a medication is known to result in deficiency, nutrient supplementation is usually recommended as an adjunct therapy.

Impaired Coagulation

Anticoagulants, which are prescribed specifically to reduce blood clotting, sometimes lead to excessive bleeding. These medications work by interfering with one of the steps involved in blood clotting, such as platelet function, vitamin K function, or the synthesis of clotting proteins. Other than anticoagulants, drugs that impair coagulation include aspirin and other nonsteroidal anti-inflammatory drugs (NSAIDs), acetaminophen, cimetidine (Tagamet), ranitidine (Zantac), and thiazide diuretics.[9] The anticoagulant effects may be augmented if several of these drugs are used simultaneously. Sometimes the slow, chronic bleeding that develops when these drugs are taken may go unnoticed until excessive blood loss has occurred.

Aplastic Anemia

Many classes of drugs are associated with **aplastic anemia**, a type of anemia that occurs when the bone marrow fails to produce adequate numbers of blood cells. The categories of drugs associated with aplastic anemia include anticonvulsants, antibiotics, antidiabetic drugs, diuretics, antithyroid drugs, and anticancer agents.[10] Aplastic anemia can also be caused by a genetic defect or may result from viral infections or exposure to toxins.

TABLE H19-1 Laboratory Tests for Evaluating Iron Deficiency and Anemia of Chronic Disease

Laboratory Test	Effect of Iron Deficiency	Effect of Chronic Disease
Red blood cell (RBC) size and number	Microcytic; reduced RBC count	Normocytic or microcytic; reduced RBC count
Serum iron	Low	Low
Serum ferritin	Low	Normal or elevated
Serum transferrin	Elevated	Low
Total iron-binding capacity	High	Low
Bone marrow iron	Low	Normal or elevated

Hemolytic Anemia

Some patients may develop hemolytic anemia as a result of drug interactions with red blood cells. For example, a drug may alter the red blood cell membrane in such a way that a component of the membrane becomes an antigen and induces an antibody response that destroys the cell.[11] Several types of antibiotics, including penicillin and cephalosporin, may cause this type of response. Withdrawal of the drug can eventually reverse the anemia, and sometimes medications are given to suppress the immune response.

In conclusion, anemia is a disorder associated with many diseases, and it may also be caused by disease treatment. When anemia occurs during illness, its causes must be investigated before it leads to complications that worsen prognosis. The medical history, blood tests, and peripheral blood smears may all help to determine the reasons why anemia has developed.

REFERENCES

1. K. S. Zuckerman, Approach to the anemias, in L. Goldman and D. Ausiello, eds., *Cecil Textbook of Medicine* (Philadelphia: Saunders, 2004), pp. 963–971.
2. Zuckerman, 2004; J. E. Ansell, Cardinal manifestations of hematologic disease, anemias, and related conditions, in J. Noble and coeditors, *Textbook of Primary Care Medicine* (St. Louis: Mosby, 2001), pp. 1027–1037.
3. T. P. Duffy, Microcytic and hypochromic anemias, in L. Goldman and D. Ausiello, eds., *Cecil Textbook of Medicine* (Philadelphia: Saunders, 2004), pp. 1003–1008.
4. C. N. Roy, D. A. Weinstein, and N. D. Andrews, 2002 E. Mead Johnson Award for Research in Pediatrics lecture: The molecular biology of the anemia of chronic disease: A hypothesis, *Pediatric Research* 53 (2003): 507–512.
5. Roy, Weinstein, and Andrews, 2003.
6. Duffy, 2004; Roy, Weinstein, and Andrews, 2003; D. A. Weinstein and coauthors, Inappropriate expression of hepcidin is associated with iron refractory anemia: Implications for the anemia of chronic disease, *Blood* 100 (2002): 3776–3781.
7. Roy, Weinstein, and Andrews, 2003.
8. S. P. Stabler and R. H. Allen, Megaloblastic anemias, in L. Goldman and D. Ausiello, eds., *Cecil Textbook of Medicine* (Philadelphia: Saunders, 2004), pp. 1050–1057.
9. M. Shuman, Hemorrhagic disorders: Abnormalities of platelet and vascular function, in L. Goldman and D. Ausiello, eds., *Cecil Textbook of Medicine* (Philadelphia: Saunders, 2004), pp. 1060–1069.
10. H. Castro-Malaspina and R. J. O'Reilly, Aplastic anemia and related disorders, in L. Goldman and D. Ausiello, eds., *Cecil Textbook of Medicine* (Philadelphia: Saunders, 2004), pp. 1044–1050.
11. A. D. Schreiber, Autoimmune and intravascular hemolytic anemias, in L. Goldman and D. Ausiello, eds., *Cecil Textbook of Medicine* (Philadelphia: Saunders, 2004), pp. 1013–1021.

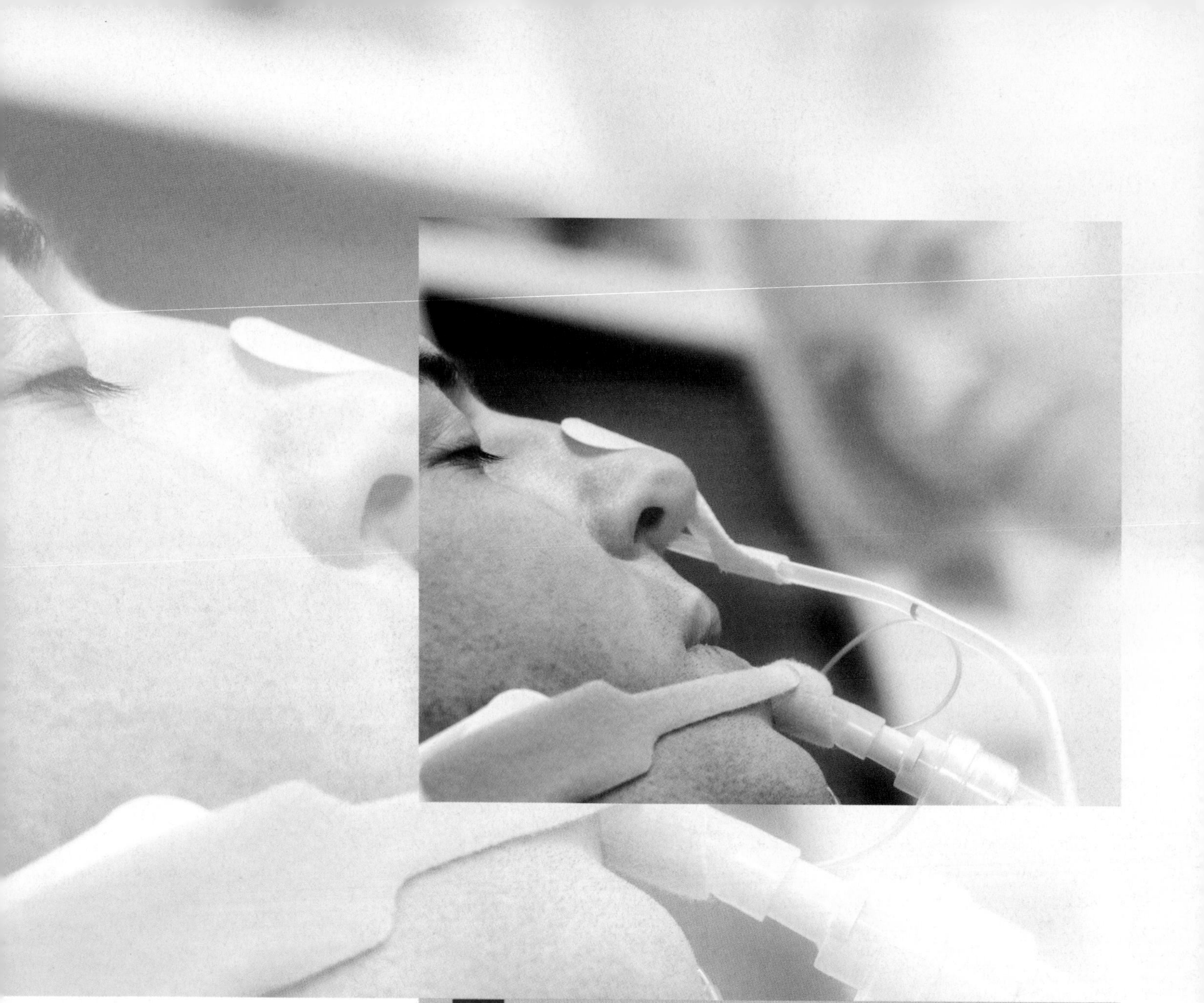

Nutrition in the Clinical Setting

Patients are often too sick to obtain the energy and nutrients they need by consuming foods. In such cases, enteral nutrition support can help many patients regain health. Because some enteral formulas are common grocery items, patients usually feel comfortable using them as oral supplements or as meal substitutes. Tube feedings, however, are unfamiliar to most people, and patients and caregivers may be resistant at first. Showing understanding and carefully explaining the procedure can help to alleviate patients' concerns.

Enteral Nutrition Support

Some illnesses may interfere with eating, digestion, or absorption to such a degree that conventional foods cannot supply the necessary nutrients. In such cases, **nutrition support**—the delivery of formulated nutrients (prepared nutrient solutions)—can meet a patient's nutritional needs. **Enteral nutrition** provides nutrients using the gastrointestinal (GI) tract. Enteral nutrition includes oral diets or supplements, but the term more often refers to the use of tube feedings, which supply nutrients directly to the stomach or intestine via a thin, flexible tube. **Parenteral nutrition,** discussed in Chapter 21, provides nutrients intravenously to patients who do not have adequate gastrointestinal function to handle enteral feedings. If the GI tract remains functional, enteral nutrition support is usually preferred, partly to avoid the expense and complications associated with intravenous feedings and partly to preserve healthy GI function.

If gastrointestinal function is normal and a poor appetite is the primary nutrition problem, enteral formulas are generally provided as an oral supplement to the usual diet. If patients cannot consume enough food or drink enough formula to meet nutrient needs, tube feedings are often used to deliver the required nutrients.

Enteral Formulas

Over 100 enteral formulas are currently marketed.[1] Most formulas can supply all of an individual's nutrient requirements when consumed in sufficient volume, a necessity for the patient who is using a tube feeding or oral liquid diet for more than a few days. Thus, an enteral formula can be considered a liquid form of a standard or modified diet.

Several enteral products are sold in pharmacies and grocery stores for home use; examples include Ensure, Boost, and Carnation Instant Breakfast. These

nutrition support: the delivery of formulated nutrients via a feeding tube or intravenous infusion.

enteral (EN-ter-al) **nutrition:** the provision of nutrients using the GI tract, including the use of tube feedings and oral diets.

parenteral (par-EN-ter-al) **nutrition:** the intravenous provision of nutrients that bypasses the GI tract.
- **par** = beside
- **entero** = intestine

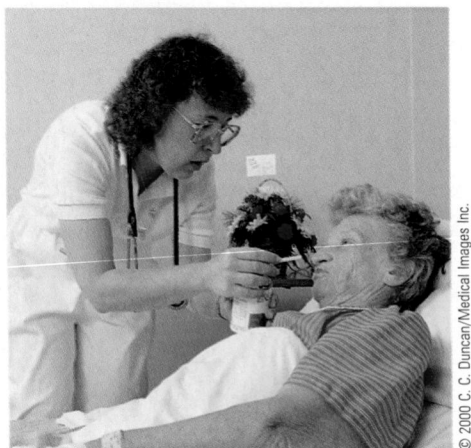

© 2000 C. Duncan/Medical Images Inc.

Patients can drink enteral formulas when they are unable to consume enough food from a conventional diet.

◆ Reminder: The *macronutrients* are carbohydrates, fats, and proteins.

products are used as dietary supplements by people who have trouble meeting nutritional needs or as convenient meal replacements by healthy individuals. The products are available in ready-to-drink liquid form or in powdered forms that must be reconstituted with water or milk.

Types of Enteral Formulas

Enteral formulas are categorized according to their macronutrient sources. ◆ **Standard formulas** usually contain intact proteins and polysaccharides, whereas **elemental formulas** contain macronutrients that have been broken down to some extent and require less digestion. **Specialized formulas** are designed to meet the specific needs of certain diseases. When an ideal formula is unavailable, a **modular formula** can be prepared in the hospital pharmacy by combining individual macronutrient preparations (called *modules*). Examples of enteral formulas are provided in Appendix K.

Standard Formulas Standard formulas, also called *polymeric formulas,* are provided to individuals who can digest and absorb nutrients without difficulty. They contain intact proteins extracted from milk or soybeans or a combination of **protein isolates** (proteins isolated from foods). The carbohydrate sources include modified starches, glucose polymers (such as maltodextrin), and sugars. A few formulas, called **blenderized formulas,** are made from whole foods and derive their protein primarily from pureed meat or poultry.

Elemental Formulas Elemental formulas, also called *hydrolyzed, chemically defined,* or *monomeric formulas,* are prescribed for patients who have compromised digestive or absorptive functions. Elemental formulas contain proteins and carbohydrates that have been partially or fully broken down to fragments that require little (if any) digestion. The formulas are often low in fat and may contain **medium-chain triglycerides (MCT)** to ease digestion and absorption. Table 20-1 compares the sources of macronutrients in standard and elemental formulas.

Specialized Formulas Specialized formulas, also called *disease-specific formulas,* are designed to meet the specific nutrient needs of patients with particular illnesses. Products have been developed for individuals with liver, kidney, and lung diseases; glucose intolerance; and metabolic stress. Disease-specific formulas are generally expensive, and their effectiveness is controversial.

Modular Formulas Modular formulas, created from individual macronutrient preparations called *modules,* are prepared for patients who require specific nutrient combinations to treat their illnesses. Vitamin and mineral preparations are

standard formulas: enteral formulas that contain mostly intact proteins and polysaccharides; also called *polymeric formulas.*

elemental formulas: enteral formulas that contain carbohydrates and proteins that are partially or fully hydrolyzed; also called *hydrolyzed, chemically defined,* or *monomeric formulas.*

specialized formulas: enteral formulas designed to meet the nutrient needs of patients with specific illnesses; also called *disease-specific formulas.*

modular formulas: enteral formulas prepared in the hospital from *modules* that contain single macronutrients; used for people with unique nutrient needs.

protein isolates: proteins that have been isolated from foods.

blenderized formulas: enteral formulas that are prepared by using a food blender to mix and puree whole foods.

medium-chain triglycerides (MCT): triglycerides that contain fatty acids that are 8 to 10 carbons in length. MCT do not require digestion and can be absorbed in the absence of lipase or bile.

TABLE 20-1	Macronutrient Sources in Standard and Elemental Formulas		
Type of Formula	Carbohydrate Sources	Protein Sources	Fat Sources
Standard formulas	Corn syrup solids Hydrolyzed cornstarch Sucrose Fructose	Intact proteins, such as casein, whey, lactalbumin, and soy protein isolates Milk protein concentrate Egg white	Vegetable oils (such as corn oil, soybean oil, and canola oil) MCT Palm kernel oil
Elemental formulas	Hydrolyzed cornstarch Maltodextrin Fructose	Hydrolyzed casein, whey, lactalbumin, or soy protein Crystalline amino acids	Vegetable oils (such as corn oil, soybean oil, and canola oil) MCT

NOTE: MCT = medium-chain triglycerides

also included in these formulas so that they can meet all of a person's nutrient needs. In some cases, one or more modules are added to other enteral formulas to adjust their nutrient composition.

Formula Characteristics

The varying nutrient and energy densities in enteral formulas allow them to meet the needs of patients while providing different volumes of fluid. The fiber content influences fecal bulk, colonic function, and blood glucose control. These properties affect the administration of tube feedings, as well as the side effects that patients may experience.

Macronutrient Composition The percentages of protein, carbohydrate, and fat vary substantially among enteral formulas. The protein content of most formulas ranges from 12 to 20 percent of total kcalories.[2] Note that protein needs are high in patients with severe metabolic stress, whereas protein restrictions are necessary for patients with kidney disease. Carbohydrate and fat provide most of the energy in enteral formulas; standard formulas generally provide 40 to 60 percent of kcalories from carbohydrate and 30 to 40 percent of kcalories from fat.[3]

Energy Density The energy density of enteral formulas ranges from 0.5 to 2.0 kcalories per milliliter of fluid. Standard formulas typically provide 1.0 to 1.2 kcalories per milliliter and are appropriate for patients with average fluid requirements. Formulas that have higher energy densities can meet energy and nutrient needs in a smaller volume of fluid and thus benefit patients who have high nutrient needs or fluid restrictions. Individuals with high fluid needs can be given a formula with low energy density or be supplied with additional water via the feeding tube or intravenously.

Fiber Content The fiber content must be taken into account when selecting an enteral formula. Fiber-containing formulas can be helpful for normalizing intestinal function, treating diarrhea or constipation, and maintaining blood glucose control. Conversely, fiber-containing formulas are avoided in patients with acute intestinal conditions, pancreatitis, or procedures involving the intestines.

Osmolality **Osmolality** refers to the moles of osmotically active solutes (*osmoles*) per kilogram of solvent. ◆ An enteral formula with an osmolality similar to that of blood serum (about 300 milliosmoles per kilogram) is an **isotonic formula**, whereas a **hypertonic formula** has an osmolality greater than that of blood serum.

Most enteral formulas have osmolalities between 300 and 700 milliosmoles per kilogram; generally, hydrolyzed formulas and nutrient-dense formulas have higher osmolalities than standard formulas. Most people are able to tolerate both isotonic and hypertonic feedings without difficulty.[4] When medications are infused along with enteral feedings, however, the osmotic load increases substantially and may contribute to the diarrhea experienced by many tube-fed patients.

◆ Osmotically active solutes affect the movement of water across biological membranes (see p. 405).

IN SUMMARY

Enteral formulas are liquid diets that can meet all of a patient's nutritional needs. Standard formulas contain intact proteins and polysaccharides and are provided to patients who can digest and absorb nutrients without difficulty; elemental formulas meet the nutrient needs of patients with limited digestive and absorptive functions. Specialized formulas are available for use in patients with specific diseases. Modular formulas, which contain individual macronutrients, can be used to modify other formulas. Formulas differ in their macronutrient composition, energy density, fiber content, and osmolality. Most people can tolerate isotonic and hypertonic formulas without difficulty.

osmolality (OZ-moe-LAL-ih-tee): the concentration of osmotically active solutes in a solution, expressed as milliosmoles (mOsm) per kilogram of solvent.

isotonic formula: a formula with an osmolality similar to that of blood serum (about 300 milliosmoles per kilogram).
- **iso** = equal
- **tono** = pressure

hypertonic formula: a formula with an osmolality greater than that of blood serum.

Enteral Nutrition in Medical Care

A person with a functioning GI tract who cannot meet nutrient needs with conventional foods alone may be a candidate for enteral nutrition support. Enteral feedings are preferred over intravenous feedings because they help to stimulate or maintain gut function, cause fewer complications, and are less costly.[5] Similarly, oral feedings are preferred to tube feedings when the person is able to drink enteral formulas, because drinking the formulas prevents the stress, complications, and expense associated with tube feedings. ◆

◆ A decision tree for selecting an appropriate feeding method is shown in Figure 21-1 on p. 688.

Oral Use of Enteral Formulas

As mentioned, enteral formulas can fully meet the nutritional needs of individuals who can consume only liquids or who require hydrolyzed nutrients. In most cases, however, patients drink enteral formulas to supplement their diets when they cannot consume enough food to meet their needs. Enteral formulas provide a reliable source of nutrients and add energy and protein to the diets of malnourished patients. Those who are weak or debilitated may also find it easier to manage formulas than meals.

When a patient drinks a formula, taste becomes an important consideration. Allowing patients to sample different products and flavors and select the ones they prefer helps to promote acceptance. The "How to" below offers additional suggestions for helping patients to accept and enjoy oral formulas.

Indications for Tube Feedings

Tube feedings are typically recommended for patients at risk of developing protein-energy malnutrition who are unable to consume adequate food or formula for at least seven days.[6] The following medical conditions may indicate the need for tube feedings:

- Severe swallowing disorders
- Impaired motility in the upper GI tract
- Gastrointestinal obstructions and **fistulas** that can be bypassed with a feeding tube
- Certain types of intestinal surgeries
- Mechanical ventilation
- Extremely high nutrient requirements

fistulas (FIST-you-luz): abnormal passages between organs or tissues (or between an internal organ and the body's surface) that permit the passage of fluids or secretions.

HOW TO Help Patients Accept Oral Formulas

People using enteral formulas are often quite ill and have poor appetites. Even when a person enjoys a formula, the taste can become monotonous in time. Hydrolyzed formulas are usually less palatable than standard formulas, and patients may find them difficult to drink. Health professionals can help by trying these suggestions:

- Let the patient sample different formulas that are appropriate for his or her needs, and use only those that the patient enjoys.
- Serve formulas attractively and remind patients to drink them. Formulas offered in

a glass on an attractive plate may be more appealing than those served from a can with an unfamiliar name.

- If a patient finds the smell of a formula unappealing, it may help to cover the top of the glass with plastic wrap or a lid, leaving just enough room for a straw.
- Provide easy access. Keep the formula close to the patient's bed where it can be reached with little effort and within sight so that the patient is reminded to drink it. Patients who are very ill may lack the motivation to reach for the formula, let alone drink it.

- Try keeping the formula in an ice bath so that it will be cool and refreshing when the patient drinks it. Check with the patient to make sure the colder temperature is suitable.
- For patients with little appetite, offer the formula in smaller amounts that are easy to tolerate, and serve it more frequently during the day.
- If the patient stops enjoying the formula, recommend different flavors or try other formulas.

- Little or no appetite for extended periods, especially if the patient is malnourished
- Mental incapacitation due to confusion, neurological disorders, or coma

Contraindications for tube feedings include severe GI bleeding, high-output fistulas, **intractable** vomiting or diarrhea, complete intestinal obstruction, and severe malabsorption.[7] In addition, some clinical studies suggest that tube feedings are not always effective in some of the patient populations in which they are routinely used; thus, the decision to use tube feedings should be considered in light of the most recent research evidence.[8]

Feeding Routes

The feeding route chosen depends on the medical condition, expected duration of tube feeding, and potential complications of a particular route. Figure 20-1 illustrates the main feeding routes, and the Glossary of Tube Feeding Routes on p. 668 describes each route.

Gastrointestinal Access When a patient is expected to be tube-fed for less than four weeks, a **nasogastric** or **nasoenteric** route is generally chosen; for these routes, the feeding tube is passed into the GI tract via the nose. The patient is frequently awake during **transnasal** (through-the-nose) placement of a feeding tube. While the patient is in a slightly upright position with head tilted, the tube is inserted into a nostril and passed into the stomach (nasogastric placement), duodenum (**nasoduodenal** placement), or jejunum (**nasojejunal** placement). If the patient is awake and alert, he or she can swallow water to ease the tube's passage. The final position of the feeding tube tip is verified by abdominal X-ray or other means. In infants, **orogastric** placement, in which the feeding tube is

intractable: not easily managed or controlled.

FIGURE 20-1 | Tube Feeding Routes

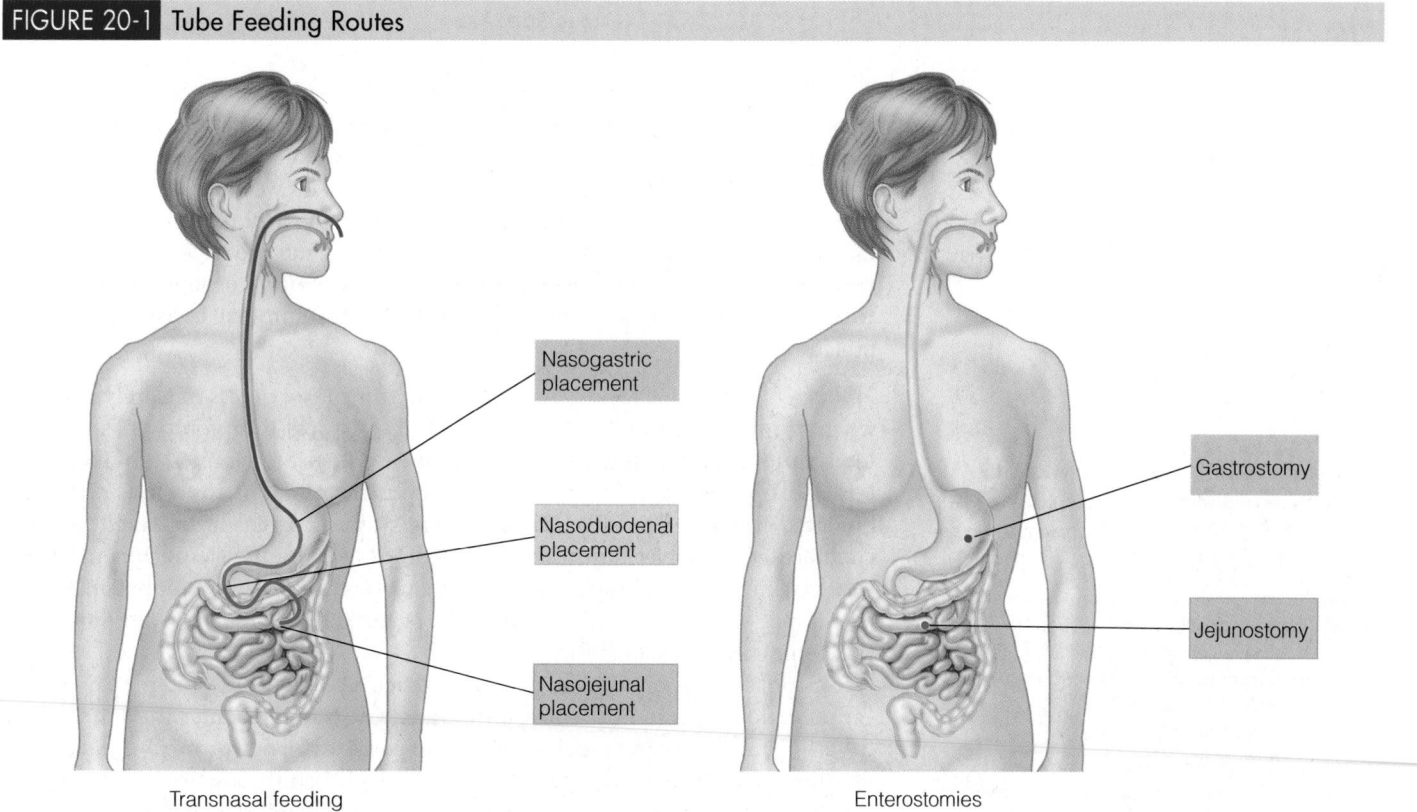

Nasogastric placement

Nasoduodenal placement

Nasojejunal placement

Gastrostomy

Jejunostomy

Transnasal feeding tube placements

Enterostomies

GLOSSARY OF TUBE FEEDING ROUTES

For each type of tube placement, the terms are listed in order from the upper to lower organs of the digestive system.

transnasal: through the nose. A *transnasal feeding tube* is one that is inserted through the nose.

- **nasogastric (NG):** tube is placed into the stomach via the nose.
- **nasoenteric:** tube is placed into the GI tract via the nose. (*Nasoenteric feedings* usually refer to *nasoduodenal* and *nasojejunal* feedings.)

nasoduodenal (ND): tube is placed into the duodenum via the nose.

nasojejunal (NJ): tube is placed into the jejunum via the nose.

orogastric: tube is placed into the stomach via the mouth. This method is often used to feed infants because a nasogastric tube can hinder the infant's breathing.

enterostomy (EN-ter-AH-stoe-mee): an opening into the GI tract through which a feeding tube can be passed.

- **gastrostomy** (gah-STRAH-stoe-mee): an opening into the stomach through which a feeding tube can be passed. A nonsurgical technique for creating a gastrostomy under local anesthesia is called *percutaneous endoscopic gastrostomy (PEG).*

- **jejunostomy** (JE-ju-NAH-stoe-mee): an opening in the jejunum through which a feeding tube can be passed. A nonsurgical technique for creating a jejunostomy is called *percutaneous endoscopic jejunostomy (PEJ).* The tube can either be guided into the jejunum via a gastrostomy or passed directly into the jejunum (*direct PEJ*).

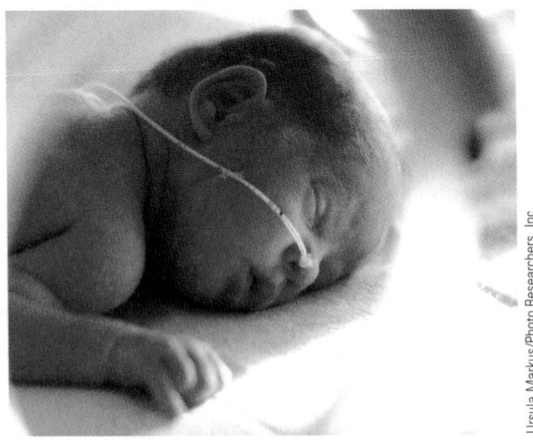

Ursula Markus/Photo Researchers, Inc.

A transnasal feeding tube accesses the GI tract via the nose.

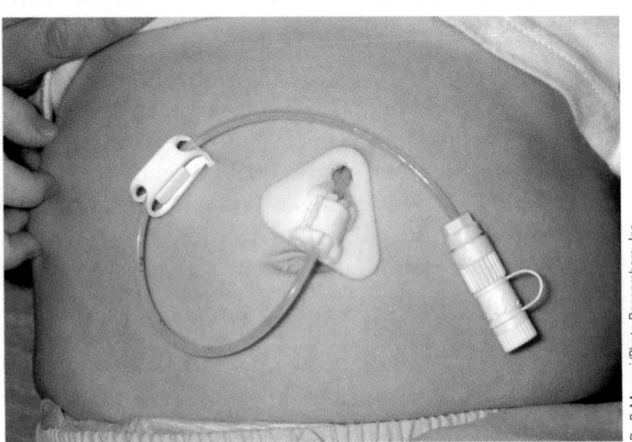

Dr. P. Marazzi/Photo Researchers, Inc.

In a gastrostomy, the feeding tube accesses the GI tract through the skin.

passed into the stomach via the mouth, is sometimes preferred over transnasal routes; this placement allows the infant to breathe more normally during feedings.

When a patient will be tube-fed for longer than four weeks or if the nasoenteric route is inaccessible due to an obstruction or other medical reasons, a direct route to the stomach or intestine may be created by passing the tube through an **enterostomy,** an opening in the stomach (**gastrostomy**) or jejunum (**jejunostomy**). An enterostomy can be made by either surgical incision or needle puncture.

Selecting a Feeding Route As mentioned, transnasal access is usually preferred when the tube feeding duration is expected to be less than four weeks, and enterostomies are often appropriate when tube feedings are planned for longer periods. Gastric feedings, such as the nasogastric and gastrostomy routes, are preferred whenever possible. These feedings are more easily tolerated and less complicated to deliver than intestinal feedings because the stomach controls the rate at which nutrients enter the intestine. Gastric feedings are not possible, however, if patients have gastric obstructions or motility disorders that interfere with the stomach's ability to empty.

Gastric feedings are often avoided in patients at high risk of **aspiration,** ◆ a common complication in which formula or GI secretions enter the lungs, often from the backflow of stomach contents. **Aspiration pneumonia,** a lung disorder that is sometimes fatal, may result. Although health practitioners frequently administer nasoenteric feedings to minimize the possibility of aspiration, studies have not consistently shown that gastric feedings are associated with increased as-

◆ Aspiration risk is high in patients with esophageal disorders, neurological diseases, and conditions that reduce consciousness or cause dementia.

aspiration: drawing in by suction or breathing; a common complication of enteral feedings in which foreign material enters the lungs, often from GI secretions or the reflux of stomach contents.

aspiration pneumonia: a lung disease resulting from the abnormal entry of foreign material; caused by either bacterial infection or irritation of the lower airways.

TABLE 20-2 Comparison of Tube Feeding Routes[a]

Insertion Method and Feeding Site	Advantages	Disadvantages
Transnasal	Does not require surgery or incisions for placement.	Easy to remove by disoriented patients; long-term use may irritate the nasal passages, throat, and esophagus.
Nasogastric	Easiest to insert and confirm placement; feedings can often be given intermittently and without an infusion pump.	Highest risk of aspiration in compromised patients (controversial).[b]
Nasoduodenal and nasojejunal	Lower risk of aspiration in compromised patients[b]; allow for enteral nutrition earlier than gastric feedings following severe stress; may allow for enteral feeding when obstruction, fistulas, or other medical conditions prevent gastric feeding.	More difficult to insert and confirm placement; feedings require an infusion pump for administration; may take longer to reach nutrition goals.
Tube enterostomies	Allow lower esophageal sphincter to remain closed, reducing the risk of aspiration[b]; more comfortable than transnasal insertion for long-term use; site is not visible under clothing.	May require general anesthesia for insertion; require incisions; greater risk of complications from the insertion procedure; greater risk of infection; may cause skin irritation around the insertion site.
Gastrostomy	Feedings can often be given intermittently and without a pump; easier to insert than a jejunostomy.	Moderate risk of aspiration in high-risk patients.[b]
Jejunostomy	Lowest risk of aspiration[b]; allows for enteral nutrition earlier following severe stress; may allow for enteral feeding when obstructions, fistulas, or medical conditions prevent gastric feeding.	Most difficult to insert; feedings require an infusion pump for administration; may take longer to reach nutrition goals.

[a]Relative to other tube feeding routes. The actual advantages and disadvantages of different insertion procedures depend on the person's medical condition.
[b]The risk of aspiration associated with the different feeding routes is under investigation.

piration risk.[9] Table 20-2 summarizes the advantages and disadvantages of the various tube feeding routes.

Feeding Tubes Feeding tubes are made from soft, flexible materials (usually silicone, polyurethane, or polyvinyl) and come in a variety of lengths and diameters. The tube selected largely depends on the patient's age and size, the feeding route, and the formula's viscosity. In many cases, the tube selected is the smallest-diameter tube through which the formula will flow without clogging.

The outer diameter of a feeding tube is measured in **French units,** in which each unit equals $^1/_3$ millimeter; thus, a "12 French" feeding tube has a 4-millimeter diameter. ◆ The inner diameter depends on the thickness of the tubing material. Double-lumen tubes are also available; these allow a single tube to be used for both intestinal feedings and **gastric decompression,** a procedure in which the stomach contents of patients with motility disorders are removed by suction.

Formula Selection

The formula is selected after careful assessment of the patient's medical problems, fluid and nutrition status, and ability to digest and absorb nutrients; some of the factors considered are shown in Figure 20-2 on p. 670. Generally, the best formula is one that meets the patient's medical and nutrient needs with the lowest risk of complications and the lowest cost. The vast majority of patients can use standard formulas. A person with a functional, but impaired, GI tract may require an elemental formula. Some nutrition-related criteria that may influence formula selection include:

- *Nutrient and energy needs.* As with patients consuming regular diets, an adjustment in macronutrient and energy intakes may be necessary for tube-fed patients. For example, patients with diabetes may need to control carbohydrate intake, critical care patients may have high protein and energy requirements, and patients with chronic kidney disease may need to limit their intake of protein and several minerals.

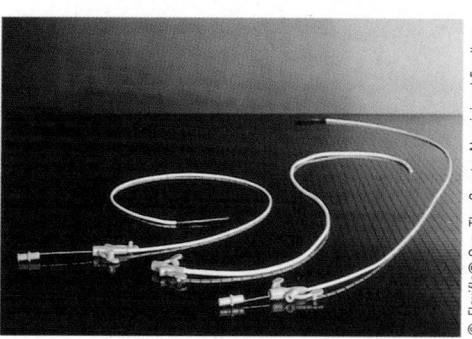

The thin wires protruding from the ends of these feeding tubes are stylets, which stiffen the tubes to ease insertion and are discarded thereafter. The Y-connector (shown here in orange) provides a port for administering water or medications without disrupting the feeding.

◆ 1 French = $^1/_3$ mm
   12 French = 12 × $^1/_3$ mm = 4 mm

French units: units of measure used to indicate the size of a feeding tube's outer diameter; 1 French unit equals $^1/_3$ millimeter.

gastric decompression: the removal of the stomach contents (swallowed saliva, stomach secretions, and gas) of patients who have motility disorders or obstructions that prevent stomach emptying.

FIGURE 20-2 Selecting a Formula

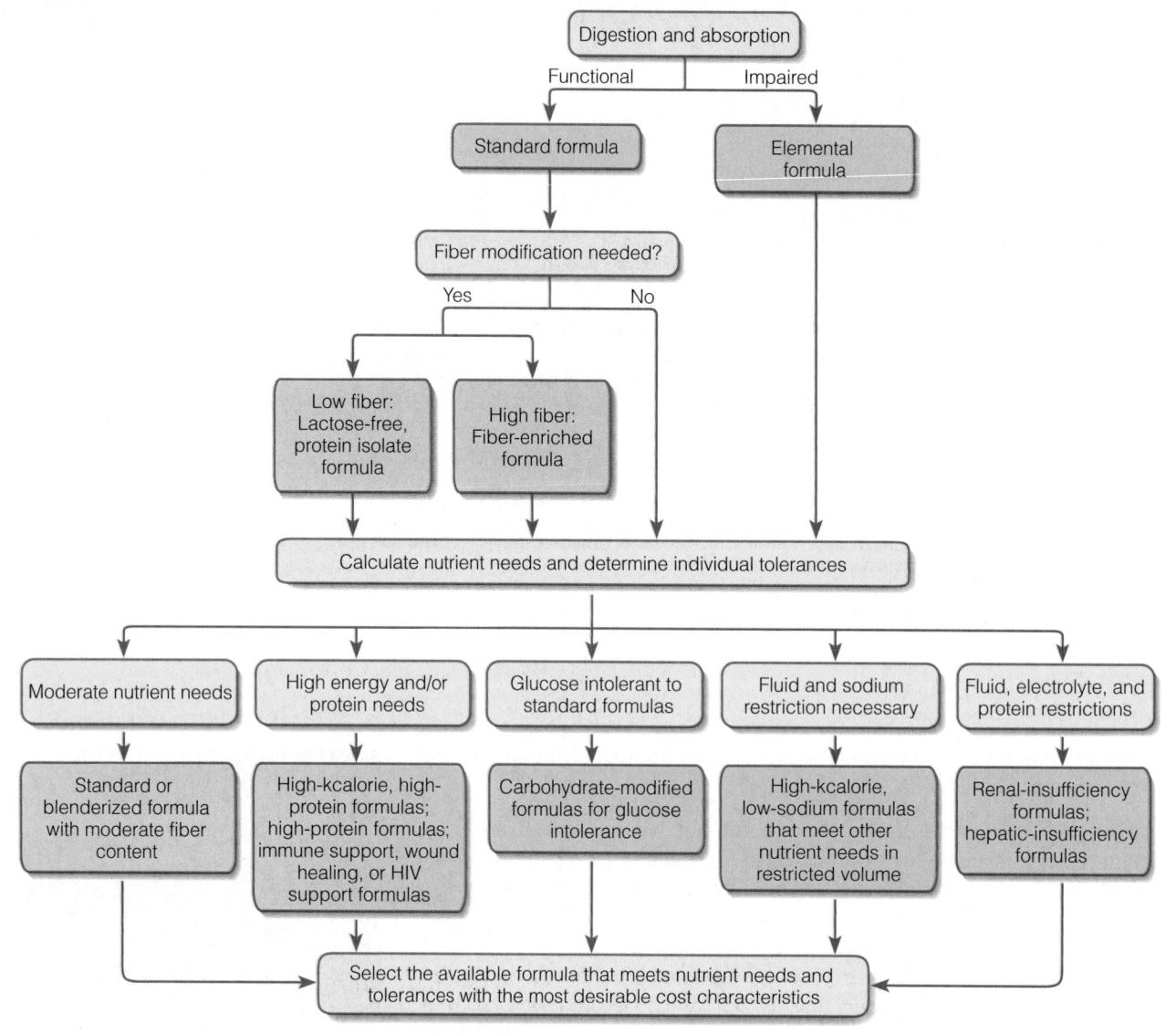

- *Fluid requirements.* High nutrient needs must be met using the volume of formula a patient can tolerate. If fluids are restricted, the formula should have adequate nutrient content and energy density to deliver the necessary nutrients in the volume prescribed.

- *The need for fiber modifications.* The choice of formulas is narrower if fiber intake needs to be high or low. Formulas that provide fiber may be helpful for managing diarrhea, constipation, or hyperglycemia in some patients; other patients may need to avoid fiber due to an increased risk of GI obstructions.[10]

- *Individual tolerances (food allergies and sensitivities).* Most formulas are lactose-free, because many patients who need enteral formulas have some degree of lactose intolerance. Many formulas are also gluten-free and can accommodate the needs of individuals with celiac disease (gluten sensitivity).

Health care facilities stock a limited number of formulas, so formula selection is limited by availability. Initially, the medical staff may make an educated guess as to the best formula based on the criteria previously mentioned, and the decision can be reappraised based on the patient's response to the formula. Note that few research studies have evaluated the effectiveness of the various specialized formulas,[11] so their additional expense may be difficult to justify.

Meeting Water Needs

Although water needs vary, ◆ many adults require about 2000 milliliters (about 2 quarts) of water daily. Fluids may be restricted in persons with kidney, liver, or heart disease. Additional water is required in patients with fever, high urine output, diarrhea, excessive sweating, severe vomiting, fistula drainage, high-output ostomies, blood loss, or open wounds.

In alert adults, thirst is often a good indicator of water needs. People who complain of thirst may be given more water unless medical orders restrict fluid intake. In the elderly, however, thirst may be slow to develop in response to dehydration. Health professionals routinely monitor patients' weight changes, record fluid intake and output, and measure urine specific gravity to evaluate hydration status. (Chapter 17 provides additional information about evaluating hydration.)

Formula Water Content The water in formulas meets a substantial portion of water needs. Standard formulas contain about 85 percent water, or about 850 milliliters of water per liter of formula. Nutrient-dense formulas contain about 69 to 72 percent water; exact amounts can be obtained from the product label or manufacturer's information sheet. In addition to the water in formulas, water can be provided by flushing water separately through the feeding tube.

Routine Flushes To prevent clogging, feeding tubes are routinely flushed with 20 to 30 milliliters of warm water before and after each feeding and about every 4 hours when feedings are continued throughout the day. The water used for routine flushes should be included when estimating fluid intakes.

◆ To estimate fluid requirements in adults and children:
- Adults: allow 30 to 40 mL/kg; 25 to 30 mL/kg in older adults.
- Children: allow 50 to 60 mL/kg.
- Infants: allow 100 to 150 mL/kg.

IN SUMMARY

Enteral formulas are provided to patients who need to consume liquid or elemental diets or who require tube feedings. A nasoenteric feeding route is preferred for short-term tube feedings, whereas enterostomies are used for longer-term feedings. Because the stomach delivers nutrients into the intestine at a controlled rate, gastric feedings are often preferred, although they are frequently avoided in patients at risk of aspiration. A chief concern in formula selection is the formula's ability to meet the patient's nutrient requirements. Formulas meet a substantial portion of the water requirements, and additional water can be provided by flushing water through the feeding tube.

Administration of Tube Feedings

After the feeding route and formula have been selected, attention turns to delivering the formula. The methods of tube feeding administration vary somewhat from one health care facility to the next. The procedures presented in the following sections are general guidelines.

Safe Handling

Individuals who are ill or malnourished often have suppressed immune systems, making them vulnerable to infection from foodborne illness. To prevent contamination, the personnel involved in preparing and delivering formulas should work in clean environments, using clean equipment and clean hands.

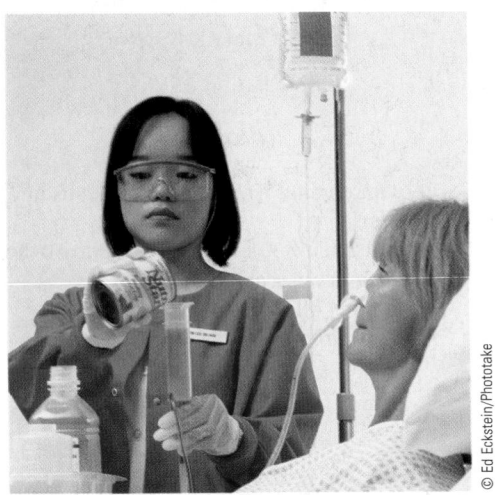

In an open feeding system, the formula is transferred from its original packaging to a feeding container.

© Ed Eckstein/Phototake

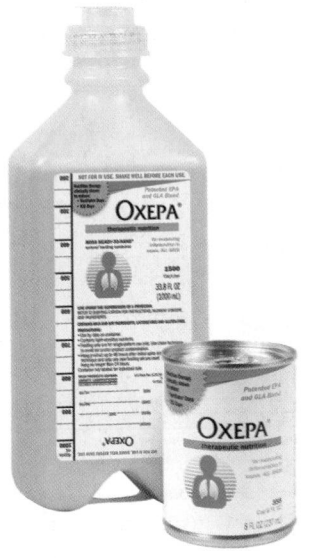

© Courtesy of Abbott Nutrition

In a closed feeding system, the formula is prepackaged in a container that can be attached to a feeding tube, such as the bottle shown on the left. The formula in the can at right can be used in an open feeding system.

◆ A gastric sample usually has a pH between 1 and 4. An intestinal sample should have a pH between 6 and 8.

open feeding system: a delivery system that requires the formula to be transferred from its original packaging to a feeding container before being administered through the feeding tube.

closed feeding system: a delivery system in which the formula comes prepackaged in a container that is ready to be attached to the feeding tube for administration.

Safety Guidelines As mentioned in earlier chapters, health care facilities have specific protocols for handling food products and formulas based on the potential hazards and critical control points in food preparation, referred to as *HACCP* (*Hazard Analysis and Critical Control Points*) systems. Personnel involved with preparing or delivering formula should be aware of the specific HACCP systems at their facility related to formula preparation and administration.

Feeding Systems Formulas are available in open feeding systems and closed feeding systems. With an **open feeding system,** the formula needs to be transferred from its original packaging to a feeding container. Examples include formulas that are packaged in cans or bottles, concentrates that need to be diluted, and powders that require reconstitution. In a **closed feeding system,** the formula is prepackaged in a container that can be connected directly to a feeding tube. Closed systems are less likely to become contaminated, require less nursing time, and can hang for longer periods of time than open systems. Although closed systems cost more initially, they may be less expensive in the long run by preventing bacterial contamination and thus avoiding the costs of treating infections.

At the Nursing Station After the formula reaches the nursing station, the nursing staff assumes responsibility for its safe handling. Hands should be carefully washed before handling formulas and feeding containers. Some facilities require that nonsterile gloves be worn whenever formulas are handled. The following steps can reduce the risk of formula contamination when using open feeding systems:

- Before opening a can of formula, clean the lid with a disposable alcohol wipe and wash the can opener with detergent and hot water. (Check HACCP protocols for details.) If you do not use the entire can at one feeding, label the can with the date and time it was opened.
- Store opened cans or mixed formulas in clean, closed containers. Refrigerate the unused portion of formula promptly.
- Discard unlabeled or improperly labeled containers and all opened containers of formula that are not used within 24 hours.

At the Bedside To reduce the risk of bacterial infections in tube-fed patients, the nurse should hang no more than an 8-hour supply of formula when using an open feeding system. The nurse should discard any formula that remains, rinse out the feeding bag and tubing, and add fresh formula to the feeding bag. A new feeding container and tubing (except for the feeding tube itself) is necessary every 24 hours.

For closed feeding systems, the hang time should be no longer than 24 to 48 hours. Contamination is more likely with the longer time periods.

Initiating and Progressing a Tube Feeding

Before starting a tube feeding, health practitioners can ease fears by fully discussing the procedure with the patient and family members, who may feel anxious about using a feeding tube. The discussion should address the reasons why tube feeding is appropriate and the benefits and risks of the procedure. The "How to" on p. 673 offers suggestions that may help ease the concerns of patients who may benefit from tube feeding.

Tube Placement Serious complications can develop if a transnasal tube is accidentally inserted into the respiratory tract or if formula or GI secretions are aspirated into the lungs. To minimize the risk of incorrect tube placement, clinicians often use X-rays to verify the position of the feeding tube before a feeding is initiated. Another technique is to test the pH of a sample of bodily fluid drawn into the feeding tube; recall that the pH of stomach fluid is much lower than the pH of fluid obtained from the intestine or respiratory tract. ◆ After the tube's placement has been confirmed, the nurse secures the tube to the patient's nose and cheek with tape and monitors the position of the tubing throughout the day.

To reduce the risk of aspiration, the patient's upper body is elevated to at least a 30- to 45-degree angle during the feeding and for 30 minutes after the feeding whenever possible. The addition of blue food coloring to formula was formerly suggested as a means of identifying aspirated formula in lung secretions; however, this practice is now discouraged because several deaths have been attributed to its use.[12]

Formula Delivery A day's nutrient needs can be met by delivering relatively large amounts of formula several times per day **(intermittent feedings)** or smaller amounts continuously throughout the day **(continuous feedings).** A patient may also start on a continuous feeding schedule and gradually transition to intermittent feeding. Each method has specific uses, advantages, and disadvantages.

Intermittent feedings are best tolerated when they are delivered into the stomach (not the intestine). Generally, a total of about 250 to 400 milliliters is delivered over 20 to 40 minutes using a gravity drip method or an infusion pump. The exact amount is determined by dividing the volume of formula required for meeting a patient's nutrient needs into several daily feedings, as shown in the "How to" on p. 674. Because of the relatively high volume of formula delivered at a time, intermittent feedings may be difficult for some patients to tolerate, and the risk of aspiration may be higher than with continuous feedings. An advantage of intermittent feedings is that they are similar to the usual pattern of eating and allow the patient freedom of movement between meals.

intermittent feedings: delivery of about 250 to 400 milliliters of formula over 20 to 40 minutes.

continuous feedings: slow delivery of formula at a constant rate over an 8- to 24-hour period.

HOW TO Help Patients Cope with Tube Feedings

The thought of being "force-fed" is frightening to many people. Some may envision thick feeding tubes or fear that the procedure will be painful. Others may associate tube feedings with disabling injury or irreversible illness. Patients may be less apprehensive once they understand the insertion procedure, the expected duration of the tube feeding, and the strategic role that nutrition plays in recovery from disease. The pointers that follow can help health professionals prepare patients for transnasal tube feedings:

- Allow the patient to see and touch the feeding tube. Seeing firsthand that the tube is soft and narrow (only about half the diameter of a pencil) often alleviates anxiety.
- Show the patient how the feeding apparatus is attached to the feeding tube, and explain how the feeding will work. For young children, use dolls or stuffed toys to demonstrate tube insertion and feeding procedures.
- Explain that the patient remains fully alert during the procedure and helps pass the tube by swallowing. A numbing solution sprayed on the back of the throat minimizes discomfort and prevents gagging during the procedure.
- Tell the patient that once the tube has been inserted, people have no problem talking, and they become accustomed to the tube's presence within a few hours. In most cases, the patient can easily swallow foods and liquids with the tube in place. If permitted, favorite foods or beverages can still be enjoyed.
- Assure the patient that the tube feeding will be temporary, if such assurance is appropriate.

A tube feeding may be frightening for some patients, but others may be relieved to know that they can receive sound nutrition without any effort. As they feel better and begin to eat again, the volume of the feeding can be reduced and then tube feeding discontinued when oral intake is adequate.

Tube feedings may cause some patients to feel that they have lost control over an important aspect of their lives. They may also feel self-conscious about how the feeding tube looks or feel awkward when moving around with the equipment. A few measures can help:

- Involve patients in the decision-making and care process whenever possible. Patients can help to arrange their daily feeding schedules and can perform some of the feeding procedures themselves.
- Show patients how to manipulate the feeding equipment so that they can get out of bed and move around.
- Encourage patients to maintain contact with friends and keep busy with the hobbies and activities they enjoy. This measure is especially important for children, teens, and those on long-term feedings.

When caring for infants and children, keep the developmental age of the child in mind, and work with parents to ensure that appropriate feeding skills are mastered. Infants can be provided with a pacifier during feedings to help maintain the associations between sucking, swallowing, and fullness. When possible, the formula can be provided by bottle to an infant, or by spoon to a child, to further develop skills.

The more complex the procedure, the easier it becomes for health care professionals to focus on the procedure and disregard a patient's emotional response. No matter how many technicalities you have to keep in mind, remember to stay focused on the person receiving your care.

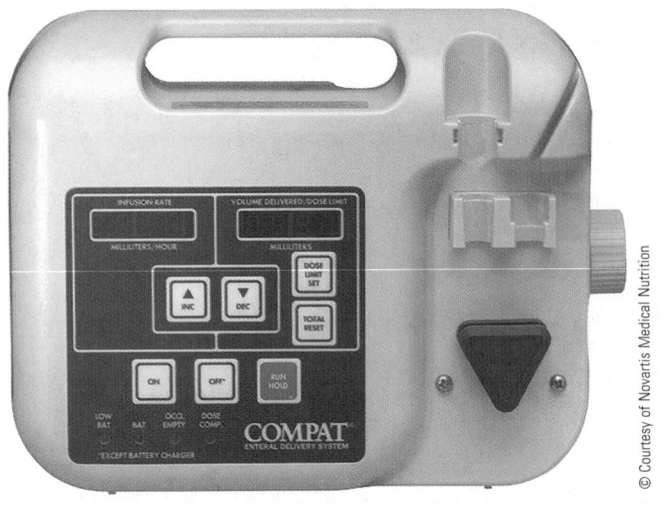

The delivery of intermittent and continuous feedings can be controlled with an infusion pump.

Rapid delivery of a large volume of formula into the stomach (250 to 500 milliliters in less than 20 minutes) is called a **bolus feeding.** This type of feeding may be given every 3 to 4 hours using a syringe. Bolus feedings can cause abdominal discomfort, nausea, and cramping in some patients, especially when the feeding is initiated. The risk of aspiration is also greater than with other methods of feeding. For these reasons, bolus feedings are used only in patients who are not critically ill.

Continuous feedings are delivered slowly and at a constant rate over a period of 8 to 24 hours. Continuous feedings are used in patients who receive intestinal feedings. This method of feeding is also recommended for critically ill patients, because delivering relatively small volumes at a time may reduce nausea, diarrhea, and possibly, the risk of aspiration. An infusion pump is required to ensure accurate and steady flow rates; consequently, the feedings can limit the patient's freedom of movement and are also more costly.

Formula Strength and Volume Formula administration techniques vary among institutions, so protocols should be reviewed carefully before working with patients. The patient's response can help in guiding formula delivery. Keep in mind that few studies have evaluated the various methods for initiating and progressing enteral feedings.

The formulas are typically provided full-strength, but they are occasionally diluted if the patient's fluid requirements are high and water needs cannot be met by other means.[13] In addition, dilution may sometimes be necessary to improve the flow of a viscous formula.

Intermittent feedings may start with 60 to 120 milliliters at the initial feeding and be increased by 30 to 60 milliliters at each feeding until the goal volume is reached. Continuous feedings may start at about 20 to 50 milliliters per hour and be raised by 10 to 25 milliliters per hour every 4 to 8 hours.[14] Concentrated formulas are often started at the slower rates. For both intermittent and continuous feedings, the delivery rate and amount of increase depend on the patient's tolerance to the formula. If the new rate is not tolerated, the rate of delivery progresses more slowly to give the person additional time to adapt. If a patient on intermittent feeding cannot tolerate the feeding, continuous feeding may be a better choice.

Checking Gastric Residuals When a patient receives a gastric feeding, the nurse regularly measures the **gastric residual volume** (the volume of formula remaining in the stomach after feeding) to ensure that the stomach is emptying properly. The gastric residual is measured by gently withdrawing the gastric contents through the feeding tube using a syringe, usually before each intermittent

> **bolus** (BOH-lus) **feeding:** delivery of about 250 to 500 milliliters of formula in less than 20 minutes.
>
> **gastric residual volume:** the volume of formula remaining in the stomach from a previous feeding.

HOW TO Determine the Formula Volumes to Administer in Tube Feedings

After selecting a formula that meets the patient's medical and nutrient needs, the clinician determines the volume of formula that meets those needs. Consider a patient who needs 2000 kcalories daily and is receiving a standard formula that provides 1.0 kcalorie per milliliter. The total volume of formula required would be 2000 milliliters per day:

$$x \text{ mL} \times 1.0 \text{ kcal/mL} = 2000 \text{ kcal}$$

$$x \text{ mL} = \frac{2000 \text{ kcal}}{1.0 \text{ kcal/mL}} = 2000 \text{ mL}$$

If the patient is to receive intermittent feedings 6 times per day, he will need about 330 milliliters of formula at each feeding:

$$2000 \text{ mL} \div 6 \text{ feedings} = 333 \text{ mL/feeding}$$

Alternatively, if he is to receive intermittent feedings 8 times per day, he will need 250 milliliters (or about 1 can of ready-to-feed formula) at each feeding:

$$2000 \text{ mL} \div 8 \text{ feedings} = 250 \text{ mL/feeding}$$

He will probably tolerate this volume of formula best if it is delivered over a 30-minute period at each feeding. If the patient is to receive the formula continuously over 24 hours, he will need about 85 milliliters of formula each hour:

$$2000 \text{ mL} \div 24 \text{ hours} = 83 \text{ mL/hr}$$

feeding and every 4 to 6 hours during continuous feedings.[15] Although opinions vary, some experts recommend that an evaluation be conducted if the gastric residual exceeds 200 milliliters and that feedings be withheld if it exceeds 500 milliliters. If the tendency to accumulate fluids persists, the physician may recommend intestinal feedings or begin drug therapy to stimulate gastric emptying.

Medication Delivery through Feeding Tubes

Patients receiving tube feedings sometimes require one or more medications that need to be delivered through feeding tubes. Because medications can interact with enteral formulas in the same ways that they interact with foods, potential diet-drug interactions must be considered. In addition, some medications may need to be exposed to the acidic stomach environment and thus cannot be administered via an intestinal feeding tube. Medications can also cause feeding tubes to clog. The "How to" below provides some guidelines about medication administration that may help to prevent complications.

Diarrhea Medications are a major cause of the diarrhea that frequently accompanies tube feedings. Diarrhea is especially associated with the administration of sorbitol-containing medications, laxatives, and some types of antibiotics.[16] The high osmolality of many liquid medications can cause diarrhea, so dilution of hypertonic medications may be helpful.

Medications and Continuous Feedings Continuous feedings are ordinarily stopped during the administration of medication so that the components of enteral formulas do not interfere with the medication's absorption. The feeding is typically halted for 15 minutes before and 15 minutes after medication delivery. Some medications may require a longer formula-free interval; for example, feedings need to be stopped for 1 to 2 hours before and after administering phenytoin, a medication that controls seizures.[17] In such cases, the formula's delivery rate needs to be increased so that the correct amount of formula can be delivered.

Tube Feeding Complications

Complications are a frequent occurrence during tube feedings. The following paragraphs discuss some of the common complications, which include gastrointestinal

---

> **HOW TO** Administer Medications to Patients Receiving Tube Feedings
>
> The pharmacist is your best resource for learning how and when medications can be administered via feeding tubes, especially when you are dealing with an unfamiliar medication. Check with the pharmacist to learn the following:
>
> - Whether a particular medication is known to be incompatible with formulas.
> - The proper timing of medication administration to avoid drug-nutrient interactions.
> - For patients using intestinal feedings, whether a medication can be absorbed without exposure to stomach acid.
> - Whether a liquid form of a medication is available, and if so, the appropriate dosage of the liquid form.
> - If only tablets are available, whether the tablets can be crushed and mixed with
>
> water. Enteric-coated and sustained-release medications should not be crushed due to the potential for adverse effects.
>
> In general, it is best to give medications by mouth instead of by tube whenever possible. In some cases, the injectable form of a medication may be the best option. For medications that must be given by feeding tube:
>
> - Do not mix medications with enteral formulas. Do not mix medications together.
> - Before administering medications, ensure that the feeding tube is placed correctly, that it is not clogged, and that the gastric residual is not excessive.
> - Position the patient in a semiupright position (30 degrees or higher) to prevent aspiration.
>
> - Flush the feeding tube with 15 to 30 milliliters of warm (body temperature) water before and after administering a medication. When more than one medication is administered, flush the tube with 5 milliliters of water after each medication is given.
> - Use liquid forms of medications whenever possible. Dilute very viscous or hypertonic liquid medications with 10 to 30 milliliters of water before administering them through the feeding tube.
> - If tablets are used, crush tablets to a fine powder and mix with 30 milliliters of warm water before administering.

problems, such as aspiration and diarrhea; mechanical problems directly related to the tube feeding process; and metabolic problems, such as biochemical alterations and nutrient deficiencies. Examples of these complications, along with preventive and corrective measures, are summarized in Table 20-3 on p. 677.

Gastrointestinal Complications Diarrhea may be caused by malabsorption problems, medications, bacterial overgrowth, malnutrition, or more rarely, hypertonic formulas. Constipation sometimes occurs due to dehydration, motility impairments, obstructions, and low-fiber intakes. Impaired gastric motility or inadequate functioning of the lower esophageal sphincter may result in aspiration of GI secretions or formula. Other GI complications include abdominal discomfort, nausea, and vomiting.

Mechanical Complications Mechanical problems include clogged feeding tubes, malfunctioning feeding pumps, and feeding tubes that become dislodged after placement. The feeding tube itself may be a physical irritant and may warrant a change to a different type of tubing or a different feeding route. Nasoenteric tube placement may cause a number of side effects, such as dry mouth from increased mouth breathing and reduced salivary secretions, blocked eustachian tubes and resultant middle ear infections, and sinus infections due to blocking of the sinus tract. Sometimes ostomies are associated with leakage of gastrointestinal secretions at the site of tube insertion.

Metabolic Complications Common metabolic complications include fluid imbalances (either dehydration or overhydration), electrolyte imbalances, and glucose intolerance. Routine blood tests may be necessary to monitor levels of potassium, phosphorus, sodium, and glucose until a patient has stabilized. Some patients may need insulin or medications to reverse hyperglycemia. Vitamin K and essential fatty acid deficiencies may result if formulas lacking these nutrients are used for a prolonged period.

Monitoring Tube Feedings Many complications of tube feeding can be prevented by choosing the most appropriate feeding route, formula, and delivery method. Attention to a patient's primary medical condition and medication use is important as well. The health practitioners responsible for the day-to-day care of the patient routinely monitor body weight, hydration status, and results of laboratory tests to detect problems before complications develop. Table 20-4 on p. 678 provides a monitoring schedule that may help with the early detection of common tube feeding problems.

Transition to Table Foods

Once the condition requiring a tube feeding resolves, the volume of formula can be tapered off as the patient gradually shifts to an oral diet. The steps in the transition depend on the patient's medical condition and the type of feeding the patient is receiving. Individuals using continuous feedings are often switched to intermittent feedings initially. In some patients, swallowing function may need to be evaluated before oral feedings begin. Patients receiving elemental formulas may begin the transition by using a standard formula, either orally or via tube feeding. If the patient has not consumed lactose for a month or longer, a low-lactose diet may be better tolerated. Oral intake should supply about two-thirds of estimated nutrient needs before the tube feeding is discontinued completely.[18] The Case Study on p. 678 allows you to consider the many factors involved in tube feedings.

TABLE 20-3 Causes and Prevention or Correction of Tube Feeding Complications

Complications	Possible Causes	Preventive/Corrective Measures
Aspiration of formula	Compromised lower esophageal sphincter, delayed gastric emptying	Use nasoenteric, gastrostomy, or jejunostomy feedings in high-risk patients; check tube placement; elevate head of bed during and for 45 minutes after feeding; check gastric residuals.
Clogged feeding tube	Formula too thick for tube	Select appropriate tube size; flush tubing with water before and after giving formula; use infusion pump to deliver thick formulas. Remedies that may help to unclog feeding tubes include flushes with warm water or solutions that contain pancreatic enzymes or bicarbonate; consult pharmacist for more options.
	Medications delivered through feeding tube	Use oral, liquid, or injectable medications whenever possible; dilute thick or sticky liquid medications with water before administering; crush tablets to a fine powder and mix with water (except enteric-coated or sustained-release medications); flush tubing with water before and after medications are given; give medications individually; do not add medications to the feeding container.
Constipation	Low-fiber formula	Provide additional fluids; use high-fiber formula.
	Lack of exercise	Encourage walking and other activities, if appropriate.
Dehydration and electrolyte imbalance	Excessive diarrhea	See items under *Diarrhea*.
	Inadequate fluid intake	Provide additional fluid.
	Carbohydrate intolerance	Use continuous drip administration of formula; monitor blood glucose; select a formula with a lower amount or different type of carbohydrate; provide a formula with a higher fat content.
	Excessive protein intake	Monitor blood electrolyte levels; reduce protein intake.
Diarrhea, cramps, abdominal distention	Bacterial contamination	Use fresh formula every 24 hours; store opened or mixed formula in a refrigerator; rinse feeding bag and tubing before adding fresh formula; change feeding apparatus every 24 hours; prepare formula with clean hands using clean equipment in a clean environment.
	Lactose intolerance	Use lactose-free formula in patients with current or potential lactose intolerance.
	Hypertonic formula	Use small volume of formula and increase volume gradually.
	Rapid formula administration	Use slow administration rate or use continuous drip feedings.
	Malnutrition/low serum albumin	Use small volume of dilute formula and increase volume and concentration gradually.
Hyperglycemia	Diabetes, hypermetabolism, drug therapy	Check blood glucose; slow administration rate; provide adequate fluids; select a formula with a lower amount or different type of carbohydrate; provide a formula with a higher fat content.
Nausea and vomiting	Obstruction	Discontinue tube feeding.
	Delayed gastric emptying	Check gastric residual; slow administration rate, use continuous drip feedings, or discontinue tube feeding.
	Intolerance to concentration or volume of formula	Use small volume of formula and increase volume and concentration gradually; use continuous drip feedings.
	Psychological reaction to tube feeding	Address patient's concerns.
Skin irritation at enterostomy site	Leakage of GI secretions and friction caused by the tube	Keep site clean; inspect area for redness, tenderness, and drainage; use protective skin cream.

NOTE: Many of the complications presented here can be caused by the patient's primary disorder or drug therapy rather than the tube feeding itself. In such a case, the corrective measure would include treatment of the disorder or a change in drug therapy. Additionally, other corrective measures that require a physician's order are not shown here.

TABLE 20-4 Monitoring Patients on Tube Feedings[a]

Before starting a new feeding:	Conduct a complete nutrition assessment. Check tube placement.
Before each intermittent feeding:	Check patient's position. Check tube placement. Check gastric residual volume. Flush feeding tube with water.
After each intermittent feeding:	Flush feeding tube with water.
Every hour:	Check infusion pump rate, when applicable.
Every 4 hours:	Check vital signs, including blood pressure, temperature, pulse, and respiration.
Every 4 to 6 hours of continuous feeding:	Check patient's position. Check gastric residual volume. Flush feeding tube with water.
Every day:	Check intake and output and hydration status. Check blood glucose; once stable, check blood glucose weekly (individuals without diabetes). Change feeding container and attached tubing. Clean feeding equipment.
Twice weekly:	Check body weight (check daily if patient is nutritionally unstable).
As necessary:	Observe patient for undesirable responses to tube feeding, such as delayed gastric emptying, nausea, vomiting, or diarrhea. Check results of laboratory tests. Check nitrogen balance.

[a]Guidelines vary among institutions. Monitoring frequency depends on the patient's medical condition. Patients beginning tube feedings and patients who are medically or nutritionally unstable need more intense monitoring.

CASE STUDY   Graphics Designer Requiring Enteral Nutrition Support

Sharyn Eschler is a 24-year-old graphics designer who suffered multiple fractures when she fell from a cliff while hiking. She has been in the hospital for 2 weeks and has no appetite. Sharyn weighed 140 pounds upon her arrival in the hospital, but she has lost 8 pounds over the course of her hospitalization. Due to the nature of her injuries, she is in traction and is immobile, although the head of her bed can be elevated 45 degrees. From the diet history, it appears that Sharyn's nutrition status was adequate prior to hospitalization. The health care team agrees that nasoduodenal tube feeding should be instituted before her nutrition status deteriorates further. The standard formula selected for the feeding is lactose-free, and Sharyn's nutrient requirements can be met with 2200 milliliters of the formula per day.

1. What steps can be taken to prepare Sharyn for tube feeding? What are some general reasons why nasoduodenal placement of the feeding tube might be preferred over nasogastric placement?
2. What parameters should be monitored to ensure that Sharyn's fluid needs are being met? How can additional fluids be given? Estimate Sharyn's fluid needs using her current weight and the fluid intake range suggested in the margin on p. 671.
3. The physician's orders specify that the feeding should be given continuously over 18 hours. Using the method shown in the "How to" on p. 674, determine an appropriate feeding rate.
4. What steps can the health care team take to prevent aspiration? Describe precautions that should be taken if Sharyn is to receive medications through the feeding tube.
5. After three days of feeding, Sharyn develops diarrhea. Check pp. 675–676 and Table 20-3 to determine the possible causes. What measures can be taken to correct the diarrhea?

IN SUMMARY

To maximize the benefits of tube feedings, formulas should be prepared and administered using food safety techniques that minimize the risk of complications. Tube placement should be verified and monitored to reduce the risks of aspiration and inadvertent placement into the respiratory tract. Depending on the feeding route and medical condition, the formula can be delivered in bolus feedings, intermittently, or continuously. Medications should be given separately and accompanied by water flushes to prevent tube clogging. Complications of tube feedings can be gastrointestinal, mechanical, or metabolic in nature.

Clinical Portfolio

CENGAGENOW™
academic.cengage.com/login

■ Appendix K provides examples of enteral formulas on the market and lists their energy and macronutrient contents. Select one standard formula and one elemental formula from Tables K-1 and K-2, respectively. For the two formulas you selected, calculate the volume of formula that would meet the energy needs of a patient who requires about 1750 kcalories daily. Use these results in answering the following questions:

a. What is the amount of protein, carbohydrate, and fat that the patient would obtain in a typical day? Determine the percentages of kcalories that come from carbohydrate and fat. Do these percentages fall within the Acceptable Macronutrient Distribution Ranges described in Chapter 1 (p. 18)?

b. Tables J-1 and J-2 show the formula volumes that would meet the Reference Daily Intakes (RDI). Would the volumes you obtained meet typical vitamin and mineral needs?

■ The administration of tube feedings requires attention to many technical details, which makes it easy to focus on the procedure rather than the patient. Imagine that your brother, sister, or a parent requires a transnasal tube feeding. How might this person react to the need for a tube feeding? How would you explain the benefits and possible problems associated with the procedure? Think about the ways you would want the health practitioner to help your relative.

NUTRITION ON THE NET

For further study of topics covered in this chapter, log on to **academic.cengage .com/nutrition/rolfes/UNCN8e**. Go to Chapter 20, then to Nutrition on the Net.

• To learn more about the appropriate uses of enteral and parenteral nutrition, visit the websites of these organizations:

American Society for Parenteral and Enteral Nutrition: **www.clinnutr.org**

Canadian Parenteral-Enteral Nutrition Association: **www.cpena.ca/home.html**

British Association for Parenteral and Enteral Nutrition: **www.bapen.org.uk/**

• To learn about home enteral nutrition, visit the website of the Oley Foundation, a national, nonprofit organization that provides information, outreach services, and emotional support for consumers of home enteral and parenteral services: **www.oley.org**

NUTRITION ASSESSMENT CHECKLIST for People Receiving Tube Feedings

Medical History
Check the medical record for medical conditions that:

- Alter nutrient needs and influence the formula selection
- Influence the selection of tube placement sites (gastric versus intestinal) and feeding routes
- Suggest the length of time that the tube feeding will be needed

Monitor the medical record for complications or risks that may influence the formula selection or delivery technique, including:

- Aspiration
- Constipation
- Fluid and electrolyte imbalances
- Diarrhea
- Hyperglycemia
- Nausea and vomiting
- Skin irritation

Medications
Check medications for those that can cause side effects similar to the adverse effects associated with the tube feeding, such as:

- Nausea and vomiting

- Diarrhea
- Constipation
- GI discomfort

For medications delivered through the feeding tube, check:

- Form of medication and possible alternatives
- Viscosity of liquid medications
- Potential for diet-drug interactions

Dietary Intake
To assess nutritional adequacy, check to see whether:

- The formula is appropriate for patient's needs
- Supplemental water is provided to meet needs
- The formula is administered as prescribed

Anthropometric Data
Measure baseline height and weight, and monitor body weight regularly. If weight is not appropriate:

- Determine whether energy needs have been correctly assessed.

- Check to see if the formula is being delivered as prescribed.
- Check for signs of dehydration or overhydration.

Laboratory Tests
Check serum and urine tests for signs of:

- Fluid and electrolyte imbalances
- Glucose intolerance
- Adequacy of protein intake (serum protein levels)
- Improvement or deterioration of the medical condition

Physical Signs
Look for physical signs of:

- Dehydration or overhydration
- Delayed gastric emptying (gastric residual volume)
- Malnutrition

STUDY QUESTIONS

CENGAGENOW™
To assess your understanding of chapter topics, take the Student Practice Test and explore the modules recommended in your Personalized Study Plan. Log on to **academic.cengage.com/login.**

These questions will help you review the chapter. You will find the answers in the discussions on the pages provided.

1. Characterize standard formulas, elemental formulas, specialized formulas, and modular formulas, and describe situations in which they are used. (pp. 664–665)

2. Discuss how macronutrient composition, energy density, fiber content, and osmolality vary in enteral formulas. (p. 665)

3. Identify reasons why oral intake of enteral formulas may be advised. Suggest ways for improving patient acceptance of formulas. (p. 666)

4. List the types of patients who may benefit from tube feedings. (pp. 666–667)

5. Describe the different tube feeding routes, and suggest reasons why each might be used. Discuss advantages and disadvantages of each. (pp. 667–669)

6. Identify measures that can help to prevent contamination of enteral formulas and equipment. (pp. 671–672)

7. Contrast the different methods of formula delivery, and discuss the possible advantages and disadvantages associated with each. Discuss how clinicians can help to relieve anxiety about tube feeding procedures. (p. 673–674)

8. Describe the problems that can occur when medications are delivered through feeding tubes. Suggest guidelines that can prevent these problems. (p. 675)

9. Discuss complications often associated with tube feedings. Summarize possible causes and some measures that can prevent or correct these complications. (pp. 676–677)

These multiple choice questions will help you prepare for an exam. Answers can be found on p. 681.

1. Which of the following statements is correct?
   a. Standard formulas contain whole proteins or protein isolates.
   b. Standard formulas contain free amino acids or small peptide chains.
   c. Modular formulas contain a mixture of proteins, carbohydrates, and fats.
   d. Elemental formulas may contain protein isolates or whole proteins.

2. *Osmolality* refers to an enteral formula's:
   a. energy density.
   b. nutrient density.
   c. fiber content.
   d. concentrations of molecules and ionic particles.

3. For a patient who is at high risk of aspiration and is not expected to be able to eat table foods for several months, an appropriate placement of a feeding tube might be:
   a. nasogastric.
   b. nasoenteric.
   c. gastrostomy.
   d. jejunostomy.

4. In selecting an appropriate enteral formula for a patient, the primary consideration is:
   a. formula osmolality.
   b. the patient's nutrient needs.
   c. availability of infusion pumps.
   d. formula cost.

5. An important measure that may prevent bacterial contamination in tube feeding formulas is:
   a. nonstop feeding of formula.
   b. using the same feeding bag and tubing each day.
   c. discarding opened containers of formula not used within 24 hours.
   d. adding formula to the feeding container before it empties completely.

6. Compared with intermittent feedings, continuous feedings:
   a. always require an infusion pump.
   b. allow greater freedom of movement.
   c. are more similar to normal patterns of eating.
   d. are associated with more GI side effects.

7. A patient needs 1800 milliliters of formula a day. If the patient is to receive formula intermittently every 4 hours, how many milliliters of formula will she need at each feeding?
   a. 225
   b. 300
   c. 400
   d. 425

8. The term that describes the volume of formula remaining in the stomach from a previous feeding is:
   a. residue.
   b. osmolar load.
   c. gastric residual.
   d. intermittent feeding.

9. The health professional using a feeding tube to deliver medications recognizes that:
   a. medications given by feeding tube generally do not cause GI complaints.
   b. medications can usually be added directly to the feeding container.
   c. enteral formulas do not interact with medications in the same way that foods do.
   d. thick or sticky liquid medications and crushed tablets can clog feeding tubes.

10. Tube feedings can gradually be discontinued when:
    a. discharge planning begins.
    b. the patient experiences hunger.
    c. the medical condition resolves.
    d. the patient is able to eat foods or drink formula in sufficient amounts.

REFERENCES

1. A. M. Malone, Enteral formula selection, in P. Charney and A. Malone, eds., *ADA Pocket Guide to Enteral Nutrition* (Chicago: American Dietetic Association, 2006), pp. 63–122.
2. M. E. Shike, Enteral feeding, in M. E. Shils and coeditors, *Modern Nutrition in Health and Disease* (Baltimore: Lippincott Williams & Wilkins, 2006), pp. 1554–1566.
3. Shike, 2006.
4. C. R. Parrish and S. McCray, Enteral feeding: Dispelling myths, *Practical Gastroenterology* 27 (September 2003): 33–50.
5. N. Gupta and R. G. Martindale, Parenteral vs. enteral nutrition, in G. Cresci, ed., *Nutrition Support for the Critically Ill Patient: A Guide to Practice* (Boca Raton, Fla.: Taylor & Francis Group, 2005), pp. 193–208.
6. J. L. Rombeau, Enteral nutrition, in L. Goldman and D. Ausiello, eds., *Cecil Medicine* (Philadelphia: Saunders, 2008), pp. 1617–1621.
7. M. Marian and P. Charney, Patient selection and indications for enteral feedings, in P. Charney and A. Malone, eds., *ADA Pocket Guide to Enteral Nutrition* (Chicago: American Dietetic Association, 2006), pp. 1–25; Shike, 2006.

8. R. L. Koretz, Do data support nutrition support? Part II. Enteral artificial nutrition, *Journal of the American Dietetic Association* 107 (2007): 1374–1380.
9. Shike, 2006; B. Taylor and J. E. Mazuski, Enteral feeding access in the critically ill, in G. Cresci, ed., *Nutrition Support for the Critically Ill Patient: A Guide to Practice* (Boca Raton, Fla.: Taylor & Francis Group, 2005), pp. 235–252.
10. A. M. Malone, Enteral formulations, in G. Cresci, ed., *Nutrition Support for the Critically Ill Patient: A Guide to Practice* (Boca Raton, Fla.: Taylor & Francis Group, 2005), pp. 253–277.
11. Shike, 2006; Malone, 2005.
12. L. Klein, Is blue dye safe as a method of detection for pulmonary aspiration? *Journal of the American Dietetic Association* 104 (2004): 1651–1652; J. P. Maloney and T. A. Ryan, Detection of aspiration in enterally

fed patients: A requiem for bedside monitors of aspiration, *Journal of Parenteral and Enteral Nutrition* 26 (2002): S34–S42.
13. C. Thompson, Initiation, advancement, and transition of enteral feedings, in P. Charney and A. Malone, eds., *ADA Pocket Guide to Enteral Nutrition* (Chicago: American Dietetic Association, 2006), pp. 123–154; C. R. Parrish, J. Krenitsky, and C. Kusenda, Enteral feeding challenges, in G. Cresci, ed., *Nutrition Support for the Critically Ill Patient: A Guide to Practice* (Boca Raton, Fla.: Taylor & Francis Group, 2005), pp. 321–340.
14. Thompson, 2006; Parrish, Krenitsky, and Kusenda, 2005.
15. M. K. Russell, Monitoring complications of enteral feedings, in P. Charney and A. Malone, eds., *ADA Pocket Guide to Enteral Nutrition* (Chicago: American Dietetic Association, 2006), pp. 155–192.
16. Russell, 2006.
17. Russell, 2006.
18. Thompson, 2006.

ANSWERS

Study Questions (multiple choice)

1. a   2. d   3. d   4. b   5. c   6. a   7. b   8. c   9. d   10. d

Inborn Errors of Metabolism

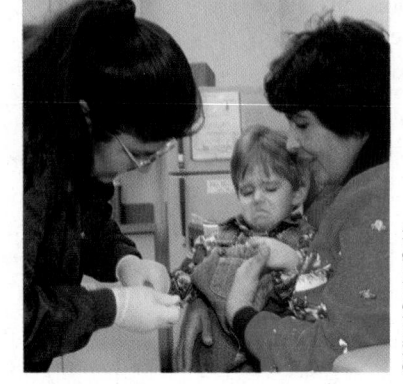

© Miguel Gandert/Corbis

Chapter 20 described the use of enteral formulas for patients who are unable to meet their nutrient needs with conventional foods. Such is the case for individuals with some inborn errors of metabolism; for them, enteral formulas play a vital role in disease management. This highlight describes some inborn errors of metabolism and discusses the role of diet in two of these disorders: phenylketonuria and galactosemia. The accompanying glossary defines terms related to inborn errors of metabolism.

Inborn Errors of Metabolism

An **inborn error of metabolism** is an inherited trait, caused by a genetic **mutation,** that results in the absence, deficiency, or malfunction of a protein that has a critical metabolic role.[1] The protein may function as an enzyme, receptor, transport protein, or structural protein. When the body fails to make a protein, the functions that depend on that protein are impaired. For example, when an enzyme is missing or malfunctioning in a metabolic pathway that typically converts compound A to compound B, compound A will accumulate and compound B will not be made. The excess of compound A and the lack of compound B may have harmful effects. Furthermore, the imbalances in one pathway may affect other pathways and ultimately cause a number of metabolic and physiologic disturbances. The severity of the inborn error's effects are ultimately related to the degree of impairment caused by the altered or missing protein.

Treatment for Inborn Errors of Metabolism

Successful treatment for an inborn error of metabolism depends on the ability to screen newborns and diagnose metabolic diseases before irreversible damage can occur. After a genetic defect is identified, family members undergo **genetic counseling** to evaluate the likelihood that they may pass on the disorder to future offspring. During counseling, couples may learn about reproductive options such as artificial insemination, in vitro fertilization, or prenatal monitoring after conception.

Medical nutrition therapy is the primary treatment for many inborn errors that involve nutrient metabolism. Once the biochemical pathway affected by a mutation is identified, a health practitioner may be able to manipulate elements of the diet to compensate for deficiencies and excesses. Dietary intervention generally involves restricting substances that cannot be properly metabolized and supplying substances that cannot be produced. Thus, dietary changes may be able to improve outcomes of some inborn errors by:

- Preventing the accumulation of toxic metabolites
- Replacing nutrients that are deficient as a result of a defective metabolic pathway
- Providing a diet that supports normal growth and development and maintains health

GLOSSARY

cystic fibrosis: an inherited disorder that affects the transport of chloride across epithelial cell membranes; primarily affects the gastrointestinal and respiratory systems.

galactosemia (ga-LACK-toe-SEE-me-ah): an inherited disorder that affects galactose metabolism. Accumulated galactose causes damage to the liver, kidneys, and brain in untreated patients.

gene therapy: treatment for inherited disorders, in which DNA sequences are introduced into the chromosomes of affected cells, prompting the cells to express the protein needed to correct the disease.

genetic counseling: support for families at risk of genetic disorders; involves diagnosis of disease, identification of inheritance patterns within the family, and review of reproductive options.

hemophilia (HE-moh-FEEL-ee-ah): inherited bleeding disorders characterized by deficiency or malfunction of plasma proteins needed for clotting blood.

inborn error of metabolism: an inherited trait (one that is present at birth) that causes the absence, deficiency, or malfunction of a protein that has a critical metabolic role.

metabolites: products of metabolism; compounds produced by a biochemical pathway.

mutation: an inheritable alteration in the DNA sequence of a gene.

phenylketonuria (FEN-il-KEY-toe-NU-ree-ah) or *PKU:* an inherited disorder that affects the conversion of the essential amino acid phenylalanine to the amino acid tyrosine.

Nondietary therapies can treat some inborn errors of metabolism, although the options are somewhat limited. In some cases, the missing protein is infused; this is the primary means of treating **hemophilia,** caused by deficiency of one of the plasma proteins needed for clotting blood. Drug therapy is the main treatment for some inborn errors, including **cystic fibrosis** (discussed in Chapter 24), which is characterized by a defect that prevents normal chloride transport across cell membranes. Future approaches may include **gene therapy,** a treatment that introduces DNA sequences into the chromosomes of affected cells, prompting the cells to express the protein needed to correct the abnormality.

The following sections of this highlight present a sampling of inborn errors that benefit primarily from medical nutrition therapy. A classic example is phenylketonuria, a metabolic disorder that affects one's ability to metabolize the essential amino acid phenylalanine.

Phenylketonuria

One of many inborn errors affecting amino acid metabolism, **phenylketonuria (PKU)** affects approximately 1 out of every 10,000 births in the United States each year.[2] The screening of newborns for PKU is one of the most common genetic tests in the United States and many other countries. The early detection and treatment of PKU have successfully prevented most of the damaging consequences of this disorder.

The Error in PKU

In PKU, the missing or defective protein is a liver enzyme that converts the essential amino acid phenylalanine to the amino acid tyrosine (see Figure H20-1). Without this enzyme, phenylalanine and its **metabolites** (metabolic products) accumulate and damage the developing nervous system. The impairment in the metabolic pathway also prevents liver synthesis of tyrosine and tyrosine-derived compounds (such as the neurotransmitter epinephrine). Under these conditions, tyrosine becomes essential: the body cannot produce tyrosine, and therefore the diet must supply it.

Although PKU's most debilitating effect is on brain development, other symptoms may manifest if the condition is untreated. Infants with PKU may have poor appetites and grow slowly. They may be irritable or have tremors or seizures. Their bodies and urine may have a musty odor. Their skin may be unusually pale, and they may develop skin rashes. In older children and adults who discontinue treatment, neurological and psychological problems are common.

Detecting PKU

PKU is not evident at birth, but diagnosis in the first few days of life and early treatment prevent its devastating effects. For this reason, newborns are screened for PKU in all 50 states.[3] A standard blood test for phenylalanine is typically conducted by heel puncture, often after the infant has consumed several meals containing protein. Abnormal results require further testing. Before widespread newborn screening, infants with PKU demonstrated developmental delays (for example, inability to crawl) by six to nine months of age. By the time parents recognized the problem, the damage was irreversible.

Medical Nutrition Therapy for PKU

The only current treatment for PKU is a diet that restricts phenylalanine and supplies tyrosine so that the blood levels of these amino acids are maintained within safe ranges. Because phenylalanine is an essential amino acid, the diet cannot exclude it completely. Children with PKU need phenylalanine to grow, but they cannot handle excesses without detrimental effects. Therefore, their diets must provide enough phenylalanine to support growth and health but not so much as

---

FIGURE H20-1 Biochemical Alterations in PKU

Normal:

Normally, the amino acid phenylalanine follows two pathways, one in the liver and the other in the kidneys. In the liver, the enzyme phenylalanine hydroxylase adds a hydroxyl group (OH) to produce the amino acid tyrosine. Tyrosine, in turn, produces melanin, the pigmented compound found in skin and brain cells; the neurotransmitters epinephrine and norepinephrine; and the hormone thyroxin. In the kidneys, enzymes convert phenylalanine to by-products that are excreted.

In the liver:

Phenylalanine $\xrightarrow{\text{Phenylalanine hydroxylase}}$ Tyrosine $\longrightarrow$ Melanin / Epinephrine / Norepinephrine / Thyroxin

In the kidneys:

Phenylalanine $\longrightarrow$ Phenylpyruvic acid (a ketone body) $\longrightarrow$ Other phenyl acids (excreted)

In PKU:

Individuals with PKU lack the liver enzyme phenylalanine hydroxylase, impairing conversion of phenylalanine to tyrosine. Phenylalanine accumulates in the liver and blood, reaching the kidneys in abnormally high concentrations. In the kidneys, an aminotransferase enzyme converts phenylalanine to the ketone body phenylpyruvic acid, which spills into the urine—thus the name phenylketonuria.

In the liver:

Phenylalanine (accumulates) $\xrightarrow{\text{Phenylalanine hydroxylase (deficient)}}$ Tyrosine (deficient)

In the kidneys:

Phenylalanine (accumulates) $\longrightarrow$ Phenylpyruvic acid (accumulates) $\longrightarrow$ Other phenyl acids (accumulate)

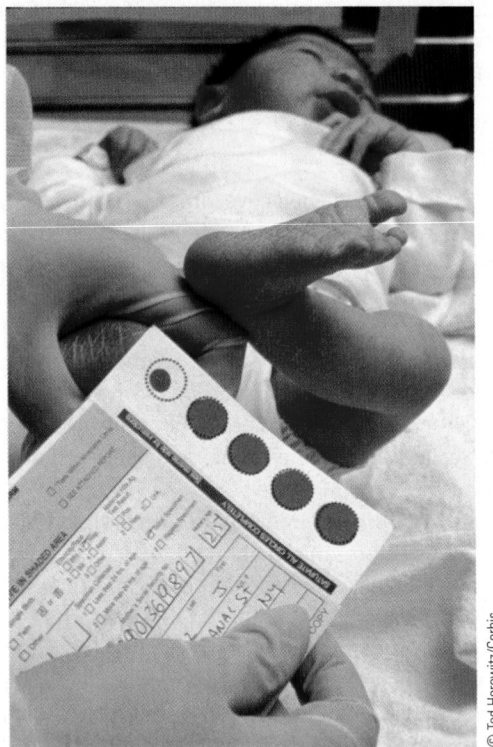

© Ted Horowitz/Corbis

A simple blood test screens newborns for PKU—
a common inborn error of metabolism.

phenylalanine, so only limited amounts are allowed. Low-protein flours and mixes are available for making low-phenylalanine breads, pasta, cakes, and cookies. Foods that do not contain phenylalanine, such as jams, jellies, and most sweeteners, can be used freely. Growth rates and nutrition status are monitored to ensure that the diet is adequate.

Parents and children may need to develop creative ways to make these diets enjoyable. The formula can be flavored or combined with fruits or juices to make smoothies or frozen juice bars. Sandwiches can include low-phenylalanine breads and fillings such as mashed bananas or avocados, shredded carrots and olives, or tomato slices with mayonnaise. Children often enjoy creating special recipes with permitted foods to make their choices more varied and to share meals with friends.

Continuing Dietary Restrictions

Lifelong adherence to a phenylalanine-restricted diet is currently recommended for all individuals with PKU. (Adults continue to use phenylalanine-free formulas, which generally provide about half of their protein and energy needs.) Elevated phenylalanine levels can adversely affect cognitive function at any age. Case studies have suggested that individuals with PKU who discontinue dietary management may have problems with attention span, concentration, and memory. It is especially important that women with PKU maintain safe phenylalanine concentrations during pregnancy. Elevated phenylalanine levels, especially during the first trimester, have been associated with mental retardation and organ malformations in the offspring of PKU mothers who have discontinued dietary treatment.[4]

Galactosemia

Galactosemia is an example of an inborn error of carbohydrate metabolism. Individuals with galactosemia are deficient in one of the enzymes needed to metabolize galactose, a sugar that is primarily found in milk products (recall that each lactose molecule contains a molecule of galactose). An accumulation of galactose can cause damage in multiple tissues. Infants with galactosemia who are given milk react with severe vomiting and liver jaundice within days of the initial feeding. Serious liver damage can develop and progress to symptomatic cirrhosis. Other complications may include kidney failure, cataracts, and brain damage. Treatment in the first weeks of life can prevent the most detrimental effects of galactose accumulation, but if treatment is delayed, the damage to the brain is irreversible.[5]

The Galactosemia Diet

The diet for galactosemia is much simpler than the diet for PKU. For one thing, galactose is not an essential nutrient. The galactosemia diet essentially eliminates galactose from the diet and does not need to provide a carefully determined amount of any nutrient, as the PKU diet does. In addition, dietary galactose is primarily obtained from lactose (the milk sugar), so the main focus of dietary treatment is the exclusion of milk and milk products. A

to cause harm. The diets must also provide tyrosine, which is an essential nutrient for individuals with PKU. To ensure that blood concentrations of phenylalanine and tyrosine are close to normal, blood tests are performed periodically, and diets are adjusted when necessary. If the dietary treatment is conscientiously followed, it can prevent the symptoms described earlier. Adults must continue to follow the PKU diet as well to prevent deterioration in brain function.

The PKU Diet

Central to the PKU diet (for all ages) is the use of an enteral formula that is phenylalanine-free yet supplies energy, amino acids, vitamins, and minerals. In infants, the phenylalanine-free formula is supplemented with measured amounts of breast milk or regular infant formula, which provide the phenylalanine that an infant needs for growth. Low-phenylalanine infant formulas are also available and are useful for infants who meet most or all of their nutrient needs by consuming formula. Formula requirements need to be recalculated periodically to accommodate the growing infant's shifting needs for protein, phenylalanine, tyrosine, and energy.

Once food consumption begins, a phenylalanine-free formula supplies the needed amino acids, and foods that contain phenylalanine are carefully monitored. All proteins contain some phenylalanine; therefore, high-protein foods such as meat, fish, poultry, milk, cheese, legumes, and nuts (including peanut butter) are omitted. Fruits, vegetables, and cereals also contain number of other foods that contain galactose in substantial amounts, such as organ meats and some legumes, fruits, and vegetables, must also be avoided or restricted. Patients receive food lists that identify the galactose content of common foods.

Infants diagnosed with galactosemia are given lactose-free formulas to meet their nutrient needs. Once a child can consume adequate amounts of regular foods, special formulas are unnecessary. However, care must be taken to ensure that the diet supplies adequate calcium.

Long-Term Complications

Although the early introduction of a galactose-restricted diet can eliminate the acute toxic effects of galactosemia, complications of the disease may develop despite an individual's compliance with diet therapy. For example, most patients experience delays in speech and language development. Ovarian failure occurs in up to 85 percent of women who have galactosemia.[6] In addition, some evidence suggests that IQ declines as a person with galactosemia ages. The reasons for these long-term complications are not fully understood.

As our scientific understanding of human genetics and biochemistry increases, more inborn errors of metabolism are being recognized. Mainstays of management for these diseases include effective diagnosis, early treatment, and control of environmental factors that cause toxicity. In some cases, dietary changes are central to treatment and can prevent serious complications. Not all inborn errors are easily treated, however. Future developments in biotechnology may someday allow medical practitioners to correct genetic errors using gene therapy.

REFERENCES

1. L. J. Elsas II, Approach to inborn errors of metabolism, in L. Goldman and D. Ausiello, eds., *Cecil Medicine* (Philadelphia: Saunders, 2008), pp. 1539–1546.
2. S. D. Cederbaum, Disorders of phenylalanine and tyrosine metabolism, in L. Goldman and D. Ausiello, eds., *Cecil Medicine* (Philadelphia: Saunders, 2008), pp. 1573–1576.
3. L. J. Elsas II and P. B. Acosta, Inherited metabolic disease: Amino acids, organic acids, and galactose, in M. E. Shils and coeditors, *Modern Nutrition in Health and Disease* (Baltimore: Lippincott Williams & Wilkins, 2006), pp. 909–959.
4. Cederbaum, 2008.
5. L. J. Elsas II, Galactosemia, in L. Goldman and D. Ausiello, eds., *Cecil Medicine* (Philadelphia: Saunders, 2008), pp. 1555–1558.
6. Elsas, Galactosemia, 2008.

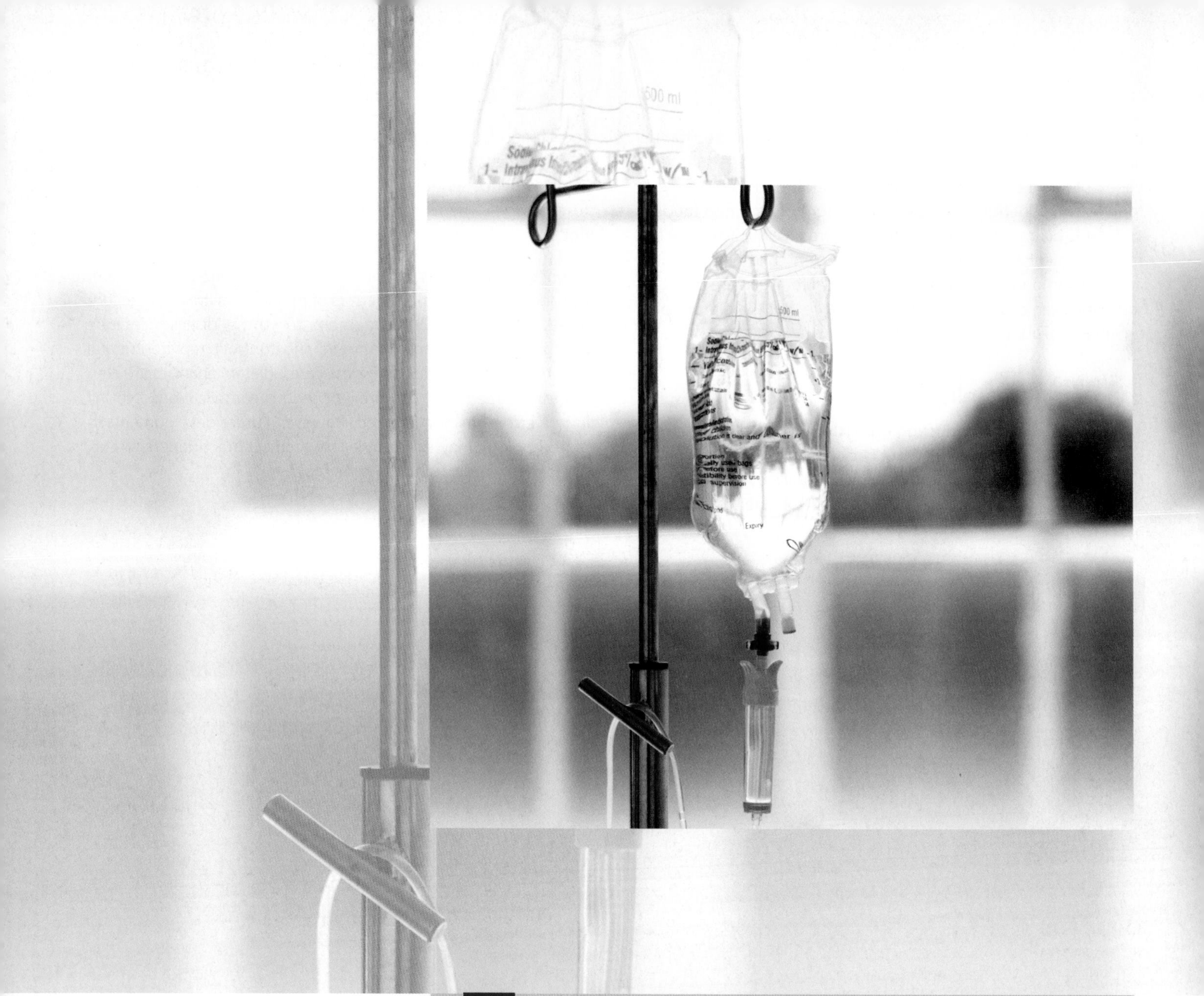

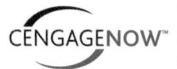

George Doyle/Getty Images

Nutrition in the Clinical Setting

The science of medical nutrition was dramatically changed in 1968 by the demonstration that all nutrient needs could be met intravenously. Since then, health practitioners have had a way to feed people who otherwise might have died from malnutrition. Although intravenous feeding techniques have advanced considerably since 1968, parenteral nutrition remains expensive and is sometimes associated with serious complications. For these reasons, health practitioners subscribe to the adage "If the GI tract works, use it."

CENGAGENOW™ The CengageNOW logo indicates an opportunity for online self-study, linking you to interactive tutorials and videos based on your level of understanding.

academic.cengage.com/login

Parenteral Nutrition Support

Chapter 20 described how enteral formulas can supplement or replace conventional foods to meet nutritional needs. Enteral formulas cannot be used when intestinal function is inadequate, however, and therefore the ability to meet nutrient needs intravenously is a lifesaving option for critically ill persons. Unfortunately, the procedure is costly and associated with a number of potentially dangerous complications. If the gastrointestinal (GI) tract is functional, enteral nutrition support is preferred, partly to avoid the expense and complications associated with intravenous feedings and partly to preserve healthy GI function. Figure 21-1 on p. 688 summarizes the decision-making process for selecting the most appropriate feeding method.

Indications for Parenteral Nutrition

As with other nutrition therapies, the decision to use parenteral nutrition is based on a thorough assessment of the patient's medical condition and nutrient needs. Generally, parenteral nutrition is indicated for patients who do not have functioning GI tracts and who are either malnourished or likely to become so. Parenteral support may also be of benefit if using the GI tract would cause harm to the patient, as when severe tissue damage in the small intestine requires bowel rest for an extended period. Thus, patients with the following conditions are often considered candidates for parenteral nutrition:

- Intestinal obstructions or fistulas
- Paralytic ileus (intestinal paralysis)
- Short-bowel syndrome (a substantial portion of the small intestine has been removed)
- Intractable vomiting or diarrhea
- Severe electrolyte, mineral, and glucose imbalances
- Severe pancreatitis

FIGURE 21-1 Selecting a Feeding Route

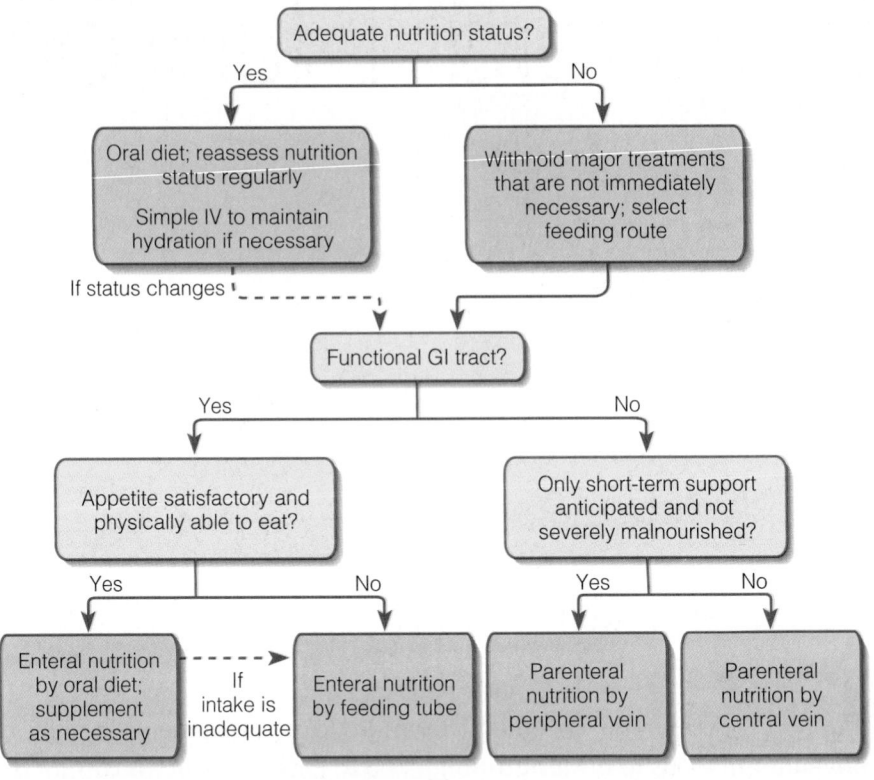

The peripheral veins can provide access to the blood for delivery of parenteral solutions.

peripheral veins: small-diameter veins that carry blood from the arms and legs.

central veins: large-diameter veins located close to the heart.

peripheral parenteral nutrition (PPN): a type of nutrition support in which intravenous feedings are delivered into peripheral veins.

osmolarity: the concentration of osmotically active particles in a solution, expressed as milliosmoles per liter (mOsm/L). *Osmolality* is an alternative expression of a solution's osmotic properties that is used in clinical practice and is expressed as milliosmoles per kilogram (mOsm/kg).

- Bone marrow transplants
- Severe malnutrition and intolerance to enteral nutrition

Some clinical studies suggest that parenteral nutrition is not always effective in the patient populations in which it is routinely used; thus, the decision to use it should be considered in light of the most recent research evidence.[1] In addition, parenteral nutrition is unlikely to be beneficial when used for periods of less than 7 to 14 days.[2]

Once the decision to use parenteral nutrition has been made, the access site must be selected. The access sites for intravenous feedings fall into two main categories: nutrients may be delivered into the **peripheral veins** located in the arms and legs or into the large-diameter **central veins** located near the heart.

Peripheral Parenteral Nutrition In **peripheral parenteral nutrition (PPN),** nutrient needs are met using only the peripheral veins. Peripheral veins can be damaged by overly concentrated solutions, however: phlebitis (inflammation of the vein) may result, characterized by redness, swelling, and tenderness at the infusion site. Therefore, the **osmolarity** of parenteral solutions is usually kept between 600 and 900 milliosmoles per liter,[3] so PPN can supply only limited amounts of energy and protein. The "How to" on p. 689 compares the expressions for *osmolarity* and *osmolality*, both of which can be used to express the osmolar concentration of a solution.

PPN is used most often in patients who require short-term nutrition support (about 7 to 10 days) and who do not have high nutrient needs or fluid restrictions. The use of PPN is not possible if the peripheral veins are too weak to tolerate the procedure. In most cases, it is necessary to rotate venous access sites to prevent inflammation.

Total Parenteral Nutrition Most patients meet their nutrient needs using the larger, central veins, where blood volume is greater and nutrient concentrations

HOW TO Express the Osmolar Concentration of a Solution

Chapter 12 described the concept of osmosis and explained how the solute concentration influences a solution's osmotic pressure. *Osmolarity* and *osmolality* are both used to express the osmolar concentration of a solution—that is, the concentration of osmotically active solutes in a solution:

- Osmolarity refers to the milliosmoles per liter of solution (mOsm/L).
- Osmolality refers to the milliosmoles per kilogram of solvent (mOsm/kg).

The *milliosmole* is a unit that represents the ions and molecules that contribute to the solution's osmotic pressure.

Whereas osmolarity refers to a volume of solution that includes the solutes of interest, osmolality refers to the solutes separately from the solvent in which they are dis-

solved. A second difference between the expressions is that osmolarity is expressed in terms of the solution's volume, whereas osmolality is expressed in terms of the solvent's weight.

Osmolarity and osmolality are both used in clinical practice, but they are derived in different ways. Osmolarity is typically calculated using equations that account for the nutrients and electrolytes in biological solutions. Osmolality is usually a measured value obtained using an *osmometer,* a common instrument in hospital laboratories.

Osmolarity and osmolality are roughly equivalent when describing dilute aqueous solutions at room or body temperature. This is because 1 liter of water weighs 1 kilogram, and the solutes contribute little to the volume of the solution.

---

do not need to be limited. Because this method can reliably meet a person's complete nutrient requirements, it is called **total parenteral nutrition (TPN).** Central veins lie close to the heart, where the large volume of blood rapidly dilutes parenteral solutions. Therefore, patients with very high nutrient needs or fluid restrictions are able to receive the nutrient-dense solutions they require. TPN is also preferred for patients who require long-term intravenous feedings.

There are several ways to access central veins. The tip of a central venous **catheter** can be placed directly into a large-diameter central vein or threaded into a central vein through a peripheral vein (see Figure 21-2). Peripheral insertion of

total parenteral nutrition (TPN): a type of nutrition support in which intravenous feedings are delivered into a central vein.

catheter: a thin tube placed within a narrow lumen (such as a blood vessel) or body cavity; can be used to infuse or withdraw fluids or keep a passage open.

FIGURE 21-2 Accessing Central Veins for Total Parenteral Nutrition

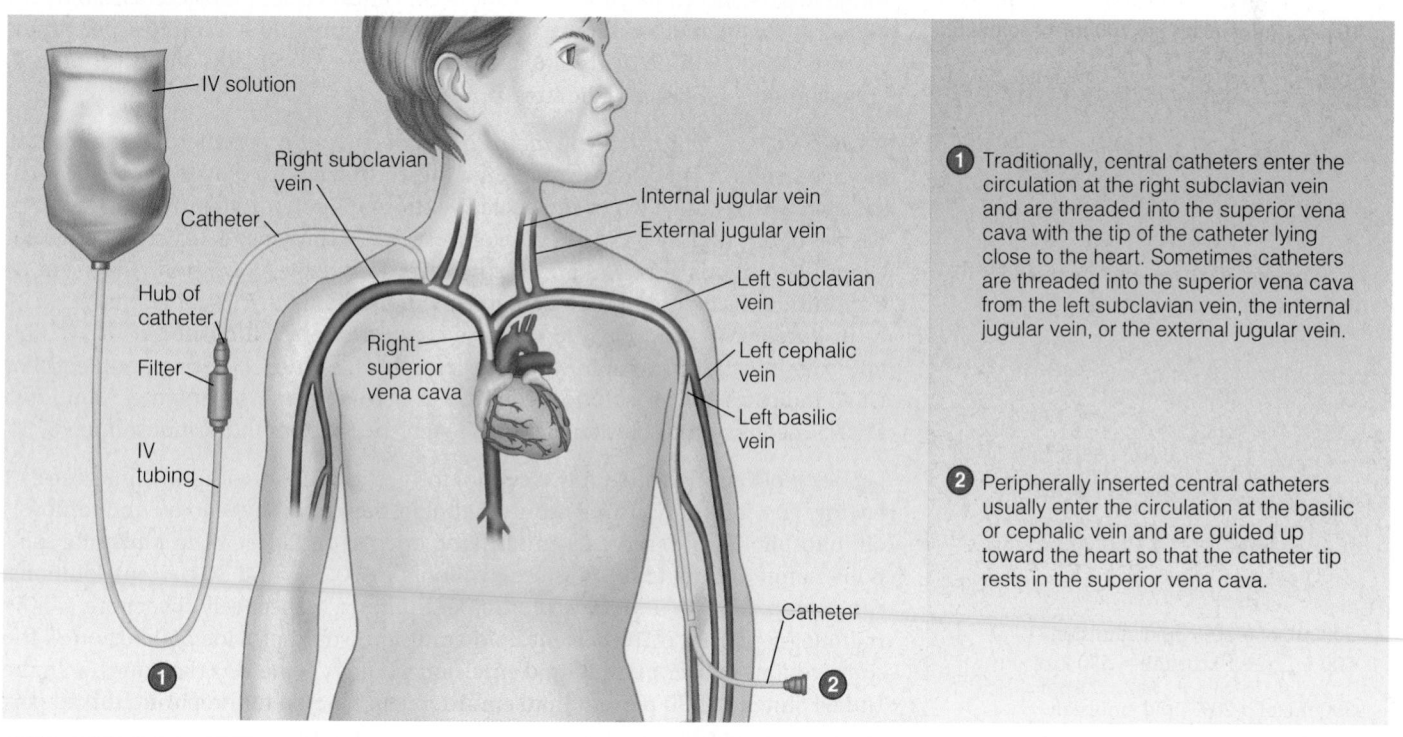

1 Traditionally, central catheters enter the circulation at the right subclavian vein and are threaded into the superior vena cava with the tip of the catheter lying close to the heart. Sometimes catheters are threaded into the superior vena cava from the left subclavian vein, the internal jugular vein, or the external jugular vein.

2 Peripherally inserted central catheters usually enter the circulation at the basilic or cephalic vein and are guided up toward the heart so that the catheter tip rests in the superior vena cava.

central catheters is less invasive and lower in cost than the direct insertion of catheters into central veins; this method is usually preferred for short-term venous access (about two months or less in duration).[4]

> ### IN SUMMARY
>
> Parenteral nutrition support delivers nutrients intravenously. It is used in patients whose GI tract is not functioning and who may readily become malnourished. Patients receiving parenteral nutrition generally have intestinal disorders or are critically ill. If nutrients are infused directly into peripheral veins (peripheral parenteral nutrition), nutrient concentrations must be limited to avoid inflammation of the veins. The infusion of nutrients into central veins (total parenteral nutrition) can supply nutrient-dense solutions and is used for long-term intravenous feedings.

Parenteral Solutions

The pharmacies located within health care institutions are often responsible for preparing parenteral solutions. This arrangement is convenient because the pharmacist can customize formulations to meet patients' nutrient needs and because the solutions have a limited shelf life. This section describes the nutrients and characteristics of parenteral solutions.

Parenteral Nutrients

Parenteral solutions provide the combinations of amino acids, carbohydrate, lipids, vitamins, and minerals that are best suited to meet patients' requirements. Because the nutrients are provided intravenously, they must be given in forms that are safe to inject directly into the bloodstream.

Amino Acids Parenteral solutions contain all of the essential amino acids and various combinations of the nonessential amino acids. Amino acid concentrations range from 3.5 to 15 percent; ◆ the more concentrated solutions are used only for TPN. Just as in regular foods, the amino acids provide 4 kcalories per gram. Disease-specific amino acid solutions are available for patients with liver failure, kidney failure, and metabolic stress.

◆ A 10 percent amino acid solution supplies 10 g of amino acids per 100 mL of solution.

Carbohydrate Glucose is the main source of energy in parenteral feedings. It is provided in the form dextrose monohydrate, in which each glucose molecule is associated with a single water molecule. Dextrose monohydrate provides 3.4 kcalories per gram, slightly less than pure glucose, which provides 4 kcalories per gram. Dextrose solutions are available in concentrations between 2.5 and 70 percent. ◆ Concentrations greater than 12.5 percent are used only in TPN solutions.[5]

In parenteral solutions, the dextrose concentration is indicated by a "D" followed by its concentration in water (W) or normal saline (NS). For example, D5 or D5W indicates that a solution contains 5 percent dextrose in water. Similarly, D5/NS means that a solution contains 5 percent dextrose in normal saline.

◆ A 10 percent dextrose solution provides 10 g of dextrose monohydrate per 100 mL of solution.

Lipids Lipid emulsions supply essential fatty acids and are a significant source of energy. The emulsions usually contain triglycerides from soybean oil and safflower oil, phospholipids to serve as emulsifying agents, and glycerol to make the solutions isotonic. Lipid emulsions are available in 10, 20, and 30 percent solutions, containing 1.1, 2.0, and 3.0 kcalories per milliliter, respectively. Therefore, a 500-milliliter container of 10 percent lipid emulsion would provide 550 kcalories; the same volume of a 20 percent lipid emulsion would provide 1000 kcalories. ◆ In the United States, the 30 percent lipid emulsion can be used for preparing mixed parenteral solutions but cannot be directly infused into patients.[6]

◆ 500 mL of a 10% lipid emulsion:
500 mL × 1.1 kcal/mL = 550 kcal

500 mL of a 20% lipid emulsion:
500 mL × 2 kcal/mL = 1000 kcal

Lipid emulsions are often provided daily and may supply 20 to 30 percent of total kcalories. Including lipids as an energy source reduces the need for energy from dextrose and lowers the risk of hyperglycemia in glucose-intolerant patients. Lipid infusions must be restricted in patients with hypertriglyceridemia, however. There is also some concern that lipid emulsions that contain excessive linoleic acid can suppress some aspects of the immune response.

Fluids and Electrolytes Daily fluid needs are estimated to be about 30 to 40 milliliters per kilogram of body weight in young adults and 25 to 30 milliliters per kilogram of body weight in older adults, averaging between 1500 and 2500 milliliters for most people. The amounts are adjusted according to daily fluid losses and the results of hydration assessment.

The electrolytes added to parenteral solutions include sodium, potassium, chloride, calcium, magnesium, and phosphorus. The amounts in parenteral solutions differ from DRI values because the nutrients are infused directly into the blood and are not influenced by absorption, as they are when consumed orally. Because electrolyte imbalances can be lethal, electrolyte management by experienced professionals is necessary whenever intravenous therapies are used. Blood tests are administered daily to monitor electrolyte levels until patients have stabilized.

The electrolyte content of parenteral solutions is expressed in *milliequivalents (mEq)*, which are units indicating the number of ionic charges provided by electrolytes. ◆ The body's fluids are neutral solutions that contain equal numbers of positive and negative charges.

Vitamins and Trace Minerals Commercial multivitamin and trace mineral preparations are routinely added to parenteral solutions. All of the vitamins are usually included, although a preparation without vitamin K is available for patients using warfarin therapy.[7] ◆ The trace minerals usually added to parenteral solutions include zinc, copper, chromium, selenium, and manganese. Iron is excluded because it alters the stability of other ingredients in parenteral mixtures; therefore, special forms of iron need to be injected separately.

Osmolarity Recall that the osmolarity of PPN solutions is limited to 900 milliosmoles per liter because peripheral veins are sensitive to high nutrient concentrations, whereas TPN solutions may be as nutrient dense as necessary. The components of a solution that contribute most to its osmolarity are amino acids, dextrose, and electrolytes: as concentrations of these nutrients increase, the osmolarity of a solution increases. Because lipids contribute little to osmolarity, lipid emulsions are used to increase the energy provided in PPN solutions. Table 21-1 on p. 692 presents a method for estimating the osmolarity of a parenteral solution.

Medications To avoid the need for a separate infusion site, medications are occasionally added directly to parenteral solutions or infused through a separate port (attached via a Y-connector). The administration of a second solution using a separate port in a catheter is called a **piggyback.** Insulin, for example, is sometimes added by piggyback to improve glucose tolerance. Heparin (an anticoagulant) may be added to prevent clotting at the catheter tip. In practice, few medications are added to parenteral solutions so that potential drug-nutrient interactions can be avoided.

Solution Preparation

The parenteral solution prescribed depends on the patient's medical condition and nutrition status and the method of venous access. Prescriptions for parenteral solutions are highly individualized and may need to be recalculated daily until the patient's condition is stable. Figure 21-3 (p. 693) provides an example of a parenteral nutrition order form.

Parenteral Formulations When a parenteral solution contains dextrose, amino acids, and lipids, it is called a **total nutrient admixture (TNA),** a **3-in-1**

A lipid emulsion gives a parenteral solution a milky white color.

◆ Milliequivalents are determined by dividing an ion's molecular weight (MW) by its number of charges. For example:
- For calcium, MW = 40, and the ion has 2 positive charges: 40 ÷ 2 = 20. Thus, 1 mEq of Ca⁺⁺ is equivalent to 20 mg of calcium.
- For sodium, MW = 23, and the ion has 1 positive charge: 23 ÷ 1 = 23. Thus, 1 mEq of Na⁺ is equivalent to 23 mg of sodium.
- *1 mEq of Ca⁺⁺ has the same number of charges as 1 mEq of Na⁺.*

◆ Reminder: The anticoagulant warfarin works by interfering with vitamin K's blood-clotting function (see Chapter 19).

piggyback: the administration of a second solution using a separate port in an intravenous catheter.

total nutrient admixture (TNA): a parenteral solution that contains dextrose, amino acids, and lipids; also called a **3-in-1 solution** or an **all-in-one solution.**

TABLE 21-1 Osmolarity Contribution of Nutrients in Parenteral Solutions

For an estimate of the osmolarity of a 1-liter parenteral solution:
- Multiply the grams of amino acids in the solution by 10.
- Multiply the grams of dextrose in the solution by 5.
- Multiply the milliequivalents of electrolytes in the solution by 2.
- Multiply the grams of lipids in the solution by 1.5.

Add the values obtained to determine the approximate osmolarity.

Example:
A liter of a TPN solution has the approximate composition shown below. Calculate the osmolarity contribution of each component and estimate the total osmolarity of the solution.

Amino acids: 40 g	Sodium: 40 mEq	Calcium: 4.8 mEq
Dextrose: 250 g	Potassium: 35 mEq	Magnesium: 8 mEq
Lipids: 40 g	Chloride: 77 mEq	Phosphate: 21 mEq

Answer:
Amino acids: 40 g × 10 = 400 mOsm/L.
Dextrose: 250 g × 5 = 1250 mOsm/L.
Electrolytes: (40 + 35 + 77 + 4.8 + 8 + 21) × 2 = 371.6 mOsm/L.
Lipids: 40 g × 1.5 = 60 mOsm/L.
Total osmolarity: 400 + 1250 + 371.6 + 60 = 2081.6 mOsm/L.

NOTE: mEq = milliequivalents; mOsm/L = milliosmoles per liter.
SOURCE: Adapted from A. M. Coulston, C. L. Rock, and E. R. Monsen, eds., *Nutrition in the Prevention and Treatment of Disease* (San Diego: Academic Press, 2001), Table 7; p. 254.

solution, or an **all-in-one solution.** A **2-in-1 solution** excludes lipids, and the lipid emulsion is administered separately, often by piggyback administration. The administration of TNA solutions is simpler because only one infusion pump is required; however, the addition of lipid emulsion to solutions reduces their stability, a major concern when TNA solutions are compounded. Generally, a TNA with a high lipid concentration is more stable than one with a low lipid concentration, because diluting the lipids reduces the concentration of the phospholipids that emulsify the solution. Lipids are often administered separately when they are not a major energy source and are used only to provide essential fatty acids. The "How to" below describes a method for calculating the macronutrient and energy content of a parenteral solution.

Nonprotein kCalorie-to-Nitrogen Ratio Some practitioners calculate the **nonprotein kcalorie-to-nitrogen ratio** to assess whether the nitrogen provided by the solution is sufficient for maintaining muscle tissue. A ratio of 150:1 to 200:1 is

2-in-1 solution: a parenteral solution that contains dextrose and amino acids, but excludes lipids.

nonprotein kcalorie-to-nitrogen ratio: a ratio between the nonprotein kcalories and nitrogen content of the diet; used to assess whether the nitrogen intake is sufficient for maintaining muscle tissue.

HOW TO Calculate the Macronutrient and Energy Content of a Parenteral Solution

Suppose a patient is receiving 1.25 liters (1250 milliliters) of a parenteral solution that contains 5 percent amino acids and 30 percent dextrose, supplemented by 250 milliliters of a 20 percent lipid emulsion daily. How many grams of protein and carbohydrate is the person receiving, and what is the total energy intake for the day?

Amino acids:

$$5\% \text{ amino acids} = \frac{5 \text{ g amino acids}}{100 \text{ mL}}$$

$$\frac{5 \text{ g amino acids}}{100 \text{ mL}} \times 1250 \text{ mL} = 62.5 \text{ g of amino acids}$$

62.5 g amino acids × 4.0 kcal/g = 250 kcal

Carbohydrate:

$$30\% \text{ dextrose} = \frac{30 \text{ g dextrose}}{100 \text{ mL}}$$

$$\frac{30 \text{ g dextrose}}{100 \text{ mL}} \times 1250 \text{ mL} = 375 \text{ g of dextrose}$$

375 g dextrose × 3.4 kcal/g = 1275 kcal

Lipids:
Recall that a 20 percent lipid emulsion provides 2.0 kcalories per milliliter. If the patient is given 250 milliliters of the emulsion:

250 mL × 2.0 kcal/mL = 500 kcal

Total energy intake:

250 kcal + 1275 kcal + 500 kcal = 2025 kcal

FIGURE 21-3 Sample Parenteral Nutrition Order Form

Physician Orders
PARENTERAL NUTRITION (PN) – ADULT

Primary Diagnosis: _____ Ht: _____ cm **Dosing Wt:** _____ kg

PN Indication: _____ Allergies _____

Instructions: This form must be completed for a new order or continuation of PN and faxed to the Pharmacy by [Insert Time] to receive same day preparation. PN administration begins at [Insert Time]. Contact the Nutrition Support Service at (XXX) XXX-XXXX for additional information.

Administration Route: CVC or PICC   *Note: Proper tip placement of the CVC or PICC must be confirmed prior to PN infusion*

Peripheral IV (PIV) *(Final PN Osmolarity ≤ _____ mOsm/L)*

Monitoring: Daily weights, Strict input & output, Bedside glucose monitoring every _____ hours

Na, K, Cl, CO_2, Glucose, BUN, Scr, Mg, PO_4 every _____

T, Bili, Alk Phos, AST, ALT, Albumin, Triglycerides, Calcium every _____

Base Solution: *Select one*	*Parenteral nutrition MUST be administered through a dedicated infusion port and filtered with a 1.2-micron in-line filter at all times. Discard any unused volume after 24 hours.*

PERIPHERAL 2-in-1	**CENTRAL 2-in-1**	**CENTRAL 3-in-1**
Dextrose _____ g	Dextrose _____ g	Dextrose _____ g
Amino Acids (*Brand _____*) _____ g	Amino Acids (*Brand _____*) _____ g	Amino Acids (*Brand _____*) _____ g
For patients with PIV and established glucose tolerance; Provides _____ kcal; Maximum Rate not to exceed _____ mL/hour	*For patients with CVC or PICC and established glucose tolerance; Provides _____ kcal; Maximum Rate not to exceed _____ mL/hour*	Fat Emulsion (*Brand _____*) _____ g *For patients with CVC or PICC and established glucose/fat emulsion tolerance; Provides _____ kcal; Maximum Rate not to exceed _____ mL/hour*
RATE & VOLUME: _____ mL/hour for _____ hours = _____ mL/day ***Must specify***		*Use of additional fat emulsion not required with 3-in-1 base solution*

or **CYCLIC INFUSION:** _____ mL/hour for _____ hours, then _____ mL/hour for _____ hours = _____ mL/day

Fat Emulsion (*Brand _____*) – *via PIV or CVC with 2-in-1 base solutions* (*Select caloric density & volume*)

10%	250 mL	Infuse at _____ mL/hour over _____ hours	Frequency _____
20%	500 mL	(*Note: infusions < 4 or > 12 hours not recommended*)	*Discard any unused volume after 12 hours.*

Additives: *(per day)*		**Normal Dosages**	**Additives:** *(per day)*
Sodium Chloride	_____ mEq	*1-2 mEq Sodium/kg/day*	**Regular Insulin** _____ units
as Acetate	_____ mEq	*pH or CO_2 dependent*	*Recommend if hyperglycemic, start*
as Phosphate	_____ mmol of PO_4	*Consider if hyperkalemic*	*with 1 unit for every 10 g of dextrose*
Potassium Chloride	_____ mEq	*1-2 mEq Potassium/kg/day*	
as Acetate	_____ mEq	*pH or CO_2 dependent*	**Pharmacy Use Only: Ca/PO_4**
as Phosphate	_____ mmol of PO_4	*20-40 mmol/day (1 mmol Phos = 1.5 mEq K)*	**Limit Checked** _____
Calcium **Gluconate**	_____ mEq	*5-15 mEq/day*	*(Note: Some brands of amino acids contain phosphate)*
Magnesium **Sulfate**	_____ mEq	*8-24 mEq/day*	
Adult **Multivitamins**	_____ mL/day	*Contains Vitamin K 150 mcg*	
Adult **Trace Elements**	_____ mL/day	*Zn ___ mg, Cu ___ mg, Mn ___ mg, Cr ___ mcg, Se ___ mcg (with normal hepatic function)*	
H_2 **Antagonist** _____	_____ mg	*____ mg/day with normal renal function*	
Other:			

Physician's Signature: _____ Pager Number: _____ Date/time: _____

Orders transcribed by: _____ Date/time: _____ Orders verified by: _____ Date/time: _____

SEND COMPLETED ORDERS TO PHARMACY

considered adequate for stable patients, whereas a ratio of 100:1 or less is preferred for critically ill patients who have difficulty maintaining muscle mass. The ratio is sometimes used as a guideline for preparing parenteral solutions, in conjunction with information related to the patient's clinical condition and nutrition status.[8] The "How to" on p. 694 shows how to calculate the nonprotein kcalorie-to-nitrogen ratio.

HOW TO Calculate the Nonprotein kCalorie-to-Nitrogen Ratio

The nonprotein kcalorie-to-nitrogen ratio of the diet is sometimes used to determine whether a patient is receiving adequate nitrogen to maintain muscle tissue. A ratio between 150:1 and 200:1 is often adequate for stable patients, whereas ratios of 100:1 and below may be necessary for patients who are critically ill.

To calculate the nonprotein kcalorie-to-nitrogen ratio, determine the energy intake from carbohydrates and lipids and compare this amount to the nitrogen intake. To determine the nitrogen intake, multiply the amino acid or protein intake by 16% (protein contains about 16% nitrogen by weight).

The "How to" on p. 692 describes a parenteral solution that provides 1275 kcalories from dextrose, 500 kcalories from lipids, and 62.5 grams of amino acids. Using these values to calculate the nonprotein kcalorie-to-nitrogen ratio:

Nonprotein kcalories:

1275 kcal (dextrose) + 500 kcal (lipids) = 1775 kcal

Nitrogen content:

62.5 g amino acids × 16% nitrogen = 62.5 g × 0.16 = 10 g

Nonprotein kcalorie-to-nitrogen ratio:

1775 ÷ 10 = 178:1

Thus, the parenteral solution in the example has a nonprotein kcalorie-to-nitrogen ratio of 178:1, which is likely to be adequate for a stable hospital patient.

Safety Concerns Intravenous feedings are similar to tube feedings in that careful attention to solution preparation and handling can minimize complications. To prevent bacterial contamination and maintain stability, parenteral solutions are compounded in the pharmacy under aseptic conditions, shielded from light, and refrigerated. Prior to infusion, the solutions are removed from the refrigerator and allowed to reach room temperature. During feedings, the solution and catheter need to be checked frequently for signs of contamination.

IN SUMMARY

Prescriptions for parenteral solutions are individualized to meet each patient's needs. The solutions are compounded in hospital pharmacies using commercial nutrient preparations and include amino acids, dextrose, electrolytes, vitamins, and trace minerals. Lipid emulsions may be included in the solution or may be administered separately. A parenteral solution that includes lipids is called a total nutrient admixture, a 3-in-1 solution, or an all-in-one solution. Few medications are added to parenteral solutions due to the potential for drug-nutrient interactions. Parenteral solutions are prepared and handled using aseptic techniques to prevent contamination.

Administering Parenteral Nutrition

◆ Reminder: A *nutrition support team* is a multidisciplinary team of health care professionals who are responsible for the provision of nutrients by tube feeding or intravenous infusion.

Parenteral nutrition is a complex treatment that requires skills from a variety of disciplines. Many hospitals organize nutrition support teams, ◆ consisting of physicians, nurses, dietitians, and pharmacists, that specialize in the provision of both intravenous and tube feedings. Members of the team may serve as advisers to other clinicians or may manage nutrition support directly. They may also have administrative responsibilities, such as receiving patients, purchasing supplies, developing guidelines, and keeping records. Figure 21-4 describes the typical roles of each member of the nutrition support team.

Insertion and Care of Intravenous Catheters

Although skilled nurses can place catheters into peripheral veins, only qualified physicians can insert catheters directly into central veins. Patients may be awake for the procedure and be given local anesthesia. Unnecessary apprehension can be avoided by explaining the procedure to the patient beforehand.

FIGURE 21-4 The Nutrition Support Team

The physician
- Diagnoses medical problems
- Performs medical procedures
- Coordinates and prescribes therapy
- Directs and supervises team
- Approves guidelines and protocols
- Consults with other physicians

The nurse
- Assesses nursing needs
- Performs direct patient care
- Explains medical procedures and treatment plans
- Instructs patients regarding medical care
- Acts as a liaison between team and nursing staff
- Coordinates discharge plans

All team members
- Review current research
- Analyze new products
- Develop guidelines
- Provide in-service training
- Monitor patients
- Correct problems
- Educate patients
- Evaluate the outcome of the care provided and cost savings
- Promote the appropriate use of nutrition support
- Improve communications among team members and between the team and other health care professionals

The dietitian
- Assesses nutrition status
- Determines patients' nutrient needs
- Recommends appropriate diet therapy
- Reevaluates patients regularly
- Instructs patients about their diets
- Acts as a liaison between the team and the dietary department

The pharmacist
- Recommends appropriate drug therapy
- Identifies drug-drug and diet-drug interactions
- Identifies drug-related complications
- Educates patients about their medications
- Acts as a liaison between the team and the pharmacy

Catheter-related problems frequently cause complications (see Table 21-2). Catheters may be improperly positioned or may dislodge after placement. Air can leak into catheters, obstructing blood flow. Catheters in peripheral veins may cause phlebitis, necessitating reinsertion at an alternate site. A catheter may become clogged from blood clotting or from a buildup of scar tissue around the catheter tip. Catheters are also a leading cause of infection: contamination may be introduced during insertion or may develop at the placement site.

To reduce the risk of complications, nurses use aseptic techniques when inserting catheters, changing tubing, or changing a dressing that covers the catheter

TABLE 21-2 Potential Complications of Parenteral Nutrition

Catheter-Related	Metabolic
Air embolism	Abnormalities in liver function
Blood clotting at catheter tip	Electrolyte imbalances
Clogging of catheter	Gallbladder disease
Dislodgment of catheter	Hyperglycemia, hypoglycemia
Improper placement	Hypertriglyceridemia
Infection, sepsis	Metabolic bone disease
Phlebitis	Nutrient deficiencies
Tissue injury	Refeeding syndrome

TABLE 21-3 Patient Monitoring during Parenteral Nutrition

Before Starting

- Perform a nutrition assessment.
- Record body height, weight, and body mass index.
- Confirm catheter placement by X-ray.
- Check laboratory tests: blood glucose, sodium, potassium, calcium, phosphorus, magnesium, bicarbonate, serum proteins, liver enzymes, triglycerides, blood urea nitrogen, creatinine, bilirubin, complete blood count.

Every 4 to 8 Hours

- Check vital signs, including body temperature.
- Check blood glucose (once stabilized, check daily).
- Inspect catheter site for signs of inflammation or infection (frequency depends on patient condition).
- Check pump infusion rate and appearance of parenteral solution and tubing.

Daily

- Replace parenteral solution and tubing.
- Check blood glucose and serum electrolytes until stabilized.
- Monitor weight changes.
- Record fluid intake and output.

Several Times Weekly (or as needed)

- Reassess nutrition status.
- Check laboratory tests to monitor blood chemistry (once stabilized, may check less frequently).

◆ Chapter 18 provides more information about the clear liquid diet.

continuous parenteral nutrition: continuous administration of parenteral solutions over a 24-hour period.

cyclic parenteral nutrition: administration of a parenteral solution over a 10- to 16-hour period.

site. Unusual bleeding or a wet dressing suggests a problem with catheter placement. A change in infusion rate may indicate a clogged catheter. Infection may be indicated by redness or swelling around the catheter site or by an unexplained fever. Routine inspections of equipment and frequent monitoring of patients' symptoms help to minimize the problems associated with catheter use.

Administration of Parenteral Solutions

Before determining the best method for administering the parenteral solution, the nutrition support team must consider the patient's clinical condition and the potential for complications.[9] The preferred protocols for initiating and advancing parenteral infusions vary among institutions. One approach is to start the infusion at a slow rate and increase the rate gradually over a 2- to 3-day period. For example, 40 milliliters per hour can be infused during the first 24 hours of administration (supplying 960 milliliters), and the rate can be increased by 1 liter per day until the goal rate is reached. Another method is to give the full volume of a nutrient-dilute solution on the first day and advance nutrient concentrations as tolerated. Some protocols suggest starting solutions at full strength unless there is a risk that the patient will become hyperglycemic. If complications are unlikely, the solution can usually be advanced to full volume and full strength by the second day of the infusion.[10]

Parenteral solutions can be infused continuously over 24 hours (**continuous parenteral nutrition**) or during 10- to 16-hour periods only (**cyclic parenteral nutrition**). Continuous feedings are given to critically ill and malnourished patients who cannot receive adequate nutrition in the shorter time periods. Cyclic feedings are often provided at night so that patients can participate in routine activities during the day. This method is especially suited to patients who require long-term parenteral support or who will be infusing parenteral solutions at home. Patients may begin with continuous feedings and transition to cyclic feedings as their condition improves.

Regular monitoring can help to prevent complications. The parenteral solution and tubing are checked daily for signs of contamination. Routine testing of glucose, lipids, and electrolyte levels helps to determine tolerance to solutions. Frequent reassessment of nutrition status may be necessary until a patient has stabilized. Rapid changes in infusion rate are discouraged in some patients due to a risk of developing hyperglycemia or hypoglycemia.[11] Table 21-3 lists some guidelines for monitoring patients undergoing intravenous feedings.

Discontinuing Intravenous Feedings

The method used for transitioning a patient to oral feedings depends on the patient's overall health and medical condition. The patient must have adequate GI function before parenteral feedings can be tapered off and enteral feedings begun. Other factors to consider include the length of time that the patient was receiving parenteral support, the type of intravenous feeding, and the follow-up treatment planned.

During the transition to oral feedings, a combination of feeding methods is often necessary. Parenteral feedings are usually tapered off at the same time that tube feedings or oral feedings are begun, such that the two feeding methods can together supply the needed nutrients. Clear liquids are generally the first foods offered and include pulp-free fruit juices, soft drinks, and clear broths; small amounts are given initially to determine tolerance. ◆ Later feedings include beverages and solid foods that are unlikely to cause discomfort. If gastrointestinal symptoms (such as nausea, vomiting, bloating, or diarrhea) develop, oral feedings are limited in size or frequency until the intestines adapt. Once about two-thirds to three-fourths of nutrient needs can be provided enterally, the intravenous feedings may be discontinued.

Transitioning to an oral diet is sometimes difficult because a person's appetite remains suppressed for several weeks after parenteral nutrition is terminated. Patients receiving continuous parenteral feedings may have better appetites during the day if they are switched to nocturnal cyclic feedings before beginning oral intakes.

Managing Metabolic Complications

As has been discussed, the catheters used for parenteral nutrition may cause a number of serious complications. This section describes some metabolic complications that may result from parenteral feedings (review Table 21-2) and some suggestions for managing them.[12]

Hyperglycemia Hyperglycemia ◆ most often occurs in patients who are glucose intolerant or undergoing severe metabolic stress. It can be prevented by providing insulin along with feedings or by restricting the amount of dextrose in a solution. Dextrose infusions are generally limited to less than 5 milligrams per kilogram of body weight per minute in critically ill adult patients so that the carbohydrate intake does not exceed the maximum glucose oxidation rate. Premature infants are especially likely to develop hyperglycemia, because their pancreas and liver are not fully functioning.

◆ For most patients receiving parenteral nutrition, blood glucose levels should not exceed 200 mg/dL.

Hypoglycemia Although uncommon, hypoglycemia sometimes occurs when feedings are interrupted or discontinued or if excessive insulin is given. In patients at risk, such as young infants, feedings may be tapered off over several hours before discontinuation. Another option is to infuse a 10 percent dextrose solution at the same time that the parenteral feedings are interrupted or stopped.[13]

Hypertriglyceridemia Hypertriglyceridemia may develop in critically ill patients who cannot tolerate the amount of lipid emulsion supplied. Patients at risk include those with severe infection, liver disease, kidney failure, hyperglycemia, and use of immunosuppressant or corticosteroid medications. If blood triglyceride levels exceed 500 milligrams per deciliter, lipid infusions should be reduced or stopped.[14]

Refeeding Syndrome Severely malnourished patients who are fed aggressively (parenterally or otherwise) may develop **refeeding syndrome,** characterized by electrolyte and fluid imbalances and hyperglycemia. These effects occur because dextrose infusions raise circulating insulin levels, which promote anabolic processes that quickly remove phosphate, potassium, and magnesium from the blood. The altered electrolyte levels can lead to fluid retention and life-threatening changes in organ systems. Heart failure and respiratory failure are possible consequences.

Refeeding syndrome generally develops within two weeks of beginning parenteral feedings.[15] The patients at highest risk are those who have experienced chronic malnutrition or substantial weight loss. Symptoms include edema, cardiac arrhythmias, muscle weakness, and confusion. To prevent refeeding syndrome, health practitioners start parenteral feedings slowly and carefully monitor electrolyte and glucose levels when malnourished patients begin receiving nutrition support.

Abnormal Liver Function Fatty liver often results from parenteral support, but it is usually corrected when the parenteral feedings are discontinued. Long-term parenteral nutrition, however, may result in chronic, irreversible liver disease that can eventually lead to liver failure. The cause of the liver abnormalities is unclear.

Liver enzyme levels are monitored weekly during parenteral support, and abnormal values are often seen within weeks of beginning the feedings. The patients at highest risk are those with pre-existing GI or liver disorders, malnutrition, and severe infection.[16] To minimize the risk, practitioners are careful to avoid giving the patient excess energy, dextrose, or lipids. Some amount of enteral intake may be encouraged to reduce the amount of parenteral support necessary. Cyclic feedings may be less problematic than continuous feedings. Note that various critical illnesses and disease treatments can also cause liver complications, so parenteral nutrition cannot be assumed to be the underlying cause.[17]

refeeding syndrome: a condition that sometimes develops when a severely malnourished person is aggressively fed; characterized by electrolyte and fluid imbalances and hyperglycemia.

Gallbladder Disease Gallbladder problems frequently develop when the GI tract is not used for long periods. When parenteral nutrition continues for more than four weeks, sludge (thickened bile) often builds up in the gallbladder and may eventually lead to gallstone formation. Prevention is sometimes possible by initiating enteral feedings before problems develop. Patients requiring long-term parenteral nutrition may be given cholecystokinin injections to cause gallbladder contraction and bile release or may have their gallbladders removed surgically.

Metabolic Bone Disease Long-term parenteral nutrition has been associated with lower bone density and bone mineralization, which may be related to altered calcium, phosphorus, magnesium, and sodium metabolism. Imbalanced intakes of vitamin D, vitamin K, and phosphorus have also been implicated. Therefore, nutrition status and bone density are routinely monitored during long-term parenteral nutrition. The ideal intervention for metabolic bone disease varies among patients; it may include dietary adjustments, nutrient supplements, medications, and physical activity.[18]

IN SUMMARY

A nutrition support team, which is made up of physicians, nurses, dietitians, and pharmacists, may administer parenteral nutrition support or serve as advisers to other clinicians. Parenteral solutions may be initiated gradually or provided at full volume and full strength in selected patients. Critically ill patients may require continuous feedings, whereas healthier patients and long-term users may prefer cyclic feedings. Catheters are frequently the cause of complications, which include improper placement or dislodgment, infection, clotting, embolism, and phlebitis. Metabolic complications include hyperglycemia and hypoglycemia; hypertriglyceridemia; fluid and electrolyte imbalances; and diseases affecting the liver, gallbladder, and bone. When the need for parenteral nutrition resolves, patients are transitioned to an enteral diet as the volume of parenteral nutrition is gradually reduced. The Case Study can be used to check your understanding of the concepts introduced in this chapter.

Nutrition Support at Home

Occasionally, a patient must continue to receive nutrition support, either tube feedings or parenteral nutrition, after a medical condition has stabilized. For such a person, home nutrition support might be an option.

The use of home nutrition support is rapidly expanding. Current technology allows for the safe administration of nutrition support in home settings, and insurance coverage often pays a substantial portion of the costs. Home health services and home infusion pharmacies can provide the equipment, enteral formulas or parenteral solutions, and services necessary for home nutrition care. Most important, patients using these services can continue to receive specialized nutrition care while leading normal lives.

Candidates for Home Nutrition Support

Individuals referred for home nutrition support usually need long-term nutrition care for chronic medical conditions. Users of home nutrition services (or their families and other caregivers) must be intellectually capable of learning the necessary procedures, monitoring the treatment, and managing complications as necessary. The home should be clean and have adequate storage for formulas or solutions and

---

CASE STUDY Geologist Requiring Parenteral Nutrition

Jerry Huang, a 27-year-old geologist with an inflammatory intestinal disease, underwent a surgical procedure in which a substantial portion of his small intestine was removed. He had received TPN prior to surgery and continued to receive it afterwards. After 10 days, tube feeding was begun, which initially delivered very small feedings.

1. List some reasons why the nutrition support team initially chose TPN to provide nutrition support to this patient. How would you explain the need for parenteral feedings to Jerry?

2. Describe the components of a typical TPN solution. Calculate the energy content of 1 liter of a solution that provides 140 grams of dextrose monohydrate, 45 grams of amino acids, and 90 milliliters of 20 percent lipid emulsion. Then calculate the nonprotein kcalorie-to-nitrogen ratio in the solution. If Jerry's energy requirement is 2100 kcalories per day, how many liters of solution will he need each day?

3. Why is it important that Jerry begin enteral feedings as soon as possible? Assuming that Jerry eventually tolerates a tube feeding, in what ways can the health care team help Jerry make the transition from parenteral feedings to tube feedings? Consider some of the physiological problems that Jerry might face when he begins eating an oral diet.

4. If Jerry is unable to meet his nutrient needs orally, he may need to continue tube feeding or TPN at home. As you read through the section on nutrition support at home, consider the factors that would make Jerry a good candidate for a home nutrition support program. Consider both the benefits of a proposed program and the problems he could encounter.

---

equipment. The costs should be clearly explained to families who cannot get insurance reimbursement. Candidates for home nutrition support include the following:

- For home enteral nutrition, individuals who have functioning GI tracts and illnesses that prevent food from reaching the digestive tract. Examples include patients with head and neck cancers and individuals with neurological impairments that cause difficulties with swallowing.

- For home parenteral nutrition, individuals who have illnesses that severely impair nutrient absorption or cause motility problems in the stomach or intestines. Examples include persons who have had large portions of their small intestine removed and those with intestinal obstructions or malabsorption conditions.

Planning Home Nutrition Care

As with the nutrition support provided in health care facilities, planning for home nutrition care involves decisions about access sites, formulas, and nutrient delivery methods. Users of home services should be involved in the decision making to ensure long-term compliance and satisfaction.

Home Enteral Nutrition Access to the GI tract is possible using either nasal tubes or enterostomies. People sometimes learn to place nasogastric tubes themselves, which may improve acceptance of the therapy. Active children and adults often prefer low-profile gastrostomy tubes, which allow them to lead a more normal lifestyle. Jejunostomy tubes may be required for some individuals but are less convenient, because the frequent feedings required for people with jejunostomies can interfere with daytime activities.

The choice of formula for home use is influenced by its cost and availability. Insurance reimbursements do not always include the cost of formula, which is considered to be a "food" product. Fo   r this reason, some people choose to prepare simple formulas at home. Blenderizing home-cooked foods is possible, but the foods need to be strained to remove particles and clumps that may obstruct the tube. Closed (ready-to-hang) feeding systems are useful for avoiding contamination risk

Portable pumps and convenient carrying cases allow people who require nutrition support at home to move about freely.

Ethical Issues in Nutrition Care

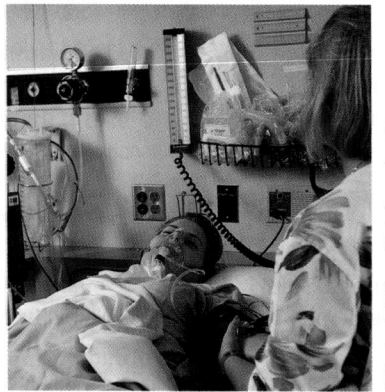

© Carolyn A. McKeone/Photo Researchers, Inc.

As with other medical technologies, the availability of specialized nutrition support forces health care professionals and members of our society to face difficult **ethical** issues. When medical treatments prolong life by merely delaying death, the lifetime that remains may be of extremely low quality. This highlight examines the ethical dilemmas that clinicians must face when dealing with patients in critical care. The glossary below defines the relevant terms.

Ethical Considerations

If providing nutrition care can do little to promote recovery, is it morally and legally appropriate to withhold or to withdraw nutrition support? Do patients and family members have the rights to make these types of decisions themselves? How important is the input of the health professional? In attempting to answer questions such as these, health professionals must consider the following ethical principles:[1]

- A patient has the right to make decisions concerning his or her own well-being **(patient autonomy)**, even if refusing treatment could result in death. It is generally accepted that a patient's preferences should take precedence over the desires of others.[2]

- A patient should be fully informed of a treatment's benefits and risks in a fair and honest manner **(disclosure).** A patient's acceptance of a treatment that has been adequately disclosed is considered **informed consent.**

- A patient must have the mental capacity to make appropriate health care decisions **(decision-making capacity).** If a patient is mentally incapable of doing so, a person designated by the patient should serve as a **surrogate** decision maker.

GLOSSARY

advance directive: written or oral instruction regarding one's preferences for medical treatment to be used in the event of becoming incapacitated.

beneficence (be-NEF-eh-sense): the act of performing beneficial services rather than harmful ones.

cardiopulmonary resuscitation (CPR): life-sustaining treatment that supplies oxygen and restores a person's ability to breathe and pump blood.

decision-making capacity: the ability to understand pertinent information and make appropriate decisions; known as *decision-making competency* within the legal system.

defibrillation: life-sustaining treatment in which an electronic device is used to shock the heart and reestablish a pattern of normal contractions. Defibrillation is used when the

heart has arrhythmias or has experienced cardiac arrest.

dialysis: life-sustaining treatment in which a patient's blood is filtered using selective diffusion through a semipermeable membrane; substitutes for kidney function.

disclosure: the act of revealing pertinent information. For example, clinicians should accurately describe proposed tests and procedures, their benefits and risks, and alternative approaches.

distributive justice: the equitable distribution of resources.

do-not-resuscitate (DNR) order: a request by a patient or surrogate to withhold cardio-pulmonary resuscitation.

durable power of attorney: a legal document (sometimes called a *health care proxy*) that gives legal authority to another (a *health care agent*) to make medical decisions in the event of incapacitation.

ethical: in accordance with accepted principles of right and wrong.

futile: medical care that will not improve the medical circumstances of a patient.

health care agent: a person given legal authority to make medical decisions for another in the event of incapacitation.

informed consent: a patient's or caregiver's agreement to undergo a treatment that has been adequately disclosed. Persons must be mentally competent in order to make the decision.

living will: a written statement that specifies the medical procedures desired or not desired in the event that a person is unable to communicate or is incapacitated; also called a *medical directive.*

maleficence (mah-LEF-eh-sense): the act of doing evil or harm.

mechanical ventilation: life-sustaining treatment in which a mechanical ventilator is used to substitute for a patient's failing lungs.

patient autonomy: a principle of self-determination, such that patients (or surrogate decision makers) are free to choose the medical interventions that are acceptable to them, even if they choose to refuse interventions that may extend their lives.

persistent vegetative state (PVS): a vegetative mental state resulting from brain injury that persists for at least one month. Individuals lose awareness and the ability to think but retain noncognitive brain functions, such as motor reflexes and normal sleep patterns.

surrogate: a substitute; a person who takes the place of another.

- The potential benefits **(beneficence)** of any treatment should outweigh its potential harm **(maleficence).**
- Health care providers must determine whether the provision of health care to one patient would unfairly limit the care of other patients **(distributive justice).**

Although these principles may seem simple and obvious, it is often difficult to determine the appropriate action to take during intensive care.[3] When clinicians and families disagree, the courts may be asked to decide.

When a patient's preferences are unknown, the medical staff is obligated to provide any and all available care that is likely to sustain the patient's life. Nutrition support and hydration are both considered life-sustaining treatments because withholding or withdrawing either can result in death. Other life-sustaining treatments include **cardiopulmonary resuscitation (CPR),** which supplies oxygen and restores a person's ability to breathe and pump blood; **defibrillation,** in which an electronic device shocks the heart and reestablishes normal contractions; **mechanical ventilation,** which substitutes for lung function; and **dialysis,** which substitutes for kidney function.

Ethical Dilemmas

Although life-sustaining treatments are readily provided to patients who have a reasonable chance of recovering from illness, it is sometimes difficult to determine the best course of action for patients who are dying or who are unlikely to regain consciousness. Under such circumstances, such treatments may be considered **futile** because they are unable to improve the outcome of disease or increase the patient's comfort and well-being. If patients or caregivers demand treatment that health practitioners have determined to be useless, a legal resolution may be required. Conversely, medical personnel may find it objectionable to withdraw life support when they know that the inevitable consequence will be the patient's death.

Legal Decisions

One of the landmark cases involving nutrition support concerned Nancy Cruzan, who suffered permanent and irreversible brain damage after a car crash in 1983, when she was 26 years of age.[4] After she had been in a **persistent vegetative state (PVS)** for five years, her parents requested permission to discontinue tube feeding, but hospital staff refused to honor the request, and the matter was taken to court. The Missouri Supreme Court determined that Nancy had never definitively stated her "right to die" wishes and that her parents were unable to make such a request for her. The court also stated that preserving life, no matter what its quality, should take precedence over all other considerations. Nancy's parents appealed the ruling, but in 1990, the U.S. Supreme Court upheld the Missouri Supreme Court in a five-to-four decision. Three witnesses were eventually found who could testify that Nancy would not desire life-sustaining treatment under the circumstances, and the court finally granted permission to remove the feeding tube. This case illustrates the importance of

having an **advance directive** (discussed in a later section) that clearly indicates one's preferences for medical treatment in the event of incapacitation.

In a more recent case that received widespread media attention, the spouse and parents of a patient in a persistent vegetative state fought a ten-year legal battle over her medical care. In 1990, at the age of 25, Terri Schiavo suffered a full cardiac arrest.[5] She initially fell into a coma, but her condition evolved into a persistent vegetative state that was considered irreversible. Despite the neurologists' diagnosis and a series of computed tomography (CT) and magnetic resonance imaging (MRI) scans showing extensive brain atrophy, her parents maintained that she was minimally conscious and could improve with rigorous treatment. Her husband, who was legally responsible for her care, insisted that she would never have wanted to be kept alive in a vegetative state. Like Nancy Cruzan, Terri had never expressed her wishes in an advance directive.

In 1998, Terri's husband filed a petition to have her feeding tube removed, and a Florida court approved the motion in February 2000. Although Terri's parents appealed, an appeals court affirmed the decision, and the Florida Supreme Court declined to review the case. In April 2001, Terri's physicians removed her feeding tube, but within days, a federal circuit court judge ordered it to be reinserted and reopened the case. Eventually, the motions filed by the parents were dismissed, and Terri's feeding tube was removed for the second time in October 2003. Within days, the Florida legislature passed a bill known as "Terri's Law" that gave the governor the authority to intervene, and Governor Jeb Bush ordered the feeding tube reinserted. A year later, Florida's Supreme Court declared Terri's Law to be unconstitutional. Although the governor appealed the decision, his appeal was rejected in January 2005. Terri's feeding tube was removed for the third time in March 2005. Despite emergency petitions by her parents and an attempt by the U.S. Congress to have her case reconsidered, the courts refused to grant a restraining order, and Terri died 13 days after her feeding tube was removed.

Religious Viewpoints

The withdrawal of nutrition support and other life-sustaining treatments may not be acceptable to persons of some religious faiths. For example, Orthodox Jews believe that the soul is present in people who are alive (even if permanently unconscious) and disallow any actions that would hasten death.[6] If a person's or family's religious beliefs are not in accord with medical recommendations, health practitioners are expected to consider the viewpoint and try to resolve the issue in some way. If practitioners are unable to comply with the wishes of a patient or caregiver, the care of the patient should be transferred elsewhere.

Advance Planning

Health care professionals should discuss the importance of advance directives with competent patients. Individuals are encouraged to discuss their medical preferences with family members and surrogate decision makers so that their wishes will be considered in the event that they become incapacitated. Advance directives are incorporated into the medical record and updated when appropriate. They take effect only if a physician determines that a patient lacks the ability to understand and make decisions about available treatments. If a person's preferences are unknown, decisions are based on a patient's best interests as determined by a caregiver or family member.[7]

Advance Directives

A person may declare preferences about medical treatments in a **living will,** sometimes called a *medical directive.* Living wills can include detailed instructions about life-sustaining procedures that a person does or does not want. Another important directive is a **durable power of attorney** (sometimes called a *health care proxy*), in which another person (a **health care agent**) is appointed to act as decision maker in the event of incapacitation. The agent should understand one's medical preferences and be absolutely trustworthy. Only one person can be designated, although one or two alternates may also be listed. If an agent is given comprehensive power to supervise care, he or she may make decisions about medical staff, health care facilities, and medical procedures.

Laws regarding advance directives vary from state to state. In some states, nutrition and hydration are not considered life-sustaining treatments, and a person's instructions about them may need to be indicated separately. Some states restrict the use of advance directives to terminal illness or disallow them if a woman is pregnant. State statutes also specify characteristics of people who may serve as health care agents and witnesses. Generally, advance directives created in one state are honored in another.

The Do-Not-Resuscitate Order

A **do-not-resuscitate (DNR) order** is frequently used to withhold CPR in the event of cardiopulmonary arrest, which occurs too suddenly for deliberate decision-making.[8] A DNR order is written in the medical record as other directives are, but it does not exclude the use of other life-prolonging measures. A DNR order is most often used in patients with serious illnesses or advanced age. Some institutions allow a physician to write a DNR order for a patient who has a poor prognosis, but the physician must inform the patient or surrogate if this is done.

Organ and Tissue Donation

End-of-life decisions invariably raise questions about a dying patient's preferences concerning organ and tissue donation. Even if a donor card has been signed, it is important to let family members know one's wishes, as the family may need to sign a consent form in order for donation to occur. Although organ donation is a difficult topic to bring up near the time of death, potential donors can be assured that their gift could greatly enhance or save the lives of others.

Ethical questions sometimes arise when organs are donated. A physician must alert an organ procurement team about a donor's existence and arrange to maintain organ functions until organs are retrieved. Treatments that maintain the viability of organs and tissues cannot be used if they may harm the donor. Sometimes the care of a donor and the needs of a potential recipient may appear to be in conflict, but the care of donors and recipients is always kept separate and performed by different physicians.

Ongoing Issues

Despite the availability of advance directives, only about 20 percent of people in the United States have completed one.[9] Furthermore, advance directives are often unavailable when intensive care decisions are made: one study found that only 57.5 percent of patient charts indicating the existence of an advance directive actually contained a copy.[10] In addition, advance directives are sometimes too general or vague to guide treatment decisions.

Physicians must often make treatment decisions before they have a chance to discuss them with patients or caregivers.[11] In many cases, life-sustaining treatments are begun without the prior knowledge of patients or their decision makers, or treatments continue even if patients want them stopped. Patients who are fully aware of treatment options and clearly state their preferences are more likely to be successful at obtaining the care they desire.

Medical decisions that are planned in advance and discussed with close friends and family can help to prevent decision-making dilemmas during emergency situations. Health practitioners should strive to provide the best information possible so that patients can consider all of the medical options available and make their preferences known to medical personnel.

REFERENCES

1. M. A. Grippi, Ethics in critical care, in A. P. Fishman and coeditors, *Fishman's Manual of Pulmonary Diseases and Disorders* (New York: McGraw-Hill, 2002), pp. 1111–1114.
2. E. J. Emanuel, Bioethics in the practice of medicine, in L. Goldman and D. Ausiello, eds., *Cecil Medicine* (Philadelphia: Saunders, 2008), pp. 6–11.
3. Emanuel, 2008.
4. J. O. Maillet, R. L. Potter, and L. Heller, Position of the American Dietetic Association: Ethical and legal issues in nutrition, hydration, and feeding, *Journal of the American Dietetic Association* 102 (2002): 716–726.
5. R. Cranford, Facts, lies, and videotapes: The permanent vegetative state and the sad case of Terri Schiavo, *Journal of Law, Medicine, and Ethics* 33 (2005): 363–372.
6. Maillet, Potter, and Heller, 2002.
7. American College of Physicians, Ethics manual, *Annals of Internal Medicine* 128 (1998): 576–594.
8. American College of Physicians, 1998.
9. Emanuel, 2008.
10. Institute of Medicine, *Approaching Death: Improving Care at the End of Life* (Washington, D.C.: National Academy Press, 1997), pp. 202–203.
11. Emanuel, 2008.

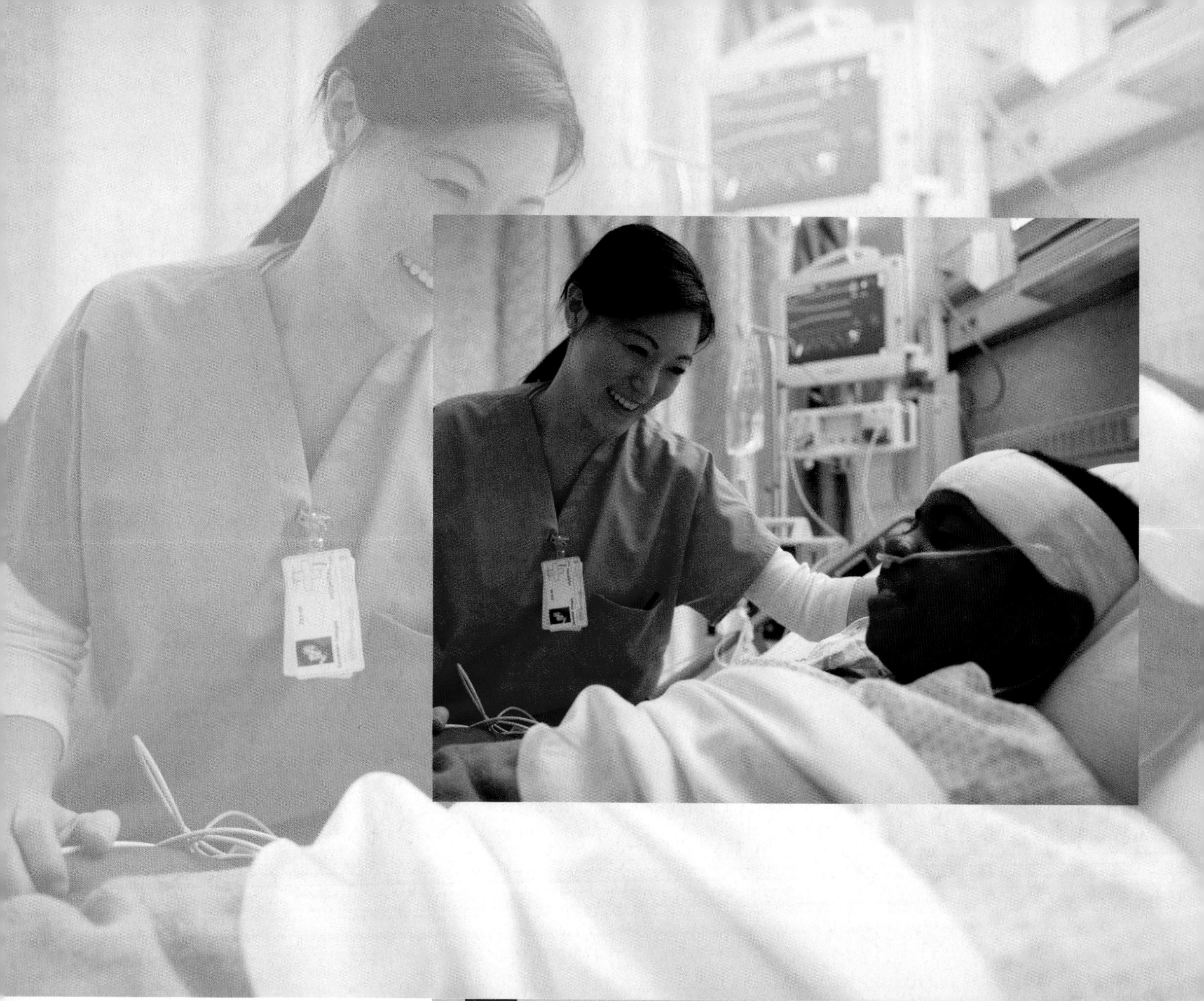

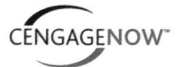

Nutrition in the Clinical Setting

The body's dramatic response to severe stress can alter metabolism enough to threaten survival. Many patients with severe stress require life support measures and intensive monitoring. Stress also raises nutritional needs considerably—increasing the risk of malnutrition even in previously healthy individuals. Providing nutrition care for these patients is not only challenging; it is often ineffective for preventing weight loss and muscle tissue losses. Despite these difficulties, the health care professional must determine the best measures to take in order to limit damage and promote recovery.

CENGAGENOW™ The CengageNOW logo indicates an opportunity for online self-study, linking you to interactive tutorials and videos based on your level of understanding.

academic.cengage.com/login

Metabolic and Respiratory Stress

This chapter addresses the nutrition care provided to patients who undergo certain types of physiological stress. **Metabolic stress,** a disruption in the body's internal chemical environment, can result from uncontrolled infections or extensive tissue damage, such as deep, penetrating wounds or multiple broken bones. As the first part of this chapter explains, the body's stress response is an attempt to restore balance, but it can have both helpful and harmful effects. Later sections of this chapter describe **respiratory stress,** which is characterized by inadequate oxygen and excessive carbon dioxide in the blood and tissues. Both metabolic and respiratory stress can lead to **hypermetabolism** (above-normal metabolic rate), **wasting** (breakdown of muscle mass and loss of strength), and in severe circumstances, life-threatening complications. The highlight following this chapter discusses the causes and consequences of **multiple organ dysfunction syndrome,** the simultaneous dysfunction of two or more organ systems, which is often fatal.

The Body's Responses to Stress and Injury

The **stress response** is the body's *nonspecific* response to a variety of stressors, such as infection, fractures, surgery, and burns. During stress, the body's metabolic processes focus on immediate survival, while functions of lesser consequence are delayed. Energy is of primary importance, and therefore the energy nutrients are mobilized from storage and made available in the blood. Heart rate and respiration (breathing rate) increase to deliver oxygen and nutrients to cells more quickly, and blood pressure rises. Meanwhile, energy is diverted from processes that are not life sustaining, such as growth, reproduction, and long-term immunity. If

metabolic stress: a disruption in the body's chemical environment due to the effects of disease or injury. Metabolic stress is characterized by changes in metabolic rate, heart rate, blood pressure, hormonal status, and nutrient metabolism.

respiratory stress: abnormal gas exchange between the air and blood, resulting in lower-than-normal oxygen levels and higher-than-normal carbon dioxide levels.

hypermetabolism: a higher-than-normal metabolic rate.

wasting: the breakdown of muscle tissue that results from disease or malnutrition.

multiple organ dysfunction syndrome: the dysfunction of two or more organ systems that develops during intensive care; often results in death.

stress response: the chemical and physical changes that occur within the body during stress.

Although severe metabolic stress can have damaging consequences, the healthy body handles minor stresses quickly and efficiently.

◆ The catecholamines, glucagon, and cortisol have actions that oppose those of insulin and are therefore referred to as *counterregulatory hormones*.

stress continues for a long period, interference with these processes begins to cause damage, possibly resulting in growth retardation and illness.

Hormonal Responses to Stress

The stress response is mediated by several hormones, which are released into the blood soon after the onset of injury (see Table 22-1).[1] The catecholamines (epinephrine and norepinephrine), often called the *fight-or-flight hormones*, stimulate heart muscle, raise blood pressure, and increase metabolic rate. Epinephrine also promotes glucagon secretion by the pancreas, prompting the release of nutrients from storage. The steroid hormone cortisol enhances protein degradation, raising amino acid levels in the blood and making amino acids available for conversion to glucose. All of these hormones have similar effects on glucose and fat metabolism, causing the breakdown of glycogen (glycogenolysis), the production of glucose from amino acids (gluconeogenesis), and the breakdown of triglycerides in adipose tissue (lipolysis). ◆ Thus, the combined effects of these hormones contribute to hyperglycemia, which often accompanies critical illness. Two other hormones induced by stress, aldosterone and antidiuretic hormone, help to maintain blood volume by stimulating the kidneys to reabsorb more sodium and water, respectively.

Cortisol's effects can be detrimental when stress is prolonged. In excess, cortisol causes the depletion of protein in muscle, bone, connective tissue, and skin. It impairs wound healing, so high cortisol levels are especially dangerous for a patient with severe injuries. Because cortisol inhibits protein synthesis, consuming more protein cannot easily reverse tissue losses. Excess cortisol also leads to insulin resistance, contributing to hyperglycemia. In addition, cortisol suppresses immune responses, increasing susceptibility to infection. Note that the pharmaceutical forms of cortisol are common anti-inflammatory medications (such as *cortisone* and *prednisone*); their long-term use can cause undesirable side effects such as muscle wasting, thinning of the skin, diabetes, and early osteoporosis.

The Inflammatory Response

Cells of the immune system mount a quick, nonspecific response to infection or tissue injury. This so-called **inflammatory response** serves to contain and destroy infectious agents (and their products) and prevent further tissue damage. The inflammatory response also triggers various events that promote healing. As in the stress response, however, there is a delicate balance between a response that protects tissues from further injury and an excessive response that can cause additional damage to tissue.

inflammatory response: a group of nonspecific immune responses to infection or injury.

TABLE 22-1	Metabolic Effects of Hormones Released during the Stress Response
Hormone	**Metabolic Effects**
Catecholamines	• Increase in metabolic rate • Glycogen breakdown in liver and muscle • Glucose production from amino acids • Release of fatty acids from adipose tissue • Glucagon secretion from pancreas
Glucagon	• Glycogen breakdown in liver • Glucose production from amino acids • Release of fatty acids from adipose tissue
Cortisol	• Protein degradation • Enhancement of glucagon's action on liver glycogen • Glucose production from amino acids • Release of fatty acids from adipose tissue
Aldosterone	• Retention of sodium
Antidiuretic hormone	• Retention of water

The Inflammatory Process The inflammatory response begins with the dilation of blood vessels that deliver blood to the site of an injury (arterioles) and the constriction of small blood vessels that carry blood away from an infected area (venules). The capillaries within the damaged tissue become more permeable, allowing some blood plasma to escape and causing local edema. These changes in blood vessels ◆ limit or prevent the spread of infection and encourage the entry of immune cells that can destroy foreign agents (see Figure 22-1). Among the first cells to arrive are the **phagocytes,** which slip through gaps between the endothelial cells that form the vessel walls. The phagocytes engulf microorganisms and destroy them with hydrolytic enzymes and reactive forms of oxygen. When inflammation becomes chronic, these normally useful products of phagocytes can damage healthy tissue.

Mediators of Inflammation Numerous chemical substances control the inflammatory process. These *mediators* are released from damaged tissue, blood vessel cells, and activated immune cells. Many of them help to regulate more than one step in the process. Some of the examples that follow were introduced in Highlight 17's discussion of immunity. Histamine, a small molecule similar to an amino acid in structure, is released from granules within **mast cells,** causing vasodilation and capillary permeability. ◆ Fragments of complement proteins ◆ trigger histamine's release from mast cells and help to recruit and activate phagocytes. Other compounds that participate in the inflammatory process include several cytokines ◆ (especially interleukin-1, interleukin-6, and tumor necrosis factor-α) and various **eicosanoids** (which are derived from dietary fatty acids). Note that most anti-inflammatory medications, including steroidal drugs (such as cortisone and prednisone) and nonsteroidal anti-inflammatory drugs (such as aspirin and ibuprofen), act by blocking eicosanoid synthesis.

Changing dietary fat sources can have subtle effects on the inflammatory process.[2] The major precursor for the eicosanoids is arachidonic acid, which derives from the omega-6 fatty acids in vegetable oils. Some omega-3 fatty acids compete with arachidonic acid and inhibit the production of the most powerful inflammatory mediators. Partially replacing vegetable oils rich in omega-6 fatty acids with food sources high in omega-3 fatty acids (such as fish oil) helps to suppress inflammation, but it is not a reliable treatment.

◆ The classic signs of inflammation that accompany altered blood flow are:
  • *Swelling*—from the accumulation of fluid at the site of injury
  • *Redness*—from the dilation of small blood vessels in the injured area
  • *Heat*—from the influx of warm arterial blood
  • *Pain*—from the pressure of edema within the damaged tissue and the actions of certain chemical mediators on pain receptors

◆ *Antihistamines* are medications taken to reduce the effects of histamine.

◆ Reminder: *Complement* is a collective term for a group of plasma proteins that assist the activities of antibodies.

◆ Reminder: *Cytokines* are hormone-like proteins that regulate immune responses.

---

phagocytes (FAG-oh-sites): white blood cells (neutrophils and macrophages) that have the ability to engulf and destroy antigens.
  • **phagein** = to eat

mast cells: cells within connective tissue that produce and release histamine.

eicosanoids (eye-KO-sa-noids): 20-carbon molecules derived from dietary fatty acids that help to regulate blood pressure, blood clotting, and other body functions. (Eicosanoids were introduced on p. 154.)
  • **eicosa** = twenty

FIGURE 22-1 The Inflammatory Process

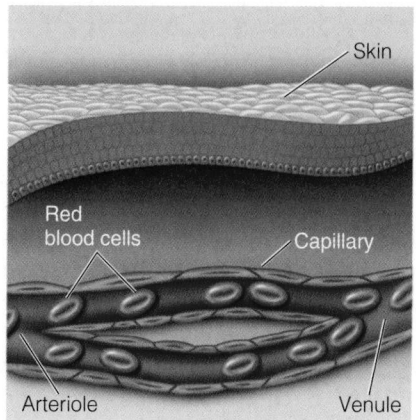

Cells lining the blood vessels lie close together, and normally do not allow the contents to cross into tissue.

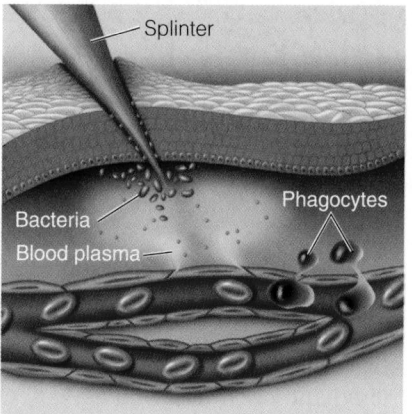

When tissues are damaged, immune cells release histamine, which dilates some blood vessels, increasing blood flow to the damaged area. Fluid leaks out of capillaries (causing swelling), and phagocytes escape between the small gaps in the blood vessel walls.

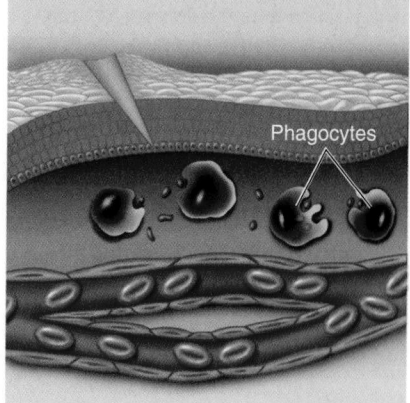

Phagocytes engulf bacteria and disable them with hydrolytic enzymes and reactive forms of oxygen.

◆ C-reactive protein, the best clinical indicator of the acute-phase response, becomes elevated during many chronic diseases.

Systemic Effects of Inflammation Cytokines released during the inflammatory process induce **systemic** effects as well as the localized effects described earlier. Within hours (or, in some cases, days) after inflammation, infection, or severe injury, the liver steps up its production of certain proteins in an effort known as the **acute-phase response.**[3] These acute-phase proteins include **C-reactive protein,** ◆ complement, blood-clotting proteins such as fibrinogen and prothrombin, and others. At the same time, plasma concentrations of albumin, iron, and zinc fall (recall from Chapter 17 that albumin levels are often measured to assess health and nutrition status). The acute-phase response is accompanied by muscle catabolism to make amino acids available for glucose production, tissue repair, and immune protein synthesis; consequently, negative nitrogen balance (and wasting) frequently results. Other clinical features include an elevated metabolic rate, increased neutrophils in the blood, lethargy, anorexia, and often, fever.

Severe inflammation that persists for more than a few days can lead to a life-threatening condition: **systemic inflammatory response syndrome (SIRS).** SIRS is a whole-body response to unresolved inflammation and is diagnosed when a patient's condition becomes severe enough to raise heart rate, respiratory rate, white blood cell counts, and/or body temperature to critical levels. If identical symptoms result from infection, the condition is called **sepsis.** Complications associated with SIRS or sepsis include fluid retention and tissue edema, low blood pressure, and impaired blood flow. If the reduction of blood flow is severe enough to deprive the body's tissues of oxygen and nutrients (a condition known as **shock**), multiple organs may fail simultaneously, as discussed in Highlight 22.

IN SUMMARY

The stress and inflammatory responses are nonspecific responses to stressors that cause infection and injury. The stress response is mediated by the catecholamine hormones, cortisol, and glucagon, which together raise nutrient levels in blood, stimulate heart rate, raise blood pressure, and increase metabolic rate. Aldosterone and antidiuretic hormone help to maintain adequate blood volume. The inflammatory process—mediated by compounds released from damaged tissues, immune cells, and blood vessels—results in systemic effects that alter nutrient metabolism, heart rate, blood pressure, body temperature, and immune cell functions. Signs of inflammation in injured tissues include swelling, redness, heat, and pain. Persistent, severe inflammation may result in shock and lead to multiple organ dysfunction.

---

systemic (sih-STEM-ic): relating to the entire body.

acute-phase response: changes in body chemistry resulting from infection, inflammation, or injury; characterized by alterations in plasma proteins.

C-reactive protein: an acute-phase protein released from the liver during acute inflammation or stress.

systemic inflammatory response syndrome (SIRS): a whole-body response to acute inflammation; characterized by raised heart and respiratory rates, abnormal white blood cell counts, and abnormal body temperature.

sepsis: an acute inflammatory response caused by infection; characterized by symptoms similar to those of SIRS.

shock: a severe reduction in blood flow that deprives the body's tissues of oxygen and nutrients; characterized by reduced blood pressure, raised heart and respiratory rates, and muscle weakness.

abscesses (AB-sess-es): accumulated pus that is surrounded by inflamed tissue.

debridement: the surgical removal of dead, damaged, or contaminated tissue resulting from burns or wounds; helps to prevent infection and hasten healing.

Nutrition Treatment of Acute Stress

As described earlier, an excessive response to metabolic stress can worsen illness and even threaten survival. Therefore, a critical care unit must manage both the acute medical condition that initiated stress and the complications that arise as a result of the stress and inflammatory responses. Immediate concerns during severe stress are to restore lost fluids and electrolytes and to remove underlying stressors. Thus, initial treatments include administering intravenous solutions to correct fluid and electrolyte imbalances, treating infections, repairing wounds, draining **abscesses** (pus), and removing dead tissue **(debridement).** After stabilization, nutrient needs can be estimated and medical nutrition therapy provided.

Determining Nutritional Requirements

The most critical metabolic changes in patients undergoing metabolic stress include hypermetabolism, negative nitrogen balance, hyperglycemia, and insulin resistance. Hypermetabolism and negative nitrogen balance can result in wasting, which may impair organ function and delay recovery. Hyperglycemia increases the risk of infection, a dangerous problem during critical illness. Thus, the principal goals of medical nutrition therapy are to provide a diet that preserves lean tissue, maintains immune defenses, and promotes healing.

Feeding an acutely stressed patient is challenging. Overfeeding increases the risks of refeeding syndrome ◆ and its associated hyperglycemia. Underfeeding may worsen negative nitrogen balance and increase lean tissue losses. Assessment of nutritional needs can be complicated, however, because fluid imbalances prevent accurate measurements of weight, and laboratory data may reflect the metabolic alterations of illness rather than the person's nutrition status.

The amounts of protein and energy to provide during acute illness are controversial and still under investigation. Research results have been mixed, in part because the wide assortment of conditions that cause metabolic stress makes each patient's situation somewhat different. Moreover, protein and energy needs can vary substantially over the course of illness. The guidelines presented here are subject to change as new findings help to resolve the complex issues related to nutrient intakes and delivery methods. In all cases, clinicians need to closely observe patients' responses to feedings and readjust nutrient intakes as necessary.

Estimating Energy Needs for Acute Stress A common method for determining the energy needs of acutely stressed individuals is to estimate resting metabolic rate (RMR) and then multiply the result by a stress factor to account for the increased energy requirements of stress and healing.[4] This method was introduced in the "How to" on p. 621 (Chapter 18); Table 18-3 (p. 620) lists some common predictive equations used for estimating RMR.

Stress factors depend on the severity of the illness and the patient's nutrition status. Generally, energy needs are increased by fever, mechanical ventilation, restlessness, and the presence of open wounds; patients with burns and infections often have the highest energy needs. For many critically ill patients, the stress factor 1.2 provides adequate energy, and higher factors may result in overfeeding.[5] Note that patients who are critically ill are usually bedridden and inactive, so the energy needed for physical activity is minimal. Table 22-2 outlines the use of stress factors during acute metabolic stress and provides examples that are appropriate for critical care patients; keep in mind that the particular values used can vary somewhat among institutions.

A number of predictive equations with "built-in" stress factors have been developed for use in critical care populations. As an example, the Ireton-Jones equation,

◆ Reminder: *Refeeding syndrome* can develop when a severely malnourished person is aggressively fed; it is associated with fluid and electrolyte imbalances and hyperglycemia.

TABLE 22-2 Estimating Energy Needs for Patients with Acute Metabolic Stress

Step 1. Estimate energy needs to support resting metabolic rate (RMR) using indirect calorimetry or a predictive equation (see Table 18-3, p. 620).

Step 2. Multiply the patient's RMR by an appropriate stress factor for acute illness (see the example in the "How to" box on p. 621). Examples of stress factors include:

- Advanced liver disease: 1.0 to 1.16
- Inflammatory bowel disease (active): 1.05 to 1.10
- Pancreatitis: 1.13 to 1.21
- Surgery: 1.2 to 1.4
- Leukemia: 1.25 to 1.35
- Mechanical ventilation: 1.32 to 1.34
- Burns (20 to 30 percent of body surface): 1.6 to 1.7
- Burns (30 to 40 percent of body surface): 1.8 to 1.9
- Burns (40 to 45 percent of body surface): 2.0

SOURCES: American Dietetic Association, *Nutrition Care Manual* (Chicago: American Dietetic Association, 2007); N. Barak, E. Wall-Alonso, and M. D. Sitrin, Evaluation of stress factors and body weight adjustments currently used to estimate energy expenditure in hospitalized patients, *Journal of Parenteral and Enteral Nutrition* 26 (2002): 231–238.

HOW TO Estimate the Energy Needs of a Critical Care Patient

Equations that estimate the energy needs of critically ill patients sometimes include "built-in" stress factors. The Ireton-Jones equation, shown here, was developed for use in critically ill populations and includes factors for traumatic injury and burns:

$$\text{Energy needs (kcal per day)} = 1925 + [5 \times \text{wt (kg)}] - [10 \times \text{age (yr)}] + (281 \times \text{sex}) + (292 \times \text{trauma}) + (851 \times \text{burn})$$

where sex is male ($\times$ 1) or female ($\times$ 0), trauma is the presence of physical injury ($\times$ 1) or not ($\times$ 0), and burn is the presence of a burn injury ($\times$ 1) or not ($\times$ 0).

Example: Erin is a 27-year-old female patient who weighs 140 pounds (63.6 kilograms). Two days ago, she was severely injured in an automobile accident and is currently being cared for in a critical care unit. She did not suffer a burn injury. Using the Ireton-Jones equation, her daily energy needs can be estimated as follows:

Energy needs (kcal per day)

$$= 1925 + (5 \times 63.6 \text{ kg}) - (10 \times 27 \text{ yr}) + (281 \times 0) + (292 \times 1) + (851 \times 0)$$

$$= 1925 + 318 - 270 + 0 + 292 + 0 = 2265 \text{ kcal}$$

SOURCES: D. Frankenfield, Prediction of resting metabolic rate in critically ill adult patients: Results of a systematic review of the evidence, *Journal of the American Dietetic Association* 107 (2007): 1552–1561; C. S. Ireton-Jones and coauthors, Equations for the estimation of energy expenditures in patients with burns with special reference to ventilatory status, *Journal of Burn Care and Rehabilitation* 13 (1992): 330–333.

---

described in the "How to" above, includes multipliers for the presence of trauma and burn injuries; this equation provides relatively accurate results in both obese and nonobese patients.[6] Several other equations in current use include factors for other pertinent variables, such as body temperature, respiratory rate, and use of mechanical ventilation.[7]

A quick method for estimating energy needs is to multiply a person's body weight by a factor appropriate for the medical condition. For example, daily energy needs for critical care patients often fall within the range of 25 to 30 kcalories per kilogram of body weight[8]; a patient weighing 160 pounds (72.7 kilograms) may therefore require between 1818 and 2181 kcalories per day. ◆ The energy intake can be started within this range and then adjusted as the patient's body weight and other determinants of nutrition status change.

◆ 72.7 kg $\times$ 25 kcal/kg = 1818 kcal
72.7 kg $\times$ 30 kcal/kg = 2181 kcal

Protein Requirements in Acute Stress To protect lean tissue, the protein intakes recommended during acute stress are higher than DRI values. ◆ In most critically ill patients, protein needs range between 1 and 2 grams per kilogram body weight per day[9]; burn patients often require between 2 and 3 grams per kilogram body weight each day due to the substantial losses of protein associated with burn wounds.[10] Even with adequate protein, however, negative nitrogen balance cannot be prevented during acute stress, because the accompanying metabolic processes encourage the degradation of body protein. The bed rest required during critical illness also contributes substantially to muscle breakdown.

◆ Reminder: The protein RDA for adults is 0.8 grams per kilogram body weight.

The amino acids glutamine and arginine are sometimes added to the diets of acutely stressed and immune-compromised patients. Several studies have suggested that glutamine supplementation may improve immune function, preserve muscle mass, and reduce mortality rates in critically ill patients.[11] Arginine supplementation has been shown to have beneficial effects on the immune responses and nitrogen balance of critically ill and postoperative patients.[12] Although glutamine and arginine are often added to enteral formulas promoted for wound healing and enhanced immunity, their use remains controversial.

Carbohydrate and Fat Intakes in Acute Stress The bulk of energy needs are supplied from carbohydrate and fat. Carbohydrate is usually the main source of energy, providing 50 to 60 percent of total energy requirements.[13] When parenteral feedings are necessary, dextrose is provided to critically ill patients at no more than 5 milligrams per kilogram body weight per minute to prevent hyperglycemia (see p. 697).

Fat provides both energy and essential fatty acids. In critically ill patients who are not at risk for hypertriglyceridemia (blood triglyceride levels are less than 300 milligrams per deciliter), fat intakes should range between 1.0 and 1.5 grams per kilogram body weight per day.[14] In patients with severe hyperglycemia, fat may supply up to 50 percent of kcalories, although high fat intakes may suppress immune function and increase the risks of developing infections and hypertriglyceridemia. Patients with blood triglyceride levels above 300 to 400 milligrams per deciliter may require fat restriction.[15]

Micronutrient Needs in Acute Stress Acutely stressed patients may have increased micronutrient needs, but specific requirements remain unknown.[16] In hypermetabolic patients, the need for B vitamins may be higher to support the increase in energy metabolism. A number of micronutrients, such as vitamin A, vitamin C, and zinc, have critical roles in immunity and wound healing, and their supplementation may speed recovery under certain circumstances. Patients with burns and tissue injuries may have increased requirements for trace minerals due to tissue losses; in several studies, supplementation of zinc, copper, and selenium improved immune responses in severely burned patients.[17]

Plasma levels of micronutrients are often altered during critical illness. The acute-phase response causes a redistribution of some micronutrients (such as zinc and iron) that lowers their blood levels; therefore, micronutrient status is sometimes difficult to interpret. Blood concentrations of trace minerals should be monitored in patients receiving parenteral nutrition support to ensure that excessive amounts are not given intravenously.

Approaches to Nutrition Care in Acute Stress

As mentioned earlier, the initial care following acute stress focuses on maintaining fluid and electrolyte balances. Simple intravenous solutions often contain dextrose, providing minimal kcalories. Once feedings begin, patients may require a combination of methods to meet their nutritional needs. If poor appetite, the medical condition, or a medical procedure (such as mechanical ventilation) interferes with food intake, nutrition support may be warranted.

As Chapter 20 explained, enteral nutrition support is preferred over parenteral nutrition in patients with normal intestinal function. However, parenteral nutrition support may be required if patients cannot achieve adequate nutrient intakes from enteral feedings. In one study of critically ill patients, only 53 percent were able to meet nutrient needs from enteral feedings alone, and energy intakes averaged 77 percent of the amounts prescribed.[18] Consequently, parenteral nutrition is sometimes used to supplement enteral feedings; for patients who are likely to become malnourished during critical illness, it may be the main source of nutrients.

Once patients transition to oral feedings, meeting protein and energy needs may be difficult, and enteral formulas are often given to supplement the diet. Many such formulas have high nutrient density, and some contain extra amounts of nutrients believed to promote healing, such as vitamin A, zinc, and the amino acids arginine and glutamine. A high-kcalorie, high-protein diet is often prescribed, although care must be taken not to overfeed patients who are at risk of developing refeeding syndrome or hyperglycemia. Nutrient needs should be reassessed frequently until the patient's condition stabilizes.

Patients with Burn Injuries

Burns are among the most severe injuries that a person may experience, and they have destructive effects on growth and health that may persist long after the

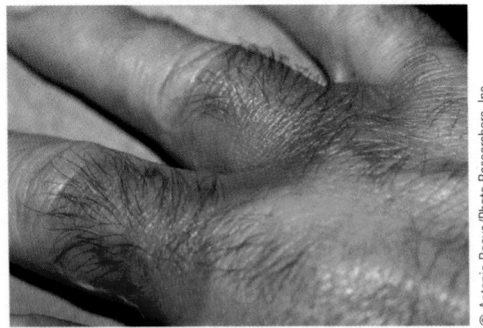

A first-degree burn injures the epidermis and is characterized by pink or reddened skin.

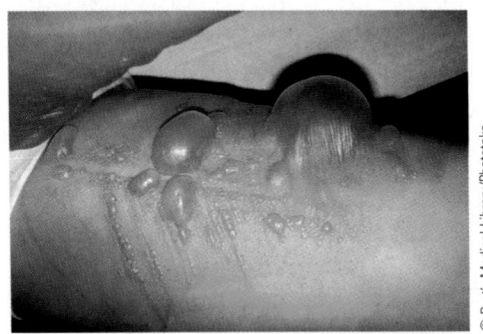

A second-degree burn damages the epidermis and a portion of the dermis and causes redness, swelling, and blistering.

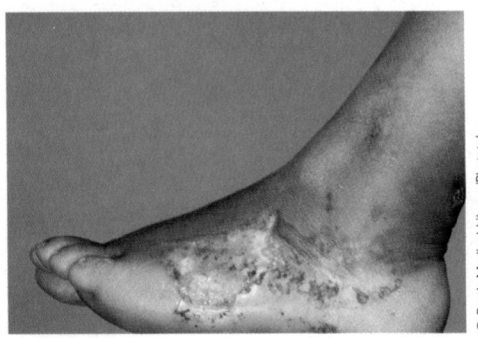

A third-degree burn destroys both the epidermis and dermis and may involve the tissues beneath the skin.

◆ Chapter 18 provides information about the high-kcalorie, high-protein diet (see pp. 624–625).

hypovolemia (HIGH-poe-voe-LEE-me-ah): low blood volume.

epidermis (eh-pih-DER-miss): the outer layer of the skin.

dermis: the connective tissue layer underneath the epidermis that contains the skin's blood vessels and nerves.

burns have healed. Causes of burns include flames or scalding water, chemical agents, electricity, and irradiation. Frequent complications include infection and **hypovolemia,** which can increase the risk of death.[19]

Burn Classification Burns are classified according to how deeply they penetrate the skin and underlying tissue. First-degree burns affect only the **epidermis** and are pink or red, dry, and painful (for example, a sunburn). Second-degree burns (also called partial-thickness burns) involve both the epidermis and a portion of the **dermis.** They are red, wet, and blistery, and extremely painful because nerve endings are exposed. Third-degree burns (also known as full-thickness burns) destroy both the epidermis and dermis and may extend into the tissues below; their appearance may be waxy white, brown and leathery, or black and charred. These burns are deep enough to destroy nerves and are therefore painless.[20]

Burn size in adults is often estimated by dividing the body into 11 parts; each part represents about 9 percent of the total body surface area (TBSA).[21] The head and neck region and each arm are equivalent to about 9 percent TBSA each; the front torso, the back torso, and each leg represent approximately 18 percent TBSA each (see the margin drawing on p. 717). The severity of a burn is based both on its thickness and on the amount of surface area involved.

Treatment for Burn Injuries Emergency measures after a burn include the removal of clothing and smoldering material from the skin. Burns caused by acid or chemical compounds must be flushed with copious amounts of water. Wounds are cleaned and debris removed. Blisters and dead tissue are debrided, if necessary. Finally, the surface is covered with topical antibacterial agents and sterile dressings. Immediate care also includes fluid replacement, as the fluid losses through burned skin can be considerable. Some burn victims need immediate oxygen support or mechanical ventilation. Pain relief medication is also required soon after injury.

Metabolic Changes in Burn Patients Burn injuries cause severe metabolic stress and the inflammatory response, as discussed earlier in this chapter, and therefore they result in hypermetabolism, tissue breakdown, and altered nutrient metabolism. With the protective skin barrier partially destroyed, burns are accompanied by losses of evaporative water and body heat. Second- and third-degree burns can cause substantial losses of protein and micronutrients. Extensive burns can disrupt gastrointestinal function.

Medical Nutrition Therapy for Burn Patients The objectives of nutrition care for burn patients are to achieve nitrogen balance and minimize tissue losses. The nutrition prescription is typically a high-kcalorie, high-protein diet.[22] ◆ Energy needs can be calculated as described previously (see Table 22-2 on p. 713 and the "How to" on p. 714), although equations that consider additional factors such as burn severity and ventilator use are sometimes used during peak recovery periods. The suggested protein intake is 2 to 3 grams per kilogram body weight or 20 to 25 percent of total kcalories in individuals with burns greater than 10 percent TBSA.[23] For less extensive burns, a protein intake of 1.2 grams per kilogram body weight is usually adequate. Adequacy of protein intake is assessed by monitoring nitrogen balance, serum proteins, and wound-healing ability. Micronutrient supplements are often provided and may include high amounts of vitamin A, vitamin C, and zinc, which are thought to support immunity and promote wound healing. Fluid needs must be monitored carefully during the recovery period; the patient's hydration status is evaluated by monitoring urine output and serum electrolyte levels.

Some patients may need to be evaluated for feeding ability; problems that may interfere with eating include burns on the face, hands, and arms; bulky dressings; frequent dressing changes; and pain medications that cause sedation. Patients who are able to eat are often offered small, frequent meals rather than large meals and are provided with oral supplements and nutrient-dense snacks to help them meet energy and protein needs. A combination of oral feedings and tube feedings is often necessary. Some burn patients develop gastroparesis or intestinal ileus (stomach or intestinal paralysis) and may require nasoenteric feedings (discussed

Mortgage Broker with a Severe Burn

David Bray, a 42-year-old mortgage broker, has been admitted to intensive care. He suffered a severe burn covering 35 percent of his body when he was trapped inside a burning building. His wife told the nurse that Mr. Bray's height is 6 feet and that he usually weighs about 175 pounds. The physician ordered lab work, including serum protein concentrations, but the results have not yet been received.

1. Identify Mr. Bray's immediate needs after the injury. Describe the initial concerns of the health care team and the measures they might take soon after Mr. Bray's arrival at the hospital.
2. Considering Mr. Bray's condition, what problems might the health care team encounter when they attempt to obtain information that can help them assess his nutrition status? What additional concerns might they have if Mr. Bray was malnourished before he experienced the burn?
3. Estimate Mr. Bray's energy and protein needs (use a protein factor of 2.5 grams per kilogram). What problems may interfere with Mr. Bray's ability to meet his nutrient needs?
4. Due to complications that developed during tube feeding, Mr. Bray was able to obtain only 65 percent of his energy requirements. What other feeding options may be considered?

Burn size can be estimated by sectioning the total body surface area (TBSA) as shown.

in Chapter 20). Parenteral support may be required if intestinal function is lacking, if complications that interfere with enteral feedings develop, or if nutrient requirements cannot be met by tube feeding alone. The accompanying Case Study reviews the nutrition care of a burn patient.

IN SUMMARY

Severe metabolic stress causes hypermetabolism and negative nitrogen balance and may result in wasting. The objectives of nutrition care during stress are to provide a diet that can preserve muscle tissue, maintain immune defenses, and promote healing. Energy intake should sustain nitrogen balance but should not result in overfeeding. Protein recommendations are increased to help prevent tissue losses and allow healing of damaged tissue. Enteral and parenteral feedings are sometimes needed to meet the high nutrient requirements of acutely stressed patients. Burn patients require fluid replacement immediately after a burn injury, and a high-kcalorie, high-protein diet during recovery.

Nutrition and Respiratory Stress

Some medical problems upset the gas exchange process between the air and blood and result in respiratory stress, which is characterized by a reduction in the blood's oxygen supply and an increase in carbon dioxide levels. In some individuals, the excessive carbon dioxide in the blood disrupts the breathing pattern and thereby interferes with food intake. Moreover, the labored breathing caused by respiratory disorders entails a higher energy cost than normal breathing does, raising energy needs and increasing carbon dioxide production further. Lung diseases make physical activity difficult and can lead to muscle wasting. Weight loss and malnutrition therefore become dangerous outcomes of some types of respiratory illnesses.

Chronic Obstructive Pulmonary Disease

Chronic obstructive pulmonary disease (COPD) refers to a group of conditions characterized by the persistent obstruction of airflow through the lungs.

chronic obstructive pulmonary disease (COPD): a group of lung diseases characterized by persistent obstructed airflow through the lungs and airways; includes chronic bronchitis and emphysema.

FIGURE 22-2 The Respiratory System

Inhaled air travels via the trachea to the bronchi and bronchioles, the major airways of the lungs. Oxygen and carbon dioxide are exchanged across the thin-walled alveoli, which are surrounded by capillaries.

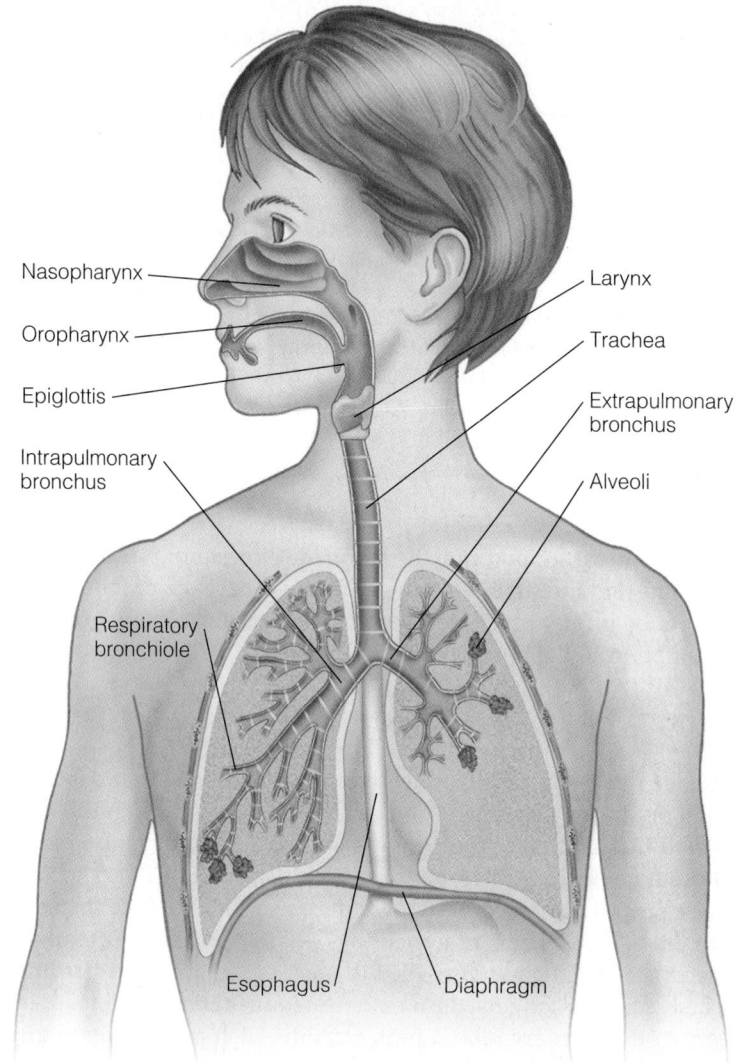

Nasopharynx

Oropharynx

Epiglottis

Intrapulmonary bronchus

Respiratory bronchiole

Larynx

Trachea

Extrapulmonary bronchus

Alveoli

Esophagus

Diaphragm

SOURCE: Based on a drawing in Carol Mattson Porth, *Pathophysiology*, 5th ed. (Lippincott Williams & Wilkins, 1998).

bronchi (BRON-key), **bronchioles** (BRON-key-oles): the main airways of the lungs. The singular form of bronchi is *bronchus*.

alveoli (al-VEE-oh-lie): air sacs in the lungs. One sac is an *alveolus*.

chronic bronchitis (bron-KYE-tis): a lung disorder characterized by persistent inflammation and excessive secretions of mucus in the main airways of the lungs; diagnosis is based on the presence of a chronic, productive cough for at least three months of the year for two successive years.

emphysema (EM-fih-ZEE-mah): a progressive lung condition characterized by the breakdown of the lungs' elastic structure and destruction of the walls of the bronchioles and alveoli, reducing the surface area involved in respiration.

dyspnea (DISP-nee-ah): shortness of breath.

Figure 22-2 illustrates the main airways (**bronchi** and **bronchioles**) and air sacs (**alveoli**) of the normal respiratory system, and Figure 22-3 shows how they are altered in COPD. The two main types of COPD are **chronic bronchitis** and **emphysema**, and many patients display features of both conditions[24]:

- *Chronic bronchitis* is characterized by persistent inflammation and excessive secretions of mucus in the main airways of the lungs, which may ultimately thicken and become too narrow for adequate mucus clearance. Chronic bronchitis is diagnosed when a chronic, productive cough persists for at least three months of the year for two consecutive years.

- *Emphysema* is characterized by the breakdown of the lungs' elastic structure and destruction of the walls of the bronchioles and alveoli, changes that significantly reduce the surface area needed for respiration. Emphysema is diagnosed on the basis of clinical signs and the results of lung function tests.

Both chronic bronchitis and emphysema are associated with abnormal levels of oxygen and carbon dioxide in the blood and shortness of breath (**dyspnea**).

FIGURE 22-3 Chronic Obstructive Pulmonary Disease

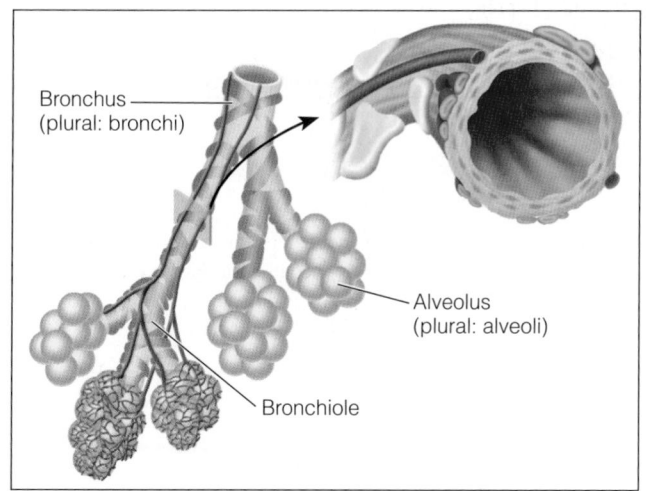

Healthy bronchi provide an open passageway for air. Healthy alveoli permit gas exchange between the air and blood.

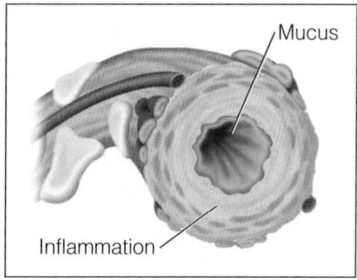

Chronic bronchitis is characterized by inflammation, excessive secretion of mucus, and narrowing of the bronchi – factors that reduce normal airflow.

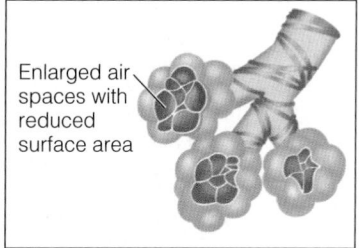

Emphysema is characterized by gradual destruction of the walls separating the alveoli and reduced lung elasticity.

COPD may eventually lead to respiratory or heart failure and ranks as the fourth leading cause of death in the United States.[25]

COPD is a debilitating condition. Generally, dyspnea worsens as the condition progresses, resulting in dramatic reductions in physical activity and quality of life. Activities of daily living such as bathing or dressing may cause exhaustion or breathlessness. Weight loss and wasting are common in the advanced stages of disease and may result from hypermetabolism, poor food intake, and the actions of various inflammatory proteins. As with other chronic illnesses, anxiety and depression are a concern, and psychological distress may reduce a COPD patient's ability to cope with the demands of treatment.

Causes of COPD Cigarette smoking is the primary risk factor in 90 percent of COPD cases[26] and is especially damaging when combined with respiratory infections or an occupational exposure to dusts or chemicals. Only a minority of smokers (about 15 percent) develop COPD, however; thus genetic susceptibility also contributes to its development. Genetic factors are especially likely in patients with early-onset COPD. Alpha-1-antitrypsin deficiency, an inherited disorder, occurs in 1 to 2 percent of patients with COPD.[27] These individuals have inadequate blood levels of a plasma protein (alpha-1 antitrypsin) that normally inhibits enzymatic breakdown of the lungs' connective tissue.

Treatment of COPD The primary objectives of COPD treatment are to prevent the disease from progressing and relieve major symptoms (dyspnea and coughing). Individuals with COPD are encouraged to quit smoking to prevent disease progression and to get vaccinated against influenza and pneumonia to avoid complications. The most frequently prescribed medications are bronchodilators, which improve airflow, and corticosteroids (anti-inflammatory medications), which help to prevent symptom recurrence; note that corticosteroids promote catabolic processes and can exacerbate the muscle loss that often accompanies COPD. For people with severe COPD, supplemental oxygen therapy (12 hours daily) can maintain normal oxygen levels in the blood and reduce mortality risk. The Diet-Drug Interactions feature on p. 720 lists nutrition-related effects of the medications used to treat COPD.

DIET-DRUG INTERACTIONS

Check this table for notable nutrition-related effects of the medications discussed in this chapter.

	Gastrointestinal Effects	Interactions with Dietary Substances	Metabolic Effects
Bronchodilators (theophylline, dyphylline)	Increased gastric acid secretion, acid reflux.	Caffeine may enhance drug effects.	
Corticosteroids (prednisone)			Glucose intolerance, sodium retention, negative nitrogen balance, appetite stimulation, weight gain, growth suppression in children.

◆ The altered sense of taste in patients with COPD may be due to chronic mouth breathing, which dries the mouth. Taste is also affected by the use of certain medications, including some bronchodilators.

Patients who need supplemental oxygen can use lightweight, portable equipment that allows them to move about freely.

© Courtesy of Airsep Corporation

Medical Nutrition Therapy for COPD The main goals of nutrition therapy for COPD are to correct the malnutrition that affects up to 60 percent of patients with COPD,[28] promote the maintenance of a healthy body weight, and prevent muscle wasting. Research studies have found that COPD patients with low body weights have higher mortality rates[29]; thus, encouraging adequate food intake is usually the main focus of the nutrition care plan. The hypermetabolism in COPD patients is due to the increased workload of the muscles involved in breathing; resting metabolic rates are about 15 percent above the values predicted by RMR equations.[30] Note that some patients with COPD may be overweight or obese, which puts an additional strain on the respiratory system; such patients may benefit from energy restriction and gradual weight reduction.

Food intake often declines as COPD progresses, although the causes of poor intake vary among patients. Dyspnea may interfere with chewing or swallowing. Appetite may be affected by medications, depression or anxiety, or altered taste perception. ◆ Physical changes in the diaphragm and lungs may reduce abdominal volume, leading to early satiety. Some patients may become too disabled to shop or prepare food or may lack adequate support at home. The health practitioner must assess the unique needs of a COPD patient before proposing a nutrition care plan.

Some patients may benefit by eating small, frequent meals rather than two or three large ones. The lower energy content of small meals reduces the carbon dioxide load,[31] plus abdominal discomfort and dyspnea may be reduced when less food is consumed. If bloating is a problem, patients should avoid foods that increase gas formation. Some individuals may eat better if they receive supplemental oxygen at mealtimes. Consuming adequate fluids should be encouraged to help prevent the secretion of overly thick mucus; however, some patients should consume liquids between meals so as not to interfere with food intake. For undernourished patients, a high-kcalorie, high-protein diet may be helpful, but excessive energy intakes may increase carbon dioxide output and increase respiratory stress. Liquid supplements are sometimes given between meals to improve weight gain or exercise endurance, but high-energy supplements (those containing more than 250 kcalories) may induce satiety and reduce energy intake at mealtime.[32]

Pulmonary Formulas Some enteral formulas available for use in pulmonary disease provide more kcalories from fat and fewer from carbohydrate than standard formulas. The ratio of carbon dioxide production to oxygen consumption in cells is lower when fat is consumed, so theoretically these formulas should lower respiratory requirements. However, research studies have not confirmed that the reduced-carbohydrate formulas improve clinical outcomes more than moderate energy intakes, so the benefits of using these pulmonary formulas are uncertain.[33]

Incorporating an Exercise Program Loss of muscle can be more readily prevented or reversed if the treatment plan includes a carefully designed exercise program.[34] With exercise, patients are likely to see improvements in their endurance and become less fearful of their physical limitations. For some patients, the combination of an exercise plan and oral supplements may be better for maintaining weight and improving muscle status than either component of treatment alone.[35] The accompanying Case Study allows you to review the nutrition care for a patient with COPD.

Respiratory Failure

In respiratory failure, the gas exchange between the air and the circulating blood is severely impaired, causing abnormal levels of tissue gases that can be life threatening. Respiratory failure can develop from chronic disease (such as COPD) or may arise suddenly (*acute* respiratory failure). Various conditions that affect lung function can be the underlying cause of failure. For example, respiratory failure may result from obstruction in the upper airways or from weakness or paralysis of the muscles involved in respiration. An embolus lodged within the lungs may prevent blood flow, or toxic substances may damage lung tissue. Surgery sometimes results in respiratory failure due to the depressive effects of anesthesia or because some abdominal procedures can affect breathing. Severe trauma and infection are common triggers of **acute respiratory distress syndrome (ARDS)**, an acute form of respiratory failure marked by extensive lung damage. ARDS is a life-threatening condition that usually requires the use of mechanical ventilation to restore normal oxygen and carbon dioxide levels.

Consequences of Respiratory Failure Impaired gas exchange results in **hypoxemia** (low oxygen levels in the blood) and **hypercapnia** (excessive carbon dioxide in the blood). An inadequate oxygen supply within tissues **(hypoxia)** inhibits cellular function and can ultimately cause cell death. Hypercapnia can lead to **acidosis,** which interferes with the functions of the central nervous system. To compensate for respiratory failure, a person breathes more rapidly, and the heart rate increases. The skin may become sweaty and develop a bluish cast **(cyanosis).** Headache, confusion, and drowsiness may occur. Severe cases of respiratory failure can cause heart arrhythmias and ultimately, coma.

Acute Respiratory Distress Syndrome (ARDS) Acute respiratory distress syndrome (ARDS) typically occurs in individuals who have no history of lung disease, and it often follows acute lung injury due to sepsis, trauma, severe pneumonia,

acute respiratory distress syndrome (ARDS): respiratory failure triggered by severe lung injury; a medical emergency that causes dyspnea and pulmonary edema and usually requires assisted (mechanical) ventilation.

hypoxemia (high-pock-SEE-me-ah): a low level of oxygen in the blood.

hypercapnia (high-per-CAP-nee-ah): excessive carbon dioxide in the blood.

hypoxia (high-POCK-see-ah): a low amount of oxygen in body tissues.

acidosis: acid accumulation in body tissues; depresses the central nervous system and can lead to disorientation and, eventually, coma.

cyanosis (sigh-ah-NOH-sis): a bluish cast in the skin due to the color of deoxygenated hemoglobin. Cyanosis is most evident in individuals with lighter, thinner skin; it is mostly seen on lips, cheeks, and ears and under the nails.

CASE STUDY Elderly Person with Emphysema

John Norback is an 82-year-old man who has emphysema that severely affects both lungs. He is 5 feet 9 inches tall and currently weighs 150 pounds, about 20 pounds less than his weight in earlier years. He lives with a daughter and son-in-law and eats meals with their family. He becomes breathless when eating and when walking around the house, and he feels tired all the time. A medical clinic recently ordered oxygen therapy for home use, but supplies have not yet arrived. Mr. Norback's daughter is concerned about her father's recent weight loss and breathlessness.

1. Assess Mr. Norback's risk of malnutrition, using information from Table 17-9 in Chapter 17 (p. 602). What factors may have contributed to his weight loss?
2. What are possible reasons for Mr. Norback's difficulty with eating? List some dietary suggestions that may help to improve his appetite and food intake. How might the use of oxygen therapy help?
3. Based on the history given, what factors may account for Mr. Norback's tiredness? What suggestions would you give Mr. Norback and his daughter regarding physical activity?

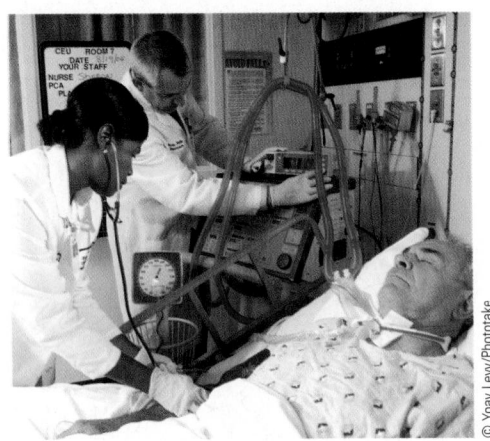

Mechanical ventilation controls the rate and amount of oxygen supplied to a patient's airways.

inhalation of smoke or toxic chemicals, or aspiration of gastric contents.[36] It can develop over 12 to 48 hours or may take several days to appear. The lungs exhibit extensive inflammation and fluid buildup (pulmonary edema) that interfere with lung ventilation and with gas exchange in the alveoli. Blood oxygen levels are typically very low. The later stages of ARDS are associated with a proliferation of lung cells, which causes fibrosis and disrupts lung structure. A dangerous complication of ARDS is a progression to multiple organ dysfunction syndrome, which is described in the highlight following this chapter.

Treatment of Respiratory Failure The treatment of respiratory failure focuses on supporting lung function and correcting the underlying disorder. Because respiratory failure can be caused by a number of different conditions, treatment plans vary considerably. In individuals with chronic lung disease, providing oxygen therapy via a face mask or nasal tubing can relieve symptoms, but patients with ARDS usually require mechanical ventilation until they are able to breathe independently. Fluids may need careful monitoring to maintain fluid balance and prevent overload; diuretics are sometimes prescribed to mobilize the fluid that has accumulated in lung tissue. Medications may be given to treat infections, keep airways open, or relieve inflammation. Complications are common in ARDS and must be forestalled to prevent multiple organ dysfunction.

Medical Nutrition Therapy for Acute Respiratory Failure Dietary recommendations are individualized according to the patient's condition. The primary concern is to supply enough energy and protein to support lung function without overtaxing the compromised respiratory system.[37] Fluid restrictions may be necessary to help reverse pulmonary edema. As usual, when nutrition support is necessary, enteral nutrition is preferred over parenteral nutrition.

Energy Needs Energy needs can be estimated by using a predictive equation to determine the RMR and adjusting the result with an appropriate stress factor. The body weight used in the equation may need to be corrected for edema, which is often present in patients with acute respiratory failure. A stress factor of 1.2 can be used initially, although needs may be higher in malnourished patients or if fever or infections are present.[38] Energy intakes above 1.5 times RMR are not recommended for patients with acute respiratory failure because excessive energy intakes generate extra carbon dioxide and increase the risk for complications.

For patients using mechanical ventilation, the stress factors 1.34 and 1.32 are often appropriate for men and women, respectively.[39] Predictive equations that include respiratory rates (in breaths per minute or liters per minute) and lung volume (in liters) are also used for determining energy needs in ventilated patients. Note that mechanical ventilation can sometimes reduce energy expenditure, because patients using ventilators are often heavily sedated and immobile, which lowers energy requirements.

Fluids Dehydration may develop due to a low fluid intake, an increase in bronchial secretions, or diuretic therapy. Although dehydration can impede the clearance of lung secretions, pulmonary edema is often present, so fluid restriction may be required to prevent the accumulation of additional fluids in lung tissue. The edema may make it difficult to assess whether a critically ill patient is maintaining weight.

Nutrition Support Patients with acute respiratory failure are often unable to eat meals and may need nutrition support. Tube feedings are usually used if the intestine is functional, and intestinal feedings are preferred over gastric feedings because they reduce the risk of aspiration. Nutrient-dense formulas (2 kcalories per milliliter) are generally recommended for patients with fluid restrictions. Formulas enriched with omega-3 fatty acids and vitamins A, C, and E have been found to be helpful in patients with ARDS.[40] If the risk of aspiration is too high to continue enteral feedings, parenteral nutrition support may be considered.

IN SUMMARY

Respiratory stress from chronic or acute disease affects body weight, muscle mass, and the normal functioning of all body tissues. Chronic obstructive pulmonary disease (COPD) is a debilitating, progressive illness that can lead to malnutrition, muscle wasting, and activity intolerance. Depending on individual needs, the goals of nutrition therapy are to improve food intake, maintain proper weight, preserve muscle tissue, and improve exercise endurance. Acute respiratory failure can develop from COPD or arise in persons with no history of lung disease. It often follows acute lung injury due to infection, trauma, or inhalation of toxic substances. Medical nutrition therapy may include nutrition support and fluid restrictions.

 Clinical Portfolio

CENGAGENOW™
academic.cengage.com/login

■ Adam is a 49-year-old male who is 6 feet 2 inches tall and has a usual body weight of 180 pounds. He was physically injured by an explosion in the chemistry lab where he works and is now in the intensive care unit. Using the method described in the "How to" on p. 714, estimate Adam's energy requirement. Estimate his protein requirement, using the factor 1.5 grams per kilogram of body weight.

■ Turning again to the case described in the first item, assume that Adam requires a tube feeding and can tolerate a standard enteral formula. Check Appendix K to find at least three formulas that the nutrition support team might select for tube feeding. Determine the volume of each formula that would be needed to meet Adam's energy and protein needs. Would this volume also meet the recommendations for vitamins and minerals?

■ Ayla is a 23-year-old law student admitted to the hospital following an automobile accident in which she broke several bones and ruptured part of her small intestine. She has been in the hospital for several weeks and has just begun eating table foods. Her brother, who was driving the vehicle, was also seriously injured and nearly lost his life. Aside from the increased nutritional needs imposed by the stress of the accident, discuss how the following factors might interfere with Ayla's ability to improve her nutrition status:

• Ayla's injuries are painful.
• Ayla's medications cause drowsiness.
• Ayla is depressed.
• Ayla is often out of her room for X-rays and other diagnostic tests when the menus and food trays arrive.
• Ayla's food intake is sometimes restricted due to the procedures she is undergoing.

How might these problems be resolved to improve Ayla's food intake?

NUTRITION ON THE NET

- To uncover additional information relevant to critical care, visit these sites: American Association of Critical-Care Nurses: **www.aacn.org** American Society for Parenteral and Enteral Nutrition: **www.clinnutr.org**

- These sites provide resources regarding burns for patients and their families: American Burn Association: **www.ameriburn.org** Burn Survivor Resource Center: **www.burnsurvivor.com**

- This nonprofit group provides comprehensive, up-to-date information for burn care professionals: **www.burnsurgery.org**

- To learn more about lung diseases, visit these sites: American Lung Association: **www.lungusa.org** Canadian Lung Association: **www.lung.ca** National Heart, Lung, and Blood Institute: **www.nhlbi.nih.gov**

NUTRITION ASSESSMENT CHECKLIST for People Undergoing Metabolic or Respiratory Stress

Medical History
Check the medical record to determine:

- Cause of stress
- Severity of stress
- Route of feeding (oral, tube feeding, or parenteral)
- Whether any organ system is compromised

For patients with COPD, check to determine:

- Degree of breathing difficulty
- Use of oxygen therapy
- Activity tolerance

Review the medical record for complications related to underfeeding or overfeeding, such as:

- Dehydration or fluid overload
- Electrolyte imbalances
- Fatty liver
- Hyperglycemia
- Hypertriglyceridemia

Medications
Record all medications and note:

- Side effects that may alter food intake or nutrition status
- Use of theophylline, in patients who may need to avoid caffeine

Dietary Intake
If the patient is not meeting nutrition goals:

- Monitor intakes to ensure that the patient is receiving the diet prescribed.
- Investigate appetite problems or difficulties with eating.
- Consider interventions to improve food intake.
- Consider the need for supplementation.
- In patients with COPD, consider problems that may hamper the patient's ability to prepare or consume foods.

Anthropometric Data
Measure baseline height and weight, and monitor daily weights. Remember that body weight can fluctuate in acutely ill patients who undergo fluid resuscitation. Once a patient's weight has stabilized:

- Reevaluate protein and energy needs.
- Consider the need to alter the energy prescription to meet weight goals.

Laboratory Tests
Laboratory tests that may be affected by stress and therefore require careful interpretation include:

- Albumin
- Transferrin

- Prealbumin
- C-reactive protein
- Serum iron and zinc
- Total lymphocyte count (white blood cell counts are often elevated)

Monitor laboratory tests for signs of:

- Dehydration or fluid overload
- Electrolyte and acid-base imbalances
- Hyperglycemia
- Hypertriglyceridemia
- Nutrient deficiencies
- Negative nitrogen balance
- Organ dysfunction or organ function that has normalized

Physical Signs
Regularly assess vital signs, including:

- Blood pressure
- Pulse
- Body temperature
- Respiration

Look for physical signs of:

- Protein-energy malnutrition
- Dehydration or fluid overload
- Nutrient deficiencies and excesses

STUDY QUESTIONS

CENGAGENOW™

These questions will help you review the chapter. You will find the answers in the discussions on the pages provided.

1. Describe the effects of the stress response on the body. Specify the main hormones involved and their metabolic effects. (pp. 709–710)

2. Discuss the main effects of the inflammatory response following infection or injury. Identify the chemical mediators involved, and explain how they help to regulate the inflammatory process. (pp. 710–712)

3. Characterize the acute-phase response, giving examples of the clinical symptoms and changes in blood chemistry that usually result. (p. 712)

4. How does metabolic stress affect nutrition status? Explain why nutrition status may be difficult to evaluate in an acutely stressed individual. (pp. 712–713)

5. Describe how energy and protein needs are estimated during acute stress. Which micronutrients are sometimes supplemented? (pp. 713–715)

6. Characterize first-, second-, and third-degree burns. What measures are taken immediately after a burn occurs? (p. 716)

7. Identify the objectives of nutrition care for burn patients. Explain how the energy and protein needs of patients with burns are estimated. What measures are taken to provide adequate nutrient intakes? (pp. 716–717)

8. What is chronic obstructive pulmonary disease (COPD)? Describe its causes and treatment. Discuss the possible effects of COPD on body composition. (pp. 717–720)

9. Describe respiratory failure and its consequences. Identify the key elements of medical treatment and medical nutrition therapy. What are potential causes of acute respiratory distress syndrome (ARDS)? (pp. 721–722)

These multiple choice questions will help you prepare for an exam. Answers can be found on p. 726.

1. Which of the following metabolic changes accompanies acute stress?
    a. reduced plasma concentrations of glucose and fatty acids
    b. reduced blood volume and blood pressure
    c. increased insulin action
    d. catabolism of protein in skeletal muscle and connective tissue

2. Tissue injury is followed by:
    a. fluid accumulation in damaged tissue.
    b. reduced blood flow to injured tissue.
    c. reduced capillary permeability.
    d. decreased body temperature.

3. What is a possible effect of replacing vegetable oils rich in omega-6 fatty acids with oils rich in omega-3 fatty acids?
    a. improvement in blood circulation
    b. suppression of inflammation
    c. protection against sepsis
    d. hypertriglyceridemia

4. The acute-phase response results in increased plasma concentrations of:
    a. albumin.
    b. iron.
    c. C-reactive protein.
    d. zinc.

5. Which of the following statements concerning protein and energy recommendations during acute metabolic stress is true?
    a. Protein and energy recommendations are similar to those for healthy people.
    b. Protein and energy recommendations are reduced because a stressed individual cannot metabolize nutrients normally.
    c. Acutely stressed individuals can benefit from as much protein and energy as can be provided.
    d. Protein and energy recommendations are high in order to minimize muscle tissue losses.

6. The immediate treatment provided to a patient with a severe burn includes:
    a. parenteral feedings.
    b. fluid replacement.
    c. blood transfusions.
    d. micronutrient supplementation.

7. The primary risk factor for COPD is:
    a. alpha-1 antitrypsin deficiency.
    b. occupational exposure to dusts or chemicals.
    c. cigarette smoking.
    d. respiratory infections.

8. A primary feature of emphysema is:
    a. obstruction within the bronchi.
    b. obstruction within the bronchioles.
    c. destruction of the walls separating the alveoli.
    d. excessive lung elasticity.

9. The weight loss and wasting that often occur in COPD can be caused by:
    a. reduced food intake.
    b. increased metabolic rate.
    c. reduced exercise tolerance.
    d. all of the above.

10. Medical nutrition therapy for a person with respiratory failure includes:
    a. careful attention to providing enough, but not too much, energy.
    b. a generous fluid intake to facilitate mucus clearance.
    c. a high-fat intake to prevent weight loss.
    d. a high-carbohydrate intake to limit carbon dioxide production.

REFERENCES

1. S. F. Lowry and J. M. Perez, The hypercata-bolic state, in M. E. Shils and coeditors, *Modern Nutrition in Health and Disease* (Baltimore: Lippincott Williams & Wilkins, 2006), pp. 1381–1400; M. I. T. D. Correia and C. T. de Almeida, Metabolic response to stress, in G. Cresci, ed., *Nutrition Support for the Critically Ill Patient: A Guide to Practice* (Boca Raton, Fla.: Taylor & Francis Group, 2005), pp. 3–13.

2. A. A. Spector, Essential fatty acids, in M. H. Stipanuk, ed., *Biochemical, Physiological, and Molecular Aspects of Human Nutrition* (St. Louis: Saunders, 2006), pp. 518–540; D. L. Waitzberg, R. S. Torrinhas, and L. De Nardi, Lipid metabolism: Comparison of stress and nonstressed states, in G. Cresci, ed., *Nutrition Support for the Critically Ill Patient: A Guide to Practice* (Boca Raton, Fla.: Taylor & Francis Group, 2005), pp. 49–67.

3. C. A. Dinarello and R. Porat, The acute phase response, in L. Goldman and D. Ausiello, eds., *Cecil Textbook of Medicine* (Philadelphia: Saunders, 2004), pp. 1733–1735.

4. D. Frankenfield, Energy requirements in the critically ill patient, in G. Cresci, ed., *Nutrition Support for the Critically Ill Patient: A Guide to Practice* (Boca Raton, Fla.: Taylor & Francis Group, 2005), pp. 83–98.

5. D. A. Schoeller, Making indirect calorimetry a gold standard for predicting energy requirements for institutionalized patients, *Journal of the American Dietetic Association* 107 (2007): 390–392; K. A. Kudsk and G. S. Sacks, Nutrition in the care of the patient with surgery, trauma, and sepsis, in M. E. Shils and coeditors, *Modern Nutrition in Health and Disease* (Baltimore: Lippincott Williams & Wilkins, 2006), pp. 1414–1435.

6. D. Frankenfield, Prediction of resting metabolic rate in critically ill adult patients: Results of a systematic review of the evidence, *Journal of the American Dietetic Association* 107 (2007): 1552–1561; C. S. Ireton-Jones and coauthors, Equations for the estimation of energy expenditures in patients with burns with special reference to ventilatory status, *Journal of Burn Care and Rehabilitation* 13 (1992): 330–333.

7. Frankenfield, 2007.

8. Kudsk and Sacks, 2006.

9. J. Lefton and P. P. Lopez, Macronutrient requirements: Carbohydrate, protein, and lipid, in G. Cresci, ed., *Nutrition Support for the Critically Ill Patient: A Guide to Practice* (Boca Raton, Fla.: Taylor & Francis Group, 2005), pp. 99–108.

10. American Dietetic Association, *Nutrition Care Manual* (Chicago: American Dietetic Association, 2007).

11. Lowry and Perez, 2006.

12. P. Furst, Protein and amino acid metabolism: Comparison of stressed and non-stressed states, in G. Cresci, ed., *Nutrition Support for the Critically Ill Patient: A Guide to Practice* (Boca Raton, Fla.: Taylor & Francis Group, 2005), pp. 27–47.

13. Kudsk and Sacks, 2006.

14. Kudsk and Sacks, 2006.

15. Lefton and Lopez, 2005.

16. K. Sriram and J. I. Cué, Micronutrient and antioxidant therapy in critically ill patients, in G. Cresci, ed., *Nutrition Support for the Critically Ill Patient: A Guide to Practice* (Boca Raton, Fla.: Taylor & Francis Group, 2005), pp. 109–123.

17. Kudsk and Sacks, 2006.

18. A.S.P.E.N. Board of Directors and The Clinical Guidelines Task Force, Guidelines for the use of parenteral and enteral nutrition in adult and pediatric patients, *Journal of Parenteral and Enteral Nutrition* 26 (2002): 1–138SA.

19. M. H. Beers and coeditors, *The Merck Manual of Diagnosis and Therapy* (Whitehouse Station, N.J.: Merck Research Laboratories, 2006), pp. 2592–2597.

20. R. H. Demling and J. D. Gates, Medical aspects of trauma and burn care, in L. Goldman and D. Ausiello, eds., *Cecil Medicine* (Philadelphia: Saunders, 2008), pp. 790–797; Beers and coeditors, 2006, pp. 2592–2597.

21. Beers and coeditors, 2006, pp. 2592–2597.

22. American Dietetic Association, 2007.

23. American Dietetic Association, 2007.

24. N. Anthonisen, Chronic obstructive pulmonary disease, in L. Goldman and D. Ausiello, eds., *Cecil Medicine* (Philadelphia: Saunders, 2008), pp. 619–627; Beers and coeditors, pp. 400–422.

25. Anthonisen, 2008.

26. T. J. Prendergast and S. J. Ruoss, Pulmonary disease, in S. J. McPhee and W. F. Ganong, eds., *Pathophysiology of Disease: An Introduction to Clinical Medicine* (New York: McGraw-Hill/Lange, 2006), pp. 218–258.

27. Beers and coeditors, 2006, pp. 400–422.

28. B. Suckling, M. M. Johnson, and R. Chin, Jr., Nutrition, respiratory function, and disease, in M. E. Shils and coeditors, *Modern Nutrition in Health and Disease* (Baltimore: Lippincott Williams & Wilkins, 2006), pp. 1462–1474.

29. Beers and coeditors, 2006, pp. 400–422; M. S. McCarthy, Pulmonary failure, in G. Cresci, ed., *Nutrition Support for the Critically Ill Patient: A Guide to Practice* (Boca Raton, Fla.: Taylor & Francis Group, 2005), pp. 481–490.

30. McCarthy, 2005.

31. A. M. Malone, Enteral formula selection: A review of selected product categories, *Practical Gastroenterology* 29 (June 2005): 44–74.

32. M. A. P. Vermeeren and coauthors, Acute effects of different nutritional supplements on symptoms and functional capacity in patients with chronic obstructive pulmonary disease, *American Journal of Clinical Nutrition* 73 (2001): 295–301.

33. McCarthy, 2005; Malone, 2005.

34. C. F. Donner and A. Patessio, Exercise in stable COPD, in T. Similowski, W. A. Whitelaw, and J.-P. Derenne, eds., *Clinical Management of Chronic Obstructive Pulmonary Disease* (New York: Marcel Dekker, 2002), pp. 731–758; R. M. Senior, Chronic obstructive pulmonary disease: Epidemiology, pathophysiology, pathogenesis, clinical course, management, and rehabilitation, in A. P. Fishman and coeditors, *Fishman's Manual of Pulmonary Diseases and Disorders* (New York: McGraw-Hill, 2002), pp. 118–141.

35. M. C. Steiner and coauthors, Nutritional enhancement of exercise performance in chronic obstructive pulmonary disease: A randomised controlled trial, *Thorax* 58 (2003): 745–751.

36. L. D. Hudson and A. S. Slutsky, Acute respiratory failure, in L. Goldman and D. Ausiello, eds., *Cecil Medicine* (Philadelphia: Saunders, 2008), pp. 723–734.

37. L. M. Bellini, Nutrition in acute respiratory failure, in A. P. Fishman and coeditors, *Fishman's Manual of Pulmonary Diseases and Disorders* (New York: McGraw-Hill, 2002), pp. 1082–1089.

38. American Dietetic Association, *Manual of Clinical Dietetics* (Chicago: American Dietetic Association, 2000), pp. 580–581.

39. N. Barak, E. Wall-Alonso, and M. D. Sitrin, Evaluation of stress factors and body weight adjustments currently used to estimate energy expenditure in hospitalized patients, *Journal of Parenteral and Enteral Nutrition* 26 (2002): 231–238.

40. Suckling, Johnson, and Chin, Jr., 2006; McCarthy, 2005.

ANSWERS

Study Questions (multiple choice)

1. d 2. a 3. b 4. c 5. d 6. b 7. c 8. c 9. d 10. a

Multiple Organ Dysfunction Syndrome

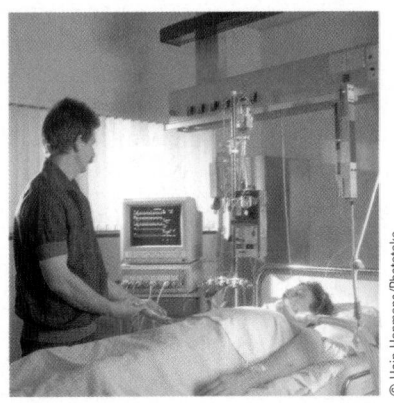

© Hein Hopmans/Phototake

Multiple organ dysfunction syndrome (MODS), also called *multiple organ failure,* is a frequent cause of death in intensive care patients. Described as the progressive dysfunction of two or more of the body's organ systems, MODS most often involves the lungs, liver, kidneys, and gastrointestinal (GI) tract. MODS is not a disease per se, but rather a late stage of severe illness or injury that results from a severe inflammatory response (discussed in Chapter 22).[1] MODS can be initiated by a number of very different critical illnesses and conditions, including acute respiratory failure, trauma, sepsis, burn injuries, extensive surgery, and pancreatitis. This highlight discusses how MODS develops, the manner in which it is treated, and the importance of its prevention.

Multiple organ dysfunction syndrome was recognized as a clinical entity only after World War II. Prior to the mid-20th century, patients with severe illnesses or multiple injuries frequently died of shock or circulatory failure. After fluid replacement and blood transfusions became standard treatments, the kidneys became the organs at highest risk, and kidney failure became the most common cause of death. Eventually, physicians learned to better support kidney function by providing appropriate electrolyte solutions and improving urine output. With improved kidney care, the lungs became the most vulnerable organ after severe injury. Improved treatment of respiratory failure eventually led to the current situation: advances in critical care allow patients to survive severe illnesses and injuries, but the body's defenses often overburden organs that were not originally injured.

Development of MODS

As discussed in Chapter 22, injury and infection cause the release of chemical mediators that have systemic (whole-body) effects. A severe, persistent inflammatory response can lead to systemic inflammatory response syndrome (SIRS), which is associated with a constellation of symptoms including fever, raised heart and respiratory rates, and abnormal white blood cell counts. SIRS is a normal adaptive response to a severe insult, but if not reversed quickly enough it can progress to shock, which is characterized by extremely low blood pressure and an inadequate blood supply for the tissues and organs of the body.[2]

As might be expected from a systemic reduction in blood availability, shock can impair numerous organ systems. The abnormal delivery of oxygen and nutrients to tissues and insufficient removal of wastes result in irreversible injury to cells and tissues. Although each organ system is affected differently, ultimately one or more organs may begin to fail. The failure of one organ may place excessive demands on another, causing the second to fail as well. The progression of SIRS to MODS reflects the inability of the body's defenses and medical treatments to counter the detrimental effects of a sustained and potent inflammatory response.

The sequence of organ dysfunction often follows a similar pattern among patients: first the lungs fail, then the liver, and finally the kidneys, GI tract, or heart.[3] Other organs or systems may also become involved, and each additional failure reduces the likelihood of survival. Table H22-1 lists the organs and systems most often involved in MODS and the potential consequences of their failure.

Factors That Influence Organ Dysfunction

The specific pathophysiology of MODS is poorly understood. Although early reports attempted to link the development of MODS directly to sepsis, sepsis is not present in all cases. Infection often results from impaired immune function and therefore is a frequent consequence of MODS, but it is not necessarily the underlying trigger of organ dysfunction. Recall from Chapter 22 that sepsis gives rise to symptoms identical to those seen in SIRS. Figure H22-1 on p. 728 illustrates the relationships among SIRS, infection, sepsis, and MODS.

TABLE H22-1	Physiological Effects of Organ or System Failure
Organ or System	**Effects of Failure**
Lungs	Inability to maintain gas exchange
Liver	Altered metabolic processes
Kidneys	Inability to regulate blood volume, maintain electrolytes, remove wastes
Heart	Low cardiac output, low blood pressure, inadequate circulation, shock
GI tract	Impaired digestion and absorption, abnormal bleeding, bacterial translocation
Immune system	Infection, sepsis
Coagulation system	Excessive bleeding or coagulation
Central nervous system	Decreased perceptions, brain injury, coma

FIGURE H22-1 Relationships among SIRS, Sepsis, and Multiple Organ Dysfunction Syndrome

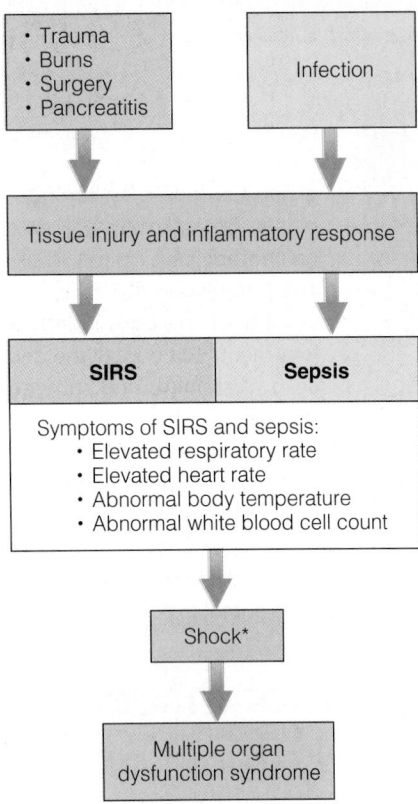

*After critical injury, shock may sometimes precede and be the cause of SIRS.

Finding the exact cause of MODS is difficult because its clinical course differs greatly among patient populations. Epidemiological studies have, however, identified a number of factors that increase risk. For example, people who develop MODS are often older, have multiple or severe injuries, and develop severe infections. Table H22-2 lists the major risk factors associated with MODS, some of which are discussed in the following sections.

TABLE H22-2 Factors That Influence Risk of Multiple Organ Dysfunction Syndrome

Age over 55 years

Prior chronic disease

Persistent SIRS

Major infection

Blood transfusions

Severity of tissue injury

Length of time between injury and arrival at hospital

Malnutrition

Age

Patients over 55 years old are several times more likely to develop MODS than are younger patients. In elderly patients, the increased risk may be due to the presence of chronic illnesses that directly affect organ function, such as heart disease, lung disease, diabetes, or liver damage. Aging also decreases the functional reserve of organs, thereby reducing an older patient's ability to deal with the additional stress that arises during critical illness.

Severity of SIRS

The length of time that SIRS persists is related to the development of MODS. In one study, patients who had SIRS that persisted for more than three days were more likely to develop MODS than patients who had SIRS for less than two days.[4]

Infection

Prolonged SIRS can suppress immune function and increase the risk of developing an infection. During hospital stays, critically ill patients often contract pneumonia—the principal infection associated with MODS. The risks of infection and sepsis greatly increase with the use of invasive catheters, which are frequently needed during intensive care to provide oxygen support, intravenous fluid resuscitation, nutrition support, and urine clearance.

Blood Transfusions

Blood transfusions are immunosuppressive and may increase a patient's risks of developing infection or sepsis. Blood transfusions frequently have adverse effects that can add further stress; they may cause acute lung injury, allergic reactions, red blood cell hemolysis (breakdown), and other complications.

Treatment for MODS

Once MODS has developed, extensive medical support is needed until the inflammatory response has abated. Unfortunately, aggressive treatments can have damaging effects of their own and may cause further injury to organs that are already weakened by illness. Health practitioners must be aware of the adverse effects of aggressive therapies and remain alert to a patient's responses to treatments. Therapies that are often used to manage MODS include:

- *Lung support.* Mechanical ventilation is used to assist injured lungs and sustain gas exchange.
- *Fluid resuscitation.* Fluids and electrolytes are supplied to restore blood volume and maintain electrolyte balance.
- *Support of heart and blood vessel function.* Medications help to sustain or increase cardiac output and maintain adequate blood pressure.
- *Kidney support.* Hemofiltration or dialysis helps to prevent the buildup of toxic metabolites in blood.

- *Protection against infection.* Antibiotic therapy may reverse or prevent infections.
- *Nutrition support.* Enteral and parenteral nutrition support provides nutrients, helps to prevent excessive wasting, and promotes recovery.

Because mortality rates for MODS are so high, prevention must be considered at the earliest stages of injury and treatment, before an excessive inflammatory response can cause further damage. Health practitioners have learned to identify the conditions that can increase organ stress whether they are due to a disease process, an inflammatory response, or an aggressive treatment that is intended to provide organ support. Although improvements in care over the past few decades have reduced some of the complications that arise during intensive care, rates of mortality from MODS have not changed. Thus a focus on prevention is critical until a better understanding of the pathophysiology of MODS is achieved, which may lead to additional therapeutic options.

REFERENCES

1. J. Parrillo, Approach to the patient with shock, in L. Goldman and D. Ausiello, eds., *Cecil Medicine* (Philadelphia: Saunders, 2008), pp. 742–750; D. Johnson and I. Mayers, Multiple organ dysfunction syndrome: A narrative review, *Canadian Journal of Anesthesia* 48 (2001): 502–509.

2. J. A. Russell, Shock syndromes related to sepsis, in L. Goldman and D. Ausiello, eds., *Cecil Medicine* (Philadelphia: Saunders, 2008), pp. 755–763.

3. P. J. Offner and E. E. Moore, Risk factors for MOF and pattern of organ failure following severe trauma, in A. E. Baue, E. Faist, and D. E. Fry, eds., *Multiple Organ Failure: Pathophysiology, Prevention, and Therapy* (New York: Springer-Verlag, 2000), pp. 30–43.

4. Offner and Moore, 2000.

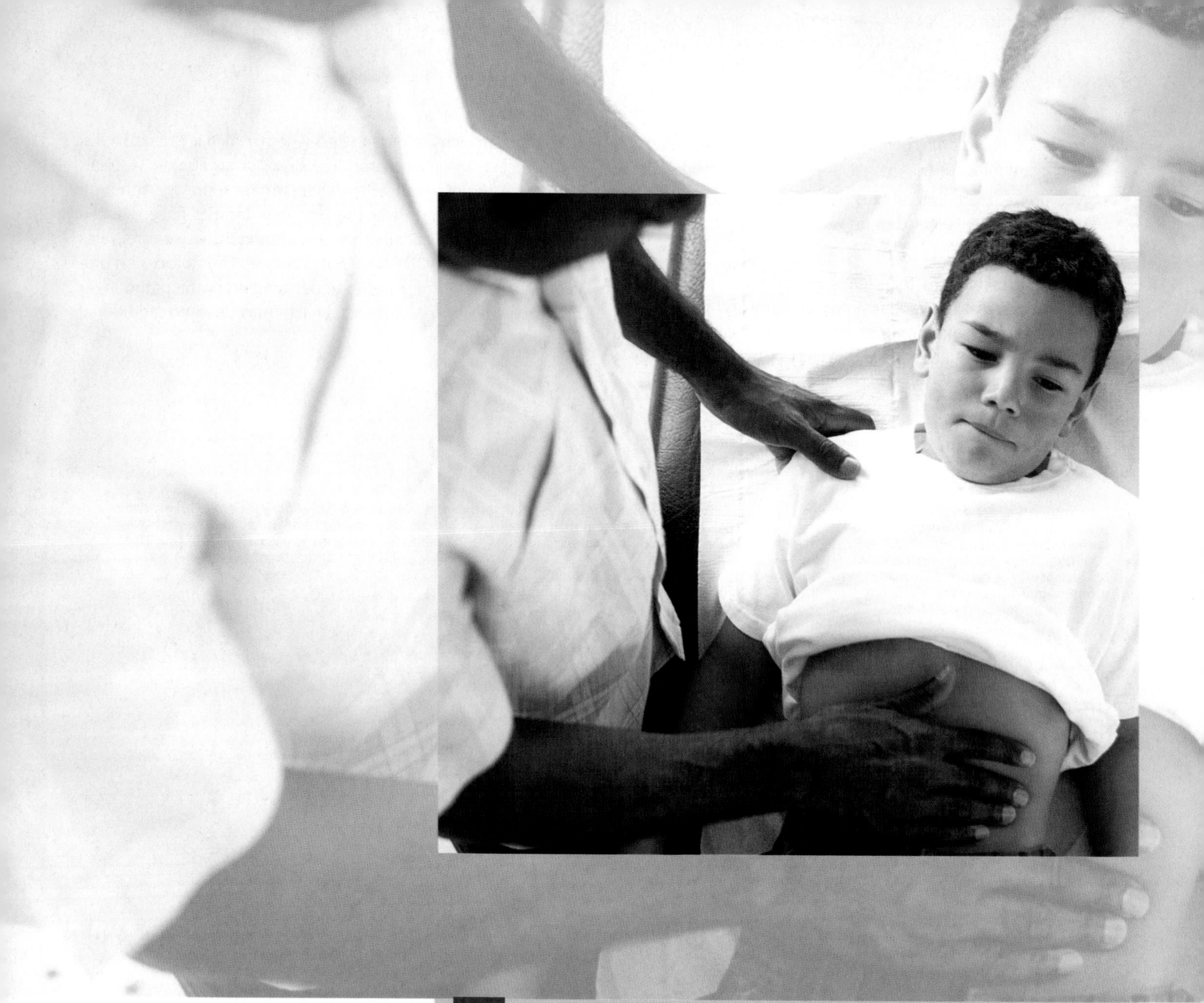

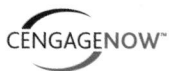

Ian Hooton/Photo Researchers, Inc.

Nutrition in the Clinical Setting

Gastrointestinal illnesses account for a significant fraction of hospital admissions and visits to health practitioners each year. Diagnosis is not always straightforward, however, as many patients with gastrointestinal complaints exhibit no physical abnormalities. Evaluation therefore requires a detailed review of a patient's symptoms and responses to dietary adjustments. Because gastrointestinal complications frequently accompany other illnesses, the medical history can sometimes uncover the underlying source of distress.

CENGAGENOW™ The CengageNOW logo indicates an opportunity for online self-study, linking you to interactive tutorials and videos based on your level of understanding.

academic.cengage.com/login

Upper Gastrointestinal Disorders

The remarkable gastrointestinal (GI) tract provides a means of delivering nutrients to the body's interior. When various medical conditions impair some of the GI tract's functions, dietary adjustments can help to ease symptoms and prevent malnutrition. This chapter discusses common upper GI tract symptoms and disorders; the next chapter describes conditions that affect the lower GI tract. Highlight 23 presents several mouth and dental problems and their associations with chronic disease.

Figure 23-1 on p. 732 illustrates the upper GI tract and reviews its functions. ◆ In the mouth, the teeth and jaw muscles work together to break down food to a consistency that is easily swallowed. Upon swallowing, a bolus of food passes through the pharynx to the esophagus, where peristaltic contractions move the bolus toward the stomach. The lower esophageal sphincter relaxes to allow the bolus to enter the stomach and then closes to prevent reflux (backward flow) of stomach contents.

◆ See Chapter 3 for a complete review of the GI tract and its functions; common digestive problems and simple self-help measures were introduced in Highlight 3.

Conditions Affecting the Esophagus

Disorders of the esophagus may be accompanied by difficulty with swallowing, a sensation of something "stuck in the throat," or pain in the chest area. This section examines the causes and treatments of the two most common problems affecting the esophagus. Dysphagia (difficulty swallowing), introduced in Chapter 18, is discussed in more detail here. Gastroesophageal reflux disease, often referred to as "heartburn," was introduced and defined in Highlight 3.

Dysphagia

The act of swallowing involves multiple processes. In the initial, or **oropharyngeal**, phase of swallowing, muscles in the mouth and tongue propel the bolus of

oropharyngeal (OR-oh-fah-ren-JEE-al): involving the mouth and pharynx.

FIGURE 23-1 The Upper GI Tract

Mouth
Chews and mixes food with saliva.

Pharynx
Directs food from mouth to esophagus.

Epiglottis
Protects airway during swallowing.

Upper esophageal sphincter
Allows passage from mouth to esophagus. Prevents backflow from esophagus.

Esophagus
Conducts food to stomach.

Lower esophageal sphincter
Allows passage from esophagus to stomach. Prevents backflow from stomach.

Stomach
Adds acid, enzymes, and fluid. Churns, mixes, and grinds food to a liquid mass.

Pyloric sphincter
Allows passage from stomach to small intestine. Prevents backflow from small intestine.

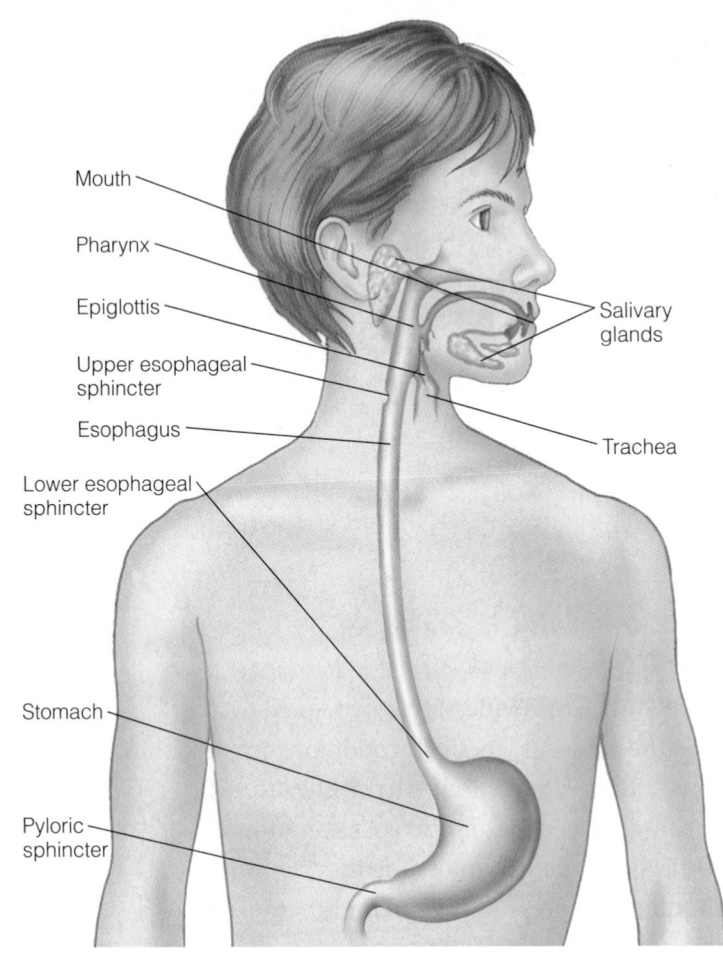

Mouth
Pharynx
Epiglottis
Upper esophageal sphincter
Esophagus
Lower esophageal sphincter
Stomach
Pyloric sphincter
Salivary glands
Trachea

Salivary glands
Secrete saliva (provides moisture and contains starch-digesting enzymes).

Trachea
Allows air to pass to and from lungs.

food through the pharynx and into the esophagus. At the same time, tissues of the soft palate prevent food from entering the nasal passages, and the epiglottis blocks the opening to the trachea to prevent aspiration of food substances or saliva into the lungs. In the second, or **esophageal**, phase of swallowing, peristalsis forces the bolus through the esophagus, and the lower esophageal sphincter relaxes to allow passage of the bolus into the stomach. Due to the many tasks involved in swallowing, dysphagia can result from a number of different physical or neurological conditions. Table 23-1 lists some potential causes of dysphagia, which are categorized according to the phase of swallowing that is impaired.[1]

Oropharyngeal Dysphagia A person with **oropharyngeal dysphagia** typically has a neuromuscular condition that upsets the swallowing reflex or impairs the necessary movements of the tongue and other oral tissues. Symptoms include an inability to initiate swallowing, coughing during or after swallowing (due to aspiration), and nasal regurgitation. Other signs include bad breath, a gurgling noise after swallowing, a hoarse or "wet" voice, or a speech disorder. Oropharyngeal dysphagia occurs frequently in elderly people and is often caused by stroke.[2]

Esophageal Dysphagia A person with **esophageal dysphagia** usually has an obstruction in the esophagus or a motility disorder; the main symptom is often the sensation of food "sticking" in the esophagus after it is swallowed. An obstruction can be caused by a **stricture** (abnormal narrowing), tumor, or compression of the esoph-

esophageal (eh-SOF-ah-JEE-al): involving the esophagus.

oropharyngeal dysphagia: an inability to transfer food from the mouth and pharynx to the esophagus; usually caused by a neurological or muscular disorder.

esophageal dysphagia: an inability to move food through the esophagus; usually caused by an obstruction or a motility disorder.

stricture: abnormal narrowing of a passageway; often due to inflammation, scarring, or a congenital abnormality.

agus by surrounding tissues. Whereas an obstruction may affect the passage of solid foods only and may not affect liquids, a motility disorder hinders the passage of both solids and liquids. **Achalasia,** the most common motility disorder, is a degenerative nerve condition affecting the esophagus; it is characterized by impaired peristalsis and incomplete relaxation of the lower esophageal sphincter when swallowing.[3]

Complications of Dysphagia Health practitioners should be alert to the various complications that can indicate a dysphagia problem. If food consumption is reduced due to dysphagia, malnutrition and weight loss may occur. Individuals who cannot swallow liquids are at increased risk of dehydration. A serious and potentially life-threatening complication associated with dysphagia is aspiration, which may cause airway obstruction, choking, or respiratory infections, including pneumonia. If a person does not have a normal cough reflex, aspiration is more difficult to diagnose and may go unnoticed.

Evaluation of Dysphagia Although the signs and symptoms of dysphagia can help a health care provider recognize the condition, diagnosing the exact cause generally requires further examination. A barium swallow study often reveals the nature of the problem. In this procedure, the patient consumes foods or liquids that contain barium (a metallic element visible on X-rays), and the swallowing process is monitored using a video X-ray technique known as videofluoroscopy. Another assessment method, endoscopy, uses a thin, flexible tube to examine the esophageal lumen directly. Peristalsis and sphincter pressure can be measured using a manometer, a flexible catheter containing multiple pressure sensors that is passed into the esophagus. A neurological examination may be needed to evaluate mental status, physical reflexes, and the cranial nerves associated with swallowing.

Nutrition Intervention for Dysphagia Modifying the physical properties of foods and beverages and using alternative feeding methods can help to compensate for swallowing difficulties. Because a wide variety of defects can cause dysphagia, finding the best diet is often a challenge. Even after careful assessment of a person's swallowing abilities, the most appropriate foods may be determined only by trial and error. A person's swallowing abilities may fluctuate over time, so the dietary plan needs frequent reassessment.

The National Dysphagia Diet, developed in 2002 by a panel of dietitians, speech and language therapists, and a food scientist, has helped to standardize the nutrition care of dysphagia patients.[4] Table 23-2 on p. 734 presents brief descriptions of the different levels of the diet and some sample meals.[5] After the appropriate dietary level is selected, it must be adjusted to suit the person's swallowing abilities and tolerances. A consultation with a swallowing expert, such as a speech and language therapist, is often necessary.

Food Properties and Preparation Foods included in dysphagia diets should have easy-to-manage textures and consistencies. Soft, cohesive foods are easier to handle than hard or crumbly foods. Moist foods are better tolerated than dry foods. Some foods within a category may be acceptable and others may not; for example, some cookies are soft and tender, whereas others are hard and brittle. Sticky or gummy foods, such as peanut butter and cream cheese, may be difficult to clear from the mouth and throat.

The textures of foods can be altered to make them easier to swallow. Foods are often pureed, mashed, ground, or minced (review Table 18-5 on p. 623). Foods that have more than one texture, such as vegetable soup or cereal with milk, are harder to handle, so ingredients may be blended to a single consistency and items such as nuts and seeds omitted.

Consuming foods that have a similar consistency can quickly become monotonous. By using commercial thickeners and food molds, pureed foods can be formed into attractive shapes. Including a variety of flavors and colors can also make a meal more appealing. The "How to" on p. 735 offers additional suggestions that can improve the acceptance of pureed and other mechanically altered foods.

TABLE 23-1	Causes of Dysphagia

Oropharyngeal Dysphagia

- Alzheimer's disease (advanced stages)
- Developmental disabilities
- Goiter
- Lou Gehrig's disease (amyotrophic lateral sclerosis)
- Multiple sclerosis
- Muscular dystrophy
- Myasthenia gravis
- Parkinson's disease
- Poliomyelitis
- Stroke

Esophageal Dysphagia

- Achalasia
- Enlarged atrium (right side of heart)
- Esophageal cancer
- Esophageal spasm
- Scleroderma
- Strictures (from inflammation, scarring, or a congenital abnormality)
- Thoracic tumor (usually lung cancer)

Courtesy Diamond Crystal Specialty Foods

Can you tell that the foods in this photo are pureed foods shaped with food molds?

achalasia (ack-ah-LAY-zhah): an esophageal disorder characterized by weakened peristalsis and impaired relaxation of the lower esophageal sphincter.
- **a** = without
- **chalasia** = relaxation

TABLE 23-2 National Dysphagia Diet

Level 1: Dysphagia Pureed

Foods should be pureed or well mashed, homogeneous, and cohesive. This diet is for patients with moderate to severe dysphagia and poor oral or chewing ability.

Sample menus:
- *Breakfast:* Cream of wheat, slurried muffins or pancakes,[a] pureed scrambled eggs, mashed bananas, fruit juice without pulp (thickened as needed), coffee or tea (if thin liquids are acceptable).
- *Lunch or dinner:* Pureed tomato soup, slurried crackers, pureed meat or poultry, broccoli soufflé, mashed potatoes with gravy, pureed carrots or green beans, smooth applesauce, pureed peaches, chocolate pudding.

Foods to avoid: Cheese (including cottage cheese), oatmeal, rice, peanut butter, fruit preserves with chunks or seeds, chunky applesauce, beverages with pulp, coarsely ground pepper, herbs.

Level 2: Dysphagia Mechanically Altered

Foods should be moist, cohesive, and soft textured and should easily form a bolus. This diet is for patients with mild to moderate dysphagia; some chewing ability is required.

Sample menus:
- *Breakfast:* Moist oatmeal, cornflakes or puffed rice cereal with milk (thickened as needed), moist pancakes or muffins (with butter, margarine, or jam), soft scrambled eggs, cottage cheese, ripe bananas or cooked fruit without skin or seeds, fruit juice (thickened as needed), coffee or tea (if thin liquids are allowed).
- *Lunch or dinner:* Soup with easy-to-chew meat and vegetables; slurried bread or crackers; minced, tender-cooked meat; well-cooked pasta with moist meatballs and meat sauce; baked potato with gravy; soft, tender-cooked vegetables (not fibrous or rubbery); canned peach slices; soft fruit pie (with bottom crust only); soft, smooth chocolate bar.

Foods to avoid: Dry foods, frankfurters, sausage, hard-cooked eggs, corn and clam chowders; sandwiches, pizza, sliced cheese, rice, potato skins, French fries, corn, broccoli, asparagus, brussels sprouts, cabbage, peanut butter, coconut, nuts and seeds, fruit with skin or seeds, canned pineapple, dried fruit, chewy candies (such as caramel or licorice).

Level 3: Dysphagia Advanced

Foods should be moist and be in bite-sized pieces when swallowed; foods with mixed textures are included. This diet is for patients with mild dysphagia and adequate chewing ability.

Sample menus:
- *Breakfast:* Cereal with milk, moist pancakes or muffins (with butter, margarine, or jam), poached or scrambled eggs, cheese, soft fresh fruit (peeled) or berries, coffee or tea (if thin liquids are tolerated).
- *Lunch or dinner:* Chicken noodle soup; moistened crackers or bread; thin-sliced tender meat; moist, soft-cooked potatoes or rice; tender-cooked vegetables; shredded lettuce with dressing; fresh peach or melon; canned fruit salad; moist chocolate chip cookie (without nuts).

Foods to avoid: Dry foods; corn and clam chowders; potato skins; corn; raw vegetables; chunky peanut butter; coconut; nuts and seeds; hard fruit (such as apples or pears); fruit with skin, seeds, or stringy textures (such as mango or pineapple); uncooked dried fruit; fruit leathers; popcorn; chewy candies (such as caramel or licorice).

Liquid Consistencies (only those tolerated are allowed in the diet)

- *Thin:* Watery fluids; may include milk, coffee, tea, juices, carbonated beverages.
- *Nectarlike:* Fluids thicker than water that can be sipped through a straw; may include buttermilk, eggnog, tomato juice.
- *Honeylike:* Fluids that can be eaten with a spoon but do not hold their shape; may include honey, tomato sauce, yogurt.
- *Spoon-thick:* Thick fluids that must be eaten with a spoon and can hold their shape; may include milk pudding, thickened applesauce.

---

[a] Slurried foods are mixed with liquid until the consistency is appropriate; they may be gelled and shaped to improve appearance.
SOURCE: American Dietetic Association, *Nutrition Care Manual* (Chicago: American Dietetic Association, 2007).

Properties of Liquids Thickened liquids are easier to swallow than thin liquids such as water or juice. Table 23-2 describes the four levels of liquid consistencies prescribed for dysphagia patients, referred to as thin, nectarlike, honeylike, and spoon-thick. To increase liquid viscosity, commercial starch thickeners are stirred into beverages and other liquid foods, such as soup broths. Some beverages can lose their appeal when thickened; for example, individuals may find thickened coffee and tea unacceptable. Furthermore, hydration is more difficult to maintain when a patient has access to only thickened beverages, which are less acceptable for quenching thirst.

Feeding Strategies for Dysphagia Depending on the nature of the swallowing problem, some patients can learn new feeding techniques to help them compensate for their disability. For example, people with oropharyngeal dysphagia can do exercises that strengthen the jaws, tongue, or larynx or learn new methods of swallowing that allow them to consume a normal diet. Changing the position of the head and neck while eating can also minimize some swallowing problems. Speech and language therapists are often responsible for teaching patients these techniques.

Gastroesophageal Reflux Disease

◆ Highlight 3 introduced gastroesophageal reflux disease and discussed strategies for preventing its recurrence.

Gastroesophageal reflux disease (GERD) ◆ is a condition of gastric reflux that causes frequent discomfort and, sometimes, tissue damage. Reflux of the stomach's acidic contents can irritate the esophagus, and small amounts may enter the mouth. Peo-

---

HOW TO	Improve Acceptance of Mechanically Altered Foods

Take a moment to think about a meal of pureed or ground foods. A typical dinner of baked chicken, potatoes, carrots, and green beans can look like mounds of differently colored mush. The foods may taste great, but a person may have little appetite before trying a first bite. To improve appetite, be creative when preparing and serving meals:

- Help to stimulate the appetite by preparing favorite foods and foods with pleasant smells. Enliven food flavors with seasonings and spices.
- Use attractive plates and silverware to improve the visual appeal of a meal. Consider colors and shapes when arranging foods on a plate; colorful garnishes can add color and eye appeal. Substitute brightly colored vegetables for white vegetables; for example, replace mashed potatoes with mashed sweet potatoes.
- Try layering ingredients so that the entrée looks like a fancy casserole. For example, recipes can resemble such popular entrées as lasagna, shepherd's pie, and moussaka.
- Shape pureed and ground foods to resemble traditional dishes. Flatten a spoonful of pureed meat to make a patty, or use small scoops so that meat resembles meatballs. Use food molds to restore slurried breads and pureed meats to their traditional shapes.

Efforts to improve the visual appearance of foods can go a long way toward helping people eat nourishing meals and maintain a healthy weight.

---

ple who suffer from GERD often refer to these symptoms as *heartburn* or *acid indigestion*. Reflux does not necessarily cause symptoms or injury—it occurs occasionally in healthy people and is a problem only if it creates complications and requires lifestyle changes or medical treatment.

Causes of GERD The lower esophageal sphincter is the main barrier to gastric reflux, so GERD can result if the sphincter muscle is weak or relaxes inappropriately. Medical conditions that interfere with the sphincter's mechanism or prevent rapid clearance of acid from the esophagus can also predispose a person to GERD.

Conditions associated with high rates of GERD include pregnancy, asthma, and **hiatal hernia,** a condition in which a portion of the stomach protrudes above the diaphragm (see Figure 23-2 on p. 736). Pregnancy is the most common predisposing condition; as many as two-thirds of pregnant women report heartburn, which often begins in the first trimester.[6] Some medications may increase the risk of reflux, as does the use of nasogastric tubes in tube feedings. Various other conditions and substances can exacerbate GERD by either weakening the sphincter or raising pressure within the stomach; Table 23-3 on p. 736 lists examples.

Consequences of GERD If gastric acid remains in the esophagus long enough to damage the esophageal lining, the resulting inflammation is called **reflux esophagitis.** Severe and chronic inflammation may lead to esophageal ulcers, with consequent bleeding. Healing and scarring of ulcerated tissue may narrow the inner diameter of the esophagus, causing esophageal stricture. A slowly progressive dysphagia for solid foods sometimes results, and swallowing occasionally becomes painful. Pulmonary disease may develop if gastric contents are aspirated into the lungs. Chronic reflux is also associated with **Barrett's esophagus,** a condition in which damaged esophageal cells are gradually replaced by cells that resemble those in gastric or intestinal tissue; such cellular changes increase the risk of developing esophageal cancer. GERD can also damage tissues in the mouth, pharynx, and larynx, resulting in eroded tooth enamel, sore throat, and laryngitis.[7]

Treatment of GERD Treatment objectives are to alleviate symptoms and facilitate the healing of damaged tissue. Severe ulcerative disease may require immediate acid-suppressing medication, whereas a mild case may be managed with dietary and lifestyle modifications. The "How to" on p. 737 offers suggestions that may help to prevent the recurrence of gastrointestinal reflux.

Medications that suppress gastric acid secretion help the healing process by reducing the damaging effects of acid on esophageal tissue. **Proton-pump inhibitors** are the most effective of the antisecretory agents and are used both for

hiatal hernia: a condition in which the upper portion of the stomach protrudes above the diaphragm; most cases are asymptomatic.

reflux esophagitis: inflammation in the esophagus related to the reflux of acidic stomach contents.

Barrett's esophagus: a condition in which esophageal cells damaged by chronic exposure to stomach acid are replaced by cells that resemble those in the stomach or small intestine, sometimes becoming cancerous.

proton-pump inhibitors: a class of drugs that inhibit the enzyme that pumps hydrogen ions (protons) into the stomach. Examples include omeprazole (Prilosec) and lansoprazole (Prevacid).

FIGURE 23-2 The Upper GI Tract, Acid Reflux, and Hiatal Hernia

Normal

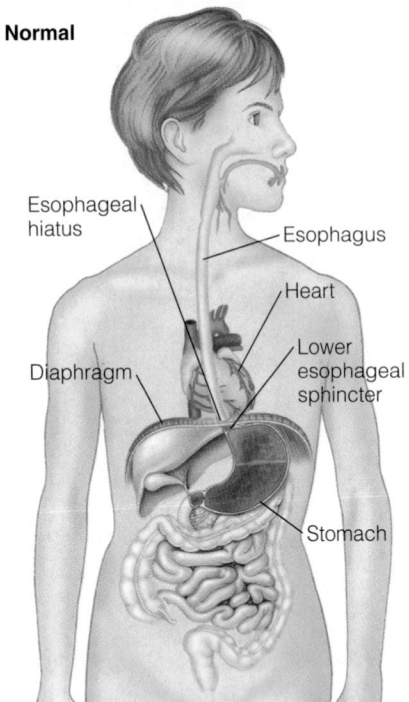

Esophageal hiatus

Esophagus

Heart

Lower esophageal sphincter

Diaphragm

Stomach

The stomach normally lies below the diaphragm, and the esophagus passes through the esophageal hiatus. The lower esophageal sphincter prevents reflux of stomach contents.

Acid reflux

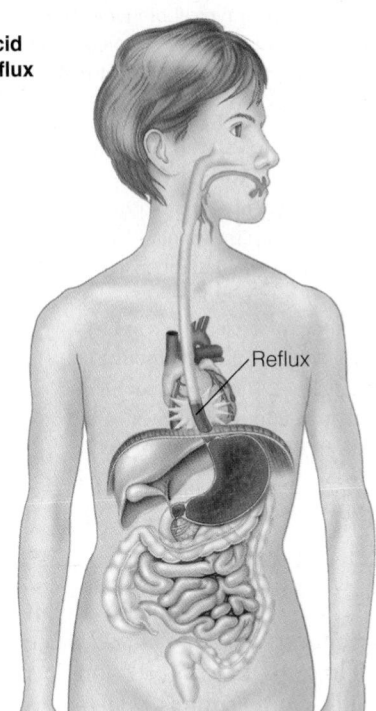

Reflux

Whenever the pressure in the stomach exceeds the pressure in the esophagus, as can occur with overeating and overdrinking, the chance of reflux increases. The resulting "heartburn" is so named because it is felt in the area of the heart.

Hiatal hernia

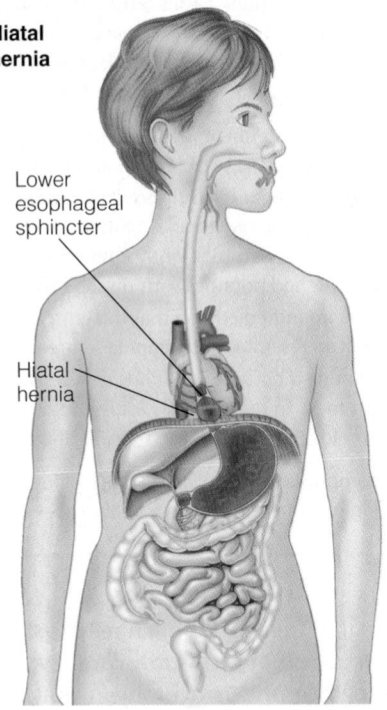

Lower esophageal sphincter

Hiatal hernia

Risk of acid reflux may increase as a consequence of a hiatal hernia. A "sliding" hiatal hernia occurs when part of the stomach, along with the lower esophageal sphincter, rises above the diaphragm.

rapid healing of esophagitis and as a maintenance treatment. Other drugs used for GERD include **histamine-2 receptor blockers** (often referred to as *H2 blockers*) and antacids, which neutralize gastric acid. Although antacids are frequently used to relieve occasional heartburn, they are not necessarily appropriate for GERD be-

TABLE 23-3 Conditions and Substances Associated with Esophageal Reflux

Conditions That Raise the Likelihood of Reflux	Substances That Weaken Lower Esophageal Sphincter Pressure
Ascites (accumulation of fluid in the abdomen)	Alcohol
Delayed gastric emptying	Anticholinergic agents
Eating large meals	Antihistamines
Lying flat after eating	Caffeine
Obesity	Calcium channel blockers
Pregnancy	Chocolate
Wearing clothes that fit tightly across the waist or abdomen	Cigarette smoking
	Diazepam
	Garlic
	High-fat foods
	Onions
	Peppermint and spearmint oils
	Progesterone
	Theophylline
	Tricyclic antidepressants

histamine-2 receptor blockers: a class of drugs that suppress acid secretion by inhibiting receptors on acid-producing cells; commonly called **H2 blockers.** Examples include cimetidine (Tagamet), ranitidine (Zantac), and famotidine (Pepcid).

HOW TO Manage Gastrointestinal Reflux Disease

Management of GERD often requires lifestyle changes to help minimize discomfort and reduce recurrence of acid reflux. Recommendations generally include the following:

- Avoid eating bedtime snacks or lying down after meals. Meals should be consumed at least two to three hours before bedtime.
- Reduce nighttime reflux by elevating the head of the bed on 6-inch blocks, inserting a foam wedge under the mattress, or propping pillows under the head and upper torso.
- Consume only small meals, and drink liquids between meals so that the stomach does not become overly distended, which can exert pressure on the lower esophageal sphincter.

- Limit foods that weaken lower esophageal sphincter pressure or increase gastric acid secretion; these include chocolate, fatty foods, spearmint and peppermint, coffee (both caffeinated and decaffeinated), and tea.
- Avoid cigarettes and alcohol; both relax the lower esophageal sphincter.
- Avoid bending over and wearing tight-fitting garments; both can cause pressure in the stomach to increase, heightening the risk of reflux.
- Advise obese individuals to lose weight if they are adequately motivated. Obesity can increase abdominal pressure.
- During periods of esophagitis, avoid foods and beverages that may irritate the

esophagus, such as citrus fruits and juices, tomato products, pepper, spicy foods, carbonated beverages, and very hot or very cold foods (depending on individual tolerances).
- Avoid using nonsteroidal anti-inflammatory drugs (NSAIDs) such as aspirin, naproxen, and ibuprofen, which can damage the esophageal mucosa.

Food tolerances among people with GERD can vary markedly. Health professionals can help patients pinpoint food intolerances by advising them to keep a record of the foods and beverages consumed, as well as any resulting symptoms.

---

cause they have only short-term effects, are associated with gastrointestinal side effects, and may cause some nutrient deficiencies when used over the long term.

Surgery may be required in severe cases of GERD that are unresponsive to medications and lifestyle changes. In one popular procedure (called *fundoplication*), the upper section of the stomach (the fundus) is gathered up around the esophagus and sewn in such a way that the esophagus and sphincter are surrounded by stomach muscle, which increases pressure within the esophagus and fortifies the sphincter muscle. Esophageal strictures are often treated by dilating the esophagus (with an inflatable balloon-like device or a fixed-size dilator) or by using surgical approaches. The Case Study below includes questions that review the usual treatments for a patient with GERD.

CASE STUDY   Accountant with GERD

Gina Rinaldi is a 49-year-old accountant who is 5 feet 4 inches tall and weighs 165 pounds. She recently underwent a complete physical examination. She told her physician that she had been feeling fairly well until she began experiencing heartburn, which has progressively become more frequent and painful. The heartburn often occurs after she eats a large meal and is particularly bad after she goes to bed at night. By directly examining the esophageal lumen using an endoscope (a thin, flexible tube equipped with an optical device), the physician found evidence of reflux esophagitis and a slight narrowing throughout the length of the esophagus.

Mrs. Rinaldi's medical history does not indicate any significant health problems. During her last physical exam, her physician advised her to stop smoking cigarettes and to lose 20 pounds, but she has not attempted to do either. The nutrition assessment reveals that Mrs. Rinaldi is feeling stressed because it is the middle of the tax season. She usually has little time for breakfast, eats a lunch of fast foods while continuing to work at her desk, and eats a large dinner at around 8:00 P.M. She generally has wine with dinner and another alcoholic beverage later in the evening.

1. Explain to Mrs. Rinaldi the meaning of the medical diagnoses *reflux esophagitis* and *esophageal stricture.*
2. From the brief history provided, list the factors and behaviors that increase Mrs. Rinaldi's risks of experiencing reflux. What recommendations can you make to help her change these behaviors?
3. What medications might the physician prescribe, and why?

IN SUMMARY

Dysphagia and gastroesophageal reflux are the most common esophageal disorders. Dysphagia can interfere with food intake and increase the risk of aspiration. Treatment may include dietary adjustments, strengthening exercises, and using different swallowing techniques. Gastroesophageal reflux disease (GERD) may lead to esophageal ulcers, inflammation, bleeding, and stricture. Treatment includes the use of acid-suppressing drugs and lifestyle changes.

Conditions Affecting the Stomach

Stomach disorders range from occasional bouts of discomfort to severe conditions that require surgery. This section begins with a discussion of **dyspepsia** (often called "indigestion"), the sensation of pain or discomfort in the upper abdomen that occurs after food consumption. More serious stomach conditions that may benefit from dietary adjustments include **gastritis** and **peptic ulcers,** which most often result from bacterial infection or from the use of medications that damage the stomach lining.

Dyspepsia

Dyspepsia refers to the general symptoms of indigestion in the upper abdominal region, which may include stomach pain, gnawing sensations, early satiety, nausea, vomiting, and bloating. These symptoms sometimes indicate the presence of more serious illnesses, such as GERD or peptic ulcer disease. Although about 25 percent of the population experiences dyspepsia, only one person in four seeks medical attention.[8]

Causes of Dyspepsia Abdominal symptoms don't always lead to a clear diagnosis, as the cause of the symptoms can be difficult to identify. Various medical conditions can cause abdominal discomfort; they include peptic ulcers, GERD, motility disorders, malabsorptive disorders (discussed in Chapter 24), gallbladder disease, and tumors in the esophagus or stomach. Chronic diseases such as diabetes mellitus, heart disease, and hypothyroidism can sometimes be accompanied by gastric symptoms. Some medications, including aspirin (and other nonsteroidal anti-inflammatory drugs), antibiotics, digitalis, and theophylline, can cause gastrointestinal distress. Dietary supplements may also be a cause; for example, iron and potassium supplements and some herbal products can cause gastrointestinal problems. Intestinal conditions such as irritable bowel syndrome or lactose intolerance may mimic dyspepsia. Although pinpointing the cause of the symptoms can be difficult, a complete examination is in order if the individual experiences weight loss, persistent vomiting, dysphagia, anemia, or bleeding, which suggest the presence of serious illness.[9]

Potential Food Intolerances Although many people attribute their symptoms to eating certain foods or spices, controlled studies have been unable to find associations between specific foods and dyspepsia. Coffee can induce symptoms in about 50 percent of dyspepsia patients, however, and also increases gastric acid production and acid reflux.[10] Spicy foods may cause some injury to the mucosal lining and exacerbate the pain from a preexisting ulcer. High-fat meals can slow gastric emptying and thereby exacerbate dyspepsia. To minimize symptoms, people with dyspepsia are sometimes advised to consume small meals with well-cooked foods that are not overly seasoned and to consume meals in a relaxed atmosphere.[11]

Bloating and Stomach Gas The feeling of bloating may be caused by excessive gas in the stomach, which accumulates when air is swallowed. Swallowing air often

dyspepsia: a feeling of pain, bloating, or discomfort in the upper abdominal area, often called **indigestion;** a symptom of illness rather than a disease itself.

- **dys** = bad; impaired
- **pepsia** = refers to digestion

gastritis: inflammation of stomach tissue.

peptic ulcers: ulcers in the gastrointestinal mucosa resulting from exposure to gastric secretions; may develop in the esophagus, stomach, or duodenum.

- **peptic** = related to digestion

accompanies gum chewing, smoking, rapid eating, drinking carbonated beverages, and using a straw. Omitting these practices generally helps to correct the problem.

Nausea and Vomiting

Nausea and vomiting accompany many illnesses and are common side effects of medications. Although occasional vomiting is not dangerous, prolonged vomiting can cause fluid and electrolyte imbalances and may require medical care. Chronic vomiting can reduce food intake and lead to malnutrition and nutrient deficiencies.

The timing of vomiting gives clues about its cause. Vomiting that occurs within an hour after a meal suggests a peptic ulcer or a psychological cause. If it occurs more than one hour after a meal, possible causes include food poisoning, an obstruction that prevents stomach emptying, or a stomach motility disorder.

Treatment of Nausea and Vomiting The main goal of treatment is to find and correct the underlying disorder. Most cases are short-lived and require no treatment. Restoring hydration may be necessary. If a medication is the cause, taking it with food may help. If the cause is unknown or the underlying disorder cannot be corrected, medications that suppress nausea and vomiting can be prescribed. People with **intractable vomiting**—vomiting that is not easily controlled—may require intravenous nutrition support.

Dietary Interventions Sometimes nausea can be prevented or improved with dietary measures. Eating and drinking slowly may be helpful, as may eating small meals that do not distend the stomach. Drinking clear, cold beverages such as carbonated drinks or fruit juices may ease symptoms. Foods that may help to reduce nausea include dry, salty foods like crackers or pretzels. Fried or spicy foods and foods with strong odors should be avoided. Foods that are cold or at room temperature may be better tolerated than hot meals. Individuals sometimes have strong food aversions when nauseated, and tolerances vary greatly.

Gastritis

Gastritis is a general term that refers to inflammation of the stomach mucosa. ◆ As shown in Table 23-4, gastritis can result from infection, chemical substances, and diseases and treatments that damage the stomach lining. Most often, gastritis results from **Helicobacter pylori** infection or the use of nonsteroidal anti-inflammatory drugs (NSAIDs), both primary causes of peptic ulcer disease as well.

If the gastric mucosa shows signs of tissue destruction, ulcers, or hemorrhaging (severe bleeding), the condition may be called **erosive gastritis,** even if inflammation is not present.[12] Erosive gastritis can be caused by substances or treatments that irritate the gastric mucosa, including alcohol and other chemical substances, viral infection, radiation treatment, and bile reflux. Gastritis that becomes chronic and is associated with tissue destruction is known as *atrophic gastritis;* ◆ it is especially prevalent among adults over 60 years of age.

Complications of Gastritis The extensive tissue damage that sometimes develops in chronic gastritis can disrupt gastric secretory functions. If hydrochloric acid secretions become abnormally low **(hypochlorhydria)** or absent **(achlorhydria),** absorption of nonheme iron and vitamin B_{12} can be impaired, and the risk of deficiency increases. Pernicious anemia, a condition characterized by the destruction of stomach cells that produce intrinsic factor, is a late complication of atrophic gastritis and a primary cause of vitamin B_{12} deficiency (see p. 343).

Dietary Interventions for Gastritis Dietary recommendations depend on an individual's symptoms. If gastritis is asymptomatic, no dietary adjustments are needed. If pain or discomfort is present, the patient should avoid irritating foods and beverages; these usually include alcohol, coffee (including decaffeinated), tea, cola beverages, spicy foods, and fatty or greasy foods. If food consumption increases pain or

---

TABLE 23-4 Potential Causes of Gastritis

Infection

- Bacterial: *Helicobacter pylori, Actinomyces israelii*
- Fungal: *Candida albicans*
- Parasitic: Cryptosporidiosis, nematode infection
- Viral: Cytomegalovirus

Chemical Substances

- Alcohol
- Cocaine
- Drugs (especially aspirin and other NSAIDs)
- Ingestion of corrosive materials

Internal (bodily) Causes

- Autoimmune
- Bile reflux
- Stress
- Systemic illness/sepsis

Miscellaneous

- Food sensitivity (allergy)
- Foreign bodies
- High salt intake
- Radiation therapy

◆ The suffix *-itis* refers to the presence of inflammation in an organ or tissue.

◆ Reminder: Atrophic gastritis was introduced in Chapter 10 (p. 343).

intractable vomiting: vomiting that is not easily managed or controlled.

Helicobacter pylori: a species of bacterium that colonizes gastric mucosa; a primary cause of gastritis and peptic ulcer disease.

erosive gastritis: erosion of the gastric mucosa, characterized by tissue destruction, ulcers, and hemorrhaging; often caused by the toxic effects of chemical substances or radiation treatment.

hypochlorhydria (HIGH-poe-clor-HIGH-dree-ah): a reduction in gastric acid secretion.

achlorhydria (AY-clor-HIGH-dree-ah): absence of gastric acid secretion.

causes nausea and vomiting, food intake should be avoided for 24 to 48 hours to rest the stomach. Nutrition support may be necessary if the patient cannot tolerate food for a prolonged period. If gastritis results in hypochlorhydria or achlorhydria, supplementation of iron and vitamin B_{12} may be warranted.

Peptic Ulcer Disease

A **peptic ulcer** is an ulceration that develops in the gastrointestinal mucosa when gastric acid and pepsin overwhelm mucosal defenses and destroy mucosal tissue. A primary factor in peptic ulcer development is *Helicobacter pylori* infection, which is present in about 80 percent of patients with duodenal ulcers and 60 percent of those with gastric ulcers.[13] ◆ Another major factor is the use of NSAIDs, which have both topical and systemic effects that can damage mucosal tissue. In rare cases, ulcers may develop from disorders that cause excessive acid secretion: one such condition is **Zollinger-Ellison syndrome,** characterized by the presence of gastrin-secreting tumors in the duodenum or pancreas. ◆ Ulcer risk can be increased by cigarette smoking, psychological stress, and genetic factors.[14]

Effects of Emotional Stress Although most ulcers are associated with *H. pylori* infection or NSAID use, about a quarter of ulcers develop in people for other reasons.[15] Emotional stress is not believed to cause ulcers per se, but it has effects on physiological processes and behaviors that may increase a person's vulnerability. Physiological effects of stress vary among individuals but may include rapid stomach emptying (which increases the acid load in the duodenum), hormonal changes that impair wound healing, and increases in acid and pepsin secretions. Stress may also lead to behavioral changes, including increased use of cigarettes, alcohol, and NSAIDs—all potential risk factors for ulcers. Thus, stress may play a contributory role in ulcer development, although its precise effects are not fully understood.

Symptoms of Peptic Ulcers Peptic ulcer symptoms vary. Some people are asymptomatic or experience only mild discomfort. Ulcer "pain" may be experienced as a hunger pain, a sensation of gnawing, or a burning pain in the stomach region. The pain or discomfort of ulcers may be relieved by food and recur several hours after a meal, especially if the ulcer is duodenal. Gastric ulcers may be aggravated by food and can cause loss of appetite and eventual weight loss. ◆ Ulcer symptoms tend to go into remission regularly and recur every few weeks or months.[16]

Complications of Peptic Ulcers Peptic ulcers are a major cause of gastrointestinal bleeding, which occurs in about 15 to 20 percent of ulcer cases.[17] Bleeding is a potential cause of death and, if severe, may indicate the need for surgical intervention. Severe bleeding is evidenced by black, tarry stool samples or, occasionally, vomit that resembles coffee grounds. Other serious complications of ulcers include perforations of the stomach or duodenum (sometimes leading directly into the peritoneal cavity) and gastric outlet obstruction due to scarring or inflammation.

Drug Therapy for Peptic Ulcers The goals of ulcer treatment are to relieve pain, promote healing, and prevent recurrence. In most cases, treatment requires using a combination of antibiotics to eradicate *H. pylori* infection and/or discontinuing the use of aspirin and other NSAIDs, which can irritate the gastric mucosa and delay healing. Antisecretory and acid-neutralizing drugs may be prescribed to relieve pain and allow healing; these include proton-pump inhibitors, H2 blockers, and antacids (as used in GERD; see the earlier discussion on pp. 735–737). Bismuth preparations (such as Pepto-Bismol) or sucralfate may help by coating the gastrointestinal lining and preventing further tissue erosion. The most frequently prescribed drug regimen is a "triple therapy" that includes two antibiotics and one other type of drug; the antibiotics used to treat *H. pylori* infection include amoxicillin, clarithromycin, metronidazole, and tetracycline.[18] See the Diet-Drug Interactions feature for nutrition-related effects of the medications used in ulcer treatment.

◆ The specific reasons why ulcers develop are not known; only 10 to 15 percent of individuals with chronic *Helicobacter pylori* infection actually develop a peptic ulcer.

◆ Reminder: *Gastrin* is a hormone that signals stomach cells to secrete hydrochloric acid.

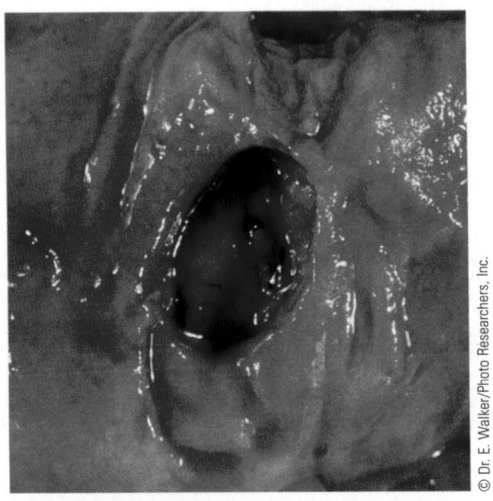

A peptic ulcer, such as the gastric ulcer shown here, damages mucosal tissue and may cause pain and bleeding.

◆ In the United States, most ulcers are duodenal ulcers. In Asian countries, gastric ulcers are more common.

peptic ulcer: an erosion in the gastrointestinal mucosa resulting from the destructive effects of gastric acid and pepsin.

Zollinger-Ellison syndrome: a condition characterized by the presence of gastrin-secreting tumors in the duodenum or pancreas.

© Dr. E. Walker/Photo Researchers, Inc.

DIET-DRUG INTERACTIONS

Check this table for notable nutrition-related effects of the medications discussed in this chapter.

	Gastrointestinal Effects	Interactions with Dietary Substances	Metabolic Effects
Antacids (aluminum hydroxide, magnesium hydroxide, calcium carbonate)	Constipation (aluminum- or calcium-containing antacids), diarrhea (magnesium-containing antacids).	May decrease iron, folate, or vitamin B_{12} absorption.	Hypophosphatemia (from excess aluminum, magnesium, or calcium).
Antibiotics (for *Helicobacter pylori* infection; may include amoxicillin, metronidazole, tetracycline)	Diarrhea (amoxicillin, tetracycline), nausea and vomiting (tetracycline), altered taste sensation (metronidazole).	Avoid alcohol with metronidazole; tetracycline decreases iron absorption and binds calcium in the GI tract, reducing absorption of both the tetracycline and the calcium.	—
Antisecretory agents (proton-pump inhibitors, H2 blockers)	Constipation, nausea and vomiting, abdominal pain (proton-pump inhibitors).	May decrease iron, folate, and vitamin B_{12} absorption.	—
Coating agents (bismuth preparations, sucralfate)	Constipation, diarrhea.	—	Hypophosphatemia, calcium retention (sucralfate).

Nutrition Care for Peptic Ulcers The goals of nutrition care are to correct nutrient deficiencies, if necessary, and encourage dietary and lifestyle practices that minimize symptoms.[19] Patients should avoid dietary items that increase acid secretion or irritate the gastrointestinal lining; examples include alcohol, coffee and other caffeine-containing beverages, chocolate, and pepper, although individual tolerances vary. Small meals may be better tolerated than large ones. Patients should avoid food consumption for at least two hours before bedtime. Cigarette smoking should be discouraged, as it can delay healing and increase the risk of ulcer recurrence.[20] There is no evidence that dietary adjustments can alter the rate of healing.[21]

IN SUMMARY

Dyspepsia refers to general symptoms of indigestion such as abdominal pain, nausea, and vomiting, which are caused by many different medical conditions. Gastritis and peptic ulcer disease are most often associated with *Helicobacter pylori* infection, which can be eradicated by antibiotic therapy. In addition, NSAID use can promote gastritis and peptic ulcer disease by damaging the mucosal lining. Extensive damage to the mucosa may reduce gastric secretions and increase the risks of developing iron and vitamin B_{12} deficiencies. The nutrition care for gastritis and peptic ulcer disease includes correcting any nutritional deficiencies that develop and eliminating dietary substances that can cause pain or discomfort.

Gastric Surgery

Gastric surgery is sometimes necessary for treating stomach cancers, some ulcer complications, and ulcers that are resistant to drug therapy. In recent years, gastric surgery has also become a popular treatment for severe obesity. Because gastric surgeries can interfere with stomach function either temporarily or permanently, patients generally need to make dietary adjustments afterward.

Stomach cancers are often treated with a **gastrectomy**, a surgical procedure that removes the diseased areas of the stomach. To suppress gastric acid secretion, a **vagotomy** is sometimes performed; this procedure severs the **vagus nerve**, which normally stimulates the cells that produce gastric acid. Because a vagotomy may impair gastric motility, it is sometimes followed by a **pyloroplasty**, which widens the pyloric sphincter to ensure drainage from the stomach to the duodenum. **Bariatric surgery,** the type of surgery that treats severe obesity, was introduced in Chapter 9 (pp. 292–293) and is discussed later in this chapter.

Gastrectomy

Figure 23-3 illustrates some typical gastrectomy procedures. In a partial gastrectomy, only part of the stomach is removed, and the remaining portion is connected to the duodenum or jejunum. In a total gastrectomy, the surgeon removes the entire stomach and connects the esophagus directly to the small intestine.

Nutrition Care after Gastrectomy The primary goals of nutrition care after a gastrectomy are to meet the nutrient needs of the postsurgical patient and promote the healing of stomach tissue. Another goal is to prevent problems that may arise due to altered stomach function. As the next section will describe, some gastric surgeries increase the risk of **dumping syndrome,** a group of symptoms that result when a large amount of food passes rapidly into the small intestine.

Following a gastrectomy, oral intake of fluids and foods is suspended until some healing has occurred, and fluids are supplied intravenously.[22] Ice chips (melted in the mouth), small sips of water, and broth are usually the first fluids given orally. Once fluids are tolerated, patients are offered liquid meals (with no sugars) at first, and they usually progress to solid foods by the fourth or fifth day after surgery. They may receive tube feedings if complications prevent a normal progression to solid foods.[23]

Dietary adjustments after gastrectomy are influenced by the size of the remaining stomach, which influences meal size, and the stomach emptying rate, which affects food tolerances. Initially, the patient is offered small meals and snacks that include only one or two food items; these foods may contain protein (fish, lean meats, and eggs), fat, and complex carbohydrates (bread, potatoes, and vegetables). Depending on the amount of food tolerated, the patient progresses to five or six small meals per day. The patient should avoid sweets and sugars, because they increase osmolarity in the small intestine and potentiate the dumping syndrome (discussed below). Some patients may need to avoid milk products due to lactose intolerance. Soluble fibers may be added to meals to help delay stomach emptying and reduce diarrhea. Although tolerances vary, patients may have difficulty with fatty foods, highly spiced foods, carbonated drinks, caffeine-containing beverages, alcohol, extremely hot or cold foods, peppermint, and chocolate. Liquids are restricted during meals due to the limited stomach capacity and because liquids can

gastrectomy (gah-STREK-ta-mee): the surgical removal of part of the stomach (partial gastrectomy) or the entire stomach (total gastrectomy).

vagotomy (vay-GOT-oh-mee): surgery that severs the vagus nerve in order to suppress gastric acid secretion. This surgery may require a follow-up *pyloroplasty* procedure to allow stomach drainage.

vagus nerve: the cranial nerve that regulates hydrochloric acid secretion and peristalsis. Effects elsewhere in the body include regulation of heart rate and bronchiole constriction.

pyloroplasty (pye-LORE-oh-PLAS-tee): surgery that enlarges the pyloric sphincter.

bariatric (BAH-ree-AH-trik) **surgery:** surgery that treats severe obesity.
• **baros** = weight

dumping syndrome: symptoms that result from the rapid emptying of an osmotic load from the stomach into the small intestine. Early symptoms include nausea, abdominal cramps, weakness, and diarrhea; later symptoms are those of hypoglycemia.

FIGURE 23-3 Typical Gastrectomy Procedures

In a gastrectomy, part or all of the stomach is surgically removed. The dashed lines show the removed section.

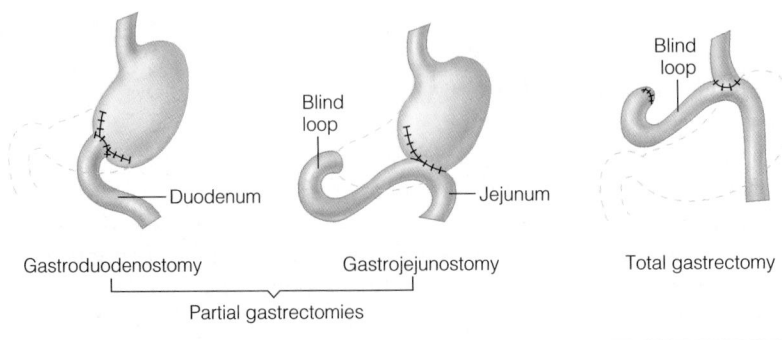

Gastroduodenostomy   Gastrojejunostomy   Total gastrectomy

Partial gastrectomies

TABLE 23-5 Postgastrectomy Diet

Food Category	Foods Recommended (as tolerated)	Foods to Limit (unless tolerated)
Meat and meat alternates	Lean tender meats, fish, poultry, shellfish, eggs, smooth nut butters	Fried, tough, or chewy meats; frankfurters and sausages; bacon; luncheon meats; dried peas and beans
Milk and milk products	Milk, plain yogurt, mild cheeses	Milk shakes, chocolate milk, fruit yogurt
Breads and cereals	Breads, crackers, bagels, pasta, and breakfast cereals made from enriched white flour (cereals should contain no added sugars)	Breads and cereals with more than 2 grams of fiber per serving; baked goods with dried fruits, nuts, or seeds; granola; frosted cereals; pastries; doughnuts
Vegetables	Tender-cooked vegetables without peels, skins, or seeds; raw lettuce	Raw vegetables (except lettuce), beets, broccoli, brussels sprouts, cabbage, cauliflower, collard and mustard greens, corn, potato skins
Fruit	Canned fruit without added sugars, bananas, melon	Canned fruits in syrup, raw fruits (except bananas and melons), dried fruits, fruit juices
Beverages	Decaffeinated coffee and tea, beverages sweetened with artificial sweeteners	Caffeinated beverages; alcoholic beverages; beverages sweetened with sugars, corn syrup, and honey

SOURCE: American Dietetic Association, *Nutrition Care Manual* (Chicago: American Dietetic Association, 2007).

increase the stomach emptying rate. Table 23-5 lists foods that are often permitted and those that are limited in postgastrectomy diets.[24]

Dumping Syndrome The dumping syndrome is characterized by a group of symptoms resulting from rapid gastric emptying. Ordinarily, the pyloric sphincter controls the rate of flow from the stomach into the duodenum. After some types of stomach surgery, the hypertonic gastric contents are no longer regulated and can rush into the small intestine more quickly after meals, causing a number of unpleasant effects. Early symptoms can occur within 30 minutes and may include nausea, vomiting, abdominal cramping, diarrhea, lightheadedness, rapid heartbeat, and others (see Table 23-6). These symptoms may be due to intestinal distention (which causes the release of an excess amount of vasoactive chemicals), a shift of fluid from blood vessels to the intestine that lowers blood volume, and an increase in peristaltic activity. Several hours later, symptoms of hypoglycemia may occur (see Table 23-6) because the unusually large spike in blood glucose following the meal (due to rapid nutrient influx and absorption) can result in an excessive insulin response.

Dietary adjustments can greatly minimize or prevent dumping syndrome. The goals are to limit the amount of food material that reaches the intestine, slow the

TABLE 23-6 Symptoms of Dumping Syndrome

Early Dumping Syndrome	Late Dumping Syndrome
Symptoms may begin within 30 minutes after eating:	Symptoms may begin 1 to 3 hours after eating:
• Abdominal fullness, cramps	• Anxiety
• Diarrhea	• Confusion, difficulty thinking
• Dizziness	• Headache
• Flushing, sweating	• Hunger
• Nausea and vomiting	• Palpitations
• Rapid heartbeat	• Sweating
• Weakness, feeling faint	• Weakness, feeling faint

◆ *Octreotide* inhibits gastrointestinal motility, thereby slowing both gastric emptying and transit time in the small intestine.

◆ Bacterial overgrowth is described in Chapter 24.

◆ Fat malabsorption reduces calcium absorption because the negatively charged fatty acids combine with calcium (which is positively charged) and prevent its absorption.

rate of gastric emptying, and reduce foods that increase hypertonicity. Therefore, meal size is limited, fluids are restricted during meals, and sugars (including milk sugar) are restricted. The "How to" below lists practical suggestions for reducing the occurrence of dumping syndrome. In some cases, drugs that inhibit gastrointestinal motility (such as octreotide) ◆ may help. The Case Study on p. 745 provides the opportunity to design a menu for a postgastrectomy patient who is at risk for dumping syndrome.

Postsurgical Complications and Nutrition Status Substantial weight loss can sometimes be an unintended consequence of gastrectomy. It may take time for a postgastrectomy patient to learn the amount of food that can be consumed without causing discomfort. The symptoms associated with meals may lead to food avoidance, weight loss, and eventually, malnutrition. Other nutrition problems that may occur after gastrectomy include the following:[25]

- *Fat malabsorption.* Fat digestion and absorption may be impaired due to the accelerated transit of food material, which prevents the normal mixing of fat with lipase and bile. If the duodenum has been removed or bypassed, less lipase is secreted into the intestine for fat digestion. Another potential problem is bacterial overgrowth in the small intestine, which interferes with bile function. ◆ Fat malabsorption can lead to deficiencies of fat-soluble vitamins and some minerals. Supplemental pancreatic enzymes are sometimes provided to improve fat digestion. Medium-chain triglycerides, which are more easily digested and absorbed, can be used to supply additional fat kcalories.

- *Bone disease.* Osteoporosis and osteomalacia are common outcomes following a gastrectomy, and they are most likely a consequence of vitamin D and calcium deficiencies.[26] The fat malabsorption discussed earlier can also cause malabsorption of vitamin D and calcium; ◆ furthermore, patients at risk of dumping syndrome may need to avoid milk products, which are among the best sources of these nutrients. Bone density should be monitored during the years following surgery, and supplementation of calcium and vitamin D is often recommended.

- *Anemia.* After gastrectomy, iron and vitamin B_{12} deficiencies are common. The reduced gastric secretions impair the absorption of both iron and vitamin B_{12}. If the duodenum has been removed or is bypassed, the risk of iron deficiency increases because the duodenum is a major site of iron absorption. Supplementation of both iron and vitamin B_{12} is usually warranted after surgery.

HOW TO Alter the Diet to Reduce Symptoms of Dumping Syndrome

Dietary adjustments can minimize or prevent symptoms of dumping syndrome. The following suggestions often help:

- Eat smaller meals to fit the reduced capacity of the stomach. Increase the number of meals consumed daily so that energy intake is adequate.
- Eat in a relaxed setting. Eat slowly and chew food thoroughly.
- Limit the amount of fluid taken with meals. Avoid consuming beverages within 45 minutes before and after meals, but be sure to include adequate fluid intake during the day to avoid dehydration.

- Avoid juices and sweetened beverages and foods that contain large amounts of sugar. Avoid carbonated beverages if they cause bloating.
- Use artificial sweeteners to sweeten beverages and desserts.
- Avoid foods and beverages that are very hot or very cold, unless tolerated.
- Include fiber-rich foods in each meal. Sometimes adding soluble fibers like pectin or guar gum to meals can help to control symptoms.
- Avoid milk and most milk products, which are high in lactose. Enzyme-treated milk

should also be avoided because the breakdown products of lactose (glucose and galactose) can also cause symptoms. Cheese may be better tolerated, because its lactose content is low. Make an effort to consume nonmilk calcium sources such as green leafy vegetables, tofu, and fish with bones.

- If symptoms of hypoglycemia continue, try including a protein-rich food in each meal.
- Lie down for 20 to 30 minutes (or longer) after eating to help slow the transit of food to the small intestine. While eating a meal, sit upright.

> **CASE STUDY**   Biology Teacher Requiring Gastric Surgery
>
> Marie Erwin, a 58-year-old biology teacher, was admitted to the hospital for gastric surgery after numerous medical treatments failed to manage her severe peptic ulcer disease. A gastrojejunostomy was performed, and after about 24 hours, Mrs. Erwin was able to take small sips of warm water. The health care team anticipates multiple nutrition-related problems and is taking measures to prevent them.
>
> 1. Review Figure 23-3 to better understand Mrs. Erwin's surgical procedure. Consider the possibilities that she might experience the following symptoms: early satiety, nausea and vomiting, weight loss, dumping syndrome, fat malabsorption, anemia, and bone disease. Explain why each of these conditions may occur.
> 2. What type of diet will the physician prescribe for Mrs. Erwin after she begins eating solid foods? Create a day's worth of menus, using foods from Table 23-5.
> 3. What advice can you give Mrs. Erwin that will help to prevent dumping syndrome? List several foods from each major food group that may cause symptoms of dumping syndrome.

Bariatric Surgery

The most popular surgical options for weight reduction, the *gastric bypass* and *gastric banding* procedures, were introduced in Chapter 9 (see Figure 9-5 on p. 293). The gastric bypass operation, ◆ which accounts for more than 70 percent of bariatric surgeries,[27] constructs a small gastric pouch that reduces stomach capacity and thereby restricts meal size. In addition, the surgeon alters the digestive route by connecting the pouch directly to the jejunum; this reduces nutrient absorption because the nutrients bypass a significant portion of the small intestine. In the gastric banding procedure, a gastric pouch is created using a fluid-filled inflatable band. Adjusting the band's fluid level can tighten or loosen the band and alter the size of the opening between the pouch and the rest of the stomach. A smaller opening reduces the speed at which the pouch is emptied and prolongs the sense of fullness after a meal. Whereas the gastric bypass operation is somewhat permanent, the gastric banding procedure is fully reversible.

◆ The gastric bypass surgery is also known as a *Roux-en-Y gastric bypass* because the reconstructed small intestine resembles the letter *Y*.

The weight loss achieved after gastric bypass and gastric banding is similar, although patients often lose weight faster after the gastric bypass procedure. Although research results vary, the average long-term (5- to 7-year) weight loss following bariatric surgery is approximately 55 percent of excess weight.[28]

Although bariatric surgeries are effective treatments for morbid obesity, patients should have realistic expectations about the amount of weight they are likely to lose, the diet they will need to follow, and the complications that may ensue. Some types of bariatric surgery can dramatically affect health and nutrition status, and patients may require lifelong management.

Dietary Guidelines after Bariatric Surgery   The gastric pouch created by surgery eventually expands to hold about $1/2$ cup of food, but its initial capacity is only a few tablespoons. Only sugar-free clear liquids and low-fat broths are given on the day following bariatric surgery.[29] Afterward, patients consume a liquid diet at first, followed by pureed foods and then solid foods; soft foods like tender meats, tender-cooked vegetables, and canned fruits are tolerated most easily. Only small portions should be consumed, because overeating can stretch the gastric pouch. Similarly, fluids must be consumed separately from meals to avoid excessive distention.

Other nutrition-related concerns include the following:

- *Protein intake.* The protein recommendation for bariatric patients is 1.5 grams of protein per kilogram of body weight; however, intakes are often lower than recommended.[30] Patients are generally instructed to eat high-protein foods before consuming other foods in a meal and to consume liquid protein supplements regularly.

HOW TO Alter Dietary Habits to Achieve and Maintain Weight Loss after Bariatric Surgery

Patients need to learn new dietary habits after bariatric surgery. The following recommendations may help:

- Chew food thoroughly, and consume only small amounts. Use a small spoon, and take small bites. Relax and enjoy the meal, taking at least 20 minutes to eat.
- Understand that at first, the appropriate portion of each food served at mealtime may be only a few spoonfuls. Learn to recognize the sensations that occur when the gastric pouch is full. Signs of fullness may include pressure in the stomach region, a slight feeling of nausea, or pain in the upper chest or shoulder.

- Learn to recognize foods that cause problems. Foods that are dry, sticky, or fibrous may be difficult to tolerate during the weeks after surgery.
- To control vomiting, try eating smaller volumes of food, eating more slowly, and avoiding foods that are known to cause difficulty. Continued vomiting may be a sign that appropriate food behaviors are not being maintained.
- Eat only at mealtimes. Snacking throughout the day can become a bad habit that causes weight to be regained.
- Avoid consuming liquids within 45 minutes of mealtime. Consume liquids between

meals only. Avoid high-kcalorie drinks such as soda, alcoholic beverages, and milk shakes. Carbonated beverages increase stomach gas and may cause bloating.

- Drink adequate fluids between meals to avoid dehydration. Most people meet a substantial portion of their fluid needs by eating foods, but a patient who has had bariatric surgery cannot do this. Therefore, fluid intakes need to be increased after surgery.
- Engage in regular physical activity. Activity is a valuable aid to weight maintenance and can help to maintain lean tissue while weight is being lost.

- *Vitamin and mineral deficiencies.* Bariatric patients have a high risk of developing nutrient deficiencies due to reduced food intake, reduced gastric secretions, and nutrient malabsorption. Supplemental vitamin B_{12}, iron, and calcium are usually recommended after surgery. In addition, a daily multivitamin/mineral supplement can ensure that patients meet their needs for other nutrients.
- *Foods to avoid.* Patients may find some foods difficult to manage; dry, doughy, or fibrous foods may cause pain or vomiting. Foods that are often problematic after bariatric surgery include sticky foods like rice, pasta, and soft breads; tough or chewy meats; and foods that have seeds, peels, or husks.
- *Dumping syndrome.* To avoid symptoms of dumping syndrome, gastric bypass patients must carefully control food portions, avoid foods high in sugars, and consume liquids between meals (review the "How to" on p. 744).

After bariatric surgery, patient education and counseling are critical for weight loss and weight management; patients also need to learn the elements of a healthy diet. The "How to" on this page includes additional dietary suggestions for patients who have undergone bariatric surgery.

Postsurgical Concerns in Gastric Bypass Surgery The complications that arise after gastric bypass surgery are similar to those that arise after gastrectomy and may include fat malabsorption, bone disease, and anemia. Rapid weight loss also increases a person's risk of developing gallbladder disease; patients at especially high risk sometimes have their gallbladders removed while undergoing bariatric surgery. After weight loss, plastic surgery may be necessary to remove extra skin, especially on the abdomen, buttocks, hips, and thighs.

IN SUMMARY

Gastric surgeries are used to treat cancer, peptic ulcer complications, and obesity. This surgery may alter stomach size or function, so dietary adjustments are required afterward. Common postsurgical complications include fat malabsorption, bone disease, anemia, and dumping syndrome. After bariatric surgery, patients must learn to consume appropriate food portions; use dietary supplements to prevent nutrient deficiencies; and choose foods that are unlikely to cause abdominal discomfort, vomiting, or dumping syndrome.

CENGAGENOW™
academic.cengage.com/login

1. The diets described in this chapter are highly individualized: a particular food may cause discomfort for one person and have no effect on another. Describe some practical ways to keep track of food intolerances.

2. Although some individuals may require a mechanically altered diet for just a few weeks, others may have medical problems that require long-term use of such diets. Consider the difference between working with a person who has had a swallowing problem for years and a person who recently had mouth surgery and is just beginning to eat again.

   ■ Explain how the needs of these individuals may differ. What nutrition-related problems may develop if a person has been following a restrictive dysphagia diet for several years?

   ■ Using Table 23-2 and the "How to" on p. 735, create a day's worth of menus for a person who requires long-term use of a pureed dysphagia diet and tolerates only liquids that have a honeylike consistency.

NUTRITION ON THE NET

For further study of topics covered in this chapter, log on to **academic.cengage .com/nutrition/rolfes/UNCN8e**. Go to Chapter 23, then to Nutrition on the Net.

- Visit the websites of these organizations to find information that is helpful both for health practitioners and patients with gastrointestinal problems:
  American College of Gastroenterology: **www.acg.gi.org**
  American Gastroenterological Association: **www.gastro.org**

National Institute of Diabetes and Digestive and Kidney Diseases, a division of the National Institutes of Health: **www2.niddk.nih.gov**

- Find more information about dysphagia at the Dysphagia Resource Center: **www.dysphagia.com**

- Learn more about *Helicobacter pylori* from the Helicobacter Foundation: **www.helico.com**

NUTRITION ASSESSMENT CHECKLIST for People with Upper GI Tract Disorders

Medical History

Check the medical history to uncover conditions or treatments that may:

- Lead to dry mouth
- Interfere with chewing or swallowing
- Lead to dyspepsia, nausea, or vomiting

Check for a medical diagnosis of:

- Gastritis or peptic ulcer
- GERD
- Hiatal hernia
- Pernicious anemia

For a patient who has undergone gastric surgery, check for the following complications:

- Anemia
- Bone disease
- Dumping syndrome
- Fat malabsorption

Medications

Record all medications and note:

- Aspirin or NSAID use in patients with gastritis or peptic ulcer disease
- Medications that may cause dry mouth
- Medications that may cause nausea and vomiting

To help alleviate nausea, suggest that medications be taken with food, when possible.

Dietary Intake

To devise an acceptable meal plan, obtain:

- An accurate and thorough record of food intake
- A record of foods that provoke symptoms of dyspepsia, nausea, GERD, gastritis, peptic ulcers, or dumping syndrome

For patients on long-term dysphagia diets, monitor:

- Appetite
- Tolerances to foods
- Variety of foods offered and regularly consumed

Anthropometric Data

Measure baseline height and weight. Address weight loss early to prevent malnutrition for patients with:

- Dysphagia or difficulty chewing
- Dyspepsia or nausea of long duration
- Malabsorption
- Dumping syndrome

Laboratory Tests

Check laboratory tests for signs of dehydration for patients with:

- Dumping syndrome
- Persistent vomiting

Check laboratory tests for nutrition-related anemia in patients with:

- Conditions that require long-term use of antisecretory medications
- Gastritis
- Previous gastric surgeries

Physical Signs

Look for physical signs of:

- Dehydration—in patients with persistent vomiting or dumping syndrome
- Iron and vitamin B_{12} deficiencies—in patients with hypochlorhydria or achlorhydria

STUDY QUESTIONS

These questions will help you review the chapter. You will find the answers in the discussions on the pages provided.

1. Provide examples of conditions that can interfere with swallowing. Describe how diets are adjusted to meet the needs of people with dysphagia. (pp. 731–734)

2. Discuss ways to provide appetizing meals for people consuming mechanically altered foods. (p. 735)

3. Identify the symptoms, causes, and complications of gastroesophageal reflux disease (GERD). What lifestyle modifications can benefit patients with GERD? (pp. 734–737)

4. What are possible causes of nausea and vomiting? Discuss interventions that may help. (p. 739)

5. Specify the common causes of gastritis and peptic ulcer disease. Explain the possible consequences of these diseases. Describe the role of diet therapy for both conditions. (pp. 739–741)

6. Describe the gastric disorders that may benefit from gastric surgery. What dietary adjustments are usually required after a gastrectomy procedure? (pp. 741–744)

7. What complications may arise after gastric surgery? Discuss the dietary interventions that may help to prevent these consequences. (pp. 743–744)

8. Describe the surgical procedures used to treat morbid obesity. Discuss the dietary recommendations for patients who have undergone bariatric surgery. (pp. 745–746)

These multiple choice questions will help you prepare for an exam. Answers can be found on p. 749.

1. If a patient with dysphagia has difficulty swallowing solids but can easily swallow liquids:
   a. the problem is probably a motility disorder.
   b. the patient most likely has achalasia.
   c. the problem is probably an esophageal obstruction.
   d. the patient may also develop oropharyngeal dysphagia.

2. The health practitioner working with a patient with dysphagia should recognize that:
   a. only pureed foods should be given to minimize the risk of aspiration.
   b. the patient can have any food that can be comfortably and safely chewed and swallowed.
   c. highly seasoned foods are often restricted.
   d. conventional diets are unable to meet total nutrient needs, and supplements are always necessary.

3. Gastroesophageal reflux disease (GERD) is:
   a. characterized by frequent backflow of the stomach's gastric secretions into the esophagus.
   b. a protuberance of a portion of the stomach above the lower esophageal sphincter.
   c. an erosion of the lining of the stomach caused by excess acid in gastric secretions.
   d. an obstruction of the lower esophagus that results in dysphagia.

4. Conditions associated with an increased risk of developing GERD include:
   a. hiatal hernia.
   b. asthma.
   c. pregnancy.
   d. all of the above.

5. For the patient with persistent vomiting, the major nutrition-related concern(s) is/are:
   a. dehydration and malnutrition.
   b. reflux esophagitis.
   c. dyspepsia.
   d. peptic ulcers.

6. Chronic gastritis frequently leads to:
   a. dumping syndrome.
   b. bone disease.
   c. iron and vitamin B_{12} deficiencies.
   d. excessive hydrochloric acid secretion.

7. The primary cause of most peptic ulcers is:
   a. consumption of spicy foods.
   b. hypochlorhydria.
   c. smoking cigarettes.
   d. *Helicobacter pylori* infection.

8. Foods discouraged for patients with gastritis or active ulcers include those that:
   a. are high in fiber.
   b. irritate the gastric mucosa.
   c. are easy to swallow.
   d. contain simple sugars.

9. People at risk of dumping syndrome should generally avoid:
   a. high-fiber foods.
   b. sweets and sugars.
   c. beverages.
   d. bread and potatoes.

10. The health practitioner assessing a patient who underwent a gastrectomy several years ago should be alert to signs of:
    a. dysphagia.
    b. GERD.
    c. anemia.
    d. gastritis.

REFERENCES

1. R. C. Orlando, Diseases of the esophagus, in L. Goldman and D. Ausiello, eds., *Cecil Medicine* (Philadelphia: Saunders, 2008), pp. 998–1009; M. H. Beers and coeditors, *The Merck Manual of Diagnosis and Therapy* (Whitehouse Station, N.J.: Merck Research Laboratories, 2006), pp. 62–183.
2. R. Terré and F. Mearin, Oropharyngeal dysphagia after the acute phase of stroke: Predictors of aspiration, *Neurogastroenterology and Motility* 18 (2006): 200–205.
3. Orlando, 2008.
4. The National Dysphagia Diet Task Force, *The National Dysphagia Diet: Standardization for Optimal Care* (Chicago: American Dietetic Association, 2002).
5. American Dietetic Association, *Nutrition Care Manual* (Chicago: American Dietetic Association, 2007).
6. J. E. Richter, Gastroesophageal reflux disease during pregnancy, *Gastroenterology Clinics of North America* 32 (2003): 235–261.
7. Orlando, 2008; Beers and coeditors, 2006.
8. N. J. Talley, Functional gastrointestinal disorders: Irritable bowel syndrome, dyspepsia, and noncardiac chest pain, in L. Goldman and D. Ausiello, eds., *Cecil Medicine* (Philadelphia: Saunders, 2008), pp. 990–998.
9. Beers and coeditors, 2006.
10. Talley, 2008.
11. S. Escott-Stump, *Nutrition and Diagnosis-Related Care* (Baltimore: Lippincott Williams & Wilkins, 2008).
12. Beers and coeditors, 2006.
13. B. Cryer and S. J. Spechler, Peptic ulcer disease, in M. Feldman, L. S. Friedman, and L. J. Brandt, eds., *Sleisenger and Fordtran's Gastrointestinal and Liver Disease* (Philadelphia: Saunders, 2006), pp. 1089–1110.
14. Y. Yuan, I. T. Padol, and R. H. Hunt, Peptic ulcer disease today, *Nature Clinical Practice Gastroenterology and Hepatology* 3 (2006): 80–89; N. W. Bunnett and V. R. Lingappa, Gastrointestinal disease, in S. J. McPhee and W. F. Ganong, eds., *Pathophysiology of Disease: An Introduction to Clinical Medicine* (New York: McGraw-Hill/Lange, 2006), pp. 338–388.
15. K. Ramakrishnan and R. C. Salinas, Peptic ulcer disease, *American Family Physician* 76 (2007): 1005–1012.
16. Beers and coeditors, 2006.
17. Ramakrishnan and Salinas, 2007.
18. D. Y. Graham and J. J. Y. Sung, *Helicobacter pylori*, in M. Feldman, L. S. Friedman, and L. J. Brandt, eds., *Sleisenger and Fordtran's Gastrointestinal and Liver Disease* (Philadelphia: Saunders, 2006), pp. 1049–1066.
19. American Dietetic Association, 2007.
20. Cryer and Spechler, 2006.
21. W. F. Stenson, The esophagus and stomach, in M. E. Shils and coeditors, *Modern Nutrition in Health and Disease* (Baltimore: Lippincott Williams & Wilkins, 2006), pp. 1179–1188.
22. P. L. Bayer, Medical nutrition therapy for upper gastrointestinal tract disorders, in L. K. Mahan and S. Escott-Stump, eds., *Krause's Food and Nutrition Therapy* (Philadelphia: Saunders, 2008), pp. 654–672.
23. American Dietetic Association, 2007.
24. American Dietetic Association, 2007.
25. C. R. Parrish, Post-gastrectomy: Managing the nutrition fall-out, *Nutrition Issues in Gastroenterology* 18 (2004): 63–75.
26. Parrish, 2004.
27. S. Klein, Obesity, in M. Feldman, L. S. Friedman, and L. J. Brandt, eds., *Sleisenger and Fordtran's Gastrointestinal and Liver Disease* (Philadelphia: Saunders, 2006), pp. 409–425.
28. P. E. O'Brien and coauthors, Systematic review of medium-term weight loss after bariatric operations, *Obesity Surgery* 16 (2006): 1032–1040.
29. American Dietetic Association, 2007.
30. American Dietetic Association, 2007.

ANSWERS

Study Questions (multiple choice)

1. c   2. b   3. a   4. d   5. a   6. c   7. d   8. b   9. b   10. c

Dental Health and Chronic Illness

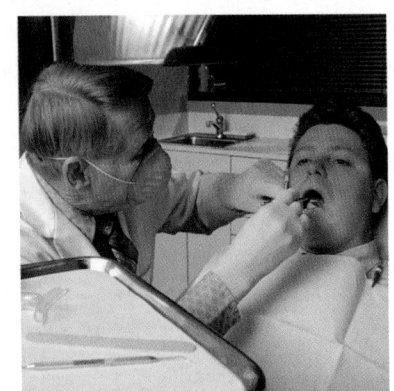

The relationships between nutrition and dental health were discussed earlier in this book. Chapter 4 described the effects of sugar and other fermentable carbohydrates on tooth decay. Chapter 13 explained how fluoride can help to prevent dental caries. Chapter 15 discussed the development of tooth decay in babies who are given bottles for prolonged periods. This highlight identifies other types of dental problems. It includes examples of how chronic illnesses may influence the development of dental diseases and, conversely, how dental diseases may increase the risks of developing chronic illnesses. Related terms are defined in the accompanying glossary.

Periodontal Disease

Recall from Chapter 4 that dental caries develops when the bacteria that reside in dental plaque ferment sugars and produce acid that dissolves tooth enamel (review Figure 4-14 on p. 119). Deposits of plaque can thicken and lead to other dental problems as well. As plaque accumulates on the tooth surface, it fills with calcium and phosphate, eventually forming **dental calculus.** Calculus may develop either at the gum surface or in the crevice between the gum and a tooth; its presence may cause further plaque retention. The buildup of plaque and calculus increases the likelihood of infection and subsequent inflammation.

Periodontal disease refers to inflammatory conditions involving the **periodontium**—the tissues that support the tooth in its bony socket. The periodontium includes the gums (called **gingiva**), other connective tissues surrounding the tooth, and the bone underneath. Inflammation of the gums, called **gingivitis,** is characterized by redness, bleeding, and swelling of gum tissue. **Periodontitis** is an inflammation of the other tissues surrounding the tooth. As plaque invades the space below the gum line, the combination of toxic bacterial by-products and the body's immune response can destroy the tissues holding a tooth in place. Left untreated, the tissues and bone of the peridontium may ultimately be destroyed, leading to permanent tooth loss.

Risk Factors

Dental plaque is the major risk factor associated with periodontal disease, and the severity of the disease is related to the amount of plaque present. Tobacco smoking is another factor, possibly because of its destructive effects on cellular immune responses.[1] The risk of developing periodontal disease is especially high if a person has a chronic illness that impairs immune status, such as diabetes mellitus or HIV infection. Other risk factors include genetic susceptibility, stress, and dental conditions that increase plaque accumulation, such as poorly aligned teeth or calculus buildup. Strategies for reducing risk focus on improving oral hygiene (proper brushing and flossing) and encouraging smoking cessation.

Signs and Symptoms

Periodontal disease typically begins with gingivitis; the gums may bleed readily from brushing or flossing and be tender and swollen. The gap between an infected gum and a tooth usually deepens, allowing food particles to get caught easily. A bad taste in the mouth or persistent bad breath is sometimes the first sign

---

GLOSSARY

dental calculus: mineralized dental plaque, often associated with inflammation and bleeding.

gingiva (jin-JYE-va, JIN-jeh-va): the gums.

gingivitis (jin-jeh-VYE-tus): inflammation of the gums; characterized by redness, swelling, and bleeding.

periodontal disease: a disease that affects the connective tissue structures that support the teeth.

periodontitis: inflammation or degeneration of the tissues that support the teeth.

periodontium: the tissues that support the teeth, including the gums, cementum (bonelike material covering the dentin layer of the tooth), periodontal ligament, and underlying bone.
• **peri** = around, surrounding
• **odont** = tooth

Sjögren's syndrome: an autoimmune disease characterized by the destruction of secretory glands, especially those that produce saliva and tears, resulting in dry mouth and dry eyes.

xerostomia: dry mouth caused by reduced salivary flow.
• **xero** = dry
• **stomia** = mouth

Reminder: Dental caries refers to tooth decay. *Dental plaque* is an accumulation of bacteria and their by-products that grow on teeth and can lead to dental caries and gum disease.

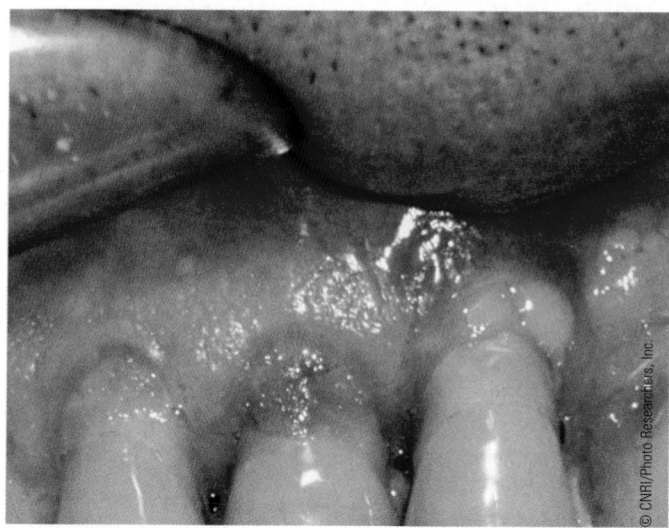

Periodontal disease destroys the tissues and bones that hold teeth in place.

TABLE H23-1 Suggestions for Managing Dry Mouth

- Take frequent sips of water or a sugarless beverage.
- Use sugarless candy or gum to help stimulate salivary flow.
- Suck on ice cubes or frozen fruit juice bars (unless their coldness causes discomfort).
- Avoid citrus juices and spicy or salty foods if they cause mouth irritation.
- Avoid dry foods like toast, chips, and crackers.
- Avoid caffeine, alcohol, and smoking, which may dry the mouth.
- Consume foods that have a high fluid content, such as soups, stews, sauces and gravies, yogurt, and pureed fruits.
- Try over-the-counter saliva substitutes (available as gels, sprays, and tablets), especially just before meals and at bedtime.
- Try rinsing the mouth with small amounts of vegetable oil or softened margarine.
- Use a humidifier during the night.
- Pay strict attention to oral hygiene, brushing and flossing at least twice daily. Try to brush immediately after each meal.
- Avoid alcohol- and detergent-containing mouthwashes, which may dry and irritate the mouth.
- If dry mouth is caused by a medication, ask your physician about possible alternatives.
- Ask your physician if using a medication to stimulate saliva secretion may be of benefit; examples include nicotinic acid tablets and pilocarpine.

of gingivitis. In severe cases, pus may surround the teeth and gums. The teeth may be sensitive and chewing painful. If bone is destroyed, the affected gums usually recede, and teeth may loosen or change position.

Treatment

Treatment of periodontal disease depends on the extent of damage. In mild cases, deep cleaning and proper oral hygiene may reverse the condition. Antimicrobial mouth rinses and topical antibiotics are often prescribed to control infection. Surgical approaches are sometimes necessary to remove plaque or calculus deposits underneath gum tissue or to replace tissues that have been destroyed.

Dry Mouth

Secretions of the salivary glands protect the teeth and the mouth's soft tissues. Saliva lubricates oral tissues and contains antimicrobial proteins to defend against bacteria and fungi. The buffers in saliva raise the mouth's pH so that tooth enamel is protected from the acid produced by caries-causing bacteria. The calcium and phosphate concentrations of saliva help to prevent dissolution of enamel. Thus saliva helps to control plaque formation, prevent infection within the mouth, and maintain tooth enamel. If salivary secretions are low or absent, the risk of developing dental caries and periodontal disease increases.[2]

Dry mouth **(xerostomia),** caused by reduced salivary flow, is a side effect of many medications and is associated with a number of diseases and disease treatments. Antihistamines, antihypertensive agents, antidepressants, decongestants, and other medications can cause dry mouth. Poorly controlled diabetes mellitus is often associated with dry mouth, as are conditions that directly affect salivary gland function, such as **Sjögren's syndrome.** Radiation therapy used to treat head and neck cancers often damages salivary glands, sometimes permanently. Mouth breathing is also a common cause of dry mouth.[3]

A reduction in salivary flow can impair health in other ways as well. Dry mouth may interfere with speech and cause bad breath. Mouth infections are more common. Chewing and swallowing are more difficult, and taste sensation is diminished. Dentures may be uncomfortable to wear, and ulcerations may develop where they contact the mouth. Dry mouth may cause a person to reduce food intake and thereby increase malnutrition risk. Table H23-1 lists suggestions that may help to manage dry mouth.

Dental Health and Chronic Illness

Maintaining dental health is sometimes challenging for a person with a chronic illness. As mentioned earlier, many medications can reduce salivary secretions, along with the immune protection that saliva provides. This section describes how several conditions may upset oral health and increase the risks of developing dental problems.

Diabetes Mellitus

For a number of reasons, periodontal disease is more prevalent among people with diabetes mellitus, especially those whose diabetes is poorly controlled. People with diabetes often have impaired immune responses and a greater susceptibility to infections. Diabetes also favors the growth of bacteria that tend to infect periodontal tissues. The damaging effects of hyperglycemia weaken the collagen structure of tissues, making them more vulnerable. In addition, people with diabetes tend to have higher plaque accumulations and dry mouth, predisposing them to periodontal disease.[4]

Because the risk of developing dental caries and oral fungal infections is greater for people with diabetes, they must pay strict attention to oral hygiene. Smoking is discouraged because it can increase periodontal disease risk nearly 10-fold in people with diabetes.[5] Health care providers should advise patients with diabetes that glucose control and routine dental care are critical to preventing periodontal disease.

Human Immunodeficiency Virus (HIV) Infection/AIDS

HIV infection is characterized by compromised immunity, and the risk of developing periodontal disease is closely linked to the extent of HIV infection. In untreated persons, fungal and viral infections are common and may cause burning in the mouth and painful ulcerations. The current use of antiretroviral therapies, however, has substantially reduced the incidence of oral infections in HIV-positive persons.[6] Those at greatest risk of developing dental disease include smokers, individuals who decline therapy, and patients in advanced stages of disease.[7] HIV-infected individuals often have dry mouth as a result of medications or salivary gland dysfunction.[8]

Oral Cancers

Radiation treatment of oral cancers can cause serious oral and dental complications.[9] The inflammation and tissue damage may be so severe that the radiation treatment may need to be halted or the intensity reduced substantially. Radiation can also reduce salivary flow, causing the problem of dry mouth described earlier. Other complications include fungal and viral infections, changes in taste sensation, and tissue and muscle scarring (which often reduces chewing ability). To minimize complications, dental care is often initiated before radiation therapy begins.

Dental Health and Disease Risk

Dental diseases may have adverse effects on health beyond their effects on teeth.[10] The bacteria that reside on dental tissues can enter the bloodstream and cause infections elsewhere in the body. Evidence supports a link between dental bacteria and other conditions, including the following:

- *Systemic inflammation.* The inflammatory process induced by periodontal disease increases levels of cytokines and other mediators that have systemic effects. Systemic inflammation may contribute to the development of certain chronic illnesses, including heart disease and diabetes.

- *Respiratory illnesses.* The teeth of hospital patients often become colonized with bacteria that cause respiratory illnesses. In one study, only patients whose teeth were colonized by respiratory pathogens ended up with pneumonia.[11]

- *Atherosclerosis and heart disease.* Bacteria associated with gingivitis can attack the cells lining the blood vessels, possibly affecting the process of atherosclerosis. In one study, researchers found a significant association between serum antibodies to periodontal bacteria and the incidence of heart disease.[12]

- *Diabetes mellitus.* The presence of periodontal disease can make it more difficult for persons with diabetes to attain glucose control.

Although this evidence is suggestive, researchers studying the effects of periodontal disease on health have yet to prove cause-and-effect relationships between dental health and other conditions. Additional studies will help to clarify the complex interactions between dental disease and chronic illnesses.

REFERENCES

1. B. Loos and coauthors, Lymphocyte numbers and function in relation to periodontitis and smoking, *Journal of Periodontology* 75 (2004): 557–564.
2. D. P. DePaola and coauthors, Nutrition and dental medicine, in M. E. Shils and coeditors, *Modern Nutrition in Health and Disease* (Baltimore: Lippincott Williams & Wilkins, 2006), pp. 1152–1178.
3. T. E. Daniels, Diseases of the mouth and salivary glands, in L. Goldman and D. Ausiello, eds., *Cecil Medicine* (Philadelphia: Saunders, 2008), pp. 2867–2874; DePaola and coauthors, 2006.
4. American Dietetic Association, Position of the American Dietetic Association: Oral health and nutrition, *Journal of the American Dietetic Association* (2007): 1418–1428; DePaola and coauthors, 2006.
5. D. C. Matthews, The relationship between diabetes and periodontal disease, *Journal of the Canadian Dental Association* 68 (2002): 161–164.
6. V. Ramirez-Amador and coauthors, The changing clinical spectrum of human immunodeficiency virus (HIV)-related oral lesions in 1,000 consecutive patients: A 12-year study in a referral center in Mexico, *Medicine (Baltimore)* 82 (2003): 39–50; J. D. Eyeson and coauthors, Oral manifestations of an HIV positive cohort in the era of highly active anti-retroviral therapy (HAART) in South London, *Journal of Oral Pathology and Medicine* 31 (2002): 169–174.
7. T. Alpagot and coauthors, Risk factors for periodontitis in HIV patients, *Journal of Periodontal Research* 39 (2004): 149–157.
8. M. Navazesh and coauthors, A 4-year longitudinal evaluation of xerostomia and salivary gland hypofunction in the Women's Interagency HIV Study participants, *Oral Surgery Oral Medicine Oral Pathology Oral Radiology and Endodontics* 95 (2003): 693–698.
9. American Dietetic Association, 2007; J. B. Epstein and coauthors, Cancer-related oral health care services and resources: A survey of oral and dental care in Canadian cancer centres, *Journal of the Canadian Dental Association* 70 (2004): 302–304.
10. Y.-T. A. Teng and coauthors, Periodontal health and systemic disorders, *Journal of the Canadian Dental Association* 68 (2002): 188–192.
11. Teng, 2002.
12. Teng, 2002.

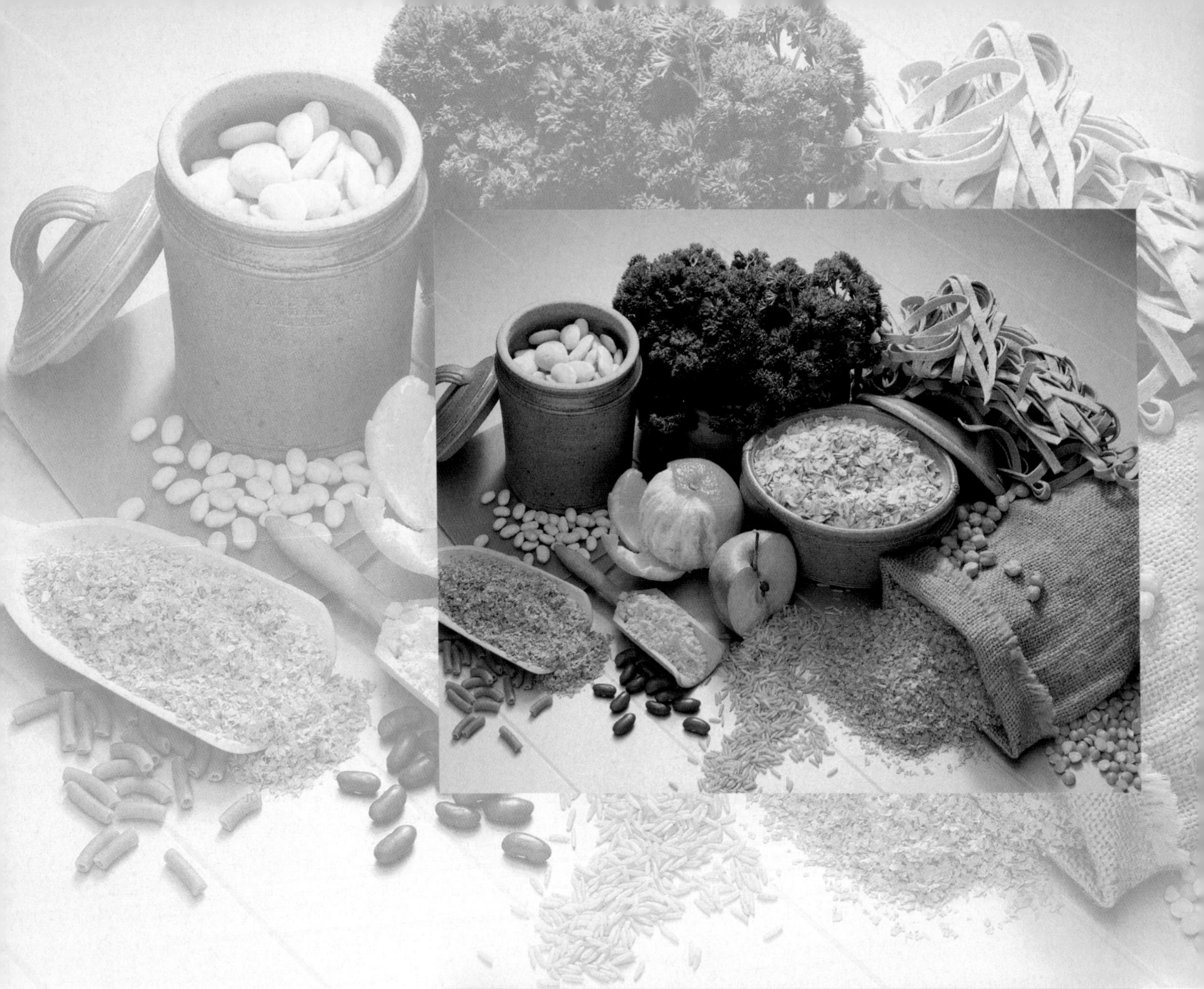

The CengageNOW logo indicates an opportunity for online self-study, linking you to interactive tutorials and videos based on your level of understanding.

academic.cengage.com/login

Nutrition in the Clinical Setting

Disorders affecting the lower gastrointestinal tract can interfere substantially with a patient's diet and lifestyle. Patients with some diseases must adhere to diets that are complicated and difficult to follow. Furthermore, foods that patients tolerate can vary considerably. In follow-up visits, health care professionals should ensure that patients understand the diet prescription and help to pinpoint difficult foods. They can also suggest ways to make restrictive diets more acceptable.

Lower Gastrointestinal Disorders

This chapter discusses medical conditions that can upset the digestive and absorptive functions of the lower gastrointestinal (GI) tract. As you may recall from Chapter 3, the lower GI tract consists of the small intestine (the duodenum, jejunum, and ileum), the large intestine, the rectum, and the anus. The digestion and absorption of nutrients occur primarily in the small intestine. The pancreas and gallbladder support these complex functions by delivering digestive secretions to the duodenum, the segment of small intestine closest to the stomach. The large intestine reabsorbs water and facilitates the excretion of waste material. Figure 24-1 on p. 756 illustrates the lower GI tract and related organs and reviews the functions of each organ, and pp. 72–74 in Chapter 3 provide additional detail.

Common Intestinal Problems

Nearly all people experience occasional intestinal problems, which usually clear up without medical treatment. Intestinal discomfort can sometimes drive a person to seek medical attention, however, and the symptoms may be evidence of a serious intestinal disorder or other illness. The most common intestinal problems and their causes and treatments are discussed below.

Constipation

A medical diagnosis of constipation is based, in part, on a defecation frequency of fewer than three bowel movements per week. Other symptoms may include the passage of hard stool and excessive straining during defecation. In some cases, however, a person's perception of constipation may be due to a mistaken notion of what constitutes "normal" bowel habits, so the person's expectations about bowel function may need to be addressed.

Constipation is much more prevalent among women than men and is a common complaint during pregnancy. The incidence of constipation increases with

FIGURE 24-1 The Lower GI Tract and Related Organs

Stomach
Adds acid, enzymes, and fluid. Churns, mixes, and grinds food to a liquid mass.

Pyloric sphincter
Allows passage from stomach to small intestine. Prevents backflow from small intestine.

Small intestine
Produces enzymes that digest energy-yielding nutrients to smaller nutrient particles. Cells absorb nutrients into blood and lymph.

Ileocecal valve (sphincter)
Allows passage from small to large intestine. Prevents backflow from large intestine and controls transit through intestine.

Large intestine (colon)
Reabsorbs water and minerals. Passes waste (fiber and some water) to rectum.

Rectum
Stores waste prior to elimination.

Anus
Holds rectum closed. Opens to allow elimination.

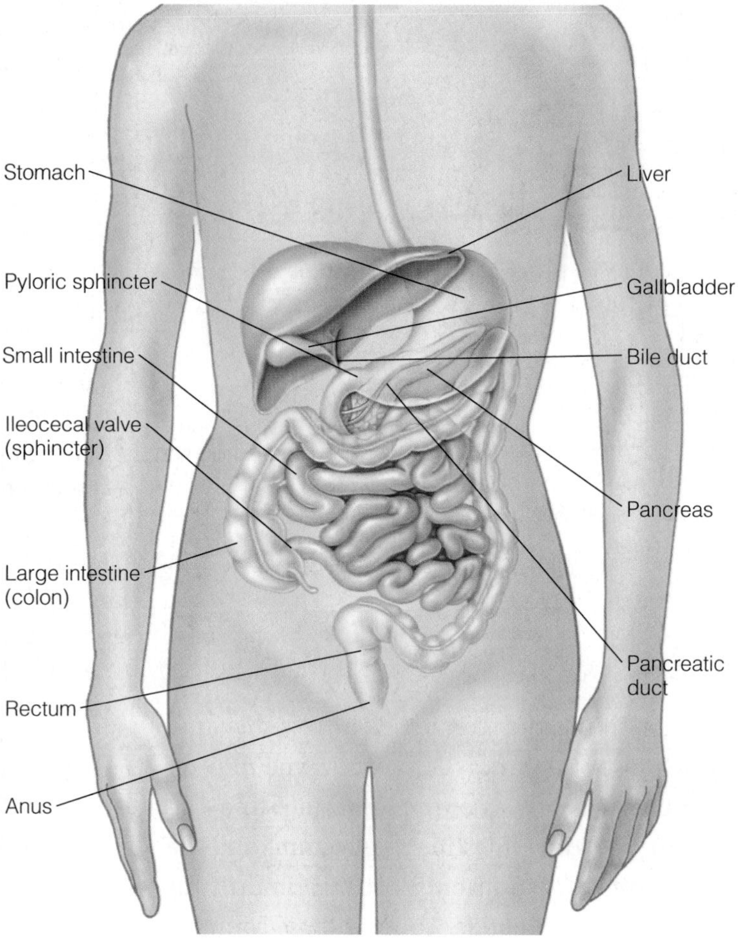

Liver
Manufactures bile salts (detergent-like substances), to help digest fats.

Gallbladder
Stores bile until needed.

Bile duct
Conducts bile from liver into small intestine.

Pancreas
Manufactures enzymes to digest energy-yielding nutrients and bicarbonate to neutralize acidic stomach contents that enter the small intestine. (Also produces insulin and glucagon.)

Pancreatic duct
Conducts pancreatic juice from pancreas into small intestine.

Labels in figure: Stomach, Pyloric sphincter, Small intestine, Ileocecal valve (sphincter), Large intestine (colon), Rectum, Anus, Liver, Gallbladder, Bile duct, Pancreas, Pancreatic duct

---

aging, although elderly individuals tend to find straining during defecation to be a greater problem than infrequent defecation.[1]

Causes of Constipation In Western societies, constipation generally correlates with low-fiber diets, low food intake, and physical inactivity. All of these factors can extend transit time, leading to increased water reabsorption within the colon and dry, hard stools that are difficult to pass. In addition, the frequency and duration of peristaltic contractions are reduced in some patients with constipation.[2]

Medical conditions often associated with constipation include diabetes mellitus, hypothyroidism, and chronic kidney disease. Neurological conditions such as multiple sclerosis, Parkinson's disease, and spinal cord injuries may cause motor problems that lead to constipation. During pregnancy, women may experience constipation when the enlarged uterus presses against the rectum and colon. Constipation is also a common side effect of several classes of medications and some dietary supplements, including opiate-containing analgesics, tricyclic antidepressants, anticonvulsants, calcium channel blockers, aluminum-containing antacids, and iron and calcium supplements.

Treatment of Constipation The primary treatment for constipation is a gradual increase in fiber intake to about 20 to 25 grams per day.[3] ◆ High-fiber diets increase

◆ The fiber DRI for women and men aged 19 to 50 years are 25 and 38 grams, respectively. Recommendations for increasing fiber intake are described on pp. 124–125, and Appendix H includes the fiber values of most common foods.

stool weights and promote a more rapid transit of materials through the colon. Foods that increase stool weight the most are wheat bran, fruits, and vegetables.[4] Bran intake can be increased by adding bran cereals and whole wheat bread to the diet or by mixing bran powder with beverages or foods. The transition to a high-fiber diet may be difficult for some people because it can increase intestinal gas, so high-fiber foods should be added gradually, as tolerated. Fiber supplements such as methylcellulose (Citrucel), psyllium (Metamucil, Fiberall), and polycarbophil (a synthetic fiber) are also effective (see Table 24-1); these supplements can be mixed with beverages and taken several times daily. Unlike other fibers, methylcellulose and polycarbophil do not increase intestinal gas.

Several other measures may also help constipation. Consuming adequate fluid prevents dehydration, which draws water from the colon to increase hydration in the rest of the body. Adding prunes or prune juice to the diet is often recommended because prunes contain compounds that have a mild laxative effect. Increasing daily exercise can help to stimulate peristalsis.

Laxatives Many laxatives can be purchased without prescription. They work by increasing stool weight, increasing the water content of the stool, or stimulating peristaltic contractions. Table 24-1 includes examples of common laxatives and describes their modes of action. Enemas and suppositories (chemicals introduced into the rectum) are also used to promote defecation; they work by distending and stimulating the rectum or by lubricating the stool.

Medical Interventions Patients with severe constipation who do not respond to dietary or laxative treatments may require medications that stimulate colonic contractions. Physical therapy and biofeedback techniques are sometimes successful in training patients to relax their pelvic muscles more effectively. Surgical interventions are a last resort and include colonic resections and colostomy operations, which are discussed later in this chapter.

High-fiber foods promote regular bowel movements.

© Andrew McClenaghan/Photo Researchers, Inc.

TABLE 24-1	Laxatives and Bulk-Forming Agents			
Laxative Type	**Active Ingredients**	**Product Examples**	**Method of Action**	**Cautions**
Fiber (bulk formers)	Methylcellulose, polycarbophil, psyllium, malt soup extract	Metamucil, Citrucel, FiberLax	Fiber supplements increase stool weight and aid in formation of soft, bulky stools. Similar effects are achieved by adding bran to the diet. For mild constipation. Safe for long-term use.	Some fiber supplements may increase flatulence. Psyllium may cause an allergic reaction.
Emollients (stool softeners)	Docusate sodium	Colace	Detergent action promotes the mixing of water with stools. Prevents formation of dry, hard stools.	Do not increase stool weight. Limited effectiveness.
Nonabsorbable sugars (osmotic laxatives)	Lactulose, sorbitol, mannitol	Cephulac	Unabsorbed sugars attract water to large intestine and promote softer stools. Must be used for several days to take effect. Safe for long-term use.	May cause flatulence and cramps. Can lose effectiveness over time.
Saline laxatives (osmotic laxatives)	Magnesium hydroxide, magnesium citrate, sodium sulfate	Milk of magnesia	Unabsorbed salts attract and retain water in large intestine and stimulate contractions.	May cause bloating and watery stools or diarrhea. Should be used with caution. Avoid using in renal patients and children.
Stimulant or irritant laxatives	Senna, bisacodyl, cascara, castor oil, aloe	Ex-Lax, Correctol, Dulcolax	Act as local irritants to colonic tissue; stimulate peristalsis and mucosal secretions. For moderate to severe constipation. Long-term use is discouraged.	Usually given only after milder treatments fail. May alter fluid and electrolyte balances. May lead to laxative dependency.

TABLE 24-2 Foods That May Increase Intestinal Gas
Apples
Beer
Broccoli
Brussels sprouts
Cabbage
Carbonated beverages
Cauliflower
Corn
Dried beans and peas
Fruit juices
Leeks
Milk products (if patients are lactose intolerant)
Onions
Peanuts
Pears
Potatoes
Turnips

◆ Reminder: The *soluble fibers* in foods are more readily fermented in the small intestine than the *insoluble fibers*.

Intestinal Gas

As mentioned in the previous section, increased intestinal gas (**flatulence**) can be an unpleasant side effect of consuming a high-fiber diet. The undigested fibers pass into the colon, where they are fermented by bacteria, which produce gas as a by-product. ◆ Other incompletely digested or poorly absorbed carbohydrates have similar effects; these include fructose, sugar alcohols (sorbitol, mannitol, and maltitol), the indigestible carbohydrates in beans (raffinose and stachyose), and some forms of resistant starch, found in grain products and potatoes. Table 24-2 lists examples of foods commonly associated with excessive gas production, although individual responses vary. Malabsorptive disorders (discussed later in this chapter) can cause considerable flatulence because the undigested nutrients pass to the colon, where they are metabolized by colonic bacteria. Swallowed air that is not expelled by belching may travel to the intestines and be a source of intestinal gas (see p. 95).

Many people blame abdominal bloating and pain on excessive gas, but these symptoms do not correlate well with increased intestinal gas.[5] In fact, most people who self-diagnose a flatulence problem have no more intestinal gas than others. Individuals who frequently experience symptoms of abdominal bloating and pain are sometimes diagnosed with irritable bowel syndrome (see pp. 774–776) or dyspepsia (discussed in Chapter 23).

Diarrhea

Diarrhea is characterized by the passage of frequent, watery stools. In most cases, it lasts for only a day or two and subsides without complication. Severe or persistent diarrhea, however, can cause dehydration and electrolyte imbalances. If chronic, it may lead to weight loss and malnutrition. Serious cases of diarrhea are often accompanied by other symptoms such as fever, cramps, dyspepsia, or bleeding, which help in diagnosing the cause.

Causes of Diarrhea Diarrhea is a complication of many different medical conditions and may also be induced by infections, medications, or dietary substances. It results from inadequate fluid reabsorption in the intestines, sometimes in conjunction with an increase in intestinal secretions.[6] Most cases of diarrhea are categorized as *osmotic* or *secretory*. In *osmotic diarrhea*, unabsorbed nutrients or other substances attract water to the colon and increase fecal water content; the usual causes include lactase deficiency, high intakes of poorly absorbed sugars (such as sorbitol, mannitol, or fructose), and ingestion of laxatives that contain magnesium or phosphates. In *secretory diarrhea*, the fluid secreted by the intestines exceeds the amount that can be reabsorbed by intestinal cells. Secretory diarrhea is often due to bacterial food poisoning but can also be caused by various chemical substances and inflammatory conditions. Motility disorders can also cause diarrhea: rapid transit within the colon shortens the contact time needed for fluid reabsorption, whereas slow transit can promote bacterial overgrowth (discussed in a later section) and thereby alter intestinal secretions.[7]

Acute cases of diarrhea start abruptly and may persist for several weeks; they are frequently caused by viral, bacterial, or protozoal infections or occur as a side effect of medications. Chronic diarrhea, which persists for a month or longer, can result from altered GI tract motility, intestinal inflammation, malabsorptive and endocrine disorders, infectious diseases, radiation treatment, and many other conditions. As mentioned in earlier chapters, diarrhea is a frequent complication of tube feedings or may occur when enteral feedings are resumed after a period of bowel rest (see Chapters 20 and 21).

Medical Treatment of Diarrhea Correcting the underlying medical disorder is the first step in treating diarrhea. For example, antibiotics are prescribed for treat-

flatulence: the condition of having excessive intestinal gas, which causes abdominal discomfort.

ing infections. If a medication causes diarrhea, a different drug may be prescribed. If certain foods are responsible, they can be omitted from the diet. Bulk-forming agents such as psyllium (Metamucil) or methylcellulose (Citrucel) can help to reduce the liquidity of the stool. If chronic diarrhea does not respond to treatment, antidiarrheal drugs may be prescribed to slow GI motility or reduce intestinal secretions. Probiotics ◆ may be beneficial for certain types of diarrhea (especially diarrhea caused by infections), but standard treatment protocols have not been developed.[8] People with severe, **intractable** diarrhea sometimes require total parenteral nutrition.

Oral Rehydration Therapy Severe diarrhea necessitates the replacement of lost fluid and electrolytes. Oral rehydration solutions can be purchased or easily mixed using water, salts, and glucose or sucrose (see the recipe in the margin). ◆ The addition of carbohydrate to the rehydration solution facilitates sodium and water absorption. Commercial sports drinks are not recommended for rehydration because their sodium contents are too low to replace losses resulting from severe diarrhea; they can be used if accompanied by salty snack foods, however.[9] When diarrhea results in extreme dehydration, intravenous solutions are used to quickly replenish fluid and electrolytes.

Nutrition Therapy for Diarrhea Because diarrhea can develop for numerous reasons, the nutrition prescription depends on the medical diagnosis and severity of the condition. The dietary treatment often recommended is a low-residue, low-fat, lactose-free diet.[10] The low-residue diet limits foods that contribute to colonic residue, such as those with significant amounts of fiber, resistant starch, lactose (in lactose-intolerant individuals), fructose, and sugar alcohols. Fructose and sugar alcohols, which are poorly absorbed, retain fluids in the colon and contribute to osmotic diarrhea. Similarly, milk products may worsen osmotic diarrhea in persons who are lactose intolerant. Avoidance of fatty foods is recommended because they can sometimes aggravate diarrhea. Gas-producing foods can increase intestinal distention and cause additional discomfort. Patients should avoid coffee and tea because caffeine stimulates GI motility and can thereby reduce water reabsorption. In the treatment of formula-fed infants, apple pectin or banana flakes are sometimes added to formulas to help thicken stool consistency. Table 24-3 lists examples of foods that may worsen diarrhea, although individual tolerances vary.

◆ *Probiotics* are live bacteria provided in foods and dietary supplements for the purpose of preventing or treating disease. Highlight 24 describes the potential health benefits of probiotics.

◆ An oral rehydration solution can be mixed from the following ingredients:
- $^1/_2$ tsp sodium chloride (table salt)
- $^1/_3$ tsp potassium chloride (salt substitute)
- $^3/_4$ tsp sodium bicarbonate (baking soda)
- $1^1/_3$ tbs sugar
- 1 qt water

intractable: not easily managed or controlled.

TABLE 24-3 Foods That May Worsen Diarrhea

Foods to Avoid	Rationale	Selected Examples
High-fiber foods	They increase colonic residue.	Breads and cereals with more than 2 g fiber per serving, fruits and vegetables with peels or skins.
Foods with indigestible carbohydrates	They contribute to osmotic diarrhea.	Artichokes, asparagus, brussels sprouts, cabbage, dried beans and peas, fruit, garlic, green beans, leeks, onions, wheat, zucchini.
Foods that contain fructose or sugar alcohols	They contribute to osmotic diarrhea.	Dried fruits, fresh fruits (except bananas), fruit juices, fructose-sweetened soft drinks, sugar-free gums and candies.
Milk products, if person is lactose intolerant	They contribute to osmotic diarrhea.	Milk and milk products.
Gas-producing foods	They increase abdominal discomfort.	Foods with poorly digested or absorbed carbohydrates (including foods listed in the three categories directly above).
Caffeine-containing beverages	They increase intestinal motility.	Coffee, tea, colas, energy drinks.

[a]Individual tolerances vary; the foods to avoid are best determined by trial and error.
SOURCES: American Dietetic Association, *Nutrition Care Manual* (Chicago: American Dietetic Association, 2007); J. S. Barrett and P. R. Gibson, Clinical ramifications of malabsorption of fructose and other short-chain carbohydrates, *Practical Gastroenterology* (August 2007): 51–65; M. H. Beers and coeditors, *The Merck Manual of Diagnosis and Therapy* (Whitehouse Station, N.J.: Merck Research Laboratories, 2006), pp. 77–80.

IN SUMMARY

The most common intestinal problems include constipation, intestinal gas, and diarrhea. Constipation accompanies a wide range of conditions but generally correlates with low-fiber diets, low food intake, and physical inactivity. Intestinal gas is largely produced by bacteria that colonize the colon and is often associated with nutrient malabsorption. Diarrhea can result from intestinal infections, malabsorption, motility disorders, medications, or dietary substances and may require oral rehydration therapy to replace fluid and electrolyte losses. Dietary modifications may help to improve bowel function and alleviate intestinal discomfort.

TABLE 24-4 Potential Causes of Malabsorption

Genetic disorders
- Enzyme deficiencies

Pancreatic disorders
- Chronic pancreatitis
- Cystic fibrosis

Intestinal disorders
- Bacterial overgrowth
- Celiac disease
- Crohn's disease
- HIV enteropathy
- Radiation enteritis

Intestinal infections
- AIDS-related infections
- Giardiasis

Liver disease (bile insufficiency)

Surgeries
- Gastric bypass surgery
- Intestinal bypass surgery
- Intestinal resection (short bowel syndrome)

resection: the surgical removal of part of an organ or body structure.

steatorrhea (stee-AT-or-REE-ah): excessive fat in the stools resulting from fat malabsorption; characterized by stools that are loose, frothy, and foul smelling due to a high fat content.
- **steat** = fat
- **rheo** = flow

soaps: chemical compounds that form between fatty acids and positively charged minerals.

Malabsorption Syndromes

To digest and absorb nutrients, we depend on normal digestive secretions and healthy intestinal mucosa. Disorders of the pancreas that cause enzyme deficiencies can impair nutrient digestion and lead to widespread malabsorption. Intestinal inflammation, certain medications, and cancer therapies can damage mucosal tissue and interfere with its absorptive functions. In some cases, the treatment of an intestinal disorder requires surgical removal of a section **(resection)** of the small intestine, leaving minimal absorptive capacity in the portion that remains. Table 24-4 lists examples of diseases and treatments that are frequently associated with malabsorption.

Malabsorption rarely involves a single nutrient. When malabsorption is caused by pancreatic enzyme deficiencies, all macronutrients—protein, carbohydrate, and fat—may be affected. If fat is malabsorbed, essential fatty acids, fat-soluble vitamins, and minerals are usually malabsorbed as well. Malabsorptive disorders and their treatments can further tax nutrition status by causing complications that alter food intake, raise nutrient needs, and incur additional nutrient losses.

Fat Malabsorption

Fat is the nutrient most frequently malabsorbed, because both digestive enzymes and bile must be present for its digestion. Thus, fat malabsorption often develops when an illness interferes with the production or secretion of either pancreatic lipase or bile. For example, pancreatitis and cystic fibrosis can decrease the secretion of pancreatic lipase, whereas severe liver disease can reduce bile availability. Fat malabsorption also results from conditions that damage the intestinal mucosa, such as inflammatory bowel diseases or radiation treatment for cancer. Motility disorders that cause rapid gastric emptying or rapid intestinal transit can also lead to fat malabsorption because they prevent the normal mixing of dietary fat with lipase and bile.

Fat malabsorption is often evidenced by **steatorrhea,** the presence of excessive fat in the stools. Steatorrhea can be evaluated by performing a 72-hour fecal collection and measuring the stool's fat content. In healthy individuals, the fecal fat excretion is generally less than 7 grams per day, although diarrhea can raise the fecal fat output to up to 14 grams per day.[11]

Consequences of Fat Malabsorption Fat malabsorption is associated with losses of food energy, essential fatty acids, fat-soluble vitamins, and some minerals (see Figure 24-2). Weight loss is possible unless the individual consumes alternative sources of energy. Deficiencies of fat-soluble vitamins and essential fatty acids are common in chronic conditions. Malabsorption of some minerals, including calcium, magnesium, and zinc, often develops because the minerals form **soaps** with unabsorbed fatty acids and bile acids. Calcium deficiency can lead to bone loss, which is further aggravated by the vitamin D deficiency that is common in cases of fat malabsorption.

FIGURE 24-2 The Consequences of Fat Malabsorption

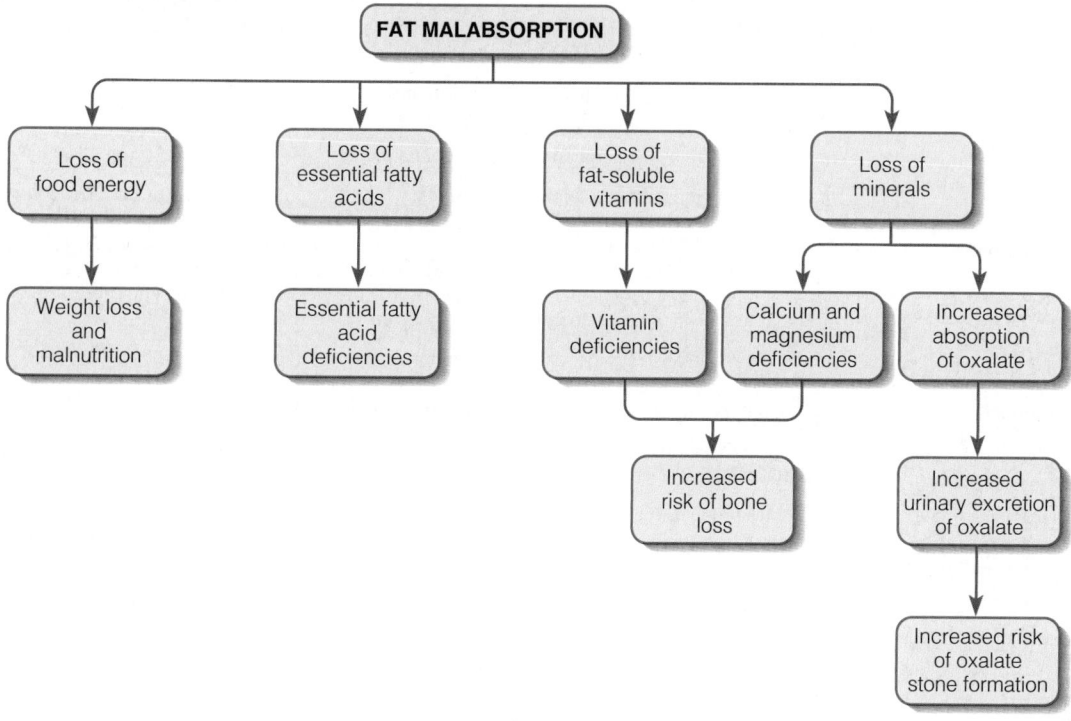

Another consequence of fat malabsorption is an increased risk of kidney stones, which are most often composed of calcium oxalate. ◆ The oxalates in foods ordinarily bind to calcium in the small intestine and are excreted in the stool. If calcium instead binds to fatty acids or bile acids, oxalates are free to be absorbed into the blood and are ultimately excreted in the urine. The risk of developing oxalate stones increases when urinary oxalate levels are high. Kidney stones are discussed further in Chapter 28.

Dietary Adjustments If steatorrhea does not improve, a fat-restricted diet may be recommended (see Table 24-5 on p. 762). The objectives of fat restriction are to relieve intestinal symptoms that are aggravated by fat intake (usually diarrhea and flatulence) and to reduce vitamin and mineral losses. Fat should not be restricted more than necessary because fat is an important source of energy. Medium-chain triglycerides (MCT), which do not require lipase or bile for digestion and absorption, can be used as an alternative source of dietary fat (although MCT oil does not provide essential fatty acids). The "How to" on p. 763 offers suggestions for following a fat-restricted diet and for using MCT oil.

Bacterial Overgrowth

Ordinarily, the stomach and small intestine are protected from **bacterial overgrowth** by gastric acid, which destroys bacteria, and by peristalsis, which flushes bacteria through the small intestine before they multiply.[12] When bacterial overgrowth does occur, it disrupts fat digestion and absorption, because excessive bacteria in the small intestine dismantle the bile acids needed for fat emulsification. Consequently, deficiencies of the fat-soluble vitamins may eventually develop. The bacteria also compete for vitamin B_{12}, impairing its absorption and increasing the risk of vitamin B_{12} deficiency. Although symptoms of bacterial overgrowth are often minor and nonspecific, severe cases may lead to chronic diarrhea, steatorrhea, abdominal discomfort, bloating, and weight loss.

◆ Reminder: *Oxalates* are plant compounds that bind with some minerals to form complexes that the body cannot absorb. They are present in green leafy vegetables such as beet greens and spinach.

bacterial overgrowth: excessive bacterial colonization of the stomach and small intestine; may be caused by low gastric acidity, altered gastrointestinal motility, mucosal damage, or contamination.

TABLE 24-5 Fat-Restricted Diet

General guidelines	*For fat restriction of 25 grams per day:* Limit meat and meat alternates to 4 ounces daily (cooked weight); limit fat equivalents to 1 per day (see Fats section below).
	For fat restriction of 50 grams per day: Limit meat and meat alternates to 6 ounces daily (cooked weight); limit fat equivalents to 3 to 5 per day.
	To raise fat content: Allow additional servings of meat or fat.
	To lower fat content: Restrict servings of meat or fat.
Meat and meat alternates	*Recommended:* Choose lean meat, fish, and poultry only. Preparation methods can include broiling, roasting, grilling, or boiling. Trim visible fat and remove poultry skin before consuming. For sandwiches, select turkey breast or other low-fat luncheon meats. Limit eggs to 2 per week or use low-fat egg substitutes. Meat alternates include tofu and dried beans and peas.
	Avoid: Pork and beans, sausage, bacon, frankfurters, spareribs, duck, goose, tuna packed in oil, fried meats.
Milk and milk products	Choose milk products that contain less than 1 gram fat per serving.
	Recommended: Fat-free milk, fat-free yogurt, fat-free sour cream substitutes, fat-free half-and-half and cream substitutes, fat-free cheeses.
	Avoid: Milk products that are not fat-free.
Breads, cereals, rice, and pasta	Choose breads, cereals, rice, and pasta dishes that contain less than 1 gram fat per serving.
	Recommended: Whole-grain breads, soda crackers, cooked cereals and most cold cereals, plain tortillas, bagels, English muffins, fat-free muffins, graham crackers, plain rice, plain noodles and pasta.
	Avoid: Biscuits, pancakes, waffles, doughnuts, granola, snack crackers that contain fat, corn chips, cornbread, fried rice, pasta sauces with added fats.
Vegetables	Choose vegetables that contain less than 1 gram fat per serving.
	Recommended: All vegetables prepared without added fats.
	Avoid: Buttered or fried vegetables, creamed vegetables, au gratin style, french-fried potatoes, olives, sauces with added fats.
Fruit	Choose fruits that contain less than 1 gram fat per serving.
	Recommended: All fruits prepared without added fats.
	Avoid: Avocado, fruit dips made with fat or coconut.
Desserts	Choose desserts that contain less than 1 gram fat per serving.
	Recommended: Sherbet, fruit ices, fruit whips, flavored gelatin, angel food cake, meringues, fat-free puddings, fat-free baked products, fat-free ice cream or frozen yogurt, fat-free candies (marshmallows, jelly beans, hard candy).
	Avoid: Cakes, cookies, pies, and pastries made with fat; puddings made with whole milk or eggs; ice cream; candies made with fat (caramel, chocolates).
Fats	Choose 1 fat equivalent daily if fat restriction is 25 grams per day, and 3 to 5 fat equivalents daily if fat restriction is 50 grams per day.
	One fat equivalent is equal to: Vegetable oil: 1 tsp. Butter, margarine: 1 tsp, or 1 tbs diet margarine. Mayonnaise: 1 tsp, or 1 tbs reduced-kcalorie mayonnaise. Salad dressing: 1 tbs, or 2 tbs reduced-kcalorie dressing. Nuts: 6 almonds or cashews, 10 peanuts, 2 tsp peanut butter, 4 halves walnuts or pecans.
Beverages	Choose beverages that contain less than 1 gram fat per serving.
	Recommended: Coffee, tea, soft drinks, juices, fat-free milk, coffee substitutes.
	Avoid: Beverages made with milk (unless fat-free milk) or added cream, chocolate milk, eggnog, milk shakes.

SOURCE: Adapted from American Dietetic Association, *Manual of Clinical Dietetics* (Chicago: American Dietetic Association, 2000), pp. 697–702.

Causes of Bacterial Overgrowth Conditions that impair intestinal motility and allow material to stagnate can increase susceptibility to bacterial overgrowth. For example, in some types of gastric surgery, a portion of the small intestine is bypassed, preventing the flow of material in the bypassed region and allowing bacteria to flourish (see the "blind loop" shown in Figure 23-3 on p. 742). Intestinal motility can also be reduced by strictures, obstructions, and diverticula (protrusions) in the small intes-

HOW TO Follow a Fat-Restricted Diet

Fat-restricted diets can be difficult to follow. Fats add flavors, aromas, and textures to foods—characteristics that make foods more enjoyable. Unlike some diets that can be introduced gradually, a fat-restricted diet is often implemented immediately, allowing little time for adaptation. These suggestions may help:

- Fat is better tolerated if provided in small portions. Divide the day's allotment into several servings that can be consumed throughout the day.
- Use variety to enhance enjoyment of meals: vary flavors, textures, colors, and seasonings.
- Look for fat-free items when grocery shopping. Incorporate fat-free ingredients when preparing favorite recipes.
- Try fat-free and low-fat condiments to improve the diet's palatability. Experiment with herbs and spices. Instead of butter, use fruit butters on toast. Use butter-flavored granules on vegetables. Replace mayonnaise on sandwiches with spicy mustard. Replace salad dressings with flavored vinegars.
- Avoid products that contain the fat substitute olestra, which may aggravate GI symptoms.

If patients are interested in using MCT oil:

- Explain that MCT products are expensive, but that the cost is sometimes covered by medical insurance.
- Advise patients to add MCT oil to the diet gradually. Diarrhea and abdominal cramps may result if too much is used at once. Tolerance to MCT oil may improve in time.
- Advise patients that MCT oil may have an unpleasant taste when used alone. Suggest using MCT oil in recipes as a substitute for regular oil. MCT oil can replace oil in salad dressings, be incorporated into sauces, and be used in cooking or baking. It can also be added to fat-free milk products to make milk shakes.
- Point out that MCT oil should not be used to fry foods because it decomposes at lower temperatures than most cooking oils.

NOTE: MCT = medium-chain triglyceride.

tine, as well as by chronic diseases such as diabetes mellitus, scleroderma, and chronic kidney disease.[13]

Reduced secretions of gastric acid can also lead to bacterial overgrowth. Possible causes include atrophic gastritis, acid-suppressing medications, and acid-reducing surgery (vagotomy) for peptic ulcer disease.

Treatment for Bacterial Overgrowth Treatment may include antibiotics to suppress bacterial growth and surgical correction of the anatomical defects that contribute to a motility disorder. Medications may be given to stimulate peristalsis. Dietary supplements are provided to correct nutrient deficiencies, especially deficiencies in the fat-soluble vitamins A, D, and E; calcium; and vitamin B_{12}.

IN SUMMARY

Malabsorption syndromes can be caused by reduced digestive secretions or damaged intestinal mucosa. Malabsorption usually affects multiple nutrients and causes complications that impair nutrition status further. Fat malabsorption, usually indicated by the development of steatorrhea, is associated with the loss of food energy and deficiencies of essential fatty acids, fat-soluble vitamins, and some minerals. Bacterial overgrowth can result from impaired peristalsis or reduced gastric acidity and typically causes malabsorption of fat and vitamin B_{12}.

Conditions Affecting the Pancreas

As mentioned previously, pancreatic disorders can lead to maldigestion and malabsorption due to the impaired secretion of digestive enzymes. ◆ This section describes several pancreatic illnesses that are characterized by widespread malabsorption.

◆ Reminder: The digestive secretions of the pancreas include bicarbonate and digestive enzymes. The bicarbonate neutralizes the acidic gastric contents that enter the duodenum, and the digestive enzymes break down protein, carbohydrate, and fat.

Pancreatitis

Pancreatitis is an inflammatory disease of the pancreas. The inflammation causes damage to pancreatic tissue and the release of active pancreatic enzymes, which may destroy the surrounding tissues. Although mild cases may subside in a few days, other cases can persist for weeks or months. Chronic pancreatitis can lead to irreversible damage to pancreatic tissue and permanent loss of function.

Acute Pancreatitis Acute pancreatitis is short-lived and does not cause permanent damage. It is most often caused by gallstones or excessive alcohol use—factors that account for more than 70 percent of acute cases.[14] Less common causes include hypertriglyceridemia (triglyceride concentrations higher than 1000 milligrams per deciliter), exposure to toxins, and the use of some medications.

Common symptoms include severe abdominal pain, nausea and vomiting, and abdominal distention. Elevated serum levels of amylase and lipase—released by damaged pancreatic tissue into the blood—help to confirm the diagnosis. In most patients, the condition resolves within a week with no complications. More severe cases may lead to renal failure, sepsis, and other complications that often require prolonged hospitalization.

Nutrition Therapy for Acute Pancreatitis Oral fluids and food are withheld until pain and tenderness have subsided, and fluids and electrolytes are supplied intravenously. After three to seven days, patients may be able to consume small amounts ($^1/_2$ to 1 cup) of fluids; either zero-kcalorie fluids or clear liquids may be provided initially.[15] If tolerated, the diet progresses from liquid feedings to solid foods, given in small, frequent feedings. Because fat stimulates the pancreas more than other nutrients, a low-fat diet may be better tolerated at first. In severe pancreatitis, jejunal tube feedings are sometimes necessary; either standard formulas or elemental formulas ◆ may be used, depending on patient tolerance.[16] Protein and energy needs are high in severe cases due to the catabolic and hypermetabolic effects of inflammation.

Chronic Pancreatitis Chronic pancreatitis is characterized by permanent damage to pancreatic tissue, resulting in impaired secretion of digestive enzymes. About 70 to 90 percent of chronic pancreatitis cases are caused by excessive alcohol consumption.[17] Unlike acute pancreatitis, chronic pancreatitis is not caused by gallstones. In children, most cases are attributable to cystic fibrosis, discussed in a later section.

In chronic pancreatitis, the abdominal pain is often severe and unrelenting and worsens with eating. Analgesics or opiate drugs are often needed for pain control. Fat maldigestion develops sooner than maldigestion of protein or carbohydrate, and steatorrhea is common in advanced cases. Food avoidance (due to pain associated with eating) and malabsorption may lead to weight loss and malnutrition. Advanced cases are associated with reductions in both insulin and glucagon secretions, and diabetes eventually develops in up to 80 percent of patients.[18]

Nutrition Therapy for Chronic Pancreatitis The objectives of nutrition therapy are to improve nutrition status, reduce malabsorption, and prevent symptom recurrence. Protein and energy needs are high in patients who have lost weight or are malnourished. Dietary supplements are often needed to correct nutrient deficiencies, which may be due to malabsorption or to the alcohol abuse that caused the disease. Patients must avoid alcohol completely because it can worsen pancreatic function.[19]

Steatorrhea is usually treated with pancreatic enzyme replacement.[20] Pancreatic enzymes are often **enteric coated** to resist the acidity of the stomach and do not dissolve until the pH is above 5.5. If nonenteric-coated preparations are used, acid-suppressing drugs may be required. Fecal fat concentrations are monitored to determine if the enzyme treatment has been effective. A low-fat diet may be necessary if steatorrhea is resistant to treatment, in which case medium-chain triglycerides can be used to replace other fat sources. The accompanying Case Study allows you to apply your knowledge of chronic pancreatitis to a clinical situation.

◆ Reminder: An *elemental formula* contains hydrolyzed nutrients that require minimal digestion and are easily absorbed.

enteric coated: refers to medications or enzyme preparations that can withstand gastric acidity and dissolve only at a higher pH.

CASE STUDY   Retired Executive with Chronic Pancreatitis

Taylor Gray is a 62-year-old man with chronic pancreatitis. He was forced to resign his position as president of an import-export company in his early 50s, when his problems with alcohol and declining health seriously impaired his abilities to run the company. His wife divorced him shortly thereafter, and he currently lives alone.

At 5 feet 11 inches tall, Mr. Gray weighs 145 pounds. He continues to experience frequent, severe abdominal pain and steatorrhea. Mr. Gray has been advised to follow a high-kcalorie, high-protein diet with no fat restrictions and uses enzyme replacement with meals, but he has difficulty eating enough food. He has not used alcohol for 8 months. Mr. Gray takes pain medications, acid-suppressing drugs, and a multivitamin daily.

1. What are the possible reasons for Mr. Gray's difficulty maintaining weight? What suggestions may help? Why was he advised to consume a high-protein diet?
2. Explain why malabsorption develops in chronic pancreatitis. Which nutrients are most likely to be affected?
3. How can Mr. Gray's physician determine whether the enzyme replacement therapy is effective? Explain why Mr. Gray must continue to use acid suppressants.
4. If Mr. Gray continues to experience steatorrhea, what measures would you suggest? What complications may develop if the steatorrhea is not controlled?

Cystic Fibrosis

Cystic fibrosis is the most common life-threatening genetic disorder among Caucasians, with an incidence of approximately 1 in 2000 to 1 in 3000 white births. In the United States, about 30,000 people are affected.[21] Cystic fibrosis is caused by a genetic mutation that disturbs the transport of chloride ions in **exocrine** glands. This defect results in thickened glandular secretions and a broad range of serious complications. Until a few decades ago, few infants born with cystic fibrosis survived to adulthood. Now, with early detection and advances in medical treatment, the average life span has extended beyond 30 years of age, with many patients surviving into their 50s.[22]

Consequences of Cystic Fibrosis Cystic fibrosis is characterized by abnormal chloride and sodium levels in exocrine secretions. The secretions are unusually viscous and can obstruct the ducts that normally allow their passage. The blockages that develop can disrupt tissue function and cause tissue damage. The major complications of cystic fibrosis involve the lungs, pancreas, ◆ and sweat glands:

- *Lung disease.* The abnormally thick mucus secretions cause obstructions in many of the small airways of the lungs. The obstructions lead to chronic coughing and persistent respiratory infections, both of which contribute to progressive inflammation in the bronchial tissues. The lung damage that eventually develops results in breathing difficulties and lower exercise tolerance. As with other obstructive airway diseases, ◆ in cystic fibrosis nutrition status becomes impaired due to hypermetabolism, the greater energy cost of labored breathing, and anorexia (loss of appetite). The chronic respiratory infections raise energy needs further.

- *Pancreatic disease.* Approximately 85 percent of cystic fibrosis patients develop thickened pancreatic secretions that obstruct the pancreatic ducts, causing digestive enzymes to accumulate in the pancreas and destroy pancreatic tissue. Fewer pancreatic enzymes reach the small intestine, leading to malabsorption of protein, fat, and fat-soluble vitamins. Other problems that may develop over time include pancreatitis, hyperglycemia (due to destruction of the insulin-producing cells), and diabetes.[23]

- *Other complications.* Because cystic fibrosis affects all exocrine secretions, complications typically develop in many other tissues or organs. Salt losses in sweat are usually excessive, increasing the risk of dehydration. Intestinal

◆ Reminder: The pancreas secretes digestive enzymes and bicarbonate into the digestive tract (*exocrine* secretions) and the hormones insulin and glucagon into the bloodstream (*endocrine* secretions).

◆ Chapter 22 describes the nutrition problems associated with chronic obstructive lung diseases.

cystic fibrosis: an inherited disease characterized by the presence of abnormally viscous exocrine secretions; often leads to respiratory illness and pancreatic insufficiency.

exocrine: pertains to external secretions, such as those of the mucous membranes or the skin. Opposite of *endocrine*, which pertains to hormonal secretions into the blood.
- **exo** = outside
- **krinein** = to secrete

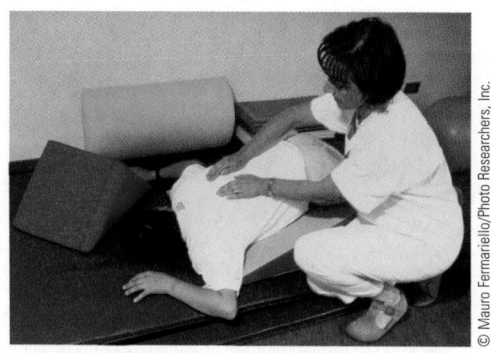

Postural drainage, a type of physical therapy used in treating cystic fibrosis, helps to clear the thick, sticky secretions that block airways and increase infection risk.

obstruction is a common symptom in newborn infants and may also occur in older patients. Gallbladder and liver diseases may result from bile duct obstructions. Abnormalities in genital tissues cause sterility in men and reduced fertility in women.

Nutrition Therapy for Cystic Fibrosis Children with cystic fibrosis are chronically undernourished, grow poorly, and have difficulty maintaining normal body weight. Their energy and protein needs are high due to increased requirements, nutrient malabsorption, and reduced food consumption. To achieve normal growth, energy intakes should be 20 to 50 percent higher than DRI values, and protein should provide 15 to 20 percent of kcalories.[24] To compensate for fat malabsorption, about 35 to 40 percent of kcalories should come from fat. A patient with cystic fibrosis is therefore encouraged to consume high-kcalorie and high-fat foods, eat frequent meals and snacks, and supplement meals with milk shakes or liquid dietary supplements. Supplemental tube feedings can help to improve nutrition status if energy intakes are inadequate.

Pancreatic enzyme replacement therapy is a central feature of cystic fibrosis treatment. Supplemental enzymes must be included with every meal or snack. For infants and small children, the contents of capsules are mixed in small amounts of liquid or a soft food (such as applesauce) and fed with a spoon. Enzyme dosages may need to be adjusted if malabsorption continues, as evidenced by poor growth or GI symptoms such as steatorrhea, intestinal gas, or abdominal pain.

The risk of nutrient deficiency depends on the degree of malabsorption. The nutrients of greatest concern include the fat-soluble vitamins, essential fatty acids, and calcium. Multivitamin and fat-soluble vitamin supplements are routinely recommended. The liberal use of table salt and salty foods is encouraged to make up for losses of sodium in sweat. The accompanying Case Study checks your understanding of the nutrition therapy for a child with cystic fibrosis.

IN SUMMARY

Chronic pancreatic disorders can lead to widespread maldigestion and malabsorption due to impaired secretion of digestive enzymes. Acute pancreatitis is short-lived and does not cause permanent damage, but it requires that food and liquids be withheld until healing has occurred. Chronic pancreatitis, most often caused by alcohol abuse, results in digestive enzyme deficiencies and requires pancreatic enzyme replacement therapy. Cystic fibrosis, a genetic disorder associated with thickened exocrine secretions, causes obstructive lung disease and pancreatic damage. Children with cystic fibrosis have high protein and energy requirements and must use pancreatic enzyme replacement therapy and dietary supplements to reverse malnutrition.

CASE STUDY   Child with Cystic Fibrosis

Julie is a 7-year-old girl diagnosed with cystic fibrosis. Symptoms of steatorrhea and failure to gain weight in infancy prompted the tests that led to the diagnosis. She is currently 45 inches tall and weighs 42 pounds. Her height for age and weight for age fall near the 10th percentile (see Appendix E). Julie eats regular foods during the day and receives additional nutrients by tube feedings delivered overnight.

1. What do the height and weight percentiles tell you about Julie's nutrition status? Why is growth failure common in children with cystic fibrosis?
2. Explain why Julie's energy needs are so much higher than normal. Describe the elements of the diet that Julie should follow to gain weight. Would her energy requirements change if she developed a respiratory infection?
3. Explain to Julie's parents how to use enzyme replacement therapy effectively.
4. Julie's parents are hoping to discontinue the nightly tube feedings. Do you think the tube feedings are necessary? Why or why not?

Conditions Affecting the Small Intestine

When the intestinal mucosa is damaged due to inflammation, infection, or other causes, malabsorption is the likely outcome. This section discusses *celiac disease* and *inflammatory bowel diseases* (intestinal conditions that can impair mucosal function) and *short bowel syndrome* (the malabsorptive condition that results when a substantial portion of the small intestine is surgically removed).

Celiac Disease

Celiac disease is an immune disorder characterized by an abnormal immune response to a protein fraction in **wheat gluten** ◆ and to related proteins in barley and rye. The reaction to gluten causes severe damage to the intestinal mucosa and subsequent malabsorption. Celiac disease may affect as many as 1 in every 133 persons in the United States.[25]

Consequences of Celiac Disease The immune reaction to gluten can cause striking changes in intestinal tissue. In affected areas, the absorptive surface appears flattened due to the shortening or absence of villi and overdeveloped crypts (see the photos). ◆ The reduction in mucosal surface area (and, therefore, in intestinal digestive enzymes) can be substantial. The damage may be restricted to the duodenum or may involve the full length of the small intestine. Individuals with severe disease may malabsorb all nutrients to some degree, especially the macronutrients, fat-soluble vitamins, electrolytes, calcium, magnesium, zinc, iron, folate, and vitamin B_{12}.[26] As a result of nutrient deficiencies, patients often develop anemia and have low bone mineral density.

Symptoms of celiac disease include GI disturbances such as diarrhea, steatorrhea, and flatulence. Because lactase deficiency can result from mucosal damage, the GI symptoms may be exacerbated by milk products. Children with celiac disease often have stunted growth and are severely underweight. Adults may develop bone disorders and have fertility problems. Individuals with celiac disease who do not eliminate gluten from the diet are at increased risk of developing intestinal and lymphatic cancers.[27]

Some gluten-sensitive individuals may have few GI symptoms but react to gluten by developing a severe rash. This condition is called **dermatitis herpetiformis** and requires dietary adjustments similar to those for celiac disease.

Nutrition Therapy for Celiac Disease The treatment for celiac disease is lifelong adherence to a gluten-free diet. Improvement in symptoms is often evident within several weeks, although mucosal healing can sometimes take years. If lactase deficiency is suspected, patients should avoid lactose-containing foods until the intestine has recovered. Dietary supplements can be used to meet micronutrient needs and reverse deficiencies.[28]

The gluten-free diet eliminates foods that contain wheat, barley, and rye (see Table 24-6 on p. 768). Because many foods contain ingredients derived from these grains, foods that are problematic are not always obvious. Even small amounts of gluten may cause symptoms in some persons, so patients should carefully check ingredient lists on food labels. Gluten-containing products that may be overlooked include beer, caramel coloring, coffee substitutes, communion wafers, imitation meats, malt syrup, medications, salad dressings, and soy sauce. Special gluten-free products can be purchased to replace common food items such as bread, pasta, and cereals. Although somewhat expensive, these foods increase food choices and allow celiac patients to enjoy foods that would otherwise be forbidden.

Although most people with celiac disease can safely consume moderate amounts of oats, oats grown in the United States may be contaminated with wheat, barley, or rye.[29] Oats are often grown in rotation with other grains and may

◆ The protein in wheat gluten that has toxic effects in celiac disease is called *gliadin*.

◆ Reminder: *Crypts* are tubular glands that lie between the intestinal villi and secrete intestinal juices into the small intestine.

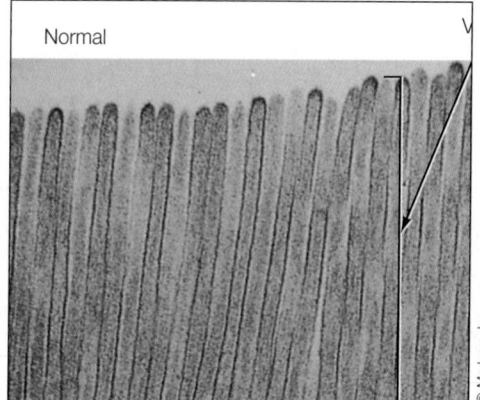

In the healthy intestine, the villi greatly increase the absorptive surface area.

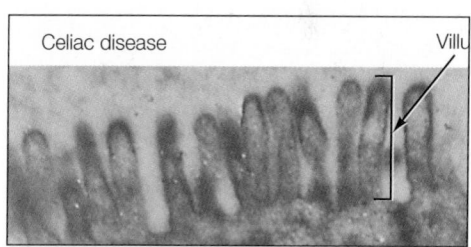

In celiac disease, the villi may be shortened or absent, resulting in substantial reductions in nutrient absorption.

celiac (SEE-lee-ack) **disease:** a condition characterized by an abnormal immune reaction to wheat gluten that causes severe intestinal damage and nutrient malabsorption; also called *gluten-sensitive enteropathy* or *celiac sprue*.

wheat gluten (GLU-ten): a family of water-insoluble proteins in wheat; includes the gliadin (GLY-ah-din) proteins that are toxic to persons with celiac disease.

dermatitis herpetiformis (DERM-ah-TYE-tis HER-peh-tih-FOR-mis): a gluten-sensitive disorder characterized by a severe skin rash.

TABLE 24-6 Gluten-Free Diet

Meat and meat alternates	• *Recommended:* Fresh, frozen, salted, and smoked meats (unless processed meats contain any prohibited grains); products made with hydrolyzed vegetable protein (HVP) or hydrolyzed plant protein (HPP); eggs; dried beans and peas; tofu. • *Questionable:* Luncheon meats, sandwich spreads, meat loaf, frozen burgers, sausage, imitation meat products, meat extenders, egg substitutes, dried egg products, dry-roasted nuts, peanut butter. • *Avoid:* Products that are breaded or prepared in cream sauces, gravies.
Milk and milk products	• *Recommended:* Milk, buttermilk, plain yogurt, cheese. • *Questionable:* Milk shakes, cheese spreads, flavored yogurt, frozen yogurt, chocolate milk. • *Avoid:* Malted milk and malted milk powders.
Breads, cereals, rice, and pasta	• *Recommended:* Breads, baked products, and cereals made with corn, rice, soy, potato starch, potato flour, hominy, buckwheat, millet, teff, sorghum, amaranth, quinoa, arrowroot, and tapioca; pasta and noodles made with grains or starches listed above; corn tacos and corn tortillas. • *Questionable:* Oatmeal and oat bran; rice crackers, rice cakes, and corn cakes. • *Avoid:* Breads, baked products, cereals, tortillas, or pastas made with wheat, rye, barley, triticale, spelt, kamut, wheat germ, wheat bran, graham flour, durum flour, wheat starch, bulgur, farina, or semolina from wheat; commercially prepared mixes for biscuits, cornbread, muffins, pancakes, or waffles; malt and malt flavoring; pretzels; matzos.
Fruits and vegetables	• *Recommended:* Any unprocessed fruits or vegetables. • *Questionable:* French fries, especially in fast-food restaurants; commercial salad dressings; fruit pie fillings; dried fruits. • *Avoid:* Scalloped potatoes (with wheat flour), creamed vegetables, vegetables dipped in batters.
Desserts	• *Recommended:* Ice cream, sherbet, egg custards, or gelatin desserts that do not contain gluten; pure baking chocolate; chocolate chips; hard candy. • *Questionable:* Icing, powdered sugar, candies, chocolate bars, marshmallows. • *Avoid:* Puddings thickened with wheat flour; ice cream or sherbets that contain gluten stabilizers; baked products or doughnuts made with wheat, rye, or barley; ice cream cones; licorice.
Beverages	• *Recommended:* Coffee; tea; cocoa; soft drinks; distilled alcoholic beverages such as rum, gin, whisky, and vodka; wine. • *Questionable:* Instant tea or coffee, coffee substitutes, chocolate drinks, hot cocoa mixes. • *Avoid:* Beer, ale, lager, malted beverages, cereal beverages (Postum), beverages that contain nondairy cream substitutes.

SOURCE: Adapted from American Dietetic Association, *Manual of Clinical Dietetics* (Chicago: American Dietetic Association, 2000), pp. 181–191.

Gluten-free products help people with celiac disease enjoy a wider variety of foods.

© Polara Studios, Inc.

Crohn's disease: an inflammatory bowel disease that usually occurs in the lower portion of the small intestine and the colon. Inflammation may pervade the entire intestinal wall.

ulcerative colitis (ko-LY-tis): an inflammatory bowel disease that involves the colon. Inflammation affects the mucosa and submucosa of the intestinal wall.

become contaminated during harvesting or processing. Although some oat millers have developed procedures that remove other grains from oats, no oats available in the United States are guaranteed to be gluten-free. Individuals who wish to try oats can be advised to limit their intakes to the amounts found to be safe (about $1/2$ cup of dry rolled oats per day) and to contact millers or product manufacturers to determine whether contamination with gluten-containing products is likely.[30]

A gluten-free diet may become monotonous unless care is taken to diversify food choices. The diet can also be a social liability by restricting food choices when individuals eat in restaurants, visit friends, or travel. Nonadherence is common when individuals are away from home.[31] Dietetic counseling may help celiac patients learn how to meet their nutrient needs and expand meal options despite dietary constraints.

Inflammatory Bowel Diseases

Inflammatory bowel diseases are chronic inflammatory disorders that damage gastrointestinal tissue. Both genetic and environmental factors are believed to contribute to the development of these diseases, but the exact triggers are unknown. Disease onset occurs most often in individuals between 15 and 25 years of age.[32]

Table 24-7 compares the two major forms of inflammatory bowel disease, **Crohn's disease** and **ulcerative colitis.** Crohn's disease usually involves the small intestine and may lead to nutrient malabsorption, whereas ulcerative colitis affects the colon, which is past the absorptive areas. Both diseases are characterized

TABLE 24-7	Comparison of Crohn's Disease and Ulcerative Colitis	
	Crohn's Disease	**Ulcerative Colitis**
Location of inflammation	Approximately 40 percent of cases involve the ileum and cecum, 30 percent are in the small intestine only, and 25 percent are in the colon.	Inflammation is confined to the rectum and colon; it begins at the rectum and spreads into the colon.
Pattern of inflammation	Discrete areas separated by normal tissue ("skip" lesions).	Continuous inflammation that begins at the rectum and ends abruptly within the colon.
Depth of damage	Damage throughout all layers of tissue; causes deep fissures that give intestinal tissue a "cobblestone" appearance.	Damage primarily in the mucosa and submucosa.
Fistulas	Common.	Usually do not occur.
Cancer risk	Increased.	Greatly increased.

by periods of active disease interspersed with periods of remission. Nutrient losses can result from tissue damage, bleeding, and diarrhea.

Complications of Crohn's Disease Crohn's disease may occur in any region of the GI tract, but most cases involve the ileum and/or large intestine. Lesions may develop in different areas in the intestine, with normal tissue separating affected regions (called "skip" lesions). During exacerbations, the inflammation can extend deeply into intestinal tissue and be accompanied by ulcerations, fissures, and fistulas. ◆ Loops of intestine may become matted together. Scar tissue eventually thickens and stiffens the intestinal wall, narrowing the lumen and sometimes causing strictures or obstructions. About 60 to 75 percent of patients require surgical resections during the course of illness, although disease often recurs in the remaining intestine.[33] Patients with Crohn's disease are at increased risk of developing intestinal cancers.

Malnutrition may result from malabsorption, nutrient losses (especially of protein) associated with the tissue damage, reduced food intake, and surgical resections that shorten the small intestine. If the ileum is affected, bile acids may become depleted. ◆ The result is malabsorption of fat, fat-soluble vitamins, calcium, magnesium, and zinc (the minerals bind to the unabsorbed fatty acids). Because the ileum is the site of vitamin B_{12} absorption, deficiency can develop unless the patient is given vitamin B_{12} injections. Anemia may result from bleeding, the inadequate absorption of nutrients involved in blood cell formation, or the metabolic effects of chronic illness (see Highlight 19). Anorexia develops due to abdominal discomfort and the effects of cytokines produced during the inflammatory process. [34]

◆ Reminder: *Fistulas* are abnormal passages between organs or tissues that permit the passage of fluids or secretions.

◆ Reminder: Most of the bile used during digestion is eventually reabsorbed in the ileum and returned to the liver.

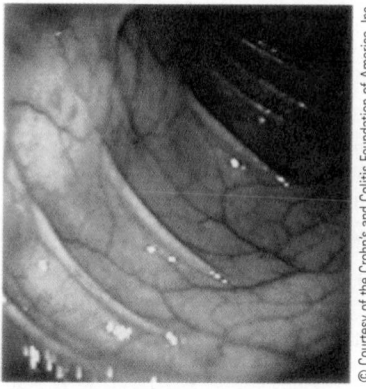

The healthy colon has a smooth surface with a visible pattern of fine blood vessels.

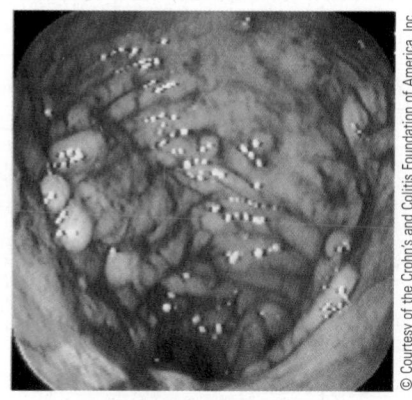

In Crohn's disease, the mucosa has a "cobblestone" appearance due to deep fissuring in the inflamed mucosal tissue.

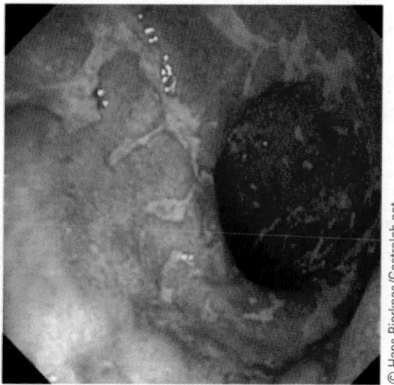

In ulcerative colitis, the colon appears inflamed and reddened, and ulcers are visible.

Complications of Ulcerative Colitis Ulcerative colitis always involves the rectum and usually extends into the colon. Inflammation is continuous along the length of intestine affected, ending abruptly at the area where healthy tissue begins. Tissue erosion or ulceration develops primarily in the mucosa and submucosa (the top two layers of intestinal tissue). During active episodes, patients have frequent, urgent bowel movements that are small in volume. Stools are often streaked with blood and contain mucus.

Although mild disease may cause few complications, weight loss, fever, and weakness are common when most of the colon is involved. Severe disease is often associated with anemia (due to blood loss), dehydration, and electrolyte imbalances. Protein losses from the inflamed tissue can be substantial. A **colectomy** (removal of the colon) is performed in 20 to 25 percent of patients and prevents future recurrence.[35] Colon cancer risk is substantially increased in ulcerative colitis patients.

Drug Treatment of Inflammatory Bowel Diseases Medications help to reduce inflammation, control symptoms, and minimize complications. The drugs prescribed include antidiarrheal agents, immunosuppressants, anti-inflammatory drugs (usually corticosteroids and salicylates), and antibiotics. Although these medications may allow the patient to achieve and maintain remission, some may cause side effects that are detrimental to nutrition status. The Diet-Drug Interactions feature below lists some nutrition-related effects of the medications used in inflammatory bowel diseases.

Nutrition Therapy for Crohn's Disease In patients with Crohn's disease, dietary measures depend on the functional status of the GI tract and the symptoms and complications that develop; thus, nutrition care is highly variable. Crohn's disease often requires aggressive dietary management because it can lead to protein-energy malnutrition (PEM), nutrient deficiencies, and growth failure in children.

During disease exacerbations, a low-residue, low-fat diet provided in small, frequent feedings can minimize stool output and reduce symptoms of malabsorption. Tube feedings may be necessary to supplement the diet or may be the sole means of providing nutrients; some patients may tolerate elemental formulas more easily than standard formulas. High-kcalorie, high-protein diets are prescribed to prevent or treat malnutrition; the amount of protein recommended may be 50 percent higher than DRI levels.[36] Liquid supplements can increase energy intake and improve weight gain. Vitamin and mineral supplements are usually required, especially if nutrient

colectomy: removal of a portion or all of the colon.

DIET-DRUG INTERACTIONS

Check this table for notable nutrition-related effects of the medications discussed in this chapter.

	Gastrointestinal Effects	Interactions with Dietary Substances	Metabolic Effects
Antidiarrheals	Constipation.	—	—
Anti-inflammatory drugs (sulfasalazine, corticosteroids)	Nausea, heartburn (sulfasalazine).	Sulfasalazine may decrease folate absorption; supplementation is recommended.	Anemia (sulfasalazine); fluid retention, hyperglycemia, hypocalcemia, hypokalemia, hypophosphatemia, increased appetite, protein catabolism (corticosteroids).
Laxatives	Diarrhea.	Mineral oil may decrease absorption of fat-soluble vitamins, but it is not often used.	Fluid and electrolyte imbalances, laxative dependency.
Pancreatic enzyme replacements	Diarrhea, nausea, stomach cramps, irritation to GI mucosa.	—	—

TABLE 24-8	Management of Symptoms and Complications in Crohn's Disease
Symptom or Complication	**Possible Dietary Measures**
Growth failure/weight loss	High-kcalorie diet
	Enteral supplements
	Elemental tube feedings
Anorexia/pain with eating	Small, frequent meals
	Enteral supplements
	If long-term (>5 to 7 days): elemental tube feedings
Malabsorption	High-kcalorie diet
	Nutrient supplementation
Steatorrhea (fat malabsorption)	Fat restriction
	Medium-chain triglycerides
	Nutrient supplementation
Diarrhea	Fluid and electrolyte replacement
	Nutrient supplementation
Lactose intolerance	Avoidance of lactose-containing foods
Nutrient deficiencies	Nutrient-dense diet
	Nutrient supplementation
Intestinal recovery	High-protein diet
	Glutamine supplementation
Strictures/fistulas	Low-fiber diet
Severe bowel obstruction/high-output fistulas/severe exacerbations of disease	Total parenteral nutrition

malabsorption is present; nutrients at risk include iron, magnesium, zinc, calcium, vitamin D, vitamin B_{12}, and folate.[37] Table 24-8 includes examples of other dietary adjustments that may be beneficial for patients with Crohn's disease.

During periods of remission, dietary restrictions are unnecessary unless complications develop. Symptoms that may interfere with adequate food intake include anorexia, pain, and diarrhea. Restricted intake of lactose, fructose, and sorbitol may improve symptoms of diarrhea or intestinal gas. Adequate fluid replacement should be encouraged in patients with diarrhea. Individuals with partial obstructions may need to restrict high-fiber foods. Although research studies suggest that supplementation with fish oil, glutamine, or probiotics may be helpful, more research is necessary to confirm these benefits.[38]

Nutrition Therapy for Ulcerative Colitis In most cases, the diet for ulcerative colitis requires few adjustments. As in Crohn's disease, the symptoms and complications that arise are managed with specific dietary measures (see Table 24-8). During disease exacerbations, the primary goals are to restore fluid and electrolyte balances and correct deficiencies that result from protein and blood losses; dietary adjustments are based on the extent of bleeding and diarrhea output. Thus, adequate protein, energy, fluid, and electrolytes need to be provided. A low-fiber diet may reduce irritation by minimizing fecal volume. If colon function becomes severely impaired, food and fluids may be withheld and fluids and electrolytes supplied intravenously until colon function is restored.[39]

Short Bowel Syndrome

The treatment of Crohn's disease, cancers of the small intestine, and other intestinal disorders may include the surgical resection of a major portion of the small intestine. **Short bowel syndrome** is the malabsorption syndrome that results when the absorptive capacity of the remaining intestine is insufficient for meeting nutritional needs. Without appropriate dietary adjustments, short bowel syndrome can result in fluid and electrolyte imbalances and multiple nutrient deficiencies. Symptoms include diarrhea, steatorrhea, dehydration, weight loss, and growth impairment in children.

short bowel syndrome: the malabsorption syndrome that follows resection of the small intestine, which results in insufficient absorptive capacity in the remaining intestine.

Consequences of Short Bowel Syndrome Figure 24-3 reviews nutrient absorption in the GI tract and describes how absorption is affected by surgical resections. Generally, up to 50 percent of the small intestine can be resected without serious nutritional consequences.[40] More extensive resections lead to generalized malabsorption, and patients may need lifelong parenteral nutrition to supplement oral intakes. Other problems that may develop include kidney stones (due to the effects of fat malabsorption on urinary oxalate levels; review p. 761) and gallstones (due to bile malabsorption and the subsequent imbalance between bile acid and cholesterol concentrations in bile; see Chapter 25). Furthermore, loss of the ileocecal valve (between the ileum and cecum) increases the likelihood that colonic bacteria will infiltrate the small intestine and cause bacterial overgrowth.

Intestinal Adaptation After an intestinal resection, the remaining intestine undergoes **intestinal adaptation,** an adaptive response that dramatically improves its absorptive efficiency. Adaptation depends on the presence of nutrients in the lumen, and therefore oral intakes are begun as soon as possible after surgery to stimulate the growth of intestinal tissue. Many patients can eventually return to a normal diet if their intestinal adaptation compensates sufficiently for the removed length of intestine.

The adaptive changes are more prominent in the ileum; thus, removal of the ileum ultimately has more severe consequences than removal of the jejunum. Over a one- to two-year period after resection of the jejunum, the ileum gradually develops taller villi and deeper crypts; it also increases in diameter and length, improving its absorptive capacity enough to replace the jejunum's absorptive functions. Loss of the ileum, however, can permanently disrupt both bile acid and vitamin B_{12} absorption. Depletion of bile acids exacerbates fat malabsorption; furthermore, the unabsorbed bile acids can worsen diarrhea because they irritate the colon walls and increase colonic secretions.[41]

Intestinal adaptation is achieved more easily if both the ileum and colon remain intact. A functional colon is beneficial because its resident bacteria metabolize unabsorbed nutrients and produce some usable nutrients, such as short-chain fatty acids, that are readily absorbed in the colon. An intact colon also helps to reduce losses of fluids and electrolytes.

intestinal adaptation: after resection, the process of intestinal recovery that leads to improved absorptive capacity.

FIGURE 24-3 Nutrient Absorption and Consequences of Intestinal Surgeries

About 90 to 95 percent of nutrient absorption takes place in the first half of the small intestine. After a resection, nutrient absorption may be reduced.

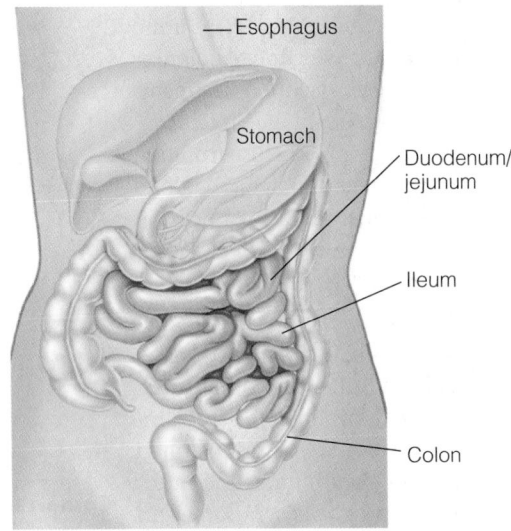

WHAT IS ABSORBED

Duodenum/jejunum
- Simple carbohydrates
- Fats
- Amino acids
- Vitamins[a]
- Minerals[a]
- Water

Ileum
- Bile salts
- Vitamin B_{12}
- Water
(Assumes absorptive function of duodenum and jejunum with adaptation)

Colon
- Water
- Electrolytes
- Short-chain fatty acids

POSSIBLE CONSEQUENCES OF RESECTION

Duodenum/jejunum
- Minimal consequences if the ileum remains intact
- Calcium and iron malabsorption if duodenum resected

Ileum
- Fat malabsorption
- Protein malabsorption
- Malabsorption of fat-soluble vitamins and vitamin B_{12}
- Reduced calcium, magnesium, and zinc absorption
- Fluid losses
- Diarrhea/steatorrhea

Colon
- Fluid and electrolyte losses
- Diarrhea
(Losses are compounded if ileum is also resected)

Labels on figure: Esophagus, Stomach, Duodenum/jejunum, Ileum, Colon

[a] The absorption of vitamins and minerals begins in the duodenum and continues throughout the length of the small intestine.

Treatment of Short Bowel Syndrome To meet their nutritional needs with enteral nutrition alone, adults need at least 40 to 80 inches (100 to 200 centimeters) of small intestine; the length required depends on the region of intestine that remains. If patients do not have adequate absorptive capacity, the usual treatment is lifelong parenteral nutrition. Intestinal transplantation is an option for patients who cannot continue parenteral nutrition because of life-threatening complications.

Nutrition Therapy for Short Bowel Syndrome Immediately after a resection, fluids and electrolytes must be supplied intravenously. In the first few weeks after surgery, the fluid losses from diarrhea can be substantial (sometimes exceeding 2 liters per day), so appropriate rehydration therapy is critical to recovery. Antidiarrheal medications may be needed to limit losses. The diarrhea gradually lessens as intestinal adaptation progresses.

Total parenteral nutrition meets nutrient needs after surgery and is gradually reduced as oral feedings increase. To promote intestinal adaptation, oral feedings may be started just a few days after surgery, after diarrhea subsides somewhat and some bowel function is restored. Initial oral intake may consist of occasional sips of clear, sugar-free liquids, with intake progressing to larger amounts of liquid formulas and then to solid foods, as tolerated. Very small, frequent feedings can utilize the remaining intestine most efficiently. Parenteral feedings are tapered and gradually discontinued once oral intakes can supply adequate nourishment. Some patients require tube feedings in addition to oral feedings to successfully meet nutrient needs.

The exact diet prescribed for short bowel syndrome depends on the portion of intestine removed, the length of intestine that remains, and whether the colon is still intact; moreover, dietary readjustments may be required as intestinal adaptation progresses.[42] A high-kcalorie intake is typically recommended to compensate for malabsorption. To provide adequate kcalories, a high-fat, low-carbohydrate diet may be recommended if the fat is sufficiently tolerated. Conversely, a high-complex-carbohydrate, low-fat diet may be suggested for patients who have an intact colon. The carbohydrate stimulates the production of short-chain fatty acids in the colon, thereby providing an additional 500 to 1000 kcalories daily, and the low fat intake can improve steatorrhea. In some cases, medium-chain triglycerides may be added as an energy source.

CASE STUDY   Economist with Short Bowel Syndrome

Judi Morel is a 28-year-old economist with an 8-year history of Crohn's disease. Judi is 5 feet 7 inches tall. Three years ago, she underwent a small bowel resection and remained free of active disease for 2 years. During that time, her symptoms subsided; she was able to tolerate most foods without any problem and gained weight. Ten months ago, Judi experienced a severe flare-up of her Crohn's disease. Since that time, she has lost 15 pounds and currently weighs 118 pounds. She has experienced severe abdominal pain and fatigue that have persisted despite aggressive medical management that included intravenous nutrition. Five days ago, Judi finally underwent another resection, which left her with 40 percent of healthy small intestine. Her colon is intact. She is experiencing extensive diarrhea.

1. Describe the manifestations of Crohn's disease, and explain why surgery is sometimes performed as part of the treatment. Describe the complications of disease that may affect nutrient needs.
2. Using the BMI table in the back of the book, check the ideal weight range for a person of Judi's height. What nutrition-related concerns are suggested by Judi's recent weight loss? What other nutrition problems did Judi probably experience as a consequence of Crohn's disease?
3. Discuss the complications that may follow an extensive intestinal resection. What factors may affect a person's ability to meet nutrient needs with an oral diet?
4. Discuss the dietary progression that is recommended following an intestinal resection. After Judi is able to eat solid foods, what factors may affect the type of diet that is recommended for her?

Dietary choices are tailored according to individual symptoms and tolerances. Some patients may be lactose intolerant, but they can lessen symptoms by consuming small amounts of lactose-containing foods throughout the day. Patients should avoid concentrated sweets, which draw fluid into the intestines, if they cause additional diarrhea. Due to the high risk of developing kidney stones as a result of calcium malabsorption, a low-oxalate diet may be recommended (see Chapter 28).

Vitamin and mineral supplements can prevent deficiencies from developing due to malabsorption. If fat is malabsorbed, patients may need supplements of fat-soluble vitamins, calcium, magnesium, and zinc. If a large portion of the ileum has been removed, vitamin B_{12} must be injected. Iron absorption is likely to be compromised if upper portions of the small intestine have been removed. Other nutrients at risk may include vitamin A, vitamin C, potassium, and selenium.[43] The Case Study on p. 773 can help you review the material on short bowel syndrome.

IN SUMMARY

Disorders of the small intestine that cause damage to mucosal tissue, such as celiac disease and Crohn's disease, often result in malabsorption. A gluten-free diet can alleviate the symptoms of celiac disease. Intestinal resection is often necessary to treat Crohn's disease. Ulcerative colitis is an inflammatory condition that affects the colon only, and severe cases can be treated by colectomy. Short bowel syndrome is a frequent consequence of major resections of the small intestine and may require permanent parenteral nutrition support. In many patients, intestinal adaptation may improve absorptive capacity after surgical resection.

Conditions Affecting the Large Intestine

The large intestine moves undigested materials to the rectum and has a central role in maintaining fluid and electrolyte balances. Its bacterial population ferments the undigested nutrients that reach the colon and produces short-chain fatty acids and some vitamins that our bodies can absorb and use. This section describes several conditions that may disrupt the normal functioning of the large intestine.

Irritable Bowel Syndrome

◆ Irritable bowel syndrome was introduced in Highlight 3.

People with **irritable bowel syndrome** ◆ experience chronic and recurring intestinal symptoms that cannot be explained by specific physical abnormalities. The symptoms may include both diarrhea and constipation, abdominal pain or discomfort, flatulence, and distention; the abdominal pain is generally aggravated by eating and relieved by defecation. In some patients, symptoms may be mild; in others, the disturbances in defecation can interfere with work and social activities and dramatically alter lifestyle and sense of well-being. Irritable bowel syndrome generally occurs in individuals between 30 and 50 years of age and affects twice as many women as men. About 30 percent of people with the disorder eventually become asymptomatic.[44]

Although the causes of irritable bowel syndrome remain elusive, people with the disorder tend to have excessive colonic responses to meals, GI hormones, and stress. Many individuals exhibit hypersensitivity to a normal degree of intestinal distention and may feel discomfort when experiencing normal meal transit or typical amounts of intestinal gas. Intestinal motility after meals may be excessive, leading to diarrhea, or be reduced, causing constipation. Symptoms often worsen during

irritable bowel syndrome: an intestinal disorder of unknown cause that affects the functioning of the lower bowel; symptoms include abdominal pain, flatulence, diarrhea, and constipation.

periods of psychological stress, and stressful life events may trigger the illness. Some evidence suggests that an infection may cause the initial GI disturbance and that tissue sensitization persists after the infection has healed.

Diagnosing irritable bowel syndrome is often difficult because its symptoms are typical of other GI disorders and laboratory tests for the condition are nonexistent. Up to 5 percent of patients with symptoms of irritable bowel syndrome are found to have celiac disease.[45] Other GI disturbances that may cause similar symptoms include bacterial overgrowth, inflammatory bowel diseases, lactose intolerance, hypothyroidism or hyperthyroidism, and intestinal cancers. Between 40 and 60 percent of patients diagnosed with irritable bowel syndrome have coexisting psychiatric illnesses, such as anxiety, depression, or panic disorder, which can exacerbate symptoms or reduce the ability to cope with the GI condition.[46]

Treatment of Irritable Bowel Syndrome Medical treatment of irritable bowel syndrome often includes dietary adjustments, stress management, and behavioral therapies. Medications may be prescribed to manage symptoms, although they are not always helpful. The drugs prescribed may include antidiarrheal agents, anticholinergics (which affect GI motility), antidepressants, and laxatives.

Nutrition Therapy for Irritable Bowel Syndrome Although dietary adjustments may be useful, measures that help one symptom can sometimes make another worse. The usual dietary advice is to increase fiber intake to 25 to 35 grams per day, which helps to reduce constipation and improve stool bulk.[47] To minimize discomfort from intestinal gas, patients should add fiber-containing foods gradually. They should avoid dietary substances that produce excessive gas unless they can tolerate these foods (review Table 24-2). If diarrhea persists, a bulking agent (psyllium) is usually effective.

A careful evaluation of the diet history can help to reveal the foods and behaviors most closely associated with intestinal discomfort. Avoidance of milk products may benefit those who are lactose intolerant. Patients should avoid caffeine and alcohol because they can exacerbate GI symptoms. Other adjustments that are often helpful include consuming small, frequent meals instead of larger ones and avoiding fatty foods. Psychological associations have a strong influence on food tolerance, so foods that patients perceive to be problematic should be discussed so that

---

CASE STUDY New College Graduate with Irritable Bowel Syndrome

Marcy Hudson is a 22-year-old recent college graduate who began her first professional job in a bank one month ago. As a college student, she occasionally experienced abdominal pain and cramping after eating. She also had frequent bouts of diarrhea and felt somewhat better after bowel movements. Once Marcy began her new job, her symptoms occurred more frequently. At first she attributed her symptoms to job stress, but when the symptoms continued for several months, she decided to see her physician. After taking a careful history and conducting tests to rule out other bowel disorders, the physician diagnosed irritable bowel syndrome. The physician prescribed bulk-forming agents and advised Marcy to keep a record of her food intake and symptoms for one week. Marcy was then referred to a dietitian for a review of her dietary record. The dietitian noticed that Marcy routinely drank several cups of coffee in the morning and had large meals for lunch and dinner. Marcy often ate out in Mexican restaurants and favored highly spiced foods and refried beans. Between meals, she snacked on low-carbohydrate foods sweetened with sugar alcohols and drank several cans of soda daily. Her dietary fiber intake, however, totaled only about 15 grams daily.

1. Describe the characteristics of irritable bowel syndrome to Marcy, and indicate the role that stress might play in her illness.
2. Explain how the record of food intake and symptoms might be helpful in devising an appropriate diet plan for Marcy. Are any of the foods in Marcy's diet likely to be aggravating her symptoms?
3. What dietary measures might benefit individuals with irritable bowel syndrome? What problems might the dietary changes cause?

the diet is not restricted unnecessarily. Review the Case Study on p. 775 to apply your knowledge about irritable bowel syndrome to a clinical situation.

Diverticular Disease of the Colon

◆ Diverticulosis was introduced in Highlight 3.

Diverticulosis ◆ refers to the presence of pebble-sized herniations (protrusions) in the intestinal wall, known as diverticula (see the photo). In Western societies, the diverticula occur most often in the sigmoid colon, the portion of the colon just above the rectum. The prevalence of diverticulosis increases with age, occurring in 50 to 80 percent of individuals over 80 years of age.[48] Most people with diverticulosis are symptom-free and remain unaware of the condition until a complication develops.

Epidemiological studies suggest that the development of diverticula is strongly influenced by the amount of dietary fiber a person consumes. Researchers have theorized that the increased stool weight and bulk associated with high-fiber diets may reduce the workload of the circular muscles that move wastes through the colon.[49] Low-fiber diets require more vigorous muscle contractions, increasing pressure within the segments immediately adjacent to the circular muscles. This increase in pressure induces small areas of intestinal tissue to balloon outward over time.

Diverticulitis Inflammation or infection sometimes develops in the area around a diverticulum. This condition, called *diverticulitis,* is the most common complication of diverticulosis, affecting 10 to 25 percent of individuals with the condition.[50] It is thought to result from hardened fecal matter that abrades the mucosal lining, causing inflammation and possibly a microperforation that leads to subsequent infection. If the infection spreads to adjacent organs, fistulas may develop. Less frequently, the infection spreads to the peritoneal cavity, causing life-threatening illness. Symptoms of diverticulitis may include persistent abdominal pain, tenderness in the affected area, fever, constipation, diarrhea, and bleeding. Anorexia, nausea, and vomiting may also occur.

Treatment for Diverticular Disease Treatment for diverticulosis is necessary only if symptoms develop. Because increasing dietary fiber may prevent disease progression and the development of intestinal symptoms, patients are advised to consume a high-fiber diet, with an emphasis on insoluble fiber sources. The fiber intake should be gradually increased to ensure tolerance. Bulk-forming agents, such as psyllium, can raise fiber intakes if food sources are insufficient. Avoiding nuts and seeds is sometimes suggested to prevent disease progression and complications, but evidence is inadequate to justify or refute the recommendation.[51]

Patients with diverticulitis may need antibiotics to treat infections and, possibly, pain-control medications. In mild cases, a clear liquid diet may be advised initially, with progression to solid foods as symptoms resolve. In more severe cases, bowel rest is necessary (oral fluids and food are withheld), and fluids are given intravenously. Oral intakes are gradually reintroduced as the condition improves, beginning with clear liquids and progressing to a restricted-fiber, low-residue diet until inflammation and bleeding subside.[52] After recovery, a high-fiber diet is recommended to prevent disease progression and symptom recurrence. Surgical interventions are sometimes necessary to treat complications of diverticulitis and may include removal of the affected portion of colon.[53]

© Hans Bjorknas/Gastrolab

Diverticulosis is charcterized by herniations in the intestinal wall, which are asymptomatic unless complicated by infection or inflammation.

IN SUMMARY

Irritable bowel syndrome and diverticular disease are common disorders affecting the colon. Irritable bowel syndrome is characterized by abdominal pain and alternating diarrhea and constipation. Although the causes are unknown, the disorder is influenced by stress and psychological factors. Diverticulosis is often asymptomatic until complications develop; its prevalence increases with advancing age and may be related to low fiber intakes. Patients with irritable bowel syndrome or diverticulosis may benefit from a high-fiber diet.

Colostomies and Ileostomies

An *ostomy* is a surgically created opening (called a **stoma**) in the abdominal wall through which dietary wastes can be eliminated. A permanent ostomy is necessary after a partial or total colectomy. A temporary ostomy is sometimes constructed so that part or all of the colon can be bypassed after injury or extensive surgery. To create the stoma, the cut end of the remaining segment of functioning intestine is brought through an opening in the abdominal wall and stitched in place so that it empties to the exterior. The stoma can be formed from a section of the colon **(colostomy)** or ileum **(ileostomy),** as shown in Figure 24-4. Conditions that may require these procedures include inflammatory bowel diseases, diverticulitis, and colorectal cancers.

To collect wastes, a disposable bag is affixed to the skin around the stoma and emptied during the day as needed. Alternatively, an interior pouch can be surgically constructed behind the stoma using intestinal tissue, and the pouch can be emptied with a catheter when convenient. Stool consistency varies according to the length of colon that is functional. If a small portion of the colon is absent or bypassed, the stools may continue to be semisolid. If the entire colon has been removed or is bypassed, absorption of fluid and electrolytes into the body is reduced substantially, and the output is liquid. Due to the difficulty in obtaining enough water to replace losses, patients with ileostomies often have low urine output and an increased risk of developing kidney stones.

Nutrition Therapy for Patients with Ostomies The nutrition care after an ostomy depends on the length of colon removed and the portion of ileum that remains, so dietary adjustments are individualized according to the surgical procedure and symptoms that develop afterward. Following surgery, the diet gradually progresses from clear liquids that are low in sugars to a normal meal plan, as tolerated. To reduce stool output, a low-residue diet is often recommended. To determine food tolerances, patients should add questionable foods to the diet one at a time and in small amounts to assess their effects; a food that causes problems can be tried again later. Appropriate fluid and electrolyte intakes should be encouraged when a large portion of the colon has been removed.

> **stoma** (STOE-ma): a surgical opening made in the abdominal wall.
>
> **colostomy** (co-LAH-stoe-me): a surgical procedure that creates a stoma using a section of the colon.
>
> **ileostomy** (ill-ee-AH-stoe-me): a surgical procedure that creates a stoma using the ileum.

FIGURE 24-4 Colostomy and Ileostomy

Colostomy

Ileostomy

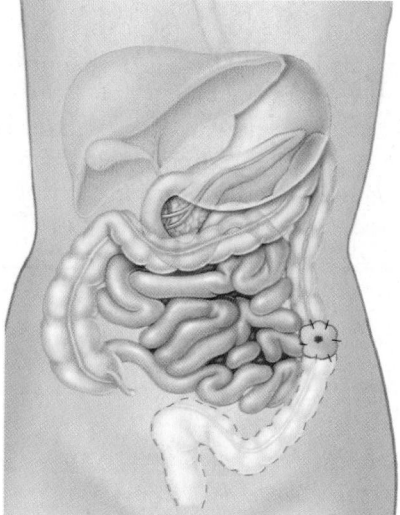

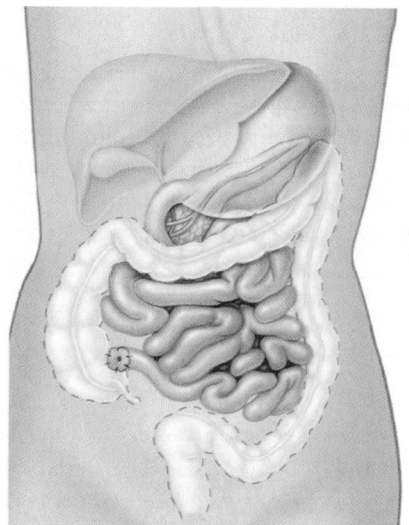

In a colostomy, a portion of the colon is removed or bypassed, and the stoma is formed from the remaining section of functional colon.

In an ileostomy, the entire colon is removed or bypassed, and the stoma is formed from the ileum.

People with ileostomies need to chew thoroughly to ensure that foods can be adequately digested and to prevent obstructions, which are a common complication due to the small diameter of the ileal lumen. Foods high in insoluble fibers are sometimes avoided because they shorten intestinal transit time, may cause obstructions, and increase stool output. Because the colon is no longer available for reabsorbing water, the diet should provide at least 8 cups of liquid daily to prevent dehydration. To ensure adequate electrolyte intakes, patients are encouraged to use salt liberally and to ingest beverages with added electrolytes (such as sports drinks and oral rehydration beverages), if necessary. If a large portion of the ileum has been removed, fat malabsorption may occur due to bile acid depletion, and vitamin B_{12} injections may be required.

Dietary concerns after colostomies depend on the length of colon remaining. Most patients have no dietary restrictions and can return to a regular diet.[54] Patient concerns may include stool odors, excessive gas production, and diarrhea. If a large portion of colon was removed, recommendations may be similar to those given to ileostomy patients.

Obstructions As mentioned, foods that are incompletely digested can cause obstructions, a primary concern of ileostomy patients. Although these patients can consume almost any food that is cut into small pieces and carefully chewed, the following foods may cause difficulty: celery, coconut, corn, dried fruit, grapes, nuts, popcorn, raw cabbage (for example, in coleslaw), and unpeeled apples.[55]

Reducing Gas and Odors Persons with ostomies are often concerned about foods that may increase gas production or cause strong odors. Foods that may cause excessive gas include those listed in Table 24-2 on p. 758; practices that increase gas formation include smoking, gum chewing, tobacco chewing, using drinking straws, and eating quickly. Foods that sometimes produce unpleasant odors include asparagus, beer, broccoli, brussels sprouts, cabbage, dried beans and peas, eggs, fish, garlic, and onions. Foods that may help to reduce odors include buttermilk, cranberry juice, parsley, and yogurt.[56]

Diarrhea Examples of foods that may aggravate diarrhea were listed in Table 24-3 on p. 759. Foods and dietary substances that may thicken stool include applesauce, banana flakes, cheese, creamy peanut butter, marshmallows, pasta, pectin, potatoes, pretzels, white bread, and white rice.[57] What works may differ for each individual, however, and is best determined by trial and error.

IN SUMMARY

Colostomies and ileostomies are surgically created openings in the abdominal wall using the colon or ileum. Fluid and electrolyte requirements are greater after an ostomy because colon function is reduced or absent. Foods that are poorly digested may cause obstructions in people with ostomies, although thorough chewing can reduce risk. These individuals should avoid foods if they provoke diarrhea or cause excessive gas or strong odors.

Clinical Portfolio

CENGAGENOW™
academic.cengage.com/login

1. A health practitioner working with a patient with a constipation problem provided him with detailed information about a high-fiber diet. At a follow-up appointment, the patient reported no change in symptoms. His food diary for that day showed that he consumed an omelet and toast for breakfast and a sandwich with juice for lunch.
    - Considering these two meals only, what additional information would help the health practitioner evaluate the man's compliance with the diet he was given?
    - Review the discussion about fiber in Chapter 4, and create a one-day menu that provides the DRI for fiber for an adult male, using the fiber values listed in Appendix H.

2. Using Table 24-5 on p. 762 as a guide, plan a day's menus for a diet containing 50 grams of fat. Take care to make the meals both palatable and nutritious. How can these menus be improved using the suggestions in the "How to" on p. 763?

3. As stated in this chapter, treatment of celiac disease is deceptively simple—eliminate wheat, barley, and rye, and possibly oats. Remaining on a gluten-free diet is more challenging than it appears, however.
    - Randomly select ten of your favorite snack and convenience foods. Take a trip to the grocery store, and check the labels of the products you selected to see if they would be allowed on a gluten-free diet. Keep in mind that the labels may not list all offending ingredients.
    - Find acceptable substitutes for the products that are not allowed, either by substituting other foods or by checking for gluten-free products in the grocery store. If you have access to the Internet, you may want to investigate websites that advertise gluten-free products to get an idea of what's available.

NUTRITION ON THE NET

For further study of topics covered in this chapter, log on to **academic.cengage .com/nutrition/rolfes/UNCN8e**. Go to Chapter 24, then to Nutrition on the Net.

- Visit the websites of these organizations to find information that is helpful both for health practitioners and patients with gastrointestinal problems:
    American College of Gastroenterology: **www.acg.gi.org**
    American Gastroenterological Association: **www.gastro.org**
    National Institute of Diabetes and Digestive and Kidney Diseases, a division of the National Institutes of Health: **www2.niddk.nih .gov**

- Find additional information about cystic fibrosis at the website of the Cystic Fibrosis Foundation: **www.cff.org/home**

- Find more information about celiac disease by visiting these websites:
    Celiac Disease Foundation: **www.celiac.org**
    Celiac Sprue Association: **www.csaceliacs.org**
    Gluten Intolerance Group: **www.gluten.net**

- Learn more about inflammatory bowel diseases at the website of the Crohn's and Colitis Foundation of America: **www.ccfa.org**

NUTRITION ASSESSMENT CHECKLIST for People with Lower GI Tract Disorders

Medical History

Check the medical record for diseases that:

- Cause chronic GI symptoms, such as irritable bowel syndrome or ulcerative colitis
- Interfere with pancreatic enzyme secretion, such as chronic pancreatitis or cystic fibrosis
- Interfere with nutrient absorption, such as Crohn's disease or celiac disease

Check for surgical procedures involving the lower GI tract, such as:

- Intestinal resections
- Ileostomy
- Colostomy

Check for the following symptoms or complications:

- Anemia
- Bacterial overgrowth
- Bone disease
- Constipation
- Diarrhea, dehydration
- Fistulas
- Lactose intolerance

- Nutrient deficiencies
- Obstructions
- Oxalate kidney stones
- Poor growth, in children
- Steatorrhea

Medications

Check for medications or dietary supplements that may:

- Cause constipation or diarrhea
- Interfere with food intake by causing nausea, vomiting, cramps, dry mouth, or drowsiness
- Alter appetite or nutrient needs

Dietary Intake

Note the following problems, and contact the dietitian if you suspect difficulties such as:

- Poor appetite or food intake
- Food intolerances
- Inadequate fiber intake, in patients with constipation
- Lactose intolerance, in patients with diarrhea
- Inadequate fluid intake

Anthropometric Data

Measure baseline height and weight. Address weight loss early to prevent malnutrition in patients with:

- Severe or persistent diarrhea
- Nutrient malabsorption

Laboratory Tests

Check laboratory tests for signs of dehydration, electrolyte imbalances, nutrient deficiencies, and anemia in patients with:

- Severe or persistent diarrhea
- Nutrient malabsorption
- Intestinal resections

Physical Signs

Look for physical signs of:

- Dehydration
- Protein-energy malnutrition
- Essential fatty acid and fat-soluble vitamin deficiencies
- Folate and vitamin B_{12} deficiencies
- Mineral deficiencies

STUDY QUESTIONS

CENGAGENOW™
To assess your understanding of chapter topics, take the Student Practice Test and explore the modules recommended in your Personalized Study Plan. Log on to **academic.cengage.com/login.**

These questions will help you review the chapter. You will find the answers in the discussions on the pages provided.

1. What measures can help to prevent and treat constipation? What are the modes of action of the laxatives used to treat constipation? (pp. 755–757)

2. Describe possible causes of diarrhea and the dietary measures that may be included during treatment. (pp. 758–759)

3. Identify conditions that can cause fat malabsorption, and discuss the primary nutrition problems that can result. (pp. 760–761)

4. Explain why bacterial overgrowth develops, and describe its effects on nutrition status. (pp. 761–763)

5. Explain how chronic pancreatitis and cystic fibrosis can result in malabsorption. In what ways are their dietary treatments similar? (pp. 763–766)

6. Discuss the cause of celiac disease, and describe the diet used in its treatment. Explain why a gluten-free diet may be difficult for patients to follow. (pp. 767–768)

7. Compare the effects of Crohn's disease and ulcerative colitis on nutrition status, and describe the dietary measures that may be required during the course of illness. (pp. 768–771)

8. Identify possible causes of short bowel syndrome. What nutrition problems often develop? Discuss the adaptive process that occurs in the remaining intestine after a portion of intestine is resected. (pp. 771–774)

9. Describe irritable bowel syndrome, and discuss the ways in which diet can be used in its treatment. (pp. 774–776)

10. Specify the factors that increase the risk of developing diverticular disease. What dietary modification is most useful in its prevention and treatment? (p. 776)

11. Describe the primary nutrition-related concerns of people who have undergone ileostomies and colostomies. (pp. 777–778)

These multiple choice questions will help you prepare for an exam. Answers can be found on p. 782.

1. The health practitioner advising an elderly patient with constipation encourages the patient to:
   a. consume a low-fat diet low in sodium.
   b. consume a high-protein diet rich in calcium.
   c. eliminate gas-forming foods from the diet.
   d. gradually add high-fiber foods to the diet.

2. Osmotic diarrhea often results from:
   a. excessive colonic contractions.
   b. excessive fluid secretion by the intestines.
   c. nutrient malabsorption.
   d. viral, bacterial, or protozoal infections.

3. Nutrition problems that may result from fat malabsorption include all of the following *except:*
   a. weight loss.
   b. essential amino acid deficiencies.
   c. bone loss.
   d. oxalate kidney stones.

4. Common nutrition problems associated with bacterial overgrowth in the stomach and small intestine include:
   a. sensitivity to gluten.
   b. fat malabsorption and vitamin B_{12} deficiency.
   c. constipation.
   d. permanent loss of digestive enzymes.

5. The majority of chronic pancreatitis cases can be attributed to:
   a. bacterial and viral infections.
   b. gallstones.
   c. excessive alcohol use.
   d. elevated triglyceride levels.

6. Chronic pancreatitis and cystic fibrosis are both treated with:
   a. intestinal resection.
   b. postural drainage.
   c. enzyme replacement therapy.
   d. stool softeners.

7. A person on a gluten-free diet must avoid products containing:
   a. wheat, barley, and rye.
   b. barley, soybeans, and corn.
   c. wheat, corn, and rice.
   d. buckwheat, rice, and millet.

8. Symptoms of irritable bowel syndrome most often include:
   a. nausea and vomiting.
   b. weight loss and malnutrition.
   c. strong odors and obstructions.
   d. constipation, diarrhea, and flatulence.

9. Diverticulosis is most often usually associated with:
   a. a low-fiber diet.
   b. inadequate exercise.
   c. intestinal surgery.
   d. a high-fiber diet.

10. After an ileostomy, the most serious concern is that:
    a. the diet is too restrictive to meet nutrient needs.
    b. waste disposal causes frequent daily interruptions.
    c. fluid restrictions prevent patients from drinking beverages freely.
    d. incompletely digested foods may cause obstructions.

REFERENCES

1. S. M. Patel and A. J. Lembo, Constipation, in M. Feldman, L. S. Friedman, and L. J. Brandt, eds., *Sleisenger and Fordtran's Gastrointestinal and Liver Disease* (Philadelphia: Saunders, 2006), pp. 221–253.
2. Patel and Lembo, 2006.
3. Patel and Lembo, 2006.
4. Standing Committee on the Scientific Evaluation of Dietary Reference Intakes, Food and Nutrition Board, Institute of Medicine, *Dietary Reference Intakes for Energy, Carbohydrate, Fiber, Fat, Fatty Acids, Cholesterol, Protein, and Amino Acids* (Washington, D.C.: National Academies Press, 2002).
5. H. Ohge and M. D. Levitt, Intestinal gas, in M. Feldman, L. S. Friedman, and L. J. Brandt, eds., *Sleisenger and Fordtran's Gastrointestinal and Liver Disease* (Philadelphia: Saunders, 2006), pp. 187–197.
6. L. R. Schiller and J. H. Sellin, Diarrhea, in M. E. Shils and coeditors, *Modern Nutrition in Health and Disease* (Baltimore: Lippincott Williams & Wilkins, 2006), pp. 159–186.
7. Schiller and Sellin, 2006.
8. M. de Vrese and P. R. Marteau, Probiotics and prebiotics: Effects on diarrhea, *American Journal of Clinical Nutrition* 137 (2007): 803S–811S.
9. Schiller and Sellin, 2006.
10. American Dietetic Association, *Nutrition Care Manual* (Chicago: American Dietetic Association, 2007).
11. C. Högenauer and H. F. Hammer, Maldigestion and malabsorption, in M. Feldman, L. S. Friedman, and L. J. Brandt, eds., *Sleisenger and Fordtran's Gastrointestinal and Liver Disease* (Philadelphia: Saunders, 2006), pp. 2199–2241.
12. S. O'Mahony and F. Shanahan, Enteric bacterial flora and bacterial overgrowth, in M. Feldman, L. S. Friedman, and L. J. Brandt, eds., *Sleisenger and Fordtran's Gastrointestinal and Liver Disease* (Philadelphia: Saunders, 2006), pp. 2243–2256.
13. O'Mahony and Shanahan, 2006.
14. C. Owyang, Pancreatitis, in L. Goldman and D. Ausiello, eds., *Cecil Medicine* (Philadelphia: Saunders, 2008), pp. 1070–1078.
15. M. Raimondo and J. S. Scolapio, Nutrition in pancreatic disorders, in M. E. Shils and coeditors, *Modern Nutrition in Health and Disease* (Baltimore: Lippincott Williams & Wilkins, 2006), pp. 1227–1234.
16. American Dietetic Association, 2007.
17. C. E. Forsmark, Chronic pancreatitis, in M. Feldman, L. S. Friedman, and L. J. Brandt, eds., *Sleisenger and Fordtran's Gastrointestinal and Liver Disease* (Philadelphia: Saunders, 2006), pp. 1271–1308.
18. Forsmark, 2006.
19. Raimondo and Scolapio, 2006.
20. Owyang, 2008.
21. M. J. Welsh, Cystic fibrosis, in L. Goldman and D. Ausiello, eds., *Cecil Medicine* (Philadelphia: Saunders, 2008), pp. 627–631.
22. D. C. Whitcomb, Hereditary, familial, and genetic disorders of the pancreas and pancreatic disorders in childhood, in M. Feldman, L. S. Friedman, and L. J. Brandt, eds., *Sleisenger and Fordtran's Gastrointestinal and Liver Disease* (Philadelphia: Saunders, 2006), pp. 1203–1240.
23. Welsh, 2008.
24. K. M. Hayek, Medical nutrition therapy for cystic fibrosis: Beyond pancreatic enzyme replacement therapy, *Journal of the American Dietetic Association* 105 (2005): 1186–1188; S. W. Powers and S. R. Patton, A comparison of nutrient intake between infants and toddlers with and without cystic fibrosis, *Journal of the American Dietetic Association* 103 (2003): 1620–1625.
25. A. Fasano and coauthors, Prevalence of celiac disease in at-risk and not-at-risk groups in the United States: A large multicenter study, *Archives of Internal Medicine* 163 (2003): 286–292.
26. C. E. Semrad and D. W. Powell, Approach to the patient with diarrhea and malabsorption, in L. Goldman and D. Ausiello, eds., *Cecil Medicine* (Philadelphia: Saunders, 2008), pp. 1019–1042.
27. J. J. Heidelbaugh and coauthors, Gastroenterology, in R. E. Rakel, ed., *Textbook of Family Medicine* (Philadelphia: Saunders, 2007), pp. 1115–1171; R. J. Farrell and C. P. Kelly, Celiac sprue and refractory sprue, in M. Feldman, L. S. Friedman, and L. J. Brandt, eds., *Sleisenger and Fordtran's Gastrointestinal and Liver Disease* (Philadelphia: Saunders, 2006), pp. 2277–2306.
28. American Dietetic Association, 2007.
29. American Dietetic Association, 2007.
30. T. Thompson, Oats and the gluten-free diet, *Journal of the American Dietetic Association* 103 (2003): 376–379.
31. A. L. Lee and J. M. Newman, Celiac diet: Its impact on quality of life, *Journal of the American Dietetic Association* 103 (2003): 1533–1535.
32. W. F. Stenson, Inflammatory bowel disease, in L. Goldman and D. Ausiello, eds., *Cecil Medicine* (Philadelphia: Saunders, 2008), pp. 1042–1050.
33. Stenson, 2008; B. E. Sands, Crohn's disease, in M. Feldman, L. S. Friedman, and L. J. Brandt, eds., *Sleisenger and Fordtran's Gastrointestinal and Liver Disease* (Philadelphia: Saunders, 2006), pp. 2459–2498.
34. A. M. Griffiths, Inflammatory bowel disease, in M. E. Shils and coeditors, *Modern Nutrition in Health and Disease* (Baltimore: Lippincott Williams & Wilkins, 2006), pp. 1209–1218.

35. Stenson, 2008.
36. American Dietetic Association, 2007.
37. American Dietetic Association, 2007.
38. American Dietetic Association, 2007.
39. Stenson, 2008.
40. A. L. Buchman, Short bowel syndrome, in M. Feldman, L. S. Friedman, and L. J. Brandt, eds., *Sleisenger and Fordtran's Gastrointestinal and Liver Disease* (Philadelphia: Saunders, 2006), pp. 2257–2276.
41. K. N. Jeejeebhoy, Short bowel syndrome, in M. E. Shils and coeditors, *Modern Nutrition in Health and Disease* (Baltimore: Lippincott Williams & Wilkins, 2006), pp. 1201–1208; Buchman, 2006.
42. Buchman, 2006; Jeejeebhoy, 2006; C. R. Parrish, The clinician's guide to short bowel syndrome, *Practical Gastroenterology* (September 2005): 67–106.
43. Buchman, 2006; Jeejeebhoy, 2006.

44. N. J. Talley, Functional gastrointestinal disorders: Irritable bowel syndrome, dyspepsia, and noncardiac chest pain, in L. Goldman and D. Ausiello, eds., *Cecil Medicine* (Philadelphia: Saunders, 2008), pp. 990–998.
45. Heidelbaugh and coauthors, 2007.
46. Talley, 2008.
47. American Dietetic Association, 2007.
48. C. Prather, Inflammatory and anatomic diseases of the intestine, peritoneum, mesentery, and omentum, in L. Goldman and D. Ausiello, eds., *Cecil Medicine* (Philadelphia: Saunders, 2008), pp. 1050–1061.
49. J. M. Fox and N. H. Stollman, Diverticular disease of the colon, in M. Feldman, L. S. Friedman, and L. J. Brandt, eds., *Sleisenger and Fordtran's Gastrointestinal and Liver Disease* (Philadelphia: Saunders, 2006), pp. 2613–2632.

50. Prather, 2008.
51. Prather, 2008.
52. American Dietetic Association, 2007.
53. Prather, 2008, American Dietetic Association, 2007; Fox and Stollman, 2006.
54. K. Willcutts, K. Scarano, and C. W. Eddins, Ostomies and fistulas: A collaborative approach, *Practical Gastroenterology* (December 2005): 63–79.
55. American Dietetic Association, 2007.
56. American Dietetic Association, 2007; Willcutts, Scarano, and Eddins, 2005.
57. American Dietetic Association, 2007; Willcutts, Scarano, and Eddins, 2005.

ANSWERS

Study Questions (multiple choice)

1. d   2. c   3. b   4. b   5. c   6. c   7. a   8. d   9. a   10. d

Probiotics and Intestinal Health

Soon after birth, the warm, nutrient-rich environment within the gastrointestinal tract is colonized by a wide variety of bacterial species. The approximately 10 trillion cells of bacteria inhabiting our bodies **(flora)** make up more than 90 percent of all our cells. Most bacterial cells reside in our colon, which harbors over 400 different species.[1] Although the exact composition of intestinal bacteria varies among individuals, the pattern within an individual tends to remain constant over time, fluctuating somewhat due to illness, antibiotic treatment, and to some extent, dietary factors. Table H24-1 lists the predominant types of bacteria that colonize the human intestines, and Table H24-2 shows how the bacterial populations vary within different regions of the GI tract.

Over the past several decades, nutritional scientists and microbiologists have tried to determine whether **probiotics**—live, **nonpathogenic** bacteria supplied in sufficient numbers to possibly benefit our health—can be useful for preventing or treating various medical conditions. Although the diseases of interest include gastrointestinal disorders, researchers have also been studying the effects of bacterial cells on cancer, immune system disorders, and other illnesses. This highlight discusses some of the research and explains some of the issues involved in selecting and consuming probiotic bacteria. The accompanying glossary defines the relevant terms.

Our Intestinal Flora

Intestinal bacteria can benefit our health in a number of different ways. First, the bacteria degrade much of our undigested or unabsorbed dietary carbohydrate, including dietary fibers, starch that is resistant to digestion, and poorly absorbed sugars and sugar alcohols. In turn, the bacteria produce some vitamins, as well as short-chain fatty acids that our cells can use as an energy source. Intestinal bacteria also stimulate our immune defenses and may prevent the overgrowth of **pathogenic** bacteria in the gastrointestinal tract. Healthy bacteria may help to prevent invasion of our tissues by pathogenic bacteria by creating a barrier on the intestinal walls.[2]

Probiotic Bacteria

For bacteria to be "probiotic"—that is, beneficial to health—they must be nonpathogenic when consumed. They must survive their transit through the digestive tract; therefore, they must be resistant to destruction by stomach acid, bile, and other digestive substances. They should be able to alter the intestinal environment in some way that is beneficial to the human host, either by producing antimicrobial substances, altering immune defenses, metabolizing undigested foodstuffs, or protecting the intestinal walls.[3]

Probiotic bacteria must be consumed in high amounts—between 100 million and 100 billion live bacteria per day—to

TABLE H24-1 Intestinal Flora

Predominant Types	Subdominant Types
Bacteroides	Enterobacteria
Bifidobacteria	Enterococci
Clostridia	*Escherichia* species
Eubacteria	*Klebsiella* species
Peptococcus species	Lactobacilli
Peptostreptococcus species	Micrococci
Ruminococcus species	Staphylococci

TABLE H24-2 Bacterial Populations in the Gastrointestinal Tract

Organ	Total Bacteria (per mL of contents)
Stomach	0 to 100
Small intestine: duodenum	0 to 1000
Small intestine: jejunum and ileum	10^5 to 10^8
Colon	10^{10} to 10^{12}

GLOSSARY

flora: the bacteria that normally reside in a person's body.

nonpathogenic: not capable of causing disease.

pathogenic: capable of causing disease.

prebiotics: nondigestible substances in foods that stimulate the growth of probiotic bacteria within the large intestine.

probiotics: live bacteria provided in foods and dietary supplements for the purpose of preventing or treating disease.

survive in sufficient numbers to influence the bacterial populations in the large intestine; a serving of yogurt usually provides these amounts. Carefully controlled studies have not found that probiotic bacteria actually *colonize* the intestine, however, as they are no longer detected in fecal or intestinal samples once ingestion of the probiotic product stops.[4] Note that only a few different types of bacteria are used in foods, and the relatively small amounts consumed cannot compete with the huge populations that normally populate our digestive tract.

Probiotic Bacteria and Disease

Although results of research studies vary, probiotic bacteria may help to prevent and treat some gastric and intestinal disorders, alter susceptibility to food allergens and alleviate some allergy symptoms, and improve the availability and digestibility of various nutrients.[5] Some evidence suggests that probiotics may help to prevent or reverse infections in the urethra and vagina.[6] Other potential benefits include improved immune responses, reduced symptoms of lactose intolerance, and reduced cancer risk.[7]

Much of the research investigating probiotics and intestinal illness has focused on the prevention and treatment of infectious diarrhea. For example, controlled trials have suggested that certain strains of probiotic bacteria may shorten the duration of diarrhea caused by rotavirus infection in infants and children, decrease the incidence of traveler's diarrhea in tourists visiting high-risk areas, and prevent the recurrence of infectious diarrhea in hospitalized patients.[8] In studies of children and adults using antibiotics, some strains of probiotic bacteria have been shown to reduce the incidence and duration of antibiotic-associated diarrhea. As another example, some studies have suggested that probiotic treatment may help to reduce the recurrence of *pouchitis,* an inflammation of the surgical pouch created in patients who have had an ileostomy or colostomy.[9]

Despite promising research results thus far, there are no clear conclusions about the appropriate probiotic doses or durations of treatment for many of these conditions.[10] Moreover, the beneficial effects of one bacterial strain cannot be extrapolated to other strains of the same species.[11] Thus, individuals who decide to consume probiotic-containing foods and supplements to benefit their health cannot be certain that the substances they use will help their condition. At best, probiotics should be considered an adjunct therapy rather than a primary treatment for an illness.

Probiotics in the Diet

Probiotics are provided mainly by fermented foods. In the United States, yogurt and acidophilus milk are produced using various species of lactobacilli and bifidobacteria, although the species are chosen for their ability to produce desirable food products rather than their potential health benefits.[12] In Europe and Asia, food products containing probiotic bacteria include yogurt, milk, ice cream, oatmeal gruel, and soft drinks.[13] Although lactobacilli are used to produce various other fermented food products, such as sauerkraut, pickles, brined olives, Korean kimchi, and sausages, the foods do not necessarily contain adequate numbers of live bacteria to benefit health.[14]

Various species of *Lactobacillus* are used in the production of fermented food products, such as the foods shown in this photo.

A number of companies market probiotic supplements, which are available in capsules, tablets, and powders. Because probiotic bacteria are living organisms, storage conditions may affect their viability—heat, moisture, and oxygen can reduce survival times—and therefore consumers should check the expiration date before purchasing a product. When a consumer group (ConsumerLab .com) tested 13 probiotic supplements, they found that 5 of the products contained substantially fewer live bacteria than was claimed on the label.[15] Thus, there is no guarantee that a dietary supplement will contain the amount of bacteria expected.

Certain nondigestible substances in food, called **prebiotics,** may stimulate the growth or activity of resident bacteria within the large intestine; prebiotics include some of the carbohydrates found in asparagus, chicory root, garlic, Jerusalem artichokes, onions, and other foods.[16] Because the intestinal bacteria that degrade these substances produce gas as a by-product, people who consume high amounts of these foods may experience more flatulence than usual.

Safety Concerns

One major concern is the possibility that probiotic bacteria may cause infection in immune-compromised individuals. Various species of probiotic bacteria, including *Lactobacillus* species, have been isolated from the infection sites of severely ill individuals who were consuming the probiotic orally.[17] The individuals most likely to be susceptible to infectious complications include patients with reduced immunity, such as people with AIDS or cancer and those undergoing organ transplantation. Care should be taken to inquire about probiotic use in these patients.

Other concerns are related to the lack of industry standards for probiotics in foods and supplements: the concentrations of probiotic bacteria in foods may vary substantially. Thus, a consumer who wishes to try probiotics would find it difficult to determine how much of a product to consume in order to achieve the desired effect.

In recent years, the contributions of our intestinal flora to health have been increasingly recognized. Preliminary research suggests that altering our bacterial populations by consuming probiotics or prebiotics may help to improve our defenses against certain illnesses. Additional studies are needed to verify the beneficial effects of probiotics and prebiotics and to develop standard protocols that can be used for treating illness.

REFERENCES

1. S. O'Mahony and F. Shanahan, Enteric bacterial flora and bacterial overgrowth, in M. Feldman, L. S. Friedman, and L. J. Brandt, eds., *Sleisenger and Fordtran's Gastrointestinal and Liver Disease* (Philadelphia: Saunders, 2006), pp. 2243–2256.
2. F. Guarner and J.-R. Malagelada, Gut flora in health and disease, *Lancet* 361 (2003): 512–519.
3. P. Winkler and coauthors, Molecular and cellular basis of microflora-host interactions, *Journal of Nutrition* 137 (2007): 756S–772S.
4. B. Corthésy, H. R. Gaskins, and A. Mercenier, Cross-talk between probiotic bacteria and the host immune system, *Journal of Nutrition* 137 (2007): 781S–790S.
5. M. de Vrese and P. R. Marteau, Probiotics and prebiotics: Effects on diarrhea, *Journal of Nutrition* 137 (2007): 803S–811S; A. C. Ouwehand, Antiallergic effects of probiotics, *Journal of Nutrition* 137 (2007): 794S–797S.
6. G. Reid and J. Burton, Use of *Lactobacillus* to prevent infection by pathogenic bacteria, *Microbes and Infection* 4 (2002): 319–324.
7. S. Parvez and coauthors, Probiotics and their fermented food products are beneficial for health, *Journal of Applied Microbiology* 100 (2006): 1171–1185.
8. de Vrese and Marteau, 2007.
9. J. J. Jones and A. E. Foxx-Orenstein, Probiotics in inflammatory bowel disease, *Practical Gastroenterology* (March 2006): 44–50.
10. E. I. Benchimol and D. R. Mack, Safety issues of probiotic ingestion, *Practical Gastroenterology* (November 2005): 23–34; R. S. Carvalho and M. Oliva-Hemker, Clinical indications for the use of probiotics in the pediatric population, *Practical Gastroenterology* (October 2005): 51–64.
11. Guarner and Malagelada, 2003.
12. K. J. Heller, Probiotic bacteria in fermented foods: Product characteristics and starter organisms, *American Journal of Clinical Nutrition* 73 (2001): 374S–379S.
13. C. Stanton and coauthors, Market potential for probiotics, *American Journal of Clinical Nutrition* 73 (2001): 476S–483S; G. Molin, Probiotics in foods not containing milk or milk constituents, with special reference to *Lactobacillus plantarum* 299v, *American Journal of Clinical Nutrition* 73 (2001): 380S–385S.
14. P. Lavermicocca and coauthors, Study of adhesion and survival of Lactobacilli and Bifidobacteria on table olives with the aim of formulating a new probiotic food, *Applied and Environmental Microbiology* 71 (2005): 4233–4240.
15. ConsumerLab.com, Product review: Probiotic supplements (including *Lactobacillus acidophilus, Bifidobacterium,* and others), available at www.consumerlab.com; site visited November 27, 2007.
16. S. Kolida and G. R. Gibson, Prebiotic capacity of inulin-type fructans, *Journal of Nutrition* 137 (2007): 2503S–2506S.
17. Benchimol and Mack, 2005; N. Ishibashi and S. Yamazaki, Probiotics and safety, *American Journal of Clinical Nutrition* 73 (2001): 465S–470S.

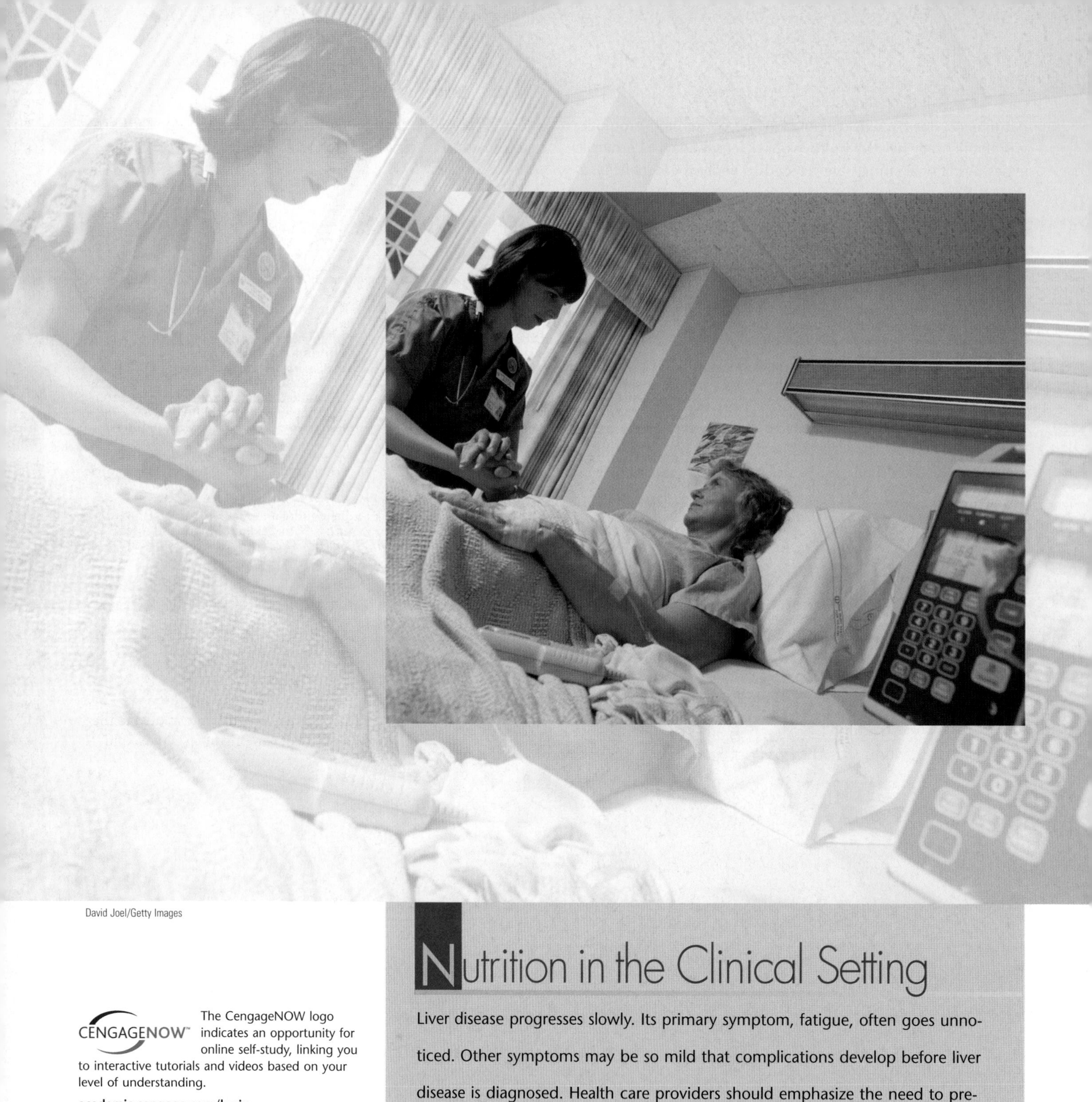

CENGAGENOW™

The CengageNOW logo indicates an opportunity for online self-study, linking you to interactive tutorials and videos based on your level of understanding.

academic.cengage.com/login

Nutrition in the Clinical Setting

Liver disease progresses slowly. Its primary symptom, fatigue, often goes unnoticed. Other symptoms may be so mild that complications develop before liver disease is diagnosed. Health care providers should emphasize the need to preserve remaining liver function, as healthy liver tissue can proliferate, improving prognosis. Preventing additional damage is the principal means of avoiding liver failure or transplantation.

Liver Disease and Gallstones

The liver is the most metabolically active organ in the body. As you may recall from Chapter 7, the liver plays a central role in processing, storing, and redistributing the nutrients provided by the meals we eat. ◆ The liver synthesizes most of the proteins that circulate in plasma, including albumin, clotting proteins, and transport proteins. As Chapter 3 described, the liver produces the bile that emulsifies fat during digestion. In addition, the liver detoxifies drugs and alcohol and processes excess nitrogen so that it can be safely excreted as urea. If the liver's numerous roles are upset by liver damage or disease, the effects on health and nutrition status can be profound.

As Figure 25-1 on p. 788 shows, the liver is ideally situated for receiving and processing the nutrients absorbed by the small intestine. The portal vein's nutrient-rich blood supplies 70 to 80 percent of the blood that enters liver tissue, whereas the rest arrives via hepatic arteries. Blood is returned to the heart by way of the hepatic vein and then circulates throughout the body. The **biliary system** of channels and ducts carries bile and other substances from the liver to the duodenum while a meal is being digested. Between meals, the bile is diverted to the gallbladder, where it is stored and concentrated until needed for a subsequent meal.

◆ Table 7-1 on p. 215 summarizes the chemical reactions of the liver that are related to the metabolism of carbohydrates, lipids, and protein.

Fatty Liver and Hepatitis

Fatty liver and hepatitis are the two most common disorders affecting the liver. Although both conditions may be mild and are usually reversible, each may progress to more serious illness and eventually cause liver damage.

biliary system: the gallbladder and ducts that deliver bile from the liver and gallbladder to the small intestine.

FIGURE 25-1 The Liver, Biliary System, and Associated Blood Vessels

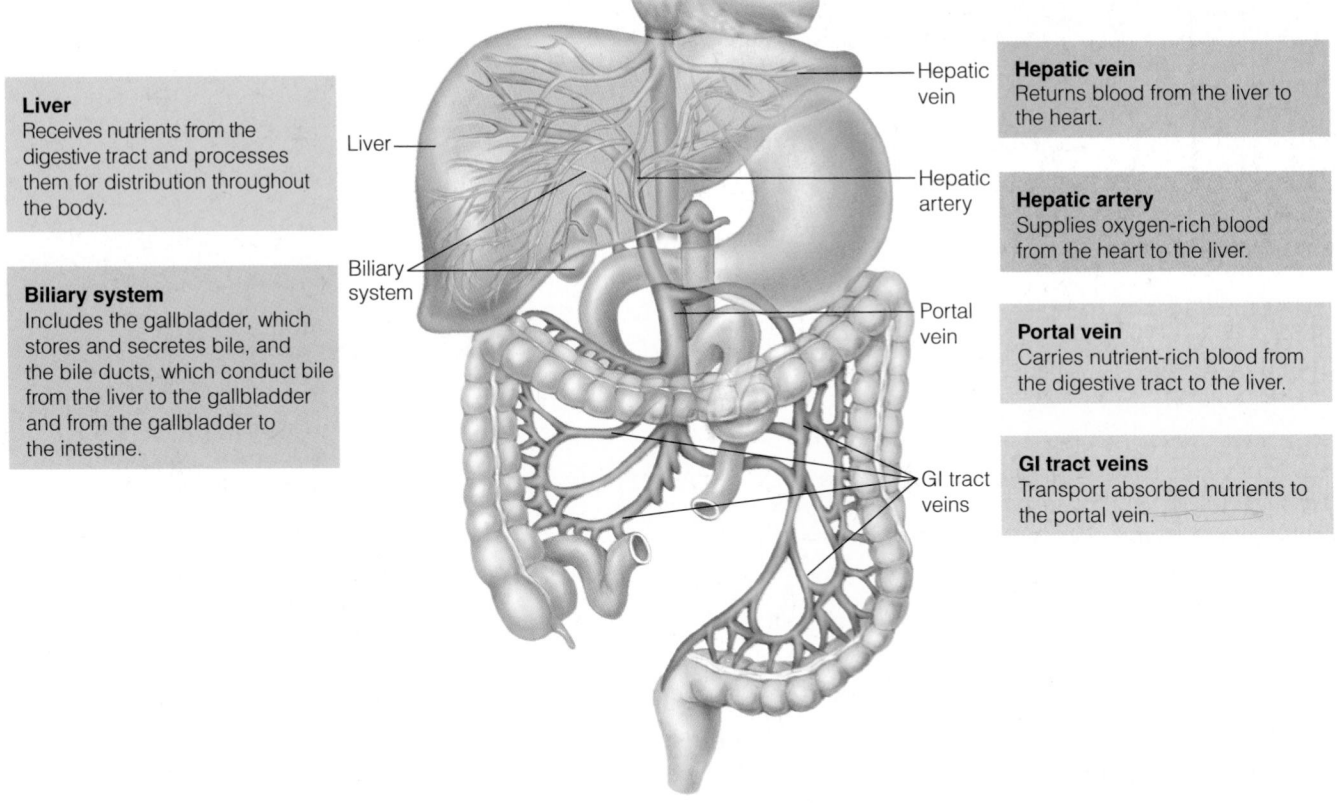

Liver
Receives nutrients from the digestive tract and processes them for distribution throughout the body.

Biliary system
Includes the gallbladder, which stores and secretes bile, and the bile ducts, which conduct bile from the liver to the gallbladder and from the gallbladder to the intestine.

Liver

Biliary system

Hepatic vein

Hepatic artery

Portal vein

GI tract veins

Hepatic vein
Returns blood from the liver to the heart.

Hepatic artery
Supplies oxygen-rich blood from the heart to the liver.

Portal vein
Carries nutrient-rich blood from the digestive tract to the liver.

GI tract veins
Transport absorbed nutrients to the portal vein.

Fatty Liver

◆ Reminder: *Fatty liver* is an accumulation of triglycerides in the liver; also called *hepatic steatosis* (STEE-ah-TOE-sis).

◆ Insulin resistance promotes fat accumulation in the liver by stimulating fatty acid synthesis, inhibiting fatty acid oxidation, and blocking VLDL secretion into the bloodstream.

◆ The liver enzymes ALT (alanine aminotransferase) and AST (aspartate aminotransferase) are involved in amino acid catabolism.

hepatomegaly (HEP-ah-toe-MEG-ah-lee): enlargement of the liver.

steatohepatitis (STEE-ah-to-HEP-ah-TIE-tis): liver inflammation that is associated with fatty liver.

Fatty liver ◆ is an accumulation of fat in liver tissue. Ordinarily, the liver's triglycerides are packaged into very-low-density lipoproteins (VLDL) and exported to the bloodstream (see Chapter 5). Although the exact reasons why fat accumulates are often unknown, fatty liver represents an imbalance between the amount of fat synthesized in the liver or picked up from the blood and the amount exported to the blood via VLDL. Fatty liver has been estimated to affect 20 percent or more of the adult population in the United States.[1]

Fatty liver is a clinical finding that is common to many conditions. It is present in the majority of patients who have alcoholic liver disease and can also result from exposure to drugs and toxic metals. As a frequent complication of insulin resistance, fatty liver often accompanies diabetes mellitus, metabolic syndrome, and obesity.[2] ◆ Other causes of fatty liver include long-term total parenteral nutrition, gastrointestinal bypass surgery, and diseases of malnutrition (such as kwashiorkor).[3]

Consequences of Fatty Liver In many individuals, fatty liver is asymptomatic and causes no harm. In other cases, it may be accompanied by liver enlargement **(hepatomegaly)**, inflammation **(steatohepatitis)**, and fatigue. If liver damage and scarring develop, fatty liver may progress to cirrhosis (discussed in a later section), liver failure, or liver cancer.[4]

Fatty liver is a common cause of abnormal liver enzyme levels in the blood. Laboratory findings may include elevated blood concentrations of the liver enzymes ALT and AST, as well as increased levels of triglycerides, cholesterol, and glucose. ◆ Table 17-10 on p. 603 provides normal ranges for these liver enzymes.

Treatment of Fatty Liver The usual treatment for fatty liver is to eliminate the factors that cause it. For example, if fatty liver is due to alcohol abuse or drug treatment, it may improve after the patient discontinues use of the substance. In patients with elevated blood lipids, fatty liver may improve after blood lipid levels are lowered. An appropriate treatment for obese or diabetic patients might be weight reduction, increased physical activity, or medications that improve insulin sensitivity. Rapid weight loss should be discouraged, however, because it may accelerate the progression of liver disease.[5] Note that lifestyle modifications are not always successful in reversing fatty liver, especially in patients who lack the usual risk factors.

Hepatitis

Hepatitis, or liver inflammation, results from damage to liver tissue. Most often, the damage is caused by infection with specific viruses, which are designated by the letters A, B, C, D, and E. Hepatitis can also be caused by excessive alcohol intake, exposure to certain drugs and toxic chemicals, and fatty liver disease. A number of herbal remedies are reported to cause hepatitis; they include chaparral, germander, ma huang, jin bu huan, kava kava, and skullcap.[6] Less common causes of hepatitis include infection with other viruses and autoimmune diseases.

Viral Hepatitis In the United States, acute hepatitis ◆ is most often caused by infection with hepatitis viruses A, B, and C (see Table 25-1).[7] Hepatitis A virus (HAV), which is spread via the fecal-oral route, is due to the contamination of foods and beverages with fecal material. Outbreaks of HAV infection are often associated with floods and other natural disasters, when inadequately treated sewage may contaminate water supplies with fecal matter. Less frequently, HAV infection is spread by consuming undercooked shellfish obtained from contaminated waters. Individuals at highest risk of HAV infection include those in lower socioeconomic groups, day care workers, illicit drug users, and travelers to regions where it is endemic.[8] In the United States, the incidence of HAV infection has declined 88 percent since 1995 as a result of routine vaccinations in children and high-risk individuals.[9] HAV infection usually resolves within a few months and does not cause chronic illness or permanent liver damage.

Over half of viral hepatitis cases are caused by infection with hepatitis B or C viruses. Hepatitis B virus (HBV) is transmitted by infected blood or needles or by sexual contact and is carried by at least 1.25 to 1.5 million persons in the United States.[10] As a precautionary measure, HBV vaccinations are recommended for health care workers, sexually active adults, users of illicit injected drugs, and newborn infants and children. Hepatitis C virus (HCV) is transmitted by blood contact as well but is less efficiently spread by sexual contact. HCV is a major cause of chronic hepatitis and is currently carried by 3.2 million individuals in the United States.[11] ◆ No vaccine is available to protect against HCV infection. Chronic HBV or HCV infection may lead to cirrhosis and liver cancer.

◆ Hepatitis is considered *acute* if it lasts less than six months; *chronic* cases are those that last six months or longer.

◆ There are fewer new cases of HCV infection than of HBV infection each year, but more HCV cases become chronic. Therefore, there are more HCV carriers than HBV carriers.

hepatitis (hep-ah-TYE-tis): inflammation of the liver.

TABLE 25-1	Features of Hepatitis Viruses			
Hepatitis Virus	% of Viral Cases[a]	Major Mode of Transmission	Chronic Disease Rate (% of cases)	Vaccination Available
A	37	Fecal-oral	None	Yes
B	45	Bloodborne, sexual transmission	2–7	Yes
C	18	Bloodborne	50–85	No

[a]Although not listed here, a small fraction of viral hepatitis cases are caused by hepatitis viruses D and E.

SOURCES: J. H. Hoofnagle, Acute viral hepatitis, in L. Goldman and D. Ausiello, eds., *Cecil Medicine* (Philadelphia: Saunders, 2008), pp. 1101–1108; Centers for Disease Control and Prevention, Surveillance for acute viral hepatitis—United States, 2005, *Morbidity and Mortality Weekly Report* 56, No. SS-3 (2007).

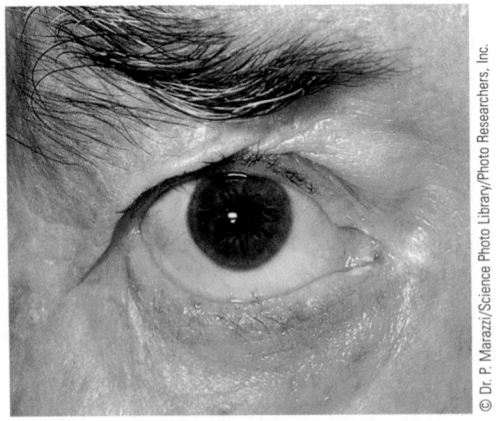

Jaundice is a yellow discoloration of the tissues that is most easily seen in the whites of the eyes.

◆ Jaundice results when liver dysfunction impairs the metabolism of bilirubin, a breakdown product of hemoglobin that is normally eliminated in bile. Accumulation of bilirubin in the bloodstream leads to yellow discoloration of tissues.

TABLE 25-2	Causes of Cirrhosis

Alcoholic liver disease

Autoimmune hepatitis

Bile duct obstructions
- Complications of gallbladder surgery
- Cystic fibrosis
- Diseases that cause bile duct injury

Drug-induced liver injury

Inherited disorders
- Galactosemia
- Glycogen storage disease
- Hemochromatosis (causes excessive liver iron)
- Wilson's disease (causes excessive liver copper)

Steatohepatitis (fatty liver disease)

Viral hepatitis
- Hepatitis B
- Hepatitis C

---

jaundice (JAWN-dis): yellow discoloration of the skin and eyes due to an accumulation of bilirubin, a breakdown product of hemoglobin that normally exits the body via bile secretions.

cirrhosis (sih-ROE-sis): an advanced stage of liver disease in which extensive scarring replaces healthy liver tissue, causing impaired liver function and liver failure.

Symptoms of Hepatitis The effects of hepatitis depend on the cause and severity of the disease. Individuals with mild or chronic hepatitis are often asymptomatic. The onset of acute hepatitis may be accompanied by fatigue, nausea, anorexia, and pain in the liver area. The liver is often slightly enlarged and tender. **Jaundice** (yellow discoloration of tissues) may develop, causing yellowing of the skin, urine, and the whites of the eyes. ◆ Other symptoms of hepatitis include fever, headache, muscle weakness, and skin rashes. Increased serum levels of the liver enzymes ALT and AST are common. Chronic hepatitis may cause complications that are typical of liver cirrhosis.

Treatment of Hepatitis Hepatitis is treated with supportive care, such as bed rest (if necessary) and an appropriate diet. Hepatitis patients should avoid substances that irritate the liver, such as alcohol and drugs or dietary supplements that cause liver damage. Hepatitis A infection usually resolves without the use of medications. Antiviral agents may be used to treat HBV and HCV infections; these medications include lamivudine and ribavirin, which block viral replication, and interferon, which both inhibits viral replication and enhances immune responses.[12] Nonviral forms of hepatitis may be treated with anti-inflammatory and immunosuppressant drugs. Hospitalization is not required for hepatitis unless other medical conditions or complications hamper recovery.

Nutrition Therapy for Hepatitis Nutrition care varies according to a patient's symptoms and nutrition status.[13] Some individuals require no dietary changes. Those with anorexia or gastrointestinal discomfort may find small, frequent meals easier to tolerate. Persons who are malnourished should consume adequate protein and energy to replenish nutrient stores; the diet should include 1.0 to 1.2 grams of protein per kilogram of body weight. Individuals with fluid retention should avoid high-sodium foods. A low-fat diet, with fat limited to less than 30 percent of total kcalories, may be necessary for those with steatorrhea. Patients with persistent vomiting may require fluid and electrolyte replacement. Liquid supplements can be helpful for improving nutrient intakes.

IN SUMMARY

Fatty liver can result from excessive alcohol intake, drug toxicity, and chronic disorders such as diabetes and obesity. Hepatitis is frequently caused by viral infection but can also result from alcohol abuse, drug toxicity, and other causes. Although fatty liver is often benign, hepatitis can become chronic and lead to cirrhosis and liver cancer. Treatment of hepatitis includes avoiding substances that cause liver damage, taking medications for viral infection, and following dietary measures that improve nutrition status.

Cirrhosis

Cirrhosis is the final phase of chronic liver disease. Long-term liver disease gradually destroys liver tissue, leading to scarring (fibrosis) in some regions and small areas of regenerated, healthy tissue in others. As the disease progresses, the scarring becomes more extensive, leaving fewer areas of healthy tissue. A cirrhotic liver is often shrunken and has an irregular, nodular appearance. Cirrhosis impairs liver function and can eventually lead to liver failure. It is the 12th leading cause of death in the United States.[14]

Table 25-2 lists some common causes of cirrhosis. In the United States, most cases are caused by alcoholic liver disease and chronic hepatitis C infection, followed by fatty liver disease and chronic hepatitis B infection.[15] Additional causes include other types of chronic hepatitis; bile duct blockages, which cause bile acids to accumulate to toxic levels in the liver; drug-induced liver injury; and inherited disorders that cause toxic substances to build up in the liver.

Consequences of Cirrhosis

About 40 percent of cirrhosis patients are asymptomatic.[16] The effects of cirrhosis may be minimal at first, as liver damage often progresses slowly. Initial symptoms are usually nonspecific and may include fatigue, weakness, anorexia, and weight loss. Later, the decline in liver function can lead to metabolic disturbances: patients may develop anemia, bruise easily, and be more susceptible to infections. If bile obstruction occurs, jaundice and fat malabsorption are likely. The physical changes in liver tissue may interfere with blood flow, causing fluid to accumulate in blood vessels and body tissues. Advanced cirrhosis can disrupt kidney and lung function. Figure 25-2 illustrates some of the clinical effects of liver cirrhosis, and later sections describe some of these complications in more detail.

Table 25-3 on p. 792 lists laboratory tests that are used to monitor the extent of liver damage. Liver enzyme levels are increased because injured liver tissue releases the enzymes into the bloodstream. Levels of bilirubin may be elevated if the liver is too damaged to process it or if bile ducts are blocked and prevent its excretion. Reduced synthesis of plasma proteins by the liver lowers albumin levels and extends blood-clotting time. Liver damage also impairs the conversion of ammonia to urea, causing ammonia levels in the blood to rise.

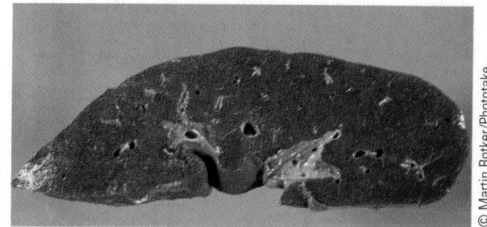

Normal liver tissue is smooth and has a regular texture.

A cirrhotic liver has an irregular, nodular appearance. The nodules represent clusters of regenerating cells within the damaged liver tissue.

FIGURE 25-2 Clinical Effects of Liver Cirrhosis

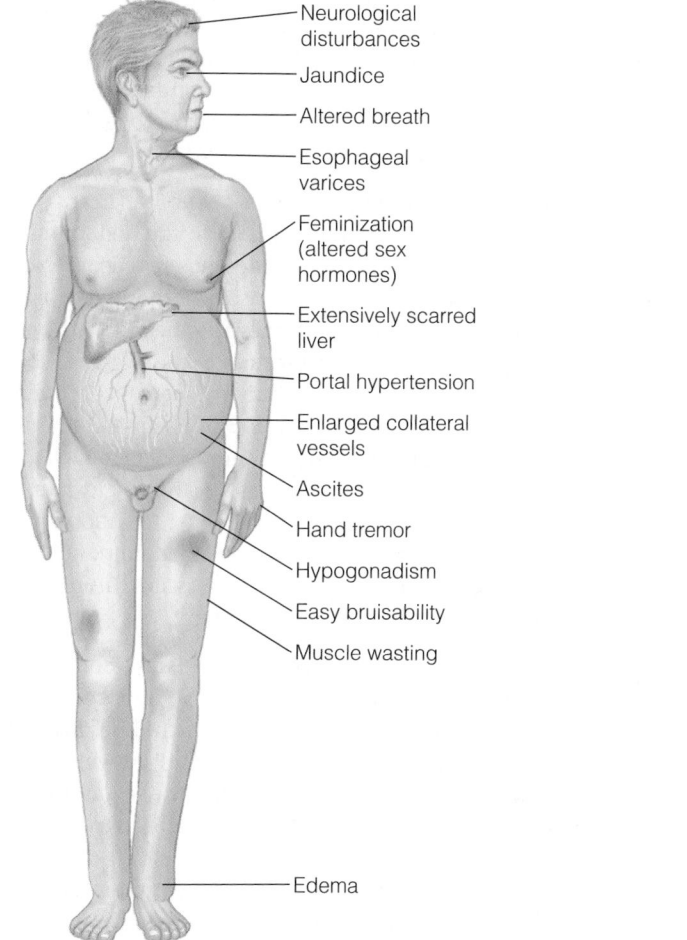

- Neurological disturbances
- Jaundice
- Altered breath
- Esophageal varices
- Feminization (altered sex hormones)
- Extensively scarred liver
- Portal hypertension
- Enlarged collateral vessels
- Ascites
- Hand tremor
- Hypogonadism
- Easy bruisability
- Muscle wasting
- Edema

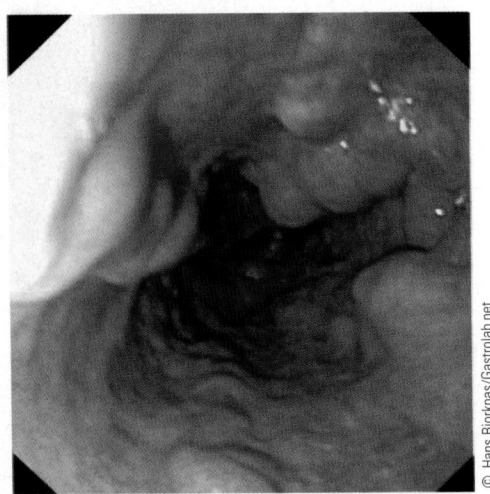

Esophageal varices, such as the one shown here, may protrude into the lumen and be vulnerable to rupture and bleeding.

◆ Reminder: The *portal vein* is the large blood vessel that carries nutrient-rich blood from the digestive tract to the liver.

portal hypertension: elevated blood pressure in the portal vein due to obstructed blood flow through the liver.

collaterals: blood vessels that enlarge to allow an alternative pathway for diverted blood.

varices (VAH-rih-seez): abnormally dilated blood vessels (singular: *varix*).

ascites (ah-SIGH-teez): an abnormal accumulation of fluid in the abdominal cavity.

sinusoids: the small, capillary-like passages that carry blood through liver tissue.

hepatic encephalopathy (en-sef-ah-LOP-ah-thie): a condition in advanced liver disease characterized by altered neurological functioning, including personality changes, reduced mental abilities, and disturbances in motor function.
  • **encephalo** = brain
  • **pathy** = disease

hepatic coma: loss of consciousness resulting from severe liver disease.

TABLE 25-3 Laboratory Tests for Evaluation of Liver Disease

Laboratory Test	Normal Ranges (serum)	Values in Liver Disease
Alanine aminotransferase (ALT)	Male: 10–40 U/L Female: 7–35 U/L	Elevated
Albumin	3.4–4.8 g/dL	Decreased
Alkaline phosphatase	25–100 U/L	Normal or elevated
Ammonia	15–45 μg N/dL	Elevated
Aspartate aminotransferase (AST)	10–30 U/L	Elevated
Bilirubin (total)	0.3–1.2 mg/dL	Elevated
Blood urea nitrogen (BUN)	6–20 mg/dL	Normal or decreased
Prothrombin time[a]	10–13 seconds	Prolonged

[a]The test for prothrombin time evaluates the clotting ability of blood.
NOTE: U/L = units per liter; N = nitrogen; dL = deciliter.

Portal Hypertension A large volume of blood normally flows through the liver. The portal vein ◆ and hepatic artery together supply approximately 1500 milliliters (about 1.5 quarts) of blood each minute to the extensive network of vessels in the liver. The scarred tissue of a cirrhotic liver impedes the flow of blood, which is mostly supplied by the portal vein. The resistance to blood flow within the liver causes a rise in blood pressure within the portal vein, called **portal hypertension.**

Collaterals and Gastroesophageal Varices When blood flow through the portal vein is impeded, the blood is diverted to the smaller blood vessels surrounding the liver. These **collaterals** develop throughout the gastrointestinal (GI) tract and in regions near the abdominal wall. As pressure builds, the collateral vessels become enlarged and engorged, forming abnormally dilated vessels called **varices** (see the photo). Esophageal and gastric varices are vulnerable to rupture because they have thin walls and often bulge into the lumen. If ruptured, they can cause massive bleeding that is sometimes fatal. The blood loss is exacerbated by the liver's reduced production of blood-clotting factors.

Ascites Within 10 years of disease onset, about 50 percent of cirrhosis patients develop **ascites,** a large accumulation of fluid in the abdominal cavity. The development of ascites indicates that liver damage has reached a critical stage, as half of patients with ascites die within 2 years.[17] Ascites is thought to be a consequence of portal hypertension, reduced albumin synthesis by the diseased liver, and altered kidney function. Portal hypertension raises pressure in the liver's **sinusoids,** forcing fluid to leak from the blood into the abdominal cavity. The fluid accumulation is exacerbated by low plasma albumin levels, because albumin helps to retain the fluid in blood vessels. The increased pressure in the portal vein triggers the release of chemical factors (such as nitric oxide) that dilate blood vessels, thereby lowering the pressure within vessels elsewhere; this activates sodium and water retention by the kidneys and leads to additional water pooling. Ascites can cause abdominal discomfort and early satiety, which contribute to malnutrition. Because ascites can raise body weight considerably, weight changes may be difficult to interpret.

Hepatic Encephalopathy Advanced liver disease sometimes leads to **hepatic encephalopathy,** a disorder characterized by abnormal neurological functioning. Symptoms of hepatic encephalopathy include changes in mental abilities, personality, and motor functions (see Table 25-4). At worst, amnesia, seizures, and **hepatic coma** may develop. Although hepatic encephalopathy is fully reversible with medical treatment, prognosis is poor when it progresses to the advanced stages.[18]

The exact causes of hepatic encephalopathy remain elusive, although elevated blood ammonia levels are thought to play a key role in its development due to ammonia's neurotoxicity.[19] Other compounds potentially toxic to brain tissue, such as

TABLE 25-4	Symptoms of Hepatic Encephalopathy	
Early Stages	Middle Stages	Later Stages
Lack of attention	Poor memory	Disorientation
Irritability, depression	Drowsiness	Amnesia
Impaired judgment	Slurred speech	Muscular rigidity
Lack of coordination	Jerking movements	Abnormal reflexes
Tremor	Sleep disorders	Delirium, stupor

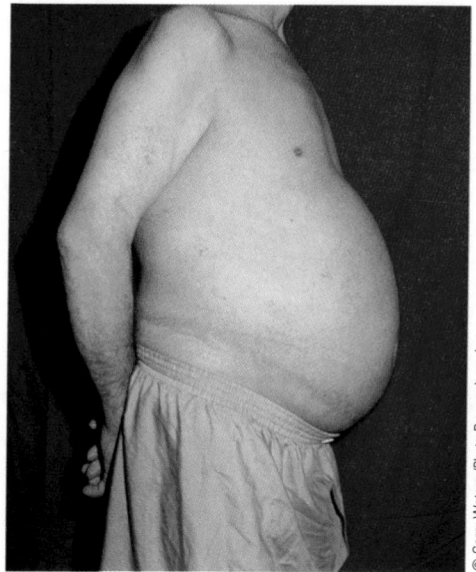

Ascites is caused by different diseases, but cirrhosis is the underlying cause in most patients with this condition.

sulfur compounds, short-chain fatty acids, and gamma-aminobutyric acid (GABA; a neurotransmitter), may accumulate in brain cells and disturb brain function. Another theory is that the brain's neurotransmitters are altered by an increased ratio of aromatic amino acids to branched-chain amino acids in brain tissue, a result of disordered amino acid metabolism in the liver. ◆ Most likely, a combination of metabolic abnormalities contributes to the disruption in neurological functioning.

Elevated Blood Ammonia Levels Much of the body's free ammonia is produced by bacterial action on unabsorbed dietary protein in the colon. Normally, the liver extracts this ammonia from portal blood and converts it to urea, which is then excreted by the kidneys. In advanced liver disease, ammonia-laden portal blood bypasses the liver by way of collateral vessels and reaches the general blood circulation, causing a substantial increase in the ammonia that reaches brain tissue. Although ammonia levels do not correlate well with the degree of neurological impairment in hepatic encephalopathy, ammonia-reducing medications can successfully reverse the neurological symptoms.[20]

Malnutrition and Wasting Most patients with cirrhosis develop protein-energy malnutrition (PEM) and experience some degree of wasting. Malnutrition is usually caused by a combination of factors (see Table 25-5). Patients may consume less food due to reduced appetite, gastrointestinal symptoms, or fatigue. Ascites often causes early satiety. If the diet is sodium restricted (to treat ascites), meals may seem monotonous or unpalatable. Fat malabsorption is common due to reduced bile flow, which leads to steatorrhea and deficiencies of the fat-soluble vitamins and some minerals. Other GI symptoms, such as diarrhea, vomiting, and gastrointestinal bleeding, contribute to additional nutrient losses. If cirrhosis is a consequence of alcohol abuse, multiple nutrient deficiencies may be present.

◆ The aromatic amino acids—phenylalanine, tyrosine, and tryptophan—have carbon rings in their side groups. The branched-chain amino acids are leucine, isoleucine, and valine; their side groups have a branched structure.

Treatment of Cirrhosis

Treatment of cirrhosis is individualized according to the severity of the illness and the complications that develop. Supportive care, including an appropriate diet and

TABLE 25-5	Possible Causes of Malnutrition in Liver Disease
Mechanism	Examples
Reduced nutrient intake	Anorexia, early satiety (due to ascites), nausea and vomiting, restrictive diets, effects of medications (including gastrointestinal disturbances and taste changes), abdominal pain, fatigue, fasting for medical procedures
Malabsorption/nutrient losses	Fat malabsorption (due to reduced bile flow), vomiting, diarrhea, gastrointestinal bleeding, effects of medications (including malabsorption and nutrient losses from diuretic use)
Altered metabolism/increased nutrient needs	Hypermetabolism, catabolism, infections/inflammation, inadequate protein synthesis, reduced nutrient storage and metabolism in the liver

avoidance of liver toxins, promotes recovery and helps to prevent further damage. Abstinence from alcohol is critical for preserving liver function and extending survival. Antiviral therapy may be prescribed to treat viral infections. Patients may be screened for life-threatening complications, such as gastroesophageal varices and liver cancer. Liver transplantation may be necessary in advanced cirrhosis.

Drugs are prescribed to treat the symptoms and complications that accompany cirrhosis. Medications for portal hypertension and varices include propranolol and octreotide, which reduce portal blood pressure, and vasopressin, which constricts blood vessels. Diuretics may help to control portal hypertension and ascites; common examples include spironolactone and furosemide. To stimulate the appetite and promote weight gain, megestrol acetate or dronabinol may be prescribed. Drug treatment of hepatic encephalopathy focuses on controlling blood ammonia levels. Lactulose, a nonabsorbable disaccharide often used as a laxative, helps to reduce ammonia production and absorption in the colon. The antibiotic neomycin is an alternative treatment for elevated ammonia that works by altering bacterial populations. The Diet-Drug Interactions feature lists potential nutritional problems associated with these medications.

Medical Nutrition Therapy for Cirrhosis

Nutrition therapy for cirrhosis is customized to each patient's needs, which vary considerably and depend on the accompanying complications. Patients with cirrhosis are generally at high risk of protein-energy malnutrition and muscle wasting; therefore, protein and energy intakes must be high enough to maintain nitrogen balance. Dietary substances that may cause additional liver injury should be avoided; examples include alcohol, certain drugs, herbal supplements, and vitamin or mineral megadoses. Table 25-6 lists the general dietary guidelines for cirrhosis, which are discussed in the sections that follow.

Energy To estimate energy requirements, health practitioners measure or calculate the resting metabolic rate (RMR) and then apply a stress factor, as described in the "How to" on p. 621 (Chapter 18). If possible, indirect calorimetry should be used to

DIET-DRUG INTERACTIONS

Check this table for notable nutrition-related effects of the medications discussed in this chapter.

	Gastrointestinal Effects	Interactions with Dietary Substances	Metabolic Effects
Appetite stimulants (megestrol acetate, dronabinol)	Nausea, vomiting, diarrhea.	—	Hyperglycemia (megestrol acetate).
Diuretics (furosemide, spironolactone[a])	Dry mouth, anorexia, decreased taste perception.	Furosemide's bioavailability is reduced when taken with food.	Fluid and electrolyte imbalances,[a] hyperglycemia (spironolactone), hyperlipidemia (spironolactone), thiamin and zinc deficiencies.
Immunosuppressants (cyclosporine, tacrolimus)	Nausea, vomiting, diarrhea, anorexia (tacrolimus).	Grapefruit juice can raise serum concentrations of these drugs to toxic levels. Cyclosporine potentiates the effects of alcohol. The bioavailability of tacrolimus is reduced when the drug is taken with food.	Electrolyte imbalances, hypertension, hyperglycemia, hyperlipidemia.
Lactulose	Diarrhea.	—	Fluid and electrolyte imbalances.

[a]*Furosemide* is a "potassium-wasting" diuretic; patients should increase intakes of potassium-rich foods. *Spironolactone* is a "potassium-sparing" diuretic; patients should avoid supplemental potassium and potassium-containing salt substitutes.

TABLE 25-6 Medical Nutrition Therapy for Liver Cirrhosis

Energy	• Energy needs may be approximately 20% above resting metabolic rate (RMR); calculate or measure RMR and apply the stress factor 1.2 initially (see Table 18–3 and the "How To" on p. 621).
	• Use estimated dry body weight for RMR calculations in patients with ascites.
	• Energy requirements may be higher in patients with infection or malnutrition. Energy requirements may be lower in patients who would benefit from weight loss.
Meal frequency	• To improve food intake, patients should consume small meals 4 to 6 times daily.
Protein	• Provide 0.8 to 1.2 g protein per kilogram of dry body weight per day to maintain nitrogen balance and prevent wasting.
Carbohydrate	• No carbohydrate restrictions unless patient has insulin resistance or diabetes.
	• For persons with insulin resistance or diabetes, provide up to 50% to 60% of kcalories from carbohydrates (mainly complex carbohydrates); carbohydrate intake should be consistent from day to day and at each meal and snack.
Fat	• No fat restrictions unless fat malabsorption is present.
	• If fat is malabsorbed, restrict fat to 30% of total kcalories or as necessary to control steatorrhea; use medium-chain triglycerides (MCT) to increase kcalories.
Sodium and fluid	• Restrict sodium as necessary to control ascites; 2000 mg sodium per day is adequate restriction in most cases.
	• If ascites is accompanied by low serum sodium levels (less than 128 mEq/L), restrict fluids to 1200 to 1500 mL per day. In severe cases (serum sodium less than 125 mEq/L), restrict fluids to 1000 to 1200 mL per day.
Vitamins and minerals	• Ensure adequate intake from diet or supplements based on individual needs.

determine RMR. For most patients with cirrhosis, the stress factor 1.2 can be used initially and adjusted as necessary.[21] Nutrient malabsorption, recent weight loss, and infection can increase energy needs. For patients with ascites, RMR calculations should use either the patient's desirable weight or an estimated dry weight (weight without ascites). A value for dry weight can be obtained after diuretic therapy or after a medical procedure that directly removes excess abdominal fluid.

Many patients with cirrhosis have difficulty consuming enough food to achieve good nutrition status. They may better tolerate four to six daily feedings than three meals per day. Liquid dietary supplements can help to improve energy intakes. The "How to" below offers additional suggestions that can help a patient meet energy needs.

HOW TO Help the Person with Cirrhosis Eat Enough Food

Individuals with cirrhosis often have difficulty consuming enough food to prevent malnutrition and its consequences. Ascites and gastrointestinal symptoms such as nausea and vomiting may interfere with food intake. Fatigue may cause disinterest in food preparation. Sodium restrictions may make foods unpalatable. To improve food intake:

• If nutrient restrictions are necessary, make sure the patient fully understands how to modify the diet so that food intake is not restricted unnecessarily. Provide lists of acceptable foods and menus. Explain how recipes can be altered so that favorite foods can still be incorporated into the diet.

• Suggest between-meal snacks during the day and a snack at bedtime. An oral supplement such as Ensure can substitute for a snack and requires no preparation. Snacks should not be consumed within two hours of meals, or they may reduce appetite at mealtime.

• If the patient has little appetite or is quickly satiated, suggest foods that are higher in

food energy, such as whole milk instead of reduced-fat milk or canned fruit that is packed in heavy syrup instead of fruit juice.

• Recommend energy boosters. Cream sauces and gravies can add kcalories to entrées. Fruit juices and fruit nectars can substitute for drinking water. The following additions can boost the energy content of meals:

• Sour cream and butter—on vegetables and potatoes

• Mayonnaise—in sandwiches and salads

• Half-and-half and light cream—in soups and on cereals

• Hard-boiled eggs—in casseroles and meat loaf

• Cheese—in salads and casseroles and melted on steamed vegetables

• Peanut butter, nut butters, and cream cheese—on crackers or celery and in milk shakes

• Chopped nuts—in salads, cooked cereals, and bakery products

Low-sodium diets are recommended for treating ascites and other medical

conditions, including kidney and heart disorders. The "How To" on p. 860 offers suggestions to help patients implement sodium restrictions. To improve the palatability of low-sodium meals:

• Suggest that patients replace the salt they use for cooking and seasoning with strong-flavored herbs and spices such as chili powder, coriander, cumin, curry powder, garlic, ginger, lemon, mint, and parsley.

• Advise patients to check food labels to learn the sodium content of the foods they eat. Similar products may be available that are lower in sodium. (Persons using potassium-sparing diuretics should be cautioned to avoid salt substitutes that replace sodium with potassium.)

Offer support and encouragement to the patient with cirrhosis. Severe weight loss is less likely to occur if dietary advice is provided before problems progress.

◆ Reminder: The protein RDA for healthy adults is 0.8 g/kg.

Protein The protein recommendation is 0.8 to 1.2 grams of protein per kilogram of body weight per day, based on desirable weight or dry weight. ◆ Patients with hepatic encephalopathy should avoid excessive protein consumption, and their protein intake should be spread throughout the day so that they consume only modest amounts at each meal. Protein restriction is not helpful, however, because inadequate protein can worsen malnutrition and wasting. In an attempt to normalize altered amino acid ratios in brain tissue and improve mental status, some health care providers may prescribe enteral formulas with reduced aromatic amino acids and added branched-chain amino acids. Clinical studies testing the use of these formulas have yielded mixed results, however, and their routine use is not currently recommended.[22]

Carbohydrate Carbohydrate provides a substantial proportion of energy needs. Many patients with cirrhosis are insulin resistant and require medications or insulin to manage their hyperglycemia. These individuals should follow the dietary guidelines for diabetes: consume mostly complex carbohydrates, and consume them at regular intervals throughout the day.

Fat Fat provides both energy and essential fatty acids. In patients with fat malabsorption, fat intake may be restricted to less than 30 percent of total kcalories or as necessary to control steatorrhea. Medium-chain triglycerides (MCT) may be used to provide additional energy, although essential fatty acids cannot be obtained from MCT oils and may need to be supplemented. Severe steatorrhea warrants supplementation of the fat-soluble vitamins, calcium, magnesium, and zinc (see Chapter 24).

◆ Table 28-1 (p. 877) and the "How to" on p. 860 provide information about following a sodium-restricted diet.

Sodium and Fluid Patients with ascites are generally advised to restrict sodium. Ascites is partly caused by the kidneys' reabsorption of sodium into the blood, which results in sodium and water retention. Therefore, the treatment usually includes both a moderate sodium restriction (to no more than 2000 milligrams of sodium per day) and diuretic therapy to promote fluid loss. ◆ Potassium intake should be monitored if a potassium-wasting diuretic (such as furosemide) is used.

Many patients find low-sodium diets unpalatable, so some health practitioners may allow a more liberal sodium intake and depend on diuretics to mobilize excess fluids. If patients do not respond to sodium restriction and diuretic therapy, fluid may be removed from the abdomen by surgical puncture (**paracentesis**) or may be diverted to the bloodstream using a catheter (**peritoneovenous shunt**).

Fluid restriction may be necessary when ascites is accompanied by a low concentration of serum sodium. If the sodium level falls below 128 milliequivalents per liter, the fluid intake should be limited to 1200 to 1500 milliliters daily; with a sodium level below 125 milliequivalents per liter, fluids should be restricted to 1000 to 1200 milliliters per day.[23]

Vitamins and Minerals Vitamin and mineral deficiencies are common in patients with cirrhosis due to the effects of illness, disease complications, or the alcohol abuse that may have induced the liver disease. Thus, multivitamin supplementation is often necessary. If steatorrhea is present, fat-soluble nutrients can be provided in a water-soluble form. Patients with esophageal varices may find it easier to ingest supplements in liquid form.

Enteral and Parenteral Nutrition Support In patients who are unable to consume enough food, tube feedings may be infused overnight as a supplement to oral intakes or may replace oral feedings entirely. Although standard formulas are often appropriate, an energy-dense, moderate-protein, low-electrolyte formula may be necessary for patients with ascites or fluid restrictions. In patients with esophageal varices, the feeding tube should be as narrow and flexible as possible to prevent rupture and bleeding. Parenteral nutrition support should be considered for patients who are unable to tolerate enteral feedings due to intestinal obstruction, gastrointestinal bleeding, or uncontrollable vomiting. To avoid excessive fluid delivery, patients with ascites typically require concentrated parenteral solutions, which are infused into central veins. The Case Study on p. 797 allows you to apply your knowledge of cirrhosis to a clinical situation.

paracentesis (pah-rah-sen-TEE-sis): a surgical puncture of a body cavity with an aspirator to draw out excess fluid.

peritoneovenous (PEH-rih-toe-NEE-oh-VEE-nus) **shunt:** a surgical passage created between the peritoneum and the jugular vein to divert fluid and relieve ascites. The peritoneum is the membrane that surrounds the abdominal cavity.

Carpenter with Cirrhosis

Marty Hamilton, a 49-year-old carpenter, has just been diagnosed with cirrhosis, which is a consequence of his alcohol abuse over the past 25 years. Although he recognizes that he has an alcohol problem and recently entered an alcohol rehabilitation program, he is still drinking. At 5 feet 8 inches tall, Mr. Hamilton, who formerly weighed 160 pounds, now weighs 130 pounds. According to family members, he is showing signs of mental deterioration, such as forgetfulness and an inability to concentrate. He is jaundiced and appears thin, although his abdomen is distended with ascites. Laboratory findings indicate elevated serum concentrations of AST, ALT, and ammonia; reduced albumin levels; and hyperglycemia.

1. Do Mr. Hamilton's laboratory values suggest liver disease? Compare the results of his laboratory tests with the values shown in Table 25-3 (p. 792).
2. From the limited information available, evaluate Mr. Hamilton's nutrition status. What medical problem makes it difficult to interpret his present weight? Describe the development of that type of problem in liver disease, and explain how the diet is usually adjusted for such a patient.
3. Calculate Mr. Hamilton's energy and protein needs. Describe the general diet you might recommend for him. What suggestions do you have for increasing his energy intake?
4. Explain the significance of Mr. Hamilton's elevated blood ammonia levels. What are some signs that would indicate that he is undergoing mental decline?
5. Describe each of the following complications of liver disease: portal hypertension, jaundice, and gastroesophageal varices. What complication may result if the esophageal varices are not treated?

IN SUMMARY

Liver cirrhosis is characterized by extensive fibrosis and permanent liver dysfunction. The primary causes of cirrhosis in the United States are hepatitis C infection and alcohol abuse. Symptoms of cirrhosis include fatigue, gastrointestinal disturbances, anorexia, and weight loss. Complications include portal hypertension, gastroesophageal varices, ascites, and hepatic encephalopathy. Treatment of cirrhosis is highly individualized and depends on the accompanying symptoms and complications. Both drug therapies and dietary adjustments are usually necessary. If warranted, the diet may need to be restricted in fat, sodium, or fluids. A person with cirrhosis often has a poor food intake and is at high risk of malnutrition.

Liver Transplantation

Acute or chronic liver disease can lead to liver failure, in which case liver transplantation is the only remaining treatment option. The most common illnesses that precede liver transplantation are chronic hepatitis C infection and alcoholic liver disease, which account for about 50 percent of liver transplant cases.[24] The 5-year survival rate among transplant recipients ranges from 58 to 81 percent, depending on the cause of illness.[25] Complications such as ascites and hepatic encephalopathy worsen the prognosis.

Nutrition Status of Transplant Patients As mentioned earlier, advanced liver disease is usually associated with malnutrition, which can increase the risk of complications following a liver transplant. Evaluating nutrition status in transplant candidates can be difficult because liver dysfunction and malnutrition often have similar metabolic effects. In addition, fluid retention can mask weight loss and alter anthropometric and laboratory values. Correcting malnutrition prior to transplant surgery can help speed recovery after the surgery.

Posttransplantation Concerns The immediate concerns following a transplant are organ rejection and infection. Immunosuppressive drugs, including prednisone, cyclosporine, and tacrolimus, help to reduce the immune responses that cause rejection, but they also raise the risk of infection. Infections are a potential cause of death following a liver transplant; therefore, antibiotics and antiviral medications are prescribed to reduce infection risk.

Immunosuppressive drugs can affect nutrition status in numerous ways. Gastrointestinal side effects include nausea, vomiting, diarrhea, abdominal pain, and mouth sores. Some medications may alter appetite and taste perception. Some of the drugs may cause hyperglycemia or outright diabetes, which may need to be controlled with insulin. Electrolyte and fluid imbalances are common. Other possible effects include hypertension, hyperlipidemias, protein catabolism, and increased osteoporosis risk.[26]

Protein and energy requirements are increased after transplantation due to the stress of surgery. High-kcalorie, high-protein snacks and enteral supplements can help the transplant patient meet postsurgical needs. Vitamin and mineral supplementation is also an integral part of nutrition care. To help transplant patients avoid developing foodborne illnesses, health practitioners can provide information about food safety measures, such as cooking meats adequately, washing fresh produce, and avoiding foods that may be contaminated. Highlight 18 (pp. 632–639) provides additional information about food safety.

IN SUMMARY

Liver transplantation has improved the long-term outlook for patients with advanced liver disease. Transplant patients are usually malnourished and may have medical problems that affect transplant success. Due to the potential for organ rejection, immunosuppressive drugs are prescribed following surgery. Use of these drugs increases the risk of infection, and the drugs have side effects that can impair nutrition status and general health.

Gallbladder Disease

As described earlier in this chapter, the gallbladder concentrates and stores the bile produced by the liver until the bile is needed for fat digestion (see Figure 25-3). Disorders that obstruct the liver's release of bile can damage the liver. More commonly, disorders of the biliary system—the gallbladder and bile ducts—result in the formation of **gallstones.** Gallstones affect an estimated 30 million people in the United States, or about 10 percent of the population.[27]

Types of Gallstones

The formation of gallstones, or **cholelithiasis**, results from excessive concentration and crystallization of the compounds in bile. Bile is a solution of bile acids, cholesterol, phospholipids, proteins, and bile pigment (bilirubin). While stored in the gallbladder, bile's concentration increases approximately 10-fold as its water content is extracted. Factors that raise bile's cholesterol concentration, promote crystal formation, or reduce gallbladder motility favor gallstone formation.[28]

Cholesterol Gallstones In about 80 percent of cases, the gallstones are composed primarily of cholesterol, although they also contain calcium salts and bilirubin. The cholesterol in bile precipitates out of solution and forms small crystals, which eventually coalesce to form stones. The stones can be as small as a pea or as large as a Ping-Pong ball. Some people tend to form many small stones, while others may form only one or two large ones.

Cholesterol gallstones often develop because the bile concentrate thickens and forms a **sludge** that cannot be easily expelled by gallbladder contraction. Biliary

gallstones: stones that form in the gallbladder from crystalline deposits of cholesterol or bilirubin.

cholelithiasis (KOH-leh-lih-THIGH-ah-sis): formation of gallstones.
- **chole** = bile
- **lithiasis** = formation of stones

sludge: literally, a semisolid mass. Biliary sludge is made up of mucus, cholesterol crystals, and bilirubin granules.

FIGURE 25-3 The Gallbladder and Bile Ducts

Gallbladder

(from the liver)

Stomach

Common hepatic duct

Cystic duct

Pancreas

Common bile duct

Duodenum

Pancreatic duct

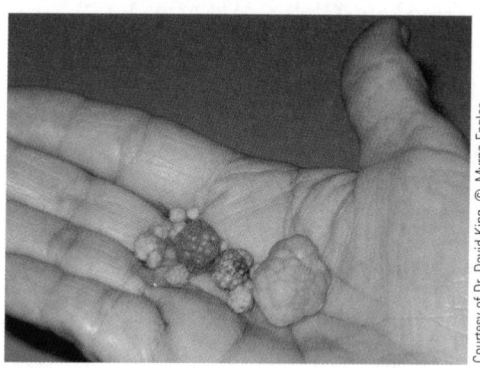

Most gallstones are made primarily of cholesterol; they can be as small as a pea or as large as a Ping-Pong ball.

Courtesy of Dr. David King. © Myrna Engler

sludge frequently develops after rapid weight loss, gastric bypass surgery, and long-term total parenteral nutrition and also occurs during pregnancy.

Pigment Gallstones Pigment stones account for 10 to 25 percent of gallstone cases in the United States, although they are much more common in Asian countries.[29] Pigment stones are primarily made up of the calcium salt of bilirubin (calcium bilirubinate). They often develop as a result of bacterial infection, which alters bilirubin and causes it to precipitate out of bile and form stones. Other cases result from excessive red blood cell breakdown, leading to an abnormal accumulation of bilirubin. Conditions associated with pigment stone formation include biliary tract infections, pancreatitis, and red blood cell disorders, such as sickle-cell anemia. Pigment stones may form in either the gallbladder or a bile duct. Unlike the crystalline cholesterol stones, pigment stones are soft and easily crushed.

Consequences of Gallstones

About 85 percent of gallstones are asymptomatic and are discovered accidentally while testing for other conditions.[30] However, many people experience an aggressive course of illness with recurring symptoms.

Gallstone Symptoms Gallstone pain usually arises when a gallstone temporarily blocks the cystic duct, which leads from the gallbladder to the common bile duct (see Figure 25-3). The pain is steady and severe and may last for several minutes or several hours. Although the pain is usually located in the upper abdomen, it may radiate to the chest or to the back. Nausea, vomiting, and bloating may also be present. Symptoms usually develop after meals, especially after eating fatty foods. Pain may also occur during the night and awaken a person from sleep.

Complications of Gallstones If a gallstone remains lodged in the cystic duct, it can obstruct bile flow to the duodenum and cause **cholecystitis**—distention and inflammation of the gallbladder. Cholecystitis can lead to infection or to more severe complications, including perforation of the gallbladder, **peritonitis,** and fistulas. If gallstones obstruct the common bile duct (see Figure 25-3), they can block bile flow from the liver and lead to jaundice or damage to liver tissue. An impacted stone within the bile ducts may lead to infection and the condition known as **bacterial cholangitis,** which causes severe pain, sepsis, and fever and is

cholecystitis (KOH-leh-sih-STY-tis): inflammation of the gallbladder, usually caused by obstruction of the cystic duct by gallstones.

peritonitis: inflammation of the peritoneal membrane, which lines the abdominal cavity.

bacterial cholangitis (KOH-lan-JYE-tis): bacterial infection involving the bile ducts.

often a medical emergency. Gallstones can block the pancreatic duct as well—a primary cause of acute pancreatitis. Due to the potential danger of these complications, individuals should seek medical attention if gallstone pain does not resolve over time or if fever, jaundice, or persistent nausea and vomiting develop.

Risk Factors for Gallstones

Gallstone risk is influenced by a number of genetic and lifestyle factors. As described in the sections that follow, the risk factors are typically related to increased cholesterol concentrations in bile or reduced gallbladder motility, which promote crystallization and subsequent gallstone formation.

Ethnicity Although the genetic factors related to gallstone formation are not yet clear, ethnicity strongly influences gallstone formation. The Pima Indians are an exceptionally high-risk population; gallstones develop in about 70 percent of adult women. Other high-risk populations include Native Americans in the United States and Canada, Scandinavians, Chileans, and Bolivians. In the United States, African Americans have a lower prevalence of gallstones than white populations.[31]

Aging Because gallstones cannot dissolve spontaneously, gallstone prevalence increases with age. Moreover, bile composition tends to change with aging: the cholesterol concentration increases while bile acids decrease, leading to a greater likelihood of cholesterol crystallization.

Gender The incidence of gallstones in women is nearly three times that in men during the reproductive years, although it falls to a similar level after menopause.[32] The reason for the gender difference is that estrogen alters cholesterol metabolism and causes an increased secretion of cholesterol into bile. The use of estrogen replacement therapy after menopause increases gallstone risk in postmenopausal women.

Pregnancy Some women experience their first gallstone symptoms during pregnancy. Gallstone risk is increased in pregnancy due to hormonal changes: higher serum estrogen levels increase the secretion of cholesterol into bile, and higher progesterone levels impair gallbladder motility.[33]

Obesity and Weight Loss Obesity is associated with increased cholesterol synthesis in the liver, leading to a greater release of cholesterol into bile. In a clinical study that investigated the incidence of gallstones among women of different weights, researchers found that the women who had a body mass index (BMI) greater than 45 had a risk of gallstone formation seven times that of the nonobese women.[34] ◆

Gallstones frequently develop as a result of rapid weight loss, occurring in about 25 percent of obese persons on very-low-kcalorie diets and in as many as half of individuals who undergo gastric bypass surgery.[35] Dieting increases the secretion of cholesterol into bile and may also decrease gallbladder motility. The oral ingestion of bile salts has been shown to reduce the risk of gallstone formation during rapid weight loss.

Other Risk Factors Long-term total parenteral nutrition usually reduces gallbladder motility, increasing the development of biliary sludge. Some medications (such as octreotide) may have similar effects. The medication clofibrate, used for heart disease, increases the cholesterol concentration of bile, promoting crystallization. High triglyceride levels in blood are also associated with increased gallstone risk, as are hyperinsulinemia, insulin resistance, and diabetes mellitus.

Treatment for Gallstones

Asymptomatic gallstones generally do not require treatment. Gallstones that cause symptoms or complications are usually treated by gallbladder surgery or by nonsur-

◆ Reminder: A healthy weight usually falls between a BMI of 18.5 and 25.0. Overweight and obesity are usually defined by BMIs above 25 and 30, respectively.

gical procedures that dissolve or fragment the stones. To minimize symptoms before the gallbladder or gallstones are removed, a low-fat diet (with less than 30 percent of total kcalories from fat) may be prescribed.[36]

Surgery Gallbladder removal, or **cholecystectomy**, is the primary treatment for patients with recurring gallstones.[37] The preferred surgical approach is a **laparoscopic** method, which relies on narrow surgical telescopes (laparoscopes) to view and perform the necessary procedures via small incisions in the abdomen. The procedure takes only one or two hours, and many patients are discharged on the same day as the surgery. In patients with complications that make organ removal difficult, open cholecystectomy may be performed. In this procedure, the surgeon cuts through the abdominal muscle and exposes the abdominal cavity, allowing direct access to the gallbladder and bile ducts. An open cholecystectomy is associated with a greater risk of infection, more pain, and a lengthier recovery time than the laparoscopic procedure.

Once the gallbladder has been removed, the common bile duct collects bile between meals and releases it into the duodenum at mealtimes. Most patients have no problems after they recover from surgery, although some may experience diarrhea, abdominal pain, and other gastrointestinal symptoms. The diarrhea may result from an increased amount of bile in the large intestine, which has a laxative effect. Abdominal pain is sometimes caused by the presence of residual stones within the common bile duct that were overlooked during surgery or that formed within the duct itself. Bile duct injuries occasionally result from the surgical procedure.

Nonsurgical Procedures Nonsurgical methods are used primarily in patients who have small cholesterol stones and transient conditions associated with gallstone formation. The gallstones can be treated by oral intake of ursodeoxycholic acid (ursodiol), a bile acid that reduces cholesterol secretion by the liver and eventually causes the cholesterol crystals in gallstones to dissolve. Ursodeoxycholic acid must be used for 6 to 12 months and is best suited for stones that are 5 millimeters (about $1/4$ inch) in diameter or smaller. Recurrence rates after dissolution are as high as 50 percent.[38]

Cholesterol gallstones can be fragmented using **shock-wave lithotripsy**, a procedure that is also used to fragment kidney stones. This technique uses high-amplitude sound waves (called shock waves) to break gallstones into pieces that are small enough to either pass into the intestine or be dissolved with ursodeoxycholic acid. Shock-wave lithotripsy can be performed only in patients with few gallstones. Success is highest in patients with solitary stones that are less than 20 millimeters ($3/4$ inch) in diameter. Recurrence of gallstones has been reported in up to 44 percent of patients using this procedure.[39]

IN SUMMARY

Gallstones are the most common disorder affecting the gallbladder. They are formed by the concentration of compounds in bile, especially cholesterol and the bile pigment bilirubin. Most people with gallstones have no symptoms. Symptomatic gallstones can cause recurring pain and gastrointestinal problems that usually appear after meals and may persist for several hours. The risk of gallstone formation increases with age and is highest in premenopausal and pregnant women and in persons who are obese or who undergo rapid weight loss. Treatments for gallstones include gallbladder removal and gallstone dissolution or fragmentation.

cholecystectomy (KOH-leh-sis-TEK-toe-mee): surgical removal of the gallbladder.

laparoscopic: pertaining to procedures that use a laparoscope for internal examination or surgery. A laparoscope is a narrow surgical telescope that is inserted into the abdominal cavity through a small incision. A video camera is usually attached so that the procedure can be viewed on a television monitor.

shock-wave lithotripsy: a nonsurgical procedure that uses high-amplitude sound waves to fragment gallstones.

Clinical Portfolio

1. Vijaya Reddy is a college student who visited relatives near her parents' birthplace in Anantapur, India, during summer vacation. Although her relatives provided boiled or purified water at their home, they occasionally took Vijaya to local restaurants, where she drank tap water. Several weeks after Vijaya returned home, she developed flulike symptoms and started feeling extremely tired. She also experienced upper abdominal pain and felt nauseous after meals. After her roommate told her that her eyes and skin appeared yellow, she knew something was definitely wrong. A physician at the student health center diagnosed hepatitis.

   ▧ Which type of hepatitis did Vijaya most likely have?

   ▧ What additional symptoms can develop? Is Vijaya's condition likely to become chronic?

   ▧ What medical treatment is suggested for Vijaya's condition? Describe the dietary modifications that may be necessary in some cases.

2. As discussed in the section on cirrhosis, many patients develop protein-energy malnutrition and wasting during the course of illness. Review Table 25-5 to find examples of problems that may lead to malnutrition. Select three nutrition or medical problems (from the "Examples" column), and discuss the complications of liver disease that may cause the problems you selected. What dietary or medical treatments can help in managing these problems?

NUTRITION ON THE NET

For further study of topics covered in this chapter, log on to **academic.cengage .com/nutrition/rolfes/UNCN8e**. Go to Chapter 25, then to Nutrition on the Net.

• To obtain additional information about liver diseases, visit the American Liver Foundation and the Canadian Liver Foundation: **www.liverfoundation.org** and **www.liver.ca**

• To find out more about resources and support for children with liver diseases and liver transplants, visit the Children's Liver Alliance: **www.liverkids.org.au**

• Learn more about hepatitis by visiting the Hepatitis Foundation International: **www.hepfi.org**

• To uncover more information about liver transplants, search the CenterSpan Transplant News Network: **www.centerspan.org**

NUTRITION ASSESSMENT CHECKLIST for People with Disorders of the Liver and Gallbladder

Medical History

Check the medical record to determine:

- Type of liver disorder
- Cause of the liver disorder
- If the patient has received a liver transplant
- If the patient has a history of gallstones

Review the medical record for complications that may alter nutritional needs, including:

- Abdominal pain
- Anemia
- Ascites
- Esophageal varices
- Hepatic encephalopathy
- Impaired kidney or lung function
- Infections
- Insulin resistance or diabetes mellitus
- Malabsorption
- Malnutrition
- Pancreatitis

Medications

In patients with liver dysfunction, the risk of diet-drug interactions is high because most drugs are metabolized in the liver. Risk of interactions is intensified for patients with:

- Ascites (medications may take a long time to reach the liver)
- Renal failure (medications are often metabolized further in the kidneys and excreted in the urine)
- Malnutrition

- Multiple prescriptions
- Long-term medication use

Dietary Intake

For patients with fatty liver, pay special attention to:

- Energy intake, if the patient is overweight or malnourished, has diabetes, or is receiving total parenteral nutrition
- Carbohydrate intake, if the patient has diabetes or is receiving total parenteral nutrition
- Alcohol abuse

For patients with hepatitis, cirrhosis, or ascites:

- Check appetite.
- Ensure that energy and nutrient intakes are adequate.
- Determine alcohol consumption.
- Determine whether sodium or fluid restriction is warranted.
- Base energy needs on desirable weight or estimated dry weight to avoid overfeeding.

Anthropometric Data

Take baseline height and weight measurements, and monitor weight regularly. For patients with ascites and edema:

- Monitor weight changes to evaluate the degree of fluid retention.
- Remember that the patient may be malnourished, and weight may be deceptively high.

Laboratory Tests

Note that albumin and serum proteins are often reduced in people with liver disease and cannot always be used as indicators of nutrition status. Review the following laboratory test results to assess liver function:

- Albumin
- Alkaline phosphatase
- ALT and AST
- Ammonia
- Bilirubin
- Prothrombin time

Check laboratory test results for complications associated with liver failure, including:

- Anemia
- Decreased renal function
- Fluid retention
- Hyperglycemia

Physical Signs

Look for physical signs of:

- Fluid retention (ascites and edema)
- PEM (muscle wasting and unintentional weight loss)
- Nutrient deficiencies

STUDY QUESTIONS

CENGAGENOW

To assess your understanding of chapter topics, take the Student Practice Test and explore the modules recommended in your Personalized Study Plan. Log on to **academic.cengage.com/login**.

These questions will help you review the chapter. You will find the answers in the discussions on the pages provided.

1. Describe fatty liver, and identify its possible causes. What consequences of fatty liver may develop? What are possible treatments? (pp. 788–789)

2. What is hepatitis, and what are its primary causes? Compare the features of infection with hepatitis virus A, B, and C. Identify nutritional concerns for patients with hepatitis. (pp. 789–790)

3. Describe the progression of liver disease to cirrhosis. What are the most common causes of cirrhosis in the United States? (pp. 790–791)

4. Discuss the consequences of cirrhosis, including its clinical effects and complications such as portal hyperten-

sion, gastroesophageal varices, ascites, and hepatic encephalopathy. What metabolic changes may result from altered liver function? (pp. 791–793)

5. How does cirrhosis affect nutrition status? Describe the dietary treatment of a patient with cirrhosis. Discuss the special dietary concerns of patients with ascites and esophageal varices. (pp. 793–797)

6. Discuss the problems that arise following liver transplantation that can affect nutrition status. What dietary modifications may be necessary? (pp. 797–798)

7. Explain how gallstones form, and describe the features of the two main types of gallstones. What complications are associated with gallstone disease? (pp. 798–800)

8. Discuss the major risk factors for gallstone disease. Describe the primary methods of treatment. (pp. 800–801)

These multiple choice questions will help you prepare for an exam. Answers can be found on p. 805.

1. Which of the following dietary strategies would be the most appropriate for reversing fatty liver associated with diabetes mellitus?
    a. following a low-protein diet
    b. following a fat-restricted diet
    c. following a fluid- and sodium-restricted diet
    d. modifying energy to achieve a desirable weight and modifying carbohydrates to attain blood glucose control

2. Which of the following statements about hepatitis is true?
    a. Chronic hepatitis can progress to cirrhosis.
    b. Whatever the cause of hepatitis, symptoms are typically severe.
    c. People with hepatitis require high-kcalorie, high-protein diets.
    d. HCV infection can be spread through contaminated foods and water.

3. Esophageal varices are a dangerous complication of liver disease primarily because they:
    a. interfere with food intake.
    b. can lead to massive bleeding.
    c. divert blood flow from the GI tract.
    d. contribute to hepatic encephalopathy.

4. A complication of liver disease that contributes to the development of ascites is:
    a. portal hypertension.
    b. rising blood ammonia levels.
    c. elevated serum albumin levels.
    d. insulin resistance.

5. A patient with cirrhosis may develop personality changes and motor dysfunction, which are signs of:
    a. jaundice.
    b. hepatic encephalopathy.
    c. hyperammonemia.
    d. hepatic coma.

6. With respect to protein intake, patients with cirrhosis should:
    a. consume no more than the protein RDA.
    b. restrict protein intake to 0.6 gram per kilogram of body weight.
    c. use formulas enriched with aromatic amino acids to meet their protein needs.
    d. maintain nitrogen balance by consuming 0.8 to 1.2 grams of protein per kilogram of body weight per day.

7. People with ascites must often restrict dietary intake of:
    a. fat.
    b. protein.
    c. sugars.
    d. sodium.

8. Dietary concerns after a liver transplant include all of the following *except:*
    a. severe protein restrictions that are difficult to adhere to.
    b. increased risk of foodborne illness.
    c. gastrointestinal side effects of medications.
    d. reduced appetite and altered taste perception from medications.

9. Regarding risk factors for gallstone disease:
    a. prevalence is much higher in men than in women.
    b. gallstone risk is increased during pregnancy.
    c. rapid weight loss can temporarily shrink gallstones.
    d. risk is generally similar among ethnic groups.

10. Nonsurgical approaches to gallstone treatment include:
    a. cholecystectomy.
    b. weight loss.
    c. dissolution and fragmentation.
    d. immunosuppressant drug therapy.

REFERENCES

1. A. M. Diehl, Alcoholic and nonalcoholic steatohepatitis, in L. Goldman and D. Ausiello, eds., *Cecil Medicine* (Philadelphia: Saunders, 2008), pp. 1135–1139.
2. M. M. Yeh and E. M. Brunt, Pathology of nonalcoholic fatty liver disease, *American Journal of Clinical Pathology* 128 (2007): 837–847; M. F. Abdelmalek and A. M. Diehl, Nonalcoholic fatty liver disease as a complication of insulin resistance, *Medical Clinics of North America* 91 (2007): 1125–1149.
3. A. E. Reid, Nonalcoholic fatty liver disease, in M. Feldman, L. S. Friedman, and L. J. Brandt, eds., *Sleisenger and Fordtran's Gastrointestinal and Liver Disease* (Philadelphia: Saunders, 2006), pp. 1793–1805.
4. Reid, 2006.
5. Diehl, 2008.
6. J. H. Lewis, Liver disease caused by anesthetics, toxins, and herbal preparations, in M. Feldman, L. S. Friedman, and L. J. Brandt, eds., *Sleisenger and Fordtran's Gastrointestinal and Liver Disease* (Philadelphia: Saunders, 2006), pp. 1793–1805.

7. Centers for Disease Control and Prevention, Surveillance for acute viral hepatitis—United States, 2005, *Morbidity and Mortality Weekly Report* 56, No. SS-3 (2007).
8. D. M. Shoemaker and coauthors, Infectious diseases, in R. E. Rakel, ed., *Textbook of Family Medicine* (Philadelphia: Saunders, 2007), pp. 317–352.
9. J. H. Hoofnagle, Acute viral hepatitis, in L. Goldman and D. Ausiello, eds., *Cecil Medicine* (Philadelphia: Saunders, 2008), pp. 1101–1108.
10. R. Perrillo and S. Nair, Hepatitis B and D, in M. Feldman, L. S. Friedman, and L. J. Brandt, eds., *Sleisenger and Fordtran's Gastrointestinal and Liver Disease* (Philadelphia: Saunders, 2006), pp. 1647–1679.
11. Centers for Disease Control and Prevention, 2007.
12. S. Safrin, Antiviral agents, in B. G. Katzung, ed., *Basic and Clinical Pharmacology* (New York: McGraw-Hill/Lange, 2007), pp. 790–818.

13. American Dietetic Association, *Nutrition Care Manual* (Chicago: American Dietetic Association, 2007).
14. G. Garcia-Tsao, Cirrhosis and its sequelae, in L. Goldman and D. Ausiello, eds., *Cecil Medicine* (Philadelphia: Saunders, 2008), pp. 1140–1147.
15. Garcia-Tsao, 2008.
16. J. J. Heidelbaugh and coauthors, Gastroenterology, in R. E. Rakel, ed., *Textbook of Family Medicine* (Philadelphia: Saunders, 2007), pp. 1115–1171.
17. B. A. Runyon, Ascites and spontaneous bacterial peritonitis, in M. Feldman, L. S. Friedman, and L. J. Brandt, eds., *Sleisenger and Fordtran's Gastrointestinal and Liver Disease* (Philadelphia: Saunders, 2006), pp. 1935–1964.
18. J. G. Fitz, Hepatic encephalopathy, hepatopulmonary syndromes, hepatorenal syndrome, and other complications of liver disease, in M. Feldman, L. S. Friedman, and L. J. Brandt, eds., *Sleisenger and Fordtran's*

Gastrointestinal and Liver Disease (Philadelphia: Saunders, 2006), pp. 1965–1991.
19. Garcia-Tsao, 2008; Fitz, 2006.
20. Fitz, 2006.
21. American Dietetic Association, 2007.
22. C. S. Lieber, Nutrition in liver disorders and the role of alcohol, in M. E. Shils and coeditors, *Modern Nutrition in Health and Disease* (Baltimore: Lippincott Williams & Wilkins, 2006), pp. 1235–1259; B. Als-Nielsen and coauthors, Branched-chain amino acids for hepatic encephalopathy (Cochrane Review), *The Cochrane Library* 3 (2004).
23. American Dietetic Association, 2007.
24. E. B. Keeffe, Hepatic failure and liver transplantation, in L. Goldman and D. Ausiello, eds., *Cecil Medicine* (Philadelphia: Saunders, 2008), pp. 1147–1152.
25. Keeffe, 2008.
26. American Dietetic Association, 2007; M. J. Weiss, V. T. Armenti, and J. M. Hasse, Drug-nutrient interactions in transplantation, in J. I. Boullata and V. T. Armenti, eds., *Handbook of Drug-Nutrient Interactions* (Totowa, N.J.: Humana Press, 2004), pp. 425–440.
27. N. H. Afdhal, Diseases of the gallbladder and bile ducts, in L. Goldman and D. Ausiello, eds., *Cecil Medicine* (Philadelphia: Saunders, 2008), pp. 1152–1161.
28. J. D. Browning and J. Sreenarasimhaiah, Gallstone disease, in M. Feldman, L. S. Friedman, and L. J. Brandt, eds., *Sleisenger and Fordtran's Gastrointestinal and Liver Disease* (Philadelphia: Saunders, 2006), pp. 1387–1418.
29. Browning and Sreenarasimhaiah, 2006.
30. Afdhal, 2008.
31. Browning and Sreenarasimhaiah, 2006.
32. Afdhal, 2008.
33. Browning and Sreenarasimhaiah, 2006.
34. Browning and Sreenarasimhaiah, 2006.
35. Browning and Sreenarasimhaiah, 2006.
36. American Dietetic Association, 2007.
37. R. E. Glasgow and S. J. Mulvihill, Treatment of gallstone disease, in M. Feldman, L. S. Friedman, and L. J. Brandt, eds., *Sleisenger and Fordtran's Gastrointestinal and Liver Disease* (Philadelphia: Saunders, 2006), pp. 1419–1442.
38. Glasgow and Mulvihill, 2006.
39. Glasgow and Mulvihill, 2006.

ANSWERS

Study Questions (multiple choice)

1. d 2. a 3. b 4. a 5. b 6. d 7. d 8. a 9. b 10. c

Food Allergies

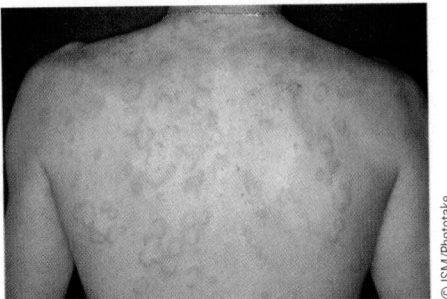

© ISM/Phototake

Some of the diseases discussed in this book involve adverse reactions to specific foods. Chapter 15 explained that such responses can be categorized either as *food allergies,* which elicit an immune response, or *food intolerances,* which are caused by other physiological processes. Celiac disease and dermatitis herpetiformis, for example, are characterized by allergic reactions to gluten, whereas lactose intolerance, a result of lactase deficiency, is a type of food intolerance. This highlight focuses on the diagnosis and treatment of food allergies, beginning with a brief review of the body's reactions to an **allergen.** The accompanying glossary defines the relevant terms.

gerous effect of allergy is **anaphylaxis,** a systemic (whole-body) reaction that may cause difficulty breathing and a dangerous fall in blood pressure, potentially leading to shock. People whose food allergies are intense enough to cause anaphylaxis are often prescribed epinephrine, which they can self-inject in an emergency.

Contact dermatitis or hives can also develop on skin after physical contact with food. In the condition known as **oral allergy syndrome,** hives, swelling, and itching are mostly confined to the lips, tongue, mouth, and throat. These symptoms usually develop following the consumption of raw fruits and vegetables.[3]

A Review of Food Allergy

A food allergy occurs when a food component, usually an incompletely digested protein, is absorbed into the blood and elicits an immune response.[1] The allergen is treated as a foreign particle that needs to be neutralized, and allergen-specific antibodies are produced to mount a defense. These antibodies are attached to specialized cells (mast cells and basophils) that release inflammatory mediators when they encounter the allergen. The mediators circulate in the blood and may trigger symptoms in the GI tract, skin, respiratory system, and circulatory system. The foods most likely to cause an allergy include eggs, fish, milk, peanuts, shellfish, soybeans, tree nuts, and wheat.[2]

Common symptoms of food allergies include skin rashes, itching, abdominal pain, vomiting, and diarrhea. **Hives** occur frequently; these raised, swollen patches of skin or mucous membranes are associated with intense itching. The most dan-

Diagnosis of Food Allergy

If a food allergy is suspected, an accurate diagnosis can help a person avoid unnecessary dietary restrictions. Parents who believe that a food allergy is causing health or behavioral problems may limit their children's food intakes, which can adversely affect growth and nutrition status.[4] A timely diagnosis can also help a person avoid accidental exposure to a food allergen.

Diagnosis often requires a thorough medical history, physical examination, and laboratory tests. The medical history can help to establish whether the symptoms are a response to a true food allergy rather than a food intolerance, foodborne illness, or food toxicity. To help pinpoint the foods that cause symptoms, patients are generally advised to keep a **food and symptom diary,** which provides a record of the foods consumed, the amounts, and the symptoms that develop. Other helpful data include the brands of foods consumed, ingredient lists of packaged foods, and the exact timing of symptom onset. If the symptoms arise several hours or days after the offending food is ingested, the exact cause of the allergy may be more difficult to identify.

Oral Food Challenges

When performed properly, food challenges are considered the gold standard for diagnosing food allergy. In an oral challenge, a food suspected to cause allergy is presented to a patient in a dose suggested by the medical history. If the test substance does not cause symptoms, the challenge is repeated to rule out a false-negative result. Ideally, food challenges are double blinded and placebo controlled: test foods are mixed into other foods or provided in capsules, and placebos are identical in appearance, taste, and texture. A food challenge can be labor intensive and cannot be performed if a patient has a history of severe anaphylaxis.

Elimination Diets

In an elimination diet, the patient omits common food allergens from the diet until symptoms subside, and then reintroduces individual foods one by one. Although foods that cause symptoms are sometimes easily identified using this method, it may be difficult to identify allergens when they are ingredients in packaged foods. Also, allergic reactions sometimes persist for some time after the allergens are removed from the diet; in these cases, an elemental formula diet (which contains no intact proteins) may be needed to stabilize the patient before foods are reintroduced.

Skin-Prick Testing

The skin-prick test evaluates the patient's responses to commercially prepared food extracts that are introduced into the skin (see the photo). Substances that cause areas of redness and swelling greater than 3 millimeters in diameter are considered possible allergens, and larger responses suggest a greater potential for allergy. Although the rate of false-positive results for skin tests is about 50

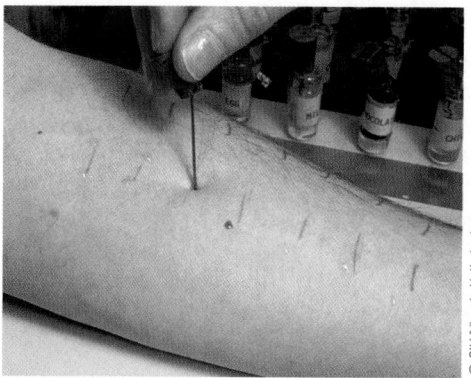

In a skin-prick test, extracts containing food allergens are placed on the skin, and the skin is pricked using a lancet or needle. This technique introduces small amounts of the allergens into the skin.

© SIU/Visuals Unlimited

percent (meaning that half of the reactions that appear to be positive are actually negative), the absence of a reaction is fairly good evidence that the test substance is not the cause of allergy.

Antibody Blood Testing

Measures of food-specific serum antibodies are useful for assessing the presence of food allergies; generally, a high antibody level suggests an increased risk of an allergic response to a food. Because a person with low antibody levels may still experience an allergic reaction, however, antibody test results need to be considered along with other methods of diagnosis.

Treatment of Food Allergy

Food allergies are treated by eliminating all dietary sources of an allergen. Successful treatment depends in part on the patient's ability to identify hidden sources of allergens in foods with multiple ingredients (see Table H25-1). Inadvertent ingestion of

TABLE 25-1	Food Avoidance in Milk, Egg, and Peanut Allergies	
Food Allergy	**Excluded Food Ingredients**	**Hidden Sources**
Milk allergy	Milk (including dried, evaporated, and condensed milks), milk solids, buttermilk, yogurt, cheese, butter, artificial butter flavor, half-and-half, cream, whipped cream, custard, pudding, ice cream, casein (or caseinates), whey, protein hydrolysates, lactalbumin, lactoferrin, lactoglobulin.	Margarine, luncheon meats, frankfurters and sausages, high-protein products (including bars, flours, and beverages), nougat candy, chocolate bars, caramel color or flavorings, coffee whiteners, bakery glazes, salad dressings, sauces. Meats sliced at a delicatessen are subject to cross-contamination from sliced cheeses.
Egg allergy	Eggs (including powdered eggs and egg substitutes), egg white, eggnog, meringue, albumin, globulin, lysozyme, ovalbumin, ovoglobulin, ovomucin, ovomucoid, ovotransferrin, ovovitellin, lecithin (some food labels may indicate that a "binder" or "emulsifier" was added).	Many baked products and baking mixes, noodles and pastas, mayonnaise, béarnaise and hollandaise sauces, breaded meats and vegetables, candies, fondants, marshmallows, frozen desserts, ice cream, custards and puddings, frankfurters and sausages, processed meats, cocoa drinks, salad dressings, bakery glazes.
Peanut allergy	Peanuts (also called ground nuts), peanut butter, peanut flour, nut pieces, mixed nuts, beer nuts, artificial nuts, mandalona nuts, peanut sauces (common in Asian cuisine), hydrolyzed vegetable protein (HVP), cold-pressed or gourmet peanut oils (may contain peanut residue).	Chocolate and candy bars, power bars, marzipan, nougat, breakfast cereals, egg rolls, satay sauce, curries, salad dressings. Cross-contamination is possible from food-processing equipment; caution is required when purchasing baked products, ice creams, candies, nut butters, and sunflower seeds.

allergens often occurs because foods become contaminated during meal preparation or food processing. Problem foods may also be consumed at restaurants, schools, and other public places, where the foods' ingredients are not always obvious. The foods that account for most allergic reactions in infants and children are cow's milk, eggs, and peanuts[5]; the dietary issues involved with these foods are described in the sections that follow.

Milk Allergy

Milk and the proteins derived from milk are common ingredients in many prepared and packaged foods, so people with milk allergies must check ingredient lists carefully. As an example, foods that are labeled "nondairy" (such as nondairy creamers) may contain the milk protein casein. In addition, individuals with milk allergies need to avoid milk from all animals due to the potential for **cross-reactivity.** Obtaining sufficient calcium and vitamin D from nonmilk sources may be difficult, and supplementation is often warranted. A milk allergy may be difficult to differentiate from lactose intolerance because both conditions can produce gastrointestinal symptoms.

Egg Allergy

Eggs and egg proteins are common ingredients in many recipes and processed foods; Table H25-1 lists terms that may be used on food labels when egg protein is present. People with egg allergy should avoid eggs from all birds to prevent cross-reactivity. Because flu vaccines are prepared using egg embryos, people with egg allergies need to check with their physicians before being vaccinated.

Peanut Allergy

Some people with peanut allergies have severe reactions, including anaphylaxis, to even the smallest quantities of peanuts. Although peanut allergy is not ordinarily associated with other nut allergies, patients may be advised to avoid all nuts due to potential contamination from food-processing equipment (see Table H25-1). Parents often fear that skin contact with peanut butter or inhalation of peanut dust may cause severe allergic reactions, but there is little evidence that this occurs.[6]

Reevaluation of Food Allergy

Due to the stringent dietary restrictions required for some food allergies, health care providers advise that patients with these allergies be reevaluated periodically so that they do not continue the restrictions unnecessarily. Most young children outgrow food allergies within three to five years, and many older children and adults also lose their allergies in time.[7] Individuals with allergies to peanuts, tree nuts, and seafood are least likely to develop tolerance. Reevaluation may require oral food challenges and skin-prick tests, although substantial caution is necessary in patients who experienced severe allergic reactions after consuming certain foods.

REFERENCES

1. H. A. Sampson, Food allergies, in M. Feldman, L. S. Friedman, and L. J. Brandt, eds., *Sleisenger and Fordtran's Gastrointestinal and Liver Disease* (Philadelphia: Saunders, 2006), pp. 427–439.
2. S. L. Taylor and S. L. Hefle, Food allergies and intolerances, in M. E. Shils and coeditors, *Modern Nutrition in Health and Disease* (Baltimore: Lippincott Williams & Wilkins, 2006), pp. 1512–1530.
3. Sampson, 2006.
4. Taylor and Hefle, 2006.
5. Sampson, 2006.
6. T. T. Perry and coauthors, Distribution of peanut allergen in the environment, *Journal of Allergy and Clinical Immunology* 113 (2004): 973–976; S. J. Simonte and coauthors, Relevance of casual contact with peanut butter in children with peanut allergy, *Journal of Allergy and Clinical Immunology* 112 (2003): 180–182.
7. L. B. Schwartz, Systemic anaphylaxis, food allergy, and insect sting allergy, in L. Goldman and D. Ausiello, eds., *Cecil Medicine* (Philadelphia: Saunders, 2008), pp. 1949–1950; Sampson, 2006.

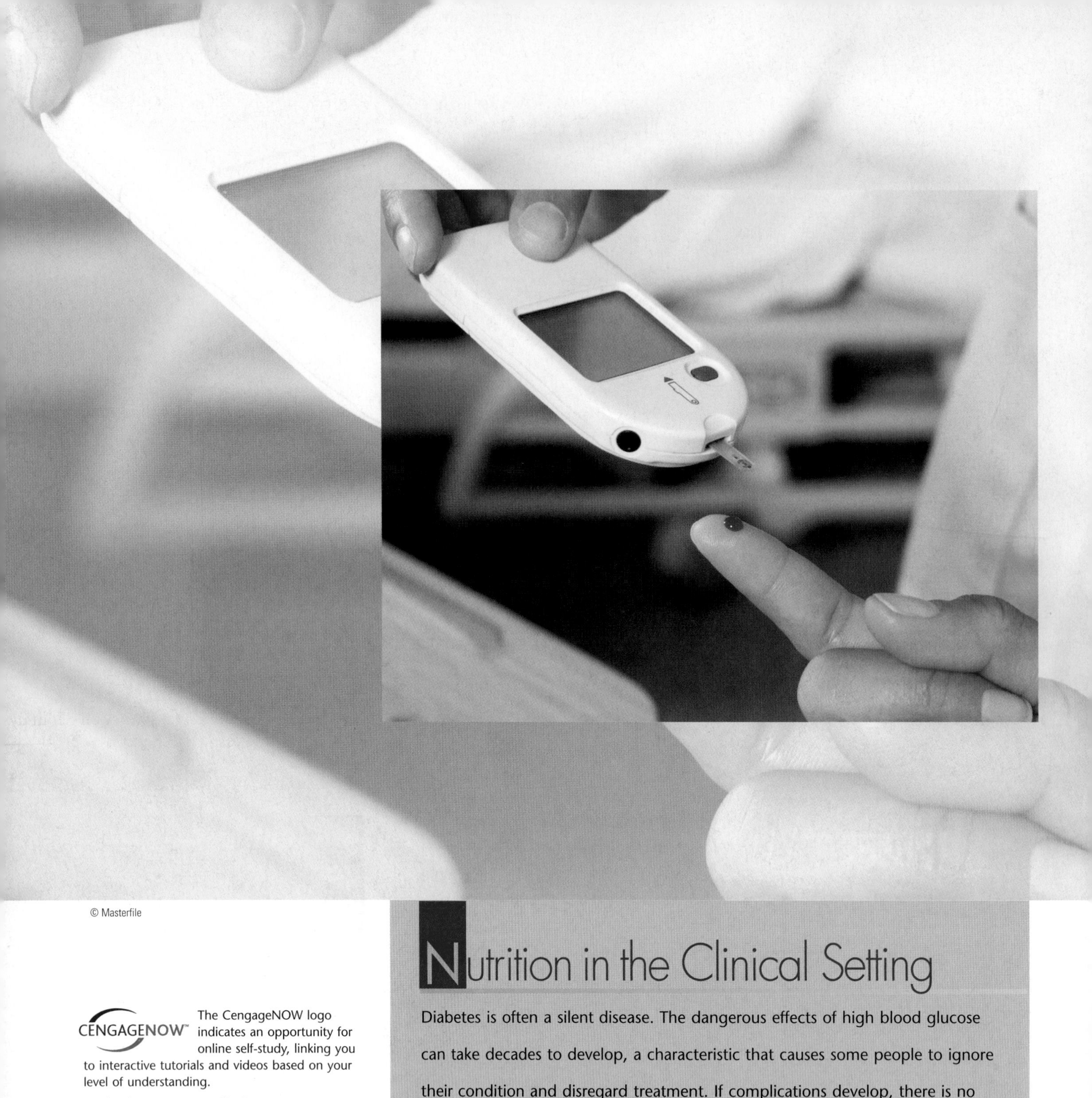

Nutrition in the Clinical Setting

Diabetes is often a silent disease. The dangerous effects of high blood glucose can take decades to develop, a characteristic that causes some people to ignore their condition and disregard treatment. If complications develop, there is no way to correct the damage to heart, kidneys, nerves, and eyes that has occurred. Because most diabetes care requires self-management, the challenge for health practitioners is to motivate patients to make the dietary and lifestyle changes that are necessary. The good news is that careful management allows individuals with diabetes to live long, healthy, and productive lives.

Diabetes Mellitus

The incidence of **diabetes mellitus** ◆ is steadily increasing in the United States and many other countries (see Figure 26-1 on p. 812). The condition affects an estimated 10.2 percent of adults aged 20 and older in the United States, or more than 30 million people.[1] About 28 percent of persons with diabetes are unaware that they have it,[2] a danger because its damaging effects often occur before symptoms develop. Diabetes ranks sixth among the leading causes of death in the United States. It also contributes to the development of other life-threatening diseases, including heart disease and kidney failure, which are discussed in the two chapters that follow. The glossary on p. 812 defines diabetes-related symptoms and complications.

◆ An unrelated condition with a similar name is *diabetes insipidus,* a pituitary disorder that causes a deficiency of antidiuretic hormone.

Overview of Diabetes Mellitus

The term *diabetes mellitus* refers to metabolic disorders characterized by elevated blood glucose concentrations and disordered insulin metabolism. People with diabetes may be unable to secrete sufficient insulin or use insulin effectively, or they may have both types of abnormalities. ◆

Normally, insulin secretions rise after food is ingested, and the insulin enables muscle and adipose cells to take up newly absorbed glucose from the blood. Insulin is also secreted between meals in smaller amounts to restrain the glucose-raising actions of glucagon, a hormone that promotes glucose production in the liver (gluconeogenesis) and the breakdown of liver glycogen. In diabetes, insulin secretion may be impaired, cells that are normally responsive to insulin may become resistant to its effects, or both. This situation leads to the reduced utilization of glucose in muscle and adipose cells and unrestrained gluconeogenesis in the liver. The result is **hyperglycemia,** a marked elevation in blood glucose levels that can ultimately cause damage to blood vessels, nerves, and tissues. Because insulin also promotes the synthesis of triglycerides and protein in body cells, a defect in insulin metabolism leads to the degradation of these nutrients, an increase

◆ Reminder: *Insulin* is a pancreatic hormone that regulates blood glucose concentrations. Its actions are countered mainly by the hormone *glucagon.*

diabetes (DYE-ah-BEE-teez) **mellitus:** a group of metabolic disorders characterized by hyperglycemia and disordered insulin metabolism.
- **diabetes** = siphon (in Greek), referring to the excessive passage of urine that is characteristic of untreated diabetes
- **mellitus** = sweet, honey-like

FIGURE 26-1 Prevalence of Diabetes among Adults in the United States

Key:

☐ Missing data	☐ 6%–6.9%
☐ <5%	☐ 7%–7.9%
☐ 5%–5.9%	■ 8+%

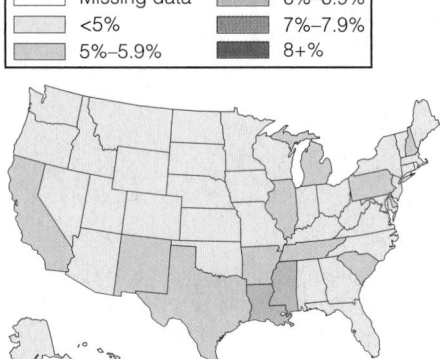

1994: 36 states had a diabetes prevalence of less than 5% and no state had a prevalence of 7% or greater.

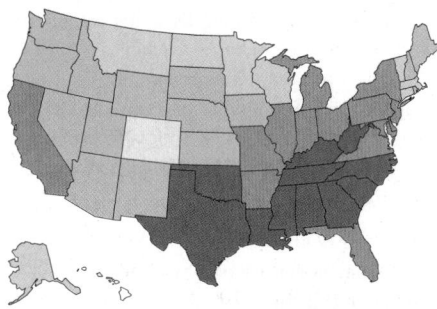

2005: Only 1 state had a diabetes prevalence of less than 5%, and 25 states had a prevalence of 7% or greater.

in fatty acid and triglyceride levels in the blood, and muscle wasting. Table 26-1 summarizes the effects of insulin insufficiency on nutrient metabolism in the body.

Symptoms of Diabetes Mellitus

Symptoms of diabetes (see Table 26-2) are usually related to the degree of hyperglycemia present. When the plasma glucose concentration rises above about 200 milligrams per deciliter (mg/dL), it exceeds the **renal threshold**, which is the concentration at which the kidneys begin to pass glucose into the urine **(glycosuria).** The presence of glucose in the urine draws additional water from the blood, increasing the amount of urine produced. Thus the symptoms that arise in diabetes typically include frequent urination **(polyuria),** dehydration, and increased thirst **(polydipsia).** Some people lose weight and have an increased appetite **(polyphagia)** as a result of the nutrient depletion that occurs when insulin is deficient. Another potential consequence of hyperglycemia is blurred vision, which is caused by the exposure of eye tissues to hyperosmolar fluids. ◆ Increased infections are common in individuals with diabetes and may be due to hyperglycemia, impaired circulation, or weakened immune function. In some cases, constant fatigue is the only symptom and may be related to altered energy metabolism, dehydration, or other effects of the disease.

Diagnosis of Diabetes Mellitus

The diagnosis of diabetes is based primarily on plasma glucose levels, which can be measured under fasting conditions or at random times during the day. ◆ In some cases, an **oral glucose tolerance test** is given: the individual ingests a 50- or 75-gram glucose load, and plasma glucose is measured at one or more time intervals following glucose ingestion. The following criteria are currently used to diagnose diabetes:

- The plasma glucose concentration of a blood sample obtained at a random time during the day (without regard to food intake) is 200 mg/dL or greater,

GLOSSARY OF DIABETES-RELATED SYMPTOMS AND COMPLICATIONS

acetone breath: a distinctive fruity odor on the breath of a person with ketosis.

claudication (CLAW-dih-KAY-shun): pain in the legs while walking; usually due to an inadequate supply of blood to muscles.

dawn phenomenon: morning hyperglycemia that is caused by the early-morning release of growth hormone, which counteracts insulin's glucose-lowering effects.

diabetic coma: a coma that occurs in uncontrolled diabetes; may be due to diabetic ketoacidosis, the hyperosmolar hyperglycemic state, or severe hypoglycemia.

diabetic nephropathy (neh-FRAH-pah-thee): damage to the kidneys that results from long-term diabetes.

diabetic neuropathy (nur-RAH-pah-thee): complications of diabetes that cause damage to nerves.

diabetic retinopathy (REH-tih-NAH-pah-thee): retinal damage that results from long-term diabetes.

gangrene: death of tissue due to a deficient blood supply and/or infection.

gastroparesis (GAS-troe-pah-REE-sis): delayed stomach emptying.

glycosuria (GLY-co-SOOR-ee-ah): an abnormal amount of glucose in urine.

hyperglycemia: elevated blood glucose concentrations. Normal fasting plasma glucose is less than 100 mg/dL. Fasting plasma glucose from 100 to 125 mg/dL suggests prediabetes; values of 126 mg/dL and above suggest diabetes.

hyperosmolar hyperglycemic state: extreme hyperglycemia associated with hyperosmolar blood, dehydration, and altered mental status; formerly called **hyperglycemic hyperosmolar nonketotic coma.**

hypoglycemia: abnormally low concentrations of blood glucose. In diabetes, hypoglycemia is treated when plasma glucose levels fall below 70 mg/dL.

ketoacidosis (KEY-toe-ass-ih-DOE-sis): an acidosis (lowering of blood pH) that results from the excessive production of ketone bodies.

ketonuria (KEY-toe-NOOR-ee-ah): the presence of ketone bodies in the urine.

macrovascular complications: disorders that affect the large blood vessels, including the coronary arteries and arteries of the limbs.

microalbuminuria: the presence of albumin (a blood protein) in the urine, a sign of diabetic nephropathy.

microvascular complications: disorders that affect the small blood vessels and capillaries, including those in the retinas and kidneys.

polydipsia (POL-ee-DIP-see-ah): excessive thirst.

polyphagia (POL-ee-FAY-jee-ah): excessive appetite or food intake.

polyuria (POL-ee-YOOR-ree-ah): excessive urine secretion.

rebound hyperglycemia: hyperglycemia that results from the release of counterregulatory hormones following nighttime hypoglycemia; also called the **Somogyi phenomenon.**

TABLE 26-1	Effects of Insulin Insufficiency on Nutrient Metabolism

Insulin normally promotes nutrient uptake after meals, as well as the synthesis of glycogen, triglycerides, and protein in liver, adipose, and muscle tissue. A defect in insulin metabolism inhibits these processes, leading to the effects shown in this table.

Nutrient	Effects of Insulin Insufficiency
Carbohydrate	• Decreased glucose uptake by muscle and adipose tissue • Decreased glycogen synthesis in muscle and liver • Increased glycogen breakdown in muscle and liver • Increased gluconeogenesis in the liver • Hyperglycemia
Fat	• Decreased triglyceride synthesis in adipose tissue • Increased triglyceride breakdown in adipose tissue • Increased fatty acid and triglyceride levels in the blood • Increased production of ketone bodies in the liver
Protein	• Decreased amino acid uptake by muscle cells • Decreased protein synthesis • Increased protein breakdown • Muscle wasting and growth retardation

TABLE 26-2	Symptoms of Diabetes Mellitus

Frequent urination (polyuria)

Dehydration, dry mouth

Increased thirst (polydipsia)

Blurred vision

Increased infections

Weight loss

Increased hunger (polyphagia)

Fatigue

and classic symptoms of diabetes (such as polyuria, polydipsia, and unexplained weight loss) are present.

- The plasma glucose concentration is 126 mg/dL or greater after a fast of at least eight hours.
- The plasma glucose concentration measured two hours after a 75-gram glucose load is 200 mg/dL or greater.

Overt symptoms of hyperglycemia help to confirm the diagnosis. Otherwise, a diagnosis of diabetes is confirmed only if a subsequent test yields similar results.

The term **prediabetes** pertains to individuals with blood glucose levels between normal and diabetic, that is, between 100 and 125 mg/dL when fasting (a condition known as *impaired fasting glucose*) or between 140 and 200 mg/dL when measured two hours after ingesting a 75-gram glucose load (a condition known as *impaired glucose tolerance*). Although people with prediabetes are usually asymptomatic, they are at increased risk of developing diabetes and cardiovascular diseases.[3] Prediabetes has been estimated to affect approximately 28 percent of adults in the United States,[4] and it is especially prevalent among those who are overweight or obese.

Types of Diabetes Mellitus

Table 26-3 on p. 814 lists features of the two main types of diabetes, type 1 and type 2 diabetes. Pregnancy can lead to abnormal glucose tolerance and the condition known as *gestational diabetes* (discussed later in this chapter), which often resolves after pregnancy but is a risk factor for type 2 diabetes. ◆ Diabetes can also be caused by medical conditions that damage the pancreas or interfere with insulin function.

Type 1 Diabetes **Type 1 diabetes** accounts for about 5 to 10 percent of diabetes cases. It is usually caused by **autoimmune** destruction of the pancreatic beta cells, which produce and secrete insulin. By the time symptoms develop, the damage to the beta cells has progressed so far that insulin must be supplied exogenously, most often by injection. Although the precise cause of the autoimmune attack is usually unknown, environmental toxins or infections are likely triggers. People with type 1 diabetes often have a genetic susceptibility for the disorder and are at increased risk of developing other autoimmune diseases.

◆ Reminder: *Osmolarity* refers to the concentration of osmotically active particles in solution. Hyperglycemia causes the body's fluids to become *hyperosmolar*, meaning that they have an abnormally high osmolarity.

◆ Normal fasting plasma glucose levels are approximately 75 to 100 mg/dL (published values vary).

◆ Gestational diabetes was introduced in Chapter 14.

renal threshold: the blood concentration of a substance that exceeds the kidneys' capacity for reabsorption, causing the substance to be passed into the urine.

oral glucose tolerance test: a test that evaluates a person's ability to tolerate an oral glucose load.

prediabetes: the condition in which blood glucose levels are higher than normal but not high enough to be diagnosed as diabetes.

type 1 diabetes: the type of diabetes that accounts for 5 to 10 percent of diabetes cases and usually results from autoimmune destruction of pancreatic beta cells.

autoimmune: an immune response directed against the body's own tissues.
- **auto** = self

TABLE 26-3 Features of Type 1 and Type 2 Diabetes

Feature	Type 1	Type 2
Prevalence in diabetic population	5% to 10% of cases	90% to 95% of cases
Age of onset	<30 years	>45 years[a]
Associated conditions	Autoimmune diseases, viral infection, inherited factors	Obesity, aging, inherited factors
Major defect	Destruction of pancreatic beta cells; insulin deficiency	Insulin resistance; insulin deficiency (relative to needs)
Insulin secretion	Little or none	Varies; may be normal, increased, or decreased
Requirement for insulin therapy	Always	Sometimes
Other names	Juvenile-onset diabetes	Adult-onset diabetes
	Insulin-dependent diabetes mellitus (IDDM)	Noninsulin-dependent diabetes mellitus (NIDDM)
	Ketosis-prone diabetes	Ketosis-resistant diabetes

[a] The incidence of type 2 diabetes is increasing in children and adolescents; in over 90% of these cases, it is associated with overweight or obesity and a family history of type 2 diabetes.

◆ Reminder: *Ketone bodies* are products of fat metabolism that are produced in the liver; they accumulate in tissues when fatty acids are released in abnormally high amounts from adipose tissue.

Type 1 diabetes usually develops during childhood or adolescence, and symptoms may appear abruptly in previously healthy children.[5] Classic symptoms are frequent urination, weight loss, and increased thirst. **Ketoacidosis**—acidosis due to excessive production of ketone bodies—is sometimes the first sign of disease. ◆ Disease onset tends to be more gradual in individuals who develop type 1 diabetes in later years. Blood tests that detect antibodies to insulin, pancreatic islet cells, and pancreatic enzymes can confirm the diagnosis and help to predict development of the disease in close relatives.

Type 2 Diabetes **Type 2 diabetes** is the most prevalent form of diabetes, accounting for 90 to 95 percent of cases, and it is often asymptomatic. The primary defect in type 2 diabetes is **insulin resistance**, a reduced sensitivity to insulin in muscle, adipose, and liver cells. To compensate, the pancreas secretes larger amounts of insulin, and plasma insulin concentrations can rise to abnormally high levels **(hyperinsulinemia).** Over time, the pancreas becomes less able to compensate for the cells' reduced sensitivity to insulin, and hyperglycemia worsens. The high demand for insulin can eventually exhaust the beta cells of the pancreas and lead to impaired insulin secretion and reduced plasma insulin concentrations. Type 2 diabetes is therefore associated with both insulin resistance and relative insulin deficiency; that is, the amount of insulin is insufficient to compensate for its diminished effect in the cells.

Although the actual causes of type 2 diabetes are unknown, the risk is substantially increased by obesity (especially abdominal obesity), aging, and physical inactivity. An estimated 80 to 90 percent of individuals with type 2 diabetes are obese, and obesity itself can directly cause some degree of insulin resistance.[6] ◆ The prevalence of type 2 diabetes increases with age and exceeds 22 percent in persons over 60 years of age; however, many of these cases remain undiagnosed.[7] Inherited factors strongly influence risk, and type 2 diabetes is more common in certain ethnic populations, including Native Americans, Hispanic Americans, Mexican Americans, African Americans, Asian Americans, and Pacific Islanders.

◆ Highlight 26 provides information about the relationship between obesity and insulin resistance.

Type 2 Diabetes in Children and Adolescents Although most cases of type 2 diabetes are diagnosed in individuals over 45 years old, children and adolescents who are overweight or have a family history of diabetes also are at increased risk. Because type 2 diabetes is frequently asymptomatic, it is generally detected in children only when high-risk groups are screened for the disease. For example, when 167 obese children of different ethnic groups were screened, prediabetes was detected in 25 percent of children between 4 and 10 years of age and in 21 percent of adolescents between 11 and 18 years of age.[8] Among the Pima Indians in Ari-

type 2 diabetes: the type of diabetes that accounts for 90 to 95 percent of diabetes cases and usually results from insulin resistance coupled with insufficient insulin secretion.

insulin resistance: reduced sensitivity to insulin in muscle, adipose, and liver cells.

hyperinsulinemia: abnormally high levels of insulin in the blood.

zona, a population with one of the highest rates of type 2 diabetes in the world, overt type 2 diabetes was reported in 2.2 percent of 10- to 14-year-old children and in 5 percent of 15- to 19-year-old teens.[9] Routine screening and prevention programs that target food intake and activity patterns can be important safeguards for preventing diabetes in children at risk.

Prevention of Type 2 Diabetes Mellitus

Clinical studies suggest that lifestyle changes can delay or prevent the incidence of type 2 diabetes in individuals at risk. In the Diabetes Prevention Program, a multicenter trial of 3234 adults with impaired glucose tolerance, dietary changes and increased physical activity led to a 58 percent reduction in diabetes incidence.[10] Based on the results of this and similar studies, guidelines for diabetes prevention include the following strategies:*[11]

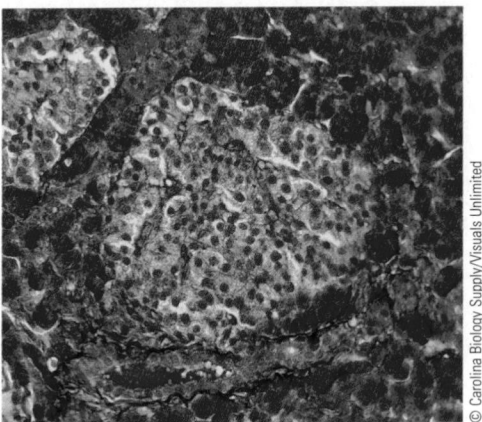

Cross sections of the pancreas reveal distinct areas known as the islets of Langerhans, which contain the alpha cells that produce glucagon and the beta cells that produce insulin.

- *Weight management.* A sustained weight loss of 5 to 10 percent of body weight is recommended for overweight and obese individuals. If weight loss cannot be achieved, healthy eating behaviors should be encouraged to prevent additional weight gain.

- *Active lifestyle.* At least 30 minutes of moderate physical activity, such as brisk walking, is recommended daily.

- *Dietary modifications.* An increased intake of whole grains and dietary fiber has been associated with a reduced risk for type 2 diabetes. Individuals who are overweight or obese should decrease their intake of dietary fat to avoid consuming excessive energy.

- *Regular monitoring.* Individuals at risk should be monitored every one or two years to check for the possible development of type 2 diabetes. If necessary, they can be provided with additional counseling, education, or resources.

Clinical trials have found that a light to moderate alcohol intake (one to two drinks per day) may reduce the risk of developing type 2 diabetes, compared with either abstinence from alcohol or heavy drinking. Alcohol's protective effect may be attributable to an increased secretion of adiponectin, an adipose hormone that improves insulin sensitivity.[12] Specific recommendations regarding alcohol intake are unavailable, however, because the risk of adverse effects from alcohol ingestion must be considered on an individual basis.

Acute Complications of Diabetes Mellitus

Untreated diabetes may result in life-threatening complications. As described earlier, insufficient insulin can result in significant disturbances in energy metabolism (review Table 26-1). Severe hyperglycemia can lead to dehydration and electrolyte imbalances. In treated diabetes, hypoglycemia (low blood glucose) is a possible complication of inappropriate management.

Diabetic Ketoacidosis in Type 1 Diabetes A severe lack of insulin causes diabetic ketoacidosis. Without insulin, glucagon's effects become more pronounced, leading to the unrestrained breakdown of the triglycerides in adipose tissue and the protein in muscle. As a result, an increased supply of fatty acids and amino acids arrives in the liver, where fatty acid oxidation and gluconeogenesis proceed unchecked. ◆ The increased rate of fatty acid oxidation results in excessive amounts of acetyl CoA and subsequent ketone body production. Ketone bodies, which are acidic, can reach dangerously high levels in the bloodstream (ketoacidosis) and spill into the urine **(ketonuria).** Blood pH typically falls below 7.30

◆ Chapter 7 provides details about these metabolic pathways.

---

* The antidiabetic medication metformin may be beneficial for preventing diabetes in high-risk individuals, such as those with obesity, a sedentary lifestyle, prediabetes, and a family history of diabetes.

◆ Reminder: *Ketosis* is an abnormal increase in the production of ketone bodies.

◆ Bicarbonate is a buffer in the blood that corrects acidosis. The acid (H^+) and bicarbonate (HCO_3^-) combine to form carbonic acid (H_2CO_3), which breaks down to water (H_2O) and carbon dioxide (CO_2). The carbon dioxide is then exhaled.

(blood pH normally ranges between 7.35 and 7.45). In diabetic ketoacidosis, blood glucose concentrations usually exceed 250 mg/dL and may rise above 1000 mg/dL in severe cases. The main features of diabetic ketoacidosis thus include severe ketosis, ◆ acidosis, and hyperglycemia.

Patients with ketoacidosis may exhibit symptoms of both acidosis and dehydration. Acidosis is partially corrected by exhalation of carbon dioxide, so rapid or deep breathing is characteristic. ◆ Polyuria and polydipsia (frequent urination and extreme thirst) can accompany the hyperglycemia, lowering blood volume and blood pressure and depleting electrolytes. In response, patients may demonstrate marked fatigue, lethargy, nausea, and vomiting. Ketone accumulation is sometimes evident by a fruity odor on a person's breath **(acetone breath).** Mental state may vary from alertness to comatose **(diabetic coma).** Diabetic coma was a frequent cause of death before insulin was routinely used to manage diabetes.

Diabetic ketoacidosis is sometimes the earliest sign that leads to diagnosis of type 1 diabetes, but more often it results from inappropriate treatment (such as missed insulin injections), illness or infection, alcohol abuse, or other physiological stressors.[13] The condition usually develops quickly, within hours or a few days. It is a medical emergency that requires insulin therapy to correct the hyperglycemia, intravenous fluid and electrolyte replacement, and in some cases, bicarbonate therapy to treat acidosis. Antibiotics may be necessary if infection is present. The mortality rate in diabetic ketoacidosis is nearly 5 percent in individuals under 40 years of age, but exceeds 20 percent in elderly individuals.[14]

Hyperosmolar Hyperglycemic State in Type 2 Diabetes The **hyperosmolar hyperglycemic state** is a condition of severe hyperglycemia, dehydration, and hyperosmolarity that develops in the absence of significant ketosis. When ketosis is present, it is much milder than in diabetic ketoacidosis because enough insulin is available in type 2 diabetes to suppress fatty acid oxidation and consequent ketone body production. Because glucagon's actions dominate, however, gluconeogenesis leads to dramatic increases in blood glucose levels, which typically exceed 600 mg/dL and may rise above 2000 mg/dL. The extreme hyperglycemia causes substantial fluid losses, leading to depleted blood volume and electrolyte imbalances. Blood plasma may become so hyperosmolar as to cause neurological abnormalities, such as abnormal reflexes, motor impairments, reduced verbal ability, and seizures; about 10 percent of patients lapse into coma.

The hyperosmolar hyperglycemic state is sometimes the first sign of type 2 diabetes in older persons. It is usually precipitated by infection, illness, or a drug treatment that impairs insulin action or secretion, and it often develops because patients are unable to recognize thirst or adequately replace fluid losses due to age, illness, sedation, or incapacity. Unlike diabetic ketoacidosis, the condition often evolves slowly, over several days or weeks; the absence of clinical signs can delay its diagnosis. Treatment includes intravenous fluid and electrolyte replacement and insulin therapy. The mortality rate approaches 15 percent, primarily because the condition occurs more often in older patients who have cardiovascular disease or other major illnesses.[15]

Hypoglycemia **Hypoglycemia,** or low blood glucose, is the most frequent complication of type 1 diabetes and may occur in type 2 diabetes as well. It arises from the inappropriate management of diabetes rather than from the disease itself, and it usually results from excessive dosages of insulin or antidiabetic drugs, prolonged exercise, skipped or delayed meals, inadequate food intake, or the consumption of alcohol without food. Symptoms of hypoglycemia include sweating, shakiness, heart palpitations, slurred speech, double vision, and irritability. Mental confusion may prevent a person from recognizing the problem and taking such corrective action as ingesting glucose tablets, juice, or candy. If hypoglycemia occurs during the night, patients may be completely unaware of its presence. Severe hypoglycemia or a delay in treatment can cause irreversible brain damage. Hypoglycemia is the most frequent cause of coma in insulin-treated patients and is believed to account for 3 to 4 percent of deaths in this population.[16]

Chronic Complications of Diabetes Mellitus

Excessive blood glucose can eventually alter cellular functions and damage cells and tissues. Glucose nonenzymatically combines with proteins, producing molecules that ultimately break down to form reactive compounds known as **advanced glycation end products (AGEs);** these AGEs can accumulate and cause damage to cells and blood vessels. Excessive glucose also promotes the production and accumulation of sorbitol, which increases oxidative stress and alters molecular structures and functions. The chronic complications of diabetes typically involve the large blood vessels **(macrovascular complications),** smaller vessels such as arterioles and capillaries **(microvascular complications),** and the nervous system **(diabetic neuropathy).** Other tissues adversely affected by diabetes include the lens of the eye and the skin; cataracts, glaucoma, and various skin disorders sometimes develop. Increased infections are common in diabetes, a possible consequence of hyperglycemia, impaired circulation, or depressed immune responses. In individuals with type 2 diabetes, complications often develop before the diabetes is diagnosed.

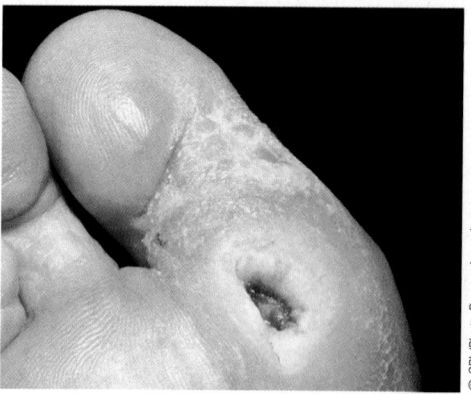

Foot ulcers are a common complication of diabetes because blood circulation is impaired (which slows healing) and nerve damage dampens foot pain (delaying recognition and treatment of cuts and bruises).

Macrovascular Complications The damage caused by diabetes accelerates the development of atherosclerosis in the coronary arteries and the arteries of the limbs. Cardiovascular diseases are the leading cause of death in people with diabetes, accounting for up to 70 percent of deaths.[17] Type 2 diabetes is often accompanied by multiple risk factors for coronary heart disease, including hypertension, abnormal blood lipids, and obesity. ◆ In addition, people with diabetes have increased tendencies for thrombosis (blood clot formation) and abnormal ventricle function, both of which can worsen the clinical course of heart disease.[18]

Impaired blood flow in the arteries of the limbs increases the risk of **claudication** (pain while walking) and contributes to the development of foot ulcers (see the photo). Left untreated, foot ulcers can lead to **gangrene** (tissue death), and some patients require foot amputation, a major cause of disability in individuals with diabetes. About 15 to 20 percent of persons with diabetes are hospitalized with foot complications during the course of illness.[19]

◆ People with type 2 diabetes frequently develop the *metabolic syndrome*, a cluster of symptoms associated with insulin resistance (including hyperglycemia, hypertension, and altered blood lipids) that substantially increase heart disease risk (see Highlight 26).

Microvascular Complications Long-term diabetes is associated with a thickening of the basement membrane of capillaries and small arterioles, which impairs the normal functioning of these blood vessels. The primary microvascular complications involve the retina of the eye and the kidneys. Diabetes is currently the leading cause of both adult blindness and kidney failure in the United States.[20]

In **diabetic retinopathy,** the weakened capillaries of the retina leak fluid, lipids, or blood, causing local edema or hemorrhaging. The defective blood flow also leads to damage and scarring within retinal tissue. New blood vessels eventually form, but they are fragile and bleed easily, releasing blood and proteins that obscure vision. The retinal changes usually occur after an individual has had diabetes for many years; for example, 80 percent of diabetes patients develop retinopathy after 15 years.[21] Diabetic retinopathy progresses most rapidly when diabetes is poorly controlled, and intensive management substantially reduces the risk.

In **diabetic nephropathy,** damage to the kidneys' specialized capillaries prevents adequate filtration of the blood, which is evidenced by abnormal protein losses in the urine **(microalbuminuria).** As the kidney damage worsens, urine production decreases and nitrogenous wastes accumulate in the blood; eventually, the individual requires dialysis (artificial filtration of blood) to survive. ◆ Because the kidneys normally regulate blood volume and blood pressure (see pp. 401–402), inadequate kidney function leads to high blood pressure in patients with nephropathy. Kidney failure eventually develops in about 30 to 35 percent of

◆ Chapter 28 provides details about the progression and treatment of chronic kidney disease.

advanced glycation end products (AGEs): reactive compounds formed after glucose combines with protein; AGEs can damage tissues and lead to diabetic complications.

patients with type 1 diabetes and 20 percent of those with type 2 diabetes.[22] As with diabetic retinopathy, intensive diabetes management can help slow the progression of kidney damage.

Diabetic Neuropathy Neuropathy, or nerve degeneration, occurs in about 50 percent of diabetes cases.[23] The extent of nerve damage depends on the severity and duration of hyperglycemia. Symptoms of neuropathy vary and may be experienced as pain or burning, numbness and tingling in the hands and feet, or loss of sensation. Pain and cramping, especially in the legs, are often severe during the night and may interrupt sleep. Neuropathy also contributes to the development of foot ulcers because cuts and bruises may go unnoticed until wounds are severe. Other manifestations of neuropathy include sweating abnormalities, sexual dysfunction, constipation, and delayed stomach emptying **(gastroparesis).**

> ### IN SUMMARY
>
> Diabetes mellitus is a chronic condition characterized by inadequate insulin secretion or impaired insulin action. In type 1 diabetes, the pancreas secretes little or no insulin, and insulin therapy is necessary for survival. Type 2 diabetes is characterized by insulin resistance coupled with relative insulin deficiency, and disease risk is increased by obesity, aging, and physical inactivity. Acute complications of diabetes include diabetic ketoacidosis, in which hyperglycemia is accompanied by ketosis and acidosis, and the hyperosmolar hyperglycemic state, characterized by severe hyperglycemia, dehydration, and possible mental impairments. Another acute complication, hypoglycemia, is usually a consequence of inappropriate disease management. Chronic complications of diabetes include macrovascular disorders such as cardiovascular diseases and peripheral vascular disease, microvascular conditions such as diabetic retinopathy and diabetic nephropathy, and diabetic neuropathy.

Treatment of Diabetes Mellitus

Diabetes is a chronic and progressive illness that requires lifelong treatment. Managing blood glucose levels is a delicate balancing act that involves meal planning, proper timing of medications, and physical exercise. Frequent adjustments in treatment are often necessary to establish good **glycemic** control. Individuals with type 1 diabetes require insulin therapy for survival. Type 2 diabetes is initially treated with diet therapy and exercise, but most patients eventually need antidiabetic medications or insulin. Diabetes management becomes even more difficult once complications develop. Although the health care team must determine the appropriate therapy, the individual with diabetes ultimately assumes much of the responsibility for treatment and therefore requires education in self-management of the disease.

Treatment Goals

The main goal of diabetes treatment is to maintain blood glucose levels within a desirable range to prevent or reduce the risk of complications. As discussed in the next section, clinical trials have demonstrated that *intensive* diabetes treatment, which keeps blood glucose levels tightly controlled, can reduce the incidence and severity of chronic complications. Therefore, maintenance of near-normal glucose levels has become the fundamental objective of all diabetes care plans. Other goals of treatment include maintaining healthy blood lipid concentrations, controlling blood pressure, and managing weight—measures that can help to prevent or delay diabetes complications as well.

glycemic (gly-SEE-mic): pertaining to blood glucose.

TABLE 26-4	Comparison of Conventional and Intensive Therapies for Type 1 Diabetes	
	Conventional Therapy	**Intensive Therapy**
Blood glucose monitoring	Monitored daily	Monitored at least three times daily
Insulin therapy	One or two daily injections; no daily adjustments	Three or more daily injections or use of external insulin pump; dosage adjusted according to results of glucose monitoring and expected carbohydrate intake
Advantages	Fewer incidences of severe hypoglycemia; less weight gain	Delayed progression of retinopathy, nephropathy, and neuropathy
Disadvantages	More rapid progression of retinopathy, nephropathy, and neuropathy	Twofold to threefold increase in severe hypoglycemia; weight gain; increased risk of becoming overweight

Benefits of Intensive Treatment Several landmark studies conducted in the 1980s and 1990s confirmed that keeping blood glucose levels as close to normal as possible offers clear advantages over less rigorous diabetes treatment. The Diabetes Control and Complications Trial was a multicenter trial that tested whether the intensive treatment of type 1 diabetes would decrease the frequency and severity of microvascular and neurological complications.[24] In this study, 1441 persons with type 1 diabetes were randomly assigned to receive either conventional or intensive therapy, as summarized in Table 26-4. The subjects were followed for an average of 6.5 years. The participants undergoing intensive therapy had delayed onset and reduced progression of retinopathy, nephropathy, and neuropathy; however, they also experienced increased incidences of severe hypoglycemia and gained more weight. A later trial, the United Kingdom Prospective Diabetes Study, found similar advantages to using intensive treatment in type 2 diabetes.*[25]

Diabetes Self-Management Education Newly diagnosed patients and their families have much to learn about diabetes and its management. Diabetes education provides an individual with the knowledge and skills necessary to implement treatment. The primary instructor is often a **Certified Diabetes Educator (CDE)**, a health care professional (often a nurse or dietitian) who has specialized knowledge about diabetes treatment and the health education process. To manage diabetes, patients need to learn about appropriate meal planning, medication administration, blood glucose monitoring, weight management, appropriate physical activity, and prevention of complications.

Evaluating Diabetes Treatment

Diabetes treatment is largely evaluated by monitoring glycemic status. Good glycemic control requires frequent home monitoring of blood glucose using a glucose meter, referred to as **self-monitoring of blood glucose.** In this procedure, a drop of blood from a finger prick is applied to a chemically treated paper strip, which is then analyzed for glucose. Glucose testing provides valuable feedback when the patient adjusts food intake, medications, and physical activity and is helpful for preventing hypoglycemia. Ideally, patients with type 1 diabetes should monitor blood glucose three or more times daily—and more frequently when therapy is adjusted. Self-monitoring of blood glucose is also useful in type 2 diabetes, although the recommended frequency depends on the specific needs of individual patients.[26]

Long-Term Glycemic Control Health care providers periodically evaluate long-term glycemic control by measuring **glycated hemoglobin** (abbreviated **HbA$_{1c}$**). The glucose in blood freely enters red blood cells and nonenzymatically attaches to hemoglobin molecules in direct proportion to the amount of glucose

Certified Diabetes Educator (CDE): a health care professional who specializes in diabetes management education. Certification is obtained from the National Certification Board for Diabetes Educators.

self-monitoring of blood glucose: home monitoring of blood glucose levels using a glucose meter.

glycated hemoglobin (HbA$_{1c}$): hemoglobin molecules to which glucose has been nonenzymatically attached; the level of HbA$_{1c}$ in blood helps to evaluate long-term glycemic control. Also called *glycosylated hemoglobin.*

---

*Intensive treatment may be inappropriate for some individuals with diabetes; examples include individuals with limited life expectancies or a history of hypoglycemia and middle-aged or older adults with previous heart disease or multiple heart disease risk factors.

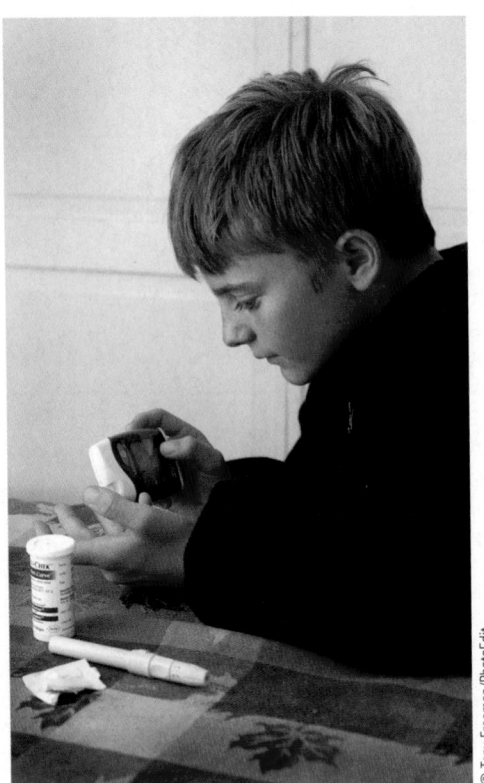

Self-monitoring of blood glucose can help individuals with diabetes learn how to maintain blood glucose levels within a desirable range.

present. Because the life span of red blood cells averages 120 days, the percentage of HbA_{1c} is a measure of glycemic control during the preceding two to three months (the average age of circulating red blood cells).[27] In people without diabetes, HbA_{1c} is typically less than 6 percent of total hemoglobin. The goal of diabetes treatment is an HbA_{1c} value under 7 percent,[28] but the concentration is often markedly higher even in people with diabetes who are maintaining near-normal blood glucose levels.

The **fructosamine test** is sometimes conducted to determine glycemic control for the preceding two-week period. This test determines the nonenzymatic glycation of serum proteins (primarily albumin), which have a shorter half-life than hemoglobin. Most often, the fructosamine test is used to evaluate recent adjustments in diabetes treatment or glycemic control during pregnancy. The test cannot be interpreted correctly in patients with kidney or liver disease.

Monitoring for Long-Term Complications Individuals with diabetes are routinely monitored for signs of long-term complications. Blood pressure is measured at each checkup. Annual lipid screening is suggested for most adult patients. Routine checks for urinary protein (microalbuminuria) help to determine if nephropathy has developed. Physical examinations generally screen for signs of retinopathy, neuropathy, and foot problems.

Ketone Testing Ketone testing checks for the development of ketoacidosis if symptoms are present or if risk has increased due to acute illness, stress, or pregnancy. Both blood and urine tests are available for home use, although the blood tests are currently more reliable.[29] Ketone testing is most useful for patients who have type 1 diabetes or gestational diabetes. Individuals with type 2 diabetes may produce excessive ketone bodies when severely stressed by infection or trauma.

Body Weight Concerns

Whereas individuals with newly diagnosed type 1 diabetes are likely to be thin, most people with type 2 diabetes are overweight or obese. Body weight and children's growth patterns are monitored to evaluate whether energy intakes are appropriate.

Body Weight in Type 1 Diabetes In general, people with type 1 diabetes are less likely to be overweight than those in the general population. However, excessive weight gain is sometimes an unwanted side effect of improved glycemic control, especially in those undergoing intensive insulin therapy. Although the cause of weight gain is unclear, it is possibly related to the insulin treatment, which may stimulate fat synthesis or induce energy intake in some way.[30] In addition, insulin treatment eliminates energy losses from glycosuria, a change that can contribute to energy excess.[31] Although patients should try to prevent excessive weight gain, concerns about weight should not discourage the use of intensive therapy, which is associated with longer life expectancy and fewer complications than occur with conventional therapy. It is also important to ensure that growing children receive sufficient energy for normal growth and development.

Body Weight in Type 2 Diabetes Because excessive body fat can worsen insulin resistance, weight loss is often recommended for those who are overweight or obese. Even moderate weight loss (10 to 20 pounds) can help to improve glycemic control, blood lipid levels, and blood pressure. Weight loss is most beneficial early in the course of diabetes, before insulin secretion has diminished. The positive effects appear to be related to kcaloric restriction rather than to weight loss itself, and improvements in blood glucose usually appear within days after a weight-loss program is initiated.[32] Clinical studies, however, have indicated that the improved glycemic control often diminishes within a year after weight loss. The reason for this outcome may be that subjects are no longer restricting kcalories and have begun to regain the lost weight.[33]

fructosamine test: a measurement of glycated serum proteins; used to analyze glycemic control over the preceding two weeks. Also known as the *glycated albumin test* or the *glycated serum protein test*.

Not all persons with type 2 diabetes are overweight or obese. Older adults and those in long-term care facilities are often underweight and may need to gain weight. Low body weight increases risks of morbidity and mortality in these individuals.

Medical Nutrition Therapy: Nutrient Recommendations

Medical nutrition therapy has a considerable influence on diabetes outcome. The appropriate dietary choices can both improve blood glucose levels and slow the progression of diabetes complications. As always, the nutrition care plan must take personal preferences and lifestyle habits into account. Dietary intakes need to be modified to accommodate growth, lifestyle changes, aging, and any complications that develop. Although all members of the diabetes care team should understand the principles of dietary treatment, a registered dietitian is best suited to design and implement the medical nutrition therapy for diabetes patients. This section presents the nutrient recommendations for diabetes. Meal-planning strategies are described later in the chapter.

Total Carbohydrate Intake The amount of carbohydrate ingested has the greatest influence on blood glucose levels after meals—the more grams of carbohydrate ingested, the greater the glycemic response. The carbohydrate recommendation is based in part on the person's metabolic needs (that is, the type of diabetes or degree of glucose tolerance) and individual preferences. In addition, the carbohydrate intake must be fairly consistent at meals and snacks to help reduce fluctuations in blood glucose levels between meals. Low-carbohydrate diets, which restrict carbohydrate intake to less than 130 grams per day, are not recommended.[34]

Carbohydrate Sources Different carbohydrate-containing foods have different effects on blood glucose levels; for example, consuming a portion of white rice may cause blood glucose to rise more than would consuming a similar portion of barley. This *glycemic effect* of foods is influenced by the type of carbohydrate in a food, the food's fiber content, the preparation method, the other foods included in a meal, and individual tolerances. The glycemic index (GI), a ranking of carbohydrate foods based on their average glycemic effect, has been compiled from the scientific literature; some individuals may find this resource helpful when making food choices. ◆ The GI is not a primary consideration when treating diabetes, however, because research studies investigating the possible benefits of low-GI diets on glycemic control have had mixed results.[35] In addition, there is considerable variability in individual responses to specific carbohydrate foods. Nonetheless, high-fiber, minimally processed foods, which typically have more moderate effects on blood glucose than do highly processed, starchy foods, are among the foods frequently recommended for persons with diabetes.

◆ Chapter 4 provides additional information about the glycemic index (see pp. 115–116). The website **www.glycemicindex.com** provides glycemic index values for a wide variety of common foods.

Fiber Fiber recommendations for individuals with diabetes are similar to those for the general population; ◆ thus, people with diabetes are encouraged to include fiber-rich foods such as legumes, whole-grain cereals, fruits, and vegetables in their diet. Although some studies have suggested that very high intakes of fiber (50 grams or more per day) may improve glycemic control, the benefits have not been consistent across studies, and many individuals may have difficulty tolerating such large amounts of fiber.[36]

◆ The fiber DRI for adult women and men ranges from 21 to 38 g; check the DRI table on the inside front cover of this text for specific values.

Sugars A common misperception is that people with diabetes need to avoid sugar and sugar-containing foods. In reality, table sugar (sucrose), made up of glucose and fructose, has a lower glycemic effect than that of starch. Because moderate consumption of sugar has not been shown to adversely affect glycemic control,[37] sugar recommendations for people with diabetes are similar to those for the general population, which suggest minimizing foods and beverages that contain added sugars. However, sugars and sugary foods must be counted as part of the daily carbohydrate allowance.

Although fructose has a minimal glycemic effect, its use as an added sweetener is not advised because excessive dietary fructose may adversely affect blood lipid levels. (Note that it is not necessary to avoid the naturally occurring fructose in fruits and vegetables.) Sugar alcohols (such as sorbitol and maltitol) have lower glycemic effects than glucose, fructose, or sucrose, but their use has not been found to significantly improve long-term glycemic control. Artificial sweeteners (such as aspartame, saccharin, and sucralose) contain no digestible carbohydrate and can be safely used in place of sugar.

Dietary Fat As mentioned earlier, people with diabetes are at high risk of developing cardiovascular diseases. Guidelines for dietary fat are similar to those for other persons at risk: saturated fat intake should be limited to less than 7 percent of total kcalories, *trans* fat intake should be minimized, and cholesterol intake should be limited to less than 200 milligrams daily.[38] Dietary strategies for cardiovascular disease are discussed further in Chapter 27.

Protein The protein intake in people with diabetes should be between 15 and 20 percent of total kcalories, which is the usual range of protein intake in the general population. Although small, short-term studies have suggested that diets with higher protein intakes may improve glycemic control, increase satiety, and help with weight loss, the long-term effects of such diets on diabetes management and complications are unknown.[39] In addition, high protein intakes are discouraged because they may be detrimental to kidney function in some individuals.

Alcohol Use in Diabetes Alcohol can be used in moderation by adults with diabetes. Guidelines are similar to those for the general population, which advise a daily limit of one drink for women and two drinks for men. ◆ However, individuals using insulin or medications that promote insulin secretion should consume food when they ingest alcoholic beverages to avoid hypoglycemia. Alcohol can cause hypoglycemia by interfering with glucose production in the liver. Conversely, excessive alcohol can worsen hyperglycemia, and it can also raise triglyceride levels in susceptible persons. People who should avoid alcohol include pregnant women and individuals with pancreatitis, advanced neuropathy, abnormally high triglyceride levels, or a history of alcohol abuse.[40]

Micronutrients Micronutrient recommendations for people with diabetes are the same as for the general population. Vitamin and mineral supplementation is not recommended unless nutrient deficiencies develop; those at risk include the elderly, pregnant or lactating women, strict vegetarians, and individuals on kcalorie-restricted diets.[41] Although some studies have suggested that supplemental chromium can improve glycemic control in type 2 diabetes, results have not been consistent. At present, chromium supplementation is not recommended for those with type 2 diabetes.

◆ Reminder: One drink is equivalent to 12 ounces of beer; 5 ounces of wine; 10 ounces of wine cooler; or 1¹/₂ ounces of 80 proof distilled spirits such as gin, rum, vodka, and whiskey.

Medical Nutrition Therapy: Meal-Planning Strategies

Dietitians provide a number of meal-planning strategies to help people with diabetes maintain glycemic control. These strategies emphasize control of carbohydrate intake and portion sizes. A regular eating pattern, with carbohydrate intake spaced evenly throughout the day, is typically recommended. Sample menus, which include commonly eaten foods, can help to illustrate general principles. Initial dietary instructions may include a discussion of the *Dietary Guidelines for Americans* or other recommendations designed for the general population (see Chapter 2), as well as guidelines for improving blood lipids and other cardiovascular risk factors. People using intensive insulin therapy must learn to coordinate insulin injections with meals and to match insulin dosages to carbohydrate intake, as discussed later.

Carbohydrate Counting Carbohydrate-counting techniques are simpler and more flexible than other menu-planning approaches and are widely used for planning diabetes diets. Carbohydrate counting works as follows: After a dietitian determines a person's nutrient and energy needs, the individual is given a daily carbohydrate allowance, often divided into a pattern of meals and snacks according to individual preferences. The carbohydrate allowance can be expressed in grams or as the number of carbohydrate portions allowed per meal (see Table 26-5 on p. 824). The user of the plan need only be concerned about meeting carbohydrate goals and can select from any of the carbohydrate-containing food groups when planning meals (see Table 26-6 on p. 824 and Figure 26-2 on p. 825). Although encouraged to make healthy food choices, the individual has the freedom to choose the foods desired at each meal without risking loss of glycemic control. Some people may also need guidance about noncarbohydrate foods to help them choose a healthy diet that improves blood lipids or energy intakes. The "How to" on pp. 824–825 provides more information about using carbohydrate counting in clinical practice.

Carbohydrate counting is taught at different levels of complexity depending on a person's needs and abilities. The basic carbohydrate-counting method just described can be helpful for most people, although it requires a consistent carbohydrate intake from day to day to match the medication or insulin regimen. Advanced carbohydrate counting allows more flexibility but is best suited for patients using intensive insulin therapy. With this method, a person can determine the specific dose of insulin needed to cover the amount of carbohydrate consumed at a meal. The person is then free to choose the types and portions of food desired without sacrificing glycemic control. Advanced carbohydrate counting requires some training and should be attempted only after an individual has mastered more basic methods.

Exchange Lists for Meal Planning The exchange list system is an alternative meal-planning method, although it is more complex and difficult for patients to learn than carbohydrate counting. This system of meal planning was introduced in Chapter 2 and is described further in Appendix G (Appendix I for Canadians). The exchange system sorts foods according to their proportions of carbohydrate, fat, and protein so that each item in a food group (or "exchange list") has a similar macronutrient and energy content (see pp. G-1–G-2). Thus any food on a list can be exchanged, or traded, for any other food on the same list without affecting the macronutrient balance in a day's meals. Although the exchange list system can be helpful for individuals who want a structured dietary plan that provides specific percentages of protein, carbohydrate, and fat, it offers no advantages for maintaining glycemic control and is less flexible than carbohydrate counting.

The exchange lists can be helpful resources for individuals using carbohydrate-counting methods because the portions in the exchange lists are interchangeable with the portions used in carbohydrate counting. For example, foods listed in the starch, fruit, and milk exchange lists are equivalent to carbohydrate "portions," as each item contains approximately 15 grams of carbohydrate (see Tables G-4, G-5, and G-6; note that the carbohydrate in the milk exchanges can be rounded up to 15 grams). In the list labeled "Sweets, Desserts, and Other Carbohydrates" (Table G-7), the number of carbohydrate portions per serving is indicated in the far-right column.

Insulin Therapy

Insulin therapy is necessary for individuals who cannot produce enough insulin to meet their metabolic needs. It is therefore required by people with type 1 diabetes and those with type 2 diabetes who cannot maintain glycemic control with antidiabetic medications, diet, and exercise. The pancreas normally secretes insulin in rel-

HOW TO Use Carbohydrate Counting in Clinical Practice

1. The first step in basic carbohydrate counting is to determine an appropriate carbohydrate intake and suitable distribution pattern; an example is shown in Table 26-5. A nutrition assessment can help to estimate a person's usual energy and carbohydrate intakes. The carbohydrate level should be acceptable to the person using the plan. Frequent monitoring of blood glucose levels can help determine whether additional carbohydrate restriction would be helpful.

The example given in Table 26-5 illustrates a meal pattern for a person consuming 2000 kcalories daily with a carbohydrate allowance of 50 percent of kcalories. This is calculated as follows:

50% × 2000 kcal = 1000 kcal of carbohydrate

$$\frac{1000 \text{ kcal carbohydrate}}{4 \text{ kcal/g carbohydrate}} = 250 \text{ g carbohydrate/day}$$

$$\frac{250 \text{ g carbohydrate}}{15 \text{ g/1 carbohydrate portion}} = 16.7 \text{ carbohydrate portions/day}$$

2. The distribution of carbohydrates among meals and snacks is based on both individual preferences and metabolic needs. In type 1 diabetes, the insulin regimen must coordinate with the individual's dietary and lifestyle choices. People using conventional insulin therapy must have a consistent carbohydrate intake from day to day to match their particular insulin prescription, whereas those using intensive therapy can alter insulin dosages when carbohydrate intakes change. People with type 2 diabetes are encouraged to develop dietary patterns that suit their lifestyle and medication schedules. For all types of diabetes, the carbohydrate recommendation may need to be altered periodically to improve blood glucose control.

3. Carbohydrate counting can be done in one of two ways:
   • Count the grams of carbohydrate provided by foods.
   • Count carbohydrate portions, expressed in terms of servings that contain approximately 15 grams each.

Success with carbohydrate counting requires knowledge about the food sources of carbohydrates and an understanding of portion control. As shown in Table 26-6, food selections that contain about 15 grams of carbohydrate are interchangeable. The portions of foods that contain 15 grams may vary substantially, however, even among foods in a single food group. Accurate carbohydrate counting often requires instruction and practice in portion control using measuring cups, spoons, and a food scale. Food lists that indicate the carbohydrate contents of common foods are available from the American Diabetes Association and the American Dietetic Association; these are helpful resources for learning carbohydrate-counting methods.

When using packaged foods, individuals should check the Nutrition Facts panel of food labels to find the carbohydrate content of a serving. If the fiber content is greater than 5 grams per serving, it should be subtracted from the *Total Carbohydrate* value, as fiber does not contribute to blood glucose. If the sugar alcohol content is greater than 5 grams per serving, half of the grams of sugar alcohol can be subtracted from the *Total Carbohydrate* value.

TABLE 26-6 Portion Sizes of Carbohydrate-Containing Foods

Food Groups with Sample Portion Sizes

Bread, cereal, rice, and pasta: 1 portion = 15 g carbohydrate.
1 slice of bread or 1 tortilla
1/2 English muffin
3/4 c unsweetened, ready-to-eat cereal
1/2 c cooked oatmeal
1/3 c cooked rice or pasta

Fruit: 1 portion = 15 g carbohydrate.
1 medium apple, orange, or peach
1 small banana
3/4 c blueberries or chopped pineapple
1/2 c apple juice or orange juice

Milk products: 1 portion = 12 g carbohydrate; may be rounded up to 15 g for ease in counting carbohydrate portions.
1 c milk (whole, low-fat, or fat-free)
1 c buttermilk
6 oz plain yogurt

Starchy vegetables: 1 portion = 15 g carbohydrate.
1 small (3-oz) potato
1/2 c canned or frozen corn
1/3 c baked beans
1 c winter squash, cubed

Sweets and desserts: Considerable variation in carbohydrate content; portions listed contain approximately 15 g.
1/2 c ice cream
2 sandwich cookies (with cream filling)
1/2 frosted cupcake
1 granola bar (1 oz)
1 tbs honey

Nonstarchy vegetables: 1 portion = 3 to 6 g carbohydrate; 3 servings are equivalent to 1 carbohydrate portion; can be disregarded if less than 3 servings are consumed.
1/2 c cooked cauliflower
1/2 c cooked cabbage, collards, or kale
1/2 c cooked okra
1/2 c diced or raw tomatoes

NOTE: Unprocessed meats, fish, and poultry contain negligible amounts of carbohydrate.

TABLE 26-5 Sample Carbohydrate Distribution for a 2000-kCalorie Diet

Meals	Carbohydrate Allowance	
	Grams	Portions[a]
Breakfast	60	4
Lunch	60	4
Afternoon snack	30	2
Dinner	75	5
Evening snack	30	2
Totals	**255 g**	**17**

NOTE: The carbohydrate allowance in this example is approximately 50% of total kcalories.
[a] 1 portion = 15 g carbohydrate = 1 portion of starchy food, milk, or fruit.

HOW TO Use Carbohydrate Counting in Clinical Practice—*continued*

4. Once they have learned the basic carbohydrate-counting method, individuals can select whatever foods they wish, as long as they do not exceed their carbohydrate goals. Figure 26-2 shows a day's menu that follows the dietary plan shown in Table 26-5. Although carbohydrate counting focuses on a single macronutrient, people using this technique should be encouraged to follow a healthy eating plan that meets other dietary objectives as well.

FIGURE 26-2 Translating Carbohydrate Portions into a Day's Meals

✳ SAMPLE MENU ✳

	Carbohydrate Portions			Carbohydrate Portions
Breakfast:			**Afternoon snack:**	
Carbohydrate goal = 4 portions or 60 g.			**Carbohydrate goal = 2 portions or 30 g.**	
3/4 c unsweetened, ready-to-eat cereal	1		2 sandwich cookies	1
1/2 c low-fat milk	1/2		1 c low-fat millk	1
1 scrambled egg	—		**Dinner:**	
1 slice whole wheat toast (with margarine or butter)	1		**Carbohydrate goal = 5 portions or 75 g.**	
6 oz orange juice	1 1/2		4 oz grilled steak	—
Coffee (without milk or sugar)	—		1 small baked potato (with margarine or butter)	1
Lunch:			Corn on cob, 1 large ear	2
Carbohydrate goal = 4 portions or 60 g.			1/2 c steamed collard greens[a]	
1 tuna salad sandwich (includes 2 slices whole-grain bread, mayonnaise)	2		1 c sliced, raw tomatoes[a]	1
6 oz yogurt (plain) with 3/4 c blueberries and artificial sweetener	2		1/2 c ice cream	1
Diet cola	—		**Evening snack:**	
			Carbohydrate goal = 2 portions or 30 g.	
			1 medium apple	1
			1 oz granola bar	1

[a]Three servings of nonstarchy vegetables are equivalent to 1 carbohydrate portion.

atively low amounts between meals and during the night (called *basal insulin*) and in much higher amounts when meals are ingested. Ideally, the insulin treatment should reproduce the natural pattern of insulin secretion as closely as possible.

Insulin Preparations The forms of insulin that are available differ by their onset of activity, timing of peak activity, and duration of effects. Figure 26-3 on p. 826 and Table 26-7 show how insulin preparations are classified: they may be rapid acting (lispro, aspart, and glulisine), short acting (regular), intermediate acting (NPH), or long acting (glargine and detemir), thereby allowing substantial flexibility in establishing a suitable insulin regimen.[42] The rapid- and short-acting insulins are used at mealtimes, whereas the intermediate- and long-acting insulins provide basal insulin for the periods between meals and during the night. Thus, mixtures of several types of insulin can produce greater glycemic control than any one type alone. Several premixed formulations are also available; several examples are listed in Table 26-7.

Most insulin is produced by recombinant DNA techniques that allow the mass production of human insulin by bacteria or yeast. The different forms of insulin are made by chemically modifying insulin's amino acid sequence or by combining insulin with special buffers or peptides that alter insulin's concentration, solubility, or duration of activity in the body.

Insulin Delivery Insulin is most often administered by **subcutaneous** injection, either self-administered or provided by caregivers. ◆ Disposable **syringes,** which can be filled from vials that contain multiple doses of insulin, are the most common devices used for injecting insulin. Another option is to use insulin pens,

◆ Because insulin is a protein, it would be destroyed by digestive processes if taken orally.

subcutaneous (sub-cue-TAY-nee-us): beneath the skin.

syringes: devices used for injecting medications. A syringe consists of a hypodermic needle attached to a hollow tube with a plunger inside.

FIGURE 26-3 Effects of Insulin Preparations

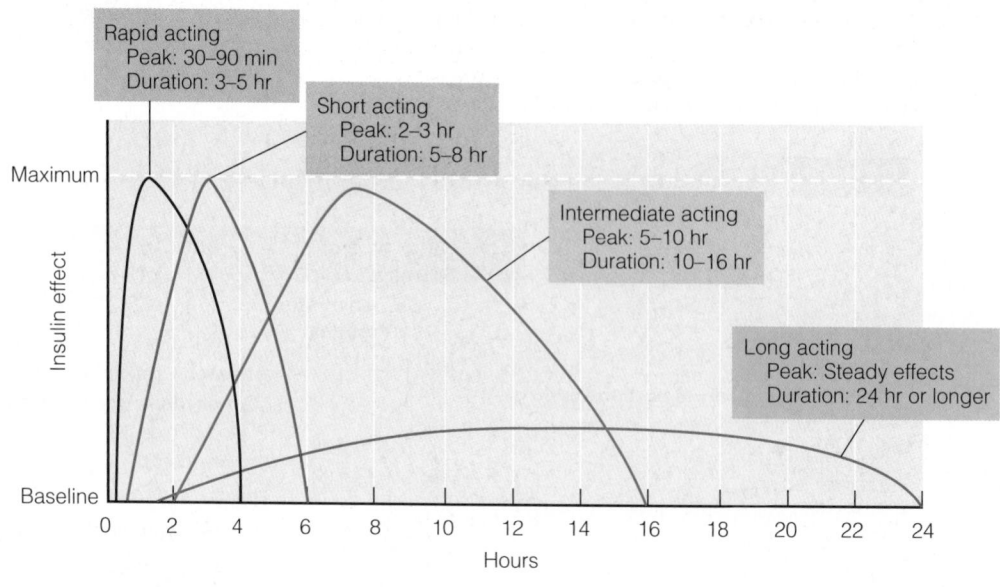

Rapid acting
Peak: 30–90 min
Duration: 3–5 hr

Short acting
Peak: 2–3 hr
Duration: 5–8 hr

Intermediate acting
Peak: 5–10 hr
Duration: 10–16 hr

Long acting
Peak: Steady effects
Duration: 24 hr or longer

Maximum

Insulin effect

Baseline

0   2   4   6   8   10   12   14   16   18   20   22   24

Hours

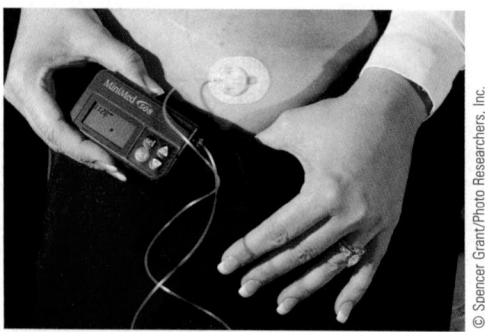

External insulin pumps deliver insulin continuously through thin, flexible tubing inserted into the skin.

© Spencer Grant/Photo Researchers, Inc.

which look like permanent marking pens; these injection devices eliminate the need to carry syringes and separate containers of insulin. Insulin pens are available as disposable pens, which contain prefilled insulin, and reusable pens, which can be fitted with prefilled cartridges and replaceable needles. To eliminate the need for multiple punctures, injection ports for insulin are sometimes inserted through the skin and left in place for several days. Another option is an insulin pump, a computerized device that can be programmed to deliver basal insulin continuously and bolus doses at mealtimes. The pump infuses insulin through thin, flexible tubing that remains in the skin. The pump can be worn under clothes, attached to a belt, or kept in a pocket.

An inhalation powder that supplied rapid-acting insulin was available for several years but is no longer marketed due to inadequate sales. Alternative forms of insulin that can be inhaled are currently under development.

Insulin Regimen for Type 1 Diabetes Type 1 diabetes is best managed with intensive insulin therapy, which involves multiple daily injections of several types of insulin or the use of an insulin pump (review Table 26-4). Usually, intermediate- or long-acting insulin meets basal insulin needs, and rapid- or short-acting insulin is injected before meals. ◆ Three or more daily injections are required for good glycemic control. Simpler regimens involve twice-daily injections of a mixture of intermediate- and short-acting insulin. Regimens that include three or more injections allow for greater flexibility in carbohydrate intake and meal tim-

◆ Rapid-acting insulin begins working within 15 minutes, so it can be injected right before a meal. Short-acting insulin requires a half-hour wait before the meal can begin.

TABLE 26-7 Insulin Preparations

Form of Insulin	Common Preparations	Onset of Action	Peak Activity	Duration of Action
Rapid acting	Lispro Aspart Glulisine	5 to 15 minutes	30 to 90 minutes	3 to 5 hours
Short acting	Regular	30 minutes	2 to 3 hours	5 to 8 hours
Intermediate acting	NPH	2 to 4 hours	5 to 10 hours	10 to 16 hours
Long acting	Glargine Detemir	1 to 2 hours	Steady effects	24 hours or longer
Insulin mixtures (with sample ratios)	NPH/regular (70:30) NPH/regular (50:50)	Variable; depends on formulation	Variable; depends on formulation	Variable; depends on formulation

ing. With fewer injections, the timing of both meals and injections must be similar from day to day to avoid periods of insulin deficiency or excess.

A person using intensive therapy must learn to accurately determine the amount of insulin to inject before each meal. The amount required depends on the premeal blood glucose level, the carbohydrate content of the meal, and the person's body weight and sensitivity to insulin. To determine insulin sensitivity, a person keeps careful records of food intake, insulin dosages, and blood glucose levels. Eventually, these records are analyzed by medical personnel to determine the appropriate **carbohydrate-to-insulin ratio** for that individual, which assists in calculating insulin doses at mealtime. Intensive therapy allows for substantial variation in food intake and lifestyle, but it requires frequent testing of blood glucose levels and a good understanding of carbohydrate counting.

After insulin therapy is initiated, persons with type 1 diabetes may experience a temporary remission of disease symptoms and a reduced need for insulin, known as the "honeymoon phase." The remission is due to a temporary improvement in pancreatic function and may last for several weeks or months.[43] It is important to anticipate this period of remission to avoid insulin excess. In all cases, diabetes eventually returns, and the patient must reinstate full insulin treatment.

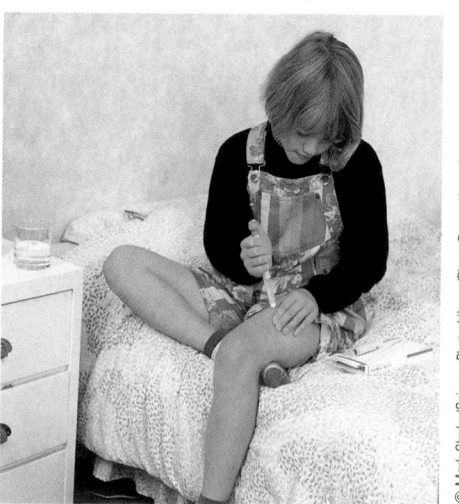

Children often become adept at administering the insulin they require.

Insulin Regimen for Type 2 Diabetes Approximately 27 percent of people diagnosed with type 2 diabetes are treated with insulin therapy.[44] Although initial treatment of type 2 diabetes may involve diet therapy, physical activity, and oral antidiabetic medications, long-term results with these treatments are often disappointing. As the disease progresses, pancreatic function worsens, and many individuals require insulin therapy to maintain glycemic control.

Many possible regimens can be used to control type 2 diabetes. Some persons may be treated with insulin alone, whereas others may use insulin in combination with other antidiabetic agents. Many patients need only one or two daily injections. Some regimens involve a mixture of rapid- and intermediate-acting insulin in the morning and an injection of intermediate- or long-acting insulin at dinner or before bedtime. In other cases, only a single injection of intermediate- or long-acting insulin may be needed at bedtime.[45] Doses and timing are adjusted according to the results of blood glucose self-monitoring.

Insulin Therapy and Hypoglycemia Hypoglycemia is the most common complication of insulin treatment, although it may also result from the use of some oral antidiabetic drugs. It most often results from intensive insulin therapy, because the attempt to attain near-normal blood glucose levels increases the risk of overtreatment.

Hypoglycemia can be corrected with the immediate intake of glucose or a carbohydrate-containing food. Usually, 15 to 20 grams of carbohydrate ◆ can relieve hypoglycemia in about 15 minutes, although patients should retest their blood glucose levels after 15 minutes in case additional treatment is necessary.[46] Foods that provide pure glucose yield a better response than foods that contain other sugars, such as sucrose or fructose. People using insulin are usually advised to carry glucose tablets or a source of carbohydrate that can be readily ingested. Those at risk of severe hypoglycemia are often given prescriptions for the hormone glucagon, which can be injected by caregivers in case of unconsciousness.

◆ Each of the following sources provides approximately 15 g of carbohydrate:
- Glucose tablets: 2 to 3 tablets
- Table sugar: 4 tsp
- Honey: 1 tbs
- Jelly beans: 15 small
- Grape juice, unsweetened: 1/2 c
- Orange juice, canned: 1/2 c

Fasting Hyperglycemia Insulin therapy sometimes must be adjusted to prevent fasting hyperglycemia, which has three possible causes. The usual cause is a waning of insulin action during the night due to insufficient insulin. A second possibility, known as the **dawn phenomenon,** occurs when blood glucose levels increase in the morning due to the early morning secretion of growth hormone, which counteracts insulin's actions. Less frequently, fasting hyperglycemia develops as a result of nighttime hypoglycemia, which causes hormonal responses that stimulate glucose production; the resulting condition is known as **rebound hyperglycemia.** Whatever the cause, fasting hyperglycemia can be treated by adjusting the dose or formulation of insulin administered in the evening.

carbohydrate-to-insulin ratio: the amount of carbohydrate that can be handled per unit of insulin. On average, every 15 grams of carbohydrate requires about 1 unit of rapid- or short-acting insulin.

Antidiabetic Drugs

Treatment of type 2 diabetes often requires the use of oral medications and injectable drugs other than insulin. These drugs can improve hyperglycemia by several modes of action: they can stimulate insulin secretion, suppress glucagon secretion, decrease insulin resistance, reduce glucose production in the liver, improve glucose utilization in tissues, delay stomach emptying, or delay carbohydrate absorption. Treatment may involve the use of a single medication (monotherapy) or a combination of several medications (combination therapy). By utilizing several mechanisms at once, combination therapy achieves more rapid and sustained glycemic control than is possible with monotherapy. Table 26-8 lists examples of antidiabetic drugs, and the Diet-Drug Interactions feature lists their nutrition-related effects. Because medications cannot replace the benefits offered by dietary modifications and physical activity, persons with diabetes should be advised to continue both.

Physical Activity and Diabetes Management

Regular physical activity can improve glycemic control considerably and is therefore a central feature of diabetes management. A regular exercise program improves insulin sensitivity, which can reduce insulin requirements. Physical activity also benefits other aspects of health, including blood lipid levels, blood pressure, body weight, and cardiovascular functioning. People with diabetes are encouraged to regularly perform both aerobic activity and resistance exercise unless contraindicated. They should undertake at least 150 minutes of moderate-intensity activity and/or 90 minutes of vigorous activity each week. In addition, they should participate in a resistance exercise program that targets all major muscle groups three times weekly.[47]

Physical Activity and Insulin Therapy People who do not have diabetes maintain blood glucose levels during physical activity because their normal hormonal responses—a fall in insulin levels and increased secretion of glucagon and epinephrine—promote glucose production in the liver. In people who use insulin, the natural hormonal balance is upset: blood glucose levels drop during activity because injected insulin promotes rapid consumption of glucose by exercising muscles and also blocks glucose synthesis by the liver. For this reason, insulin should not be injected immediately before exercise, because it can lead to hypoglycemia. Conversely, a complete lack of insulin contributes to hyperglycemia, because liver glucose production is unchecked.

TABLE 26-8 Antidiabetic Drugs

Drug Category	Common Examples	Mode of Action
Alpha-glucosidase inhibitors	Acarbose, miglitol	Delay carbohydrate absorption
Amylin analogs (injected)	Pramlintide	Suppress glucagon secretion, delay stomach emptying, suppress appetite
Biguanides	Metformin	Inhibit liver glucose production, improve glucose utilization
D-phenylalanine derivatives	Nateglinide	Stimulate insulin secretion by the pancreas
DPP-4 inhibitors	Sitagliptin	Improve insulin secretion, suppress glucagon secretion, delay stomach emptying
Incretin mimetics (injected)	Exenatide	Improve insulin secretion, suppress glucagon secretion, delay stomach emptying
Meglitinides	Repaglinide	Stimulate insulin secretion by the pancreas
Sulfonylureas	Chlorpropamide, glipizide, glyburide, tolbutamide	Stimulate insulin secretion by the pancreas
Thiazolidinediones	Pioglitazone, rosiglitazone	Decrease insulin resistance

DIET-DRUG INTERACTIONS

Check this table for notable nutrition-related effects of the medications discussed in this chapter.

	Gastrointestinal Effects	Interactions with Dietary Substances	Metabolic Effects
Sulfonylureas	Nausea, vomiting, cramps, diarrhea.	Avoid using with alcohol due to a toxic reaction that causes flushing, throbbing head and neck pain, shortness of breath, palpitations, and sweating. Avoid using with dietary supplements that contain ginseng, garlic, fenugreek, coriander, celery, as they may increase risk of hypoglycemia.	Hypoglycemia, weight gain, allergic skin reactions.
Biguanides (metformin)	Abdominal pain, nausea, vomiting, gas, cramps, diarrhea, metallic taste, anorexia.	—	Asymptomatic vitamin B_{12} deficiency.
Thiazolidinediones	—	—	Weight gain, fluid retention, edema, anemia.
Alpha-glucosidase inhibitors	Abdominal pain, nausea, gas, cramps, diarrhea.	—	Elevated liver enzymes, hyperbilirubinemia.

People with type 1 diabetes must carefully adjust food intake and insulin therapy to prevent hypoglycemia during physical activity. They should check their blood glucose levels both before and after an activity. Insulin doses that precede exercise often need to be reduced substantially. If blood glucose is below 100 mg/dL, the individual should consume carbohydrate before beginning the activity. Additional carbohydrate may be needed during or after prolonged activity or even several hours after the activity is completed. The individual should avoid strenuous exercise if fasting blood glucose levels are 250 mg/dL or higher and should avoid all types of physical activity when blood glucose levels are 300 mg/dL or higher or ketosis is present.[48] Predicting the precise adjustments in diet and insulin becomes easier as a person gains experience in maintaining glycemic control.

Physical Activity in Type 2 Diabetes Regular physical activity can improve the metabolic outcomes associated with type 2 diabetes, including problems with blood glucose, blood lipids, and blood pressure. People with type 2 diabetes are often overweight and sedentary, however, and many develop complications during the course of disease. Before an exercise program is planned, a medical evaluation should screen for problems that may be aggravated by certain activities. Complications involving the heart and blood vessels, eyes, kidneys, feet, and nervous system may limit the types of activity recommended.

Only mild or moderate exercise may be prescribed at first. For obese, inactive persons, a short walk at a comfortable pace may be the first activity suggested. Persons with retinopathy should avoid heavy lifting or straining, which may raise blood pressure and damage eye tissue. Persons with nephropathy often have reduced capacity for physical activity, and strenuous exercise is discouraged. Peripheral neuropathy precludes repetitive weight-bearing exercises such as jogging and step exercises, because these activities may lead to foot ulcerations. The use of protective foot gear, such as gel soles or air midsoles and socks that prevent blisters, can help to prevent foot trauma. To prevent dehydration,

STUDY QUESTIONS

These questions will help you review the chapter. You will find the answers in the discussions on the pages provided.

1. Describe the symptoms that develop as a consequence of hyperglycemia. How is diabetes diagnosed? (pp. 811–813)

2. Compare the features of the two main types of diabetes. Describe gestational diabetes. (pp. 813–814)

3. Discuss the acute complications that may arise in uncontrolled diabetes. (pp. 815–817)

4. Describe the macrovascular and microvascular complications that develop from prolonged exposure to high blood glucose concentrations. Discuss the problems associated with diabetic neuropathy. (pp. 817–818)

5. What are the goals of medical and nutrition therapy for people with diabetes? Explain how diabetes treatment is evaluated. (pp. 818–820)

6. Discuss the dietary recommendations for people with diabetes. Describe the meal-planning strategies they can use to control carbohydrate intakes. (pp. 821–825)

7. Describe the insulin regimens for type 1 and type 2 diabetes. Explain how insulin therapy is coordinated with food intake and physical activity. (pp. 826–829)

8. Discuss the modes of action of the various antidiabetic drugs. (p. 828)

9. What are the risks of poorly controlled diabetes during pregnancy? Describe the general recommendations for diabetic women who become pregnant. Describe the dietary adjustments that may be necessary for women with gestational diabetes. (pp. 830–832)

These multiple choice questions will help you review for an exam. Answers can be found on p. 835.

1. Which of the following is characteristic of type 1 diabetes?
   a. Abdominal obesity increases risk.
   b. The pancreas makes little or no insulin.
   c. It is the predominant form of diabetes.
   d. It often arises during pregnancy.

2. Which of the following describes type 2 diabetes?
   a. It is usually an autoimmune disease.
   b. The pancreas makes little or no insulin.
   c. Diabetic ketoacidosis is a common complication.
   d. Chronic complications may develop before it is diagnosed.

3. The chronic complications associated with all types of diabetes result from:
   a. altered kidney function.
   b. infections that deplete nutrient reserves.
   c. weight gain and hypertension.
   d. damage to blood vessels and nerves.

4. Long-term glycemic control is usually evaluated by:
   a. self-monitoring of blood glucose.
   b. testing urinary ketone levels.
   c. measuring glycated hemoglobin.
   d. testing urinary protein levels (microalbuminuria).

5. Regarding dietary carbohydrate, a patient with diabetes should be most concerned about:
   a. consuming the correct quantity of carbohydrate at each meal or snack.
   b. consuming the correct proportion of sugars, starches, and fiber in meals.
   c. avoiding added sugars and kcaloric sweeteners.
   d. choosing meals with ideal proportions of protein, carbohydrate, and fat.

6. Which of the following is true regarding the general use of alcohol in diabetes?
   a. A serving of alcohol is considered part of the carbohydrate allowance.
   b. Alcohol contributes to hyperglycemia and should be avoided completely.
   c. Alcohol can cause hypoglycemia and should therefore be consumed with food if patients use insulin or medications that stimulate insulin secretion.
   d. Patients can use alcohol in unlimited quantities unless they are pregnant.

7. The meal-planning strategy best suited to all people with diabetes is:
   a. carbohydrate counting.
   b. the exchange list system.
   c. following menus and recipes provided by a registered dietitian.
   d. any approach that best helps the patient control blood glucose levels.

8. A patient using intensive insulin therapy is likely to follow a regimen that involves:
   a. twice-daily injections that combine short-, intermediate-, and long-acting insulin in each injection.
   b. a mixture of intermediate- and long-acting insulin injected between meals.
   c. multiple daily injections that supply basal insulin and precise insulin doses at each meal.
   d. the use of both insulin and oral antidiabetic agents.

9. In a person who has previously maintained good glycemic control, hyperglycemia can be precipitated by:
   a. infections or illnesses.
   b. chronic alcohol ingestion.
   c. undertreatment of hypoglycemia.
   d. prolonged exercise.

10. Women with pregnancies complicated by diabetes:
    a. generally benefit from larger meals and a snack at bedtime.
    b. often need less carbohydrate at breakfast.
    c. need more carbohydrate than women with diabetes who are not pregnant.
    d. need more kcalories to support the pregnancy than women without diabetes.

REFERENCES

1. National Center for Health Statistics, *Health, United States, 2007—With Chartbook on Trends in the Health of Americans* (Hyattsville, Md.: 2007), p. 248.
2. National Center for Health Statistics, 2007.
3. American Diabetes Association, Diagnosis and classification of diabetes mellitus, *Diabetes Care* 30 (2007): S42–S47.
4. W. Rosamond and coauthors, Heart disease and stroke statistics—2007 update: A report from the American Heart Association Statistics Committee and Stroke Statistics Subcommittee, *Circulation* 115 (2007): e69–e171.
5. S. E. Inzucchi and R. S. Sherwin, Type 1 diabetes mellitus, in L. Goldman and D. Ausiello, eds., *Cecil Medicine* (Philadelphia: Saunders, 2008), pp. 1727–1747.
6. *Diabetes Mellitus: A Guide to Patient Care* (Philadelphia: Lippincott Williams & Wilkins, 2007), p. 12.
7. National Center for Health Statistics, 2007.
8. F. R. Kaufman, Diabetes management in children and adolescents, in A. Peters Harmel and R. Mathur, eds., *Davidson's Diabetes Mellitus: Diagnosis and Treatment* (Philadelphia: Saunders, 2004), pp. 299–321.
9. Kaufman, 2004; P. A. Tartaranni and C. Bogardus, Obesity and diabetes mellitus, in D. Porte, Jr., R. S. Sherwin, and A. Baron, eds., *Ellenberg and Rifkin's Diabetes Mellitus* (New York: McGraw-Hill, 2003), pp. 401–413.
10. *Diabetes Mellitus: A Guide to Patient Care*, 2007, pp. 37–49; J. Wylie-Rosett and L. M. Delahanty, Diabetes prevention, in T. A. Ross, J. L. Boucher, and B. S. O'Connell, eds., *American Dietetic Association Guide to Diabetes: Medical Nutrition Therapy and Education* (Chicago: American Dietetic Association, 2005), pp. 49–58.
11. *Diabetes Mellitus: A Guide to Patient Care*, 2007, pp. 37–49; Wylie-Rosett and Delahanty, 2005.
12. J. W. Beulens and coauthors, Effect of moderate alcohol consumption on adipokines and insulin sensitivity in lean and overweight men: A diet intervention study, *European Journal of Clinical Nutrition* (2007): epub 6 June 2007 (DOI 10.1038/sj.ejcn .1602821); J. W. Beulens and coauthors, Alcohol consumption and risk of type 2 diabetes among older women, *Diabetes Care* 28 (2005): 2933–2938.
13. Inzucchi and Sherwin, Type 1 diabetes mellitus, 2008.
14. U. Masharani and M. S. German, Pancreatic hormones and diabetes mellitus, in D. G. Gardner and D. Shoback, eds., *Greenspan's Basic and Clinical Endocrinology* (New York: McGraw-Hill/Lange, 2007), pp. 661–747.
15. Masharani and German, 2007.
16. Inzucchi and Sherwin, Type 1 diabetes mellitus, 2008.
17. S. E. Inzucchi and R. S. Sherwin, Type 2 diabetes mellitus, in L. Goldman and D. Ausiello, eds., *Cecil Medicine* (Philadelphia: Saunders, 2008), pp. 1748–1760.

18. L. H. Young and D. A. Chyun, Heart disease in patients with diabetes, in D. Porte, Jr., R. S. Sherwin, and A. Baron, eds., *Ellenberg and Rifkin's Diabetes Mellitus* (New York: McGraw-Hill, 2003), pp. 823–844.
19. R. G. Frykberg, Diabetic foot ulcers: Pathogenesis and management, *American Family Physician* 66 (2002): 1655–1662.
20. Inzucchi and Sherwin, Type 1 diabetes mellitus, 2008.
21. Masharani and German, 2007.
22. Inzucchi and Sherwin, Type 1 diabetes mellitus, 2008.
23. Inzucchi and Sherwin, Type 1 diabetes mellitus, 2008.
24. Diabetes Control and Complications Trial Research Group, The effect of intensive treatment of diabetes on the development and progression of long-term complications in insulin-dependent diabetes mellitus, *New England Journal of Medicine* 329 (1993): 977–986.
25. American Diabetes Association, Implications of the United Kingdom Prospective Diabetes Study, *Diabetes Care* 21 (1998): 2180–2184.
26. American Diabetes Association, Standards of medical care in diabetes—2007, *Diabetes Care* 30 (2007): S4–S41.
27. Masharani and German, 2007.
28. American Diabetes Association, Standards of medical care in diabetes, 2007.
29. A. Fischl, Monitoring, in T. A. Ross, J. L. Boucher, and B. S. O'Connell, eds., *American Dietetic Association Guide to Diabetes: Medical Nutrition Therapy and Education* (Chicago: American Dietetic Association, 2005), pp. 106–115.
30. M. Ryan and coauthors, Is a failure to recognize an increase in food intake a key to understanding insulin-induced weight gain? *Diabetes Care* epub 17 December 2007 (DOI 10.2337/dc07-1171); A. N. Jacob and coauthors, Potential causes of weight gain in type 1 diabetes mellitus, *Diabetes, Obesity and Metabolism* 8 (2006): 404–411.
31. A. Daly, Use of insulin and weight gain: Optimizing diabetes nutrition therapy, *Journal of the American Dietetic Association* 107 (2007): 1386–1393.
32. B. S. Baliga, Z. Bloomgarden, and C. Nonas, Medical nutrition therapy for patients with type-2 diabetes, in J. I. Mechanick and E. M. Brett, eds., *Nutritional Strategies for the Diabetic and Prediabetic Patient* (Boca Raton, Fla.: CRC Press, 2006), pp. 81–103; D. D. Hensrud, Dietary treatment and long-term weight loss and maintenance in type 2 diabetes, *Obesity Research* 9 (2001): 348S–353S.
33. Hensrud, 2001.
34. American Diabetes Association, Nutrition recommendations and interventions for diabetes, *Diabetes Care* 30 (2007): S48–S65.
35. American Diabetes Association, Nutrition recommendations and interventions for diabetes, *Diabetes Care* 31 (2008): S61–S78; T. M. S. Wolever and coauthors, The Canadian Trial of Carbohydrates in Diabetes

(CCD), a 1-y controlled trial of low-glycemic-index dietary carbohydrate in type 2 diabetes: No effect on glycated hemoglobin but reduction in C-reactive protein, *American Journal of Clinical Nutrition* 87 (2008): 114–125.
36. American Diabetes Association, Nutrition recommendations and interventions for diabetes, 2007.
37. American Diabetes Association, Nutrition recommendations and interventions for diabetes, 2007.
38. American Diabetes Association, Nutrition recommendations and interventions for diabetes, 2007.
39. American Diabetes Association, Nutrition recommendations and interventions for diabetes, 2007.
40. American Diabetes Association, Nutrition recommendations and interventions for diabetes, 2007.
41. American Diabetes Association, Nutrition recommendations and interventions for diabetes, 2007.
42. M. S. Nolte and J. H. Karam, Pancreatic hormones and antidiabetic drugs, in B. G. Katzung, ed., *Basic and Clinical Pharmacology* (New York: McGraw-Hill/Lange, 2007), pp. 683–705; Masharani and German, 2007.
43. Inzucchi and Sherwin, Type 1 diabetes mellitus, 2008.
44. C. E. Koro and coauthors, Glycemic control from 1988 to 2000 among U.S. adults diagnosed with type 2 diabetes, *Diabetes Care* 27 (2004): 17–20.
45. Inzucchi and Sherwin, Type 2 diabetes mellitus, 2008.
46. American Diabetes Association, Standards of medical care in diabetes, 2007.
47. American Diabetes Association, Standards of medical care in diabetes, 2007.
48. Inzucchi and Sherwin, Type 1 diabetes mellitus, 2008; *Diabetes Mellitus: A Guide to Patient Care*, 2007, p. 74.
49. American Diabetes Association, Physical activity/exercise and diabetes, *Diabetes Care* 27 (2004): S58–S62.
50. *Diabetes Mellitus: A Guide to Patient Care*, 2007, pp. 248–249.
51. Inzucchi and Sherwin, Type 2 diabetes mellitus, 2008; American Diabetes Association, Gestational diabetes mellitus, *Diabetes Care* 27 (2004): S88–S90.
52. American Diabetes Association, Preconception care of women with diabetes, *Diabetes Care* 27 (2004): S76–S78.
53. Z. Hussain and L. Jovanovic, Nutritional strategies in pregestational, gestational, and postpartum diabetic patients, in J. I. Mechanick and E. M. Brett, eds., *Nutritional Strategies for the Diabetic and Prediabetic Patient* (Boca Raton, Fla.: CRC Press, 2006), pp. 133–148.
54. O. Langer, Oral antidiabetic drugs in pregnancy: The other alternative, *Diabetes Spectrum* 20 (2007): 101–105.

ANSWERS

Study Questions (multiple choice)

1. b 2. d 3. d 4. c 5. a 6. c 7. d 8. c 9. a 10. b

The Metabolic Syndrome

Chapter 26 described how insulin resistance—a reduced sensitivity to insulin in muscle, adipose, and liver cells—can contribute to hyperglycemia and hyperinsulinemia and, eventually, to type 2 diabetes. Insulin resistance is also a central feature of several other conditions, including the **metabolic syndrome,** a condition that raises the risk of developing cardiovascular diseases (CVD) and type 2 diabetes. The metabolic syndrome is a cluster of at least three of the following: hyperglycemia, obesity, hypertriglyceridemia (elevated blood triglycerides), reduced HDL cholesterol levels, and hypertension (high blood pressure). This highlight describes how the metabolic syndrome is diagnosed, how and why it might develop, its consequences, and current treatment approaches. The accompanying glossary defines the relevant terms.

Prevalence of the Metabolic Syndrome

Table H26-1 lists the laboratory values used to identify the metabolic syndrome, which currently affects an estimated 29 percent of the adult population in the United States.[1] As Figure H26-1 shows, the prevalence of the metabolic syndrome increases with age. Risk also varies among ethnic groups: Hispanic Americans have the highest incidence in the United States, with an overall prevalence of 36 percent.[2] Although the precise cause of the metabolic syndrome is not known, the close relationship between abdominal obesity and insulin resistance suggests that the current obesity crisis in the United States may be partly responsible for the high prevalence of the condition.

Obesity and the Metabolic Syndrome

Excessive abdominal fat induces a number of metabolic changes that lead to insulin resistance, which then leads to hyperglycemia and other abnormalities. The following sections explore several of these relationships.

Effects of Obesity on Insulin Action

Adipose cells that reside in the abdominal region are more metabolically active than adipose cells elsewhere.[3] Within these cells, triglycerides break down more rapidly, increasing fatty acid levels in the blood. The higher fatty acid concentrations inhibit the actions of insulin receptors, the proteins that recognize and bind insulin at cell surfaces.[4] Unless the pancreas can secrete enough

TABLE H26-1	Features of the Metabolic Syndrome

Metabolic syndrome is diagnosed when a person has three or more of the following symptoms.

Symptom	Diagnostic Criteria
Hyperglycemia	Fasting plasma glucose ≥100 mg/dL
Abdominal obesity	Waist circumference >40" in men, >35" in women
Hypertriglyceridemia	≥150 mg/dL
Reduced HDL cholesterol	<40 mg/dL in men, <50 mg/dL in women
Hypertension	≥130/85 mm Hg

FIGURE H26-1 Prevalence of Metabolic Syndrome in the U.S. Population

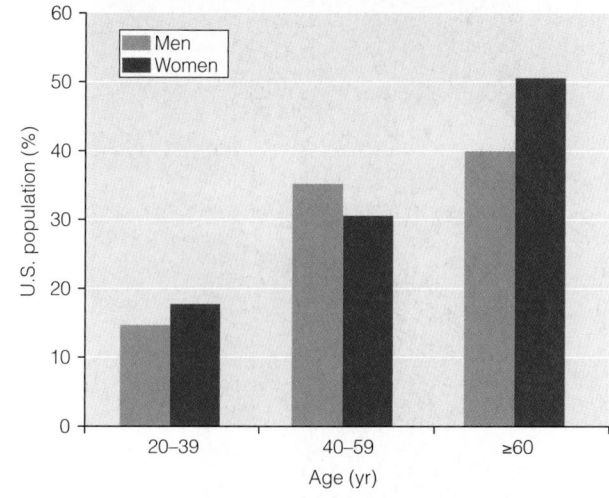

Metabolic Syndrome and CVD Risk

As mentioned, individuals with the metabolic syndrome are at increased risk of developing cardiovascular diseases. The disorders that characterize the metabolic syndrome—obesity, lipid abnormalities, and hypertension—are all independent risk factors for CVD. In addition, both insulin resistance and elevated lipoprotein levels can cause damage to blood vessels, promoting inflammation and accelerating the progression of atherosclerosis.[9] Blood vessel inflammation induces liver secretion of **fibrinogen,** a protein that promotes blood clot formation (recall that blood clotting in arteries can lead to a heart attack or stroke). C-reactive protein, which is elevated by both inflammation and obesity, inhibits **nitric oxide** production by blood vessel cells, an effect that impairs blood vessel activity and also promotes blood clotting.[10] Another procoagulant factor—**plasminogen activator inhibitor-1**—is overproduced as a consequence of both obesity and hyperinsulinemia. Thus, the combined effect of these multiple abnormalities can worsen atherosclerosis and increase the risks of developing heart attack and stroke.

Treatment of the Metabolic Syndrome

The metabolic syndrome is primarily treated with dietary and lifestyle changes, with the goal of correcting abnormalities that increase CVD risk.[11] In most individuals, a combination of weight loss and physical activity can improve insulin resistance, blood pressure, and blood lipid levels. Additional dietary strategies depend on a patient's specific symptoms. If dietary and lifestyle changes are not successful, medications may be prescribed. Because effective treatment requires lifelong commitment, health care providers should work with patients to develop a treatment plan that they are willing to adopt.

Dietary Management

Weight reduction is often recommended for obese individuals, and even a small weight loss (10 to 20 pounds) can improve symptoms. Many people find it difficult to achieve and maintain weight loss, however, and should be encouraged to make other dietary changes that can improve their health. In individuals with hypertriglyceridemia, the general recommendation is to reduce intake of added sugars and refined grain products (soda, juices, white bread, sweetened cereal, and desserts) and increase servings of whole grains and foods high in fiber (whole wheat bread, oatmeal, legumes, fruits, and vegetables).[12] In some people, carbohydrate restriction may help to reduce blood triglyceride levels and improve hyperglycemia.[13] Including fish in the diet each week may also improve triglyceride levels. Individuals with hypertension are encouraged to reduce sodium intake and increase consumption of fruits and vegetables and low-fat milk products.

insulin to compensate, glucose uptake from the blood is reduced, contributing to hyperglycemia.

Obesity can also alter production of the hormones and proteins made in adipose cells.[5] For example, obesity causes a reduced secretion of **adiponectin,** a hormone that improves insulin sensitivity. Conversely, **resistin,** a hormone that contributes to insulin resistance, is released in greater amounts. Enlarged adipose cells also boost their production of certain cytokines that induce the synthesis of liver proteins that promote inflammation and blood coagulation.[6] People who are obese often have elevated levels of C-reactive protein, a marker of inflammation linked to an increased risk of CVD.

Obesity and Hypertension

Obesity increases the risk of developing high blood pressure, a common component of the metabolic syndrome. Both insulin resistance and hyperinsulinemia may be implicated in raising blood pressure.[7] Insulin resistance interferes with the normal relaxation and dilation of blood vessels. Hyperinsulinemia promotes reabsorption of sodium by the kidneys, resulting in fluid retention and increased blood volume. These effects contribute to an increase in blood pressure.

Obesity and Hypertriglyceridemia

Abdominal obesity is frequently associated with blood lipid abnormalities.[8] Increases in body weight are linked with higher triglyceride and LDL cholesterol levels and lower HDL cholesterol levels. As a result of obesity, adipose cells are less responsive to insulin and release more fatty acids into the bloodstream. At the same time, they are less able to extract and store triglycerides from chylomicrons and VLDL. To keep up with the greater influx of fatty acids, the liver must accelerate its production of VLDL, and hypertriglyceridemia develops.

A diet low in saturated fat, *trans* fats, and cholesterol can help to reduce LDL cholesterol levels. Chapter 27 includes additional information about dietary modifications that can reduce CVD risk.

Physical Activity

Regular physical activity helps with weight management and may also improve blood lipid concentrations, hypertension, and insulin resistance—all changes that can reduce the risk of developing CVD. A regular exercise program can also prevent or delay the onset of diabetes in persons at risk.[14] A program that includes both aerobic exercise and strength training is best. A minimum of 30 minutes of moderate aerobic activity (brisk walking, jogging, or cycling) daily is suggested, although longer periods (1 hour daily) are recommended for weight control. A sedentary lifestyle can worsen the progression of metabolic syndrome and should be discouraged.

Drug Therapy

If dietary and lifestyle changes are unsuccessful, medications may be prescribed to correct hypertriglyceridemia and hypertension (Chapter 27 provides details). Insulin resistance is not routinely treated with drug therapy in nondiabetic patients due to insufficient evidence that the medications can benefit individuals with the metabolic syndrome.[15]

As explained in this highlight, the metabolic syndrome consists of a cluster of related disorders that increase the risk for developing CVD. Whereas the common features of the metabolic syndrome are independent risk factors for CVD, in combination they may raise risk twofold to threefold. Treatment of the metabolic syndrome emphasizes dietary and lifestyle changes. The following chapter provides additional information about the dietary and lifestyle changes that can reduce CVD risk.

© Rolf Bruderer/Corbis

Regular exercise can reduce the risks of developing the metabolic syndrome, cardiovascular diseases, and type 2 diabetes.

REFERENCES

1. E. S. Ford, W. H. Giles, and A. H. Mokdad, Increasing prevalence of the metabolic syndrome among U.S. adults, *Diabetes Care* 27 (2004): 2444–2449.
2. Z. T. Bloomgarden, American Association of Clinical Endocrinologists (AACE) consensus conference on the insulin resistance syndrome, *Diabetes Care* 26 (2003): 1297–1303.
3. R. F. Kushner and J. L. Roth, Nutritional strategies for patients with obesity and the metabolic syndrome, in J. I. Mechanick and E. M. Brett, eds., *Nutritional Strategies for the Diabetic and Prediabetic Patient* (Boca Raton, Fla.: CRC Press, 2006), pp. 55–80; D. E. Moller and K. D. Kaufman, Metabolic syndrome: A clinical and molecular perspective, *Annual Review of Medicine* 56 (2005): 45–62.
4. G. A. Bray and C. M. Champagne, Obesity and the metabolic syndrome: Implications for dietetics practitioners, *Journal of the American Dietetic Association* 104 (2004): 86–89.
5. M. D. Jensen, Obesity, in L. Goldman and D. Ausiello, eds., *Cecil Medicine* (Philadelphia: Saunders, 2008), pp. 1643–1652; Kushner and Roth, 2006.
6. Bray and Champagne, 2004; S. M. Grundy, Inflammation, hypertension, and the metabolic syndrome, *Journal of the American Medical Association* 290 (2003): 3000–3002.
7. Grundy, 2003.
8. Jensen, 2008.
9. Moller and Kaufman, 2005.
10. S. M. Grundy and coauthors, Clinical management of metabolic syndrome: Report of the American Heart Association/National Heart, Lung, and Blood Institute/American Diabetes Association Conference on Scientific Issues Related to Management, *Circulation* 109 (2004): 551–556.
11. Kushner and Roth, 2006; D. Deen, Metabolic syndrome: Time for action, *American Family Physician* 69 (2004): 2875–2882.
12. Grundy and coauthors, 2004.
13. Kushner and Roth, 2006.
14. *Diabetes Mellitus: A Guide to Patient Care* (Philadelphia: Lippincott Williams & Wilkins, 2007), pp. 37–49.
15. Grundy and coauthors, 2004.

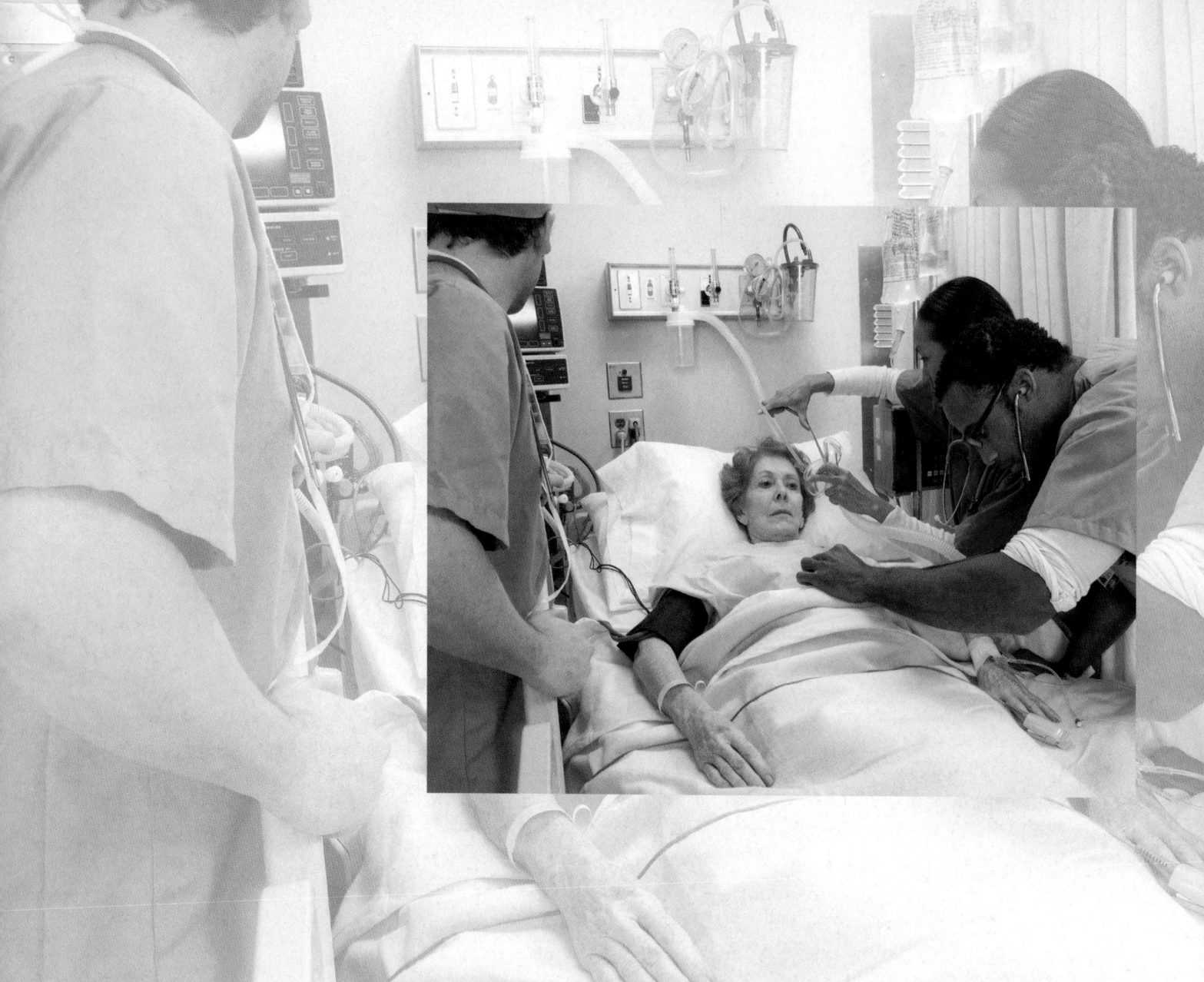

Nutrition in the Clinical Setting

Each heartbeat sends oxygen-rich blood to the body's tissues. When the functions of the heart and blood vessels are disturbed, as is common in cardiovascular diseases, the disrupted blood supply hinders the ability of cells to carry out their metabolic functions. At first, people with cardiovascular disease may not realize that their weakness, fatigue, or shortness of breath are symptoms of a cardiovascular condition. When their condition worsens, however, the complications can be disabling and interfere with many aspects of daily life.

Cardiovascular Diseases

Cardiovascular disease (CVD) is a general term describing diseases of the heart and blood vessels. ◆ **Coronary heart disease (CHD),** the most common form of CVD, is caused by atherosclerosis in the coronary arteries that supply blood to the heart muscle. If atherosclerosis interferes with blood flow in these arteries, the resulting deprivation of oxygen and nutrients can destroy heart tissue and cause a **myocardial infarction (MI)**—a **heart attack**. When the blood supply to brain tissue is blocked, a **stroke** occurs. Both heart attack and stroke may result in dis-ablement or death. This chapter describes these and other cardiovascular disorders. Figure 27-1 on p. 842 shows the percentages of deaths resulting from all types of CVD. The glossary on p. 842 defines some common terms related to CVD.

Cardiovascular disease is responsible for approximately 36 percent of deaths in the United States, claiming more lives than the next four lead-ing causes of death combined.[1] Although many people assume that heart conditions are men's diseases, more women than men die each year from the various types of CVD. Furthermore, CVD is a global health issue; it is the leading cause of death in Europe and contributes to over 29 percent of deaths worldwide.[2]

◆ Cardiovascular disease was introduced in Chapter 5 (see p. 157) and discussed further in Highlight 15.

Atherosclerosis

In atherosclerosis, ◆ sometimes called "hardening of the arteries," the artery walls become progressively thickened due to an accumulation of fatty deposits, smooth muscle cells, and fibrous connective tissue, collectively known as **plaque.** Plaque can exist in a stable form that does not cause complications or an unstable form called **vulnerable plaque.** Vulnerable plaque has only a thin, fibrous barrier be-

◆ Atherosclerosis is the most common form of *arteriosclerosis,* a more general term for arte-rial diseases that are characterized by abnor-mally thickened walls and lost elasticity.

FIGURE 27-1 Percentage Breakdown of Deaths from Cardiovascular Diseases in the United States, 2004

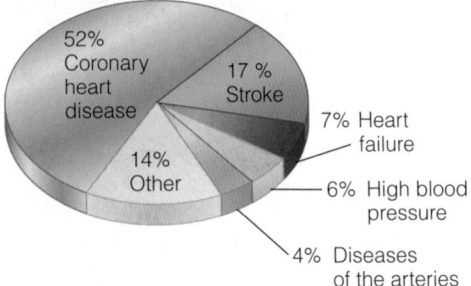

SOURCE: W. Rosamond and coauthors, Heart disease and stroke statistics—2008 update: A report from the American Heart Association Statistics Committee and Stroke Statistics Subcommittee, *Circulation* 117 (2008): epub 17 Dec 2007 (DOI 10.1161/CIRCULATIONAHA.107.187998); site visited January 2, 2008.

tween its lipid-rich core and the arterial lumen.[3] It is highly susceptible to rupture, which then promotes blood clot formation (**thrombosis**) within the artery.

Consequences of Atherosclerosis

As atherosclerosis worsens, it can eventually narrow the lumen of an artery and interfere with blood flow. If the plaque ruptures and results in thrombosis, the blood clot (**thrombus**) may enlarge in time and ultimately obstruct blood flow. A portion of a clot can also break free (**embolus**) and travel through the circulatory system until it lodges in a narrowed artery and shuts off blood flow to the surrounding tissue (**embolism**). Most complications of atherosclerosis result from the deficiency of blood and oxygen within the tissue served by an artery (**ischemia**).

Atherosclerosis can affect almost any organ or tissue in the body, and accordingly is a major cause of disablement or death. Obstructed blood flow in the coronary arteries can cause pain or discomfort in the chest and surrounding regions (**angina pectoris**) or lead to a heart attack. Obstructed blood flow to the brain can cause injury or destruction to brain tissue, or a stroke. Inadequate blood circulation in the legs can cause fatigue and pain while walking, known as **intermittent claudication.** Blockage of the arteries that supply the kidneys can result in kidney disease or even acute kidney failure.

Atherosclerosis is the most common cause of an **aneurysm**—an abnormal bulge in a weakened blood vessel. Plaque can weaken the blood vessel wall, and eventually the pressure of blood flow can cause the weakened region to stretch and balloon outward. Aneurysms can rupture and lead to massive bleeding and death, particularly when a large vessel such as the aorta is affected. In the arteries of the brain, an aneurysm may lead to bleeding within the brain, a coma, or a stroke.

Development of Atherosclerosis

Atherosclerosis begins to develop as early as childhood or adolescence and usually progresses over several decades before symptoms develop. It initially arises in response to minimal but chronic injuries that damage the inner arterial wall. The first lesions tend to develop in regions where the arteries branch or bend because

GLOSSARY OF TERMS RELATED TO CARDIOVASCULAR DISEASE

aneurysm (AN-you-rih-zum): an abnormal enlargement or bulging of a blood vessel (usually an artery) caused by damage to or weakness in the blood vessel wall.

angina (an-JYE-nah or AN-ji-nah) **pectoris:** a condition caused by ischemia in the heart muscle that results in discomfort or dull pain in the chest region. The pain often radiates to the left shoulder and arm or to the back, neck, and lower jaw.

cardiovascular disease (CVD): a general term describing diseases of the heart and blood vessels.

• **cardio** = heart

• **vascular** = blood vessels

coronary heart disease (CHD): a chronic, progressive disease characterized by obstructed blood flow in the coronary arteries; also called *coronary artery disease*.

embolism (EM-boh-lizm): the obstruction of a blood vessel by an embolus, causing sudden tissue death.

• **embol** = to insert, plug

embolus (EM-boh-lus): an abnormal particle, such as a blood clot or air bubble, that travels in the blood.

intermittent claudication (claw-dih-KAY-shun): severe pain and weakness in the legs (especially the calves) that is caused by inadequate blood supply to the muscles; it usually occurs with walking and subsides during rest.

ischemia (iss-KEE-mee-a): inadequate blood supply within tissues due to obstructed blood flow in the arteries.

myocardial (MY-oh-CAR-dee-al) **infarction** (in-FARK-shun), or **MI:** death of heart muscle caused by a sudden reduction in coronary blood flow; also called a *heart attack* or *cardiac arrest*.

• **myo** = muscle

• **cardial** = heart

• **infarct** = tissue death

plaque (PLACK): an accumulation of fatty deposits, smooth muscle cells, and fibrous connective tissue in blood vessels.

stroke: a sudden injury to brain tissue resulting from impaired blood flow through an artery that supplies blood to the brain; also called a *cerebrovascular accident*.

• **cerebro** = brain

thrombosis (throm-BOH-sis): the formation or presence of a blood clot in blood vessels. A *coronary thrombosis* occurs in a coronary artery, and a *cerebral thrombosis* occurs in an artery that supplies blood to the brain.

• **thrombo** = clot

thrombus: a blood clot formed within a blood vessel that remains attached to its place of origin.

vulnerable plaque: a form of plaque, susceptible to rupture, that is lipid-rich and has only a thin, fibrous barrier between the arterial lumen and the plaque's lipid core.

the blood flow is disturbed in those areas (see Figure H15-1 in Highlight 15, p. 555). The damage to the artery attracts monocytes, ◆ T cells, and platelets to the region. The monocytes slip under the artery's thin layer of **endothelial cells** and engulf LDL (low-density lipoprotein) cholesterol, ◆ becoming **foam cells;** these fat-laden foam cells collect along the artery wall, creating regions known as *fatty streaks* (see Figure 27-2). Next, the smooth muscle cells in the arterial tissue begin to divide, ingest LDL particles, and produce fibrous connective tissue. The thickening plaque accumulates calcium, and the cholesterol within the lipid core may crystallize and harden. The processes involved in atherosclerosis occur in response to cytokines and other signaling molecules produced by the white blood cells and endothelial cells.[4]

As atherosclerosis progresses, the artery may expand outward to accommodate the plaque, such that a decrease in the lumen diameter does not occur.[5] In other cases, the accumulating plaque causes narrowing, rather than expansion, of the artery. Although arteries that expand are less likely to interfere with blood flow, they usually are associated with the unstable vulnerable plaque, which is more likely to rupture, induce clotting, and increase the risk of heart attack or stroke. The arteries that accommodate plaque only by narrowing may impede blood flow, but they generally have a more stable plaque structure, with a lower lipid content and a thicker barrier between the plaque and arterial lumen. The discovery of variations in plaque anatomy may help to explain why some CVD treatments can dramatically reduce the risk of heart attack and stroke even when plaque volume and the lumen diameter do not change.[6]

Causes of Atherosclerosis

The reasons why atherosclerosis develops and progresses are complex. Generally, the factors that initiate atherosclerosis either cause direct damage to the artery

◆ Reminder: *Monocytes* are phagocytic white blood cells that circulate in the blood. Once they enter tissues, they become macrophages (see Highlight 17).

◆ Reminder: LDL (low-density lipoproteins) transport cholesterol in the blood.

endothelial cells: the type of cells that line the blood vessels, lymphatic vessels, and body cavities.

foam cells: swollen cells in the artery wall that accumulate lipids.

FIGURE 27-2 Stages of Plaque Progression

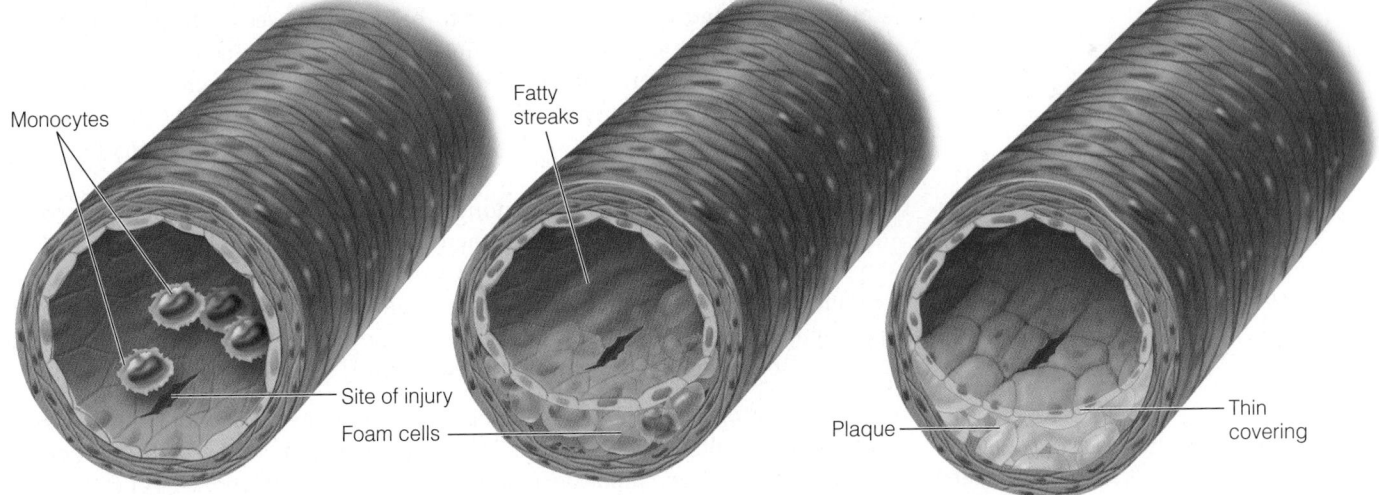

Monocytes—phagocytic white blood cells—circulate in the bloodstream and respond to injury on the artery wall.

Monocytes slip under blood vessel cells and engulf LDL cholesterol, becoming foam cells. The thin layers of foam cells that develop on artery walls are known as *fatty streaks*.

A fatty streak thickens and forms plaque as it accumulates additional lipids, smooth muscle cells, connective tissue, and cellular debris.

The artery may expand to accommodate plaque. When this occurs, the plaque that develops often contains a large lipid core with a thin, fibrous covering and is vulnerable to rupture and thrombosis.

wall or allow lipid materials to penetrate its surface. The factors that promote the progression of atherosclerosis and the development of complications typically induce plaque rupture and blood coagulation.

Inflammation and Infection Plaque formation is an inflammatory response to an injury on the artery wall, so the body's immune system is directly involved in its development.[7] ◆ There is also evidence that a persistent infection within the body may contribute to plaque formation. A number of different bacterial and viral antigens have been identified in plaque.[8]

Shear Stress/Hypertension The stress of blood flow along the artery walls—called shear stress—can cause mechanical damage within the arteries. The resulting inflammatory response can initiate the development of plaque.[9] Hypertension (high blood pressure) intensifies the stress of blood flow on the arterial walls and also induces the production of proinflammatory cytokines in endothelial cells.[10]

Cigarette Smoking Chemicals in smoke (including nicotine) are toxic to endothelial cells, and the resulting damage can initiate plaque development. Smoking also promotes blood clotting, raises LDL cholesterol, and lowers HDL (high-density lipoprotein) cholesterol—effects that can cause atherosclerosis to progress. Another effect of smoking is vasoconstriction, which increases the risk of heart attack and stroke in arteries already narrowed by atherosclerosis. Passive smoking can cause these effects as well.[11]

Elevated LDL and VLDL High blood levels of low-density lipoproteins and very-low-density lipoproteins (VLDL) ◆ promote atherosclerosis, particularly if the lipoproteins are oxidized.[12] Oxidized LDL and VLDL are actively taken up and retained in the artery wall. Oxidation may result from free-radical generation by macrophages and endothelial cells or various enzyme reactions. HDL help prevent the oxidation of LDL and also remove cholesterol from circulation, so low levels of HDL contribute to the development of atherosclerosis. It is not yet known whether the oxidation of dietary lipids contributes significantly to the levels of oxidized LDL and VLDL in the blood. ◆

There are distinct subtypes of LDL that vary in size and density. LDL size is inversely associated with heart disease risk: the smallest, most dense LDL are the most **atherogenic** (and more likely to become oxidized), whereas larger, less dense LDL are less atherogenic.[13] People who have small, dense LDL frequently have elevated VLDL and low HDL levels as well. This lipoprotein profile, which is especially prevalent in individuals with metabolic syndrome and type 2 diabetes, is associated with an approximately threefold increased risk of coronary heart disease.[14]

Elevated concentrations of a variant form of LDL called *lipoprotein(a)* have been found to speed the progression of atherosclerosis and to raise the risk of various types of CVD.[15] Lipoprotein(a) levels are primarily genetically determined and are influenced to only a minor degree by age and environmental factors. Although pharmacological doses of a form of niacin may lower lipoprotein(a) levels, the benefit of treating lipoprotein(a) is not yet clear.[16]

Diabetes Mellitus Diabetes can initiate and accelerate the development of atherosclerosis in multiple ways. Chronic hyperglycemia leads to the production of advanced glycation end products (AGEs) ◆, which damage the proteins of endothelial cells and thereby upset the functioning of endothelial tissue. AGEs also promote inflammation and cause oxidative stress.[17] By other mechanisms, diabetes increases tendencies for vasoconstriction, plaque rupture, and blood clotting.[18]

Age and Gender Advancing age is strongly associated with atherosclerosis due to the cumulative exposure to risk factors and the degeneration of arterial cells with age. Aging becomes a significant risk factor for men aged 45 or older and for women aged 55 or older. After menopause, women's risk increases because the loss of estrogen reduces arterial defenses against atherosclerosis and causes a rise in LDL cholesterol levels.[19] ◆ Levels of the amino acid homocysteine, which may

◆ C-reactive protein, an acute-phase protein secreted during the inflammatory response, is associated with increased heart disease risk (see Chapter 22).

◆ Reminder: VLDL transport triglycerides in the blood. In clinical practice, VLDL levels are commonly referred to as *blood triglycerides*.

◆ Disorders characterized by abnormal levels of blood lipids are called *hyperlipidemias* or *dyslipidemias*.

◆ Reminder: *Advanced glycation end products* are reactive compounds formed after glucose nonenzymatically attaches to proteins.

◆ Estrogen replacement therapy after menopause has mixed effects on heart disease risk; it can improve endothelial function, lower LDL, and raise HDL, but it can also promote blood clotting.

atherogenic: able to initiate or promote atherosclerosis.

damage artery walls and promote blood clotting, rise with age and are generally higher in men; however, researchers have not determined whether homocysteine is a cause or an effect of the disease process.[20] ◆

IN SUMMARY

Atherosclerosis, which is characterized by the buildup of arterial plaque, can lead to complications such as angina pectoris, heart attack, stroke, intermittent claudication, kidney disease, and aneurysms. Plaque is caused by factors that damage the artery wall and promote blood coagulation. Leading causes of plaque formation and progression include inflammation, shear stress, hypertension, cigarette smoking, elevated LDL and VLDL cholesterol levels, reduced HDL cholesterol levels, diabetes, and aging.

◆ Reminder: Blood homocysteine levels are influenced by intakes of folate, vitamin B_{12}, and vitamin B_6 (see Chapter 10).

Coronary Heart Disease (CHD)

Coronary heart disease (CHD), also called *coronary artery disease,* is the most common type of cardiovascular disease and the leading cause of death in the United States.[21] As mentioned, CHD is characterized by impaired blood flow through the coronary arteries, which may lead to angina pectoris, heart attack, or sudden death. CHD is usually caused by atherosclerosis but occasionally results from **spasm** or inflammatory conditions that cause narrowing of the coronary arteries. Although atherosclerosis can advance enough to fully block an artery, most heart attacks occur with less than 50 percent blockage.[22]

The lifetime risk of developing CHD is 49 percent for men and 32 percent for women.[23] Women typically develop CHD about 10 years later in life than men do, and their incidences of serious complications like heart attack and sudden death lag behind men's by about 20 years.[24] In both genders, CHD can lie dormant for years: over half of sudden deaths from CHD occur without prior symptoms in both men and women.[25]

Symptoms of Coronary Heart Disease

Symptoms of CHD usually arise only after many years of disease progression. In angina pectoris and heart attacks, the pain or discomfort usually occurs in the chest region and may be perceived as a feeling of heaviness, pressure, or squeezing. Pain may radiate to the left arm, shoulders, back, throat, jaws, or teeth. In angina pectoris, the symptoms are often triggered by exertion, persist for several minutes, and subside with rest. In a heart attack, the pain may be severe, last longer, and occur without exertion. Other symptoms of CHD include shortness of breath, nausea, vomiting, sweating, lightheadedness, weakness, and anxiety. In some cases, a feeling of indigestion or lower abdominal pain may occur.[26]

Evaluating Risk for Coronary Heart Disease

Because CHD develops over many years, prevention should begin well before symptoms appear. Population studies have suggested that about 90 percent of people with CHD have at least one of the four classic risk factors: smoking, high LDL cholesterol, high blood pressure, and diabetes.[27] These and other major risk factors that can be modified by changes in diet and lifestyle are listed in Table 27-1; only age, gender, and family history cannot be modified.

CHD Risk Assessment Risk assessment requires several key laboratory measures (see Table 27-2 on p. 846) and a thorough medical history. A complete lipoprotein profile (also called a *blood lipid profile*) includes measures of total cholesterol, LDL

TABLE 27-1 Risk Factors for CHD

Major Risk Factors for CHD (not modifiable)

- Increasing age
- Male gender
- Family history of premature heart disease

Major Risk Factors for CHD (modifiable)

- High blood LDL cholesterol
- Low blood HDL cholesterol
- High blood pressure (hypertension)
- Diabetes
- Obesity (especially abdominal obesity)
- Physical inactivity
- Cigarette smoking
- An "atherogenic" diet (high in saturated fats and low in vegetables, fruits, and whole grains)

NOTE: Risk factors highlighted in color have relationships with diet.
SOURCE: Expert Panel on Detection, Evaluation, and Treatment of High Blood Cholesterol in Adults (Adult Treatment Panel III), *Third Report of the National Cholesterol Education Program (NCEP),* NIH publication no. 02-5215 (Bethesda, Md.: National Heart, Lung, and Blood Institute, 2002), pp. II-15–II-20.

spasm: a sudden, forceful, and involuntary muscle contraction.

◆ Reminder: Polyunsaturated fats are more susceptible to oxidation than saturated fats, although the clinical significance of consuming high amounts is still unknown.

◆ Reminder: Polyunsaturated fatty acids are precursors for the eicosanoids, which mediate inflammation (see p. 711).

Polyunsaturated and Monounsaturated Fat As described in the previous section, replacing saturated fat with either monounsaturated or polyunsaturated fat helps to lower LDL levels. A switch to polyunsaturated fat tends to have the greater effect, but it also promotes a slight reduction in HDL cholesterol.[31] Other concerns are that high intakes of polyunsaturated fat may contribute to oxidative stress ◆ or increase inflammation within the body. ◆ Therefore, TLC guidelines limit polyunsaturated fat to 10 percent of total kcalories, whereas up to 20 percent of kcalories from monounsaturated fat are allowed. Keep in mind that most polyunsaturated fat in the diet consists of omega-6 fatty acids, such as linoleic acid; omega-3 fatty acids may have beneficial effects on heart disease risk, as described in a later section.

Total Fat For people whose fat intake includes substantial saturated fat, limiting total fat may indirectly reduce saturated fat. Therefore, the TLC recommendation for total fat is an intake of 25 to 35 percent of kcalories. People with the metabolic syndrome, who typically have elevated blood triglycerides, may benefit from a fat intake at the upper end of this range (30 to 35 percent) so that their carbohydrate intakes are not excessive. Fat intakes higher than 35 percent of kcalories are discouraged because they may promote weight gain in some people.

Dietary Cholesterol A high cholesterol intake can raise LDL levels, and reducing dietary cholesterol lowers LDL cholesterol in most people. The TLC recommendation is a cholesterol intake of less than 200 milligrams per day. Currently, the daily cholesterol intake in the United States averages about 273 milligrams, although it is higher in men (347 milligrams) than in women (235 milligrams).[32] Eggs contribute about one-third of the cholesterol in the American diet, followed by meats, milk, and cheese.

Trans Fat *Trans*-fatty acids raise LDL cholesterol levels, and when they replace saturated fats in the diet (as when stick margarine replaces butter), they may also cause a decline in HDL cholesterol. *Trans* fats may also raise CHD risk by altering blood vessel function, promoting inflammation, and reducing LDL size.[33] The TLC recommendation is to keep *trans* fat intake as low as possible.

Most sources of *trans* fats are foods made with partially hydrogenated vegetable oils; examples include baked goods such as crackers, cookies, and doughnuts; snack foods such as potato chips and corn chips; and fried foods such as French fries and fried chicken. In the past few years, many food manufacturers have been reducing their use of hydrogenated oils and reformulating products so that they contain minimal amounts of *trans* fats. Thus, margarine, vegetable shortening, and many baked goods are now available with little or no *trans* fat.

Soluble Fibers As Chapter 4 explained, a diet rich in soluble, viscous fibers can reduce LDL cholesterol levels by inhibiting cholesterol and bile absorption in the small intestine and reducing cholesterol synthesis in the liver (see p. 122). An extra 5 to 10 grams of soluble fiber daily is associated with a 5 percent reduction in LDL cholesterol. Dietary sources of soluble fibers include oats, barley, legumes, and fruits. The soluble fiber from psyllium seed husks, frequently used to treat constipation, is effective for lowering cholesterol levels when used as a dietary supplement.

◆ Plant sterols are extracted from soybeans and pine tree oils, and they are then hydrogenated to produce the plant stanols that are added to commercial products.

Plant Sterols and Stanols Foods or supplements that contain significant amounts of plant sterols or plant stanols can help to lower LDL cholesterol levels. ◆ Plant sterols and stanols are added to various food products, such as margarine or cheese, or supplied in dietary supplements. These plant compounds work by interfering with cholesterol and bile absorption. Clinical trials have shown that a bit more than 1 tablespoon of margarine daily (containing about 2 grams of plant sterols) can lower LDL cholesterol by up to 15 percent without lowering HDL cholesterol.[34]

◆ Some people are more sensitive to sodium intakes than others, as discussed in a later section (pp. 858–859)

Sodium and Potassium Intakes Excessive sodium in the diet can raise blood pressure, ◆ whereas potassium can lower blood pressure. A low-sodium diet that contains generous amounts of fruits and vegetables, low-fat milk products, nuts, and whole grains has been found to substantially reduce blood pressure, largely due to the diet's content of potassium and several other minerals that have blood pressure–lowering effects. This diet (the *DASH Eating Plan*) and other factors that influence blood pressure are discussed in a later section (see pp. 858–860).

Fish and Omega-3 Fatty Acids The omega-3 fatty acids in fish, known as EPA and DHA, ◆ may benefit people who have had a heart attack by suppressing inflammation, reducing blood clotting, stabilizing heart rhythm, and lowering triglyceride levels. In addition, including fish in the diet can reduce CHD risk because fish is low in saturated fat and often replaces entrées that contain animal fats. The American Heart Association recommends consuming two servings of fish per week, with an emphasis on fatty fish.[35] Fish oil supplements (providing 1 gram of EPA and DHA daily) may be helpful for individuals with documented CHD, and they are particularly useful for treating individuals with elevated triglyceride levels (see p. 854).[36] Chapter 5 provides additional information about the omega-3 fatty acids and fish oil supplements.

The 18-carbon omega-3 fatty acids found in flaxseed and other land plants have lesser or different effects than the omega-3 fatty acids from marine sources. Although some evidence suggests that moderate increases in these plant sources of omega-3 fatty acids may improve CHD risk, additional research is needed to confirm their benefits.[37]

Alcohol Moderate consumption of alcohol—from beer, wine, or liquor—has favorable effects on HDL cholesterol levels, atherosclerosis, inflammation, blood-clotting activity, and insulin resistance.[38] Alcohol use is also inversely related to the incidence of heart attack. These benefits are most apparent in men and women who are at least 45 and 55 years old, respectively. Of note, only low or moderate amounts of alcohol—no more than one drink daily for women and two for men—have been found to lower CHD risk, and higher intakes are associated with higher mortality rates. One "drink" is equivalent to 12 ounces of beer, 5 ounces of wine, 10 ounces of wine cooler, or 1¹/₂ ounces of 80 proof distilled spirits such as gin, rum, vodka, and whiskey.

For some people, alcohol's negative effects can offset any health advantages. Alcohol consumption is associated with cancers of the gastrointestinal (GI) tract and several other cancers, including liver cancer and breast cancer.[39] Alcohol is destructive to the liver and male reproductive system, and high alcohol intakes can elevate blood pressure and triglyceride levels. Moreover, up to 10 percent of adults misuse alcohol.[40] For these reasons, nondrinkers are not encouraged to start drinking in an effort to decrease their risk for CHD.

Regular Physical Activity Regular aerobic activity reverses a number of risk factors for CHD: it can lower triglycerides, raise HDL, lower blood pressure, promote weight loss, improve insulin sensitivity, strengthen heart muscle, and increase coronary artery size and tone. ◆ Aerobic activities that use large muscle groups have the greatest benefits; such activities include brisk walking, running, swimming, cycling, stair-stepping, and cross-country skiing. Alternatives for busy people include heavy house cleaning, lawn mowing, raking leaves, and walking to and from work.

Research studies have found that the most active persons have CHD rates that are about half the rates of those who are the least active.[41] The American Heart Association recommends that all adults participate in at least 30 minutes of moderate-intensity physical activity on most days of the week, whereas 60 minutes of physical activity is suggested for adults who are attempting or maintaining weight loss.[42] If preferred, physical activity can be divided into several sessions during the day. Note that vigorous activity increases the risk of heart attack and sudden death in individuals with diagnosed heart disease, so sedentary adults are advised to increase their activity levels gradually.[43]

Smoking Cessation Cigarette smoking is a major risk factor for CHD, as well as for other types of cardiovascular disease. ◆ Compounds in smoke damage blood vessel cells, cause chronic inflammation, decrease the oxygen-carrying capacity of

© Ronnie Kaufman/Corbis

Regular aerobic exercise can strengthen the cardiovascular system, promote weight loss, reduce blood pressure, and improve blood glucose and lipid levels.

◆ Reminder: *EPA* and *DHA* are abbreviations for eicosapentaenoic acid and docosahexaenoic acid, respectively, the 20- and 22-carbon polyunsaturated fatty acids found in fish. See pp. 154 and 174 for a review of these fatty acids.

◆ Resistance exercise can also help to reverse some CHD risk factors, but its overall effect on CHD risk is uncertain.

◆ Cigar and pipe smoking can also increase the risk of CHD, but the risk may be lower because the smoke is less likely to be inhaled.

Feeding Disabilities

Chapter 27 referred to difficulties following a stroke that can interfere with the ability to eat independently. This highlight discusses the problems faced by individuals who must cope with disabilities that interfere with the process of eating, such as those that interfere with chewing and swallowing. These obstacles can arise at any time during a person's life and from any number of causes. An infant may be born with a physical impairment such as cleft palate; an adolescent may lose motor control following injuries sustained in an automobile accident; an older adult may struggle with the pain of arthritis or the mental deterioration of dementia. Table H27-1 lists some of the conditions that may lead to feeding problems.

Effects of Disabilities on Nutrition Status

Eating and drinking require a considerable number of individual coordinated motions. Consider an infant learning the skills required for feeding: each step—sitting, grasping cups and utensils, bringing food to the mouth, biting, chewing, and swallowing—requires coordinated movements. An injury or disability that interferes with any of these movements can lead to feeding problems and inadequate food intake. Total food intake is often significantly reduced when individuals with inefficient motor function take a long time to eat.[1] Difficulties that affect procurement of food, such as the inability to drive, walk, or carry groceries, can also lower food intake and lead to malnutrition and weight loss.

TABLE H27-1 Conditions That May Lead to Feeding Problems

The following conditions may lead to feeding problems by interfering with a person's ability to suck, bite, chew, swallow, or coordinate hand-to-mouth movements.

• Accidents	• Language, visual, or hearing impairment
• Amputations	• Multiple sclerosis
• Arthritis	• Muscle weakness
• Birth defects	• Muscular dystrophy
• Cerebral palsy	• Neuromotor dysfunction
• Cleft palate	• Parkinson's disease
• Down syndrome	• Polio
• Head injuries	• Spinal cord injuries
• Huntington's chorea	• Stroke

Energy Requirements

Disabilities may either increase or decrease energy needs. Those that affect muscle tension and mobility can reduce physical activity and, consequently, energy requirements. Other disabilities, such as certain forms of cerebral palsy, cause involuntary muscle activity that raises energy requirements.[2] Loss of a limb due to amputation reduces energy needs in proportion to the weight and metabolism represented by the missing limb, but energy needs may be greater if an individual increases activity to compensate for the loss, such as by propelling a wheelchair. Because the effects of disabilities are often unpredictable, the health care practitioner may find it difficult to assess energy requirements until weight gain or loss has occurred.

Overweight and obesity often accompany conditions that limit mobility or result in short stature; examples include Down syndrome (see p. 498) and spina bifida (see p. 481). Obesity may also develop because the family or caregiver provides an inappropriate amount of food, sometimes out of sympathy for the individual who has a disability.[3] In these cases, the health practitioner may need to counsel the family or caregiver about appropriate food choices and portion sizes.

Effects of Disease Symptoms and Medications

Physical symptoms of disease sometimes interfere with eating and nutrition status. Examples include nausea, frequent coughing or choking, difficulty breathing, and gastroesophageal reflux. Individuals with speech and hearing problems may have a difficult time communicating with caregivers about thirst and hunger. Mobility problems can lead to bone demineralization and pressure sores.

Conditions that require the use of multiple medications can also have a significant impact on nutrition status (see Chapter 19).[4] Medications may increase or decrease appetite, interfere with nutrient metabolism, or have gastrointestinal effects that cause pain or discomfort with eating.

Social Concerns

Because mealtimes are a critical time for social interaction, individuals with feeding problems may encounter emotional and social problems if they are unable to participate. Children may fail to develop social skills, whereas adults may miss the social stimulation that mealtimes provide. Individuals should be encouraged to sit with family and friends during meals so that they are not deprived of the social and cultural aspects of eating.

Independent Eating for People with Disabilities

The evaluation and treatment of feeding problems often involve the joint efforts of health care professionals from a variety of disciplines, including dietitians, nurses, occupational and physical therapists, speech-language pathologists, and dentists. Together, these professionals evaluate each patient's dietary needs and assess abilities to chew, sip, swallow, grasp utensils, use utensils to pick up foods, and bring foods from the plate to the mouth. A speech-language pathologist most often evaluates chewing and swallowing abilities and trains patients to use lips, tongue, and throat for eating and speaking. An occupational therapist can demonstrate alternate feeding strategies, including changes in body position that improve feeding, techniques for handling utensils and food, and use of special feeding devices.

Adaptive feeding equipment can help patients with feeding disabilities gain independence.

© Courtesy of Sammons Preston/Patterson Medical Products, Inc.

Feeding Strategies

Direct observation of a patient during mealtimes allows health care professionals to assess current eating behaviors, demonstrate feeding techniques, monitor the patient's and caregiver's understanding of the techniques, and evaluate how well the care plan is working. To illustrate, consider a child with a feeding problem caused by hypersensitivity to oral stimulation. The health care professional may start by teaching the caregiver to gently and playfully stroke the child's face with a hand, washcloth, or soft toy. Once the child tolerates touch on less sensitive areas of the face, the health care professional may encourage the caregiver to slowly begin to rub the child's lips, gums, palate, and tongue. With time, the child may be better able to tolerate the presence of food in the mouth. Examples of other strategies that can help feeding problems are listed in Table H27-2.

Adaptive Feeding Equipment

Adaptive feeding devices can make a remarkable difference in a person's ability to eat independently. Figure H27-1 on p. 870 shows a few of the many special feeding devices that are available and describes their uses. Other examples of adaptive equipment include specialized chairs to improve posture, bolsters inserted under arms to improve elbow stability, and raised trays or eating surfaces to simplify hand-to-mouth movements.[5]

Sometimes, despite the best efforts of all involved, a patient is unable to consume enough food by mouth. In these cases, tube feedings can help to improve nutrition status. Tube feedings are also recommended for patients who have severe dysphagia (difficulty swallowing) or aspiration pneumonia.[6]

A Note for Caregivers

The responsibility of caring for a person with a feeding problem can frequently overwhelm a caregiver. Caring for a person with disabilities requires time and patience—and many new therapies to be learned and administered. The caregiver may spend many

---

TABLE H27-2 Interventions for Feeding-Related Problems

Inability to Suck
- Use squeeze bottles, which do not require sucking, to express liquids into the mouth.
- Place a spoon on the center of the tongue and apply downward pressure to stimulate sucking.
- Apply rhythmic, slow strokes on the tongue to alter the tongue position and improve the sucking response.

Inability to Chew
- Place foods between gums and teeth to promote chewing.
- Improve chewing skills with foods of different textures; for example, fruit leathers stimulate jaw movements but dissolve quickly enough to minimize choking.
- Provide soft foods that require minimal chewing or are easily chewed.

Inability to Swallow
- Provide thickened liquids, pureed foods, and moist foods that form boluses easily.
- Provide cold formulas, frozen fruit juice bars, and ice; cold substances promote swallowing movements by the tongue and soft palate.
- Make sure the patient's jaw and lips are closed to facilitate swallowing action.
- Correct posture and head position if they interfere with swallowing ability.

Inability to Grasp or Coordinate Movements
- Provide utensils that have modified handles, or are smaller or larger as necessary.
- Encourage the use of hands for feeding if utensils are difficult to maneuver.
- Provide plates with food guards to prevent spilling.
- Supply clothing protection.

Impaired Vision
- Place foods (meats, vegetables) in similar locations on the plate at each meal.
- Provide plates with food guards to prevent spilling.

SOURCES: J. Case-Smith and R. Humphry, Feeding and oral motor skills, in J. Case-Smith, A. S. Allen, and P. N. Pratt, eds., *Occupational Therapy for Children* (St. Louis: Mosby–Year Book, 1996), pp. 430–460; S. Escott-Stump, *Nutrition and Diagnosis-Related Care* (Baltimore: Lippincott Williams & Wilkins, 2002), pp. 64–65.

FIGURE H27-1 Examples of Adaptive Feeding Devices

Utensils

Rocker knife

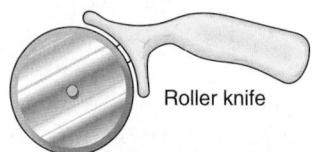

Roller knife

People with only one arm or hand may have difficulty cutting foods and may appreciate using a *rocker knife* or a *roller knife*.

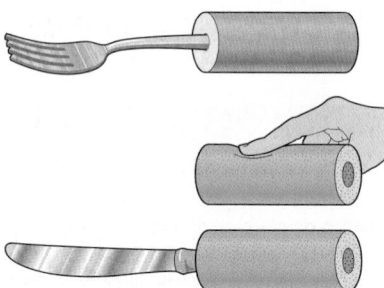

People with a limited range of motion can feed themselves better when they use *flatware with built-up handles*.

People with extreme muscle weakness may be able to eat with a *utensil holder*.

For people with tremors, spasticity, and uneven jerky movements, *weighted utensils* can aid the feeding process.

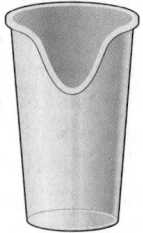

Battery-powered feeding machines enable people with severe limitations to eat with less assistance from others.

Plates

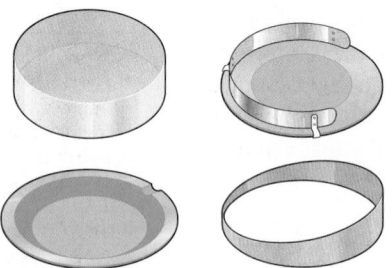

People who have limited dexterity and difficulty maneuvering food find *scoop dishes* or *food guards* useful.

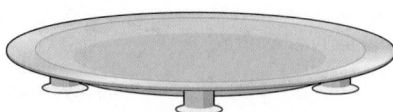

People with uncontrolled or excessive movements might move dishes around while eating and may benefit from using *unbreakable dishes with suction cups*.

Cups

People with limited neck motion can use a *cutout plastic cup*.

Two-handed cups enable people with moderate muscle weakness to lift a cup with two hands.

People with uncontrolled or excessive movements might prefer to drink liquids from a *covered cup* or glass with a *slotted opening* or *spout*.

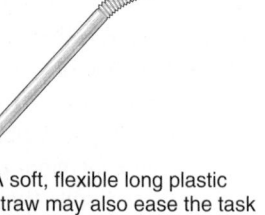

A soft, flexible long plastic straw may also ease the task of drinking.

hours preparing special foods, monitoring the use of adaptive feeding equipment, and helping with feedings. Moreover, a person with disabilities may need help with other tasks as well, and all may require a considerable amount of time. In many cases, a caregiver receives little or no assistance. These conditions may lead to strained interactions between caregiver and patient and cause frustration and depression.[7] Psychologists can offer counseling to patients or caregivers to help them adjust, and all members of the health care team can offer emotional support and practical suggestions to ease caregivers' responsibilities and frustrations.

Successful therapy for people with feeding disabilities requires the involvement of many health care professionals and depends on accurate identification of impaired feeding skills and determination of appropriate interventions. Ideally, with training, people with disabilities attain total independence—they are able to prepare, serve, and eat nutritionally adequate food daily without help. In some cases, these goals can be met with the help of caregivers. The combined efforts of the health care team can support both patients and caregivers in enhancing quality of life and in achieving independence to the greatest degree possible.

REFERENCES

1. E. B. Fung and coauthors, Feeding dysfunction is associated with poor growth and health status in children with cerebral palsy, *Journal of the American Dietetic Association* 102 (2002): 361–368.
2. Position of the American Dietetic Association: Providing nutrition services for infants, children, and adults with developmental disabilities and special health care needs, *Journal of the American Dietetic Association* 104 (2004): 97–107.
3. H. H. Cloud, Expanding roles for dietitians working with persons with developmental disabilities, *Journal of the American Dietetic Association* 97 (1997): 129–130.
4. Position of the American Dietetic Association, 2004.
5. J. Case-Smith and R. Humphry, Feeding and oral motor skills, in J. Case-Smith, A. S. Allen, and P. N. Pratt, eds., *Occupational Therapy for Children* (St. Louis: Mosby–Year Book, 1996), pp. 430–460.
6. Position of the American Dietetic Association, 2004.
7. Fung and coauthors, 2002.

AJPhoto/Photo Researchers, Inc.

Nutrition in the Clinical Setting

Each bean-shaped kidney is only about the size of a fist, yet the kidneys carry out many critical functions. Among other tasks, the kidneys shoulder much of the responsibility for maintaining the body's chemical balance. If the kidneys fail to function, toxic compounds build up in the blood, causing a wide range of symptoms and life-threatening complications. Unfortunately, acute kidney diseases have high mortality rates, and chronic kidney disease is underdiagnosed and undertreated, as symptoms do not arise until the later stages. Health practitioners must learn to recognize and treat renal diseases early, before kidney damage progresses and causes irreversible illness.

Renal Diseases

The two kidneys sit just above the waist on each side of the spinal column. As part of the urinary system (see Figure 28-1 on p. 874), they are responsible for filtering blood and removing excess fluid and wastes for elimination in urine. Many body functions, including the work of the heart, blood vessels, enzymes, and cell membranes, depend on the kidneys to maintain normal fluid volume and composition. Because the kidneys are so proficient at this task, disturbances in body fluids that result from food intake, physical activity, and metabolism are normally corrected within hours. In addition, the kidneys perform a number of other metabolic roles, as discussed in the following section. Thus, **renal** disorders not only result in fluid and electrolyte imbalances, but can have widespread effects on health.

Functions of the Kidneys

The functional unit of the kidneys is the **nephron,** introduced on p. 402 (see Figure 12-2). Within each nephron, the **glomerulus,** a ball-shaped tuft of capillaries, serves as a gateway through which the components of blood must pass to form **filtrate.**◆ The glomerulus and surrounding **Bowman's capsule** function like a sieve, retaining blood cells and plasma proteins in blood while allowing fluid and small solutes to enter the nephron's system of **tubules.** As the filtrate passes through the tubules, its composition continuously changes as some of its components are reabsorbed and returned to the body via capillaries surrounding each tubule. Eventually, the remaining filtrate enters a **collecting duct** shared by several nephrons, and additional water is reabsorbed to form the final urine product. ◆ The urine travels through the ureters to the bladder for temporary storage (review Figure 28-1). By filtering the blood and forming urine, the kidneys regulate the extracellular fluid volume and osmolarity, electrolyte concentrations, and acid-base balance. They also excrete metabolic waste products such as urea and creatinine, as well as various drugs and toxins.

renal (REE-nal): pertaining to the kidneys.

nephron (NEF-ron): the functional unit of the kidneys, consisting of a glomerulus and tubules.

• **nephros** = kidney

glomerulus (gloh-MEHR-yoo-lus): a tuft of capillaries within the nephron that filters water and solutes from blood as urine production begins (plural: *glomeruli*).

filtrate: the substances that pass through the glomerulus and travel through the nephron's tubules, eventually forming urine.

Bowman's (BOE-minz) **capsule:** a cuplike component of the nephron that surrounds the glomerulus and collects the filtrate that is passed to the tubules.

tubules: tubelike structures of the nephron that process filtrate during urine production. The tubules are surrounded by capillaries that reabsorb substances retained by tubule cells.

collecting duct: the last portion of a nephron's tubule, where the final concentration of urine occurs. One collecting duct is shared by several nephrons.

FIGURE 28-1 The Kidneys and Urinary Tract

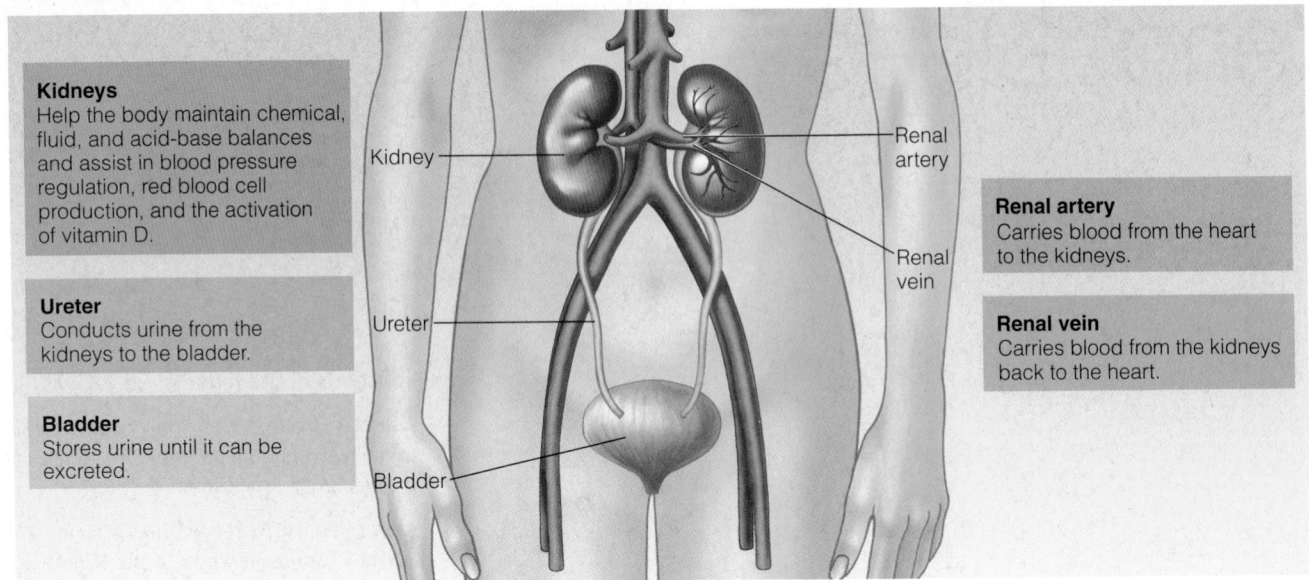

Kidneys
Help the body maintain chemical, fluid, and acid-base balances and assist in blood pressure regulation, red blood cell production, and the activation of vitamin D.

Ureter
Conducts urine from the kidneys to the bladder.

Bladder
Stores urine until it can be excreted.

Renal artery
Carries blood from the heart to the kidneys.

Renal vein
Carries blood from the kidneys back to the heart.

Kidney

Renal artery

Renal vein

Ureter

Bladder

◆ The rate at which the kidneys form filtrate is known as the *glomerular filtration rate,* discussed later in this chapter.

◆ About 99 percent of the substances in filtrate, including water, are reabsorbed, leaving only 1 to 2 liters of urine to be excreted daily.

In addition to their remarkable role in maintaining homeostasis, the kidneys have other critical roles:

• The kidneys help to regulate blood pressure by secreting the enzyme renin. Renin catalyzes the formation of angiotensin, a potent vasoconstrictor that narrows the diameters of arterioles and thereby raises blood pressure (for details, review p. 401 and Figure 12-3 on p. 403). Angiotensin also stimulates the release of aldosterone, an adrenal hormone that triggers the kidneys to reabsorb more sodium. Sodium reabsorption promotes water retention and increases plasma volume, which raises blood pressure.

• The kidneys produce the hormone **erythropoietin,** which stimulates the production of red blood cells in the bone marrow (see Highlight 19 for details).

• The kidneys convert vitamin D to its active form, 1,25-dihydroxyvitamin D_3 (see Figure 11-9 on p. 377), thereby playing a primary role in calcium regulation and bone formation.

Subsequent sections of this chapter explain how renal diseases can interfere with these kidney functions and severely disrupt health.

IN SUMMARY

The kidneys are responsible for filtering the blood and removing wastes for excretion in urine. By adjusting the blood's volume and composition, the kidneys help to maintain homeostasis within the body. Other kidney functions include the production of enzymes and hormones that regulate blood pressure, stimulate red blood cell production, and activate vitamin D.

erythropoietin (eh-RITH-ro-POY-eh-tin): a hormone made by the kidneys that stimulates red blood cell production.

nephrotic (neh-FROT-ik) **syndrome:** a syndrome associated with kidney disorders that cause urinary protein losses exceeding 3.0 to 3.5 g/day; symptoms include low serum albumin, elevated blood lipids, and edema.

proteinuria (PRO-teen-NOO-ree-ah): loss of protein, mostly albumin, in the urine; also known as *albuminuria.*

The Nephrotic Syndrome

The **nephrotic syndrome** is not a specific disease; rather, the term refers to kidney disorders that cause urinary protein losses (**proteinuria**) exceeding 3.0 to 3.5 grams per day.[1] Although the nephrotic syndrome can occur at any age, it is seen most often in children between 1$^1/_2$ and 4 years of age.[2] The condition arises when damage to the glomeruli increases their permeability to plasma proteins, allowing protein to escape into the urine. Although various diseases involving the glomeruli are often the cause, the nephrotic syndrome may also be a complication of diabetes mellitus, immunological and hereditary disorders, infections (involving the kidneys or elsewhere in the body), chemical damage (from medications or illicit drugs), and some cancers. Along with proteinuria, typical clinical findings include low serum albumin levels, edema, elevated blood lipids, and blood coagulation disorders. The nephrotic syndrome can sometimes progress to renal failure.

Consequences of the Nephrotic Syndrome

In the nephrotic syndrome, urinary protein losses generally average about 8 grams daily, but additional protein is lost due to protein catabolism within the kidney tubules.[3] The liver attempts to compensate for these losses by increasing its synthesis of certain plasma proteins, but the excessive production of some of the proteins causes additional complications.

Edema Because albumin is the most abundant plasma protein, it is also the protein with the most significant loss in urine. Consequently, its blood level is markedly reduced, which contributes to a fluid shift from blood plasma to the interstitial spaces and, thus, edema. ◆ Defective sodium excretion also contributes to edema: the kidney tubules reabsorb sodium in greater amounts than usual, causing sodium and water retention within the body.[4]

◆ Reminder: Plasma proteins, such as albumin, help to maintain fluid balance within the blood.

Risk of Cardiovascular Disease People with the nephrotic syndrome frequently have elevated levels of low-density lipoproteins (LDL), very-low-density lipoproteins (VLDL), and the more damaging LDL variant known as lipoprotein(a). The impaired clearance of VLDL from blood is largely due to reduced lipoprotein lipase on the blood vessel walls.[5] ◆ The risk of blood clotting is increased in the nephrotic syndrome due to urinary losses of proteins that inhibit blood clotting and elevated levels of plasma proteins that favor clotting. These factors raise the risks of developing heart disease and stroke.[6]

◆ Reminder: *Lipoprotein lipase* is the enzyme that hydrolyzes the triglycerides in lipoproteins.

Other Effects of the Nephrotic Syndrome The proteins lost in urine include immunoglobulins (antibodies) and vitamin D–binding protein. Depletion of immunoglobulins increases susceptibility to infection. Loss of vitamin D–binding protein results in lower vitamin D and calcium levels and increases the risk of rickets in children. Patients with the nephrotic syndrome frequently develop protein-energy malnutrition (PEM) and muscle wasting from the continued proteinuria. Figure 28-2 on p. 876 summarizes the effects of urinary protein losses in the nephrotic syndrome.

Treatment of the Nephrotic Syndrome

Medical treatment of the nephrotic syndrome requires diagnosis and management of the underlying disorder responsible for the proteinuria. Complications are managed with medications and medical nutrition therapy. The drugs prescribed may include diuretics, ACE inhibitors (which reduce protein losses), lipid-lowering drugs, anticoagulants, anti-inflammatory drugs (usually corticosteroids), and immunosuppressants (such as cyclosporine).[7] Nutrition therapy helps to alleviate edema, prevent PEM, and slow the progression of atherosclerosis.

Protein and Energy Meeting protein and energy needs helps to minimize losses of muscle tissue. However, high-protein diets are not advised because they can exacerbate urinary protein losses and result in further damage to the kidneys.[8] Instead, the protein intake should fall between 0.8 and 1.0 gram per kilogram of body weight per day; at least half of the protein consumed should be from high-quality sources, such as milk products, meat, fish, poultry, eggs, and soy products. An adequate energy intake (about 35 kcalories per kilogram of body weight daily)

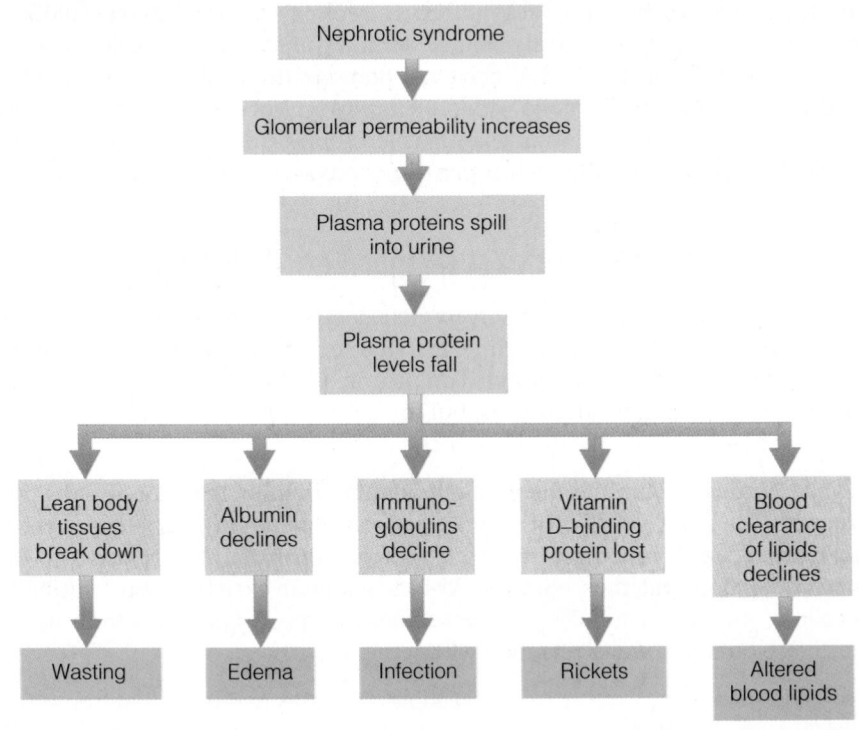

FIGURE 28-2 Consequences of Urinary Protein Losses in the Nephrotic Syndrome

sustains weight and spares protein. Weight loss or infections suggest the need for additional kcalories.

Fat As Chapter 27 explained, a diet low in saturated fat, *trans* fats, cholesterol, and refined sugars helps to control elevated blood lipids. Thus, patients with the nephrotic syndrome should limit saturated fat intake to 7 percent of total kcalories, avoid foods that contain *trans* fats, and limit cholesterol intake to 200 milligrams per day. Dietary measures are usually inadequate for controlling blood lipids, however, so physicians usually prescribe lipid-lowering medications as well.

Sodium and Potassium Sodium restriction helps to control edema; therefore, sodium intake is often limited to 1 to 2 grams daily.[9] Table 28-1 provides guidelines for following a diet restricted to 2 grams of sodium. If diuretics prescribed for the edema cause potassium losses, patients are encouraged to select foods rich in potassium (see Chapter 12, pp. 414–415).

Vitamins and Minerals Patients with the nephrotic syndrome may require vitamin D and calcium supplementation to help prevent bone loss and rickets. Multivitamin/mineral supplementation is often advised to avoid additional nutrient deficiencies; nutrients at risk include vitamin B_6, vitamin B_{12}, folate, iron, copper, and zinc.[10] ◆

◆ Nutrient deficiencies may develop if the carrier proteins for nutrients are lost in the urine.

IN SUMMARY

The nephrotic syndrome is characterized by urinary protein losses exceeding 3.0 to 3.5 grams per day. Complications include edema, lipid abnormalities, blood coagulation disorders, reduced immunity, rickets, and PEM. Medications treat the underlying condition and help to manage complications. The diet should provide sufficient protein and energy to maintain health, but patients should avoid consuming excess protein. Other dietary adjustments may be needed to correct edema, lipid disorders, and nutrient deficiencies.

TABLE 28-1 Sodium-Restricted Diet

General Guidelines

About 75% of sodium in a typical diet comes from processed foods, about 10% from unprocessed natural foods, and about 15% from table salt. With this in mind:
- Choose fresh foods and foods frozen or canned without added salt.
- Avoid adding salt to foods while cooking.
- Avoid adding salt to foods at the table.
- When eating out, ask that meals be prepared without salt.

Sodium in Foods

All foods contain sodium, but some contain more than others. Use the information about the average sodium contents of foods to tailor the diet to individual preferences.

Food Group	Serving Size	Sodium (mg) per Serving
Fresh meats, poultry, freshwater fish; low-sodium canned meats and fish; low-sodium peanut butter and cheese; unsalted cottage cheese, soybeans, and textured vegetable protein	1 oz	20
Regular fat-free, low-fat, and whole milk and yogurt	8 oz	120
Eggs	1	60
Fresh artichokes, beets, carrots, and celery; beet, collard, dandelion, mustard, and turnip greens	$1/2$ c	50
Regular canned vegetables	$1/2$ c	300
Regular white and whole-grain bread	1 slice	150
Butter and margarine	1 tsp	50
Salt	$1/2$ tsp	1000

Other Foods

These foods can be used freely or with some limits with respect to sodium, although some of these foods may need to be restricted for weight or blood lipid control:
- All fruits and fruit juices
- Low-sodium canned or frozen vegetables without added salt except those listed above; low-sodium vegetable juices
- Low-sodium bread and bread products; puffed rice and wheat and shredded wheat cereals; rice; pasta
- Soups, casseroles, and recipes made with allowed foods and ingredients
- Unsalted butter, margarine, nuts, and gravy; low-sodium mayonnaise and salad dressing; shortening
- Low-sodium catsup, mustard, Tabasco sauce, and other condiments; low-sodium baking powder

These foods and dishes prepared with them are high in sodium and should be avoided:
- Cured, canned, salted, or smoked meats, poultry, and fish such as bacon, luncheon meats, corned beef, kosher meats, and canned tuna and salmon; imitation fish products, salted textured vegetable protein, peanut butter, and nuts
- Buttermilk, regular cheeses
- Maraschino cherries; crystallized or glazed fruits; dried fruits with sodium sulfite added
- Pickles, pickled vegetables, sauerkraut, regular vegetable juices
- Instant and quick-cooking hot cereals; commercial bread products made from self-rising flour or cornmeal; salted snack foods
- Salt pork and bacon; commercial salad dressing; olives; regular gravy; catsup, baking powder, soy sauce, bouillon

A Sample Diet Restricted to 2 Grams Sodium

Using the preceding information, many diet plans that meet individual needs are possible. Using the guidelines for a heart-healthy diet, a typical plan for a day might look like this:

Food Group	Sodium (mg)
Meat, 6 oz (6 × 20 mg)	120
Milk, 3 c (3 × 120 mg)	360
Fruit, 3 servings	Negligible
Vegetables, $1/2$ c vegetables with some sodium (1 × 50)	50
Vegetables, other vegetables and legumes, 2 servings	Negligible
Whole-grain bread, 4 slices (4 × 150 mg)	600
Salted margarine, 6 servings (6 × 50 mg)	300
Total	1430

Individuals can use the remainder of the sodium allowance for whatever foods they choose. The sodium content of other foods can be determined by reading food labels or using food composition tables. An individual may choose to use some ($1/4$ tsp) table salt or a favorite food that contains sodium.

Acute Renal Failure

In **acute renal failure,** also known as *acute kidney injury,* kidney function deteriorates rapidly, over hours or days. The loss of kidney function reduces urine output and allows nitrogenous wastes to build up in blood. The degree of renal dysfunction varies from mild to severe. With prompt treatment, acute renal failure is often reversible, although mortality rates are high, ranging from 36 to 86 percent.[11]

Causes of Acute Renal Failure

Various disorders can lead to acute renal failure, and it often develops as a consequence of severe illness, injury, or surgery. To aid in diagnosis and treatment, its causes are commonly classified as prerenal, intrinsic, or postrenal. *Prerenal* factors are those that cause a sudden reduction in blood flow to the kidneys; they often involve a severe stressor such as heart failure, shock, or blood loss. Factors that damage kidney tissue, such as infections, toxins, drugs, or direct trauma, are classified as *intrinsic* causes of acute renal failure. *Postrenal* factors are those that prevent excretion of urine due to urinary tract obstructions. Table 28-2 provides examples of specific disorders that may cause acute renal failure.

Consequences of Acute Renal Failure

A decline in renal function alters the composition of blood and urine. The kidneys may become unable to regulate the levels of electrolytes, acid, and nitrogenous wastes in the blood. Urine may be diminished in quantity or absent, leading to fluid retention. Diagnosis is often a complex task because the clinical effects can be subtle and vary according to the underlying cause of disease.

Fluid and Electrolyte Imbalances About half of patients with acute renal failure experience **oliguria,** producing less than 400 milliliters of urine per day.[12] ◆ The reduced excretion of electrolytes results in sodium retention and elevated levels of potassium, phosphate, and magnesium in the blood. Elevated potassium **(hyperkalemia)** is of particular concern, because potassium imbalances can alter heart rate and lead to heart failure. Elevated serum phosphate levels **(hyperphosphatemia)** ◆ promote an increased secretion of parathyroid hormone, leading to losses in bone calcium. Due to the sodium retention and reduced urine production, edema is a common symptom of acute renal failure and may be apparent as puffiness in the face and hands and swelling of the feet and ankles.

◆ Normal urine volume exceeds 800 milliliters per day, which is equivalent to about 27 fluid ounces, or almost 3$^1/_2$ cups.

◆ To measure serum phosphate, the phosphorus content of the blood is analyzed; thus the terms *serum phosphate* and *serum phosphorus* are often used interchangeably.

acute renal failure: abrupt loss of kidney function over a period of hours or days; also known as *acute kidney injury.*

oliguria (OL-lih-GOO-ree-ah): an abnormally low amount of urine, often less than 400 mL/day.

hyperkalemia (HIGH-per-ka-LEE-me-ah): elevated serum potassium levels.

hyperphosphatemia (HIGH-per-fos-fa-TEE-me-ah): elevated serum phosphate levels.

TABLE 28-2	Causes of Acute Renal Failure	
Prerenal Factors (60% to 70% of cases)	**Intrinsic Factors** (25% to 40% of cases)	**Postrenal Factors** (5% to 10% of cases)
• **Low blood volume or pressure:** Hemorrhage, burns, sepsis or shock, anaphylactic reactions, nephrotic syndrome, gastrointestinal losses, diuretics, antihypertensive medications	• **Vascular disorders:** Sickle-cell disease, diabetes mellitus, transfusion reactions	• **Obstructions (ureter or bladder):** Strictures, tumors, stones, trauma
• **Renal artery disorders:** Blood clots or emboli, stenosis, aneurysm, trauma	• **Obstructions (within kidney):** Inflammation, tumors, stones, scar tissue	• **Prostate disorders:** Cancer or hyperplasia
• **Heart disorders:** Heart failure, heart attack, arrhythmias	• **Renal injury:** Infections, environmental contaminants, drugs, medications, *Escherichia coli* food poisoning	• **Renal vein thrombosis**
		• **Bladder disorders:** Neurological conditions, bladder rupture
		• **Pregnancy**

Uremia As a result of impaired kidney function, nitrogen-containing waste products—blood urea nitrogen (BUN), creatinine, and uric acid—accumulate in the blood. ◆ Moreover, the tissue catabolism that accompanies illness generates more nitrogenous waste than usual. The clinical outcome, called **uremia**, includes symptoms such as fatigue, lethargy, confusion, headache, anorexia, a metallic taste in the mouth, nausea and vomiting, and diarrhea. In more serious cases, elevated blood pressure, rapid heartbeat, seizures, and delirium or coma may occur. It is sometimes difficult to distinguish the symptoms of uremia from those of the underlying illness.

◆ A progressive rise in BUN or creatinine suggests the presence of acute renal failure. Refer to Table 17-10 on p. 603 for normal laboratory values.

Treatment of Acute Renal Failure

Treatment of acute renal failure involves a combination of drug therapy, **dialysis,** ◆ and medical nutrition therapy to restore fluid and electrolyte balances and minimize blood concentrations of toxic waste products. Correcting the underlying illness is necessary to prevent further damage to the kidneys.

In oliguric patients (those with reduced urine production), recovery from renal failure sometimes begins with a period of **diuresis,** in which large amounts of fluid are excreted. Because tubular function is minimal at this stage, electrolytes may not be sufficiently reabsorbed; consequently, both fluid depletion and electrolyte imbalances become a concern. Patients with this pattern of recovery (generally those whose renal failure is due to tubular injury) must be monitored closely in case they require fluid and electrolyte replacement.[13]

◆ Highlight 28 describes common dialysis procedures, including *continuous renal replacement therapy*, the approach usually used for treating acute renal failure.

Drug Treatment in Acute Renal Failure Because kidney function is required for drug excretion, patients may need to use lower doses of their usual medications to compensate for limited urine output. Conversely, dialysis treatment may increase losses of some drugs, and doses may need to be increased. Drugs that are **nephrotoxic** (including some antibiotics and nonsteroidal anti-inflammatory drugs) must be avoided until kidney function improves.

The medications prescribed for acute renal failure depend on the underlying cause of illness and the complications that develop. For patients with edema, diuretics may be used to mobilize fluids; furosemide (Lasix) is the usual choice. Immunosuppressants may be prescribed for patients with inflammatory conditions. Patients with hyperkalemia are given potassium-exchange resins that bind potassium ions in the gastrointestinal (GI) tract, ensuring the elimination of potassium in the stool. Rapid correction of hyperkalemia requires the use of insulin, which causes a temporary shift of extracellular potassium into the cells. (Glucose must be supplied along with insulin to prevent hypoglycemia.) If acidosis is present, bicarbonate may be administered orally or intravenously.[14]

Protein and Energy Although its effects are highly variable, acute renal failure is typically a catabolic condition associated with hypermetabolism and muscle wasting. Thus, sufficient protein and energy must be ingested to preserve muscle mass. Initially, the patient may be provided with 35 kcalories per kilogram of body weight per day, while body weight is monitored to ensure that energy intake is adequate. If available, indirect calorimetry provides the best estimate of energy needs.

Protein contributes nitrogen, increasing the kidneys' workload, but intake should be sufficient to prevent negative nitrogen balance and additional wasting. Protein recommendations are influenced by kidney function, the degree of catabolism, and the use of dialysis (dialysis removes nitrogenous wastes).[15] Protein restriction to about 0.6 to 0.8 grams per kilogram of body weight per day may be necessary for patients with limited kidney function who are not treated with dialysis. Higher intakes (up to 2.0 grams per kilogram daily) may be recommended if kidney function improves or the treatment includes dialysis. Patients who are catabolic or septic typically have high protein needs, but they require dialysis to accommodate the additional nitrogen load.

uremia (you-REE-me-ah): the abnormal accumulation of nitrogen-containing substances, especially urea, in the blood; also called *azotemia* (AZE-oh-TEE-me-ah).

dialysis (dye-AH-lih-sis): a treatment that removes wastes and excess fluid from the blood after the kidneys have stopped functioning. The most common types of dialysis are *hemodialysis* and *peritoneal dialysis* (see Highlight 28).

diuresis (DYE-uh-REE-sis): increased urine production.

nephrotoxic: toxic to the kidneys.

Fluids Health practitioners can assess fluid status by monitoring weight fluctuations, blood pressure, pulse rates, and appearance of the skin and mucous membranes. Another method is to measure serum sodium concentrations: a low level of sodium often indicates excessive fluid intake, and a high level suggests inadequate intake.

Fluid balance must be restored in patients who are either overhydrated or dehydrated. Thereafter, fluid needs can be estimated by measuring urine output and adding about 500 milliliters to account for the water lost from skin, lungs, and perspiration. An individual with fever, vomiting, or diarrhea requires additional fluid. Patients undergoing dialysis can ingest fluids more freely.[16]

Electrolytes Serum electrolyte levels are monitored closely to determine appropriate electrolyte intakes. Generally, the potassium intake should be restricted to 2000 to 3000 milligrams daily. A phosphorus restriction may be necessary if hyperphosphatemia is present. In patients with oliguria, sodium may be restricted to between 1100 and 3300 milligrams per day; the sodium intake can be adjusted to match urinary sodium losses unless edema is present.[17] Patients undergoing dialysis can consume electrolytes more freely. As mentioned previously, oliguric patients who experience diuresis at the beginning of the recovery period may need electrolyte replacement to compensate for urinary losses.

Enteral and Parenteral Nutrition Some patients need enteral or parenteral nutrition support to obtain adequate energy. Enteral nutrition (tube feeding) is generally preferred over parenteral nutrition because it is less likely to cause infection and sepsis. Enteral formulas for patients with renal failure are more kcalorically dense and have lower protein and electrolyte concentrations than standard formulas. Total parenteral nutrition is necessary only if patients are severely malnourished or cannot consume food for more than 14 days.

IN SUMMARY

Acute renal failure is characterized by a rapid loss in kidney function, causing a buildup of fluid, electrolytes, and nitrogenous wastes in the blood. Acute renal failure may be caused by prerenal, intrinsic, or postrenal factors. Consequences may include oliguria, hyperkalemia, hyperphosphatemia, and uremia. If hyperkalemia develops, it can alter heart rate and lead to heart failure. Acute renal failure is treated with medications, dialysis, and dietary modifications. The accompanying Case Study checks your understanding of acute renal failure.

Chronic Kidney Disease

Unlike acute renal failure, in which kidney function declines suddenly and rapidly, **chronic kidney disease** ◆ is characterized by gradual, irreversible deterioration. Because the kidneys have a large functional reserve, ◆ the disease typically progresses over many years without causing symptoms. Patients are typically diagnosed late in the course of illness, after most kidney function has been lost.[18]

The most common causes of chronic kidney disease are diabetes mellitus and hypertension, which are estimated to cause 45 and 27 percent of cases, respectively.[19] Other conditions that lead to chronic kidney disease include inflammatory, immunological, and hereditary diseases that directly involve the kidneys. **Polycystic kidney disease,** characterized by the formation of multiple cysts in both kidneys, is a common hereditary condition that accounts for about 5 to 10 percent of chronic kidney disease cases.[20] Chronic kidney disease sometimes follows acute renal failure.

◆ Chronic kidney disease is also known as *chronic renal failure.*

◆ The kidneys' ability to function despite loss of nephrons is referred to as *renal reserve.*

chronic kidney disease: a kidney disease characterized by gradual, irreversible deterioration of the kidneys; also called **chronic renal failure.**

polycystic kidney disease: a hereditary disorder characterized by the formation of multiple cysts in the kidneys.

TABLE 28-3 Clinical Effects of Chronic Kidney Disease

Early Stages

- Anorexia
- Fatigue
- Headache
- Hypertension
- Itching
- Kidney inflammation or nephrotic syndrome
- Nausea and vomiting
- Proteinuria, hematuria (blood in urine)

Advanced Stages

- Anemia, bleeding tendency
- Cardiovascular disease
- Confusion, mental impairments
- Electrolyte abnormalities
- Fluid retention
- Hormonal abnormalities
- Metabolic acidosis
- Peripheral neuropathy
- Protein-energy malnutrition
- Reduced immunity
- Renal osteodystrophy

CASE STUDY — Store Manager with Acute Renal Failure

Catherine Garber is a 42-year-old store manager admitted to the hospital's intensive care unit. She was first seen in the emergency room with severe edema, headache, nausea and vomiting, and a rapid heart rate. She reported an inability to pass more than minimal amounts of urine in the past two days. Her son, who drove her to the emergency room, reported that she had missed work for several days and seemed confused and unusually tired. Laboratory tests revealed elevated serum creatinine, BUN, and potassium levels. After learning from her medical history that Mrs. Garber had begun taking penicillin earlier in the week, the physician diagnosed acute renal failure, probably caused by a reaction to the medication. Mrs. Garber is 5 feet 3 inches tall and weighs 125 pounds.

1. Describe the probable reason for Mrs. Garber's inability to produce urine. Is her reaction to penicillin considered a prerenal, intrinsic, or postrenal cause of renal failure? Give examples of other medical problems that can cause acute renal failure.
2. What medications can the physician prescribe to treat Mrs. Garber's edema and hyperkalemia? What recommendation is likely regarding her continued use of penicillin?
3. What concerns should be kept in mind when determining Mrs. Garber's energy, protein, fluid, and electrolyte needs during acute renal failure? How would dialysis treatment alter recommendations?
4. After treatment begins, Mrs. Garber suddenly begins producing copious amounts of urine. How should this development alter dietary treatment?

As you read through the discussion of chronic kidney disease, consider how Mrs. Garber's diet should change if her kidney problems become chronic.

Consequences of Chronic Kidney Disease

In the early stages of chronic kidney disease, the nephrons compensate by enlarging so that they can handle the extra workload. As the nephrons deteriorate, however, there is additional work for the remaining nephrons. The overburdened nephrons continue to degenerate until finally the kidneys are unable to function adequately, resulting in kidney failure. Once the extent of kidney damage necessitates active treatment—either dialysis or a kidney transplant—the condition is classified as **end-stage renal disease (ESRD).** Without intervention at this stage, an individual cannot survive. Table 28-3 lists the common clinical effects of the early and advanced stages of chronic kidney disease. Symptoms of chronic kidney disease may not appear until over 75 percent of kidney function is lost.[21]

Renal disease is evaluated using the **glomerular filtration rate (GFR),** the rate at which the kidneys form filtrate. The GFR can be estimated using predictive equations that are based on serum creatinine levels, ◆ age, gender, race, and body size. Table 28-4 shows how chronic kidney disease is classified according to estimated GFR. Other laboratory measures used to assess kidney function include urinary protein levels, BUN, and the ratio of albumin to creatinine in a urine sample.[22]

◆ Reminder: *Creatinine* is a waste product of creatine, a nitrogen-containing compound in muscle cells.

TABLE 28-4 Evaluation of Chronic Kidney Disease

Stage of Disease	Description	GFR[a] (mL/min per 1.73 m²)
1	Kidney damage with normal or increased GFR	≥90
2	Kidney damage with mildly decreased GFR	60–89
3	Moderately decreased GFR	30–59
4	Severely decreased GFR	15–29
5	Kidney failure	<15 (or undergoing dialysis)

[a]Glomerular filtration rate, or GFR, is estimated from the Modification of Diet in Renal Disease study equation and is based on age, gender, race, and calibration for serum creatinine. Normal GFR is approximately 125 mL/min.
SOURCE: A. S. Levey and coauthors, National Kidney Foundation practice guidelines for chronic kidney disease: Evaluation, classification, and stratification, *Annals of Internal Medicine* 139 (2003): 137–147.

end-stage renal disease (ESRD): an advanced stage of chronic kidney disease in which dialysis or a kidney transplant is necessary to sustain life.

glomerular filtration rate (GFR): the rate at which filtrate is formed within the kidneys, normally approximately 125 mL/min.

◆ Reminder: *Aldosterone* promotes sodium (and therefore water) retention and potassium excretion.

◆ Reminder: *Parathyroid hormone* helps to regulate serum concentrations of calcium and phosphorus. Elevated parathyroid hormone stimulates bone turnover and the release of calcium from bone into blood.

Altered Electrolytes and Hormones Fluid and electrolyte disturbances may not develop until the third or fourth stage of chronic kidney disease (see Table 28-4).[23] As GFR falls, the increased activity by the remaining nephrons is often sufficient to maintain electrolyte excretion. A number of hormonal adaptations also help to regulate electrolyte levels, but these changes may cause complications of their own. The increased secretion of aldosterone ◆ helps to prevent increases in serum potassium but contributes to the development of hypertension (in patients who were not previously hypertensive). Increased secretion of parathyroid hormone ◆ helps to prevent elevations in serum phosphorus but contributes to bone loss and the development of **renal osteodystrophy,** a bone disorder common in renal patients. Electrolyte imbalances are likely when GFR becomes extremely low (less than 5 milliliters per minute), when hormonal adaptations are inadequate, or when intakes of water and electrolytes are either very restricted or excessive.

Because the kidneys are responsible for maintaining acid-base balance, acidosis often develops in chronic kidney disease. Although usually mild, the acidosis exacerbates renal bone disease because compounds in bone (for example, protein and phosphates) are released to buffer the acid in blood.

Uremic Syndrome Uremia usually develops during the final stages of chronic renal failure, when GFR is below 15 milliliters per minute and BUN exceeds 60 milligrams per deciliter.[24] The many symptoms and complications that develop during this stage of illness are collectively known as the **uremic syndrome.** The uremia itself can cause subtle mental dysfunctions and neuromuscular changes such as muscle cramping, twitching, and restless leg syndrome. Other complications include:

- *Impaired hormone synthesis.* Diseased kidneys are unable to produce erythropoietin, causing anemia. Reduced production of active vitamin D contributes to bone disease.

- *Impaired hormone degradation.* Imbalances develop in hormones involved in growth, reproduction, fluid balance, blood glucose regulation, and nutrient metabolism.

- *Bleeding abnormalities.* Defects in platelet function and clotting factors prolong bleeding time and contribute to bruising, gastrointestinal bleeding, and anemia.

- *Increased cardiovascular disease risk.* Chronic kidney disease worsens risk factors such as hypertension, insulin resistance, and homocysteinemia. Elevated parathyroid hormone levels lead to calcification of blood vessels and heart tissue. Patients are at increased risk of stroke, heart attack, and heart failure.

- *Reduced immunity.* Patients with uremia have poor immune responses and are at high risk of developing infections, which are a frequent cause of death.

Protein-Energy Malnutrition Patients with chronic kidney disease often develop PEM and wasting. Clinical studies have suggested that renal patients may have inadequate protein and energy intakes, even during the early stages of disease.[25] Anorexia is thought to contribute to poor food intake and may result from hormonal disturbances, nausea and vomiting, restrictive diets, uremia, and medications. Nutrient losses also contribute to malnutrition and may be a consequence of vomiting, diarrhea, gastrointestinal bleeding, and dialysis. In addition, many of the illnesses that lead to chronic kidney disease can induce a catabolic state that contributes to protein losses.[26]

renal osteodystrophy: a bone disorder that develops in patients with chronic kidney disease as a consequence of increased secretion of parathyroid hormone, reduced serum calcium, acidosis, and impaired vitamin D activation by the kidneys.

uremic syndrome: the cluster of symptoms associated with a GFR below 15 mL/min, including uremia, anemia, bone disease, hormonal imbalances, bleeding impairment, increased cardiovascular disease risk, and reduced immunity.

Treatment of Chronic Kidney Disease

The goals of treatment for patients with chronic kidney disease are to slow disease progression and prevent or alleviate symptoms. Dietary measures help to prevent

PEM and weight loss. Once kidney disease reaches the final stages, dialysis or a kidney transplant is necessary to sustain life.

Drug Therapy for Chronic Kidney Disease Drug therapies help to control some of the complications associated with chronic kidney disease. Treatment of hypertension is critical for slowing disease progression and reducing cardiovascular disease risk; thus antihypertensive drugs are usually prescribed (see Chapter 27). Some antihypertensive drugs (such as ACE inhibitors) can reduce proteinuria, helping to prevent additional kidney damage. Anemia is usually treated by injection or intravenous administration of erythropoietin (epoetin). Other common drug treatments include the administration of phosphate binders (taken with food) to reduce serum phosphorus levels, sodium bicarbonate to reverse acidosis, and cholesterol-lowering medications. Supplementation with active vitamin D (called *calcitriol)* helps to raise serum calcium and reduce parathyroid hormone levels.

Dialysis Dialysis replaces kidney function by removing excess fluid and wastes from the blood. In **hemodialysis,** the blood is circulated through a **dialyzer** (artificial kidney), where it is bathed by a **dialysate,** a solution that selectively removes fluid and wastes. In **peritoneal dialysis,** the dialysate is infused into a person's peritoneal cavity, and blood is filtered by the peritoneum (the membrane that surrounds the abdominal cavity). After several hours, the dialysate is drained, removing unneeded fluid and wastes. Highlight 28 provides additional information about dialysis.

Medical Nutrition Therapy for Chronic Kidney Disease The patient's diet strongly influences disease progression, the development of complications, and serum levels of nitrogenous wastes and electrolytes. Because the dietary measures for chronic kidney disease are complex and nutrient needs change frequently during the course of illness, a dietitian who specializes in renal disease is best suited to provide medical nutrition therapy. Table 28-5 summarizes the dietary guidelines for patients in the different stages of chronic kidney disease. The predialysis guidelines apply to patients in stages 1 through 4; by stage 5, either hemodialysis or peritoneal dialysis is necessary. Because patients' needs can vary considerably, actual recommendations should be based on the results of a nutrition assessment.

hemodialysis (HE-moe-dye-AL-ih-sis): a treatment that removes fluids and wastes from the blood by passing the blood through a dialyzer.

dialyzer (DYE-ah-LYE-zer): a machine used in hemodialysis to filter the blood; also called an *artificial kidney.*

dialysate (dye-AL-ih-sate): the solution used in dialysis to draw wastes and fluids from the blood.

peritoneal (PEH-rih-toe-NEE-al) **dialysis:** a treatment that removes fluids and wastes from the blood by using the peritoneal membrane as a filter.

TABLE 28-5 Dietary Recommendations for Chronic Kidney Disease

Nutrient	Predialysis	Hemodialysis	Peritoneal Dialysis
Energy[a] (kcal/kg body weight)	35 for <60 years old 30–35 for ≥60 years old	35 for <60 years old 30–35 for ≥60 years old	35 for <60 years old 30–35 for ≥60 years old (total kcalories should include those absorbed from the dialysate)
Protein (g/kg body weight)	0.60–0.75 (≥50% high-quality proteins)	≥1.2 (≥50% high-quality proteins)	≥1.2–1.3 (≥50% high-quality proteins)
Fat	As necessary to maintain a healthy lipid profile	As necessary to maintain a healthy lipid profile	As necessary to maintain a healthy lipid profile
Fluid (mL/day)	Unrestricted if urine output is normal	1000 plus urine output	As necessary to maintain fluid balance
Sodium (mg/day)	1000–3000	1000–3000	2000–4000
Potassium (mg/day)	Unrestricted unless hyperkalemia is present	2000–3000; adjust according to serum potassium levels	3000–4000; adjust according to serum potassium levels
Calcium (mg/day)	1000–1500	≤2000 from diet and medications	≤2000 from diet and medications
Phosphorus (mg/day)	800–1000 if serum phosphorus or parathyroid hormone is elevated	800–1000 if serum phosphorus or parathyroid hormone is elevated	800–1000 if serum phosphorus or parathyroid hormone is elevated

[a]Values listed apply to adults; recommendations for children should not fall below DRI levels.
SOURCE: National Kidney Foundation

◆ Reminder: Foods with high energy density contain a high number of kcalories per unit weight; these foods are generally high in fat and low in water content.

◆ The dialysate contains glucose in order to draw fluid from the blood to the peritoneal cavity by osmosis.

◆ The protein RDA for adults is 0.8 g/kg body weight.

◆ Reminder: Most salad dressings and mayonnaise products are made with polyunsaturated or monounsaturated vegetable oils.

In a renal diet, at least half of the protein consumed should be from high-quality sources such as eggs, milk, meat, poultry, and fish.

Energy The energy intake should be high enough to allow patients to maintain a healthy weight and to prevent wasting. Foods and beverages with high energy density are typically recommended. ◆ Malnourished patients may require oral supplements or tube feedings to maintain weight.

Patients undergoing peritoneal dialysis can absorb a substantial amount of glucose from the dialysate, which can contribute as many as 800 kcalories daily. ◆ These kcalories must be included in estimates of energy intake. Weight gain is sometimes a problem when peritoneal dialysis continues for a long period.

Protein A low-protein diet is usually prescribed to help slow the progression of kidney disease. During the predialysis period, the recommended protein intake is 0.60 to 0.75 grams per kilogram of body weight per day.[27] ◆ Low-protein diets produce fewer nitrogenous wastes and therefore reduce the risk of uremia. In addition, low-protein diets supply less phosphorus than high-protein diets, reducing the risks associated with hyperphosphatemia. Low-protein breads, pastas, and other grain-based products are commercially available to help renal patients improve their energy intakes without increasing protein consumption.

Because renal patients often develop PEM, however, their diet must provide enough protein to meet needs and prevent wasting. Therefore, at least 50 percent of the protein consumed should be from high-quality protein sources (such as eggs, milk products, meat, poultry, fish, and soybeans) to ensure that the patient consumes adequate amounts of the essential amino acids.

Because of the high risk of wasting and compliance difficulties associated with low-protein diets, some dietitians suggest that patients consume higher amounts of protein to preserve health.[28] Another option is to encourage patients to undergo a resistance training program, which can improve protein utilization, nitrogen retention, and muscle strength despite adherence to a low-protein diet.[29] Once dialysis has begun, protein restrictions can be relaxed, because dialysis removes nitrogenous wastes and results in some amino acid losses as well.

Lipids Patients with chronic kidney disease are at increased risk of coronary heart disease and are therefore advised to restrict their intakes of saturated fat, *trans* fat, and cholesterol to help control elevated blood lipids. Although patients are often encouraged to consume high-fat foods to improve their energy intakes, the foods they select should provide mostly unsaturated fats; good choices include nuts and seeds, oil-based salad dressings, mayonnaise, ◆ avocados, and soybean products (Highlight 5 provides additional suggestions).

Sodium and Fluids As kidney disease progresses, the patient excretes less urine and cannot handle normal intakes of sodium and fluids. Recommendations depend on the total urine output, changes in body weight and blood pressure, and serum sodium levels. A rise in body weight and blood pressure suggests that the person is retaining sodium and fluid; conversely, declines in these measurements indicate fluid loss. Most people with kidney disease tend to retain sodium and may benefit from mild restriction; less frequently, a patient may have a salt-wasting condition that requires additional dietary sodium.

Fluids are not restricted until urine output decreases. For a person who is neither dehydrated nor overhydrated, the daily fluid intake should match the daily urine output. (Obligatory water losses—from skin and lungs—are replaced from the water contained in solid foods.) Once a person is on dialysis, sodium and fluid intakes should be controlled so that only about 2 pounds of water weight is gained daily—this excess fluid is then removed during the next dialysis treatment.[30] Patients on fluid-restricted diets should be advised that foods such as flavored gelatin, soups, fruit ices, frozen fruit juice bars, and ice milk contribute to the fluid allowance.

Potassium Before dialysis treatments begin, most renal patients can handle typical intakes of potassium. Potassium restrictions are usually necessary only in patients with elevated potassium levels. Individuals with diabetic nephropathy are at high risk of hyperkalemia and may need to limit dietary potassium during the early stages of disease. Conversely, potassium supplementation may be necessary for persons using potassium-wasting diuretics.

Dialysis patients must control potassium intakes to prevent hyperkalemia or, more rarely, **hypokalemia.** Restriction is usually necessary for people treated with hemodialysis, whereas those undergoing peritoneal dialysis can consume potassium more freely. Recommended intakes are based on serum potassium levels, renal function, medications, and the dialysis procedure used.

All fresh foods provide potassium, but some fruits and vegetables contain such high amounts that their regular use is discouraged in patients at risk of hyperkalemia; examples include avocados, bananas, beets, brussels sprouts, chard, dates, dried beans and peas, dried fruit, honeydew melon, nectarines, oranges, potatoes, spinach, sweet potatoes, tomatoes, winter squash, and zucchini.[31] Foods in other food groups that are high in potassium include milk and milk products, molasses, and nuts; patients with potassium restrictions must limit these foods as well. Individuals with renal failure should be cautioned that salt substitutes and other low-sodium products often contain potassium chloride, which people on a potassium-restricted diet should avoid. Appendix H provides additional information about the potassium content of common foods.

People on a renal diet can consume most fruits and vegetables in limited amounts.

Calcium, Phosphorus, and Vitamin D To prevent bone disease, calcium and phosphorus intakes may need adjustment, even during the early stages of kidney disease. Laboratory values usually help to guide dietary recommendations for these nutrients. Serum calcium levels must be monitored to guard against **hypercalcemia,** which can develop in response to simultaneous calcium and vitamin D supplementation. Elevated serum phosphorus levels indicate the need for dietary phosphorus restriction. Vitamin D supplementation is standard treatment for many renal patients, ◆ but the amount prescribed depends on the serum levels of calcium, phosphorus, and parathyroid hormone.

High-protein foods are also high in phosphorus, so the protein-restricted diets consumed by predialysis patients curb phosphorus intakes as well. After dialysis treatments begin and protein intakes are liberalized, phosphate binders (taken with meals) become essential for phosphorus control. Because foods that are rich in calcium (such as milk and milk products) are usually high in phosphorus and are therefore restricted, patients must rely on calcium supplements to meet their calcium needs.

◆ Reminder: Diseased kidneys are unable to produce activated vitamin D, which normally regulates calcium absorption and helps to maintain serum calcium levels.

Vitamins and Minerals The restrictive renal diet interferes with vitamin and mineral intakes, increasing the risk of deficiencies. In addition, patients treated with dialysis lose water-soluble vitamins and some trace minerals into the dialysate. Dietary supplements for renal patients typically supply generous amounts of folate and vitamin B_6—1 milligram and 10 milligrams per day, respectively—along with recommended amounts of the other water-soluble vitamins. Supplemental vitamin C should be limited to 100 milligrams per day, because excessive intakes can contribute to kidney stone formation in individuals at risk (see p. 889). Supplements containing vitamins A and E are not recommended because these vitamins sometimes accumulate in patients with renal disease.[32]

Iron deficiency is common in hemodialysis patients and may be due to inadequate erythropoietin, gastrointestinal bleeding, reduced iron absorption, or blood losses associated with the dialysis treatment.[33] Intravenous administration of iron, in conjunction with erythropoietin therapy, is more effective than oral iron supplementation for improving iron status.

Enteral and Parenteral Nutrition Enteral and parenteral nutrition support can provide nutrients to renal patients who cannot consume adequate amounts of food. The enteral formulas suitable for chronic kidney disease are more kcalorically dense and have lower protein and electrolyte concentrations than standard formulas. **Intradialytic parenteral nutrition** is an option for supplying supplemental nutrients to dialysis patients; this technique combines parenteral

hypokalemia (HIGH-po-ka-LEE-me-ah): low serum potassium levels.

hypercalcemia (HIGH-per-kal-SEE-me-ah): elevated serum calcium levels.

intradialytic parenteral nutrition: the infusion of nutrients during hemodialysis, often providing amino acids, dextrose, lipids, and some trace minerals.

feedings with hemodialysis treatments. An advantage of this approach is that the volume of parenteral solution infused can be simultaneously removed (recall that fluid intake is controlled in dialysis patients). However, clinical studies have not shown intradialytic parenteral nutrition to be more successful than oral supplementation in improving nutrition status or mortality rates in malnourished dialysis patients.[34]

Dietary Compliance Adhering to a renal diet is probably the most difficult aspect of treatment for patients with renal disease. These patients often require extensive counseling once multiple dietary restrictions become necessary. Depending on the stage of illness and the patient's laboratory values, the renal diet may limit protein, fluids, sodium, potassium, and phosphorus, thereby affecting food selections from all major food groups. Because these diets have so many restrictions, patient compliance is often a problem.[35] The "How to" below provides suggestions to help patients comply with renal diets. The accompanying Case Study allows you to apply your knowledge about chronic kidney disease and hemodialysis.

Kidney Transplants

A preferred alternative to dialysis in patients with end-stage renal disease is kidney transplantation.[36] A successful kidney transplant restores kidney function, allows a more liberal diet, and frees the patient from routine dialysis. Given the choice, many patients would prefer transplants, but the demand for suitable kidneys far exceeds the supply. Other barriers to transplantation include advanced age, poor health, financial difficulties, and abnormalities of the urinary tract. Fewer than 20 percent of patients who develop end-stage renal disease receive a kidney transplant.[37]

Immunosuppressive Drug Therapy To prevent tissue rejection following transplant surgery, patients require high doses of immunosuppressive drugs such as corticosteroids, cyclosporine, tacrolimus, sirolimus, and azathioprine. These drugs have multiple effects that can alter nutrition status, including nausea, vomiting, diarrhea, glucose intolerance, altered blood lipids, fluid retention, hypertension,

---

HOW TO Help Patients Comply with a Renal Diet

Patients with renal disease and their caregivers face considerable challenges as they learn to manage a renal diet. The following suggestions may help:

1. *To keep track of fluid intake:*
   - Fill a container with an amount of water equal to your total fluid allowance. Each time you use a liquid food or beverage, discard an equivalent amount of water from the container. The amount remaining in the container will show you how much fluid you have left for the day.
   - Be sure to save enough fluid to take medications.

2. *To help control thirst:*
   - Chew gum or suck hard candy.
   - Freeze beverages to a semisolid state so that they take longer to consume.
   - Add lemon juice or crumpled mint leaves to water to make it more refreshing.
   - Gargle with refrigerated mouthwash.

3. *To increase the energy content of meals:*
   - Add extra margarine or butter to rice, noodles, breads, crackers, and cooked vegetables. Add extra salad dressing or oil to salads.
   - Add nondairy whipped toppings to desserts.
   - Include fried foods in your diet.

4. *To include more of your favorite vegetables in meals:*
   - Consult your dietitian or nurse to learn whether you can safely use the process of leaching to remove some of the potassium from vegetables.
   - To leach potassium from vegetables: Cut the vegetables into $1/8$-inch slices and rinse. Soak the vegetables in a large amount of warm water for two hours—about ten parts of water to one part of vegetables. Cook the vegetables using five parts of water to one part of vegetables.

5. *To prevent the diet from becoming monotonous:*
   - Experiment with new combinations of allowed foods.
   - Substitute nondairy products for milk products. Nondairy products, which are lower in protein, phosphorus, and potassium, can substitute for milk and add energy to the diet.
   - Add flavor to foods by seasoning with garlic, onion, chili powder, curry powder, oregano, mint, basil, parsley, pepper, or lemon juice.
   - Consult a dietitian or nurse when you want to eat restricted foods. Many restricted foods can be used occasionally and in small amounts if the menu is carefully adjusted.

CASE STUDY Banker with Chronic Kidney Disease

Thomas Stone is a 55-year-old banker who developed chronic kidney disease as a result of hypertension. His condition was discovered several years ago, when routine laboratory tests revealed elevated serum creatinine and BUN levels. Since then, he has been taking antihypertensive medications and restricting dietary sodium; he reported difficulty following the low-protein diet that was also prescribed. Mr. Stone recently visited his doctor with complaints of low urine output and reduced sensation in his hands and feet. He also reported feeling drowsy at work and mentioned that he was bruising more than usual. The examination revealed a 9-pound weight gain since his last visit and swelling in his ankles and feet. Tests revealed that his GFR had fallen to 10 milliliters per minute. Mr. Stone is 5 feet 8 inches tall and normally weighs 160 pounds.

1. Explain how chronic kidney disease progresses. What happens to GFR, serum creatinine levels, and BUN as renal function declines?
2. Describe the clinical effects you would expect during the final stage of disease, when kidney failure develops. Explain the significance of each of Mr. Stone's physical complaints.
3. Explain why a low-sodium, low-protein diet was prescribed for Mr. Stone at a former visit. What energy and protein intakes were probably recommended at that time?
4. The physician determines that Mr. Stone's kidney disease has reached the final stage and prescribes hemodialysis. How will dialysis alter Mr. Stone's diet? Calculate his new protein recommendation, and compare it to the amount of protein recommended before dialysis. What other changes in nutrient intake may be necessary?

and infection. Because immunosuppressive drug therapy increases the risk of foodborne infection, food safety guidelines should be provided to patients and caregivers. The Diet-Drug Interactions feature summarizes the nutrition-related effects of immunosuppressants and other drugs mentioned in this chapter.

DIET-DRUG INTERACTIONS

Check this table for notable nutrition-related effects of the medications discussed in this chapter.

	Gastrointestinal Effects	Interactions with Dietary Substances	Metabolic Effects
Anti-inflammatory/ immunosuppressive drugs			
Corticosteroids	—	—	Glucose intolerance, sodium retention, negative nitrogen balance, appetite stimulation, weight gain, growth suppression in children.
Cyclosporine	—	Grapefruit juice raises drug levels and increases risk of toxicity. Avoid potassium supplements and salt substitutes with potassium. St. John's wort can alter drug efficacy.	Hyperkalemia, hypomagnesia, hyperglycemia, hyperlipidemia.
Phosphate binders (calcium containing)	Constipation.	—	Electrolyte imbalances.
Potassium-exchange resins (sodium polystyrene sulfonate)	—	—	Fluid retention, hypokalemia, hypocalcemia.
Potassium citrate	Nausea, vomiting, stomach pain, diarrhea.	—	Hyperkalemia.

TABLE 28-6	Dietary Guidelines Following a Kidney Transplant

- **Energy:** Initially, 30–35 kcal/kg body weight per day; reduce to 25–30 kcal/kg body weight per day to maintain healthy body weight, and adjust as necessary.
- **Protein:** Initially, 1.3 to 1.5 g/kg body weight per day; reduce to 1.0 g/kg per day after 6–8 weeks.
- **Carbohydrate:** For hyperglycemia, maintain consistent carbohydrate intake at meals and snacks.
- **Fat:** Limit saturated fat (to <10% of total kcalories) and cholesterol (to <300 mg/day) to help control serum lipids.
- **Sodium:** Generally unrestricted. Restrict to 2–4 g/day if fluid retention and hypertension are present.
- **Potassium:** Generally unrestricted. Adjust according to serum potassium levels if necessary.
- **Calcium:** 1200 mg/day to minimize bone loss associated with drug therapy.
- **Phosphorus:** Unrestricted.
- **Fluid:** Unrestricted.

SOURCE: J. A. Beto and V. K. Bansal, Medical nutrition therapy in chronic kidney failure: Integrating clinical practice guidelines, *Journal of the American Dietetic Association* 104 (2004): 404–409.

Nutrition Therapy after Kidney Transplant Protein and energy requirements increase after surgery due to stress and the catabolic effects of drug therapy. Once recovery is under way, the side effects of drugs can strongly influence dietary treatment. Typical dietary modifications are shown in Table 28-6.[38] Hyperglycemia (due to drug treatment) may be improved by controlling carbohydrate intake, although oral medications or insulin therapy may also be required. Because blood lipids are frequently elevated, patients should limit saturated fat and cholesterol intakes. Sodium, potassium, and phosphorus intakes are often liberalized following a transplant, but serum electrolyte levels must be monitored closely because some drug therapies can cause hyperkalemia or hypophosphatemia. Calcium supplementation is advised due to urinary calcium losses associated with corticosteroids (which are often used as immunosuppressants). To avoid foodborne illness, patients should avoid raw or undercooked meat, fish, poultry, and eggs; unpasteurized milk and juices; cheese made with unpasteurized milk; fresh bean sprouts; and food that is spoiled or moldy.[39]

IN SUMMARY

Chronic kidney disease causes gradual loss of kidney function and often results from long-standing diabetes mellitus or hypertension. Depending on the stage of chronic kidney disease, its complications may include fluid and electrolyte disturbances, hypertension, renal osteodystrophy, mental impairments, bleeding abnormalities, anemia, increased risk for cardiovascular disease, and reduced immunity. Treatment can slow disease progression and correct complications; it includes drug therapies, dialysis, and medical nutrition therapy. Dietary measures usually feature a low-protein diet, controlled fluid and sodium intakes, phosphorus restrictions, and calcium and vitamin D supplementation; potassium restrictions are usually necessary after dialysis treatment begins. Kidney transplantation can restore renal function and liberalize dietary restrictions.

Kidney Stones

Approximately 12 percent of men and 6 percent of women in the United States develop one or more **kidney stones** during their lifetimes.[40] A kidney stone is a crystalline mass that forms within the urinary tract. Although stones are often asymptomatic, their passage can cause severe pain or block the urinary tract. Stones tend to recur but can be prevented with dietary measures and medical treatment.

kidney stones: crystalline masses that form in the urinary tract; also called *renal calculi* and *nephrolithiasis*.

Formation of Kidney Stones

Kidney stones develop when stone constituents become concentrated in urine, allowing crystals to form and grow. About 70 percent of kidney stones are made up primarily of calcium oxalate. Less commonly, stones are composed of uric acid, the amino acid cystine, calcium phosphate, or magnesium ammonium phosphate (the latter are known as *struvite* stones). Factors that predispose an individual to stone formation include the following:

- *Dehydration* or *low urine volume,* which promotes the crystallization of minerals and other compounds in urine.

- *Obstruction,* which prevents the flow of urine and encourages salt precipitation.

- *Urine acidity,* which affects the dissolution of urinary constituents. Some stones form more readily in acidic urine, whereas others form in alkaline urine.

- *Metabolic factors,* which affect the presence of compounds that either promote or inhibit crystal growth.

- *Renal disease,* which is associated with calcification of tissues and phosphate accumulation.

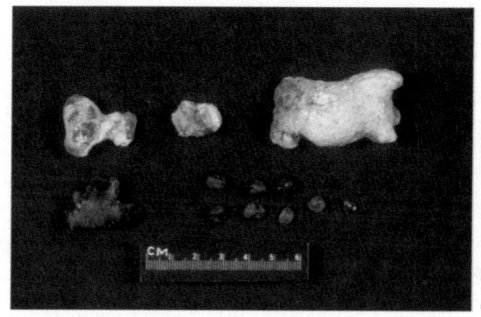

The most common type of kidney stone is composed of calcium oxalate crystals, as shown here.

The most common types of kidney stones are described in this section.

Calcium Oxalate Stones The most common abnormality in people with calcium oxalate stones is **hypercalciuria** (elevated urinary calcium levels). Hypercalciuria can result from excessive calcium absorption, impaired calcium reabsorption in kidney tubules, or elevated serum levels of parathyroid hormone or vitamin D. However, some people with calcium oxalate stones excrete normal amounts of calcium in the urine, and the reason they form stones is unknown.

Elevated urinary oxalate levels, or **hyperoxaluria,** also promote the formation of calcium oxalate crystals. Oxalate is a normal product of metabolism that readily binds to calcium. Hyperoxaluria may reflect an increase in the body's synthesis of oxalate or increased absorption from dietary sources. ◆ People who form calcium oxalate stones are advised to reduce their dietary intake of oxalate (see Table 28-7 on p. 890) and to avoid supplementation with vitamin C, which degrades to oxalate in the body.[41]

◆ Reminder: Fat malabsorption promotes oxalate absorption, thereby increasing the risk of forming calcium oxalate stones (see Chapter 24).

Uric Acid Stones Uric acid stones develop when the urine is abnormally acidic, contains excessive uric acid, or both. These stones are frequently associated with **gout,** a metabolic disorder characterized by elevated uric acid levels in the blood and urine. Other diseases that increase the risk of uric acid stones include leukemia, lymphoma, and glycogen storage disease: these conditions are associated with the overproduction of **purines,** which degrade to uric acid in the body. A diet rich in purines also contributes to high uric acid levels (see Table 28-8 on p. 890).

Cystine and Struvite Stones Cystine stones can form in people with the inherited disorder **cystinuria,** in which the renal tubules are unable to reabsorb the amino acid cystine. The abnormality results in abnormally high concentrations of cystine in the urine, leading to subsequent crystallization and stone formation. **Struvite** stones, composed primarily of magnesium ammonium phosphate, form in alkaline urine as a result of the bacterial degradation of urea to ammonia. The ammonia elevates urinary pH to a level that favors struvite formation. Struvite stones can accompany chronic urinary infections or disorders that interfere with urinary flow.

hypercalciuria (HIGH-per-kal-see-YOO-ree-ah): elevated urinary calcium levels.

hyperoxaluria (HIGH-per-ox-ah-LOO-ree-ah): elevated urinary oxalate levels.

gout (GOWT): a metabolic disorder characterized by elevated uric acid levels in the blood and urine and the deposition of uric acid in and around the joints, causing acute joint inflammation.

purines (PYOO-reens): products of nucleotide metabolism that degrade to uric acid.

cystinuria (SIS-tin-NOO-ree-ah): an inherited disorder characterized by elevated urinary excretion of several amino acids, including cystine.

struvite (STROO-vite): crystals of magnesium ammonium phosphate.

renal colic: the intense pain that occurs when a kidney stone passes through the ureter.

Consequences of Kidney Stones

In most cases, kidney stones do not pose serious medical problems. Small stones can readily pass through the ureters and out of the body with minimal treatment.

Renal Colic A stone passing through the ureter can produce severe, stabbing pain, called **renal colic.** Generally, the pain begins in the back and intensifies as

TABLE 28-7 Foods High in Oxalate

Vegetables	Fruits	Other
Beans, green and wax	Blackberries	Chocolate and chocolate beverages*
Beets*	Blueberries	Cocoa
Celery	Currants, red	Coffee
Chard, Swiss	Gooseberries	Draft beer
Collard greens	Grapes, Concord	Fruitcake
Dandelion greens	Lemon peel	Grits
Eggplant	Lime peel	Nuts, nut butters*
Endive	Orange peel	Peanut butter*
Escarole	Raspberries	Pepper
Leeks	Rhubarb*	Soybean crackers
Legumes	Strawberries*	Tea*
Okra		Tofu
Parsley		Wheat bran*
Potatoes, sweet		Wheat germ
Spinach*		
Squash, summer		

NOTE: The oxalate content of many foods has not been analyzed, and even fewer studies have been conducted to determine which foods raise urinary oxalate levels. * The foods marked with an asterisk have been documented to raise urinary oxalate levels and should be avoided by people who form calcium stones.

TABLE 28-8 Foods High in Purines

Organ Meats	Meat and Meat Products	Seafood
Brains	Game meat	Anchovies
Kidney	Gravies	Herring
Liver	Meat extracts	Mackerel
Sweetbreads		Sardines
		Scallops

SOURCE: J. A. T. Pennington, *Bowes and Church's Food Values of Portions Commonly Used* (Philadelphia: J. B. Lippincott, 1994), p. 387.

the stone travels toward the bladder (review Figure 28-1 on p. 874). The pain can be severe enough to cause nausea and vomiting and sometimes requires medication. When the stone reaches the bladder, the pain abruptly stops. Blood may appear in the urine **(hematuria)** as a result of damage to the kidney or ureter lining.

Urinary Tract Complications Depending on the location of the stone, symptoms may include urination urgency, frequent urination, or inability to urinate. Stones that are unable to pass through the ureter can cause a urinary tract obstruction and possibly lead to infection.

hematuria (HE-mah-TOO-ree-ah): blood in the urine.

Prevention and Treatment of Kidney Stones

Solutes are less likely to crystallize and form stones in dilute urine. Therefore, people who form kidney stones are advised to drink 12 to 16 cups of fluids daily in order to maintain urine volumes of least $2^{1}/_{2}$ liters per day.[42] Additional fluid may be needed in hot weather or if an individual is extremely active. Acceptable fluid sources include tea, coffee, wine, and beer, but apple and grapefruit juices should be limited because they may increase the risk of stones.[43]

Calcium Oxalate Stones Dietary measures and drug treatments aim to reduce urinary calcium and oxalate levels. Furthermore, the uric acid content of urine must be kept low, because uric acid reduces the solubility of calcium oxalate.[44] Thiazide diuretics are a mainstay of drug therapy and help to reduce urinary calcium by enhancing calcium reabsorption in the kidney tubules.[45] Other medications include cholestyramine, which reduces oxalate absorption, and allopurinol, which reduces uric acid production in the body. Potassium citrate can inhibit the formation and growth of crystals but may cause stomach upset and diarrhea.

Medical nutrition therapy includes adjustments in calcium, oxalate, protein, and sodium intakes.[46] Patients should consume adequate calcium from food sources (800 to 1200 milligrams per day) because dietary calcium combines with oxalate in the intestines, reducing oxalate absorption and helping to control hyperoxaluria. ◆ Conversely, low-calcium diets promote oxalate absorption and higher urinary oxalate levels. Foods high in oxalate should be restricted, because dietary oxalate contributes to urinary oxalate content (review Table 28-7). High protein and sodium intakes increase urinary calcium excretion, so moderate protein consumption (about 0.8 to 1.0 gram per kilogram of body weight per day) and sodium restriction (no more than 3450 milligrams daily) are also advised. Vitamin C intake should be limited to the DRI (75 and 90 milligrams per day for women and men, respectively).

Uric Acid Stones Drug treatments for uric acid stones include allopurinol to reduce urinary uric acid and potassium citrate to reduce urine acidity. Diets restricted in purines may also help to control urinary uric acid levels (review Table 28-8). Because all meats, poultry, fish, and shellfish contain considerable amounts of purines, strict dietary control over a long period may be difficult to achieve. In addition, the benefits of purine restriction are unknown.

Cystine and Struvite Stones High fluid intakes may prevent the formation of cystine stones in some patients, whereas other individuals require drug therapy to reduce cystine production in the body. Medications frequently prescribed include penicillamine and tiopronin, which reduce cystine levels, and potassium citrate, which reduces urine acidity.

Preventing urinary tract infections is an important strategy for preventing struvite stones. Patients with these stones may require antibiotic therapy to prevent further stone formation.

© James Darell/Getty Images

Drinking plenty of water throughout the day is the most important measure for preventing kidney stones.

◆ Because calcium supplements can elevate urinary calcium levels, they are not as helpful as food sources of calcium.

IN SUMMARY

Kidney stones form when stone constituents—calcium oxalate, calcium phosphate, uric acid, cystine, or magnesium ammonium phosphate—crystallize in urine. Complications include renal colic, difficulty with urination, and obstruction. Kidney stones may be prevented by maintaining urine volumes of at least 2^1/$_2$ liters daily. Other dietary measures include consumption of enough calcium to control oxalate absorption, dietary oxalate and purine restrictions, a moderate protein intake, and sodium restriction.

CENGAGENOW™
academic.cengage.com/login

Clinical Portfolio

1. A person with chronic kidney disease may need multiple medications to control disease progression and treat symptoms and complications. For people with diabetes and hyperlipidemias who develop chronic kidney disease, medications might include insulin, oral hypoglycemic drugs, antihypertensives, diuretics, lipid-lowering medications, and phosphate binders. Review the nutrition-related side effects of these medications. Describe the ways in which these medications may make it harder for people to maintain nutrition status.

2. Because the diet for chronic kidney disease is so restrictive, patients find it difficult to manage and maintain over the long term. Review the suggestions in the "How to" on p. 886. Can you think of additional suggestions that may help? List ideas that may help patients adjust to each of the different aspects of their renal diets.

NUTRITION ON THE NET

For further study of topics covered in this chapter, log on to **academic.cengage.com/nutrition/rolfes/UNCN8e**. Go to Chapter 28, then to Nutrition on the Net.

- To search for specific topics related to kidney diseases, dialysis, and kidney transplants, visit these sites:
Kidney Foundation of Canada: **www.kidney.ca**
National Institute of Diabetes and Digestive and Kidney Diseases: **www2.niddk.nih.gov**
National Kidney Foundation: **www.kidney.org**
Renalnet Kidney Information Clearinghouse: **www.renalnet.org**

- To find materials for patients with kidney diseases, visit the American Association of Kidney Patients: **www.aakp.org**

- To find more information about kidney stones, visit the Oxalosis and Hyperoxaluria Foundation: **www.ohf.org**

- To see photographs of kidney stones, visit the website of the Louis C. Herring and Company Laboratory: **www.herringlab.com**

NUTRITION ASSESSMENT CHECKLIST for People with Renal Disorders

Medical History

Check the medical record to determine:

- Degree of renal function
- Cause of the nephrotic syndrome or kidney failure
- Type of dialysis, if appropriate
- Whether the patient has received a kidney transplant
- Type of kidney stone

Review the medical record for complications that may alter nutritional needs:

- Anemia
- Diabetes mellitus
- Edema or oliguria
- Hyperlipidemia
- Hypertension
- Metabolic stress or infection
- Protein-energy malnutrition

Medications

Assess risks for medication-related malnutrition related to:

- Long-term use of medications
- Multiple medication use, especially if medications affect nutrition status

For all patients with renal diseases, note:

- Whether medications or supplements contain electrolytes that must be controlled
- Use of drugs or herbs that may be toxic to the kidneys

Dietary Intake

For patients with the nephrotic syndrome, kidney failure, or kidney transplants, assess intakes of:

- Protein and energy
- Fluid
- Vitamins, especially vitamin D
- Minerals, especially calcium, phosphorus, iron, and electrolytes

For patients with kidney stones or a history of kidney stones:

- Stress the need to drink plenty of fluids throughout the day.
- Assess intake of calcium, oxalate, sodium, protein, purines, or vitamin C, as appropriate for the type of stone.

Anthropometric Data

Take accurate baseline height and weight measurements. Keep in mind that:

- Fluid retention due to the nephrotic syndrome or kidney failure can mask malnutrition.
- For dialysis patients, the weight measured immediately after the dialysis treatment (called the *dry weight*) most accurately reflects the person's true weight. Rapid weight gain between dialysis treatments reflects fluid retention. If fluid retention is excessive, review fluid intake to determine if the patient understands and is complying with diet recommendations.

Laboratory Tests

Note that serum protein levels are often low in patients with nephrotic syndrome or kidney failure. Review the following laboratory test results to assess the degree of renal function and response to treatments:

- Blood urea nitrogen (BUN)
- Creatinine
- Glomerular filtration rate (GFR)
- Serum electrolytes
- Urinary protein

Check laboratory test results for complications associated with kidney disease, including:

- Anemia
- Hyperglycemia
- Hyperlipidemia
- Hyperparathyroidism (related to bone disease)

Physical Signs

For patients with nephrotic syndrome or kidney failure, look for physical signs of:

- Bone disease
- Dehydration or fluid retention
- Hyperkalemia
- Iron deficiency
- Uremia

STUDY QUESTIONS

These questions will help you review the chapter. You will find the answers in the discussions on the pages provided.

1. Describe the kidneys' role in maintaining homeostasis. Discuss other functions of the kidneys. (pp. 873–874)

2. Define the nephrotic syndrome, and describe the consequences that can develop. Discuss the elements of dietary treatment recommended for the nephrotic syndrome. (pp. 874–876)

3. Describe acute renal failure, and list possible causes. Discuss how its consequences can disrupt health. Describe the medical treatment of patients with acute renal failure and the elements of nutrition therapy. (pp. 878–880)

4. Explain how chronic kidney disease differs from acute renal failure. What changes occur as chronic kidney disease progresses? Discuss the symptoms and complications associated with the uremic syndrome. (pp. 880–882)

5. Identify the objectives of treatment for chronic kidney disease, and discuss the role of dialysis. Describe how dietary recommendations change during the course of illness. (pp. 882–883)

6. What are the fluid and electrolyte (sodium, potassium, and phosphorus) recommendations for patients with chronic kidney disease? What adjustments are needed in vitamin and mineral intakes, and why? (pp. 884–885)

7. Explain why renal patients often have difficulty adhering to a renal diet. Discuss ways to help patients comply with recommendations. (p. 886)

8. Discuss the nutrient needs of a kidney transplant patient. How can immunosuppressive drug therapy affect nutrition status? (pp. 886–888)

9. Identify factors that affect kidney stone formation. Describe the composition of the most common types of kidney stones. Discuss dietary adjustments that may help to prevent kidney stone recurrence. (pp. 888–891)

These multiple choice questions will help you prepare for an exam. Answers can be found on p. 895.

1. Which of the following is *not* a function of the kidneys?
   a. activation of vitamin K
   b. maintenance of acid-base balance
   c. elimination of metabolic waste products
   d. maintenance of fluid and electrolyte balances

2. The nephrotic syndrome frequently results in:
   a. the uremic syndrome.
   b. oliguria.
   c. edema.
   d. renal colic.

3. Dietary recommendations for patients with the nephrotic syndrome include:
   a. a high-protein intake.
   b. sodium restriction.
   c. potassium and phosphorus restrictions.
   d. fluid restriction.

4. Hyperkalemia is often treated by:
   a. eliminating potassium from the diet.
   b. using diuretics to increase potassium losses.
   c. increasing fluid consumption.
   d. using potassium-exchange resins, which bind potassium in the GI tract.

5. Fluid requirements for oliguric patients are estimated by adding about ____ milliliters to the volume of urine output.
   a. 100
   b. 300
   c. 500
   d. 750

6. The most common cause of chronic kidney disease is:
   a. diabetes mellitus.
   b. hypertension.
   c. autoimmune disease.
   d. exposure to toxins.

7. A person with chronic kidney disease, who has been following a renal diet for several years, begins hemodialysis treatment. An appropriate dietary adjustment would be to:
   a. reduce protein intake.
   b. consume protein more liberally.
   c. increase intakes of sodium and water.
   d. consume potassium and phosphorus more liberally.

8. Which of the following nutrients may be unintentionally restricted when a patient restricts phosphorus intake?
   a. fluid
   b. calcium
   c. potassium
   d. sodium

9. Most kidney stones are made primarily from:
   a. struvite.
   b. uric acid.
   c. calcium oxalate.
   d. cystine.

10. Treatment for all kidney stones includes:
   a. dietary oxalate restriction.
   b. dietary protein restriction.
   c. vitamin C supplementation.
   d. a fluid intake that maintains a urine volume of at least $2^1/_2$ liters per day.

REFERENCES

1. G. B. Appel, Glomerular disorders and nephrotic syndromes, in L. Goldman and D. Ausiello, eds., *Cecil Medicine* (Philadelphia: Saunders, 2008), pp. 866–876.
2. M. H. Beers and coeditors, *The Merck Manual of Diagnosis and Therapy* (Whitehouse Station, N.J.: Merck Research Laboratories, 2006), pp. 2004–2006.
3. Appel, 2008.
4. J. Goddard and coauthors, Kidney and urinary tract disease, in N. A. Boon, N. R. Colledge, and B. R. Walker, eds., *Davidson's Principles and Practice of Medicine* (Philadelphia: Churchill Livingstone/Elsevier, 2006), pp. 455–518.
5. G. C. Shearer and G. A. Kaysen, Endothelial bound lipoprotein lipase (LpL) depletion in hypoalbuminemia results from decreased endothelial binding, not decreased secretion, *Kidney International* 70 (2006): 647–653.
6. Goddard and coauthors, 2006.
7. Appel, 2008; J. A. Charlesworth, D. M. Gracey, and B. A. Pussell, Adult nephrotic syndrome: Non-specific strategies for treatment, *Nephrology* 13 (2008): 45–50.
8. American Dietetic Association, *Nutrition Care Manual* (Chicago: American Dietetic Association, 2008).
9. American Dietetic Association, 2008.
10. American Dietetic Association, 2008; G. M. Podda and coauthors, Abnormalities of

homocysteine and B vitamins in the nephrotic syndrome, *Thrombosis Research* 120 (2007): 647–652.
11. B. A. Molitoris, Acute kidney injury, in L. Goldman and D. Ausiello, eds., *Cecil Medicine* (Philadelphia: Saunders, 2008), pp. 862–866.
12. C. L. Edelstein and R. W. Schrier, Acute renal failure: Pathogenesis, diagnosis, and management, in R. W. Schrier, ed., *Renal and Electrolyte Disorders* (Philadelphia: Lippincott Williams & Wilkins, 2003), pp. 401–455.
13. Goddard and coauthors, 2006.
14. Molitoris, 2008; Goddard and coauthors, 2006.
15. American Dietetic Association, 2008.
16. American Dietetic Association, 2008.
17. American Dietetic Association, 2008.
18. R. G. Luke, Chronic renal failure, in L. Goldman and D. Ausiello, eds., *Cecil Textbook of Medicine* (Philadelphia: Saunders, 2004), pp. 708–716.
19. W. E. Mitch, Chronic kidney disease, in L. Goldman and D. Ausiello, eds., *Cecil Medicine* (Philadelphia: Saunders, 2008), pp. 921–930.
20. Beers and coeditors, 2006, pp. 1978–1979.
21. Goddard and coauthors, 2006.
22. Mitch, 2008.
23. Mitch, 2008; Beers and coeditors, 2006, pp. 1978–1979.

24. Luke, 2004.
25. R. Mehrotra and J. D. Kopple, Nutritional management of maintenance dialysis patients: Why aren't we doing better? *Annual Review of Nutrition* 21 (2001): 343–379.
26. J. D. Kopple, Nutrition, diet, and the kidney, in M. E. Shils and coeditors, *Modern Nutrition in Health and Disease* (Baltimore, Md.: Lippincott Williams & Wilkins, 2006), pp. 1475–1511.
27. American Dietetic Association, 2008.
28. J. A. Beto and V. K. Bansal, Medical nutrition therapy in chronic kidney disease: Integrating clinical practice guidelines, *Journal of the American Dietetic Association* 104 (2004): 404–409.
29. C. Castaneda and coauthors, Resistance training to counteract the catabolism of a low-protein diet in patients with chronic renal insufficiency, *Annals of Internal Medicine* 135 (2001): 965–976.
30. Beto and Bansal, 2004.
31. American Dietetic Association, 2008.
32. Beto and Bansal, 2004.
33. N. Tolkoff-Rubin, Treatment of irreversible renal failure, in L. Goldman and D. Ausiello, eds., *Cecil Medicine* (Philadelphia: Saunders, 2008), pp. 936–947.
34. N. J. Cano and coauthors, Intradialytic parenteral nutrition does not improve survival in malnourished hemodialysis

patients: A 2-year multicenter, prospective, randomized study, *Journal of the American Society of Nephrology* 18 (2007): 2583–2591; L. B. Pupim and coauthors, Intradialytic oral nutrition improves protein homeostasis in chronic hemodialysis patients with deranged nutritional status, *Journal of the American Society of Nephrology* 17 (2006): 149–157.

35. C. L. Durose and coauthors, Knowledge of dietary restrictions and the medical consequences of noncompliance by patients on hemodialysis are not predictive of dietary compliance, *Journal of the American Dietetic Association* 104 (2004): 35–41.

36. Tolkoff-Rubin, 2008.
37. W. Wang and L. Chan, Chronic renal failure: Manifestations and pathogenesis, in R. W. Schrier, ed., *Renal and Electrolyte Disorders* (Philadelphia: Lippincott Williams & Wilkins, 2003), pp. 456–497.
38. Beto and Bansal, 2004.
39. American Dietetic Association, 2008.
40. G. C. Curhan, Nephrolithiasis, in L. Goldman and D. Ausiello, eds., *Cecil Medicine* (Philadelphia: Saunders, 2008), pp. 897–903.
41. L. K. Massey, M. Liebman, and S. A. Kynast-Gales, Ascorbate increases human oxaluria and kidney stone risk, *Journal of Nutrition* 135 (2005): 1673–1677.
42. American Dietetic Association, 2008.
43. S. Escott-Stump, *Nutrition and Diagnosis-Related Care* (Baltimore, Md.: Lippincott Williams & Wilkins, 2008), pp. 800–804.
44. F. L. Coe, A. Evan, and E. Worcester, Kidney stone disease, *Journal of Clinical Investigation* 115 (2005): 2598–2608.
45. Curhan, 2008.
46. American Dietetic Association, 2008.

ANSWERS

Study Questions (multiple choice)

1. a 2. c 3. b 4. d 5. c 6. a 7. b 8. b 9. c 10. d

Dialysis

Although there is no perfect substitute for one's own kidneys, dialysis offers a life-sustaining treatment option for people with chronic kidney disease who develop renal failure. Dialysis can serve as a permanent treatment or as a temporary measure to sustain life until a suitable kidney donor can be found. Dialysis can also restore fluid and electrolyte balances in patients with acute renal failure. Clinicians who routinely work with renal patients should understand how dialysis procedures work. This highlight describes the process of dialysis and outlines the different types of procedures used. The accompanying glossary defines the relevant terms.

The Basics of Dialysis

As described in this section, dialysis removes excess fluids and wastes from the blood by employing the processes of **diffusion, osmosis,** and **ultrafiltration** (see Figure H28-1). The dialysate, a solution similar in composition to normal blood plasma, is delivered to a compartment beside a **semipermeable membrane;** the person's blood flows along the other side of the membrane. The semipermeable membrane acts like a filter: small molecules such as urea and glucose can pass through microscopic pores in the membrane, whereas large molecules are unable to cross.

In *hemodialysis,* the tiny tubes that carry blood through the dialyzer are made of materials that serve as semipermeable membranes. In *peritoneal dialysis,* the body's peritoneal membrane, rich with blood vessels, is used to filter blood.

Removal of Solutes

The chemical composition of the dialysate affects the movement of solutes across the semipermeable membrane. When the concentration of a substance is lower in the dialysate than in the blood, the substance—provided it can cross the membrane—will diffuse out of the blood. For example, the goal is to remove as much as possible of the waste product urea from the blood, so the dialysate contains no urea. For many other solutes, the dialysate is adjusted so that only excesses will be removed. Potassium can be removed from the blood, for example, by providing a dialysate that has a lower concentration of potassium than is found in the person's blood. The dialysate must contain some potassium, however; otherwise the blood potassium would fall too low.

The dialysate can also be used to add needed components back into the blood. For a person with acidosis, for example, bases such as bicarbonate are added to the dialysate; the bases then move by diffusion into the blood to alleviate the acidosis.

Removal of Fluid

Because albumin and other plasma proteins are so adept at retaining fluids in blood, osmosis alone is not an efficient process for removing fluid. In hemodialysis, a **pressure gradient** is created between the blood and the dialysate. Most modern dialyzers produce *positive* pressure in the blood compartment and *negative* pressure in the dialysate compartment, establishing a pressure

GLOSSARY

continuous ambulatory peritoneal dialysis (CAPD): the most common method of peritoneal dialysis; involves frequent exchanges of dialysate, which remains in the peritoneal cavity throughout the day.

continuous renal replacement therapy (CRRT): a slow, continuous method of removing solutes and/or fluids from blood by gently pumping blood across a filtration membrane over a prolonged time period.

diffusion: movement of solutes from an area of high concentration to one of low concentration.

hemofiltration: removal of fluid and solutes by pumping blood across a membrane; no osmotic gradients are created during the process.

oncotic pressure: the pressure exerted by fluid on one side of a membrane as a result of osmosis.

osmosis: movement of water across a membrane toward the side where solutes are more concentrated.

peritonitis: inflammation of the peritoneal membrane.

pressure gradient: the change in pressure over a given distance. In dialysis, a pressure gradient is created between the blood and the dialysate.

semipermeable membrane: a membrane that allows some particles to pass through, but not others.

ultrafiltration: removal of fluids and solutes from blood by using pressure to transfer the blood across a semipermeable membrane.

urea kinetic modeling: a method of determining the adequacy of dialysis treatment by calculating the urea clearance from blood.

FIGURE H28-1 | Diffusion, Osmosis, and Ultrafiltration

Diffusion

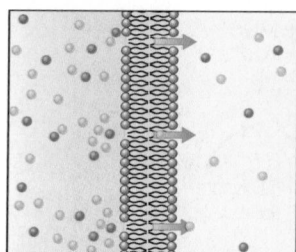

Small molecules (electrolytes and waste products) move from an area of high concentration to an area of low concentration by diffusion.

Osmosis

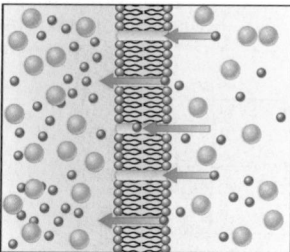

Water moves from an area of high water concentration to an area of low water concentration. In other words, water moves toward the side where solutes are more concentrated.

Ultrafiltration

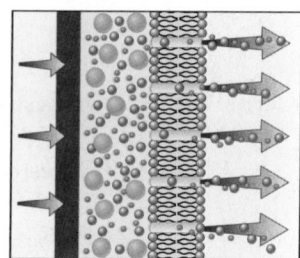

Pressure squeezes water and small molecules through the pores of a semipermeable membrane during ultrafiltration.

gradient that "pushes" water (and accompanying solutes) through the pores of the membrane.[1] This process, called ultrafiltration, relies on pumps to establish an appropriate flow rate between the blood and the dialysate.

Evaluation of Dialysis Treatment

A number of methods have been devised for gauging the adequacy of dialysis treatment. The most common method is **urea kinetic modeling,** a technique that evaluates the amount of urea cleared from the blood. The formula used most often is Kt/V, where K is the amount of urea cleared, t is the time spent on dialysis, and V is the blood volume. The value obtained indicates whether the patient has undergone sufficient dialysis; the goal is a Kt/V result of approximately 1.2. Because technical data (such as dialyzer clearance data, blood flow rate, and dialysate flow rate) need to be incorporated into the calculation, the computation is usually done by computer analysis. Current treatment guidelines recommend that hemodialysis adequacy be evaluated at least monthly, or more often if problems develop or patients are noncompliant.[2]

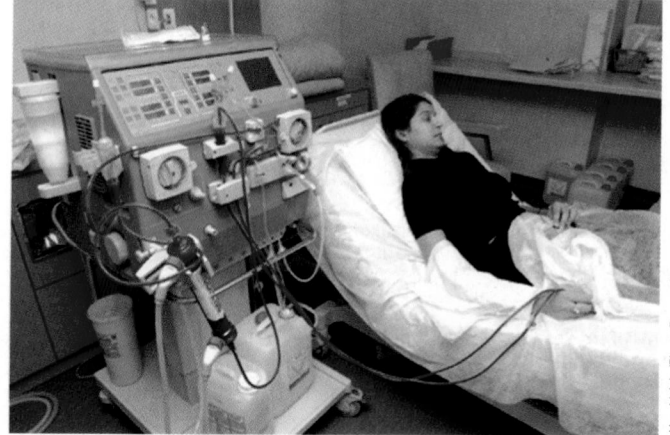

During hemodialysis, blood passes through a dialyzer, where wastes are extracted, and the cleansed blood is returned to the body.

Hpa-Voisin/Photo Researchers

Types of Dialysis

Three approaches are currently used to remove fluids and wastes from the body: hemodialysis, peritoneal dialysis, and continuous renal replacement therapy. The latter procedure is used only to treat acute renal failure.

Hemodialysis

As described previously, hemodialysis utilizes a dialyzer to cleanse the patient's blood. Although dialyzers vary in efficiency, the treatment usually lasts 3 to 4 hours and is required at least 3 times weekly. Some studies suggest that patients undergoing daily hemodialysis for briefer periods (2 to $2^1/_2$ hours) may tolerate dialysis treatment better and have fewer complications, but this approach has not been widely adopted.[3] Most patients visit dialysis centers to obtain treatment; home hemodialysis programs are available, but only about 2 percent of patients use them.

Although lifesaving, hemodialysis is associated with a substantial number of complications.[4] Problems at the vascular access site include infections and blood clotting. Hypotension can develop while blood is circulated through the dialyzer. Muscle cramping often occurs during the procedure, especially in the hands, legs, and feet. Blood losses can worsen anemia, which is already severe in two-thirds of patients beginning hemodialysis treatment.[5] Patients may also experience headaches, weakness, nausea, vomiting, restlessness, and agitation.[6]

Peritoneal Dialysis

In peritoneal dialysis, the peritoneal membrane surrounding the abdominal organs serves as a semipermeable membrane. The dialysate is infused into a catheter that empties into the peritoneal space—the space within the abdomen near the intestines (see Figure H28-2 on p. 898). In the most common procedure, **continuous ambulatory peritoneal dialysis (CAPD),** the dialysate remains in the peritoneal cavity for 4 to 6 hours, after which it is drained and replaced with fresh dialysate (about 2 to 3 liters in adults). Generally, the dialysate solution is exchanged four times daily and requires only about 30 minutes to drain and replace.

Because a pressure gradient cannot be created in the peritoneal cavity, as it can in a dialyzer, the glucose concentration in the dialysate must be high enough to create enough **oncotic pressure** to draw fluid from the blood. As indicated in Chapter 28, a substantial amount of glucose can be absorbed into the patient's blood and may contribute to weight gain over time. The high glucose load may also cause hyperglycemia and hypertriglyceridemia in some patients.

Peritoneal dialysis offers a number of advantages over hemodialysis: vascular access is not required, dietary restrictions are fewer, and the procedure can be scheduled when convenient.

FIGURE H28-2 Peritoneal Dialysis

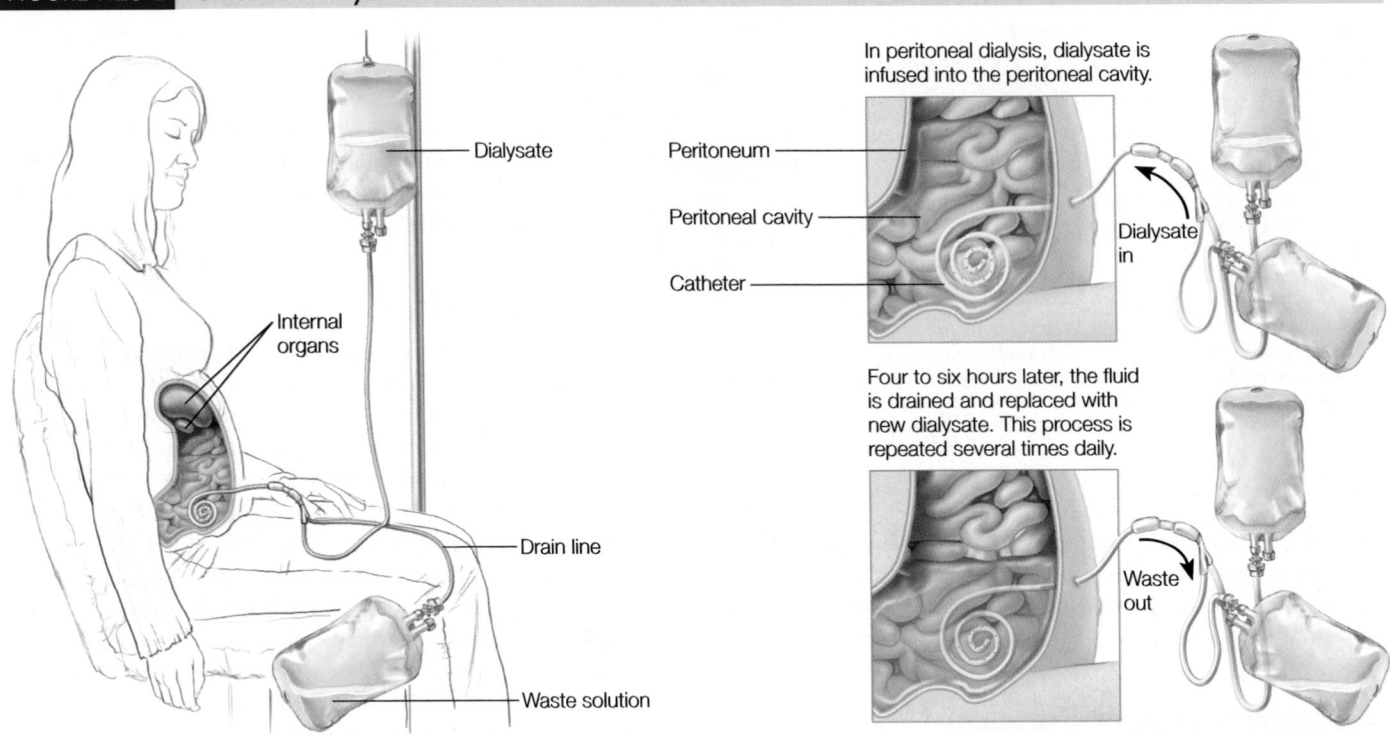

In peritoneal dialysis, dialysate is infused into the peritoneal cavity.

- Dialysate
- Internal organs
- Drain line
- Waste solution

- Peritoneum
- Peritoneal cavity
- Catheter
- Dialysate in

Four to six hours later, the fluid is drained and replaced with new dialysate. This process is repeated several times daily.

- Waste out

The most common complication is infection, which can occur at the catheter site or within the peritoneal cavity **(peritonitis).** Other problems that may arise include blood clotting in the catheter, catheter migration, and abdominal hernia due to the dialysate volume.

Continuous Renal Replacement Therapy

In people with acute renal failure, **continuous renal replacement therapy (CRRT)** removes fluids and wastes. CRRT utilizes the process of **hemofiltration,** in which blood is gently pumped across a filtration membrane over a prolonged time period. (This process differs from dialysis treatments that rely on the diffusion of wastes across a membrane into the dialysate.) Either a pump or the patient's own blood pressure moves the blood across the membrane. The procedure can be used to remove fluids, solutes, or both. Some patients require fluid replacement during the procedure to maintain adequate blood volume, so hydration status must be closely monitored.

The use of CRRT is advantageous in acute care situations because it corrects imbalances without causing sudden shifts in blood volume, which are poorly tolerated in acute care patients. In addition, replacement fluids can include parenteral feedings without upsetting fluid balance. Complications include clotting problems, damage to arteries, and inadequate blood flow rates in hypotensive patients.

Dialysis and CRRT help to remove the wastes and fluids that are normally removed by healthy kidneys. Although these procedures cannot restore the kidneys' hormonal functions, they provide a lifesaving means of alleviating symptoms of uremia, hypertension, and edema.

REFERENCES

1. C. F. Gutch, Principles of hemodialysis, in C. F. Gutch, M. H. Stoner, and A. L. Corea, eds., *Review of Hemodialysis for Nurses and Dialysis Personnel* (St. Louis: Mosby, 1999), pp. 35–45.
2. National Kidney Foundation, K/DOQI clinical practice guidelines for hemodialysis adequacy: Update 2006, available at www.kidney.org/PROFESSIONALS/kdoqi/guideline_upHD_PD_VA/hd_guide2.htm; site visited January 18, 2008.
3. A. Pierratos, New approaches to hemodialysis, *Annual Review of Medicine* 55 (2004): 179–189.
4. N. Tolkoff-Rubin, Treatment of irreversible renal failure, in L. Goldman and D. Ausiello, eds., *Cecil Medicine* (Philadelphia: Saunders, 2008), pp. 936–947.
5. Tolkoff-Rubin, 2008.
6. Gutch, 1999.

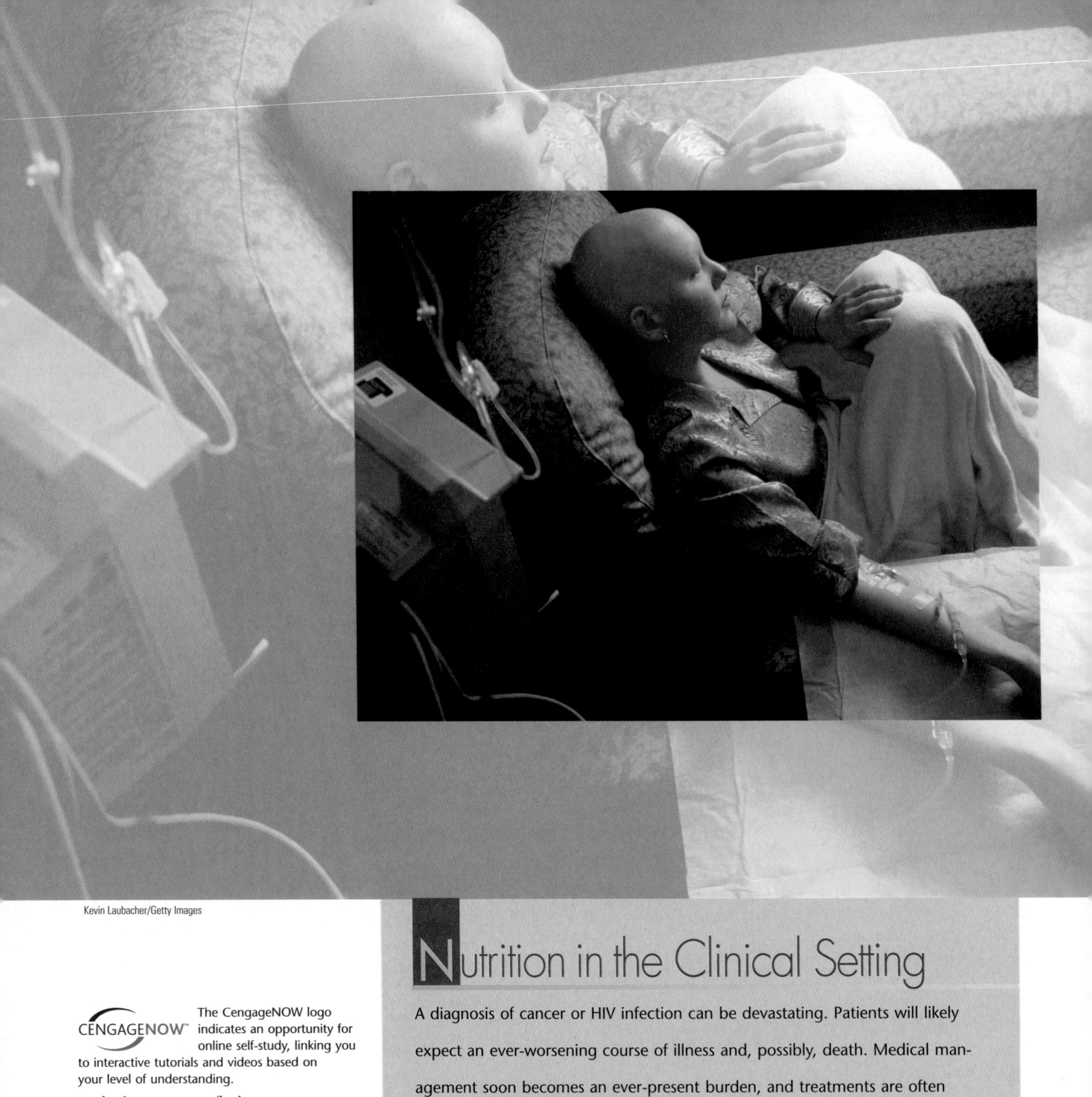

Kevin Laubacher/Getty Images

Nutrition in the Clinical Setting

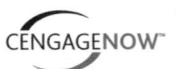

A diagnosis of cancer or HIV infection can be devastating. Patients will likely expect an ever-worsening course of illness and, possibly, death. Medical management soon becomes an ever-present burden, and treatments are often unpleasant. For both illnesses, however, extraordinary therapeutic advances have been made. Treatment options have expanded, and patients have benefited from vast improvements in quality of life. The health practitioner's knowledge and empathy are the patient's most important resources—and an important source of hope.

CENGAGENOW™ The CengageNOW logo indicates an opportunity for online self-study, linking you to interactive tutorials and videos based on your level of understanding.

academic.cengage.com/login

Cancer and HIV Infection

Although **cancers** and **HIV (human immunodeficiency virus)** infections are distinct disorders, from a nutritional standpoint they share some similarities. Both disorders have debilitating effects that influence nutritional needs, and both can lead to severe wasting in advanced cases. These illnesses require medical nutrition therapy that is highly individualized based on the symptoms manifested and the organ systems involved.

Cancer

Cancer, the growth of **malignant** tissue, ranks just below cardiovascular disease as a cause of death in the United States. Cancer is not a single disorder, however; there are many different kinds of malignant growths. The different types of cancer have different characteristics, occur in different locations in the body, take different courses, and require different treatments. ◆ Whereas an isolated, nonspreading type of skin cancer may be removed in a physician's office with no effect on nutrition status, advanced cancers—especially those of the gastrointestinal (GI) tract and pancreas—can seriously impair nutrition status.

How Cancer Develops

The development of cancer, called **carcinogenesis,** often proceeds slowly and continues for several decades. A cancer arises from mutations in the genes that control cell division in a single cell.[1] These mutations may promote cellular growth, interfere with growth restraint, or prevent cellular death. The affected cell thereby loses its built-in capacity for halting cell division and produces daughter cells with the same genetic defects. As the abnormal mass of cells, called a **tumor** (or *neoplasm*), grows, ◆ blood vessels form to supply the tumor with the nutrients it needs to support its growth. The tumor can disrupt the functioning of the normal tissue around it, and some tumor cells may **metastasize,** spreading to another region in the body. In leukemia (cancer affecting the white blood cells), the abnormal cells do not form a tumor but rather accumulate in the blood and other tissues. Figure 29-1 on p. 902 illustrates the steps in cancer development.

cancers: malignant growths or tumors that result from abnormal and uncontrolled cell division.

HIV (human immunodeficiency virus): the virus that causes acquired immune deficiency syndrome (AIDS). HIV destroys immune cells and progressively impedes the body's ability to fight infections and certain cancers.

malignant (ma-LIG-nent): describes a cancerous cell or tumor, which can injure healthy tissue and spread cancer to other regions of the body.

carcinogenesis (CAR-sin-oh-JEN-eh-sis): the process of cancer development.

tumor: an abnormal tissue mass that has no physiological function; also called a *neoplasm* (NEE-oh-plazm).

metastasize (meh-TAS-tah-size): the spread of cancer cells from one part of the body to another.

FIGURE 29-1 Cancer Development

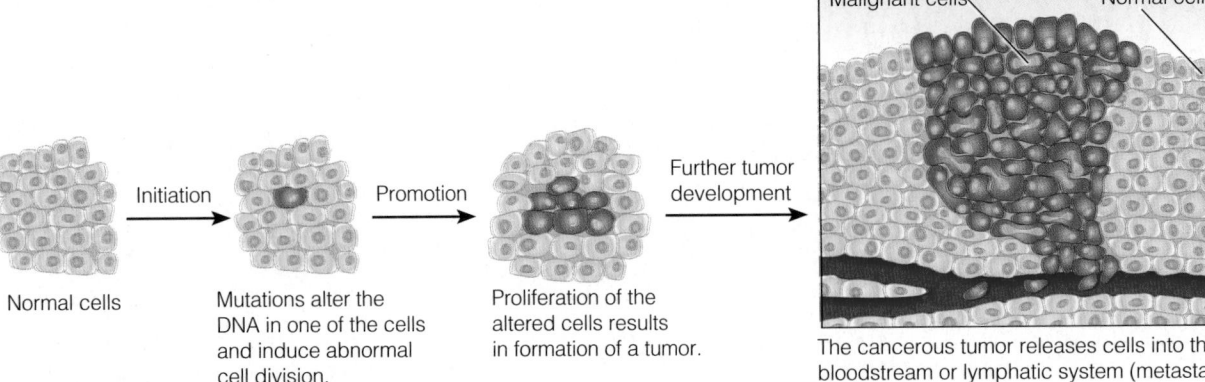

Normal cells → Initiation → Mutations alter the DNA in one of the cells and induce abnormal cell division. → Promotion → Proliferation of the altered cells results in formation of a tumor. → Further tumor development → The cancerous tumor releases cells into the bloodstream or lymphatic system (metastasis).

Malignant cells / Normal cells

◆ Cancers are classified by the tissues or cells from which they develop:
  • *Adenomas* (ADD-eh-NO-muz) arise from glandular tissues.
  • *Carcinomas* (CAR-sih-NO-muz) arise from epithelial tissues.
  • *Gliomas* (gly-OH-muz) arise from glial cells of the central nervous system.
  • *Leukemias* (loo-KEY-mee-uz) arise from white blood cell precursors.
  • *Lymphomas* (lim-FOE-muz) arise from lymphoid tissues.
  • *Melanomas* (MEL-ah-NO-muz) arise from pigmented skin cells.
  • *Myelomas* (MY-ah-LOE-muz) arise from plasma cells in the bone marrow.
  • *Sarcomas* (sar-KO-muz) arise from connective tissues, such as muscle or bone.

◆ An abnormal mass of cells that is noncancerous is called a *benign* tumor.

The reasons why cancers develop are numerous and varied. Vulnerability to cancer is sometimes inherited, as when a person is born with a genetic defect that alters DNA structure, function, or repair. Certain metabolic processes may initiate carcinogenesis, as when phagocytes (immune cells) produce oxidants that cause DNA damage, or when chronic inflammation increases the rate of cell division, increasing the risk of a damaging mutation. More often, cancers are caused by interactions between a person's genes and environmental agents. Exposure to cancer-causing substances, or **carcinogens,** may either induce genetic mutations that lead to cancer or promote proliferation of cancerous cells. Table 29-1 provides examples of environmental factors that increase cancer risk.

TABLE 29-1 Environmental Factors That Increase Cancer Risk

Environmental Factors	Cancer Sites
Aflatoxins (toxins in moldy peanuts or grains)	Liver
Alcohol[a]	Mouth, pharynx, larynx, esophagus, colon, rectum, liver, breast
Asbestos[b]	Lung, pleura, peritoneum
Chromium (hexavalent) compounds	Nasal cavity, lung
Estrogen-progesterone replacement therapy	Breast
Immunosuppressive medications	Lymphoid tissues, liver
Infection with *Helicobacter pylori*	Stomach
Infection with hepatitis B and hepatitis C viruses	Liver
Infection with human papillomavirus (HPV)	Cervix
Ionizing radiation (X-rays, radioactive isotopes, and other sources)	White blood cells (leukemia), esophagus, stomach, colon, thyroid, lung, bladder, breast
Tobacco[a]	Nasal cavity, lung, mouth, pharynx, larynx, esophagus, stomach, colon, rectum, liver, pancreas, kidney, renal pelvis, bladder
Ultraviolet radiation (sun exposure)	Skin

[a]A combined exposure to alcohol and tobacco multiplies the risks of developing cancers of the oral cavity, pharynx, larynx, and esophagus.
[b]Risk is greatly increased in cigarette smokers.
SOURCES: M. J. Thun, Epidemiology of cancer, in L. Goldman and D. Ausiello, eds., *Cecil Medicine* (Philadelphia: Saunders, 2008), pp. 1335–1340; World Cancer Research Fund/American Institute for Cancer Research, *Food, Nutrition, Physical Activity, and the Prevention of Cancer: A Global Perspective* (Washington, D.C.: American Institute for Cancer Research, 2007), pp. 157–171.

carcinogens (CAR-sin-oh-jenz or car-SIN-oh-jenz): substances that can cause cancer (the adjective is *carcinogenic*).

Nutrition and Cancer Risk

Like other environmental factors, diet and lifestyle strongly influence cancer risk. Certain food components may directly damage DNA, alter the metabolism of carcinogens by liver enzymes, or inhibit the formation of carcinogens in the body.[2] In addition, energy balance and growth rates affect the rate of cell division and consequently influence the rates at which mutations form and are replicated. Table 29-2 lists examples of nutrition-related factors that may increase or decrease the risk of developing cancer.

Nutrition and Increased Cancer Risk As shown in Table 29-2, obesity is a risk factor for a number of different cancers, including some relatively common cancers such as colon cancer and postmenopausal breast cancer. Obesity increases cancer risk, in part, by altering levels of hormones that influence cell growth, such as the sex hormones, insulin, and several kinds of growth factors. For example, in the case of breast cancer in postmenopausal women, the hormone estrogen is likely involved: obese women have higher estrogen levels than do lean women, because adipose tissue produces estrogen.

Although studies in animals have suggested that high-fat diets can promote tumor growth, studies of humans have not proved that the effects of fat are independent of the effects of energy intake and physical activity.[3] Evidence from population studies is mixed: high-fat diets often, but not always, correlate with high cancer rates. Within single populations, cancer rates do not reliably reflect fat intakes. In addition, the type of fat consumed may be critical: studies of colon and rectal cancers implicate animal fats but not vegetable fat, and a number of studies suggest that consuming fatty fish may be protective.[4]

TABLE 29-2 Nutrition-Related Factors That Influence Cancer Risk

Nutrition-Related Factors[a]	Cancer Sites
Factors that increase cancer risk:	
Obesity	Esophagus, colon, rectum, pancreas, gallbladder, kidney, breast (postmenopausal), endometrium
Red meat, processed meats	Colon, rectum
Salted and salt-preserved foods	Stomach
Beta-carotene supplements	Lung[b]
High-calcium diets (over 1500 mg daily)	Prostate
Low level of physical activity[c]	Colon, breast (postmenopausal), endometrium
Factors that decrease cancer risk:	
Fruits and nonstarchy vegetables	Lung, mouth, pharynx, larynx, esophagus, stomach
Carotenoid-containing foods	Lung, mouth, pharynx, larynx, esophagus
Tomato products	Prostate
Allium vegetables (onion, garlic)	Stomach, colon, rectum
Vitamin C–containing foods	Esophagus
Folate-containing foods	Pancreas
Fiber-containing foods	Colon, rectum
Milk and calcium supplements	Colon, rectum
High level of physical activity[c]	Colon, breast (postmenopausal), endometrium

[a]Altered cancer risk is associated with high intakes of the dietary substances listed. The cancer sites associated with alcohol are included in Table 29-1.
[b]Cancer risk is increased in tobacco smokers and may not apply to other groups.
[c]Physical activity may influence cancer risk by altering body fatness, intestinal transit time, insulin sensitivity, hormone levels, enzyme activities, and immune responses.
SOURCE: World Cancer Research Fund/American Institute for Cancer Research, *Food, Nutrition, Physical Activity, and the Prevention of Cancer: A Global Perspective* (Washington, D.C.: American Institute for Cancer Research, 2007).

Cruciferous vegetables, such as cauliflower, broccoli, and brussels sprouts, contain nutrients and phytochemicals that inhibit cancer development.

Food preparation methods are responsible for producing certain types of carcinogens. Cooking meat, poultry, and fish at high temperatures causes carcinogens to form in these foods.[5] Carcinogens also accompany the smoke that adheres to food during grilling, and they are present in the charred surfaces of grilled meat and fish. However, the cancer risk from eating such foods is unclear, because the biological actions of these carcinogens are modulated by other dietary components, including compounds in vegetables and other plant foods. In several population studies, consumption of well-cooked meats was linked to cancers of the stomach, colon, breast, and prostate.[6]

Nutrition and Decreased Cancer Risk A considerable number of human studies have found a link between the consumption of fruits and vegetables and reduced incidences of certain cancers (review Table 29-2). Fruits and vegetables contain both nutrients and phytochemicals with antioxidant activity, and these substances may prevent or reduce the oxidative reactions in cells that cause DNA damage. Phytochemicals may also help to inhibit carcinogen production in the body, enhance immune functions that protect against cancer development, and promote enzyme reactions that inactivate carcinogens.[7] In addition, certain fruits and vegetables provide the B vitamin folate, which plays roles in DNA synthesis and repair; thus, inadequate folate intakes may allow DNA damage to accumulate.

Although research reports in the 1970s and 1980s suggested that a fiber-rich diet could protect against colon cancer, recent studies have cast doubt on the earlier analyses.[8] The earlier studies depended on the ability of colon cancer patients to recall the foods they had consumed during the preceding years, whereas more recent studies—considered more reliable—tracked the subjects' health behaviors and cancer outcomes for extended periods (10 to 20 years). Moreover, some studies that had found fiber to be protective did not analyze factors such as physical activity, smoking, or folate intake, all of which can influence cancer outcome. A fiber-rich diet may be protective, in part, because high-fiber foods usually contain high levels of nutrients and phytochemicals that are protective against cancer. Table 29-3 summarizes the dietary and lifestyle practices that may help to reduce the risk of developing cancer.

TABLE 29-3 Recommendations for Reducing Cancer Risk

Maintain a healthy body weight.
- Balance energy intake with appropriate physical activity.
- Avoid weight gain and increases in waist circumference throughout adulthood.
- If overweight or obese, achieve a healthy body weight.

Be physically active.
- For adults: engage in moderate to vigorous activity for 30 minutes on at least 5 days of the week; 45 to 60 minutes is preferable.
- For children and adolescents: engage in moderate to vigorous activity for 60 minutes on at least 5 days of the week.

Choose a healthy diet that emphasizes plant sources.
- Consume five or more servings of a variety of vegetables and fruits daily.
- Choose whole-grain products instead of processed (refined) grains.
- Limit consumption of red meats and processed meats.
- Avoid salt-preserved and salty foods.
- Avoid moldy grains and legumes.

Limit consumption of alcoholic beverages.
- For women: drink no more than one drink daily.
- For men: drink no more than two drinks daily.

Avoid using tobacco in any form.

SOURCES: World Cancer Research Fund/American Institute for Cancer Research, *Food, Nutrition, Physical Activity, and the Prevention of Cancer: A Global Perspective* (Washington, D.C.: American Institute for Cancer Research, 2007); L. H. Kushi and coauthors, American Cancer Society guidelines on nutrition and physical activity for cancer prevention: Reducing the risk of cancer with healthy food choices and physical activity, *CA: A Cancer Journal for Clinicians* 56 (2006): 254–281.

Consequences of Cancer

Once cancer develops, its consequences depend on the location of the tumor, its severity, and the treatment. The complications that develop are often due to the tumor's impingement on surrounding tissues. Nonspecific effects of cancer include anorexia, lethargy, weight loss, night sweats, and fever.[9] During the early stages, many cancers produce no symptoms, and the person may be unaware of the threat to health.

Wasting Associated with Cancer Anorexia, muscle wasting, weight loss, and fatigue typify **cancer cachexia,** which eventually affects over 80 percent of people with terminal cancer.[10] Weight loss is often evident at the time that cancer is diagnosed, and severe malnutrition, typically seen in the later stages of cancer, is the ultimate cause of death in many cases. Without adequate energy and nutrients, the body is poorly equipped to maintain organ function, support immune defenses, and mend damaged tissues. An involuntary weight loss of more than 10 percent, which indicates significant malnutrition, is cause for concern.[11]

Many factors play a role in the wasting associated with cancer. Cytokines, ◆ released by both tumor cells and immune cells, induce a catabolic state. The combined effects of a poor appetite, accelerated and abnormal metabolism, and the diversion of nutrients to support tumor growth result in a lower supply of energy and nutrients at a time when demands are high. Appetite and food intake are further disturbed by the effects of treatments and medications prescribed for cancer patients.

◆ The cytokines that induce cachexia include tumor necrosis factor-α, interleukin-1, interleukin-6, and γ-interferon.

Metabolic Changes The metabolic changes that arise in cancer exacerbate the wasting described in the previous section.[12] Cancer patients exhibit an increased rate of protein turnover, but reduced muscle protein synthesis. ◆ Gluconeogenesis increases, further straining the body's supply of protein (recall that muscle supplies the amino acids used in glucose production). Triglyceride breakdown increases, elevating serum lipids. Many patients develop insulin resistance. These metabolic abnormalities help to explain why people with cancer fail to regain lean tissue or maintain healthy body weights even when they are consuming adequate energy and nutrients.

◆ Reminder: *Protein turnover* refers to the continuous degradation and synthesis of the body's proteins.

Anorexia and Reduced Food Intake Anorexia is a major contributor to the wasting associated with cancer. Some factors that contribute to anorexia or otherwise reduce food intake include:

- *Chronic nausea and early satiety.* People with cancer frequently experience nausea and a premature feeling of fullness after eating small amounts of food.

- *Fatigue.* People with cancer often tire easily and lack the energy to prepare and eat meals. Once cachexia develops, these tasks become even more difficult.

- *Pain.* People in pain may have little interest in eating, particularly if eating makes the pain worse.

- *Mental stress.* A cancer diagnosis can cause distress, anxiety, and depression, all of which may reduce appetite. Facing and undergoing cancer treatments induces additional psychological stress.

- *Effects of cancer therapies.* Therapies for cancer (including medications, chemotherapy, radiation therapy, surgery, and bone marrow transplants) can reduce food intake by causing nausea, vomiting, altered taste perceptions, food aversions, inflammation of the mouth and esophagus, dry mouth, mouth sores, difficulty swallowing, intestinal cramping, diarrhea, and constipation.

- *Gastrointestinal obstructions.* A tumor may partially or completely obstruct a portion of the GI tract, causing complications such as nausea and vomiting, early satiety, delayed gastric emptying, and bacterial overgrowth. Some patients with obstructions are unable to tolerate oral diets.

cancer cachexia (ka-KEK-see-ah): a wasting syndrome associated with cancer that is characterized by anorexia, muscle wasting, weight loss, and fatigue.

TABLE 29-4 Nutrition-Related Side Effects of Cancer Surgeries

Head and Neck Surgeries

Difficulty with chewing/swallowing

Inability to chew/swallow

Esophageal Resection

Diarrhea

Fistula formation

Reduced gastric acid secretion

Reduced gastric motility

Steatorrhea (fat malabsorption)

Stenosis (constriction)

Gastric Resection

Dumping syndrome

General malabsorption

Hypoglycemia

Lack of gastric acid

Vitamin B$_{12}$ malabsorption

Intestinal Resection

Blind loop syndrome

Diarrhea

Fluid and electrolyte imbalances

Hyperoxaluria

Malabsorption

Steatorrhea

Pancreatic Resection

Diabetes mellitus

Malabsorption

◆ One drug that inhibits cell division is *methotrexate,* which closely resembles the B vitamin folate (see Figure 19-1 on p. 651). Folate is required for cell division because it is needed for DNA synthesis. Methotrexate works by blocking activity of the enzyme that converts folate to its active form.

---

chemotherapy: the use of drugs to arrest or destroy cancer cells; these drugs are called *antineoplastic agents.*

radiation therapy: the use of X-rays, gamma rays, or atomic particles to destroy cancer cells.

radiation enteritis: inflammation of intestinal tissue caused by radiation therapy.

bone marrow transplant: a procedure that replaces bone marrow that has been destroyed by cancer treatment; it is also used to treat certain types of cancers and blood disorders. Also called *hematopoietic stem cell transplantation.*

tissue rejection: destruction of donor tissue by the recipient's immune system, which recognizes the donor cells as foreign.

Treatments for Cancer

The primary medical treatments for cancer—surgery, chemotherapy, radiation therapy, or any combination of the three—aim to remove cancer cells, prevent further tumor growth, and alleviate symptoms.[13] The likelihood of effective treatment is highest with early detection and intervention. Because treatment decisions are difficult and cancer therapies have considerable side effects, patients rely on health care providers to help them make informed decisions.

Surgery Surgery is performed to remove tumors, determine the extent of cancer, and protect nearby tissues. Often, surgery must be followed by other cancer treatments to prevent growth of new tumors. The acute metabolic stress caused by surgery raises protein and energy needs and can exacerbate wasting. Surgery also contributes to pain, fatigue, and anorexia, all of which can reduce food intake at a time when nutritional needs are substantial. Blood loss contributes to nutrient losses and further exacerbates malnutrition. Some surgeries can have long-term effects on nutrition status (see Table 29-4).

Chemotherapy **Chemotherapy** relies on the use of drugs to treat cancer; it is used to inhibit tumor growth, shrink tumors before surgery, and prevent or eradicate metastasis. Some cancer drugs interfere with the process of cell division; ◆ others sterilize cells that are in a resting phase and are not actively dividing. Ideally, chemotherapy would wipe out cancer cells without destroying healthy ones. Unfortunately, most of these drugs have toxic effects on normal cells as well and are especially damaging to rapidly dividing cells, such as those of the GI tract, skin, and bone marrow. Some of the newer drugs are able to target properties specific to cancer cells and are better tolerated by the body's tissues. Table 29-5 includes a summary of the nutrition-related side effects that may result from chemotherapy.

Radiation Therapy **Radiation therapy** treats cancer by bombarding cancer cells with X-rays, gamma rays, or various atomic particles. These treatments induce the formation of reactive oxygen species, ◆ such as superoxide and hydroxyl radicals, which can damage cellular DNA and cause cell death. Newer techniques are able to focus radiation directly at tumors and minimize damage to nearby tissues. An advantage of radiation therapy over surgery is that it can shrink tumors while preserving organ structure and function. Compared with chemotherapy, radiation therapy is better able to target specific regions of the body, rather than involving all body cells. Nonetheless, radiation therapy can damage healthy tissues and sometimes has long-term detrimental effects on nutrition status. Radiation to the head and neck area can damage the salivary glands and taste buds, causing inflammation, dry mouth, and a reduced sense of taste; in severe cases, damage may be permanent. Radiation treatment in the lower abdominal area can cause **radiation enteritis,** an inflammatory condition of the small intestine that causes nausea, vomiting, malabsorption, and diarrhea. Table 29-5 includes additional side effects of radiation treatment that affect nutrition status.

Bone Marrow Transplant A **bone marrow transplant** replaces bone marrow that has been destroyed by chemotherapy or radiation therapy, and it is one of the primary treatments for leukemia, lymphomas, and multiple myeloma.[14] If possible, bone marrow cells are collected from the patient before chemotherapy or radiation treatment begins so that it is not necessary to find a separate donor. If another person's cells are used, the patient must take immunosuppressant drugs to prevent **tissue rejection.**

The treatments that bone marrow transplant patients undergo have a substantial impact on their food intake and nutrition status. The chemotherapy or radiation therapy preceding the transplant and the immunosuppressant drugs required afterward can impair immune function substantially and increase the risk of foodborne illness. Other common complications include anorexia, dry mouth, altered taste sensations, inflamed mucous membranes, malabsorption, nausea, vomiting, and diarrhea. Patients are often unable to consume adequate food and may require nutrition support, as described in a later section.

TABLE 29-5	Nutrition-Related Side Effects of Chemotherapy and Radiation Therapy		
	Reduced Nutrient Intake	**Accelerated Nutrient Losses**	**Altered Metabolism**
Chemotherapy	Abdominal pain	Diarrhea	Fluid and electrolyte imbalances
	Anorexia	Intestinal ulcers	Hyperglycemia
	Mouth ulcers	Malabsorption	Interference with vitamins or other metabolites
	Nausea	Vomiting	Negative nitrogen and calcium balances
	Taste alterations		Secondary effects of malnutrition, infection, or tissue damage (inflammation)
	Vomiting		
Radiation	Anorexia	Blood loss from intestine and bladder	Fluid and electrolyte imbalances as a consequence of vomiting, diarrhea, or malabsorption
	Damage to teeth and jaws	Diarrhea	Secondary effects of malnutrition, infection, or tissue damage (inflammation)
	Dysphagia	Fistulas	
	Esophagitis	Intestinal obstructions	
	Mouth ulcers	Malabsorption	
	Nausea	Radiation enteritis	
	Reduced salivary secretions	Vomiting	
	Taste alterations		
	Thick salivary secretions		
	Vomiting		

Medications to Combat Anorexia and Wasting To help cancer patients combat anorexia, medications may be prescribed to stimulate the appetite and promote weight gain. One of the most effective medications, megestrol acetate (Megace), is a synthetic compound similar in structure to the hormone progesterone. Dronabinol (Marinol), which resembles the psychoactive ingredient in marijuana, stimulates the appetite at doses that have minimal mental effects. Under investigation are medications that may help to restore lean tissue, such as anabolic steroids, growth hormone, and insulin-like growth factor.[15]

Alternative Therapies Many patients turn to *complementary and alternative medicine (CAM)* ◆ to assist them in their fight against cancer. Patients may turn to CAM because they wish to gain more control over treatment or because they are concerned about the effectiveness of conventional approaches. Although few abandon conventional medicine, up to 80 percent of cancer patients combine one or more CAM approaches with standard treatment.[16] Many patients do not discuss their use of CAM with physicians.

Dietary supplements and herbal remedies are among the most frequently used CAM therapies. Although many supplements can be used without risk, some may have adverse effects or interfere with conventional treatments. Use of the herbal remedy St. John's wort, for example, can reduce the effectiveness of some anticancer drugs.[17] As another example, some studies suggest that antioxidant supplements interfere with chemotherapy and radiation treatments.[18] Clinical trials of several popular supplements are in progress to learn more about their potential effects and interactions with treatments.

◆ Reminder: Reactive oxygen species and their effects on cells were described in Highlight 11.

◆ *Complementary and alternative medicine (CAM)* refers to health care practices that have not been proved to be effective and consequently are not included as part of conventional treatment. Highlight 29 provides additional information about CAM.

Medical Nutrition Therapy for Cancer

The objectives of medical nutrition therapy for cancer patients are to minimize loss of weight and muscle tissue, correct nutrient deficiencies, and provide a diet that patients can tolerate and enjoy despite the complications of illness. Appropriate nutrition care helps patients preserve their strength and improves recovery after stressful cancer treatments. Moreover, malnourished cancer patients develop more complications and have shorter survival times than patients who maintain good nutrition status.[19]

HOW TO Increase kCalories and Protein in Meals

To increase the energy content of a meal, try these suggestions:

- *Butter or margarine.* Melt on pasta, potatoes, rice, and cooked vegetables. Add to hot cereals, casseroles, and soups. Spread liberally on bread, crackers, and rolls.
- *Mayonnaise.* Add to pasta, tuna, and potato salads. Use as a dressing for raw or cooked vegetables.
- *Cream cheese.* Spread on raw vegetables, toast, and crackers. Mix into chopped fruit. Use as a spread in sandwiches made with luncheon meats.
- *Half-and-half and cream.* Replace milk or water with half-and-half or cream in soups, sauces, hot chocolate, desserts, mashed potatoes, and cold and cooked cereals.
- *Nuts.* Add chopped nuts to pasta dishes, stir-fried vegetables, fruit salads, and green salads. Use nut meats in baked products.
- *Beverages.* Replace water and nonkaloric beverages with sweetened drinks, fruit juices, and milk shakes.

These suggestions can help to add protein to a meal:

- *Powdered milk (use full-fat milk powder, if available).* Add to recipes that include milk. Dissolve extra milk powder into milk-containing beverages. Stir into hot cereals, potato dishes, casseroles, and sauces. Add to scrambled eggs, hamburger, and meat loaf.
- *Cheese.* Melt on burgers, meat loaf, cooked vegetables, scrambled eggs, casseroles, and potatoes. Add cottage cheese to casseroles, egg dishes, pasta recipes, and salad dressings. Grate hard cheeses and sprinkle on soups, salads, and cooked vegetable dishes.
- *Eggs.* Add raw eggs when preparing casseroles, meatballs, and hamburgers. Add chopped hard-cooked eggs to salads, vegetable dishes, sandwich fillings, and pasta and potato salads.
- *Meats.* Add meat pieces to soups, egg dishes, casseroles, bean dishes, and pasta sauces. Add minced meats to vegetable dishes. Add chunks of cooked chicken or turkey to salads.

Because there are many forms of cancer and a variety of potential treatments, nutritional needs among cancer patients vary considerably. Furthermore, a person's needs may change at different stages of illness. Patients should be screened for malnutrition when cancer is diagnosed and be reassessed during the treatment and recovery periods.

Protein and Energy For patients at risk of weight loss and wasting, protein and energy needs are considerable. Protein recommendations range from 1.0 to 1.2 grams per kilogram of body weight for nonstressed patients, 1.2 to 1.5 grams per kilogram for those undergoing treatment, and 1.5 to 2.5 grams per kilogram for patients with substantial protein losses or cachexia.[20] Energy needs may be 25 to 35 kcalories per kilogram of body weight, depending on the patient's current weight, activity level, degree of metabolic stress, and energy needs for weight regain and tissue repair. Health practitioners should regularly monitor patients' weight changes and adjust intake recommendations as necessary. Patients who cannot eat adequate food may be able to meet their needs by supplementing the diet with nutrient-dense formulas. The "How to" above provides suggestions that can help to increase the energy and protein content of meals. ◆

◆ Reminder: The high-kcalorie, high-protein diet, which is appropriate for some individuals with cancer, was described in Chapter 18 (see p. 624).

Although weight loss is a problem for many cancer patients, breast cancer patients often gain weight.[21] The weight gain occurs during the first two years after breast cancer diagnosis and is associated with an increase in total body fat. By discussing weight maintenance soon after diagnosis and encouraging physical activity, health practitioners can help patients avoid unnecessary weight gain.

Managing Symptoms and Complications A thorough nutrition assessment often uncovers specific problems or symptoms that interfere with food consumption. Table 29-6 lists dietary considerations related to cancers affecting different sites in the body. The "How to" on p. 910 outlines a variety of dietary strategies that may improve food intake and alleviate symptoms. Patients' responses to these strategies may vary considerably, and in some cases a number of adjustments may be necessary.

Enteral and Parenteral Nutrition Support Nutrition support is used in limited situations during cancer treatment. Generally, tube feedings and parenteral nutrition are provided to patients who have long-term or permanent gastrointestinal impairment or are experiencing complications that interfere with food intake.[22] For example, many patients undergoing radiation therapy for head and neck cancers require long-term tube feeding and may need to continue tube feedings at home. ◆ Par-

◆ Irradiation to the head and neck regions often causes dysphagia and mouth sores.

TABLE 29-6	Dietary Considerations for Specific Cancers	
Cancer Sites	**Common Complications**[a]	**Possible Dietary Measures**
Brain and nervous system	Chewing and swallowing difficulties, difficulty feeding oneself	Mechanically altered diet, use of adaptive feeding devices (see Highlight 27)
Head and neck[b]	Swallowing difficulty, aspiration, inflamed mucosa, dry mouth, altered taste sensation	Tube feeding, mechanically altered diet
Esophagus	Swallowing difficulty, obstruction, acid reflux, inflamed mucosa	Tube feeding, mechanically altered diet
Stomach	Anorexia, delayed stomach emptying, early satiety, dumping syndrome, malabsorption	Tube feeding (for obstruction or unmanageable dumping syndrome); postgastrectomy diet; small, frequent meals; limited sugars and insoluble fibers (see Chapter 23)
Intestine	Fluid and electrolyte imbalances, altered bowel function, malabsorption, lactose intolerance, inflamed mucosa, bacterial overgrowth, short bowel syndrome (if resected), obstruction	Tube feeding or total parenteral nutrition for obstruction, enteritis, or short bowel syndrome; fat- and lactose-restricted diet (see Chapter 24)
Pancreas	Malabsorption, bile insufficiency, hyperglycemia	Fat-restricted diet; enzyme replacement (see Chapter 24); small, frequent meals; carbohydrate-controlled diet (Chapter 26)

[a] Actual complications depend on the specific methods used for treating the cancer.
[b] Includes cancers of the pharynx, larynx, salivary glands, and oral and nasal cavities.

enteral nutrition is reserved for patients who have inadequate GI function, such as individuals with chronic radiation enteritis. Whenever possible, enteral nutrition is strongly preferred over parenteral nutrition, to preserve GI function and avoid infection.

Nutrition Therapy for Bone Marrow Transplant Patients Patients who undergo bone marrow transplants may require total parenteral nutrition (TPN) before and after the transplant, as the GI tract is often severely damaged by the chemotherapy or radiation treatment required beforehand. When GI function returns, the patient can begin consuming small amounts of food along with TPN. As oral intake improves, TPN is gradually tapered. Some patients may require a high-kcalorie, high-protein diet to reverse malnutrition. Because recipients of bone marrow transplants are severely immunocompromised, they should be instructed to follow safe food-handling practices to minimize the risk of foodborne illness (see Highlight 18). In addition, they must avoid foods that are likely to contain unsafe levels of bacteria, such as raw fruits and vegetables; undercooked meat, poultry, and eggs; leftover luncheon meats and meat spreads; blue-veined cheeses; unpasteurized dairy products, juices, honey, and beer; and foods from salad bars or street vendors.[23]

IN SUMMARY

Cancer arises from mutations in the genes that control cell division. Some dietary substances promote carcinogenesis, while others may help to prevent cancer. Cancer's effects on nutrition status depend on the type of cancer a person has, its severity, and the methods used to treat the cancer. Cancer cachexia is a frequent complication of cancer and may be related to anorexia, altered metabolism, and responses to treatment. Medical treatments for cancer include surgery, chemotherapy, and radiation therapy, which remove cancer cells, prevent tumor growth, and alleviate symptoms. Medical nutrition therapy for cancer patients aims to minimize weight loss and wasting, correct deficiencies, and manage complications that impair food intake. The accompanying Case Study on p. 911 allows you to apply information about nutrition and cancer to a clinical situation.

HOW TO Help Patients Handle Food-Related Problems

In people with cancer or HIV infections, various complications can interfere with eating. Health practitioners can try to identify the specific problems that patients are having and offer appropriate solutions. Not every suggestion will work for each patient; encourage patients to experiment and find the strategies that work best.

I just don't have an appetite.
- Eat small meals and snacks at regular times each day.
- Eat the largest meal at the time of day when you feel the best.
- Include nutrient-dense foods in meals, and consume them before other foods.
- Indulge in favorite foods throughout the day. Serve foods attractively.
- Avoid drinking large amounts of liquids before or with meals.
- Eat in a pleasant and relaxed environment. Eat with family and friends when possible.
- Listen to your favorite music or enjoy a program on TV while you eat.
- Take a walk before you eat.

I am too tired to fix meals and eat.
- Let family members and friends prepare food for you.
- Obtain foods that are easy to prepare and easy to eat, such as sandwiches, frozen dinners, take-out meals from restaurants, instant breakfast drinks, liquid formulas, and energy bars.

Foods just don't taste right.
- Brush your teeth or use mouthwash before you eat.
- Consume foods chilled or at room temperature.
- Choose eggs, fish, poultry, and milk products instead of meats.
- Experiment with sauces, seasonings, herbs, and spices to improve food's flavor.
- Use plastic, rather than metal, eating utensils.
- Save your favorite foods for times when you are not feeling nauseated.

I am nauseated a lot of the time, and sometimes I need to vomit.
- Consume liquids throughout the day to replace fluids.
- If you become nauseated from chemotherapy treatments, avoid eating for at least two hours before treatments.
- Consume smaller meals, and eat slowly.
- Avoid foods and meals that have strong odors or are fatty, greasy, or gas forming.

I am having problems chewing and swallowing food.
- Experiment with food consistencies to find the ones you can manage best. Thin liquids, dry foods, and sticky foods (such as peanut butter) are often difficult to swallow.
- Add sauces and gravies to dry foods.
- Drink fluids with meals to ease chewing and swallowing.
- Try using a straw to drink liquids.
- Tilt your head forward and backward to see if you can swallow more easily when your head is positioned differently.

I have sores in my mouth, and they hurt when I eat.
- Use cold or frozen foods; they are often soothing.
- Try soft foods such as ice cream, milk shakes, bananas, applesauce, mashed potatoes, cottage cheese, and macaroni and cheese.
- Avoid foods that irritate mouth sores, such as citrus fruits and juices, tomatoes and tomato-based products, spicy foods, foods that are very salty, foods with seeds (such as poppy seeds and sesame seeds) that can scrape the sores, and coarse foods such as raw vegetables and toast.
- Ask your doctor about using a local anesthetic solution such as lidocaine before eating to reduce pain.
- Use a straw for drinking liquids in order to bypass the sores.

My mouth is really dry.
- Rinse your mouth with warm saltwater or mouthwash frequently. Avoid using mouthwash that contains alcohol.
- Drink small amounts of liquid frequently between meals.
- Ask your doctor or pharmacist about medications that can help dry mouth.
- Use sour candy or gum to stimulate the flow of saliva.
- Add broth, sauces, gravies, mayonnaise, butter, or margarine to dry foods.
- Make sure you brush your teeth and floss regularly to prevent cavities and oral infections.

I am having trouble with diarrhea.
- Drink plenty of fluids. Salty broths and soups, diluted fruit juices, and sports drinks are good choices. For severe diarrhea, try oral rehydration formulas that are commercially prepared.
- Avoid foods and beverages that increase gas, such as legumes, onions, vegetables of the cabbage family, foods that contain sorbitol or mannitol, and carbonated beverages.
- Try using lactase enzyme replacements when you use milk products in case you are experiencing lactose intolerance. Yogurt and aged cheeses may be easier to tolerate than milk and fresh cheeses.
- Avoid high-fat foods if you are fat intolerant.
- Avoid caffeine.
- Eat smaller meals, and eat more frequently.
- Check with your doctor about using digestive enzyme replacements if you have had diarrhea for a long time.

I am having trouble with constipation.
- Drink plenty of fluids. Try warm fluids, especially in the morning.
- Eat whole-grain breads and cereals, nuts, fresh fruits and vegetables, prunes, and prune juice. Avoid refined carbohydrate foods such as white bread, white rice, and pasta.
- Engage in physical activity regularly.

Public Relations Consultant with Cancer

Janet Woodhouse is a 58-year-old public relations consultant who was recently diagnosed with colon cancer after a routine colonoscopy, a procedure in which the colon is examined using a flexible tube attached to an optical device. Mrs. Woodhouse is scheduled to have surgery to remove the segment of colon that contains the tumor and to determine if the cancer has spread to the surrounding lymph nodes and, possibly, other organs. The nurse completing the nutrition assessment finds that Mrs. Woodhouse is 5 feet 5 inches tall and weighs 178 pounds. The patient spends most of the day sitting and has little time to engage in recreational exercise. Her diet is high in fat and typically includes red meat at both lunch and dinner. She eats two or three servings of fruits and vegetables each day, although she does not like green leafy vegetables very much. She rarely drinks milk or consumes milk products.

1. Review Table 29-2 on p. 903, and describe the factors in Mrs. Woodhouse's diet and lifestyle that may have contributed to the development of colon cancer.
2. What symptoms and complications may arise after colon surgery and impair nutrition status? If the cancer team decides that Mrs. Woodhouse needs follow-up chemotherapy, how might the chemotherapy affect her nutrition status?
3. If Mrs. Woodhouse is unresponsive to treatment and her cancer progresses, she may develop cancer cachexia. Describe this syndrome, its causes, and its consequences.
4. Provide suggestions that may help Mrs. Woodhouse handle the following problems should they develop: poor appetite, fatigue, taste alterations, nausea and vomiting, chewing and swallowing difficulties, mouth sores, dry mouth, diarrhea, constipation, and weight loss.

HIV Infection

Possibly, the most infamous infectious disease today is **acquired immune deficiency syndrome (AIDS).** AIDS develops from infection with HIV (human immunodeficiency virus), which attacks the immune system and disables a person's defenses against other diseases, including infections and certain cancers. Then these diseases—which would cause few, if any, symptoms in people with healthy immune systems—destroy health and life.

The HIV/AIDS epidemic continues to sweep across countries, especially in sub-Saharan Africa. Table 29-7 shows its impact worldwide and in North America. For many years the destructive effects of HIV infection seemed unstoppable, but in the mid to late 1990s the death rate from AIDS began to decline in the United States, and the progression from HIV infection to AIDS slowed dramatically. AIDS still has no cure, but remarkable progress has been made in understanding and treating HIV infection.

Without a cure for AIDS, the best course is prevention. HIV is most often sexually transmitted and can be spread by direct contact with contaminated body fluids, such as blood, semen, vaginal secretions, and breast milk. Because many people remain symptom-free during the early stages of infection, they may not realize that they can pass the infection to others. To reduce the spread of HIV infection, individuals at risk (see Table 29-8) are encouraged to undergo testing. A blood test can usually detect HIV antibodies within several months after exposure and, often, after 1 or 2 weeks. An estimated 25 percent of persons in the United States who have HIV infection are unaware that they are infected.[24]

Consequences of HIV Infection

HIV infection destroys immune cells that have a protein called CD4 on their surfaces. The cells most affected are the **helper T cells,** ◆ also called *CD4+ T cells* because the presence of CD4 is a primary characteristic. HIV is able to enter the helper T cells and induce them to produce additional copies of the virus, thus perpetuating

TABLE 29-7 The HIV and AIDS Epidemic at a Glance, 2007

Stage of Epidemic	World	North America
Individuals living with HIV infection or AIDS	33,200,000	1,300,000
Individuals newly infected with HIV	2,500,000	46,000
AIDS deaths	2,100,000	21,000

SOURCE: Joint United Nations Programme on HIV/AIDS and World Health Organization, *AIDS epidemic update: December 2007,* available at http://data.unaids.org/pub/EPISlides/2007/2007_epiupdate_en.pdf; site visited February 3, 2008.

TABLE 29-8 Risk Factors for HIV Infection

- History of receiving blood transfusions or clotting factors between 1978 and 1985
- Infant born to a mother with HIV infection
- Intravenous drug use in which syringes are shared among users
- Sexual contact with multiple partners
- Sexual contact with intravenous drug users, prostitutes, or individuals with a history of HIV or other sexually transmitted diseases
- Unsafe sexual practices

◆ *T cells* are lymphocytes that develop in the thymus gland. The other lymphocytes are the *B cells* (which develop in bone marrow) and *natural killer cells.*

acquired immune deficiency syndrome (AIDS): the late stage of illness caused by infection with the human immunodeficiency virus (HIV); characterized by severe damage to immune function.

helper T cells: lymphocytes that have a specific protein called CD4 on their surfaces and therefore are also known as *CD4+ T cells;* the cells most affected in HIV infection.

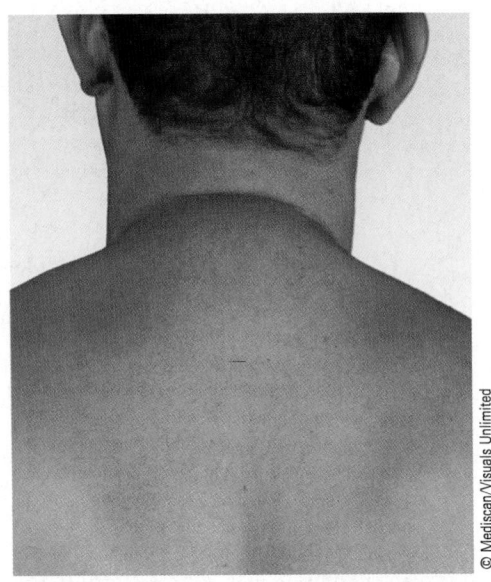

© Mediscan/Visuals Unlimited

HIV-lipodystrophy is sometimes evident by the accumulation of fatty tissue at the base of the neck, referred to as *buffalo hump*.

and exacerbating the infection. Other cells that have the CD4 protein (and are infected by HIV) include tissue macrophages, blood monocytes, and certain cells of the central nervous system.[25] Early symptoms of HIV infection are nonspecific and may include fever, sore throat, malaise, swollen lymph nodes, skin rashes, muscle and joint pain, and diarrhea. After these symptoms subside, many people remain symptom-free for 5 to 10 years or even longer. If the HIV infection is not treated, however, the depletion of T cells eventually increases the person's susceptibility to **opportunistic infections**—that is, infections caused by microorganisms that normally do not cause disease in healthy individuals.

The term *AIDS* applies to the advanced stages of HIV infection, in which the inability to fight illness allows a number of serious diseases and complications to develop; such **AIDS-defining illnesses** include severe infections, certain cancers, and wasting of muscle tissue. Health practitioners evaluate disease progression by measuring the concentrations of helper T cells and circulating virus (called the *viral load*) and by monitoring clinical symptoms. Although current drug therapies can dramatically slow the progression of HIV infection, the drugs' side effects may make it difficult for patients to adhere to treatments, as discussed in several of the following sections.

Lipodystrophy Some of the drug treatments that suppress HIV infection cause abnormalities in glucose and fat metabolism, which affect an estimated 25 to 50 percent of these patients. These complications, collectively known as the **HIV-lipodystrophy syndrome,** include body fat redistribution, abnormal blood lipid levels, and insulin resistance.[26] Patients tend to accumulate abdominal fat and lose fat from the face, arms, and legs; thus, they appear to be thin except for a "pot belly." Also observed are breast enlargement (in both men and women), fat accumulation at the base of the neck (called a **buffalo hump**), and benign growths composed of fat tissue (called **lipomas**). The changes in body composition are often disfiguring and may cause physical discomfort; moreover, patients often develop hypertriglyceridemia, elevated LDL (low-density lipoprotein) cholesterol levels, low HDL (high-density lipoprotein) cholesterol levels, glucose intolerance, and hyperinsulinemia. The reasons for the development of lipodystrophy are unknown.

Weight Loss and Wasting Even with effective treatment of HIV infection, weight loss and wasting are ongoing problems for HIV-infected patients.[27] The Centers for Disease Control and Prevention defines *AIDS-related wasting syndrome* as a 10 percent weight loss within a 6-month period, accompanied by diarrhea or fever without a known cause for more than 30 days. The wasting has been linked with accelerated disease progression, reduced strength, and fatigue. In the later stages of AIDS, wasting is severe and increases the risk of death. Much as in cancer, the wasting associated with HIV infection has many causes: anorexia and inadequate food intake, altered metabolism, malabsorption, chronic diarrhea, and diet-drug interactions.

Anorexia and Reduced Food Intake As mentioned, inadequate food intake is a key factor in the development of wasting. Poor food intake may result from various factors, including the following:

- *Emotional distress, pain, and fatigue.* The physical and social problems that accompany chronic illness may cause fear, anxiety, and depression, which contribute to anorexia. Pain and fatigue, which may be associated with some disease complications, can cause anorexia and difficulty with eating.

- *Oral infections.* The oral infections associated with HIV infection can cause discomfort and interfere with food consumption. Common infections include thrush and herpes simplex virus infection. **Thrush** can cause mouth pain, dysphagia (difficulty swallowing), and altered taste sensation; infection with **herpes simplex virus** may cause painful lesions around the lips and in the mouth.

- *Respiratory disorders.* Respiratory infections, including pneumonia and tuberculosis, are common in people with HIV infection. Symptoms often include chest pain, shortness of breath, and cough, which interfere with eating and contribute to anorexia.

opportunistic infections: infections from microorganisms that normally do not cause disease in healthy people, but are damaging to persons with compromised immune function.

AIDS-defining illnesses: diseases and complications associated with the later stages of an HIV infection, including wasting, recurrent bacterial pneumonia, opportunistic infections, and certain cancers.

HIV-lipodystrophy (LIP-oh-DIS-tro-fee) **syndrome:** a group of abnormalities in fat and glucose metabolism that may result from drug treatments for HIV infection; changes include body fat redistribution, abnormal blood lipid levels, and insulin resistance. The accumulation of abdominal fat is sometimes called *protease paunch.*

buffalo hump: the accumulation of fatty tissue at the base of the neck.

lipomas (lih-POE-muz): benign tumors composed of fatty tissue.

thrush: a fungal infection of the mouth and throat, most often caused by *Candida albicans.*

herpes simplex virus: a common virus that can cause blisterlike lesions on the lips and in the mouth.

- *Cancer.* As described earlier in this chapter, cancer leads to anorexia for numerous reasons. In addition, **Kaposi's sarcoma**, a type of cancer frequently associated with HIV infection, can cause lesions in the mouth and throat that make eating painful.

- *Medications.* The medications given to treat HIV infection, other infections, and cancer often cause anorexia, nausea and vomiting, altered taste sensation, food aversions, and diarrhea.

GI Tract Complications Complications of HIV infection involving the GI tract may result from opportunistic infections, the HIV infection itself, and medications.[28] In addition to the oral infections described previously, infections may develop in the esophagus, stomach, and intestines. Moreover, advanced AIDS is often accompanied by characteristic changes in the small intestinal lining: the villi appear shortened and flattened, and the absorptive area is substantially reduced. ◆ These changes can cause malabsorption, steatorrhea, and diarrhea.

As described earlier, many patients are unable to tolerate the medications used to suppress HIV and develop nausea, vomiting, and diarrhea. Furthermore, the medications that treat the viral, parasitic, and fungal infections in the GI tract contribute to bacterial overgrowth. Thus HIV-infected patients face an extremely high risk of malnutrition due to the combination of intestinal discomfort, bacterial overgrowth, malabsorption, and nutrient losses from vomiting, steatorrhea, and diarrhea.

Neurological Complications Neurological complications may be a consequence of HIV infection, immune suppression, or cancers and infections that target brain tissue.[29] Clinical features include mild to severe dementia; muscle weakness and gait disturbances; and pain, numbness, and tingling in the legs and feet. Neurological impairments are usually more pronounced in the advanced stages of AIDS.

Other Complications Patients with HIV infection can develop anemia due to chronic inflammation, ◆ nutrient malabsorption, blood loss, disturbances in bone marrow function, or medication side effects. HIV infection may also lead to skin disorders (rashes, infections, and cancers), eye diseases (retinal infection and detachment), kidney diseases (nephrotic syndrome and chronic kidney disease), and coronary heart disease.[30]

Treatments for HIV Infection

Although there is no cure for HIV infection, treatments can help to slow its progression, reduce complications, and alleviate pain. The standard treatment for suppressing HIV infection, called *highly active antiretroviral therapy (HAART)*, combines three or more antiretroviral drugs.[31] Table 29-9 on p. 914 lists the major drug categories included in antiretroviral therapy and describes the drugs' modes of action. These antiretroviral agents have multiple adverse effects that make their long-term use difficult to tolerate. In addition to the GI effects discussed previously, side effects include skin rashes, headache, anemia, tingling and numbness, hepatitis, pancreatitis, and kidney stones. Thus, although HAART has improved life span and quality of life for many patients, the drug regimens are difficult to adhere to and cause complications that require continual management. The Diet-Drug Interactions feature on p. 914 summarizes the nutrition-related effects of some of the antiretroviral agents and other drugs mentioned in this chapter.

Control of Lipodystrophy Treatment strategies for lipodystrophy are under investigation. Both aerobic activity and resistance training help to reduce abdominal fat, although some patients opt for cosmetic surgery.[32] Patients may be given alternative antiretroviral drugs to alleviate symptoms. Medications may be prescribed to treat abnormal blood lipid levels and insulin resistance.

Control of Anorexia and Wasting Anabolic hormones, appetite stimulants, and regular physical activity have been successful in reversing weight loss and increasing muscle mass in HIV-infected patients.[33] Testosterone and human growth

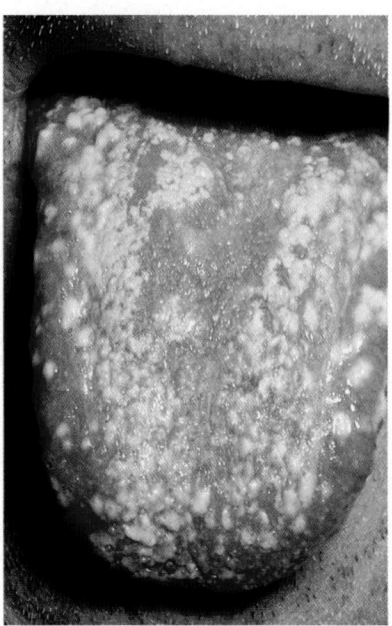

The oral infection *thrush* is easily identified by the characteristic milky white patches that appear on the tongue.

◆ The AIDS-related abnormalities in the intestinal mucosa are sometimes referred to as *HIV enteropathy* (EN-ter-OP-ah-thy).

◆ Reminder: The *anemia of chronic disease* often develops during chronic illness and is characterized by altered iron distribution in tissues and reduced synthesis of red blood cells, among other abnormalities (See Highlight 19).

Kaposi's (cap-OH-seez) **sarcoma:** a common cancer in HIV-infected persons that is characterized by lesions in the skin, lungs, and GI tract.

TABLE 29-9 Antiretroviral Drugs for Treatment of HIV Infection

Category	Examples	Mode of Action
CCR5 antagonist	Maraviroc	CCR5 antagonists prevent HIV from entering cells by blocking a membrane receptor on the host cell.
Fusion inhibitor	Enfuvirtide	Fusion inhibitors prevent HIV from entering cells by binding a viral protein needed for its entry.
Integrase inhibitor	Raltegravir	Integrase inhibitors impair the function of HIV's integrase enzyme, which incorporates viral DNA into the host cell's genome.
Non-nucleoside reverse transcriptase inhibitor (NNRTI)	Delavirdine, efavirenz, nevirapine	NNRTI bind active sites on HIV's reverse transcriptase enzyme, blocking the ability of HIV to produce DNA copies of its genetic material.
Nucleoside reverse transcriptase inhibitor (NRTI)	Didanosine, lamivudine, zidovudine (AZT)	As analogs of the nucleosides needed for DNA synthesis, NRTI impair the ability of HIV's reverse transcriptase enzyme to produce usable copies of DNA.
Protease inhibitor (PI)	Saquinavir, ritonavir, indinavir	PI inhibit HIV's protease enzyme, which cleaves HIV's gene products into usable structural proteins.

SOURCES: U.S. Department of Health and Human Services, Panel on Antiretroviral Guidelines for Adults and Adolescents, *Guidelines for the use of antiretroviral agents in HIV-1-infected adults and adolescents,* January 29, 2008, pp. 1–128, available at www.aidsinfo.nih.gov/ContentFiles/AdultandAdolescentGL.pdf; site visited February 4, 2008; S. Safrin, Antiviral agents, in B. G. Katzung, ed., *Basic and Clinical Pharmacology* (New York: Lange/McGraw-Hill, 2007), pp. 790–818.

DIET-DRUG INTERACTIONS

Check this table for notable nutrition-related effects of the medications discussed in this chapter.

	Gastrointestinal Effects	Interactions with Dietary Substances	Metabolic Effects
Appetite stimulants (megestrol acetate, dronabinol)	Nausea, vomiting, diarrhea.	—	Hyperglycemia (megestrol acetate).
Didanosine	Nausea, vomiting, dry mouth, altered taste perception, anorexia, constipation.	Avoid alcohol and aluminum- and magnesium-containing antacids.	Pancreatitis.
Enfuvirtide	Nausea, vomiting, anorexia, diarrhea, constipation.	—	Pancreatitis, increased blood triglycerides.
Methotrexate	Nausea, vomiting, diarrhea, reduced absorption of vitamin B_{12} and calcium.	Milk may reduce methotrexate absorption if the milk and methotrexate are ingested together.	Increased serum uric acid levels, anemia, liver toxicity.
Ritonavir	Nausea, vomiting, altered taste perception, anorexia, diarrhea.	—	Pancreatitis, diabetes, reduced blood levels of copper and zinc; increased levels of triglycerides, liver enzymes, creatine kinase, and uric acid.
Zidovudine (AZT)	Nausea, vomiting, altered taste perception, anorexia, mouth sores, constipation.	—	Anemia, reduced blood levels of copper and zinc.

NOTE: Other antiretroviral drugs that treat HIV infection have gastrointestinal and metabolic side effects; only a few are listed here as examples.

hormone have demonstrated positive effects on nitrogen balance and lean tissue content, especially in combination with resistance training. A regular program of resistance exercise improves muscle mass and strength and corrects some of the metabolic abnormalities (altered blood lipids and insulin resistance) that are common in HIV-infected patients. The medications megestrol acetate and dronabinol (described on p. 907) are sometimes prescribed to stimulate appetite and improve weight gain, although much of the weight increase is attributable to a gain of fat rather than lean tissue.[34]

Alternative Therapies Like cancer patients, people with HIV infection and AIDS are frequently tempted to try unconventional methods of treatment. Although many alternative therapies are harmless, they can be expensive at a time when financial security is of concern. Monitoring patients' use of dietary supplements is essential to reduce the possibility of nutrient-drug and herb-drug interactions.

Medical Nutrition Therapy for HIV Infection

Nutrition assessment and counseling should begin as soon as a patient is diagnosed with HIV infection. The initial assessment provides baseline data with which to monitor progress throughout the course of the disease, and it should include an evaluation of body weight and body composition. Follow-up measurements may indicate the need to adjust dietary recommendations and drug therapies.

Weight Maintenance A primary objective of nutrition therapy is to help the patient with HIV infection maintain weight and muscle tissue.[35] Health practitioners should attempt to determine the factors that interfere with the patient's food intake and physical activity, as well as offer suggestions that may help to prevent future weight problems. Some individuals may benefit from a high-kcalorie, high-protein diet. The addition of nutrient-dense snacks, protein or energy bars, and oral supplements can improve intakes. Liquid formulas may be useful for the person who is too tired to eat or prepare meals. If food consumption is difficult, small, frequent feedings may be better tolerated than several large meals. The "How to" on p. 908 provides suggestions for adding energy and protein to the diet.

Vitamins and Minerals Vitamin and mineral needs of people with HIV infections are highly variable, and little information is available concerning specific needs. Because nutrient deficiencies are likely to result from reduced food intake, malabsorption, diet-drug interactions, and nutrient losses, multivitamin-mineral supplements are often recommended. Patients should be cautioned to maintain intakes that are close to DRI recommendations, however, as some studies suggest that micronutrient supplements may have some harmful effects in individuals on antiretroviral drug therapies.[36]

Metabolic Complications As mentioned, patients with HIV infection who are using antiretroviral drugs frequently develop insulin resistance and elevated triglyceride and LDL (low-density lipoprotein) cholesterol levels. Treating these problems often requires both medications and dietary adjustments. Patients should be advised to achieve or maintain a desirable weight, replace saturated fats with monounsaturated and polyunsaturated fats, increase fiber intake, and limit intakes of *trans*-fatty acids, cholesterol, added sugars, and alcohol. Regular physical activity can improve both insulin resistance and blood lipid levels. ◆ If problems persist, alternative antiretroviral medications may be prescribed in an attempt to improve the metabolic abnormalities.

Symptom Management The discomfort associated with antiretroviral therapy, opportunistic GI infections, and malabsorption may make food consumption difficult, and problems such as vomiting and diarrhea contribute to fluid and electrolyte losses. The "How to" on p. 910 describes measures that can improve food and fluid intakes and alleviate discomfort in individuals with these problems.

Resistance training can help a person with HIV infection maintain muscle mass and strength.

◆ Additional suggestions for managing insulin resistance and hyperlipidemias are available in Chapters 26 and 27, respectively.

Food Safety The depressed immunity of people with HIV infections places them at extremely high risk of developing foodborne infections. Health practitioners should caution patients about their high susceptibility to foodborne illness and provide detailed instructions about the safe handling and preparation of foods (see Highlight 18). Water can also be a source of foodborne illness and is a common cause of **cryptosporidiosis** in HIV-infected individuals. Because water quality varies throughout the United States, patients should consult their local health departments to determine whether the local tap water is safe to drink. If not, or to take additional safety measures, they should boil drinking water for one minute. Some, but not all, types of filtered and bottled waters are safe.

Enteral and Parenteral Nutrition Support In later stages of illness, people with HIV infections may be unable to consume enough food and may need aggressive nutrition support. Tube feedings are preferred whenever the GI tract is functional; they can be provided at night to supplement oral diets consumed during the day. Parenteral nutrition is reserved for patients who are unable to tolerate enteral nutrition, such as those with GI obstructions that prevent food intake. For individuals with severe malabsorption, orally administered hydrolyzed formulas containing medium-chain triglycerides may be as effective as parenteral nutrition for reversing weight loss and wasting. For either type of feeding, careful measures are necessary to avoid bacterial contamination of nutrient formulas and feeding equipment.

IN SUMMARY

By attacking immune cells, HIV causes progressive damage to immune function and may eventually lead to AIDS. Improved drug therapies have slowed the progression of HIV infection; however, these drugs may promote the HIV-lipodystrophy syndrome, characterized by body fat redistribution, abnormal lipid levels, and insulin resistance. HIV infection is often associated with weight loss and wasting, anorexia, and various complications that affect food intake. Dietary adjustments, resistance training, and medications can help patients maintain their weight and prevent wasting. People with HIV infection must pay strict attention to food safety guidelines to prevent foodborne illnesses. The Case Study provides an opportunity to review the nutritional concerns of a person with HIV infection.

CASE STUDY Financial Planner with HIV Infection

Three years ago, Darrell Meckler, a 34-year-old financial planner, sought medical help when he began feeling run-down and developed a painful white fungal infection over his mouth and tongue. The presence of thrush, recent weight loss, and anemia alerted Mr. Meckler's physician to the possibility of an HIV infection. When Mr. Meckler tested positive for HIV, he and his family and friends were devastated by the news, but those close to him have remained supportive. During the three years since Mr. Meckler began antiretroviral drug therapy, he has maintained his weight but has also developed lipodystrophy and hypertriglyceridemia. Mr. Meckler is 6 feet tall and currently weighs 185 pounds. He occasionally develops diarrhea and sometimes anorexia.

1. Describe lipodystrophy, and discuss its typical pattern in people who have an HIV infection. What adjustments in treatment and lifestyle may be helpful for Mr. Meckler?
2. Describe an appropriate diet for Mr. Meckler. What strategies may improve his problems with diarrhea and anorexia? Suggest reasons why diarrhea and anorexia may develop in people with HIV infections.
3. Explain why an HIV infection can lead to wasting as the disease progresses to the later stages. What recommendations may be helpful for maintaining weight and health if wasting becomes a problem?

cryptosporidiosis (KRIP-toe-spor-ih-dee-OH-sis): a foodborne illness caused by the parasite *Cryptosporidium parvum*.

CENGAGENOW™
academic.cengage.com/login

Clinical Portfolio

1. Consider the nutrition problems that may develop in a 36-year-old woman with a malignant brain tumor that affects her ability to move the right side of her body (including the tongue) and to speak coherently. She is taking a pain medication that makes her nauseated and sleepy. Her expected survival time is only about six months.

   ▦ If she is right-handed, how might her impairment interfere with eating? What suggestions do you have for overcoming this problem?

   ▦ How might her nutrition status be affected by her inability to communicate effectively? What suggestions may help?

   ▦ In what ways might the pain medication she is taking affect her nutrition status?

2. Various types of chronic conditions can lead to weight loss and wasting. For some of these conditions, such as Crohn's disease or celiac disease (Chapter 24), diet is a cornerstone of treatment. For others, such as cancer and HIV infection, nutrition plays a supportive role. What determines whether nutrition plays a primary role or a supportive role in the treatment of disease?

NUTRITION ON THE NET

For further study of topics covered in this chapter, log on to **academic.cengage .com/nutrition/rolfes/UNCN8e**. Go to Chapter 29, then to Nutrition on the Net.

- To learn more about cancer, including risk factors, prevention, screening, detection, treatments (including nutrition), and support networks, visit these sites:
  American Association for Cancer Research: **www.aacr.org**
  American Cancer Society: **www.cancer.org**
  American Institute for Cancer Research: **www.aicr.org**
  National Cancer Institute: **www.cancer.gov**

- To find additional information about HIV infection and AIDS, visit these sites:
  AIDS Education Global Information System: **www.aegis.com**

AIDSinfo, an information service provided by the U.S. Department of Health and Human Services:
aidsinfo.nih.gov
The Body: **www.thebody.com**
UCSF Center for HIV Information: **hivinsite.ucsf.edu**

- To review information about safe food handling, visit the FDA's Center for Food Safety and Applied Nutrition:
  vm.cfsan.fda.gov

NUTRITION ASSESSMENT CHECKLIST for People with Cancer or HIV Infections

Medical History

Check the medical record to determine:

- Type and stage of cancer
- Stage of HIV infection

Review the medical record for complications that may alter medical nutrition therapy, including:

- Altered organ function
- Altered taste perception
- Anorexia
- Dry mouth and oral infections
- GI symptoms and infections
- Hyperlipidemias
- Insulin resistance
- Malnutrition and wasting

Medications

For patients with cancer or HIV infections:

- Check medications to identify potential diet-drug interactions.
- Recommend the use of antinauseants at mealtime, if needed.
- Ask about the use of dietary supplements, including herbal remedies.

For cancer patients who require chemotherapy:

- Recommend strategies to prevent food aversions.
- Offer suggestions for managing drug-related complications.

For HIV-infected patients using antiretroviral drug therapy:

- Remind patients that some drugs are better absorbed with foods and that others must be taken on an empty stomach.
- Help patients work out a medication schedule that suits their lifestyle and is timed appropriately in regard to food intake.
- Offer suggestions for managing drug-related complications.

Dietary Intake

For patients with poor food intakes and weight loss:

- Determine the reasons for reduced food intake.
- Offer appropriate suggestions to improve food intake.
- Provide interventions before weight loss progresses too far.

For patients with HIV infections who experience weight gain, elevated triglyceride or LDL cholesterol levels, or hyperglycemia:

- Assess the diet for energy, total fat, types of fat, carbohydrates, fiber, and sugars.
- For patients with hyperlipidemias, recommend a diet low in saturated fat, *trans*-fatty acids, and sugars.
- For patients with hyperglycemia, recommend a consistent carbohydrate intake that emphasizes complex carbohydrates.
- Recommend regular physical activity for weight control and for improving blood lipid levels and insulin resistance.

Anthropometric Data

Take baseline height and weight measurements, monitor weight regularly, and suggest dietary adjustments for weight maintenance, if necessary. Remember that body composition may change without affecting body weight. Perform baseline and periodic body composition measurements in HIV-infected patients who are using antiretroviral drug therapy.

Laboratory Tests

Note that albumin and other serum proteins may be reduced in patients with cancer or HIV infections, especially in those experiencing wasting. Check laboratory tests for indications of:

- Anemia
- Dehydration
- Elevated LDL cholesterol levels
- Elevated triglyceride levels
- Hyperglycemia

For patients with HIV infections, evaluate disease progression by checking:

- Helper T cell counts
- Viral load

Physical Signs

Look for physical signs of:

- Dehydration (especially for patients with fever, vomiting, or diarrhea)
- Kaposi's sarcoma
- Oral infections
- Protein-energy malnutrition and wasting

STUDY QUESTIONS

CENGAGENOW™

To assess your understanding of chapter topics, take the Student Practice Test and explore the modules recommended in your Personalized Study Plan. Log on to **academic.cengage.com/login.**

These questions will help you review the chapter. You will find the answers in the discussions on the pages provided.

1. Describe the process of tumor formation. What factors contribute to cancer development? Discuss the dietary factors that may increase or decrease the risk of cancer. (pp. 901–904)

2. What is cancer cachexia? What factors promote its development? (p. 905)

3. Explain how cancer and its treatments can cause alterations in food intake and metabolism, and possibly lead to malnutrition. (p. 905)

4. Discuss the elements of medical nutrition therapy for cancer, as well as strategies that can improve food intake. (pp. 907–910)

5. Explain how HIV is transmitted, and list risk factors associated with an HIV infection. (p. 911)

6. Describe the possible consequences of HIV infection, such as reduced immunity, HIV-lipodystrophy syndrome, wasting, GI complications, neurological complications, and anemia. Explain why an HIV infection often results in anorexia and reduced food intake. (pp. 911–913)

7. Describe how an HIV infection is treated, and discuss the potential complications associated with treatment. Discuss the features of medical nutrition therapy for HIV-infected and AIDS patients. (pp. 913–916)

8. Why are people with HIV infections highly susceptible to foodborne illness? Describe some measures that can be taken to prevent foodborne illness. (p. 916)

These multiple choice questions will help you prepare for an exam. Answers can be found on p. 920.

1. Which of these dietary substances may help to protect against cancer?
    a. alcohol
    b. well-cooked meats, poultry, and fish
    c. animal fats
    d. phytochemicals from fruits and vegetables

2. The metabolic changes that often accompany cancer include all of the following *except:*
    a. increased triglyceride breakdown.
    b. increased protein turnover.
    c. increased muscle protein synthesis.
    d. insulin resistance.

3. An advantage of radiation therapy over chemotherapy is that:
    a. radiation is not damaging to rapidly dividing cells.
    b. irradiation's side effects do not include malnutrition.
    c. radiation can be directed toward the regions affected by cancer.
    d. the radiation used is too weak to damage GI tissues.

4. Although many cancer patients lose weight, which type of cancer is often associated with weight gain?
    a. kidney cancer
    b. breast cancer
    c. colon cancer
    d. lung cancer

5. Oral diets after bone marrow transplants may restrict:
    a. fiber.
    b. carbohydrates.
    c. high-protein foods.
    d. raw fruits and vegetables.

6. HIV can enter and destroy these immune cells:
    a. B cells
    b. helper T cells
    c. natural killer cells
    d. neutrophils

7. HIV-lipodystrophy syndrome may result in all of these changes *except:*
    a. increased abdominal fat.
    b. increased fat in the arms and legs.
    c. fat accumulation at the base of the neck.
    d. hypertriglyceridemia.

8. Mouth sores in people with HIV infections are most frequently due to:
    a. oral infections.
    b. dehydration.
    c. nutrient deficiency.
    d. foodborne illnesses.

9. Megestrol acetate and dronabinol are:
    a. medications used to promote weight gain.
    b. protease inhibitors that fight HIV infection.
    c. medications that treat common opportunistic infections.
    d. anabolic hormones that promote gain of muscle tissue.

10. To prevent cryptosporidiosis, a person with HIV infection may need to:
    a. wash hands carefully before meals.
    b. avoid consuming undercooked meat, poultry, and eggs.
    c. consume a high-kcalorie, high-protein diet.
    d. boil drinking water for one minute.

REFERENCES

1. R. S. K. Chaganti, Genetics of cancer, in L. Goldman and D. Ausiello, eds., *Cecil Medicine* (Philadelphia: Saunders, 2008), pp. 1340–1344.
2. W. C. Willett and E. Giovannucci, Epidemiology of diet and cancer risk, in M. E. Shils and coeditors, *Modern Nutrition in Health and Disease* (Philadelphia: Lippincott Williams & Wilkins, 2006), pp. 1267–1279.
3. Willett and Giovannucci, 2006.
4. World Cancer Research Fund/American Institute for Cancer Research, *Food, Nutrition, Physical Activity, and the Prevention of Cancer: A Global Perspective* (Washington, D.C.: American Institute for Cancer Research, 2007), pp. 280–288.
5. R. J. Turesky, Formation and biochemistry of carcinogenic heterocyclic aromatic amines in cooked meats, *Toxicology Letters* 168 (2007): 219–227; T. Sugimura and coauthors, Heterocyclic amines: Mutagens/carcinogens produced during cooking of meat and fish, *Cancer Science* 95 (2004): 290–299.
6. S. Koutros and coauthors, Meat and meat mutagens and risk of prostate cancer in the

agricultural health study, *Cancer Epidemiology, Biomarkers, and Prevention* 17 (2008): 80–87; World Cancer Research Fund/American Institute for Cancer Research, pp. 116–128; Turesky, 2007.
7. R. H. Liu, Potential synergy of phytochemicals in cancer prevention: Mechanism of action, *Journal of Nutrition* 134 (2004): 3479S–3485S.
8. Willett and Giovannucci, 2006.
9. H. S. Rugo, Paraneoplastic syndromes and other non-neoplastic effects of cancer, in L. Goldman and D. Ausiello, eds., *Cecil Medicine* (Philadelphia: Saunders, 2008), pp. 1353–1362.
10. American Dietetic Association, *Nutrition Care Manual* (Chicago: American Dietetic Association, 2008).
11. M. Schattner and M. Shike, Nutrition support of the patient with cancer, in M. E. Shils and coeditors, *Modern Nutrition in Health and Disease* (Philadelphia: Lippincott Williams & Wilkins, 2006), pp. 1290–1313.
12. T. Agustsson and coauthors, Mechanism of increased lipolysis in cancer cachexia, *Cancer Research* 67 (2007): 5531–5537; L. G.

Melstrom and coauthors, Mechanisms of skeletal muscle degradation and its therapy in cancer cachexia, *Histology and Histopathology* 22 (2007): 805–814; A. Lelbach, G. Muzes, and J. Feher, Current perspectives of catabolic mediators of cancer cachexia, *Medical Science Monitor* 13 (2007): RA168–173.
13. M. C. Perry, Principles of cancer therapy, in L. Goldman and D. Ausiello, eds., *Cecil Medicine* (Philadelphia: Saunders, 2008), pp. 1370–1387.
14. J. M. Vose and S. Z. Pavletic, Hematopoietic stem cell transplantation, in L. Goldman and D. Ausiello, eds., *Cecil Medicine* (Philadelphia: Saunders, 2008), pp. 1328–1332.
15. Schattner and Shike, 2006.
16. C. S. Roberts, Patient-physician communication regarding use of complementary therapies during cancer treatment, *Journal of Psychosocial Oncology* 23 (2005): 35–60.
17. S. Marchetti and coauthors, Concise review: Clinical relevance of drug–drug and herb–drug interactions mediated by the ABC transporter ABCB1 (MDR1, P-glycoprotein), *The Oncologist* 12 (2007): 927–941;

I. Meijerman, J. H. Beijnen, and J. H. M. Schellens, Herb-drug interactions in oncology: Focus on mechanisms of induction, *The Oncologist* 11 (2006): 742–752.

18. B. Bruemmer and coauthors, The association between vitamin C and vitamin E supplement use before hematopoietic stem cell transplant and outcomes to two years, *Journal of the American Dietetic Association* 103 (2003): 982–990; H. E. Seifried and coauthors, The antioxidant conundrum in cancer, *Cancer Research* 63 (2003): 4295–4298.

19. Schattner and Shike, 2006.

20. American Dietetic Association, 2008.

21. N. Saquib and coauthors, Weight gain and recovery of pre-cancer weight after breast cancer treatments: Evidence from the women's healthy eating and living (WHEL) study, *Breast Cancer Research and Treatment* 105 (2007): 177–186; G. Makari-Judson, C. H. Judson, and W. C. Mertens, Longitudinal patterns of weight gain after breast cancer diagnosis: Observations beyond the first year, *Breast Journal* 13 (2007): 258–265.

22. Schattner and Shike, 2006.

23. American Dietetic Association, 2008.

24. UNAIDS/WHO, *AIDS Epidemic Update: December 2006,* available at http://data.unaids.org/pub/EpiReport/2006/2006_Epi-Update_en.pdf; site visited February 3, 2008.

25. G. M. Shaw, Biology of human immunodeficiency viruses, in L. Goldman and D. Ausiello, eds., *Cecil Medicine* (Philadelphia: Saunders, 2008), pp. 2557–2561.

26. P. Koutkia and S. Grinspoon, HIV-associated lipodystrophy: Pathogenesis, prognosis, treatment, and controversies, *Annual Review of Medicine* 55 (2004): 303–317.

27. S. Grinspoon and K. Mulligan, Weight loss and wasting in patients infected with human immunodeficiency virus, *Clinical Infectious Diseases* 36 (2003): S69–S78.

28. J. G. Bartlett, Gastrointestinal manifestations of human immunodeficiency virus and acquired immunodeficiency syndrome, in L. Goldman and D. Ausiello, eds., *Cecil Medicine* (Philadelphia: Saunders, 2008), pp. 2582–2585.

29. J. R. Berger and A. Nath, Neurologic complications of human immunodeficiency virus infection, in L. Goldman and D. Ausiello, eds., *Cecil Medicine* (Philadelphia: Saunders, 2008), pp. 2607–2611.

30. E. G. L. Wilkins, Human immunodeficiency virus infection and the human acquired immunodeficiency syndrome, in N. A. Boon, N. R. Colledge, and B. R. Walker, eds., *Davidson's Principles and Practice of Medicine* (Philadelphia: Churchill Livingstone/Elsevier, 2006), pp. 377–402.

31. Panel on Antiretroviral Guidelines for Adults and Adolescents, Department of Health and Human Services, *Guidelines for the Use of Antiretroviral Agents in HIV-1-Infected Adults and Adolescents,* January 29, 2008, pp. 1–128, available at www.aidsinfo.nih.gov/ContentFiles/AdultandAdolescentGL.pdf; site visited February 4, 2008; S. Safrin, Antiviral agents, in B. G. Katzung, ed., *Basic and Clinical Pharmacology* (New York: Lange/McGraw-Hill, 2007), pp. 790–818.

32. Koutkia and Grinspoon, 2004.

33. Grinspoon and Mulligan, 2003.

34. K. Mulligan and coauthors, Testosterone supplementation of megestrol therapy does not enhance lean tissue accrual in men with human immunodeficiency virus-associated weight loss: A randomized, double-blind, placebo-controlled, multicenter trial, *Journal of Clinical Endocrinology and Metabolism* 92 (2007): 563–570.

35. C. Fields-Gardner and coauthors, Position of the American Dietetic Association and Dietitians of Canada: Nutrition intervention in the care of persons with human immunodeficiency virus infection, *Journal of the American Dietetic Association* 104 (2004): 1425–1441; J. Nerad and coauthors, General nutrition management in patients infected with human immunodeficiency virus, *Clinical Infectious Diseases* 36 (2003): S52–S62.

36. P. K. Drain and coauthors, Micronutrients in HIV-positive persons receiving highly active antiretroviral therapy, *American Journal of Clinical Nutrition* 85 (2007): 333–345; Fields-Gardner and coauthors, 2004.

ANSWERS

Study Questions (multiple choice)

1. d   2. c   3. c   4. b   5. d   6. b   7. b   8. a   9. a   10. d

Complementary and Alternative Medicine

© Art Montes De Oca/Taxi/Getty Images

The medical treatments described in the clinical chapters are based upon current scientific understanding of human physiology and biochemistry and are generally supported by well-conducted clinical research. In **conventional medicine,** if a treatment is tested and found to be ineffective, it is eventually abandoned. If a novel therapy is demonstrated by clinical research to be effective and the benefits of using it outweigh its risks, it is incorporated into mainstream medical practice.[1] This highlight examines therapies that have *not* been scientifically validated and are therefore not currently accepted by conventional medical professionals; these therapies fall into a category called **complementary and alternative medicine (CAM).** The accompanying glossary defines related terms.

An estimated 36 percent of adults in the United States use some form of CAM (excluding the use of prayer).[2] CAM is most prevalent among people with chronic, debilitating diseases; for example, 84 percent of AIDS patients reportedly use CAM.[3] Many patients use CAM as an adjunct to conventional medicine—often for symptoms or illnesses that are not sufficiently helped by conventional treatments. CAM therapies remain popular despite the dearth of evidence demonstrating their effectiveness. Reasons for their popularity include consumers' growing interest in self-help measures, the noninvasive nature of many CAM therapies, and the positive interactions consumers have with CAM practitioners.[4]

In response to the enormous popularity of CAM in the United States, in 1998 Congress established the **National Center for Complementary and Alternative Medicine (NCCAM),** which is now one of the 27 institutes that make up the National Institutes of Health (NIH). The missions of NCCAM are to investigate complementary and alternative therapies by funding well-designed scientific studies and to provide authoritative information for consumers and health professionals. If enough evidence is found to support the use of a complementary or alternative therapy, it will likely become a treatment regularly offered by conventional health practitioners.[5]

Defining Complementary and Alternative Medicine

CAM includes a huge variety of approaches, philosophies, and treatments that have not been proven effective for treating disease; some of them are described in the Glossary of Alternative Therapies (p. 923). When these therapies are used in place of conventional medicine, they are called *alternative;* when used together with conventional medicine, they are called *complementary.* The term *alternative* may be misleading in that it inappropriately implies that unproven methods of treatment are valid alternatives to conventional treatments.

Due to substantial consumer interest, health care professionals are finding it necessary to learn about CAM therapies so that they can better communicate with patients regarding their medical care and advise them when an alternative approach conflicts with standard therapy or presents a danger to health. To provide medical students with objective information about CAM, half of U.S. medical schools now offer elective courses about alternative forms of treatment.[6] Physicians who practice **integrative medicine** refer patients for complementary therapies while continuing to provide standard treatments.

GLOSSARY

complementary and alternative medicine (CAM): diverse medical and health care systems, practices, and products that currently are not considered part of conventional medicine; also called *unconventional* or *unorthodox therapies.*

- *Complementary medicine* refers to unconventional therapies that are used *in addition to,* and not simply as a replacement for, conventional medicine.

- *Alternative medicine* refers to unconventional therapies that are used *in place of* conventional medicine.

conventional medicine: diagnosis and treatment of diseases as practiced by a doctor of medicine (M.D.) or doctor of osteopathy (D.O.) and assisted by allied health professionals such as registered nurses, pharmacists, and physical therapists; also called *Western, mainstream,* or *orthodox medicine.*

integrative medicine: medical care that combines mainstream medical treatments and referrals to practitioners of CAM therapies.

National Center for Complementary and Alternative Medicine (NCCAM): a federal agency that researches and provides information about complementary and alternative therapies.

Overview of CAM Therapies

CAM encompasses any and all therapies that are not normally part of conventional medicine. Consequently, the list of CAM approaches includes hundreds of advertised therapies purchased and used by consumers. Unfortunately, CAM has become a marketing buzzword and is used by unscrupulous sellers of worthless treatments.

The NCCAM has classified CAM therapies as shown in Table H29-1 and defined in the Glossary of Alternative Therapies on p. 923. Several popular examples are described in this highlight. Other examples are discussed on the NCCAM website **(http://nccam.nih.gov/health).**

Alternative Medical Systems

Alternative medical systems are based on beliefs that lack the scientific basis of the theories underlying conventional medicine. Virtually all of these alternative systems were developed well over 100 years ago, before our bodies' biochemical and physiological processes were well understood. The alternative forms of diagnoses and treatments may appeal to consumers because the interventions are nontechnical and seem nonthreatening. In general, however, the alternative theories and practices remain rooted in the past and have not been updated to include our current knowledge.

Naturopathic Medicine

Naturopathic medicine proposes that a person's natural "life force" can foster self-healing. This life force is allegedly stimulated by certain health-promoting factors and suppressed by excesses and deficiencies. Naturopaths believe that ill health results from an internal disruption rather than from external disease-causing agents. Naturopathic therapies aim to enhance the natural healing powers of the body and may include special diets or fasting, herbal remedies and other dietary supplements, acupuncture, homeopathy, massage, and various other interventions.

Homeopathic Medicine

Homeopathic medicine is based on the dubious theory that "like cures like." Homeopaths believe that a substance that causes a particular set of symptoms can be used to cure a disease that has similar symptoms. Homeopathic remedies are usually substantially diluted in the belief that dilution increases potency, and most remedies are so extremely diluted that the original substance is no longer present. Homeopaths theorize that even though their remedies no longer contain a diluted substance, they still have powerful healing effects because the water structure is somehow altered during the dilution process used to prepare homeopathic medicines. This theory, however, conflicts with scientific understanding of water structure and properties.

Traditional Chinese Medicine

Traditional Chinese medicine (TCM) includes a large number of folk practices that originated in China. TCM is based on the theory that the body has pathways (called *meridians*) that conduct energy (called *qi;* pronounced "chee"). The interrupted flow of qi is believed to cause illness. TCM practices allegedly improve the flow of qi and include acupuncture, qi gong, herbal remedies, dietary practices, and massage. (Acupuncture and qi gong are described in a later section on energy therapies.) Ironically, TCM is used by relatively few in the Chinese population, as Chinese physicians have largely adopted the Western approach to managing illness.[7]

TABLE H29-1 Examples of Complementary and Alternative Medicine

Alternative Medical Systems

- Naturopathic medicine
- Homeopathic medicine
- Traditional Chinese medicine
- Ayurveda

Mind-Body Interventions

- Meditation
- Faith healing (prayer)
- Mental healing (including hypnotherapy)
- Music, art, and dance therapy

Biologically Based Therapies

- Dietary supplements
- Foods and special diets
- Herbal products
- Hormones
- Aromatherapy

Manipulative and Body-Based Methods

- Chiropractic
- Massage therapy
- Osteopathic manipulation
- Reflexology

Energy Therapies

- Biofield therapies (including therapeutic touch, acupuncture, and qi gong)
- Bioelectrical therapies (including electrical and magnetic fields).

Mind-Body Interventions

Mind-body therapies attempt to improve a person's sense of psychological or spiritual well-being despite the presence of illness. The treatments are also used in the hope of reducing stress, dealing with pain, or lowering blood pressure. Some of these therapies have been incorporated into mainstream medicine for stress reduction or relaxation. For example, **biofeedback** training, in which individuals learn to monitor skin temperature, muscle tension, or brain wave activity while practicing relaxation techniques, is frequently taught by behavioral medicine specialists to

GLOSSARY OF ALTERNATIVE THERAPIES

acupuncture (AK-you-PUNK-chur): a therapy that involves inserting thin needles into the skin at specific anatomical points, allegedly to correct disruptions in the flow of energy within the body.

aromatherapy: inhalation of oil extracts from plants to cure illness or enhance health.

ayurveda: a traditional medical system from India that promotes the use of diet, herbs, meditation, massage, and yoga for preventing and treating illness.

bioelectrical or **bioelectro-magnetic therapies:** therapies that involve the unconventional use of electric or magnetic fields to cure illness.

biofeedback: a technique in which individuals are trained to gain voluntary control of certain physiological processes, such as skin temperature or brain wave activity, to help reduce stress and anxiety.

biofield therapies: healing methods based on the belief that illnesses can be healed by manipulating energy fields that purportedly surround and penetrate the body. Examples include *acupuncture, qi gong,* and *therapeutic touch.*

chiropractic (KYE-roh-PRAK-tic): an alternative medical system based on the unproven theory that spinal manipulation can restore health.
- A *subluxation* is a misaligned vertebra or other spinal alteration that may cause illness.
- *Adjustment* is the manipulative therapy practiced by chiropractors.

faith healing: the use of prayer or belief in divine intervention to promote healing.

homeopathic (HO-mee-oh-PATH-ic) **medicine:** a practice based on the theory that "like cures like"; that is, substances believed to cause certain symptoms are prescribed for curing the same symptoms, but are given in extremely diluted amounts.
- **homeo** = like
- **pathos** = suffering

hypnotherapy: a technique that uses hypnosis and the power of suggestion to improve health behaviors, relieve pain, and promote healing.

imagery: the use of mental images of things or events to aid relaxation or promote self-healing.

massage therapy: manual manipulation of muscles to reduce tension, increase blood circulation, improve joint mobility, and promote healing of injuries.

meditation: a self-directed technique of calming the mind and relaxing the body.

naturopathic (NAY-chur-oh-PATH-ic) **medicine:** an approach to medical care using practices alleged to enhance the body's natural healing abilities. Treatments may include a variety of alternative therapies including dietary supplements, herbal remedies, exercise, and homeopathy.

osteopathic (OS-tee-oh-PATH-ic) **manipulation:** a CAM technique performed by a doctor of osteopathy (D.O., or osteopath) that includes deep tissue massage and manipulation of the joints, spine, and soft tissues. A D.O. is a fully trained and licensed medical physician, although osteopathic manipulation has not been proved to be an effective treatment.

qi gong (chee-GUNG): a Chinese system that combines movement, meditation, and breathing techniques and allegedly cures illness by enhancing the flow of qi (energy) within the body.

reflexology: a technique that applies pressure or massage on areas of the hands or feet to allegedly cure disease or relieve pain in other areas of the body; sometimes called *zone therapy.*

therapeutic touch: a technique of passing hands over a patient to purportedly identify energy imbalances and transfer healing power from therapist to patient; also called *laying on of hands.*

traditional Chinese medicine (TCM): an approach to medical care based on the concept that illness can be cured by enhancing the flow of qi (energy) within a person's body. Treatments may include herbal therapies, physical exercises, meditation, acupuncture, and remedial massage.

---

help patients reduce stress or anxiety. Other techniques to reduce stress and promote relaxation include **meditation,** art and music therapy, and prayer.

The clinical applications of other mind-body therapies are far more questionable. An example is guided **imagery,** in which a person tries to reverse the disease process (for example, shrink a tumor) by using mental pictures. Another example is the use of **faith healing** in place of proven conventional treatments to cure disease.

Biologically Based Therapies

Biological therapies include the use of natural products, such as vitamin supplements, herbal and plant extracts, and special foods. The most popular biological therapy is the use of herbal products, which was discussed in Chapter 19. An overview of some other popular biologically based treatments follows.

Hormones

Some hormones or hormonelike products that are derived from foods are considered dietary supplements and can be sold over

Biofeedback training is a stress reduction and relaxation technique.

© Cindy Charles/PhotoEdit, Inc.

the counter. Because the FDA does not regulate these products, there is no way of knowing whether they are safe or effective. Moreover, the amount of active ingredient in a dose, as listed on the label, may not be accurate, and the potential hazards of using these products are not known. One example of a hormone that is available over the counter is melatonin, a hormone made by the pineal gland and alleged to reverse sleep disorders and prevent jet lag. Another example is the adrenal hormone DHEA (dehydroepiandrosterone), which is promoted to enhance immunity, increase muscle mass, improve memory, and defend against aging.

Glucosamine-Chondroitin Supplements

The use of glucosamine and chondroitin supplements is an example of a CAM therapy that is being considered for adoption by mainstream medicine, depending on the outcome of studies of their safety and effectiveness. Glucosamine and chondroitin are produced in the body and help to maintain joint cartilage. Although study results have been mixed, some clinical trials have found that glucosamine and chondroitin supplements reduced moderate to severe symptoms of osteoarthritis better than a placebo; these findings have prompted some physicians to suggest using these supplements for pain relief.[8] Recent studies have cast doubt on the earlier findings, however, and several trials are still in progress.[9]

Aromatherapy

Aromatherapy is the practice of inhaling aromatic substances derived from plants, called *essential oils*. Aromatherapy allegedly improves health and enhances natural healing processes. Popular examples of essential oils include those from eucalyptus, lavender, peppermint, rosemary, and lemon.

Manipulative and Body-Based Methods

Manipulative interventions include physical touch, forceful movement of different parts of the body, and the application of pressure. Some practitioners maintain that special energy fields are also manipulated during the physical treatment and that proper energy flow induces healing, as described in the later section on energy therapies.

Chiropractic

Chiropractic theory proposes that keeping the nervous system free from obstruction allows the body to heal itself, because the healing process stems from the brain and is conducted via the spinal cord and nerves to all parts of the body. Chiropractors claim to diagnose illnesses by detecting subluxations in the spine, which are variously described as misaligned vertebrae or pinched nerves that allegedly cause subtle interferences within the nervous system. The main treatment is the adjustment, a manual manipulation that is said to correct a subluxation and restore the body's natural healing ability. Although spinal manipulation has mainly been found to be helpful for improving back pain, most chiropractors still assert that chiropractic can cure disease rather than simply relieve symptoms.[10] For example, many chiropractors promote spinal manipulation to treat infectious diseases, prevent cancer, and regulate menstrual periods, even though the nervous system and spinal alignment do not play roles in the pathology of these conditions.

Massage Therapy

Massage therapy is the manipulation of muscle and connective tissue to improve muscle function, reduce pain, or promote relaxation. Massage therapists may also apply heat or cold and give advice about exercises that may improve muscle tone and range of motion. Massage is often integrated into conventional physical therapy, although some massage therapists may incorrectly suggest that massage is a valid treatment for a wide range of medical conditions.

Energy Therapies

Two categories of therapies involve the alleged curative power of "energy." **Biofield therapies** are said to influence the energy that surrounds or pervades the human body, and their proponents claim that an energy therapy can strengthen or restore a person's "energy flow" and induce healing. Acupuncture, qi gong, and therapeutic touch are among the therapies that subscribe to these theories. Note that CAM adherents often use the term *energy* unscientifically and that there is no objective evidence of this sort of energy flow. **Bioelectrical** or **bioelectromagnetic therapies** use electric or magnetic fields to allegedly promote healing; for example, magnets have been marketed with claims that they can improve circulation, reduce inflammation, and speed recovery from injuries.

Acupuncture

Acupuncture, a component of traditional Chinese medicine, is based on the theory that disease is caused by the disrupted flow of qi through the body. Acupuncture allegedly corrects such disruptions and restores health. The practice involves the shallow insertion of stainless steel needles into the skin at designated points on the body, sometimes accompanied by a low-frequency current to produce greater stimulation.

Qi Gong

Qi gong is another therapy originating in China that is said to improve the flow of qi within the body. Qi gong masters allegedly cure disease by releasing energy from their body and passing it to the person being treated. Self-help practices include deep breathing, certain types of physical exercise, and concentration and relaxation techniques.

Therapeutic Touch

Therapeutic touch is based on the premise that the "healing force" of a practitioner can be used to cure disease. Practitioners claim to identify and correct energy imbalances by passing their hands above a patient's body and transferring "excess energy" to the patient.

Is CAM Safe and Effective?

As mentioned earlier, CAM treatments are generally excluded from mainstream medical practice because there is no evidence proving that they are effective for treating the diseases and medical conditions for which they are used. Many consumers think otherwise and seem satisfied that these treatments "work." How is this dichotomy to be explained?

Does CAM Work?

Surveys suggest that consumers perceive their visits to CAM therapists as far more pleasant than their visits to conventional health practitioners. CAM therapists spend more time with patients, are more attentive, and use less invasive interventions.[11] Self-help measures are encouraged, so the consumer has more control over the treatment. The therapies appear to be more "natural" and to have fewer side effects. Possible explanations for "cures" include the following:

- A person may seem cured because of misdiagnosis; that is, the condition diagnosed by the CAM practitioner may not have actually existed.

- The condition may have been self-limiting, or it may have gone into temporary remission after the treatment.

- Undue credit may be inappropriately assigned to the CAM therapy when the improvement was actually due to a previous or concurrent conventional treatment.

- The placebo effect may have had an influence on the course of disease.

The central question remains: Do the CAM therapies merely make people feel better, or do they really get better? This question can be answered only by well-controlled research studies.

Potential Hazards of CAM

One of the attractions of alternative therapies is the assumption that they are safe. Recall, however, the concerns associated with the use of herbal products discussed in Chapter 19, which include the potential toxicity of herbal ingredients, product contamination or adulteration, and interactions with conventional medications. Between 1990 and 1999, the FDA recalled more than 100 dietary supplements due to hazards associated with their use.[12]

Another concern is that use of CAM therapies may delay the use of reliable treatments that have demonstrable benefits.[13] Various reports have described people with treatable medical conditions who suffered permanent disability or death when they were misdiagnosed or improperly treated by CAM practitioners. For example, a rare but well-known risk of spinal cord injury or stroke is associated with a type of cervical manipulation performed by chiropractors.[14] Unfortunately, because most CAM therapies are not regulated or monitored, there are no accurate estimates of their adverse effects.

Working with Patients Who Use CAM

CAM therapies may have consequences that influence the course of a disease and its treatment. Accordingly, it is important that health practitioners routinely inquire about the use of CAM therapies and educate patients about the hazards of postponing or stopping conventional treatment.[15] Patients should also be told about potential interactions between conventional treatments and CAM therapies. Some patients may want to learn about differences between evidence-based medical practices and untested CAM theories and may be interested in the integrative medicine options available.

All alternative therapies have one characteristic in common: their effectiveness is, for the most part, unproven. As mentioned previously, patients often choose alternative therapies because of positive interactions with CAM practitioners. Thus, all health care practitioners should realize that empathizing with patients may go a long way toward winning their trust and improving their compliance with therapy. In addition, health practitioners need to regularly use reliable, objective resources to update their knowledge about unconventional practices so that they can knowledgeably discuss these options with patients.

REFERENCES

1. S. E. Straus, Complementary and alternative medicine, in L. Goldman and D. Ausiello, eds., *Cecil Medicine* (Philadelphia: Saunders, 2008), pp. 206–209.
2. P. M. Barnes and E. Powell-Griner, Complementary and alternative medicine use among adults: United States, 2002, *Advance Data from Vital and Health Statistics* 343 (2004): 1–19.
3. J. D. Berman and S. E. Straus, Implementing a research agenda for complementary and alternative medicine, *Annual Review of Medicine* 55 (2004): 239–254.
4. E. Ernst, The role of complementary and alternative medicine, *British Medical Journal* 321 (2000): 1133–1135.
5. Straus, 2008.
6. B. Barzansky, H. S. Jonas, and S. I. Etzel, Educational programs in U.S. medical schools, 1999–2000, *Journal of the American Medical Association* 284 (2000): 1114–1120.
7. D. Normile, The new face of Chinese medicine, *Science* 299 (2003): 188–190.

8. O. Bruyere and J. Y. Reginster, Glucosamine and chondroitin sulfate as therapeutic agents for knee and hip osteoarthritis, *Drugs and Aging* 24 (2007): 573–580.
9. R. M. Rozendaal and coauthors, Effect of glucosamine sulfate on hip osteoarthritis, *Annals of Internal Medicine* 148 (2008): 268–277; S. Reichenbach and coauthors, Meta-analysis: Chondroitin for osteoarthritis of the knee or hip, *Annals of Internal Medicine* 146 (2007): 580–590.
10. American Medical Association, *Alternative Medicine*, report 12 of the Council on Scientific Affairs, A-97 (Chicago: American Medical Association, 1997), available at www.ama-assn.org/ama/pub/category/13638.html; site visited January 29, 2008.
11. B. Barrett and coauthors, What complementary and alternative medicine practitioners say about health and health care, *Annals of Family Medicine* 2 (2004): 253–259; American Medical Association, 1997.
12. Berman and Straus, 2004.

13. Straus, 2008.
14. W.- L. Chen and coauthors, Vertebral artery dissection and cerebellar infarction following chiropractic manipulation, *Emergency Medical Journal* 23 (2006): e1.doi:10.1136/emj.2004.015636; W. S. Smith and coauthors, Spinal manipulative therapy is an independent risk factor for vertebral artery dissection, *Neurology* 13 (2003):1424–1428; R. Dziewas and coauthors, Cervical artery dissection—Clinical features, risk factors, therapy and outcome in 126 patients, *Journal of Neurology* 250 (2003): 1179–1184.
15. American Medical Association, 1997.

Appendixes

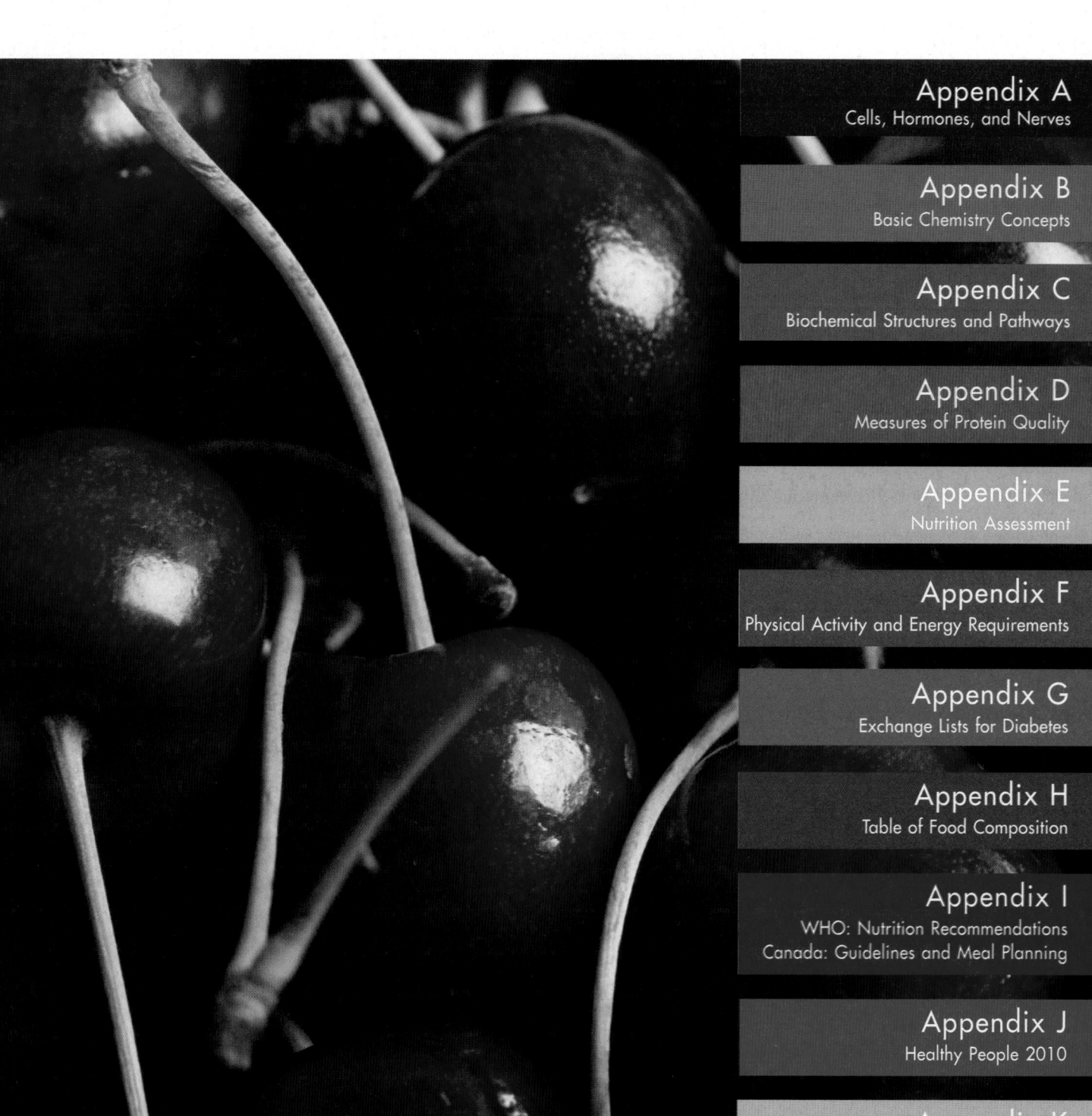

Appendix

A

GLOSSARY OF CELL STRUCTURES

cell: the basic structural unit of all living things.

cell membrane: the thin layer of tissue that surrounds the cell and encloses its contents; made primarily of lipid and protein.

chromosomes: a set of structures within the nucleus of every cell that contains the cell's genetic material, DNA, associated with other materials (primarily proteins).

cytoplasm (SIGH-toh-plazm): the cell contents, except for the nucleus.
- **cyto** = cell
- **plasm** = a form

cytosol: the fluid of cytoplasm; contains water, ions, nutrients, and enzymes.

endoplasmic reticulum (en-doh-PLAZ-mic reh-TIC-you-lum): a complex network of intracellular membranes. The **rough endoplasmic reticulum** is dotted with ribosomes, where protein synthesis takes place. The **smooth endoplasmic reticulum** bears no ribosomes.
- **endo** = inside
- **plasm** = the cytoplasm

Golgi (GOAL-gee) **apparatus:** a set of membranes within the cell where secretory materials are packaged for export.

lysosomes (LYE-so-zomes): cellular organelles; membrane-enclosed sacs of degradative enzymes.
- **lysis** = dissolution

mitochondria (my-toh-KON-dree-uh); singular **mitochondrion:** the cellular organelles responsible for producing ATP aerobically; made of membranes (lipid and protein) with enzymes mounted on them.
- **mitos** = thread (referring to their slender shape)
- **chondros** = cartilage (referring to their external appearance)

nucleus: a major membrane-enclosed body within every cell, which contains the cell's genetic material, DNA, embedded in chromosomes.
- **nucleus** = a kernel

organelles: subcellular structures such as ribosomes, mitochondria, and lysosomes.
- **organelle** = little organ

Cells, Hormones, and Nerves

This appendix is offered as an optional chapter for readers who want to enhance their understanding of how the body coordinates its activities. It presents a brief summary of the structure and function of the body's basic working unit (the cell) and of the body's two major regulatory systems (the hormonal system and the nervous system).

The Cell

The body's organs are made up of millions of cells and of materials produced by them. Each **cell** is specialized to perform its organ's functions, but all cells have common structures (see the accompanying glossary and Figure A-1). Every cell is contained within a **cell membrane.** The cell membrane assists in moving materials into and out of the cell, and some of its special proteins act as "pumps" (described in Chapter 6). Some features of cell membranes, such as microvilli (Chapter 3), permit cells to interact with other cells and with their environments in highly specific ways.

Inside the membrane lies the **cytoplasm,** which is filled with **cytosol,** or cell "fluid." The cytoplasm contains much more than just fluid, though. It is a highly organized system of fibers, tubes, membranes, particles, and subcellular **organelles** as complex as a city. These parts intercommunicate, manufacture and exchange materials, package and prepare materials for export, and maintain and repair themselves.

Within each cell is another membrane-enclosed body, the **nucleus.** Inside the nucleus are the **chromosomes,** which contain the genetic material, DNA. The DNA encodes all the instructions for carrying out the cell's activities. The role of DNA in coding for cell proteins is summarized in Figure 6-7 on p. 188. Chapter 6 also describes the variety of proteins produced by cells and the ways they perform the body's work.

Among the organelles within a cell are ribosomes, mitochondria, and lysosomes. Figure 6-7 briefly refers to the **ribosomes;** they assemble amino acids into proteins, following directions conveyed to them by RNA.

The **mitochondria** are made of intricately folded membranes that bear thousands of highly organized sets of enzymes on their inner and outer surfaces. Mitochondria are crucial to energy metabolism (described in Chapter 7) and muscles conditioned to work aerobically are packed with them. Their presence is implied whenever the TCA cycle and electron transport chain are mentioned because the mitochondria house the needed enzymes.*

The **lysosomes** are membranes that enclose degradative enzymes. When a cell needs to self-destruct or to digest materials in its surroundings, its lysosomes free their enzymes. Lysosomes are active when tissue repair or remodeling is taking place—for example, in cleaning up infections, healing wounds, shaping embryonic organs, and remodeling bones.

Besides these and other cellular organelles, the cell's cytoplasm contains a highly organized system of membranes, the **endoplasmic reticulum.** The ribosomes may either float free in the cytoplasm or be mounted on these membranes. A membranous surface dotted with ribosomes looks speckled under the microscope and is called "rough" endoplasmic reticulum; such a surface without ribosomes is called "smooth." Some intracellular membranes are organized into tubules that collect cellular materials, merge with the cell membrane, and discharge their contents to the outside of

---

*For the reactions of glycolysis, the TCA cycle, and the electron transport chain, see Chapter 7 and Appendix C. The reactions of glycolysis take place in the cytoplasm; the conversion of pyruvate to acetyl CoA takes place in the mitochondria, as do the TCA cycle and electron transport chain reactions. The mitochondria then release carbon dioxide, water, and ATP as their end products.

FIGURE A-1 The Structure of a Typical Cell

The cell shown might be one in a gland (such as the pancreas) that produces secretory products (enzymes) for export (to the intestine). The rough endoplasmic reticulum with its ribosomes produces the enzymes; the smooth reticulum conducts them to the Golgi region; the Golgi membranes merge with the cell membrane, where the enzymes can be released into the extracellular fluid.

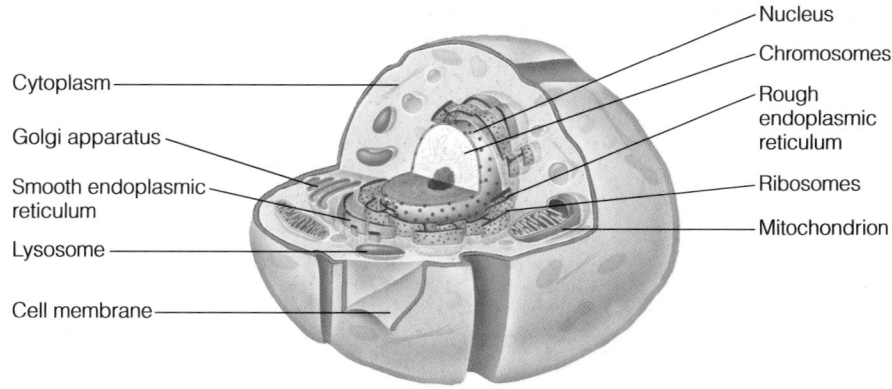

- The study of hormones and their effects is **endocrinology.**

- The **pituitary gland** in the brain has two parts—the **anterior** (front) and the **posterior** (hind).

---

ribosomes (RYE-boh-zomes): protein-making organelles in cells; composed of RNA and protein.
- **ribo** = containing the sugar ribose (in RNA)
- **some** = body

---

the cell; these membrane systems are named the **Golgi apparatus,** after the scientist who first described them. The rough and smooth endoplasmic reticula and the Golgi apparatus are continuous with one another, so secretions produced deep in the interior of the cell can be efficiently transported to the outside and released. These and other cell structures enable cells to perform the multitudes of functions for which they are specialized.

The actions of cells are coordinated by both hormones and nerves, as the next sections show. Among the types of cellular organelles are receptors for the hormones delivering instructions that originate elsewhere in the body. Some hormones penetrate the cell and its nucleus and attach to receptors on chromosomes, where they activate certain genes to initiate, stop, speed up, or slow down synthesis of certain proteins as needed. Other hormones attach to receptors on the cell surface and transmit their messages from there. The hormones ◆ are described in the next section; the nerves, in the one following.

The Hormones

A chemical compound—a **hormone**—originates in a gland and travels in the bloodstream. The hormone flows everywhere in the body, but only its target organs respond to it, because only they possess the receptors to receive it.

The hormones, the glands they originate in, and their target organs and effects are described in this section. Many of the hormones you might be interested in are included, but only a few are discussed in detail. Figure A-2 (p. A-4) identifies the glands that produce the hormones, and the accompanying glossary defines the hormones discussed in this section.

Hormones of the Pituitary Gland and Hypothalamus

The anterior pituitary gland ◆ produces the following hormones, each of which acts on one or more target organs and elicits a characteristic response:

- **Adrenocorticotropin (ACTH)** acts on the adrenal cortex, promoting the production and release of its hormones.

- **Thyroid-stimulating hormone (TSH)** acts on the thyroid gland, promoting the production and release of thyroid hormones.

- **Growth hormone (GH)** or **somatotropin** acts on all tissues, promoting growth, fat breakdown, and the formation of antibodies.

GLOSSARY OF HORMONES

adrenocorticotropin (ad-REE-noh-KORE-tee-koh-TROP-in) or **ACTH**: a hormone, so named because it stimulates (trope) the adrenal cortex. The adrenal gland, like the pituitary, has two parts, in this case an outer portion (cortex) and an inner core (medulla). The realease of ACTH is mediated by **corticotropin-releasing hormone (CRH).**

aldosterone: a hormone from the adrenal gland involved in blood pressure regulation.
- **aldo** = aldehyde

angiotensin: a hormone involved in blood pressure regulation that is activated by **renin** (REN-in), an enzyme from the kidneys.
- **angio** = blood vessels
- **tensin** = pressure
- **ren** = kidneys

antidiuretic hormone (ADH): the hormone that prevents water loss in urine (also called **vasopressin**).
- **anti** = against
- **di** = through
- **ure** = urine
- **vaso** = blood vessels
- **pressin** = pressure

calcitonin (KAL-see-TOH-nin): a hormone secreted by the thyroid gland that regulates (tones) calcium metabolism.

erythropoietin (eh-RITH-ro-POY-eh-tin): a hormone that stimulates red blood cell production.
- **erythro** = red (blood cell)
- **poiesis** = creating (like poetry)

estrogens: hormones responsible for the menstrual cycle and other female characteristics.
- **oestrus** = the egg-making cycle
- **gen** = gives rise to

FIGURE A-2 The Endocrine System

These organs and glands release hormones that regulate body processes. An *endocrine gland* secretes its product directly into *(endo)* the blood; for example, the pancreas cells that produce insulin. An *exocrine gland* secretes its product(s) out *(exo)* to an epithelial surface either directly or through a duct; the sweat glands of the skin and the enzyme-producing glands of the pancreas are both examples. The pancreas is therefore both an endocrine and an exocrine gland.

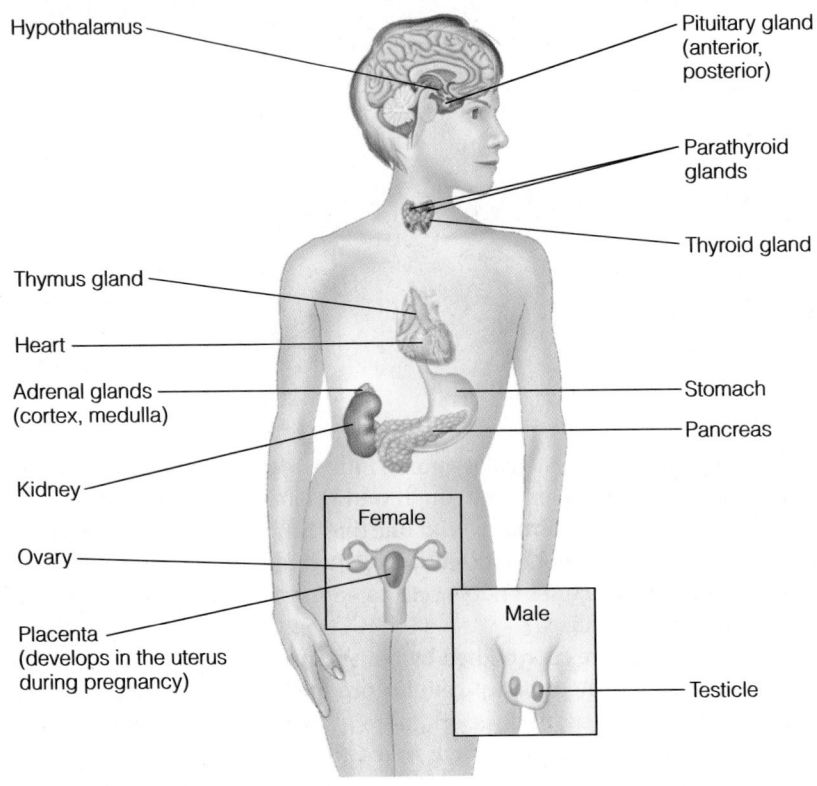

Hypothalamus — Pituitary gland (anterior, posterior)

Parathyroid glands

Thyroid gland

Thymus gland

Heart

Adrenal glands (cortex, medulla)

Stomach

Pancreas

Kidney

Ovary

Placenta (develops in the uterus during pregnancy)

Female

Male

Testicle

◆ Hormones that are turned off by their own effects are said to be regulated by **negative feedback.**

follicle-stimulating hormone (FSH): a hormone that stimulates maturation of the ovarian follicles in females and the production of sperm in males. (The ovarian follicles are part of the female reproductive system where the eggs are produced.) The release of FSH is mediated by **follicle-stimulating hormone releasing hormone (FSH–RH).**

glucocorticoids: hormones from the adrenal cortex that affect the body's management of glucose.
- **gluco** = glucose
- **corticoid** = from the cortex

growth hormone (GH): a hormone secreted by the pituitary that regulates the cell division and protein synthesis needed for normal growth (also called **somatotropin**). The release of GH is mediated by **GH-releasing hormone (GHRH)** and **GH-inhibiting hormone (GHIH).**

hormone: a chemical messenger. Hormones are secreted by a variety of endocrine glands in response to altered conditions in the body. Each hormone travels to one or more specific target tissues or organs, where it elicits a specific response to maintain homeostasis.

luteinizing (LOO-tee-in-EYE-zing) hormone (LH): a hormone that stimulates ovulation and the development of the corpus luteum (the small tissue that develops from a ruptured ovarian follicle and secretes hormones); so called because the follicle turns yellow as it matures. In men, LH stimulates testosterone secretion. The release of LH is mediated by **luteinizing hormone–releasing hormone (LH–RH).**
- **lutein** = a yellow pigment

- **Follicle-stimulating hormone (FSH)** acts on the ovaries in the female, promoting their maturation, and on the testicles in the male, promoting sperm formation.
- **Luteinizing hormone (LH)** also acts on the ovaries, stimulating their maturation, the production and release of progesterone and estrogens, and ovulation; and on the testicles, promoting the production and release of testosterone.
- **Prolactin,** secreted in the female during pregnancy and lactation, acts on the mammary glands to stimulate their growth and the production of milk.

Each of these hormones has one or more signals that turn it on and another (or others) that turns it off. ◆ Among the controlling signals are several hormones from the hypothalamus:

- **Corticotropin-releasing hormone (CRH),** which promotes release of ACTH, is turned on by stress and turned off by ACTH when enough has been released.
- **TSH-releasing hormone (TRH),** which promotes release of TSH, is turned on by large meals or low body temperature.
- **GH-releasing hormone (GHRH),** which stimulates the release of growth hormone, is turned on by insulin.
- **GH-inhibiting hormone (GHIH or somatostatin),** which inhibits the release of GH and interferes with the release of TSH, is turned on by hypoglycemia and/or physical activity and is rapidly destroyed by body tissues so that it does not accumulate.

- **FSH/LH–releasing hormone (FSH/LH–RH)** is turned on in the female by nerve messages or low estrogen and in the male by low testosterone.
- **Prolactin-inhibiting hormone (PIH)** is turned on by high prolactin levels and off by estrogen, testosterone, and suckling (by way of nerve messages).

Let's examine some of these controls. PIH, for example, responds to high prolactin levels (remember, prolactin promotes milk production). High prolactin levels ensure that milk is made and—by calling forth PIH—ensure that prolactin levels don't get too high. But when the infant is suckling—and creating a demand for milk—PIH is not allowed to work (suckling turns off PIH). The consequence: prolactin remains high, and milk production continues. Demand from the infant thus directly adjusts the supply of milk. The need is met through the interaction of the nerves and hormones.

As another example, consider CRH. Stress, perceived in the brain and relayed to the hypothalamus, switches on CRH. On arriving at the pituitary, CRH switches on ACTH. Then ACTH acts on its target organ, the adrenal cortex, which responds by producing and releasing stress hormones. The stress hormones trigger a cascade of events involving every body cell and many other hormones.

The numerous steps required to set the stress response in motion make it possible for the body to fine-tune the response; control can be exerted at each step. These two examples illustrate what the body can do in response to two different stimuli—producing milk in response to an infant's need and gearing up for action in an emergency.

The posterior pituitary gland produces two hormones, each of which acts on one or more target cells and elicits a characteristic response:

- **Antidiuretic hormone (ADH)**, or **vasopressin**, acts on the arteries, promoting their contraction, and on the kidneys, preventing water excretion. ADH is turned on whenever the blood volume is low, the blood pressure is low, or the salt concentration of the blood is high (see Chapter 12). It is turned off by the return of these conditions to normal.
- **Oxytocin** acts during late pregnancy on the uterus, inducing contractions, and during lactation on the mammary glands, causing milk ejection. Oxytocin is produced in response to reduced progesterone levels, suckling, or the stretching of the cervix.

Hormones That Regulate Energy Metabolism

Hormones produced by a number of different glands have effects on energy metabolism:

- Insulin from the pancreas beta cells is turned on by many stimuli, including raised blood glucose. It acts on cells to increase glucose and amino acid uptake into them and to promote the secretion of GHRH.
- Glucagon from the pancreas alpha cells responds to low blood glucose and acts on the liver to promote the breakdown of glycogen to glucose, the conversion of amino acids to glucose, and the release of glucose into the blood.
- Thyroxine from the thyroid gland responds to TSH and acts on many cells to increase their metabolic rate, growth, and heat production.
- Norepinephrine and epinephrine ◆ from the adrenal medulla respond to stimulation by sympathetic nerves and produce reactions in many cells that facilitate the body's readiness for fight or flight: increased heart activity, blood vessel constriction, breakdown of glycogen and glucose, raised blood glucose levels, and fat breakdown. Norepinephrine and epinephrine also influence the secretion of the many hormones from the hypothalamus that exert control on the body's other systems.
- Growth hormone (GH) from the anterior pituitary (already mentioned).
- **Glucocorticoids** from the adrenal cortex become active during times of stress and carbohydrate metabolism.

◆ Norepinephrine and epinephrine were formerly called **noradrenalin** and **adrenalin**, respectively.

oxytocin (OCK-see-TOH-sin): a hormone that stimulates the mammary glands to eject milk during lactation and the uterus to contract during childbirth.
- **oxy** = quick
- **tocin** = childbirth

progesterone: the hormone of gestation (pregnancy).
- **pro** = promoting
- **gest** = gestation (pregnancy)
- **sterone** = a steroid hormone

prolactin (proh-LAK-tin): a hormone so named because it promotes (*pro*) the production of milk (*lacto*). The release of prolactin is mediated by **prolactin-inhibiting hormone (PIH)**.

relaxin: the hormone of late pregnancy.

somatostatin (GHIH): a hormone that inhibits the release of growth hormone; the opposite of **somatotropin (GH)**.
- **somato** = body
- **stat** = keep the same
- **tropin** = make more

testosterone: a steroid hormone from the testicles, or testes. The steroids, as explained in Chapter 5, are chemically related to, and some are derived from, the lipid cholesterol.
- **sterone** = a steroid hormone

thyroid-stimulating hormone (TSH): a hormone secreted by the pituitary that stimulates the thyroid gland to secrete its hormones—thyroxine and triiodothyronine. The release of TSH is mediated by **TSH-releasing hormone (TRH)**.

TABLE B-1 Chemical Symbols for the Elements

Key:
- Elements found in energy-yielding nutrients, vitamins, and water
- Major minerals
- Trace minerals

Number of Protons (Atomic Number)	Element	Number of Electrons in Outer Shell	Number of Protons (Atomic Number)	Element	Number of Electrons in Outer Shell
1	Hydrogen (H)	1			
2	Helium (He)	2	57	Lanthanum (La)	2
3	Lithium (Li)	1	58	Cerium (Ce)	2
4	Beryllium (Be)	2	58	Cerium (Ce)	2
5	Boron (B)	3	58	Cerium (Ce)	2
6	Carbon (C)	4	60	Neodymium (Nd)	2
7	Nitrogen (N)	5	61	Promethium (Pm)	2
8	Oxygen (O)	6	62	Samarium (Sm)	2
9	Fluorine (F)	7	63	Europium (Eu)	2
10	Neon (Ne)	8	64	Gadolinium (Gd)	2
11	Sodium (Na)	1	65	Terbium (Tb)	2
12	Magnesium (Mg)	2	66	Dysprosium (Dy)	2
13	Aluminum (Al)	3	67	Holmium (Ho)	2
14	Silicon (Si)	4	68	Erbium (Er)	2
15	Phosphorus (P)	5	69	Thulium (Tm)	2
16	Sulfur (S)	6	70	Ytterbium (Yb)	2
17	Chlorine (Cl)	7	71	Lutetium (Lu)	2
18	Argon (Ar)	8	72	Hafnium (Hf)	2
19	Potassium (K)	1	73	Tantalum (Ta)	2
20	Calcium (Ca)	2	74	Tungsten (W)	2
21	Scandium (Sc)	2	75	Rhenium (Re)	2
22	Titanium (Ti)	2	76	Osmium (Os)	2
23	Vanadium (V)	2	77	Iridium (Ir)	2
24	Chromium (Cr)	1	78	Platinum (Pt)	1
25	Manganese (Mn)	2	79	Gold (Au)	1
26	Iron (Fe)	2	80	Mercury (Hg)	2
27	Cobalt (Co)	2	81	Thallium (Tl)	3
28	Nickel (Ni)	2	82	Lead (Pb)	4
29	Copper (Cu)	1	83	Bismuth (Bi)	5
30	Zinc (Zn)	2	84	Polonium (Po)	6
31	Gallium (Ga)	3	85	Astatine (At)	7
32	Germanium (Ge)	4	86	Radon (Rn)	8
33	Arsenic (As)	5	87	Francium (Fr)	1
34	Selenium (Se)	6	88	Radium (Ra)	2
35	Bromine (Br)	7	89	Actinium (Ac)	2
36	Krypton (Kr)	8	90	Thorium (Th)	2
37	Rubidium (Rb)	1	91	Protactinium (Pa)	2
38	Strontium (Sr)	2	92	Uranium (U)	2
39	Yttrium (Y)	2	93	Neptunium (Np)	2
40	Zirconium (Zr)	2	94	Plutonium (Pu)	2
41	Niobium (Nb)	1	95	Americium (Am)	2
42	Molybdenum (Mo)	1	96	Curium (Cm)	2
43	Technetium (Tc)	1	97	Berkelium (Bk)	2
44	Ruthenium (Ru)	1	98	Californium (Cf)	2
45	Rhodium (Rh)	1	99	Einsteinium (Es)	2
46	Palladium (Pd)	—	100	Fermium (Fm)	2
47	Silver (Ag)	1	101	Mendelevium (Md)	2
48	Cadmium (Cd)	2	102	Nobelium (No)	2
49	Indium (In)	3	103	Lawrencium (Lr)	2
50	Tin (Sn)	4	104	Rutherfordium (Rf)	2
51	Antimony (Sb)	5	105	Dubnium (Db)	2
52	Tellurium (Te)	6	106	Seaborgium (Sg)	2
53	Iodine (I)	7	107	Bohrium (Bh)	2
54	Xenon (Xe)	8	108	Hassium (Hs)	2
55	Cesium (Cs)	1	109	Meitnerium (Mt)	2
56	Barium (Ba)	2	110	Darmstadtium (Ds)	2

Besides hydrogen, the atoms most common in living things are carbon (C), nitrogen (N), and oxygen (O), whose atomic numbers are 6, 7, and 8, respectively. Their structures are more complicated than that of hydrogen, but each of them possesses the same number of electrons as there are protons in the nucleus. These electrons are found in orbits, or shells (shown below).

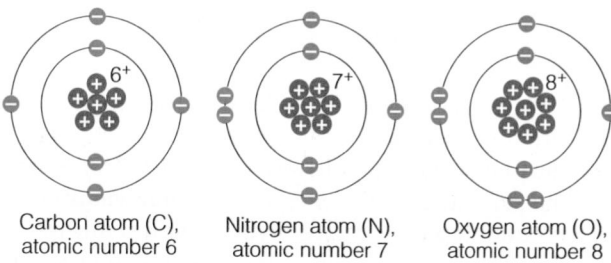

Carbon atom (C),
atomic number 6

Nitrogen atom (N),
atomic number 7

Oxygen atom (O),
atomic number 8

In these and all diagrams of atoms that follow, only the protons and electrons are shown. The neutrons, which contribute only to atomic weight, not to charge, are omitted.

The most important structural feature of an atom for determining its chemical behavior is the number of electrons in its outermost shell. The first, or innermost, shell is full when it is occupied by two electrons; so an atom with two or more electrons has a filled first shell. When the first shell is full, electrons begin to fill the second shell.

The second shell is completely full when it has eight electrons. A substance that has a full outer shell tends not to enter into chemical reactions. Atomic number 10, neon, is a chemically inert substance because its outer shell is complete. Fluorine, atomic number 9, has a great tendency to draw an electron from other substances to complete its outer shell, and thus it is highly reactive. Carbon has a half-full outer shell, which helps explain its great versatility; it can combine with other elements in a variety of ways to form a large number of compounds.

Atoms seek to reach a state of maximum stability or of lowest energy in the same way that a ball will roll down a hill until it reaches the lowest place. An atom achieves a state of maximum stability:

- By gaining or losing electrons to either fill or empty its outer shell.

- By sharing its electrons with other atoms and thereby completing its outer shell.

The number of electrons determines how the atom will chemically react with other atoms. The atomic number, not the weight, is what gives an atom its chemical nature.

Chemical Bonding

Atoms often complete their outer shells by sharing electrons with other atoms. In order to complete its outer shell, a carbon atom requires four electrons. A hydrogen atom requires one. Thus, when a carbon atom shares electrons with four hydrogen atoms, each completes its outer shell (as shown in the next column). Electron sharing binds the atoms together and satisfies the conditions of maximum stability for the molecule. The outer shell of each atom is complete, since hydrogen effectively has the required two electrons in its first (outer)

shell, and carbon has eight electrons in its second (outer) shell; and the molecule is electrically neutral, with a total of ten protons and ten electrons.

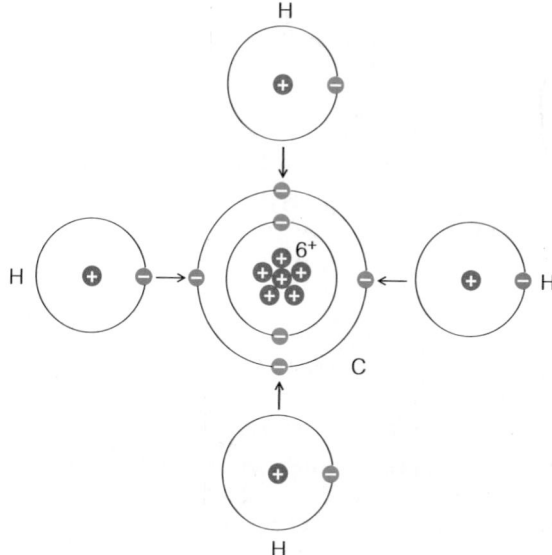

When a carbon atom shares electrons with four hydrogen atoms, a methane molecule is made.

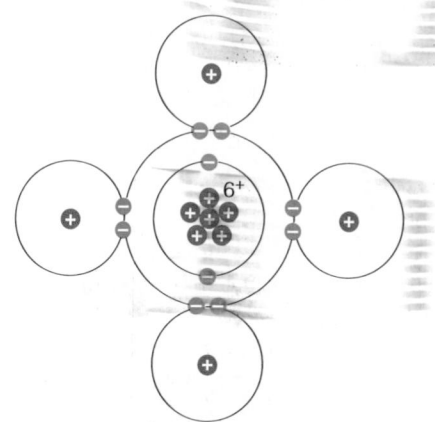

The chemical formula for methane is CH_4. Note that by sharing electrons, every atom achieves a filled outer shell.

Bonds that involve the sharing of electrons, like the bonds between carbon and the four hydrogens, are the most stable kind of association that atoms can form with one another. These bonds are called covalent bonds, and the resulting combination of atoms are called molecules. A single pair of shared electrons forms a single bond. A simplified way to represent a single bond is with a single line. Thus the structure of methane (CH_4) could be represented like this:

$$H-\overset{\displaystyle H}{\underset{\displaystyle H}{C}}-H$$

Methane (CH_4)

Similarly, one nitrogen atom and three hydrogen atoms can share electrons to form one molecule of ammonia (NH_3):

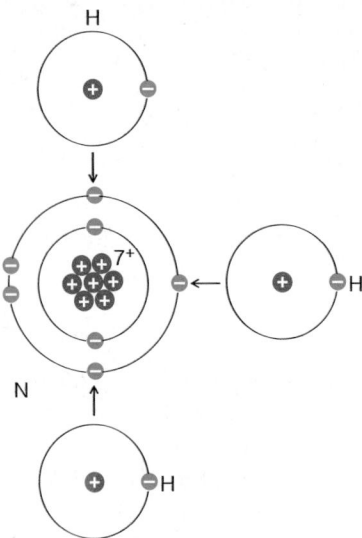

When a nitrogen atom shares electrons with three hydrogen atoms, an ammonia molecule is made.

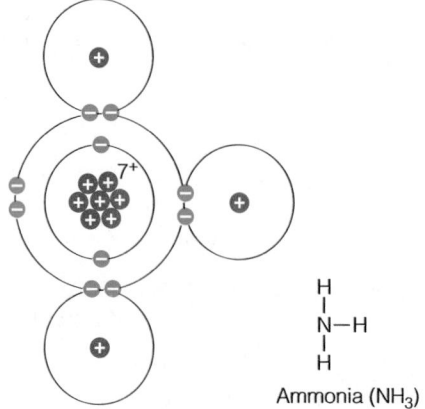

Ammonia (NH₃)

H
|
N—H
|
H

The chemical formula for ammonia is NH₃. Count the electrons in each atom's outer shell to confirm that it is filled.

One oxygen atom may be bonded to two hydrogen atoms to form one molecule of water (H₂O):

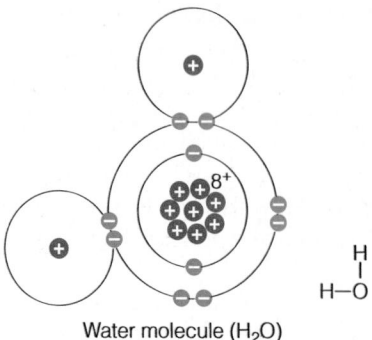

Water molecule (H₂O)

H
|
H—O

When two oxygen atoms form a molecule of oxygen, they must share two pairs of electrons. This double bond may be represented as two single lines:

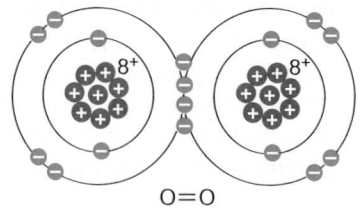

O=O

Oxygen molecule (O₂)

Small atoms form the tightest, most stable bonds. H, O, N, and C are the smallest atoms capable of forming one, two, three, and four electron-pair bonds respectively. This is the basis for the statement in Chapter 4 that in drawings of compounds containing these atoms, hydrogen must always have one, oxygen two, nitrogen three, and carbon four bonds radiating to other atoms:

$$H- \qquad -O- \qquad -\overset{|}{N}- \qquad -\overset{|}{\underset{|}{C}}-$$

The stability of the associations between these small atoms and the versatility with which they can combine make them very common in living things. Interestingly, all cells, whether they come from animals, plants, or bacteria, contain the same elements in very nearly the same proportions. The elements commonly found in living things are shown in Table B-2.

TABLE B-2	Elemental Composition of the Human Body	
Element	Chemical Symbol	By Weight (%)
Oxygen	O	65
Carbon	C	18
Hydrogen	H	10
Nitrogen	N	3
Calcium	Ca	1.5
Phosphorus	P	1.0
Potassium	K	0.4
Sulfur	S	0.3
Sodium	Na	0.2
Chloride	Cl	0.1
Magnesium	Mg	0.1
Total		99.6[a]

[a]The remaining 0.40 percent by weight is contributed by the trace elements: chromium (Cr), copper (Cu), zinc (Zn), selenium (Se), molybdenum (Mo), fluorine (F), iodine (I), manganese (Mn), and iron (Fe). Cells may also contain variable traces of some of the following: boron (B), cobalt (Co), lithium (Li), strontium (Sr), aluminum (Al), silicon (Si), lead (Pb), vanadium (V), arsenic (As), bromine (Br), and others.

Appendix B

Formation of Ions

An atom such as sodium (Na, atomic number 11) cannot easily fill its outer shell by sharing. Sodium possesses a filled first shell of two electrons and a filled second shell of eight; there is only one electron in its outermost shell:

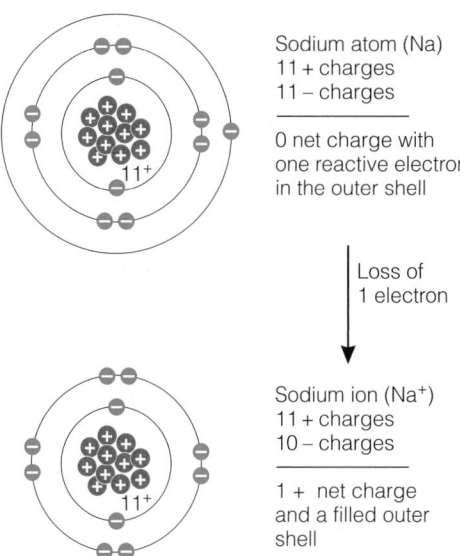

Sodium atom (Na)
11 + charges
11 – charges

0 net charge with one reactive electron in the outer shell

Loss of 1 electron

Sodium ion (Na$^+$)
11 + charges
10 – charges

1 + net charge and a filled outer shell

If sodium loses this electron, it satisfies one condition for stability: a filled outer shell (now its second shell counts as the outer shell). However, it is not electrically neutral. It has 11 protons (positive) and only 10 electrons (negative). It therefore has a net positive charge. An atom or molecule that has lost or gained one or more electrons and so is electrically charged is called an ion.

An atom such as chlorine (Cl, atomic number 17), with seven electrons in its outermost shell, can share electrons to fill its outer shell, or it can gain one electron to complete its outer shell and thus give it a negative charge:

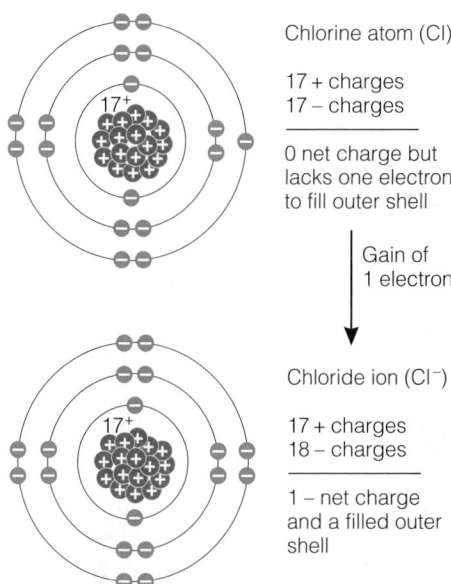

Chlorine atom (Cl)

17 + charges
17 – charges

0 net charge but lacks one electron to fill outer shell

Gain of 1 electron

Chloride ion (Cl$^-$)

17 + charges
18 – charges

1 – net charge and a filled outer shell

A positively charged ion such as sodium ion (Na$^+$) is called a cation; a negatively charged ion such as a chloride ion (Cl$^-$) is called an anion. Cations and anions attract one another to form salts:

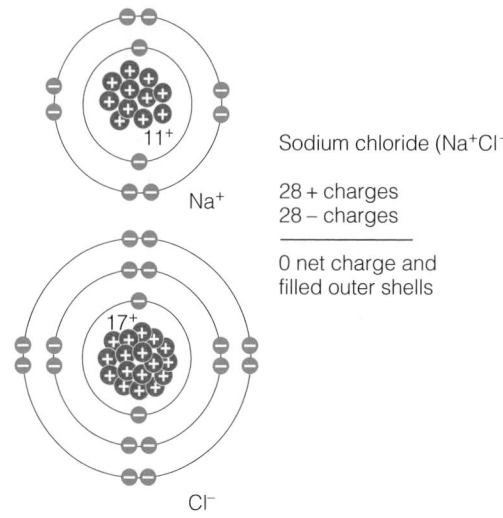

Sodium chloride (Na$^+$Cl$^-$)

28 + charges
28 – charges

0 net charge and filled outer shells

With all its electrons, sodium is a shiny, highly reactive metal; chlorine is the poisonous greenish yellow gas that was used in World War I. But after sodium and chlorine have transferred electrons, they form the stable white salt familiar to you as table salt, or sodium chloride (Na$^+$Cl$^-$). The dramatic difference illustrates how profoundly the electron arrangement can influence the nature of a substance. The wide distribution of salt in nature attests to the stability of the union between the ions. Each meets the other's needs (a good marriage).

When dry, salt exists as crystals; its ions are stacked very regularly into a lattice, with positive and negative ions alternating in a three-dimensional checkerboard structure. In water, however, the salt quickly dissolves, and its ions separate from one another, forming an electrolyte solution in which they move about freely. Covalently bonded molecules rarely dissociate like this in a water solution. The most common exception is when they behave like acids and release H$^+$ ions, as discussed in the next section.

An ion can also be a group of atoms bound together in such a way that the group has a net charge and enters into reactions as a single unit. Many such groups are active in the fluids of the body. The bicarbonate ion is composed of five atoms—one H, one C, and three Os—and has a net charge of -1 (HCO$_3^-$). Another important ion of this type is a phosphate ion with one H, one P, and four O, and a net charge of -2 (HPO$_4^{-2}$).

Whereas many elements have only one configuration in the outer shell and thus only one way to bond with other elements, some elements have the possibility of varied configurations. Iron is such an element. Under some conditions iron loses two electrons, and under other circumstances it loses three. If iron loses two electrons, it then has a net charge of +2, and we call it ferrous iron (Fe^{++}). If it donates three electrons to another atom, it becomes the +3 ion, or ferric iron (Fe^{+++}).

Ferrous iron (Fe^{++}) (had 2 outer-shell electrons but has lost them)	Ferric iron (Fe^{+++}) (had 3 outer-shell electrons but has lost them)
26 + charges	26 + charges
24 − charges	23 − charges
2 + net charge	3 + net charge

Remember that a positive charge on an ion means that negative charges—electrons—have been lost and not that positive charges have been added to the nucleus.

Water, Acids, and Bases

Water

The water molecule is electrically neutral, having equal numbers of protons and electrons. When a hydrogen atom shares its electron with oxygen, however, that electron will spend most of its time closer to the positively charged oxygen nucleus. This leaves the positive proton (nucleus of the hydrogen atom) exposed on the outer part of the water molecule. We know, too, that the two hydrogens both bond toward the same side of the oxygen. These two facts explain why water molecules are polar: they have regions of more positive and more negative charge.

Polar molecules like water are drawn to one another by the attractive forces between the positive polar areas of one and the negative poles of another. These attractive forces, sometimes known as polar bonds or hydrogen bonds, occur among many molecules and also within the different parts of single large molecules. Although very weak in comparison with covalent bonds, polar bonds may occur in such abundance that they become exceedingly important in determining the structure of such large molecules as proteins and DNA.

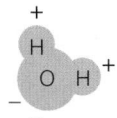

This diagram of the polar water molecule shows displacement of electrons toward the O nucleus; thus the negative region is near the O and the positive regions are near the H atoms.

Water molecules have a slight tendency to ionize, separating into positive (H^+) and negative (OH^-) ions. In pure water, a small but constant number of these ions is present, and the number of positive ions exactly equals the number of negative ions.

Acid

An acid is a substance that releases H^+ ions (protons) in a water solution. Hydrochloric acid (HCl^-) is such a substance because it dissociates in a water solution into H^+ and Cl^- ions.

Acetic acid is also an acid because it dissociates in water to acetate ions and free H^+:

$$H-\overset{\overset{\displaystyle H}{|}}{\underset{\underset{\displaystyle H}{|}}{C}}-\overset{\overset{\displaystyle O}{\|}}{C}-O-H \longrightarrow H-\overset{\overset{\displaystyle H}{|}}{\underset{\underset{\displaystyle H}{|}}{C}}-\overset{\overset{\displaystyle O}{\|}}{C}-O^- + H^+$$

Acetic acid dissociates into an acetate ion and a hydrogen ion.

The more H^+ ions released, the stronger the acid.

pH

Chemists define degrees of acidity by means of the pH scale, which runs from 0 to 14. The pH expresses the concentration of H^+ ions: a pH of 1 is extremely acidic, 7 is neutral, and 13 is very basic. There is a tenfold difference in the concentration of H^+ ions between points on this scale. A solution with pH 3, for example, has ten times as many H^+ ions as a solution with pH 4. At pH 7, the concentrations of free H^+ and OH^- are exactly the same—1/10,000,000 moles per liter (1027 moles per liter).* At pH 4, the concentration of free H^+ ions is 1/10,000 (1024) moles per liter. This is a higher concentration of H^+ ions, and the solution is therefore acidic. Figure 3-6 on p. 77 presents the pH scale.

Bases

A base is a substance that can combine with H^+ ions, thus reducing the acidity of a solution. The compound ammonia is such a substance. The ammonia molecule has two electrons that are not shared with any other atom; a hydrogen ion (H^+) is just a naked proton with no shell of electrons at all. The proton readily combines with the ammonia molecule to form an ammonium ion; thus a free proton is withdrawn from the solution and no longer contributes to its acidity. Many compounds containing nitrogen are important bases in living systems. Acids and bases neutralize each other to produce substances that are neither acid nor base.

$$:\overset{\overset{\displaystyle H}{|}}{\underset{\underset{\displaystyle H}{|}}{N}}-H + H^+ \longrightarrow H-\overset{\overset{\displaystyle H}{|}}{\underset{\underset{\displaystyle H}{|}}{N^+}}-H$$

Ammonia captures a hydrogen ion from water. The two dots here represent the two electrons not shared with another atom. These dots are ordinarily not shown in chemical structure drawings. Compare this drawing with the earlier diagram of an ammonia molecule (p. B-4).

Chemical Reactions

A chemical reaction, or chemical change, results in the breakdown of substances and the formation of new ones. Almost all such reactions involve a change in the bonding of atoms. Old bonds are broken, and new ones are formed. The nuclei of atoms are never involved in chemical reactions—only their

---

*A mole is a certain number (about 6×10^{23}) of molecules. The pH of a solution is defined as the negative logarithm of the hydrogen ion concentration of the solution. Thus, if the concentration is 10^{-2} (moles per liter), the pH is 2; if 10^{-8}, the pH is 8; and so on.

Diagrams:

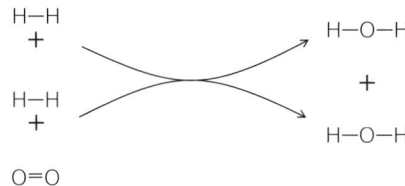

2 Hydrogen molecules

1 Oxygen molecule

2 Water molecules

Structures:

H—H
+
                              H—O—H
                              +
H—H
+   H—O—H

O=O

Formulas:

$$2H_2 + O_2 \longrightarrow 2H_2O$$

Hydrogen and oxygen react to form water.

outer-shell electrons take part. At the end of a chemical reaction, the number of atoms of each type is always the same as at the beginning. For example, two hydrogen molecules ($2H_2$) can react with one oxygen molecule (O_2) to form two water molecules ($2H_2O$). In this reaction two substances (hydrogen and oxygen) disappear, and a new one (water) is formed, but at the end of the reaction there are still four H atoms and two O atoms, just as there were at the beginning. Because the atoms are now linked in a different way, their characteristics or properties have changed.

In many instances chemical reactions involve not the re-linking of molecules but the exchanging of electrons or protons among them. In such reactions the molecule that gains one or more electrons (or loses one or more hydrogen ions) is said to be reduced; the molecule that loses electrons (or gains

protons) is oxidized. A hydrogen ion is equivalent to a proton. Oxidation and reduction reactions take place simultaneously because an electron or proton that is lost by one molecule is accepted by another. The addition of an atom of oxygen is also oxidation because oxygen (with six electrons in the outer shell) accepts two electrons in becoming bonded. Oxidation, then, is loss of electrons, gain of protons, or addition of oxygen (with six electrons); reduction is the opposite—gain of electrons, loss of protons, or loss of oxygen. The addition of hydrogen atoms to oxygen to form water can thus be described as the reduction of oxygen *or* the oxidation of hydrogen.

If a reaction results in a net increase in the energy of a compound, it is called an endergonic, or "uphill," reaction (energy, *erg,* is added into, *endo,* the compound). An example is the chief result of photosynthesis, the making of sugar in a plant from carbon dioxide and water using the energy of sunlight. Conversely, the oxidation of sugar to carbon dioxide and water is an exergonic, or "downhill," reaction because the end products have less energy than the starting products. Oftentimes, but not always, reduction reactions are endergonic, resulting in an increase in the energy of the products. Oxidation reactions often, but not always, are exergonic.

Chemical reactions tend to occur spontaneously if the end products are in a lower energy state and therefore are more stable than the reacting compounds. These reactions often give off energy in the form of heat as they occur. The generation of heat by wood burning in a fireplace and the maintenance of human body warmth both depend on energy-yielding chemical reactions. These downhill reactions occur easily, although they may require some activation energy to get them started, just as a ball requires a push to start rolling.

Uphill reactions, in which the products contain more energy than the reacting compounds started with, do not occur until an energy source is provided. An example of such an energy source is the sunlight used in photosynthesis, where carbon dioxide and water (low-energy compounds) are combined to form the sugar glucose (a higher-energy compound). Another example is the use of the energy in glucose to combine two low-energy compounds in the body into the high-energy

Energy change as reaction occurs

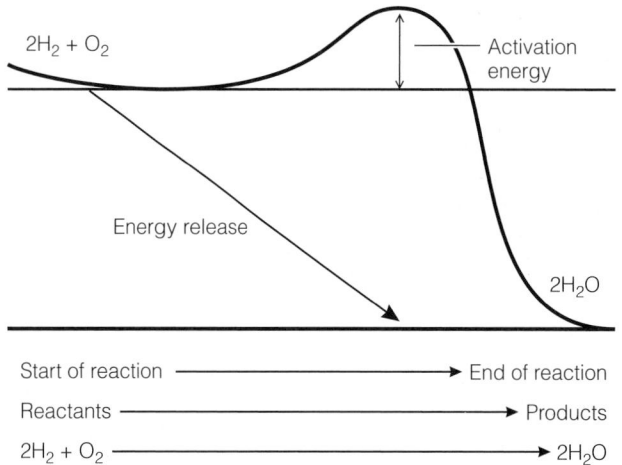

compound ATP (see Chapter 7). The energy in ATP may be used to power many other energy-requiring, uphill reactions. Clearly, any of many different molecules can be used as a temporary storage place for energy.

Neither downhill nor uphill reactions occur until something sets them off (activation) or until a path is provided for them to follow. The body uses enzymes as a means of providing paths and controlling chemical reactions (see Chapter 6). By controlling the availability and the action of its enzymes, the cells can "decide" which chemical reactions to prevent and which to promote.

Formation of Free Radicals

Normally, when a chemical reaction takes place, bonds break and re-form with some redistribution of atoms and rearrangement of bonds to form new, stable compounds. Normally, bonds don't split in such a way as to leave a molecule with an odd, unpaired electron. When they do, free radicals are formed. Free radicals are highly unstable and quickly react with other compounds, forming more free radicals in a chain reaction. A cascade may ensue in which many highly reactive radicals are generated, resulting finally in the disruption of a living structure such as a cell membrane.

Free radicals are formed. The dots represent single electrons that are available for sharing (the atom needs another electron to fill its outer shell).

| Free radical | Compound with weak bond (perhaps an unsaturated fatty acid) | New stable compound (water or an alcohol) | Free radical |

Free radicals destroy biological compounds. The free radical attacks a weak bond in a biological compound, disrupting it and forming a new stable molecule and another free radical. This free radical can attack another biological compound, and so on.

Oxidation of some compounds can be induced by air at room temperature in the presence of light. Such reactions are thought to take place through the formation of compounds called peroxides:

Peroxides:

H—O—O—H	Hydrogen peroxide
R—O—O—H	Hydroperoxides (R is any carbon chain with appropriate numbers of H)
R—O—O—R	Peroxide

Some peroxides readily disintegrate into free radicals, initiating chain reactions like those just described.

Free radicals are of special interest in nutrition because the antioxidant properties of vitamins C and E as well as beta-carotene and the mineral selenium are thought to protect against the destructive effects of these free radicals (see Highlight 11). For example, vitamin E on the surface of the lungs reacts with, and is destroyed by, free radicals, thus preventing the radicals from reaching underlying cells and oxidizing the lipids in their membranes.

Biochemical Structures and Pathways

The diagrams of nutrients presented here are meant to enhance your understanding of the most important organic molecules in the human diet. Following the diagrams of nutrients are sections on the major metabolic pathways mentioned in Chapter 7—glycolysis, fatty acid oxidation, amino acid degradation, the TCA cycle, and the electron transport chain—and a description of how alcohol interferes with these pathways. Discussions of the urea cycle and the formation of ketone bodies complete the appendix.

CONTENTS

C Appendix

Carbohydrates

Monosaccharides

Glucose (alpha form). The ring would be at right angles to the plane of the paper. The bonds directed upward are above the plane; those directed downward are below the plane. This molecule is considered an alpha form because the OH on carbon 1 points downward.

Glucose (beta form). The OH on carbon 1 points upward.
Fructose, galactose: see Chapter 4.

Glucose (alpha form) shorthand notation. This notation, in which the carbons in the ring and single hydrogens have been eliminated, will be used throughout this appendix.

Disaccharides

Glucose   Glucose

Maltose.

Galactose

Glucose

Lactose (alpha form).

Glucose   Fructose

Sucrose.

Polysaccharides

As described in Chapter 4, starch, glycogen, and cellulose are all long chains of glucose molecules covalently linked together.

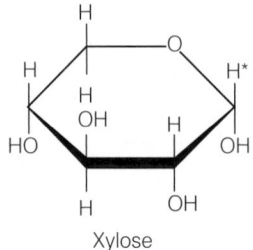

Amylose (unbranched starch)

Amylopectin (branched starch)

Starch. Two kinds of covalent linkages occur between glucose molecules in starch, giving rise to two kinds of chains. Amylose is composed of straight chains, with carbon 1 of one glucose linked to carbon 4 of the next (α-1,4 linkage). Amylopectin is made up of straight chains like amylose but has occasional branches arising where the carbon 6 of a glucose is also linked to the carbon 1 of another glucose (α-1,6 linkage).

Glycogen. The structure of glycogen is like amylopectin but with many more branches.

Cellulose. Like starch and glycogen, cellulose is also made of chains of glucose units, but there is an important difference: in cellulose, the OH on carbon 1 is in the beta position (see p. C-1). When carbon 1 of one glucose is linked to carbon 4 of the next, it forms a β-1,4 linkage, which cannot be broken by digestive enzymes in the human GI tract.

Fibers, such as hemicelluloses, consist of long chains of various monosaccharides.

Monosaccharides common in the backbone chain of hemicelluloses:

Xylose

Mannose

Galactose

---

*These structures are shown in the alpha form with the H on the carbon pointing upward and the OH pointing downward, but they may also appear in the beta form with the H pointing downward and the OH upward.

Monosaccharides common in the side chains of hemicelluloses:

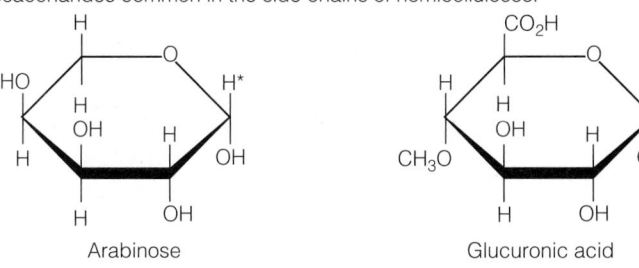

Arabinose Glucuronic acid Galactose

Hemicelluloses. The most common hemicelluloses are composed of a backbone chain of xylose, mannose, and galactose, with branching side chains of arabinose, glucuronic acid, and galactose.

Lipids

TABLE C-1	Saturated Fatty Acids Found in Natural Fats		
Saturated Fatty Acids	Chemical Formulas	Number of Carbons	Major Food Sources
Butyric	C_3H_7COOH	4	Butterfat
Caproic	$C_5H_{11}COOH$	6	Butterfat
Caprylic	$C_7H_{15}COOH$	8	Coconut oil
Capric	$C_9H_{19}COOH$	10	Palm oil
Lauric	$C_{11}H_{23}COOH$	12	Coconut oil, palm oil
Myristic[a]	$C_{13}H_{27}COOH$	14	Coconut oil, palm oil
Palmitic[a]	$C_{15}H_{31}COOH$	16	Palm oil
Stearic[a]	$C_{17}H_{35}COOH$	18	Most animal fats
Arachidic	$C_{19}H_{39}COOH$	20	Peanut oil
Behenic	$C_{21}H_{43}COOH$	22	Seeds
Lignoceric	$C_{23}H_{47}COOH$	24	Peanut oil

[a]Most common saturated fatty acids.

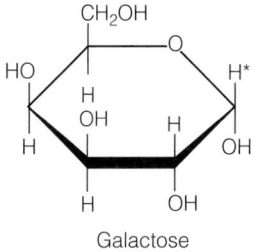

TABLE C-2	Unsaturated Fatty Acids Found in Natural Fats					
Unsaturated Fatty Acids	Chemical Formulas	Number of Carbons	Number of Double Bonds	Standard Notation[a]	Omega Notation[b]	Major Food Sources
Palmitoleic	$C_{15}H_{29}COOH$	16	1	16:1;9	16:1ω7	Seafood, beef
Oleic	$C_{17}H_{33}COOH$	18	1	18:1;9	18:1ω9	Olive oil, canola oil
Linoleic	$C_{17}H_{31}COOH$	18	2	18:2;9,12	18:2ω6	Sunflower oil, safflower oil
Linolenic	$C_{17}H_{29}COOH$	18	3	18:3;9,12,15	18:3ω3	Soybean oil, canola oil
Arachidonic	$C_{19}H_{31}COOH$	20	4	20:4;5,8,11,14	20:4ω6	Eggs, most animal fats
Eicosapentaenoic	$C_{19}H_{29}COOH$	20	5	20:5;5,8,11,14,17	20:5ω3	Seafood
Docosahexaenoic	$C_{21}H_{31}COOH$	22	6	22:6;4,7,10,13,16,19	22:6ω3	Seafood

NOTE: A fatty acid has two ends; designated the methyl (CH_3) end and the carboxyl, or acid (COOH), end.

[a]Standard chemistry notation begins counting carbons at the acid end. The number of carbons the fatty acid contains comes first, followed by a colon and another number that indicates the number of double bonds; next comes a semicolon followed by a number or numbers indicating the positions of the double bonds. Thus the notation for linoleic acid, an 18-carbon fatty acid with two double bonds between carbons 9 and 10 and between carbons 12 and 13, is 18:2;9,12.

[b]Because fatty acid chains are lengthened by adding carbons at the acid end of the chain, chemists use the omega system of notation to ease the task of identifying them. The omega system begins counting carbons at the methyl end. The number of carbons the fatty acid contains comes first, followed by a colon and the number of double bonds; next come the omega symbol (ω) and a number indicating the position of the double bond nearest the methyl end. Thus linoleic acid with its first double bond at the sixth carbon from the methyl end would be noted 18:2ω6 in the omega system.

Protein: Amino Acids

The common amino acids may be classified into the seven groups listed on the next page. Amino acids marked with an asterisk (*) are essential.

1. Amino acids with aliphatic side chains, which consist of hydrogen and carbon atoms (hydrocarbons):

Glycine (Gly)

Alanine (Ala)

Valine* (Val)

Leucine* (Leu)

Isoleucine* (Ile)

2. Amino acids with hydroxyl (OH) side chains:

Serine (Ser)

Threonine* (Thr)

3. Amino acids with side chains containing acidic groups or their amides, which contain the group NH_2:

Aspartic acid (Asp)

Glutamic acid (Glu)

Asparagine (Asn)

Glutamine (Gln)

4. Amino acids with basic side chains:

Lysine* (Lys)

Arginine (Arg)

Histidine* (His)

5. Amino acids with aromatic side chains, which are characterized by the presence of at least one ring structure:

Phenylalanine* (Phe)

Tyrosine (Tyr)

Tryptophan* (Trp)

6. Amino acids with side chains containing sulfur atoms:

Cysteine (Cys)

Methionine* (Met)

7. Imino acid:

Proline (Pro)

Proline has the same chemical structure as the other amino acids, but its amino group has given up a hydrogen to form a ring.

Vitamins and Coenzymes

Vitamin A: retinol. This molecule is the alcohol form of vitamin A.

Vitamin A: retinal. This molecule is the aldehyde form of vitamin A.

Vitamin A: retinoic acid. This molecule is the acid form of vitamin A.

Vitamin A precursor: beta-carotene. This molecule is the carotenoid with the most vitamin A activity.

Thiamin. This molecule is part of the coenzyme thiamin pyrophosphate (TPP).

Thiamin pyrophosphate (TPP). TPP is a coenzyme that includes the thiamin molecule as part of its structure.

Riboflavin. This molecule is a part of two coenzymes—flavin mononucleotide (FMN) and flavin adenine dinucleotide (FAD).

Flavin mononucleotide (FMN). FMN is a coenzyme that includes the riboflavin molecule as part of its structure.

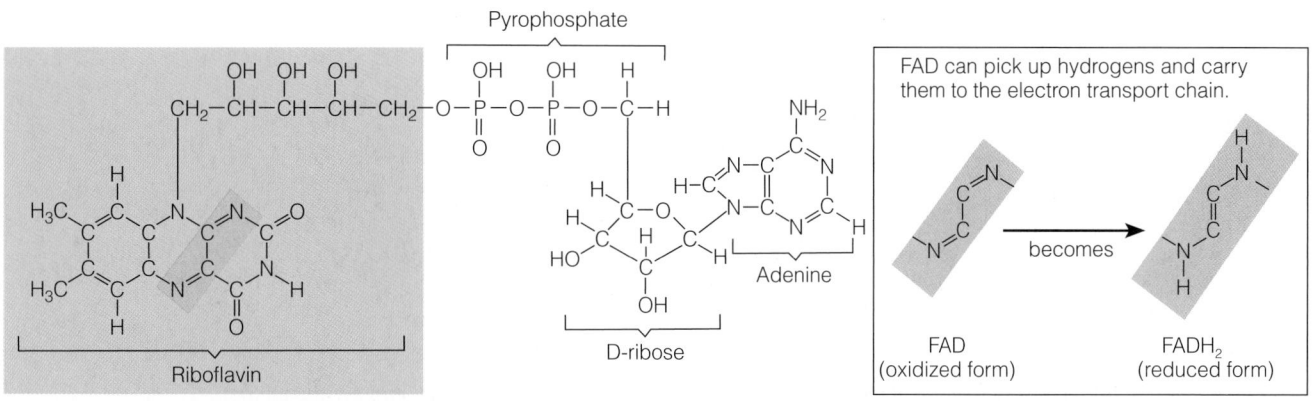

Flavin adenine dinucleotide (FAD). FAD is a coenzyme that includes the riboflavin molecule as part of its structure.

Niacin (nicotinic acid and nicotinamide). These molecules are a part of two coenzymes—nicotinamide adenine dinucleotide (NAD^+) and nicotinamide adenine dinucleotide phosphate ($NADP^+$).

Nicotinamide adenine dinucleotide (NAD^+) and nicotinamide adenine dinucleotide phosphate ($NADP^+$). NADP has the same structure as NAD but with a phosphate group attached to the O instead of the H.

Reduced NAD^+ (NADH). When NAD^+ is reduced by the addition of H^+ and two electrons, it becomes the coenzyme NADH. (The dots on the H entering this reaction represent electrons—see Appendix B.)

Vitamin B_6 (a general name for three compounds—pyridoxine, pyridoxal, and pyridoxamine). These molecules are a part of two coenzymes—pyridoxal phosphate and pyridoxamine phosphate.

Pyridoxal phosphate (PLP) and pyridoxamine phosphate. These coenzymes include vitamin B_6 as part of their structures.

Vitamin B_{12} (cyanocobalamin). The arrows in this diagram indicate that the spare electron pairs on the nitrogens attract them to the cobalt.

Folate (folacin or folic acid). This molecule consists of a double ring combined with a single ring and at least one glutamate (a nonessential amino acid marked in the box). Folate's biologically active form is tetrahydrofolate.

Tetrahydrofolate. This active coenzyme form of folate has four added hydrogens. An intermediate form, dihydrofolate, has two added hydrogens.

Pantothenic acid. This molecule is part of coenzyme A (CoA).

Coenzyme A (CoA). Coenzyme A is a coenzyme that includes pantothenic acid as part of its structure.

Biotin.

Vitamin C. Two hydrogen atoms with their electrons are lost when ascorbic acid is oxidized and gained when it is reduced again.

Ascorbic acid (reduced form)

Dehydroascorbic acid (oxidized form)

7-dehydrocholesterol

Carbon #7

Ultraviolet light on the skin

Vitamin D₃ (also called cholecalciterol or calciol)

Hydroxylation in the liver

25-hydroxy-vitamin D₃ (also called calcidiol)

Carbon #25

Hydroxylation in the kidneys

1,25-dihydroxy-vitamin D₃ (also called calcitrol)

Carbon #1

Vitamin D. The synthesis of active vitamin D begins with 7-dehydrocholesterol. (The carbon atoms at which changes occur are numbered.)

Vitamin E (alpha-tocopherol). The number and position of the methyl groups (CH₃) bonded to the ring structure differentiate among the tocopherols.

Tocotrienols contain double bonds here.

Vitamin K. Naturally occurring compounds with vitamin K activity include phylloquinones (from plants) and menaquinones (from bacteria).

Menadione. This synthetic compound has the same activity as natural vitamin K.

Adenosine triphosphate (ATP), the energy carrier. The cleavage point marks the bond that is broken when ATP splits to become ADP + P.

Adenosine diphosphate (ADP).

Glycolysis

Figure C-1 depicts the events of glycolysis. The following text describes key steps as numbered on the figure.

FIGURE C-1 | Glycolysis

Notice that galactose and fructose enter at different places but continue on the same pathway.

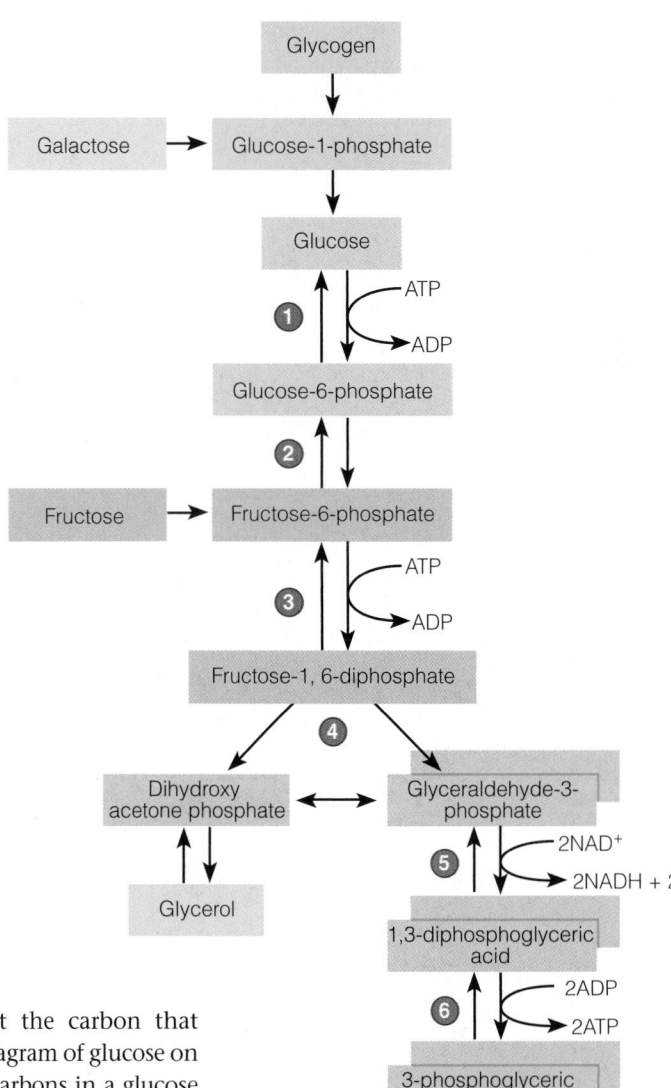

1. A phosphate is attached to glucose at the carbon that chemists call number 6 (review the first diagram of glucose on p. C-1 to see how chemists number the carbons in a glucose molecule). The product is called, logically enough, glucose-6-phosphate. One ATP molecule is used to accomplish this.

2. Glucose-6-phosphate is rearranged by an enzyme.

3. A phosphate is added in another reaction that uses another molecule of ATP. The product this time is fructose-1,6-diphosphate. At this point the six-carbon sugar has a phosphate group on its first and sixth carbons and is ready to break apart.

4. When fructose-1,6-diphosphate breaks in half, the two three-carbon compounds are not identical. Each has a phosphate group attached, but only glyceraldehyde-3-phosphate converts directly to pyruvate. The other compound, however, converts easily to glyceraldehyde-3-phosphate.

5. In the next step, enough energy is released to convert NAD^+ to $NADH + H^+$.

6. In two of the following steps ATP is regenerated.

Remember that in effect two molecules of glyceraldehyde-3-phosphate are produced from glucose; therefore, four ATP molecules are generated from each glucose molecule. Two ATP were needed to get the sequence started, so the net gain at this point is two ATP and two molecules of $NADH + H^+$. As you will see later, each $NADH + H^+$ moves to the electron transport chain to unload its hydrogens onto oxygen, producing more ATP.

Fatty Acid Oxidation

Figure C-2 presents fatty acid oxidation. The sequence is as follows.

1. The fatty acid is activated by combining with coenzyme A (CoA). In this reaction, ATP loses two phosphorus atoms (PP, or pyrophosphate) and becomes AMP (adenosine monophosphate)—the equivalent of a loss of two ATP.

2. In the next reaction, two H with their electrons are removed and transferred to FAD, forming $FADH_2$.

3. In a later reaction, two H are removed and go to NAD^+ (forming $NADH + H^+$).

4. The fatty acid is cleaved at the "beta" carbon, the second carbon from the carboxyl (COOH) end. This break results in a fatty acid that is two carbons shorter than the previous one and a two-carbon molecule of acetyl CoA. At the same time, another CoA is attached to the fatty acid, thus activating it for its turn through the series of reactions.

5. The sequence is repeated with each cycle producing an acetyl CoA and a shorter fatty acid until only a 2-carbon fatty acid remains—acetyl CoA.

In the example shown in Figure C-2, palmitic acid (a 16-carbon fatty acid) will go through this series of reactions seven times, using the equivalent of two ATP for the initial activation and generating seven $FADH_2$, seven $NADH + H^+$, and eight acetyl CoA. As you will see later, each of the seven $FADH_2$ will enter the electron transport chain to unload its hydrogens onto oxygen, yielding two ATP (for a total of 14). Similarly, each $NADH + H^+$ will enter the electron transport chain to unload its hydrogens onto oxygen, yielding three ATP (for a total of 21). Thus the oxidation of a 16-carbon fatty acid uses 2 ATP and generates 35 ATP. When the eight acetyl CoA enter the TCA cycle, even more ATP will be generated, as a later section describes.

Amino Acid Degradation

The first step in amino acid degradation is the removal of the nitrogen-containing amino group through either deamination (Figure 7-14 on p. 226) or transamination (Figure 7-15 on p. 226) reactions. Then the remaining carbon skeletons may enter the metabolic pathways at different places, as shown in Figure C-3.

The TCA Cycle

The tricarboxylic acid, or TCA, cycle is the set of reactions that break down acetyl CoA to carbon dioxide and hydrogens. To link glycolysis to the TCA cycle, pyruvate enters the mitochondrion, loses a carbon group, and bonds with a molecule of CoA to become acetyl CoA. The TCA cycle uses any substance that can be converted to acetyl CoA directly or indirectly through pyruvate.

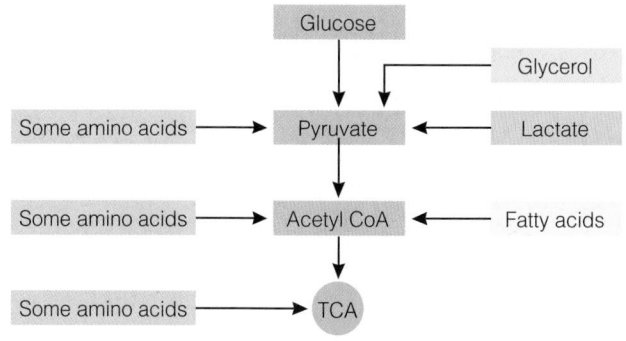

The step from pyruvate to acetyl CoA is complex. We have included only those substances that will help you understand

FIGURE C-2 · Fatty Acid Oxidation

Palmitic acid (16C)
→ ATP
CoA → ① → AMP + PP
Activated palmitic acid
→ FAD
② → $FADH_2$
→ H_2O
→ NAD^+
③ → $NADH + H^+$
CoA → ④
⑤
Activated myristic acid (14C) + Acetyl CoA (2C)

FIGURE C-3 Amino Acid Degradation

After losing their amino groups, carbon skeletons can be converted to one of seven molecules that can enter the TCA cycle (presented in Figure C-4).

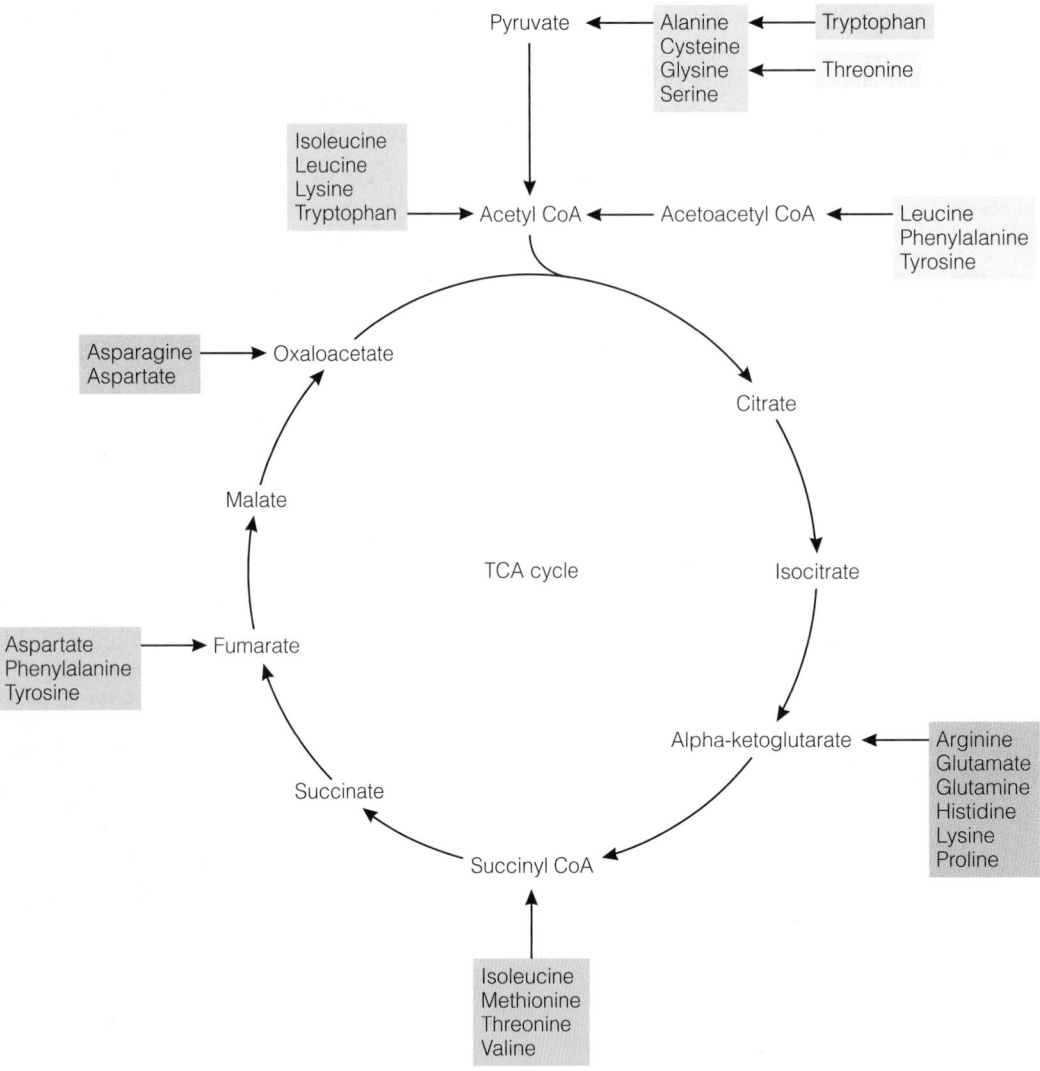

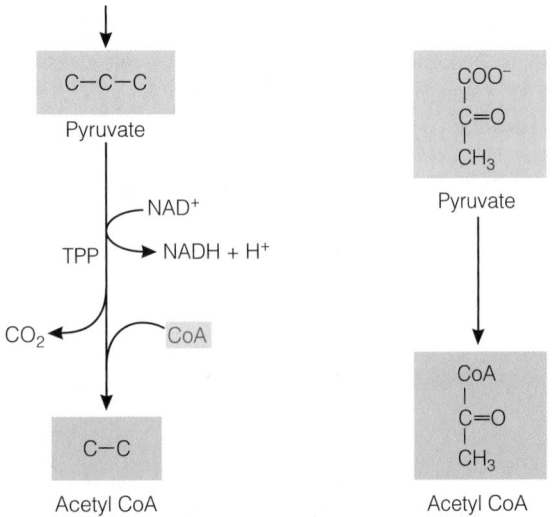

The step from pyruvate to acetyl CoA. (TPP and NAD are coenzymes containing the B vitamins thiamin and niacin, respectively.)

the transfer of energy from the nutrients. Pyruvate loses a carbon to carbon dioxide and is attached to a molecule of CoA. In the process, NAD$^+$ picks up two hydrogens with their associated electrons, becoming NADH + H$^+$.

Let's follow the steps of the TCA cycle (see the corresponding numbers in Figure C-4).

1. The two-carbon acetyl CoA combines with a four-carbon compound, oxaloacetate. The CoA comes off, and the product is a six-carbon compound, citrate.

2. The atoms of citrate are rearranged to form isocitrate.

3. Now two H (with their two electrons) are removed from the isocitrate. One H becomes attached to the NAD$^+$ with the two electrons; the other H is released as H$^+$. Thus NAD$^+$ becomes NADH + H$^+$. (Remember this NADH + H$^+$, but let's follow the carbons first.) A carbon is combined with two oxygens, forming carbon dioxide (which diffuses away into the blood and is exhaled). What is left is the five-carbon compound alpha-ketoglutarate.

FIGURE C-4 The TCA Cycle

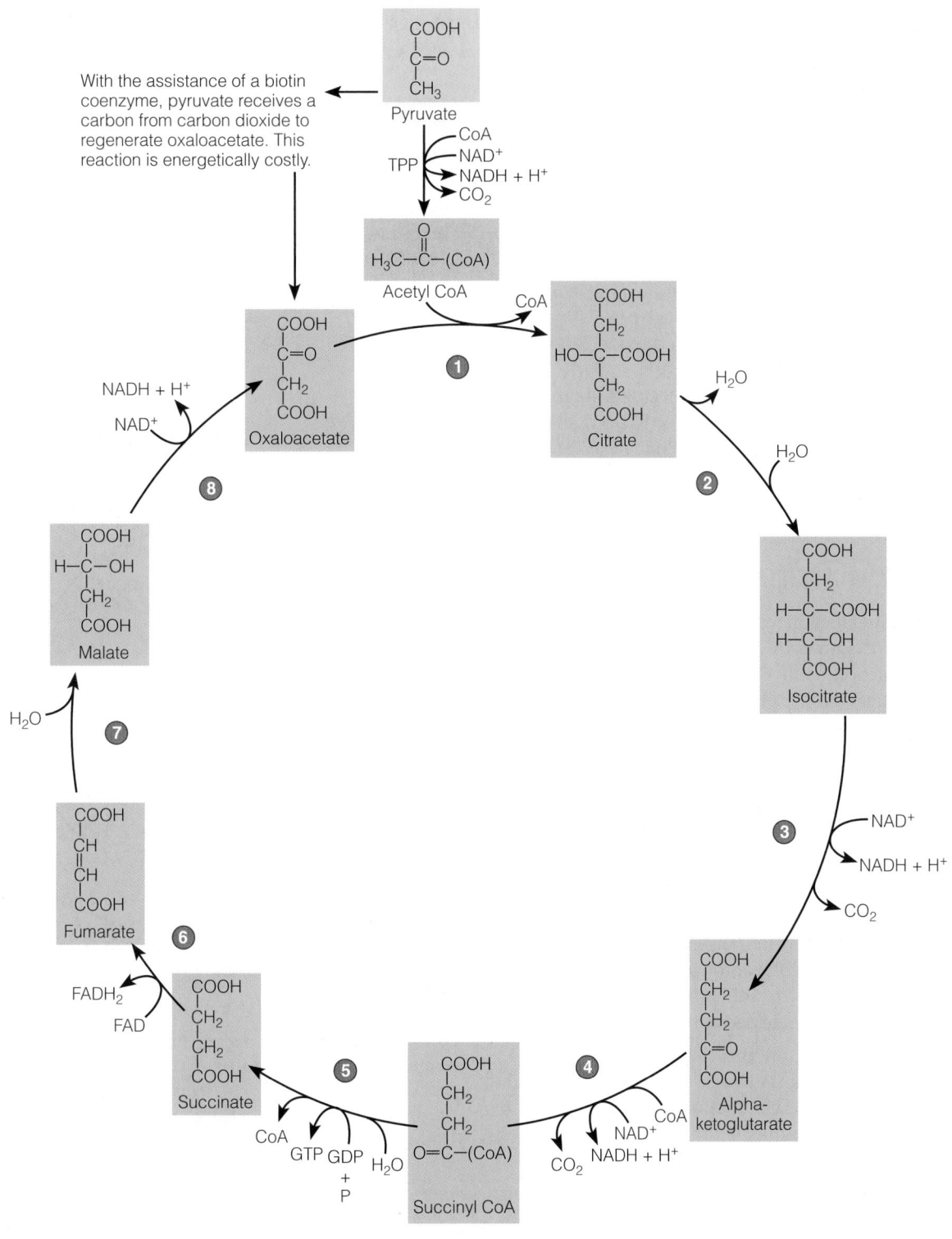

4. Now two compounds interact with alpha-ketoglutarate —a molecule of CoA and a molecule of NAD⁺. In this complex reaction, a carbon and two oxygens are removed (forming carbon dioxide); two hydrogens are removed and go to NAD⁺ (forming NADH + H⁺); and the remaining four-carbon compound is attached to the CoA, forming succinyl

CoA. (Remember this NADH + H⁺ also. You will see later what happens to it.)

5. Now two molecules react with succinyl CoA—a molecule called GDP and one of phosphate (P). The CoA comes off, the GDP and P combine to form the high-energy compound GTP (similar to ATP), and succinate remains. (Remember this GTP.)

6. In the next reaction, two H with their electrons are removed from succinate and are transferred to a molecule of FAD (a coenzyme like NAD+) to form FADH₂. The product that remains is fumarate. (Remember this FADH₂.)

7. Next a molecule of water is added to fumarate, forming malate.

8. A molecule of NAD+ reacts with the malate; two H with their associated electrons are removed from the malate and form NADH + H+. The product that remains is the four-carbon compound oxaloacetate. (Remember this NADH + H+.)

We are back where we started. The oxaloacetate formed in this process can combine with another molecule of acetyl CoA (step 1), and the cycle can begin again, as shown in Figure C-4.

So far, we have seen two carbons brought in with acetyl CoA and two carbons ending up in carbon dioxide. But where are the energy and the ATP we promised?

A review of the eight steps of the TCA cycle shows that the compounds NADH + H+ (three molecules), FADH₂, and GTP capture energy originally found in acetyl CoA. To see how this energy ends up in ATP, we must follow the electrons further—into the electron transport chain.

The Electron Transport Chain

The six reactions described here are those of the electron transport chain, which is shown in Figure C-5. Since oxygen is required for these reactions, and ADP and P are combined to form ATP in several of them (ADP is phosphorylated), these reactions are also called oxidative phosphorylation.

An important concept to remember at this point is that an electron is not a fixed amount of energy. The electrons that bond the H to NAD+ in NADH have a relatively large amount of energy. In the series of reactions that follow, they release this energy in small amounts, until at the end they are attached (with H) to oxygen (O) to make water (H₂O). In some of the steps, the energy they release is captured into ATP in coupled reactions.

1. In the first step of the electron transport chain, NADH reacts with a molecule called a flavoprotein, losing its electrons (and their H). The products are NAD+ and reduced flavoprotein. A little energy is released as heat in this reaction.

2. The flavoprotein passes on the electrons to a molecule called coenzyme Q. Again they release some energy as heat, but ADP and P bond together and form ATP, storing much of the energy. This is a coupled reaction: ADP + P → ATP.

3. Coenzyme Q passes the electrons to cytochrome *b*. Again the electrons release energy.

4. Cytochrome *b* passes the electrons to cytochrome *c* in a coupled reaction in which ATP is formed: ADP + P → ATP.

5. Cytochrome *c* passes the electrons to cytochrome *a*.

6. Cytochrome *a* passes them (with their H) to an atom of oxygen (O), forming water (H₂O). This is a coupled reaction in which ATP is formed: ADP + P → ATP.

As Figure C-5 shows, each time NADH is oxidized (loses its electrons) by this means, the energy it releases is captured into three ATP molecules. When the electrons are passed on to water at the end, they are much lower in energy than they were originally. This completes the story of the electrons from NADH.

As for FADH₂, its electrons enter the electron transport chain at coenzyme Q. From coenzyme Q to water, ATP is generated in only two steps. Therefore, FADH₂ coming out of the TCA cycle yields just two ATP molecules.

One energy-receiving compound of the TCA cycle (GTP) does not enter the electron transport chain but gives its energy directly to ADP in a simple phosphorylation reaction. This reaction yields one ATP.

It is now possible to draw up a balance sheet of glucose metabolism (see Table C-3). Glycolysis has yielded 4 NADH + H+ and 4 ATP molecules and has spent 2 ATP. The 2 acetyl CoA

FIGURE C-5 The Electron Transport Chain

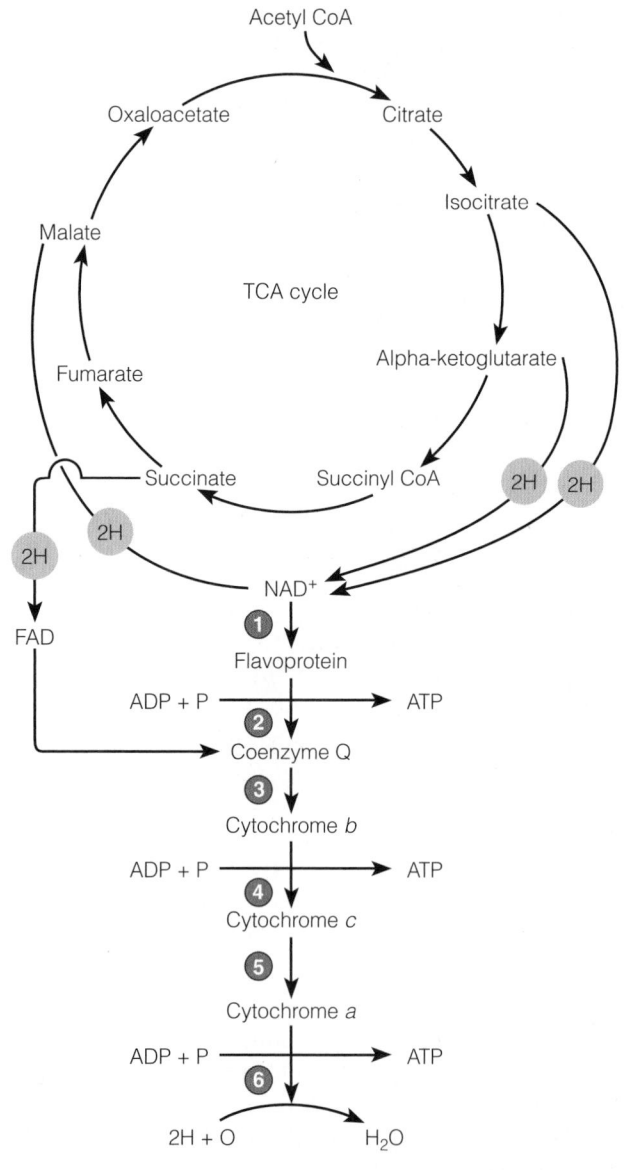

going through the TCA cycle have yielded 6 NADH + H$^+$, 2 FADH$_2$, and 2 GTP molecules. After the NADH + H$^+$ and FADH$_2$ have gone through the electron transport chain, there are 28 ATP. Added to these are the 4 ATP from glycolysis and the 2 ATP from GTP, making the total 34 ATP generated from one molecule of glucose. After the expense of 2 ATP is subtracted, there is a net gain of 32 ATP.*

A similar balance sheet from the complete breakdown of one 16-carbon fatty acid would show a net gain of 129 ATP. As mentioned earlier, 35 ATP were generated from the seven FADH$_2$ and seven NADH + H$^+$ produced during fatty acid oxidation. The eight acetyl CoA produced will each generate 12 ATP as they go through the TCA cycle and the electron transport chain, for a total of 96 more ATP. After subtracting the 2 ATP needed to activate the fatty acid initially, the net yield from one 16-carbon fatty acid: 35 + 96 − 2 = 129 ATP.

These calculations help explain why fat yields more energy (measured as kcalories) per gram than carbohydrate or protein. The more hydrogen atoms a fuel contains, the more ATP will be generated during oxidation. The 16-carbon fatty acid molecule, with its 32 hydrogen atoms, generates 129 ATP, whereas glucose, with its 12 hydrogen atoms, yields only 32 ATP.

The TCA cycle and the electron transport chain are the body's major means of capturing the energy from nutrients in ATP molecules. Other means, such as anaerobic glycolysis, contribute energy quickly, but the aerobic processes are the most efficient. Biologists and chemists understand much more about these processes than has been presented here.

Alcohol's Interference with Energy Metabolism

Highlight 7 provides an overview of how alcohol interferes with energy metabolism. With an understanding of the TCA cycle, a few more details may be appreciated. During alcohol metabolism, the enzyme alcohol dehydrogenase oxidizes alcohol to acetaldehyde while it simultaneously reduces a molecule of NAD$^+$ to NADH + H$^+$. The related enzyme acetaldehyde dehydrogenase reduces another NAD$^+$ to NADH + H$^+$ while it oxidizes acetaldehyde to acetyl CoA, the compound that enters the TCA cycle to generate energy. Thus, whenever alcohol is being metabolized in the body, NAD$^+$ diminishes, and NADH + H$^+$ accumulates. Chemists say that the body's "redox state" is altered, because NAD$^+$ can oxidize, and NADH + H$^+$ can reduce, many other body compounds. During alcohol metabolism, NAD$^+$ becomes unavailable for the multitude of reactions for which it is required.

TABLE C-3	Balance Sheet for Glucose Metabolism	
		ATP
Glycolysis:	4 ATP − 2 ATP	2
1 glucose to 2 pyruvate	2 NADH + H$^+$	3-5[a]
2 pyruvate to 2 acetyl CoA	2 NADH + H$^+$	5
TCA cycle and electron transport chain:		
2 isocitrate	2 NADH + H$^+$	5
2 alpha-ketoglutarate	2 NADH + H$^+$	5
2 succinyl CoA	2 GTP	2
2 succinate	2 FADH$_2$	3
2 malate	2 NADH + H$^+$	5
Total ATP collected from one molecule glucose:		30–32

[a]Each NADH + H$^+$ from glycolysis can yield 1.5 or 2.5 ATP. See the accompanying text.

As the previous sections just explained, for glucose to be completely metabolized, the TCA cycle must be operating, and NAD$^+$ must be present. If these conditions are not met (and when alcohol is present, they may not be), the pathway will be blocked, and traffic will back up—or an alternate route will be taken. Think about this as you follow the pathway shown in Figure C-6.

In each step of alcohol metabolism in which NAD$^+$ is converted to NADH + H$^+$, hydrogen ions accumulate, resulting in a dangerous shift of the acid-base balance toward acid (Chapter 12 explains acid-base balance). The accumulation of NADH + H$^+$ slows TCA cycle activity, so pyruvate and acetyl CoA build up. This condition favors the conversion of pyruvate to lactate, which serves as a temporary storage place for hydrogens from NADH + H$^+$. The conversion of pyruvate to lactate restores some NAD$^+$, but a lactate buildup has serious consequences of its own. It adds to the body's acid burden and interferes with the excretion of uric acid, causing goutlike symptoms. Molecules of acetyl CoA become building blocks for fatty acids or ketone bodies. The making of ketone bodies consumes acetyl CoA and generates NAD$^+$; but some ketone bodies are acids, so they push the acid-base balance further toward acid.

Thus alcohol cascades through the metabolic pathways, wreaking havoc along the way. These consequences have physical effects, which Highlight 7 describes.

The Urea Cycle

Chapter 7 sums up the process by which waste nitrogen is eliminated from the body by stating that ammonia molecules combine with carbon dioxide to produce urea. This is true, but it is not the whole story. Urea is produced in a multistep process within the cells of the liver.

---

*The total may sometimes be 30 ATP. The NADH + H$^+$ generated in the cytoplasm during glycolysis pass their electrons on to shuttle molecules, which move them into the mitochondria. One shuttle, malate, contributes its electrons to the electron transport chain before the first site of ATP synthesis, yielding 5 ATP. Another, glycerol phosphate, adds its electrons into the chain beyond that first site, yielding 3 ATP. Thus sometimes 5, and sometimes 3, ATP result from the NADH + H$^+$ that arise from glycolysis. The amount depends on the cell.

GLOSSARY

amino acid scoring: a measure of protein quality assessed by comparing a protein's amino acid pattern with that of a reference protein; sometimes called **chemical scoring.**

biological value (BV): a measure of protein quality assessed by measuring the amount of protein nitrogen that is retained from a given amount of protein nitrogen absorbed.

net protein utilization (NPU): a measure of protein quality assessed by measuring the amount of protein nitrogen that is retained from a given amount of protein nitrogen eaten.

PDCAAS (protein digestibility–corrected amino acid score): a measure of protein quality assessed by comparing the amino acid score of a food protein with the amino acid requirements of preschool-age children and then correcting for the true digestibility of the protein; recommended by the FAO/WHO and used to establish protein quality of foods for Daily Value percentages on food labels.

protein efficiency ratio (PER): a measure of protein quality assessed by determining how well a given protein supports weight gain in growing rats; used to establish the protein quality for infant formulas and baby foods.

Measures of Protein Quality

In a world where food is scarce and many people's diets contain marginal or inadequate amounts of protein, it is important to know which foods contain the highest-quality protein. Chapter 6 describes protein quality, and this appendix presents different measures researchers use to assess the quality of a food protein. The accompanying glossary defines related terms.

Amino Acid Scoring

Amino acid scoring evaluates a protein's quality by determining its amino acid composition and comparing it with that of a reference protein. The advantages of amino acid scoring are that it is simple and inexpensive, it easily identifies the limiting amino acid, and it can be used to score mixtures of different proportions of two or more proteins mathematically without having to make up a mixture and test it. Its chief weaknesses are that it fails to estimate the digestibility of a protein, which may strongly affect the protein's quality; it relies on a chemical procedure in which certain amino acids may be destroyed, making the pattern that is analyzed inaccurate; and it is blind to other features of the protein (such as the presence of substances that may inhibit the digestion or utilization of the protein) that would only be revealed by a test in living animals.

Table D-1 (p. D-1) shows the reference pattern for the nine essential amino acids. To interpret the table, read, "For every 3210 units of essential amino acids, 145 must be histidine, 340 must be isoleucine, 540 must be leucine," and so on. To compare a test protein with the reference protein, the experimenter first obtains a chemical analysis of the test protein's amino acids. Then, taking 3210 units of the amino acids, the experimenter compares the amount of each amino acid to the amount found in 3210 units of essential amino acids in egg protein. For example, suppose the test protein contained (per 3210 units) 360 units of isoleucine; 500 units of leucine; 350 of lysine; and for each of the other amino acids, more units than egg protein contains. The two amino acids that are low are leucine (500 as compared with 540 in egg) and lysine (350 versus 440 in egg). The ratio, amino acid in the test protein divided by amino acid in egg, is 500/540 (or about 0.93) for leucine and 350/440 (or about 0.80) for lysine. Lysine is the limiting amino acid (the one that falls shortest compared with egg). If the protein's limiting amino acid is 80 percent of the amount found in the reference protein, it receives a score of 80.

PDCAAS

The **protein digestibility–corrected amino acid score,** or **PDCASS,** compares the amino acid composition of a protein with human amino acid requirements and corrects for digestibility. First the protein's amino acid composition is determined, and then it is compared against the amino acid requirements of preschool-age children. This comparison reveals the most limiting amino acid—the one that falls shortest compared with the reference. If a food protein's limiting amino acid is 70 percent of the amount found in the reference protein, it receives a score of 70. The amino acid score is multiplied by the food's protein digestibility percentage to determine the PDCAAS. The box on p. D-2 provides an example of how to calculate the PDCAAS, and Table D-2 (p. D-1) lists the PDCAAS values of selected foods.

Biological Value

The **biological value (BV)** of a protein measures its efficiency in supporting the body's needs. In a test of biological value, two nitrogen balance studies are done. In the first, no protein is fed, and nitrogen (N) excretions in the urine and feces are measured. It is assumed that under these conditions, N lost in the urine is the amount the body always necessarily loses by filtration into the urine each day, regardless of what protein is fed (endogenous N). The N lost in the feces (called metabolic N) is the amount the body invariably loses into the intestine each day, whether or not food protein is fed. (To help you remember the terms: endogenous N is "urinary N on a zero-protein diet"; metabolic N is "fecal N on a zero-protein diet.")

In the second study, an amount of protein slightly below the requirement is fed. Intake and losses are measured; then the BV is derived using this formula:

$$BV = \frac{N \text{ retained}}{N \text{ absorbed}} \times 100$$

The denominator of this equation expresses the amount of nitrogen *absorbed*: food N minus fecal N (excluding the metabolic N the body would lose in the feces anyway, even without food). The numerator expresses the amount of N *retained* from the N absorbed: absorbed N (as in the denominator) minus the N excreted in the urine (excluding the endogenous N the body would lose in the urine anyway, even without food). The more nitrogen retained, the higher the protein quality. (Recall that when an essential amino acid is missing, protein synthesis stops, and the remaining amino acids are deaminated and the nitrogen excreted.)

Egg protein has a BV of 100, indicating that 100 percent of the nitrogen absorbed is retained. Supplied in adequate quantity, a protein with a BV of 70 or greater can support human growth as long as energy intake is adequate. Table D-3 presents the BV for selected foods.

This method has the advantages of being based on experiments with human beings (it can be done with animals, too, of course) and of measuring actual nitrogen retention. But it is also cumbersome, expensive, and often impractical, and it is based on several assumptions that may not be valid. For example, the physiology, normal environment, or typical food intake of the subjects used for testing may not be similar to those for whom the test protein may ultimately be used. For another example, the retention of protein in the body does not necessarily mean that it is being well utilized. Considerable exchange of protein among tissues (protein turnover) occurs, but is hidden from view when only N intake and output are measured. The test of biological value wouldn't detect if one tissue were shorted.

Net Protein Utilization

Like BV, **net protein utilization (NPU)** measures how efficiently a protein is used by the body and involves two balance studies. The difference is that NPU measures retention of food nitrogen rather than food nitrogen absorbed (as in BV). The formula for NPU is:

$$NPU = \frac{N \text{ retained}}{N \text{ intake}} \times 100$$

The numerator is the same as for BV, but the denominator represents food N intake only—not N absorbed.

This method offers advantages similar to those of BV determinations and is used more frequently, with animals as the test subjects. A drawback is that if a low NPU is obtained, the test results offer no help in distinguishing between two possible causes: a poor amino acid composition of the test protein or poor digestibility. There is also a limit to the extent to which animal test results can be assumed to be applicable to human beings.

TABLE D-1 A Reference Pattern for Amino Acid Scoring of Proteins

Essential Amino Acids	Reference Protein—Whole Egg (mg amino acid/g nitrogen)
Histidine	145
Isoleucine	340
Leucine	540
Lysine	440
Methionine + cystine[a]	355
Phenylalanine + tyrosine[b]	580
Threonine	294
Tryptophan	106
Valine	410
Total	3210

[a]Methionine is essential and is also used to make cystine. Thus the methionine requirement is lower if cystine is supplied.
[b]Phenylalanine is essential and is also used to make tyrosine if not enough of the latter is available. Thus the phenylalanine requirement is lower if tyrosine is also supplied.

TABLE D-2 PDCAAS Values of Selected Foods

Casein (milk protein)	1.00
Egg white	1.00
Soybean (isolate)	.99
Beef	.92
Pea flour	.69
Kidney beans (canned)	.68
Chickpeas (canned)	.66
Pinto beans (canned)	.66
Rolled oats	.57
Lentils (canned)	.52
Peanut meal	.52
Whole wheat	.40

NOTE: 1.0 is the maximum PDCAAS a food protein can receive.

TABLE D-3 Biological Values (BV) of Selected Foods

Egg	100
Milk	93
Beef	75
Fish	75
Corn	72

NOTE: 100 is the maximum BV a food protein can receive.

HOW TO Measure Protein Quality Using PDCAAS

To calculate the PDCAAS (protein digestibility–corrected amino acid score), researchers first determine the amino acid profile of the test protein (in this example, pinto beans). The second column of the table below presents the essential amino acid profile for pinto beans. The third column presents the amino acid reference pattern.

To determine how well the food protein meets human needs, researchers calculate the ratio by dividing the second column by the third column (for example, 30 ÷ 18 = 1.67). The amino acid with the lowest ratio is the most limiting amino acid—in this case, methionine. Its ratio is the amino acid score for the protein—in this case, 0.84.

The amino acid score alone, however, does not account for digestibility. Protein digestibility, as determined by rat studies, yields a value of 79 percent for pinto beans. Together, the amino acid score and the digestibility value determine the PDCAAS:

$$\text{PDCAAS} =$$
protein digestibility × amino acid score
PDCAAS for pinto beans =
$$0.79 \times 0.84 = 0.66$$

Thus the PDCAAS for pinto beans is 0.66. Table D-2 lists the PDCAAS values of selected foods.

The PDCAAS is used to determine the % Daily Value on food labels. To calculate the % Daily Value for protein for canned pinto beans, multiply the number of grams of protein in a standard serving (in the case of pinto beans, 7 grams per ½ cup) by the PDCAAS:

$$7 \text{ g} \times 0.66 = 4.62$$

This value is then divided by the recommended standard for protein (for children over age four and adults, 50 grams):

$$4.62 \div 50 = 0.09 \text{ (or 9\%)}$$

The food label for this can of pinto beans would declare that one serving provides 7 grams protein, and if the label included a % Daily Value for protein (which is optional), the value would be 9 percent.

Essential Amino Acids	Amino Acid Profile of Pinto Beans (mg/g protein)	Amino Acid Reference Pattern (mg/g protein)	Amino Acid Score
Histidine	30.0	18	1.67
Isoleucine	42.5	25	1.70
Leucine	80.4	55	1.46
Lysine	69.0	51	1.35
Methionine (+ cystine)	21.1	25	0.84
Phenylalanine (+ tyrosine)	90.5	47	1.93
Threonine	43.7	27	1.62
Tryptophan	8.8	7	1.26
Valine	50.1	32	1.57

TABLE D-4 Protein Efficiency Ratio (PER) Values of Selected Proteins

Casein (milk)	2.8
Soy	2.4
Glutein (wheat)	0.4

Protein Efficiency Ratio

The **protein efficiency ratio (PER)** measures the weight gain of a growing animal and compares it to the animal's protein intake. Until recently, the PER was generally accepted in the United States and Canada as the official method for assessing protein quality, and it is still used to evaluate proteins for infants.

Young rats are fed a measured amount of protein and weighed periodically as they grow. The PER is expressed as:

$$\text{PER} = \frac{\text{weight gain (g)}}{\text{protein intake (g)}}$$

This method has the virtues of economy and simplicity, but it also has many drawbacks. The experiments are time-consuming; the amino acid needs of rats are not the same as those of human beings; and the amino acid needs for growth are not the same as for the maintenance of adult animals (growing animals need more lysine, for example). Table D-4 presents PER values for selected foods.

Nutrition Assessment: Supplemental Information

Chapter 17 described data from nutrition assessments that help health professionals evaluate patients' nutrition status and nutrient needs. This appendix provides additional information that may be useful for complete assessments.

Growth Charts

Health professionals generally evaluate physical development by monitoring the growth rate of a child and comparing this rate with those on standard charts. Standard charts compare length or height to age, weight to age, weight to length, head circumference to age, and body mass index (BMI) ◆ to age. Although individual growth patterns may vary, a child's growth curve will generally stay at about the same percentile throughout childhood. In children whose growth has been retarded, nutrition rehabilitation will ideally induce height and weight to increase to higher percentiles. In overweight children, the goal is for weight to remain stable as height increases, until weight becomes appropriate for height.

To evaluate growth in infants, an assessor uses charts such as those in Figures E-1 (A and B) through E-3 (A and B) on pp. E-2 to E-4. ◆ The assessor follows these steps to plot a weight measurement on a percentile graph:

- Select the appropriate chart based on age and gender.
- Locate the child's age along the horizontal axis on the bottom of the chart.
- Locate the child's weight in pounds or kilograms along the vertical axis.
- Mark the chart where the age and weight lines intersect, and read off the percentile.

For other measures, the assessor follows a similar procedure, using the appropriate chart. (When length is measured, use the chart for birth to 36 months; when height is measured, use the chart for 2 to 20 years.) Once all of the measures are plotted on growth percentile charts, a skilled clinician can begin to interpret the data. Ideally, the height, weight, and head circumference should be in roughly the same percentile.

Percentile charts divide the measures of a population into 100 equal divisions. Thus half of the population falls above the 50th percentile, and half falls below. The use of percentile measures allows for comparisons among people of the same age and gender. For example, a six-month-old female infant whose weight is at the 75th percentile weighs more than 75 percent of the female infants her age.

Head circumference is generally measured in children under two years of age. Since the brain grows rapidly before birth and during early infancy, extreme and chronic malnutrition during these times can impair brain development, curtailing the number of brain cells and the size of head circumference. Nonnutritional factors, such as certain disorders and genetic variation, can also influence head circumference.

CONTENTS

◆ Reminder: The *body mass index (BMI)* is an index of a person's weight in relation to height, determined by dividing the weight in kilograms by the square of the height in meters:

$$BMI = \frac{Weight\ (kg)}{Height\ (m)^2}$$

◆ Chapter 15 presents BMI charts for children and adolescents.

Appendix
E

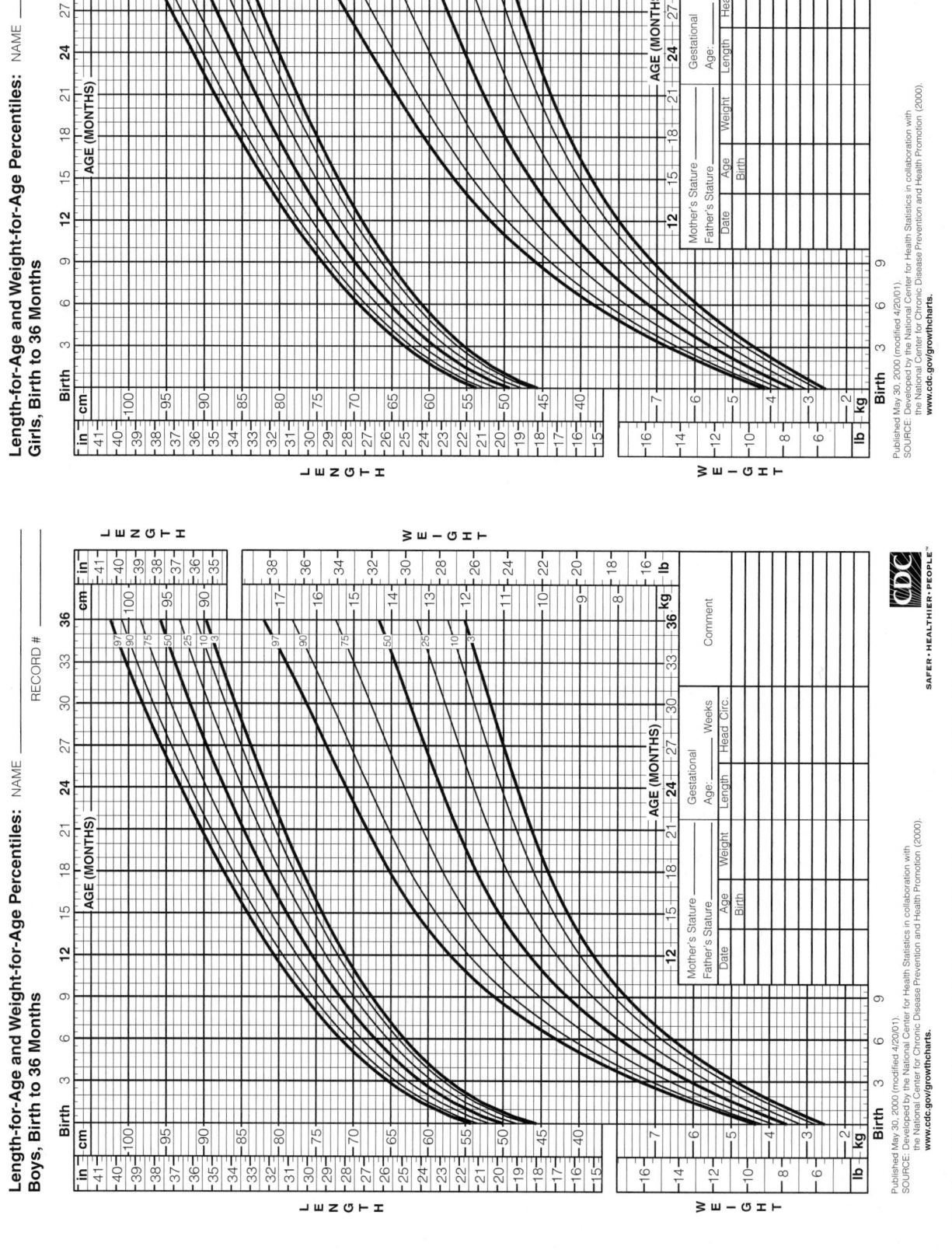

FIGURE E-1A Length-for-Age and Weight-for-Age Percentiles: Boys, Birth to 36 Months

FIGURE E-1B Length-for-Age and Weight-for-Age Percentiles: Girls, Birth to 36 Months

FIGURE E-2A Head Circumference-for-Age and Weight-for-Length Percentiles: Boys, Birth to 36 Months

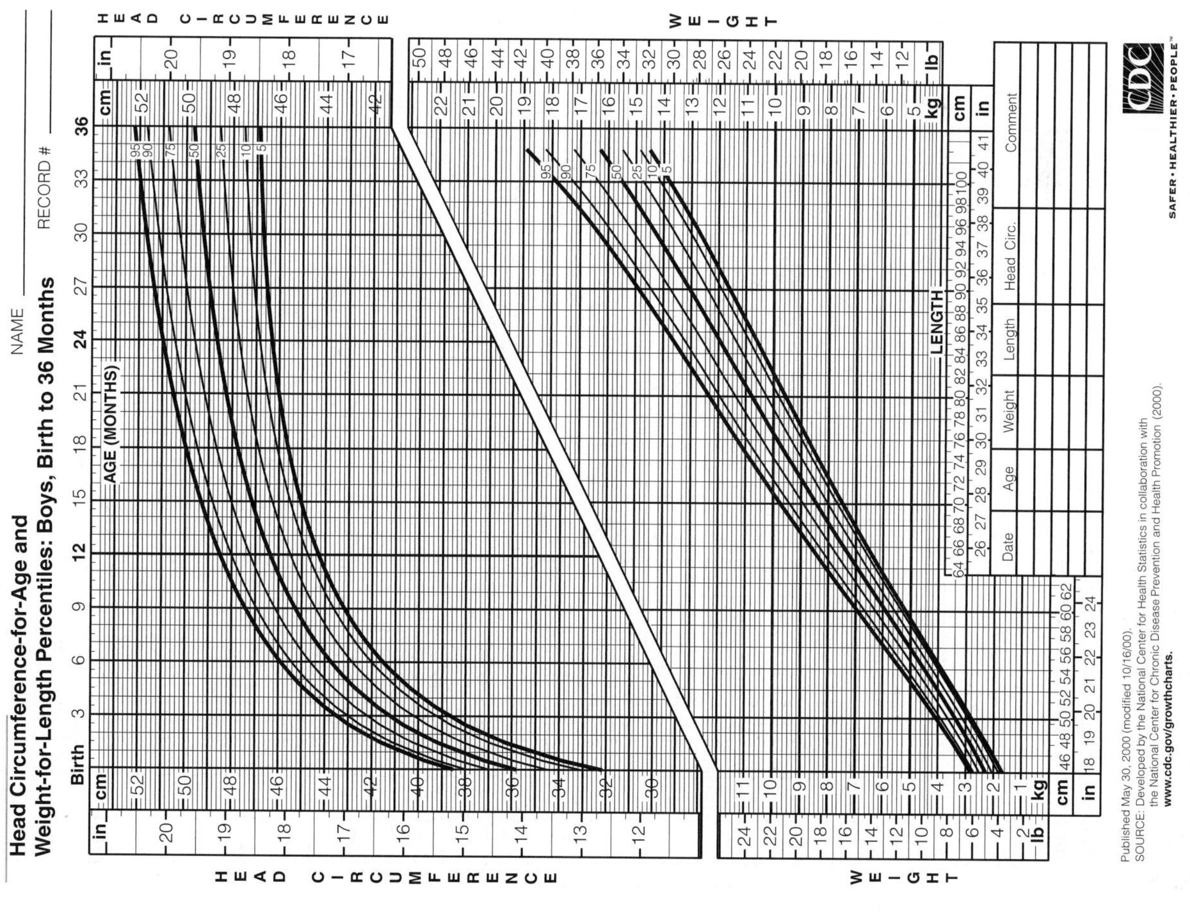
Head Circumference-for-Age and
Weight-for-Length Percentiles: Boys, Birth to 36 Months
NAME
RECORD #
AGE (MONTHS)
LENGTH
Date
Age
Weight
Length
Head Circ.
Comment
Published May 30, 2000 (modified 10/16/00).
SOURCE: Developed by the National Center for Health Statistics in collaboration with
the National Center for Chronic Disease Prevention and Health Promotion (2000).
www.cdc.gov/growthcharts.
CDC
SAFER • HEALTHIER • PEOPLE™

FIGURE E-2B Head Circumference-for-Age and Weight-for-Length Percentiles: Girls, Birth to 36 Months

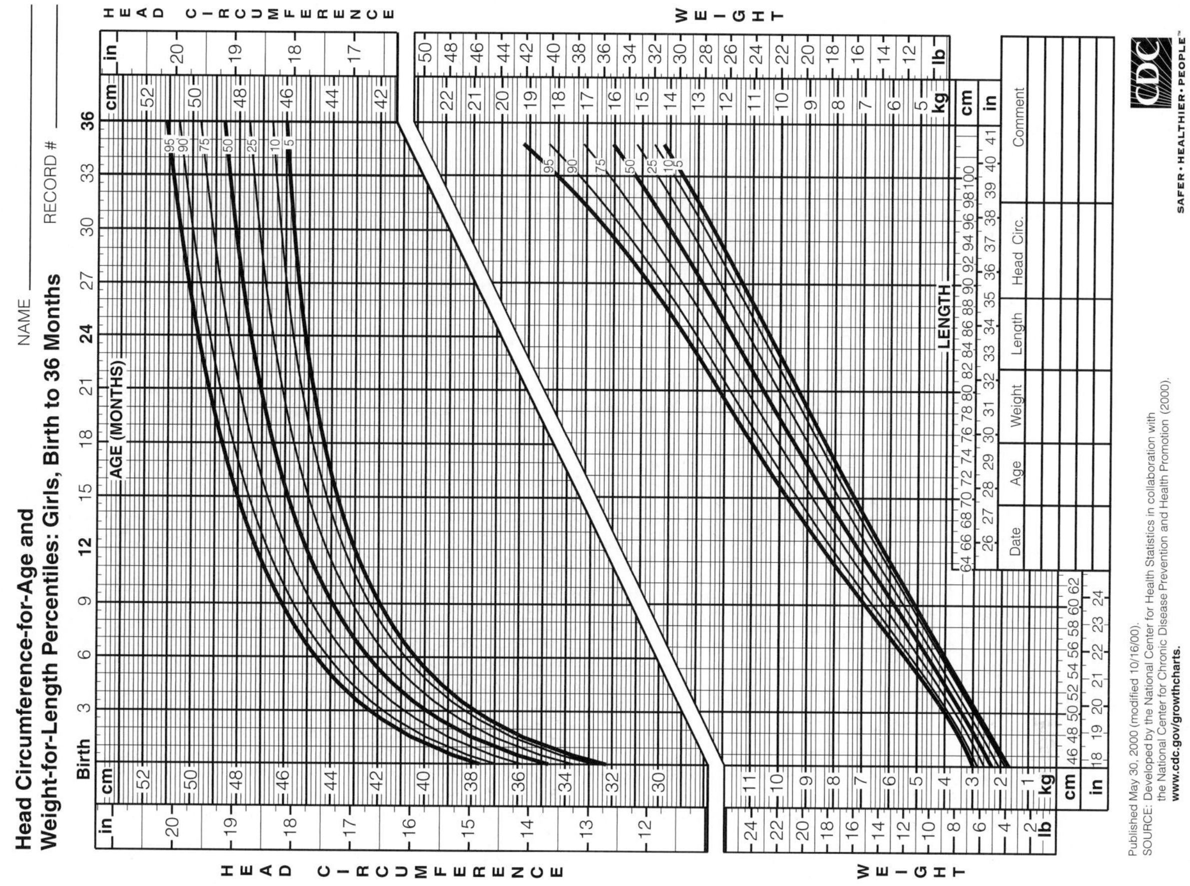
Head Circumference-for-Age and
Weight-for-Length Percentiles: Girls, Birth to 36 Months
NAME
RECORD #
AGE (MONTHS)
LENGTH
Date
Age
Weight
Length
Head Circ.
Comment
Published May 30, 2000 (modified 10/16/00).
SOURCE: Developed by the National Center for Health Statistics in collaboration with
the National Center for Chronic Disease Prevention and Health Promotion (2000).
www.cdc.gov/growthcharts.
CDC
SAFER • HEALTHIER • PEOPLE™

FIGURE E-3A Stature-for-Age and Weight-for-Age Percentiles: Boys, 2 to 20 Years

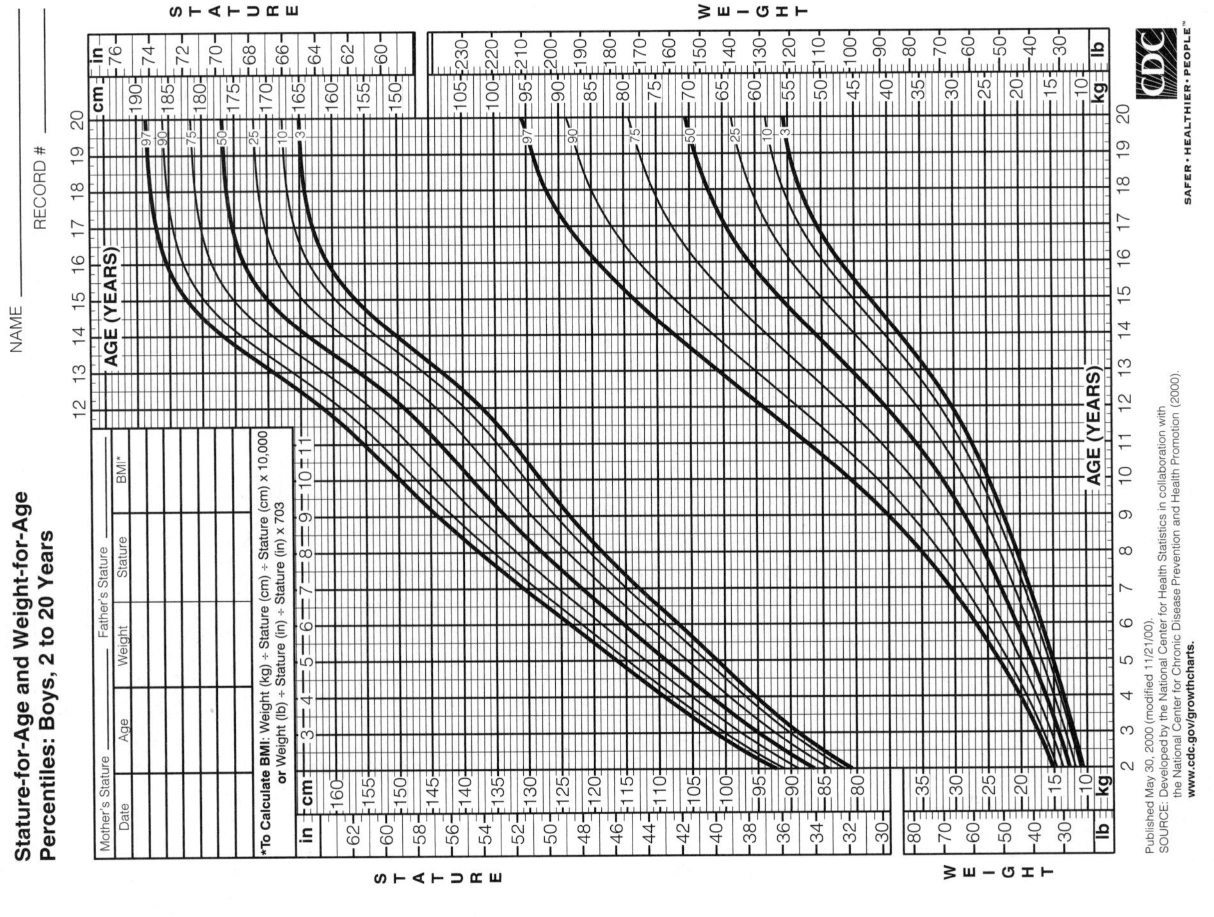
Stature-for-Age and Weight-for-Age Percentiles: Boys, 2 to 20 Years
NAME
RECORD #
Mother's Stature
Father's Stature
Date
Age
Weight
Stature
BMI*
*To Calculate BMI: Weight (kg) ÷ Stature (cm) ÷ Stature (cm) x 10,000
or Weight (lb) ÷ Stature (in) ÷ Stature (in) x 703
AGE (YEARS)
STATURE
WEIGHT
cm
in
kg
lb
Published May 30, 2000 (modified 11/21/00).
SOURCE: Developed by the National Center for Health Statistics in collaboration with
the National Center for Chronic Disease Prevention and Health Promotion (2000).
www.cdc.gov/growthcharts.
CDC
SAFER • HEALTHIER • PEOPLE™

FIGURE E-3B Stature-for-Age and Weight-for-Age Percentiles: Girls, 2 to 20 Years

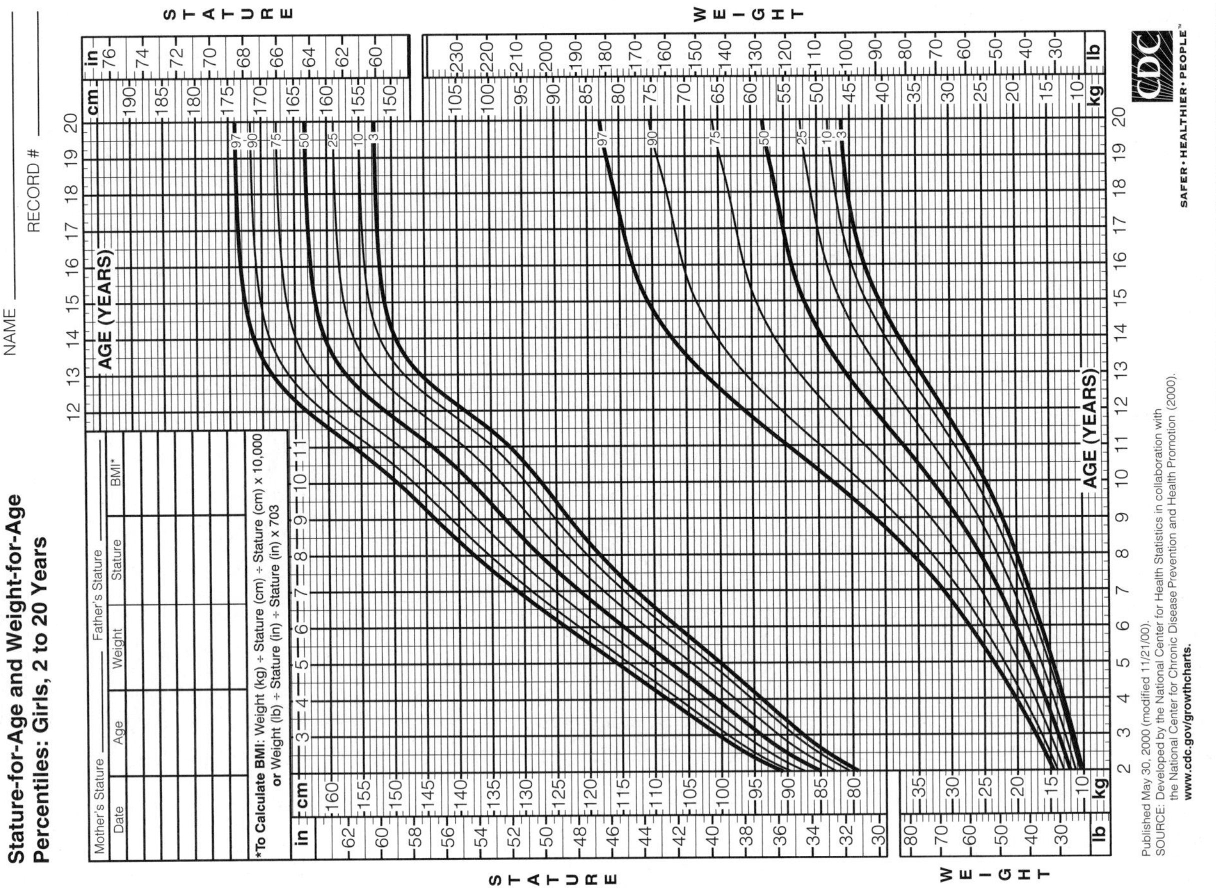
Stature-for-Age and Weight-for-Age Percentiles: Girls, 2 to 20 Years
NAME
RECORD #
Mother's Stature
Father's Stature
Date
Age
Weight
Stature
BMI*
*To Calculate BMI: Weight (kg) ÷ Stature (cm) ÷ Stature (cm) x 10,000
or Weight (lb) ÷ Stature (in) ÷ Stature (in) x 703
AGE (YEARS)
STATURE
WEIGHT
cm
in
kg
lb
Published May 30, 2000 (modified 11/21/00).
SOURCE: Developed by the National Center for Health Statistics in collaboration with
the National Center for Chronic Disease Prevention and Health Promotion (2000).
www.cdc.gov/growthcharts.
CDC
SAFER • HEALTHIER • PEOPLE™

Measures of Body Fat and Lean Tissue

Significant weight changes in both children and adults can reflect overnutrition or undernutrition with respect to energy and protein. To estimate the degree to which fat stores or lean tissues are affected by malnutrition, several anthropometric measurements are useful.

Skinfold Measures Skinfold measures provide a good estimate of total body fat and a fair assessment of the fat's location. Most body fat lies directly beneath the skin, and the thickness of this subcutaneous fat correlates with total body fat. In some parts of the body, such as the back and the back of the arm over the triceps muscle, this fat is loosely attached. ◆ As illustrated in Figure E-4, an assessor can measure the thickness of the fat with calipers that apply a fixed amount of pressure. If a person gains body fat, the skinfold increases proportionately; if the person loses fat, it decreases. Measurements taken from central-body sites better reflect changes in fatness than those taken from upper sites (arm and back). Because subcutaneous fat may be thicker in one area than in another, skinfold measurements are often taken at three or four different places on the body (including upper-, central-, and lower-body sites); the sum of these measures is then compared to standard values. In some situations, the triceps skinfold measurement alone may be used because it is easily accessible. Triceps skinfold measures greater than 15 millimeters in men or 25 millimeters in women suggest excessive body fat.

Waist Circumference Chapter 8 described how fat distribution correlates with health risks and mentioned that the waist circumference is a valuable indicator of

◆ Common sites for skinfold measures:
- Triceps
- Biceps
- Subscapular (below shoulder blade)
- Suprailiac (above hip bone)
- Abdomen
- Upper thigh

FIGURE E-4 How to Measure the Triceps Skinfold

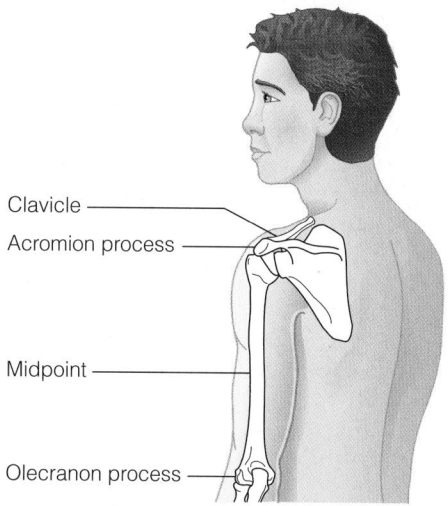
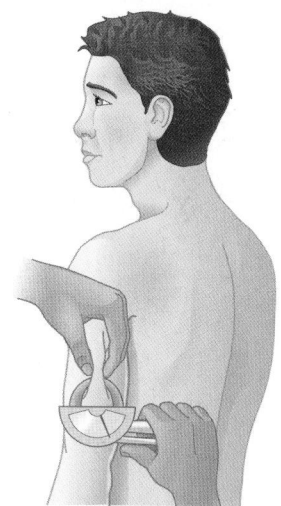

Clavicle
Acromion process
Midpoint
Olecranon process

A. Find the midpoint of the arm:
1. Ask the subject to bend his or her arm at the elbow and lay the hand across the stomach. (If he or she is right-handed, measure the left arm, and vice versa.)
2. Feel the shoulder to locate the acromion process. It helps to slide your fingers along the clavicle to find the acromion process. The olecranon process is the tip of the elbow.
3. Place a measuring tape from the acromion process to the tip of the elbow. Divide this measurement by 2, and mark the midpoint of the arm with a pen.

B. Measure the skinfold:
1. Ask the subject to let his or her arm hang loosely to the side.
2. Grasp a fold of skin and subcutaneous fat between the thumb and forefinger slightly above the midpoint mark. Gently pull the skin away from the underlying muscle. (This step takes a lot of practice. If you want to be sure you don't have muscle as well as fat, ask the subject to contract and relax the muscle. You should be able to feel if you are pinching muscle.)

3. Place the calipers over the skinfold at the midpoint mark, and read the measurement to the nearest 1.0 millimeter in two to three seconds. (If using plastic calipers, align pressure lines, and read the measurement to the nearest 1.0 millimeter in two to three seconds.)
4. Repeat steps 2 and 3 twice more. Add the three readings, and then divide by 3 to find the average.

abdominal fat. To measure waist circumference, the assessor places a nonstretchable tape around the person's body, crossing just above the upper hip bones and making sure that the tape remains on a level horizontal plane on all sides (see Figure E-5). The tape is tightened slightly, but without compressing the skin.

Waist-to-Hip Ratio The waist-to-hip ratio assesses abdominal obesity, but it offers no advantage over the waist circumference alone. To calculate the waist-to-hip ratio, divide the waistline measurement by the hip measurement. ◆ In general, women with a waist-to-hip ratio of 0.8 or greater and men with a waist-to-hip ratio of 0.9 or greater have an increased risk of developing diabetes and cardiovascular diseases.

Hydrodensitometry To estimate body density using hydrodensitometry, the person is weighed twice—first on land and then again when submerged in water. Underwater weighing usually generates a good estimate of body fat and is useful in research, although the technique has drawbacks: it requires bulky, expensive, and nonportable equipment. Furthermore, submerging some people in water (especially those who are very young, very old, ill, or fearful) is difficult and not well tolerated.

Bioelectrical Impedance To measure body fat using the bioelectrical impedance technique, a very-low-intensity electrical current is briefly sent through the body by way of electrodes placed on the wrist and ankle. Fat impedes the flow of electricity; thus the magnitude of the current is influenced by the body fat content. Recent food intake and hydration status can influence results. As with other anthropometric techniques, bioelectrical impedance requires standardized procedures and calibrated instruments.

◆ The calculation of waist-to-hip ratio in a woman with a 28-inch waist and 38-inch hips is 28 ÷ 38 = 0.74.

| FIGURE E-5 | How to Measure Waist Circumference |

Place the measuring tape around the waist just above the bony crest of the hip. The tape runs parallel to the floor and is snug (but does not compress the skin). The measurement is taken at the end of normal expiration.

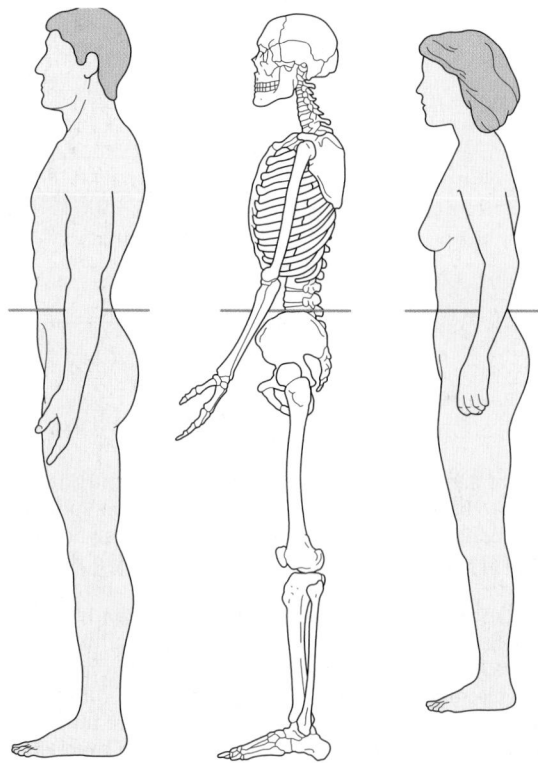

SOURCE: National Institutes of Health Obesity Education Initiative, *Clinical Guidelines on the Identification, Evaluation, and Treatment of Overweight and Obesity in Adults* (Washington, D.C.: U.S. Department of Health and Human Services, 1998), p. 59.

TABLE E-1   Methods of Estimating Body Fat Content and Distribution

Method	Cost	Ease of Use	Accuracy	Measures Fat Distribution
Height and weight	Low	Easy	High	No
Skinfolds	Low	Easy	Low	Yes
Circumferences	Low	Easy	Moderate	Yes
Ultrasound	Moderate	Moderate	Moderate	Yes
Hydrodensitometry	Low	Moderate	High	No
Heavy water tritiated	Moderate	Moderate	High	No
Deuterium oxide, or heavy oxygen	High	Moderate	High	No
Potassium isotope (^{40}K)	Very high	Difficult	High	No
Total body electrical conductivity (TOBEC)	High	Moderate	High	No
Bioelectrical impedance (BIA)	Moderate	Easy	High	No
Dual energy X-ray absorptiometry (DEXA)	High	Easy	High	No
Computed tomography (CT)	Very high	Difficult	High	Yes
Magnetic resonance imaging (MRI)	Very high	Difficult	High	Yes

SOURCE: Adapted with permisssion from G. A. Bray, a handout presented at the North American Association for the Study of Obesity and Emory University School of Medicine Conference on Obesity Update: Pathophysiology, Clinical Consequences, and Therapeutic Options, Atlanta, Georgia, August 31–September 2, 1992.

Clinicians use many other methods to estimate body fat and its distribution. Each has advantages and disadvantages, as Table E-1 summarizes.

Nutritional Anemias

Anemia, a symptom of a wide variety of nutrition- and nonnutrition-related disorders, is characterized by a reduced oxygen-carrying capacity of blood. Iron, folate, and vitamin B_{12} deficiencies—caused by inadequate intake, poor absorption, or abnormal metabolism of these nutrients—are the most common causes of nutritional anemias. Table E-2 on p. E-8 lists laboratory tests that distinguish among the various nutrition-related anemias. Some nonnutrition-related causes of anemia include massive blood loss, infections, hereditary blood disorders such as sickle-cell anemia, and chronic liver or kidney disease.

Assessment of Iron Status

Chapter 13 described the progression of iron deficiency in detail, ◆ as well as the roles of some of the proteins involved in iron metabolism. This section describes the various tests that assess iron status, and Table E-3 on p. E-8 provides acceptable values. Although other tests are more specific for detecting the early stages of iron deficiency, hemoglobin and hematocrit are most often used to detect iron-deficiency anemia because they are inexpensive and easily measured.

Serum Ferritin In the initial stage of iron deficiency, iron stores diminish. Iron is stored in the protein ferritin, which is located in the liver, spleen, and bone marrow. Serum ferritin values provide a noninvasive estimate of iron stores, because the ferritin levels in blood reflect the amounts stored in the tissues. Serum ferritin is not a reliable indicator of iron deficiency, however, because its concentrations are increased by infection, inflammation, alcohol consumption, and liver disease.

Serum Iron and Total Iron-Binding Capacity (TIBC) Early stages of iron deficiency are characterized by reduced levels of serum iron, which represent the amount

◆ Reminder: Iron deficiency progresses as follows:
1. Iron stores diminish
2. Transport iron decreases
3. Hemoglobin production falls

TABLE E-2 Laboratory Tests Useful for Evaluating Nutrition-Related Anemias

Test or Test Result	What It Reflects
For Anemia (general)	
Hemoglobin (Hg)	Total amount of hemoglobin in the red blood cells (RBC)
Hematocrit (Hct)	Percentage of RBC in the total blood volume
Red blood cell (RBC) count	Number of RBC
Mean corpuscular volume (MCV)	RBC size; helps to determine if anemia is microcytic (iron deficiency) or macrocytic (folate or vitamin B_{12} deficiency)
Mean corpuscular hemoglobin concentration (MCHC)	Hemoglobin concentration within the average RBC; helps to determine if anemia is hypochromic (iron deficiency) or normochromic (folate or vitamin B_{12} deficiency)
Bone marrow aspiration	The manufacture of blood cells in different developmental states
For Iron-Deficiency Anemia	
↓ Serum ferritin	Early deficiency state with depleted iron stores
↓ Transferrin saturation	Progressing deficiency state with diminished transport iron
↑ Erythrocyte protoporphyrin	Later deficiency state with limited hemoglobin production
For Folate-Deficiency Anemia	
↓ Serum folate	Progressing deficiency state
↓ RBC folate	Later deficiency state
For Vitamin B_{12}–Deficiency Anemia	
↓ Serum vitamin B_{12}	Progressing deficiency state
Schilling test	Absorption of vitamin B_{12}

TABLE E-3 Criteria for Assessing Iron Status

Laboratory Test	Acceptable Values	Effect of Iron Deficiency
Serum ferritin	Male: 20–250 ng/mL Female: 10–120 ng/mL	Lower than normal
Serum iron	Male: 60–175 µg/dL Female: 50–170 µg/dL	Lower than normal
Total iron-binding capacity	250–450 µg/dL	Higher than normal
Transferrin saturation	Male: 20–50% Female: 15–50%	Lower than normal
Erythrocyte protoporphyrin	< 70 µg/dL red blood cells	Higher than normal
Hemoglobin (Hb)	Male: 13.5–17.5 g/dL Female: 12.0–16.0 g/dL	Lower than normal
Hematocrit (Hct)	Male: 39–49% Female: 35–45%	Lower than normal
Mean corpuscular volume (MCV)	80–100 fL	Lower than normal

NOTE: ng = nanogram, µg = microgram, dL = deciliter, nmol = nanomoles, fL=femtoliter
SOURCES: L. Goldman and D. Ausiello, coeditors, *Cecil Medicine* (Philadelphia: Saunders, 2008), pp. 2983–2991; R. J. Wood and A. G. Ronnenberg, Iron, in M. E. Shils and coeditors, *Modern Nutrition in Health and Disease* (Baltimore: Lippincott Williams & Wilkins, 2006), pp. 248–270.

of iron bound to transferrin, the iron transport protein. Total iron-binding capacity (TIBC) is a measure of the total amount of iron that the transferrin in blood can carry; thus, it is an indirect measure of the transferrin content of blood. During iron deficiency, the liver produces more transferrin in an effort to increase iron transport capacity, and therefore iron depletion is characterized by an increase in TIBC. TIBC reflects liver function as well as changes in iron metabolism.

Transferrin Saturation The percentage of transferrin that is saturated with iron is an indirect measure derived from the serum iron and total iron-binding capacity measures, as follows:

$$\%\text{Transferrin} = \frac{\text{serum iron}}{\text{total iron-binding capacity}} \times 100$$

During iron deficiency, transferrin saturation decreases. The transferrin saturation value is a useful indicator of iron status because it includes information about both the iron and transferrin content of blood.

Erythrocyte Protoporphyrin The iron-containing molecule in hemoglobin is heme, which is formed from iron and protoporphyrin. Protoporphyrin accumulates in the blood when iron supplies are inadequate for the formation of heme. However, levels of protoporphyrin may also be increased when hemoglobin synthesis is impaired for other reasons, such as lead poisoning or inflammation.

Hemoglobin When iron stores become depleted, hemoglobin production is impaired, and symptoms of anemia may eventually develop. Hemoglobin's usefulness in evaluating iron status is limited, however, because hemoglobin concentrations drop fairly late in the development of iron deficiency, and other nutrient deficiencies and medical conditions can also alter hemoglobin concentrations.

Hematocrit The hematocrit is the percentage of the total blood volume occupied by red blood cells. To measure the hematocrit, a clinician spins the blood samples in a centrifuge to separate the red blood cells from the plasma. Low values indicate a reduced number or size of red blood cells. Although this test is not specific for iron status, it can help to detect the presence of iron-deficiency anemia.

Mean Corpuscular Volume (MCV) The hematocrit value divided by the red blood cell count provides a measure of the average size of a red blood cell, referred to as the mean corpuscular volume (MCV). Such a measure helps to classify the type of anemia that is present. In iron deficiency, the red blood cells are smaller than average (microcytic cells).

Assessment of Folate and Vitamin B_{12} Status

Folate deficiency and vitamin B_{12} deficiency present a similar clinical picture—an anemia characterized by abnormally large, misshapen, and immature red blood cells (megaloblastic cells). Distinguishing between folate and vitamin B_{12} deficiency is essential, however, because their treatments differ. Giving folate to a person with vitamin B_{12} deficiency improves many of the test results indicative of vitamin B_{12} deficiency, but this would be a dangerous treatment because vitamin B_{12} deficiency causes nerve damage that folate cannot correct. Thus, inappropriate folate administration masks vitamin B_{12}–deficiency anemia, and nerve damage worsens. For this reason, it is critical to determine whether the anemia results from a folate deficiency or from a vitamin B_{12} deficiency. Several of the following assessment measures help to make this distinction.

Mean Corpuscular Volume (MCV) As previously mentioned, MCV is a measure of red blood cell size. In folate and vitamin B_{12} deficiencies, the red blood cells are larger than average, or macrocytic. Macrocytic cells are not necessarily indicative of nutrient deficiency, however, as they may also result from a high alcohol intake, liver disease, and various medications.

Serum Folate and Vitamin B_{12} Levels Analyses of serum folate and vitamin B_{12} levels are usually among the first tests conducted to determine the cause of macrocytic red blood cells. The presence of low serum levels of either nutrient is consistent with a deficiency of that nutrient, whereas adequate levels can help to rule out deficiency. Folate levels are not a specific measure of folate status, however; they may increase after folate consumption, and decrease due to alcohol consumption, pregnancy, or use of anticonvulsants. The folate levels in red blood cells (called erythrocyte folate) correlate well with folate stores and can help to diagnose folate deficiency, but the more reliable testing methods are not widely available. Table E-4 shows the acceptable ranges for these tests.

Methylmalonic Acid and Homocysteine Levels To determine whether a nutrient deficiency is present, clinicians can measure the levels of substances that accumulate when the functions of that nutrient are impaired. For example, blood levels of the amino acid homocysteine are usually increased by both folate and vitamin B_{12} deficiency, because both nutrients are needed for its metabolism. Methylmalonic acid, a breakdown product of several amino acids, requires vitamin B_{12} for its metabolism; hence, its levels accumulate in serum as a result of vitamin B_{12} deficiency. Because methylmalonic acid levels are not influenced by folate status, this measure is useful in distinguishing between folate and vitamin B_{12} deficiency.

Schilling Test As Chapter 10 explained, vitamin B_{12} deficiency most often results from malabsorption, not poor intake. The Schilling test helps to diagnose malabsorption of vitamin B_{12}. After the patient takes an oral dose of radioactive vitamin B_{12}, a urine test determines whether the vitamin B_{12} was absorbed. More extensive testing can determine the cause of malabsorption.

Cautions about Nutrition Assessment

The tests outlined in this appendix yield information that becomes meaningful only when conducted and interpreted by a skilled clinician. Potential sources of error may be introduced at any step, from the collection of samples to the analysis and reporting of data. Equipment must be regularly calibrated to ensure accuracy of measurements. In addition, the assessor must keep in mind that few tests may be specific to the nutrient of interest alone, and lab results may reflect physiological processes other than the ones being tested. Furthermore, many tests are not sensitive enough to detect the early stages of deficiency; thus follow-up testing is often necessary to identify a nutrition problem.

TABLE E-4	Criteria for Assessing Folate and Vitamin B_{12} Status	
Laboratory Test	**Acceptable Range**	**Effect of Folate or Vitamin B_{12} Deficiency**
Serum folate	3–16 ng/mL	Reduced in folate deficiency
Erythrocyte folate	140–628 ng/mL packed cells	Reduced in folate deficiency
Serum vitamin B_{12}	200–835 pg/mL	Reduced in vitamin B_{12} deficiency
Serum methylmalonic acid	70–270 nmol/L	Increased in vitamin B_{12} deficiency
Serum homocysteine	5–-14 μmol/L	Increased in folate or vitamin B_{12} deficiency

NOTE: ng = nanogram, pg = picogram, nmol = nanomole, μmol = micromoles
SOURCE: L. Goldman and D. Ausiello, coeditors, *Cecil Medicine* (Philadelphia: Saunders, 2008), p. 1238 and pp. 2983–2991.

Physical Activity and Energy Requirements

Chapter 8 described how to calculate estimated energy requirements (EER) for adults by using an equation that accounts for gender, age, weight, height, and physical activity level. Table F-1 presents additional equations to determine the EER for infants, children, adolescents, and pregnant and lactating women.

This appendix helps you determine the correct physical activity (PA) factor to use in the equations, either by calculating the physical activity level or by estimating it. For those who prefer to bypass these steps, the appendix presents tables that provide a shortcut to estimating total energy expenditure.*

Calculating Physical Activity Level

To calculate your physical activity level, record all of your activities for a typical 24-hour day, noting the type of activity, the level of intensity, and the duration. Then, using a copy of Table F-2, find your activity in the first column (or an activity that is reasonably similar) and multiply the number of minutes spent on that activity by the factor in the third column. Put your answer in the last column and total the accumulated values for the day. Now add the subtotal of the last column to 1.1 (to account for basal energy and the thermic effect of food) as shown. This score indicates your physical activity level. Using Table F-3, find the PA factor for your age and gender that correlates with your physical activity level and use it in the energy equations presented in Table F-1.

Estimating Physical Activity Level

As an alternative to recording your activities for a day, you can use the third column of Table F-3 to decide if your daily activity is sedentary, low active, active, or very active. Find the PA factor for your age and gender that correlates with your typical physical activity level and use it in the energy equations presented in Table F-1.

Using a Shortcut to Estimate Total Energy Expenditure

The DRI Committee has developed estimates of total energy expenditure based on the equations for adults presented in Table F-1. These estimates are presented in Table F-4 for women and Table F-5 for men. You can use these tables to estimate your energy requirement—that is, the number of kcalories needed to maintain your current body weight. On the table appropriate for your gender, find your height in meters (or inches) in the left-hand column. Then follow the row across to find your weight in kilograms (or pounds). (If you can't find your exact height and weight, choose a value between the two closest ones.) Look down the column to find the number of kcalories that corresponds to your activity level.

Importantly, the values given in the tables are for 30-year-old people. Women 19 to 29 should add 7 kcalories per day for each year below age 30; older women should subtract 7 kcalories per day for each year above age 30. Similarly, men 19 to 29 should add 10 kcalories per day for each year below age 30; older men should subtract 10 kcalories per day for each year above age 30.

---

*This appendix, including the tables, is adapted from Committee on Dietary Reference Intakes, *Dietary Reference Intakes for Energy, Carbohydrate, Fiber, Fat, Fatty Acids, Cholesterol, Protein, and Amino Acids* (Washington, D.C.: National Academies Press, 2002/2005).

TABLE F-4   Total Energy Expenditure (TEE in kCalories per Day) for Women 30 Years of Age[a] at Various Levels of Activity and Various Heights and Weights

Heights m (in)	Physical Activity Level	Weight[b] kg (lb)					
1.45 (57)		38.9 (86)	45.2 (100)	52.6 (116)	63.1 (139)	73.6 (162)	84.1 (185)
				kCalories			
	Sedentary	1564	1623	1698	1813	1927	2042
	Low active	1734	1800	1912	2043	2174	2304
	Active	1946	2021	2112	2257	2403	2548
	Very active	2201	2287	2387	2553	2719	2886
1.50 (59)		41.6 (92)	48.4 (107)	56.3 (124)	67.5 (149)	78.8 (174)	90.0 (198)
				kCalories			
	Sedentary	1625	1689	1771	1894	2017	2139
	Low active	1803	1874	1996	2136	2276	2415
	Active	2025	2105	2205	2360	2516	2672
	Very active	2291	2382	2493	2671	2849	3027
1.55 (61)		44.4 (98)	51.7 (114)	60.1 (132)	72.1 (159)	84.1 (185)	96.1 (212)
				kCalories			
	Sedentary	1688	1756	1846	1977	2108	2239
	Low active	1873	1949	2081	2230	2380	2529
	Active	2104	2190	2299	2466	2632	2798
	Very active	2382	2480	2601	2791	2981	3171
1.60 (63)		47.4 (104)	55.0 (121)	64.0 (141)	76.8 (169)	89.6 (197)	102.4 (226)
				kCalories			
	Sedentary	1752	1824	1922	2061	2201	2340
	Low active	1944	2025	2168	2327	2486	2645
	Active	2185	2276	2396	2573	2750	2927
	Very active	2474	2578	2712	2914	3116	3318
1.65 (65)		50.4 (111)	58.5 (129)	68.1 (150)	81.7 (180)	95.3 (210)	108.9 (240)
				kCalories			
	Sedentary	1816	1893	1999	2148	2296	2444
	Low active	2016	2102	2556	2425	2594	2763
	Active	2267	2364	2494	2682	2871	3059
	Very active	2567	2678	2824	3039	3254	3469
1.70 (67)		53.5 (118)	62.1 (137)	72.3 (159)	86.7 (191)	101.2 (223)	115.6 (255)
				kCalories			
	Sedentary	1881	1963	2078	2235	2393	2550
	Low active	2090	2180	2345	2525	2705	2884
	Active	2350	2453	2594	2794	2994	3194
	Very active	2662	2780	2938	3166	3395	3623
1.75 (69)		56.7 (125)	65.8 (145)	76.6 (169)	91.9 (202)	107.2 (236)	122.5 (270)
				kCalories			
	Sedentary	1948	2034	2158	2325	2492	2659
	Low active	2164	2260	2437	2627	2817	3007
	Active	2434	2543	2695	2907	3119	3331
	Very active	2758	2883	3054	3296	3538	3780
1.80 (71)		59.9 (132)	69.7 (154)	81.0 (178)	97.2 (214)	113.4 (250)	129.6 (285)
				kCalories			
	Sedentary	2015	2106	2239	2416	2593	2769
	Low active	2239	2341	2529	2731	2932	3133
	Active	2519	2634	2799	3023	3247	3472
	Very active	2855	2987	3172	3428	3684	3940

(continued)

[a] For each year below 30, add 7 kcalories/day to TEE. For each year above 30, subtract 7 kcalories/day from TEE.
[b] These columns represent a BMI of 18.5, 22.5, 25, 30, 35, and 40, respectively.

TABLE F-4 Total Energy Expenditure (TEE in kCalories per Day) for Women 30 Years of Age[a] at Various Levels of Activity and Various Heights and Weights—continued

Heights m (in)	Physical Activity Level	Weight[b] kg (lb)					
1.85 (73)		63.3 (139)	73.6 (162)	85.6 (189)	102.7 (226)	119.8 (264)	136.9 (302)
		kCalories					
	Sedentary	2083	2179	2322	2509	2695	2882
	Low active	2315	2422	2624	2836	3049	3262
	Active	2605	2727	2904	3141	3378	3615
	Very active	2954	3093	3292	3562	3833	4103
1.90 (75)		66.8 (147)	77.6 (171)	90.3 (199)	108.3 (239)	126.4 (278)	144.4 (318)
		kCalories					
	Sedentary	2151	2253	2406	2603	2800	2996
	Low active	2392	2505	2720	2944	3168	3393
	Active	2693	2821	3011	3261	3511	3760
	Very active	3053	3200	3414	3699	3984	4270
1.95 (77)		70.3 (155)	81.8 (180)	95.1 (209)	114.1 (251)	133.1 (293)	152.1 (335)
		kCalories					
	Sedentary	2221	2328	2492	2699	2906	3113
	Low active	2470	2589	2817	3053	3290	3526
	Active	2781	2917	3119	3383	3646	3909
	Very active	3154	3309	3538	3838	4139	4439

[a]For each year below 30, add 7 kcalories/day to TEE. For each year above 30, subtract 7 kcalories/day from TEE.
[b]These columns represent a BMI of 18.5, 22.5, 25, 30, 35, and 40, respectively.

TABLE F-5 Total Energy Expenditure (TEE in kCalories per Day) for Men 30 Years of Age[a] at Various Levels of Activity and Various Heights and Weights

Heights m (in)	Physical Activity Level	Weight[b] kg (lb)					
1.45 (57)		38.9 (86)	47.3 (100)	52.6 (116)	63.1 (139)	73.6 (163)	84.1 (185)
		kCalories					
	Sedentary	1777	1911	2048	2198	2347	2496
	Low active	1931	2080	2225	2393	2560	2727
	Active	2127	2295	2447	2636	2826	3015
	Very active	2450	2648	2845	3075	3305	3535
1.50 (59)		41.6 (92)	50.6 (107)	56.3 (124)	67.5 (149)	78.8 (174)	90.0 (198)
		kCalories					
	Sedentary	1848	1991	2126	2286	2445	2605
	Low active	2009	2168	2312	2491	2670	2849
	Active	2215	2394	2545	2748	2951	3154
	Very active	2554	2766	2965	3211	3457	3703
1.55 (61)		44.4 (98)	54.1 (114)	60.1 (132)	72.1 (159)	84.1 (185)	96.1 (212)
		kCalories					
	Sedentary	1919	2072	2205	2376	2546	2717
	Low active	2089	2259	2401	2592	2783	2974
	Active	2305	2496	2646	2862	3079	3296
	Very active	2660	2887	3087	3349	3612	3875

(continued)

[a]For each year below 30, add 10 kcalories/day to TEE. For each year above 30, subtract 10 kcalories/day from TEE.
[b]These columns represent a BMI of 18.5, 22.5, 25, 30, 35, and 40, respectively.

| TABLE F-5 | Total Energy Expenditure (TEE in kCalories per Day) for Men 30 Years of Age[a] at Various Levels of Activity and Various Heights and Weights—continued |

Heights m (in)	Physical Activity Level	Weight[b] kg (lb)					
1.60 (63)		47.4 (104)	57.6 (121)	64.0 (141)	76.8 (169)	89.6 (197)	102.4 (226)
		kCalories					
	Sedentary	1993	2156	2286	2468	2650	2831
	Low active	2171	2351	2492	2695	2899	3102
	Active	2397	2601	2749	2980	3210	3441
	Very active	2769	3010	3211	3491	3771	4051
1.65 (65)		50.4 (111)	61.3 (129)	68.1 (150)	81.7 (180)	95.3 (210)	108.9 (240)
		kCalories					
	Sedentary	2068	2241	2369	2562	2756	2949
	Low active	2254	2446	2585	2801	3017	3234
	Active	2490	2707	2854	3099	3345	3590
	Very active	2880	3136	3339	3637	3934	4232
1.70 (67)		53.5 (118)	65.0 (137)	72.3 (159)	86.7 (191)	101.2 (223)	115.6 (255)
		kCalories					
	Sedentary	2144	2328	2454	2659	2864	3069
	Low active	2338	2542	2679	2909	3139	3369
	Active	2586	2816	2961	3222	3483	3743
	Very active	2992	3265	3469	3785	4101	4417
1.75 (69)		56.7 (125)	68.9 (145)	76.6 (169)	91.9 (202)	107.2 (236)	122.5 (270)
		kCalories					
	Sedentary	2222	2416	2540	2757	2975	3192
	Low active	2425	2641	2776	3020	3263	3507
	Active	2683	2927	3071	3347	3623	3900
	Very active	3108	3396	3602	3937	4272	4607
1.80 (71)		59.9 (132)	72.9 (154)	81.0 (178)	97.2 (214)	113.4 (250)	129.6 (285)
		kCalories					
	Sedentary	2301	2507	2628	2858	3088	3318
	Low active	2513	2741	2875	3132	3390	3648
	Active	2782	3040	3183	3475	3767	4060
	Very active	3225	3530	3738	4092	4447	4801
1.85 (73)		63.3 (139)	77.0 (162)	85.6 (189)	102.7 (226)	119.8 (264)	136.9 (302)
		kCalories					
	Sedentary	2382	2599	2718	2961	3204	3447
	Low active	2602	2844	2976	3248	3520	3792
	Active	2883	3155	3297	3606	3915	4223
	Very active	3344	3667	3877	4251	4625	4999
1.90 (75)		66.8 (147)	81.2 (171)	90.3 (199)	108.3 (239)	126.4 (278)	144.4 (318)
		kCalories					
	Sedentary	2464	2693	2810	3066	3322	3579
	Low active	2693	2948	3078	3365	3652	3939
	Active	2986	3273	3414	3739	4065	4390
	Very active	3466	3806	4018	4413	4807	5202
1.95 (77)		70.3 (155)	85.6 (180)	95.1 (209)	114.1 (251)	133.1 (293)	152.1 (335)
		kCalories					
	Sedentary	2547	2789	2903	3173	3443	3713
	Low active	2786	3055	3183	3485	3788	4090
	Active	3090	3393	3533	3875	4218	4561
	Very active	3590	3948	4162	4578	4993	5409

[a] For each year below 30, add 10 kcalories/day to TEE. For each year above 30, subtract 10 kcalories/day from TEE.
[b] These columns represent a BMI of 18.5, 22.5, 25, 30, 35, and 40, respectively.

Exchange Lists for Diabetes

Chapter 2 introduced the exchange system, and this appendix provides details from the *2008 Choose Your Foods: Exchange Lists for Diabetes.* Appendix I presents Canada's meal-planning system.

Exchange lists can help people with diabetes to manage their blood glucose levels by controlling the amount and kinds of carbohydrates they consume. These lists can also help in planning diets for weight management by controlling kcalorie and fat intake.

The Exchange System

The exchange system sorts foods into groups by their proportions of carbohydrate, fat, and protein (Table G-1 on p. G-2). These groups may be organized into several exchange lists of foods (Tables G-2 through G-12 on pp. G-3–G-16). For example, the carbohydrate group includes these exchange lists:

- Starch
- Fruits
- Milk (fat-free, reduced-fat, and whole)
- Sweets, Desserts, and Other Carbohydrates
- Nonstarchy Vegetables

Then any food on a list can be "exchanged" for any other on that same list. Another group for alcohol has been included as a reminder that these beverages often deliver substantial carbohydrate and kcalories, and therefore warrant their own list.

Serving Sizes

The serving sizes have been carefully adjusted and defined so that a serving of any food on a given list provides roughly the same amount of carbohydrate, fat, and protein, and, therefore, total energy. Any food on a list can thus be exchanged, or traded, for any other food on the same list without significantly affecting the diet's energy-nutrient balance or total kcalories. For example, a person may select 17 small grapes or $^1/_2$ large grapefruit as one fruit exchange, and either choice would provide roughly 15 grams of carbohydrate and 60 kcalories. A whole grapefruit, however, would count as 2 fruit exchanges.

To apply the system successfully, users must become familiar with the specified serving sizes. A convenient way to remember the serving sizes and energy values is to keep in mind a typical item from each list (review Table G-1).

The Foods on the Lists

Foods do not always appear on the exchange list where you might first expect to find them. They are grouped according to their energy-nutrient contents rather than by their source (such as milks), their outward appearance, or their vitamin and mineral contents. For example, cheeses are grouped with meats (not milk) because, like meats, cheeses contribute energy from protein and fat but provide negligible carbohydrate.

For similar reasons, starchy vegetables such as corn, green peas, and potatoes are found on the Starch list with breads and cereals, not with the vegetables. Likewise, bacon is grouped with the fats and oils, not with the meats.

Diet planners learn to view mixtures of foods, such as casseroles and soups, as combinations of foods from different exchange lists. They also learn to interpret food labels with the exchange system in mind.

G
Appendix

Controlling Energy, Fat, and Sodium

The exchange lists help people control their energy intakes by paying close attention to serving sizes. People wanting to lose weight can limit foods from the Sweets, Desserts, and Other Carbohydrates and Fats lists, and they might choose to avoid the Alcohol list altogether. The Free Foods list provide low-kcalorie choices.

By assigning items like bacon to the Fats list, the exchange lists alert consumers to foods that are unexpectedly high in fat. Even the Starch list specifies which grain products contain added fat (such as biscuits, cornbread, and waffles) by marking them with a symbol to indicate added fat (the symbols are explained in the table keys). In addition, the exchange lists encourage users to think of fat-free milk as milk and of whole milk as milk with added fat, and to think of lean meats as meats and of medium-fat and high-fat meats as meats with added fat. To that end, foods on the milk and meat lists are separated into categories based on their fat contents (review Table G-1). The Milk list is subdivided for fat-free, reduced-fat, and whole; the meat list is subdivided for lean, medium-fat, and high-fat. The meat list also includes plant-based proteins, which tend to be rich in fiber. Notice that many of these foods (p. G-11) bear the symbol for "high fiber."

People wanting to control the sodium in their diets can begin by eliminating any foods bearing the "high sodium" symbol. In most cases, the symbol identifies foods that, in one serving, provide 480 milligrams or more of sodium. Foods on the "Combination Foods" or "Fast Foods" lists that bear the symbol provide more than 600 milligrams of sodium. Other foods may also contribute substantially to sodium (consult Chapter 12 for details).

TABLE G-1 The Food Lists

Lists	Typical Item/Portion Size	Carbohydrate (g)	Protein (g)	Fat (g)	Energy[a] (kcal)
Carbohydrates					
Starch[b]	1 slice bread	15	0–3	0–1	80
Fruits	1 small apple	15	—	—	60
Milk					
Fat-free, low-fat, 1%	1 c fat-free milk	12	8	0–3	100
Reduced-fat, 2%	1 c reduced-fat milk	12	8	5	120
Whole	1 c whole milk	12	8	8	160
Sweets, desserts, and other carbohydrates[c]	2 small cookies	15	varies	varies	varies
Nonstarchy vegetables	½ c cooked carrots	5	2	—	25
Meat and Meat Substitutes					
Lean	1 oz chicken (no skin)	—	7	0–3	45
Medium-fat	1 oz ground beef	—	7	4–7	75
High-fat	1 oz pork sausage	—	7	8+	100
Plant-based proteins	½ c tofu	varies	7	varies	varies
Fats	1 tsp butter	—	—	5	45
Alcohol	12 oz beer	varies	—	—	100

[a]The energy value for each exchange list represents an approximate average for the group and does not reflect the precise number of grams of carbohydrate, protein, and fat. For example, a slice of bread contains 15 grams of carbohydrate (60 kcalories), 3 grams protein (12 kcalories), and a little fat—rounded to 80 kcalories for ease in calculating. A half-cup of vegetables (not including starchy vegetables) contains 5 grams carbohydrate (20 kcalories) and 2 grams protein (8 more), which has been rounded down to 25 kcalories.

[b]The Starch list includes cereals, grains, breads, crackers, snacks, starchy vegetables (such as corn, peas, and potatoes), and legumes (dried beans, peas, and lentils).

[c]The Sweets, Desserts, and Other Carbohydrates list includes foods that contain added sugars and fats such as sodas, candy, cakes, cookies, doughnuts, ice cream, pudding, syrup, and frozen yogurt.

Appendix G

Planning a Healthy Diet

To obtain a daily variety of foods that provide healthful amounts of carbohydrate, protein, and fat, as well as vitamins, minerals, and fiber, the meal plan for adults and teenagers should include at least:

- 2 to 3 servings of nonstarchy vegetables
- 2 servings of fruits
- 6 servings of grains (at least 3 of whole grains), beans, and starchy vegetables
- 2 servings of low-fat or fat-free milk
- about 6 ounces of meat or meat substitutes
- *small* amounts of fat and sugar

The actual amounts are determined by age, gender, activity levels, and other factors that influence energy needs. Refer to Chapter 8 as you read through these sections to get an idea of how exchange lists can be useful in planning a diet.

TABLE G-2	Starch

The Starch list includes bread, cereals and grains, starchy vegetables, crackers and snacks, and legumes (dried beans, peas, and lentils).
1 starch choice = 15 g carbohydrate, 0–3 g protein, 0–1 g fat, and 80 kcal.
Note: In general, one starch exchange is $^1/_2$ c cooked cereal, grain, or starchy vegetable; $^1/_3$ c cooked rice or pasta; 1 oz of bread product; $^3/_4$ oz to 1 oz of most snack foods.

Bread

Food	Serving Size
Bagel, large (about 4 oz)	$^1/_4$ (1 oz)
▽ Biscuit, $2^1/_2$ inches across	1
Bread	
☺ reduced-kcalorie	2 slices ($1^1/_2$ oz)
white, whole-grain, pumpernickel, rye, unfrosted raisin	1 slice (1 oz)
Chapatti, small, 6 inches across	1
▽ Cornbread, $1^3/_4$ inch cube	1 ($1^1/_2$ oz)
English muffin	$^1/_2$
Hot dog bun or hamburger bun	$^1/_2$ (1 oz)
Naan, 8 inches by 2 inches	$^1/_4$
Pancake, 4 inches across, $^1/_4$ inch thick	1
Pita, 6 inches across	$^1/_2$
Roll, plain, small	1 (1 oz)
▽ Stuffing, bread	$^1/_3$ cup
▽ Taco shell, 5 inches across	2
Tortilla, corn, 6 inches across	1
Tortilla, flour, 6 inches across	1
Tortilla, flour, 10 inches across	$^1/_3$
▽ Waffle, 4-inch square or 4 inches across	1

Cereals and Grains

Food	Serving Size
Barley, cooked	$^1/_3$ cup
Bran, dry	
☺ oat	$^1/_4$ cup
☺ wheat	$^1/_2$ cup
☺ Bulgur (cooked)	$^1/_2$ cup
Cereals	
☺ bran	$^1/_2$ cup
cooked (oats, oatmeal)	$^1/_2$ cup
puffed	$1^1/_2$ cups
shredded wheat, plain	$^1/_2$ cup
sugar-coated	$^1/_2$ cup
unsweetened, ready-to-eat	$^3/_4$ cup
Couscous	$^1/_3$ cup
Granola	
low-fat	$^1/_4$ cup
▽ regular	$^1/_4$ cup
Grits, cooked	$^1/_2$ cup
Kasha	$^1/_2$ cup
Millet, cooked	$^1/_3$ cup
Muesli	$^1/_4$ cup
Pasta, cooked	$^1/_3$ cup

KEY

☺ = More than 3 grams of dietary fiber per serving.

▽ = Extra fat, or prepared with added fat. (Count as 1 starch + 1 fat.)

▯ = 480 milligrams or more of sodium per serving.

TABLE G-2   Starch—*continued*

Cereals and Grains

Food	Serving Size
Polenta, cooked	$1/3$ cup
Quinoa, cooked	$1/3$ cup
Rice, white or brown, cooked	$1/3$ cup
Tabbouleh (tabouli), prepared	$1/2$ cup
Wheat germ, dry	3 Tbsp
Wild rice, cooked	$1/2$ cup

Starchy Vegetables

Food	Serving Size
Cassava	$1/3$ cup
Corn	$1/2$ cup
on cob, large	$1/2$ cob (5 oz)
☺ Hominy, canned	$3/4$ cup
☺ Mixed vegetables with corn, peas,	
or pasta	1 cup
☺ Parsnips	$1/2$ cup
☺ Peas, green	$1/2$ cup
Plantain, ripe	$1/3$ cup
Potato	
baked with skin	$1/4$ large (3 oz)
boiled, all kinds	$1/2$ cup or $1/2$ medium (3 oz)
▽ mashed, with milk and fat	$1/2$ cup
French fried (oven-baked)[a]	1 cup (2 oz)
☺ Pumpkin, canned, no sugar added	1 cup
Spaghetti/pasta sauce	$1/2$ cup
☺ Squash, winter (acorn, butternut)	1 cup
☺ Succotash	$1/2$ cup
Yam, sweet potato, plain	$1/2$ cup

Crackers and Snacks[b]

Food	Serving Size
Animal crackers	8
Crackers	
▽ round-butter type	6
saltine-type	6
▽ sandwich-style, cheese or peanut	
butter filling	3
▽ whole-wheat regular	2–5 ($3/4$ oz)
☺ whole-wheat lower fat or crispbreads	2–5 ($3/4$ oz)

Crackers and Snacks[b]

Food	Serving Size
Graham cracker, $2^1/2$-inch square	3
Matzoh	$3/4$ oz
Melba toast, about 2-inch by 4-inch piece	4
Oyster crackers	20
Popcorn	3 cups
▽ ☺ with butter	3 cups
☺ no fat added	3 cups
☺ lower fat	3 cups
Pretzels	$3/4$ oz
Rice cakes, 4 inches across	2
Snack chips	
fat-free or baked (tortilla, potato),	
baked pita chips	15–20 ($3/4$ oz)
▽ regular (tortilla, potato)	9–13 ($3/4$ oz)

Beans, Peas, and Lentils[c]

The choices on this list count as 1 starch + 1 lean meat.

Food	Serving Size
☺ Baked beans	$1/3$ cup
☺ Beans, cooked (black, garbanzo, kidney, lima, navy, pinto, white)	$1/2$ cup
☺ Lentils, cooked (brown, green, yellow)	$1/2$ cup
☺ Peas, cooked (black-eyed, split)	$1/2$ cup
🧂 ☺ Refried beans, canned	$1/2$ cup

KEY

☺ = More than 3 grams of dietary fiber per serving.

▽ = Extra fat, or prepared with added fat. (Count as 1 starch + 1 fat.)

🧂 = 480 milligrams or more of sodium per serving.

[a]Restaurant-style French fries are on the Fast Foods list.
[b]For other snacks, see the Sweets, Desserts, and Other Carbohydrates list. For a quick estimate of serving size, an open handful is equal to about 1 cup or 1 to 2 ounces of snack food.
[c]Beans, peas, and lentils are also found on the Meat and Meat Substitutes list.

TABLE G-3 Fruits

Fruit[a]

The Fruits list includes fresh, frozen, canned, and dried fruits and fruit juices. 1 fruit choice = 15 g carbohydrate, 0 g protein, 0 g fat, and 60 kcal.
Note: In general, one fruit exchange is $1/2$ c canned or fresh fruit or unsweetened fruit juice; 1 small fresh fruit (4 oz); 2 Tbsp dried fruit.

Food	Serving Size	Food	Serving Size
Apple, unpeeled, small	1 (4 oz)	Nectarine, small	1 (5 oz)
Apples, dried	4 rings	☺ Orange, small	1 ($6^1/2$ oz)
Applesauce, unsweetened	$1/2$ cup	Papaya	$1/2$ or 1 cup cubed (8 oz)
Apricots		Peaches	
canned	$1/2$ cup	canned	$1/2$ cup
dried	8 halves	fresh, medium	1 (6 oz)
☺ fresh	4 whole ($5^1/2$ oz)	Pears	
Banana, extra small	1 (4 oz)	canned	$1/2$ cup
☺ Blackberries	$3/4$ cup	fresh, large	$1/2$ (4 oz)
Blueberries	$3/4$ cup	Pineapple	
Cantaloupe, small	$1/3$ melon or 1 cup cubed (11 oz)	canned	$1/2$ cup
		fresh	$3/4$ cup
Cherries		Plums	
sweet, canned	$1/2$ cup	canned	$1/2$ cup
sweet fresh	12 (3 oz)	dried (prunes)	3
Dates	3	small	2 (5 oz)
Dried fruits (blueberries, cherries, cranberries, mixed fruit, raisins)	2 Tbsp	☺ Raspberries	1 cup
Figs		☺ Strawberries	$1^1/4$ cup whole berries
dried	$1^1/2$	☺ Tangerines, small	2 (8 oz)
☺ fresh	$1^1/2$ large or 2 medium ($3^1/2$ oz)	Watermelon	1 slice or $1^1/4$ cups cubes ($13^1/2$ oz)
Fruit cocktail	$1/2$ cup		
Grapefruit			
large	$1/2$ (11 oz)		
sections, canned	$3/4$ cup		
Grapes, small	17 (3 oz)		
Honeydew melon	1 slice or 1 cup cubed (10 oz)		
☺ Kiwi	1 ($3^1/2$ oz)		
Mandarin oranges, canned	$3/4$ cup		
Mango, small	$1/2$ ($5^1/2$ oz) or $1/2$ cup		

Fruit Juice

Food	Serving Size
Apple juice/cider	$1/2$ cup
Fruit juice blends, 100% juice	$1/3$ cup
Grape juice	$1/3$ cup
Grapefruit juice	$1/2$ cup
Orange juice	$1/2$ cup
Pineapple juice	$1/2$ cup
Prune juice	$1/3$ cup

KEY

☺ = More than 3 grams of dietary fiber per serving.

▽ = Extra fat, or prepared with added fat.

▤ = 480 milligrams or more of sodium per serving.

[a]The weight listed includes skin, core, seeds, and rind.

TABLE G-4	Milk

The Milk list groups milks and yogurts based on the amount of fat they have (fat-free/low-fat, reduced-fat, and whole). Cheeses are found on the Meat and Meat Substitutes list and cream and other dairy fats are found on the Fats list.

Note: In general, one milk choice is 1 cup (8 fluid ounces or $^1/_2$ pint) milk or yogurt.

Milk and Yogurts

Food	Serving Size
Fat-free or low-fat (1%)	
1 fat-free/low-fat milk choice = 12 g carbohydrate, 8 g protein, 0–3 g fat, and 100 kcal.	
Milk, buttermilk, acidophilus milk, Lactaid	1 cup
Evaporated milk	$^1/_2$ cup
Yogurt, plain or flavored with an artificial sweetener	$^2/_3$ cup (6 oz)
Reduced-fat (2%)	
1 reduced-fat milk choice = 12 g carbohydrate, 8 g protein, 5 g fat, and 120 kcal.	
Milk, acidophilus milk, kefir, Lactaid	1 cup
Yogurt, plain	$^2/_3$ cup (6 oz)
Whole	
1 whole milk choice = 12 g carbohydrate, 8 g protein, 8 g fat, and 160 kcal.	
Milk, buttermilk, goat's milk	1 cup
Evaporated milk	$^1/_2$ cup
Yogurt, plain	8 oz

Dairy-Like Foods

Food	Serving Size	Count as
Chocolate milk		
fat-free	1 cup	1 fat-free milk + 1 carbohydrate
whole	1 cup	1 whole milk + 1 carbohydrate
Eggnog, whole milk	$^1/_2$ cup	1 carbohydrate + 2 fats
Rice drink		
flavored, low-fat	1 cup	2 carbohydrates
plain, fat-free	1 cup	1 carbohydrate
Smoothies, flavored, regular	10 oz	1 fat-free milk + $2^1/_2$ carbohydrates
Soy milk		
light	1 cup	1 carbohydrate + $^1/_2$ fat
regular, plain	1 cup	1 carbohydrate + 1 fat
Yogurt		
and juice blends	1 cup	1 fat-free milk + 1 carbohydrate
low carbohydrate (less than 6 grams carbohydrate per choice)	$^2/_3$ cup (6 oz)	$^1/_2$ fat-free milk
with fruit, low-fat	$^2/_3$ cup (6 oz)	1 fat-free milk + 1 carbohydrate

TABLE G-5 Sweets, Desserts, and Other Carbohydrates

1 other carbohydrate choice = 15 g carbohydrate, variable grams protein, variable grams fat, and variable kcalories.

Note: In general, one choice from this list can substitute for foods on the Starch, Fruits, or Milk lists.

Beverages, Soda, and Energy/Sports Drinks

Food	Serving Size	Count as
Cranberry juice cocktail	$1/2$ cup	1 carbohydrate
Energy drink	1 can (8.3 oz)	2 carbohydrates
Fruit drink or lemonade	1 cup (8 oz)	2 carbohydrates
Hot chocolate		
regular	1 envelope added to 8 oz water	1 carbohydrate + 1 fat
sugar-free or light	1 envelope added to 8 oz water	1 carbohydrate
Soft drink (soda), regular	1 can (12 oz)	$2^1/_2$ carbohydrates
Sports drink	1 cup (8 oz)	1 carbohydrate

Brownies, Cake, Cookies, Gelatin, Pie, and Pudding

Food	Serving Size	Count as
Brownie, small, unfrosted	$1^1/_4$-inch square, $^7/_8$ inch high (about 1 oz)	1 carbohydrate + 1 fat
Cake		
angel food, unfrosted	$1/_{12}$ of cake (about 2 oz)	2 carbohydrates
frosted	2-inch square (about 2 oz)	2 carbohydrates + 1 fat
unfrosted	2-inch square (about 2 oz)	1 carbohydrate + 1 fat
Cookies		
chocolate chip	2 cookies ($2^1/_4$ inches across)	1 carbohydrate + 2 fats
gingersnap	3 cookies	1 carbohydrate
sandwich, with crème filling	2 small (about $^2/_3$ oz)	1 carbohydrate + 1 fat
sugar-free	3 small or 1 large ($^3/_4$–1 oz)	1 carbohydrate + 1–2 fats
vanilla wafer	5 cookies	1 carbohydrate + 1 fat
Cupcake, frosted	1 small (about $1^3/_4$ oz)	2 carbohydrates + 1–$1^1/_2$ fats
Fruit cobbler	$1/2$ cup ($3^1/_2$ oz)	3 carbohydrates + 1 fat
Gelatin, regular	$1/2$ cup	1 carbohydrate
Pie		
commercially prepared fruit, 2 crusts	$1/_6$ of 8-inch pie	3 carbohydrates + 2 fats
pumpkin or custard	$1/_8$ of 8-inch pie	$1^1/_2$ carbohydrates + $1^1/_2$ fats
Pudding		
regular (made with reduced-fat milk)	$1/2$ cup	2 carbohydrates
sugar-free or sugar- and fat-free (made with fat-free milk)	$1/2$ cup	1 carbohydrate

Candy, Spreads, Sweets, Sweeteners, Syrups, and Toppings

Food	Serving Size	Count as
Candy bar, chocolate/peanut	2 "fun size" bars (1 oz)	$1^1/_2$ carbohydrates + $1^1/_2$ fats
Candy, hard	3 pieces	1 carbohydrate
Chocolate "kisses"	5 pieces	1 carbohydrate + 1 fat
Coffee creamer		
dry, flavored	4 tsp	$1/2$ carbohydrate + $1/2$ fat
liquid, flavored	2 Tbsp	1 carbohydrate

(continued)

TABLE G-5	Sweets, Desserts, and Other Carbohydrates—*continued*

Candy, Spreads, Sweets, Sweeteners, Syrups, and Toppings

Food	Serving Size	Count as
Fruit snacks, chewy (pureed fruit concentrate)	1 roll ($^3/_4$ oz)	1 carbohydrate
Fruit spreads, 100% fruit	$1^1/_2$ Tbsp	1 carbohydrate
Honey	1 Tbsp	1 carbohydrate
Jam or jelly, regular	1 Tbsp	1 carbohydrate
Sugar	1 Tbsp	1 carbohydrate
Syrup		
chocolate	2 Tbsp	2 carbohydrates
light (pancake type)	2 Tbsp	1 carbohydrate
regular (pancake type)	1 Tbsp	1 carbohydrate

Condiments and Sauces[a]

Food	Serving Size	Count as
Barbeque sauce	3 Tbsp	1 carbohydrate
Cranberry sauce, jellied	$^1/_4$ cup	$1^1/_2$ carbohydrates
🧂 Gravy, canned or bottled	$^1/_2$ cup	$^1/_2$ carbohydrate + $^1/_2$ fat
Salad dressing, fat-free, low-fat, cream-based	3 Tbsp	1 carbohydrate
Sweet and sour sauce	3 Tbsp	1 carbohydrate

Doughnuts, Muffins, Pastries, and Sweet Breads

Food	Serving Size	Count as
Banana nut bread	1-inch slice (1 oz)	2 carbohydrates + 1 fat
Doughnut		
cake, plain	1 medium ($1^1/_2$ oz)	$1^1/_2$ carbohydrates + 2 fats
yeast type, glazed	$3^3/_4$ inches across (2 oz)	2 carbohydrates + 2 fats
Muffin (4 oz)	$^1/_4$ muffin (1 oz)	1 carbohydrate + $^1/_2$ fat
Sweet roll or Danish	1 ($2^1/_2$ oz)	$2^1/_2$ carbohydrates + 2 fats

Frozen Bars, Frozen Desserts, Frozen Yogurt, and Ice Cream

Food	Serving Size	Count as
Frozen pops	1	$^1/_2$ carbohydrate
Fruit juice bars, frozen, 100% juice	1 bar (3 oz)	1 carbohydrate
Ice cream		
fat-free	$^1/_2$ cup	$1^1/_2$ carbohydrates
light	$^1/_2$ cup	1 carbohydrate + 1 fat
no sugar added	$^1/_2$ cup	1 carbohydrate + 1 fat
regular	$^1/_2$ cup	1 carbohydrate + 2 fats
Sherbet, sorbet	$^1/_2$ cup	2 carbohydrates
Yogurt, frozen		
fat-free	$^1/_3$ cup	1 carbohydrate
regular	$^1/_2$ cup	1 carbohydrate + 0–1 fat

KEY

🧂 = 480 milligrams or more of sodium per serving.

[a] You can also check the Fats list and Free Foods list for other condiments.

TABLE G-5 Sweets, Desserts, and Other Carbohydrates—*continued*

Granola Bars, Meal Replacement Bars/Shakes, and Trail Mix

Food	Serving Size	Count as
Granola or snack bar, regular or low-fat	1 bar (1 oz)	$1^1/_2$ carbohydrates
Meal replacement bar	1 bar ($1^1/_3$ oz)	$1^1/_2$ carbohydrates + 0–1 fat
Meal replacement bar	1 bar (2 oz)	2 carbohydrates + 1 fat
Meal replacement shake, reduced kcalorie	1 can (10–11 oz)	$1^1/_2$ carbohydrates + 0–1 fat
Trail mix		
candy/nut-based	1 oz	1 carbohydrate + 2 fats
dried fruit-based	1 oz	1 carbohydrate + 1 fat

TABLE G-6 Nonstarchy Vegetables

The Nonstarchy Vegetables list includes vegetables that have few grams of carbohydrates or kcalories; starchy vegetables are found on the Starch list. 1 nonstarchy vegetable choice = 5 g carbohydrate, 2 g protein, 0 g fat, and 25 kcal.
Note: In general, one nonstarchy vegetable choice is $^1/_2$ cup cooked vegetables or vegetable juice or 1 cup raw vegetables. Count 3 cups of raw vegetables or $1^1/_2$ cups of cooked vegetables as one carbohydrate choice.

Nonstarchy Vegetables[a]

Amaranth or Chinese spinach
Artichoke
Artichoke hearts
Asparagus
Baby corn
Bamboo shoots
Beans (green, wax, Italian)
Bean sprouts
Beets
🜨 Borscht
Broccoli
☺ Brussels sprouts
Cabbage (green, bok choy, Chinese)
☺ Carrots
Cauliflower
Celery
☺ Chayote
Coleslaw, packaged, no dressing
Cucumber
Eggplant
Gourds (bitter, bottle, luffa, bitter melon)
Green onions or scallions
Greens (collard, kale, mustard, turnip)
Hearts of palm
Jicama

Kohlrabi
Leeks
Mixed vegetables (without corn, peas, or pasta)
Mung bean sprouts
Mushrooms, all kinds, fresh
Okra
Onions
Oriental radish or daikon
Pea pods
☺ Peppers (all varieties)
Radishes
Rutabaga
🜨 Sauerkraut
Soybean sprouts
Spinach
Squash (summer, crookneck, zucchini)
Sugar pea snaps
☺ Swiss chard
Tomato
Tomatoes, canned
🜨 Tomato sauce
🜨 Tomato/vegetable juice
Turnips
Water chestnuts
Yard-long beans

KEY

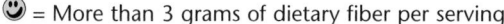

☺ = More than 3 grams of dietary fiber per serving.

🜨 = 480 milligrams or more of sodium per serving.

[a]Salad greens (like chicory, endive, escarole, lettuce, romaine, spinach, arugula, radicchio, watercress) are on the Free Foods list.

TABLE G-7   Meat and Meat Substitutes

The Meat and Meat Substitutes list groups foods based on the amount of fat they have (lean meat, medium-fat meat, high-fat meat, and plant-based proteins).

Lean Meats and Meat Substitutes

1 lean meat choice = 0 g carbohydrate, 7 g protein, 0–3 g fat, and 100 kcal.

Food	Amount
Beef: Select or Choice grades trimmed of fat: ground round, roast (chuck, rib, rump), round, sirloin, steak (cubed, flank, porterhouse, T-bone), tenderloin	1 oz
Beef jerky	1 oz
Cheeses with 3 grams of fat or less per oz	1 oz
Cottage cheese	$^1/_4$ cup
Egg substitutes, plain	$^1/_4$ cup
Egg whites	2
Fish, fresh or frozen, plain: catfish, cod, flounder, haddock, halibut, orange roughy, salmon, tilapia, trout, tuna	1 oz
Fish, smoked: herring or salmon (lox)	1 oz
Game: buffalo, ostrich, rabbit, venison	1 oz
Hot dog with 3 grams of fat or less per oz (8 dogs per 14 oz package) *Note: May be high in carbohydrate.*	1
Lamb: chop, leg, or roast	1 oz
Organ meats: heart, kidney, liver *Note: May be high in cholesterol.*	1 oz
Oysters, fresh or frozen	6 medium
Pork, lean	
Canadian bacon	1 oz
rib or loin chop/roast, ham, tenderloin	1 oz
Poultry, without skin: Cornish hen, chicken, domestic duck or goose (well-drained of fat), turkey	1 oz
Processed sandwich meats with 3 grams of fat or less per oz: chipped beef, deli thin-sliced meats, turkey ham, turkey kielbasa, turkey pastrami	1 oz
Salmon, canned	1 oz
Sardines, canned	2 medium
Sausage with 3 grams of fat or less per oz	1 oz
Shellfish: clams, crab, imitation shellfish, lobster, scallops, shrimp	1 oz
Tuna, canned in water or oil, drained	1 oz
Veal, lean chop, roast	1 oz

Medium-Fat Meat and Meat Substitutes

1 medium-fat meat choice = 0 g carbohydrate, 7 g protein, 4–7 g fat, and 130 kcal.

Food	Amount
Beef: corned beef, ground beef, meatloaf, Prime grades trimmed of fat (prime rib), short ribs, tongue	1 oz
Cheeses with 4–7 grams of fat per oz: feta, mozzarella, pasteurized processed cheese spread, reduced-fat cheeses, string	1 oz
Egg *Note: High in cholesterol, so limit to 3 per week.*	1
Fish, any fried product	1 oz
Lamb: ground, rib roast	1 oz
Pork: cutlet, shoulder roast	1 oz
Poultry: chicken with skin; dove, pheasant, wild duck, or goose; fried chicken; ground turkey	1 oz
Ricotta cheese	2 oz or $^1/_4$ cup
Sausage with 4–7 grams of fat per oz	1 oz
Veal, cutlet (no breading)	1 oz

High-Fat Meat and Meat Substitutes

1 high-fat meat choice = 0 g carbohydrate, 7 g protein, 8+ g fat, and 150 kcal. These foods are high in saturated fat, cholesterol, and kcalories and may raise blood cholesterol levels if eaten on a regular basis. Try to eat 3 or fewer servings from this group per week.

Food	Amount
Bacon	
pork	2 slices (16 slices per lb or 1 oz each, before cooking)
turkey	3 slices ($^1/_2$ oz each before cooking)
Cheese, regular: American, bleu, brie, cheddar, hard goat, Monterey jack, queso, and Swiss	1 oz
Hot dog: beef, pork, or combination (10 per lb-sized package)	1
Hot dog: turkey or chicken (10 per lb-sized package)	1
Pork: ground, sausage, spareribs	1 oz
Processed sandwich meats with 8 grams of fat or more per oz: bologna, pastrami, hard salami	1 oz
Sausage with 8 grams fat or more per oz: bratwurst, chorizo, Italian, knockwurst, Polish, smoked, summer	1 oz

(continued)

TABLE G-7	Meats and Meat Substitutes—*continued*

Plant-Based Proteins

1 plant-based protein choice = variable grams carbohydrate, 7g protein, variable grams fat, and variable kcalories. Because carbohydrate content varies among plant-based proteins, you should read the food label.

Food	Serving Size	Count as
"Bacon" strips, soy-based	3 strips	1 medium-fat meat
☺ Baked beans	$1/3$ cup	1 starch + 1 lean meat
☺ Beans, cooked: black, garbanzo, kidney, lima, navy, pinto, white[a]	$1/2$ cup	1 starch + 1 lean meat
☺ "Beef" or "sausage" crumbles, soy-based	2 oz	$1/2$ carbohydrate + 1 lean meat
"Chicken" nuggets, soy-based	2 nuggets ($1^1/2$ oz)	$1/2$ carbohydrate + 1 medium-fat meat
☺ Edamame	$1/2$ cup	$1/2$ carbohydrate + 1 lean meat
Falafel (spiced chickpea and wheat patties)	3 patties (about 2 inches across)	1 carbohydrate + 1 high-fat meat
Hog dog, soy-based	1 ($1^1/2$ oz)	$1/2$ carbohydrate + 1 lean meat
☺ Hummus	$1/3$ cup	1 carbohydrate + 1 high-fat meat
☺ Lentils, brown, green, or yellow	$1/2$ cup	1 carbohydrate + 1 lean meat
☺ Meatless burger, soy-based	3 oz	$1/2$ carbohydrate + 2 lean meats
☺ Meatless burger, vegetable- and starch-based	1 patty (about $2^1/2$ oz)	1 carbohydrate + 2 lean meats
Nut spreads: almond butter, cashew butter, peanut butter, soy nut butter	1 Tbsp	1 high-fat meat
☺ Peas, cooked: black-eyed and split peas	$1/2$ cup	1 starch + 1 lean meat
🧂 ☺ Refried beans, canned	$1/2$ cup	1 starch + 1 lean meat
"Sausage" patties, soy-based	1 ($1^1/2$ oz)	1 medium-fat meat
Soy nuts, unsalted	$3/4$ oz	$1/2$ carbohydrate + 1 medium-fat meat
Tempeh	$1/4$ cup	1 medium-fat meat
Tofu	4 oz ($1/2$ cup)	1 medium-fat meat
Tofu, light	4 oz ($1/2$ cup)	1 lean meat

KEY

☺ = More than 3 grams of dietary fiber per serving.

▽ = Extra fat, or prepared with added fat. (Add an additional fat choice to this food.)

🧂 = 480 milligrams or more of sodium per serving (based on the sodium content of a typical 3-oz serving of meat, unless 1 or 2 oz is the normal serving size).

[a]Beans, peas, and lentils are also found on the Starch list; nut butters in smaller amounts are found in the Fats list.

TABLE G-8 | Fats

Fats and oils have mixtures of unsaturated (polyunsaturated and monounsaturated) and saturated fats. Foods on the Fats list are grouped together based on the major type of fat they contain. 1 fat choice = 0 g carbohydrate, 0 g protein, 5 g fat, and 45 kcal.

Note: In general, one fat exchange is 1 teaspoon of regular margarine, vegetable oil, or butter; 1 tablespoon of regular salad dressing.

When used in large amounts, bacon and peanut butter are counted as high-fat meat choices (see Meat and Meat Substitutes list). Fat-free salad dressings are found on the Sweets, Desserts, and Other Carbohydrates list. Fat-free products such as margarines, salad dressings, mayonnaise, sour cream, and cream cheese are found on the Free Foods list.

Monounsaturated Fats

Food	Serving Size
Avocado, medium	2 Tbsp (1 oz)
Nut butters (*trans* fat-free): almond butter, cashew butter, peanut butter (smooth or crunchy)	1½ tsp
Nuts	
almonds	6 nuts
Brazil	2 nuts
cashews	6 nuts
filberts (hazelnuts)	5 nuts
macadamia	3 nuts
mixed (50% peanuts)	6 nuts
peanuts	10 nuts
pecans	4 halves
pistachios	16 nuts
Oil: canola, olive, peanut	1 tsp
Olives	
black (ripe)	8 large
green, stuffed	10 large

Polyunsaturated Fats

Food	Serving Size
Margarine: lower-fat spread (30%–50% vegetable oil, *trans* fat-free)	1 Tbsp
Margarine: stick, tub (*trans* fat-free) or squeeze (*trans* fat-free)	1 tsp
Mayonnaise	
reduced-fat	1 Tbsp
regular	1 tsp
Mayonnaise-style salad dressing	
reduced-fat	1 Tbsp
regular	2 tsp
Nuts	
Pignolia (pine nuts)	1 Tbsp
walnuts, English	4 halves
Oil: corn, cottonseed, flaxseed, grape seed, safflower, soybean, sunflower	1 tsp
Oil: made from soybean and canola oil—Enova	1 tsp
Plant stanol esters	
light	1 Tbsp
regular	2 tsp

Polyunsaturated Fats

Food	Serving Size
Salad dressing	
🧂 reduced-fat	2 Tbsp
Note: May be high in carbohydrate.	
🧂 regular	1 Tbsp
Seeds	
flaxseed, whole	1 Tbsp
pumpkin, sunflower	1 Tbsp
sesame seeds	1 Tbsp
Tahini or sesame paste	2 tsp

Saturated Fats

Food	Serving Size
Bacon, cooked, regular or turkey	1 slice
Butter	
reduced-fat	1 Tbsp
stick	1 tsp
whipped	2 tsp
Butter blends made with oil	
reduced-fat or light	1 Tbsp
regular	1½ tsp
Chitterlings, boiled	2 Tbsp (½ oz)
Coconut, sweetened, shredded	2 Tbsp
Coconut milk	
light	⅓ cup
regular	1½ Tbsp
Cream	
half and half	2 Tbsp
heavy	1 Tbsp
light	1½ Tbsp
whipped	2 Tbsp
whipped, pressurized	¼ cup
Cream cheese	
reduced-fat	1½ Tbsp (¾ oz)
regular	1 Tbsp (½ oz)
Lard	1 tsp
Oil: coconut, palm, palm kernel	1 tsp
Salt pork	¼ oz
Shortening, solid	1 tsp
Sour cream	
reduced-fat or light	3 Tbsp
regular	2 Tbsp

KEY

🧂 = 480 milligrams or more of sodium per serving.

TABLE G-9 Free Foods

A "free" food is any food or drink choice that has less than 20 kcalories and 5 grams or less of carbohydrate per serving.

- Most foods on this list should be limited to 3 servings (as listed here) per day. Spread out the servings throughout the day. If you eat all 3 servings at once, it could raise your blood glucose level.
- Food and drink choices listed here without a serving size can be eaten whenever you like.

Low Carbohydrate Foods

Food	Serving Size
Cabbage, raw	1/2 cup
Candy, hard (regular or sugar-free)	1 piece
Carrots, cauliflower, or green beans, cooked	1/4 cup
Cranberries, sweetened with sugar substitute	1/2 cup
Cucumber, sliced	1/2 cup
Gelatin	
dessert, sugar-free	
unflavored	
Gum	
Jam or jelly, light or no sugar added	2 tsp
Rhubarb, sweetened with sugar substitute	1/2 cup
Salad greens	
Sugar substitutes (artificial sweeteners)	
Syrup, sugar-free	2 Tbsp

Modified Fat Foods with Carbohydrate

Food	Serving Size
Cream cheese, fat-free	1 Tbsp (1/2 oz)
Creamers	
nondairy, liquid	1 Tbsp
nondairy, powdered	2 tsp
Margarine spread	
fat-free	1 Tbsp
reduced-fat	1 tsp
Mayonnaise	
fat-free	1 Tbsp
reduced-fat	1 tsp
Mayonnaise-style salad dressing	
fat-free	1 Tbsp
reduced-fat	1 tsp
Salad dressing	
fat-free or low-fat	1 Tbsp
fat-free, Italian	2 Tbsp
Sour cream, fat-free or reduced-fat	1 Tbsp
Whipped topping	
light or fat-free	2 Tbsp
regular	1 Tbsp

Condiments

Food	Serving Size
Barbecue sauce	2 tsp
Catsup (ketchup)	1 Tbsp
Honey mustard	1 Tbsp
Horseradish	

Condiments

Food	Serving Size
Lemon juice	
Miso	1 1/2 tsp
Mustard	
Parmesan cheese, freshly grated	1 Tbsp
Pickle relish	1 Tbsp
Pickles	
dill	1 1/2 medium
sweet, bread and butter	2 slices
sweet, gherkin	3/4 oz
Salsa	1/4 cup
Soy sauce, light or regular	1 Tbsp
Sweet and sour sauce	2 tsp
Sweet chili sauce	2 tsp
Taco sauce	1 Tbsp
Vinegar	
Yogurt, any type	2 Tbsp

Drinks/Mixes

Any food on the list—without a serving size listed—can be consumed in any moderate amount.

- Bouillon, broth, consommé
- Bouillon or broth, low-sodium
- Carbonated or mineral water
- Club soda
- Cocoa powder, unsweetened (1 Tbsp)
- Coffee, unsweetened or with sugar substitute
- Diet soft drinks, sugar-free
- Drink mixes, sugar-free
- Tea, unsweetened or with sugar substitute
- Tonic water, diet
- Water
- Water, flavored, carbohydrate free

Seasonings

Any food on this list can be consumed in any moderate amount.
- Flavoring extracts (for example, vanilla, almond, peppermint)
- Garlic
- Herbs, fresh or dried
- Nonstick cooking spray
- Pimento
- Spices
- Hot pepper sauce
- Wine, used in cooking
- Worcestershire sauce

KEY

= 480 milligrams or more of sodium per serving.

TABLE G-10   Combination Foods

Many foods are eaten in various combinations, such as casseroles. Because "combination" foods do not fit into any one choice list, this list of choices provides some typical combination foods.

Entrees

Food	Serving Size	Count as
🔲 Casserole type (tuna noodle, lasagna, spaghetti with meatballs, chili with beans, macaroni and cheese)	1 cup (8 oz)	2 carbohydrates + 2 medium-fat meats
🔲 Stews (beef/other meats and vegetables)	1 cup (8 oz)	1 carbohydrate + 1 medium-fat meat + 0–3 fats
Tuna salad or chicken salad	$^1/_2$ cup ($3^1/_2$ oz)	$^1/_2$ carbohydrate + 2 lean meats + 1 fat

Frozen Meals/Entrees

Food	Serving Size	Count as
🔲 ☻ Burrito (beef and bean)	1 (5 oz)	3 carbohydrates + 1 lean meat + 2 fats
🔲 Dinner-type meal	generally 14–17 oz	3 carbohydrates + 3 medium-fat meats + 3 fats
🔲 Entrée or meal with less than 340 kcalories	about 8–11 oz	2–3 carbohydrates + 1–2 lean meats
Pizza		
🔲 cheese/vegetarian, thin crust	$^1/_4$ of a 12 inch ($4^1/_2$–5 oz)	2 carbohydrates + 2 medium-fat meats
🔲 meat topping, thin crust	$^1/_4$ of a 12 inch (5 oz)	2 carbohydrates + 2 medium-fat meats + $1^1/_2$ fats
🔲 Pocket sandwich	1 ($4^1/_2$ oz)	3 carbohydrates + 1 lean meat + 1–2 fats
🔲 Pot pie	1 (7 oz)	$2^1/_2$ carbohydrates + 1 medium-fat meat + 3 fats

Salads (Deli-Style)

Food	Serving Size	Count as
Coleslaw	$^1/_2$ cup	1 carbohydrate + $1^1/_2$ fats
Macaroni/pasta salad	$^1/_2$ cup	2 carbohydrates + 3 fats
🔲 Potato salad	$^1/_2$ cup	$1^1/_2$–2 carbohydrates + 1–2 fats

Soups

Food	Serving Size	Count as
🔲 Bean, lentil, or split pea	1 cup	1 carbohydrate + 1 lean meat
🔲 Chowder (made with milk)	1 cup (8 oz)	1 carbohydrate + 1 lean meat + $1^1/_2$ fats
🔲 Cream (made with water)	1 cup (8 oz)	1 carbohydrate + 1 fat
🔲 Instant	6 oz prepared	1 carbohydrate
🔲 with beans or lentils	8 oz prepared	$2^1/_2$ carbohydrates + 1 lean meat
🔲 Miso soup	1 cup	$^1/_2$ carbohydrate + 1 fat
🔲 Oriental noodle	1 cup	2 carbohydrates + 2 fats
Rice (congee)	1 cup	1 carbohydrate
🔲 Tomato (made with water)	1 cup (8 oz)	1 carbohydrate
🔲 Vegetable beef, chicken noodle, or other broth-type	1 cup (8 oz)	1 carbohydrate

KEY

☻ = More than 3 grams of dietary fiber per serving.

▽ = Extra fat, or prepared with added fat.

🔲 = 600 milligrams or more of sodium per serving (for combination food main dishes/meals).

TABLE G-11 Fast Foods

The choices in the Fast Foods list are not specific fast-food meals or items, but are estimates based on popular foods. Ask the restaurant or check its website for nutrition information about your favorite fast foods.

Breakfast Sandwiches

Food	Serving Size	Count as
Egg, cheese, meat, English muffin	1 sandwich	2 carbohydrates + 2 medium-fat meats
Sausage biscuit sandwich	1 sandwich	2 carbohydrates + 2 high-fat meats + $3^1/_2$ fats

Main Dishes/Entrees

Food	Serving Size	Count as
Burrito (beef and beans)	1 (about 8 oz)	3 carbohydrates + 3 medium-fat meats + 3 fats
Chicken breast, breaded and fried	1 (about 5 oz)	1 carbohydrate + 4 medium-fat meats
Chicken drumstick, breaded and fried	1 (about 2 oz)	2 medium-fat meats
Chicken nuggets	6 (about $3^1/_2$ oz)	1 carbohydrate + 2 medium-fat meats + 1 fat
Chicken thigh, breaded and fried	1 (about 4 oz)	$^1/_2$ carbohydrate + 3 medium-fat meats + $1^1/_2$ fats
Chicken wings, hot	6 (5 oz)	5 medium-fat meats + $1^1/_2$ fats

Oriental

Food	Serving Size	Count as
Beef/chicken/shrimp with vegetables in sauce	1 cup (about 5 oz)	1 carbohydrate + 1 lean meat + 1 fat
Egg roll, meat	1 (about 3 oz)	1 carbohydrate + 1 lean meat + 1 fat
Fried rice, meatless	$^1/_2$ cup	$1^1/_2$ carbohydrates + $1^1/_2$ fats
Meat and sweet sauce (orange chicken)	1 cup	3 carbohydrates + 3 medium-fat meats + 2 fats
Noodles and vegetables in sauce (chow mein, lo mein)	1 cup	2 carbohydrates + 1 fat

Pizza

Food	Serving Size	Count as
Pizza		
cheese, pepperoni, regular crust	$^1/_8$ of a 14 inch (about 4 oz)	$2^1/_2$ carbohydrates + 1 medium-fat meat + $1^1/_2$ fats
cheese/vegetarian, thin crust	$^1/_4$ of a 12 inch (about 6 oz)	$2^1/_2$ carbohydrates + 2 medium-fat meats + $1^1/_2$ fats

Sandwiches

Food	Serving Size	Count as
Chicken sandwich, grilled	1	3 carbohydrates + 4 lean meats
Chicken sandwich, crispy	1	$3^1/_2$ carbohydrates + 3 medium-fat meats + 1 fat
Fish sandwich with tartar sauce	1	$2^1/_2$ carbohydrates + 2 medium-fat meats + 2 fats
Hamburger		
large with cheese	1	$2^1/_2$ carbohydrates + 4 medium-fat meats + 1 fat
regular	1	2 carbohydrates + 1 medium-fat meat + 1 fat
Hot dog with bun	1	1 carbohydrate + 1 high-fat meat + 1 fat
Submarine sandwich		
less than 6 grams fat	6-inch sub	3 carbohydrates + 2 lean meats
regular	6-inch sub	$3^1/_2$ carbohydrates + 2 medium-fat meats + 1 fat
Taco, hard or soft shell (meat and cheese)	1 small	1 carbohydrate + 1 medium-fat meat + $1^1/_2$ fats

KEY

☻ = More than 3 grams of dietary fiber per serving.

▽ = Extra fat, or prepared with added fat.

🧂 = 600 milligrams or more of sodium per serving (for fast-food main dishes/meals).

(continued)

TABLE G-11 Fast Foods—continued

Salads

Food	Serving Size	Count as
🧂 😊 Salad, main dish (grilled chicken type, no dressing or croutons)		1 carbohydrate + 4 lean meats
Salad, side, no dressing or cheese	Small (about 5 oz)	1 vegetable

Sides/Appetizers

Food	Serving Size	Count as
🔻 French fries, restaurant style	small	3 carbohydrates + 3 fats
	medium	4 carbohydrates + 4 fats
	large	5 carbohydrates + 6 fats
🧂 Nachos with cheese	small (about 4$1/2$ oz)	2$1/2$ carbohydrates + 4 fats
🧂 Onion rings	1 serving (about 3 oz)	2$1/2$ carbohydrates + 3 fats

Desserts

Food	Serving Size	Count as
Milkshake, any flavor	12 oz	6 carbohydrates + 2 fats
Soft-serve ice cream cone	1 small	2$1/2$ carbohydrates + 1 fat

KEY

😊 = More than 3 grams of dietary fiber per serving.

🔻 = Extra fat, or prepared with added fat.

🧂 = 600 milligrams or more of sodium per serving (for fast-food main dishes/meals).

TABLE G-12 Alcohol

1 alcohol equivalent = variable grams carbohydrate, 0 g protein, 0 g fat, and 100 kcal.

Note: In general, one alcohol choice ($1/2$ oz absolute alcohol) has about 100 kcalories. For those who choose to drink alcohol, guidelines suggest limiting alcohol intake to 1 drink or less per day for women, and 2 drinks or less per day for men. To reduce your risk of low blood glucose (hypoglycemia), especially if you take insulin or a diabetes pill that increases insulin, always drink alcohol with food. While alcohol, by itself, does not directly affect blood glucose, be aware of the carbohydrate (for example, in mixed drinks, beer, and wine) that may raise your blood glucose.

Alcoholic Beverage	Serving Size	Count as
Beer		
light (4.2%)	12 fl oz	1 alcohol equivalent + $1/2$ carbohydrate
regular (4.9%)	12 fl oz	1 alcohol equivalent + 1 carbohydrate
Distilled spirits: vodka, rum, gin, whiskey		
80 or 86 proof	1$1/2$ fl oz	1 alcohol equivalent
Liqueur, coffee (53 proof)	1 fl oz	1 alcohol equivalent + 1 carbohydrate
Sake	1 fl oz	$1/2$ alcohol equivalent
Wine		
dessert (sherry)	3$1/2$ fl oz	1 alcohol equivalent + 1 carbohydrate
dry, red or white (10%)	5 fl oz	1 alcohol equivalent

Table of Food Composition

This edition of the table of food composition includes a wide variety of foods. It is updated with each edition to reflect current nutrient data for foods, to remove outdated foods, and to add foods that are new to the marketplace.* The nutrient database for this appendix is compiled from a variety of sources, including the USDA Standard Release database and manufacturers' data. The USDA database provides data for a wider variety of foods and nutrients than other sources. Because laboratory analysis for each nutrient can be quite costly, manufacturers tend to provide data only for those nutrients mandated on food labels. Consequently, data for their foods are often incomplete; any missing information on this table is designated as a dash. Keep in mind that a dash means only that the information is unknown and should not be interpreted as a zero. A zero means that the nutrient is not present in the food.

Whenever using nutrient data, remember that many factors influence the nutrient contents of foods. These factors include the mineral content of the soil, the diet fed to the animal or the fertilizer used on the plant, the season of harvest, the method of processing, the length and method of storage, the method of cooking, the method of analysis, and the moisture content of the sample analyzed. With so many influencing factors, users should view nutrient data as a close approximation of the actual amount.

For updates, corrections, and a list of more than 8000 foods and codes found in the diet analysis software that accompanies this text, visit www.thomsonedu.com/nutrition and click on Diet Analysis Plus.

- *Fats* Total fats, as well as the breakdown of total fats to saturated, mono-unsaturated, polyunsaturated, and *trans* fats, are listed in the table. The fatty acids seldom add up to the total in part due to rounding but also because values are derived from a variety of laboratories.

- *Trans Fats* *Trans* fat data has been listed in the table. Because food manufacturers have only been required to report *trans* fats on food labels since January 2006, much of the data is incomplete. Missing *trans* fat data is designated with a dash. As additional *trans* fat data becomes available, the table will be updated.

- *Vitamin A and Vitamin E* In keeping with the 2001 RDA for vitamin A, this appendix presents data for vitamin A in micrograms (µg) RAE. Similarly, because the 2000 RDA for vitamin E is based only on the alpha-tocopherol form of vitamin E, this appendix reports vitamin E data in milligrams (mg) alpha-tocopherol, listed on the table as Vit E (mg α).

- *Bioavailability* Keep in mind that the availability of nutrients from foods depends not only on the quantity provided by a food, but also on the amount absorbed and used by the body—the bioavailability. The bioavailability of folate from fortified foods, for example, is greater than from naturally occurring sources. Similarly, the body can make niacin from the amino acid tryptophan, but niacin values in this table (and most databases) report preformed niacin only. Chapter 10 provides conversion factors and additional details.

- *Using the Table* The foods and beverages in this table are organized into several categories, which are listed at the head of each right-hand page. Page numbers are provided, and each group is color-coded to make it easier to find individual foods.

- *Caffeine Sources* Caffeine occurs in several plants, including the familiar coffee bean, the tea leaf, and the cocoa bean from which chocolate is made.

---

*This food composition table has been prepared by Wadsworth Publishing Company. The nutritional data are supplied by Axxya Systems.

Most human societies use caffeine regularly, most often in beverages, for its stimulant effect and flavor. Caffeine contents of beverages vary depending on the plants they are made from, the climates and soils where the plants are grown, the grind or cut size, the method and duration of brewing, and the amounts served. The accompanying table shows that, in general, a cup of coffee contains the most caffeine; a cup of tea, less than half as much; and cocoa or chocolate, less still. As for cola beverages, they are made from kola nuts, which contain caffeine, but most of their caffeine is added, using the purified compound obtained from decaffeinated coffee beans. The FDA lists caffeine as a multipurpose GRAS substance ◆ that may be added to foods and beverages. Drug manufacturers use caffeine in many products.

◆ Reminder: A GRAS substance is one that is "generally recognized as safe."

TABLE   Caffeine Content of Selected Beverages, Foods, and Medications

Beverages and Foods	Serving Size	Average (mg)
Coffee		
Brewed	8 oz	95
Decaffeinated	8 oz	2
Instant	8 oz	64
Tea		
Brewed, green	8 oz	30
Brewed, herbal	8 oz	0
Brewed, leaf or bag	8 oz	47
Instant	8 oz	26
Lipton Brisk iced tea	12 oz	7
Nestea Cool iced tea	12 oz	12
Snapple iced tea (all flavors)	16 oz	42
Soft drinks		
A & W Creme Soda	12 oz	29
Barq's Root Beer	12 oz	18
Coca-Cola	12 oz	30
Dr. Pepper, Mr. Pibb, Sunkist Orange	12 oz	36
A&W Root Beer, club soda, Fresca, ginger ale, 7-Up, Sierra Mist, Sprite, Squirt, tonic water, caffeine-free soft drinks	12 oz	0
Mello Yello	12 oz	51
Mountain Dew	12 oz	45
Pepsi	12 oz	32
Energy drinks		
Amp	8.4 oz	70
Aqua Blast	.5 L	90
Aqua Java	.5 L	55
E Maxx	8.4 oz	74
Java Water	.5 L	125
KMX	8.4 oz	33
Krank	.5 L	100
Red Bull	8.3 oz	67
Red Devil	8.4 oz	42
Sobe Adrenaline Rush	8.3 oz	77
Sobe No Fear	16 oz	141
Water Joe	.5 L	65

Beverages and Foods	Serving Size	Average (mg)
Other beverages		
Chocolate milk or hot cocoa	8 oz	5
Starbucks Frappuccino Mocha	9.5 oz	72
Starbucks Frappuccino Vanilla	9.5 oz	64
Yoohoo chocolate drink	9 oz	3
Candies		
Baker's chocolate	1 oz	26
Dark chocolate covered coffee beans	1 oz	235
Dark chocolate, semisweet	1 oz	18
Milk chocolate	1 oz	6
Milk chocolate covered coffee beans	1 oz	224
White chocolate	1 oz	0
Foods		
Frozen yogurt, Ben & Jerry's coffee fudge	1 cup	85
Frozen yogurt, Häagen-Dazs coffee	1 cup	40
Ice cream, Starbucks coffee	1 cup	50
Ice cream, Starbucks Frappuccino bar	1 bar	15
Yogurt, Dannon coffee flavored	1 cup	45

Drugs[a]	Serving Size	Average (mg)
Cold remedies		
Coryban-D, Dristan	1 tablet	30
Diuretics		
Aqua-Ban	1 tablet	100
Pre-Mens Forte	1 tablet	100
Pain relievers		
Anacin, BC Fast Pain Reliever	1 tablet	32
Excedrin, Midol, Midol Max Strength	1 tablet	65
Stimulants		
Awake, NoDoz	1 tablet	100
Awake Maximum Strength, Caffedrine, NoDoz Maximum Strength, Stay Awake, Vivarin	1 tablet	200
Weight-control aids		
Dexatrim	1 tablet	200

[a] A pharmacologically active dose of caffeine is defined as 200 milligrams.

NOTE: The FDA suggests a maximum of 65 milligrams per 12-ounce cola beverage but does not regulate the caffeine contents of other beverages. Because products change, contact the manufacturer for an update on products you use regularly.

Source: Adapted from USDA database Release 18 (**http://www.nal.usda.gov/fnic/foodcomp/Data/**), Caffeine content of foods and drugs, Center for Science and the Public Interest (**www.cspinet.org/new/cafchart.htm**), and R. R. McCusker, B. A. Goldberger, and E. J. Cone, Caffeine content of energy drinks, carbonated sodas, and other beverages, *Journal of Analytical Toxicology* 30 (2006): 112–114.

H Appendix

TABLE H–1

Food Composition

(DA+ code is for Wadsworth Diet Analysis program)   (For purposes of calculations, use "0" for t, <1, <.1, <.01, etc.)

DA + Code	Food Description	Quantity	Measure	Wt (g)	H₂O (g)	Ener (kcal)	Prot (g)	Carb (g)	Fiber (g)	Fat (g)	Fat Breakdown (g) Sat	Mono	Poly	Trans
	BREADS, BAKED GOODS, CAKES, COOKIES, CRACKERS, CHIPS, PIES													
	Bagels													
8534	Cinnamon & raisin	1	item(s)	71	23	195	7	39	2	1	0.19	0.12	0.48	—
4910	Enriched, all varieties	1	item(s)	71	23	195	7	38	2	1	0.16	0.09	0.49	0
4911	Plain, enriched, toasted	1	item(s)	66	18	195	7	38	2	1	0.16	0.09	0.49	0
8538	Oat bran	1	item(s)	71	23	181	8	38	3	1	0.14	0.18	0.35	—
12079	Whole grain	1	item(s)	85	—	170	9	35	6	2.5	0	—	—	0
	Biscuits													
25008	Biscuits	1	item(s)	41	16	121	3	16	1	5	1.40	1.41	1.82	0
16729	Scone	1	item(s)	42	11	149	4	19	1	6	2.01	2.55	1.26	—
25166	Wheat biscuits	1	item(s)	55	21	162	4	22	1	7	1.90	1.92	2.51	0
	Bread													
325	Boston brown, canned	1	slice(s)	45	21	88	2	19	2	1	0.13	0.09	0.25	—
8716	Bread sticks, plain	4	item(s)	24	1	99	3	16	1	2	0.34	0.86	0.87	—
25176	Cornbread	1	piece(s)	55	26	141	5	18	1	5	2.09	1.44	1.50	0
327	Cracked wheat	1	slice(s)	25	9	65	2	12	1	1	0.23	0.48	0.17	—
9079	Croutons, plain	¼	cup(s)	8	<1	31	1	6	<1	<1	0.11	0.23	0.10	—
8582	Egg	1	slice(s)	40	14	115	4	19	1	2	0.64	0.92	0.44	—
8585	Egg, toasted	1	slice(s)	37	10	117	4	19	1	2	0.60	1.11	0.43	—
329	French	1	slice(s)	25	9	69	2	13	1	1	0.16	0.30	0.17	—
8591	French, toasted	1	slice(s)	23	7	69	2	13	1	1	0.16	0.30	0.17	—
8597	Indian fry	1	item(s)	90	24	296	6	48	2	9	2.08	3.59	2.33	—
332	Italian	1	slice(s)	30	11	81	3	15	1	1	0.26	0.24	0.42	—
1393	Mixed grain	1	slice(s)	26	10	65	3	12	2	1	0.21	0.40	0.24	—
8604	Mixed grain, toasted	1	slice(s)	24	8	65	3	12	2	1	0.21	0.40	0.24	—
8605	Oat bran	1	slice(s)	30	13	71	3	12	1	1	0.21	0.48	0.51	—
8608	Oat bran, toasted	1	slice(s)	27	10	70	3	12	1	1	0.21	0.47	0.50	—
8609	Oatmeal	1	slice(s)	27	10	73	2	13	1	1	0.19	0.43	0.46	—
8613	Oatmeal, toasted	1	slice(s)	25	8	73	2	13	1	1	0.19	0.43	0.46	—
1409	Pita	1	item(s)	60	19	165	5	33	1	1	0.10	0.06	0.32	—
7905	Pita, whole wheat	1	item(s)	64	20	170	6	35	5	2	0.26	0.22	0.68	—
338	Pumpernickel	1	slice(s)	32	12	80	3	15	2	1	0.14	0.30	0.40	—
334	Raisin, enriched	1	slice(s)	26	9	71	2	14	1	1	0.28	0.60	0.18	—
8625	Raisin, toasted	1	slice(s)	24	7	71	2	14	1	1	0.28	0.60	0.18	—
10168	Rice, white	1	slice(s)	42	—	140	1	21	1	6	0.50	—	—	0
8653	Rye	1	slice(s)	32	12	83	3	15	2	1	0.20	0.42	0.26	—
8654	Rye, toasted	1	slice(s)	29	9	82	3	15	2	1	0.20	0.42	0.25	—
336	Rye, light	1	slice(s)	25	9	65	2	12	2	1	0.20	0.30	0.30	—
8588	Sourdough	1	slice(s)	25	9	69	2	13	1	1	0.16	0.30	0.17	—
8592	Sourdough, toasted	1	slice(s)	23	7	69	2	13	1	1	0.16	0.30	0.17	—
491	Submarine or hoagie roll	1	item(s)	135	41	400	11	72	4	8	1.80	3.00	2.20	—
8596	Vienna, toasted	1	slice(s)	23	7	69	2	13	1	1	0.16	0.30	0.17	—
8670	Wheat	1	slice(s)	25	9	65	2	12	1	1	0.22	0.43	0.23	—
8671	Wheat, toasted	1	slice(s)	23	7	65	2	12	1	1	0.22	0.43	0.23	—
340	White	1	slice(s)	25	9	67	2	13	1	1	0.18	0.17	0.34	—
1395	Whole wheat	1	slice(s)	46	15	128	4	24	3	2	0.37	0.53	1.35	—
	Cakes													
386	Angel food, from mix	1	slice(s)	50	16	129	3	29	<1	<1	0.02	0.01	0.06	—
8772	Butter pound, ready to eat, commercially prepared	1	slice(s)	75	18	291	4	37	<1	15	8.67	4.43	0.80	—
8737	Carrot, cream cheese frosting, from mix	1	slice(s)	111	23	484	5	52	1	29	5.43	7.24	15.10	—
4931	Chocolate, chocolate icing, commercially prepared	1	slice(s)	64	15	235	3	35	2	10	3.05	5.61	1.18	—
8756	Chocolate, from mix	1	slice(s)	95	23	340	5	51	2	14	5.16	5.74	2.62	—
393	Devil's food cupcake, chocolate frosting	1	item(s)	35	8	120	2	20	1	4	1.80	1.60	0.60	—
8757	Fruitcake, ready to eat, commercially prepared	1	piece(s)	43	11	139	1	26	2	4	0.45	1.81	1.43	—
1397	Pineapple upside down, from mix	1	slice(s)	115	37	367	4	58	1	14	3.35	5.97	3.77	—
411	Sponge, from mix	1	slice(s)	63	19	187	5	36	<1	3	0.82	0.99	0.41	—
8817	White, coconut frosting, from mix	1	slice(s)	112	23	399	5	71	1	12	4.36	4.14	2.42	—
8819	Yellow, chocolate frosting, ready to eat, commercially prepared	1	slice(s)	64	14	243	2	35	1	11	2.98	6.14	1.35	—
8822	Yellow, vanilla frosting, ready to eat, commercially prepared	1	slice(s)	64	14	239	2	38	<1	9	1.52	3.91	3.30	—
	Snack cakes													
8791	Chocolate snack cake, creme filled, w/frosting	1	item(s)	50	10	188	2	30	<1	7	1.43	2.85	2.62	—
25010	Cinnamon coffee cake	1	piece(s)	72	23	231	4	36	1	8	2.19	2.65	2.99	0

PAGE KEY: H–2 = Breads/Baked Goods   H–6 = Cereal/Rice/Pasta   H–10 = Fruit   H–14 = Vegetables/Legumes   H–24 = Nuts/Seeds   H–26 = Vegetarian   H–28 = Dairy   H–34 = Eggs   H–34 = Seafood   H–36 = Meats   H–40 = Poultry   H–40 = Processed meats   H–42 = Beverages   H–46 = Fats/Oils   H–48 = Sweets   H–50 = Spices/Condiments/Sauces   H–52 = Mixed foods/Soups/Sandwiches   H–58 = Fast food   H–74 = Convenience meals   H–76 = Baby foods

Chol (mg)	Calc (mg)	Iron (mg)	Magn (mg)	Pota (mg)	Sodi (mg)	Zinc (mg)	Vit A (µg)	Thia (mg)	Vit E (mg α)	Ribo (mg)	Niac (mg)	Vit B$_6$ (mg)	Fola (µg)	Vit C (mg)	Vit B$_{12}$ (µg)	Sele (µg)
0	13	2.70	20	105	229	0.80	15	0.27	0.22	0.20	2.19	0.04	79	<1	0	22
0	53	2.53	21	72	379	0.62	0	0.38	0.07	0.22	3.24	0.04	75	0	0	23
0	53	2.52	20	72	379	0.62	0	0.31	0.08	0.20	2.91	0.03	64	0	0	23
0	9	2.19	22	82	360	0.64	1	0.24	0.23	0.24	2.10	0.03	70	<1	0	24
0	200	1.08	120	0	200	4.5	0	0.44	—	0.5	8	0.6	—	0	1.79	0
<1	33	1.01	6	37	205	0.27	9	0.13	0.01	0.12	1.08	0.01	26	0	<.1	7
49	80	1.31	7	48	288	0.29	—	0.15	0.43	0.16	1.20	0.03	8	<.1	<1	—
<1	57	1.22	16	81	321	0.42	12	0.16	0.01	0.13	1.49	0.03	29	<.1	<.1	12
<1	32	0.95	28	143	284	0.23	11	0.01	0.14	0.05	0.50	0.04	5	0	<.1	10
0	5	1.03	8	30	158	0.21	0	0.14	0.24	0.13	1.27	0.02	39	0	0	9
21	88	1.01	10	59	209	0.57	38	0.13	0.33	0.16	0.98	0.04	36	2	<1	6
0	11	0.70	13	44	135	0.31	0	0.09	—	0.06	0.92	0.08	15	0	<.1	6
0	6	0.31	2	9	52	0.07	0	0.05	—	0.02	0.41	0.00	10	0	0	3
20	37	1.22	8	46	197	0.32	25	0.18	0.10	0.17	1.94	0.03	42	0	<.1	12
21	38	1.24	8	47	200	0.32	26	0.14	0.11	0.16	1.77	0.02	36	0	<.1	12
0	19	0.63	7	28	152	0.22	0	0.13	0.08	0.08	1.19	0.01	37	0	0	8
0	19	0.63	7	28	152	0.22	0	0.10	0.07	0.07	1.07	0.01	22	0	0	8
0	210	3.24	14	67	626	0.45	0	0.39	—	0.27	3.27	0.02	67	0	0	21
0	23	0.88	8	33	175	0.26	0	0.14	0.09	0.09	1.31	0.01	57	0	0	8
0	24	0.90	14	53	127	0.33	0	0.11	0.09	0.09	1.13	0.09	31	<.1	<.1	8
0	24	0.90	14	53	127	0.33	0	0.08	0.08	0.08	1.02	0.08	28	<.1	<.1	8
0	20	0.94	11	44	122	0.27	1	0.15	0.13	0.10	1.45	0.02	24	0	0	9
0	19	0.93	9	33	121	0.28	1	0.12	0.13	0.09	1.29	0.01	19	0	0	9
0	18	0.73	10	38	162	0.28	1	0.11	0.13	0.06	0.85	0.02	17	0	<.1	7
0	18	0.74	10	39	163	0.28	1	0.09	0.13	0.06	0.77	0.02	13	<.1	<.1	7
0	52	1.57	16	72	322	0.50	0	0.36	0.18	0.20	2.78	0.02	64	0	0	16
0	10	1.96	44	109	340	0.97	0	0.22	0.39	0.05	1.82	0.17	22	0	0	28
0	22	0.92	17	67	215	0.47	0	0.10	0.13	0.10	0.99	0.04	30	0	0	8
0	17	0.75	7	59	101	0.19	0	0.09	0.07	0.10	0.90	0.02	28	<.1	0	5
0	17	0.76	7	59	102	0.19	0	0.07	0.07	0.09	0.81	0.02	24	<.1	0	5
0	40	1.08	—	45	160	—	0	0.23	—	0.14	1.20	—	40	0	—	—
0	23	0.91	13	53	211	0.36	0	0.14	0.11	0.11	1.22	0.02	35	<1	0	10
0	23	0.90	12	53	210	0.36	0	0.11	0.11	0.10	1.09	0.02	30	<1	0	10
0	20	0.70	4	51	175	0.18	0	0.10	—	0.08	0.80	0.01	5	0	<.1	8
0	19	0.63	7	28	152	0.22	0	0.13	0.08	0.08	1.19	0.01	37	0	0	8
0	19	0.63	7	28	152	0.22	0	0.10	0.07	0.07	1.07	0.01	22	0	0	8
0	100	3.80	—	128	683	—	0	0.54	—	0.33	4.50	0.05	—	0	—	42
0	19	0.63	7	28	152	0.22	0	0.10	0.07	0.07	1.07	0.01	22	0	0	8
0	26	0.83	12	50	133	0.26	0	0.10	0.07	0.07	1.03	0.02	23	0	0	8
0	26	0.83	12	50	132	0.26	0	0.08	0.07	0.06	0.93	0.02	19	0	0	8
0	38	0.94	6	25	170	0.19	0	0.11	0.05	0.08	1.10	0.02	28	0	0	4
0	15	1.43	37	144	159	0.69	0	0.14	0.35	0.10	1.83	0.09	30	0	0	18
0	42	0.12	4	68	255	0.07	0	0.05	0.00	0.10	0.09	0.00	10	0	<.1	8
166	26	1.04	8	89	299	0.35	112	0.10	—	0.98	0.03	31	0	<1	7	—
60	28	1.39	20	124	273	0.54	—	0.15	—	0.17	1.13	0.08	13	1	<1	—
27	28	1.41	22	128	214	0.44	—	0.02	—	0.09	0.37	0.03	11	<.1	<.1	2
55	57	1.53	30	133	299	0.66	38	0.13	—	0.20	1.08	0.04	26	<1	<1	11
19	21	0.70	—	46	92	—	—	0.04	—	0.05	0.30	—	2	0	—	2
2	14	0.89	7	66	116	0.12	3	0.02	0.39	0.04	0.34	0.02	9	<1	<.1	1
25	138	1.70	15	129	367	0.36	71	0.18	—	0.18	1.37	0.04	30	1	<.1	11
107	26	1.00	6	89	144	0.37	49	0.10	—	0.19	0.76	0.04	25	0	<1	12
1	101	1.30	13	111	318	0.37	13	0.14	0.13	0.21	1.19	0.03	35	<1	<.1	12
35	24	1.33	19	114	216	0.40	21	0.08	—	0.10	0.80	0.02	14	0	<1	2
35	40	0.68	4	34	220	0.16	12	0.06	—	0.04	0.32	0.02	17	0	<.1	4
9	37	1.68	21	61	213	0.26	3	0.11	1.09	0.15	1.21	0.01	120	0	<.1	1
26	50	1.46	10	81	277	0.38	35	0.14	0.23	0.16	1.17	0.02	30	<.1	<1	10

TABLE H–1
Food Composition

(DA+ code is for Wadsworth Diet Analysis program)   (For purposes of calculations, use "0" for t, <1, <.1, <.01, etc.)

DA + Code	Food Description	Quantity	Measure	Wt (g)	H₂O (g)	Ener (kcal)	Prot (g)	Carb (g)	Fiber (g)	Fat (g)	Sat	Mono	Poly	Trans
	BREADS, BAKED GOODS, CAKES, COOKIES, CRACKERS, CHIPS, PIES—Continued													
16777	Funnel cake	1	item(s)	90	37	278	7	29	1	14	2.77	4.46	6.33	—
8794	Sponge snack cake, creme filled	1	item(s)	43	9	155	1	27	<1	5	1.09	1.73	1.40	—
	Snacks, chips, pretzels													
29428	Bagel chips, plain	3	item(s)	29	—	130	3	19	1	5	0.50	—	—	—
29429	Bagel chips, toasted onion	3	item(s)	29	—	130	4	20	1	5	0.50	—	—	—
38192	Chex traditional snack mix	1	cup(s)	46	—	198	3	33	2	6	0.76	—	—	—
654	Potato chips, salted	20	item(s)	28	1	152	2	15	1	10	3.11	2.79	3.46	—
8816	Potato chips, unsalted	20	item(s)	28	1	152	2	15	1	10	3.11	2.79	3.46	—
4641	Tortilla chips, plain	6	item(s)	28	1	142	2	18	2	7	1.43	4.39	1.03	—
5096	Pretzels, plain, hard, twists	5	item(s)	30	1	114	3	24	1	1	0.23	0.41	0.37	—
4632	Pretzels, whole wheat	1	ounce(s)	28	1	103	3	23	2	1	0.16	0.29	0.24	—
	Cookies													
8859	Animal crackers	12	piece(s)	30	0	134	2	22	<1	4	1.03	2.29	0.56	—
8876	Brownie, prepared from mix	1	item(s)	24	3	112	1	12	1	7	1.76	2.60	2.26	—
25207	Chocolate chip cookies	1	item(s)	30	4	140	2	16	1	8	2.09	3.26	2.09	0
8915	Chocolate sandwich cookie, extra creme filling	1	item(s)	13	<1	65	<1	9	<1	3	0.50	1.39	1.22	1.10
14145	Fig Newtons	1	item(s)	16	—	55	1	10	1	1	0.50	0.50	0.00	0.50
8920	Fortune cookie	1	item(s)	8	1	30	<1	7	<1	<1	0.05	0.11	0.04	—
25208	Oatmeal cookies	1	item(s)	69	12	234	6	45	3	4	0.70	1.28	1.85	0
25213	Peanut butter cookies	1	item(s)	35	4	163	4	17	1	9	1.65	4.72	2.43	0
33095	Sugar cookies	1	item(s)	16	4	61	1	7	<1	3	0.63	1.27	0.87	0
9002	Vanilla sandwich cookie, creme filling	1	item(s)	10	<1	48	<1	7	<1	2	0.30	0.84	0.76	—
	Crackers													
9008	Cheese crackers (mini)	30	item(s)	30	1	151	3	17	1	8	2.81	3.63	0.74	—
9010	Cheese crackers (mini), low salt	30	item(s)	30	1	151	3	17	1	8	2.82	2.70	1.44	—
9012	Cheese cracker sandwich w/peanut butter	4	item(s)	28	1	139	3	16	1	7	1.23	3.64	1.43	—
8928	Honey graham crackers	4	item(s)	28	1	118	2	22	1	3	0.43	1.14	1.07	—
9016	Matzo crackers, plain	1	item(s)	28	1	112	3	24	1	<1	0.06	0.04	0.17	—
9024	Melba toast	3	item(s)	15	1	59	2	11	1	<1	0.07	0.12	0.19	—
14189	Ritz crackers	5	item(s)	16	<1	80	1	10	1	4	0.50	1.50	0.00	—
9014	Rye crispbread crackers	1	item(s)	10	1	37	1	8	2	<1	0.01	0.02	0.06	—
9028	Rye melba toast	3	item(s)	15	1	58	2	12	1	1	0.07	0.14	0.20	—
9040	Rye wafer	1	item(s)	11	1	37	1	9	3	<.1	0.01	0.02	0.04	—
432	Saltine crackers	5	item(s)	15	1	65	1	11	<1	2	0.44	0.96	0.25	0.54
9046	Saltine crackers, low salt	5	item(s)	15	1	65	1	11	<1	2	0.44	0.96	0.25	—
9048	Snack crackers, round	10	item(s)	30	1	151	2	18	<1	8	1.13	3.19	2.86	—
9050	Snack crackers, round, low salt	10	item(s)	30	1	151	2	18	<1	8	1.13	3.19	2.86	—
9052	Snack cracker sandwich, cheese filling	4	item(s)	28	1	134	3	17	1	6	1.72	3.15	0.72	—
9054	Snack cracker sandwich, peanut butter filling	4	item(s)	28	1	138	3	16	1	7	1.38	3.86	1.30	—
9044	Soda crackers	5	item(s)	15	1	65	1	11	<1	2	0.44	0.96	0.25	0.54
9055	Wheat crackers	10	item(s)	30	1	142	3	19	1	6	1.55	3.43	0.84	—
9057	Wheat crackers, low salt	10	item(s)	30	1	142	3	19	1	6	1.55	3.43	0.84	—
9059	Wheat cracker sandwich, cheese filling	4	item(s)	28	1	139	3	16	1	7	1.16	2.90	2.57	—
9061	Wheat cracker sandwich, peanut butter filling	4	item(s)	28	1	139	4	15	1	7	1.29	3.29	2.48	—
9022	Whole wheat crackers	7	item(s)	28	1	124	2	19	3	5	0.95	1.65	1.85	—
	Pastry													
16754	Apple fritter	1	item(s)	17	6	62	1	6	<1	4	0.87	1.69	1.13	—
5118	Cinnamon sweet roll w/icing, from refrigerator dough	1	item(s)	30	7	109	2	17	1	4	1.00	2.23	0.52	—
4945	Croissant, butter	1	item(s)	57	13	231	5	26	1	12	6.59	3.15	0.62	—
9096	Danish pastry, nut	1	item(s)	65	13	280	5	30	1	16	3.78	8.90	2.78	—
4947	Doughnut, cake	1	item(s)	47	10	198	2	23	1	11	1.70	4.37	3.70	—
9105	Doughnut, cake, chocolate glazed	1	item(s)	42	7	175	2	24	1	8	2.16	4.74	1.04	—
9115	Doughnut, creme filling	1	item(s)	85	32	307	5	26	1	21	4.62	10.27	2.62	—
437	Doughnut, glazed	1	item(s)	60	15	242	4	27	1	14	3.49	7.72	1.74	—
9117	Doughnut, jelly filling	1	item(s)	85	30	289	5	33	1	16	4.12	8.69	2.02	—
10617	Toaster pastry, brown sugar cinnamon	1	item(s)	50	5	210	3	35	1	6	1.00	4.00	1.00	—
30928	Toaster pastry, cream cheese	1	item(s)	54	—	200	3	23	1	11	3.50	—	—	—
	Muffins													
25015	Blueberry	1	item(s)	63	30	160	3	23	1	6	0.87	1.48	3.25	0
4997	Bran, from mix	1	item(s)	50	18	138	3	23	2	5	1.18	2.34	0.72	—
9189	Corn, ready to eat	1	item(s)	57	19	174	3	29	2	5	0.77	1.20	1.83	—

PAGE KEY: H–2 = Breads/Baked Goods H–6 = Cereal/Rice/Pasta H–10 = Fruit H–14 = Vegetables/Legumes H–24 = Nuts/Seeds H–26 = Vegetarian H–28 = Dairy H–34 = Eggs H–34 = Seafood H–36 = Meats H–40 = Poultry H–40 = Processed meats H–42 = Beverages H–46 = Fats/Oils H–48 = Sweets H–50 = Spices/Condiments/Sauces H–52 = Mixed foods/Soups/Sandwiches H–58 = Fast food H–74 = Convenience meals H–76 = Baby foods

Chol (mg)	Calc (mg)	Iron (mg)	Magn (mg)	Pota (mg)	Sodi (mg)	Zinc (mg)	Vit A (µg)	Thia (mg)	Vit E (mg α)	Ribo (mg)	Niac (mg)	Vit B$_6$ (mg)	Fola (µg)	Vit C (mg)	Vit B$_{12}$ (µg)	Sele (µg)
63	128	1.86	18	154	273	0.64	—	0.24	1.55	0.32	1.86	0.05	14	<1	<1	—
7	19	0.55	3	37	155	0.12	2	0.07	0.50	0.06	0.52	0.01	17	<.1	<.1	1
0	0	0.72	—	45	70	—	0	—	—	—	—	—	—	0	0	—
0	0	0.72	—	50	300	—	0	—	—	—	—	—	—	0	0	—
0	0	0.55	0	76	623	0.00	0	0.09	—	0.05	1.22	0.00	12	0	—	—
0	7	0.46	19	362	169	0.31	0	0.05	1.91	0.06	1.09	0.19	13	9	0	2
0	7	0.46	19	362	2	0.31	0	0.05	2.59	0.06	1.09	0.19	13	9	0	2
0	44	0.43	25	56	150	0.43	1	0.02	1	0.05	0.36	0.08	3	0	0	2
0	11	1.30	11	44	515	0.26	0	0.14	—	0.19	1.58	0.03	51	0	0	2
0	8	0.76	9	122	58	0.18	0	0.12	—	0.08	1.86	0.08	15	<1	0	—
0	13	0.82	5	30	1118	0.19	—	0.10	0.04	0.09	1.04	0.00	50	0	<.1	—
18	14	0.44	13	42	82	0.23	42	0.03	—	0.05	0.24	0.02	7	<.1	<.1	3
13	11	0.70	12	62	109	0.24	27	0.07	0.54	0.06	0.82	0.02	16	<.1	<.1	4
0	3	0.37	4	16	64	0.08	0	0.01	0.25	0.02	0.20	0.00	6	0	<.1	<1
0	5	0.36	—	40	60	—	4	0.03	—	0.04	0.22	—	—	<1		<1
<1	1	0.12	1	3	22	0.01	<.1	0.01	0.00	0.01	0.15	0.00	5	0	<.1	<1
<.1	26	1.94	49	177	311	1.43	48	0.23	0.23	0.12	1.24	0.09	30	<1	<.1	17
13	28	0.67	22	104	157	0.46	51	0.08	0.74	0.09	1.81	0.05	21	<.1	<.1	5
18	5	0.32	2	13	50	0.08	31	0.04	0.28	0.04	0.28	0.01	8	<.1	<.1	3
0	3	0.22	1	9	35	0.04	0	0.03	0.16	0.02	0.27	0.00	5	0	0	<1
4	45	1.43	11	44	299	0.34	9	0.17	0.66	0.13	1.40	0.17	46	0	<1	3
4	45	1.44	11	32	137	0.33	—	0.18	—	0.12	1.41	0.18	8	0	<1	—
0	14	0.76	16	61	199	0.29	0	0.15	0.16	0.08	1.63	0.04	26	0	<.1	2
0	7	1.04	8	38	169	0.23	0	0.06	0.09	0.09	1.15	0.02	13	0	0	3
0	4	0.90	7	32	1	0.19	0	0.11	0.02	0.08	1.11	0.03	5	0	0	10
0	14	0.56	9	30	124	0.30	0	0.06	0.06	0.04	0.62	0.01	19	0	0	5
0	20	0.72	3	10	135	0.23	—	0.07	—	0.04	0.45	0.01	10	1	0	—
0	3	0.24	8	32	26	0.24	0	0.02	0.08	0.01	0.10	0.02	5	0	0	4
0	12	0.55	6	29	135	0.20	0	0.07	—	0.04	0.71	0.01	13	0	0	6
0	4	0.65	13	54	87	0.31	0	0.05	0.09	0.03	0.17	0.03	5	<.1	0	3
0	18	0.81	4	19	195	0.12	0	0.08	0.15	0.07	0.79	0.01	19	0	0	2
0	18	0.81	4	109	95	0.12	0	0.08	0.02	0.07	0.79	0.01	19	0	0	3
0	36	1.08	8	40	254	0.20	0	0.12	0.61	0.10	1.21	0.02	27	0	0	2
0	36	1.08	8	107	112	0.20	0	0.12	0.61	0.10	1.21	0.02	27	0	0	2
1	72	0.67	10	120	392	0.17	5	0.12	0.06	0.19	1.05	0.01	28	<.1	<.1	6
0	23	0.78	15	60	201	0.32	0	0.14	0.58	0.08	1.71	0.04	24	0	<.1	3
0	18	0.81	4	19	195	0.12	0	0.08	0.15	0.07	0.79	0.01	19	0	0	2
0	15	1.32	19	55	239	0.48	0	0.15	0.15	0.10	1.49	0.04	35	0	0	2
0	15	1.32	19	61	85	0.48	0	0.15	0.15	0.10	1.49	0.04	15	0	0	10
2	57	0.73	15	86	256	0.24	5	0.10	—	0.12	0.89	0.07	18	<1	<.1	7
0	48	0.75	11	83	226	0.23	0	0.11	—	0.08	1.65	0.04	20	0	0	6
0	14	0.86	28	83	185	0.60	0	0.06	0.24	0.03	1.27	0.05	8	0	0	4
14	9	0.25	2	24	7	0.09	—	0.03	0.07	0.04	0.23	0.01	2	<1	<.1	—
0	10	0.80	4	19	250	0.10	—	0.12	—	0.07	1.09	0.01	14	<.1	<.1	—
38	21	1.16	9	67	424	0.43	101	0.22	—	0.14	1.25	0.03	35	<1	<.1	13
30	61	1.17	21	62	236	0.57	6	0.14	0.53	0.16	1.50	0.07	54	1	<1	9
17	21	0.92	9	60	257	0.26	—	0.10	—	0.11	0.87	0.03	22	<.1	<1	0
24	89	0.95	14	45	143	0.24	5	0.02	0.09	0.03	0.20	0.01	19	<.1	<.1	2
20	21	1.56	17	68	263	0.68	9	0.29	0.25	0.13	1.91	0.06	60	0	<1	9
4	26	0.36	13	65	205	0.46	2	0.53	—	0.04	0.39	0.03	13	<.1	<.1	5
22	21	1.50	17	67	249	0.64	14	0.27	0.37	0.12	1.82	0.09	58	0	<1	11
0	0	1.80	—	70	190	—	—	0.15	—	0.17	2.00	0.20	40	0	0	—
15	0	1.08	—	—	230	—	—	—	—	—	—	—	—	0	0	—
20	50	1.15	7	56	288	0.39	20	0.14	0.76	0.15	1.14	0.03	29	<1	<1	9
34	16	1.27	29	74	234	0.57	—	0.10	—	0.12	1.44	0.09	33	0	<.1	—
15	42	1.60	18	39	297	0.31	30	0.16	0.46	0.19	1.16	0.05	46	0	<.1	9

TABLE H-1
Food Composition

(DA+ code is for Wadsworth Diet Analysis program)   (For purposes of calculations, use "0" for t, <1, <.1, <.01, etc.)

DA + Code	Food Description	Quantity	Measure	Wt (g)	H₂O (g)	Ener (kcal)	Prot (g)	Carb (g)	Fiber (g)	Fat (g)	Sat	Mono	Poly	Trans
												Fat Breakdown (g)		

BREADS, BAKED GOODS, CAKES, COOKIES, CRACKERS, CHIPS, PIES—Continued

DA + Code	Food Description	Quantity	Measure	Wt (g)	H₂O (g)	Ener (kcal)	Prot (g)	Carb (g)	Fiber (g)	Fat (g)	Sat	Mono	Poly	Trans
9121	English muffin, plain, enriched	1	item(s)	57	24	134	4	26	2	1	0.15	0.17	0.51	—
29582	English, toasted	1	item(s)	50	19	128	4	25	1	1	0.14	0.16	0.48	—
9145	English, wheat	1	item(s)	57	24	127	5	26	3	1	0.16	0.16	0.48	—
	Granola bars													
38161	Kudos milk chocolate w/fruit & nuts	1	item(s)	28	—	90	2	15	1	3	1.00	—	—	—
38196	Nature Valley banana nut crunchy	1	item(s)	21	—	95	2	14	1	4	0.50	—	—	—
38187	Nature Valley fruit n nut trail mix	1	item(s)	35	—	140	3	25	2	4	0.50	—	—	—
1383	Plain, hard	1	item(s)	25	1	115	2	16	1	5	0.58	1.07	2.95	—
4606	Plain, soft	1	item(s)	28	2	126	2	19	1	5	2.06	1.08	1.51	—
	Pies													
454	Apple pie, from home recipe	1	slice(s)	155	73	411	4	58	2	19	4.73	8.36	5.17	—
470	Pecan pie, from home recipe	1	slice(s)	122	24	503	6	64	0	27	4.87	13.64	6.97	—
472	Pumpkin pie, from home recipe	1	slice(s)	155	91	316	7	41	0	14	4.92	5.73	2.81	—
9007	Pie crust, frozen, ready to bake, enriched, baked	1	slice(s)	16	2	82	1	8	<1	5	1.69	2.51	0.65	—
5052	Pie crust, prepared w/water, baked	1	slice(s)	20	2	100	1	10	<1	6	1.54	3.46	0.77	—
	Rolls													
8555	Crescent dinner roll	1	item(s)	28	10	80	2	14	1	1	0.34	0.70	0.25	—
489	Hamburger roll or bun, plain	1	item(s)	43	15	120	4	21	1	2	0.47	0.48	0.85	—
490	Hard roll	1	item(s)	57	18	167	6	30	1	2	0.35	0.65	0.98	—
5127	Kaiser roll	1	item(s)	57	18	167	6	30	1	2	0.35	0.65	0.98	—
5130	Whole wheat roll or bun	1	item(s)	28	9	76	2	15	2	1	0.24	0.34	0.62	—
	Sport bars													
37026	Balance original chocolate	1	item(s)	50	—	200	14	22	1	6	3.50	—	—	—
37024	Balance original peanut butter	1	item(s)	50	—	200	14	22	1	6	2.50	—	—	—
36580	Clif Bar chocolate brownie energy bar	1	item(s)	68	—	240	10	41	6	4	1.00	—	—	—
36583	Clif Bar crunchy peanut butter energy bar	1	item(s)	68	—	240	12	39	5	5	0.50	—	—	—
36584	Clif Luna tropical crisp energy bar	1	item(s)	48	—	180	10	24	2	5	3.50	0.00	0.00	—
12005	Powerbar apple cinnamon	1	item(s)	65	—	230	10	45	3	3	0.50	1.50	0.50	—
16078	Powerbar banana	1	item(s)	65	—	230	9	45	3	2	0.50	1.00	0.50	—
16080	Powerbar chocolate	1	item(s)	65	—	230	10	45	3	2	0.50	0.50	1.00	—
16079	Powerbar mocha	1	item(s)	65	—	230	10	45	3	3	1.00	1.00	0.50	—
	Tortillas													
1391	Corn tortillas, soft	1	item(s)	26	11	58	1	12	1	1	0.09	0.17	0.29	—
1669	Flour tortilla	1	item(s)	32	9	104	3	18	1	2	0.56	1.21	0.34	—
1390	Taco shells, hard	1	item(s)	13	1	62	1	8	1	3	0.43	1.19	1.13	—
	Pancakes, waffles													
8926	Pancakes, blueberry, from recipe	3	item(s)	114	61	253	7	33	1	10	2.26	2.64	4.74	—
5037	Pancakes, from mix w/egg & milk	3	item(s)	114	60	249	9	33	2	9	2.33	2.36	3.33	—
9219	Waffle, plain, frozen, toasted	2	item(s)	66	28	174	4	27	2	5	0.95	2.12	1.84	—
500	Waffle, plain, from recipe	1	item(s)	75	32	218	6	25	2	11	2.14	2.64	5.08	—
30311	Waffle, 100% whole grain	1	item(s)	75	32	201	7	25	2	8	2.35	3.38	2.06	—
	CEREAL, FLOUR, GRAIN, PASTA, NOODLES, POPCORN													
	Grain													
2861	Amaranth, dry	½	cup(s)	98	10	365	14	65	15	6	1.62	1.40	2.82	—
1953	Barley, pearled, cooked	½	cup(s)	79	54	97	2	22	3	<1	0.07	0.04	0.17	—
1956	Buckwheat groats, cooked, roasted	½	cup(s)	84	64	77	3	17	2	1	0.11	0.16	0.16	—
1957	Bulgur, cooked	½	cup(s)	91	71	76	3	17	4	<1	0.04	0.03	0.09	—
1963	Couscous, cooked	½	cup(s)	79	57	88	3	18	1	<1	0.02	0.02	0.05	—
1967	Millet, cooked	½	cup(s)	120	86	143	4	28	2	1	0.21	0.22	0.61	—
1969	Oat bran, dry	½	cup(s)	47	3	116	8	31	7	3	0.62	1.12	1.30	—
1972	Quinoa, dry	½	cup(s)	85	8	318	11	59	5	5	0.50	1.30	1.99	—
	Rice													
129	Brown, long grain, cooked	½	cup(s)	98	71	108	3	22	2	1	0.18	0.32	0.31	—
2863	Brown, medium grain, cooked	½	cup(s)	98	71	109	2	23	2	<1	0.16	0.29	0.28	—
37488	Jasmine, saffroned, cooked	½	cup(s)	280	—	340	8	78	0	0	0.00	—	—	0
30280	Pilaf, cooked	½	cup(s)	103	74	129	2	22	1	3	0.67	1.61	0.95	—
28066	Spanish, cooked	½	cup(s)	120	3	25	2	1	<1	<1	0.33	0.07	18.31	0
2867	White glutinous, cooked	½	cup(s)	87	67	84	2	18	1	<1	0.03	0.06	0.06	—
482	White, instant long grain, enriched, boiled	½	cup(s)	83	63	81	2	18	<1	<1	0.04	0.04	0.04	—
484	White, long grain, boiled	½	cup(s)	79	54	103	2	22	<1	<1	0.06	0.07	0.06	—
486	White, long grain, enriched, parboiled, cooked	½	cup(s)	88	63	100	2	22	<1	<1	0.06	0.07	0.06	—
1194	Wild brown, cooked	½	cup(s)	82	61	83	3	17	1	<1	0.04	0.04	0.17	—

PAGE KEY: H–2 = Breads/Baked Goods   H–6 = Cereal/Rice/Pasta   H–10 = Fruit   H–14 = Vegetables/Legumes   H–24 = Nuts/Seeds   H–26 = Vegetarian
H–28 = Dairy   H–34 = Eggs   H–34 = Seafood   H–36 = Meats   H–40 = Poultry   H–40 = Processed meats   H–42 = Beverages   H–46 = Fats/Oils
H–48 = Sweets   H–50 = Spices/Condiments/Sauces   H–52 = Mixed foods/Soups/Sandwiches   H–58 = Fast food   H–74 = Convenience meals   H–76 = Baby foods

Chol (mg)	Calc (mg)	Iron (mg)	Magn (mg)	Pota (mg)	Sodi (mg)	Zinc (mg)	Vit A (µg)	Thia (mg)	Vit E (mg α)	Ribo (mg)	Niac (mg)	Vit B$_6$ (mg)	Fola (µg)	Vit C (mg)	Vit B$_{12}$ (µg)	Sele (µg)
0	30	1.43	12	75	264	0.40	0	0.25	—	0.16	2.21	0.02	42	0	<.1	—
0	95	1.36	11	72	252	0.38	0	0.19	0.17	0.14	1.90	0.02	15	<.1	<.1	—
0	101	1.64	21	106	218	0.61	0	0.25	0.26	0.17	1.91	0.05	36	0	0	17
0	200	0.36	—	—	60	—	0	—	—	—	—	—	—	0	0	—
0	10	0.54	—	60	80	—	0	—	—	—	—	—	—	0	—	—
0	0	0.00	—	—	95	—	0	—	—	—	—	—	—	0	—	—
0	15	0.72	24	82	72	0.50	2	0.06	—	0.03	0.39	0.02	6	<1	0	4
<1	30	0.73	21	92	79	0.43	0	0.08	—	0.05	0.15	0.03	7	0	<1	5
0	11	1.74	11	122	327	0.29	17	0.23	—	0.17	1.91	0.05	37	3	0	12
106	39	1.81	32	162	320	1.24	100	0.23	—	0.22	1.03	0.07	32	<1	<1	15
65	146	1.97	29	288	349	0.71	660	0.14	—	0.31	1.21	0.07	33	3	<1	11
0	3	0.36	3	18	104	0.05	0	0.04	0.42	0.06	0.39	0.01	9	0	<.1	<1
0	12	0.43	3	12	146	0.08	0	0.06	—	0.04	0.47	0.01	20	0	0	—
0	39	0.89	6	39	157	0.17	0	0.14	0.02	0.09	1.10	0.01	—	0	<.1	—
0	59	1.43	9	40	206	0.28	0	0.17	0.03	0.14	1.79	0.03	48	0	<.1	8
0	54	1.87	15	62	310	0.54	0	0.27	0.24	0.19	2.42	0.02	54	0	0	22
0	54	1.87	15	62	310	0.54	0	0.27	—	0.19	2.42	0.02	54	0	0	22
0	30	0.69	24	78	136	0.57	0	0.07	—	0.04	1.05	0.06	9	0	0	14
3	100	4.50	40	160	180	3.75	—	0.38	—	0.43	5.00	0.50	100	60	2	18
3	100	4.50	40	130	230	3.75	—	0.38	—	0.43	5.00	0.50	100	60	2	18
0	250	5.40	120	260	150	3.75	—	0.38	—	0.26	4.00	0.40	80	60	1	18
0	250	5.40	120	300	290	3.75	—	0.38	—	0.34	6.00	0.40	100	60	1	14
0	350	6.30	140	120	135	5.25	—	1.50	—	1.70	20.00	2.00	400	60	6	25
0	300	6.30	140	110	90	5.25	0	1.50	—	1.70	20.00	2.00	400	60	6	—
0	300	6.30	140	200	90	5.25	0	1.50	—	1.70	20.00	2.00	400	60	6	—
0	300	6.30	140	150	90	5.25	0	1.50	—	1.70	20.00	2.00	400	60	6	—
0	300	6.30	140	150	90	5.25	0	1.50	—	1.70	20.00	2.00	400	60	6	—
0	46	0.36	17	40	42	0.24	0	0.03	0.07	0.02	0.39	0.06	26	0	0	1
0	40	1.06	8	42	153	0.23	0	0.17	0.06	0.09	1.14	0.02	33	0	0	7
0	21	0.33	14	24	49	0.19	0	0.03	0.22	0.01	0.18	0.04	17	0	0	2
64	235	1.96	18	157	470	0.62	57	0.22	—	0.31	1.74	0.06	41	3	<1	16
81	245	1.48	25	227	576	0.86	82	0.23	—	0.36	1.40	0.12	105	1	<1	7
16	153	2.95	15	84	519	0.38	253	0.25	0.65	0.31	2.93	0.59	36	0	2	11
52	191	1.73	14	119	383	0.50	49	0.19	—	0.26	1.55	0.04	51	<1	<1	35
71	196	1.56	30	173	374	0.85	—	0.15	0.32	0.25	1.47	0.09	14	<1	<1	—
0	149	7.40	259	357	20	3.10	0	0.08	—	0.20	1.25	0.22	48	4	0	—
0	9	1.04	17	73	2	0.64	0	0.07	0.01	0.05	1.62	0.09	13	0	0	7
0	6	0.67	43	74	3	0.51	0	0.03	0.08	0.03	0.79	0.06	12	0	0	2
0	9	0.87	29	62	5	0.52	0	0.05	0.01	0.03	0.91	0.08	16	0	0	1
0	6	0.30	6	46	4	0.20	0	0.05	0.10	0.02	0.77	0.04	12	0	0	22
0	4	0.76	53	74	2	1.09	0	0.13	0.02	0.10	1.60	0.13	23	0	0	1
0	27	2.54	110	266	2	1.46	0	0.55	0.47	0.10	0.44	0.08	24	0	0	21
0	51	7.86	179	629	18	2.81	0	0.17	—	0.34	2.49	0.19	42	0	0	—
0	10	0.41	42	42	5	0.61	0	0.09	0.03	0.02	1.49	0.14	4	0	0	10
0	10	0.51	43	77	1	0.6	0	0.09	—	0.01	1.29	0.14	4	0	0	38
0	—	2.16	—	—	780	—	—	—	—	—	—	—	—	—	—	—
0	13	1.16	9	55	403	0.38	—	0.13	0.28	0.02	1.24	0.06	4	<1	<.1	—
1	47	0.78	48	1	13	0.13	<1	0.03	0.06	0.19	8.71	0.14	<.1	7	<.1	9
0	2	0.12	4	9	4	0.36	0	0.02	0.03	0.01	0.25	0.02	1	0	0	5
0	7	0.52	4	3	2	0.20	0	0.06	0.01	0.04	0.73	0.01	58	0	0	3
0	8	0.95	9	28	1	0.39	0	0.13	0.03	0.01	1.17	0.07	46	0	0	6
0	17	0.99	11	32	3	0.27	0	0.22	0.01	0.02	1.23	0.02	67	0	0	7
0	2	0.49	26	83	2	1.09	0	0.04	—	0.07	1.05	0.11	21	0	0	<1

TABLE H–1
Food Composition
(DA+ code is for Wadsworth Diet Analysis program)　　(For purposes of calculations, use "0" for t, <1, <.1, <.01, etc.)

DA + Code	Food Description	Quantity	Measure	Wt (g)	H₂O (g)	Ener (kcal)	Prot (g)	Carb (g)	Fiber (g)	Fat (g)	Sat	Mono	Poly	Trans
	CEREAL, FLOUR, GRAIN, PASTA, NOODLES, POPCORN—Continued													
	Flour & grain fractions													
505	All purpose flour, self rising, enriched	½	cup(s)	63	7	221	6	46	2	1	0.10	0.05	0.26	—
503	All purpose flour, white, bleached, enriched	½	cup(s)	63	7	228	6	48	2	1	0.10	0.05	0.26	—
1643	Barley flour	½	cup(s)	56	6	198	4	45	2	1	0.16	0.10	0.38	—
383	Buckwheat flour, whole groat	½	cup(s)	60	7	201	8	42	6	2	0.41	0.57	0.57	—
504	Cake wheat flour, enriched	½	cup(s)	55	7	197	4	43	1	<1	0.07	0.04	0.21	—
426	Cornmeal, degermed, enriched	½	cup(s)	69	8	253	6	54	5	1	0.16	0.28	0.49	—
424	Cornmeal, yellow whole grain	½	cup(s)	61	6	221	5	47	4	2	0.31	0.58	1.00	—
1644	Masa corn flour, enriched	½	cup(s)	57	5	208	5	43	5	2	0.30	0.57	0.98	—
1976	Rice flour, brown	½	cup(s)	79	9	287	6	60	4	2	0.44	0.80	0.79	—
1645	Rice flour, white	½	cup(s)	79	9	289	5	63	2	1	0.30	0.35	0.30	—
1978	Rye flour, dark	½	cup(s)	64	7	207	9	44	14	2	0.20	0.21	0.77	—
1980	Semolina, enriched	½	cup(s)	84	11	301	11	61	3	1	0.13	0.10	0.36	—
2827	Soy flour, raw	½	cup(s)	43	2	186	15	15	4	9	1.27	1.94	4.96	—
1990	Wheat germ, crude	2	tablespoon(s)	14	2	52	3	7	2	1	0.24	0.20	0.86	—
506	Whole wheat flour	½	cup(s)	60	6	203	8	44	7	1	0.19	0.14	0.47	—
	Breakfast bars													
39230	Atkins Morning Start apple crisp	1	item(s)	37	—	170	11	12	6	9	4.00	—	—	—
10574	Health Valley fat free apple	1	item(s)	38	—	110	2	26	3	0	0.00	0.00	0.00	0
10647	Nutri-Grain blueberry cereal bar	1	item(s)	37	5	140	2	27	1	3	0.50	2.00	0.50	—
10648	Nutri-Grain raspberry cereal bar	1	item(s)	37	5	140	2	27	1	3	0.50	2.00	0.50	—
10649	Nutri-Grain strawberry cereal bar	1	item(s)	37	5	140	2	27	1	3	0.50	2.00	0.50	—
	Breakfast cereals, hot													
363	Corn grits, white, regular & quick, enriched, cooked w/water & salt	½	cup(s)	121	103	71	2	16	<1	<1	0.03	0.06	0.10	—
8636	Corn grits, yellow, regular & quick, enriched, cooked w/salt	½	cup(s)	121	103	71	2	16	<1	<1	0.03	0.06	0.10	—
1260	Cream of Wheat, instant, prepared	½	cup(s)	121	106	61	2	13	<1	<.1	0.01	0.01	0.04	0
365	Farina, enriched, cooked w/water & salt	½	cup(s)	117	102	56	2	12	<1	<.1	0.01	0.01	0.03	—
8657	Oatmeal, cooked w/water	½	cup(s)	117	100	74	3	13	2	1	0.19	0.37	0.44	—
5500	Oatmeal, maple & brown sugar, instant, prepared	1	item(s)	198	150	200	5	40	2	2	0.42	0.74	0.85	—
5510	Oatmeal, ready to serve, packet	1	item(s)	186	158	112	4	20	3	2	0.38	0.66	0.76	—
	Breakfast cereals, ready to eat													
1197	All-Bran	1	cup(s)	62	2	160	8	46	20	2	0.00	0.00	1.00	0
1200	All-Bran Buds	1	cup(s)	91	3	212	6	73	42	3	—	—	—	0
1199	Apple Jacks	1	cup(s)	33	1	130	1	30	1	1	—	—	—	0
13633	Bran Flakes, Post	1	cup(s)	40	1	133	4	32	7	1	0.00	0.00	0.71	—
1204	Cap'n Crunch	1	cup(s)	36	1	144	2	30	1	2	0.53	0.39	0.27	—
1205	Cap'n Crunch Crunchberries w/wildberry colors	1	cup(s)	35	1	139	2	29	1	2	0.49	0.39	0.28	—
1206	Cheerios	1	cup(s)	30	1	110	3	22	3	2	0.00	0.50	0.50	—
3415	Cocoa Puffs	1	cup(s)	30	1	120	1	26	0	1	—	—	—	—
1207	Cocoa Rice Krispies	1	cup(s)	41	1	160	1	36	1	1	0.67	0.00	0.00	—
5522	Complete wheat bran flakes	1	cup(s)	39	1	120	4	31	7	1	—	—	—	0
1211	Corn Flakes	1	cup(s)	28	1	100	2	24	1	0	0.00	0.00	0.00	0
1247	Corn Pops	1	cup(s)	31	1	120	1	28	0	0	0.00	0.00	0.00	0
1937	Cracklin' Oat Bran	1	cup(s)	65	0	266	5	47	7	9	2.70	4.70	1.33	—
1220	Froot Loops	1	cup(s)	32	1	120	1	28	1	1	0.50	0.00	0.00	—
38214	Frosted Cheerios	1	cup(s)	30	—	120	2	25	1	1	0.00	0.00	0.00	—
372	Frosted Flakes	1	cup(s)	41	1	160	1	37	1	0	0.00	0.00	0.00	0
38215	Frosted Mini Chex	1	cup(s)	40	—	146	1	36	0	0	0.00	0.00	0.00	0
10268	Frosted Mini-Wheats	5	item(s)	51	3	180	5	41	5	1	0.00	0.00	0.50	0
38216	Frosted Wheaties	1	cup(s)	40	—	146	1	36	<1	0	0.00	0.00	0.00	0
1223	Granola, prepared	½	cup(s)	61	0	299	9	32	5	15	2.76	4.7	6.53	—
13334	Granola, Quaker 100% natural, oats & honey	½	cup(s)	48	0	219	5	31	3	9	3.83	4.0	1.19	—
13335	Granola, Quaker 100% natural, oats, honey & raisins	½	cup(s)	51	0	225	5	34	3	9	3.57	3.80	1.10	—
2415	Honey Bunches of Oats honey roasted	1	cup(s)	40	1	160	3	33	1	2	0.67	1.20	0.13	—
1227	Honey Nut Cheerios	1	cup(s)	30	1	120	3	24	2	2	0.00	0.50	0.00	—
2424	Honeycomb	1	cup(s)	22	<1	83	2	20	<1	<1	0.00	—	—	—
10286	Kashi puffed	1	cup(s)	25	—	70	3	13	2	1	0.00	—	—	—
1231	Kix	1	cup(s)	23	<1	90	2	20	1	<1	0.00	0.00	0.00	—
30569	Life	1	cup(s)	43	2	160	4	33	3	2	0.35	0.64	0.61	—
1233	Lucky Charms	1	cup(s)	30	1	120	2	25	1	1	0.00	0.00	0.00	—

PAGE KEY: H–2 = Breads/Baked Goods H–6 = Cereal/Rice/Pasta H–10 = Fruit H–14 = Vegetables/Legumes H–24 = Nuts/Seeds H–26 = Vegetarian
H–28 = Dairy H–34 = Eggs H–34 = Seafood H–36 = Meats H–40 = Poultry H–40 = Processed meats H–42 = Beverages H–46 = Fats/Oils
H–48 = Sweets H–50 = Spices/Condiments/Sauces H–52 = Mixed foods/Soups/Sandwiches H–58 = Fast food H–74 = Convenience meals H–76 = Baby foods

Chol (mg)	Calc (mg)	Iron (mg)	Magn (mg)	Pota (mg)	Sodi (mg)	Zinc (mg)	Vit A (µg)	Thia (mg)	Vit E (mg α)	Ribo (mg)	Niac (mg)	Vit B6 (mg)	Fola (µg)	Vit C (mg)	Vit B12 (µg)	Sele (µg)
0	211	2.92	12	78	794	0.39	0	0.42	0.03	0.26	3.65	0.03	123	0	0	22
0	9	2.90	14	67	1	0.44	0	0.49	0.04	0.31	3.69	0.03	114	0	0	21
0	16	0.71	45	186	4	1.05	0	0.07	—	0.03	2.57	0.16	13	0	0	2
0	25	2.44	151	346	7	1.87	0	0.25	0.19	0.11	3.69	0.35	32	0	0	3
0	8	3.99	9	57	1	0.34	0	0.49	0.01	0.23	3.70	0.02	101	0	0	3
0	3	2.85	28	112	2	0.50	8	0.49	0.10	0.28	3.47	0.18	161	0	0	5
0	4	2.10	77	175	21	1.11	7	0.23	0.26	0.12	2.22	0.19	15	0	0	9
0	80	4.11	63	170	3	1.01	0	0.81	0.09	0.43	5.61	0.21	133	0	0	9
0	9	1.56	88	228	6	1.94	0	0.35	0.95	0.06	5.01	0.58	13	0	0	—
0	8	0.28	28	60	0	0.63	0	0.11	0.09	0.02	2.05	0.34	3	0	0	12
0	36	4.13	159	467	1	3.60	1	0.20	0.90	0.16	2.73	0.28	38	0	0	23
0	14	3.64	39	155	1	0.88	0	0.68	0.22	0.48	5.00	0.09	153	0	0	75
0	88	2.71	183	1070	6	1.67	3	0.25	0.93	0.49	1.84	0.20	147	0	0	3
0	6	0.90	34	128	2	1.77	0	0.27	—	0.07	0.98	0.19	40	0	0	11
0	20	2.33	83	243	3	1.76	0	0.27	0.49	0.13	3.82	0.20	26	0	0	42
0	200	—	—	90	70	—	—	0.23	—	0.26	3.00	—	—	9	—	—
0	0	0.72	—	160	25	—	—	0.09	—	0.03	0.40	—	—	1	—	—
0	200	1.80	8	75	110	1.50	—	0.38	—	0.43	5.00	0.50	40	0	0	—
0	200	1.80	8	70	110	1.50	—	0.38	—	0.43	5.00	0.50	40	0	0	—
0	200	1.80	8	55	110	1.50	—	0.38	—	0.43	5.00	0.50	40	0	0	—
0	4	0.73	6	25	270	0.08	0	0.10	0.02	0.07	0.87	0.03	40	0	0	4
0	4	0.73	6	25	270	0.08	2	0.10	0.02	0.07	0.87	0.03	40	0	0	3
0	27	8.60	2	17	1	0.10	0	0.07	—	0.04	0.60	0.01	357	0	0	—
0	5	0.58	2	15	383	0.09	0	0.07	0.01	0.05	0.57	0.01	40	0	0	11
0	9	0.80	28	66	1	0.57	0	0.13	0.12	0.02	0.15	0.02	5	0	0	9
0	26	6.84	50	126	404	1.04	0	1.02	—	0.05	1.57	0.31	30	0	0	11
0	21	3.96	45	112	241	0.93	0	0.60	—	0.05	0.78	0.19	19	0	0	4
0	300	9.00	200	700	160	3.00	300	0.75	—	0.85	10.00	4.00	800	12	12	6
0	0	13.64	182	909	606	4.55	455	1.14	—	1.29	15.15	6.06	1212	18	18	26
0	0	4.50	8	35	150	1.50	150	0.38	—	0.43	5.00	0.50	100	15	2	2
0	0	10.77	80	253	293	2.00	—	0.50	—	0.57	6.65	0.67	133	0	2	—
0	5	6.00	20	72	269	4.99	3	0.51	—	0.57	6.66	0.67	133	0	2	7
<.1	7	6.14	19	71	242	5.12	2	0.51	—	0.57	6.66	0.67	133	<.1	0	7
0	100	8.10	40	95	280	3.75	150	0.38	—	0.43	5.00	0.50	200	6	2	11
0	100	4.50	8	50	170	3.75	0	0.38	—	0.43	5.00	0.50	100	6	2	2
0	53	5.99	11	67	253	2.00	200	0.50	—	0.57	6.65	0.67	133	20	2	6
0	0	23.94	53	226	279	19.95	299	2.00	—	2.26	26.60	2.66	532	80	8	4
0	0	8.10	3	25	200	0.17	150	0.38	—	0.43	5.00	0.50	100	6	2	1
0	0	1.80	2	25	120	1.50	150	0.38	—	0.43	5.00	0.50	100	6	2	2
0	27	2.38	80	293	186	2.00	299	0.49	—	0.56	6.65	0.67	218	20	2	14
0	0	4.50	8	35	150	1.50	150	0.38	—	0.43	5.00	0.50	100	15	2	2
0	100	4.50	16	55	210	3.75	—	0.38	—	0.43	5.00	0.50	100	6	2	—
0	0	5.99	4	27	200	0.21	200	0.50	—	0.57	6.65	0.67	133	8	2	2
0	133	11.97	—	33	266	3.99	—	0.50	—	0.57	6.65	0.67	266	8	2	—
0	0	15.30	60	170	5	1.50	—	0.38	—	0.43	5.00	0.50	100	0	2	2
0	133	10.77	0	47	266	9.98	—	1.00	—	1.13	13.30	1.33	532	8	4	—
0	48	2.59	107	328	13	2.5	2	0.44	3.59	0.17	1.29	0.18	51	1	0	17
1	61	1.21	51	225	20	1.04	1	0.12	—	0.11	0.81	0.07	17	<1	0.1	8
1	59	1.24	49	250	19	0.99	<1	0.12	—	0.11	0.8	0.07	16	<1	0.1	9
0	0	3.59	21	67	253	0.40	—	0.50	—	0.57	6.65	0.67	133	0	2	—
0	100	4.50	24	95	270	3.75	—	0.38	—	0.43	5.00	0.50	200	6	2	7
0	0	2.03	6	26	165	1.13	—	0.28	—	0.32	3.74	0.37	75	0	1	—
0	0	0.72	—	35	0	—	0	0.03	—	0.03	0.80	0.00	—	0	—	—
0	113	6.08	6	26	203	2.81	113	0.28	—	0.32	3.75	0.38	150	5	1	5
0	124	11.92	41	121	218	5.32	1	0.53	—	0.60	7.10	0.70	142	0	0	11
0	100	4.50	16	60	210	3.75	—	0.38	—	0.43	5.00	0.50	200	6	2	6

TABLE H–1
Food Composition

(DA+ code is for Wadsworth Diet Analysis program)   (For purposes of calculations, use "0" for t, <1, <.1, <.01, etc.)

DA + Code	Food Description	Quantity	Measure	Wt (g)	H₂O (g)	Ener (kcal)	Prot (g)	Carb (g)	Fiber (g)	Fat (g)	Sat	Fat Breakdown (g) Mono	Poly	Trans
	CEREAL, FLOUR, GRAIN, PASTA, NOODLES, POPCORN—Continued													
1201	Multi-Bran Chex	1	cup(s)	58	1	200	4	49	7	2	0.00	0.00	0.00	0
38220	Multi Grain Cheerios	1	cup(s)	30	—	110	3	24	3	1	0.00	0.00	0.00	—
1238	Nutri-Grain golden wheat	1	cup(s)	40	—	133	4	31	5	1	0.00	0.00	0.67	—
1241	Product 19	1	cup(s)	30	1	100	2	25	1	0	0.00	0.00	0.00	0
32432	Puffed rice, fortified	1	cup(s)	14	<1	56	1	13	<1	<.1	0.02	—	—	—
32433	Puffed wheat, fortified	1	cup(s)	12	0	44	2	10	<1	<1	0.02	—	—	—
2420	Raisin Bran	1	cup(s)	59	5	190	4	47	8	1	0.00	0.10	0.36	—
1244	Rice Chex	1	cup(s)	25	1	96	2	22	<1	0	0.00	0.00	0.00	0
1245	Rice Krispies	1	cup(s)	26	1	96	2	23	0	0	0.00	0.00	0.00	0
5593	Shredded Wheat	1	cup(s)	25	1	88	3	20	3	1	0.04	0.01	0.10	—
1248	Smacks	1	cup(s)	36	1	133	3	32	1	1	0.00	0.00	0.00	—
1246	Special K	1	cup(s)	31	1	110	7	22	1	0	0.00	0.00	0.00	0
3428	Total, corn flakes	1	cup(s)	23	1	83	2	18	1	0	0.00	0.00	0.00	0
1253	Total whole grain	1	cup(s)	40	1	146	3	31	4	1	0.00	0.00	0.00	—
1254	Trix	1	cup(s)	30	1	120	1	27	1	1	0.00	0.00	0.00	—
382	Wheat germ, toasted	2	tablespoon(s)	14	0	54	4	7	2	1	0.25	0.21	0.93	—
1257	Wheaties	1	cup(s)	30	1	110	3	24	3	1	0.00	0.00	0.00	
	Pasta, noodles													
449	Chinese chow mein noodles, cooked	½	cup(s)	23	<1	119	2	13	1	7	0.99	1.73	3.90	—
1995	Corn pasta, cooked	½	cup(s)	70	48	88	2	20	3	1	0.07	0.13	0.23	—
448	Egg noodles, enriched, cooked	½	cup(s)	80	55	106	4	20	1	1	0.25	0.34	0.33	0.02
440	Macaroni, enriched, cooked	½	cup(s)	70	46	99	3	20	1	<1	0.07	0.06	0.19	—
1996	Pasta, plain, fresh-refrigerated, cooked	½	cup(s)	64	44	84	3	16	0	1	0.10	0.08	0.27	—
1725	Ramen noodles, cooked	½	cup(s)	114	95	104	3	15	1	4	0.19	0.22	0.21	—
2878	Soba noodles, cooked	½	cup(s)	95	69	94	5	20	0	<.1	0.02	0.02	0.03	—
2879	Somen noodles, cooked	½	cup(s)	88	60	115	4	24	0	<1	0.02	0.02	0.06	—
493	Spaghetti, al dente, cooked	½	cup(s)	65	42	95	4	20	1	1	0.05	0.05	0.15	—
2884	Spaghetti, whole wheat, cooked	½	cup(s)	70	47	87	4	19	3	<1	0.07	0.05	0.15	—
1563	Spinach egg noodles, enriched, cooked	½	cup(s)	80	55	105	4	19	2	1	0.29	0.39	0.28	—
2000	Tricolor vegetable macaroni, enriched, cooked	½	cup(s)	67	46	86	3	18	3	<.1	0.01	0.01	0.03	—
	Popcorn													
476	Air popped	1	cup(s)	8	<1	31	1	6	1	<1	0.05	0.09	0.15	—
4619	Caramel	1	cup(s)	35	1	152	1	28	2	5	1.27	1.01	1.58	—
4620	Cheese flavored	1	cup(s)	37	1	196	3	19	4	12	2.38	3.61	5.72	—
477	Popped in oil	1	cup(s)	33	1	165	3	19	3	9	1.61	2.70	4.43	—
	FRUIT AND FRUIT JUICES													
	Apples													
223	Raw medium, w/peel	1	item(s)	138	118	72	<1	19	3	<1	0.04	0.01	0.07	—
224	Slices	½	cup(s)	55	47	29	<1	8	1	<.1	0.02	0.00	0.03	—
946	Slices w/o skin, boiled	½	cup(s)	85	73	45	<1	12	2	<1	0.05	0.01	0.09	—
948	Dried, sulfured	½	cup(s)	22	7	52	<1	14	2	<.1	0.01	0.00	0.02	—
952	Juice, from frozen concentrate	½	cup(s)	120	105	56	<1	14	<1	<1	0.02	0.00	0.04	—
225	Juice, unsweetened, canned	½	cup(s)	124	109	58	<.1	14	<1	<1	0.02	0.01	0.04	—
226	Applesauce, sweetened, canned	½	cup(s)	128	101	97	<1	25	2	<1	0.04	0.01	0.07	—
227	Applesauce, unsweetened, canned	½	cup(s)	122	108	52	<1	14	1	<.1	0.01	0.00	0.02	—
38492	Crabapples	1	item(s)	35	28	27	<1	7	1	<1	0.02	0.00	0.03	—
	Apricot													
228	Fresh w/o pits	4	item(s)	140	121	67	2	16	3	1	0.04	0.24	0.11	—
230	Halves, dried, sulfured	¼	cup(s)	33	10	79	1	21	2	<1	0.01	0.02	0.02	—
229	Halves w/skin, canned in heavy syrup	½	cup(s)	129	100	107	1	28	2	<1	0.01	0.04	0.02	—
	Avocado													
233	California, whole, w/o skin or pit	1	item(s)	170	123	284	3	15	12	26	3.59	16.61	3.42	—
234	Florida, whole, w/o skin or pit	1	item(s)	304	240	365	7	24	17	31	5.90	16.70	5.00	—
2998	Pureed	⅛	cup(s)	29	21	46	1	2	2	4	0.61	2.82	0.52	—
	Banana													
235	Fresh whole, w/o peel	1	item(s)	118	88	105	1	27	3	<1	0.13	0.04	0.09	—
4580	Dried chips	¼	cup(s)	55	2	287	1	32	4	19	16.00	1.08	0.35	—
	Blackberries													
237	Raw	½	cup(s)	72	63	31	1	7	4	<1	0.01	0.03	0.20	—
958	Unsweetened, frozen	½	cup(s)	76	62	48	1	12	4	<1	0.01	0.03	0.18	—
	Blueberries													
238	Raw	½	cup(s)	72	61	41	1	10	2	<1	0.02	0.03	0.11	—
959	Canned in heavy syrup	½	cup(s)	128	98	113	1	28	2	<1	0.03	0.06	0.18	—
960	Unsweetened, frozen	½	cup(s)	78	67	40	1	10	2	1	0.04	0.07	0.22	—

Chol (mg)	Calc (mg)	Iron (mg)	Magn (mg)	Pota (mg)	Sodi (mg)	Zinc (mg)	Vit A (μg)	Thia (mg)	Vit E (mg α)	Ribo (mg)	Niac (mg)	Vit B$_6$ (mg)	Fola (μg)	Vit C (mg)	Vit B$_{12}$ (μg)	Sele (μg)
0	100	16.20	60	220	390	3.75	158	0.38	—	0.03	5.00	0.50	100	6	2	5
0	100	18.00	24	85	200	15.00	—	1.50	—	1.70	20.00	2.00	400	15	6	—
0	0	1.46	32	146	279	4.99	0	0.50	—	0.57	6.65	0.67	133	20	2	9
0	0	18.00	16	50	210	15.00	225	1.50	—	1.70	20.00	2.00	400	60	6	4
0	1	4.44	4	16	<1	0.14	0	0.36	—	0.25	4.94	0.01	3	0	0	1
0	3	3.8	17	42	<1	0.28	0	0.31	—	0.21	4.23	0.02	4	0	0	15
0	20	10.80	80	340	300	2.25	—	0.53	—	0.60	7.00	0.70	140	0	2	—
0	80	7.20	7	28	232	3.00	—	0.30	—	0.34	4.00	0.40	160	5	1	1
0	0	1.44	13	32	256	0.48	120	0.30	—	0.34	4.80	0.40	80	5	1	4
0	10	1.08	31	92	2	0.70	0	0.07	—	0.06	1.77	0.10	12	0	0	1
0	0	0.48	11	53	67	0.40	200	0.50	—	0.57	6.65	0.67	133	8	2	17
0	0	8.70	16	60	220	0.90	225	0.53	—	0.60	7.00	2.00	400	15	6	7
0	750	13.50	0	23	158	11.25	113	1.13	22.50	1.28	15.00	1.50	300	45	5	1
0	1330	23.94	32	120	253	19.95	200	2.00	31.24	2.26	26.60	2.66	532	80	8	2
0	100	4.50	0	15	190	3.75	150	0.38	—	0.43	5.00	0.50	100	6	2	6
0	6	1.28	45	134	<1	2.35	0	0.23	—	0.11	0.78	0.13	50	<1	0	9
0	0	8.10	32	110	220	7.50	150	0.75	2.26	0.85	10.00	1.00	200	6	3	1
0	5	1.06	12	27	99	0.32	0	0.13	—	0.09	1.34	0.02	20	0	0	10
0	1	0.18	25	22	0	0.44	2	0.04	0.78	0.02	0.39	0.04	4	0	0	2
26	10	1.27	15	22	6	0.49	5	0.15	0.14	0.07	1.19	0.03	51	0	<.1	17
0	5	0.98	13	22	1	0.37	0	0.14	0.04	0.07	1.17	0.02	54	0	0	15
21	4	0.73	12	15	4	0.36	4	0.13	—	0.10	0.63	0.02	41	0	<.1	—
18	9	0.89	9	34	415	0.31	—	0.08	—	0.05	0.71	0.03	4	<.1	<.1	—
0	4	0.45	9	33	57	0.11	0	0.09	—	0.02	0.48	0.04	7	0	0	—
0	7	0.46	2	25	141	0.19	0	0.02	—	0.03	0.09	0.01	2	0	0	—
0	7	1.00	12	52	1	0.35	0	0.12	0.04	0.07	0.90	0.04	8	0	0	40
0	11	0.74	21	31	2	0.57	0	0.08	0.21	0.03	0.49	0.06	4	0	0	18
26	15	0.87	19	30	10	0.50	4	0.20	0.46	0.10	1.18	0.09	51	0	<1	17
0	7	0.33	13	21	4	0.29	3	0.08	0.06	0.04	0.72	0.02	44	0	0	13
0	1	0.22	11	24	<1	0.28	1	0.02	0.02	0.02	0.16	0.02	2	0	0	1
2	15	0.61	12	38	73	0.20	1	0.02	0.42	0.02	0.77	0.01	2	0	<.1	1
4	42	0.83	34	97	331	0.75	14	0.05	—	0.09	0.54	0.09	4	<1	<1	4
0	3	0.92	36	74	292	0.87	3	0.04	—	0.04	0.51	0.07	6	<.1	0	2
0	8	0.17	7	148	1	0.06	4	0.02	—	0.04	0.13	0.06	4	6	0	0
0	3	0.07	3	59	1	0.02	2	0.01	—	0.01	0.05	0.02	2	3	0	0
0	4	0.16	3	75	1	0.03	2	0.01	0.04	0.01	0.08	0.04	1	<1	0	<1
0	3	0.30	3	97	19	0.04	0	0.00	0.11	0.03	0.20	0.03	0	1	0	<1
0	7	0.31	6	151	8	0.05	0	0.00	0.01	0.02	0.05	0.04	0	1	0	<1
0	9	0.46	4	148	4	0.04	0	0.03	0.01	0.02	0.12	0.04	0	1	0	<1
0	5	0.45	4	78	4	0.05	1	0.02	0.27	0.04	0.24	0.03	1	2	0	<1
0	4	0.15	4	92	2	0.04	1	0.02	0.26	0.03	0.23	0.03	1	1	0	<1
0	6	0.13	2	68	<1	—	0	0.01	—	0.01	0.04	—	2	3	0	—
0	18	0.55	14	363	1	0.28	134	0.04	1.25	0.06	0.84	0.08	13	14	0	<1
0	18	0.88	11	383	3	0.13	59	0.00	1.43	0.02	0.85	0.05	3	<1	0	1
0	12	0.39	9	181	5	0.14	80	0.03	0.77	0.03	0.49	0.07	3	4	0	<1
0	22	1.00	49	861	14	1.12	104	0.12	3.35	0.24	3.24	0.47	105	15	0	1
0	30	0.50	73	1067	6	1.20	185	0.00	0.09	0.10	2.00	0.20	106	53	0	0
0	3	0.16	8	139	2	0.18	2	0.02	0.60	0.04	0.50	0.07	17	3	0	<1
0	6	0.31	32	422	1	0.18	4	0.04	0.12	0.09	0.78	0.43	24	10	0	1
0	10	0.69	42	296	3	0.41	2	0.05	0.13	0.01	0.39	0.14	8	3	0	1
0	21	0.45	14	117	1	0.38	8	0.01	0.84	0.02	0.47	0.02	18	15	0	<1
0	22	0.60	17	106	1	0.19	5	0.02	0.88	0.03	0.91	0.05	26	2	0	<1
0	4	0.20	4	55	1	0.12	2	0.03	0.41	0.03	0.30	0.04	4	7	0	<.1
0	6	0.42	5	51	4	0.09	3	0.04	0.49	0.07	0.14	0.05	3	1	0	<1
0	6	0.14	4	42	1	0.06	2	0.03	0.37	0.03	0.41	0.05	6	2	0	0

TABLE H–1

Food Composition

(DA+ code is for Wadsworth Diet Analysis program)   (For purposes of calculations, use "0" for t, <1, <.1, <.01, etc.)

DA + Code	Food Description	Quantity	Measure	Wt (g)	H₂O (g)	Ener (kcal)	Prot (g)	Carb (g)	Fiber (g)	Fat (g)	Sat	Mono	Poly	Trans
												Fat Breakdown (g)		
	FRUIT AND FRUIT JUICES—Continued													
	Boysenberries													
961	Canned in heavy syrup	½	cup(s)	128	98	113	1	29	3	<1	0.01	0.02	0.09	—
962	Unsweetened, frozen	½	cup(s)	66	57	33	1	8	3	<1	0.01	0.02	0.10	—
35576	**Breadfruit**	1	item(s)	384	271	396	4	104	17	1	0.00	0.00	0.00	—
	Cherries													
3000	Sour red, raw	½	cup(s)	78	67	39	1	9	1	<1	0.05	0.06	0.07	—
967	Sour red, canned in water	½	cup(s)	122	110	44	1	11	1	<1	0.03	0.03	0.04	—
240	Sweet, raw	½	cup(s)	73	60	46	1	12	2	<1	0.03	0.03	0.04	—
3004	Sweet, canned in heavy syrup	½	cup(s)	127	98	105	1	27	2	<1	0.04	0.05	0.06	—
969	Sweet, canned in water	½	cup(s)	124	108	57	1	15	2	<1	0.03	0.04	0.05	—
	Cranberries													
3007	Chopped, raw	½	cup(s)	55	48	25	<1	7	3	<.1	0.01	0.01	0.03	—
1638	Cranberry juice cocktail	½	cup(s)	127	108	72	0	18	<1	<1	0.01	0.02	0.06	—
241	Cranberry juice cocktail, low calorie, w/saccharin	½	cup(s)	127	120	24	<.1	6	0	<.1	0.00	0.00	0.00	—
1717	Cranberry apple juice drink	½	cup(s)	123	100	87	<.1	22	<1	<.1	0.00	0.00	0.00	—
242	Cranberry sauce, sweetened, canned	¼	cup(s)	69	42	105	<1	27	1	<1	0.01	0.01	0.05	—
	Dates													
244	Domestic, chopped	¼	cup(s)	45	0	126	1	33	4	<1	0.01	0.01	0	—
243	Domestic, whole	¼	cup(s)	45	0	126	1	33	4	<1	0.01	0.01	0	—
	Figs													
973	Raw, medium	2	item(s)	101	80	74	1	19	3	<1	0.06	0.07	0.14	—
975	Canned in heavy syrup	½	cup(s)	130	99	114	<1	30	3	<1	0.03	0.03	0.06	—
974	Canned in water	½	cup(s)	124	106	66	<1	17	3	<1	0.02	0.03	0.06	—
	Fruit cocktail & salad													
245	Fruit cocktail, canned in heavy syrup	½	cup(s)	124	100	91	<1	23	1	<.1	0.01	0.02	0.04	—
978	Fruit cocktail, canned in juice	½	cup(s)	119	104	55	1	14	1	<.1	0.00	0.00	0.00	—
977	Fruit cocktail, canned in water	½	cup(s)	119	108	38	<1	10	1	<.1	0.01	0.01	0.02	—
979	Fruit salad, canned in water	½	cup(s)	123	112	37	<1	10	1	<.1	0.01	0.02	0.03	—
	Gooseberries													
981	Raw	½	cup(s)	75	66	33	1	8	3	<1	0.03	0.04	0.24	—
982	Canned in light syrup	½	cup(s)	126	101	92	1	24	3	<1	0.02	0.02	0.14	—
	Grapefruit													
3022	Raw, pink or red	½	cup(s)	115	101	48	1	12	2	<1	0.02	0.02	0.04	—
247	Raw, white	½	item(s)	118	107	39	1	10	1	<1	0.02	0.02	0.03	—
251	Juice, pink, sweetened, canned	½	cup(s)	125	109	58	1	14	<1	<1	0.02	0.02	0.03	—
249	Juice, white	½	cup(s)	124	111	48	1	11	<1	<1	0.02	0.02	0.03	—
248	Sections, canned in light syrup	½	cup(s)	127	106	76	1	20	1	<1	0.02	0.02	0.03	—
983	Sections, canned in water	½	cup(s)	122	110	44	1	11	<1	<1	0.02	0.02	0.03	—
	Grapes													
255	American, slip skin	½	cup(s)	46	37	31	<1	8	<1	<1	0.05	0.01	0.05	—
256	European, red or green, adherent skin	½	cup(s)	80	61	55	1	14	1	<1	0.04	0.01	0.04	—
259	Juice, sweetened, added vitamin C, from frozen concentrate	½	cup(s)	125	109	64	<1	16	<1	<1	0.04	0.01	0.03	—
3159	Juice drink, canned	½	cup(s)	125	109	63	<1	16	0	0	0.00	0.000	0.00	—
3060	Raisins, seeded, packed	¼	cup(s)	41	7	122	1	32	3	<1	0.07	0.01	0.07	—
987	**Guava, raw**	1	item(s)	90	77	46	1	11	5	1	0.15	0.05	0.23	—
35593	**Guava, strawberry**	1	item(s)	6	5	4	<.1	1	<1	<.1	0.01	0.00	0.02	—
3027	**Jackfruit**	½	cup(s)	83	61	78	1	20	1	<1	0.05	0.04	0.07	—
8458	**Kiwi fruit**	1	item(s)	77	63	53	1	11	3	1	0.02	0.03	0.19	—
	Lemon													
992	Raw	1	item(s)	108	94	22	1	12	5	<1	0.04	0.01	0.10	—
262	Juice	1	tablespoon(s)	15	14	4	<.1	1	<.1	0	0.00	0.00	0.00	—
993	Peel	1	teaspoon(s)	2	2	1	<.1	<1	<1	<.1	0.00	0.00	0.00	—
	Lime													
994	Raw	1	item(s)	67	61	15	<1	6	2	<.1	0.01	0.01	0.02	—
269	Juice	1	tablespoon(s)	15	14	4	<.1	1	<.1	<.1	0.00	0.00	0.00	—
995	**Loganberries, frozen**	½	cup(s)	74	62	40	1	10	4	<1	0.01	0.02	0.13	—
	Mandarin orange													
1038	Canned in juice	½	cup(s)	125	111	46	1	12	1	<.1	0.00	0.01	0.01	—
1039	Canned in light syrup	½	cup(s)	126	105	77	1	20	1	<1	0.02	0.02	0.03	—
999	**Mango**	½	item(s)	104	85	67	1	18	2	<1	0.07	0.10	0.05	—
1005	**Nectarine, raw, sliced**	½	cup(s)	69	60	30	1	7	1	<1	0.02	0.06	0.08	—
	Melons													
271	Cantaloupe	½	cup(s)	80	72	27	1	7	1	<1	0.04	0.00	0.07	—
1000	Casaba melon	½	cup(s)	85	78	24	1	6	1	<.1	0.02	0.00	0.03	—

PAGE KEY: H–2 = Breads/Baked Goods H–6 = Cereal/Rice/Pasta H–10 = Fruit H–14 = Vegetables/Legumes H–24 = Nuts/Seeds H–26 = Vegetarian
H–28 = Dairy H–34 = Eggs H–34 = Seafood H–36 = Meats H–40 = Poultry H–40 = Processed meats H–42 = Beverages H–46 = Fats/Oils
H–48 = Sweets H–50 = Spices/Condiments/Sauces H–52 = Mixed foods/Soups/Sandwiches H–58 = Fast food H–74 = Convenience meals H–76 = Baby foods

Chol (mg)	Calc (mg)	Iron (mg)	Magn (mg)	Pota (mg)	Sodi (mg)	Zinc (mg)	Vit A (µg)	Thia (mg)	Vit E (mg α)	Ribo (mg)	Niac (mg)	Vit B$_6$ (mg)	Fola (µg)	Vit C (mg)	Vit B$_{12}$ (µg)	Sele (µg)
0	23	0.55	14	115	4	0.24	3	0.03	—	0.04	0.29	0.05	44	8	0	1
0	18	0.56	11	92	1	0.15	2	0.03	0.57	0.02	0.51	0.04	42	2	0	<1
0	65	2.07	96	1882	8	0.46	8	0.42	—	0.12	3.46	0.00	54	111	0	2
0	12	0.25	7	134	2	0.08	50	0.02	0.05	0.03	0.31	0.03	6	8	0	0
0	13	1.67	7	120	9	0.09	46	0.02	0.28	0.05	0.22	0.05	10	3	0	0
0	9	0.26	8	161	0	0.05	2	0.02	0.05	0.02	0.11	0.04	3	5	0	0
0	11	0.44	11	183	4	0.13	10	0.03	0.29	0.05	0.50	0.04	5	5	0	0
0	14	0.45	11	162	1	0.10	10	0.03	0.29	0.05	0.51	0.04	5	3	0	0
0	4	0.14	3	47	1	0.06	2	0.01	0.66	0.01	0.06	0.03	1	7	0	<.1
0	4	0.19	3	23	3	0.09	0	0.01	0.28	0.01	0.04	0.02	0	45	0	<1
0	11	0.05	3	32	4	0.03	0	0.00	0.06	0.00	0.01	0.00	0	41	0	0
0	6	0.15	2	34	9	0.22	0	0.01	0.15	0.02	0.07	0.03	0	39	0	0
0	3	0.15	2	18	20	0.03	1	0.01	0.57	0.01	0.07	0.01	1	1	0	<1
0	17	0.45	19	292	1	0.12	1	0.02	0.02	0.02	0.56	0.07	9	<1	0	1
0	17	0.45	19	292	1	0.12	1	0.02	0.02	0.02	0.56	0.07	9	<1	0	1
0	35	0.37	17	233	1	0.15	7	0.06	0.11	0.05	0.40	0.11	6	2	0	<1
0	35	0.36	13	128	1	0.14	3	0.03	0.16	0.05	0.55	0.09	3	1	0	<1
0	35	0.36	12	128	1	0.15	2	0.03	0.10	0.05	0.55	0.09	2	1	0	<1
0	7	0.36	6	109	7	0.10	12	0.02	0.50	0.02	0.46	0.06	4	2	0	1
0	9	0.25	8	113	5	0.11	18	0.01	0.47	0.02	0.48	0.06	4	3	0	1
0	6	0.30	8	111	5	0.11	15	0.02	0.47	0.01	0.43	0.06	4	2	0	1
0	9	0.37	6	96	4	0.10	27	0.02	—	0.03	0.46	0.04	4	2	0	1
0	19	0.23	8	149	1	0.09	11	0.03	0.28	0.02	0.23	0.06	5	21	0	<1
0	20	0.42	8	97	3	0.14	9	0.03	—	0.07	0.19	0.02	4	13	0	1
0	25	0.09	10	155	0	0.08	30	0.05	0.15	0.03	0.23	0.06	15	36	0	<1
0	14	0.07	11	175	0	0.08	2	0.04	0.15	0.02	0.32	0.05	12	39	0	2
0	10	0.45	13	203	3	0.08	0	0.05	0.05	0.03	0.40	0.03	13	34	0	<1
0	11	0.25	15	200	1	0.06	2	0.05	0.27	0.02	0.25	0.05	12	47	0	<1
0	18	0.51	13	164	3	0.10	0	0.05	0.11	0.03	0.31	0.03	11	27	0	1
0	18	0.50	12	161	2	0.11	0	0.05	0.11	0.03	0.30	0.02	11	27	0	1
0	6	0.13	2	88	1	0.02	2	0.04	0.09	0.03	0.14	0.05	2	2	0	<.1
0	8	0.29	6	153	2	0.06	6	0.06	0.15	0.06	0.15	0.07	2	9	0	<.1
0	5	0.13	5	26	3	0.05	0	0.02	0.00	0.03	0.16	0.05	1	30	0	<1
0	4	0.13	4	41	1	0.03	0	0.01	0.00	0.02	0.09	0.02	1	20	0	<1
0	12	1.07	12	340	12	0.07	0	0.05	—	0.08	0.46	0.08	1	2	0	<1
0	18	0.28	9	256	3	0.21	28	0.05	0.66	0.05	1.08	0.13	13	165	0	1
0	1	0.01	1	18	2	—	—	0.00	—	0.00	0.04	0.00	—	2	0	—
0	28	0.50	31	251	2	0.35	12	0.02	—	0.09	0.33	0.09	12	6	0	<1
0	30	0.38	14	251	2	0.10	4	—	—	0.02	0.25	0.05	<.1	74	0	—
0	66	0.76	13	157	3	0.11	2	0.05	—	0.04	0.22	0.12	—	83	0	1
0	1	0.00	1	19	<1	0.01	<1	0.00	0.02	0.00	0.02	0.01	2	7	0	<.1
0	3	0.02	<1	3	<1	0.01	<.1	0.00	0.00	0.00	0.01	0.00	<1	3	0	<.1
0	9	0.06	5	78	1	0.05	1	0.02	0.15	0.01	0.10	0.03	7	20	0	<.1
0	1	0.00	1	17	<1	0.01	<1	0.00	0.03	0.00	0.02	0.01	1	5	0	<.1
0	19	0.47	15	107	1	0.25	1	0.04	0.64	0.02	0.62	0.05	19	11	0	<1
0	14	0.34	14	166	6	0.63	54	0.10	0.12	0.04	0.55	0.05	6	43	0	<1
0	9	0.47	10	98	8	0.30	53	0.07	0.13	0.06	0.56	0.05	6	25	0	1
0	10	0.13	9	161	2	0.04	39	0.06	1.16	0.06	0.60	0.14	14	29	0	1
0	4	0.19	6	139	0	0.12	12	0.02	0.53	0.02	0.78	0.02	3	4	0	0
0	7	0.17	10	215	13	0.14	136	0.03	0.04	0.02	0.59	0.06	17	30	0	<1
0	9	0.29	9	155	8	0.06	0	0.01	0.04	0.03	0.20	0.14	7	19	0	<1

H Appendix

TABLE H–1

Food Composition   (DA+ code is for Wadsworth Diet Analysis program)   (For purposes of calculations, use "0" for t, <1, <.1, <.01, etc.)

DA + Code	Food Description	Quantity	Measure	Wt (g)	H₂O (g)	Ener (kcal)	Prot (g)	Carb (g)	Fiber (g)	Fat (g)	Sat	Mono	Poly	Trans
	FRUIT AND FRUIT JUICES—Continued													
272	Honeydew	½	cup(s)	89	80	32	<1	8	1	<1	0.03	0.00	0.05	—
318	Watermelon	½	cup(s)	77	71	23	<1	6	<1	<1	0.01	0.03	0.04	—
	Orange													
273	Raw	1	item(s)	131	114	62	1	15	3	<1	0.02	0.03	0.03	—
3040	Peel	1	teaspoon(s)	2	1	2	<.1	1	<1	<.1	0.00	0.00	0.00	—
274	Sections	½	cup(s)	90	78	43	1	11	2	<1	0.01	0.02	0.02	—
275	Juice	½	cup(s)	124	109	56	1	13	<1	<1	0.03	0.04	0.05	—
29630	Juice, fresh squeezed	½	cup(s)	124	109	56	1	13	<1	<1	0.03	0.04	0.05	—
14414	Juice w/calcium & extra vitamin C	½	cup(s)	125	109	55	1	13	<1	0	0.00	0.00	0.00	—
278	Juice, unsweetened, from frozen concentrate	½	cup(s)	125	110	56	1	13	<1	<.1	0.01	0.01	0.01	—
	Papaya													
282	Raw	½	cup(s)	70	62	27	<1	7	1	<.1	0.03	0.03	0.02	—
16830	Dried, strips	2	item(s)	46	12	119	2	30	5	<1	0.13	0.12	0.09	—
35640	**Passion fruit, purple**	1	item(s)	18	13	17	<1	4	3	<1	0.00	0.00	0.00	—
	Peach													
283	Raw, medium	1	item(s)	98	87	38	1	9	1	<1	0.02	0.07	0.08	—
285	Halves, canned in heavy syrup	½	cup(s)	131	104	97	1	26	2	<1	0.01	0.05	0.06	—
286	Halves, canned in water	½	cup(s)	122	114	29	1	7	2	<.1	0.01	0.03	0.03	—
290	Slices, sweetened, frozen	½	cup(s)	125	93	118	1	30	2	<1	0.02	0.06	0.08	—
	Pear													
291	Raw	1	item(s)	166	139	96	1	26	5	<1	0.01	0.04	0.05	—
8672	Asian	1	item(s)	122	108	51	1	13	4	<1	0.01	0.06	0.07	—
293	Danjou	1	item(s)	200	168	120	1	30	5	1	0.00	0.20	0.20	—
294	Halves, canned in heavy syrup	½	cup(s)	133	107	98	<1	25	2	<1	0.01	0.04	0.04	—
1012	Halves, canned in juice	½	cup(s)	124	107	62	<1	16	2	<.1	0.00	0.02	0.02	—
1017	**Persimmon**	1	item(s)	25	16	32	<1	8	0	<1	0.01	0.02	0.02	—
	Pineapple													
295	Raw, diced	½	cup(s)	78	67	37	<1	10	1	<.1	0.01	0.01	0.03	—
3053	Canned in extra heavy syrup	½	cup(s)	130	101	108	<1	28	1	<1	0.01	0.02	0.05	—
1019	Canned in juice	½	cup(s)	125	104	75	1	20	1	<.1	0.01	0.01	0.04	—
296	Canned in light syrup	½	cup(s)	126	108	66	<1	17	1	<1	0.01	0.02	0.05	—
1018	Canned in water	½	cup(s)	123	112	39	1	10	1	<1	0.01	0.01	0.04	—
299	Juice, unsweetened, canned	½	cup(s)	125	107	70	<1	17	<1	<1	0.01	0.01	0.04	—
1024	**Plantain, cooked**	½	cup(s)	77	52	89	1	24	2	<1	0.05	0.01	0.03	—
300	**Plum, raw, large**	1	item(s)	83	72	38	1	9	1	<1	0.01	0.11	0.04	—
1027	**Pomegranate**	1	item(s)	154	125	105	1	26	1	<1	0.06	0.07	0.10	—
	Prunes													
5644	Dried	2	item(s)	17	5	40	<1	11	1	<.1	0.01	0.06	0.02	—
305	Dried, stewed	½	cup(s)	119	86	128	1	33	4	<1	0.00	0.15	0.04	—
306	Juice, canned	1	cup(s)	256	208	182	2	45	3	<.1	0.01	0.05	0.02	—
	Raisins, *see* grapes													
	Raspberries													
309	Raw	½	cup(s)	62	53	32	1	7	4	<1	0.01	0.04	0.23	—
310	Red, sweetened, frozen	½	cup(s)	125	91	129	1	33	6	<1	0.01	0.02	0.11	—
311	**Rhubarb, cooked with sugar**	½	cup(s)	120	82	140	1	38	3	<.1	0.00	0.00	0.05	—
	Strawberries													
313	Raw	½	cup(s)	72	65	23	<1	6	1	<1	0.01	0.03	0.11	—
315	Sweetened, frozen, thawed	½	cup(s)	128	100	99	1	27	2	<1	0.01	0.02	0.09	—
16828	**Tangelo**	1	item(s)	95	82	45	1	11	2	<1	0.01	0.02	0.02	—
	Tangerine													
316	Raw	1	item(s)	84	74	37	1	9	2	<1	0.02	0.03	0.03	—
1040	Juice	½	cup(s)	124	110	53	1	12	<1	<1	0.03	0.04	0.05	—
	VEGETABLES, LEGUMES													
	Amaranth													
1042	Leaves, raw	1	cup(s)	28	26	6	1	1	0	<.1	0.03	0.02	0.04	—
1043	Leaves, boiled, drained	½	cup(s)	66	60	14	1	3	0	<1	0.03	0.03	0.05	—
8683	**Arugula leaves, raw**	1	cup(s)	20	18	5	1	1	<1	<1	0.02	0.01	0.06	—
	Artichoke													
1044	Boiled, drained	1	item(s)	120	101	60	4	13	6	<1	0.04	0.01	0.08	—
2885	Hearts, boiled, drained	½	cup(s)	84	71	42	3	9	5	<1	0.03	0.00	0.06	—
	Asparagus													
566	Boiled, drained	½	cup(s)	90	83	20	2	4	2	<1	0.06	0	0.12	—
568	Canned, drained	½	cup(s)	121	114	23	3	3	2	1	0.18	0.03	0.34	—
565	Tips, frozen, boiled, drained	½	cup(s)	90	82	25	3	4	1	<1	0.09	0.01	0.17	—

PAGE KEY: H–2 = Breads/Baked Goods H–6 = Cereal/Rice/Pasta H–10 = Fruit H–14 = Vegetables/Legumes H–24 = Nuts/Seeds H–26 = Vegetarian
H–28 = Dairy H–34 = Eggs H–34 = Seafood H–36 = Meats H–40 = Poultry H–40 = Processed meats H–42 = Beverages H–46 = Fats/Oils
H–48 = Sweets H–50 = Spices/Condiments/Sauces H–52 = Mixed foods/Soups/Sandwiches H–58 = Fast food H–74 = Convenience meals H–76 = Baby foods

Chol (mg)	Calc (mg)	Iron (mg)	Magn (mg)	Pota (mg)	Sodi (mg)	Zinc (mg)	Vit A (µg)	Thia (mg)	Vit E (mg α)	Ribo (mg)	Niac (mg)	Vit B$_6$ (mg)	Fola (µg)	Vit C (mg)	Vit B$_{12}$ (µg)	Sele (µg)
0	5	0.15	9	203	16	0.08	3	0.03	0.02	0.01	0.37	0.08	17	16	0	1
0	5	0.19	8	86	1	0.08	22	0.03	0.04	0.02	0.14	0.03	2	6	0	<1
0	52	0.13	13	237	0	0.09	14	0.11	0.24	0.05	0.37	0.08	39	70	0	1
0	3	0.02	<1	4	<.1	0.01	<1	0.00	0.00	0.00	0.02	0.00	1	3	0	<.1
0	36	0.09	9	164	0	0.06	10	0.08	0.16	0.04	0.26	0.05	27	48	0	<1
0	14	0.25	14	248	1	0.06	12	0.11	0.05	0.04	0.50	0.05	37	62	0	<1
0	14	0.25	14	248	1	0.06	—	0.11	0.05	0.04	0.50	0.05	38	62	0	—
0	176	—	—	226	0	—	5	0.08	—	—	0.40	0.06	30	54	0	—
0	11	0.12	12	237	1	0.06	6	0.10	0.25	0.02	0.25	0.05	55	48	0	<1
0	17	0.07	7	180	2	0.05	39	0.02	0.51	0.02	0.24	0.01	27	43	0	<1
0	73	0.30	30	783	9	0.21	—	0.06	2.22	0.09	0.93	0.05	58	38	0	—
0	2	0.29	5	63	5	—	—	0.00	—	0.02	0.27	—	3	5	0	<1
0	6	0.25	9	186	0	0.17	16	0.02	0.72	0.03	0.79	0.02	4	6	0	<.1
0	4	0.35	7	121	8	0.12	22	0.01	0.64	0.03	0.80	0.02	4	4	0	<1
0	2	0.39	6	121	4	0.11	33	0.01	0.60	0.02	0.64	0.02	4	4	0	<1
0	4	0.46	6	163	8	0.06	18	0.02	0.77	0.04	0.82	0.02	4	118	0	1
0	15	0.28	12	198	2	0.17	2	0.02	0.20	0.04	0.26	0.05	12	7	0	<1
0	5	0.00	10	148	0	0.02	0	0.01	0.15	0.01	0.27	0.03	10	5	0	<1
0	22	0.50	12	250	0	0.24	—	0.04	1.00	0.08	0.20	0.04	15	8	0	1
0	7	0.29	5	86	7	0.11	0	0.01	0.11	0.03	0.32	0.02	1	1	0	0
0	11	0.36	9	119	5	0.11	0	0.01	0.10	0.01	0.25	0.02	1	2	0	0
0	7	0.63	—	78	<1	—	—	—	—	—	—	—	—	17	0	0
0	10	0.22	9	89	1	0.08	2	0.06	0.02	0.02	0.38	0.09	12	28	0	<.1
0	18	0.49	20	133	1	0.14	1	0.12	—	0.03	0.37	0.10	7	9	0	—
0	17	0.35	17	152	1	0.12	2	0.12	0.01	0.02	0.35	0.09	6	12	0	<1
0	18	0.49	20	132	1	0.15	3	0.11	0.01	0.03	0.37	0.09	6	9	0	1
0	18	0.49	22	156	1	0.15	2	0.11	0.01	0.03	0.37	0.09	6	9	0	<1
0	21	0.33	16	168	1	0.14	0	0.07	0.03	0.03	0.32	0.12	29	13	0	<1
0	2	0.45	25	358	4	0.10	35	0.04	0.10	0.04	0.58	0.18	20	8	0	1
0	5	0.14	6	130	0	0.08	14	0.02	0.21	0.02	0.34	0.02	4	8	0	0
0	5	0.46	5	399	5	0.18	8	0.05	0.92	0.05	0.46	0.16	9	9	0	1
0	9	0.42	8	125	1	0.09	17	0.01	0.00	0.03	0.33	0.04	1	1	0	<1
0	23	0.46	21	383	1	0.19	37	0.00	0.23	0.12	0.85	0.23	0	3	0	<1
0	31	3.02	36	707	10	0.54	0	0.04	0.31	0.18	2.01	0.56	0	10	0	2
0	15	0.42	14	93	1	0.26	1	0.02	0.54	0.02	0.37	0.03	13	16	0	<1
0	19	0.81	16	143	1	0.23	4	0.02	0.90	0.06	0.29	0.04	33	21	0	<1
0	174	0.25	16	115	1	—	—	0.02	—	0.03	0.25	—	—	4	0	—
0	12	0.30	9	110	1	0.10	1	0.02	0.21	0.02	0.28	0.03	17	42	0	<1
0	14	0.60	8	125	1	0.06	1	0.02	0.31	0.10	0.37	0.04	5	50	0	1
0	38	0.10	10	172	0	0.07	—	0.08	0.17	0.04	0.27	0.06	29	51	0	—
0	12	0.08	10	132	1	0.20	29	0.09	0.17	0.02	0.13	0.06	17	26	0	<1
0	22	0.25	10	220	1	0.04	16	0.07	0.16	0.02	0.12	0.05	6	38	0	<1
0	60	0.65	15	171	6	0.25	0	0.01	—	0.04	0.18	0.05	24	12	0	<1
0	138	1.49	36	423	14	0.58	92	0.01	—	0.09	0.37	0.12	38	27	0	1
0	32	0.29	9	74	5	0.09	24	0.01	0.09	0.02	0.06	0.01	19	3	0	<.1
0	54	1.55	72	425	114	0.59	11	0.08	0.23	0.08	1.20	0.13	61	12	0	<1
0	38	1.08	50	297	80	0.41	8	0.05	0.16	0.06	0.84	0.09	43	8	0	<1
0	21	0.81	12	202	13	0.54	49	0.14	1.35	0.12	0.97	0.07	134	7	0	5
0	19	0.73	12	208	347	0.48	50	0.07	0.38	0.12	1.15	0.13	116	22	0	2
0	21	0.58	12	196	4	0.50	—	0.06	1.08	0.09	0.93	0.02	121	22	0	4

TABLE H–1
Food Composition

(DA+ code is for Wadsworth Diet Analysis program) (For purposes of calculations, use "0" for t, <1, <.1, <.01, etc.)

DA + Code	Food Description	Quantity	Measure	Wt (g)	H₂O (g)	Ener (kcal)	Prot (g)	Carb (g)	Fiber (g)	Fat (g)	Sat	Mono	Poly	Trans
	VEGETABLES, LEGUMES —Continued													
	Bamboo shoots													
1048	Boiled, drained	½	cup(s)	60	58	7	1	1	1	<1	0.03	0.00	0.06	—
1049	Canned, drained	½	cup(s)	65	62	12	1	2	1	<1	0.06	0.01	0.12	—
	Beans													
1801	Adzuki beans, boiled	½	cup(s)	115	76	147	9	28	8	<1	0.04	—	—	—
511	Baked beans w/franks, canned	½	cup(s)	129	89	182	9	20	9	8	3.02	3.64	1.07	—
512	Baked beans w/pork in tomato sauce, canned	½	cup(s)	127	92	124	7	25	6	1	0.50	0.56	0.17	—
513	Baked beans w/pork in sweet sauce, canned	½	cup(s)	127	89	140	7	27	7	2	0.71	0.80	0.24	0
1805	Black beans, boiled	½	cup(s)	86	57	114	8	20	7	<1	0.12	0.04	0.20	—
14597	Chickpeas, garbanzo beans, or bengal gram, boiled	½	cup(s)	82	49	134	7	22	6	2	0.22	0.48	0.95	—
569	Fordhook lima beans, frozen, boiled, drained	½	cup(s)	85	62	88	5	16	5	<1	0.07	0.02	0.14	—
1806	French beans, boiled	½	cup(s)	89	59	114	6	21	8	1	0.07	0.05	0.40	—
2773	Great northern beans, boiled	½	cup(s)	89	61	104	7	19	6	<1	0.12	0.02	0.17	—
2736	Hyacinth beans, boiled, drained	½	cup(s)	44	38	22	1	4	0	<1	0.05	0.06	0.00	—
515	Lima beans, boiled, drained	½	cup(s)	85	57	105	6	20	5	<1	0.06	0.02	0.13	—
570	Lima beans, baby, frozen, boiled, drained	½	cup(s)	90	65	95	6	18	5	<1	0.06	0.02	0.13	—
579	Mung beans, sprouted, boiled, drained	½	cup(s)	62	62	13	1	3	<1	<.1	0.02	0.00	0.02	—
510	Navy beans, boiled	½	cup(s)	91	57	129	8	24	6	1	0.13	0.05	0.22	0
32816	Pinto beans, boiled, drained, no salt added	½	cup(s)	114	106	25	2	5	0	<1	0.04	0.03	0.21	—
1052	Pinto beans, frozen, boiled, drained	½	cup(s)	47	27	76	4	15	4	<1	0.03	0.02	0.13	—
514	Red kidney beans, canned	½	cup(s)	128	99	109	7	20	8	<1	0.06	0.03	0.24	—
1810	Refried beans, canned	½	cup(s)	127	96	119	7	20	7	2	0.60	0.71	0.19	—
1053	Shell beans, canned	½	cup(s)	123	111	37	2	8	4	<1	0.03	0.02	0.13	—
1670	Soybeans, boiled	½	cup(s)	86	54	149	14	9	5	8	1.12	1.70	4.36	—
1108	Soybeans, green, boiled, drained	½	cup(s)	90	62	127	11	10	4	6	0.67	1.09	2.71	—
1807	White beans, small, boiled	½	cup(s)	90	57	127	8	23	9	1	0.15	0.05	0.25	—
574	Green string beans, canned, fat added in cooking	½	cup(s)	93	63	41	1	4	2	3	0.51	1.23	0.75	—
575	Yellow snap, string or wax beans, boiled, drained	½	cup(s)	62	56	22	1	5	2	<1	0.04	0.00	0.09	—
576	Yellow snap, string or wax beans, frozen, boiled, drained	½	cup(s)	68	62	19	1	4	2	<1	0.02	0.00	0.05	—
	Beets													
580	Whole, boiled, drained	2	item(s)	100	87	44	2	10	2	<1	0.03	0.04	0.06	—
581	Sliced, boiled, drained	½	cup(s)	85	74	37	1	8	2	<1	0.02	0.03	0.05	—
583	Sliced, canned, drained	½	cup(s)	85	77	26	1	6	1	<1	0.02	0.02	0.04	—
2730	Pickled, canned with liquid	½	cup(s)	114	93	74	1	18	3	<1	0.01	0.02	0.03	—
584	Beet greens, boiled, drained	½	cup(s)	72	64	19	2	4	2	<1	0.02	0.03	0.05	—
585	**Cowpeas or black-eyed peas, boiled, drained**	½	cup(s)	83	60	80	3	17	4	<1	0.07	0.02	0.13	—
	Broccoli													
587	Raw, chopped	½	cup(s)	44	39	15	1	3	1	<1	0.02	0.00	0.02	—
588	Chopped, boiled, drained	½	cup(s)	78	70	27	2	6	3	<1	0.06	0.03	0.13	—
590	Frozen, chopped, boiled, drained	½	cup(s)	92	83	26	3	5	3	<1	0.02	0.01	0.05	—
16848	**Broccoflower, raw, chopped**	½	cup(s)	32	29	10	1	2	1	<.1	0.01	0.01	0.04	—
	Brussels sprouts													
591	Boiled, drained	½	cup(s)	78	69	28	2	6	2	<1	0.08	0.03	0.20	—
592	Frozen, boiled, drained	½	cup(s)	78	67	33	3	6	3	<1	0.06	0.02	0.16	—
	Cabbage													
594	Raw, shredded	1	cup(s)	70	65	17	1	4	2	<.1	0.01	0.01	0.04	—
595	Boiled, drained, no salt added	1	cup(s)	150	140	33	2	7	3	1	0.08	0.05	0.29	—
35611	Chinese (pak choi or bok choy), boiled w/salt, drained	1	cup(s)	170	162	20	3	3	2	<1	0.04	0.02	0.13	—
16869	Kim chee	1	cup(s)	150	138	31	2	6	2	<1	0.04	0.02	0.15	—
596	Red, shredded, raw	1	cup(s)	70	63	22	1	5	1	<1	0.02	0.01	0.09	—
597	Savoy, shredded, raw	1	cup(s)	70	64	19	1	4	2	<.1	0.01	0.00	0.03	—
11710	**Capers**	1	teaspoon(s)	5	—	0	0	0	0	0	0.00	0.00	0.00	0
	Carrots													
600	Raw	½	cup(s)	61	54	25	1	6	2	<1	0.02	0.01	0.06	0
8691	Raw, baby	8	item(s)	80	72	28	1	7	1	<1	0.02	0.01	0.05	0
601	Agrated	½	cup(s)	55	49	23	1	5	2	<1	0.02	0.01	0.06	0
602	Sliced, boiled, drained	½	cup(s)	78	70	27	<1	6	2	<1	0.02	0	0.08	—

PAGE KEY: H–2 = Breads/Baked Goods H–6 = Cereal/Rice/Pasta H–10 = Fruit H–14 = Vegetables/Legumes H–24 = Nuts/Seeds H–26 = Vegetarian H–28 = Dairy H–34 = Eggs H–34 = Seafood H–36 = Meats H–40 = Poultry H–40 = Processed meats H–42 = Beverages H–46 = Fats/Oils H–48 = Sweets H–50 = Spices/Condiments/Sauces H–52 = Mixed foods/Soups/Sandwiches H–58 = Fast food H–74 = Convenience meals H–76 = Baby foods

Chol (mg)	Calc (mg)	Iron (mg)	Magn (mg)	Pota (mg)	Sodi (mg)	Zinc (mg)	Vit A (µg)	Thia (mg)	Vit E (mg α)	Ribo (mg)	Niac (mg)	Vit B$_6$ (mg)	Fola (µg)	Vit C (mg)	Vit B$_{12}$ (µg)	Sele (µg)
0	7	0.14	2	320	2	0.28	0	0.01	—	0.03	0.18	0.06	1	0	0	<1
0	5	0.21	3	52	5	0.43	1	0.02	0.41	0.02	0.09	0.09	2	1	0	<1
0	32	2.30	60	612	9	2.04	0	0.13	—	0.07	0.82	0.11	139	0	0	1
8	62	2.22	36	302	553	2.40	5	0.07	0.59	0.07	1.16	0.06	39	3	0	8
9	71	4.15	44	380	557	7.41	5	0.07	0.13	0.06	0.63	0.09	29	4	0	6
9	77	2.10	43	336	425	1.90	1	0.06	0.04	0.08	0.44	0.11	47	4	0	6
0	23	1.81	60	305	1	0.96	0	0.21	—	0.05	0.43	0.06	128	0	0	1
0	40	2.37	39	239	6	1.25	1	0.10	0.29	0.05	0.43	0.11	141	1	0	3
0	26	1.55	36	258	59	0.63	9	0.06	0.25	0.05	0.91	0.10	18	11	0	1
0	56	0.96	50	327	5	0.57	0	0.12	—	0.05	0.48	0.09	66	1	0	1
0	60	1.89	44	346	2	0.78	0	0.14	—	0.05	0.60	0.10	90	1	0	4
0	18	0.33	18	114	1	0.17	3	0.02	—	0.04	0.21	0.01	20	2	0	1
0	27	2.08	63	485	14	0.67	16	0.12	0.12	0.08	0.88	0.16	22	9	0	2
0	25	1.76	50	370	26	0.50	7	0.06	0.58	0.05	0.69	0.10	14	5	0	2
0	7	0.40	9	63	6	0.29	1	0.03	0.04	0.06	0.50	0.03	18	7	0	<1
0	64	2.26	54	335	1	0.96	0	0.18	0.01	0.06	0.48	0.15	127	1	0	5
0	17	0.75	20	111	58	0.19	0	0.08	—	0.07	0.82	0.06	146	7	0	1
0	24	1.27	25	304	39	0.32	0	0.13	—	0.05	0.30	0.09	16	<1	0	1
0	31	1.61	36	329	436	0.70	0	0.13	0.77	0.11	0.58	0.03	65	1	0	2
10	44	2.10	42	338	378	1.48	0	0.03	0.00	0.02	0.40	0.18	14	8	0	2
0	36	1.21	18	134	409	0.33	13	0.04	0.04	0.07	0.25	0.06	22	4	0	1
0	88	4.42	74	443	1	0.99	0	0.13	0.30	0.25	0.34	0.20	46	1	0	6
0	131	2.25	54	485	13	0.82	7	0.23	—	0.14	1.13	0.05	100	15	0	1
0	65	2.54	61	414	2	0.98	0	0.21	—	0.05	0.24	0.11	123	0	0	1
0	24	0.81	12	100	266	0.26	129	0.01	0.40	0.05	0.18	0.03	—	4	0.00	—
0	29	0.80	16	187	2	0.22	5	0.05	0.28	0.06	0.38	0.03	21	6	0	<1
0	33	0.59	16	85	6	0.32	7	0.02	0.24	0.06	0.26	0.04	16	3	0	<1
0	16	0.79	23	305	77	0.35	2	0.03	0.04	0.04	0.33	0.07	80	4	0	1
0	14	0.67	20	259	65	0.30	2	0.02	0.03	0.03	0.28	0.06	68	3	0	1
0	13	1.55	14	126	165	0.18	1	0.01	0.03	0.03	0.13	0.05	26	3	0	<1
0	12	0.47	17	168	300	0.30	1	0.01	—	0.05	0.28	0.06	31	3	0	1
0	82	1.37	49	654	174	0.36	276	0.08	1.30	0.21	0.36	0.10	10	18	0	1
0	106	0.92	43	345	3	0.84	65	0.08	0.18	0.12	1.15	0.05	105	1.81	0	2
0	21	0.32	9	139	15	0.18	15	0.03	0.34	0.05	0.28	0.08	28	39	0	1
0	31	0.52	16	229	32	0.35	76	0.05	1.13	0.10	0.43	0.16	84	51	0	1
0	30	0.56	12	131	10	0.26	52	0.05	1.21	0.07	0.42	0.12	52	37	0	1
0	11	0.23	6	96	7	0.20	0	0.03	0.01	0.03	0.23	0.07	18	28	0	—
0	28	0.94	16	247	16	0.26	30	0.08	0.34	0.06	0.47	0.14	47	48	0	1
0	20	0.37	14	225	12	0.19	36	0.08	0.40	0.09	0.42	0.22	78	35	0	<1
0	33	0.41	11	172	13	0.13	6	0.04	0.10	0.03	0.21	0.07	30	23	0	1
0	47	0.26	12	146	12	0.14	11	0.09	0.18	0.08	0.42	0.17	30	30	0	1
0	158	1.77	19	631	459	0.29	360	0.05	0.15	0.11	0.73	0.28	70	44	0	1
0	145	1.28	27	375	995	0.36	—	0.07	0.08	0.10	0.75	0.34	88	80	0	—
0	32	0.56	11	170	19	0.15	39	0.04	0.12	0.05	0.29	0.15	13	40	0	<1
0	25	0.28	20	161	20	0.19	35	0.05	—	0.02	0.21	0.13	56	22	0	1
0	—	—	—	—	105	—	—	—	—	—	—	—	—	—	0	—
0	20	0.18	7	195	42	0.15	367	0.04	0.40	0.04	0.60	0.08	12	4	0	<1
0	26	0.71	8	190	62	0.14	552	0.02	—	0.03	0.44	0.08	26	7	0	1
0	18	0.17	7	177	38	0.13	333	0.04	0.36	0.03	0.54	0.08	11	3	0	<1
0	23	0.26	7.8	183	45	0.15	671	0.05	0.80	0.03	0.5	0.11	11	2.8	0	<1

TABLE H–1

Food Composition   (DA+ code is for Wadsworth Diet Analysis program)   (For purposes of calculations, use "0" for t, <1, <.1, <.01, etc.)

DA + Code	Food Description	Quantity	Measure	Wt (g)	H₂O (g)	Ener (kcal)	Prot (g)	Carb (g)	Fiber (g)	Fat (g)	Sat	Mono	Poly	Trans
	VEGETABLES, LEGUMES—Continued													
1055	Juice, canned	½	cup(s)	123	109	49	1	11	1	<1	0.03	0.01	0.09	—
32725	**Cassava or manioc**	½	cup(s)	103	61	165	1	39	2	<1	0.08	0.08	0.05	—
	Cauliflower													
605	Raw, chopped,	½	cup(s)	50	46	13	1	3	1	<1	0.02	0.01	0.05	—
606	Boiled, drained	½	cup(s)	62	58	14	1	3	2	<1	0.04	0.02	0.13	—
607	Frozen, boiled, drained	½	cup(s)	90	85	17	1	3	2	<1	0.03	0.01	0.09	—
	Celery													
609	Diced	½	cup(s)	60	58	8	<1	2	1	<1	0.03	0.02	0.05	—
608	Stalk	2	item(s)	80	76	11	1	2	1	<1	0.03	0.03	0.06	—
	Chard													
1056	Swiss chard, raw	1	cup(s)	36	33	7	1	1	1	<.1	0.01	0.01	0.03	—
1057	Swiss chard, boiled, drained	½	cup(s)	88	81	18	2	4	2	<.1	0.01	0.01	0.02	—
	Collard greens													
610	Boiled, drained	½	cup(s)	95	87	25	2	5	3	<1	0.04	0.02	0.16	—
611	Frozen, chopped, boiled, drained	½	cup(s)	85	75	31	3	6	2	<1	0.05	0.02	0.18	—
	Corn													
29614	Yellow corn, fresh, cooked	1	item(s)	100	72	107	3	25	3	1	0.19	0.37	0.59	—
612	Yellow sweet corn, boiled, drained	½	cup(s)	82	57	89	3	21	2	1	0.16	0.31	0.49	—
614	Yellow sweet corn, frozen, boiled, drained	½	cup(s)	82	63	66	2	16	2	1	0.08	0.16	0.26	—
615	Yellow creamed sweet corn, canned	½	cup(s)	128	101	92	2	23	2	1	0.08	0.16	0.25	—
618	**Cucumber**	¼	item(s)	75	72	11	<1	3	<1	<.1	0.03	0.00	0.04	—
16870	**Cucumber, kim chee**	½	cup(s)	75	68	16	1	4	1	<.1	0.02	0.00	0.03	—
	Dandelion greens													
2734	Raw	1	cup(s)	55	47	25	1	5	2	<1	0.09	0.01	0.17	—
620	Chopped, boiled, drained	½	cup(s)	53	47	17	1	3	2	<1	0.08	0.01	0.14	—
1066	**Eggplant, boiled, drained**	½	cup(s)	48	43	17	<1	4	1	<1	0.02	0.01	0.04	—
621	**Endive or escarole, chopped, raw**	1	cup(s)	53	49	9	1	2	2	<1	0.03	0.00	0.05	—
8784	**Jicama or yambean**	½	cup(s)	65	59	25	<1	6	3	<.1	0.01	0.00	0.03	—
	Kale													
29313	Raw	1	cup(s)	67	57	34	2	7	1	<1	0.06	0.03	0.23	—
623	Frozen, chopped, boiled, drained	½	cup(s)	65	59	20	2	3	1	<1	0.04	0.02	0.15	—
	Kohlrabi													
1071	Raw	1	cup(s)	135	123	36	2	8	5	<1	0.02	0.01	0.06	—
1072	Boiled, drained	½	cup(s)	83	74	24	1	6	1	<.1	0.01	0.01	0.04	—
	Leeks													
1073	Raw	1	cup(s)	89	74	54	1	13	2	<1	0.04	0.00	0.15	—
1074	Boiled, drained	½	cup(s)	52	47	16	<1	4	1	<1	0.01	0.00	0.06	—
	Lentils													
522	Boiled	½	cup(s)	99	69	115	9	20	8	<1	0.05	0.06	0.17	—
1075	Sprouted	1	cup(s)	77	52	82	7	17	0	<1	0.04	0.08	0.17	—
	Lettuce													
624	Butterhead, boston, or bibb	1	cup(s)	55	53	7	1	1	1	<1	0.02	0.00	0.06	—
625	Butterhead leaves	11	piece(s)	83	79	11	1	2	1	<1	0.02	0.00	0.10	—
626	Iceberg	1	cup(s)	55	53	6	<1	1	1	<.1	0.01	0.00	0.03	—
628	Iceberg, chopped	1	cup(s)	55	53	6	<1	1	1	<.1	0.01	0.00	0.03	—
629	Looseleaf	1	cup(s)	56	54	8	1	2	1	<.1	0.01	0.00	0.05	—
1665	Romaine, shredded	1	cup(s)	56	53	10	1	2	1	<1	0.02	0.01	0.09	—
	Mushrooms													
15585	Crimini (about 6)	3	ounce(s)	85	28	4	3	2	0	0	0.00	0.00	0	0
8700	Enoki	30	item(s)	90	80	31	2	6	2	<1	0.04	0.01	0.14	—
630	Mushrooms, raw	½	cup(s)	35	32	8	1	1	<1	<1	0.02	0.00	0.05	—
1079	Mushrooms, boiled, drained	½	cup(s)	78	71	22	2	4	2	<1	0.05	0.01	0.14	—
1080	Mushrooms, canned, drained	½	cup(s)	78	71	20	1	4	2	<1	0.03	0.00	0.09	—
15587	Portobello, grilled	1	item(s)	85	30	3	4	3	0	0	0.00	0.00	0	0
2743	Shiitake, cooked	½	cup(s)	73	61	40	1	10	2	<1	0.04	0.05	0.02	—
	Mustard greens													
29319	Raw	1	cup(s)	56	51	15	2	3	2	<1	0.01	0.05	0.02	—
2744	Frozen, boiled, drained	½	cup(s)	75	70	14	2	2	2	<1	0.01	0.08	0.04	—
	Okra													
632	Sliced, boiled, drained	½	cup(s)	80	74	18	1	4	2	<1	0.04	0.02	0.04	—
32742	Frozen, boiled, drained, no salt added	½	cup(s)	92	84	26	2	5	3	<1	0.07	0.05	0.07	—
16866	Batter coated, fried	11	piece(s)	83	55	160	2	13	2	11	1.50	2.80	6.37	—
	Onions													
633	Raw, chopped	½	cup(s)	80	71	34	1	8	1	<.1	0.02	0.02	0.05	—
635	Chopped, boiled, drained	½	cup(s)	106	93	47	1	11	1	<1	0.03	0.03	0.08	—

PAGE KEY: H–2 = Breads/Baked Goods   H–6 = Cereal/Rice/Pasta   H–10 = Fruit   H–14 = Vegetables/Legumes   H–24 = Nuts/Seeds   H–26 = Vegetarian
H–28 = Dairy   H–34 = Eggs   H–34 = Seafood   H–36 = Meats   H–40 = Poultry   H–40 = Processed meats   H–42 = Beverages   H–46 = Fats/Oils
H–48 = Sweets   H–50 = Spices/Condiments/Sauces   H–52 = Mixed foods/Soups/Sandwiches   H–58 = Fast food   H–74 = Convenience meals   H–76 = Baby foods

Chol (mg)	Calc (mg)	Iron (mg)	Magn (mg)	Pota (mg)	Sodi (mg)	Zinc (mg)	Vit A (µg)	Thia (mg)	Vit E (mg α)	Ribo (mg)	Niac (mg)	Vit B$_6$ (mg)	Fola (µg)	Vit C (mg)	Vit B$_{12}$ (µg)	Sele (µg)
0	30	0.57	17	359	36	0.22	1176	0.11	1.43	0.07	0.47	0.27	5	10	0	1
0	16	0.28	22	279	14	0.35	1	0.09	0.20	0.05	0.88	0.09	28	21	0	1
0	11	0.22	8	152	15	0.14	1	0.03	0.04	0.03	0.26	0.11	29	23	0	<1
0	10	0.20	6	88	9	0.11	1	0.03	0.04	0.03	0.25	0.11	27	27	0	<1
0	15	0.37	8	125	16	0.12	0	0.03	0.05	0.05	0.28	0.08	37	28	0	<1
0	24	0.12	7	157	48	0.08	13	0.01	0.16	0.03	0.19	0.04	22	2	0	<1
0	32	0.16	9	208	64	0.10	18	0.02	0.22	0.05	0.26	0.06	29	2	0	<1
0	18	0.65	29	136	77	0.13	110	0.01	0.68	0.03	0.14	0.04	5	11	0	<1
0	51	1.98	75	480	157	0.29	268	0.03	1.65	0.08	0.32	0.07	8	16	0	1
0	133	1.10	19	110	15	0.22	386	0.04	0.84	0.10	0.55	0.12	88	17	0	<1
0	179	0.95	26	213	43	0.23	489	0.04	1.06	0.10	0.54	0.10	65	22	0	1
0	2	0.6	32	248	242	0.47	22	0.21	0.09	0.07	1.6	0.05	—	6	0	—
0	2	0.50	26	204	14	0.39	11	0.18	0.07	0.06	1.32	0.05	38	5	0	<1
0	2	0.39	23	191	1	0.52	8	0.02	0.06	0.05	1.08	0.08	29	3	0	1
0	4	0.49	22	172	365	0.68	5	0.03	0.09	0.07	1.23	0.08	58	6	0	1
0	12	0.21	10	111	2	0.15	4	0.02	0.02	0.02	0.07	0.03	5	2	0	<1
0	7	3.62	6	88	766	0.38	—	0.02	0.36	0.02	0.35	0.08	17	3	0	—
0	103	1.71	20	219	42	0.23	137	0.11	2.65	0.14	0.45	0.14	15	19	0	<1
0	74	0.95	13	122	23	0.15	260	0.07	1.79	0.09	0.27	0.08	7	9	0	<1
0	3	0.12	5	59	<1	0.06	1	0.04	0.20	0.01	0.29	0.04	7	1	0	<.1
0	27	0.44	8	165	12	0.41	57	0.04	0.23	0.04	0.21	0.01	75	3	0	<1
0	8	0.39	8	98	3	0.10	1	0.01	0.30	0.02	0.13	0.03	8	13	0	<1
0	90	1.14	23	299	29	0.29	515	0.07	—	0.09	0.67	0.18	19	80	0	1
0	90	0.61	12	209	10	0.12	478	0.03	0.60	0.07	0.44	0.06	9	16	0	1
0	32	0.54	26	473	27	0.04	3	0.07	0.65	0.03	0.54	0.20	22	84	0	1
0	21	0.33	16	281	17	0.26	2	0.03	0.43	0.02	0.32	0.13	10	45	0	1
0	53	1.87	25	160	18	0.11	74	0.05	0.82	0.03	0.36	0.21	57	11	0	1
0	16	0.57	7	45	5	0.03	1	0.01	—	0.01	0.10	0.06	13	2	0	<1
0	19	3.30	36	365	2	1.26	0	0.17	0.11	0.07	1.05	0.18	179	1	0	3
0	19	2.47	28	248	8	1.16	2	0.18	—	0.10	0.87	0.15	77	13	0	<1
0	19	0.69	7	132	3	0.11	92	0.03	0.10	0.03	0.20	0.05	40	2	0	<1
0	29	1.02	11	196	4	0.17	137	0.05	0.15	0.05	0.29	0.07	60	3	0	<1
0	11	0.19	4	84	5	0.09	9	0.02	0.10	0.01	0.07	0.03	31	2	0	<1
0	11	0.19	4	84	5	0.09	9	0.02	0.10	0.01	0.07	0.03	31	2	0	<1
0	20	0.48	7	109	16	0.10	208	0.04	0.16	0.05	0.21	0.05	21	10	0	<1
0	19	0.55	8	139	5	0.13	163	0.04	0.07	0.04	0.18	0.04	77	14	0	<1
0	1	—	—	33	—	0	—	—	—	—	—	—	—	0	0	—
0	1	0.80	14	343	3	0.51	0	0.08	0.01	0.09	3.28	0.04	27	11	0	14
0	1	0.18	3	110	1	0.18	0	0.03	0.00	0.15	1.35	0.04	6	1	<.1	3
0	5	1.36	9	278	2	0.68	0	0.06	0.01	0.23	3.48	0.07	14	3	0	9
0	9	0.62	12	101	332	0.56	0	0.07	0.01	0.02	1.24	0.05	9	0	0	3
40	<1	—	—	10	—	0	—	—	—	—	—	—	—	0	0	—
0	2	0.32	10	85	3	0.96	0	0.03	0.01	0.12	1.09	0.12	15	<1	0	18
0	58	0.82	18	199	14	0.11	295	0.05	1.13	0.06	0.45	0.10	105	39	0	1
0	76	0.84	10	104	19	0.15	266	0.03	1.01	0.04	0.19	0.08	53	10	0	<1
0	62	0.22	29	108	5	0.34	11	0.11	0.22	0.04	0.70	0.15	37	13	0	<1
0	88	0.62	47	215	3	0.57	16	0.09	0.29	0.11	0.72	0.04	134	11	0	1
2	54	1.13	32	170	110	0.44	—	0.16	1.51	0.13	1.29	0.11	34	9	<.1	—
0	18	0.15	8	115	2	0.13	0	0.04	0.02	0.02	0.07	0.12	15	5	0	<1
0	23	0.26	12	177	3	0.22	0	0.04	0.02	0.02	0.18	0.14	16	6	0	1

TABLE H–1

Food Composition (DA+ code is for Wadsworth Diet Analysis program)   (For purposes of calculations, use "0" for t, <1, <.1, <.01, etc.)

DA + Code	Food Description	Quantity	Measure	Wt (g)	H₂O (g)	Ener (kcal)	Prot (g)	Carb (g)	Fiber (g)	Fat (g)	Sat	Mono	Poly	Trans
	VEGETABLES, LEGUMES—Continued													
2748	Frozen, boiled, drained	½	cup(s)	106	98	30	1	7	2	<1	0.02	0.01	0.04	—
16850	Red onions, sliced, raw	½	cup(s)	58	52	22	1	5	1	<.1	0.02	0.01	0.04	—
636	Scallions, green or spring onions	2	item(s)	30	27	10	1	2	1	<.1	0.01	0.01	0.02	—
1081	Onion rings, breaded & pan fried, frozen, heated	11	item(s)	78	22	318	4	30	1	21	6.70	8.49	3.99	—
16860	**Palm hearts, cooked**	½	cup(s)	73	51	75	2	19	1	<1	0.03	0.00	0.07	—
637	**Parsley, chopped**	1	tablespoon(s)	4	3	1	<1	<1	<1	<.1	0.01	0.01	0.00	—
638	**Parsnips, sliced, boiled, drained**	½	cup(s)	78	63	55	1	13	3	<1	0.04	0.09	0.04	—
	Peas													
639	Green peas, canned, drained	½	cup(s)	85	69	59	4	11	3	<1	0.05	0.03	0.14	—
641	Green peas, frozen, boiled, drained	½	cup(s)	80	64	62	4	11	4	<1	0.04	0.02	0.10	—
35694	Pea pods, boiled w/salt, drained	½	cup(s)	80	71	34	3	6	2	<1	0.04	0.02	0.08	—
1082	Peas & carrots, canned w/liquid	½	cup(s)	128	112	48	3	11	3	<1	0.06	0.03	0.16	—
1083	Peas & carrots, frozen, boiled, drained	½	cup(s)	80	69	38	2	8	2	<1	0.06	0.03	0.16	—
640	Snow or sugar peas, raw	½	cup(s)	32	28	13	1	2	1	<.1	0.01	0.01	0.03	—
2750	Snow or sugar peas, frozen, boiled, drained	½	cup(s)	80	69	42	3	7	2	<1	0.06	0.03	0.13	—
29324	Split peas, sprouted	½	cup(s)	60	37	77	5	17	0	<1	0.07	0.04	0.20	—
	Peppers													
643	Green bell or sweet, raw	½	cup(s)	75	70	15	1	3	1	<1	0.04	0.01	0.05	—
644	Green bell or sweet, boiled, drained	½	cup(s)	68	62	19	1	5	1	<1	0.02	0.01	0.07	—
1664	Green hot chili	1	item(s)	45	39	18	1	4	1	<.1	0.01	0.00	0.05	—
1663	Green hot chili, canned w/liquid	½	cup(s)	68	63	14	1	3	1	<.1	0.01	0.00	0.04	—
1086	Jalapeno, canned w/liquid	½	cup(s)	68	60	18	1	3	2	1	0.07	0.04	0.35	—
8703	Yellow bell or sweet	1	item(s)	186	171	50	2	12	2	<1	0.06	0.03	0.21	—
1087	**Poi**	½	cup(s)	122	87	136	<1	33	<1	<1	0.04	0.00	0.07	—
	Potatoes													
5791	Baked, flesh & skin	1	item(s)	202	144	220	5	51	4	<1	0.05	0.00	0.09	—
645	Baked, flesh only	½	cup(s)	61	46	57	1	13	1	<.1	0.02	0.00	0.03	—
1088	Baked, skin only	1	item(s)	58	27	115	2	27	5	<.1	0.02	0.00	0.02	—
5794	Boiled, drained, skin & flesh	1	item(s)	150	116	129	3	30	2	<1	0.04	0.00	0.06	—
647	Boiled, flesh only	½	cup(s)	78	60	67	1	16	1	<.1	0.02	0.00	0.03	—
5795	Boiled in skin, drained, flesh only	1	item(s)	136	105	118	3	27	2	<1	0.04	0.00	0.06	—
2759	Microwaved	1	item(s)	202	146	212	5	49	5	<1	0.05	0.00	0.09	—
5804	Microwaved, skin only	1	item(s)	58	37	77	3	17	4	<.1	0.02	0.00	0.02	—
2760	Microwaved in skin, flesh only	½	cup(s)	78	57	78	2	18	1	<.1	0.02	0.00	0.03	—
1089	Au gratin, prepared w/butter	½	cup(s)	123	91	162	6	14	2	9	5.80	2.63	0.34	
1090	Au gratin mix, prepared w/water, whole milk, & butter	½	cup(s)	114	90	106	3	15	1	5	2.94	1.34	0.15	
648	French fried, deep fried, prepared from raw	14	item(s)	70	32	190	3	24	2	10	1.93	4.21	2.97	—
649	French fried, frozen, heated	14	item(s)	70	40	140	2	22	2	5	0.88	3.33	0.55	—
1091	Hashed brown	½	cup(s)	78	37	207	2	27	2	10	1.11	3.13	2.78	—
653	Mashed, from dehydrated granules w/milk, water, & margarine	½	cup(s)	105	80	122	2	17	1	5	1.27	2.05	1.41	—
652	Mashed, w/margarine & whole milk	½	cup(s)	105	79	119	2	18	2	4	1.05	1.83	1.27	—
1097	Potato puffs, frozen, heated	½	cup(s)	64	34	142	2	20	2	7	3.26	2.79	0.51	—
1093	Scalloped, prepared w/butter	½	cup(s)	123	99	105	4	13	2	5	2.76	1.27	0.20	—
1094	Scalloped mix, prepared w/water, whole milk, & butter	½	cup(s)	114	90	106	2	15	1	5	2.99	1.38	0.22	—
	Pumpkin													
1773	Boiled, drained	½	cup(s)	123	115	25	1	6	1	<.1	0.05	0.01	0.00	—
656	Canned	½	cup(s)	123	110	42	1	10	4	<1	0.18	0.05	0.02	—
	Radicchio									<.1				
2498	Raw	1	cup(s)	40	37	9	1	2	<1	<1	0.02	0.00	0.00	—
8731	Raw, leaves	10	item(s)	80	75	18	1	4	1	<1	0.05	0.01	0.09	—
657	**Radishes**	6	item(s)	27	26	4	<1	1	<1	<.1	0.01	0.00	0.01	—
1099	**Rutabaga, boiled, drained**	½	cup(s)	85	76	33	1	7	2	<1	0.02	0.02	0.08	—
658	**Sauerkraut, canned**	½	cup(s)	114	105	22	1	5	3	<1	0.04	0.01	0.07	—
	Seaweed													
1102	Kelp	½	cup(s)	41	33	17	1	4	1	<1	0.10	0.04	0.02	—
1104	Spirulina, dried	½	cup(s)	8	<1	22	4	2	<1	1	0.20	0.05	0.16	—
1106	**Shallots**	3	tablespoon(s)	30	24	22	1	5	0	<.1	0.01	0.00	0.01	—
	Soybeans													
1670	Boiled	½	cup(s)	86	<1	149	14	9	5	8	1.11	1.7	4.35	—
2825	Dry roasted	½	cup(s)	86	1	388	34	28	7	19	2.69	4.11	10.50	—
2824	Roasted, salted	½	cup(s)	86	2	405	30	29	15	22	3.16	4.82	12.33	—

PAGE KEY: H–2 = Breads/Baked Goods   H–6 = Cereal/Rice/Pasta   H–10 = Fruit   H–14 = Vegetables/Legumes   H–24 = Nuts/Seeds   H–26 = Vegetarian
H–28 = Dairy   H–34 = Eggs   H–34 = Seafood   H–36 = Meats   H–40 = Poultry   H–40 = Processed meats   H–42 = Beverages   H–46 = Fats/Oils
H–48 = Sweets   H–50 = Spices/Condiments/Sauces   H–52 = Mixed foods/Soups/Sandwiches   H–58 = Fast food   H–74 = Convenience meals   H–76 = Baby foods

Chol (mg)	Calc (mg)	Iron (mg)	Magn (mg)	Pota (mg)	Sodi (mg)	Zinc (mg)	Vit A (µg)	Thia (mg)	Vit E (mg α)	Ribo (mg)	Niac (mg)	Vit B$_6$ (mg)	Fola (µg)	Vit C (mg)	Vit B$_{12}$ (µg)	Sele (µg)
0	17	0.32	6	115	13	0.07	0	0.02	0.01	0.03	0.15	0.07	14	3	0	<1
0	11	0.13	6	90	2	0.11	0	0.02	0.01	0.01	0.09	0.07	11	4	0	—
0	22	0.44	6	83	5	0.12	15	0.02	0.17	0.02	0.16	0.02	19	6	0	<1
0	24	1.32	15	101	293	0.33	9	0.22	—	0.11	2.82	0.06	52	1	0	3
0	13	1.23	7	1318	10	2.72	—	0.03	0.37	0.13	0.62	0.53	15	5	0	—
0	5	0.24	2	21	2	0.04	16	0.00	0.03	0.00	0.05	0.00	6	5	0	<.1
0	29	0.45	23	286	8	0.20	0	0.06	0.78	0.04	0.56	0.07	45	10	0	1
0	17	0.81	14	147	214	0.60	23	0.10	0.03	0.07	0.62	0.05	37	8	0	1
0	19	1.22	18	88	58	0.54	84	0.23	0.02	0.08	1.18	0.09	47	8	0	1
0	34	1.58	21	192	192	0.30	43	0.10	0.31	0.06	0.43	0.12	23	38	0	1
0	29	0.96	18	128	332	0.74	368	0.09	—	0.07	0.74	0.11	23	8	0	1
0	18	0.75	13	126	54	0.36	374	0.18	0.42	0.05	0.92	0.07	21	6	0	1
0	14	0.66	8	63	1	0.09	17	0.05	0.12	0.03	0.19	0.05	13	19	0	<1
0	47	1.92	22	174	4	0.39	53	0.05	0.38	0.10	0.45	0.14	28	18	0	1
0	22	1.36	34	229	12	0.63	5	0.14	—	0.09	1.85	0.16	86	6	0	<1
0	7	0.25	7	130	2	0.10	13	0.04	0.28	0.02	0.36	0.17	8	60	0	0
0	6	0.31	7	113	1	0.08	10	0.04	0.36	0.02	0.32	0.16	11	51	0	<1
0	8	0.54	11	153	3	0.14	27	0.04	0.31	0.04	0.43	0.13	10	109	0	<1
0	5	0.34	10	127	798	0.12	24	0.01	0.47	0.03	0.54	0.10	7	46	0	<1
0	16	1.28	10	131	1136	0.23	58	0.03	0.47	0.03	0.27	0.13	10	7	0	<1
0	20	0.86	22	394	4	0.32	19	0.05	—	0.05	1.66	0.31	48	341	0	1
0	19	1.07	29	223	15	0.27	4	0.16	2.80	0.05	1.34	0.33	26	5	0	1
0	20	2.75	55	844	16	0.65	0	0.22	—	0.07	3.32	0.70	22	26	0	2
0	3	0.21	15	239	3	0.18	0	0.06	0.02	0.01	0.85	0.18	5	8	0	<1
0	20	4.08	25	332	12	0.28	1	0.07	0.02	0.06	1.78	0.36	13	8	0	<1
0	13	1.27	34	572	7	0.47	0	0.15	—	0.03	2.13	0.44	15	18	0	—
0	6	0.24	16	256	4	0.21	0	0.08	0.01	0.01	1.02	0.21	7	6	0	<1
0	7	0.42	30	515	5	0.41	0	0.14	—	0.03	1.96	0.41	14	18	0	<1
0	22	2.50	55	903	16	0.73	0	0.24	—	0.06	3.46	0.69	24	31	0	1
0	27	3.45	21	377	9	0.30	0	0.04	—	0.04	1.29	0.29	10	9	0	<1
0	4	0.32	20	321	5	0.26	0	0.10	—	0.02	1.27	0.25	9	12	0	<1
28	146	0.78	25	485	530	0.85	78	0.08	—	0.14	1.22	0.21	13	12	0	3
17	94	0.36	17	249	499	0.27	59	0.02	—	0.09	1.07	0.05	8	4	0	3
0	9	1.02	28	731	8	0.53	0	0.10	0.09	0.05	1.90	0.33	13	21	0	—
0	6	0.87	15	293	21	0.28	0	0.08	0.08	0.02	1.46	0.22	8	7	0	<1
0	11	0.43	27	449	267	0.37	0	0.13	0.12	0.01	1.80	0.37	12	10	0	<1
2	34	0.22	21	163	181	0.25	49	0.09	0.54	0.09	0.91	0.17	8	7	<1	6
1	21	0.27	20	342	350	0.32	43	0.10	0.44	0.05	1.23	0.26	9	11	<.1	1
0	19	1.00	12	243	477	0.19	0	0.13	0.15	0.05	1.38	0.15	11	4	0	<1
15	70	0.70	23	463	410	0.49	39	0.08	—	0.11	1.29	0.22	13	13	0	2
13	41	0.43	16	231	388	0.28	40	0.02	—	0.06	1.17	0.05	11	4	0	2
0	18	0.70	11	282	1	0.28	306	0.04	0.98	0.10	0.51	0.05	11	6	0	<1
0	32	1.70	28	252	6	0.21	953	0.03	1.30	0.07	0.45	0.07	15	5	0	<1
0	8	0.23	5	121	9	0.25	<1	0.01	0.90	0.01	0.10	0.02	24	3	0	<1
0	15	0.46	10	242	18	0.50	1	0.01	1.81	0.02	0.20	0.05	48	6	0	1
0	7	0.09	3	63	11	0.08	0	0.00	0.00	0.01	0.07	0.02	7	4	0	<1
0	41	0.45	20	277	17	0.30	0	0.07	0.27	0.03	0.61	0.09	13	16	0	1
0	34	1.67	15	193	751	0.22	1	0.02	0.11	0.02	0.16	0.15	27	17	0	1
0	68	1.16	49	36	94	0.50	2	0.02	0.35	0.06	0.19	0.00	73	1	0	<1
0	9	2.14	15	102	79	0.15	2	0.18	0.38	0.28	0.96	0.03	7	1	0	1
0	11	0.36	6	100	4	0.12	18	0.02	—	0.01	0.06	0.10	10	2	0	<1
0	88	4.42	74	443	0.86	0.98	0.86	0.13	0.30	0.24	0.34	0.2	46	1	0	6
0	120	3.40	196	1173	2	4.10	0	0.37	—	0.65	0.91	0.19	176	4	0	17
0	119	3.35	125	1264	140	2.70	9	0.09	0.78	0.12	1.21	0.18	181	2	0	16

Appendix **H**

TABLE H–1

Food Composition

(DA+ code is for Wadsworth Diet Analysis program)   (For purposes of calculations, use "0" for t, <1, <.1, <.01, etc.)

DA + Code	Food Description	Quantity	Measure	Wt (g)	H₂O (g)	Ener (kcal)	Prot (g)	Carb (g)	Fiber (g)	Fat (g)	Sat	Mono	Poly	Trans
	VEGETABLES, LEGUMES—Continued											**Fat Breakdown (g)**		
30282	Soup (miso)	1	cup(s)	240	218	85	6	8	2	3	0.59	1.05	1.47	—
8739	Sprouted, stir fried	3	ounce(s)	85	57	106	11	8	1	6	0.84	1.37	3.41	—
	Soy products													
1813	Soy milk	1	cup(s)	240	214	118	9	11	3	5	0.51	0.78	2.00	—
2838	Tofu, dried, frozen (koyadofu)	3	ounce(s)	85	5	408	41	12	6	26	3.73	5.70	14.57	—
13844	Tofu, extra firm	3	ounce(s)	79	—	80	8	2	1	4	0.50	0.87	2.60	—
13843	Tofu, firm	3	ounce(s)	79	—	80	8	2	1	4	0.50	0.87	2.17	—
1816	Tofu, firm, w/calcium sulfate & magnesium chloride (nigari)	3	ounce(s)	85	72	65	7	3	<1	4	0.54	0.83	2.14	—
1817	Tofu, fried	3	ounce(s)	85	43	230	15	9	3	17	2.48	3.79	9.69	—
13841	Tofu, silken	3	ounce(s)	91	—	30	6	0	1	1	0.50	0.51	1.52	—
13842	Tofu, soft	3	ounce(s)	91	—	30	6	1	1	1	0.50	1.00	2.00	—
1671	Tofu, soft, w/calcium sulfate & magnesium chloride (nigari)	3	ounce(s)	85	73	52	6	2	<1	3	0.45	0.69	1.76	—
	Spinach													
659	Raw, chopped	1	cup(s)	30	27	7	1	1	1	<1	0.02	0.00	0.05	—
663	Canned, drained	½	cup(s)	108	100	25	3	4	3	1	0.09	0.02	0.23	—
660	Chopped, boiled, drained	½	cup(s)	90	82	21	3	3	2	<1	0.04	0.01	0.10	—
661	Chopped, frozen, boiled, drained	½	cup(s)	95	84	30	4	5	4	<1	0.09	0.00	0.20	—
662	Leaf, frozen, boiled, drained	½	cup(s)	95	84	30	4	5	4	<1	0.09	0.00	0.20	—
8470	Trimmed leaves	1	cup(s)	32	27	3	1	<.1	3	<.1	—	—	—	—
	Squash													
1662	Acorn, baked	½	cup(s)	103	85	57	1	15	5	<1	0.03	0.01	0.06	—
29702	Acorn, boiled, mashed	½	cup(s)	123	110	42	1	11	3	<.1	0.02	0.01	0.04	—
1661	Butternut, baked	½	cup(s)	103	90	41	1	11	3	<.1	0.02	0.01	0.04	—
29451	Butternut, frozen, boiled	½	cup(s)	132	116	51	2	13	2	<.1	0.02	0.00	0.04	—
32773	Butternut, frozen, boiled, mashed, no salt added	½	cup(s)	122	<1	47	1	12	0	<.1	0.02	0.00	0.03	—
29700	Crookneck & straightneck, boiled, drained	½	cup(s)	90	84	18	1	4	1	<1	0.05	0.02	0.11	—
29703	Hubbard, baked	v	cup(s)	103	87	51	3	11	0	1	0.13	0.05	0.27	—
1660	Hubbard, boiled, mashed	½	cup(s)	118	107	35	2	8	3	<1	0.09	0.03	0.18	—
29704	Spaghetti, boiled, drained, or baked	½	cup(s)	78	72	21	1	5	1	<1	0.05	0.02	0.10	—
664	Summer, all varieties, sliced, boiled, drained	½	cup(s)	90	84	18	1	4	1	<1	0.06	0.02	0.12	—
665	Winter, all varieties, baked, mashed	½	cup(s)	103	91	38	1	9	3	<1	0.13	0.05	0.27	—
1112	Zucchini, boiled, drained	½	cup(s)	90	85	14	1	4	1	<.1	0.01	0.00	0.02	—
1113	Zucchini, frozen, boiled, drained	½	cup(s)	113	107	19	1	4	1	<1	0.03	0.01	0.06	—
	Sweet potatoes													
666	Baked, peeled	½	cup(s)	100	76	90	2	21	3	<1	0.03	0.00	0.06	—
667	Boiled, mashed	½	cup(s)	166	133	126	2	29	4	<1	0.05	0.00	0.10	—
668	Candied, home recipe	½	cup(s)	84	56	115	1	23	2	3	1.13	0.53	0.12	—
670	Canned, vacuum pack	½	cup(s)	100	76	91	2	21	2	<1	0.04	0.01	0.09	—
2765	Frozen, baked	½	cup(s)	88	65	88	2	21	2	<1	0.02	0.00	0.05	—
1136	Yams, baked or boiled, drained	½	cup(s)	68	48	79	1	19	3	<.1	0.02	0.00	0.04	—
32785	**Taro shoots, cooked, no salt added**	½	cup(s)	70	67	10	1	2	0	<.1	0.01	0.00	0.02	—
	Tomatillo													
8774	Raw	2	item(s)	68	62	22	1	4	1	1	0.09	0.11	0.28	—
8777	Raw, chopped	½	cup(s)	66	60	21	1	4	1	1	0.09	0.10	0.28	—
	Tomato													
671	Fresh, ripe, red	1	item(s)	123	116	22	1	5	1	<1	0.05	0.06	0.16	—
16846	Fresh, cherry	5	item(s)	85	80	18	1	4	1	<1	0.03	0.04	0.11	—
3952	Diced, red	½	cup(s)	90	85	16	1	4	1	<1	0.04	0.05	0.12	—
1118	Boiled, red	½	cup(s)	120	113	22	1	5	1	<1	0.02	0.02	0.05	—
675	Juice, canned	½	cup(s)	122	115	21	1	5	<1	<.1	0.01	0.01	0.03	—
75	Juice, no salt added	½	cup(s)	122	115	21	1	5	<1	<.1	0.01	0.01	0.03	—
1699	Paste, canned	2	tablespoon(s)	33	24	27	1	6	1	<1	0.04	0.03	0.07	—
1700	Puree, canned	¼	cup(s)	63	55	24	1	6	1	<1	0.02	0.02	0.05	—
1125	Sauce, canned	¼	cup(s)	61	55	20	1	5	1	<1	0.02	0.02	0.06	—
1120	Stewed, canned, red	½	cup(s)	128	117	33	1	8	1	<1	0.03	0.04	0.10	—
8778	Sun dried	½	cup(s)	27	4	70	4	15	3	1	0.12	0.13	0.30	—
8783	Sun dried in oil, drained	¼	cup(s)	28	15	59	1	6	2	4	0.52	2.38	0.57	—
	Turnips													
677	Turnips, cubed, boiled, drained	½	cup(s)	78	73	17	1	4	2	<.1	0.01	0.00	0.03	—
678	Turnip greens, chopped, boiled, drained	½	cup(s)	72	67	14	1	3	3	<1	0.04	0.01	0.07	—
679	Turnip greens, frozen, chopped, boiled, drained	½	cup(s)	82	74	24	3	4	3	<1	0.08	0.02	0.14	—

PAGE KEY: H–2 = Breads/Baked Goods H–6 = Cereal/Rice/Pasta H–10 = Fruit H–14 = Vegetables/Legumes H–24 = Nuts/Seeds H–26 = Vegetarian H–28 = Dairy H–34 = Eggs H–34 = Seafood H–36 = Meats H–40 = Poultry H–40 = Processed meats H–42 = Beverages H–46 = Fats/Oils H–48 = Sweets H–50 = Spices/Condiments/Sauces H–52 = Mixed foods/Soups/Sandwiches H–58 = Fast food H–74 = Convenience meals H–76 = Baby foods

Chol (mg)	Calc (mg)	Iron (mg)	Magn (mg)	Pota (mg)	Sodi (mg)	Zinc (mg)	Vit A (µg)	Thia (mg)	Vit E (mg α)	Ribo (mg)	Niac (mg)	Vit B6 (mg)	Fola (µg)	Vit C (mg)	Vit B12 (µg)	Sele (µg)
0	64	1.89	37	361	988	0.87	—	0.06	0.96	0.16	2.61	0.17	57	4	<1	—
0	70	0.34	82	482	12	1.79	1	0.36	—	0.16	0.94	0.14	108	10	0	1
0	10	1.39	46	338	29	0.55	5	0.39	3.24	0.17	0.35	0.10	5	0	0	3
0	310	8.28	50	17	5	4.17	22	0.42	—	0.27	1.01	0.24	78	1	0	46
0	60	1.08	78	—	0	—	0	—	0.03	—	—	—	—	0	0	—
0	60	1.08	52	—	0	—	0	—	—	—	—	—	—	0	0	—
0	138	1.23	39	150	7	0.85	1	0.07	—	0.08	0	0.05	28	<1	0	8
0	316	4.14	51	124	14	1.69	1	0.14	0.03	0.04	0.09	0.08	23	0	0	24
0	300	0.73	35	—	65	—	0	—	—	—	—	—	—	0	2	—
0	300	0.72	33	—	65	—	0	—	—	—	—	—	—	0	2	—
0	94	0.94	23	102	7	0.54	1	0.03	0.01	0.03	0.45	0.04	37	<1	0	8
0	30	0.81	24	167	24	0.16	141	0.02	0.61	0.06	0.22	0.06	58	8	0	<1
0	138	2.49	82	375	29	0.50	531	0.02	2.10	0.15	0.42	0.11	106	16	0	2
0	122	3.21	78	419	63	0.68	472	0.09	1.87	0.21	0.44	0.22	131	9	0	1
0	145	1.86	78	287	92	0.47	573	0.07	3.36	0.17	0.42	0.13	115	2	0	5
0	145	1.86	78	287	92	0.47	573	0.07	3.36	0.17	0.42	0.13	115	2	0	5
0	25	2.13	25	134	38	0.18	—	0.03	—	0.06	0.18	0.07	<.1	8	0	—
0	45	0.95	44	448	4	0.17	22	0.17	—	0.01	0.90	0.20	19	11	0	1
0	32	0.69	32	322	4	0.13	50	0.12	—	0.01	0.65	0.14	13	8	0	<1
0	42	0.62	30	291	4	0.13	572	0.07	1.32	0.02	0.99	0.13	19	15	0	1
0	25	0.77	12	176	3	0.16	—	0.07	—	0.05	0.61	0.09	22	5	0	1
0	23	0.70	11	162	2	0.14	406	0.05	—	0.05	0.56	0.08	19	4	0	1
0	18	0.41	18	184	2	0.25	29	0.04	—	0.03	0.39	0.09	20	7	0	<1
0	17	0.48	23	367	8	0.15	310	0.08	—	0.05	0.57	0.18	16	10	0	1
0	12	0.33	15	253	6	0.12	236	0.05	0.14	0.03	0.39	0.12	12	8	0	<1
0	16	0.26	9	91	14	0.16	5	0.03	0.09	0.02	0.63	0.08	6	3	0	<1
0	24	0.32	22	173	1	0.35	10	0.04	0.13	0.04	0.46	0.06	18	5	0	<1
0	23	0.45	13	448	1	0.23	268	0.02	0.12	0.07	0.51	0.17	21	10	0	<1
0	12	0.32	20	228	3	0.16	50	0.04	0.11	0.04	0.39	0.07	15	4	0	<1
0	19	0.54	15	219	2	0.23	11	0.05	0.14	0.05	0.44	0.05	9	4	0	<1
0	38	0.69	27	475	36	0.32	961	1.45	0.71	0.11	1.49	0.29	6	20	0	<1
0	45	1.20	30	382	45	0.33	1310	0.09	1.56	0.08	0.89	0.27	10	21	0	<1
7	22	0.95	9	159	59	0.13	176	0.02	—	0.04	0.33	0.03	9	6	0	1
0	22	0.89	22	312	53	0.18	399	0.04	1.00	0.06	0.74	0.19	17	26	0	1
0	31	0.48	18	332	7	0.26	722	0.06	0.68	0.05	0.49	0.16	19	8	0	1
0	10	0.36	12	458	5	0.14	4	0.06	0.26	0.02	0.38	0.16	11	8	0	<1
0	10	0.29	6	241	1	0.38	2	0.03	—	0.04	0.57	0.08	2	13	0	1
0	5	0.42	14	182	1	0.15	4	0.03	0.26	0.02	1.26	0.04	5	8	0	<1
0	5	0.41	13	177	1	0.15	4	0.03	0.25	0.02	1.22	0.04	5	8	0	<1
0	12	0.33	14	292	6	0.20	76	0.04	0.66	0.02	0.73	0.09	18	16	0	0
0	4	0.37	9	189	8	0.07	53	0.05	0.46	0.03	0.53	0.07	—	16	0	—
0	9	0.24	10	213	5	0.15	38	0.03	0.49	0.02	0.53	0.07	14	11	0	0
0	13	0.82	11	262	13	0.17	29	0.04	0.67	0.03	0.64	0.09	16	27	0	1
0	12	0.52	13	279	328	0.18	28	0.06	0.39	0.04	0.82	0.14	24	22	0	<1
0	12	0.52	13	279	12	0.18	28	0.06	0.39	0.04	0.82	0.14	24	22	0	<1
0	12	0.98	14	333	259	0.21	25	0.02	1.41	0.05	1.01	0.07	4	7	0	2
0	11	1.11	14	274	249	0.23	16	0.02	1.23	0.05	0.92	0.08	7	7	0	3
0	8	0.62	10	203	321	0.12	10	0.01	1.27	0.04	0.60	0.06	6	4	0	<1
0	43	1.70	15	264	282	0.22	11	0.06	1.06	0.04	0.91	0.02	6	10	0	1
0	30	2.45	52	925	566	0.54	12	0.14	0.00	0.13	2.44	0.09	18	11	0	1
0	13	0.74	22	430	73	0.21	18	0.05	—	0.11	1.00	0.09	6	28	0	1
0	26	0.14	7	138	12	0.09	0	0.02	0.02	0.02	0.23	0.05	7	9	0	<1
0	99	0.58	16	146	21	0.10	274	0.03	1.35	0.05	0.30	0.13	85	20	0	1
0	125	1.59	21	184	12	0.34	441	0.04	2.18	0.06	0.38	0.05	32	18	0	1

TABLE H–1

Food Composition

(DA+ code is for Wadsworth Diet Analysis program)   (For purposes of calculations, use "0" for t, <1, <.1, <.01, etc.)

DA + Code	Food Description	Quantity	Measure	Wt (g)	H₂O (g)	Ener (kcal)	Prot (g)	Carb (g)	Fiber (g)	Fat (g)	Sat	Mono	Poly	Trans
	VEGETABLES, LEGUMES—Continued													
	Vegetables, mixed													
1132	Canned, drained	½	cup(s)	82	71	40	2	8	2	<1	0.04	0.01	0.10	—
680	Frozen, boiled, drained	½	cup(s)	91	76	59	3	12	4	<1	0.03	0.01	0.07	—
7489	Vegetable juice, V8 100%	½	cup(s)	120	113	25	1	5	1	0	0.00	0.00	0.00	0
7490	Vegetable juice, V8 low sodium	½	cup(s)	120	113	25	0	7	1	0	0.00	0.00	0.00	0
7491	Vegetable juice, V8 spicy hot	½	cup(s)	120	113	25	1	5	1	0	0.00	0.00	0.00	0
	Water chestnuts													
31073	Sliced, drained	½	cup(s)	75	70	20	<1	5	1	0	0.00	0.00	0.00	0
31087	Whole	½	cup(s)	75	70	20	<1	5	1	0	0.00	0.00	0.00	0
1135	**Watercress**	1	cup(s)	34	32	4	1	<1	<1	<.1	0.01	0.00	0.01	—
	NUTS, SEEDS, AND PRODUCTS													
	Almonds													
32886	Blanched	¼	cup(s)	36	2	211	8	7	4	18	1.41	11.70	4.37	—
32887	Dry roasted, no salt added	¼	cup(s)	35	1	206	8	7	4	18	1.40	11.61	4.36	—
29724	Dry roasted, salted	¼	cup(s)	35	1	206	8	7	4	18	1.40	11.61	4.36	—
29725	Oil roasted, salted	¼	cup(s)	39	1	238	8	7	4	22	1.65	13.66	5.31	—
508	Slivered	¼	cup(s)	34	2	195	7	7	4	17	1.31	10.85	4.12	—
1137	Almond butter, no salt added	1	tablespoon(s)	16	<1	101	2	3	1	9	0.90	6.14	1.98	—
32940	Almond butter, salt added	1	tablespoon(s)	16	<1	101	2	3	1	9	0.90	6.14	1.98	—
1138	**Beechnuts, dried**	¼	cup(s)	57	4	327	4	19	5	28	3.25	12.43	11.41	—
517	**Brazil nuts, unblanched, dried**	¼	cup(s)	35	1	230	5	4	3	23	5.30	8.59	7.20	—
1166	**Breadfruit seeds, roasted**	¼	cup(s)	57	28	118	4	23	3	2	0.41	0.20	0.82	—
1139	**Butternuts, dried**	¼	cup(s)	30	1	184	7	4	1	17	0.39	3.13	12.82	—
	Cashews													
1140	Dry roasted	¼	cup(s)	34	1	197	5	11	1	16	3.14	9.36	2.68	—
518	Oil roasted	¼	cup(s)	33	1	189	5	10	1	16	2.76	8.42	2.78	—
32889	Cashew butter, no salt added	1	tablespoon(s)	16	<1	94	3	4	<1	8	1.56	4.66	1.34	—
32931	Cashew butter, salt added	1	tablespoon(s)	16	<1	94	3	4	<1	8	1.56	4.66	1.34	—
	Coconut													
32896	Dried, not sweetened	¼	cup(s)	60	2	393	4	14	10	38	34.06	1.63	0.42	—
1153	Dried, shredded, sweetened	¼	cup(s)	24	3	122	1	12	1	9	7.68	0.37	0.09	—
520	Shredded	¼	cup(s)	21	10	75	1	3	2	7	6.27	0.30	0.08	—
	Chestnuts													
1152	Chinese, roasted	¼	cup(s)	57	23	136	3	30	0	1	0.10	0.35	0.17	—
32895	European, boiled & steamed	¼	cup(s)	57	39	74	1	16	0	1	0.15	0.27	0.31	—
32911	European, roasted	¼	cup(s)	57	23	139	2	30	3	1	0.23	0.43	0.49	—
32922	Japanese, boiled & steamed	¼	cup(s)	57	49	32	<1	7	0	<1	0.02	0.06	0.03	—
32923	Japanese, roasted	¼	cup(s)	57	28	114	2	26	0	<1	0.07	0.24	0.12	—
4958	**Flaxseeds or linseeds**	¼	cup(s)	57	5	276	11	19	16	19	1.79	3.85	12.54	—
32904	**Ginkgo nuts, dried**	¼	cup(s)	57	7	197	6	41	0	1	0.22	0.42	0.42	—
	Hazelnuts or filberts													
32901	Blanched	¼	cup(s)	57	3	357	8	10	6	35	2.65	27.32	3.15	—
32902	Dry roasted, no salt added	¼	cup(s)	57	1	366	9	10	5	35	2.56	26.43	4.80	—
1156	**Hickorynuts, dried**	¼	cup(s)	30	1	197	4	5	2	19	2.11	9.78	6.57	—
	Macadamias													
1157	Raw	¼	cup(s)	34	<1	241	3	5	3	25	4.04	19.72	0.50	—
32905	Dry roasted, no salt added	¼	cup(s)	34	1	241	3	4	3	25	4.00	19.86	0.50	—
32932	Dry roasted, salt added	¼	cup(s)	34	1	240	3	4	3	25	4.00	19.86	0.50	—
	Mixed nuts													
1159	With peanuts, dry roasted	¼	cup(s)	34	1	203	6	9	3	18	2.36	10.75	3.69	—
32933	With peanuts, dry roasted, salt added	¼	cup(s)	34	1	203	6	9	3	18	2.36	10.75	3.69	—
32906	Without peanuts, oil roasted, no salt added	¼	cup(s)	36	1	221	6	8	2	20	3.27	11.93	4.12	—
	Peanuts													
2807	Dry roasted	¼	cup(s)	37	0	214	9	8	3	18	2.51	8.99	5.72	—
2806	Dry roasted, salted	¼	cup(s)	37	0	214	9	8	3	18	2.51	8.99	5.72	—
1763	Oil roasted, salted	¼	cup(s)	36	0	216	10	5	3	19	3.12	9.33	5.49	—
2804	Raw	¼	cup(s)	37	2	207	9	6	3	18	2.49	8.92	5.68	—
1884	Peanut butter, chunky	1	tablespoon(s)	16	<1	94	4	3	1	8	1.53	3.77	2.27	—
30303	Peanut butter, low sodium	1	tablespoon(s)	16	<1	95	4	3	1	8	1.66	3.88	2.21	—
30305	Peanut butter, reduced fat	1	tablespoon(s)	18	<1	94	5	6	1	6	1.33	2.91	1.85	—
524	Peanut butter, smooth	1	tablespoon(s)	16	<1	96	4	3	1	8	1.60	3.96	2.38	—
	Pecans													
32907	Dry roasted, no salt added	¼	cup(s)	57	1	403	5	8	5	42	3.56	24.92	11.66	—

PAGE KEY: H–2 = Breads/Baked Goods   H–6 = Cereal/Rice/Pasta   H–10 = Fruit   H–14 = Vegetables/Legumes   H–24 = Nuts/Seeds   H–26 = Vegetarian
H–28 = Dairy   H–34 = Eggs   H–34 = Seafood   H–36 = Meats   H–40 = Poultry   H–40 = Processed meats   H–42 = Beverages   H–46 = Fats/Oils
H–48 = Sweets   H–50 = Spices/Condiments/Sauces   H–52 = Mixed foods/Soups/Sandwiches   H–58 = Fast food   H–74 = Convenience meals   H–76 = Baby foods

Chol (mg)	Calc (mg)	Iron (mg)	Magn (mg)	Pota (mg)	Sodi (mg)	Zinc (mg)	Vit A (µg)	Thia (mg)	Vit E (mg α)	Ribo (mg)	Niac (mg)	Vit B₆ (mg)	Fola (µg)	Vit C (mg)	Vit B₁₂ (µg)	Sele (µg)
0	22	0.86	13	237	121	0.33	474	0.04	0.28	0.04	0.47	0.06	20	4	0	<1
0	23	0.75	20	154	32	0.45	195	0.06	0.40	0.11	0.77	0.07	17	3	0	<1
0	20	0.54	13	270	310	0.24	50	0.05	—	0.03	0.87	0.17	—	30	0	—
0	20	0.36	—	420	70	—	63	0.02	—	0.02	0.75	—	—	30	0	—
0	20	0.36	13	255	370	0.24	50	0.05	—	0.03	0.88	0.17	—	18	0	—
0	7	0.23	—	—	6	—	0	—	—	—	—	—	—	2	—	—
0	7	0.23	—	—	6	—	0	—	—	—	—	—	—	2	—	—
0	41	0.07	7	112	14	0.04	80	0.03	0.34	0.04	0.07	0.04	3	15	0	<1
0	78	1.35	100	249	10	1.13	0	0.07	8.96	0.20	1.33	0.04	11	0	0	1
0	92	1.56	99	257	<1	1.22	0	0.03	8.97	0.30	1.33	0.04	11	0	0	1
0	92	1.56	99	257	117	1.22	0	0.03	8.97	0.30	1.33	0.04	11	0	0	1
0	114	1.44	108	274	133	1.20	0	0.04	10.19	0.31	1.44	0.05	11	0	0	1
0	84	1.45	93	246	<1	1.13	0	0.08	8.73	0.27	1.32	0.04	10	0	0	1
0	43	0.59	48	121	2	0.49	0	0.02	—	0.10	0.46	0.01	10	<1	0	—
0	43	0.59	48	121	72	0.49	0	0.02	—	0.10	0.46	0.01	10	<1	0	1
0	1	1.40	0	578	22	0.20	0	0.17	—	0.21	0.50	0.39	64	9	0	4
0	56	0.85	132	231	1	1.42	0	0.22	2.01	0.01	0.10	0.04	8	<1	0	671
0	49	0.51	35	615	16	0.59	9	0.23	—	0.14	4.20	0.24	34	4	0	8
0	16	1.21	71	126	<1	0.94	2	0.11	—	0.04	0.31	0.17	20	1	0	5
0	15	2.06	89	194	5	1.92	0	0.07	0.32	0.07	0.48	0.09	24	0	0	4
0	14	1.97	89	205	4	1.74	0	0.12	0.30	0.07	0.56	0.10	8	<.1	0	7
0	7	0.80	41	87	2	0.83	0	0.05	—	0.03	0.26	0.04	11	0	0	2
0	7	0.80	41	87	98	0.83	0	0.05	0.15	0.03	0.26	0.04	11	0	0	2
0	15	1.98	54	323	22	1.20	0	0.04	0.26	0.06	0.36	0.18	5	1	0	11
0	4	0.47	12	82	64	0.44	0	0.01	0.10	0.00	0.12	0.07	2	<1	0	4
0	3	0.51	7	75	4	0.23	0	0.01	0.05	0.00	0.11	0.01	5	1	0	2
0	11	0.85	51	271	2	0.53	0	0.09	—	0.05	0.85	0.25	41	22	0	4
0	26	0.98	31	405	15	0.14	1	0.08	—	0.06	0.41	0.13	22	15	0	—
0	16	0.52	19	336	1	0.32	1	0.14	0.28	0.10	0.76	0.28	40	15	0	1
0	6	0.30	10	67	3	0.23	1	0.07	—	0.03	0.31	0.06	10	5	0	—
0	20	1.19	36	242	11	0.81	2	0.26	—	—	0.40	0.24	33	16	0	—
0	111	3.48	203	381	19	2.34	0	0.10	—	0.09	0.78	0.52	156	1	0	3
0	11	0.91	30	566	7	0.38	31	0.24	—	0.10	6.65	0.36	60	17	0	—
0	84	1.87	91	373	0	1.25	1	0.27	9.92	0.06	0.88	0.33	44	1	0	2
0	70	2.48	98	428	0	1.42	2	0.19	8.66	0.07	1.16	0.35	50	2	0	2
0	18	0.64	52	131	<1	1.29	2	0.26	—	0.04	0.27	0.06	12	1	0	2
0	28	1.24	44	123	2	0.44	0	0.40	0.18	0.05	0.83	0.09	4	<1	0	1
0	23	0.89	40	122	1	0.43	0	0.24	0.19	0.03	0.76	0.12	3	<1	0	1
0	23	0.89	40	122	89	0.43	0	0.24	0.19	0.03	0.76	0.12	3	<1	0	4
0	24	1.27	77	204	4	1.30	<1	0.07	—	0.07	1.61	0.10	17	<1	0	1
0	24	1.27	77	204	229	1.30	0	0.07	3.75	0.07	1.61	0.10	17	<1	0	3
0	38	0.93	90	196	4	1.68	<1	0.18	—	0.17	0.71	0.06	20	<1	0	—
0	20	0.82	64	240	2	1.20	0	0.15	2.56	0.03	4.93	0.09	53	0	0	3
0	20	0.82	64	240	297	1.20	0	0.15	2.89	0.03	4.93	0.09	53	0	0	3
0	22	0.54	63	261	115	1.18	0	0.03	2.50	0.03	4.97	0.16	43	<1	0	1
0	34	1.67	61	257	7	1.19	0	0.23	3.04	0.05	4.40	0.13	88	0	0	3
0	8	0.33	31	101	75	0.52	0	0.02	1.01	0.02	2.19	0.07	15	0	0	1
0	6	0.29	25	107	3	0.47	0	0.01	1.23	0.01	2.14	0.07	12	0	0	—
0	6	0.34	31	120	97	0.50	0	0.05	1.20	0.01	2.63	0.06	11	0	0	—
0	8	0.30	28	88	80	0.47	0	0.01	1.44	0.02	2.14	0.07	12	0	0	1
0	41	1.59	75	240	1	2.87	4	0.26	0.74	0.06	0.66	0.11	9	<1	0	2

TABLE H–1
Food Composition

(DA+ code is for Wadsworth Diet Analysis program)   (For purposes of calculations, use "0" for t, <1, <.1, <.01, etc.)

DA + Code	Food Description	Quantity	Measure	Wt (g)	H₂O (g)	Ener (kcal)	Prot (g)	Carb (g)	Fiber (g)	Fat (g)	Sat	Mono	Poly	Trans
	DAIRY—Continued													
900	Roquefort	1	ounce(s)	28	11	103	6	1	0	9	5.39	2.37	0.37	—
21	Swiss	1	ounce(s)	28	10	106	8	2	0	8	4.98	2.04	0.27	—
	Imitation cheese													
7998	Shredded imitation cheddar	¼	cup(s)	28	—	90	5	2	0	7	1.50	—	—	—
8028	Shredded imitation mozzarella	¼	cup(s)	28	—	80	6	1	0	6	1.00	—	—	—
	Cottage Cheese													
9	Low fat, 1% fat	½	cup(s)	113	93	81	14	3	0	1	0.73	0.33	0.04	—
8	Low fat, 2% fat	½	cup(s)	113	90	102	16	4	0	2	1.38	0.62	0.07	—
	Cream cheese													
11	Cream cheese	2	tablespoon(s)	29	16	101	2	1	0	10	6.37	2.85	0.37	—
17366	Fat free cream cheese	2	tablespoon(s)	30	23	29	4	2	0	<1	0.27	0.10	0.02	—
10438	Tofutti Better Than Cream Cheese	2	tablespoon(s)	30	—	80	1	1	0	8	2.00	—	6.00	
	Processed cheese													
22	American cheese, processed	1	ounce(s)	28	11	106	6	<1	0	9	5.58	2.54	0.28	—
24	American cheese food, processed	1	ounce(s)	28	12	94	5	2	0	7	4.23	2.05	0.31	—
25	American cheese spread, processed	1	ounce(s)	28	14	82	5	2	0	6	3.78	1.77	0.18	—
9110	Kraft deluxe singles pasteurized process American cheese	1	ounce(s)	28	—	110	5	1	0	9	6.00	—	—	—
23	Swiss cheese, processed	1	ounce(s)	28	12	95	7	1	0	7	4.55	2.00	0.18	
	Soy cheese													
10430	Nu Tofu cheddar flavored cheese alternative	1	ounce(s)	28	—	70	6	1	0	4	0.50	2.50	1.00	
10435	Nu Tofu mozzarella flavored cheese alternative	1	ounce(s)	28	—	70	6	2	0	4	0.50	2.50	1.00	
	Cream													
26	Half & half	1	tablespoon(s)	15	12	20	<1	1	0	2	1.07	0.50	0.06	—
28	Light coffee or table, liquid	1	tablespoon(s)	15	11	29	<1	1	0	3	1.80	0.84	0.11	—
30	Light whipping cream, liquid	1	tablespoon(s)	15	10	44	<1	<1	0	5	2.90	1.36	0.13	—
32	Heavy whipping cream, liquid	1	tablespoon(s)	15	9	52	<1	<1	0	6	3.45	1.60	0.21	—
34	Whipped cream topping, pressurized	1	tablespoon(s)	4	2	10	<1	<1	0	1	0.52	0.24	0.03	—
	Sour cream													
36	Sour cream	2	tablespoon(s)	24	17	51	1	1	0	5	3.13	1.45	0.19	—
30556	Fat free sour cream	2	tablespoon(s)	32	26	24	1	5	0	0	0.00	0.00	0.00	0
	Imitation cream													
3659	Coffeemate nondairy creamer, liquid	1	tablespoon(s)	16	—	20	0	2	0	1	0.00	0.50	0.00	—
40	Cream substitute, powder	1	teaspoon(s)	2	<.1	11	<.1	1	0	1	0.65	0.02	0.00	—
35972	Nondairy coffee whitener, liquid, frozen	1	tablespoon(s)	16	12	22	<1	2	0	2	0.31	1.20	0.00	—
35975	Nondairy dessert topping, pressurized	1	tablespoon(s)	5	3	12	<.1	1	0	1	0.88	0.09	0.01	—
35976	Nondairy dessert topping, frozen	1	tablespoon(s)	5	3	16	<.1	1	0	1	1.09	0.08	0.03	—
904	Imitation sour cream	2	tablespoon(s)	24	17	50	1	2	0	5	4.27	0.14	0.01	—
	Fluid milk													
57	Fat free, nonfat, or skim	1	cup(s)	245	223	83	8	12	0	<1	0.29	0.12	0.02	—
58	Fat free, nonfat, or skim, w/nonfat milk solids	1	cup(s)	245	221	91	9	12	0	1	0.40	0.16	0.02	—
54	Low fat, 1%	1	cup(s)	244	219	102	8	12	0	2	1.54	0.68	0.09	—
55	Low fat, 1%, w/nonfat milk solids	1	cup(s)	245	220	105	9	12	0	2	1.48	0.69	0.09	—
60	Low fat buttermilk	1	cup(s)	245	221	98	8	12	0	2	1.34	0.62	0.08	—
51	Reduced fat, 2%	1	cup(s)	244	218	122	8	11	0	5	2.35	2.04	0.17	—
52	Reduced fat, 2%, w/nonfat milk solids	1	cup(s)	245	218	125	9	12	0	5	2.93	1.36	0.17	—
50	Whole, 3.3%	1	cup(s)	244	216	146	8	11	0	8	4.55	1.98	0.48	—
	Canned													
61	Whole evaporated	2	tablespoon(s)	32	23	42	2	3	0	2	1.45	0.74	0.08	—
62	Fat free, nonfat, or skim evaporated	2	tablespoon(s)	32	25	25	2	4	0	<.1	0.04	0.02	0.00	—
63	Sweetened condensed	2	tablespoon(s)	38	10	123	3	21	0	3	2.10	0.93	0.13	—
	Dried Milk													
64	Dried buttermilk	¼	cup(s)	30	1	118	10	15	0	2	1.09	0.51	0.07	—
65	Instant nonfat dry milk w/added vitamin A	¼	cup(s)	17	1	63	6	9	0	<1	0.08	0.03	0.00	—
5234	Skim milk powder	¼	cup(s)	18	1	64	6	9	0	<1	0.08	0.03	0.01	—
907	Whole dry milk	¼	cup(s)	32	1	161	9	12	0	9	5.43	2.57	0.22	—
909	**Goat milk**	1	cup(s)	244	212	168	9	11	0	10	6.51	2.71	0.36	—
	Chocolate milk													
69	Low fat	1	cup(s)	250	211	158	8	26	1	3	1.54	0.75	0.09	—
68	Reduced fat	1	cup(s)	250	209	180	8	26	1	5	3.10	1.47	0.18	—
67	Whole milk	1	cup(s)	250	206	208	8	26	2	8	5.26	2.48	0.31	—
33156	Chocolate syrup, fortified, prepared w/milk	1	cup(s)	263	220	197	8	24	<1	8	5.22	2.44	0.31	—

PAGE KEY: H–2 = Breads/Baked Goods H–6 = Cereal/Rice/Pasta H–10 = Fruit H–14 = Vegetables/Legumes H–24 = Nuts/Seeds H–26 = Vegetarian
H–28 = Dairy H–34 = Eggs H–34 = Seafood H–36 = Meats H–40 = Poultry H–40 = Processed meats H–42 = Beverages H–46 = Fats/Oils
H–48 = Sweets H–50 = Spices/Condiments/Sauces H–52 = Mixed foods/Soups/Sandwiches H–58 = Fast food H–74 = Convenience meals H–76 = Baby foods

Chol (mg)	Calc (mg)	Iron (mg)	Magn (mg)	Pota (mg)	Sodi (mg)	Zinc (mg)	Vit A (µg)	Thia (mg)	Vit E (mg α)	Ribo (mg)	Niac (mg)	Vit B$_6$ (mg)	Fola (µg)	Vit C (mg)	Vit B$_{12}$ (µg)	Sele (µg)
25	185	0.16	8	25	507	0.58	82	0.01	—	0.16	0.21	0.03	14	0	<1	4
26	221	0.06	11	22	54	1.22	62	0.02	0.11	0.08	0.03	0.02	2	0	1	5
0	150	0.00	—	—	420	—	—	—	—	—	—	—	—	0	—	—
0	150	0.00	8	—	320	1.20	—	0.00	—	0.26	0.00	0.00	40	0	<1	—
5	69	0.16	6	97	459	0.43	12	0.02	0.01	0.19	0.14	0.08	14	0	1	10
9	78	0.18	7	108	459	0.47	24	0.03	0.02	0.21	0.16	0.09	15	0	1	12
32	23	0.35	2	35	86	0.16	106	0.00	0.09	0.06	0.03	0.01	4	0	<1	1
2	56	0.05	4	49	164	0.26	84	0.02	0.00	0.05	0.05	0.02	11	0	<1	1
0	0	0.00	—	—	135	—	0	—	—	—	—	—	—	0	—	—
27	156	0.05	8	48	422	0.81	72	0.01	0.08	0.10	0.02	0.02	2	0	<1	4
23	162	0.16	9	83	359	0.91	57	0.02	0.06	0.15	0.05	0.02	2	0	<1	5
16	160	0.09	8	69	382	0.74	49	0.01	0.05	0.12	0.04	0.03	2	0	<1	3
25	150	0.00	0	25	450	0.90	84	—	—	0.10	—	—	—	0	<1	—
24	219	0.17	8	61	388	1.02	56	0.00	0.10	0.08	0.01	0.01	2	0	<1	5
0	200	0.36	—	—	190	—	—	—	—	—	—	—	—	0	—	—
0	150	0.36	—	—	190	—	—	—	—	—	—	—	—	0	—	—
6	16	0.01	2	20	6	0.08	15	0.01	0.05	0.02	0.01	0.01	<1	<1	<.1	<1
10	14	0.01	1	18	6	0.04	27	0.00	0.08	0.02	0.01	0.00	<1	<1	<.1	<.1
17	10	0.00	1	15	5	0.04	42	0.00	0.13	0.02	0.01	0.00	1	<.1	<.1	<.1
21	10	0.00	1	11	6	0.03	62	0.00	0.16	0.02	0.01	0.00	1	<.1	<.1	<.1
3	4	0.00	<1	6	5	0.01	7	0.00	0.02	0.00	0.00	0.00	<1	0	<.1	<.1
11	28	0.01	3	35	13	0.06	42	0.01	0.14	0.04	0.02	0.00	3	<1	<.1	1
3	40	0.00	3	41	45	0.16	—	0.01	0.00	0.05	0.02	0.01	4	0	<.1	—
0	0	0.00	—	30	0	—	0	0.02	—	0.02	0.20	—	—	0	—	—
0	<1	0.02	<.1	16	4	0.01	<.1	0.00	0.01	0.00	0.00	0.00	0	0	0	<.1
0	1	0.00	<.1	30	13	0.00	—	0.00	—	0.00	0.00	0.00	0	0	0	<1
0	<1	0.00	<.1	1	3	0.00	—	0.00	—	0.00	0.00	0.00	0	0	0	<.1
0	<1	0.01	<.1	1	1	0.00	—	0.00	—	0.00	0.00	0.00	0	0	0	<1
0	1	0.09	1	39	24	0.28	0	0.00	0.18	0.00	0.00	0.00	0	0	0	1
5	223	1.23	22	238	108	2.08	149	0.11	0.02	0.45	0.23	0.09	12	0	1	8
5	316	0.12	37	419	130	1.00	149	0.10	0.00	0.43	0.22	0.11	12	2	1	5
12	264	0.85	27	290	122	2.12	142	0.05	0.02	0.45	0.23	0.09	12	0	1	8
10	314	0.12	34	397	127	0.98	145	0.10	—	0.42	0.22	0.11	12	2	1	6
10	284	0.12	27	370	257	1.03	17	0.08	0.12	0.38	0.14	0.08	12	2	1	5
20	271	0.24	27	342	115	1.17	134	0.10	0.07	0.45	0.22	0.09	12	<1	1	6
20	314	0.12	34	397	127	0.98	137	0.10	—	0.42	0.22	0.11	12	2	1	6
24	246	0.07	24	325	105	0.93	68	0.11	0.15	0.45	0.26	0.09	12	0	1	9
9	82	0.06	8	95	33	0.24	20	0.01	0.04	0.10	0.06	0.02	3	1	<.1	1
1	93	0.09	9	106	37	0.29	38	0.01	0.00	0.10	0.06	0.02	3	<1	<.1	1
13	109	0.07	10	142	49	0.36	28	0.03	0.06	0.16	0.08	0.02	4	1	<1	6
21	360	0.09	33	484	157	1.22	15	0.12	0.03	0.48	0.27	0.10	14	2	1	6
3	215	0.05	20	298	96	0.77	124	0.07	0.00	0.30	0.16	0.06	9	1	1	5
3	222	0.06	21	307	99	0.79	0	0.07	—	0.31	0.16	0.06	9	1	1	5
31	296	0.15	28	431	120	1.08	83	0.09	0.16	0.39	0.21	0.10	12	3	1	5
27	327	0.12	34	498	122	0.73	139	0.12	0.17	0.34	0.68	0.11	2	3	<1	3
8	288	0.60	33	425	153	1.03	145	0.10	0.05	0.42	0.32	0.10	13	2	1	5
18	285	0.60	33	423	150	1.03	138	0.09	0.10	0.41	0.32	0.10	13	2	1	5
30	280	0.60	33	418	150	1.03	65	0.09	0.15	0.41	0.31	0.10	13	2	1	5
34	292	2.68	32	460	147	0.92	—	0.09	—	0.55	6.53	0.11	13	2	1	5

TABLE H–1
Food Composition

(DA+ code is for Wadsworth Diet Analysis program)   (For purposes of calculations, use "0" for t, <1, <.1, <.01, etc.)

DA + Code	Food Description	Quantity	Measure	Wt (g)	H₂O (g)	Ener (kcal)	Prot (g)	Carb (g)	Fiber (g)	Fat (g)	Sat	Mono	Poly	Trans
	DAIRY—Continued													
908	Cocoa, hot, prepared w/milk	1	cup(s)	250	206	193	9	27	3	6	3.58	1.69	0.09	0.18
33184	Cocoa mix with aspartame, added sodium & vitamin A, no added calcium or phosphorus, prepared with water	1	cup(s)	192	177	56	2	10	1	<1	0.00	0.15	0.01	—
70	**Eggnog**	1	cup(s)	254	189	343	10	34	0	19	11.29	5.67	0.86	—
	Breakfast drinks													
10093	Carnation Instant Breakfast classic chocolate malt, prepared w/skim milk, no sugar added	1	cup(s)	243	—	142	11	21	<1	1	0.89	—	—	—
10091	Carnation Instant Breakfast strawberry creme, prepared w/skim milk	1	cup(s)	273	—	220	13	39	0	<1	0.40	—	—	—
10094	Carnation Instant Breakfast strawberry creme, prepared w/skim milk, no sugar added	1	cup(s)	243	—	134	12	21	0	<1	0.45	—	—	—
10092	Carnation Instant Breakfast vanilla creme, prepared w/skim milk, no sugar added	1	cup(s)	273	—	220	13	39	0	<1	0.40	—	—	—
1417	Ovaltine rich chocolate flavor, prepared w/skim milk	1	cup(s)	243	—	134	12	21	0	<1	0.45	—	—	—
8539	**Malted milk, chocolate mix, fortified, prepared w/milk**	1	cup(s)	265	216	223	9	29	1	9	4.95	2.17	0.54	—
	Milkshakes													
73	Chocolate	1	cup(s)	227	164	270	7	48	1	6	3.81	1.77	0.23	—
74	Vanilla	1	cup(s)	227	169	254	9	40	0	7	4.28	1.98	0.26	—
	Ice cream													
4776	Chocolate	½	cup(s)	66	37	143	3	19	1	7	4.49	2.12	0.27	—
16514	Chocolate, soft serve	½	cup(s)	87	50	177	3	24	1	8	5.17	2.43	0.31	—
12137	Chocolate fudge, fat free no sugar added	½	cup(s)	71	—	100	4	22	0	0	0.00	0.00	0.00	0
82	Light vanilla	½	cup(s)	66	42	109	4	18	<1	3	1.71	0.57	0.10	—
78	Light vanilla, soft serve	½	cup(s)	86	60	108	4	19	0	2	1.40	0.65	0.09	—
16523	Sherbet, all flavors	½	cup(s)	97	64	133	1	29	<1	2	1.12	0.51	0.08	—
4778	Strawberry	½	cup(s)	66	40	127	2	18	1	6	3.43	—	—	—
76	Vanilla	½	cup(s)	66	40	133	2	16	<1	7	4.48	1.96	0.30	—
12146	Vanilla chocolate swirl, fat free, no sugar added	½	cup(s)	71	—	100	4	20	0	0	0.00	0.00	0.00	0
	Soy desserts													
10694	Tofutti low fat vanilla fudge nondairy frozen dessert	½	cup(s)	70	—	120	2	24	0	2	1.00	—	—	—
15721	Tofutti premium chocolate supreme nondairy frozen dessert	½	cup(s)	60	—	180	3	18	0	11	2.00	—	—	—
15720	Tofutti premium vanilla nondairy frozen dessert	½	cup(s)	60	—	190	2	20	0	11	2.00	—	—	—
	Ice milk													
16516	Flavored, not chocolate	½	cup(s)	66	45	91	2	15	0	3	1.72	0.81	0.11	—
16517	Chocolate	½	cup(s)	66	43	95	3	17	<1	2	1.29	0.61	0.08	—
	Pudding													
25032	Chocolate	½	cup(s)	144	110	154	5	23	1	5	2.78	1.94	0.23	0
1923	Chocolate, sugar free, prepared w/2% milk	½	cup(s)	133	—	100	5	14	<1	3	1.50	—	—	—
1722	Rice	½	cup(s)	113	73	175	6	26	1	6	1.99	2.14	0.88	—
4747	Tapioca, ready to eat	1	item(s)	142	105	169	3	28	<1	5	0.85	2.24	1.93	—
25031	Vanilla	½	cup(s)	136	110	116	5	17	<.1	3	1.31	1.21	0.16	0
1924	Vanilla, sugar free, prepared w/2% milk	½	cup(s)	133	90	4	12	<1	2	2	10	150	0.00	—
	Frozen yogurt													
4785	Chocolate, soft serve	½	cup(s)	72	46	115	3	18	2	4	2.61	1.26	0.16	—
1747	Fruit varieties	½	cup(s)	113	80	144	3	24	0	4	2.63	1.11	0.11	—
4786	Vanilla, soft serve	½	cup(s)	72	47	117	3	17	0	4	2.46	1.14	0.15	—
	Milk substitutes													
	Lactose free													
16081	Fat free calcium fortified milk	1	cup(s)	240	—	90	9	13	0	0	0.00	—	—	0
36486	Low fat milk	1	cup(s)	240	—	110	8	13	0	3	1.50	—	—	—
36487	Reduced fat milk	1	cup(s)	240	—	130	8	13	0	5	3.00	—	—	—
36488	Whole milk	1	cup(s)	240	—	160	8	12	0	9	5.00	—	—	—
	Rice													
10083	Rice Dream carob rice beverage	1	cup(s)	240	—	150	1	32	0	3	0.00	—	—	—
10087	Rice Dream vanilla enriched rice beverage	1	cup(s)	240	—	130	1	28	0	2	0.00	—	—	—
17089	Rice Dream original rice beverage, enriched	1	cup(s)	240	—	120	1	25	0	2	0.00	—	—	—

PAGE KEY: H–2 = Breads/Baked Goods H–6 = Cereal/Rice/Pasta H–10 = Fruit H–14 = Vegetables/Legumes H–24 = Nuts/Seeds H–26 = Vegetarian
H–28 = Dairy H–34 = Eggs H–34 = Seafood H–36 = Meats H–40 = Poultry H–40 = Processed meats H–42 = Beverages H–46 = Fats/Oils
H–48 = Sweets H–50 = Spices/Condiments/Sauces H–52 = Mixed foods/Soups/Sandwiches H–58 = Fast food H–74 = Convenience meals H–76 = Baby foods

Chol (mg)	Calc (mg)	Iron (mg)	Magn (mg)	Pota (mg)	Sodi (mg)	Zinc (mg)	Vit A (µg)	Thia (mg)	Vit E (mg α)	Ribo (mg)	Niac (mg)	Vit B$_6$ (mg)	Fola (µg)	Vit C (mg)	Vit B$_{12}$ (µg)	Sele (µg)
20	263	1.20	58	493	110	1.58	128	0.10	0.08	0.46	0.33	0.10	13	1	1	7
<1	90	0.75	33	405	171	0.52	27	0.04	0.06	0.21	0.16	0.05	2	<1	<1	2
150	330	0.51	48	419	137	1.17	114	0.09	0.51	0.48	0.27	0.13	3	4	1	11
9	445	4.01	89	632	196	3.38	—	0.35	—	0.45	4.45	0.45	4	27	1	8
9	500	4.47	100	638	360	3.75	—	0.38	—	0.51	5.08	0.48	100	30	1	9
9	445	4.01	89	570	187	3.38	—	0.33	—	0.45	4.45	0.45	89	27	1	8
9	500	4.50	100	630	240	3.75	—	0.38	—	0.51	5.00	0.50	100	30	2	9
9	445	4.01	89	570	187	3.38	—	0.33	—	0.45	4.45	0.45	89	27	1	8
27	339	3.76	45	578	231	1.17	904	0.76	0.16	1.32	11.08	1.01	19	32	1	12
25	299	0.70	36	508	252	1.09	41	0.11	0.11	0.50	0.28	0.06	11	0	1	4
27	331	0.23	27	415	215	0.88	57	0.07	0.11	0.44	0.33	0.10	16	0	1	5
22	72	0.61	19	164	50	0.38	78	0.03	0.20	0.13	0.15	0.04	11	<1	<1	2
22	103	0.33	19	192	44	0.48	—	0.04	0.22	0.13	0.11	0.03	5	1	<1	—
0	80	0.36	—	—	60	—	—	—	—	—	—	—	—	0	—	—
17	77	0.05	9	137	49	0.48	91	0.02	0.08	0.11	0.06	0.02	3	<1	<1	1
10	135	0.05	12	190	60	0.46	25	0.04	0.05	0.17	0.10	0.04	5	1	<1	3
5	52	0.14	8	93	44	0.46	—	0.02	0.03	0.07	0.09	0.03	4	4	<1	—
19	79	0.14	9	124	40	0.22	63	0.03	—	0.17	0.11	0.03	8	5	<1	1
29	84	0.06	9	131	53	0.46	78	0.03	0.20	0.16	0.08	0.03	3	<1	<1	1
0	80	0.00	50	0					—							
0	0	0.00	—	8	90	—	0	—	—	—	—	—	—	0	—	—
0	0	0.00	—	7	180	—	0	—	—	—	—	—	—	0	—	—
0	0	0.00	—	2	210	—	0	—	—	—	—	—	—	0	—	—
9	91	0.07	10	138	56	0.29	—	0.04	0.06	0.17	0.06	0.04	4	1	<1	—
6	94	0.17	13	155	41	0.38	—	0.03	0.05	0.12	0.09	0.03	4	<1	<1	—
35	138	1.04	29	211	135	1.07	73	0.04	0.00	0.25	0.18	0.03	7	<1	<1	5
10	150	0.72	—	330	310	—	—	0.06	—	0.26	—	—	—	0	—	—
71	130	1.21	21	250	253	0.61	—	0.10	0.06	0.26	0.73	0.08	14	1	<1	—
1	119	0.33	11	136	226	0.38	0	0.03	0.43	0.14	0.44	0.03	4	1	<1	2
35	133	0.25	14	173	134	0.63	73	0.03	0.00	0.24	0.11	0.03	6	<1	<1	5
—	190	380	—	—	<1	—	—	0.17	—	—	—	—	0	—	—	—
4	106	0.90	19	188	71	0.35	32	0.03	—	0.15	0.22	0.05	8	<1	<1	2
15	113	0.52	11	176	71	0.32	—	0.05	0.10	0.20	0.08	0.05	5	1	<.1	—
1	103	0.22	10	152	63	0.30	42	0.03	0.08	0.16	0.21	0.06	4	1	<1	2
3	500	0.00	—	—	130	—	100	—	—	—	—	—	—	0	—	—
15	300	0.00	—	—	125	—	100	—	—	—	—	—	—	0	—	—
20	300	0.00	—	—	125	—	98	—	—	—	—	—	—	0	—	—
35	300	0.00	—	—	125	—	58	—	—	—	—	—	—	0	—	—
0	20	0.72	—	—	100	—	—	—	—	—	—	—	—	1	—	—
0	300	0.00	—	—	90	—	—	—	—	—	—	—	—	0	2	—
0	300	0.00	13	60	90	0.24	—	0.07	—	0.00	0.84	0.08	—	0	2	—

TABLE H–1

Food Composition

(DA+ code is for Wadsworth Diet Analysis program)   (For purposes of calculations, use "0" for t, <1, <.1, <.01, etc.)

DA + Code	Food Description	Quantity	Measure	Wt (g)	H₂O (g)	Ener (kcal)	Prot (g)	Carb (g)	Fiber (g)	Fat (g)	Sat	Mono	Poly	Trans
	SEAFOOD—Continued													
1594	Broiled or baked w/butter	3	ounce(s)	85	54	155	23	0	0	6	1.16	2.29	2.33	—
2938	Coho, farmed, raw	3	ounce(s)	85	60	136	18	0	0	7	1.54	2.83	1.58	—
154	**Sardines, Atlantic, with bones, canned in oil**	2	item(s)	24	14	50	6	0	0	3	0.36	0.92	1.23	—
	Scallops													
155	Mixed species, breaded, fried	3	item(s)	47	27	100	8	5	0	5	1.24	2.09	1.32	—
1599	Steamed	3	ounce(s)	85	65	90	14	2	0	3	—	—	—	—
1839	**Snapper, mixed species, cooked, dry heat**	3	ounce(s)	85	60	109	22	0	0	1	0.31	0.27	0.50	—
	Squid													
1868	Mixed species, fried	3	ounce(s)	85	55	149	15	7	0	6	1.60	2.34	1.82	—
16617	Steamed or boiled	3	ounce(s)	85	63	90	15	3	0	1	0.35	0.11	0.51	—
1570	**Striped bass, cooked, dry heat**	3	ounce(s)	85	62	105	19	0	0	3	0.55	0.72	0.85	—
1601	**Sturgeon, steamed**	3	ounce(s)	85	59	111	17	0	0	4	0.97	2.04	0.73	—
1840	**Surimi, formed**	3	ounce(s)	85	65	84	13	6	0	1	0.16	0.13	0.38	—
1842	**Swordfish, cooked, dry heat**	3	ounce(s)	85	58	132	22	0	0	4	1.20	1.68	1.00	—
1846	**Tuna, yellowfin or ahi, raw**	3	ounce(s)	85	60	92	20	0	0	1	0.20	0.13	0.24	—
	Tuna, canned													
159	Light, canned in oil, drained	2	ounce(s)	57	34	113	17	0	0	5	0.87	1.68	1.64	—
355	Light, canned in water, drained	2	ounce(s)	57	42	66	14	0	0	<1	0.13	0.09	0.19	—
33211	Light, no salt, canned in oil, drained	2	ounce(s)	57	34	112	17	0	0	5	0.87	1.67	1.64	—
33212	Light, no salt, canned in water, drained	2	ounce(s)	57	43	66	14	0	0	<1	0.13	0.09	0.19	—
2961	White, canned in oil, drained	2	ounce(s)	57	36	105	15	0	0	5	0.73	1.85	1.69	—
351	White, canned in water, drained	2	ounce(s)	57	41	73	13	0	0	2	0.45	0.44	0.63	—
33213	White, no salt, canned in oil, drained	2	ounce(s)	57	36	105	15	0	0	5	0.94	1.41	1.92	—
33214	White, no salt, canned in water, drained	2	ounce(s)	57	42	73	13	0	0	2	0.45	0.44	0.63	—
	Yellowtail													
2970	Mixed species, raw	2	ounce(s)	57	42	83	13	0	0	3	0.73	1.13	0.81	—
8548	Mixed species, cooked, dry heat	3	ounce(s)	85	57	159	25	0	0	6	1.44	2.21	1.52	—
	Shellfish, meat only													
1857	Abalone, mixed species, fried	3	ounce(s)	85	51	161	17	9	0	6	1.40	2.33	1.42	—
16618	Abalone, steamed or poached	3	ounce(s)	85	41	177	29	10	0	1	0.25	0.18	0.18	—
	Crab													
1851	Blue crab, canned	2	ounce(s)	57	43	56	12	0	0	1	0.14	0.12	0.25	—
1852	Blue crab, cooked, moist heat	3	ounce(s)	85	66	87	17	0	0	2	0.19	0.24	0.58	—
8562	Dungeness crab, cooked, moist heat	3	ounce(s)	85	62	94	19	1	0	1	0.14	0.18	0.35	—
1860	**Clams, cooked, moist heat**	3	ounce(s)	85	54	126	22	4	0	2	0.16	0.15	0.47	—
1853	**Crayfish, farmed, cooked, moist heat**	3	ounce(s)	85	69	74	15	0	0	1	0.18	0.21	0.35	—
	Oysters													
8720	Baked or broiled	3	ounce(s)	85	69	90	6	3	0	6	1.38	2.18	1.88	—
152	Eastern, farmed, raw	3	ounce(s)	85	73	50	4	5	0	1	0.38	0.13	0.50	—
8715	Eastern, wild, cooked, moist heat	3	ounce(s)	85	60	116	12	7	0	4	1.31	0.53	1.65	—
8584	Pacific, cooked, moist heat	3	ounce(s)	85	55	139	16	8	0	4	0.87	0.66	1.52	—
1865	Pacific, raw	3	ounce(s)	85	70	69	8	4	0	2	0.43	0.30	0.76	—
1854	**Lobster, northern, cooked, moist heat**	3	ounce(s)	85	65	83	17	1	0	1	0.09	0.14	0.08	—
1862	**Mussels, blue, cooked, moist heat**	3	ounce(s)	85	52	146	20	6	0	4	0.72	0.86	1.03	—
	Shrimp													
1855	Mixed species, cooked, moist heat	3	ounce(s)	85	66	84	18	0	0	1	0.25	0.17	0.37	—
158	Mixed species, breaded, fried	3	ounce(s)	85	45	206	18	10	<1	10	1.77	3.24	4.32	—
	BEEF, LAMB, PORK													
	Beef													
4450	Breakfast strips, cooked	2	slice(s)	23	0	101	7	<1	0	8	3.24	3.8	0.35	—
174	Corned, canned	3	ounce(s)	85	49	213	23	0	0	13	5.25	5.07	0.54	—
33147	Cured, thin sliced	2	ounce(s)	57	31	87	18	2	0	1	0.54	0.48	0.04	—
4581	Jerky	1	ounce(s)	28	0	116	9	3	1	7	3.08	3.21	0.28	—
	Ground													
4411	Extra lean, broiled, well	3	ounce(s)	85	46	225	24	0	0	13	5.28	5.88	0.50	—
4417	Lean, broiled, medium	3	ounce(s)	85	47	231	21	0	0	16	6.16	6.87	0.59	—
4418	Lean, broiled, well	3	ounce(s)	85	45	238	24	0	0	15	5.89	6.56	0.56	—
4423	Regular, broiled, medium	3	ounce(s)	85	46	246	20	0	0	18	6.91	7.70	0.65	—
	Rib													
4183	Rib, whole, lean & fat, ¼" fat, roasted	3	ounce(s)	85	39	320	19	0	0	27	10.71	11.42	0.94	—
	Roast													
4264	Bottom round, lean & fat, ¼" fat, braised	3	ounce(s)	85	44	241	24	0	0	15	5.71	6.63	0.58	—

PAGE KEY: H–2 = Breads/Baked Goods   H–6 = Cereal/Rice/Pasta   H–10 = Fruit   H–14 = Vegetables/Legumes   H–24 = Nuts/Seeds   H–26 = Vegetarian
H–28 = Dairy   H–34 = Eggs   H–34 = Seafood   H–36 = Meats   H–40 = Poultry   H–40 = Processed meats   H–42 = Beverages   H–46 = Fats/Oils
H–48 = Sweets   H–50 = Spices/Condiments/Sauces   H–52 = Mixed foods/Soups/Sandwiches   H–58 = Fast food   H–74 = Convenience meals   H–76 = Baby foods

Chol (mg)	Calc (mg)	Iron (mg)	Magn (mg)	Pota (mg)	Sodi (mg)	Zinc (mg)	Vit A (µg)	Thia (mg)	Vit E (mg α)	Ribo (mg)	Niac (mg)	Vit B$_6$ (mg)	Fola (µg)	Vit C (mg)	Vit B$_{12}$ (µg)	Sele (µg)
40	15	1.02	27	377	99	0.56	—	0.14	1.15	0.05	8.33	0.19	4	2	2	41
43	10	0.29	26	383	40	0.37	48	0.08	—	0.09	5.79	0.56	11	1	2	11
34	108	0.70	9	95	121	0.31	16	0.01	0.49	0.05	1.25	0.04	3	0	2	13
28	20	0.38	27	155	216	0.49	10	0.01	—	0.05	0.69	0.06	23	1	1	13
27	21	0.22	—	238	366	—	—	—	0.16	—	—	—	—	2	—	—
40	34	0.20	31	444	48	0.37	30	0.05	—	0.00	0.29	0.39	5	1	3	42
221	33	0.86	32	237	260	1.48	9	0.05	—	0.39	2.21	0.05	12	4	1	44
227	31	0.63	29	192	356	1.49	—	0.02	1.17	0.32	1.70	0.04	4	3	1	—
88	16	0.92	43	279	75	0.43	26	0.10	—	0.03	2.17	0.29	9	0	4	40
63	11	0.59	30	239	389	0.36	—	0.07	0.53	0.07	8.31	0.19	14	0	2	—
26	8	0.22	37	95	122	0.28	17	0.02	0.54	0.02	0.19	0.03	2	0	1	24
43	5	0.88	29	314	98	1.25	35	0.04	—	0.10	10.02	0.32	2	1	2	52
38	14	0.62	43	378	31	0.44	15	0.37	0.43	0.04	8.33	0.77	2	1	<1	31
10	7	0.79	18	118	202	0.51	13	0.02	0.50	0.07	7.06	0.06	3	0	1	43
17	6	0.87	15	134	192	0.44	10	0.02	0.19	0.04	7.53	0.20	2	0	2	46
10	7	0.79	18	117	28	0.51	13	0.02	—	0.07	7.03	0.06	3	0	1	43
17	6	0.87	15	134	28	0.44	10	0.02	—	0.04	7.53	0.20	2	0	2	46
18	2	0.37	19	189	225	0.27	3	0.01	1.30	0.04	6.63	0.24	3	0	1	34
24	8	0.55	19	134	214	0.27	3	0.00	0.48	0.02	3.29	0.12	1	0	1	37
18	2	0.37	19	189	28	0.27	14	0.01	—	0.04	6.63	0.24	3	0	1	34
24	8	0.55	19	134	28	0.27	3	0.00	—	0.02	3.29	0.12	1	0	1	37
31	13	0.28	17	238	22	0.29	16	0.08	—	0.02	3.86	0.09	2	2	1	21
60	25	0.53	32	457	43	0.56	26	0.14	—	0.04	7.41	0.15	3	2	1	40
80	31	3.23	48	241	502	0.81	2	0.19	—	0.11	1.62	0.13	12	2	1	44
143	50	4.85	69	295	980	1.38	—	0.29	6.74	0.13	1.90	0.22	6	3	1	—
50	57	0.48	22	212	189	2.28	1	0.05	1.04	0.05	0.78	0.09	24	2	<1	18
85	88	0.77	28	275	237	3.59	2	0.09	1.56	0.04	2.81	0.15	43	3	6	34
65	50	0.37	49	347	321	4.65	26	0.05	—	0.17	3.08	0.15	36	3	9	40
57	78	23.77	15	534	95	2.32	145	0.13	—	0.36	2.85	0.09	25	19	84	54
116	43	0.94	28	202	82	1.26	13	0.04	—	0.07	1.42	0.11	9	<1	3	29
42	37	5.30	38	126	418	72.22	60	0.07	0.99	0.06	1.04	0.05	8	3	15	—
21	37	4.91	28	105	151	32.23	7	0.09	—	0.06	1.08	0.05	15	4	14	54
89	77	10.19	81	239	359	154.37	46	0.16	—	0.15	2.11	0.10	12	5	30	61
85	14	7.82	37	257	180	28.25	124	0.11	0.72	0.38	3.08	0.08	13	11	24	131
43	7	4.35	19	143	90	14.14	69	0.06	—	0.20	1.71	0.04	9	7	14	65
61	52	0.33	30	299	323	2.48	22	0.01	0.85	0.06	0.91	0.07	9	0	3	36
48	28	5.71	31	228	314	2.27	77	0.26	—	0.36	2.55	0.09	65	12	20	76
166	33	2.63	29	155	190	1.33	58	0.03	1.17	0.03	2.20	0.11	3	2	1	34
150	57	1.07	34	191	292	1.17	48	0.10	—	0.11	2.60	0.08	20	1	2	35
27	2	0.70	6	93	509	1.43	0	0.02	0.07	0.05	1.46	0.07	2	0	1	6
73	10	1.77	12	116	855	3.03	0	0.02	0.13	0.12	2.07	0.11	8	0	1	36
45	3	1.58	11	140	1582	2.49	0	0.03	0.00	0.12	1.85	0.16	5	0	1	13
14	6	1.53	14	170	628	2.30	0	0.04	0.14	0.04	0.49	0.05	38	0	<1	3
84	8	2.35	21	314	70	5.47	0	0.06	—	0.27	4.97	0.27	9	0	2	19
74	9	1.79	18	256	65	4.56	0	0.04	—	0.18	4.39	0.22	8	0	2	25
86	10	2.08	20	297	76	5.27	0	0.05	—	0.20	5.07	0.26	9	0	2	22
77	9	2.07	17	248	71	4.40	0	0.03	—	0.16	4.90	0.23	8	0	2	16
72	9	1.96	16	252	54	4.45	0	0.06	—	0.14	2.86	0.20	6	0	2	19
82	5	2.65	19	240	43	4.17	0	0.06	0.17	0.20	3.17	0.28	9	0	2	27

TABLE H–1

Food Composition

(DA+ code is for Wadsworth Diet Analysis program)   (For purposes of calculations, use "0" for t, <1, <.1, <.01, etc.)

DA + Code	Food Description	Quantity	Measure	Wt (g)	H₂O (g)	Ener (kcal)	Prot (g)	Carb (g)	Fiber (g)	Fat (g)	Sat	Mono	Poly	Trans
	BEEF, LAMB, PORK—Continued													
169	Bottom round, separable lean, ¼" fat, roasted	3	ounce(s)	85	57	161	24	0	0	6	2.13	2.83	0.24	—
4147	Chuck, arm pot roast, lean & fat, ¼" fat, braised	3	ounce(s)	85	41	282	23	0	0	20	7.97	8.68	0.77	—
4161	Chuck, blade roast, lean & fat, ¼" fat, braised	3	ounce(s)	85	40	293	23	0	0	22	8.70	9.44	0.78	—
5853	Chuck, blade roast, separable lean, ¼" trim, pot roasted	3	ounce(s)	85	39	209	27	0	0	10	3.94	4.37	0.33	—
4295	Eye of round, lean, ¼" fat, roasted	3	ounce(s)	85	55	149	25	0	0	5	1.76	2.06	0.15	—
4285	Eye of round, lean & fat, ¼" fat, roasted	3	ounce(s)	85	51	195	23	0	0	11	4.23	4.66	0.39	—
	Steak													
1757	Rib, small end, lean, ¼" fat, broiled	3	ounce(s)	85	49	188	24	0	0	10	3.84	4.01	0.27	—
4349	Short loin, T-bone steak, lean, ¼" fat, broiled	3	ounce(s)	85	52	174	23	0	0	9	3.05	4.23	0.26	—
4348	Short loin, T-bone steak, lean & fat, ¼" fat, broiled	3	ounce(s)	85	43	274	19	0	0	21	8.29	9.58	0.75	—
4360	Top loin, prime, lean & fat, ¼" fat, broiled	3	ounce(s)	85	43	275	22	0	0	20	8.16	8.61	0.73	—
	Variety													
188	Liver, pan fried	3	ounce(s)	85	53	149	23	4	0	4	1.27	0.56	0.49	0.17
4447	Tongue, simmered	3	ounce(s)	85	49	236	16	0	0	19	6.91	8.59	0.56	0.71
	Lamb													
	Chop													
3275	Loin, domestic, lean & fat, ¼" fat, broiled	3	ounce(s)	85	44	269	21	0	0	20	8.36	8.25	1.43	—
3287	Shoulder, arm, domestic, lean & fat, ¼" fat, braised	3	ounce(s)	85	38	294	26	0	0	20	8.39	8.65	1.45	—
3290	Shoulder, arm, domestic, lean, ¼" fat, braised	3	ounce(s)	85	42	237	30	0	0	12	4.28	5.24	0.78	—
	Leg													
3264	Domestic, lean & fat, ¼" fat, cooked	3	ounce(s)	85	46	250	21	0	0	18	7.51	7.50	1.28	—
	Rib													
183	Domestic, lean, ¼" fat, broiled	3	ounce(s)	85	50	200	24	0	0	11	3.95	4.43	1.00	—
182	Domestic, lean & fat, ¼" fat, broiled	3	ounce(s)	85	40	307	19	0	0	25	10.80	10.30	2.01	—
	Shoulder													
187	Arm & blade, domestic, choice, lean, ¼" fat, roasted	3	ounce(s)	85	54	173	21	0	0	9	3.47	3.71	0.81	—
186	Arm & blade, domestic, choice, lean & fat, ¼" fat, roasted	3	ounce(s)	85	48	235	19	0	0	17	7.17	6.94	1.38	—
	Variety													
3375	Brain, pan fried	3	ounce(s)	85	52	232	14	0	0	19	4.82	3.42	1.94	—
3406	Tongue, braised	3	ounce(s)	85	49	234	18	0	0	17	6.66	8.50	1.06	—
	Pork													
	Cured													
161	Bacon, cured, broiled, pan fried or roasted	2	slice(s)	13	2	68	5	<1	0	5	1.73	2.33	0.57	0
29229	Bacon, Canadian style, cured	2	ounce(s)	57	38	89	12	1	0	4	1.26	1.79	0.36	—
35422	Breakfast strips, cured, cooked	3	slice(s)	34	9	156	10	<1	0	12	4.34	5.58	1.92	—
16561	Ham, smoked or cured, lean, cooked	1	slice(s)	42	28	66	11	0	0	2	0.77	1.06	0.27	—
189	Ham, cured, boneless, 11% fat, roasted	3	ounce(s)	85	55	151	19	0	0	8	2.65	3.77	1.20	—
1316	Ham, cured, extra lean, 5% fat, roasted	3	ounce(s)	85	58	123	18	1	0	5	1.54	2.23	0.46	—
29215	Ham, cured, extra lean, 4% fat, canned	2	ounce(s)	57	42	68	10	0	0	3	0.86	1.25	0.22	—
	Chop													
32671	Loin, blade, lean & fat, pan fried	3	ounce(s)	85	42	291	18	0	0	24	8.65	9.97	2.64	—
32672	Loin, center cut, lean & fat, pan fried	3	ounce(s)	85	45	236	25	0	0	14	5.11	6.00	1.62	—
32682	Loin, center rib, boneless, lean & fat, braised	3	ounce(s)	85	49	217	22	0	0	13	5.21	6.13	1.12	—
32603	Loin, center rib, lean, broiled	3	ounce(s)	85	48	186	26	0	0	8	2.94	3.78	0.53	—
32481	Loin, whole, lean, braised	3	ounce(s)	85	52	174	24	0	0	8	2.87	3.54	0.60	—
32478	Loin, whole, lean & fat, braised	3	ounce(s)	85	50	203	23	0	0	12	4.35	5.15	1.00	—
	Leg or ham													
32471	Rump portion, lean & fat, roasted	3	ounce(s)	85	48	214	25	0	0	12	4.47	5.42	1.17	—
32468	Whole, lean & fat, roasted	3	ounce(s)	85	47	232	23	0	0	15	5.50	6.70	1.43	—
	Ribs													
32696	Loin, country style, lean, roasted	3	ounce(s)	85	49	210	23	0	0	13	4.52	5.49	0.94	—
32693	Loin, country style, lean & fat, roasted	3	ounce(s)	85	43	279	20	0	0	22	7.83	9.36	1.71	—
	Shoulder													
32629	Arm picnic, lean, roasted	3	ounce(s)	85	51	194	23	0	0	11	3.66	5.09	1.02	—

PAGE KEY: H–2 = Breads/Baked Goods H–6 = Cereal/Rice/Pasta H–10 = Fruit H–14 = Vegetables/Legumes H–24 = Nuts/Seeds H–26 = Vegetarian
H–28 = Dairy H–34 = Eggs H–34 = Seafood H–36 = Meats H–40 = Poultry H–40 = Processed meats H–42 = Beverages H–46 = Fats/Oils
H–48 = Sweets H–50 = Spices/Condiments/Sauces H–52 = Mixed foods/Soups/Sandwiches H–58 = Fast food H–74 = Convenience meals H–76 = Baby foods

Chol (mg)	Calc (mg)	Iron (mg)	Magn (mg)	Pota (mg)	Sodi (mg)	Zinc (mg)	Vit A (µg)	Thia (mg)	Vit E (mg α)	Ribo (mg)	Niac (mg)	Vit B$_6$ (mg)	Fola (µg)	Vit C (mg)	Vit B$_{12}$ (µg)	Sele (µg)
66	4	2.66	24	332	56	3.92	0	0.06	—	0.20	3.45	0.31	10	0	2	23
84	9	2.64	16	209	51	5.81	0	0.06	0.19	0.20	2.70	0.24	8	0	3	21
88	11	2.64	16	196	54	7.07	0	0.06	0.15	0.20	2.06	0.22	4	0	2	21
74	11	3.12	20	224	60	8.72	0	0.06	—	0.23	0	0.24	—	0	2	23
59	4	1.66	23	336	53	4.03	0	0.08	—	0.14	3.19	0.32	6	0	2	23
61	5	1.56	20	308	50	3.69	0	0.07	0.15	0.14	2.97	0.30	6	0	2	22
68	11	2.18	23	335	59	5.94	0	0.09	0.12	0.19	4.08	0.34	7	0	3	19
50	5	3.11	22	278	65	4.34	0	0.09	0.12	0.21	3.94	0.33	7	0	2	9
58	7	2.56	18	234	58	3.56	0	0.08	0.19	0.18	3.29	0.28	6	0	2	10
67	8	1.89	20	294	54	3.85	0	0.07	—	0.15	3.96	0.31	6	0	2	19
324	5	5.24	19	298	65	4.45	6582	0.15	0.39	2.91	14.85	0.87	221	1	71	28
112	4	2.22	13	156	55	3.48	0	0.02	0.25	0.25	2.97	0.13	6	1	3	11
85	17	1.54	20	278	65	2.96	0	0.09	0.11	0.21	6.04	0.11	15	0	2	23
102	21	2.03	22	260	61	5.17	0	0.06	0.13	0.21	5.66	0.09	15	0	2	32
103	22	2.30	25	287	65	6.21	0	0.06	0.15	0.23	5.38	0.11	19	0	2	32
82	14	1.60	20	264	61	3.79	0	0.09	0.12	0.21	5.66	0.11	15	0	2	22
77	14	1.88	25	266	72	4.48	0	0.09	0.15	0.21	5.57	0.13	18	0	2	26
84	16	1.60	20	230	65	3.40	0	0.08	0.10	0.19	5.95	0.09	12	0	2	20
74	16	1.81	21	225	58	5.13	0	0.08	0.15	0.22	4.90	0.13	21	0	2	24
78	17	1.67	20	213	56	4.45	0	0.08	0.12	0.20	5.23	0.11	18	0	2	22
2128	18	1.73	19	304	133	1.70	0	0.14	—	0.31	3.87	0.20	6	20	20	10
161	9	2.24	14	134	57	2.54	0	0.07	—	0.36	3.14	0.14	3	6	5	24
14	1	0.18	4	71	291	0.44	1	0.05	0.04	0.03	1.40	0.04	<1	0	<1	8
28	5	0.39	10	195	799	0.79	0	0.43	0.12	0.10	3.53	0.22	2	0	<1	14
36	5	0.67	9	158	714	1.25	0	0.25	0.09	0.13	2.58	0.12	1	0	1	8
23	3	0.40	9	133	557	1.08	0	0.29	0.11	0.11	2.11	0.20	2	0	<1	—
50	7	1.14	19	348	1275	2.10	0	0.62	0.26	0.28	5.23	0.26	3	0	1	17
45	7	1.26	12	244	1023	2.45	0	0.64	0.21	0.17	3.42	0.34	3	0	1	17
22	3	0.53	10	206	712	1.09	0	0.47	0.10	0.13	3.01	0.26	3	0	<1	8
72	26	0.75	18	282	57	2.71	3	0.53	0.17	0.25	3.36	0.29	3	1	1	30
78	23	0.77	25	361	68	1.96	2	0.97	0.21	0.26	4.76	0.40	5	1	1	33
62	4	0.78	14	329	34	1.76	2	0.45	0.21	0.21	3.67	0.26	3	<1	<1	28
69	26	0.70	24	357	55	2.02	2	0.95	0.25	0.28	5.25	0.40	3	<1	1	40
67	15	0.96	17	329	43	2.11	2	0.56	0.18	0.23	3.90	0.33	3	1	<1	41
68	18	0.91	16	318	41	2.02	2	0.54	0.20	0.22	3.76	0.31	3	1	<1	39
82	10	0.89	23	318	53	2.40	3	0.64	0.19	0.28	3.96	0.27	3	<1	1	40
80	12	0.86	19	299	51	2.52	3	0.54	0.19	0.27	3.89	0.34	9	<1	1	39
79	25	1.10	20	297	25	3.24	2	0.49	—	0.29	3.97	0.37	4	<1	1	36
78	21	0.90	20	293	44	2.01	3	0.76	—	0.29	3.67	0.38	4	<1	1	32
81	8	1.21	17	299	68	3.46	2	0.49	—	0.30	3.67	0.35	4	<1	1	33

TABLE H–1

Food Composition

(DA+ code is for Wadsworth Diet Analysis program)   (For purposes of calculations, use "0" for t, <1, <.1, <.01, etc.)

DA + Code	Food Description	Quantity	Measure	Wt (g)	H₂O (g)	Ener (kcal)	Prot (g)	Carb (g)	Fiber (g)	Fat (g)	Sat	Mono	Poly	Trans
	BEEF, LAMB, PORK—Continued													
32626	Arm picnic, lean & fat, roasted	3	ounce(s)	85	44	270	20	0	0	20	7.47	9.12	2.00	—
	Rabbit													
3366	Domesticated, roasted	3	ounce(s)	85	52	167	25	0	0	7	2.04	1.84	1.33	—
3367	Domesticated, stewed	3	ounce(s)	85	50	175	26	0	0	7	2.13	1.93	1.39	—
	Veal													
3391	Liver, braised	3	ounce(s)	85	51	163	24	3	0	5	1.69	0.97	0.88	0.26
3319	Rib, lean only, roasted	3	ounce(s)	85	55	150	22	0	0	6	1.77	2.26	0.57	—
1732	**Deer or venison, roasted**	3	ounce(s)	85	55	134	26	0	0	3	1.06	0.75	0.53	—
	POULTRY													
	Chicken													
29562	Flaked, canned	2	ounce(s)	57	37	97	10	<1	0	6	1.62	2.32	1.29	—
	Fried													
29632	Breast, meat only, breaded, baked or fried	3	ounce(s)	85	44	193	25	7	<1	7	1.62	2.66	1.73	—
35327	Broiler breast, meat only, fried	3	ounce(s)	85	51	159	28	<1	0	4	1.10	1.46	0.91	—
36413	Broiler breast, meat & skin, flour coated, fried	3	ounce(s)	85	48	189	27	1	<.1	8	2.08	2.98	1.67	—
35389	Broiler drumstick, meat only, fried	3	ounce(s)	85	53	166	24	0	0	7	1.81	2.50	1.68	—
36414	Broiler drumstick, meat & skin, flour coated, fried	3	ounce(s)	85	48	208	23	1	<.1	12	3.11	4.61	2.75	—
35406	Broiler leg, meat only, fried	3	ounce(s)	85	52	177	24	1	0	8	2.12	2.92	1.89	—
35484	Broiler wing, meat only, fried	3	ounce(s)	85	51	179	26	0	0	8	2.13	2.62	1.76	—
29580	Patty, fillet, or tenders, breaded, cooked	3	ounce(s)	85	42	241	14	13	<1	15	4.62	7.25	1.87	—
	Roasted, meat only													
35409	Broiler chicken leg	3	ounce(s)	85	55	162	23	0	0	7	1.95	2.59	1.68	—
35486	Broiler chicken wing	3	ounce(s)	85	53	173	26	0	0	7	1.92	2.22	1.51	—
35138	Roasting chicken, dark meat	3	ounce(s)	85	57	151	20	0	0	7	2.07	2.82	1.70	—
35136	Roasting chicken, light meat	3	ounce(s)	85	58	130	23	0	0	3	0.92	1.29	0.79	—
35132	Roasting chicken	3	ounce(s)	85	57	142	21	0	0	6	1.54	2.13	1.28	—
	Stewed													
3174	Meat only, stewed	3	ounce(s)	85	48	150	23	0	0	6	1.56	2.03	1.3	—
1268	Gizzard, simmered	3	ounce(s)	85	58	124	26	0	0	2	0.57	0.45	0.30	0.11
1270	Liver, simmered	3	ounce(s)	85	57	142	21	1	0	6	1.75	1.20	1.08	0.08
	Duck													
1286	Domesticated, meat & skin, roasted	3	ounce(s)	85	44	286	16	0	0	24	8.22	10.97	3.10	—
1287	Domesticated, meat only, roasted	3	ounce(s)	85	55	171	20	0	0	10	3.54	3.15	1.22	—
	Goose													
35507	Domesticated, meat & skin, roasted	3	ounce(s)	85	44	259	21	0	0	19	5.84	8.72	2.14	—
35524	Domesticated, meat only, roasted	3	ounce(s)	85	49	202	25	0	0	11	3.88	3.69	1.31	—
1297	Liver pâté, smoked, canned	4	tablespoon(s)	52	19	240	6	2	0	23	7.51	13.32	0.44	—
	Turkey													
3256	Ground turkey, cooked	3	ounce(s)	85	51	200	23	0	0	11	2.88	4.16	2.75	—
222	Roasted, fryer roaster breast, meat only	3	ounce(s)	85	58	115	26	0	0	1	0.20	0.11	0.17	—
219	Roasted, dark meat, meat only	3	ounce(s)	85	54	159	24	0	0	6	2.06	1.39	1.84	—
220	Roasted, light meat, meat only	3	ounce(s)	85	56	133	25	0	0	3	0.88	0.48	0.73	—
3263	Patty, batter coated, breaded, fried	1	item(s)	94	47	266	13	15	<1	17	4.41	7.02	4.43	—
1302	Turkey roll, light meat	2	slice(s)	57	41	83	11	<1	0	4	1.15	1.42	0.99	—
1303	Turkey roll, light & dark meat	2	slice(s)	57	40	84	10	1	0	4	1.16	1.30	1.01	—
	PROCESSED MEATS													
	Beef													
1331	Corned beef loaf, jellied, sliced	2	slice(s)	57	39	87	13	0	0	3	1.47	1.52	0.18	—
	Bologna													
13458	Made w/chicken, pork, & beef	1	slice(s)	28	15	90	3	1	0	8	3.00	4.05	1.10	—
13461	Light, made w/pork, chicken, & beef	1	slice(s)	28	18	60	3	2	0	4	1.50	2.04	0.43	—
13459	Beef	1	slice(s)	28	15	90	3	1	0	8	3.50	4.26	0.31	—
13565	Turkey	1	slice(s)	28	19	50	3	1	0	4	1.00	1.09	0.98	—
	Chicken													
13562	Oven roasted white chicken	1	slice(s)	28	20	40	4	1	0	3	0.50	—	—	—
	Ham													
13581	Honey glazed, traditional carved	2	slice(s)	45	—	50	8	1	0	2	0.50	0.68	0.18	—
13777	Deli sliced cooked	1	slice(s)	28	—	30	5	1	0	1	0.50	0.39	0.11	—
13778	Deli sliced honey	1	slice(s)	28	—	35	5	1	0	1	0.50	0.39	0.11	—
8614	**Pork & beef mortadella, sliced**	2	slice(s)	46	24	143	8	1	0	12	4.37	5.23	1.44	—

PAGE KEY: H–2 = Breads/Baked Goods   H–6 = Cereal/Rice/Pasta   H–10 = Fruit   H–14 = Vegetables/Legumes   H–24 = Nuts/Seeds   H–26 = Vegetarian
H–28 = Dairy   H–34 = Eggs   H–34 = Seafood   H–36 = Meats   H–40 = Poultry   H–40 = Processed meats   H–42 = Beverages   H–46 = Fats/Oils
H–48 = Sweets   H–50 = Spices/Condiments/Sauces   H–52 = Mixed foods/Soups/Sandwiches   H–58 = Fast food   H–74 = Convenience meals   H–76 = Baby foods

Chol (mg)	Calc (mg)	Iron (mg)	Magn (mg)	Pota (mg)	Sodi (mg)	Zinc (mg)	Vit A (µg)	Thia (mg)	Vit E (mg α)	Ribo (mg)	Niac (mg)	Vit B$_6$ (mg)	Fola (µg)	Vit C (mg)	Vit B$_{12}$ (µg)	Sele (µg)
80	16	1.00	14	276	60	2.93	2	0.44	—	0.26	3.33	0.30	3	<1	1	29
70	16	1.93	18	326	40	1.93	0	0.08	—	0.18	7.17	0.40	9	0	7	33
73	17	2.01	17	255	31	2.01	0	0.05	0.37	0.14	6.09	0.29	8	0	6	33
434	5	4.34	17	280	66	9.55	17973	0.15	0.58	2.43	11.18	0.78	281	1	72	16
98	10	0.82	20	264	82	3.82	0	0.05	0.31	0.25	6.38	0.23	12	0	1	9
95	6	3.80	20	285	46	2.34	0	0.15	—	0.51	5.70	—	—	0	—	11
35	8	0.90	7	148	410	0.8	19	0	—	0.07	3.6	0.19	—	0	<1	—
67	19	1.05	25	223	450	0.84	—	0.08	—	0.10	10.98	0.47	4	0	<1	—
77	14	0.97	26	235	67	0.92	—	0.07	—	0.11	12.57	0.54	3	0	<1	22
76	14	1.01	26	220	65	0.94	—	0.07	—	0.11	11.69	0.49	5	0	<1	20
80	10	1.12	20	212	82	2.74	—	0.07	—	0.20	5.23	0.33	8	0	<1	17
77	10	1.14	20	195	76	2.46	—	0.07	—	0.19	5.13	0.30	9	0	<1	16
84	11	1.19	21	216	82	2.53	—	0.07	—	0.21	5.69	0.33	8	0	<1	16
71	13	0.97	18	177	77	1.80	—	0.04	—	0.11	6.16	0.50	3	0	<1	22
51	14	1.06	17	209	452	0.88	—	0.08	—	0.12	5.71	0.26	9	<1	<1	—
80	10	1.11	20	206	77	2.43	—	0.06	—	0.20	5.37	0.32	7	0	<1	19
72	14	0.99	18	179	78	1.82	—	0.04	—	0.11	6.22	0.50	3	0	<1	21
64	9	1.13	17	191	81	1.81	14	0.05	—	0.16	4.88	0.26	6	0	<1	17
64	11	0.92	20	201	43	0.66	7	0.05	0.23	0.08	8.90	0.46	3	0	<1	22
64	10	1.03	18	195	64	1.29	10	0.05	—	0.13	6.70	0.35	4	0	<1	21
71	12	0.99	18	153	60	1.69	13	0.04	0.23	0.13	5.19	0.22	5	0	<1	18
315	14	2.71	3	152	48	3.76	0	0.02	0.17	0.18	2.65	0.06	4	0	1	35
479	9	9.89	21	224	65	3.38	3384	0.25	0.70	1.69	9.39	0.64	491	24	14	70
71	9	2.30	14	173	50	1.58	54	0.15	0.59	0.23	4.10	0.15	5	0	<1	17
76	10	2.30	17	214	55	2.21	20	0.22	0.59	0.40	4.34	0.21	9	0	<1	19
77	11	2.41	19	280	60	2.23	18	0.07	—	0.28	3.55	0.32	2	0	<1	19
82	12	2.44	21	330	65	2.70	10	0.08	—	0.33	3.47	0.40	10	0	<1	22
78	36	2.86	7	72	362	0.48	521	0.05	—	0.16	1.31	0.03	31	0	5	23
87	21	1.64	20	230	91	2.43	0	0.05	0.29	0.14	4.10	0.33	6	0	<1	32
71	10	1.30	25	248	44	1.48	0	0.04	0.08	0.11	6.37	0.48	5	0	<1	27
72	27	1.98	20	247	67	3.79	0	0.05	0.54	0.21	3.10	0.31	8	0	<1	35
59	16	1.15	24	259	54	1.73	0	0.05	0.08	0.11	5.81	0.46	5	0	<1	27
58	13	2.07	14	259	752	1.35	10	0.09	1.18	0.18	2.16	0.19	26	0	<1	19
24	23	0.73	9	142	277	0.88	0	0.05	0.07	0.13	3.97	0.18	2	0	<1	13
31	18	0.77	10	153	332	1.13	0	0.05	0.19	0.16	2.72	0.15	3	0	<1	17
27	6	1.16	6	57	540	2.32	0	0.00	—	0.06	1.00	0.07	5	0	1	10
30	0	0.36	6	43	290	0.40	0	—	—	—	—	—	—	0	—	—
15	0	0.36	6	46	310	0.45	0	—	—	—	—	—	—	0	—	—
20	0	0.36	4	47	310	0.57	0	0.01	—	0.03	0.68	0.05	4	0	<1	—
20	40	0.36	6	43	270	0.52	0	—	—	—	—	—	—	0	—	—
15	0	0.36	7	85	350	0.32	0	—	—	—	—	—	—	0	—	—
25	0	0.72	—	—	560	—	0	—	—	—	—	—	—	0	—	—
15	0	0.00	—	—	240	—	0	—	—	—	—	—	—	0	—	—
15	0	0.00	—	—	240	—	0	—	—	—	—	—	—	0	—	—
26	8	0.64	5	75	573	0.97	0	0.05	0.10	0.07	1.23	0.06	1	0	1	10

TABLE H–1
Food Composition

(DA+ code is for Wadsworth Diet Analysis program)   (For purposes of calculations, use "0" for t, <1, <.1, <.01, etc.)

DA + Code	Food Description	Quantity	Measure	Wt (g)	H₂O (g)	Ener (kcal)	Prot (g)	Carb (g)	Fiber (g)	Fat (g)	Sat	Mono	Poly	Trans
	PROCESSED MEATS—Continued													
1323	**Pork olive loaf**	2	slice(s)	57	33	133	7	5	0	9	3.32	4.47	1.10	—
1324	**Pork pickle & pimento loaf**	2	slice(s)	57	32	149	7	3	0	12	4.45	5.45	1.47	—
	Sausages & frankfurters													
37296	Beerwurst beef beer salami (bierwurst)	1	slice(s)	29	17	74	4	1	0	6	2.50	2.69	0.21	—
37257	Beerwurst pork beer salami	1	slice(s)	21	13	50	3	<1	0	4	1.32	1.89	0.50	—
35338	Berliner, pork & beef	1	ounce(s)	28	17	65	4	1	0	5	1.72	2.27	0.45	—
37299	Braunschweiger pork liver sausage	1	slice(s)	15	0	51.34	1.97	0.34	0	4.48	1.52	2.08	0.52	—
37298	Bratwurst pork, cooked	1	piece(s)	74	42	181	10	2	0	14	5.15	6.73	1.51	—
1329	Cheesefurter or cheese smokie, beef & pork	1	item(s)	43	23	141	6	1	0	12	4.52	5.89	1.30	—
1330	Chorizo, beef & pork	2	ounce(s)	57	18	258	14	1	0	22	8.15	10.43	1.96	—
8600	Frankfurter, beef	1	item(s)	45	23	149	5	2	0	13	5.26	6.44	0.53	—
202	Frankfurter, beef & pork	1	item(s)	57	32	174	7	1	1	16	6.14	7.79	1.56	—
1293	Frankfurter, chicken	1	item(s)	45	26	116	6	3	0	9	2.49	3.82	1.82	—
3261	Frankfurter, turkey	1	item(s)	45	28	102	6	1	0	8	2.65	2.51	2.25	—
37275	Italian sausage, pork, cooked	1	item(s)	68	34	220	14	1	0	17	6.14	8.13	2.23	—
37307	Kielbasa, kolbassa, pork & beef	2⅛	ounce(s)	61	37	135	10	2	0	9	3.40	4.44	1.06	—
1333	Knockwurst or knackwurst, beef & pork	2	ounce(s)	57	31	174	6	2	0	16	5.79	7.26	1.66	—
37285	Pepperoni, beef & pork	1	slice(s)	11	3	55	2	<1	0	5	1.77	2.32	0.48	—
37313	Polish sausage, pork	2	slice(s)	57	31	163	8	2	—	14	4.91	6.42	1.46	—
206	Salami, beef, cooked, sliced	2	slice(s)	46	28	119	6	1	0	10	4.54	4.90	0.48	—
37272	Salami, pork, dry or hard	1	slice(s)	13	5	52	3	<1	0	4	1.52	2.05	0.48	—
3262	Salami, turkey	2	slice(s)	57	31	125	8	11	<.1	5	1.98	1.80	1.43	0
7162	Sausage, breakfast, turkey	2½	ounce(s)	100	67	190	17	<1	0	13	3.90	6.23	3.33	0
8620	Smoked sausage, beef & pork	2	ounce(s)	57	31	181	7	1	0	16	5.54	6.94	2.23	0
8619	Smoked, sausage, pork	2	ounce(s)	57	22	221	13	1	0	18	6.42	8.30	2.13	—
37273	Smoked, sausage, pork link	1	piece(s)	76	30	295	17	2	—	24	8.58	11.09	2.85	—
1336	Summer sausage, thuringer, or cervelat, beef & pork	2	ounce(s)	57	29	190	9	<1	0	17	6.82	7.35	0.68	—
37294	Vienna sausage, cocktail, beef & pork, canned	1	piece(s)	16	10	45	2	<1	0	4	1.49	2.01	0.27	—
	Spreads													
32419	Pork & beef sandwich spread	4	tablespoon(s)	60	36	141	5	7	<1	10	3.59	4.57	1.54	—
1318	Ham salad spread	¼	cup(s)	60	38	130	5	6	0	9	3.04	4.32	1.62	—
	Turkey													
16049	Breast, hickory smoked, slices	1	slice(s)	56	—	50	11	1	0	0	0.00	0.00	0.00	0
13606	Breast, hickory smoked fat free	1	slice(s)	28	—	25	4	1	0	0	0.00	0.00	0.00	0
16047	Breast, honey roasted, slices	1	slice(s)	56	—	60	11	2	0	0	0.00	0.00	0.00	0
16048	Breast, oven roasted, slices	1	slice(s)	56	—	50	11	1	0	0	0.00	0.00	0.00	0
13583	Breast, traditional carved	2	slice(s)	45	—	40	9	0	0	1	0.00	0.07	0.14	—
13604	Breast, oven roasted, fat free	1	slice(s)	28	—	25	4	1	0	0	0.00	0.00	0.00	0
13567	Turkey ham, 10% water added	1	slice(s)	28	20	35	5	0	0	1	0.00	0.22	0.31	—
13596	Turkey pastrami	2	ounce(s)	56	—	70	11	1	0	2	1.00	—	—	—
13597	Turkey salami	2	ounce(s)	56	—	120	8	1	0	9	2.50	2.92	2.30	—
	BEVERAGES													
	Alcoholic													
	Beer													
866	Ale, mild	12	fluid ounce(s)	360	332	148	1	13	1	0	0.00	0.00	0.00	—
686	Beer	12	fluid ounce(s)	356	336	118	1	6	<1	<1	0.00	0.00	0.00	0
869	Beer, light	12	fluid ounce(s)	354	337	99	1	5	0	0	0.00	0.00	0.00	0
16886	Beer, nonalcoholic	12	fluid ounce(s)	360	353	32	1	5	0	0	0.00	0.00	0.00	0
31608	Budweiser beer	12	fluid ounce(s)	355	328	143	1	11	0	0	0.00	0.00	0.00	0
31609	Bud Light beer	12	fluid ounce(s)	355	335	110	1	7	0	0	0.00	0.00	0.00	0
31613	Michelob Beer	12	fluid ounce(s)	355	323	155	1	13	0	0	0.00	0.00	0.00	0
31614	Michelob Light beer	12	fluid ounce(s)	355	330	134	1	12	0	0	0.00	0.00	0.00	0
	Gin, rum, vodka, whiskey													
687	Distilled alcohol, 80 proof	1	fluid ounce(s)	28	19	64	0	0	0	0	0.00	0.00	0.00	0
688	Distilled alcohol, 86 proof	1	fluid ounce(s)	28	18	70	0	<.1	0	0	0.00	0.00	0.00	0
689	Distilled alcohol, 90 proof	1	fluid ounce(s)	28	17	73	0	0	0	0	0.00	0.00	0.00	0
856	Distilled alcohol, 94 proof	1	fluid ounce(s)	28	17	76	0	0	0	0	0.00	0.00	0.00	0
857	Distilled alcohol, 100 proof	1	fluid ounce(s)	28	16	82	0	0	0	0	0.00	0.00	0.00	0
	Liqueurs													
3142	Coffee liqueur, 63 proof	1	fluid ounce(s)	35	14	107	<.1	11	0	<1	0.04	0.01	0.04	—
33187	Coffee liqueur, 53 proof	1	fluid ounce(s)	35	11	117	<.1	16	0	<1	0.04	0.01	0.04	—

Appendix H

PAGE KEY: H–2 = Breads/Baked Goods H–6 = Cereal/Rice/Pasta H–10 = Fruit H–14 = Vegetables/Legumes H–24 = Nuts/Seeds H–26 = Vegetarian
H–28 = Dairy H–34 = Eggs H–34 = Seafood H–36 = Meats H–40 = Poultry H–40 = Processed meats H–42 = Beverages H–46 = Fats/Oils
H–48 = Sweets H–50 = Spices/Condiments/Sauces H–52 = Mixed foods/Soups/Sandwiches H–58 = Fast food H–74 = Convenience meals H–76 = Baby foods

Chol (mg)	Calc (mg)	Iron (mg)	Magn (mg)	Pota (mg)	Sodi (mg)	Zinc (mg)	Vit A (µg)	Thia (mg)	Vit E (mg α)	Ribo (mg)	Niac (mg)	Vit B_6 (mg)	Fola (µg)	Vit C (mg)	Vit B_{12} (µg)	Sele (µg)
22	62	0.31	11	169	843	0.78	34	0.17	0.14	0.15	1.04	0.13	1	0	1	9
21	54	0.58	10	193	789	0.80	12	0.17	0.24	0.14	1.17	0.11	3	0	1	8
18	3	0.44	4	67	265	0.71	0	0.02	—	0.04	0.99	0.05	1	0	1	5
12	2	0.16	3	53	261	0.36	0	0.12	—	0.04	0.69	0.07	1	0	<1	—
13	3	0.33	4	80	368	0.70	0	0.11	—	0.06	0.88	0.06	1	0	1	4
23.69	1.36	1.42	1.67	27.49	131.54	0.42	641.01	0.03	—	0.23	1.27	0.05	—	0	3.05	8.81
44	33	0.96	11	157	412	1.70	0	0.37	—	0.14	2.37	0.16	1	1	1	16
29	25	0.46	6	89	465	0.97	20	0.11	0.00	0.07	1.25	0.06	1	0	1	7
50	5	0.90	10	226	700	1.93	0	0.36	0.12	0.17	2.91	0.30	1	0	1	12
24	6	0.68	6	70	513	1.11	0	0.02	0.09	0.07	1.07	0.04	2	0	1	4
29	6	0.66	6	95	638	1.05	10	0.11	0.14	0.07	1.50	0.07	2	0	1	8
45	43	0.90	5	38	617	0.47	18	0.03	0.10	0.05	1.39	0.14	2	0	<1	8
48	48	0.83	6	81	642	1.40	0	0.02	0.28	0.08	1.86	0.10	4	0	<1	7
53	16	1.02	12	207	627	1.62	0	0.42	—	0.16	2.83	0.22	3	1	1	15
41	27	0.88	10	169	566	1.23	0	0.14	—	0.13	1.75	0.11	3	0	1	11
34	6	0.37	6	113	527	0.94	0	0.19	—	0.08	1.55	0.10	1	0	1	8
9	1	0.15	2	38	224	0.28	0	0.04	—	0.03	0.55	0.03	<1	0	<1	—
40	7	0.82	8	102	546	1.10	0	0.29	—	0.08	1.96	0.11	1	1	1	10
33	3	1.01	6	86	524	0.81	0	0.05	0.09	0.09	1.49	0.08	1	0	1	7
10	2	0.17	3	48	289	0.54	0	0.12	—	0.04	0.72	0.07	<1	0	<1	3
45	42	0.87	15	225	616	1.76	1	0.24	0.14	0.17	2.26	0.24	6	12	1	11
92	57	2.20	18	188	665	2.07	0	0.04	0.00	0.12	3.55	0.29	5	1	<1	—
33	7	0.43	7	101	517	0.71	7	0.11	0.07	0.06	1.67	0.09	1	0	<1	0
39	17	0.66	11	191	851	1.60	0	0.40	0.14	0.15	2.57	0.20	3	1	1	12
52	23	0.88	14	255	1137	2.14	0	0.53	—	0.20	3.43	0.27	4	0	1	16
43	7	1.44	8	154	704	1.45	0	0.09	0.12	0.19	2.44	0.15	1	0	3	12
8	2	0.14	1	16	152	0.26	0	0.01	—	0.02	0.26	0.02	1	0	<1	3
23	7	0.47	5	66	608	0.61	16	0.10	1.04	0.08	1.04	0.07	1	0	1	6
22	5	0.35	6	90	547	0.66	0	0.26	1.04	0.07	1.26	0.09	1	0	<1	11
25	0	0.72	—	—	730	—	0	—	—	—	—	—	—	0	—	—
10	0	0.00	—	—	300	—	0	—	—	—	—	—	—	0	—	—
20	0	0.72	—	—	640	—	0	—	—	—	—	—	—	0	—	—
20	0	0.72	—	—	620	—	0	—	—	—	—	—	—	0	—	—
20	0	0.72	—	—	540	—	0	—	—	—	—	—	—	0	—	—
10	0	0.00	—	—	330	—	0	—	—	—	—	—	—	0	—	—
20	0	0.36	6	81	310	0.73	0	—	—	—	—	—	—	0	—	—
40	0	0.72	—	—	590	—	0	—	—	—	—	—	—	0	—	—
50	40	0.72	—	—	500	—	0	—	—	—	—	—	—	0	—	—
0	18	0.11	—	—	18	—	0	0.02	0.00	0.10	1.63	—	—	0	<.1	—
0	18	0.07	21	89	14	0.04	0	0.02	0.00	0.09	1.61	0.18	21	0	<.1	2
0	18	0.14	18	64	11	0.11	0	0.03	0.00	0.11	1.39	0.12	14	0	<.1	2
0	25	0.04	32	90	18	0.04	—	0.02	0.00	0.10	1.63	0.18	22	0	<.1	—
0	18	0.11	21	89	9	0.07	0	0.02	0.00	0.09	1.61	0.18	21	0	<.1	4
0	18	0.14	18	64	9	0.11	0	0.03	0.00	0.11	1.39	0.12	15	0	<.1	4
0	18	0.11	21	89	9	0.07	0	0.02	0.00	0.09	1.61	0.18	21	0	<.1	4
0	18	0.14	18	64	9	0.11	0	0.03	0.00	0.11	1.39	0.12	15	0	<.1	4
0	0	0.01	0	1	<1	0.01	0	0.00	0.00	0.00	0.00	0.00	0	0	0	0
0	0	0.01	0	1	<1	0.01	0	0.00	0.00	0.00	0.00	0.00	0	0	0	0
0	0	0.01	0	1	<1	0.01	0	0.00	0.00	0.00	0.00	0.00	0	0	0	0
0	0	0.01	0	1	<1	0.01	0	0.00	0.00	0.00	0.00	0.00	0	0	0	0
0	0	0.01	0	1	<1	0.01	0	0.00	0.00	0.00	0.00	0.00	0	0	0	0
0	<1	0.02	1	10	3	0.01	0	0.00	—	0.00	0.05	0.00	0	0	0	<1
0	<1	0.02	1	10	3	0.01	0	0.00	0.00	0.00	0.05	0.00	0	0	0	<1

TABLE H–1

Food Composition

(DA+ code is for Wadsworth Diet Analysis program)   (For purposes of calculations, use "0" for t, <1, <.1, <.01, etc.)

DA + Code	Food Description	Quantity	Measure	Wt (g)	H₂O (g)	Ener (kcal)	Prot (g)	Carb (g)	Fiber (g)	Fat (g)	Fat Breakdown (g)			
											Sat	Mono	Poly	Trans
	BEVERAGES—Continued													
736	Cordials, 54 proof	1	fluid ounce(s)	30	9	106	<.1	13	0	<.1	0.02	0.01	0.04	—
	Wine													
858	Champagne, domestic	5	fluid ounce(s)	150	—	105	<1	4	0	0	0.00	0.00	0.00	0
861	Red wine, California	5	fluid ounce(s)	150	133	125	<1	4	0	0	0.00	0.00	0.00	0
690	Sweet dessert wine	5	fluid ounce(s)	150	106	240	<1	21	0	0	0.00	0.00	0.00	0
1481	White wine	5	fluid ounce(s)	148	132	100	<1	1	0	0	0.00	0.00	0.00	0
1811	Wine cooler	10	fluid ounce(s)	300	270	150	<1	18	<.1	<.1	0.01	0.00	0.02	—
	Carbonated													
692	Club soda	12	fluid ounce(s)	355	355	0	0	0	0	0	0.00	0.00	0.00	0
12010	Coca-Cola Classic cola soda	12	fluid ounce(s)	360	—	146	0	41	0	0	0.00	0.00	0.00	0
12031	Coke diet cola soda	12	fluid ounce(s)	360	—	2	0	<1	0	0	0.00	0.00	0.00	0
693	Cola	12	fluid ounce(s)	426	380	179	<1	46	0	0	0.00	0.00	0.00	—
9522	Cola soda, decaffeinated	12	fluid ounce(s)	372	331	156	<1	40	0	0	0.00	0.00	0.00	0
1415	Cola, low calorie w/aspartame	12	fluid ounce(s)	355	354	4	<1	<1	0	0	0.00	0.00	0.00	0
9524	Cola, decaffeinated, low calorie w/aspartame	12	fluid ounce(s)	355	354	4	<1	<1	0	0	0.00	0.00	0.00	0
1412	Cream soda	12	fluid ounce(s)	371	321	189	0	49	0	0	0.00	0.00	0.00	0
31899	Diet 7 Up	12	fluid ounce(s)	360	—	0	0	0	0	0	0.00	0.00	0.00	0
695	Ginger ale	12	fluid ounce(s)	366	334	124	0	32	0	0	0.00	0.00	0.00	0
694	Grape soda	12	fluid ounce(s)	372	330	160	0	42	0	0	0.00	0.00	0.00	0
1876	Lemon lime soda	12	fluid ounce(s)	368	330	147	0	38	0	0	0.00	0.00	0.00	—
29392	Mountain Dew diet soda	12	fluid ounce(s)	360	—	0	0	0	0	0	0.00	0.00	0.00	0
29391	Mountain Dew soda	12	fluid ounce(s)	360	—	170	0	46	0	0	0.00	0.00	0.00	0
3145	Orange soda	12	fluid ounce(s)	372	326	179	0	46	0	0	0.00	0.00	0.00	0
1414	Pepper-type soda	12	fluid ounce(s)	368	329	151	0	38	0	<1	0.26	0.00	0.00	
2391	Pepper-type or cola soda, low calorie w/saccharin	12	fluid ounce(s)	355	354	0	0	<1	0	0	0.00	0.00	0.00	0
29389	Pepsi diet cola soda	12	fluid ounce(s)	360	—	0	0	0	0	0	0.00	0.00	0.00	0
29388	Pepsi regular cola soda	12	fluid ounce(s)	360	—	150	0	41	0	0	0.00	0.00	0.00	0
696	Root beer	12	fluid ounce(s)	370	330	152	0	39	0	0	0.00	0.00	0.00	0
31898	7 Up	12	fluid ounce(s)	360	—	240	0	59	0	0	0.00	0.00	0.00	0
12034	Sprite diet soda	12	fluid ounce(s)	360	—	4	0	0	0	0	0.00	0.00	0.00	0
12044	Sprite soda	12	fluid ounce(s)	360	—	144	0	39	0	0	0.00	0.00	0.00	0
	Coffee													
731	Brewed	8	fluid ounce(s)	237	236	9	<1	0	0	0	0.00	0.00	0.00	0
9520	Brewed, decaffeinated	8	fluid ounce(s)	237	235	5	<1	1	0	0	0.00	0.00	0.00	0
16882	Cappuccino	8	fluid ounce(s)	240	224	78	4	6	<1	4	2.53	1.18	0.15	—
16883	Cappuccino, decaffeinated	8	fluid ounce(s)	240	224	78	4	6	<1	4	2.53	1.18	0.15	—
16880	Espresso	8	fluid ounce(s)	237	235	5	<1	1	0	0	0.00	0.00	0.00	0
16881	Espresso, decaffeinated	8	fluid ounce(s)	237	235	5	<1	1	0	0	0.00	0.00	0.00	0
732	Instant, prepared	8	fluid ounce(s)	239	237	5	<1	1	0	0	0.00	0.00	0.00	0
	Fruit drinks													
29357	Crystal Light low calorie lemonade drink	8	fluid ounce(s)	240	—	5	0	0	0	0	0.00	0.00	0.00	0
6012	Fruit punch drink w/added vitamin C, canned	8	fluid ounce(s)	276	242	129	0	33	<1	<.1	0.01	0.01	0.01	0
260	Grape drink, canned	8	fluid ounce(s)	250	221	113	<.1	29	0	0	0.00	0.00	0.00	0
266	Lemonade, from frozen concentrate	8	fluid ounce(s)	248	213	131	<1	34	<1	<1	0.02	0.00	0.04	—
268	Limeade, from frozen concentrate	8	fluid ounce(s)	247	220	104	<.1	26	0	<.1	0.00	0.00	0.00	—
31143	Gatorade Thirst Quencher, all flavors	8	fluid ounce(s)	240	—	50	0	14	0	0	0.00	0.00	0.00	0
17372	Kool-Aid (lemonade/punch/fruit drink)	8	fluid ounce(s)	248	220	108	<1	28	<1	<.1	0.01	0.01	0.02	—
17225	Kool-Aid sugar free, low calorie tropical punch mix, prepared	8	fluid ounce(s)	240	—	5	0	0	0	0	0.00	0.00	0.00	0
14266	Odwalla strawberry 'c' monster fruit drink	8	fluid ounce(s)	240	—	150	2	34	1	1	0.00	—	—	0
10080	Odwalla strawberry lemonade quencher	8	fluid ounce(s)	240	—	120	1	28	1	0	0.00	0.00	0.00	0
10099	Snapple fruit punch	8	fluid ounce(s)	240	—	110	0	29	0	0	0.00	0.00	0.00	0
10096	Snapple kiwi strawberry	8	fluid ounce(s)	240	211	110	0	28	0	0	0.00	0.00	0.00	0
	Slim Fast ready to drink shake													
16056	Dark chocolate fudge	11	fluid ounce(s)	325	—	220	10	42	5	3	1.00	1.50	0.50	—
16054	French vanilla	11	fluid ounce(s)	325	—	220	10	40	5	3	0.50	1.50	0.50	—
16055	Strawberries n cream	11	fluid ounce(s)	325	—	220	10	40	5	3	0.50	1.50	0.50	—
	Tea													
733	Tea, prepared	8	fluid ounce(s)	237	236	2	0	1	0	0	0.00	0.00	0.01	0
33179	Decaffeinated, prepared	8	fluid ounce(s)	237	236	2	0	1	0	0	0.00	0.00	0.01	0
1877	Herbal, prepared	8	fluid ounce(s)	237	236	2	0	<1	0	0	0.00	0.00	0.01	0
734	Instant tea mix, unsweetened, prepared	8	fluid ounce(s)	237	236	2	<.1	<1	0	0	0.00	0.00	0.00	0

Appendix H

PAGE KEY: H–2 = Breads/Baked Goods   H–6 = Cereal/Rice/Pasta   H–10 = Fruit   H–14 = Vegetables/Legumes   H–24 = Nuts/Seeds   H–26 = Vegetarian
H–28 = Dairy   H–34 = Eggs   H–34 = Seafood   H–36 = Meats   H–40 = Poultry   H–40 = Processed meats   H–42 = Beverages   H–46 = Fats/Oils
H–48 = Sweets   H–50 = Spices/Condiments/Sauces   H–52 = Mixed foods/Soups/Sandwiches   H–58 = Fast food   H–74 = Convenience meals   H–76 = Baby foods

Chol (mg)	Calc (mg)	Iron (mg)	Magn (mg)	Pota (mg)	Sodi (mg)	Zinc (mg)	Vit A (µg)	Thia (mg)	Vit E (mg α)	Ribo (mg)	Niac (mg)	Vit B$_6$ (mg)	Fola (µg)	Vit C (mg)	Vit B$_{12}$ (µg)	Sele (µg)
0	<1	0.02	<1	5	2	0.01	0	0.00	0.00	0.00	0.02	0.00	0	0	0	—
0	—	—	—	—	—	—	—	—	—	—	0.00	—	—	—	0	—
0	12	1.43	16	171	15	0.15	0	0.02	0.00	0.04	0.12	0.05	1	0	<.1	—
0	12	0.36	14	138	14	0.11	0	0.03	0.00	0.03	0.32	0.00	0	0	0	1
0	13	0.47	15	118	7	0.10	0	0.01	—	0.01	0.10	0.02	0	0	0	<1
0	17	0.81	16	135	25	0.17		0.01	0.03	0.02	0.13	0.04	4	5	<.1	—
0	18	0.04	4	7	75	0.36	0	0.00	0.00	0.00	0.00	0.00	0	0	0	0
0	—	—	—	0	50	—	0	—	—	—	—	—	—	0	—	—
0	—	—	—	18	42	—	0	—	—	—	—	—	—	0	—	—
0	13	0.09	4	4	17	0.04	0	0.00	0.00	0.00	0.00	0.00	0	0	0	<1
0	11	0.07	4	4	15	0.04	0	0.00	0.00	0.00	0.00	0.00	0	0	0	<1
0	11	0.11	4	21	18	0.00	0	0.02	0.00	0.08	0.00	0.00	0	0	0	0
0	14	0.11	4	0	21	0.28	0	0.02	0.00	0.08	0.00	0.00	0	0	0	<1
0	19	0.19	4	4	44	0.26	0	0.00	0.00	0.00	0.00	0.00	0	0	0	0
0	—	—	—	116	53	—	—	—	0.00	—	—	—	—	0	—	—
0	11	0.66	4	4	26	0.18	0	0.00	0.00	0.00	0.00	0.00	0	0	0	<1
0	11	0.30	4	4	56	0.26	0	0.00	0.00	0.00	0.00	0.00	0	0	0	0
0	7	0.26	4	4	41	0.18	0	0.00	0.00	0.00	0.06	0.00	0	0	0	0
0	—	—	—	70	35	—	—	—	—	—	—	—	—	0	—	—
0	—	—	—	0	70	—	—	—	—	—	—	—	—	0	—	—
0	19	0.22	4	7	45	0.37	0	0.00	—	0.00	0.00	0.00	0	0	0	0
0	11	0.15	0	4	37	0.15	0	0.00	—	0.00	0.00	0	0	0	<1	
0	14	0.07	4	14	57	0.11	0	0.00	0.00	0.00	0.00	0.00	0	0	0	<1
0	—	—	—	30	35	—	—	—	—	—	—	—	—	0	—	—
0	—	—	—	0	35	—	—	—	—	—	—	—	—	0	—	—
0	18	0.18	4	4	48	0.26	0	0.00	0.00	0.00	0.00	0.00	0	0	0	<1
0	—	—	—	0	113	—	—	—	—	—	—	—	—	0	—	—
0	—	—	—	110	36	—	0	—	—	—	—	—	—	0	—	—
0	—	—	—	0	71	—	0	—	—	—	—	—	—	0	—	—
0	2	0.02	5	114	2	0.02	0	0.00	0.02	0.12	0.00	0.00	5	0	0	0
0	5	0.12	12	128	5	0.05	0	0.00	0.00	0.00	0.53	0.00	<1	0	0	0
17	152	0.26	22	250	62	0.50	—	0.04	0.10	0.20	0.37	0.05	5	1	<1	0
17	152	0.26	22	250	62	0.50	—	0.04	0.10	0.20	0.37	0.05	5	1	<1	—
0	5	0.12	12	128	5	0.05	0	0.00	0.05	0.00	0.53	0.00	<1	0	0	—
0	5	0.12	12	128	5	0.05	0	0.00	0.05	0.00	0.53	0.00	<1	0	0	—
0	10	0.10	7	72	5	0.02	0	0.00	0.00	0.00	0.56	0.00	0	0	0	<1
0	0	0.00	—	160	20	—	0	—	—	—	—	—	—	0	—	—
0	22	0.58	6	69	61	0.33	—	0.06	0.00	0.06	0.06	0.00	4	99	0	0
0	5	0.45	3	30	15	0.30	0	0.00	0.00	0.01	0.03	0.01	0	85	0	<1
0	10	0.52	5	50	7	0.07	0	0.02	0.02	0.07	0.05	0.02	2	13	0	<1
0	7	0.02	2	22	5	0.02	0	0.00	0.00	0.01	0.02	0.01	2	6	0	<1
0	10	0.18	—	30	110	—	—	—	—	—	—	—	—	1	—	—
0	14	0.46	5	50	31	0.20	—	0.04	—	0.05	0.05	0.01	4	42	0	1
0	0	0.00	—	10	10	—	0	—	—	—	—	—	—	6	—	—
0	20	1.44	—	330	40	—	—	—	—	—	—	—	—	600	0	—
0	20	0.00	—	70	30	—	0	—	—	—	—	—	—	60	0	—
0	0	0.00	—	20	10	—	0	—	—	—	—	—	—	0	0	—
0	0	0.00	—	40	10	—	0	—	—	—	—	—	—	0	0	—
5	400	2.70	140	600	220	2.25	—	0.53	—	0.60	7.00	0.70	120	60	2	18
5	400	2.70	140	600	220	2.25	—	0.53	—	0.60	7.00	0.70	120	60	2	18
5	400	2.70	140	600	220	2.25	—	0.53	—	0.60	7.00	0.70	120	60	2	18
0	0	0.05	7	88	7	0.05	0	0.00	0.00	0.03	0.00	0.00	12	0	0	0
0	0	0.05	7	88	7	0.05	0	0.00	0.00	0.03	0.00	0.00	12	0	0	0
0	5	0.19	2	21	2	0.09	0	0.02	0.00	0.01	0.00	0.00	0	0	0	0
0	7	0.05	5	47	7	0.02	0	0.00	0.00	0.00	0.09	0.00	0	0	0	0

TABLE H-1
Food Composition

(DA+ code is for Wadsworth Diet Analysis program)   (For purposes of calculations, use "0" for t, <1, <.1, <.01, etc.)

DA + Code	Food Description	Quantity	Measure	Wt (g)	H₂O (g)	Ener (kcal)	Prot (g)	Carb (g)	Fiber (g)	Fat (g)	Sat	Mono	Poly	Trans
	BEVERAGES—Continued													
735	Instant lemon flavored tea mix w/sugar, prepared	8	fluid ounce(s)	259	236	88	<1	22	0	<.1	0.01	0.00	0.02	—
	Water													
1413	Mineral water, carbonated	8	fluid ounce(s)	237	237	0	0	0	0	0	0.00	0.00	0.00	0
33183	Poland spring water, bottled	8	fluid ounce(s)	237	237	0	0	0	0	0	0.00	0.00	0.00	0
1	Tap water	8	fluid ounce(s)	237	237	0	0	0	0	0	0.00	0.00	0.00	—
1879	Tonic water	8	fluid ounce(s)	244	222	83	0	21	0	0	0.00	0.00	0.00	0
	FATS AND OILS													
	Butter													
104	Butter	1	tablespoon(s)	15	2	108	<1	<.1	0	12	6.13	5.00	0.43	—
921	Unsalted	1	tablespoon(s)	15	3	108	<1	<.1	0	12	7.71	3.15	0.46	—
107	Whipped	1	tablespoon(s)	11	2	82	<.1	<.1	0	9	5.76	2.67	0.34	—
944	Whipped, unsalted	1	tablespoon(s)	11	2	82	<.1	<.1	0	9	5.76	2.67	0.34	—
2522	Butter Buds, dry butter substitute	1	teaspoon(s)	2	—	8	0	2	0	0	0.00	0.00	0.00	0
	Fats, cooking													
2671	Beef tallow, semisolid	1	tablespoon(s)	13	0	115	0	0	0	13	6.37	5.35	0.51	—
922	Chicken fat	1	tablespoon(s)	13	<.1	115	0	0	0	13	3.81	5.72	2.68	—
5454	Household shortening w/vegetable oil	1	tablespoon(s)	13	0	115	0	0	0	13	3.39	5.56	2.75	2.20
111	Lard	1	tablespoon(s)	13	0	114	0	0	0	13	4.94	5.68	1.41	—
	Margarine													
114	Margarine	1	tablespoon(s)	14	2	101	<1	<1	0	11	2.23	5.05	3.58	—
116	Soft	1	tablespoon(s)	14	2	101	<1	<.1	0	11	1.95	4.02	4.88	—
117	Soft, unsalted	1	tablespoon(s)	14	3	101	<1	<1	0	11	1.95	5.26	3.62	—
928	Unsalted	1	tablespoon(s)	14	3	101	<.1	<.1	0	11	2.12	5.17	3.53	—
119	Whipped	1	tablespoon(s)	9	1	64	<.1	<.1	0	7	1.17	3.25	2.51	—
	Spreads													
16164	I Can't Believe It's Not Butter! whipped spread	1	tablespoon(s)	14	4	60	0	0	0	7	1.50	1.50	2.50	—
16157	Promise vegetable oil spread, stick	1	tablespoon(s)	14	4	90	0	0	0	10	2.50	2.00	4.00	—
	Oils													
2681	Canola	1	tablespoon(s)	14	0	120	0	0	0	14	0.97	8.01	4.03	—
120	Corn	1	tablespoon(s)	14	0	120	0	0	0	14	1.73	3.29	7.98	0.04
122	Olive	1	tablespoon(s)	14	0	119	0	0	0	14	1.82	9.98	1.35	—
124	Peanut	1	tablespoon(s)	14	0	119	0	0	0	14	2.28	6.24	4.32	—
2693	Safflower	1	tablespoon(s)	14	0	120	0	0	0	14	0.84	10.15	1.95	—
923	Sesame	1	tablespoon(s)	14	0	120	0	0	0	14	1.93	5.40	5.67	—
130	Soybean w/cottonseed oil	1	tablespoon(s)	14	0	120	0	0	0	14	2.45	4.01	6.54	—
128	Soybean, hydrogenated	1	tablespoon(s)	14	0	120	0	0	0	14	2.03	5.85	5.11	—
2700	Sunflower	1	tablespoon(s)	14	0	120	0	0	0	14	1.77	6.28	4.95	—
357	**Pam original no stick cooking spray**	1	serving(s)	0	—	0	0	0	0	0	0.00	0.00	0.00	—
	Salad dressing													
132	Blue cheese	2	tablespoon(s)	31	10	154	1	2	0	16	3.03	3.76	8.51	
133	Blue cheese, low calorie	2	tablespoon(s)	32	25	32	2	1	0	2	0.82	0.57	0.78	—
1764	Caesar	2	tablespoon(s)	30	10	158	<1	1	<.1	17	2.64	4.05	9.86	—
29654	Creamy, reduced calorie, fat free, cholesterol free, sour cream and/or buttermilk & oil	2	tablespoon(s)	32	24	34	<1	6	0	1	0.16	0.21	0.46	—
29617	Creamy, reduced calorie, sour cream and/or buttermilk & oil	2	tablespoon(s)	30	22	48	<1	2	0	4	0.63	0.98	2.40	—
134	French	2	tablespoon(s)	31	11	143	<1	5	0	14	1.76	2.63	6.56	—
135	French, low fat	2	tablespoon(s)	33	18	76	<1	10	<1	4	0.36	1.92	1.64	—
136	Italian	2	tablespoon(s)	29	17	86	<1	3	0	8	1.32	1.86	3.80	—
137	Italian, diet	2	tablespoon(s)	30	25	23	<1	1	0	2	0.14	0.66	0.51	—
139	Mayonnaise type	2	tablespoon(s)	29	12	115	<1	7	0	10	1.44	2.65	5.29	—
942	Oil & vinegar	2	tablespoon(s)	31	15	140	0	1	0	16	2.84	4.62	7.52	—
1765	Ranch	2	tablespoon(s)	30	12	146	<1	2	<.1	16	2.32	3.85	8.92	—
3666	Ranch, reduced calorie	2	tablespoon(s)	30	21	62	<1	2	<.1	6	1.13	1.79	2.89	—
940	Russian	2	tablespoon(s)	31	11	151	<1	3	0	16	2.23	3.61	9.00	—
939	Russian, low calorie	2	tablespoon(s)	33	21	46	<1	9	<.1	1	0.20	0.29	0.75	—
941	Sesame seed	2	tablespoon(s)	31	12	136	1	3	<1	14	1.90	3.64	7.68	—
142	Thousand island	2	tablespoon(s)	31	15	115	<1	5	<1	11	1.59	2.46	5.68	—
143	Thousand island, low calorie	2	tablespoon(s)	31	19	62	<1	7	<1	4	0.23	1.98	0.82	—
	Sandwich spreads													
138	Mayonnaise w/soybean oil	1	tablespoon(s)	14	2	99	<1	1	0	11	1.64	2.70	5.89	0.04

Appendix H

Chol (mg)	Calc (mg)	Iron (mg)	Magn (mg)	Pota (mg)	Sodi (mg)	Zinc (mg)	Vit A (µg)	Thia (mg)	Vit E (mg α)	Ribo (mg)	Niac (mg)	Vit B6 (mg)	Fola (µg)	Vit C (mg)	Vit B12 (µg)	Sele (µg)
0	5	0.05	5	49	8	0.03	0	0.00	0.00	0.04	0.09	0.01	0	<1	0	<1
0	33	0.00	0	0	2	0.00	0	0.00	—	0.00	0.00	0.00	0	0	0	0
0	2	0.02	2	0	2	0.00	0	0.00	—	0.00	0.00	0.00	0	0	0	0
0	4.74	0.00	2.37	0	4.74	0	0	0	0.57	0	0	0	0	0	0	0
0	2	0.02	0	0	10	0.24	0	0.00	0.00	0.00	0.00	0.00	0	0	0	0
32	4	0.00	<1	4	86	0.01	103	0.00	0.35	0.01	0.01	0.00	<1	0	<.1	<1
32	4	0.00	<1	4	2	0.01	103	0.00	0.35	0.01	0.01	0.00	<1	0	<.1	<1
25	3	0.02	<1	3	94	0.01	78	0.00	0.26	0.00	0.00	0.00	<1	0	<.1	<1
25	3	0.02	<1	3	1	0.01	—	0.00	0.26	0.00	0.01	0.00	<1	0	<.1	—
0	0	0.00	0	2	70	0.00	0	0.00	0.00	0.00	0.00	0.00	<1	0	0	—
14	0	0.00	0	0	0	0.00	0	0.00	0.35	0.01	0.00	0.00	0	0	0	<.1
11	0	0.00	0	0	0	0.00	0	0.00	0.35	0.00	0.00	0.00	0	0	0	<.1
0	0	0.00	0	0	0	0.00	0	0.00	—	0.00	0.00	0.00	0	0	0	—
12	0	0.00	0	0	0	0.01	0	0.00	0.08	0.00	0.00	0.00	0	0	0	<.1
0	4	0.01	<1	6	133	0.00	115	0.00	1.27	0.01	0.00	0.00	<1	<.1	<.1	0
0	4	0.00	<1	5	152	0.00	103	0.00	0.99	0.00	0.00	0.00	<1	<.1	<.1	0
0	4	0.00	<1	5	4	0.00	103	0.00	1.23	0.00	0.00	0.00	<1	<.1	<.1	0
0	2	0.00	<1	4	<1	0.00	115	0.00	1.80	0.00	0.00	0.00	<1	<.1	<.1	0
0	2	0.00	<1	3	97	0.00	—	0.00	0.45	0.00	0.00	0.00	<.1	<.1	<.1	—
0	10	0.18	—	4	70	—	—		1.65	0.00	0.00	0.00	—	1	—	—
0	10	0.18	—	9	90	—	—		0.00	—	0.00	0.00	—	1	—	—
0	0	0.00	0	0	0	0.00	0	0.00	2.33	0.00	0.00	0.00	0	0	0	0
0	0	0.00	0	0	0	0.00	0	0.00	1.94	0.00	0.00	0.00	0	0	0	0
0	<1	0.09	0	<1	<1	0.00	0	0.00	1.94	0.00	0.00	0.00	0	0	0	0
0	0	0.00	0	0	0	0.00	0	0.00	2.12	0.00	0.00	0.00	0	0	0	0
0	0	0.00	0	0	0	0.00	0	0.00	4.64	0.00	0.00	0.00	0	0	0	0
0	0	0.00	0	0	0	0.00	0	0.00	0.19	0.00	0.00	0.00	0	0	0	0
0	0	0.00	0	0	0	0.00	0	0.00	1.65	0.00	0.00	0.00	0	0	0	0
0	0	0.00	0	0	0	0.00	0	0.00	1.10	0.00	0.00	0.00	0	0	0	0
0	0	0.00	0	0	0	0.00	0	0.00	—	0.00	0.00	0.00	0	0	0	0
0	0	0.00	—	0	0	—	0	—	0.00	—	—	—	—	0	0	—
5	25	0.06	0	11	335	0.08	21	0.00	1.84	0.03	0.03	0.01	9	1	<.1	<1
<1	28	0.16	2	2	384	0.08	—	0.01	0.08	0.03	0.02	0.01	1	<.1	<.1	—
1	7	0.05	1	9	323	0.03	—	0.00	1.57	0.00	0.01	0.00	1	0	<.1	—
0	12	0.08	2	43	320	0.06	0	0.00	0.21	0.02	0.01	0.01	1	0	0	—
0	2	0.04	1	11	307	0.01	—	0.00	0.72	0.00	0.01	0.01	4	<1	<.1	—
0	7	0.25	2	21	261	0.09	7	0.01	1.56	0.02	0.06	0.00	0	0	<.1	0
0	4	0.28	3	35	262	0.07	9	0.01	0.10	0.02	0.15	0.02	1	0	0	1
0	2	0.19	1	14	486	0.04	1	0.00	1.47	0.01	0.00	0.02	0	0	0	1
2	3	0.20	1	26	410	0.06	<1	0.00	0.06	0.00	0.00	0.02	0	0	0	2
8	4	0.06	1	3	209	0.05	19	0.00	0.61	0.01	0.00	0.00	2	0	<.1	<1
0	0	0.00	0	2	<1	0.00	0	0.00	1.44	0.00	0.00	0.00	0	0	0	0
1	4	0.03	1	8	354	0.01	—	0.00	1.85	0.00	0.00	0.00	<1	<.1	<.1	—
<1	5	0.01	1	8	414	0.02	—	0.00	0.73	0.01	0.01	0.00	<1	<.1	<.1	—
6	6	0.18	1	48	266	0.13	5	0.02	1.02	0.02	0.18	0.01	3	2	<.1	<1
2	6	0.20	1	51	283	0.03	1	0.00	0.13	0.00	0.00	0.00	1	2	<.1	1
0	6	0.18	0	48	306	0.03	1	0.00	1.53	0.00	0.00	0.00	0	0	0	<1
8	5	0.37	2	33	269	0.08	3	0.45	1.25	0.02	0.13	0.00	0	0	0	<1
<1	5	0.28	2	62	254	0.06	5	0.01	0.31	0.01	0.13	0.00	0	0	0	0
5	2	0.07	<1	5	78	0.02	12	0.00	0.72	0.00	0.00	0.08	1	0	<.1	<1

TABLE H–1

Food Composition

(DA+ code is for Wadsworth Diet Analysis program) (For purposes of calculations, use "0" for t, <1, <.1, <.01, etc.)

DA + Code	Food Description	Quantity	Measure	Wt (g)	H₂O (g)	Ener (kcal)	Prot (g)	Carb (g)	Fiber (g)	Fat (g)	Fat Breakdown (g)			
											Sat	Mono	Poly	Trans
	FATS AND OILS—Continued													
2708	Mayonnaise w/soybean & safflower oils	1	tablespoon(s)	14	0	98.94	0.15	0.37	0	10.95	1.18	1.79	7.59	—
140	Mayonnaise, low calorie	1	tablespoon(s)	16	10	37	<.1	3	0	3	0.53	0.72	1.70	—
141	Tartar sauce	2	tablespoon(s)	28	9	144	<1	4	<.1	14	2.14	4.13	7.57	—
	SWEETS													
4799	**Butterscotch or caramel topping**	2	tablespoon(s)	41	13	103	1	27	<1	<.1	0.05	0.01	0.00	—
	Candy													
1786	Almond Joy candy bar	1	item(s)	49	5	240	2	29	2	13	9.00	3.63	0.74	0
1785	Bit-o-Honey candy	6	item(s)	40	2	170	1	34	0	3	2.00	0	20	—
33375	Butterscotch candy	2	piece(s)	12	1	47	<.1	11	0	<1	0.25	0.10	0.01	—
1701	Chewing gum, stick	1	item(s)	3	<.1	7	0	2	<.1	<.1	0.00	0.00	0.00	—
33378	Chocolate fudge w/nuts, prepared	2	piece(s)	38	3	175	2	26	1	7	2.29	1.41	2.81	—
1787	Jelly beans	15	item(s)	43	3	159	0	40	<.1	<.1	0.00	0.00	0.00	—
1784	Kit Kat wafer bar	1	item(s)	42	1	220	3	27	1	11	7.00	3.53	0.34	0
4674	Krackel candy bar	1	item(s)	41	1	220	3	26	1	11	6.00	3.94	0.37	0
4934	Licorice	4	piece(s)	44	7	147	1	34	1	1	0.18	0.07	0.00	—
1780	Life Savers candy	1	item(s)	2	—	8	0	2	0	<.1	0.00	—	—	0
1790	Lollipop	1	item(s)	28	—	108	0	28	0	0	0.00	0.00	0.00	0
4679	M & Ms peanut chocolate candy, small bag	1	item(s)	49	1	250	5	30	2	13	5.00	5.42	2.07	—
1781	M & Ms plain chocolate candy, small bag	1	item(s)	48	1	240	2	34	1	10	6.00	3.30	0.30	—
4673	Milk chocolate bar	1	item(s)	91	1	483	8	53	2	28	16.69	7.20	0.63	—
1783	Milky Way bar	1	item(s)	58	4	270	2	41	1	10	5.00	3.50	0.35	—
1788	Peanut brittle	1½	ounce(s)	43	<1	206	3	30	1	8	1.76	3.43	1.94	—
1789	Reese's peanut butter cups	2	piece(s)	45	1	250	5	25	1	14	5.00	6.17	2.34	0
4689	Reese's pieces candy, small bag	1	item(s)	46	1	230	6	26	1	11	7.00	0.97	0.46	0
33399	Semisweet chocolate candy, made w/butter	½	ounce(s)	14	<.1	68	1	9	1	4	2.49	1.41	0.13	—
1782	Snickers bar	1	item(s)	59	3	280	4	35	1	14	5.00	6.13	2.89	—
4694	Special Dark chocolate bar	1	item(s)	41	<1	220	2	24	3	13	8.00	4.59	0.41	0
4695	Starburst fruit chews, original fruits	1	package	59	4	240	0	48	0	5	1.00	2.10	1.83	—
4698	Taffy	3	piece(s)	45	2	169	<.1	41	0	1	0.92	0.43	0.05	—
4699	Three Musketeers bar	1	item(s)	60	4	260	2	46	1	8	4.50	2.59	0.27	—
4702	Twix caramel cookie bars	2	item(s)	58	2	280	3	37	1	14	5.00	7.75	0.49	—
4705	York peppermint pattie	1	item(s)	42	4	170	1	34	1	3	2.00	1.32	0.12	0
	Frosting, icing													
4760	Chocolate frosting, ready to eat	2	tablespoon(s)	28	5	112	<1	18	<1	5	1.55	2.54	0.60	—
4771	Creamy vanilla frosting, ready to eat	2	tablespoon(s)	28	4	118	0	19	<.1	5	0.84	1.37	2.24	0
17291	Dec-a-Cake variety pack candy decoration	1	teaspoon(s)	4	—	15	0	3	0	1	0.00	—	—	—
536	White icing	2	tablespoon(s)	40	3	163	<1	32	0	4	0.86	2.07	1.19	—
	Gelatin													
13697	Gelatin snack, all flavors	1	item(s)	99	97	70	1	17	0	0	0.00	0.00	0.00	0
2616	Mixed fruit gelatin mix, sugar free, low calorie, prepared	½	cup(s)	121	—	10	1	0	0	0	0.00	0.00	0.00	0
548	**Honey**	1	tablespoon(s)	21	4	64	<.1	17	<.1	0	0.00	0.00	0.00	0
	Jams, Jellies													
23054	Jams, jellies, preserves, all flavors	1	tablespoon(s)	20	<.1	56	<.1	14	<1	<.1	0.00	0.01	0.00	—
23278	Jams, jellies, preserves, all flavors, low sugar	1	tablespoon(s)	18	<.1	25	<.1	6	<1	<.1	0.00	0.01	0.02	—
545	**Marshmallows**	4	item(s)	29	5	92	1	23	<.1	<.1	0.02	0.02	0.01	—
4800	**Marshmallow cream topping**	2	tablespoon(s)	28	6	91	<1	22	<.1	<.1	0.02	0.02	0.01	—
555	**Molasses**	1	tablespoon(s)	20	4	58	0	15	0	<.1	0.00	0.01	0.01	—
4780	**Popsicle or ice pop**	1	item(s)	59	47	42	0	11	0	0	0.00	0.00	0.00	—
	Sugar													
559	Brown, packed	1	teaspoon(s)	5	<.1	17	0	4	0	0	0.00	0.00	0.00	0
563	Powdered, sifted	⅓	cup(s)	33	<.1	130	0	33	0	<.1	0.01	0.01	0.02	—
561	White granulated	1	teaspoon(s)	4	<.1	15	0	4	0	0	0.00	0.00	0.00	—
	Sugar Substitute													
1760	Equal sweetener, packet	1	item(s)	1	<.1	4	<.1	1	0	0	0.00	0.00	0.00	—
13029	Splenda granular no calorie sweetener	1	teaspoon(s)	1	—	2	0	1	0	0	0.00	0.00	0.00	—
1759	Sweet n Low sugar substitute, packet	1	item(s)	1	<.1	4	0	1	0	0	0.00	0.00	0.00	—
	Syrup													
3148	Chocolate	2	tablespoon(s)	38	12	105	1	24	1	<1	0.19	0.11	0.01	—
29676	Maple	¼	cup(s)	80	26	209	0	54	0	<1	0.03	0.05	0.08	—
4795	Pancake	¼	cup(s)	80	30	187	0	49	1	0	0.00	0.00	0.00	0

PAGE KEY: H–2 = Breads/Baked Goods H–6 = Cereal/Rice/Pasta H–10 = Fruit H–14 = Vegetables/Legumes H–24 = Nuts/Seeds H–26 = Vegetarian
H–28 = Dairy H–34 = Eggs H–34 = Seafood H–36 = Meats H–40 = Poultry H–40 = Processed meats H–42 = Beverages H–46 = Fats/Oils
H–48 = Sweets H–50 = Spices/Condiments/Sauces H–52 = Mixed foods/Soups/Sandwiches H–58 = Fast food H–74 = Convenience meals H–76 = Baby foods

Chol (mg)	Calc (mg)	Iron (mg)	Magn (mg)	Pota (mg)	Sodi (mg)	Zinc (mg)	Vit A (µg)	Thia (mg)	Vit E (mg α)	Ribo (mg)	Niac (mg)	Vit B$_6$ (mg)	Fola (µg)	Vit C (mg)	Vit B$_{12}$ (µg)	Sele (µg)
8.14	2.48	0.06	0.13	4.69	78.38	0.01	11.59	0	3.04	0	0	0.07	1.1	0	0.03	0.22
4	<.1	0.00	<.1	2	80	0.02	0	0.00	0.32	0.00	0.00	0.00	0	0	0	—
11	6	0.21	1	10	200	0.05	—	0.00	0.97	0.00	0.01	0.08	2	<1	<.1	—
<1	22	0.08	3	34	143	0.08	11	0.00	—	0.04	0.02	0.01	1	<1	<.1	0
3	20	0.36	33	138	70	0.40	0	0.02	—	0.08	0.24	—	—	0	—	—
0.00	—	—	85	—	0	—	—	—	—	—	—	0	—	—		
1	<1	0.00	<1	<1	47	0.00	3	0.00	0.01	0.00	0.00	0.00	0	0	0	<.1
0	0	0.00	0	<.1	<.1	0.00	0	0.00	0.00	0.00	0.00	0.00	0	0	0	<.1
5	21	0.75	21	68	16	0.54	14	0.03	0.10	0.04	0.12	0.03	6	<.1	<.1	1
0	1	0.06	1	16	21	0.02	0	0.00	0.00	0.00	0.00	0.00	0	0	0	<1
3	40	0.36	16	126	25	0.52	8	0.07	—	0.23	1.07	0.05	60	0	<.1	2
3	60	0.37	—	169	80	—	0	—	—	—	—	—	—	0	—	—
0	3	0.13	3	28	109	0.07	0	0.01	0.08	0.02	0.04	0.00	0	0	0	—
0	<1	0.04	—	0	1	—	0	0.00	—	0.00	0.00	—	—	0	—	0
0	0	0.00	—	—	11	—	0	0.00	—	0.00	0.00	—	—	0	—	1
5	40	0.36	36	171	25	1.13	15	0.03	—	0.07	1.60	0.04	17	1	<.1	2
5	40	0.36	20	127	30	0.46	15	0.03	—	0.07	0.11	0.01	3	1	<1	1
22	228	0.83	61	399	92	1.00	20	0.06	—	0.26	0.15	0.10	11	2	<1	—
5	60	0.18	20	140	95	0.41	15	0.02	—	0.07	0.20	0.03	6	1	<1	3
5	11	0.52	18	71	189	0.37	17	0.06	1.09	0.02	1.13	0.03	20	0	<.1	1
3	20	0.36	40	233	140	0.82	7	0.11	—	0.08	2.08	0.07	25	0	<.1	2
0	40	0.00	20	182	90	0.35	25	0.04	—	0.07	1.31	0.03	13	0	<.1	1
3	5	0.44	16	52	2	0.23	<1	0.01	—	0.01	0.06	0.01	<1	0	0	<1
5	40	0.36	42	—	140	1.38	15	0.03	—	0.07	1.60	0.05	23	1	<.1	3
3	0	0.72	46	136	0	0.60	0	0.01	—	0.03	0.16	0.01	1	0	0	1
0	10	0.18	1	1	0	0.00	—	0.00	—	0.00	0.00	0.00	0	30	0	<1
4	1	0.03	<1	2	40	0.02	—	0.00	—	0.01	0.01	0.00	0	0	<.1	—
5	20	0.36	18	80	110	0.33	14	0.02	—	0.03	0.20	0.01	0	1	<1	2
5	40	0.36	18	117	115	0.45	15	0.09	—	0.13	0.69	0.02	14	1	<1	1
0	0	0.36	25	71	10	0.31	0	0.01	—	0.04	0.34	0.01	2	0	<.1	—
0	2	0.40	6	55	51	0.08	0	0.00	0.44	0.00	0.03	0.00	<1	0	0	<1
0	1	0.04	<1	10	52	0.02	0	0.00	0.43	0.08	0.06	0.00	2	0	0	<.1
0	0	0.00	—	—	15	—	0	—	—	—	—	—	—	0	—	—
<1	5	0.02	—	7	92	—	—	0.00	0.33	0.01	0.00	—	—	<.1	—	—
0	0	0.00	—	0	40	—	0	—	—	—	—	—	—	0	—	—
0	0	0.00	0	0	50	0.00	0	0.00	0.00	0.00	0.00	0.00	0	0	0	—
0	1	0.09	<1	11	1	0.05	0	0.00	0.00	0.01	0.03	0.01	<1	<1	0	<1
0	4	0.10	1	15	6	0.01	0.00	0.00	0.00	0.02	0.01	0.00	2.20	1.76	0.00	—
0	2	0.05	1	19	<1	0.02	0.76	0.00	0.01	0.01	0.03	0.01	—	4.93	0.00	—
0	1	0.07	1	1	23	0.01	0	0.00	0.00	0.00	0.02	0.00	<1	0	0	<1
0	1	0.06	1	1	23	0.01	0	0.00	0.00	0.00	0.02	0.00	<1	0	0	1
0	41	0.94	48	293	7	0.06	0	0.01	0.00	0.00	0.19	0.13	0	0	0	4
0	0	0.00	1	2	7	0.01	0	0.00	0.00	0.00	0.00	0.00	0	0	0	0
0	4	0.09	1	16	2	0.01	0	0.00	0.00	0.00	0.00	0.00	<.1	0	0	<.1
0	<1	0.01	0	1	<1	0.00	0	0.00	0.00	0.01	0.00	0.00	0	0	0	<1
0	<.1	0.00	0	<.1	0	0.00	0	0.00	0.00	0.00	0.00	0.00	0	0	0	<.1
0	0	0.00	0	0	0	0.00	0	0.00	0.00	0.00	0.00	0.00	0	0	0	0
0	10	0.18	—	—	<1	—	—	0.02	0.00	0.02	0.20	—	—	1	0	—
0	0	0.00	0	—	0	0.00	0	0.00	0.00	0.00	0.00	0.00	0	0	0	0
0	5	0.79	24	84	27	0.27	0	0.00	0.00	0.02	0.12	0.00	1	<.1	0	1
0	54	0.96	11	163	7	3.33	0	0.00	0.00	0.01	0.02	0.00	5	0	0	<1
0	2	0.02	2	12	66	0.06	0	0.00	0.00	0.01	0.01	0.00	0	0	0	0

TABLE H–1
Food Composition

(DA+ code is for Wadsworth Diet Analysis program)   (For purposes of calculations, use "0" for t, <1, <.1, <.01, etc.)

DA + Code	Food Description	Quantity	Measure	Wt (g)	H₂O (g)	Ener (kcal)	Prot (g)	Carb (g)	Fiber (g)	Fat (g)	Sat	Mono	Poly	Trans
	SPICES, CONDIMENTS, SAUCES													
	Spices													
807	Allspice, ground	1	teaspoon(s)	2	<1	5	<1	1	<1	<1	0.05	0.01	0.04	—
1171	Anise seeds	1	teaspoon(s)	2	<1	7	<1	1	<1	<1	0.01	0.21	0.07	—
729	Baker's yeast active	1	teaspoon(s)	4	<1	12	2	2	1	<1	0.02	0.10	0.00	—
683	Baking powder, double acting, w/phosphate	1	teaspoon(s)	5	<1	2	<.1	1	<.1	0	0.00	0.00	0.00	0
1611	Baking soda	1	teaspoon(s)	5	<.1	0	0	0	0	0	0.00	0.00	0.00	0
8552	Basil	1	teaspoon(s)	1	1	<1	<.1	<.1	<.1	<.1	0.00	0.00	0.00	—
34959	Basil, fresh	1	piece(s)	1	<1	<1	<.1	<.1	<.1	<.1	0.00	0.00	0.00	—
808	Basil, ground	1	teaspoon(s)	1	<.1	4	<1	1	1	<.1	0.00	0.01	0.03	—
809	Bay leaf	1	teaspoon(s)	1	<.1	2	<1	<1	<1	<.1	0.01	0.01	0.01	—
11720	Betel leaves	1	ounce(s)	28	—	17	2	2	0	<.1	—	—	—	—
818	Black pepper	1	teaspoon(s)	2	<1	5	<1	1	1	<.1	0.02	0.02	0.02	—
730	Brewer's yeast	1	teaspoon(s)	3	<1	8	1	1	1	0	0.00	0.00	0.00	0
35417	Capers	1	teaspoon(s)	4	—	2	0	0	0	0	0.00	0.00	0.00	—
1172	Caraway seeds	1	teaspoon(s)	2	<1	7	<1	1	1	<1	0.01	0.15	0.07	—
819	Cayenne pepper	1	teaspoon(s)	2	<1	6	<1	1	<1	<1	0.06	0.05	0.15	—
1173	Celery seeds	1	teaspoon(s)	2	<1	8	<1	1	<1	1	0.04	0.32	0.07	—
1174	Chervil, dried	1	teaspoon(s)	1	<.1	1	<1	<1	<.1	<.1	0.00	0.01	0.01	—
810	Chili powder	1	teaspoon(s)	3	<1	8	<1	1	1	<1	0.08	0.09	0.19	—
8553	Chives, chopped	1	teaspoon(s)	1	1	<1	<.1	<.1	<.1	<.1	0.00	0.00	0.00	—
8556	Cilantro	1	teaspoon(s)	2	1	<1	<.1	<.1	<.1	<.1	0.00	0.00	0.00	—
811	Cinnamon, ground	1	teaspoon(s)	2	<1	6	<.1	2	1	<.1	0.01	0.01	0.01	—
812	Cloves, ground	1	teaspoon(s)	2	<1	7	<1	1	1	<1	0.11	0.03	0.15	—
1175	Coriander leaf, dried	1	teaspoon(s)	1	<.1	2	<1	<1	<.1	<.1	0.00	0.01	0.00	—
1176	Coriander seeds	1	teaspoon(s)	2	<1	5	<1	1	1	<1	0.02	0.24	0.03	—
1706	Cornstarch	1	tablespoon(s)	8	1	30	<.1	7	<.1	<.1	0.00	0.00	0.00	—
11729	Cumin, ground	1	teaspoon(s)	5	—	11	<1	1	1	<1	—	—	—	—
1177	Cumin seeds	1	teaspoon(s)	2	<1	8	<1	1	<1	<1	0.03	0.29	0.07	—
1178	Curry powder	1	teaspoon(s)	2	<1	7	<1	1	1	<1	0.04	0.11	0.05	—
1179	Dill seeds	1	teaspoon(s)	2	<1	6	<1	1	<1	<1	0.02	0.20	0.02	—
1180	Dill weed, dried	1	teaspoon(s)	1	<.1	3	<1	1	<1	<.1	0.00	0.01	0.00	—
34949	Dill weed, fresh	5	piece(s)	1	1	<1	<.1	<.1	<.1	<.1	0.00	0.01	0.00	—
4949	Fennel leaves, fresh	1	teaspoon(s)	1	1	<1	<.1	<.1	0	<.1	0.00	0.00	0.00	—
1181	Fennel seeds	1	teaspoon(s)	2	<1	7	<1	1	1	<1	0.01	0.20	0.03	—
1182	Fenugreek seeds	1	teaspoon(s)	4	<1	12	1	2	1	<1	0.05	—	—	—
11733	Garam masala, powder	1	ounce(s)	28	—	107	4	13	0	4	—	—	—	—
1067	Garlic clove	1	item(s)	3	2	4	<1	1	<.1	<.1	0.00	0.00	0.01	—
813	Garlic powder	1	teaspoon(s)	3	<1	9	<1	2	<1	<.1	0.00	0.00	0.01	—
1183	Ginger, ground	1	teaspoon(s)	2	<1	6	<1	1	<1	<1	0.03	0.02	0.02	—
1068	Ginger root	2	teaspoon(s)	4	3	3	<.1	1	<.1	<.1	0.01	0.01	0.01	—
35497	Leeks, bulb & lower leaf, freeze-dried	¼	cup(s)	1	<.1	3	<1	1	<.1	<.1	0.00	0.00	0.01	—
1184	Mace, ground	1	teaspoon(s)	2	<1	8	<1	1	<1	1	0.16	0.19	0.07	—
1185	Marjoram, dried	1	teaspoon(s)	1	<.1	2	<.1	<1	<1	<.1	0.00	0.01	0.03	—
1186	Mustard seeds, yellow	1	teaspoon(s)	3	<1	15	1	1	<1	1	0.05	0.65	0.18	—
814	Nutmeg, ground	1	teaspoon(s)	2	<1	12	<1	1	<1	1	0.57	0.07	0.01	—
2747	Onion flakes, dehydrated	1	teaspoon(s)	2	<.1	6	<1	1	<1	<.1	0.00	0.00	0.00	—
1187	Onion powder	1	teaspoon(s)	2	<1	7	<1	2	<1	<.1	0.00	0.00	0.01	—
815	Oregano, ground	1	teaspoon(s)	2	<1	5	<1	1	1	<1	0.04	0.01	0.08	—
816	Paprika	1	teaspoon(s)	2	<1	6	<1	1	1	<1	0.04	0.03	0.17	—
817	Parsley, dried	1	teaspoon(s)	0	<.1	1	<.1	<1	<.1	<.1	0.00	0.01	0.00	—
1189	Poppy seeds	1	teaspoon(s)	3	<1	15	1	1	1	1	0.14	0.18	0.86	—
1190	Poultry seasoning	1	teaspoon(s)	2	<1	5	<1	1	<1	<1	0.05	0.02	0.03	—
1191	Pumpkin pie spice, powder	1	teaspoon(s)	2	<1	6	<.1	1	<1	<1	0.11	0.02	0.01	—
1192	Rosemary, dried	1	teaspoon(s)	1	<1	4	<.1	1	1	<1	0.09	0.04	0.03	—
11723	Rosemary, fresh	1	teaspoon(s)	1	<1	1	<.1	<1	<.1	<.1	0.02	0.01	0.01	—
2722	Saffron powder	1	teaspoon(s)	1	<.1	2	<.1	<1	<.1	<.1	0.01	0.00	0.01	—
11724	Sage	1	ounce(s)	28	—	34	1	4	0	1	—	—	—	—
1193	Sage, ground	1	teaspoon(s)	1	<.1	2	<.1	<1	<1	<.1	0.05	0.01	0.01	—
822	Salt, table	¼	teaspoon(s)	2	<.1	0	0	0	0	0	0.00	0.00	0.00	0
30189	Salt substitute	¼	teaspoon(s)	1	—	<.1	0	<.1	0	0	0.00	0.00	0.00	0
30190	Salt substitute, seasoned	¼	teaspoon(s)	1	—	1	<.1	<1	0	<.1	0.00			
1194	Savory, ground	1	teaspoon(s)	1	<1	4	<.1	1	1	<.1	0.05	—	—	—
820	Sesame seed kernels, toasted	1	teaspoon(s)	3	<1	15	<1	1	<1	1	0.18	0.49	0.57	—
11725	Sorrel	1	tablespoon(s)	9	—	2	<1	<1	<.1	<.1	0.00	—	—	—

PAGE KEY: H–2 = Breads/Baked Goods H–6 = Cereal/Rice/Pasta H–10 = Fruit H–14 = Vegetables/Legumes H–24 = Nuts/Seeds H–26 = Vegetarian
H–28 = Dairy H–34 = Eggs H–34 = Seafood H–36 = Meats H–40 = Poultry H–40 = Processed meats H–42 = Beverages H–46 = Fats/Oils
H–48 = Sweets H–50 = Spices/Condiments/Sauces H–52 = Mixed foods/Soups/Sandwiches H–58 = Fast food H–74 = Convenience meals H–76 = Baby foods

Chol (mg)	Calc (mg)	Iron (mg)	Magn (mg)	Pota (mg)	Sodi (mg)	Zinc (mg)	Vit A (µg)	Thia (mg)	Vit E (mg α)	Ribo (mg)	Niac (mg)	Vit B$_6$ (mg)	Fola (µg)	Vit C (mg)	Vit B$_{12}$ (µg)	Sele (µg)
0	13	0.13	3	20	1	0.02	1	0.00	—	0.00	0.05	0.00	1	1	0	<.1
0	14	0.78	4	30	<1	0.11	<1	0.01	—	0.01	0.06	0.01	<1	<1	0	<1
0	3	0.66	4	80	2	0.26	0	0.09	0.00	0.22	1.59	0.06	94	<.1	<.1	1
0	339	0.52	2	<1	363	0.00	0	0.00	0.00	0.00	0.00	0.00	0	0	0	<.1
0	0	0.00	0	0	1259	0.00	0	0.00	0.00	0.00	0.00	0.00	0	0	0	<.1
0	1	0.03	1	4	<.1	0.01	2	0.00	—	0.00	0.01	0.00	1	<1	0	<.1
0	1	—	<1	2	<.1	0.00	—	0.00	—	0.00	0.01	0.00	<1	—	0	<.1
0	30	0.59	6	48	<1	0.08	7	0.00	0.10	0.00	0.10	0.03	4	1	0	<.1
0	5	0.26	1	3	<1	0.02	2	0.00	—	0.00	0.01	0.01	1	<1	0	<.1
0	110	2.29	—	156	2	—	—	0.04	—	0.07	0.20	—	—	1	0	—
0	9	0.61	4	26	1	0.03	<1	0.00	0.02	0.01	0.02	0.01	<1	<1	0	<.1
0	6	0.47	6	51	3	0.21	0	0.42	—	0.11	1.00	0.07	104	0	0	0
0	0	0.00	—	—	140	—	0	—	—	—	—	—	—	0	—	—
0	14	0.34	5	28	<1	0.12	<1	0.01	0.05	0.01	0.08	0.01	<1	<1	0	<1
0	3	0.14	3	36	1	0.04	37	0.01	0.54	0.02	0.16	0.04	2	1	0	<1
0	35	0.90	9	28	3	0.14	<.1	0.01	0.02	0.01	0.06	0.02	<1	<1	0	<1
0	8	0.19	1	28	<1	0.05	2	0.00	—	0.00	0.03	0.01	2	<1	0	<1
0	7	0.37	4	50	26	0.07	39	0.01	—	0.02	0.21	0.10	3	2	0	<1
0	1	0.02	<1	3	<.1	0.01	2	0.00	0.76	0.00	0.01	0.00	1	1	0	<.1
0	1	0.03	<1	8	1	0.00	—	0.00	—	0.00	0.02	0.00	1	1	0	<.1
0	28	0.88	1	12	1	0.05	<1	0.00	0.02	0.00	0.03	0.01	1	1	0	<.1
0	14	0.18	6	23	5	0.02	1	0.00	0.18	0.01	0.03	0.01	2	2	0	<1
0	7	0.25	4	27	1	0.03	2	0.01	—	0.01	0.06	0.00	2	3	0	<1
0	13	0.29	6	23	1	0.08	0	0.00	—	0.01	0.04	—	0	<1	0	<1
0	<1	0.04	<1	<1	1	0.00	0	0.00	0.00	0.00	0.00	0.00	0	0	0	<1
0	20	—	—	44	5	—	—	—	—	—	—	—	—	—	—	—
0	20	1.39	8	38	4	0.10	1	0.01	0.07	0.01	0.10	0.01	<1	<1	0	<1
0	10	0.59	5	31	1	0.08	1	0.01	0.44	0.01	0.07	0.02	3	<1	0	<1
0	32	0.34	5	25	<1	0.11	<.1	0.01	—	0.01	0.06	0.01	<1	<1	0	<1
0	18	0.49	5	33	2	0.03	3	0.00	—	0.00	0.03	0.02	2	1	0	0
0	2	—	1	7	1	0.01	—	0.00	—	0.00	0.02	0.00	2	—	0	—
0	1	0.03	—	4	<.1	—	—	0.00	—	0.00	0.01	0.00	—	<1	0	—
0	24	0.37	8	34	2	0.07	<1	0.01	—	0.01	0.12	0.01	—	<1	0	0
0	7	1.24	7	28	2	0.09	<1	0.01	—	0.01	0.06	0.02	2	<1	0	<1
0	215	9.25	94	411	28	1.07	—	0.10	—	0.09	0.71	—	0	0	0	—
0	5	0.05	1	12	1	0.03	0	0.01	0.00	0.00	0.02	0.04	<.1	1	0	<1
0	2	0.08	2	31	1	0.07	0	0.01	0.02	0.00	0.02	0.08	<.1	1	0	1
0	2	0.21	3	24	1	0.08	<1	0.00	0.32	0.00	0.09	0.02	1	<1	0	1
0	1	0.02	2	17	1	0.01	0	0.00	0.01	0.00	0.03	0.01	<1	<1	0	<.1
0	3	0.06	1	19	<1	0.01	<1	0.01	—	0.00	0.03	0.01	3	1	0	<.1
0	4	0.24	3	8	1	0.04	1	0.01	—	0.01	0.02	0.00	1	<1	0	<.1
0	12	0.50	2	9	<1	0.02	2	0.00	0.01	0.00	0.02	0.01	2	<1	0	<.1
0	17	0.33	10	23	<1	0.19	<.1	0.02	0.10	0.01	0.26	0.01	3	<.1	0	4
0	4	0.07	4	8	<1	0.05	<1	0.01	0.00	0.00	0.03	0.00	2	<1	0	<.1
0	4	0.03	2	27	<1	0.03	<.1	0.00	0.00	0.00	0.02	0.03	3	1	0	<.1
0	8	0.05	3	20	1	0.05	0	0.01	0.01	0.00	0.01	0.03	3	<1	0	<.1
0	24	0.66	4	25	<1	0.07	5	0.01	0.28	0.00	0.09	0.02	4	1	0	<.1
0	4	0.50	4	49	1	0.09	55	0.01	0.63	0.04	0.32	0.08	2	1	0	<.1
0	4	0.29	1	11	1	0.01	2	0.00	0.02	0.00	0.02	0.00	1	<1	0	<.1
0	41	0.26	9	20	1	0.29	0	0.02	0.03	0.00	0.03	0.01	2	<.1	0	<.1
0	15	0.53	3	10	<1	0.05	2	0.00	0.03	0.00	0.04	0.02	2	<1	0	<1
0	12	0.34	2	11	1	0.04	<1	0.00	0.02	0.00	0.04	0.01	1	<1	0	<.1
0	15	0.35	3	11	1	0.04	2	0.01	—	0.01	0.01	0.02	4	1	0	<.1
0	2	0.05	1	5	<1	0.01	1	0.00	—	0.00	0.01	0.00	1	<1	0	—
0	1	0.08	2	12	1	0.01	<1	0.00	—	0.00	0.01	0.01	1	1	0	<.1
0	170	—	45	110	1	0.48	—	0.03	—	—	—	—	—	—	0	—
0	12	0.20	3	7	<.1	0.03	2	0.01	0.05	0.00	0.04	0.02	2	<1	0	<.1
0	<1	0.00	<.1	<1	581	0.00	0	0.00	0.00	0.00	0.00	0.00	0	0	0	<.1
0	7	0.00	<.1	604	<.1	—	0	0.00	—	—	—	—	—	0	—	—
0	0	0	476	<1	—	0	—	—	—	—	—	—	—	0	—	—
0	30	0.53	5	15	<1	0.06	4	0.01	—	—	0.06	0.03	—	1	0	<.1
0	4	0.21	9	11	1	0.28	<.1	0.03	0.01	0.01	0.15	0.00	3	0	0	<.1
0	—	—	—	—	<1	—	—	—	—	—	—	—	—	—	—	—

TABLE H–1

Food Composition

(DA+ code is for Wadsworth Diet Analysis program)   (For purposes of calculations, use "0" for t, <1, <.1, <.01, etc.)

DA + Code	Food Description	Quantity	Measure	Wt (g)	H₂O (g)	Ener (kcal)	Prot (g)	Carb (g)	Fiber (g)	Fat (g)	Sat	Mono	Poly	Trans
	SPICES, CONDIMENTS, SAUCES—Continued													
11721	Spearmint	1	teaspoon(s)	2	2	1	<.1	<1	<1	<.1	0.00	0.00	0.01	—
35498	Sweet green peppers, freeze-dried	¼	cup(s)	2	<.1	5	<1	1	<1	<.1	0.01	0.00	0.03	—
11726	Tamarind leaves	1	ounce(s)	28	—	33	2	5	0	1	—	—	—	—
11727	Tarragon	1	ounce(s)	28	—	14	1	2	0	<1	—	—	—	—
1195	Tarragon, ground	1	teaspoon(s)	2	<1	5	<1	1	<1	<1	0.03	0.01	0.06	—
11728	Thyme, fresh	1	teaspoon(s)	1	1	1	<.1	<1	<1	<.1	0.00	0.00	0.00	—
821	Thyme, ground	1	teaspoon(s)	1	<1	4	<1	1	1	<1	0.04	0.01	0.02	—
1196	Turmeric, ground	1	teaspoon(s)	2	<1	8	<1	1	<1	<1	0.07	0.04	0.05	—
11995	Wasabi	1	tablespoon(s)	14	11	11	1	2	<1	<.1	—	—	—	—
1188	White pepper	1	teaspoon(s)	2	<1	7	<1	2	1	<.1	0.02	0.02	0.01	—
	Condiments													
674	Catsup or ketchup	1	tablespoon(s)	15	11	14	<1	4	<1	<.1	0.01	0.01	0.04	—
703	Dill pickle	1	ounce(s)	28	26	5	<1	1	<1	<.1	0.01	0.00	0.02	—
1641	Horseradish sauce, prepared	1	teaspoon(s)	5	3	10	<1	<1	<.1	1	0.59	0.28	0.04	—
140	Mayonnaise, low calorie	1	tablespoon(s)	16	10	37	<.1	3	0	3	0.53	0.72	1.70	—
138	Mayonnaise w/soybean oil	1	tablespoon(s)	14	2	99	<1	1	0	11	1.64	2.70	5.89	0.04
1682	Mustard, brown	1	teaspoon(s)	5	4	5	<1	<1	<.1	<1	—	—	—	—
700	Mustard, yellow	1	teaspoon(s)	5	4	3	<1	<1	<1	<1	0.01	0.11	0.03	—
706	Sweet pickle relish	1	tablespoon(s)	15	9	20	<.1	5	<1	<.1	0.01	0.03	0.02	—
141	Tartar sauce	2	tablespoon(s)	28	9	144	<1	4	<.1	14	2.14	4.13	7.57	—
	Sauces													
685	Barbecue sauce	2	tablespoon(s)	31	25	23	1	4	<1	1	0.08	0.24	0.21	—
834	Cheese sauce	¼	cup(s)	70	49	121	5	5	<1	9	4.19	2.67	1.81	—
32123	Chili enchilada sauce, green	2	tablespoon(s)	57	53	15	1	3	1	<1	0.04	0.04	0.13	0
32122	Chili enchilada sauce, red	2	tablespoon(s)	32	24	27	1	5	2	1	0.08	0.05	0.43	0
29688	Hoisin sauce	1	tablespoon(s)	16	7	35	1	7	<1	1	0.09	0.15	0.27	—
16670	Mole poblano sauce	½	cup(s)	133	103	155	5	11	2	11	2.67	5.15	2.91	—
29689	Oyster sauce	1	tablespoon(s)	16	13	8	<1	2	<.1	<.1	0.01	0.01	0.01	—
1655	Pepper sauce or tabasco	1	teaspoon(s)	5	5	1	<.1	<.1	<.1	<.1	0.01	0.00	0.02	—
347	Salsa	2	tablespoon(s)	16	14	4	<1	1	<1	<.1	0.00	0.00	0.02	—
841	Soy sauce	1	tablespoon(s)	18	13	10	1	2	0	<.1	0.00	0.00	0.01	—
839	Sweet & sour sauce	2	tablespoon(s)	39	30	37	<.1	9	<.1	<.1	0.00	0.00	0.00	—
1613	Teriyaki sauce	1	tablespoon(s)	18	12	15	1	3	<.1	0	0.00	0.00	0.00	0
25294	Tomato sauce	½	cup(s)	112	100	46	2	8	2	1	0.18	0.29	0.72	0
728	White sauce, medium	¼	cup(s)	63	47	92	2	6	<1	7	1.78	2.78	1.79	—
1654	Worcestershire sauce	1	teaspoon(s)	6	4	4	0	1	0	0	0.00	0.00	0.00	0
	Vinegar													
30853	Balsamic	1	tablespoon(s)	15	—	10	0	2	0	0	0.00	0.00	0.00	0
727	Cider	1	tablespoon(s)	15	14	2	0	1	0	0	0.00	0.00	0.00	0
1673	Distilled	1	tablespoon(s)	15	14	2	0	1	0	0	0.00	0.00	0.00	0
15439	Tarragon	1	tablespoon(s)	16	—	0	0	0	0	0	0.00	0.00	0.00	0
	MIXED FOODS, SOUPS, SANDWICHES													
	Mixed Dishes													
16652	Almond chicken	1	cup(s)	242	186	280	22	16	3	15	1.91	6.07	5.62	—
25224	Barbecued chicken	2	piece(s)	177	100	325	27	15	<1	17	4.63	6.78	3.71	0
25227	Bean burrito	1	item(s)	149	82	327	17	33	6	15	8.30	4.73	0.85	0
9516	Beef & vegetable fajita	1	item(s)	223	144	397	23	35	3	18	5.50	7.53	3.45	—
16796	Beef or pork egg roll	2	item(s)	128	85	227	10	19	1	12	2.88	5.96	2.64	—
177	Beef stew w/vegetables, prepared	1	cup(s)	245	201	220	16	15	3	11	4.40	4.50	0.50	—
30233	Beef stroganoff w/noodles	1	cup(s)	256	190	343	20	23	2	19	7.37	5.62	4.47	—
16651	Cashew chicken	1	cup(s)	242	131	644	43	17	3	46	7.75	20.83	14.47	—
475	Cheese pizza	2	slice(s)	126	60	281	15	41	0	6	3.08	1.98	0.98	—
30330	Cheese quesadilla	1	item(s)	54	19	183	6	18	1	10	3.49	3.42	2.16	—
215	Chicken & noodles, prepared	1	cup(s)	240	170	365	22	26	1	18	5.10	7.10	3.90	—
30239	Chicken & vegetables w/broccoli, onion, bamboo shoots in soy based sauce	1	cup(s)	162	112	287	22	6	1	19	5.13	7.65	4.68	—
25093	Chicken cacciatore	1	cup(s)	230	166	266	28	5	1	14	3.98	5.78	3.11	0
28020	Chicken fried turkey steak	3	ounce(s)	85	48	122	13	12	1	2	0.59	0.37	0.78	—
218	Chicken pot pie	1	cup(s)	252	154	542	23	42	3	31	9.79	12.52	7.03	—
30240	Chicken teriyaki	1	cup(s)	244	163	339	51	13	1	7	1.78	2.03	1.71	—
25119	Chicken waldorf salad	½	cup(s)	100	68	178	14	6	1	11	1.76	3.18	5.05	0
25099	Chili con carne	¾	cup(s)	215	175	197	14	21	7	7	2.55	2.83	0.54	0
1062	Coleslaw	¾	cup(s)	90	73	62	1	11	1	2	0.35	0.64	1.22	—
1896	Combination pizza, w/meat & vegetables	2	slice(s)	158	75	368	26	43	5	11	3.07	5.09	1.83	—

Appendix H

PAGE KEY: H–2 = Breads/Baked Goods H–6 = Cereal/Rice/Pasta H–10 = Fruit H–14 = Vegetables/Legumes H–24 = Nuts/Seeds H–26 = Vegetarian
H–28 = Dairy H–34 = Eggs H–34 = Seafood H–36 = Meats H–40 = Poultry H–40 = Processed meats H–42 = Beverages H–46 = Fats/Oils
H–48 = Sweets H–50 = Spices/Condiments/Sauces H–52 = Mixed foods/Soups/Sandwiches H–58 = Fast food H–74 = Convenience meals H–76 = Baby foods

Chol (mg)	Calc (mg)	Iron (mg)	Magn (mg)	Pota (mg)	Sodi (mg)	Zinc (mg)	Vit A (µg)	Thia (mg)	Vit E (mg α)	Ribo (mg)	Niac (mg)	Vit B$_6$ (mg)	Fola (µg)	Vit C (mg)	Vit B$_{12}$ (µg)	Sele (µg)
0	4	0.23	1	9	1	0.02	4	0.00	—	0.00	0.02	0.00	2	<1	0	—
0	2	0.17	3	51	3	0.04	3	0.02	0.06	0.02	0.12	0.04	4	30	0	<.1
0	85	1.48	20	—	—	—	—	0.07	—	0.03	1.16	—	—	1	0	—
0	48	—	14	128	3	0.17	—	0.04	—	—	—	—	—	1	0	—
0	18	0.52	6	48	1	0.06	3	0.00	0.10	0.02	0.14	0.04	4	1	0	<.1
0	3	0.14	1	5	<.1	0.01	2	0.00	—	0.00	0.01	0.00	<1	1	0	—
0	26	1.73	3	11	1	0.09	3	0.01	—	0.01	0.07	0.01	4	1	0	<.1
0	4	0.91	4	56	1	0.10	0	0.00	—	0.01	0.11	0.04	1	1	0	<.1
0	13	0.11	—	—	—	—	—	0.02	—	0.01	0.07	—	—	11	0	—
0	6	0.34	2	2	<1	0.03	0	0.00	0.10	0.00	0.01	0.00	<1	1	0	<.1
0	3	0.08	3	57	167	0.04	7	0.00	0.22	0.07	0.23	0.02	2	2	0	<.1
0	3	0.15	3	33	363	0.04	3	0.00	0.03	0.01	0.02	0.00	<1	1	0	0
2	5	0.00	1	7	15	0.01	—	0.00	0.03	0.01	0.00	0.00	1	<.1	<.1	—
4	<.1	0.00	<.1	2	80	0.02	0	0.00	0.32	0.00	0.00	0.00	0	0	0	—
5	2	0.07	<1	5	78	0.02	12	0.00	0.72	0.00	0.00	0.08	1	0	<.1	<1
0	6	0.09	1	7	68	0.02	0	0.00	0.09	0.00	0.01	0.00	<1	<.1	0	—
0	4	0.09	2	8	56	0.03	<1	0.00	0.01	0.00	0.02	0.00	<1	<1	0	2
0	<1	0.13	1	4	122	0.02	1	0.00	0.06	0.00	0.03	0.00	<1	<1	0	0
11	6	0.21	1	10	200	0.05	—	0.00	0.97	0.00	0.01	0.08	2	<1	<.1	—
0	6	0.28	6	54	255	0.06	<1	0.01	0.01	0.01	0.28	0.02	1	2	0	<1
20	128	0.15	6	21	578	0.68	56	0.00	—	0.08	0.02	0.01	3	<1	<.1	2
0	5	0.36	9	126	62	0.11	—	0.03	0.00	0.02	0.63	0.06	6	44	0	0
0	7	1.05	11	231	114	0.15	—	0.02	0.00	0.22	0.61	0.34	7	<1	0	<1
<1	5	0.16	4	19	258	0.05	0	0.00	0.04	0.03	0.19	0.01	4	<.1	0	<1
1	37	1.51	57	283	305	0.95	—	0.07	1.72	0.09	1.82	0.09	14	5	<.1	—
0	5	0.03	1	9	437	0.01	0	0.00	0.00	0.02	0.24	0.00	2	<.1	<.1	1
0	1	0.06	1	6	32	0.01	4	0.00	—	0.00	0.01	0.01	<.1	<1	0	—
0	5	0.16	2	34	69	0.04	5	0.01	0.19	0.01	0.13	0.02	3	2	0	<.1
0	3	0.36	6	32	1029	0.07	0	0.01	0.00	0.02	0.61	0.03	3	0	0	—
0	5	0.20	1	8	98	0.01	0	0.00	—	0.01	0.12	0.04	<1	0	0	—
0	5	0.31	11	41	690	0.02	0	0.01	0.00	0.01	0.23	0.02	4	0	0	<1
0	21	1.08	19	431	199	0.30	48	0.05	0.39	0.05	1.18	0.13	15	15	0	1
4	74	0.21	9	98	221	0.26	—	0.04	—	0.12	0.25	0.03	3	1	<1	—
0	6	0.30	1	45	56	0.01	—	0.00	0.00	0.01	0.04	0.00	0	1	0	—
0	0	0.00	—	—	0	—	0	—	—	—	—	—	—	0	—	—
0	1	0.09	3	15	<1	0.00	0	0.00	0.00	0.00	0.00	0.00	0	0	0	<.1
0	1	0.09	0	2	<1	0.00	0	0.00	0.00	0.00	0.00	0.00	0	0	0	5
0	0	0.00	—	0	0	—	0	—	—	—	—	—	—	0	0	—
40	69	1.97	60	549	526	1.62	—	0.09	4.11	0.20	9.48	0.44	26	7	<1	—
120	26	1.64	31	387	477	2.69	69	0.07	0.01	0.37	6.92	0.39	15	5	<1	19
38	331	2.95	45	384	514	1.92	119	0.24	0.01	0.29	1.82	0.15	115	4	<1	18
45	84	3.74	37	476	757	3.51	—	0.39	0.80	0.30	5.37	0.38	23	27	2	—
74	30	1.66	20	248	547	0.91	—	0.32	1.28	0.25	2.55	0.19	20	4	<1	—
71	29	2.90	—	613	292	—	—	0.15	0.51	0.17	4.70	—	—	17	<.1	15
74	70	3.26	37	393	818	3.66	—	0.21	1.25	0.31	3.80	0.21	17	1	2	—
96	74	2.92	94	640	1355	2.24	—	0.23	4.11	0.22	19.76	0.88	64	11	<1	—
19	233	1.16	32	219	672	1.63	147	0.37	—	0.33	4.96	0.09	69	3	1	27
13	132	1.21	13	77	230	0.64	—	0.13	0.43	0.14	1.09	0.04	6	15	<.1	—
103	26	2.20	—	149	600	—	—	0.05	—	0.17	4.30	—	—	0	—	29
84	22	1.38	29	344	962	1.70	—	0.08	1.12	0.17	7.90	0.32	13	8	<1	—
103	45	2.21	37	444	451	2.01	53	0.10	0.00	0.21	9.20	0.54	15	8	<1	22
27	69	1.34	19	197	139	1.08	5	0.15	0.00	0.18	3.46	0.22	21	<1	<1	16
69	64	3.38	38	393	651	1.93	607	0.40	1.06	0.40	7.24	0.24	31	11	<1	—
157	52	3.27	67	589	3209	3.75	—	0.15	0.59	0.37	16.69	0.89	23	6	1	—
42	20	0.78	24	197	246	1.13	21	0.04	0.62	0.10	4.05	0.25	15	2	<1	11
27	43	3.16	50	646	865	2.44	25	0.13	0.02	0.23	3.01	0.18	56	10	1	10
7	41	0.53	9	163	21	0.18	48	0.06	—	0.06	0.24	0.11	24	29	0	1
41	202	3.07	36	357	765	2.23	117	0.43	—	0.35	3.92	0.19	65	3	1	22

TABLE H–1
Food Composition

(DA+ code is for Wadsworth Diet Analysis program)   (For purposes of calculations, use "0" for t, <1, <.1, <.01, etc.)

DA + Code	Food Description	Quantity	Measure	Wt (g)	H₂O (g)	Ener (kcal)	Prot (g)	Carb (g)	Fiber (g)	Fat (g)	Sat	Mono	Poly	Trans
	MIXED FOODS, SOUPS, SANDWICHES—Continued													
1574	Crab cakes, from blue crab	1	item(s)	60	43	93	12	<1	0	5	0.89	1.69	1.36	—
32144	Enchiladas w/green chili sauce (enchiladas verdes)	1	item(s)	144	104	207	9	18	3	12	6.35	3.65	0.96	0
2793	Falafel patty	3	item(s)	51	18	170	7	16	0	9	1.22	5.19	2.12	—
28546	Fettuccine alfredo	1	cup(s)	222	81	247	11	42	1	3	1.61	0.79	0.43	0
32146	Flautas	3	item(s)	162	78	438	25	36	4	22	8.22	8.80	2.29	—
29629	Fried rice w/meat or poultry	1	cup(s)	198	129	329	12	41	1	12	2.27	3.53	5.69	—
16649	General tso chicken	1	cup(s)	146	91	293	19	16	1	17	3.98	6.27	5.27	—
1826	Green salad	¾	cup(s)	104	99	17	1	3	2	<.1	0.01	0.00	0.04	—
1814	Hummus	½	cup(s)	123	80	218	6	25	5	11	1.38	6.04	2.56	—
16650	Kung pao chicken	1	cup(s)	162	88	431	29	11	2	31	5.19	13.95	9.69	—
16622	Lamb curry	1	cup(s)	236	188	256	28	3	1	14	3.93	4.92	3.35	—
25253	Lasagna w/ground beef	1	cup(s)	237	157	288	18	22	2	15	7.47	4.84	0.84	0
442	Macaroni & cheese	1	cup(s)	200	122	393	15	40	1	19	8.18	6.72	2.66	—
25105	Meat loaf	1	slice(s)	115	85	244	17	7	<1	16	6.15	6.89	0.83	0
16646	Moo shi pork	1	cup(s)	151	77	512	19	5	1	46	6.84	14.80	22.07	—
16788	Nachos w/beef, beans, cheese, tomatoes, & onions	7	item(s)	551	284	1496	40	119	19	99	22.34	40.19	30.69	—
1668	Pepperoni pizza	2	slice(s)	142	66	362	20	40	1	14	4.47	6.28	2.33	—
655	Potato salad	½	cup(s)	125	95	179	3	14	2	10	1.79	3.10	4.67	—
29637	Ravioli, meat filled, w/tomato or meat sauce, canned	1	cup(s)	251	196	220	9	38	2	4	1.58	1.49	0.41	—
25109	Salisbury steaks w/mushroom sauce	1	serving(s)	135	102	251	17	9	1	15	5.98	6.67	0.76	0
16637	Shrimp creole w/rice	1	cup(s)	243	176	311	27	28	1	9	1.83	3.79	2.88	—
497	Spaghetti & meat balls w/tomato sauce, prepared	1	cup(s)	248	174	330	19	39	3	12	3.90	4.40	2.20	—
28585	Spicy thai noodles (pad thai)	8	ounce(s)	231	74	222	9	36	3	6	0.83	3.33	1.83	0
33073	Stir fried pork & vegetables w/rice	1	cup(s)	235	173	349	15	34	2	16	5.55	6.87	2.62	0
28588	Stuffed shells	2½	item(s)	299	189	292	18	33	3	10	3.81	3.57	1.62	0
16821	Sushi w/egg in seaweed	6	piece(s)	156	117	190	9	20	<1	8	2.09	3.02	1.55	—
16819	Sushi w/vegetables & fish	6	piece(s)	156	102	217	8	44	2	1	0.16	0.14	0.20	—
16820	Sushi w/vegetables in seaweed	6	piece(s)	156	110	182	3	41	1	<1	0.10	0.11	0.11	—
25266	Sweet & sour pork	¾	cup(s)	249	206	264	29	17	1	8	2.59	3.51	1.48	0
16824	Tabouli, tabbouleh, or tabuli	1	cup(s)	160	124	199	3	16	4	15	2.04	10.83	1.37	—
25276	Three bean salad	½	cup(s)	99	82	95	2	10	3	6	0.76	1.41	3.48	0
160	Tuna salad	½	cup(s)	103	65	192	16	10	0	9	1.58	2.96	4.23	0
25241	Turkey & noodles	1	cup(s)	319	228	271	24	21	1	9	2.39	3.48	2.27	0
16794	Vegetable egg roll	2	item(s)	128	90	202	5	20	2	12	2.46	5.71	2.65	—
16818	Vegetable sushi, no fish	6	piece(s)	156	99	225	5	50	2	<1	0.11	0.10	0.14	—
	Sandwiches													
1744	Bacon, lettuce & tomato w/mayonnaise	1	item(s)	164	97	349	11	34	2	19	4.54	7.22	6.07	—
30287	Bologna & cheese w/margarine	1	item(s)	111	46	350	13	28	1	20	8.55	8.40	2.28	—
30286	Bologna w/margarine	1	item(s)	83	34	256	7	26	1	13	4.08	6.31	2.07	—
16546	Cheese	1	item(s)	83	31	262	10	27	1	13	5.59	4.77	1.67	—
8789	Cheeseburger, large, plain	1	item(s)	185	72	609	30	47	0	33	14.84	12.74	2.44	—
8624	Cheeseburger, large, w/bacon, vegetables, & condiments	1	item(s)	195	85	608	32	37	2	37	16.24	14.49	2.71	—
1745	Club w/bacon, chicken, tomato, lettuce, & mayonnaise	1	item(s)	246	137	555	31	48	3	26	5.94	—	—	—
1908	Cold cut submarine w/cheese & vegetables	1	item(s)	228	132	456	22	51	2	19	6.81	8.23	2.28	—
30247	Corned beef	1	item(s)	130	75	268	19	25	2	10	3.75	3.96	0.80	—
25283	Egg salad	1	item(s)	126	72	278	10	29	1	13	2.96	3.97	4.79	—
16686	Fried egg	1	item(s)	96	50	226	10	26	1	9	2.29	3.51	1.64	—
16547	Grilled cheese	1	item(s)	83	27	292	10	27	1	16	6.22	6.29	2.54	—
16659	Gyro w/onion & tomato	1	item(s)	105	67	170	12	21	1	4	1.53	1.41	0.43	—
1906	Ham & cheese	1	item(s)	146	74	352	21	33	2	15	6.44	6.74	1.38	—
31890	Ham w/mayonnaise	1	item(s)	112	55	282	14	27	1	13	3.06	5.04	3.79	—
756	Hamburger, double patty, large, w/condiments & vegetables	1	item(s)	226	121	540	34	40	0	27	10.52	10.33	2.80	—
8793	Hamburger, large, plain	1	item(s)	137	58	426	23	32	2	23	8.38	9.88	2.14	—
8795	Hamburger, large, w/vegetables & condiments	1	item(s)	218	121	512	26	40	3	27	10.42	11.42	2.20	—
25134	Hot chicken salad	1	item(s)	98	49	239	16	23	1	9	2.83	2.61	2.76	0
1411	Hot dog w/bun, plain	1	item(s)	98	53	242	10	18	2	15	5.11	6.85	1.71	—
25133	Hot turkey salad	1	item(s)	98	50	221	16	23	1	7	2.23	1.76	2.28	0
30249	Pastrami	1	item(s)	134	71	331	14	27	2	18	6.18	8.74	1.02	—

PAGE KEY: H–2 = Breads/Baked Goods   H–6 = Cereal/Rice/Pasta   H–10 = Fruit   H–14 = Vegetables/Legumes   H–24 = Nuts/Seeds   H–26 = Vegetarian
H–28 = Dairy   H–34 = Eggs   H–34 = Seafood   H–36 = Meats   H–40 = Poultry   H–40 = Processed meats   H–42 = Beverages   H–46 = Fats/Oils
H–48 = Sweets   H–50 = Spices/Condiments/Sauces   H–52 = Mixed foods/Soups/Sandwiches   H–58 = Fast food   H–74 = Convenience meals   H–76 = Baby foods

Chol (mg)	Calc (mg)	Iron (mg)	Magn (mg)	Pota (mg)	Sodi (mg)	Zinc (mg)	Vit A (µg)	Thia (mg)	Vit E (mg α)	Ribo (mg)	Niac (mg)	Vit B$_6$ (mg)	Fola (µg)	Vit C (mg)	Vit B$_{12}$ (µg)	Sele (µg)
90	63	0.65	20	194	198	2.45	34	0.05	—	0.05	1.74	0.10	32	2	4	24
27	266	1.08	38	251	276	1.27	—	0.07	0.03	0.16	1.28	0.18	45	59	<1	6
0	28	1.74	42	298	150	0.77	1	0.07	—	0.08	0.53	0.06	47	1	0	1
9	153	1.88	32	123	386	1.48	51	0.35	0.00	0.34	2.60	0.06	103	1	<1	35
73	146	2.66	61	223	886	3.44	0	0.10	0.10	0.17	3.00	0.27	96	0	1	37
102	36	2.66	31	182	821	1.42	—	0.30	1.60	0.19	3.51	0.24	24	3	<1	—
65	27	1.49	24	250	906	1.40	—	0.10	1.62	0.19	6.28	0.28	17	12	<1	—
0	13	0.65	11	178	27	0.22	59	0.03	—	0.05	0.57	0.08	38	24	0	<1
0	60	1.93	36	213	298	1.34	0	0.11	0.92	0.06	0.49	0.49	73	10	0	3
64	49	1.96	63	428	907	1.50	—	0.15	4.32	0.15	13.23	0.59	43	8	<1	—
89	36	2.97	40	495	495	6.62	—	0.09	1.30	0.28	8.05	0.20	27	1	3	—
68	222	2.33	40	437	493	2.81	108	0.19	0.22	0.29	3.02	0.20	50	10	1	22
30	323	2.26	42	263	800	1.95	327	0.25	0.72	0.40	2.18	0.10	12	<1	<1	—
85	54	2.09	21	278	423	3.55	27	0.08	0.00	0.29	3.77	0.13	20	<1	2	17
172	30	1.45	26	330	1078	1.83	—	0.50	5.39	0.38	2.90	0.31	22	8	1	—
82	699	6.71	205	1067	1611	7.55	—	0.31	7.71	0.50	5.62	0.85	59	14	1	—
28	129	1.87	17	305	534	1.04	105	0.27	—	0.47	6.09	0.11	74	3	<1	26
85	24	0.81	19	318	661	0.39	40	0.10	—	0.08	1.11	0.18	9	13	0	5
17	28	2.04	23	337	1354	1.19	—	0.22	0.70	0.20	2.88	0.14	17	22	<1	—
60	64	2.21	23	282	370	3.66	27	0.11	0.00	0.30	4.00	0.13	22	<1	2	17
181	101	4.44	64	439	381	1.73	—	0.29	2.07	0.10	4.77	0.22	12	18	1	—
89	124	3.70	—	665	1009	—	82	0.25	—	0.30	4.00	—	22	—	22
37	32	1.58	50	187	598	1.08	38	0.18	0.36	0.13	1.88	0.17	44	22	<.1	3
46	39	2.65	32	394	574	2.07	80	0.51	0.38	0.20	5.07	0.30	102	18	<1	23
35	241	3.18	63	462	543	1.68	280	0.32	0.00	0.36	4.64	0.30	109	15	<1	36
217	42	1.63	18	128	527	0.98	—	0.12	0.67	0.29	1.33	0.13	29	2	<1	—
11	24	2.18	25	204	340	0.79	—	0.26	0.25	0.07	2.77	0.15	14	4	<1	—
0	20	1.54	20	99	153	0.70	—	0.20	0.12	0.04	1.86	0.14	10	2	0	—
74	41	1.78	35	622	624	2.53	64	0.80	0.20	0.37	6.69	0.65	14	10	1	50
0	29	1.25	36	246	799	0.48	—	0.08	2.43	0.05	1.14	0.11	31	29	0	—
0	26	0.96	15	144	224	0.31	12	0.04	0.89	0.06	0.26	0.06	31	9	0	3
13	17	1.03	19	182	412	0.57	25	0.03	0.00	0.07	6.87	0.08	8	2	1	42
77	60	2.69	33	379	576	2.64	108	0.23	0.29	0.32	6.40	0.30	60	1	1	34
60	29	1.61	18	193	548	0.51	—	0.16	1.28	0.21	1.59	0.10	27	6	<1	—
0	23	2.40	23	158	369	0.84	—	0.28	0.16	0.06	2.44	0.13	15	4	0	—
20	76	2.54	27	328	837	0.98	—	0.39	1.16	0.27	3.81	0.20	31	15	<1	—
35	221	2.18	24	185	940	1.68	—	0.30	0.56	0.33	2.77	0.12	21	<.1	1	—
16	60	1.96	15	112	598	0.85	—	0.29	0.50	0.21	2.73	0.08	19	<.1	<1	—
19	216	1.75	20	135	655	1.14	—	0.25	0.47	0.29	2.04	0.07	19	<.1	<1	—
96	91	5.46	39	644	1589	5.55	185	0.48	—	0.57	11.17	0.28	74	0	3	39
111	162	4.74	45	332	1043	6.83	82	0.31	—	0.41	6.63	0.31	86	2	2	33
72	116	4.05	47	463	855	1.65	—	0.61	1.53	0.44	11.92	0.59	48	9	1	—
36	189	2.51	68	394	1651	2.58	71	1.00	—	0.80	5.49	0.14	87	12	1	31
46	67	2.67	20	187	1177	2.24	—	0.24	0.21	0.25	3.23	0.10	22	2	1	—
217	107	2.60	18	147	494	0.94	94	0.26	0.13	0.43	2.27	0.16	82	1	1	24
207	80	2.25	17	120	433	0.85	—	0.27	0.66	0.41	2.06	0.10	34	0	<1	—
19	219	1.76	21	137	696	1.15	—	0.19	0.72	0.28	1.86	0.06	13	<.1	<1	—
34	46	1.85	21	209	272	2.30	—	0.24	0.26	0.21	3.14	0.13	18	4	1	—
58	130	3.24	16	291	771	1.37	96	0.31	0.29	0.48	2.69	0.20	76	3	1	23
36	59	2.10	23	245	1033	1.50	—	0.71	0.50	0.31	4.89	0.26	19	0	<1	—
122	102	5.85	50	570	791	5.67	5	0.36	—	0.38	7.57	0.54	77	1	4	26
71	74	3.58	27	267	474	4.11	0	0.29	—	0.29	6.25	0.23	60	0	2	27
87	96	4.93	44	480	824	4.88	24	0.41	—	0.37	7.28	0.33	83	3	2	34
39	114	1.93	20	150	470	1.22	28	0.20	0.28	0.23	4.93	0.20	54	<1	<1	17
44	24	2.31	13	143	670	1.98	0	0.24	—	0.27	3.65	0.05	48	<.1	1	26
37	113	2.04	22	167	459	1.09	23	0.19	0.29	0.21	4.36	0.23	54	<1	<1	20
51	68	2.64	23	243	1335	2.69	—	0.29	0.27	0.27	4.77	0.13	21	2	1	—

TABLE H–1
Food Composition

(DA+ code is for Wadsworth Diet Analysis program)   (For purposes of calculations, use "0" for t, <1, <.1, <.01, etc.)

DA + Code	Food Description	Quantity	Measure	Wt (g)	H₂O (g)	Ener (kcal)	Prot (g)	Carb (g)	Fiber (g)	Fat (g)	Sat	Mono	Poly	Trans
	MIXED FOODS, SOUPS, SANDWICHES—Continued													
16701	Peanut butter	1	item(s)	93	24	344	13	37	3	17	3.55	8.16	4.58	—
30306	Peanut butter & jelly	1	item(s)	93	24	330	11	42	3	15	3.00	6.87	3.82	—
1910	Roast beef, plain	1	item(s)	139	68	346	22	33	1	14	3.61	6.80	1.71	—
1909	Roast beef submarine w/mayonnaise & vegetables	1	item(s)	216	127	410	29	44	—	13	7.09	1.84	2.61	—
1907	Steak w/mayonnaise & vegetables	1	item(s)	204	104	459	30	52	2	14	3.81	5.34	3.35	—
25288	Tuna salad	1	item(s)	179	102	414	24	29	2	22	3.61	5.46	11.43	—
31891	Turkey w/mayonnaise	1	item(s)	143	75	330	29	26	1	11	2.61	3.25	4.40	—
30283	Turkey submarine w/cheese, lettuce, tomato, & mayonnaise	1	item(s)	277	156	583	37	51	3	25	7.15	8.03	7.81	—
	Soups													
25296	Bean	1	cup(s)	301	253	191	14	29	6	2	0.67	0.83	0.53	0
711	Bean with pork, condensed, prepared w/water	1	cup(s)	265	223	180	8	24	9	6	1.59	2.28	1.91	—
713	Beef noodle, condensed, prepared w/water	1	cup(s)	244	224	83	5	9	1	3	1.15	1.24	0.49	—
825	Cheese, condensed, prepared w/milk	1	cup(s)	251	207	231	9	16	1	15	9.11	4.09	0.45	—
826	Chicken broth, condensed, prepared w/water	1	cup(s)	244	234	39	5	1	0	1	0.39	0.59	0.27	—
25297	Chicken noodle	1	cup(s)	286	258	117	11	11	1	3	0.78	1.10	0.66	—
827	Chicken noodle, condensed, prepared w/water	1	cup(s)	241	222	75	4	9	1	2	0.65	1.11	0.55	—
724	Chicken noodle, dehydrated, prepared w/water	1	cup(s)	252	237	58	2	9	<1	1	0.31	0.52	0.39	—
823	Cream of asparagus, condensed, prepared w/milk	1	cup(s)	248	213	161	6	16	1	8	3.32	2.08	2.23	—
824	Cream of celery, condensed, prepared w/milk	1	cup(s)	248	214	164	6	15	1	10	3.94	2.46	2.65	—
708	Cream of chicken, condensed, prepared w/milk	1	cup(s)	248	210	191	7	15	<1	11	4.64	4.46	1.64	—
715	Cream of chicken, condensed, prepared w/water	1	cup(s)	244	221	117	3	9	<1	7	2.07	3.27	1.49	—
709	Cream of mushroom, condensed, prepared w/milk	1	cup(s)	248	210	203	6	15	<1	14	5.13	2.98	4.61	—
716	Cream of mushroom, condensed, prepared w/water	1	cup(s)	244	220	129	2	9	<1	9	2.44	1.71	4.22	—
25298	Cream of vegetable	1	cup(s)	285	251	165	7	15	2	9	1.56	4.62	1.92	—
16689	Egg drop	1	cup(s)	244	229	73	8	1	0	4	1.15	1.52	0.59	—
25138	Golden squash	1	cup(s)	258	224	144	8	21	2	4	0.84	2.18	0.88	0
16663	Hot & sour	1	cup(s)	244	210	161	15	5	1	8	2.72	3.40	1.20	—
28054	Lentil chowder	1	cup(s)	229	188	150	11	27	12	<1	0.09	0.08	0.22	0
28560	Macaroni & bean	1	cup(s)	229	129	136	6	21	5	3	0.48	2.06	0.59	0
714	Manhattan clam chowder, condensed, prepared w/water	1	cup(s)	244	224	78	2	12	1	2	0.38	0.38	1.29	—
28561	Minestrone	1	cup(s)	230	177	99	4	16	5	2	0.32	1.30	0.43	0
717	Minestrone, condensed, prepared w/water	1	cup(s)	241	220	82	4	11	1	3	0.55	0.70	1.11	—
28038	Mushroom & wild rice	1	cup(s)	230	188	81	4	12	2	<1	0.05	0.02	0.15	0
828	New England clam chowder, condensed, prepared w/milk	1	cup(s)	248	211	164	9	17	1	7	2.95	2.26	1.09	—
28036	New England style clam chowder	1	cup(s)	229	207	83	3	15	2	<1	0.08	0.03	0.05	0
28566	Old country pasta	1	cup(s)	228	164	135	6	20	3	3	1.17	1.60	0.63	0
725	Onion, dehydrated, prepared w/water	1	cup(s)	246	237	27	1	5	1	1	0.12	0.32	0.07	—
16667	Shrimp gumbo	1	cup(s)	244	206	171	10	19	3	7	1.34	3.02	2.05	—
28037	Southwestern corn chowder	1	cup(s)	229	202	102	5	18	2	<1	0.12	0.12	0.20	0
25140	Split pea	1	cup(s)	165	117	85	4	19	2	<1	0.07	0.03	0.18	0
718	Split pea with ham, condensed, prepared w/water	1	cup(s)	253	207	190	10	28	2	4	1.77	1.80	0.63	—
710	Tomato, condensed, prepared w/milk	1	cup(s)	248	210	161	6	22	3	6	2.90	1.61	1.12	—
719	Tomato, condensed, prepared w/water	1	cup(s)	244	220	85	2	17	<1	2	0.37	0.44	0.95	—
726	Tomato vegetable, dehydrated, prepared w/water	1	cup(s)	253	237	56	2	10	1	1	0.38	0.30	0.08	—
28595	Turkey noodle	1	cup(s)	228	203	106	8	14	2	2	0.27	1.06	0.67	0
28051	Turkey vegetable	1	cup(s)	227	203	98	11	8	2	1	0.32	0.17	0.30	0
720	Vegetable beef, condensed, prepared w/water	1	cup(s)	244	224	78	6	10	<1	2	0.85	0.81	0.12	—
28598	Vegetable gumbo	1	cup(s)	229	168	153	4	26	3	4	0.61	2.93	0.56	0
25141	Vegetable	1	cup(s)	252	225	96	5	20	4	—	0.06	0.04	0.16	0
721	Vegetarian vegetable, condensed, prepared w/water	1	cup(s)	241	223	72	2	12	—	2	0.29	0.82	0.72	—

PAGE KEY: H–2 = Breads/Baked Goods   H–6 = Cereal/Rice/Pasta   H–10 = Fruit   H–14 = Vegetables/Legumes   H–24 = Nuts/Seeds   H–26 = Vegetarian
H–28 = Dairy   H–34 = Eggs   H–34 = Seafood   H–36 = Meats   H–40 = Poultry   H–40 = Processed meats   H–42 = Beverages   H–46 = Fats/Oils
H–48 = Sweets   H–50 = Spices/Condiments/Sauces   H–52 = Mixed foods/Soups/Sandwiches   H–58 = Fast food   H–74 = Convenience meals   H–76 = Baby foods

Chol (mg)	Calc (mg)	Iron (mg)	Magn (mg)	Pota (mg)	Sodi (mg)	Zinc (mg)	Vit A (µg)	Thia (mg)	Vit E (mg α)	Ribo (mg)	Niac (mg)	Vit B$_6$ (mg)	Fola (µg)	Vit C (mg)	Vit B$_{12}$ (µg)	Sele (µg)
1	80	2.47	62	272	479	1.25	0	0.33	2.39	0.25	6.46	0.17	43	0	<.1	—
1	68	2.11	53	239	409	1.06	—	0.27	2.02	0.21	5.45	0.15	37	<1	<.1	—
51	54	4.23	31	316	792	3.39	11	0.38	—	0.31	5.87	0.26	57	2	1	29
73	41	2.81	67	330	845	4.38	30	0.41	—	0.41	5.96	0.32	71	6	2	26
73	92	5.16	49	524	798	4.53	20	0.41	—	0.37	7.30	0.37	90	6	2	42
53	100	3.29	35	302	795	1.08	46	0.26	0.35	0.26	12.29	0.48	70	1	2	71
69	78	3.10	34	315	490	2.94	—	0.30	0.74	0.33	6.64	0.46	24	0	<1	—
70	324	3.88	51	552	2408	2.66	—	0.53	1.19	0.49	12.50	0.54	46	5	2	—
5	80	3.08	61	590	690	1.41	26	0.27	0.03	0.15	3.61	0.23	139	3	<1	8
3	85	2.15	48	421	996	1.09	48	0.09	0.80	0.03	0.59	0.04	34	2	<.1	8
5	15	1.10	5	100	952	1.54	7	0.07	0.68	0.06	1.07	0.04	20	<1	<1	7
48	289	0.80	20	341	1019	0.68	359	0.06	—	0.33	0.50	0.08	10	1	<1	7
0	10	0.51	2	210	776	0.24	0	0.01	0.05	0.07	3.35	0.02	5	0	<1	0
24	26	1.34	16	335	776	0.77	49	0.15	0.02	0.16	5.57	0.13	40	1	<1	10
7	17	0.77	5	55	1106	0.39	36	0.05	0.10	0.06	1.39	0.03	22	<1	<1	6
10	5	0.50	8	33	577	0.20	3	0.20	0.13	0.08	1.09	0.03	18	0	<.1	10
22	174	0.87	20	360	1042	0.92	62	0.10	—	0.28	0.88	0.06	30	4	<1	8
32	186	0.69	22	310	1009	0.20	114	0.07	—	0.25	0.44	0.06	7	1	<1	5
27	181	0.67	17	273	1047	0.67	179	0.07	—	0.26	0.92	0.07	7	1	1	8
10	34	0.61	2	88	986	0.63	163	0.03	—	0.06	0.82	0.02	2	<1	<.1	7
20	179	0.60	20	270	918	0.64	35	0.08	1.24	0.28	0.91	0.06	10	2	<1	4
2	46	0.51	5	100	881	0.59	15	0.05	0.95	0.09	0.72	0.01	5	1	<.1	1
1	68	1.38	17	312	784	0.74	100	0.12	1.06	0.20	3.27	0.12	37	10	<1	5
103	21	0.75	5	220	729	0.48	—	0.02	0.29	0.19	3.03	0.05	15	0	<1	—
4	203	1.63	39	412	500	1.72	454	0.17	0.53	0.38	1.15	0.15	32	10	1	8
34	29	1.89	29	382	1561	1.51	—	0.27	0.12	0.25	4.97	0.20	13	1	<1	—
<1	47	4.07	55	590	26	1.44	163	0.21	0.06	0.12	1.69	0.30	164	13	0	3
<1	64	1.86	32	254	489	0.46	174	0.15	0.35	0.13	1.36	0.09	59	7	0	9
2	27	1.63	12	188	578	0.98	56	0.03	0.34	0.04	0.82	0.10	10	4	4	9
0	68	1.76	31	273	423	0.38	138	0.10	0.23	0.10	0.69	0.07	47	12	0	4
2	34	0.92	7	313	911	0.75	118	0.05	—	0.04	0.94	0.10	36	1	0	8
0	27	1.08	26	332	267	0.87	4	0.06	0.07	0.21	2.97	0.14	18	4	<.1	4
22	186	1.49	22	300	992	0.79	57	0.07	0.45	0.24	1.03	0.13	10	3	10	13
2	69	1.29	26	430	236	0.66	34	0.07	0.02	0.12	1.02	0.20	17	12	3	4
6	51	2.32	47	434	319	0.69	114	0.20	0.01	0.15	2.42	0.23	65	17	<.1	9
0	12	0.15	5	64	849	0.05	0	0.03	0.00	0.06	0.48	0.00	2	<1	0	2
51	99	2.34	51	515	515	0.93	—	0.19	1.90	0.10	2.54	0.19	59	26	<1	—
1	65	1.10	24	374	200	0.73	46	0.08	0.09	0.14	1.65	0.22	27	37	<1	2
0	30	1.25	33	352	608	0.57	112	0.12	0.00	0.09	1.67	0.21	61	9	0	<1
8	23	2.28	48	400	1007	1.32	23	0.15	—	0.08	1.47	0.07	3	2	<1	8
17	159	1.81	22	449	744	0.30	64	0.13	1.24	0.25	1.52	0.16	17	68	<1	2
0	12	1.76	7	264	695	0.24	29	0.09	2.32	0.05	1.42	0.11	15	66	0	<1
0	8	0.63	20	104	1146	0.18	10	0.06	0.35	0.05	0.79	0.05	10	6	0	5
24	27	1.40	22	200	372	0.67	81	0.20	0.02	0.11	2.68	0.15	45	5	<1	13
20	36	1.30	22	383	328	0.90	110	0.08	0.01	0.09	3.33	0.27	21	10	<1	9
5	17	1.12	5	173	791	1.54	95	0.04	0.37	0.05	1.03	0.08	10	2	<1	4
0	52	1.90	35	313	471	0.56	15	0.17	0.58	0.07	1.59	0.16	51	18	0	4
0	41	2.45	38	688	674	0.78	118	0.12	0.00	0.13	2.37	0.27	33	23	0	5
0	22	1.08	7	210	822	0.46	116	0.05	—	0.05	0.92	0.06	10	1	0	4

TABLE H–1

Food Composition   (DA+ code is for Wadsworth Diet Analysis program)   (For purposes of calculations, use "0" for t, <1, <.1, <.01, etc.)

DA + Code	Food Description	Quantity	Measure	Wt (g)	H₂O (g)	Ener (kcal)	Prot (g)	Carb (g)	Fiber (g)	Fat (g)	Sat	Mono	Poly	Trans
	FAST FOOD													
	Arby's													
36094	Au jus sauce	1	serving(s)	85	—	5	<1	1	<.1	<.1	0.02	—	—	—
751	Beef 'n cheddar sandwich	1	item(s)	198	—	480	23	43	2	24	8.00	—	—	—
9279	Cheddar curly fries	1	serving(s)	170	—	460	6	54	4	24	6.00	—	—	—
36131	Chocolate shake	1	serving(s)	397	—	480	10	84	0	16	8.00	—	—	—
36045	Curly fries, large	1	serving(s)	198	—	620	8	78	7	30	7.00	—	—	—
36044	Curly fries, medium	1	serving(s)	128	—	400	5	50	4	20	5.00	—	—	—
9265	Fish fillet sandwich	1	item(s)	220	—	529	23	50	2	27	7.00	9.20	10.60	—
752	Ham 'n cheese sandwich	1	item(s)	170	—	340	23	35	1	13	4.50	—	—	—
36048	Homestyle fries, large	1	serving(s)	213	—	560	6	79	6	24	6.00	—	—	—
36047	Homestyle fries, medium	1	serving(s)	142	—	370	4	53	4	16	4.00	—	—	—
33465	Homestyle fries, small	1	serving(s)	113	—	300	3	42	3	13	3.50	—	—	—
9267	Italian sub sandwich	1	item(s)	312	—	780	29	49	3	53	15.00	—	—	—
36041	Market Fresh grilled chicken caesar salad w/o dressing	1	serving(s)	338	—	230	33	8	3	8	3.50	—	—	—
9291	Roast beef deluxe sandwich, light	1	item(s)	182	—	296	18	33	6	10	3.00	5.00	2.00	—
9251	Roast beef sandwich, giant	1	item(s)	228	—	480	32	41	3	23	10.00	—	—	—
9249	Roast beef sandwich, junior	1	item(s)	129	—	310	16	34	2	13	4.50	—	—	—
750	Roast beef sandwich, regular	1	item(s)	157	—	350	21	34	2	16	6.00	—	—	—
2009	Roast beef sandwich, super	1	item(s)	245	—	470	22	47	3	23	7.00	—	—	—
9269	Roast beef sub sandwich	1	item(s)	334	—	760	35	47	3	48	16.00	—	—	—
9295	Roast chicken deluxe sandwich, light	1	item(s)	194	—	260	23	33	3	5	1.00	—	—	—
9293	Roast turkey deluxe sandwich, light	1	item(s)	194	—	260	23	33	3	5	0.50	—	—	—
36132	Strawberry shake	1	serving(s)	397	—	500	11	87	0	13	8.00	—	—	—
9273	Turkey sub sandwich	1	item(s)	306	—	630	26	51	2	37	9.00	—	—	—
36130	Vanilla shake	1	serving(s)	397	—	470	10	83	0	15	7.00	—	—	—
	Auntie Anne's													
35371	Cheese dipping sauce	1	serving(s)	35	—	100	3	4	0	8	4.00	—	—	—
35353	Cinnamon sugar soft pretzel	1	item(s)	120	—	350	9	74	2	2	0.00	—	—	—
35354	Cinnamon sugar soft pretzel w/butter	1	item(s)	120	—	450	8	83	3	9	5.00	—	—	—
35372	Marinara dipping sauce	1	serving(s)	35	—	10	0	4	0	0	0.00	0.00	0.00	0
35357	Original soft pretzel	1	item(s)	120	—	340	10	72	3	1	0.00	—	—	—
35358	Original soft pretzel w/butter	1	item(s)	120	—	370	10	72	3	4	2.00	—	—	—
35359	Parmesan herb soft pretzel	1	item(s)	120	—	390	11	74	4	5	2.50	—	—	—
35360	Parmesan herb soft pretzel w/butter	1	item(s)	120	—	440	10	72	9	13	7.00	—	—	—
35361	Sesame soft pretzel	1	item(s)	120	—	350	11	63	3	6	1.00	—	—	—
35362	Sesame soft pretzel w/butter	1	item(s)	120	—	410	12	64	7	12	4.00	—	—	—
35364	Sour cream & onion soft pretzel	1	item(s)	120	—	310	9	66	2	1	0.00	—	—	—
35366	Sour cream & onion soft pretzel w/butter	1	item(s)	120	—	340	9	66	2	5	3.00	—	—	—
35373	Sweet mustard dipping sauce	1	serving(s)	35	—	60	1	8	0	2	1.00	—	—	—
35367	Whole wheat soft pretzel	1	item(s)	120	—	350	11	72	7	2	0.00	—	—	—
35368	Whole wheat soft pretzel w/butter	1	item(s)	120	—	370	11	72	7	5	1.50	—	—	—
	Boston Market													
34975	Bbq baked beans	¾	cup(s)	201	—	270	8	48	12	5	2.00	—	—	—
34976	Black beans & rice	1	cup(s)	227	—	300	8	45	5	10	1.50	—	—	—
34978	Butternut squash	¾	cup(s)	193	—	150	2	25	6	6	4.00	—	—	—
35006	Caesar side salad	1	serving(s)	119	—	300	5	13	1	26	4.50	—	—	—
34979	Chicken gravy	1	ounce(s)	28	—	15	0	2	0	1	0.00	—	—	—
34973	Chicken pot pie	1	item(s)	425	—	750	26	57	2	46	14.00	—	—	—
35007	Cole slaw	¾	cup(s)	184	—	300	2	30	3	19	3.00	—	—	—
35057	Cornbread	1	item(s)	68	—	200	3	33	1	6	1.50	—	—	—
35008	Cranberry walnut relish	¾	cup(s)	210	—	350	3	75	3	5	0.00	—	—	—
34980	Creamed spinach	¾	cup(s)	181	—	260	9	11	2	20	13.00	—	—	—
34981	Glazed carrots	¾	cup(s)	153	—	280	1	35	4	15	3.00	—	—	—
34983	Green bean casserole	¾	cup(s)	170	—	80	1	9	2	5	1.50	—	—	—
34982	Green beans	¾	cup(s)	85	—	70	1	6	2	4	0.50	—	—	—
34967	Half chicken, w/skin	1	item(s)	277	—	590	70	4	0	33	10.00	—	—	—
34984	Homestyle mashed potatoes	¾	cup(s)	173	—	210	4	30	2	9	5.00	—	—	—
34985	Homestyle mashed potatoes & gravy	1	cup(s)	201	—	230	4	32	3	9	5.00	—	—	—
34969	Honey glazed ham	5	ounce(s)	142	—	210	24	10	0	8	3.00	—	—	—
34988	Hot cinnamon apples	¾	cup(s)	181	—	250	0	56	3	5	0.50	—	—	—
34989	Macaroni & cheese	¾	cup(s)	192	—	280	13	33	1	11	6.00	—	—	—
34970	Meatloaf	5	ounce(s)	142	—	282	20	15	1	17	7.28	—	—	—
35012	Old-fashioned potato salad	¾	cup(s)	150	—	200	3	22	2	12	2.00	—	—	—

PAGE KEY: H–2 = Breads/Baked Goods H–6 = Cereal/Rice/Pasta H–10 = Fruit H–14 = Vegetables/Legumes H–24 = Nuts/Seeds H–26 = Vegetarian
H–28 = Dairy H–34 = Eggs H–34 = Seafood H–36 = Meats H–40 = Poultry H–40 = Processed meats H–42 = Beverages H–46 = Fats/Oils
H–48 = Sweets H–50 = Spices/Condiments/Sauces H–52 = Mixed foods/Soups/Sandwiches H–58 = Fast food H–74 = Convenience meals H–76 = Baby foods

Chol (mg)	Calc (mg)	Iron (mg)	Magn (mg)	Pota (mg)	Sodi (mg)	Zinc (mg)	Vit A (µg)	Thia (mg)	Vit E (mg α)	Ribo (mg)	Niac (mg)	Vit B$_6$ (mg)	Fola (µg)	Vit C (mg)	Vit B$_{12}$ (µg)	Sele (µg)
0	0	0.00	—	—	386	—	0	—	—	—	—	—	—	0	—	—
90	100	3.60	—	—	1240	—	0	—	—	—	—	—	—	1	—	—
5	60	1.80	—	—	1290	—	0	—	—	—	—	—	—	15	—	—
45	500	0.72	—	—	370	—	38	—	—	—	—	—	—	2	—	—
0	0	2.70	—	—	1540	—	0	—	—	—	—	—	—	21	—	—
0	0	1.80	—	—	990	—	0	—	—	—	—	—	—	15	—	—
43	90	3.78	—	450	864	—	10	0.35	—	0.31	5.60	—	—	1	—	—
90	150	2.70	—	—	1450	—	20	—	—	—	—	—	—	1	—	—
0	0	1.80	—	—	1070	—	0	—	—	—	—	—	—	30	—	—
0	0	1.08	—	—	710	—	0	—	—	—	—	—	—	21	—	—
0	0	0.72	—	—	570	—	0	—	—	—	—	—	—	15	—	—
120	250	2.70	—	—	2440	—	—	—	—	—	—	—	—	2	—	—
80	200	1.80	—	—	920	—	—	—	—	—	—	—	—	42	—	—
42	130	4.50	—	392	826	—	40	0.27	—	0.49	8.40	—	—	8	—	—
110	60	5.40	—	—	1440	—	0	—	—	—	—	—	—	0	—	—
70	60	2.70	—	—	740	—	0	—	—	—	—	—	—	0	—	—
85	60	3.60	—	—	950	—	0	—	—	—	—	—	—	0	—	—
85	80	3.60	—	—	1130	—	40	—	—	—	—	—	—	1	—	—
130	300	4.50	—	—	2230	—	40	—	—	—	—	—	—	4	—	—
40	100	2.70	—	—	1010	—	—	—	—	—	—	—	—	2	—	—
40	80	1.80	—	—	980	—	—	—	—	—	—	—	—	1	—	—
15	350	0.36	—	—	340	—	36	—	—	—	—	—	—	1	—	—
100	200	0.36	—	—	2170	—	—	—	—	—	—	—	—	2	—	—
45	500	1.08	—	—	360	—	39	—	—	—	—	—	—	2	—	—
10	100	0.00	—	—	510	—	—	—	—	—	—	—	—	0	—	—
0	20	1.98	—	—	410	—	0	—	—	—	—	—	—	0	—	—
25	30	2.34	—	—	430	—	—	—	—	—	—	—	—	0	—	—
0	0	0.00	—	—	180	—	0	—	—	—	—	—	—	0	—	—
0	30	2.34	—	—	900	—	0	—	—	—	—	—	—	0	—	—
10	30	2.16	—	—	930	—	—	—	—	—	—	—	—	0	—	—
10	80	1.80	—	—	780	—	—	—	—	—	—	—	—	1	—	—
30	60	1.80	—	—	660	—	—	—	—	—	—	—	—	1	—	—
0	20	2.88	—	—	840	—	0	—	—	—	—	—	—	0	—	—
15	20	2.70	—	—	860	—	—	—	—	—	—	—	—	0	—	—
0	30	1.98	—	—	920	—	—	—	—	—	—	—	—	0	—	—
10	40	2.16	—	—	930	—	—	—	—	—	—	—	—	0	—	—
40	0	0.00	—	—	120	—	0	—	—	—	—	—	—	0	—	—
0	30	1.98	—	—	1100	—	0	—	—	—	—	—	—	0	—	—
10	30	2.34	—	—	1120	—	—	—	—	—	—	—	—	0	—	—
0	100	3.60	—	—	540	—	42	—	—	—	—	—	—	6	—	—
0	40	1.80	—	—	1050	—	0	—	—	—	—	—	—	4	—	—
20	80	1.08	—	—	560	—	1150	—	—	—	—	—	—	30	—	—
15	100	0.72	—	—	690	—	—	—	—	—	—	—	—	9	—	—
0	0	0.00	—	—	180	—	0	—	—	—	—	—	—	0	—	—
110	40	4.50	—	—	1530	—	—	—	—	—	—	—	—	1	—	—
20	60	0.72	—	—	540	—	108	—	—	—	—	—	—	36	—	—
25	0	1.08	—	—	390	—	0	—	—	—	—	—	—	0	—	—
0	0	5.40	—	—	0	—	0	—	—	—	—	—	—	0	—	—
55	250	2.70	—	—	740	—	—	—	—	—	—	—	—	9	—	—
0	40	1.08	—	—	80	—	1000	—	—	—	—	—	—	1	—	—
5	20	0.72	—	—	670	—	—	—	—	—	—	—	—	2	—	—
0	40	0.36	—	—	250	—	30	—	—	—	—	—	—	5	—	—
290	0	2.70	—	—	1010	—	0	—	—	—	—	—	—	0	—	—
25	40	0.36	—	—	590	—	53	—	—	—	—	—	—	15	—	—
25	60	0.36	—	—	780	—	—	—	—	—	—	—	—	15	—	—
75	0	1.08	—	—	1460	—	0	—	—	—	—	—	—	0	—	—
0	20	0.36	—	—	45	—	—	—	—	—	—	—	—	0	—	—
30	300	1.44	—	—	890	—	—	—	—	—	—	—	—	0	—	—
68	91	2.46	—	—	592	—	—	—	—	—	—	—	—	1	—	—
15	60	1.08	—	—	450	—	0	—	—	—	—	—	—	6	—	—

TABLE H–1

Food Composition

(DA+ code is for Wadsworth Diet Analysis program)   (For purposes of calculations, use "0" for t, <1, <.1, <.01, etc.)

DA + Code	Food Description	Quantity	Measure	Wt (g)	H₂O (g)	Ener (kcal)	Prot (g)	Carb (g)	Fiber (g)	Fat (g)	Sat	Mono	Poly	Trans
	FAST FOOD—Continued													
34965	Quarter chicken, dark meat, no skin	1	item(s)	95	—	190	22	1	0	10	3.00	—	—	—
34966	Quarter chicken, dark meat, w/skin	1	item(s)	125	—	320	30	2	0	21	6.00	—	—	—
34963	Quarter chicken, white meat, no skin or wing	1	item(s)	140	—	170	33	2	0	4	1.00	—	—	—
34964	Quarter chicken, white meat, w/skin & wing	1	item(s)	152	—	280	40	2	0	12	3.50	—	—	—
34993	Rice pilaf	1	cup(s)	137	—	140	2	24	1	4	0.50	—	—	—
34968	Rotisserie turkey breast, skinless	5	ounce(s)	142	—	170	36	3	0	1	0.00	—	—	—
34998	Savory stuffing	1	cup(s)	132	—	190	4	27	2	8	1.50	—	—	—
34999	Squash casserole	¾	cup(s)	187	—	330	7	20	3	24	13.00	—	—	—
35003	Steamed vegetables	1	cup(s)	102	—	30	2	6	2	0	0.00	—	—	0
35004	Sweet potato casserole	¾	cup(s)	181	—	280	3	39	2	13	4.50	—	—	—
35005	Whole kernel corn	¾	cup(s)	146	—	180	5	30	2	4	0.50	—	—	—
	Burger King													
29731	Biscuit with sausage, egg, & cheese	1	item(s)	189	—	650	20	38	1	46	14.00	—	—	1
3739	BK Broiler chicken sandwich	1	item(s)	258	—	550	30	52	3	25	5.00	—	—	—
14249	Cheeseburger	1	item(s)	133	—	360	19	31	2	17	8.00	—	—	0.50
14251	Chicken sandwich	1	item(s)	224	—	660	25	53	3	39	8.00	—	—	2.20
3808	Chicken Tenders, 8 pieces	1	serving(s)	123	—	340	22	20	1	19	5.00	—	—	3.50
14259	Chocolate shake, small	1	item(s)	333	—	620	12	72	2	32	21.00	—	—	0
29732	Croissanwich w/sausage & cheese	1	item(s)	107	—	420	14	23	1	31	11.00	—	—	2
14261	Croissanwich w/sausage, egg, & cheese	1	item(s)	157	—	520	19	24	1	39	14.00	—	—	1.93
3809	Double cheeseburger	1	item(s)	189	—	540	32	32	2	31	15.00	—	—	1.50
14244	Double Whopper	1	item(s)	374	—	980	52	52	4	62	22.00	—	—	2
14245	Double Whopper w/cheese	1	item(s)	399	—	1070	57	53	4	70	27.00	—	—	2.50
14250	Fish Fillet sandwich	1	item(s)	185	—	520	18	44	2	30	8.00	—	—	1.12
14255	French fries, medium, salted	1	item(s)	117	—	360	4	46	4	18	5.00	—	—	4.50
14262	French toast sticks	1	serving(s)	112	—	390	6	46	2	20	4.50	—	—	4.50
14248	Hamburger	1	item(s)	121	—	310	17	31	2	13	5.00	—	—	0.50
14263	Hash brown rounds, small	1	serving(s)	75	—	230	2	23	2	15	4.00	—	—	5.0
14256	Onion rings, medium	1	serving(s)	91	—	320	4	40	3	16	4.00	—	—	3.50
39000	Tendercrisp chicken sandwich	1	item(s)	310	—	810	28	72	6	47	8.00	—	—	4.28
14258	Vanilla shake, small	1	item(s)	305	—	560	11	56	1	32	21.00	—	—	0
1736	Whopper	1	item(s)	291	—	710	31	52	4	43	13.00	—	—	1
14243	Whopper w/cheese	1	item(s)	316	—	800	36	53	4	50	18.00	—	—	2
	Carl's Jr													
10801	Carl's Catch fish sandwich	1	item(s)	201	—	530	18	55	2	28	7.00	—	1.89	—
10862	Carl's Famous Star hamburger	1	item(s)	254	—	590	24	50	3	32	9.00	—	—	—
10866	Charboiled chicken salad-to-go	1	item(s)	350	—	200	25	12	4	7	3.00	—	1.02	—
10855	Charboiled Sante Fe chicken sandwich	1	item(s)	220	—	540	28	37	2	31	8.00	—	—	—
10790	Chicken stars (6 pieces)	6	item(s)	90	—	260	13	14	1	16	4.50	—	1.71	—
34864	Chocolate shake, small	1	item(s)	595	—	530	14	96	0	10	7.00	—	—	—
10797	Crisscut fries	1	serving(s)	139	—	410	5	43	4	24	5.00	—	—	—
10799	Double western bacon cheeseburger	1	item(s)	308	—	920	51	65	3	50	21.00	—	6.55	—
34855	Famous bacon cheeseburger	1	item(s)	279	—	700	31	51	3	41	13.00	—	—	—
14238	French fries, small	1	serving(s)	92	—	290	5	37	3	14	3.00	—	—	—
10798	French toast dips w/o syrup	1	serving(s)	105	—	370	6	42	1	20	2.50	—	1.35	—
34856	Hamburger	1	item(s)	119	—	280	14	36	1	9	3.50	—	—	—
10802	Onion rings	1	serving(s)	127	—	430	7	53	3	22	5.00	—	0.84	—
38925	Six Dollar burger	1	item(s)	539	—	1000	39	72	6	82	25.00	—	—	—
34858	Spicy chicken sandwich	1	item(s)	198	—	480	14	47	2	26	5.00	—	—	—
34867	Strawberry shake, small	1	item(s)	595	—	510	14	91	0	10	7.00	—	—	—
10865	Super Star hamburger	1	item(s)	345	—	790	41	51	3	47	15.00	—	—	—
10818	Vanilla shake, small	1	item(s)	595	—	470	15	78	0	11	7.00	—	—	—
10770	Western bacon cheeseburger	1	item(s)	225	—	660	31	64	3	30	12.00	—	4.85	—
	Chick Fil-A													
38746	Biscuit w/bacon, egg, & cheese	1	item(s)	155	—	430	16	38	1	24	9.00	—	—	2.85
38747	Biscuit w/egg	1	item(s)	135	—	340	11	38	1	16	4.50	—	—	3
38748	Biscuit w/egg & cheese	1	item(s)	148	—	390	13	38	1	21	7.00	—	—	2.98
38753	Biscuit w/gravy	1	item(s)	191	—	310	5	44	1	13	3.50	—	—	3.98
38752	Biscuit w/sausage, egg, & cheese	1	item(s)	189	—	540	18	43	1	33	13.00	—	—	2.67
38741	Biscuit, plain	1	item(s)	78	—	260	4	38	1	11	2.50	—	—	2.97
38771	Carrot & raisin salad	1	item(s)	91	—	130	1	22	2	5	1.00	—	—	0
38761	Chargrilled chicken cool wrap	1	item(s)	245	—	380	29	54	3	6	3.00	—	—	0
38766	Chargrilled chicken garden salad	1	item(s)	275	—	180	22	9	3	6	3.00	—	—	0
38758	Chargrilled chicken sandwich	1	item(s)	157	—	280	26	30	1	7	1.50	—	—	0

PAGE KEY: H–2 = Breads/Baked Goods   H–6 = Cereal/Rice/Pasta   H–10 = Fruit   H–14 = Vegetables/Legumes   H–24 = Nuts/Seeds   H–26 = Vegetarian
H–28 = Dairy   H–34 = Eggs   H–34 = Seafood   H–36 = Meats   H–40 = Poultry   H–40 = Processed meats   H–42 = Beverages   H–46 = Fats/Oils
H–48 = Sweets   H–50 = Spices/Condiments/Sauces   H–52 = Mixed foods/Soups/Sandwiches   H–58 = Fast food   H–74 = Convenience meals   H–76 = Baby foods

Chol (mg)	Calc (mg)	Iron (mg)	Magn (mg)	Pota (mg)	Sodi (mg)	Zinc (mg)	Vit A (µg)	Thia (mg)	Vit E (mg α)	Ribo (mg)	Niac (mg)	Vit B$_6$ (mg)	Fola (µg)	Vit C (mg)	Vit B$_{12}$ (µg)	Sele (µg)
115	0	1.08	—	—	440	—	0	—	—	—	—	—	—	0	—	—
155	0	1.80	—	—	500	—	0	—	—	—	—	—	—	0	—	—
85	0	0.72	—	—	480	—	0	—	—	—	—	—	—	0	—	—
135	0	1.08	—	—	510	—	0	—	—	—	—	—	—	0	—	—
0	20	1.08	—	—	520	—	—	—	—	—	—	—	—	4	—	—
100	20	1.80	—	—	850	—	0	—	—	—	—	—	—	0	—	—
5	40	1.44	—	—	620	—	—	—	—	—	—	—	—	2	—	—
70	200	0.72	—	—	1110	—	—	—	—	—	—	—	—	5	—	—
0	40	0.35	—	—	135	—	389	—	—	—	—	—	—	18	—	—
10	40	1.08	—	—	190	—	—	—	—	—	—	—	—	9	—	—
0	0	0.36	—	—	170	—	20	—	—	—	—	—	—	5	—	—
190	150	2.70	—	—	1600	—	90	—	—	—	—	—	—	0	—	—
105	60	3.60	—	—	1110	—	—	0.46	—	0.23	10.50	—	—	6	—	—
50	150	3.60	—	—	790	—	63	0.25	—	0.32	4.18	—	—	1	—	—
70	80	2.70	—	—	1330	—	—	0.47	—	0.30	9.59	—	—	0	—	—
50	20	0.72	—	—	840	—	—	0.14	—	0.12	10.93	—	—	0	—	—
95	350	1.08	—	—	310	—	42	0.11	—	0.56	0.24	—	—	0	—	—
45	100	3.60	—	—	840	—	—	—	—	—	—	—	—	0	—	—
210	300	4.50	—	—	1090	—	140	0.36	—	0.42	4.35	—	—	0	—	—
100	250	4.50	—	—	1050	—	100	0.26	—	0.45	6.37	—	—	1	—	—
160	150	9.00	—	—	1070	—	—	0.40	—	0.60	11.08	—	—	9	—	—
185	300	9.00	—	—	1500	—	—	0.40	—	0.67	11.07	—	—	9	—	—
55	150	2.70	—	—	840	—	14	—	—	—	—	—	—	1	—	—
0	20	0.72	—	—	640	—	0	0.16	—	0.48	2.32	—	—	9	—	—
0	60	1.80	—	—	440	—	0	0.19	—	0.22	2.86	—	—	0	—	—
40	76	3.60	—	—	580	—	9	0.25	—	0.29	4.26	—	—	1	—	—
0	0	0.36	—	—	450	—	0	0.11	—	0.07	2.11	—	—	1	—	—
0	97	0.00	—	—	460	—	0	0.14	—	0.09	2.33	—	—	0	—	—
60	80	4.50	—	—	1800	—	—	—	—	—	—	—	—	9	—	—
95	300	0.36	—	—	220	—	39	0.11	—	0.64	0.22	—	—	0	—	—
85	150	6.30	—	—	980	—	52	0.39	—	0.44	7.33	—	—	9	—	—
110	250	6.30	—	—	1420	—	157	0.39	—	0.51	7.31	—	—	9	—	—
80	150	1.80	—	—	1030	—	60	—	—	—	—	—	—	2	—	—
70	100	4.50	—	—	910	—	—	—	—	—	—	—	—	6	—	—
75	150	1.80	—	—	440	—	—	—	—	—	—	—	—	5	—	—
95	200	2.70	—	—	1210	—	—	—	—	—	—	—	—	6	—	—
40	20	1.08	—	—	480	—	0	—	—	—	—	—	—	0	—	—
45	600	1.08	—	—	350	—	0	—	—	—	—	—	—	0	—	—
0	20	1.80	—	—	950	—	0	—	—	—	—	—	—	12	—	—
155	300	7.20	—	—	1770	—	—	—	—	—	—	—	—	1	—	—
95	200	5.40	—	—	1310	—	102	—	—	—	—	—	—	6	—	—
0	0	1.08	—	—	180	—	0	—	—	—	—	—	—	21	—	—
0	40	1.08	—	—	430	—	0	0.26	—	0.24	2.00	—	—	0	—	—
35	80	2.70	—	—	480	—	0	—	—	—	—	—	—	1	—	—
0	20	0.72	—	—	700	—	0	—	—	—	—	—	—	4	—	—
135	350	5.40	—	—	1690	—	—	—	—	—	—	—	—	21	—	—
40	100	2.70	—	—	1220	—	—	—	—	—	—	—	—	6	—	—
45	600	0.00	—	—	330	—	0	—	—	—	—	—	—	0	—	—
130	100	7.20	—	—	980	—	—	—	—	—	—	—	—	9	—	—
50	600	0.00	—	—	350	—	0	—	—	—	—	—	—	0	—	—
85	200	5.40	—	—	1410	—	40	—	—	—	—	—	—	1	—	—
265	150	3.60	—	—	1070	—	—	—	—	—	—	—	—	0	—	—
245	80	2.70	—	—	740	—	—	—	—	—	—	—	—	0	—	—
260	150	2.70	—	—	960	—	—	—	—	—	—	—	—	0	—	—
5	60	1.80	—	—	930	—	—	—	—	—	—	—	—	0	—	—
280	150	3.60	—	—	1030	—	—	—	—	—	—	—	—	0	—	—
0	60	1.80	—	—	670	—	0	—	—	—	—	—	—	0	—	—
0	20	0.36	—	—	90	—	—	—	—	—	—	—	—	4	—	—
70	200	2.70	—	—	1060	—	—	—	—	—	—	—	—	6	—	—
70	150	0.72	—	—	660	—	—	—	—	—	—	—	—	30	—	—
70	80	1.80	—	—	980	—	0	—	—	—	—	—	—	2	—	—

TABLE H-1

Food Composition

(DA+ code is for Wadsworth Diet Analysis program)   (For purposes of calculations, use "0" for t, <1, <.1, <.01, etc.)

DA + Code	Food Description	Quantity	Measure	Wt (g)	H₂O (g)	Ener (kcal)	Prot (g)	Carb (g)	Fiber (g)	Fat (g)	Sat	Mono	Poly	Trans
	FAST FOOD—Continued													
38759	Chargrilled deluxe chicken sandwich	1	item(s)	195	—	290	27	31	2	7	1.50	—	—	0
38742	Chicken biscuit	1	item(s)	137	—	400	16	43	2	18	4.50	—	—	2.83
38743	Chicken biscuit w/cheese	1	item(s)	151	—	450	19	43	2	23	7.00	—	—	2.85
38762	Chicken caesar wrap	1	item(s)	227	—	460	36	52	2	10	6.00	—	—	0
38757	Chicken deluxe sandwich	1	item(s)	208	—	420	28	39	2	16	3.50	—	—	0
38764	Chicken salad sandwich	1	item(s)	153	—	350	20	32	5	15	3.00	—	—	0
38756	Chicken sandwich	1	item(s)	170	—	410	28	38	1	15	3.50	—	—	0
38768	Chick-n-Strip salad	1	item(s)	331	—	390	34	22	4	18	5.00	—	—	0
38763	Chick-n-Strips	4	item(s)	127	—	290	29	14	1	13	2.50	—	—	0
38770	Coleslaw	1	item(s)	105	—	210	1	14	2	17	2.50	—	—	0
38755	Hash browns	1	serving(s)	84	—	170	2	20	2	9	4.50	—	—	1
38765	Hearty breast of soup	1	cup(s)	241	—	140	8	18	1	4	1.00	—	—	0
38778	Icedream, small cone	1	item(s)	135	—	160	4	28	0	4	2.00	—	—	0
38774	Icedream, small cup	1	serving(s)	213	—	230	5	38	0	6	3.50	—	—	0
38775	Lemonade	1	cup(s)	255	—	170	0	41	0	1	0.00	—	—	0
38776	Lemonade, diet	1	cup(s)	255	—	25	0	5	0	0	0.00	0.00	0.00	0
38777	Nuggets	8	item(s)	113	—	260	26	12	1	12	2.50	—	—	0
38769	Side salad	1	item(s)	108	—	60	3	4	2	3	1.50	—	—	0
38767	Southwest chargrilled salad	1	item(s)	303	—	240	22	17	5	8	3.50	—	—	0
38772	Waffle potato fries, small, salted	1	serving(s)	85	—	280	3	37	5	14	5.00	—	—	1.50
	Cinnabon													
39569	Caramel Pecanbon	1	item(s)	272	—	1100	16	141	8	56	10.00	—	—	5
39572	Caramellata Chill w/whipped cream	16	fluid ounce(s)	480	—	406	10	61	0	14	8.00	—	—	—
39571	Cinnapoppers	1	serving(s)	74	—	368	4	41	2	21	11.00	—	—	1
39567	Classic roll	1	item(s)	221	—	813	15	117	4	32	8.00	—	—	5
39568	Minibon	1	item(s)	92	—	339	6	49	2	13	3.00	—	—	2
39573	Mochalatta chill w/whipped cream	16	fluid ounce(s)	480	—	362	9	55	0	13	8.00	—	—	—
39570	Stix	5	item(s)	85	—	379	6	41	1	21	6.00	—	—	4
	Dairy Queen													
1466	Banana split	1	item(s)	369	—	510	8	96	3	12	8.00	3.00	0.50	0
38552	Brownie Earthquake	1	serving(s)	304	—	740	10	112	0	27	16.00	—	—	3
38561	Chocolate chip cookie dough blizzard, small	1	item(s)	319	—	720	12	105	0	28	14.00	—	—	2.50
1464	Chocolate malt, small	1	item(s)	418	—	650	15	111	0	16	10.00	—	—	0.50
38541	Chocolate shake, small	1	item(s)	397	—	560	13	93	1	15	10.00	—	—	0.50
17257	Chocolate soft serve	½	cup(s)	94	—	150	4	22	0	5	3.50	—	—	0
1463	Chocolate sundae, small	1	item(s)	163	—	280	5	49	0	7	4.50	1.00	1.00	0
1462	Dipped cone, small	1	item(s)	156	—	340	6	42	1	17	9.00	4.00	3.00	1
38555	Oreo cookies blizzard, small	1	item(s)	283	—	570	11	83	1	21	10.00	—	—	2.50
38547	Royal Treats Peanut Buster parfait	1	item(s)	305	—	730	16	99	2	31	17.00	—	—	0
17256	Vanilla soft serve	½	cup(s)	94	—	140	3	22	0	5	3.00	—	—	0
	Domino's													
31606	Barbeque wings	1	item(s)	25	—	50	6	2	<1	2	0.65	—	—	—
31604	Breadsticks	1	item(s)	37	—	116	3	18	1	4	0.79	—	—	—
37551	Buffalo chicken kickers	1	item(s)	24	14	47	4	3	<1	2	0.39	—	—	—
37548	Cinnastix	1	item(s)	32	8	122	2	15	1	6	1.15	—	—	—
	Classic hand tossed pizza													
31573	America's favorite feast, 12"	2	slice(s)	205	99	508	22	57	4	22	9.20	—	—	—
31574	America's favorite feast, 14"	2	slice(s)	283	138	697	30	79	5	30	12.70	—	—	—
37543	Bacon cheeseburger feast, 12"	2	slice(s)	198	60	549	25	55	3	26	11.62	—	—	—
37545	Bacon cheeseburger feast, 14"	2	slice(s)	275	121	762	35	75	4	36	16.10	—	—	—
37546	Barbeque feast, 12"	2	slice(s)	192	85	506	22	62	3	20	9.08	—	—	—
37547	Barbeque feast, 14"	2	slice(s)	262	115	691	30	85	4	27	12.24	—	—	—
31569	Cheese, 12"	2	slice(s)	159	—	375	15	55	3	11	4.81	—	—	—
31570	Cheese, 14"	2	slice(s)	219	—	516	21	75	4	15	6.72	—	—	—
37538	Deluxe feast, 12"	2	slice(s)	201	102	465	20	57	3	18	7.66	—	—	—
37540	Deluxe feast, 14"	2	slice(s)	273	138	627	26	78	5	24	10.20	—	—	—
31685	Deluxe, 12"	2	slice(s)	213	—	465	20	57	3	18	7.65	—	—	—
31694	Deluxe, 14"	2	slice(s)	273	—	627	26	78	5	24	10.20	—	—	—
31686	Extravaganzza, 12"	2	slice(s)	245	127	576	27	59	4	27	11.56	—	—	—
31695	Extravaganzza, 14"	2	slice(s)	329	171	773	36	88	5	36	15.42	—	—	—
31575	Hawaiian feast, 12"	2	slice(s)	204	105	450	21	58	3	16	7.20	—	—	—
31576	Hawaiian feast, 14"	2	slice(s)	283	147	623	29	80	5	22	10.09	—	—	—
31687	Meatzza, 12"	2	slice(s)	213	—	560	26	57	3	26	11.40	—	—	—
31696	Meatzza, 14"	2	slice(s)	293	139	753	35	78	5	34	15.24	—	—	—

PAGE KEY: H–2 = Breads/Baked Goods H–6 = Cereal/Rice/Pasta H–10 = Fruit H–14 = Vegetables/Legumes H–24 = Nuts/Seeds H–26 = Vegetarian
H–28 = Dairy H–34 = Eggs H–34 = Seafood H–36 = Meats H–40 = Poultry H–40 = Processed meats H–42 = Beverages H–46 = Fats/Oils
H–48 = Sweets H–50 = Spices/Condiments/Sauces H–52 = Mixed foods/Soups/Sandwiches H–58 = Fast food H–74 = Convenience meals H–76 = Baby foods

Chol (mg)	Calc (mg)	Iron (mg)	Magn (mg)	Pota (mg)	Sodi (mg)	Zinc (mg)	Vit A (µg)	Thia (mg)	Vit E (mg α)	Ribo (mg)	Niac (mg)	Vit B$_6$ (mg)	Fola (µg)	Vit C (mg)	Vit B$_{12}$ (µg)	Sele (µg)
70	80	1.80	—	—	990	—	—	—	—	—	—	—	—	5	—	—
30	60	2.70	—	—	1200	—	0	—	—	—	—	—	—	0	—	—
45	150	2.70	—	—	1430	—	—	—	—	—	—	—	—	0	—	—
80	500	2.70	—	—	1390	—	—	—	—	—	—	—	—	1	—	—
60	100	2.70	—	—	1300	—	—	—	—	—	—	—	—	2	—	—
65	150	1.80	—	—	880	—	—	—	—	—	—	—	—	0	—	—
60	100	2.70	—	—	1300	—	—	—	—	—	—	—	—	0	—	—
80	200	0.36	—	—	860	—	—	—	—	—	—	—	—	30	—	—
65	20	0.36	—	—	730	—	—	—	—	—	—	—	—	1	—	—
20	40	0.36	—	—	180	—	—	—	—	—	—	—	—	27	—	—
10	0	0.72	—	—	350	—	—	—	—	—	—	—	—	0	—	—
25	40	1.08	—	—	900	—	—	—	—	—	—	—	—	0	—	—
15	100	0.36	—	—	80	—	—	—	—	—	—	—	—	0	—	—
25	150	0.00	—	—	100	—	—	—	—	—	—	—	—	0	—	—
0	0	0.36	—	—	10	—	0	—	—	—	—	—	—	15	—	—
0	0	0.36	—	—	5	—	0	—	—	—	—	—	—	15	—	—
70	40	1.08	—	—	1090	—	0	—	—	—	—	—	—	0	—	—
10	100	0.00	—	—	75	—	—	—	—	—	—	—	—	15	—	—
60	200	1.08	—	—	770	—	—	—	—	—	—	—	—	24	—	—
15	20	0.00	—	—	105	—	0	—	—	—	—	—	—	21	—	—
63	—	—	—	—	600	—	—	—	—	—	—	—	—	—	—	—
46	—	—	—	—	187	—	—	—	—	—	—	—	—	—	—	—
62	—	—	—	—	104	—	—	—	—	—	—	—	—	—	—	—
67	—	—	—	—	801	—	—	—	—	—	—	—	—	—	—	—
27	—	—	—	—	337	—	—	—	—	—	—	—	—	—	—	—
46	100	0.00	—	—	252	—	—	—	—	—	—	—	—	0	—	—
16	—	—	—	—	413	—	—	—	—	—	—	—	—	—	—	—
30	250	1.80	—	860	180	—	—	0.15	—	0.60	0.20	—	—	15	—	—
50	250	1.80	—	—	350	—	—	—	—	—	—	—	—	0	—	—
50	350	2.70	—	—	370	—	—	—	—	—	—	—	—	1	—	—
55	450	1.80	—	—	370	—	—	—	—	—	—	—	—	2	—	—
50	450	1.44	—	—	280	—	—	0.12	—	—	—	—	—	2	—	—
15	100	0.72	—	—	75	—	—	—	—	—	—	—	—	0	—	—
20	200	1.08	—	278	140	—	—	0.06	—	0.24	0.20	—	—	0	—	—
20	200	1.08	—	290	130	—	—	0.06	—	0.26	0.20	—	—	1	—	—
40	350	2.70	—	—	430	—	—	—	—	—	—	—	—	1	—	—
35	300	1.80	—	—	400	—	—	—	—	—	—	—	—	1	—	—
15	150	0.72	—	—	70	—	150	—	—	—	—	—	—	0	—	—
26	6	0.32	—	—	175	—	—	—	—	—	—	—	—	<.1	—	—
0	<.1	0.87	—	—	152	—	—	—	—	—	—	—	—	6	—	—
9	3	0.00	—	—	163	—	—	—	—	—	—	—	—	0	—	—
0	6	0.70	—	—	110	—	—	—	—	—	—	—	—	<.1	—	—
49	202	3.70	—	—	1221	—	—	—	—	—	—	—	—	1	—	—
68	281	5.10	—	—	1685	—	—	—	—	—	—	—	—	1	—	—
60	293	3.56	—	—	1274	—	—	—	—	—	—	—	—	0	—	—
84	395	4.96	—	—	1809	—	—	—	—	—	—	—	—	0	—	—
46	—	—	—	—	1206	—	—	—	—	—	—	—	—	—	—	—
63	393	4.42	—	—	1672	—	—	—	—	—	—	—	—	2	—	—
23	187	2.99	—	—	776	—	131	—	—	—	—	—	—	0	—	—
32	261	4.13	—	—	1080	—	184	—	—	—	—	—	—	0	—	—
40	199	3.56	—	—	1063	—	—	—	—	—	—	—	—	1	—	—
53	276	4.84	—	—	1432	—	—	—	—	—	—	—	—	2	—	—
40	199	3.56	—	—	1063	—	—	—	—	—	—	—	—	1	—	—
53	276	4.85	—	—	1432	—	—	—	—	—	—	—	—	2	—	—
60	290	4.08	—	—	1348	—	—	—	—	—	—	—	—	1	—	—
89	403	5.48	—	—	1780	—	—	—	—	—	—	—	—	2	—	—
41	274	3.30	—	—	1102	—	—	—	—	—	—	—	—	2	—	—
57	384	4.57	—	—	1544	—	—	—	—	—	—	—	—	3	—	—
344	282	3.71	—	—	1463	—	—	—	—	—	—	—	—	<1	—	—
85	393	5.04	—	—	1947	—	—	—	—	—	—	—	—	<1	—	—

TABLE H-1

Food Composition

(DA+ code is for Wadsworth Diet Analysis program)   (For purposes of calculations, use "0" for t, <1, <.1, <.01, etc.)

DA + Code	Food Description	Quantity	Measure	Wt (g)	H₂O (g)	Ener (kcal)	Prot (g)	Carb (g)	Fiber (g)	Fat (g)	Sat	Mono	Poly	Trans
	FAST FOOD—Continued													
31571	Pepperoni feast, extra pepperoni & cheese, 12"	2	slice(s)	196	87	534	24	56	3	25	10.92	—	—	—
31572	Pepperoni feast, extra pepperoni & cheese, 14"	2	slice(s)	270	121	732	33	77	4	34	15.00	—	—	—
31577	Vegi feast, 12"	2	slice(s)	203	107	439	19	57	4	16	7.09	—	—	—
31578	Vegi feast, 14"	2	slice(s)	278	147	304	27	78	5	22	9.89	—	—	—
37549	Dot cinnamon	1	item(s)	28	8	99	2	15	1	4	0.68	—	—	—
31605	Double cheesy bread	1	item(s)	35	11	123	4	13	1	6	2.06	—	—	—
31607	Hot wings	1	item(s)	25	—	45	5	1	<1	2	0.65	—	—	—
	Thin crust pizza													
31583	America's favorite, 12"	¼	item(s)	159	—	408	19	34	2	23	9.77	—	—	—
31584	America's favorite, 14"	¼	item(s)	202	—	557	26	47	3	31	13.19	—	—	—
31579	Cheese, 12"	¼	item(s)	106	—	273	12	31	2	12	9.37	—	—	—
31580	Cheese, 14"	¼	item(s)	148	—	382	17	43	2	17	6.72	—	—	—
31688	Deluxe, 12"	¼	item(s)	159	—	363	16	34	2	19	7.64	—	—	—
31697	Deluxe, 14"	¼	item(s)	202	—	494	22	47	3	25	10.20	—	—	—
31689	Extravaganzza, 12"	¼	item(s)	159	—	425	20	34	3	24	9.41	—	—	—
31698	Extravaganzza, 14"	¼	item(s)	202	—	571	27	48	4	31	12.44	—	—	—
31585	Hawaiian, 12"	¼	item(s)	159	—	349	18	35	2	16	7.20	—	—	—
31586	Hawaiian, 14"	¼	item(s)	202	—	489	25	48	3	23	10.09	—	—	—
31690	Meatzza, 12"	¼	item(s)	159	—	458	23	33	2	27	11.39	—	—	—
31699	Meatzza, 14"	¼	item(s)	202	—	619	31	46	3	36	15.24	—	—	—
31581	Pepperoni, extra pepperoni & cheese 12"	¼	item(s)	159	—	420	20	32	2	24	10.46	—	—	—
31582	Pepperoni, extra pepperoni & cheese 14"	¼	item(s)	202	—	586	28	45	3	34	14.55	—	—	—
31587	Vegi, 12"	¼	item(s)	159	—	338	16	34	3	17	7.08	—	—	—
31588	Vegi, 14"	¼	item(s)	202	—	471	22	47	3	23	9.89	—	—	—
	Ultimate deep dish pizza													
31596	America's favorite, 12"	2	slice(s)	235	—	617	26	59	4	33	12.88	—	—	—
31702	America's favorite, 14"	2	slice(s)	311	—	851	36	84	5	44	17.35	—	—	—
31590	Cheese, 12"	2	slice(s)	181	—	482	19	56	3	22	7.91	—	—	—
31591	Cheese, 14"	2	slice(s)	257	—	677	26	80	5	30	10.88	—	—	—
31589	Cheese, 6"	1	item(s)	215	—	598	23	68	4	28	9.94	—	—	—
31691	Deluxe, 12"	2	slice(s)	235	—	527	23	59	4	29	10.75	—	—	—
31700	Deluxe, 14"	2	slice(s)	311	—	788	31	84	5	38	14.36	—	—	—
31692	Extravaganzza, 12"	2	slice(s)	235	—	635	27	59	4	34	12.52	—	—	—
31701	Extravaganzza, 14"	2	slice(s)	311	—	866	36	85	6	45	16.60	—	—	—
31599	Hawaiian, 12"	2	slice(s)	235	—	558	24	60	4	26	10.31	—	—	—
31600	Hawaiian, 14"	2	slice(s)	311	—	784	35	85	5	36	14.25	—	—	—
31693	Meatzza, 12"	2	slice(s)	235	—	667	30	58	4	37	14.50	—	—	—
31703	Meatzza, 14"	2	slice(s)	311	—	914	40	83	5	49	19.40	—	—	—
31593	Pepperoni, extra pepperoni & cheese 12"	2	slice(s)	235	—	629	26	57	4	34	13.57	—	—	—
31594	Pepperoni, extra pepperoni & cheese 14"	2	slice(s)	311	—	880	37	82	5	47	18.71	—	—	—
31602	Vegi, 12"	2	slice(s)	235	—	547	22	59	4	26	10.19	—	—	—
31603	Vegi, 14"	2	slice(s)	311	—	765	32	84	6	36	14.05	—	—	—
31598	With ham & pineapple tidbits, 6"	1	item(s)	430	—	619	25	70	4	28	10.19	—	—	—
31595	With Italian sausage, 6"	1	item(s)	430	—	642	25	70	4	31	11.33	—	—	—
31592	With pepperoni, 6"	1	item(s)	430	—	647	25	69	4	32	11.70	—	—	—
31601	With vegetables, 6"	1	item(s)	430	—	619	23	71	5	29	10.11	—	—	—
	In-n-Out Burger													
34374	Cheeseburger	1	item(s)	268	—	480	22	39	3	27	10.00	—	—	—
34391	Cheeseburger w/mustard & ketchup	1	item(s)	268	—	400	22	41	3	18	9.00	—	—	—
34390	Cheeseburger, lettuce leaves instead of buns	1	item(s)	300	—	330	18	11	2	25	9.00	—	—	—
34377	Chocolate shake	1	item(s)	425	—	690	9	83	0	36	24.00	—	—	—
34375	Double-Double cheeseburger	1	item(s)	328	—	670	37	40	3	41	18.00	—	—	—
34393	Double-Double cheeseburger w/mustard & ketchup	1	item(s)	328	—	590	37	42	3	32	17.00	—	—	—
34392	Double-Double cheeseburger, lettuce leaves instead of buns	1	item(s)	361	—	520	33	11	2	39	17.00	—	—	—
34376	French fries	1	item(s)	125	—	400	7	54	2	18	5.00	—	—	—
34373	Hamburger	1	item(s)	243	—	390	16	39	3	19	5.00	—	—	—
34389	Hamburger w/mustard & ketchup	1	item(s)	243	—	310	16	41	3	10	4.00	—	—	—
34388	Hamburger, lettuce leaves instead of buns	1	item(s)	275	—	240	12	10	2	17	4.50	—	—	—

PAGE KEY: H–2 = Breads/Baked Goods   H–6 = Cereal/Rice/Pasta   H–10 = Fruit   H–14 = Vegetables/Legumes   H–24 = Nuts/Seeds   H–26 = Vegetarian   H–28 = Dairy   H–34 = Eggs   H–34 = Seafood   H–36 = Meats   H–40 = Poultry   H–40 = Processed meats   H–42 = Beverages   H–46 = Fats/Oils   H–48 = Sweets   H–50 = Spices/Condiments/Sauces   H–52 = Mixed foods/Soups/Sandwiches   H–58 = Fast food   H–74 = Convenience meals   H–76 = Baby foods

Chol (mg)	Calc (mg)	Iron (mg)	Magn (mg)	Pota (mg)	Sodi (mg)	Zinc (mg)	Vit A (µg)	Thia (mg)	Vit E (mg α)	Ribo (mg)	Niac (mg)	Vit B$_6$ (mg)	Fola (µg)	Vit C (mg)	Vit B$_{12}$ (µg)	Sele (µg)
57	279	3.36	—	—	1349	—	155	—	—	—	—	—	—	<1	—	—
78	390	4.66	—	—	1855	—	233	—	—	—	—	—	—	<1	—	—
34	279	3.44	—	—	987	—	—	—	—	—	—	—	—	1	—	—
47	389	4.71	—	—	1369	—	—	—	—	—	—	—	—	2	—	—
0	6	0.59	—	—	86	—	—	—	—	—	—	—	—	<.1	—	—
6	47	0.66	—	—	164	—	—	—	—	—	—	—	—	<1	—	—
26	5	0.30	—	—	354	—	—	—	—	—	—	—	—	1	—	—
51	318	1.52	—	—	1285	—	—	—	—	—	—	—	—	<1	—	—
69	444	2.07	—	—	1751	—	—	—	—	—	—	—	—	1	—	—
23	225	0.97	—	—	835	—	125	—	—	—	—	—	—	0	—	—
32	315	1.36	—	—	1172	—	175	—	—	—	—	—	—	0	—	—
40	237	1.54	—	—	1123	—	—	—	—	—	—	—	—	1	—	—
53	330	2.08	—	—	1523	—	—	—	—	—	—	—	—	2	—	—
53	245	1.95	—	—	1408	—	—	—	—	—	—	—	—	1	—	—
69	340	2.59	—	—	1871	—	—	—	—	—	—	—	—	2	—	—
41	312	1.28	—	—	1162	—	—	—	—	—	—	—	—	2	—	—
57	437	1.80	—	—	1635	—	—	—	—	—	—	—	—	3	—	—
64	320	1.69	—	—	1523	—	—	—	—	—	—	—	—	<1	—	—
454	446	2.27	—	—	2039	—	—	—	—	—	—	—	—	<1	—	—
54	316	1.34	—	—	1362	—	162	—	—	—	—	—	—	<1	—	—
76	442	1.87	—	—	1900	—	227	—	—	—	—	—	—	<1	—	—
34	317	1.42	—	—	1047	—	—	—	—	—	—	—	—	1	—	—
47	442	1.94	—	—	1460	—	—	—	—	—	—	—	—	2	—	—
58	334	4.43	—	—	1573	—	—	—	—	—	—	—	—	1	—	—
78	464	6.24	—	—	2155	—	—	—	—	—	—	—	—	1	—	—
30	241	3.88	—	—	1123	—	151	—	—	—	—	—	—	<1	—	—
41	335	5.53	—	—	1575	—	210	—	—	—	—	—	—	1	—	—
36	295	4.67	—	—	1341	—	174	—	—	—	—	—	—	1	—	—
47	253	4.45	—	—	1410	—	—	—	—	—	—	—	—	2	—	—
62	349	6.25	—	—	1927	—	—	—	—	—	—	—	—	2	—	—
60	261	4.86	—	—	1696	—	—	—	—	—	—	—	—	2	—	—
78	359	6.76	—	—	2275	—	—	—	—	—	—	—	—	2	—	—
48	328	4.19	—	—	1449	—	—	—	—	—	—	—	—	2	—	—
67	457	5.97	—	—	2039	—	—	—	—	—	—	—	—	3	—	—
379	336	4.60	—	—	1810	—	—	—	—	—	—	—	—	1	—	—
501	466	6.44	—	—	2443	—	—	—	—	—	—	—	—	1	—	—
61	332	4.25	—	—	1650	—	187	—	—	—	—	—	—	1	—	—
85	462	6.04	—	—	2304	—	260	—	—	—	—	—	—	1	—	—
41	333	4.33	—	—	1334	—	—	—	—	—	—	—	—	2	—	—
57	462	6.11	—	—	1864	—	—	—	—	—	—	—	—	2	—	—
43	298	4.84	—	—	1498	—	—	—	—	—	—	—	—	1	—	—
45	302	4.89	—	—	1478	—	—	—	—	—	—	—	—	1	—	—
47	299	4.81	—	—	1524	—	168	—	—	—	—	—	—	1	—	—
36	307	5.10	—	—	1472	—	—	—	—	—	—	—	—	5	—	—
60	200	3.60	—	—	1000	—	188	—	—	—	—	—	—	15	—	—
55	200	3.60	—	—	1080	—	182	—	—	—	—	—	—	15	—	—
60	200	1.08	—	—	720	—	—	—	—	—	—	—	—	18	—	—
95	300	0.72	—	—	350	—	143	—	—	—	—	—	—	0	—	—
120	350	5.40	—	—	1430	—	184	—	—	—	—	—	—	15	—	—
115	350	5.40	—	—	1510	—	229	—	—	—	—	—	—	15	—	—
120	350	1.08	—	—	1160	—	275	—	—	—	—	—	—	18	—	—
0	20	1.80	—	—	245	—	0	—	—	—	—	—	—	0	—	—
40	40	3.60	—	—	640	—	50	—	—	—	—	—	—	15	—	—
35	40	3.60	—	—	720	—	75	—	—	—	—	—	—	15	—	—
40	40	1.08	—	—	370	—	—	—	—	—	—	—	—	18	—	—

TABLE H–1

Food Composition   (DA+ code is for Wadsworth Diet Analysis program)   (For purposes of calculations, use "0" for t, <1, <.1, <.01, etc.)

DA + Code	Food Description	Quantity	Measure	Wt (g)	H₂O (g)	Ener (kcal)	Prot (g)	Carb (g)	Fiber (g)	Fat (g)	Sat	Mono	Poly	Trans
	FAST FOOD—Continued													
34379	Strawberry shake	1	item(s)	425	—	690	8	91	2	33	22.00	—	—	—
34378	Vanilla shake	1	item(s)	425	—	680	9	78	2	37	25.00	—	—	—
	Jack in the Box													
30392	Bacon ultimate cheeseburger	1	item(s)	353	—	1120	52	59	2	55	28.00	—	—	3.13
1740	Breakfast Jack	1	item(s)	133	—	310	14	34	1	14	5.00	—	—	0
14074	Cheeseburger	1	item(s)	116	—	300	14	31	2	13	6.00	—	—	0.89
14106	Chicken breast pieces	5	piece(s)	150	—	360	27	24	1	17	3.00	—	—	4.48
37241	Chicken club salad	1	item(s)	535	—	310	28	15	5	16	6.00	—	—	0
14111	Chocolate ice cream shake	1	item(s)	315	—	660	11	89	1	29	18.00	—	—	1
14075	Double cheeseburger	1	item(s)	155	—	410	20	32	1	22	11.00	—	—	—
14098	French fries, jumbo	1	serving(s)	142	—	410	4	55	4	20	4.50	—	—	5.34
14099	French fries, super scoop	1	serving(s)	198	—	580	6	77	6	28	6.00	—	—	7.07
14073	Hamburger	1	item(s)	104	—	250	12	30	2	9	3.50	—	—	0.88
14090	Hash browns	1	serving(s)	57	—	150	1	13	2	10	2.50	—	—	3
14072	Jack's Spicy Chicken sandwich	1	item(s)	253	—	580	24	53	3	31	6.00	—	—	2.81
1468	Jumbo Jack hamburger	1	item(s)	269	—	600	22	58	3	31	11.00	—	—	1.55
1469	Jumbo Jack hamburger w/cheese	1	item(s)	294	—	690	26	60	3	38	16.00	—	—	1.55
1470	Onion rings	1	serving(s)	119	—	500	6	51	3	30	5.00	—	—	10
33141	Sausage, egg, & cheese biscuit	1	item(s)	223	—	760	25	33	2	60	20.00	—	—	5.72
14095	Seasoned curly fries	1	serving(s)	125	—	400	6	45	5	23	5.00	—	—	7
14077	Sourdough Jack	1	item(s)	244	—	700	30	36	3	49	16.00	—	—	2.98
37249	Southwest chicken salad	1	serving(s)	598	—	340	28	31	9	13	6.00	—	—	0
14112	Strawberry ice cream shake	1	item(s)	313	—	640	10	84	0	28	18.00	—	—	1
14078	Ultimate cheeseburger	1	item(s)	328	—	990	41	59	2	66	28.00	—	—	3.05
14110	Vanilla ice cream shake	1	item(s)	285	—	570	12	65	0	29	18.00	—	—	1
	Jamba Juice													
31646	Banana berry smoothie	24	fluid ounce(s)	719	—	470	5	112	5	2	0.50	—	—	—
31647	Caribbean passion smoothie	24	fluid ounce(s)	730	—	440	4	102	4	2	1.00	—	—	—
38422	Carrot juice	16	fluid ounce(s)	472	—	100	3	23	0	1	0.00	—	—	—
31648	Chocolate mood smoothie	24	fluid ounce(s)	612	—	690	16	142	2	8	4.50	—	—	—
31649	Citrus squeeze smoothie	24	fluid ounce(s)	729	—	450	4	105	5	2	1.00	—	—	—
31650	Coffee mood smoothie	24	fluid ounce(s)	560	—	596	13	121	1	6	4.00	—	—	—
31651	Coldbuster smoothie	24	fluid ounce(s)	724	—	430	5	100	5	3	1.00	—	—	—
31652	Cranberry craze smoothie	24	fluid ounce(s)	731	—	420	6	97	4	2	1.00	—	—	—
31654	Jamba powerboost smoothie	24	fluid ounce(s)	730	—	440	6	103	7	2	0.00	—	—	—
38423	Lemonade	16	fluid ounce(s)	483	—	300	1	75	0	0	0.00	0.00	0.00	0
31656	Lime sublime smoothie	24	fluid ounce(s)	721	—	450	3	104	6	2	1.00	—	—	—
31657	Mango-a-go-go smoothie	24	fluid ounce(s)	739	—	500	4	117	4	2	1.00	—	—	—
38424	Orange juice, freshly squeezed	16	fluid ounce(s)	496	—	220	3	52	1	1	0.00	—	—	—
38426	Orange/carrot juice	16	fluid ounce(s)	484	—	160	3	37	0	1	0.00	—	—	—
31660	Orange-a-peel smoothie	24	fluid ounce(s)	726	—	440	9	102	5	1	0.00	—	—	—
31665	Protein berry pizzaz smoothie	24	fluid ounce(s)	710	—	440	20	92	6	2	0.00	—	—	—
31667	Raspberry refresher smoothie	24	fluid ounce(s)	636	—	442	3	101	8	3	0.90	—	—	—
31668	Razzmatazz smoothie	24	fluid ounce(s)	730	—	480	3	112	4	2	1.00	—	—	—
31669	Strawberries wild smoothie	24	fluid ounce(s)	725	—	450	6	105	4	0	0.00	—	—	—
38421	Strawberry tsunami smoothie	24	fluid ounce(s)	740	—	530	4	128	4	2	1.00	—	—	—
38427	Vibrant C juice	16	fluid ounce(s)	448	—	210	2	50	1	0	0.00	0.00	0.00	0
38428	Wheatgrass juice, freshly squeezed	1	ounce(s)	32	—	5	1	1	0	0	0.00	0.00	0.00	0
	Kentucky Fried Chicken (KFC)													
31850	BBQ baked beans	1	serving(s)	156	—	190	6	33	6	3	1.00	—	—	0.29
31853	Biscuit	1	item(s)	56	—	180	4	20	1	10	2.50	—	—	3.44
31851	Coleslaw	1	serving(s)	142	—	232	2	26	3	14	2.00	—	—	0.27
31842	Colonel's Crispy Strips	3	item(s)	150	—	340	28	20	0	16	4.50	—	—	4.47
31849	Corn on the cob	1	item(s)	162	—	150	5	35	2	2	0.00	—	—	0
3761	Extra Crispy chicken, breast	1	item(s)	162	—	470	34	19	0	28	8.00	—	—	4.50
3762	Extra Crispy chicken, drumstick	1	item(s)	60	—	160	12	5	0	10	2.50	—	—	1.50
3763	Extra Crispy chicken, thigh	1	item(s)	114	—	370	21	12	0	26	7.00	—	—	3
3764	Extra Crispy chicken, whole wing	1	item(s)	52	—	190	10	10	0	12	3.50	—	—	2
31833	Honey BBQ wing pieces	6	item(s)	189	—	607	33	33	1	38	10.00	—	—	5.42
10810	Hot & spicy chicken, breast	1	item(s)	179	—	450	33	20	0	27	8.00	—	—	0
10813	Hot & spicy chicken, drumstick	1	item(s)	60	—	140	13	4	0	9	2.50	—	—	0
10811	Hot & spicy chicken, thigh	1	item(s)	128	—	390	22	14	0	28	8.00	—	—	0
10812	Hot & spicy chicken, whole wing	1	item(s)	55	—	180	11	9	0	11	3.00	—	—	0
10859	Hot wings pieces	6	piece(s)	135	—	471	27	18	2	33	8.00	—	—	4.03
31848	Macaroni & cheese	1	serving(s)	153	—	180	7	21	2	8	3.00	—	—	2.81

PAGE KEY: H–2 = Breads/Baked Goods   H–6 = Cereal/Rice/Pasta   H–10 = Fruit   H–14 = Vegetables/Legumes   H–24 = Nuts/Seeds   H–26 = Vegetarian
H–28 = Dairy   H–34 = Eggs   H–34 = Seafood   H–36 = Meats   H–40 = Poultry   H–40 = Processed meats   H–42 = Beverages   H–46 = Fats/Oils
H–48 = Sweets   H–50 = Spices/Condiments/Sauces   H–52 = Mixed foods/Soups/Sandwiches   H–58 = Fast food   H–74 = Convenience meals   H–76 = Baby foods

Chol (mg)	Calc (mg)	Iron (mg)	Magn (mg)	Pota (mg)	Sodi (mg)	Zinc (mg)	Vit A (µg)	Thia (mg)	Vit E (mg α)	Ribo (mg)	Niac (mg)	Vit B$_6$ (mg)	Fola (µg)	Vit C (mg)	Vit B$_{12}$ (µg)	Sele (µg)
85	250	0.00	—	—	280	—	134	—	—	—	—	—	—	0	—	—
90	300	0.00	—	—	390	—	145	—	—	—	—	—	—	0	—	—
160	300	7.20	—	600	2260	—	—	—	—	—	—	—	—	1	—	—
210	150	3.60	—	210	770	—	—	—	—	—	—	—	—	4	—	—
40	150	3.60	—	180	840	—	40	—	—	—	—	—	—	0	—	—
80	20	1.80	—	430	970	—	—	—	—	—	—	—	—	1	—	—
65	300	3.60	—	1010	890	—	—	—	—	—	—	—	—	54	—	—
110	350	0.36	—	720	270	—	215	—	—	—	—	—	—	0	—	—
70	250	4.50	—	280	920	—	—	—	—	—	—	—	—	1	—	—
0	20	1.08	—	550	690	—	0	—	—	—	—	—	—	6	—	—
0	20	1.44	—	770	960	—	0	—	—	—	—	—	—	9	—	—
30	100	3.60	—	155	610	—	0	—	—	—	—	—	—	0	—	—
0	10	0.18	—	190	230	—	0	—	—	—	—	—	—	0	—	—
60	150	1.80	—	470	950	—	—	—	—	—	—	—	—	9	—	—
45	164	4.92	—	390	980	—	—	—	—	—	—	—	—	10	—	—
75	250	4.50	—	420	1360	—	—	—	—	—	—	—	—	9	—	—
0	40	2.70	—	140	420	—	40	—	—	—	—	—	—	18	—	—
280	100	2.70	—	240	1390	—	—	—	—	—	—	—	—	0	—	—
0	40	1.80	—	580	890	—	—	—	—	—	—	—	—	0	—	—
80	200	4.50	—	450	1220	—	—	—	—	—	—	—	—	9	—	—
60	300	4.50	—	1020	920	—	—	—	—	—	—	—	—	48	—	—
110	350	0.00	—	610	220	—	202	—	—	—	—	—	—	0	—	—
130	300	7.20	—	480	1670	—	—	—	—	—	—	—	—	1	—	—
115	400	0.00	—	630	220	—	218	—	—	—	—	—	—	0	—	—
5	200	1.08	32	1000	85	0.30	—	0.06	0.32	0.26	1.20	0.40	33	15	0	0
5	100	1.80	24	810	60	0.30	—	0.09	0.64	0.26	5.00	0.50	100	78	0	1
0	150	2.70	80	1030	250	0.90	0	0.53	—	0.26	5.00	0.70	80	18	0	6
25	500	1.08	32	760	280	0.60	0	0.09	0.00	0.85	0.40	0.08	9	6	1	4
5	150	1.80	60	1150	50	0.30	—	0.30	0.40	0.26	1.90	0.40	100	168	0	1
28	455	0.30	49	634	429	1.50	—	0.10	0.16	0.60	0.30	0.10	18	7	1	3
5	100	1.08	60	1240	35	15.00	—	0.38	17.71	0.34	3.00	0.40	122	1302	0	1
5	250	1.44	16	500	90	0.30	—	0.03	0.64	0.26	5.00	0.50	100	54	0	1
0	1100	1.44	480	1110	40	15.00	—	5.25	17.71	5.78	66.00	6.80	640	294	10	70
0	20	0.00	8	200	10	0.00	0	0.03	0.00	0.17	14.00	1.80	320	36	0	0
5	150	1.80	32	660	75	0.60	—	0.12	0.32	0.26	7.00	0.80	160	66	<1	1
5	100	1.08	24	800	60	0.30	—	0.15	1.61	0.26	5.00	0.70	120	72	0	1
0	60	1.08	60	990	0	0.30	0	0.45	—	0.14	2.00	0.20	160	246	0	0
0	100	1.80	60	1010	125	0.60	0	0.45	—	0.26	3.00	0.50	120	132	0	3
0	250	1.80	60	1350	100	0.30	—	0.38	0.64	0.43	3.00	0.40	140	240	0	1
0	1100	2.62	39	650	240	0.58	0	0.09	0.31	0.10	1.55	0.40	58	60	0	4
3	104	2.20	56	806	47	0.80	—	0.10	0.40	0.30	1.60	0.40	43	35	<1	1
5	150	1.80	32	790	70	0.60	—	0.09	0.32	0.26	6.00	0.90	160	60	0	1
0	250	1.80	32	1020	115	0.30	—	0.03	0.32	0.34	1.20	0.20	32	60	0	1
5	100	1.08	24	480	10	0.30	0	0.06	—	0.34	14.00	1.80	320	90	0	1
0	20	1.08	40	720	0	0.30	0	0.30	—	0.10	1.60	0.40	80	678	0	0
0	0	1.80	8	80	0	0.00	0	0.03	—	0.03	0.40	0.04	16	4	0	3
5	80	1.80	—	—	760	—	—	—	—	—	—	—	—	1	—	—
0	20	1.08	—	—	560	—	—	—	—	—	—	—	—	1	—	—
8	30	0.18	—	—	284	—	65	—	—	—	—	—	—	34	—	—
70	10	0.72	—	—	1140	—	—	—	—	—	—	—	—	1	—	—
0	10	0.18	—	—	20	—	10	—	—	—	—	—	—	4	—	—
135	19	1.44	—	—	1230	—	—	—	—	—	—	—	—	1	—	—
70	9	0.65	—	—	415	—	—	—	—	—	—	—	—	1	—	—
120	19	1.04	—	—	710	—	—	—	—	—	—	—	—	1	—	—
55	9	0.34	—	—	390	—	—	—	—	—	—	—	—	1	—	—
193	40	1.44	—	—	1145	—	—	—	—	—	—	—	—	5	—	—
130	10	1.07	—	—	1450	—	—	—	—	—	—	—	—	1	—	—
65	20	0.68	—	—	380	—	—	—	—	—	—	—	—	1	—	—
125	10	1.44	—	—	1240	—	—	—	—	—	—	—	—	1	—	—
60	10	0.72	—	—	420	—	—	—	—	—	—	—	—	1	—	—
150	40	1.44	—	—	1230	—	—	—	—	—	—	—	—	1	—	—
10	150	0.18	—	—	860	—	350	—	—	—	—	—	—	1	—	—

TABLE H–1

Food Composition   (DA+ code is for Wadsworth Diet Analysis program)   (For purposes of calculations, use "0" for t, <1, <.1, <.01, etc.)

DA + Code	Food Description	Quantity	Measure	Wt (g)	H₂O (g)	Ener (kcal)	Prot (g)	Carb (g)	Fiber (g)	Fat (g)	Sat	Mono	Poly	Trans
	FAST FOOD—Continued													
31847	Mashed potatoes with gravy	1	serving(s)	136	—	120	1	17	2	6	1.00	—	—	0.50
10825	Original Recipe chicken, breast	1	item(s)	161	—	370	40	11	0	19	6.00	—	—	2.50
10826	Original Recipe chicken, drumstick	1	item(s)	59	—	140	14	4	0	8	2.00	—	—	1
10827	Original Recipe chicken, thigh	1	item(s)	126	—	360	22	12	0	25	7.00	—	—	1.50
10828	Original Recipe chicken, whole wing	1	item(s)	47	—	145	11	5	0	9	2.50	—	—	1
3760	Original Recipe chicken sandwich w/sauce	1	item(s)	200	—	450	29	33	2	22	5.00	—	—	—
31834	Original Recipe chicken sandwich w/o sauce	1	item(s)	187	—	360	29	21	1	13	3.50	—	—	—
31852	Potato salad	1	serving(s)	160	—	230	4	23	3	14	2.00	—	—	0.31
10845	Potato wedges	1	serving(s)	156	—	376	6	53	5	15	4.20	—	—	6.12
10853	Rotisserie Gold chicken, breast & wing w/skin	4	ounce(s)	114	—	218	26	1	0	12	3.51	—	—	—
10851	Rotisserie Gold chicken, thigh & leg w/skin	4	ounce(s)	114	—	260	23	1	0	18	5.15			
10852	Rotisserie Gold chicken, thigh & leg w/o skin	4	ounce(s)	117	—	217	27	0	0	12	3.50	—	—	—
31843	Spicy Crispy Strips	3	item(s)	115	—	335	25	23	1	15	4.00	—	—	—
10854	Tender Roast chicken, breast w/o skin	1	item(s)	118	—	169	31	1	0	4	1.20	—	—	—
	Long John Silver													
39392	Baked cod	1	serving(s)	101	—	120	22	1	0	5	1.00	—	—	—
3777	Batter dipped fish sandwich	1	item(s)	177	—	440	17	48	3	20	5.00	—	—	—
37568	Battered fish	1	item(s)	92	—	230	11	16	0	13	4.00	—	—	—
37569	Breaded clams	1	serving(s)	85	—	240	8	22	1	13	2.00	—	—	—
39404	Clam chowder	1	item(s)	227	—	220	9	23	0	10	4.00	—	—	—
39398	Cocktail sauce	1	ounce(s)	28	—	25	0	6	0	0	0.00	0.00	0.00	0
3770	Coleslaw	1	serving(s)	113	—	200	1	15	4	15	2.50	1.76	4.10	—
39394	Crunchy shrimp basket	21	item(s)	114	—	340	12	32	2	19	5.00	—	—	—
39400	French fries, large	1	item(s)	142	—	390	4	56	5	17	4.00	—	—	—
3774	Fries regular	1	serving(s)	85	—	230	3	34	3	10	2.50	7.40	5.10	—
3779	Hushpuppy	1	piece(s)	23	—	60	1	9	1	3	0.50	—	—	—
3781	Shrimp batter-dipped	1	piece(s)	14	—	45	2	3	0	3	1.00	—	—	—
39399	Tartar sauce	1	ounce(s)	28	—	100	0	4	0	9	1.50	—	—	—
39395	Ultimate fish sandwich	1	item(s)	199	—	500	20	48	3	25	8.00	—	—	—
	McDonald's													
2247	Barbecue sauce	1	serving(s)	28	—	45	0	10	0	0	0.00	0.00	0.00	0
737	Big Mac hamburger	1	item(s)	216	—	590	24	47	3	34	11.00	—	—	1.48
738	Cheeseburger	1	item(s)	121	—	330	15	36	2	14	6.00	—	—	1.02
29775	Chicken McGrill sandwich	1	item(s)	213	—	400	25	37	2	17	3.00	—	—	0
3792	Chicken McNuggets	4	item(s)	72	—	210	10	12	1	13	2.50	—	—	1.13
1873	Chicken McNuggets	6	item(s)	108	—	310	15	18	2	20	4.00	—	—	1.69
73	Chocolate milkshake	8	fluid ounce(s)	227	164	270	7	48	1	6	3.81	1.77	0.23	—
29774	Crispy chicken sandwich	1	item(s)	219	—	500	22	46	2	26	4.50	—	—	1.50
743	Egg McMuffin	1	item(s)	138	—	300	18	29	2	12	4.50	—	—	0.42
742	Filet-o-fish sandwich	1	item(s)	156	—	470	15	45	1	26	5.00	—	—	1.11
2257	French fries, large	1	serving(s)	176	—	540	8	68	6	26	4.50	—	—	6.18
1872	French fries, small	1	serving(s)	68	—	210	3	26	2	10	1.50	—	—	2.30
2244	French fries, super size	1	serving(s)	198	—	610	9	77	7	29	5.00	—	—	—
33822	Fruit n' yogurt parfait	1	item(s)	338	—	380	10	76	2	5	2.00	—	—	0.18
2251	Garden salad	1	item(s)	177	—	35	2	7	3	0	0.00	0.00	0.00	0
739	Hamburger	1	item(s)	107	—	280	12	35	2	10	4.00	—	—	0.51
2003	Hash browns	1	item(s)	53	—	130	1	14	1	8	1.50	—	—	2
2249	Honey sauce	1	item(s)	14	—	45	0	12	0	0	0.00	0.00	0.00	—
33816	McSalad Shaker chef salad	1	item(s)	206	—	150	17	5	2	8	3.50	—	—	—
33817	McSalad Shaker garden salad	1	item(s)	149	—	100	7	4	2	6	3.00	—	—	—
33818	McSalad Shaker grilled chicken caesar salad	1	item(s)	163	—	100	17	3	2	3	1.50	—	—	—
38396	Newman's Own cobb salad dressing	1	item(s)	59	—	120	1	9	0	9	1.50	—	—	0.01
38397	Newman's Own creamy caesar salad dressing	1	item(s)	59	—	190	2	4	0	18	3.50	—	—	0.29
38398	Newman's Own low fat balsamic vinaigrette salad dressing	1	item(s)	44	—	40	0	4	0	3	0.00	—	—	0.01
38399	Newman's Own ranch salad dressing	1	item(s)	59	—	290	1	4	0	30	4.50	—	—	0.22
1874	Plain hotcakes w/syrup & margarine	3	item(s)	228	—	600	9	104	0	17	3.00	—	—	4
740	Quarter Pounder hamburger	1	item(s)	172	—	430	23	37	2	21	8.00	—	—	1.01
741	Quarter Pounder hamburger w/cheese	1	item(s)	200	—	530	28	38	2	30	13.00	—	—	1.51
2005	Sausage McMuffin w/egg	1	item(s)	164	—	450	20	29	2	28	10.00	—	—	0.59

PAGE KEY: H–2 = Breads/Baked Goods H–6 = Cereal/Rice/Pasta H–10 = Fruit H–14 = Vegetables/Legumes H–24 = Nuts/Seeds H–26 = Vegetarian
H–28 = Dairy H–34 = Eggs H–34 = Seafood H–36 = Meats H–40 = Poultry H–40 = Processed meats H–42 = Beverages H–46 = Fats/Oils
H–48 = Sweets H–50 = Spices/Condiments/Sauces H–52 = Mixed foods/Soups/Sandwiches H–58 = Fast food H–74 = Convenience meals H–76 = Baby foods

Chol (mg)	Calc (mg)	Iron (mg)	Magn (mg)	Pota (mg)	Sodi (mg)	Zinc (mg)	Vit A (µg)	Thia (mg)	Vit E (mg α)	Ribo (mg)	Niac (mg)	Vit B$_6$ (mg)	Fola (µg)	Vit C (mg)	Vit B$_{12}$ (µg)	Sele (µg)
1	10	0.36	—	—	440	—	—	—	—	—	—	—	—	1	—	—
145	20	1.14	—	—	1145	—	—	—	—	—	—	—	—	1	—	—
75	10	0.70	—	—	440	—	—	—	—	—	—	—	—	1	—	—
165	10	1.00	—	—	1060	—	—	—	—	—	—	—	—	1	—	—
60	10	0.36	—	—	370	—	—	—	—	—	—	—	—	1	—	—
70	40	1.80	—	—	940	—	—	—	—	—	—	—	—	1	—	—
60	40	1.80	—	—	890	—	—	—	—	—	—	—	—	1	—	—
15	20	2.70	—	—	540	—	100	—	—	—	—	—	—	1	—	—
4	36	1.55	—	—	1323	—	—	—	—	—	—	—	—	8	—	—
102	7	0.12	—	—	718	—	—	—	—	—	—	—	—	1	—	—
127	8	0.14	—	—	764	—	—	—	—	—	—	—	—	1	—	—
128	10	0.18	—	—	772	—	—	—	—	—	—	—	—	1	—	—
70	20	0.90	—	—	1140	—	—	—	—	—	—	—	—	1	—	—
112	10	0.18	—	—	797	—	—	—	—	—	—	—	—	1	—	—
90	20	0.72	—	—	240	—	—	—	—	—	—	—	—	0	—	—
35	60	3.60	—	—	1120	—	—	—	—	—	—	—	—	9	—	—
30	20	1.80	—	—	700	—	—	—	—	—	—	—	—	5	—	—
10	20	1.08	—	—	1110	—	—	—	—	—	—	—	—	0	—	—
25	150	0.72	—	—	810	—	—	—	—	—	—	—	—	0	—	—
0	0	0.00	—	—	250	—	—	—	—	—	—	—	—	0	—	—
20	40	0.36	—	223	340	0.70	34	0.07	—	0.08	2.35	—	—	18	—	—
105	500	1.80	—	—	720	—	—	—	—	—	—	—	—	1	—	—
0	0	0.00	—	—	580	—	—	—	—	—	—	—	—	24	—	—
0	0	0.00	—	370	350	0.30	—	0.09	—	0.02	1.60	—	—	15	—	—
0	20	0.36	—	—	200	—	—	—	—	—	—	—	—	0	—	—
15	0	0.00	—	—	125	—	—	—	—	—	—	—	—	1	—	—
15	0	0.00	—	—	250	—	—	—	—	—	—	—	—	0	—	—
50	150	3.60	—	—	1310	—	—	—	—	—	—	—	—	9	—	—
0	10	0.18	—	45	250	—	3	—	—	—	—	—	—	4	—	—
85	300	4.50	—	430	1090	—	60	—	—	—	—	—	—	4	—	—
45	250	2.70	—	250	830	—	60	—	—	—	—	—	—	2	—	—
60	200	2.70	—	440	890	—	—	—	—	—	—	—	—	6	—	—
35	20	0.72	—	180	460	—	—	—	—	—	—	—	—	1	—	—
50	20	0.72	—	260	680	—	—	—	—	—	—	—	—	1	—	—
25	299	0.70	36	508	252	1.09	41	0.11	0.11	0.50	0.28	0.06	11	0	1	4
50	200	2.70	—	400	1100	—	—	—	—	—	—	—	—	6	—	—
235	300	2.70	—	210	830	—	—	—	0.72	—	—	—	—	1	—	—
50	200	1.80	—	280	890	—	40	—	—	—	—	—	—	1	—	—
0	20	1.44	—	1210	350	—	—	—	—	—	—	—	—	21	—	—
0	10	0.36	—	470	135	—	—	—	—	—	—	—	—	9	—	—
0	20	1.44	—	1370	390	—	—	—	—	—	—	—	—	24	—	—
15	300	1.80	—	550	240	—	—	—	—	—	—	—	—	24	—	—
0	40	1.09	—	410	20	—	—	—	—	—	—	—	—	24	—	—
30	200	2.70	—	230	590	—	5	—	—	—	—	—	—	2	—	—
0	10	0.36	—	210	330	—	—	—	—	—	—	—	—	2	—	—
0	10	0.18	—	7	0	—	—	—	—	—	—	—	—	1	—	—
95	150	1.44	—	360	740	—	323	—	—	—	—	—	—	15	—	—
75	150	1.08	—	290	120	—	273	—	—	—	—	—	—	15	—	—
40	100	1.08	—	420	240	—	—	—	—	—	—	—	—	12	—	—
10	40	0.18	—	13	440	—	—	—	0.00	—	—	—	—	1	—	—
20	60	0.18	—	16	500	—	—	—	15.40	—	—	—	—	1	—	—
0	10	0.18	—	9	730	—	—	—	0.00	—	—	—	—	2	—	—
20	40	0.18	—	64	530	—	—	—	—	—	—	—	—	1	—	—
20	100	4.50	—	280	770	—	—	—	—	—	—	—	—	1	—	—
70	200	4.50	—	370	840	—	10	—	—	—	—	—	—	2	—	—
95	350	4.50	—	420	1310	—	100	—	—	—	—	—	—	2	—	—
255	300	2.70	—	260	930	—	115	—	0.72	—	—	—	—	1	—	—

TABLE H–1

Food Composition

(DA+ code is for Wadsworth Diet Analysis program)   (For purposes of calculations, use "0" for t, <1, <.1, <.01, etc.)

DA + Code	Food Description	Quantity	Measure	Wt (g)	H₂O (g)	Ener (kcal)	Prot (g)	Carb (g)	Fiber (g)	Fat (g)	Fat Breakdown (g)			
											Sat	Mono	Poly	Trans
	FAST FOOD—Continued													
3163	Strawberry milkshake	8	fluid ounce(s)	226	168	256	8	43	1	6	3.93	—	—	—
74	Vanilla milkshake	8	fluid ounce(s)	227	169	254	9	40	0	7	4.28	1.98	0.26	—
	Pizza Hut													
39009	Hot chicken wings	2	item(s)	57	—	110	11	1	0	6	2.00	—	—	0.25
14025	Meat Lovers hand tossed pizza	1	slice(s)	125	—	320	16	30	2	15	7.00	—	—	0.53
14026	Meat Lovers pan pizza	1	slice(s)	130	—	360	16	29	2	20	7.00	—	—	0.53
31009	Meat Lovers stuffed crust pizza	1	slice(s)	188	—	500	25	44	3	25	11.00	—	—	1.11
14024	Meat Lovers thin 'n crispy pizza	1	slice(s)	112	—	310	15	22	2	18	8.00	—	—	0.57
14031	Pepperoni Lovers hand tossed pizza	1	slice(s)	114	—	300	15	30	2	14	7.00	—	—	0.50
14032	Pepperoni Lovers pan pizza	1	slice(s)	119	—	350	15	29	2	19	8.00	—	—	0.50
31011	Pepperoni Lovers stuffed crust pizza	1	slice(s)	171	—	480	23	44	3	24	11.00	—	—	1.05
14030	Pepperoni Lovers thin 'n crispy pizza	1	slice(s)	94	—	270	13	22	2	14	7.00	—	—	0.51
10834	Personal Pan pepperoni pizza	1	slice(s)	59	—	150	7	18	—	6	2.50	—	—	0.97
10842	Personal Pan supreme pizza	1	slice(s)	73	—	170	8	19	1	7	3.00	—	—	0.95
39013	Personal Pan Veggie Lovers pizza	1	slice(s)	69	—	150	6	19	1	6	2.00	—	—	0.50
14028	Veggie Lovers hand tossed pizza	1	slice(s)	120	—	220	10	31	2	6	3.00	—	—	0.25
14029	Veggie Lovers pan pizza	1	slice(s)	125	—	260	10	31	2	12	4.00	—	—	0.26
31010	Veggie Lovers stuffed crust pizza	1	slice(s)	181	—	370	17	45	3	14	7.00	—	—	0.53
14027	Veggie Lovers thin 'n crispy pizza	1	slice(s)	110	—	190	8	23	2	7	3.00	—	—	0.54
39012	Wing blue cheese dipping sauce	1	item(s)	43	—	230	2	2	0	24	5.00	—	—	1
39011	Wing ranch dipping sauce	1	item(s)	43	—	210	1	4	0	22	3.50	—	—	0.50
	Starbucks													
38042	Apple cider, tall steamed	12	fluid ounce(s)	360	—	180	0	45	0	0	0.00	0.00	0.00	0
38052	Cappuccino, tall	12	fluid ounce(s)	360	—	120	7	10	0	6	4.00	—	—	—
38053	Cappuccino, tall nonfat	12	fluid ounce(s)	360	—	80	7	11	0	0	0.00	0.00	0.00	0
38054	Cappuccino, tall soy milk	12	fluid ounce(s)	360	—	100	5	13	1	3	0.00	—	—	—
38059	Cinnamon spice mocha, tall nonfat w/o whipped cream	12	fluid ounce(s)	360	—	170	11	32	0	0	0.50	0.00	0.00	0
38057	Cinnamon spice mocha, tall w/whipped cream	12	fluid ounce(s)	360	—	320	10	31	0	17	11.00	—	—	—
38051	Espresso, single shot	1	fluid ounce(s)	30	—	5	0	1	0	0	0.00	0.00	0.00	—
38088	Flavored syrup, 1 pump	1	serving(s)	10	—	20	0	5	0	0	0.00	0.00	0.00	—
32562	Frappuccino coffee drink, lite mocha	9½	fluid ounce(s)	281	—	100	7	12	3	3	2.00	—	—	0
38079	Frappuccino, grande chocolate malt	16	fluid ounce(s)	480	—	470	15	87	2	10	3.50	—	—	—
38075	Frappuccino, grande mocha malt	12	fluid ounce(s)	360	—	430	14	91	1	7	4.00	—	—	—
32561	Frappuccino low fat coffee drink, all flavors	9½	fluid ounce(s)	281	—	190	6	39	0	3	2.00	—	—	—
38067	Frappuccino, tall caramel	12	fluid ounce(s)	360	—	210	4	43	0	3	1.50	—	—	—
38078	Frappuccino, tall chocolate	12	fluid ounce(s)	360	—	290	13	52	1	5	1.00	—	—	—
38069	Frappuccino, tall chocolate brownie	12	fluid ounce(s)	360	—	270	5	51	1	7	4.50	—	—	—
38070	Frappuccino, tall coffee	12	fluid ounce(s)	360	—	190	4	38	0	3	1.50	—	—	—
38071	Frappuccino, tall espresso	12	fluid ounce(s)	360	—	160	4	33	0	2	1.50	—	—	—
38073	Frappuccino, mocha	12	fluid ounce(s)	360	—	220	5	44	0	3	1.50	—	—	—
38072	Frappuccino, tall mocha coconut	12	fluid ounce(s)	360	—	300	5	58	2	7	5.00	—	—	—
38080	Frappuccino, tall vanilla	12	fluid ounce(s)	360	—	260	11	47	0	4	1.00	—	—	—
38074	Frappuccino, tall white chocolate	12	fluid ounce(s)	360	—	240	5	48	0	4	2.50	—	—	—
33111	Latte, tall w/nonfat milk	12	fluid ounce(s)	360	335	123	12	17	0	1	0.40	0.16	0.02	0
33112	Latte, tall w/whole milk	12	fluid ounce(s)	360	325	212	11	17	0	11	6.90	3.24	0.42	—
33109	Macchiato, tall caramel w/nonfat milk	12	fluid ounce(s)	360	—	140	7	27	0	1	0.40	—	—	—
33110	Macchiato, tall caramel w/whole milk	12	fluid ounce(s)	360	—	190	6	27	0	7	4.00	—	—	—
33107	Mocha coffee drink, tall nonfat, w/o whipped cream	12	fluid ounce(s)	360	—	180	12	33	1	2	1.50	0.68	0.08	—
38089	Mocha syrup	1	serving(s)	17	—	25	1	6	0	1	0.00	—	—	—
33108	Mocha, tall w/whole milk	12	fluid ounce(s)	360	—	340	12	33	1	20	12.00	3.48	0.44	—
38084	Tazo chai black tea, tall	12	fluid ounce(s)	360	—	210	6	36	0	5	3.50	—	—	—
38083	Tazo chai black tea, tall nonfat	12	fluid ounce(s)	360	—	170	6	37	0	0	0.00	0.00	0.00	—
38087	Tazo chai black tea, tall soy milk	12	fluid ounce(s)	360	—	190	4	39	1	2	0.00	—	—	—
38063	Tazo chai creme frappuccino, tall	12	fluid ounce(s)	360	—	280	11	51	0	4	1.00	—	—	—
38076	Tazo iced tea, tall	12	fluid ounce(s)	360	—	60	0	16	0	0	0.00	0.00	0.00	0
38077	Tazo tea, grande lemonade	16	fluid ounce(s)	480	—	120	0	31	0	0	0.00	0.00	0.00	0
38065	Tazoberry creme frappuccino, tall	12	fluid ounce(s)	360	—	240	4	54	1	1	0.00	—	—	—
38066	Tazoberry frappuccino, tall	12	fluid ounce(s)	360	—	140	1	36	1	0	0.00	0.00	0.00	0
38045	Vanilla creme steamed nonfat milk, tall w/whipped cream	12	fluid ounce(s)	360	—	180	12	32	0	0	0.00	0.00	0.00	—
38046	Vanilla creme steamed soy milk, tall w/whipped cream	12	fluid ounce(s)	360	—	300	8	37	1	12	6.00	—	—	—

PAGE KEY: H–2 = Breads/Baked Goods H–6 = Cereal/Rice/Pasta H–10 = Fruit H–14 = Vegetables/Legumes H–24 = Nuts/Seeds H–26 = Vegetarian
H–28 = Dairy H–34 = Eggs H–34 = Seafood H–36 = Meats H–40 = Poultry H–40 = Processed meats H–42 = Beverages H–46 = Fats/Oils
H–48 = Sweets H–50 = Spices/Condiments/Sauces H–52 = Mixed foods/Soups/Sandwiches H–58 = Fast food H–74 = Convenience meals H–76 = Baby foods

Chol (mg)	Calc (mg)	Iron (mg)	Magn (mg)	Pota (mg)	Sodi (mg)	Zinc (mg)	Vit A (µg)	Thia (mg)	Vit E (mg α)	Ribo (mg)	Niac (mg)	Vit B$_6$ (mg)	Fola (µg)	Vit C (mg)	Vit B$_{12}$ (µg)	Sele (µg)
25	256	0.25	29	412	188	0.82	59	0.10	—	0.44	0.40	0.10	7	2	1	5
27	331	0.23	27	415	215	0.88	57	0.07	0.11	0.44	0.33	0.10	16	0	1	5
70	0	0.36	—	—	450	—	—	—	—	—	—	—	—	0	—	—
40	150	1.80	—	—	830	—	—	—	—	—	—	—	—	6	—	—
40	150	2.70	—	—	810	—	—	—	—	—	—	—	—	6	—	—
65	250	2.70	—	—	1450	—	—	—	—	—	—	—	—	9	—	—
45	150	1.80	—	—	880	—	—	—	—	—	—	—	—	9	—	—
40	200	1.80	—	—	730	—	58	—	—	—	—	—	—	2	—	—
40	200	2.70	—	—	710	—	58	—	—	—	—	—	—	2	—	—
65	300	2.70	—	—	1300	—	—	—	—	—	—	—	—	4	—	—
40	200	1.44	—	—	700	—	58	—	—	—	—	—	—	2	—	—
15	80	1.44	—	—	340	—	38	—	—	—	—	—	—	1	—	—
15	80	1.86	—	—	400	—	—	—	—	—	—	—	—	4	—	—
10	80	1.80	—	—	280	—	—	—	—	—	—	—	—	4	—	—
15	150	1.80	—	—	490	—	—	—	—	—	—	—	—	9	—	—
15	150	2.70	—	—	470	—	—	—	—	—	—	—	—	9	—	—
35	250	2.70	—	—	980	—	—	—	—	—	—	—	—	12	—	—
15	150	1.44	—	—	480	—	—	—	—	—	—	—	—	12	—	—
25	20	0.00	—	—	550	—	0	—	—	—	—	—	—	0	—	—
10	0	0.00	—	—	340	—	0	—	—	—	—	—	—	0	—	—
0	0	1.08	—	—	15	—	0	—	—	—	—	—	—	0	0	—
25	250	0.00	—	—	95	—	0	—	—	—	—	—	—	1	0	—
3	200	0.00	—	—	100	—	0	—	—	—	—	—	—	0	0	—
0	250	0.72	—	—	75	—	0	—	—	—	—	—	—	0	0	—
5	300	0.72	—	—	150	—	0	—	—	—	—	—	—	0	0	—
70	350	1.08	—	—	140	—	0	—	—	—	—	—	—	2	0	—
0	0	0.00	—	—	0	—	0	—	—	—	—	—	—	0	0	—
0	0	0.00	—	—	0	—	0	—	—	—	—	—	—	0	0	—
13	200	1.08	—	—	80	—	—	—	—	—	—	—	—	0	—	—
15	250	2.70	—	—	420	—	0	—	—	—	—	—	—	12	0	—
20	250	1.08	—	—	390	—	0	—	—	—	—	—	—	0	0	—
12	220	0.00	—	—	110	—	—	—	—	—	—	—	—	0	—	—
10	150	0.00	—	—	180	—	0	—	—	—	—	—	—	0	0	—
3	400	1.80	—	—	300	—	0	—	—	—	—	—	—	5	0	—
10	150	1.44	—	—	220	—	0	—	—	—	—	—	—	0	0	—
10	150	0.00	—	—	180	—	0	—	—	—	—	—	—	0	0	—
10	100	0.00	—	—	160	—	0	—	—	—	—	—	—	0	0	—
10	150	0.72	—	—	180	—	0	—	—	—	—	—	—	0	0	—
10	150	1.08	—	—	220	—	0	—	—	—	—	—	—	0	0	—
3	400	0.00	—	—	280	—	0	—	—	—	—	—	—	4	0	—
10	150	0.00	—	—	210	—	0	—	—	—	—	—	—	0	0	—
6	420	0.18	40	—	174	1.35	—	0.12	—	0.47	0.36	0.14	18	4	1	—
46	400	0.18	47	254	165	1.28	—	0.13	—	0.54	0.35	0.14	17	3	1	—
25	250	0.36	—	—	110	—	—	—	—	—	—	—	—	2	—	—
25	200	0.36	—	—	105	—	—	—	—	—	—	—	—	1	—	—
5	350	2.70	—	—	150	—	—	—	—	—	—	—	—	2	—	—
0	0	0.72	—	—	0	—	0	—	—	—	—	—	—	0	0	—
47	300	0.18	—	—	169	—	—	—	—	—	—	—	—	2	—	—
20	200	0.36	—	—	85	—	0	—	—	—	—	—	—	1	0	—
5	200	0.36	—	—	95	—	0	—	—	—	—	—	—	0	0	—
0	200	0.72	—	—	70	—	0	—	—	—	—	—	—	0	0	—
3	400	0.00	—	—	280	—	0	—	—	—	—	—	—	4	0	—
0	0	0.00	—	—	0	—	0	—	—	—	—	—	—	0	0	—
0	0	0.00	—	—	15	—	0	—	—	—	—	—	—	5	0	—
0	150	0.00	—	—	125	—	0	—	—	—	—	—	—	1	0	—
0	0	0.00	—	—	30	—	0	—	—	—	—	—	—	0	0	—
5	350	0.00	—	—	170	—	0	—	—	—	—	—	—	0	0	—
30	400	1.44	—	—	130	—	0	—	—	—	—	—	—	0	0	—

TABLE H–1

Food Composition

(DA+ code is for Wadsworth Diet Analysis program)   (For purposes of calculations, use "0" for t, <1, <.1, <.01, etc.)

DA + Code	Food Description	Quantity	Measure	Wt (g)	H₂O (g)	Ener (kcal)	Prot (g)	Carb (g)	Fiber (g)	Fat (g)	Sat	Mono	Poly	Trans
	FAST FOOD—Continued													
38044	Vanilla creme steamed whole milk, tall w/whipped cream	12	fluid ounce(s)	360	—	340	10	31	0	18	12.00	—	—	—
38090	Whipped cream	1	serving(s)	27	—	100	0	2	0	9	6.00	—	—	—
38062	White chocolate mocha, tall nonfat w/o whipped cream	12	fluid ounce(s)	360	—	260	12	45	0	4	3.00	—	—	—
38061	White chocolate mocha, tall w/whipped cream	12	fluid ounce(s)	360	—	410	11	44	0	20	13.00	—	—	—
38048	White hot chocolate, tall w/o whipped cream	12	fluid ounce(s)	360	—	300	15	51	0	5	3.50	—	—	—
38047	White hot chocolate, tall w/whipped cream	12	fluid ounce(s)	360	—	460	13	50	0	22	15.00	—	—	—
38050	White hot chocolate soy milk, tall w/whipped cream	12	fluid ounce(s)	360	—	420	11	56	1	16	9.00	—	—	—
	Subway													
34023	Asiago caesar chicken wrap	1	item(s)	244	—	413	22	47	2	15	3.00	—	—	0
38622	Atkins-friendly chicken bacon ranch wrap	1	item(s)	213	—	480	40	19	11	27	9.00	—	—	0
38623	Atkins-friendly turkey bacon melt wrap	1	item(s)	199	—	430	32	22	12	25	9.00	—	—	0
34029	Bacon & egg breakfast sandwich	1	item(s)	127	—	302	14	29	1	15	4.00	—	—	0
32045	Chocolate chip cookie	1	item(s)	48	—	209	3	29	1	10	3.50	—	—	1.07
32048	Chocolate chip M&M cookie	1	item(s)	48	—	210	2	29	1	10	3.00	—	—	2.67
32049	Chocolate chunk cookie	1	item(s)	48	—	210	2	30	1	10	3.00	—	—	2.67
4024	Classic Italian B.M.T. sandwich, 6", white bread	1	item(s)	250	—	453	21	40	3	24	8.00	—	—	0
16397	Club salad	1	item(s)	323	—	145	17	12	3	4	1.00	—	—	0
3422	Club sandwich, 6", white bread	1	item(s)	253	—	294	22	40	3	5	1.50	—	—	0
4030	Cold cut trio sandwich, 6", white bread	1	item(s)	254	—	415	19	40	3	20	7.00	—	—	0
34030	Ham & egg breakfast sandwich	1	item(s)	147	—	291	15	30	1	12	3.00	—	—	0
3885	Ham sandwich, 6", white bread	1	item(s)	219	—	261	17	39	3	5	1.50	—	—	0
34026	Honey mustard melt sandwich, 6", Italian bread	1	item(s)	258	—	373	23	47	3	11	5.00	—	—	—
34027	Horseradish roast beef sandwich, 6", Italian bread	1	item(s)	230	—	401	18	42	3	17	3.00	—	—	—
4651	Meatball sandwich, 6", white bread	1	item(s)	284	—	501	23	46	4	25	10.00	—	—	0.75
15839	Melt sandwich, 6", white bread	1	item(s)	256	—	380	23	41	3	15	5.00	—	—	—
32046	Oatmeal raisin cookie	1	item(s)	48	—	197	3	29	1	8	2.00	—	—	2.67
32047	Peanut butter cookie	1	item(s)	48	—	220	3	26	1	12	3.00	—	—	1.07
3957	Roast beef sandwich, 6", white bread	1	item(s)	220	—	264	18	39	3	5	1.00	—	—	0
16403	Roasted chicken breast salad	1	item(s)	304	—	137	16	12	3	3	0.50	—	—	—
16378	Roasted chicken breast sandwich, 6", white bread	1	item(s)	234	—	311	25	40	3	6	1.50	—	—	0
34028	Southwest steak & cheese sandwich, 6", Italian bread	1	item(s)	255	—	412	23	42	4	18	6.00	—	—	—
4032	Spicy italian sandwich, 6", white bread	1	item(s)	213	—	458	19	42	2	24	9.00	—	—	0
4031	Steak & cheese sandwich, 6", white bread	1	item(s)	253	—	362	23	41	4	13	4.50	—	—	0
34024	Steak & cheese wrap	1	item(s)	245	—	353	22	46	3	9	4.00	—	—	—
32050	Sugar cookie	1	item(s)	48	—	222	2	28	1	12	3.00	—	—	3.73
16402	Tuna salad	1	item(s)	314	—	238	13	11	3	16	4.00	—	—	—
15844	Tuna sandwich, 6", white bread	1	item(s)	252	—	419	18	39	3	21	5.00	—	—	—
15834	Turkey breast & ham sandwich, 6", white bread	1	item(s)	229	—	267	18	40	3	5	1.00	—	—	0
34025	Turkey breast & bacon wrap	1	item(s)	228	—	318	19	45	2	7	2.50	—	—	—
16376	Turkey breast sandwich, 6", white bread	1	item(s)	220	—	254	16	39	3	4	1.00	—	—	0
16375	Veggie delite, 6", white bread	1	item(s)	163	—	200	7	37	3	3	0.50	—	—	0
32051	White macadamia nut cookie	1	item(s)	48	—	221	2	27	1	12	3.00	—	—	1.07
	Taco Bell													
29906	7-layer burrito	1	item(s)	283	—	530	18	67	10	22	8.00	—	—	3
744	Bean burrito	1	item(s)	198	—	370	14	55	8	10	3.50	—	—	2
749	Beef burrito supreme	1	item(s)	248	—	440	18	51	7	18	8.00	—	—	2
33417	Beef chalupa supreme	1	item(s)	153	—	390	14	31	3	24	10.00	—	—	3
29910	Beef gordita supreme	1	item(s)	153	—	310	14	30	3	16	7.00	—	—	0.50
2014	Beef soft taco	1	item(s)	99	—	210	10	21	2	10	4.50	—	—	1
10860	Beef soft taco supreme	1	item(s)	134	—	260	11	22	3	14	7.00	—	—	1
2018	Big beef burrito supreme	1	item(s)	291	—	510	23	52	11	23	9.00	6.55	1.61	—
14467	Big chicken burrito supreme	1	item(s)	255	—	460	27	50	3	17	6.00	—	—	—
34472	Chicken burrito supreme	1	item(s)	248	—	410	21	50	5	14	6.00	—	—	2
33418	Chicken chalupa supreme	1	item(s)	153	—	370	17	30	1	20	8.00	—	—	3

PAGE KEY: H–2 = Breads/Baked Goods   H–6 = Cereal/Rice/Pasta   H–10 = Fruit   H–14 = Vegetables/Legumes   H–24 = Nuts/Seeds   H–26 = Vegetarian
H–28 = Dairy   H–34 = Eggs   H–34 = Seafood   H–36 = Meats   H–40 = Poultry   H–40 = Processed meats   H–42 = Beverages   H–46 = Fats/Oils
H–48 = Sweets   H–50 = Spices/Condiments/Sauces   H–52 = Mixed foods/Soups/Sandwiches   H–58 = Fast food   H–74 = Convenience meals   H–76 = Baby foods

Chol (mg)	Calc (mg)	Iron (mg)	Magn (mg)	Pota (mg)	Sodi (mg)	Zinc (mg)	Vit A (µg)	Thia (mg)	Vit E (mg α)	Ribo (mg)	Niac (mg)	Vit B6 (mg)	Fola (µg)	Vit C (mg)	Vit B12 (µg)	Sele (µg)
75	40	0.00	—	—	160	—	0	—	—	—	—	—	—	2	0	—
40	0	0.00	—	—	10	—	0	—	—	—	—	—	—	0	0	—
5	400	0.00	—	—	210	—	0	—	—	—	—	—	—	0	0	—
70	400	0.00	—	—	210	—	0	—	—	—	—	—	—	2	0	—
10	450	0.00	—	—	250	—	0	—	—	—	—	—	—	0	0	—
75	500	0.00	—	—	250	—	0	—	—	—	—	—	—	4	0	—
35	500	1.44	—	—	210	—	0	—	—	—	—	—	—	0	0	—
46	40	2.70	—	—	1320	—	—	—	—	—	—	—	—	15	—	—
90	350	2.70	—	—	1340	—	—	—	—	—	—	—	—	7	—	—
65	300	2.70	—	—	1650	—	—	—	—	—	—	—	—	5	—	—
185	60	1.80	—	—	480	—	—	—	—	—	—	—	—	15	—	—
12	0	1.00	—	—	135	—	0	—	—	—	—	—	—	0	—	—
13	0	1.00	—	—	135	—	0	—	—	—	—	—	—	0	—	—
12	0	1.00	—	—	150	—	0	—	—	—	—	—	—	0	—	—
56	100	2.70	—	—	1740	—	—	—	—	—	—	—	—	24	—	—
30	40	1.80	—	—	1070	—	—	—	—	—	—	—	—	30	—	—
30	40	3.60	—	—	1250	—	60	—	—	—	—	—	—	24	—	—
57	150	3.60	—	—	1670	—	100	—	—	—	—	—	—	24	—	—
189	60	2.70	—	—	700	—	67	—	—	—	—	—	—	15	—	—
25	40	2.70	—	—	1260	—	—	—	—	—	—	—	—	24	—	—
41	100	2.70	—	—	1570	—	—	—	—	—	—	—	—	24	—	—
27	40	3.60	—	—	880	—	—	—	—	—	—	—	—	24	—	—
56	100	3.60	—	—	1350	—	—	—	—	—	—	—	—	24	—	—
41	100	2.70	—	—	1690	—	—	—	—	—	—	—	—	24	—	—
14	0	1.00	—	—	180	—	0	—	—	—	—	—	—	0	—	—
0	0	1.00	—	—	200	—	0	—	—	—	—	—	—	0	—	—
20	40	3.60	—	—	840	—	60	—	—	—	—	—	—	24	—	—
36	40	1.08	—	—	730	—	—	—	—	—	—	—	—	30	—	—
48	60	3.60	—	—	880	—	—	—	—	—	—	—	—	24	—	—
44	100	6.30	—	—	1120	—	—	—	—	—	—	—	—	24	—	—
57	30	3.00	—	—	1498	—	—	—	—	—	—	—	—	13	—	—
37	100	6.30	—	—	1200	—	—	—	—	—	—	—	—	24	—	—
37	150	7.20	—	—	1400	—	—	—	—	—	—	—	—	15	—	—
18	0	1.00	—	—	170	—	0	—	—	—	—	—	—	0	—	—
42	100	1.08	—	—	880	—	177	—	—	—	—	—	—	30	—	—
42	100	2.70	—	—	1180	—	100	—	—	—	—	—	—	24	—	—
23	40	2.70	—	—	1210	—	—	—	—	—	—	—	—	24	—	—
24	60	2.70	—	—	1490	—	—	—	—	—	—	—	—	15	—	—
15	40	2.70	—	—	1000	—	—	—	—	—	—	—	—	24	—	—
0	40	1.80	—	—	500	—	—	—	—	—	—	—	—	24	—	—
13	0	1.00	—	—	140	—	0	—	—	—	—	—	—	0	—	—
25	300	3.59	—	—	1360	—	—	—	—	—	—	—	—	5	—	—
10	200	2.69	—	—	1200	—	53	—	—	—	—	—	—	5	—	—
40	200	2.70	—	—	1330	—	351	—	—	—	—	—	—	9	—	—
40	150	1.80	—	—	600	—	—	—	—	—	—	—	—	5	—	—
35	150	2.70	—	—	590	—	—	—	—	—	—	—	—	5	—	—
25	100	1.80	—	—	620	—	44	—	—	—	—	—	—	2	—	—
40	150	1.80	—	—	630	—	73	—	—	—	—	—	—	5	—	—
60	150	2.70	—	493	1500	—	877	—	—	0.07	—	—	—	5	—	—
70	101	1.46	—	—	1200	—	—	—	—	—	—	—	—	2	—	—
45	200	2.70	—	—	1270	—	—	—	—	—	—	—	—	9	—	—
45	100	1.08	—	—	530	—	—	—	—	—	—	—	—	5	—	—

TABLE H–1

Food Composition

(DA+ code is for Wadsworth Diet Analysis program)   (For purposes of calculations, use "0" for t, <1, <.1, <.01, etc.)

DA + Code	Food Description	Quantity	Measure	Wt (g)	H₂O (g)	Ener (kcal)	Prot (g)	Carb (g)	Fiber (g)	Fat (g)	Sat	Mono	Poly	Trans
	FAST FOOD—Continued													
29900	Chicken fajita wrap supreme	1	item(s)	255	—	510	20	53	3	24	7.76	—	—	—
29895	Choco taco ice cream dessert	1	item(s)	113	—	310	3	37	1	17	10.00	—	—	—
10794	Cinnamon twists	1	serving(s)	35	—	160	0	28	0	5	1.00	—	—	1.50
14465	Grilled chicken burrito	1	item(s)	198	—	390	19	49	3	13	4.00	—	—	—
29911	Grilled chicken gordita supreme	1	item(s)	153	—	290	17	28	2	12	5.00	—	—	0
14463	Grilled chicken soft taco	1	item(s)	99	—	190	14	19	0	6	2.50	—	—	—
29912	Grilled steak gordita supreme	1	item(s)	153	—	290	16	28	2	13	6.00	—	—	0.50
29904	Grilled steak soft taco	1	item(s)	127	—	280	12	21	1	17	4.50	—	—	1
29905	Grilled steak soft taco supreme	1	item(s)	135	—	240	15	20	2	11	5.00	—	—	—
2021	Mexican pizza	1	serving(s)	216	—	550	21	46	7	31	11.00	—	—	5
2011	Nachos	1	serving(s)	99	—	320	5	33	2	19	4.50	—	—	5
2012	Nachos bellgrande	1	serving(s)	308	—	780	20	80	12	43	13.00	—	—	10
34473	Steak burrito supreme	1	item(s)	248	—	420	19	50	6	16	7.00	—	—	2
33419	Steak chalupa supreme	1	item(s)	153	—	370	15	29	2	22	8.00	—	—	3
29899	Steak fajita wrap supreme	1	item(s)	255	—	510	21	52	3	25	8.00	—	—	—
747	Taco	1	item(s)	78	—	170	8	13	3	10	4.00	—	—	0.50
2015	Taco salad w/salsa, with shell	1	serving(s)	533	—	790	31	73	13	42	15.00	—	—	8.75
14459	Taco supreme	1	item(s)	113	—	220	9	14	3	14	7.00	—	—	1
748	Tostada	1	item(s)	170	—	250	11	29	7	10	4.00	—	—	1.50
29901	Veggie fajita wrap supreme	1	item(s)	255	—	470	11	55	3	22	7.00	—	—	—
	CONVENIENCE MEALS													
	Banquet													
29961	Barbeque chicken meal	1	item(s)	281	—	330	16	37	2	13	3.00	—	—	—
14788	Boneless white fried chicken meal	1	item(s)	234	—	490	14	49	2	27	7.00	—	—	—
29960	Fish sticks meal	1	item(s)	187	—	270	13	31	3	10	3.00	—	—	—
29957	Lasagna with meat sauce meal	1	item(s)	312	—	320	15	46	7	9	4.00	—	—	—
14777	Macaroni & cheese meal	1	item(s)	340	—	420	15	57	5	14	8.00	—	—	—
1741	Meatloaf meal	1	item(s)	269	—	240	14	20	4	11	4.00	—	—	—
39418	Pepperoni pizza meal	1	item(s)	191	—	480	11	56	5	23	8.00	—	—	—
33759	Roasted white turkey meal	1	item(s)	255	—	230	14	30	5	6	2.00	—	—	—
1743	Salisbury steak meal	1	item(s)	269	197	380	12	28	3	24	12.00	—	—	—
	Budget Gourmet													
1914	Cheese manicotti w/meat sauce	1	item(s)	284	194	420	18	38	4	22	11.00	6.00	1.34	
1915	Chicken w/fettucini	1	item(s)	284	—	380	20	33	3	19	10.00	—	—	
3986	Light beef stroganoff	1	item(s)	248	177	290	20	32	3	7	4.00	—	—	
3996	Light sirloin of beef in herb sauce	1	item(s)	269	214	260	19	30	5	7	4.00	2.30	0.31	
3987	Light vegetable lasagna	1	item(s)	298	227	290	15	36	5	9	1.79	0.89	0.60	
	Healthy Choice													
36979	Bowls chicken teriyaki with rice	1	item(s)	298	—	330	19	50	5	6	2.00	2.00	2.00	—
9425	Cheese French bread pizza	1	item(s)	170	—	360	20	57	5	5	1.50	—	—	
9306	Chicken enchilada suprema meal	1	item(s)	320	252	360	13	59	8	7	3.00	2.00	2.00	
9316	Lemon pepper fish meal	1	item(s)	303	—	280	11	49	5	5	2.00	1.00	2.00	
9322	Traditional salisbury steak meal	1	item(s)	354	250	360	23	45	5	9	3.50	4.00	1.00	
9359	Traditional turkey breasts meal	1	item(s)	298	—	330	21	50	4	5	2.00	1.50	1.50	
9451	Zucchini lasagna	1	item(s)	383	—	280	13	47	5	4	2.50	—	—	
	Stouffers													
2363	Cheese enchiladas with mexican rice	1	serving(s)	276	—	370	12	48	5	14	5.00	—	—	—
2313	Cheese French bread pizza	1	serving(s)	294	—	370	14	43	3	16	6.00	—	—	—
11138	Cheese manicotti w/tomato sauce	1	item(s)	255	—	330	17	35	3	13	8.00	—	—	—
2366	Chicken pot pie	1	item(s)	284	—	740	23	56	4	47	18.00	12.41	10.48	—
11116	Homestyle baked chicken breast w/mashed potatoes & gravy	1	item(s)	252	—	260	19	21	1	11	3.00	—	—	—
11146	Homestyle beef pot roast & potatoes	1	item(s)	252	—	270	16	25	3	12	4.50	—	—	—
11152	Homestyle roast turkey breast w/stuffing & mashed potatoes	1	item(s)	273	—	300	16	34	2	11	3.00	—	—	—
11043	Lean Cuisine Cafe Classics baked chicken & whipped potatoes w/stuffing	1	item(s)	227	—	240	17	33	3	5	1.50	1.50	1.00	0
11046	Lean Cuisine Cafe Classics honey mustard chicken	1	item(s)	213	—	260	18	37	1	4	1.50	1.00	1.00	0
360	Lean Cuisine Everyday Favorites chicken chow mein w/rice	1	item(s)	255	—	210	12	33	2	3	1.00	1.00	0.50	0
9467	Lean Cuisine Everyday Favorites fettucini alfredo	1	item(s)	262	—	280	13	40	2	7	3.50	2.00	1.00	0
11055	Lean Cuisine Everyday Favorites lasagna w/meat sauce	1	item(s)	291	—	300	19	41	3	8	4.00	2.00	0.50	0
9479	Lean Cuisine French bread deluxe pizza	1	item(s)	174	—	330	18	44	3	9	3.50	1.50	1.00	0

PAGE KEY: H–2 = Breads/Baked Goods H–6 = Cereal/Rice/Pasta H–10 = Fruit H–14 = Vegetables/Legumes H–24 = Nuts/Seeds H–26 = Vegetarian
H–28 = Dairy H–34 = Eggs H–34 = Seafood H–36 = Meats H–40 = Poultry H–40 = Processed meats H–42 = Beverages H–46 = Fats/Oils
H–48 = Sweets H–50 = Spices/Condiments/Sauces H–52 = Mixed foods/Soups/Sandwiches H–58 = Fast food H–74 = Convenience meals H–76 = Baby foods

Chol (mg)	Calc (mg)	Iron (mg)	Magn (mg)	Pota (mg)	Sodi (mg)	Zinc (mg)	Vit A (µg)	Thia (mg)	Vit E (mg α)	Ribo (mg)	Niac (mg)	Vit B6 (mg)	Fola (µg)	Vit C (mg)	Vit B12 (µg)	Sele (µg)
57	165	1.52	—	—	1182	—	—	—	—	—	—	—	—	7	—	—
20	60	0.72	—	—	100	—	—	—	—	—	—	—	—	0	—	—
0	0	0.37	—	—	150	—	0	—	—	—	—	—	—	0	—	—
40	151	1.44	—	—	1240	—	—	—	—	—	—	—	—	2	—	—
45	100	1.80	—	—	530	—	—	—	—	—	—	—	—	5	—	—
30	100	1.08	—	—	550	—	15	—	—	—	—	—	—	1	—	—
35	100	2.70	—	—	520	—	—	—	—	—	—	—	—	4	—	—
30	100	1.44	—	—	650	—	29	—	—	—	—	—	—	4	—	—
35	100	1.08	—	—	510	—	29	—	—	—	—	—	—	4	—	—
45	350	3.60	—	—	1030	—	—	—	—	—	—	—	—	6	—	—
4	80	0.72	—	—	530	—	0	—	—	—	—	—	—	0	—	—
35	200	2.70	—	—	1300	—	162	—	—	—	—	—	—	6	—	—
35	200	2.70	—	—	1260	—	789	—	—	—	—	—	—	9	—	—
35	100	1.44	—	—	520	—	—	—	—	—	—	—	—	4	—	—
50	150	1.80	—	—	1200	—	—	—	—	—	—	—	—	6	—	—
25	60	1.08	—	—	350	—	44	—	—	—	—	—	—	2	—	—
65	400	6.23	—	—	1670	—	—	—	—	—	—	—	—	21	—	—
40	80	1.44	—	—	360	—	73	—	—	—	—	—	—	5	—	—
15	150	1.44	—	—	710	—	281	—	—	—	—	—	—	5	—	—
30	150	1.44	—	—	990	—	—	—	—	—	—	—	—	6	—	—
50	40	1.08	—	—	1210	—	0	—	—	—	—	—	—	5	—	—
65	60	1.08	—	—	1150	—	—	—	—	—	—	—	—	0	—	—
30	60	1.44	—	—	690	—	—	—	—	—	—	—	—	2	—	—
20	100	2.70	—	—	1170	—	—	—	—	—	—	—	—	0	—	—
20	150	1.44	—	—	1330	—	0	—	—	—	—	—	—	0	—	—
30	0	1.80	—	—	1040	—	0	—	—	—	—	—	—	0	—	—
35	150	1.80	—	—	870	—	0	—	—	—	—	—	—	0	—	—
25	60	1.80	—	—	1070	—	—	—	—	—	—	—	—	4	—	—
60	40	1.44	—	—	1140	—	0	—	—	—	—	—	—	0	—	—
85	300	2.70	45	484	810	2.29	—	0.45	—	0.51	4.00	0.23	31	0	1	—
85	100	2.70	—	—	810	—	—	0.15	—	0.43	6.00	—	—	0	—	—
35	40	1.80	39	280	580	4.71	—	0.17	—	0.37	4.28	0.27	19	2	3	—
30	40	1.80	58	540	850	4.81	—	0.16	—	0.29	5.53	0.37	38	6	2	—
15	283	3.03	79	420	780	1.39	—	0.22	—	0.45	3.13	0.32	75	59	<1	—
40	20	0.72	—	—	600	—	—	—	—	—	—	—	—	15	—	—
10	350	3.60	—	—	600	—	—	—	—	—	—	—	—	12	—	—
30	40	1.44	—	—	580	—	—	—	—	—	—	—	—	4	—	—
30	40	0.36	—	—	580	—	—	—	—	—	—	—	—	30	—	—
45	80	2.70	—	—	580	—	—	—	—	—	—	—	—	21	—	—
35	40	1.44	—	—	600	—	—	—	—	—	—	—	—	0	—	—
10	200	1.80	—	—	310	—	—	—	—	—	—	—	—	0	—	—
25	200	1.44	—	360	890	—	—	—	—	—	—	—	—	12	—	—
15	200	1.80	—	240	880	—	—	—	—	—	—	—	—	0	—	—
40	350	1.08	—	430	810	—	—	—	—	—	—	—	—	1	—	—
65	150	2.70	—	—	1170	—	—	—	—	—	—	—	—	2	—	—
50	20	0.72	—	500	760	—	0	—	—	—	—	—	—	0	—	—
35	20	1.80	—	790	820	—	—	—	—	—	—	—	—	6	—	—
35	40	0.72	—	450	1190	—	0	—	—	—	—	—	—	0	—	—
30	80	0.72	—	480	690	—	—	—	—	—	—	—	—	0	—	—
35	60	0.36	—	370	640	—	—	—	—	—	—	—	—	0	—	—
30	20	0.36	—	310	620	—	—	—	—	—	—	—	—	0	—	—
20	200	0.36	—	260	670	—	0	—	—	—	—	—	—	0	—	—
30	200	1.08	—	590	650	—	—	—	—	—	—	—	—	5	—	—
20	100	1.80	—	390	630	—	—	—	—	—	—	—	—	9	—	—

TABLE H–1

Food Composition

(DA+ code is for Wadsworth Diet Analysis program)   (For purposes of calculations, use "0" for t, <1, <.1, <.01, etc.)

DA + Code	Food Description	Quantity	Measure	Wt (g)	H₂O (g)	Ener (kcal)	Prot (g)	Carb (g)	Fiber (g)	Fat (g)	Sat	Fat Breakdown (g)		
												Mono	Poly	*Trans*
	CONVENIENCE MEALS—Continued													
	Weight Watchers													
11164	Smart Ones chicken enchiladas suiza entree	1	serving(s)	255	—	270	15	33	2	9	3.50	—	—	—
11155	Smart Ones garden lasagna entree	1	item(s)	312	—	270	14	36	5	7	3.50	—	—	—
11187	Smart Ones pepperoni pizza	1	item(s)	158	—	390	23	46	4	12	4.00	—	—	—
31514	Smart Ones spicy penne pasta & ricotta	1	item(s)	289	—	280	11	45	4	6	2.00	—	—	—
31512	Smart Ones spicy szechuan style vegetables & chicken	1	item(s)	255	—	220	11	39	3	2	0.50	—	—	—
	BABY FOODS													
787	Apple juice	4	fluid ounce(s)	127	112	60	0	15	<1	<1	0.02	0.00	0.04	—
778	Applesauce, strained	4	tablespoon(s)	64	55	31	<1	8	1	<1	0.02	0.01	0.04	—
779	Bananas w/tapioca, strained	4	tablespoon(s)	60	50	34	<1	9	1	<.1	0.02	0.01	0.01	—
604	Carrots, strained	4	tablespoon(s)	56	52	15	<1	3	1	<.1	0.01	0.00	0.03	—
770	Chicken noodle dinner, strained	4	tablespoon(s)	64	55	42	2	6	1	1	0.38	0.55	0.30	—
801	Green beans, strained	4	tablespoon(s)	60	0.05	15	0.77	3.53	1.13	0.05	0.01	0	0.03	—
910	Human milk, mature	2	fluid ounce(s)	62	54	43	1	4	0	3	1.24	1.02	0.31	—
760	Mixed cereal, prepared w/whole milk	4	ounce(s)	114	85	128	5	18	1	4	2.19	1.25	0.43	—
772	Mixed vegetable dinner, strained	2	ounce(s)	57	50	23	1	5	1	<.1	0.00	0.00	0.06	—
762	Rice cereal, prepared w/whole milk	4	ounce(s)	114	85	131	4	19	<1	4	2.64	1.02	0.16	—
758	Teething biscuits	1	item(s)	11	1	43	1	8	<1	<1	0.17	0.16	0.09	—

PAGE KEY: H–2 = Breads/Baked Goods H–6 = Cereal/Rice/Pasta H–10 = Fruit H–14 = Vegetables/Legumes H–24 = Nuts/Seeds H–26 = Vegetarian
H–28 = Dairy H–34 = Eggs H–34 = Seafood H–36 = Meats H–40 = Poultry H–40 = Processed meats H–42 = Beverages H–46 = Fats/Oils
H–48 = Sweets H–50 = Spices/Condiments/Sauces H–52 = Mixed foods/Soups/Sandwiches H–58 = Fast food H–74 = Convenience meals H–76 = Baby foods

Chol (mg)	Calc (mg)	Iron (mg)	Magn (mg)	Pota (mg)	Sodi (mg)	Zinc (mg)	Vit A (µg)	Thia (mg)	Vit E (mg α)	Ribo (mg)	Niac (mg)	Vit B₆ (mg)	Fola (µg)	Vit C (mg)	Vit B₁₂ (µg)	Sele (µg)
50	250	1.08	—	—	660	—	—	—	—	—	—	—	—	4	—	—
30	350	1.80	—	—	610	—	—	—	—	—	—	—	—	6	—	—
45	450	1.80	—	320	650	—	55	—	—	—	—	—	—	5	—	—
5	150	2.70	—	250	400	—	—	—	—	—	—	—	—	6	—	—
10	150	1.80	—	—	730	—	—	—	—	—	—	—	—	2	—	—
0	5	0.72	4	115	4	0.04	1	0.01	0.76	0.02	0.11	0.04	0	73	0	<1
0	3	0.14	2	45	1	0.01	1	0.01	0.38	0.02	0.04	0.02	1	25	0	<1
0	3	0.12	6	53	5	0.04	1	0.01	0.36	0.02	0.11	0.07	4	10	0	<1
0	12	0.21	5	110	21	0.08	321	0.01	0.29	0.02	0.26	0.04	8	3	0	<1
10	17	0.41	9	89	15	0.35	70	0.03	0.13	0.04	0.46	0.04	7	<.1	<.1	2
0	23.39	0.44	14.39	94.8	1.2	0.12	27	0.01	0.31	0.05	0.2	0.02	21	3.11	0	0.18
9	20	0.02	2	31	10	0.10	38	0.01	0.05	0.02	0.11	0.01	3	3	<.1	1
12	250	11.85	31	226	53	0.81	28	0.49	—	0.66	6.56	0.07	12	1	<.1	—
0	12	0.19	6	69	5	0.09	77	0.01	—	0.02	0.29	0.04	5	2	0	<1
12	272	13.85	51	216	52	0.73	25	0.53	—	0.57	5.91	0.13	9	1	<.1	4
0	29	0.39	4	36	40	0.10	3	0.03	0.03	0.06	0.48	0.01	5	1	<.1	3

WHO: Nutrition Recommendations
Canada: Guidelines and Meal Planning

This appendix presents nutrition recommendations from the World Health Organization (WHO) and details for Canadians on the *Eating Well with Canada's Food Guide* and the *Beyond the Basics* meal planning system.

Nutrition Recommendations from WHO

The World Health Organization (WHO) has assessed the relationships between diet and the development of chronic diseases. Its recommendations include:

- Energy: sufficient to support growth, physical activity, and a healthy body weight (BMI between 18.5 and 24.9) and to avoid weight gain greater than 11 pounds (5 kilograms) during adult life
- Total fat: 15 to 30 percent of total energy
- Saturated fatty acids: <10 percent of total energy
- Polyunsaturated fatty acids: 6 to 10 percent of total energy
- Omega-6 polyunsaturated fatty acids: 5 to 8 percent of total energy
- Omega-3 polyunsaturated fatty acids: 1 to 2 percent of total energy
- *Trans*-fatty acids: <1 percent of total energy
- Total carbohydrate: 55 to 75 percent of total energy
- Sugars: <10 percent of total energy
- Protein: 10 to 15 percent of total energy
- Cholesterol: <300 mg per day
- Salt (sodium): <5 g salt per day (<2 g sodium per day), appropriately iodized
- Fruits and vegetables: ≥400 g per day (about 1 pound)
- Total dietary fiber: >25 g per day from foods
- Physical activity: one hour of moderate-intensity activity, such as walking, on most days of the week

Eating Well with Canada's Food Guide

Figure I-1 presents the 2007 *Eating Well with Canada's Food Guide,* which interprets Canada's *Guidelines for Healthy Eating* (see Table 2-2 on p. 40) for consumers and recommends a range of servings to consume daily from each of the four food groups. Additional publications, which are available from Health Canada ◆ through its website, provide many more details.

◆ Search for "Canada's food guide" at Health Canada: **www.hc-sc.gc.ca**

FIGURE I-1   *Eating Well with Canada's Food Guide*

Health Canada   Santé Canada   Your health and safety... our priority.   Votre santé et votre sécurité... notre priorité.

Eating Well with Canada's Food Guide

Canada

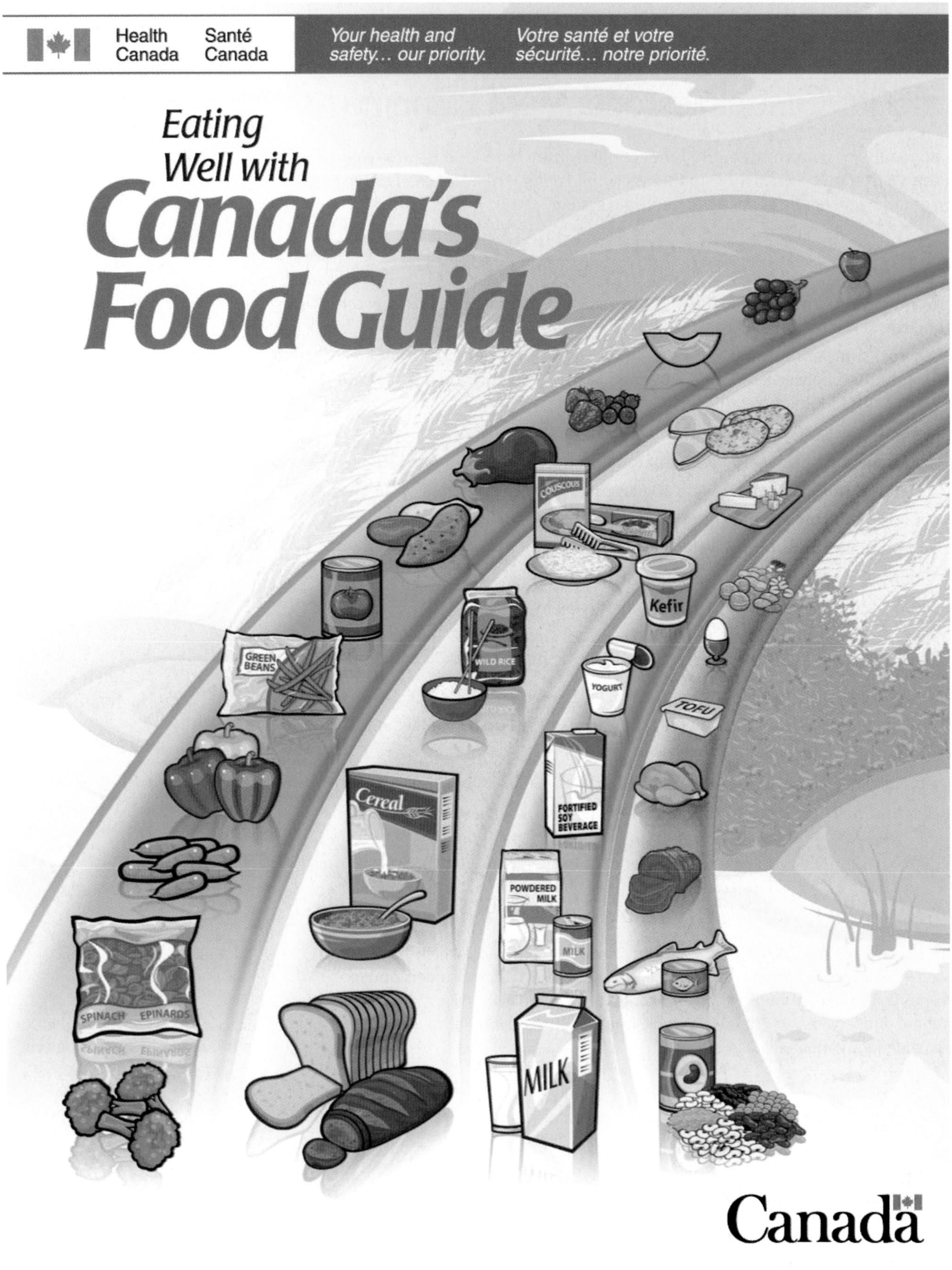

FIGURE I-1   *Eating Well with Canada's Food Guide—continued*

Recommended Number of **Food Guide Servings** per Day

	Children			Teens		Adults			
Age in Years	2-3	4-8	9-13	14-18		19-50		51+	
Sex	Girls and Boys			Females	Males	Females	Males	Females	Males
Vegetables and Fruit	4	5	6	7	8	7-8	8-10	7	7
Grain Products	3	4	6	6	7	6-7	8	6	7
Milk and Alternatives	2	2	3-4	3-4	3-4	2	2	3	3
Meat and Alternatives	1	1	1-2	2	3	2	3	2	3

The chart above shows how many Food Guide Servings you need from each of the four food groups every day.

Having the amount and type of food recommended and following the tips in *Canada's Food Guide* will help:

• Meet your needs for vitamins, minerals and other nutrients.

• Reduce your risk of obesity, type 2 diabetes, heart disease, certain types of cancer and osteoporosis.

• Contribute to your overall health and vitality.

FIGURE I-1   *Eating Well with Canada's Food Guide—continued*

What is One Food Guide Serving?
Look at the examples below.

Fresh, frozen or canned vegetables
125 mL (½ cup)

Leafy vegetables
Cooked: 125 mL (½ cup)
Raw: 250 mL (1 cup)

Fresh, frozen or canned fruits
1 fruit or 125 mL (½ cup)

100% Juice
125 mL (½ cup)

Bread
1 slice (35 g)

Bagel
½ bagel (45 g)

Flat breads
½ pita or ½ tortilla (35 g)

Cooked rice, bulgur or quinoa
125 mL (½ cup)

Cereal
Cold: 30 g
Hot: 175 mL (¾ cup)

Cooked pasta or couscous
125 mL (½ cup)

Milk or powdered milk (reconstituted)
250 mL (1 cup)

Canned milk (evaporated)
125 mL (½ cup)

Fortified soy beverage
250 mL (1 cup)

Yogurt
175 g
(¾ cup)

Kefir
175 g
(¾ cup)

Cheese
50 g (1 ½ oz.)

Cooked fish, shellfish, poultry, lean meat
75 g (2 ½ oz.)/125 mL (½ cup)

Cooked legumes
175 mL (¾ cup)

Tofu
150 g or
175 mL (¾ cup)

Eggs
2 eggs

Peanut or nut butters
30 mL (2 Tbsp)

Shelled nuts and seeds
60 mL (¼ cup)

Oils and Fats

- Include a small amount – 30 to 45 mL (2 to 3 Tbsp) – of unsaturated fat each day. This includes oil used for cooking, salad dressings, margarine and mayonnaise.
- Use vegetable oils such as canola, olive and soybean.
- Choose soft margarines that are low in saturated and trans fats.
- Limit butter, hard margarine, lard and shortening.

FIGURE I-1   *Eating Well with Canada's Food Guide—continued*

Make each Food Guide Serving count...
wherever you are – at home, at school, at work or when eating out!

▸ **Eat at least one dark green and one orange vegetable each day.**
- Go for dark green vegetables such as broccoli, romaine lettuce and spinach.
- Go for orange vegetables such as carrots, sweet potatoes and winter squash.

▸ **Choose vegetables and fruit prepared with little or no added fat, sugar or salt.**
- Enjoy vegetables steamed, baked or stir-fried instead of deep-fried.

▸ **Have vegetables and fruit more often than juice.**

▸ **Make at least half of your grain products whole grain each day.**
- Eat a variety of whole grains such as barley, brown rice, oats, quinoa and wild rice.
- Enjoy whole grain breads, oatmeal or whole wheat pasta.

▸ **Choose grain products that are lower in fat, sugar or salt.**
- Compare the Nutrition Facts table on labels to make wise choices.
- Enjoy the true taste of grain products. When adding sauces or spreads, use small amounts.

▸ **Drink skim, 1%, or 2% milk each day.**
- Have 500 mL (2 cups) of milk every day for adequate vitamin D.
- Drink fortified soy beverages if you do not drink milk.

▸ **Select lower fat milk alternatives.**
- Compare the Nutrition Facts table on yogurts or cheeses to make wise choices.

▸ **Have meat alternatives such as beans, lentils and tofu often.**

▸ **Eat at least two Food Guide Servings of fish each week.***
- Choose fish such as char, herring, mackerel, salmon, sardines and trout.

▸ **Select lean meat and alternatives prepared with little or no added fat or salt.**
- Trim the visible fat from meats. Remove the skin on poultry.
- Use cooking methods such as roasting, baking or poaching that require little or no added fat.
- If you eat luncheon meats, sausages or prepackaged meats, choose those lower in salt (sodium) and fat.

Enjoy a variety of foods from the four food groups.

Satisfy your thirst with water!

Drink water regularly. It's a calorie-free way to quench your thirst. Drink more water in hot weather or when you are very active.

* Health Canada provides advice for limiting exposure to mercury from certain types of fish. Refer to www.healthcanada.gc.ca for the latest information.

Appendix I

FIGURE I-1 *Eating Well with Canada's Food Guide—continued*

Advice for different ages and stages...

Children

Following *Canada's Food Guide* helps children grow and thrive.

Young children have small appetites and need calories for growth and development.

- Serve small nutritious meals and snacks each day.

- Do not restrict nutritious foods because of their fat content. Offer a variety of foods from the four food groups.

- Most of all... be a good role model.

Women of childbearing age

All women who could become pregnant and those who are pregnant or breastfeeding need a multivitamin containing **folic acid** every day. Pregnant women need to ensure that their multivitamin also contains **iron**. A health care professional can help you find the multivitamin that's right for you.

Pregnant and breastfeeding women need more calories. Include an extra 2 to 3 Food Guide Servings each day.

Here are two examples:
- Have fruit and yogurt for a snack, or

- Have an extra slice of toast at breakfast and an extra glass of milk at supper.

Men and women over 50

The need for **vitamin D** increases after the age of 50.

In addition to following *Canada's Food Guide*, everyone over the age of 50 should take a daily vitamin D supplement of 10 µg (400 IU).

How do I count Food Guide Servings in a meal?

Here is an example:

Vegetable and beef stir-fry with rice, a glass of milk and an apple for dessert		
250 mL (1 cup) mixed broccoli, carrot and sweet red pepper	=	2 **Vegetables and Fruit** Food Guide Servings
75 g (2 ½ oz.) lean beef	=	1 **Meat and Alternatives** Food Guide Serving
250 mL (1 cup) brown rice	=	2 **Grain Products** Food Guide Servings
5 mL (1 tsp) canola oil	=	part of your **Oils and Fats** intake for the day
250 mL (1 cup) 1% milk	=	1 **Milk and Alternatives** Food Guide Serving
1 apple	=	1 **Vegetables and Fruit** Food Guide Serving

FIGURE I-1   *Eating Well with Canada's Food Guide—continued*

Eat well and be active today and every day!

The benefits of eating well and being active include:

- Better overall health.
- Lower risk of disease.
- A healthy body weight.
- Feeling and looking better.
- More energy.
- Stronger muscles and bones.

Be active

To be active every day is a step towards better health and a healthy body weight.

Canada's Physical Activity Guide recommends building 30 to 60 minutes of moderate physical activity into daily life for adults and at least 90 minutes a day for children and youth. You don't have to do it all at once. Add it up in periods of at least 10 minutes at a time for adults and five minutes at a time for children and youth.

Start slowly and build up.

Eat well

Another important step towards better health and a healthy body weight is to follow *Canada's Food Guide* by:

- Eating the recommended amount and type of food each day.
- Limiting foods and beverages high in calories, fat, sugar or salt (sodium) such as cakes and pastries, chocolate and candies, cookies and granola bars, doughnuts and muffins, ice cream and frozen desserts, french fries, potato chips, nachos and other salty snacks, alcohol, fruit flavoured drinks, soft drinks, sports and energy drinks, and sweetened hot or cold drinks.

Read the label

- Compare the Nutrition Facts table on food labels to choose products that contain less fat, saturated fat, trans fat, sugar and sodium.
- Keep in mind that the calories and nutrients listed are for the amount of food found at the top of the Nutrition Facts table.

Nutrition Facts
Per 0 mL (0 g)

Amount	% Daily Value
Calories 0	
Fat 0 g	0 %
Saturates 0 g	0 %
+ Trans 0 g	
Cholesterol 0 mg	
Sodium 0 mg	0 %
Carbohydrate 0 g	0 %
Fibre 0 g	0 %
Sugars 0 g	
Protein 0 g	

Vitamin A	0 %	Vitamin C	0 %
Calcium	0 %	Iron	0 %

Limit trans fat

When a Nutrition Facts table is not available, ask for nutrition information to choose foods lower in trans and saturated fats.

Take a step today...

- ✓ Have breakfast every day. It may help control your hunger later in the day.
- ✓ Walk wherever you can – get off the bus early, use the stairs.
- ✓ Benefit from eating vegetables and fruit at all meals and as snacks.
- ✓ Spend less time being inactive such as watching TV or playing computer games.
- ✓ Request nutrition information about menu items when eating out to help you make healthier choices.
- ✓ Enjoy eating with family and friends!
- ✓ Take time to eat and savour every bite!

For more information, interactive tools, or additional copies visit Canada's Food Guide on-line at:
www.healthcanada.gc.ca/foodguide

or contact:
Publications
Health Canada
Ottawa, Ontario K1A 0K9
E-Mail: publications@hc-sc.gc.ca
Tel.: 1-866-225-0709
Fax: (613) 941-5366
TTY: 1-800-267-1245

Également disponible en français sous le titre : Bien manger avec le Guide alimentaire canadien

This publication can be made available on request on diskette, large print, audio-cassette and braille.

chyme (KIME): the semiliquid mass of partly digested food expelled by the stomach into the duodenum.

cirrhosis (sih-ROE-sis): an advanced stage of liver disease in which extensive scarring replaces healthy liver tissue, causing impaired liver function and liver failure; often associated with alcoholism.

claudication (CLAW-dih-KAY-shun): pain in the legs while walking; usually due to an inadequate supply of blood to muscles.

clear liquid diet: a diet that consists of foods that are liquid at body temperature, require minimal digestion, and contribute limited residue (undigested material) in the colon.

clinically severe obesity: a BMI of 40 or greater or a BMI of 35 or greater with additional medical problems. A less preferred term used to describe the same condition is *morbid obesity*.

closed feeding system: a delivery system in which the formula comes prepackaged in a container that is ready to be attached to the feeding tube for administration.

CoA (coh-AY): coenzyme A; the coenzyme derived from the B vitamin pantothenic acid and central to energy metabolism.

coenzymes: complex organic molecules that work with enzymes to facilitate the enzymes' activity. Many coenzymes have B vitamins as part of their structures (Figure 10-1 on p. 327 in Chapter 10 illustrates coenzyme action).

colectomy: removal of a portion or all of the colon.

colitis (ko-LYE-tis): inflammation of the colon.

collagen (KOL-ah-jen): the protein from which connective tissues such as scars, tendons, ligaments, and the foundations of bones and teeth are made.

collaterals: blood vessels that enlarge to allow an alternative pathway for diverted blood.

collecting duct: the last portion of a nephron's tubule, where the final concentration of urine occurs. One collecting duct is shared by several nephrons.

colonic irrigation: the popular, but potentially harmful practice of "washing" the large intestine with a powerful enema machine.

colostomy (co-LAH-stoe-me): a surgical procedure that creates a stoma using a section of the colon.

colostrum (ko-LAHS-trum): a milklike secretion from the breast, present during the first day or so after delivery before milk appears; rich in protective factors.

complement: a group of plasma proteins that assist the activities of antibodies.

complementary and alternative medicine (CAM): diverse medical and health care systems, practices, and products that currently are not considered part of conventional medicine; also called *unconventional* or *unorthodox therapies*.

- *Complementary medicine* refers to unconventional therapies that are used *in addition to,* and not simply as a replacement for, conventional medicine.

- *Alternative medicine* refers to unconventional therapies that are used *in place of* conventional medicine.

complementary proteins: two or more dietary proteins whose amino acid assortments complement each other in such a way that the essential amino acids missing from one are supplied by the other.

complex carbohydrates (starches and fibers): polysaccharides composed of straight or branched chains of monosaccharides.

compound: a substance composed of two or more different atoms—for example, water (H_2O).

conception: the union of the male sperm and the female ovum; fertilization.

condensation: a chemical reaction in which two reactants combine to yield a larger product.

conditionally essential amino acid: an amino acid that is normally nonessential, but must be supplied by the diet in special circumstances when the need for it exceeds the body's ability to produce it.

confectioners' sugar: finely powdered sucrose, 99.9% pure.

congregate meals: nutrition programs that provide food for the elderly in conveniently located settings such as community centers.

conjugated linoleic acid: a collective term for several fatty acids that have the same chemical formula as linoleic acid (18 carbons, two double bonds) but with different configurations.

constipation: the condition of having infrequent or difficult bowel movements.

contamination iron: iron found in foods as the result of contamination by inorganic iron salts from iron cookware, iron-containing soils, and the like.

continuous ambulatory peritoneal dialysis (CAPD): the most common method of peritoneal dialysis; involves frequent exchanges of dialysate, which remains in the peritoneal cavity throughout the day.

continuous feedings: slow delivery of formula at a constant rate over an 8- to 24-hour period.

continuous parenteral nutrition: continuous administration of parenteral solutions over a 24-hour period.

continuous renal replacement therapy (CRRT): a slow, continuous method of removing solutes and/or fluids from blood by gently pumping blood across a filtration membrane over a prolonged time period.

control group: a group of individuals similar in all possible respects to the experimental group except for the treatment. Ideally, the control group receives a placebo while the experimental group receives a real treatment.

conventional medicine: diagnosis and treatment of diseases as practiced by a doctor of medicine (M.D.) or doctor of osteopathy (D.O.) and assisted by allied health professionals such as registered nurses, pharmacists, and physical therapists; also called *Western, mainstream,* or *orthodox medicine*.

Cori cycle: the path from muscle glycogen to glucose to pyruvate to lactate (which travels to the liver) to glucose (which can travel back to the muscle) to glycogen; named after the scientist who elucidated this pathway.

corn sweeteners: corn syrup and sugars derived from corn.

corn syrup: a syrup made from cornstarch that has been treated with acid, high temperatures, and enzymes that produce glucose, maltose, and dextrins. See also *high-fructose corn syrup (HFCS)*.

cornea (KOR-nee-uh): the transparent membrane covering the outside of the eye.

coronary heart disease (CHD): a chronic, progressive disease characterized by obstructed blood flow in the coronary arteries; also called *coronary artery disease*.

correlation (CORE-ee-LAY-shun): the simultaneous increase, decrease, or change in two variables. If A increases as B increases, or if A decreases as B decreases, the correlation is **positive**. (This does not mean that A causes B or vice versa.) If A increases as B decreases, or if A decreases as B increases, the correlation is **negative**. (This does not mean that A prevents B or vice versa.) Some third factor may account for both A and B.

correspondence schools: schools that offer courses and degrees by mail. Some correspondence schools are accredited; others are not.

cortical bone: the very dense bone tissue that forms the outer shell surrounding trabecular bone and comprises the shaft of a long bone.

coupled reactions: pairs of chemical reactions in which some of the energy released from the breakdown of one compound is used to create a bond in the formation of another compound.

covert (KOH-vert): hidden, as if under covers.

cretinism (CREE-tin-ism): a congenital disease characterized by mental and physical retardation and commonly caused by maternal iodine deficiency during pregnancy.

critical pathways: coordinated programs of treatment that merge the care plans of different health practitioners; also called *clinical pathways*.

critical periods: finite periods during development in which certain events occur that will have irreversible effects on later developmental stages; usually a period of rapid cell division.

Crohn's disease: an inflammatory bowel disease that usually occurs in the lower portion of the small intestine and the colon. Inflammation may pervade the entire intestinal wall.

cross-contamination: the contamination of food by bacteria that occurs when the food comes into contact with surfaces previously touched by raw meat, poultry, or seafood.

cross-reactivity: an antibody reaction involving an antigen other than the one that induced the antibody's formation.

cryptosporidiosis (KRIP-toe-spor-ih-dee-OH-sis): a foodborne illness caused by the parasite *Cryptosporidium parvum.*

crypts (KRIPTS): tubular glands that lie between the intestinal villi and secrete intestinal juices into the small intestine.

cyanosis (sigh-ah-NOH-sis): a bluish cast in the skin due to the color of deoxygenated hemoglobin. Cyanosis is most evident in individuals with lighter, thinner skin; it is mostly seen on lips, cheeks, and ears and under the nails.

cyclamate (SIGH-kla-mate): an artificial sweetener that is being considered for approval in the United States and is available in Canada as a tabletop sweetener, but not as an additive.

cyclic parenteral nutrition: administration of a parenteral solution over a 10- to 16-hour period.

cystic fibrosis: an inherited disorder that affects the transport of chloride across epithelial cell membranes; primarily affects the gastrointestinal and respiratory systems.

cystinuria (SIS-tin-NOO-ree-ah): an inherited disorder characterized by elevated urinary excretion of several amino acids, including cystine.

cytokines (SIGH-toe-kines): signaling proteins produced by the body's cells; those produced by white blood cells regulate immune cell development and immune responses.

cytoplasm (SIGH-toh-plazm): the cell contents, except for the nucleus.

cytosol: the fluid of cytoplasm; contains water, ions, nutrients, and enzymes.

D

Daily Values (DV): reference values developed by the FDA specifically for use on food labels.

dawn phenomenon: morning hyperglycemia that is caused by the early-morning release of growth hormone, which counteracts insulin's glucose-lowering effects.

deamination (dee-AM-ih-NAY-shun): removal of the amino (NH_2) group from a compound such as an amino acid.

debridement: the surgical removal of dead, damaged, or contaminated tissue resulting from burns or wounds; helps to prevent infection and hasten healing.

decision-making capacity: the ability to understand pertinent information and make appropriate decisions; known as *decision-making competency* within the legal system.

defecate (DEF-uh-cate): to move the bowels and eliminate waste.

defibrillation: life-sustaining treatment in which an electronic device is used to shock the heart and reestablish a pattern of normal contractions. Defibrillation is used when the heart has arrhythmias or has experienced cardiac arrest.

deficient: the amount of a nutrient below which almost all healthy people can be expected, over time, to experience deficiency symptoms.

dehydration: the condition in which body water output exceeds water input. Symptoms include thirst, dry skin and mucous membranes, rapid heartbeat, low blood pressure, and weakness.

denaturation (dee-NAY-chur-AY-shun): the change in a protein's shape and consequent loss of its function brought about by heat, agitation, acid, base, alcohol, heavy metals, or other agents.

dental calculus: mineralized dental plaque, often associated with inflammation and bleeding.

dental caries: decay of teeth.

dental plaque: a gummy mass of bacteria that grows on teeth and can lead to dental caries and gum disease.

dermatitis herpetiformis (DERM-ah-TYE-tis HER-peh-tih-FOR-mis): a gluten-sensitive disorder characterized by a severe skin rash.

dermis: the connective tissue layer underneath the epidermis that contains the skin's blood vessels and nerves.

dextrose: an older name for glucose.

diabetes (DYE-ah-BEE-teez) **mellitus:** a group of metabolic disorders characterized by hyperglycemia resulting from insufficient or ineffective insulin.

diabetic coma: a coma that occurs in uncontrolled diabetes; may be due to diabetic ketoacidosis, the hyperosmolar hyperglycemic state, or severe hypoglycemia.

diabetic nephropathy (neh-FRAH-pah-thee): damage to the kidneys that results from long-term diabetes.

diabetic neuropathy (nur-RAH-pah-thee): complications of diabetes that cause damage to nerves.

diabetic retinopathy (REH-tih-NAH-pah-thee): retinal damage that results from long-term diabetes.

dialysate (dye-AL-ih-sate): the solution used in dialysis to draw wastes and fluids from the blood.

dialysis (dye-AH-lih-sis): a treatment that removes wastes and excess fluid from the blood after the kidneys have stopped functioning. The most common types of dialysis are *hemodialysis* and *peritoneal dialysis* (see Highlight 28).

dialyzer (DYE-ah-LYE-zer): a machine used in hemodialysis to filter the blood; also called an *artificial kidney.*

diarrhea: the frequent passage of watery bowel movements.

diet: the foods and beverages a person eats and drinks.

diet manual: a resource that specifies the foods allowed and restricted in modified diets and provides sample menus.

diet orders: specific instructions regarding dietary management; also called *diet prescriptions.*

dietary antioxidants: substances typically found in foods that significantly decrease the adverse effects of free radicals on normal functions in the body.

dietary fibers: in plant foods, the *nonstarch polysaccharides* that are not digested by human digestive enzymes, although some are digested by GI tract bacteria. Dietary fibers include cellulose, hemicelluloses, pectins, gums, and mucilages and the nonpolysaccharides lignins, cutins, and tannins.

dietary folate equivalents (DFE): the amount of folate available to the body from naturally occurring sources, fortified foods, and supplements, accounting for differences in the bioavailability from each source.

Dietary Reference Intakes (DRI): a set of nutrient intake values for healthy people in the United States and Canada. These values are used for planning and assessing diets and include:
- Estimated Average Requirements (EAR)
- Recommended Dietary Allowance (RDA)
- Adequate Intakes (AI)
- Tolerable Upper Intake Levels (UL)

dietetic technician: a person who has completed a minimum of an associate's degree from an accredited university or college and an approved dietetic technician program that includes a supervised practice experience. See also *dietetic technician, registered (DTR).*

dietetic technician, registered (DTR): a dietetic technician who has passed a national examination and maintains registration through continuing professional education.

dietitian: a person trained in nutrition, food science, and diet planning. See also *registered dietitian.*

diffusion: movement of solutes from an area of high concentration to one of low concentration.

digestion: the process by which food is broken down into absorbable units.

digestive enzymes: proteins found in digestive juices that act on food substances, causing them to break down into simpler compounds.

digestive system: all the organs and glands associated with the ingestion and digestion of food.

dipeptide (dye-PEP-tide): two amino acids bonded together.

disaccharides (dye-SACK-uh-rides): pairs of monosaccharides linked together. See Appendix C for the chemical structures of the disaccharides.

disclosure: the act of revealing pertinent information. For example, clinicians should accurately describe proposed tests and procedures, their benefits and risks, and alternative approaches.

discretionary kcalorie allowance: the kcalories remaining in a person's energy allowance after consuming enough nutrient-dense foods to meet all nutrient needs for a day.

disordered eating: eating behaviors that are neither normal nor healthy, including restrained eating, fasting, binge eating, and purging.

dissociates (dis-SO-see-aites): physically separates.

distilled liquor or **hard liquor:** an alcoholic beverage made by fermenting and distilling grains; sometimes called *distilled spirits.*

distilled water: water that has been vaporized and recondensed, leaving it free of dissolved minerals.

distributive justice: the equitable distribution of resources.

diuresis (DYE-uh-REE-sis): increased urine production.

diverticula (dye-ver-TIC-you-la): sacs or pouches that develop in the weakened areas of the intestinal wall (like bulges in an inner tube where the tire wall is weak).

diverticulitis (DYE-ver-tic-you-LYE-tis): infected or inflamed diverticula.

diverticulosis (DYE-ver-tic-you-LOH-sis): the condition of having diverticula. About one in every six people in Western countries develops diverticulosis in middle or later life.

DNA (deoxyribonucleic acid): the double helix molecules of which genes are made.

do-not-resuscitate (DNR) order: a request by a patient or surrogate to withhold cardiopulmonary resuscitation.

docosahexaenoic (DOE-cossa-HEXA-ee-NO-ick) **acid (DHA):** an omega-3 polyunsaturated fatty acid with 22 carbons and six double bonds; present in fish and synthesized in limited amounts in the body from linolenic acid.

dolomite: a compound of minerals (calcium magnesium carbonate) found in limestone and marble. Dolomite is powdered and is sold as a calcium-magnesium supplement. However, it may be contaminated with toxic minerals, is not well absorbed, and interacts adversely with absorption of other esssential minerals.

double-blind experiment: an experiment in which neither the subjects nor the researchers know which subjects are members of the experimental group and which are serving as control subjects, until after the experiment is over.

Down syndrome: a genetic abnormality that causes mental retardation, short stature, and flattened facial features.

drink: a dose of any alcoholic beverage that delivers $\frac{1}{2}$ oz of pure ethanol:
- 5 oz of wine
- 10 oz of wine cooler
- 12 oz of beer
- 1$\frac{1}{2}$ oz of hard liquor (80 proof whiskey, scotch, rum, or vodka)

drug: a substance that can modify one or more of the body's functions.

DTR: *see dietetic technician, registered.*

dumping syndrome: symptoms that result from the rapid emptying of an osmotic load from the stomach into the small intestine. Early symptoms include nausea, abdominal cramps, weakness, and diarrhea; later symptoms are those of hypoglycemia.

duodenum (doo-oh-DEEN-um, or doo-ODD-num): the top portion of the small intestine (about "12 fingers' breadth" long in ancient terminology).

durable power of attorney: a legal document (sometimes called a *health care proxy*) that gives legal authority to another (a *health care agent*) to make medical decisions in the event of incapacitation.

dysentery (DISS-en-terry): an infection of the digestive tract that causes diarrhea.

dyspepsia: a feeling of pain, bloating, or discomfort in the upper abdominal area, often called **indigestion;** a symptom of illness rather than a disease itself.

dysphagia (dis-FAY-jah): difficulty in swallowing.

dyspnea (DISP-nee-ah): shortness of breath.

E

eating disorders: disturbances in eating behavior that jeopardize a person's physical or psychological health.

eclampsia (eh-KLAMP-see-ah): a severe stage of preeclampsia characterized by convulsions.

edema (eh-DEEM-uh): the swelling of body tissue caused by excessive amounts of fluid in the interstitial spaces; seen in protein deficiency (among other conditions).

eicosanoids (eye-COSS-uh-noyds): derivatives of 20-carbon fatty acids; biologically active compounds that help to regulate blood pressure, blood clotting, and other body functions. They include *prostaglandins* (PROS-tah-GLAND-ins), *thromboxanes* (throm-BOX-ains), and *leukotrienes* (LOO-ko-TRY-eens).

eicosapentaenoic (EYE-cossa-PENTA-ee-NO-ick) **acid (EPA):** an omega-3 polyunsaturated fatty acid with 20 carbons and five double bonds; present in fish and synthesized in limited amounts in the body from linolenic acid.

electrolyte solutions: solutions that can conduct electricity.

electrolytes: salts that dissolve in water and dissociate into charged particles called ions.

electron transport chain: the final pathway in energy metabolism that transports electrons from hydrogen to oxygen and captures the energy released in the bonds of ATP.

element: a substance composed of atoms that are alike—for example, iron (Fe).

elemental formulas: enteral formulas that contain carbohydrates and proteins that are partially or fully hydrolyzed; also called *hydrolyzed, chemically defined,* or *monomeric formulas.*

embolism (EM-boh-lizm): the obstruction of a blood vessel by an embolus, causing sudden tissue death.

embolus (EM-boh-lus): an abnormal particle, such as a blood clot or air bubble, that travels in the blood.

embryo (EM-bree-oh): the developing infant from two to eight weeks after conception.

emergency shelters: facilities that are used to provide temporary housing.

emetic (em-ETT-ic): an agent that causes vomiting.

emphysema (EM-fih-ZEE-mah): a progressive lung condition characterized by the breakdown of the lungs' elastic structure and destruction of the walls of the bronchioles and alveoli, reducing the surface area involved in respiration.

empty-kcalorie foods: a popular term used to denote foods that contribute energy but lack protein, vitamins, and minerals.

emulsifier (ee-MUL-sih-fire): a substance with both water-soluble and fat-soluble portions that promotes the mixing of oils and fats in a watery solution.

end-stage renal disease (ESRD): an advanced stage of chronic kidney disease in which dialysis or a kidney transplant is necessary to sustain life.

endoplasmic reticulum (en-doh-PLAZ-mic reh-TIC-you-lum): a complex network of intracellular membranes. The rough endoplasmic reticulum is dotted with ribosomes, where protein synthesis takes place. The smooth endoplasmic reticulum bears no ribosomes.

endothelial cells: the type of cells that line the blood vessels, lymphatic vessels, and body cavities.

enemas: solutions inserted into the rectum and colon to stimulate a bowel movement and empty the lower large intestine.

energy: the capacity to do work. The energy in food is chemical energy. The body can convert this chemical energy to mechanical, electrical, or heat energy.

energy density: a measure of the energy a food provides relative to the amount of food (kcalories per gram).

energy-yielding nutrients: the nutrients that break down to yield energy the body can use:
• Carbohydrate
• Fat
• Protein

enriched: the addition to a food of nutrients that were lost during processing so that the food will meet a specified standard.

enteral (EN-ter-al) nutrition: the provision of nutrients using the GI tract, including the use of tube feedings and oral diets.

enteric coated: refers to medications or enzyme preparations that can withstand gastric acidity and dissolve only at a higher pH.

enteropancreatic (EN-ter-oh-PAN-kree-AT-ik) circulation: the circulatory route from the pancreas to the intestine and back to the pancreas.

enterostomy (EN-ter-AH-stoe-mee): an opening into the GI tract through which a feeding tube can be passed.

enzymes: proteins that facilitate chemical reactions without being changed in the process; protein catalysts.

epidemic (ep-ih-DEM-ick): the appearance of a disease (usually infectious) or condition that attacks many people at the same time in the same region.

epidermis (eh-pih-DER-miss): the outer layer of the skin.

epigenetics: the study of heritable changes in gene function that occur without a change in the DNA sequence.

epiglottis (epp-ih-GLOTT-iss): cartilage in the throat that guards the entrance to the trachea and prevents fluid or food from entering it when a person swallows.

epinephrine (EP-ih-NEFF-rin): a hormone of the adrenal gland that modulates the stress response; formerly called **adrenaline.** When administered by injection, epinephrine counteracts anaphylactic shock by opening the airways and maintaining heartbeat and blood pressure.

epithelial (ep-i-THEE-lee-ul) cells: cells on the surface of the skin and mucous membranes.

epithelial tissue: the layer of the body that serves as a selective barrier between the body's interior and the environment. (Examples are the cornea of the eyes, the skin, the respiratory lining of the lungs, and the lining of the digestive tract.)

erosive gastritis: erosion of the gastric mucosa, characterized by tissue destruction, ulcers, and hemorrhaging; often caused by the toxic effects of chemical substances or radiation treatment.

erythrocyte (eh-RITH-ro-cite) hemolysis (he-MOLL-uh-sis): the breaking open of red blood cells (erythrocytes); a symptom of vitamin E–deficiency disease in human beings.

erythrocyte protoporphyrin (PRO-toe-PORE-fe-rin): a precursor to hemoglobin.

erythropoiesis (eh-RIH-throh-poy-EE-sis): production of red blood cells within the bone marrow.

erythropoietin (eh-RITH-ro-POY-eh-tin): a hormone made by the kidneys that stimulates red blood cell production.

esophageal (eh-SOF-ah-JEE-al): involving the esophagus.

esophageal dysphagia: an inability to move food through the esophagus; usually caused by an obstruction or a motility disorder.

esophageal (ee-SOF-ah-GEE-al) sphincter: a sphincter muscle at the upper or lower end of the esophagus. The lower esophageal sphincter is also called the *cardiac sphincter.*

esophagus (ee-SOFF-ah-gus): the food pipe; the conduit from the mouth to the stomach.

essential amino acids: amino acids that the body cannot synthesize in amounts sufficient to meet physiological needs (see Table 6-1 on p. 182).

essential fatty acids: fatty acids needed by the body but not made by it in amounts sufficient to meet physiological needs.

essential nutrients: nutrients a person must obtain from food because the body cannot make them for itself in sufficient quantity to meet physiological needs; also called **indispensable nutrients.** About 40 nutrients are currently known to be essential for human beings.

Estimated Average Requirement (EAR): the average daily amount of a nutrient that will maintain a specific biochemical or physiological function in half the healthy people of a given age and gender group.

Estimated Energy Requirement (EER): the average dietary energy intake that maintains energy balance and good health in a person of a given age, gender, weight, height, and level of physical activity.

estrogens: hormones responsible for the menstrual cycle and other female characteristics.

ethanol: a particular type of alcohol found in beer, wine, and distilled liquor; also called *ethyl alcohol* (see Figure H7-1). Ethanol is the most widely used—and abused—drug in our society. It is also the only legal, nonprescription drug that produces euphoria.

ethical: in accordance with accepted principles of right and wrong.

exchange lists: diet-planning tools that organize foods by their proportions of carbohydrate, fat, and protein. Foods on any single list can be used interchangeably.

exocrine: pertains to external secretions, such as those of the mucous membranes or the skin. Opposite of endocrine, which pertains to hormonal secretions into the blood.

experimental group: a group of individuals similar in all possible respects to the control group except for the treatment. The experimental group receives the real treatment.

extra lean: less than 5 g of fat, 2 g of saturated fat and *trans* fat combined, and 95 mg of cholesterol per serving and per 100 g of meat, poultry, and seafood.

extracellular fluid: fluid outside the cells. Extracellular fluid includes two main components—the interstitial fluid and plasma. Extracellular fluid accounts for approximately one-third of the body's water.

F

fad diets: popular eating plans that promise quick weight loss. Most fad diets severely limit certain foods or overemphasize others (for example, never eat potatoes or pasta or eat cabbage soup daily).

faith healing: the use of prayer or belief in divine intervention to promote healing.

false negative: a test result indicating that a condition is not present (negative) when in fact it is present (therefore false).

false positive: a test result indicating that a condition is present (positive) when in fact it is not (therefore false).

fat replacers: ingredients that replace some or all of the functions of fat and may or may not provide energy.

fat-free: less than 0.5 g of fat per serving (and no added fat or oil); synonyms include "zero-fat," "no-fat," and "nonfat."

fats: lipids that are solid at room temperature (77°F or 25°C).

fatty acid: an organic compound composed of a carbon chain with hydrogens attached and an acid group (COOH) at one end and a methyl group (CH3) at the other end.

fatty acid oxidation: the metabolic breakdown of fatty acids to acetyl CoA; also called **beta oxidation.**

fatty liver: an early stage of liver deterioration seen in several diseases, including kwashiorkor and alcoholic liver disease. Fatty liver is characterized by an accumulation of fat in the liver cells.

fatty streaks: accumulations of cholesterol and other lipids along the walls of the arteries.

FDA (Food and Drug Administration): a part of the Department of Health and Human Services' Public Health Service that is responsible for ensuring the safety and wholesomeness of all dietary supplements and food processed and sold in interstate commerce except meat, poultry, and eggs (which are under the jurisdiction of the USDA); inspecting food plants and imported foods; and setting standards for food composition and product labeling.

female athlete triad: a potentially fatal combination of three medical problems—disordered eating, amenorrhea, and osteoporosis.

fermentable: the extent to which bacteria in the GI tract can break down fibers to fragments that the body can use.

indigestion: incomplete or uncomfortable digestion, usually accompanied by pain, nausea, vomiting, heartburn, intestinal gas, or belching.

inflammation: a nonspecific response to injury or infection; a type of innate immune response.

inflammatory response: a group of nonspecific immune responses to infection or injury.

informed consent: a patient's or caregiver's agreement to undergo a treatment that has been adequately disclosed. Persons must be mentally competent in order to make the decision.

innate immunity: immunity that is present at birth, unchanging throughout life, and nonspecific for particular antigens; also called **natural immunity.**

inorganic: not containing carbon or pertaining to living things.

inositol (in-OSS-ih-tall): a nonessential nutrient that can be made in the body from glucose. Inositol is a part of cell membrane structures.

insoluble fibers: indigestible food components that do not dissolve in water. Examples include the tough, fibrous structures found in the strings of celery and the skins of corn kernels.

insulin (IN-suh-lin): a hormone secreted by special cells in the pancreas in response to (among other things) increased blood glucose concentration. The primary role of insulin is to control the transport of glucose from the bloodstream into the muscle and fat cells.

insulin resistance: the condition in which a normal amount of insulin produces a subnormal effect in muscle, adipose, and liver cells, resulting in an elevated fasting glucose; a metabolic consequence of obesity that precedes type 2 diabetes.

integrative medicine: medical care that combines mainstream medical treatments and referrals to practitioners of CAM therapies.

intermittent claudication (claw-dih-KAY-shun): severe pain and weakness in the legs (especially the calves) that is caused by inadequate blood supply to the muscles; it usually occurs with walking and subsides during rest.

intermittent feedings: delivery of about 250 to 400 milliliters of formula over 20 to 40 minutes.

Internet (the net): a worldwide network of millions of computers linked together to share information.

interstitial (IN-ter-STISH-al) **fluid:** fluid between the cells (intercellular), usually high in sodium and chloride. Interstitial fluid is a large component of extracellular fluid.

intestinal adaptation: after resection, the process of intestinal recovery that leads to improved absorptive capacity.

intra-abdominal fat: fat stored within the abdominal cavity in association with the internal abdominal organs, as opposed to the fat stored directly under the skin (subcutaneous fat).

intracellular fluid: fluid within the cells, usually high in potassium and phosphate. Intracellular fluid accounts for approximately two-thirds of the body's water.

intractable: not easily managed or controlled.

intractable vomiting: vomiting that is not easily managed or controlled.

intradialytic parenteral nutrition: the infusion of nutrients during hemodialysis, often providing amino acids, dextrose, lipids, and some trace minerals.

intravenous feedings: the provision of nutrients through a vein, bypassing the intestine; also called **parenteral nutrition.**

intrinsic factor: a glycoprotein (a protein with short polysaccharide chains attached) secreted by the stomach cells that binds with vitamin B_{12} in the small intestine to aid in the absorption of vitamin B_{12}.

invert sugar: a mixture of glucose and fructose formed by the hydrolysis of sucrose in a chemical process; sold only in liquid form and sweeter than sucrose. Invert sugar is used as a food additive to help preserve freshness and prevent shrinkage.

ions (EYE-uns): atoms or molecules that have gained or lost electrons and therefore have electrical charges. Examples include the positively charged sodium ion (Na^+) and the negatively charged chloride ion (Cl^-). For a closer look at ions, see Appendix B.

iron deficiency: the state of having depleted iron stores.

iron overload: toxicity from excess iron.

iron-deficiency anemia: severe depletion of iron stores that results in low hemoglobin and small, pale red blood cells. Anemias that impair hemoglobin synthesis are **microcytic** (small cell).

irritable bowel syndrome: an intestinal disorder of unknown cause that affects the functioning of the lower bowel; symptoms include abdominal pain, flatulence, diarrhea, and constipation.

ischemia (iss-KEE-mee-a): inadequate blood supply within tissues due to obstructed blood flow in the arteries.

ischemic strokes: strokes caused by the obstruction of blood flow to brain tissue.

isotonic formula: a formula with an osmolality similar to that of blood serum (about 300 milliosmoles per kilogram).

J

jaundice (JAWN-dis): yellow discoloration of the skin and eyes due to an accumulation of bilirubin, a breakdown product of hemoglobin that normally exits the body via bile secretions.

jejunostomy (JE-ju-NAH-stoe-mee): an opening in the jejunum through which a feeding tube can be passed. A nonsurgical technique for creating a jejunostomy is called *percutaneous endoscopic jejunostomy (PEJ)*. The tube can either be guided into the jejunum via a gastrostomy or passed directly into the jejunum *(direct PEJ)*.

jejunum (je-JOON-um): the first two-fifths of the small intestine beyond the duodenum.

K

Kaposi's (cap-OH-seez) **sarcoma:** a common cancer in HIV-infected persons that is characterized by lesions in the skin, lungs, and GI tract.

kcalorie (energy) control: management of food energy intake.

kcalorie counts: the determination of food energy (and often, protein) consumed by patients for one or more days.

kcalorie-free: fewer than 5 kcal per serving.

kefir (keh-FUR): a fermented milk created by adding *Lactobacillus acidophilus* and other bacteria that break down lactose to glucose and galactose, producing a sweet, lactose-free product.

keratin (KARE-uh-tin): a water-insoluble protein; the normal protein of hair and nails.

keratinization: accumulation of keratin in a tissue; a sign of vitamin A deficiency.

keratomalacia (KARE-ah-toe-ma-LAY-shuh): softening of the cornea that leads to irreversible blindness; seen in severe vitamin A deficiency.

keto (KEY-toe) **acid:** an organic acid that contains a carbonyl group (C=O).

ketoacidosis (KEY-toe-ass-ih-DOE-sis): an acidosis (lowering of blood pH) that results from the excessive production of ketone bodies.

ketone (KEE-tone) **bodies:** the product of the incomplete breakdown of fat when glucose is not available in the cells.

ketonuria (KEY-toe-NOOR-ee-ah): the presence of ketone bodies in the urine.

ketosis (kee-TOE-sis): an undesirably high concentration of ketone bodies in the blood and urine.

kidney stones: crystalline masses that form in the urinary tract; also called *renal calculi* and *nephrolithiasis*.

kwashiorkor (kwash-ee-OR-core, kwash-ee-or-CORE): a form of PEM that results either from inadequate protein intake or, more commonly, from infections.

L

lactadherin (lack-tad-HAIR-in): a protein in breast milk that attacks diarrhea-causing viruses.

lactase: an enzyme that hydrolyzes lactose.

lactase deficiency: a lack of the enzyme required to digest the disaccharide lactose into its component monosaccharides (glucose and galactose).

lactate: a 3-carbon compound produced from pyruvate during anaerobic metabolism.

lactation: production and secretion of breast milk for the purpose of nourishing an infant.

lacto-ovo-vegetarians: people who include milk, milk products, and eggs, but exclude meat, poultry, fish, and seafood from their diets.

lactoferrin (lack-toh-FERR-in): a protein in breast milk that binds iron and keeps it from supporting the growth of the infant's intestinal bacteria.

lactose (LAK-tose): a disaccharide composed of glucose and galactose; commonly known as milk sugar.

lactose intolerance: a condition that results from inability to digest the milk sugar lactose; characterized by bloating, gas, abdominal discomfort, and diarrhea. Lactose intolerance differs from milk allergy, which is caused by an immune reaction to the protein in milk.

lactovegetarians: people who include milk and milk products, but exclude meat, poultry, fish, seafood, and eggs from their diets.

laparoscopic: pertaining to procedures that use a laparoscope for internal examination or surgery. A laparoscope is a narrow surgical telescope that is inserted into the body through a small incision. A video camera is usually attached so that the procedure can be viewed on a television monitor.

large intestine or **colon** (COAL-un): the lower portion of intestine that completes the digestive process. Its segments are the ascending colon, the transverse colon, the descending colon, and the sigmoid colon.

larynx: the upper part of the air passageway that contains the vocal cords; also called the voice box (see Figure H3-1).

laxatives: substances that loosen the bowels and thereby prevent or treat constipation.

LDL (low-density lipoprotein): the type of lipoprotein derived from very-low-density lipoproteins (VLDL) as VLDL triglycerides are removed and broken down; composed primarily of cholesterol.

lean: less than 10 g of fat, 4.5 g of saturated fat and *trans* fat combined, and 95 mg of cholesterol per serving and per 100 g of meat, poultry, and seafood.

lean body mass: the body minus its fat content.

lecithin (LESS-uh-thin): one of the phospholipids. Both nature and the food industry use lecithin as an emulsifier to combine water-soluble and fat-soluble ingredients that do not ordinarily mix, such as water and oil.

legumes (lay-GYOOMS, or LEG-yooms): plants of the bean and pea family, with seeds that are rich in protein compared with other plant-derived foods.

leptin: a protein produced by fat cells under direction of the *ob* gene that decreases appetite and increases energy expenditure; sometimes called the **ob** protein.

less: at least 25% less of a given nutrient or kcalories than the comparison food (see individual nutrients); synonyms include "fewer" and "reduced."

less cholesterol: 25% or less cholesterol than the comparison food (reflecting a reduction of at least 20 mg per serving), and 2 g or less saturated fat and *trans* fat combined per serving.

less fat: 25% or less fat than the comparison food.

less saturated fat: 25% or less saturated fat and *trans* fat combined than the comparison food.

let-down reflex: the reflex that forces milk to the front of the breast when the infant begins to nurse.

leukocytes: blood cells that function in immunity; also called *white blood cells.*

levulose: an older name for fructose.

license to practice: permission under state or federal law, granted on meeting specified criteria, to use a certain title (such as dietitian) and offer certain services. **Licensed dietitians** may use the initials **LD** after their names.

life expectancy: the average number of years lived by people in a given society.

life span: the maximum number of years of life attainable by a member of a species.

light or **lite:** one-third fewer kcalories than the comparison food; 50% or less of the fat or sodium than the comparison food; any use of the term other than as defined must specify what it is referring to (for example, "light in color" or "light in texture").

lignans: phytochemicals present in flaxseed, but not in flax oil, that are converted to phytosterols by intestinal bacteria and are under study as possible anticancer agents.

limiting amino acid: the essential amino acid found in the shortest supply relative to the amounts needed for protein synthesis in the body. Four amino acids are most likely to be limiting:
• Lysine
• Methionine
• Threonine
• Tryptophan

linoleic (lin-oh-LAY-ick) **acid:** an essential fatty acid with 18 carbons and two double bonds.

linolenic (lin-oh-LEN-ick) **acid:** an essential fatty acid with 18 carbons and three double bonds.

lipids: a family of compounds that includes triglycerides, phospholipids, and sterols. Lipids are characterized by their insolubility in water. (Lipids also include the fat-soluble vitamins, described in Chapter 11.)

lipomas (lih-POE-muz): benign tumors composed of fatty tissue.

lipoprotein lipase (LPL): an enzyme that hydrolyzes triglycerides passing by in the bloodstream and directs their parts into the cells, where they can be metabolized for energy or reassembled for storage.

lipoproteins (LIP-oh-PRO-teenz): clusters of lipids associated with proteins that serve as transport vehicles for lipids in the lymph and blood.

listeriosis: an infection caused by eating food contaminated with the bacterium *Listeria monocytogenes,* which can be killed by pasteurization and cooking but can survive at refrigerated temperatures; certain ready-to-eat foods, such as hot dogs and deli meats, may become contaminated after cooking or processing, but before packaging.

liver: the organ that manufactures bile. (The liver's many other functions are described in Chapter 7.)

living will: a written statement that specifies the medical procedures desired or not desired in the event that a person is unable to communicate or is incapacitated; also called a *medical directive.*

longevity: long duration of life.

low: an amount that would allow frequent consumption of a food without exceeding the Daily Value for the nutrient. A food that is naturally low in a nutrient may make such a claim, but only as it applies to all similar foods (for example, "fresh cauliflower, a low-sodium food"); synonyms include "little," "few," and "low source of."

low birthweight (LBW): a birthweight of 5½ lb (2500 g) or less; indicates probable poor health in the newborn and poor nutrition status in the mother during pregnancy, before pregnancy, or both. Normal birthweight for a full-term baby is 6½ to 8¾ lb (about 3000 to 4000 g).

low cholesterol: 20 mg or less cholesterol per serving and 2 g or less saturated fat and *trans* fat combined per serving.

low fat: 3 g or less fat per serving.

low kcalorie: 40 kcal or less per serving.

low saturated fat: 1 g or less saturated fat and less than 0.5 g of *trans* fat per serving.

low sodium: 140 mg or less per serving.

low-residue diet: a diet low in fiber and other food constituents that contribute to colonic residue.

low-risk pregnancy: a pregnancy characterized by indicators that make a normal outcome likely.

lumen (LOO-men): the space within a vessel, such as the intestine.

lutein (LOO-teen): a plant pigment of yellow hue; a phytochemical believed to play roles in eye functioning and health.

luteinizing (LOO-tee-in-EYE-zing) **hormone (LH):** a hormone that stimulates ovulation and the development of the corpus luteum (the small tissue that develops from a ruptured ovarian follicle and secretes hormones); so called because the follicle turns yellow as it matures. In men, LH stimulates testosterone secretion. The release of LH is mediated by **luteinizing hormone–releasing hormone (LH–RH).**

lycopene (LYE-koh-peen): a pigment responsible for the red color of tomatoes and other red-hued vegetables; a phytochemical that may act as an antioxidant in the body.

lymph (LIMF): a clear yellowish fluid that is similar to blood except that it contains no red blood cells or platelets. Lymph from the GI tract transports fat and fat-soluble vitamins to the bloodstream via lymphatic vessels.

lymphatic (lim-FAT-ic) **system:** a loosely organized system of vessels and ducts that convey fluids toward the heart. The GI part of the lymphatic system carries the products of fat digestion into the bloodstream.

lymphatic vessels: vessels through which lymph travels.

lymphocytes (LIM-foe-sites): white blood cells that recognize specific antigens and therefore function in adaptive immunity; include *T cells* and *B cells*.

lymphoid tissues: tissues that have roles in immunity.

lysosomes (LYE-so-zomes): cellular organelles; membrane-enclosed sacs of degradative enzymes.

lysozyme (LYE-so-zyme): an enzyme with antibacterial properties; found in immune cells and body secretions such as tears, saliva, and sweat.

M

macrobiotic diets: extremely restrictive diets limited to a few grains and vegetables; based on metaphysical beliefs and not on nutrition. A macrobiotic diet might consist of brown rice, miso soup, and sea vegetables, for example.

macrocytic anemia: anemia characterized by large red blood cells, as occurs in folate and vitamin B_{12} deficiency; also called **megaloblastic anemia.**

macrophages (MAK-roe-fay-jez): monocytes that have left circulation and settled in a tissue, where they serve as scavengers and activate the immune response.

macrosomia (MAK-roh-SOH-mee-ah): the condition of having an abnormally large body; in infants, refers to birth weights of 4000 g (8 lb 13 oz) and above.

macrovascular complications: disorders that affect the large blood vessels, including the coronary arteries and arteries of the limbs.

macular (MACK-you-lar) **degeneration:** deterioration of the macular area of the eye that can lead to loss of central vision and eventual blindness. The **macula** is a small, oval, yellowish region in the center of the retina that provides the sharp, straight-ahead vision so critical to reading and driving.

magnesium: a cation within the body's cells, active in many enzyme systems.

major minerals: essential mineral nutrients found in the human body in amounts larger than 5 g; sometimes called **macrominerals.**

maleficence (mah-LEF-eh-sense): the act of doing evil or harm.

malignant (ma-LIG-nent): describes a cancerous cell or tumor, which can injure healthy tissue and spread cancer to other regions of the body.

malnutrition: any condition caused by excess or deficient food energy or nutrient intake or by an imbalance of nutrients.

maltase: an enzyme that hydrolyzes maltose

maltose (MAWL-tose): a disaccharide composed of two glucose units; sometimes known as *malt sugar.*

mammary glands: glands of the female breast that secrete milk.

maple sugar: a sugar (mostly sucrose) purified from the concentrated sap of the sugar maple tree.

marasmus (ma-RAZ-mus): a form of PEM that results from a severe deprivation, or impaired absorption, of energy, protein, vitamins, and minerals.

massage therapy: manual manipulation of muscles to reduce tension, increase blood circulation, improve joint mobility, and promote healing of injuries.

mast cells: cells within connective tissue that produce and release histamine.

matrix (MAY-tricks): the basic substance that gives form to a developing structure; in the body, the formative cells from which teeth and bones grow.

matter: anything that takes up space and has mass.

Meals on Wheels: a nutrition program that delivers food for the elderly to their homes.

meat replacements: products formulated to look and taste like meat, fish, or poultry; usually made of textured vegetable protein.

mechanical ventilation: life-sustaining treatment in which a mechanical ventilator is used to substitute for a patient's failing lungs.

medical nutrition therapy: nutrition care provided by a registered dietitian; includes assessing nutrition status, diagnosing nutrition problems, and providing nutrition care.

meditation: a self-directed technique of calming the mind and relaxing the body.

medium-chain triglycerides (MCT): triglycerides that contain fatty acids that are 8 to 10 carbons in length. MCT do not require digestion and can be absorbed in the absence of lipase or bile.

MEOS or **microsomal** (my-krow-SO-mal) **ethanol-oxidizing system:** a system of enzymes in the liver that oxidize not only alcohol but also several classes of drugs.

metabolic stress: a disruption in the body's chemical environment due to the effects of disease or injury. Metabolic stress is characterized by changes in metabolic rate, heart rate, blood pressure, hormonal status, and nutrient metabolism.

metabolic syndrome: a cluster of interrelated clinical symptoms, including obesity, insulin resistance, high blood pressure, and abnormal blood lipids, which together increase cardiovascular disease risk twofold to threefold; also known as *syndrome X* or *insulin resistance syndrome.*

metabolism: the sum total of all the chemical reactions that go on in living cells. Energy metabolism includes all the reactions by which the body obtains and expends the energy from food.

metabolites: products of metabolism; compounds produced by a biochemical pathway.

metalloenzymes (meh-TAL-oh-EN-zimes): enzymes that contain one or more minerals as part of their structures.

metallothionein (meh-TAL-oh-THIGH-oh-neen): a sulfur-rich protein that avidly binds with and transports metals such as zinc.

metastasize (meh-TAS-tah-size): the spread of cancer cells from one part of the body to another.

MFP factor: a peptide released during the digestion of **m**eat, **f**ish, and **p**oultry that enhances nonheme iron absorption.

micelles (MY-cells): tiny spherical complexes of emulsified fat that arise during digestion; most contain bile salts and the products of lipid digestion, including fatty acids, monoglycerides, and cholesterol.

microalbuminuria: the presence of albumin (a blood protein) in the urine, a sign of diabetic nephropathy.

microarray technology: research tools that analyze the expression of thousands of genes simultaneously and search for particular gene changes associated with a disease. DNA microarrays are also called *DNA chips.*

microcytic anemia: anemia characterized by small, hypochromic (pale) red blood cells, as occurs in iron deficiency.

microvascular complications: disorders that affect the small blood vessels and capillaries, including those in the retinas and kidneys.

microvilli (MY-cro-VILL-ee, MY-cro-VILL-eye): tiny, hairlike projections on each cell of every villus that can trap nutrient particles and transport them into the cells; singular **microvillus.**

milk anemia: iron-deficiency anemia that develops when an excessive milk intake displaces iron-rich foods from the diet.

milliequivalents (mEq): the concentration of electrolytes in a volume of solution. Milli-equivalents are a useful measure when considering ions because the number of charges reveals characteristics about the solution that are not evident when the concentration is expressed in terms of weight.

mineral oil: a purified liquid derived from petroleum and used to treat constipation.

mineral water: water from a spring or well that typically contains 250 to 500 parts per million (ppm) of minerals. Minerals give water a distinctive flavor. Many mineral waters are high in sodium.

mineralization: the process in which calcium, phosphorus, and other minerals crystallize on the collagen matrix of a growing bone, hardening the bone.

minerals: inorganic elements. Some minerals are essential nutrients required in small amounts by the body for health.

misinformation: false or misleading information.

mitochondria (my-toh-KON-dree-uh); singular **mitochondrion:** the cellular organelles responsible for producing ATP aerobically; made of membranes (lipid and protein) with enzymes mounted on them.

moderation: in relation to alcohol consumption, not more than two drinks a day for the average-size man and not more than one drink a day for the average-size woman.

moderation (dietary): providing enough but not too much of a substance.

modified diet: a diet that is altered by changing food consistency or nutrient content or by including or eliminating specific foods; also called a *therapeutic diet.*

modular formulas: enteral formulas prepared in the hospital from *modules* that contain single macronutrients; used for people with unique nutrient needs.

molasses: the thick brown syrup produced during sugar refining. Molasses retains residual sugar and other by-products and a few minerals; blackstrap molasses contains significant amounts of calcium and iron.

molecule: two or more atoms of the same or different elements joined by chemical bonds. Examples are molecules of the element oxygen, composed of two oxygen atoms (O_2), and molecules of the compound water, composed of two hydrogen atoms and one oxygen atom (H_2O).

molybdenum (mo-LIB-duh-num): a trace element.

monocytes (MON-oh-sites): cells released from the bone marrow that move into tissues and mature into macrophages.

monoglycerides: molecules of glycerol with one fatty acid attached. A molecule of glyc-erol with two fatty acids attached is a **diglyceride.**

monosaccharides (mon-oh-SACK-uh-rides): carbohydrates of the general formula $C_nH_{2n}O_n$ that typically form a single ring. See Appendix C for the chemical structures of the monosaccharides.

monounsaturated fatty acid (MUFA): a fatty acid that lacks two hydrogen atoms and has one double bond between carbons—for example, oleic acid. A **monounsaturated fat** is composed of triglycerides in which most of the fatty acids are monounsaturated.

more: at least 10% more of the Daily Value for a given nutrient than the comparison food; synonyms include "added" and "extra."

mouth: the oral cavity containing the tongue and teeth.

mucous (MYOO-kus) **membranes:** the membranes, composed of mucus-secreting cells, that line the surfaces of body tissues.

mucus (MYOO-kus): a slippery substance secreted by cells of the GI lining (and other body linings) that protects the cells from exposure to digestive juices (and other destructive agents). The lining of the GI tract with its coat of mucus is a **mucous membrane.** (The noun is **mucus**; the adjective is **mucous.**)

multiple organ dysfunction syndrome: the dysfunction of two or more organ systems that develops during intensive care; often results in death.

muscle dysmorphia (dis-MORE-fee-ah): a psychiatric disorder characterized by a preoccupation with building body mass.

muscular dystrophy (DIS-tro-fee): a hereditary disease in which the muscles gradually weaken. Its most debilitating effects arise in the lungs.

mutation: an inheritable alteration in the DNA sequence of a gene.

myocardial (MY-oh-CAR-dee-al) **infarction** (in-FARK-shun), or **MI:** death of heart muscle caused by a sudden reduction in coronary blood flow; also called a *heart attack* or *cardiac arrest.*

myoglobin: the oxygen-holding protein of the muscle cells.

N

NAD (nicotinamide adenine dinucleotide): the main coenzyme form of the vitamin niacin. Its reduced form is NADH.

narcotic (nar-KOT-ic): a drug that dulls the senses, induces sleep, and becomes addictive with prolonged use.

nasoduodenal (ND): tube is placed into the duodenum via the nose.

nasoenteric: tube is placed into the GI tract via the nose. (*Nasoenteric feedings* usually refer to *nasoduodenal* and *nasojejunal* feedings.)

nasogastric (NG): tube is placed into the stomach via the nose.

nasojejunal (NJ): tube is placed into the jejunum via the nose.

National Center for Complementary and Alternative Medicine (NCCAM): a federal agency that researches and provides information about complementary and alternative therapies.

natural killer cells: lymphocytes that confer nonspecific immunity by destroying a wide array of viruses and tumor cells.

natural water: water obtained from a spring or well that is certified to be safe and sanitary. The mineral content may not be changed, but the water may be treated in other ways such as with ozone or by filtration.

naturopathic (NAY-chur-oh-PATH-ic) **medicine:** an approach to medical care using practices alleged to enhance the body's natural healing abilities. Treatments may include a variety of alternative therapies including dietary supplements, herbal remedies, exercise, and homeopathy.

neotame (NEE-oh-tame): an artificial sweetener composed of two amino acids (phenylalanine and aspartic acid); approved for use in the United States.

nephron (NEF-ron): the functional unit of the kidneys, consisting of a glomerulus and tubules.

nephrotic (neh-FROT-ik) **syndrome:** a syndrome associated with kidney disorders that cause urinary protein losses exceeding 3.0 to 3.5 g/day; symptoms include low serum albumin, elevated blood lipids, and edema.

nephrotoxic: toxic to the kidneys.

net protein utilization (NPU): a measure of protein quality assessed by measuring the amount of protein nitrogen that is retained from a given amount of protein nitrogen eaten.

neural tube: the embryonic tissue that forms the brain and spinal cord.

neural tube defects: malformations of the brain, spinal cord, or both during embryonic development that often result in lifelong disability or death.

neurofibrillary tangles: snarls of the thread-like strands that extend from the nerve cells, commonly found in the brains of people with Alzheimer's dementia.

neurons: nerve cells; the structural and functional units of the nervous system. Neurons initiate and conduct nerve impulse transmissions.

neuropeptide Y: a chemical produced in the brain that stimulates appetite, diminishes energy expenditure, and increases fat storage.

neurotransmitters: chemicals that are released at the end of a nerve cell when a nerve impulse arrives there. They diffuse

renal threshold: the blood concentration of a substance that exceeds the kidneys' capacity for reabsorption, causing the substance to be passed into the urine.

renin (REN-in): an enzyme from the kidneys that activates angiotensin.

replication (REP-lih-KAY-shun): repeating an experiment and getting the same results. The skeptical scientist, on hearing of a new, exciting finding, will ask, "Has it been replicated yet?" If it hasn't, the scientist will withhold judgment regarding the finding's validity.

requirement: the lowest continuing intake of a nutrient that will maintain a specified criterion of adequacy.

resection: the surgical removal of part of an organ or body structure.

resistant starches: starches that escape digestion and absorption in the small intestine of healthy people.

resistin (re-ZIST-in): a hormone produced by adipose cells that induces insulin resistance.

respiratory stress: abnormal gas exchange between the air and blood, resulting in lower-than-normal oxygen levels and higher-than-normal carbon dioxide levels.

resting metabolic rate (RMR): similar to the basal metabolic rate (BMR), a measure of the energy use of a person at rest in a comfortable setting, but with less stringent criteria for recent food intake and physical activity. Consequently, the RMR is slightly higher than the BMR.

reticulocytes: immature red blood cells released into blood by the bone marrow.

retina (RET-in-uh): the layer of light-sensitive nerve cells lining the back of the inside of the eye; consists of rods and cones.

retinoids (RET-ih-noyds): chemically related compounds with biological activity similar to that of retinol; metabolites of retinol.

retinol activity equivalents (RAE): a measure of vitamin A activity; the amount of retinol that the body will derive from a food containing preformed retinol or its precursor beta-carotene.

retinol-binding protein (RBP): the specific protein responsible for transporting retinol.

rheumatoid (ROO-ma-toyd) **arthritis:** a disease of the immune system involving painful inflammation of the joints and related structures.

rhodopsin (ro-DOP-sin): a light-sensitive pigment of the retina; contains the retinal form of vitamin A and the protein opsin.

riboflavin (RYE-boh-flay-vin): a B vitamin. The coenzyme forms are **FMN (flavin mononucleotide)** and **FAD (flavin adenine dinucleotide).**

rickets: the vitamin D–deficiency disease in children characterized by inadequate mineralization of bone (manifested in bowed legs or knock-knees, outward-bowed chest, and knobs on ribs). A rare type of rickets, not caused by vitamin D deficiency, is known as *vitamin D–refractory rickets.*

risk factor: a condition or behavior associated with an elevated frequency of a disease but not proved to be causal. Leading risk factors for chronic diseases include obesity, cigarette smoking, high blood pressure, high blood cholesterol, physical inactivity, and a diet high in saturated fats and low in vegetables, fruits, and whole grains.

RNA (ribonucleic acid): a compound similar to DNA, but RNA is a single strand with a ribose sugar instead of a deoxyribose sugar and uracil instead of thymine as one of its bases.

S

saccharin (SAK-ah-ren): an artificial sweetener that has been approved for use in the United States. In Canada, approval for use in foods and beverages is pending; currently available only in pharmacies and only as a tabletop sweetener, not as an additive.

saliva: the secretion of the salivary glands. Its principal enzyme begins carbohydrate digestion.

salivary glands: exocrine glands that secrete saliva into the mouth.

salt: a compound composed of a positive ion other than H^+ and a negative ion other than OH^-. An example is sodium chloride $(Na^+ Cl^-)$.

salt sensitivity: a characteristic of individuals who respond to a high salt intake with an increase in blood pressure or to a low salt intake with a decrease in blood pressure.

sarcopenia (SAR-koh-PEE-nee-ah): loss of skeletal muscle mass, strength, and quality.

satiating: having the power to suppress hunger and inhibit eating.

satiation (say-she-AY-shun): the feeling of satisfaction and fullness that occurs during a meal and halts eating. Satiation determines how much food is consumed during a meal.

satiety (sah-TIE-eh-tee): the feeling of fullness and satisfaction that occurs after a meal and inhibits eating until the next meal. Satiety determines how much time passes between meals.

saturated fat-free: less than 0.5 g of saturated fat and 0.5 g of *trans* fat per serving.

saturated fatty acid: a fatty acid carrying the maximum possible number of hydrogen atoms—for example, stearic acid. A saturated fat is composed of triglycerides in which most of the fatty acids are saturated.

scurvy: the vitamin C–deficiency disease.

secondary deficiency: a nutrient deficiency caused by something other than an inadequate intake such as a disease condition or drug interaction that reduces absorption, accelerates use, hastens excretion, or destroys the nutrient.

secondary hypertension: hypertension that results from a known physiological abnormality.

secretin (see-CREET-in): a hormone produced by cells in the duodenum wall. Target organ: the pancreas. Response: secretion of bicarbonate-rich pancreatic juice.

segmentation (SEG-men-TAY-shun): a periodic squeezing or partitioning of the intestine at intervals along its length by its circular muscles.

selective menus: menus that provide choices in some or all menu categories.

selenium (se-LEEN-ee-um): a trace element.

self-monitoring of blood glucose: home monitoring of blood glucose levels using a glucose meter.

semipermeable membrane: a membrane that allows some particles to pass through, but not others.

semiselective menus: menus that combine aspects of both selective and nonselective menus.

senile dementia: the loss of brain function beyond the normal loss of physical adeptness and memory that occurs with aging.

senile plaques: clumps of the protein fragment beta-amyloid on the nerve cells, commonly found in the brains of people with Alzheimer's dementia.

sepsis: an acute inflammatory response caused by infection; characterized by symptoms similar to those of SIRS.

serotonin (SER-oh-TONE-in): a neurotransmitter important in sleep regulation, appetite control, intestinal motility, obsessive-compulsive behaviors, and mood disorders. Serotonin is synthesized in the body from the amino acid tryptophan with the help of vitamin B_6.

set point: the point at which controls are set (for example, on a thermostat). The set-point theory that relates to body weight proposes that the body tends to maintain a certain weight by means of its own internal controls.

shock: a severe reduction in blood flow that deprives the body's tissues of oxygen and nutrients; characterized by reduced blood pressure, raised heart and respiratory rates, and muscle weakness.

shock-wave lithotripsy: a nonsurgical procedure that uses high-amplitude sound waves to fragment gallstones.

short bowel syndrome: the malabsorption syndrome that follows resection of the small intestine, which results in insufficient absorptive capacity in the remaining intestine.

sibutramine (sigh-BYOO-tra-mean): a drug used in the treatment of obesity that slows the reabsorption of serotonin in the brain, thus suppressing appetite and creating a feeling of fullness.

sickle-cell anemia: a hereditary form of anemia characterized by abnormal sickle- or crescent-shaped red blood cells. Sickled cells interfere with oxygen transport and blood flow. Symptoms are precipitated by dehydration and insufficient oxygen (as may occur at high altitudes) and include hemolytic anemia (red blood cells burst), fever, and severe pain in the joints and abdomen.

simple carbohydrates (sugars): monosaccharides and disaccharides.

sinusoids: the small, capillary-like passages that carry blood through liver tissue.

Sjögren's syndrome: an autoimmune disease characterized by the destruction of secretory glands, especially those that produce saliva and tears, resulting in dry mouth and dry eyes.

sludge: literally, a semisolid mass. Biliary sludge is made up of mucus, cholesterol crystals, and bilirubin granules.

small intestine: a 10-foot length of small-diameter intestine that is the major site of digestion of food and absorption of nutrients. Its segments are the duodenum, jejunum, and ileum.

soaps: chemical compounds that form between fatty acids and positively charged minerals.

sodium: the principal cation in the extracellular fluids of the body; critical to the maintenance of fluid balance, nerve impulse transmissions, and muscle contractions.

sodium-free and **salt-free:** less than 5 mg of sodium per serving.

soft water: water with a high sodium or potassium content.

soluble fibers: indigestible food components that dissolve in water to form a gel. An example is pectin from fruit, which is used to thicken jellies.

solutes (SOLL-yutes): the substances that are dissolved in a solution. The number of molecules in a given volume of fluid is the **solute concentration.**

somatic (so-MAT-ick) **nervous system:** the division of the nervous system that controls the voluntary muscles, as distinguished from the autonomic nervous system, which controls involuntary functions.

somatostatin (GHIH): a hormone that inhibits the release of growth hormone; the opposite of somatotropin (GH).

soup kitchens: programs that provide prepared meals to be eaten on site.

spasm: a sudden, forceful, and involuntary muscle contraction.

specialized formulas: enteral formulas designed to meet the nutrient needs of patients with specific illnesses; also called *disease-specific formulas.*

sperm: the male reproductive cell, capable of fertilizing an ovum.

sphincter (SFINK-ter): a circular muscle surrounding, and able to close, a body opening. Sphincters are found at specific points along the GI tract and regulate the flow of food particles.

spina (SPY-nah) **bifida** (BIFF-ih-dah): one of the most common types of neural tube defects; characterized by the incomplete closure of the spinal cord and its bony encasement.

spring water: water originating from an underground spring or well. It may be bubbly (carbonated), or "flat" or "still," meaning not carbonated. Brand names such as "Spring Pure" do not necessarily mean that the water comes from a spring.

standard diet: a diet that includes all foods and meets the nutrient needs of healthy people; also called a *regular diet.*

standard formulas: enteral formulas that contain mostly intact proteins and polysaccharides; also called *polymeric formulas.*

starches: plant polysaccharides composed of glucose.

steatohepatitis (STEE-ah-to-HEP-ah-TIE-tis): liver inflammation that is associated with fatty liver.

steatorrhea (stee-AT-or-REE-ah): excessive fat in the stools resulting from fat malabsorption; characterized by stools that are loose, frothy, and foul smelling due to a high fat content.

sterile: free of microorganisms, such as bacteria.

sterols (STARE-ols or STEER-ols): compounds containing a four ring carbon structure with any of a variety of side chains attached.

stevia (STEE-vee-ah): a South American shrub whose leaves are used as a sweetener; sold in the United States as a dietary supplement that provides sweetness without kcalories.

stoma (STOE-ma): a surgical opening made in the abdominal wall.

stomach: a muscular, elastic, saclike portion of the digestive tract that grinds and churns swallowed food, mixing it with acid and enzymes to form chyme.

stools: waste matter discharged from the colon; also called feces (FEE-seez).

stress: any threat to a person's well-being; a demand placed on the body to adapt.

stress fractures: bone damage or breaks caused by stress on bone surfaces during exercise.

stress response: the chemical and physical changes that occur within the body during stress.

stressors: environmental elements, physical or psychological, that cause stress.

stricture: abnormal narrowing of a passageway; often due to inflammation, scarring, or a congenital abnormality.

stroke: a sudden injury to brain tissue resulting from impaired blood flow through an artery that supplies blood to the brain; also called a *cerebrovascular accident.*

structure-function claims: statements that characterize the relationship between a nutri-

ent or other substance in a food and its role in the body.

struvite (STROO-vite): crystals of magnesium ammonium phosphate.

subclavian (sub-KLAY-vee-an) **vein:** the vein that provides passageway from the lymphatic system to the vascular system.

subclinical deficiency: a deficiency in the early stages, before the outward signs have appeared.

subcutaneous (sub-cue-TAY-nee-us): beneath the skin.

subjects: the people or animals participating in a research project.

successful weight-loss maintenance: achieving a weight loss of at least 10% of initial body weight and maintaining the loss for at least one year.

sucralose (SUE-kra-lose): an artificial sweetener approved for use in the United States and Canada.

sucrase: an enzyme that hydrolyzes sucrose

sucrose (SUE-krose): a disaccharide composed of glucose and fructose; commonly known as table sugar, beet sugar, or cane sugar. Sucrose also occurs in many fruits and some vegetables and grains.

sudden infant death syndrome (SIDS): the unexpected and unexplained death of an apparently well infant; the most common cause of death of infants between the second week and the end of the first year of life; also called *crib death.*

sugar replacers: sugarlike compounds that can be derived from fruits or commercially produced from dextrose; also called **sugar alcohols** or **polyols.** Sugar alcohols are absorbed more slowly than other sugars and metabolized differently in the human body; they are not readily utilized by ordinary mouth bacteria. Examples are **maltitol, mannitol, sorbitol, xylitol, isomalt,** and **lactitol.**

sugar-free: less than 0.5 g of sugar per serving.

sulfate: the oxidized form of sulfur.

sulfur: a mineral present in the body as part of some proteins.

supplement: any pill, capsule, tablet, liquid, or powder that contains vitamins, minerals, herbs, or amino acids; intended to increase dietary intake of these substances.

surrogate: a substitute; a person who takes the place of another.

sushi: vinegar-flavored rice and seafood, typically wrapped in seaweed and stuffed with colorful vegetables. Some sushi is stuffed with raw fish; other varieties contain cooked seafood.

syringes: devices used for injecting medications. A syringe consists of a hypodermic needle attached to a hollow tube with a plunger inside.

systemic (sih-STEM-ic): relating to the entire body.

systemic inflammatory response syndrome (SIRS): a whole-body response to acute inflammation; characterized by raised heart and respiratory rates, abnormal white blood cell counts, and abnormal body temperature.

T

T cell: a lymphocyte that attacks antigens; functions in cell-mediated immunity.

tagatose (TAG-ah-tose): a monosaccharide structurally similar to fructose that is incompletely absorbed and thus provides only 1.5 kcalories per gram; approved for use as a "generally recognized as safe" ingredient.

TCA cycle or **tricarboxylic** (try-car-box-ILL-ick) **acid cycle:** a series of metabolic reactions that break down molecules of acetyl CoA to carbon dioxide and hydrogen atoms; also called the **Kreb's cycle** after the biochemist who elucidated its reactions.

tempeh (TEM-pay): a fermented soybean food, rich in protein and fiber.

teratogenic (ter-AT-oh-jen-ik): causing abnormal fetal development and birth defects.

testosterone: a steroid hormone from the testicles, or testes. The steroids, as explained in Chapter 5, are chemically related to, and some are derived from, the lipid cholesterol.

textured vegetable protein: processed soybean protein used in vegetarian products such as soy burgers; see also *meat replacements.*

theory: a tentative explanation that integrates many and diverse findings to further the understanding of a defined topic.

therapeutic touch: a technique of passing hands over a patient to purportedly identify energy imbalances and transfer healing power from therapist to patient; also called *laying on of hands.*

thermic effect of food (TEF): an estimation of the energy required to process food (digest, absorb, transport, metabolize, and store ingested nutrients); also called the **specific dynamic effect (SDE)** of food or the **specific dynamic activity (SDA)** of food. The sum of the TEF and any increase in the metabolic rate due to overeating is known as **diet-induced thermogenesis (OIT).**

thermogenesis: the generation of heat; used in physiology and nutrition studies as an index of how much energy the body is expending.

thiamin (THIGH-ah-min): a B vitamin. The coenzyme form is **TPP (thiamin pyrophosphate).**

thirst: a conscious desire to drink.

thoracic (thor-ASS-ic) **duct:** the main lymphatic vessel that collects lymph and drains into the left subclavian vein.

thrombosis (throm-BOH-sis): the formation or presence of a blood clot in blood vessels. A *coronary thrombosis* occurs in a coronary artery, and a *cerebral thrombosis* occurs in an artery that supplies blood to the brain.

thrombus: a blood clot formed within a blood vessel that remains attached to its place of origin.

thrush: a fungal infection of the mouth and throat, most often caused by *Candida albicans.*

thyroid-stimulating hormone (TSH): a hormone secreted by the pituitary that stimulates the thyroid gland to secrete its hormones—thyroxine and triiodothyronine. The release of TSH is mediated by TSH-releasing hormone (TRH).

tissue rejection: destruction of donor tissue by the recipient's immune system, which recognizes the donor cells as foreign.

tocopherol (tuh-KOFF-er-ol): a general term for several chemically related compounds, one of which has vitamin E activity. (See Appendix C for chemical structures.)

tofu (TOE-foo): a curd made from soybeans, rich in protein and often fortified with calcium; used in many Asian and vegetarian dishes in place of meat.

Tolerable Upper Intake Level (UL): the maximum daily amount of a nutrient that appears safe for most healthy people and beyond which there is an increased risk of adverse health effects.

tolerance level: the maximum amount of residue permitted in a food when a pesticide is used according to the label directions.

total nutrient admixture (TNA): a parenteral solution that contains dextrose, amino acids, and lipids; also called a **3-in-1 solution** or an **all-in-one solution.**

total parenteral nutrition (TPN): a type of nutrition support in which intravenous feedings are delivered into a central vein.

trabecular (tra-BECK-you-lar) **bone:** the lacy inner structure of calcium crystals that supports the bone's structure and provides a calcium storage bank.

trace minerals: essential mineral nutrients found in the human body in amounts smaller than 5 g; sometimes called **microminerals.**

trachea (TRAKE-ee-uh): the air passageway from the larynx to the lungs; also called the *windpipe.*

traditional Chinese medicine (TCM): an approach to medical care based on the concept that illness can be cured by enhancing the flow of qi (energy) within a person's body. Treatments may include herbal therapies, physical exercises, meditation, acupuncture, and remedial massage.

trans fat-free: less than 0.5 g of *trans* fat and less than 0.5 g of saturated fat per serving.

trans-fatty acids: fatty acids with hydrogens on opposite sides of the double bond.

transamination (TRANS-am-ih-NAY-shun): the transfer of an amino group from one amino acid to a keto acid, producing a new nonessential amino acid and a new keto acid.

transferrin (trans-FAIR-in): the iron transport protein.

transient hypertension of pregnancy: high blood pressure that develops in the second half of pregnancy and resolves after childbirth, usually without affecting the outcome of the pregnancy.

transient ischemic attacks (TIAs): brief ischemic strokes that cause short-term neurological symptoms.

transnasal: through the nose. A *transnasal feeding tube* is one that is inserted through the nose.

triglycerides (try-GLISS-er-rides): the chief form of fat in the diet and the major storage form of fat in the body; composed of a molecule of glycerol with three fatty acids attached; also called *triacylglycerols* (try-ay-seel-GLISS-er-ols).

tripeptide: three amino acids bonded together.

tube feedings: liquid formulas delivered through a tube placed in the stomach or intestine.

tubules: tubelike structures of the nephron that process filtrate during urine production. The tubules are surrounded by capillaries that reabsorb substances retained by tubule cells.

tumor: an abnormal tissue mass that has no physiological function; also called a *neoplasm* (NEE-oh-plazm).

turbinado (ter-bih-NOD-oh) **sugar:** sugar produced using the same refining process as white sugar, but without the bleaching and anti-caking treatment. Traces of molasses give turbinado its sandy color.

type 1 diabetes: the type of diabetes that accounts for 5 to 10% of diabetes cases and usually results from autoimmune destruction of pancreatic beta cells and failure to produce insulin.

type 2 diabetes: the type of diabetes that accounts for 90 to 95% of diabetes cases and usually results from insulin resistance coupled with insufficient insulin secretion.

type I osteoporosis: osteoporosis characterized by rapid bone losses, primarily of trabecular bone.

type II osteoporosis: osteoporosis characterized by gradual losses of both trabecular and cortical bone.

U

ulcer: a lesion of the skin or mucous membranes characterized by inflammation and damaged tissues. See also *peptic ulcer.*

ulcerative colitis (ko-LY-tis): an inflammatory bowel disease that involves the colon. Inflammation affects the mucosa and submucosa of the intestinal wall.

ultrafiltration: removal of fluids and solutes from blood by using pressure to transfer the blood across a semipermeable membrane.

umbilical (um-BILL-ih-cul) **cord:** the ropelike structure through which the fetus's veins and arteries reach the placenta; the route of nourishment and oxygen to the fetus and the route of waste disposal from the fetus. The scar in the middle of the abdomen that marks the former attachment of the umbilical cord is the **umbilicus** (um-BILL-ih-cus), commonly known as the "belly button."

undernutrition: deficient energy or nutrients.

underweight: body weight below some standard of acceptable weight that is usually defined in relation to height (such as BMI); BMI below 18.5.

unsaturated fatty acid: a fatty acid that lacks hydrogen atoms and has at least one double bond between carbons (includes monounsaturated and polyunsaturated fatty acids). An **unsaturated fat** is composed of triglycerides in which most of the fatty acids are unsaturated.

unspecified eating disorders: eating disorders that do not meet the defined criteria for specific eating disorders.

urea (you-REE-uh): the principal nitrogen-excretion product of protein metabolism. Two ammonia fragments are combined with carbon dioxide to form urea.

urea kinetic modeling: a method of determining the adequacy of dialysis treatment by calculating the urea clearance from blood.

uremia (you-REE-me-ah): the abnormal accumulation of nitrogen-containing substances, especially urea, in the blood; also called *azotemia* (AZE-oh-TEE-me-ah).

uremic syndrome: the cluster of symptoms associated with a GFR below 15 mL/min, including uremia, anemia, bone disease, hormonal imbalances, bleeding impairment, increased cardiovascular disease risk, and reduced immunity.

uterus (YOU-ter-us): the muscular organ within which the infant develops before birth.

V

vagotomy (vay-GOT-oh-mee): surgery that severs the vagus nerve in order to suppress gastric acid secretion. This surgery may require a follow-up *pyloroplasty* procedure to allow stomach drainage.

vagus nerve: the cranial nerve that regulates hydrochloric acid secretion and peristalsis. Effects elsewhere in the body include regulation of heart rate and bronchiole constriction.

validity (va-lid-ih-tee): having the quality of being founded on fact or evidence.

variables: factors that change. A variable may depend on another variable (for example, a child's height depends on his age), or it may be independent (for example, a child's height does not depend on the color of her eyes). Sometimes both variables correlate with a third variable (a child's height and eye color both depend on genetics).

varices (VAH-rih-seez): abnormally dilated blood vessels (singular: *varix*).

variety (dietary): eating a wide selection of foods within and among the major food groups.

vasoconstrictor (VAS-oh-kon-STRIK-tor): a substance that constricts or narrows the blood vessels.

vegans (VEE-gans): people who exclude all animal-derived foods (including meat, poultry, fish, eggs, and dairy products) from their diets; also called **pure vegetarians, strict vegetarians,** or **total vegetarians.**

vegetarians: a general term used to describe people who exclude meat, poultry, fish, or other animal-derived foods from their diets.

veins (VANES): vessels that carry blood to the heart.

very low sodium: 35 mg or less per serving.

villi (VILL-ee, or VILL-eye): fingerlike projections from the folds of the small intestine; singular **villus.**

viscous: a gel-like consistency.

vitamin A: all naturally occurring compounds with the biological activity of retinol (RET-ih-nol), the alcohol form of vitamin A.

vitamin A activity: a term referring to both the active forms of vitamin A and the precursor forms in foods without distinguishing between them.

vitamin B$_6$: a family of compounds—pyridoxal, pyridoxine, and pyridoxamine. The primary active coenzyme form is **PLP (pyridoxal phosphate).**

vitamin B$_{12}$: a B vitamin characterized by the presence of cobalt (see Figure 13-12, p. 462). The active forms of coenzyme B$_{12}$ are **methylcobalamin** and **deoxyadenosylcobalamin.**

vitamins: organic, essential nutrients required in small amounts by the body for health.

VLDL (very-low-density lipoprotein): the type of lipoprotein made primarily by liver cells to transport lipids to various tissues in the body; composed primarily of triglycerides.

vomiting: expulsion of the contents of the stomach up through the esophagus to the mouth.

vulnerable plaque: a form of plaque, susceptible to rupture, that is lipid-rich and has only a thin, fibrous barrier between the arterial lumen and the plaque's lipid core.

W

waist circumference: an anthropometric measurement used to assess a person's abdominal fat.

wasting: the gradual atrophy (loss) of body tissues; associated with protein-energy malnutrition or chronic illness.

water balance: the balance between water intake and output (losses).

water intoxication: the rare condition in which body water contents are too high in all body fluid compartments.

wean: to gradually replace breast milk with infant formula or other foods appropriate to an infant's diet.

websites: Internet resources composed of text and graphic files, each with a unique URL (Uniform Resource Locator) that names the site (for example, www.usda.gov).

weight management: maintaining body weight in a healthy range by preventing gradual weight gain over time and losing weight if overweight.

well water: water drawn from ground water by tapping into an aquifer.

Wernicke-Korsakoff (VER-nee-key KORE-sah-kof) **syndrome:** a neurological disorder typically associated with chronic alcoholism and caused by a deficiency of the B vitamin thiamin; also called *alcohol-related dementia.*

wheat gluten (GLU-ten): a family of water-insoluble proteins in wheat; includes the gliadin (GLY-ah-din) proteins that are toxic to persons with celiac disease.

whey protein: a by-product of cheese production; falsely promoted as increasing muscle mass. Whey is the watery part of milk that separates from the curds.

white sugar: pure sucrose or "table sugar," produced by dissolving, concentrating, and recrystallizing raw sugar.

whole grain: a grain milled in its entirety (all but the husk), not refined.

wine: an alcoholic beverage made by fermenting grape juice.

World Wide Web (the web, commonly abbreviated **www**): a graphical subset of the Internet.

X

xanthophylls (ZAN-tho-fills): pigments found in plants; responsible for the color changes seen in autumn leaves.

xerophthalmia (zer-off-THAL-mee-uh): progressive blindness caused by severe vitamin A deficiency.

xerosis (zee-ROW-sis): abnormal drying of the skin and mucous membranes; a sign of vitamin A deficiency.

xerostomia: dry mouth caused by reduced salivary flow.

Y

yogurt: milk product that results from the fermentation of lactic acid in milk by *Lactobacillus bulgaricus* and *Streptococcus thermophilus.*

Z

Zollinger-Ellison syndrome: a condition characterized by the presence of gastrin-secreting tumors in the duodenum or pancreas.

zygote (ZY-goat): the product of the union of ovum and sperm; so-called for the first two weeks after fertilization.

Index

Aids to Calculation

Many mathematical problems have been worked out in the "How to" sections of the text and practice problems have been provided in the "Nutrition Calculations" sections at the end of some chapters. These pages offer additional help and examples.

Conversions

A conversion factor is a fraction that converts a measurement expressed in one unit to another unit—for example, from pounds to kilograms or from feet to meters. To create a conversion factor, an equality (such as 1 kilogram = 2.2 pounds) is expressed as a fraction:

$$\frac{1 \text{ kg}}{2.2 \text{ lb}} \text{ and } \frac{2.2 \text{ lb}}{1 \text{ kg}}$$

To convert the units of a measurement, use the fraction with the desired unit in the numerator.

Example 1: Convert a weight of 130 pounds to kilograms. Multiply 130 pounds by the conversion factor that includes both pounds and kilograms, with the desired unit (kilograms) in the numerator:

$$130 \text{ lb} \times \frac{1 \text{ kg}}{2.2 \text{ lb}} = \frac{130 \text{ kg}}{2.2} = 59 \text{ kg}$$

Alternatively, to convert a measurement from one unit of measure to another, multiply the given measurement by the appropriate equivalent found in the accompanying table of weights and measures.

Example 2: Convert 64 fluid ounces to liters. Locate the equivalent measure from the table (1 ounce = 0.03 liter) and multiply the number of ounces by 0.03:

$$64 \text{ oz} \times 0.03 \text{ oz/L} = 1.9 \text{ L}$$

Percentages

A percentage is a fraction whose denominator is 100. For example:

$$50\% = \frac{50}{100}$$

Like other fractions, percentages are used to express a portion of a quantity. Fractions whose denominators are numbers other than 100 can be converted to percentages by first dividing the numerator by the denominator and then multiplying the result by 100.

Example 3: Express $^5/_8$ as a percent.

$$\frac{5}{8} = 5 \div 8 = 0.625$$

$$0.625 \times 100 = 62.5\%$$

The following examples show how to calculate specific percentages.

Example 4: Suppose your energy intake for the day is 2000 kcalories (kcal) and your recommended energy intake is 2400 kcalories. What percent of the recommended energy intake did you consume?

Divide your intake by the recommended intake.
2000 kcal (intake) ÷ 2400 kcal (recommended) = 0.83

Multiply by 100 to express the decimal as a percent.
0.83 × 100 = 83%

Example 5: Suppose a man's intake of vitamin C is 120 milligrams and his RDA is 90 milligrams. What percent of the RDA for vitamin C did he consume?

Divide the intake by the recommended intake.
120 mg (intake) ÷ 90 mg (RDA) = 1.33

Multiply by 100 to express the decimal as a percent.
1.33 × 100 = 133%

Example 6: Dietary recommendations suggest that carbohydrates provide 45 to 65 percent of the day's energy intake. If your energy intake is 2000 kcalories, how much carbohydrate should you eat?

Because this question has a range of acceptable answers, work the problem twice. First, use 45% to find the least amount you should eat.

Divide 45 by 100 to convert to a decimal.
45 ÷ 100 = 0.45

Multiply kcalories by 0.45.
2000 kcal × 0.45 = 900 kcal

Divide kcalories by 4 to convert carbohydrate kcal to grams.
900 kcal ÷ 4 kcal/g = 225 g

Now repeat the process using 65% to find the maximum number of grams of carbohydrates you should eat.

Divide 65 by 100 to convert it to a decimal.
65 ÷ 100 = 0.65

Multiply kcalories by 0.65.
2000 kcal × 0.65 = 1300 kcal

Divide kcalories by 4 to convert carbohydrate kcal to grams.
1300 kcal ÷ 4 kcal/g = 325 g

If you plan for between 45% and 65% of your 2000-kcalorie intake to be from carbohydrates, you should eat between 225 grams and 325 grams of carbohydrates.

Weights and Measures

LENGTH

1 centimeter (cm) = 0.39 inches (in)
1 foot (ft) = 30 centimeters (cm)
1 inch (in) = 2.54 centimeters (cm)
1 meter (m) = 39.37 inches (in)

WEIGHT

1 gram (g) = 0.001 kilogram (kg)
    = 1000 milligram (mg)
    = .035 ounce (oz)
1 kilogram (kg) = 1000 grams (g)
    = 2.2 pounds (lb)
1 microgram (µg) = 0.001 milligram (mg)
1 milligram (mg) = 0.001 gram (g)
    = 1000 microgram (µg)
1 ounce (oz) = 28 grams (g)
    = 0.03 kilograms (kg)
1 pound (lb) = 454 grams (g)
    = 0.45 kilograms (kg)
    = 16 ounces (oz)

VOLUME

1 cup = 16 tablespoons (tbs or T)
    = 0.25 liter (L)
    = 236 milliliters (mL, commonly rounded to 250 mL)
    = 8 ounces (oz)
1 liter (L) = 33.8 fluid ounces (fl oz)
    = 0.26 gallons (gal)
    = 2.1 pints (pt)
    = 1.06 quarts (qt)
    = 1000 milliliters (mL)
1 milliliter (mL) = 0.001 liter (L)
    = 0.03 fluid ounces (fl oz)
1 ounce (oz) = 0.03 liter (L)
    = 30 milliliters (mL)
1 pint (pt) = 2 cups (c)
    = 0.47 liters (L)
    = 16 ounces (oz)
1 quart (qt) = 4 cups (c)
    = 0.95 liters (L)
    = 32 ounces (oz)
1 tablespoon (tbs or T) = 3 teaspoons (tsp)
    =15 milliliters (mL)
1 teaspoon (tsp) = 5 milliliters (mL)
1 gallon (gal) = 16 cups (c)
    = 3.8 liters (L)
    = 128 ounces (oz)

ENERGY

1 millijoule (mJ) = 240 kcalories (kcal)
1 kilojoule (kJ) = 0.24 kcalories (kcal)
1 kcalorie (kcal) = 4.2 kilojoule (kJ)
1 g alcohol = 7 kcal = 29 kJ
1 g carbohydrate = 4 kcal = 17 kJ
1 g fat = 9 kcal = 37 kJ
1 g protein = 4 kcal = 17 kJ

TEMPERATURE

To change from Fahrenheit (°F) to Celsius (°C), subtract 32 from the Fahrenheit measure and then multiply that result by 0.56.

To change from Celsius (°C) to Fahrenheit (°F), multiply the Celsius measure by 1.8 and add 32 to that result.

A comparison of some useful temperatures is given below.

	Celsius	Fahrenheit
Boiling point	100°C	212°F
Body temperature	37°C	98.6°F
Freezing point	0°C	32°F

Daily Values for Food Labels

The Daily Values are standard values developed by the Food and Drug Administration (FDA) for use on food labels. The values are based on 2000 kcalories a day for adults and children over 4 years old. Chapter 2 provides more details.

Nutrient	Amount
Protein[a]	50 g
Thiamin	1.5 mg
Riboflavin	1.7 mg
Niacin	20 mg NE
Biotin	300 µg
Pantothenic acid	10 mg
Vitamin B_6	2 mg
Folate	400 µg
Vitamin B_{12}	6 µg
Vitamin C	60 mg
Vitamin A	5000 IU[b]
Vitamin D	400 IU[b]
Vitamin E	30 IU[b]
Vitamin K	80 µg
Calcium	1000 mg
Iron	18 mg
Zinc	15 mg
Iodine	150 µg
Copper	2 mg
Chromium	120 µg
Selenium	70 µg
Molybdenum	75 µg
Manganese	2 mg
Chloride	3400 mg
Magnesium	400 mg
Phosphorus	1000 mg

[a] The Daily Values for protein vary for different groups of people: pregnant women, 60 g; nursing mothers, 65 g; infants under 1 year, 14 g; children 1 to 4 years, 16 g.
[b] Equivalent values for nutrients expressed as IU are: vitamin A, 1500 RAE (assumes a mixture of 40% retinol and 60% beta-carotene); vitamin D, 10 µg; vitamin E, 20 mg.

Food Component	Amount	Calculation Factors
Fat	65 g	30% of kcalories
Saturated fat	20 g	10% of kcalories
Cholesterol	300 mg	Same regardless of kcalories
Carbohydrate (total)	300 g	60% of kcalories
Fiber	25 g	11.5 g per 1000 kcalories
Protein	50 g	10% of kcalories
Sodium	2400 mg	Same regardless of kcalories
Potassium	3500 mg	Same regardless of kcalories

GLOSSARY OF NUTRIENT MEASURES

kcal: kcalories; a unit by which energy is measured (Chapter 1 provides more details).

g: grams; a unit of weight equivalent to about 0.03 ounces.

mg: milligrams; one-thousandth of a gram.

µg: micrograms; one-millionth of a gram.

IU: international units; an old measure of vitamin activity determined by biological methods (as opposed to new measures that are determined by direct chemical analyses). Many fortified foods and supplements use IU on their labels.
- For vitamin A, 1 IU = 0.3 µg retinol, 3.6 µg β-carotene, or 7.2 µg other vitamin A carotenoids
- For vitamin D, 1 IU = 0.02 µg cholecalciferol
- For vitamin E, 1 IU = 0.67 natural α-tocopherol (other conversion factors are used for different forms of vitamin E)

mg NE: milligrams niacin equivalents; a measure of niacin activity (Chapter 10 provides more details).
- 1 NE = 1 mg niacin
  = 60 mg tryptophan (an amino acid)

µg DFE: micrograms dietary folate equivalents; a measure of folate activity (Chapter 10 provides more details).
- 1 µg DFE = 1 µg food folate
  = 0.6 µg fortified food or supplement folate taken with food
  = 0.5 µg supplement folate taken on an empty stomach

µg RAE: micrograms retinol activity equivalents; a measure of vitamin A activity (Chapter 11 provides more details).
- 1 µg RAE = 1 µg retinol
  = 12 µg β-carotene
  = 24 µg other vitamin A carotenoids

mmol: millimoles; one-thousanth of a mole, the molecular weight of a substance. To convert mmol to mg, multiply by the atomic weight of the substance.
- For sodium, mmol × 23 = mg Na
- For chloride, mmol × 35.5 = mg Cl
- For sodium chloride, mmol × 58.5 = mg NaCl

1 What are the requirements of my writing project?

If I am writing to fulfill an assignment, do I understand that assignment? If I am writing on my own, do I have definite expectations of what I will accomplish?

2 As I proceed in this project, what do I need to know?

Do I have a good understanding of my subject, or do I need more information? Have I considered the possible audiences who might read my writing?

3 What hypothesis can I use as my working purpose?

How many different hypotheses can I formulate about my subject? Which of them seems to direct and control my information in the most effective manner?

4 What purpose have I discovered for this writing project?

Has my purpose changed as I learned more about my subject and audience? If so, in what ways? Have I discovered, by working with a hypothesis or hypotheses, what I want to do in my writing?

5 What is my thesis?

How can I state my main idea about my subject in a thesis sentence? Does my thesis limit the scope of my writing to what I can demonstrate in the available space? Does it focus my writing on one specific assertion? Does it make an exact statement about what my writing intends to do?

1 What is my general impression of my writing?

Do I find my writing clear, unambiguous, and likely to engage my readers? Have I carried out my purpose at every level; that is, am I satisfied that the *how* of my writing — its attitude, organization, and language — conveys the *what* of my ideas?

2 What tone have I established in my writing?

Is my tone informative, affective, or a blend of both? How much distance have I maintained between myself and my readers? Is my tone appropriate for my subject and audience? Is it maintained consistently?

3 How can I characterize the overall style of my writing?

Have I written this essay in a moderate style, opting for more or less colloquialism or formality as my purpose requires? Does my purpose in fact require me to be overtly colloquial or formal?

4 Are my sentences well constructed and easy on the ear?

Have I written sentences varying in length and style so that they hold my readers' interest? Have I avoided the choppiness that comes from too many basic or loosely coordinated sentences? Have I avoided the density that comes from too many complicated sentences with multiple subordinate clauses?

5 Have I used words as effectively as possible?

Do the connotations and denotations of my words support my purpose? Have I avoided unnecessary formalities and slang? Is my language specific and, when appropriate, vivid? Have I inadvertently mixed metaphors? Have I used imagery successfully to *heighten* effects, not merely to strive after them?

WRITING
WITH A
PURPOSE

Joseph F. Trimmer

Ball State University

James M. McCrimmon

WRITING WITH A PURPOSE

Ninth Edition

HOUGHTON MIFFLIN COMPANY   Boston

Dallas   Geneva, Illinois   Palo Alto   Princeton, New Jersey

Printed in the U.S.A.

Library of Congress Catalog Card Number: 87–80258

ISBN: 0–395–35777–2

  BCDEFGHIJ–VH–9543210–898

ILLUSTRATION CREDITS

Cover photograph by James Scherer. Sculpture by José de
Rivera, *Construction 195;* The Currier Gallery of Art, Manchester, New Hampshire. Gift of Grace Borgenicht.

We are grateful to the following individuals for permission
to reproduce their photographs on the opening pages of the
chapters in this text:

Chapter 1:   Michael deCamp/The Image Bank
Chapter 2:   Lou Jones
Chapter 3:   Barbara Kasten
Chapter 4:   Lou Jones
Chapter 5:   Todd Gipstein/Photo Researchers
Chapter 6:   Harold Sund/The Image Bank
Chapter 7:   H. Wendler/The Image Bank
Chapter 8:   Lou Jones
Chapter 9:   Ed Carlin/The Picture Cube
Chapter 10: David Witbeck/The Picture Cube
Chapter 11: Mel Ingber/Photo Researchers
Chapter 12: Ellis Herwig/The Picture Cube
Chapter 13: Judith Sedwick/The Picture Cube
Chapter 14: Herb Snitzer/Stock, Boston

TEXT CREDITS

"Colloquial" and "Style" (definitions) and "Contract"
(entry facsimile), Copyright © 1981 by Houghton Mifflin Company. Reprinted by permission of *The American Heritage Dictionary of the English Language.*

Bill Barich, "The Crazy Life," November 3, 1986.
Reprinted by permission; © 1986 Bill Barich. Originally in *The
New Yorker.*

Samuel Beckett, "Breath" reprinted from *The Collected
Shorter Plays of Samuel Beckett* reprinted by permission of
Grove Press Inc. and Faber & Faber Ltd. Copyright © 1970 by
Samuel Beckett.

Copyright page continues on page 509.

For Carol

BRIEF CONTENTS

Note: See also complete Table of Contents on page ix.

CONTENTS

ix

PREFACE

INTEGRATING PURPOSE AND PROCESS

Writing with a Purpose has always been distinguished by its emphasis on the role of purpose in the writing process, its comprehensive coverage of the materials and problems basic to the introductory writing course, and its effective use of examples and exercises to illustrate how writers make decisions that produce successful writing. Although the revisions embodied in the Ninth Edition retain and reinforce these traditional features of the text, they also introduce and incorporate the best of contemporary theory and practice in the teaching of writing. The result is a blend of familiar and new material invigorated by fresh approaches and examples and enlivened by student writing that evolves through the various stages of the writing process.

PART ONE: THE WRITING PROCESS

Overview

Part One, "The Writing Process," covers all aspects of composing from planning through revising. Chapter 1, which provides an overview of this process, details the variety of approaches writers use to complete their tasks and discusses those three activities common to every writing situation: selecting your subject, analyzing your audience, and determining your purpose. *Purpose* receives an expanded definition that is carefully reinforced throughout the text as the principal touchstone by which writers measure their progress through the writing process.

Planning

Drafting

Revising

Writing in progress: Expressive to investigative

The remainder of Part One, enriched by new examples and exercises, focuses on the three stages of the writing process — planning, drafting, and revising. Chapter 2 offers multiple planning strategies, some new to this edition, to demonstrate how students can discover and evaluate their thinking in writing. Chapter 3 presents methods to arrange and assess the material discovered in planning to guide the creation of a discovery draft and then a more successful second draft. Chapter 4 defines the revising process, demonstrates methods for revising an essay, and then provides an extended case study of revising from discovery draft to final draft. All the chapters in Part One are unified by recurring student writing-in-progress that illustrates a range of projects from expressive writing about personal experience to investigative writing about academic subjects.

PART TWO: THE EXPRESSION OF IDEAS

Writing skills

Professional writers: Comments on writing

Discovering structure

Composing an argument

Writing in progress: Thematic focus

Professional writers: Essays

Part Two (Chapters 5 through 10) has been extensively revised. The chapters are as thorough and as substantive in covering the discrete skills of effective writing — methods of development, argument, paragraphs, sentences, diction, tone and style — as were those in the Eighth Edition. But each chapter has been totally reconceived so as to place each skill within the composing process. This is done partly through the use of quotations from personal interviews with professional writers who explain their own method for using the skill, and partly through the work of student writers who try to employ the skill in their own writing process. In particular, Chapter 5, "Common Methods of Development," and Chapter 6, "Argument," have been completely rewritten to illustrate how a student writer works with the patterns of exposition and the structures of argument to plan, draft, and revise a compelling piece of writing. The writing examples and exercises in each chapter have been selected to cluster around a different disciplinary theme to provide readers with the opportunity to discover provocative comparisons. The review exercises include an extended piece of writing from the professional writer interviewed, tying the chapter together by supplying a "final draft" where the skill discussed can be examined and analyzed.

PART THREE: SPECIAL ASSIGNMENTS

The essay examination

The critical essay

The research paper

Student research paper

Part Three, "Special Assignments," has also been completely revised to demonstrate how the essay exam, the critical essay, and the research paper evolve from a writing process. Student writers, guided by the advice of professional writers, work through the special stages of each assignment. Chapter 11 illustrates how to evaluate exam questions, to abstract key concepts and phrases, and to use this material to develop ideas and reach conclusions to achieve successful essay exam answers. Chapter 12 demonstrates how the use of specific reading and writing strategies enables one student to compose a critical essay that integrates his personal response to a literary work with his understanding of the basic elements of literature.

Chapters 13 and 14 provide extensive analysis and illustrations of the many steps embedded in *planning* and *writing* the research paper. Chapter 13 demonstrates how to select a subject for research; how to select, assess, and analyze sources; and how to use current research tools, such as the computer search, to locate information. Chapter 14 presents the methods by which information composed by other researchers can be incorporated into and help advance the student's own research paper. Special attention is given to the purpose and procedures of quoting, documenting, and listing sources. The whole range of planning and writing activities required to produce a successful research paper is illustrated by one student's progress through the process. Her paper, "Light and Literacy: The Windows of Chartres," concludes the chapter and is fully annotated so that readers can assess the decisions made during the writing process and the methods used to embody those decisions in an appropriate format.

APPENDIX: WRITING WITH A WORD PROCESSOR

Planning/Drafting Revising/Editing

An Appendix, new to this edition, covers the special applications of the word processor to the writing process of student writers. In addition to suggestions for adapting the planning, drafting, and revising strategies in the text for use on the computer, "Writing with a Word Processor" supplies additional strategies uniquely suited to work on the computer as well as advice and cautions useful for those just learning or expanding their use of this technology.

SPECIAL FEATURES

Integrated art program

A range of writing assignments

Two special features of the Ninth Edition are the art program and the writing assignments. The artwork that opens each chapter was selected to serve as a metaphor for many of the ideas discussed in the chapter. The rest of the artwork, placed throughout the chapters, illustrates terms or themes or serves as the focus for exercises and writing assignments. The writing assignments that conclude each chapter exhibit the range of the writing in the text — expressive, informative, persuasive, investigative, and interpretive — and encourage students to begin writing full essays as early as the first chapter. The final assignment in each chapter asks students to interpret the visual metaphor that begins the chapter, requiring them to think and write again about the purpose of each chapter in *Writing with a Purpose.*

HANDBOOK OF GRAMMAR AND USAGE

The Handbook of Grammar and Usage has been thoroughly revised and reorganized for this edition. A new and easier to use coding system, new explanations and examples, and new and more abundant exercises make the Handbook a more useful tool for classroom or individual review on a systematic basis or in response to occasional need or questions. (The text is also available in a brief version without the handbook for instructors who do not use a handbook or who use a separate handbook.)

ANCILLARIES

For the student

Accompanying the main text and new to this edition is *Resources for Writing with a Purpose,* by Brock Dethier of the University of New Hampshire, a set of writing exercises and assignments keyed to the writing process and to the writing strategies discussed in the text and related to readings on subjects from academic disciplines ranging from business to medicine, from contemporary history to film studies. In combination, the writing assignments and readings encourage students to think and write more frequently and with greater focus on topics of current interest to which they will be pleased to find that they have something to contribute.

For the instructor

The instructor's manual accompanying the text, *Teaching with a Purpose,* by Alice Gillam-Scott of the University of Wisconsin–Milwaukee, features an introductory essay for new instructors that discusses the benefits of various classroom paradigms, comments on all aspects of the Ninth Edition, and provides answers for all the exercises. In addition, the author provides practical advice throughout *Teaching with a Purpose* for using the text in traditional composition classrooms, with peer tutors and small groups, and in computer-assisted classrooms; she provides additional activities for each of these classroom configurations, as well, and an annotated bibliography of the theoretical literature on which the text is based.

Other supporting materials

Other ancillaries available include:

- The Writing Process Workshop, a set of transparencies that provides an overview of the writing process by showing one student's writing in progress as she works through one idea from planning to final draft.

- Grade Performance Analyzer (GPA), a computerized grade book.

- Works-in-Progress: Houghton Mifflin College Word Processing (WIP), a word processing program for students' use.

- Instructor's Support Package, two sets of exercises to accompany the Handbook of Grammar and Usage and three sets of Diagnostic Tests (also available in a computerized version).

Please contact your regional sales office for additional information on these and other items.

ACKNOWLEDGMENTS

It is a pleasure to acknowledge those who have helped shape this edition of *Writing with a Purpose.*

For his counsel throughout the planning, drafting, and revising of the Ninth Edition, thanks to James M. McCrimmon.

For his help in undertaking the substantial revision and reorganization of the Handbook of Grammar and Usage, my thanks to Robert Perrin, Indiana State University.

For her help in composing the instructor's guide, *Teaching with a Purpose,* my thanks to Alice Gillam-Scott, University of Wisconsin–Milwaukee.

For their extensive comments on the plans for the Ninth Edition and for their criticism of the evolution of those plans through several drafts of the manuscript, I would like to thank the following people: Katherine Adams, Loyola University, New Orleans; John L. Adams, University of New Orleans; Mary Helen Adams, Hopkinsville Community College; Richard P. Batteiger, Oklahoma State University; Mary Bly, University of California, Davis; Daniel Brislane, Saint Bonaventure University; Joan Buckley, Concordia College; Pam Buesking, Indiana University at Kokomo; Patricia Card, Hawaii Pacific College; Fay E. Chandler, Pasadena City College; William Leon Coburn, University of Nevada; Margaret

Cole, Arapahoe Community College; Don Richard Cox, University of Tennessee; Martha Day, Richard Bland College of the College of William and Mary; Sarah Dye, Elgin Community College; Francis Ferrance, Victor Valley College; Christopher Gould, University of North Carolina at Wilmington; John Hellman, The Ohio State University at Lima; Anne Herrington, University of Massachusetts at Amherst; Michael J. Hogan, The University of New Mexico; John Hollow, Ohio University; Roger B. Horn, Charles County Community College; Tony J. Howard, Collin County Community College; Nadene A. Keene, Indiana University at Kokomo; Leon Keens, Maple Woods Community College; James B. King, Hillsdale College; George Klawitter, Viterbro College; Nancy G. Little Wright, Motlow State Community College; Larry Longerbeam, Cleveland State University; Frankie Loving, Jackson State University; Thomas E. Martinez, Villanova University; Shirlee A. McGuire, Olivet Nazarene University; David Middleton, Trinity University; Deane Minahan, University of Wisconsin–Superior; Maxine F. Moore, Johnson C. Smith University; Toni J. Morris, University of Indianapolis; Walter Mullen, Mississippi Gold Coast Junior College; Carol A. Perrin, Hawaii Pacific College; Nancy Posselt, Midlands Technical College; Susan Pratt, Loyola University, New Orleans; T. J. Ray, University of Mississippi; Jane Robinson, St. Petersburg Junior College; William Scarpaci, Rock Valley College; J. W. Smith, Henderson County Junior College; Betty J. Steele, Averett College; Kathleen Steele, DePauw University; Edward Stone, University of Arkansas at Monticello; R. Gregory Sutcliffe, Lehigh County Community College; Ethel F. Taylor, North Carolina A&T University; Lawana Trout, Central State University (Oklahoma); Brena B. Walker, Anderson College; Dorothy Wells, Southern University in New Orleans; John Whiting, Orange County Community College; Paul A. Wood, Villanova University; and Liliana Zancu, Millersville University.

For their willingness to talk at length about the craft they practice with such precision and intelligence, I would like to thank Calvin Trillin, John McPhee, Anna Quindlen, Bill Barich, Richard Selzer, Annie Dillard, and Patricia Hampl.

For their eagerness to participate in a variety of writing experiments, I would like to thank the students in my composition classes. In particular, I am indebted to the inventive, humorous, and informative observations of the student authors whose writing gives this edition its unique voice: Wally Armstrong, Robert Scheffel, Christy Kovacs, Joanne Malbone, Ray Peterson, Jenny Miller, Susan Reidenback, Paul Singer, Ellen Haack, Rod Meyers, Jane Graham, Matt Fisher, Richard Gant, and Laura Burney.

Special thanks also go to Rai Peterson and Dana Bausbach for their help in researching this text, and to Kathy Love and Karen Taylor for their assistance in preparing the manuscript.

J.F.T.

WRITING WITH A PURPOSE

PART ONE
THE WRITING PROCESS

1
TOWARD PURPOSEFUL WRITING

Some people seem to find writing easy. They sit down and write, work until they are finished, and turn out a first draft that is so good that it is their final draft. Everyone has heard stories about students who dash off perfect term papers the night before an assignment is due. And many people know the story of how Jack Kerouac wrote his novel *On the Road* — composing it in fourteen days and typing it on a roll of Teletype paper to avoid changing pages. But such ease of composition is rare. Most people feel nervous — or downright alarmed — when they begin to write. They are paralyzed by the blank legal pad, typewriter paper, or word processor screen. The right words and ideas seem suddenly elusive, the task overwhelming.

All writers — students writing papers, business people drafting memos, journalists assembling news stories, novelists composing lengthy works of fiction — know these basic frustrations. Writing is hard work. But writing is also opportunity: to express something about yourself, to explore and explain ideas, to assess the claims of other people. To make good use of these opportunities, you need to develop the confidence to overcome the frustrations of writing.

You can gain this confidence by writing and learning from the work — your own and that of others. In fact, experienced writers have learned some lessons you may find helpful. Consider what Ernest Hemingway says about his early writing experiences in Paris in the 1920s.

> It was wonderful to walk down the long flights of stairs knowing that I'd had good luck working. I always worked until I had something done and I always stopped when I knew what was going to happen next. That way I could be sure of going on the next day. But sometimes when I was starting a new story and I could not get it going, I would sit in front of the fire and squeeze the peel of the little oranges into the edge of the flame and watch the sputter of blue that they made. I would stand and look out over the roofs of

Paris and think, "Do not worry. You have always written before and you will write now. All you have to do is write one true sentence. Write the truest sentence that you know." So finally I would write one true sentence, and then go on from there. It was easy then because there was always one true sentence that I knew or had seen or had heard someone say. If I started to write elaborately, or like someone introducing or presenting something, I found that I could cut that scrollwork or ornament out and throw it away and start with the first true simple declarative sentence I had written. Up in that room I decided that I would write one story about each thing that I knew about. I was trying to do this all the time I was writing, and it was good and severe discipline. (Ernest Hemingway, *A Moveable Feast*)

Hemingway's words suggest that experienced writers possess the following knowledge and attitudes:

1. *Experienced writers have faith in their writing habits.* Because they have written successfully before, experienced writers believe they will write successfully again. They put this belief at risk each time they struggle to solve the problems of a new task, but although their confidence may fluctuate, they do not lose faith. They believe the habits that have worked before — writing in a special environment, maintaining a disciplined schedule, and using familiar tools — will work again.

2. *Experienced writers understand the stages in their writing process.* Because they have written often, experienced writers can identify predictable stages in their writing process. They know why each stage contains smaller steps, how each stage connects to the next, and when it is necessary to repeat one or all stages to produce a successful piece of writing. As they move forward from stage to stage, or backward to repeat a stage, they are confident that they are making progress because they are identifying and solving problems.

3. *Experienced writers rely on the basic elements in the writing situation to guide them.* Because they have had to write under many conditions and for many occasions, experienced writers have learned to focus on the constants in every writing situation. They may talk from time to time about "inspiration" or "the Muse," but they do not depend on such mysterious forces. Effective writing emerges from effective decision making, and effective decisions are made best when writers understand their *subject, audience,* and *purpose.* To enhance their understanding of these basic elements, experienced writers use practical guidelines to help them think and write more effectively.

One of the primary aims of this book is to offer you experience, and thereby confidence, in writing. In the following chapters, you will learn and practice the strategies professional and student writers apply to their craft. This chapter will discuss the writer's environment and habits, the three-stage division of the writing process, and the central elements — subject, audience, and purpose — that every writer must understand to proceed through the process.

THE WRITER'S ENVIRONMENT AND HABITS

From kindergarten to college, you have probably produced dozens of stories, book reports, and themes on a wide variety of subjects. But if you are like many student writers, you may feel you haven't learned much from this experience. The assignments were so different, the writing experiences so inconsistent, and the results so predictable that you may still feel like a beginner. Indeed, most beginning writers are so preoccupied with completing an assignment that they rarely think about how they did it. Asked to describe their writing process, students usually begin by identifying their writing *habits* — describing the conditions and tools they believe they need whenever they write.

Some of these habits are formed by chance. If you produced a paper you were proud of by secluding yourself in a quiet room, inscribing neatly shaped words in a spiral notebook with a soft-lead pencil, then you may be convinced that you need isolation, silence, and simple tools to perform successfully. But you might have produced that same paper seated at the kitchen table, surrounded by

© J. D. Sloan/The Picture Cube

family conversation, banging out your sentences on an electric typewriter. If so, you might believe that to write effectively you require a comfortable environment, reassuring noises, and efficient machinery. Indeed, after a time some writers look upon their writing habits as rituals, procedures to be followed faithfully each time they write. They wear the same flannel shirt, choose the same background music, or sharpen an entire box of pencils before they begin.

Although many writing habits are formed by coincidence and perpetuated by ritual, most come about almost unconsciously and conform with a writer's other personal habits. If, like Felix (the finicky character in Neil Simon's comedy *The Odd Couple*), you must have a carefully organized environment to accomplish any task — from finding your socks to preparing dinner for guests — then you will no doubt develop precise writing habits. You will write regularly, at approximately the same time each day, and for about the same amount of time, in a serene atmosphere. On the other hand, if like Felix's opposite, Oscar, you can tolerate disorder, then your writing habits are likely to reflect your preference for flexibility. Within reason (and within sight of your deadline), you will probably write in bursts of energy: at different hours of the day and night, for varying stretches of time, and in all sorts of places. Neither style is better than the other. What matters is that you find the conditions and tools that work best for you.

■■■■■ *Exercise*

Most people have developed special rituals for writing. Using the following questions as a guide, compile a list of your writing habits.

1. What experiences or people helped you form your writing habits?
2. How do your writing habits resemble your other work habits? Make a list of similarities and differences.
3. What type of physical environment do you need to write effectively? What happens to you when you are forced to write in a "hostile" environment?
4. What kind of tools do you prefer? What happens to you when you have to work with "alien" tools?
5. When do you write best — early in the morning or late in the evening? How long can you write at one sitting?

THE STAGES OF THE WRITING PROCESS

Whatever your writing habits, they are simply the enabling conditions that allow you to begin and pursue your writing process. They are the physical and psychological scenery for the central action — the intellectual procedure you use to move through the stages of composition. This book divides the writing process into three stages: *planning, drafting,* and *revising.* They will be discussed briefly in this chapter and at greater length, with examples, in Chapters 2, 3 and 4.

Planning

Planning is an orderly procedure that brings about a desired result. As the first stage in the writing process, *planning is a series of strategies designed to find and formulate information in writing.* When you begin a writing project, you need to discover what is possible, locate and explore a variety of subjects, and invent alternative ways to think and write about each subject. You need to consider all ideas, however mundane or unsettling, to create and shape the substance of your text. In Chapter 2 you will learn several planning strategies for generating information you can transform into a first draft.

Drafting

Drafting is a procedure for executing a preliminary sketch. As the second stage in the writing process, *drafting is a series of strategies designed to organize and develop a sustained piece of writing.* Once planning has enabled you to identify several subjects and encouraged you to gather information on those subjects from different perspectives, you need to select one subject, organize your information into meaningful clusters, and then discover the links that connect those clusters. In Chapter 3 you will learn how to use drafting techniques to produce a preliminary text. Chapters 5 through 10 will give you additional techniques for composing your preliminary draft.

Revising

Revising is a procedure for improving work in progress. As the third and final stage in the writing process, *revising is a series of strategies designed to reexamine and reevaluate the choices that have created a piece of writing.* After you have completed your preliminary draft, you need to stand back from your text and decide whether to embark on *global revision* — a complete re-creation of the world of your writing — or to begin *local revision* — a concerted effort to perfect the smaller elements in your writing. In Chapter 4, you will learn how to use global revision to rethink, reenvision, and rewrite your work. In Chapters 7 through 10 you will learn strategies for making the small rearrangements and subtle refinements that will help you achieve your most effective writing.

Working Within the Process

The division of the writing process into three stages is deceptive. The discussion of *planning, drafting,* and *revising* suggests that writing proceeds in a linear sequence in which you complete all the activities in one stage and then move to the next. But writing is a complex intellectual activity that may demand a flexible, recursive sequence. You may have to repeat the activities in one stage several times before you are ready to move to the next, or you may have to loop back to an earlier stage before you can go forward again. For example, you may have to try numerous planning strategies until you have generated enough information to work with. Or in the middle of drafting, you may have to return to planning because you discover that the relationships you thought you saw in your material are not there after all or cannot be supported. When you revise, you may find that your essay doesn't explain certain assumptions or illustrate some assertions, so that you must return to planning or drafting. Indeed, although planning, drafting, and revising are distinct activities, at any point in the writing process you are likely to be doing all three at once.

Experienced writers seem to perform within the process in different ways. Some spend an enormous amount of time planning every detail before they write; others prefer to dispense with planning and discover their direction in drafting or revising. For example, the American humorist James Thurber once acknowledged that he and one of his collaborators worked quite differently when writing a play:

> Elliot Nugent . . . is a careful constructor. When we were working on *The Male Animal* together, he was constantly concerned with plotting the play. He could plot the thing from back to front — what was going to happen here, what sort of situation would end the first-act curtain and so forth. I can't work that way. Nugent would say, "Well, Thurber, we've got our problem, we've got all these people in the living room. Now what are we going to do with them?" I'd say that I didn't know and couldn't tell him until I'd sat down at my typewriter and found out. I don't believe the writer should know too much where he's going. . . . (James Thurber, *Writers at Work: The Paris Review Interviews*)

Even experienced writers with established routines for producing a particular kind of work admit that each project inevitably presents new problems. Virginia Woolf planned, drafted, and revised some of her novels with great speed, but she was bewildered by her inability to repeat the process with other novels:

> . . . blundering on at *The Waves*. I write two pages of arrant nonsense, after straining; I write variations of every sentence; compromises; bad shots; possibilities; till my writing book is like a lunatic's dream. Then I trust to inspiration on re-reading; and pencil them into some sense. Still I am not satisfied. . . . I press to my centre. I don't care if it all is scratched out . . . and then, if nothing comes of it — anyhow I have examined the possibilities. But I wish I enjoyed it more. I don't have it in my head all day like the *Lighthouse* and *Orlando*. (Virginia Woolf, "Boxing Day" 1929, *A Writer's Diary*)

Writers may discover problems when they are asked to work in a different context. For example, writers who feel comfortable telling stories about their personal experience may encounter unexpected twists and turns in their writing process when they are asked to write about the lives of other people, explain a historical event, or analyze the arguments in an intellectual controversy. For example, Calvin Trillin, one of our most versatile writers, admits that his writing process changes dramatically when he shifts from writing investigative reports to writing humorous essays or weekly columns.

> In my reporting pieces, I worry a lot about structure. Everything is there — in interviews, clippings, documents — but I don't know how to get it all in. I think that's why I do what we call around the house the vomit-out. I just start writing — to see how much I've got, how it might unfold, and what I've got to do to get through to the end. In my columns and humor pieces, I usually don't know the end or even the middle. I might start with a joke, but I don't know where it's going, so I fiddle along, polishing each paragraph, hoping something will tell me what to write next. (Personal interview)

This wide range of variations suggests that what appears to be a simple three-stage process may be a confused and often contradictory procedure. But experienced writers know that confusion and contradiction may be inevitable but will be temporary in the evolution of most pieces of writing. Confusion occurs when you know too little about your writing project; contradiction occurs when you think too little about what you know. The secret to solving both problems is to consult the constants in every writing context.

MAKING DECISIONS IN THE WRITING PROCESS

As you write, you discover that you are constantly making decisions. Some of these decisions are complex, as when you are trying to shape ideas; others are simple, as when you are trying to select words. But each decision, large or small, affects every other decision you make so that you are continuously adjusting and readjusting your writing to be sure it is consistent, coherent, and clear. You can test the effectiveness of your decisions by measuring them against this dictum: in every writing situation, a writer is trying to communicate a *subject* to an *audience* for a *purpose*.

Initially, think of these three elements as *prompters,* as ways to consider what you want to write about and how you want to write about it. Later, as you move through planning to drafting and revising, think of them as *touchstones,* as ways to assess what you set out to accomplish. But mainly think of them as *guidelines,* as ways to control every decision you make throughout the writing process, from formulating ideas to refining sentences.

SELECTING YOUR SUBJECT

Many student writers complain that their biggest problem is finding a subject. Sometimes that problem seems less complicated because the subject is named in the writing assignment. But assignments vary in how they are worded, what they assume, and what they expect. For example, you may be asked to discuss two characters in a play. This assignment does not identify a subject; it merely identifies an area in which a subject can be found. Another version of that assignment might ask you to compare and contrast the way two characters make compromises. This assignment identifies a subject but assumes you know how to work with a specific form (the comparison and contrast essay) and expects you to produce specific information (two ways of defining and dealing with compromise). In other words, the second assignment selects your subject, but you must still select and develop the subject matter of your essay.

When you have a free choice of subjects, your problem may appear more complicated. No one is helping you find or focus your subject. On the other hand, no one is telling you what to do or how to do it. You are free to make your own decisions. For example, you may decide to compare two characters in your neighborhood or to analyze how your favorite character (you) faces the difficulty of making compromises.

Whether you are responding to an assignment or creating one, you need to take certain steps to find a suitable subject. First, select a subject you know or can learn something about. The more you know about a subject, the more likely you are to make it your own, shaping it according to your own perspective. If, in addition, you select a subject, such as television, that is familiar to most of your readers, you will know that you share an area of common knowledge that allows you more freedom to explore *your* observations, ideas, and values. Second, select a subject you can restrict. A subject such as television is really a broad general category that contains an unlimited supply of smaller, more specific subjects. The more you can restrict your subject, the more likely you are to control your investigation, identify vivid illustrations, and maintain a unified focus.

For example, the general category *television* can be divided into subtopics, each of which can be restricted further into specific subjects.

situation comedies ⟶	women in sitcoms
detective series ⟶	the role of luck in solving television crime
news broadcasts ⟶	the image of trustworthiness projected by news anchors
sports broadcasts ⟶	the language of "color" commentators
quiz shows ⟶	the symbolism of set designs on quiz shows
music videos ⟶	nostalgia in music videos

Although the subjects on the right could profit from further restriction, they illustrate why selecting a specific subject simplifies your writing task: it helps you focus on and form judgments about a concrete topic.

Finally, before you select a subject, you need to ask yourself three questions: Is it significant? Is it interesting? Is is manageable? A significant subject need not be ponderous or solemn. In fact, many significant subjects grow out of ordinary observations rather than from grandiose declarations. But you do need to decide whether a specific subject raises important issues (the reliability of television journalism) or appeals to the common experience of your readers (the evocative power of music). Similarly, an interesting subject need not be dazzling or spectacular, but it does need to capture your attention. If it bores you, it will surely bore

"Write about dogs!"

Drawing by Booth; © 1976 The New Yorker Magazine Inc.

your readers. You need to decide why a specific subject fascinates you (why you are attracted to quiz shows that give away money) and how you can make the subject more intriguing for your readers (how private detectives manipulate their friends in high places to gather evidence). A manageable subject is neither so limited that you can exhaust it in a few pages (the language of one sports broadcaster) nor so vast that lengthy articles or books would be required to discuss it adequately (the evolution of female stereotypes in situation comedies).

Ultimately, you must develop your own methods for answering these questions as you examine the choice of subjects available and as you consider your audience and purpose. When you compare subjects in the context of the complete writing situation, you will naturally prefer some to others. To decide which one will enable you to produce the most suitable subject, measure it against the criteria set forth in the following guidelines:

Guidelines for Selecting Your Subject

1. *What do I know about my subject?*
   Do I know about my subject in some depth, or do I need to learn more about it? What are the sources of my knowledge — direct experience, observation, reading? How does my knowledge give me a special or unusual perspective on my subject?

2. *What is the focus of my subject?*
   Is my subject too general? How can I restrict it to a more specific subject that I can develop in greater detail?

3. *What is significant about my subject?*
   What issues of general importance does it raise? What fresh insight can I contribute to my readers' thinking on this issue?

4. *What is interesting about my subject?*
   Why is this subject interesting to me? How can I interest my readers in it?

5. *Is my subject manageable?*
   Can I write about my subject in a particular form, within a certain number of pages? Do I feel in control of my subject or confused by it? If my subject is too complicated or too simplistic, how can I make it more manageable?

ANALYZING YOUR AUDIENCE

Most inexperienced writers assume that their audience is their writing teacher. But writing teachers, like writing assignments, often vary in what they teach, what they assume, and what they expect. Such variation has often prompted inexperienced writers to define their writing tasks as "trying to figure out what the teacher wants." This definition is both naive and sophisticated. On the one hand, it suggests that the sole purpose of any writing assignment is to satisfy the whims of another individual. On the other hand, it suggests that when writers analyze the knowledge, assumptions, and expectations of their readers they develop a clearer perception of their subject and purpose. To make this analysis effective, however, writers must remember that they are writing for multiple audiences.

The most immediate audience is *you.* You write not only to convey your ideas to others but also to clarify them for yourself. To think of yourself as an audience, however, you must stop thinking like a writer and begin thinking like a reader. This change in perspective offers advantages, for you are the reader you know best. You are also a fairly representative reader because you share broad concerns and interests with other people. If you feel that your writing is clear, lively, and informed, other readers will probably feel that way too. If you sense that your text is confused or incomplete, your audience will be likewise disappointed.

The main drawback to considering yourself as audience is your inclination to deceive yourself. You want every sentence and paragraph to be perfect, but you know how much time and energy you invested in composing them, and that effort may blur your judgment. You may accept bad writing from yourself, even though you wouldn't accept it from someone else. For that reason you need a second audience. These readers—usually your friends, classmates, or teachers—are your most attentive audience. They help you choose your subject, coach you through the various stages of the writing process, and counsel you about how to improve your sentences and paragraphs. As you write, you must certainly anticipate detailed advice from these readers. But you must remember that writing teachers and even peers are essentially collaborators. They know what you have considered, cut, and corrected. The more they help you, the more eager they are to commend your writing as it approaches their standards of acceptability.

Your most significant audience consists of readers who do not know how much time and energy you invested in your writing or care about how many choices you considered and rejected. These readers want writing that tells them something interesting or important, and they are put off by writing that is tedious or trivial. It is this wider audience that you (and your collaborators) must consider as you work through the writing process.

At times this audience seems vague, and you may wonder how you can direct your writing to it if you do not know any of its distinguishing features. In those cases, it may be helpful to imagine a reader — an attentive, sensible, reasonably informed person who will give you an objective reading as long as you do not waste his or her time. This reader, specifically imagined though often termed the "general reader," "the universal reader," or "the common reader," is essentially a fiction, but a helpful fiction. Imagine an important person whom you respect and whose respect you want; your writing will benefit from the objectivity and sincerity with which you address this reader.

Many times, however, especially as you learn more about your subject, you discover a more specific audience for your writing. You may sometimes discern a number of specific audiences, in which case you will ultimately have to choose among them. Consider the following example of how you might identify and analyze several audiences.

Your knowledge of contemporary music suggests that music videos might make an interesting subject, but you may find it difficult to focus this knowledge. You decide to identify your audience, and after some deliberation you see that you have at least three possible audiences: (1) those who love music videos — MTV addicts who relish the creative nuances in each new film; (2) those who hate music videos — MTV critics who despise the commercial manipulation in every three-minute tape; and (3) those who are indifferent to music videos — a

Cyndi Lauper in an MTV production

© Allan Tannenbaum/SYGMA

group whose members have never heard of MTV or who have seen only an occasional clip.

Now that you have identified your audiences, analyze the distinctive features of each group. What do they know? What do they think they know? What do they need to know? The more you know about each group, the more you will be able to direct your writing to their assumptions and expectations. If you know the contemporary music scene, you will have little difficulty analyzing MTV's devotees and detractors. You have heard the addicts applaud the imaginative performances, costumes, plots, and technical innovations that they believe enrich their appreciation of the music. Similarly, you have heard the critics complain that videos present contrived performances, outrageous costumes, pointless plots, and haphazard trick photography — antics that they say destroy their appreciation of music by reducing it to a single, often bizarre interpretation of each song.

At first you may have difficulty with the third group because those readers have no preconceptions of music videos either as a new art form or as grotesque commercials. In some ways, readers in the third group are like the "general reader" — thoughtful, discerning people who are willing to read about music videos if you can convince them that the subject is worth their attention.

As you consider addressing each of these audiences, you should be able to restrict your subject to anticipate their interests. For example, as you think about the addict, you might decide to describe the various innovative techniques video makers have used to render one of the many recurring themes in contemporary music. As you think about the critic, you might decide to portray the plight of performers who must redirect their creative and financial resources toward producing effective videos rather than inspiring music. As you think about the neutral readers, you may decide that although they may have no interest in music videos (or even in contemporary music) they may be intrigued by a demonstration of how this particular art form has influenced another art form, feature films.

Although this sort of audience analysis helps you visualize a group of readers, it does not help you decide which group is most suitable for your essay. If you target one group, you may fall into the trap of allowing its preferences to determine the direction of your writing. If you try to accommodate all three groups, you may waver indecisively among them, so that your writing never finds any direction. Your decision about audience, like your decision about subject, has to be made in the context of the complete writing situation. Both decisions are ultimately related to your discovery of purpose — what you want to do in your essay. In the next several pages, you will learn how purpose guides your decisions in writing. But first, look at the following guidelines for analyzing your audience:

Guidelines for Analyzing Your Audience

1. *Who are the readers that will be most interested in my writing?*
   What is their probable age, sex, education, economic status, and social position? What values, assumptions, and prejudices characterize their general attitudes toward life?

2. *What do my readers know or think they know about my subject?*
   What is the probable source of their knowledge — direct experience, observation, reading, rumor? Will my readers react positively or negatively toward my subject?

3. *Why will my readers read my writing?*
   If they know a great deal about my subject, what will they expect to learn from reading my essay? If they know only a few things about my subject, what will they expect to be told about it? Will they expect to be entertained, informed, or persuaded?

4. *How can I interest my readers in my subject?*
   If they are hostile toward it, how can I convince them to give my writing a fair reading? If they are sympathetic, how can I fulfill and enhance their expectations? If they are neutral, how can I catch and hold their attention?

5. *How can I help my readers read my writing?*
   What kind of organizational pattern will help them see its purpose? What kind of guideposts and transitional markers will they need to follow this pattern? What (and how many) examples will they need to understand my general statements?

DETERMINING YOUR PURPOSE

The central idea of this book is that writers write most effectively when they are "writing with a purpose." Inexperienced writers occasionally have difficulty writing with *a* purpose, because they see many purposes: to complete the assignment, to earn a good grade, to publish their writing. These "purposes" are outside the writing situation, but they certainly influence the way you think about your purpose. For example, if you want a good grade, you will define your purpose in terms of your teacher's writing assignment. If you want to publish your essay, then you will define your purpose in terms of a given publication's statements about its editorial policies.

When *purpose* is considered as an element inside the writing situation, the term has a specific meaning: *purpose is the overall design that governs what writers do in their writing.* Writers who have determined their purpose know what kind of information they need, how they want to organize and develop it, and why they think it is important. In effect, purpose directs and controls all the decisions writers make. It is both the *what* of that process and the *how* — that is, the specific subject the writer selects and the strategies, from establishing organization to refining style, the writer uses to communicate the subject most effectively.

The difficulty with this definition is that finding a purpose to guide you through the writing process *is* the purpose of the writing process. Writing is both a procedure for *discovering* what you know and a procedure for *demonstrating* what you know. For that reason, you must maintain a double vision of your purpose. You must think of it as a preliminary objective that helps illuminate the decisions you have to make. You must also think of it as a final assertion that helps you implement what you intend to do in your writing.

Forming a Working Purpose: The Hypothesis

As you begin writing, you start to acquire a general sense of your purpose from your preliminary decisions about subject and audience. Suppose again that you choose to write about music videos. You might approach this subject in many ways, but you have to start somewhere, so you decide to focus on their intent. This preliminary decision will help you gather and sort information, but as you assess it, you may see that you do not know what you want to do with it. You do not have a purpose; you have only a general notion that suggests a number of perspectives. Any one of these perspectives might provide an interesting angle from which to investigate your subject, and you can use any one of them to form a *hypothesis* — a working purpose.

Forming a hypothesis is a major step in determining your purpose. Sometimes you come to your writing certain of your hypothesis: you know from the outset what you want to prove and what you need to prove it. More often, you need to consider various possibilities; to convey something meaningful in your writing, something that bears your own mark, you need to keep an open mind and explore your options fully. Eventually, however, you must choose one hypothesis that you think most accurately establishes what you want to say about your subject and how you want to say it.

How do you know which hypothesis to choose? There is no easy answer to this question. The answer ultimately emerges from your temperament, experiences, and interests, and also from the requirements of the writing situation — whether you are writing for yourself or on assignment. Sometimes you can make the choice intuitively as you proceed; in thinking about your subject and audience, you see at once the perspective you want to adopt and how it will direct your writing. At other times you may find it helpful to write out various hypotheses and then consider their relative effectiveness. Which will be most interesting to write about? Which best expresses your way of looking at things? With which can you make the strongest case or most compelling assertions?

For example, as you look at the information you have gathered about the intent of music videos, you might see several hypotheses:

- Videos are merely advertising gimmicks.
- Videos are an imaginative new art form.
- Videos are a simple pictorial translation of a song.
- Videos reinterpret music in complex visual images.

Any one of these or other hypotheses might guide you to the next phase of your writing. You must decide which one suits you best, and your decision must take into account the nature of the subject and your intended audience.

Testing Your Hypothesis: The Discovery Draft

After you have chosen your hypothesis, you need to determine whether this preliminary statement of purpose really provides the direction and control you need to produce an effective piece of writing. You can test your hypothesis by writing a first, or *discovery,* draft. Sometimes your discovery draft demonstrates that your hypothesis works. More often, however, as you continue in the writing process, you discover new information or unforeseen complications that cause you to modify your original hypothesis. In other cases, you discover that you simply cannot prove what your hypothesis suggested you might be able to prove.

Whatever you discover about your hypothesis, you must proceed in writing. If your discovery draft reveals that your hypothesis represents what you want to prove and needs only slight modification, then change your perspective somewhat or find the additional information so that you can modify it. If, on the other hand, your discovery draft demonstrates that your hypothesis lacks conviction, or that you do not have (and you suspect you cannot get) the information you need to make your case, then choose another hypothesis that more accurately reflects your intentions.

For example, if you choose a hypothesis that music videos are simply advertising gimmicks and start your discovery draft, you may realize that even though you have gathered information to illustrate the effect of videos on record sales, you don't find the evidence overwhelming or particularly interesting. Since contemporary music has always mixed commercial and artistic intentions, it comes as no surprise that music videos exhibit a similar mixture. You see flaws in your reasoning, holes in your essay. If you remember the conclusions of a recent opinion poll revealing that most consumers prefer to hear a record before they see the video or an interview with a famous film director who acknowledged his admiration for the aesthetic innovations in videos, you have uncovered additional evidence that compromises your hypothesis or makes it less compelling. You need to modify that hypothesis or, more likely, find another.

Purpose and Thesis

Whether you proceed with your original hypothesis, modify it, or choose another, you must eventually arrive at a final decision about your purpose. You make that decision during revision, when you know what you want to do and how you want to do it. Once you have established your purpose, you can make or refine other decisions — about your organization, examples, and style. One way to express your purpose is to state your thesis. A *thesis* is a sentence that establishes your writing commitment by stating the main idea you are going to develop. The thesis sentence usually appears in the first paragraph of your essay. Although the thesis is often called a purpose statement, thesis and purpose are not precisely the same thing. Your purpose is both contained in and larger than your thesis; it is all the discoveries and decisions you have used to create that thesis and all the strategies you will use to demonstrate it in a sustained and successful piece of writing.

As the preceding paragraph indicates, your thesis is usually a written statement. It makes a restricted, unified, and precise assertion about your subject — an assertion that can be developed in the amount of space you have, that treats only one idea, and that is open to only one interpretation. Deriving a thesis from the information you have about your subject and stating it well are key steps in the writing process; for that reason they will be discussed in more detail in Chapter 3. For now, however, remember that your thesis is stated after you have determined your purpose and that your thesis expresses your main idea about your subject. For example, if you finally determine that you want to write not about the commercial purposes of videos but rather about the kind of commercial music that benefits most by transformation into videos, your thesis might be stated in this way: "The most innovative videos are created from music whose lyrics do not prescribe a specific narrative line."

In certain writing situations you do not need to write out your thesis. This does not mean that your writing in such situations lacks purpose, only that you do not need to compose a purpose statement. Narrating a story or describing an object, for example, does not usually require an explicit thesis; the thesis is suggested, or implied.

In many ways the difference between a hypothesis (a working purpose) and a thesis (a final assertion) explains why you can speculate about your purpose before you write but you can specify your purpose only after you have written. This connection between your writing process and your writing purpose requires you to pause frequently to consult the criteria set forth in the following guidelines:

Guidelines for Determining Your Purpose

1. *What are the requirements of my writing project?*
   If I am writing to fulfill an assignment, do I understand that assignment? If I am writing on my own, do I have definite expectations of what I will accomplish?

2. *As I proceed in this project, what do I need to know?*
   Do I have a good understanding of my subject, or do I need more information? Have I considered the possible audiences who might read my writing?

3. *What hypothesis can I use as my working purpose?*
   How many different hypotheses can I formulate about my subject? Which of them seems to direct and control my information in the most effective manner?

4. *What purpose have I discovered for this writing project?*
   Has my purpose changed as I learned more about my subject and audience? If so, in what ways? Have I discovered, by working with a hypothesis or hypotheses, what I want to do in my writing?

5. *What is my thesis?*
   How can I state my main idea about my subject in a thesis sentence? Does my thesis limit the scope of my writing to what I can demonstrate in the available space? Does it focus my writing on one specific assertion? Does it make an exact statement about what my writing intends to do?

COORDINATING DECISIONS IN THE WRITING PROCESS

You need to understand how each of the three elements — subject, audience, and purpose — helps you measure your progress through the writing process. But you must not assume that these guidelines lock you into the fixed sequence of selecting your subject, analyzing your audience, and then determining your purpose. Subject, audience, and purpose are too interconnected for you to follow such a simple procedure. Indeed, they resemble the elements in a complex chemical formula. You can isolate the elements in the formula, but you cannot understand what they create until you combine them. And each time you alter one element, however slightly, you change the character of the other elements, inventing another network of relationships and a new formula. Thus to make informed and purposeful decisions about your writing, you must not only understand the separate contributions of subject, audience, and purpose, but also learn to coordinate the complex relationship among them.

■■■■■ *Exercise*

Describe the conceptions of subject, audience, and purpose implied in each original hypothesis, and then explain how each new decision will affect the three elements. Write a new hypothesis that reflects the changes.

1. *Original hypothesis:*   Videos have had an enormous influence on contemporary cinema.
   New decision: You decide to discuss earlier films featuring music, such as *Fantasia,* that influenced videos.

2. *Original hypothesis:*   Videos are shown almost exclusively on cable television.
   New decision: You decide to address your essay to someone trying to decide whether to subscribe to cable television.

3. *Original hypothesis:*   Videos enrich our understanding and appreciation of a song's lyrics.
   New decision: You decide to prove that film makers (not songwriters or musicians) are in charge of creating videos.

—————— *Review Exercise* ■■■■■■■■■■■

To review what you have learned in this chapter, read the following essay in which Wally, a student writer, describes his experiences painting for commission. Then discuss the questions that follow.

Brandon's Clown

Few of those words of wisdom that are passed   1
from father to son are followed. Most are simply acknowledged and forgotten. My father's advice about my adolescent love for painting was simple and direct: "Son, you have a special talent. Be smart. Use it to make money." With his words guiding me, I took my love to the marketplace. I began accepting commissions, painting fantasies for those who didn't have the skill or desire to paint their own.

Creating artwork for a client, I soon discov- 2
ered, demands much more than talent and a lust for money. A friend of mine in Art History 110 was offered $175 to paint a mural in the lobby of her home-town bank. Her only direction from the bank manager was to paint "something rural

looking." Obviously, rural looking can be anything from cows lapping at a sunlit pond to combines ripping the heads from a field of golden wheat. Somewhere in that wide spectrum was the painting the manager wanted. Such flexibility forces the artist to play mind reader or simply follow his own desires and hope the client likes the result. A less common, perhaps even less desirable commission is the verbal blueprint. The client knows exactly what he wants. "I want a portrait," he says, "that makes me look like General Patton. I want two American flags in the background and a Doberman at my side." Although there is some security in knowing what the client expects, most artists aren't desperate enough to follow such instructions.

Once the actual work begins, the question   3
that most frequently plagues the artist is "Why am I doing this?" — especially when the commission calls for an uncharacteristic shift in creative expression. I am known on campus as "the guy who paints screaming faces," a tag I am

quite happy with. I am proud of the dark maelstroms of anxiety I can create on my canvas. For hours after leaving the easel, I will walk about with my shoulders stooped and head down, frowning, lost in reflections on man's inability to overcome despair.

Naturally, my first commission was for a **4** nursery painting. A young mother expecting her first child wanted an oil painting to hang in a newly remodeled nursery. Mutual friends had informed her that I was willing, able, and cheap. My father's advice prompted me to take the job and I soon found myself discussing infant tastes with the mother-to-be, a pleasant young woman whose only exposure to art was probably the Sunday funnies.

"You know, something happy and cheery," **5** she said between snatches of dialogue on *General Hospital*. "The usual stuff for a kid's room."

"Is there any kind of theme to your nur- **6** sery?" I asked hesitantly.

"Well, it's blue," she said. "If it's a boy, **7** we're going to call him Brandon. Something like that would be nice."

I followed her finger to a baby stroller, dec- **8** orated with circus animals. "Oh," I said, as the room began to darken and spin. "I see."

For the next several weeks, I had night- **9** mares of pink giraffes and smiling zebras parading across my empty canvas. I spent hours hunched over my sketchbook, filling page after page with oddly shaped animal skeletons. Before long, the book looked like a Disneyland mortuary. The entire project seemed so pointless that I was exhausted with frustration. How would I satisfy my cheerful client and remain true to my creative vision?

Resolved to compromise my pride for a **10** buck, I began gathering my paints, brushes and canvas with all the enthusiasm of the only girl in the eighth grade not chosen for the pompom squad. I reviewed my sketches hoping to find a subject and suddenly spotted a quick drawing I had made of a clown. Eureka! The answer seemed obvious: a clown. Of course. Clowns had screaming faces, even though they concealed their real pain behind their painted masks. I would mask my true intentions by painting a clown with a slightly sorrowful expression lurk-

ing behind his grease paint. He would hold the strings of two balloons, suggesting comparisons to a man behind prison bars.

So I began to paint. Every morning before **11** class, I ran to my studio to reappraise the work I had done the night before. The glistening, bright arts were so inviting. Algebra could wait. I picked out my favorite brush and dove back into the painting, detailing areas I had sketched before bedtime, making the highlights lighter and the shadows darker. Gradually, the clown's face began to emerge from the swirl of color. "Brandon's Clown" was my best work to date!

Feeling every bit like the proud parent **12** myself, I carried the finished canvas home from my college studio to unveil it for my mother. As I held it up for her reaction, I was certain she would gush with glee and call the neighbors. But after a few moments of silence, she said "It's black."

My defenses flew to Red Alert. "It's sup- **13** posed to be black. How can you emphasize light and shadow without using black?"

Leaving Mom to her *Reader's Digest,* I **14** drove to my client's house in hopes of a better reception. Occasionally I glanced at the painting lying next to me on the seat. Perhaps it *was* a bit dark — for a nursery painting, that is. The expression on the clown's face suddenly seemed harsh and scolding. How could defenseless little Brandon lie in his crib and stare at this all day long?

"How do you like your painting, Mrs. **15** Hobbs?" I asked meekly. For some reason, the canvas which had seemed so perfect that morning in my studio now looked totally out of place in Brandon's nursery. A beautiful, tiny wooden rocking horse swayed peacefully on the dresser top. An 8 × 10 hospital glossy of little Brandon, looking fresh and innocent, hung on the wall. And in the corner, nearly lost in the bundle of blankets, was Brandon himself, fast asleep and completely unaware of this ugly black thing that was about to disturb the harmony of his world.

"It's fine. Will you take a check?" his **16** mother asked as she hung the painting right above Brandon's head.

As I left holding my first commission, I did **17** not think about my father's advice or my artistic reputation. I thought only of poor Brandon. I will not be surprised if, twenty years from now, a deranged young man stops me on the street and smashes a clown painting over my head. I probably deserve worse.

Questions for Discussion

1. What kind of work habits does Wally follow when he paints?

2. In what ways does the process of painting *Brandon's Clown* resemble the three-stage writing process?

3. What is the subject of the essay "Brandon's Clown"? What general subjects are introduced in paragraphs 1 and 2? What specific subject is illustrated by the rest of the essay?

4. Who is the audience for this essay? What does Wally assume about the knowledge and expectations of his readers? How do his opening paragraphs help catch and hold his readers' attention?

5. What is the purpose of this essay? Does Wally imply his thesis or announce it in a direct statement? If it is implied, formulate his thesis. If it is stated, identify his thesis sentence.

Brandon's Clown

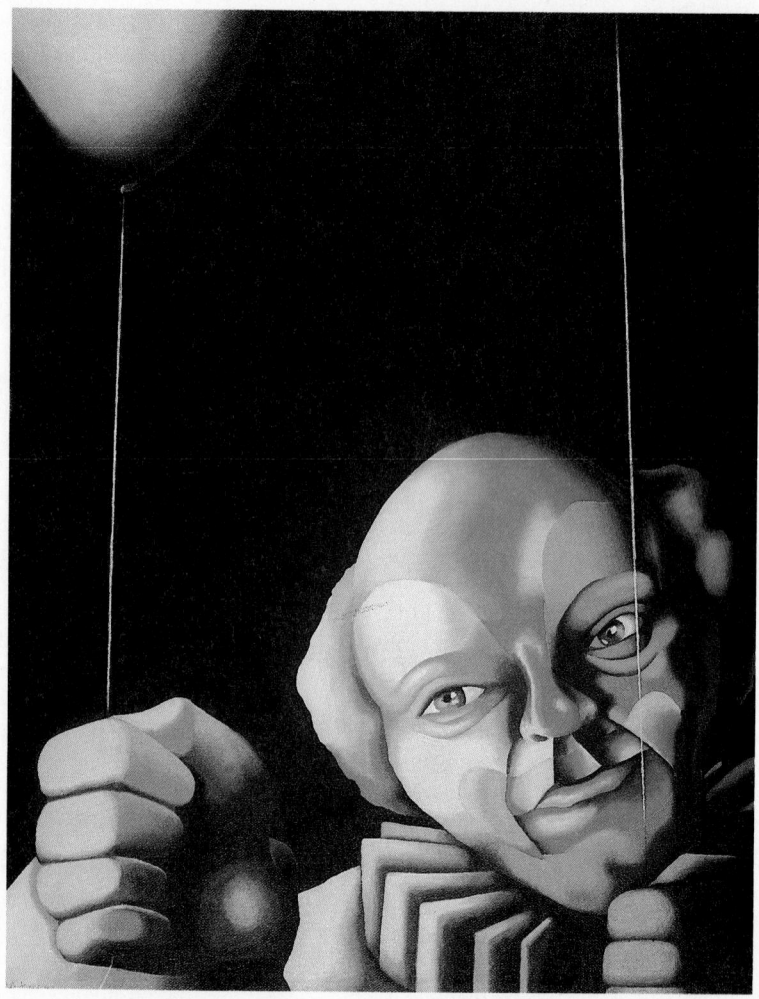

"Brandon's Clown," original painting by Wallace L. Armstrong

Assignments

1. Using the questions about writing habits on page 6, interview a classmate about his or her writing habits. After you have compiled your information, write a brief description of this person as a writer.

2. After completing the exercise on page 20, select a subject, analyze an audience, and determine a purpose for your own essay on music videos.

3. Try to remember some words of wisdom that your parents or a trusted friend passed on to you as guidelines for giving your life a sense of purpose. Then describe an incident in which you tried to follow those guidelines.

4. Examine the details in Wally's painting, *Brandon's Clown*. In what ways does the painting resemble the description of it in the essay "Brandon's Clown"? Imagine you are Brandon and explain how the presence of this painting in your room has influenced your life.

5. The photograph on page 2 was selected as a metaphor representing an important aspect of this chapter. Examine the photograph and determine the possible meanings of this metaphor as you discern them in the photograph. Then write an essay explaining how the metaphor has been reinforced or revised by the ideas in this chapter. You may want to suggest another metaphor that illustrates your understanding of these ideas.

2
PLANNING

All writers have trouble getting started. In Chapter 1, Ernest Hemingway was quoted as saying that when he had trouble starting a new story he would squeeze orange peels into the fire, gaze out the window, and worry about whether he would ever write again. Activities like these are not a waste of time; some may even be necessary if writers are to think their way into a writing project. Random thinking may seem pointless, but it often helps writers break away from outside distractions and begin to focus on their writing. Random thinking is a waste of time, however, if it encourages writers to wait for the perfect moment to begin working. Those who wait for the perfect moment usually do a lot of waiting and very little writing. Eventually, all writers must do what Hemingway recalls he did: give themselves a strong pep talk. "Stop worrying! Start writing!"

The best way to start writing is to begin planning. Inexperienced writers define planning as essentially a thinking activity. First, they plan inside their heads what they want to say, and then they copy their thoughts onto a piece of paper. Unfortunately, they discover that such planning usually produces two kinds of failure: (1) they cannot think through everything they want to say before they write, and (2) they cannot simply transfer their thinking to writing. In fact, planning is primarily a writing activity, as most experienced writers can testify. Although they admit that they do some planning before they write, they insist that they do their most productive planning after they have begun writing. For them, planning is not so much thinking *and* writing as it is *thinking-in-writing*. Read what the poet William Stafford has to say about the relationship between thinking and writing:

> When I write, I like to have an interval before me when I am not likely to be interrupted. For me, this means usually the early morning, before others are awake. I get pen and paper, take a glance out the window (often it is dark out there), and wait. It is like fishing. But I do not wait very long, for there is always a nibble — and this is where receptivity comes in. To get started I will accept

anything that occurs to me. Something always occurs, of course, to any of us.
We can't keep from thinking. Maybe I have to settle for an immediate impression:
it's cold, or hot, or dark, or bright, or in between! Or — well, the possibilities
are endless. If I put down something, that thing will help the next thing come,
and I'm off. If I let the process go on, things will occur to me that were not at
all in my mind when I started. These things, odd or trivial as they may be, are
somehow connected. And if I let them string out, surprising things will happen.
(William Stafford, "A Way of Writing," *Writing the Australian Crawl*)

SOURCES AND STRATEGIES

As the first stage in the writing process, planning helps you uncover,
explore, and evaluate a topic. Whether you are assigned a topic by your
teacher or are free, like Stafford, to accept any topic that occurs to you,
planning helps you locate and produce information in writing. Hemingway, you
will recall, found his "one true sentence" in one of *three* sources: (1) something
he knew (memory); (2) something he had seen (observation); or (3) something he
learned through someone else (research).

Those three sources contain an abundance of material for writing. To tap
them, however, you will need to employ one of several thinking-in-writing strategies. The strategies not only identify familiar (or forgotten) information, but also
create, as Stafford suggests, surprising new information. The strategies are flexible enough that you can apply them to all three sources or use them during other
stages of the writing process. You may discover, for example, that a strategy
designed to tap your memory — such as freewriting — works equally well as a
method for recording observations, conducting research, or solving a problem
you encounter in drafting or revising.

As you experiment with a particular strategy, try to answer two questions:
(1) How much information can I produce about this topic in writing? (2) How can
I use this information to create or refine an interesting piece of writing? You can
answer the first after you have worked with one or more thinking-in-writing
strategies. You can answer the second by evaluating your information against the
guidelines provided in Chapter 1 for selecting your subject, analyzing your audience, and determining your purpose.

USING MEMORY IN WRITING

Your own past is one of your best sources of information. Since childhood you have been accumulating memories about people, places, and
things; about growing up, falling down, leaving home, staying put — the
list is endless. These memories often pop into your head when you least expect
them. A face in a crowd reminds you of an old friend. A whiff of suntan lotion
finds you playing volleyball at the beach. The sound of a word — *Schwinn* —
recalls the freedom of peddling your first bike. When you begin to write, how-

ever, you cannot remember on impulse. You must remember on purpose, searching your past for information you may want to use. Select a prompt — a word, an image, or a question — to unlock your memory and focus your attention. Next, experiment with one or more of the following writing strategies designed to stimulate your memory: *listing, clustering, freewriting,* and *keeping a journal.*

Listing

Listing is a simple way to produce information in writing. If you are creating your own assignment, you can make a list of potential subjects. If you are responding to directions, you can list what you already know or need to know about the subject of the assignment. As you restrict your subject, or identify its various subdivisions, you can make additional lists to help you recall more precise information. Begin by giving your list a title, a prompt that will evoke events, impressions, and ideas. Write the title at the top of a new page in your notebook, and then, working down the page as quickly as possible, list any word or phrase that comes to mind. Don't stop to edit, organize, or evaluate the items in your list. Simply spill them down the page in whatever form they occur to you. Sometimes you will work in explosive bursts, producing ideas that instantly generate related ideas. At other times you will pause, probing for a new direction as you move from one isolated idea to another. Don't be discouraged by these occasional pauses or by the strange notions you jot down to keep the list going. The pauses may identify dead ends or topics that require further consideration, and the strange notions may provoke unexpected insights into your subject.

One of your richest memory sources is probably your family. As he thought about his family, Robert focused on his relationship with his father. Using as a prompt the word *Dad,* he produced the following list in a ten-minute writing session:

Dad

P.K. — preacher's kid, defensive, embarrassed, rebellious, special
friends' fathers had normal jobs — no explanations
worked at bank, factory — on weekends, slept late, played golf
Dad works at church — and at home
<u>Saturday</u> "Keep quiet! Dad's working on sermon"
Typing in study. No phone calls. Turn down T.V.
Skips lunch — dinner, glassy-eyed, distant
<u>Sunday Morning</u> — gone before breakfast — dress rehearsal
Mother checks our shoes, knots ties

Marches us to first pew — everybody looking

Organ, choir, congregation — processional

Dad in robe, carrying Bible — distinguished, important

Sunday voice — scripture, sermon (fidgety, sleepy)

Prayer — "Our Heavenly Father" — power, majesty

Heavenly father, my father — same words, same rules

Sunday afternoons — Sabbath — Keep it Holy

No football, no movies, no fun — goodie-two-shoes

Career day at school — doctor, lawyer, police chief

"What's your father do?"

Nothing. He's a minister."

Evaluating Lists   After completing your list, examine your information carefully. What subject dominates your list? Can you identify other subjects? What subjects do you want to develop in greater detail? Underline or star the items that seem most promising. Circle and connect those that seem to go together. Pick a word or phrase and start another list. You may decide that you can use most of the information on your list, or you may discover that only a few fragments or sentences are worth pursuing. But one of those sentences may be what Hemingway called a "true sentence," a sentence that points the way to your subject, audience, and purpose.

Robert's list focuses immediately on his attitude toward his father's job. The initials *P.K.* evoke emotions he cannot easily explain. His friends' fathers have *normal* jobs, but his Dad's job is different. The weekend schedule clarifies that difference and suggests the possibility of a specific subject that is significant, interesting, and manageable. It also suggests at least three groups of readers (normal people, preachers' kids, and preachers) who might find the subject intriguing. But, most important, the list hints at several hypotheses about Robert's perception of his father's work that he may want to explore in greater detail, either through another list or through one of the other planning strategies described in this section. When he has collected enough information, he will sort through his sketchy hypotheses and select one to guide his writing during drafting.

Clustering

At first, clustering may seem an elaborate way to make a list. But the differences, though subtle, are significant. In listing, you place a word at the top of the page and work downward in a linear sequence. If a particular phrase strikes a responsive chord, you are likely to follow that line of thought, neglecting others. By numbering the items or counting the lines on a page, you may unconsciously limit your thinking.

By contrast, in clustering you place a circled word or phrase in the middle of the page and add ideas to it, moving freely all over the page. As ideas occur to you, draw lines outward from the nucleus, jotting down and circling a word or phrase; keep moving outward, naming and circling related ideas, details, and examples. When a sequence reaches an end, return to the nucleus and begin another and another. If you think of new ideas or illustrations for previous sequences, skip back to that cluster and add the information. Circling each item encourages you to see it as a whole, and as each circle spawns additional circles, you will see not fragments but a web of relationships. This pattern of radiating lines and budding clusters may be impossible for another person to interpret, but for the creator it provides a wonderful shorthand for possible writing projects.

When Christy began to search her memory for writing material, she recalled her mother's warning about college: "It's a place to work hard, to make something of yourself. It's not a place to play." For Christy, the word *play* suddenly possessed evocative power, and she created the following cluster.

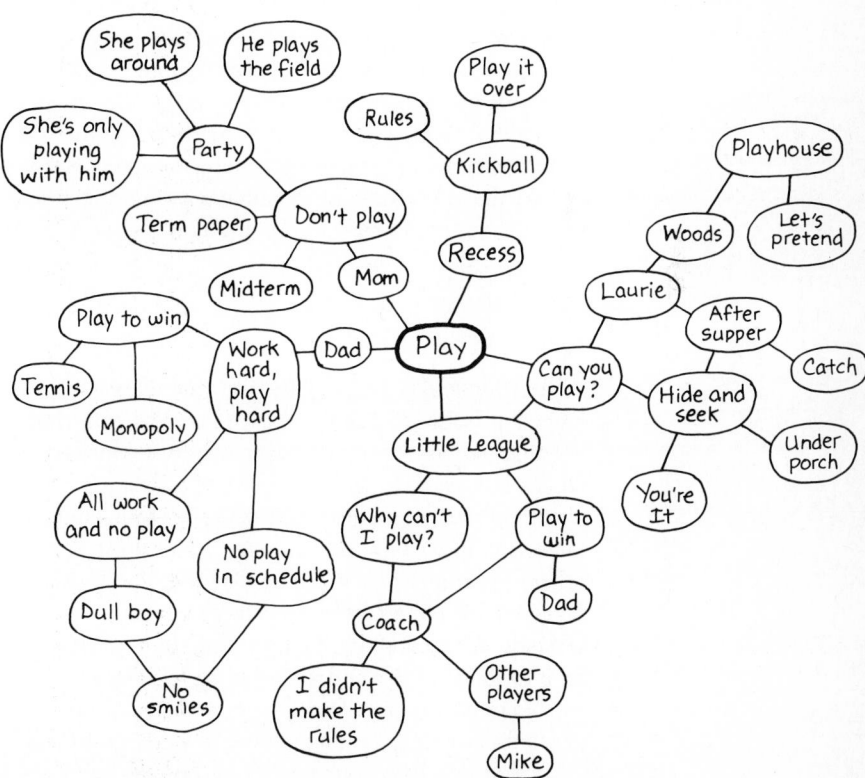

Evaluating Clusters   To assess your cluster, focus on the main circle. Does it still function as the center for the smaller clusters it produced? If it does not hold the system together, examine your cluster for a new center. Look at the movement from generalization to detail within each major cluster. Is the topic being restricted or expanded? Mark the clusters that enclose an individual world worthy of separate investigation. What threads weave all or several of these clusters into an integrated pattern? Cross out clusters that do not seem to illuminate the subject, attract an audience, or reflect your emerging purpose.

Christy's cluster reveals five smaller worlds in the universe of play:

- Can you play?
- Recess
- Little League
- Dad
- Mom

Each of these worlds contains specific subjects. The first three point to childhood memories and so would appeal to the common experience of many readers. The others refer to adult attitudes toward work and play that might engage a variety of readers concerned about the place of those activities in their lives. But as Christy played around with the word *play,* she was attracted to the relationship between rules and play. She was particularly fascinated by the contrast between the world of "Can you play?" where she and Laurie could create and change the rules for their own enjoyment and the world of "Dad" where everyone had to play by the rules and nobody smiled. Once she decides how to transform this insight into a hypothesis, she may be able to include examples from other sections of her cluster. Or she may discard the whole effort and, given her new perspective, reconsider her mother's warning in another cluster or thinking-in-writing strategy.

Freewriting

Like listing and clustering, freewriting helps you write down as quickly as possible what you can remember. But freewriting differs from those strategies by encouraging you to remember in coherent blocks and to write in phrases and sentences. Freewriting is *free,* however, because you don't have to worry about writing perfect sentences. You write for a sustained period of time (usually ten to fifteen minutes) without stopping. Under these conditions, you don't have time to reflect on where you are going, where you have been, or how anything looks. You certainly don't have time to correct misspelled words, cross out awkward phrases, or reread what you have written. Once you are finished, you will have time to look it over and abstract useful ideas.

You can practice freewriting as an unfocused or focused activity. In unfocused freewriting, you simply begin writing, transcribing onto the page the first ideas that pop into your head and then allowing your mind to wander to other subjects. The chief advantage of this kind of activity is that it forces you to write *immediately,* by-passing the agony of selecting the perfect topic. Its chief disadvantage is that it often forces you to write about the *immediate,* the impulses and events that have just occurred. But while you are filling up the page, you may discover a provocative word or forgotten episode that leads to a more complex and deeply textured pattern of ideas.

Joanne, for example, completed the following unfocused freewriting during the confusion of the first day of her writing class:

> *Here I go again. Summer's gone. September stomach back. Worry about clothes, teachers, grades. I mean how many times did I check and double-check my schedule, and then — wrong room, wrong class. Everybody's looking at blotches on my neck. Act like you're the only one who doesn't know. If you get lucky, finding this class — surprise, write an essay. Freewriting, fat chance. First days are always a lie. I mean what do you people want? Solve the polynomial. Remember the <u>Maine</u>. Find the unknown. Never could figure out Bridgeman (Latin) — never learned to decline. My little revolt. He liked it — seemed to matter. Weird. Wonder if this matters? What are we supposed to be doing anyway? Why not "What I did on my summer vacation" or better yet "What I didn't do on my summer vacation"?*

It is often difficult to evaluate unfocused freewriting because the sentences slide from subject to subject, never stopping long enough to develop. Underline words or phrases that seem significant or suspicious. Do they deal with different topics? Do they have a common theme? Does one evoke a particularly powerful memory? Answering these questions may help you discover information for a focused freewriting exercise. Joanne was surprised that she remembered Mr. Bridgeman's Latin class. She decided to reconstruct the episode in a focused freewriting, using "my little revolt" as a prompt.

My Little Revolt

> *"I'm a tough teacher. This is a tough class. Half of you will be gone in two weeks. The rest of you will love Latin." Tough love! Bridgeman paces behind desk like a caged panther. Knit shirts, no ties — pumps iron. Made Linda Thomas cry three days straight; then she was gone. Everybody got it. Even Tommy Hirschen, Latin ace. "Miss Reinhardt. Go to board. Decline <u>fēmina</u>." Stomach lurches. Don't cry in front of boys. Already sniffling — bad cold, no Kleenex. "Begin!" Nominative, genitive, dative . . . accusative? . . . Damn! Play with chalk. Read back of eraser. "Miss Reinhardt?" "I'm thinking." Thinking I need a Kleenex. Nose dripping, eyes watering. Should have stayed home. "Cease" — favorite word. Never stop or that will be all. "Mr. Hirschen. Finish declining <u>fēmina</u>." Blotches on neck. Nose running all the way to desk.*

Ask Mary Keller for Kleenex — too loud. "Cease." Heart pounds. "What is it now, Miss Reinhardt?" "I need a Kleenex." Long sigh — "Is there a kind soul who can give Miss Reinhardt a Kleenex?" Book bags, purses, hands — six tissues. Don't look up. Stack them slowly in pile. Reshuffle them. Make him wait. Peel off top tissue and blow. Crumple in wad. Walk slowly toward his desk. He's still waiting. Drumming pencil. Drop wad in basket. Look him in the eye — "You may continue." Turn around . . . Hirschen's mouth open. Keller's in shock. Note on my desk. "You're dead." Bell. "Miss Reinhardt. May I see you before you leave?" Class crunches out. Hirschen whispers, "Bridgeman kills." Don't cry no matter what. We're alone. He smiles . . . "Congratulations. I've been waiting a long time for somebody to do something like that." Like what? And that was it. Still can't figure it out. Never told anybody. Let 'em fight their own battles with Bridgeman. Never scared of him again. Never loved Latin. Tough luck, Bridgeman.

Evaluating Freewriting   Like Joanne, you will need to make a number of decisions about your subject, audience, and purpose, if you want to write a longer, more finished essay based on your freewriting. Reread both of Joanne's freewritings and then answer the following questions:

1. How does Joanne's decision to focus on her revolt restrict her subject? What does she find significant and interesting about this incident? What aspects has she managed effectively? What aspects need further development?

2. How might an assessment of her audience help Joanne transform her planning into a successful draft? What specific sections of her freewriting attract the sympathy of her readers? What additional information will they need to understand her revolt?

3. What is Joanne trying to prove about her subject? Is she attempting to prove more than one thing? Make a list of hypotheses she could use to clarify her intentions.

Keeping a Journal

Keeping a journal is a way of keeping track of your thinking-in-writing. Your first step is to purchase a notebook. It can be loose-leaf, if you like adding and rearranging pages; spiral-bound, if you like choosing from a variety of sizes and shapes; or bound, if you like preserving your words in a permanent volume. Whatever it looks like, write in it on a regular basis — preferably every day.

Although you are free to write whatever you want in a journal, avoid turning it into a diary that simply recounts your daily schedule. A journal records the activities of your mind. As a source book for writing projects, you can use it to collect and create images, copy snatches of conversation or sentences from your reading, or comment on events and ideas. Because you are keeping your journal for yourself, you need not be particular about how you record this information.

You can compile lists, draw clusters, freewrite, or even compose preliminary drafts — any activity that helps you think in writing.

Keeping a journal has two advantages: it encourages you to *take risks* and to *take stock.* A journal provides a no-fault environment where you can try out all sorts of writing experiments, from personal essays to dramatic dialogues. You can play memory games like naming the boundaries of your childhood neighborhood, or you can try to honestly explore the emotions you felt during your first success or the reasons you never told anyone your deepest secret. Some of those experiences may seem hazy as you try to reconstruct them in your mind, but as you write down specific words and phrases, you will be surprised by how much you remember.

A journal is also a good place to take stock of yourself as a writer. As you make lists or compose paragraphs, comment in your journal about your writing. If you are having trouble focusing your subject, interview yourself on paper about your problem. What is the *key issue* in your subject? What do you really want to say about it? What is preventing you from saying it? Does everybody see this issue the way you do? Your answers to those questions may help you restrict your subject, anticipate your readers, and underline your purpose.

In his journal, Ray registers his anger about having to move. But the occasion gives him an opportunity to reflect on the relationship between his possessions and his past.

> 6/12   *Got my notice today. Pay increase or move. Ticks me off. Can't pay more than I am now. Gotta move. Three times in three years. Packing is terrible. Finding boxes. Lugging stuff around. Trying to decide what to pitch. New place always smaller. No storage. Renters don't have attics. Unless they live in them. Jerry lives in an attic, but he can only stand up straight in the middle of the room. Most apartments have closets. Where have all the attics gone? Maybe I could write about that. Not big question. I'm no architect. Just some attics I have known. Trouble is, I've only known one.*
>
> *Grandparents had a great attic at the farm. Every time we visited, Grandma would lead us up the stairs above the ceiling to play. Strange smell — especially on rainy days. Dust on everything — not dirt — powdery, gray dust. Made me sneeze. We would make up all kinds of games. Little window looked over top of trees. Captain's deck on pirate ship, command central for Star Ship Enterprise. Dress up in old clothes. Grandma would come back after awhile. Show us dress Aunt Marilyn was married in, Black Stallion books Dad read when he was a kid, pillow with Marine Corps insignia on it that Uncle George sent from Korea. Everything had something to do with us — uncles, cousins, Dad when he was young. Like the Waltons. Writing about it makes me feel good. Sound like that guy who narrated the Waltons, remembering when things stayed the same.*
>
> *Dad's house — our house — didn't have an attic. Crawl space covered with Corning Pink Insulation. Energy efficient. Kept some stuff in*

the basement. Tools, fishing gear — his stuff. What happened to our stuff? Old toys, my crutches when I broke my leg, Little League trophies, merit badges, photo albums. What happened to our stuff? Divorced, burned, dumped. Sent to Goodwill. Garage sale. Sold our past. I could write about that. Always sad when I drive by those sales — rummage. Is that a noun or a verb? People's stuff out on the street. If you don't have any stuff, it's hard to remember.

That's why I hate to move. What will I do with my stuff? Posters, books, records, junk in junk drawers. Walk around apartment picking up stuff, remembering all sorts of crazy things. Ray's museum. What happens when the movers come? Big guys — Sid, Burt stitched on uniforms. Start tossing stuff around. "Hey, Buddy, this stuff go?" Yes, that <u>stuff</u> goes! STUFF, STUFF, STUFF. I want all my STUFF. I could <u>explain</u> why, but there's so much stuff. Stuff it for now. Write about it after the move when I'll have more and less stuff.

Evaluating Journal Entries   Ray is making good use of his journal. He starts with a piece of news — as he might in a diary — but shifts quickly to consider the plight of people who have no place to store their possessions. One question — "Where have all the attics gone?" — suggests the possibility of a writing project. As he evaluates the subject, he quickly restricts it to "attics I have known" and then realizes that he has only known one — his grandparents' attic. Nevertheless, this focus evokes pleasant memories from his early childhood, when the family was together. As he writes about that time, he identifies with the narrator of an old television show, *The Waltons,* which recreated an unchanging past.

These memories lead Ray to an important insight. His family did not have an attic, a fact that may explain why his possessions were misplaced during his parents' divorce, given away, or sold at a garage sale. This speculation suggests another subject (garage sales), points to a possible hypothesis ("If you don't have any stuff, it's hard to remember"), and brings Ray back to his immediate dilemma ("What will I do with my stuff?").

At this point in his planning, Ray is simply testing the possibilities of several *subjects* — attics, garage sales, moving, stuff. He has not thought very much about *audience,* but he can assume that all his readers have faced or will face the problems he explores in his journal. His planning has suggested, however, several hypotheses that might explain why possessions are so important — why he wants to keep his stuff. Although he has used his journal to take stock of these potential purposes, he decides to postpone making any final decision about which one to use. Perhaps the trauma of moving will give him new insights that will enable him to evaluate his subject, audience, and purpose more effectively.

■■■■■ *Exercises*

1. Pick a word from the list below (or choose a word of your own) and then make a list or draw a cluster. When you have finished, evaluate your infor-

mation for potential topics. Which ones do you find most interesting, surprising, or unsettling?

shy	ice cream	germs
hair	stereo	check
whoops	weekends	obligations
proof	bookcase	clutter

2. Choose a topic for a focused freewriting exercise. Limit yourself to fifteen minutes. Select one of the topics you identified in the list or cluster you created in question 1 (for example, the rewards of ice cream, the fear of germs) or a topic you have heard mentioned in a class or discussed among your friends.

3. Comment in your journal on all or any of the following:
   a. Reading assignments from other courses that pose interesting questions or solve difficult problems
   b. Letters from friends and family that trigger memories
   c. Situations that cause you happiness or stress
   d. Striking images you see — in the media or on your daily rounds
   e. Individuals — contemporary, historic, or fiction — who interest you
   f. Major news stories of the day that may affect your life

USING OBSERVATION IN WRITING

Like most people, you spend much of your life as a casual observer, acknowledging but rarely examining the blur of daily experience. You have learned to *look for* the things you must see — the assignment on the syllabus, the stop sign at the intersection. Occasionally you will *look at* some person or some thing new or unusual — a new student in the front row, a remodeled store in the old building around the corner. But for the most part, you *look through* the world the way you look through a magazine, flipping pages without any particular purpose. When you begin to write, however, you must train yourself to look for interesting subjects, significant features, and telling details to help you compose vivid verbal pictures. Two observation strategies are *mapping* and *speculating*.

Mapping

Mapping is a strategy for studying a subject during an extended period of observation — usually twenty to thirty minutes. If you are convinced you have no artistic talent, or if you have not sketched anything in a long time, you may hesitate to experiment with this strategy. But mapping, like clustering, is not designed to produce an object for admiration; it is simply another strategy for thinking on paper. Others may be fascinated by your map, but only you can interpret the significance of its lines, figures, and legends.

You can draw three kinds of maps; depending on your subject and purpose, you may decide to use one, two, or all three types.

- *Spatial maps* resemble a traditional map or blueprint. The map maker looks down on a static scene and outlines and defines the spatial relationship among its various parts.

- *Activity charts* use the basic outline provided by a spatial map to document the kind and amount of action that occurs in the scene. Usually, the map maker uses a variety of dotted lines to illustrate where action originates and concentrates.

- *Figure drawings* focus on your subject in two- or three-dimensional detail. Drawing this kind of map requires more time and talent, but usually shows you more of your subject's unique features than any other method.

Regardless of the format you choose, follow the same mapping procedure. First, buy a large sketch pad (at least 18 by 24 inches) and a box of pencils with good erasers. Second, select a subject to map. You may decide to sit in a familiar spot and wait for a subject to appear, wander around until you bump into something or somebody interesting, or choose a specific subject that you suspect will

"He's closer, but not better."

Drawing by Sempé; © 1984 The New Yorker Magazine, Inc.

prove worth your time and attention. Third, select a lookout spot. Lookout spots function like a photographer's view finder, helping you frame and center the subject you want to observe. Every subject suggests dozens of possible lookout spots, each providing a slightly different perspective. Experiment with several locations, and then select the best one for your subject and purpose.

Once you are situated, hold your hands upright in front of your eyes to form an imaginary picture frame. Expand and contract this frame to see how much you want to include within the borders of your map. Then draw your subject, outlining and identifying the physical features, major characters, and energy centers that attract your eye. Finally, fill in the details, and use annotations to correct any distortions of emphasis or proportion. Study your subject in the way a stage designer examines a scene to create a theatrical set. Locate and list everything you would need to reconstruct this set somewhere else — backdrops, characters, costumes, furniture, props.

One assignment in Jenny's writing class called for her to observe a local business and compose its profile. Armed with her sketch pad, she ventured into town in search of a suitable subject. She considered several possibilities — a deli, a Laundromat, and a clothing store — but settled on the Varsity Barbershop, a hangout for male students, faculty, and various members of the community. Slightly apprehensive about invading this predominantly male world, she explained her assignment and was warmly accepted by the barbers, who began joking with one another about how famous Jenny's map would make their shop. She selected a lookout spot and spent almost an hour drawing the maps on pages 38 and 39.

Evaluating Maps   Before leaving your lookout spot, examine your map as you would a painting. Form a general impression of the lines, figures, and shapes you have made. Next, move your eyes slowly across your map (side to side, top to bottom, center to periphery), examining the details in each section. If you see details that require further explanation, ask your subject to identify movements or to label props so that you can add more precise annotations. This information will help you illuminate your subject and evoke its texture for your readers once you begin describing it. Later, look for the magnetic center — the subject that pushes itself into the foreground of your drawing. Why does it attract your attention? How does it connect to or contrast with the subjects it forces into the background? Propose several hypotheses about your map that will help you interpret your observations in writing.

Jenny began by constructing a spatial map of the Varsity Barbershop. She established her lookout spot in a chair beneath the price sign and blocked out the major areas in the shop (barber chairs, waiting chairs, cash register); then she added important details (barber tools, hair supplies, magazines). But as she evaluated this map she realized that she was interested more in the action in the shop than in its props and scenery. Doing an activity chart shows her that the owner, Everett, is the central character in the shop's daily drama. Finding him an appealing *subject,* she does a detailed character drawing. Her observations show that men of all ages are customers; thus, her *audience* could be men (who are familiar with the rituals of a barbershop) or women (who are interested in those rituals or suspect there may be similarities between the activities in barbershops and those in beauty salons). As Jenny focuses on Everett, however, she thinks less about cutting hair and more about *him.* He is the energy center, the individual who dominates and orchestrates the conversation in the shop. She is close to for-

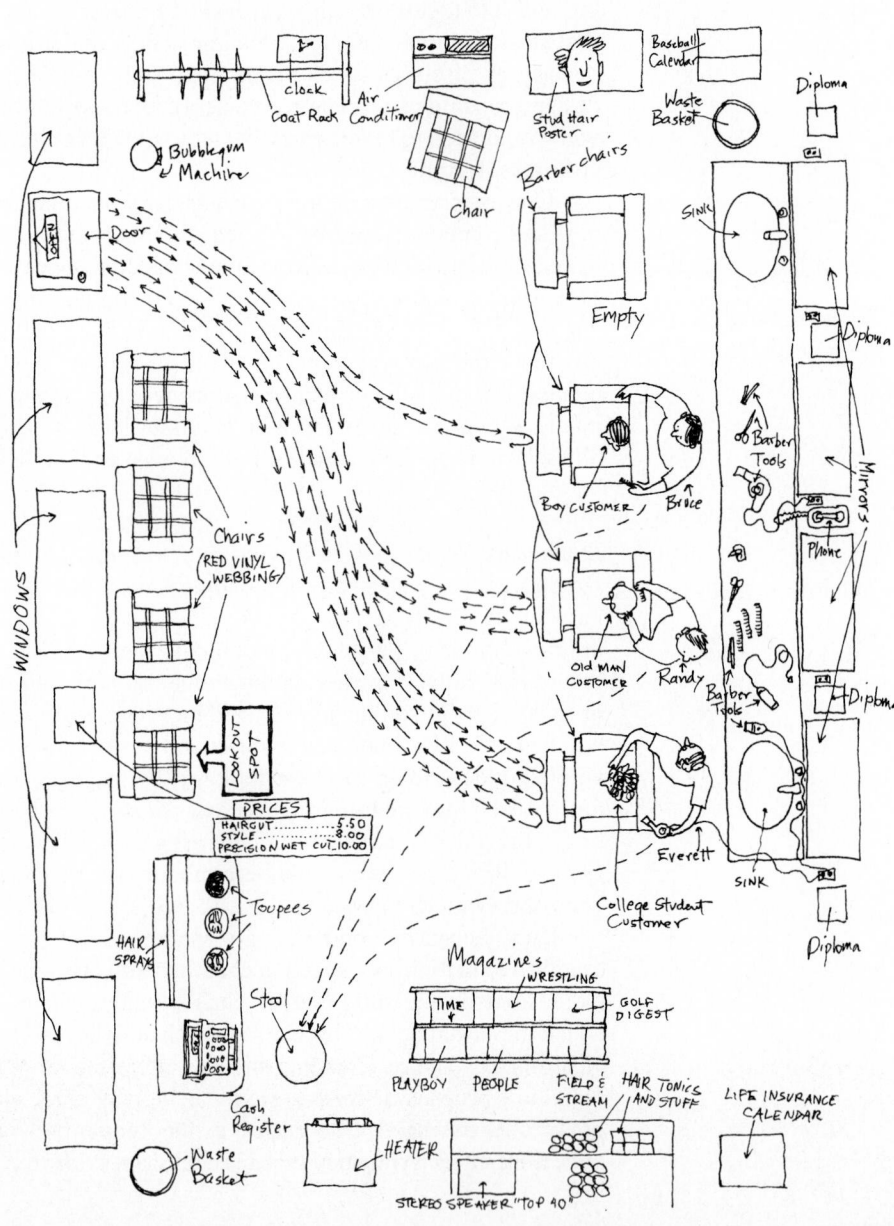

mulating a *hypothesis* to try out in a draft that reflects this growing sense of focus: "A single personality (Everett) gives a place (Varsity Barbershop) its atmosphere and character."

Speculating

Speculating will help you produce different interpretations of what you have observed. The word *speculate* suggests an imaginative, hypothetical kind of seeing: you study your subject, mull over its possible meanings, and make several conjectures about its significance. Your subject might be an observed object, a scene, an event, or anything else you would like to think about, such as an idea or concept (for example, the American educational system or truth) or a character from a play or novel. As a thinking-in-writing strategy, speculating encourages you to expand your planning by considering your subject from three different perspectives[1]:

▓ *Speculate about your subject as an object.*
  This perspective suggests that you see your subject at rest, as a static object or scene that at a given moment has a fixed identity.

▓ *Speculate about your subject as an action.*
  This perspective allows you to see your subject in motion, as a dynamic process that changes form as it progresses.

▓ *Speculate about your subject as a network.*
  This perspective enables you to see your subject as a series of connections, as a complex system of relationships that extend and enhance its importance.

As you think about your subject from each of these perspectives, note your speculations in your journal. If one perspective does not produce provocative information, move on to the next. Write what you see and make guesses about what you don't see. Don't try to write perfect sentences; you are still *planning*. You are looking for something interesting and significant about your subject that you can reveal to an audience for a purpose.

Speculating both generates information (if you have at least some notion of your subject) and focuses and interprets information you have already collected. For example, here are Jenny's speculations about what she discovered from mapping the barbershop:

Varsity Barbershop as an Object

The Varsity Barbershop has been a fixture in town for almost forty years. Probably never changed in all that time. A permanent set. If it had a checkerboard and an overhead fan, it could be Floyd's Barbershop on the old Andy Griffith television series. There are four brown leather and chrome barber chairs facing the windows, and four red, webbed vinyl and aluminum waiting chairs facing the barber chairs. Behind three of the barber chairs stand the three barbers, waiting for somebody to come in. They watch the people who walk down the street. Sometimes they sit in their own chairs. They don't talk much when there are no customers. Mannequins in a store window. Behind the chairs are large mirrors and a series of counters and sinks where they keep their tools. On the walls between the mirrors are their diplomas from barber school. Near the door are a coat rack, an air conditioner, and a large poster of this terrific-

---

1. Based on the tagmemic theory of invention in Richard E. Young, Alton L. Becker, and Kenneth L. Pike, *Rhetoric: Discovery and Change* (New York: Harcourt Brace Jovanovich, 1970).

looking stud with wonderful hair. Near the cash register are hair supplies, including a few sample toupees, and a magazine rack — Field and Stream, Golf Digest, and Playboy — typical male reading. You can hardly see the price sign. When you sit in the waiting chair, it's above your head. When you are in the barber's chair, you can see it. As a fixed object, there's not much to see here. Drab. Dull. The shop comes to life when a customer walks in.

Varsity Barbershop as Action

There are so many things to see when the shop is busy. Above everything else there's the murmur of conversation. Listen carefully and you can pick out individual sounds — scissors snipping, razors trimming, and the vacuum whirring. But first you come in: "Hi, how are you?" Everybody gets a hello from every barber, but the biggest hello comes from Everett. Then you have to find a chair, pick out a magazine, and wait your turn. Sometimes you talk to the barbers, especially Everett, while you wait, but mostly you wait. Which barber you get depends on rotation — "Next?" Some people say "Go ahead" to somebody else. "I'm waiting for Everett." Sometimes there is a long waiting time. Sometimes you're next. A few people just come in to talk or say "Hi" to Everett. Always crowded at 5:00 p.m. on Fridays. Once you get in the chair, you talk about how you want it cut. After a few jokes. Everett, who wears one of the "rugs" he sells, starts cutting your hair — combs, scissors, electric trimmers. He may turn you around a couple of times so you can see what he's doing in the mirror, but usually he waits to show you the result when he's done. All the time he talks. To you, to the other customers, to the other barbers. Randy is his straight man. Feeds him lines so he can get Everett to tell a joke. Bruce isn't as busy as Randy and Everett — he's been at the shop only a year. He picks up the broom and sweeps the hair clippings from under the chairs and pushes them down the floor to the wastebasket. When it's not busy, Randy likes to play the stereo. Works as a drummer on weekends. "Where you playing this weekend, Randy?" "Moose." When you're done, you pay at the cash register and say "Goodbye."

Varsity Barbershop as Network

Varsity Barbershop is definitely a network. Everett connects everybody with everybody. The shop is in the middle of town, across from the bus stop and the bank. Everett owns some property nearby so he knows all about real estate. Also knows all the other shopkeepers. Probably knows their parents. When he goes out for coffee, he talks to everybody. Gets the lowdown, the latest. Comes back to the shop loaded with news about who's planning a sale, who's hiring, and who's selling out. Everybody who comes into the shop has news. The mayor, bankers, policemen, bus drivers, students, professors, kids — everybody has something to tell Everett. And Everett passes it on. Insurance man named Jerry comes into the shop. Just to sell Everett some life insurance. Everett says he cashed in his old policy and invested it in money market. "Oh, that Everett. He's a smart one," says Randy. Jerry takes bait. Soon they are shouting. "Everybody's got to have insurance," says Jerry. "Not really," says Everett. "You're crazy," shouts Jerry. "Why? I got money. I'm taken care of." "I'm talking about the average person," says Jerry. "Oh well, then," Randy interrupts, "you're not talking about Everett." Everybody laughs. Bruce says that when the Brinks truck can't get through to the bank, "they just drive over here and ask Everett for a pile of hundreds." Everybody laughs again. Old gossip about Everett and his millions. Everett must know everybody in town. And when you think about how many students have gotten haircuts here in forty years, he probably knows half the state. During alumni week, the old grads come back to talk to Everett. He gets more news. Next week he'll pass it on. The Varsity Barbershop is an information switchboard, and Everett is the chief operator.

Evaluating Speculations   Jenny's planning activities have yielded a wealth of information about the Varsity Barbershop. Before she drafts an extended profile of this local business, she needs to reconsider all her information in terms of her potential subject, audience, and purpose. Reexamine her mapping and speculating exercises; then write out your answers to the following questions. Your instructor may want you to discuss your answers in class.

1. How many specific subjects can you identify in the three speculations?

2. Which speculation produces the most interesting perspective on the barbershop? How does that perspective define a significant and manageable subject?

3. What potential audience could Jenny assume for her profile? What different audiences might she assume for each of her three speculations?

4. What hypotheses do you find in Jenny's mapping and speculating exercises?

5. What purpose might Jenny choose for writing an essay about a small community barbershop? Compose a thesis sentence she could use to state the purpose of her essay.

USING RESEARCH IN WRITING   After exploring your memories and examining your observations, you may decide that you need more information about your subject to compose an effective piece of writing. Memory and observation, after all, define your subject in *your* terms — the way you remember or see it. To broaden and deepen your understanding, you must reach beyond the limits of your personal experience to determine how others have perceived your subject. Almost any subject you choose (worst flavors of ice cream, best hiking trails, most unusual junk mail) has been researched by other writers. Sometimes their research reflects and reinforces your view of the subject. More often it enlarges, confuses, or even contradicts your conclusions. You may want to reject this unsettling information and reaffirm the world as you see it. But you should resist this temptation, suspending your judgment until you have given the evidence a fair hearing. After such consideration, you will need to decide whether to dismiss this evidence entirely, accept the assertions that challenge your weakest assumptions, or integrate all its findings into your expanding interpretation of your subject.

In Chapters 13 and 14 you will learn how to follow the formal procedures for planning and writing the research paper, which is a more complex piece of writing than the simple expository writing discussed here. In this chapter you will learn how to practice two informal planning strategies designed to gather information in writing: *interviewing* and *reading*.

Interviewing

Interviewing someone is the most direct way to investigate a subject, but the process is never so simple or spontaneous as it appears on television. To use the strategy effectively, you must learn how to *prepare* for an interview, how to *manage* and *record* an interview, and how to *evaluate* the results.

How to Prepare for an Interview Begin by compiling a list of people who may know something about your subject. The list could include *experts* (people who have studied your subject), *participants* (people who have *lived* your subject), and *information brokers* (people who may know nothing about your subject but know how to put you in touch with people who do). Contact the people on your list and ask for an interview, stating in general terms what you are trying to find out and why you think they may be able to help. Ask to talk to them for a specific amount of time — no longer than an hour — and don't overstay your welcome. As you schedule your interviews, anticipate how your subjects may respond to your questions. Are they likely to be friendly, hostile, or merely puzzled when you arrive? Sometimes you can anticipate their attitudes because you know them or because you have heard them express their opinions on your subject. At other times you can anticipate their attitudes by imagining yourself in their position. How would you feel about being interviewed on this subject? What topics would you want to be asked about? What topics would you want to avoid? What questions would you have about the interviewer's motives and methods?

Next, ask yourself what you want to learn from each subject. To feel prepared and to avoid getting sidetracked, write out a list of questions following the journalist's formula: Who? What? Where? When? Why? and How? These questions will help you organize what you want to learn. Most interviews are dynamic, disorganized conversations. For that reason, you must remind yourself that your questions are likely to provoke surprising answers that will inspire new questions. The best interviewers prepare to discover things they need to know and plan to uncover things they had not expected to learn.

How to Manage and Record an Interview   Interviews can be formal occasions in which the interviewer acts like an attorney grilling witnesses and the interviewee acts like a jittery suspect. Or interviews can be informal occasions — pleasant and productive conversations among friends. Sometimes you cannot control the mood of an interview because the person you are talking to is suspicious before you arrive. But most of the time, you can put yourself and your subject at ease by following a few basic tips:

- Don't feel that you must apologize for your interview. You may want to say something about how much you appreciate your subject's willingness to talk to you. Such remarks are part of the etiquette of interviewing. But someone who has agreed to be interviewed is probably flattered that you are there and has already made time for you on the day's schedule.

- Tape recorders can supply a valuable record of your conversation, but they make some people uncomfortable. Try purchasing a small notebook. Keep it out of sight as much as possible, and write in it as little as possible. When you do write, jot down words and phrases rather than complete sentences. Keep your eyes on the person you are talking to, not on your notebook.

- Begin your interview by talking about interesting and safe topics. If you are conducting the interview in the subject's home or office, ask about some object or photograph you have noticed in the room. Even if you don't ask about these props, try to list them in your notebook because they may help you interpret what your subject thinks is important in his or her life.

- Don't tell your subject everything *you* want to know before your subject tells you what he or she knows. You are trying to learn what other people think, not prove what you think. The best strategy is to encourage the interviewee to act like an expert: ask your subject to help you understand the topic at hand by answering your questions.

- Use prepared questions only when the conversation drifts far away from the designated topic. For the most part, allow the conversation to develop naturally. Before you begin to end the interview, however, review your prepared questions as a final check.

- Save two questions for the end of the interview: (1) What should I have asked that I didn't ask? (2) Whom else do I need to interview (or what do I need to read) to understand the topic we've been discussing?

How to Evaluate an Interview   Once you have completed the interview, return to your room and immediately reconstruct the conversation in writing. Describe the atmosphere of the room where the interview took place, the appearance of your subject, and the varying attitudes (eager, evasive, expansive) he or she expressed during the discussion. Transform your words and phrases into complete sentences. If your notes seem incomplete or if you want to be sure you are quoting your subject accurately, call the person to double-check your information. Such follow-up calls are not invitations to censor. People often want to revise what they said during an interview, but only you can decide whether their second thoughts should be yours. Your follow-up call is simply an effort to make the first version complete and accurate.

After you have reproduced the interview in writing, talk to yourself in your journal about what you learned. Did your subject provide useful answers to your prepared questions? What answers confirmed your assessment of his or her biases? What answers surprised you? What new questions did you ask? If you were to draft an essay based on this interview, could you identify a subject, anticipate an audience, and formulate a hypothesis? This last assessment suggests the degree to which even the most informal interview can propel you into further planning activities. Once you try to evaluate what other people have told you, you will discover that you are making lists of questions you should have asked, people you want to talk to next, and material you think you'd better read.

Susan decided to write about the set design in Arthur Miller's *Death of a Salesman* for a short research paper for her literature class. The University Theater was scheduled to produce the play during the next semester, and she decided to interview the theater's set designer.

10/25   Decided on my research paper — "Set Design in Arthur Miller's Death of a Salesman*." Theater is producing it next semester. John knows the set designer — D. C. Schwanger. Says he's easy to talk to — real creative. Got an interview Thursday at 1:30. Wonder what he'll do with* Salesman*? Better think up some questions so I don't look dumb.*

1. *Who designed the original set for* Salesman*?*
2. *What kind of problems will you have to solve in designing the set?*
3. *Where will you find props for the play — trophy, tape recorder, flute?*
4. *When will you have to start building the set to complete it on time?*
5. *How will you create the set for the play — follow Miller's instructions or your own intuitions?*
6. *Why do some playwrights provide more information than others about the significance of the set?*

10/28   Notes from Interview with D. C. Schwanger
Just back from Schwanger interview. First question blew my topic. Since Jo Mielziner's prize-winning design for the original production, nobody has done much with Salesman*. Schwanger says all you need is three rooms for the house and open space (apron) for other scenes —*

Timothy Sawyer and Dawn Davis in a scene from Hedda Gabler

© 1987 Michael Romanos/Photo courtesy Nickerson Theatre, Inc.

restaurant, hotel, garden. He's more interested in sets he's designing for Hedda Gabler *and* Waiting for Godot. *Need to write down what happened.*

Sign above office door: "The Miracle Worker." Room filled with props (mostly swords), drawings of sets (<u>renderings</u>), awards (Outstanding Creative Achievement). Next to Schwanger's desk is a drafting table stacked with blueprints (<u>working drawings</u>). Sign behind the table: "It's a small world, but I wouldn't want to paint it." Behind his desk Schwanger looks small, unimpressive — until he starts talking about his work. Real dynamo.

Designing set for one play, building set for another, tinkering with another in performance — complicated schedule. Set designer has to be organized. Reads play, researches how other designers staged it, then looks at history books to see what a living room should look like in a specific time (<u>Hedda</u>). Next collaborates with director about initial ideas — sometimes there is disagreement, but director is always in charge. Next step, makes a rendering or scale model to illustrate things like space, mood, symbolism. Then the working drawings — 15–20 pages of blueprints for each play. Designer has to be architect, historian, artist, draftsman, engineer, carpenter, plumber, electrician, upholsterer. Turns drawings over to shop foreman for construction but may have to change everything in rehearsal. Too much furniture. Lights change color of paint. Audience can't see props.

New challenge with every play. Historical plays like <u>Hedda</u> you have to be accurate. Modern plays like <u>Godot</u> you have to be creative.

John Bottoms (left)
and Mark Linn-Baker
(right) *in a scene from*
Waiting for Godot

© Richard Feldman/Photo courtesy American Repertory Theatre

*Strange — less set — i.e. Godot — frees you to be more creative with
what you have — tree. Each play requires you to learn something new.
With Equus, weld metal masks. With Miracle Worker, get water to the
pump.*

*Schwanger blew my paper, but fired me up about the whole process.
Volunteered to work for him next semester. I could still do Salesman, but
I'd rather pick another play or maybe write about several plays that
present different design problems. Wonder if I could write about the job
of set designer and use different plays as examples? Audience doesn't
know how complicated a job it is. Just sit there and watch the actors.
Never think about plumbing or electricity unless it doesn't work. Maybe I
could interview some other designers. Read textbooks or something
by Mielziner.*

Susan's interview produced much information on the world of set design. Her
original topic, "Set Design in Arthur Miller's *Death of a Salesman,*" is still feasible, but it no longer seems interesting or significant. Read through her notes, and
then try to answer these questions:

1. To what extent does Susan's list of questions *prepare* her for her interview?

2. How does she *manage* Schwanger's replies to her questions?

3. Why does the interview change the focus of Susan's research paper?

4. What kind of audience does she anticipate for the new essay she is planning?

5. What hypothesis is she beginning to formulate about this new subject?

Reading

Reading what other people have written about your subject is probably the most common strategy for gathering information for writing. But merely consuming the words of other people will not help you compose your own. You must be a critical consumer — *selecting, analyzing,* and *evaluating* what you read for one purpose: to help you write. In Chapters 13 and 14, you will perform these activities according to the more formal requirements of the research paper: checking bibliographies, quoting evidence, documenting sources. In this chapter, you will use reading as an informal procedure for exploring different approaches to your subject.

The easiest way to keep track of your reading is to write about it in your journal. Think of reading as another form of interviewing — a way of talking to people who have already thought and written about your subject. So, just as you did in interviewing, start by making a list of material you want to read. Be sure to look for variety in your reading material: skim magazines in a bookstore; consult with a local expert on your subject; or talk to a reference librarian. Once you start reading, each selection will refer you to others; then your problem will be deciding when you have read enough. There is no way to make that decision — you can *always* read more — but usually you know that you have read enough when the same names, ideas, and issues begin to pop up in everything you read.

On most subjects, reading material falls into three broad categories: eyewitness accounts (printed in popular magazines), expert studies (documented in technical journals or official reports), and advocates' opinions (expressed in editorials or promotional pamphlets). Although these categories suggest the kind of material you need to read, they may not help you classify a specific selection. Indeed, many selections combine the features of all three categories. An editorial, for example, may cite personal anecdotes and statistical data to advance its argument. To sort out the material on your list, you need a set of questions to help you interpret what you are reading. These questions are similar to the guidelines you follow as you write, but in this case you are deciding how another writer, writing in a specific format and context, selects a subject, analyzes an audience, and determines a purpose:

1. Who is the author? What makes the author an expert on the subject? What might have formed the author's point of view?

2. Where and when was the article or book published? You may know little about the author, but you may know something about the point of view of the magazine (*Ms., Scientific American*) and whether the date of publication makes the information current and reliable.

3. What is the author's general subject? What is the restricted subject? What does this perspective contribute to your understanding of *your* subject? Does it confirm your perception, raise new and confusing questions, or lead you into detours and dead ends?

4. To whom is the writing addressed? What attitudes does the author assume in

his or her audience? What knowledge does he or she expect these readers to possess or want to obtain?

5. What is the author's purpose? Does the author state a specific thesis or imply one? How does the thesis, stated or implied, control the direction of the writing?

These questions not only help you select the most intriguing reading material from your list, but they also help you analyze and evaluate it. Keep your journal with you so you can write about your reading as you *preview, read,* and *review.*

▪ *Preview:* Before you read each selection, try to anticipate the kinds of information and the point of view you will find. Annotate each entry on your list, indicating what you expect to find when you read it.

▪ *Read:* As you read each selection, write a few brief notes in your journal. Comment on the major points, copy memorable sentences or important data, and write out short answers to your reading questions. You will also want to write any questions the author raises that you have not considered.

▪ *Review:* Put the material aside and write about the most vivid and thought-provoking things you remember. Thumb back through your reading notes. What did you discover that contributes to your view of the subject? Evaluate what the selection said and what it didn't say.

Paul follows this three-part procedure as he does preliminary reading for an essay on energy alternatives. He begins by reading about the most obvious alternatives — solar energy, nuclear energy, and synthetic fuels — but eventually he restricts his subject to an offbeat but plausible option: garbage. Next, he compiles a short reading list, noting his expectations for each selection.

Reading List on Garbage

Preview

1. *"Where Will All the Garbage Go?" Atlantic (March 1985). Author is an environmentalist — wrote a book about small farms. Drawing for article shows island of Manhattan shaped like garbage scow. World Trade Towers serve as smokestacks. Probably focus on New York, but relate problems to general question in title. Who reads Atlantic?*

2. *Handbook of Solid Waste Disposal (1977). Long, technical book by three professors of civil engineering. One chapter on "Energy Recovery" written for experts. Charts, tables, drawings. Survey of research, case studies, bibliography, appendix on cost estimates for building a plant. Information may be dated — mostly from late 60s and early 70s.*

3. *"The Sweet Smell of Profits from Trash," Fortune (April 1985). Don't know author (Leinster), but mag. usually focuses on business. Articles grouped under "Corporate Performance," "Money and Money-*

*makers," etc. This article under "Looking Ahead." Written for poten-
tial investors. "Investor's Snapshot" lists sales, profit, and price per
share for the two companies featured in article.*

Finally Paul reads and reviews each selection, evaluating what the article contributes to his own planning.

Reading notes

1. *"Where Will All the Garbage Go?" Focuses on Fresh Kill — (strange
name) — world's largest landfill. Staten Island. 11,000 tons of garbage each day. Title good question. Burning garbage hot political
issue. Smoke/dioxins. "Incidence of cancer." Barry Commoner
against plant, then changes his mind if they promise to enforce emis-
sion standards. Right Barry! Don't you watch "60 Minutes"? Public
attitude — "out of sight, out of mind." How can Fresh Kill be out of
sight? Mountain peaks over 500 feet — highest on eastern seaboard
south of Maine. Ends with story about a "Maintenance Artist." Builds
enormous sculpture of garbage truck out of mirrored glass. Arranges
"dance" for tugboats and garbage scows. Says, "If we don't blow
ourselves up . . . biggest problem will be what to do with the
garbage."*

Review

*Most striking image — mountains of garbage at Fresh Kill. Incredible
name. 11,000 tons a day. How does it all get there? Who works there?
What do people who live nearby think of mountain range? How would
they like a big garbage-burning plant? Smoke? Article deals with problem,
not solution. No one seems to have a solution. "Garbage always wins."*

Reading notes

2. *Handbook of Solid Waste Disposal. Several plans. Combustion of
prepared waste. What is it? Uses only residential, not industrial waste.
Does that solve problem? Describes burning waste to produce a waste
product that can be burned. Problem — when all this "prepared"
waste burns, it produces a heavy accumulation of new waste called
"bottom ash." Corrosive. Pollution not significantly higher than plants
burning other fuels. Great news! Another system uses "mixed municipal waste." Problem — additional recovery system for by-products.
Charts show what materials can be reclaimed and their worth. Blueprints of plants. 1,000 tons a day. Looks like something out of Star
Wars. Six pages of charts on the chemicals that come from stackgas.
Never say whether any of them can hurt you.*

Review

*Waste — garbage, refuse, trash, residential waste, industrial waste, mixed
municipal waste. I could write my whole essay on these words. The diagrams for the plants look really complicated. Rube Goldberg cartoons.
Charts are complete, but the numbers are neutral. Never say anything
about environmental impact. Engineers. Talk about economic feasibility
and technical problems. Should probably reread this chapter. Know one
thing, 1,000 tons a day won't dent Fresh Kill.*

Reading notes

3. *"The Sweet Smell of Profits from Trash." Description of a ground-breaking ceremony for $80 million garbage plant in Florida. Growth industry. When* <u>*properly*</u> *run, no problem. Monaco has a plant "topped by a public tennis court with a restaurant inside, a mere 500 yards from Prince Rainier's palace." Garbage in Monaco? Prince strolls over for a "power" lunch at the plant. Spotters watch for dangerous garbage — cars, appliances, old war souvenirs (land mines and grenades). Big claws — like in penny arcades — grab garbage for fire. When it burns, you can hear the "occasional zing of aerosol cans" bouncing off furnace walls. Problems — tax reforms could prevent depreciation and investment credits. Two compounds in smoke cause cancer. Talks to Kiwanis, Chamber of Commerce, etc. Shows them a glass of cloudy water from a pond near a landfill. Says, "I'll smell the smoke if you drink this water." Projected profits for the two corporations that will dominate this "burgeoning industry."*

Review

Monaco example makes plant seem harmless. Kiwanis example reduces complex problem to simple either/or. All the money talk suggests these guys are serious. Will probably battle lobbyists from Sierra Club. Artist was right. Garbage is a big problem. Engineers are right. Should develop technology to convert garbage into energy. Business executives also right. See profit potential in all this garbage. Still, everything seems shaky.

Evaluating Reading   Paul has come a long way since he decided to restrict his paper to garbage. His view of the subject has widened so that he doubts his ability to cover it. He is astonished at how much his readers will have to know to understand the deceptively simple plan to convert garbage into energy. But as he rereads this material and searches for additional information, he discovers a working hypothesis: "Converting garbage into energy creates several kinds of problems." When he composes his discovery draft, Paul will have to refine this hypothesis into a thesis.

▰▰▰ *Exercises*

1. Think of a decision that will require you to gather information. For example, you may need to decide where to go on your next spring break or which stereo components to buy to upgrade your system. Gather information by interviewing someone knowledgeable about the subject: a travel agent, stereo equipment salesperson, etc. Prepare for, conduct, and evaluate the interview.

2. Choose a subject about which you want or need to do some research, such as the risks of strenuous exercise or the usefulness of personal computers. In your journal, compile and preview a short, varied reading list about your subject. When you have completed your reading, evaluate each source in your journal.

A FINAL WORD ABOUT PLANNING

The most important thing to learn about planning is to know when to stop. You probably know people who spend their lives making plans. If they have to study for a test, they make lists of what to study, map out a schedule for studying each item on the list, and then talk about their study plans to everyone else taking the test. Somewhere in the midst of all their planning, they lose sight of the project — the test. Planning as a part of the writing process has a specific and limited purpose: thinking-in-writing. Once you have completed this thinking, you must come to some conclusion about what you have written. You must make choices about your *subject, audience,* and *purpose.* And you must be willing to try out those choices in a fuller, more sustained piece of writing. You must begin drafting.

Planning Exercises

1. The following observations about keeping a journal were made by three important writers: Henry David Thoreau, nineteenth-century American philosopher and essayist; Virginia Woolf, twentieth-century British novelist and essayist; and Joan Didion, twentieth-century American novelist and essayist. Read each passage carefully, then discuss all three in terms of the following questions:
   a. How would you characterize the attitude of each writer toward keeping a journal?
   b. Do you see any similarities among the three attitudes? Any differences?
   c. Which writer's remarks best approximate how you feel about keeping a journal? Explain your answer.

To set down such choice experiences that my own writings may inspire me and at last I may make wholes of parts. Certainly it is a distinct profession to rescue from oblivion and to fix the sentiments and thoughts which visit all men more or less generally, that the contemplation of the unfinished picture may suggest its harmonious completion. Associate reverently and as much as you can with your loftiest thoughts. Each thought that is welcomed and recorded is a nest egg, by the side of which more will be laid. Thoughts accidentally thrown together become a frame in which more may be developed and exhibited. Perhaps this is the main value of a habit of writing, of keeping a journal — that so we remember our best hours and stimulate ourselves. My thoughts are my company. They have a certain individuality and separate existence, aye, personality. Having by chance recorded a few disconnected thoughts and then brought them into juxtaposition, they suggest a whole new field in which it was possible to labor and to think. Thought begat thought. (Henry David Thoreau, *Journal,* 1852)

. . . I have just re-read my year's diary and am much struck by the rapid haphazard gallop at which it swings along, sometimes indeed jerking almost intolerably over the cobbles. Still if it were not written rather faster than the fastest type-writing, if I stopped and took thought, it would never be written at all; and the advantage of the method is that it sweeps up accidentally several stray matters which I should exclude if I hesitated, but which are the diamonds of the dustheap. . . . (Virginia Woolf, January 20, 1919, *A Writer's Diary*)

So the point of my keeping a notebook has never been, nor is it now, to have an accurate factual record of what I have been doing or thinking. That would be a different impulse entirely, an instinct for reality which I sometimes envy but do not possess. At no point have I ever been able successfully to keep a diary; my approach

*A nuclear power plant
at San Onofre,
California*

© 1986 Peter Menzel/Stock Boston

to daily life ranges from the grossly negligent to the merely absent, and on those few occasions when I have tried dutifully to record a day's events, boredom has so overcome me that the results are mysterious at best. What is this business about "shopping, typing piece, dinner with E, depressed"? Shopping for what? Typing what piece? Who is E? Was this "E" depressed, or was I depressed? Who cares? . . .

How it felt to me: that is getting closer to the truth about a notebook. I sometimes delude myself about why I keep a notebook, imagine that some thrifty virtue derives from preserving everything observed. See enough and write it down, I tell myself, and then some morning when the world seems drained of wonder, some day when I am only going through the motions of doing what I am supposed to do, which is write — on that bankrupt morning I will simply open my notebook and there it will all be, a forgotten account with accumulated interest, paid passage back to the world out there: . . . I imag-

ine, in other words, that the notebook is about other people. But of course it is not. . . . (Joan Didion, "On Keeping a Notebook," *Slouching Towards Bethlehem*)

2. Examine the photograph above according to this procedure:
   a. Describe your general impression of the picture.
   b. Identify the major landmarks, activity centers, and focal points.
   c. Observe and record the specific details in each section of the picture.
   d. Formulate a hypothesis about this picture that will allow you to communicate a subject to an audience.

3. Read the following essay, based on interviews with professional travel photographers. In what ways are their various *planning* strategies similar to the thinking-in-writing strategies described in this chapter? Which of these strategies might have produced the photographs on pages 54 and 55?

A chance shot:
A Russian woman at a
May Day celebration in
Red Square, Moscow

© 1980 Burt Glinn/Magnum Photos, Inc.

The Traveling Photographer / Robert S. Winkler

The ways to a good travel photograph are as varied as the combinations of personality and experience that can exist behind the camera. The photographers represented in this special issue recognize that; asked for their comments about their work, they agree on many matters and differ greatly on others, but the most important thing they share, aside from talent, is a willing acceptance of any photographic approach that has, as its end, a meaningful picture. It can only be encouraging for other traveling photographers to see that there is more than one way to skin a cat.

Perhaps in no area do these photographers differ more than in what they do to prepare for their trips. For Pete Turner, who specializes in color and design — making his living from advertising photography but taking his pleasure from travel photographs made for himself — thorough preparation is the rule.

*A planned shot:
A dramatically lit night
view of the Sphinx in
Cairo, Egypt*

© 1980 Pete Turner/The Image Bank

"I usually have a target or a plan — something in mind that I'm going to photograph. For example, I might go to Kenya with the intention of photographing Mount Kilimanjaro. I'd try to make arrangements to be there at the right time of year — not in the rainy season — and I'd think about the best light to photograph in. My primary objective would be to photograph at the best time and in the best light.

"My bible is the *Pan Am World Guidebook*. You can stick it in your suitcase. It covers most

of the countries in the world and tells you the number of days of rain or sunshine each month, the temperature, the electric current, the places of interest. I always refer to that first because I think the two most important factors when you're photographing anywhere are temperature and rainfall. After that, I get any photographic books I can and study the images. I like to know what's been done before so that I can create my own vision. If I were visiting the Sphinx and the Pyramids, I wouldn't need to do research because I've been there a number of times, but if I'd never been there, it would be useful to know that they're lit up at night. Without a plan, travel photography can be a disaster. You can't go just hoping to find something."

And yet Jay Maisel, a photographer whose work crosses several photographic disciplines and who therefore refuses to be categorized ("I'm interested in light"), believes the less preparation the better. But his approach to travel photography rests on more than just a faint hope.

"You don't really know what you're looking for before you leave. Obviously, it's the juxtaposition of some kind of color and shape and light that catches you, but there's no way to say what it's going to be in advance. I try not to anticipate anything. Especially in travel photography, there's a tendency to take the pictures you've seen before you've been there. I try to avoid that by going out as clean and as stupid as I can be. I don't do a lot of research; I try not to see too much of a place before I get there. What you don't want to do is to start saying, 'I'll do that picture, but a little better.' You try to be as fresh as you can."

Burt Glinn also avoids looking at other photographers' pictures of an area he will visit, "because they keep hanging up in your mind, and pretty soon you start taking pictures of other people's pictures. The only way you get things fresh is by going there and being surprised. Now, that doesn't mean you shouldn't first steep yourself in the essential character of the country. But if you're a photographer going to Russia, instead of looking at the book I did on Russia, you would do better to read Tolstoy or Chekhov to understand what the Russians are like." . . .

The most important part of taking pictures — recognizing the image and getting it on film — is also the most difficult. As Fred Maroon defines the problem:

"When you're hunting for images, most of the time you're saying no to yourself. You always seem to see the best pictures when you're on a dead run — looking out of the corner of your eye, you see off in the distance this marvelous thing, and you're not prepared to shoot it. The

worst thing a photographer can do is say, 'I'll get that later.' It's never the same. On the other hand, you can see a situation where some things are right but certain essential factors are missing — the lighting might be bad, and you're a slave to the light. But when the right circumstances develop — a big snowfall or a fog, or if you get up at four in the morning and reach that spot when there are no people or traffic or other things obstructing the view, and if the quality of light gives it some sort of magic — it can become exquisite, and then the picture is there."

Aiming always for the target he's chosen before a trip, Pete Turner at the same time keeps his mind and his eyes open to "photographs of opportunity" that might occur along the way. Echoing Fred Maroon, he says that if a place you're planning to photograph is particularly attractive to tourists, the best time to photograph is dawn.

"Those places get too crowded during the day. If you want a real sense of the place, a sense of experiencing something fresh for the first time, get there at first light. It's calm, it's quiet, and the light is great in the morning. Dawn is the best time to photograph, period. If it were up to me, I'd just do sunsets; I hate getting up. I never was an early person, and I have to make myself get up to do those shots, but it's worth it." . . .

John Lewis Stage takes a psychological approach. He may travel around for a day or two before he starts to shoot, examining how he feels about a place.

"The photographs that grab me are the ones that transmit into visual form what I feel emotionally. That could be anything from a portrait to a scenic to a still life, as long as it has the feel of the area as I see it. Even if you're an amateur, and you go to the Greek Islands, and you just love that sunbaked feeling — or anything else that gives you pleasure — then you ought to try and identify what it *really* is that gives you pleasure, and then it's surprising how automatically you start grasping those things visually if you're any kind of a photographer. Photography is no big mystery, but it's really quite individual. And thank goodness for that."

Writing Assignments

The following assignments encourage effective planning by suggesting that you mix sources and strategies.

1. Find some photographs of yourself in an old family album or, better yet, some essays you wrote in elementary or high school. Observe the photos or read the essays carefully, and then make a *list* or *cluster* in your journal, trying to remember "how it felt to be me at that age."

2. Interview one of your grandparents, one of your family's oldest friends, or someone you know in your home town. Pick somebody unlike yourself, someone who grew up in a different time and place. Instead of following the interview procedures discussed on pages 43–47, ask your subject to *map* his or her childhood neighborhood. As your subject sketches this world, ask him or her to do some *speculating* about the objects, processes, and networks suggested by the map.

3. You have just read about a contest called "Fantasies Unlimited." According to the rules of the contest, you are to describe the distinctive features of a place you would like to visit. The contest winners will receive enough money to visit the place for two weeks. Plan your essay by observing (photographs, maps), interviewing (travel agents, tourists, former residents), and reading (encyclopedia articles, travel essays, newspaper articles, guide books).

4. Investigate a center of activity in your community — a grocery store, a gas station, a bank. First, *freewrite* about your attitude toward the place. Next, go there and *map* it, paying attention to how people interact with the environment and which people attract and direct the flow of energy. Then, *interview* the people who inhabit the place — cashiers, mechanics, tellers. Ask them what they know about the place and how its procedures might be improved. Finally, *read* a variety of material on any controversial issues facing this center — competition from generic goods, hidden costs of gasoline credit cards, declining interest rates. As you work with each strategy, note in your *journal* how your initial attitudes toward the center were reinforced, refocused, or revised.

5. The photograph on page 24 was selected as a metaphor representing an important aspect of this chapter. Examine the photograph and determine the possible meanings of this metaphor as you discern them in the photograph. Then write an essay explaining how the metaphor has been reinforced or revised by the ideas discussed in this chapter. You may want to suggest another metaphor to illustrate your understanding of these ideas.

3
DRAFTING

Through drafting, you determine whether the information you discovered in planning can be shaped into successful writing. In Chapter 2 you learned to apply thinking-in-writing strategies to explore likely topics and to assess the potential of each topic by using the guidelines for selecting a subject, analyzing an audience, and determining a purpose. Now you will learn to work the most promising of these topics into a draft.

Although drafting requires you to make specific choices about your subject, audience, and purpose, you should resist the temptation to think of these choices as final. Inexperienced writers often assume that composing the first draft is virtually the last stage in the writing process. They believe that, with a little quick repair work, they can convert first drafts into final drafts. Experienced writers, on the other hand, presume that the first draft is merely the first attempt to produce a sustained piece of writing. Planning allows them to examine possible topics; drafting enables them to experiment with possible arrangements of a topic. They expect this experiment to lead to new discoveries, some of which emerge in the first draft, but most of which will emerge in a completely new draft. Indeed, experienced writers try several drafts. With each one, they come closer to what they want to say and how they want to say it.

In the following passage, May Sarton compares the art of flower arranging to the art of writing, indirectly commenting on the experienced writer's view of drafting.

> After breakfast I spend an hour or more arranging and rearranging seven or eight bunches of flowers for the house. There are flowers indoors here all the year around — in winter, bowls of narcissus, geraniums brought in from the windowboxes in the autumn, cut flowers from a local florist when all else fails. But from late May on I have a variety to play with, and the joy becomes arduous and complex. Arranging flowers is like writing in that it is an art of choice. Not everything can be used of the rich material that rushes forward demanding utterance. And just as one tries one word after

another, puts a phrase together only to tear it apart, so one arranges flowers. It is engrossing work, and needs a fresh eye and a steady hand. When you think the thing is finished, it may suddenly topple over, or look too crowded after all, or a little meager. It needs one more note of bright pink, or it needs white. White in a bunch of flowers does a little of what black does in a painting, I have found. It acts as a catalyst for all the colors. After that first hour I have used up my "seeing energy" for a while, just as, after three hours at my desk, the edge begins to go, the critical edge. (May Sarton, *Plant Dreaming Deep*)

It may be a relief to learn that drafting allows you additional opportunities to discover your subject, audience, and purpose, but you cannot lapse into inactivity. You must continue to exercise the "art of choice," evaluating your information, arranging and rearranging it, until you compose a coherent draft. This chapter presents four approaches to creating a first draft: making a *scratch outline*, drafting a *hypothesis*, composing a *discovery draft*, and constructing a *descriptive outline*.

THE SCRATCH OUTLINE

To make a scratch outline you examine your material for revealing relationships and then determine a preliminary organization. The scratch outline does not impose a format; you arrange your material into whatever pattern seems helpful. As you struggle with your first draft, you will learn more about your subject, audience, and purpose; this new information will lead to more precise outlines for any subsequent drafts. But the pattern you create in your scratch outline will form the middle portion (*body*) of your draft. Later your preliminary hypothesis about this pattern will lead to an introduction that makes a statement about your subject and why you are writing about it.

Begin by assembling whatever lists, maps, reading notes, and so on that you produced during planning. Then, read through your materials several times, looking for recurrent patterns and any ideas or phrases that stand out. Use any system you like — marking key words, using numbers or circles — to identify the related information you discover. Once you have established these groups of information, establish connections among them and arrange them into a meaningful pattern, copying it onto another sheet of paper. Don't discard your planning materials; you may need to consult them again.

Sometimes you will discover that your planning has already evoked a simple pattern, such as Joanne's narration about Mr. Bridgeman's Latin class or Jenny's analysis of the activity at the Varsity Barbershop. More often, your information suggests several contradictory or inconsistent methods of arrangement or none at all. Such confusion usually indicates that your information is too meager to work with (in which case you will have to do some more planning) or that your information is too varied to work with (in which case you will have to make some tough choices). All the cherished items you found during planning will rush forward in drafting, "demanding utterance." But if you try to give everything equal attention, your essay is likely to "topple over." Some information may be significant enough to form a major division in your outline, but other information may be less significant, useful only as minor subdivisions or illustrative examples. Still other information — wonderful, colorful, delightful information — may have to be eliminated because it does not fit into your scheme. As you arrange and re-arrange various sections of your scratch outline, information may even change status: significant information may become unimportant and be eliminated; insignificant information may suddenly become important. All the choices you make ultimately depend on one question: "What arrangement best enables me to communicate my *subject* to an *audience* for a *purpose?*"

Ellen explored several possible subjects during planning (her favorite movies, the kinds of people she saw at the health food store, the books she purchased for her personal library), but each time she tried out a thinking-in-writing strategy, she found herself focusing on her friend Barry and how often he offered elaborate explanations for his actions. Ellen decided that Barry's behavior was an interesting subject and might lead to a significant piece of writing. Using *Barry* as her prompt, she produced the following list.

Barry

Drives me nuts

Type A — always busy, can't seem to relax

Discipline — routine, schedules, lists

Right way to do everything — always giving advice

Making granola — the right spoon, bowl

No junk food — Big Macs, doughnuts

Health nut — what's good for me

Salad bars, yogurt (?)

Sneaks ice cream — Heavenly Hash

Burns it off with exercise — but no jogging

Shin splints, tendons, new study

Rebounder — bounces to old time rock'n'roll

Says he likes opera and ballet — improves his taste

Never listens, falls asleep — the sleep of the tasteless

Books — looks at diet books, buys classics

Impress himself, never reads — dust — "I'll get around to it someday"

Movies — Siskel and Ebert vs. Revenge of the Nerds

Always has his reasons — getting out of the house

Naps, wine, avoiding work

Good for me

Ellen is ready to begin drafting, but she needs to find a pattern in her information. Reading through her list, she begins to circle and number certain topics. She uses arrows to suggest how something from one part of her list belongs in another and question marks and asterisks to indicate problems she needs to solve or explore in greater detail.

Barry

? Drives me nuts

1. Reasons

1. Type A — always busy, can't seem to relax (3.)

1. Discipline — routine, schedules, lists

1. Right way to do everything — * *always giving advice*

2. Food

2. Making granola — the right spoon, bowl

2. No junk food — Big Macs, doughnuts

2. Health nut — what's good for me (1.)

2. Salad bars, yogurt (?)

2. Sneaks ice cream — Heavenly Hash

3. Exercise

3. Burns it off with exercise — but no jogging

3. Shin splints, tendons, new study (1.)

3. Rebounder — bounces to old time rock'n'roll (4.)

4. Culture

4. Says he likes opera and ballet — improves his taste (1.*)

4. Never listens, falls asleep — the sleep of the tasteless

4. Books — looks at diet books, buys classics (2.)

4. Impress himself, never reads — dust — "I'll get around to it someday" (1.*)

4. Movies — Siskel and Ebert vs. Revenge of the Nerds

1. Always has his reasons — getting out of the house

3. Naps, wine, avoiding work (2.) (3.)

* *1. Good for me*

By circling and numbering, Ellen identifies and labels several information groups. Some groups evolve from the pattern of associations on the list (*food*); others form incomplete but potentially useful categories (*exercise*); still others are so cluttered that they will require additional planning to make them precise (*culture*). Many of the items on Ellen's list could prompt the creation of new lists ("Type A," "Sneaks ice cream," "sleep of the tasteless"). But Ellen decides that a phrase appearing intermittently throughout her list (*good for me*) forms the most promising topic for her first draft, one that might organize the information in her other groups. She continues her thinking-in-writing in a scratch outline.

Good for Me

1. *Food*
   you are what you eat
   health food — granola, complex carbohydrates
   junk food — Whoppers, doughnuts
   cancer — new studies — yogurt (?), salad bars (?)

2. *Exercise*
   keep in shape
   aerobics — jogging, swimming, rebounder (heart)
   isometrics — pumping iron, violent sports (competition, stress)
   serious injuries (shin splints), marathons — dehydration

3. *Culture*
   improve your mind
   novels, classics, history, philosophy
   romances, trash, junk magazines ("inquiring minds want to know")
   big tomes — fall asleep
   some reading better than none

Through her scratch outline, Ellen has discovered some new information ("you are what you eat," "pumping iron," "junk magazines") and has deleted other information that does not seem to fit into her plan ("movies," "wine," "opera"). Her outline has helped her restrict her subject to three categories of things *good for me*. Although still sketchy, they suggest an emerging pattern that is potentially significant, interesting, and manageable. Some potential audiences have also emerged: people interested in food, exercise, or culture, and people interested in what is "good for me." Before she can convert this outline into an essay, however, Ellen needs to draft a preliminary hypothesis.

DRAFTING A HYPOTHESIS

A hypothesis states a possible purpose for your writing. Unlike the thesis, which formally asserts what you will prove in your essay, the hypothesis expresses a tentative purpose. You should certainly try to make this statement restricted, unified, and precise, but you may discover that writing your first draft modifies rather than demonstrates your purpose. Indeed, once you have completed your first draft, you may decide to revise your purpose completely. The first draft is like a laboratory experiment: it gives you an opportunity to test one possible explanation (*hypothesis*) for the pattern you derived from your planning.

To draft a hypothesis, read through your scratch outline and then write out answers to these questions:

1. What do I expect to accomplish in this writing project?
2. What is my attitude toward the material I have gathered?
3. What hypotheses could I prove by writing about it?
4. Which hypothesis seems the most restricted, unified, and precise?

Your answers to the first two questions may be lengthy and provisional since you are trying to clarify your aims and attitudes. You will probably not use this information in any direct way in your essay, but you may use it to determine your point of view toward your material and the ultimate effect you hope to achieve. Your answers to the second two questions should be simple and direct. These questions are designed to produce conditional theses, one of which you can use in your introduction to help keep your writing on track and to alert your readers to the main idea you are trying to develop.

When Ellen started to draft her hypothesis, she was confused about what she wanted to prove about Barry's preoccupation with what is "good for me." Her answers to the first two questions exhibit her uncertainty, while her answers to the second two illustrate her attempts to resolve her confusion by proposing several provisional theses.

1. *What do I want to accomplish with this project? It all started with Barry. Drives me nuts with his explanations. Everything is either good or bad. I guess I wanted to show him that things aren't that simple. There are lots of Barrys in the world. Just like parents, always telling you what's good for you — body, mind, and soul. That's what I tried to do with my categories — food, exercise, and culture. It's not exact, but it might work. The interesting things in each group are the ones that don't fit or that change. What used to be good for you now gives you cancer. Maybe I could make something out of that.*

2. *I think my attitude toward this material is weird. I started out thinking about Barry. He's so compulsive he's funny. Maybe not so funny. People who think things have to be good or bad are always changing their minds or cooking up new explanations. They could be devious, dangerous. How do other people react to the Barrys of the world? Maybe they're just like me. I mean I'm always trying to figure out whether he's good for me. I change my mind everyday — weird.*

3. *Possible hypotheses:*
   a. Everybody wants to know what's "good for me."

    b. Most people try to divide the world into two categories — what's good for me and what's bad for me.

    c. The most interesting things in the world cannot be classified as absolutely good or bad.

4. *I'm still not sure any of these will work, but if I had to pick one, I'd probably go with c.*

As Ellen tries to draft her hypothesis, she struggles to clarify her subject, anticipate her readers, and define her purpose. She knows that this material is interesting and that she is raising significant issues that will appeal to most readers. Her scratch outline is simple enough to guide her through a first draft. As she tinkers with various ways to word her hypothesis, she feels confident enough to begin writing.

THE DISCOVERY DRAFT

Novelist Dorothy Canfield Fisher once compared writing the first draft to skiing down a steep slope she wasn't sure she was clever enough to manage. Although you have compiled a large body of information during planning, organized that information into a scratch outline, and drafted several hypotheses about its significance, you cannot stand at the top of the hill

forever. You must push off and see if your preparation has made you clever enough to manage the long white slope of blank paper in front of you.

This first draft is called a *discovery draft* because you should expect as you write to discover something new about the subject, audience, and purpose of your essay. Some of what you discover will be disappointing. Sections of your essay that you felt certain were complete suddenly seem sketchy. Connections that you saw between sections disappear or appear in forms you had not anticipated. And individual sections may not prove useful because certain items cannot be developed in any detail, duplicate other items, or detract from your working hypothesis.

Most discoveries — even the negative ones — help you learn more about what you want to say and how you want to say it. As you convert notes into sentences and group sentences into paragraphs, your writing will talk to you — telling you things you had forgotten, making unexpected connections, pointing toward things you need to find out. Sections of your scratch outline will expand and contract before your eyes; your hypothesis may reshape itself into a more subtle statement. Your discovery draft gives you something to work with — a text, a core of information that you can rearrange or refine in a subsequent draft.

Ellen's discovery draft illustrates some of these changes as she struggles to say what she means.

Good for Me

Most people divide the world into two categories: (1) things that are good for me and (2) things that are bad for me. This system works as a general rule, but the most interesting things do not fit easily into one of the two categories.

First, there is food. Everybody knows that raw vegetables, whole grain cereals, and fresh fruit are nutritious, and that French fries, doughnuts, and Whoppers are junk. But what about yogurt? Most advertisements claim it is a health food, but a recent study points out that one 8-ounce serving of flavored yogurt contains 9 teaspoons of sugar. And what about salad bars? They are supposed to be good for you — better than a slab of red meat or a basket of deep-fried shrimp — but recent studies charge that dangerous chemicals are sprayed on salad bar vegetables to keep them looking good. What's supposed to be good for you might give you tooth decay or cancer.

Next, there is exercise. Doctors are always telling you to stay fit. Exercise, particularly aerobic exercise like jogging, is supposed to be good for you because it controls your weight, lowers your cholesterol, and strengthens the efficiency of your heart muscle. But there is a trade-off. Jogging can cause serious injuries such as sprained ankles, pulled tendons, and shin splints. And there is a recent study that says severe physical exertion may actually accelerate the causes of coronary heart disease. One doctor says "jog," another says "don't," and another says "it all depends." *How can people know for certain whether jogging will be "good for me"?*

Finally, there is reading. Teachers, parents, and politicians are always saying that reading improves your mind. But what kind of reading leads to improvement? Books that are supposed to help you — diet books, or pop-psychology books — don't really improve your mind. They are stacked next to the candy near the cash register, and everybody knows they are trash. But some of the books that are supposed to be good for you — the classics — many people find too long and boring. If they can't finish the books that are supposed to be good for them, how can people improve their minds?

The world is really too complex for a simple good for me/bad for me system. Take ice cream, for example. On one level it is bad for me because it contains large amounts of sugar and cholesterol. On another level, however, it is really good for me. After a long day of dieting, exercising, and reading, what could be better for me than a few scoops of Heavenly Hash?

THE DESCRIPTIVE OUTLINE

A descriptive outline helps you assess what you have accomplished during drafting. In it you report what you have done with the discovery draft and speculate about the composition of your next draft.

To construct a descriptive outline, place your discovery draft on one side of your desk and some blank paper on the other. Do not look at your scratch outline; your objective is to make a new outline that describes your draft. Counting the introductory paragraph as *1*, number in sequence each paragraph, and then list those numbers on your blank piece of paper. For each number, write down as briefly as possible *(a)* what each paragraph *says* and *(b)* what each paragraph *does*.

There is a subtle but significant difference between what a paragraph *says* and what it *does*. When you identify what a paragraph *says*, you are concerned with subject matter, with the major topic discussed in the paragraph. When you identify what a paragraph *does*, you are concerned with writing strategies, with the development of each paragraph and its function within the larger design of the essay. A paragraph can do many things: it can tell a story, describe a scene, list examples, compare evidence. You will learn more about what paragraphs can do individually and collectively in Chapters 5, 6, and 7. For the moment, use your own words to describe what each paragraph in your discovery draft *says* and *does*.

After you complete the descriptive outline, read through your discovery draft again. Then reread your planning and drafting material. Now return to your descriptive outline. What kind of draft does it describe? Are you inspired or disappointed by what you see? Does the outline reveal an interesting progression of ideas? Or does it show predictable patterns and insufficient development?

All this rereading should prompt you to draw some important conclusions about what you have already achieved and what you still need to do. Being as honest as you can, list at the bottom of your descriptive outline, your conclusions about the effectiveness of your essay's introduction; the interest, completeness, and coherence of its body; and the appropriateness of its summation. Ellen's conclusions show that she realized that she could do a number of things to improve her essay. After you have studied her outline, reread her discovery draft (pp. 66–67); then proceed to her conclusions.

1. a. *People divide world into two categories.*
   b. *Introduces hypothesis: most interesting items don't fit easily into one of two categories.*

2. a. *Although most people know good from bad, some items cause confusion.*
   b. *Identifies two examples of food whose classification has been challenged by recent studies.*

3. a. *Although exercise is considered good for you, it can be dangerous.*
   b. *Compares benefits of exercise with possible risk factors.*

4. a. *Reading improves your mind — if you read.*
   b. *Compares not reading trash with not reading classics.*

5. a. *The world is too complex for a simple system of good/bad.*
   b. *Illustrates how ice cream can be both bad and good.*

Conclusions

1. *Everything goes wrong in first sentence when I replace* Barry *with* People. *Sounds more scholarly, but I'm not really interested anymore. Anybody can write about people; I want to write about Barry.*

2. *The hypothesis is probably all right. What does* interesting *mean? It would be more interesting if it introduced Barry's explanations.*

3. *Three sections — food, exercise, and reading — are really predictable. This is good. This is bad. Here's a problem. If I used Barry, I could include more examples under each category — or create some new categories — and show how Barry tries to deal with the problems.*

4. *The conclusion is good, but it doesn't sound like me. Heavenly Hash is really Barry talking.*

5. *Mainly, I need to get back to Barry and develop more examples from his point of view.*

Ellen's descriptive outline has helped her identify what she has done right, what she has done wrong, and what she needs to do next. Her conclusions form an agenda for the next stage in the writing process — revising. But before Ellen revises her discovery draft, she must transform her hypothesis into an effective thesis and consider the organization of her next draft by constructing a formal outline.

COMPOSING AN EFFECTIVE THESIS

A thesis asserts the main idea you will develop in your writing. In a sense, it summarizes your ideas about your subject and suggests your point of view toward it. You cannot make such an assertion with any confidence until you understand your purpose. You will seldom succeed if you try to compose your thesis first and then write an essay that meets its specifications. It is more efficient to work by testing various hypotheses in drafts, selecting the one that best controls your material, and refining it into a thesis.

An effective thesis, then, derives from and makes a compelling statement about your writing. But once you have selected a thesis, you must word it properly if it is to be effective.

Making Your Thesis Restricted, Unified, and Precise

Restricted, unified, and *precise* theses were defined in Chapter 1. The definitions are worth amplifying here, for the more fully your thesis reflects these qualities, the better it will control your writing in your remaining drafts.

To be *restricted* a thesis must limit the scope of an essay to what can be discussed in detail in the space available. A thesis such as "The United States has serious pollution problems" might be suitable for a long magazine article, but trying to treat it in a three-page essay would force you to make statements so broad that your readers would find them superficial and uninformative. A better thesis about pollution might be one of the following:

- The government has not been sufficiently aggressive in enforcing the regulations that control the disposal of chemical wastes.

- In Toledo, industrial expansion has resulted in severe air and water pollution.

- Widespread use of agricultural pesticides threatens the survival of certain species of wildlife.

A good thesis is *unified* if it expresses only one idea. The following thesis contains not one idea but three: "The use of drugs has increased significantly in the last fifteen years. Hard drugs are admittedly dangerous, but there is considerable disagreement about marijuana." This thesis commits the writer to three topics: (1) the increase in the use of drugs, (2) the dangerous effects of hard drugs, and (3) the controversy about marijuana. Each of these topics could form the thesis of a separate essay. To try to deal with all three would almost surely result in an unfocused essay consisting of three unrelated and underdeveloped sections.

Lack of unity most often arises when a thesis contains two or more coordinate parts. The thesis "Compared with other languages, English has a relatively simple grammar, but its spelling is confusing" could lead to separate treatments of grammar and spelling. If you wanted to relate the two topics — for example, by contrasting the ease of learning grammar with the difficulty of learning spelling — the relationship has to be implied in the thesis: "In learning English, foreigners usually have less trouble with grammar than with spelling." If your chief interest is spelling, it would be safer to ignore grammar: "Foreigners have a hard time with English spelling." Sometimes one part of a two-part thesis can be embedded in the other. For example, "The amateur ideal of the Olympic Games is being threatened; professionalism is on the increase" can be rewritten as "Increased professionalism threatens the amateur ideal of the Olympic Games."

Compare the lack of unity in the statements on the left with the reworded versions on the right:

Not Unified	*Unified*
Many of the silent letters in English were once pronounced. The pronunciation has changed, but the old spelling was standardized.	Many of the silent letters in English words are a result of the standardization of spelling while pronunciation was still changing.
The nuclear bomb has immense destructive power, and there is no adequate defense against it.	There is no adequate defense against the immense destructive power of the nuclear bomb.
Baseball players have achieved a new independence, and there is nothing the owners can do about the situation.	Baseball owners can do nothing about the new independence of their players.

Finally, a thesis is *precise* when it can have only one interpretation. For example, "My home town is one of the most unusual in the state" does not indicate the content of your essay, because *unusual* is vague and can mean many things. Readers will want to know in what way the town is unusual. If they have to read the whole essay to find out, the thesis does not help them. Moreover, because its wording is vague, the thesis does not help you see what you need to develop in your essay.

Words such as *unusual, colorful, inspiring,* and *interesting* are too vague for a thesis. So are metaphors. The thesis "Where instructors are concerned, all that glitters is not gold" may seem clever, but what does it mean? That the best scholars are not always the best teachers? That instructors who are good classroom performers do not always help students master the subject? Or does it mean something else? The precise meaning of a thesis should be immediately clear. Metaphors may be effective in the text of your essay, but they can be troublesome in your thesis.

Throughout her drafting, Ellen has had difficulty composing an effective thesis. Her first hypothesis is imprecise because it relies on the vague word *inter-*

esting: "The most interesting things in the world cannot be classified as absolutely good or bad." A second, even more imprecise version appears in the first paragraph of her discovery draft: "This system works as a general rule, but the most interesting things do not fit easily into one of the two categories." This version also suggests two subjects — a system and things that do not fit into it — which could destroy the unity of her essay. In her conclusions to her descriptive outline, Ellen recognizes the problem word and hints that she might be able to draft a more effective thesis if she restricted it to Barry's difficulties explaining what is "good for me." This insight allows Ellen to restate her thesis. She might begin writing another draft immediately, but she decides instead to reevaluate her material and organization by constructing a formal outline.

▉▉▉ *Exercise*

Some of the following statements would make acceptable theses; others, because they lack restriction, unity, or precision, would not. Explain your reasons for rejecting those that are unacceptable.

1. Foods with high fiber content are very important for your diet.

2. It's easy to see the beginning of things and harder to see the ends.

3. Social historians agree that the American Dream is no more than the snows of yesteryear.

4. Although the average person thinks of gorillas as ferocious, chest-beating monsters, they are actually gentle creatures who live at peace with other animals.

5. Jane Fonda's workout tapes have enjoyed an enormous success, and they have proved that self-help videos are now a major consumer item.

CONSTRUCTING A FORMAL OUTLINE

A formal outline can serve as an additional writing tool, helping you to discover the need for more information and enabling you to organize a more precise design before you begin another draft. It breaks your topic into major units, marked by Roman numerals, and subdivides these into minor units, marked by capital letters; the next subdivision is marked by Arabic numerals, and a still smaller subdivision is marked by lower-case letters. One warning about this format: if you make any subdivision in your outline, you must have at least *two* subdivisions. You cannot divide something into only one part.

If it seems helpful, give your outline a title. Then begin to construct the formal outline by laying out the major, Roman-numeral headings first; then break each Roman-numeral heading into capital-letter entries, and so on, completing each level of division before starting on the next lower level. This procedure keeps you in control of your outline: you will not distort your organization by developing some headings too much and others too little. As you work your way through each division, you may discover new structural patterns that will require you to revise all the headings in your outline. Such discoveries will help you draft a more consistent outline that will reveal a more coherent pattern.

When Ellen looks at her two previous outlines, she decides that she wants her formal outline to follow the same order. When she starts to work her way through each division, however, she starts reclaiming from her original list the information that applies directly to Barry. As a result, her organization — and outline — change dramatically. She drops one major division (*exercise*) and adds two new divisions (*movies* and *wine*). In the lower subdivisions, she rearranges information ("ice cream" is moved from the conclusion to the first major division) and creates new information (all the items on movies and wine).

It is not always necessary to construct a formal outline to produce a new draft. Your discovery draft, descriptive outline, and reformulated thesis may give you all the direction you need. But Ellen's formal outline shows why it is such a powerful tool.

A Man With All Reasons

Thesis: *Barry admits that he has difficulty explaining what's "good for me."*

I. *Barry has difficulty defending his eating habits*
   A. *He knows the foods that are nutritious*
   B. *He knows the foods that are junk*
   C. *He is attracted to foods that are both*
      1. *Yogurt has become a problem*
      2. *Ice cream presents a dilemma*

II. *Barry has difficulty explaining his reading habits*
   A. *He knows how to select good books*
      1. *He reads reviews*
      2. *He talks to friends*
      3. *He consults lists*
   B. *He resists his fascination for self-help books*
   C. *He buys classics*
      1. *He believes in the value of owning such books*
      2. *He never reads them*

III. *Barry has difficulty justifying his choice of movies*
   A. *He attends culturally acclaimed movies*
   B. *He also attends "junk" movies*
   C. *He contrives elaborate justifications to explain such inconsistency*

IV. *Barry has difficulty rationalizing his preference for wine*
   A. *Wine enhances his appetite*

    B. *Wine relaxes his activity*
    C. *Wine is served to French and Italian children*
    D. *Wine is a health risk*
       1. *Its high sugar content could induce diabetes*
       2. *Its high alcohol content could induce chemical dependency*

Evaluating Outlines

Ellen can evaluate the usefulness of her formal outline as a stage in her writing process and as a completed product. For the first assessment, she compares her new outline with her previous writing on Barry's "good for me" justifications.

1. How much of the new outline derives from the original planning?

2. How many of the conclusions following the descriptive outline have been incorporated into the outline?

3. How effectively does the outline indicate possibilities for controlling *subject, audience,* and *purpose* in a final draft?

To make the *second* evaluation, Ellen considers the following criteria:

1. *Is the thesis satisfactory?* Because the thesis controls the whole outline, a faulty thesis invites trouble all along the way. A vigorous checking of the thesis is therefore the first and most important step in testing an outline.

"Barry admits that he has difficulty explaining what's 'good for me' " is certainly more restricted, unified, and precise than Ellen's earlier hypotheses. It seems to control the whole outline, but does it say everything Ellen wants to say about her purpose? Ellen considers how she might open up her thesis to assert more about the information she is trying to present. She thinks, too, about why she feels this information is significant.

2. *Is the relationship among the parts clear and consistent?* In a good outline it should be clear how each main heading relates to the thesis and how each subdivision helps develop its main heading. If there is any doubt about the relation of any heading to the thesis, that heading is either poorly stated or indicates an inconsistency in the outline.

Ellen's decision to focus on Barry's explanations has helped her establish a clearer connection among the four major divisions of her outline. Her subdivisions also seem to follow a clear and more consistent pattern. She examines the outline carefully, however, for potential trouble spots that might disrupt the organization of her next draft.

3. *Does the order of the parts provide an effective progression?* Just as the sentences within each paragraph must follow a logical order, so must the parts of an outline. If any of the parts is out of order, the disorder will be magnified in the essay.

Ellen's major divisions — *food, reading, movies,* and *wine* — suggest a progression different from the divisions of her earlier draft — *food, exercise,* and *reading.* She examines this new progression to determine if it is logical and then does the same with the material she has grouped under her capital letters and Arabic numerals.

4. *Is the outline complete?* This is not one question but two: (a) Are all the major units of the subject represented? (b) Is each major unit subdivided far enough to guide the development of the essay? The first question is especially important for essays that classify something or explain how something is done: all classes or steps must be included. The second question depends on the scope of the essay. For short papers, the outline may not need to go beyond the main headings. For longer papers, the outline needs to be developed in greater detail so that you can balance and control the information you group under each heading.

Ellen's four major divisions do not represent every issue on which Barry has an opinion. She has omitted Barry's feelings about what kind of exercise is "good for me" to focus on the more interesting new subjects — movies and wine. She needs to consider whether these changes make her outline more or less complete. The other subdivisions will require the same consideration to ensure the completeness of her essay.

5. *Can each entry be developed in detail?* Each entry in the outline should be developed when the essay is written. There is no rigid rule about how much development each entry should receive. Sometimes a single entry will require two or three paragraphs; occasionally several minor entries may be dealt with in a single paragraph.

In her discovery draft, Ellen devoted one paragraph to each of her three major divisions. She must consider whether her formal outline suggests the same coverage. The secondary divisions in Ellen's outline (those marked with capital letters) appeared singly as sentences in her discovery draft. Ellen must consider whether these, and the entries she marks with Arabic numerals, contain enough information to develop into separate paragraphs or will remain single, supporting sentences.

A FINAL WORD ABOUT DRAFTING

The most important thing to learn about drafting is to expect frustration. You certainly know people who give up on a project if their effort fails to measure up. They think that additional effort is a punishment for failing to succeed on the first try. But writing, like any other valuable work, requires effort. For that reason, think of drafting as an opportunity rather than an ordeal. Remember May Sarton's comparison of arranging flowers and drafting a piece of writing. Both processes are arduous and complex, but they are also a joy. They require a fresh eye and a steady hand and, after awhile, they use up your seeing

"The Flower Girl," detail of a painting by Charles Cromwell Ingham (1796–1863). Oil on canvas (1846). The Metropolitan Museum of Art, Gift of William Church Osborn, 1902.

energy. In Chapter 4, you will study strategies that will help you recover that energy. Revising helps you *see* your material *again;* it helps you find a catalyst to restore your creative ingenuity and critical edge.

Drafting Exercises

1. Read Ellen's final draft, "A Man with All Reasons," and then use a descriptive outline to assess her achievement. In your conclusions, consider whether this draft is better than or merely different from her discovery draft on pages 66–67. What further revisions might she make?

A Man with All Reasons

Barry's in the kitchen making granola. His long, narrow frame moves deftly around the work area, pausing to scan the recipe book on the counter or search the shelves for the ingredients — wheat germ, bran, rolled oats, sunflower seeds, raisins, honey — he plans to mix in a large, gray earthenware bowl. While he works, he hums a vaguely recognizable version of "Blue Moon," drumming on the top of the counter or the edge of the bowl with a wooden spoon. He's happy. Making granola always has this effect on him. He doesn't like to make granola — it's messy and time-consuming. He doesn't like to eat granola — it's too crunchy without milk and too mushy with it. But he likes to contemplate his reasons for eating granola—"it's good for me." 1

Barry moves through the world like he moves through the kitchen, carefully sorting everything into two categories: (1) things that are "good for me" and (2) things that are "not good for me." Although he is attracted to the simplicity of this system, he often has difficulty explaining what's "good for me." 2

Since he is a health nut, Barry seems most adept at sorting food. His first category includes nutritious foods such as raw vegetables, whole grains, and fish, while his second includes junk food such as French fries, doughnuts, and Whoppers. But not all food can be classified so easily. For example, Barry is currently in a quandary over yogurt. For years, he ranked it close to granola as a major "good for me." But recent reports on the sugar content of flavored yogurt have made its position shaky. Ice cream presents another dilemma. It contains too much sugar and cholesterol to be "good for me," but Barry 3

loves it. Whenever he does something that is "good for me," he likes to reward himself with a large bowl of Heavenly Hash — which is obviously, "not good for me."

Barry is also dedicated to improving his mind and is constantly looking for books that will be "good for me." He reads reviews, talks to his more literary friends, and consults an old mimeographed list of "150 Great Novels Every Well-Educated Person Should Read" that was given to him by his high school English teacher. Armed with such advice, he can work his way through a bookstore, sorting titles into his two categories. The cash register provides the ultimate test, however. He may be attracted to _Jane Fonda's Workout Book,_ justifying his interest by saying that a strenuous exercise program would be "good for me." But despite his attraction to Jane and his dedication to health, he does not buy the book; exercise books are trash. Instead, he purchases _Milton Gross' Complete Stories of the Great Operas,_ explaining that knowing more about opera would be "good for me." Unfortunately, the book gathers dust next to one of his other carefully chosen purchases, H. G. Wells' _The Outline of History._ Although he has never opened these books, he believes he will get around to them eventually because they are "good for me." 4

Movies present a similar difficulty. Barry believes that only critically acclaimed films about significant subjects are "good for me." Following the critics' advice, he went to see _My Dinner With Andre,_ only to fall asleep during the first course. He blinked awake during the credits, saying the nap had been "good for me." On the other hand, he went to see _The Revenge of the Nerds,_ knowing that the critics did not think it would be "good for me." He managed to stifle his laughter and maintain his disdain throughout the film. But walking to the car he announced that getting out of the house had been "good for me." 5

And finally there's wine. Barry assigns wine a high position in his ranking of things "good for 6 me," and he has his reasons. Foremost among these is wine's appetite-enhancing properties. He is slightly underweight and views anything that encourages him to eat as "good for me." Wine is also a relaxant. Barry leans toward the Type A personality and so considers anything that helps him slow down as "good for me." And then there are the French and Italian children. He has read that French and Italian parents give even very young children wine on a regular basis. Parents couldn't possibly give their own children something that isn't good for them. He worries about wine, however, because sugar and alcohol are not "good for me."

Recently, as he was fretting about the onset 7 of diabetes or alcoholism, he visited a friend in the hospital. When his friend's dinner tray arrived, he spotted a plastic glass of clear liquid between the Jello and the peas.

"Do they serve wine *here*? he asked. 8

"Yeh, pretty nice, huh?" the friend replied, 9 not realizing the cause of Barry's sudden elation.

"See, I told you it was good for me." 10

2. Read the following essay by Suzanne Britt Jordan. State her thesis and then construct a formal outline for her essay. What is her purpose in this essay? (Remember, purpose and thesis are not identical.)

The Lean and Hungry Look

Caesar was right. Thin people need watching. 1 I've been watching them for most of my adult life, and I don't like what I see. When these narrow fellows spring at me, I quiver to my toes. Thin people come in all personalities, most of them menacing. You've got your "together" thin person, your mechanical thin person, your condescending thin person, your tsk-tsk thin person, your efficiency-expert thin person. All of them are dangerous.

In the first place, thin people aren't fun. 2 They don't know how to goof off, at least in the best, fat sense of the word. They've always got to be adoing. Give them a coffee break, and they'll jog around the block. Supply them with a quiet evening at home, and they'll fix the screen

door and lick S&H green stamps. They say things like "there aren't enough hours in the day." Fat people never say that. Fat people think the day is too damn long already.

Thin people make me tired. They've got 3 speedy little metabolisms that cause them to bustle briskly. They're forever rubbing their bony hands together and eying new problems to "tackle." I like to surround myself with sluggish, inert, easygoing fat people, the kind who believe that if you clean it up today, it'll just get dirty again tomorrow.

Some people say the business about the 4 jolly fat person is a myth, that all of us chubbies are neurotic, sick, sad people. I disagree. Fat people may not be chortling all day long, but they're a hell of a lot *nicer* than the wizened and shriveled. Thin people turn surly, mean and hard at a young age because they never learn the value of a hot-fudge sundae for easing tension. Thin people don't like gooey soft things because they themselves are neither gooey nor soft. They are crunchy and dull, like carrots. They go straight to the heart of the matter while fat people let things stay all blurry and hazy and vague, the way things actually are. Thin people want to face the truth. Fat people know there is no truth. One of my thin friends is always staring at complex, unsolvable problems and saying, "The key thing is . . ." Fat people never say that. They know there isn't any such thing as the key thing about anything.

Thin people believe in logic. Fat people see 5 all sides. The sides fat people see are rounded blobs, usually gray, always nebulous and truly not worth worrying about. But the thin person persists. "If you consume more calories than you burn," says one of my thin friends, "you will gain weight. It's that simple." Fat people always grin when they hear statements like that. They know better.

Fat people realize that life is illogical and 6 unfair. They know very well that God is not in his heaven and all is not right with the world. If God was up there, fat people could have two doughnuts and a big orange drink anytime they wanted it.

Thin people have a long list of logical things 7
they are always spouting off to me. They hold
up one finger at a time as they reel off these
things, so I won't lose track. They speak slowly
as if to a young child. The list is long and full of
holes. It contains tidbits like "get a grip on yourself," "cigarettes kill," "cholesterol clogs," "fit
as a fiddle," "ducks in a row," "organize" and
"sound fiscal management." Phrases like that.

They think these 2,000-point plans lead to 8
happiness. Fat people know happiness is elusive
at best and even if they could get the kind thin
people talk about, they wouldn't want it. Wisely,
fat people see that such programs are too dull,
too hard, too off the mark. They are never better
than a whole cheesecake.

Fat people know all about the mystery of 9
life. They are the ones acquainted with the night,
with luck, with fate, with playing it by ear. One
thin person I know once suggested that we
arrange all the parts of a jigsaw puzzle into
groups according to size, shape and color. He
figured this would cut the time needed to complete the puzzle by at least 50 per cent. I said I
wouldn't do it. One, I like to muddle through.
Two, what good would it do to finish early?
Three, the jigsaw puzzle isn't the important
thing. The important thing is the fun of four people (one thin person included) sitting around a
card table, working a jigsaw puzzle. My thin
friend had no use for my list. Instead of joining
us, he went outside and mulched the boxwoods.
The three remaining fat people finished the puz-

zle and made chocolate, double-fudge brownies
to celebrate.

The main problem with thin people is they 10
oppress. Their good intentions, bony torsos, tight
ships, neat corners, cerebral machinations and
pat solutions look like dark clouds over the
loose, comfortable, spread-out, soft world of the
fat. Long after fat people have removed their
coats and shoes and put their feet up on the coffee table, thin people are still sitting on the edge
of the sofa, looking neat as a pin, discussing
rutabagas. Fat people are heavily into fits of
laughter, slapping their thighs and whooping it
up, while thin people are still politely waiting
for the punch line.

Thin people are downers. They like math 11
and morality and reasoned evaluation of the limitations of human beings. They have their skinny
little acts together. They expound, prognose,
probe and prick.

Fat people are convivial. They will like you 12
even if you're irregular and have acne. They
will come up with a good reason why you never
wrote the great American novel. They will cry
in your beer with you. They will put your name
in the pot. They will let you off the hook. Fat
people will gab, giggle, guffaw, gallumph, gyrate
and gossip. They are generous, giving and gal-
lant. They are gluttonous and goodly and great.
What you want when you're down is soft and
jiggly, not muscled and stable. Fat people know
this. Fat people have plenty of room. Fat people
will take you in.

Writing Assignments

1. Select any item on Ellen's list (p. 61) that interests
you, such as "Drives me nuts," "getting out of the
house," "Rebounder." Use this item to start your
own list and then group your information into a
scratch outline. Next formulate several hypotheses and determine which one seems the most significant, interesting, and manageable.

2. Make a list of people you know who can be characterized by their intense interest in one particular

activity, such as buying record albums, eating
chocolate, taking chances, or saving money. Use
various planning strategies to discover how you
can compose an effective thesis about the casual,
careful, and compulsive members of this group.

3. Use a phrase such as *makes me laugh* or *drives me
nuts* to describe a recurrent situation that amuses
or annoys you. Generate information with a planning strategy such as clustering or freewriting and

work the material into a discovery draft. Compose a descriptive outline to determine what the draft says and does.

4. Reread the information you gathered for writing assignment 4 in Chapter 2 (p. 57). Make a scratch outline, draft several hypotheses, and then select one to guide you through a discovery draft. Analyze your results by constructing a descriptive outline.

5. The photograph on page 58 was selected as a met-metaphor representing an important aspect of this chapter. Examine the photograph and determine the possible meanings of this metaphor as you discern them in the photograph. Then write an essay explaining how the metaphor has been reinforced or revised by the ideas discussed in this chapter. You may want to suggest another metaphor to illustrate your understanding of these ideas.

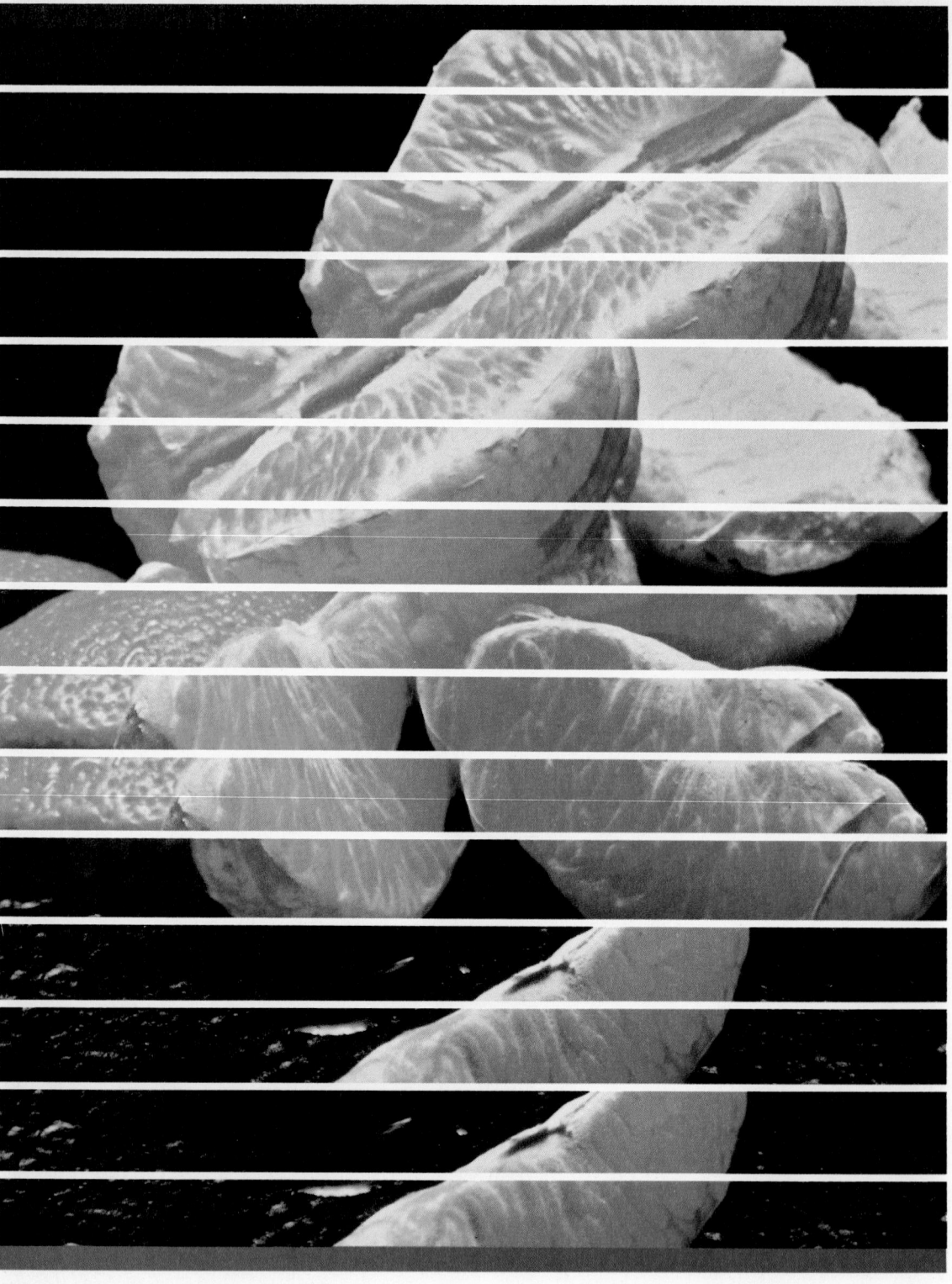

4
REVISING

Revising is the process of *seeing again,* of discovering a new vision for the writing you produced in planning and drafting. In a sense, you have been revising from the moment you began writing. As you experimented with thinking-in-writing strategies in planning, you revised your decisions about subject, audience, and purpose. As you tried to embody those decisions in drafting, you revised again when you saw a better thesis or a more effective outline. Now that you are ready to revise, you must try to gain a new perspective on writing you have already looked at several times. One thing is certain: you will not see much that is new if you think of revising as simply taking one last look.

Nothing distinguishes inexperienced writers from experienced writers more dramatically than the way they look at revision. For inexperienced writers, revision means *fixing* the first draft. They expect to fix a few overlooked errors by scratching out a sentence, rearranging a few phrases, or substituting one word for another. For experienced writers, revision means *creating* the final draft. They know they have overlooked many trouble spots, but before they begin perfecting and polishing, they want to be sure they are looking at the final text.

Revision is a two-stage process. During the first stage (the focus of this chapter), you use various reading strategies to help you rethink, reorder, or rewrite substantial portions of your draft. When you are satisfied with this *global revision,* you can focus your attention on the second stage, *local revision,* and begin repairing individual paragraphs, sentences, phrases, and words. (Strategies for implementing local revision are presented in Chapters 7 through 10.) And work on this second stage may suggest a more effective method to revise a section or even the entirety of your essay.

Novelist Eudora Welty discusses her attitude toward revision as a continuous process:

> My ideal way to write . . . is to write the whole first draft through in one sitting, then work as long as it takes on revisions, and then

write the final version all in one, so that in the end the whole thing amounts to one long sustained effort. . . . [Using a typewriter] helps give me the feeling of making my work objective. I can correct better if I see it in typescript. After that, I revise with scissors and pins. Pasting is too slow, and you can't undo it, but with pins you can move things from anywhere to anywhere, and that's what I really love doing — putting things in their best and proper place, revealing things at the time when they matter most. Often I shift things from the very beginning to the very end. . . .

[Writing] is so much an inward thing that reading the proofs later can be a real shock. . . . [T]here's . . . a strange moment with every book when I move from the position of writer to the position of reader, and I suddenly see my words with the eyes of the cold public. It gives me a terrible sense of exposure, as if I'd gotten sunburned. (Eudora Welty, *Writers at Work: The Paris Review Interviews*)

LOOKING TO REVISE

Once you consider revising a creative rather than a cleaning-up activity, you will look at yourself, your readers, and your text with more objectivity.

How You Look at It

Revision is always hard work, especially since you have already invested considerable effort in your writing. But try to look objectively at what you have accomplished. Rethink, don't merely glance over, every aspect of your writing, from your most abstract assumptions to your most concrete assertions. Do not fall in love with your words, no matter how clever or inspiring they seem: your real concern is effectiveness not eloquence. Like Welty, you should use revision to achieve coherence and clarity — by moving "things from anywhere to anywhere, . . . putting things in their best and proper place, revealing things at the time when they matter most."

How Someone Else Looks at It

When you are ready to revise, you may welcome the perspective of another reader — a teacher, editor, or friend. Your readers may ask about the larger elements in your writing (its subject, audience, or purpose) or its smaller elements (sentence structure, punctuation, or diction); they may talk about their general impression (whether your writing was interesting, funny, or dull). As you revise, you must consider all reactions to your work — even if they seem threatening or trivial. You must consider what your readers saw, why they liked or disliked what they saw, and whether and how you can use *their* observations to strengthen *your* vision of your writing. Various readers may give you confusing or contradictory advice, but remember that ultimately *you* are responsible for deciding how to revise your text.

How It Looks

Once you start revising, your writing will often look unfamiliar. Sometimes this can be misleading. If you have neatly retyped your messy, handwritten draft, it may now look compact and complete, as if cosmetic surgery had somehow solved your writing problems. But you must look beneath this seductive surface to uncover problems that still require rethinking and rewriting.

More often, the strangeness of your writing will lead more directly to new insights. Sometimes the passage of time will make your writing seem unfamiliar. It has been incubating in your unconscious while you have been doing other things, and now, after a few days — or even a few hours — as you reestablish contact with your writing, you may see faults that were not apparent before. Perhaps you have not restricted your subject, anticipated your audience, or defined your purpose effectively. This new perspective compels you to look for solutions that will make your writing stronger and sharper.

If your discovery draft produced new information and new questions, they may have altered your perspective. Following these new directions may cause fundamental changes in your writing.

But mainly your writing will look unfamiliar when you revise because you have changed position. In *writing,* your task was to create ideas; in *revising,* your task is to judge the effectiveness of those ideas. And just as you used a set of guidelines for writing, so now you may use a set of guidelines for reading and evaluating what you have written.

READING TO REVISE When you are reading to revise, you are still actively involved in the writing process. You are *not* proofreading. Proofreading presumes the existence a completed manuscript. When you proofread, you are merely checking your writing one last time to prove that you have not mangled sentences, misplaced punctuation, or misspelled words. Reading to revise, in contrast, assumes the existence of an evolving manuscript. When you read to revise, you are trying to identify the strengths and weaknesses of your subject. You are sharpening your perception of what your audience knows (what *is* on the page) and speculating about what your audience needs to know (what *is not* on the page). You are determining whether your purpose controls your essay and the connections among its parts.

Each of the following reading strategies suggests that you read your writing as though you were someone else — a fictitious person in a special situation who is reading for a particular reason. Each strategy will highlight a different facet of your work. Taken together, they will lead to a complete revision.

Reading for Subject

Imagine you are seated in the waiting room of your dentist's office, flipping through several magazines, looking for something to read to pass the time. What subjects attract and sustain your attention? Now pick up your writing and skim through it as quickly as possible. Then put it aside and, maintaining your identify as an impatient patient, jot down your general reactions using the following question as a guide:

1. Why was I attracted to this essay? How did the title or first few sentences — the lead — convince me that the essay was worth reading?

2. What is the specific subject of the essay? Did the essay focus on the subject immediately, or did I have to read a lot of preliminary material or prolonged digressions?

3. What is significant about the subject? Is it a subject I like reading about, one that I need to know about, or one that I ought to think about?

4. What makes the subject interesting — the attitude of the writer, the nature of the subject, or the way the subject is presented?

5. Does the essay seem the right length? Is it long enough to answer all my questions yet short enough to keep my mind from wandering?

Reading for Audience

Now imagine you are seated in a large banquet hall listening to an after-dinner speech. You cannot avoid listening to the speech, so you decide to determine how well the speaker has anticipated the needs and expectations of the audience. One way to simulate this situation is to read your writing aloud to anyone who will listen — the important thing is to *hear* your writing. An excellent way to accomplish this objective is to read into a tape recorder; then, with your manuscript in hand, play back the "speech" with you as its audience. As you listen, use the following questions to identify passages that need revision. Stop the recorder each time your annotations become too detailed to keep up with the tape. Once you have copied down your comments, push the button and resume listening to your writing being read.

1. What kinds of people does the speaker expect to find in the audience? Does the speech acknowledge their values, assumptions, and prejudices?

2. What role does the speaker invite the members of the audience to take (for example, that of dedicated, discerning people)?

3. What are the members of the audience likely to know about the subject of the speech?

4. Does the speaker ask questions that the audience would ask about the sub-

ject? Does the speaker answer those questions when they need to be answered? Does the speaker anticipate challenging or hostile questions?

5. Does the speaker help the audience focus on the subject and follow the development of the parts? Where might the audience get bored, confused, or annoyed? Does the end echo and fulfill the promise of the beginning?

Reading for Purpose

Imagine you are seated in an attorney's office about to sign a contract that will have an enormous impact on your life. You must determine what the thesis of this contract promises and whether the various sections and subsections deliver on that promise. Read your writing slowly and deliberately, underlining your thesis and tracing its connection to each major topic in your essay. If there are sections that need to be rewritten, rearranged, or deleted, now is the time for renegotiation. The following guidelines will remind you of the purpose of the "contract" and call your attention to how that purpose is effected through its various "clauses":

1. Does the essay rest on hidden and undocumented assumptions? How can these assumptions be introduced into the wording of the essay?

2. What is the purpose of the essay? Is that purpose expressed openly, or must it be inferred from the text?

3. What is the thesis of the essay? What specific promise does it make to its readers? Is the thesis sufficiently restricted, unified, and precise that it can be demonstrated?

4. Does the body of the essay fulfill the promise of the thesis? Is there a direct, logical, and dramatic connection among the various parts?

5. Is each part of the essay sufficiently developed with evidence that is germane, reliable, and verifiable? Does new evidence need to be introduced to clarify the thesis?

REVISION AGENDA   The detailed analysis produced by answering the reading-to-revise questions will help you establish a *revision agenda,* a plan for rethinking, rearranging, and rewriting the next draft of your essay. You can prepare this agenda in at least three ways:

▪ You may find that subject, audience, and purpose so overlap that you cannot undertake three separate readings. For you, the easiest procedure may be to keep the three imaginary readers in your head simultaneously and, after one "combined" reading, to prepare a single revision agenda.

░ You may prefer considering subject, audience, and purpose singly and thoroughly and so read through the text three times, answering the sets of questions that accompany each reading. After completing all three readings, you prepare one revision agenda.

░ A complex or difficult-to-formulate subject may invite three complete revisions. After you "read for subject," you prepare a revision agenda and then a new draft. You use this text to "read for audience" and the text generated then to "read for purpose."

You will have to decide which procedure helps you produce the most effective revision agenda and the most polished revisions. Sometimes the nature of the writing project determines which procedure you use. If you have to write under pressure on a fairly straightforward subject, then you may be able to work with one revision agenda. If, on the other hand, you have time to allow a complex subject to evolve, then you may want to generate several revision agendas to systematically identify and solve the problems in your writing.

Revision is an *intuitive* process: a sudden impulse inspires you to relocate a paragraph or realign the parts of a sentence. And it is a *recursive* process: you are constantly stepping back from your writing to see the big picture, moving forward to touch up some small detail, and stepping back again to see how the altered detail changes the composition of the whole. But revision is also a *logical* process. In any piece of writing, you will discover large problems of form and focus and smaller problems of syntax and diction. You may be tempted to fix the simple problems first and the more difficult problems later. But the logical way to proceed is to work on the difficult problems first, for in solving them you may eliminate the simple problems or at least discover an efficient method for dealing with them.

Organize your revision agenda by asking yourself three questions:

1. What did I try to do in this draft?

2. What are its strengths and weaknesses?

3. What revisions do I want to make in my next draft?

Answer the third question by writing yourself prescriptions containing action verbs — for example, "Collapse section on . . ."; "Expand paragraphs on . . ."; "Reword sentences in . . . ." Group the large problems (subject, audience, purpose) at the top of the list and the smaller problems (style, mechanics, usage) at the bottom.

REVISING:
A CASE STUDY
Early in his writing course, Rod decided to write about his high school trip to Washington, D.C. His junior class had spent almost a year preparing for the trip, traveled miles across country in a crowded bus, and then crammed what should have been a month of sightseeing into four days of

frantic, nonstop touring. Rod learned so much during this year-long experience that he had difficulty sorting out all that he remembered, saw, and read before and after his trip. He decided to attack the problem slowly and systematically in several drafts — writing to discover what he remembered, reading to fill the gaps in his information, then trying a new perspective in a new draft. Most writing assignments do not require you to revise as many times as Rod does, but Rod, an architecture major, is intrigued by design problems. He wants his writing to capture the essence of his Washington experience. His shaping and reshaping of his material through the term provides a case study of how one writer rethinks, revises, and rewrites his work.

Rod began by planning. In a listing exercise, he remembered the preparations made for the "trip of his life." In a freewriting exercise, he remembered the people, buildings, monuments, and heat of Washington in June. His scratch outline used three major headings: *Preparation, Trip,* and *D.C.* He used this working hypothesis to begin his discovery draft: everyone cooperated through months of preparation to make our trip to Washington, D.C., a success.

Our Trip to D.C.
(Discovery Draft)

In my junior year in high school, my class sold magazine subscriptions to make money for a trip to Washington, D.C. Everything we did that year seemed to focus on Washington. In first period (econ), we had market reports and strategy sessions — we had to make our quotas or there would be no trip. In second period (history) we learned the history of Washington. In third period (art), we learned about the architect's design for the city, the famous museums, and all the monuments. I took really good notes in this class because I wanted to be a world-famous architect when I graduated from college. I still have that notebook and my notes actually make sense. Art class was the high point of my education for Washington.

By lunch, everybody had had enough — although I vaguely remember cutesy names on the cafeteria menu such as Lincoln Dogs and Congressional Burgers. In the afternoon, we learned about how the government worked — the different branches and agencies — in civics, but I can't remember learning anything about D.C. in math, science, and gym.

At home there was no relief. Mom and Dad were always showing me articles and stuff about Washington, and my little sister pasted a picture of the Washington Monument on my art notebook. Even Uncle Fred — the Midwest's most fanatic Bears' fan — sent me a list of the Redskins' draft choices.

The trip was a real bummer. All the junior class, plus our counselors, Mr. and Mrs. Franklin (my art teacher), and our driver Dave crammed into this bus with lousy air conditioning and headed across Indiana, Ohio, West Virginia, Pennsylvania, and Virginia to D.C. It was awful hot and kids were always asking Dave for a pit stop. When he stopped, somebody was always late getting back to the bus, so Mr. Franklin was always walking up and down the aisle counting heads before Dave could pull out.

We had four days in Washington to see the whole thing. We stayed in a motel in Virginia, but every morning Dave drove us into downtown D.C. On the first day we saw our congressman, the Capitol, the Supreme Court, the Library of Congress, and then we stood in line a long time to see the White House. On the second and third days, we visited the museums on the Mall. My favorite was the Air and Space Museum, especially the film about flying. After we were through at one museum, we would run to the next one just to be in an air-conditioned building.

Even though we drove by the monuments every day, we didn't visit them until the last day. The Jefferson and Lincoln were interesting and you could see them quickly, but you had to stand in line — a long, long line — to see the Washington Monument. But I liked that monument the best. Ever since we hit town, I kept seeing it everywhere. I'd walk around any corner and there it would be. It seemed big and yet so simple beside all those other buildings. All the postcards I sent home had pictures of the Washington Monument on them. Whenever I think of our trip, I think of that monument standing in the middle of downtown Washington like a big paperweight.

Revision Agenda

1. What did I want to do in this draft?
   Describe our junior class trip to D.C.

2. What are its strengths and weaknesses?
   Need more focused hypothesis. Scratch outline had three parts, but it doesn't work. Takes too long to get to D.C. Everything jammed together at the end — like the trip. I do like part about Washington Monument — paperweight.

3. What revisions do I want to make in my next draft?
   a. Cut material on school and trip.
   b. Focus on monument — add information — tour guides, souvenir shop, people in line.
   c. Describe view from top — impressive.
   d. Rewrite introductory paragraph, thesis?
   e. Cut out high school slang — "bummer," "lousy," "stuff."

Revising for Subject

Rod's trip to Washington seemed a specific subject, but he now realizes that it contains many small, unevenly developed subjects. He decides to focus on the

Washington Monument because it represents everything he learned about and saw in Washington. He begins to think about what kind of information he can use to develop his subject in detail and how he can introduce and interpret his subject for his readers. He writes a new hypothesis — "I was impressed by my visit to the Washington Monument" — and then tries to implement his revision agenda in a second draft.

VISITING THE WASHINGTON MONUMENT
(2nd Draft)

During my junior year in high school, my class sold magazine subscriptions to make money for a trip to Washington, D.C. Our stay was brief, but we saw dozens of historic sites. I was certainly impressed with the large government buildings and the elaborate architecture of the monuments, especially the Jefferson and Lincoln Memorials, but for simple beauty nothing could compare with the Washington Monument.

The afternoon was sunny and hot as our bus drove into the parking lot near the Washington Monument. We jockeyed for window positions to take photographs and let out tremendous whoops to let everyone know we had arrived. Our driver parked the bus, and we squirmed in our seats for another five minutes while Mrs. Franklin, our art teacher, told us to take notes on what the guides said and reminded us that young ladies and gentlemen stand quietly in line. With that said, we stampeded off the bus and raced through the maze of cars and buses toward the monument.

The line was over two hundred yards long and snaked along the sidewalk like a Chinese dragon. It ended, oddly enough, with a middle-aged couple from Indiana.

"Are you kids from Indiana?" the lady asked. "We seen your bus pull in."

"We sure are," I answered. "We sold magazine subscriptions to get here." She seemed impressed. She treated us to a detailed account of their vacation to date, naming every battlefield, landmark, memorial, bridge, souvenir shop, rest area, and McDonald's they had visited. She placed special emphasis on the Jack Daniels Distillery in Tennessee.

"I wish this line would get movin'. It's hotter than blazes out here," grumbled her husband. He removed his Funk Hybrids cap and wiped the sweat off his forehead with a red handkerchief. "We got's plenty of buildin's to see before tonight." Before his wife could tell us about how quickly the line moved at Jack Daniels, we spotted a group of girls in front of us. My friends and I spent the next quarter of an hour trying to figure out how we could meet them. But the girls ignored our leers, so we concentrated on spraying our own girls with our recently purchased "Greenie-Meanie" squirt guns.

Finally we stood at the base of the monument and craned our necks to see the top. It was bigger than I ever imagined, and I thought about the guys who had to build the thing, particularly those guys who had to do the peak.

When we finally moved inside, we were greeted by a tour guide (actually a park ranger) who told us that the Monument was 555 feet, 5-1/8 inches high, that it weighs 90,854 tons, and that it cost $1,187,710.31 to build. While I was writing these numbers in my notebook, he explained that we could not walk up the 898 steps to the top, but would have to take a 13-second elevator ride. I was pleased in a way, because it was too hot to walk, but if it were cooler it would be fun to walk up to the top and count those steps.

Inside the elevator another park ranger told us historical facts about the monument. He only had a few seconds. He called it an obelisk and told us that it took 105 years to build because they had trouble raising money and figuring out where to put it. The ground around the monument used to be a swamp or something.

At the top another ranger walked us around the observation deck. There was a magnificent view out of each window. To the east, you could see down the Mall to Capitol Hill. To the west, we could see the Lincoln Memorial. To the south, you could see the Pentagon straight ahead and the Jefferson. To the north you could see the National Cathedral on the hill and the White House below. Everybody was taking pictures, pointing to places we had been, and trying to see our bus in the parking lot and our motel across the river. I asked the ranger why there were bars on the windows. He said they installed them during the Depression after three people had committed suicide by jumping out one of the windows.

After awhile we came down and walked over to the souvenir shop where we loaded up on pamphlets, postcards, and little models of the monument. Our heads were groggy with information and our bodies were sticky with sweat, so we walked back to the bus, spraying an occasional victim with our "Greenie-Meanies." But strangely enough, I was in no mood for horsing around. I stopped and looked back at the stark, jutting obelisk. I was impressed. The Washington Monument, like the man it honors, is a rugged individual.

Revision Agenda

1. *What did I try to do in this essay?*
   Describe the Washington Monument.

2. *What are its strengths and weaknesses?*
   Best part — when I get inside and describe what I see from the top. Also, conclusion (last few sentences) — simple, rugged design. Still sound like high school kid — "Greenie-Meanies." Funny couple from Indiana didn't know the difference between a monument and a distillery.

3. *What revisions do I want to make in my next draft?*

a. *Tell readers something important about monument.*
b. *Cut out all the personal stuff — waiting in line, girls.*
c. *Look over notes from art class, tour guides, souvenir shop.*
d. *Write a straight history of monument.*
e. *Fix introduction, thesis — make it more objective.*
f. *Imagine I'm a historian, tour guide so that I'll talk like one — serious.*
g. *Check vocabulary — no kid stuff.*

Revising for Audience

Rod is beginning to recognize that he knows more about his subject than what he remembers about standing in line to observe it. He is also beginning to recognize the difference between *telling* what happened and *using* what happened to create a subject and develop a thesis. But his major reaction to his second draft is his discovery of audience. He wants to separate himself from the high school kids who don't care what they are seeing and the couple who don't understand what they are seeing. He wants to tell both groups that there is more to the monument than a place to find girls or yet another building to see. Its history is special and significant. Rod thinks that he has gathered enough information from his art teacher, the tour guides, and the souvenir shop to tell that history in some detail. This decision, made to enlighten his audience, also enlarges his subject and redefines his purpose.

THE STORY BEHIND THE WASHINGTON MONUMENT
(3rd Draft)

Since people first spoke, they have probably had a few good things to say about one another. Some people receive "high fives," some just congratulations. Others live in our books, and a very few are important enough to have monuments built in their honor. When we look at those stone and steel structures, we often do not consider the time and work that went into their construction. The Washington Monument, for example, took 105 years to build and cost nearly 1.2 million dollars. Its construction was delayed repeatedly due to lack of funds and lack of interest. Once it was even stolen. The story behind the monument is in many ways as interesting as the story behind Washington himself.

The idea of building a monument to honor George Washington was discussed for several decades before it was finally examined seriously. In 1783, the Continental Congress voted to erect a bronze equestrian statue depicting the father of our country. This idea was not mentioned again in 1799, when Congress proposed instead a pyramid-type mausoleum. Congress failed to provide the funds, estimated at twenty-five cents per citizen. Only faint rumblings about the monument were heard during the next thirty years.

Senator Henry Clay of Kentucky revived the notion in 1832 and managed to allocate money for a twenty-ton statue of a toga-clad Washington, but this half-naked sculpture did not satisfy a group of local citizens who preferred the 150-foot shaft already erected in Washington's honor in Baltimore. The following year these people formed the Washington National Monument Society, and, during the next three years, collected $28,000 and sponsored a design competition. Robert Mills drew the winning plan, a 600-foot obelisk on a colonnade base, 250 feet in diameter, known as the National Pantheon. This temple was to contain statues of the signers of the Declaration of Independence, Revolutionary War heroes, and a colossal George Washington.

The laying of the cornerstone was scheduled for Washington's birthday in 1848, but Congress was slow in approving the site, and the ceremony was rescheduled for July 4. Fifteen thousand people, including President Polk and Congressman Lincoln, watched Benjamin B. French lay the cornerstone with the same silver trowel that Washington had used to begin construction of the Capitol in 1793.

Although the cornerstone was laid, construction of the monument was delayed by a political group called the Know-

Nothings. In 1855, they infiltrated the Monument Society, held a secret election, and took over the leadership of the Society, thereby "stealing" the monument. As a result, Congress promptly withdrew its pledge of $200,000 and the Society was discredited. Over the next three years, the Know-Nothings collected $285.09 and added thirteen rows of inferior marble to the monument.

By 1858 the Society was back in reliable hands, but the monument remained a 176-foot stump until 1876 when the U.S. Army Corps of Engineers began removing the work of the Know-Nothings. On July 5 Congress promised $200,000, and a completion date of October 19, 1881, was announced. Meanwhile, the U.S. Minister to Italy, George P. Marsh, suggested that the monument be simplified, leaving a stark, classical obelisk. The Society agreed and on August 7, 1880, President Hayes laid the second cornerstone. Four year later the tip was set in place. On Washington's next birthday, the monument was officially dedicated; on October 9, 1888, it was opened to the public.

Despite lack of funds and lack of interest, the Washington Monument finally grew to be the tallest masonry structure in the world. And despite its simple appearance, it is one of the most visited memorials in the country. It is a tribute to both the father of our country and our country itself.

Revision Agenda

1. *What did I try to do in this draft?*
   Write a simple history of monument — how it got started, the different designs, and all the obstacles people had to overcome to build it.

2. *What are its strengths and weaknesses?*
   Good information on history. Could have used more statistics and stupid things people have done there (tour guide). Sound too much like tour guide — a dull one. Dan says only good part is Know-Nothings. He's not a terrific audience — likes reruns of "Flintstones."

> *Still, he's right. It gets dull in places. Problem may be thesis — two subjects?*

3. *What revisions do I want to make in my next draft?*
   a. *Cut some of the historical information. Like me, most people are probably more interested in zany stories.*
   b. *Revise thesis to focus on history of <u>monument</u>, not monument and man — don't know much about <u>George</u>.*
   c. *Hook reader with opening — everybody knows the facts, but here are some things you don't know.*
   d. *Get my experience (tourist) back into essay. Might confuse things, but may help readers see monument.*
   e. *Keep good paragraphs — background information makes them clear and developed — but fix them so readers see connection to thesis.*
   f. *Make style more upbeat — too many facts, too dull.*

Revising for Purpose

By trying to write a pure information essay rather than a personal experience essay, Rod has created three problems. First, because he has allowed his information to control his essay, he sounds — to himself, at least — like a dull tour guide. Second, he has inadvertently allowed himself to write a thesis that promises a comparison between the monument and the man. Rod doesn't have the information to write the second half of this comparison — nor does he want to. His purpose is to discuss the history of the monument. Third, he has eliminated himself from his essay. In trying to inform his readers about the significance of the monument, he has prevented them from seeing, as he did, its impressive, simple beauty. The historical information is interesting, but Rod's attitude toward it (a mixture of admiration and amusement) does not really come through.

To tell the story behind the Washington Monument, Rod has to select and arrange his information so that it reflects *his* attitude, not that of a tour guide. By focusing on the unusual history of the monument, Rod formulates a more restricted, unified, and precise thesis, and eliminating the comparison between the man and the monument makes his subject more manageable. Rod is finally confident that his new purpose will create an interesting subject that will appeal to his readers.

REVISING THE WASHINGTON MONUMENT
(4th Draft)

The facts are simple: The Washington Monument stands 555 feet 5-1/8 inches. Its base is 55 feet 1-1/2 square inches. It is faced with 9,613 marble slabs, weighs 90,854 tons, can withstand a 145-mph gale, and cost $1,187,710.31. But these statistics aren't the whole story. In the two centuries since the idea was first mentioned, the Washington Monument has been revised, moved, stolen, ignored, held hostage, and almost blown up.

It began in 1783 when the Continental Congress voted to erect a bronze equestrian statue of George dressed in Roman garb, wielding a spear. Sixteen years later, Congress decided to honor George with a pyramid. Three decades later somebody thought George looked best dressed in a toga. This last idea was actually turned into a twenty-ton statue. But George looked a little cold in his toga. Not satisfied with a half-naked George, the good people of Washington formed the Washington National Monument Society in 1833 to see if they could create a fitting tribute for their hero.

In three years, the Society collected $28,000, and in 1836 they held a contest to select the best design. There were a lot of crazy designs: one looked like a Gothic cathedral, another like an Egyptian sphinx. But the committee selected a drawing by Robert Mills that called for a 600-foot obelisk rising out of a colonnade base, 250 feet in diameter, to be known as the National Pantheon. This temple was to contain statues of the signers of the Declaration of Independence, Revolutionary War heroes, and a colossal statue of George. But to cut costs, the monument was stripped of all its ornamentation and shortened slightly to conform to the classical proportions of an obelisk. For years, however, artists included the National Pantheon in their drawings.

George's monument was not only modified, it was also moved. Pierre Charles L'Enfant, the architect who designed the city, originally envisioned the monument in the intersection of an axis running west from the Capitol and south from the

White House. This spot proved too marshy to support much weight. If they had put George there, he would have been a leaning or sinking tower. So they moved him one hundred yards south.

Finally, in 1848, Benjamin B. French, Grand Master of the capital's Lodge of Free and Accepted Masons, laid the cornerstone with the same silver trowel George had used to begin construction of the Capitol. Fifteen thousand people attended the ceremony, including Martha's grandson, President Polk, and Congressman Abraham Lincoln. The odd thing is that nobody remembers which stone is the cornerstone.

The Society continued to raise money for George's monument. It also collected a number of carved stones from various states, Indian tribes, and foreign countries. Greece sent a block from the Parthenon; Egypt donated a stone from the Alexandrian Library; France sent a piece of Napoleon's tomb; and Italy sent a hunk of lava from Mt. Vesuvius. All these different stones (190 in all) decorate the monument — with the exception of one. On March 6, 1854, a stone sent by Pope Pius IX from the Temple of Concord was stolen. A political group called the Know-Nothings was credited with the heist, but the stone was never found.

In 1858, these Know-Nothings added another page to the crazy history of the monument by infiltrating the Society, electing themselves as officers, and essentially ''stealing'' the monument. Congress promptly withdrew its $200,000 pledge. In their three-year reign, the Know-Nothings collected $285.09 and added thirteen rows of inferior marble to the monument.

Twenty-five years later, when serious construction finally resumed, Col. Casey of the U.S. Army Corps of Engineers discovered that the rope used to lift supplies to the top of the monument had been pulled down. He solved this problem by killing a pigeon. He tied a long, thin wire to the pigeon's leg, which was attached to a rope. Then he let the bird loose inside the monument and fired a shotgun. The pigeon flew out of the monument's top carrying the wire and rope with it. But as soon as it cleared the monument, it was blown out of the sky by a sharpshooter. The wire was retrieved, and the rope pulled through.

George's memorial was opened to the public in 1888 and has been involved in some bizarre stunts ever since. In 1892 ''Gabby'' Street caught a baseball thrown out of one of the windows at the top of the monument. In 1915, Dr. Aldredo Warsaw, a New York baritone, sang two verses of a ballad while friends dangled him from one of the windows. During the Depression, three people committed suicide by jumping out the windows. (Iron bars have since been installed.) In 1966, 76-year-old Edna Rousseau made 307 round-trip climbs on the 898 steps. In 1982, a man urging the cause of nuclear disarmament held the monument hostage and threatened to blow it up if people didn't listen to him. He was shot by a police sharp-shooter, and his bomb was discovered to be a fake.

There's obviously more to George's monument than meets the eye. Of course, you don't have to know the inside story to like the monument. Most people who see a picture of the monument on a calendar or a postcard like what they see. But to truly understand and appreciate the monument, you do need to know the inside story. When you visit George's monument on some hot summer afternoon and have to stand in line for over two hours for a thirteen-second elevator ride to the top, you need to know how long it took the planners and builders to reach the top: all those false starts, crazy designs, and engineering obstacles that had to be overcome to produce this beautifully simple structure. When you see people in the line who don't know why they are there or who are just fooling around as they would at Disneyland, you need to know that the two-hundred-year history of the monument is filled with bewildered and misguided people. And when you have finished your visit, you need to take time to appreciate what you have seen. Turn and look at it one last time. There it stands — a solitary, rugged individual, indifferent to the confusion below. It is a peaceful and impressive sight.

Revision Agenda

1. *What did I try to do in this draft?*
   Tell the eventful history of the Washington Monument — make the big paperweight more human and interesting for my reader.
2. *What are its strengths and weaknesses?*
   I see mainly strengths. Thesis names topics of most of my paragraphs. Other paragraphs — stones and pigeon — add to the spirit of thesis. Should I call Washington "Old George"?
3. *What revisions do I want to make in my next draft?*
   a. *Not many — do I need another draft?*
   b. *Fix up a few sentences and find some better words — esp. verbs.*
   c. *Decide what to do about "the George problem."*
   d. *Mainly, I think I'm done. What did the guy who laid the last stone in the monument say when he was done?*

Consider the following questions about Rod's four revision agendas.

1. How does Rod's *subject* change from his discovery to his final draft?
2. How does Rod's concern about *audience* affect his decision making?
3. What is Rod's *purpose* for each draft? How does he try to embody that purpose in a thesis statement? To what extent does each thesis control the decisions he makes in writing?
4. How effectively does Rod follow Eudora Welty's notion of revising? For example, how does he move things around in each draft — "putting things in their best and proper place, revealing things at the time when they matter most . . . [shifting] things from the very beginning to the very end"?
5. Rod is obviously happiest with his last draft. Which draft do you like best? Explain your answer.

A FINAL WORD ABOUT REVISING

There is no final word about revising. You can revise endlessly, re-arranging information, rewriting paragraphs, substituting new words. The more you look the more you will see. But at some point, revising becomes rationalizing — an excuse for idle tinkering. The test of global revision is whether it produces significant improvement. Sometimes additional revision actually destroys good writing, replacing spontaneous, original insights with self-conscious, overwrought commentary. Like Rod, you have to know when to say "I'm done." Strategies for local revision to help you polish and perfect your last draft are contained in Part Two.

Revision Exercises

1. Each of the four designs shown on page 101 was proposed to honor George Washington. Compare and contrast the three preliminary sketches with the photograph of the actual monument. In partic-ular, examine the detail and annotations on each of the sketches. Then answer these questions:
   a. What *subject* is suggested by each structure? What clues help you identify the subject? Which designs suggest more than one subject? What are they?
   b. What kind of *audience* does each designer imagine for his structure? What kind of values and tastes does the designer attribute to the audience who would be impressed with his design?
   c. What is the *purpose* of a national monument to George Washington? Which of these designs fulfills that purpose most effectively? Explain your answer.

2. Compare and contrast the following descriptions of George Washington.
   a. How does each author define his *subject*? What kind of details does each select to illuminate his portrait?
   b. What assumptions does each writer make about the needs and expectations of his *audience*?
   c. What is each writer's *purpose* in composing his portrait of George Washington?
   d. Which portrait is the most interesting? Which one is the most believable? Explain your answers.

Everybody likes a boy who is strong and manly, and that was what George Washington was. The boy who, while in his early teens, could tame an unbroken colt, firmly keep his seat until he had mastered the wild and plunging thoroughbred, the boy who could "down" the best wrestlers in the county, who could throw a stone clear across the Rappahannock, toss bars and pitch quoits better than any man or boy about him, and sight and fire a rifle held with one hand only; the boy who could always be trusted to keep his promises, tell the truth, and do as he was bid without asking why, was a boy who could be at once bold and brave, good and gentle, sturdy and strong, wise and cautious. . . .

He would get dreadfully "mad" with other boys sometimes, and he was so strong that if he had been at all bad, he might have been what is called a bully. But, even when he was a small boy, he had learned to control his temper; and this he always did throughout his useful life, losing it so seldom and only when there was every excuse for his getting "mad," that we can set this splendid habit of self-control as one of the things that made him great and noble. . . .

Even as a boy, you see, George Washington had what we call the qualities of mind and brain, the courage, the caution and the determi-nation to succeed that made him, in after years, a leader of men and the chieftain of America. (Elbridge S. Brooks, *The True Story of George Washington*)

Sketch 1

Sketch 2

Sketch 3

Sketches courtesy National Archives. Photo © John Aikins 1986/Uniphoto

©Ellis Herwig/The Picture Cube

If George Washington were alive today, what a shining mark he would be for the whole camorra of uplifters, forward-lookers and professional patriots! He was the Rockefeller of his time, the richest man in the United States, a promoter of stock companies, a land-grabber, an exploiter of mines and timber. He was a bitter opponent of foreign entanglements, and denounced their evils in harsh, specific terms. He had a liking for forthright and pugnacious men, and a contempt for lawyers, schoolmasters and all other such obscurantists. He was not pious. He drank whis-key whenever he felt chilly, and kept a jug of it handy. He knew far more profanity than Scripture, and used and enjoyed it more. He had no belief in the infallible wisdom of the common people, but regarded them as inflammatory dolts, and tried to save the Republic from them. He advocated no sure cure for all the sorrows of the world, and doubted that such a panacea existed. (H. L. Mencken, "Pater Patriae")

. . . when Washington became president he wanted to be styled His Mightiness. The Senate was agreeable. The House of Representatives was not, and referred the other house to the Constitution which speaks of the chief executive as, simply, the president. In fact, the Speaker — the droll Mr. Muhlenberg — went so far as to suggest that perhaps the General would like to be known as "His High and Mightiness." Muhlenberg's mild pleasantry was not well received by the greatest man in the world who would very much have enjoyed, I suspect, being king had he not lacked a son, a prince of Virginia, to succeed him. (Gore Vidal, *Burr*)

Washington was asked to be heroic, but heroically restrained. Not only did he (and he alone) play a crucial role at *every* stage of the revolution — in the war, the postwar period, the passage of the Constitution, and the establishment of a working government. He had to show a complete virtue, a complementarity of competing traits, that was asked of no other. There may

have been greater warriors than he — but Arnold proved venal, Gates petty, Charles Lee unstable. Others might *use* armies better; but Washington best grasped that the problem was to *create* an army, to keep it in existence, by embodying its cause. There were better philosophers of the republican ideal. But the assurance of Washington's example swayed more people than did their arguments. Jefferson and Hamilton offered brilliant analyses of events, in the first administration; but Washington steered a steadier course than either could. He did not come to describe heroism, but to enact it. . . .

Washington's disinterested service, in war and peace, in power and in retirement, was not flawless; but it was so rounded and balanced that, for people with a less classical ideal of rule, it looks *soporifically* perfect. (Garry Wills, *Cincinnatus: George Washington and the Enlightenment*)

Writing Assignments

1. Evaluate your school's symbol or mascot in terms of its appropriateness to the school's spirit and achievements. Construct a revision agenda, based on the questions on pages 84 and 85 on subject, audience, and purpose; then write a brief proposal for a new symbol or mascot.

2. Transform something that you have written exclusively from the point of view of your personal experience — an extended journal entry, a letter you haven't yet mailed to a friend or to your family, an essay — into a more objective, informative piece of writing. Try to avoid the two problems that Rod discovered when he tried this assignment in his third draft : (a) his information took over his writing, and (b) he lost his unique perspective on his information — his purpose for presenting it.

3. Find a magazine article that is directed toward readers interested in historic preservation. Good sources are *American Heritage, Historic Preservation,* and *Smithsonian.* Read the article carefully, and write a brief analysis of its audience. Then profile a completely different audience — for

example, people interested in economic progress — and write down some ways in which the subject might be made appealing to them. Finally, draw up a careful revision agenda indicating how you would revise the article for the new audience.

4. Return to the essay that you began planning in writing assignment 4 of Chapter 2 (p. 57) and drafted in writing assignment 4 of Chapter 3 (p. 78). Reread the list of conclusions you drew up when you analyzed it; then — looking to revise subject, audience, and purpose — reread the essay itself. Draw up a revision agenda and then revise your essay.

5. The photograph on page 80 was selected as a metaphor representing an important aspect of this chapter. Examine the photograph and determine the possible meanings of this metaphor as you discern them in the photograph. Then write an essay explaining how the metaphor has been reinforced or revised by the ideas discussed in this chapter. You may want to suggest another metaphor to illustrate your understanding of these ideas.

PART TWO
THE EXPRESSION OF IDEAS

5
COMMON METHODS OF DEVELOPMENT

In Part One you learned how working through the stages of the writing process enables you to discover your subject, audience, and purpose. You also learned how these discoveries suggest various structures for shaping your writing. Once Joanne decided to write about Mr. Bridgeman, she began telling a story about her "little revolt" in his Latin class. Once Ellen decided to explain Barry's preoccupation with "what's good for me," she began to think of illustrative examples that would engage her audience. And once Rod decided to write about the Washington Monument, he reformulated his purpose to better analyze the various stages in the monument's unusual history. In each case, the method of development — the structure of the writing — emerged from material that the writer discovered in planning, drafting, and revising.

In Part Two you will learn how and when to use various structures to express your ideas more effectively. Essayist John McPhee reminds us that "everything in writing is a structure within a structure within a structure down to the simple sentence, which, of course, is also a structure" (personal interview). As you work your way through particular writing assignments, you need to decide whether to impose a structure on your information or to look for a structure within your information. There are occasions when you may need to employ a formal structure. For example, Part Three, "Special Assignments," discusses ways of complying with very specific instructions for developing and structuring essay examinations, critical essays, and research papers. Much of your writing, however, will evolve through such a complicated process of decision making that a formal structure chosen at an early stage is likely to change later.

McPhee explains that selecting a method (or methods) of developing his writing is probably the most important decision he makes, but one he can make only after immersing himself in the writing process:

> The structure of a piece of writing arises from within the material I have col-
> lected. When I go out to gather information on a subject, I am observing, asking
> questions, constantly scribbling notes. Then I go to the library, read, and make
> more notes. I end up with this great pile of miscellaneous notes that don't reveal
> their important relationships. It's at this point that I go over my notes again and
> again, looking for a couple of things that might illuminate each other. When I
> find them, I think, "Well, that makes sense. I could put these two things side by
> side." So I put them there. Once these two go together, gradually, slowly, the
> other segments work themselves out until I have a beginning, middle, and end.
> It's sort of the way the structure of some kind of mineral might get itself together.
> (Personal interview)

Like McPhee, you will uncover various patterns for developing your ideas
throughout the writing process. In *planning,* these patterns often emerge as
answers to the basic questions you might ask about any body of information:
What is it? How does it work? Why does it matter? These questions are like the
different lenses you attach to your camera: each lens gives you a different picture
of your subject. If you decide to write about women artists, for example, you
might formulate a series of questions such as: Why are comparatively few women
ranked among the world's great artists? What historical forces may have discouraged women from becoming artists? Who is in charge of the "ranking" process, and how does it work? Do the ways women look at the world differ from the
ways men look at it, and if so what implications does this difference have for
women's art and its reception? As you can see, each question not only shifts your
perspective on your subject, but also suggests a different method for developing
your information about it.

If planning gives you the opportunity to envision your subject from a variety
of perspectives, then *drafting* encourages you to develop the pattern (or patterns)
that appear to you most effective for demonstrating your purpose. In some writing projects, a pattern may seem to emerge naturally from your planning. If you
decide to write about your observation of a game of lacrosse, your choice seems
obvious: to tell what happened. In attempting this, however, you may need to
answer other questions about this unfamiliar sport: What do the field and equipment look like? How is it played? How is it similar to or different from other
sports? Developing this new information may require you to develop new patterns that challenge your ability to work with them in ways that advance your
purpose without distorting your structure.

Logically, this judgment is best made during *revising.* As you look over your
draft, you will need to make two decisions. First, you must decide whether individual segments or patterns of information develop or distort your original structure. The history of lacrosse — its creation by Iroquois Indians, its discovery by
French explorers, and its development by Canadians — is an interesting body of
information, but it may need to be reshaped, relocated, or even eliminated in
revision to preserve the purpose of your original design: to tell what happened.

Second, you must decide whether your original design, often mirroring the method by which you discovered the structure of your essay, is still the best method for presenting your information to your audience. Instead of using this organic structure, telling what happened, you may decide that you can best express your ideas by choosing a more formal structure, comparing lacrosse to games with which your readers are already familiar such as soccer or hockey.

Whatever you decide, you need to understand each pattern's purpose, techniques, and effect if you are going to use it successfully to develop a paragraph, a section of your essay, or your whole essay. The remainder of this chapter presents and analyzes the most common methods of development, except for argumentation, which, because of its complexity, is given more extended explanation in Chapter 6.

WHAT HAPPENED? NARRATION

Narration tells a story to make a point. It can be used in an anecdotal, abbreviated way to introduce or illustrate a complicated subject or in an extended way to provide a detailed, personal account of "what happened." An effective narrative has a *plot* — a meaningful and dramatic sequence of action — which may or may not follow the order in which events actually occurred and usually focuses on some tension or conflict within the writer, between the writer and others, or between the writer and the environment.

A narrative depends on pace to effect its purpose. *Pace* means the speed at which events are narrated. Sometimes you need to slow the pace to describe one aspect of an event in great detail; at other times several related events can be quickly summarized in a few sentences. By making such decisions about pace, you control the development of your narrative to dramatize its purpose for your readers.

A narrative requires a *point of view,* that is, establishing the person and position of the narrator. If you tell the story as "I" or "we," you are writing in the *first person;* if you recount what "he," "she," or "they" did, you are writing in the *third person.* In addition to identity, *person* refers to the attitude and personality of the narrator. *Position* is the narrator's closeness to the action in both space and time: the narrator may be a participant or an observer and may be telling the story as it happens, shortly thereafter, or much later.

In the following narrative excerpt, humorist Garrison Keillor recalls the drama and disappointment of an early romance.

Introduces characters and context for narration

Focuses on one scene (the kiss)

When I left Lake Wobegon, Donna Bunsen and I promised each other we'd read the same books that summer as a token of our love, which we sealed with a kiss in her basement. She wore white shorts and a blue blouse with white stars. She poured a cup of Clorox bleach in the washing machine, and then we kissed. In books, men and women "embraced passionately," but I didn't know how much passion to use, so I put my

Summarizes a process (selecting reading list)

arms around her and held my lips to hers and rubbed her lovely back, under the wings. Our reading list was ten books, five picked by her and five by me, and we made a reading schedule so that, although apart, we would have the same things on our minds at the same time and would think of each other. We each picked the loftiest books we knew of, such as Plato's *Republic, War and Peace, The Imitation of Christ,* the *Bhaga-vad-Gita, The Art of Loving,* to have great thoughts to share all summer

Distance from scene speeds up pace and produces several conflicts (does not read books, does not send letter, does not reply to cards)

as we read, but I didn't get far; my copy of Plato sat in my suitcase, and I fished it out only to feel guilty for letting her down so badly. I wrote her a letter about love, studded with Plato quotes picked out of Bartlett's, but didn't mail it, it was so shameless and false. She sent me two post-cards from the Black Hills, and in the second she asked, "Do you still love me?" I did, but evidently not enough to read those books and

Passage of time resolves conflicts (Donna marries)

Conclusion evokes significance of narration

become someone worthy of love, so I didn't reply. Two years later she married a guy who sold steel supermarket shelving, and they moved to San Diego. I think of her lovingly every time I use Clorox. Half a cup is enough to bring it all back. (Garrison Keillor, *Lake Wobegon Days*)

Like Keillor, Jane wants to write about a significant experience in her life, her first whale watch. Although she was only at sea for a few hours, she was overwhelmed by events, information, and insight. She decides that one way to explore her ideas is to follow the sequence of events in a narrative.

1. How does Jane establish the position and attitude of the narrator?

2. What possible conflicts does she introduce to dramatize her narration?

3. What techniques does she use to slow down the pace of her narration?

4. What event brings the narration to its climax?

5. How does the conclusion resolve the conflicts and confirm the purpose of the narration?

It was mist upon mist. The Portuguese Princess plowed out of Province-town harbor toward Stellwagen Bank where the humpbacks had been feeding for several months. I was seated forward, so that the gasoline fumes would not combine with the rolling swell to make me sick. Noth-ing was going to spoil this long-anticipated adventure. My insistence chased away the mist, and we chugged along in the hot sun for about an hour. Some people tried to talk above the sound of the engine: "I've watched all the Cousteau specials." "Have you seen the skeletons at the Museum of Natural History?" Others checked their camera lenses or scanned the horizon through binoculars. "Thar she blows," yelled our tour guide. "Off the port bow. There!" I followed his finger, straining to see what he saw in the dark blue sea. Nothing. Then, suddenly, against the skyline, I saw it. A definite vertical spray. Then a pause. Then another. A whale. At last, a whale.

▬▬ *Writing Assignments*

1. Freewrite in your journal about an important promise you have broken. Then expand your freewriting into a narrative, telling how you came to make the promise and why you failed to keep it.

2. List the events that contributed to your achievement of a long-anticipated goal. Then compose a narrative in which you help your readers appreciate your accomplishment by emphasizing the conflicts, large and small, that you had to overcome.

©1983 Paula M. Lerner/The Picture Cube

WHAT DOES IT LOOK LIKE? DESCRIPTION

Description presents a verbal portrait of a person, place, or thing. It can be used to enrich other forms of writing or as a dominant strategy for developing a picture of a subject. A successful description does not depend solely on visual effects, however, but attempts to identify the subject's significant features by evoking all the senses. Impressions and observations are arranged in an appropriate pattern designed to capture both *detail* and *wholeness*. Specific, vivid details make your readers see what you see, and arranging those details in an appropriate sequence helps your readers understand why your subject is interesting and significant. Description is not merely a catalogue of facts or a collection of ornaments; like narration, it must make a point.

In the following passage, reprinted from her autobiography *Journey Around My Room,* the poet Louise Bogan describes the preparations for a new season.

Establishes major point of description ("Everyone knew what he had to face.")

The whole town, late in October, felt the cold coming on; in bleak afternoons the lights came out early in the frame houses; lights showed clearly across the river in the chill dusk in houses and in the mill. Everyone knew what he had to face. After the blaze of summer that had parched paint and shingle, winter was closing in to freeze wood and stone to the

Evokes senses: touch *(cold),* smell *(plaster),* sight *(piano, curtains, books, pictures),* taste *(tea),* sound *(voices, leaves)*

core. The whole house, in winter, turned as cold as a tomb. The upper rooms smelled of cold plaster and cold wood. The parlor was shut; the piano stood shut and freezing against the wall; the lace curtains fell in starched frigid folds down to the cold grain of the carpet. The little padded books on the table, the lace doilies under them, the painted china vases, and the big pictures hanging against the big pattern of the wallpaper all looked distant, desolate, and to no purpose when the door was opened into the room's icy air. The life of the house went on in the sitting room and in the kitchen, for in both these rooms there were stoves. If

Arranges detail in a sequence (public world outside to private world inside)

my mother happened to be on good terms with Mrs. Parsons or with Mrs. Gardner, who lived across the street, they would come to visit her in the early twilight, on those days when the lamps were lit at four o'clock. My mother made tea and the women sat talking in low secret voices beside the kitchen table. I sat in the sitting room, and heard their voices and the sound of dry leaves blowing along the walk at the side of the house. (Louise Bogan, "The Neighborhood," *Journey Around My Room: The Autobiography of Louise Bogan*)

Jane saw and photographed so many whales on her voyage that she has trouble sorting out her impressions. She decides that a description of a whale will concentrate and enrich her narrative. She describes the last whale she saw — the one that saw her.

1. *How does Jane introduce the main point of her description?*

2. *What senses does she evoke to help her readers see what she sees?*

3. *What specific details does she use to describe the finback?*

4. *How does she arrange these details so that they create a dominant impression?*

5. *What details in Jane's description justify her concluding sentence?*

The biologist hushed us. A finback, the world's second largest whale, had been spotted swimming toward the boat. He was way off course from the usual pattern of his species. Finbacks are shy, solitary creatures and, because they are still hunted, usually stay clear of boats. The boat, its engine cut, rolled on four-foot swells. The smell of oil and gas hung in the air, making my head ache and my throat tighten as I tried to swallow. Soon, we saw him — a sea-green monolith, silhouetted several feet beneath the water. His color was uneven. The small distinguishing dorsal fin was speckled with white and gray spots. He was silent. We were silent. He swam purposefully toward the boat. Then under. More silence. I crossed to the other side in time to see his huge greenish head explode through the surface. One watery eye looked back at me before he arched his body and sounded. We did not see him again, but we knew we had seen something rare.

▬▬▬ *Writing Assignments*

1. List the rituals you and your family perform as you prepare for a new season. Then use specific sensory details to create a lively portrait of these ordinary procedures.

2. Draw a cluster that documents your reaction to your first encounter with an extraordinary machine (helicopter) or animal (racehorse); then expand your planning by reading several technical or scientific descriptions of your subject. Weave this factual detail into your impressionistic description of what you saw.

HOW DO YOU DO IT?
PROCESS ANALYSIS

A *process* is a sequence of actions or changes that bring about a result. The development of the human embryo from conception to birth is one process; the procedure by which citizens of the United States elect the president is another. To *analyze* a process effectively, you must know it thoroughly or learn as much as you can about it, and then you must divide it into steps or stages. A careful division will help you see the best way to describe the process: several small steps may be combined into one large step; certain steps may be suspended while others are completed; still other steps may require special emphasis so that they are not overlooked or reversed. After you have established an effective pattern of steps, you must explain each step, acknowledging how it connects to the next and what special knowledge or tools will be needed to complete it.

You must also make a careful assessment of your audience to determine whether you want to provide them with information about *how something works* or *happens* or to give them instructions on *how to do something.* In the first case, your readers may be interested in the stage and operation of a particular process even though they have no direct involvement in it. In the second case, they will want to know enough about the process so they can actually perform it.

In the following selection, George Orwell analyzes how the process of coal mining works.

Introduces process to be analyzed (acknowledges his own limitations as an expert)

Compares unfamiliar process to familiar one

Explains earlier procedure

Analyzes step 1: describes special tools, details smaller steps within step 1

Acknowledges negative effects

Analyzes step 2: describes special tools, details smaller steps within step 2

When you have been down two or three pits you begin to get some grasp of the processes that are going on underground. (I ought to say, by the way, that I know nothing whatever about the technical side of mining: I am merely describing what I have seen.) Coal lies in thin seams between enormous layers of rock, so that essentially the process of getting it out is like scooping the central layer from a Neapolitan ice. In the old days the miners used to cut straight into the coal with pick and crowbar — a very slow job because coal, when lying in its virgin state, is almost as hard as rock. Nowadays the preliminary work is done by an electrically-driven coal-cutter, which in principle is an immensely tough and powerful band-saw, running horizontally instead of vertically, with teeth a couple of inches long and half an inch or an inch thick. It can move backwards or forwards on its own power, and the men operating it can rotate it this way and that. Incidentally it makes one of the most awful noises I have ever heard, and sends forth clouds of coal dust which make it impossible to see more than two or three feet and almost impossible to breathe. The machine travels along the coal face cutting into the base of the coal and undermining it to the depth of five feet or five feet and a half; after this it is comparatively easy to extract the coal to the depth to which it has been undermined. Where it is "difficult getting," however, it has also to be loosened with explosives. A man with an electric drill, like a rather smaller version of the drills used in street-mending, bores holes at intervals in the coal, inserts blasting powder, plugs it with clay, goes round the corner if there is one handy (he is supposed to retire to twenty-five yards distance) and touches off the charge with an electric

Advises about possible dangers

current. This is not intended to bring the coal out, only to loosen it. Occasionally, of course, the charge is too powerful, and then it not only brings the coal out but brings the roof down as well. (George Orwell, from *The Road to Wigan Pier*)

One of the real dangers Jane faced when she went on the whale watch was her susceptibility to seasickness. She had been at sea enough and sick enough to qualify as an expert on the illness. In her draft, she composes a section in which she provides instructions about how to avoid the whale watcher's curse.

1. How does Jane establish the need for an effective procedure?

2. How many steps does she include in her pre-boarding precautions?

3. How does she use past experience to establish her credentials as an expert?

4. What additional steps does she take during the actual whale watch?

5. How does she present the negative effects of not following this procedure?

A person who suffers from motion sickness and loves whale watching has a problem. Whales do not fit under microscopes. They must be watched at sea, atop rolling four-foot swells. I have solved this problem by developing a procedure to combat the sickness that attacks me when "I go down to the sea in boats." Several hours before boarding, I eat a mild meal: toast, tea, mashed potatoes. An hour later, I take two Dramamine tablets. Then, I dress warmly, knowing that I will not be able to go below when I get cold. When I board the ship, I stay away from the engine, which spews gas and oil fumes into the air. Once underway, I stand forward, always to windward, so that I can feel the cold, salty breeze on my face. When we finally encounter some whales, I try to avoid looking too quickly from whale to whale. Finally, on the way back to port, I avoid the green tourist who has not taken these precautions. The next "Thar she blows" might be her.

Writing Assignments

1. Observe the steps in an unfamiliar process. Divide the process into major stages and then smaller steps. Before writing your analysis, interview an expert who knows all the little tricks and hidden steps that must be performed accurately if the process is to produce successful results.

2. Analyze a process that confuses, intimidates, or terrifies most people. Use personal experience to demystify the process: "I did this and survived to tell the tale. So can you."

HOW IS IT SIMILAR TO OR DIFFERENT FROM SOMETHING ELSE? COMPARISON

A *comparison* systematically analyzes and evaluates the similarities of two or more things. (A *contrast* is a comparison that emphasizes differences rather than similarities.) An effective comparison demonstrates one of three general purposes: two things thought to be different are shown to be quite similar; two things thought to be similar are shown to be quite different; or two things, although comparable, are shown to be not equal (that is, one is shown to be better than the other). To demonstrate one of these purposes, you will need to develop your comparison according to either the divided (*A + B*) pattern or the alternating (*A/B + A/B*) pattern.

The Divided Pattern of Comparison — *A + B*

The *divided pattern*, the most common of the two strategies, divides the comparison into two sections, the first devoted to a discussion of *A* and the second to a discussion of *B*. Linking the examples in *A* to those in *B* — for example, by making three points about *A* and three similar points about *B* — unifies the two contrasting parts. The points should be in the same sequence and, where possible, paired points should be treated in the same amount of space. Although such exact pairings are not always necessary, in working out your purpose you should demonstrate that *A* and *B* are inextricably bound.

The two main parts of an *A + B* comparison should develop cumulatively. A simple description of two unlike houses, for example, becomes a contrast only when its thesis points out the significance of the differences. In some cases, as when one side of the contrast is so well known to readers that it need not be stated, the comparison is *implied*. If you are contrasting American football (*A*) with English rugby (*B*) for an American audience, the details of *A* are already in the mind of your audience and need only be mentioned in passing to explain the significant details of *B*.

From an interview with Richard, the marine biologist on the *Portuguese Princess,* and from her reading, Jane gathers information on two of the world's largest creatures — the diplodocus and the blue whale. She then develops this information in a divided pattern that contrasts their enormous size with their gentle behavior.

Thesis contrasts perception with reality

A. *Diplodocus*
  1. *Enormous size*
  2. *Small brain*
  3. *Gentle behavior*
  4. *Eating habits*

B. *Blue whale*
  1. *Enormous size*
  2. *Large brain*
  3. *Gentle behavior*
  4. *Eating habits*

The creatures that seem to threaten man most with their size and power have often been quite gentle. They have banded together to care for their young, to protect one another, and to provide each other with company. Consider the big, dumb diplodocus — his huge body laboring with a slow, steady drumbeat through the prehistoric forest. Although his body was enormous, his brain was small, so small that it did not generate enough power to control his weighty tail. He needed a separate ganglion to control that end of his body. His size promised danger, but Big D spent most of his day grazing harmlessly on vegetation with his pals. When a meat-eating enemy approached, he lumbered into knee-deep water and munched on a soggy plant.

Similarly, the blue whale, the largest creature ever to exist, is as gentle as a golden retriever. One hundred feet long, he glides soundlessly through the ocean, his enormous head concealing his fifteen-pound brain. Like Big D, Big Blue was once a pack animal, but his numbers have been so depleted by hunters that he is rarely seen alone, much less in groups. He feeds on krill and other microscopic marine life. Instead of teeth, he has baleen, long strands of tissue that hang from his upper jaw. He sifts his food through this tissue by swimming, mouth open, through tons of water. When he closes his mouth, the water strains back through the baleen, and he swallows the remaining food in gigantic gulps.

The Alternating Pattern of Comparison — A/B + A/B

The *alternating pattern* develops your material through matched pairs of *A* and *B* details, expressed either in the same paragraph or in the same sentence. The divided pattern is perhaps easier to organize and control, particularly in short essays, but unless you connect the two subjects with a clear thesis, you may discover that you have written two separate essays. The alternating pattern requires you to organize your material more precisely, especially in a longer essay, but the pattern is often more interesting and accessible for your reader, because the point-by-point development can be written in balanced sentences that reinforce the comparison with every pair of matched details. Gretel Ehrlich uses the alternating pattern in the following passage to compare the romanticized conception (*A*) and the real character (*B*) of the American cowboy.

1. *How does Ehrlich use the Marlboro man to introduce the two items in her comparison?*

2. *How does her comment on romanticizing the cowboy establish the thesis of her comparison?*

3. *How does she use parallel sentence structure to emphasize the contrast between A and B?*

4. *How does the contrast between toughness and "toughing it out" serve as an apt summary of this contrast?*

5. *What does the comparison of cowboys with rocks contribute to her thesis?*

When I'm in New York but feeling lonely for Wyoming I look for the Marlboro ads in the subway. What I'm aching to see is horseflesh, the glint of a spur, a line of distant mountains, brimming creeks, and a reminder of the ranchers and cowboys I've ridden with for the last eight years. But the men I see in those posters with their stern, humorless looks remind me of no one I know here. In our hellbent earnestness to romanticize the cowboy we've ironically disesteemed his true character. If he's "strong and silent" it's because there's probably no one to talk to. If he "rides away into the sunset" it's because he's been on horseback since four in the morning moving cattle and he's trying, fifteen hours later, to get home to his family. If he's "a rugged individualist" he's also part of a team: ranch work is teamwork and even the glorified open-range cowboys of the 1880s rode up and down the Chisholm Trail in the company of twenty or thirty other riders. Instead of the macho, trigger-happy man our culture has perversely wanted him to be, the cowboy is more apt to be convivial, quirky, and softhearted. To be "tough" on a ranch has nothing to do with conquests and displays of power. More often than not, circumstances — like the colt he's riding or an unexpected blizzard — are overpowering him. It's not toughness but "toughing it out" that counts. In other words, this macho, cultural artifact the cowboy has become is simply a man who possesses resilience, patience, and an instinct for survival. "Cowboys are just like a pile of rocks — everything happens to them. They get climbed on, kicked, rained and snowed on, scuffed up by wind. Their job is 'just to take it,' " one old timer told me. (Gretel Ehrlich, *The Solace of Open Spaces*)

Writing Assignments

1. Speculate in your journal on some of the similarities you have observed between two apparently different activities — eating and sleeping, talking and listening, reading and running. List the points of your comparison; decide on an appropriate arrangement of the points; and then develop your comparison according to one of the two patterns.

2. Read through several popular magazines looking for common comparisons: for example, the heart is like a pump (medicine), an election is a popularity contest or a horse race (politics), a war is like a game (the military), football is show business (sports). Develop an essay in which you demonstrate how one of these comparisons demeans rather than defines the "true character" of its principal subject.

WHAT KIND OF SUBDIVISIONS DOES IT CONTAIN? CLASSIFICATION

Classification organizes information into groups and categories. An effective classification begins by defining a subject and then dividing it into major categories based on a common trait. The categories are then arranged in a sequence that shows that the division is *consistent* (the same principle is used to classify each category), *complete* (no major categories are omitted), and *significant* (the categories and subcategories are arranged in an order that demonstrates some purpose).

Classification, like comparison, builds on your readers' expectations for precision, balance, and order. Call your readers' attention to the principle you have used to classify each category, or clearly imply it, and devote approximately the same amount of space to each category. Finally, arrange your categories and subpoints clearly and logically so that your reader is able to follow your system. Mortimer Adler employs these strategies in classifying book owners.

Establishes three classes of book owners.

Devotes approximately same amount of space in each category to how owners use books

Arranges categories in logical sequence, concluding with the person who really owns books

There are three kinds of book owners. The first has all the standard sets and best-sellers — unread, untouched. (This deluded individual owns woodpulp and ink, not books.) The second has a great many books — a few of them read through, most of them dipped into, but all of them as clean and shiny as the day they were bought. (This person would probably like to make books his own, but is restrained by a false respect for their physical appearance.) The third has a few books or many — every one of them dog-eared and dilapidated, shaken and loosened by continual use, marked and scribbled in from front to back. (This man owns books.) (Mortimer J. Adler, "How to Mark a Book," *Saturday Review of Literature*)

Jane was fascinated by the many types of whales. The books she read emphasized their physical differences, and Richard, who had studied many individual whales, helped her understand their behavioral idiosyncrasies. During planning, Jane tries to combine her sources to classify whales according to their "personalities."

1. How does Jane attempt to establish the completeness of her classification?

2. How does she use identifying labels to distinguish among her three categories?

Although there are over one hundred kinds of whales in the world (and even more names to designate each kind), most whales can be classified into three major categories. The gregarious whales are outgoing, curious, and friendly toward people. This category includes extroverts, like the humpback, who have baleen and feed off microscopic marine life, and acrobats, like the porpoises, who have teeth and eat small fish. The

3. How does she avoid the apparent contradiction that baleen and teeth introduce into her system?

4. What kind of evidence does she use to illustrate the "personalities" of each category?

5. What is significant about the sequence in which she presents her categories?

aggressive whales are formidable, ornery, and dangerous. This category includes killer whales, who hunt in packs, stalking all sorts of marine life from penguins to baby blue whales, and sperm whales, who hunt alone, devouring such impressive prey as octopi, and who have been known to ram ships. The shy whales are solitary, enigmatic, and unpredictable. This category includes whales like the finback, who seem indifferent to other whales as well as to people, and the blue whale, who has been hunted to near-extinction by those eager to kill the world's largest creature.

Writing Assignments

1. Interview friends or members of your family about their experiences with tools. Establish at least three categories that will enable you to classify people according to how they use their tools. Compose a thesis statement that asserts the significance of your categories.

2. Read about an established method for classifying information such as categories for rating films or the procedure for naming plants. Then devise a principle that will enable you to simplify the system. In your classification, demonstrate why your new system is more efficient.

HOW WOULD YOU CHARACTERIZE IT? DEFINITION

Definition provides a necessary explanation of a word or concept. The explanation may be a simple substitution of a familiar for an unfamiliar word, as when you substitute *cancer* for *carcinoma*. It may be the addition of a phrase, as when you follow *vintage* with "the yield of wine or grapes from a particular vineyard or district during one season." It may be a single sentence: "In the theater, a *prompter* is a person who provides cues for the actors or singers on stage." Or the definition may consist of one or more paragraphs, or even a whole essay, in which you explain your subject, such as *tradition* or *excellence*, in some depth.

Definitions may be short, stipulative, or extended. *Short* definitions, like the definition of *carcinoma, vintage,* and *prompter,* provide brief explanations similar to those in a dictionary. *Stipulative* definitions specify the exact meaning or sense of a word as you intend to use it in your writing. Some words may have favorable or unfavorable connotations, depending on the audience. In a political election, for example, a candidate, on being called a *liberal*, might reply, "Yes, I am a liberal, but only in the true sense of that word. It originally meant *free*, and I believe in the freedom of individuals to think, speak, and act according to their consciences." The candidate is emphasizing one meaning and excluding others. Such definitions enable your readers to see the purpose of your words. The definition of *purpose* on page 16 is another example of a stipulative definition.

Extended definitions, usually longer pieces of writing, explain a writer's view of a subject, sometimes using short or stipulative definitions, but going far beyond both. The writer adds to, modifies, and illustrates the basic definition.

Any patterns of development may be used: comparison of a word's synonyms, for example, or analysis of changes in its meaning. Poet Laurie Lee combines different strategies in his extended definition of *charm*.

Thesis: defines by simple substitution: "Charm is . . ."

Charm is the ultimate weapon, the supreme seduction, against which there are few defences. If you've got it, you need almost nothing else, neither money, looks, nor pedigree. It's a gift, only given to give away, and the more used the more there is. It is also a climate of behaviour set for perpetual summer and thermostatically controlled by taste and tact. . . .

Defines the "process" by which it is acquired

You recognize charm by the feeling you get in its presence. You know who has it. But can you get it, too? Properly, you can't, because it's a quickness of spirit, an originality of touch you have to be born with. Or it's something that grows naturally out of another quality, like the simple desire to make people happy. Certainly, charm is not a question of learning palpable tricks, like wrinkling your nose, or having a laugh in your voice, or gaily tossing your hair out of your dancing eyes and twisting your mouth into succulent love-knots. Such signs, to the nervous, are ominous warnings which may well send him streaking for cover. On the other hand, there is an antenna, a built-in awareness of others, which most people have, and which care can nourish.

Defines negatively: "charm is not . . ."

Defines by classifying qualities of charm

But in a study of charm, what else does one look for? Apart from the ability to listen — rarest of all human virtues and most difficult to sustain without vagueness — apart from warmth, sensitivity, and the power to please, what else is there visible? A generosity, I suppose, which makes no demands, a transaction which strikes no bargains, which doesn't hold itself back till you've filled up a test-card making it clear that you're worth the trouble. Charm can't withhold, but spends itself willingly on young and old alike, on the poor, the ugly, the dim, the boring, on the last fat man in the corner. It reveals itself also in a sense of ease, in casual but perfect manners, and often in a physical grace which springs less from an accident of youth than from a confident serenity of mind. Any person with this is more than just a popular fellow, he is also a social healer. (Laurie Lee, "Charm," *I Can't Stay Long*)

Defines by describing its effect on the person who possesses it

During Jane's whale-watching experience, she encountered many unusual words. She had heard some of them before (*breeching, spouting*) but did not appreciate their true significance until she saw the actions they identified. Others she heard for the first time on the *Portuguese Princess* (*krill, baleen*), grasping their meaning as Richard explained them. But one term, *gam*, was used without explanation. In reviewing her notes, the word popped up again, prompting her to research and meditate on its meaning in the following definition.

1. *How does Jane introduce the problem of definition?*
2. *How do the quotations from Melville add credibility to her definition?*

What is a gam? When Herman Melville discussed the word in Moby Dick, he was certain he was the first writer to define it: "you might wear out your index finger running up and down the columns of dictionaries and never find the word." But he was equally certain that the word needed an official definition because it "has now for many years been in constant use among fifteen thousand true born Yankees" (Herman

©South Sea Whaling. From a colored woodcut by Duncal. The Bettman Archive, Inc.

3. How does she use substitution to clarify the precise meaning of the word?

4. How does the speculation about its origin extend the definition?

5. How do her closing examples illustrate this meaning?

Melville, "The Gam," Moby Dick). Perhaps because of Melville, the word now appears in most dictionaries. Gam, like its equally unusual synonyms pod and school, refers to a herd of whales. By extension, gam refers to a meeting of two or more whaling ships at sea. Its precise origin is unknown, but some speculate that the word derives from gammon, a little-used word for talk or chatter. And that's what whales and whalers do when they gam. They socialize. Whales seem to enjoy each other's company as they feed. And whalers, once they meet at sea, stop the hunt, visit between ships, and settle down to food and conversation.

Writing Assignments

1. Select a common word, such as *stress* or *class*, that seems to have many subtle meanings. Freewrite a few simple definitions: "Class is . . . ." Next, try a few negative definitions: "Class is not . . . ." Then research the word in a dictionary to determine its precise meanings and in a thesaurus to discover words with similar meanings. Finally, expand your planning into an extended definition.

2. Select an unusual word that puzzled you when you first heard or read it, and interview several "authorities" about what the word might mean. Check the word in several dictionaries (including a dictionary of slang), and then use your planning and personal experience to prepare your own definition.

WHY DID IT HAPPEN? CAUSAL ANALYSIS

Causal analysis, like process analysis, details a sequence of steps that produce a result. But rather than describing these steps, causal analysis examines them for *causes* and *effects.* Such an analysis can be developed in three ways: by describing an action or event and then demonstrating its effect; by describing an action or event and then determining its cause; or by examining two related actions or events and proving a cause and effect connection between them.

In each case, you must be careful not to exaggerate or oversimplify the cause and effect relationship. You may mistake coincidence for cause, or you may identify one cause as *the* direct cause when any number of complex causes (working independently or in combination) could have produced the same effect. In the following passage Alan Devoe analyzes the many causes of hibernation.

Suggests variety of causes:
1. cold
2. diminishing food supply
3. increased darkness
4. silence

Illustrates that different causes produce same effect in different animals

Supplies another cause:
5. instinctual behavior

Presents new effect for investigators: hibernators often do not sleep at all in captivity

The season of hibernating begins quite early for some of the creatures of outdoors. It is not alone the cold which causes it; there are a multiplicity of other factors — diminishing food supply; increased darkness as the fall days shorten; silence — frequently decisive. Any or all of these may be the signal for entrance into the Long Sleep, depending upon the habits and make-up of the particular creature. Among the skunks, it is usually the coming of the cold that sends them, torpid, to their root-lined underground burrows; but many other mammals (for instance, ground squirrels) begin to grow drowsy when the fall sun is still warm on their furry backs and the food supply is not at all diminished. This ground-squirrel kind of hibernating, independent of the weather and the food supply, may be an old race habit, an instinctual behavior pattern like the unaccountable migrations of certain birds. Weather, food, inheritance, darkness — all of these obscurely play their parts in bringing on the annual subsidence into what one biologist has called "the little death." Investigation of the causes will need a good many years before they can be understood, for in captivity, where observation is more easy than in the wild, the hibernators often do not sleep at all. (Alan Devoe, *Lives Around Us*)

Her first whale watch had a profound effect on Jane, and as she rea more about whale watching, she realized that many other people had had similarly significant experiences. In fact, she speculated that whale watching had produced three important effects.

1. How does Jane account for the origin of whale watching?

2. What events enabled it to become a popular activity?

3. What have been the three effects this popularity?

Since its inception over a decade ago, whale watching has become a major industry. It began as a scientific enterprise as marine biologists, concerned about extinction, attempted to chart the feeding and migrating patterns of whales. The biologists soon became so adept at sighting whales and so enthusiastic about what they saw that they offered the experience to the public, producing three significant effects. First, whale watching has become an enormously popular recreational experience. The exhilarating voyage out to sea, the spectacular acrobatics of the whales, and the sensation of seeing something extraordinary rivals the

4. *How does Jane arrange these effects in a significant sequence?*

5. *Why is the third effect potentially the most important?*

best rides at Disneyland. Second, whale watching has increased public awareness about preservation. Once people have seen the power, grace, and agility of whales, they can never again be indifferent to their wholesale slaughter. Third, whale watching in America has become so profitable that entrepreneurs are trying to convince the nations that still hunt whales, such as Japan and the Soviet Union, that there is more money to be made in watching whales than in killing them.

▬▬▬ *Writing Assignments*

1. Analyze the possible causes for one of your common problems, such as occasional insomnia or headaches. Read a general reference work, such as *The Columbia Medical Encyclopedia,* to identify some probable causes. Then interview others who are afflicted by the same problem to see how they explain its causes. Assess each cause in a brief essay, and include those variations in the pattern (such as extended periods of slumber) that cannot be explained by your analysis.

2. Speculate on the possible effects that an event such as one of the following might have on your life:
   a. Purchasing an expensive car
   b. Studying in England for a semester
   c. Trying out for a part in a play
   d. Working for a political candidate

───────── *Review Exercises* ▬▬▬▬

1. Jane's final essay, "Watching Whales," incorporates much of the writing she developed during planning and drafting. Read her essay and review her writing above. What patterns has she reshaped, relocated, or eliminated during revising? Is the overall structure of her essay the most effective method for accomplishing her purpose? Explain your answer.

Watching Whales

The problem with whales is that they don't fit   1
under a microscope. As a small girl, I walked around and around the large skeleton of a blue whale on the first floor of the American Museum of Natural History, trying to compare its size to the skeletons of the brontosaurus and diplodocus I had seen on the fourth floor. Like those creatures from another time, whales challenged and

held my imagination. I could never see enough of them. The drawings in textbooks outlined their shape but reduced their enormous size to a few inches. The photographs in magazines suggested their magnificence but usually focused on their parts — the head rising to breech, the tail arching to dive. Last summer, hoping to see more, I booked passage on the Portuguese Princess out of Provincetown, Massachusetts, for my first whale watch.

A motley crowd assembled on MacMillan   2
Pier in the early morning mist. Seasoned whale watchers checked their binoculars, cameras, and lens cases. Casual tourists chased after wayward children, counted their supply of Dramamine, and looked anxiously at the Princess as it creaked and groaned against it moorings. Various members of the crew, dressed in T-shirts and cutoffs, peered into the engine hole or stud-

ied the sighting charts stacked on the elevated table behind the wheel house. The haunting sounds of whale songs were piped through the mist by the boat's loud-speaker system. Soon, Richard, a marine biologist, dressed like the crew but wearing a Red Sox cap, walked out of the mist and jumped aboard. He checked the charts for a few minutes and then picked up a clipboard to call the seventy names on his new manifest.

I crossed the gangplank and took a position 3 forward, away from the engine, knowing that the combination of gasoline fumes and rolling seas could make me sick. The engine sputtered to life, drowning out the whale songs and the nervous chatter of passengers. As the Princess plowed out of the harbor, I looked back to watch Provincetown disappear. The streets crowded with gawking summer people blurred, and the silhouette of the old fishing village — shacks resting on huge pilings, seagulls diving behind weathered boats — lingered briefly on the horizon.

"We're headed for Stellwagen Bank," Rich- 4 ard announced from his perch behind the wheel house. "This is an area of shallow water and undersea crags that attracts microscopic marine life, plankton, and krill, the staple diet of many whales, especially the humpbacks. Humpbacks are baleen whales. They do not hunt, kill, and devour. Instead of teeth, they have baleen, long strands of tissue that hang from their upper jaws. They sift their food through this tissue by swimming, mouth open, through tons of water. When they close their mouths, the water strains back through the baleen, and they swallow the remaining food in gigantic gulps."

The mist had cleared, the sun burned my 5 neck, and a cold, salty spray occasionally splashed my face as the Princess headed into the wind for the next half hour. The passengers talked, gestured, and looked for signs. "I've watched all the Cousteau specials," a New Hampshire bride was telling me when Richard yelled, "Thar she blows! Off the port bow! There!" We followed his finger, focusing binoculars, twisting telephoto lenses, straining to see what he had seen in the dark blue swells.

Against the skyline, I saw it. A definite vertical spray. Then a pause. Then another.

I was actually seeing two sprays, one from 6 each of the whale's spouts or nostrils. Whales are mammals and so, like us, cannot breathe under water. They store air in their lungs before they dive and then "blow" it out their spouts when they surface. This whale was diving and blowing steadily, heading away from Stellwagen Bank. Richard tried to identify its flukes, the large flippers at the end of its tail. The patterns and notches on a whale's flukes are like fingerprints, enabling expert whale watchers to identify individual whales. But this whale, swimming intently toward some destination, was too far away for even Richard to see its colors.

The Princess churned on for another thirty 7 minutes. Suddenly, the captain killed the engine. "We're here." And so were they. As we drifted, several sets of feeding humpbacks swam into view. A mother, her calf beside her, dived, surfaced, and waved her flukes in a perfect vertical before diving again. Her calf, eager to imitate, tried the same maneuver over and over, but his tail and flukes flopped sideways at the last minute like a sloppy cartwheel. At our stern, an obstreperous male leaped totally out of the water and then fell sideways with an impressive clap. I could see his huge, vulnerable underbelly and the white undersides of his long flippers. When he burst out of the water, he held his flippers at a slight angle to his side. As he descended, he moved them gracefully away from his body like the arms of a ballerina finishing a pirouette.

After an hour of dashing around the deck, 8 elbowing my way to the rail, and snapping pictures of everything I could see, I thought I had seen enough. Other excursion boats and private craft had slowly encircled the feeding whales. Shortwave radios crackled: "Look to the west. A male is breeching." "Over to the south. Two bachelors are gamming." "In the center. The mother and calf are resting." The noise prompted thoughts of other messages delivered on distant seas: "Finback, south, southwest. Lower the boats. Aim the harpoon." Richard had apparently seen enough as well. The engine erupted, and the Princess began to turn. I looked forward, trying to sort it all out — the graceful, powerful gentleness of the whales; the curious, careless intrusiveness of the whale watchers. Why couldn't we simply watch? Why did people have to hunt, kill, and destroy?

The engine stopped. Richard's voice came 9 over the loud-speaker in an excited, controlled whisper. "There's a finback heading toward us. He's way off course. Finbacks are shy and solitary. They're still hunted and so usually stay clear of boats. If we're quiet, we may see something rare." Seventy watchers became still. The Princess rolled on four-foot swells. Soon we saw him — a sea-green monolith, his huge bulk silhouetted several feet beneath the surface. His distinguishing fin was mottled in patterns of white and gray. He was swimming purposefully toward us. We were obviously in his path, but not in his way. He was silent. We were silent. Closer, closer, and then under. More silence. I rushed to the other side. His green head exploded through the surface twenty feet away. One large, watery eye looked back at me before he arched his body into the sea. He swam in a straight line for a couple of hundred yards, blew, and then sounded.

Although we watched eagerly for him to 10 surface again, we looked in vain. Soon we began to look at one another. We smiled, but we did not talk. Neither did Richard. No explanation was necessary. We had watched a whale. A whale had watched us. Finally, the engine coughed, and the Portuguese Princess resumed her journey back to Provincetown.

2. Consider John McPhee's comments on structure on pages 107 and 108 and this essay. What is the predominant pattern of the essay? Does this pattern seem to emerge from the material or to be imposed upon it? Explain your answer. Identify the other patterns of development evident in this essay, making annotations like those accompanying the examples in the chapter.

Under the Snow

When my third daughter was an infant, I could 1 place her against my shoulder and she would stick there like velvet. Only her eyes jumped from place to place. In a breeze, her bright-red

hair might stir, but she would not. Even then, there was profundity in her repose.

When my fourth daughter was an infant, I 2 wondered if her veins were full of ants. Placing her against a shoulder was a risk both to her and to the shoulder. Impulsively, constantly, everything about her moved. Her head seemed about to revolve as it followed the bestirring world.

These memories became very much alive 3 some months ago when — one after another — I had bear cubs under my vest. Weighing three, four, 5.6 pounds, they were wild bears, and for an hour or so had been taken from their dens in Pennsylvania. They were about two months old, with fine short brown hair. When they were made to stand alone, to be photographed in the mouth of a den, they shivered. Instinctively, a person would be moved to hold them. Picked up by the scruff of the neck, they splayed their paws like kittens and screamed like baby bears. The cry of a baby bear is muted, like a human infant's heard from her crib down the hall. The first cub I placed on my shoulder stayed there like a piece of velvet. The shivering stopped. Her bright-blue eyes looked about, not seeing much of anything. My hand, cupped against her back, all but encompassed her rib cage, which was warm and calm. I covered her to the shoulders with a flap of down vest and zipped up my parka to hold her in place.

I was there by invitation, an indirect result 4 of work I had been doing nearby. Would I be busy on March 14th? If there had been a conflict— if, say, I had been invited to lunch on that day with the Queen of Scotland and the King of Spain — I would have gone to the cubs. The first den was a rock cavity in a lichen-covered sandstone outcrop near the top of a slope, a couple of hundred yards from a road in Hawley. It was on posted property of the Scrub Oak Hunting Club — dry hardwood forest underlain by laurel and patches of snow — in the northern Pocono woods. Up in the sky was Buck Alt. Not long ago, he was a dairy farmer, and now he was working for the Keystone State, with directional antennae on his wing struts angled in the direction of bears. Many bears in Pennsylvania have radios around their necks as a result of the summer trapping work of Alt's son Gary, who is a wildlife biologist. In winter, Buck Alt flies the country listening to the radio, crissing and crossing until the bears come on. They come on stronger the closer to them he flies. The transmitters are not omnidirectional. Suddenly, the sound cuts out. Buck looks down, chooses a landmark, approaches it again, on another vector. Gradually, he works his way in, until he is flying in ever tighter circles above the bear. He marks a map. He is accurate within two acres. The plane he flies is a Super Cub.

The den could have served as a set for a 5 Passion play. It was a small chamber, open on one side, with a rock across its entrance. Between the freestanding rock and the back of the cave was room for one large bear, and she was curled in a corner on a bed of leaves, her broad head plainly visible from the outside, her cubs invisible between the rock and a soft place, chuckling, suckling, in the wintertime tropics of their own mammalian heaven. Invisible they were, yes, but by no means inaudible. What biologists call chuckling sounded like starlings in a tree.

People walking in woods sometimes come 6 close enough to a den to cause the mother to get up and run off, unmindful of her reputation as a fearless defender of cubs. The cubs stop chuckling and begin to cry: possibly three, four cubs — a ward of mewling bears. The people hear the crying. They find the den and see the cubs. Sometimes they pick them up and carry them away, reporting to the state that they have saved the lives of bear cubs abandoned by their mother. Wherever and whenever this occurs, Gary Alt collects the cubs. After ten years of bear trapping and biological study, Alt has equipped so many sows with radios that he has been able to conduct a foster-mother program with an amazingly high rate of success. A mother in hibernation will readily accept a foster cub. If the need to place an orphan arises somewhat later, when mothers and their cubs are out and around, a sow will kill an alien cub as soon as she smells it. Alt has overcome this problem by stuffing sows' noses with Vicks VapoRub. One way or another, he has found new families for forty-seven orphaned cubs. Forty-six have sur-

vived. The other, which had become accustomed over three weeks to feedings and caresses by human hands, was not content in a foster den, crawled outside, and died in the snow.

With a hypodermic jab stick, Alt now 7 drugged the mother, putting her to sleep for the duration of the visit. From deeps of shining fur, he fished out cubs. One. Two. A third. A fourth. Five! The fifth was a foster daughter brought earlier in the winter from two hundred miles away. Three of the four others were male — a ratio consistent with the heavy preponderance of males that Alt's studies have shown through the years. To various onlookers he handed the cubs for safekeeping while he and several assistants carried the mother into the open and weighed her with block and tackle. To protect her eyes, Alt had blindfolded her with a red bandanna. They carried her upside down, being extremely careful lest they scrape and damage her nipples. She weighed two hundred and nineteen pounds. Alt had caught her and weighed her some months before. In the den, she had lost ninety pounds. When she was four years old, she had had four cubs; two years later, four more cubs; and now, after two more years, four cubs. He knows all that about her, he had caught her so many times. He referred to her as Daisy. Daisy was as nothing compared with Vanessa, who was sleeping off the winter somewhere else. In ten seasons, Vanessa had given birth to twenty-three cubs and had lost none. The growth and reproductive rates of black bears are greater in Pennsylvania than anywhere else. Black bears in Pennsylvania grow more rapidly than grizzlies in Montana. Eastern black bears are generally much larger than Western ones. A seven-hundred-pound bear is unusual but not rare in Pennsylvania. Alt once caught a big boar [a large male bear] like that who had a thirty-seven-inch neck and was a hair under seven feet long.

This bear, nose to tail, measured five feet 8 five. Alt said, "That's a nice long sow." For weighing the cubs, he had a small nylon stuff sack. He stuffed it with bear and hung it on a scale. Two months before, when the cubs were born, each would have weighed approximately half a pound — less than a newborn porcupine. Now the cubs weighed 3.4, 4.1, 4.4, 4.6, 5.6 —

cute little numbers with soft tan noses and erec-
tile pyramid ears. Bears have sex in June and
July, but the mother's system holds the fertilized
egg away from the uterus until November, when
implantation occurs. Fetal development lasts
scarcely six weeks. Therefore, the creatures who
live upon the hibernating mother are so small
that everyone survives.

The orphan, less winsome than the others,   9
looked like a chocolate-covered possum. I kept
her under my vest. She seemed content there
and scarcely moved. In time, I exchanged her
for 5.6 — the big boy in the litter. Lifted by the
scruff and held in the air, he bawled, flashed his
claws, and curled his lips like a woofing boar. I
stuffed him under the vest, where he shut up
and nuzzled. His claws were already more than
half an inch long. Alt said that the family would
come out of the den in a few weeks but that
much of the spring would go by before the cubs
gained weight. The difference would be that they
were no longer malleable and ductile. They
would become pugnacious and scratchy, not to
say vicious, and would chew up the hand that
caressed them. He said, "If you have an enemy,
give him a bear cub."

Six men carried the mother back to the den, 10
the red bandanna still tied around her eyes. Alt
repacked her into the rock. "We like to return
her to the den as close as possible to the way
we found her," he said. Someone remarked that
one biologist can work a coon, while an army is
needed to deal with a bear. An army seemed to
be present. Twelve people had followed Alt to
the den. Some days, the group around him is
four times as large. Alt, who is in his thirties,
was wearing a visored khaki cap with a blue-
and-gold keystone on the forehead, and a khaki
cardigan under a khaki jump suit. A lithe and
light-bodied man with tinted glasses and a blond
mustache, he looked like a lieutenant in the
Ardennes Forest. Included in the retinue were
two reporters and a news photographer. Alt
encourages media attention, the better to soften
the image of the bears. He says, "People fear
bears more than they need to, and respect them
not enough." Over the next twenty days, he had
scheduled four hundred visitors — state sena-
tors, representatives, commissioners, television

reporters, word processors, biologists, friends —
to go along on his rounds of dens. Days before,
he and the denned bears had been hosts to the
BBC. The Brits wanted snow. God was having
none of it. The BBC brought in the snow.

In the course of the day, we made a brief   11
tour of dens that for the time being stood vacant.
Most were rock cavities. They had been used
before, and in all likelihood would be used again.
Bears in winter in the Pocono Plateau are like
chocolate chips in a cookie. The bears seldom
go back to the same den two years running, and
they often change dens in the course of a winter.
In a forty-five-hundred-acre housing development called Hemlock Farms are twenty-three
dens known to be in current use and countless
others awaiting new tenants. Alt showed one
that was within fifteen feet of the intersection of
East Spur Court and Pommel Drive. He said that
when a sow with two cubs was in there he had
seen deer browsing by the outcrop and ignorant
dogs stopping off to lift a leg. Hemlock Farms is
expensive, and full of cantilevered cypress and
unencumbered glass. Houses perch on high flat
rock. Now and again, there are bears in the rock
— in, say, a floor-through cavity just under the
porch. The owners are from New York. Alt does
not always tell them that their property is zoned
for bears. Once, when he did so, a "FOR SALE"
sign went up within two weeks.

Not far away is Interstate 84. Flying over it   12
one day, Buck Alt heard an oddly intermittent
signal. Instead of breaking off once and cleanly,
it broke off many times. Crossing back over, he
heard it again. Soon he was in a tight turn, now
hearing something, now nothing, in a pattern
that did not suggest anything he had heard
before. It did, however, suggest the interstate.
Where a big green sign says, "MILFORD 11, PORT
JERVIS 20," Gary hunted around and found the
bear. He took us now to see the den. We went
down a steep slope at the side of the highway
and, crouching, peered into a culvert. It was
about fifty yards long. There was a disc of daylight at the opposite end. Thirty inches in diameter, it was a perfect place to stash a body, and
that is what the bear thought, too. On Gary's
first visit, the disc of daylight had not been visible. The bear had denned under the eastbound

lanes. She had given birth to three cubs. Soon after he found her, heavy rains were predicted. He hauled the family out and off to a vacant den. The cubs weighed less than a pound. Two days later, water a foot deep was racing through the culvert.

Under High Knob, in remote undeveloped 13 forest about six hundred meters above sea level, a slope falling away in an easterly direction contained a classic excavated den: a small entrance leading into an intimate ovate cavern, with a depression in the center for a bed — in all, about twenty-four cubic feet, the size of a refrigerator-freezer. The den had not been occupied in several seasons, but Rob Buss, a district game protector who works regularly with Gary Alt, had been around to check it three days before and had shined his flashlight into a darkness stuffed with fur. Meanwhile, six inches of fresh snow had fallen on High Knob, and now Alt and his team, making preparations a short distance from the den, scooped up snow in their arms and filled a big sack. They had nets of nylon mesh. There was a fifty-fifty likelihood of yearling bears in the den. Mothers keep cubs until their second spring. When a biologist comes along and provokes the occupants to emerge, there is no way to predict how many will appear. Sometimes they keep coming and coming, like clowns from a compact car. As a bear emerges, it walks into the nylon mesh. A drawstring closes. At the same time, the den entrance is stuffed with a bag of snow. That stops the others. After the first bear has been dealt with, Alt removes the sack of snow. Out comes another bear. A yearling weighs about eighty pounds, and may move so fast that it runs over someone on the biological team and stands on top of him sniffing at his ears. Or her ears. Janice Gruttadauria, a research assistant, is a part of the team. Bear after bear, the procedure is repeated until the bag of snow is pulled away and nothing comes out. That is when Alt asks Rob Buss to go inside and see if anything is there.

Now, moving close to the entrance, Alt 14 spread a tarp on the snow, lay down on it, turned on a five-cell flashlight, and put his head inside the den. The beam played over thick black fur and came to rest on a tiny foot. The sack of snow would not be needed. After drugging the mother with a jab stick, he joined her in the den. The entrance was so narrow he had to shrug his shoulders to get in. He shoved the sleeping mother, head first, out of the darkness and into the light.

While she was away, I shrugged my own 15 shoulders and had a look inside. The den smelled of earth but not of bear. The walls were dripping with roots. The water and protein metabolism of hibernating black bears has been explored by the Mayo Clinic as a research model for, among other things, human endurance on long flights through space and medical situations closer to home, such as the maintenance of anephric human beings who are awaiting kidney transplants.

Outside, each in turn, the cubs were put in 16 the stuff sack — a male and a female. The female weighed four pounds. Greedily, I reached for her when Alt took her out of the bag. I planted her on my shoulder while I wrote down facts about her mother: weight, a hundred and ninety-two pounds; length, fifty-eight inches; some toes missing; severe frostbite from a bygone winter evidenced along the edges of the ears.

Eventually, with all weighing and tagging 17 complete, it was time to go. Alt went into the den. Soon he called out that he was ready for the mother. It would be a tight fit. Feet first, she was shoved in, like a safe-deposit box. Inside, Alt tugged at her in close embrace, and the two of them gradually revolved until she was at the back and their positions had reversed. He shaped her like a doughnut — her accustomed den position. The cubs go in the center. The male was handed in to him. Now he was asking for the female. For a moment, I glanced around as if looking to see who had her. The thought crossed my mind that if I bolted and ran far enough and fast enough I could flag a passing car and keep her. Then I pulled her from under the flap of my vest and handed her away.

Alt and others covered the entrance with 18 laurel boughs, and covered the boughs with snow. They camouflaged the den, but that was not the purpose. Practicing wildlife management to a fare-thee-well, Alt wanted the den to be even darker than it had been before; this would

6
ARGUMENT

Argument, in a sense, underlies all writing. In expressing your ideas, no matter what method or combination of methods you use, you are "arguing" some point about your subject. For example, in Chapter 5, each of Jane's writing ventures attempts to demonstrate a particular thesis: whales can be understood best when seen at sea; whales may seem threatening, but are really gentle; whale watching has helped increase public support for whale preservation. Jane uses these informal arguments to advance her primary purpose — explaining her impressions of a whale watch.

Formal arguments, too, commonly employ expository patterns to convey information. But since their primary purpose is to convince others — to accept a proposal, to challenge a situation, to support some cause — formal arguments must be developed according to rules of evidence and logical reasoning. These demands may seem formidable, but in many ways they are simply extensions of what you have learned about those elements that govern the writing process.

You begin composing an argument by planning, by investigating a variety of sources so that you can discover a subject. Next you organize your argument into a draft so that it most effectively and fully expresses your subject to persuade your audience. Finally, you examine your text to revise those features that obscure the clarity or weaken the credibility of your purpose.

As you move through these stages, you will expand your appreciation of other arguments (particularly those of the opposition), develop your authority to present those features of your argument that matter most, and increase your understanding of those aspects of your argument or your opponent's argument that might never be resolved or that might be negotiated.

INVESTIGATING THE ARGUMENT: PLANNING

In planning an argument you *select a subject, collect and use evidence,* and *consider the opposition.*

Selecting a Subject

Not every subject can be treated in a formal argument. Common sense tells you that there is no point in arguing about *facts.* You may like to argue about when the Red Sox last won the World Series or how many times Katharine Hepburn has won an Oscar, but such trivial disputes can be settled quickly by looking up the correct answer in an almanac. You may like to argue that chocolate chocolate chip is the best flavor of ice cream in the world or that all politicians are corrupt — but these assertions are merely *opinions,* based on preference (or prejudice) rather than on evidence.

A formal argument must focus on a subject that can be debated. Such subjects are open to interpretation because opposing opinions can be supported with evidence and the audience is free to consider the opposing claims and to choose sides.

As you think about possible subjects for an argument, review "Guidelines for Selecting Your Subject" in Chapter 1 (p. 12). Your subject should be one you know something about, consider interesting and significant, and can restrict to a manageable size. When you develop your subject into an argument, you will need to view it as a controversy with at least two contending sides. What you know about a subject (or what you find out about it) will lead you to take a side. But to convince your readers that yours is the right side (or, at least, the better side), you will need to support your position.

Exercise

Matt is an older student who, after working for several years in a factory, has returned to college to study pre-law. In his business law course, he is asked to write an argument about a current controversy in the work place. Read his list of potential subjects and identify those that are (a) facts, (b) opinions, (c) open to debate. Explain your choices.

1. Unions have ruined free enterprise in America.

2. Random drug testing of workers is an invasion of privacy.

3. Health-care costs for alcoholics and their families are double the costs for nonalcoholics and their families.

4. Workers on the day shift are more conscientious than those on the night shift.

5. Less than 10 percent of the upper management of America's major corporations belongs to a minority group.

6. The recreational use of drugs and alcohol outside the work place does not affect workers' performance on the job.

7. Drug tests are too unreliable to form the basis of company policy.

Collecting and Using Evidence

Once you have selected a subject, start gathering evidence. You might begin by making lists, drawing clusters, or doing some freewriting in your journal. Such activities will help you identify what you already know (or think you know) about your subject, but eventually you will need to expand and deepen your knowledge by conducting some research. Anna Quindlen, a columnist for the *New York Times,* suggests that an effective investigation requires several planning strategies:

> I begin by reading, reading, reading. I have to understand my subject before I interview people, otherwise I won't understand the terms they use or the issues they raise. Also, I have to understand where they're coming from. You need to watch out for those red flags, those people on one side of an argument who distort issues by making highly inflammatory statements. Sometimes you can catch their bias when you ask other people to comment on their remarks. So much of reporting is bouncing from one person to another, trying to get all sides of the story. You never get it all. You simply run out of time. (Personal interview)

Whatever planning strategy you use, follow Quindlen's advice and collect information from all sides of your subject, not just your side. The most common kinds of evidence are *facts, judgments,* and *testimony.*

Facts   Facts are a valuable ally in building an argument because they cannot be debated. It is a fact that the stock market crashed on October 29, 1929. It is a fact that the Dow Jones Industrial Average first exceeded 2,000 on January 8, 1987. But not all facts are so clear-cut; some may require explanation. And some statements that look like facts may not be facts. A stock analyst who announces a company's projected earnings for the next five years is making an estimate, not a statement of fact.

Judgments   Judgments are conclusions inferred from facts. Unlike opinions, judgments lend credibility to an argument because they result from careful reasoning. A doctor considering a patient's symptoms reaches a tentative diagnosis of either tuberculosis or a tumor. If a laboratory test eliminates tuberculosis, then the patient probably has a tumor that is either malignant or benign, questions that can be settled by surgery. The doctor's judgment emerges from the following procedure:

1. *The patient's symptoms are caused by either tuberculosis or a tumor.*
   This diagnosis is a judgment based on knowledge of the symptoms of or facts about the two conditions.

2. *The patient does not have tuberculosis.*
   This is a judgment based on the results of a laboratory test, that is, a fact.

3. *The patient has a tumor.*
   This judgment is inferred from the facts of 1 and 2.

4. *The patient does not have cancer.*
This judgment is determined after surgery, which reveals the fact that the tumor is benign.

Testimony Testimony affirms or asserts facts. A person who has had direct experience (an *eyewitness*) or who has developed expertise in a subject (an *expert witness*) can provide testimony based on fact, judgment, or both. An eyewitness is asked to report facts, as when an observer reports seeing a man drown in a strong current. An expert witness is asked to study facts and render a judgment, as when a coroner reports that an autopsy has shown the the victim did not drown but died of a heart attack.

Both kinds of testimony can constitute powerful evidence. Eyewitness testimony provides authenticity. Expert testimony provides authority. Each has its limitations, however. An eyewitness is not always trustworthy: eyewitness testimony can be distorted by faulty observation or biased opinion. An expert witness is not infallible or always unbiased: expert testimony, though often difficult for the nonexpert to challenge, can be disputed by other experts employing a different method of investigation. Each type of testimony can be abused. An eyewitness account of an event may be convincing, but it should not be used to draw parallels to unrelated events. And expert credentials in one field, whatever eminence they convey, do not automatically carry over to other (even related) fields.

Using the Evidence Evaluate your evidence by determining whether it is *pertinent, verifiable,* and *reliable.* A stock analyst who uses the success of the polio vaccine as a reason for investing in a drug company researching a vaccine for the common cold is not presenting evidence *pertinent* to the argument. A historian who claims that Amelia Earhart's flying ability was impaired by Alzheimer's disease is using an argument that is not *verifiable.* And an attorney who builds a case on the eyewitness testimony of a person who has been arrested several times for public intoxication is not using the most *reliable* evidence. (See "Evaluating Your Sources," Chapter 13).

As you collect your evidence, be careful to identify and record its source. Your readers will want to know where you found the information important to your argument. If you cannot identify your source, your reader may question your knowledgeability. If your information comes from dubious sources, your readers may doubt your credibility. Identify information from printed and other sources in the text of your paper. (See "Documenting Your Sources," Chapter 13.)

■■■■ *Exercise*

Matt has decided to write about drug testing in the work place. During his last year at the factory, the subject became a hot issue in contract negotiations: management proposed random drug testing on the job; the union opposed all forms of drug testing. As he investigated the subject, Matt gathered evidence from many sources. Examine these items from his notes to determine for each: (a) whether it is a fact, a judgment, or eyewitness or expert testimony, and (b) whether it is likely to be *pertinent, verifiable,* and *reliable* evidence.

1. *"According to the Research Triangle Institute, a respected North Carolina business-sponsored research organization, drug abuse cost the United States economy $60 billion in 1983, or nearly 30% more than the $47 billion estimated for 1980."* (Janice Castro, "Battling the Enemy Within," Time *17 March 1986: 53.)*

2. *"Right now, if the foreman doesn't like somebody, he can ask him to take a drug test — claiming probable cause. That makes the guy look suspicious. Even though Fred's test came back clean, everybody said, 'Well, there must be something in it.' That's not fair."* (Factory worker, personal interview, 10 November 1986.)

3. *"Employers have always had rules to govern the work place. You can't be late or absent without just cause. There are some capital offenses that are just cause for firing. You can't be intoxicated on the job. Your work is impaired, you jeopardize your co-workers, and you destroy the reliability of the company's product. Testing for drug abuse is just another way to ensure safety in the work place."* (Chief counsel at local factory, personal interview, 15 November 1986.)

4. The Centers for Disease Control reports that in the thirteen laboratories they studied, there was *"up to 67% error rate in false positive identification of drugs."* (Hugh J. Hansen, "Crisis in Drug Testing," Journal of American Medical Association, *April 1985: 2382.)*

5. *"An estimated 45% of the Fortune 500 corporations will be involved in drug testing next year."* (Fern Schumer Chapman, "The Ruckus Over Medical Testing," Fortune, *19 August 1985: 58.)*

6. *"Drug abuse is hard to detect. There's no odor. People seem to act normally. Management wants to use this urine test as a cheap way to see if they have a problem. What if they do? Are they going to invest in a rehabilitation program or just fire people?"* (Local union leader, personal interview, 12 November 1986.)

7. *"It costs over $10,000 to hire and train even an entry level employee. For the average employee this cost exceeds $25,000. Before I terminate an employee for drug abuse I need to be confident that he or she is really a drug abuser. I can't afford to fire a productive employee on the basis of a test that isn't much better than flipping a coin."* (Lewis L. Maltby, "An Employer's Perspective," The Drug Testing Debate: Remedy or Reaction, *Washington, D.C.: American Civil Liberties Union (1986): 4.)*

████████

Considering the Opposition

If you have selected a debatable subject and collected a broad range of evidence, you have seen that controversies are tricky. You may remain convinced that your side is the right side; you may be tempted, given your discovery of new information, to change sides; or you may feel confused as you try to balance the evidence on both sides of the debate. To complete your investigation and test your original position (or to write your way out of your current confusion), you must give serious consideration to the opposition.

Assume the identity of your opposition and draft an argument entitled "Their Side." This planning exercise will enable you to understand the important evidence that makes your opponents' position credible — evidence that you may have to concede when you present your side of the argument. The exercise will also help you discover the weak spots in their argument — points that you will have to disguise when you write "Their Side," but points that you can dispute when you write "My Side." And finally, this exercise will encourage you to identify the points, often overlooked in the heat of argument, on which both sides agree.

If you are going to compose an effective "Their Side," you have to be *knowledgeable* and *fair.* You must present your opponents' best argument in such a way that they will recognize it as a thorough and accurate statement of their position. If you set up a "straw man," presenting your opponents' weakest evidence in the most unfavorable light with the intention of knocking it down, the exercise will not be useful to you. Knowledge of your opponents' best argument and a sincere effort to avoid bias will help you avoid an unfair, unproductive effort. Common signs of such unfairness are *distortion, slanting,* and *quoting out of context.*

Distortion   You distort evidence if you intentionally exaggerate your opponents' views. Councilman Jones supports Planned Parenthood because the agency provides valuable information on birth control. Councilman Smith misrepresents his opponent's position by claiming that his support of Planned Parenthood constitutes an endorsement of extreme forms of birth control such as abortion.

Slanting   If you select facts favorable to your position and suppress those unfavorable to it, you are slanting the evidence. A business executive who says there can be no real poverty in a country where the average annual income is $10,000 ignores two facts: (1) the average includes incomes of $1 million or more, and (2) a great many incomes fall far below the average.

Quoting Out of Context   If you remove words deliberately (or carelessly) from their original context and reuse them in a new context that changes their meaning, you are quoting out of context. In reviewing a play, a critic writes: "The plot is fascinating in a strange way: you keep waiting for something to happen, but nothing does. The characters never come close to greatness, and the few witty lines are out of place among the predictable dialogue." An advertisement based on this review quotes out of context if it reads: "Fascinating plot . . . characters close to greatness . . . witty lines."

■ *Exercise*

When Matt began his investigation, he was opposed to drug testing in the work place. As a former factory worker, he resisted management's inclination to establish arbitrary rules to control employees. As a pre-law student, he was sensitive to policies that disregarded the rights of the individual. But the more research he collected on drug testing, the more he came to acknowledge the legitimate claims of the other side. Although he still opposed drug testing, he tried to assess the power of the opposition by composing a "Their Side" exercise. Discuss whether *those in favor of drug testing* would see Matt's presentation of their position as knowledgeable and fair: (1) What types of evidence does he present? (2) Does this evidence seem to be the best evidence — most *pertinent, verifiable, reliable?* (3) In what sections of the essay might Matt be accused of misrepresenting the opposing position?

Their Side

On January 4, 1987, an Amtrak train traveling north of Baltimore collided with three Conrail engines, causing the worst accident in Amtrak's history. Sixteen passengers were killed, 170 were injured, and five locomotives from the two trains were destroyed. Investigators have since discovered that the crew operating the Conrail engines had been using marijuana.

The Amtrak disaster near Baltimore is not an isolated example. Every day news stories report truck drivers in drug-related accidents, air-traffic controllers impaired on the job by drug abuse, and assembly-line workers selling and using drugs in the work place. Dr. Howard Franbrel, medical director at Rockwell, estimates that "20% to 25% of the workers at the plant responsible for the final assembly of the space shuttle were high on drugs, alcohol or both" (Castro 53). The Research Triangle Institute in North Carolina reports that "drug abuse cost the United States economy $60 billion in 1983, or nearly 30% more than the $47 billion estimated for 1980" (Castro 53).

Employers have an obligation to protect the safety of workers, the efficiency of the work place, and the reliability of their products. Workers on drugs are less productive, more likely to harm themselves or their co-workers, and more likely to steal from their company. Because they are more susceptible to illness, they drive up the premiums of health insurance. And because their errors create flawed products or unreliable services, they increase the probability of lawsuits and the cost of liability insurance.

Fortunately there is a solution to this growing problem: drug testing. Drug testing, particularly urine analysis, is now being used by a large number of corporations to protect the safety of their employees and the investment of their stockholders. An estimated 45% of Fortune 500 companies will be involved in drug testing by next year (Chapman 57).

There are three basic methods of drug testing currently in use: pre-screening, probable cause testing, and random/massive testing. Most employers require a physical examination for all job applicants. The additional cost of a pre-employment urine analysis is minimal and well worth the expense since employers can thus identify drug users before they are hired, trained, and become a problem.

The probable cause method is used when a worker's performance provides sufficient reason to justify a test. Drug abuse is more difficult to detect than alcohol abuse: there is no obvious smell, slurred speech, or red eyes. But there are signs of possible drug abuse: excessive absenteeism, inconsistent work performance, deterioration of personal appearance, severe financial problems, and increased trips to the rest room.

In random or massive testing any part or all of the work force is tested to spot abusers. This is the most costly and controversial of the three methods. Weekly or monthly testing of an entire work force is an enormous investment, but given the pervasiveness of drugs in our society such an investment may prevent a costly accident. There are those who argue that without probable cause, such testing is an invasion of workers' privacy. But if workers are drug free, then requiring such a test is no more an invasion of privacy than requiring protective masks or helmets. If workers are users, then they need to be placed in a rehabilitation program before they endanger themselves or their co-workers.

Employers are in business to make money. To do so, they need a dependable work force. In 1982, an Illinois Central Gulf freight train carrying hazardous chemicals derailed at Livingston, Louisiana, because the engineer, affected by drugs, had fallen asleep. In 1984, a spectacular rear-end collision of two trains near Newcastle, Wyoming, occurred while six of the twelve crew members were using drugs. In July 1984, two Amtrak trains collided in Queens, killing one and injuring 125; traces of marijuana and cocaine were later found in the urine of the track signal-operator. To prevent disasters like these and the one near Baltimore, American industry must adopt a more aggressive drug-testing program.

ORGANIZING THE ARGUMENT: DRAFTING

Once you have completed your preliminary investigation, you need to develop, organize, and draft your argument, using information you learned in Chapters 3, 4, and 5. Anna Quindlen points out that organizing an effective argument is terribly important and terribly difficult:

Structure is a slippery subject. I usually like to lead with an anecdote, particularly if I can use a quote that works like a punch in the stomach. Also, up front, in what I call my *nut paragraph,* I like to spell out the issues I've discovered. "Where are we going? Where have we been? Have we gone too far?" Then I'll get into an "on the one hand, on the other hand" pattern where I present the sides in the debate. But in a true controversy, the debate is taking place on so many levels — emotional, ethical, social — that I can't choose sides. I switch back and forth and usually wind up somewhere in the middle. (Personal interview)

The final shape of an argument depends, as Quindlen suggests, on a series of interrelated decisions. To make *informed* decisions, you need to *analyze the audience, arrange the evidence,* and *monitor the appeals.*

Analyzing the Audience

In preceding chapters you have learned that knowing your audience helps you to discover and demonstrate your purpose. This is particularly true in argument because you are attempting to convince your audience to believe in your purpose. To do this you must assess what potential audiences are likely to know about your subject, what they believe is important in the controversy, and what kind of evidence will influence their judgment.

Use the advice below, on considering the audience for argument, in conjunction with the "Guidelines for Analyzing Your Audience" in Chapter 1 (p. 15).

Identify Specific Readers You need to identify the specific groups of readers who form the potential audience for your argument. Some may seem in the thick of things — friendly or hostile to your position. Others may seem on the edge — skeptical about any proposal or merely uncommitted. Unfortunately, the sides in controversies are rarely that clear-cut, and your potential audience usually contains many people with uncertain or conflicting views. Identify these specific groups of potential readers for your argument, and acknowledge their conflicting opinions and anticipate their various objections in your draft.

Identify with Your Readers The "Their Side" exercise helped you see the debate from the perspective of your opponent. Now, as you identify other readers who have only partial or passing interest in the controversy, place yourself in those positions and assess the issues from those viewpoints. Who are the friendlies? What do they know? Who are the skeptics? What makes them resist? Who are the uncommitted? What will make them decide? Such imaginative identification helps you establish the points of conflict and areas of agreement among the various readers in your audience.

Establish Your Identity for Your Readers In drafting your argument, you inevitably establish an identity (or a *persona*) that reveals to your readers your attitude toward your subject. You should be alert to strategies that strengthen your bond with your readers and avoid tactics that distance you from them. As the "Their Side" exercise helped show, the most effective arguments consider all sides of a controversy. In arguing, your identity should be that of the "reasonable citizen," who follows a balanced discussion with a thoughtfully rendered judgment. You must consider your readers, too, to be reasonable citizens. However indignant you are about the situation you are writing about, you cannot be indignant with your readers — not if you want their agreement. You can concede or refute certain points, but you cannot flatter or talk down to your readers, or neglect to provide logical proof of your argument — not if you want their respect. You are trying to establish a partnership of mutual inquiry. Anything that distorts your identity in that partnership will destroy your credibility.

Provide Your Readers With Appropriate Evidence   If you respect your readers, you will not ask them to accept unsupported assertions. A business executive would not ask her partners to accept her unsupported word that investing in another company will be profitable. She would provide detailed evidence for their review and would address the particular concerns (management, labor, debt, productivity, competition) of individual partners with appropriate, specific evidence. She may believe the investment is sound, but if she wants to convince her partners, she must supply the right kind of evidence. As a writer of argument, you have an obligation to spell out in detail why you think your readers should accept your conclusion. And you must provide the specific kinds of evidence your readers will find compelling.

Draft an Argument Your Readers Can Read   You are asking your readers to agree with you, so make their job as easy as possible. Highly complex arguments, technical terminology, abstract diction, confusing statistics — all these make communication difficult. You want to convince your readers that you are knowledgeable, but you do not want to clutter your argument with arcane or unnecessary information that confuses and annoys them. Anticipate your readers' needs and try to fulfill them to achieve the response you seek.

▮▮▮▮ *Exercise*

As Matt thinks about drafting his argument about drug testing in the work place, he identifies the following specific readers as potential members of his audience. Examine his list. Has he overlooked any group that has a stake in the controversy? Identify readers who might belong to more than one group. Then discuss how each group of readers sees the issue.

Employer	Competitors (other employers)
Worker (drug free)	Health insurance claims agent
Worker (abuser)	Potential employee
Drug rehabilitation counselor	Stockholder

Medical laboratory analyst

Union negotiator

Company attorney

Consumer

Liability insurance executive

American Civil Liberties Union representative

Arranging the Evidence

Since every controversy creates its own problems and possibilities, no one method of arrangement will always work best. Sometimes you may even have to combine methods to make your case. To make an informed decision, you need to consider how you might adapt your evidence to the three common methods of arrangement: *induction, deduction,* and *accommodation.*

Induction   Often called the scientific method, induction begins by presenting specific evidence and then moves to a general conclusion. This arrangement reflects the history of your investigation. You began your research with a hypothesis about what you wanted to find out and where you needed to look. You collected a cross section of examples until a pattern emerged. At this point, you made what scientists call an "inductive leap": you determined that although you had not collected every example, you had examined enough examples to risk proposing a probable conclusion.

To incorporate this arrangement into a draft, begin by posing the question that prompted your research:

Question

Why is our company losing so many valuable computer operators to other companies?

Then arrange your individual pieces of evidence in such a way that they help your readers see the pattern you discovered. You need not list all the false leads or blind alleys you encountered along the way, unless they changed your perspective or confirmed your judgment.

Evidence

1. Most computer operators are women who have preschool children. (Provide examples.)
2. A nearby day-care center used by employees has closed because it lost federal funding. (Provide examples.)
3. Other day-care centers in the area are inconvenient and understaffed. (Provide examples.)
4. Other companies provide on-site day care for children of employees. (Provide examples.)
5. On-site day care is beneficial to the emotional well-being of both preschool children and their mothers, because of the possibility of contact during the workday.

Finally, present the conclusion that seems warranted by the evidence you have presented.

Conclusion

Therefore, our company needs to provide on-site day care to retain valuable employees.

Deduction   Usually identified with classical reasoning, deduction begins with a general statement or *major premise* that, when restricted by a *minor premise,* leads to a specific conclusion. Unlike induction, which in theory makes an assertion only in its conclusion, deduction does make initial assertions (based on evidence) from which a conclusion is derived.

To use this organization in your draft, consider the pattern of this three-step syllogism.

Major premise

Retention of computer operators who have preschool children is promoted by on-site day care.

Minor premise

Our company wants to retain computer operators who have preschool children.

Conclusion

Our company should establish on-site day-care centers.

To gain your audience's acceptance of your major and minor premises, you must support each assertion with specific evidence. Demonstrate that retaining computer operators who have preschool children is promoted by on-site day-care centers and that "our company" wants to retain computer operators who have preschool children. If your readers then accept your premises, they are logically committed to accept your conclusion.

Accommodation   Sometimes called "nonthreatening argument," accommodation arranges evidence so that all parties believe their position has received a fair hearing. Induction reveals how a chain of evidence leads to a conclusion. Deduction demonstrates why certain premises demand a single conclusion. Although both procedures work effectively in specific situations, they occasionally defeat your purpose. Readers may feel trapped by the relentless march of your argument; though unable to refute your logic, they are still unwilling to listen to reason. Accommodation takes this into account. Instead of trying to win the argument, you try to improve communication and increase understanding.

To employ this strategy, begin by composing an objective description of the controversy: women computer operators who have preschool children are leaving the company. Then draft a complete and accurate statement of the contending positions, supplying the evidence that makes each position credible.

Corporation board

We need a qualified work force, but we are not in business to provide social services. (Provide evidence.)

Fellow workers (single, male, etc.)

We understand their problem, but providing an on-site day-care center is giving expensive, preferential treatment to a small segment of the work force. (Provide evidence.)

We need better computer operators if we are going to compete, and we will provide what is necessary to hire them. (Provide evidence.)

Next show where and why you and the various parties agree: the corporation should not be in the day-care business; providing a center is preferential treatment; women computer operators have the right to market their skills in a competitive market. Then present your own opinion, explaining where it differs from other positions and why it deserves serious consideration: we have invested a large amount of money in training our work force; child care is an appropriate investment in view of the long-term contribution these people will provide for the corporation. Finally, present a proposal that might resolve the issue in a way that recognizes the interests of all concerned: suggest that the corporation help fund the nearby day-care center that was previously supported by government money.

■■■ *Exercise*

Matt decides to make a scratch outline to organize his evidence *against* drug testing. What method of arrangement does he use? What sections of his outline suggest the potential for combining several methods?

Drug Testing in the Work Place

I. *Drug testing is unreliable*
   Cheap screening tests
   Lab error — "sink tests" (dumping samples)
   "Black market" for urine samples
   "Chain of custody" for urine samples
   Cross reactivity — false positives from other chemicals
   Expensive confirmatory tests

II. *Drug testing is counterproductive*
   Decline in workers' morale
   Potential for discrimination
   Constant surveillance — time consuming
   Worker paranoia — worry about behavior, possibility of false positives
   Supervisors play cop — loss of rapport, cooperation with workers
   Status of productive but rehabilitated worker?
   Class bias — rehabilitate executive; fire janitor

III. *Drug testing is an invasion of employees' privacy*
   Presumption of guilt
   Humiliating test — objective observer
   Confidentiality of test results — liability?
   Control life style outside work place
   Dangerous precedent — genetic testing for future health risks (long-term illness, insurance premiums)

███████████

Monitoring the Appeals

In organizing your draft you must keep track of how you are using the three basic appeals of argument: the *emotional* appeal, the *ethical* appeal, and the *logical* appeal. These three appeals are rarely fully separate; they all weave in and out of virtually every argument. But to control their effects to your advantage, you must know when and why you are using them.

The Emotional Appeal   Readers feel as well as think, and to be thoroughly convinced they must be emotionally as well as intellectually engaged by your argument. Some people think that the emotional appeal is suspect; because it relies on the feelings, instincts, and opinions of readers, they link it to the devious manipulations of advertising or politics. The emotional appeal is often used to stampede an audience into thoughtless action, but such abuses do not negate its value. The emotional appeal should never replace more rational appeals, but it can be an effective strategy for convincing your readers that they need to pay attention to your arguments.

The greatest strength of the emotional appeal is also its greatest weakness. Dramatic examples, presented in concrete images and connotative language, personalize a problem and produce powerful emotions. Some examples produce predictable emotions: an abandoned puppy or a lonely old woman evokes pity; a senseless accident or recurring incompetence evokes anger; a smiling family or a heroic deed evokes delight. Some examples, however, produce unpredictable results, and their dramatic presentation often works against your purpose. It would be difficult to predict, for instance, how all your readers would respond to a working mother's fears for and guilt about her preschool child in day care. Some might pity her, while others might disdain her inability to solve her problems. Since controversial issues attract a range of passions, use the emotional appeal with care.

The Ethical Appeal   The character (or *ethos*) of the writer, not the writer's morality, is the basis of the ethical appeal. It suggests that the writer is someone to be trusted, a claim that emerges from a demonstration of competence as an authority on the subject under discussion. Readers trust a writer such as Anna Quindlen and are inclined to agree with her arguments because she has established a reputation for informed, reasonable, and reliable writing about controversial subjects.

You can incorporate the ethical appeal into your argument either by citing authorities, such as Quindlen, who have conducted thorough investigations of your subject, or by following the example of authorities in your competent treatment of evidence. There are two potential dangers with the ethical appeal. First, you cannot win the trust of your readers by citing as an authority in one field someone who is an authority in another. Lee Iacocca is an established authority

on building cars, but he should not be cited as an authority on nuclear disarmament. And second, you cannot convince your readers that you are knowledgeable if you present your argument exclusively in personal terms. Your own experience in trying to find reliable, convenient day-care services may make your argument for providing on-site day care more forceful, but you need to consider the experience of other people to establish your ethical appeal.

The Logical Appeal   The rational methods used to develop an argument constitute a logical appeal. Some people think that the forceful use of logic, like the

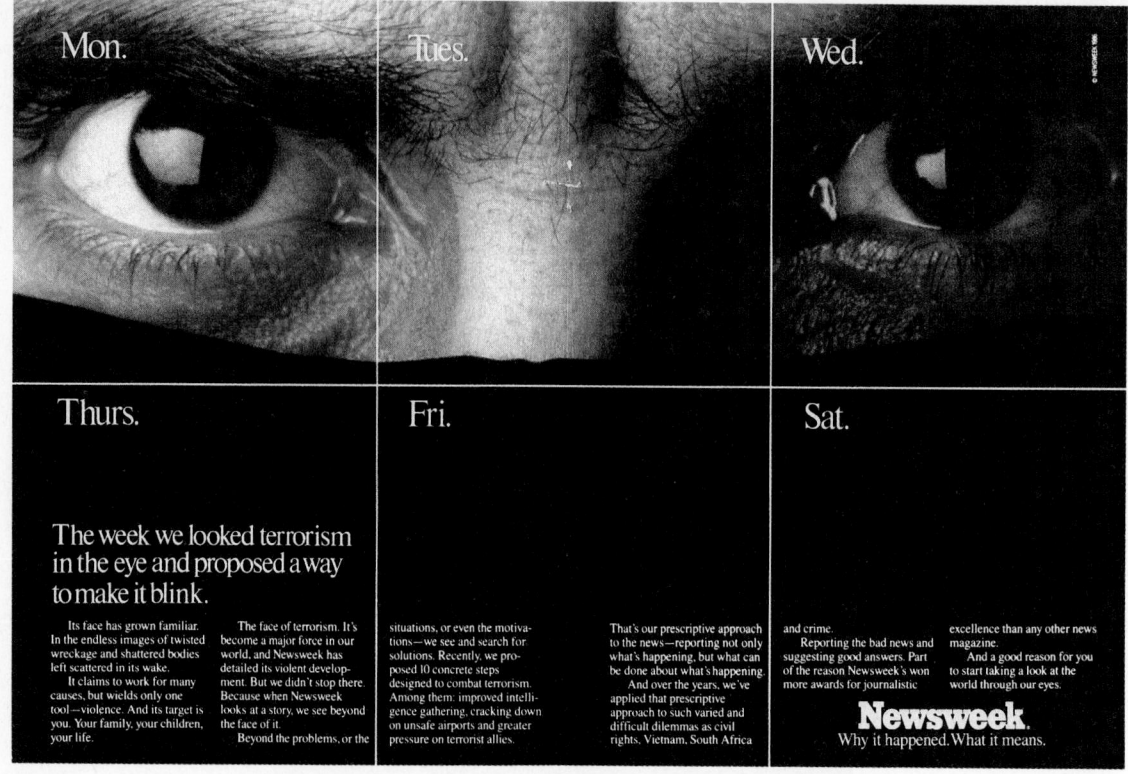

Mon.   Tues.   Wed.

Thurs.   Fri.   Sat.

The week we looked terrorism in the eye and proposed a way to make it blink.

Its face has grown familiar. In the endless images of twisted wreckage and shattered bodies left scattered in its wake.

It claims to work for many causes, but wields only one tool—violence. And its target is you. Your family, your children, your life.

The face of terrorism. It's become a major force in our world, and Newsweek has detailed its violent development. But we didn't stop there. Because when Newsweek looks at a story, we see beyond the face of it.

Beyond the problems, or the situations, or even the motivations—we see and search for solutions. Recently, we proposed 10 concrete steps designed to combat terrorism. Among them: improved intelligence gathering, cracking down on unsafe airports and greater pressure on terrorist allies.

That's our prescriptive approach to the news—reporting not only what's happening, but what can be done about what's happening.

And over the years, we've applied that prescriptive approach to such varied and difficult dilemmas as civil rights, Vietnam, South Africa

and crime.

Reporting the bad news and suggesting good answers. Part of the reason Newsweek's won more awards for journalistic

excellence than any other news magazine.

And a good reason for you to start taking a look at the world through our eyes.

Newsweek.
Why it happened. What it means.

precise use of facts, makes an argument absolutely true. But controversies contain many truths, no one of which can be graded simply true or false. By using the logical appeal, you acknowledge that arguments are conducted in a world of probability, not certainty; by offering a series of reasonable observations and conclusions, you establish the most reliable case.

The logical appeal is widely used and accepted in argument. Establishing the relationships that bind your evidence to your proposition engages your readers' reasoning power, and an appeal to their intelligence and common sense is likely to win their assent. But the logical appeal is not infallible. Its limit is in acknowledging limits: How much evidence is enough? There is no simple answer to this question. The amount of evidence required to convince fellow workers that your company should provide on-site day care may not be sufficient to persuade the company's board of directors. On the other hand, too much evidence, however methodically analyzed, may win the argument but lose your audience. Without emotional or ethical appeal, your "reasonable" presentation may be put aside in favor of more urgent issues. Accurate and cogent reasoning is the basis for any sound argument, but the logical appeal, like the emotional and ethical appeals, must be monitored carefully during drafting to make sure it is accomplishing your purpose.

◼ *Exercise*

In his ongoing investigation of drug testing in the work place, Matt has collected a great deal of evidence, considered the position of his opponents, and analyzed the claims of his various readers. Now he wants to turn off these other voices and rediscover his own: he is ready to write "My Side" of the argument. As you read his discovery draft, identify the various appeals he has used to accomplish his purpose. Note the places that were probably strengthened by his decision to write "Their Side" first. Finally, point to the sections that would have been weaker had he written "My Side" first.

My Side

The increased public awareness of drug abuse in our society has created a sense of public alarm about the potential dangers of drug abuse in the work place. Although such concern is justifiable, particularly when it affects the safety of workers and consumers, most companies are trying to solve this complex problem with a simple urine test. Whether it is used to weed out job applicants or identify suspected abusers on the job, urine testing is a misguided attempt to find a quick fix. The test is not only unreliable and counterproductive, it is a serious invasion of workers' privacy.

The biggest argument for urine testing is that it is a cheap and efficient way to spot abusers. It is certainly cheap. Many test kits cost less than ten dollars. But they are hardly reliable. In order to perform an effective drug test, urine samples must be taken according to strict procedures, to guarantee the right kind of sample and to protect against sample switching from the rapidly developing "black market" for clean urine. When the test is taken at the job site, this "chain of custody" is often broken, increasing the opportunity for error and fraud.

The laboratories that analyze these samples are also suspect. Many of them have been set up to make a fast buck off the high demand for drug testing. The Centers for Disease Control discovered that in the thirteen laboratories they studied, there was "up to 67% error rate in false positive identification of drugs" (Hansen 2382). Some of these labs reported false negatives 100% of the time when they tested for certain drugs. Dr. Joe Boone, chief of CDC's clinical chemistry and toxicology section, reports that "if these labs would have dumped the samples down the sink and tossed a coin, they would have come up with the same reliability in their test results" (Chapman 57).

To even approach reliability, an initial urine screening needs to be checked by a second, more expensive "confirmatory test." This test attempts to offset the problem of "cross-reactivity." According to David Greenblatt, chief of clinical pharmacy at Tufts New England Medical Center, "many chemicals in the body, such as caffeine, cough syrup, or antiasthmatic medication can throw off findings" (Chapman 59). Employers who do not invest in a confirmatory testing procedure will never know whether job applicants or employees who tested positive were abusing drugs or following doctor's orders.

Questions about the reliability of drug testing contribute to a decline in workers' morale. Rather than ensuring productivity, urine testing has created a climate of paranoia. Workers distrust the motives of their employers and worry about the potential for discrimination. If they become too emotional, make too many trips to the rest room, or behave in any way that is viewed as suspicious, they could be asked to take a urine test. Even if the results are negative, they are "on record" as former suspects. Supervisors, who should focus their attention on workers' productivity, are not responsible for worker surveillance. They may know little about the symptoms of drug abuse, but they are expected to look for telltale signs. By playing super-cop, they destroy their rapport with their workers.

A more serious threat to workers' morale emerges from the debate about the purpose of drug testing. Does the company want to pacify its stockholders, protect its workers, or identify abusers? More to the point, what is the fate of those who are identified as abusers? Are they replaced or rehabilitated? If the latter, who pays for the rehabilitation? Workers' morale is certainly not bolstered by rumors that executives are counseled in rehabilitation programs paid for by company insurance, whereas workers are quickly fired.

The most significant objection to urine testing, particularly random/massive testing, is that it is an invasion of workers' privacy. The test may seem simple enough, like other tests in a routine physical, but the legal and ethical consequences are hardly routine. First, the institution of the test raises serious concerns regarding an employer's right to control an employee's activities off duty. Second, the test presumes the guilt of all workers until the "results" establish their innocence. Third, these highly debatable results may ruin the career of a valuable employee who has no way to appeal a decision or reform his or her behavior. And finally, such testing sets a dangerous precedent. There is already significant corporate support for genetic testing — a test that determines an employee's predisposition toward such "costly" diseases as diabetes and cancer (Chapman 60).

There is reason for the public to be alarmed about drug abuse in the work place, but the evidence on urine testing suggests that the public should be equally alarmed about the abuse of drug testing in the work place.

**ELIMINATING
FALLACIES: REVISING**
Once you have organized your argument, reexamine your draft to see if it accomplishes your purpose. Revision, as you learned in Chapter 4, can mean *global revision* — the rethinking, reordering, and rewriting of your text — or *local revision* often necessitated by a small problem: a faulty thesis, an undeveloped paragraph, or an unclear sentence. In argument, many local problems are the result of errors in reasoning, known as *fallacies*. Some fallacies are unintentional, created in the haste of composition. Others are intentional, created to deceive readers. In either case, they oversimplify or distort evidence, thus making an argument unreliable. Read the following list as a guide to finding and eliminating fallacies from your writing during revision, or to detecting them in the writing of others.

Hasty generalizations are conclusions drawn from inadequate or atypical evidence. Suppose there are twelve women and ten men in your section of business law. On the final exam, the four highest scores were made by men and the four lowest scores by women. On the basis of this evidence, you conclude that women are unqualified for business law. There is no justification for that generalization because twelve women in one section are not typical of all women, and twelve is hardly an adequate sample to justify such a sweeping conclusion. To avoid hasty generalizations, make sure you provide sufficient and appropriate evidence to support your conclusions.

Post hoc, ergo propter hoc (Latin for "after this, therefore because of this") asserts that one event caused another because it preceded it. Lowering the drinking age in your state should not be claimed as the direct cause of the increased crime rate. The first event may have contributed in some way to the other, but the two events may be merely coincidental. Since direct cause and effect relationships are difficult to establish, do not overstate your claims for events that may have many plausible explanations

Begging the question occurs when part of what has to be proved is assumed to be true. When a lawyer seeks to use as evidence the results of a polygraph test because such tests verify the truth, he begs this question: Are polygraph tests reliable? Sometimes begging the question can produce a *circular argument* that restates in different words what you are trying to prove: "Polygraph tests do not provide reliable evidence because their results cannot be trusted." As you revise, check each of your claims to make sure you advance your argument with evidence; do not assume a claim is credible because you have used "new and improved" language.

Either-or states a position in such categorical terms — either A or B — that it ignores (or denies) the possibility of other alternatives. An economist who argues that "the United States must either adopt a new industrial policy or face financial ruin" does not see (or prefers not to see) that there are many possible ways to solve our economic difficulties and that financial ruin does not inevitably follow from the failure to adopt one policy. If you have conducted a thorough investigation, you know that there are more than two sides to your subject.

■ **Ad hominem** (Latin for "to the man") distorts an argument by attacking the character of the opponent to arouse the emotions or prejudices of the audience. This strategy attempts to discredit an opponent by using labels (or stereotypes) with highly unfavorable connotations: "Senator Hoover is an avowed feminist. Her antidiscrimination bill just expresses her radical political views." As you have seen, an expert witness's trustworthiness is always on trial. If Senator Hoover's opinions are not supported by evidence, then her credibility should be discounted. But as long as she presents evidence for her case, you should resist calling her names and address her argument.

■ **Red herring** is a smoked fish that has a reddish color and strong odor. A hunter may drag a red herring across a trail to distract the hounds and trick them into following a new scent. In argument, a *red herring* is an unrelated issue introduced to divert attention from the real issue: "As long as we are discussing whether women should receive equal pay, we should also discuss whether women still want to retain preferred treatment on social occasions, as when men open doors for them and pay for their meals. It seems to me that women want equal and preferred treatment at the same time." Such diversions confuse the issue by introducing irrelevancies. Economic equality and social courtesy are two different issues.

■ **Faulty analogy** attempts to argue that because two things are alike in some ways they are alike in all ways. Attending school may be analogous to working at a factory in that both require long hours to complete specific tasks for a supervisor. But school work and factory work are not the same. An analogy can be useful in helping your readers understand how something complex or unknown is similiar to something simple and familiar. But such illustrations should not be confused with proof. Arguments by analogy are particularly suspect when an insignificant resemblance is proposed as evidence for a more relevant (but unsupportable) comparison.

■■■■ *Exercises*

1. In checking through his planning notes and preliminary drafts, Matt found several assertions that looked like fallacies. In the following list, identify the probable fallacies and then discuss how they might be revised to avoid oversimplifying or distorting the evidence.
   a. Every Monday morning he shows up at work looking tired and acting moody. He must be using drugs on the weekend.
   b. We must face the problem of drug abuse with an aggressive testing program or face the possibility of a serious industrial accident.
   c. The union leadership is against drug testing. But most of those guys dress

like back-alley thugs. What else would you expect from people who look like they belong on "Miami Vice"?

d. Requiring urine tests is no more an invasion of workers' privacy than requiring protective masks or helmets.

e. What would happen if the executive vice president was arrested for drunk driving? Executives often have drinks when they conduct business over long lunches.

f. Several corporations have started talking about genetic testing. In a few years, all workers in the United States will be tested to see if they have a predisposition toward "costly" diseases like diabetes or cancer.

g. Drug testing violates workers' civil rights by invading their privacy.

2. Review Matt's two drafts, "Their Side" (pp. 137–38) and "My Side" (pp. 147–48), examining each for fallacies. Speculate on how Matt could eliminate those fallacies to make the same point in a more reliable and effective way.

Review Exercises

1. Study the advertisements on pages 145 and 146 and consider the following questions.
   a. What information in this chapter do these advertisements reflect?
   b. Which of the appeals on pages 144–47 dominate the texts of the ads? Which dominate the photographs? Do the photographs confirm or contradict the messages in the texts?
   c. Does the choice to use black and white in the ad on page 145 further or hinder its purpose? Does the use of color in the ad on page 146 contribute to or detract from its effect?

2. When Matt finished his "My Side" draft, he knew that he had presented only one side of a complex issue. Combining "Their Side" with "My Side," he produced this final draft of his argument. Compare Matt's three drafts to determine how each arranges its evidence, uses the three appeals, and fulfills its purpose.

Drug Abuse in the Work Place

In the last decade drug abuse has reached epidemic proportions in our country — destroying the lives of our young people, exhausting the energies of our legal system, and eroding our fundamental values. Nowhere is the problem more alarming than in the work place, where drug abuse costs the American economy in excess of $60 billion a year. All of us — employers, workers, and consumers — want to cure this plague that threatens the health of our culture. Unfortunately, there is as much controversy about the cure as there is about the disease.

Employers, rightfully concerned about the investments of their stockholders, the safety of their workers, and the credibility of their goods and services, favor various forms of urine testing to identify drug abusers. Many employers now include a urine test as part of the routine pre-employment physical. Such a test allows them to spot abusers before they are hired, trained, and become problems. Other employers use urine testing when a worker's behavior suggests that drug abuse is a reasonable explanation for impaired performance. Such testing enables employers to restore the productivity of their work force and to encourage rehabilitation of abusers. Employers whose workers' tasks directly affect the safety of fellow workers or consumers, or whose workers are engaged in sensitive security positions, may employ massive or random testing to guarantee the integrity of the work place.

Workers are concerned about protecting the integrity of the work place, but they are equally concerned about protecting their own integrity. The initial urine test now being administered by most employers is highly unreliable. False results (positive and negative) can be produced by careless procedures in taking the sample or by inaccurate procedures in testing the sample. In fact, the Centers for Disease Control reported that in the thirteen laboratories they studied, there was "up to 67% error rate in false positive identification[s]" (Hansen 2382). The reliability of these tests is compromised by the problem of "cross-reactivity" — chemicals in the body such as caffeine or cough syrup that can corrupt the results.

If urine testing is judged unreliable, workers are justifiably concerned about the potential for discrimination. Uninformed supervisors may misread their behavior and ask for a drug test. If the results are falsely positive, they are uncertain about their right to appeal. If the results are negative, they are uncertain about their status as "former suspects." And if the results are positive, they are uncertain of their fate. Is the employer interested in replacing or rehabilitating employees?

This question is at the heart of the controversy. Putting aside the issues of efficiency, reliability, and discrimination, and the most serious issue of invasion of workers' privacy, all of us need to be clear about the purpose of drug testing. Employers need a reliable work force; workers want a safe environment; and consumers expect dependable goods and services. But is drug testing the best way to satisfy these needs and expectations?

Pre-employment drug screening does not guarantee that workers have never used drugs or, once employed, will not become drug abusers. It simply provides highly questionable evidence about one sample of a worker's urine. Testing for probable cause may prevent or determine the cause of some accidents, but it does not solve the larger problem of ensuring a drug-free work place. And massive or random testing of a large work force (like the testing of "suspicious workers") may identify a few abusers, but it could also discredit innocent employees. The whole procedure, rather than increasing productivity, destroys the mutual trust between employer and workers so essential to quality performance.

In order to solve the complex problem of drug abuse in the work place, we need more than a simple urine test. What we need is a comprehensive drug policy developed by all those concerned. Under such a policy, drug testing might still be used, but employers would make a concerted effort to establish and monitor the credentials of urinanalysis laboratories. They would also spell out in detail the purpose and procedures of all drug testing. Job applicants would be told in advance that drug screening is required and informed about the results. Workers on the job would help establish the specific criteria for impairment. If they fail to meet those criteria and then fail the test, they would have the right to ask for a series of confirmatory tests. In all cases, especially those that involved massive or random testing, the employer would protect the confidentiality of test results and provide an agreed-upon counseling procedure for those who do not pass.

More important than mere drug testing, however, is an enlightened employee assistance program. Such a program would have as its primary goals the well-being and long-term productivity of employees. For example, employers would be more concerned about obtaining detailed references about an applicant's qualifications and work history than in collecting a urine sample. On the job, workers and supervisors would be given extensive educational programs about the causes and consequences of drug abuse. If detailed medical examinations confirm drug abuse, workers would be given sick leave and assigned to a designated drug treatment program. For those who refuse treatment or who become multiple offenders, a predetermined administrative action is justified. But for those who enter rehabilitation, the prospects are good. Companies that provide such treatment report a 70% to 75% success record (Castro 57).

Lewis L. Maltby, Vice President, Drexelbrook Controls, Inc., describes the kind of consensus policy that should cure drug abuse in the work place: "We can attack drug abuse with drug testing. It's quick, it's easy, and it's cheap. It just doesn't work. It gives us inaccurate and irrelevant information and undermines the trust of good employees who resent being ordered to pee in a bottle when they've done nothing wrong. Or, we can take the time to learn about our employees, watch their job performance, and help them when it starts to slip. It's time consuming, difficult, and expensive. But it works. Not just in preventing work place drug abuse, but in creating a committed and productive work place " (Maltby 4).

3. Compare Anna Quindlen's comments on organizing an argument (pp. 138–39) with her essay below. How does she present evidence for the contending positions? How does she introduce and support her own opinions? How does her accommodation of the opposing positions make a compelling argument?

Death Penalty's False Promise: An Eye for an Eye / *Anna Quindlen*

Ted Bundy and I go back a long way, to a time when there was a series of unsolved murders in Washington State known only as the Ted murders. Like a lot of reporters, I'm something of a crime buff. But the Washington Ted murders — and the ones that followed in Utah, Colorado and finally in Florida, where Ted Bundy was convicted and sentenced to die — fascinated me because I could see myself as one of the victims. I looked at the studio photographs of young women with long hair, pierced ears, easy smiles, and I read the descriptions: polite, friendly, quick to help, eager to please. I thought about being approached by a handsome young man asking for help, and I knew if I had been in the wrong place at the wrong time I would have been a goner.

By the time Ted finished up in Florida, law enforcement authorities suspected he had murdered dozens of young women. He and the death penalty seemed made for each other.

The death penalty and I, on the other hand, seem to have nothing in common. But Ted Bundy has made me think about it all over again, now that the outlines of my 60's liberalism have been filled in with a decade as a reporter covering some of the worst back alleys in New York City and three years as a mother who, like most, would lay down her life for her kids.

Simply put, I am opposed to the death penalty. I would tell that to any judge or lawyer undertaking the voir dire of jury candidates in a state in which the death penalty can be imposed. That is why I would be excused from such a jury. In a rational, completely cerebral way, I think the killing of one human being as punishment for the killing of another makes no sense and is inherently immoral.

But whenever my response to an important subject is rational and completely cerebral, I know there is something wrong with it — and so it is here. I have always been governed by my gut, and my gut says I am hypocritical about the death penalty. That is, I do not in theory think that Ted Bundy, or others like him, should be put to death. But if my daughter had been the one clubbed to death as she slept in a Tallahassee sorority house, and if the bite mark left in her buttocks had been one of the prime pieces of evidence against the young man charged with her murder, I would with the greatest pleasure kill him myself.

The State of Florida will not permit the parents of Bundy's victims to do that, and, in a way, that is the problem with an emotional response to capital punishment. The only reason for a death penalty is to exact retribution. Is there anyone who really thinks that it is a deterrent, that there are considerable numbers of criminals out there who think twice about committing crimes because of the sentence involved? The ones I have met in the course of my professional duties have either sneered at the justice system, where they can exchange one charge for another with more ease than they could return a shirt to a clothing store, or they have simply believed that it is the other guy who will get caught, get convicted, get the stiffest sentence. Of course, the death penalty would act as a deterrent by eliminating recidivism, but then so would life without parole, albeit at greater taxpayer expense.

I don't believe deterrence is what most proponents seek from the death penalty anyhow. Our most profound emotional response is to want criminals to suffer as their victims did. When a man is accused of throwing a child from a high-rise terrace, my emotional — some might say hysterical — response is that he should be given an opportunity to see how endless the seconds are from the 31st story to the ground. In a civilized society that will never happen. And so what many people want from the death penalty, they will never get.

Death is death, you may say, and you would be right. But anyone who has seen someone die suddenly of a heart attack and someone else slip slowly into the clutches of cancer knows that there are gradations of dying.

I watched a television re-enactment one night of an execution by lethal injection. It was well done; it was horrible. The methodical approach, people standing around the gurney waiting, made it more awful. One moment there was a man in a prone position; the next moment that man was gone. On another night I watched a television movie about a little boy named Adam Walsh, who disappeared from a shopping center in Florida. There was a re-enactment of Adam's parents coming to New York, where they appeared on morning talk shows begging for their son's return, and in their hotel room, where they received a call from the police saying that Adam had been found: not all of Adam, actually, just his severed head, discovered in the waters of a Florida canal. There is nothing anyone could do that is bad enough for an adult who took a 6-year-old boy away from his parents, perhaps tortured, then murdered, him and cut off his head. Nothing at all. Lethal injection? The electric chair? Bah.

And so I come back to the position that the death penalty is wrong, not only because it consists of stooping to the level of the killers, but also because it is not what it seems. Just before Ted Bundy's most recent execution date was postponed, pending further appeals, the father of his last known victim, a 12-year-old girl, said what almost every father in his situation must feel. "I wish they'd bring him back to Lake City," said Tom Leach of the town where Kimberly Leach lived and died, "and let us all have at him." But the death penalty doesn't let us all have at him in the way Mr. Leach seems to mean. What he wants is for something as horrifying as what happened to his child to happen to Ted Bundy. And that is impossible.

Writing Assignments

1. Select a controversial public issue with which you have had some direct personal experience. Read several articles that present "their side." Give the other side a thorough hearing, paying particular attention to the assumptions and judgments that support its reasoning. Assume the identity of your opponent and write an "annual report to the stockholders" explaining "your" position.

2. Select the same issue as you did for assignment 1 (or a similar issue), and interview several friends who share your views on the subject. Then write a "letter to the editor" in which you present your/our side of the issue. Arrange your evidence and adjust your tone to anticipate the reactions of your audience.

3. Write an analysis of Anna Quindlen's "Death Penalty's False Promise: An Eye for an Eye." Consider information and opinions you have read, heard, or seen on television in defending or refuting Quindlen's argument that the death penalty offers a "false promise" because it cannot give society what it wants — or needs.

4. The photograph on page 130 was selected as a metaphor representing an important aspect of this chapter. Examine the photograph and determine the possible meanings of this metaphor as you discern them in the photograph. Then write an essay explaining how the metaphor has been reinforced or revised by the ideas discussed in this chapter. You may want to suggest another metaphor to illustrate your understanding of these ideas.

7

PARAGRAPHS: UNITS OF DEVELOPMENT

A paragraph is a set of related sentences that express or develop a topic. A paragraph is usually part of an extended piece of writing, although in some situations you may need only one paragraph to fulfill your purpose. In narration or description, a new paragraph usually signals a shift in time, scene, or speaker. In exposition or argument, a new paragraph adds information or announces another point supporting your thesis. This chapter discusses two kinds of paragraphs: *topical paragraphs,* those that develop a topic or idea; and *special paragraphs,* those that introduce or conclude a piece of writing or provide a transition between major parts.

Paragraphs serve several purposes for you and your readers. You can use them to divide your subject into manageable units of information. By grouping ideas into paragraphs, you show the relationship of ideas to one another and their significance to your overall purpose. You can also use paragraphs to control emphasis. By placing a paragraph in a particular position, you demonstrate the relative importance of an idea in your essay. Finally, you can use paragraphs to establish rhythm. By interrupting a series of long paragraphs with a short paragraph, or by creating a series of brief paragraphs, you establish and vary cadence in your writing.

Readers use your paragraphs to grasp key points and follow your reasoning. Extended, uninterrupted passages tax their attention. They expect and need to see the regularly spaced indentations that indicate a new paragraph. This visual cue tells them that they have completed one topic and are about to take up another. Because the new paragraph promises new information, they refocus their attention to see how this change alters or advances your essay.

Like so many other writing procedures, making paragraphs depends on your subject, your audience, your purpose — and your own composing preferences. Some writers like to plan every detail of a project before they begin, assigning each subdivision of their subject to a specific paragraph. Such a procedure usually produces a blueprint for an orderly essay, but it may box the writer into a pattern that forces connections and discourages discoveries. Other writers prefer a more open-ended procedure. They know what they want to accomplish but realize that when they start writing they will discover ideas and associations that will call for paragraphs they could not have anticipated. Bill Barich, a regular contributor to *The New Yorker,* admits that he prefers this more organic approach to creating paragraphs.

> When I am planning an essay or a piece of reporting, I usually have a sense of the beginning and ending, but my paragraphs tend to arrange themselves during the writing. The process is instinctive and often surprising. I have general ideas grouped together in my notes — facts and information I want to include. But I don't know where I will put them until the piece begins to unfold. Information gathers around a center and becomes a paragraph. It's a thought unit, an energy unit. When the energy runs out, I move on, establishing a rhythm that tells me how much information I can use in the next paragraph.
>
> This sounds a little disorderly, but for me a piece becomes lifeless if I know from the beginning what goes where. I am less interested in getting the information in the right place the first time through than I am in finding the rhythm that will carry the information. I like the structure to be organic, to flow naturally. Once I have finished writing, I can check for things like sequence and fullness of statement. Maybe there is something in my notes that I can insert in a phrase or use in a new paragraph. I may push myself to shorten everything, just to see what that does to the energy of the work. I'll shift, compress, and strip away extraneous material, hoping that what remains is diamond bright. (Personal interview)

Whether you chart every paragraph before you write or form paragraphs intuitively as you compose, once you have made them you must be sure, as Barich suggests, that they meet certain criteria. The following pages discuss the four characteristics of effective topical paragraphs, the primary functions of special paragraphs, and strategies for revising paragraphs.

THE REQUIREMENTS OF TOPICAL PARAGRAPHS

An effective topical paragraph must meet four requirements. First, it must discuss one topic only; that is, its statements and illustrations must display a *unity* of subject matter, often expressed in a topic sentence. Second, it must say all that your readers need to know about the topic; it must be *complete* enough to do what it is intended to do. Third, the sentences within the paragraph must exhibit an *order* that your readers can recognize and follow. Fourth, the sentences within the paragraph must display *coherence,* allowing readers to move easily from one sentence to the next without feeling that there are gaps in the sequence of ideas or points.

© Cliff Feulner/The Image Bank

Unity

Paragraph unity requires the consistent development of the idea in your paragraph. To achieve this, each sentence must show a clear connection to the topic. In the following paragraph, the first sentence introduces Middletown, the topic that the paragraph intends to develop. Each succeeding sentence casts light on that topic; all the sentences together focus on the distinguishing characteristics of Middletown. This close relationship among individual sentences give the paragraph unity.

(1) Middletown is not an actual place but an assortment of principles — the democratic ideal, the golden mean, the common citizen — central to our sense of the American experience. (2) We have located it in a middle landscape, somewhere between the desolation of the wilderness and the debauchery of the city, and we have envisioned it as a middle-sized community, large enough to provide everything its citizens want and small enough to preserve everything they need. (3) Our culture contains many Middletowns. (4) Although their names change — Winesburg, Main Street, Mayberry — their image remains the same. (5) Their courthouse squares, tree-lined streets, and warm-hearted people represent our desire for stability in a century of bewildering change.

Any sentence that digresses or drifts away from the topic blurs the focus of the paragraph and obscures your purpose. Consider the following paragraph.

(1) The legend of the Old South has a certain timeless beauty. (2) On the veranda of stately mansions, courtly gentlemen and charming ladies talked quietly of family, land, and cotton. (3) Even the slaves, who worked the fields, were said to be

content. (4) The poor whites were barely visible until the vigilante movement after the Civil War. (5) They formed secret societies such as the Klan to terrorize the black community and to acquire political power. (6) Political power led to economic power, and by the turn of the century the poor whites were no longer poor.

The first three sentences deal with the legend of the Old South, and the reader assumes that its "timeless beauty" will be the topic of the rest of the paragraph. In sentence 4, however, the writer shifts to poor whites, a topic that is "barely visible" in the legend. The remaining sentences build on the allusion to the vigilante movement and shift the focus of the paragraph to the Klan and its use of violence to acquire political and economic power. Most readers can follow this digression, but as they proceed past the first few sentences, they become less sure about the main topic of the paragraph and more certain that the paragraph lacks unity.

▆▆▆ *Exercise*

Read the following paragraph, checking for unity by marking shifts in focus. Each sentence should clearly refer to the topic of Oregon as an ideal place to live. Suggest revisions where you think they are needed.

(1) With its seaside beaches, snowcapped mountains, and extensive forests, Oregon is an ideal place to live. (2) Its magnificent forests alone, covering more than 30 million acres, make it a natural paradise. (3) The National Park Service protects 17 million acres in national forests, but the rest is used for Oregon's lumber industry. (4) Loggers "harvest" trees with chain saws and then send their "crop" to the mill on trucks. (5) Many sawmills that used to employ 100 or more people are now closed because of the decline in the lumber business. (6) But loggers are not like steelworkers. (7) They don't go on relief and wait for their factory to reopen. (8) Self-reliant, eccentric, and impatient, loggers simply move to other parts of the country to look for work.

▆▆▆▆ The Topic Sentence

A *topic sentence* is a statement that presents the main idea to be developed in the paragraph. It is often a single sentence, although sometimes you will need two sentences to state the topic. By beginning a paragraph with a topic sentence, you immediately tell your readers what main idea you are going to develop. In turn, they expect that the remaining sentences in your paragraph will elaborate on that idea. The following example shows how an opening topic sentence controls a paragraph.

There is no simple formula for describing the intricate logic of the Pawnee people's lives. One thing is clear — that no one is caught within the social code. Against the backdrop of his natural environment, each individual stands as his own person. The Old World design for the human personality does not apply to this New World Man. The Pawnee child was born into a community from the beginning, and he never acquired the notion that he was closed in "within four walls." He was literally trained to feel that the world around him was his home — *kahuraru,* the universe, meaning literally the inside land, and that his house was a small model of it. The infinite cosmos was his constant source of strength and his ultimate progenitor, and there was no reason why he should hesitate to set out alone and explore the wide world, even though years should pass before he returned. Not only was he not confined within four walls but he was not closed in with a permanent group of people. The special concern of his mother did not mean that he was so closely embedded with her emotionally that he was not able to move about. (Gene Weltfish, *The Lost Universe*)

Gene Weltfish begins his paragraph with a simple declarative sentence: "There is no simple formula for describing the intricate logic of the Pawnee people's lives." His purpose is to explain the complexity of that logic, and the remaining sentences in the paragraph advance that purpose by showing how the Pawnees imagined the universe and their place in it.

Not all topical paragraphs begin with topic sentences. If you begin composing and don't discover the main point of your paragraph until you reach the end, you must decide whether to move the topic sentence to the beginning, place it somewhere else, or leave it where you found it — at the end. In most cases, you will probably want to place it at the beginning to cue your readers about the significance of the sentences to come. On a few occasions, you may want to place it in mid-paragraph, where it serves as a commentary on the sentences that come before or after it.

In the following paragraph, Richard Rodriguiz uses his first sentence to make a flat statement of fact: two nuns come to visit. He uses his second sentence to explain the reason for their visit: they suggest that the family speak English more often. But he defers his actual statement of topic until the third sentence, where he confirms the significance of the preceding sentences and points a direction for the rest of the paragraph.

I remember when, 20 years ago, two grammar-school nuns visited my childhood home. They had come to suggest — with more tact than was necessary, because my parents accepted without question the church's authority — that we make a greater effort to speak as much English around the house as possible. The nuns realized that my brothers and I led solitary lives largely because we were barely able to comprehend English in a school where we were the only Spanish-speaking students. My mother and father complied as best they could. Heroically, they gave up speaking to us in Spanish — the language that formed so much of the family's sense of intimacy in an alien world — and began to speak a broken English. Instead of Spanish sounds, I began hearing sounds that were new, harder, less friendly. More important, I was encouraged to respond in English. (Richard Rodriguiz, "On Becoming a Chicano," *Saturday Review*)

Sometimes you will want readers to follow the path that led to your discovery. On such occasions, build your paragraph toward a topic sentence that provides an appropriate conclusion. Ralph Ellison uses this technique skillfully in the following paragraph.

> To live in Harlem is to dwell in the very bowels of the city; it is to pass a labyrinthine existence among streets that explode monotonously skyward with the spires and crosses of churches and clutter under foot with garbage and decay. Harlem is a ruin — many of its ordinary aspects (its crimes, its casual violence, its crumbling buildings with littered area-ways, ill-smelling halls and vermin-invaded rooms) are indistinguishable from the distorted images that appear in dreams, and which, like muggers haunting a lonely hall, quiver in the waking mind with hidden and threatening significance. Yet this is no dream but the reality of well over four hundred thousand Americans; a reality which for many defines and colors the world. Overcrowded and exploited politically and economically, Harlem is the scene and symbol of the Negro's perpetual alienation in the land of his birth. (Ralph Ellison, "Harlem Is Nowhere," *Shadow and Act*)

Not every paragraph needs a topic sentence. Sometimes your readers will be able to infer your purpose from the way you express your thoughts. On such occasions a topic sentence would be gratuitous, out of place. In the following paragraph, for example, E. B. White, one of America's master stylists, evokes a summer place without composing a formal topic sentence.

> Summertime, oh, summertime, pattern of life indelible, the fade-proof lake, the woods unshatterable, the pasture with the sweetfern and the juniper forever and ever, summer without end; this was the background, and the life along the shore was the design, the cottages with their innocent and tranquil design, their tiny docks with the flagpole and the American flag floating against the white clouds in the blue sky, the little paths over the roots of the trees leading from camp to camp and the paths leading back to the outhouses and the can of lime for sprinkling, and at the souvenir counters at the store the miniature birch-bark canoes and the postcards that showed things looking a little better than they looked. This was the American family at play, escaping the city heat, wondering whether the newcomers in the camp at the head of the cove were "common" or "nice," wondering whether it was true that the people who drove up for Sunday dinner at the farmhouse were turned away because there wasn't enough chicken. (E. B. White, "Once More to the Lake," *Essays of E. B. White*)

As a rule, you should reserve paragraphs without topic sentences for special occasions. When used for effect, they can be powerful and memorable, but you risk misleading your readers and yourself if you routinely decline to write topic sentences.

Remember, your readers are most disoriented and most alert at the beginning of a paragraph. They look to your first sentences for help. They expect that you will tell them how to interpret the new group of sentences they are about to read. Although you don't have to place your topic sentence at the beginning of a paragraph, or even use a topic sentence, be aware that your readers will scrutinize your first sentence for a clue to the unity of your paragraph. If that opening sentence does not declare the topic of your paragraph, it must at least point your readers toward the sentence or group of sentences that does identify or evoke the main topic.

Completeness

Completeness, the second major requirement of an effective paragraph, is relative. The amount of explanation an idea requires depends on the amount your readers need. You must decide this based on your knowledge of your subject and of your audience. Too much information can overwhelm readers, too little can annoy them. Inexperienced writers tend to supply too little detail. Consider this example.

> American Sign Language is a language unto itself. It has its own rules that determine the meaning and significance of each gesture. Experts familiar with these rules can use their hands to communicate extremely complex thoughts.

If that is all the writer is going to say about American Sign Language, this paragraph is incomplete for most readers because the first two sentences merely state the topic. This paragraph would only "satisfy" readers who were already familiar with the intricacies of American Sign Language — but such readers would not even need this brief paragraph. For those who do need it, the writer must provide further explanation:

> American Sign Language — ASL — is a language unto itself, with its own syntax and grammar. Adjectives follow nouns, as in Romance languages. In sign, one says "house blue," establishing a picture of what is being described and then embellishing on that. Many sign language "sentences" begin with a time element and then proceed with what happened, thereby conjugating the verbs. The movement of the shoulders, the speed of the hands, the facial expression, the number of repetitions of a sign, combine with the actual signs to give meaning to the language. Signing is precise. The casual gestures hearing people make when talking have no meaning in sign language. Hearing people who do learn sign usually practice "signed English," a word-for-word coding of English into signs, but that translation sorely limits the language. In some hands, signing is an art equal to an actor's rendering of Shakespeare. It is not just swoops and swirls but an enormous variety of expression, just as a great actor's delivery is completely different from some ham's idea of haughty speeches. (Lou Ann Walker, *A Loss for Words*)

Detailed information is necessary to make the meaning of the short paragraph clear. Of course, you can begin with a short paragraph if your purpose is to state the idea and then develop the topic in greater detail in a subsequent paragraph. In that case, the second paragraph would complete the first.

Here is another example of an incomplete paragraph:

> A child understands far more than we suspect. She may not understand words too well, but she senses what they mean by how they are spoken.

If the writer stops there, the reader is left with only a topic sentence and a brief statement that barely begins to develop the main idea. The reader needs to know how a child can "sense" the meaning of words she does not understand. The complete version of the paragraph supplies the answer:

There is also a sense in which a child understands far more than we suspect. Because a child doesn't understand words too well (and also because his nervous system is not yet deadened by years spent as a lawyer, accountant, advertising executive, or professor of philosophy), a child attends not only to what we say but to everything about us as we say it — tone of voice, gesture, facial expression, bodily tensions, and so on. A child attends to a conversation between grown-ups with the same amazing absorption. Indeed, a child listening is, I hope, like a good psychiatrist listening — or like a good semanticist listening — because she watches not only the words but also the nonverbal events to which words bear, in all too many cases, so uncertain a relationship. Therefore a child is in some matters quite difficult to fool, especially on the subject of one's true attitude toward her. For this reason many parents, without knowing it, are to a greater or lesser degree in the situation of the worried mother who said to the psychiatrist to whom she brought her child, "I tell her a dozen times a day that I love her, but the brat still hates me. Why, doctor?" (S. I. Hayakawa, "Words and Children," *Through the Conversation Barrier*)

The additions to the paragraph clarify the process children use to understand words: they pay attention to tone of voice, gestures, facial expressions, and so on. The final example illustrates that children see beyond the words we use and understand what we really mean.

These examples demonstrate that you need to spell out the implications of a topic sentence with facts, illustrations, explanations — whatever is needed. Unless you give your readers the necessary information, they will have difficulty grasping your purpose. You can easily flesh out incomplete paragraphs once you realize that every generalization must be developed with supporting details.

▬▬▬ *Exercise*

The following paragraphs are incomplete. Complete them by adding several examples or a sustained illustration from your own experience.

1. You cannot and should not try to eliminate all anger from your life. If you react mildly to everything, you will often suppress your true feelings. You must learn to recognize situations where expressing anger is counterproductive.

2. Most families have a private set of signs that enable them to communicate with one another without having to say a word. Family members use these signs in public to warn each other about a potential problem.

▬▬▬ Order

The third requirement for effective paragraphs is consistent *order.* You saw in the paragraph on the Old South (pp. 159–160) that when sentences point in various directions readers are likely to have trouble following the writer's line of reasoning.

Order in a paragraph is like organization in an essay, but because paragraphs are smaller in scope, it may be easier to consider order as *direction of movement*. Four directional patterns in expository paragraphs are from *general to particular,* from *particular to general,* from *whole to parts,* and from *question to answer* or *effect to cause.*

General to Particular   A common pattern in expository paragraphs moves from a general statement, often a topic sentence, to specific explanations or illustrations of that statement. The purpose of the paragraph is to help the reader understand the general statement. That meaning becomes increasingly clear as the paragraph progresses. You saw this kind of clarification in the paragraphs on American Sign Language and children's ability to make sense of nonverbal language. Here is another example.

> It is easy to produce examples of the many ways in which Americans attempt to minimize, circumvent, or deny the interdependence upon which all human societies are based. We seek a private house, a private means of transportation, a private garden, a private laundry, self-service stores, and do-it-yourself skills of every kind. An enormous technology seems to have set itself the task of making it unnecessary for one human being ever to ask anything of another in the course of going about his daily business. Even within the family Americans are unique in their feeling that each member should have a separate room, and even a separate telephone, television, and car, when economically possible. We seek more and more privacy, and feel more and more alienated and lonely when we get it. What accidental contacts we do have, furthermore, seem more intrusive, not only because they are unsought but because they are unconnected with any familiar pattern of interdependence. (Philip Slater, *The Pursuit of Loneliness*)

Particular to General   A particular-to-general pattern reverses the preceding pattern. It begins with specific information and leads to a general conclusion, as in this example.

> We look at old family photographs in which we stand next to black, boxy Fords and are wearing period costumes, and we do not gaze fascinated because there we are young again, or there we are standing, as we never will again in life, next to our mother. We stare and drift because there we are ... historical. It is the dress, the black car that dazzle us now and draw us beyond our mother's bright arms which once caught us. We reach into the attractive impersonality of something more significant than ourselves. (Patricia Hampl, *A Romantic Education*)

Had Hampl chosen, she could have followed a general-to-particular order by beginning with her last sentence and then illustrating its significance with details about family photographs.

The direction of movement within an individual paragraph is often determined by the direction of movement within an extended passage. As Bill Barich points out, paragraphs develop a rhythm for carrying various kinds of information. In the paragraphs that precede the quoted example, Hampl uses several general-to-particular patterns to discuss the search for a family past. In the exam-

The Bettmann Archive, Inc.

ple, she reverses the order to emphasize the particulars — family photographs. She returns to the general-to-particular order in the next paragraph, when she explores the impersonality of history.

Whole to Parts   Sometimes the purpose of a paragraph is to show the parts or divisions of a topic, as in this example.

> There are medium friends, and pretty good friends, and very good friends indeed, and these friendships are defined by their level of intimacy. And what we'll reveal at each of these levels of intimacy is calibrated with care. We might tell a medium friend, for example, that yesterday we had a fight with our husband. And we might tell a pretty good friend that this fight with our husband made us so mad that we slept on the couch. And we might tell a very good friend that the reason we got so mad in that fight that we slept on the couch had something to do with that girl who works in his office. But it's only to our very best friends that we're willing to tell all, to tell what's going on with that girl in his office. (Judith Viorst, "Friends, Good Friends — and Such Good Friends," *Redbook*)

This order is also called *partitive* or *enumerative*. The opening statement announces the divisions of the topic, often indicating their number; the rest of the paragraph identifies and defines each of the parts. The partitive or enumerative paragraph is often used in argument, either to introduce the issues to be considered or to summarize a discussion. In exposition, it is used to introduce the categories to be analyzed. Such paragraphs are usually less detailed than Viorst's paragraph because succeeding paragraphs supply information about the categories.

Question to Answer, Effect to Cause   A paragraph may begin with a question and give the answer or with an effect and explain the cause. Such a paragraph may have no specific topic sentence beyond the opening question or statement of the effect. This paragraph begins with a question, which it proceeds to answer.

> So what are we to do, those of us whose habit and pleasure and doom is our tendency, as a Georgia lady put it, to "fly off at every other whipstitch?" Think in terms of movable feasts, for a start. Live here, wherever here may be, as if we were going to belong here for the rest of our lives. Learn to hallow whatever ground we happen to stand on or land on. Like medieval knights who took their tapestries along on Crusades, like modern Afghanis with their yurts, we must pack such totems and icons as we can to make short-term quarters feel like home. Pillows, small rugs, watercolors can dispel much of the chilling anonymity of a sublet apartment or motel room. When we can, we should live in rooms with stoves or fireplaces or anyway candlelight. The ancient saying still is true: Extinguished hearth, extinguished family. Round tables help, too, and as a friend of mine once put it, so do "too many comfortable chairs, with surfaces to put feet on, arranged so as to encourage a maximum of eye-contact." Such rooms inspire good talk, of which good clans can never have enough. (Jane Howard, *Families*)

The next paragraph begins with an effect and then considers some of the causes of that effect.

> There is something very wrong with an institution that so often disintegrates at the very point it is supposed to be the most useful. It is obviously absurd that we should marry, have children, get divorced, and then start the whole thing over in an even more difficult and complicated way. The "reconstituted" families of remarriage give statistical stability to the big picture. But the clue to what is wrong with the big picture lies in the "transition period to new family life," which is likely to be three or four years when a woman is caring for very small children. Nearly half the children born today will spend a significant part of their lives in a single-parent home. Raising a child or — worse — children, alone, is the wrong way to do it. It is too hard, not on the children — if recent studies mean anything — but on the mother. Of American women divorced and separated, only 4 percent receive alimony, and only 23 percent with children receive child support. Women earn, on the average, fifty-nine cents for every dollar men earn. And of all female-headed families, 41.8 percent live below the poverty level. (Jane O'Reilly, "But Who Will Take Care of the Children?")

Summary of Main Orders of Movement Within Paragraphs

General to particular	Opening general statement or topic sentence followed by illustration or details of explanation or proof. The paragraph may conclude with a restatement of the topic sentence.
Particular to general	From a series of detailed statements to a conclusion drawn from them. If there is a topic sentence, it occurs at or near the end of the paragraph.
Whole to parts	Paragraph begins with an introductory statement about the number of parts and then explains each part: often a first, second, third order.
Question to answer, or *effect to cause*	Paragraph begins with question or effect, then answers the question or shows the cause.

Coherence

A paragraph is coherent when the sentences are woven together in such a way that readers move easily from one sentence to the next and read the paragraph as an integrated discussion rather than as a series of separate sentences.

If you have a sense of purpose when you begin writing, then you are not likely to have trouble with coherence. Lack of coherence often results if you think about your topic one sentence at a time: you write one sentence, stop, think a minute, write a second sentence, stop, and continue in a series of spurts and pauses. Paragraphs written in this way are likely to lack coherence because your ideas do not flow from one sentence to the next and continuity is lost.

A paragraph can exhibit unity, completeness, and order but still lack coherence. The writer of the following paragraph is composing a paper to describe the ordeals of the immigration experience. Here he tries to illustrate the processing procedure at Ellis Island in New York Harbor, the official place of entry for most European immigrants from 1892 until 1943.

> The immigrants were herded off the boat onto Ellis Island. Inspectors talking in a strange language pushed them into a building where metal railings divided them into lines. They waited as the doctors examined them for diseases and defects. Some were separated from their families and sent to other parts of the building. The rest moved on to the next test. They were asked about their relatives, their politics, their work, their money. The questions were confusing, and they were never sure they gave the right answers.

Although this paragraph moves in an orderly sequence, it requires revision. Consider the following points:

▪ What is the purpose of the paragraph? You know from the description of the paper that the writer wants to describe the ordeal of immigration. The information does illustrate this experience, but its significance would be clearer if the paragraph possessed a topic sentence that bound the individual sentences together. Without such a sentence, a reader may not see the writer's purpose in describing the activity on Ellis Island.

▪ What is the relationship between this paragraph and the one that preceded it? The reader is told that the immigrants were on a boat, but there is no introductory statement that explains the transition from boat to island.

▪ Although the repetition of *they* provides some coherence, the sentence structure becomes monotonous, emphasizing the paragraph's disjointedness.

Now consider this revised version.

Although the immigrants endured many physical hardships during the crossing, they were not prepared for the psychological ordeal of Ellis Island. They had come so far, and now they realized that they could be sent back. Inspectors shouting strange words herded them into a massive building where metal railings divided them into lines. There they waited in silent humiliation as one doctor after another poked at their bodies, looking for diseases and defects. Occasional screams followed family members who were pulled out of line and led off to another part of the building. The rest moved on to the next test. Interpreters asked them about their relatives, their politics, their work, their money. Every question seemed like a trap. If they told the truth, they might be turned away. If they lied, they might be caught. If they said nothing, they might be marked dumb and sent back where they came from.

The revision achieved the following improvements:

▪ The topic sentence states the purpose of the paragraph, gives significance to all the details that follow, and connects this paragraph to preceding paragraphs.

▪ The more detailed and evocative explanation of the psychological ordeal of the inspection process helps bind the sentences together more effectively than in the unsatisfactory version, where sentences were connected only by a common subject.

▪ The concluding sentences dramatize the anxiety of those waiting for clearance and sum up the content of the whole paragraph.

This contrast of the original and revised paragraphs shows how an unsatisfactory piece of writing can better express the main idea by providing the links that reveal the relationship among the sentences. Other connective devices that increase paragraph coherence are *pronouns, contrast, repetitive structure,* and *transitional markers.*

Coherence Through Pronoun Reference   Because it refers to an antecedent, a pronoun points back (or forward) to create a simple, natural connection. Notice in the following paragraph how the pronoun *they* links the whole paragraph to the antecedent *emigrants*. Pronoun repetition reinforces the purpose of this paragraph: to illustrate that emigrants "shared certain characteristics."

> Most of the emigrants shared certain characteristics as a group: they were men and women who had already made one or more moves before in a restless search for better lands. They were children of parents who themselves had moved to new lands. If ever a people could be said to have been "prepared" for the adventure of the Overland Trail, it would have to be these men and women. They possessed the assortment of skills needed to make the journey and to start again. They had owned land before, had cleared land before, and were prepared to clear and own land again. And they were young. Most of the population that moved across half the continent were between sixteen and thirty-five years of age. (Lillian Schlissel, *Women's Diaries of Westward Journey*)

Coherence Through Repetitive Structure   Purposeless repetition should be avoided, but deliberate repetition of key words, phrases, or sentence patterns can make sentences flow into a coherent paragraph. In the following example, every sentence after the first has the same structure and the same opening words, "There is nothing." This kind of repetition (discussed as *parallel structure* in Chapter 8) ties the sentences together in a coherent development of the topic sentence.

> America, the richest and most powerful nation in the world, can well lead the way in this revolution of values. There is nothing to prevent us from paying adequate wages to schoolteachers, social workers and other servants of the public to insure that we have the best available personnel in these positions which are charged with the responsibility of guiding our future generations. There is nothing but a lack of social vision to prevent us from paying an adequate wage to every American citizen whether he be a hospital worker, laundry worker, maid or day laborer. There is nothing except shortsightedness to prevent us from guaranteeing an annual minimum — and *livable* — income for every American family. There is nothing, except a tragic death wish, to prevent us from reordering our priorities, so the pursuit of peace will take precedence over the pursuit of war. There is nothing to keep us from remolding a recalcitrant status quo with bruised hands until we have fashioned it into a brotherhood. (Martin Luther King, Jr., *Where Do We Go From Here: Chaos or Community?*)

Coherence Through Contrasted Elements   When a topic sentence calls for a comparison or contrast, the pairing of the contrasted elements enhances coherence. In the following paragraph, the middle sentences illustrate the contrast between the lives of Grant and Lee — a contrast that is announced in the topic sentence and summarized in the concluding sentence.

> So Grant and Lee were in complete contrast, representing two diametrically opposed elements in American life. Grant was the modern man emerging; beyond him, ready to come on the stage, was the great age of steel and machinery, of crowded cities and a restless, burgeoning vitality. Lee might have

ridden down from the old age of chivalry, lance in hand, silken banner fluttering over his head. Each man was the perfect champion of his cause, drawing both his strengths and his weaknesses from the people he led. (Bruce Catton, "Grant and Lee: A Study in Contrasts")

Coherence Through Transitional Markers   *Transitional markers* are words or phrases often placed at or near the beginning of a sentence or clause to signal the relationship between the new sentence or clause and the one before it. The most common markers are the conjunctions *and, or, nor, but,* and *for.* Others — sometimes called *transitional connectives* — indicate the direction a new sentence is about to take and prepare the reader for what is to follow. The most common transitional connectives are used as follows:

- To introduce an illustration: for example, for instance, to illustrate

- To add another phrase of the same idea: second, in the second place, then, furthermore, next, moreover, in addition, similarly, again, also, finally

- To point out a contrast or qualification: on the other hand, nevertheless, despite this fact, on the contrary, still, however, conversely, instead

- To indicate a conclusion or result: therefore, in conclusion, to sum up, consequently, as a result, accordingly, in other words

Coherence Through Connection Between Paragraphs   Coherence is necessary not only within a paragraph but also between the paragraphs of an essay, so that your readers can see how each paragraph is related to those that precede or follow it. The following passage begins with an assertion. At the end of the first paragraph, the writer notes a change, thus providing the transitional link to the next paragraphs and giving coherence to the whole passage.

Those of us who grew up in the fifties believed in the permanence of our American-history textbooks. To us as children, those texts were the truth of things: they were American history. It was not just that we read them before we understood that not everything that is printed is the truth, or the whole truth. It was that they, much more than other books, had the demeanor and trappings of authority. They were weighty volumes. They spoke in measured cadences: imperturbable, humorless, and as distant as Chinese emperors. Our teachers treated them with respect, and we paid them abject homage by memorizing a chapter a week. But now the textbook histories have changed, some of them to such an extent that an adult would find them unrecognizable.

One current junior-high-school American history begins with a story about a Negro cowboy called George McJunkin. It appears that when McJunkin was riding down a lonely trail in New Mexico one cold spring morning in 1925 he discovered a mound containing bones and stone implements, which scientists later proved belonged to an Indian civilization ten thousand years old. The book goes on to say that scientists now believe there were people in the Americas at least twenty thousand years ago. It discusses the Aztec, Mayan, and Incan civilizations and the meaning of the word "culture" before introducing the European explorers.

Another history text — this one for the fifth grade — begins with the story of how Henry B. Gonzalez, who is a member of Congress from Texas, learned about his own nationality. When he was ten years old, his teacher told him he was an American because he was born in the United States. His grandmother, however, said, "The cat was born in the oven. Does that make him bread?" After reporting that Mr. Gonzalez eventually went to college and law school, the book explains that "the melting pot idea hasn't worked out as some thought it would," and that now "some people say that the people of the United States are more like a salad bowl than a melting pot." (Frances Fitzgerald, *America Revised*)

███ *Exercise*

Read the following passage excerpted from "The Crazy Life," Bill Barich's article on youth gangs in Los Angeles. The passage is presented without its original paragraphing. Reread Barich's comments on paragraphing (p. 158); review the four requirements of effective topical paragraphs; and then discuss how you would break this passage into manageable units.

The landmark work in the sociology of gangs is Frederic M. Thrasher's "The Gang," which was published in 1927. Thrasher was a founder of the Chicago school, a methodology that stressed the importance of interviews and direct observation. In pursuing his study, he observed more than a thousand Illinois gangs before arriving at his well-known theory that gangs are largely a phenomenon of immigrant communities. According to Thrasher, they represent an ethnic group in transition, waiting out its adolescence until it can be assimilated into the mainstream. The underlying assumption is that the attractions of a so-called "normal" life — a job, a family, a house in the suburbs — far outweigh the attractions of a life of crime. Over the years, Thrasher's ideas would be repeated, with variations, in many other studies, monographs, and books, and they still echo in current sociological theory, coloring the way youth-gang members are perceived, making them seem distant, opposite, always somewhat less than human. In Los Angeles County, there are Hispanic youth gangs whose histories go back almost a century, involving three and sometimes four generations of men. There are black gangs of such size, sophistication, and economic well-being that they put many small corporations to shame. There are gangs of Chinese teen-agers who run gambling emporiums as skillfully as old Vegas hands. When immigrants, legal or illegal, come to Southern California, their children form gangs — Korean, Vietnamese, Filipino, Honduran, Salvadoran, Nicaraguan, Guatemalan. There are Samoan gangs in Los Angeles County, and gangs from Tonga, and they feud with each other just as their ancestors did on the islands. Increasingly, in affluent suburban towns, there are gangs of white teen-agers, kids from decent homes, who — the saying goes — "have everything," and still take to the streets. (Bill Barich, "The Crazy Life," *The New Yorker*)

© Rafael Macia/Photo Researchers, Inc.

SPECIAL PARAGRAPHS So far in this chapter you have been examining topical paragraphs, the main paragraphs in an essay, which develop your topic or some aspect of it. Special paragraphs are used to *introduce* or *conclude* an essay and to *mark transitions* from one unit to another.

Introductory Paragraphs

Readers want to know what you are writing about and whether they will find your subject interesting and significant. Your introductory paragraph leads into your essay, giving your readers a preview. Most introductions contain attention-getting statements that engage the readers' interest or statements that suggest the organization or indicate the scope, focus, or thesis of your essay. Notice how Bill Barich blends these in his introductory paragraph to "The Crazy Life."

The first time I met Manuel Velazquez, he greeted me awkwardly, unable to shake hands. He had cut himself on a broken bottle while crawling around in a tunnel to read a new graffito, and a doctor at a local clinic had sewed him up with seventeen stitches and then wrapped the wound in cotton, gauze, and tape. Manuel was stoical about his injury, seeing it as an unfortunate but perhaps necessary consequence of his job, which is to keep teen-age gang members in the San Fernando Valley, in California, from killing one another in wars. In Los Angeles County, of which the valley is a part, there are an estimated fifty thousand youth-gang members, and about three hundred of them are expected to be murdered this year. In the old days of youth-gang warfare, the days of "Blackboard Jungle" and "West Side Story," a boy might arm himself with a knife or a homemade zip gun, but now in times of trouble he has access to .357 magnum pistols, hunting rifles with pinpoint scopes, and Uzi semi-automatics from Israel. (Bill Barich, "The Crazy Life")

You can utilize many strategies in your introductory paragraph to suggest your intentions. The writers whose essays you have examined in previous chapters have used the following techniques:

Direct Statement   Matt uses direct statement to introduce two of his essays on drug abuse in the work place. In one version, he begins with a general statement about the effect of drug abuse, restricts this statement to drug abuse in the work place, and then offers further clarification (see Chapter 6, p. 151):

> In the last decade drug abuse has reached epidemic proportions in our country — destroying the lives of our young people, exhausting the energies of our legal system, and eroding our fundamental values. Nowhere is the problem more alarming than in the work place, where drug abuse costs the American economy in excess of $60 billion a year. All of us — employers, workers, and consumers — want to cure this plague that threatens the health of our culture. Unfortunately, there is as much controversy about the cure as there is about the disease.

In his "My Side" essay Matt begins with three statements about the problem of drug abuse in the work place (see Chapter 6, p. 147). These narrow the focus of the paragraph but build toward a direct statement of his thesis — that drug testing is unreliable, counterproductive, and an invasion of workers' privacy.

> The increased public awareness of drug abuse in our society has created a sense of public alarm about the potential dangers of drug abuse in the work place. Although such concern is justifiable, particularly when it affects the safety of workers and consumers, most companies are trying to solve this complex problem with a simple urine test. Whether it is used to weed out job applicants or identify suspected abusers on the job, urine testing is a misguided attempt to find a quick fix. The test is not only unreliable and counterproductive, it is a serious invasion of worker's privacy.

Factual Information   Rod experimented with several opening paragraphs before deciding on this opening statement: "The facts are simple." He follows this with a barrage of facts about the Washington Monument that sets up the "hook" of the monument's dramatic story (see Chapter 4, pp. 95–96).

The facts are simple: The Washington Monument stands 555 feet 5⅛ inches. Its base is 55 feet 1½ square inches. It is faced with 9,613 marble slabs, weighs 90,854 tons, can withstand a 145-mph gale, and cost $1,187,710.31. But these statistics aren't the whole story. In the two centuries since the idea was first mentioned, the Washington Monument has been revised, moved, stolen, ignored, held hostage, and almost blown up.

Quotation   Wally uses his father's "words of wisdom" to set up his essay about painting *Brandon's Clown* (see Chapter 1, p. 20). The quotation guides the reader through the narration and serves as an ironic commentary on Wally's subject and purpose.

Few of those words of wisdom that are passed from father to son are followed. Most are simply acknowledged and forgotten. My father's advice about my adolescent love for painting was simple and direct: "Son, you have a special talent. Be smart. Use it to make money." With his words guiding me, I took my love to the marketplace. I began accepting commissions, painting fantasies for those who didn't have the skill or desire to paint their own.

Dramatic Episode   Ellen's description of Barry making granola dramatizes the unique features of Barry's personality (see Chapter 3, pp. 75–76). She concludes the episode with his infamous words, "it's good for me" — words with which Barry, as the reader soon discovers, justifies everything.

Barry's in the kitchen making granola. His long, narrow frame moves deftly around the work area, pausing to scan the recipe book on the counter or search the shelves for the ingredients — wheat germ, bran, rolled oats, sunflower seeds, raisins, honey — he plans to mix in a large, gray earthenware bowl. While he works, he hums a vaguely recognizable version of "Blue Moon," drumming on the top of the counter or the edge of the bowl with a wooden spoon. He's happy. Making granola always has this effect on him. He doesn't like to make granola — it's messy and time-consuming. He doesn't like to eat granola — it's too crunchy without milk and too mushy with it. But he likes to contemplate his reasons for eating granola — "it's good for me."

Anecdote   After her attention-getting first sentence, Jane uses an anecdote about her trip to the American Museum of Natural History to introduce her reasons for her whale-watching trip (see Chapter 5, p. 122). The purpose of her allusions to textbooks and magazines is to arouse the curiosity of her readers, making them want to "see more."

The problem with whales is that they don't fit under a microscope. As a small girl, I walked around and around the large skeleton of a blue whale on the first floor of the American Museum of Natural History, trying to compare its size to the skeletons of the brontosaurus and diplodocus I had seen on the fourth floor. Like those creatures from another time, whales challenged and held my imagination. I could never see enough of them. The drawings in textbooks outlined their shape but reduced their enormous size to a few inches. The photographs in magazines suggested their magnificence but usually focused on their parts — the head rising to breech, the tail arching to dive. Last summer, hoping to see more, I booked passage on the *Portuguese Princess* out of Provincetown, Massachusetts, for my first whale watch.

Transitional Paragraphs

A transitional paragraph signals a change in content. It tells your readers that they have finished one main unit and are moving to the next, or it tells them that they are moving from a general explanation to specific examples or applications. A transitional paragraph is often as brief as one sentence:

> So much for the parents. We come now to the children.
>
> Let us see how this theory works in practice.
>
> And this brings us to the final ordeal of the crossing: Ellis Island.
>
> A few examples will make this explanation clear.

Sometimes you will need to supplement such signals with a concise summary of what has been covered or with a statement of what is to come, as John McPhee did in this paragraph from "Under the Snow" (see Chapter 5, p. 124).

> These memories became very much alive some months ago when — one after another — I had bear cubs under my vest. Weighing three, four, 5.6 pounds, they were wild bears, and for an hour or so had been taken from their dens in Pennsylvania. They were about two months old, with fine short brown hair. When they were made to stand alone, to be photographed in the mouth of a den, they shivered. Instinctively, a person would be moved to hold them. Picked up by the scruff of the neck, they splayed their paws like kittens and screamed like baby bears. The cry of a baby bear is muted, like a human infant's heard from her crib down the hall. The first cub I placed on my shoulder stayed there like a piece of velvet. The shivering stopped. Her bright-blue eyes looked about, not seeing much of anything. My hand, cupped against her back, all but encompassed her rib cage, which was warm and calm. I covered her to the shoulders with a flap of down vest and zipped up my parka to hold her in place.

Concluding Paragraphs

Not every essay needs a concluding paragraph. If you have adequately demonstrated your thesis, nothing more is necessary. You do not need to add a paragraph that mechanically restates the obvious: "Thus I have shown. . . ."

A good concluding paragraph does not necessarily sum up the ideas in an essay. Bill Barich concludes "The Crazy Life" with a quotation from Manuel, the man he introduced at the beginning of the essay. This paragraph echoes the introduction, but it also opens up the essay by suggesting that Manuel will always have more work to do.

"You can't get depressed about it," he said. "You know, with the people on my team, I tell them they have to make their job fun — even though it's a crummy job. It's got lots of negatives, but lots of positives, too. Once you start an input into somebody's life, you begin to influence them, you break the monotony. Some kids, they don't know any other kind of life. What we do, it's like throwing a wrench inside an engine. It screws up the structure. Suddenly, you're a part of their lives. That's good to feel. Sometimes it's like I can almost control what's going on. It's like having a sixth sense. Like seeing into the future and controlling it. That's when you know you're doing your job." (Bill Barich, "The Crazy Life," *The New Yorker*)

An effective concluding paragraph leaves the reader with a sense of completeness, a conviction that the point has been made, that nothing else needs to be said; it contributes something significant to the essay that could not have been accomplished by a "Thus I have shown . . ." conclusion.

The following examples, taken from essays in previous chapters, show various strategies for concluding paragraphs.

Restatement and Recommendation   Rod summarizes the main points he has made about the inside story of the Washington Monument and goes on to suggest how that information can help you appreciate this "peaceful and impressive sight" (see Chapter 4, pp. 98–99).

There's obviously more to George's monument than meets the eye. Of course, you don't have to know the inside story to like the monument. Most people who see a picture of the monument on a calendar or a postcard like what they see. But to truly understand and appreciate the monument, you do need to know the inside story. When you visit George's monument on some hot summer afternoon and have to stand in line for over two hours for a thirteen-second elevator ride to the top, you need to know how long it took the planners and builders to reach the top: all those false starts, crazy designs, and engineering obstacles that had to be overcome to produce this beautifully simple structure. When you see people in the line who don't know why they are there or who are just fooling around as they would at Disneyland, you need to know that the two-hundred-year history of the monument is filled with bewildered and misguided people. And when you have finished your visit, you need to take time to appreciate what you have seen. Turn and look at it one last time. There it stands — a solitary, rugged individual, indifferent to the confusion below. It is a peaceful and impressive sight.

Prediction   Wally begins his conclusion by drawing on the information he has discussed in other paragraphs — his father's advice, his artistic reputation. But he concludes by predicting that in twenty years Brandon may pay him back for his creation (see Chapter 1, pp. 21–22).

As I left holding my first commission, I did not think about my father's advice or my artistic reputation. I thought only of poor Brandon. I will not be surprised if, twenty years from now, a deranged young man stops me on the street and smashes a clown painting over my head. I probably deserve worse.

Resolution   Jane presents a chronological wind-up to her narrative and a final, climactic insight into her experience (see Chapter 5, p. 124).

> Although we watched eagerly for him to surface again, we looked in vain. Soon we began to look at one another. We smiled, but we did not talk. Neither did Richard. No explanation was necessary. We had watched a whale. A whale had watched us. Finally, the engine coughed, and the *Portuguese Princess* resumed her journey back to Provincetown.

Quotation   Sometimes a single line of dialogue makes an effective ending, especially when the words resonate like Ellen's (see Chapter 3, p. 76):

> "See, I told you it was good for me."

At other times, a carefully chosen quotation from an expert can strengthen the authority of your conclusion. Matt uses this strategy to make his final points about drug testing in the work place (see Chapter 6, pp. 152–53).

> Lewis L. Maltby, Vice President, Drexelbrook Controls, Inc., describes the kind of consensus policy that should cure drug abuse in the work place: "We can attack drug abuse with drug testing. It's quick, it's easy, and it's cheap. It just doesn't work. It gives us inaccurate and irrelevant information and undermines the trust of good employees who resent being ordered to pee in a bottle when they've done nothing wrong. Or, we can take the time to learn about our employees, watch their job performance, and help them when it starts to slip. It's time consuming, difficult, and expensive. But it works. Not just in preventing work-place drug abuse, but in creating a committed and productive work place" (Maltby 4).

REVISING TOPICAL PARAGRAPHS

As you learned in Chapter 4, there are two types of revision: *global revision,* employed while you are still creating your final draft, and *local revision,* in which you fine-tune your final draft. Local revision of paragraphs ensures that each paragraph in your essay not only works externally, to advance the purpose of your essay, but also works internally to create a sensible sequence of thought. The discussion of topical paragraphs in this chapter has shown you some ways to think about local revision, especially in the case of the paragraph about Ellis Island (pp. 168–69). To identify problems in your own paragraphs, you must view your paragraphs as your readers will see them. In other words, after you have refined your subject, audience, and purpose through global revision, and after you have established a reasonably firm organization, you must give your work a sentence-by-sentence reading, concentrating on the relation of sentences to each other and to the topic sentence.

One helpful technique in such close reading is creating a descriptive outline (see Chapter 3). As you recall, in a descriptive outline you describe what each paragraph in an essay says (that is, its main idea) and does (that is, how the paragraph supports the main idea). You should be able to state succinctly the main idea of every topical paragraph you write. If in reading a paragraph you find that you cannot succinctly state its main idea, you probably need to reconsider your topic sentence — or create one in the first place. If you find that you cannot determine what a paragraph does, the problem may be that it does nothing: it may fail to support the topic sentence or fail to add up to any particular meaning.

You need then to consider the paragraph's internal qualities: unity, completeness, order, and coherence. A paragraph that does not exhibit each of these qualities will not effectively advance your purpose. It will leave questions unanswered, details unsupported, transitions incomplete; and your readers will receive less from your writing than you intended. You must question both the meaning and function of each sentence in your paragraph and the overall effect of all the sentences.

In the following example a writer reports a series of impressions.

> (1) On Bourbon Street in the French Quarter of New Orleans there are a 50-cent peepshow and a theater that shows pornographic movies. (2) Pictures painted by talented artists are for sale in a shop down the street. (3) Canned music blares through doors held partly open by hustlers of strip joints. (4) There is a concert hall with no doors and no admission fee, where the crowd is entertained by jazz musicians. (5) In some places there are "dancing girls" who just walk across the stage and do "bumps" in what is supposed to be a dancer routine. (6) Sometimes there is a young woman who dances gracefully. (7) She has mastered the techniques that the better burlesques made popular in earlier years.

This paragraph does not lack detail; it is a series of details that try to capture the confusion of Bourbon Street. But the paragraph is all detail and no pattern: it has no unifying idea, no recognizable order, and no semblance of coherence.

The details in the paragraph fall into two groups: sentences 1, 3, and 5 show an unfavorable impression of Bourbon Street; sentences 2, 4, 6, and 7 suggest a more favorable impression. By classifying the details in this way, you can discern that the writer probably intends to contrast the two impressions. This purpose stated in a topic sentence is: "Bourbon Street in the French Quarter of New Orleans is a contrast of vulgarity and art." With that topic sentence as the controlling idea of the paragraph, you can revise its order in one of two ways:

- Use an $A + B$ contrast in which all the details suggesting vulgarity (A) are placed in the first half of the paragraph and all those suggesting art (B) are placed in the second half, with a transitional marker (*but* or *on the other hand*) to mark the change.
- Use an $A/B + A/B$ contrast in which matched details of vulgarity and art alternate within each sentence.

For this paragraph, the second arrangement is better because it is truer to the constant contrasts the writer is trying to describe. Here is a revised version.

> Bourbon Street in the French Quarter of New Orleans is a contrast of vulgarity and art. Just a few doors down the street from a 50-cent peepshow and a theater that presents pornographic movies is a shop displaying for sale paintings by talented artists. At strip joints canned music blares out from doors kept ajar by hustlers seeking to entice passers-by; yet not far away is a concert hall with no admission fee, where musicians play first-rate jazz. Even the "dancing girls" offer a sharp contrast: most limit themselves to a slow walk across the stage, interrupted by exaggerated "bumps"; but a few gracefully demonstrate the techniques that once made burlesque at its best an art form.

REVISING SPECIAL PARAGRAPHS

Because introductory, transitional, and concluding paragraphs are used for special purposes, they require a somewhat different approach in revision. In judging how well they do what they are meant to do, you should focus less on method than on effect.

Be sure that your introductory paragraph clearly states the thesis you intend to develop and that the thesis is advantageously positioned within the paragraph. Or, if you have used an attention-getting device, you need to look carefully at this "hook" to see that it entices rather than confuses your readers.

Check to see that your opening paragraph fits your final draft. An introductory paragraph written before the final draft was completed may no longer be appropriate. Does your opening seem misleading, flat, or simply uninspiring? View it with an open mind, and do not hesitate to revise it completely if your other revisions make a different opening necessary.

As you revise, pay particular attention to transitional paragraphs, the structural seams of your essay. Significant revisions elsewhere may require you to revise your transitional paragraphs or to delete them and use a concluding sentence or an opening phrase in existing paragraphs. Or you may discover that changes in your essay require you to add a transitional paragraph to clarify the movement from one unit to another. Such decisions require you to reexamine the connections between your ideas to be certain that you have chosen the most effective strategy for communicating those connections to your readers.

When you revise a concluding paragraph, consider whether your essay needs a formal conclusion. You do not want to belabor the obvious. If your conclusion merely repeats what you have already said, delete it. If you decide that your essay does need a conclusion, consider the impression you want to leave in closing. You will sometimes discover that other revisions — both global and local — have weakened the effectiveness of your original conclusion. After revising the other parts of your essay, you will know what you must do to round off your work effectively, perhaps even memorably.

————— *Review Exercises*

Identify the problems in the following paragraphs and then revise each paragraph.

1. The front porch was once a door sill. It became a square platform large enough to hold two chairs. It lengthened and began to expand around the sides of the house. At one time, it was long and narrow, just wide enough for a row of chairs. People sat there watching and talking about their neighbors. Porches were a status symbol of economic prosperity and social prominence. Architects complained that they were too ornate. Cars created dust and fumes, so families went inside. The car also gave young lovers more privacy than the porch swing. Air conditioning made it unnecessary to go outside for air. People built backyard patios so the family could have some privacy.

2. It is unusual for one family to live in the same community for more than two generations. The average American family moves once every five years. Americans want to look for their roots. Many have complex ties to different parts of the world and are proud of their heritage. It's hard to trace the effects

of immigration. Most families are scattered all over the country. Parents or grandparents who used to bring everyone together for Christmas spend December in trailer parks in Florida.

3. The Gateway Arch in St. Louis is a 630-foot stainless-steel arch. It marks the most prominent point of embarkation for those who traveled into the new territories. A Museum of Westward Expansion, near the arch, commemorates the frontier experience. Inside the steel structure is a contraption like a Ferris wheel that travels to the apex of the arch. The small windows provide a thirty-mile view in any direction. On the Mississippi River, tied to a dock near the arch, is a fast-food restaurant designed to look like a riverboat.

4. The people whose ancestors built the Great Wall of China made one of the most widely acclaimed contributions to American history. Over thirteen thousand Chinese built the western half of the transcontinental railroad. Digging tunnels through mountains and track beds across deserts, they laid 10 miles of track a day. On May 10, 1869, when a golden spike was driven at Promontory Point, Utah, no Chinese were visible. Americans resented the strength and skill of the "little yellow men." Anti-Chinese resentment was so high that Congress passed a series of Exclusion Acts that prohibited Chinese immigration. These laws even were repealed in 1943, and Chinese were admitted under a strict quota system.

Writing Assignments

1. Your family, like most, probably has a relative about whom the rest tell stories. Using several planning strategies, gather information about this person. Group this material under different headings, and then arrange those headings into a sequence that presents a coherent profile of your family's favorite hero or villain.

2. Select a minority group that you are unfamiliar with — Amish, Eskimos, Libertarians — and then read several articles about how that group has been treated in the United States. Draft several topic sentences that advance an explanation for this treatment. Then compose an essay by developing those sentences into topical paragraphs. Experiment with each paragraph by placing the topic sentence in different positions.

3. Read Bill Barich's complete essay "The Crazy Life," in *The New Yorker* (November 3, 1986: 97–130). Select an extended passage. Identify the movement within each paragraph and from paragraph to paragraph. Write an essay analyzing how Barich uses his paragraphs to help his readers grasp the purpose of his essay.

4. The photograph on page 156 was selected as a metaphor representing an important aspect of this chapter. Examine the photograph and determine the possible meanings of this metaphor as you discern them in the photograph. Then write an essay explaining how the metaphor has been reinforced or revised by the ideas discussed in this chapter. You may want to suggest another metaphor to illustrate your understanding of these ideas.

8
SENTENCES: PATTERNS OF EXPRESSION

As you saw in Chapter 7, the sentences in a well-written paragraph are not isolated statements; they exist in complex interrelationships with each other. Although traditionally called units of composition, sentences are units chiefly in the grammatical sense that each sentence has its own subject and predicate and is not part of another sentence. Beyond these basic attributes, however, sentences vary widely in style — and therefore in the effect they create. By varying the arrangement of words, phrases, and clauses in sentences, writers reveal emphasis, shade meaning, and create various kinds of movement in their writing.

Richard Selzer, surgeon and writer, composes sentences in two stages:

> The *first* propels me across the page with an armful of language. It's like the technique of the abstract painter. He will make a brush stroke on a bare canvas. That stroke leads to another and the combination to a third until the canvas is done and he realizes what he has made. The *second* is the fully conscious tinkering that goes on. The artist is always blacking out parts of his canvas and correcting small details until the painting is done in his mind. So one of my essays isn't finished until I cross out half of it and revise its individual parts. I try to respond to the marvellous elasticity of the language by constantly opening up or compressing my sentences. When I am done, they possess a distinctive rhythm and resonance. I can pick them out immediately on any page of prose. They're like my signature. (Personal interview)

This chapter will take a close look at how writers such as Selzer express their purpose by expanding, combining, and revising their sentences.

EXPANDING AND COMBINING SENTENCES

Like a brush stroke, a basic sentence moves across the page in a simple line. It consists of a subject and a predicate. The predicate may be a complete verb or a verb that needs something to complete it. For example, the sentence "The doctor smiled" consists of the subject *doctor* and the complete verb *smiled*. But in the following two sentences the verb requires completion:

> The doctor + *examined* + the patient. [Subject, verb, and object]
>
> The doctor + *was* + courteous. [Subject, verb, and complement]

A *basic sentence*, then, is a main clause consisting of subject and verb and any object or complement required to complete the verb.

Any element in a basic sentence may be modified by adjectives, adverbs, phrases, or clauses that describe or limit the words being modified. Or the whole main clause rather than one of its elements may be modified. For example, in the

"Athletic Contest" (1915), by Max Weber (1881–1961). Oil on canvas. The Metropolitan Museum of Art, George A. Hearn Fund, 1967.

sentence "If you can't do it, I'll ask Dr. Helena," the introductory *if* clause modifies the whole main clause and is called a *sentence modifier*.

In the following examples the basic sentence is underscored; the modifiers are italicized and are connected by arrows to the words they modify.

Many students *enrolled in the pre-med curriculum* find *organic* chemistry

extremely difficult.

To prepare for a test they study *for hours* with *incredible* concentration.

Some, *recognizing their limitations in math,* hire *private* tutors *to help them*

prepare for each exam.

Others study alone, *because they are driven to master the material in seclusion.*

If their study habits predict how they will behave as doctors, some will consult

with colleagues, while others will insist on making their own diagnosis.

As those examples show, modifiers may be single words or phrases (*many, private, for hours*) or subordinate clauses (*because they are driven to master the material in seclusion*). The modifiers may come before, after, or within the main clause. In the last two sentences, the subordinate clauses modify the whole main clause and are therefore sentence modifiers.

■■■ *Exercise*

Distinguish between basic sentences and modifying words or phrases in the following passage from Richard Selzer's "An Absence of Windows." First underline the basic sentence; then circle the modifiers.

(1) Part of my surgical training was spent in a rural hospital in eastern Connecticut. (2) The building was situated on the slope of a modest hill. (3) Behind it, cows grazed in a pasture. (4) The operating theater occupied the fourth, the ultimate floor, wherefrom high windows looked down upon the scene. (5) To glance up from our work and see lovely cattle about theirs, calmed the frenzy of the most temperamental of prima donnas. (6) Intuition tells me that our patients had fewer wound infections and made speedier recoveries than those operated upon in the airless sealed boxes where we now strive. (7) Certainly the surgeons were of gentler stripe.

Expanding Sentences by Modification

In effective writing, details communicate specific information and hold the reader's interest. Any sentence can be enriched by modification. Notice how much detail is added by the italicized modifiers in the following sentences.

> My father kept his own books, *in a desk calendar that recorded in his fine Spencerian handwriting the names of the patients he had seen each day, each name followed by the amount he charged, and that number followed by the amount received.* (Lewis Thomas, "Amity Street," *The Youngest Science: Notes of a Medicine Watcher*)

> I debated whether I should major in zoology *so early in the term before I knew whether my high school science courses had prepared me for the difficult challenges of the college curriculum.*

> The drug companies, *usually operating through private physicians with access to the prisons,* can obtain healthy human subjects *living in conditions that are difficult, if not impossible, to duplicate elsewhere.* (Jessica Mitford, *Kind and Unusual Punishment*)

> In this portion of the chart, *after all the histories are taken, after the chest has been thumped and the spleen has been fingered, after the white cells have been counted and the potassium surveyed,* the doctor can be *seen to abandon the position of recorder and assume that of* natural scientist. (Gerald Weissman, "The Chart of the Novel," *The Woods Hole Cantata: Essays on Science and Society*)

In sentences like these, the effect comes not from the main clause, which provides no details, but from the specific information provided by the modifiers. You can easily see that this is so by isolating the main clause from the modifiers:

My father kept his own books.

I debated whether I should major in zoology.

The drug companies can obtain healthy human subjects.

In this portion of the chart the doctor can be natural scientist.

Effective modifiers such as the ones shown in the full sentences above have two qualities. First, they are not tacked on as afterthoughts but are essential parts of the writers' purposes. A significant portion of Thomas' recollection of his father's books focuses on the shape of the handwriting and the pattern of the accounts; the student's debate about majoring in zoology is occasioned by her doubts about her high school preparation in science; Weissman's description of a doctor's writing process is based on the details he has to record. In each case the modifying details are necessary to express the writer's complete thought.

Second, effective modification is grounded in observation or experience. Weissman knows from experience how to keep a patient's chart. Mitford has researched her subject and knows about the doctors' cooperation and the conditions that made prisoners ideal subjects for experiments. In each example, the writer has used modification to get his or her observations into the sentences.

Suppose you were asked to expand these three sentences by modification:

I am uncomfortable in my doctor's office.

My doctor always seems impatient.

Doctors study the human body.

To complete the assignment, you need to know the context in which the writer is working and his or her purpose for composing the sentences. Without that knowledge, asking you to expand them is asking you to supply a context and purpose. By drawing on your own experience, you can describe why doctors offices make you uncomfortable, what makes your doctor appear impatient, and how doctors study the human body. You might expand the first sentence like this:

I am uncomfortable in my doctor's office because while I sit in the posh chairs of the waiting room I imagine that my body is being secretly afflicted with all sorts of life-threatening diseases.

■ *Exercises*

1. Drawing on your own experience, expand the other two basic sentences above to make them fuller expressions of the ideas they suggest to you.

2. Study Thomas Eakins' painting (p. 208) of Dr. Samuel Gross operating in his clinic. Draft a basic sentence that conveys your impression of the painting. Devise a cluster of specific details and use it to expand your basic sentence into a richer and more complete sentence.

Combining Sentences by Coordination

Coordination is combining or joining similar elements into pairs or series. As in the following pairs of examples, sentences can be combined by using a common subject or predicate and compounding the remaining element.

Medical technology contributes to the high cost of health care. Medical liability insurance is also extremely expensive.

Compound subject/
Common predicate

Medical technology and medical liability insurance both contribute to the high cost of health care.

Doctors must complete their charts under pressure. They also must carefully think through their comments and notations.

Compound predicate/
Common subject

Doctors completing their charts must carefully think through their comments and notations and must write under pressure.

▬▬ *Exercise*

By combining parts through coordination, compress the sentences in each group into a single sentence.

1. Many young people cannot afford college tuition.
   The expense of living away from home also keeps them from going to college.

2. Coronary heart disease is the major cause of death in this country.
   Cancer also causes many deaths.

3. Journalism is an overcrowded field just now.
   Currently there are more licensed secondary school teachers then there are positions.
   The legal profession is overcrowded.

4. My doctor warned me about trying to lose weight too fast.
   My coach reminded me of the danger involved.
   My mother told me the same thing.

▬▬▬ Using Parallel Structures

When two or more coordinate elements have the same form, they are said to have *parallel structure*. Such structures may be unnoticed in any sentence, but when the parallelism is conspicuous, the whole sentence may be called a *parallel sentence*. Thus, "He was without a job, without money, without opportunity, and without hope" is a parallel sentence in which the four phrases have the same form (they all start with *without*) and the same grammatical function (they all complete the verb *was*).

Consider these contrasted sentences:

I am in favor of equal economic rights for women. Women should be able to compete with men for jobs for which they are both qualified. The pay should be the same for the same jobs. There should be the same opportunities for advancement.	I am in favor of equal economic rights for women: the right to compete with men for jobs for which they are both qualified, the right to get the same pay for the same job, and the right to equal opportunities for advancement.

Both versions assert the same three rights. The version on the left advances the assertions in four sentences, each using a different subject. The version on the right states all three rights in one sentence and focuses the reader's attention by the repetition of the phrase "the right to." The parallel structure of the second version gives it the unity, coherence, and emphasis that the first version lacks.

Parallel elements may be single words, phrases, clauses, or sentences (contributing to a paragraph with parallel structure); they may act as subjects, objects, verbs, or adverbial or adjectival modifiers. But for elements in a pair or a series to be parallel, all members must have the same form and serve the same grammatical function. You cannot coordinate nouns with adjectives, verbs with infinitives, or phrases with clauses. Any attempt to do so disrupts parallelism, disappoints your readers, and produces awkwardness.

In each of the following sentences, the italicized element disrupts parallelism by switching from one grammatical form to another.

> The children were laughing, squealing, and *danced*. [To restore parallelism, *danced* should be changed to *dancing*.]

> My two ambitions are to become a doctor and *having* enough money to give my children a good education. [Parallelism requires that *to become* must be matched with *to have*.]

> My parents taught me such things as honesty, faith, *to be fair*, and *having patience*. [Parallelism with the two nouns would be sustained by replacing the infinitive and participial phrases with the nouns *fairness* and *patience*.]

The following diagrams show the similarity in form and function of various parallel structures.

- *Parallel predicates*

He ⎯⎯⎯ ⎡ walked past the information desk,
⎢ followed the signs to obstetrics,
⎢ selected a seat in the waiting room, and
⎣ watched the other expectant fathers pace the floor.

- *Prepositional phrases* in a series

Fat people know all about the mystery of life. They are the ones acquainted ⎯⎯⎯ ⎡ with the night,
⎢ with luck,
⎢ with fate,
⎣ with playing it by ear.

(Suzanne Britt Jordan)

- *Participial phrases* in a series

This was the American family at play ⎯⎯⎯ ⎡ escaping the city heat,
⎢ wondering whether the newcomers in the camp at the head of the cove were "common" or "nice,"
⎣ wondering whether it was true that the people who drove up for Sunday dinner at the farmhouse were turned away because there wasn't enough chicken.

(E. B. White)

▬▬ *Exercises*

1. Identify and diagram as shown above the parallel elements in the following sentences.
   a. We hold these truths to be self-evident: that all men are created equal, that they are endowed by their Creator with certain unalienable Rights. . . .
   b. Berton Roueche's narratives of "medical detection" are full of patients with unusual symptoms, of laboratory technicians with specialized knowledge, and of doctors with extraordinary diagnostic powers.

2. The notes that follow list the distinguishing characteristics of people who exhibit Type A and Type B behavior. Organize the items into parallel lists, and then write a parallel sentence coordinating the information in the two series.

Type A Behavior	*Type B Behavior*
Obsessed with deadlines	Evokes serenity in others
Intense need to win at all costs	Enjoys creating images and metaphors
Conversation dominated by numbers	Secure sense of self-esteem
Harshly critical of others	Rarely wears a watch

3. Create a list of notes in the brain suggested by the notes on the computer. Then compose a parallel sentence contrasting the advantages of the brain over the computer as a thinking "machine."

Computer	*Brain*
Circuits	
Blank disk	
Storage	
Program	

When parallelism is extended through a paragraph, each sentence becomes an element in a series. An example of a series of parallel sentences appears on page 170, in the paragraph by Dr. Martin Luther King, in which each sentence begins with the phrase "There is nothing." Such deliberate repetition binds individual sentences into a coherent paragraph. Consider the features of conspicuous parallelism in the following paragraph.

> H. G. Wells continues to be a biographer's dream and book reviewer's waltz. His life stretches very nearly from Appomattox to Hiroshima. He was one of the world's great storytellers, the father of modern science fiction, an autobiographic novelist of scandalous proportions, a proselytizer for world peace through brain power, an unsurpassed popular historian, a journalist and inexhaustible pamphleteer, the friend and worthy adversary of great men and the lover of numerous and intelligent women. (R. Z. Sheppard, *Time*)

Combining Sentences by Subordination

Subordination reduces one sentence to a subordinate clause or to a phrase that becomes part of another sentence. Consider the following basic sentences.

We left early. We had work to do.

The second of these sentences may be embedded in the first as a subordinate clause.

We left early *because we had work to do.*

In the next example, a basic sentence is reduced to a phrase.

The man was evidently in great pain. He was taken to a hospital.

The man, *evidently in great pain,* was taken to the hospital.

As those examples show, when two sentences are combined by subordination, the information in one sentence is embedded in the other. Using subordination, a writer can pack a great deal of information into a sentence and still emphasize what is most important. This does not mean that short, basic sentences are always inappropriate. Often they are useful to create variety and achieve emphasis. But consistent use of basic sentences in a sustained piece of writing can seem repetitive and dull.

In the following example, a complex sentence is created through the subordination of four basic sentences.

1. Last weekend I saw a science fiction film.
2. Three friends went with me.
3. The film focused on the experiments of a mad doctor.
4. He altered his patients' lives by manipulating their dreams.

Last weekend three friends and I saw a science fiction film in which a mad doctor altered his patients' lives by manipulating their dreams.

The original four sentences contain 32 words. Using subordination to embed sentences 2, 3, and 4 in sentence 1 creates a denser sentence in which the same information is expressed economically, in 24 words, and the monotony of the original sentences is eliminated.

■■■■ *Exercise*

Practice creating denser sentences through subordination by reducing each of the following sets of sentences to a single sentence without omitting any of the information.

1. Scientists use guinea pigs in their laboratory experiments.
   They inject them with a disease.
   They observe their behavior.
   They dissect them.
   They examine the effect of the disease on their organs.

2. Michelangelo studied anatomy.
   He dissected cadavers.
   Such gruesome work helped him understand human bones and muscles.
   His sculptures celebrate the human body.

3. X-rays can penetrate the human body.
   They produce images on photographic film.
   Shadows on the picture reveal changes in body tissue.

4. The early history of medicine is filled with guesswork.
   Doctors did not know what caused disease.
   They developed cures through trial and error.
   Most of the cures did not work.
   Cures that worked were considered magical.

5. People went to spas to restore their health.
   Most spas were located in beautiful settings.
   They featured special mineral waters.
   The waters were supposed to purge the body of disease.
   These watering spots developed into vacation resorts.

Combining for a Purpose

When you combine basic sentences into denser structures through coordination and subordination, you should choose the sentence pattern that best suits your purpose. If you are writing a paper on the latest developments in frozen foods for a nutrition class, you might collect the following information.

1. Frozen food has long been part of the American way of life.

2. Frozen dinners once consisted of such dull items as lima beans, gooey potatoes, and mystery meat.

3. Currently available frozen-dinner entrées include beef bourguignon with glazed carrots, asparagus crepes with Mornay sauce, and vegetarian lasagna.

4. Some frozen-food companies have even developed gourmet dietetic dinners.

5. Americans have fast-paced, busy life styles these days.

6. They are also more concerned than before about eating healthful, interesting, and balanced meals.

Strung together in that order, these sentences have little meaning. They trace a sequence, to be sure, but what is significant about that sequence and what are its implications? The answer, of course, depends on your purpose. If you have a purpose, you can combine the sentences through coordination or subordination to achieve focus and establish relationships among the facts.

For example, if you want to propose a cause and effect relationship between the change in American life styles and the change in frozen food, you can reduce sentence 1 to a subordinate clause and embed it in sentence 2.

> Frozen dinners, long part of the American way of life, once consisted of such dull items as lima beans, gooey potatoes, and mystery meat.

Through a similar process you can combine sentences 5 and 6 and 3 and 4.

> To satisfy busy Americans' concern for healthy, interesting, and balanced meals, today's frozen-food companies have developed more sophisticated fare: beef bourguignon with glazed carrots, asparagus crepes with Mornay sauce, vegetarian lasagna, and even gourmet dietetic dinners.

With a coordinating conjunction you can combine the two sentences into one.

> Frozen dinners, long part of the American way of life, once consisted of such dull items as lima beans, gooey potatoes, and mystery meat, but to satisfy busy Americans' concern for healthy, interesting, and balanced meals, today's food companies have developed more sophisticated fare: beef bourguignon with glazed carrots, asparagus crepes with Mornay sauce, vegetarian lasagna, and even gourmet dietetic dinners.

But if you want to demonstrate that those who once found frozen dinners unappetizing now have a reason to try them, you can combine the original sentences for this purpose.

> Busy Americans once identified frozen dinners with such dull items as lima beans, gooey potatoes, and mystery meat, but today those same consumers are tempted by a new variety of healthy, interesting, and balanced frozen meals: beef bourguignon with glazed carrots, asparagus crepes with Mornay sauce, vegetarian lasagna, and even dietetic dinners.

The last version focuses on the customer rather than on the company, but the focus is still achieved by subordination and coordination.

The three devices you have been studying — modification, coordination, and subordination — allow you to combine and present information in complex sentences. Organizing ideas into an effective form is just as important in a sentence as it is in a paragraph or essay. In a series of simple sentences, each idea is stated separately; the statements lack coherence, and relationships among the ideas are not clear. Moreover, such sentences lack emphasis; all information appears to be of equal importance. By subordinating the less important to the more important, or by showing the equality of ideas through coordination, you can revise such sentences so that they exhibit variety, clarity, and emphasis.

■■■■ *Exercise*

For each of the following clusters, combine as many of the items as you need (you may not need or want to use them all) into a sentence that fulfills each of the two purposes specified for each cluster. Combine any remaining items into a second, supporting sentence.

Cluster A

1. Clothes reveal an individual's personality and attitudes.
2. Clothes like blue jeans and T-shirts make a statement.
3. Dark blue suits and white laboratory jackets also make a statement.
4. Certain occasions, situations, or jobs require appropriate attire.
5. "Appropriate" attire is often formal: a dress, suit, tie.
6. People should be free to dress as they please.
7. In many professional situations, people judge others by their appearance.

Purpose 1 You want to assert that people should be free to dress as they please.
Purpose 2 You want to argue that people should dress appropriately in professional situations.

Cluster B

1. Hospitals are often the setting for television soap operas.
2. Characters spend more time discussing their personal problems than their professional responsibilities.
3. Doctors are seen making rounds; nurses are seen keeping charts.
4. Soap operas rarely show medical procedures.
5. Characters conveniently develop illnesses to complicate the plot.
6. Doctors use vague medical terminology to discuss cases.

Purpose 1 You want to demonstrate that soap operas show superficial treatment of illness.
Purpose 2 You want to illustrate that soap operas use hospitals as a convenient setting for dramatic conflict.

Cluster C

1. Many of today's sports injuries are treated with arthroscopic surgery.
2. Major surgery requires doctors to cut through the muscle to find the damaged area.
3. Such surgery requires many months of rehabilitation.
4. An arthroscopic surgeon punctures the muscle with a catheter carrying a microscopic television camera and another catheter carrying appropriate surgical tools.
5. He locates the injury with the camera and uses the tool to repair it.
6. Players can often return to action within several weeks.
7. Any injury requires sufficient recuperation.
8. Reinjury of damaged areas can result in disability.

Purpose 1 You want to argue that arthroscopic surgery is a major medical breakthrough.
Purpose 2 You want to suggest that the availability of arthroscopic surgery can give players a false sense of security.

TYPES OF SENTENCES AND THEIR EFFECTS

In this section you will learn to use to good effect three important sentence types: the *balanced* sentence, the *periodic* sentence, and the *cumulative* sentence.

The Balanced Sentence

In a balanced sentence, two coordinate but contrasting structures are set off against each other like the weights on a balance scale. In each of the following sentences the underlined parts balance each other.

> Many are called but few are chosen. (Matthew 22:14)

> I came to bury Caesar, not to praise him. (William Shakespeare, *Julius Caesar*)

> Where I used to suture the tissue of the body together, now I suture words together. (Richard Selzer)

When you read a balanced sentence aloud, you tend to pause between the balanced parts. That pause is often marked by a coordinating conjunction (*but, or, nor, yet, and*), sometimes by *not* (as in the second sentence), and sometimes by punctuation alone (as in the third sentence). Whatever the marker, it serves as a fulcrum, the point at which the contrasted parts balance against each other, as the following diagram illustrates.

| but |
| When a man dies on shore, his body remains with his friends, and the "mourners go about the streets," | when a man falls overboard at sea and is lost, there is a suddenness in the event . . . which gives it an air of awful mystery. |
| | (Richard Henry Dana) |

The balanced structure points up the contrast in thought. It is effective when two subjects are to be contrasted within the same sentence.

Exercise

Compose a sentence about each item in each pair. Then combine the two sentences into a balanced sentence.

jogging	swimming	pain	stress
doctors	dentists	painting	photograph
smokers	nonsmokers	mural	miniature

Periodic and Cumulative Sentences

A *periodic* sentence builds to a climactic statement in its final main clause. The writer withholds the main idea until the end of the sentence. Here is an example:

> Just before I went away to college, my father took me aside, as I had expected, and said, as I had not expected, "Now, Son, if a strange woman comes up to you on a street corner and offers to take your watch around the corner and have it engraved, don't do it." (Eric Lax)

The father's remarks lead up to the advice given in final main clause—"don't do it"—which provides a climax, like the punch line of a joke. All the rest of the sentence has been a preparation for that statement.

Cumulative sentences, also called *loose sentences,* reverse the order of periodic sentences. Instead of withholding the main idea until the last clause, the writer states it immediately and then adds examples and details. Compare the style and effect of the cumulative and periodic sentence in these two examples.

> I fought migraine then, ignored the warnings it sent, went to school and later to work in spite of it, sat through lectures in Middle English and presentations to advertisers with involuntary tears running down the right side of my face, threw up in washrooms, stumbled home by instinct, emptied ice trays onto my bed and tried to freeze the pain in my right temple, wished only for a neurosurgeon who would do a lobotomy on house call, and cussed my imagination. (Joan Didion, "In Bed," *The White Album*)

> For doctors, who confront death when they go to work in the morning as routinely as other people deal with balance sheets and computer print outs, and for me, to whom a chest x-ray or a blood test will never again be a simple routine procedure, it is particularly important to face the fact of death squarely, to talk about it with one another. (Alice Trillin, "Of Dragons and Garden Peas: A Cancer Patient Talks to Doctors," *The New England Journal of Medicine*)

Both writers could have reversed the order of their sentences. Didion could have listed her agonies first, building to her main clause, "I fought migraine." But by announcing her struggle first and then providing the details, she exhausts the main idea and forces her reader to experience some of the exhaustion she feels. Similarly, Trillin could have begun with a clause that states her main idea — that doctors and patients talk squarely to each other about death — and then added the details to illustrate her assertion. But by suspending her assertion until she establishes the contrasting perspectives of doctors and patients, she gains the reader's sympathy and interest. In each example the style and effect of the sentence advance the writer's purpose.

▆▆▆ *Exercise*

Write a periodic sentence describing the most frustrating, fulfilling, or comic day you have had recently. Use your opening clauses to build suspense and reveal your final assertion in the last clause. Then reverse this order and write a cumulative sentence. Make your assertion first and then accumulate detailed examples. Which pattern is more effective? Explain your answer.

REVISING SENTENCES

You have been examining the structure of different types of sentences and techniques for increasing the density of sentences through modification, coordination, and subordination. The process of shaping ideas into sentences is a learning process, and as you grope toward a satisfactory statement of your meaning, you will try out different sentence structures, revising your sentences while you write them. But in this section you will be trying, through local revision, to make your sentences more effective expressions of the ideas you intend them to convey, to improve sentences that have already been written — to "tinker," as Selzer suggests, with your completed draft — by revising for *clarity, emphasis, economy,* and *variety.*

Revising for Clarity

This section is concerned only with revising confusing sentence structure, even though lack of clarity can result from faulty grammar or punctuation, misleading pronoun reference, or vague or ambiguous wording, as well. Unclear sentence structure sometimes occurs when a writer tries to pack too much information into one sentence. The following sentence illustrates this problem.

> Last month while I was visiting the federal buildings in Washington on a guided tour, we went to the National Art Gallery, where we had been for an hour when the rest of the group was ready to move on to the Treasury Building and I told a friend with the group that I wanted to stay in the Art Gallery a while longer and I would rejoin the group about a half an hour later, but I never did, even though I moved more quickly than I wanted to from room to room, not having seen after about four more hours all that I wanted to see.

As written, this sentence of 108 words consists of three main clauses and eight subordinate clauses. This involved structure is hard going for both writer and reader. The goal of revision should be to simplify the structure by reducing the number of clauses. This can be done by (1) distributing the clauses into two or more sentences or (2) omitting material irrelevant to the writer's purpose.

Here is a revision that follows the first strategy.

> While I was visiting the National Art Gallery with a group tour last month, I decided to stay longer when the group left after an hour, and so I told a friend that I would rejoin the group at the Treasury Building in about half an hour. I moved from room to room much more quickly than I wanted to, but after four more hours I had not seen all I wanted to see. I never did rejoin the tour group that day.

This revision distributes all the original material into three sentences and makes the passage easier to read. In addition, the revision uses 25 fewer words, a reduction of over 23 percent.

Here is a revision that follows the second strategy, cutting the original drastically by leaving out irrelevant material.

> While visiting the National Art Gallery with a tour group last month, I stayed four hours after the group left. Even then I did not see all that I wanted to.

This version reduces the original eleven clauses to four and compresses the 108-word sentence to 31 words in two sentences.

Both revisions are clearer than the original. The first revision is minor, since it makes little change in content. The second is major, since it both selects and reorganizes the content. In addition to these revisions, others are possible. Try a few variations to see which you prefer.

Notice that the revisions above reduce the amount of information in the original sentence. This may seem to contradict what was said earlier about combining sentences to increase their density. But in fact there is no contradiction. Some sentences should be combined to achieve greater density; others should be separated into several sentences to achieve greater clarity. The decision to combine or to separate, to enrich or to simplify, depends on your material and your best judgment about your audience and purpose.

■■■ *Exercise*

Simplify the structure of the following sentences to make them easier to read.

1. A controversy has centered on a commonly used herbicide called 2-4-5-T, with producers insisting that it is not harmful to humans but with independent researchers saying there is evidence that it can contribute to miscarriages and development of cancer, among other problems, in people exposed to it, and the controversy has been highly publicized recently because the herbicide is similar in its chemical make-up to the defoliant used by the American military in Vietnam that is now suspected of causing cancer and other serious diseases in people who were exposed to it there.

2. For centuries artists have known that the paint that they keep in their studios and that they allow to collect on their clothing, and that they are constantly touching and breathing, so that it penetrates their skin and gets into their blood stream, contains lead, which can poison them, causing them to have convulsive cramps, fatigue, and making them look sickly by comparison to the portraits of the healthy subjects they paint.

3. In the movies, artists are often portrayed as tormented and anguished people who live in poverty and squalor that causes them to contract all sorts of strange diseases that make them suffer so that they become more sensitive than normal people, which enables them to create beautiful art in the midst of their illness, even though they have to die for their art and only become famous once they are dead.

Revising for Emphasis

Emphasis reflects your purpose and helps you convey that purpose to your readers. Among the numerous ways available to express any idea, the most effective are those that underscore your purpose — that best achieve the effect you have in mind. You can create purposeful emphasis by means of *emphatic word order* (including *climactic order*), *emphatic repetition,* and *emphatic voice.*

Emphatic Word Order   To employ emphatic word order, you must know what you want to emphasize and which positions in a sentence provide the most emphasis. In an English sentence the positions of greatest emphasis are the beginning and the end. The most important material should be placed in those positions; less important material, in mid-sentence. Unimportant details piled up at the end of a sentence get more emphasis than they deserve and make your readers feel that the sentence is "running out of gas."

Notice the difference between the following statements.

Unemphatic Order	*Emphatic Order*
From 1904 to 1914, Americans built the 50-mile Panama Canal, which caused over five thousand workers to die of malaria in the jungles of Central America.	Over five thousand workers died of malaria in the jungles of Central America when Americans labored from 1904 to 1914 to build the 50-mile Panama Canal.

The most important information in this sentence is that over five thousand workers died of malaria to build a 50-mile canal — about one hundred workers per mile. The version on the left puts the number of deaths and the size of the canal in the least important position, lessening the impact. The version on the right puts this information where it will get the most emphasis, relegating the less important information to the middle of the statement.

████ *Exercise*

Revise the order of each sentence to emphasize the points you think most important.

1. He said that the United Nations had failed in its chief function, to preserve peace, although it had done much of which it could be proud in the areas of health and literacy and was still performing valuable services in other areas.

2. It is entirely possible that longevity depends on the luck of your genes, I sometimes think.

3. A proposal that has caused much discussion about our national health policy is the one about insurance for catastrophic illness that is now before Congress.

4. A problem that is important to our environment, noise pollution, is one that we have only recently given much attention to.

5. Dr. Albert Schweitzer, even when he was ninety, worked in his clinic all day and played his piano at night because he believed that a sense of purpose and creativity was the best medicine for any illness he might have.

Climactic Order   Climactic word order achieves emphasis by building to a major idea. The effect of a periodic sentence depends on climactic order, but climax may be used in other sentences as well. The following example contrasts anticlimactic and climactic order. Study both versions of the sentence to determine what changes were made in revision.

Anticlimactic Order	*Climactic Order*
Near the end of *A Separate Peace,* Dr. Stanpole says to Gene that he must tell him that his best friend Finny is dead, the sort of news that the doctor fears the boys of Gene's generation will hear much of.	Near the end of *A Separate Peace,* Dr. Stanpole tells Gene that he must give him the sort of news he fears the boys of Gene's generation will hear much of, that Gene's best friend Finny is dead.

Emphatic Repetition   Unintentional repetition generally weakens a sentence, as the following examples show:

The disappointing results were all the more disappointing because we were sure that the experiment would be a success, and so were disappointed in the results.

The writer who wrote the novel that won the award for the best novel of the year did not attend the awards ceremony.

Deliberate repetition, by contrast, can produce a desired emphasis. You have seen how the repetition of words can help unify a paragraph and how the use of parallel and balanced structures creates coherence. The following examples show the effective use of deliberate repetition.

If we *write* about our scientific *observations,* our *observations* inevitably color our *writing.* (Leigh Hafrey, "Write About What You Know: Big Bang or Grecian Urn")

And this hell was simply, that he had never in his life owned anything — *not his* wife, *not his* house, *not his* child — which could *not,* at any instant, be taken from him by the power of white people. (James Baldwin)

Emphatic Voice   Although it is a commonplace that writers should use verbs in the active rather than in the passive voice, this sound advice is worth repeating. The active voice creates natural, vigorous sentences; the passive voice encourages awkward shifts in structure and anemic, evasive wordiness.

Weak Passive	*More Emphatic Active Voice*
Fantasies of the self are created by modern artists.	Modern artists create fantasies of the self.
Real people afflicted by real diseases have been helped by medical science.	Medical science helps real people afflicted by real diseases.

▰ *Exercise*

Revise the following sentences by changing passive verbs to the active voice.

1. Changing the criteria for what people consider ugly has been proposed by critics as the goal of much modern art.

2. It must surely be recognized by the American Hospital Association that such costs cannot be afforded by many families.

3. The critic said that the winners in the photographic competition would be announced by her within three days.

4. Once the danger was gone, the safety precautions that had been so carefully observed by us were abandoned.

But in some cases, the passive voice may provide greater emphasis than the active voice. Passive constructions can be used to by-pass a sentence's grammatical subject in order to emphasize a more important element, such as a significant action or object of an action. For example:

Active Voice	*Passive Voice*
A person may not smoke in this section of the plane.	Smoking is prohibited in this section of the plane.
The doctor performed the emergency surgery under battery-operated lights.	The emergency surgery was performed under battery-operated lights.

Choosing active or passive voice is like any other choice you make in the writing process. You must judge by results: Which form provides the emphasis you want? But, to avoid the awkward or ungrammatical sentences that can result from misuse of the passive, use the active voice unless the passive advances your purpose more effectively.

Revising for Economy

Economical prose achieves an equivalence between the number of words used and the amount of meaning they convey. A sentence is not economical because it is short or wordy because it is long. Consider these two statements.

I should like to make it entirely clear to one and all that neither I nor any of my associates or fellow workers had anything at all to do in any way, shape, or form with this illicit and legally unjustifiable act that has been committed.	I want to make it clear to everyone that neither I nor any of my associates had anything to do with this illegal act.

The version on the left takes 46 words to say what is more clearly said on the right in 24. The 22 additional words do not add significant information: they merely make reading more difficult and annoy the reader by useless repetition.

Now contrast the following statements.

His defense is not believable.	His defense is not believable: at points it is contradicted by the unanimous testimony of other witnesses, and it offers no proof that that testimony is false; it ignores significant facts about which there can be no dispute, or evades them by saying that he does not recollect them; it contains inconsistencies that he is unable to resolve, even when specifically asked to do so.

The version on the right contains over twelve times more words than the one on the left, and its greater length is justified by the greater information it provides. Both versions express the same judgment, but the second presents the reasons for that judgment. If the writer believes these reasons must be stated, it would be foolish to omit them simply to compose a shorter sentence. Decisions about economy must always be made in relation to meaning and purpose.

Wordiness — the failure to achieve economy — is a common writing problem. Essays can be wordy because of scanty planning, corrected through global revision, or a monotonous style, corrected through better use of coordination and subordination. To eliminate wordiness *within* a sentence entails local revision. The two most common methods are deleting useless words and phrases and substituting more economical expressions for wordy ones.

Cutting Out Useless Words and Phrases

~~It seems unnecessary to point out that~~ the purpose of chemistry lab is to give students ~~the kind of~~ practical experience ~~they need~~ in testing chemical formulas.

~~I would say in response to your question that~~ the task of the art teacher is to help students ~~develop the ability~~ to understand the function of shape, line, color, and light.

Picasso ~~was an artist who~~ took ~~everyday~~ common objects such as a ~~bicycle~~ seat and the handlebars of an old bicycle and ~~through the process of his imagination~~ transformed them into ~~a work of art called~~ *Bull's Head.*

Substituting an Economical Expression for a Wordy One

^{Contemporary researchers in}
~~The forward looking thinking of those working in the area of~~ coronary
^{emphasize}
artery disease ~~tends to place a great deal of emphasis on~~ the potential of

laser technology.

^{an international}
The Armory Show, ~~a famous~~ exhibition of painting and sculpture ~~that~~
^{held}
~~brought together the works of artists from throughout the world and~~

~~displayed them~~ in the Armory of the 69th Regiment in New York City in
^{introduced}
1913, ~~enabled~~ the American public to ~~see for the first time artistic~~

~~movements such as~~ Cubism and Expressionism.

^{Over}   ^{ago}   ^{argued that}
~~It has been more than~~ thirty years ~~since~~ C. P. Snow ~~presented his~~
^{our}
~~argument that the~~ world ~~we live in~~ is divided into *two cultures* — one ~~of~~
^{dedicated to}
~~these cultures was preoccupied with~~ the humanities; the other ~~was~~
^{to}
~~interested only in~~ the sciences.

▰ *Exercise*

Eliminate wordiness in the following sentences.

1. As far as the average citizen is concerned, it is probable that most people are not greatly concerned with the latest critical reaction to the new fads in the world of art.

2. When we studied defense mechanisms, which we did in our psychology class, I discovered that I use most of the mechanisms that are discussed in the textbook.

3. Concerning the question of whether men are stronger than women, it seems to me that the answer is variable, depending on how one interprets the word *stronger*.

4. When, after much careful and painstaking study of the many and various problems involved, the administrators in charge of the different phases of the operation of the hospital made the decision to build a rehabilitation center for patients with all sorts of disabilities, a completely new staff of doctors, nurses, and physical therapists had to be hired.

Revising for Variety

Variety is a characteristic not of single sentences but of a succession of sentences, and it is best seen in a paragraph. But variety is achieved through modification, coordination, subordination, and changes in word order as shown in the following example.

1. Maxwell Perkins was born in 1884 and died in 1947.

2. He worked for Charles Scribner's Sons for thirty-seven years.

3. He was the head editor for Scribner's for the last twenty of those thirty-seven years.

4. He was almost certainly the most important American editor in the first half of the twentieth century.

5. He worked closely with Thomas Wolfe, Scott Fitzgerald, and Ernest Hemingway.

6. He also worked closely with a number of other well-known writers.

The sentences are of similar length (10, 9, 15, 17, 11, and 11 words, respectively) and structure (subject + predicate). Their lack of variety becomes monotonous. Now contrast the same passage revised for variety.

> Maxwell Perkins (1884–1947), head editor of Charles Scribner's Sons for the last twenty of his thirty-seven years with that company, was almost certainly the most important American editor in the first half of the twentieth century. Among the many well-known writers with whom he worked closely were Thomas Wolfe, Scott Fitzgerald, and Ernest Hemingway.

The revision combines the original into two sentences of 37 and 18 words, respectively, and results in greater economy (57 words instead of 73), greater density, and more variety. The following operations produced the revision.

■ Sentences 1, 2, 3, and 4 of the original were combined into a new sentence by: (a) making *Maxwell Perkins* the subject of the new sentence; (b) placing his dates in parentheses; (c) reducing sentences 2 and 3 to a phrase in apposition with the subject of the new sentence; and (d) making sentence 4 the complement of the new sentence.

■ Sentences 5 and 6 of the original were combined into a new sentence by: (a) having them share a common verb, *were*; and (b) making sentence 6 the subject of *were* and sentence 5 the complement.

■■■■ *Exercise*

Consider the possible revisions listed below the paragraph about Nathaniel Hawthorne's story "The Birthmark." Decide which of those procedures you want to use to revise the paragraph. You do not need to use them all; use those that will give you the best paragraph.

> (1) Nathaniel Hawthorne's "The Birthmark" is one of his most famous short stories. (2) It is essentially a story about the limits of science and human perfectibility. (3) Aylmer has conducted many previous experiments in the attempt to improve nature, but all of them have failed. (4) Even so, he decides to perfect his wife Georgiana's beauty by removing a tiny birthmark on her cheek. (5) At the beginning of his experiment, Aylmer is confident he will succeed. (6) He secludes Georgiana in a private chamber. (7) Then he doctors her with strange medicines concocted in his laboratory. (8) He soon discovers that the birthmark is stronger than he thought. (9) He tries to avoid another failure by giving Georgiana an extremely powerful potion. (10) At the climax of the story, Aylmer sees his experiment succeed. (11) Unfortunately, it does so at a price. (12) Georgiana loses her birthmark. (13) She also loses her life.

Consider these revisions:

■ Combine 1 and 2 by omitting everything after *is* in 1 and adding everything after *essentially* in 2. Write the revised sentence.

■ Reduce most of 3 to a subordinate clause, "whose previous attempts to improve nature have failed," and insert it after *Aylmer* in 3 and before *decides* in 4. Write the revised sentence.

■ Leave 5 as it is.

■ Combine 6, 7, and 8 into a parallel structure using *but* to separate 7 and 8. Write the revised sentence.

■ Change the pattern of 9 by beginning the sentence with an introductory phrase, "To avoid another failure," and converting "by risking" to "he decides to risk." Write the revised sentence.

■ Join 10 and 11 by adding a dash after *succeed* and then inserting the phrase "at a price." Write the revised sentence.

■ Combine 12 and 13 with coordinating conjunction *and*. Write the revised sentence.

Using any of those revised sentences, or any revisions of your own, rewrite the complete paragraph.

Here are three pieces of advice on sentence variety.

1. *Don't overdo it.* It is neither necessary nor effective to give every sentence a different structure. Within a paragraph, try to have *some* variety in the pattern of your sentences. Most of your sentences will probably be basic sentences containing about 20 words. Individual sentences will range from 10 words or fewer to 30 words or more and will include balanced, periodic, or cumulative structures.

2. *Postpone revising for variety until you have written your first draft.* As Richard Selzer suggests, the process of *tinkering,* of opening up and compressing sentences, takes place after you have a number of sentences on the page. As you rework your completed sentences, read them aloud, listening for the variations in sound and rhythm that Selzer says are essential to effective writing. You can even *see* recurring structures in sentences, just as Barich sees unvaried patterns in paragraphs, by noticing that they all occupy about the same number of lines.

3. *Be aware of the effect that sentence length has on your readers.* In general, long sentences slow down reading, and short ones speed it up. Short sentences are often effective as topic sentences because they state the general idea simply; longer sentences are often needed to develop the idea. Short sentences are excellent for communicating a series of actions, emotions, or impressions; longer sentences are more appropriate for analysis and explanation. Short sentences are closer to the rhythms of speech and are therefore suitable when you want to adopt an intimate tone and a conversational style. The more formal the tone and style, the more likely you are to use long and complex sentences.

Keep in mind that those statements are relative to the particular material and perspective you are trying to present to your readers. As in all writing, your choice should reflect your purpose.

—————————————— *Review Exercises* ▐

1. Edit the following paragraph into a more effective statement. First read the paragraph as a whole. Then go over it sentence by sentence, making whatever changes you think desirable. Finally, rewrite the paragraph in its revised form.

One of the conceptions not founded in fact that many people have about abstract painters is that they don't possess the skills necessary to draw a landscape or a face. In fact, most abstract painters have developed the ability to be extremely adept at drawing objects and people with photographic accuracy. Instead of painting such realistic portraits, however, they design thickly woven textures of paint that enable them to create more universal symbols that express aspects of all human experience. Some of these symbol are called "biomorphic" because they appear to resemble organic forms or fragments of human anatomy sandwiched into densely packed spatial landscapes. These symbols are often interpreted to express the abstract painter's sense of fragmentation in modern society.

2. Revise these sentences into an effective, coherent paragraph on surgeons, using all the information provided.
   a. Surgeons are revered as medical priests.
   b. They are cloaked and masked in green vestments.
   c. They have acquired a secret knowledge.
   d. They are given absolute authority over life and death.
   e. They cut the human body with special tools.
   f. They eliminate disease.
   g. They stitch the body together.
   h. They wash their hands after the ritual.
   i. The people they restore to health view the process as a miracle.

3. Review Richard Selzer's comments on composing sentences on page 183. Then read the excerpt from his essay "The Knife," below. Determine how he has created a "distinctive rhythm and resonance" in his sentences by using different patterns of expression.

The Knife

One holds the knife as one holds the bow of a cello or a tulip — by the stem. Not palmed nor gripped nor grasped, but lightly, with the tips of the fingers. The knife is not for pressing. It is for drawing across the field of skin. Like a slender fish, it waits, at the ready, then, go! It darts, followed by a fine wake of red. The flesh parts, falling away to yellow globules of fat. Even now, after so many times, I still marvel at its power — cold, gleaming, silent. More, I am still struck with a kind of dread that it is I in whose hand the blade travels, that my hand is its vehicle, that yet again this terrible steel-bellied thing and I have conspired for a most unnatural purpose, the laying open of the body of a human being.

A stillness settles in my heart and is carried to my hand. It is the quietude of resolve layered over fear. And it is this resolve that lowers us, my knife and me, deeper and deeper into the person beneath. It is an entry into the body that is nothing like a caress; still, it is among the gentlest of acts. Then stroke and stroke again, and we are joined by other instruments, hemostats and forceps, until the wound blooms with strange flowers whose looped handles fall to the side in steely array.

There is sound, the tight click of clamps fixing teeth into severed blood vessels, the snuffle and gargle of the suction machine clearing the field of blood for the next stroke, the litany of monosyllables with which one prays his way down and in: *clamp, sponge, suture, tie, cut.* And there is color. The green of the cloth, the white of the sponges, the red and yellow of the body. Beneath the fat lies the fascia, the tough fibrous sheet encasing the muscles. It must be sliced and the red beef of the muscles separated. Now there are retractors to hold apart the wound. Hands move together, part, weave. We are fully engaged, like children absorbed in a game or the craftsmen of some place like Damascus.

Deeper still. The peritoneum, pink and gleaming and membranous, bulges into the wound. It is grasped with forceps, and opened. For the first time we can see into the cavity of the abdomen. Such a primitive place. One expects to find drawings of buffalo on the walls. The sense of trespassing is keener now, heightened by the world's light illuminating the organs, their secret colors revealed — maroon and salmon and yellow. The vista is sweetly vulnerable at this moment, a kind of welcoming. An arc of the liver shines high and on the right, like a dark sun. It laps over the pink sweep of the stomach, from whose lower border the gauzy omentum is draped, and through which veil one sees, sinuous, slow as just-fed snakes, the indolent coils of the intestine.

You turn aside to wash your gloves. It is a ritual cleansing. One enters this temple double washed. Here is man as microcosm, representing in all his parts the earth, perhaps the universe.

I must confess that the priestliness of my profession has ever been impressed on me. In the beginning there are vows, taken with all solemnity. Then there is the endless harsh novitiate of training, much fatigue, much sacrifice.

"The Gross Clinic," by Thomas Eakins (1844–1916). Oil on canvas (1875).
From the Jefferson Medical College of Thomas Jefferson University, Philadelphia.

At last one emerges as celebrant, standing close to the truth lying curtained in the Ark of the body. Not surplice and cassock but mask and gown are your regalia. You hold no chalice, but a knife. There is no wine, no wafer. There are only the facts of blood and flesh.

And if the surgeon is like a poet, then the scars you have made on countless bodies are like verses into the fashioning of which you have poured your soul. I think that if years later I were to see the trace from an old incision of mine, I should know it at once, as one recognizes his pet expressions.

But mostly you are a traveler in a dangerous country, advancing into the moist and jungly cleft your hands have made. Eyes and ears are shuttered from the land you left behind; mind empties itself of all other thought. You are the root of groping fingers. It is a fine hour for the fingers, their sense of touch so enhanced. The blind must know this feeling. Oh, there is risk everywhere. One goes lightly. The spleen. No! No! Do not touch the spleen that lurks below the left leaf of the diaphragm, a manta ray in a coral cave, its bloody tongue protruding. One poke and it might rupture, exploding with sudden hemorrhage. The filmy omentum must not be torn, the intestine scraped or denuded. The hand finds the liver, palms it, fingers running along its sharp lower edge, admiring. Here are the twin mounds of the kidneys, the apron of the omentum hanging in front of the intestinal coils. One lifts it aside and the fingers dip among the loops, searching, mapping territory, establishing boundaries. Deeper still, and the womb is touched, then held like a small muscular bottle — the womb and its earlike appendages, the ovaries. How they do nestle in the cup of a man's hand, their power all dormant. They are frailty itself.

There is a hush in the room. Speech stops. The hands of others, assistants and nurses, are still. Only the voice of the patient's respiration remains. It is the rhythm of a quiet sea, the sound of waiting. Then you speak, slowly, the terse entries of a Himalayan climber reporting back.

"The stomach is okay. Greater curvature clean. No sign of ulcer. Pylorus, duodenum fine. Now comes the gall-bladder. No stones. Right kidney, left, all right. Liver . . . uh-oh."

Your speech lowers to a whisper, falters, stops for a long, long moment, then picks up again at the end of a sigh that comes through your mask like a last exhalation.

"Three big hard ones in the left lobe, one on the right. Metastatic deposits. Bad, bad. Where's the primary? Got to be coming from somewhere."

The arm shifts direction and the fingers drop lower and lower into the pelvis — the body impaled now upon the arm of the surgeon to the hilt of the elbow.

"Here it is."

The voice goes flat, all business now.

"Tumor in the sigmoid colon, wrapped all around it, pretty tight. We'll take out a sleeve of the bowel. No colostomy. Not that, anyway. But, God, there's a lot of it down there, Here, you take a feel."

You step back from the table, and lean into a sterile basin of water, resting on stiff arms, while the others locate the cancer.

Writing Assignments

1. Compose a freewriting exercise about your most memorable experience with illness. Follow Richard Selzer's advice and simply make one "brush stroke" after another until you have "painted" a fairly complete portrait of the experience. Then "doctor" your sentences — expanding and combining them to revise your freewriting into a finished essay.

2. Read several articles or chapters in books about a particular twentieth-century painter. Then write an essay explaining how the painter used certain techniques — applying paint with fingers, brushes, spray devices, or palette knives, or pouring paint directly from cans — to create the overall effect of one of his or her famous paintings.

3. Richard Selzer uses many comparisons in the passage above from "The Knife" to illustrate the process he is describing. Select one comparison and write an extended analysis that justifies the two points in the comparison. For example, in what ways is the artistry of the surgeon similar to the artistry of the writer?

4. The photograph on page 182 was selected as a metaphor representing an important aspect of this chapter. Examine the photograph and determine the possible meanings of this metaphor as you discern them in the photograph. Then write an essay explaining how the metaphor has been reinforced or revised by the ideas discussed in this chapter. You may want to suggest another metaphor to illustrate your understanding of these ideas.

9
DICTION: THE CHOICE OF WORDS

As you think your way through a sentence, you inevitably search for the best words to convey your thoughts. Sometimes, especially when you are quite clear about what you want to say, the words come so easily that you are hardly aware of choosing them. At other times, especially when you are trying to discover what you want to say, you find yourself scratching out one choice after another as you search for the exact word to express your meaning. Such revisions are not necessarily a sign of indecision. The best writers worry constantly about diction — the selection and use of words for effective communication. Perhaps they are the best writers partly because they take pains to choose the best word.

Words are not right or wrong in themselves. What makes a particular word right is the effect it creates in the context of your sentence or paragraph. Annie Dillard, celebrated for her evocative descriptions of nature, explains how she chooses words.

> I learn words by learning worlds. Any writer does that out of simple curiosity. When I choose words, I think about their effect — of course. I like to create a rich prose surface that pommels the reader with verbs and images. I think of them as jabs. Jab, jab, jab, left. Jab, jab, jab, right. That's the vigor I want. (Personal interview)

To create vigor in your writing and to advance your purpose, you must, like Dillard, learn the words that represent the "worlds" you want to write about and learn to use words for their effect. You must learn the denotations and connotations of words.

DENOTATION AND CONNOTATION

The most familiar use of words is to name things — plants, people, oceans, stars. When words are used in this way, the things they refer to are called their *denotations*. The word *molecule* most commonly denotes small structures of atoms. The denotation of *Mars* is the fourth planet from the sun. The denotation is a word's explicit meaning.

But some words acquire connotations as well as denotations. A connotation is an *implicit* meaning, an implied or suggested attitude that is not stated outright. When you label theories about life on Mars "improbable" or "preposterous," you are not only describing them, you are also expressing, and inviting your readers to share, an attitude toward them.

In each of these sentences, the writer implies a different attitude toward similar events.

■ My wife asked me why I was *slashing* the shrubbery. I told her I was merely *pruning* it.

■ The difference between *childish pranks* and acts of *vandalism* depends on whose child does the mischief.

■ Although most bathers thought the high surf looked *threatening,* a few thought it looked *challenging.*

■ When our team of scientists traveled across the crater to *collect samples,* they encountered another team *stealing evidence.*

The contrast in each example is not between denotations and connotations but between favorable and unfavorable connotations. *Slashing* and *pruning* refer to similar actions; but the first implies destructive recklessness, and the second suggests careful cutting.

The words you choose should support your purpose. If you wish to report objectively, select words that suggest a neutral attitude: *collect samples,* not *stealing evidence.* If you wish to convey a tolerant or approving attitude, use words that invite a tolerant or approving response: *pruning, childish pranks, challenging.* If you wish to suggest disapproval, select words with unfavorable connotations: *slashing, act of vandalism, threatening.*

■■■■■ *Exercise*

In the following sentences the blank may be filled with any of the words in parentheses, but each choice creates different connotations. Discuss how each choice affects the writer's intention and the reader's interpretation of the sentence.

1. The roses _____ the trellis. (climbed, adorned, strangled)

2. She was a _____ reader. (compulsive, critical, perceptive)

3. The children were _____. (sleepy, exhausted, weary)

4. The reef _____ beneath the surface. (appeared, loomed, glimmered)

5. The comet _____ across the night sky. (shot, blinked, blazed)

THREE QUALITIES OF EFFECTIVE DICTION

Choice of diction is always made with reference to a particular sentence and to the total context of your writing. For this reason no dictionary or thesaurus will give you *the* right word. A dictionary presents a word's various meanings, and a thesaurus provides a list of synonyms, words with slightly varying meanings. You must decide which word and meaning meet your needs. To make this decision, consider the qualities of effective diction: *appropriateness, specificity,* and *imagery.*

Appropriateness

Words are appropriate when they are suited to your subject, audience, and purpose. Imagine an astronomer reporting the discovery of a new star to a convention of scientists and then to the viewers of a morning television show. The subject is the same, but the audience and purpose are so different that the speaker alters the content, manner, and language of the report. Choices to accommodate audience and purpose affect not only diction, the subject of this chapter, but tone and style, the subjects of the next chapter. Diction, tone, and style are alike in requiring you to make important decisions about the degree of formality appropriate for a given context. Jeans and a T-shirt are not appropriate for a formal dance; an evening gown or tuxedo is conspicuously inappropriate for the classroom; in the same way, some words inappropriate in some situations are perfectly acceptable in others. The best way to understand this distinction is to consider four types of words: *learned, popular, colloquial,* and *slang.*

Learned and Popular Words   Most words in English, as in other languages, are common to the speech of educated and uneducated speakers alike. These words are the basic elements of everyday communication. They are called *popular words* because they belong to the whole populace.

By contrast, there are words that you read more often than you hear, write more often than you speak — words used more widely by educated than by uneducated people, and more likely to be used on formal than on informal occasions. Such words are called *learned words.*

The following list contrasts pairs of popular and learned words that have similar denotations.

Popular	*Learned*	*Popular*	*Learned*
agree	concur	help	succor
begin	commence	move easily	facilitate
clear	lucid	secret	esoteric
disagree	remonstrate	think	cogitate
end	terminate	wordy	verbose

Colloquialisms   The term *colloquial* is defined by *The American Heritage Dictionary* as "characteristic of or appropriate to the spoken language or writing that seeks its effect; informal in diction or style of expression." Colloquialisms are not "incorrect" or "bad" English. They are the kinds of words people, educated and uneducated alike, use when they are speaking together informally. Their deliberate use in writing conveys the impression of direct and intimate conversation. To achieve this effect you might use contractions (*don't, wasn't, hasn't*) or clipped words (*taxi, phone*). Other typical colloquialisms are:

awfully (*for* very)	fix (*for* predicament)	movie (*for* film)
back of (*for* behind)	it's me	over with (*for* completed)
cute	kind of (*for* somewhat)	peeve (*for* annoy)
exam	a lot of; lots of	plenty (*as an adverb*)
expect (*for* suppose)	mad (*for* angry)	sure (*for* certainly)

Slang   The *Oxford English Dictionary* defines slang as "language of a highly colloquial type." Notice that the adjective used is *colloquial*, not *vulgar* or *incorrect*. Slang is used by everyone. The appropriateness of slang, however, depends on the occasion. A college president would probably avoid slang in a public speech but might use it at an informal gathering.

Slang satisfies a desire for novelty of expression. It is often borrowed from the vocabularies of particular occupations or activities: *input* (computer technology), *on the beam* (aerial navigation), *behind the eight ball* (pool), *dunk* (basketball). Some slang words are from the "insider" languages of those who commit crimes or take drugs: *sting, torch, grass, stoned*. Much slang is borrowed from the popular vocabulary and given new meanings: *flipped, split, cool, soul, rap, high, trip, wheels, vibes, bread*. Some slang proves so imaginative and useful that it becomes part of the popular vocabulary, but most quickly becomes dated and obscure and loses its impact.

The scale below shows the four types of diction within the range of formality.

Learned   Popular   Colloquial   Slang

Most formal   *Least formal*

In most of your writing, words from the middle of this scale will be appropriate. Unfortunately some inexperienced writers think that formality is a virtue and that big, fancy words are more impressive than short, common ones. If writers cannot maintain an appropriate level of formality, their diction becomes strained and inconsistent.

The following passage provides a humorous illustration of such inconsistency. In this scene from George Bernard Shaw's play *Pygmalion* (from which the musical *My Fair Lady* was created), Liza Doolittle, a cockney flower girl who is being taught by Professor Higgins to speak like a lady, meets her first test at a small party at the home of Mrs. Higgins, the professor's mother. Notice the contrast between Liza's first speech and her last.

Mrs. Higgins:	Will it rain, do you think?
Liza:	The shallow depression in the west of these islands is likely to move slowly in an easterly direction. There are no indications of any great change in the barometrical situation.
Freddy:	Ha! Ha! How awfully funny!
Liza:	What is wrong with that, young man? I bet I got it right.
Freddy:	Killing!
Mrs. Eynsford Hill:	I'm sure I hope it won't turn cold. There's so much influenza about. It runs right through our whole family regularly every spring.
Liza:	My aunt died of influenza: so they said. . . . But it's my belief they done the old woman in. . . . Why should *she* die of influenza? She come through diphtheria right enough the year before. I saw her with my own eyes. Fairly blue with it, she was. They all thought she was dead; but my father he kept ladling gin down her throat 'til she came to so sudden that she bit the bowl off the spoon. . . . What call would a woman with that strength in her have to die of influenza? What become of her new straw hat that should have come to me? Somebody pinched it; and what I say is, them as pinched it done her in.

The obvious switch from formal to highly colloquial speech is justified by Shaw's purpose, which is to show Liza in a transitional stage at which she cannot yet consistently maintain the pose of being a well-educated young woman. Liza does not see that her learned comment on the weather is inappropriate in this situation, and therefore she has no idea why Freddy is laughing. When the subject changes to influenza, she forgets she is supposed to be a lady and reverts to her natural speech — which is much more expressive and colorful than her phony formality. Her inconsistency is amusing, as Shaw meant it to be.

But in the compositions of inexperienced writers, most inconsitencies are not intended for humorous effect. They slip in when the writer is not in control of *how* to say *what* he or she wants to say. The writer may start off like Liza, hoping to make a good impression, but the writer's natural voice asserts itself and the result is neither formal nor informal diction but an embarrassing mixture. Try to choose words consistently appropriate to your purpose throughout the writing process, and, of course, in revision, stay alert for unintentional shifts in diction.

■ *Exercise*

The following paragraph does not maintain a consistent level of diction. Identify the words and phrases that seem too formal or too informal. Then rewrite the paragraph, substituting more appropriate diction to make the language of the paragraph consistent.

In my perusal of the morning paper, I often pause to take a gander at my horoscope. This stuff is supposed to be figured out on a chart of the heavens, which manifests the positions of the sun, moon, and the signs of the zodiac at the honest to goodness time and location of your birth. These configurations are then juxta-posed to the twelve hours of the celestial sphere. The signs are presumed to hold sway over certain parts of the body, and the houses are supposed to tell you what's happening in the various conditions of life. The degree of influence attributed to these houses depends on a bunch of factors. Sometimes my horo-scope predicts the orb of my daily activities with confounding accuracy. But most of the time it's just hogwash.

Specificity

General and *specific* are opposite terms. Words are general when they refer not to individual things but to groups or classes: *mother, flowers, hurricane.* Words are specific when they refer to individual persons, objects, or events: *Joe's mother, the flowers in the vase near the window, Hurricane Allen.* A general term may be made more specific with a modifier that restricts the reference to a particular member of the group or class.

Specific and *general* are also relative terms: a word may seem specific in one context and general in another, as this diagram shows.

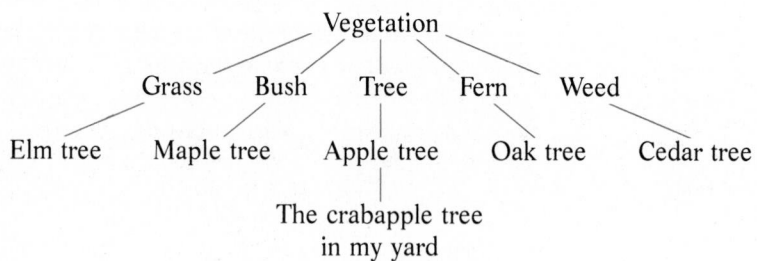

Most general	Vegetation
Less general	Grass Bush Tree Fern Weed
Still less general	Elm tree Maple tree Apple tree Oak tree Cedar tree
Specific	The crabapple tree in my yard

Exercise

For each set of terms, show the gradation from general to specific. Place the most general term at the left and the most specific at the right, as in this example:

matter, food, fruit, citrus fruit, orange

1. Labrador retriever, quadruped, bird dog, animal, dog
2. protons, molecule, electrons, atom, nucleus
3. bush, rosebush, plant, decorative bush, Tropicana rosebush
4. Jupiter, Milky Way, sun, solar system, galaxy
5. boxer, heavyweight, Joe Lewis, athlete

Your context determines whether a specific or a general word is required. Some purposes require generalities. A president's inaugural address, for exam-ple, does not deal with specifics; it states general policies and intentions. The best policy is to be as specific as the situation permits. Notice how the specific lan-

guage in the examples on the right communicates meaning that is no[t]
by the general diction on the left.

A drop of water contains particles.	A drop of water contains microsco[pic] strands of algae and multicellular aquatic organisms.
Saturn has rings.	Saturn is surrounded by a vast swarm of meteoric fragments revolving in various orbits.

The term *concrete* is used to describe some kinds of specific diction. *Concrete* is the opposite of *abstract*. Concrete words refer to particular things or qualities that can be perceived by your senses: details of appearance, sounds, smells, textures, tastes. Abstract words refer to qualities shared by many people or things: newness, width, size, shape, value, joy, anger. Abstract qualities cannot be perceived directly by observation; they are concepts that you infer from what you see.

In the following paragraph, Annie Dillard uses concrete, sensory detail to illustrate her abstract opening sentence. The image of successive flocks of red-winged blackbirds materializing from the dense, green foliage of the Osage orange tree remains in the mind in a way impossible for concepts such as *nature* or *revelation*.

> For nature does reveal as well as conceal; now-you-don't-see-it, now-you-do. For a week this September migrating red-winged blackbirds were feeding heavily down by Tinker Creek at the back of the house. One day I went out to investigate the racket; I walked up to a tree, an Osage orange, and a hundred birds flew away. They simply materialized out of the tree. I saw a tree, then a whisk of color, then a tree again. I walked closer and another hundred blackbirds took flight. Not a branch, not a twig budged: the birds were apparently weightless as well as invisible. Or, it was as if the leaves of the Osage orange had been freed from a spell in the form of red-winged blackbirds; they flew from the tree, caught my eye in the sky, and vanished. When I looked again at the tree, the leaves had reassembled as if nothing had happened. Finally I walked directly to the trunk of the tree and a final hundred, the real diehards, appeared, spread, and vanished. How could so many hide in the tree without my seeing them? The Osage orange, unruffled, looked just as it had looked from the house, when three hundred red-winged blackbirds cried from its crown. I looked upstream where they flew, and they were gone. Searching, I couldn't spot one. I wandered upstream to force them to play their hand, but they'd crossed the creek and scattered. One show to a customer. These appearances catch at my throat; they are the free gifts, the bright coppers at the roots of trees. (Annie Dillard, "Sight Into Insight," *Pilgrim at Tinker Creek*)

Words that refer to sensory experiences — to what you see, hear, touch, taste, and smell — call up sensory images (particularly when they are embedded in strong verbs) and create the "rich prose surface" Dillard uses to "jab" her reader. The following list gives additional examples (some words fit into more than one sensory category).

■ *Touch*   chill, clammy, cold, grainy, gritty, jarring, knobby, moist, numb, rough, smooth, sting, tingle

■ *Taste*   bland, bitter, brackish, metallic, minty, peppery, salty, sour, spicy, sweet

■ *Smell*   acrid, fetid, greasy, musky, musty, pungent, putrid, rancid, rank, reek, stench

■ *Sound*   bellow, blare, buzz, chime, clang, clatter, clink, crackle, crash, creak, gurgle, hiss, hum, murmur, pop, purr, rattle, rustle, screech, snap, squeak, whine, whisper

■ *Sight*   blaze, bleary, bloody, chalky, dappled, ebony, flame, flicker, florid, foggy, gaudy, glare, glitter, glossy, grimy, haze, inky, leaden, muddy, pallid, sallow, shadow, smudged, streak, tawny

■■■ *Exercise*

1. Read the following passage and note the words or phrases that seem most concrete. Then copy the passage over without those words or phrases. In what ways is the meaning of the second version diminished?

   A single knoll rises out of the plain in Oklahoma, north and west of the Wichita Range. For my people, the Kiowas, it is an old landmark, and they gave it the name Rainy Mountain. The hardest weather in the world is there. Winter brings blizzards, hot tornadic winds arise in the spring, and in summer the prairie is an anvil's edge. The grass turns brittle and brown, and it cracks beneath your feet. There are green belts along the rivers and creeks, linear groves of hickory and pecan, willow and witch hazel. At a distance in July or August the steaming foliage seems almost to writhe in fire. Great green and yellow grasshoppers are everywhere in the tall grass, popping up like corn to sting the flesh, and tortoises crawl about on the red earth, going nowhere in the plenty of time. Loneliness is an aspect of the land. All things in the plain are isolate; there is no confusion of objects in the eye, but *one* hill or *one* tree or *one* man. To look upon that landscape in the early morning, with the sun at your back, is to lose the sense of proportion. Your imagination comes to life, and this, you think, is where Creation was begun. (N. Scott Momaday, "The Way to Rainy Mountain")

2. The following contrasted statements deal with the same subject. From each pair, choose the statement that you think is more concrete. Explain the specific reasons for your choice.

   In the past, girls in rural communities had no facilities for bathing except those offered by some neighboring stream. In such circumstances a bathing suit was not always a necessity, but if one was worn it was likely to consist of nothing more than some discarded article of clothing tailored to fit the occasion.

   Forty years ago, if the farmer's daughter went swimming she swam in the crick below the pasture, and if she wore a bathing suit, which was not as customary as you may think, it was likely to be a pair of her brother's outgrown overalls trimmed with scissors as her discretion might suggest.

"Magic Moment," by Lori Malott, 1985. Watercolor on handmade watercolor paper. Exhibited from 1985 to 1987 in Kansas Landscape Show. Painting owned by Dr. Rita Stucky, Topeka, Kansas.

Suddenly I felt something on the biceps of my right arm — a queer light touch, clinging for an instant, and then the smooth glide of its body. I could feel the muscles of the snake's body slowly contract and relax. At last I saw a flat, V-shaped head, with two glistening, black, protruding buttons. A thin, pointed, sickening yellow tongue slipped out, then in, accompanied by a sound like that of escaping steam.

Suddenly I felt the snake moving over my arm. I felt the contraction of its muscles as it moved. Then I saw its ugly head and its evil-looking eyes. All the time its tongue kept moving in and out, making a kind of hissing noise.

Imagery

Imagery has two general meanings when applied to diction: the images or pictures that concrete words sometimes suggest and figures of speech such as similes and metaphors. The first meaning includes the pictorial quality of phrases such as *an anvil's edge, green belts, popping up like corn.* In this section you will learn about the second meaning — the use of figurative language.

The chief element in all figures of speech is an imaginative comparison in which dissimilar things are described as having a meaningful similarity. The writer, by thus linking the unfamiliar with the familiar, creates a context in which the reader may more easily or clearly understand a new aspect of the subject or new information and ideas. Here is an example:

> The moon was a ghostly galleon tossed upon cloudy seas. (Alfred Noyes)

The line of poetry compares the moon to a sailing ship. Now in most ways the moon is quite unlike a ship. But as the poet watches it alternately emerging from behind the clouds and disappearing into them again, he thinks of a ship alternately emerging and disappearing from view as it rides the troughs and crests of the waves. In his imagination the moon is being tossed by the clouds as a ship is tossed by the waves.

The most commonly used figures of speech are *simile, metaphor, analogy, personification,* and *allusion.* Each figure makes a comparison, but each has its own characteristic form and use.

Simile   A simile compares two things — *A* and *B* — by asserting that one is like the other. A simile usually contains the word *like, as,* or *so* and is used to transfer to *A* the qualities or feelings associated with *B.* Thus, when Annie Dillard in "Total Eclipse" describes a solar eclipse, she imagines that the sky (*A*) — or more precisely the moon — functions like the lens cover (*B*) on a camera, covering the lens and shutting out all light:

> The sky snapped over the sun like a lens cover.

Here are some other similes:

> Insects in the first frosts of autumn all run down like little clocks. (Loren Eisely, "How Flowers Changed the World," *The Star Thrower*)

> When, as you approach, [the iguanas] swish away, there is a flash of azure, green and purple over the stones, the colour seems to be standing behind them in the air, like a comet's luminous tail. (Isak Dinesen, *Out of Africa*)

> Floating on one's back is like riding between two skies. (Edward Hoagland, "Summer Pond")

> His face was as blank as a pan of uncooked dough. (William Faulkner)

> Laverne wasn't too bad a dancer, but the other one, old Marty, . . . was like dragging the Statue of Liberty around the floor. (J. D. Salinger)

Metaphor   A metaphor compares two things by identifying one with the other. It does not say that *A* is like *B* but instead states that *A* is *B.* Lewis Thomas is using metaphor when he suggests that the sky encloses the earth just as a membrane surrounds an organ or cell.

> Aloft, floating free beneath the moist, gleaming membrane of bright blue sky, is the rising earth, the only exuberant thing in this part of the cosmos. (Lewis Thomas, "The World's Biggest Membrane," *The Lives of a Cell*)

Here are other metaphors:

> Time is but a stream I go a-fishing in. (Henry David Thoreau, *Walden*)

> Suddenly the whole room broke into a sea of shouting, as they saw me rise. Waves of rejoicing swept the place. (Langston Hughes, *The Big Sea*)

> Each rich nation is a lifeboat full of comparatively rich people. In the ocean outside each lifeboat swim the poor of the world who want to get in, or at best to share some of the wealth. (Garrett Hardin, "Lifeboat Ethics: The Case Against Helping the Poor")

Many words and phrases no longer thought of as figures of speech were originally metaphors or similes. "At bay" originally described a hunted animal cornered by pursuers and forced to turn and fight the baying hounds. "Crestfallen" first described a cock that had been humbled in a cockfight. Many other expressions can be analyzed as metaphors although you no longer think of them as figures of speech — expressions such as the "mouth" of a river, the "face" of a clock, the "brow" of a hill. Such expressions are often called, metaphorically, *dead* or *frozen metaphors*. They are so common in the language that it is hard to write a paragraph without them, but you should try to avoid those (such as "rosy" red, "dirt cheap," and "face the music") that have become clichés.

Analogy   An analogy is an extended metaphor that, through several sentences or paragraphs, explains an abstract idea or seeks to persuade readers that because two things are alike, a conclusion drawn from one suggests a similar conclusion from the other. In the following passage, Robert Jastrow uses analogy to explain the shape of our galaxy.

> The Galaxy is flattened by its rotating motion into the shape of a disk whose thickness is roughly one-fifth of its diameter. Most of the stars in the Galaxy are in this disk, although some are located outside it. A relatively small, spherical cluster of stars, called the nucleus of the Galaxy, bulges out of the disk at the center. The center structure resembles a double sombrero with the gigantic nucleus as the crown and the disk as the brim. The sun is located in the brim of the sombrero about three-fifths of the way out from the center of the edge. When we look into the sky in the direction of the disk we see so many stars that they are not visible as separate points of light, but blend together into a luminous band stretching across the sky. This band is called the Milky Way. (Robert Jastrow, "The Size of Things," *Red Giants and White Dwarfs*)

Personification   Personification is a figure of speech by which inanimate objects or abstractions are given human or animal characteristics. Thus winds are said to "roar" or "bite"; flames "eat hungrily" at a burning house and many even "devour" it; a tree may "bow" before a gale or in fair weather "stretch" its

Spiral Galaxy NGC, 2997 (type Sc), in Antlia. Photo by David F. Malin,
© 1980 Anglo-Australian Telescope Board.

branches; truth or virtue emerges "triumphant"; and justice is "blind." In these examples, the writer imagines a resemblance between the actions observed and the actions of an animal or person.

Such implied comparisons are often effective, but they should be used with restraint. If they seem exaggerated ("the waves roared their threat to the listening clouds while the palm trees nodded their approval"), your reader is likely to reject them as far-fetched or as an unsuccessful attempt to be "literary."

Allusion   An allusion is a comparison between a historical, literary, or mythological event or person and the subject under discussion. When someone is said to possess Machiavellian guile, he or she is being associated with the crafty manipulation of power outlined by Niccolò Machiavelli in his Renaissance treatise on political power, *The Prince*. Or when a political scandal is called another Teapot Dome or another Watergate, it is being likened to the most notorious political scandals in United States history.

A successful allusion provides a flash of wit or insight and gives your readers the pleasure of recognition. But you must be reasonably sure that an allusion is suited to your audience; if your readers do not recognize or understand the allusion, they will be confused and not appreciate its effect. Calling writing a research paper a "Sisyphean task" is amusing only if your audience remembers that Sisyphus was condemned forever to roll a huge stone up a steep hill in Hades, only to have it roll back down again just as it neared the top.

■■■ *Exercises*

1. In the following selections, underline effective examples of specificity and imagery.

When wild ducks or wild geese migrate in their season, a strange tide rises in the territories over which they sweep. As if magnetized by the great triangular flight, the barnyard fowl leap a foot or two into the air and try to fly, . . . and a vestige of savagery quickens their blood. All the ducks on the farm are transformed for an instant into migrant birds, and into those hard little heads, till now filled with humble images of pools and worms and barnyards, there swims a sense of continental expanse, of the breadth of seas and the salt taste of the ocean wind. The duck totters to the right and left in its wire enclosure, gripped by a sudden passion to perform the impossible and a sudden love whose object is a mystery. (Antoine de Saint Exupéry, *Wind, Sand and Stars*)

The word is terracide. As in homicide, or genocide. Except it's terra. Land.

It is not committed with guns and knives, but with great, relentless bulldozers and thundering dump trucks, with giant shovels like mythological creatures, their girdered necks lifting massive steel mouths high above the tallest trees. And with dynamite. They cut and blast and rip apart mountains to reach the minerals inside, and when they have finished there is nothing left but naked hills, ugly monuments to waste, stripped of everything that once held them in place, cut off from the top and sides and dug out from the inside and then left, restless, to slide down on houses and wash off into rivers and streams, rendering the land unlivable and the water for miles downstream undrinkable.

Terracide. Or, if you prefer, strip-mining. (Skip Rozin, "People of the Ruined Hills," *Audubon*)

"On nights like that," Raymond Chandler once wrote about the Santa Ana, "every booze party ends in a fight. Meek little wives feel the edge of the carving knife and study their husbands' necks. Anything can happen." That was the kind of wind it was. I did not know then that there was any basis for the effect it had on all of us, but it turns out to be another of those cases in which science bears out folk wisdom. The Santa Ana, which is named for one of the canyons it rushes through, is a *foehn* wind, like the *foehn* of Austria and Switzerland and the *hamsin* of Israel. There are a number of persistent malevolent winds, perhaps the best known of which are the mistral of France and the Mediterranean sirocco, but a *foehn* wind has distinct characteristics: it occurs on the leeward slope of a mountain range and, although the air begins as a cold mass, it is warmed as it comes down the mountain and appears finally as a hot dry wind. Whenever and wherever a *foehn* blows, doctors hear about headaches and nausea and allergies, about "nervousness," about "depression." In Los Angeles some teachers do not attempt to conduct formal classes during a Santa Ana, because the children become unmanageable. In Switzerland the suicide rate goes up during the *foehn,* and in the courts of some Swiss cantons the wind is considered a mitigating circumstance for crime. Surgeons are said to watch the wind, because blood does not clot normally during a *foehn.* A few years ago an Israeli physicist discovered that not only during such winds, but for the ten or twelve hours which precede them, the air carries an unusually high ratio of positive to negative ions. No one seems to know exactly why that should be; some talk about friction and others suggest solar disturbances. In any case the positive ions are there, and what an excess of positive ions does, in the simplest terms, is make people unhappy. One cannot get much more mechanistic than that. (Joan Didion, "Los Angeles Notebook," *Slouching Towards Bethlehem*)

2. Evaluate the effectiveness of the following figures of speech from the student writing used in previous chapters. Discuss how each figure contributes to the writer's purpose.

   a. Wally, "Brandon's Clown" (Chapter 1):

   Before long, the book looked like a Disneyland mortuary.

   b. Rod, "Our Trip to D.C." (Discovery draft, Chapter 4):

   Whenever I think of our trip, I think of that monument standing in the middle of downtown Washington like a big paperweight.

   c. Jane, "Watching Whales" (Chapter 5):

   When he burst out of the water, he held his flippers at a slight angle to his side. As he descended, he moved them gracefully away from his body like the arms of a ballerina finishing a pirouette.

   d. Matt, "Drug Abuse in the Work Place" (Chapter 6):

   All of us — employers, workers, consumers — want to cure this plague that threatens the health of our culture.

REVISING DICTION

In considering the qualities of effective diction, you evaluated ways of expressing ideas, and your evaluation included a great deal of revision. In this stage of local revision, you need to change your emphasis from expressing yourself effectively to rooting out the ineffective. Work through your text slowly, looking for words that don't fully express your purpose. Four major weaknesses to watch for are *vagueness, jargon, triteness,* and *ineffective imagery.*

Eliminating Vagueness

Words are vague when, in context, they do not convey a single, specific meaning to your readers. Consider this sentence:

   I could tell by the funny look on her face that she was mad.

Funny and *mad* can have quite specific meanings, but not in that sentence. What does *funny* mean here — a purposeful attempt to create laughter? What does *mad* mean — "insane," as it might in another sentence? Certainly not. "Angry," then, or "annoyed" or "irritated" or "offended"? A reader cannot be sure. The writer can remove doubt by using more specific diction.

I could tell by the *way her face stiffened* that she was *offended.*

Words like *funny* and *mad* belong to a group called *utility words.* Such words, as their name implies, are useful. In ordinary conversation, which does not usually permit or require deliberate choice and offers little chance for revision, utility words are common and often pass unnoticed. In writing, they may be adequate if the context limits them to a single, clear interpretation. But because their meaning is often left vague, they should be used with caution. The following list shows some of the most common utility words:

affair	funny	marvelous	regular
awful	gadget	matter	silly
business	glamorous	nature	situation
cute	goods	neat	stuff
fantastic	gorgeous	nice	terrible
fierce	great	outfit	terrific
fine	line	peculiar	weird
freak	lovely	pretty	wonderful

Usually the simplest way to clarify a vague utility word or phrase is to substitute a specific word or phrase, as in the following examples:

It was a ~~peculiar~~ statement. *(puzzling)*

The news is ~~terrible.~~ *(alarming)*

The weather will be ~~fantastic.~~ *(clear, sunny, and in the low 80s.)*

Vagueness is not limited to unclear utility words. Any word or phrase that is more general than the intended meaning should be revised. The substitutions in these sentences make the information more specific.

The class was discussing ~~an essay.~~ *(Annie Dillard's "Total Eclipse.")*

Professor Jones is studying ~~erosion.~~ *(how this winter's storms eroded the sand dunes on Cape Cod.)*

■ *Exercises*

1. Assume that the italicized utility words are not made clear by context. Substitute more specific diction to give a precise meaning.

   a. She is a doctor, but I don't know what her *line* is.

   b. Our *organization* thought the *matter* was *peculiar*.

   c. One *aspect* of the *proposition* is its effect on prices.

   d. What a *terrific* surprise to meet so many *cool* people at the same *affair*.

   e. The price they are charging for gasoline is *something else*.

2. Circle the more specific expression in each parenthetical set.

   The whole surface of the ice was *(a chaos – full)* of movement. It looked like an enormous *(mass – jigsaw puzzle)* stretching away to infinity and being *(pushed – crunched)* together by some invisible but irresistible force. The impression of its *(titanic – great)* power was heightened by the unhurried deliberateness of the motion. Whenever two thick *(pieces – floes)* came together, their edges *(met – butted)* and *(moved – ground)* against one another for a time. Then, when neither of them showed signs of yielding, they rose *(uncertainly – quiveringly)*, driven by the *(implacable – tremendous)* power behind them. Sometimes they would stop *(altogether – abruptly)* as the unseen forces affecting the ice appeared mysteriously to lose interest. More frequently, though, the two floes — often ten feet thick or more — would continue to rise, *(rearing up – tenting up)* until one or both of them toppled over, creating a pressure ridge.

Eliminating Jargon

Jargon originally meant meaningless chatter. Later it meant the specialized language of a group or profession, as in *habeas corpus* (law) and *cursor* (computer technology). There is no reason not to use learned and technical terms for audiences and situations for which they are appropriate, but to use them unnecessarily when addressing a general audience is a violation of the basic rule that your style should fit your purpose and audience. Jargon in informal writing is pretentious and frustrating to your audience. The following contrast shows inappropriate jargon that results in vague, indirect writing. The version on the left comes from the King James translation of the Bible; the version on the right is George Orwell's "translation" of the same material into modern jargon.

I returned and saw under the sun, that the race is not to the swift, nor the battle to the strong, neither yet bread to the wise, nor yet riches to men of understanding, nor yet favor to men of skill; but time and chance happeneth to them all.	Objective considerations of contemporary phenomena compel the conclusion that success or failure in competitive activities exhibits no tendency to be commensurate with innate capacity, but that a considerable element of the unpredictable must invariably be taken into account.

Although the Biblical version first appeared in 1611, it makes more sense and is easier to read than the "translation," which smothers simplicity and clarity under a blanket of vague, polysyllabic words. As Orwell points out:

The first contains forty-nine words but only sixty syllables, and all of its words are those of everyday life. The second contains thirty-eight words of ninety syllables; eighteen of its words are from Latin roots, and one from Greek. The first sentence contains six vivid images, and only one phrase ("time and chance") that could be called vague. The second contains not a single fresh, arresting phrase, and in spite of its ninety syllables it gives only a shortened version of the meaning contained in the first. ("Politics and the English Language," *Shooting an Elephant and Other Essays*)

Jargon has three chief characteristics:

- Highly abstract, often technical, diction that shows a fondness for learned rather than popular words: *have the capability to* for *can, maximize productivity* for *increase production,* and *utilization of mechanical equipment* for *use of machinery.*

- Excessive use of the passive voice. If machines break down, they "are found to be functionally impaired." If a plan does not work, "its objectives were not realized." If management failed to consider the effects of certain changes on the workers, the error is reported like this: "With respect to employee reactions, management seems to have been inadequately advised."

- Conspicuous wordiness, as illustrated in the examples given above.

Jargon combines inappropriateness, vagueness, and wordiness into one consistently unintelligible style. Writers who lapse into jargon do so because they believe that ordinary language is not good enough. Like Liza Doolittle, they are trying to make a good impression. But the best way to make a good impression with your writing is to have something to say and to say it clearly.

The only way to revise jargon is to get rid of it. Leaving it in your writing forces your readers to do mentally the rewriting that you should have done, with the difficulty that they may not be sure what you meant.

Notice how the following passage creates ambiguity for the reader.

(1) Rigorous comprehension and innermost instincts are the features of collaborative inquiry that spelunkers must engage when they delve into the unexplored terrain of a new cave. (2) They must scrutinize the formation of the rocks and the current of the river, establish their directional movement with the assistance of a compass, and maintain a precise topographical record so that they can re-negotiate their expedition to its inception. (3) But since they have no established record, they must still commit their full confidence to their instincts when they reach an intersection. At that juncture they must contemplate whether traversing one passageway will bring them into contact with a chamber enlarged by a high ceiling or with a corridor that narrows into a cul-de-sac.

The writer probably consulted a thesaurus to dress up the language of that needlessly "learned" paragraph. If you rewrite the paragraph, reducing each sentence to a summary of what the writer was trying to say, you get this:

1. Spelunkers use knowledge and instincts to explore a new cave.

2. They study rock formations and river currents; they use a compass; and they need to find their way out.

3. Since they have no maps, at intersections they must rely on their instincts.

4. They may turn into big chambers or dead ends.

If you join the revised sentences, you get a paragraph that says all the original was trying to say, but says it more clearly and in fewer words — 51 words instead of 119. You might get a still better revision by tinkering with the sentence structure, adding or subtracting words where appropriate:

When they explore a new cave, spelunkers rely on educated guesswork. They study the rock formation and river current, chart their movement with a compass, and mark their trail so that they can find their way out. But since they have no map, they must trust their instincts at each intersection. They may select a passage that turns into a vaulted chamber or a dead end.

The revised paragraph says nothing that was not said in the revised sentences. But it sharpens the focus by beginning with a clear topic sentence and then

Drawing by M. Stevens; © 1985 The New Yorker Magazine, Inc.

explaining that sentence. The pomposities of the original version have been eliminated, and the meaning is conveyed in half the space.

▪▪▪▪ *Exercise*

Using the procedure shown above, revise this paragraph.

> Last month as I was perambulating through the stacks, I happened to encounter a book filled with photographs of volcanoes. Not yet having realized my latent interest in such geological formations, I was surprised at the manner in which this book enthralled my attention. The color displays of the spectacular discharge of Washington's St. Helens and the florid lava flows of Hawaii's Kilauea I found to be particularly disquieting. It was these photographs which led me to the decision to espouse volcanoes as the subject of my research paper.

▬▬▬▬ Eliminating Triteness

The terms *trite, hackneyed,* and *cliché* are used to describe expressions, once colorful and apt, that have been used so often that they have lost their freshness and force. Like outdated slang, trite expressions once evoked images and conveyed a sense of discovery through language. Striking when new, such phrases become meaningless when they are overused. Note these examples:

apple of her eye	hook, line, and sinker
birds of a feather	lock, stock, and barrel
black sheep	mountains out of molehills
blind as a bat	sober as a judge
budding genius	teeth like pearls
diamond in the rough	thick as thieves
fly in the ointment	water over the dam

Trite diction blocks thought. Writers who use ready-made phrases instead of creating their own soon find it difficult to think beyond stereotyped expressions: any change in personnel becomes a "shakeup"; all hopes become "fond," "foolish," or "forlorn"; defeats are "crushing"; changes in the existing system are "noble experiments" or "dangerous departures"; unexpected occurrences are "bolts from the blue"; and people who "sow wild oats" always have to "pay the piper" even though they are "as poor as church mice."

When you spot triteness in your writing, you should remove it. Unfortunately, what an experienced writer considers trite may seem original to an inexperienced writer. The way to learn to detect and avoid triteness in your writing is to learn to recognize triteness in the speech and writing of others. Notice the italicized clichés in this paragraph.

> *Money doesn't grow on trees,* and how well I've learned that. What a *rude awakening* when I realized that the odd change I used to ask for at home was not available at school. I had thought my allowance was enormous but, before I

knew it, it had *trickled through my fingers*. College has taught me that *"A penny saved is a penny earned."* I have learned to live within my allowance and even to *save something for a rainy day*. What I have learned in college is *not all in the books*. I have learned to *shoulder my responsibilities*.

How many of those phrases do you recognize as prefabricated units that can be inserted into any sentence? If you can identify triteness in another writer's work, you can identify it in your own.

▬▬▬ *Exercises*

1. Politics, television, popular music, and sports seem to encourage the use of trite expressions. Although some writers interpret these subjects with originality and insight, most seem to type their texts on a cliché machine. Examine the following paragraph to determine whether it provides any insights into football or simply repeats clichés.

   Wherever the gridiron game is played in the United States — on a sandlot, a high school field, or in a college or professional stadium — the players learn through the school of hard knocks the invaluable lesson that only by the men's blending together like birds of a feather can the team win. It is a lesson they do not forget on the gridiron. Off the field, they duly remember it. In society, the former player does not look upon himself as a lone wolf on the prowl who has the right to do his own thing — that is, to observe only his individual social laws. He knows he is a part of the big picture and must conduct himself as such. He realizes that only by playing as a team man can he do his share in making society what it should be — the protector and benefactor of all. The man who has been willing to make the sacrifice to play football knows that teamwork is essential in this modern day and age and that every citizen must pull his weight in the boat if the nation is to prosper. So he has little difficulty in adjusting to his roles in family life and in the world of business and to his duties as a citizen in the total scheme of things. In short, his football training helps make him a better citizen and person, better able to play the big game of life.

2. Rewrite that paragraph to eliminate the major objections made about it during class discussion.

3. Choose as a general subject a current issue in the world of politics, television, or popular music. Restrict that general subject to a topic that you think is important to students, and write a letter to the campus newspaper expressing your opinion. Then reread your letter to catch any trite language you may have used.

▬▬▬▬▬
Eliminating Ineffective Imagery

An effective figure of speech can make your writing concrete; a figure that is trite, far-fetched, obscure, or confused mars your writing and distracts your readers. *Mixed metaphors* — metaphors that try to combine two or more incompatible images in a single figure — are especially ineffective. Consider the following:

The businessman decided to test the political water by throwing his hat into the ring as a trial balloon.

This sentence mixes three overused, equally unimaginative images. "Testing the water" has a literal meaning in cooking and bathing. "Throwing one's hat into the ring" was the conventional way of issuing a challenge in the days of bare-knuckle prize fights. Trial balloons were once used to determine wind direction and velocity. Each of those images can be used in a political context, but fitting all three into one consistent figure of speech is impossible. The writer of the sentence should abandon metaphor and simply say, "The businessman decided to gauge his chances of being nominated by announcing his intention to run."

A poor figure of speech is worse than none at all. Imagery is ruined if the picture it conveys is preposterous or dull. If the figure you intended to strengthen your writing instead weakens it, delete or revise it. Test every figure of speech by *seeing* the picture it presents. If you visualize what your words are saying, then you will see and eliminate ineffective or mixed images during revision.

■■■■ *Exercise*

These sentences try unsuccessfully to combine discordant elements in the same figure of speech. Visualize the images they suggest, and then revise each sentence by creating an acceptable figure or restating the idea without a figure.

1. A host of music fans flocked into the arena like an avalanche.

2. The president's ill-advised action has thrown the ship of state into low gear, and unless the members of Congress wipe out party lines and carry the ball as a team, it may take months to get the country back on the track.

3. When I try to focus my microscope, I do something that throws a monkey wrench into the ointment.

4. NASA expressed confidence that the hearing would allow the real facts of the case to come out in the wash.

Review Exercise ■■■■■■■■■■■

Compare Annie Dillard's comments (p. 211) on choosing and using words with her diction in this excerpt from her collection *Teaching a Stone to Talk*. Select several specific examples where her word choice exhibits the three qualities of effective diction — appropriateness, specificity, and imagery. How does each example advance her purpose?

Total Eclipse

It began with no ado. It was odd that such a well-advertised public event should have no starting gun, no overture, no introductory speaker. I should have known right then that I was out of my depth. Without pause or preamble, silent as orbits, a piece of the sun went away. We looked at it through welders' goggles. A piece of the sun was missing; in its place we saw empty sky.

I had seen a partial eclipse in 1970. A partial eclipse is very interesting. It bears almost no relation to a total eclipse. Seeing a partial eclipse bears the same relation to seeing a total eclipse as kissing a man does to marrying him, or as flying in an airplane does to falling out of an airplane. Although the one experience precedes the other, it in no way prepares you for it. During a partial eclipse the sky does not darken — not even when 94 percent of the sun is hidden. Nor does the sun, seen colorless through protective devices, seem terribly strange. We have all seen a sliver of light in the sky; we have all seen the crescent moon by day. However, during a partial eclipse the air does indeed get cold, precisely as if someone were standing between you and the fire. And blackbirds do fly back to their roosts. I had seen a partial eclipse before, and here was another.

What you see in an eclipse is entirely different from what you know. It is especially different for those of us whose grasp of astronomy is so frail that, given a flashlight, a grapefruit, two oranges, and fifteen years, we still could not figure out which way to set the clocks for Daylight Saving Time. Usually it is a bit of a trick to keep your knowledge from blinding you. But during an eclipse it is easy. What you see is much more convincing than any wild-eyed theory you may know.

You may read that the moon has something to do with eclipses. I have never seen the moon yet. You do not see the moon. So near the sun, it is as completely invisible as the stars are by day. What you see before your eyes is the sun going through phases. It gets narrower and narrower, as the waning moon does, and, like the ordinary moon, it travels alone in the simple sky. The sky is of course background. It does not appear to eat the sun; it is far behind the sun. The sun simply shaves away; gradually, you see less sun and more sky.

The sky's blue was deepening, but there was no darkness. The sun was a wide crescent, like a segment of tangerine. The wind freshened and blew steadily over the hill. The eastern hill across the highway grew dusky and sharp. The towns and orchards in the valley to the south were dissolving into the blue light. Only the thin river held a trickle of sun.

Now the sky to the west deepened to indigo, a color never seen. A dark sky usually loses color. This was a saturated, deep indigo, up in the air. Stuck up into that unworldly sky was the cone of Mount Adams, and the alpenglow was upon it. The alpenglow is that red light of sunset which holds out on snowy mountaintops long after the valleys and tablelands are dimmed. "Look at Mount Adams," I said, and that was the last sane moment I remember.

I turned back to the sun. It was going. The sun was going, and the world was wrong. The grasses were wrong; they were platinum. Their every detail of stem, head, and blade shone lightless and artificially distinct as an art photographer's platinum print. This color has never been seen on earth. The hues were metallic; their finish was matte. The hillside was a nineteenth-century tinted photograph from which the tints had faded. All the people you see in the photograph, distinct and detailed as their faces look, are now dead. The sky was navy blue. My hands were silver. All the distant hills' grasses were finespun metal which the wind laid down. I was watching a faded color print of a movie filmed in the Middle Ages; I was standing in it, by some mistake. I was standing in a movie of hillside grasses filmed in the Middle Ages. I missed my own century, the people I knew, and the real light of day.

I looked at Gary. He was in the film. Everything was lost. He was a platinum print, a dead artist's version of life. I saw on his skull the darkness of night mixed with the colors of day. My mind was going out; my eyes were receding the way galaxies recede to the rim of space. Gary was light-years away, gesturing inside a circle of darkness, down the wrong end of a telescope. He smiled as if he saw me; the stringy crinkles around his eyes moved. The sight of him, familiar and wrong, was something I was remembering from centuries hence, from the other side of death: yes, *that* is the way he used to look, when we were living. When it was our generation's turn to be alive. I could not hear him; the wind was too loud. Behind him the sun was going. We had all started down a chute of time. At first it was pleasant; now there was no stopping it. Gary was chuting away across space, moving and talking and catching my eye, chuting down the long corridor of separation. The skin on his face moved like thin bronze plating that would peel.

The grass at our feet was wild barley. It was the wild einkorn wheat which grew on the hilly flanks of the Zagros Mountains, above the

Euphrates valley, above the valley of the river we called *River*. We harvested the grass with stone sickles, I remember. We found the grasses on the hillsides; we built our shelter beside them and cut them down. That is how he used to look then, that one, moving and living and catching my eye, with the sky so dark behind him, and the wind blowing. God save our life.

From all the hills came screams. A piece of sky beside the crescent sun was detaching. It was a loosened circle of evening sky, suddenly lighted from the back. It was an abrupt black body out of nowhere; it was a flat disk; it was almost over the sun. That is when there were screams. At once this disk of sky slid over the sun like a lid. The sky snapped over the sun like a lens cover. The hatch in the brain slammed. Abruptly it was dark night, on the land and in the sky. In the night sky was a tiny ring of light. The hole where the sun belongs is very small. A thin ring of light marked its place. There was no sound. The eyes dried, the arteries drained, the lungs hushed. There was no world. We were the world's dead people rotating and orbiting around and around, embedded in the planet's crust, while the earth rolled down. Our minds were light-years distant, forgetful of almost everything. Only an extraordinary act of will could recall to us our former, living selves and our contexts in matter and time. We had, it seems, loved the planet and loved our lives, but could no longer remember the way of them. We got the light wrong. In the sky was something that should not be there. In the black sky was a ring of light. It was a thin ring, an old, thin silver wedding band, an old, worn ring. It was an old wedding band in the sky, or a morsel of bone. There were stars. It was all over.

Writing Assignments

1. Mark Twain once said "The difference between the right word and the nearly right word is the difference between lightning and a lightning bug." Compose a narrative essay about the consequences of your inability to select the right word(s) to explain your behavior on an important occasion.

2. Select a troublesome term that you have encountered in your reading about science — such as *entropy, refraction,* or *void.* Look up its history in the *Oxford English Dictionary* and some of its synonyms in *Roget's Thesaurus.* Then compose an extended definition of the word, pointing out its denotative and connotative meanings.

3. Compose a causal analysis in which you explain why extraordinary events such as a solar eclipse seem to produce extreme psychological reactions. You may want to read Annie Dillard's complete essay, "Total Eclipse," in *Teaching a Stone to Talk,* for her reflections on the subject.

4. The photograph on page 210 was selected as a metaphor representing an important aspect of this chapter. Examine the photograph and determine the possible meanings of this metaphor as you discern them in the photograph. Then write an essay explaining how the metaphor has been reinforced or revised by the ideas discussed in this chapter. You may want to suggest another metaphor to illustrate your understanding of these ideas.

10
TONE AND STYLE

In Chapters 8 and 9, you examined sentence structure and diction separately, but you must also consider their joint impact on the tone and style of a piece of writing. You are generally familiar with the words tone and style, of course. When someone says, "Don't speak to me in that tone," or, "The style of that book is difficult," you know what is meant. In this chapter, however, *tone* and *style* are technical concepts that you need to understand and control to improve your writing.

As you work your way through the stages of the writing process, you try to sort out what you think about your material and how you want your readers to think about it. For example, Ellen's initial planning (in Chapter 3) captured much of Barry's quirkiness, but her scratch outline reduced his behavior to abstractions (*food, exercise,* and *culture*). This led her to eliminate Barry and adopt an academic tone for her working hypothesis and discovery draft. She discovered this problem through her descriptive outline and concluded that her draft sounded scholarly but wasn't very interesting. By returning to Barry in her new outline and subsequent draft, she achieved a more entertaining tone and engaging style and revealed Barry's quirkiness to her readers.

Essayist Patricia Hampl, like Ellen, finds that the right tone is essential for an effective writing style.

> The issue of tone is central to me because I want my readers to know that another human being is speaking to them. I've never had much interest in the objective tone of the godlike narrator. It's supposed to give your writing a kind of neutral authority, but I don't think it exists. Even God has a personality. There's the God of light. The God of mercy. The philosophical issue at stake here is how to maintain the middle position, the position of a human being who possesses both authority and reliability. I try to create that balance by mixing a formal tone with a more conversational tone. You can deal with only so much operatic truth. Then you have to acknowledge the popcorn stand and the peanut gallery. (Personal interview)

TONE   Range of Tone: Informative to Affective

As Hampl suggests, writers can adopt an objective or a subjective tone. You adopt the first when your purpose is to give your readers authoritative information. You adopt the second when your purpose is to affect or influence your readers in some way. The following two excerpts demonstrate these two extremes of tone.

> The United States was the first nation in history so many of whose citizens could go so far simply in quest of fun and culture. The size of this phenomenon made international travel, for the first time, a major element in world trade, a new problem for the American economy and for American balance of payments, and a new opportunity for the destination countries. In 1970 the Department of Commerce estimated that the expenditures of American travelers overseas had reached $2 billion each year. In the United States, economists began to count foreign travel as a major import, and other countries began to plan for tourism by Americans as a principal export. (Daniel Boorstin, *The Americans: The Democratic Experience*)

Notice the following characteristics of this passage:

- Boorstin is principally concerned with giving readers information about his subject. He assumes that they will be interested in the historic and economic significance of travel.

- Boorstin's writing is informative and objective — he does not show his personal feelings. He supports his judgment about America's preoccupation with travel by citing experts: the Department of Commerce and economists.

- As a result of Boorstin's purpose and strategies, his tone is informative, factual, and impersonal.

Contrast Boorstin's passage with this one:

> I don't know where we got the idea we have to go away for a vacation. I suppose the travel industry sold it to us. The travel business is the second largest industry in the United States. It's always trying to get us to go someplace *else* to spend our money when most of us don't have any trouble spending it right here where we are. The industry tries to make us feel cheap if we don't go on an expensive trip.
>   Well, I've got my plans all made for my vacation. I'm not going *anywhere:* How do you like that, travel industry? Show me all the luxurious accommodations you want, tempt me with pictures of bikini-clad girls with windswept hair on pearl-white beaches — I'm not going. I'm staying put is what I'm doing. We've been tourists, and none of us likes being a tourist, so this summer I'll be somewhere I've never been on vacation — right where I live. (Andy Rooney, *A Few Minutes With Andy Rooney*)

This sample is quite different from Boorstin's.

■ Rooney's purpose is to offer his opinion, not to give information. He does tell his readers that the travel industry is the second largest in the United States, but he tells them little more about it — except how it makes him feel.

■ Rooney's writing is affective and *subjective*. Rooney writes almost exclusively about how *he* feels about the subject of vacations: "I've got my plans," "I'm not going *anywhere*," "I'm staying put." His language reveals his defiant attitude: "How do you like that, travel industry?"

■ As a result of Rooney's purpose and strategies, his tone is affective, opinionated, and personal.

These two passages represent the extremes of tone on a scale ranging from informative to affective.

Informative ———————————————————————— Affective
(Boorstin passage)   (Rooney passage)

Between the extremes, you could place other samples that show varying degrees of objectivity and subjectivity; those in the middle would be equally balanced between the two.

Each time you write, you will have to decide how informative or affective your writing should be. Certain assignments — laboratory reports, summaries of books or events, essay examinations — usually require an informative tone. Persuasive essays, intended to convince a reader to believe or do something, are usually affective. But, as Hampl indicates, a balance of informative and affective elements produces the most flexible and natural tone. The balance in your essays will depend on your subject and purpose. Rod's final essay on the Washington Monument (in Chapter 4) is informative because it provides facts about the monument's history, but it is also affective because Rod tries to show his readers why he feels that there is "more to George's monument than meets the eye."

Inexperienced writers sometimes believe that the way to be objective is to avoid writing in the first person. However, when you are writing about your own experience, ideas, and feelings, you need to use the pronoun *I*. Moreover, a third-person account is not necessarily an objective treatment of the subject. The statement "I saw the car run through a red light" is more objective than "The crazy fool drove through a red light," even though the first sentence contains *I* and the second does not.

■■■■ *Exercise*

Where on the scale above would you place the following passages? Examine each passage closely so that you can support your judgment by reference to specific details.

Passage 1

Why is it almost impossible to gaze directly at the Grand Canyon under these circumstances and see it for what it is — as one picks up a strange object from one's back yard and gazes directly at it? It is almost impossible because the Grand Canyon, the thing as it is, has been appropriated by the symbolic complex which has already been formed in the sightseer's mind. Seeing the canyon under approved circumstances is seeing the symbolic complex head on. The thing is no longer the thing as it confronted the Spaniard; it is rather that which has already been formulated — by picture postcard, geography book, tourist folders, and the words *Grand Canyon*. As a result of this preformulation, the source of the sightseer's pleasure undergoes a shift. Where the wonder and delight of the Spaniard arose from his penetration of the thing itself, from a progressive discovery of depths, patterns, colors, shadows, etc., now the sightseer measures his satisfaction *by the degree to which the canyon conforms to the preformed complex*. If it does so, if it looks just like the postcard, he is pleased; he might even say, "Why it is every bit as beautiful as a picture postcard!" He feels he has not been cheated. But if it does not conform, if the colors are somber, he will not be able to see it directly; he will only be conscious of the disparity between what it is and what it is supposed to be. He will say later that he was unlucky in not being there at the right time. The highest point, the term of the sightseer's satisfaction, is not the sovereign discovery of the thing before him; it is rather the measuring up of the thing to the criterion of the preformed symbolic complex. (Walker Percy, "The Loss of the Creature," *The Message in the Bottle*)

Passage 2

Everything comes onto the island: nothing much goes off, even by evaporation. Once it was a gateway to a New World, now it is a portal chiefly to itself. Manhattan long ago abandoned its melting-pot function. Nobody even tries to Americanize the Lebanese or the Lithuanians now, and indeed the ethnic enclaves of the island seem to me to become more potently ethnic each time I visit the place. Nothing could be much more Italian than the Festival of St. Anthony of Padua down on Mulberry Street, when the families of Little Italy stroll here and there through their estate, pausing often to greet volatile contemporaries and sometimes munching the soft-shelled crabs which, spread-eagled on slices of bread like zoological specimens, are offered loudly for sale by street vendors. Harlem has become almost a private city in itself, no longer to be slummed through by whities after dinner, while Manhattan's Chinatown is as good a place as anywhere in the world to test your skill at that universal challenge, trying to make a Chinese waiter smile. (Jan Morris, "Manhattan: The Islanders," *Destinations: Essays From Rolling Stone*)

Passage 3

The next day when I was already in full flight — aboard a northward bound train — I could not have accounted, if it had been demanded of me, for all the varied forces that were making me reject the culture that had molded and shaped me. I was leaving without a qualm, without a single backward glance. The face of the South that I had known was hostile and forbidding, and yet out of all the conflicts and the curses, the blows and the anger, the tension and the terror, I had somehow gotten the idea that life could be different, could be lived in a fuller and richer manner. As had happened when I had fled the orphan home, I was now running more away from something than toward something. But that did not matter to me. My mood was: I've got to get away; I can't stay here. (Richard Wright, *Black Boy*)

© Gary Ladd

Distance

Another element of tone is the impression of distance between writer and reader.

Consider a professor lecturing to a large class. She is separated from her listeners by her position on a platform; she cannot speak to each student in the audience individually. The situation requires her to speak more slowly, more loudly, and more formally than if she were conferring with a student in her office. The lecture room produces both physical and stylistic distance. Consider the distance in a statement in a printed syllabus that "Students are expected to hand in assignments on the date stipulated" and an instructor's after class remark, "Joe, you've got to get your papers in on time." In the first statement the writer is impersonal and remote; in the second the speaker is personal and close.

The following selections illustrate the difference between writing that tries to get close to the reader and writing that addresses the reader from a distance.

Example 1

Did you ever wonder, looking at a map or whirling a globe, how the intricate shape of continents, shorelines, rivers and mountains came to be mapped, and who designed the system of parallels and meridians? Or how people found out, in the first place, that the earth is shaped like a globe? If we had not been told so I am not ever sure that you or I would have found it out for ourselves. (Erwin Raisz, *Mapping the World*)

Here the distance between writer and reader is very slight. By using the second person, Raisz gives the impression of speaking personally to each reader. His conversational and questioning tone and his diction suggest that he identifies with his readers.

Example 2

The map predates the book (even a fairly ordinary map may contain several books' worth of information). It is the oldest means of information storage, and can present the most subtle facts with great clarity. It is a masterly form of compression, a way of miniaturizing a country or society. Most hill-climbers and perhaps all mountaineers know the thrill at a certain altitude of looking down and recognizing the landscape that is indistinguishable from a map. The only pleasure I take in flying in a jet plane is the experience of matching a coastline or the contour of a river to the corresponding map in my memory. A map can do many things, but I think its chief use is in lessening our fear of foreign parts and helping us anticipate the problems of dislocation. Maps give the world coherence. It seems to me one of man's supreme achievements that he knew the precise shape of every continent and practically every river-vein on earth long before he was able to gaze at them whole from the window of a rocketship. (Paul Theroux, "Mapping the World," *Sunrise With Seamonsters*)

This paragraph is addressed not to any particular reader but to all readers. Theroux does not address his readers as "you" or try to appeal to their special interests. He is more interested in what he has to say than in his audience. His tone establishes greater distance between writer and reader than existed in the passage by Raisz.

The impression of distance comes chiefly from sentence structure and diction, the linguistic bases of tone, discussed below under "Style." Remember that the tone of your writing depends in part on decisions you make about the distance, or degree of separation, you want to maintain between yourself and your readers. The decision is not arbitrary; it is related to decisions about your attitude toward your subject, and it should be consistent with your purpose.

STYLE *The American Heritage Dictionary* defines style as "the way in which something is said or done, as distinguished from its substance." In writing, *style* refers to "how something is said" — the way a writer arranges sentences and words; *substance* means "what is said" — the ideas or message that the writer wishes to convey. The difficulty with the dictionary definition is that it assumes that what is said can somehow be examined apart from how it is said. In writing, however, the *what* and the *how* are virtually indistinguishable. The most subtle changes in sentences or diction can change the ideas or message. A simple, though more accurate definition, then, would be "Style is the way something is written."

How do you describe the way something is written? Consider the following passage from *The Adventures of Huckleberry Finn,* by Mark Twain. Huck, staying with the Grangerfords after his raft has been wrecked, has been reading a "poem" written by the youngest daughter, now dead. The poem, intended to be a sad poem about a young man who fell into a well and drowned, is so maudlin and so badly written that it is funny. Huck thinks it is very good. After reading it, he says

> If Emmeline Grangerford could make poetry like that before she was fourteen, there ain't no telling what she could a done by-and-by. Buck [her younger brother] said she could rattle off poetry like nothing. She didn't ever have to stop and think. He said she would slap down a line, and if she couldn't find anything to rhyme with it she would just scratch it out and slap down another one, and go ahead. She warn't particular; she could write about anything you choose to give her to write about just so it was sadful. Every time a man died, or a woman died, or a child died, she would be on hand with her "tribute" before he was cold. She called them tributes. The neighbors said it was the doctor first, then Emmeline, then the undertaker — the undertaker never got in ahead of Emmeline but once, and then she hung fire on a rhyme for the dead person's name, which was Whistler. She warn't ever the same, after that; she never complained, but she kinder pined away and did not live long.

Notice the following things about this passage:

- Mark Twain is having fun with the subject. He knows the poem is sentimental to the point of being ridiculous, and he expects the reader to see that too.

- Huck himself is serious. Huck, the narrator, says what a boy like him, but not a man like Twain, would say. Even though Twain is the writer, the *voice* you hear is that of Huck Finn. Twain creates an ironical situation in which everything Huck says about Emmeline's poetry confirms a different judgment in his readers.

- The diction is appropriate to the speaker. The humorous effect would be lost if Huck were made to speak like a sophisticated adult. He must speak in his own voice.

- All these elements are so interrelated that a change in any one of them would spoil the total effect.

Even this brief sample suggests that when you write something you must consider your attitude toward your subject, your relationship with your readers, and the language you use to express your ideas. The tone of your writing reveals the decisions you have made about the first two. The third is expressed in the choices you make as you develop your purpose. As you have seen throughout this text, a clear sense of purpose guides you through all the choices you make from planning through revision. If these choices are consistent, your writing will exhibit a distinctive *style* that will express your attitude toward and embody your understanding of your purpose.

Range of Style: Diction, Sentence Structure, and Tone

Style, of course, rests finally on language. Look closely at the sentence structure, diction, and tone of the following passages to see what generalizations you can make about them.

Example 1

> Airplanes are invariably scheduled to depart at such times as 7:54, 9:21 or 11:37. This extreme specificity has the effect on the novice of instilling in him the twin beliefs that he will be *arriving* at 10:08, 1:43 or 4:22, and that he should get to the airport on time. These beliefs are not only erroneous but actually unhealthy, and could easily be dispelled by an attempt on the part of the airlines toward greater realism. Understandably, they may be reluctant to make such a radical change all at once. In an effort to make the transition easier I offer the following graduated alternatives to "Flight 477 to Minneapolis will depart at 8:03 P.M.":
> a. Flight 477 to Minneapolis will depart oh, let's say, eightish.
> b. Flight 477 to Minneapolis will depart around eight, eight-thirty.
> c. Flight 477 to Minneapolis will depart while it's still dark.
> d. Flight 477 to Minneapolis will depart before the paperback is out.
> (Fran Leibowitz, "From Leibowitz' Travel Hints," *Social Studies*)

Analysis of Example 1: Sentence Structure

The nine sentences in this passage average only 17 words in length, and almost half of them are set apart in a list of brief announcements. Most of the sentences consist of one or two independent clauses. Only the second sentence has a slightly inverted construction. The rest of the sentences follow a subject-verb-object order.

Diction

In the whole passage there are proportionately few words more than two syllables long. *Invariably, specificity* and *graduated* are the most notable. Except for *novice* and *erroneous* there are no learned words. There are a few contractions (*let's, it's*) and one example of slang (*eightish*).

Tone

Leibowitz's attitude toward her subject is humorous and ironic. Supposedly informative — after all, she is offering advice — her attitude is really affective because she wants to exploit her readers' exasperation over airline schedules. The distance between writer and reader is slight. Leibowitz writes as a fellow sufferer commenting on a common frustration. You hear the voice of a world-weary crank, slightly angry about a situation over which she has no control.

Summary

With simple sentences and diction, Leibowitz establishes her tone and creates a style that achieves ease and clarity. The diction and the pace of her sentences are those of informal conversation, creating an impression of talking in print. In Chapter 9 the word *colloquial* was used to describe diction of this sort. That term can apply to a style, as well. A colloquial style is not common in college writing, but it is by no means inappropriate when it is used — as it is here — to fulfill the writer's purpose.

Now contrast the next example with Leibowitz's writing. Read the following passage aloud, slowly, to hear how it sounds.

Example 2

What a fierce weird pleasure to lie in my berth at night in the luxurious palace-car, drawn by the mighty Baldwin — embodying, and filling me, too, full of the swiftest motion, and most resistless strength! It is late, perhaps midnight or after — distances join'd like magic — as we speed through Harrisburg, Columbus, Indianapolis. The element of danger adds zest to it all. On we go, rumbling and flashing, with our loud whinnies thrown out from time to time, or trumpet-blasts, into the darkness. Passing the homes of men, the farms, barns, cattle — the silent villages. And the car itself, the sleeper, with curtains drawn and lights turn'd down — in the berths the slumberers, many of them women and children — as on, on, on, we fly like lightning through the night — how strangely sound and sweet they sleep! (They say the French Voltaire in his time designated the grand opera and a ship of war the most signal illustrations of the growth of humanity's and art's advance beyond primitive barbarism. Perhaps if the witty philosopher were here these days, and went in the same car with perfect bedding and feed from New York to San Francisco, he would shift his type and sample to one of our American sleepers.) (Walt Whitman, *Specimen Days*)

Analysis of Example 2: Sentence Structure

The paragraph has only eight sentences. Although there are two short sentences (one a fragment), most of the sentences are long. The average sentence length is 26 words, against 17 in the Leibowitz passage; the last three sentences are 42, 32, and 39 words long. The sentences are not simple. Subjects and verbs are often separated — sometimes widely — by modifiers. Several of the sen-

Lithograph by Currier and Ives, 1884: The Bettmann Archive, Inc.

tences exhibit extremely complicated punctuation. Throughout the paragraph, parallel and periodic structures are used for rhythmic and other effects.

Diction

About 27 percent of the words in the passage are more than two syllables long, and it contains a smaller proportion of monosyllables than the Leibowitz example. Whitman's diction is sprinkled with learned words — *resistless, slumberers, barbarism.* Several phrases have a lofty, poetic ring — "embodying, and filling me, too, full of the swiftest motion, and most resistless strength!"

Tone

Whitman is celebrating the mythic adventure of train travel, and his attitude toward his subject is clearly subjective. The distance between writer and reader is great; the emphasis is on the subject, not on the reader. The affective tone is impersonal, dignified, eloquent.

Summary

As in the Leibowitz example, all the components blend into a consistent style. But what a different style! Whitman aims at eloquence, not ease or familiarity. You would not use this style to give directions, explain a process, answer an examination, or report a story in a newspaper. Whitman's style shares many of the characteristics of nineteenth-century prose, characteristics still found in prose seeking similar effects. It could be called a *grand* style, but the usual name for it is *formal.*

Example 3

In those days differences in vegetation between a main highway and a side road were even greater than now. There weren't many state highways but the best of them stretched straight ahead with mathematical perfection, even though just of gravel. These rights-of-way were usually wide and from the road edge to the boundary fences were mowed pretty much as now, though road building had not been preceded by earth moving. As a result, there were more native plants in among the grass and fewer of those cosmopolitan tramps which take so readily to disturbed habitats. But the side roads, those fascinating side roads, they are mostly gone and there is nothing in the modern road system to compare with them. The wheel track wound here and there, depending upon the slope and the vegetation and the character of the land. Sometimes it was straight for a little way; frequently it wobbled. There was often grass in the roadway outside the actual wheel tracks; shrubs like sumac and elderberry pressed so close to the road that you could smell them as you drove by and children snatched at the flowers. Accommodating drivers of the local stage-line learned to snip off small twigs with a snap of the buggy whip and present them to lady passengers. (Edgar Anderson, "Horse-and-Buggy Countryside," *Landscape*)

Analysis of Example 3: Sentence Structure

This paragraph has nine sentences, all following the standard pattern: subject-verb-object with modifiers and subordinate clauses appearing in predictable places. There are no conspicuous periodic sentences. Sentences do vary in length, from 11 to 38 words, but most fall between 19 and 26, averaging 21 (between the 17-word Leibowitz average and the 26-word Whitman average).

Diction

In this 214-word passage, only 15 words (or about 7 percent) are more than two syllables long. There are a few learned words such as *cosmopolitan* and *habitat,* but they are counterbalanced by popular words such as *snip* and *snap.*

Tone

Anderson's attitude toward his subject is both informative and affective. He wants to inform his readers about country roads, and he wants to describe how it must have looked and felt to ride down one of these roads in a horse-and-buggy. His choice of details — children snatching flowers, stage drivers snipping off flowers with their buggy whips — helps establish his informal, familiar tone. The distance between writer and reader is greater than in the first example but less than the second example.

Summary

    Anderson's style is neither so simple as Leibowitz's nor so involved as Whitman's. It has none of Leibowitz's chattiness and none of the poetry or exaltation that pervades Whitman's writing. Its balance of sentence length, diction, and tone produces a *moderate* style.

Formal	Moderate	Colloquial
∧	∧	∧
(Whitman passage)	*(Anderson passage)*	*(Leibowitz passage)*

"When I say 'Please pass the butter,' why do you say 'Hey, no problem'?"

The diagram suggests three things.

▨ The scale indicates degree of formality, from most to least; it does not measure degree of excellence. No one style is better than another. All three are standard styles, and each is appropriate in some situations — although the moderate style is appropriate for most writing situations.

▨ Each classification embodies a range. A particular sample may be more or less formal than others. For example, a legal contract is more formal than Whitman's paragraph.

▨ There is no clear division between styles. The overlapping in the diagram is intended to suggest that the moderate style has such a broad range that it may include some formal and colloquial elements. Such inclusions, however, must be consistent with the writer's purpose. Edgar Anderson's use of *habitat*, for example, was justified because his purpose was to identify the characteristics of a particular landscape.

See the table on page 247 for a summary of the three styles.

▬▬ *Exercise*

Analyze the sentence structure, diction, and tone of each of the following passages, using supporting details. Then rate each passage on the scale provided.

Most Formal				Most Colloquial

Competitors in conquest have overlooked the vital soul of Africa herself, from which emanates the true resistance to conquest. The soul is not dead, but silent, the wisdom not lacking, but of such simplicity as to be counted non-existent in the tinker's mind of modern civilization. Africa is of an ancient age and the blood of many of her peoples is as venerable and as chaste as truth. What upstart race, sprung from some recent, callow century to arm itself with steel and boastfulness, can match in purity the blood of a single Masai Murani whose heritage may have stemmed not far from Eden? It is not the weed that is corrupt; roots of the weed sucked first life from the genesis of earth and hold the essence of it still. Always the weed returns; the cultured plant retreats before it. Racial purity, true aristocracy, devolve not from edict, nor from rote, but from the preservation of kinship with the elemental forces and purposes of life whose understanding is not farther beyond the mind of a Native shepherd than beyond the cultured fumblings of a mortar-board intelligence. (Beryl Markham, *West with the Night*)

We were sitting in a restaurant called Orchid Garden, in the Wanchai district, beginning our first meal in Hong Kong, and I had just sampled something called fish-brain soup. I was about to comment. Alice was looking a bit anxious. She was concerned, I think, that over the years I might have created a vision of Hong Kong in my mind that could not be matched by the reality — like some harried businessman who finally arrives in what he has pictured as the remote, otherworldly peace of a Tahiti beach only to be hustled by a couple of hip beach-

Summary of the Formal, Moderate, and Colloquial Styles

	Formal	Moderate	Colloquial
Sentences	Relatively long and involved; likely to make considerable use of parallel, balanced, and periodic structures; fragments rare.	Of medium length, averaging between 15 and 25 words; mostly standard structure but with some use of parallelism and balanced and periodic sentences; fragments occasional.	Short, simple structures; mainly subject-verb-object order; almost no use of balanced or periodic sentences; fragments common.
Diction	Extensive vocabulary, some use of learned and abstract words; no slang; almost no contractions or clipped words.	Ranges from learned to colloquial but mostly popular words; both abstract and concrete diction; occasional contractions and clipped words; may contain some inconspicuous slang.	Diction limited to popular and colloquial words, frequent contractions, and clipped words; frequent use of utility words; more slang than in moderate style.
Tone	Always a serious attitude toward an important subject; may be either subjective or objective and informative or affective; no attempt to establish closeness with reader, who is almost never addressed as "you"; personality of the writer often inconspicuous; whole tone usually dignified and impersonal.	Attitude toward subject may be serious or light, objective or subjective, informative or affective; relationship with reader close but seldom intimate; writer often refers to himself or herself as "I" and to reader as "you"; but the range of moderate style is so broad that it can vary from semi-formal to semi-colloquial.	Attitude toward subject may be serious or light but is usually subjective; close, usually intimate, relation with reader, who is nearly always addressed as "you"; whole tone is that of informal conversation.
Uses	A restricted style used chiefly for scholarly or technical writing for experts, or for essays and speeches that aim at eloquence or inspiration; a distinguished style, but not one for everyday use or practical affairs.	The broadest and most usable style for expository and argumentative writing and for all but the most formal of public speeches; the prevailing style in nontechnical books and magazines, in newspaper reports and editorials, in college lectures and discussions, in all student writing except some fiction.	Light, chatty writing as in letters to close friends; a restricted style that is inappropriate to most college writing except fiction.

umbrella salesmen wearing "Souvenir of Fort Lauderdale" T-shirts. Even before we had a meal, she must have noticed my surprise at discovering that most of the other visitors in Hong Kong seemed to be there for purposes other than eating. That's the sort of thing that can put a visionary off his stride. How would the obsessed mountain climber feel if he arrived in Nepal after years of fantasizing about a clamber up the Himalayas, and found that most of the other tourists had come to observe the jute harvest? It appeared that just about everyone else had come to Hong Kong to shop. Hong Kong has dozens of vast shopping malls — floor after floor of shops run by cheerfully competitive merchants who knock off 10 percent at the hint of a frown and have never heard of sales tax. There are restaurants in some of the shopping malls, but most of the visitors seemed too busy shopping to eat. It was obvious that they would have come to Hong Kong even if it had been one of those British colonies where the natives have been taught to observe the queen's birthday by boiling brussels sprouts for an extra month. That very morning, in the lobby of a hotel, we had noticed a couple in late middle age suddenly drawing close to share some whispered intimacy in what Alice, the romantic, took to be a scene of enduring affection until one of the softly spoken phrases reached her ears — "customs declaration." (Calvin Trillin, "Hong Kong Dream," *Third Helping*)

If you want to get something done, here is a professional secret: don't try to rush it by laying down the law that it *must be done* within a given time, but come back, each day if need be, and ask again, and do it with a smile. While evidences of irritation are fatal, persistence is never resented if it is clothed in good manners and good temper. When you can't wait any longer, the thing to do is to take an unfair advantage of your Brazilian friend — appeal to him to arrange matters for you because, first, you are in trouble with your principals; second, you are embarrassed by your own failure; third, you are obliged to leave by a given boat and will be humiliated if matters have not been arranged — anything to put it on a personal basis. When you do this your Brazilian is lost. He feels that you have been reasonable and patient and that he cannot throw you to the wolves. And in arranging matters for you he probably has to make a series of appeals more or less similar to yours. (Hugh Gibson, *Rio*)

Some Practical Advice About Style

You have examined the elements of style in passages by other writers. Now you are ready to apply what you have learned to your own writing. As you revise your writing, keep these stylistic considerations in mind.

1. *Let your purpose be your guide.* A clear sense of purpose controls all the choices you make at every stage in writing. Style results from that control.

2. *Generally, choose a moderate style.* There is nothing wrong with a formal or a colloquial style when it is appropriate. Unnecessary formality, however, often leads to pretentiousness and wordiness. The writer tries too hard to be impressive or literary when it would be enough to clearly express his purpose. The colloquial style, if used for serious treatment of a serious subject, can undermine the writer's purpose. The best policy, as Hampl suggests, is to maintain the middle position, "the position of a human being who possesses both authority and reliability." You can then mix objective and subjective — informative and affective — as your purpose demands.

3. *Keep your style consistent.* Probably the worst stylistic defect is inconsistency. An inconsistent style is not the same as a moderate style in which formal and colloquial elements are balanced for a purpose. Writing that is totally inconsistent has no discoverable purpose and therefore no discoverable style. The

inconsistencies in tone and diction that often occur in individual paragraphs, sentences, or words can be removed in revision.

a. *Inconsistency in tone.* Conspicuous inconsistency in tone is likely to jar a reader. It is most obvious when colloquial elements appear in a formal style or formal elements in a colloquial style. Because a moderate style can range from semi-formal to semi-colloquial, it can tolerate usages that would be conspicuous in the extreme styles.

b. *Inconsistency in diction.* As you begin planning and drafting your essay, you gradually commit yourself to a recognizable approach to both your subject and your reader. Your choice of words will either contribute to stylistic consistency or obscure the pattern in your writing. You are most likely to confuse readers by choosing words that are close but not close enough to your meaning. Do you want to say that your traveling companions are *insensitive, naive,* or *undiscriminating?* Your purpose determines the "best" choice of words.

4. *Try to see your writing as your readers will see it.* This advice may be the hardest to follow. People assume that what is clear to them will be clear to others, but common experience demonstrates that this assumption is not always true. In everyday conversation, a frequently asked question is "What do you mean by that?" Because your readers will not have the opportunity to ask you that question, try to anticipate their need for clarity and completeness.

5. *Be as specific as you can.* Writing is a difficult medium because it is abstract. The word *apple* is more abstract than any apple you ever ate, because it leaves out your actual experience with apples — their shape, size, color, texture, and taste. The problem all writers face is how to make abstractions concrete. The two common solutions are to (a) illustrate the meaning of general statements with examples, and (b) choose words that are specific.

6. *Revise for style.* You revise for style when you undertake global revision. Your style is dependent on your purpose, so if you revise your purpose, you must revise your style. Likewise, if you revise your subject or audience, you must make new decisions about sentence structure, diction, and tone. For example, when Rod changed the focus of his essay on the Washington Monument and directed it toward a wider audience, he modified his style to avoid colloquial terms and breezy references.

Local revision involves the nuts and bolts of style. In fact, the changes you make in local revision will probably make the most improvement in your style. Follow your local revision with one last reading — just for style. Read the text aloud or, better yet, listen to it read aloud, and concentrate on its overall effect. Imagine yourself listening to a band or a singer performing a piece of music. As you listen to your own writing, concentrate on the style of the piece and keep alert for any false notes.

The following guidelines provide some basic questions to ask yourself when you revise for style.

Guidelines for Revising Your Style

1. *What is my general impression of my writing?*
   Do I find my writing clear, unambiguous, and likely to engage my readers? Have I carried out my purpose at every level; that is, am I satisfied that the *how* of my writing — its attitude, organization, and language — conveys the *what* of my ideas?

2. *What tone have I established in my writing?*
   Is my tone informative, affective, or a blend of both? How much distance have I maintained between myself and my readers? Is my tone appropriate for my subject and audience? Is it maintained consistently?

3. *How can I characterize the overall style of my writing?*
   Have I written this essay in a moderate style, opting for more or less colloquialism or formality as my purpose requires? Does my purpose in fact require me to be overtly colloquial or formal?

4. *Are my sentences well constructed and easy on the ear?*
   Have I written sentences varying in length and style so that they hold my readers' interest? Have I avoided the choppiness that comes from too many basic or loosely coordinated sentences? Have I avoided the density that comes from too many complicated sentences with multiple subordinate clauses?

5. *Have I used words as effectively as possible?*
   Do the connotations and denotations of my words support my purpose? Have I avoided unnecessary formalities and slang? Is my language specific and, when appropriate, vivid? Have I inadvertently mixed metaphors? Have I used imagery successfully to *heighten* effects, not merely to strive after them?

Once you have completed all your revisions and have typed your final draft, you should proofread your writing very carefully. Proofreading is a close reading of the final version to catch errors in grammar, spelling, and punctuation, as well as typographical errors, that survived your revisions or crept into the final copy. Proofreading should be done slowly, preferably aloud. If possible, you should allow some time to elapse between your final typing and your proofreading. In that way you are more likely to read with a fresh eye.

Review Exercise

Read this excerpt from Patricia Hampl's memoir *A Romantic Education*, keeping in mind the questions posed in "Guidelines for Revising Your Style" above. Then write an evaluation of Hampl's style, using as examples specific details relating to each of the five topics in the guidelines. Finally, compare your analysis with Hampl's comments on tone and style at the beginning of this chapter.

Prague

In May 1975, during the spring music festival that opens every year with a performance of Smetana's "Ma Vlast" ("My Homeland"), I went to Prague for the first time. The lilacs were in bloom everywhere: the various lavenders of the French and Persian lilac, and the more unusual — except in Prague — double white lilac. Huge flat red banners with yellow lettering were hoisted everywhere too, draped across homely suburban factories and from the subtle rose and mustard baroque buildings of Staré Město (the Old City).

The banners were in honor of the thirtieth anniversary of the liberation of Prague by the Soviet Army in May 1945. 30 *Let,* "30 years," it said everywhere, even on the visa stamp in my passport. Many offices and stores had photographs in their windows, blowups from 1945 showing Russian soldiers accepting spring bouquets from shy little girls, Russian soldiers waving from tanks to happy crowds. For the first time, I was in a city where the end of the Second World War was really celebrated, where history was close at hand. Prague was the first Continental European city I had seen (I had come from London) and it was almost weirdly intact, not modern. On the plane from America to London I had reminded myself that London would be modern; *my* England was so much a product of the nineteenth-century novels and poetry I'd been reading all my life that I knew I would be shocked to see automobiles. And in fact, nothing could prepare me for the slump I felt in London: I wanted the city of Becky Sharpe, of Daniel Deronda, even of Clarissa Dalloway, not the London of Frommer's guidebooks and the thrill of finding lunch for 5 pence.

Prague stopped my tourism flat. The weight of its history and the beauty of its architecture came to me first as an awareness of dirt, a sort of ancient grime I had never seen before. It bewitched me, that dirt, caught in the corners of baroque moldings and decorative cornices, and especially I loved the dusty filth of the long, grave windows at sunset when the light flared against the tall oblongs and caused them to look gilded.

I had arrived in a river city, just as I had left one in St. Paul. But the difference . . . On the right bank of the Vltava (in German, the Moldau) the buildings were old — to me. Some of them were truly old, churches and wine cellars and squares dating from the Middle Ages. But the real look, especially of the residential and shopping areas of Nové Město (New City — new since the fourteenth century when Charles IV founded it), was art nouveau, highly decorative, the Bohemian version of the Victorian. Across the river, in Malá Strana (Small Side — Prague's Left Bank), the city became most intensely itself, however; it rose baroquely up, villa by villa, palace crushed to palace, gardens crumbling and climbing, to the castle that ran like a great crown above it on a bluff.

The city silenced me. It was just as well I didn't know the language and was traveling alone. There was nothing for me to say. I was here to look.

My original intention in going to Prague was simple: to see the place my grandparents had come from, to hear the language they had spoken. I knew Prague was Kafka's city, I knew Rilke had been born here, and I had read his *Letters to a Young Poet* many times. I was a young poet myself. But my visit wasn't for them. Mine was the return of a third generation American, the sort of journey that is so inexplicable to the second generation: "What are you going to *do* there?" my father asked me before I left Minnesota.

That spring, the lines at the Prague Čedok office (the government agency that runs tourism in Czechoslovakia) were dotted here and there with young Americans looking for family villages. The young couple in front of me in the line had come from Cleveland. The man was asking a young travel agent, who was dressed in a jeans skirt and wore nail polish the color of an eggplant, how he and his wife could get to a village whose name he couldn't manage to pronounce.

"There does not seem to be such a place," she told him. She couldn't find a name on the map with a spelling that corresponded to the one the young man had brought from Cleveland

Charles Bridge, Prague

© Bob Krist

on the piece of paper he was holding out to his wife. ("It's *his* family," she said to me. "I'm Irish.")

They decided to set out, anyway, for a village in Moravia that had a similar spelling. "I guess that's the place," the young man said, without much conviction.

I asked my father's question: "What are you going to *do* there?"

"Look around," he said. "Maybe somebody" — he meant a relative — "will be there."

My own slip of paper, which I'd brought

from St. Paul, had the name of my grandmother's village ("spelling approximate," my cousin had written next to the name when he gave it to me), which was supposed to be near Třeboň, a small town in southern Bohemia. On the map, Třeboň was set among lakes (like Minneapolis, I had thought); here, the guidebook said, "the famous carp" were caught.

Suddenly, just then, as my turn came up, I had no heart for the approximate name of the village, for the famous carp, the kind of journey the Cleveland couple had set for themselves.

SMETANOVO NÁBŘEŽÍ
STARÉ MĚSTO - PRAHA 1

Prague coffeehouse

("We're going to do the same thing in Ireland," the wife said.) I stepped out of the line, crumpled up my piece of paper, and left it in an ashtray. The absurdity of trying to get to Třeboň, and from there to wherever this village with the approximate spelling was supposed to be, lay on me like a plank. I felt like a student who drops out of medical school a semester before graduation; I was almost there and, suddenly, it didn't matter, I didn't want what I'd been seeking. Apparently I wanted something else.

"Do you want to get something to eat?" I asked the couple from Cleveland.

But they didn't have time. "We have to split," she said.

"Yeah," said her husband. "Its a long way. . . ."

I put Třeboň out of my mind and spent the rest of the week walking aimlessly around Prague. If I had answered my father's question — what was I *doing* — I would have said I was sitting in coffeehouses, in between long, aimless walks. A long trip for a cup of coffee, but I was listless, suddenly lacking curiosity and I felt, as I sat in the Slavia next to the big windows that provide one of the best views of the Hradčany in the city, that simply by staring out the window I was doing my bit: there it was, the castle, and I was looking at it. I had fallen on the breast of the Middle European coffeehouse and I was content among the putty- and dove-colored clothes, the pensioners stirring away the hours, the tables of university students studying and writing their papers, the luscious waste of time, the gossip whose ardor I sensed in the bent heads, lifted eyebrows — because of course I couldn't understand a word.

In Vienna the coffeehouses — not all of them, thank God — lose their leases to McDonald's and fast-food chains. But in Prague the colors just fade and become more, rather than less, what they were. The coffeehouse is deeply attached to the idea of conversation, the exchange of ideas, and therefore, to a political society. In *The Agony of Czechoslovakia '38/'68,* Kurt Weisskopf remembers the Prague coffeehouses before the Second World War.

You were expected to patronize the coffeehouse of your group, your profession, your political party. Crooks frequented the Golden Goose, or the Black Rose in the centre of Prague. Snobs went to the Savarin, whores and their prospective clients to the Lind or Julis; commercial travelers occupied the front part of Cafe Boulevard, while the rear was the traditional meeting place of Stalinists and Trotskyites, glaring at each other as they sat around separate marble tables. The rich went to the Urban, "progressive" intellectuals to the Metro. Abstract painters met at the Union, and surrealists at the Manes where they argued with impressionists. You were still served by the black-coated waiters if you did not belong, but so contemptuously that you realized how unwelcome your presence was. The papers in their bamboo frames and the magazines in their folders, an essential part of the Central European coffeehouse service, were regrettably not available to intruders. If you went to the wrong coffeehouse you were just frozen out.

But if you fitted in, socially, politically, philosophically, artistically or professionally, well, then the headwaiter and the manager treated you almost as a relative and you were even deemed worthy of credit.

"Rudolf, switch on the light over the Communists, they can't read their papers properly," old Loebl, manager of the Edison, used to instruct the headwaiter. "And see that the Anarchists get more iced water." Once you had ordered your coffee you were entitled to free glasses of iced water brought regularly by the trayload; this was called a "swimming pool."

The water was still brought, as I sat at the marble table of the Slavia, though not by the trayload. Perhaps there were political conversations, even arguments: I couldn't tell. But the newspapers were the official ones of the Communist Party, including, in English, the *Daily Worker*. When I asked someone I'd struck up a conversation with, who was quoted as a source on American news, I was told, "Gus Hall," the president of the American Communist Party.

"As a typical American?" I asked incredulously. "As the voice of the people," I was told. It struck me as funny, not sinister, although later I realized I was annoyed.

I walked around Prague, hardly caring if I hit the right tourist spots, missing baroque gems, I suppose, getting lost, leaving the hotel without a map as if I had no destination. I just walked, stopping at coffeehouses, smoking unfiltered cigarettes and looking out from the blue wreath around me to other deep-drawing smokers. Everyone seemed to have time to sit, to smoke as if smoking were breathing, to stare into the vacancy of private thought, if their thoughts were private. I was in the thirties, I'd finally arrived in my parents' decade, the men's soft caps, the dove colors of Depression pictures, the acquiescence to circumstance, the ruined quality. For the first time I recognized the truth of beauty: that it is brokenness, it is on its knees. I sat and watched it and smoked (I don't smoke but I found myself buying cigarettes), smoked a blue relation between those coffeehouses and me. For this sadness turned out to be, to me, beautiful. Or rather, the missing quality of beauty, whatever makes it approachable, became apparent in Prague. I could sit, merely breathing, and be part of it. I was beautiful — at last. And I didn't care — at last. I stumbled through the ancient streets, stopped in the smoke-grimed coffeehouses and added my signature of ash, anonymous and yet entirely satisfied. I had ceased even to be a reverse immigrant — I sought no one, no sign of my family or any ethnic heritage that might be mine. I was, simply, in the most beautiful place I had ever seen, and it was grimy and sad and broken. I was relieved of some weight, the odd burden of happiness and unblemished joy of the adored child — or perhaps I was free of beauty itself as an abstract concept. I didn't think about it and didn't bother to wonder. I sat and smoked; I walked and got lost and didn't care because I couldn't get lost. I hardly understood that I was happy: my happiness consisted of encountering sadness. I simply felt *accurate*.

_____ *Writing Assignments* ▐▬▬▬▬▬▬▬▬▬▬▬▬▬▬

1. Select a place you have visited on several occasions and compare and contrast how two means of travel — walking, driving a car, riding in a boat, bicycling, flying in an airplane — presented different images of the place. As you plan and draft your essay, consider why you prefer one mode of travel to the other.

2. In "Girl Talk–Boy Talk," *Science 85* (January/February 1985), John Pfeiffer reports that researcher "Sally McConnell-Ginet of Cornell finds that women's voices are more colorful — they vary their pitch and change pitch more frequently than do men's voices. In one experiment, women immediately assumed a monotone style when asked to imitate men's speech. McConnell-Ginet regards speaking tunefully as an effective strategy for getting and holding attention, a strategy used more often by women than men." Use McConnell-Ginet's theory and speculation as a starting point for developing your own analysis of the way in which gender, class, or profession may affect the tone of a person's speech.

3. Reread Robert Winkler's essay "The Traveling Photographer" (pp. 53–55). Compare the attitudes toward travel presented in the interviews in the essay with the ideas expressed in Walker Percy's description of the Grand Canyon (pp. 237–38) and Patricia Hampl's description of Prague (starting on p. 251). Then analyze how each author's traveling style is reflected in his or her writing style.

4. The photograph on page 234 was selected as a metaphor representing an important aspect of this chapter. Examine the photograph and determine the possible meanings of this metaphor as you discern them in the photograph. Then write an essay explaining how the metaphor has been reinforced or revised by the ideas discussed in this chapter. You may want to suggest another metaphor to illustrate your understanding of these ideas.

PART THREE
SPECIAL ASSIGNMENTS

11
THE
ESSAY
EXAMINATION

The essay examination is one of the most practical of all academic writing assignments. By asking you to compose an answer to a specific question, with a limited amount of time for organizing your thoughts, it calls on most of the skills you have developed as a writer. As historian T. H. White points out in his book *In Search of History,* "One must grasp the question quickly; answer hard, with minimum verbiage; and do it all against a speeding clock."

Instructors frequently complain that students produce their worst writing on essay examinations. Of course, the "speeding clock" makes it impossible for you to work slowly and carefully through the stages of planning, drafting, and revising your responses. And certainly, the pressure of taking an examination does not encourage stylistic polish. But in spite of these limitations, you can still plan what you want to say, develop your purpose into an adequate essay, and reserve some time to review and proofread your answer. If you simply begin writing without using those strategies, you are likely to produce an essay that is irrelevant, inadequate, unclear, and even self-contradictory.

If you practice the major principles of purposeful writing discussed in Parts One and Two of this book, you will be able to improve the quality of your essay answers. Of course, you will not learn the subject matter of your examinations in this chapter. But many weak responses to essay examination questions are caused not by ignorance of the subject matter but by carelessness, haste, or panic. The recommendations in this chapter will help you avoid such pitfalls.

UPI/Bettmann Newsphotos

READ THE QUESTION CAREFULLY

Before you begin to answer any part of an examination, read the question carefully to see what it asks you to write about and how it asks you to write about it. If you misinterpret the question, your answer may be inadequate, even if it shows detailed knowledge of the subject and is otherwise well written. Essay questions are generally carefully written to identify a specific subject and indicate a specific approach. So before you begin to write, ask yourself, "What subject does this question require me to write about and how does it require me to write about it?" Look for key words — such as *evaluate, trace,* or *contrast* — that indicate the appropriate organization for your information. If you are asked to *analyze* a passage, a summary or a paraphrase will not satisfy the requirement. If you are asked to *compare* two characters in a play, a description of each character may not develop the comparison. *Never begin to write until you have a clear idea of the form and content of the answer required.*

Study the following question and the two answers to it to see the difference in answers derived from a careful and from a careless reading of the question.

Examination question

FDR and JFK had different conceptions of the presidential press conference. Using these two photographs as evidence, describe and contrast their conceptions.

Photo courtesy The John F. Kennedy Library, Boston. Photo No. AR 6703B.

Answer 1

FDR and JFK had different conceptions of the presidential press conference. FDR saw it as a private discussion with a small group of friendly reporters; JFK saw it as a public performance in front of a large audience that included reporters and the American people. In the two photographs, these contrasting views can be seen in each president's choice of setting, characters, and activity.

FDR chose the Oval Office for his intimate press conferences. The informality of this setting is illustrated by the small group of reporters standing close to the president, the bouquet of flowers on his desk, and the good will evident in the president's smile. By contrast, JFK chose the auditorium at the State Department for his more imposing press conferences. The formality of this setting is illustrated by the way reporters are separated from the president by theater seating, by the elevated stage and rostrum, and by the microphones on the lectern. Instead of the casual bouquet of flowers, the flags and official seals of the president suggest that this press conference is a ceremonial occasion.

FDR invited thirty or so reporters to his weekly press conferences. He thought of them as "family" and was eager to talk with them about various issues of public policy. Because all of them were print journalists, he provided them with information that they could use to write their stories. JFK held open news conferences, and usually the four-hundred-seat auditorium was filled to overflowing. He thought of reporters as "friendly adversaries" and enjoyed making news by sparring with them on live television. In the first photo, FDR seems prepared to answer questions by himself. In the second, JFK is flanked by a group of advisors who can help him if he is asked a complicated question.

The drama captured in the FDR photo suggests the activity of a family council. FDR allowed each reporter to ask questions, but he conducted his conferences like after-dinner discussions, lecturing leisurely to reporters about the state of the economy or foreign policy and dictating when they could or could not quote him directly. JFK called on reporters, but since he staged his press conferences as thirty-minute television shows, he had to be selective about which questions he acknowledged and which answers he developed in detail. In the first photo, FDR seems like a beneficent father who has stopped discussion to pose for a family photograph. In the second, JFK seems like a matinee idol performing not just for the reporters in the room but for the millions of people watching on television.

This is an excellent response to the examiner's question. The first paragraph asserts a basic contrast in the two presidents' conception of the press conference that is based on supporting evidence from the photographs. The three paragraphs that follow use the alternating pattern to contrast FDR's and JFK's choice of, respectively, setting, characters, and dramatic activity. Each detail develops the contrast announced in the opening paragraph. There are no digressions. And each paragraph concludes with a brief summary that reaffirms the basic contrast. The writer knows what the question requires. That determines her purpose and controls what she has to say.

Answer 2

The photograph of FDR's press conference is very crowded. About twenty people, not all of them smiling, are trying to squeeze themselves into the shot. It is not clear from the photograph that all these people are reporters — only one seems to be writing. All seem aware of the photographer, who has apparently asked them to stop their discussion and pose. The president's desk is covered, but not cluttered, with paper. Other objects on his desk are an inkstand, a telephone, and a large bouquet of flowers. The president is seated in a large, comfortable chair. Because FDR was crippled by polio, photographers agreed to maintain the illusion of his good health by taking his picture only when he was seated. His smile suggests that the press conference has been productive.

The photograph of JFK's press conference is also crowded, but most of the reporters are seated in (or standing around the rim of) an auditorium, while the president stands on a platform behind a large lectern. Some of his advisors appear to be seated on his left. Near the double doors to his right are two television cameras. Behind the windows on the second floor is the control center. Barely visible behind the glass are the technicians who control the light and sound for the president's performance and broadcast it to the American public. The president seems to be making a thoughtful reply to an important question. Most of the reporters in the room are either listening to his response or writing significant quotes in their notebooks.

Even though it shows good observation of details, answer 2 is unsatisfactory, chiefly because it does not respond directly to the question. The purpose imposed by the directions was to use the two photographs to *describe and contrast* the different conceptions FDR and JFK had of the presidential press conference.

Answer 2 describes each photograph separately but pays almost no attention to the contrast the descriptions were to demonstrate. The writer digresses by mentioning FDR's polio, which is irrelevant to the description of the press conference. This answer lacks a thesis; it tells readers what each press conference looks like but says nothing about the two presidents' differing conceptions of them.

THINK OUT YOUR ANSWER BEFORE WRITING

Plan your answer before you begin to write it. Because in an essay examination you will have almost no chance for composing a discovery draft or for extensive revising, your answer must be drafted correctly the first time. Most examination questions attempt to restrict your subject, specify a method of development and organization, and suggest a thesis. Thinking about the implications of the information given in the question itself will bring to mind explanatory or illustrative details that you can use in your answer. For many questions, it is wise to start, as you learned in Chapter 3, by making a scratch outline to organize the information you want to use and then formulating a thesis statement or, in the case of brief answers, a topic sentence.

Answer 1 below shows a carefully planned response to this question:

Examination question

Explain how the Arctic region supports various forms of plant and animal life.

The student thinks over the question and frames a topic sentence:

> *The seasonal changes in the permafrost enable the Arctic region to support various forms of plant and animal life.*

The topic sentence (or thesis) shows that the writer understands that the question is asking him to discuss *how* a specific characteristic of the Arctic region — the seasonal changes in the permafrost — supports various forms of plant and animal life. His topic sentence will require him to define *permafrost,* to explain how seasonal changes affect it, and to demonstrate how these changes support plant and animal life. Notice how his answer satisfies the requirements of both the question and his own topic sentence.

Answer 1

> *The seasonal changes in the permafrost enable the Arctic region to support various forms of plant and animal life. During most of the year the ground is covered with snow, and permafrost freezes the sublayers of the soil to depths ranging from 50 to 4,800 feet. But during the spring months the snow melts and the top layer of the permafrost thaws into a swampy marsh. This layer is not deep enough to support trees, whose roots need from 4 to 8 feet of unfrozen soil to grow, but it does support a wide variety of stunted vegetation, composed mostly of mosses, lichens, and grasses. These plant communities, some featuring brightly colored flowers, support the life of insects, migrating birds, and grazing mammals such as reindeer.*

This answer is an excellent example of purposeful writing in a paragraph: topic sentence, followed by supporting detail, followed by a concluding sentence that expands the reference to "plant and animal life." Because the writer planned

© George Calef/Masterfile

his whole answer from beginning to end, he controls his paragraph and demonstrates his understanding of the content behind the question.

Answer 2 does not show the same careful thought about the requirements of the question or the use of an appropriate topic sentence to control the answer. Instead the student plunges into a summary of the facts without considering how they relate to the question she is supposed to be answering.

Answer 2

> *The Arctic region is able to support a wide range of plant and animal life. Although most of the area is virtually treeless, it does support specialized vegetation such as grasses and sedges. These plants have brightly colored flowers but thrive in swampy soil that makes walking extremely difficult. They provide habitation for many insects, including flies and mosquitoes that make what appears to be a beautiful meadow unbearable for most travelers. The most common mammals in this region are lemming, reindeer, Arctic fox, and wolf. Birds, mainly hawks, also migrate to the Arctic region during the summer months.*

This paragraph does not answer the question. It catalogues the variety of plant and animal life in the Arctic region but fails to analyze the factor (the spring thaw of the top layer of the permafrost) that enables this life to thrive. The answer also digresses by its allusions to the difficulties encountered by travelers.

Failure to read the question carefully enough to see what it asks and failure to plan your answer are related faults. If you know the subject, careful reading of the question will suggest an answer, and planning the answer will give you a check against the wording in the question. If you miss the first step, you will probably miss the second also.

WRITE A COMPLETE ANSWER

Unless the directions specify a short answer, do not write a one- or two-sentence response in an essay examination. Be sure to distinguish between a short-answer test and an essay examination. A short-answer test tests your ability to recall facts, and each question can usually be answered in one or two minutes. Usually there are from twenty to thirty such questions in a fifty-minute quiz. An essay examination, by contrast, tests your ability to interpret facts — to select and organize information that supports a thesis. Because such an answer requires more extended writing, the examiner assumes that you may need as much as fifty minutes to compose each answer. Sometimes the directions specify how much time to allow for each answer, but if they do not, the number of questions in the test indicates approximately how much time you should devote to each answer.

A complete answer is one that deals with the subject as fully as possible within the time limits. An answer that is complete for a short-answer test will be inadequate for an essay examination. For example, with a few additions, the second sentence of answer 1 contrasting FDR's and JFK's conceptions of press conferences would be a complete answer for a short-answer test.

> *FDR saw [the presidential press conference] as a private discussion with a small group of friendly reporters; JFK saw it as a public performance in front of a large audience that included reporters and the American people.*

But this would be an inadequate answer for an essay examination because it lacks the detailed analysis of the two photographs required by the question.

The answers below further illustrate the difference between a complete and an incomplete answer — to the following essay examination question.

Examination question

What lessons did American military strategists learn from the French experience in Vietnam?

Answer 1

American military strategists learned little from the French experience in Vietnam because, like the French, they failed to understand the culture of the country and the dedication of the enemy. For example, the Americans ignored the lesson of the French "agroville" experiment when they adopted the Strategic Hamlet program. Both programs violated ancient Vietnamese traditions by removing peasants from their sacred villages and herding them into secure but alien fortresses. In each case, many peasants sneaked back to their homes, and those who remained nursed a growing hostility toward those who were trying to save them.

Similarly, the U.S. did not understand the lesson of Dien Bien Phu. The French assumed that a massive increase in highly trained forces could over-power a poorly armed enemy. At Dien Bien Phu, however, the Vietminh used superhuman effort to haul their small supply of large guns through dense jungles to trap and capture the superior French forces. Like the French, American mili-tary and political leaders failed to learn that increases in personnel and weapons were insufficient to dispose of a dedicated army fighting for its homeland.

Answer 2

The United States never made the error of Dien Bien Phu, where the French troops were trapped in a valley by enemy artillery. But, by and large, the United States repeated the central French mistake of failing to understand the country it was trying to save from communism.

Answer 2 is incomplete for two reasons: first, it does not mention the real lesson of Dien Bien Phu (that a small, dedicated army can outmaneuver large forces), and second, it fails to illustrate how the Americans repeated the central French mistake of ignoring the culture of the country it was trying to help. Although the second answer may be adequate for a short-answer examination, it is not developed in enough detail for an essay examination.

Completeness in an essay examination resembles completeness in para-graphs (discussed in Chapter 7). The topic sentence of a paragraph or the thesis of an essay is necessarily a general statement. To make that statement clear and convincing, you must develop it with specific details. This is especially true when an examination question, such as the one on Vietnam, calls for a judgment. Your judgment is only an assertion until you support it with evidence. If the writer of answer 1 on Vietnam had stopped with the assertion in his first sentence, he would have written an incomplete answer. His explanatory details provide the evidence that makes his answer complete.

DO NOT PAD YOUR ANSWER

Padding an answer with needless repetition or irrelevant detail is more likely to hurt than to help. A padded answer suggests that you are trying to conceal a lack of knowledge. Graders are not easily persuaded that an answer is good just because it is long. They are more likely to be annoyed at hav-ing to sort through all the padding to find the information that might be relevant to the question. It is your responsibility to select, present, and develop the essen-tial information called for in the question.

The answer on page 267 is a padded response to the question of what lessons American strategists learned from the French experience in Vietnam. The grader's comments identify the two major weaknesses of the answer — failure to explain the lesson learned and the useless repetition of the content of the first paragraph in the second. The thesis statement reveals that the writer was not prepared to answer the question. She said all she had to say in the first paragraph, but she seemed to feel that saying it over again, from a slightly different perspective, would somehow make her answer acceptable. Through the use of marginal comments and editorial suggestions, the grader calls attention to these deficiencies.

What specific lesson did the United States learn from the French defeat at Dien Bien Phu?

> *vague*
> *The Americans learned a great deal from the French war with the revolutionary army called the Vietminh, headed by the communist leader Ho Chi Minh. Throughout their war with these guerrillas, the French continuously increased their troop commitments to higher and higher numbers because they believed that more soldiers and more guns could bring about a speedy victory against a small and poorly armed enemy. At Dien Bien Phu, the French parachuted over 6,000 crack troops into a valley to begin an all-out assault on the Vietminh. But they did not realize that the enemy had worked its way through the jungles to occupy the high ridges above the valley where it had a superior military position.*

This ¶ merely restates the ideas in ¶ 1 and further delays your response to the question about what Americans may have learned from this battle.

> *The capture of the French troops at Dien Bien Phu is an example of how important military mistakes are often made by miscalculation of the enemy's strategy. The French assumed that they could confront the enemy in a traditional ground war. But because the Vietminh was a guerrilla army, it never allowed the French the opportunity to engage in this kind of military action. Instead, it hauled its guns to the top of a high ridge and surprised the French.*

Your problem seems to begin with your vague thesis sentence. You seem uncertain about what the United States learned from the French, so you simply present (rather than interpret) the facts of Dien Bien Phu.

PROOFREAD YOUR ANSWER

Reserve some time at the end of the examination to reread and correct your answer. You will not have time to write a second draft, but you may have enough time for a quick review and revision. First, reread the question. Then, as you read your answer, determine whether you supplied all the required information. Sometimes a subtle change in diction or sentence structure will clarify your answer. At other times you may need to add a sentence or paragraph to develop or redirect your answer. Use a caret (∧) to indicate where such insertions belong, and then place the additional information in a numbered box in the margin or at the end of your paper. Help your instructor follow these revisions by writing a brief note near the insertion. If the writer of the essay above had reserved time for such revisions, she might have sharpened her thesis, added a few sentences about what the United States learned from Dien Bien Phu, and deleted the repetitious second paragraph.

Review Exercise

Examine each of the following paired answers to questions on literature, psychology, and cultural geography. Decide which is the better answer, and be prepared to explain the reasons for your choice. Even if you are unfamiliar with the subjects, you should be able to decide which answer satisfies the requirements of the question.

Question 1

Just before he dies, Laertes says to Hamlet, "Mine and my father's death come not on thee, nor thine on me." In view of the facts of the play, how do you interpret this statement?

Answer 1

Laertes' statement fits some of the facts but not all of them and is best understood as a request to let bygones be bygones. True, Hamlet is not responsible for Laertes' death, because Hamlet thought he was engaging in a friendly bout with blunted swords. When he picked up Laertes' sword in the mix-up, he did not know it was poisoned. Since Laertes deliberately put the poison there, he was responsible for both Hamlet's death and his own. Hamlet killed Polonius by mistake, thinking that the person behind the curtain was the king. To that extent it was an accidental killing, but a killing nevertheless. I think Laertes' statement is intended not as a literal description of the facts but as a reconciliation speech. I interpret the statement to mean: "We have both been the victims of the king's treachery. Forgive me for your death, as I forgive you for mine and my father's."

Answer 2

Laertes returns from France and learns that his father has been killed by Hamlet. He is almost mad with grief and rage, and in a stormy scene with the king he demands revenge. He and the king conspire to arrange a duel between Laertes and Hamlet in which Laertes will use a poisoned sword. The duel takes place after Ophelia's funeral, and Laertes cuts Hamlet with the poisoned sword. Then, in a scuffle, their swords are knocked from their hands, and Hamlet picks up Laertes' poisoned sword and wounds him. Meanwhile the king has put poison in a goblet of wine he intends for Hamlet, but the queen drinks the wine by mistake. When Hamlet sees that she is dying, he kills the king; then both Hamlet and Laertes die.

Question 2

Explain the chief differences between neurosis and psychosis.

Answer 1

The chief differences between neurosis and psychosis are the extent to which a person is alienated from reality and his or her chances of making a workable adjustment to normal living. The boundary between the two cannot be precisely drawn; therefore the differences are best illustrated at their extremes.

A person suffering from neurosis may feel serious anxieties but still be able to handle the ordinary activities of daily living. For example, a woman may have a phobia about being left alone with a red-headed man because a male with red hair once assaulted her. But as long as she avoids that particular situation, she is able to fulfill her domestic and business responsibilities. Through psychiatric counseling she may learn to understand the cause of her phobia and either get rid of it or control it. Fears of heights and of crowds are examples of other neuroses. They are not central to the way one organizes one's life and can be alleviated either by counseling or by avoidance of situations in which the neurotic response is likely to occur.

A psychotic person is so divorced from reality that in severe cases, like paranoid schizophrenia, he or she lives in a private world that has little relation to the real one. A man who thinks he is Moses and feels a divinely granted right to punish those who break any of the Ten Commandments has reorganized experience around a delusion that makes life bearable for him. His delusion is necessary to his continued existence. In a sense he has found a therapy that works for him. He will resist psychiatric help because he thinks he no longer has any problem: it is the sinners who have problems. Such a person may be helped to some degree by specialized, institutional care, but the chances of a complete recovery are slim.

Between these extremes are conditions clas-sifiable as either neurosis or psychosis. In such cases psychiatrists may disagree in their diagnoses.

Answer 2

Because there is some neurosis in all of us, we all utilize defense mechanisms against our frus-trations. For that matter, a psychotic person may also use such defenses, but is less likely to be aware of what he or she is doing. We may repress our frustrations — simply refuse to think about them. Or we may defend by consciously developing characteristics that are the opposite of those we disapprove of. For example, a per-son who is troubled by a tendency toward gree-diness may force himself or herself to generously give away prized possessions. We can also escape frustration and low self-esteem by projec-tion of our faults, blaming them on other peo-ple. Or we can use rationalization by devising excuses to justify our behavior. Finally, we may save our pride by fantasizing. That is, a young man may imagine that the girl who has declined to date him is cheering wildly in a basketball gymnasium when he sets a new scoring record and that she will be waiting for him with bated breath at the locker-room door.

Neurotic people may need the help that defense mechanisms can give them, but exces-sive use of such devices can make a problem worse. And if a neurotic condition becomes so serious that the person is out of touch with the real world and locked into his or her private world, that person is psychotic. As I think back over my answer, it seems to me that a psychotic person would be more likely to use some of these mechanisms than others.

Question 3

Contrast the different concepts of *border* that are illustrated by these two maps of North America.

Answer 1

These two maps provide a dramatic contrast between the geopolitical and theoretical con-cepts of border. The first map designates the official borders of the United States, which have been formed by three methods. First, there are natural borders. Rivers such as the Ohio and

Mississippi, lakes such as Michigan and Erie, and mountains such as the Smokies and Sierra Nevada provide natural ways to separate various regions of the country from one another. Sec-ond, the map documents historical borders that were negotiated as the result of specific military conflicts — for example, the border separating Canada and Maine was negotiated to end the Aroostook War (1842), and the border separat-ing Mexico and Texas was negotiated to end the Mexican War (1848). And third, this map illustrates political borders that were formed by the process of achieving statehood. After the formation of the thirteen original colonies, large blocks of land, particularly west of the Missis-sippi, were shaped by local politicians and fed-eral agencies into territories that eventually became states. The borders of these structures — for example, Colorado and Oklahoma — were often drawn without reference to the con-tours of the land or the interests of the inhabitants.

By contrast, the second map illustrates Joel Garreau's theory that the United States can be understood only by reshaping its borders into "the nine nations of North America." By ignor-ing the principles used to establish the borders of the first map, Garreau is able to speculate about how peoples' attitudes toward issues such as economics and ecology give rise to more nat-ural alliances — and hence more meaningful borders — than those on the geopolitical map. For example, Garreau says that the neatly boxed borders of Colorado make no sense. The wheat farmers in the eastern part of the state feel a more natural alliance with farmers in "The Breadbasket" than with the industrialists in Denver who want to mine the resources of "The Empty Quarter." Similarly, the thin strip along the Pacific coast contains not three states, but two nations. What separates them, Garreau argues, is their attitude toward water. "Ecoto-pia" is the only place in the West with enough water, whereas "MexAmerica" must suck water from other places to support its dry, sunny climate.

Answer 2

The first map illustrates the international borders separating the United States from

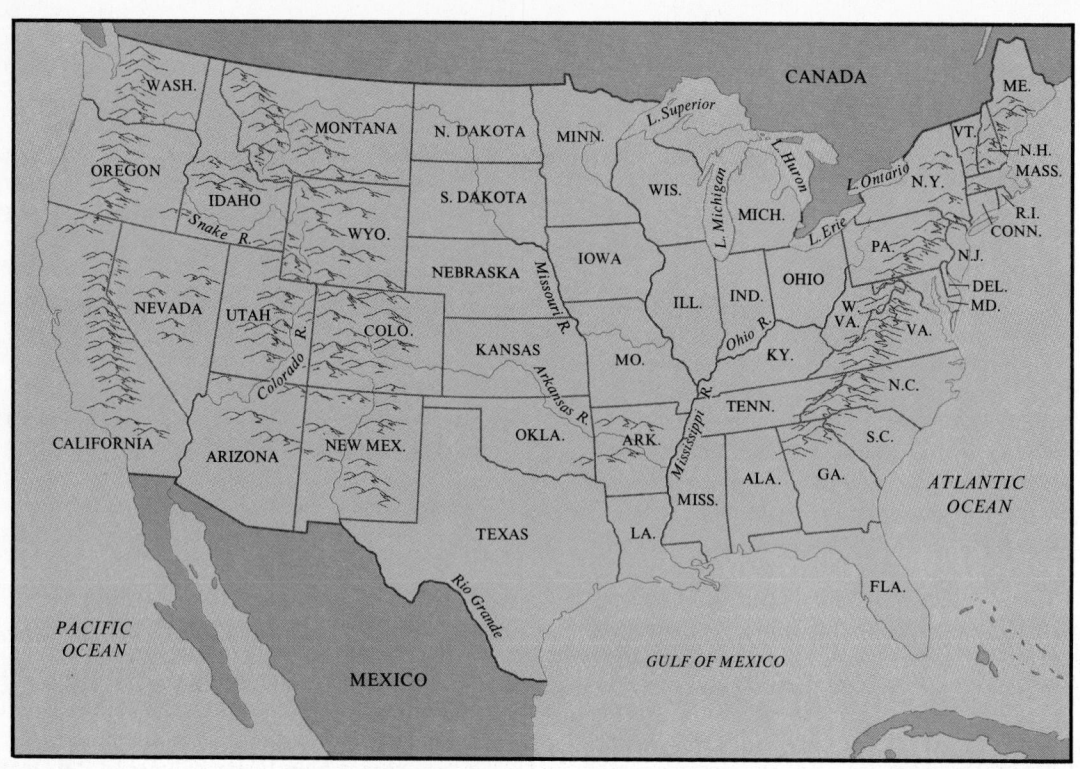

Map by Graf-Tech, Inc.

Canada and Mexico, and the national borders separating the states. Some of these borders are irregular because they are formed by the shoreline of the ocean or the banks of a river. Other borders are extremely regular because they were drawn by surveyors who were marking out plots of land for the United States government. For example, the borders of the western states look like rectangular blocks, whereas the borders of the eastern states look like oddly shaped pieces of a puzzle.

The second map does not acknowledge the borders that separate the United States from Canada and Mexico, or the borders that separate the various states from one another. It illustrates Joel Garreau's nine "nations" of North America. Each nation has an identifying name, symbol, and border. Some of these nations are vaguely recognizable as traditional regions in the United States. For example, the borders of "The Foundry" enclose most of the industrial Northeast, whereas the borders of "Dixie" enclose most of the Old South. But most of the map rearranges the traditional borders of North America. For example, "The Empty Quarter" includes most of the underdeveloped land of the United States and Canada, and "MexAmerica" joins southern California, the Southwest, and Mexico.

Review Exercise

Using any essay examination that you have previously written, preferably an unsatisfactory one, review the wording of the question to determine the assumptions the examiner made about *what* you knew and the instructions he or she gave you about *how* to compose your answer. Reread your answer. Did you (1) read the question carefully, (2) think out your answer before you wrote, (3) compose a complete answer, (4) avoid

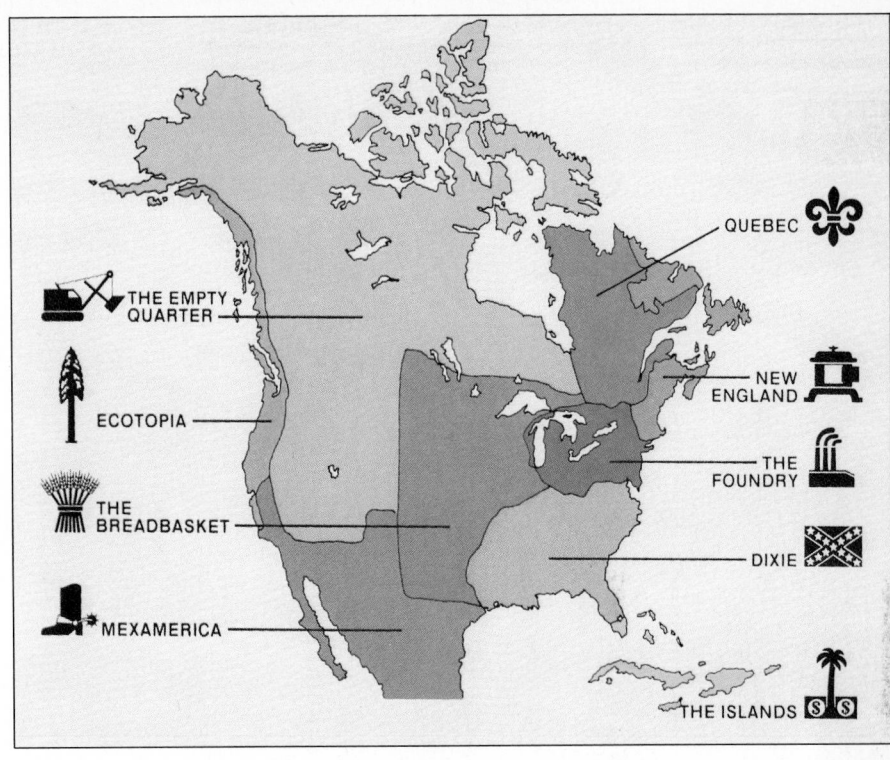

© Robert Anthony, Inc. 1987. From *Nine Nations of North America*, by Joel Garreau, © Houghton Mifflin Co.

padding your answer, and (5) proofread your answer? Try making a descriptive outline of your essay, showing what each paragraph *says* and *does*. Such an outline may help you see how you misread the question or mismanaged the presentation of your answer. Using your instructor's comments as a guide, revise your essay so that it provides a more effective answer to the question.

Writing Assignments

1. In a brief essay, compare and contrast your writing process and study habits when you develop an essay through several drafts and when you are forced to write under time pressure.

2. Choose one of the essay questions presented in this chapter and, after conducting some research, write a longer and more detailed response than those appearing in the chapter.

3. Rewrite one of the essay questions in this chapter so that it no longer contains the key words on organizing your information — for example, change the question on presidential press conferences (p. 260) to "Look at these two photographs and discuss FDR's and JFK's press conferences." Or rewrite a "discuss" question you have been given in one of your classes so that it contains key words that help you develop an answer. Then, using the two versions of the question as examples, write an essay in which you define the characteristics of an effective essay examination question.

4. The photograph on page 258 was selected as a metaphor representing an important aspect of this chapter. Examine the photograph and determine the possible meanings of this metaphor as you discern them in the photograph. Then write an essay explaining how the metaphor has been reinforced or revised by the ideas discussed in this chapter. You may want to suggest another metaphor to illustrate your understanding of these ideas.

12
THE CRITICAL ESSAY

The critical essay helps readers understand a subject. The word *critical* in this context does not mean to find fault. Its Greek root means "to separate, discern, or choose." Thus, the critical essay attempts to enhance the reader's understanding of a subject by analyzing its parts and interpreting its meaning. It may deal with any subject worthy of serious study — a painting, a building, a social movement. In most writing classes, however, the critical essay assignment focuses on literature.

Although procedures for composing a critical essay resemble procedures followed in writing other essays, the success of the critical essay depends on another process — the informed reading of imaginative literature. Reading literature is different from reading a newspaper or magazine. In reading those texts, you may be a "speed-reader," grasping the main point early, skimming through the middle, and glancing at the end to see how it turns out. Such habits can cause trouble when you read a poem or a play.

As novelist William Gass points out, reading a work of literature is a "slow, old-fashioned . . . complicated, profound, silent, still, very private, very solitary yet civilizing activity" ("Of Speed-Readers and Lip-Movers"). To engage in that sort of reading, you need to distinguish among three literary genres — *fiction, poetry,* and *drama,* each with its own history, conventions, and subcategories — and recognize five elements common to nearly every literary work — *plot, characterization, setting, point of view,* and *theme.* These elements may be combined with such subtlety that you may not see them as separate concepts, or one or two of them may be given special attention. Some works may require you to consider additional elements. In much poetry, for example, you will need to consider rhyme, sound patterns, meter, and form. In considering dramatic productions, you will need to evaluate actors, costumes, lighting, and sets. But in most cases, your ability to recognize the interaction of the five basic elements will enable you to complete the informed reading essential for writing a critical essay.

BASIC ELEMENTS OF LITERATURE

Plot

The plot is a coherent, unified, and meaningful sequence of events that forms the beginning, middle, and end of a work of literature. The author must begin by revealing where and when the events take place, who the characters are, and what situation has brought them together. This introductory material is termed the *exposition.* In some cases, as in a brief poem, the exposition may be short because the author expects the readers to understand immediately the circumstances and setting. In other cases, such as in a complex drama, the playwright may need to provide a lengthy exposition to help the audience understand the events that are about to occur.

The middle section of the plot begins when a new set of complications develops to disrupt the existing order. These complications are almost always the result of *conflict* — conflicts among characters, conflicts between characters and their environment, conflicts in the thoughts, desires, or choices of a single character. In some works, all three types of conflict occur together. As conflict intensifies, a moment of crisis, called the *climax,* is reached. The climax usually marks the end of the major action of the plot. All that remains is the revelation of the consequences of whatever occurred during the climax. This final clarification or unraveling of the plot is called the *dénouement,* from a French word that means to untangle knots.

All plots have a beginning, middle, and end, but they do not always follow a chronological or predictable sequence. Some begin in the middle of the action or near the end. In such instances, the author may use a *flashback* to take the reader back in time to witness a scene that explains the current action. Early in a work, an author may use *foreshadowing* to suggest how the action might be resolved. Occasionally such hints turn out to be false clues, and the plot resolves itself in an unexpected way. This reversal of expectations is called *situational irony.*

Characterization

The method by which an author creates, reveals, and develops characters is called *characterization.* An author may describe characters directly, telling the reader what people look like, how they behave, and what they think; or an author may reveal characters indirectly, suggesting their appearance, personality, and values through their words and deeds or through the words and deeds of others.

A literary work usually focuses on a single character. In a poem, this character is often the speaker, who reveals thoughts or describes events. In fiction or drama, this character, the *protagonist,* is often opposed by another character or characters, the *antagonist.* The antagonist need not be a person; it might be the environment, society, or some aspect of the protagonist's personality.

Central characters who change in some significant way as a result of the conflicts they must resolve are often called *dynamic.* Characters who remain unchanged by the experiences they encounter are called *static*. Another way to distinguish between characters is to use the terms *round* and *flat*. A round character, usually the central character, is a fully developed, complex, often contradictory personality. A flat character, usually a minor character, possesses no depth or complexity but is so predictable in words and deeds that it is often called a stock character, or *stereotype*.

In thinking about literary characters, your main concern is to interpret their behavior. When a character makes choices or fails to make choices, you need to explain his or her *motivation*. In some cases, motivation can be explained by acknowledging that the character is responding to external factors such as social expectations. In other cases, motivation can be explained only by assessing internal factors such as psychological perception. A mixture of social and psychological factors motivates the most complex characters. When characters reveal their inability to understand their own motivation, making choices that the reader knows are uninformed or incorrect, they create a situation known as *dramatic irony*.

Setting

In simple terms, setting is the place, time, and social context established by a work of literature. Sometimes an author's choice of a physical place — a living room in a play, for example — seems relatively insignificant. More often, an author chooses a specific physical location — a farm in the rural South, a ghetto in the urban North — because it embodies the conflict in the plot and the choices available to the characters. Similarly, some authors seem to be indifferent to the constraints of time. A poet may evoke a mood in a kind of timeless present; a novelist may establish a story in a particular historical period. To interpret a literary work effectively, you need to determine the author's purpose in ignoring or exploiting restrictions of place and time.

The combination of place and time creates the *social context* for a literary work. Certain stories, poems, and plays evolve from the assumptions, rituals, and shared beliefs that shape the characters and their world. Understanding this context may help you explain an author's *tone*, or attitude toward the world he or she is describing, or an author's use of *symbols*, the evocation of artifacts, images, and ideas that illuminate the meaning of the work.

Point of View

Literally, point of view is the position one occupies in viewing an object. Applied to literature, the phrase refers to both *position* (the narrator's or speaker's proximity to the action in time and space) and *person* (the narrator's or speaker's personality and attitude). The term *narrator* is usually reserved for novels and short stories; the term *speaker* is used for poems. Neither term suits drama, because in a play the action unfolds as each character speaks.

In determining how to reveal the action and ideas in a literary work, an author may choose from four basic points of view.

Third-person omniscient   The narrator, usually assumed to be the author, tells the story. He or she can move at will through time, across space, and into the minds of each character to tell the reader anything that is necessary to understand the story.

Third-person limited omniscient   Although the author is still the narrator, he or she gives up total omniscience and limits the point of view to the experience and perception of one character in the story. Instead of knowing everything, the reader knows only what this one character knows or is able to learn.

First person   The author selects one character to tell the story or express an idea. The character may be involved in the action or reflect on it from afar. This character may tell about events as they are happening or recall events many years after they have taken place.

Dramatic   The author presents the external action of the story directly, as if it were being acted on stage or filmed by a movie camera. The author does not attempt to comment on or interpret the character's actions, thoughts, or feelings. All the reader knows about the action is inferred from the character's public words and deeds.

Through your ability to identify a point of view, you can understand other literary elements, such as *tone* and *style*. In Chapter 10, you learned how to use tone and style to enhance the purpose of your writing. In a work of imaginative literature, the tone and style of the narrator or speaker also further the author's purpose. The tone may be straightforward or ironic, bitter or enthusiastic. The style may be colloquial or formal, simple or complex. In each case the author makes choices that shape the meaning of the work. Your ability to assess the significance of these choices will help you interpret a work more effectively.

Theme

In addition to showing characters in conflict, a work of literature expresses several ideas about human experience. Different readers, working from different perspectives, may respond to the same work in dramatically different ways. Their disagreement prompts further discussion and analysis, encouraging additional reading of a work to find support for a particular interpretation.

To discover and state the theme that seems to integrate your perceptions and responses to a literary work, look for general statements about human experience. These statements are sometimes presented by the omniscient narrator and sometimes by individual characters. In the case of statements by a character, you need to examine the statement in light of your understanding of events. A character who makes a generalization that seems directly opposite to the author's intended meaning may be using *verbal irony*.

If you cannot find any direct statement of theme, ask yourself what the central characters have learned from their experience, or ask yourself what you have learned, from reading of this experience, that the characters have somehow failed to perceive.

You may also gain insight into theme by looking at how authors express their meaning through literary devices such as symbol and allusion. A *symbol* is a person, act, or thing that has both literal significance and metaphorical meaning. A symbol often pulls together several complex, interrelated ideas. An author may use one symbol or a series of related symbols to suggest a theme. An *allusion* suggests a thematic connection between some aspect of the story and something similar in literature, history, or myth. Such references suggest significant comparisons or ironic contrasts that illuminate the theme.

GUIDELINES FOR READING LITERATURE

Now that you have some knowledge of the basic elements that interact within any work of literature, you need to approach the reading of a specific work by following some practical reading strategies. Some of these strategies resemble those discussed in Chapter 2 (pp. 48–49). Others are especially suited to reading literature.

Preview

It is often possible to learn a few important things about a literary work before you read it.

- Begin with a careful consideration of the title, which, as the first clue to the world you are about to enter, may reveal significant information about what you are about to discover. After you have finished reading, you should reconsider the title to see if your first impression was accurate. In most instances, you will discover that the title anticipated what you found and provides a focus for your interpretation.

- Next, consider what you know about the author. If you have read other works by the same author, you may be familiar with her or his themes or techniques. Read whatever introduction or headnote is provided; it may contain important biographical information. If such information is not provided, you may want to use a standard reference guide such as *Who's Who* to find out something about the author's life and work.

- Finally, consider the genre of the literary work you are about to read. Recalling the basic differences between genres may give you some initial insight into the text you are about to read.

Read

As you read a literary text, do not be a passive observer, simply absorbing words and turning pages. Be an active reader, analyzing characters and speculating about their behavior.

- As you read, annotate the work. Use any convenient method. Underline words or sentences that seem important, mark transitions, or star statements that may help you once you begin to compose your interpretation.

- Ask yourself questions about what you are reading. If you are confused about a character's motives, reactions, or observations, take time to write out your questions in the margin of the work. Even though further reading may answer your questions, the act of writing them down encourages you to analyze what seems to be happening and why.

- As you read, and particularly when you reread a work, mark the places in the text where the five basic elements are most apparent — for example, divisions within the plot, clues to character motivation, shifts in point of view.

Review

Once you have finished reading a work, you will want to reread it to analyze how it achieves its effect and contemplate the ways it might be interpreted. A poem with a difficult meter or that remains opaque after a first reading, or a story whose ending surprised you, may send you back to the text for an immediate rereading, causing you to postpone using the following planning strategies. Often, however, using these strategies immediately after a first reading produces unexpected insights and interesting questions to guide your second reading.

- Make a list of the ideas and images that come to mind as you think about the work. You need not organize this information according to any particular pattern. Consult your memory to determine what aspects of the story made the strongest impression.

- Using the freewriting techniques you learned in Chapter 2, compose a more extended response to the story, using statements or questions or both. Your purpose is to write out your immediate reaction to the work. Did you like it or not? More importantly, what confused you about the characters and the way they attempted to resolve their problems?

- Compose (or consult) a series of questions about the five basic elements in the work. Some anthologies (or instructors) provide such questions. These questions, like yours, are composed by active readers who are attempting to achieve an informed reading of the work. You may discover that you have already answered some of them in your first reading. Other questions, however, may send you back to the work to look more closely at specific elements of plot, characterization, setting, point of view, or theme.

Now use these guidelines to read and draft your initial responses to the following short story. A headnote introduces the author, and study questions are provided in the exercise on pages 287–88.

Everyday Use   *for your grandmama*

by Alice Walker

ALICE WALKER was born in 1944 in Eatonton, Georgia. After graduating from Sarah Lawrence College, she became active in the civil rights movement, helping to register voters in Georgia, teaching in the Head Start program in Mississippi, and working on the staff of the New York City welfare department. In subsequent years she developed her own writing career while working as writer-in-residence at several universities. She has written a biography for children, *Langston Hughes, American Poet* (1973), edited an important literary anthology, *I Love Myself When I'm Laughing . . . and Then Again When I'm Looking Mean and Impressive: A Zora Neale Hurston Reader* (1979), and compiled a collection of her essays, *In Search of Our Mothers' Gardens* (1983). These works reveal her interest in the themes of sexism and racism, themes embodied in her three widely acclaimed novels: *The Third Life of Grange Copeland* (1970), *Meridian* (1976), and especially *The Color Purple* (1982), recently made into a motion picture. Her stories, collected in *In Love and Trouble: Stories of Black Women* (1973) and *You Can't Keep a Good Woman Down* (1981), also examine the complex experiences of black women.

I will wait for her in the yard that Maggie and I made so clean and wavy yesterday afternoon. A yard like this is more comfortable than most people know. It is not just a yard. It is like an extended living room. When the hard clay is swept clean as a floor and the fine sand around the edges lined with tiny, irregular grooves anyone can come and sit and look up into the elm tree and wait for the breezes that never come inside the house.

Maggie will be nervous until after her sister goes: she will stand hopelessly in corners homely and ashamed of the burn scars down her arms and legs, eyeing her sister with a mixture of envy and awe. She thinks her sister has held life always in the palm of one hand, that "no" is a word the world never learned to say to her.

You've no doubt seen those TV shows where the child who has "made it" is confronted, as a surprise, by her own mother and father, tottering in weakly from backstage. (A pleasant surprise, of course: What would they do if parent and child came on the show only to curse out and insult each other?) On TV mother and child embrace and smile into each other's faces. Sometimes the mother and father weep, the child wraps them in her arms and leans across the table to tell how she would not have made it without their help. I have seen these programs.

Sometimes I dream a dream in which Dee and I are suddenly brought together on a TV program of this sort. Out of a dark and soft-seated limousine I am ushered into a bright room filled with many people. There I meet a smiling, gray, sporty man like Johnny Carson who shakes my hand and tells me what a fine girl I have. Then we are on the stage and Dee is embracing me with tears in her eyes. She pins on my dress a large orchid, even though she has told me once that she thinks orchids are tacky flowers.

In real life I am a large, big-boned woman with rough, man-working hands. In the winter I wear flannel nightgowns to bed and overalls during the day. I can kill and clean a hog as mercilessly as a man. My fat keeps me hot in zero weather. I can work all day, breaking ice to get water for washing. I can eat pork liver cooked over the open fire minutes after it comes steaming from the hog. One winter I knocked a bull calf straight in the brain between the eyes with a sledge hammer and had the meat hung up to chill before nightfall. But of course all this does not show on television. I am the way my daughter would want me to be: a hundred pounds lighter, my skin like an uncooked barley pancake. My hair glistens in the hot bright lights. Johnny Carson has much to do to keep up with my quick and witty tongue.

But that is a mistake. I know even before I wake up. Who ever knew a Johnson with a quick tongue? Who can even imagine me looking a strange white man in the eye? It seems to me I have talked to them always with one foot raised in flight, with my head turned in whichever way is farthest from them. Dee, though. She would always look anyone in the eye. Hesitation was no part of her nature.

"How do I look, Mama?" Maggie says, showing just enough of her thin body enveloped in pink skirt and red blouse for me to know she's there, almost hidden by the door.

"Come out into the yard," I say.

Have you ever seen a lame animal, perhaps a dog run over by some careless person rich enough to own a car, sidle up to someone who is ignorant enough to be kind to him? That is the way my Maggie walks. She has been like this, chin on chest, eyes on ground, feet in shuffle, ever since the fire that burned the other house to the ground.

Dee is lighter than Maggie, with nicer hair and a fuller figure. She's a woman now, though sometimes I forget. How long ago was it that the other house burned? Ten, twelve years? Sometimes I can still hear the flames and feel Maggie's arm sticking to me, her hair smoking and her dress falling off her in little black papery flakes. Her eyes seemed stretched open, blazed open by the flames reflected in them. And Dee. I see her standing off under the sweet gum tree she used to dig gum out of; a look of concentration on her face as she watched the last dingy gray board of the house fall in toward the red-hot brick chimney. Why don't you do a dance around the ashes? I'd wanted to ask her. She had hated the house that much.

I used to think she hated Maggie, too. But that was before we raised the money, the church and me, to send her to Augusta to school. She used to read to us without pity; forcing words, lies, other folks' habits, whole lives upon us two, sitting trapped and ignorant underneath her voice. She washed us in a river of make-believe, burned us with a lot of knowledge we didn't necessarily need to know. Pressed us to her with the serious way she read, to shove us away at just the moment, like dimwits, we seemed about to understand.

Dee wanted nice things. A yellow organdy dress to wear to her graduation from high school; black pumps to match a green suit she'd made from an old suit somebody gave me. She was determined to stare down any disaster in her efforts. Her eyelids would not flicker for minutes at a time. Often I fought off the temptation to shake her. At sixteen she had a style of her own: and knew what style was.

I never had an education myself. After second grade the school was closed down. Don't ask me why: in 1927 colored asked fewer questions than they do now. Sometimes Maggie reads to me. She stumbles along good-naturedly but can't see well. She knows she is not bright. Like good looks and money, quickness passed her by. She will marry John Thomas (who has mossy teeth in an earnest face) and then I'll be free to sit here and I guess just sing church songs to myself. Although I never was a good singer. Never could carry a tune. I was always better at a man's job. I used to love to milk till I was hoofed in the side in '49. Cows are soothing and slow and don't bother you, unless you try to milk them the wrong way.

I have deliberately turned my back on the house. It is three rooms, just like the one that burned, except the roof is tin; they don't make shingle roofs any more. There are no real windows, just some holes cut in the sides, like the port-holes in a ship, but not round and not square, with rawhide holding the shutters up on the outside. This house is in a pasture, too, like the other one. No doubt when Dee sees it she will want to tear it down. She wrote me once that no matter where we "choose" to live, she will manage to come see us. But she will never bring her friends. Maggie and I thought about this and Maggie asked me, "Mama, when did Dee ever *have* any friends?"

She had a few. Furtive boys in pink shirts hanging about on washday after school. Nervous girls who never laughed. Impressed with her they worshiped the well-turned phrase, the cute shape, the scalding humor that erupted like bubbles in lye. She read to them.

When she was courting Jimmy T she didn't have much time to pay to us, but turned all her faultfinding power on him. He *flew* to marry a cheap gal from a family of ignorant flashy people. She hardly had time to recompose herself.

When she comes I will meet — but there they are!

Maggie attempts to make a dash for the house, in her shuffling way, but I stay her with my hand, "Come back here," I say. And she stops and tries to dig a well in the sand with her toe.

It is hard to see them clearly through the strong sun. But even the first glimpse of leg out of the car tells me it is Dee. Her feet were always neat-looking, as if God himself had shaped them with a certain style. From the other side of the car comes a short, stocky man. Hair is all over his head a foot long and hanging from his chin like a kinky mule tail. I hear Maggie suck in her breath. "Uhnnnh," is what it sounds like. Like when you see the wriggling end of a snake just in front of your foot on the road. "Uhnnnh."

Dee next. A dress down to the ground, in this hot weather. A dress so loud it hurts my eyes. There are yellows and oranges enough to throw back the light of the sun. I feel my whole face warming from the heat waves it throws out.

Earrings, too, gold and hanging down to her shoulders. Bracelets dangling and making noises when she moves her arm up to shake the folds of the dress out of her armpits. The dress is loose and flows, and as she walks closer, I like it. I hear Maggie go "Uhnnnh" again. It is her sister's hair. It stands straight up like the wool on a sheep. It is black as night and around the edges are two long pig-tails that rope about like small lizards disappearing behind her ears.

"Wa-su-zo-Tean-o!" she says, coming on in that gliding way the dress makes her move. The short stocky fellow with the hair to his navel is all grinning and he follows up with "Asalamalakim, my mother and sister!" He moves to hug Maggie but she falls back, right up against the back of my chair. I feel her trembling there and when I look up I see the perspiration falling off her chin.

"Don't get up," says Dee. Since I am stout it takes something of a push. You can see me trying to move a second or two before I make it. She turns, showing white heels through her sandals, and goes back to the car. Out she peeks next with a Polaroid. She stoops down quickly and lines up picture after picture of me sitting there in front of the house with Maggie cowering behind me. She never takes a shot without making sure the house is included. When a cow comes nibbling around the edge of the yard she snaps it and me and Maggie *and* the house. Then she puts the Polaroid in the back seat of the car, and comes up and kisses me on the forehead.

Meanwhile Asalamalakim is going through the motions with Maggie's hand. Maggie's hand is as limp as a fish, and probably as cold, despite the sweat, and she keeps trying to pull it back. It looks like Asalamalakim wants to shake hands but wants to do it fancy. Or maybe he don't know how people shake hands. Anyhow, he soon gives up on Maggie.

"Well," I say. "Dee."

"No, Mama," she says. "Not 'Dee,' Wangero Leewanika Kemanjo!"

"What happened to 'Dee'?" I wanted to know.

"She's dead," Wangero said. "I couldn't bear it any longer being named after the people who oppress me."

"You know as well as me you was named after your aunt Dicie," I said. Dicie is my sister. She named Dee. We called her "Big Dee" after Dee was born.

"But who was *she* named after?" asked Wangero.

"I guess after Grandma Dee," I said.

"And who was she named after?" asked Wangero.

"Her mother," I said, and saw Wangero was getting tired. "That's about as far back as I can trace it," I said. Though, in fact, I probably could have carried it back beyond the Civil War through the branches.

"Well," said Asalamalakim, "there you are."

"Uhnnnh," I heard Maggie say.

"There I was not," I said, "before 'Dicie' cropped up in our family, so why should I try to trace it that far back?"

He just stood there grinning, looking down on me like somebody inspecting a Model A car. Every once in a while he and Wangero sent eye signals over my head.

"How do you pronounce this name?" I asked.

"You don't have to call me by it if you don't want to," said Wangero.

"Why shouldn't I?" I asked. "If that's what you want us to call you, we'll call you."

"I know it might sound awkward at first," said Wangero.

"I'll get used to it," I said. "Ream it out again."

Well, soon we got the name out of the way. Asalamalakim had a name twice as long and three times as hard. After I tripped over it two or three times he told me to just call him Hakim-a-barber. I wanted to ask him was he a barber, but I didn't really think he was, so I didn't ask.

"You must belong to those beef-cattle peoples down the road," I said. They said "Asalamalakim" when they met you, too, but they didn't shake hands. Always too busy: feeding the cattle, fixing the fences, putting up salt-lick shelters, throwing down hay. When the white folks poisoned some of the herd the men stayed up all night with rifles in their hands. I walked a mile and a half just to see the sight.

Hakim-a-barber said, "I accept some of their doctrines, but farming and raising cattle is not my style." (They didn't tell me, and I didn't ask, whether Wangero [Dee] had really gone and married him.)

We sat down to eat and right away he said he didn't eat collards and pork was unclean. Wangero, though, went on through the chitlins and corn bread, the greens and everything else. She talked a blue streak over the sweet potatoes. Everything delighted her. Even the fact that we still used the benches her daddy made for the table when we couldn't afford to buy chairs.

"Oh, Mama!" she cried. Then turned to Hakim-a-barber. "I never knew how lovely these benches are. You can feel the rump prints," she said, running her hands underneath her and along the bench. Then she gave a sigh and her hand closed over Grandma Dee's butter dish. "That's it!" she said. "I knew there was something I wanted to ask you if I could have." She jumped up from the table and went over in the corner where the churn stood, the milk in its clabber by now. She looked at the churn and looked at it.

"This churn top is what I need," she said. "Didn't Uncle Buddy whittle it out of a tree you all used to have?"

"Yes," I said.

"Uh huh," she said happily. "And I want the dasher, too."

"Uncle Buddy whittle that, too?" asked the barber.

Dee (Wangero) looked at me.

"Aunt Dee's first husband whittled the dash," said Maggie so low you almost couldn't hear her. "His name was Henry, but they called him Stash."

"Maggie's brain is like an elephant's," Wangero said, laughing. "I can use the churn top as a centerpiece for the alcove table," she said, sliding a plate over the churn, "and I'll think of something artistic to do with the dasher."

When she finished wrapping the dasher the handle stuck out. I took it for a moment in my hands. You didn't even have to look close to see where hands pushing the dasher up and down to make butter had left a kind of sink in the wood. In fact, there were a lot of small sinks; you could see where thumbs and fingers had sunk into the wood. It was beautiful light yellow wood, from a tree that grew in the yard where Big Dee and Stash had lived.

After dinner Dee (Wangero) went to the trunk at the foot of my bed and started rifling through it. Maggie hung back in the kitchen over the dishpan. Out came Wangero with two quilts. They had been pieced by Grandma Dee and then Big Dee and me had hung them on the quilt frames on the front porch and quilted them. One was in the Lone Star pattern. The other was Walk Around the Moun-

Courtesy The Shelburne Museum, Shelburne, Vermont

tain. In both of them were scraps of dresses Grandma Dee had worn fifty and more years ago. Bits and pieces of Grandpa Jarrell's Paisley shirts. And one teeny faded blue piece, about the size of a penny matchbox, that was from Great Grandpa Ezra's uniform that he wore in the Civil War.

"Mama," Wangero said sweet as a bird. "Can I have these old quilts?"

I heard something fall in the kitchen, and a minute later the kitchen door slammed.

"Why don't you take one or two of the others?" I asked. "These old things was just done by me and Big Dee from some tops your grandma pieced before she died."

"No," said Wangero, "I don't want those. They are stitched around the borders by machine."

"That's make them last better," I said.

"That's not the point," said Wangero. "These are all pieces of dresses Grandma used to wear. She did all this stitching by hand. Imagine!" She held the quilts securely in her arms, stroking them.

"Some of the pieces, like those lavender ones, come from old clothes her mother handed down to her," I said, moving up to touch the quilts. Dee (Wangero) moved back just enough so that I couldn't reach the quilts. They already belonged to her.

"Imagine!" she breathed again, clutching them closely to her bosom.

"The truth is," I said, "I promised to give them quilts to Maggie, for when she marries John Thomas."

She gasped like a bee had stung her.

"Maggie can't appreciate these quilts!" she said. "She'd probably be backward enough to put them to everyday use."

"I reckon she would," I said. "God knows I been saving 'em for long enough with nobody using 'em. I hope she will!" I didn't want to bring up how I had offered Dee (Wangero) a quilt when she went away to college. Then she had told me they were old-fashioned, out of style.

"But they're *priceless*!" she was saying now, furiously; for she has a temper. "Maggie would put them on the bed and in five years they'd be in rags. Less than that!"

"She can always make some more," I said. "Maggie knows how to quilt."

Dee (Wangero) looked at me with hatred. "You just will not understand. The point is these quilts, *these* quilts!"

"Well," I said, stumped. "What would *you* do with them?"

"Hang them," she said. As if that was the only thing you *could* do with quilts.

Maggie by now was standing in the door. I could almost hear the sound her feet made as they scraped over each other.

"She can have them, Mama," she said, like somebody used to never winning anything, or having anything reserved for her. "I can 'member Grandma Dee without the quilts."

I looked at her hard. She had filled her bottom lip with checkerberry snuff and it gave her face a kind of dopey, hangdog look. It was Grandma Dee and Big Dee who taught her how to quilt herself. She stood there with her scarred hands hidden in the folds of her skirt. She looked at her sister with something like fear but she wasn't mad at her. This was Maggie's portion. This was the way she knew God to work.

When I looked at her like that something hit me in the top of my head and ran down to the soles of my feet. Just like when I'm in church and the spirit of God touches me and I get happy and shout. I did something I never had done before: hugged Maggie to me, then dragged her on into the room, snatched the quilts out of Miss Wangero's hands and dumped them into Maggie's lap. Maggie just sat there on my bed with her mouth open.

"Take one or two of the others," I said to Dee.

But she turned without a word and went out to Hakim-a-barber.

"You just don't understand," she said, as Maggie and I came out to the car.

"What don't I understand?" I wanted to know.

"Your heritage," she said. And then she turned to Maggie, kissed her, and said, "You ought to try to make something of yourself, too, Maggie. It's really a new day for us. But from the way you and Mama still live you'd never know it."

She put on some sunglasses that hid everything above the tip of her nose and her chin.

Maggie smiled; maybe at the sunglasses. But a real smile, not scared. After we watched the car dust settle I asked Maggie to bring me a dip of snuff. And then the two of us sat there just enjoying, until it was time to go in the house and go to bed.

PLANNING THE CRITICAL ESSAY An informed reading of a literary work is the most important stage in planning the critical essay. But to transform your reading into writing, you need to try out some of the thinking-in-writing strategies suggested as "review" activities (pp. 278–79). For example, after Richard read "Everyday Use," he made a list of the characters, scenes, and images that he remembered most vividly.

> *Mama — fat, proud, smart — sledge hammer*
>
> *Johnny Carson reunion vs. real reunion*
>
> *Maggie — fire, ugly, afraid of Dee, knows how to make a quilt*
>
> *Dee — educated, ashamed of family (Wangero), Hakim-a-barber*
>
> *Butter dish, churn*
>
> *Quilts — bits and pieces, suddenly fashionable*
>
> *Hang on the wall*
>
> *Mama's choice*

After looking over this list, Richard decided to focus his attention on Dee's attempt to take the quilts. In a freewriting exercise, he explored his reaction to this event:

> *I was really furious with Dee, or Wangero, or whatever her name is. She didn't want those quilts when she was in college. Out of style. Now that they're "in," she wants to hang one in her apartment. Probably put spotlights on them with a brass plaque explaining how priceless they are. Good thing Mama saw through her. Maggie would have given them to her. She can always make others. But that's not the point. It's the principle. They're Maggie's quilts. Shouldn't have to give them up.*

Not everybody reacts to a particular literary work in the same way. For example, Julia's freewriting exercise expresses a slightly different interpretation of Dee's behavior.

> *Dee is not completely at fault for the way she acts. She's what her family wanted her to become — educated. Mama and Maggie think she's famous. Johnny Carson. Every family has somebody that outgrows it. Dee lives in a different world. Different values. Car, dress, bracelets, boyfriend. Wangero. Trying to discover roots. Knows what style is. People should respect heritage. Try to make something out of themselves.*

By contrast to Richard's emotional reaction to the story, Julia's response seems thoughtful and objective. Her analysis may explain Dee's behavior, but it does not account for Mama's decision to give the quilts to Maggie. Richard, on the other hand, may need to think through the reasons for his anger, but his assessment provides a more complete picture of the story's events. Obviously, Julia and Richard have a great deal to discuss with each other about "Everyday Use." Each response adds some insight to the other, prompting both writers to rethink their views and perhaps return to the story to look for evidence that supports their positions.

To test their reactions to the story, they need to consider the five basic elements contributing to the story's effect. One way is to compose and answer questions about these elements or attempt to answer questions provided by their instructor.

■■■■■ *Exercise*

Answer the following questions about how the five basic elements interact in "Everyday Use." Respond to each question in three or four sentences. For some questions, you may need to reread sections of the story before you compose your answer.

Questions About Plot

1. How does Mama's dream about a family reunion on television introduce the conflicts in the story?
2. How does Mama's decision to give the quilts to Maggie mark the climax of the story?
3. How do Wangero's comments about heritage and self-development bring the story to an ironic conclusion?

Questions About Characterization

1. How does Mama's description of her working ability establish her character?
2. How do Maggie's scars explain her lack of self-esteem?
3. How do Dee's attire, boyfriend, an new name justify Mama's comment that Dee "know what style was"?

Questions About Setting

1. How do Mama's opening and closing comments about the yard evoke the physical and social setting of the story?
2. Why does Mama suspect that Dee will want to "tear down" the house?
3. Why does Wangero photograph the house and pay so much attention to the benches, churn, and quilts?

Questions About Point of View

1. Why is it appropriate for Mama to tell this story? For example, how might the story change if it were told by Maggie or Dee?
2. What is Mama's attitude toward Maggie's accident and Dee's education?

3. What attitude toward her story does Walker suggest by her dedication, *"for your grandmama"*?

Questions About Theme

1. How does the conflict about the quilts symbolize the themes of tradition and style?

2. How do you interpret Wangero's comment that her sister should "make something" of herself?

3. How do you interpret Mama's observation that "Maggie knows how to quilt"?

DRAFTING THE CRITICAL ESSAY

After you have worked your way through several thinking-in-writing strategies, you should have learned enough about a particular literary work to draft your essay.

Selecting Your Subject

Sort through the subject ideas you uncovered during planning. Even a subject assigned by your instructor and restricted to a specific aspect of a story, poem, or play should enable you to integrate your planning material into a unified essay. If your instructor asks you to select your own subject, be sure to restrict it to a specific aspect of the work. You cannot discuss everything you know about "Everyday Use," for example, but you can focus your attention on an aspect of the story that will enable you to compose a thoughtful and thorough analysis.

Analyzing Your Audience

You must assume that your audience for a critical essay is a group of informed readers. This means that your readers have read the story you are about to interpret and have probably formed their own opinions about its meaning and significance. For that reason, you do not need to summarize the plot. But you do need to analyze relevant aspects of the plot (and even quote certain passages) so that your readers can see how you arrived at your interpretation.

Determining Your Purpose

Restricting your subject will help determine your purpose, but remember that you need to prove something about your subject. In a critical essay on a literary work, this means proving that your particular interpretation is supported by a careful examination of the text. To put your purpose into operation, compose a thesis that embodies the interpretation you are trying to prove.

As Richard began drafting his essay on "Everyday Use," he decided that he wanted to restrict his subject to the conflicting attitudes toward the word *heritage* that he discovered in the story. He began by making a scratch outline grouping words and details that reveal how Mama and Dee might use the word.

1. *Mama*
   House-yard-homestead
   Hard work, hog killing
   Family — children, all the Dees (Civil War)
   Church songs
   Skills — whittling, quilts
   Saving for "everyday use"
   Can always make another

2. *Dee*
   School — discovers other habits (heritages)
   Wanted nice things — style (vs. heritage)
   Rediscovers heritage: Afro, pigtails, jewelry
   New name (oppression vs. tradition)
   Photographs — artifacts, something artistic
   Skills — admired, appreciated, priceless
   New Day

Making this simple division lets Richard see the significance of some aspects of the story that were not on his original list or in his freewriting exercise and lets him draft a preliminary hypothesis about this information.

> *In "Everyday Use," Alice Walker presents two attitudes toward heritage: Mama's and Maggie's is something used everyday, and Wangero's is a matter of what is in style.*

As he thinks about this hypothesis, Richard decides to use his knowledge of the basic elements of literature to provide a more critical perspective on his material. He revises his thesis to focus on *how* Alice Walker develops this *theme* through *characterization*.

> *In "Everyday Use," Alice Walker presents two attitudes toward heritage: Mama's and Maggie's is something used everyday, and Wangero's is a matter of what is in style.*

The advantage of this thesis is that it sets up a structure for Richard's essay. He can contrast the two attitudes toward heritage as he contrasts the characters in three separate sections: (a) *appearance*, (b) *actions*, and (c) *appreciation of family possessions*. His use of the various stages of the reading and writing process enables him to compose the following draft.

The Meaning of Heritage in "Everyday Use"

Alice Walker's short story "Everyday Use" is about the conflict between a mother and a daughter over their heritage. Heritage for Mama and her daughter Maggie is a matter of everyday living, or "everyday use" as the title of the story suggests. Mama and Maggie are not conscious of their heritage because it is so much a part of their lives. For Mama's daughter Dee, however, heritage is a matter of style, a fashionable obsession with one's roots. Walker develops these contrasting attitudes through her characters' appearance, actions, and appreciation of family possessions.

The description of the characters in the story introduces the conflicting attitudes toward heritage. Mama admits that she is fat and manly. Her hands are calloused and rough from a lifetime of hard work. Maggie is described as homely, wearing a pink skirt and red blouse, and bearing scars from a fire that burned down the family home. Dee, however, is beautiful and stylish, wearing a striking, brightly colored African dress, earrings, bracelets, sunglasses, and hair in the full-bodied African style that, according to Mama, "stands straight up like the wool on a sheep." Even Dee's feet are pretty, "as if God himself had shaped them with a certain style."

The characters' actions further develop the theme of heritage. By telling us that she can butcher hogs "as mercilessly as a man" or break ice "to get water for washing," Mama suggests that she possesses necessary survival skills. She is at home in her world, sweeping the front yard as though it were "an extended living room." And she prepares and eats chitlins, corn bread, and collards because they are inexpensive and readily available. Although she is less assertive, Maggie displays similar domestic skills, particularly her ability to make quilts. Both women live their heritage. Dee, however, thinks of herself as outside this world. When she was a girl, she used to read things to her family that they "didn't necessarily need to know," washing them "in a river of make-believe." She wanted to dress nicely and impress others with a "well-turned phrase." Now that she has become educated and fascinated by African culture, she changes her name to Wangero Leewanika Kemanjo. She does not want to bear the name of the "people who oppress me," even though, ironically, her name has a long history in the family. Unaware of this inconsistency, Wangero returns home with her Muslim boyfriend, Hakim-a-barber, to take pictures of her mother and sister with the house and cow, probably to show her friends the "down home" aura of the homestead.

The characters' appreciation of certain family possessions intensifies the difference between the two notions of heritage. Dee-Wangero rediscovers Mama's wooden benches: "I never knew how lovely these benches are. You can feel the rump prints." She also views the churn top and the dasher as quaint because they were created by a primitive skill, whittling, and because she can use them to create something else, such as an artistic centerpiece for her alcove table. And finally, she is captivated by the quilts that Grandma Dee and Big Dee stitched together out of bits and pieces of family clothing from as far back as the Civil War. She wants to hang them on a wall as she would priceless paintings. To Mama and Maggie, however, these objects are indispensable to their everyday living. The churn top, dasher, benches, and quilts are for "everyday use," not for stylish decoration.

At the climax of the story, Maggie is tempted to give her quilts to Dee-Wangero. Mama has been saving them for Maggie to use when she starts her own home, but Maggie says that she can remember "Grandma Dee without the quilts." Mama acknowledges that Maggie "can always make some more," and then suddenly she feels "something hit me in the top of my head and ran down to the soles of my feet." She hugged Maggie, "snatched the quilts out of Miss Wangero's hands and dumped them into Maggie's lap." This decision leaves Maggie speechless and Dee-Wangero momentarily annoyed. But clearly Mama and Maggie understand their heritage better than Dee-Wangero thinks they do. The final scene of the story reveals the difference between an artificial and a real heritage. Dee-Wangero puts on her sunglasses and rides away in the dust, while Mama and Maggie finish the day by sitting there "just enjoying."

REVISING THE CRITICAL ESSAY

To assess what you have accomplished in your draft, use the strategies for global and local revision that you have already learned to clarify the internal focus and development of your text. To complete this revision process, however, reread the literary work, using your draft as a guide. If your draft helps you understand the interaction of the basic elements in the work, then you have probably composed a fairly complete essay. But if, as you reread the work, you discover important features of plot, characterization, setting, point of view, or theme that you have overlooked or underestimated, then you may need to revise your essay. Sometimes such discoveries will require you to rethink the wording of your thesis or the transitions between the major divisions of your paper. At other times, they will remind you that by adding a detail, modifying an assertion, or quoting a particular passage, you can make a more compelling case for your interpretation.

■■■■■ *Exercise*

Reread Alice Walker's "Everyday Use." Then reread Richard's essay "The Meaning of Heritage in 'Everyday Use.'" Analyze his introductory paragraph to determine how he has restricted his subject, what assumptions he has made about his audience, and how he has worded his thesis statement. Then check the unity, development, and coherence of each of his paragraphs. Finally, write Richard a note commenting on how his essay affected your understanding of Walker's story. Point first to the aspects of his essay that were particularly useful, but also suggest revisions (large or small) that you feel he may want to consider in rethinking his critical interpretation of the story.

_____ *Writing Assignments* ■■■■■■

1. Examine Edwin Arlington Robinson's poem "Mr. Flood's Party" by using the reading strategies described in Guidelines for Reading Literature (pp. 277–79). Then read the student essay "Old Eben and Mr. Flood," and write a brief essay that includes answers to the following questions:
   a. In what ways did your response differ from the student's response?
   b. What aspects of the poem did you overlook in your reading?
   c. What aspects did she overlook in her writing?

Mr. Flood's Party | *Edwin Arlington Robinson*

Old Eben Flood, climbing alone one night
Over the hill between the town below
And the foresaken upland hermitage
That held as much as he should ever know
On earth again of home, paused warily.   5
The road was his with not a native near;
And Eben, having leisure, said aloud,
For no man else in Tilbury Town to hear:

"Well, Mr. Flood, we have the harvest moon
Again, and we may not have many more;   10
The bird is on the wing, the poet says,
And you and I have said it here before.
Drink to the bird." He raised up to the light
The jug that he had gone so far to fill,
And answered huskily: "Well, Mr. Flood,   15
Since you propose it, I believe I will."

Alone, as if enduring to the end
A valiant armor of scarred hopes outworn,
He stood there in the middle of the road
Like Roland's ghost winding a silent horn.   20
Below him, in the town among the trees,
Where friends of other days had honored him,
A phantom salutation of the dead
Rang thinly till old Eben's eyes were dim.

Then, as a mother lays her sleeping child   25
Down tenderly, fearing it may awake,
He set the jug down slowly at his feet
With trembling care, knowing that most
    things break;
And only when assured that on firm earth
It stood, as the uncertain lives of men   30
Assuredly did not, he paced away,
And with his hand extended paused again:

"Well, Mr. Flood, we have not met like this
In a long time; and many a change has come
To both of us, I fear, since last it was   35
We had a drop together. Welcome home!"
Convivially returning with himself,
Again he raised the jug up to the light;
And with an acquiescent quaver said:
"Well, Mr. Flood, if you insist, I might.   40

"Only a very little, Mr. Flood—
For auld lang syne. No more, sir; that will do."
So, for the time, apparently it did,
And Eben evidently thought so too;
For soon amid the silver loneliness   45
Of night he lifted up his voice and sang,
Secure, with only two moons listening,
Until the whole harmonious landscape rang—

"For auld lang syne." The weary throat
  gave out,
The last word wavered, and the song was done. 50
He raised again the jug regretfully
And shook his head, and was again alone.
There was not much that was ahead of him,
And there was nothing in the town below—
Where strangers would have shut the   55
  many doors
That many friends had opened long ago.

Old Eben and Mr. Flood | *Student Essay*

"Mr. Flood's Party" embraces two separate
worlds. The first world is one of reality, consisting of Eben's consciousness of his present state,
in which he is lonely and desolate; the other
world is one of illusion, made up of Mr. Flood's
dreams of his fellowship with friends of bygone
days. The distinction between the poem's two
worlds is emphasized by the figure of the "two
moons." One moon is real, but the other moon
is just a part of Mr. Flood's illusion.

Robinson begins the poem by placing Eben
in the world of reality. In the first words of the
first line, "Old Eben Flood, climbing alone," the
reader gets a hint of Eben's age and solitude. In
the next line, the phrase "Over the hill" continues to express the concept of age. Eben's
position "between the town below/And the for-
saken upland hermitage" strengthens the idea of
loneliness. Here, he is caught in the middle. In
the words of the last stanza, "there was nothing
in the town below," and yet "There was not
much that was ahead of him." The world of
reality continues for Eben in the second stanza.
In addressing himself, Eben recognizes his age
in the passage: "we have the harvest moon/
Again, and we may not have many more." He
also acknowledges the movement of time in the
statement, "The bird is on the wing." On this
note old Eben takes a drink, and his world of
reality begins to fade.

In the third stanza, Eben is still aware of his
state of solitude, but this awareness has taken
on a heroic quality. He is described as "enduring
to the end/A valiant armor of scarred hopes outworn." This is a very noble image of Mr. Flood,
who has merely outlasted his expectations and
outlived his time. "Like Roland's ghost winding
a silent horn," so Mr. Flood calls for the help of
his comrades by raising the jug to his mouth. In
the world of illusion, Flood's friends answer the
call. At the end of the third stanza, these "friends
of other days" who "had honored him" greet
Mr. Flood in a "phantom salutation of the dead."
Old Eben's desolation is forgotten in this illusion.
He is living in a dream of the past through his
imagined reunion with his deceased friends.
Feeling secure in the company of his party, Mr.
Flood sings "For auld lang syne."

Robinson then informs the reader that "The
weary throat gave out,/The last word wavered,
and the song was done." Likewise, Mr. Flood's
world of illusion fades, and the party is over.

As the party ends, the world of reality
returns. Eben, as if he were waking from a
dream, "shook his head, and was again alone."
In the line that states, "There was not much that
was ahead of him," Eben recognizes that there
is neither much left of his life, nor much left in
his life. The words of the last three lines sum up
Eben's situation: "there was nothing in the town
below — /Where strangers would have shut the
many doors/That many friends had opened long
ago." In the realm of reality, Eben knows that
the world has changed. He also knows that all
of his contemporaries are gone. As the poem
ends, Eben returns to being a lonely old man
with time passing rapidly by him.

In conclusion, the name Eben Flood is most
appropriate for the main character in this poem.
The name can be broken down into "ebb and
flood." The flood describes the high tide of
Eben's life. This is the period in which Eben had
friends and hopes. At that time, life held much
in store for him. The image of the flood also
describes the flood of memories from his past,
which pour in the flood from the jug. All of Mr.
Flood's dreams of a bygone era comprise the
world of illusion. The ebb describes the decline
of Eben's life and the low level of his present
existence. The ebb also represents the world of
reality, in which Mr. Flood must face up to loneliness, desolation, and age. Eben may always
want to live in the flood of illusion, but he must
always return to the ebb of reality. After all, time
and tide wait for no man.

2. After examining the two poems below, following the Guidelines for Reading Literature, compose a critical essay on this topic: *Compare the tension between romantic love and everyday reality in Richard Wilbur's "A Late Aubade" and Adrienne Rich's "Living in Sin."*

A Late Aubade | *Richard Wilbur*

You could be sitting now in a carrel
Turning some liver-spotted page,
Or rising in an elevator-cage
Toward Ladies' Apparel.

You could be planting a raucous bed   5
Of salvia, in rubber gloves,
Or lunching through a screed of someone's loves
With pitying head,

Or making some unhappy setter
Heel, or listening to a bleak   10
Lecture on Schoenberg's serial technique.
Isn't this better?

Think of all the time you are not
Wasting, and would not care to waste,
Such things, thank God, not being to your taste.   15
Think what a lot

Of time, by woman's reckoning,
You've saved, and so may spend on this,
You who had rather lie in bed and kiss
Than anything.   20

It's almost noon, you say? If so,
Time flies, and I need not rehearse
The rosebuds-theme of centuries of verse.
If you *must* go,

Wait for a while, then slip downstairs   25
And bring us up some chilled white wine,
And some blue cheese, and crackers, and
    some fine
Ruddy-skinned pears.

Living in Sin | *Adrienne Rich*

She had thought the studio would keep itself;
no dust upon the furniture of love.
Half heresy, to wish the taps less vocal,
the panes relieved of grime. A plate of pears,
a piano with a Persian shawl, a cat   5
stalking the picturesque amusing mouse
had risen at his urging.
Not that at five each separate stair would writhe
under the milkman's tramp; that morning light
so coldly would delineate the scraps   10
of last night's cheese and three sepulchral
    bottles;
that on the kitchen shelf among the saucers
a pair of beetle-eyes would fix her own—
envoy from some village in the moldings . . .
Meanwhile, he, with a yawn,   15
sounded a dozen notes upon the keyboard,
declared it out of tune, shrugged at the mirror,
rubbed at his beard, went out for cigarettes;
while she, jeered by the minor demons,
pulled back the sheets and made the bed
    and found   20
a towel to dust the table-top,
and let the coffee-pot boil over on the stove.
By evening she was back in love again,
though not so wholly but throughout the night
she woke sometimes to feel the daylight coming 25
like a relentless milkman up the stairs.

3. Read Samuel Beckett's "play" *Breath*. In what sense does it illustrate the basic elements of literature? In what sense is it meant to be performed? How would a theater audience respond to such a performance? Consider these questions as you write a critical essay interpreting the theme of *Breath*.

Breath / *Samuel Beckett*

Curtain

1. Minimum light on stage littered with miscellaneous rubbish. Hold about five seconds.

2. Faint brief cry and immediately inspiration and slow increase of light together reaching maximum together in about ten seconds. Silence and hold about five seconds.

3. Expiration and slow decrease of light together reaching minimum together (light as in 1) in about ten seconds and immediately cry as before. Silence and hold about five seconds.

Curtain

Rubbish No verticals, all scattered and lying.

Cry Instant of recorded vagitus. Important that two cries be identical, switching on and off strictly synchronized light and breath.

Breath Amplified recording.

Light If 0 = dark and 10 = bright, light should move from about 3 minimum to 6 maximum and back.

4. The photograph on page 272 was selected as a metaphor representing an important aspect of this chapter. Examine the photograph and determine the possible meanings of this metaphor as you discern them in the photograph. Then write an essay explaining how the metaphor has been reinforced or revised by the ideas discussed in the chapter. You may want to suggest another metaphor to illustrate your understanding of these ideas.

13
PLANNING THE RESEARCH PAPER

You will probably write a number of research papers (also called library or term papers) during your formal education, papers that will figure prominently in your course grades but also provide you with an opportunity to discover new ways of thinking and writing. Noted historian Barbara Tuchman remembers researching her undergraduate honors thesis as "the single most formative experience of my career. . . . It was not a tutor or a teacher or a fellow student or a great book or the shining example of some famous lecturer. . . . It was the stacks at Widener [library]. They were *my* Archimedes' bathtub, my burning bush, my dish of mold where I found my personal penicillin" (Barbara Tuchman, "In Search of History," *Practicing History*).

The assignment to write a research paper is similar to other writing assignments in that you must discover information to fulfill a specific purpose. But it differs from other writing assignments, in that your major source of information is not memory, observation or informal reading (as in most personal essays), your textbook or lecture notes (as in the essay examination), or one or more literary texts (as in the critical essay) but — as Tuchman suggests — the books, articles, and documents housed in your university library. To locate the information you need and then to use it in your paper require skills in thinking, reading, and writing that you do not draw upon in other assignments.

Chapter 13 introduces the stages you must work through in *planning* a research paper:

■ Understanding the assignment
■ Making a schedule
■ Selecting a subject
■ Finding sources
■ Evaluating sources
■ Taking notes
■ Filling gaps

Chapter 14 takes up the stages involved in *writing* the paper. In these chapters, you will follow Laura, a student writer, as she plans, drafts, and revises her paper on the purpose of the stained-glass windows in the cathedral at Chartres. Laura's final paper, "Light and Literacy: The Stained-Glass Windows at Chartres," appears, fully annotated, at the end of Chapter 14 so you can see how she made important decisions about its content and how she implemented those decisions.

UNDERSTANDING THE ASSIGNMENT

Before you begin working, you need to determine the kind of paper you are to write. There are two basic kinds of research papers. In one you are expected to compile a *survey;* in the other you are expected to conduct an *argument.*

The Survey

The survey is a factual review of what other researchers have written about a subject. When you select a subject for a survey, focus on an issue or problem that has provoked extensive commentary or controversy, such as the causes of acid rain, the effects of gun control, the merits of educational reform. Imagine that your readers are curious about your subject but uncommitted to any particular position. They expect you to examine all sides of the subject objectively and to document your sources accurately so that they can read more about your subject. Your purpose is not to present your own argument about the subject but to identify and summarize the major arguments of others.

The Argument

The argument presents your analysis of a subject that has been researched by others; you interpret the information you uncovered in your research. You work from the perspective you have chosen, and you devise your own method of organizing and analyzing sources. Imagine that your readers are curious and uncommitted but ready to be convinced by a compelling argument. They expect you to acknowledge opinions that do not support your own, but they also expect you to present a forceful analysis, citing the proper authorities to support your viewpoint. Your purpose is not to compile a neutral summary of what others have written but to make your own contribution to a growing body of knowledge.

MAKING A SCHEDULE

Your instructor will specify when your paper is due and may require you to submit your work in stages so that both of you can track your progress. If your instructor does not provide a timetable, make one yourself. Start with the deadline and then work backward through the process, assigning a specific amount of time to each stage. Be cautious; allow yourself plenty of time to complete each activity. Be conscientious; work in the library for a certain number of hours each week. And be pragmatic; produce some kind of written material (journal entries, note cards, drafts) at the end of each stage. Post your schedule in a prominent place and consult it often. A schedule for a research paper might look like this:

Time	Activity	Written Product
Week 1	Study assignment. Make out schedule. Use journal to assess subject, audience, and purpose. Pick general subject. Read background material.	Schedule; journal entries; general subject; notes on background reading
Week 2	Select a specific subject. Formulate several hypotheses. Begin compiling bibliography of possible sources.	Specific subject; several hypotheses; source cards
Week 3	Locate and evaluate possible sources. Begin reading and taking notes.	Note cards
Week 4	Restrict subject. Analyze most valuable sources. Identify gaps in research.	Restricted-subject note cards; new source cards
Week 5	Locate and read additional sources. Take notes.	New note cards
Week 6	Select hypothesis. Develop outline.	Hypothesis; outline
Week 7	Write first draft. Prepare revision agenda for next draft. New outline.	First draft; revision agenda; new outline
Week 8	Write final draft. Check quotations. Complete documentation. Compile "works cited" list. Type and proofread final manuscript.	Completed assignment

Even if you start planning your paper the day you receive your assignment and follow a schedule like the one above, inevitably you will have to adjust your timetable as your work proceeds. A good rule of thumb is to add two weeks to your schedule for unexpected difficulties. You may need this time for situations like these:

- *Some things take more time than you planned.* Because the article you must read by Friday has to be ordered through interlibrary loan, you have to wait two weeks to complete your background reading.

- *Some stages prove more difficult than you expected.* Because your search strategy turns up only a few sources that deal directly with your subject, you have to find new sources or a new angle on your subject.

- *Deadlines on your schedule have to be adjusted, making it more difficult to meet subsequent deadlines.* Difficulties composing the final draft cut into time set aside for typing and proofreading.

Writers who start promptly, map out a reasonable timetable, make allowances for setbacks, and work efficiently can produce a research paper on time. Those who leave everything to the last minute will discover too late that they cannot throw together a satisfactory paper overnight. You need to live with your project for several weeks, reading and assimilating sources. A realistic schedule is a written reminder of your goal and encourages you to work at a steady pace, committing your discoveries and ideas to writing as soon as possible.

SELECTING A SUBJECT

Selecting a subject for a research paper is like staking a prospector's claim. You *hope* the claim will produce gold, but you won't *know* until you begin digging. Some instructors, therefore, ask students to select a subject from a pretested list. If your instructor instead asks you to select your own subject, assess potential subjects according to the following criteria of successful research subjects. (Review the Guidelines for Selecting a Subject discussed in Chapter 1.)

1. *Select a subject you can research.* This may seem an obvious requirement, but many subjects cannot be researched.
    a. *Some subjects are too autobiographical.* A paper that draws primarily on your own experience — "Growing Up in Oklahoma City" — does not require you to search for information in other sources.
    b. *Some subjects are too subjective.* No amount of research will resolve a question of personal taste, such as "Which is the better poet — Yeats or Eliot?"
    c. *Some subjects are too restricted.* A mechanical process — "How to Operate a VCR" — that can be explained by only one source does not require significant research.
    d. *Some subjects are too current.* Events that produce today's headlines — "Scandal at the State House" — have not been studied in sufficient depth for you to find enough information about them.
    e. *Some subjects are too specialized.* A subject such as "Reactions of German-American Pacifists to the Great Sioux Massacre of 1864" cannot be researched if your library does not own or have access to the required special documents.
2. *Select a subject you can restrict.* Before you begin your research, you may worry that you will not be able to find enough information. Once you begin, however, you are likely to find that nearly every source reveals new aspects of your subject. Instead of feeling overwhelmed, take control of your subject and reduce it to a manageable size. Two factors will help you:
    a. *The time you scheduled for planning your paper.* Whether you select a subject that you already know something about or one that is new to you, be realistic about how much you can learn in the time available. A subject

such as heart disease, for example, will lead to more sources in the card catalogue alone than you will have time to read, analyze, and understand. Restrict your subject to a specific aspect of heart disease, such as one of its suspected causes. Restricting your subject even further, to one method used to control a cause of heart disease, might lead to an even better focus for your paper.

b. *The space available to develop your paper.* If your assignment restricts the length of your paper, be realistic about how much you can cover in the specified number of pages. A subject such as word processors, for example, will reveal more sources than can be listed in ten pages, let alone usefully developed into an argument. Restricting your subject to changes in American business correspondence or even changes in *one* American business's correspondence with its customers will be more manageable.

3. *Select a subject you can live with.* Writing a research paper requires you to work with one subject for a long time. If this subject does not fascinate you, if you do not care about the questions it poses or the answers you can provide, you will become bored, and your planning will be careless and your writing uninspired. Be sure to select a subject that holds your interest.

4. *Select a subject that will appeal to other readers.* Although the immediate audience for your research paper is your instructor, imagine at least two other audiences. One is the authors you have come to know while doing your research. In a sense, you are carrying on an extended conversation with these writers about a new direction in an area familiar to them. The other audience is the intelligent "general reader," who is always interested in new information or new approaches. You are asking this reader to consider your thesis. As you select your subject, strike a balance between the expectations of these two audiences. Previous researchers should consider your work substantive, not trendy; the general reader should find it innovative, not shopworn.

5. *Select a subject you can prove something about.* If you are writing an *argument*, not a *survey*, the purpose of your paper is to use sources to support a thesis. You must be sure to select a subject that will yield a thesis, not a summary of other researchers' arguments. Your subject must be focused, so that you can control your evidence to support your argument, and complex, so that you can develop and sustain your argument throughout your paper.

▬▬▬ *Exercise*

One way to discover a subject is to explore, in your journal, what you already know about certain general subjects. Laura, a library science major, tries this as she thinks about possible subjects for a research paper in her English class. Read her journal entry and then discuss how she has applied the above criteria for selecting a subject.

Research paper assigned today. Need a subject. Johnson says we can write about our major. <u>Library Science</u>. How to catalogue books? How to preserve old books? How to keep books from being stolen? All too technical. Maybe something about driving bookmobile last summer — seeing kids, parents, old folks stocking up on summer reading. Not sure where that would go. Making books available to the people?

Maybe something about book I read waiting for customers — picture book on stained-glass windows. How they are made, restored. One section had floor plan for windows at famous cathedrals. Windows at Chartres. Dull from outside, but inside — a slide show. Light changes outside, colors change inside like a big kaleidoscope. Each window supposed to tell a story. Panels look like Sunday comics. Could people <u>read</u> them? Dark Ages. Big subject. What's my subject? Chartres' windows? Still pretty big. Over 175 windows. Needs more focus. Maybe look at those slides Jill brought back from Chartres last summer — a good place to begin.

FINDING SOURCES

Once you have selected your general subject, begin the formal process of researching it in the library. Most libraries, like most cities, try to help visitors by providing tours, publishing maps and directories, and hiring guides (librarians) to work at specific locations throughout the building. Because no two libraries are organized exactly alike, even the most experienced researchers depend on this kind of assistance to help them work in an unfamiliar library.

You will spare yourself considerable frustration at the outset if you realize that your best allies are librarians. No matter how difficult or ridiculous your question seems, the librarians have probably heard it or one like it before. And if they cannot give you an answer, they can show you where and how to look for one. Indeed, they can direct you to many more sources than are mentioned in this chapter. You have already considered how to select a subject. Librarians will help you formulate a *search strategy* — a systematic procedure for finding information on your subject. The diagram on page 303 shows you the sequence of steps in a search strategy.

Background Information

Read several overviews of your general subject so that you can learn about its history, major themes, and principal figures. Such background information can be found in general and specialized encyclopedias located in the reference collection of the library.

Search strategy

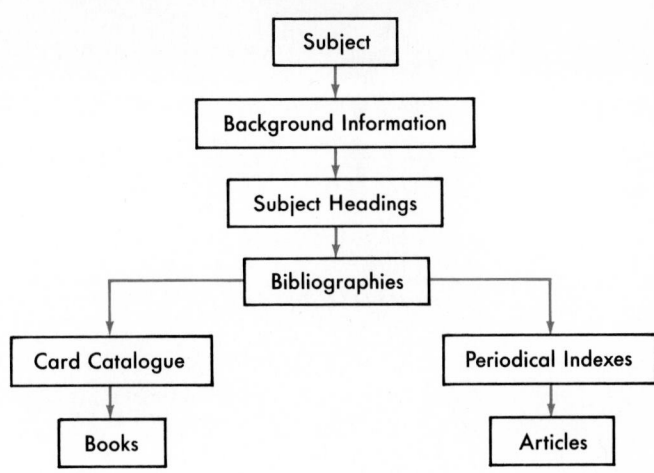

General Encyclopedias   These reference works, written for a general audience, cover a wide variety of subjects. The entries, generally arranged in alphabetical order, are not technical or scholarly, but they often contain a brief list of sources that do treat the subject in depth. You can find the main entry about your general subject by looking in the appropriate volume. To locate every reference to your subject in the encyclopedia, you can consult the index, which is usually the last volume. The latter method enables you to see how your subject is subdivided and cross-referenced, and reading the additional entries may give you a perspective that will help you restrict or revise your subject. Standard encyclopedias include

Collier's Encyclopedia
The New Columbia Encyclopedia
Encyclopedia Americana
The New Encyclopedia Britannica
World Book

Specialized Encyclopedias   These books are usually devoted to one or two disciplines or to specific subject areas. They provide a narrow, detailed coverage of particular topics. The entries tend to be more technical than those in a general encyclopedia, with longer lists of sources. Because specialized encyclopedias are available on a wide range of academic and general interest subjects, you will need to use the card catalogue to locate them. Look up your general subject and then look for the subheading "Dictionaries." The *Encyclopedia of Philosophy*, for example, is listed under the subject heading "Philosophy — Dictionaries." The following brief list of specialized encyclopedias shows the variety of material available.

© Ron Sherman/Bruce Coleman, Inc.

The International Encyclopedia of Higher Education
McGraw-Hill Encyclopedia of Science and Technology
Rock On: The Illustrated Encyclopedia of Rock 'n' Roll

Biographical Sources   These works give brief accounts of notable figures, listing such information as family history, educational background, major accomplishments, and significant publications. A short list of additional sources is often included. If your background reading points to one or two people significant in the development of your general subject, you should research these figures in a biographical source. Some examples are listed below.

Biography Index
Dictionary of American Biography
The International Who's Who

███   **Exercise**

Laura uses her journal to make notes on the information she discovered in encyclopedias about Chartres and stained glass. Read her notes and consider how her background reading helps Laura shape her subject. Note what information enriches, confuses, or expands her original subject.

Over ten headings on stained glass in Americana *index. Some look promising —* Gothic art *— others too contemporary —* Tiffany. *About ten headings on Chartres focusing on every aspect of the cathedral — doors, paintings, sculpture, spires, stained-glass windows. Entry on Chartres deals with history of church — site, destruction (various causes), constant rebuilding. Mentions "warm glow of light" inside and features brief discussion of "Blue Virgin" window — famous because of the blue glass used for her tunic — cathedral's relic. Cross-reference under Chartres to* stained glass *— ten-page entry by Jane Hayward, The Met. Deals with technique of making windows — design, types of glass, cutting, leading, setting. Also deals with history of glass staining (early Gothic to twentieth century). Importance of Abbot Suger, who commissioned windows for St. Denis cathedral. Good discussion of glassmakers' workshops and the iconographic plans for various cathedrals. Choosing right colors a problem — needed to make figures "legible." Not much on Chartres, but good bibliography.*

Specialized encyclopedias not all that useful. The Encyclopedia of Glass *has only four pages on stained-glass windows. Mostly covers making glass and glassware and art objects.* Encyclopedia of World Art *has 24 pages, plus bibliography, on stained-glass windows. Major discussion is on evolution of various techniques — such as "grisaille," a method used for altering tone of colors and outlining details. Problem of "authorship" — relationship between designer and glazier (the craftsman who paints glass and assembles windows). "Chartres' cycle remains the greatest moment, the apogee, of medieval painting on glass." Discussion too focused on the* technique *of Chartres' glassmakers, problems of restoration. The church was rebuilt many times — though some older windows preserved. No discussion of content, the "legendary scenes" in the windows. How did people understand all the symbols? Did they already know the stories? How? They didn't know how to read. Need to connect stories in windows with problem of literacy. McLuhan's* Gutenberg Galaxy. *Maybe he has some stuff on how (what?) people learned before they could read.*

Subject Headings

When you have done some background reading, you will discover that there are many ways to classify, subdivide, and cross-reference your topic. This cluster of categories should tell you three things:

1. You must restrict your general subject (in Laura's case, stained-glass windows) to the specific subject you want to write about (the stained-glass windows at Chartres).
2. You may find your specific subject listed under several headings (Chartres, stained-glass windows, Gothic cathedrals, Cathedrals in art).

3. You may have to read material that does not deal directly with your subject in order to discover a perspective that will organize your information (medieval culture, oral traditions, the history of literacy, illuminated books and manuscripts).

The best way to discover subject headings related to your subject is to consult the reference book entitled *Library of Congress Subject Headings*. This guide, usually located near the card catalogue, shows how your subject is classified and cross-referenced in the card catalogue. Sometimes you will have to use some ingenuity to determine how your subject might be listed. For example, Laura found no subject headings under *Chartres* or *stained-glass windows*. But a cross-reference to *glass painting and staining* produced the headings *art, church decoration and ornament,* and *Gothic glass painting and staining* that might have sources on her subject.

Throughout your research you will revise your subject headings. Some will be eliminated because they fail to turn up sources that focus on your subject. Others will be added because they lead to sources that give you valuable insights. In fact, you may have to strike out in a new direction to find the angle that gives the various parts of your subject the greatest coherence.

For example, Laura added to her initial headings, *Chartres* and *stained-glass windows, medieval culture* and *literacy,* and considered but discarded *glass-staining techniques* and *stained-glass window restoration.* Although Laura's subject may seem to be getting out of control, Laura is actually developing several hypotheses about the relationship between Chartres' windows and literacy. Her hypotheses will help her shape her draft.

Bibliographies

When you have compiled a preliminary list of subject headings, you are ready to begin building your bibliography. You begin with a *working bibliography,* a list of books and articles you intend to consult as you plan your paper. (When you finish writing your research paper, you will have a *final bibliography,* a typewritten list of "works cited.") Consult a reference guide to reference books, such as the American Library Association's *Guide to Reference Books* (10th ed., 1986), compiled by Eugene P. Sheehy. This guide will reveal whether someone has already published a specialized bibliography on your subject. Specialized bibliographies not only save you time but also ensure that you consult the most significant sources. Such bibliographies are often *annotated,* providing a brief description (and perhaps an evaluation) of each book and article listed. Specialized bibliographies, however, may not cover recent works. A bibliography published in 1950, for example, leaves you to track down elsewhere the research done in the last four decades.

Three other ways to locate specialized bibliographies are to:

- Check the *Bibliography Index,* a reference work that lists articles, parts of books, and pamphlets devoted in whole or in part to bibliographies
- Check the card catalogue under your subject and subject headings for the subheading "Bibliography." For example:

Cathedrals—Bibliography

Glass—Bibliography

Literacy—Bibliography

- Check the catalogue card under your subject heading to see if it contains a notation indicating that a source includes a bibliography. (See the example below.)

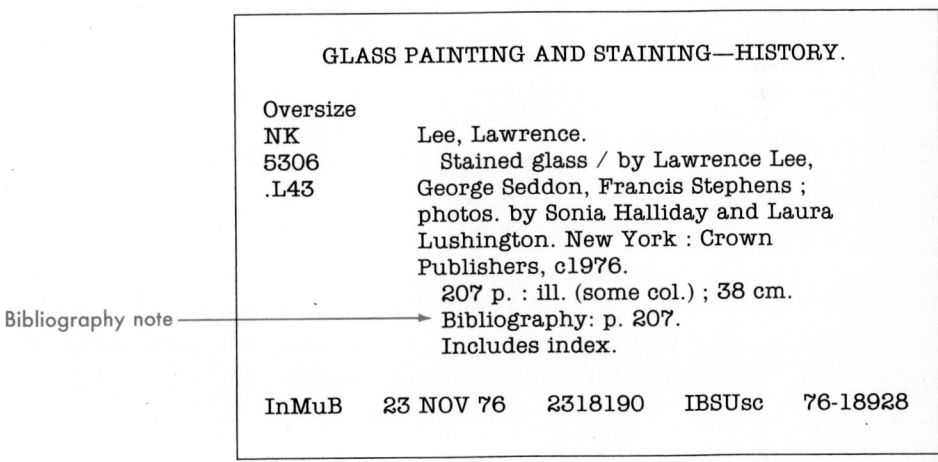

GLASS PAINTING AND STAINING—HISTORY.

Oversize
NK   Lee, Lawrence.
5306   Stained glass / by Lawrence Lee,
.L43   George Seddon, Francis Stephens ;
                   photos. by Sonia Halliday and Laura
                   Lushington. New York : Crown
                   Publishers, c1976.
                       207 p. : ill. (some col.) ; 38 cm.

Bibliography note ————————————→ Bibliography: p. 207.
                   Includes index.

InMuB   23 NOV 76   2318190   IBSUsc   76-18928

Card Catalogue

After locating a specialized bibliography or determining that none is available, begin constructing your working bibliography with the two major research tools in the library — the *card catalogue* and the *periodical indexes.* The card catalogue, arranged alphabetically by subject, author, and title, lists books and other material available in the library.

Author card

If you know the author of a book about your subject, or suspect that a certain author might have written about the subject, look in the card catalogue for the *author* (or *main*) *card.* It will give you not only the title but much other useful information:

Author card

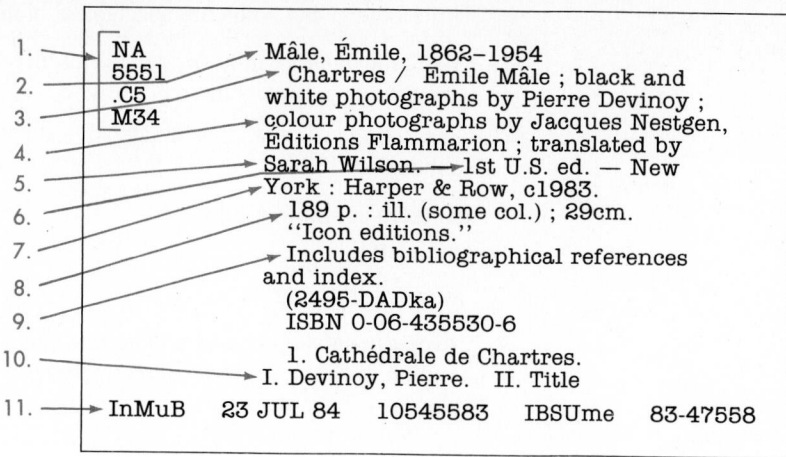

1. *Call number and location symbol (if applicable).* The letters and numbers, in the exact order in which they appear on the card, constitute the call number, which indicates the book's location on the library's shelves.

2. *Author's name.* The last name appears first, followed by additional data, such as birth and death dates.

3. *Title.* The title is given as it appears on the title page of the book.

4. *Photographer's or illustrator's name.* If the book contains special photographs or illustrations, the names of the artists are often listed.

5. *Translator's name.* If the text has been translated from another language, the name of the translator appears on the card.

6. *Edition statement.* The card identifies the particular edition held by the library.

7. *Imprint.* The imprint information includes the place of publication, the publisher, and the year of publication.

8. *Number of pages; size of book; and maps, charts, or illustrations.* Numbers in small Roman numerals (xii) are sometimes used to indicate the number of pages of front matter.

9. *Special features.* These may include the bibliography, appendixes, or index.

10. *Tracings.* These list the subject headings under which the book is catalogued. Tracings are most useful for locating other subject areas that might pertain to your subject.

11. *Codes.* Codes provide librarians with computer cataloguing information; this information is *not* pertinent to your research or for locating the book.

Title card

If you know the title of the book but have forgotten the author, or if you suspect that one of your subject headings might be included in a book title, look for a *title card.*

Title card

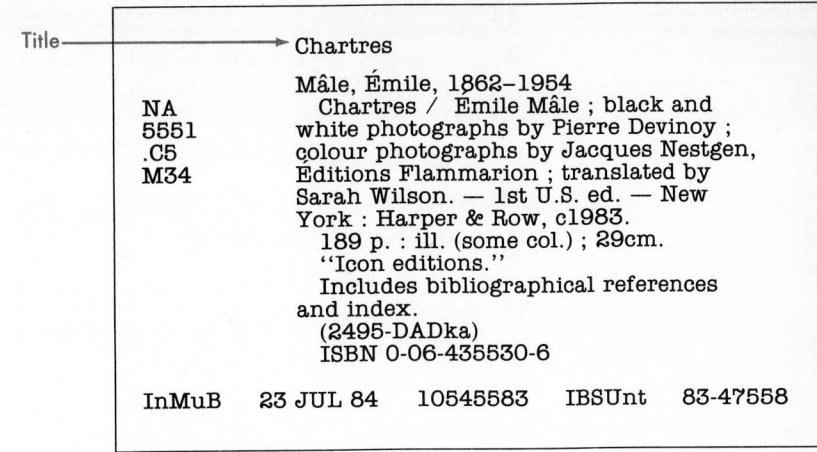

Title ————————→ Chartres

NA
5551
.C5
M34

Mâle, Émile, 1862–1954
    Chartres / Émile Mâle ; black and
white photographs by Pierre Devinoy ;
colour photographs by Jacques Nestgen,
Éditions Flammarion ; translated by
Sarah Wilson. — 1st U.S. ed. — New
York : Harper & Row, c1983.
    189 p. : ill. (some col.) ; 29cm.
    "Icon editions."
    Includes bibliographical references
and index.
    (2495-DADka)
    ISBN 0-06-435530-6

InMuB   23 JUL 84   10545583   IBSUnt   83-47558

Subject card

If you know only your subject or several related subject areas (this will be the case if you haven't found a specialized bibliography of your subject), look for *subject cards.*

Subject card

General
subject
heading

(See
tracings
on author
card.)

————→ CATHÉDRALE DE CHARTRES.

NA
5551
.C5
M34

Mâle, Émile, 1862–1954
    Chartres / Émile Mâle ; black and
white photographs by Pierre Devinoy ;
colour photographs by Jacques Nestgen,
Éditions Flammarion ; translated by
Sarah Wilson. — 1st U.S. ed. — New
York : Harper & Row, c1983.
    189 p. : ill. (some col.) ; 29cm.
    "Icon editions."
    Includes bibliographical references
and index.
    (2495-DADka)
    ISBN 0-06-435530-6

After you have located a book in the card catalogue, locate it on the shelf. Most libraries now arrange books according to the *Library of Congress system,* although some libraries retain the *Dewey decimal system* or are converting to the Library of Congress system and so have part of their collection classified by one system and part by the other.

Library of Congress system

A	General Works	**M**	Music
B	Philosophy, Psychology, Religion	**N**	Fine Arts
C–D	History and Topography (except America)	**P**	Language and Literature
E–F	America	**Q**	Science
G	Geography, Anthropology, Sports and Games	**R**	Medicine
H	Social Sciences	**S**	Agriculture, Forestry
J	Political Science	**T**	Engineering and Technology
K	Law	**U**	Military Science
L	Education	**V**	Naval Science
		Z	Bibliography and Library Science

In the Dewey decimal system, materials are arranged by numbers that represent subject areas.

Dewey decimal system

000–009	General works	**700–799**	The arts. Fine and decorative arts
100–199	Philosophy and related disciplines	**800–899**	Literature (belles-lettres)
200–299	Religion	**900–999**	General geography and history and their auxiliaries
300–399	Social sciences		
400–499	Language		
500–599	Pure sciences		
600–699	Technology (applied sciences)		

Knowing which classification system your library uses will help you find a specific book. But because in both systems books are shelved by subject, browsing through the shelves near where you found a particular book you were looking for may reveal other intriguing, relevant titles that you failed to notice in the card catalogue. For example, browsing on the shelf where she found Émile Mâle's *Chartres,* Laura found many books on the Gothic cathedral. Similarly, by browsing in the general area designated "Fine Arts (N)," she found some books on stained-glass that she had overlooked in the card catalogue that were shelved under NK, the designation for Art and Industry, Decoration, and Antiques.

Periodical Indexes

Periodicals (magazines and journals) exist on almost every conceivable subject. Their value in research is their currency and specificity, and back issues can show the ways your subject has been treated in the past. Indexes to periodicals are of two types, general and subject.

General indexes list articles from hundreds of nontechnical and nonscholarly magazines. Individual articles are classified by author and by subject, and most broad topics are subdivided. The most useful general index is *Reader's Guide to Periodical Literature,* which is issued monthly and bound into annual volumes.

Subject indexes list articles on specific subjects. The articles tend to be more detailed and scholarly than those listed in the *Reader's Guide.* Here is a brief sampling of the subject indexes available:

Art Index   *Humanities Index*

Business Periodicals Index   *Music Index*

Education Index   *Social Science Index*

Reader's Guide

Glass painting and staining
See also
Merry Go Round Stained Glass (Firm)

Albinas Elskus explores the possibilities of vitreous paint. R. Kehlmann. il por *Am Craft* 45:10–14 D '85/Ja '86

Antiques: the double image [Chinese mirror painting] J. A. Cuadrado. il *Archit Dig* 42:78–83 Ja '85

Farewell to toxic fumes [stained glass studio] G. Proust. *Mother Earth News* 96:37 N/D '85

The legacy of Johan Thorn Prikker [ecclesiastical stained glass] R. Kehlmann. bibl f il *Am Craft* 45:26–31 Ap/My '85

Light, color, glass [work of L. Hovey] C. W. Ward. il pors *Americana* 13:42–5 Jl/Ag '85

Toluca's glass-covered garden. R. A. Camp. il *Americas* 37:60–3 Mr/Ap '85

Conservation and restoration
The restoration of medieval stained glass. G. Frenzel. il *Sci Am* 252:126–9+ My '85

Glass painting and staining, Miniature
More decorative screens [miniature stained glass screen] S. S. Bain. il *Creat Crafts Miniat* 9:31–2 D '85

Humanities Index

ORAL tradition
See also
Folklore—Transmission

Kisra legend as oral history. M. H. Stewart. In: J Afric Hist Stud 13 no 1:51–70 '80

Memories with no historian: tr. by J. Wicke and D. Moschenberg. G. Duby. Yale Fr Stud no59:7–16 '80

Mother of the muses: in praise of memory. C. C. Park. Am Scholar 50:55–71 Wint '80/81

Narrative techniques and the oral tradition in The scarlet letter. J. G. Bayer. Am Lit 52:250–63 My '80

Swatchel omi: Punch and Judy and the oral tradition. R. Leach. il Theatre Q 9:66–76 Wint '80

Le territoire de la mythologie. M. Detienne. Class Philol 75:97–111 Ap '80

* { Varieties and consequences of medieval literacy and illiteracy. F. H. Bäuml. Speculum 55:237–65 Ap '80

Wu Tzu-hsü pien-wen and its sources: part II. D. Johnson. Harv J Asiatic Stud 40:465–505 D '80

Art Index

GLASS painting and staining, Alsatian
Vitraux de la nef de Niederhaslach. C. Block. Bull Mon 131 no 1:63 '73

GLASS painting and staining, British
Musée des beaux-arts de Strasbourg: deux cartons de vitraux par Edward Burne-Jones. J. Christian. bibliog f il R du Louvre 22 no6:525–30 '72

Saving Canterbury's medieval glass. M. Caviness. il Country Life 152:739–40 S 28 '72

Stained glass of Lavers and Barraud. M. Harrison. il (pt col) Connoisseur 183:194–9 Jl '73

Windows of King's college chapel, Cambridge, by H. G. Wayment
Review by L. Grodecki. Gaz Beaux Arts s6 v80:sup6–7 O '72
Review by G. W. Thomas. Connoisseur 181:138 O '72

Work of C. E. Kempe. R. Mander. il pors Apollo ns 97:160–3 F '73

See also
Glass painting and staining, Victorian

GLASS painting and staining, Chinese
European painting on glass found in rural Ontario. W. Z. Nixon. il Orient Arts ns 18 no2:171–2 Summer '72

GLASS painting and staining, Flemish
Deux panneaux du musée de Cleveland. F. Perrot. Bull Mon 130 no 1:69–70 '72

GLASS painting and staining, French
Corpus vitrearum medii aevi: les vitraux de l'église Saint-Ouen de Rouen, by J. Lafond and others
Review by J. Hayward. Art Bull 55:293–6 Je '73

Églises et vitraux de la région de Pont-Audemer, by M. Baudot and J. Lafond
Review by F. Perrot. Bull Mon 130 no 1:87–8 '72

Un fragment de la vitrerie de la Sainte-Chapelle de Paris à la cathédrale de Cantorbéry; Les vitraux de Saint-Vincent de Rouen. F. Perrot. Bull Mon 131 no 1:62 '73

Redating of the thirteenth-century grisaille windows of Chartres cathedral. M. P. Lillich. bibliog f il plan Gesta 11 no 1:11–18 '72

Vitrail: vitrerie du bas-côté sud de la cathédrale de Strasbourg. C. Block. Bull Mon 130 no4:365 '72

Vitraux héraldiques venant du château d'Écouen. F. Perrot. bibliog f il R du Louvre 23 no2:77–82 '73

Most periodical indexes are located in the reference area of your library. When you have selected the best indexes for your subject, look under your subject headings to find major listings and then check any subheadings or cross-references for additional information. When you look at a specific entry, you will see that it contains (1) the title of the article, (2) the name of the author, (3) the abbreviated title of the periodical, (4) the volume number of the periodical, (5) page references, and (6) date of publication. Copy all this information on to your source card (see "Taking Notes," pp. 316–22). You may also want to make a note of the title and volume of the index in case you need to recheck the information. Abbreviations used in the entries are spelled out in the *front* of most indexes.

Laura checked *Reader's Guide, Art Index,* and *Humanities Index* for periodical sources. She discovered that for her purposes the listings in *Reader's Guide* were too general, focusing on subjects such as restoration and decoration. By contrast, the listings in *Art Index* were too specialized, often written in languages other than English and focusing on problems such as authorship and dating. Initially, her search through *Humanities Index* proved even more frustrating. She began with *literacy* as a subject heading, was referred to *illiteracy* as a heading, and then cross-referenced to *oral tradition.* Under this heading, however, she found an important article, F. H. Bäuml's "Varieties and Consequences of Medieval Literacy and Illiteracy." (This entry is marked with an asterisk in the excerpt from *Humanities Index* shown on page 311.)

Newspapers, Documents, Microforms, and Computer Searches

In addition to the card catalogue and the periodical indexes, four other tools will help you find information on your subject.

Newspapers   Many major newspapers, such as the *New York Times, The Times* (London), and the *Washington Post,* publish their own indexes. Newspapers give day-to-day accounts of events and provide details often omitted in later, more general discussions. Consulting them will help you identify important factual information about your subject and enable you to trace its historical development.

Government Documents and Publications   The United States government publishes an enormous amount of information on a wide variety of subjects. These publications are housed in a separate area of the library and catalogued according to a separate classification system. There are indexes to this information, such as *American Statistics Index* and the *Monthly Catalogue of United States Government Publications,* but to use the collection effectively you will need help from someone familiar with government documents. A documents librarian or a member of your library's reference staff will be able to provide this guidance.

Microforms Most libraries do not have space for all the documents they wish to keep. Thus many periodicals and other materials are photographically reduced in size and stored on *microfilm* (reels of film) or *microfiche* (sheets of film). With the aid of mechanical viewers you can enlarge and read these documents easily or make photocopies of the enlarged images. Librarians can tell you which books and periodicals are stored in microform and show you how to operate the microreaders.

Computer Searches An increasing number of libraries make use of computers to store and access information. The Library of Congress in Washington, D.C., for example, is entering its entire card catalogue into a computer. Instead of flipping through cards, users must now search for sources at computer terminals.

Information from many fields is stored in data bases that can be accessed through a computer. Medical information, government statistics, stock-market figures, law cases, abstracts of educational articles, and bibliographies are among the types of material that computers search quickly, thoroughly, and accurately. Sometimes the results can be printed out while you wait; but if you require a large amount of material you may have to wait a few days.

If your library offers computer searches, consult with the librarian about your subject and subject headings to determine if a computer search is appropriate. If so, the librarian will use your subject headings to search data bases for books, articles, and other documents. Computer searches may cost as little as ten dollars; some cost a good deal more. Your librarian can advise you whether, given your subject, you should invest in the service.

EVALUATING SOURCES

To compile a working bibliography of books and articles that might pertain to your subject, you must decide which items are likely to prove most useful. The best way to discover this is through careful reading, but you will *never* have time to read every possible source. If Laura attempts to read everything she finds on *stained-glass windows* and *Chartres,* her reading will consume all her research and writing time. If she decides to read in addition all the books and articles she can find on *medieval culture* and *literacy,* she will spend her life trying to keep up with an ever-growing reading list.

All researchers need guidelines and short cuts to help them make intelligent guesses about the potential value of the sources they uncover. Following such guidelines will help you eliminate some sources immediately, discard others after determining that they do not focus on your subject, and concentrate on the sources or parts of sources that will make the most significant contribution to your research.

Guidelines for Evaluating Sources

1. *The source should be relevant.*

   Whether a particular source is relevant is not always apparent. When you first begin your research, your lack of perspective on your subject may make every source seem potentially relevant. Sometimes the titles of articles and books may be misleading or vague, leading you to examine a work unrelated to your subject, or to assume that a work is too general or theoretical when actually it focuses on an essential aspect of your subject. Finally, your reading will occasionally change the status of some sources. What seemed irrelevant to yesterday's perspective on your subject may suddenly seem crucial to today's more informed definition of your purpose. The key to assessing the relevance of a source is to restrict your subject. The sooner you limit your subject, the sooner you can determine the relevance of a particular book or article.

2. *The source should be current.*

   You want to be sure that the information in your sources is reliable and up-to-date. A paper on the latest cures for cancer should not rely on an article published in 1945. On the other hand, if you are analyzing the public's attitude toward cancer at different times, the 1945 article might be relevant. Not all old works are dated, however. Experts in many subjects acknowledge standard or "classic" books or articles that have advanced major interpretations. You should read those that pertain to your subject.

3. *The source should be comprehensive.*

   Some sources will focus on an extremely narrow aspect of your subject; others will cover its every feature and many related topics as well. Always begin your reading with the most comprehensive source, because it will probably provide the essential information contained in the more specialized sources, and you may not have to read the second source.

4. *The source should direct you to other sources.*

   Catalogue cards will show whether a book contains a bibliography. Skimming will reveal whether an article contains extensive notes. The most helpful notes include annotations about the sources cited. Books and articles that describe and evaluate other sources help you decide whether you want to read sources you have already found and point you toward sources you have overlooked.

Short Cuts for Evaluating Sources

Locate Annotated Bibliographies If you are lucky enough to find an annotated bibliography on your subject or if the notes in an article contain extensive annotations, you can determine quickly whether the sources they describe are worth reading.

Read Book Reviews   If you want to determine whether a particular book is reliable, see how it was reviewed when it was first published. Reference guides, such as the *Book Review Digest* and the *New York Times Index,* contain either summaries or references to book reviews that should help you evaluate the book's content and critical reception.

Obtain the Advice of Experts   Many people on your campus or in your community are experts on certain subjects. A quick phone call or visit to these people can help you identify the "must-reads" or classic treatments of your subject. They can also direct you to annotated bibliographies and special indexes. And, finally, they can refer you to unusual sources that you would not find by following a normal search strategy. Such books and articles, often dealing with other subjects, may introduce you to new ideas or new methods of interpretation.

Review the Table of Contents   To determine the way a book develops its major ideas, study the table of contents. The chapter titles and subheadings work like an outline, giving you a general sense of the author's understanding and organization of the subject.

Read the Introduction   To discover the particular focus of a source, read its introduction — the preface and often the first chapter of a book, the first few paragraphs of an article. A few pages is usually enough to detect the author's thesis and decide whether it is relevant and the source valuable.

Browse Through the Index   An index works like a miniature card catalogue, helping you see whether a source has information on your subject, how it classifies and cross-references your subject, how much information it devotes to each of your subject's features, and precisely where that information is located.

████ *Exercise*

The following titles come from Laura's working bibliography on the stained-glass windows at Chartres. Using the Guidelines for Evaluating Sources (p. 314), what guesses would you make about their usefulness for Laura's project? Number the entries to indicate the order in which you think they should be consulted.

Adams, Henry, *Mont-Saint-Michel and Chartres.* Boston: Houghton, 1905.

Arwas, Victor. *Glass: Art Nouveau to Art Deco.* New York: Rizzoli, 1977.

Haile, H. G. "Luther and Literacy." *Publications of the Modern Language Association* 91 (1976): 816–28.

Hall, Edward T. *The Silent Language.* New York: Doubleday, 1959.

Harrigan, P. J. "Education and Society in Modern France." *Canadian Journal of History* 14 (1979): 442–48.

McLuhan, Marshall, *The Gutenberg Galaxy: The Making of Typographic Man.* Toronto: U of Toronto P, 1962.

Marchini, Giuseppe. *Italian Stained Glass Windows.* New York: Abrams, 1956.

Panofsky, Erwin. *Gothic Architecture and Scholasticism.* New York: Meridian, 1957.

Singler, J. V. "The Psychology of Literacy." *Language* 59 (1983): 893–901.

Wulf, Maurice De. *Philosophy and Civilization in the Middle Ages.* Princeton: Princeton UP, 1922.

TAKING NOTES   As you find your sources and select those that you suspect will prove most useful, start making source cards and taking notes.

Source Cards

Fill out a 3″ × 5″ source card for every item you intend to read. The card will help you keep track of each source from the beginning of your research (when you locate the source in the card catalogue or periodical index) to the point at which you type up the results of your research (when you enter the source in your list of "works cited").

Two sample source cards, the first for a book, the second for a periodical, appear on page 317.

Two pieces of information on the source card will help you during the research process. The call number (at the left) enables you to locate the book in the library. The source number (on the right) enables you to code your note cards so that you can identify the source without recopying the bibliographic information. Note the bibliographic information on your source cards in the format you would use on a list of "Works Cited" (see Listing Sources, pp. 345–53). Complete and accurate source cards will save you time as you research and write your paper, so it is worthwhile to make a card for any material that survives your initial evaluation, even though you are not yet sure which sources will surface in your paper. When you are ready to type your "Works Cited" list, simply arrange in alphabetical order the cards for the sources you have used and type the appropriate information for each entry from the card.

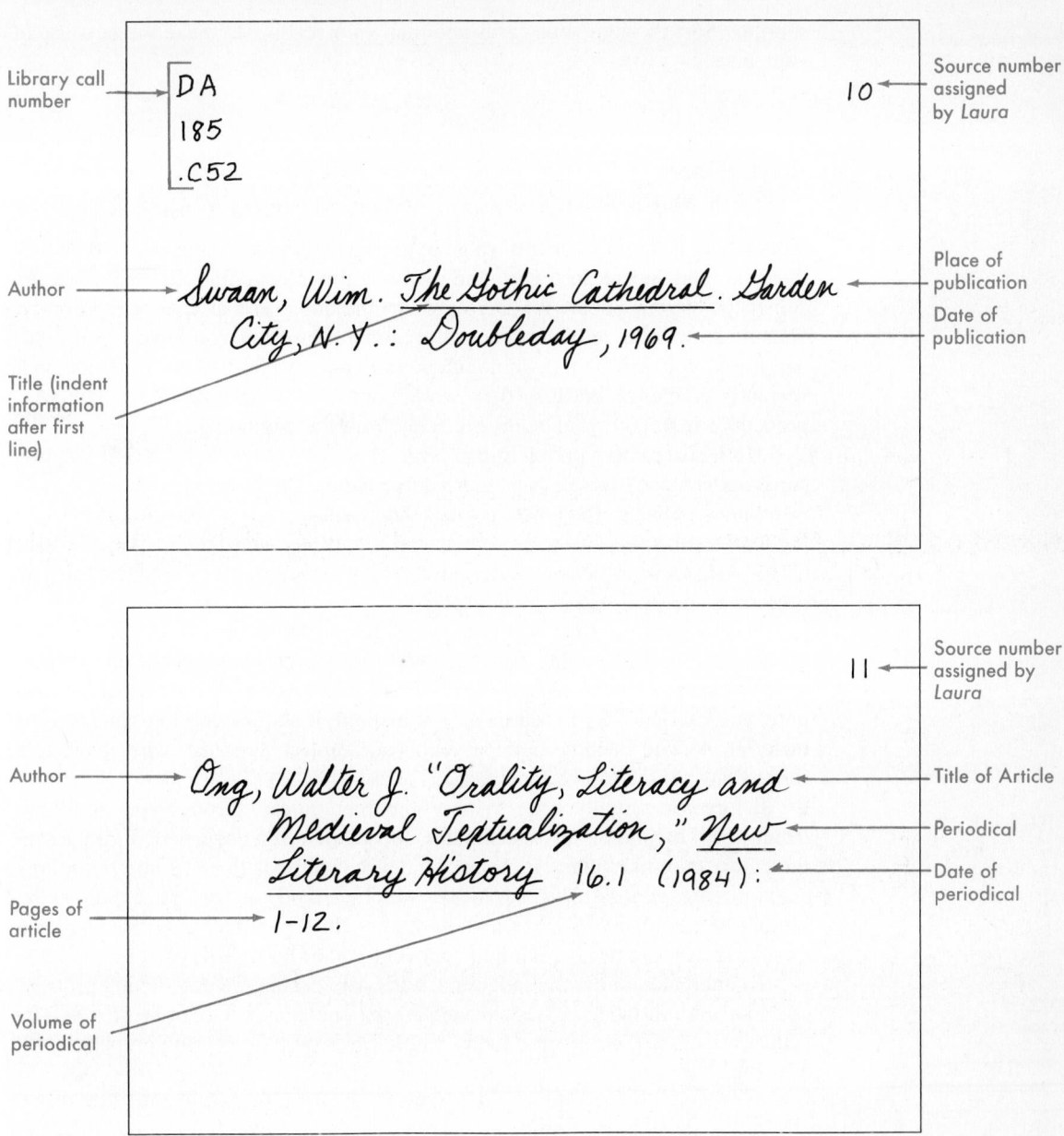

Library call number — DA 185 .C52

Source number assigned by *Laura* — 10

Author — Swaan, Wim. *The Gothic Cathedral*. Garden City, N.Y.: Doubleday, 1969.

Title (indent information after first line)

Place of publication

Date of publication

Source number assigned by *Laura* — 11

Author — Ong, Walter J. "Orality, Literacy and Medieval Textualization," *New Literary History* 16.1 (1984):

Pages of article — 1–12.

Volume of periodical

Title of Article

Periodical

Date of periodical

After you have made source cards, use the evaluation guidelines to establish some priorities for your reading. What sources seem most crucial? What sources will take the most time to read? What sources are likely to help you find other sources? Sort through your cards and plan your reading to make the best use of your time.

Note Cards

Note taking is the most critical stage in the research process because it demands that you read, select, interpret, and evaluate the information that will form the substance of your paper. When you return the books and articles to the library, your notes will be your only record of your research. If you have taken notes carelessly, you will be in trouble when you begin writing. Many students inadvertently plagiarize because they work from inaccurate note cards. The wise procedure is to take your notes precisely from the beginning.

Most researchers prefer to use large cards (4″ × 6″ or 5″ × 7″) for note cards and small cards (3″ × 5″) for source cards. The two sizes make it easy to distinguish between the types of cards, and the large cards provide more space for information. Just as your source cards may become part of your "Works Cited" list, so each of your note cards may produce one of the ideas you will present in the final text of your paper.

Make a separate card for every piece of important information you discover in each source. Identify the note card with the source-card number, the author, and the pages where you found the information. By using separate cards for each note, you can shuffle your cards as you begin to look for ways to organize your material. As you become familiar with your subject, you may wish to write a subheading on the top of each card.

Before you actually write down a note, read quickly through your source to determine if it contains any worthwhile information. If it does, then during a second, more careful reading, use one of the three methods to make notes: *quoting, summarizing, paraphrasing.* No matter which method you use, leave room at the bottom of each card for your own comment. In this space, consider how you might use the note or how it might connect to other note cards you have written.

To understand how to read, react, and write during the note-taking process, consider the following passage from Barbara Tuchman's *A Distant Mirror: The Calamitous 14th Century* and Laura's note cards on it, appearing on pages 319, 320, and 321.

> The average layman acquired knowledge mainly by ear, through public sermons, mystery plays, and the recital of narrative poems, ballads, and tales, but during Enguerrand's lifetime, reading by educated nobles and upper bourgeois increased with the increased availability of manuscripts. Books of universal knowledge, mostly dating from the 13th century and written in (or translated from the Latin into) French and other vernaculars for the use of the layman, were literary staples familiar in every country over several centuries. A 14th century man drew also

on the Bible, romances, bestiaries, satires, books of astronomy, geography, universal history, church history, rhetoric, law, medicine, alchemy, falconry, hunting, fighting, music, and any number of special subjects. Allegory was the guiding concept. Every incident in the Old Testament was considered to pre-figure in allegory what was to come in the New. Everything in nature concealed an allegorical meaning relating to some aspect of Christian doctrine. Allegorical figures — Greed, Reason, Courtesy, Love, False-Seeming, Do-Well, Fair Welcome, Evil Rumor — peopled the tales and political treatises.

Quoting Sources

Quoting an author's text word for word is the easiest way to record information, but you should use this method selectively. Quote only the passages that deal

Source-card number	12
Subject heading	*Oral Tradition*
Author	Tuchman
Pages	59-60

Quotation marks

"The average layman acquired knowledge mainly by ear, through public sermons, mystery plays and the recital of narrative poems, ballads and tales...."

Laura's comment — Good description of oral culture. Introduction?

directly with your subject in particularly memorable language. When you write a quotation on your note card, place quotation marks at the beginning and end of the passage to indicate that you are quoting. If you decide to leave out part of the text, use *three* ellipsis points (. . .) to indicate that you have omitted words from the middle of a sentence and *four* ellipsis points (. . . .) to indicate that you have omitted one or more sentences. (See the discussion of ellipsis points in Section

28 of the Handbook of Grammar and Usage.) Finally, proofread your note against the original to confirm that every word and mark of punctuation is in its proper place.

Summarizing Sources

Summarizing an author's views is an effective way to record the essence of a text. Use this method when the author states a thesis or analyzes evidence in a way that either anticipates or contradicts your emerging perception of your argument. Use key words or two or three phrases to briefly restate the principal points in the argument.

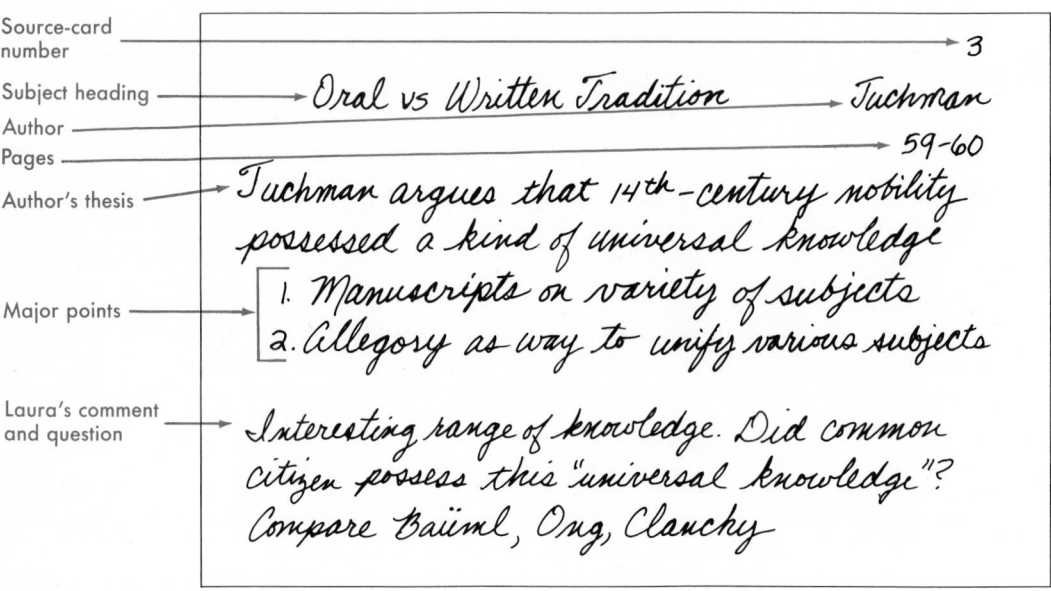

Source-card number — 3

Subject heading — *Oral vs Written Tradition*

Author — *Tuchman*

Pages — 59–60

Author's thesis — *Tuchman argues that 14th-century nobility possessed a kind of universal knowledge*

Major points —
1. *Manuscripts on variety of subjects*
2. *Allegory as way to unify various subjects*

Laura's comment and question — *Interesting range of knowledge. Did common citizen possess this "universal knowledge"? Compare Bäuml, Ong, Clanchy*

Paraphrasing Sources

Paraphrasing an author's views is both the most useful and the most misused method of note taking. It is *not* simply a casual way for you to reproduce the author's views nearly word for word without using quotation marks. Rather, paraphrasing requires you to think through what the author has said and then restate the information *in your own words*. To accomplish this objective you must understand what the author has said and then reformulate her or his opinion without adding or deleting significant information and without distorting the intent of the original passage. The paraphrase combines the advantages of quoting and summarizing because it allows you to reproduce the essence of the author's argument and adapt it to the flow of your own argument.

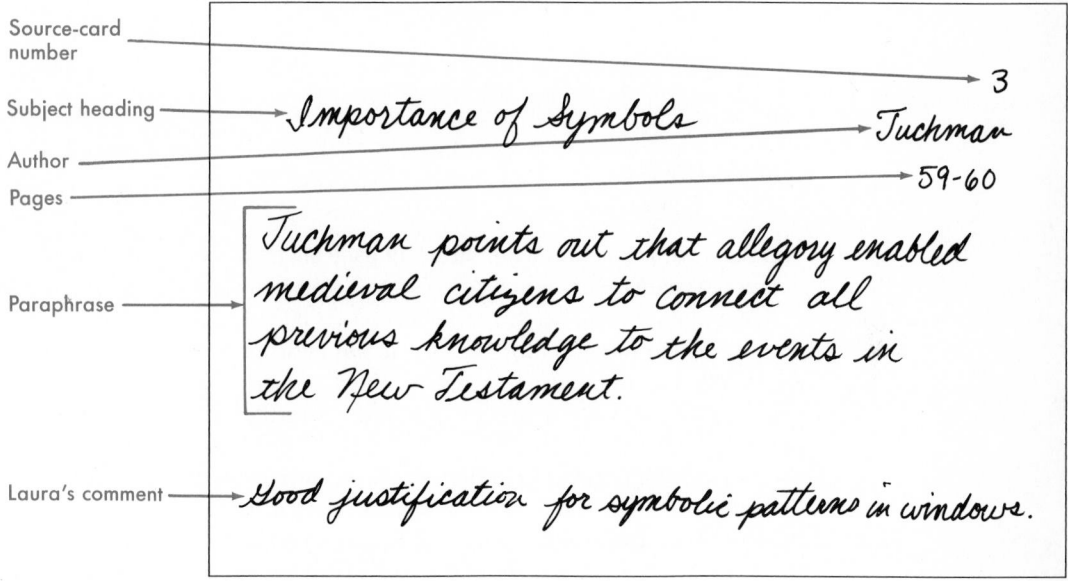

Source-card number

Subject heading

Author

Pages

Paraphrase

Laura's comment

3

Importance of Symbols → *Juchman*
→ *59-60*

Juchman points out that allegory enabled medieval citizens to connect all previous knowledge to the events in the New Testament.

Good justification for symbolic patterns in windows.

■■■ *Exercise*

The following passage is from Kenneth Clark's *Civilisation: A Personal View* (New York: Harper, 1969): 59–60. Read it carefully and then, following the procedures discussed above, make out three note cards. On the first, quote a few important sentences. On the second, summarize the major points of the passage. On the third, paraphrase Clark's thesis. At the bottom of each card, state the significance of the note and how you might use it in a paper.

> So much has been written about the Gothic style that one feels inclined to take it for granted. But it remains one of the most remarkable of human achievements. Since the first expression of civilised life in architecture, say the pyramid of Sakara, man had thought of buildings as a weight on the ground. He had accepted their material nature and although he had tried to make them transcend it by means of proportion or by the colour of precious marbles, he had always found himself limited by problems of stability and weight. In the end it kept him down to the earth. Now by the devices of the Gothic style — the shaft with its cluster of columns, passing without interruption into the vault and the pointed

arch — he could make stone seem weightless: the weightless expression of his spirit.

By the same means he could surround his space with glass. Suger said that he did this in order to get more light, but he found that these areas of glass could be made into an ideal means of impressing and instructing the faithful — far better than wall-painting because with a resonance, an effect on the senses, that the matte surface of a wall-painting could never have. 'Man may rise to the contemplation of the divine through the senses.' Well, nowhere else, I think, is this saying of the old pseudo–St Denis so wonderfully illustrated as it is in Chartres Cathedral. As one looks at the painted glass which completely surrounds one, it seems almost to set up a vibration in the air. . . .

Photocopying

Photocopying has had as great an impact on research as have microtexts and computer data bases. When you discover a particularly valuable source, copying all of it (if it is an article) or major portions of it (if it is a book) may seem to be the most efficient thing to do. You can cut out appropriate passages and paste them onto individual note cards, thus ensuring that you have precise quotations. You can even review the pertinent information in the original when you begin to write.

But if you decide to photocopy every source you uncover, you will waste a lot of time, paper, and money. The purpose of reading and note taking is to extract the essence of an article or book, to identify the information most pertinent to your argument. Photocopying is not an alternative to evaluating your sources. Compiling a large stack of paper only postpones your examination, reading, and assessment of the material.

FILLING GAPS

Planning the research paper is like planning any other paper. As you evaluate your information, you discover that you need more or different kinds of material. As you conduct your research, you constantly refine and restrict your subject. By the time you finish taking notes on the sources you first discovered and the sources you have uncovered along the way, you know where the holes in your research are, what gaps you have to fill if you are to present a thorough and coherent argument.

The research process, like the writing process, is recursive. You must loop back to the beginning to devise additional subject headings and check specialized bibliographies and other references to find the sources to fill the gaps in your information. When you begin writing your paper, you may have to make yet another loop to discover information that forges missing connections; but because you have worked through the process once and know what you are looking for, these further efforts should be quick and to the point.

"By God, for a minute there it suddenly all made sense!"

Drawing by Gahan Wilson; © 1986 The New Yorker Magazine, Inc.

Writing Assignments

1. Recount your own experience of trying to find something in a library. You may want to treat the experience as a comedy of errors, in which you find yourself hopelessly lost and confused, or as a detective story, in which you find clues and track down your source.

2. Read several of Barbara Tuchman's essays about her own research in *Practicing History* (New York: Knopf, 1981). Using her experience with one project as evidence, analyze the lessons she learned as she tried to solve several research problems.

3. Skim through Kenneth Clark's *Civilisation: A Personal View*. This book, originally a script for a tel-

evision series on PBS, told the story of Western civilization through dramatic examples of art and architecture, like Chartres. Select another of Clark's examples and devise a potential subject heading. Use your library's resources to compile a working bibliography.

4. The photograph on page 296 was selected as a metaphor representing an important aspect of this chapter. Examine the photograph and determine the possible meanings of this metaphor as you discern them in the photograph. Then write an essay explaining how the metaphor has been reinforced or revised by the ideas discussed in this chapter. You may want to suggest another metaphor to illustrate your understanding of these ideas.

14 WRITING THE RESEARCH PAPER

For the research paper as for any writing project, eventually you must stop planning and start writing. The large stack of note cards you have prepared points to two accomplishments: (1) you have learned a systematic procedure for gathering information (a procedure you can apply to other subjects), and (2) you have learned a great deal about the research that has already been published on your subject. But, as Barbara Tuchman suggests, most people are "overimpressed by research. People are always saying to me in awed tones, 'Think of all the *research* you must have done!' as if this were the hard part. It is not; writing, being a creative process, is much harder and takes twice as long" (Barbara Tuchman, "Problems in Writing the Biography of General Stilwell," *Practicing History*).

Your stack of note cards will guide you through the hard part of research — writing your paper. But they can also present two problems: (1) you have probably gathered more information than you can use in one paper, and (2) you have become so immersed in what other researchers have written about your subject that you may no longer know what *you* want to write about it. The most effective way to overcome these difficulties is to settle down and write a first draft.

This chapter will discuss the basic steps in writing that first draft. It will also show you some of Laura's initial difficulties as she tried to get her first draft started. And further, it will give you detailed advice about how to quote and document sources in the final draft. Read Laura's final draft and accompanying annotations at the end of the chapter to see how she resolved her difficulties and integrated her research into her writing.

The major topics discussed in this chapter are

- Organizing a preliminary outline
- Developing a thesis
- Writing the first draft
- Creating the introduction
- Quoting sources
- Documenting sources
- Listing sources
- Typing the final draft
- Laura's research paper

ORGANIZING A PRELIMINARY OUTLINE

Before drafting your paper, try out several *preliminary* outlines to determine the most effective pattern for your information. After you have completed your first draft, use a *descriptive* outline (or a revision agenda) to evaluate what you have written. Once you are ready to write the final draft of your research paper, prepare a *formal* outline to guide your composing process. (You may want to review the types, formats, and uses of each of these outlines in Chapter 3.) Your instructor may require you to submit your formal outline with your paper as a kind of "table of contents." (See Laura's formal outline, pp. 356–57.)

To organize a preliminary outline, read through your note cards, matching up those that deal with specific subdivisions of your subject. If you have composed subheadings and written them at the top of each card, your large stack of cards will organize itself quickly into a series of smaller stacks. Each stack could become a major section or minor subdivision on your outline. If you discover that several stacks focus on similar aspects of your subject, contain too few cards to develop a significant subdivision, or deal with subheadings no longer appropriate to the emerging design of your paper, eliminate the irrelevant note cards or re-label them so that they fit under a more appropriate subheading.

Once you have established your categories, arrange them into various patterns to find the one that seems most effective. (You may wish to review the patterns in Chapter 5, "Common Methods of Development.") Some of your subject headings may suggest a simple pattern (a *description* of the stained-glass windows at Chartres); others may point to a more complicated pattern. For example, Laura's research led her to study the history of stained-glass windows, the design of the stained-glass windows at Chartres, and the evolution of oral tradition in a preliterate society. Because she suspected there was a complex cause and effect relationship among these factors, she considered writing a *causal analysis*. But she was perplexed about how to arrange her material. Should she analyze causes and then effects? Which factors (stained-glass windows, Chartres' windows, the oral tradition) were causes and which were effects? She decided that she didn't know enough about these factors to analyze their interrelationship, and she realized that even if she did have sufficient knowledge, she would need more space and time than her assignment allowed to compose such an analysis. She sorted her cards again and discovered a historical sequence that she could use to organize her preliminary outline.

Preliminary Outline: The Windows at Chartres

1. *Reasons for building medieval cathedrals*
   religious
   educational
   economic

2. *Building Chartres cathedral*
   pagan history
   relic
   cult of the Virgin

3. *Chartres' various destructions*
   marauders
   fires — the great fire of 1194

4. *The great rebuilding, 1194–1260*
   survival of relic
   cult of the carts
   donations by guilds

5. *The windows*
   Biblical stories
   Christian allegory
   secular history
   local history

A preliminary outline reveals at a glance the kind of paper your sources will allow you to write. As Laura looks at her outline, she sees that her planning has led to several interesting decisions. Her reading about the history of stained-glass windows had taken her far afield of her primary subject, Chartres' windows. She read several general histories of the medieval cathedral and many specific histories of Chartres, which gave rise to ideas that dominate her preliminary outline, relegating the subject of her paper, "The Windows at Chartres," to section 5, "The Windows," only. The subheadings in this section look promising, however, because they remind Laura of what prompted her original interest in the windows — their ability to "tell" stories. Laura suspects that if she reviews her research on illiteracy and the oral tradition, she may discover some insights that will help her revise her preliminary outline so that it demonstrates a thesis.

DEVELOPING A THESIS

Developing a thesis is the most difficult task in composing the research paper. Your note cards, even when they are sorted and arranged into stacks, represent the voices of authority. It is easy to be intimidated by your own research, easy to let your sources speak for you. But remember, your aim is to advance your own argument — to discuss and analyze a topic from your own point of view. Like all other writing assignments, the research paper requires you to write with a purpose.

To find that purpose, review the comments you wrote at the bottom of your note cards. If you consistently commented on or posed questions about the ideas in your notes, then you have already established a degree of independence from your resources. As you reread your comments, look for common denominators in your thinking. What fascinated you about your subject? What connections did you see among the various sources you read?

Once you have identified some common themes, convert each into a hypothesis and try to match it to your preliminary outline. Some of your hypotheses may match up easily, leading to assertions that explain each of your subdivisions and pointing toward a thesis. Others may require you to revise the subdivisions or expand the range of categories you need to consider. Still others may suggest that you reorganize your outline completely. The particular thesis you try to advance in your first draft depends on a number of factors — your personal preference, your understanding of the information you have gathered, and your confidence in your ability to demonstrate what you propose to prove.

As Laura looked back through her comments, she discovered that two ideas expressed by other writers clarified what she wanted to prove in her own research. Marshall McLuhan's observation that cathedrals in the Middle Ages served as the "books of the people" gave Laura a rationale for her project — analyzing how one cathedral, Chartres, functioned as a kind of information center. But McLuhan did not really explain how the cathedral performed this function. Kenneth Clark's report that in the medieval church the stained-glass windows were designed to "impress and instruct the faithful" provided her with more specific evidence about the educational purpose of stained-glass windows.

A friend who had visited the cathedral had given Laura a set of slides of its windows, and she began to study specific windows to see what they attempted to teach. She found that the four categories she had listed in section 5 of the preliminary outline — Bible stories, Christian allegory, secular history, and local history — were comprehensive. In fact, she found four windows that not only illustrated these four categories but also illustrated the historical information she had included in the other sections of her preliminary outline. Laura then formulated three hypotheses about the stained-glass windows at Chartres.

- The history of Chartres cathedral is illustrated in four of its stained-glass windows.

- Examining four of the stained-glass windows at Chartres reveals the scale and detail of education available at this famous medieval cathedral.

■ Although most people in the Middle Ages could not read verbal histories of their culture, they could read visual histories such as those found in four stained-glass windows at Chartres.

■ *Exercise*

Discuss how each of Laura's hypotheses matches up with her preliminary outline (p. 327). In what ways will each hypothesis require Laura to change the major headings and subdivisions of her outline? What information will she have to add, eliminate, or relabel?

WRITING THE FIRST DRAFT

Writing the first draft of your research paper is like writing the first draft of any other paper — it is a discovery exercise. You have to discover whether your planning will enable you to communicate a subject to an audience for a purpose. Some of your discoveries will seem familiar because they are common in every writing situation: the information that seemed so complete in your notes now seems sketchy; and the overall purpose of your paper, so clear in your mind when you began, now strikes you as confused or inconsistent. Other discoveries may prove unsettling because they seem unique to the research paper: your first draft may follow your preliminary outline and support your thesis; but it may seem stiff, mechanical, and dull.

For most inexperienced writers, first-draft dullness derives from the inability to compose a simple, straightforward introduction that asserts a thesis and the inability to weave quoted material gracefully and naturally into the body of the paper. The following sections will illustrate and discuss some methods for solving these problems.

CREATING THE INTRODUCTION

Do not be surprised if you have to struggle with the introduction to your research paper. You have learned a great deal about your subject and are eager to display that learning in the body of your paper. But in order to "get there," you have to write an introduction that establishes the focus of your subject, attracts the attention of your readers, and asserts the purpose of your paper. Some writers cannot write such an introduction until they have discovered precisely what they are going to say in the body of the paper. For that reason, they prefer to write the introduction after they have drafted the rest of the paper — when they know exactly what they want to introduce. Other writers cannot develop the body of the paper until they have defined its exact direction. For that reason, they draft several versions of the introduction, hoping to learn what they want to introduce. Either method will require you to make a series of adjustments — some large, some small — to the introduction and to the body when you revise the final draft.

Laura rejects her first hypothesis as too flat and limited. Her attempts to work out the implications of her other two hypotheses and to find a purpose for her paper produced two different introductions. Her revision agendas reveal her dissatisfaction with them because each failed to establish the subject and clarify her thesis. Both introductions and revision agendas are shown below.

FIRST DRAFT: INTRODUCTION

The educational purpose of the medieval church is much in evidence at the great cathedral of Chartres. Towering above the wheat fields of northern France, it testifies to the presence and prominence of Christianity in medieval life. To appreciate the full significance of Chartres, however, one must enter the cathedral and meditate on the more than 175 stained-glass windows that illuminate its interior. These spectacular patterns of color were created by unknown twelfth- and thirteenth-century artists. Each window forms a text, exhibiting the characters, legends, and symbols of a complex story. "Reading" four of them reveals the scale and detail of religious education available at Chartres.

Revision Agenda

1. *What did I try to do in this draft?*
   I tried to illustrate how cathedrals such as Chartres used stained-glass windows for educational purposes.

2. *What are its strengths and weaknesses?*
   I focus immediately on Chartres and get to the four windows I plan to "read" by the end of the first paragraph. But nobody knows why people need this kind of education. Nothing on illiteracy. No comparison between the way we learn now (reading books) and the various ways people learned in Middle Ages (e.g., looking at windows).

3. *What revisions do I want to make in my next draft?*
   a. *Establish context of preliterate culture.*
   b. *Explain that illiteracy did not mean ignorance.*
   c. *Use McLuhan's quote to connect discussion of illiteracy to analysis of Chartres.*
   d. *Focus on four windows at Chartres as examples.*

SECOND DRAFT: INTRODUCTION

The information explosion that continues to bombard modern people simply could not exist in a preliterate society. In the medieval world, the ability to learn was limited to what people could see, hear, and remember (Ong 1). And perception and memory are, after all, highly fallible. But an illiterate during the Middle Ages was not necessarily ignorant or uneducated. One could learn orally or depend on the literacy of another (Bäuml 242). In fact, everyone in the Middle Ages did not need to read and write. People were well acquainted with vernacular stories and the content of the Bible, in spite of their illiteracy (Bäuml 247). Our methods of constructing arguments, telling stories, and presenting complex information grew out of the oral communication strategies employed during the Middle Ages. The medieval church, with its sermons, schools, and iconography, was the place where such strategies flourished. Marshall McLuhan points out that "the medieval cathedrals were the 'books of the people' " (108). Thus, although most people could not read verbal texts, they could read visual texts such as the stained-glass windows at Chartres.

Revision Agenda

1. What did I try to do in this draft?
   I tried to connect the discussion of illiteracy in the Middle Ages to educational purpose of stained-glass windows at Chartres.

2. What are its strengths and weaknesses?
   Ong and Bäuml help me define illiteracy in the Middle Ages. They also help me demonstrate that illiteracy did not mean ignorance. McLuhan's quote connects discussion of illiteracy to educational purpose of cathedral. Discussion seems wordy (ironic). Instead of introducing windows at Chartres, introduction seems to hide them. Last sentence seems like afterthought rather than thesis. Suggests I am going to analyze all the windows rather than just four.

3. *What revisions do I want to make in my next draft?*
   a. *Make a clearer presentation of preliterate culture — maybe use some specific examples.*
   b. *Explain difference between verbal illiteracy and visual literacy.*
   c. *More dramatic presentation of Chartres and its windows as place where illiterate could learn. Maybe use material from first draft. Maybe I'll need more than one paragraph to set up thesis.*
   d. *Use thesis to assert what paper will prove by examining four windows.*
   e. *Show illustrations of four windows — check with Johnson about how to do it.*

When she completed her second revision agenda, Laura was confident that her second hypothesis had led to a thesis and purpose for her paper. She reworked her introduction and drafted the body of the paper, encountering a number of tricky problems. For example, she had to decide how to weave her source material into her text and how to prepare and mount the illustrations. You will see how Laura handled these problems when you read her paper and the annotations accompanying it.

■ *Exercise*

Compare Laura's two draft introductions (pp. 330, 331) with the introduction that she uses in her final paper (pp. 358–59). How do her revision agendas help her expand her introduction and sharpen her thesis? What specific decisions seem to produce the biggest changes in her final draft?

QUOTING SOURCES

The most persistent challenge posed by the research paper is deciding *when* and *where* to cite your sources to support your argument. For every division on your outline, you have a stack of cards that contains the words and ideas of other writers. If you use this material, you must acknowledge the source of your information (see Documenting Sources, pp. 338–45, and Listing Sources, pp. 345–53). Many inexperienced writers, however, do more than merely use their sources; they allow their sources to write the paper for them. Excessive quoting distorts the balance between your writing and the writing of others and makes your paper seem to be a scrapbook of other people's opinions. And it can disrupt the flow of your argument by introducing ideas and images that may not deal directly with your thesis.

To avoid these problems, you should be selective when you quote. Because each quotation creates a special effect, ask yourself these questions when you are deciding whether to quote a passage.

1. *Will the substance of the passage make a significant contribution to my subject?* Sometimes a passage may seem significant because it provides exten-

sive evidence for its conclusions. But you may be able to make the same point more effectively by summarizing or paraphrasing rather than quoting the passage.

2. *Will the phrasing of the passage seem memorable to my readers?* You do not want to blur the effect of a quotation by quoting too much material. Nor do you want to waste the effect of a quotation by quoting uninspired or unintelligible writing. You should quote only key sentences or phrases — those that convey the author's meaning in especially vivid language.

3. *Will the reputation of the author give credibility to my argument?* The mere mention of certain "experts" produces controversy or distorts your argument. If the authority of your sources is suspect, there is no point in quoting them.

When you determine that you want to use a particular quotation, you have to decide how to incorporate it into your own writing. There are several methods for quoting material; the one you choose depends on why you are quoting the passage and how much you intend to quote.

Introducing Quotations

All quotations must be placed within quotation marks or set off from your text. They must be documented with a parenthetical reference that includes the exact page numbers from which the quotation was taken and, if necessary, the author's name or another word or phrase that identifies the source. A lead-in phrase or sentence identifying the person you are quoting and the reason he or she is being quoted helps your readers follow your reasoning and prepares them for the special effect conveyed by a quotation. There are at least three ways to introduce a quotation.

Here is the most common method of introducing a quotation:

M. T. Clanchy, noted medieval historian, argues in From Memory to Written Record: England, 1066–1307 that "reliance on literacy can be narrowing, since it restricts communication to those who have learned its techniques" (8).

The person to be quoted (M. T. Clanchy), his expertise (medieval historian), and the source of the quotation *(From Memory to Written Record: England, 1066–1307)* are all identified. But the purpose for which Clanchy is being quoted is not explained; presumably, this passage would appear in a paragraph in which the context or reason for the quotation has been established. Introductory phrases such as the one above help your readers see how the argument you have stated is advanced by an authority.

In the second form of introducing a quotation, the person being quoted (medieval historian M. T. Clanchy, author of *From Memory to Written Record: England, 1066–1307)* and the reason he is being quoted (he explains the effects of our dependence on written texts) are both identified in a sentence that concludes with a colon; the quotation follows, as in this example:

> Medieval historian M. T. Clanchy explains the effects of our dependence on written texts in From Memory to Written Record: England, 1066–1307: "reliance on literacy can be narrowing, since it restricts communication to those who have learned its techniques" (8).

The third method of identifying a quotation relies on the assumption that you have already introduced the writer's full name and credentials earlier in the paper:

> "Reliance on literacy can be narrowing," Clanchy points out, "since it restricts communication to those who have learned its techniques" (8).

Compare Laura's treatment of the Clanchy passage in her paper on page 370.

Length of Quotations

The length (and look) of your quotation will determine its effect. A brief, pointed quotation, generally worked into the syntax of your sentence, is often the best way to advance your argument. But on rare occasions (perhaps no more than two or three times in a ten-page research paper), you may want to quote a long passage that expresses the main ideas you are trying to present.

Long quotations (four or more lines of prose; three or more lines of poetry) are usually introduced by a colon or comma, set off from your text by triple-spacing above and below, and indented ten spaces from the left margin. This special placement identifies the passage as a quotation, so do not enclose it in quotation marks. In such block quotations, the final period goes *before* rather than *after* the parenthetical reference. Here is an example.

> In The Gothic Cathedral, Otto von Simon explains the importance of Christian relics such as the Holy Tunic:

The religion of medieval man was a communication
with a sacred reality that was invisible, yet immediately and continuously present. The veneration of
saints and their relics, and the repercussions that
this cult exerted upon nearly every phase of medieval
life . . . are unintelligible unless the immediacy of
this relationship with the supernatural is properly
understood. (10)

Short quotations (less than four lines of prose, less than three lines of poetry)
are usually run in with your text, unless they deserve special emphasis. You can
introduce a short quotation by one of the three methods described earlier. Or, as
Laura does in the example below, you can work brief phrases from your source
into the syntax of your sentence, using quotation marks and identifying the source
with a parenthetical reference. See how Laura uses this passage in her paper on
pages 362–64.

The presence of the Virgin's Holy Tunic not only provided a
sacred link to the supernatural world, but also produced a significant effect ''upon nearly every phase of medieval life'' (von
Simon 10).

Integrating Quotations

Sometimes, to make a quoted passage fit smoothly into the flow of your sentences, you will have to use *ellipsis* and *brackets*. Use ellipsis points when you
want to omit part of the quoted passage to make it conform to your sentence. Use
three points (. . .) to indicate omission of material within the sentence. Use four
points (. . . .) to indicate the omission of a whole sentence or more. Use brackets
when you need to add your own words to a quotation to make the passage complete or grammatically correct or, on occasion, when you need to make an editorial comment. (See Sections 27 and 28, on brackets and ellipsis, respectively, in
the Handbook of Grammar and Usage.) For example, Laura wants to integrate
phrases from the quotation below from Henry Adams' *Mont-Saint-Michel and
Chartres* into her own sentence.

Above the signature, in the first panel, the Emperor Constantine is seen, asleep,
in Constantinople, on an elaborate bed, while an angel is giving him the order to
seek aid from Charlemagne against the Saracens. Charlemagne appears, in full

armour of the year 1200, on horseback. Then Charlemagne, sainted, wearing his halo, converses with two bishops on the subject of a crusade for the rescue of Constantine. In the next scene, he arrives at the gates of Constantinople where Constantine receives him. The fifth picture is most interesting; Charlemagne has advanced with his knights and attacks the Saracens; the Franks wear coats-of-mail, and carry long, pointed shields; the infidels carry round shields; Charlemagne, wearing a crown, strikes off with one blow of his sword the head of a Saracen emir; but the battle is desperate; the chargers are at full gallop, and a Saracen is striking at Charlemagne with his battle-axe. After the victory has been won, the Emperor Constantine rewards Charlemagne by the priceless gift of three chasses or reliquaries, containing a piece of the true Cross; the Suaire or grave-cloth of the Saviour; and a tunic of the Virgin. Charlemagne then returns to France, and in the next medallion presents the three chasses and the crown of the Saracen king to the church at Aix, which to a French audience meant the Abbey of Saint-Denis. This scene closes the first volume of the story.

Notice how Laura uses ellipsis and brackets to incorporate Adams' phrasing into her own.

In Mont-Saint-Michel and Chartres, Henry Adams points out that the window begins with the Emperor asleep in Constantinople, as an angel gives him orders to "seek aid from Charlemagne against the Saracens. . . . Then Charlemagne . . . wearing his halo, converses with two bishops [about] . . . the rescue of Constantine" (156).

See page 362 to see how Laura works this passage into the context of her paper.

Using Summary and Paraphrase

Often the most efficient way to work your sources into your own writing is to summarize or paraphrase them. As you remember from Chapter 13, a summary states the thesis or outlines the principal points of an author's argument; a paraphrase is a restatement of the author's ideas in *your own words* (see pp. 320–21). Because the words of a summary or paraphrase are yours, they do not have to be enclosed by quotation marks. But because the ideas come from someone else, you do need to cite the source in your text and document the passage with a parenthetical reference.

The following passage is quoted from Painton Cowen's *Rose Windows*.

The School at Chartres seems to have elaborated a concept of evolution, drawn from all available knowledge, Christian and pagan alike, which embodied the Old Testament, number, geometry, nature, the Cosmos, Divine Love, and the

New Testament. It is essentially the Logos, the Word, of St. John's Gospel (John 1:1) seen in the light of the latest knowledge of the age.

Laura summarizes this passage in a paraphrase. See how she works this material into her paper on page 367.

As Painton Cowen points out, the School of Chartres was interested in how the word of God evolved through the ages and through all areas of knowledge, Christian and pagan, Old Testament and New Testament (14).

Plagiarism

Plagiarism is the use of someone else's writing without giving proper credit — or perhaps without giving any credit at all — to the writer of the original. Whether plagiarism is intentional or unintentional, it is a serious offense that can be easily avoided by adhering scrupulously to the following advice. You should document your sources whenever you

- Use a direct quotation
- Copy a table, chart, or other diagram
- Construct a table from data provided by others
- Summarize or paraphrase a passage in your own words
- Present specific examples, figures, or factual information that are taken from a specific source and used to explain or support your judgments

The following excerpt from Émile Mâle's *Chartres* and the examples of a student's use of it illustrate the problem of plagiarism.

Original Version

Very often images of protective divinities, called "Mothers," were found in the vicinity of sacred springs of the Gallo-Roman period. There were generally three figures in such images, but often a single figure could be found: a seated female figure with a child in her lap, bearing a striking resemblance to the Virgin of the medieval period.

Version A

Images of protective divinities, called "mothers," were found near the sacred springs. There were three figures in such images, but often a single figure could be found: a seated female figure with a child in her lap, bearing a striking resemblance to the Virgin of the medieval period.

That is plagiarism in its worst form. Because the writer of Version A does not indicate in the text or in a parenthetical reference that the words and ideas belong to Mâle, her readers will believe the words are hers. She has stolen the words and ideas and attempted to cover the theft by changing or omitting an occasional word.

Version B

> Émile Mâle points out that images of protective divinities, called "Mothers," were found in the vicinity of sacred springs of the Gallo-Roman period. There were generally three figures in such images, but often a single figure could be found: a seated female figure with a child in her lap, bearing a striking resemblance to the Virgin of the medieval period (8).

Version B is also plagiarism, even though the writer acknowledges his source and documents the passage with a parenthetical reference. Obviously the writer has copied the original word for word, yet he has supplied no quotation marks to indicate the extent of the borrowing. As written and documented, the passage masquerades as a paraphrase, when in fact it is a direct quotation.

Version C

> Émile Mâle explains that images of protective gods called "Mothers" were often found near such springs, one of whom, "a seated female figure with a child in her lap, [bears] a striking resemblance to the Virgin of the medieval period" (8).

Laura's version (Version C) represents one, although not the only, satisfactory way of handling this source material. Laura has identified her source at the beginning of the sentence, letting her readers know who is being quoted. She then paraphrases most of the material in her own words, placing within quotation marks the parts of the original she wants to quote and using brackets to maintain the grammatical integrity of her own sentence. Finally, she provides a parenthetical reference. By following this procedure, Laura has made perfectly clear which words are hers and which belong to Mâle. See how she works this sentence into her paper on pages 359–60.

DOCUMENTING SOURCES

The purpose of documenting each source with a parenthetical reference is twofold: (1) to avoid the appearance of representing somebody else's work as your own and (2) to refer your readers to your list of "works cited," where they will find a complete citation on your source. Although there is general agreement about the purpose of documentation, different fields of knowledge, periodicals, and publishers prefer different styles.

The two most commonly used styles are those recommended by the Modern Language Association in its *MLA Handbook for Writers of Research Papers* (1984) and by the American Psychological Association in its *Publication Manual of the American Psychological Association* (1983). The two styles differ:

MLA Style	*APA Style*
Sources are documented by parenthetical reference to author's last name and to page number. No punctuation separates these elements, and no abbreviation for *page* or *pages* is used.	Sources are documented by parenthetical reference to author's last name, publication date, and page number. Commas separate these elements, and the abbreviations *p.* and *pp.* are used.

Here is the same source documented according to each style:

MLA Style

Although the common languages of the Middle Ages were "widely spoken, [they] were rarely written or read" (Bäuml 237–38).

WORKS CITED

Bäuml, Franz H. "Varieties and Consequences of Medieval Literacy and Illiteracy." Speculum 55 (1980): 237–65.

APA Style

Although the common languages of the Middle Ages were "widely spoken, [they] were rarely written or read" (Bäuml, 1980, pp. 237–38).

REFERENCE

Bäuml, F. H. (1980). Varieties and consequences of medieval literacy and illiteracy. Speculum, 55, 237–65.

Different departments in your college or university may require a particular style; your instructors will tell you which documentation style they require. If you are writing a research paper in the social sciences, your instructor is likely to require APA style. If you are writing a research paper in the humanities, your instructor is likely to require MLA style. Because this chapter is designed to help you write a research paper in a composition class, its examples, including Laura's research paper, follow MLA style.

■■■■■■

Sample Citations

Frequently you will need to cite sources that are not so straightforward as the Bäuml citation — for example, a book written by more than one author or several works by the same author. In such cases, follow the recommendations below. Each example of a citation is followed by the entry that would appear in the "works cited" list.

Citing One Work by an Author of Two or More Works   If you are citing two or more titles by the same author, place a comma after the author's last name, add a shortened version of the title of the work, and then supply the relevant page numbers. Another solution is to cite the author's last name and title in your sentence and then add the page numbers in a parenthetical reference.

> Once society reaches a certain stage of industrial growth, it will shift its energies to the production of services (Toffler, Future 221).

> Toffler argues in The Third Wave that society has gone through two eras (agricultural and industrial) and is now entering another — the information age (26).

WORKS CITED

Toffler, Alvin. Future Shock. New York: Random, 1970.

---. The Third Wave. New York: Morrow, 1980.

Citing One Work by an Author Who Has the Same Last Name as Another Author in Your List of Cited Works   When you are citing two or more authors with the same last name, avoid confusion by supplying each author's first name in the parenthetical reference or in your sentence. In the list of cited works, the two authors should be alphabetized by their first names.

> Critics have often debated the usefulness of the psychological approach to literary interpretation (Frederick Hoffman 317).

> Daniel Hoffman argues that folklore and myth provide valuable insights for the literary critic (9–15).

WORKS CITED

Hoffman, Daniel G. <u>Form and Fable in American Fiction</u>. New

York: Oxford, 1961.

Hoffman, Frederick J. <u>Freudianism and the Literary Mind</u>.

Baton Rouge: Louisiana State UP, 1945.

Citing a Work by More Than One Author If you are citing a book by two authors, you have the option of naming them in your sentence or of putting their names in a parenthetical reference. If you are citing a book by three or more authors, you should probably place their names in a parenthetical reference to sustain the readability of your sentence. The authorship of a work by three or more authors can be given in a shortened form by using the first author's last name and "et al." (an abbreviation for the Latin phrase *et alia,* meaning "and others").

Boller and Story interpret the Declaration of Independence

as Thomas Jefferson's attempt to list America's grievances

against England (58).

Other historians view the Declaration of Independence as

Jefferson's attempt to formulate the principles of America's

political philosophy (Norton et al. 124).

WORKS CITED

Boller, Paul, and Ronald Story. <u>A More Perfect Union: Docu-</u>

<u>ments in U.S. History</u>. Boston: Houghton, 1984.

Norton, Mary Beth, et al. <u>A People and a Nation: A History of</u>

<u>the United States</u>. Boston: Houghton, 1984.

Citing a Multivolume Work If you are citing one volume from a multivolume work, indicate in your parenthetical reference the specific volume you used.

William Faulkner's initial reluctance to travel to Stockholm

to receive the Nobel Prize produced considerable consternation

in the American embassy (Blotner 2: 1347).

WORKS CITED

Blotner, Joseph. <u>Faulkner: A Biography</u>. 2 vols. New York:

Random, 1974.

Citing a Work by Title   If you are citing a source for which no author is named, use a shortened version of the title — or the title itself, if it is short — in either the text citation or a parenthetical reference. If you shorten the title, be sure to begin with the word by which the source is alphabetized in the list of cited works.

The recent exhibit of nineteenth-century patent models at

the Cooper-Hewitt Museum featured plans for such inventions

as the Rotating Blast-Producing Chair, and Improved, Creeping

Doll, and the Life Preserving Coffin: In Doubtful Cases of Actual

Death ("Notes").

Notice that page numbers are omitted from the parenthetical reference when a one-page article is cited.

WORKS CITED

"Notes and Comments: The Talk of the Town." <u>The New Yorker</u>.

16 July 1984: 23.

Citing an Illustration   Place the illustrative material as close as possible to the part of the text it illustrates. The illustration — a photograph, advertisement, map, drawing, or graph — should be labeled "Figure" (often abbreviated "Fig."), assigned an Arabic number, and, if appropriate, given a caption and complete citation. In your text, use a parenthetical reference to guide the reader to the illustration: "(see Fig. 1)." Place all necessary information below the illustration:

Fig. 1. "Your Gateway to Tomorrow." Advertisement. <u>Money</u>

Mar. 1983: 1–2.

Citing Literary Works   Because some literary works — novels, plays, poems — are available in several editions, MLA recommends that you give more information than just a page number so that readers who are not using the same edition as you are can locate in their books the passage you are citing. After the page number, add a semicolon and other appropriate information, using lower-case abbreviations such as *pt., sec.,* and *ch.* (for *part, section,* and *chapter*).

Although Flaubert sees Madame Bovary for what she is —
a silly, romantic woman — he insists that "none of us can ever
express the exact measure of his needs or his thoughts or his
sorrows" and that all of us "long to make music that will melt
the stars" (216; pt. 2, ch. 12).

WORKS CITED

Flaubert, Gustave. Madame Bovary: Patterns of Provincial Life.

Trans. Francis Steegmuller. New York: Random-Modern

Library, 1957.

When you cite classic verse plays and poems, omit all page numbers and
document by division(s) and line(s), using periods to separate the various
numbers. You can also use appropriate abbreviations to designate certain well-
known works. For example, "*Od.* 8.326" refers to book 8, line 326, of Homer's
Odyssey. Do not use the abbreviations "l" or "ll" to designate lines because they
can be confused with numbers. Once you have established in your text which
numbers indicate lines, you may omit the words *line* and *lines* and simply use the
numbers.

Also, as shown in the *Odyssey* citation just given, use Arabic numbers rather
than Roman numerals for division and page numbers. Some teachers prefer
Roman numerals for designating acts and scenes in plays (for example, "*Macbeth*
III.iv"), but if your instructor does not insist on them, use Arabic numbers with
appropriate abbreviations to cite famous plays: "*Mac.* 3.4."

Citing More Than One Work in a Single Parenthetical Reference   If you need
to mention two or more works in a single parenthetical reference, document each
reference according to the normal pattern, but use semicolons to separate the
citations.

(Oleson 59; Trimble 85; Hylton 63)

WORKS CITED

Hylton, Marion Willard. "On the Trail of Pollen: Momaday's

House Made of Dawn." Critique 14.2 (1972): 60–69.

Oleson, Carole. "The Remembered Earth: Momaday's House

Made of Dawn." South Dakota Review 2 (1973): 59–78.

Trimble, Martha Scott. N. Scott Momaday. Western Writers

Series. Boise, ID: Boise State College, 1973.

Although the MLA style provides this procedure for documenting multiple citations within parenthetical references, that kind of documentation is often the result of "scholarly" padding and may be disruptive for readers. If multiple citations are absolutely necessary, MLA recommends that they be placed in a bibliographical endnote or footnote.

Using Notes With Parenthetical References

A *superscript numeral* (a number raised above the line) placed at an appropriate place in the text — usually at the end of a sentence — signals a note. The note itself, identified by a matching number, may appear at the end of the text (as an endnote) or at the bottom of the page on which its superscript appears (as a footnote).

In MLA style, notes (preferably endnotes) are reserved for two specific purposes.

Notes Containing Additional Commentary

Thurber's reputation continued to grow until the 1950s,

when he was forced to give up drawing because of his

blindness.[1]

[1]Thurber's older brother accidentally shot him in the eye

with an arrow when they were children, causing the immediate

loss of that eye. He gradually lost the sight of the other eye

because of complications from the accident and a cataract.

Notes Listing or Evaluating Sources or Referring to Additional Sources

The argument that American policy in Vietnam was on the

whole morally justified has come under attack from many

quarters.[2]

[2]For a useful sampling of opinion, see Buckley 20; Draper

32; Nardin and Slater 437.

Notice that the sources cited in note 2 are documented like parenthetical references. Complete citations would be given in the "works cited" list.

WORKS CITED

Buckley, Kevin. "Vietnam: The Defense Case." New York Times
    7 Dec. 1978: 19–24.

Draper, Theodore. "Ghosts of Vietnam." Dissent 26 (1979):
    30–41.

Nardin, Terry, and Jerome Slater. "Vietnam Revisited." World
    Politics 33 (1981): 436–48.

LISTING SOURCES: SAMPLE ENTRIES

In the MLA documentation style readers can locate complete information about your sources only in a list of cited works. The list goes at the end of your paper and, as its title "Works Cited" suggests, contains only the sources you have *cited* in your paper. Occasionally, your instructor may require a list of the works you *consulted*. Such a list would include not only the sources you cited in your paper but also the sources you consulted while conducting your research. (If you have questions about the kind of list you are to prepare, ask your instructor.)

Even though the list of cited works appears at the end of your paper, it must be compiled *before* you begin writing. The bibliographic information on your source cards will eventually constitute your "works cited" list. In order to create the list, you must alphabetize your source cards, being careful that each entry is complete according to the appropriate format. As you write, you may need to add or delete source cards. Be sure to identify your sources clearly and accurately in your text and to provide complete bibliographic information about each one in your list of cited works.

When you type your final list, follow the instructions given below. For an illustration of this format, see Laura's "Works Cited" list on pages 372–73.

1. Paginate the "Works Cited" list as a continuation of your text. If the conclusion of your paper appears on page 9, begin your list on page 10, unless there is an intervening page of endnotes.

2. List all entries in alphabetical order according to the last name of the author.

3. Double-space between successive lines of an entry and between entries.

4. After the first line of an entry, indent successive lines five spaces.

5. If you are listing more than one work by the same author, alphabetize the works according to title (excluding any initial articles — *a, an, the*). Instead

of repeating the author's name with each citation, for the second and additional works, type three hyphens and a period, skip two spaces, and then give the title:

Lanham, Richard A. <u>Literacy and the Survival of Humanism</u>.

   New Haven: Yale, 1983.

---. <u>Style: An Anti-Textbook</u>. New Haven: Yale, 1974.

The form of each entry in your "Works Cited" list will vary according to the type of source you are citing. The major variations are illustrated below. If you need additional information, consult *MLA Handbook for Writers of Research Papers* (1984).

Books

Citations for books have three main parts — author, title, and facts of publication. Separate each part with a period followed by two spaces. (The first sample entry is described completely; significant variations in subsequent entries are noted in marginal annotations.)

A Book by a Single Author or Agency

Tuchman, Barbara W. <u>A Distant Mirror: The Calamitous 14th</u>

   <u>Century</u>. New York: Knopf, 1978.

- The author's last name comes before the given name or initial to facilitate alphabetizing. Use the name exactly as it appears on the title page of the sources.

- If the book is the work of an agency (committee, organization, or department) instead of an individual, the name of the agency takes the place of the author's name.

- If no author or agency is given, the citation begins with and is alphabetized by the title of the source.

- The title and subtitle of the book are underlined.

- The place of publication, the publisher, and the date of publication are named in that order and are punctuated and spaced as in the preceding Tuchman citation. A colon separates the place of publication from the name of the publishing company, and a comma separates the publisher's name from the date.

■ If more than one place of publication is given on the title page, mention only the first.

■ If the place of publication might be unfamiliar or unclear to your readers, add an abbreviation identifying the appropriate state or country: Cambridge, MA.

■ Shorten the publisher's name, as long as the shortened form is easily identifiable: *Houghton Mifflin* can be *Houghton; Harvard University Press* can be *Harvard UP.*

■ When you cannot locate one or more pieces of information concerning the facts of publication, use these abbreviations in the appropriate positions.

No place: n.p.
No publisher: n.p.
No date: n.d.

A Book by Two or Three Authors

Ashby, Eric, and Mary Anderson. The Rise of the Student Estate in Britain. Cambridge, MA: Harvard UP, 1970.

Lee, Lawrence, George Seddon, and Frances Stephens. Stained Glass. New York: Crown, 1976.

A Book by Three or More Authors

Sheridan, Marion C., et al. The Motion Picture and the Teaching of English. New York: Appleton, 1965.

A Book with an Editor

Kuhn, Thomas, ed. The Essential Tension: Selected Studies in Scientific Tradition and Change. Chicago: U of Chicago P, 1977.

A Book with an Author and an Editor

Emphasis on author

Ginsberg, Allen. Journals: Early Fifties, Early Sixties. Ed. Gordon Ball. New York: Grove, 1977.

Emphasis on editor

Ball, Gordon, ed. Journals: Early Fifties, Early Sixties. By Allen Ginsberg. New York: Grove, 1977.

Works in an Anthology

Citation of one work only

Tyler, Anne. "Still Just Writing." The Writer on Her Work. Ed. Janet Sternberg. New York: Norton, 1980. 3–16.

Basis for multiple citations

Sternberg, Janet, ed. The Writer on Her Work. New York: Norton, 1980.

Multiple citations

Walker, Alice. "One Child of One's Own: A Meaningful Digression Within the Works." Sternberg 121–40.

Walker, Margaret. "On Being Female, Black, and Free." Sternberg 95–106.

An Article in an Alphabetically Arranged Reference Book

"Graham, Martha." Who's Who of American Women. 13th ed. 1983–1984.

Frequently revised work

Hayward, Jane. "Stained Glass." Encyclopedia Americana 1983 ed.

A Multivolume Work

Citing all volumes

Blotner, Joseph. Faulkner: A Biography. 2 vols. New York: Random, 1974.

Citing one volume

Blotner, Joseph. Faulkner: A Biography. New York: Random, 1974. Vol. 2.

An Edition Other Than the First

Bailey, Sydney. British Parliamentary Research. 3rd ed. Boston: Houghton, 1971.

An Introduction, Preface, Foreword, or Afterword

Bernstein, Carl. Afterword. Poison Penmanship: The Gentle Art of Muckraking. By Jessica Mitford. New York: Random, 1979. 275–77.

A Book in a Series

Longley, John L., Jr. Robert Penn Warren. Southern Writers Series 2. Austin, TX: Steck, 1969.

A Republished Book

> Malamud, Bernard. <u>The Natural</u>. 1952. New York: Avon, 1980.

Published Proceedings of a Conference

> Shusterman, Alan J., ed. <u>Capitalizing on Ideas: New Alliances</u>
>> <u>for Business</u>. Proceedings from a Conference of Indiana
>> Business Leaders. 10–11 April 1983. Indianapolis: Indiana
>> Committee for the Humanities, 1983.

A Translation

> Mâle, Émile. <u>Chartres</u>. Trans. Sara Wilson. New York: Harper,
>> 1983.

A Book with a Title Usually Italicized in Its Title

> Miller, James E., Jr. <u>A Critical Guide to</u> Leaves of Grass. Chi-
>> cago: U of Chicago P, 1957.

Articles

Citations for articles in periodicals, like citations for books, contain three parts: author, title, and facts of publication. But articles include more complicated facts of publication such as the periodical title, the volume number, year of publication, and inclusive page numbers. You can usually find this information on the first page of the article and on the cover or title page of the periodical. The first entry is discussed completely; significant variations in subsequent entries are noted in marginal annotations.

> Fulwiler, Toby. "How Well Does Writing Across the Curriculum
>> Work?" <u>College English</u> 46 (1984): 112–25.

- Cite author (last name first).
- Place title of article within quotation marks.
- Underline title of periodical.
- Place volume number after title of periodical.
- Enclose year of publication within parentheses.

■ Use colon to separate date of periodical from inclusive page numbers of article.

■ If a periodical pages its issues continuously through an annual volume, as does *College English*, give only the volume number, not the issue number.

An Article in a Journal That Pages Each Issue Separately or Uses Only Issue Numbers

Bird, Harry. "Some Aspects of Prejudice in the Roman World." University of Windsor Review 10.1 (1975): 64–75.

An Article from a Monthly or Bimonthly Periodical

Jacobs, Jane. "The Dynamic of Decline." Atlantic April 1984: 98–114.

An Article from a Weekly or Biweekly Periodical

Arlen, Michael J. "Onward and Upward With the Arts: Thirty Seconds." The New Yorker 15 Oct. 1979: 55–146.

Article from a Daily Newspaper

Article with by-line

Whited, Charles. "The Priceless Treasure of the Marqueses." Miami Herald 15 July 1973: 1.

Separately paginated sections

"Culture Shock: Williamsburg and Disney World, Back to Back." New York Times 21 Sept. 1975: sec. 10:1.

Newspaper title that does not include city name

"Oliver North Faces Congress." Union Star [Schenectady, NY] 7 July 1987: 1.

Editorial, Letter to Editor, Review

"From Good News to Bad." Editorial. Washington Post 16 July 1984: 10.

Coldwater, Charles F., MD. Letter. The Muncie Star 17 June 1987: 4.

Griswold, Charles L., Jr. "Soul Food." Rev. of Statecraft as Soulcraft: What Government Does, by George F. Will. American Scholar 53 (1984): 401–06.

An Article Whose Title Contains a Quotation or a Title Within Quotation Marks

> Carpenter, Lynette. "The Daring Gift in Ellen Glasgow's 'Dare's Gift.' " Studies in Short Fiction 21 (1984): 95–102.

An Abstract from Dissertation Abstracts (DA) or Dissertation Abstracts International (DAI)

> Creek, Mardena Bridges. "Myth, Wound, Accommodation: American Literary Response to the War in Vietnam." DAI 43 (1982): 3593A. Ball State U.

Other Sources

You may sometimes have to list sources other than books and articles. In fact, MLA changed the name of its concluding list of sources from "Bibliography" (literally "description of books") to "Works Cited" because of the variety of non-book sources that may be cited. The particular treatment of "other sources" depends on what information is available and what information needs to be included to enable readers to locate the same material themselves. Here are some sample entries with brief annotations.

Government Documents

GPO stands for Government Printing Office.

> United States. Federal Communications Commission. Investigation of the Telephone Industry in the United States. 76th Cong., 1st sess. H. Doc. 340. Washington: GPO, 1939.

Computer Software

Cite:
1. type of computer it can be used on
2. amount of memory required
3. operating system
4. program format

> Volkswriter Delux. Computer software. Lifetree Software, 1983. IBM, 128K, PC-DOS 2.0, disk.

> WordPerfect. Computer software. WordPerfect Corporation, 1986. IBM, 256K, PC-DOS 2.0, disk.

Films, Radio and Television Programs

> Julia. Dir. Fred Zimmerman. With Jane Fonda, Vanessa Redgrave, and Jason Robards, Jr. TCF, 1977.

> "If God Ever Listened: A Portrait of Alice Walker." Horizons.

Prod. Jane Rosenthal. Natl. Public Radio. WBST, Muncie, IN. 3 March 1984.

"The Campaign." Middletown. Created by Peter Davis. Dir. Tom Cohen. PBS. WQED, Pittsburgh. 24 March 1982.

Plays and Concerts

Emphasis on performance

The Real Thing. By Tom Stoppard. Dir. Mike Nichols. With Jeremy Irons and Glenn Close. Plymouth Theatre, New York. 4 June 1984.

Emphasis on conductor

Zuckerman, Pinchas, cond. St. Paul Chamber Orch. Concert. Symphony Hall, Boston. 19 Nov. 1983.

Recordings

Emphasis on composer

Mozart, Wolfgang A. Cosi Fan Tutte. With Kiri Te Kanawa, Frederica von Stade, David Rendall, and Philippe Huttenlocher. Cond. Alain Lombard. Strasbourg Philharmonic Orch. RCA, SRL3-2629, 1978.

Emphasis on performer

McGarrigle, Kate and Anna. Love Over and Over. Polygram, 810 042-1 Y-1, 1982.

Works of Art

Botticelli, Sandor. Giuliano de' Medici. Samuel H. Kress Collection. The National Gallery of Art, Washington.

de Rivera, José. Construction 195. The Currier Gallery of Art, Manchester, NH.

Maps and Charts

Sonoma and Napa Counties. Map. San Francisco. California State Automobile Association, 1984.

Published and Unpublished Letters

Published letter

Fitzgerald, F. Scott. "To Ernest Hemingway." 1 June 1934. The

Letters of F. Scott Fitzgerald. Ed. Andrew Turnbull. New

York: Scribner's, 1963. 308–10.

Unpublished letter   Stowe, Harriet Beecher. Letter to George Eliot. 25 May 1869.

Berg Collection. New York Public Library.

Interviews

Published interview   Ellison, Ralph. Interview. "Indivisible Man." Atlantic. With

James Alan McPherson. 226 (1970): 45–60.

Unpublished interview   McPhee, John. Personal interview. 4 November 1986.

TYPING THE FINAL DRAFT

When you type the final draft of your research paper, follow these general specifications.

1. Use white, 20-pound bond, 8½″ × 11″ paper.

2. Use a pica typewriter or high quality printer.

3. Double-space the text throughout — *including* quotations, notes, and "works cited" list.

4. Maintain margins of 1 inch at the top and the bottom and on both sides of each page.

5. Indent 5 spaces at the beginning of each paragraph.

6. Leave 2 spaces after periods and other terminal marks of punctuation.

7. Leave 1 space after commas and other internal marks of punctuation.

8. Arrange the information on the title page so that it is balanced on the page. Center the title one third of the way down the page. *Do not underline your own title.* After the title, leave several line spaces and then type your name. At the bottom of the title page, centered on separate lines, list (a) the course and section numbers, (b) your instructor's name, and (c) the date.

9. A formal outline, if required, follows the title page and serves as the table of contents for the final paper. Use topics or full sentences, and follow the outline formats discussed in Chapter 3. Type the title of your paper at the top of the first page of the outline; then type your thesis, triple-spaced below the title. Do not list your introduction or conclusion in your outline.

10. Type your title at the top of the first page of your paper. Triple space to the first sentence of your paper.

11. Number the pages of your manuscript consecutively, placing the page number in the upper-right corner. Begin numbering on the second page

(page 2) of your paper; do *not* number the title page, the formal outline, or the first page of the text of your paper. Number all text pages after the first, including any that contain illustrations or endnotes and the "works cited" list.

12. Label any illustrations. Position the caption 2 line-spaces below the figure, and align the caption with the left side of the figure. Mount each illustration on bond paper using rubber cement or dry-mount tissue.

When you have finished typing, proofread every page carefully — including the title page, the formal outline, captions, endnotes, and "Works Cited" list. Make a photocopy of your paper. If your instructor has asked you to hand in your note cards with your paper, arrange them in a logical order, put a rubber band around them, and place them in an envelope.

RESEARCH PAPER: STUDENT MODEL

Reading the final draft of Laura's research paper will demonstrate several valuable lessons. You will see that an informed writer can make any subject, however unfamiliar, significant and interesting. You will see how a carefully crafted, compelling essay evolves from the complex and often confusing attempts to discover a subject, audience, and purpose. And you will see how Laura uses the specific techniques of analyzing, paraphrasing, quoting, and documenting sources to advance her purpose.

Consider, too, as you read, the intriguing relation Laura establishes between visual and textual literacy. How does this theme contribute to your perspective as a writer and as a reader of *Writing with a Purpose?*

Title, double-spaced

Light and Literacy:

The Windows of Chartres

Name on separate line

Laura Burney

Double-space:
Course, section

English 104, Section 3

Instructor

Mr. Johnson

Date submitted

May 12, 1987

Light and Literacy: The Windows of Chartres

Thesis: *Compare Laura's final thesis with her preliminary hypotheses on page 327. Notice that Laura uses three paragraphs to establish a context for her thesis.*

Outline: *The outline is placed between title page and the text, and its pages are not numbered.*

Thesis: Reading four of these radiant "texts" illustrates how Chartres' stained-glass windows provided the faithful with visual instruction in various aspects of their history.

 I. The "Blue Virgin" window fuses pre-Christian and Christian history.

   A. Pagan legends prefigure medieval Virgin.

   B. The "Blue Virgin" window illustrates Virgin's importance to Christian history.

 II. The "Charlemagne" window associates secular and church history.

   A. Charlemagne receives Virgin's Holy Tunic for victory in Crusades.

   B. Charlemagne's grandson, Charles the Bald, gives Holy Tunic to Chartres.

   C. Relic accounts for fame of Chartres.

III. The "Cult of the Carts" window documents local history.

   A. The cathedral is destroyed, but the relic survives.

   B. The Great Rebuilding involves the whole medieval community.

   C. The Cult of the Carts reminds citizens of their achievement.

   D. School of Chartres helps interpret windows.

Outline: *Compare Laura's formal outline with her preliminary outline on page 327. Her first four headings evolve from her thesis statement. Her last heading evolves from her discussion of windows and leads to her conclusion. Notice that neither her introduction (paragraphs 1–3) nor her conclusion (paragraph 17) appears on her outline.*

IV. The "Good Samaritan" window teaches theory of Biblical history.

    A. The lower half of window renders New Testament parable.

    B. The upper half illustrates Old Testament account of the Fall.

    C. Juxtaposition reveals hidden connection between two Biblical stories.

V. The windows exhibit problems of visual and verbal literacy.

    A. The windows at Chartres are extremely difficult to read.

    B. Our reliance on verbal literacy makes it difficult to appreciate other forms of literacy.

    C. Medieval world was dependent on oral and visual literacy.

Title: *Laura uses her title to prepare readers for her thesis (visual literacy) and her specific subject (the windows at Chartres).*

Light and Literacy:

The Windows at Chartres

It is probably difficult for anyone who can read this sentence to imagine life in the Middle Ages. Most of the people who lived in Europe between the fourth and fifteenth centuries were illiterate. "Their common languages, though widely spoken, were rarely written or read" (Bäuml 227–28). A few clergy and nobles understood Latin, "the language in which reading and writing were taught" (Bäuml 228). And they circulated manuscripts (written in or translated from Latin by scribes) that contained versions of the Bible, romances, histories, and essays on various subjects (Tuchman 60). But for even the most important citizens of this society, illiteracy was likely to be the status quo (Haile 817).

Illiteracy, however, did not mean ignorance. Medieval citizens learned by looking, listening, and remembering (Ong 1). In noble halls and common taverns, they could hear the epic poems and narrative ballads that recorded their cultural history. And in medieval churches, the information centers of their world, they could hear public sermons, watch mystery plays, and gaze at the artwork that illustrated how Christian doctrine explained their lives. As Marshall McLuhan points out, "the medieval cathedrals were the 'books of the people' "(108).

Introduction: *Laura establishes context for thesis. (See revision agendas on pp. 330 and 331.) Notice how she uses the McLuhan quote as transition from general discussion of medieval literacy to description of Chartres and the statement of her thesis.*

2

3

Nowhere is the educational purpose of the medieval church more in evidence than at the great cathedral of Chartres. Towering above the wheat fields of northern France, its double spires, flying buttresses, and ornate sculpture testify to the presence and prominence of Christianity in medieval life. But this imposing architecture serves as a mere frame for the more than 175 stained-glass windows that filter God's light into the cathedral's interior. These spectacular patterns of color, created by twelfth- and thirteenth-century artists (many of them now unknown), exhibit the characters, legends, and symbols of Christian culture. Reading four of these radiant "texts" illustrates how Chartres' stained-glass windows provided the faithful with visual instruction in various aspects of their history.

The images in Notre Dame de la Belle Verriera, or the "Blue Virgin" fuse the pre-Christian and Christian history of the cathedral (see Figure 1). According to legend, the first of Chartres' many Christian churches was built on the site of a pagan well whose waters were reputed to have miraculous powers (Swaan 118–19). Émile Mâle explains that images of protective gods called "Mothers" were often found near such springs, one of whom, "a seated female figure with a child in her lap, [bears] a striking resemblance to the Virgin

4

Thesis: *Laura not only announces her purpose but also prepares her readers for the four-part division of paper.*

Citing an illustration: *Laura uses a parenthetical reference to cite Figure 1 and inserts the illustration close to her discussion of it in the text.*

3

of the medieval period" (8). Another unusual legend tells how, long before the birth of Mary, a pagan king from the Chartres area was inspired to have "a sculpture made of a Virgin bearing a child with this prophetic inscription: Virgini pariturae — 'to the Virgin about to give birth' " (Mâle 8). These legends add significance to yet another: "The Blessed Virgin herself was said to have written, in Hebrew, to the martyrs who evangelized Chartres, agreeing to her coronation as the church's queen" (Lee 64).

The window portrays a seven-foot figure of the Virgin wearing a crown, seated on a rich throne, and surrounded by adoring angels. The Christ child sits in her lap, his right hand raised in blessing, his left hand holding the Scriptures. Above the Virgin descends the image of the Holy Ghost, in the form of a dove, signifying one of the most miraculous events in Christian history, the Annunciation. The most striking colors in the window radiate from the many folds of the Virgin's blue robe, which she was reputed to have worn on this occasion. Medieval citizens looking at the glass image of this robe knew its significance, for the robe itself was sealed in a vault in the west nave of the cathedral — Chartres' one great relic and the reason for its renown.

5

Introduction of Quotation: *Laura works Mâle quote into her own sentence and introduces Lee quote with a sentence that suggests its significance.*

4

Figure 1. "The Blue Virgin."

5

The association of the Virgin's Holy Tunic with the
secular history of the Middle Ages is portrayed in the
"Charlemagne" window (see Figure 2). In <u>Mont-Saint-
Michel and Chartres</u>, Henry Adams points out that the
window begins with the Emperor Constantine asleep
in Constantinople, as an angel gives him orders "to
seek aid from Charlemagne against the Saracens. . . .
Then Charlemagne . . . wearing his halo, converses
with two bishops [about] . . . the rescue of Constantine" (156). The battle scenes of this Crusade are
represented in several panels, with Charlemagne and
his knights wearing coats of mail, while the infidels
are identified by round shields. In one scene, Charlemagne beheads a Saracen with one swing of his sword.
After the victory, Constantine presents Charlemagne
with three gifts: a piece of the cross, the shroud of
Christ, and the Holy Tunic of the Virgin (Adams 156).

    This legend was familiar to those who gazed at the
window, for not only was Charlemagne the epic hero
of the Middle Ages, but also his grandson, Charles the
Bald, in 876, gave Chartres the Holy Tunic, making it
the center for the worship of the Virgin. Scholars,
philosophers, and pilgrims mused that "the cathedral
appeared to be her dwelling place on earth" (Mâle
9–10). As von Simon points out, the presence

6

7

*Integrating Quotation:
Laura uses ellipsis and
brackets to work portions
of Adams' text into
her own.*

*Paraphrase: Laura
summarizes a long passage
from Adams in a
paraphrase.*

6

Figure 2. "The Charlemagne Window."

7

**Introduction of
Quotation:** *Laura names
the author of her source in
her sentence and reduces
parenthetical reference to
a page number.*

of the Virgin's Holy Tunic not only provided a sacred link to the supernatural world, but also produced a significant effect "upon nearly every phase of medieval life" (10). Indeed, the acquisition of the Holy Tunic enabled Chartres to become an economic as well as an educational center. Swaan suggests that "it is no coincidence that the four great annual fairs with which the prosperity of Chartres was linked, coincide with the four Feasts of the Virgin" (120).

8

In 1194 Chartres' association with the cult of the Virgin seemed finished when a fire reduced most of the cathedral to ashes. But the Holy Tunic survived, as did a major portion of the "Blue Virgin" window. Swaan reports that these events were read as miracles: "The fact that the Virgin had suffered her 'palace' to be destroyed was . . . interpreted not as a sign that she had abandoned Chartres but rather, . . . 'that she desired a new and even more splendid church to be built in her honor' " (119).

9

Longer Quotation: *Laura
quotes this long but
memorable passage to
describe the events
illustrated in the "Cult of
the Carts" window.*

The rebuilding of Chartres, which started almost immediately but took 66 years to complete, awakened the imagination of the medieval world. Kenneth Clark provides a memorable written account of an earlier stage of the cathedral's construction:

8

In the year 1144, they say, when the towers
seemed to be rising as if by magic, the faithful
harnessed themselves to the carts which were
bringing stone, and dragged them from the
quarry to the cathedral. The enthusiasm
spread throughout France. Men and women
came from far away carrying heavy burdens
of provisions for the workmen — wine, oil,
corn. Amongst them were lords and ladies,
pulling carts with the rest. There was perfect
discipline, and a most profound silence. (56)

An equally memorable portrait of the cathedral's
reconstruction can be seen in the window known as
the "Cult of the Carts" (see Figure 3). Two heavy
carts — one loaded with sacks, the other with a huge
barrel — are pulled by men toward a statue of the
Virgin that is surrounded by devoted pilgrims. Above
the statue, looking down from her heavenly throne,
the Virgin watches approvingly the re-creation of her
"favorite Earthly abode" (Clark 58).

    This window provides an important lesson in local
history. Mâle explains that the window is a "rare
instance of the commemoration of an almost
contemporary event, which perhaps the artist himself
witnessed" (171). Moreover, the window serves as a

10

11

9

Figure 3. "The Cult of the Carts."

10

forceful reminder that virtually every stratum of
medieval society was involved in the rebuilding
campaign. Many of the windows created for the new
cathedral were donated by local guilds: carpenters,
bakers, butchers, tanners (Swaan 122). The "Cult of
the Cart" window reminded the local citizens of their
achievement and inspired other Christians to similar
acts of generosity.

    As the new cathedral reached completion in 1260,
Chartres asserted its pre-eminence as a center of
learning. Clergy, scholars, and philosophers congre-
gated beneath the towering spires, forming one of the
most important schools in the age. As Painton Cowen
points out, the School at Chartres was interested
in how the word of God evolved through the ages and
through all areas of knowledge, Christian and pagan,
Old Testament and New Testament (14). Although
the School encouraged the study of ancient manuscripts among the literate, it also promoted the
"reading" of the spectacular "texts" that "enlightened" the cathedral. Sermons, tours, and religious
festivals were designed to help medieval people interpret the windows' portrayal of Biblical history from
Creation to Second Coming to Final Judgment.

12

11

13

Quotation and
Paraphrase: *Laura quotes
Mâle to advance her thesis
about the teaching
function of the window.*

Then she paraphrases
*Mâle to help her interpret
details of the window.*

A vivid example of this Christian teaching is
illustrated in the "Good Samaritan" window (see
Figure 4). Mâle suggests that the "learned men of the
Middle Ages saw [the parable], as we do, as a lesson of
charity, but beyond this they perceived further
secrets" (170). The lower part of the window renders
the familiar parable. A traveler sets out from
Jerusalem, is attacked by thieves, and robbed of his
belongings, including his tunic. Although he is naked
and injured, the priest and the Levite pass by him. The
Good Samaritan, however, possessing "the
countenance of Christ," stops, binds the man's
wounds, puts him on his own horse, and then leads
him to an inn (Mâle 171).

14

The upper half of the window places the parable in
a larger context. It shows the creation of Adam and
Eve, the story of their sin, culminating in their fall
from grace, banishment from the Garden of Eden,
condemnation to a life of labor and pain, and the
murder of their son Abel by his brother Cain.
According to Mâle, the juxtaposition of these two
stories instructs the viewer to see their "hidden"
connection:

Longer Quotation: *Laura
uses long quotation from
Mâle to explain the
"hidden" connection
between two Biblical
stories in the window.*

> The traveler is man: he is attacked by a group
> of thieves, that is to say by all the sins

12

Figure 4. "The Good Samaritan."

13

together, who fall upon him and take away his
tunic, symbol of his immortality. Such was
Adam, condemned, after his sin, along with
his descendants to labor and death. The priest
and the Levite represent the Law of Moses and
the Old Testament incapable of the salvation
of sick humanity. The Good Samaritan is
Christ himself, who binds the wounds that
Moses was unable to heal and who leads man
into an inn, in other words the Church. This
view explains why the history of the Fall has
its place in the parable and why the figure of
Christ dominates the whole composition. (171)

The "Good Samaritan" window, like all the
windows at Chartres, is extremely complex, leading
many to question its original effectiveness in teaching
the faithful. Clark admits that "it is quite hard to find
out what is going on in the various windows, even
when one goes around with a crib prepared by some
learned student of iconography" (60). But perhaps our
difficulty in reading the windows at Chartres comes
not so much from their visual complexity as from our
reliance on the verbal complexity of language.

M. T. Clanchy argues that our dependence on
literacy narrows our ability to imagine and appreciate

Using Authorities: Laura uses Clark quotation to acknowledge the possible limitations in her thesis.

Using Authorities: Laura counterbalances Clark quotation with Clanchy's discussion of medieval literacy.

15

16

14

other forms of communication (8). In the medieval world, people relied on what they saw and heard. Most important transactions were sealed with a symbolic object "to impress the event on the memory of all those present" (203). Many medieval manuscripts were addressed to "all those seeing and hearing these letters" (202). Indeed, as Clanchy points out, most medieval manuscripts seem more preoccupied with pictorial illuminations than with functional communication (227).

Conclusion: *Clanchy's discussion leads Laura to speculate that the stained-glass windows performed a vital function in a world that depended on oral and visual literacy.*

17

Seen from this perspective, the windows at Chartres (and other medieval cathedrals) seem to mark the midway point between oral and verbal literacy. Pilgrims at Chartres had either heard the legendary stories they saw celebrated in the windows or were eager to have those windows interpreted by some knowledgeable storyteller. In either case, they left the cathedral with a renewed reverence for the world they knew or an expanded vision of the world beyond. In the Middle Ages, "pictures" such as those of the Virgin, Charlemagne, the Cult of the Carts, and the Good Samaritan were worth a thousand words. It is only in our time that they take a thousand words.

15

Works Cited

Adams, Henry. Mont-Saint-Michel and Chartres.
     Boston: Houghton, 1905.

Bäuml, Franz H. "Varieties and Consequences of
     Medieval Literacy and Illiteracy." Speculum 55
     (1980): 237–65.

Clanchy, M. T. From Memory to Written Record:
     England, 1066–1307. Cambridge, MA: Harvard UP,
     1979.

Clark, Kenneth. Civilisation: A Personal View. New
     York: Harper, 1970.

Cowen, Painton. Rose Windows. San Francisco:
     Chronicle, 1979.

Haile, H. G. "Luther and Literacy." Publications of the
     Modern Language Association 91 (1976): 816–28.

Lee, Lawrence, George Seddon, and Francis Stephens.
     Stained Glass. New York: Crown, 1976.

Mâle Émile. Chartres. Trans. Sara Wilson. New York:
     Harper, 1983.

McLuhan, H. Marshall. The Gutenberg Galaxy: The
     Making of Typographic Man. Toronto: U of
     Toronto P, 1962.

Ong, Walter J. "Orality, Literacy and Medieval
     Textualization." New History 16.1 (1984): 1–12.

Swaan, Wim. The Gothic Cathedral. Garden City, NY:
     Doubleday, 1969.

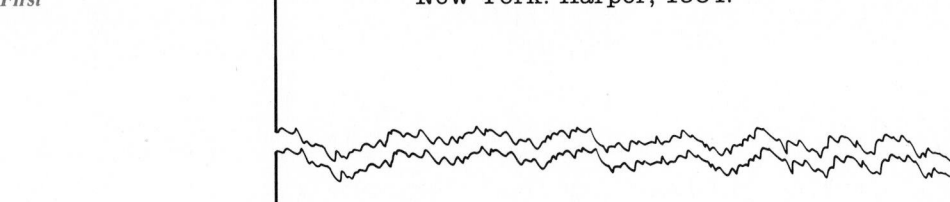

16

Tuchman, Barbara W. <u>A Distant Mirror: The Calamitous 14th Century</u>. New York: Knopf, 1978.

von Simon, Otto Georg. <u>The Gothic Cathedral</u>. 2nd ed. New York: Harper, 1964.

Entry for Book in an Edition Other Than the First

Writing Assignments

1. Revise Laura's research paper. Begin by trying to condense her introduction to one paragraph and giving her thesis a stronger argumentative edge. Then expand the four sections to accommodate your research on and reactions to the windows.

2. Research and then write a paper on some aspect of another method for communicating information in a preliterate culture. A good overview of this subject can be found in Charles A. Ferguson and Shirley Brice Heath, eds., *Language in the USA* (Cambridge: Cambridge UP, 1981).

3. Research the problem of how information is communicated in an aliterate culture — that is, in a culture where people can read but don't. You may

wish to write your paper on television or, even more specifically, on music videos. Begin with Marshall McLuhan, *Understanding Media: The Extensions of Man* (New York: McGraw, 1964), and Marie Winn, *The Plug-In Drug* (New York: Viking, 1977).

4. The photograph on page 324 was selected as a metaphor representing an important aspect of this chapter. Examine the photograph and determine the possible meanings of this metaphor as you discern them in the photograph. Then write an essay explaining how the metaphor has been reinforced or revised by the ideas discussed in the chapter. You may want to suggest another metaphor to illustrate your understanding of these ideas.

Appendix
WRITING WITH A WORD PROCESSOR

Writing with a word processor expands opportunities for fully exploring the writing process and thus the possibility of more fully expressing your purpose. When personal computers first gained popularity in the late 1970s, they were most commonly used for computer games. Recently, however, word processing has become the primary use of personal computers. Like a growing number of people, you can use word processors for all kinds of personal, academic, and professional writing. Your use of a word processor during your school years serves both short-term and long-term purposes: The word processor is a writing tool that can improve your academic writing by making planning, drafting, and revising easier; and the ability to use a word processor develops technical skills that will be helpful in any profession you enter.

THE WORD PROCESSOR AS A WRITING TOOL

You may have friends who talk about writing with a word processor with the enthusiasm of the newly converted, giving glowing testimonials about what this new tool has done for their writing. Although such testimonials can be boring to the uninitiated, your friends' enthusiasm is well based. The microcomputer can increase your efficiency and effectiveness, as well as your enjoyment of writing. Make no mistake: the word processor does not miraculously transform writing into a quick and simple task. After all, it cannot do for you the hard thinking that good writing requires. It can, however, encourage experimental planning and drafting and extensive revision through its capacity to save and change texts easily.

Because writing is such a personal activity, how you choose to utilize this tool is an individual matter. Some of you, especially those who have been composing on a typewriter, may very quickly begin to do all your writing on the computer — planning, drafting, and revising. Others may continue to plan and produce the first draft by hand and turn to the word processor only for second drafts and subsequent revisions. Still others may initially use the word processor like a typewriter, entering and editing only the final draft. Just as there is no one right way to write, so too there is no one right way to use the computer for writing. Over time, however, you will probably find yourself taking increasing advantage of the computer's capabilities.

As you contemplate using a word processor for the first time, you may share a common fear of many novice computer writers. Writers new to the word processor often fear that they will lose their work because of a mistaken command, a damaged disk, or a computer "crash." Three precautions can reduce the risk of such misfortunes:

- Save the writing you are working on frequently.
- After every day's work, make back-up copies of your disk.
- Print "hard" copies of your work regularly (that is, print out your work on paper).

USING THE WORD PROCESSOR FOR PLANNING

Some people argue that the word processor is useful for planning because typing on an electronic keyboard more closely matches the pace of thought than does writing with a pen or pencil or typing on a manual or even electric typewriter. Others suggest that writing on a word processor is like talking to oneself or thinking aloud. In any case, for many, the word processor seems to encourage the "play" and experimentation with ideas and language that are the heart of planning.

Nearly all the thinking-in-writing strategies mentioned in Chapter 2 adapt readily to planning with a word processor. *Listing,* for example, can be done as easily on screen as by hand. Simply type your prompt at the top of your screen and begin listing all the ideas, associations, and images that it brings to mind. To evaluate your list, review it and highlight promising items with an asterisk, underlining, or boldface. As suggested in Chapter 2, you might then pick one of these promising ideas and begin another list or do some freewriting, using the new idea as your prompt. After that activity, you may wish to return to your original list and insert some of the new ideas that emerged. To do this you will need to save your initial lists and other plans; you can do this by keeping them at the beginning of the file for your upcoming assignment, in a separate file, or in print. Since these steps are easy to do, you should get into the habit of saving your planning so that you do not prematurely delete plans to which you later find you need to refer.

Clustering cannot be done on screen in the same manner as it can on paper, but with some computers you can use graphics to represent idea associations. For example, you might create a flow chart to show how one idea leads to others. Or you might put your prompt in the middle of the screen and produce clusters of ideas in each of the four corners, later drawing arrows or lines on your hard copy to show the connections among the ideas in the four clusters.

Freewriting can be done easily with the word processor. In fact, the exploratory nature of freewriting makes it an ideal strategy for generating ideas and at the same time practicing computer-assisted writing. You just begin "fishing," as poet William Stafford puts it, for whatever happens along. You should try to write without stopping for ten to fifteen minutes to get the most out of this activity. To push your freewriting a bit further, try this strategy. After ten to fifteen minutes of freewriting, stop, reread what you've written, and look for a focus. This might be a surprising phrase, a problem, an intriguing idea, or a remembered incident. Move down on the screen a few lines and type this focus as a prompt for another ten to fifteen minutes of focused freewriting.

Another computer-assisted freewriting strategy is what the composition researchers Sheridan Blau and Stephen Marcus call "invisible writing." Simply turn down the brightness control on the monitor, so that your screen appears blank. You then begin typing whatever comes to mind or, if you are using a prompt, whatever it triggers. You may find this technique liberating because you will not be distracted by what you've just written; such forced concentration on what you wish to say next can produce creative results. After you've done this for ten or fifteen minutes, turn the brightness control back up, read, save, and print what you've written. As recommended in Chapter 2, it is wise to conclude a freewriting session by reviewing and evaluating your work for interesting ideas and passages. These passages can be identified for later use by asterisks placed at the beginning and end of the passage, underlining, boldface, or brackets, or they may be transferred to a file reserved for them. You can also annotate promising passages by hand on your hard copy.

Keeping an electronic journal, an alternative to keeping a journal in a notebook, is a planning activity particularly suited to the word processor. Although an electronic journal cannot be carried around like a notebook, it offers several other advantages: it is easier to store; it is less likely to get lost (especially if you work on your word processor at home or in your dorm room and don't have to carry your disks with you); it allows you to transfer chunks of your journal writing directly into a draft; and since it is typed, it is easier to read. There are, however, several guidelines you should follow. First, be sure that an electronic journal is acceptable to your teacher. Next, use a separate disk for your journal entries. And finally, create a file on your journal disk to serve as a table of contents. Call this file "Contents"; and before or after you have written an entry, record in your contents file the date and a phrase indicating the focus of the entry. You can also keep track of your entries by making a separate file for each. Or, if you are keeping different kinds of entries — assigned freewriting entries, reading notes, self-selected freewriting entries — you may wish to create separate "contents" files for each type. If your teacher wants you to submit selected entries or the whole journal, you will need to print the required material.

If you are using your journal to evoke and store potential essay topics, you may wish to reread and evaluate your journal entries from time to time, abstracting promising entries and transferring them to another file kept for such theme-starters. If you wish to move blocks of your journal directly into the text of a paper, check your computer manual for instructions on "block moves."

The two observation strategies mentioned in Chapter 2 are not equally adaptable to the word processor. *Mapping* is adaptable only to computers that have graphics capabilities. *Speculating* is adaptable to any computer. Mapping requires you to translate your sketches or notes about your subject's relationships, activity patterns, and figures into computer maps and sketches. If you have graphics or "painting" software and know how to use it, this translation process can promote further insights into and new perspectives on your subject.

Speculating is a particularly rich planning strategy since it allows you to explore your subject from several perspectives. You can speculate on a word processor by doing a series of three focused freewritings or by using a grid. Some computer programs will create multi-celled grids. In other cases, you yourself might create the grid, using computer graphics. A speculation grid for Jenny's subject in Chapter 2, the Varsity Barbershop, might look like this:

	UNIQUE FEATURES	VARIATIONS POSSIBLE	PLACE IN SYSTEM
OBJECT	Established for 40 yrs. Old-fashioned barbershop décor; chairs, diplomas, men's mags.	Only a few modern touches—poster of a young man with "mod" hairdo, air-conditioner.	Like small-town barbershops everywhere, could be Floyd's on Andy Griffith Show. Male meeting place.
ACTION	Scissors snipping, razors trimming. Main action—male conversation. Everett at center.	Activity level varies, often busy. Not much talk when there are no customers, Randy plays radio.	Center of community activity. Communications switchboard.
NETWORK	Located in middle of town. Everett in middle of everything—knows everybody, respected, other business interests.	Everett's role in community has developed over time. Place and its role will change when Everett retires.	Place where men gather to exchange news, gossip, political opinions, financial advice. Everett a town elder.

Interviewing and *reading,* the two research activities mentioned in Chapter 2, will yield information that can be stored in computer files. Translating your interview notes or transcribing your interview tape into computer files works best if you do it soon after the interview. Thus, the translation process becomes an opportunity to interpret and speculate about the data you have just collected. After recording your interview notes in the actual sequence in which you took them, you may want to copy this file. You can then preserve the file containing the original version of the interview while using the duplicate file to experiment with your subject's comments, moving them around and arranging them to best reveal the patterns and themes that emerged during the interview.

Whether you are keeping informal reading notes or working on a research paper, the word processor is a useful tool. If you type the bibliographic information about each of your sources at the beginning of your notes for that source, you will be able to keep track of that information and transfer it, if necessary, into your bibliography or "works cited" list. Some word-processing programs can take your bibliographic information and format and alphabetize it, saving you a lot of time.

Below the bibliographic information, enter a summary of the reading, paraphrases of key ideas, direct quotations, and your own comments and questions. This method of keeping notes has many advantages. You can insert additional comments when you review your notes; you can move paraphrases, quotations, or comments directly into the text of your paper; and you can merge related readings into a single file that represents a subtopic or theme. A further benefit is that such notes are less likely to be lost than those done on note cards.

In addition to transferring and expanding traditional planning techniques with the word processor, you may wish to use or your teacher may encourage you to use various invention (or planning) programs. These programs prompt you to explore your topic by asking you various questions — speculating questions similar to those found in Chapter 2; the reporter's six questions (*who? what? when? where? why?* and *how?*), mentioned in Chapter 2; and questions that ask for definition, comparison, and so on, similar to the questions on structure in Chapter 5. Some new software, called idea processors or thought processors, offers outlining programs as a planning device. These programs can be useful and can make the planning process more gamelike; however, most are programmed versions of the planning strategies presented in Chapter 2 (which can be used on word-processing software), and so are not essential to productive planning.

As noted at the end of Chapter 2, at some point you must stop planning and begin drafting. Because of the ease of electronic word processing, you may be more tempted than usual to continue freewriting and note taking, putting off the more difficult and threatening business of drafting. The following suggestions may help you overcome this tendency. Set specific time limits on your planning, and ease into drafting by viewing it as an extension of your freewriting. Even if you think you have more planning to do — and you always will think so, for most writers end up planning and replanning throughout the writing process — begin drafting. Focused freewriting might aid in this transition. After your listing, clustering, or freewriting, summarize the gist of an interesting train of thought, or form a hypothesis and begin drafting with the same sense of freedom you felt during freewriting. After all, you have nothing to lose. Computer-produced drafts can be endlessly reshaped without complete retyping.

USING THE WORD PROCESSOR FOR DRAFTING

The fluid nature of screen text counteracts a first-draft block for some writers. However, if you find yourself reluctant to begin drafting, even on a word processor, you might try one of these strategies:

- You may wish to try *invisible writing,* described above, to produce your first draft. This technique, by preventing you from becoming preoccupied with exact wording and correctness, frees you to concentrate on developing your ideas.

- Another strategy for overcoming a first-draft block is similar to freewriting. Write nonstop for a certain period of time, say thirty minutes, with the aim of producing a complete representation of your ideas. Such a draft, discussed in Chapter 3 as a *discovery draft,* is deliberately tentative.

- Another strategy is to draft a *scratch outline* and then compose your draft in chunks. Your first goal, for example, might be to draft one of the topical paragraphs. This strategy breaks the job of drafting into small segments and allows you to compose your draft in any sequence you wish. If the introductory paragraph has you stumped, begin with the first or second topical paragraph. A word processor allows you to easily reorder the chunks later.

The word processor can help you get started with your draft, and it can also help you keep going. When you get stuck, shift to freewriting or listing. This can get you over the various blocks and on to other sections where your ideas are more forthcoming. If you use brackets or asterisks to mark the places at which you feel blocked, you can easily locate those places with the "find" function on your computer when you return to drafting. Another way to get through a problem section is to make a bracketed note to yourself to work on it later and then to go on. Still another idea is to draft several versions of the problem section and decide later which works best.

If your word-processing software includes a thesaurus, you will want to use it if you are having trouble finding the right word. All you have to do is type a word that comes close to the one you want but can't quite think of. The thesaurus will suggest several alternatives. Often the discovery of the right word is exactly what you need to spur you on to further composing.

A final way in which the word processor can be useful is by helping you to evaluate your drafts. After completing your first draft, triple-space your text, read through it (or have a friend read it), and make notes for later revision. Some word processors have a "split screen" or a "window" function that allows you to annotate your screen text. In each case, a second screen is superimposed on some portion of the screen holding your text. The "split screen" function divides the screen into two sections; the "window" function opens a "window" in one corner of your screen. In both cases, you can control the portion of the screen devoted to your original text and the portion devoted to revision. You might use these functions to make on-screen comments or to compose a descriptive outline (explained in Chapter 3) of your draft. Since these functions work by merging two files, the draft and comments or descriptive outline will print out separately.

In addition to using on-screen evaluation strategies, you should always evaluate a hard copy of your draft. On a screen you can see only a very small portion of your text (usually 24 lines) at a time, so it can be difficult to keep in mind the content of preceding and following pages. Reading screen text does not offer the same complete perspective as reading the easily turned pages of a printed copy. Some writers find it helpful to move back and forth from printed text to screen text, reading and annotating the former, then making the desired changes on the latter.

Some word processing programs will provide a word count, a useful drafting aid that will help you avoid length problems in drafts for assignments that impose a specific number of words. Programs may also offer an abbreviation feature. If your subject involves a name, title, or word that you use repeatedly — such as *photosynthesis* or *Feodor Dostoevski* — you can avoid typing and retyping all those letters (and the risk of misspellings) by pressing a special key followed by a *P,* a *D,* an asterisk, or whatever symbol you choose: the word or words will appear automatically.

USING THE WORD PROCESSOR FOR REVISING

Probably the most touted benefit of computer-assisted writing is its encouragement of revision. But to some novice computer writers, the easy revision of screen texts leads to preoccupation with word changes and mechanical corrections. Substantial revision, unlike editing and proofreading, involves reseeing and reshaping the ideas in a draft. In Chapter 4 this process is called *global revision*. In the past, you may have avoided or minimized this kind of revision because it meant changing, recopying, or retyping large portions of the draft you had worked so hard to produce. But with the computer, you can make substantial changes — adding paragraphs and sentences, deleting sentences and paragraphs, shifting and reordering paragraphs and larger sections — without recopying or retyping the entire paper. The revision process becomes physically easier and far less tedious.

On-screen revision, which should be coupled with the evaluation of printed drafts, can be divided into three categories: (1) writer-initiated activities, (2) responses to peer or teacher feedback, (3) software-generated activities.

Some writer-initiated strategies have already been mentioned — using the "split screen" function to annotate a draft, marking blocks of text for later revision, and composing several versions of a particular section. Since such on-screen revisions involve reorganizing, reworking, and experimenting with the draft, you may wish to copy the original draft, so that you can retrieve portions of it if you decide you prefer them to any of your subsequent revisions.

The easiest kind of revision is addition or deletion of text. Perhaps in evaluating your draft you decide that a certain term needs a definition, a generalization needs support, or an example needs elaboration. All you need to do is position the cursor at the point where you want to insert material and begin typing. (Some programs require that you use an "insert" function to do this; if you don't, as you type you will cover and eliminate material that you may wish to keep.) If you decide to eliminate any portion of your draft, delete the material word by word or mark the entire block of text and delete it all at once.

Manipulation of blocks of text is a revision strategy especially suited to the word processor. Easy to accomplish, block and cut-and-paste moves allow you to reorder sentences, paragraphs, or larger sections to see what effect these shifts have on the overall effectiveness and meaning of your text. Sometimes, for example, you may want to try your conclusion as an introduction or to reorder your topical paragraphs. You may decide that an example you used in one paragraph would be more effective in another. Word processors enable you to reorganize your draft with ease and yet to return readily to your earlier version if you saved your original draft.

As suggested in Chapter 4, another important guide to global revision is the feedback you receive from others. For computer-produced drafts this feedback can be offered by computer as well as on a hard copy. A peer editor, friend, or teacher who reads your draft on screen and wishes to offer comments may do so in several ways. The reader may insert his or her comments right into the text in bracketed caps, like this: [CAN YOU ILLUSTRATE HOW THIS WOULD WORK?] Or, if your draft is triple-spaced, the reader's comments may be inserted between the lines. If you find such inserted comments intrusive, the reader may place his or her comments at the end of your draft (you can read, print, and then erase them) or in a separate file kept for comments; comments from different sources can be kept in the same file or in separate files. Some teachers arrange for class members to respond to one another's drafts through electronic mail or electonic bulletin boards. If your reader will be responding to a hard copy, be sure that you triple-space or leave wide right or left margins for comments.

There are computer programs especially designed for aiding revision. You may find that the software you received with your computer includes a text-feedback program. Sometimes these programs are available in computer-assisted classrooms or campus writing labs. Although some of them are designed strictly for editing and proofreading, others (such as revision prompters, revision demonstrators, and speech synthesizers) are intended to help with global revision. Revision prompters can represent your text to you in various ways; for example, the program may show you a paragraph from your draft and ask you whether it has a clear focus. Revision demonstrators show you examples of an expert writer revising a text with a word processor. Speech synthesizers read your draft aloud to you, allowing you the different perspective of being a listener rather than a reader. Because the voice is flat and robotic, this oral reading can give you an objective view of your paper's coherence and meaning, though at times the mechanical voice can be difficult to understand.

Some of the strategies useful for global revision — annotating drafts, manipulating sentences and paragraphs, employing text feedback programs — are also useful for local revision. As discussed in Chapters 7 through 10, local revision involves fine-tuning individual paragraphs, sentences, phrases, and words. You may, for example, wish to experiment by reordering the sentences in your topical paragraphs or by drafting several versions of your introductory paragraph to vary tone or emphasis. Or ask a peer reader to mark your draft, using one of the methods discussed above, to show needed transitions or problems with paragraph unity, completeness, order, or coherence.

Similarly, you may experiment with sentence structure and diction, drafting several versions of a sentence or listing several possible word choices, or you may wish to ask a peer reader to annotate your draft for awkward or ineffective sentences or diction. Style-checking programs summarize the sentence types in your draft (the number or percentage of simple, compound, or compound-complex sentences), sometimes suggesting the need to vary sentence structure. Also, some style checkers calculate the number of words per sentence and flag wordy constructions. Such feedback on sentence types, constructions, and length may be helpful, but it must be evaluated in terms of your subject, audience, and purpose. As mentioned elsewhere, the computer thesaurus aids in word choice, although here, too, you will want to evaluate the computer's advice or suggestions in light of your subject or purpose.

Reading computer drafts aloud should be part of your local revision process. The computer's encouragement of constant tinkering can cause you to overlook wordiness and overly long sentences. Your ear is the best guide for catching such infelicities and for determining whether your tone and style are consonant with your purpose.

It is important to print each successive draft for evaluation and revision. The on-screen methods for revision complement but do not replace reading and annotating the hard copy. There are several reasons why this is true. First, it is hard to evaluate the overall organization and coherence of a paper by reading on screen. The slow process of scrolling from page to page to read screen text makes it difficult to remember what you have written. When you read a printed copy, you can refresh your memory by quickly referring to earlier pages. Second, as you compose or enter your paper on the computer, you inevitably make many changes. Since you are not forced to recopy or retype the entire paper to incorporate these changes, it can be difficult to see how they affect the paper as a whole. So, even though you may be impressed with the professional look of your computer-produced paper and hesitant to read it for revision and mark it up, resist the temptation to forgo this step. Another copy that is just as professional looking can easily be produced later, and this one will have the benefit of careful, thorough review and revision.

USING THE WORD PROCESSOR FOR EDITING AND PROOFREADING

The last stage of work on any assignment is editing and proofreading. Some software programs offer editing aids, but you should develop your own strategies for editing and proofreading with a word processor.

First, here is some general advice. When you read screen text for editing purposes, it is easy to miss mechanical and grammatical errors. Because the text looks so neat and you already know what's on the page, your eyes may skip over words and fail to catch errors. Computer-type errors such as double letters and undeleted words — the computer's counterpart to typos — are also easy to miss. Sometimes, as you make quick changes, you may forget to delete a word or phrase, and a sentence like this results: "The following essays *explore examine* causes and effects." At other times, you may fail to add an intended word. In any case, you need to read through your screen text carefully. First read each page line by line; then move the cursor through the page word by word. When you find errors, delete the incorrect form and add the correct one. Double-check by proofreading a hard copy.

The "find/replace" function is a specific word processor editing and proofreading tool that can be used in several ways to check and correct your text. If you discover that you have misspelled a word throughout your paper, you can use the "find/replace" function to correct all the misspellings of that word. You can also use the "find/replace" feature if you wish to replace one word with another throughout your paper. For example, you might start writing about a personal experience using a pseudonym (for example, Sarah) instead of your real name (Jane). If you later decide to use your real name, you can change all the Sarahs to Janes in seconds by employing the "find/replace" feature.

When you are drafting or revising, you may wish to mark with asterisks all words or sentences about which you have questions of diction, style, or grammatical correctness. Instead of interrupting your writing to check your handbook or dictionary, you can use the "find/replace" function to return to each marked spot for later editing. You can also use this feature to find all the periods in your text if you wish to check for sentence completeness, all the commas if you wish to check for correct usage, or all instances of *is* and *are* if you wish to see whether you can replace them with stronger verbs or make passive constructions active.

Peer editors can also offer proofreading and editing advice by marking your text in certain ways — placing asterisks on either side of a misspelled word [*recieve*], a *P* at the beginning of each sentence with a punctuation error, and so on. Using the "find/replace" function, you can locate the errors flagged by your peer editor.

You may be familiar with some programmed editing and proofreading aids — for example, spell and style checkers, thesauruses, and error detectors. Although these text-feedback programs are helpful, they must be used with caution and should accompany, not replace, self-editing and proofreading strategies.

▪ *Spell checkers* check the correctness of the spelling of each word in your text by comparing it to the contents of an online dictionary. However, the computer does not recognize the *contexts* of the words you use and so will not pick up errors like these: *compute* for *computer* or *there* for *their*.

- *Style checkers* vary from program to program. Generally, they locate potential problem areas, such as passive constructions, wordy phrases, and clichés, and may suggest changes. Some style checkers give you a statistical analysis of the number of words in your essay, average word length, readability level, number and types of sentences. The advice provided by all style programs is generic and should not be followed blindly. They are meant as guides to editing, not as instructions for editing. Since computers cannot understand the context in which a certain word or phrase appears, they cannot determine the most appropriate word or phrase. Only you, perhaps with the advice of a teacher or a peer editor, can do that.

- A programmed *thesaurus* can suggest various alternative word choices, but you must decide which word expresses the meaning you intend.

- *Error detectors* are very new and relatively untried. If you use one, remember only you can determine whether the sentence fragment detected by the program is an intentional fragment that you wish to keep for stylistic effect or whether it is an error that should be corrected.

Remember, no matter how neat your printed copy may look, it still needs a careful proofreading.

In addition to the uses of the word processor mentioned here, there are other electronic aids for special writing tasks. If you are taking a business or technical writing course, you can set up certain formats (for example, tables, graphs, spreadsheets) for frequent or repeated use. A number of aids are available for writers of research papers: computer data searches, programs to translate your bibliographic entries and notes into various formats (including MLA and APA); programs to index items in a long paper.

The computer is a powerful writing tool and one that promises to continue to grow in usefulness and capacity. Exciting new advances are under development. In the not-too-distant future, for example, voice-reproduction word processors will work like dictating machines, entering the text as you speak your thoughts aloud. Already, writers working on joint projects in different cities use telephone connections to send their drafts to one another in seconds.

Powerful though they are, however, computers are not the writer's most valuable tool. Even with advances in artificial intelligence, the computer's simulation of human thought is crude and rigid. An example of an advanced computer's translation from one language to another illustrates the computer's limitations. The English saying "The spirit is willing, but the flesh is weak," when translated by computer into Russian and back into English, becomes "The vodka is good, but the meat is rotten." The complexity and resourcefulness of the human mind, the richness of human language, and the boundlessness of the human imagination are, and will remain, the writer's most powerful tools.

HANDBOOK OF GRAMMAR AND USAGE

THE EVOLUTION
OF ENGLISH

The language that Americans speak and write is descended from the language spoken by the English, Scottish, and Irish immigrants who founded the British colonies in America. Their language, in turn, was descended from the languages of Germanic tribes who, during the fifth and sixth centuries, invaded Britain and settled there. One of these tribes, the Angles, later became known as the Englisc (English) and gave their name to a country and a language, both of which they shared with other peoples — the Saxons, the Jutes, and, later, the Danes and the Normans.

The language that has come down to us from that Anglo-Saxon beginning has undergone great changes. Modern college students find Chaucer's fourteenth-century English something of a puzzle. And before Chaucer — well, judge for yourself. Here is the opening of the Lord's Prayer as it was written in the ninth, fourteenth, and seventeenth centuries, respectively:

Old English	Middle English	Modern English
Fæder ūre þū þe eart on heofonum, sī þīn nama gehālgod. Tōbecume þīn rīce, Gewurþe ðīn willa on eorðan swā swā on heofonum.	Oure fadir that art in heuenes, halwid be thi name; thi kyngdom cumme to; be thi wille don as in heuen and in erthe.	Our Father which art in heaven, Hallowed be thy name. Thy kingdom come. Thy will be done on earth as it is in heaven.

A contrast of these three versions offers a brief but revealing impression of the changes that occurred in the language during eight hundred years, and these differences would seem even greater if we could reproduce also the changes in sound that took place. For example, Old English ū and ī were pronounced like the *oo* in *boot* and the *e* in *me* respectively, so that *ūre* was pronounced "oo'ruh" and *sī*, "see."

In grammar the major change has been the simplification of grammatical forms. Old English (700–1100) was a highly *inflected* language, one that made grammatical distinctions by changes in the form of a word. For example, nouns were declined in five cases (nominative, genitive, dative, accusative, and instrumental) as well as in singular and plural numbers. Adjectives and the definite article were declined to agree with the nouns they modified. Here is the declension, in the singular only, of "the good man" with the approximate pronunciation enclosed in quotation marks at the right:

Case	Declension	Pronunciation
N. (*man* as subject)	sē gōda mann	"say goada man"
G. ("of the good man" or "the good man's")	ðaes gōdan mannes	"thas goadan mannes"
D. ("to the good man")	ðǣm gōdan menn	"tham goadan men"
A. (*man* as object)	ðone gōdan mann	"thonna goadan man"
I. ("by the good man")	ðȳ gōdan menn	"thee goadan men"

In Modern English the article and the adjective are not declined at all. The noun retains the genitive case and has singular and plural forms. We distinguish between subject and object by word order, and we have replaced the dative and instrumental endings by the prepositions *to* and *by*. As a result, the whole declensional system has been greatly simplified. Verbs still show considerable inflection, though much less than in Old English.

Along with this simplification of grammatical forms went a great increase in vocabulary as new words were introduced through association with foreign cultures. During the eighth and ninth centuries, Scandinavian raiders settled along the coast of England and brought into the language some fourteen hundred place names and about one thousand common words. In 1066 the Normans conquered England, and for three hundred years their French language dominated the court and the affairs in which the nobility was most involved — government, army, law, church, art, architecture, fashions, and recreation. Between 1100 and 1500 over ten thousand French words were absorbed into the language. During the fourteenth, fifteenth, and sixteenth centuries, English writers borrowed heavily from Latin. Language historians estimate that more than half of the present English vocabulary came from Latin, either directly or through one of the Romance languages, especially French. And as the English-speaking countries grew in political, economic, and cultural importance, their language borrowed from all over the world the words it needed to name the things and ideas that Anglo-Americans were acquiring. Today the vocabulary of the English language is international in origin, as the following list illustrates:

algebra (Arabic)	dollar (German)	polo (Tibetan)
amen (Hebrew)	flannel (Welsh)	silk (Chinese)
bantam (Javanese)	garage (French)	shampoo (Hindi)
boor (Dutch)	garbage (Italian)	ski (Norwegian)
caravan (Persian)	inertia (Latin)	tag (Swedish)
cashew (Portuguese)	kimono (Japanese)	toboggan (American Indian)
chorus (Greek)	leprechaun (Old Irish)	vodka (Russian)
coffee (Turkish)	polka (Polish)	whiskey (Gaelic)

UNDERSTANDING SENTENCE ELEMENTS

The ability to *compose* effective sentences does not depend on an ability to describe sentences grammatically or linguistically. Rather, most proficient writers use intuitive language skills, developed over a number of years, when they compose sentences. Yet when writers understand sentence elements and structure, they are better able to *revise* sentences to achieve a clear form and purpose. To the end of effective sentence revision, the following sections offer a brief overview of basic sentence elements.

1   RECOGNIZE THE BASIC ELEMENTS OF SENTENCES

Although many sentences are complicated word structures, all sentences, even the most complicated, are built from a few basic elements: *subjects* (S), *verbs* (V), *objects* (O), and *complements* (C). These elements work together to express a central idea that may be further developed or refined by other elements: *modifiers* (M) and *conjunctions* (Conj).

The verb with its objects, complements, and modifiers is known as the *predicate* of the sentence. The predicate describes the action performed by the subject or the state of being of the subject. Subject and predicate are the two main parts of a simple sentence.

     S   V
The lawyer wrote. [Subject + Verb.]

     S   V   O
The lawyer wrote the brief. [Subject + Verb + Object.]

       S   V   O Conj   V O
The trial lawyer hurriedly wrote the Hernandez brief but then carefully revised it. [Subject + Verb + Object + Conjunction + Verb + Object. Modifiers used throughout.]

These examples show that a sentence composed of the basic elements can be made more specific and informative through expansion.

1a   *Subjects identify the people, places, things, ideas, qualities, or conditions that act, are acted upon, or are described in a sentence.*

Nouns and pronouns are the most common subjects, but phrases (groups of words without verbs) or clauses (groups of words with subjects and verbs) may also be subjects.

⌐S⌐ V   C
To win is her objective. [Subject *(phrase)* + Verb + Complement.]

⌐————————S————————⌐ V   C
What President Aquino wants most is political stability. [Subject *(clause)* + Verb + Complement.]

1b   **Verbs express action (*select, walk*) or a state of being (*seem, is*). Verbs consist of single words (*develop*) or groups of words (*might have developed*).**

A verb that requires an object to complete its meaning is a ***transitive verb***. A verb that does not require an object to complete its meaning is an ***intransitive verb***. Notice that some verbs can be either transitive or intransitive.

⌐————S————⌐ Trans V   O
President Roosevelt **ordered** the evacuation.

                    Intrans V
After two years, Senator Harris **resigned**.

             Trans V   O
The building inspector **examined** the wiring.

     Intrans V
Pandas **eat** voraciously.

A verb connecting a complement to a subject is a ***linking verb***.

The child **seemed** frightened.

After years of study, Fred **became** an aerospace engineer.

1c   **Objects are nouns or pronouns that complete the ideas expressed by subjects and transitive verbs.**

Direct objects answer the questions *what?* or *whom?* ***Indirect objects*** answer the questions *to whom?* or *for whom?*

          ⌐——S——⌐ V   D. O.
With great care, Dr. Rodriguez completed the **report**. [*Report* tells *what* Dr. Rodriguez completed.]

          ⌐—S—⌐ V   I. O.   D. O.
Dr. Rodriguez sent the **immunologist** the **report**. [*Report* tells *what* Dr. Rodriguez sent; *immunologist* explains *to whom* he sent it.]

1d
Complements are adjectives or nouns that complete the ideas in a sentence by modifying the subject (*predicate adjective*) or by renaming the subject (*predicate noun*).

Complements are joined to the subjects of sentences by linking verbs, such as *am, are, is, was, were, become, get, feel, look,* and *seem.*

Throughout the competition, Warren remained **optimistic**. [Predicate adjective modifies the subject *Warren.*]

In the end, Warren was first **runner-up**. [Predicate noun renames the subject *Warren.*]

1e
Modifiers (typically adjectives, adverbs, and prepositional phrases) describe or limit subjects, verbs, objects, complements, or other modifiers.

Modifiers alter the meanings of other words by answering one of these questions: *what kind? which one? how many? whose? how? when? where? how often?* or *to what extent?*

Long speeches are unacceptable. [What kind of speeches?]

Those four-wheelers are dangerous. [Which four-wheelers?]

We received **sixty-seven** applications. [How many applications?]

The sub-committee shared **its** findings. [Whose findings?]

The immigrant **slowly** completed the form. [How did he complete it?]

After the tennis match, we celebrated. [When did we celebrate?]

Leave the carton **in the mailroom.** [Where should it be left?]

Michael called his doctor **frequently.** [How often did he call?]

The glassblower **very** skillfully formed the stem. [What degree of skill did the glassblower use?]

1f
Conjunctions join and relate two or more words, phrases, or clauses in a sentence.

Coordinating conjunctions (*and, but, for, nor, or, so,* and *yet*) link equivalent sentence elements.

John Kander **and** Fred Ebb collaborated on several major musicals. [Conjunction links two subjects.]

The burglars gained access to the vault **yet** left its contents intact. [Conjunction links two verbs.]

He fought unenthusiastically **but** skillfully. [Conjunction links two modifiers.]

I will not do it, **nor** will I recommend anyone else who might. [Conjunction links two clauses.]

Correlative conjunctions, such as *both . . . and, either . . . or, neither . . . nor,* and *not only . . . but also,* work in pairs and also link equivalent sentence elements.

Both Senator Robins **and** Representative Hershell received contributions from the tobacco industry. [Conjunction links two subjects.]

Marion will go **either** to Butler University to study pharmacology **or** to Indiana University to study dentistry. [Conjunction links two modifiers.]

Subordinating conjunctions, such as *after, although, because, even if, so that, until,* and *when,* join clauses but subordinate one clause to another. The subordinate clause, introduced by the subordinating conjunction, can be positioned at the beginning, in the middle, or at the end of a sentence.

Because she was outspoken on the subject of women writers, Virginia Woolf has become a central figure in feminist criticism. [Subordinate clause first.]

Virginia Woolf has become a central figure in feminist criticism **because she was outspoken on the subject of women writers.** [Subordinate clause last.]

Virginia Woolf, **because she was outspoken on the subject of women writers,** has become a central figure in feminist criticism. [Subordinate clause embedded.]

■ *Exercise*

Identify each of the underlined words or phrases as a subject (S), verb (V), object (O), complement (C), modifier (M), or conjunction (Conj). (Consider proper names as single elements but consider all other words separately.)

1. Although the 1981 baseball strike lasted seven weeks, the 1985 baseball strike lasted only three days.

2. Cadets at West Point are considered members of the regular Army.

3. Jason sent me an application for Duke University, in hopes that I too would apply for admission.

4. In 1967, a fire <u>aboard</u> Apollo 1 killed <u>Virgil Grissom</u>, <u>Edward White</u>, <u>and</u> <u>Roger Chaffee</u>.

5. <u>General Washington</u> <u>commissioned</u> seven <u>ships</u> to fight against the <u>British</u> navy.

6. <u>Universal Studios</u> <u>and</u> 20th Century-Fox <u>produced</u> the <u>five</u> motion pictures with the highest revenues.

7. <u>The Iliad and The Odyssey</u>, composed by the <u>Greek</u> poet Homer, are <u>main-</u> <u>stays</u> of <u>most humanities</u> curricula.

8. <u>Thomas à Becket</u> <u>was</u> the <u>archbishop of Canterbury</u> during the reign of Henry II.

9. <u>In spite of</u> recent declines in sales, <u>General Motors</u>, <u>Ford</u>, <u>and Chrysler</u> are still among the fifteen largest corporations in the United States.

10. <u>U.S. Grant</u> <u>was</u> an effective <u>general</u> <u>but</u> an <u>ineffectual president</u>.

2 RECOGNIZE BASIC SENTENCE PATTERNS

There are four basic sentence patterns: simple, compound, complex, and compound-complex.

2a

A *simple sentence* contains one independent clause, that is, a subject and verb that can stand alone as a grammatically complete sentence.

The subject and verb of a simple sentence may appear in compound form. A simple sentence need not be simplistic, but it does present a single idea. In fact, simple sentences, because they present ideas clearly, are useful for creating emphasis.

    S   V   C
Hurricanes are frightening.

    S   S   V
Hurricanes and other tropical storms are both frightening and dangerous. [Though more complicated than the preceding example, this is a simple sentence because it has one (compound) subject, and one predicate; the sentence also includes one compound complement.]

2b

A *compound sentence* contains two or more independent clauses joined by a comma and a coordinating conjunction *(and, but, for, nor, or, so, and yet)* or by a semicolon alone.

A compound sentence presents a balanced relationship between the clauses that are joined, thus emphasizing that the ideas in the sentence are of equal importance.

> We moved to the Gulf coast to escape the cold Ohio winters, but then we were terrified by tropical storms. [Two independent clauses joined by a conjunction.]

> We moved to Florida in 1978; however, we stayed only five years and then returned to Ohio. [Two independent clauses joined by a semicolon and a conjunctive adverb.]

2c

A *complex sentence* contains at least two clauses: one independent and one or more subordinate clauses.

Subordinating conjunctions, such as *although, because, since, when,* and *while,* and relative pronouns, such as *that, what, which* and *who,* join the clauses in a complex sentence. One clause, and thus the interrelationship of the ideas in the sentence, is emphasized over the others.

> S V   S   V   C
> Although it rains in the midmornings, the afternoons are generally sunny.
> [Subordinate clause, then independent clause.]

> S   V   O   S   V ┌──────O──────
> We enjoyed our stay in Florida, even though we knew that we could not
> ──────┐
> remain. [Independent clause, then subordinate clause.]

2d

A *compound-complex sentence* contains three or more clauses—at least two independent clauses and one subordinate clause.

A compound-complex sentence establishes a complicated relationship among a series of ideas.

> S   V   S   V   O
> While we lived in Florida, we survived four minor hurricanes without injuries or
> S   V   O
> property damage, but we never developed the nonchalance of native Floridians.
> [Subordinate clause, then independent clause, then another independent clause.]

███████ *Exercise*

Identify the following sentences as simple, compound, complex, or compound complex.

1. Lightning is a discharge of electricity between two clouds or between a cloud and the Earth.

2. Because of deaths during the war with the Soviet Union and because of massive emigration to Iran and Pakistan, Afghanistan's population has shrunk by one third in the last decade.

3. According to 1985 statistics, Northern Ireland has the highest unemployment rate in Europe.

4. The original purpose of the Crusades was to take Christianity to the non-Christian "infidels," but the holy wars also served to enrich trade and the arts in Europe.

5. Although the 1986 Tax Reform Act is supposed to be revenue neutral, it provides too many loopholes for selected businesses.

6. Mongolia is located in eastern Asia, between Siberia and China, and is slightly larger than Alaska.

7. After his notorious raid on Harpers Ferry in 1859, John Brown was captured and hanged; in the years that followed, his name became a symbol of ineffectual militant protest.

8. Although *bona fide* means "in good faith" in Latin, it is commonly used today to mean "genuine."

9. Because the costs of American materials and labor are high, sales of American-made shoes have plummeted, and sales of imports from Brazil and South Korea have risen.

10. Most Americans assume that the U.S. Navy is our oldest maritime service, yet the U.S. Coast Guard was established in 1790, eight years before the Navy.

3

EXPAND AND VARY SENTENCE PATTERNS

As writers move from drafting sentences to revising them, they make decisions about diction, about the placement of phrases and clauses, and about the structural patterns of sentences. The larger issue — sentence patterns — is important because sentence construction (and reconstruction) often determines how effectively writing communicates with readers.

3a

Use Coordination

When isolated, successive sentences present ideas that together establish an important, parallel relationship, those sentences can be effectively combined by coordination. *Coordination* is the joining of simple sentences to form compound sentences, and it is also the meshing or combining of related sentence elements through pairing or seriation.

President Reagan initially denied that the United States had traded arms for hos-

tages. *but t*The Tower Commission subsequently revealed otherwise. [Two sentences effectively joined by a coordinating conjunction.]

Oliver North's *and John Poindexter's* activities were largely unsupervised. ~~John Poindexter's activities~~

~~were also unsupervised.~~ [Combining the sentences — by using a compound subject — improves the emphasis.]

3b

Use Subordination

When isolated, successive sentences present related ideas. The sentences can be combined through subordination. *Subordination* is the joining of simple sentences to form complex or compound-complex sentences, but it is also the embedding of words, phrases, or clauses in the structure of the most important sentence.

Mark Chagall ~~was~~ a native of Russia. He emigrated to France. *Where* ~~While in France,~~ he

produced vibrant and dreamlike paintings. [Three independent clauses compressed into one complex sentence.]

Because Chagall's images are often childlike. *u*Undiscerning observers sometimes find his

work simplistic. *but o*Observant critics see an ingenious use of childhood perception in

his work. [Three independent clauses reworked into a compound-complex sentence that reveals a cause and effect relationship among the ideas.]

Subordination is a useful means of indicating emphasis and bringing variety to sentences. Consider subordination the groundwork of a mature and effective sentence style and work to use it to achieve your purpose.

▮▮▮▮ *Exercise*

Use coordination and subordination to combine the following pairs of sentences.

1. Four men from the United States have won the Olympic figure skating title. Only one, Dick Button, won the title twice.

2. *Amadeus* popularized the works of Mozart. The plot of the play and film is historically inaccurate.

3. Ten percent of home-study lawyers pass the California Bar Exam. Sixty percent of law-school-trained lawyers pass.

4. Alfred Smith was the first Catholic to run for president. He lost by a wide margin to Herbert Hoover in 1928.

5. Trademarks are usually specialized symbols, products, or company names. They can also be individual words and letters.

▮▮▮▮ *Exercise*

The following paragraph contains simple sentences only. Use coordination and subordination to combine sentences and produce an effective and varied paragraph.

The Beatles were the most successful pop group of all time. They began playing in Liverpool, England. The group had four members. They were John Lennon, Paul McCartney, George Harrison, and Ringo Starr. Their early music was characterized by a simple rhythm-and-blues style. It also had simple harmonies and lyrics. The Beatles' early hits included "She Loves You," "Please Please Me," and "I Want to Hold Your Hand." These simple songs attracted worldwide interest. John Lennon and Paul McCartney later wrote complicated songs. The lyrics became imaginative and philosophical. The music itself became varied. It was also complex. They experimented with new instruments and recording equipment. Their later work included the albums *Rubber Soul, The White Album,* and *Abbey Road.* The Beatles' most sophisticated work was the album *Sergeant Pepper's Lonely Hearts Club Band.* It contained some of the group's most memorable songs. Each member of the band developed separate interests. The group disbanded in 1980.

▬▬▬ *Understanding Sentence Elements: Review Exercises*

In the following paragraph, label each of the underlined elements as a subject (S), Verb (V), Object (O), complement (C), modifier (M), or conjunction (Conj).

Our Sun is a sphere of superheated gas. Hydrogen atoms at its core fuse, creating the atomic reactions that produce both light and energy. Scientists estimate that the temperature of the Sun's core is twenty million degrees centigrade, while the surface temperature is approximately six thousand degrees centigrade. The diameter of the Sun is roughly 850,000 miles. These figures suggest that our Sun is neither a hot nor a large star when compared with others in our solar system.

The following paragraph contains simple sentences only. Use coordination and subordination to combine sentences to make this paragraph effective. Then label your sentences by type: simple (S), compound (C), Complex (CX), and compound-complex (CCX).

The Black Death devastated Europe between 1348 and 1666. The disease was brought to Europe through Italy. Traders carried it from the Black Sea area. The epidemic of 1348 killed one-fourth of the population of Europe. The disease was carried by fleas. This was unknown in the fourteenth and seventeenth centuries. The fleas lived on rats. The disease raged, subsided, and reemerged for three hundred years. The worst epidemic in England occurred in 1665. Entire towns and villages were wiped out. London's population decreased by one-tenth. The population was approximately 450,000. People were terrified. They tried all kinds of cures. They didn't understand the nature of the disease. The cures failed. Samuel Pepys wrote about the Black Death in his *Diary.* Daniel DeFoe wrote a fictional account titled *Journal of the Plague Year.*

WRITING LOGICAL AND EFFECTIVE SENTENCES

4   ## SENTENCE SENSE (SS)

Creating a uniform impression in your writing is important, especially at the sentence level, because readers need consistency to understand fully your ideas. Unnecessary shifts in point of view, tense, mood, and voice are distracting and may be confusing.

4a   ### Use a consistent point of view.

Point of view is the standpoint from which you present information. If you select the *first-person* point of view, then your subject *(I, we)* is the speaker. If you select the **second person,** then your subject *(you)* is being spoken to. If you select the **third person,** then your subject *(he, she, it, they)* is being spoken about. Selecting the appropriate point of view and maintaining it are important for clarity.

Swimming instructors must be patient if they work with children. ~~You~~ *They* must acknowledge that some children have never swum before, and ~~you~~ *they* must acclimate children to the water. Instructors must be willing to work slowly and teach skills gradually. [Elimination of shift from third person *(swimming instructors)* to second person *(you)*.]

4b   ### Use verb tenses consistently.

Verb tenses signal chronological relationships among ideas. Unnecessary shifts in tense will confuse your readers.

After filling out a tentative class schedule, the student goes to see his or her advisor for approval. The advisor examine*s* the schedule to see that requirements ~~were~~ *are* met, and the student submits the signed computer form. [The shifts from present tense *(goes)* to the past tense *(examined)* to the present tense *(submits)* make these sentences confusing. It is unclear whether they describe a typically repeated sequence, for which present tense would be appropriate, or a completed sequence, for which past tense would be suitable.]

4c   **Use a consistent mood.**

Three moods indicate how you view the actions or conditions you are describing in your sentences. The ***indicative mood*** is used to make statements of fact or opinion and ask questions. The ***imperative mood*** is used to express commands. The ***subjunctive mood*** is used to indicate doubt, conditional situations, statements contrary to fact, and wishes. In most writing, mood should remain consistent.

He suggested that people

Betting on a horse race, Mr. McMillan explained, can be risky. ~~Bet~~ only money that

they

~~you~~ can afford to lose. [Elimination of confusing shift from the indicative to the imperative. The first sentence *described;* the second *advised.*]

4d   **Use a consistent voice.**

Active voice and passive voice create different kinds of emphasis in sentences. They should not be mixed in successive sentences that describe the same subject. (See Section 5, "Active and Passive Sentences.")

The research assistants carefully compile the results of the questionnaire. First, they

they record

record the sex, age, race, and religion of each respondent. Then, profession,

income, and education ~~were recorded~~. Next, the assistants note responses to indi-

vidual interpretive questions. [Elimination of shift into the passive voice.]

████ *Exercise*

Revise each set of sentences so that it constitutes a paragraph that is consistent in mood, point of view, voice, and tense.

1. Department of Energy spokespersons have suggested that Americans save
   energy in small but important ways. They suggest walking rather than driving, coordinating short trips, and driving at slower speeds. Turn off lights
   when you are not in a room. Wash only full loads. Lower your thermostats.

2. The U.S. Fish and Wildlife Service now restricts bird hunting in most areas.
   Hunters can no longer use lead shot, since it poisons birds that are wounded
   but not killed. You must also restrict hunting to specified seasons, and

hunters must limit the number of birds they kill. Penalties are also severe if you are caught violating these protective laws.

3. The National Forest Service has made timberlands available to private logging companies. New logging roads are built, destroying the forest floor. Trees are removed, and fish and wildlife are threatened. Irreparable damage is being done.

4. In the last few years, industrial pollution of water has declined. The Clean Water Act has given government agencies the right to assign stiff fines to plants and foundries in violation of existing pollution standards. These companies then had to correct the problem or risk further fines. Most industries adapted to these procedures.

5. Spokespeople for these agencies and services address important issues at arranged press conferences. Facts are given, and violators are identified. The American people are given information by these spokespeople to help them understand these national concerns.

5 ACTIVE AND PASSIVE SENTENCES (ACT & PAS)

Active sentences emphasize the people or things responsible for actions and conditions. Passive sentences focus on people or things that are acted upon. What would be the object in an active sentence is used as the subject in a passive sentence, and a form of the verb *to be* is used with the main verb. As a result, passive sentences are always slightly longer than active sentences. Although most readers and writers prefer active sentences, you should select the sentence pattern that most effectively matches your purpose.

> *Active:* Congress approved a multi-billion-dollar highway improvement bill. [Congress is emphasized.]
>
> *Passive:* A multi-billion-dollar highway improvement bill was approved by Congress. [The highway improvement bill is emphasized.]

5a Use Active Sentences Most of the Time

Use active sentences to indicate who takes responsibility for actions and events.

Dr. Taylor misdiagnosed and mistreated Jeremy's respiratory problem.

Use active sentences to create emphasis.

A tornado in Texas destroyed property worth over $50 million.

Use active sentences for economy of expression.

Active: Lionel Richie sang the national anthem at the opening game of the season. [13 words]

Passive: The national anthem was sung by Lionel Richie at the opening game of the season. [15 words]

5b Use Passive Sentences Selectively

Use passive sentences when the people who are responsible for actions are not known.

The superintendent's window was broken sometime over the weekend.

Use passive sentences to emphasize the receiver of the action instead of those responsible.

Van Gogh's *Sunflowers* was sold for $39.9 million.

Use passive sentences to emphasize actions that are more important than specific people who might be reponsible.

Lasers are currently being used to treat medical problems as diverse as cancer, cataracts, and varicose veins.

▓▓▓ *Exercise*

The following sentences are written in the passive voice. Rewrite those that would be more effective in the active voice.

1. Twenty people were killed when a car bomb exploded in Teheran.

2. Over $800 million had been deposited in personal Swiss bank accounts by Ferdinand Marcos, the ousted president of the Philippines.

3. Sanctions against South Africa were approved by the Senate, despite President Reagan's urgings.

4. Details of the nuclear accident at Chernobyl were withheld by Soviet officials for several days.

5. Franklin D. Roosevelt was elected to four terms as president.

6   **MAINTAIN PARALLELISM AMONG SENTENCE ELEMENTS**

Parallelism in a sentence requires that similar ideas be presented in similar form and that elements that are similar in function appear in similar grammatical form. Parallelism is an important principle of both grammar and style.

Grammatically, sentence elements linked by coordinating or correlative conjunctions should be similar in form: a clause should be followed by a clause, a phrase by a phrase, a noun by a noun, a verb by a verb of the same tense, and so on. (See Consistent Verb Tenses, Section 4b.)

Stylistically, parallelism creates balance and emphasis. It can, therefore, be used to create desired effects.

6a   **Maintain parallelism with coordinating conjunctions.**

The evangelist ended the service with a hymn and ~~calling on~~ *with a call for* sinners to repent. [Two prepositional phrases separated by a coordinating conjunction.]

6b   **Maintain parallelism with correlative conjunctions.**

The owners of VCRs can either tape films from network broadcasts or ~~can~~ rent films from video clubs. [Correlative conjunctions followed by two verbs without auxiliaries.]

6c   **Repeat key words to clarify a parallel construction.**

The commission has the power to investigate, *to* conciliate, *to* hold hearings, *to* subpoena witnesses, *to* issue cease-and-desist commands, *to* order reinstatements, and *to* direct hiring. [Parallelism emphasized by repeating *to*.]

■■■ *Exercise*

Revise the following sentences to eliminate faulty parallelism.

1. The narrator of *Invisible Man* was idealistic, intelligent, and tried to advance the cause of black people.

2. Holden Caufield, the main character of *Catcher in the Rye,* rejected hypocrisy in other people but was ignoring his own hypocrisy.

3. Thornton Wilder won Pulitzer prizes not only for his plays *Our Town* and *The Skin of Our Teeth* but also he won for his novel *The Bridge of San Luis Rey.*

4. The stories of Flannery O'Connor allow readers to examine unusual characters, to explore psychological motivations, and consider macabre situations.

5. Willy Loman could neither understand his own problems nor could he accept the help of friends.

7   WORD ORDER (WD OR)

7a   Inversions

The common order of words in sentences can be briefly summarized:

- Subjects precede verbs.
- Verbs precede objects or complements.
- Indirect objects precede direct objects.
- Adjectives precede the words they modify.
- Adverbs usually follow verbs they modify, but they precede adjectives or other adverbs.
- Prepositional phrases follow the words they modify.
- Independent clauses often precede subordinate clauses, although three variations are common: (1) clauses used as adjectives follow the words they modify, (2) clauses used as adverbs often precede the independent clause, and (3) clauses used as nouns occupy the subject or object positions.
- Closely related material is best kept as close together as possible.

Although these principles usually govern word order in sentences, any element of a sentence may be moved to create emphasis or interest. Variations of common word order, however, should produce neither awkward nor unidiomatic writing.

Common Order
The chancellor **quickly and superficially** responded to the interviewer's questions.

Inverted Order
Quickly and superficially, the Chancellor responded to the interviewer's questions.

The team doesn't stand a chance **without Terrance.**	**Without Terrance,** the team doesn't stand a chance.
The computer terminals were installed **at last.**	**At last,** the computer terminals were installed.
The company will pay relocation expenses **if employees are transferred.**	**If employees are transferred,** the company will pay relocation expenses.
Jessica said, "I can only attend the Art Institute if I receive a scholarship."	"I can only attend the Art Institute," **Jessica said,** "if I receive a scholarship."

7b   Emphatic Order

To achieve a desired effect in a sentence, writers can vary the location of key information. In a typical sentence, information placed near the beginning or end will be emphasized. Placing important information in an independent clause, instead of in a subordinate clause or phrase, strengthens emphasis.

Place important information first or last; do not bury it in the middle.

Unemphatic: On 14 April 1865, **Abraham Lincoln was shot** at Ford's Theatre.

Emphatic: **Abraham Lincoln was shot** at Ford's Theatre on 14 April 1865.

Emphatic: At Ford's Theatre on 14 April 1865, **Abraham Lincoln was shot.**

Unemphatic: That novel, **as far as I know,** was my biggest commercial failure.

Emphatic: **As far as I know,** that novel was my biggest commercial failure.

Unemphatic: She is innocent, **in my opinion.**

Emphatic: **In my opinion,** she is innocent.

Place key information in independent clauses, not in subordinate clauses or phrases.

Unemphatic: He fell from the roof, **thus breaking his neck.**

Emphatic: He fell from the roof and **broke his neck.**

▬▬ *Exercise*

Revise the following sentences to improve awkward or unemphatic word order.

1. The Supreme Court refused to consider the appeal, according to the late news last night.

2. The major evidence had been acquired during a search without a proper warrant, thus resulting in a dismissal of the case.

3. The evidence shows, the prosecuting attorney suggested, that Marshall Tireman is guilty of stealing industrial secrets.

4. The judge agreed to admit the video tape as evidence after the defense attorney made a special appeal.

5. The lawyers, even, had not expected such a large settlement in the case.

8   POSITION MODIFIERS CAREFULLY (MOD)

A modifier must clearly relate to a word in a sentence and explain, describe, define, or limit the word to which it relates. When a modifier is not positioned properly, the modification can be both awkward and confusing.

8a   **Long modifiers should not separate a subject and verb or a verb and its complement.**

Although modifiers may be placed between a subject and verb or between a verb and its complement, such positioning often makes a sentence difficult to read and interpret. Reposition the modifiers so that they do not break the flow of the sentence.

The renovation, *because of fund-raising activities and because of competitive bidding by major contracting firms,* was delayed.

The final bid was, *even though it was thousands lower than the initial bids,* still too high.

8b   **Avoid dangling modifiers.**

Opening modifiers that do not modify the subject of a sentence are said to dangle — hence, the name **dangling modifier.** To correct such an error, either revise the independent clause so that the introductory phrase can logically modify the subject, or revise the introductory phrase to make it a subordinate clause.

To qualify for the award, ~~the committee requires that~~ candidates *must* have sixty class hours and a 3.50 GPA.

While *I was* waiting for my date in the lobby, two men in tuxedos got into a violent argument.

8c   **Avoid squinting modifiers.**

A *squinting modifier* seems to modify the word before it *and* the word after it. Reposition the modifier to clarify the meaning, or use *that* to eliminate confusion.

that
The reporter said before noon she would finish the article.

The reporter said before noon she would finish the article.

8d   **Avoid split infinitives.**

A *split infinitive* occurs when a modifier falls between *to* and the primary verb. Writers and readers disagree about whether split infinitives are grammatically or stylistically acceptable. To be on the safe side, reposition the modifier.

Darren began to furiously pack his luggage to try to make the nine o'clock flight.

███ *Exercise*

Eliminate ambiguities in the following sentences by changing the position of misleading modifiers.

1. At one time his parents said he had been an engineering student.

2. The stage set, based on original paintings and engravings from the eighteenth century, was breathtaking.

3. The car was in the garage that he wrecked.

4. Marc promised on his way home to pick me up.

5. They talked about going on a second honeymoon but never did.

6. My brother hung the painting in the hallway that I gave him for his birthday.

7. The short story was, because of its convoluted sentences and obscure imagery, almost incomprehensible.

8. There is a panel discussion tonight about drug addiction in the student lounge.

9. I thought of writing often but never did.

10. Reading the personal letters of famous people is a way to usefully and completely understand their reactions to public situations.

9

COMPARISONS (CP)

When you include comparisons in your sentences, consider your diction carefully to ensure that the ideas are clear and complete.

9a

Include all the words needed to make a comparison clear and complete.

Flying to Chicago is more convenient than ~~taking~~ a train.

Levis are more popular than any ~~other~~ jeans.

9b

Do not write an implied comparison.

An implied comparison presents only part of the necessary context. The words *better, less, more,* and *worse* and words formed with the suffix *-er* signal the need for fully stated comparisons; use *than* and explain the comparison completely.

The house on Elm Street is better suited to our needs ~~than the others we've seen~~.

The orchestra's performance of Beethoven's *Ninth Symphony* was much worse ~~than expected~~.

▆▆▆ *Exercise*

Revise the following sentences to make their comparisons logical and complete.

1. Once Carla began taking her medication regularly, she felt much better.

2. Having had a two-hour practice session, the students were no longer as confused.

3. Taking a taxi or riding the subway is certainly more convenient than a car.

4. Reeboks are more popular than any tennis shoe.

5. Revising a paper is much easier using a word processor.

10

CONCISENESS (CON)

Writing that is concise expresses ideas in as few words as possible; it is free of needless repetition and useless words. To make your writing concise, eliminate words, phrases, and clauses that do not further your purpose.

10a   **Do not repeat words needlessly.**

~~The car~~ *We* were looking for ~~was~~ a car for highway travel. [12 words reduced to 9.]

10b   **Do not repeat ideas that are already understood.**

~~The~~ frown ~~on~~ Todd's ~~face~~ suggested that he was ~~depressingly~~ saddened by his interview. [14 words reduced to 10.]

10c   **Eliminate expletive constructions whenever possible.**

Expletives, such as *it is, there is, there are, here is, here are,* and so on, add words to sentences without clarifying meaning.

~~There were~~ *J* three cars *were* involved in the accident. [8 words reduced to 7.]

10d   **Write active sentences whenever possible. (See Active and Passive Sentences, Section 5.)**

My Aunt Ruth made The prize-winning quilt ~~was made by my Aunt Ruth.~~ [9 words reduced to 7.]

10e   **Replace wordy phrases with brief expressions.**

I ~~am of the opinion~~ *think* that we should resubmit the insurance claim. [12 words reduced to 9.]

10f   **Replace forms of the verb *to be* with stronger verbs.**

Counselors ~~are responsible for~~ *must* complet~~ing~~ *e* the transcript portion of the applications. [11 words reduced to 9.]

10g   **When possible, replace nonrestrictive clauses with appositives.**

Nonrestrictive clauses, clauses that provide useful or interesting but nonessential information, can often be replaced with appositives, simple words or phrases that provide definitions for other words or phrases in a sentence. To save words and tighten and clarify your writing, consider using appositives in place of nonrestrictive clauses.

Sandra Day O'Connor, ~~who was~~ the first woman appointed to the Supreme Court, assumed her duties in August 1981. [19 words reduced to 17 through substitution of an appositive for a nonrestrictive clause.]

■■■■ *Exercise*

Make the following wordy sentences concise. Note the number of words saved through revision.

1. There should be two waiters to serve every ten people at the banquet, or there will be unnecessary delays occurring. [20 words reduced to _____.]

2. After the violent eruption of Nevada de Ruiz, relief agencies joined together in their efforts to help the unfortunate victims. [20 words reduced to _____.]

3. Wynton Marsalis, who plays both classical and jazz trumpet, scorns pop music. [12 words reduced to _____.]

4. At this point in time, we should prepare for spring floods, in the event that the Wabash River will crest as it did last year. [25 words reduced to _____.]

5. Finalists in the oratory competition will be evaluated by seven judges. [11 words reduced to _____.]

6. The original prototype for the Ford Mustang is on display at the Ford Museum in Detroit, Michigan. [17 words reduced to _____.]

7. A house made of brick is more costly but more maintenance free than a house made of wood. [18 words reduced to _____.]

8. Secret Service agents are responsible for protecting the current president, past presidents, and their families. [16 words reduced to _____.]

9. In the humble opinion of this writer, Academy Awards present indications of popularity rather than quality. [16 words reduced to _____.]

10. The real truth is that there is no money available to support and maintain the scholarship. [16 words reduced to _____.]

▰▰ *Exercise*

Revise the following paragraph to make it concise. Try a number of strategies and notice how much the paragraph improves when you eliminate unnecessary words and bloated phrases.

Prior to beginning the search for gainful employment, gather together necessary and essential information and materials. Assemble a list of your experiences in educational institutions and in the work place and be sure to include the months or years involved in each situation. Prepare a résumé that includes facts and information about yourself, personally, and about yourself, academically and professionally. Make sure that there are clear sections in the résumé to cover each of these important and crucial topics. Proofread the final copy of the résumé in order to be aware of and correct any errors or mistakes. Then photocopy the résumé so that you still have at your disposal a copy of the résumé for future reference.

▰▰ *Writing Logical and Effective Sentences: Review Exercises*

Revise the following sets of sentences to create logical and effective sentences. Identify the kinds of problems that required correction.

1. Human figures were elongated and were rendered in sallow yellows and greens by the Spanish artist El Greco.

2. Cubism is, with its emphasis on presenting the surfaces of all objects — both living and inanimate — in abstract geometric forms, alien to many people's artistic sensibilities.

3. Although his work was not popular during his lifetime, Van Gogh paints with bold colors and exaggerated forms. Modern collectors have valued his work since his death.

4. It is clear that there are only a few major pop art paintings of lasting aesthetic value. There are many others that are simply cultural curiosities.

5. When one sees the work of Rembrandt in a well-lighted gallery, you will be

413

impressed by the rich texture of his work and the subtle variations in his gold and brown tones.

6. Neoclassical artists of the eighteenth century objected to the visual excesses of Baroque and Rococo art and imitate the symmetry and simple forms of Greek and Roman art.

7. Picasso's versatility as a sculptor is evident in his ability to skillfully and ingeniously use "junk" in his welded works.

8. Da Vinci, Raphael, David, Rembrandt, Van Gogh, Monet, and Picasso would surely be included if one was to make a list of major European painters.

9. New York's Chrysler Building — with its use of zigzag forms, angular metal ornamentation, and strong vertical lines — is an exemplary model of Art Deco architecture.

10. Up until the middle of our current century, most prominent and important painters and sculptors from the United States of America trained and went to school in the countries of Europe.

11. Once, painters worked almost exclusively on wood panels or plaster walls. Then stretched canvas was used. Today, wood is being used again by many artists.

12. The Louvre in Paris houses more major works of art than any museum.

13. Prior to viewing a major exhibition, I would offer encouragement to inexperienced and untrained viewers to peruse or skim the catalogue prepared to accompany the exhibition.

14. To create what he described as an unconscious interpretation of reality, paint was splattered on canvas by Jackson Pollock.

15. Stressing the dreamlike, the unusual, and the bizarre, we found Surrealistic art unsettling.

WRITING GRAMMATICAL SENTENCES

11   **ELIMINATE SENTENCE FRAGMENTS (FRAG)**

A *sentence fragment* is a group of words presented as if it were a complete sentence — with a capital letter at the beginning and a period at the end. A sentence fragment, however, lacks a subject or a verb or both and does not express a complete thought. Eliminate sentence fragments in one of four ways, depending on the type of fragment.

11a   **Add a subject when necessary, or join the fragment to another sentence.**

Charlie Chaplin was a multitalented man. *He w*Wrote, directed, and starred in his own

films.

Charlie Chaplin *was* a multitalented man, *w*Wrote, directed, and starred in his own

films.

11b   **Add a verb when necessary, or join the fragment to another sentence.**

Grigori Rasputin, a Russian monk in Czar Nicholas' court, He was assassinated in

1916.

11c   **Omit the subordinating conjunction, or connect the fragment to an independent clause.**

Mother Teresa
~~Because she~~ tirelessly helped the poor in Calcutta. *She* ~~Mother Teresa~~ was awarded the

Nobel Peace Prize in 1979.

11d   **Attach a phrase to a related sentence.**

Leonard Bernstein received wide acclaim on Broadway, *n*Notably for the score of

West Side Story.

415

████ *Exercise*

Eliminate each fragment by making it into a sentence or by combining it with a sentence.

1. The *Robert E. Lee,* a renovated river boat that now operates as a restaurant. It is an excellent place to eat.

2. We made our way up the mountain trail with much difficulty. Slipping on rocks and snagging our clothes in the underbrush.

3. Chad has only one ambition. To play the violin in a major symphony.

4. Many people dread one part of medical exams more than any other. Having a blood sample taken.

5. In a political speech, candidates should appeal to the entire audience. Not just to those who believe as they do.

6. Even though the cost of automobile insurance is high. Repairs on damaged cars are even more exorbitant.

7. Having come this far. We must see the matter through.

8. Whatever challenge the office presents. I believe our new member of Congress will meet it successfully.

9. When the chairperson stated, "I will not compromise on any issue on which I have taken a stand." I began to question her judgment.

10. Rita Moreno has won all major performance awards. An Oscar, an Emmy, a Grammy, and a Tony.

12   ELIMINATE FUSED SENTENCES (FS) AND COMMA SPLICES (CS)

A *fused sentence* (also called a *run-on sentence*) results when no punctuation or coordinating conjunction separates two or more independent clauses. A *comma splice* results when two or more independent clauses are joined with only a comma. Eliminate these sentence errors in one of four ways.

12a   **Use a period to separate independent clauses, forming two sentences.**

Lorraine Hansberry was the first black female playwright of importance, she wrote

A Raisin in the Sun.

12b   **Use a semicolon to separate independent clauses and form a compound sentence.**

Through flying, Charles Lindbergh gained his notoriety; Amelia Earhart lost her life.

12c   **Insert a coordinating conjunction between independent clauses to form a compound sentence.**

Helen Keller was both blind and deaf, *but* she was a skillful author and lecturer.

12d   **Use a subordinating conjunction to put the less important idea in a subordinate clause and form a complex sentence.**

although
Paul Revere is known to most people as a Revolutionary War patriot he is known to

collectors as a silversmith and engraver.

Be especially sensitive to the use of conjunctive adverbs, such as *consequently, however, moreover, nevértheless,* and *therefore.* They do not link clauses grammatically. Misinterpreting their function in sentences is a common cause of comma splices.

Oscar Wilde fancied himself a poet and critic, however, he is most remembered as a playwright and wit.

████ *Exercise*

Correct the following fused sentences and comma splices.

1. The comma splice can confuse readers, it is usually less troublesome, however, than the fused sentence.

2. Members of the Drama Guild have rehearsed carefully for tonight's show, the director feels certain it will be a success.

3. The war is over the fighting is not.

4. The air traffic controller made the best decision he could at the time, looking back, he saw what he should have done differently.

5. It is too late to sign up for the proficiency exam this term, however, students can sign up for next term's exam.

6. Pay attention to the instructions you must follow them exactly.

7. Much has been done the Civil Liberties Union believes that much more needs to be done.

8. Stockholders don't have to liquidate their assets this week, all they need to do is sign papers of intent.

9. Clean-up is scheduled for Monday, Tuesday, and Wednesday the plant closes on Friday.

10. No conclusive evidence has been uncovered, the commissioners will meet again tomorrow.

13   AGREEMENT (AGR)

Agreement in grammar refers to the correspondence of key sentence elements in number, person, and gender. Two kinds of agreement are grammatically important in most sentences: subject-verb agreement and pronoun-antecedent agreement.

13a   Subject-Verb Agreement

In simplest terms, a singular subject requires a singular verb, and a plural subject requires a plural verb. A number of troublesome constructions can cause confusion, however, and require consideration.

When subjects are joined by *and*, use a plural verb.

Although each of the subjects may be singular, the compounding makes a plural verb necessary.

O'Connor and Reinquest **speak** articulately for the dissenters. [Plural verb with compound subject.]

A fool and his money **are** soon parted.

When subjects are joined by *or, nor, but, either . . . or, neither . . . nor, or not only . . . but also,* use a verb that agrees with the subject that is nearer to the verb.

Either Weixlmann or Stein **is** my choice for president. [Singular verb with two singular subjects.]

The coach or the co-captains **supervise** the practices each day. [Plural verb agrees with *co-captains,* the nearer subject.]

Neither Lewis, his two partners, nor their lawyers **were** at the press conference. [Plural verb agrees with *lawyers,* the nearer subject.]

Either Jean or you **are** to accept the award for the entire cast. [Verb agrees with *you,* the nearer subject.]

When this rule produces an awkward though correct sentence, consider revising the sentence.

You ~~or he is~~ *two are* the leading contender*s*.

When a subject is followed by a phrase containing a noun that differs in number or person from the subject, use a verb that agrees with the subject, not with the noun in the phrase.

The attitude of these men **is** decidedly hostile. [Singular verb agrees with *attitude,* the singular subject.]

The ballots with her name **have** been recalled. [Plural verb agrees with *ballots,* the plural subject.]

When an indefinite pronoun, such as *anybody, anyone, each, either, everybody, neither, nobody,* and *someone,* is used as a subject, use a singular verb.

Ultimately, someone **has** to accept responsibility.

Anybody who wants to **has** the right to attend the hearing.

Everyone **has** the same chance.

When a collective noun is used as a subject, use a singular verb or a plural verb to clarify the meaning.

When a collective noun emphasizes the unity of a group, use a singular verb. When a collective noun emphasizes group members as individuals, use a plural verb.

The clergy **is** grossly underpaid. [Singular verb because whole group is meant.]

The clergy **are** using their pulpits to speak out against oppression. [Plural verb because individual members are meant.]

When an expletive construction, such as *here is, here are, there is,* and *there are,* is used as both subject and verb, match the verb to the noun that follows.

Here **is** your receipt. [Singular verb with singular noun *receipt.*]

Here **are** the copies you requested. [Plural verb with plural noun *copies.*]

There **is** no excuse for such behavior. [Singular verb with singular noun *excuse.*]

There **are** several solutions to the city's problems. [Plural verb with plural noun *solutions.*]

The verb in a relative clause introduced by *who, which,* or *that* agrees in number with the pronoun's antecedent.

Jessica is one performer who **acts** with restraint. [Singular verb with singular antecedent *performer.*]

Philip Roth writes books that **illustrate** the absurdities of modern life. [Plural verb with plural antecedent *books.*]

When a compound subject is preceded by *each* or *every,* use a singular verb.

Each and *every* indicate that persons or things are being considered individually.

Each boy and girl **takes** shop and home economics.

Every basket of peaches and flat of strawberries **was** sold.

When a subject is followed by a predicate noun that differs in number from the subject, the verb agrees with the subject, not with the complement.

Although predicate nouns restate the subject of the sentence, their word forms do not always agree in number; that is, the predicate noun and the subject may not be both singular or both plural. Use the subject, not the predicate noun, to determine the appropriate subject-verb agreement.

Her chief source of enjoyment **is** books. [Singular verb with singular subject *source.*]

Books **are** her chief source of enjoyment. [Plural verb with plural subject *books.*]

When a plural noun has a singular meaning, use a singular verb.

Some subjects may initially appear to be plural, although they are singular. *Electronics, mathematics, semantics,* and *geriatrics* appear to be in plural form but are names of individual fields of study. Expressions such as *gin and tonic* and *ham and eggs* are also singular, because they name a single drink and a single dish.

No news **is** good news.

Scotch and soda **is** not as popular as it once was.

When fractions, measurements, money, time, weight, and volume are considered as single units, use singular verbs.

Three days **is** too long to wait.

Jerrid feels that 165 pounds **is** his ideal weight.

Twenty-two percent **is** the accepted rate for credit-card financing.

With titles of individual works, even those containing plural words, use a singular verb.

All the King's Men **is** an enlightening political novel.

Dorothy Parker's "Good Souls" **is** about congenial, often exploited people.

Words used as words take a singular verb.

Amateur athletes is used to describe participants as varied as Little League pitchers and endorsement-rich track-and-field stars.

13b   ## Pronoun-Antecedent Agreement

Pronouns must agree with their antecedents (the nouns or pronouns to which they refer) in number and person. A singular pronoun must be used with a singular antecedent; a plural pronoun must be used with a plural antecedent. (See also Case, Section 14.)

The workers received **their** wages. [Plural third-person pronoun with plural third-person antecedent *workers*.]

The DC 10 changed **its** course and landed at Cincinnati. [Singular third-person pronoun with singular third-person antecedent *DC 10*.]

Singular pronouns must also agree in gender with their antecedents. A masculine pronoun must be used with a masculine antecedent; a feminine pronoun must be used with a feminine antecedent; and a neuter pronoun must be used with a neuter antecedent. (See also Case, Section 14.)

Masculine:	he	him	his	himself
Feminine:	she	her	hers	herself
Neuter:	it	it	its	itself

The generic use of masculine pronouns is no longer universally acceptable. Use both masculine and feminine pronouns when an antecedent could be either male or female *(he or she, his or hers.)* Alternatively, use plural, genderless antecedents and pronouns whenever possible.

Each teacher must submit **his or her** annual report by March 15. [Singular masculine and feminine pronouns with male or female antecedent *teacher*.]

Teachers must submit **their** annual reports by March 15. [Plural pronoun with plural antecedent *teachers*.]

These principles of pronoun-antecedent agreement apply consistently to all situations, but a number of troublesome constructions require special consideration.

When the antecedents *each, either, neither,* and *none* are followed by a phrase that contains a plural noun, use a singular pronoun.

Although the noun in the phrase may be plural, *each, either, neither* and *none* refer to elements individually. Consequently, the pronoun must be singular.

Neither of the boys would accept the responsibility for **his** actions. [Singular pronoun with *neither* as antecedent.]

Either of these women may lose **her** position. [Singular pronoun with *either* as antecedent.]

When *everybody, each, either, everyone, neither, nobody,* and a person are antecedents, use a singular pronoun.

Although in context these words may imply plurality, the word forms are singular and therefore singular pronouns are required. Do not use masculine pronouns generically to refer to these genderless singular antecedents. Use both masculine and feminine forms, or alternatively, substitute plural antecedents and pronouns for the singular forms.

Nobody had **his or her** work completed on time. [Singular pronouns with singular antecedent *nobody.*]

The committee members had not completed **their** work on time. [A plural pronoun with plural antecedent *members.*]

Collective nouns used as antecedents take singular or plural pronouns depending on the meaning of the sentence.

A collective noun that identifies the group as a single unit takes a singular pronoun. A collective noun that identifies the individual members of a group takes a plural pronoun.

The judge reprimanded the jury for **its** disregard of the evidence. [Singular pronoun because reference is to group as a whole.]

At the request of the defense attorney, the jury were polled and **their** individual verdicts recorded. [Plural pronoun because reference is to group members individually.]

When an antecedent is a person, use *who, whom,* or *that* to introduce qualifying phrases or clauses.

This is the architect **who** planned the civic center.

The interior designer **whom** we selected was unavailable.

The landscaper **that** worked on our property has moved.

When an antecedent is an object or concept, use *which* or *that* to introduce qualifying phrases or clauses.

Here is the package **that** she left behind.

The package, **which** she left behind, could not later be found.

When an antecedent is an animal, use *that* to introduce qualifying phrases or clauses.

Secretariat is the horse **that** you're speaking of.

■■■■ *Exercise*

Circle the correct form in parentheses.

1. Neither she nor her sons (was, were) present at the reading of the will.

2. The jury (is, are) expected to reach a verdict before midnight.

3. Each of the children is expected to bring (his, her, his or her, their) own art supplies.

4. The horse (that, who) won the Kentucky Derby went on to win the Preakness and the Belmont.

5. The team lost (its, their) first game of the season, but (it, they) won the next five games.

6. Every one of the actors who auditioned (was, were) exceptionally talented.

7. There (is, are) both food and firewood in the cabin.

8. Students (which, who) maintain grade-point averages of 3.50 or better are eligible for alumni scholarships.

9. None of the applicants presented (himself, herself, himself or herself, themselves) well in the interview.

10. Thirty hours a week (is, are) a heavy work schedule, especially if you are taking two classes.

14   **CASE (CA)**

Case is the form or position of a noun or pronoun that indicates its relation to other words in a sentence. English has three cases: **subjective, objective,** and **possessive.** In general, a noun or pronoun is in the subjective case when it acts as a subject, in the objective case when it acts as an object, and in the possessive case when it modifies a noun as in *"his* bicycle," "the *boy's* dog," *"their* future."

English nouns, pronouns, and adjectives once all showed case by changing their forms. In modern English, word order and idiomatic constructions have largely replaced case endings. Only pronouns — and chiefly the personal pronouns — still make any considerable use of case forms.

Personal pronouns change form dramatically to indicate case.

		Subjective	**Objective**	**Possessive**
Singular	*1st person*	I	me	my, mine
	2nd person	you	you	your, yours
	3rd person	he, she, it	him, her, it	his, hers, its
Plural	*1st person*	we	us	our, ours
	2nd person	you	you	your, yours
	3rd person	they	them	their, theirs

The indefinite or relative pronoun *who* also changes form to indicate case.

Subjective	**Objective**	**Possessive**
who	whom	whose
whoever	whomever	

The case of a pronoun is determined by the pronoun's function in its own clause. Pronouns used as subjects or predicate nouns, that is, nouns that follow linking verbs and restate the subject, are in the subjective case. Pronouns used as direct objects, as indirect objects, or as objects of prepositions are in the objective case. Pronouns that modify a noun or pronoun or that precede and modify a gerund are in the objective case. Use the following guidelines to select the appropriate case.

14a Uses of the Subjective Case of Personal Pronouns

▧ As the subject of a verb:

I think that **we** missed the flight.

▧ As the complement of the verb *to be:*

I'm sure it was **she.**

▧ As the appositive (restatement) of a subject or predicate noun:

The surveyors, Mr. James and **he,** plotted the acreage.

14b

Uses of the Objective Case of Personal Pronouns

▨ As the direct object or indirect object of a verb:

Mother likes **her** best.

Todd gave **us** the concert tickets.

▨ As the object of a preposition:

Sara directed the salesman to **him** and **me.**

▨ As the appositive (restatement) of a direct or indirect object:

My sister and I gave them, Mrs. Lester and **her,** nothing but trouble.

▨ As the subject of an infinitive:

I want **them** to take my place.

▨ As the object of an infinitive:

Don't expect to see **her** or **me** at a classical concert.

14c

Uses of the Possessive Case of Personal Pronouns

▨ As a modifier of a noun or pronoun:

These are **my** four children, and those are **his** three.

▨ As a modifier of a gerund:

Her skiing improved rapidly.

What's wrong with **my** buying new equipment?

14d

Distinguishing Between "We" and "Us" Used with a Noun in Apposition

The subjective case form is *we;* the objective case form is *us.* Select the pronoun that would be correct if the noun were omitted.

We tenants must file formal complaints against the management firm. [Subjective case for subject of the sentence.]

Their inattentiveness has given **us** tenants little recourse. [Objective case for indirect object.]

14e

Personal Pronouns with "Than" or "As"

The case of a pronoun following *than* or *as* in a comparison often causes difficulty. In an elliptical (incompletely expressed) construction, use the case that would be appropriate if all the words were expressed.

He is at least as capable as **she.** [Subjective case because *she* is the subject of the unexpressed verb *is.*]

The crowd liked Navratolova better than **them.** [Objective case because *them* would be the object of the verb if the comparison were expressed completely; *better than it liked them.*]

14f   Uses of the Subjective Case of the Relative Pronoun "Who"

■ As the subject of a clause:

Ralph Nader is a consumer advocate **who** gets media attention easily.

■ As the subject of a clause stated as a question:

Who donated the carpets?

14g   Uses of "Whom" — The Objective Case of the Relative Pronoun "Who"

■ As the object of a verb:

Professor Frayne is a man **whom** we admire.

■ As the object of a verb in a question:

Whom should we notify?

14h   Distinguishing Between "Whoever" and "Whomever"

The subjective case form is *whoever;* the objective case form is *whomever.* Be aware that even when a subordinate clause functions as an *object,* a pronoun that functions as the *subject* of the clause belongs in the subjective case.

Invite **whoever** will come. [Subjective case because *whoever* is the subject of *will come.*]

The committee will approve the appointment of **whomever** we select. [Objective case because *whomever* is the object of the preposition *of.*]

▇▇▇▇ *Exercise*

Revise the following sentences to correct any errors in the use of case. Some of the sentences need no correction.

1. The police suspected Boris Kraykov's associates, but he is more likely to be

   responsible than them.

2. Jim, not me, must make the recommendation.

3. Us gun collectors must be aware of people's objecting to firearms.

4. Reverend Wehrenberg is the person to whom we will go for advice.

5. They gave the finalists, Sandi and he, an enthusiastic round of applause.

6. Whoever we appoint to the council must be willing to present our case with conviction.

7. There is really no excuse for him refusing to comment.

8. The comments were directed to we two, you and I.

9. Carol is at least three years older than him.

10. Sonia will have to train whoever accepts the job.

15 VERB TENSES (T)

Verb tenses indicate the time of the action or state of being expressed. Most verbs in English have four principal parts and change in a predictable way to form the six basic tenses and the six progressive tenses.

Present-tense form: walk

Present participle: walking

Past-tense form: walked

Past participle: walked

Present tense: walk, walks

Past tense: walked

Future tense: will walk, shall walk

Present perfect tense: have walked, has walked

Past perfect tense: had walked

Future perfect tense: will have walked, shall have walked

Present progressive tense: am walking, are walking, is walking

Past progressive tense: was walking, were walking

Future progressive tense: will be walking, shall be walking

Present perfect progressive tense: have been walking, has been walking

Past perfect progressive tense: had been walking

Future perfect progressive tense: will have been walking, shall have been walking

Irregular verbs form their past tenses and their past participles through changes in spelling or word form that must be memorized. The following list contains the principal parts of the most common or troublesome verbs.

Present tense	Present participle	Past tense	Past participle
am, is, are	being	was, were	been
bear	bearing	bore	borne
beat	beating	beat	beaten
begin	beginning	began	begun
bite	biting	bit	bitten
blow	blowing	blew	blown
break	breaking	broke	broken
bring	bringing	brought	brought
burst	bursting	burst	burst
cast	casting	cast	cast
choose	choosing	chose	chosen
come	coming	came	came
deal	dealing	dealt	dealt
do	doing	did	done
draw	drawing	drew	drawn
drink	drinking	drank	drunk
eat	eating	ate	eaten
fall	falling	fell	fallen
fly	flying	flew	flown
forbid	forbidding	forbade	forbidden
forsake	forsaking	forsook	forsaken
freeze	freezing	froze	frozen
give	giving	gave	given
go	going	went	gone
grow	growing	grew	grown
hang*	hanging	hung	hung
have	having	had	had
know	knowing	knew	known
lay	laying	laid	laid
lie	lying	lay	lain
ride	riding	rode	ridden
ring	ringing	rang	rung
rise	rising	rose	risen
run	running	ran	run
see	seeing	saw	seen
shake	shaking	shook	shaken

---

* The verb *to hang,* used in the sense of "to execute," is regular: *hang, hanged, hanged.*

Present tense	Present participle	Past tense	Past participle
shoe	shoeing	shod	shod
shrink	shrinking	shrank (shrunk)	shrunk
sing	singing	sang (sung)	sung
sink	sinking	sank (sunk)	sunk
sit	sitting	sat	sat
slay	slaying	slew	slain
slink	slinking	slunk	slunk
speak	speaking	spoke	spoken
spin	spinning	spun	spun
spring	springing	sprang (sprung)	sprung
steal	stealing	stole	stolen
strive	striving	strove	striven
swear	swearing	swore	sworn
swim	swimming	swam	swum
take	taking	took	taken
teach	teaching	taught	taught
tear	tearing	tore	torn
throw	throwing	threw	thrown
wear	wearing	wore	worn
weave	weaving	wove	woven
win	winning	won	won
write	writing	wrote	written

15a   **Use the present tense to describe habitual action or actions that occur or conditions that exist in the present.**

Pamela **listens** to classical music when she writes papers.

They **are** exhausted.

15b   **Use the present tense to express general truths and scientific principles.**

The earth **tilts** slightly on its axis.

15c   **Use the present tense to describe or discuss artistic works, paintings, sculpture, etc., and literary works, novels, plays, poems, etc.**

Polonius **offers** Laertes platitudes, not advice.

Picasso's use of black and white in his large-scale painting *Guernica* establishes a stark visual equivalent for the starkness of the artist's message.

15d   **Use the past tense to describe the completed actions or conditions that existed in the past.**

Abolitionists openly **opposed** slavery, often at personal risk.

They **were** exhausted.

15e   **Use the future tense to describe actions that will occur or conditions that will exist in the future.**

The Congress **will reconvene** after a brief recess.

15f   **Use the present perfect tense to describe actions that started or conditions that existed at an unspecified time in the past and continue in the present.**

For years, Mary Tyler Moore **has been** a spokesperson for the American Diabetes Association.

15g   **Use the past perfect tense to describe actions that started or conditions that existed before a specific time in the past.**

In large part, Czar Nicholas **had ignored** the turmoil that preceded the Russian Revolution.

15h   **Use the future perfect tense to describe actions that will be completed or conditions that will exist before a specific time in the future.**

Natalie **will have submitted** her dissertation before the school year ends.

15i   **Use the progressive tenses to express ongoing actions that occur in the present, past, or future.**

I **am learning** to ski.

Sasha **had been planning** to attend the theater opening.

Rebecca **will be working** as a receptionist this summer.

15j   **Use present participles to express action that coincides with the action described by the main verb.**

Sensing that media coverage of the takeover would be negative, Albertson decided to cancel the press conference.

15k   **Use past participles and perfect participles to express actions that occurred, or to describe conditions that existed, before the action or condition described by the main verb.**

Shocked by the disparaging comments, Senator Roberson left the hearing.

Having completed her work, Sybil sat down to read.

15l   **Generally, use the past tense or past perfect tense in a subordinate clause when the verb in the independent clause is in the past or past perfect tense.**

This combination of past tenses is used to place one past action in a temporal or other relation with another past action.

Virginia Woolf **worked** in isolation because she **needed** quiet to concentrate well.

After he **had purchased** tickets for the World Series, Karl **was** unable to use them.

15m

When the verb in the independent clause is in the present, future, present perfect, or future perfect tense, use any tense in the subordinate clause that will make the meaning of the sentence clear.

In *Camelot,* Lancelot **thinks** that he **will succeed** at every venture.

Because twenty-four-carat gold **is** soft, detailed design work **will wear** away over time.

▰ *Exercise*

Select the appropriate verb tenses in the following sentences. Be ready to explain your choices.

1. Rain (is, was) water that (condenses, condensed) around dust particles and (falls, fell) to earth.

2. Normally the incidence of heart-worm disease (increases, increased) each year, but last year it (decreases, decreased).

3. On a bi-monthly basis, the Citizen's Action Coalition (sends, sent) a newsletter to its supporters.

4. Next fall, tuition at American universities (rises, will rise) to keep pace with inflation.

5. Becky Sharp (is, was) the main character of William Thackeray's *Vanity Fair,* an episodic novel published in 1847.

6. Isaac Singer (has written, had written) all of his stories in Yiddish, but they (are, were) immediately translated into English.

7. By the end of this season, we (will play, will have played) in thirty games and two tournaments.

8. (Serving, Having served) on *Time* magazine's Board of Economic Advisors, Alan Greenspan (is, was) a likely figure to head the Federal Reserve Board.

9. Because Da Vinci (experiments, experimented) with a variety of interesting pigments, many of his works (are, were) deteriorating.

10. (Opening, Having opened) the bomb casing with great care, the explosives expert (disconnects, disconnected) the timing mechanism.

16 ADJECTIVES AND ADVERBS (ADJ & ADV)

Adjectives and adverbs are both modifiers, but they serve separate purposes in sentences. Adjectives modify nouns and pronouns. Adverbs modify verbs, adjectives, and other adverbs; they may also modify phrases and clauses.

The *-ly* ending identifies many words as adverbs. However, it is not foolproof. Some adjectives end in *-ly (heavenly, lovely, leisurely)*, and many common adverbs *(very, then, always, here, now)* do not end in *-ly*. To avoid faulty modification, be certain to use adjectives only with nouns and pronouns and adverbs with verbs, adjectives, other adverbs, or whole phrases and clauses.

16a Use adjectives to modify nouns and pronouns.

His **thoughtful** assessments are always **welcome**.

They are **dependable**.

16b Use adverbs to modify verbs, adjectives, and other adverbs.

She **carefully** selected the flowers.

She was **especially** careful when she chose the roses.

She **very** carefully examined the buds and leaves on each stem.

16c Recognize the distinct uses of troublesome adjective and adverb pairs.

The following two adjective/adverb pairs, and others you may have had trouble with in your writing, should be carefully used.

Bad/Badly: Use *bad*, the adjective form, to modify nouns and pronouns, even in conjunction with sensory verbs such as appear, look, taste, and so on. Use *badly*, the adverb form, only to modify a verb.

Lendl made a series of **bad** volleys during the third match.

His prospects may seem **bad**, but they really aren't.

Although Jimmy Stewart sang several great Cole Porter songs in films, he acknowledged that he sang them **badly**.

Good/Well: The word *good,* an adjective, always modifies a noun or a pronoun. The word *well* can function as either an adverb or an adjective. As an adverb meaning "satisfactorily," *well* could modify a verb, an adjective, or another adverb. As an adjective meaning "healthy," *well* could only modify a noun or pronoun.

The lasagna smells *good.* [Adjective.]

Your point is *well* taken. [Adverb.]

Mrs. Biagi says that she feels *well* today. [Adjective.]

▮▮▮ *Exercise*

Revise the following sentences to correct faulty modification.

1. If you move quiet and slow, you can sometimes see small wildlife in this area.
2. Miss Haversham, eccentric and oppressive, treated Pip bad.
3. Competitive cyclists must react calm and quick when they need to make repairs during tournaments.
4. Make sure that the knots are tied tight and secure, or the rocking of the waves may break the boat loose from the wharf.
5. When receiving chemotherapy treatments, most patients don't feel good.

▮▮▮ *Writing Grammatical Sentences: Review Exercise*

Revise the following sentences to make them grammatical. Identify the problem in each sentence that made revision necessary.

1. Beginning in 1901, Nobel Prizes have been awarded to people who have made major contributions in the areas of peace, literature, physics, chemistry, and physiology or medicine, contributions in economics have been recognized since 1969.

2. A committee representing Yale University and the Bollingen Foundation presents their $5,000 award for poetry every two years.

3. The 1985 World Hunger Media Award was given to Bob Geldof, the rock musician who most people recognize as the organizer of the Live Aid concerts.

4. *Gödel, Escher, Bach: An Eternal Golden Braid,* Douglas R. Hofstadter's Pulitzer Prize–winning book, established philosophical and theoretical links between physics, art, and music.

5. Kennedy Center Honors have recognized the innovative work of a number of choreographers. George Balanchine, Martha Graham, Agnes de Mille, and Jerome Robbins, among others.

6. Although the musical *The Mystery of Edwin Drood* won five major Tony Awards in 1986, it has fared bad on overall ticket sales.

7. George W. Beadie and Edward L. Tatum, both of the United States, received Nobel Prizes in Physiology for their discovery that genes transmitted hereditary characteristics.

8. Each year, the Randolph Caldecott Medal, awarded by the American Library Association, recognizes whomever has produced the best illustrated book for children.

9. "We Are the World," the title cut from the album of the same name. Won Grammy Awards in 1985 for record of the year, song of the year, pop group of the year, and video–short form.

10. Henry Kissinger was Secretary of State from 1973 to 1977, under Nixon and Ford, he has received the Nobel Peace Prize (1973), the Presidential Medal of Freedom (1977), and the Medal of Liberty (n.d.).

11. Emory Holloway, Walter Jackson Bate, Justin Kaplan, Lawrence Thompson, Louis Sheaffer, and Richard W. B. Lewis have all won Pulitzer Prizes for

biographies of major writers. In spite of the awards, however, their books are more recognized by name than them.

12. MacArthur Foundation Fellowships boast awards of $164,000 to $300,000, spread over five years, these fellowships free recipients to pursue their interests.

13. The Enrico Fermi Award is given to scientists who demonstrated an "exceptional and altogether outstanding" body of work in the field of atomic energy.

14. Milos Forman has won Academy Awards for directing two highly distinct films, *One Flew Over the Cuckoo's Nest,* a black comedy about a mental ward, and *Amadeus,* a selectively retold biography of Mozart, also won Oscars as best film of the year.

15. The Columbia University Graduate School of Journalism won a George Foster Peabody Award for Broadcasting in 1985. Their collection *Seminars on Media and Society* were particularly acknowledged.

CHOOSING EFFECTIVE DICTION

Effective diction is the choice of words that best communicate your purpose to your audience. Your diction must be tailored to fit the specific context of your sentences and paragraphs and of your paper as a whole. Your choices of effective diction depend on your *subject, audience,* and *purpose.*

17   USING A DICTIONARY (DICT)

No writer should work without a dictionary. Unabridged dictionaries, like the *Oxford English Dictionary,* usually found in the reference rooms of libraries, contain vast numbers of words and lengthy, thorough definitions. They are useful when you need to find highly detailed information like full word histories or when you need to find the definition of an arcane word. Most of your needs, however, will be met by a standard desk-sized, collegiate dictionary such as

- *The American Heritage Dictionary of the English Language*
- *The Random House Dictionary of the English Language: College Edition*
- *Webster's New Collegiate Dictionary*
- *Webster's New World Dictionary of the American Language: College Edition*

Those dictionaries, and most other standard-sized collegiate dictionaries, provide a wide variety of general and useful information:

- A brief history of lexicography (the preparation and study of dictionaries)
- A brief history of the language
- An explanatory diagram of a sample word entry, with a key to the abbreviations that are used in the definitions
- Explanations and definitions of and guidance on matters of grammar and usage
- The dictionary of words (the main portion of the reference book)
- Lists of standard abbreviations
- Biographical entries (providing brief notes on important people)
- Geographical entries (providing brief notes on important places)

■ Lists of foreign words and phrases

■ Comparative tables of alphabets, calendars, and currencies

■ Tables of measurements (both American and metric)

■ Lists of signs and symbols in common use

■ Maps and illustrations

■ Pronunciation guides

■ General guides to writing (business letter forms, forms of address, manuscript preparation guidelines, and so on)

Dictionaries provide much more than definitions; they offer useful information about many subjects related to writing. Most often, however, you will turn to a dictionary to find information about specific words. To make your use of a dictionary efficient and productive, familiarize yourself with its general pattern of presenting information about words.

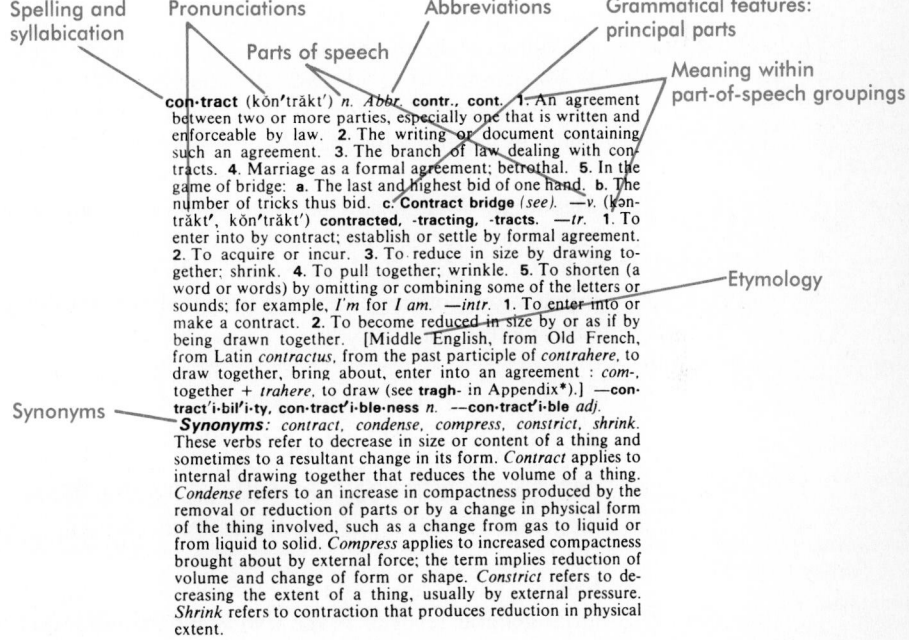

Spelling and syllabication

Pronunciations

Parts of speech

Abbreviations

Grammatical features: principal parts

Meaning within part-of-speech groupings

Etymology

Synonyms

Spelling and Syllabication   Entries begin with the word spelled out and divided into syllables, usually marked with dots: **con·tract.** When a word has several acceptable spellings, each of them will be listed, but the most common spelling will be listed first. If alternative spellings are listed as *Am* (American) and *Brit* (British), use the American spelling in your writing.

437

Pronunciation The pronunciation (usually enclosed in parentheses) follows, presented in a simplified phonetic transcription. The markings are matched to a pronunciation guide, which is usually printed at the bottom of each page. When words have more than one syllable, accent marks ('') indicate which syllables receive primary or secondary stress.

Part of Speech Abbreviations such as *n., v.,* and *adj.* indicate how a word can be used in a sentence. When a word can be used in various ways, the definitions are divided by parts of speech with, for instance, all meanings of the noun grouped together, followed by all meanings of the verb.

Abbreviations If a word is commonly abbreviated, some dictionaries include the abbreviation in the entry; other dictionaries list abbreviations in a special section.

Grammatical Features Distinctive grammatical features are noted in the entry. The principal parts of a verb are included, information that is especially helpful with irregular verbs — for example, *go, went, gone, going, goes.* Plural forms are provided for nouns that form plurals irregularly — for example, *mouse, mice.* When comparative and superlative degrees of adjectives and adverbs are formed by adding *-er* or *-est,* for example, *good, better, best* or *many, more, most,* dictionaries generally list them.

Meanings Within part-of-speech groups, meanings are most commonly presented in order of their frequency of use. Some dictionaries, however, present definitions in historical order (from oldest to newest) or hierarchical order (from primary to secondary). Read the front matter of the dictionary you are using to see what pattern it follows.

Etymology When word origins are known, they are provided, sometimes in abbreviated form (*Gk.* for Greek, *ME* for Middle English, and so on). Thus *contract* is one of a large class of words that came into English from Latin by way of Old French. It is made up of the prefix *con-* (from the Latin *com-,* meaning "together") and the root word *tract* (from the Latin *trahere,* meaning "to draw").

Labels Labels are used to identify words according to a number of specific criteria: general level of usage (Nonstandard, Informal, Slang, Dial., and so on), regional usage (Brit., Southern), and usage within an area of specialization (Law, Med., Computer Sci.).

Synonyms and Antonyms Many dictionaries include brief lists of synonyms for defined words (often introduced by the abbreviation *syns.*), to illustrate comparable word choices. Some dictionaries include antonyms as well (introduced by the abbreviation *ant.*). Check the front matter of your dictionary to see whether these aids are used and where they are placed in the entries.

18c

▄▄▄ *Exercise*

Use the following questions to familiarize yourself with your own collegiate dictionary. You will have to use all parts of the dictionary to find your answers — the front matter, entries, and appended materials. Keep in mind that your responses may vary slightly from those of people using other dictionaries.

1. If you have to hyphenate *maleficence* at the end of a line, where could you appropriately place a hyphen before finishing the word on the next line?

2. How are the following words pronounced: *acclimate, banal, data, impotent, Wagnerian*?

3. Which is the preferred spelling, *aesthetics* or *esthetics*? Is the word listed under both spellings or only under one?

18d

4. What synonyms does your dictionary list for *ghastly, lure, puzzle, single,* and *yield*?

5. What is the British meaning of the word *torch*? When *torch* is used as slang, what does it mean? What idiomatic expression uses the word *torch*?

6. What is a *schlemiel*? What is the origin of the word?

7. How many meanings are recorded for the word *vulgar*? By what pattern are the definitions arranged?

8. As what parts of speech can the word *square* be used? In what order do the parts of speech appear in the entry?

18e

9. What are the plural forms of *alumna, fungus, graffito, hippopotamus,* and *medium*?

10. What do the abbreviations *AAUW, EST, FNMA, MCAT,* and *VISTA* stand for?

11. In what years were *Marian Anderson, D. W. Griffith, Marie Antoinette, Alfred Nobel, George H. Ruth,* and *Mary Cassatt* born?

12. In what countries are *Addis Ababa, Caracas, Kuala Lumpur, Mecca,* and *Sarajevo* located?

grown, are imposing on their independence, and returning children create storm and strife when they return to the nest. But at this point in time, adults living with their parents is becoming more commonplace, and many people will have to learn to live with the situation.

▰▰▰ *Choosing Effective Diction: Review Exercise*

Revise the following sentences to improve their diction.

1. An effective piece of transcribed discourse must be easily decipherable.

2. Like a sponge absorbs water, the dancer listened to and analyzed the comments of her choreographer.

3. In the final analysis, the institution of higher learning's operation was not cost effective.

4. The NSC's clandestine operations in Central America and the Middle East were less than effective.

5. The aggressive salesperson amassed a sizable commission.

6. It goes without saying that in this day and age oral communication is crucial for getting ahead in this dog-eat-dog world.

7. Sometime or other everyone will have difficulties with interpersonal relations.

8. The principal's principal objection was not to the principal the students presented but to their method of implementing it.

9. The Herculean task of planning the reunion fell on the shoulders of the two organizers.

10. The microscopic listening devices implanted in the concrete structural members of the U.S. Embassy in Moscow would make surreptitious monitoring of conversations possible.

OBSERVING THE RULES OF PUNCTUATION

Marks of punctuation serve specific purposes in sentences. They show where thoughts end, where ideas are separated, and where pauses occur. Punctuate sentences according to the principles noted here. Remember that using punctuation unnecessarily is as confusing as omitting it when it is required.

19 PERIODS, QUESTION MARKS, AND EXCLAMATION POINTS (./?/!)

Three marks of punctuation end sentences: the **period (.)**, the **question mark (?)**, and the **exclamation point (!)**. Primarily indicators of completely stated thoughts, they serve, especially the period, a few additional purposes as well.

19a Accurate Use of Periods

Use a period to end a sentence that makes a statement.

Solar power is not yet a widely used energy source.

Use a period at the end of a question that is a courteously stated request or a command.

Will you please hand in your papers now.

Give your forms to the secretary when you've finished.

Use a period to end a sentence that contains an indirect question.

The landlord asked if we understood the terms of the lease. [The sentence reports that a question was asked; it does not directly pose the question.]

Use a period with most abbreviations.

Although a period in an abbreviation indicates that letters have been omitted, some standard abbreviations do not require periods. Consult a dictionary for guidance. (See also Section 35, "Abbreviations.")

With periods	Without periods
Mr.	FCC (government agency)
M.D.	IL (state)
Trans.	PBS (television network)

Use a period before a decimal point and with dollars and cents.

Production standards vary by only .14 millimeter.
Pi equals 3.14159.
The price was reduced to $39.95.

19b · Accurate Use of Question Marks

Use a question mark after a direct question.

The need for a question mark is usually indicated by inverted word order: part or all of the verb in the independent clause precedes the subject of the clause. In some instances, however, intent transforms a statement into a question.

Can we assume that the order has been shipped? [_Can,_ part of the verb, precedes the subject _we._]

You mean he's ill? [Intent, not word order, indicates that the sentence is a question; the question mark, though optional, confirms the writer's intent.]

Use a question mark in parentheses to indicate uncertainty about the accuracy of dates or numbers or other facts.

Modern scholars question whether Homer, a Greek poet of the ninth century (?) B.C., was the sole author of the famous epics attributed to him, _The Iliad_ and _The Odyssey._

It is not good usage to indicate possibly inaccurate words or to indicate irony by using a question mark in parentheses; changes in diction or sentence structure are more effective means of achieving these ends.

19c · Accurate Use of Exclamation Points

Use an exclamation point only to express strong emotion or to indicate unusual emphasis.

Be quiet!

Don't just stand there. Do something!

In most writing, exclamation points are not necessary or appropriate. Use them selectively.

20 COMMAS (,)

The comma is used to make the internal structure of a sentence clear. It does so in three general ways: (1) by separating elements that might otherwise be confused, (2) by setting off interrupting constructions, and (3) by marking words that are out of normal order.

20a Use commas to separate three or more coordinate items in a series.

Using a comma before the conjunction (the word that joins the items in the series) is always correct and will avoid possible confusion.

> Her favorite novelists were Melville, Lawrence, and Faulkner. [Commas separating nouns.]

> We considered displaying the statue in three places: in the lobby, in the president's office, and in the reception room. [Commas separating phrases.]

> Jack designed the set, Ira did the flat painting, and Margo did the detailed painting. [Commas separating independent clauses.]

20b Use commas between coordinate adjectives or adverbs that are not joined by a conjunction but that modify the same word individually.

When each of several adjectives modifies a noun individually, or when each of several adverbs modifies a verb, adjective, or adverb individually, commas should separate the modifiers. No comma separates the last modifier from the word it modifies. (See also Section 21d.)

> It was a dark, drizzly, depressing day. [Each adjective individually modifies *day.*]

> Rick slowly, methodically rechecked his documentation. [Each adverb individually modifies *rechecked.*]

20c In a compound sentence, use a comma before the coordinating conjunction that links the independent clauses.

This usage prevents the subject of the second clause from being misread as an additional object in the first clause. When there is no danger of a confused reading, the comma may be omitted.

> Because of financial difficulties, the farmer sold his tractor and his plows, and his land remained uncultivated in the spring. [Without the comma, *land* could be misread as another direct object in a series with *tractor* and *plows.*]

> T. S. Eliot's poetry is highly regarded but his drama is not. [No comma necessary because no confusion is likely.]

20d   **Use one comma or a pair of commas to set off a conjunctive adverb.**

Conjunctive adverbs, such as *however, moreover, therefore, consequently,* and *nevertheless,* establish logical connections between sentences. Usually they provide a transition between two statements, and they come near or at the beginning of the second statement. If no confusion will result, the comma or commas may be omitted, but using commas in these cases will always be correct.

> The warehouse was severely damaged by fire. Subsequently, the property was sold at a loss.

> Recent advances in medical research have brought hope to victims of AIDS. Some medical experts, however, feel the optimism is premature.

> Streamlining corporate management helps companies operate more smoothly; moreover, it can save on operating costs.

20e   **Use a comma after an introductory subordinate clause in a complex or compound-complex sentence.**

> Because OPEC's prices were high in the 1970s, American drivers began to conserve gasoline, showing that for once they were responsive to government recommendations.

> Although Middle English is somewhat difficult to read, the rewards of reading Chaucer make learning his language worth the effort.

Use a comma after an introductory infinitive or participial phrase.

> To prepare for her language proficiency exam, Pam skimmed five study guides.

> Moving cautiously through the rubble, the insurance agent made notes for the damage report.

20f   **Use a comma after introductory elements that function as adjectives or adverbs, unless the phrase is short and the meaning of the sentence is clear without the comma.**

A prepositional phrase at the beginning of a sentence that answers the questions *when, where,* or *under what conditions* is functioning like an adverb. Since such phrases modify the entire sentences of which they are a part, they should be followed by commas.

> After four weeks of intensive work, Jason finished the first draft of his master's thesis.

> After classes the five of us met to play basketball. [Short phrase does not require a comma.]

Use a comma to set off an introductory adverb that modifies an entire sentence.

Finally, attending conferences and workshops is an important way to meet other professionals.

20g **Use one comma or a pair of commas to set off a nonrestrictive clause or phrase.**

A ***nonrestrictive clause*** or ***phrase*** does not limit a class to a particular group or individual but modifies the whole class. It supplies additional information but can be omitted from a sentence without substantially altering the sentence's meaning.

A ***restrictive clause*** or ***phrase*** specifies a particular member or members of a group. It supplies information that is necessary to the meaning of the sentence. Restrictive clauses or phrases are not set off by commas.

The audio designer, **who creates sound effects for a play,** is an important member of a theater staff. [*Nonrestrictive clause* could be omitted without altering sentence meaning.]

New recruits, **who may join after finishing high school,** must make numerous adjustments before they are acclimated to military life. [*Nonrestrictive clause* could be omitted without altering sentence meaning.]

The audio designer **who worked on *Equus*** should be fired. [*Restrictive clause* identifies a specific audio designer.]

All soldiers **who complete basic training** will be assigned to duty within three weeks. [*Restrictive clause* identifies a specific group of soldiers.]

20h **Use one comma or a pair of commas to set off nonrestrictive appositives.**

An ***appositive*** — a word, phrase, or clause that renames a word or group of words in a sentence — can be nonrestrictive or restrictive. Nonrestrictive appositives provide nonessential information and are set off by commas. Restrictive appositives provide essential information and are not set off by commas. Appositives are grammatically equivalent to the noun or pronoun they rename.

PBS, **a nonprofit broadcasting network,** relies on corporate donations to cover most of its operating expenses. [*Nonrestrictive appositive* could be omitted without altering sentence meaning.]

The superstation **TBS** broadcast the first colorized versions of many American film classics. [*Restrictive appositive (TBS)* identifies a specific station.]

20i **Use one or a pair of commas to set off contrasted elements.**

Young children react best to positive comments, not negative ones.

South Korea, not Japan, has the highest literacy rate in Asia.

20j **Use commas to set off the words *yes* and *no*, mild interjections (*well, okay,* and so on) that begin sentences, and words in direct address.**

Yes, James Joyce's *Ulysses* is a difficult novel to read.

I suspect, my fellow Americans, that we are the victims of a hoax.

Sarah, would you please share your interpretation of the poem?

20k　Use commas to separate directly quoted material from explanatory expressions.

Expressions that signal direct quotations, such as "he said," "she replied," and so on, vary widely in form and position (they may be positioned at the beginning, in the middle, or at the end of a quotation). Wherever such expressions occur, they must be separated from the quotation, most often by using commas.

Reverend Tobias said, "State lotteries are nothing but state-sanctioned gambling."

"Lotteries, however, provide revenues that can be used to support education," **Representative Fulwiller noted.**

"If we ignore the lottery as a way of increasing revenues," **he added,** "our state's finances will continue to suffer."

20l　Use commas with numbers containing four or more digits, dates, addresses, place names, and titles and academic degrees, according to the conventions shown below.

▪ *Numbers:* Place a comma after every three digits, moving from right to left.

1,399　　　2,776,100

▪ *Dates:* In month-day-year order, a comma separates the day and year. If a date including month, day, and year appears in the middle of a sentence, a comma also follows the year.

　No commas are required in the day-month-year order or when only the month and year are used.

George Washington was born on February 22, 1732.

On October 30, 1905, Tsar Nicholas issued the October Manifesto, guaranteeing individual liberties.

Sixty percent of Hiroshima was destroyed by the atomic bomb dropped on 6 August 1945.

Hitler annexed Austria in March 1938.

▪ *Addresses:* When an address is written in a sentence, separate each element with a comma. If the address appears in the middle of a sentence, a comma must follow the last element.

She moved to **719 Maple Avenue, Cleveland, Ohio,** shortly after the Thanksgiving holidays.

The Convention Center in **Landover, Maryland,** was the site of the "Welcome Home" concert honoring Vietnam veterans.

The Olympics in **Munich, West Germany,** was plagued by terrorism.

▓ *Titles and academic degrees:* Use commas to set off these nonrestrictive elements.

At Honors Convocation, Rebecca Kingsley, **professor emerita,** presented the scholarship that bears her name.

William Leeds, **M.D.,** serves on the Marion County Health Board.

▓▓▓ *Exercise*

Supply periods, question marks, exclamation points, and commas in the following sentences. Make sure that a rule guides your placement of punctuation.

1. CBS NBC and ABC America's largest networks are now advertising programs regularly on small independent cable networks

2. The novel originally priced at $2595 did not sell well but sales increased when the price was reduced to $1795

3. Address women as Ms unless you are certain that they prefer Mrs or Miss

4. Should we send our order to the Chicago Illinois distribution center or to the Atlanta Georgia center

5. Angered that her glares did not quiet the jabbering child the old woman finally shouted "Shut up"

6. The tour guide concerned that he adapt himself to the visitors' preferences asked if they wanted to spend more time in the chapel

7. The San Francisco earthquake of April 18 1906 measure 8.3 on the Richter scale but the March 2 1933 earthquake in Japan measured 8.9

8. Much to my surprise the word *calf* is used to describe young cattle elephants antelopes rhinoceroses hippopotamuses and whales

9. Even though he was working without State Department authorization Rev Jesse Jackson secured the release of Robert Goodman Jr from Syria

10. Did you know that the West Indian island Jamaica is smaller (4244 square miles) than Connecticut

11. Because she was aware of prejudice against women Amadine Aurore Dupin published her novels under the name George Sand

12. After four months of work the restorers gave up their attempts to salvage the Venetian fresco

13. Peonies irises roses and day lilies are among Americans' favorite perennial not annual flowers

14. To be competitive in a declining market American auto manufacturers slashed interest rates and offered special rebates

15. The Tyrannosaurus Rex with teeth that measured six inches long was the fiercest of the meat-eating dinosaurs

16. Passengers who need special assistance are always asked to board airplanes before other travelers

17. Ironically taking out a mortgage is considered more stressful than having a foreclosure on a mortgage

18. Many taxpayers choose to use the "short form"; however taxpayers who wish to itemize deductions must use the "long form"

19. The mineral calcium is needed to develop and maintain bones and teeth but it is lacking in many diets

20. The film *Gandhi* begins with the leader's assassination on January 30 1948 and then recounts his life in a long flashback

21   UNNECESSARY COMMAS ⊙

Too many commas in sentences can be as confusing as too few. To avoid excessive use of commas, observe the following rules.

21a
Do not use a comma before a coordinating conjunction that joins only two words, phrases, or dependent clauses.

Isak Dineson married a Danish baron, and subsequently moved to Africa. [No comma with compound verb *married and moved*.]

I asked for advice first from my classmates, and then from Professor Bakerman. [No comma between two prepositional phrases joined by *and*.]

21b
Do not use a comma between subjects, verbs, and complements unless specific rules require that commas be used.

In the simplest sentences, no commas should break the subject-verb-complement pattern. When other information is added — appositives, nonrestrictive clauses, coordinate modifiers — commas may be necessary, but only as required by specific comma rules.

The angry soprano, walked out of the rehearsal. [No comma with subject-verb pattern.]

The angry soprano, **unhappy with the conductor,** walked out of the rehearsal. [Nonrestrictive appositive set off by commas.]

Our university's pole vaulter was, a strong contender for the title. [No comma with verb-complement pattern.]

Our university's pole vaulter was, **by general agreement,** a strong contender for the title. [Parenthetical comments, in this case a nonrestrictive prepositional phrase, require commas; see Section 20g.]

21c
Do not use a comma before the first or after the last item in a series.

Carla began attending exercise classes, to build her stamina, to lose weight, and to tone her muscles. [Comma would interrupt verb-complement pattern.]

Time, U.S. News and World Report, and *Newsweek,* are the most popular weekly magazines in America. [Comma would interrupt subject-verb pattern.]

21d
Do not use commas to separate adjectives or adverbs that cumulatively modify the same word.

When adjectives or adverbs work together to create meaning, they should *not* be separated by commas.

Four, small, red candles burned on the mantelpiece. [No commas because *red* modifies *candles, small* modifies *red candles,* and *four* modifies *small red candles.*]

21e
Do not use a comma between an adjective or an adverb and the word it modifies.]

An especially, talented, pianist opened the recital. [No commas because the adverb *especially* modifies the adjective *talented,* which modifies the noun *pianist.*]

21f   **Do not use commas to set off restrictive elements in sentences.**

Barbra Streisand's song∕ "Evergreen∕" won an Oscar in 1977. [No commas because song title is necessary for sentence clarity. (Streisand had another Oscar-winning song in a different year.)]

21g   **Do not use a comma before an expression in parentheses.**

When a comma is necessary with a parenthetical expression, it should follow the closing parenthesis.

In hopes of graduating early, Brian took six English classes∕ (English 307, 320, 337, 339, 412, and 445) but he could not manage the reading.

21h   **Do not use a comma before either an indirect or a direct quotation introduced by *that.***

Marcos said∕ that he will someday return to the Philippines.

But: Marcos said, "I will someday return to the Philippines."

Wasn't it Winston Churchill who said that "An iron curtain has descended across the Continent"?

21i   **Do not place a comma after either a question mark or an exclamation point in a direct quotation.**

Question marks and exclamation points replace the commas that are frequently required with direct quotations.

"Will we never recover from the wounds created by the Vietnam War?∕" asked Representative Martin.

■ *Exercise*

Remove unnecessary commas from the following sentences. Be ready to explain why each comma you delete is not needed.

1. Sandy Koufax was named Most Valuable Player of the World Series in 1963, and in 1965.

2. Two American cities, (Chicago, and New York City) each employ more than ten thousand police officers.

3. Four, very, small cars can park in the spaces normally allotted to three full-sized cars.

4. The geriatrician said, the symptoms suggest that Uncle Rupert probably has Alzheimer's disease.

5. Chicago's O'Hare International Airport, is the busiest airport in the United States.

6. The aging movie theater, which was once the small city's pride, needed extensive, expensive, renovation.

7. The playwright, Sophocles, is known for perfecting the form, of classical Greek tragedy.

8. Since early 1982, the copper penny has been gradually replaced, by a copper-plated zinc coin.

9. "How can we expect students, who have never taken calculus, to perform well on this portion of the exam?," Professor Carino asked.

10. California, Pennsylvania, Illinois, Michigan, and Ohio, each distributes over $1.5 million annually in unemployment benefits.

22 **SEMICOLONS (;)**

The semicolon most often functions like a period, separating independent clauses. Though it can in some specialized instances replace a comma, the semicolon should not be used routinely as a substitute for a comma.

22a **Use a semicolon to join closely related independent clauses that are not connected by a coordinating conjunction.**

In this usage, the semicolon most clearly functions like a period. As a result, make certain that each clause is, in fact, independent.

Take care of the children; let the adults take care of themselves.

22b **Use a semicolon to join independent clauses that are linked by a conjunctive adverb, or separate the clauses with a period.**

The defense attorney's closing statements were brilliantly presented; however, the facts of the case favored the prosecution.

22c **Use a semicolon to separate three or more items in a series when one or more of the items contain internal commas.**

In their essays, students commented on *The Fire Next Time,* an essay-novel by James Baldwin; *Soul on Ice,* a polemic by Eldridge Cleaver; and *Anger and Beyond,* a collection of critical essays edited by Herbert Hill.

22d Do not use a semicolon in place of a comma with a subordinate clause.

Although revisions of the tax code will eliminate many deductions, most Americans will benefit from a reduction in their overall tax rate.

22e Do not introduce a list or a clarifying phrase with a semicolon.

The colon (:) and the dash (—) are traditionally used to introduce a list or a clarification. The semicolon is not interchangeable with these marks of punctuation.

A number of long-distance services vied for consumers' business, AT&T, MCI, and Sprint.

23 COLONS (:)

The colon is a formal way to introduce a list or a clarification. The colon means "Note what follows." Use colons selectively to add clarity to your writing.

23a Use a colon to introduce a list.

The items in the series should never be direct objects, predicate nouns, predicate adjectives, or objects of prepositions. An independent clause must precede the colon.
As noted in 22e above, a dash may be used in place of the colon.

American theater and film have produced a number of notable acting families: Barrymore, Bridges, Fairbanks, Fonda, and Sheen.

23b Selectively use a colon between two independent clauses when the second explains the first.

The second clause may start with either a lower-case letter (as shown here) or with a capital letter.

Except for differences in subject matter, the rules of grammar are like the laws of chemistry: they are generalizations describing accepted principles of operation.

23c Use a colon to emphasize an appositive that comes at the end of a sentence.

Marlowe and Shakespeare introduced the dramatic use of blank verse: unrhymed iambic pentameter.

23d | **Use a colon in place of a comma to introduce or emphasize a long quotation.**

> Churchill concluded an eloquent speech with this visionary statement: "Out of the depths of sorrow and sacrifice will be born again the glory of mankind."

23e | **Use a colon between numerals designating hours and minutes, after formal salutations in formal or official correspondence, between titles and subtitles, between chapter and verse in Biblical citations, and between city and publisher in works-cited entries.**

> See "Documenting Sources" in Chapter 14.

> The speaker was scheduled to arrive on a 9:40 A.M. flight from Los Angeles.

> Dear Mr. Harper: Dear Professor Smithson:

> *Some Sort of Epic Grandeur: The Life of F. Scott Fitzgerald*

> Isaiah 12:2–4

> Cambridge: Harvard UP

> Boston: Houghton Mifflin

23f | **Do not use a colon between a verb and its complement or between a preposition and its object.**

> Jerrid's favorite restaurants are: Richard's Townhouse, The Broken Blossom, and Fernucchi's.

■■■ *Exercise*

Revise these sentences, using semicolons or colons.

1. Two lizards found in southwestern United States and northern Mexico are venomous. They are the Gila monster and the Mexican bearded lizard.

2. The Democratic party considered five cities for its national convention. Those cities were Dallas, Texas, Chicago, Illinois, Atlanta, Georgia, Washington, D.C., and Los Angeles, California.

3. Mark Spitz was an outstanding Olympic competitor. However, he was only an adequate Olympic commentator.

4. Infertility counselor Roselle Shubin made this epigrammatic comment on parenthood, "There is more to being a mother than giving birth, and more to being a father than impregnating a woman."

26

26a

26b

26c

457

If the whole sentence in which a quotation (or an exclamation) appears is a question (or an exclamation) but the quotation is not, place the question mark (or exclamation point) after the quotation marks.

I can't remember who wrote the ballad "What'll I Do?"

Were we supposed to read "What It Takes to Be a Leader"?

26d　**Use single quotation marks ('/') around material that would be enclosed by full quotation marks if it were not already within a quotation.**

Jeremy said, "Although 'Dover Beach' is one of Matthew Arnold's greatest poems, it is not one of my favorites."

27　# BRACKETS ([/])

Brackets are used to enclose an editorial or a clarifying explanation or comment inserted into a direct quotation.

Richardson commented, "Using both systems [the U.S. Customary System and the International Metric System] has caused considerable confusion for American consumers and has put U.S. industries at a trade disadvantage."

28　# ELLIPSIS (. . .)

Ellipsis, three spaced periods, is used to indicate the omission of one or more words from a quotation. The three periods, or points, that form the ellipsis are considered a unit. When the omission comes at the end of a sentence, a fourth point must be added as end punctuation. When a comma is required, it follows the ellipsis.

28a　**Use ellipsis points to show where words have been omitted from a direct quotation.**

Omit extraneous material — like parenthetical expressions or unnecessary clarifications — but do not leave out material if the omission changes the meaning of the original text.

Original Quotation	**Elliptical Quotation**
Lincoln's anti-slavery views, clarified in the Republican party platform of 1860, served to alienate not only the residents of southern states but also southern sympathizers in the North.	According to Walter Holtmire, "Lincoln's anti-slavery views . . . served to alienate not only the residents of southern states but also southern sympathizers in the North."

28b   **Use ellipsis points very selectively to indicate hesitation, a trailing off of thought, or an incomplete statement.**

The deathbed scene in James Brooks' *Terms of Endearment* is . . . manipulative.

▬▬ *Observing the Rules of Punctuation: Review Exercise*

Correct the punctuation errors in the following sentences. Be ready to identify the rules that guided your work.

1. The Internal Revenue Service IRS is responsible for administering the tax laws passed by Congress

2. From 1791 to 1862 the US government relied on tariffs to generate income however in 1862 Congress enacted the first income tax law to pay for the debts of the Civil War

3. Did you know that income taxes were not universally instituted until 1913

4. Following the 1986 tax law only a few major deductions will be allowed mortgage payments state and local taxes medical expenses and charitable contributions

5. Nevada South Dakota Texas Washington and Wyoming these states do not impose a corporate tax based on net income

6. The IRS operates from its various headquarters one national office in Washington seven regional offices sixty-three district offices and ten service centers and processes roughly 200 million returns annually

7. Beginning in 1943 taxes were withheld from wages a plan that increased the number of people who equitably paid taxes

8. According to new tax laws corporate rates will drop from 46 percent to 34 percent however a minimum tax will also be imposed to prevent major companies from paying no taxes

9. The nation's first sales taxes enacted in 1812 affected consumers of only four kinds of commodities gold silver jewelry and watches

10. Did you know that it was Benjamin Franklin who said But in this world nothing can be said to be certain except death and taxes

11. James Otis spoke the sentence that became a catch-phrase of the American Revolution Taxation without representation is tyranny

12. In 1985 the IRS collected a total of $742,871,541,000 in taxes a figure so large it is hard to envision

13. Most Americans do not object to paying taxes many however object to how the tax money is spent

14. Various tables Schedule X Y or Z are used to compute the taxes of people with incomes of more than $50000

15. Taxpayers who wish to appeal a tax charge must follow four steps 1 discuss the charge with a local appeal's office 2 submit a written protest 3 wait for a judgment 4 pay the charge or file yet another appeal with the District Claims Court

OBSERVING THE RULES OF MECHANICS

Mechanical errors in the use of capitalization, italics, quotation marks, apostrophes, hyphenation, number style, abbreviations, and spelling distract readers from the content of your writing. Heeding certain rules as you prepare the final copy of a paper will help ensure that your reader's attention is focused not on preventable mechanical errors but on what you have to say.

29 CAPITALIZATION (CAP)

29a Capitalize the first word of every sentence and of every line in conventional poetry.

Using a word processor saves time during revision.

Standardized tests present a major problem for minorities: they do not allow for cultural differences.

Whenas in silks my Julia goes
Then, then, methinks, how sweetly flows
That liquefaction of her clothes. . . .
    — Robert Herrick,
      "Upon Julia's Clothes"

29b Distinguish between *proper names*, which require capitalization, and *common names*, which do not.

I left the assignment in my professor's mailbox.

But: I left the assignment in Professor Sheldon's mailbox.

I worked in summer stock to gain experience.

But: I worked at the Summer Festival Theater to gain experience.

29c In titles, capitalize the first and last words and all important words.

As a general rule, capitalize all nouns, pronouns, verbs, adjectives, and adverbs. In addition, capitalize all prepositions and conjunctions of four or more letters. In works with two-part titles, the first word of the subtitle is also capitalized, regardless of the word's length.

Zora Neale Hurston's *Their Eyes Were Watching God*

Derrick Ashford's "Eighteenth-Century Comedy on the Modern Stage"

Marya Mannes' "TV Advertising: The Splitting Image"

29d **Capitalize the names of people, races, nationalities, languages, and places, whether used as nouns or as adjectives.**

In American usage, the terms *black, white,* and *native American* are not capitalized.

Thomas Paine	Caucasian
Malaysia	Danish customs
Nairobi	Greek festival

29e **Capitalize the names of historical and cultural periods; historical, political, and cultural events; and documents.**

the Age of Reason	the Emancipation Proclamation
the Romantic Movement	Elizabethan drama
the Battle of Hastings	Prohibition

29f **Capitalize the names of days, months, and secular and religious holidays.**

Thursday	the Fourth of July
September	Hanukkah

29g **Capitalize the names of businesses and other organizations and government agencies and offices.**

General Electric Company	National Rifle Association
Phi Delta Kappa	United States Senate

29h **Capitalize the names of schools, colleges, and universities, academic departments, specific courses, and degrees, but not general references.**

Amherst College [But *college classes.*]

Thomas Jefferson High School [But *high school teachers.*]

Department of Education [But *department meeting.*]

Sociology 245 [But *sociology course.*]

Bachelor of Arts [But *baccalaureate degree.*]

29i   **Capitalize the names of religions and their followers and religious terms for sacred persons, books, and events.**

God   Islamic law

Buddha   Christian traditions

the Koran   the Immaculate Conception

29j   **Capitalize titles that precede proper names.**

Doctor Erica Weinburg   Prime Minister Winston Churchill

Secretary of State Henry Kissinger   President Abraham Lincoln

Professor Leon Edel

A title used alone or following a proper name is not capitalized.

A college professor is the catalyst in a classroom.

Jimmy Carter, the former president, has contributed even in retirement to the benefit of the country.

29k   **Capitalize nouns designating family relationships only when they are used as proper names or when they precede proper names.**

After arthroscopic surgery, Mother's chances of leading a normal life improved.

I arrived at the apartment before Uncle Will got home from work.

Do not capitalize common nouns that name family relationships, even when the noun is preceded by a personal pronoun in the possessive case.

After arthroscopic surgery, my mother's life improved.

I arrived at the apartment before my uncle got home from work.

Never underestimate the influence of brothers and sisters.

29l   **Capitalize A.M. and P.M. and A.D. and B.C.; capitalize the call letters of radio and television stations; capitalize abbreviated forms of business, organization, and document names. (Periods may or may not be required, according to convention; consult a dictionary.)**

705 B.C.   ERA

A.D. 1066   WPFR radio

4:30 A.M.   KTVI television

4:30 P.M.   IBM

29m   **Capitalize the first word of a direct quotation if the quotation is a complete sentence or an interjection that can stand alone.**

Mr. Bennett remarked, "Though ornate by present standards, Baroque sculpture remains aesthetically pleasing."

Carla, surprised by the harsh criticism, responded, "Oh."

■■■■ **Exercise**

Supply capitalization in the following sentences, noting the rule that guides each correction.

1. the elizabethan period, a cultural and aesthetic awakening in england, began roughly a century after the italian renaissance.

2. yom kippur, the holiest jewish holiday, was observed on monday, october 13, this year.

3. although my mother and aunt beatrice are both normally critical television viewers, they both love *the young and the restless.*

4. an mba from harvard is an excellent passport to a lucrative job on wall street or with a fortune 500 firm.

5. students in french secondary schools are expected to learn english as well as one other foreign language.

6. george p. shultz, the secretary of state, travels extensively in europe and the middle east for the department of state.

7. the reverend thomas r. fitzgerald serves as president of st. louis university, a catholic university enrolling over eleven thousand students.

8. who was it who said, "i cried all the way to the bank"?

9. the abbreviation ira could refer to the irish republican army or the international reading association.

10. dorothy parker once described katharine hepburn's performance in a play with this caustic sentence: "she ran the whole gamut of emotions from a to b."

30   ITALICS (ITAL)

In print, italics, slanted type, is used to give words distinction or emphasis. In handwritten or typed manuscript, the same effects are achieved using underlining.

30a   **Italicize the titles of books, periodicals, newspapers, pamphlets, plays, films, television series (but not individual programs), radio programs, long poems, long musical compositions, record albums, paintings, and sculpture.**

Marcel Proust's *Swann's Way* (book)

the *Washington Post* (newspaper)

NCTE's *Essentials of English* (pamphlet)

Oscar Wilde's *The Importance of Being Earnest* (play)

Sidney Pollack's *Out of Africa* (film)

Norman Lear's *All in the Family* (television program)

Casey Kasem's *Top-Twenty Count Down* (radio program)

Walt Whitman's *Leaves of Grass* (long poem)

Tchaikovsky's *The Nutcracker Suite* (long musical composition)

Paul Simon's *Graceland* (album)

Pablo Picasso's *Three Musicians* (painting)

George Segal's *Girl in Doorway* (sculpture)

30b   **Italicize the names of individual ships, trains, airplanes, and spacecraft.**

Jacques Cousteau's *Calypso* (ship)

the *Orient Express* (train)

Air Force One (airplane)

Voyager 1 (spacecraft)

30c   **Italicize foreign words and phrases.**

Confined to a hospital bed, Rachel had to vote *in absentia*.

Many foreign terms have been assimilated into standard usage and, as a result, do not require italics. Consult a dictionary for guidance.

30d   **Italicize words used as words, letters used as letters, numbers used as numbers, and symbols used as symbols.**

French contains two variations of *you*, one formal and one informal.

The letters *a* and *e* used to be printed *æ*.

Make sure that you distinguish your *1*'s from your *7*'s.

The ampersand, *&*, is unacceptable in formal prose.

30e   **Italicize words selectively for emphasis.**

In formal prose, this is not considered good usage.

Jason played the song *twenty-three* times in a row!

31   QUOTATION MARKS (" / ")

Quotations marks are most commonly used to set off direct quotations, but they have a mechanical use as well.

31a   Use quotation marks with the titles of brief works or parts of complete works.

"Winning Hearts Through Minds" in *Time* (article)

Ernest Hemingway's "Old Man at the Bridge" (short story)

Gerard Manley Hopkins' "God's Grandeur" (poem)

Stephen Jay Gould's "Darwinism Defined: The Difference Between Fact and Theory" (essay)

Manhattan Transfer's "Tuxedo Junction" (song)

"The Spirit of Scholarship" in Richard Altick's *The Art of Literary Research* (chapter in a book)

"Captain Tuttle" from *M*A*S*H* (episode of television program)

■ *Exercise*

Insert italics (underlining) and quotation marks in the following sentences. Remember to place them accurately in relation to other punctuation.

1. Paul Conrad won Pulitzer prizes for editorial cartooning when he worked for two different publications: the Denver Post and the Los Angeles Times.

2. Grammies for best song and best album went to Tina Turner for What's Love Got to Do With It? and Private Dancer, respectively.

3. To demonstrate aerodynamic possibilities, engineers developed the Gossamer Albatross, an airplane propelled by peddling.

4. Blattella germanica is an eloquent sounding term to use when you mean cockroach!

5. Stuart called here sixteen times while you were gone this weekend.

6. On your final charts, please write female and male rather than ♀ and ♂.

7. A View to a Death, a pivotal chapter in Golding's novel Lord of the Flies, offers a vision of primitive, ritualistic execution.

8. In an article titled A Man With Titanic Vision, Discover magazine honored Bob Ballard as its 1986 Scientist of the Year.

9. Many American musicals have plays as their source, among them Hello, Dolly (The Matchmaker), My Fair Lady (Pygmalion), and Cabaret (I Am a Camera).

10. Kurtz, a character in Conrad's novel The Heart of Darkness, has reemerged in T. S. Eliot's Hollow Men, a brief poem, and Apocalypse Now, a long film.

32 APOSTROPHE (')

The apostrophe has three general uses: to indicate the possessive case of nouns and some pronouns, to indicate the omission of letters and numbers, to indicate the plural of letters, numbers, and words used as words.

32a Use an apostrophe and an s to form the possessive of a noun or an indefinite pronoun that does not end with an s.

Oprah Winfrey's guests

somebody's car

the men's dressing room

a month's rental fee

mother-in-law's comments

32b Use only an apostrophe, without an s, when a noun already ends with an s.

scientists' projections

Diana Ross' first film

32c To show joint possession, add an apostrophe and an s to the last name in the group. To show individual possession, add an apostrophe and an s to each name in the group.

Lerner and Lowe's musical reputation [Joint possession.]

Shakespeare's and Marlowe's dramatic innovations [Individual possession.]

32d Use an apostrophe to show the omission of letters and numbers.

| couldn't | could not | there's | there is |
| I'll | I will (*or* I shall) | the '84 Olympics | the 1984 Olympics |

32e   **Use an apostrophe and an *s* to form the plurals of numbers, letters, and words used as words.**

Kirsten received four *10*'s during the final round of competition.

Her *s*'s look like *8*'s.

His writing was cluttered with *very*'s, *really*'s, and *especially*'s.

33   **HYPHENATION (-)**

Hyphens are used for two purposes: to divide a multisyllable word at the end of a line and to join two or more words of a compound.

33a   **Hyphenate a word that must continue on a new line.**

Place hyphens between the syllables of words that must be divided (if necessary, consult a dictionary to see where a break may be made). Do not break one-syllable words or leave fewer than three letters at the end or beginning of any line. Do not hyphenate proper names.

The hurricane battered the coastline, damaging property in four cities.

33b   **Hyphenate compound words according to convention; consult a dictionary.**

Some compound words are hyphenated; others are "closed up," written as one word; still others are written as two words.

Hyphenated	Closed Up	Two Words
sister-in-law	applesauce	wedding ring
razzle-dazzle	blackboard	living room
master-at-arms	landowner	free fall

33c   **Hyphenate words that precede a noun and combine to modify it.**

heart-to-heart talk

off-the-cuff comments

never-ending problems

When the modifiers follow the noun, no hyphens are necessary.

The comments were off the cuff.

His problems were never ending.

33d **Hyphenate compound numbers ranging from twenty-one to ninety-nine; hyphenate fractions.**

a test score of sixty-seven

two-thirds of the voters

33e **Hyphenate words using the prefixes *all-*, *ex-*, and *self-*; hyphenate words using the suffix *-elect*.**

all-consuming pride

ex-wife

self-motivated student

secretary-elect of the city council

33f **Hyphenate a compound consisting of a prefix and a proper noun.**

anti-Iranian

un-American

pre-Enlightenment

post-Modern

33g **Hyphenate words formed with a prefix if the unhyphenated form would be a homonym, a word with the same spelling but a different meaning.**

re-cover (to cover again)   recover (to regain)
re-lease (to lease again)   release (to let go)

34 **NUMBERS (N)**

34a **Spell out numbers that can be expressed in one or two words.**

thirty-two source cards   twenty-six thousand dollars
five million voters   fourteen hundred miles

34b **Use digits for numbers that would require three or more words if spelled out.**

319 graduates (*not* three hundred and nineteen)

101 pages (*not* one hundred and one)

34c **Spell out a number that begins a sentence, or revise the sentence so that it does not open with a number.**

Not: 412 art dealers attended the convention.

But: Four hundred and twelve art dealers attended the convention.

Or: There were 412 art dealers at the convention.

34d

Use digits for addresses, dates, divisions of books and plays, dollars and cents, identification numbers, percentages, scores, and times.

> 2300 North 12th Street (address)
>
> 11 January 1912 or January 11, 1912 (dates)
>
> 700 B.C., A.D. 700 (dates)
>
> Chapter 17, Volume 2, Act 1, Scene 2 (divisions of books and plays)
>
> $7.95, $1,230,000, $4.9 million (dollar amounts)
>
> 314-77-2248, UTC 41 69490 (identification numbers)
>
> 81 percent, 3 percent (percentages)
>
> 117 to 93 (scores)
>
> 6:10 A.M., 2:35 P.M. (times)

35

ABBREVIATIONS (AB)

Abbreviations, shortened forms of words, should be used sparingly. Any abbreviations you use should be familiar to your readers, and they must be appropriate to the writing context. To double-check conventional abbreviations, consult a dictionary.

35a

Abbreviate titles that precede or follow people's names.

> William D. Grenville, Jr.   Judith Haverford, Ph.D.
>
> Ms. Abigail Hample   Rebecca Blair, M.D.
>
> Dr. Terence McDonald   the Rev. Stephen Pierson

35b

Use standard abbreviations and acronyms for names of organizations, corporations, and countries. Many of these abbreviations do not require periods.

> *Organizations:* AFL-CIO, YMCA, FBI
>
> *Corporations:* GTE, NBC, GM
>
> *Countries:* USA (*or* U.S.A.), USSR (*or* U.S.S.R.), UK (*or* U.K.)

35c

Before using an abbreviation that might be unfamiliar to some readers, spell out the complete name or term at its first appearance.

> Idaho State University — located in Pocatello, Idaho — is a state-supported university serving over three thousand students. ISU offers especially strong programs in health-related professions.

35d   **Use standard abbreviations with times, dates, and specific numbers. Use the dollar sign with specific amounts.**

Note the placement of the abbreviations *A.M., P.M., B.C.,* and *A.D.*

$1,245.78	500 B.C.
11:20 A.M.	A.D. 496
4:15 P.M.	part no. 339 (*or* No.)

35e   **Use standard abbreviations in works-cited entries.**

See "Documenting Sources" in Chapter 14.

Abbreviation	Meaning
ed.	editor, edition
et al.	and others
rev.	revised
rpt.	reprint
trans.	translator, translated by
vol.	volume

35f   **Use Latin abbreviations sparingly in prose.**

Latin Abbreviation	English Equivalent
c.f. (*confer*)	compare
e.g. (*exempli gratia*)	for example
et al. (*et alii*)	and others
etc. (*et cetera*)	and others, and so on
i.e. (*id est*)	that is
n.b. (*nota bene*)	note well

35g   **In most writing, do not abbreviate business designations (unless the company does), units of measurement, days of the week, months, courses of instruction, divisions of books and plays, geographical names (except in addresses), or personal names.**

Not	But
Co., Inc.	Company, Incorporated
lb., tbs.	pound, tablespoon
Fri., Oct.	Friday, October
psych., Eng.	psychology, English
chap., vol.	chapter, volume
L.A., VA	Los Angeles, Virginia
Wm., Robt.	William, Robert

■■■ *Exercise*

Insert necessary apostrophes and hyphens in the following sentences. In addition, correct the number style and forms of abbreviations.

1. 6 members of the committee returned the questionnaire, refusing to comment on MSUs drug testing program.

2. The suicide that ends Act four of *Hedda Gabler* shocked many narrow minded critics.

3. Doctor Connelly, a graduate of U. of TX at Austin, spoke to our faculty on Oct. 15 1986.

4. To attract first rate teachers to our public schools, we will have to increase teachers salaries.

5. Stephen Sondheims *Sunday in the Park With George* presents a neo Impressionist view of human relations and art.

6. Karin was delighted to receive 2 *8*s and 2 *9*s on her performance until she realized that *15*s were possible.

7. The post Civil War period was a time of exploitation and manipulation in the South.

8. The Rams won the game thirty-six to seven, having passed for one hundred and fifty four yards and having made seventy four percent of the games interceptions.

9. My father in laws Social Security check ($24967) covers slightly over two thirds of his monthly expenses.

10. The president elect of the N.C.T.E. felt that her work would require working with a state of the art computer and printer, so she bought the pair.

■ *Exercise*

Revise the following paragraph so that it is mechanically correct and consistent with conventional usage. Errors in capitalization, italics, quotation marks, apostrophes, hyphenation, number style, and abbreviations are present in the paragraph.

1600 Pennsylvania Ave., Washington, District of Columbia, is perhaps the *most famous* address in the U.S. At that site is the "White House," the residence of the President and his family. The design for the Original house was selected by Pres. Washington and Pierre L'Enfant, the french born designer of the city, and the cornerstone was set on Oct. 13, 1972. In 1814, the building was razed during a Battle of the war of 1812, and in subsequent years the interior of the structure had to be rebuilt. But the original design was heavily modified. As Mrs. John N. Pearce notes in The White House: An Historic Guide Ever changing personalities and styles of living and building have inspired the continuing metamorphosis that has marked the history of the White House. Over the years, the "White House" has served as both the official and the private residence of the first family. The 1st floor's rooms are used for public functions like Receptions and State Dinners, and its expansive, public rooms are decorated with such famous artwork as Gilbert Stuarts portrait George Washington. The limited access rooms on the second and third floors are used by the presidents family and friends. In all, the "White House" has one hundred and thirty-two rooms.

36   SPELLING (SP)

Errors in spelling, like other mechanical errors, interfere with communication because readers notice and are distracted by them. To avoid such distractions in your writing, develop good spelling habits. If you are a naturally good speller, then you have only to refresh your memory of common spelling rules. If you are a weak speller, then you need to hone your spelling skills.

Do not worry about spelling while you are planning and working on early drafts of your papers. If you interrupt your writing to check the spelling of a word, you might lose a thought that you cannot recapture. Instead make a mark near the word in question — a checkmark in the margin, the abbreviation *sp.* above the word, or a circle around the word — and continue your writing. Then, when your final draft is complete, look up the spelling of the words you have marked. This approach — checking spelling during revision — will allow you to write with a sense of continuity and attend to potential problems before typing your final copy.

36a

Review Spelling Rules

Form plurals according to the pattern of the singular word.

■ When words end in a consonant plus *o*, add *-es*.

fresco	frescoes
motto	mottoes
tomato	tomatoes

Exceptions

auto	autos
dynamo	dynamos
piano	pianos

■ When words end in a vowel plus *o*, add *-s*.

cameo	cameos
radio	radios
studio	studios

■ When words end in a consonant plus *y*, change the *y* to *i* and add *-es*.

daisy	daisies
remedy	remedies
victory	victories

■ When words end in a vowel plus *y*, add *-s*.

attorney	attorneys
key	keys
survey	surveys

■ When words end in *s*, *ss*, *sh*, *ch*, *x*, or *z*, add *-es*.

bonus	bonuses	match	matches
overpass	overpasses	tax	taxes
wish	wishes	buzz	buzzes

■ When words that are proper names end in *y,* add *-s.*

the Bellamys

the three Marys

two Germanys

Add prefixes, such as *dis-, mis-, non-, pre-, re-, un-,* and others, without altering the spelling of the root word.

similar	dissimilar
spell	misspell
restrictive	nonrestrictive
historic	prehistoric
capture	recapture
natural	unnatural

Add suffixes according to the spelling of both the root word and the suffix.

■ When words end with a silent *e* and the suffix begins with a consonant, retain the *e.*

achieve	achievement
definite	definitely
refine	refinement

Exceptions

argue	argument
awe	awful
true	truly

■ When words end with a silent *e* and the suffix begins with a vowel, drop the *e.*

accommodate	accommodating
grieve	grievance
size	sizable
tolerate	tolerating

■ When words end with a silent *e* that is preceded by a "soft" *c* or *g* and the suffix begins with a vowel, retain the *e.*

notice	noticeable	singe	singeing
trace	traceable	outrage	outrageous
change	changeable		

▓ When a one-syllable word ends with a single consonant and contains only one vowel and the suffix begins with a vowel, double the final consonant.

blot	blotted
clip	clipping
fit	fitting
skip	skipper
stop	stopping
trip	tripped

Distinguish between words spelled with _ie_ and _ei_.

The order of the vowels _ie_ and _ei_ is explained in this familiar poem:

Write _i_ before _e_
Except after _c_
Or when sounded like _ay_
As in _neighbor_ and _weigh_.

These are some exceptions to this rule:

counterfeit	leisure
either	neither
foreign	seizure
forfeit	sovereign
height	weird

36b

Improving Your Spelling Skills

Use a full-sized dictionary to check meanings and spellings of words that are often mistaken for each other.

A standard dictionary provides definitions that will help you distinguish between _affect_ and _effect, elicit_ and _illicit,_ and other confusing word pairs.

Use a spelling dictionary for easy reference when you know a word's meaning but are unsure of its spelling.

Spelling dictionaries contain lists of commonly used words with markings to indicate syllable breaks. These special dictionaries are helpful as quick references.

Concentrate on the most troublesome parts of easily misspelled words.

Give particular attention to the parts of words that lead to spelling errors — usually a single syllable or small cluster of letters.

accidentally	desperate
separate	secretary
maintenance	

Keep a record of words you have misspelled in your writing.

Most people have individual sets of words that they regularly use and regularly misspell. Use a note card, a sheet of paper, or a small notebook to record your personal list of troublesome words.

Carefully check the spelling of technical terms.

When your writing requires specialized language, verify the spelling of technical terms because their spelling is often tricky.

Use a spelling program if you are using a word-processing program.

If you use a word processor to prepare your papers, take advantage of software that can check spelling. Although spelling programs are not without problems, they can be quite helpful.

Consult the following list of frequently misspelled words.

abbreviate	aggravate	analysis
absence	aggression	analyze
absurd	airplane	annual
accelerate	alleviate	antecedent
accidentally	alley	anxiety
accommodate	allotted	apartment
accomplish	allowed	apparatus
according	ally	apparent
accumulate	although	appearance
accustom	always	appropriate
achievement	amateur	arctic
acoustics	ambiguous	argument
acquaintance	ammunition	arising
acquitted	among	arithmetic
across	amount	arouse
address	analogous	arranging

Continued on the next page.

images of generals Lee and Grant, as often as not based on the idealized statuary of the Franklin Mints Civil War Chess Set. Yet few of us have seen the civil war through the disturbing Psychological perspective of Stephen Cranes The Red Badge of Courage. Few have acknowledged in any real way that the reconstruction depressed the southern economy, gave rise to the ku klux klan, and failed to solve the ideological problems that continued to divide the country long after the deaths of six-hundred and fifty thousand soldiers.

A GLOSSARY OF CONTEMPORARY USAGE

This glossary identifies words and constructions that sometimes require attention in composition classes. Some of the entries are pairs of words that are quite different in meaning yet similar enough in spelling to be confused (see **principal, principle**). Some, such as the use of *without* as a synonym for *unless,* are nonstandard usages that are not acceptable in college writing (see **without = unless**). Some are informal constructions that may be appropriate in some situations but not in others (see **guess**).

The judgments recorded here about usage are based on the Usage Notes contained in *The American Heritage Dictionary,* supplemented by other sources. Because these authorities do not always agree, it has sometimes been necessary for the authors of this textbook to decide which judgments to accept. In coming to decisions, we have attempted to represent a consensus, but readers should be aware that on disputed items the judgments recorded in this glossary are finally those of the authors.

Because dictionaries do not always distinguish between formal, informal, and colloquial usage, it has seemed useful to indicate whether particular usages would be appropriate in college writing. The usefulness of this advice, however, depends on an understanding of its limitations. In any choice of usage, the decision depends less on what dictionaries or textbooks say than on what is consistent with the purpose and style of the writing. The student and instructor, who alone have the context of the paper before them, are in the best position to answer that question. All that the glossary can do is provide a background from which particular decisions can be made. The general assumption in the glossary is that the predominant style in college writing is moderate rather than formal or colloquial. This assumption implies that calling a usage informal in no way suggests that it is less desirable than a formal usage.

ad   *Ad* is the clipped form of *advertisement*. The full form is preferable in a formal style, especially in letters of application. The appropriateness of *ad* in college writing depends on the style of the paper.

adapt, adopt   *Adapt* means "to adjust to meet requirements": "The human body can adapt itself to all sorts of environments"; "It will take a skillful writer to adapt this novel for the movies." *Adopt* means "to take as one's own" ("He immediately adopted the idea") or — in parliamentary procedure — "to accept as law" ("The motion was adopted").

advice, advise   The first form is a noun, the second a verb: "I was advised to ignore his advice."

affect, effect   Both words may be used as nouns, but *effect,* meaning "result," is usually the word wanted: "His speech had an unfortunate effect"; "The treatments had no effect on me." The noun *affect* is a technical term in psychology. Although both words may be used as verbs, *affect* is the more common. As a verb, *affect* means "impress" or "influence": "His advice affected my decision"; "Does music affect you that way?" As a verb, *effect* is rarely required in college writing but may be used to mean "carry out" or "accomplish": "The pilot effected his mission"; "The lawyer effected a settlement."

affective, effective   See **affect, effect.** The common adjective is *effective* ("an effective argument"), meaning "having an effect." The use of *affective* is largely confined to technical discussions of psychology and semantics, in which it is roughly equivalent to "emotional." In this textbook, *affective* is used to describe a tone that is chiefly concerned with creating attitudes in the reader (see pp. 236–37).

aggravate   *Aggravate* may mean either "to make worse" ("His remarks aggravated the dispute") or "to annoy or exasperate" ("Her manners aggravate me"). Both are standard English, but there is still some objection to the second usage. If you mean *annoy, exasperate,* or *provoke,* it would be safer to use whichever of those words best expresses your meaning.

ain't   Except to record nonstandard speech, the use of *ain't* is not acceptable in college writing.

all together, altogether   Distinguish between the phrase ("They were all together at last") and the adverb ("He is altogether to blame"). *All together* means "all in one place"; *altogether* means "entirely" or "wholly."

allow   When used to mean "permit" ("No smoking is allowed on the premises"), *allow* is acceptable. Its use to mean "think" ("He allowed it could be done") is nonstandard and is not acceptable in college writing.

allusion, illusion   An *allusion* is a reference: "The poem contains several allusions to Greek mythology." An *illusion* is an erroneous mental image: "Rouge on pallid skin gives an illusion of health."

alright   A common variant spelling of *all right,* but there is still considerable objection to it. *All right* is the preferred spelling.

among, between   See **between, among.**

amount, number   *Amount* suggests bulk or weight: "We collected a considerable amount of scrap iron." *Number* is used for items that can be counted: "He has a large number of friends"; "There are a number of letters to be answered."

an Variant of the indefinite article *a*. Used instead of *a* when the word that follows begins with a vowel sound: "an apple," "an easy victory," "an honest opinion," "an hour," "an unknown person." When the word that follows begins with a consonant, or with a *y* sound or a pronounced *h*, the article used should be *a*: "a yell," "a unit," "a history," "a house." Such constructions as "a apple," "a hour" are nonstandard. The use of *an* before *historical* is an older usage that is now dying out.

and/or Many people object to *and/or* in college writing because the expression is associated with legal and commercial writing. Generally avoid it.

angle The use of *angle* to mean "point of view" ("Let's look at it from a new angle") is acceptable. In the sense of personal interest ("What's your angle?"), it is slang.

anxious = eager *Anxious* should not be used in college writing to mean "eager," as in "Gretel is anxious to see her gift." *Eager* is the preferred word in this context.

any = all The use of *any* to mean "all," as in "He is the best qualified of any applicant," is not acceptable. Say "He is the best qualified of all the applicants," or simply "He is the best-qualified applicant."

any = any other The use of *any* to mean "any other" ("The knife she bought cost more than any in the store") should be avoided in college writing. In this context, use *any other*.

anyone = all The singular *anyone* should not be used in writing to mean "all." In "She is the most talented musician of anyone I have met here," drop "of anyone."

anywheres A nonstandard variant of *anywhere*. It is not acceptable in college writing.

apt = likely *Apt* is always appropriate when it means "quick to learn" ("He is an apt student") or "suited to its purpose" ("an apt comment"). It is also appropriate when a predictable characteristic is being spoken of ("When he becomes excited he is apt to tremble"). In other situations the use of *apt* to mean "likely" ("She is apt to leave you"; "He is apt to resent it") may be too colloquial for college writing.

as = because *As* is less effective than *because* in showing causal relation between main and subordinate clauses. Since *as* has other meanings, it may in certain contexts be confusing. For example, in "As I was going home, I decided to telephone," *as* may mean "while" or "because." If there is any possibility of confusion, use either *because* or *while* — whichever is appropriate.

as = that The use of *as* to introduce a noun clause ("I don't know as I would agree to that") is colloquial. In college writing, use *that* or *whether*.

as to, with respect to = about   Although *as to* and *with respect to* are standard usage, many writers avoid these phrases because they sound stilted: "I am not concerned as to your cousin's reaction." Here *about* would be more appropriate than either *as to* or *with respect to;* "I am not concerned about your cousin's reaction."

at   Avoid the redundant *at* in such sentences as "Where were you at?" and "Where do you live at?"

author   *Author* is not fully accepted as a verb. "To write a play" is preferable to "to author a play."

awful, awfully   The real objection to *awful* is that it is worked to death. Instead of being reserved for situations in which it means "awe inspiring," it is used excessively as a utility word (see p. 225). Use both *awful* and *awfully* sparingly.

bad = badly   The ordinary uses of *bad* as an adjective cause no difficulty. As a predicate adjective ("An hour after dinner, I began to feel bad"), it is sometimes confused with the adverb *badly*. After the verbs *look, feel,* and *seem,* the adjective is preferred. Say: "It looks bad for our side," "I feel bad about the quarrel," "Our predicament seemed bad this morning." But do not use *bad* when an adverb is required, as in "He played badly," "a badly torn suit."

bank on = rely on   In college writing *rely on* is generally preferred.

being as = because   The use of *being as* for "because" or "since" in such sentences as "Being as I am an American, I believe in democracy" is nonstandard. Say "Because I am an American, I believe in democracy."

between, among   In general, use *between* in constructions involving two people or objects and *among* in constructions involving more than two: "We had less than a dollar between the two of us"; "We had only a dollar among the three of us." The general distinction, however, should be modified when insistence on it would be unidiomatic. For example, *between* is the accepted form in the following examples:

He is in the enviable position of having to choose between three equally attractive young women.

A settlement was arranged between the four partners.

Just between us girls . . . (when any number of "girls" is involved)

between you and I   Both pronouns are objects of the preposition *between* and so should be in the objective case: "between you and me."

bi-, semi-   *Bi-* means "two": "The budget for the biennium was adopted." *Semi-* means "half of": "semicircle." *Bi-* is sometimes used to mean "twice in." A bimonthly paper, for example, may be published twice a month, not once every two months, but this usage is ambiguous; *semimonthly* is preferred.

but that, but what   In such a statement as "I don't doubt but that you are correct," *but* is unnecessary. Omit it. "I don't doubt but what . . ." is also unacceptable. Delete *but what* and write *that.*

can = may   The distinction that *can* is used to indicate ability and *may* to indicate permission ("If I can do the work, may I have the job?") is not generally observed in informal usage. Either form is acceptable in college writing.

cannot help but   In college writing, the form without *but* is preferred: "I cannot help being angry." (Not: "I cannot help but be angry.")

can't hardly   A confusion between *cannot* and *can hardly.* The construction is unacceptable in college writing. Use *cannot, can't,* or *can hardly.*

capital, capitol   Unless you are referring to a government building, use *capital.* The building in which the U.S. Congress meets is always capitalized ("the Capitol"). For the various meanings of *capital,* consult your dictionary.

censor, censure   Both words come from a Latin verb meaning "to set a value on" or "judge." *Censor* is used to mean "appraise" in the sense of evaluating a book or a letter to see if it may be released ("All outgoing mail had to be censored") and is often used as a synonym for *delete* or *cut out* ("That part of the message was censored").

 Censure as a verb means "to evaluate adversely" or "to find fault with"; as a noun, it means "disapproval," "rebuke": "The editorial writers censured the speech"; "Such an attitude will invoke public censure."

center around   "Center on" is the preferred form.

cite, sight, site   *Cite* means "to refer to": "He cited chapter and verse." *Sight* means "spectacle" or "view": "The garden was a beautiful sight." *Site* means "location": "This is the site of the new plant."

compare, contrast   *Compare* can imply either differences or similarities; *contrast* always implies differences. *Compare* can be followed by either *to* or *with.* The verb *contrast* is usually followed by *with.*

Compared to her mother, she's a beauty.

I hope my accomplishments can be compared with those of my predecessor.

His grades this term contrast conspicuously with the ones he received last term.

complement, compliment   Both words can be used as nouns and verbs. *Complement* speaks of completion: "the complement of a verb"; "a full complement of soldiers to serve as an honor guard"; "Susan's hat complements the rest of her outfit tastefully." *Compliment* is associated with praise: "The instructor complimented us for writing good papers."

complement of *to be*   The choice between "It is I" and "It's me" is a choice not between standard and nonstandard usage but between formal and colloquial styles. This choice seldom has to be made in college writing, since the expression, in whatever form it is used, is essentially a spoken rather than a written sentence. Its use in writing occurs chiefly in dialogue, and then the form chosen should be appropriate to the speaker.

The use of the objective case with the third person ("That was her") is less common and should be avoided in college writing except when dialogue requires it.

continual, continuous Both words refer to a continued action, but *continual* implies repeated action ("continual interruptions," "continual disagreements"), whereas *continuous* implies that the action never ceases ("continuous pain," "a continuous buzzing in the ears").

could of = could have Although *could of* and *could have* often sound alike in speech, *of* is not acceptable for *have* in college writing. In writing, *could of, should of, would of, might of,* and *must of* are nonstandard.

council, counsel *Council* is a noun meaning "a deliberative body": "a town council," "a student council." *Counsel* can be either a noun meaning "advice" or a verb meaning "to advise": "to seek a lawyer's counsel," "to counsel a person in trouble." A person who offers counsel is a *counselor:* "Because of his low grades Quint made an appointment with his academic counselor."

credible, creditable, credulous All three words come from a Latin verb meaning "to believe," but they are not synonyms. *Credible* means "believable" ("His story is credible"); *creditable* means "commendable" ("John did a creditable job on the committee") or "acceptable for credit" ("The project is creditable toward the course requirements"); *credulous* means "gullible" ("Only a most credulous person could believe such an incredible story").

cute A word used colloquially to indicate the general notion of "attractive" or "pleasing." Its overuse shows lack of discrimination. A more specific term is often preferable.

His daughter is cute.   (lovely? petite? pleasant? charming?)

That is a cute trick.   (clever? surprising?)

He has a cute accent.   (pleasant? refreshingly unusual?)

She is a little too cute for me.   (affected? juvenile? clever?)

data is Because *data* is the Latin plural of *datum,* it logically requires a plural verb and always takes a plural verb in scientific writing: "These data have been double-checked." In popular usage and in computer-related contexts, *datum* is almost never used and *data* is treated as a singular noun and given a singular subject: "The data has been double-checked." Either *data are* or *data is* may be used in popular writing, but only *data are* is acceptable in scientific writing.

debut *Debut* is a noun meaning "first public appearance." It is not acceptable as a transitive verb ("The Little Theater will debut its new play tonight") or as an intransitive verb ("Cory Martin will debut in the new play").

decent, descent   A decent person is one who behaves well, without crudeness and perhaps with kindness and generosity. *Decent* can mean "satisfactory" ("a decent grade," "a decent living standard"). *Descent* means "a passage downward"; a descent may be either literal ("their descent into the canyon") or figurative ("hereditary descent of children from their parents," "descent of English from a hypothetical language called Indo-European").

desert, dessert   The noun *desert* means "an uncultivated and uninhabited area"; it may be dry and sandy. *Desert* can be an adjective: "a desert island." The verb *desert* means "to abandon." A *dessert* is a sweet food served as the last course at the noon or evening meal.

different from, different than   Although both *different from* and *different than* are common American usages, the preferred idiom is *different from*.

disinterested, uninterested   The distinction between these words is that *disinterested* means "unbiased" and *uninterested* means "apathetic" or "not interested." A disinterested critic is one who comes to a book with no prejudices or prior judgments of its worth; an uninterested critic is one who cannot get interested in the book. Dictionaries disagree about whether this distinction is still valid in contemporary usage and sometimes treat the words as synonyms. But in college writing the distinction is generally observed.

don't   *Don't* is a contraction of "do not," as *doesn't* is a contraction of "does not." It can be used in any college writing in which contractions are appropriate. But it cannot be used with a singular subject. "He don't" and "it don't" are nonstandard usages.

double negative   The use of two negative words within the same construction. In certain forms ("I am not unwilling to go") the double negative is educated usage for an affirmative statement; in other forms ("He hasn't got no money") the double negative is nonstandard usage. The observation that "two negatives make an affirmative" in English usage is a half-truth based on a false analogy with mathematics. "He hasn't got no money" is unacceptable in college writing, not because two negatives make an affirmative, but because it is nonstandard usage.

economic, economical   *Economic* refers to the science of economics or to business in general: "This is an economic law"; "Economic conditions are improving." *Economical* means "inexpensive" or "thrifty": "That is an economical purchase"; "He is economical to the point of miserliness."

effect, affect   See **affect, effect.**

effective, affective   See **affective, effective.**

either   Used to designate one of two things: "Both hats are becoming; I would be perfectly satisfied with either." The use of *either* when more than two things are involved ("There are three ways of working the problem; either way will give the right answer") is a disputed usage. When more than two things are involved, it is better to use *any* or *any one* instead of *either:* "There are three ways of working the problem; any one of them will give the right answer."

elicit, illicit   The first word means "to draw out" ("We could elicit no response from them"); the second means "not permitted" or "unlawful" ("an illicit sale of drugs").

emigrant, immigrant   An emigrant is a person who moves *out of* a country; an immigrant is one who moves *into* a country. Thus, refugees from Central America and elsewhere who settle in the United States are emigrants from their native countries and immigrants here. A similar distinction holds for the verbs *emigrate* and *immigrate.*

eminent, imminent   *Eminent* means "prominent, outstanding": "an eminent scientist." *Imminent* means "ready to happen" or "near in time": "War seems imminent."

enormity, enormous, enormousness   *Enormous* refers to unusual size or measure; synonyms are *huge, vast, immense:* "an enormous fish," "an enormous effort." *Enormousness* is a noun with the same connotations of size and can be applied to either good or bad effects: "The enormousness of their contribution is only beginning to be recognized"; "The enormousness of the lie almost made it believable." But *enormity* is used only for evil acts of great dimension: "The enormity of Hitler's crimes against the Jews shows what can happen when power, passion, and prejudice are all united in one human being."

enthused   *Enthused* is colloquial for *enthusiastic:* "The probability of winning has caused them to be very enthused about the campaign." In college writing use *enthusiastic.*

equally as   In such sentences as "He was equally as good as his brother," the *equally* is unnecessary. Simply write "He was as good as his brother."

etc.   An abbreviation for the Latin *et cetera,* which means "and others," "and so forth." It should be used only when the style justifies abbreviations and then only after several items in a series have been identified: "The data sheet required the usual personal information: age, height, weight, marital status, etc." An announcement of a painting contest that stated, "Entries will be judged on the basis of use of color, etc.," does not tell contestants very much about the standards by which their work is to be judged. Avoid the redundant *and* before *etc.*

expect = suppose or suspect   The use of *expect* for *suppose* or *suspect* is colloquial. In college writing use *suppose* or *suspect:* "I suppose you have written to him"; "I suspect that we have made a mistake."

fact   Distinguish between facts and statements of fact. A fact is something that exists or existed. A fact is neither true nor false; it just *is.* A statement of fact, or a factual statement, may be true or false, depending on whether it does or does not report the facts accurately.

Avoid padding a sentence with "a fact that," as in "It is a fact that all the public opinion polls predicted Truman's defeat in the 1948 election." The first five words of that sentence add no meaning. Similarly, "His guilt is admitted" says all, in fewer words, that is said by "The fact of his guilt is admitted."

famous, notorious *Famous* is a complimentary and *notorious* an uncomplimentary adjective. Well-known people of good repute are famous; those of bad repute are notorious, or infamous.

farther, further The distinction that *farther* indicates distance and *further* degree is not unanimously supported by usage studies. But to mean "in addition," only *further* is used: "Further assistance will be required."

feature = imagine The use of *feature* to mean "give prominence to," as in "This issue of the magazine features an article on juvenile delinquency," is established standard usage and is appropriate in college writing. But this acceptance does not justify the slang use of *feature,* meaning "imagine," in such expressions as "Can you feature that?" "Feature me in a dress suit," "I can't feature him as a nurse."

fewer = less *Fewer* refers to quantities that can be counted individually: "fewer male than female employees." *Less* is used for collective quantities that are not counted individually ("less corn this year than last") and for abstract characteristics ("less determination than enthusiasm").

field *Field,* in the sense of "an area of study or endeavor," is an overused word that often creates redundance: "He is majoring in the field of physics"; "Her new job is in the field of public relations." Delete "the field of" in each of these sentences.

fine = very well The colloquial use of *fine* to mean "very well" ("He is doing fine in his new position") is probably too informal for most college writing.

flaunt = flout Using *flaunt* as a synonym for *flout* confuses two different words. *Flaunt* means "to show off": "She has a habit of flaunting her knowledge to intimidate her friends." *Flout* means "to scorn or show contempt for": "He is better at flouting opposing arguments than at understanding them." In the right context either word can be effective, but the two words are not synonyms and cannot be used interchangeably.

fortuitous, fortunate *Fortuitous* means "by chance," "not planned": "Our meeting was fortuitous; we had never heard of each other before." Do not confuse *fortuitous* with *fortunate,* as the writer of this sentence has done: "My introduction to Professor Kraus was fortuitous for me; today she hired me as her student assistant." *Fortunate* would be the appropriate word here.

funny Often used in conversation as a utility word that has no precise meaning but may be clear enough in its context. It is generaly too vague for college writing. Decide in what sense the subject is "funny" and use a more precise term to convey that sense. (See the discussion of vagueness on pp. 224–25.)

get   A utility word. *The American Heritage Dictionary* lists thirty-six meanings for the individual word and more than sixty uses in idiomatic expressions. Most of these uses are acceptable in college writing. But unless the style is deliberately colloquial, avoid slang uses of *get* meaning "to cause harm to" ("She'll get me for that"), "to cause a negative reaction to" ("His bad manners really get me"), "to gain the favor of" ("He tried to get in with his boss"), and "to become up-to-date" ("Get in the swing of things").

good   The use of *good* as an adverb ("He talks good"; "She played pretty good") is not acceptable. The accepted adverbial form is *well*. The use of *good* as a predicate adjective after verbs of hearing, feeling, seeing, smelling, tasting, and the like is standard. See **bad**.

good and   Used colloquially as an intensive in such expressions as "good and late," "good and ready," "good and tired." The more formal the style, the less appropriate these intensives are. In college writing use them sparingly, if at all.

guess   The use of *guess* to mean "believe," "suppose," or "think" ("I guess I can be there on time") is accepted by all studies on which this glossary is based. There is objection to its use in formal college writing, but it should be acceptable in an informal style.

had (hadn't) ought   Nonstandard for *ought* and *ought not*. Not acceptable in college writing.

hanged, hung   Alternative past participles of *hang*. For referring to an execution, *hanged* is preferred; in other senses, *hung* is preferred.

he or she, she or he   Traditionally the masculine form *(he, his, him)* of the personal pronoun has been used to refer to an individual who could be either male or female: "The writer should revise his draft until he accomplishes his purpose." Substituting pronouns that refer to both males and females in the group, such as *he or she,* or *she or he,* corrects the implicit sexism in the traditional usage but sometimes sounds awkward. An alternative is to use plural forms: "Writers should revise their drafts until they accomplish their purpose."

hopefully   Opinion is divided about the acceptability of attaching this adverb loosely to a sentence and using it to mean "I hope": "Hopefully, the plane will arrive on schedule." This usage is gaining acceptance, but there is still strong objection to it. In college writing the safe decision is to avoid it.

idea   In addition to its formal meaning of "conception," *idea* has acquired so many supplementary meanings that it must be recognized as a utility word. Some of its meanings are illustrated in the following sentences:

The idea (thesis) of the book is simple.

The idea (proposal) she suggested is a radical one.

I got the idea (impression) that he is unhappy.

It is my idea (belief, opinion) that they are both wrong.

My idea (intention) is to leave early.

The overuse of *idea,* like the overuse of any utility word, makes for vagueness. Whenever possible, use a more precise synonym.

illicit, elicit   See **elicit, illicit.**

illusion, allusion   See **allusion, illusion.**

immigrant, emigrant   See **emigrant, immigrant.**

imminent, eminent   See **eminent, imminent.**

imply, infer   The traditional difference between these two words is that *imply* refers to what a statement means, usually to a meaning not specifically stated but suggested in the original statement, whereas *infer* is used for a listener's or reader's judgment or inference based on the statement. For example: "I thought that the weather report implied that the day would be quite pretty and sunny, but Marlene inferred that it meant we'd better take umbrellas." The dictionaries are not unanimous in supporting this distinction, but in your writing it will be better not to use *imply* as a synonym for *infer.*

individual   Although the use of *individual* to mean "person" ("He is an energetic individual") is accepted by the dictionaries, college instructors frequently disapprove of this use, probably because it is overdone in college writing. There is no objection to the adjective *individual,* meaning "single," "separate" ("The instructor tries to give us individual attention").

inferior than   Possibly a confusion between "inferior to" and "worse than." Use *inferior to:* "Today's workmanship is inferior to that of a few years ago."

ingenious, ingenuous   *Ingenious* means "clever" in the sense of "original": "an ingenious solution." *Ingenuous* means "without sophistication," "innocent": "Her ingenuous confession disarmed those who had been suspicious of her motives."

inside of, outside of   *Inside of* and *outside of* generally should not be used as compound prepositions. In place of the compound prepositions in "The display is inside of the auditorium" and "The pickets were waiting outside of the gate," write "inside the auditorium" and "outside the gate."

   Inside of is acceptable in most college writing when it means "in less than": "I'll be there inside of an hour." The more formal term is *within.*

   Both *inside of* and *outside of* are appropriate when *inside* or *outside* is a noun followed by an *of* phrase: "The inside of the house is quite attractive"; "He painted the outside of his boat dark green."

in terms of   An imprecise and greatly overused expression. Instead of "In terms of philosophy, we are opposed to his position" and "In terms of our previous experience with the company, we refuse to purchase its products," write "Philosophically, we are opposed to his position" and "Because of our previous experience with the company, we refuse to purchase its products."

497

irregardless   A nonstandard variant of *regardless*. Do not use it.

irrelevant, irreverent   *Irrelevant* means "having no relation to" or "lacking pertinence": "That may be true, but it is quite irrelevant." *Irreverent* means "without reverence": "Such conduct in church is irreverent."

it's me   This construction is essentially a spoken one. Except in dialogue, it rarely occurs in writing. Its use in educated speech is thoroughly established. The formal expression is "It is I."

-ize   The suffix *-ize* is used to change nouns and adjectives into verbs: *civilize, criticize, sterilize.* This practice is often overused, particularly in government and business. Avoid such pretentious and unnecessary jargon as *finalize, prioritize,* and *theorize.*

judicial, judicious   Judicial decisions are related to the administering of justice, often by judges or juries. A judicious person is one who demonstrates good judgment: "A judicious person would not have allowed the young boys to shoot the rapids alone."

kind of, sort of   Use a singular noun and a singular verb with these phrases: "That kind of person is always troublesome"; "This sort of attitude is deplorable." If the sense of the sentence calls for the plural *kinds* or *sorts,* use a plural noun and a plural verb: "These kinds of services are essential." In questions introduced by *what* or *which,* the singular *kind* or *sort* can be followed by a plural noun and verb: "What kind of shells are these?"

　　　The use of *a* or *an* after *kind of* ("That kind of a person is always troublesome") is usually not appropriate in college writing.

kind (sort) of = somewhat   This usage ("I feel kind of tired"; "He looked sort of foolish") is colloquial. The style of the writing will determine its appropriateness in a paper.

latter   *Latter* refers to the second of two. It should not be used to refer to the last of three or more nouns. Instead of *latter* in "Michigan, Alabama, and Notre Dame have had strong football teams for years, and yet the latter has only recently begun to accept invitations to play in bowl games," write *last* or *last-named,* or simply repeat *Notre Dame.*

lay, lie   *Lay* is a transitive verb (principal parts: *lay, laid, laid*) that means "put" or "place"; it is nearly always followed by a direct object: "She lay the magazine on the table, hiding the mail I laid there this morning." *Lie* is an intransitive verb (principal parts: *lie, lay, lain*) that means "recline" or "be situated" and does not take an object. "I lay awake all night until I decided I had lain there long enough."

leave = let   The use of *leave* for the imperative verb *let* ("Leave us face it") is not acceptable in college writing. Write "Let us face it." But *let* and *leave* are interchangeable when a noun or pronoun and then *alone* follow: "Let me alone"; "Leave me alone."

less   See **fewer.**

liable = likely   Instructors sometimes object to the use of *liable* to mean "likely," as in "It is liable to rain," "He is liable to hit you." *Liable* is used more precisely to mean "subject to" or "exposed to" or "answerable for": "He is liable to arrest"; "You will be liable for damages."

like = as, as though   The use of *like* as a conjunction ("He talks like you do"; "It looks like it will be my turn next") is colloquial. It is not appropriate in a formal style, and many people object to it in an informal style. The safest procedure is to avoid using *like* as a conjunction in college writing.

literally, figuratively   *Literally* means "word for word," "following the letter," or "in the strict sense." *Figuratively* is its opposite and means "metaphorically." In informal speech, this distinction is often blurred when *literally* is used to mean *nearly:* "She literally blew her top." Avoid this usage by maintaining the word's true meaning: "To give employees a work vacation means literally to fire them."

loath, loathe   *Loath* is an adjective meaning "reluctant," "unwilling" ("I am loath to do that"; "He is loath to risk so great an investment") and is pronounced to rhyme with *both*. *Loathe* is a verb meaning "dislike strongly" ("I loathe teas"; "She loathes an unkempt man") and is pronounced to rhyme with *clothe*.

loose, lose   The confusion of these words frequently causes misspelling. *Loose* is most common as an adjective: "a loose button," "The dog is loose." *Lose* is always used as a verb: "You are going to lose your money."

luxuriant, luxurious   These words come from the same root but have quite different meanings. *Luxuriant* means "abundant" and is used principally to describe growing things: "luxuriant vegetation," "a luxuriant head of hair." *Luxurious* means "luxury-loving" or "characterized by luxury": "He finds it difficult to maintain so luxurious a life style on so modest an income"; "The furnishings of the clubhouse were luxurious."

mad = angry or annoyed   Using *mad* to mean "angry" is colloquial: "My girl is mad at me"; "His insinuations make me mad." More precise terms — *angry, annoyed, irritated, provoked, vexed* — are generally more appropriate in college writing. *Mad* is, of course, appropriately used to mean "insane."

majority, plurality   Candidates are elected by a *majority* when they get more than half of the votes cast. A *plurality* is the margin of victory that the winning candidate has over the leading opponent, whether the winner has a majority or not.

mean = unkind, disagreeable, vicious   Using *mean* to convey the sense "unkind," "disagreeable," "vicious" ("It was mean of me to do that"; "He was in a mean mood"; "That dog looks mean") is a colloquial use. It is appropriate in most college writing, but since using *mean* loosely sometimes results in vagueness, consider using one of the suggested alternatives to provide a sharper statement.

medium, media, medias   *Medium,* not *media,* is the singular form: "The daily newspaper is still an important medium of communication." *Media* is plural: "Figuratively, the electronic media have created a smaller world." *Medias* is not an acceptable form for the plural of *medium.*

might of   See **could of.**

mighty = very   *Mighty* is not appropriate in most college writing as a substitute for *very.* Avoid such constructions as "He gave a mighty good speech."

moral, morale   Roughly, *moral* refers to conduct and *morale* refers to state of mind. A moral man is one who conducts himself according to standards for goodness. People are said to have good morale when they are cheerful, cooperative, and not too much concerned with their own worries.

most = almost   The use of *most* as a synonym for *almost* ("I am most always hungry an hour before mealtime") is colloquial. In college writing *almost* would be preferred in such a sentence.

must (adj. and n.)   The use of *must* as an adjective ("This book is must reading for anyone who wants to understand Russia") and as a noun ("It is reported that the President will classify this proposal as a must") is accepted as established usage by the dictionaries.

must of   See **could of.**

myself = I, me   *Myself* should not be used for *I* or *me.* Avoid such constructions as "John and myself will go." *Myself* is acceptably used as an intensifier ("I saw it myself"; "I myself will go with you") and as a reflexive object ("I hate myself"; "I can't convince myself that he is right.")

nauseous = nauseated   *Nauseous* does not mean "experiencing nausea"; *nauseated* has that meaning: "The thought of making a speech caused her to feel nauseated." *Nauseous* means "causing nausea" or "repulsive": "nauseous odor," "nauseous television program."

nice   A utility word much overused in college writing. Avoid excessive use of *nice* and, whenever possible, choose a more precise synonym.

That's a nice dress.   [*attractive? becoming? fashionable? well-made?*]

She's a nice person.   [*agreeable? charming? friendly? well-mannered?*]

not all that   The use of *not all that interested* to mean "not much interested" is generally not acceptable in college writing.

off, off of = from   Neither *off* nor *off of* should be used to mean "from." Write "Jack bought the old car from a stranger," not "off a stranger" or "off of a stranger."

OK, O.K.   Its use in business to mean "endorse" is generally accepted: "The manager OK'd the request." In college writing *OK* is a utility word and is subject to the general precaution concerning all such words: do not overuse it, especially in contexts in which a more specific term would give more efficient communication. For example, contrast the vagueness of *OK* in the first sentence with the discriminated meanings in the second and third sentences:

The mechanic said the tires were OK.

The mechanic said the tread on the tires was still good.

The mechanic said the pressure in the tires was satisfactory.

one   See **you.**

only   The position of *only* in such sentences as "I only need three dollars" and "If only Mother would write!" is sometimes condemned on the grounds of possible ambiguity. In practice, the context usually rules out any but the intended interpretation, but a change in the word order would result in more appropriate emphasis: "I need only three dollars"; "If Mother would only write!"

on the part of   The phrase *on the part of* ("There will be some objection on the part of the students"; "On the part of business people, there will be some concern about taxes") often contributes to wordiness. Simply say: "The students will object," "Business people will be concerned about taxes."

party = person   The use of *party* to mean "person" is appropriate in legal documents and the responses of telephone operators, but these are special uses. Generally avoid this use in college writing.

per, a   "You will be remunerated at the rate of forty dollars per diem" and "The troops advanced three miles per day through the heavy snow" show established use of *per* for *a*. But usually "forty dollars a day" and "three miles a day" would be more natural expressions in college writing.

percent, percentage   *Percent* (alternative form, *per cent*) is used when a specific portion is named: "five percent of the expenses." *Percentage* is used when no number is given: "a small percentage of the expenses." When *percent* or *percentage* is part of a subject, the noun or pronoun of the *of* phrase that follows determines the number of the verb: "Forty percent of the wheat is his"; "A large percentage of her customers pay promptly."

personal, personnel   *Personal* means "of a person": "a personal opinion," "a personal matter." *Personnel* refers to the people in an organization, especially employees: "Administrative personnel will not be affected."

phenomenon, phenomena   *Phenomenon* is singular; *phenomena* is plural: "This is a striking phenomenon." "Many new phenomena have been discovered with the radio telescope."

plenty   The use of *plenty* as a noun ("There is plenty of room") is always acceptable. Its use as an adverb ("It was plenty good") is not appropriate in college writing.

practical, practicable   Avoid interchanging the two words. *Practical* means "useful, not theoretical"; *practicable* means "feasible, but not necessarily proved successful": "The designers are usually practical, but these new blueprints do not seem practicable."

501

première   *Première* is acceptable as a noun ("The première for the play was held in a small off-Broadway theater"), but do not use it as a verb ("The play premièred in a small off-Broadway theater"). Write "The play opened . . ."

preposition (ending sentence with)   A preposition should not appear at the end of a sentence if its presence there draws undue attention to it or creates an awkward construction, as in "They are the people whom we made the inquiries yesterday about." But there is nothing wrong with writing a preposition at the end of a sentence to achieve an idiomatic construction: "Isn't that the man you are looking for?"

principal, principle   The basic meaning of *principal* is "chief" or "most important." It is used in this sense both as a noun and as an adjective: "the principal of a school," "the principal point." It is also used to refer to a capital sum of money, as contrasted with interest on the money: "He can live on the interest without touching the principal." *Principle* is used only as a noun and means "rule," "law," or "controlling idea": "the principle of 'one man, one vote' "; "Cheating is against my principles."

proceed, precede   To *proceed* is to "go forward"; to *precede* means "to go ahead of": "The blockers preceded the runner as the football team proceeded toward the goal line."

prophecy, prophesy   *Prophecy* is always used as a noun ("The prophecy came true"); *prophesy* is always a verb ("He prophesied another war").

proved, proven   When used as past participles, both forms are standard English, but the preferred form is *proved:* "Having proved the first point, we moved to the second." *Proven* is preferred when the word is used primarily as an adjective: "She is a proven contender for the championship."

quote   The clipped form for *quotation* ("a quote from *Walden*") is not acceptable in most college writing. The verb *quote* ("to quote Thoreau") is acceptable in all styles.

raise, rise   *Raise* is a transitive verb, taking an object, meaning to cause something to move up; *rise* is an intransitive verb meaning to go up (on its own): "I raised the window in the kitchen while I waited for the bread dough to rise."

rarely ever, seldom ever   The *ever* is redundant. Instead of saying "He is rarely ever late" and "She is seldom ever angry," write "He is rarely late" and "She is seldom angry."

real = really (very)   The use of *real* to mean "really" or "very" ("It is a real difficult assignment") is a colloquial usage. It is acceptable only in a paper whose style is deliberately colloquial.

reason . . . because   The construction is redundant: "The reason he couldn't complete his essay is because he lost his note cards." Substitute *that* for *because:* "The reason he couldn't complete his essay is that he lost his note cards." Better yet, simply eliminate *the reason* and *is:* "He couldn't complete his essay because he lost his note cards."

refer back   A confusion between *look back* and *refer*. This usage is objected to in college writing on the ground that since the *re-* of *refer* means "back," *refer back* is redundant. *Refer back* is acceptable when it means "refer again" ("The bill was referred back to the committee"); otherwise, use *refer* ("Let me refer you to page 17").

regarding, in regard to, with regard to   These are overused and stuffy substitutes for the following simple terms: *on, about,* or *concerning:* "The attorney spoke to you about the testimony."

respectfully, respectively   *Respectfully* means "with respect": "respectfully submitted." *Respectively* means roughly "each in turn": "These three papers were graded respectively A, C, and B."

right (adv.)   The use of *right* as an adverb is established in such sentences as "He went right home" and "It served her right." Its use to mean *very* ("I was right glad to meet him") is colloquial and should be used in college writing only when the style is colloquial.

right, rite   A *rite* is a ceremony or ritual. This word should not be confused with the various uses of *right*.

said (adj.)   The use of *said* as an adjective ("said documents," "said offense") is restricted to legal phraseology. Do not use it in college writing.

same as = just as   The preferred idiom is "just as": "He acted just as I thought he would."

same, such   Avoid using *same* or *such* as a substitute for *it, this, that, them*. Instead of "I am returning the book because I do not care for same" and "Most people are fond of athletics of all sorts, but I have no use for such," say "I am returning the book because I do not care for it" and "Unlike most people, I am not fond of athletics."

scarcely   In such sentences as "There wasn't scarcely enough" and "We haven't scarcely time," the use of *scarcely* plus a negative creates an unacceptable double negative. Say: "There was scarcely enough" and "We scarcely have time."

scarcely . . . than   The use of *scarcely . . . than* ("I had scarcely met her than she began to denounce her husband") is a confusion between "no sooner . . . than" and "scarcely . . . when." Say "I had no sooner met her than she began to denounce her husband" or "I had scarcely met her when she began to denounce her husband."

seasonable, seasonal   *Seasonable* and its adverb form *seasonably* mean "appropriate(ly) to the season": "She was seasonably dressed for a late-fall football game"; "A seasonable frost convinced us that the persimmons were just right for eating." *Seasonal* means "caused by a season": "increased absenteeism because of seasonal influenza," "flooding caused by seasonal thaws."

-selfs   The plural of *self* is *selves*. Such a usage as "They hurt themselfs" is non-standard and is not acceptable in college writing.

semi-   See **bi-, semi-**.

sensual, sensuous   *Sensual* has unfavorable connotations and means "catering to the gratification of physical desires": "Always concerned with satisfying his sexual lust and his craving for drink and rich food, the old baron led a totally sensual existence." *Sensuous* has generally favorable connotations and refers to pleasures experienced through the senses: "the sensuous comfort of a warm bath," "the sensuous imagery of the poem."

set, sit   These two verbs are commonly confused. *Set* meaning "to put or place" is a transitive verb and takes an object. *Sit* meaning "to be seated" is an intransitive verb. "You can set your books on the desk and then sit in that chair."

shall, will   In American usage the dominant practice is to use *will* in the second and third persons to express either futurity or determination and to use either *will* or *shall* in the first person.

In addition, *shall* is used in statements of law ("Congress shall have the power to ..."), in military commands ("The regiment shall proceed as directed"), and in formal directives ("All branch offices shall report weekly to the home office").

should, would   These words are used as the past forms of *shall* and *will* respectively and follow the same pattern (see **shall, will**): "I would [should] be glad to see him tomorrow"; "He would welcome your ideas on the subject"; "We would [should] never consent to such an arrangement." They are also used to convert *shall* or *will* in direct discourse into indirect discourse:

Direct Discourse	**Indirect Discourse**
"Shall I try to arrange it?" he asked.	He asked if he should try to arrange it.
I said, "They will need money."	I said that they would need money.

should of   See **could of**.

sight, site, cite   See **cite, sight, site**.

so (conj.)   The use of *so* as a connective ("The salesperson refused to exchange the merchandise; so we went to the manager") is thoroughly respectable, but its overuse in college writing is objectionable. There are other good transitional connectives — *accordingly, for that reason, on that account, therefore* — that could be used to relieve the monotony of a series of *so*'s. Occasional use of subordination ("When the salesperson refused to exchange the merchandise, we went to the manager") also brings variety to the style.

some   The use of *some* as an adjective of indeterminate number ("Some friends of yours were here") is acceptable in all levels of writing. Its use as an intensive ("That was some meal!") or as an adverb ("She cried some after you left"; "This draft is some better than the first one") should be avoided in college writing.

sort of   See **kind of.**

stationary, stationery   *Stationary* means "fixed" or "unchanging": "The battle front is now stationary." *Stationery* means "writing paper": "a box of stationery." Associate the *e* in *stationery* with the *e*'s in *letter.*

suit, suite   The common word is *suit:* "a suit of clothes"; "Follow suit, play a diamond"; "Suit yourself." *Suite,* pronounced "sweet," means "retinue" ("The President and his suite have arrived") or "set" or "collection" ("a suite of rooms," "a suite of furniture"). When *suite* refers to furniture, an alternative pronunciation is "suit."

sure = certainly   Using *sure* in the sense of "certainly" ("I am sure annoyed"; "Sure, I will go with you") is colloquial. Unless the style justifies colloquial usage, use *certainly* or *surely.*

terrific   Used at a formal level to mean "terrifying" ("a terrific epidemic") and at a colloquial level as an intensive ("a terrific party," "a terrific pain"). Overuse of the word at the colloquial level has made it almost meaningless.

than, then   *Than* is a conjunction used in comparison; *then* is an adverb indicating time. Do not confuse the two: "I would rather write in the morning than in the afternoon. My thinking seems to be clearer then."

that, which, who   *That* refers to persons or things, *which* refers to things, and *who* refers to persons. *That* introduces a restrictive clause; *which* usually introduces a nonrestrictive clause: "John argued that he was not prepared to take the exam. The exam, which had been scheduled for some time, could not be changed." "Anyone who was not ready would have to take the test anyway."

there, their, they're   Although these words are pronounced alike, they have different meanings. *There* indicates place: "Look at that dog over there." *Their* indicates possession: "I am sure it is their dog." *They're* is a contraction of "they are": "They're probably not home."

thusly   Not an acceptable variant of *thus.*

tough   The use of *tough* to mean "difficult" ("a tough assignment," "a tough decision") and "hard fought" ("a tough game") is accepted without qualification by reputable dictionaries. But its use to mean "unfortunate," "bad" ("The fifteen-yard penalty was a tough break for the team"; "That's tough") is colloquial and should be used only in a paper written in a colloquial style.

troop, troupe   Both words come from the same root and share the original meaning, "herd." In modern usage *troop* can refer to soldiers and *troupe* to actors: "a troop of cavalry," "a troop of scouts," "a troupe of circus performers," "a troupe of entertainers."

try and   *Try to* is the preferred idiom. Use "I will try to do it" in preference to "I will try and do it."

type = type of   *Type* is not acceptable as a variant form of *type of*. In "That type engine isn't being manufactured anymore," add *of* after *type*.

uninterested, disinterested   See **disinterested**.

unique   The formal meaning of *unique* is "sole" or "only" or "being the only one of its kind": "Adam was unique in being the only man who never had a mother." The use of *unique* to mean "rare" or "unusual" ("Americans watched their television sets anxiously as astronauts in the early moon landings had the unique experience of walking on the moon") has long been popular, but some people still object to this usage. The use of *unique* to mean merely "uncommon" ("a unique sweater") is generally frowned upon. *Unique* should not be modified by adverbs that express degree: *very, more, most, rather.*

up   The adverb *up* is idiomatically used in many verb-adverb combinations that act as verbs — *break up, clean up, fill up, get up, tear up.* Avoid the unnecessary or awkward separation of *up* from the verb with which it is combined, since such a separation makes *up* look at first like an adverb modifying the verb rather than an adverb combining with the verb in an idiomatic expression. For example, "They held the cashier up" and "She made her face up" are awkward. Say "They held up the cashier," "She made up her face."

use to   The *d* in *used to* is often not pronounced; it is elided before the *t* in *to*. The resulting pronunciation leads to the written expression *use to*. But the acceptable written phrase is *used to:* "I am used to the noise"; "He used to do all the grocery shopping."

very   A common intensive, but avoid its overuse.

wait on   *Wait on* means "serve": "A clerk will be here in a moment to wait on you." The use of *wait on* to mean "wait for" ("I'll wait on you if you won't be long") is a colloquialism to which there is some objection. Use *wait for:* "I'll wait for you if you won't be long."

want in, out, off   The use of *want* followed by *in, out,* or *off* ("The dog wants in"; "I want out of here"; "I want off now") is colloquial. In college writing supply an infinitive after the verb: "The dog wants to come in."

want to = ought to, should   Using *want to* as a synonym for *should* ("They want to be careful or they will be in trouble") is colloquial. *Ought to* or *should* is preferred in college writing.

where . . . at, to   The use of *at* or *to* after *where* ("Where was he at?" "Where are you going to?") is redundant. Simply write "Where was he?" and "Where are you going?"

whose, who's   *Whose* is the possessive of *who; who's* is a contraction of *who is.* "In the play, John is the character whose son leaves town. Who's going to try out for that part?"

will, shall See **shall, will.**

-wise Avoid adding the suffix *-wise,* meaning "concerning," to nouns to form such combinations as *budgetwise, jobwise, tastewise.* Some combined forms with *-wise* are thoroughly established *(clockwise, otherwise, sidewise, weatherwise),* but the fad of coining new compounds with this suffix is generally best avoided.

without = unless *Without* is not accepted as a conjunction meaning "unless." In "There will be no homecoming festivities without student government sponsors them," substitute *unless* for *without.*

with respect to See **as to.**

worst way When *in the worst way* means "very much" ("They wanted to go in the worst way"), it is too informal for college writing.

would, should See **should, would.**

would of See **could of.**

would have = had *Would* is the past-tense form of *will,* but its overuse in student writing often results in awkwardness, especially, but not only, when it is used as a substitute for *had.* Contrast the following sentences:

Awkward	Revised
If they *would have done* that earlier, there *would have been* no trouble.	If they *had done* that earlier, there *would have been* no trouble.
	or:
	Had they *done* that earlier, there *would have been* no trouble.
We *would want* some assurance that they *would accept* before we *would make* such a proposal.	We *would want* some assurance of their acceptance before we *made* such a proposal.

In general, avoid the repetition of *would have* in the same sentence.

you = one The use of *you* as an indefinite pronoun instead of the formal *one* is characteristic of an informal style. If you adopt *you* in an informal paper, be sure that this impersonal use will be recognized by your readers; otherwise, they are likely to interpret a general statement as a personal remark addressed specifically to them. Generally avoid shifting from *one* to *you* within a sentence (see p. 401).

yourself *Yourself* is appropriately used as an intensifier ("You yourself told me that") and as a reflexive object ("You are blaming yourself too much"). But usages such as the following are not acceptable: "Marian and yourself must shoulder the responsibility" and "The instructions were intended for Kate and yourself." In these two sentences, replace *yourself* with *you.* The plural form is *yourselves,* not *yourselfs.*

INDEX
of Authors and Titles

This index contains the names of all authors and titles quoted in the text. Names of student writers and titles of student writing are indicated by an asterisk (*); they are also indexed in greater detail in the main index.

INDEX

Authors and titles of all works quoted — both professional and student — will be found in the Index of Authors and Titles. Because student writers and writing play a critical role in many chapters in the text, they are also indexed in further detail in this index.

(Continued)